ENCYCLOPEDIA OF PHYSICS

CHIEF EDITOR

S. FLÜGGE

VOLUME VIa/2

MECHANICS OF SOLIDS II

EDITOR

C. TRUESDELL

WITH 25 FIGURES

SPRINGER-VERLAG
BERLIN · HEIDELBERG · NEW YORK
1972

HANDBUCH DER PHYSIK

HERAUSGEGEBEN VON

S. FLÜGGE

BAND VIa/2

FESTKÖRPERMECHANIK II

BANDHERAUSGEBER

C. TRUESDELL

MIT 25 FIGUREN

SPRINGER-VERLAG
BERLIN · HEIDELBERG · NEW YORK
1972

ISBN 3-540-05535-5 Springer-Verlag Berlin Heidelberg New York
ISBN 0-387-05535-5 Springer-Verlag New York Heidelberg Berlin

 Library of Congress Catalog Card Number A 56-2942. Printed in Germany. Satz, Druck und Bindearbeiten: Universitätsdruckerei H. Stürtz AG, Würzburg.

Contents.

The Linear Theory of Elasticity.

By

MORTON E. GURTIN.

With 18 Figures.

Dedicated to ELI STERNBERG.

A. Introduction.

1. Background. Nature of this treatise. Linear elasticity is one of the more successful theories of mathematical physics. Its pragmatic success in describing the small deformations of many materials is uncontested. The origins of the three-dimensional theory go back to the beginning of the 19th century and the derivation of the basic equations by CAUCHY, NAVIER, and POISSON. The theoretical development of the subject continued at a brisk pace until the early 20th century with the work of BELTRAMI, BETTI, BOUSSINESQ, KELVIN, KIRCHHOFF, LAMÉ, SAINT-VENANT, SOMIGLIANA, STOKES, and others. These authors established the basic theorems of the theory, namely compatibility, reciprocity, and uniqueness, and deduced important general solutions of the underlying field equations. In the 20th century the emphasis shifted to the solution of boundary-value problems, and the theory itself remained relatively dormant until the middle of the century when new results appeared concerning, among other things, Saint-Venant's principle, stress functions, variational principles, and uniqueness.

It is the purpose of this treatise to give an exhaustive presentation of the linear theory of elasticity.[1] Since this volume contains two articles by FICHERA concerning existence theorems, that subject will not be discussed here.

I have tried to maintain the level of rigor now customary in pure mathematics. However, in order to ease the burden on the reader, many theorems are stated with hypotheses more stringent than necessary.

Acknowledgement. I would like to acknowledge my debt to my friend and teacher, ELI STERNBERG, who showed me in his lectures[2] that it is possible to present the linear theory in a concise and rational form—a form palatable to both engineers and mathematicians. Portions of this treatise are based on STERNBERG's unpublished lecture notes; I have tried to indicate when such is the case. I would like to express my deep gratitude to D. CARLSON, G. FICHERA, R. HUILGOL, E. STERNBERG, and C. TRUESDELL for their valuable detailed criticisms of the manuscript. I would also like to thank G. BENTHIEN, W. A. DAY, J. ERICKSEN, R. KNOPS, M. OLIVER, G. DE LA PENHA, T. RALSTON, L. SOLOMON, E. WALSH, L. WHEELER, and W. WILLIAMS for valuable comments, and H. ZIEGLER for generously sending me a copy of

[1] Specific applications are not taken up in this article. They will be treated in a sequel by L. SOLOMON, *Some Classic Problems of Elasticity,* to appear in the Springer Tracts in Natural Philosophy.

[2] At Brown University in 1959–1961.

PRANGE'S 1916 Habilitation Dissertation. Most of the historical research for this treatise was carried out at the Physical Sciences Library of Brown University; without the continued support and hospitality of the staff of that great library this research would not have been possible. Finally, let me express my gratitude to the U.S. National Science Foundation for their support through a research grant to Carnegie-Mellon University.

2. Terminology and general scheme of notation. I have departed radically from the customary notation in order to present the theory in what I believe to be a form most easily understood by someone not prejudiced by a past acquaintance with the subject. Direct notation, rather than cartesian or general coordinates, is utilized throughout. I do not use what is commonly called "dyadic notation"; most of the notions used, e.g. vector, linear transformation, tensor product, can be found in a modern text in linear algebra.

General scheme of notation.

Italic boldface minuscules $\boldsymbol{a}, \boldsymbol{b}, \boldsymbol{u}, \boldsymbol{v}$ ...: vectors and vector fields; $\boldsymbol{x}, \boldsymbol{y}, \boldsymbol{z}, \boldsymbol{\xi}$, ...: points of space.

Italic boldface majuscules $\boldsymbol{A}, \boldsymbol{B}$, ...: (second-order) tensors or tensor fields.

Italic lightface letters A, a, α, Φ, ...: scalars or scalar fields.

C, K: fourth-order tensors.

Sans-serif boldface majuscules M, N ... (except C, K): four-tensors or fields with such values.

Sans-serif boldface minuscules u, $\boldsymbol{\xi}$, ...: four-vectors or four-vector fields.

Italic lightface majuscules B, D, Σ, ...: regions in euclidean space.

$\mathcal{G}$: a group of second order tensors.

Script majuscules $\mathcal{C}$, $\mathcal{S}$, ... (except $\mathcal{G}$): surfaces in euclidean space.

Italic indices i, j, ...: tensorial indices with the range (1, 2, 3).

Greek indices α, β, ...: tensorial indices with the range (1, 2).

Index of frequently used symbols. Only symbols used frequently are listed. It has not been possible to adhere rigidly to these notations, so that sometimes within a single section these same letters are used for quantities other than those listed below.

Symbol	Name	Place of definition or first occurrence
$\boldsymbol{A}(\boldsymbol{m})$	Acoustic tensor	
$\boldsymbol{A}$	Beltrami stress function	
B	Body	
C	Elasticity tensor	
$D_{\boldsymbol{l}}$	Set of points of application of system $\boldsymbol{l}$ of concentrated loads	
$\boldsymbol{E}$	Strain tensor	
$\mathring{\boldsymbol{E}}$	Traceless part of strain tensor	
$\overline{\boldsymbol{E}}(B)$	Mean strain	
$\mathcal{E}$	Three-dimensional euclidean space	
$\mathcal{E}^{(4)}$	$=\mathcal{E}\times(-\infty, \infty)=$space-time	

Symbol	Name	Place of definition or first occurrence
$\mathscr{G}_{\boldsymbol{x}}$	Symmetry group for the material at $\boldsymbol{x}$	
K	Kinetic energy	
K	Compliance tensor	
M	Stress-momentum tensor	
$\mathbf{0}$	Origin, zero vector, zero tensor	
P	Part of B	
$\boldsymbol{Q}$	Orthogonal tensor	
R	Plane region	
$\boldsymbol{S}$	Stress tensor	
$\mathring{\boldsymbol{S}}$	Traceless part of stress tensor	
$\bar{\boldsymbol{S}}(B)$	Mean stress	
$\mathscr{S}_1, \mathscr{S}_2$	Complementary subsets of ∂B	
$U\{\boldsymbol{E}\}$	Strain energy	
$\mathscr{U}$	Total energy	
$\mathscr{V}$	Vector space associated with $\mathscr{E}$	
$\mathscr{V}^{(4)}$	$=\mathscr{V}\times(-\infty, \infty)$	
$\boldsymbol{W}$	Rotation tensor	
$\mathscr{W}$	Singular surface	
$\boldsymbol{a}$	Amplitude of wave	
$\boldsymbol{b}$	Body force	
$\boldsymbol{c}$	Centroid of B	
c	Speed of propagation, also the constant $16\pi\mu(1-\nu)$	
c_1	Irrotational velocity	
c_2	Isochoric velocity	
$\boldsymbol{e}_i$	Orthonormal basis	
$\boldsymbol{f}$	Pseudo body force field	
$\mathscr{f}$	System of forces	
i	$\sqrt{-1}$, also function with values $i(t)=t$	
k	Modulus of compression	
$\boldsymbol{l}$	System of concentrated loads	
$\boldsymbol{m}$	Direction of propagation	
$\boldsymbol{n}$	Outward unit normal vector on ∂B	
p	Pressure	
$\boldsymbol{p}$	Position vector from the origin $\mathbf{0}$	
$\boldsymbol{p}_{\boldsymbol{c}}$	Position vector from the centroid $\boldsymbol{c}$	
$\mathscr{p}$	Admissible process, elastic process	
r	$=\lvert\boldsymbol{x}-\mathbf{0}\rvert$	
$\boldsymbol{s}$	Surface traction	
$\hat{\boldsymbol{s}}$	Prescribed surface traction	
$\mathscr{s}$	Admissible state, elastic state	
$\mathscr{s}_{\boldsymbol{y}}[\boldsymbol{l}]$	Kelvin state corresponding to a concentrated load $\boldsymbol{l}$ at $\boldsymbol{y}$	
$\mathscr{s}_{\boldsymbol{y}}^{i}$	Unit Kelvin state corresponding to the unit load $\boldsymbol{e}_i$ at $\boldsymbol{y}$	
$\bar{\mathscr{s}}_{\boldsymbol{y}}^{ij}$	Unit doublet states at $\boldsymbol{y}$	
$\mathring{\mathscr{s}}_{\boldsymbol{y}}$	Center of compression at $\boldsymbol{y}$	
$\mathring{\mathscr{s}}_{\boldsymbol{y}}^{i}$	Center of rotation at $\boldsymbol{y}$ parallel to the x_i-axis	

Symbol	Name	Place of definition or first occurrence
t	Time	
$\boldsymbol{u}$	Displacement vector	
$\hat{\boldsymbol{u}}$	Prescribed displacement on boundary	
$\boldsymbol{u}_0$	Initial displacement	
$\boldsymbol{v}_0$	Initial velocity	
$v(B)$	Volume of B	
$\boldsymbol{w}$	Rigid displacement	
$\boldsymbol{x}, \boldsymbol{y}, \boldsymbol{z}$	Points in space	
x_i	Cartesian components of $\boldsymbol{x}$	
z	Complex variable	
$\Theta\{\mathfrak{s}\}$	Functional in Hellinger-Prange-Reissner principle	
$\Lambda\{\mathfrak{s}\}$	Functional in Hu-Washizu principle	
$\Sigma_\delta(\boldsymbol{y})$	Open ball with radius δ and center at $\boldsymbol{y}$	
$\Phi\{\mathfrak{s}\}$	Functional in principle of minimum potential energy	
$\Psi\{\mathfrak{s}\}$	Functional in principle of minimum complementary energy	
β	Young's modulus	
δ_{ij}	Kronecker's delta	
$\delta v(B)$	Volume change	
ε	Internal energy density	
ε_{ijk}	Three-dimensional alternator	
$\varepsilon_{\alpha\beta}$	Two-dimensional alternator	
λ	Lamé modulus	
μ	Shear modulus	
μ_M	Maximum elastic modulus	
μ_m	Minimum elastic modulus	
ν	Poisson's ratio	
$\boldsymbol{\xi}$	Point in space-time	
ϱ	Density	
φ	Scalar field in Boussinesq-Papkovitch-Neuber solution, Airy stress function, scalar field in Lamé solution	
$\boldsymbol{\psi}$	Vector field in Boussinesq-Papkovitch-Neuber solution, vector field in Lamé solution	
$\boldsymbol{\omega}$	Rotation vector	
$\mathbf{1}$	Unit tensor	
sym	Symmetric part of a tensor	
skw	Skew part of a tensor	
tr	Trace of a tensor	
$\otimes$	Tensor product of two vectors	
∇	Gradient	
$\nabla_{(4)}$	Gradient in space-time	
$\hat{\nabla}$	Symmetric gradient	
curl	Curl	
div	Divergence	
$\mathrm{div}_{(4)}$	Divergence in space-time	
Δ	Laplacian	

Symbol	Name	Place of definition or first occurrence
$\square_1$, $\square_2$	Wave operators	
[ ]	Jump in a function	
da	Element of area	
dv	Element of volume	
*	Convolution	
#	Tensor product convolution	

B. Mathematical preliminaries.

I. Tensor analysis.

3. Points. Vectors. Second-order tensors. The space under consideration is always a ***three-dimensional euclidean point space*** $\mathscr{E}$. The term "***point***" will be reserved for elements of $\mathscr{E}$, the term "***vector***" for elements of the associated vector space $\mathscr{V}$. The ***inner product*** of two vectors $\boldsymbol{u}$ and $\boldsymbol{v}$ will be designated by $\boldsymbol{u}\cdot\boldsymbol{v}$. A ***cartesian coordinate frame*** consists of an orthonormal basis $\{\boldsymbol{e}_i\}=\{\boldsymbol{e}_1, \boldsymbol{e}_2, \boldsymbol{e}_3\}$ together with a point $\boldsymbol{0}$ called the ***origin***. The symbol $\boldsymbol{p}$ will *always* denote the ***position-vector field*** on $\mathscr{E}$ defined by

$$\boldsymbol{p}(\boldsymbol{x})=\boldsymbol{x}-\boldsymbol{0}.$$

We assume once and for all that a single, fixed cartesian coordinate frame is given. If $\boldsymbol{u}$ is a vector and $\boldsymbol{x}$ is a point, then u_i and x_i denote their (cartesian) components:

$$u_i=\boldsymbol{u}\cdot\boldsymbol{e}_i, \qquad x_i=(\boldsymbol{x}-\boldsymbol{0})\cdot\boldsymbol{e}_i.$$

We will occasionally use indicial notation; thus subscripts are assumed to range over the integers 1, 2, 3, and summation over repeated subscripts is implied:

$$\boldsymbol{u}\cdot\boldsymbol{v}=u_i v_i=\sum_{i=1}^{3} u_i v_i.$$

We denote the ***vector product*** of two vectors $\boldsymbol{u}$ and $\boldsymbol{v}$ by $\boldsymbol{u}\times\boldsymbol{v}$. In components

$$(\boldsymbol{u}\times\boldsymbol{v})_i=\varepsilon_{ijk}u_j v_k,$$

where ε_{ijk} is the *alternator*:

$$\varepsilon_{ijk}=\begin{cases}+1 \text{ if } (i,j,k) \text{ is an even permutation of } (1,2,3)\\ -1 \text{ if } (i,j,k) \text{ is an odd permutation of } (1,2,3)\\ 0 \text{ if } (i,j,k) \text{ is not a permutation of } (1,2,3).\end{cases}$$

We will frequently use the identity

$$\varepsilon_{ijk}\varepsilon_{ipq}=\delta_{jp}\delta_{kq}-\delta_{jq}\delta_{kp},$$

where δ_{ij} is the *Kronecker delta*:

$$\delta_{ij}=\begin{cases}1 & \text{if } i=j\\ 0 & \text{if } i\neq j.\end{cases}$$

For convenience, we use the term "(second-order) tensor" as a synonym for "linear transformation from $\mathscr{V}$ into $\mathscr{V}$". Thus a ***tensor*** $\boldsymbol{S}$ is a linear mapping that assigns to each vector $\boldsymbol{v}$ a vector

$$\boldsymbol{u}=\boldsymbol{S}\boldsymbol{v}.$$

The (cartesian) components S_{ij} of $\boldsymbol{S}$ are defined by

$$S_{ij}=\boldsymbol{e}_i\cdot\boldsymbol{S}\boldsymbol{e}_j,$$

so that $\boldsymbol{u}=\boldsymbol{S}\boldsymbol{v}$ is equivalent to

$$u_i=S_{ij}v_j.$$

We denote the ***identity tensor*** by $\mathbf{1}$ and the ***zero tensor*** by $\mathbf{0}$:

$$\mathbf{1}\boldsymbol{v}=\boldsymbol{v} \quad \text{and} \quad \mathbf{0}\boldsymbol{v}=\mathbf{0} \quad \text{for every vector } \boldsymbol{v}.$$

Clearly, the components of $\mathbf{1}$ are δ_{ij}. The ***product*** $\boldsymbol{S}\boldsymbol{T}$ of two tensors is defined by composition:

$$(\boldsymbol{S}\boldsymbol{T})(\boldsymbol{v})=\boldsymbol{S}(\boldsymbol{T}(\boldsymbol{v})) \quad \text{for every vector } \boldsymbol{v};$$

hence

$$(\boldsymbol{S}\boldsymbol{T})_{ij}=S_{ik}T_{kj}.$$

We write $\boldsymbol{S}^T$ for the ***transpose*** of $\boldsymbol{S}$; it is the unique tensor with the following property:

$$\boldsymbol{S}\boldsymbol{u}\cdot\boldsymbol{v}=\boldsymbol{u}\cdot\boldsymbol{S}^T\boldsymbol{v} \quad \text{for all vectors } \boldsymbol{u} \text{ and } \boldsymbol{v}.$$

We call $\boldsymbol{S}$ ***symmetric*** if $\boldsymbol{S}=\boldsymbol{S}^T$, ***skew*** if $\boldsymbol{S}=-\boldsymbol{S}^T$. Since $S_{ij}=S_{ji}^T$, it follows that $\boldsymbol{S}$ is symmetric or skew according as $S_{ij}=S_{ji}$ or $S_{ij}=-S_{ji}$. Every tensor $\boldsymbol{S}$ admits the unique decomposition

$$\boldsymbol{S}=\operatorname{sym}\boldsymbol{S}+\operatorname{skw}\boldsymbol{S},$$

where sym $\boldsymbol{S}$ is symmetric and skw $\boldsymbol{S}$ is skew; in fact,

$$\operatorname{sym}\boldsymbol{S}=\tfrac{1}{2}(\boldsymbol{S}+\boldsymbol{S}^T), \qquad \operatorname{skw}\boldsymbol{S}=\tfrac{1}{2}(\boldsymbol{S}-\boldsymbol{S}^T).$$

We call sym $\boldsymbol{S}$ the ***symmetric part*** and skw $\boldsymbol{S}$ the ***skew part*** of $\boldsymbol{S}$.

There is a one-to-one correspondence between vectors and skew tensors: given any skew tensor $\boldsymbol{W}$, there exists a unique vector $\boldsymbol{\omega}$ such that

$$\boldsymbol{W}\boldsymbol{u}=\boldsymbol{\omega}\times\boldsymbol{u} \quad \text{for every vector } \boldsymbol{u};$$

indeed,

$$\omega_i=-\tfrac{1}{2}\varepsilon_{ijk}W_{jk}.$$

We call $\boldsymbol{\omega}$ the ***axial vector*** corresponding to $\boldsymbol{W}$. Conversely, given a vector $\boldsymbol{\omega}$, there exists a unique skew tensor $\boldsymbol{W}$ such that the above relation holds; in fact,

$$W_{ij}=-\varepsilon_{ijk}\omega_k.$$

We write tr $\boldsymbol{S}$ for the ***trace*** and det $\boldsymbol{S}$ for the ***determinant*** of $\boldsymbol{S}$. In terms of the components S_{ij} of $\boldsymbol{S}$:

$$\operatorname{tr}\boldsymbol{S}=S_{ii},$$

and det $\boldsymbol{S}$ is the determinant of the matrix

$$\begin{bmatrix} S_{11} & S_{12} & S_{13} \\ S_{21} & S_{22} & S_{23} \\ S_{31} & S_{32} & S_{33} \end{bmatrix}.$$

Given any two tensors $\boldsymbol{S}$ and $\boldsymbol{T}$,

$$\operatorname{tr}(\boldsymbol{S}\boldsymbol{T}) = \operatorname{tr}(\boldsymbol{T}\boldsymbol{S}),$$
$$\det(\boldsymbol{S}\boldsymbol{T}) = (\det \boldsymbol{S})(\det \boldsymbol{T}).$$

A tensor $\boldsymbol{Q}$ is ***orthogonal*** provided

$$\boldsymbol{Q}^T\boldsymbol{Q} = \boldsymbol{Q}\boldsymbol{Q}^T = \boldsymbol{1}.$$

The set of all orthogonal tensors forms a group called the ***orthogonal group***; the set of all orthogonal tensors with positive determinant forms a subgroup called the ***proper orthogonal group***.

If $\{\boldsymbol{e}_i\}$ is an orthonormal basis and

$$\boldsymbol{e}_i' = \boldsymbol{Q}\boldsymbol{e}_i$$

with $\boldsymbol{Q}$ orthogonal, then $\{\boldsymbol{e}_i'\}$ is orthonormal. Conversely, if $\{\boldsymbol{e}_i\}$ and $\{\boldsymbol{e}_i'\}$ are orthonormal, then there exists a unique orthogonal tensor $\boldsymbol{Q}$ such that $\boldsymbol{e}_i' = \boldsymbol{Q}\boldsymbol{e}_i$. If such is the case, the components w_i' of a vector $\boldsymbol{w}$ with respect to $\{\boldsymbol{e}_i'\}$ obey the following law of transformation of coordinates:

$$w_i' = Q_{ji}\, w_j.$$

Here Q_{ij} and w_i are the components of $\boldsymbol{Q}$ and $\boldsymbol{w}$ with respect to $\{\boldsymbol{e}_i\}$. Likewise, for a tensor $\boldsymbol{S}$,

$$S_{ij}' = Q_{ki}\, Q_{lj}\, S_{kl}.$$

The ***tensor product*** $\boldsymbol{a}\otimes\boldsymbol{b}$ of two vectors $\boldsymbol{a}$ and $\boldsymbol{b}$ is the *tensor* that assigns to each vector $\boldsymbol{u}$ the vector $\boldsymbol{a}(\boldsymbol{b}\cdot\boldsymbol{u})$:

$$(\boldsymbol{a}\otimes\boldsymbol{b})(\boldsymbol{u}) = \boldsymbol{a}(\boldsymbol{b}\cdot\boldsymbol{u}) \quad \text{for every vector } \boldsymbol{u}.$$

In components

$$(\boldsymbol{a}\otimes\boldsymbol{b})_{ij} = a_i\, b_j,$$

and it follows that

$$\operatorname{tr}(\boldsymbol{a}\otimes\boldsymbol{b}) = \boldsymbol{a}\cdot\boldsymbol{b}.$$

Further, we conclude from the above definition that the negative of the vector product $\boldsymbol{a}\times\boldsymbol{b}$ is equal to twice the axial vector corresponding to the skew part of $\boldsymbol{a}\otimes\boldsymbol{b}$.

The ***inner product*** $\boldsymbol{S}\cdot\boldsymbol{T}$ of two second-order tensors is defined by

$$\boldsymbol{S}\cdot\boldsymbol{T} = \operatorname{tr}(\boldsymbol{S}^T\boldsymbol{T}) = S_{ij}\, T_{ij},$$

while

$$|\boldsymbol{S}| = \sqrt{\boldsymbol{S}\cdot\boldsymbol{S}}$$

is called the ***magnitude*** of $\boldsymbol{S}$.

Given any tensor $\boldsymbol{S}$ and any pair of vectors $\boldsymbol{a}$ and $\boldsymbol{b}$,

$$\boldsymbol{a}\cdot\boldsymbol{S}\boldsymbol{b} = \boldsymbol{S}\cdot(\boldsymbol{a}\otimes\boldsymbol{b}).$$

This identity implies that

$$(\boldsymbol{e}_i\otimes\boldsymbol{e}_j)\cdot(\boldsymbol{e}_k\otimes\boldsymbol{e}_l) = \boldsymbol{e}_k\cdot[(\boldsymbol{e}_i\otimes\boldsymbol{e}_j)\boldsymbol{e}_l] = \boldsymbol{e}_k\cdot[\boldsymbol{e}_i(\boldsymbol{e}_j\cdot\boldsymbol{e}_l)];$$

therefore, since $\{\boldsymbol{e}_i\}$ is orthonormal,

$$(\boldsymbol{e}_i\otimes\boldsymbol{e}_j)\cdot(\boldsymbol{e}_k\otimes\boldsymbol{e}_l) = \delta_{ik}\,\delta_{jl},$$

and the nine tensors $\boldsymbol{e}_i \otimes \boldsymbol{e}_j$ are orthonormal. Moreover,

$$(S_{ij}\,\boldsymbol{e}_i \otimes \boldsymbol{e}_j)\,\boldsymbol{v} = \boldsymbol{e}_i\, S_{ij}(\boldsymbol{v}\cdot\boldsymbol{e}_j) = \boldsymbol{e}_i\, S_{ij}\, v_j = \boldsymbol{S}\boldsymbol{v},$$

and thus

$$\boldsymbol{S} = S_{ij}\,\boldsymbol{e}_i \otimes \boldsymbol{e}_j.$$

Consequently, the nine tensors $\boldsymbol{e}_i \otimes \boldsymbol{e}_j$ span the set of all tensors, and we have the following theorem:

(1) *The set of all* $\left\{\begin{matrix}\textit{tensors}\\ \textit{symmetric tensors}\\ \textit{skew tensors}\end{matrix}\right\}$ *is a vector space of dimension* $\left\{\begin{matrix}9\\6\\3\end{matrix}\right\}$, *and the* $\left\{\begin{matrix}9\\6\\3\end{matrix}\right\}$ *tensors* $\left\{\begin{matrix}\boldsymbol{e}_i \otimes \boldsymbol{e}_j\\ \sqrt{2}\,\mathrm{sym}\,\boldsymbol{e}_i \otimes \boldsymbol{e}_j\\ \sqrt{2}\,\mathrm{skw}\,\boldsymbol{e}_i \otimes \boldsymbol{e}_j,\ i<j\end{matrix}\right\}$ *form an orthonormal basis.*

We say that a scalar λ is a ***principal*** or ***characteristic value*** of a tensor $\boldsymbol{S}$ if there exists a unit vector $\boldsymbol{n}$ such that

$$\boldsymbol{S}\boldsymbol{n} = \lambda\boldsymbol{n},$$

in which case we call $\boldsymbol{n}$ a ***principal direction*** corresponding to λ. The ***characteristic space*** for $\boldsymbol{S}$ corresponding to λ is the subspace of $\mathscr{V}$ consisting of *all* vectors $\boldsymbol{u}$ that satisfy the equation

$$\boldsymbol{S}\boldsymbol{u} = \lambda\boldsymbol{u}.$$

We now record, without proof, the

(2) Spectral theorem.[1] *Let* $\boldsymbol{S}$ *be a symmetric tensor. Then there exists an orthonormal basis* $\boldsymbol{n}_1, \boldsymbol{n}_2, \boldsymbol{n}_3$ *and three (not necessarily distinct) principal values* $\lambda_1, \lambda_2, \lambda_3$ *of* $\boldsymbol{S}$ *such that*

$$\boldsymbol{S}\boldsymbol{n}_i = \lambda_i \boldsymbol{n}_i \quad \text{(no sum)} \tag{a}$$

and

$$\boldsymbol{S} = \sum_{i=1}^{3} \lambda_i\, \boldsymbol{n}_i \otimes \boldsymbol{n}_i. \tag{b}$$

Conversely, if $\boldsymbol{S}$ *admits the representation* (b) *with* $\{\boldsymbol{n}_i\}$ *orthonormal, then* (a) *holds. The relation* (b) *is called a* **spectral decomposition** *of* $\boldsymbol{S}$.

(i) *If* λ_1, λ_2, *and* λ_3 *are distinct, then the characteristic spaces for* $\boldsymbol{S}$ *are the line spanned by* $\boldsymbol{n}_1$, *the line spanned by* $\boldsymbol{n}_2$, *and the line spanned by* $\boldsymbol{n}_3$.

(ii) *If* $\lambda_2 = \lambda_3$, *then* (b) *reduces to*

$$\boldsymbol{S} = \lambda_1 \boldsymbol{n}_1 \otimes \boldsymbol{n}_1 + \lambda_2 (\boldsymbol{1} - \boldsymbol{n}_1 \otimes \boldsymbol{n}_1). \tag{c}$$

Conversely, if (c) *holds with* $\lambda_1 \neq \lambda_2$; *then* λ_1 *and* λ_2 *are the only distinct principal values of* $\boldsymbol{S}$, *and* $\boldsymbol{S}$ *has two distinct characteristic spaces: the line spanned by* $\boldsymbol{n}_1$ *and the plane perpendicular to* $\boldsymbol{n}_1$.

(iii) $\lambda_1 = \lambda_2 = \lambda_3 = \lambda$ *if and only if*

$$\boldsymbol{S} = \lambda\boldsymbol{1}, \tag{d}$$

in which case the entire vector space $\mathscr{V}$ *is the only characteristic space for* $\boldsymbol{S}$.

[1] See, e.g., HALMOS [1958, *9*], § 79.

We say that two tensor $\boldsymbol{S}$ and $\boldsymbol{Q}$ ***commute*** if

$$\boldsymbol{S}\boldsymbol{Q}=\boldsymbol{Q}\boldsymbol{S}.$$

(3) Commutation theorem.[1] *A symmetric tensor* $\boldsymbol{S}$ *commutes with an orthogonal tensor* $\boldsymbol{Q}$ *if and only if* $\boldsymbol{Q}$ *leaves each characteristic space of* $\boldsymbol{S}$ *invariant,* i.e. *if and only if* $\boldsymbol{Q}$ *maps each characteristic space into itself.*

A tensor $\boldsymbol{P}$ is a ***perpendicular projection*** if $\boldsymbol{P}$ is *symmetric* and

$$\boldsymbol{P}^2=\boldsymbol{P}.$$

If $\boldsymbol{P}$ is a perpendicular projection, then

$$(\mathbf{1}-\boldsymbol{P})(\mathbf{1}-\boldsymbol{P})=\mathbf{1}-2\boldsymbol{P}+\boldsymbol{P}^2=\mathbf{1}-\boldsymbol{P},$$

and therefore $\mathbf{1}-\boldsymbol{P}$ is also a perpendicular projection. Since

$$(\mathbf{1}-\boldsymbol{P})\,\boldsymbol{P}=\boldsymbol{P}(\mathbf{1}-\boldsymbol{P})=\boldsymbol{P}-\boldsymbol{P}^2=\mathbf{0},$$

it follows that

$$\boldsymbol{P}\boldsymbol{u}\cdot(\mathbf{1}-\boldsymbol{P})\,\boldsymbol{v}=\boldsymbol{u}\cdot\boldsymbol{P}(\mathbf{1}-\boldsymbol{P})\,\boldsymbol{v}=0;$$

therefore the range of $\boldsymbol{P}$ is orthogonal to the range of $\mathbf{1}-\boldsymbol{P}$.

Two simple examples of perpendicular projections are $\mathbf{1}$ and $\mathbf{0}$. Two other examples are $\boldsymbol{n}\otimes\boldsymbol{n}$ and $\mathbf{1}-\boldsymbol{n}\otimes\boldsymbol{n}$, where $\boldsymbol{n}$ is a unit vector. The first associates with any vector its component in the direction of $\boldsymbol{n}$, the second associates with any vector its component in the plane perpendicular to $\boldsymbol{n}$. The next theorem shows that every perpendicular projection is of one of the above four types.

(4) Structure of perpendicular projections. *If* $\boldsymbol{P}$ *is a perpendicular projection, then* $\boldsymbol{P}$ *admits one of the following four representations:*

$$\mathbf{1},\ \mathbf{0},\ \boldsymbol{n}\otimes\boldsymbol{n},\ \mathbf{1}-\boldsymbol{n}\otimes\boldsymbol{n},$$

where $\boldsymbol{n}$ *is a unit vector.*

Proof. Let λ be a principal value and $\boldsymbol{n}$ a principal direction of $\boldsymbol{P}$. Then $\boldsymbol{P}\boldsymbol{n}=\lambda\boldsymbol{n}$, and hence $\boldsymbol{P}^2\boldsymbol{n}=\lambda\boldsymbol{P}\boldsymbol{n}$, which implies $\boldsymbol{P}\boldsymbol{n}=\lambda\boldsymbol{P}\boldsymbol{n}$. Thus each principal value of $\boldsymbol{P}$ is either zero or one. If all principal values are zero, then $\boldsymbol{P}=\mathbf{0}$; if all are unity, then $\boldsymbol{P}=\mathbf{1}$. If exactly one principal value is unity, then we conclude from the spectral theorem that $\boldsymbol{P}=\boldsymbol{n}\otimes\boldsymbol{n}$ with $\boldsymbol{n}$ a unit vector. If exactly two are unity, then

$$\boldsymbol{P}=\boldsymbol{m}_1\otimes\boldsymbol{m}_1+\boldsymbol{m}_2\otimes\boldsymbol{m}_2,$$

where $\boldsymbol{m}_1$ and $\boldsymbol{m}_2$ are the corresponding principal directions, and it is a simple matter to verify that this representation is equivalent to

$$\boldsymbol{P}=\mathbf{1}-\boldsymbol{n}\otimes\boldsymbol{n},$$

where $\boldsymbol{n}=\boldsymbol{m}_1\times\boldsymbol{m}_2$. □

Notice that, in view of *(4)*, the spectral decompositions (a)–(d) in *(2)* are all expansions in terms of perpendicular projections.

We shall identify fourth-order tensors with linear transformations on the space of all (second-order) tensors. Thus a ***fourth-order tensor*** $\mathsf{C}[\cdot]$ is a linear mapping that assigns to each second-order tensor $\boldsymbol{E}$ a second-order tensor

$$\boldsymbol{S}=\mathsf{C}[\boldsymbol{E}].$$

[1] This theorem is a corollary to Theorem 2, § 43 and Theorem 3, § 79 of Halmos [1958, *9*].

The components of $\mathbf{C}[\cdot]$ are defined by

$$C_{ijkl} = \boldsymbol{e}_i \cdot \mathbf{C}[\boldsymbol{e}_k \otimes \boldsymbol{e}_l]\, \boldsymbol{e}_j = (\boldsymbol{e}_i \otimes \boldsymbol{e}_j) \cdot \mathbf{C}[\boldsymbol{e}_k \otimes \boldsymbol{e}_l],$$

so that the relation $\boldsymbol{S} = \mathbf{C}[\boldsymbol{E}]$ becomes

$$S_{ij} = C_{ijkl} E_{kl}.$$

If C'_{ijkl} are the components of $\mathbf{C}[\cdot]$ with respect to a second orthonormal basis $\{\boldsymbol{e}'_1, \boldsymbol{e}'_2, \boldsymbol{e}'_3\}$ with $\boldsymbol{e}'_i = \boldsymbol{Q}\boldsymbol{e}_i$, then

$$C'_{ijkl} = Q_{mi} Q_{nj} Q_{pk} Q_{ql} C_{mnpq}.$$

The ***transpose*** $\mathbf{C}^T$ of a fourth-order tensor $\mathbf{C}$ is the unique fourth-order tensor with the following property:

$\boldsymbol{E} \cdot \mathbf{C}[\boldsymbol{F}] = \mathbf{C}^T[\boldsymbol{E}] \cdot \boldsymbol{F}$ for all (second-order) tensors $\boldsymbol{E}$ and $\boldsymbol{F}$.

In components $C^T_{ijkl} = C_{klij}$. The ***magnitude*** $|\mathbf{C}|$ of $\mathbf{C}$ is defined by

$$|\mathbf{C}| = \sup\{|\mathbf{C}[\boldsymbol{E}]| : |\boldsymbol{E}| = 1\}.$$

Clearly,

$$|\mathbf{C}[\boldsymbol{E}]| \leqq |\mathbf{C}|\, |\boldsymbol{E}|.$$

We will occasionally have to deal separately with the components u_1, u_2 in the x_1, x_2-plane of a vector $\boldsymbol{u}$. For this reason, Greek subscripts will always range over the integers 1, 2 and summation from 1 to 2 over repeated Greek subscripts is implied, i.e.,

$$u_\alpha v_\alpha = \sum_{\alpha=1}^{2} u_\alpha v_\alpha.$$

The *two-dimensional alternator* will be designated by $\varepsilon_{\alpha\beta}$; thus

$$\varepsilon_{12} = -\varepsilon_{21} = 1, \quad \varepsilon_{11} = \varepsilon_{22} = 0.$$

4. Scalar fields. Vector fields. Tensor fields. Let D be an open set in $\mathscr{E}$. By a ***scalar, vector***, or ***tensor field*** on D we mean a function φ that assigns to each point $\boldsymbol{x} \in D$, respectively, a scalar, vector, or tensor $\varphi(\boldsymbol{x})$. We say that a *scalar* field φ is ***differentiable*** at $\boldsymbol{x}$ if there exists a vector $\boldsymbol{w}$ with the following property:

$$\varphi(\boldsymbol{y}) - \varphi(\boldsymbol{x}) = \boldsymbol{w} \cdot (\boldsymbol{y} - \boldsymbol{x}) + o(|\boldsymbol{y} - \boldsymbol{x}|) \quad \text{as} \quad \boldsymbol{y} \to \boldsymbol{x}.$$

If such a vector exists, it is unique; we write $\boldsymbol{w} = \nabla\varphi(\boldsymbol{x})$ and call $\nabla\varphi(\boldsymbol{x})$ the ***gradient*** of φ at $\boldsymbol{x}$. The partial derivatives of φ at $\boldsymbol{x}$, written $\varphi_{,i}(\boldsymbol{x})$, are defined by

$$\varphi_{,i}(\boldsymbol{x}) = \frac{\partial\varphi(\boldsymbol{x})}{\partial x_i} = \nabla\varphi(\boldsymbol{x}) \cdot \boldsymbol{e}_i = [\nabla\varphi(\boldsymbol{x})]_i.$$

More generally, let $\mathscr{W}$ be a finite-dimensional inner product space, and let Ψ be a mapping from D into $\mathscr{W}$. For example, $\mathscr{W}$ may be the translation space $\mathscr{V}$ corresponding to $\mathscr{E}$ or the space of all tensors. Then Ψ is differentiable at $\boldsymbol{x} \in D$ if there exists a linear function $\mathscr{L}[\cdot]$ from $\mathscr{V}$ into $\mathscr{W}$ such that

$$\Psi(\boldsymbol{y}) - \Psi(\boldsymbol{x}) = \mathscr{L}[\boldsymbol{y} - \boldsymbol{x}] + o(|\boldsymbol{y} - \boldsymbol{x}|) \quad \text{as} \quad \boldsymbol{y} \to \boldsymbol{x}.$$

If $\mathscr{L}$ exists, it is unique; we write $\mathscr{L} = \nabla\Psi(\boldsymbol{x})$ and call $\nabla\Psi(\boldsymbol{x})$ the ***gradient*** of Ψ at $\boldsymbol{x}$. It is not difficult to show that the linear transformation $\nabla\Psi(\boldsymbol{x})$ can be computed using the formula

$$\nabla\Psi(\boldsymbol{x})[\boldsymbol{v}] = \frac{d}{d\alpha} \Psi(\boldsymbol{x} + \alpha\boldsymbol{v})\Big|_{\alpha=0},$$

which holds for every vector $\boldsymbol{v} \in \mathscr{V}$.

We say that Ψ is of ***class*** C^1 on D, or ***smooth*** on D, if Ψ is differentiable at every point of D and $\nabla\Psi$ is continuous on D. The gradient of $\nabla\Psi$ is denoted by $\nabla^{(2)}\Psi$. Continuing in this manner, we say that Ψ is of ***class*** C^N on D if it is class C^{N-1} and its $(N-1)$th gradient $\nabla^{(N-1)}\Psi$ is smooth. Clearly, the gradient of a smooth scalar field is a vector field, while the gradient of a smooth vector field is a tensor field. We say that Ψ is of ***class*** C^N on $\bar{D}$ if Ψ is of class C^N on D and for each $n\in\{0, 1, \ldots, N\}$ $\nabla^{(n)}\Psi$ has a continuous extension to $\bar{D}$. (In this case we also write $\nabla^{(n)}\Psi$ for the extended function.)

A field Ψ is ***analytic*** on D if given any $\boldsymbol{x}$ in D, Ψ can be represented by a power series in some neighborhood of $\boldsymbol{x}$. Of course, if Ψ is analytic, then Ψ is of class C^∞.

Let $\boldsymbol{u}$ be a vector field on D, and suppose that $\boldsymbol{u}$ is differentiable at a point $\boldsymbol{x}\in D$. Then the ***divergence*** of $\boldsymbol{u}$ at $\boldsymbol{x}$ is the scalar

$$\operatorname{div}\boldsymbol{u}(\boldsymbol{x}) = \operatorname{tr}\nabla\boldsymbol{u}(\boldsymbol{x}),$$

and the partial derivatives of $\boldsymbol{u}$ are defined by

$$u_{i,j}(\boldsymbol{x}) = \frac{\partial u_i(\boldsymbol{x})}{\partial x_j} = \boldsymbol{e}_i\cdot\nabla\boldsymbol{u}(\boldsymbol{x})\,\boldsymbol{e}_j = [\nabla\boldsymbol{u}(\boldsymbol{x})]_{ij};$$

hence

$$\operatorname{div}\boldsymbol{u}(\boldsymbol{x}) = u_{i,i}(\boldsymbol{x}).$$

The ***curl*** of $\boldsymbol{u}$ at $\boldsymbol{x}$, denoted $\operatorname{curl}\boldsymbol{u}(\boldsymbol{x})$, is defined to be twice the axial vector corresponding to the skew part of $\nabla\boldsymbol{u}(\boldsymbol{x})$; i.e. $\operatorname{curl}\boldsymbol{u}(\boldsymbol{x})$ is the unique vector with the property:

$$[\nabla\boldsymbol{u}(\boldsymbol{x}) - \nabla\boldsymbol{u}(\boldsymbol{x})^T]\,\boldsymbol{a} = (\operatorname{curl}\boldsymbol{u}(\boldsymbol{x}))\times\boldsymbol{a}$$

for every vector $\boldsymbol{a}$. In components

$$[\operatorname{curl}\boldsymbol{u}(\boldsymbol{x})]_i = \varepsilon_{ijk}u_{k,j}(\boldsymbol{x}).$$

We write $\hat{\nabla}\boldsymbol{u}(\boldsymbol{x})$ for the ***symmetric gradient*** of $\boldsymbol{u}$:

$$\hat{\nabla}\boldsymbol{u}(\boldsymbol{x}) = \operatorname{sym}\nabla\boldsymbol{u}(\boldsymbol{x}) = \tfrac{1}{2}\{\nabla\boldsymbol{u}(\boldsymbol{x}) + \nabla\boldsymbol{u}(\boldsymbol{x})^T\}.$$

Let $\boldsymbol{S}$ be a tensor field on D, and suppose that $\boldsymbol{S}$ is differentiable at $\boldsymbol{x}$. Then the tensor field $\boldsymbol{S}^T$ is also differentiable at $\boldsymbol{x}$; the ***divergence*** of $\boldsymbol{S}$ at $\boldsymbol{x}$, written $\operatorname{div}\boldsymbol{S}(\boldsymbol{x})$, is the unique vector with the property:

$$[\operatorname{div}\boldsymbol{S}(\boldsymbol{x})]\cdot\boldsymbol{a} = \operatorname{div}[\boldsymbol{S}^T(\boldsymbol{x})\,\boldsymbol{a}]$$

for every fixed vector $\boldsymbol{a}$. In the same manner, we define the ***curl*** of $\boldsymbol{S}$ at $\boldsymbol{x}$, written $\operatorname{curl}\boldsymbol{S}(\boldsymbol{x})$, to be the unique tensor with the property:

$$[\operatorname{curl}\boldsymbol{S}(\boldsymbol{x})]\,\boldsymbol{a} = \operatorname{curl}[\boldsymbol{S}^T(\boldsymbol{x})\,\boldsymbol{a}]$$

for every $\boldsymbol{a}$. The partial derivatives of $\boldsymbol{S}$ are given by

$$S_{ij,k}(\boldsymbol{x}) = \frac{\partial S_{ij}(\boldsymbol{x})}{\partial x_k} = \boldsymbol{e}_i\cdot[\nabla\boldsymbol{S}(\boldsymbol{x})\,\boldsymbol{e}_k]\,\boldsymbol{e}_j;$$

thus

$$[\operatorname{div}\boldsymbol{S}(\boldsymbol{x})]_i = S_{ij,j}(\boldsymbol{x}),$$
$$[\operatorname{curl}\boldsymbol{S}(\boldsymbol{x})]_{ij} = \varepsilon_{ipk}S_{jk,p}(\boldsymbol{x}).$$

Let φ be a differentiable scalar field, and suppose that $\nabla\varphi$ is differentiable at $\boldsymbol{x}$. Then we define the ***Laplacian*** of φ at $\boldsymbol{x}$ by

$$\Delta\varphi(\boldsymbol{x}) = \operatorname{div}\nabla\varphi(\boldsymbol{x}).$$

We define the ***Laplacian*** of a vector field $\boldsymbol{u}$ with analogous properties in the same manner:

$$\Delta \boldsymbol{u}(\boldsymbol{x}) = \operatorname{div} \nabla \boldsymbol{u}(\boldsymbol{x}).$$

Clearly,

$$\Delta \varphi(\boldsymbol{x}) = \varphi_{,ii}(\boldsymbol{x}),$$
$$[\Delta \boldsymbol{u}(\boldsymbol{x})]_i = u_{i,jj}(\boldsymbol{x}).$$

Finally, the ***Laplacian*** $\Delta \boldsymbol{S}(\boldsymbol{x})$ of a sufficiently smooth tensor field $\boldsymbol{S}$ is the unique tensor with the property:

$$[\Delta \boldsymbol{S}(\boldsymbol{x})]\boldsymbol{a} = \Delta[\boldsymbol{S}(\boldsymbol{x})\boldsymbol{a}];$$

in components

$$[\Delta \boldsymbol{S}(\boldsymbol{x})]_{ij} = S_{ij,kk}(\boldsymbol{x}).$$

Of future use are the following

(1) Identities. *Let φ be a scalar field, $\boldsymbol{u}$ a vector field, and $\boldsymbol{S}$ a tensor field, all of class C^2 on D. Then*

$$\operatorname{curl} \nabla\varphi = \boldsymbol{0}, \tag{1}$$

$$\operatorname{div} \operatorname{curl} \boldsymbol{u} = 0, \tag{2}$$

$$\operatorname{curl} \operatorname{curl} \boldsymbol{u} = \nabla \operatorname{div} \boldsymbol{u} - \Delta \boldsymbol{u}, \tag{3}$$

$$\operatorname{curl} \nabla \boldsymbol{u} = \boldsymbol{0}, \tag{4}$$

$$\operatorname{curl}(\nabla \boldsymbol{u}^T) = \nabla \operatorname{curl} \boldsymbol{u}, \tag{5}$$

$$\nabla \boldsymbol{u} = -\nabla \boldsymbol{u}^T \Rightarrow \nabla\nabla \boldsymbol{u} = \boldsymbol{0}, \tag{6}$$

$$\operatorname{div} \operatorname{curl} \boldsymbol{S} = \operatorname{curl} \operatorname{div} \boldsymbol{S}^T, \tag{7}$$

$$\operatorname{div}(\operatorname{curl} \boldsymbol{S})^T = \boldsymbol{0}, \tag{8}$$

$$(\operatorname{curl} \operatorname{curl} \boldsymbol{S})^T = \operatorname{curl} \operatorname{curl} \boldsymbol{S}^T, \tag{9}$$

$$\operatorname{curl}(\varphi \boldsymbol{1}) = -[\operatorname{curl}(\varphi \boldsymbol{1})]^T, \tag{10}$$

$$\operatorname{div}(\boldsymbol{S}^T \boldsymbol{u}) = \boldsymbol{u} \cdot \operatorname{div} \boldsymbol{S} + \boldsymbol{S} \cdot \nabla \boldsymbol{u}. \tag{11}$$

If $\boldsymbol{S}$ is symmetric,

$$\operatorname{tr}(\operatorname{curl} \boldsymbol{S}) = 0, \tag{12}$$

$$\operatorname{curl} \operatorname{curl} \boldsymbol{S} = -\Delta \boldsymbol{S} + 2\hat{\nabla} \operatorname{div} \boldsymbol{S} - \nabla\nabla(\operatorname{tr} \boldsymbol{S}) + \boldsymbol{1}[\Delta(\operatorname{tr})\boldsymbol{S} - \operatorname{div} \operatorname{div} \boldsymbol{S}]. \tag{13}$$

If $\boldsymbol{S}$ is symmetric and $\boldsymbol{S} = \boldsymbol{G} - \boldsymbol{1} \operatorname{tr} \boldsymbol{G}$, then

$$\operatorname{curl} \operatorname{curl} \boldsymbol{S} = -\Delta \boldsymbol{G} + 2\hat{\nabla} \operatorname{div} \boldsymbol{G} - \boldsymbol{1} \operatorname{div} \operatorname{div} \boldsymbol{G}. \tag{14}$$

If $\boldsymbol{S}$ is skew and $\boldsymbol{u}$ is its axial vector, then

$$\operatorname{curl} \boldsymbol{S} = \boldsymbol{1} \operatorname{div} \boldsymbol{u} - \nabla \boldsymbol{u}. \tag{15}$$

II. Elements of potential theory.

5. The body B. The subsurfaces $\mathscr{S}_1$ and $\mathscr{S}_2$ of ∂B. Given a region D in $\mathscr{E}$, we write $\overline{D}$ for its closure, $\mathring{D}$ for its interior, and ∂D for its boundary. We will consistently write $\Sigma_\varrho(\boldsymbol{y})$ for the ***open ball*** in $\mathscr{E}$ of radius ϱ and center at $\boldsymbol{y}$.

Let D be an open region in $\mathscr{E}$ or in the x_1, x_2-plane. We say that D is ***simply-connected*** if every closed curve in D can be continuously deformed to a point without leaving D (i.e., if given any continuous function $\boldsymbol{g}: [0,1] \to D$ satisfying $\boldsymbol{g}(0) = \boldsymbol{g}(1)$ there exists a continuous function $\boldsymbol{f}$: $[0,1] \times [0,1] \to D$ and a point $\boldsymbol{y} \in D$ such that $\boldsymbol{f}(s, 0) = \boldsymbol{g}(s)$ and $\boldsymbol{f}(s, 1) = \boldsymbol{y}$ for every $s \in [0, 1]$).

The following definitions will enable us to define concisely the notion of a regular region of space.[1] A *regular arc* is a set $\mathscr{a}$ in $\mathscr{E}$ which, for some orientation of the cartesian coordinate system, admits a representation

$$\mathscr{a} = \{\boldsymbol{x}: x_2 = f(x_1),\ x_3 = g(x_1),\ a \leqq x_1 \leqq b\},$$

where $a < b$ and f and g are smooth on $[a, b]$. A *regular curve* is a set $\mathscr{c}$ in $\mathscr{E}$ consisting of a finite number of regular arcs arranged in order, and such that the terminal point of each arc (other than the last) is the initial point of the following arc. The arcs have no other points in common, except that the terminal point of the last arc may be the initial point of the first arc, in which case $\mathscr{c}$ is a *closed regular curve*. A *regular surface element* is a (non-empty) set $\mathscr{R} \subset \mathscr{E}$ which, for some orientation of the cartesian coordinate system, admits a representation

$$\mathscr{R} = \{\boldsymbol{x}: x_3 = f(x_1, x_2),\ (x_1, x_2) \in R\},$$

where R is a compact connected region in the x_1, x_2-plane, ∂R is a closed regular curve, and f is smooth on R. The union of a finite number of regular surface elements is called a ***regular surface*** provided:

(i) the intersection of any two of the elements is either empty, a single point which is a vertex for both, or a single regular arc which is an edge for both;

(ii) the intersection of any collection of three or more elements consists at most of vertices;

(iii) any two of the elements are the first and last of a chain such that each has an edge in common with the next;

(iv) all elements having a vertex in common form a chain such that each has an edge, terminating in that vertex, in common with the next; the last may, or may not, have an edge in common with the first.

The term *edge* here refers to one of the (finite number of) regular arcs comprising the boundary of a regular surface element, while a *vertex* is a point at which two edges meet. If all the edges of a regular surface belong each to two of its surface elements, the surface is a ***closed regular surface.*** Note that a regular surface (and hence a closed regular surface) is necessarily both *connected* and *bounded*. Thus, e.g., the boundary of the spherical shell

$$\{\boldsymbol{x}: \alpha \leqq |\boldsymbol{x} - \boldsymbol{x}_0| \leqq \beta\} \qquad (\alpha < \beta)$$

is not a regular surface, but rather the union of two closed regular surfaces.

Let B be an open set in $\mathscr{E}$. We say that B is a ***bounded regular region*** if B is the interior of a closed and bounded region $\overline{B}$ in $\mathscr{E}$ whose boundary ∂B is the union of a finite number of non-intersecting closed regular surfaces. Note that the boundary of a bounded regular region may have corners and edges. We will occasionally deal with infinite domains. We say that B is an ***unbounded regular region*** if B is unbounded, but $B \cap \Sigma_\varrho(\boldsymbol{0})$ is a bounded regular region

[1] All of the definitions contained in this paragraph are from Kellogg [1929, *1*], Chap. IV, § 8.

for all sufficiently large ϱ. Note that the boundary of such a region may extend to infinity. An unbounded regular region whose boundary is finite is called an ***exterior domain***.

Throughout this article, unless specified to the contrary, B ***will denote a bounded regular region***. We will refer to B as the **body**. A (not necessarily proper) bounded regular subregion P of B will be referred to as a ***part*** of B.

We shall designate the ***outward unit normal*** to ∂B at $\boldsymbol{x} \in \partial B$ by $\boldsymbol{n}(\boldsymbol{x})$. If the function $\boldsymbol{n}$ is continuous at $\boldsymbol{x}$, then $\boldsymbol{x}$ will be referred to as a ***regular point*** of ∂B. We say that $\mathscr{S} \subset \partial B$ is a ***regular subsurface*** of ∂B if the intersection of $\mathscr{S}$ with each of the closed regular surfaces comprising ∂B is itself a regular surface. The symbols $\mathscr{S}_1$ and $\mathscr{S}_2$ will always denote ***complementary regular subsurfaces*** of ∂B; i.e. regular subsurfaces of ∂B with the following property:

$$\partial B = \mathscr{S}_1 \cup \mathscr{S}_2, \qquad \mathring{\mathscr{S}}_1 \cap \mathring{\mathscr{S}}_2 = \emptyset,$$

where $\mathring{\mathscr{S}}_\alpha$ is the (relative) interior of $\mathscr{S}_\alpha$. Note that since $\mathscr{S}_\alpha$ is regular, $\mathscr{S}_\alpha = \overline{\mathscr{S}}_\alpha$, where $\overline{\mathscr{S}}_\alpha$ is the (relative) closure of $\mathscr{S}_\alpha$.

Let B be bounded or unbounded. A partition for B is a finite collection $B_1, B_2, \ldots, B_N$ of mutually disjoint regular subregions of B such that

$$\overline{B} = \bigcup_{n=1}^{N} \overline{B}_n.$$

A function Ψ is ***piecewise continuous*** on $\overline{B}$ if there exists a partition $B_1, B_2, \ldots, B_N$ of B such that for each B_n the restriction of Ψ to B_n is bounded on $\overline{B}_n$ and continuous on $\overline{B}_n - D$, where D is a (possibly empty) *finite* subset of B_n. Ψ is ***piecewise smooth*** on $\overline{B}$ if $\nabla\Psi$ exists almost everywhere on B and is piecewise continuous on $\overline{B}$.

Piecewise continuity on regular surfaces is defined in a strictly analogous manner. We say that a function Ψ on $\mathscr{S}_\alpha$ is ***piecewise regular*** if Ψ is piecewise continuous on $\mathscr{S}_\alpha$ and each regular point of $\mathscr{S}_\alpha$ is a point of continuity of Ψ. Suppose that Ψ and $\hat{\Psi}$ are piecewise regular on $\mathscr{S}_\alpha$. Then we write $\Psi = \hat{\Psi}$ on $\mathscr{S}_\alpha$ if $\Psi(\boldsymbol{x}) = \hat{\Psi}(\boldsymbol{x})$ at each regular point $\boldsymbol{x} \in \mathscr{S}_\alpha$.

We say that B is ***properly regular***[1] provided the following hypotheses are satisfied:

(i) B is a bounded regular region;

(ii) for each $\boldsymbol{x}_0 \in \partial B$ there exists a neighborhood N of $\boldsymbol{x}_0$ and a piecewise smooth homeomorphism $\boldsymbol{h}$ of $\overline{N} \cap \overline{B}$ onto the solid hemisphere

$$\{\boldsymbol{x}\colon x_3 \geqq 0,\ x_1^2 + x_2^2 + x_3^2 \leqq 1\}$$

such that $\boldsymbol{h}$ maps $\overline{N} \cap \partial B$ onto the set $\{\boldsymbol{x}\colon x_3 = 0,\ x_1^2 + x_2^2 \leqq 1\}$;

(iii) $\det \nabla \boldsymbol{h}$ is positive and bounded away from zero.

We say that B is ***star-shaped*** if there exists a point $\boldsymbol{x}_0 \in B$ such that the line segment connecting $\boldsymbol{x}_0$ and any point $\boldsymbol{x} \in \partial B$ intersects ∂B only at $\boldsymbol{x}$. We state without proof the following property of star-shaped regions:

(1) *Let B be star-shaped. Then there exists a point $\boldsymbol{x}_0 \in B$ such that*

$$(\boldsymbol{x} - \boldsymbol{x}_0) \cdot \boldsymbol{n}(\boldsymbol{x}) > 0$$

at every regular point $\boldsymbol{x} \in \partial B$.

[1] This definition is due to FICHERA [1971, *1*], § 2.

Let (x_1, x_2, x_3) denote cartesian coordinates. We say that B is x_i-***convex*** (i fixed) if every line segment parallel to the x_i-axis lies entirely in B whenever its end points belong to B.

Consider the following projections of B:

$$B_{12}=\{(x_1, x_2): \boldsymbol{x}\in B\},$$
$$B_3=\{x_3: \boldsymbol{x}\in B\},$$

and for each $x_3\in B_3$, let

$$B_{12}(x_3)=\{(x_1, x_2): (x_1, x_2, x_3)\in B\}.$$

Note that

$$B_{12}=\bigcup_{x_3\in B_3} B_{12}(x_3).$$

(2) *Lemma*. *Let $M\geqq 1$ and $N\geqq 1$ be fixed integers, and consider the following hypotheses on B:*

(i) *B is x_3-convex.*

(ii) *$B_{12}(x_3)$ is simply-connected for each $x_3\in B_3$.*

(iii) *There exist scalar fields $\hat{x}_1$ and $\hat{x}_2$ of class C^M on $\overline{B}_3$ and C^N on B_3 such that the curve*

$$\mathscr{C}=\{(x_1, x_2, x_3): x_1=\hat{x}_1(x_3),\ x_2=\hat{x}_2(x_3),\ x_3\in B_3\}$$

is contained in B.

(iv) *There exists a scalar field $\hat{x}_3$ of class C^M on $\overline{B}_{12}$ and C^N on B_{12} such that the surface*

$$\mathscr{S}=\{(x_1, x_2, x_3): x_3=\hat{x}_3(x_1, x_2),\ (x_1, x_2)\in B_{12}\}$$

is contained in B.

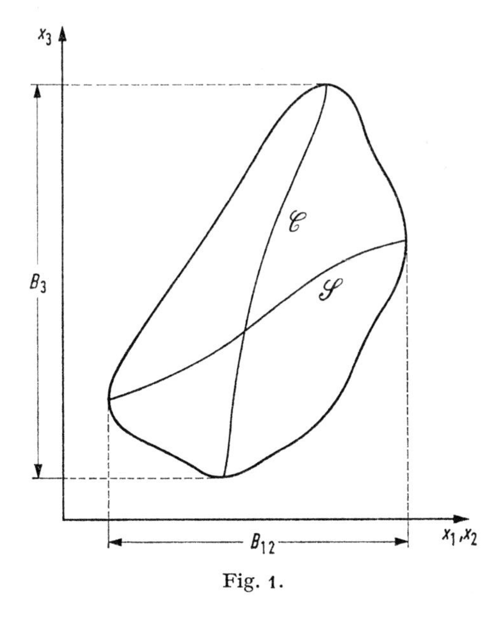

Fig. 1.

Let $\mathscr{F}(M, N)$ denote the set of all scalar fields of class C^M on $\overline{B}$ and class C^N on B. Then the following two assertions are true:

(a) *If* (i) *and* (iv) *hold, and if* $\lambda \in \mathscr{F}(M, N)$, *then there exists a field* $\omega \in \mathscr{F}(M, N)$ *such that*

$$\omega_{,3} = \lambda.$$

(b) *If* (ii) *and* (iii) *hold, and if* $v_\alpha \in \mathscr{F}(M, N)$ *and satisfies*

$$v_{1,2} - v_{2,1} = 0,$$

then there exists a field $\varphi \in \mathscr{F}(M, N)$ *such that*

$$v_\alpha = \varphi_{,\alpha}.$$

Proof. In view of (i)–(iv), the functions

$$\varphi(x_1, x_2, x_3) = \int_{\Gamma(x_1, x_2, x_3)} v_\alpha(x_1', x_2', x_3)\, dx_\alpha',$$

$$\omega(x_1, x_2, x_3) = \int_{\hat{x}_3(x_1, x_2)}^{x_3} \lambda(x_1, x_2, x_3')\, dx_3',$$

where $\Gamma(x_1, x_2, x_3)$ is any smooth path in $B_{12}(x_3)$ connecting the points $(\hat{x}_1(x_3), \hat{x}_2(x_3))$ and (x_1, x_2), have the desired properties. □

Let $\mathscr{S} \subset \partial B$. We say that B is ***convex with respect to*** $\mathscr{S}$ if the line segment connecting any pair of points $\hat{\boldsymbol{x}}$ and $\tilde{\boldsymbol{x}}$ of $\mathscr{S}$ intersects ∂B only at $\hat{\boldsymbol{x}}$ and $\tilde{\boldsymbol{x}}$. Note that this implies, in particular, that $\mathscr{S}$ is convex, though B need not be.

6. The divergence theorem. Stokes' theorem.

(1) Divergence theorem. *Let B be a bounded or unbounded regular region. Let φ be a scalar field, $\boldsymbol{u}$ a vector field, and $\boldsymbol{T}$ a tensor field, and let φ, $\boldsymbol{u}$ and $\boldsymbol{T}$ be continuous on $\overline{B}$, differentiable almost everywhere on B, and of bounded support. Then*

$$\int_{\partial B} \varphi \boldsymbol{n}\, da = \int_B \nabla \varphi\, dv,$$

$$\int_{\partial B} \boldsymbol{u} \otimes \boldsymbol{n}\, da = \int_B \nabla \boldsymbol{u}\, dv,$$

$$\int_{\partial B} \boldsymbol{u} \cdot \boldsymbol{n}\, da = \int_B \operatorname{div} \boldsymbol{u}\, dv,$$

$$\int_{\partial B} \boldsymbol{n} \times \boldsymbol{u}\, da = \int_B \operatorname{curl} \boldsymbol{u}\, dv,$$

$$\int_{\partial B} \boldsymbol{T}\boldsymbol{n}\, da = \int_B \operatorname{div} \boldsymbol{T}\, dv,$$

whenever the integrand on the right is piecewise continuous on $\overline{B}$.

Proof. The results concerning φ and $\boldsymbol{u}$ are classical and will not be verified here.[1] To establish the last relation, let $\boldsymbol{b}$ be a vector. Then

$$\boldsymbol{b} \cdot \int_{\partial B} \boldsymbol{T}\boldsymbol{n}\, da = \int_{\partial B} \boldsymbol{b} \cdot \boldsymbol{T}\boldsymbol{n}\, da = \int_{\partial B} (\boldsymbol{T}^T \boldsymbol{b}) \cdot \boldsymbol{n}\, da$$
$$= \int_B \operatorname{div} (\boldsymbol{T}^T \boldsymbol{b})\, dv = \int_B (\operatorname{div} \boldsymbol{T}) \cdot \boldsymbol{b}\, dv = \boldsymbol{b} \cdot \int_B \operatorname{div} \boldsymbol{T}\, dv,$$

which implies the desired result since $\boldsymbol{b}$ is arbitrary. □

[1] See, e.g., KELLOGG [1929, *1*].

(2) Stokes' theorem. *Let $\boldsymbol{u}$ be a smooth vector field and $\boldsymbol{T}$ a smooth tensor field on $\overline{B}$. Then given any closed regular surface $\mathscr{S}$ in $\overline{B}$,*

$$\int_{\mathscr{S}} (\operatorname{curl} \boldsymbol{u}) \cdot \boldsymbol{n}\, da = 0,$$

$$\int_{\mathscr{S}} (\operatorname{curl} \boldsymbol{T})^T \boldsymbol{n}\, da = \boldsymbol{0}.$$

Proof. The relation concerning $\boldsymbol{u}$ is well known.[1] To establish the second result, let $\boldsymbol{b}$ be a vector, and let $\boldsymbol{u} = \boldsymbol{T}^T \boldsymbol{b}$. Then

$$\boldsymbol{b} \cdot \int_{\mathscr{S}} (\operatorname{curl} \boldsymbol{T})^T \boldsymbol{n}\, da = \int_{\mathscr{S}} \boldsymbol{b} \cdot (\operatorname{curl} \boldsymbol{T})^T \boldsymbol{n}\, da = \int_{\mathscr{S}} \boldsymbol{n} \cdot (\operatorname{curl} \boldsymbol{T})\, \boldsymbol{b}\, da$$
$$= \int_{\mathscr{S}} \boldsymbol{n} \cdot \operatorname{curl} (\boldsymbol{T}^T \boldsymbol{b})\, da = \int_{\mathscr{S}} (\operatorname{curl} \boldsymbol{u}) \cdot \boldsymbol{n}\, da = 0,$$

which implies the desired result, since $\boldsymbol{b}$ is arbitrary. □

(3) Theorem on irrotational fields. *Assume that B is simply-connected.*

(i) *Let $\boldsymbol{u}$ be a class C^N $(N \geqq 1)$ vector field on B that satisfies*

$$\operatorname{curl} \boldsymbol{u} = \boldsymbol{0}.$$

Then there exists a scalar field φ of class C^{N+1} on B such that

$$\boldsymbol{u} = \nabla \varphi.$$

(ii) *Let $\boldsymbol{T}$ be a class C^N tensor field on B that satisfies*

$$\operatorname{curl} \boldsymbol{T} = \boldsymbol{0}.$$

Then there exists a vector field $\boldsymbol{u}$ of class C^{N+1} on B such that

$$\boldsymbol{T} = \nabla \boldsymbol{u}.$$

(iii) *Let $\boldsymbol{T}$ obey the hypotheses of* (ii), *and, in addition, assume that*

$$\operatorname{tr} \boldsymbol{T} = 0.$$

Then there exists a skew tensor field $\boldsymbol{W}$ of class C^{N+1} on B such that

$$\boldsymbol{T} = \operatorname{curl} \boldsymbol{W}.$$

Proof. Part (i) is well known.[2] Indeed, the function φ defined by

$$\varphi(\boldsymbol{x}) = \int_{\boldsymbol{x}_0}^{\boldsymbol{x}} \boldsymbol{u}(\boldsymbol{y}) \cdot d\boldsymbol{y}$$

has the desired properties. Here $\boldsymbol{x}_0$ is a fixed point of B, and the integral is taken along any curve in B connecting $\boldsymbol{x}_0$ and $\boldsymbol{x}$.

To establish (ii) let

$$\boldsymbol{t}_i = \boldsymbol{T}^T \boldsymbol{e}_i. \tag{a}$$

Then

$$\operatorname{curl} \boldsymbol{t}_i = \operatorname{curl} (\boldsymbol{T}^T \boldsymbol{e}_i) = (\operatorname{curl} \boldsymbol{T})\, \boldsymbol{e}_i = \boldsymbol{0},$$

and by (i) there exist scalar fields u_i such that

$$\boldsymbol{t}_i = \nabla u_i. \tag{b}$$

[1] See, e.g., Courant [1957, *4*], Chap. 5.
[2] See, e.g., Phillips [1933, *3*], § 45.

Let $\boldsymbol{u} = u_i \boldsymbol{e}_i$. Then, since $u_i = \boldsymbol{u} \cdot \boldsymbol{e}_i$, it follows that

$$\nabla u_i = (\nabla \boldsymbol{u})^T \boldsymbol{e}_i, \tag{c}$$

and (a)–(c) imply

$$\boldsymbol{T} = \nabla \boldsymbol{u}.$$

To prove (iii) let $\boldsymbol{u}$ be the vector field established in (ii), and let $\boldsymbol{W}$ be the *negative* of the skew tensor field corresponding to $\boldsymbol{u}$. Then

$$\operatorname{div} \boldsymbol{u} = 0,$$

since $\operatorname{tr} \boldsymbol{T} = 0$, and we conclude from (15) of ***(4.1)*** that

$$\operatorname{curl} \boldsymbol{W} = \nabla \boldsymbol{u}. \quad \square$$

(4) Theorem on solenoidal fields. *Assume that ∂B is of class C^3. Let $\mathscr{F}$ denote the set of all fields of class C^2 on $\bar{B}$ and class C^3 on B.*

(i)[1] *Let $\boldsymbol{u} \in \mathscr{F}$ be a vector field, and suppose that*

$$\int_{\mathscr{S}} \boldsymbol{u} \cdot \boldsymbol{n} \, da = 0$$

for every closed regular surface $\mathscr{S} \subset B$. Then there exists a vector field $\boldsymbol{w} \in \mathscr{F}$ such that

$$\boldsymbol{u} = \operatorname{curl} \boldsymbol{w}.$$

(ii) *Let $\boldsymbol{T} \in \mathscr{F}$ be a tensor field, and suppose that*

$$\int_{\mathscr{S}} \boldsymbol{T}^T \boldsymbol{n} \, da = \boldsymbol{0}$$

for every closed regular surface $\mathscr{S} \subset B$. Then there exists a tensor field $\boldsymbol{W} \in \mathscr{F}$ such that

$$\boldsymbol{T} = \operatorname{curl} \boldsymbol{W}.$$

Proof. We omit the proof of (i).[2] To establish (ii) let

$$\boldsymbol{t}_i = \boldsymbol{T} \boldsymbol{e}_i.$$

Then

$$\int_{\mathscr{S}} \boldsymbol{t}_i \cdot \boldsymbol{n} \, da = \boldsymbol{e}_i \cdot \int_{\mathscr{S}} \boldsymbol{T}^T \boldsymbol{n} \, da = 0,$$

and we conclude from (i) that there exist vector fields $\boldsymbol{w}_i$ such that

$$\boldsymbol{t}_i = \operatorname{curl} \boldsymbol{w}_i.$$

Let

$$\boldsymbol{W} = \boldsymbol{e}_i \otimes \boldsymbol{w}_i$$

and note that for any vector $\boldsymbol{a}$,

$$\boldsymbol{W}^T \boldsymbol{a} = a_i \boldsymbol{w}_i.$$

Thus

$$\boldsymbol{T} \boldsymbol{a} = a_i \boldsymbol{T} \boldsymbol{e}_i = \operatorname{curl} (a_i \boldsymbol{w}_i) = \operatorname{curl} (\boldsymbol{W}^T \boldsymbol{a}) = (\operatorname{curl} \boldsymbol{W}) \, \boldsymbol{a},$$

---

[1] The usual statement of this theorem (see, e.g., PHILLIPS [1933, *3*], § 49, COURANT [1957, *4*], pp. 404–406), which asserts that $\operatorname{div} \boldsymbol{u} = 0$ implies $\boldsymbol{u} = \operatorname{curl} \boldsymbol{w}$, is correct only when ∂B consists of a *single* closed surface. In this instance $\operatorname{div} \boldsymbol{u} = 0$ is equivalent to the assertion that

$$\int_{\mathscr{S}} \boldsymbol{u} \cdot \boldsymbol{n} \, da = 0 \tag{*}$$

for every closed regular surface $\mathscr{S} \subset B$. However, if ∂B consists of two or more closed surfaces (e.g. when B is a spherical shell) there exist divergence-free vector fields in $\mathscr{F}$ that do not satisfy (*). Note that by Stokes' theorem, (*) holds whenever $\boldsymbol{u}$ is the curl of a vector field.

[2] See LICHTENSTEIN [1929, *2*], pp. 101–106.

and hence

$$\boldsymbol{T} = \operatorname{curl} \boldsymbol{W}. \quad \square$$

(5) Properties of the Newtonian potential.[1] *Let φ be a scalar field that is continuous on $\bar{B}$ and of class C^N $(N \geqq 1)$ on B, and let*

$$\psi(\boldsymbol{x}) = -\frac{1}{4\pi} \int\limits_B \frac{\varphi(\boldsymbol{y})}{|\boldsymbol{x}-\boldsymbol{y}|} \, dv_{\boldsymbol{y}}$$

for every $\boldsymbol{x} \in B$. Then ψ is of class C^{N+1} on B and

$$\Delta \psi = \varphi.$$

Further, an analogous result holds for vector and tensor fields.

(6) Properties of the logarithmic potential.[2] *Let R be a measurable region in the x_1, x_2-plane. Let φ be a scalar field that is continuous on $\bar{R}$ and of class C^N $(N \geqq 1)$ on R, and let*

$$\psi(x_1, x_2) = -\frac{1}{2\pi} \int\limits_R \varphi(\xi_1, \xi_2) \log\left(\frac{1}{\sqrt{(x_\alpha - \xi_\alpha)(x_\alpha - \xi_\alpha)}}\right) d\xi_1 \, d\xi_2$$

for every $(x_1, x_2) \in R$. Then ψ is of class C^{N+1} on R and

$$\Delta \psi = \varphi.$$

(7) Helmholtz's theorem. *Let $\boldsymbol{u}$ be a vector field that is continuous on $\bar{B}$ and of class C^N $(N \geqq 1)$ on B. Then there exists a scalar field φ and a vector field $\boldsymbol{w}$, both of class C^N on B, such that*

$$\boldsymbol{u} = \nabla \varphi + \operatorname{curl} \boldsymbol{w},$$

$$\operatorname{div} \boldsymbol{w} = 0.$$

Proof. By ***(5)*** there exists a class C^{N+1} vector field $\boldsymbol{v}$ on B such that

$$\boldsymbol{u} = \Delta \boldsymbol{v}.$$

By (3) of ***(4.1)***,

$$\boldsymbol{u} = \nabla \operatorname{div} \boldsymbol{v} - \operatorname{curl} \operatorname{curl} \boldsymbol{v};$$

thus if we let

$$\varphi = \operatorname{div} \boldsymbol{v}, \qquad \boldsymbol{w} = -\operatorname{curl} \boldsymbol{v},$$

the desired conclusion follows. $\square$

7. The fundamental lemma. Rellich's lemma. For $\boldsymbol{x}_0$ a point of $\mathscr{E}$ and h a positive scalar, let $\Phi_h(\boldsymbol{x}_0)$ denote the set of all class C^∞ scalar fields φ on $\mathscr{E}$ with the following properties:

(a) $\varphi > 0$ on the open ball $\Sigma_h(\boldsymbol{x}_0)$ of radius h and center at $\boldsymbol{x}_0$;

(b) $\varphi = 0$ on $\mathscr{E} - \Sigma_h(\boldsymbol{x}_0)$.

For $h > 0$ the set $\Phi_h(\boldsymbol{x}_0)$ is not empty. Indeed, the function φ defined by

$$\varphi(\boldsymbol{x}) = \begin{cases} e^{\frac{-1}{h^2 - (\boldsymbol{x} - \boldsymbol{x}_0)^2}}, & \boldsymbol{x} \in \Sigma_h(\boldsymbol{x}_0) \\ 0, & \boldsymbol{x} \notin \Sigma_h(\boldsymbol{x}_0) \end{cases}$$

belongs to $\Phi_h(\boldsymbol{x}_0)$.

[1] See e.g., Kellogg [1929, *1*], p. 156; Courant [1962, *5*], p. 249.

[2] See, e.g., Kellogg [1929, *1*], p. 174; Courant [1962, *5*], p. 250.

Let $\mathcal{S}$ be a subset of ∂B. We say that a function f on $\bar{B}$ ***vanishes near*** $\mathcal{S}$ provided there exists a neighborhood N of $\mathcal{S}$ such that $f=0$ on $N \cap \bar{B}$.

We are now in a position to establish a version of the fundamental lemma of the calculus of variations that is sufficiently general for our use.

(1) Fundamental lemma. *Let $\mathcal{W}$ be a finite-dimensional inner product space. Let $\boldsymbol{w}: \bar{B} \to \mathcal{W}$ be continuous and satisfy*

$$\int_B \boldsymbol{w} \cdot \boldsymbol{v} \, dv = 0$$

for every class C^∞ function $\boldsymbol{v}: \bar{B} \to \mathcal{W}$ that vanishes near ∂B. Then

$$\boldsymbol{w} = \boldsymbol{0} \quad \text{on } \bar{B}.$$

Proof. Let $\mathbf{e}_1, \mathbf{e}_2, \ldots, \mathbf{e}_n$ be an orthonormal basis in $\mathcal{W}$, and let

$$\boldsymbol{w} = \sum_{i=1}^{n} w_i \, \mathbf{e}_i.$$

Assume that for some $\boldsymbol{x}_0 \in B$ and some integer k, $w_k(\boldsymbol{x}_0) \neq 0$. Then there exists an $h > 0$ such that $\overline{\Sigma_h(\boldsymbol{x}_0)} \subset B$ and $w_k > 0$ on $\Sigma_h(\boldsymbol{x}_0)$ or $w_k < 0$ on $\Sigma_h(\boldsymbol{x}_0)$. Assume, without loss in generality, that the former holds. If we let

$$\boldsymbol{v} = \varphi \, \mathbf{e}_k,$$

where $\varphi \in \Phi_h(\boldsymbol{x}_0)$, then $\boldsymbol{v}$ is of class C^∞ on $\bar{B}$ and vanishes near ∂B, and by properties (a) and (b) of φ,

$$\int_B \boldsymbol{w} \cdot \boldsymbol{v} \, dv = \int_B \varphi \, \boldsymbol{w} \cdot \mathbf{e}_k \, dv = \int_{\Sigma_h(\boldsymbol{x}_0)} \varphi w_k \, dv > 0.$$

But by hypothesis,

$$\int_B \boldsymbol{w} \cdot \boldsymbol{v} \, dv = 0.$$

Thus $\boldsymbol{w} = \boldsymbol{0}$ on B. But $\boldsymbol{w}$ is continuous on $\bar{B}$. Thus $\boldsymbol{w} = \boldsymbol{0}$ on $\bar{B}$. □

We now state, without proof,

(2) Rellich's lemma.[1] *Let B be properly regular. Further, let $\{\boldsymbol{u}_n\}$ be a sequence of continuous and piecewise smooth vector fields on $\bar{B}$, and suppose that there exist constants M_1 and M_2 such that*

$$\int_B |\boldsymbol{u}_n|^2 \, dv \leq M_1, \quad \int_B |\nabla \boldsymbol{u}_n|^2 \leq M_2, \quad n = 1, 2, \ldots.$$

Then there exists a subsequence $\{\boldsymbol{u}_{n_k}\}$ such that

$$\lim_{j,k \to \infty} \int_B |\boldsymbol{u}_{n_j} - \boldsymbol{u}_{n_k}|^2 \, dv = 0.$$

8. Harmonic and biharmonic fields. Let D be an open region in $\mathcal{E}$, and let Ψ be a scalar, vector, or tensor field on D. We say that Ψ is ***harmonic*** on D if Ψ is of class C^2 on D and

$$\Delta \Psi = 0.$$

We say that Ψ is ***biharmonic*** on D if Ψ is of class C^4 on D and

$$\Delta \Delta \Psi = 0.$$

We now state, without proof, some well known theorems concerning harmonic and biharmonic fields.

[1] RELLICH [1930, *4*]. See also, FICHERA [1971, *1*], Theorem 2.IV.

(1) Every harmonic or biharmonic field is analytic.

(2) Harnack's convergence theorem.[1] *Let Ψ^δ be harmonic on D for each δ $(0<\delta<\delta_0)$, and let*

$$\Psi^\delta \to \Psi \quad as \quad \delta \to 0$$

uniformly on each closed subregion of D. Then Ψ is harmonic on D, and for each fixed integer n,

$$\nabla^{(n)} \Psi^\delta \to \nabla^{(n)} \Psi \quad as \quad \delta \to 0$$

uniformly on every closed subregion of D.

(3) Mean value theorem. *Let φ and ψ be scalar, vector, or tensor fields on D with φ harmonic and ψ biharmonic. Then given any ball $\Sigma_\varrho = \Sigma_\varrho(\boldsymbol{y})$ in D:*

$$\varphi(\boldsymbol{y}) = \frac{1}{4\pi\varrho^2} \int_{\partial\Sigma_\rho} \varphi \, da, \tag{1}$$

$$\varphi(\boldsymbol{y}) = \frac{3}{4\pi\varrho^3} \int_{\Sigma_\rho} \varphi \, dv, \tag{2}$$

$$\psi(\boldsymbol{y}) + \frac{\varrho^2}{6} \Delta\psi(\boldsymbol{y}) = \frac{1}{4\pi\varrho^2} \int_{\partial\Sigma_\rho} \psi \, da, \tag{3}$$ [2]

$$\psi(\boldsymbol{y}) = \frac{3}{8\pi} \left[\frac{5}{\varrho^3} \int_{\Sigma_\rho} \psi \, dv - \frac{1}{\varrho^2} \int_{\partial\Sigma_\rho} \psi \, da \right]. \tag{4}$$ [2]

We now summarize some results concerning spherical harmonics.[3] Let (r, θ, γ) be spherical coordinates, related to the cartesian coordinates (x_1, x_2, x_3) through the mapping

$$x_1 = r \sin\theta \cos\gamma, \quad x_2 = r \sin\theta \sin\gamma, \quad x_3 = r\cos\theta,$$
$$0 \leqq r < \infty, \quad 0 \leqq \theta \leqq \pi, \quad 0 \leqq \gamma < 2\pi.$$

Then the general ***scalar solid spherical harmonic*** of degree k admits the representation

$$H^{(k)}(r, \theta, \gamma) = S^{(k)}(\theta, \gamma)\, r^k \qquad (k = 0, \pm 1, \pm 2, \ldots),$$

where $S^{(k)}(\theta, \gamma)$ is the general *surface spherical harmonic* of the same degree:

$$S^{(k)}(\theta, \gamma) = \sum_{n=0}^{|k|} [a_k^{(n)} \cos n\gamma + b_k^{(n)} \sin n\gamma]\, P_k^{(n)}(\cos\theta), \qquad b_k^{(0)} = 0.$$

Here $P_k^{(n)}$ designates the associated Legendre function of the first kind, of degree k and order n, while $a_k^{(0)}, a_k^{(n)}, b_k^{(n)}$ $(n = 1, 2, \ldots, |k|)$, for fixed k, are $2k+1$ arbitrary constants. If k and n are non-negative integers,

$$P_k^{(n)}(\xi) = (1-\xi^2)^{\frac{1}{2}n} \frac{d^n P_k(\xi)}{d\xi^n}, \qquad P_k(\xi) = \frac{1}{2^k k!} \frac{d^k(\xi^2-1)^k}{d\xi^k},$$

where P_k is the Legendre polynomial of degree k. Consequently,

$$P_k^{(0)}(\xi) = P_k(\xi), \qquad P_k^{(n)}(\xi) = 0 \qquad (n > k \geqq 0).$$

In addition, we have the recursion relations

$$P_{-k-1}^{(n)}(\xi) = P_k^{(n)}(\xi),$$

---

[1] See, for example, KELLOGG [1929, *1*], p. 249.

[2] NICOLESCO [1936, *3*]. See also DIAZ and PAYNE [1958, *6*].

[3] See, for example, POINCARÉ [1899, *1*], KELLOGG [1929, *1*], and HOBSON [1931, *4*].

which are valid for unrestricted k and n. As is evident from the above relations, *every surface harmonic of degree* $-1-k$ *is a surface harmonic of degree* k. Next, we recall that any two surface harmonics of distinct degree are orthogonal over the unit sphere $r=1$:

$$[S^{(k)}, S^{(m)}] \equiv \int_0^{2\pi}\int_0^{\pi} S^{(k)}(\theta,\gamma)\, S^{(m)}(\theta,\gamma) \sin\theta \, d\theta \, d\gamma = 0 \qquad (k \neq m,\ k \neq -m-1).$$

A ***vector solid spherical harmonic*** $\boldsymbol{h}^{(k)}$ of degree k is a vector-valued function with the property that given any vector $\boldsymbol{e}$, the function $\boldsymbol{e}\cdot\boldsymbol{h}^{(k)}$ is a scalar solid spherical harmonic of degree k.

The solid harmonics of degree k are harmonic functions, which are homogeneous of degree k with respect to the cartesian coordinates (x_1, x_2, x_3); they are homogeneous polynomials of degree k in the x_i if k is a non-negative integer. Further, if $H^{(k)}$ is a scalar solid harmonic of degree k, then $\nabla H^{(k)}$ is a vector solid harmonic of degree $k-1$ and

$$\boldsymbol{p}\cdot\nabla H^{(k)} = k H^{(k)},$$

where $\boldsymbol{p}(\boldsymbol{x}) = \boldsymbol{x} - \boldsymbol{0}$. Similarly,

$$(\nabla \boldsymbol{h}^{(k)})\,\boldsymbol{p} = k\boldsymbol{h}^{(k)}$$

for a vector solid harmonic $\boldsymbol{h}^{(k)}$ of degree k.

Turning from solid spherical harmonics to general harmonic functions, we cite the following theorem, in which $\Sigma(\infty)$ is the following *neighborhood of infinity*:

$$\Sigma(\infty) = \{\boldsymbol{x}: r_0 < |\boldsymbol{x}-\boldsymbol{0}| < \infty\}.$$

Here $r_0 > 0$ is fixed.

(4) Representation theorem for harmonic fields in a neighborhood of infinity. *Let* φ *be a harmonic scalar field on* $\Sigma(\infty)$. *Then*:

(i) φ *admits the representation*

$$\varphi = \sum_{k=-\infty}^{\infty} H^{(k)},$$

where the $H^{(k)}$ *are uniquely determined solid spherical harmonics of degree* k, *and the infinite series is uniformly convergent in every closed subregion of* $\Sigma(\infty)$;

(ii) φ *has derivatives of all orders, series representations of which may be obtained by performing the corresponding termwise differentiations, the resulting expansions being also uniformly convergent in every closed subregion of* $\Sigma(\infty)$;

(iii) *if* n *is a fixed integer, the three statements*

(α) $\varphi(\boldsymbol{x}) = O(r^{n-1})$,

(β) $\varphi(\boldsymbol{x}) = o(r^n)$,

(γ) $H^{(k)} = 0$ for $k \geqq n$,

are equivalent and imply

(δ) $\nabla\varphi(\boldsymbol{x}) = O(r^{n-2})$.

An analogous result holds for harmonic vector fields on $\Sigma(\infty)$.

We now state and prove a theorem which is the counterpart for biharmonic functions of *(4)*.

(5) Representation theorem for biharmonic fields in a neighborhood of infinity.[1] *Let* ψ *be a biharmonic scalar field on* $\Sigma(\infty)$. *Then*:

[1] PICONE [1936, *4*]. See also GURTIN and STERNBERG [1961, *11*].

(i) *ψ admits the representation*

$$\psi = \sum_{k=-\infty}^{\infty} G^{(k)} + r^2 \sum_{k=-\infty}^{\infty} H^{(k)},$$

where $G^{(k)}$ and $H^{(k)}$ are solid spherical harmonics of degree k, and both infinite series are uniformly convergent in every closed subregion of $\Sigma(\infty)$;

(ii) *ψ has derivatives of all orders, series representations of which may be obtained by performing the corresponding termwise differentiations, the resulting expansions being also uniformly convergent in every closed subregion of $\Sigma(\infty)$;*

(iii) *if n is a fixed integer, the three statements*

(α) $\psi(\boldsymbol{x}) = O(r^{n-1})$,

(β) $\psi(\boldsymbol{x}) = o(r^n)$,

(γ) $G^{(k)} = H^{(k-2)} = 0$ *for* $k \geqq n$,

are equivalent and imply

(δ) $\nabla\psi(\boldsymbol{x}) = O(r^{n-2})$.

An analogous result holds for biharmonic vector fields on $\Sigma(\infty)$.

Proof. We show first that ψ admits Almansi's representation

$$\psi = \varphi + r^2 \Phi,$$

where φ and Φ are harmonic on $\Sigma(\infty)$. A completeness proof for this representation, applicable to a region that is star-shaped with respect to $\mathbf{0}$ in its interior, is indicated by FRANK and v. MISES.[1] To establish the completeness of this representation in the present circumstances, it is sufficient to exhibit a harmonic function Φ that satisfies

$$\Delta(\psi - r^2 \Phi) = 0,$$

or equivalently,

$$4r \frac{\partial \Phi}{\partial r} + 6\Phi = \Delta\psi. \tag{a}$$

Since, by hypothesis, $\Delta\psi$ is harmonic, we know from *(4)* that it admits the expansion

$$\Delta\psi(r, \theta, \gamma) = \sum_{k=-\infty}^{\infty} \hat{S}^{(k)}(\theta, \gamma)\, r^k, \tag{b}$$

where the $\hat{S}^{(k)}(\theta, \gamma)$ are surface harmonics of degree k and the infinite series has the convergence properties asserted in parts (i) and (ii) of *(4)*. Now consider Φ defined by

$$\Phi(r, \theta, \gamma) = \sum_{k=-\infty}^{\infty} S^{(k)}(\theta, \gamma)\, r^k, \qquad S^{(k)} = \frac{\hat{S}^{(k)}}{2(2k+3)}.$$

Clearly, Φ is harmonic. Further, this function also satisfies (a), as is confirmed with the aid of (b).

On applying (i) and (ii) of *(4)* to φ and Φ, (i) and (ii) of the present theorem follow at once. With a view toward establishing (iii), we observe that (β) is immediate from (α). We show next that (β) implies (γ). Let $s^{(k)}(\theta, \gamma)$ and $S^{(k)}(\theta, \gamma)$ be the respective surface harmonics corresponding to the solid harmonics $G^{(k)}$ and $H^{(k)}$. We now multiply the right-hand member of the equation in (i) by $r^{-n} s^{(m)}(\theta, \gamma)$ $(m = 0, \pm 1, \pm 2, \ldots)$ and integrate termwise over $0 \leqq \theta \leqq \pi$,

[1] [1943, *1*], p. 848.

$0 \leqq \gamma \leqq 2\pi$. Bearing in mind the convergence properties asserted in (i) as well as the orthogonality relations, we find that the integral

$$I_{mn}(r) \equiv r^{-n} \int_0^{2\pi} \int_0^{\pi} \psi(r, \theta, \gamma)\, s^{(m)}(\theta, \gamma) \sin\theta \, d\theta \, d\gamma, \tag{c}$$

for all sufficiently large values of r, is given by

$$\begin{aligned} I_{mn}(r) = {} & r^{m-n}[s^{(m)}, s^{(m)}] + r^{-n-m+1}[s^{(-m-1)}, s^{(m)}] \\ & + r^{m-n+2}[S^{(m)}, s^{(m)}] + r^{-n-m+1}[S^{(-m-1)}, s^{(m)}]. \end{aligned} \tag{d}$$

It follows from (β) and (c) that $I_{mn}(r)$ tends to zero as $r \to \infty$ for fixed m and n; consequently, the same is true of the right-hand member of (d). Noting that the coefficients of the inner products appearing in (d) are four *distinct* powers of r, we conclude that $[s^{(m)}, s^{(m)}] = 0$ if $m - n \geqq 0$. Thus $s^{(m)} \equiv 0$, which implies that $G^{(m)} \equiv 0$ for $m \geqq n$. In a similar manner we verify that $H^{(m)} \equiv 0$ for $m \geqq n-2$. Hence (β) is a sufficient condition for (γ).

We have to show further that (γ) implies (α). This is readily seen to be true by observing that each of the series in (i) represents a field that is harmonic on $\Sigma(\infty)$, and by invoking part (iii) of ***(4)***. Finally, (γ) implies (δ) because of (ii) of the present theorem and (iii) of ***(4)***. Theorem ***(5)*** has thus been proved in its entirety. □

The following well-known proposition on divergence-free and curl-free vector fields will be extremely useful.

(6) *Let* $\boldsymbol{v}$ *be a smooth vector field on* B *that satisfies*

$$\operatorname{div} \boldsymbol{v} = 0, \qquad \operatorname{curl} \boldsymbol{v} = \boldsymbol{0}.$$

Then $\boldsymbol{v}$ *is harmonic.*

Proof. Let D be an arbitrary open ball in B. By hypothesis and (i) of ***(6.3)***, there exists a class C^2 scalar field φ on D such that

$$\boldsymbol{v} = \nabla \varphi.$$

Since $\boldsymbol{v}$ is divergence-free, φ is harmonic, and this, in turn, implies that $\boldsymbol{v}$ is harmonic. □

III. Functions of position and time.

9. Class $C^{M,N}$. By a ***time-interval*** T we mean an interval of the form $(-\infty, t_0)$, $(0, t_0)$, or $[0, t_0)$, where $t_0 > 0$ may be infinity. We frequently deal with functions of position and time having as their domain of definition the cartesian product $D \times T$ of a set D in $\mathscr{E}$ and a time-interval T. In this physical context $\boldsymbol{x}$ will always stand for a point of D and t will denote the time. If Ψ is a function on $D \times T$, we write $\Psi(\cdot, t)$ for the subsidiary mapping of D obtained by holding t fixed and $\Psi(\boldsymbol{x}, \cdot)$ for the mapping of T obtained by holding $\boldsymbol{x}$ fixed. We say that such a function Ψ is ***continuous (smooth) in time*** if $\Psi(\boldsymbol{x}, \cdot)$ is continuous (smooth) on T for each $\boldsymbol{x} \in D$. If $D \times T$ is open, we write $\nabla^{(n)} \Psi(\boldsymbol{x}, t)$ for the n-th gradient of Ψ with respect to $\boldsymbol{x}$ holding t fixed and $\Psi^{(n)}(\boldsymbol{x}, t)$ for the n-th derivative of Ψ with respect to t holding $\boldsymbol{x}$ fixed. Ordinarily, we write $\dot{\Psi}$, $\ddot{\Psi}$ instead of $\Psi^{(1)}$, $\Psi^{(2)}$.

Let M and N be non-negative integers. We say that Ψ is of ***class*** $C^{M,N}$ on $B \times (0, t_0)$ if Ψ is continuous on $B \times (0, t_0)$ and the functions

$$\nabla^{(m)} \Psi^{(n)}, \quad m \in \{0, 1, \ldots, M\}, \quad n \in \{0, 1, \ldots, N\}, \qquad m + n \leqq \max\{M, N\},$$

exist and are continuous on $B \times (0, t_0)$. We say that Ψ is of ***class*** $C^{M,N}$ on $\bar{B} \times [0, t_0)$ if Ψ is of class $C^{M,N}$ on $B \times (0, t_0)$ and for each $m \in \{0, 1, \ldots, M\}$, $n \in \{0, 1, \ldots, N\}$, $\nabla^{(m)} \Psi^{(n)}$ has a continuous extension to $\bar{B} \times [0, t_0)$ (in this case we also write $\nabla^{(m)} \Psi^{(n)}$ for the extended function). Finally, we write C^N for $C^{N,N}$.

We say that a function Ψ on $\mathscr{S}_\alpha \times [0, t_0)$ is ***piecewise regular*** if Ψ is piecewise continuous on $\mathscr{S}_\alpha \times [0, t_0)$ and $\Psi(\cdot, t)$ is piecewise regular on $\mathscr{S}_\alpha$ for $0 \leqq t < t_0$.

10. Convolutions. Let $T = [0, t_0)$ or $(-\infty, \infty)$, and let φ and ψ be scalar fields on $D \times T$ that are continuous in time. Then the ***convolution*** $\varphi * \psi$ of φ and ψ is the function on $D \times T$ defined by

$$\varphi * \psi(\boldsymbol{x}, t) = \int_0^t \varphi(\boldsymbol{x}, t - \tau)\, \psi(\boldsymbol{x}, \tau)\, d\tau .$$

Let Ψ be a function on $D \times [0, \infty)$. We say that Ψ has a ***Laplace transform*** $\overline{\Psi}$ if there exists a real number $\eta_0 \geqq 0$ such that for every $\eta \in [\eta_0, \infty)$ the integral

$$\overline{\Psi}(\boldsymbol{x}, \eta) = \int_0^\infty e^{-\eta t}\, \Psi(\boldsymbol{x}, t)\, dt$$

converges *uniformly* on D.

(1) Properties of the convolution. *Let φ, ψ, and ω be scalar fields on $D \times [0, t_0)$ that are continuous in time. Then*

(i) $\varphi * \psi = \psi * \varphi$;

(ii) $(\varphi * \psi) * \omega = \varphi * (\psi * \omega) = \varphi * \psi * \omega$;

(iii) $\varphi * (\psi + \omega) = \varphi * \psi + \varphi * \omega$;

(iv) *for $\boldsymbol{x} \in D$, if $\varphi(\boldsymbol{x}, \cdot) > 0$ on $(0, t_0)$, then*

$$\varphi * \psi(\boldsymbol{x}, \cdot) = 0 \;\Rightarrow\; \psi(\boldsymbol{x}, \cdot) = 0;$$

(v) *for $\boldsymbol{x} \in D$, if $t_0 = \infty$, then*

$$\varphi * \psi(\boldsymbol{x}, \cdot) = 0 \;\Rightarrow\; \varphi(\boldsymbol{x}, \cdot) = 0 \quad \text{or} \quad \psi(\boldsymbol{x}, \cdot) = 0;$$

(vi) *if φ is smooth in time, then*

$$\dot{\overbrace{\varphi * \psi}} = \dot{\varphi} * \psi + \varphi(\cdot, 0)\, \psi;$$

(vii) *if $t_0 = \infty$ and φ, ψ possess Laplace transforms, then so also does $\varphi * \psi$, and*

$$\overline{\varphi * \psi} = \overline{\varphi}\, \overline{\psi}$$

on $D \times [\eta_0, \infty)$ for some $\eta_0 \geqq 0$.

Proof. The results (i)–(iii), (v), and (vii) are well known,[1] while (vi) follows directly upon differentiating $\varphi * \psi$. To establish (iv) let $\varphi = \varphi(\boldsymbol{x}, \cdot)$, $\psi = \psi(\boldsymbol{x}, \cdot)$, assume that ψ does not vanish identically on $[0, t_0)$, and let

$$t_1 = \inf \{t \in [0, t_0) : \psi(t) \neq 0\}.$$

Since ψ is continuous, there exists a number t_2 with $t_1 < t_2 < t_0$ such that $\psi > 0$ on (t_1, t_2) or $\psi < 0$ on (t_1, t_2). In either case, since $\varphi > 0$, it follows that

$$\int_0^{t_2} \varphi(t_2 - \tau)\, \psi(\tau)\, d\tau \neq 0,$$

and we have a contradiction. □

[1] Properties (i)–(iii) are established in Chap. I of Mikusinski [1959, *10*]. Titchmarsh's theorem (v) is proved in Chap. II of the same book.

Let $\mathscr{W}$ be a finite-dimensional inner product space, e.g. $\mathscr{W}=\mathscr{V}$ or $\mathscr{W}=$ the set of all tensors, and let $\boldsymbol{\Phi}$ and $\boldsymbol{\Psi}$ be $\mathscr{W}$-valued fields on $D\times T$ that are continuous in time. Then we write $\varphi*\boldsymbol{\Psi}$ for the $\mathscr{W}$-valued field on $D\times T$ defined by

$$\varphi*\boldsymbol{\Psi}(\boldsymbol{x},t)=\int_0^t\varphi(\boldsymbol{x},t-\tau)\,\boldsymbol{\Psi}(\boldsymbol{x},\tau)\,d\tau,$$

and $\boldsymbol{\Phi}*\boldsymbol{\Psi}$ for the *scalar field* on $D\times T$ defined by

$$\boldsymbol{\Phi}*\boldsymbol{\Psi}(\boldsymbol{x},t)=\int_0^t\boldsymbol{\Phi}(\boldsymbol{x},t-\tau)\cdot\boldsymbol{\Psi}(\boldsymbol{x},\tau)\,d\tau.$$

Thus if φ is a scalar field, $\boldsymbol{u}$ and $\boldsymbol{v}$ vector fields, and $\boldsymbol{S}$ and $\boldsymbol{T}$ tensor fields, all defined on $D\times T$, then $\varphi*\boldsymbol{u}$ is a vector field, $\varphi*\boldsymbol{S}$ is a tensor field, and $\boldsymbol{u}*\boldsymbol{v}$ and $\boldsymbol{S}*\boldsymbol{T}$ are *scalar* fields; in components

$$[\varphi*\boldsymbol{u}]_i=\varphi*u_i,\qquad[\varphi*\boldsymbol{S}]_{ij}=\varphi*S_{ij},$$
$$\boldsymbol{u}*\boldsymbol{v}=u_i*v_i,\qquad\boldsymbol{S}*\boldsymbol{T}=S_{ij}*T_{ij}.$$

Finally, we write $\boldsymbol{S}*\boldsymbol{u}$ for the *vector* field

$$\boldsymbol{S}*\boldsymbol{u}(\boldsymbol{x},t)=\int_0^t\boldsymbol{S}(\boldsymbol{x},t-\tau)\,\boldsymbol{u}(\boldsymbol{x},\tau)\,d\tau;$$

then

$$[\boldsymbol{S}*\boldsymbol{u}]_i=S_{ij}*u_j.$$

Properties analogous to (i)–(iv) of ***(1)*** obviously hold for the convolutions defined above.

Let $\boldsymbol{l}$ and $\boldsymbol{k}$ be continuous *vector* functions on $(-\infty,\infty)$. Then $\boldsymbol{l}\#\boldsymbol{k}$ is for the *tensor* function on this interval defined by

$$\boldsymbol{l}\#\boldsymbol{k}(t)=\int_0^t\boldsymbol{l}(t-\tau)\otimes\boldsymbol{k}(\tau)\,d\tau.$$

In components

$$[\boldsymbol{l}\#\boldsymbol{k}]_{ij}=l_i*k_j.$$

For convenience, we write $\boldsymbol{l}_\alpha$ for the function on $(-\infty,\infty)$ defined by

$$\boldsymbol{l}_\alpha(t)=\boldsymbol{l}(t-\alpha).$$

Further, we let $\mathscr{L}$ denote the set of all *smooth* vector functions on $(-\infty,\infty)$ that *vanish* on $(-\infty,0)$.

(2) *Let* $\boldsymbol{l},\boldsymbol{k}\in\mathscr{L}$ *and let* $\alpha>0$. *Then*

$$\boldsymbol{l}_\alpha\#\boldsymbol{k}=\boldsymbol{l}\#\boldsymbol{k}_\alpha,$$
$$\dot{\boldsymbol{l}}_\alpha\#\boldsymbol{k}=\boldsymbol{l}\#\dot{\boldsymbol{k}}_\alpha.$$

Proof. Since

$$\left.\begin{aligned}\boldsymbol{l}=\boldsymbol{k}=\boldsymbol{0}\quad&\text{on}\quad(-\infty,0],\\ \boldsymbol{l}_\alpha=\boldsymbol{k}_\alpha=\dot{\boldsymbol{l}}_\alpha=\dot{\boldsymbol{k}}_\alpha=\boldsymbol{0}\quad&\text{on}\quad(-\infty,\alpha],\end{aligned}\right\}\tag{a}$$

the above relations hold trivially on $(-\infty,0]$. Thus choose $t\in(0,\infty)$. Then

$$\boldsymbol{l}_\alpha\#\boldsymbol{k}(t)=\int_0^t\boldsymbol{l}(t-\tau-\alpha)\otimes\boldsymbol{k}(\tau)\,d\tau.$$

Thus, letting $\tau = \lambda - \alpha$, we arrive at

$$\boldsymbol{l}_\alpha \# \boldsymbol{k}(t) = \int_\alpha^{t+\alpha} \boldsymbol{l}(t-\lambda) \otimes \boldsymbol{k}_\alpha(\lambda)\, d\lambda,$$

and, in view of (a),

$$\boldsymbol{l}_\alpha \# \boldsymbol{k}(t) = \int_0^t \boldsymbol{l}(t-\lambda) \otimes \boldsymbol{k}_\alpha(\lambda)\, d\lambda = \boldsymbol{l} \# \boldsymbol{k}_\alpha(t).$$

Next,

$$\dot{\boldsymbol{l}}_\alpha \# \boldsymbol{k}(t) = \dot{\boldsymbol{l}} \# \boldsymbol{k}_\alpha(t) = \int_0^t \dot{\boldsymbol{l}}(t-\tau) \otimes \boldsymbol{k}_\alpha(\tau)\, d\tau. \tag{b}$$

Since

$$\dot{\boldsymbol{l}}(t-\tau) = -\frac{d}{d\tau}\boldsymbol{l}(t-\tau),$$

if we integrate (b) by parts, we arrive at

$$\dot{\boldsymbol{l}}_\alpha \# \boldsymbol{k}(t) = -\left[\boldsymbol{l}(t-\tau) \otimes \boldsymbol{k}_\alpha(\tau)\right]_{\tau=0}^{\tau=t} + \int_0^t \boldsymbol{l}(t-\tau) \otimes \dot{\boldsymbol{k}}_\alpha(\tau)\, d\tau. \tag{c}$$

Because of (a), (c) yields the second relation in *(2)*. □

11. Space-time. We call the four-dimensional point space

$$\mathscr{E}^{(4)} = \mathscr{E} \times (-\infty, \infty)$$

space-time; in our physical context a point $\boldsymbol{\xi} = (\boldsymbol{x}, t) \in \mathscr{E}^{(4)}$ consists of a point $\boldsymbol{x}$ of space and a time t. The translation space associated with $\mathscr{E}^{(4)}$ is

$$\mathscr{V}^{(4)} = \mathscr{V} \times (-\infty, \infty);$$

elements of $\mathscr{V}^{(4)}$ will be referred to as ***four-vectors***. As is natural, we write

$$\mathsf{a} \cdot \mathsf{b} = \boldsymbol{a} \cdot \boldsymbol{b} + \alpha\beta$$

for the *inner product* of two four-vectors $\mathsf{a} = (\boldsymbol{a}, \alpha)$ and $\mathsf{b} = (\boldsymbol{b}, \beta)$.

A ***four-tensor*** M is a linear transformation from $\mathscr{V}^{(4)}$ into $\mathscr{V}^{(4)}$. Thus M assigns to each four-vector $(\boldsymbol{a}, \alpha)$ a four-vector

$$(\boldsymbol{b}, \beta) = \mathsf{M}(\boldsymbol{a}, \alpha).$$

Given a four-tensor M, there exists a unique tensor $\boldsymbol{M}$ (on $\mathscr{V}$), unique vectors $\boldsymbol{m}, \overline{\boldsymbol{m}} \in \mathscr{V}$, and a unique scalar λ such that for every four-vector $(\boldsymbol{a}, \alpha)$,

$$\mathsf{M}(\boldsymbol{a}, 0) = (\boldsymbol{M}\boldsymbol{a}, \boldsymbol{m} \cdot \boldsymbol{a}),$$
$$\mathsf{M}(0, \alpha) = (\overline{\boldsymbol{m}}\alpha, \lambda\alpha).$$

Thus

$$\mathsf{M}(\boldsymbol{a}, \alpha) = (\boldsymbol{M}\boldsymbol{a} + \overline{\boldsymbol{m}}\alpha, \boldsymbol{m} \cdot \boldsymbol{a} + \lambda\alpha).$$

We call the array

$$\begin{bmatrix} \boldsymbol{M} & \overline{\boldsymbol{m}} \\ \boldsymbol{m} & \lambda \end{bmatrix} \tag{a}$$

the ***space-time partition*** of M.

As before, we define the transpose M^T of M to be the unique four-tensor with the property:

$$\mathsf{M}\mathsf{a} \cdot \mathsf{b} = \mathsf{a} \cdot \mathsf{M}^T \mathsf{b} \quad \text{for all four-vectors } \mathsf{a} \text{ and } \mathsf{b}.$$

It is not difficult to show that the space-time partition of $\mathbf{M}^T$ is

$$\begin{bmatrix} \boldsymbol{M}^T & \boldsymbol{m} \\ \overline{\boldsymbol{m}} & \lambda \end{bmatrix}.$$

Let φ be a smooth scalar field on an open set D in $\mathscr{E}^{(4)}$. Then the gradient in $\mathscr{E}^{(4)}$ of φ is the four-vector field

$$\nabla_{(4)}\,\varphi = (\nabla\varphi, \dot{\varphi}).$$

Let $\mathbf{u} = (\boldsymbol{u}, \mu)$ be a smooth four-vector field and $\mathbf{M}$ a smooth four-tensor field on D (i.e. the values of $\mathbf{M}$ are four-tensors). We define the divergences of $\mathbf{u}$ and $\mathbf{M}$ by the relations:

$$\operatorname{div}_{(4)} \mathbf{u} = \operatorname{div} \boldsymbol{u} + \dot{\mu},$$

$$(\operatorname{div}_{(4)} \mathbf{M}) \cdot \mathbf{a} = \operatorname{div}_{(4)} (\mathbf{M}^T \mathbf{a}) \qquad \text{for every } \mathbf{a} \in \mathscr{V}^{(4)}.$$

For future use, we now record the following

(1) Identity. *Let* $\mathbf{M}$ *be a smooth four-tensor field on an open set* D *in* $\mathscr{E}^{(4)}$, *and let* (a) *be the space-time partition of* $\mathbf{M}$. *Then*

$$\operatorname{div}_{(4)} \mathbf{M} = (\operatorname{div} \boldsymbol{M} + \dot{\overline{\boldsymbol{m}}}, \operatorname{div} \boldsymbol{m} + \dot{\lambda}).$$

Proof. Since

$$\mathbf{M}^T \mathbf{a} = (\boldsymbol{M}^T \boldsymbol{a} + \boldsymbol{m}\alpha, \overline{\boldsymbol{m}} \cdot \boldsymbol{a} + \lambda\alpha),$$

it follows that

$$\begin{aligned} \operatorname{div}_{(4)} (\mathbf{M}^T \mathbf{a}) &= \operatorname{div} (\boldsymbol{M}^T \boldsymbol{a}) + \alpha \operatorname{div} \boldsymbol{m} + \boldsymbol{a} \cdot \dot{\overline{\boldsymbol{m}}} + \alpha \dot{\lambda} \\ &= (\operatorname{div} \boldsymbol{M} + \dot{\overline{\boldsymbol{m}}}, \operatorname{div} \boldsymbol{m} + \dot{\lambda}) \cdot (\boldsymbol{a}, \alpha), \end{aligned}$$

which implies the desired identity. □

(2) Divergence theorem in space-time. *Let* D *be a bounded regular region in* $\mathscr{E}^{(4)}$. *Let* $\mathbf{u}$ *and* $\mathbf{M}$ *be a four-vector field and a four-tensor field, respectively, both continuous on* $\overline{D}$ *and smooth on* D. *Then*

$$\int_{\partial D} \mathbf{u} \cdot \mathbf{n}\, da = \int_D \operatorname{div}_{(4)} \mathbf{u}\, dv,$$

$$\int_{\partial D} \mathbf{M}\mathbf{n}\, da = \int_D \operatorname{div}_{(4)} \mathbf{M}\, dv,$$

whenever the integrand on the right is continuous on $\overline{D}$. *Here* $\mathbf{n}$ *is the outward unit normal to* ∂D.

Proof. The first result is simply the classical divergence theorem in $\mathscr{E}^{(4)}$; the second can be established using the procedure given in the proof of ***(6.1)***. □

C. Formulation of the linear theory of elasticity.

I. Kinematics.

12. Finite deformations. Infinitesimal deformations. In this section we motivate some of the notions of the infinitesimal theory. When we present the linear theory we will give the axioms and definitions that form its foundation; while these are *motivated* by viewing the linear theory as a first-order approximation to the finite theory, *the linear theory itself is independent of these considerations and stands on its own as a completely consistent mathematical theory.*

Consider a body[1] identified with the region B it occupies in a fixed reference configuration.[2] A *deformation*[3] of B is a smooth homeomorphism $\boldsymbol{\varkappa}$ of B onto a region $\boldsymbol{\varkappa}(B)$ in $\mathscr{E}$ with $\det \nabla\boldsymbol{\varkappa} > 0$. The point $\boldsymbol{\varkappa}(\boldsymbol{x})$ is the place occupied by the material point $\boldsymbol{x}$ in the deformation $\boldsymbol{\varkappa}$, while

$$\boldsymbol{u}(\boldsymbol{x}) = \boldsymbol{\varkappa}(\boldsymbol{x}) - \boldsymbol{x} \tag{a}$$

is the *displacement* of $\boldsymbol{x}$ (see Fig. 2). The tensor fields

$$\boldsymbol{F} = \nabla\boldsymbol{\varkappa} \tag{b}$$

and $\nabla\boldsymbol{u}$ are called, respectively, the *deformation gradient* and the *displacement gradient*. By (a) and (b),

$$\nabla\boldsymbol{u} = \boldsymbol{F} - \boldsymbol{1}. \tag{c}$$

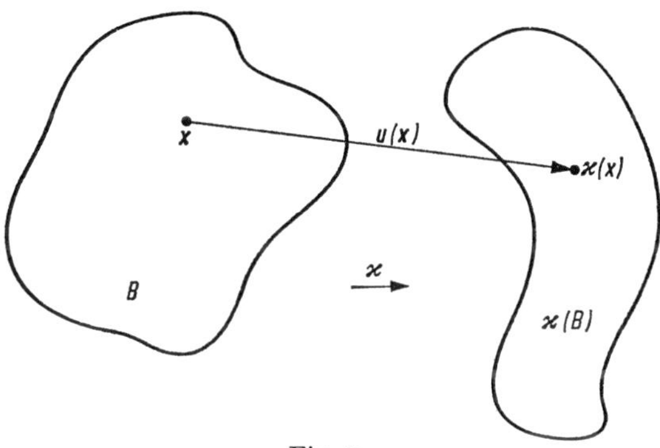

Fig. 2.

The importance of the concept of strain cannot be established by a study of kinematics alone; its relevance becomes clear when one studies the restrictions placed on constitutive assumptions by material-frame indifference.[4] Many different measures of strain appear in the literature; however all of the properly invariant choices are, in a certain sense, equivalent.[5] The most useful for our purposes is the *finite strain tensor* $\boldsymbol{D}$ defined by

$$\boldsymbol{D} = \tfrac{1}{2}(\boldsymbol{F}^T\boldsymbol{F} - \boldsymbol{1}). \tag{d}$$ [6]

Of importance in the linear theory is the *infinitesimal strain tensor*

$$\boldsymbol{E} = \tfrac{1}{2}(\nabla\boldsymbol{u} + \nabla\boldsymbol{u}^T); \tag{e}$$

by (c),

$$\boldsymbol{D} = \boldsymbol{E} + \tfrac{1}{2}\nabla\boldsymbol{u}^T\nabla\boldsymbol{u}. \tag{f}$$

The infinitesimal theory models physical situations in which the displacement $\boldsymbol{u}$ and the displacement gradient $\nabla\boldsymbol{u}$ are, in some sense, small. In view of (e) and (f), $\boldsymbol{D}$ and $\boldsymbol{E}$ can be considered functions of $\nabla\boldsymbol{u}$. Writing

$$\varepsilon = |\nabla\boldsymbol{u}|, \tag{g}$$

---

[1] See, e.g., TRUESDELL and NOLL [1965, *22*], § 15.

[2] See, e.g., TRUESDELL and NOLL [1965, *22*], § 21.

[3] TRUESDELL and TOUPIN [1960, *17*], §§ 13–58 give a thorough discussion of finite deformations.

[4] See, e.g., TRUESDELL and NOLL [1965, *22*], §§ 19, 29.

[5] See, e.g., TRUESDELL and TOUPIN [1960, *17*], § 32.

[6] The tensor $\boldsymbol{C} = \boldsymbol{F}^T\boldsymbol{F} = 2\boldsymbol{D} + \boldsymbol{1}$ is the right Cauchy-Green strain tensor (TRUESDELL and NOLL [1965, *22*], § 23).

we conclude from (f) that *to within an error of $O(\varepsilon^2)$ as $\varepsilon \to 0$, the finite strain tensor* $\boldsymbol{D}$ *and the infinitesimal strain tensor* $\boldsymbol{E}$ *coincide.*

A *finite rigid deformation* is a deformation of the form:

$$\boldsymbol{\varkappa}(\boldsymbol{x}) = \boldsymbol{y}_0 + \boldsymbol{Q}[\boldsymbol{x} - \boldsymbol{x}_0],$$

where $\boldsymbol{x}_0$ and $\boldsymbol{y}_0$ are fixed points of $\mathscr{E}$ and $\boldsymbol{Q}$ is an *orthogonal* tensor. In this instance

$$\boldsymbol{F} = \boldsymbol{Q} \quad \text{and} \quad \nabla \boldsymbol{u} = \boldsymbol{Q} - \boldsymbol{1} \tag{h}$$

are constants and

$$\boldsymbol{u}(\boldsymbol{x}) = \boldsymbol{u}_0 + \nabla \boldsymbol{u}[\boldsymbol{x} - \boldsymbol{x}_0], \tag{i}$$

where $\boldsymbol{u}_0 = \boldsymbol{y}_0 - \boldsymbol{x}_0$. Moreover, since $\boldsymbol{Q}^T \boldsymbol{Q} = \boldsymbol{1}$, (d), (e), and (h), imply

$$\begin{aligned} \boldsymbol{D} &= \boldsymbol{0}, \\ \boldsymbol{E} &= \tfrac{1}{2}(\boldsymbol{Q} + \boldsymbol{Q}^T) - \boldsymbol{1}. \end{aligned} \tag{j}$$

Thus in a rigid deformation the finite strain tensor $\boldsymbol{D}$ vanishes, but $\boldsymbol{E}$ does not. For this reason $\boldsymbol{E}$ is not used as a measure of strain in finite elasticity. By (f) and (j),

$$\boldsymbol{E} = -\tfrac{1}{2} \nabla \boldsymbol{u}^T \nabla \boldsymbol{u},$$

and we conclude from (g) that *in a rigid deformation to within an error of $O(\varepsilon^2)$ the infinitesimal strain tensor vanishes.* Thus

$$\nabla \boldsymbol{u} = -\nabla \boldsymbol{u}^T + O(\varepsilon^2);$$

i.e., to within an error of $O(\varepsilon^2)$ the displacement gradient is *skew*. This motivates our defining an *infinitesimal rigid displacement* to be a field of the form

$$\boldsymbol{u}(\boldsymbol{x}) = \boldsymbol{u}_0 + \boldsymbol{W}[\boldsymbol{x} - \boldsymbol{x}_0]$$

with $\boldsymbol{W}$ skew.

The *volume change* δV in the deformation $\boldsymbol{\varkappa}$ is given by

$$\delta V = \int_{\boldsymbol{\varkappa}(B)} dv - \int_B dv.$$

Since the Jacobian of the mapping $\boldsymbol{\varkappa}$ is $\det \nabla \boldsymbol{\varkappa} = \det \boldsymbol{F}$,

$$\int_{\boldsymbol{\varkappa}(B)} dv = \int_B \det \boldsymbol{F}\, dv;$$

thus

$$\delta V = \int_B (\det \boldsymbol{F} - 1)\, dv,$$

so that $\det \boldsymbol{F} - 1$ represents the volume change per unit volume in the deformation. In view of (c), we can consider $\boldsymbol{F}$ a function of $\nabla \boldsymbol{u}$. A simple analysis based on (c) and (g) implies that as $\varepsilon \to 0$

$$\det \boldsymbol{F} = \det(\boldsymbol{1} + \nabla \boldsymbol{u}) = 1 + \operatorname{tr} \nabla \boldsymbol{u} + O(\varepsilon^2),$$

or equivalently,

$$\det \boldsymbol{F} - 1 = \operatorname{div} \boldsymbol{u} + O(\varepsilon^2);$$

thus to within an error of $O(\varepsilon^2)$ the volume change per unit volume is equal to $\operatorname{div} \boldsymbol{u}$. For this reason we call the number

$$\delta v = \int_B \operatorname{div} \boldsymbol{u}\, dv = \int_{\partial B} \boldsymbol{u} \cdot \boldsymbol{n}\, da$$

the *infinitesimal volume change.*

13. Properties of displacement fields. Strain. In this section we study properties of (infinitesimal) displacement fields. Most of the definitions are motivated by the results of the preceding section.

A ***displacement field*** $\boldsymbol{u}$ is a class C^2 vector field over B; its value $\boldsymbol{u}(\boldsymbol{x})$ at a point $\boldsymbol{x} \in B$ is the *displacement* of $\boldsymbol{x}$. The symmetric part

$$\boldsymbol{E} = \tfrac{1}{2}(\nabla \boldsymbol{u} + \nabla \boldsymbol{u}^T)$$

of the displacement gradient, $\nabla \boldsymbol{u}$, is the (infinitesimal) ***strain field***, and the above equation relating $\boldsymbol{E}$ to $\boldsymbol{u}$ is called the ***strain-displacement relation***. The (infinitesimal) ***rotation field*** $\boldsymbol{W}$ is the skew part of $\nabla \boldsymbol{u}$, i.e.

$$\boldsymbol{W} = \tfrac{1}{2}(\nabla \boldsymbol{u} - \nabla \boldsymbol{u}^T),$$

while

$$\boldsymbol{\omega} = \tfrac{1}{2} \operatorname{curl} \boldsymbol{u}$$

is the (infinitesimal) ***rotation vector***. Thus

$$\nabla \boldsymbol{u} = \boldsymbol{E} + \boldsymbol{W},$$

and $\boldsymbol{\omega}$ is the axial vector of $\boldsymbol{W}$; i.e. for any vector $\boldsymbol{a}$,

$$\boldsymbol{W}\boldsymbol{a} = \boldsymbol{\omega} \times \boldsymbol{a}.$$

We call

$$\operatorname{div} \boldsymbol{u} = \operatorname{tr} \boldsymbol{E}$$

the ***dilatation***. The (infinitesimal) ***volume change*** $\delta v(P)$ of a part P of B due to a continuous displacement field $\boldsymbol{u}$ on $\bar{B}$ is defined by

$$\delta v(P) = \int_{\partial P} \boldsymbol{u} \cdot \boldsymbol{n} \, da,$$

and we say that $\boldsymbol{u}$ is ***isochoric*** if $\delta v(P) = 0$ for every P. By the divergence theorem,

$$\delta v(P) = \int_P \operatorname{div} \boldsymbol{u} \, dv = \int_P \operatorname{tr} \boldsymbol{E} \, dv;$$

thus $\boldsymbol{u}$ is isochoric if and only if

$$\operatorname{tr} \boldsymbol{E} \equiv 0$$

on B.

An (infinitesimal) ***rigid displacement field*** is a displacement field $\boldsymbol{u}$ of the form

$$\boldsymbol{u}(\boldsymbol{x}) = \boldsymbol{u}_0 + \boldsymbol{W}_0[\boldsymbol{x} - \boldsymbol{x}_0],$$

where $\boldsymbol{x}_0$ is a point, $\boldsymbol{u}_0$ is a vector, and $\boldsymbol{W}_0$ is a *skew* tensor. In this instance

$$\nabla \boldsymbol{u}(\boldsymbol{x}) = \boldsymbol{W}_0 = -\boldsymbol{W}_0^T \quad \text{and} \quad \boldsymbol{E}(\boldsymbol{x}) = \boldsymbol{0}$$

for every $\boldsymbol{x}$ in B. Of course, $\boldsymbol{u}$ may also be written in the form

$$\boldsymbol{u}(\boldsymbol{x}) = \boldsymbol{u}_0 + \boldsymbol{\omega}_0 \times [\boldsymbol{x} - \boldsymbol{x}_0],$$

where $\boldsymbol{\omega}_0$ is the axial vector of $\boldsymbol{W}_0$.

(1) Characterization of rigid displacements. *Let $\boldsymbol{u}$ be a displacement field. Then the following three statements are equivalent.*

(i) *$\boldsymbol{u}$ is a rigid displacement field.*

(ii) *The strain field corresponding to $\boldsymbol{u}$ vanishes on B.*

(iii) *$\boldsymbol{u}$ has the* ***projection property*** *on B: for every pair of points $\boldsymbol{x}, \boldsymbol{y} \in B$*

$$[\boldsymbol{u}(\boldsymbol{x}) - \boldsymbol{u}(\boldsymbol{y})] \cdot [\boldsymbol{x} - \boldsymbol{y}] = 0.$$

Proof. We will show that (i) ⇒ (iii) ⇒ (ii) ⇒ (i). If $\boldsymbol{u}$ is rigid, then

$$\boldsymbol{u}(\boldsymbol{x}) - \boldsymbol{u}(\boldsymbol{y}) = \boldsymbol{W}_0[\boldsymbol{x} - \boldsymbol{y}]$$

with $\boldsymbol{W}_0$ skew; hence

$$(\boldsymbol{x} - \boldsymbol{y}) \cdot [\boldsymbol{u}(\boldsymbol{x}) - \boldsymbol{u}(\boldsymbol{y})] = (\boldsymbol{x} - \boldsymbol{y}) \cdot \boldsymbol{W}_0[\boldsymbol{x} - \boldsymbol{y}] = 0$$

and (iii) holds. Next, assume that $\boldsymbol{u}$ has the projection property. Then differentiating the relation (iii) with respect to $\boldsymbol{x}$, we obtain

$$\nabla\boldsymbol{u}(\boldsymbol{x})^T [\boldsymbol{x} - \boldsymbol{y}] + \boldsymbol{u}(\boldsymbol{x}) - \boldsymbol{u}(\boldsymbol{y}) = \boldsymbol{0}.$$

Differentiating this equation with respect to $\boldsymbol{y}$ and evaluating the result at $\boldsymbol{y} = \boldsymbol{x}$, we arrive at

$$-\nabla\boldsymbol{u}(\boldsymbol{x})^T - \nabla\boldsymbol{u}(\boldsymbol{x}) = \boldsymbol{0},$$

which implies (ii). Now assume that (ii) holds. By (6) of *(4.1)*, $\nabla\nabla\boldsymbol{u} = \boldsymbol{0}$. Thus

$$\boldsymbol{u}(\boldsymbol{x}) = \boldsymbol{u}_0 + \boldsymbol{W}_0[\boldsymbol{x} - \boldsymbol{x}_0],$$

and (ii) implies that $\boldsymbol{W}_0$ is skew. □

An immediate consequence of *(1)* is the following important result.

(2) Kirchhoff's theorem[1]. *If two displacement fields $\boldsymbol{u}$ and $\boldsymbol{u}'$ correspond to the same strain field, then*

$$\boldsymbol{u} = \boldsymbol{u}' + \boldsymbol{w},$$

where $\boldsymbol{w}$ is a rigid displacement field.

The next theorem shows that (iii) in *(1)* implies (i), even if we drop the assumption that $\boldsymbol{u}$ be of class C^2, but, as we shall see, the proof is much more difficult.

(3)[2] *Let F be a non-coplanar set of points in $\mathscr{E}$, and let $\boldsymbol{u}$ be a vector field on F. Then $\boldsymbol{u}$ is a rigid displacement field if and only if $\boldsymbol{u}$ has the projection property.*

Proof. Clearly, if $\boldsymbol{u}$ is rigid, it has the projection property. To establish the converse assertion we assume that

$$[\boldsymbol{u}(\boldsymbol{x}) - \boldsymbol{u}(\boldsymbol{y})] \cdot (\boldsymbol{x} - \boldsymbol{y}) = 0$$

for every pair of points $\boldsymbol{x}, \boldsymbol{y} \in F$. Let $\boldsymbol{z} \in F$, and let $F_{\boldsymbol{z}}$ be the set of all vectors from $\boldsymbol{z}$ to points of F:

$$F_{\boldsymbol{z}} = \{\boldsymbol{v} : \boldsymbol{v} = \boldsymbol{x} - \boldsymbol{z}, \boldsymbol{x} \in F\}.$$

Let $\boldsymbol{g}$ be the vector field on $F_{\boldsymbol{z}}$ defined by

$$\boldsymbol{g}(\boldsymbol{v}) = \boldsymbol{u}(\boldsymbol{x}) - \boldsymbol{u}(\boldsymbol{z}), \qquad \boldsymbol{v} = \boldsymbol{x} - \boldsymbol{z}, \qquad \boldsymbol{x} \in F. \tag{a}$$

Then $\boldsymbol{g}(\boldsymbol{0}) = \boldsymbol{0}$, and $\boldsymbol{g}$ has the projection property:

$$[\boldsymbol{g}(\boldsymbol{v}) - \boldsymbol{g}(\boldsymbol{w})] \cdot (\boldsymbol{v} - \boldsymbol{w}) = 0 \tag{b}$$

for every pair of vectors $\boldsymbol{v}, \boldsymbol{w} \in F_{\boldsymbol{z}}$. Taking $\boldsymbol{w} = \boldsymbol{0}$ we conclude that

$$\boldsymbol{g}(\boldsymbol{v}) \cdot \boldsymbol{v} = 0$$

[1] [1859, *1*].
[2] Cf. NIELSEN [1935, *5*], Chap. 3.

for every $\boldsymbol{v} \in F_{\boldsymbol{z}}$, and this fact when combined with (b) implies

$$\boldsymbol{g}(\boldsymbol{v}) \cdot \boldsymbol{w} = -\boldsymbol{g}(\boldsymbol{w}) \cdot \boldsymbol{v} \tag{c}$$

for all $\boldsymbol{v}, \boldsymbol{w} \in F_{\boldsymbol{z}}$.

Next, since the points of F do not line in a plane, $F_{\boldsymbol{z}}$ spans the entire vector space $\mathscr{V}$. Let $\boldsymbol{w}_1, \boldsymbol{w}_2, \boldsymbol{w}_3$ be linearly independent vectors in $F_{\boldsymbol{z}}$, and let $\hat{\boldsymbol{g}}$ be the function on $\mathscr{V}$ defined by

$$\hat{\boldsymbol{g}}(\boldsymbol{v}) = v^i \boldsymbol{g}(\boldsymbol{w}_i), \tag{d}$$

where v^i are the components of $\boldsymbol{v}$ relative to $\boldsymbol{w}_i$, i.e. $\boldsymbol{v} = v^i \boldsymbol{w}_i$. Then $\hat{\boldsymbol{g}}$ is *linear*. We now show that $\boldsymbol{g}$ is the restriction of $\hat{\boldsymbol{g}}$ to $F_{\boldsymbol{z}}$. Indeed, (c) and (d) imply that for $\boldsymbol{v}, \boldsymbol{k} \in F_{\boldsymbol{z}}$

$$\hat{\boldsymbol{g}}(\boldsymbol{v}) \cdot \boldsymbol{k} = v^i \boldsymbol{g}(\boldsymbol{w}_i) \cdot \boldsymbol{k} = -v^i \boldsymbol{g}(\boldsymbol{k}) \cdot \boldsymbol{w}_i = -\boldsymbol{v} \cdot \boldsymbol{g}(\boldsymbol{k}) = \boldsymbol{g}(\boldsymbol{v}) \cdot \boldsymbol{k},$$

and since this relation must hold for every $\boldsymbol{k} \in F_{\boldsymbol{z}}$,

$$\hat{\boldsymbol{g}}(\boldsymbol{v}) = \boldsymbol{g}(\boldsymbol{v}) \quad \text{for all } \boldsymbol{v} \in F_{\boldsymbol{z}}. \tag{e}$$

Since $\hat{\boldsymbol{g}}$ is linear, we may write

$$\hat{\boldsymbol{g}}(\boldsymbol{v}) = \boldsymbol{W}\boldsymbol{v}, \tag{f}$$

where $\boldsymbol{W}$ is a tensor. Further, using (c) and (d), it is a simple matter to verify that

$$\hat{\boldsymbol{g}}(\boldsymbol{v}) \cdot \boldsymbol{w} = -\hat{\boldsymbol{g}}(\boldsymbol{w}) \cdot \boldsymbol{v}$$

for all $\boldsymbol{v}, \boldsymbol{w} \in \mathscr{V}$. Thus $\boldsymbol{W}$ is skew. Finally, (a), (e), and (f) imply

$$\boldsymbol{u}(\boldsymbol{x}) = \boldsymbol{u}(\boldsymbol{z}) + \boldsymbol{W}[\boldsymbol{x} - \boldsymbol{z}]$$

for every $\boldsymbol{x} \in F$. □

A ***homogeneous displacement field*** is a displacement field of the form

$$\boldsymbol{u}(\boldsymbol{x}) = \boldsymbol{u}_0 + \boldsymbol{A}[\boldsymbol{x} - \boldsymbol{x}_0],$$

where the point $\boldsymbol{x}_0$, the vector $\boldsymbol{u}_0$, and the tensor $\boldsymbol{A}$ are independent of $\boldsymbol{x}$. Such a field is determined, to within a rigid displacement, by the strain field $\boldsymbol{E}$, which is *constant* and equal to the symmetric part of $\boldsymbol{A}$. If $\boldsymbol{u}_0 = \boldsymbol{0}$ and $\boldsymbol{A}$ is symmetric, i.e. if

$$\boldsymbol{u}(\boldsymbol{x}) = \boldsymbol{E}[\boldsymbol{x} - \boldsymbol{x}_0],$$

then $\boldsymbol{u}$ is called a ***pure strain*** from $\boldsymbol{x}_0$.

(4) Let $\boldsymbol{u}$ be a homogeneous displacement field. Then $\boldsymbol{u}$ admits the decomposition

$$\boldsymbol{u} = \boldsymbol{w} + \hat{\boldsymbol{u}},$$

where $\boldsymbol{w}$ is a rigid displacement field and $\hat{\boldsymbol{u}}$ is a pure strain from an arbitrary point $\boldsymbol{x}_0$.

Proof. Let

$$\boldsymbol{u}(\boldsymbol{x}) = \boldsymbol{u}_0 + \boldsymbol{A}[\boldsymbol{x} - \boldsymbol{y}_0],$$

$$\boldsymbol{E} = \tfrac{1}{2}(\boldsymbol{A} + \boldsymbol{A}^T), \quad \boldsymbol{W} = \tfrac{1}{2}(\boldsymbol{A} - \boldsymbol{A}^T).$$

Then

$$\boldsymbol{u} = \boldsymbol{w} + \hat{\boldsymbol{u}},$$

where

$$\boldsymbol{w}(\boldsymbol{x}) = \boldsymbol{w}_0 + \boldsymbol{W}[\boldsymbol{x} - \boldsymbol{x}_0], \quad \boldsymbol{w}_0 = \boldsymbol{u}_0 + \boldsymbol{A}[\boldsymbol{x}_0 - \boldsymbol{y}_0],$$

$$\hat{\boldsymbol{u}}(\boldsymbol{x}) = \boldsymbol{E}[\boldsymbol{x} - \boldsymbol{x}_0]. \quad □$$

Let $\boldsymbol{x}_0$ be a given point, and let

$$\boldsymbol{p}_0(\boldsymbol{x}) = \boldsymbol{x} - \boldsymbol{x}_0.$$

Some examples of pure strains from the point $\boldsymbol{x}_0$ are:

(a) ***simple extension*** of amount e in the direction $\boldsymbol{n}$, where $|\boldsymbol{n}| = 1$:

$$\boldsymbol{u} = e(\boldsymbol{n} \cdot \boldsymbol{p}_0)\, \boldsymbol{n},$$
$$\boldsymbol{E} = e\, \boldsymbol{n} \otimes \boldsymbol{n};$$

(b) ***uniform dilatation*** of amount e:

$$\boldsymbol{u} = e\, \boldsymbol{p}_0$$
$$\boldsymbol{E} = e\, \mathbf{1};$$

(c) ***simple shear*** of amount $\varkappa$ with respect to the direction pair $(\boldsymbol{m}, \boldsymbol{n})$, where $\boldsymbol{m}$ and $\boldsymbol{n}$ are perpendicular unit vectors:

$$\boldsymbol{u} = \varkappa[(\boldsymbol{m} \cdot \boldsymbol{p}_0)\, \boldsymbol{n} + (\boldsymbol{n} \cdot \boldsymbol{p}_0)\, \boldsymbol{m}],$$
$$\boldsymbol{E} = 2\varkappa\, \mathrm{sym}\,(\boldsymbol{m} \otimes \boldsymbol{n}) = \varkappa\, [\boldsymbol{m} \otimes \boldsymbol{n} + \boldsymbol{n} \otimes \boldsymbol{m}].$$

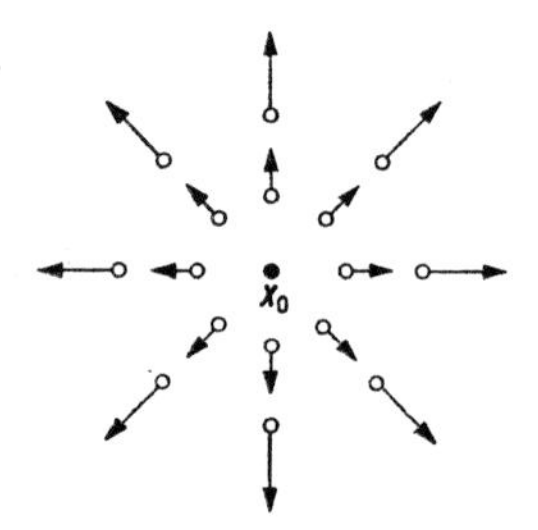

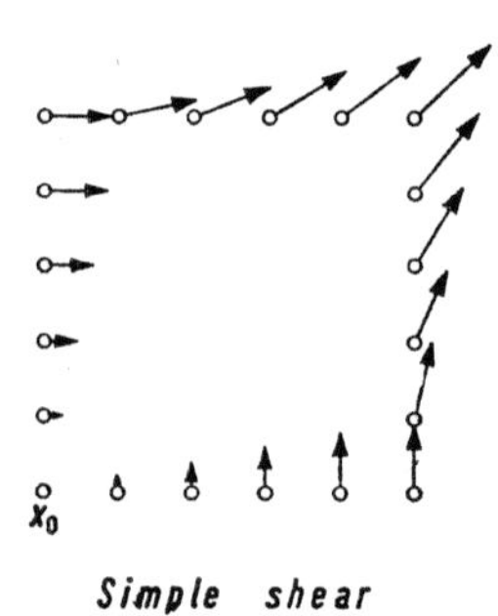

Fig. 3. Examples of pure strains from $\boldsymbol{x}_0$.

The displacement fields corresponding to (a), (b), and (c) are shown in Fig. 3. The matrix relative to an orthonormal basis $\{\boldsymbol{n}, \boldsymbol{e}_2, \boldsymbol{e}_3\}$ of $\boldsymbol{E}$ in the simple extension given in (a) is

$$[\boldsymbol{E}] = \begin{bmatrix} e & 0 & 0 \\ 0 & 0 & 0 \\ 0 & 0 & 0 \end{bmatrix},$$

and the matrix of $\boldsymbol{E}$ relative to an orthonormal basis $\{\boldsymbol{m}, \boldsymbol{n}, \boldsymbol{e}_3\}$ in the simple shear given in (c) is

$$[\boldsymbol{E}] = \begin{bmatrix} 0 & \varkappa & 0 \\ \varkappa & 0 & 0 \\ 0 & 0 & 0 \end{bmatrix}.$$

Notice that the homogeneous displacement field

$$\bar{\boldsymbol{u}} = 2\varkappa (\boldsymbol{m} \cdot \boldsymbol{p}_0)\, \boldsymbol{n}$$

has the same strain as the simple shear defined in (c) and hence differs from it by a rigid displacement field.

For a pure strain the ratio of the volume change $\delta v(B)$ of the body to its total volume $v(B)$ is $\operatorname{tr} \boldsymbol{E}$. Thus for a simple extension of amount e,

$$\frac{\delta v(B)}{v(B)} = e,$$

for a uniform dilatation of amount e,

$$\frac{\delta v(B)}{v(B)} = 3e,$$

and for any simple shear,

$$\frac{\delta v(B)}{v(B} = 0.$$

We call the principal values e_1, e_2, e_3 of a strain tensor $\boldsymbol{E}$ ***principal strains.*** By ***(3.2)***, $\boldsymbol{E}$ admits the spectral decomposition

$$\boldsymbol{E} = \sum_{i=1}^{3} e_i\, \boldsymbol{n}_i \otimes \boldsymbol{n}_i,$$

where $\boldsymbol{n}_i$ is a principal direction corresponding to e_i. We now use this decomposition to show that every pure strain can be accomplished in two ways: by three simple extensions in mutually perpendicular directions; by a uniform dilatation followed by an isochoric pure strain.

(5) Decomposition theorem for pure strains. *Let* $\boldsymbol{u}$ *be a pure strain from* $\boldsymbol{x}_0$. *Then* $\boldsymbol{u}$ *admits the following two decompositions*:

(i) $\boldsymbol{u} = \boldsymbol{u}_1 + \boldsymbol{u}_2 + \boldsymbol{u}_3$,

where $\boldsymbol{u}_1, \boldsymbol{u}_2$, *and* $\boldsymbol{u}_3$ *are simple extensions in mutually perpendicular directions from* $\boldsymbol{x}_0$;

(ii) $\boldsymbol{u} = \boldsymbol{u}_d + \boldsymbol{u}_c$,

where $\boldsymbol{u}_d$ *is a uniform dilatation from* $\boldsymbol{x}_0$, *while* $\boldsymbol{u}_c$ *is an isochoric pure strain from* $\boldsymbol{x}_0$.

Proof. Let $\boldsymbol{E}$ be the strain tensor corresponding to $\boldsymbol{u}$. It follows from the spectral decomposition for $\boldsymbol{E}$ that

$$\boldsymbol{E} = \boldsymbol{E}_1 + \boldsymbol{E}_2 + \boldsymbol{E}_3,$$

where

$$E_i = e_i \boldsymbol{n}_i \otimes \boldsymbol{n}_i \quad \text{(no sum)}.$$

Thus if we let

$$\boldsymbol{u}_i = \boldsymbol{E}_i \boldsymbol{p}_0, \qquad i = 1, 2, 3,$$

where

$$\boldsymbol{p}_0(\boldsymbol{x}) = \boldsymbol{x} - \boldsymbol{x}_0,$$

the decomposition (i) follows. On the other hand, if we let

$$\boldsymbol{u}_d = \tfrac{1}{3}(\operatorname{tr} \boldsymbol{E})\, \boldsymbol{p}_0,$$

$$\boldsymbol{u}_c = [\boldsymbol{E} - \tfrac{1}{3}(\operatorname{tr} \boldsymbol{E})\, \mathbf{1}]\, \boldsymbol{p}_0,$$

then $\boldsymbol{u}_d$ is a uniform dilatation, $\boldsymbol{u}_c$ is an isochoric pure strain, and

$$\boldsymbol{u}_d + \boldsymbol{u}_c = \boldsymbol{E}\boldsymbol{p}_0 = \boldsymbol{u}. \quad \square$$

(6) Decomposition theorem for simple shears. *Let $\boldsymbol{u}$ be a simple shear of amount $\varkappa$ with respect to the direction pair $(\boldsymbol{m}, \boldsymbol{n})$. Then $\boldsymbol{u}$ admits the decomposition*

$$\boldsymbol{u} = \boldsymbol{u}^+ + \boldsymbol{u}^-,$$

where $\boldsymbol{u}^\pm$ is a simple extension of amount $\pm\varkappa$ in the direction

$$\frac{1}{\sqrt{2}}(\boldsymbol{m} \pm \boldsymbol{n}).$$

Proof. Let $\boldsymbol{E}$ be the strain tensor corresponding to $\boldsymbol{u}$. Then

$$\boldsymbol{E} = \varkappa(\boldsymbol{m} \otimes \boldsymbol{n} + \boldsymbol{n} \otimes \boldsymbol{m}),$$

and it follows that λ is a principal value and $\boldsymbol{a}$ a principal vector of $\boldsymbol{E}$ if and only if

$$\varkappa(\boldsymbol{n} \cdot \boldsymbol{a})\, \boldsymbol{m} + \varkappa(\boldsymbol{m} \cdot \boldsymbol{a})\, \boldsymbol{n} = \lambda \boldsymbol{a}.$$

It is clear from this relation that the principal values of $\boldsymbol{E}$ are $0, +\varkappa, -\varkappa$, the corresponding principal directions

$$\boldsymbol{m} \times \boldsymbol{n}, \quad \frac{1}{\sqrt{2}}(\boldsymbol{m} + \boldsymbol{n}), \quad \frac{1}{\sqrt{2}}(\boldsymbol{m} - \boldsymbol{n}).$$

Thus we conclude from (i) of *(5)* and its proof that the decomposition given in *(6)* holds. $\square$

Trivially, every simple shear is isochoric. The next theorem asserts that every isochoric pure strain is the sum of simple shears.

(7) Decomposition theorem for isochoric pure strains. *Let $\boldsymbol{u}$ be an isochoric pure strain from $\boldsymbol{x}_0$. Then there exists an orthonormal basis $\{\boldsymbol{n}_1, \boldsymbol{n}_2, \boldsymbol{n}_3\}$ such that $\boldsymbol{u}$ admits the decomposition*

$$\boldsymbol{u} = \boldsymbol{u}_1 + \boldsymbol{u}_2 + \boldsymbol{u}_3,$$

where $\boldsymbol{u}_1$, $\boldsymbol{u}_2$, and $\boldsymbol{u}_3$ are simple shears from $\boldsymbol{x}_0$ with respect to the direction pairs $(\boldsymbol{n}_2, \boldsymbol{n}_3)$, $(\boldsymbol{n}_3, \boldsymbol{n}_1)$, and $(\boldsymbol{n}_1, \boldsymbol{n}_2)$, respectively.

Proof.[1] Let $\boldsymbol{E}$ be the strain tensor corresponding to $\boldsymbol{u}$. If $\boldsymbol{E} = \boldsymbol{0}$ the theorem is trivial; thus assume $\boldsymbol{E} \neq \boldsymbol{0}$. Clearly, it suffices to show that there exists an

[1] This proof was furnished by J. LEW (private communication) in 1968.

orthonormal basis $\{\boldsymbol{n}_1, \boldsymbol{n}_2, \boldsymbol{n}_3\}$ such that the matrix of $\boldsymbol{E}$ relative to this basis has zero entries on its diagonal:

$$\boldsymbol{n}_1 \cdot \boldsymbol{E}\boldsymbol{n}_1 = \boldsymbol{n}_2 \cdot \boldsymbol{E}\boldsymbol{n}_2 = \boldsymbol{n}_3 \cdot \boldsymbol{E}\boldsymbol{n}_3 = 0.$$

To prove this it is enough to establish the existence of non-zero vectors $\boldsymbol{\alpha}$, $\boldsymbol{\beta}$ such that

$$\boldsymbol{\alpha} \cdot \boldsymbol{\beta} = 0, \tag{a}$$

$$\boldsymbol{\alpha} \cdot \boldsymbol{E}\boldsymbol{\alpha} = \boldsymbol{\beta} \cdot \boldsymbol{E}\boldsymbol{\beta} = 0; \tag{b}$$

for if $\{\boldsymbol{n}_1, \boldsymbol{n}_2, \boldsymbol{n}_3\}$ is an orthonormal basis with

$$\boldsymbol{n}_1 = \frac{\boldsymbol{\alpha}}{|\boldsymbol{\alpha}|}, \qquad \boldsymbol{n}_2 = \frac{\boldsymbol{\beta}}{|\boldsymbol{\beta}|},$$

then (b) yields

$$\boldsymbol{n}_1 \cdot \boldsymbol{E}\boldsymbol{n}_1 = \boldsymbol{n}_2 \cdot \boldsymbol{E}\boldsymbol{n}_2 = 0,$$

and the fact that $\operatorname{tr} \boldsymbol{E} = 0$ implies

$$\boldsymbol{n}_3 \cdot \boldsymbol{E}\boldsymbol{n}_3 = 0.$$

Let $\boldsymbol{m}_1, \boldsymbol{m}_2, \boldsymbol{m}_3$ be principal directions for $\boldsymbol{E}$, and let $e_1, -e_2, -e_3$ be the corresponding principal strains. Since $\operatorname{tr} \boldsymbol{E} = 0$, we may assume, without loss in generality, that

$$e_1 \geqq e_2 \geqq e_3 \geqq 0$$

(if two of the principal strains are positive and one negative consider $-\boldsymbol{E}$ rather than $\boldsymbol{E}$). We shall now seek vectors $\boldsymbol{\alpha}$, $\boldsymbol{\beta}$ which satisfy not only (a) and (b), but also the condition

$$\beta_1 = \alpha_1, \qquad \beta_2 = \alpha_2, \qquad \beta_3 = -\alpha_3, \tag{c}$$

where α_i and β_i are the components of $\boldsymbol{\alpha}$ and $\boldsymbol{\beta}$ relative to the orthonormal basis $\boldsymbol{m}_1, \boldsymbol{m}_2, \boldsymbol{m}_3$. In view of (c), conditions (a) and (b) reduce to the system:

$$\alpha_1^2 + \alpha_2^2 - \alpha_3^2 = 0$$

$$e_1 \alpha_1^2 - e_2 \alpha_2^2 - e_3 \alpha_3^2 = 0.$$

It is easily verified that this system has the solution

$$\alpha_1^2 = e_2 + e_3,$$

$$\alpha_2^2 = e_1 - e_3,$$

$$\alpha_3^2 = e_1 + e_2,$$

and since $\boldsymbol{E} \neq \boldsymbol{0}$, the vectors $\boldsymbol{\alpha}$ and $\boldsymbol{\beta}$ so defined will be real and non-zero. □

We now consider displacement fields that are not necessarily homogeneous. Given a continuous strain field $\boldsymbol{E}$ on $\overline{B}$, we call the symmetric tensor

$$\overline{\boldsymbol{E}}(B) = \frac{1}{v(B)} \int_B \boldsymbol{E} \, dv$$

the ***mean strain***. Clearly,

$$\frac{\delta v(B)}{v(B)} = \operatorname{tr} \overline{\boldsymbol{E}}(B);$$

thus the volume change can be computed once the mean strain is known.

(8) Mean strain theorem. *Let* $\boldsymbol{u}$ *be a displacement field, let* $\boldsymbol{E}$ *be the corresponding strain field, and suppose that* $\boldsymbol{E}$ *and* $\boldsymbol{u}$ *are continuous on* $\overline{B}$. *Then the mean strain* $\overline{\boldsymbol{E}}(B)$ *depends only on the boundary values of* $\boldsymbol{u}$ *and is given by*

$$\overline{\boldsymbol{E}}(B) = \frac{1}{v(B)} \int_{\partial B} \operatorname{sym} \boldsymbol{u} \otimes \boldsymbol{n} \, da.$$

Proof. By the divergence theorem ***(6.1)***,

$$\int_{\partial B} (\boldsymbol{u} \otimes \boldsymbol{n} + \boldsymbol{n} \otimes \boldsymbol{u}) \, da = \int_B (\nabla \boldsymbol{u} + \nabla \boldsymbol{u}^T) \, dv = 2 \int_B \boldsymbol{E} \, dv. \quad \square$$

The next theorem, which is due to KORN, gives restrictions under which the L^2 norm of the gradient of $\boldsymbol{u}$ (i.e. the Dirichlet norm of $\boldsymbol{u}$) is bounded by a constant times the L^2 norm of the strain field.

(9) Korn's inequality.[1] *Let* $\boldsymbol{u}$ *be a class* C^2 *displacement field on* $\overline{B}$, *and assume that one of the following two hypotheses hold:*

(α) $\boldsymbol{u} = \boldsymbol{0}$ *on* ∂B;

(β) B *is properly regular, and either* $\boldsymbol{u} = \boldsymbol{0}$ *on a non-empty regular subsurface of* ∂B, *or*

$$\int_B \boldsymbol{W} \, dv = \boldsymbol{0},$$

where $\boldsymbol{W}$ *is the rotation field corresponding to* $\boldsymbol{u}$.

Then

$$\int_B |\nabla \boldsymbol{u}|^2 \, dv \leqq K \int_B |\boldsymbol{E}|^2 \, dv,$$

where $\boldsymbol{E}$ *is the strain field corresponding to* $\boldsymbol{u}$, *and* K *is a constant depending only on* B.

Proof. We will establish Korn's inequality only for the case in which (α) holds. The proof under (β), which is quite difficult, can be found in the article by FICHERA.[2]

Thus assume (α) holds. It follows from the definitions of $\boldsymbol{E}$ and $\boldsymbol{W}$ that

$$|\boldsymbol{E}|^2 = \tfrac{1}{2}(|\nabla \boldsymbol{u}|^2 + \nabla \boldsymbol{u} \cdot \nabla \boldsymbol{u}^T),$$
$$|\boldsymbol{W}|^2 = \tfrac{1}{2}(|\nabla \boldsymbol{u}|^2 - \nabla \boldsymbol{u} \cdot \nabla \boldsymbol{u}^T);$$

hence

$$|\boldsymbol{E}|^2 - |\boldsymbol{W}|^2 = \nabla \boldsymbol{u} \cdot \nabla \boldsymbol{u}^T. \tag{a}$$

Further, we have the identity

$$\nabla \boldsymbol{u} \cdot \nabla \boldsymbol{u}^T = \operatorname{div} [(\nabla \boldsymbol{u}) \, \boldsymbol{u} - (\operatorname{div} \boldsymbol{u}) \, \boldsymbol{u}] + (\operatorname{div} \boldsymbol{u})^2. \tag{b}$$

By (a), (b), the divergence theorem, and (α),

$$\int_B |\boldsymbol{E}|^2 \, dv - \int_B |\boldsymbol{W}|^2 \, dv = \int_B (\operatorname{div} \boldsymbol{u})^2 \, dv,$$

which implies

$$\int_B |\boldsymbol{W}|^2 \, dv \leqq \int_B |\boldsymbol{E}|^2 \, dv. \tag{c}$$

---

[1] KORN [1906, *4*], [1908, *1*], [1909, *3*]. Alternative proofs of Korn's inequality were given by FRIEDRICHS [1947, *1*], EIDUS [1951, *5*] (see also MIKHLIN [1952, *2*], §§ 40–42), GOBERT [1962, *6*], and FICHERA [1971, *1*]. Numerical values of K for various types of regions can be found in the papers by BERNSTEIN and TOUPIN [1960, *1*], PAYNE and WEINBERGER [1961, *17*], and DAFERMOS [1968, *3*].

[2] [1971, *1*], § 12.

Thus we conclude from (c) and the identity

$$|\nabla \boldsymbol{u}|^2 = |\boldsymbol{E}|^2 + |\boldsymbol{W}|^2$$

that

$$\int_B |\nabla \boldsymbol{u}|^2 \, dv \leqq 2 \int_B |\boldsymbol{E}|^2 \, dv;$$

therefore Korn's inequality holds with $K=2$. □

Let $\boldsymbol{u}$ be a displacement field on B. Then $\boldsymbol{u}$ is ***plane*** if for some choice of the coordinate frame[1]

$$u_\alpha = u_\alpha(x_1, x_2), \qquad u_3 = 0.$$

In this instance the domain of $\boldsymbol{u}$ can be identified with the following region R in the x_1, x_2-plane:

$$R = \{(x_1, x_2) : \boldsymbol{x} \in B\}.$$

Clearly, the corresponding strain field $\boldsymbol{E}$ is a function of (x_1, x_2) only and satisfies

$$E_{\alpha\beta} = \tfrac{1}{2}(u_{\alpha,\beta} + u_{\beta,\alpha}),$$
$$E_{13} = E_{23} = E_{33} = 0.$$

It is a simple matter to verify that a plane displacement field is rigid if and only if

$$u_\alpha(x) = a_\alpha + \omega\, \varepsilon_{\alpha\beta}\, x_\beta$$

where $x = (x_1, x_2)$ and a_α and ω are constants. We call a two-dimensional field u_α of this form a ***plane rigid displacement*** and a complex function of the form

$$u(z) = a_1 + i a_2 - i\omega z, \qquad z = x_1 + i x_2,$$

a ***complex rigid displacement***. Clearly, a complex function u is a complex rigid displacement if and only if its real and imaginary parts u_1 and u_2 are the components of a plane rigid displacement.

14. Compatibility. Given an arbitrary strain field $\boldsymbol{E}$, the strain displacement relation

$$\boldsymbol{E} = \tfrac{1}{2}(\nabla \boldsymbol{u} + \nabla \boldsymbol{u}^T)$$

constitutes a linear first-order partial differential equation for the displacement field $\boldsymbol{u}$. The uniqueness question appropriate to this equation was settled in the last section; we proved that any two solutions differ at most by a rigid displacement. The question of existence is far less trivial. We will show that a *necessary* condition for the existence of a displacement field is that the strain field satisfy a certain compatibility relation, and that this relation is also *sufficient* when the body is simply-connected.

The following proposition, which is of interest in itself, supplies the first step in the derivation of the equation of compatibility.

(1) *The strain field* $\boldsymbol{E}$ *and the rotation vector* $\boldsymbol{\omega}$ *corresponding to a class* C^2 *displacement field satisfy*

$$\operatorname{curl} \boldsymbol{E} = \nabla \boldsymbol{\omega}.$$

Proof. If we apply the curl operator to the strain-displacement relation and use (4) and (5) of ***(4.1)***, we find that

$$\operatorname{curl} \boldsymbol{E} = \tfrac{1}{2} \operatorname{curl}(\nabla \boldsymbol{u} + \nabla \boldsymbol{u}^T) = \tfrac{1}{2} \nabla \operatorname{curl} \boldsymbol{u} = \nabla \boldsymbol{\omega}. \quad \square$$

[1] Recall our agreement that Greek subscripts range over the integers 1, 2.

If we take the curl of the relation in ***(1)*** and use (4) of ***(4.1)***, we are immediately led to the conclusion that a necessary condition for the existence of a displacement field is that $\boldsymbol{E}$ satisfy the following ***equation of compatibility***:

$$\operatorname{curl}\operatorname{curl}\boldsymbol{E}=\mathbf{0}.$$

This result is summarized in the first portion of the next theorem; the second portion asserts the sufficiency of the compatibility relation when the body is simply-connected.

(2) Compatibility theorem.[1] *The strain field* $\mathbf{E}$ *corresponding to a class* C^3 *displacement field satisfies the equation of compatibility.*

Conversely, let B *be simply-connected, and let* $\mathbf{E}$ *be a class* C^N $(N\geqq 2)$ *symmetric tensor field on* B *that satisfies the equation of compatibility. Then there exists a displacement field* $\boldsymbol{u}$ *of class* C^{N+1} *on* B *such that* $\mathbf{E}$ *and* $\boldsymbol{u}$ *satisfy the strain-displacement relation.*

Proof. We have only to establish the converse assertion.[2] Thus let B be simply-connected, and assume that $\operatorname{curl}\operatorname{curl}\boldsymbol{E}=\mathbf{0}$. Let

$$\boldsymbol{A}=\operatorname{curl}\boldsymbol{E}. \tag{a}$$

Then

$$\operatorname{curl}\boldsymbol{A}=\mathbf{0}, \tag{b}$$

and since $\boldsymbol{E}$ is symmetric, (a) and (12) of ***(4.1)*** imply that

$$\operatorname{tr}\boldsymbol{A}=0. \tag{c}$$

By (b), (c), and (iii) of ***(6.3)***, there exists a class C^N *skew* tensor field $\boldsymbol{W}$ such that

$$\boldsymbol{A}=-\operatorname{curl}\boldsymbol{W}, \tag{d}$$

and (a), (d) imply that

$$\operatorname{curl}(\boldsymbol{E}+\boldsymbol{W})=\mathbf{0};$$

thus, by (ii) of ***(6.3)***, there exists a class C^{N+1} vector field $\boldsymbol{u}$ on B such that

$$\boldsymbol{E}+\boldsymbol{W}=\nabla\boldsymbol{u},$$

and taking the symmetric part of both sides of this equation, we arrive at the strain-displacement relation. □

In components the equation of compatibility takes the form

$$\varepsilon_{ijk}\,\varepsilon_{lmn}\,E_{jm,kn}=0,$$

or equivalently,

$$2E_{12,12}=E_{11,22}+E_{22,11},$$
$$E_{11,23}=(-E_{23,1}+E_{31,2}+E_{12,3})_{,1}, \quad \text{etc.}$$

Under the hypotheses of ***(2)*** it is possible to give an explicit formula which may be useful in computing a displacement field $\boldsymbol{u}$ corresponding to the strain

[1] Some of the basic ideas underlying this theorem are due to KIRCHHOFF [1859, *1*], who deduced three of the six equations of compatibility and indicated a procedure for determining the displacement when the strain is known. The complete equation of compatibility was first derived by SAINT-VENANT [1864, *1*], who asserted its sufficiency. The first rigorous proof of sufficiency was given by BELTRAMI [1886, *1*], [1889, *1*]. Cf. BOUSSINESQ [1871, *1*], KIRCHHOFF [1876, *1*], PADOVA [1889, *2*], E. and F. COSSERAT [1896, *1*], ABRAHAM [1901, *1*]. Explicit forms of the equation of compatibility in curvilinear coordinates were given by ODQVIST [1937, *4*], BLINCHIKOV [1938, *1*], and VLASOV [1944, *3*].

[2] This portion of the proof is simply a coordinate-free version of BELTRAMI's [1886, *1*] argument.

field $\boldsymbol{E}$. Let $\boldsymbol{x}_0$ be a fixed point in B. Then for each $\boldsymbol{x} \in B$ the line integral

$$\boldsymbol{u}(\boldsymbol{x}) = \int_{\boldsymbol{x}_0}^{\boldsymbol{x}} \boldsymbol{U}(\boldsymbol{y}, \boldsymbol{x})\, d\boldsymbol{y},$$

where

$$U_{ij}(\boldsymbol{y}, \boldsymbol{x}) = E_{ij}(\boldsymbol{y}) + (x_k - y_k)\, [E_{ij,k}(\boldsymbol{y}) - E_{kj,i}(\boldsymbol{y})],$$

is independent of the path in B from $\boldsymbol{x}_0$ to $\boldsymbol{x}$, and the function $\boldsymbol{u}$ so defined is a displacement field corresponding to the strain field $\boldsymbol{E}$.[1]

An alternative form of the equation of compatibility is given in the following proposition.

(3) *Let* $\mathbf{E}$ *be a class* C^2 *symmetric tensor field on* B. *Then* $\mathbf{E}$ *satisfies the equation of compatibility if and only if*

$$\Delta \boldsymbol{E} + \nabla\nabla(\operatorname{tr} \boldsymbol{E}) - 2\hat{\nabla} \operatorname{div} \boldsymbol{E} = \mathbf{0}.$$

Proof. The proof follows from the identities:

$$\operatorname{curl} \operatorname{curl} \boldsymbol{E} = -\Delta \boldsymbol{E} - \nabla\nabla(\operatorname{tr} \boldsymbol{E}) + 2\hat{\nabla} \operatorname{div} \boldsymbol{E} + \mathbf{1}[\Delta(\operatorname{tr} \boldsymbol{E}) - \operatorname{div} \operatorname{div} \boldsymbol{E}], \quad \text{(a)}$$

$$\operatorname{tr}[\Delta \boldsymbol{E} + \nabla\nabla(\operatorname{tr} \boldsymbol{E}) - 2\nabla \operatorname{div} \boldsymbol{E}] = 2[\Delta(\operatorname{tr} \boldsymbol{E}) - \operatorname{div} \operatorname{div} \boldsymbol{E}]. \quad \text{(b)}$$

Indeed, let $\operatorname{curl} \operatorname{curl} \boldsymbol{E} = \mathbf{0}$. Then, if we take the trace of (a), we conclude, with the aid of (b), that

$$\Delta(\operatorname{tr} \boldsymbol{E}) - \operatorname{div} \operatorname{div} \boldsymbol{E} = 0, \quad \text{(c)}$$

and this result, in view of (a), implies the relation in ***(3)***. Conversely, if that relation holds, then (b) implies (c), and (c) implies $\operatorname{curl} \operatorname{curl} \boldsymbol{E} = \mathbf{0}$. □

The following proposition will be quite useful.

(4) *Assume that* B *is simply-connected. Let* $\boldsymbol{u}$ *be a displacement field, and assume that the corresponding strain field* $\mathbf{E}$ *is of class* C^2 *on* B *and of class* C^N $(N \geq 1)$ *on* $\bar{B}$. *Then* $\boldsymbol{u}$ *is of class* C^{N+1} *on* $\bar{B}$.

Proof. Let $\boldsymbol{u}'$ be the displacement field generated by $\boldsymbol{E}$ using the procedure given in the proof of ***(2)***. By ***(13.2)***, $\boldsymbol{u}$ and $\boldsymbol{u}'$ differ by a rigid displacement field; hence it suffices to prove that $\boldsymbol{u}'$ is of class C^{N+1} on $\bar{B}$. Since $\operatorname{curl} \boldsymbol{E}$ is of class C^{N-1} on $\bar{B}$, the function $\boldsymbol{W}$ in (d) in the proof of ***(2)*** is of class C^N on $\bar{B}$, and since

$$\boldsymbol{E} + \boldsymbol{W} = \nabla \boldsymbol{u}',$$

$\boldsymbol{u}'$ is of class C^{N+1} on $\bar{B}$. □

(5) Compatibility theorem for plane displacements. *Assume that* R *is a simply-connected open region in the* x_1, x_2*-plane. Let* u_α *be a class* C^3 *field on* R, *and let*

$$E_{\alpha\beta} = \tfrac{1}{2}(u_{\alpha,\beta} + u_{\beta,\alpha}). \quad \text{(i)}$$

Then

$$2E_{12,12} = E_{11,22} + E_{22,11}. \quad \text{(ii)}$$

Conversely, let $E_{\alpha\beta}\,(=E_{\beta\alpha})$ *be a class* C^N $(N \geq 2)$ *field on* R *that satisfies* (ii). *Then there exists a class* C^{N+1} *field* u_α *on* R *such that* (i) *holds.*

[1] This explicit solution is due to CESÀRO [1906, *2*]. In this connection, see also VOLTERRA [1907, *4*], SOKOLNIKOFF [1956, *12*], BOLEY and WEINER [1960, *3*].

Proof. That (i) implies (ii) follows upon direct substitution. To prove the converse assertion assume that (ii) holds. Then

$$(E_{11,2}-E_{12,1})_{,2}=(E_{12,2}-E_{22,1})_{,1}.$$

Thus there exists a function φ on R of class C^N such that

$$E_{11,2}-E_{12,1}=\varphi_{,1}, \quad E_{12,2}-E_{22,1}=\varphi_{,2},$$

or equivalently,

$$E_{11,2}=(E_{12}+\varphi)_{,1}, \quad E_{22,1}=(E_{12}-\varphi)_{,2}.$$

Therefore there exist class C^{N+1} functions u_α on R such that

$$E_{11}=u_{1,1}, \qquad E_{22}=u_{2,2},$$
$$E_{12}+\varphi=u_{1,2}, \quad E_{12}-\varphi=u_{2,1}.$$

The first two equations are simply (i) for the case $\alpha=\beta$. On the other hand, since $E_{\alpha\beta}=E_{\beta\alpha}$, the last two equations, when added together, yield (i) for $\alpha\neq\beta$. □

Using the two-dimensional alternator $\varepsilon_{\alpha\beta}$, we can write (ii) in the form

$$\varepsilon_{\alpha\beta}\,\varepsilon_{\gamma\delta}\,E_{\alpha\gamma,\beta\delta}=0,$$

which more closely resembles its three-dimensional counterpart.

We now return to the general three-dimensional theory. A ***Volterra dislocation***[1] on B is a vector field $\boldsymbol{u}$ with the following properties:

(i) $\boldsymbol{u}$ is a class C^2 field on $B-\mathscr{S}$, where $\mathscr{S}$ is a regular surface in B.

(ii) $\boldsymbol{u}$ is not continuous across $\mathscr{S}$.

(iii) the strain field $\boldsymbol{E}$ corresponding to $\boldsymbol{u}$ is continuous across $\mathscr{S}$, and the extension (by continuity) of $\boldsymbol{E}$ to B is of class C^2.

(6) *There does not exist a Volterra dislocation on B if B is simply-connected.*

Proof. Let B be simply-connected. Assume there exists a Volterra dislocation $\boldsymbol{u}$ on B, and let $\boldsymbol{E}$ be the corresponding strain field. By the compatibility theorem ***(2)***, there exists a class C^3 displacement field $\boldsymbol{u}'$ on B such that $\boldsymbol{u}'$ and $\boldsymbol{E}$ satisfy the strain displacement relation. Thus $\boldsymbol{u}$ and $\boldsymbol{u}'$ correspond to the same strain field on $B-\mathscr{S}$, and ***(13.2)*** implies

$$\boldsymbol{u}=\boldsymbol{u}'+\boldsymbol{w} \quad \text{on } B-\mathscr{S},$$

where $\boldsymbol{w}$ is a rigid displacement field. But $\boldsymbol{u}'$ and $\boldsymbol{w}$ are continuous across $\mathscr{S}$; thus $\boldsymbol{u}$ must be continuous across $\mathscr{S}$, which contradicts (ii). □

II. Balance of momentum. The equations of motion and equilibrium.

15. Balance of momentum. Stress. In this section we determine consequences of the laws of momentum balance when the underlying motion is *infinitesimal*. The relationship between these results and the corresponding results for finite motions will be discussed in Sect. 16.

[1] VOLTERRA [1907, *4*], generalizing an idea of WEINGARTEN [1901, *7*] and TIMPE [1905, *2*]. See also BURGERS [1939, *1*].

We assume given on $\bar{B}$ a continuous strictly positive function ϱ called the ***density***; the mass of any part P of B is then

$$\int_P \varrho\, dv.$$

Let $(0, t_0)$ denote a fixed interval of time. A ***motion*** of the body is a class C^2 vector field $\boldsymbol{u}$ on $B\times(0, t_0)$. The vector $\boldsymbol{u}(\boldsymbol{x}, t)$ is the *displacement* of $\boldsymbol{x}$ at time t, while the fields $\dot{\boldsymbol{u}}$, $\ddot{\boldsymbol{u}}$,

$$\boldsymbol{E} = \tfrac{1}{2}(\nabla\boldsymbol{u} + \nabla\boldsymbol{u}^T),$$

and $\dot{\boldsymbol{E}}$ are the *velocity, acceleration, strain,* and *strain-rate.* We say that a motion is ***admissible*** if $\boldsymbol{u}$, $\dot{\boldsymbol{u}}$, $\ddot{\boldsymbol{u}}$, $\boldsymbol{E}$, and $\dot{\boldsymbol{E}}$ are continuous on $\bar{B}\times[0, t_0)$. Given an admissible motion $\boldsymbol{u}$ and a part P of B,

$$\boldsymbol{l}(P) = \int_P \dot{\boldsymbol{u}}\,\varrho\, dv$$

is the *linear momentum* of P, and

$$\boldsymbol{h}(P) = \int_P \boldsymbol{p}\times\dot{\boldsymbol{u}}\,\varrho\, dv$$

is the *angular momentum* (about the origin $\boldsymbol{0}$) of P. Note that, for P fixed, $\boldsymbol{l}(P)$ and $\boldsymbol{h}(P)$ are smooth functions of time on $[0, t_0)$; in fact,

$$\dot{\boldsymbol{l}}(P) = \int_P \ddot{\boldsymbol{u}}\,\varrho\, dv, \qquad \dot{\boldsymbol{h}}(P) = \int_P \boldsymbol{p}\times\ddot{\boldsymbol{u}}\,\varrho\, dv.$$

A ***system of forces*** $\mathscr{f}$ for the body is defined by assigning to each $(\boldsymbol{x}, t)\in\bar{B}\times[0, t_0)$ a vector $\boldsymbol{b}(\boldsymbol{x}, t)$ and, for each unit vector $\boldsymbol{n}$, a vector $\boldsymbol{s}_{\boldsymbol{n}}(\boldsymbol{x}, t)$ such that:

(i) $\boldsymbol{s}_{\boldsymbol{n}}$ is continuous on $\bar{B}\times[0, t_0)$ and of class $C^{1,0}$ on $B\times(0, t_0)$;

(ii) $\boldsymbol{b}$ is continuous on $\bar{B}\times[0, t_0)$.

We call $\boldsymbol{s}_{\boldsymbol{n}}(\boldsymbol{x}, t)$ the ***stress vector*** at $(\boldsymbol{x}, t)$. Let $\mathscr{S}$ be an oriented regular surface in B with unit normal $\boldsymbol{n}$ (Fig. 4). Then $\boldsymbol{s}_{\boldsymbol{n}(\boldsymbol{x})}(\boldsymbol{x}, t)$ is the force per unit

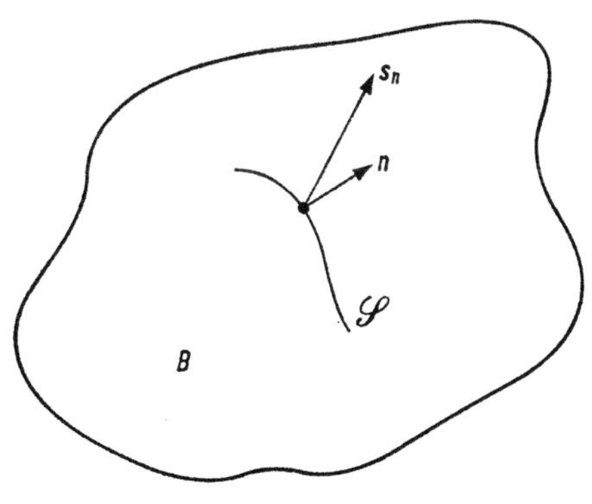

Fig. 4.

area at $\boldsymbol{x}$ exerted *by* the portion of B on the side of $\mathscr{S}$ toward which $\boldsymbol{n}(\boldsymbol{x})$ points *on* the portion of B on the other side; thus

$$\int_{\mathscr{S}} \boldsymbol{s}_{\boldsymbol{n}}\, da \equiv \int_{\mathscr{S}} \boldsymbol{s}_{\boldsymbol{n}(\boldsymbol{x})}(\boldsymbol{x}, t)\, da_{\boldsymbol{x}}$$

and

$$\int_{\mathscr{S}} \boldsymbol{p}\times\boldsymbol{s_n}\,da \equiv \int_{\mathscr{S}} \boldsymbol{p}(\boldsymbol{x})\times\boldsymbol{s}_{\boldsymbol{n}(\boldsymbol{x})}(\boldsymbol{x},t)\,da_{\boldsymbol{x}}$$

represent the total force and moment across $\mathscr{S}$. The same consideration also applies when $\boldsymbol{x}$ is located on the boundary of B and $\boldsymbol{n}$ is the outward unit normal to ∂B at $\boldsymbol{x}$; in this case $\boldsymbol{s_n}(\boldsymbol{x},t)$ is called the ***surface traction*** at $(\boldsymbol{x},t)$. The vector $\boldsymbol{b}(\boldsymbol{x},t)$ is the ***body force*** at $(\boldsymbol{x},t)$; it represents the force per unit volume exerted on the point $\boldsymbol{x}$ by bodies exterior to B. The *total force* $\boldsymbol{f}(P)$ on a part P is the total surface force exerted across ∂P plus the total body force exerted on P by the external world:

$$\boldsymbol{f}(P) = \int_{\partial P} \boldsymbol{s_n}\,da + \int_P \boldsymbol{b}\,dv.$$

Analogously, the *total moment* $\boldsymbol{m}(P)$ on P (about $\boldsymbol{0}$) is given by

$$\boldsymbol{m}(P) = \int_{\partial P} \boldsymbol{p}\times\boldsymbol{s_n}\,da + \int_P \boldsymbol{p}\times\boldsymbol{b}\,dv.$$

An ordered array $[\boldsymbol{u},\mathscr{f}]$, where $\boldsymbol{u}$ is an admissible motion and $\mathscr{f}$ a system of forces, is called a ***dynamical process***[1] if it obeys the following postulate: for every part P of B

$$\boldsymbol{f}(P) = \dot{\boldsymbol{l}}(P) \tag{m$_1$}$$

and

$$\boldsymbol{m}(P) = \dot{\boldsymbol{h}}(P). \tag{m$_2$}$$

These two relations constitute the laws of ***balance of linear and angular momentum***;[2] (m_1) is the requirement that the total force on P be equal to the rate of change of linear momentum, (m_2) that the total moment be equal to the rate of change of angular momentum. Note that by (m_1), the relation (m_2) holds for every choice of the origin $\boldsymbol{0}$ provided it holds for one such choice. Clearly, (m_1) and (m_2) can be written in the alternative forms:

$$\int_{\partial P} \boldsymbol{s_n}\,da + \int_P \boldsymbol{b}\,dv = \int_P \ddot{\boldsymbol{u}}\varrho\,dv, \tag{m$_1'$}$$

$$\int_{\partial P} \boldsymbol{p}\times\boldsymbol{s_n}\,da + \int_P \boldsymbol{p}\times\boldsymbol{b}\,dv = \int_P \boldsymbol{p}\times\ddot{\boldsymbol{u}}\varrho\,dv. \tag{m$_2'$}$$

For future use, we note that (m_2) is equivalent to the relation:

$$\operatorname{skw}\left\{\int_{\partial P} \boldsymbol{p}\otimes\boldsymbol{s_n}\,da + \int_P \boldsymbol{p}\otimes\boldsymbol{b}\,dv\right\} = \operatorname{skw}\int_P \boldsymbol{p}\otimes\ddot{\boldsymbol{u}}\varrho\,dv. \tag{m$_2''$}$$

The next theorem is one of the major results of continuum mechanics.

(1) Cauchy-Poisson theorem.[3] *Let* $\boldsymbol{u}$ *be an admissible motion and* $\mathscr{f}$ *a system of forces. Then* $[\boldsymbol{u},\mathscr{f}]$ *is a dynamical process if and only if the following two conditions are satisfied:*

[1] For the results of this section weaker definitions would suffice for an admissible motion, a system of forces, and a dynamical process. The strong definitions given here allow us to use these definitions without change in later sections.

[2] Cf. the discussion given by TRUESDELL and TOUPIN [1960, *17*], § 196.

[3] CAUCHY [1823, *1*], [1827, *1*] proved that balance of linear and angular momentum implies (i) and (ii) in (1), while POISSON [1829, *2*] established the converse assertion. In a sense, the essential ideas are implied or presumed in memoirs written by FRESNEL in 1822 [1868, *1*], but his work rests heavily on the constitutive assumptions of linear elasticity. In a still more limited sense the scalar counterpart of (i) is foreshadowed in a work on heat conduction written by FOURIER in 1814 [1822, *1*], which is even more involved with special constitutive relations. For a discussion of the history of this theorem, the reader is referred to TRUESDELL and TOUPIN [1960, *17*]. GURTIN, MIZEL, and WILLIAMS [1968, *6*] established the existence of the stress tensor under somewhat weaker hypotheses.

(i) *there exists a class* $C^{1,0}$ *symmetric tensor field* $\boldsymbol{S}$ *on* $B\times(0,t_0)$, *called the* ***stress field***, *such that for each unit vector* $\boldsymbol{n}$,

$$\boldsymbol{s_n}=\boldsymbol{S}\boldsymbol{n};$$

(ii) $\boldsymbol{u}$, $\boldsymbol{S}$, *and* $\boldsymbol{b}$ *satisfy the* ***equation of motion***:

$$\operatorname{div}\boldsymbol{S}+\boldsymbol{b}=\varrho\ddot{\boldsymbol{u}}.$$

The proof of this theorem is based on two lemmas. The first is usually referred to as the law of action and reaction.

(2) Cauchy's reciprocal theorem. *Let* $[\boldsymbol{u},\mathscr{f}]$ *be a dynamical process. Then given any unit vector* $\boldsymbol{n}$,

$$\boldsymbol{s_n}=-\boldsymbol{s_{-n}}.$$

Proof. Since B is bounded, the properties of ϱ, $\boldsymbol{u}$, and $\boldsymbol{b}$ imply that

$$k(t)=\sup_{\boldsymbol{x}\in B}|\boldsymbol{b}(\boldsymbol{x},t)-\varrho(\boldsymbol{x})\,\ddot{\boldsymbol{u}}(\boldsymbol{x},t)|$$

is finite for $0\leqq t<t_0$. Thus we conclude from (m_1') that

$$\Big|\int_{\partial P}\boldsymbol{s_n}\,da\Big|\leqq k\,v(P) \tag{a}$$

on $[0,t_0)$, where $v(P)$ is the volume of P.

Now choose a point $\boldsymbol{x}_0\in B$, a time $t\in(0,t_0)$, and a unit vector $\boldsymbol{m}$. For convenience we shall suppress the argument t in what follows. Let P_ε be a rectangular parallelepiped contained in B with center at $\boldsymbol{x}_0$ and sides parallel to $\boldsymbol{m}$ (see Fig. 5).

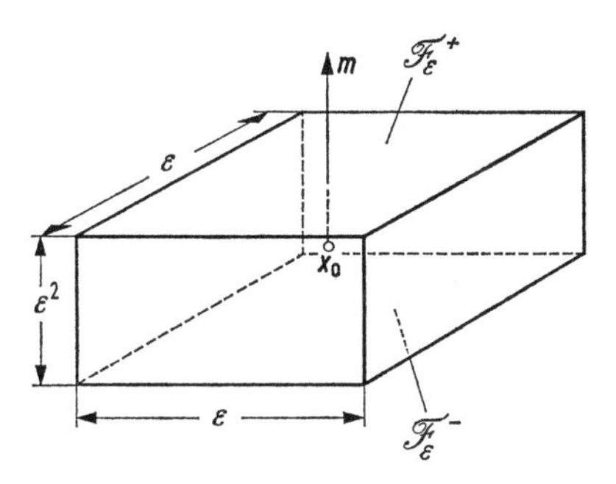

Fig. 5.

Suppose further that the top and bottom faces $\mathscr{F}_\varepsilon^+$ and $\mathscr{F}_\varepsilon^-$ with exterior normals $\boldsymbol{m}$ and $-\boldsymbol{m}$ are squares of length ε, and that the height of P_ε is ε^2. Let $\mathscr{C}_\varepsilon$ denote the union of the four side faces. Then

$$\partial P_\varepsilon=\mathscr{F}_\varepsilon^+\cup\mathscr{F}_\varepsilon^-\cup\mathscr{C}_\varepsilon \tag{b}$$

and

$$v(P_\varepsilon)=\varepsilon^4,\qquad a(\mathscr{F}_\varepsilon^\pm)=\varepsilon^2,\qquad a(\mathscr{C}_\varepsilon)=4\varepsilon^3, \tag{c}$$

where, for any surface $\mathscr{S}$, $a(\mathscr{S})$ denotes the area of $\mathscr{S}$. By (a) and (c),

$$\frac{1}{\varepsilon^2}\int_{\partial P_\varepsilon}\boldsymbol{s_n}\,da\to 0\quad\text{as}\quad\varepsilon\to 0. \tag{d}$$

Further, since $\boldsymbol{s}_{\boldsymbol{l}}$ is continuous on B for each fixed unit vector $\boldsymbol{l}$, we conclude from (c) that

$$\frac{1}{\varepsilon^2}\int\limits_{\mathscr{F}_\varepsilon^\pm} \boldsymbol{s}_{\pm\boldsymbol{m}}\, da \to \boldsymbol{s}_{\pm\boldsymbol{m}}(\boldsymbol{x}_0) \quad \text{and} \quad \frac{1}{\varepsilon^2}\int\limits_{\mathscr{C}_\varepsilon} \boldsymbol{s}_{\boldsymbol{n}}\, da \to 0 \quad \text{as} \quad \varepsilon \to 0. \tag{e}$$

By (b), (d), and (e),

$$\boldsymbol{s}_{\boldsymbol{m}}(\boldsymbol{x}_0) + \boldsymbol{s}_{-\boldsymbol{m}}(\boldsymbol{x}_0) = 0,$$

which completes the proof, since $\boldsymbol{m}$ and $\boldsymbol{x}_0$ are arbitrary. □

(3) Lemma. *Let* $\boldsymbol{S}$ *be a class* $C^{1,0}$ *tensor field on* $B\times(0,t_0)$ *with* $\boldsymbol{S}$ *and* $\operatorname{div}\boldsymbol{S}$ *continuous on* $\bar{B}\times[0,t_0)$. *Then given any part* P *of* B,

$$\int\limits_{\partial P} \boldsymbol{p}\times(\boldsymbol{S}\boldsymbol{n})\, da = \int\limits_{P} \boldsymbol{p}\times\operatorname{div}\boldsymbol{S}\, dv + 2\int\limits_{P} \boldsymbol{\sigma}\, dv$$

on $[0,t_0)$, *where* $\boldsymbol{\sigma}$ *is the axial vector corresponding to the skew part of* $\boldsymbol{S}$.

Proof. Let

$$\boldsymbol{j} = \int\limits_{\partial P} \boldsymbol{p}\times(\boldsymbol{S}\boldsymbol{n})\, da.$$

Then given any vector $\boldsymbol{e}$,

$$\boldsymbol{j}\cdot\boldsymbol{e} = \int\limits_{\partial P} (\boldsymbol{e}\times\boldsymbol{p})\cdot\boldsymbol{S}\boldsymbol{n}\, da = \int\limits_{\partial P} \boldsymbol{n}\cdot\boldsymbol{S}^T(\boldsymbol{e}\times\boldsymbol{p})\, da = \int\limits_{P} \operatorname{div}\{\boldsymbol{S}^T(\boldsymbol{e}\times\boldsymbol{p})\}\, dv,$$

and hence (11) of *(4.1)* implies

$$\boldsymbol{j}\cdot\boldsymbol{e} = \int\limits_{P} (\boldsymbol{e}\times\boldsymbol{p})\cdot\operatorname{div}\boldsymbol{S}\, dv + \int\limits_{P} \boldsymbol{S}\cdot\nabla(\boldsymbol{e}\times\boldsymbol{p})\, dv. \tag{a}$$

Next,

$$\begin{aligned} (\boldsymbol{e}\times\boldsymbol{p})\cdot\operatorname{div}\boldsymbol{S} &= \boldsymbol{e}\cdot(\boldsymbol{p}\times\operatorname{div}\boldsymbol{S}), \\ \boldsymbol{S}\cdot\nabla(\boldsymbol{e}\times\boldsymbol{p}) &= 2\boldsymbol{e}\cdot\boldsymbol{\sigma}; \end{aligned} \tag{b}$$

since $\boldsymbol{e}$ is arbitrary, (a) and (b) imply

$$\boldsymbol{j} = \int\limits_{P} \boldsymbol{p}\times\operatorname{div}\boldsymbol{S}\, dv + 2\int\limits_{P} \boldsymbol{\sigma}\, dv. \quad \square$$

Proof of (1). Assume first that (i) and (ii) of *(1)* hold. Then it follows from the properties of $\boldsymbol{u}$ and $\boldsymbol{s}_{\boldsymbol{n}}$ that $\boldsymbol{S}$ and $\operatorname{div}\boldsymbol{S}$ are continuous on $\bar{B}\times[0,t_0)$. Thus given any part P of B, (ii) and the divergence theorem imply (m_1'). Next, since $\boldsymbol{S}$ is

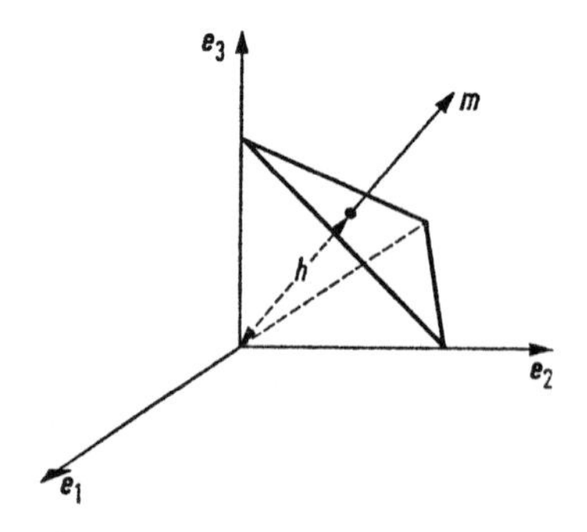

Fig. 6. Tetrahedron for the case in which $\boldsymbol{k}_1 = \boldsymbol{e}_1$, $\boldsymbol{k}_2 = -\boldsymbol{e}_2$, $\boldsymbol{k}_3 = -\boldsymbol{e}_3$.

symmetric, its axial vector $\boldsymbol{\sigma}$ vanishes, and **(3)** together with (i) and (ii) of **(1)** imply

$$\int_{\partial P} \boldsymbol{p}\times\boldsymbol{s_n}\, da=\int_P \boldsymbol{p}\times\operatorname{div}\boldsymbol{S}\, dv=\int_P \boldsymbol{p}\times(\varrho\ddot{\boldsymbol{u}}-\boldsymbol{b})\, dv,$$

which is (m_2').

Conversely, assume that $[\boldsymbol{u}, \mathscr{S}]$ is a dynamical process. Let $\boldsymbol{m}$ be a unit vector and assume that $\boldsymbol{m}\neq\pm\boldsymbol{e}_i$ for any base vector $\boldsymbol{e}_i$ of the orthonormal basis $\{\boldsymbol{e}_1, \boldsymbol{e}_2, \boldsymbol{e}_3\}$. Choose a point $\boldsymbol{x}\in B$, and consider the tetrahedron $P(h)$ in B whose sides $\pi(h)$ and $\pi_i(h)$ have outward unit normals $\boldsymbol{m}$ and $\boldsymbol{k}_i$, where

$$\boldsymbol{k}_i=-[\operatorname{sgn}(\boldsymbol{e}_i\cdot\boldsymbol{m})]\,\boldsymbol{e}_i \quad \text{(no sum)},$$

and whose faces $\pi_i(h)$ intersect at $\boldsymbol{x}$ (see Fig. 6). Let the area of $\pi(h)$ be $a(h)$, so that the area of $\pi_i(h)$ is $-a(h)\,\boldsymbol{m}\cdot\boldsymbol{k}_i$. Since $\boldsymbol{s_n}$ is continuous on $B\times(0, t_0)$ for each fixed $\boldsymbol{n}$, we conclude that

$$\frac{1}{a(h)}\int_{\pi(h)} \boldsymbol{s_m}\, da\to\boldsymbol{s_m}(\boldsymbol{x})$$

and, by **(2)**, that

$$\frac{1}{a(h)}\int_{\pi_i(h)} \boldsymbol{s}_{\boldsymbol{k}_i}\, da\to-(\boldsymbol{m}\cdot\boldsymbol{k}_i)\,\boldsymbol{s}_{\boldsymbol{k}_i}(\boldsymbol{x})=-(\boldsymbol{m}\cdot\boldsymbol{e}_i)\,\boldsymbol{s}_{\boldsymbol{e}_i}(\boldsymbol{x}) \quad \text{(no sum)} \tag{a}$$

as $h\to 0$, where we have suppressed the argument t. On the other hand, it follows from the inequality (a) in the proof of **(2)** that

$$\frac{1}{a(h)}\int_{\partial P(h)} \boldsymbol{s_n}\, da\to 0 \quad \text{as} \quad h\to 0. \tag{b}$$

Thus, since $\partial P(h)$ is the union of $\pi(h)$ and all three $\pi_i(h)$, (a) and (b) imply

$$\boldsymbol{s_m}(\boldsymbol{x})=(\boldsymbol{m}\cdot\boldsymbol{e}_i)\,\boldsymbol{s}_{\boldsymbol{e}_i}(\boldsymbol{x})=[\boldsymbol{s}_{\boldsymbol{e}_i}(\boldsymbol{x})\otimes\boldsymbol{e}_i]\,\boldsymbol{m}. \tag{c}$$

Our derivation of (c) required that $\boldsymbol{m}\neq\pm\boldsymbol{e}_i$ hold for every i. However, it follows from **(2)** that the first equation of (c) is also valid when $\boldsymbol{m}=\pm\boldsymbol{e}_i$; thus (c) holds for every unit vector $\boldsymbol{m}$. Now let $\boldsymbol{S}$ be the tensor field on $B\times(0, t_0)$ defined by

$$\boldsymbol{S}=\boldsymbol{s}_{\boldsymbol{e}_i}\otimes\boldsymbol{e}_i.$$

Then

$$\boldsymbol{s_n}=\boldsymbol{S}\boldsymbol{n} \tag{d}$$

for every unit vector $\boldsymbol{n}$. Further, $\boldsymbol{S}$ is of class $C^{1,0}$ on $B\times(0, t_0)$ and continuous on $\bar{B}\times[0, t_0)$, because $\boldsymbol{s_n}$ has these properties for every $\boldsymbol{n}$.

Next, if we apply (m_1') to an arbitrary part P (with $\bar{P}\subset B$) and use (d) and the divergence theorem we conclude that

$$\operatorname{div}\boldsymbol{S}+\boldsymbol{b}=\varrho\ddot{\boldsymbol{u}}. \tag{e}$$

Thus to complete the proof we have only to show that $\boldsymbol{S}$ is symmetric. By (m_2'), (d), and **(3)**,

$$\int_P \boldsymbol{\sigma}\, dv=0$$

for every part P with $\bar{P}\subset B$. Thus $\boldsymbol{\sigma}=\boldsymbol{0}$ on $B\times(0, t_0)$, which yields the symmetry of $\boldsymbol{S}$. □

We now give an alternative proof of the existence of the stress tensor.[1] Let $[\boldsymbol{u}, \boldsymbol{f}]$ be a dynamical process, and let $(\boldsymbol{x}_0, t) \in B \times [0, t_0)$. It suffices to show that the mapping $\boldsymbol{n} \to \boldsymbol{s}_{\boldsymbol{n}}(\boldsymbol{x}_0, t)$ is the restriction (to the set of unit vectors) of a linear function on $\mathscr{V}$. For convenience, we write

$$\boldsymbol{s}(\boldsymbol{x}, \boldsymbol{n}) = \boldsymbol{s}_{\boldsymbol{n}}(\boldsymbol{x}, t);$$

then for any $\boldsymbol{x} \in B$ we can extend the function $\boldsymbol{s}(\boldsymbol{x}, \cdot)$ to all of $\mathscr{V}$ as follows:

$$\begin{aligned} \boldsymbol{s}(\boldsymbol{x}, \boldsymbol{v}) &= |\boldsymbol{v}|\, \boldsymbol{s}\left(\boldsymbol{x}, \frac{\boldsymbol{v}}{|\boldsymbol{v}|}\right), \quad \boldsymbol{v} \neq \boldsymbol{0}, \\ \boldsymbol{s}(\boldsymbol{x}, \boldsymbol{0}) &= \boldsymbol{0}. \end{aligned} \tag{a}$$

Let α be a scalar. If $\alpha > 0$, then

$$\boldsymbol{s}(\alpha \boldsymbol{v}) = |\alpha \boldsymbol{v}|\, \boldsymbol{s}\left(\frac{\alpha \boldsymbol{v}}{|\alpha \boldsymbol{v}|}\right) = \alpha |\boldsymbol{v}|\, \boldsymbol{s}\left(\frac{\boldsymbol{v}}{|\boldsymbol{v}|}\right) = \alpha\, \boldsymbol{s}(\boldsymbol{v}), \tag{b}$$

where we have omitted the argument $\boldsymbol{x}$. If $\alpha < 0$, then (b) and Cauchy's reciprocal theorem *(2)* yield

$$\boldsymbol{s}(\alpha \boldsymbol{v}) = \boldsymbol{s}(|\alpha|(-\boldsymbol{v})) = |\alpha|\, \boldsymbol{s}(-\boldsymbol{v}) = \alpha\, \boldsymbol{s}(\boldsymbol{v}).$$

Thus $\boldsymbol{s}(\boldsymbol{x}, \cdot)$ is *homogeneous.*

To show that $\boldsymbol{s}(\boldsymbol{x}, \cdot)$ is *additive* we first note that

$$\boldsymbol{s}(\boldsymbol{x}, \boldsymbol{w}_1 + \boldsymbol{w}_2) = \boldsymbol{s}(\boldsymbol{x}, \boldsymbol{w}_1) + \boldsymbol{s}(\boldsymbol{x}, \boldsymbol{w}_2)$$

whenever $\boldsymbol{w}_1$ and $\boldsymbol{w}_2$ are linearly dependent. Suppose then that $\boldsymbol{w}_1$ and $\boldsymbol{w}_2$ are linearly independent. Fix $\varepsilon > 0$ and consider π_1, the plane through $\boldsymbol{x}_0$ with normal $\boldsymbol{w}_1$; π_2, the plane through $\boldsymbol{x}_0$ with normal $\boldsymbol{w}_2$; and π_3, the plane through $\boldsymbol{x}_0 + \varepsilon \boldsymbol{w}_3$ with normal $\boldsymbol{w}_3$, where

$$\boldsymbol{w}_3 = -(\boldsymbol{w}_1 + \boldsymbol{w}_2). \tag{c}$$

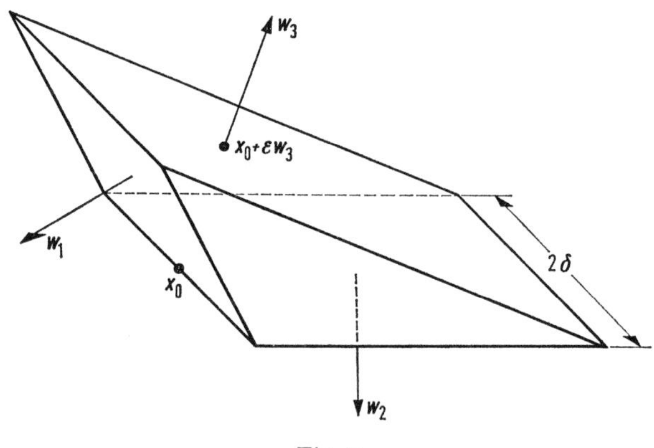

Fig. 7.

Consider the solid $\mathscr{A} = \mathscr{A}(\varepsilon)$ bounded by these three planes and two planes parallel to both $\boldsymbol{w}_1$ and $\boldsymbol{w}_2$ and a distance δ from $\boldsymbol{x}_0$ (see Fig. 7). Let ε and δ be sufficiently small that $\mathscr{A}$ is a part of B. Then

$$\partial \mathscr{A} = \bigcup_{i=1}^{5} \mathscr{W}_i,$$

[1] This proof was furnished by W. NOLL (private communication) in 1967.

where $\mathscr{W}_i$ is contained in π_i $(i=1, 2, 3)$, and $\mathscr{W}_4$ and $\mathscr{W}_5$ are parallel faces. Moreover,

$$a_i = \frac{|\boldsymbol{w}_i|}{|\boldsymbol{w}_3|} a_3 \quad (i=1, 2),$$

$$a_3 = O(\varepsilon) \quad \text{as} \quad \varepsilon \to 0,$$

$$v(\mathscr{A}) = \frac{\varepsilon}{2} |\boldsymbol{w}_3| a_3 = 2\delta\, a_4 = 2\delta\, a_5,$$

where a_i is the area of $\mathscr{W}_i$. Thus, by the continuity of $\mathbf{s}_{\boldsymbol{n}}$,

$$\boldsymbol{c} \equiv \frac{|\boldsymbol{w}_3|}{a_3} \int\limits_{\partial\mathscr{A}} \mathbf{s}_{\boldsymbol{n}}\, da = \sum_{i=1}^{3} \frac{|\boldsymbol{w}_i|}{a_i} \int\limits_{\mathscr{W}_i} \mathbf{s}\left(\boldsymbol{x}, \frac{\boldsymbol{w}_i}{|\boldsymbol{w}_i|}\right) da_{\boldsymbol{x}} + O(\varepsilon) \quad \text{as} \quad \varepsilon \to 0,$$

and (a) implies

$$\boldsymbol{c} = \sum_{i=1}^{3} \mathbf{s}(\boldsymbol{x}_0, \boldsymbol{w}_i) + o(1) \quad \text{as} \quad \varepsilon \to 0.$$

On the other hand, we conclude from estimate (a) in the proof of ***(2)*** that

$$\boldsymbol{c} = O(\varepsilon) \quad \text{as} \quad \varepsilon \to 0.$$

The last two results yield

$$\sum_{i=1}^{3} \mathbf{s}(\boldsymbol{x}_0, \boldsymbol{w}_i) = \mathbf{0};$$

since $\mathbf{s}(\boldsymbol{x}, \cdot)$ is homogeneous, this relation and (c) imply that $\mathbf{s}(\boldsymbol{x}_0, \cdot)$ is additive. Thus $\mathbf{s}(\boldsymbol{x}_0, \cdot)$ is *linear*, and, since $\boldsymbol{x}_0 \in B$ was arbitrarily chosen, NOLL's proof is complete.

The field $\boldsymbol{S}$ of the Cauchy-Poisson theorem ***(1)*** has the following properties:

(I) $\boldsymbol{S}$ is a *symmetric* tensor field of class $C^{1,0}$ on $B \times (0, t_0)$;

(II) $\boldsymbol{S}$ and div $\boldsymbol{S}$ are continuous on $\overline{B} \times [0, t_0)$.

The continuity of div $\boldsymbol{S}$ on $\overline{B} \times [0, t_0)$ follows from the equation of motion and the continuity of ϱ, $\ddot{\boldsymbol{u}}$, and $\boldsymbol{b}$ on this set. A field $\boldsymbol{S}$ with properties (I) and (II) will be called a ***time-dependent admissible stress field***. In view of the Cauchy-Poisson theorem, the specification of a dynamical process is equivalent to specifying an ordered array $[\boldsymbol{u}, \boldsymbol{S}, \boldsymbol{b}]$, where

(α) $\boldsymbol{u}$ is an admissible motion;

(β) $\boldsymbol{S}$ is a time-dependent admissible stress field;

(γ) $\boldsymbol{b}$ is a continuous vector field on $\overline{B} \times [0, t_0)$;

(δ) $\boldsymbol{u}$, $\boldsymbol{S}$, and $\boldsymbol{b}$ satisfy the equation of motion

$$\operatorname{div} \boldsymbol{S} + \boldsymbol{b} = \varrho \ddot{\boldsymbol{u}}.$$

Henceforth, a ***dynamical process*** *is an ordered array* $[\boldsymbol{u}, \boldsymbol{S}, \boldsymbol{b}]$ *with properties* (α)–(δ).

If $\boldsymbol{u}$ is independent of time, the equation of motion reduces to the ***equation of equilibrium***

$$\operatorname{div} \boldsymbol{S} + \boldsymbol{b} = \mathbf{0}.$$

Let $\boldsymbol{S}$ be a stress tensor at a point $\boldsymbol{x}$ of the body. If

$$\boldsymbol{S}\boldsymbol{n} = s\boldsymbol{n},$$

then the scalar s is called a ***principal stress*** at $\boldsymbol{x}$ and the unit vector $\boldsymbol{n}$ is called a ***principal direction*** (of stress). Since $\boldsymbol{S}$ is symmetric, there exist three mutually perpendicular principal directions $\boldsymbol{n}_1, \boldsymbol{n}_2, \boldsymbol{n}_3$ and three corresponding principal stresses s_1, s_2, s_3.

If a symmetric stress field $\boldsymbol{S}$ is constant, it automatically satisfies the equation of equilibrium (with vanishing body forces). Examples of such stress fields are:

(a) ***pure tension*** (or compression) with tensile stress σ in the direction $\boldsymbol{n}$, where $|\boldsymbol{n}| = 1$:

$$\boldsymbol{S} = \sigma \boldsymbol{n} \otimes \boldsymbol{n};$$

(b) ***uniform pressure*** p:

$$\boldsymbol{S} = -p\boldsymbol{1};$$

(c) ***pure shear*** with shear stress τ relative to the direction pair $\boldsymbol{m}, \boldsymbol{n}$, where $\boldsymbol{m}$ and $\boldsymbol{n}$ are orthogonal unit vectors:

$$\boldsymbol{S} = 2\tau \operatorname{sym}(\boldsymbol{m} \otimes \boldsymbol{n}).$$

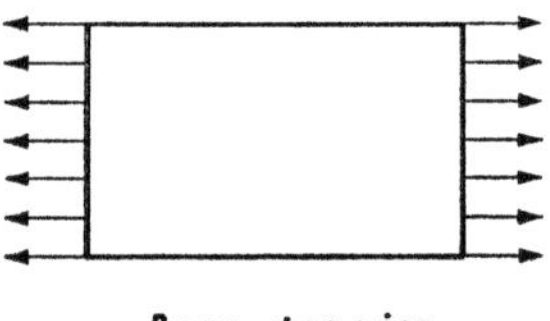

Pure tension

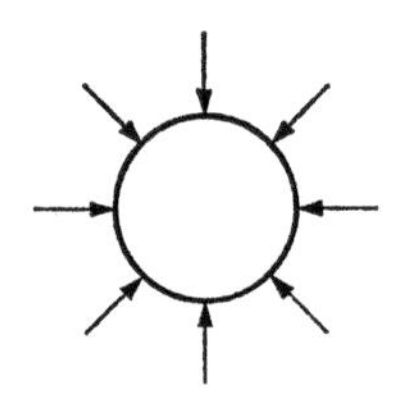

Uniform pressure

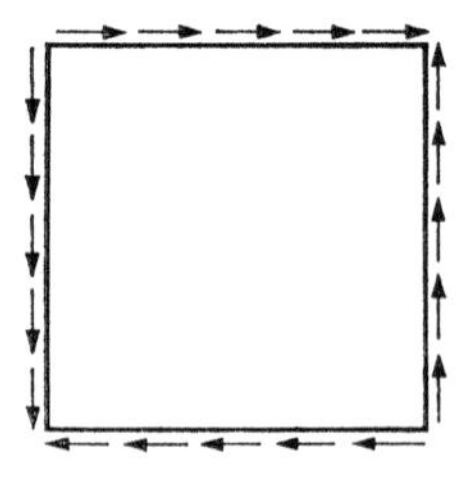

Pure shear

Fig. 8.

The matrix of the pure tension (a) relative to an orthonormal basis $\{\boldsymbol{n}, \boldsymbol{e}_2, \boldsymbol{e}_3\}$ is

$$[\boldsymbol{S}] = \begin{bmatrix} \sigma & 0 & 0 \\ 0 & 0 & 0 \\ 0 & 0 & 0 \end{bmatrix},$$

and the matrix of the pure shear (c) relative to an orthonormal basis $\{\boldsymbol{m}, \boldsymbol{n}, \boldsymbol{e}_3\}$ is

$$[\boldsymbol{S}] = \begin{bmatrix} 0 & \tau & 0 \\ \tau & 0 & 0 \\ 0 & 0 & 0 \end{bmatrix}.$$

Given a unit vector $\boldsymbol{k}$, the stress vector $\boldsymbol{s}_{\boldsymbol{k}}$ for each of the above is:

(a) $\boldsymbol{s}_{\boldsymbol{k}} = \sigma(\boldsymbol{k} \cdot \boldsymbol{n})\, \boldsymbol{n}$,

(b) $\boldsymbol{s}_{\boldsymbol{k}} = -p\boldsymbol{k}$,

(c) $\boldsymbol{s}_{\boldsymbol{k}} = \tau[(\boldsymbol{m} \cdot \boldsymbol{k})\, \boldsymbol{n} + (\boldsymbol{n} \cdot \boldsymbol{k})\, \boldsymbol{m}]$.

These three surface force fields are shown in Fig. 8.

16. Balance of momentum for finite motions.[1] In this section we show that the definitions and results of the previous section are, in a certain sense, consistent with the finite theory provided the deformations are small. We again remind the reader that this section is only to help motivate the infinitesimal theory; the remainder of the article is *independent* of the considerations given here.

Consider a *one-parameter family* $\boldsymbol{\kappa}_t$, $0 < t < t_0$, *of deformations of* B, the time t being the parameter. The point $\boldsymbol{\kappa}(\boldsymbol{x}, t) \equiv \boldsymbol{\kappa}_t(\boldsymbol{x})$ is the place occupied by $\boldsymbol{x}$ at time t, while

$$\boldsymbol{u}(\boldsymbol{x}, t) = \boldsymbol{\kappa}(\boldsymbol{x}, t) - \boldsymbol{x} \tag{a}$$

is the *displacement* of $\boldsymbol{x}$ at time t.

The *linear momentum* $\boldsymbol{l}(P)$ of a part P is the same in the finite theory as in the linear theory; i.e.

$$\boldsymbol{l}(P) = \int_P \dot{\boldsymbol{u}}\varrho\, dv, \tag{b}$$

where ϱ is the density in the reference configuration. The *angular momentum*, however, is now equal to

$$\boldsymbol{h}^*(P) = \int_P (\boldsymbol{\kappa} - 0) \times \dot{\boldsymbol{u}}\varrho\, dv. \tag{c}$$

By (a), $\dot{\boldsymbol{\kappa}} = \dot{\boldsymbol{u}}$; thus, since $\dot{\boldsymbol{u}} \times \dot{\boldsymbol{u}} = \boldsymbol{0}$,

$$\dot{\boldsymbol{h}}^*(P) = \int_P (\boldsymbol{\kappa} - 0) \times \ddot{\boldsymbol{u}}\varrho\, dv. \tag{d}$$

In the finite theory we can specify the surface forces and body forces per unit area and volume in the deformed configuration or per unit area and volume in the reference configuration. The former specification leads to the Cauchy stress tensor; the latter, which is most convenient for our purposes, leads to the Piola-Kirchhoff stress tensor. A system of forces $\mathscr{f}$ is defined exactly as in the infinitesimal theory with the quantities $\boldsymbol{s}_{\boldsymbol{n}}$ and $\boldsymbol{b}$ now interpreted as follows: given any closed regular surface $\mathscr{S}$ in $\bar{B}$ [rather than $\boldsymbol{\kappa}(\bar{B}, t)$] with unit outward normal $\boldsymbol{n}$ at $\boldsymbol{x} \in \mathscr{S}$, $\boldsymbol{s}_{\boldsymbol{n}}(\boldsymbol{x}, t)$ is the force *per unit area in the reference configuration*

[1] See, for example, Truesdell and Toupin [1960, *17*], §§ 196, 200–210, and Truesdell and Noll [1965, *22*], § 43 A.

exerted at the material point $\boldsymbol{x}$ by the material outside of $\mathscr{S}$ on the material inside; $\boldsymbol{b}(\boldsymbol{x}, t)$ is the force *per unit volume in the reference configuration* exerted on the material point $\boldsymbol{x}$ by the external world. The *force* $\boldsymbol{f}(P)$ on a part P is the same as in the linear theory:

$$\boldsymbol{f}(P) = \int_{\partial P} \boldsymbol{s_n}\, da + \int_P \boldsymbol{b}\, dv; \tag{e}$$

however, the *moment* $\boldsymbol{m}^*(P)$ on P is now given by:

$$\boldsymbol{m}^*(P) = \int_{\partial P} (\boldsymbol{\kappa} - 0) \times \boldsymbol{s_n}\, da + \int_P (\boldsymbol{\kappa} - 0) \times \boldsymbol{b}\, dv. \tag{f}$$

In terms of these quantities, the *laws of balance of linear and angular momentum* take the forms

$$\boldsymbol{f}(P) = \dot{\boldsymbol{l}}(P),$$

$$\boldsymbol{m}^*(P) = \dot{\boldsymbol{h}}^*(P),$$

or equivalently,

$$\int_{\partial P} \boldsymbol{s_n}\, da + \int_P \boldsymbol{b}\, dv = \int_P \varrho \ddot{\boldsymbol{u}}\, dv, \tag{g}$$

$$\int_{\partial P} (\boldsymbol{\kappa} - 0) \times \boldsymbol{s_n}\, da + \int_P (\boldsymbol{\kappa} - 0) \times \boldsymbol{b}\, dv = \int_P (\boldsymbol{\kappa} - 0) \times \varrho \ddot{\boldsymbol{u}}\, dv. \tag{h}$$

Now let

$$\varepsilon = \sup_{\substack{\boldsymbol{x} \in B \\ t \in (0, t_0)}} |\boldsymbol{u}(\boldsymbol{x}, t)|. \tag{i}$$

By (a),

$$\boldsymbol{\kappa}(\boldsymbol{x}, t) - 0 = \boldsymbol{u}(\boldsymbol{x}, t) + \boldsymbol{x} - 0 = \boldsymbol{u}(\boldsymbol{x}, t) + \boldsymbol{p}(\boldsymbol{x}), \tag{j}$$

and if we consider the *field* $\boldsymbol{\kappa}$ as a function of the *field* $\boldsymbol{u}$, then (i) and (j) yield

$$\boldsymbol{\kappa} - 0 = \boldsymbol{p} + O(\varepsilon) \quad \text{as} \quad \varepsilon \to 0. \tag{k}$$

Using (k) and (d), we can write (c), (f), and (h) as follows:

$$\left.\begin{aligned} \boldsymbol{h}^*(P) &= \int_P [\boldsymbol{p} + O(\varepsilon)] \times \dot{\boldsymbol{u}} \varrho\, dv, \\ \boldsymbol{m}^*(P) &= \int_{\partial P} [\boldsymbol{p} + O(\varepsilon)] \times \boldsymbol{s_n}\, da + \int_P [\boldsymbol{p} \times O(\varepsilon)] \times \boldsymbol{b}\, dv, \end{aligned}\right\} \tag{l}$$

$$\int_{\partial P} [\boldsymbol{p} + O(\varepsilon)] \times \boldsymbol{s_n}\, da + \int_P [\boldsymbol{p} + O(\varepsilon)] \times \boldsymbol{b}\, dv = \int_P [\boldsymbol{p} + O(\varepsilon)] \times \varrho \ddot{\boldsymbol{u}}\, dv. \tag{m}$$

These relations reduce to the corresponding relations in the linear theory provided we neglect the terms written $O(\varepsilon)$.

We can summarize the results established thus far as follows: *In terms of surface forces and body forces measured per unit area and volume in the reference configuration, the linear momentum of any part P and the force on P are the same as in the linear theory, and (hence) so also is the law of balance of linear momentum. The angular momentum and the moment reduce to the corresponding quantities in the linear theory provided we neglect the terms written $O(\varepsilon)$ in* (l); *therefore when these terms are neglected the law of balance of angular momentum reduces to the corresponding law in the infinitesimal theory.*

It is a simple matter to establish the consequences of balance of momentum for the finite theory. In fact, the analog of ***(15.1)*** may be stated roughly as follows: There exists a tensor field $\boldsymbol{S}$, called the *Piola-Kirchhoff stress tensor*, such that

$$\boldsymbol{s_n} = \boldsymbol{S}\boldsymbol{n}, \qquad \text{div}\,\boldsymbol{S} + \boldsymbol{b} = \varrho\ddot{\boldsymbol{u}}, \tag{n}$$

$$\boldsymbol{S}\boldsymbol{F}^T = \boldsymbol{F}\boldsymbol{S}^T, \tag{o}$$

where

$$\boldsymbol{F} = \nabla\boldsymbol{\kappa}$$

is the deformation gradient. The Eqs. (n) are identical to their counterparts in the infinitesimal theory. However, in the linear theory the stress field $\boldsymbol{S}$ is symmetric; in contrast, it is clear from (o) that in the finite theory $\boldsymbol{S}$ will in general be *non-symmetric*. If we let

$$\delta = |\nabla\boldsymbol{u}|, \tag{p}$$

then, since

$$\boldsymbol{F} = \boldsymbol{1} + \nabla\boldsymbol{u}, \tag{q}$$

(o) implies

$$\boldsymbol{S}(\boldsymbol{1} + O(\delta)) = \boldsymbol{S}^T(\boldsymbol{1} + O(\delta)).$$

Thus $\boldsymbol{S}$ *is symmetric to within terms of* $O(\delta)$ *as* $\delta \to 0$. This fact further demonstrates the consistency of the linear theory within the finite theory.

The *Cauchy stress tensor* may be defined by the relation

$$\boldsymbol{T} = (\det\boldsymbol{F})^{-1}\,\boldsymbol{S}\boldsymbol{F}^T. \tag{r}$$

It follows[1] from this definition that for any part P

$$\int_{\partial P} \boldsymbol{S}\boldsymbol{n}\,da = \int_{\partial\boldsymbol{\kappa}_t(P)} \boldsymbol{T}\boldsymbol{\nu}\,da,$$

where $\boldsymbol{\nu}$ is the outward unit normal to $\partial\boldsymbol{\kappa}_t(P)$; here $\boldsymbol{\kappa}_t(P)$ is the region of space occupied by P in the deformation $\boldsymbol{\kappa}_t$, i.e.

$$\boldsymbol{\kappa}_t(P) = \{\boldsymbol{y}: \boldsymbol{y} = \boldsymbol{\kappa}_t(\boldsymbol{x}),\ \boldsymbol{x} \in P\}.$$

Thus $\boldsymbol{T}\boldsymbol{\nu}$ is the force per unit area in the *deformed* configuration. Clearly, (r) and (o) imply that $\boldsymbol{T}$ is *symmetric*. Further, as

$$\det\boldsymbol{F} = 1 + O(\delta),$$

it follows from (p) and (q) that

$$\boldsymbol{T} = \boldsymbol{S}(1 + O(\delta)).$$

Thus *to within terms of* $O(\delta)$ *the Piola-Kirchhoff stress tensor and the Cauchy stress tensor coincide.*

17. General solutions of the equations of equilibrium. In the absence of body forces the equations of equilibrium take the form:

$$\text{div}\,\boldsymbol{S} = \boldsymbol{0}, \qquad \boldsymbol{S} = \boldsymbol{S}^T.$$

If $\boldsymbol{A}$ is a field over B, and if $\mathscr{L}$ is a differential operator with values that are symmetric tensor fields, then

$$\boldsymbol{S} = \mathscr{L}(\boldsymbol{A})$$

will be a solution of the equations of equilibrium provided

$$\text{div}\,\mathscr{L}(\boldsymbol{A}) = \boldsymbol{0}.$$

[1] See, for example, TRUESDELL and NOLL [1965, *22*], § 43 A.

A field $\boldsymbol{A}$ with this property is called a ***stress function***. The first solution of the equations of equilibrium in terms of a tress function was AIRY'S[1] two-dimensional solution in terms of a single scalar field. Three-dimensional generalizations of AIRY'S stress function were obtained by MAXWELL[2] and MORERA[3], who established two alternative solutions, each involving a triplet of scalar stress functions. BELTRAMI observed that the solutions due to MAXWELL and MORERA may be regarded as special cases of

(1) Beltrami's solution.[4] *Let* $\boldsymbol{A}$ *be a class* C^3 *symmetric tensor field on* B, *and let*

$$\boldsymbol{S} = \operatorname{curl} \operatorname{curl} \boldsymbol{A}.$$

Then

$$\operatorname{div} \boldsymbol{S} = \boldsymbol{0}, \qquad \boldsymbol{S} = \boldsymbol{S}^T.$$

Proof. The symmetry of $\boldsymbol{S}$ follows from (9) of ***(4.1)*** and the symmetry of $\boldsymbol{A}$. On the other hand, by (7) of ***(4.1)***,

$$\operatorname{div} \boldsymbol{S} = \operatorname{div} \operatorname{curl} \operatorname{curl} \boldsymbol{A} = \operatorname{curl} \operatorname{div} (\operatorname{curl} \boldsymbol{A})^T,$$

and thus we conclude from (8) of ***(4.1)*** that

$$\operatorname{div} \boldsymbol{S} = \boldsymbol{0}. \quad \square$$

Consider now a fixed cartesian coordinate frame. Then Beltrami's solution takes the form:

$$S_{ij} = \varepsilon_{imn}\, \varepsilon_{jpq}\, A_{mp,nq}.$$

Further, letting $[\boldsymbol{A}]$ denote the matrix of $\boldsymbol{A}$ relative to this frame, we have for the following special choices of $\boldsymbol{A}$:

(a) ***Airy's solution.***

$$[\boldsymbol{A}] = \begin{bmatrix} 0 & 0 & 0 \\ 0 & 0 & 0 \\ 0 & 0 & \varphi \end{bmatrix},$$

$$S_{11} = \varphi_{,22}, \quad S_{22} = \varphi_{,11}, \quad S_{12} = -\varphi_{,12}, \quad S_{13} = S_{23} = S_{33} = 0;$$

(b) ***Maxwell's solution.***

$$[\boldsymbol{A}] = \begin{bmatrix} a_1 & 0 & 0 \\ 0 & a_2 & 0 \\ 0 & 0 & a_3 \end{bmatrix},$$

$$S_{11} = a_{2,33} + a_{3,22}, \quad S_{12} = -a_{3,12}, \quad \text{etc.};$$

[1] AIRY [1863, *1*].

[2] MAXWELL [1868, *2*], [1870, *1*].

[3] MORERA [1892, *4*].

[4] BELTRAMI [1892, *1*]. This solution has been erroneously attributed to FINZI by STERNBERG [1960, *12*], to GWYTHER and FINZI by TRUESDELL [1959, *16*] and TRUESDELL and TOUPIN [1960, *17*], and to BLOKH by LUR'E [1955, *10*]. Related contributions and additional references can be found in the work of MORERA [1892, *5*], KLEIN and WIEGHARDT [1905, *1*], LOVE [1906, *5*], GWYTHER [1912, *2*], [1913, *1*], FINZI [1934, *2*], SOBRERO [1935, *7*]. KUZMIN [1945, *2*], WEBER [1948, *7*], FINZI and PASTORI [1949, *1*], KRUTKOV [1949, *5*], PERETTI [1949, *8*], BLOKH [1950, *3*], [1961, *2*], MORINAGA and NÔNO [1950, *9*], FILONENKO-BORODICH [1951, *7*, *8*], [1957, *6*], ARZHANYH [1953, *2*], SCHAEFER [1953, *18*], [1955, *12*], [1959, *12*], KRÖNER [1954, *11*], [1955, *7*, *8*], LANGHAAR and STIPPES [1954, *12*], ORNSTEIN [1954, *17*], GÜNTHER [1954, *7*], MARGUERRE [1955, *11*], DORN and SCHILD [1956, *1*], BRDIČKA [1957, *3*], RIEDER [1960, *9*], [1964, *17*, *18*, *19*], MINAGAWA [1962, *11*], [1965, *14*], GURTIN [1963, *9*], KAWATATE [1963, *16*], PASTORI [1963, *21*], INOV and VVEDENSKII [1964, *10*], NÔNO [1964, *15*], CARLSON [1966, *6*], [1967, *4*], STIPPES [1966, *25*], [1967, *14*], SCHULER and FOSDICK [1967, *13*].

(c) ***Morera's solution.***

$$[\boldsymbol{A}]=\begin{bmatrix} 0 & \omega_3 & \omega_2 \\ \omega_3 & 0 & \omega_1 \\ \omega_2 & \omega_1 & 0 \end{bmatrix}$$

$$S_{11}=-2\omega_{1,23}, \quad S_{12}=(\omega_{1,1}+\omega_{2,2}-\omega_{3,3})_{,3}, \quad \text{etc.}$$

If we define a tensor field $\boldsymbol{G}$ on B through

$$\boldsymbol{G}=\boldsymbol{A}-\tfrac{1}{2}\boldsymbol{1}\operatorname{tr}\boldsymbol{A},$$

then

$$\boldsymbol{A}=\boldsymbol{G}-\boldsymbol{1}\operatorname{tr}\boldsymbol{G},$$

and it follows from (14) of ***(4.1)*** that Beltrami's solution ***(1)*** can be written in the alternative form[1]

$$\boldsymbol{S}=-\Delta\boldsymbol{G}+2\hat{\nabla}\operatorname{div}\boldsymbol{G}-\boldsymbol{1}\operatorname{div}\operatorname{div}\boldsymbol{G}.$$

By a ***self-equilibrated stress field*** we mean a smooth symmetric tensor field $\boldsymbol{S}$ on B whose resultant force and moment vanish on each closed regular surface $\mathscr{S}$ in B:

$$\int_{\mathscr{S}}\boldsymbol{S}\boldsymbol{n}\,da=\int_{\mathscr{S}}\boldsymbol{p}\times(\boldsymbol{S}\boldsymbol{n})\,da=\boldsymbol{0}.$$

Clearly, every self-equilibrated stress field $\boldsymbol{S}$ satisfies

$$\operatorname{div}\boldsymbol{S}=\boldsymbol{0}.$$

On the other hand, if the boundary of B consists of a *single* closed surface, then any closed regular surface $\mathscr{S}$ in B encloses only points of B, and it follows from the divergence theorem that every divergence-free symmetric tensor field will be self-equilibrated. The assumption that ∂B consist of a single closed surface is necessary for the validity of this converse assertion. Indeed, let B be the spherical shell

$$B=\{\boldsymbol{x}:\alpha<|\boldsymbol{x}-\boldsymbol{x}_0|<\beta\}.$$

Then the stress field $\boldsymbol{S}$ defined on B by

$$\boldsymbol{S}(\boldsymbol{x})=\frac{\boldsymbol{a}\cdot(\boldsymbol{x}-\boldsymbol{x}_0)}{|\boldsymbol{x}-\boldsymbol{x}_0|^5}(\boldsymbol{x}-\boldsymbol{x}_0)\otimes(\boldsymbol{x}-\boldsymbol{x}_0) \qquad (\boldsymbol{a}\neq\boldsymbol{0})$$

satisfies

$$\operatorname{div}\boldsymbol{S}=\boldsymbol{0}, \quad \boldsymbol{S}=\boldsymbol{S}^T,$$

but its resultant force on any spherical surface

$$\mathscr{S}=\{\boldsymbol{x}:|\boldsymbol{x}-\boldsymbol{x}_0|=\gamma\} \qquad (\alpha<\gamma<\beta)$$

is not zero. In fact,

$$\int_{\mathscr{S}}\boldsymbol{S}\boldsymbol{n}\,da=-\frac{4\pi}{3}\boldsymbol{a}.$$

This result when combined with the next theorem shows that *the Beltrami solution and hence the solutions of* MAXWELL *and* MORERA *are, in general, not complete*; i.e. there exist stress fields that do not admit a representation as a Beltrami solution.

(2)[2] *Beltrami's solution* ***(1)*** *always yields a self-equilibrated stress field.*

---

[1] This version of Beltrami's solution is due to SCHAEFER [1953, *18*].

[2] RIEDER [1960, *9*]. See also GURTIN [1963, *9*].

Proof. Let

$$\boldsymbol{S} = \operatorname{curl} \operatorname{curl} \boldsymbol{A}$$

with $\boldsymbol{A}$ symmetric and of class C^3 on B, and let

$$\boldsymbol{W} = \operatorname{curl} \boldsymbol{A}. \tag{a}$$

Then by the symmetry of $\boldsymbol{S} = \operatorname{curl} \operatorname{curl} \boldsymbol{A}$ together with Stokes' theorem ***(6.2)***,

$$\int_{\mathscr{S}} \boldsymbol{S}\boldsymbol{n}\, da = \int_{\mathscr{S}} \boldsymbol{S}^T \boldsymbol{n}\, da = \int_{\mathscr{S}} (\operatorname{curl} \boldsymbol{W})^T \boldsymbol{n}\, da = 0$$

for every closed regular surface $\mathscr{S}$ in B. To see that the resultant moment also vanishes on $\mathscr{S}$, consider the identity

$$\boldsymbol{p} \times (\operatorname{curl} \boldsymbol{W})^T \boldsymbol{n} = [\operatorname{curl}(\boldsymbol{P}\boldsymbol{W})]^T \boldsymbol{n} + \boldsymbol{W}^T \boldsymbol{n} - (\operatorname{tr} \boldsymbol{W})\, \boldsymbol{n}, \tag{b}$$

which holds for any tensor field $\boldsymbol{W}$ and any vector $\boldsymbol{n}$, where $\boldsymbol{P}$ is the skew tensor corresponding to the position vector $\boldsymbol{p}$. In view of the symmetry of $\boldsymbol{A}$, (a) and (12) of ***(4.1)*** imply that

$$\operatorname{tr} \boldsymbol{W} = 0.$$

Hence (a), (b), and Stokes' theorem ***(6.2)*** yield

$$\begin{aligned} \int_{\mathscr{S}} \boldsymbol{p} \times (\boldsymbol{S}\boldsymbol{n})\, da &= \int_{\mathscr{S}} \boldsymbol{p} \times (\boldsymbol{S}^T \boldsymbol{n})\, da = \int_{\mathscr{S}} \boldsymbol{p} \times (\operatorname{curl} \boldsymbol{W})^T \boldsymbol{n}\, da \\ &= \int_{\mathscr{S}} [\operatorname{curl}(\boldsymbol{P}\boldsymbol{W})]^T \boldsymbol{n}\, da + \int_{\mathscr{S}} (\operatorname{curl} \boldsymbol{A})^T \boldsymbol{n}\, da = \boldsymbol{0}. \quad \square \end{aligned}$$

Thus *stress fields that are not self-equilibrated cannot be represented as Beltrami solutions.* The next theorem shows that Beltrami's solution is complete provided we restrict our attention to self-equilibrated stress fields. In order to state this theorem concisely, let $\mathscr{F}$ denote the set of all fields of class C^2 on $\bar{B}$ and class C^3 on B.

(3) Completeness of the Beltrami solution.[1] *Assume that the boundary of B is of class C^3. Let $\boldsymbol{S} \in \mathscr{F}$ be a self-equilibrated stress field. Then there exists a symmetric tensor field $\boldsymbol{A} \in \mathscr{F}$ such that*

$$\boldsymbol{S} = \operatorname{curl} \operatorname{curl} \boldsymbol{A}.$$

Proof. Since $\boldsymbol{S}$ is self-equilibrated (and hence symmetric),

$$\int_{\mathscr{S}} \boldsymbol{S}^T \boldsymbol{n}\, da = \boldsymbol{0}, \tag{a}$$

$$\int_{\mathscr{S}} \boldsymbol{p} \times (\boldsymbol{S}^T \boldsymbol{n})\, da = \boldsymbol{0}, \tag{b}$$

for every closed regular surface $\mathscr{S}$ in B. By (a) and (ii) of ***(6.4)***, there exists a tensor field $\boldsymbol{W} \in \mathscr{F}$ such that

$$\boldsymbol{S} = \operatorname{curl} \boldsymbol{W}. \tag{c}$$

[1] GURTIN [1963, *9*]. Completeness proofs were given previously by MORINAGA and NÔNO [1950, *9*], ORNSTEIN [1954, *17*], GÜNTHER [1954, *7*], and DORN and SCHILD [1956, *1*]. However, since the necessary restriction that $\boldsymbol{S}$ be self-equilibrated is absent in all of these investigations, they can at most be valid for a region whose boundary consists of a *single* closed surface. GURTIN'S [1963, *9*] proof is based on a minor variant of GÜNTHER'S [1954, *7*] argument and is quite cumbersome. The elegant proof given here is due to CARLSON [1967, *4*] (see also CARLSON [1966, *6*]).

It follows from (b), (c), the identity (b) given in the proof of ***(2)***, and Stokes' theorem ***(6.2)*** that

$$\int_{\mathscr{S}} [\boldsymbol{W} - (\operatorname{tr} \boldsymbol{W})\, \mathbf{1}]^T \boldsymbol{n}\, da = \mathbf{0}$$

for every closed regular surface $\mathscr{S}$ in B. Thus, by (ii) of ***(6.4)***, there exists a tensor field $\boldsymbol{V} \in \mathscr{F}$ such that

$$\boldsymbol{W} - (\operatorname{tr} \boldsymbol{W})\, \mathbf{1} = \operatorname{curl} \boldsymbol{V};$$

hence (c) implies

$$\boldsymbol{S} = \operatorname{curl}\operatorname{curl} \boldsymbol{V} + \operatorname{curl}[(\operatorname{tr} \boldsymbol{W})\, \mathbf{1}]. \tag{d}$$

If we let

$$\boldsymbol{A} = \operatorname{sym} \boldsymbol{V},$$

then (9) and (10) of ***(4.1)*** yield the relations

$$\operatorname{sym}\{\operatorname{curl}\operatorname{curl} \boldsymbol{V}\} = \operatorname{curl}\operatorname{curl} \boldsymbol{A},$$
$$\operatorname{sym}\{\operatorname{curl}[(\operatorname{tr} \boldsymbol{W})\, \mathbf{1}]\} = \mathbf{0}.$$

Thus, since $\boldsymbol{S}$ is symmetric, if we take the symmetric part of both sides of (d), we arrive at

$$\boldsymbol{S} = \operatorname{curl}\operatorname{curl} \boldsymbol{A}. \quad \square$$

If the boundary of B consists of a *single* closed surface, then every symmetric divergence-free stress field is self-equilibrated, and we have the following direct corollary of ***(3)***.

(4) *Assume that ∂B consists of a single class C^3 closed regular surface. Let $\boldsymbol{S} \in \mathscr{F}$ be a symmetric tensor field that satisfies*

$$\operatorname{div} \boldsymbol{S} = \mathbf{0}.$$

Then there exists a symmetric tensor field $\boldsymbol{A} \in \mathscr{F}$ such that

$$\boldsymbol{S} = \operatorname{curl}\operatorname{curl} \boldsymbol{A}.$$

The next proposition, which is a direct consequence of the compatibility theorem ***(14.2)***, shows that two tensor fields are Beltrami representations of the same stress field if and only if they differ by a strain field.

(5)[1] *Let*

$$\boldsymbol{S} = \operatorname{curl}\operatorname{curl} \boldsymbol{A}, \tag{a}$$

where $\boldsymbol{A}$ is a class C^2 tensor field on B. Further, let

$$\begin{aligned} \boldsymbol{H} &= \boldsymbol{A} + \boldsymbol{E}, \\ \boldsymbol{E} &= \tfrac{1}{2}(\nabla \boldsymbol{u} + \nabla \boldsymbol{u}^T), \end{aligned} \tag{b}$$

where $\boldsymbol{u}$ is a class C^3 vector field on B. Then

$$\boldsymbol{S} = \operatorname{curl}\operatorname{curl} \boldsymbol{H}. \tag{c}$$

Conversely, let B be simply-connected, and let $\boldsymbol{A}$ and $\boldsymbol{H}$ be class C^2 symmetric tensor fields on B that satisfy (a) *and* (c). *Then there exists a class C^3 vector field $\boldsymbol{u}$ on B such that* (b) *holds.*

It is clear from ***(5)*** that to prove the completeness of Maxwell's solution for situations in which Beltrami's solution is complete it suffices to establish the

[1] Finzi [1934, 2].

existence of a class C^3 displacement field $\boldsymbol{u}$ that satisfies the three equations

$$\tfrac{1}{2}(u_{i,j}+u_{j,i}) = -A_{ij} \quad (i \neq j). \tag{I}$$

On the other hand, the completeness of Morera's solution will follow if we can find a class C^3 solution $\boldsymbol{u}$ of the equations

$$u_{i,i} = A_{ii} \quad \text{(no sum)}. \tag{II}$$

Here $\boldsymbol{A}$ is a given class C^2 symmetric tensor field on B. Under hypotheses (i) and (iv) of ***(5.2)*** with $M=1$ and $N=3$, as well as analogous hypotheses obtained from (i) and (iv) by cyclic permutations of the integers (1, 2, 3), the system (II) has a solution. Therefore under these hypotheses *a stress field* $\boldsymbol{S}$ *admits a Morera representation whenever it admits a Beltrami representation.*

To the author's knowledge, there is, at present, no general existence theorem appropriate to the system (I).

Theorems ***(2)*** and ***(3)*** assert that the Beltrami solution is complete if and only if we restrict our attention to self-equilibrated stress fields. The next two results show that it is possible to modify Beltrami's representation so as to form a completely general solution of the equation of equilibrium.

(6) Beltrami-Schaefer solution.[1] *Let* $\boldsymbol{A}$ *be a class* C^3 *symmetric tensor field and* $\boldsymbol{h}$ *a harmonic field on* B, *and let*

$$\boldsymbol{S} = \operatorname{curl}\operatorname{curl}\boldsymbol{A} + 2\hat{\nabla}\boldsymbol{h} - \mathbf{1}\operatorname{div}\boldsymbol{h}.$$

Then

$$\operatorname{div}\boldsymbol{S} = \mathbf{0}, \qquad \boldsymbol{S} = \boldsymbol{S}^T.$$

Proof. The proof follows at once from ***(1)*** and the identity

$$\operatorname{div}[\nabla\boldsymbol{h} + (\nabla\boldsymbol{h})^T - \mathbf{1}\operatorname{div}\boldsymbol{h}] = \Delta\boldsymbol{h}. \quad \square$$

(7) Completeness of the Beltrami-Schaefer solution.[2] *Let* $\boldsymbol{S}$ *be a symmetric tensor field that is continuous on* $\bar{B}$ *and class* C^2 *on* B, *and suppose that*

$$\operatorname{div}\boldsymbol{S} = \mathbf{0}.$$

Then there exists a class C^3 *symmetric tensor field* $\boldsymbol{A}$ *and a harmonic vector field* $\boldsymbol{h}$ *on* B *such that*

$$\boldsymbol{S} = \operatorname{curl}\operatorname{curl}\boldsymbol{A} + 2\hat{\nabla}\boldsymbol{h} - \mathbf{1}\operatorname{div}\boldsymbol{h}.$$

Proof. By ***(6.5)*** there exists a class C^3 symmetric tensor field $\boldsymbol{G}$ on B such that

$$\boldsymbol{S} = -\Delta\boldsymbol{G}.$$

Thus

$$\boldsymbol{S} = -\Delta\boldsymbol{G} + 2\hat{\nabla}\operatorname{div}\boldsymbol{G} - \mathbf{1}\operatorname{div}\operatorname{div}\boldsymbol{G} - 2\hat{\nabla}\operatorname{div}\boldsymbol{G} + \mathbf{1}\operatorname{div}\operatorname{div}\boldsymbol{G},$$

and letting

$$\boldsymbol{A} = \boldsymbol{G} - \mathbf{1}\operatorname{tr}\boldsymbol{G},$$
$$\boldsymbol{h} = -\operatorname{div}\boldsymbol{G},$$

---

[1] SCHAEFER [1953, *18*].

[2] GURTIN [1963, *9*] introduced the following solution:

$$\boldsymbol{S} = \operatorname{curl}\operatorname{curl}\boldsymbol{A} + 2\Delta\hat{\nabla}\boldsymbol{b} - \nabla\nabla\operatorname{div}\boldsymbol{b}, \tag{*}$$

where $\boldsymbol{A}$ is symmetric and $\boldsymbol{b}$ biharmonic, and established its completeness. Shortly thereafter SCHAEFER noticed that the completeness of (*) implies the completeness of the Beltrami-Schaefer solution (private communication). The direct completeness proof given above appears here in print for the first time.

we see from (14) of *(4.1)* that

$$\boldsymbol{S} = \operatorname{curl}\operatorname{curl}\boldsymbol{A} + 2\hat{\nabla}\boldsymbol{h} - \mathbf{1}\operatorname{div}\boldsymbol{h}.$$

Further, the fact that $\boldsymbol{S}$ is divergence-free implies that $\boldsymbol{h}$ is harmonic. □

Let $\boldsymbol{S}$ satisfy the hypotheses of *(7)*, and let

$$\boldsymbol{S}_e = \operatorname{curl}\operatorname{curl}\boldsymbol{A},$$

$$\boldsymbol{S}_h = 2\hat{\nabla}\boldsymbol{h} - \mathbf{1}\operatorname{div}\boldsymbol{h},$$

where $\boldsymbol{A}$ and $\boldsymbol{h}$ are the fields established in *(7)*. By *(2)* $\boldsymbol{S}_e$ is self-equilibrated, while *(6)* implies that $\boldsymbol{S}_h$ is divergence-free and symmetric. Further, since $\boldsymbol{h}$ is harmonic, so is $\boldsymbol{S}_h$. Thus we have the

(8) Decomposition theorem for stress fields.[1] *Let $\boldsymbol{S}$ satisfy the hypotheses of (7). Then*

$$\boldsymbol{S} = \boldsymbol{S}_e + \boldsymbol{S}_h,$$

where $\boldsymbol{S}_e$ is a self-equilibrated stress field, and $\boldsymbol{S}_h$ is a harmonic tensor field that satisfies

$$\operatorname{div}\boldsymbol{S}_h = \mathbf{0}, \qquad \boldsymbol{S}_h = \boldsymbol{S}_h^T.$$

Finally, it is clear from the proof of *(6)* that[2]

$$\boldsymbol{S} = 2\hat{\nabla}\boldsymbol{h} - \mathbf{1}\operatorname{div}\boldsymbol{h},$$

$$\Delta\boldsymbol{h} = -\boldsymbol{b}$$

furnishes a *particular solution* of the inhomogeneous equations of equilibrium

$$\operatorname{div}\boldsymbol{S} + \boldsymbol{b} = \mathbf{0}, \qquad \boldsymbol{S} = \boldsymbol{S}^T.$$

18. Consequences of the equation of equilibrium. In this section we study properties of solutions to the *equation of equilibrium*

$$\operatorname{div}\boldsymbol{S} + \boldsymbol{b} = \mathbf{0},$$

assuming throughout that a *continuous* body force field $\boldsymbol{b}$ is prescribed on $\bar{B}$.

For convenience, we introduce the following two definitions. A vector field $\boldsymbol{u}$ is an ***admissible displacement field*** provided

(i)[3] $\boldsymbol{u}$ is of class C^2 on B;

(ii) $\boldsymbol{u}$ and $\hat{\nabla}\boldsymbol{u}$ are continuous on $\bar{B}$.

On the other hand, an ***admissible stress field*** is a symmetric tensor field $\boldsymbol{S}$ with the following properties:

(i) $\boldsymbol{S}$ is smooth on B;

(ii) $\boldsymbol{S}$ and $\operatorname{div}\boldsymbol{S}$ are continuous on $\bar{B}$.

The associated ***surface force field*** is then the vector field $\boldsymbol{s}$ defined at every regular point $\boldsymbol{x} \in \partial B$ by

$$\boldsymbol{s}(\boldsymbol{x}) = \boldsymbol{S}(\boldsymbol{x})\,\boldsymbol{n}(\boldsymbol{x}),$$

where $\boldsymbol{n}(\boldsymbol{x})$ is the outward unit normal vector to ∂B at $\boldsymbol{x}$.

---

[1] Cf. GURTIN [1963, *9*], Theorem 4.4.

[2] SCHAEFER [1953, *18*].

[3] Actually, for the results established in this section, this requirement may be replaced by the assumption that $\boldsymbol{u}$ be smooth on B.

The following lemma will be extremely useful throughout this article.

(1) Lemma. *Let* $\boldsymbol{S}$ *be an admissible stress field and* $\boldsymbol{u}$ *an admissible displacement field. Then*

$$\int_{\partial B} \mathbf{s} \cdot \boldsymbol{u}\, da = \int_B \boldsymbol{u} \cdot \operatorname{div} \boldsymbol{S}\, dv + \int_B \boldsymbol{S} \cdot \hat{\nabla} \boldsymbol{u}\, dv .$$

Proof. By the divergence theorem, (11) of *(4.1)*, and the symmetry of $\boldsymbol{S}$,

$$\int_{\partial B} \boldsymbol{u} \cdot \boldsymbol{S}\boldsymbol{n}\, da = \int_{\partial B} (\boldsymbol{S}\boldsymbol{u}) \cdot \boldsymbol{n}\, da = \int_B \operatorname{div}(\boldsymbol{S}\boldsymbol{u})\, dv = \int_B \boldsymbol{u} \cdot \operatorname{div} \boldsymbol{S}\, dv + \int_B \boldsymbol{S} \cdot \nabla \boldsymbol{u}\, dv .$$

The symmetry of $\boldsymbol{S}$ also requires that

$$\boldsymbol{S} \cdot \nabla \boldsymbol{u} = \boldsymbol{S} \cdot \nabla \boldsymbol{u}^T = \boldsymbol{S} \cdot \hat{\nabla} \boldsymbol{u},$$

and the proof is complete. □

If $\boldsymbol{S}$ is a solution of the equation of equilibrium, then $\operatorname{div} \boldsymbol{S} = -\boldsymbol{b}$, and *(1)* has the following important corollary.

(2) Theorem of work expended.[1] *Let* $\boldsymbol{S}$ *be an admissible stress field and* $\boldsymbol{u}$ *an admissible displacement field, and suppose that* $\boldsymbol{S}$ *satisfies the equation of equilibrium. Then*

$$\int_{\partial B} \mathbf{s} \cdot \boldsymbol{u}\, da + \int_B \boldsymbol{b} \cdot \boldsymbol{u}\, dv = \int_B \boldsymbol{S} \cdot \boldsymbol{E}\, dv ,$$

where $\boldsymbol{E}$ *is the strain field corresponding to* $\boldsymbol{u}$.

This theorem asserts that the work done[2] by the surface and body forces over the displacement field $\boldsymbol{u}$ is equal to the work done by the stress field over the strain field corresponding to $\boldsymbol{u}$. If $\boldsymbol{u}$ is a rigid displacement field, then $\hat{\nabla} \boldsymbol{u} = \boldsymbol{0}$, and the above relation reduces to

$$\int_{\partial B} \mathbf{s} \cdot \boldsymbol{u}\, da + \int_B \boldsymbol{b} \cdot \boldsymbol{u}\, dv = 0.$$

Thus *the work expended over a rigid displacement field vanishes.*

Our next theorem shows that, conversely, the balance laws for forces and moments are implied by the requirement that the work done by the surface and body forces be zero over every rigid displacement.

(3) Piola's theorem.[3] *Let* $\mathbf{s}$ *be an integrable vector field on* ∂B. *Then*

$$\int_{\partial B} \mathbf{s} \cdot \boldsymbol{w}\, da + \int_B \boldsymbol{b} \cdot \boldsymbol{w}\, dv = 0$$

---

[1] Cf. Theorem *(28.3)*.

[2] We follow LOVE [1927, *3*], § 120, in referring to, e.g.,

$$\int_{\partial B} \mathbf{s} \cdot \boldsymbol{u}\, da \tag{a}$$

as the work done by the surface force $\mathbf{s}$ over the displacement $\boldsymbol{u}$. In the dynamical theory the work is given by

$$\int_{t_0}^{t_1} \int_{\partial B} \mathbf{s}_t \cdot \dot{\boldsymbol{u}}_t\, da\, dt, \tag{b}$$

where $[t_0, t_1]$ is the interval of time during which the time-dependent surface force $\mathbf{s}_t$ acts, and where $\dot{\boldsymbol{u}}_t$ is the velocity of the time-dependent displacement field $\boldsymbol{u}_t$. Formally, (b) reduces to (a) if we assume that the spatial fields $\mathbf{s}$ and $\boldsymbol{u}$ are applied suddenly, i.e., if $\mathbf{s}_t(\boldsymbol{x}) = \mathbf{s}(x)\, h(t-\lambda)$ and $\boldsymbol{u}_t(\boldsymbol{x}) = \boldsymbol{u}(\boldsymbol{x})\, h(t-\lambda)$, where h is the Heaviside unit step function and $\lambda \in (t_0, t_1)$.

[3] PIOLA [1833, *2*], [1848, *1*]. See also TRUESDELL and TOUPIN [1960, *17*], § 232.

for every rigid displacement $\boldsymbol{w}$ *if and only if*

$$\int_{\partial B} \boldsymbol{s}\, da + \int_B \boldsymbol{b}\, dv = \boldsymbol{0},$$

$$\int_{\partial B} \boldsymbol{p} \times \boldsymbol{s}\, da + \int_B \boldsymbol{p} \times \boldsymbol{b}\, dv = \boldsymbol{0}.$$

Proof. Let

$$\boldsymbol{w} = \boldsymbol{u}_0 + \boldsymbol{W}_0 \boldsymbol{p}$$

with $\boldsymbol{W}_0$ skew. In view of the identity

$$\boldsymbol{a} \cdot \boldsymbol{W}_0 \boldsymbol{e} = \boldsymbol{W}_0 \cdot (\boldsymbol{a} \otimes \boldsymbol{e}),$$

which holds for all vectors $\boldsymbol{a}$ and $\boldsymbol{e}$,

$$\begin{aligned} \Lambda(\boldsymbol{w}) &\equiv \int_{\partial B} \boldsymbol{s} \cdot \boldsymbol{w}\, da + \int_B \boldsymbol{b} \cdot \boldsymbol{w}\, dv \\ &= \boldsymbol{u}_0 \cdot \left\{ \int_{\partial B} \boldsymbol{s}\, da + \int_B \boldsymbol{b}\, dv \right\} + \boldsymbol{W}_0 \cdot \left\{ \int_{\partial B} \boldsymbol{s} \otimes \boldsymbol{p}\, da + \int_B \boldsymbol{b} \otimes \boldsymbol{p}\, dv \right\}. \end{aligned}$$

Thus $\Lambda(\boldsymbol{w}) = 0$ for every rigid displacement $\boldsymbol{w}$ (or equivalently, for every vector $\boldsymbol{u}_0$ and every *skew* tensor $\boldsymbol{W}_0$) if and only if $\boldsymbol{s}$ and $\boldsymbol{b}$ satisfy balance of forces and, in addition,

$$\operatorname{skw} \left\{ \int_{\partial B} \boldsymbol{p} \otimes \boldsymbol{s}\, da + \int_B \boldsymbol{p} \otimes \boldsymbol{b}\, dv \right\} = \boldsymbol{0}.$$

This completes the proof, since the above relation is equivalent to balance of moments. □

The theorem of work expended was generalized by SIGNORINI[1] in the following manner.

(4) Signorini's theorem. *Let* $\boldsymbol{S}$ *be an admissible stress field that satisfies the equation of equilibrium, and let* $\boldsymbol{w}$ *be a smooth vector field on* $\bar{B}$. *Then*

$$\int_{\partial B} \boldsymbol{w} \otimes \boldsymbol{s}\, da + \int_B \boldsymbol{w} \otimes \boldsymbol{b}\, dv = \int_B (\nabla \boldsymbol{w})\, \boldsymbol{S}\, dv.$$

Proof. By the divergence theorem,

$$\int_{\partial B} \boldsymbol{w} \otimes \boldsymbol{s}\, da = \int_{\partial B} \boldsymbol{w} \otimes (\boldsymbol{S}\boldsymbol{n})\, da = \int_B \boldsymbol{w} \otimes (\operatorname{div} \boldsymbol{S})\, dv + \int_B (\nabla \boldsymbol{w})\, \boldsymbol{S}^T\, dv.$$

This completes the proof, since $\boldsymbol{S}^T = \boldsymbol{S}$ and $\operatorname{div} \boldsymbol{S} = -\boldsymbol{b}$. □

Since

$$\hat{\nabla} \boldsymbol{w} = \tfrac{1}{2}(\nabla \boldsymbol{w} + \nabla \boldsymbol{w}^T),$$

the symmetry of $\boldsymbol{S}$ yields

$$\operatorname{tr}[(\nabla \boldsymbol{w})\, \boldsymbol{S}] = \boldsymbol{S} \cdot \nabla \boldsymbol{w} = \boldsymbol{S} \cdot \hat{\nabla} \boldsymbol{w}.$$

Thus if we take the trace of SIGNORINI's relation, we arrive at the theorem of work expended.

Given an admissible stress field $\boldsymbol{S}$, we call the symmetric tensor

$$\bar{\boldsymbol{S}}(B) = \frac{1}{v(B)} \int_B \boldsymbol{S}\, dv$$

the ***mean stress***. Here $v(B)$ is the volume of B.

[1] SIGNORINI [1933, *4*]. Actually, SIGNORINI's theorem is slightly more general.

(5) Mean stress theorem.[1] *The mean stress corresponding to an admissible stress field that satisfies the equation of equilibrium depends only on the associated surface traction and body force fields and is given by*

$$\bar{\mathbf{S}}(B) = \frac{1}{v(B)}\left[\int_{\partial B} \mathbf{p} \otimes \mathbf{s}\, da + \int_{B} \mathbf{p} \otimes \mathbf{b}\, dv\right].$$

Proof. We simply take $\mathbf{w}$ in SIGNORINI's theorem equal to $\mathbf{p}$ and use the identity $\nabla \mathbf{p} = \mathbf{1}$. □

The next three theorems give alternative characterizations of the equations of equilibrium and compatibility.

(6) *Let* $\mathbf{S}$ *be an admissible stress field, and suppose that*

$$\int_B \mathbf{S} \cdot \nabla \mathbf{u}\, dv = \int_B \mathbf{b} \cdot \mathbf{u}\, dv$$

for every class C^∞ *vector field* $\mathbf{u}$ *on* $\bar{B}$ *that vanishes near* ∂B. *Then*

$$\operatorname{div} \mathbf{S} + \mathbf{b} = \mathbf{0}.$$

Proof. Let $\mathbf{u}$ be a class C^∞ vector field on $\bar{B}$ that vanishes near ∂B. By the symmetry of $\mathbf{S}$, $\mathbf{S} \cdot \nabla \mathbf{u} = \mathbf{S} \cdot \hat{\nabla} \mathbf{u}$; thus we conclude from ***(1)*** that the integral identity in ***(6)*** must reduce to

$$\int_B (\operatorname{div} \mathbf{S} + \mathbf{b}) \cdot \mathbf{u}\, dv = 0.$$

Since $\mathbf{u}$ was arbitrarily chosen, we conclude from the fundamental lemma ***(7.1)*** that $\operatorname{div} \mathbf{S} + \mathbf{b} = \mathbf{0}$. □

(7) *Let* $\mathbf{S}$ *be an admissible stress field, and suppose that*

$$\int_B \mathbf{S} \cdot \mathbf{E}\, dv = 0$$

for every class C^∞ *symmetric tensor field* $\mathbf{E}$ *on* $\bar{B}$ *that vanishes near* ∂B *and satisfies the equation of compatibility. Then*

$$\operatorname{div} \mathbf{S} = \mathbf{0}.$$

Proof. Let $\mathbf{u}$ be a class C^∞ vector field on $\bar{B}$ that vanishes near ∂B, and let $\mathbf{E}$ be the corresponding strain field. Then $\mathbf{E}$ satisfies the equation of compatibility and vanishes near ∂B; hence, by hypothesis,

$$\int_B \mathbf{S} \cdot \nabla \mathbf{u}\, dv = 0,$$

and the proof follows from ***(6)***. □

(8) Donati's theorem.[2] *Let* $\mathbf{E}$ *be a class* C^2 *symmetric tensor field on* B. *Further, suppose that*[3]

$$\int_B \mathbf{S} \cdot \mathbf{E}\, dv = 0$$

---

[1] CHREE [1892, *2*], p. 336. See also SIGNORINI [1932, *5*].

[2] DONATI [1890, *1*], [1894, *2*]. See also COTTERILL [1865, *1*], SOUTHWELL [1936, *5*], [1938, *5*], LOCATELLI [1940, *3, 4*], MORIGUTI [1948, *5*], KLYUSHNIKOV [1954, *10*], WASHIZU [1958, *20*], BLOKH [1962, *2*], STICKFORTH [1964, *21*], [1965, *17*], and TONTI [1967, *16, 17*].

[3] Notice that $\mathbf{S} \cdot \mathbf{E}$ is continuous on $\bar{B}$, since $\mathbf{S}$ vanishes near ∂B.

for every class C^∞ symmetric tensor field $\boldsymbol{S}$ on $\bar{B}$ that vanishes near ∂B and satisfies

$$\operatorname{div} \boldsymbol{S} = \boldsymbol{0}.$$

Then $\boldsymbol{E}$ satisfies the equation of compatibility.

Proof. Let $\boldsymbol{A}$ be a symmetric tensor field of class C^∞ on $\bar{B}$ that vanishes near ∂B, and let

$$\boldsymbol{S} = \operatorname{curl} \operatorname{curl} \boldsymbol{A}.$$

Then $\boldsymbol{S}$ vanishes near ∂B, and by *(17.1)*, $\boldsymbol{S}$ is symmetric and divergence-free. Thus

$$\int_B \boldsymbol{E} \cdot (\operatorname{curl} \operatorname{curl} \boldsymbol{A})\, dv = 0.$$

If we apply the divergence theorem twice and use the fact that $\boldsymbol{A}$ and all of its derivatives vanish on ∂B, we find that

$$\int_B \boldsymbol{E} \cdot (\operatorname{curl} \operatorname{curl} \boldsymbol{A})\, dv = \int_B (\operatorname{curl} \boldsymbol{E})^T \cdot (\operatorname{curl} \boldsymbol{A})\, dv = \int_B (\operatorname{curl} \operatorname{curl} \boldsymbol{E}) \cdot \boldsymbol{A}\, dv.$$

Thus

$$\int_B (\operatorname{curl} \operatorname{curl} \boldsymbol{E}) \cdot \boldsymbol{A}\, dv = 0.$$

Since $\operatorname{curl} \operatorname{curl} \boldsymbol{E}$ is symmetric, and since the above relation must hold for every such function $\boldsymbol{A}$, it follows from the fundamental lemma *(7.1)* that $\boldsymbol{E}$ satisfies the equation of compatibility. □

Let $\mathbf{s}$ be an integrable vector field whose domain is a regular subsurface $\mathscr{S}$ of ∂B, let $\boldsymbol{b}$ be an integrable vector field on B, and let $\boldsymbol{l}$ be a vector field whose domain is a finite point set $\{\xi_1, \xi_2, \ldots, \xi_N\}$ of $\bar{B}$. Further, let $\boldsymbol{l}_n = \boldsymbol{l}(\xi_n)$. The function $\mathbf{s}$ can be thought of as a surface traction field on $\mathscr{S}$, $\boldsymbol{b}$ as a body force field, and $\boldsymbol{l}_n$ as a concentrated load applied at ξ_n. We say that $\mathbf{s}$, $\boldsymbol{b}$, and $\boldsymbol{l}$ are in ***equilibrium*** if

$$\begin{aligned} &\int_{\mathscr{S}} \mathbf{s}\, da + \int_B \boldsymbol{b}\, dv + \sum_{n=1}^{N} \boldsymbol{l}_n = \boldsymbol{0}, \\ &\int_{\mathscr{S}} \mathbf{p} \times \mathbf{s}\, da + \int_B \mathbf{p} \times \boldsymbol{b}\, dv + \sum_{n=1}^{N} (\xi_n - 0) \times \boldsymbol{l}_n = \boldsymbol{0}. \end{aligned} \tag{E}$$

A somewhat more stringent condition is that of astatic equilibrium, in which the system of forces is required to remain in equilibrium when rotated through an arbitrary angle. More precisely, $\mathbf{s}$, $\boldsymbol{b}$ and $\boldsymbol{l}$ are in ***astatic equilibrium*** if $\boldsymbol{Q}\mathbf{s}$, $\boldsymbol{Q}\boldsymbol{b}$, and $\boldsymbol{Q}\boldsymbol{l}$ are in equilibrium for every orthogonal tensor $\boldsymbol{Q}$, i.e. if

$$\begin{aligned} &\int_{\mathscr{S}} \boldsymbol{Q}\mathbf{s}\, da + \int_B \boldsymbol{Q}\boldsymbol{b}\, dv + \sum_{n=1}^{N} \boldsymbol{Q}\boldsymbol{l}_n = \boldsymbol{0}, \\ &\int_{\mathscr{S}} \mathbf{p} \times \boldsymbol{Q}\mathbf{s}\, da + \int_B \mathbf{p} \times \boldsymbol{Q}\boldsymbol{b}\, dv + \sum_{n=1}^{N} (\xi_n - 0) \times \boldsymbol{Q}\boldsymbol{l}_n = \boldsymbol{0}, \end{aligned} \tag{A}$$

for every orthogonal $\boldsymbol{Q}$.

(9) Conditions for astatic equilibrium.[1] *Let $\mathbf{s}$, $\boldsymbol{b}$, and $\boldsymbol{l}$ be as defined above. Then $\mathbf{s}$, $\boldsymbol{b}$, and $\boldsymbol{l}$ are in astatic equilibrium if and only if the following two conditions hold:*

(i) *$\mathbf{s}$, $\boldsymbol{b}$, and $\boldsymbol{l}$ are in equilibrium;*

(ii) $\int_{\mathscr{S}} \mathbf{p} \otimes \mathbf{s}\, da + \int_B \mathbf{p} \otimes \boldsymbol{b}\, dv + \sum_{n=1}^{N} (\xi_n - 0) \otimes \boldsymbol{l}_n = \boldsymbol{0}.$

[1] In classical statics, in which the loads are all concentrated, this result has long been known (see, e.g., Routh [1908, *3*], p. 313). The extension to distributed loads is due to Berg [1969, *1*].

Proof.[1] Assume that $\boldsymbol{s}$, $\boldsymbol{b}$, and $\boldsymbol{l}$ are in astatic equilibrium. Then (i) follows from (A) with $\boldsymbol{Q}=\mathbf{1}$. Let $\boldsymbol{T}$ denote the left-hand side of (ii). By (E_2), skw $\boldsymbol{T}=\mathbf{0}$; thus $\boldsymbol{T}$ is symmetric. Next, by (A_2),

$$\mathbf{0}=\operatorname{skw}\left\{\int_{\mathscr{S}}\boldsymbol{p}\otimes\boldsymbol{Q}\boldsymbol{s}\,da+\int_{B}\boldsymbol{p}\otimes\boldsymbol{Q}\boldsymbol{b}\,dv+\sum_{n=1}^{N}(\xi_n-0)\otimes\boldsymbol{Q}\boldsymbol{l}_n\right\};$$

thus, since $\boldsymbol{a}\otimes\boldsymbol{Q}\boldsymbol{c}=(\boldsymbol{a}\otimes\boldsymbol{c})\,\boldsymbol{Q}^T$ for all vectors $\boldsymbol{a}$ and $\boldsymbol{c}$, $\operatorname{skw}(\boldsymbol{T}\boldsymbol{Q}^T)=\mathbf{0}$. Consequently,

$$\boldsymbol{T}\boldsymbol{Q}^T=\boldsymbol{Q}\boldsymbol{T}. \tag{a}$$

Let

$$\boldsymbol{T}=\sum_{i=1}^{3}\lambda_i\,\boldsymbol{n}_i\otimes\boldsymbol{n}_i \tag{b}$$

be the spectral decomposition of $\boldsymbol{T}$ (in the sense of the spectral theorem *(3.2)*). By (a) and (b),

$$\sum_{i=1}^{3}\lambda_i\,\boldsymbol{n}_i\otimes\boldsymbol{Q}\boldsymbol{n}_i=\sum_{i=1}^{3}\lambda_i(\boldsymbol{Q}\boldsymbol{n}_i)\otimes\boldsymbol{n}_i, \tag{c}$$

and this relation must hold for every orthogonal tensor $\boldsymbol{Q}$. Let $\boldsymbol{Q}$ be the orthogonal tensor with the following properties:

$$\boldsymbol{Q}\boldsymbol{n}_1=\boldsymbol{n}_2,\qquad \boldsymbol{Q}\boldsymbol{n}_2=\boldsymbol{n}_3,\qquad \boldsymbol{Q}\boldsymbol{n}_3=\boldsymbol{n}_1. \tag{d}$$

If we operate with both sides of (c) on $\boldsymbol{n}_3$ and use (d) and the fact that $\boldsymbol{n}_i\cdot\boldsymbol{n}_j=\delta_{ij}$, we find that

$$\lambda_2\,\boldsymbol{n}_2=\lambda_3\,\boldsymbol{n}_1,$$

which implies that $\lambda_2=\lambda_3=0$. Similarly, $\lambda_1=0$. Thus $\boldsymbol{T}=\mathbf{0}$ and (ii) holds.

Assume now that (i) and (ii) hold. Let $\boldsymbol{T}$ be as before. Then $\boldsymbol{T}=\mathbf{0}$ and

$$\mathbf{0}=\boldsymbol{T}\boldsymbol{Q}^T=\int_{\mathscr{S}}\boldsymbol{p}\otimes\boldsymbol{Q}\boldsymbol{s}\,da+\int_{B}\boldsymbol{p}\otimes\boldsymbol{Q}\boldsymbol{b}\,dv+\sum_{n=1}^{N}(\xi_n-0)\otimes\boldsymbol{Q}\boldsymbol{l}_n. \tag{e}$$

If we take the skew part of (e), we are led, at once, to (A_2). On the other hand, (A_1) is an immediate consequence of (E_1). □

In view of *(9)*, the mean stress theorem *(5)* has the following interesting consequence.

(10)[2] *The mean stress corresponding to an admissible stress field that satisfies the equation of equilibrium vanishes if and only if the associated surface traction and body force fields are in astatic equilibrium.*

19. Consequences of the equation of motion. In this section we study properties of solutions to the equation of motion

$$\operatorname{div}\boldsymbol{S}+\boldsymbol{b}=\varrho\ddot{\boldsymbol{u}}$$

on a given time interval $[0, t_0)$, assuming throughout that a *continuous* and *strictly positive* density field ϱ is prescribed on $\bar{B}$.

Given a dynamical process[3] $[\boldsymbol{u}, \boldsymbol{S}, \boldsymbol{b}]$, we call

$$K=\tfrac{1}{2}\int_{B}\varrho\dot{\boldsymbol{u}}^2\,dv$$

---

[1] This proof differs from that given by BERG [1969, *1*].

[2] SIGNORINI [1932, *5*].

[3] See p. 49.

the ***kinetic energy*** and

$$\int_B \boldsymbol{S} \cdot \dot{\boldsymbol{E}}\, dv$$

the *stress power;* here $\boldsymbol{E}$ is the strain field associated with $\boldsymbol{u}$.

(1) Theorem of power expended.[1] *Let* $[\boldsymbol{u}, \boldsymbol{S}, \boldsymbol{b}]$ *be a dynamical process. Then*

$$\int_{\partial B} \boldsymbol{s} \cdot \dot{\boldsymbol{u}}\, da + \int_B \boldsymbol{b} \cdot \dot{\boldsymbol{u}}\, dv = \int_B \boldsymbol{S} \cdot \dot{\boldsymbol{E}}\, dv + \dot{K}.$$

Proof. The proof follows from the theorem of work expended ***(18.2)*** with $\boldsymbol{b}$ replaced by $\boldsymbol{b} - \varrho \ddot{\boldsymbol{u}}$ and $\boldsymbol{u}$ replaced by $\dot{\boldsymbol{u}}$. □

Theorem ***(1)*** asserts that *the rate of working of the surface tractions and body forces equals the stress power plus the rate of change of kinetic energy.* If

$$\dot{\boldsymbol{u}}(\cdot, 0) = \dot{\boldsymbol{u}}(\cdot, \tau),$$

then, by ***(1)***,

$$\int_0^\tau \int_{\partial B} \boldsymbol{s} \cdot \dot{\boldsymbol{u}}\, da\, dt + \int_0^\tau \int_B \boldsymbol{b} \cdot \dot{\boldsymbol{u}}\, dv\, dt = \int_0^\tau \int_B \boldsymbol{S} \cdot \dot{\boldsymbol{E}}\, dv\, dt;$$

i.e., when the initial and terminal velocity fields coincide, the work done by the external forces is equal to the work done by the stress field.

As we shall see later, in a homogeneous deformation of a homogeneous body the stress and strain fields $\boldsymbol{S}$ and $\boldsymbol{E}$ are independent of position. In this instance $\boldsymbol{b} = \varrho \ddot{\boldsymbol{u}}$ and ***(1)*** implies

$$\boldsymbol{S} \cdot \dot{\boldsymbol{E}} = \frac{1}{v(B)} \int_{\partial B} \boldsymbol{s} \cdot \dot{\boldsymbol{u}}\, da;$$

i.e., the stress power density is equal to the rate of working of the surface tractions per unit volume.[2]

The next theorem will be of use when we discuss variational principles for elastodynamics; it shows that the equation of motion and the appropriate initial conditions are together equivalent to a single integral-differential equation.

Let $\boldsymbol{i}$ be the function on $[0, t_0)$ defined by

$$\boldsymbol{i}(t) = t.$$

(2)[3] *Let* $\boldsymbol{b}$ *be a continuous vector field on* $B \times [0, t_0)$, *let* $\boldsymbol{u}$ *be an admissible motion, and let* $\boldsymbol{S}$ *be a time-dependent admissible stress field. Further, let* $\boldsymbol{u}_0$ *and* $\boldsymbol{v}_0$ *be vector fields on* $\bar{B}$, *and let* $\boldsymbol{f}$ *be defined on* $\bar{B} \times [0, t_0)$ *by*

$$\boldsymbol{f}(\boldsymbol{x}, t) = \boldsymbol{i} * \boldsymbol{b}(\boldsymbol{x}, t) + \varrho(\boldsymbol{x})\left[\boldsymbol{u}_0(\boldsymbol{x}) + t\boldsymbol{v}_0(\boldsymbol{x})\right].$$

Then $[\boldsymbol{u}, \boldsymbol{S}, \boldsymbol{b}]$ *is a dynamical process consistent with the initial conditions*

$$\boldsymbol{u}(\cdot, 0) = \boldsymbol{u}_0, \qquad \dot{\boldsymbol{u}}(\cdot, 0) = \boldsymbol{v}_0$$

if and only if

$$\boldsymbol{i} * \operatorname{div} \boldsymbol{S} + \boldsymbol{f} = \varrho \boldsymbol{u}. \tag{a}$$

---

[1] STOKES [1851, *2*], UMOV [1874, *2*]. Cf. also v. MISES [1909, *1*], TRUESDELL and TOUPIN [1960, *17*], § 217.

[2] Cf. TRUESDELL and NOLL [1965, *22*], Eq. (18.7).

[3] GURTIN [1964, *8*], Theorem 3.1.

Proof. If the equation of motion and the initial conditions hold, then

$$i*[\operatorname{div} \boldsymbol{S}+\boldsymbol{b}](\boldsymbol{x},t)=\varrho(\boldsymbol{x})\int_0^t (t-\tau)\,\ddot{\boldsymbol{u}}(\boldsymbol{x},\tau)\,d\tau$$
$$=\varrho(\boldsymbol{x})\,\boldsymbol{u}(\boldsymbol{x},t)-\varrho(\boldsymbol{x})\,[t\boldsymbol{v}_0(\boldsymbol{x})+\boldsymbol{u}_0(\boldsymbol{x})],$$

which implies (a). The converse is easily established by reversing the above argument. □

The next lemma, which is a counterpart of **(18.1)**, will be extremely useful in establishing variational principles for elastodynamics.

(3) Lemma. *Let $\boldsymbol{w}$ be an admissible motion and $\boldsymbol{S}$ a time-dependent admissible stress field. Then*

$$\int_{\partial B} \boldsymbol{s}*\boldsymbol{w}\,da=\int_B \boldsymbol{w}*\operatorname{div}\boldsymbol{S}\,dv+\int_B \boldsymbol{S}*\hat{\nabla}\boldsymbol{w}\,dv.$$

Proof. By **(18.1)**,

$$\int_{\partial B} \boldsymbol{s}*\boldsymbol{w}\,da=\int_0^t\int_{\partial B} \boldsymbol{s}(\boldsymbol{x},t-\tau)\cdot\boldsymbol{w}(\boldsymbol{x},\tau)\,da\,d\tau$$
$$=\int_0^t\int_B [\operatorname{div}\boldsymbol{S}(\boldsymbol{x},t-\tau)\cdot\boldsymbol{w}(\boldsymbol{x},\tau)+\boldsymbol{S}(\boldsymbol{x},t-\tau)\cdot\hat{\nabla}\boldsymbol{w}(\boldsymbol{x},\tau)]\,dv\,d\tau$$
$$=\int_B \operatorname{div} S*\boldsymbol{w}\,dv+\int_B \boldsymbol{S}*\hat{\nabla}\boldsymbol{w}\,dv. \quad \square$$

When $\boldsymbol{u}_0$ and $\boldsymbol{v}_0$ are the initial values of $\boldsymbol{u}$ and $\dot{\boldsymbol{u}}$, we call the function $\boldsymbol{f}$ defined in **(2)** the **pseudo body force field**. Our next theorem gives an analog of the theorem of work expended **(18.2)**.

(4) *Let $[\boldsymbol{u},\boldsymbol{S},\boldsymbol{b}]$ be a dynamical process, let $\boldsymbol{f}$ be the corresponding pseudo body force field, and let $\boldsymbol{w}$ be an admissible motion. Then*

$$i*\int_{\partial B}\boldsymbol{s}*\boldsymbol{w}\,da+\int_B \boldsymbol{f}*\boldsymbol{w}\,dv=i*\int_B \boldsymbol{S}*\hat{\nabla}\boldsymbol{w}\,dv+\int_B \varrho\boldsymbol{u}*\boldsymbol{w}\,dv.$$

In particular, when $\boldsymbol{w}=\boldsymbol{u}$,

$$i*\int_{\partial B}\boldsymbol{s}*\boldsymbol{u}\,da+\int_B \boldsymbol{f}*\boldsymbol{u}\,dv=i*\int_B \boldsymbol{S}*\boldsymbol{E}\,dv+\int_B \varrho\boldsymbol{u}*\boldsymbol{u}\,dv.$$

Proof. By **(2)** and **(3)**,

$$i*\int_{\partial B}\boldsymbol{s}*\boldsymbol{w}\,da=\int_B (i*\operatorname{div}\boldsymbol{S})*\boldsymbol{w}\,dv+i*\int_B \boldsymbol{S}*\hat{\nabla}\boldsymbol{w}\,dv$$
$$=\int_B (\varrho\boldsymbol{u}-\boldsymbol{f})*\boldsymbol{w}\,dv+i*\int_B \boldsymbol{S}*\hat{\nabla}\boldsymbol{w}\,dv. \quad \square$$

(5) Reciprocal theorem.[1] *Let $[\boldsymbol{u},\boldsymbol{S},\boldsymbol{b}]$ and $[\tilde{\boldsymbol{u}},\tilde{\boldsymbol{S}},\tilde{\boldsymbol{b}}]$ be dynamical processes, and let $\boldsymbol{f}$ and $\tilde{\boldsymbol{f}}$ be the corresponding pseudo body force fields. Then*

$$\begin{aligned}&i*\int_{\partial B}\boldsymbol{s}*\tilde{\boldsymbol{u}}\,da+\int_B \boldsymbol{f}*\tilde{\boldsymbol{u}}\,dv-i*\int_B \boldsymbol{S}*\tilde{\boldsymbol{E}}\,dv\\ &\quad=i*\int_{\partial B}\tilde{\boldsymbol{s}}*\boldsymbol{u}\,da+\int_B \tilde{\boldsymbol{f}}*\boldsymbol{u}\,dv-i*\int_B \tilde{\boldsymbol{S}}*\boldsymbol{E}\,dv,\end{aligned} \tag{a}$$

$$\begin{aligned}&\int_{\partial B}\boldsymbol{s}*\tilde{\boldsymbol{u}}\,da+\int_B \boldsymbol{b}*\tilde{\boldsymbol{u}}\,dv-\int_B \boldsymbol{S}*\tilde{\boldsymbol{E}}\,dv+\int_B \varrho(\boldsymbol{u}_0\cdot\dot{\tilde{\boldsymbol{u}}}+\boldsymbol{v}_0\cdot\tilde{\boldsymbol{u}})\,dv\\ &\quad=\int_{\partial B}\tilde{\boldsymbol{s}}*\boldsymbol{u}\,da+\int_B \tilde{\boldsymbol{b}}*\boldsymbol{u}\,dv-\int_B \tilde{\boldsymbol{S}}*\boldsymbol{E}\,dv+\int_B \varrho(\tilde{\boldsymbol{u}}_0\cdot\dot{\boldsymbol{u}}+\tilde{\boldsymbol{v}}_0\cdot\boldsymbol{u})\,dv.\end{aligned} \tag{b}$$

---

[1] This theorem, which appears for the first time here, is based on an idea of GRAFFI [1947, *2*], [1954, *6*], [1963, *8*] (see Sect. 61 of this article).

Thus if $\boldsymbol{u}_0 = \boldsymbol{v}_0 = \tilde{\boldsymbol{u}}_0 = \tilde{\boldsymbol{v}}_0 = \mathbf{0}$, *then*

$$\int_{\partial B} \mathbf{s} * \tilde{\boldsymbol{u}}\, da + \int_B \boldsymbol{b} * \tilde{\boldsymbol{u}}\, dv - \int_B \boldsymbol{S} * \tilde{\boldsymbol{E}}\, dv = \int_{\partial B} \tilde{\mathbf{s}} * \boldsymbol{u}\, da + \int_B \tilde{\boldsymbol{b}} * \boldsymbol{u}\, dv - \int_B \tilde{\boldsymbol{S}} * \boldsymbol{E}\, dv. \quad \text{(c)}$$

Proof. The relation (a) follows from ***(4)*** and the fact that

$$\int_B \varrho \boldsymbol{u} * \tilde{\boldsymbol{u}}\, dv = \int_B \varrho \tilde{\boldsymbol{u}} * \boldsymbol{u}\, dv.$$

To verify (b) we note first that (vi) of ***(10.1)*** implies

$$\overset{\cdot\cdot}{\overline{i * \psi}} = \psi. \quad \text{(d)}$$

Thus, in view of the definition of $\boldsymbol{f}$,

$$\overset{\cdot\cdot}{\overline{\boldsymbol{f} * \tilde{\boldsymbol{u}}}} = \boldsymbol{b} * \tilde{\boldsymbol{u}} + \varrho (\boldsymbol{u}_0 \cdot \dot{\tilde{\boldsymbol{u}}} + \boldsymbol{v}_0 \cdot \tilde{\boldsymbol{u}}). \quad \text{(e)}$$

Therefore if we differentiate (a) twice with respect to t using (d), (e), and the analog of (e) for $\tilde{\boldsymbol{f}}$ and $\boldsymbol{u}$, we arrive at (b). □

Given a time-dependent admissible stress field $\boldsymbol{S}$ and an admissible motion $\boldsymbol{u}$, let $\mathbf{M}$ denote the *four-tensor* field with space-time partition

$$\begin{bmatrix} \boldsymbol{S} & -\varrho \dot{\boldsymbol{u}} \\ \mathbf{0} & 0 \end{bmatrix};$$

we call $\mathbf{M}$ the associated ***stress-momentum field***. The importance of this field is clear from the following theorem.

(6) Balance of momentum in space-time. *Let* $\boldsymbol{b}$ *be a continuous vector field on* $B \times [0, t_0)$. *Further, let* $\boldsymbol{S}$ *be a time-dependent admissible stress field, let* $\boldsymbol{u}$ *be an admissible motion, and let* $\mathbf{M}$ *be the associated stress-momentum field. Then* $[\boldsymbol{u}, \boldsymbol{S}, \boldsymbol{b}]$ *is a dynamical process if and only if*

$$\int_{\partial D} \mathbf{M}\mathbf{n}\, da + \int_D \mathbf{b}\, dv = \mathbf{0}$$

for every regular region D with $\overline{D}$ *in* $B \times (0, t_0)$. *Here* $\mathbf{n}$ *is the outward unit normal to* ∂D *and*

$$\mathbf{b} = (\boldsymbol{b}, 0).$$

Proof. By the divergence theorem ***(11.2)***, the above relation is equivalent to

$$\int_D (\operatorname{div}_{(4)} \mathbf{M} + \mathbf{b})\, dv = \mathbf{0}.$$

This equation holds for all regular D if and only if

$$\operatorname{div}_{(4)} \mathbf{M} + \mathbf{b} = \mathbf{0},$$

or equivalently, by the identity ***(11.1)***,

$$\operatorname{div} \boldsymbol{S} - \varrho \ddot{\boldsymbol{u}} + \boldsymbol{b} = \mathbf{0}. \quad □$$

III. The constitutive relation for linearly elastic materials.

20. The elasticity tensor. The laws of momentum balance hold in every dynamical process, no matter of what material the body is composed. A specific material is defined by a constitutive relation which places restrictions on the class of processes the body may undergo. Here we will be concerned exclusively with linearly elastic materials. For such materials the stress at any time and point in a dynamical process is a linear function of the displacement gradient at the *same* time and place. Moreover, while this function may depend on the point under consideration, it is independent of the time.

Therefore we say that the body is ***linearly elastic*** if for each $\boldsymbol{x} \in B$ there exists a linear transformation $\mathsf{C}_{\boldsymbol{x}}$ from the space of all tensors into the space of all *symmetric* tensors such that[1]

$$\boldsymbol{S}(\boldsymbol{x}) = \mathsf{C}_{\boldsymbol{x}}[\nabla \boldsymbol{u}(\boldsymbol{x})].$$

We call $\mathsf{C}_{\boldsymbol{x}}$ the ***elasticity tensor*** for $\boldsymbol{x}$ and the function C on B with values $\mathsf{C}_{\boldsymbol{x}}$ the ***elasticity field***. In general, $\mathsf{C}_{\boldsymbol{x}}$ depends on $\boldsymbol{x}$; if, however, $\mathsf{C}_{\boldsymbol{x}}$ and the density $\varrho(\boldsymbol{x})$ are independent of $\boldsymbol{x}$, we say that B is ***homogeneous***. We *postulate that a rigid displacement field results in zero stress at each point.*[2] Let $\boldsymbol{W}$ be an arbitrary skew tensor. Then the field

$$\boldsymbol{u}(\boldsymbol{x}) = \boldsymbol{W}[\boldsymbol{x} - \boldsymbol{x}_0]$$

is a rigid displacement field with

$$\nabla \boldsymbol{u}(\boldsymbol{x}) \equiv \boldsymbol{W};$$

thus we conclude from the above postulate that

$$\mathsf{C}_{\boldsymbol{x}}[\boldsymbol{W}] = \boldsymbol{0}.$$

Since this result must hold for every $\boldsymbol{x} \in B$ and for every skew tensor $\boldsymbol{W}$,

$$\mathsf{C}_{\boldsymbol{x}}[\boldsymbol{A}] = \mathsf{C}_{\boldsymbol{x}}[\operatorname{sym} \boldsymbol{A}]$$

for every tensor $\boldsymbol{A}$. Therefore $\mathsf{C}_{\boldsymbol{x}}$ is completely determined by its restriction to the space of all symmetric tensors. In particular, our original constitutive relation reduces to

$$\boldsymbol{S}(\boldsymbol{x}) = \mathsf{C}_{\boldsymbol{x}}[\boldsymbol{E}(\boldsymbol{x})],$$

where $\boldsymbol{E}$ is the strain field corresponding to $\nabla \boldsymbol{u}$. Since the space of all symmetric tensors has dimension 6, the matrix (relative to any basis) of the restriction of $\mathsf{C}_{\boldsymbol{x}}$ to this space is 6×6.

For the moment let us write $\mathsf{C} = \mathsf{C}_{\boldsymbol{x}}$. Consider an orthonormal basis $\{\boldsymbol{e}_1, \boldsymbol{e}_2, \boldsymbol{e}_3\}$, and let C_{ijkl} denote the components of C relative to this basis:

$$C_{ijkl} = (\boldsymbol{e}_i \otimes \boldsymbol{e}_j) \cdot \mathsf{C}[\boldsymbol{e}_k \otimes \boldsymbol{e}_l].$$

Then $\boldsymbol{S} = \mathsf{C}[\boldsymbol{E}]$ takes the form

$$S_{ij} = C_{ijkl} E_{kl}.$$

Moreover, since the values of C are symmetric, and since the value of C on any tensor $\boldsymbol{A}$ equals its value on the symmetric part of $\boldsymbol{A}$,

$$C_{ijkl} = (\operatorname{sym} \boldsymbol{e}_i \otimes \boldsymbol{e}_j) \cdot \mathsf{C}[\operatorname{sym} \boldsymbol{e}_k \otimes \boldsymbol{e}_l];$$

hence

$$C_{ijkl} = C_{jikl} = C_{ijlk}.$$

We call the 36 numbers C_{ijkl} ***elasticities***.

---

[1] Thus we have made the tacit assumption that the stress is zero when the displacement gradient vanishes; or equivalently, that the residual stress in the body is zero. At first sight it appears that such an assumption rules out elastic fluids, for which there is always a residual pressure. However, if the residual stress is a constant pressure, and if we interpret $\boldsymbol{S}(\boldsymbol{x})$ as the actual stress minus the residual stress, then all of our results are valid without change, provided, of course, we interpret the corresponding surface tractions accordingly.

[2] POISSON 1829 [1831, *1*]. See also CAUCHY [1829, *1*]. This axiom is discussed in detail by TRUESDELL and NOLL [1965, *22*], §§ 19, 19A.

(1) Properties of the elasticity tensor. *For every* $\boldsymbol{x} \in B$:

(i) $\mathsf{C}_{\boldsymbol{x}}[\boldsymbol{A}] = \mathsf{C}_{\boldsymbol{x}}[\operatorname{sym} \boldsymbol{A}]$ *for every tensor* $\boldsymbol{A}$;

(ii) $\boldsymbol{D} \cdot \mathsf{C}_{\boldsymbol{x}}[\boldsymbol{A}] = (\operatorname{sym} \boldsymbol{D}) \cdot \mathsf{C}_{\boldsymbol{x}}[\operatorname{sym} \boldsymbol{A}]$ *for every pair of tensors* $\boldsymbol{A}$ *and* $\boldsymbol{D}$;

(iii) $C_{ijkl} = (\operatorname{sym} \boldsymbol{e}_i \otimes \boldsymbol{e}_j) \cdot \mathsf{C}_{\boldsymbol{x}}[\operatorname{sym} \boldsymbol{e}_k \otimes \boldsymbol{e}_l]$, *so that*

$$C_{ijkl} = C_{jikl} = C_{ijlk}.$$

Proof. Properties (i) and (iii) have already been established. Property (ii) follows from (i) and the fact that the values of $\mathsf{C}_{\boldsymbol{x}}$ are symmetric tensors. □

If the elasticity tensor is invertible,[1] then its inverse

$$\mathsf{K}_{\boldsymbol{x}} = \mathsf{C}_{\boldsymbol{x}}^{-1}$$

is called the ***compliance tensor***; it defines a relation

$$\boldsymbol{E}(\boldsymbol{x}) = \mathsf{K}_{\boldsymbol{x}}[\boldsymbol{S}(\boldsymbol{x})]$$

between the strain $\boldsymbol{E}(\boldsymbol{x})$ and the stress $\boldsymbol{S}(\boldsymbol{x})$ at $\boldsymbol{x}$.

The following definitions will be important in what follows. Consider a fixed point $\boldsymbol{x}$ of the body, and let $\mathsf{C} = \mathsf{C}_{\boldsymbol{x}}$ denote the elasticity tensor for $\boldsymbol{x}$. We say that C is ***symmetric*** if

$$\boldsymbol{A} \cdot \mathsf{C}[\boldsymbol{B}] = \boldsymbol{B} \cdot \mathsf{C}[\boldsymbol{A}]$$

for every pair of symmetric tensors $\boldsymbol{A}$ and $\boldsymbol{B}$, ***positive semi-definite*** if

$$\boldsymbol{A} \cdot \mathsf{C}[\boldsymbol{A}] \geqq 0$$

for every symmetric tensor $\boldsymbol{A}$, and ***positive definite*** if

$$\boldsymbol{A} \cdot \mathsf{C}[\boldsymbol{A}] > 0$$

for every non-zero *symmetric* tensor $\boldsymbol{A}$. Clearly, C is symmetric if and only if its components obey

$$C_{ijkl} = C_{klij},$$

so that in this instance C has 21, rather than 36, distinct elasticities. Note also that C is invertible whenever it is positive definite.

For the remainder of the article *we assume given an elasticity field* C *on* B.

21. Material symmetry. Assume that the material is linearly elastic in the sense of the stress-strain relation

$$\boldsymbol{S} = \mathsf{C}_{\boldsymbol{x}}[\boldsymbol{E}]$$

discussed in the previous section. In order to specify the symmetry properties of the material, we introduce the notion of a symmetry group. Our first step will be to motivate this concept.

Let $\boldsymbol{x}$ be a *fixed* point in B, let $\boldsymbol{u}$ and $\boldsymbol{u}^*$ be smooth vector fields whose domain is a neighborhood of $\boldsymbol{x}$, and suppose that $\boldsymbol{u}^*$ is related to $\boldsymbol{u}$ by an orthogonal tensor $\boldsymbol{Q}$:

$$\boldsymbol{u}^*(\boldsymbol{y}^*) = \boldsymbol{Q}\boldsymbol{u}(\boldsymbol{y}) \quad \text{whenever} \quad \boldsymbol{y}^* = \boldsymbol{Q}(\boldsymbol{y} - \boldsymbol{x}) + \boldsymbol{x}. \tag{a}$$

[1] By this statement we mean: "if the *restriction* of the elasticity tensor to the space of all symmetric tensors is invertible." The elasticity tensor, when regarded as a linear transformation on the space of all tensors, cannot be invertible since its value on every skew tensor is zero.

Roughly speaking, (a) asserts that if the point $\boldsymbol{y}^*$ is obtained from $\boldsymbol{y}$ by a rotation $\boldsymbol{Q}$ about $\boldsymbol{x}$, then the displacement $\boldsymbol{u}^*(\boldsymbol{y}^*)$ is simply the displacement $\boldsymbol{u}(\boldsymbol{y})$ rotated by $\boldsymbol{Q}$. If for every such $\boldsymbol{u}$ and $\boldsymbol{u}^*$ the corresponding stresses

$$\boldsymbol{S} = \mathbf{C}_{\boldsymbol{x}}[\boldsymbol{E}(\boldsymbol{x})], \quad \boldsymbol{S}^* = \mathbf{C}_{\boldsymbol{x}}[\boldsymbol{E}^*(\boldsymbol{x})] \tag{b}$$

satisfy

$$\boldsymbol{S}^* = \boldsymbol{Q}\boldsymbol{S}\boldsymbol{Q}^T, \tag{c}$$

then we call $\boldsymbol{Q}$ a *symmetry transformation* for the material at $\boldsymbol{x}$. Condition (c) is slightly more transparent when phrased in terms of the surface force vectors $\boldsymbol{s}_{\boldsymbol{n}} = \boldsymbol{S}\boldsymbol{n}$, $\boldsymbol{s}_{\boldsymbol{n}}^* = \boldsymbol{S}^*\boldsymbol{n}$; indeed, (c) is equivalent to the requirement that

$$\boldsymbol{s}_{\boldsymbol{n}^*}^* = \boldsymbol{Q}\,\boldsymbol{s}_{\boldsymbol{n}} \quad \text{whenever} \quad \boldsymbol{n}^* = \boldsymbol{Q}\boldsymbol{n}$$

(see Fig. 9).

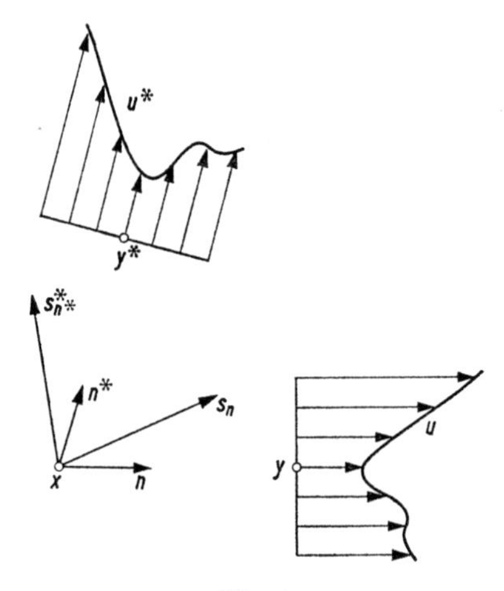

Fig. 9.

By (a)

$$\nabla \boldsymbol{u}^*(\boldsymbol{x}) = \boldsymbol{Q}\,\nabla \boldsymbol{u}(\boldsymbol{x})\,\boldsymbol{Q}^T,$$

and hence the corresponding strain tensors are related by

$$\boldsymbol{E}^*(\boldsymbol{x}) = \boldsymbol{Q}\boldsymbol{E}(\boldsymbol{x})\,\boldsymbol{Q}^T. \tag{d}$$

It follows from (b), (c), and (d) that a sufficient condition for an orthogonal tensor $\boldsymbol{Q}$ to be a symmetry transformation is that

$$\boldsymbol{Q}\mathbf{C}_{\boldsymbol{x}}[\boldsymbol{E}]\,\boldsymbol{Q}^T = \mathbf{C}_{\boldsymbol{x}}[\boldsymbol{Q}\boldsymbol{E}\boldsymbol{Q}^T] \tag{e}$$

for every symmetric tensor $\boldsymbol{E}$. That this condition is also necessary can be easily verified using displacement fields which are pure strains from $\boldsymbol{x}$.

Thus we have motivated the following definition:[1] the ***symmetry group*** $\mathscr{G}_{\boldsymbol{x}}$ for the material at $\boldsymbol{x}$ is the set of all *orthogonal* tensors $\boldsymbol{Q}$ that obey (e) for every symmetric tensor $\boldsymbol{E}$.

[1] Our definition is consistent with the notion of symmetry group (isotropy group) as employed in general non-linear continuum mechanics (see, e.g., TRUESDELL and NOLL [1965, *22*], § 31). Since the symmetry group of a solid relative to an undistorted configuration is a subgroup of the orthogonal group, our treatment is sufficiently general to include elastic solids of arbitrary symmetry. While we do not rule out non-orthogonal symmetry transformations, we do not study them. The treatment of more general symmetry groups is essential to a systematic study of fluid crystals (see, e.g., TRUESDELL and NOLL [1965, *22*], § 33 bis.).

To show that the above definition is consistent, i.e. that $\mathcal{G}_{\boldsymbol{x}}$ really is a group, it is sufficient to verify that

$$\boldsymbol{Q} \in \mathcal{G}_{\boldsymbol{x}} \Rightarrow \boldsymbol{Q}^{-1} \in \mathcal{G}_{\boldsymbol{x}},$$
$$\boldsymbol{Q}, \boldsymbol{P} \in \mathcal{G}_{\boldsymbol{x}} \Rightarrow \boldsymbol{Q}\boldsymbol{P} \in \mathcal{G}_{\boldsymbol{x}}.$$

Note first that (e) can be written in the form

$$\mathsf{C}_{\boldsymbol{x}}[\boldsymbol{E}] = \boldsymbol{Q}^T \mathsf{C}_{\boldsymbol{x}}[\boldsymbol{Q}\boldsymbol{E}\boldsymbol{Q}^T]\boldsymbol{Q}. \tag{f}$$

If $\boldsymbol{Q} \in \mathcal{G}_{\boldsymbol{x}}$, then, since

$$\boldsymbol{Q}\boldsymbol{Q}^T = \mathbf{1},$$

(f) applied to the symmetric tensor $\boldsymbol{Q}^T\boldsymbol{E}\boldsymbol{Q}$ yields

$$\mathsf{C}_{\boldsymbol{x}}[\boldsymbol{Q}^T\boldsymbol{E}\boldsymbol{Q}] = \boldsymbol{Q}^T \mathsf{C}_{\boldsymbol{x}}[\boldsymbol{Q}(\boldsymbol{Q}^T\boldsymbol{E}\boldsymbol{Q})\boldsymbol{Q}^T]\boldsymbol{Q} = \boldsymbol{Q}^T \mathsf{C}_{\boldsymbol{x}}[\boldsymbol{E}]\boldsymbol{Q}.$$

Thus we conclude from (e) that $\boldsymbol{Q}^T = \boldsymbol{Q}^{-1}$ belongs to $\mathcal{G}_{\boldsymbol{x}}$. Next, if $\boldsymbol{Q}, \boldsymbol{P} \in \mathcal{G}_{\boldsymbol{x}}$, then two applications of (f) yield

$$\begin{aligned}\mathsf{C}_{\boldsymbol{x}}[\boldsymbol{E}] &= \boldsymbol{P}^T \mathsf{C}_{\boldsymbol{x}}[\boldsymbol{P}\boldsymbol{E}\boldsymbol{P}^T]\boldsymbol{P} = \boldsymbol{P}^T\boldsymbol{Q}^T \mathsf{C}_{\boldsymbol{x}}[\boldsymbol{Q}(\boldsymbol{P}\boldsymbol{E}\boldsymbol{P}^T)\boldsymbol{Q}^T]\boldsymbol{Q}\boldsymbol{P}\\ &= (\boldsymbol{Q}\boldsymbol{P})^T \mathsf{C}_{\boldsymbol{x}}[(\boldsymbol{Q}\boldsymbol{P})\boldsymbol{E}(\boldsymbol{Q}\boldsymbol{P})^T]\boldsymbol{Q}\boldsymbol{P};\end{aligned}$$

thus $\boldsymbol{Q}\boldsymbol{P} \in \mathcal{G}_{\boldsymbol{x}}$. Therefore $\mathcal{G}_{\boldsymbol{x}}$ is, indeed, a group.

(1) Let $\mathsf{C}_{\boldsymbol{x}}$ be invertible, and let $\mathsf{K}_{\boldsymbol{x}}$ be the corresponding compliance tensor. Then

$$\boldsymbol{Q}\mathsf{K}_{\boldsymbol{x}}[\boldsymbol{S}]\boldsymbol{Q}^T = \mathsf{K}_{\boldsymbol{x}}[\boldsymbol{Q}\boldsymbol{S}\boldsymbol{Q}^T]$$

for every $\boldsymbol{Q} \in \mathcal{G}_{\boldsymbol{x}}$ and every symmetric tensor $\boldsymbol{S}$.

Proof. Choose $\boldsymbol{S}$ arbitrarily, and let $\boldsymbol{E} = \mathsf{K}_{\boldsymbol{x}}[\boldsymbol{S}]$. Then $\boldsymbol{S} = \mathsf{C}_{\boldsymbol{x}}[\boldsymbol{E}]$ and

$$\boldsymbol{Q}\boldsymbol{S}\boldsymbol{Q}^T = \mathsf{C}_{\boldsymbol{x}}[\boldsymbol{Q}\boldsymbol{E}\boldsymbol{Q}^T];$$

therefore

$$\boldsymbol{Q}\boldsymbol{E}\boldsymbol{Q}^T = \mathsf{K}_{\boldsymbol{x}}[\boldsymbol{Q}\boldsymbol{S}\boldsymbol{Q}^T],$$

or equivalently,

$$\boldsymbol{Q}\mathsf{K}_{\boldsymbol{x}}[\boldsymbol{S}]\boldsymbol{Q}^T = \mathsf{K}_{\boldsymbol{x}}[\boldsymbol{Q}\boldsymbol{S}\boldsymbol{Q}^T]. \quad \square$$

We say that the material at $\boldsymbol{x}$ is ***isotropic*** if the symmetry group $\mathcal{G}_{\boldsymbol{x}}$ equals the orthogonal group, ***anisotropic*** if $\mathcal{G}_{\boldsymbol{x}}$ is a *proper* subgroup of the orthogonal group. Clearly, $\mathcal{G}_{\boldsymbol{x}}$ always contains the two-element group $\{-\mathbf{1}, \mathbf{1}\}$ as a subgroup. In fact, $\mathcal{G}_{\boldsymbol{x}}$ is the direct product of this two-element group and a group $\mathcal{G}_{\boldsymbol{x}}^+$ which consists only of proper orthogonal transformations; hence the symmetry of the material is completely characterized by the group $\mathcal{G}_{\boldsymbol{x}}^+$. In particular, the material at $\boldsymbol{x}$ is isotropic or anisotropic according as $\mathcal{G}_{\boldsymbol{x}}^+$ equals or is a proper subgroup of the proper orthogonal group. Although there are an infinite number of subgroups of the proper orthogonal group, twelve of them seem to exhaust the kinds of symmetries occurring in theories proposed up to now as being appropriate to describe the behavior of real anisotropic elastic materials.

The first eleven of these subgroups of the proper orthogonal group correspond to the thirty-two crystal classes.[1] We denote these subgroups by $\mathscr{C}_1, \ldots, \mathscr{C}_{11}$. Table 1, which is due to COLEMAN and NOLL,[2] lists the generators of each of these groups; in this table $\{\boldsymbol{k}, \boldsymbol{l}, \boldsymbol{m}\}$ denotes a right-handed orthonormal basis, $\boldsymbol{q} \equiv \sqrt{\tfrac{1}{3}}(\boldsymbol{k} + \boldsymbol{l} + \boldsymbol{m})$, and $\boldsymbol{R}_{\boldsymbol{e}}^{\varphi}$ is the orthogonal tensor corresponding to a right-

[1] The thirty-two crystal classes are discussed by SCHOENFLIESS [1891, *1*], VOIGT [1910, *1*], FLINT [1948, *4*], DANA and HURLBUT [1959, *2*], SMITH and RIVLIN [1958, *16*], COLEMAN and NOLL [1964, *5*], TRUESDELL and NOLL [1965, *22*], BURCKHARDT [1966, *5*].

[2] [1964, *5*].

handed rotation through the angle φ, $0<\varphi<2\pi$, about an axis in the direction of the unit vector $\boldsymbol{e}$.

The last type of anisotropy, called ***transverse isotropy*** (with respect to a direction $\boldsymbol{m}$), is characterized by the group $\mathscr{C}_{12}$ consisting of $\mathbf{1}$ and the rotations $R_{\boldsymbol{m}}^{\varphi}$, $0<\varphi<2\pi$. Transverse isotropy is appropriate to real materials having a laminated or a bundled structure.

A material is called ***orthotropic*** if its symmetry group contains reflections on three mutually perpendicular planes. Such a triple of reflections is $-\boldsymbol{R}_{\boldsymbol{k}}^{\pi}$, $-\boldsymbol{R}_{\boldsymbol{l}}^{\pi}$, $-\boldsymbol{R}_{\boldsymbol{m}}^{\pi}$. Since $\boldsymbol{R}_{\boldsymbol{k}}^{\pi}\boldsymbol{R}_{\boldsymbol{l}}^{\pi}=\boldsymbol{R}_{\boldsymbol{m}}^{\pi}$ and $(\boldsymbol{R}_{\boldsymbol{k}}^{\frac{1}{2}\pi})^2=\boldsymbol{R}_{\boldsymbol{k}}^{\pi}$, it follows that the groups $\mathscr{C}_3$, $\mathscr{C}_5$, $\mathscr{C}_6$, and $\mathscr{C}_7$ correspond to orthotropic materials.

Table 1.

Crystal class	Group	Generators of $\mathscr{C}_n$	Order of $\mathscr{C}_n$
Triclinic system all classes	$\mathscr{C}_1$	$\mathbf{1}$	1
Monoclinic system all classes	$\mathscr{C}_2$	$\boldsymbol{R}_{\boldsymbol{m}}^{\pi}$	2
Rhombic system all classes	$\mathscr{C}_3$	$\boldsymbol{R}_{\boldsymbol{k}}^{\pi}$, $\boldsymbol{R}_{\boldsymbol{l}}^{\pi}$	4
Tetragonal system tetragonal-disphenoidal tetragonal-pyramidal tetragonal-dipyramidal	$\mathscr{C}_4$	$\boldsymbol{R}_{\boldsymbol{m}}^{\frac{1}{2}\pi}$	4
tetragonal-scalenohedral ditetragonal-pyramidal tetragonal-trapezohedral ditetragonal-dipyramidal	$\mathscr{C}_5$	$\boldsymbol{R}_{\boldsymbol{m}}^{\frac{1}{2}\pi}$, $\boldsymbol{R}_{\boldsymbol{k}}^{\pi}$	8
Cubic system tetartoidal diploidal	$\mathscr{C}_6$	$\boldsymbol{R}_{\boldsymbol{k}}^{\pi}$, $\boldsymbol{R}_{\boldsymbol{l}}^{\pi}$, $\boldsymbol{R}_{\boldsymbol{q}}^{\frac{2}{3}\pi}$	12
hexatetrahedral gyroidal hexoctahedral	$\mathscr{C}_7$	$\boldsymbol{R}_{\boldsymbol{k}}^{\frac{1}{2}\pi}$, $\boldsymbol{R}_{\boldsymbol{l}}^{\frac{1}{2}\pi}$, $\boldsymbol{R}_{\boldsymbol{m}}^{\frac{1}{2}\pi}$	24
Hexagonal system trigonal-pyramidal rhombohedral	$\mathscr{C}_8$	$\boldsymbol{R}_{\boldsymbol{m}}^{\frac{2}{3}\pi}$	3
ditrigonal-pyramidal trigonal-trapezohedral hexagonal-scalenohedral	$\mathscr{C}_9$	$\boldsymbol{R}_{\boldsymbol{k}}^{\pi}$, $\boldsymbol{R}_{\boldsymbol{m}}^{\frac{2}{3}\pi}$	6
trigonal-dipyramidal hexagonal-pyramidal hexagonal-dipyramidal	$\mathscr{C}_{10}$	$\boldsymbol{R}_{\boldsymbol{m}}^{\frac{1}{3}\pi}$	6
ditrigonal-dipyramidal dihexagonal-pyramidal hexagonal-trapezohedral dihexagonal-dipyramidal	$\mathscr{C}_{11}$	$\boldsymbol{R}_{\boldsymbol{k}}^{\pi}$, $\boldsymbol{R}_{\boldsymbol{m}}^{\frac{1}{3}\pi}$	12

We say that the material at $\boldsymbol{x}$ has *$\mathscr{C}_n$-symmetry* if $\mathscr{C}_n$ is a subgroup of $\mathscr{G}_{\boldsymbol{x}}$. We call two groups $\mathscr{C}_m$ and $\mathscr{C}_n$ *equivalent* if given *any* symmetric elasticity tensor $\mathbf{C}_{\boldsymbol{x}}$,

$$\mathscr{C}_m\subset\mathscr{G}_{\boldsymbol{x}}\Leftrightarrow\mathscr{C}_n\subset\mathscr{G}_{\boldsymbol{x}}.$$

A group $\mathscr{C}_n$ is *distinct* if there is no $\mathscr{C}_m$ $(m\neq n)$ equivalent to $\mathscr{C}_n$. It will be clear from the results of Sect. 26 that $\mathscr{C}_1$, $\mathscr{C}_2$, $\mathscr{C}_3$, $\mathscr{C}_4$, $\mathscr{C}_5$, $\mathscr{C}_8$, and $\mathscr{C}_9$ are distinct; $\mathscr{C}_6$ and $\mathscr{C}_7$ are equivalent; $\mathscr{C}_{10}$, $\mathscr{C}_{11}$, and $\mathscr{C}_{12}$ are equivalent. By definition, the symmetry group $\mathscr{G}_{\boldsymbol{x}}$ is a *maximal* group of symmetry transformations; thus the groups

$\mathscr{C}_6, \mathscr{C}_7, \mathscr{C}_{10}, \mathscr{C}_{11}, \mathscr{C}_{12}$ cannot be symmetry groups, since invariance with respect to one of these, $\mathscr{C}_n$ say, implies invariance with respect to another group which contains elements not in $\mathscr{C}_n$.

A unit vector $\boldsymbol{e}$ is called an ***axis of symmetry*** (for the material at $\boldsymbol{x}$) if

$$\boldsymbol{Q}\boldsymbol{e}=\boldsymbol{e}$$

for some $\boldsymbol{Q}\in\mathscr{G}_{\boldsymbol{x}}$ with $\boldsymbol{Q}\neq\mathbf{1}$. Thus $\boldsymbol{e}$ is an axis of symmetry if and only if one of the symmetry transformations is a rotation about the axis spanned by $\boldsymbol{e}$; i.e., if and only if $\boldsymbol{R}_{\boldsymbol{e}}^{\varphi}\in\mathscr{G}_{\boldsymbol{x}}$ for some φ $(0<\varphi<2\pi)$.

Consider now the stress $\boldsymbol{S}$ due to a uniform dilatation $\boldsymbol{E}=e\mathbf{1}$:

$$\boldsymbol{S}=\mathsf{C}[e\mathbf{1}],$$

where $\mathsf{C}=\mathsf{C}_{\boldsymbol{x}}$. Then

$$\boldsymbol{Q}\boldsymbol{S}\boldsymbol{Q}^T=\mathsf{C}[e\boldsymbol{Q}\mathbf{1}\boldsymbol{Q}^T]=\mathsf{C}[e\mathbf{1}]=\boldsymbol{S}$$

or

$$\boldsymbol{Q}\boldsymbol{S}=\boldsymbol{S}\boldsymbol{Q}$$

for every $\boldsymbol{Q}\in\mathscr{G}_{\boldsymbol{x}}$; i.e. $\boldsymbol{S}$ commutes with every $\boldsymbol{Q}$ in $\mathscr{G}_{\boldsymbol{x}}$. But by the commutation theorem *(3.3)*, a symmetric tensor $\boldsymbol{S}$ commutes with an orthogonal tensor $\boldsymbol{Q}$ if and only if $\boldsymbol{Q}$ leaves each of the characteristic spaces of $\boldsymbol{S}$ invariant. The only subspace of $\mathscr{V}$ invariant under all orthogonal transformations is the full vector space $\mathscr{V}$ itself. Thus if the material at $\boldsymbol{x}$ is isotropic, then every vector in $\mathscr{V}$ is a principal direction of $\boldsymbol{S}$. Therefore it follows from the spectral theorem *(3.2)* that the response to a uniform dilatation is a uniform pressure, i.e.

$$\mathsf{C}[e\mathbf{1}]=\alpha\mathbf{1}.$$

Now assume that the material at $\boldsymbol{x}$ is anisotropic and again consider the stress $\boldsymbol{S}=\mathsf{C}[e\mathbf{1}]$ due to a uniform dilatation. If $\varphi\neq\pi$, the only spaces left invariant by the rotation $\boldsymbol{R}_{\boldsymbol{e}}^{\varphi}$ are the one-dimensional space spanned by $\boldsymbol{e}$, the two-dimensional space of all vectors perpendicular to $\boldsymbol{e}$, and the entire space $\mathscr{V}$. The rotation $\boldsymbol{R}_{\boldsymbol{e}}^{\pi}$ leaves invariant, in addition, each one-dimensional space spanned by a vector perpendicular to $\boldsymbol{e}$. These facts, the spectral theorem *(3.2)*, and the commutation theorem *(3.3)* enable one to establish easily the results collected in the following table.[1]

Table 2.

Type of symmetry	Restriction on the stress $\boldsymbol{S}$ due to a uniform dilatation $\boldsymbol{E}=e\mathbf{1}$
Triclinic system ($\mathscr{C}_1$)	no restriction
Monoclinic system ($\mathscr{C}_2$)	$\boldsymbol{m}$ is a principal direction of $\boldsymbol{S}$
Rhombic system ($\mathscr{C}_3$)	$\boldsymbol{k}, \boldsymbol{l}, \boldsymbol{m}$ are principal directions of $\boldsymbol{S}$
Tetragonal system ($\mathscr{C}_4, \mathscr{C}_5$) Hexagonal system ($\mathscr{C}_8, \mathscr{C}_9, \mathscr{C}_{10}, \mathscr{C}_{11}$) Transverse isotropy ($\mathscr{C}_{12}$)	$\boldsymbol{S}=\alpha\mathbf{1}+\beta\boldsymbol{m}\otimes\boldsymbol{m}$
Cubic system ($\mathscr{C}_6, \mathscr{C}_7$) Isotropy	$\boldsymbol{S}=\alpha\mathbf{1}$

[1] Coleman and Noll [1964, *5*].

Of future use is the following

(2) Identity. *Given any two vectors $\boldsymbol{a}$, $\boldsymbol{b}$, let*

$$A(\boldsymbol{a}, \boldsymbol{b}) = \tfrac{1}{2}(\boldsymbol{a} \otimes \boldsymbol{b} + \boldsymbol{b} \otimes \boldsymbol{a}).$$

Then

$$\mathbf{C}[A(\boldsymbol{m}, \boldsymbol{n})] \cdot A(\boldsymbol{a}, \boldsymbol{b}) = \mathbf{C}[A(\boldsymbol{Q}\boldsymbol{m}, \boldsymbol{Q}\boldsymbol{n})] \cdot A(\boldsymbol{Q}\boldsymbol{a}, \boldsymbol{Q}\boldsymbol{b})$$

for all vectors $\boldsymbol{a}$, $\boldsymbol{b}$, $\boldsymbol{m}$, $\boldsymbol{n}$ and every $\boldsymbol{Q} \in \mathscr{G}_{\boldsymbol{x}}$. Here $\mathbf{C} = \mathbf{C}_{\boldsymbol{x}}$.

Proof. Since

$$\boldsymbol{Q}(\boldsymbol{m} \otimes \boldsymbol{n}) \boldsymbol{Q}^T = (\boldsymbol{Q}\boldsymbol{m}) \otimes (\boldsymbol{Q}\boldsymbol{n}),$$

it follows that

$$\boldsymbol{Q} A(\boldsymbol{m}, \boldsymbol{n}) \boldsymbol{Q}^T = A(\boldsymbol{Q}\boldsymbol{m}, \boldsymbol{Q}\boldsymbol{n}),$$

and thus

$$\boldsymbol{Q}\mathbf{C}[A(\boldsymbol{m}, \boldsymbol{n})] \boldsymbol{Q}^T = \mathbf{C}[A(\boldsymbol{Q}\boldsymbol{m}, \boldsymbol{Q}\boldsymbol{n})] \tag{a}$$

for every $\boldsymbol{Q} \in \mathscr{G}_{\boldsymbol{x}}$. Next, in view of the identity $\boldsymbol{T} \cdot (\boldsymbol{u} \otimes \boldsymbol{v}) = \boldsymbol{u} \cdot \boldsymbol{T}\boldsymbol{v}$,

$$(\boldsymbol{Q}\boldsymbol{B}\boldsymbol{Q}^T) \cdot (\boldsymbol{Q}\boldsymbol{a} \otimes \boldsymbol{Q}\boldsymbol{b}) = \boldsymbol{Q}\boldsymbol{a} \cdot \boldsymbol{Q}\boldsymbol{B}\boldsymbol{Q}^T \boldsymbol{Q}\boldsymbol{b} = \boldsymbol{a} \cdot \boldsymbol{B}\boldsymbol{b} = \boldsymbol{B} \cdot (\boldsymbol{a} \otimes \boldsymbol{b});$$

therefore if we let $\boldsymbol{B} = \mathbf{C}[A(\boldsymbol{m}, \boldsymbol{n})]$ and use (a) and (ii) of ***(20.1)***, we arrive at the desired identity. □

In view of (iii) of ***(20.1)***,

$$C_{ijkl} = A(\boldsymbol{e}_i, \boldsymbol{e}_j) \cdot \mathbf{C}_{\boldsymbol{x}}[A(\boldsymbol{e}_k, \boldsymbol{e}_l)],$$

and we conclude from the identity ***(2)*** that for an isotropic material the components of $\mathbf{C}_{\boldsymbol{x}}$ are independent of the orthonormal basis $\{\boldsymbol{e}_1, \boldsymbol{e}_2, \boldsymbol{e}_3\}$.

22. Isotropic materials. In this section we shall determine the restrictions placed on the constitutive relation

$$\boldsymbol{S} = \mathbf{C}_{\boldsymbol{x}}[\boldsymbol{E}]$$

when the material at $\boldsymbol{x}$ is isotropic; i.e. when the symmetry group $\mathscr{G}_{\boldsymbol{x}}$ equals the full orthogonal group. In a later section we will discuss the form of the elasticity tensor for various anisotropic materials.

(1) Theorem on the response of isotropic materials. *Assume that the material at $\boldsymbol{x}$ is isotropic. Then:*

(i) *The response to a uniform dilatation is a uniform pressure, i.e. there exists a constant k, called the* **modulus of compression**, *such that*

$$\mathbf{C}_{\boldsymbol{x}}[\mathbf{1}] = 3k\mathbf{1}.$$

(ii) *The response to a simple shear is a pure shear; i.e. there exists a constant μ, called the* **shear modulus**,[1] *such that*

$$\mathbf{C}_{\boldsymbol{x}}[\operatorname{sym}(\boldsymbol{m} \otimes \boldsymbol{n})] = 2\mu \operatorname{sym}(\boldsymbol{m} \otimes \boldsymbol{n})$$

whenever $\boldsymbol{m}$ and $\boldsymbol{n}$ are orthogonal unit vectors.

(iii) *If $\boldsymbol{E}_0$ is a traceless symmetric tensor, then*

$$\mathbf{C}_{\boldsymbol{x}}[\boldsymbol{E}_0] = 2\mu \boldsymbol{E}_0.$$

Proof. Assertion (i) was established in the previous chapter (see the remarks leading to Table 2).

To establish (ii) let

$$A(\boldsymbol{m}, \boldsymbol{n}) = \operatorname{sym}(\boldsymbol{m} \otimes \boldsymbol{n}),$$

[1] The shear modulus is often denoted by G.

where $\boldsymbol{m}$ and $\boldsymbol{n}$ are orthogonal unit vectors, let $\boldsymbol{e}=\boldsymbol{m}\times\boldsymbol{n}$, and choose $\boldsymbol{Q}$ such that

$$\boldsymbol{Q}\boldsymbol{m}=-\boldsymbol{m}, \quad \boldsymbol{Q}\boldsymbol{n}=-\boldsymbol{n}, \quad \boldsymbol{Q}\boldsymbol{e}=\boldsymbol{e},$$

i.e. $\boldsymbol{Q}$ is a rotation about $\boldsymbol{e}$ of 180°. Then, writing $\mathbf{C}=\mathbf{C}_{\boldsymbol{x}}$, we conclude from the identity *(21.2)* with $\boldsymbol{a}=\boldsymbol{m}$ and $\boldsymbol{b}=\boldsymbol{e}$ that

$$\mathbf{C}[\boldsymbol{A}(\boldsymbol{m},\boldsymbol{n})]\cdot\boldsymbol{A}(\boldsymbol{m},\boldsymbol{e})=-\mathbf{C}[\boldsymbol{A}(\boldsymbol{m},\boldsymbol{n})]\cdot\boldsymbol{A}(\boldsymbol{m},\boldsymbol{e}),$$

and hence

$$\mathbf{C}[\boldsymbol{A}(\boldsymbol{m},\boldsymbol{n})]\cdot\boldsymbol{A}(\boldsymbol{m},\boldsymbol{e})=0. \tag{a$_1$}$$

Similarly,

$$\mathbf{C}[\boldsymbol{A}(\boldsymbol{m},\boldsymbol{n})]\cdot\boldsymbol{A}(\boldsymbol{n},\boldsymbol{e})=0. \tag{a$_2$}$$

Next, if we choose $\boldsymbol{Q}$ such that

$$\boldsymbol{Q}\boldsymbol{m}=\boldsymbol{m}, \quad \boldsymbol{Q}\boldsymbol{n}=-\boldsymbol{n}, \quad \boldsymbol{Q}\boldsymbol{e}=\boldsymbol{e},$$

and take $\boldsymbol{a}=\boldsymbol{b}=\boldsymbol{m}, \boldsymbol{n}$, or $\boldsymbol{e}$, we conclude from *(21.2)* that

$$\mathbf{C}[\boldsymbol{A}(\boldsymbol{m},\boldsymbol{n})]\cdot\boldsymbol{A}(\boldsymbol{a},\boldsymbol{a})=0 \quad \text{for} \quad \boldsymbol{a}=\boldsymbol{m},\boldsymbol{n}, \text{ or } \boldsymbol{e}. \tag{a$_3$}$$

Now let

$$\mu=\mathbf{C}[\boldsymbol{A}(\boldsymbol{m},\boldsymbol{n})]\cdot\boldsymbol{A}(\boldsymbol{m},\boldsymbol{n}). \tag{b}$$

In view of *(3.1)*, the six tensors $\sqrt{2}\boldsymbol{A}(\boldsymbol{m},\boldsymbol{n})$, $\sqrt{2}\boldsymbol{A}(\boldsymbol{n},\boldsymbol{e})$, $\sqrt{2}\boldsymbol{A}(\boldsymbol{m},\boldsymbol{e})$, $\sqrt{2}\boldsymbol{A}(\boldsymbol{m},\boldsymbol{m})$, $\sqrt{2}\boldsymbol{A}(\boldsymbol{n},\boldsymbol{n})$, $\sqrt{2}\boldsymbol{A}(\boldsymbol{e},\boldsymbol{e})$ form an orthonormal basis for the space of all symmetric tensors. By (a) and (b), the only non-zero component of $\mathbf{C}[\boldsymbol{A}(\boldsymbol{m},\boldsymbol{n})]$ relative to this basis is the component $\sqrt{2}\mu$ relative to $\sqrt{2}\boldsymbol{A}(\boldsymbol{m},\boldsymbol{n})$; thus

$$\mathbf{C}[\boldsymbol{A}(\boldsymbol{m},\boldsymbol{n})]=2\mu\boldsymbol{A}(\boldsymbol{m},\boldsymbol{n}). \tag{c}$$

Now let $\boldsymbol{u}$ and $\boldsymbol{v}$ be orthogonal unit vectors and choose $\boldsymbol{Q}$ such that

$$\boldsymbol{Q}\boldsymbol{m}=\boldsymbol{u}, \quad \boldsymbol{Q}\boldsymbol{n}=\boldsymbol{v}.$$

Then *(21.2)* with $\boldsymbol{a}=\boldsymbol{m}$, $\boldsymbol{b}=\boldsymbol{n}$ implies

$$\mathbf{C}[\boldsymbol{A}(\boldsymbol{m},\boldsymbol{n})]\cdot\boldsymbol{A}(\boldsymbol{m},\boldsymbol{n})=\mathbf{C}[\boldsymbol{A}(\boldsymbol{u},\boldsymbol{v})]\cdot\boldsymbol{A}(\boldsymbol{u},\boldsymbol{v}),$$

and thus the scalar μ in (c) is independent of the choice of $\boldsymbol{m}$ and $\boldsymbol{n}$.

To establish (iii) let $\boldsymbol{E}_0$ be a traceless symmetric tensor. By *(13.7)* there exist scalars $\varkappa_1, \varkappa_2, \varkappa_3$ and an orthonormal basis $\boldsymbol{e}_1, \boldsymbol{e}_2, \boldsymbol{e}_3$ such that

$$\boldsymbol{E}_0=\varkappa_1\boldsymbol{A}(\boldsymbol{e}_1,\boldsymbol{e}_2)+\varkappa_2\boldsymbol{A}(\boldsymbol{e}_2,\boldsymbol{e}_3)+\varkappa_3\boldsymbol{A}(\boldsymbol{e}_3,\boldsymbol{e}_1);$$

thus (ii) and the linearity of $\mathbf{C}$ imply

$$\mathbf{C}_{\boldsymbol{x}}[\boldsymbol{E}_0]=2\mu\boldsymbol{E}_0. \quad \square$$

(2) Theorem on the elasticity tensor for isotropic materials. *The material at $\boldsymbol{x}$ is isotropic if and only if the elasticity tensor $\mathsf{C}=\mathsf{C}_{\boldsymbol{x}}$ has the form*[1]

$$\mathsf{C}[\boldsymbol{E}]=2\mu\,\boldsymbol{E}+\lambda(\operatorname{tr}\boldsymbol{E})\,\boldsymbol{1}$$

for every symmetric tensor $\boldsymbol{E}$; the scalars $\mu=\mu(\boldsymbol{x})$ and $\lambda=\lambda(\boldsymbol{x})$ are called the ***Lamé moduli.***[2] *Moreover, μ is the shear modulus and $\lambda=k-\frac{2}{3}\mu$, where k is the modulus of compression.*

Proof. We first assume that

$$\mathsf{C}[\boldsymbol{E}]=2\mu\,\boldsymbol{E}+\lambda(\operatorname{tr}\boldsymbol{E})\,\boldsymbol{1}.$$

Let $\boldsymbol{Q}$ be orthogonal. Then

$$\operatorname{tr}(\boldsymbol{Q}\boldsymbol{E}\boldsymbol{Q}^T)=\operatorname{tr}(\boldsymbol{Q}^T\boldsymbol{Q}\boldsymbol{E})=\operatorname{tr}\boldsymbol{E},$$

and therefore

$$\begin{aligned}\boldsymbol{Q}\mathsf{C}[\boldsymbol{E}]\boldsymbol{Q}^T&=2\mu\,\boldsymbol{Q}\boldsymbol{E}\boldsymbol{Q}^T+\lambda(\operatorname{tr}\boldsymbol{E})\,\boldsymbol{Q}\boldsymbol{1}\boldsymbol{Q}^T\\&=2\mu\,\boldsymbol{Q}\boldsymbol{E}\boldsymbol{Q}^T+\lambda[\operatorname{tr}(\boldsymbol{Q}\boldsymbol{E}\boldsymbol{Q}^T)]\,\boldsymbol{1}=\mathsf{C}[\boldsymbol{Q}\boldsymbol{E}\boldsymbol{Q}^T].\end{aligned}$$

Thus the material at $\boldsymbol{x}$ is isotropic.

Conversely, assume that the material at $\boldsymbol{x}$ is isotropic. Let $\boldsymbol{E}$ be an arbitrary symmetric tensor. Then $\boldsymbol{E}$ admits the decomposition

$$\boldsymbol{E}=\boldsymbol{E}_0+\tfrac{1}{3}(\operatorname{tr}\boldsymbol{E})\,\boldsymbol{1},\qquad \operatorname{tr}\boldsymbol{E}_0=0,$$

and (i) and (iii) of ***(1)*** yield

$$\mathsf{C}[\boldsymbol{E}]=2\mu\,\boldsymbol{E}_0+k(\operatorname{tr}\boldsymbol{E})\,\boldsymbol{1}.$$

But since $\boldsymbol{E}_0=\boldsymbol{E}-\frac{1}{3}(\operatorname{tr}\boldsymbol{E})\,\boldsymbol{1}$, if we let $\lambda=k-\frac{2}{3}\mu$, we arrive at the desired result. □

Note that the isotropic constitutive relation

$$\boldsymbol{S}=2\mu\,\boldsymbol{E}+\lambda(\operatorname{tr}\boldsymbol{E})\,\boldsymbol{1}$$

takes the form

$$S_{ij}=2\mu\,E_{ij}+\lambda E_{kk}\,\delta_{ij}$$

when referred to cartesian coordinates.

When the material is isotropic, ***(1)*** implies that C maps multiples of the identity into multiples of the identity and traceless tensors into traceless tensors. The next proposition shows that those are the only two tensor spaces that are invariant under C.

(3) Characteristic values of the elasticity tensor. *Assume that the material at $\boldsymbol{x}$ is isotropic with $2\mu\neq3k$. Then the linear transformation*[3] *$\mathsf{C}=\mathsf{C}_{\boldsymbol{x}}$ has exactly two characteristic values: $3k$ and 2μ. The characteristic space corresponding to $3k$ is the set of all tensors of the form $\alpha\boldsymbol{1}$, where α is a scalar: the characteristic space corresponding to 2μ is the set of all traceless symmetric tensors.*

Proof. By (i) and (iii) of ***(1)***, 2μ and $3k=2\mu+3\lambda$ are characteristic values of C. Further, if $\mathsf{C}[\boldsymbol{E}]=\beta\boldsymbol{E}$, then ***(2)*** implies that

$$(2\mu-\beta)\,\boldsymbol{E}+\lambda(\operatorname{tr}\boldsymbol{E})\,\boldsymbol{1}=\boldsymbol{0}; \tag{a}$$

thus (for $\boldsymbol{E}\neq\boldsymbol{0}$)

$$\beta=2\mu\Leftrightarrow\operatorname{tr}\boldsymbol{E}=0. \tag{b}$$

[1] The basic ideas underlying this constitutive relation were presented by FRESNEL [1866, *1*] in 1829. This form of the relation with $\mu=\lambda$ was derived from a molecular model by NAVIER [1823, *2*], [1827, *2*] and POISSON [1829, *2*]. The general isotropic relation is due to CAUCHY [1823, *1*], [1828, *1*], [1829, *1*], [1830, *1*]. Cf. POISSON [1831, *1*].

[2] LAMÉ [1852, *2*].

[3] Here C is considered as a linear transformation on the space of *symmetric* tensors.

(Note that $\lambda \neq 0$, since $3k \neq 2\mu$.) On the other hand, if $\operatorname{tr} \boldsymbol{E} \neq 0$, then (a) yields

$$\beta = 2\mu + 3\lambda \Leftrightarrow \boldsymbol{E} = \tfrac{1}{3}(\operatorname{tr} \boldsymbol{E})\, \mathbf{1}. \tag{c}$$

Therefore the two spaces specified in *(3)* are characteristic spaces of C. Taking the trace of (a), we see that

$$(2\mu + 3\lambda - \beta) \operatorname{tr} \boldsymbol{E} = 0.$$

Thus either $\beta = 2\mu + 3\lambda$ or $\operatorname{tr} \boldsymbol{E} = 0$. If $\operatorname{tr} \boldsymbol{E} = 0$, then (a) implies that $\beta = 2\mu$. Thus $2\mu + 3\lambda$ and 2μ are the only characteristic values of C. □

Note that if $2\mu = 3k$, so that $\lambda = 0$, then every symmetric tensor is a characteristic vector for C.

Assume now that the material at $\boldsymbol{x}$ is isotropic, so that

$$\boldsymbol{S} = 2\mu \boldsymbol{E} + \lambda (\operatorname{tr} \boldsymbol{E})\, \mathbf{1}. \tag{I}$$

Clearly,

$$\operatorname{tr} \boldsymbol{S} = (2\mu + 3\lambda) \operatorname{tr} \boldsymbol{E}.$$

Thus if we assume that $\mu \neq 0$, $2\mu + 3\lambda \neq 0$, then

$$\boldsymbol{E} = \frac{1}{2\mu} \boldsymbol{S} - \frac{\lambda}{2\mu(2\mu + 3\lambda)} (\operatorname{tr} \boldsymbol{S})\, \mathbf{1},$$

and we have the following result.

(4) *If the material at $\boldsymbol{x}$ is isotropic, and if $\mu \neq 0$, $2\mu + 3\lambda \neq 0$, then C is invertible, and the compliance tensor $\mathsf{K} = \mathsf{C}^{-1}$ is given by*

$$\mathsf{K}[\boldsymbol{S}] = \frac{1}{2\mu} \left[\boldsymbol{S} - \frac{\lambda}{(2\mu + 3\lambda)} (\operatorname{tr} \boldsymbol{S})\, \mathbf{1}\right]$$

for every symmetric tensor $\boldsymbol{S}$.

Let $\boldsymbol{e}_1$, $\boldsymbol{e}_2$, $\boldsymbol{e}_3$ be an orthonormal basis and consider the following special choices for $\boldsymbol{S}$ in the relation $\boldsymbol{E} = \mathsf{K}[\boldsymbol{S}]$:

(a) ***Pure shear.*** Let the matrix $[\boldsymbol{S}]$ of $\boldsymbol{S}$ relative to this basis have the form

$$[\boldsymbol{S}] = \begin{bmatrix} 0 & \tau & 0 \\ \tau & 0 & 0 \\ 0 & 0 & 0 \end{bmatrix}.$$

Then

$$[\boldsymbol{E}] = \begin{bmatrix} 0 & \varkappa & 0 \\ \varkappa & 0 & 0 \\ 0 & 0 & 0 \end{bmatrix}$$

with

$$\varkappa = \frac{1}{2\mu} \tau;$$

i.e. $\boldsymbol{E}$ is a simple shear of amount

$$\varkappa = \frac{1}{2\mu} \tau$$

with respect to the direction pair $(\boldsymbol{e}_1, \boldsymbol{e}_2)$. Conversely, if $\boldsymbol{E}$ is a simple shear of amount $\varkappa$, $\boldsymbol{S}$ is a pure shear with shear stress $\tau = 2\mu\varkappa$.

(b) ***Uniform pressure.*** Let $\boldsymbol{S}$ correspond to a uniform pressure of amount p; i.e. let

$$\boldsymbol{S} = -p\mathbf{1}.$$

Then $\boldsymbol{E}$ is a uniform dilatation of amount

$$e = -\frac{1}{3k} p$$

with

$$k = \tfrac{2}{3}\mu + \lambda .$$

Clearly,

$$p = -k(\operatorname{tr}\boldsymbol{E}).$$

Conversely, $\boldsymbol{S}$ is a uniform pressure whenever $\boldsymbol{E}$ is a uniform dilatation.

(c) ***Pure tension.*** Let $\boldsymbol{S}$ correspond to a pure tension with tensile stress σ in the $\boldsymbol{e}_1$ direction:

$$[\boldsymbol{S}] = \begin{bmatrix} \sigma & 0 & 0 \\ 0 & 0 & 0 \\ 0 & 0 & 0 \end{bmatrix}.$$

Then $[\boldsymbol{E}]$ has the form

$$[\boldsymbol{E}] = \begin{bmatrix} e & 0 & 0 \\ 0 & l & 0 \\ 0 & 0 & l \end{bmatrix},$$

with

$$e = \frac{1}{\beta}\sigma,$$

$$l = -\frac{\nu}{\beta}\sigma = -\nu e,$$

and

$$\beta = \frac{\mu(2\mu + 3\lambda)}{\mu + \lambda},$$

$$\nu = \frac{\lambda}{2(\mu + \lambda)}.$$

The modulus β is obtained by dividing the tensile stress σ by the longitudinal strain e produced by it. It is known as ***Young's modulus.***[1] The modulus ν is the ratio of the lateral contraction to the longitudinal strain of a bar under pure tension. It is known as ***Poissons' ratio.***[2] Table 3 gives the relations between the various moduli.

If we write $\mathring{\boldsymbol{E}}$ and $\mathring{\boldsymbol{S}}$ for the traceless parts of $\boldsymbol{E}$ and $\boldsymbol{S}$, i.e.

$$\mathring{\boldsymbol{E}} = \boldsymbol{E} - \tfrac{1}{3}(\operatorname{tr}\boldsymbol{E})\,\boldsymbol{1}, \quad \mathring{\boldsymbol{S}} = \boldsymbol{S} - \tfrac{1}{3}(\operatorname{tr}\boldsymbol{S})\,\boldsymbol{1},$$

then the isotropic constitutive relation (I) is equivalent to the following pair of relations[3]

$$\mathring{\boldsymbol{S}} = 2\mu\,\mathring{\boldsymbol{E}},$$

$$\operatorname{tr}\boldsymbol{S} = k\operatorname{tr}\boldsymbol{E}.$$

Another important form for the isotropic stress-strain relation is the one taken by the inverted relation *(4)* when Young's modulus β and Poisson's ratio ν are used:

$$\boldsymbol{E} = \frac{1}{\beta}\left[(1+\nu)\,\boldsymbol{S} - \nu(\operatorname{tr}\boldsymbol{S})\,\boldsymbol{1}\right].$$

---

[1] YOUNG neither defined nor used the modulus named after him. In fact, it was introduced and discussed by EULER in [1780, *1*] (cf. TRUESDELL [1960, *15*], [1960, *16*], pp. 402–403) and, according to TRUESDELL [1959, *17*], was used by EULER as early as 1727. In most texts Young's modulus is designated by E.

[2] POISSON [1829, *2*]. Poisson's ratio is often denoted by σ.

[3] STOKES [1845, *1*] introduced the moduli μ and k and noted that the response of an isotropic elastic body is completely determined by its response to shearing and to uniform compression.

Table 3.

	λ	μ	β	ν	k
λ, μ			$\frac{\mu(3\lambda+2\mu)}{\lambda+\mu}$	$\frac{\lambda}{2(\lambda+\mu)}$	$\frac{3\lambda+2\mu}{3}$
λ, β		$\frac{(\beta-3\lambda)+\sqrt{(\beta-3\lambda)^2+8\lambda\beta}}{4}$		$\frac{-(\beta+\lambda)+\sqrt{(\beta+\lambda)^2+8\lambda^2}}{4\lambda}$	$\frac{(3\lambda+\beta)+\sqrt{(3\lambda+\beta)^2-4\lambda\beta}}{6}$
λ, ν		$\frac{\lambda(1-2\nu)}{2\nu}$	$\frac{\lambda(1+\nu)(1-2\nu)}{\nu}$		$\frac{\lambda(1+\nu)}{3\nu}$
λ, k		$\frac{3(k-\lambda)}{2}$	$\frac{9k(k-\lambda)}{3k-\lambda}$	$\frac{\lambda}{3k-\lambda}$	
μ, β	$\frac{(2\mu-\beta)\mu}{\beta-3\mu}$			$\frac{\beta-2\mu}{2\mu}$	$\frac{\mu\beta}{3(3\mu-\beta)}$
μ, ν	$\frac{2\mu\nu}{(1-2\nu)}$		$2\mu(1+\nu)$		$\frac{2\mu(1+\nu)}{3(1-2\nu)}$
μ, k	$\frac{3k-2\mu}{3}$		$\frac{9k\mu}{3k+\mu}$	$\frac{1}{2}\left[\frac{3k-2\mu}{3k+\mu}\right]$	
β, ν	$\frac{\nu\beta}{(1+\nu)(1-2\nu)}$	$\frac{\beta}{2(1+\nu)}$			$\frac{\beta}{3(1-2\nu)}$
β, k	$\frac{3k(3k-\beta)}{9k-\beta}$	$\frac{3\beta k}{9k-\beta}$		$\frac{1}{2}\left[\frac{3k-\beta}{3k}\right]$	
ν, k	$\frac{3k\nu}{1+\nu}$	$\frac{3k(1-2\nu)}{2(1+\nu)}$	$3k(1-2\nu)$		

Since an elastic solid should increase its length when pulled, should decrease its volume when acted on by a pressure, and should respond to a positive shearing stress by a positive shearing strain, we would expect that

$$\beta>0, \quad k>0, \quad \mu>0.$$

Also, a pure tensile stress should produce a contraction in the direction perpendicular to it; thus

$$\nu>0.$$

Even though the above inequalities are physically well motivated, we will not assume that they hold, for in many circumstances other (somewhat weaker) assumptions are more natural. Further, these weaker hypotheses are important to the study of infinitesimal deformations superposed on finite deformations.

Finally, we remark that GAZIS, TADJBAKHSH, and TOUPIN[1] have established a method of determining the isotropic material whose elasticity tensor is nearest (in a sense they make precise) to a given elasticity tensor of arbitrary symmetry.

23. The constitutive assumption for finite elasticity. In this section we will show that the linear constitutive assumption discussed previously is, in a certain sense, consistent with the finite theory of elasticity provided we assume infinitesimal deformations and a stress-free reference configuration.

In the finite theory the stress $\boldsymbol{S}$ at any given material point $\boldsymbol{x}$ is a function of the deformation gradient $\boldsymbol{F}$ at $\boldsymbol{x}$:

$$\boldsymbol{S}=\tilde{\boldsymbol{S}}(\boldsymbol{F}). \tag{a}$$

The function $\tilde{\boldsymbol{S}}$ is the *response function* for the material point $\boldsymbol{x}$.

In view of (o) in Sect. 16, we assume that

$$\tilde{\boldsymbol{S}}(\boldsymbol{F})\boldsymbol{F}^T=\boldsymbol{F}\tilde{\boldsymbol{S}}(\boldsymbol{F})^T \tag{b}$$

for every invertible tensor $\boldsymbol{F}$; this insures that the values of $\tilde{\boldsymbol{S}}$ are consistent with the law of balance of angular momentum. In addition, we assume that $\tilde{\boldsymbol{S}}$ is compatible with the *principle of material frame indifference*; this is the requirement that the constitutive relation (a) be independent of the observer.[2] During a change in observer $\boldsymbol{F}$ transforms into $\boldsymbol{Q}\boldsymbol{F}$ and $\boldsymbol{S}$ into $\boldsymbol{Q}\boldsymbol{S}$, where $\boldsymbol{Q}$ is the orthogonal tensor corresponding to this change.[3] Since $\tilde{\boldsymbol{S}}$ must be invariant under all such changes, it must satisfy[4]

$$\tilde{\boldsymbol{S}}(\boldsymbol{F})=\boldsymbol{Q}^T\tilde{\boldsymbol{S}}(\boldsymbol{Q}\boldsymbol{F}) \tag{c}$$

for every orthogonal tensor $\boldsymbol{Q}$ and every $\boldsymbol{F}$. By the polar decomposition theorem,[5] $\boldsymbol{F}$ admits the decomposition

$$\boldsymbol{F}=\boldsymbol{R}\boldsymbol{U}, \tag{d}$$

where $\boldsymbol{R}$ is orthogonal and

$$\boldsymbol{U}=(\boldsymbol{F}^T\boldsymbol{F})^{\frac{1}{2}}. \tag{e}$$

Choosing $\boldsymbol{Q}$ in (c) equal to $\boldsymbol{R}^T$, we arrive at the relation

$$\boldsymbol{S}=\boldsymbol{R}\tilde{\boldsymbol{S}}(\boldsymbol{U}),$$

or equivalently, using (d),

$$\boldsymbol{S}=\boldsymbol{F}\boldsymbol{U}^{-1}\tilde{\boldsymbol{S}}(\boldsymbol{U}). \tag{f}$$

If we define a new function $\hat{\boldsymbol{S}}$ by the relation

$$\hat{\boldsymbol{S}}(\boldsymbol{D})=\boldsymbol{U}^{-1}\tilde{\boldsymbol{S}}(\boldsymbol{U}),$$

where $\boldsymbol{D}$ is the finite strain tensor

$$\boldsymbol{D}=\tfrac{1}{2}(\boldsymbol{U}^2-\boldsymbol{1})=\tfrac{1}{2}(\boldsymbol{F}^T\boldsymbol{F}-\boldsymbol{1}),$$

---

[1] [1963, *7*].

[2] See, e.g., TRUESDELL and NOLL [1965, *22*], §§ 19, 19A.

[3] See, e.g., TRUESDELL and NOLL [1965, *22*], § 19. The transformation rule for $\boldsymbol{F}$ is an immediate consequence of their Eq. $(19.2)_1$. The rule for the Piola-Kirchhoff stress $\boldsymbol{S}$ follows from the law of transformation for the Cauchy stress $\boldsymbol{T}$ [their Eq. $(19.2)_2$] and (r) of Sect. 16 relating $\boldsymbol{S}$ to $\boldsymbol{T}$ and $\boldsymbol{F}$.

[4] Cf. TRUESDELL and NOLL [1965, *22*], Eq. (143.9).

[5] See, e.g., HALMOS [1958, 9], § 83.

then (f) reduces to

$$\boldsymbol{S}=\boldsymbol{F}\hat{\boldsymbol{S}}(\boldsymbol{D}). \tag{g}$$

On the other hand, it is not difficult to verify that the constitutive relation (g) is consistent with (c). Further, (g) satisfies (b) if and only if

$$\boldsymbol{F}\hat{\boldsymbol{S}}(\boldsymbol{D})\,\boldsymbol{F}^T=\boldsymbol{F}\hat{\boldsymbol{S}}(\boldsymbol{D})^T\,\boldsymbol{F}^T,$$

or, since $\boldsymbol{F}$ is invertible,

$$\hat{\boldsymbol{S}}(\boldsymbol{D})=\hat{\boldsymbol{S}}(\boldsymbol{D})^T, \tag{h}$$

so that the values of $\hat{\boldsymbol{S}}$ are symmetric tensors.

It is important to note that the concept of strain arises naturally in the study of restrictions placed on constitutive relations by material frame-indifference. It is possible to derive other constitutive relations involving other strain measures; we chose one involving $\boldsymbol{D}$ because it simplifies the discussion of infinitesimal deformations.

Let $\hat{\boldsymbol{S}}$ be differentiable at $\mathbf{0}$. Then

$$\hat{\boldsymbol{S}}(\boldsymbol{D})=\hat{\boldsymbol{S}}(0)+\mathsf{C}[\boldsymbol{D}]+o(|\boldsymbol{D}|)$$

as $|\boldsymbol{D}|\to 0$, where C is a linear transformation which, by (h), maps symmetric tensors into symmetric tensors. When $\boldsymbol{F}=\mathbf{1}$, $\boldsymbol{D}=\mathbf{0}$; thus $\hat{\boldsymbol{S}}(\mathbf{0})$ represents the *residual stress in the reference configuration*. We now assume that this *residual stress vanishes*;[1] hence

$$\hat{\boldsymbol{S}}(\boldsymbol{D})=\mathsf{C}[\boldsymbol{D}]+o(|\boldsymbol{D}|). \tag{i}$$

Let

$$\delta=|\nabla\boldsymbol{u}|.$$

Then (c), (e), and (f) of Sect. 12 yield

$$\boldsymbol{F}=\mathbf{1}+O(\delta),$$
$$\boldsymbol{D}=\boldsymbol{E}+O(\delta^2)=O(\delta)$$

as $\delta\to 0$, and we conclude from (g) and (i) that

$$\boldsymbol{S}=\mathsf{C}[\boldsymbol{E}]+o(\delta)\quad\text{as}\quad\delta\to 0.$$

Therefore *if we assume that the residual stress in the reference configuration vanishes, then to within terms of $o(\delta)$ as $\delta\to 0$ the stress tensor is a linear function of the infinitesimal strain tensor. Further, since $\boldsymbol{E}=\mathbf{0}$ when $\nabla\boldsymbol{u}$ is skew, to within terms of $o(\delta)$ the stress tensor vanishes in an infinitesimal rigid displacement.*

These remarks should help to motivate the constitutive assumptions underlying the infinitesimal theory.

24. Work theorems. Stored energy. In this section we establish certain conditions, involving the notions of work and energy, which are necessary and sufficient for the elasticity tensor to be symmetric and positive definite. Thus we consider a fixed point $\boldsymbol{x}$ of the body, and let $\mathsf{C}=\mathsf{C}_{\boldsymbol{x}}$ denote the elasticity tensor for $\boldsymbol{x}$.

[1] This assumption, while basic to the theory presented here, is not satisfied, e.g., by an elastic fluid.

By a ***strain path*** Γ_E connecting E_0 and E_1 we mean a one-parameter family $E(\alpha)$, $\alpha_0 \leqq \alpha \leqq \alpha_1$, of symmetric tensors such that $E(\cdot)$ is continuous and piecewise smooth and

$$E(\alpha_0) = E_0, \quad E(\alpha_1) = E_1.$$

We say that Γ_E is ***closed*** if

$$E_0 = E_1.$$

Clearly, the strain path Γ_E determines a one-parameter family of stress tensors

$$S(\alpha) = \mathsf{C}[E(\alpha)], \quad \alpha_0 \leqq \alpha \leqq \alpha_1;$$

the ***work done along*** Γ_E is defined to be the scalar

$$w(\Gamma_E) = \int_{\alpha_0}^{\alpha_1} S(\alpha) \cdot \frac{dE(\alpha)}{d\alpha}\, d\alpha.$$

In view of the discussion following *(19.1)*, we have the following interpretation of $w(\Gamma_E)$: Consider a homogeneous elastic body with elasticity tensor C; $w(\Gamma_E)$ is the work done (per unit volume) by the surface tractions in a time-dependent homogeneous deformation with strain E.

By a ***stored energy function***[1] we mean a smooth scalar-valued function ε on the space of symmetric tensors such that $\varepsilon(0) = 0$ and[2]

$$\mathsf{C}[E] = \nabla_E\, \varepsilon(E)$$

for every symmetric tensor E. An immediate consequence of this definition is the following trivial proposition.

(1) If a stored energy function exists, it is unique, and

$$w(\Gamma_E) = \varepsilon(E_1) - \varepsilon(E_0)$$

for every strain path Γ_E connecting E_0 and E_1.

The next theorem supplies several alternative criteria for the existence of a stored energy function.

(2) Existence of a stored energy function.[3] *The following four statements are equivalent.*

(i) *The work done along each closed strain path is non-negative.*

(ii) *The work done along each closed strain path vanishes.*

(iii) C *is symmetric.*

(iv) *The function ε defined on the space of symmetric tensors by*

$$\varepsilon(E) = \tfrac{1}{2} E \cdot \mathsf{C}[E]$$

is a stored energy function.

Proof. We will show that (i) ⇒ (ii) ⇒ (iii) ⇒ (iv) ⇒ (i). Assume that (i) holds. Let Γ_E be an arbitrary closed path, and let Γ_{E^*} denote the closed path defined by

$$E^*(\alpha) = E(\alpha_0 + \alpha_1 - \alpha), \quad \alpha_0 \leqq \alpha \leqq \alpha_1.$$

---

[1] The notion of a stored energy function is due to GREEN [1839, *1*], [1841, *1*].

[2] The gradient $\nabla_E\, \varepsilon(E)$ of ε at E is defined in a manner strictly analogous to the definition given on p. 10 of the gradient of a scalar field on $\mathscr{E}$. Note that $\nabla_E\, \varepsilon(E)$ is a symmetric tensor; in components

$$[\nabla_E\, \varepsilon(E)]_{ij} = \frac{\partial \varepsilon(E)}{\partial E_{ij}}.$$

[3] GREEN [1839, *1*], [1841, *1*] proved that (iii) ⇔ (iv), UDESCHINI [1943, *3*] that (i) ⇔ (iv).

Then a simple computation tells us that

$$w(\Gamma_{\boldsymbol{E}}) = -w(\Gamma_{\boldsymbol{E}*}).$$

But (i) implies that $w(\Gamma_{\boldsymbol{E}})$ and $W(\Gamma_{\boldsymbol{E}*})$ must both be non-negative. Hence $w(\Gamma_{\boldsymbol{E}}) = 0$, which is (ii). To see that (ii) $\Rightarrow$ (iii) let $\boldsymbol{A}$ and $\boldsymbol{B}$ be symmetric tensors, and let $\Gamma_{\boldsymbol{E}}$ be the closed strain path defined by

$$\boldsymbol{E}(\alpha) = (\cos\alpha - 1)\,\boldsymbol{A} + (\sin\alpha)\,\boldsymbol{B}, \qquad 0 \leqq \alpha \leqq 2\pi.$$

Then

$$\begin{aligned} \mathbf{C}[\boldsymbol{E}(\alpha)] \cdot \frac{d\boldsymbol{E}(\alpha)}{d\alpha} &= (\sin\alpha - \sin\alpha\cos\alpha)\,\boldsymbol{A}\cdot\mathbf{C}[\boldsymbol{A}] \\ &\quad + (\cos^2\alpha - \cos\alpha)\,\boldsymbol{B}\cdot\mathbf{C}[\boldsymbol{A}] \\ &\quad - (\sin^2\alpha)\,\boldsymbol{A}\cdot\mathbf{C}[\boldsymbol{B}] + (\sin\alpha\cos\alpha)\,\boldsymbol{B}\cdot\mathbf{C}[\boldsymbol{B}] \end{aligned}$$

and

$$w(\Gamma_{\boldsymbol{E}}) = \int_0^{2\pi} \mathbf{C}[\boldsymbol{E}(\alpha)] \cdot \frac{d\boldsymbol{E}(\alpha)}{d\alpha}\, d\alpha = \pi(\boldsymbol{B}\cdot\mathbf{C}[\boldsymbol{A}] - \boldsymbol{A}\cdot\mathbf{C}[\boldsymbol{B}]).$$

Hence (ii) $\Rightarrow$ (iii). Next, define

$$\varepsilon(\boldsymbol{E}) = \tfrac{1}{2}\boldsymbol{E}\cdot\mathbf{C}[\boldsymbol{E}].$$

Since the gradient $\nabla_{\boldsymbol{E}}\varepsilon(\boldsymbol{E})$ can be computed using the formula[1]

$$\nabla_{\boldsymbol{E}}\varepsilon(\boldsymbol{E})\cdot\boldsymbol{A} = \frac{d}{d\lambda}\varepsilon(\boldsymbol{E}+\lambda\boldsymbol{A})\big|_{\lambda=0},$$

it follows that

$$\begin{aligned} \nabla_{\boldsymbol{E}}\varepsilon(\boldsymbol{E})\cdot\boldsymbol{A} &= \frac{1}{2}\frac{d}{d\lambda}\{\boldsymbol{E}\cdot\mathbf{C}[\boldsymbol{E}] + \lambda\boldsymbol{E}\cdot\mathbf{C}[\boldsymbol{A}] + \lambda\boldsymbol{A}\cdot\mathbf{C}[\boldsymbol{E}] + \lambda^2\boldsymbol{A}\cdot\mathbf{C}[\boldsymbol{A}]\}_{\lambda=0} \\ &= \frac{1}{2}\{\boldsymbol{E}\cdot\mathbf{C}[\boldsymbol{A}] + \boldsymbol{A}\cdot\mathbf{C}[\boldsymbol{E}]\}. \end{aligned}$$

Thus (iii) implies

$$\nabla_{\boldsymbol{E}}\varepsilon(\boldsymbol{E})\cdot\boldsymbol{A} = \boldsymbol{A}\cdot\mathbf{C}[\boldsymbol{E}]$$

for every symmetric tensor $\boldsymbol{A}$, which yields (iv). Finally, by *(1)*, (iv) implies (i). □

If $\mathbf{C}$ is symmetric, a stored energy function ε exists, and the constitutive equation

$$\boldsymbol{S} = \mathbf{C}[\boldsymbol{E}]$$

may be written in the form

$$\boldsymbol{S} = \nabla_{\boldsymbol{E}}\varepsilon(\boldsymbol{E}).$$

Let $\mathbf{C}$ be invertible, and let

$$\hat{\varepsilon}(\boldsymbol{A}) = \tfrac{1}{2}\boldsymbol{A}\cdot\mathbf{K}[\boldsymbol{A}]$$

for every symmetric tensor $\boldsymbol{A}$, where $\mathbf{K}$ is the compliance tensor. In view of the symmetry of $\mathbf{K}$ (which follows from the symmetry of $\mathbf{C}$),

$$\nabla_{\boldsymbol{S}}\hat{\varepsilon}(\boldsymbol{S}) = \mathbf{K}[\boldsymbol{S}],$$

and the above constitutive relation takes the alternative form

$$\boldsymbol{E} = \nabla_{\boldsymbol{S}}\hat{\varepsilon}(\boldsymbol{S}).$$

Moreover,

$$\hat{\varepsilon}(\boldsymbol{S}) = \tfrac{1}{2}\mathbf{C}[\boldsymbol{E}]\cdot\mathbf{K}[\mathbf{C}[\boldsymbol{E}]],$$

[1] Cf. p. 10.

and, since K is the inverse of C,

$$\hat{\varepsilon}(\boldsymbol{S}) = \varepsilon(\boldsymbol{E}).$$

The next two theorems give a physical motivation for some of our future assumptions concerning the elasticity tensor.

***(3)** The work done along each strain path starting from $\boldsymbol{E}=\boldsymbol{0}$ is non-negative if and only if C is symmetric and positive semi-definite.*

Proof. Assume that the work done along each strain path starting from $\boldsymbol{0}$ is non-negative. Let $\Gamma_{\boldsymbol{E}}$ be an arbitrary closed strain path starting from $\boldsymbol{E}_0$, and let

$$\boldsymbol{E}^*(\alpha) = \boldsymbol{E}(\alpha) - \boldsymbol{E}_0, \qquad \alpha_0 \leqq \alpha \leqq \alpha_1.$$

Then $\Gamma_{\boldsymbol{E}^*}$ is a closed strain path starting from $\boldsymbol{0}$ and

$$0 \leqq \int_{\alpha_0}^{\alpha_1} \mathsf{C}[\boldsymbol{E}^*(\alpha)] \cdot \frac{d\boldsymbol{E}^*(\alpha)}{d\alpha}\, d\alpha.$$

Thus, in view of the linearity of C,

$$0 \leqq \int_{\alpha_0}^{\alpha_1} \mathsf{C}[\boldsymbol{E}(\alpha)] \cdot \frac{d\boldsymbol{E}(\alpha)}{d\alpha}\, d\alpha - \mathsf{C}[\boldsymbol{E}_0] \cdot \{\boldsymbol{E}(\alpha_1) - \boldsymbol{E}(\alpha_0)\}.$$

But $\boldsymbol{E}(\alpha_1) = \boldsymbol{E}(\alpha_0)$, since $\Gamma_{\boldsymbol{E}}$ is closed. Thus the work done along each closed strain path is non-negative. Thus we may conclude from ***(2)*** that C is symmetric and, in addition, that there exists a stored energy function. Let $\boldsymbol{A}$ be a symmetric tensor, and let $\Gamma_{\boldsymbol{E}}$ be the strain path defined by

$$\boldsymbol{E}(\alpha) = \alpha \boldsymbol{A}, \qquad 0 \leqq \alpha \leqq 1.$$

Then by ***(1)*** and (iv) of ***(2)***,

$$0 \leqq w(\Gamma_{\boldsymbol{E}}) = \varepsilon(\boldsymbol{A}) - \varepsilon(\boldsymbol{0}) = \tfrac{1}{2}\boldsymbol{A} \cdot \mathsf{C}[\boldsymbol{A}];$$

hence C is positive semi-definite. Conversely, if C is symmetric, then ***(1)*** and ***(2)*** imply that given any path $\Gamma_{\boldsymbol{E}}$ connecting $\boldsymbol{0}$ and $\boldsymbol{E}_1$,

$$w(\Gamma_{\boldsymbol{E}}) = \tfrac{1}{2}\boldsymbol{E}_1 \cdot \mathsf{C}[\boldsymbol{E}_1].$$

Thus if C is positive semi-definite, $w(\Gamma_{\boldsymbol{E}}) \geqq 0$. □

***(4)** The work done along each non-closed strain path starting from $\boldsymbol{E}=\boldsymbol{0}$ is strictly positive if and only if C is symmetric and positive definite.*

Proof. Assume that the former assertion holds. Let $\Gamma_{\boldsymbol{E}}$ defined by $\boldsymbol{E}(\alpha)$, $\alpha_0 \leqq \alpha \leqq \alpha_1$, be an arbitrary *closed* strain path with

$$\boldsymbol{E}(\alpha_0) = \boldsymbol{E}(\alpha_1) = \boldsymbol{0}.$$

If $\boldsymbol{E}(\alpha) = \boldsymbol{0}$ for all α, then $w(\Gamma_{\boldsymbol{E}}) = 0$. Assume $\boldsymbol{E}(\alpha) \neq \boldsymbol{0}$ for some α, and let

$$\bar{\alpha} = \sup\{\alpha : \boldsymbol{E}(\alpha) \neq \boldsymbol{0}\}.$$

Then

$$w(\Gamma_{\boldsymbol{E}}) = \int_{\alpha_0}^{\alpha_1} \mathsf{C}[\boldsymbol{E}(\alpha)] \cdot \frac{d\boldsymbol{E}(\alpha)}{d\alpha}\, d\alpha = \lim_{\alpha^* \uparrow \bar{\alpha}} \int_{\alpha_0}^{\alpha^*} \mathsf{C}[\boldsymbol{E}(\alpha)] \cdot \frac{d\boldsymbol{E}(\alpha)}{d\alpha}\, d\alpha.$$

Since $\boldsymbol{E}(\alpha)$ $(\alpha_0 \leqq \alpha \leqq \alpha^*)$ for $\alpha^* < \bar{\alpha}$ but sufficiently close to $\bar{\alpha}$ defines a *non-closed* strain path starting from $\mathbf{0}$,

$$\int_{\alpha_0}^{\alpha^*} \mathsf{C}[\boldsymbol{E}(\alpha)] \cdot \frac{d\boldsymbol{E}(\alpha)}{d\alpha}\, d\alpha > 0;$$

hence

$$w(\Gamma_{\boldsymbol{E}}) \geqq 0.$$

Thus the work done along each strain path starting from $\mathbf{0}$ is non-negative, and we may conclude from ***(3)*** that C is symmetric. The remainder of the argument is the same as the proof of the analogous portions of ***(3)***. □

(5) *If the material is isotropic, then* C *is symmetric,*

$$\varepsilon(\boldsymbol{E}) = \mu|\boldsymbol{E}|^2 + \frac{\lambda}{2}(\mathrm{tr}\,\boldsymbol{E})^2,$$

and the following five statements are equivalent:

(i) C *is positive definite;*

(ii) $\mu > 0,\ 2\mu + 3\lambda > 0;$

(iii) $\mu > 0,\ k > 0;$

(iv) $\mu > 0,\ -1 < \nu < \frac{1}{2};$

(v) $\beta > 0,\ -1 < \nu < \frac{1}{2}.$

Proof. The symmetry of C and the above formula for $\varepsilon(\boldsymbol{E}) = \frac{1}{2}\boldsymbol{E} \cdot \mathsf{C}[\boldsymbol{E}]$ follow from ***(22.2)***. That (i)⇔(iii) is an immediate consequence of ***(22.3)*** and the sentence following its proof. Finally, the equivalence of (ii)–(v) follows from the relations given in Table 3 on p. 79. □

Let

$$\boldsymbol{E} = \mathring{\boldsymbol{E}} + \tfrac{1}{3}(\mathrm{tr}\,\boldsymbol{E})\,\mathbf{1} \quad (\mathrm{tr}\,\mathring{\boldsymbol{E}} = 0).$$

Then

$$|\boldsymbol{E}|^2 = |\mathring{\boldsymbol{E}}|^2 + \tfrac{1}{3}(\mathrm{tr}\,\boldsymbol{E})^2$$

and

$$\varepsilon(\boldsymbol{E}) = \mu|\mathring{\boldsymbol{E}}|^2 + \frac{k}{2}(\mathrm{tr}\,\boldsymbol{E})^2.$$

This alternative form for ε yields a direct verification of the implication (i)⇔(iii) in ***(5)***.

At the end of the last section we gave physically plausible arguments to support the inequalities: $\beta > 0$, $k > 0$, $\mu > 0$, $\nu > 0$. It follows from ***(5)*** that these inequalities are more than sufficient to insure positive definiteness. On the other hand, that C is positive definite implies $\beta > 0$, $k > 0$, $\mu > 0$, but *does not imply* $\nu > 0$.

Assume that C is symmetric and positive definite. Then the characteristic values of C (considered as a linear transformation on the six-dimensional space of all symmetric tensors) are all strictly positive. We call the largest characteristic value the ***maximum elastic modulus***, the smallest the ***minimum elastic modulus***.[1] The following trivial, but useful, proposition is a direct consequence of this definition and ***(22.3)***.

(6) *Let* C *be symmetric and positive definite and denote the maximum and minimum elastic moduli by* μ_M *and* μ_m, *respectively. Then*

$$\mu_m|\boldsymbol{E}|^2 \leqq \boldsymbol{E} \cdot \mathsf{C}[\boldsymbol{E}] \leqq \mu_M|\boldsymbol{E}|^2$$

[1] Cf. TOUPIN [1965, *21*].

for every symmetric tensor $\mathbf{E}$. *If, in addition, the material is isotropic with strictly positive Lamé moduli* μ *and* λ, *then*

$$\mu_M = 2\mu + 3\lambda, \qquad \mu_m = 2\mu.$$

Of course, if $\mu > 0$ and $\lambda = 0$, then $\mu_m = \mu_M = 2\mu$; if $\mu > 0$ and $\lambda < 0$ (but $2\mu + 3\lambda > 0$ so that $\mathbf{C}$ is positive definite), then $\mu_M = 2\mu$ and $\mu_m = 2\mu + 3\lambda$.

25. Strong ellipticity. In this section we relate various restrictions on the elasticity tensor $\mathbf{C} = \mathbf{C}_{\boldsymbol{x}}$ at a given point $\boldsymbol{x} \in B$.

We say that $\mathbf{C}$ is ***strongly elliptic*** if

$$\boldsymbol{A} \cdot \mathbf{C}[\boldsymbol{A}] > 0$$

whenever $\boldsymbol{A}$ is of the form

$$\boldsymbol{A} = \boldsymbol{a} \otimes \boldsymbol{b}, \qquad \boldsymbol{a} \neq \mathbf{0}, \qquad \boldsymbol{b} \neq \mathbf{0}.$$

This condition is of importance in discussing uniqueness and also in the study of wave propagation.

By ***(20.1)***

$$\boldsymbol{A} \cdot \mathbf{C}[\boldsymbol{A}] = (\operatorname{sym} \boldsymbol{A}) \cdot \mathbf{C}[\operatorname{sym} \boldsymbol{A}];$$

thus

$$\mathbf{C} \text{ positive definite} \Rightarrow \mathbf{C} \text{ strongly elliptic.}$$

The fact that the converse assertion is not true is apparent from the next proposition.

(1) *Suppose that the material is isotropic. Then*

$$\mathbf{C} \text{ strongly elliptic} \Leftrightarrow \mu > 0,\ \lambda + 2\mu > 0,$$
$$-\mathbf{C} \text{ strongly elliptic} \Leftrightarrow \mu < 0,\ \lambda + 2\mu < 0,$$

where μ *and* λ *are the Lamé moduli corresponding to* $\mathbf{C}$. *In terms of Poisson's ratio* ν,

$$\mathbf{C} \text{ strongly elliptic} \Leftrightarrow \mu > 0,\ \nu \notin [\tfrac{1}{2}, 1],$$
$$-\mathbf{C} \text{ strongly elliptic} \Leftrightarrow \mu < 0,\ \nu \notin [\tfrac{1}{2}, 1],$$

provided $\lambda + \mu \neq 0$.

Proof. By (i) of ***(20.1)*** and ***(22.2)***,

$$\begin{aligned}(\boldsymbol{a} \otimes \boldsymbol{b}) \cdot \mathbf{C}[\boldsymbol{a} \otimes \boldsymbol{b}] &= (\boldsymbol{a} \otimes \boldsymbol{b}) \cdot \mathbf{C}[\operatorname{sym} \boldsymbol{a} \otimes \boldsymbol{b}] \\ &= (\boldsymbol{a} \otimes \boldsymbol{b}) \cdot [\mu(\boldsymbol{a} \otimes \boldsymbol{b} + \boldsymbol{b} \otimes \boldsymbol{a}) + \lambda(\boldsymbol{a} \cdot \boldsymbol{b})\, \mathbf{1}] \\ &= \mu \boldsymbol{a}^2 \boldsymbol{b}^2 + (\lambda + \mu)(\boldsymbol{a} \cdot \boldsymbol{b})^2 \\ &= \mu[\boldsymbol{a}^2 \boldsymbol{b}^2 - (\boldsymbol{a} \cdot \boldsymbol{b})^2] + (\lambda + 2\mu)(\boldsymbol{a} \cdot \boldsymbol{b})^2.\end{aligned}$$

Since $(\boldsymbol{a} \cdot \boldsymbol{b})^2 \leqq \boldsymbol{a}^2 \boldsymbol{b}^2$, $\mu > 0$ and $\lambda + 2\mu > 0$ imply that $\mathbf{C}$ is strongly elliptic. Conversely, if $\mathbf{C}$ is strongly elliptic, then $\lambda + 2\mu > 0$ follows by taking $\boldsymbol{a} = \boldsymbol{b}$, $\mu > 0$ by taking $\boldsymbol{a}$ orthogonal to $\boldsymbol{b}$. The assertion concerning $-\mathbf{C}$ follows in the same manner. Finally, the last two results are consequences of the relation:

$$\nu = \frac{\lambda}{2(\lambda + \mu)}. \qquad \square$$

These results, as well as some of the results established in *(24.5)*, are given in the following table:

Table 4.

Condition	Restrictions on the moduli
Positive definite	$\mu>0,\ 2\mu+3\lambda>0$ or $\mu>0,\ \nu\in(-1,\frac{1}{2})$ or $\beta>0,\ \nu\in(-1,\frac{1}{2})$
Strongly elliptic	$\mu>0,\ \lambda+2\mu>0$ or $\mu>0,\ \nu\notin[\frac{1}{2},1]$

26. Anisotropic materials. In this section we state the restrictions that material symmetry places on the elasticity tensor $\mathbf{C}=\mathbf{C}_{\boldsymbol{x}}$ at a given point $\boldsymbol{x}$. Let $\{\boldsymbol{e}_i\}$ be a given orthonormal basis with $\boldsymbol{e}_3=\boldsymbol{e}_1\times\boldsymbol{e}_2$, and let C_{ijkl} denote the components of $\mathbf{C}$ relative to this basis:

$$C_{ijkl}=C_{jikl}=C_{ijlk}=(\operatorname{sym}\boldsymbol{e}_i\otimes\boldsymbol{e}_j)\cdot\mathbf{C}[\operatorname{sym}\boldsymbol{e}_k\otimes\boldsymbol{e}_l].$$

We assume throughout this section that a stored energy function exists, so that $\mathbf{C}$ is *symmetric*; thus

$$C_{ijkl}=C_{klij}.$$

For the *triclinic system* ($\mathscr{C}_1$-symmetry) there are no restrictions placed on $\mathbf{C}$ by material symmetry.

For the *monoclinic system* ($\mathscr{C}_2$-symmetry) let the group $\mathscr{C}_2$ be generated by $\boldsymbol{R}_{\boldsymbol{e}_3}^{\pi}$. By identity *(21.2)*,

$$C_{ijkl}=\operatorname{sym}(\boldsymbol{Q}\boldsymbol{e}_i\otimes\boldsymbol{Q}\boldsymbol{e}_j)\cdot\mathbf{C}[\operatorname{sym}(\boldsymbol{Q}\boldsymbol{e}_k\otimes\boldsymbol{Q}\boldsymbol{e}_l)]$$

for every $\boldsymbol{Q}\in\mathscr{G}_{\boldsymbol{x}}$. If we take $\boldsymbol{Q}=-\boldsymbol{R}_{\boldsymbol{e}_3}^{\pi}$, then

$$\boldsymbol{Q}\boldsymbol{e}_1=\boldsymbol{e}_1,\quad \boldsymbol{Q}\boldsymbol{e}_2=\boldsymbol{e}_2,\quad \boldsymbol{Q}\boldsymbol{e}_3=-\boldsymbol{e}_3,$$

and the above identity implies

$$C_{1123}=C_{1131}=C_{2223}=C_{2231}=C_{3323}=C_{3331}=C_{2312}=C_{3112}=0.$$

In the same manner, results for the other symmetries discussed in Sect. 21 can be deduced.[1] We now list these results, which are due to VOIGT.[2] For convenience, the 21 elasticities will be tabulated as follows:

$$\begin{array}{cccccc}
C_{1111} & C_{1122} & C_{1133} & C_{1123} & C_{1131} & C_{1112}\\
 & C_{2222} & C_{2233} & C_{2223} & C_{2231} & C_{2212}\\
 & & C_{3333} & C_{3323} & C_{3331} & C_{3312}\\
 & & & C_{2323} & C_{2331} & C_{2312}\\
 & & & & C_{3131} & C_{3112}\\
 & & & & & C_{1212}
\end{array}$$

[1] They can also be deduced using the results of SMITH and RIVLIN [1958, *16*], SIROTIN [1960, *11*].

[2] VOIGT [1882, *1*], [1887, *3*], [1900, *6*], [1910, *1*], § 287. See also KIRCHHOFF [1876, *1*], MINNIGERODE [1884, *1*], NEUMANN [1885, *3*], LOVE [1927, *3*], AUERBACH [1927, *1*], GECKELER [1928, *2*]. A discussion of methods used to measure the elastic constants of anisotropic materials, as well as a tabulation of experimental values of these constants for various materials, is given by HEARMON [1946, 2].

Monoclinic system (all classes) $\mathscr{C}_2$ generated by $\boldsymbol{R}_{\boldsymbol{e}_3}^{\pi}$ (13 elasticities)

$$\begin{matrix} C_{1111} & C_{1122} & C_{1133} & 0 & 0 & C_{1112} \\ & C_{2222} & C_{2233} & 0 & 0 & C_{2212} \\ & & C_{3333} & 0 & 0 & C_{3312} \\ & & & C_{2323} & C_{2331} & 0 \\ & & & & C_{3131} & 0 \\ & & & & & C_{1212} \end{matrix}$$

Rhombic system (all classes) $\mathscr{C}_3$ generated by $\boldsymbol{R}_{\boldsymbol{e}_1}^{\pi}$, $\boldsymbol{R}_{\boldsymbol{e}_2}^{\pi}$ (9 elasticities)

$$\begin{matrix} C_{1111} & C_{1122} & C_{1133} & 0 & 0 & 0 \\ & C_{2222} & C_{2233} & 0 & 0 & 0 \\ & & C_{3333} & 0 & 0 & 0 \\ & & & C_{2323} & 0 & 0 \\ & & & & C_{3131} & 0 \\ & & & & & C_{1212} \end{matrix}$$

Tetragonal system (tetragonal-disphenoidal, tetragonal-pyramidal, tetragonal-dipyramidal) $\mathscr{C}_4$ generated by $\boldsymbol{R}_{\boldsymbol{e}_3}^{\pi/2}$ (7 elasticities)

$$\begin{matrix} C_{1111} & C_{1122} & C_{1133} & 0 & 0 & C_{1112} \\ & C_{1111} & C_{1133} & 0 & 0 & -C_{1112} \\ & & C_{3333} & 0 & 0 & 0 \\ & & & C_{2323} & 0 & 0 \\ & & & & C_{2323} & 0 \\ & & & & & C_{1212} \end{matrix}$$

Tetragonal system (tetragonal-scalenohedral, ditetragonal-pyramidal, tetragonal-trapezohedral, ditetragonal-dipyramidal) $\mathscr{C}_5$ generated by $\boldsymbol{R}_{\boldsymbol{e}_3}^{\pi/2}$, $\boldsymbol{R}_{\boldsymbol{e}_1}^{\pi}$ (6 elasticities)

$$\begin{matrix} C_{1111} & C_{1122} & C_{1133} & 0 & 0 & 0 \\ & C_{1111} & C_{1133} & 0 & 0 & 0 \\ & & C_{3333} & 0 & 0 & 0 \\ & & & C_{2323} & 0 & 0 \\ & & & & C_{2323} & 0 \\ & & & & & C_{1212} \end{matrix}$$

Cubic system (tetartoidal, diploidal) $\mathscr{C}_6$ generated by

$$\boldsymbol{R}_{\boldsymbol{e}_1}^{\pi},\ \boldsymbol{R}_{\boldsymbol{e}_2}^{\pi},\ \boldsymbol{R}_{\boldsymbol{q}}^{2\pi/3}, \qquad \boldsymbol{q} = \sqrt{\tfrac{1}{3}}\,(\boldsymbol{e}_1 + \boldsymbol{e}_2 + \boldsymbol{e}_3);$$

(hexatetrahedral, gyroidal, hexoctahedral) $\mathscr{C}_7$ generated by $\boldsymbol{R}_{\boldsymbol{e}_1}^{\pi/2}$, $\boldsymbol{R}_{\boldsymbol{e}_2}^{\pi/2}$, $\boldsymbol{R}_{\boldsymbol{e}_3}^{\pi/2}$ (3 elasticities)

$$\begin{matrix} C_{1111} & C_{1122} & C_{1122} & 0 & 0 & 0 \\ & C_{1111} & C_{1122} & 0 & 0 & 0 \\ & & C_{1111} & 0 & 0 & 0 \\ & & & C_{2323} & 0 & 0 \\ & & & & C_{2323} & 0 \\ & & & & & C_{2323} \end{matrix}$$

Hexagonal system (trigonal-pyramidal, rhombohedral) $\mathscr{C}_8$ generated by $\boldsymbol{R}_{\boldsymbol{e}_3}^{2\pi/3}$ (7 elasticities)

$$\begin{array}{cccccc}
C_{1111} & C_{1122} & C_{1133} & C_{1123} & C_{1131} & 0 \\
 & C_{1111} & C_{1133} & -C_{1123} & -C_{1131} & 0 \\
 & & C_{3333} & 0 & 0 & 0 \\
 & & & C_{2323} & 0 & -C_{1131} \\
 & & & & C_{2323} & C_{1123} \\
 & & & & & \tfrac{1}{2}(C_{1111}-C_{1122})
\end{array}$$

Hexagonal system (ditrigonal-pyramidal, trigonal-trapezohedral, hexagonal-scalenohedral) $\mathscr{C}_9$ generated by $\boldsymbol{R}_{\boldsymbol{e}_3}^{2\pi/3}$, $\boldsymbol{R}_{\boldsymbol{e}_1}^{\pi}$ (6 elasticities)

$$\begin{array}{cccccc}
C_{1111} & C_{1122} & C_{1133} & C_{1123} & 0 & 0 \\
 & C_{1111} & C_{1133} & -C_{1123} & 0 & 0 \\
 & & C_{3333} & 0 & 0 & 0 \\
 & & & C_{2323} & 0 & 0 \\
 & & & & C_{2323} & C_{1123} \\
 & & & & & \tfrac{1}{2}(C_{1111}-C_{1122})
\end{array}$$

Hexagonal system (trigonal-dipyramidal, hexagonal-pyramidal, hexagonal-dipyramidal) $\mathscr{C}_{10}$ generated by $\boldsymbol{R}_{\boldsymbol{e}_3}^{\pi/3}$; (ditrigonal-dipyramidal, dihexagonal-pyramidal, hexagonal-trapezohedral, dihexagonal-dipyramidal) $\mathscr{C}_{11}$ generated by $\boldsymbol{R}_{\boldsymbol{e}_3}^{\pi/3}$, $\boldsymbol{R}_{\boldsymbol{e}_2}^{\pi}$; and ***transverse isotropy*** $\mathscr{C}_{12}$ generated by $\boldsymbol{R}_{\boldsymbol{e}_3}^{\varphi}$, $0<\varphi<2\pi$ (5 elasticities)

$$\begin{array}{cccccc}
C_{1111} & C_{1122} & C_{1133} & 0 & 0 & 0 \\
 & C_{1111} & C_{1133} & 0 & 0 & 0 \\
 & & C_{3333} & 0 & 0 & 0 \\
 & & & C_{2323} & 0 & 0 \\
 & & & & C_{2323} & 0 \\
 & & & & & \tfrac{1}{2}(C_{1111}-C_{1122})
\end{array}$$

In view of *(21.1)*, if the elasticity tensor $\mathbf{C}$ is invertible, then the components of the compliance tensor $\mathbf{K}=\mathbf{C}^{-1}$ will suffer restrictions exactly analogous to the ones tabulated above.

D. Elastostatics.

I. The fundamental field equations. Elastic states. Work and energy.

27. The fundamental system of field equations. The fundamental system of field equations for the time-independent behavior of a linear elastic body consists of the ***strain-displacement relation***

$$\boldsymbol{E}=\hat{\nabla}\boldsymbol{u}=\tfrac{1}{2}(\nabla\boldsymbol{u}+\nabla\boldsymbol{u}^T),$$

the ***stress-strain relation***

$$\boldsymbol{S}=\mathbf{C}[\boldsymbol{E}],$$

and the ***equation of equilibrium***[1]

$$\operatorname{div} \boldsymbol{S} + \boldsymbol{b} = \boldsymbol{0}.$$

Here, $\boldsymbol{u}$, $\boldsymbol{E}$, $\boldsymbol{S}$, and $\boldsymbol{b}$ are the displacement, strain, stress, and body force fields, while $\mathbf{C}$ is the elasticity field.

By (i) of ***(20.1)***,

$$\mathbf{C}[\widehat{\nabla \boldsymbol{u}}] = \mathbf{C}[\nabla \boldsymbol{u}];$$

thus when the displacement field is sufficiently smooth, the above equations imply the ***displacement equation of equilibrium***

$$\operatorname{div} \mathbf{C}[\nabla \boldsymbol{u}] + \boldsymbol{b} = \boldsymbol{0}.$$

Conversely, if $\boldsymbol{u}$ satisfies the displacement equation of equilibrium, and if $\boldsymbol{E}$ and $\boldsymbol{S}$ are *defined* by the strain-displacement and stress-strain relations, then the stress equation of equilibrium is satisfied. In components the displacement equation of equilibrium has the form

$$(C_{ijkl}\, u_{k,l})_{,j} + b_i = 0.$$

We assume for the remainder of this section that the material is isotropic. Then

$$\boldsymbol{S} = 2\mu \boldsymbol{E} + \lambda (\operatorname{tr} \boldsymbol{E})\, \mathbf{1},$$

where μ and λ are the Lamé moduli. Of course, μ and λ are functions of position $\boldsymbol{x}$ when the body is inhomogeneous. Therefore, since

$$\begin{aligned}
\operatorname{div}(\mu \nabla \boldsymbol{u}) &= \mu \Delta \boldsymbol{u} + (\nabla \boldsymbol{u}) \nabla \mu, \\
\operatorname{div}(\mu \nabla \boldsymbol{u}^T) &= \mu \nabla \operatorname{div} \boldsymbol{u} + (\nabla \boldsymbol{u})^T \nabla \mu, \\
\operatorname{div}[\lambda (\operatorname{div} \boldsymbol{u})\, \mathbf{1}] &= \lambda \nabla \operatorname{div} \boldsymbol{u} + (\operatorname{div} \boldsymbol{u})\, \nabla \lambda,
\end{aligned}$$

the displacement equation of equilibrium takes the form

$$\mu \Delta \boldsymbol{u} + (\lambda + \mu) \nabla \operatorname{div} \boldsymbol{u} + 2 (\widehat{\nabla \boldsymbol{u}})\, \nabla \mu + (\operatorname{div} \boldsymbol{u})\, \nabla \lambda + \boldsymbol{b} = \boldsymbol{0}.$$

Suppose now that the body is homogeneous. Then $\nabla \mu = \nabla \lambda = \boldsymbol{0}$, and the above equation reduces to ***Navier's equation***[2]

$$\mu \Delta \boldsymbol{u} + (\lambda + \mu) \nabla \operatorname{div} \boldsymbol{u} + \boldsymbol{b} = \boldsymbol{0}, \tag{a}$$

or equivalently,

$$\Delta \boldsymbol{u} + \frac{1}{1 - 2\nu} \nabla \operatorname{div} \boldsymbol{u} + \frac{1}{\mu} \boldsymbol{b} = \boldsymbol{0},$$

in terms of Poisson's ratio

$$\nu = \frac{\lambda}{2(\lambda + \mu)}.$$

Next, by (3) of ***(4.1)***,

$$\Delta \boldsymbol{u} = \nabla \operatorname{div} \boldsymbol{u} - \operatorname{curl} \operatorname{curl} \boldsymbol{u},$$

[1] Since the values of $\mathbf{C}_{\boldsymbol{x}}[\cdot]$ are assumed to be symmetric tensors, the requirement that $\boldsymbol{S} = \boldsymbol{S}^T$ is automatically satisfied.

[2] This relation was first derived by NAVIER [1823, *2*], [1827, *2*] in 1821. NAVIER's work, which is based on a molecular model, is limited to materials for which $\mu = \lambda$. The general relation involving two elastic constants first appears in the work of CAUCHY [1828, *1*]. Cf. POISSON [1829, *2*], LAMÉ and CLAPEYRON [1833, *1*], STOKES [1845, *1*], LAMÉ [1852, *2*], § 26. The analogous relation for cubic symmetry was given by ALBRECHT [1951, *1*].

and thus (a) yields[1]

$$(\lambda+2\mu)\,\nabla \operatorname{div}\boldsymbol{u}-\mu \operatorname{curl}\operatorname{curl}\boldsymbol{u}+\boldsymbol{b}=\boldsymbol{0}.$$

Operating on (a), first with the divergence and then with the curl, we find that

$$\begin{aligned}(\lambda+2\mu)\Delta \operatorname{div}\boldsymbol{u}&=-\operatorname{div}\boldsymbol{b},\\ \mu\Delta \operatorname{curl}\boldsymbol{u}&=-\operatorname{curl}\boldsymbol{b}.\end{aligned} \qquad \text{(b)}^{2}$$

If $(\lambda+2\mu)\neq 0$, $\mu\neq 0$, and div $\boldsymbol{b}$ and curl $\boldsymbol{b}$ vanish, then

$$\Delta \operatorname{div}\boldsymbol{u}=0, \qquad \Delta \operatorname{curl}\boldsymbol{u}=\boldsymbol{0}.$$

Thus, in this instance, the divergence and curl of $\boldsymbol{u}$ are harmonic fields. Trivially, these relations can be written in the form

$$\operatorname{div}\Delta\boldsymbol{u}=0, \qquad \operatorname{curl}\Delta\boldsymbol{u}=\boldsymbol{0};$$

and since a vector field with vanishing divergence and curl is harmonic, we have the following important result:[3]

$$\operatorname{div}\boldsymbol{b}=0, \qquad \operatorname{curl}\boldsymbol{b}=\boldsymbol{0}\Rightarrow \Delta\Delta\boldsymbol{u}=\boldsymbol{0}.$$

Thus if the body force field $\boldsymbol{b}$ is divergence-free and curl-free, the displacement field is biharmonic and hence analytic on B. To prove this we have tacitly assumed that $\boldsymbol{u}$ is of class C^4; in Sect. 42 we will show that these results remain valid when $\boldsymbol{u}$ is assumed to be only of class C^2.

Another interesting consequence of the displacement equation of equilibrium may be derived as follows: Let

$$\boldsymbol{v}=\mu\boldsymbol{u}+\tfrac{1}{2}(\lambda+\mu)\,\boldsymbol{p}\operatorname{div}\boldsymbol{u}, \qquad \boldsymbol{p}(\boldsymbol{x})=\boldsymbol{x}-\boldsymbol{0}.$$

Then

$$\Delta\boldsymbol{v}=\mu\Delta\boldsymbol{u}+(\lambda+\mu)\{\nabla\operatorname{div}\boldsymbol{u}+\tfrac{1}{2}\boldsymbol{p}\Delta\operatorname{div}\boldsymbol{u}\},$$

and (a) and (b) imply

$$\Delta\boldsymbol{v}=-\boldsymbol{b}-\frac{\lambda+\mu}{2(\lambda+2\mu)}\,\boldsymbol{p}\operatorname{div}\boldsymbol{b}.$$

Thus we arrive at the following result of TEDONE:[4] if the body force field vanishes, then

$$\Delta\boldsymbol{v}=\boldsymbol{0}.$$

Assume, for the time being, that the body force field $\boldsymbol{b}=\boldsymbol{0}$. A field $\boldsymbol{u}$ that satisfies Navier's equation (a) for all values of μ and λ is said to be *universal*. Since a field of this type is independent of the elastic constants μ and λ, it is a possible displacement field for *all* homogeneous and isotropic elastic materials. Clearly, $\boldsymbol{u}$ is universal if and only if[5]

$$\Delta\boldsymbol{u}=\boldsymbol{0},$$

$$\nabla\operatorname{div}\boldsymbol{u}=\boldsymbol{0};$$

---

[1] LAMÉ and CLAPEYRON [1833, *1*], LAMÉ [1852, *2*], § 26.

[2] The first of these relations is due to CAUCHY [1828, *1*]. See also LAMÉ and CLAPEYRON [1833, *1*], LAMÉ [1852, *2*], § 27.

[3] LAMÉ [1852, *2*], p. 70.

[4] [1903, *4*], [1904, *4*]. See also LICHTENSTEIN [1924, *1*].

[5] TRUESDELL [1966, *28*], p. 117.

i.e., if and only if $\boldsymbol{u}$ is harmonic and has constant dilatation. Important examples of universal displacement fields are furnished by the Saint-Venant torsion solution.[1]

If $\mu \neq 0$, $\nu \neq -1$, the stress-strain relation may be inverted to give

$$\boldsymbol{E} = \frac{1}{2\mu}\left[\boldsymbol{S} - \frac{\nu}{1+\nu}(\mathrm{tr}\,\boldsymbol{S})\,\mathbf{1}\right].$$

By *(14.3)* we can write the equation of compatibility in the form

$$\mathscr{L}(\boldsymbol{E}) = \Delta\boldsymbol{E} + \nabla\nabla(\mathrm{tr}\,\boldsymbol{E}) - 2\hat{\nabla}\,\mathrm{div}\,\boldsymbol{E} = \mathbf{0}. \tag{c}$$

Therefore

$$\mathbf{0} = \mathscr{L}(2\mu\,\boldsymbol{E}) = \mathscr{L}(\boldsymbol{S}) - \frac{\nu}{1+\nu}\mathscr{L}[(\mathrm{tr}\,\boldsymbol{S})\,\mathbf{1}]. \tag{d}$$

But

$$\mathscr{L}[(\mathrm{tr}\,\boldsymbol{S})\,\mathbf{1}] = \Delta(\mathrm{tr}\,\boldsymbol{S})\,\mathbf{1} + \nabla\nabla(\mathrm{tr}\,\boldsymbol{S}), \tag{e}$$

and by (c) with $\boldsymbol{E}$ replaced by $\boldsymbol{S}$ in conjunction with the equation of equilibrium,

$$\mathscr{L}(\boldsymbol{S}) = \Delta\boldsymbol{S} + \nabla\nabla(\mathrm{tr}\,\boldsymbol{S}) + 2\hat{\nabla}\boldsymbol{b}. \tag{f}$$

Eqs. (d)–(f) imply that

$$\Delta\boldsymbol{S} + \frac{1}{1+\nu}\nabla\nabla(\mathrm{tr}\,\boldsymbol{S}) - \frac{\nu}{1+\nu}\Delta(\mathrm{tr}\,\boldsymbol{S})\,\mathbf{1} + 2\hat{\nabla}\boldsymbol{b} = \mathbf{0}, \tag{g}$$

and taking the trace of this equation, we conclude that

$$\frac{1-\nu}{1+\nu}\Delta(\mathrm{tr}\,\boldsymbol{S}) = -\mathrm{div}\,\boldsymbol{b}. \tag{h}$$

Eqs. (g) and (h) yield the ***stress equation of compatibility***:[2]

$$\Delta\boldsymbol{S} + \frac{1}{1+\nu}\nabla\nabla(\mathrm{tr}\,\boldsymbol{S}) + 2\hat{\nabla}\boldsymbol{b} + \frac{\nu}{1-\nu}(\mathrm{div}\,\boldsymbol{b})\,\mathbf{1} = \mathbf{0},$$

which in the absence of body forces takes the simple form

$$\Delta\boldsymbol{S} + \frac{1}{1+\nu}\nabla\nabla(\mathrm{tr}\,\boldsymbol{S}) = \mathbf{0}.$$

On the other hand, if $\boldsymbol{S}$ is a symmetric tensor field on B that satisfies the equation of equilibrium and the stress equation of compatibility, and if $\boldsymbol{E}$ is defined through the inverted stress-strain relation, then $\boldsymbol{E}$ satisfies the equation of compatibility. Thus if B is simply-connected, there exists a displacement field $\boldsymbol{u}$ that satisfies the strain-displacement relation.

In view of the stress-strain and strain-displacement relations,

$$\boldsymbol{S} = \mu(\nabla\boldsymbol{u} + \nabla\boldsymbol{u}^T) + \lambda(\mathrm{div}\,\boldsymbol{u})\,\mathbf{1},$$

and the corresponding surface traction on the boundary ∂B is given by

$$\boldsymbol{s} = \boldsymbol{S}\boldsymbol{n} = \mu(\nabla\boldsymbol{u} + \nabla\boldsymbol{u}^T)\,\boldsymbol{n} + \lambda(\mathrm{div}\,\boldsymbol{u})\,\boldsymbol{n}.$$

---

[1] See, e.g., LOVE [1927, *3*], Chap. 14; SOKOLNIKOFF [1956, *12*], Chap. *4*; SOLOMON [1968, *12*], Chap. 5.

[2] This equation was obtained by BELTRAMI [1892, *1*] for $\boldsymbol{b} = \mathbf{0}$ and by DONATI [1894, *2*] and MICHELL [1900, *3*] for the general case. It is usually referred to as the Beltrami-Michell equation of compatibility although DONATI's paper appeared six years before MICHELL's. The appropriate generalization of this equation for transversely isotropic materials is given by MOISIL [1952, *3*], for cubic materials by ALBRECHT [1951, *1*].

Let $\partial \boldsymbol{u}/\partial \boldsymbol{n}$ be defined by

$$\frac{\partial \boldsymbol{u}}{\partial \boldsymbol{n}} = (\nabla \boldsymbol{u})\, \boldsymbol{n};$$

then

$$\mathbf{s} = 2\mu \frac{\partial \boldsymbol{u}}{\partial \boldsymbol{n}} + \mu (\nabla \boldsymbol{u}^T - \nabla \boldsymbol{u})\, \boldsymbol{n} + \lambda (\operatorname{div} \boldsymbol{u})\, \boldsymbol{n}.$$

But

$$(\nabla \boldsymbol{u}^T - \nabla \boldsymbol{u})\, \boldsymbol{n} = \boldsymbol{n} \times \operatorname{curl} \boldsymbol{u};$$

thus we have the following interesting formula for the surface traction:[1]

$$\mathbf{s} = 2\mu \frac{\partial \boldsymbol{u}}{\partial \boldsymbol{n}} + \mu \boldsymbol{n} \times \operatorname{curl} \boldsymbol{u} + \lambda (\operatorname{div} \boldsymbol{u})\, \boldsymbol{n}.$$

We close this section by recording the strain-displacement relations and the equations of equilibrium in rectangular, cylindrical, and polar coordinates, using physical components throughout.[2]

(a) ***Cartesian coordinates*** (x, y, z). The strain-displacement relations reduce to

$$E_{xx} = \frac{\partial u_x}{\partial x}, \quad E_{yy} = \frac{\partial u_y}{\partial y}, \quad E_{zz} = \frac{\partial u_z}{\partial z},$$

$$E_{xy} = \frac{1}{2}\left(\frac{\partial u_x}{\partial y} + \frac{\partial u_y}{\partial x}\right), \quad E_{yz} = \frac{1}{2}\left(\frac{\partial u_y}{\partial z} + \frac{\partial u_z}{\partial y}\right),$$

$$E_{zx} = \frac{1}{2}\left(\frac{\partial u_z}{\partial x} + \frac{\partial u_x}{\partial z}\right),$$

and the equations of equilibrium are given by

$$\frac{\partial S_{xx}}{\partial x} + \frac{\partial S_{xy}}{\partial y} + \frac{\partial S_{xz}}{\partial z} + b_x = 0,$$

$$\frac{\partial S_{xy}}{\partial x} + \frac{\partial S_{yy}}{\partial y} + \frac{\partial S_{yz}}{\partial z} + b_y = 0,$$

$$\frac{\partial S_{xz}}{\partial x} + \frac{\partial S_{yz}}{\partial y} + \frac{\partial S_{zz}}{\partial z} + b_z = 0.$$

(b) ***Cylindrical coordinates*** (r, θ, z). The strain-displacement relations have the form

$$E_{rr} = \frac{\partial u_r}{\partial r}, \quad E_{\theta\theta} = \frac{1}{r}\frac{\partial u_\theta}{\partial \theta} + \frac{u_r}{r}, \quad E_{zz} = \frac{\partial u_z}{\partial z},$$

$$E_{r\theta} = \frac{1}{2}\left(\frac{1}{r}\frac{\partial u_r}{\partial \theta} + \frac{\partial u_\theta}{\partial r} - \frac{u_\theta}{r}\right),$$

$$E_{rz} = \frac{1}{2}\left(\frac{\partial u_r}{\partial z} + \frac{\partial u_z}{\partial r}\right), \quad E_{\theta z} = \frac{1}{2}\left(\frac{\partial u_\theta}{\partial z} + \frac{1}{r}\frac{\partial u_z}{\partial \theta}\right),$$

while the equations of equilibrium reduce to

$$\frac{\partial S_{rr}}{\partial r} + \frac{1}{r}\frac{\partial S_{r\theta}}{\partial \theta} + \frac{\partial S_{rz}}{\partial z} + \frac{S_{rr} - S_{\theta\theta}}{r} + b_r = 0,$$

$$\frac{\partial S_{r\theta}}{\partial r} + \frac{1}{r}\frac{\partial S_{\theta\theta}}{\partial \theta} + \frac{\partial S_{\theta z}}{\partial z} + \frac{2}{r} S_{r\theta} + b_\theta = 0,$$

$$\frac{\partial S_{rz}}{\partial r} + \frac{1}{r}\frac{\partial S_{\theta z}}{\partial \theta} + \frac{\partial S_{zz}}{\partial z} + \frac{1}{r} S_{rz} + b_z = 0.$$

---

[1] Betti [1872, *1*], § 10, Eq. (53). See also Somigliana [1889, *3*], p. 38, Korn [1927, *2*], p. 13, Kupradze [1963, *17*, *18*], p. 9.

[2] See, e.g., Ericksen [1960, *6*], §§ 11–13.

(c) ***Spherical coordinates*** (r, θ, γ). The strain-displacement relations become

$$E_{rr} = \frac{\partial u_r}{\partial r}, \qquad E_{\theta\theta} = \frac{1}{r}\frac{\partial u_\theta}{\partial \theta} + \frac{u_r}{r},$$

$$E_{\gamma\gamma} = \frac{1}{r\sin\theta}\frac{\partial u_\gamma}{\partial \gamma} + \frac{u_r}{r} + \frac{u_\theta \cot\theta}{r},$$

$$E_{r\gamma} = \frac{1}{2}\left(\frac{1}{r\sin\theta}\frac{\partial u_r}{\partial \gamma} - \frac{u_\gamma}{r} + \frac{\partial u_\gamma}{\partial r}\right),$$

$$E_{r\theta} = \frac{1}{2}\left(\frac{1}{r}\frac{\partial u_r}{\partial \theta} - \frac{u_\theta}{r} + \frac{\partial u_\theta}{\partial r}\right),$$

$$E_{\gamma\theta} = \frac{1}{2}\left(\frac{1}{r}\frac{\partial u_\gamma}{\partial \theta} - \frac{u_\gamma \cot\theta}{r} + \frac{1}{r\sin\theta}\frac{\partial u_\theta}{\partial \gamma}\right),$$

and the equations of equilibrium are given by

$$\frac{\partial S_{rr}}{\partial r} + \frac{1}{r\sin\theta}\frac{\partial S_{r\gamma}}{\partial \gamma} + \frac{1}{r}\frac{\partial S_{r\theta}}{\partial \theta} + \frac{2S_{rr} - S_{\gamma\gamma} - S_{\theta\theta} + S_{r\theta}\cot\theta}{r} + b_r = 0,$$

$$\frac{\partial S_{r\gamma}}{\partial r} + \frac{1}{r\sin\theta}\frac{\partial S_{\gamma\gamma}}{\partial \gamma} + \frac{1}{r}\frac{\partial S_{\gamma\theta}}{\partial \theta} + \frac{3S_{r\gamma} + 2S_{\gamma\theta}\cot\theta}{r} + b_\gamma = 0_2$$

$$\frac{\partial S_{r\gamma}}{\partial r} + \frac{1}{r\sin\theta}\frac{\partial S_{\gamma\theta}}{\partial \gamma} + \frac{1}{r}\frac{\partial S_{\theta\theta}}{\partial \theta} + \frac{3S_{r\theta} + (S_{\theta\theta} - S_{\gamma\gamma})\cot\theta}{r} + b_\theta = 0.$$

28. Elastic states. Work and energy. Throughout this section we assume given a *continuous* elasticity field $\mathbf{C}$ on $\bar{B}$. We define the ***strain energy*** $U\{\boldsymbol{E}\}$ corresponding to a continuous strain field $\boldsymbol{E}$ on $\bar{B}$ by

$$U\{\boldsymbol{E}\} = \tfrac{1}{2}\int_B \boldsymbol{E}\cdot\mathbf{C}[\boldsymbol{E}]\,dv.$$

Note that when a stored energy function ε exists, its integral over B gives the strain energy:

$$U\{\boldsymbol{E}\} = \int_B \varepsilon(\boldsymbol{E})\,dv.$$

For future use we now record the following

(1) Lemma. *Let* $\mathbf{C}$ *be symmetric, and let* $\boldsymbol{E}$ *and* $\tilde{\boldsymbol{E}}$ *be continuous symmetric tensor fields on* $\bar{B}$. *Then*

$$U\{\boldsymbol{E}+\tilde{\boldsymbol{E}}\} = U\{\boldsymbol{E}\} + U\{\tilde{\boldsymbol{E}}\} + \int_B \boldsymbol{E}\cdot\mathbf{C}[\tilde{\boldsymbol{E}}]\,dv.$$

Proof. Since $\mathbf{C}$ is symmetric,

$$\tilde{\boldsymbol{E}}\cdot\mathbf{C}[\boldsymbol{E}] = \boldsymbol{E}\cdot\mathbf{C}[\tilde{\boldsymbol{E}}].$$

Thus

$$(\boldsymbol{E}+\tilde{\boldsymbol{E}})\cdot\mathbf{C}[\boldsymbol{E}+\tilde{\boldsymbol{E}}] = \boldsymbol{E}\cdot\mathbf{C}[\boldsymbol{E}] + \tilde{\boldsymbol{E}}\cdot\mathbf{C}[\tilde{\boldsymbol{E}}] + 2\boldsymbol{E}\cdot\mathbf{C}[\tilde{\boldsymbol{E}}],$$

which implies the desired result. □

By an ***admissible state*** we mean an ordered array $\mathscr{s} = [\boldsymbol{u}, \boldsymbol{E}, \boldsymbol{S}]$ with the following properties:

(i) $\boldsymbol{u}$ is an admissible displacement field;[1]

(ii) $\boldsymbol{E}$ is a continuous symmetric tensor field on $\bar{B}$;

(iii) $\boldsymbol{S}$ is an admissible stress field.[1]

[1] See p. 59.

Note that the fields $\boldsymbol{u}, \boldsymbol{E}$, and $\boldsymbol{S}$ need not be related. Clearly, the set of all admissible states is a vector space provided we define addition and scalar multiplication in the natural manner:

$$[\boldsymbol{u}, \boldsymbol{E}, \boldsymbol{S}] + [\tilde{\boldsymbol{u}}, \tilde{\boldsymbol{E}}, \tilde{\boldsymbol{S}}] = [\boldsymbol{u}+\tilde{\boldsymbol{u}}, \boldsymbol{E}+\tilde{\boldsymbol{E}}, \boldsymbol{S}+\tilde{\boldsymbol{S}}],$$

$$\alpha[\boldsymbol{u}, \boldsymbol{E}, \boldsymbol{S}] = [\alpha\boldsymbol{u}, \alpha\boldsymbol{E}, \alpha\boldsymbol{S}].$$

We say that $\mathfrak{s} = [\boldsymbol{u}, \boldsymbol{E}, \boldsymbol{S}]$ is an ***elastic state***[1] (on $\bar{B}$) corresponding to the body force field $\boldsymbol{b}$ if $\mathfrak{s}$ is an admissible state and

$$\boldsymbol{E} = \tfrac{1}{2}(\nabla\boldsymbol{u} + \nabla\boldsymbol{u}^T),$$
$$\boldsymbol{S} = \mathbf{C}[\boldsymbol{E}],$$
$$\operatorname{div}\boldsymbol{S} + \boldsymbol{b} = \mathbf{0}.$$

The corresponding ***surface traction*** $\boldsymbol{s}$ is then defined at every regular point of ∂B by

$$\boldsymbol{s}(\boldsymbol{x}) = \boldsymbol{S}(\boldsymbol{x})\,\boldsymbol{n}(\boldsymbol{x}),$$

where $\boldsymbol{n}(\boldsymbol{x})$ is the outward unit normal to ∂B at $\boldsymbol{x}$. We call the pair $[\boldsymbol{b}, \boldsymbol{s}]$ the ***external force system*** for $\mathfrak{s}$.

When we discuss the basic singular solutions of elastostatics, we will frequently deal with elastic states whose domain of definition is a set of the form $D = \bar{B} - F$, where F is a *finite* subset of $\bar{B}$. We say that $\mathfrak{s} = [\boldsymbol{u}, \boldsymbol{E}, \boldsymbol{S}]$ is an elastic state on D corresponding to $\boldsymbol{b}$ if $\boldsymbol{u}, \boldsymbol{E}, \boldsymbol{S}$, and $\boldsymbol{b}$ are functions on D and given any closed regular subregion $P \subset D$, the restriction of $\mathfrak{s}$ to P is an elastic state on P corresponding to the restriction of $\boldsymbol{b}$ to P.

When we omit mention of the domain of definition of an elastic state, it will always be understood to be $\bar{B}$.

(2) Principle of superposition for elastic states. *If* $[\boldsymbol{u}, \boldsymbol{E}, \boldsymbol{S}]$ *and* $[\tilde{\boldsymbol{u}}, \tilde{\boldsymbol{E}}, \tilde{\boldsymbol{S}}]$ *are elastic states corresponding to the external force systems* $[\boldsymbol{b}, \boldsymbol{s}]$ *and* $[\tilde{\boldsymbol{b}}, \tilde{\boldsymbol{s}}]$, *respectively, and if* α *and* τ *are scalars, then* $\alpha[\boldsymbol{u}, \boldsymbol{E}, \boldsymbol{S}] + \tau[\tilde{\boldsymbol{u}}, \tilde{\boldsymbol{E}}, \tilde{\boldsymbol{S}}]$ *is an elastic state corresponding to the external force system* $\alpha[\boldsymbol{b}, \boldsymbol{s}] + \tau[\tilde{\boldsymbol{b}}, \tilde{\boldsymbol{s}}]$, *where*

$$\alpha[\boldsymbol{b}, \boldsymbol{s}] + \tau[\tilde{\boldsymbol{b}}, \tilde{\boldsymbol{s}}] = [\alpha\boldsymbol{b} + \tau\tilde{\boldsymbol{b}}, \alpha\boldsymbol{s} + \tau\tilde{\boldsymbol{s}}].$$

A direct corollary of ***(18.2)*** is the

(3) Theorem of work and energy.[2] *Let* $[\boldsymbol{u}, \boldsymbol{E}, \boldsymbol{S}]$ *be an elastic state corresponding to the external force system* $[\boldsymbol{b}, \boldsymbol{s}]$. *Then*

$$\int_{\partial B} \boldsymbol{s}\cdot\boldsymbol{u}\,da + \int_B \boldsymbol{b}\cdot\boldsymbol{u}\,dv = 2\,U\{\boldsymbol{E}\}.$$

The quantity on the left-hand side of this equation is the work done by the external forces; ***(3)*** asserts that this work is equal to twice the strain energy.

(4) Theorem of positive work. *Let the elasticity field be positive definite. Then given any elastic state, the work done by the external forces is non-negative and vanishes only when the displacement field is rigid.*

Proof. If $\mathbf{C}$ is positive definite, then $U\{\boldsymbol{E}\} \geqq 0$, and the work done by the external forces is non-negative. If $U\{\boldsymbol{E}\} = 0$, then $\boldsymbol{E}\cdot\mathbf{C}[\boldsymbol{E}]$ must vanish on B. Since $\mathbf{C}$ is positive definite, this implies $\boldsymbol{E} \equiv \mathbf{0}$; hence by ***(13.1)*** the corresponding displacement field is rigid. □

[1] This notion is due to STERNBERG and EUBANKS [1955, *13*].

[2] LAMÉ [1852, *2*].

The following direct corollary to ***(4)*** will be used in Sect. 32 to establish an important uniqueness theorem for elastostatics.

(5) *Let the elasticity field be positive definite. Further, let* $[\boldsymbol{u}, \boldsymbol{E}, \boldsymbol{S}]$ *be an elastic state corresponding to vanishing body forces, and suppose that the surface traction* $\boldsymbol{s}$ *satisfies*

$$\boldsymbol{s} \cdot \boldsymbol{u} = 0 \quad \text{on } \partial B.$$

Then $\boldsymbol{u}$ *is a rigid displacement field and* $\boldsymbol{E} \equiv \boldsymbol{S} \equiv \boldsymbol{0}$.

The next theorem, which is extremely important, shows that the L^2 norm of the displacement gradient is bounded by the strain energy.

(6) Alternative form of Korn's inequality.[1] *Assume that* C *is symmetric, positive definite, and continuous on* $\bar{B}$. *Let* $\boldsymbol{u}$ *be a class* C^2 *displacement field on* $\bar{B}$, *and assume that either* (α) *or* (β) *of* ***(13.9)*** *holds. Then*

$$\int_B |\nabla \boldsymbol{u}|^2 \, dv \leqq K_0 \, U\{\boldsymbol{E}\},$$

where $\boldsymbol{E}$ *is the associated strain field and* K_0 *is a constant depending only on* C *and* B.

Proof. Let $\mu_m(\boldsymbol{x})$ denote the minimum elastic modulus for $\mathsf{C}_{\boldsymbol{x}}$.[2] By hypothesis, $\mu_m(\boldsymbol{x})$ is bounded from below on $\bar{B}$. Hence

$$\tau = \sup\left\{\frac{1}{\mu_m(\boldsymbol{x})} : \boldsymbol{x} \in B\right\} < \infty,$$

and we conclude from ***(24.6)*** that

$$|\boldsymbol{E}|^2 \leqq \tau \boldsymbol{E} \cdot \mathsf{C}_{\boldsymbol{x}}[\boldsymbol{E}]$$

for every $\boldsymbol{x} \in B$ and every symmetric tensor $\boldsymbol{E}$. This inequality and ***(13.9)*** yield the desired result. □

II. The reciprocal theorem. Mean strain theorems.

29. Mean strain and mean stress theorems. Volume change. In this section we establish some results concerning the mean values

$$\bar{\boldsymbol{S}}(B) = \frac{1}{v(B)} \int_B \boldsymbol{S} \, dv, \quad \bar{\boldsymbol{E}}(B) = \frac{1}{v(B)} \int_B \boldsymbol{E} \, dv$$

of the stress and strain fields when the body B is *homogeneous.*

(1) Second mean stress theorem. *The mean stress* $\bar{\boldsymbol{S}}$ *corresponding to an elastic state* $[\boldsymbol{u}, \boldsymbol{E}, \boldsymbol{S}]$ *depends only on the boundary values of* $\boldsymbol{u}$ *and is given by*

$$\bar{\boldsymbol{S}}(B) = \frac{1}{v(B)} \mathsf{C}\left[\int_{\partial B} \boldsymbol{u} \otimes \boldsymbol{n} \, da\right].$$

Thus for an isotropic body,

$$\bar{\boldsymbol{S}}(B) = \frac{1}{v(B)} \int_{\partial B} \{\mu(\boldsymbol{u} \otimes \boldsymbol{n} + \boldsymbol{n} \otimes \boldsymbol{u}) + \lambda(\boldsymbol{u} \cdot \boldsymbol{n}) \, \boldsymbol{1}\} \, da.$$

Proof. Since B is homogeneous, C is independent of $\boldsymbol{x}$ and

$$\bar{\boldsymbol{S}}(B) = \frac{1}{v(B)} \int_B \mathsf{C}[\boldsymbol{E}] \, dv = \mathsf{C}\left[\frac{1}{v(B)} \int_B \boldsymbol{E} \, dv\right] = \mathsf{C}[\bar{\boldsymbol{E}}(B)].$$

[1] See footnote 1 on p. 38. See also HLAVÁČEK and NEČAS [1970, *2*].

[2] See p. 85.

The first of the relations in *(1)* is an immediate consequence of this result, the mean strain theorem *(13.8)*, and (i) of *(20.1)*, while the second follows from *(22.2)*. □

The above theorem shows that the mean stress is zero when the surface displacements vanish, independent of the value of the body force field.

(2) Second mean strain theorem.[1] *Suppose that the elasticity tensor* C *is invertible with* $\mathsf{K}=\mathsf{C}^{-1}$. *Let* $[\boldsymbol{u}, \boldsymbol{E}, \boldsymbol{S}]$ *be an elastic state corresponding to the external force system* $[\boldsymbol{b}, \boldsymbol{s}]$. *Then the mean strain depends only on the external force system and is given by*

$$\overline{\boldsymbol{E}}(B)=\frac{1}{v(B)}\mathsf{K}\Big[\int_{\partial B}\boldsymbol{p}\otimes\boldsymbol{s}\,da+\int_{B}\boldsymbol{p}\otimes\boldsymbol{b}\,dv\Big].$$

Thus for an isotropic body,

$$\overline{\boldsymbol{E}}(B)=\frac{1}{v(B)}\left\{\frac{1}{2\mu}\Big(\int_{\partial B}\boldsymbol{p}\otimes\boldsymbol{s}\,da+\int_{B}\boldsymbol{p}\otimes\boldsymbol{b}\,dv\Big)\right.$$
$$\left.-\frac{\lambda}{2\mu(2\mu+3\lambda)}\Big(\int_{\partial B}\boldsymbol{p}\cdot\boldsymbol{s}\,da+\int_{B}\boldsymbol{p}\cdot\boldsymbol{b}\,dv\Big)\boldsymbol{1}\right\}.$$

Proof. The first result follows from the mean stress theorem *(18.5)*, since

$$\overline{\boldsymbol{E}}(B)=\mathsf{K}[\overline{\boldsymbol{S}}(B)];$$

the second result follows from *(22.4)*. □

Note that balance of forces

$$\int_{\partial B}\boldsymbol{s}\,da+\int_{B}\boldsymbol{b}\,dv=\boldsymbol{0}$$

insures that the above formulae are independent of the choice of the origin $\boldsymbol{0}$, while balance of moments in the form

$$\operatorname{skw}\Big[\int_{\partial B}\boldsymbol{p}\otimes\boldsymbol{s}\,da+\int_{B}\boldsymbol{p}\otimes\boldsymbol{b}\,dv\Big]=\boldsymbol{0}$$

insures that the right-hand side of each of the expressions in *(2)* is symmetric.

(3) Volume change theorem.[2] *Suppose that the elasticity tensor* C *is symmetric and invertible. Let* $[\boldsymbol{u}, \boldsymbol{E}, \boldsymbol{S}]$ *be an elastic state corresponding to the external force system* $[\boldsymbol{b}, \boldsymbol{s}]$. *Then the associated volume change* $\delta v(B)$ *depends only on the external force system and is given by*

$$\delta v(B)=\mathsf{K}[\boldsymbol{1}]\cdot\Big\{\int_{\partial B}\boldsymbol{p}\otimes\boldsymbol{s}\,da+\int_{B}\boldsymbol{p}\otimes\boldsymbol{b}\,dv\Big\},$$

where $\mathsf{K}=\mathsf{C}^{-1}$ *is the compliance tensor. If, in addition, the body is isotropic, then*

$$\delta v(B)=\frac{1}{3k}\Big\{\int_{\partial B}\boldsymbol{p}\cdot\boldsymbol{s}\,da+\int_{B}\boldsymbol{p}\cdot\boldsymbol{b}\,dv\Big\},$$

where k *is the modulus of compression.*

[1] BETTI [1872, *1*], § 6 for the case in which $\boldsymbol{b}=\boldsymbol{0}$. The terms involving the body force were added by CHREE [1892, *2*], who claimed that his results were derived independently of BETTI's. See also LOVE [1927, *3*], § 123; BLAND [1953, *5*].

[2] BETTI [1872, *1*], § 6; CHREE [1892, *2*].

Proof. Let

$$\boldsymbol{D} = \int_{\partial B} \boldsymbol{p} \otimes \boldsymbol{s}\, da + \int_{B} \boldsymbol{p} \otimes \boldsymbol{b}\, dv.$$

Given any tensor $\boldsymbol{A}$,

$$\operatorname{tr} \boldsymbol{A} = \mathbf{1} \cdot \boldsymbol{A}.$$

Thus the remark made in the paragraph preceding *(13.8)*, *(2)*, and the symmetry of K imply that

$$\delta v(B) = v(B) \operatorname{tr} \overline{\boldsymbol{E}}(B) = \operatorname{tr} \mathsf{K}[\boldsymbol{D}] = \mathbf{1} \cdot \mathsf{K}[\boldsymbol{D}] = \mathsf{K}[\mathbf{1}] \cdot \boldsymbol{D},$$

which is the first result. For an isotropic material, *(22.4)* and Table 3 on p. 79 imply that

$$\mathsf{K}[\mathbf{1}] = \frac{1}{3k} \mathbf{1};$$

hence the second formula for $\delta v(B)$ follows from the first and the identity

$$\mathbf{1} \cdot (\boldsymbol{a} \otimes \boldsymbol{c}) = \operatorname{tr}(\boldsymbol{a} \otimes \boldsymbol{c}) = \boldsymbol{a} \cdot \boldsymbol{c}. \quad \square$$

Assume that the body is homogeneous and isotropic. Then, as is clear from *(3)*, a necessary and sufficient condition that there be no change in volume is that

$$\int_{\partial B} \boldsymbol{p} \cdot \boldsymbol{s}\, da + \int_{B} \boldsymbol{p} \cdot \boldsymbol{b}\, dv = 0.$$

Thus a *solid sphere or a spherical shell under the action of surface tractions that are tangential to the boundary will not undergo a change in volume.*

Recall from Sect. 18 that the force system $[\boldsymbol{b}, \boldsymbol{s}]$ is in astatic equilibrium if

$$\int_{\partial B} \boldsymbol{Q}\boldsymbol{s}\, da + \int_{B} \boldsymbol{Q}\boldsymbol{b}\, dv = \mathbf{0},$$

$$\int_{\partial B} \boldsymbol{p} \times \boldsymbol{Q}\boldsymbol{s}\, da + \int_{B} \boldsymbol{p} \times \boldsymbol{Q}\boldsymbol{b}\, dv = \mathbf{0},$$

for every orthogonal tensor $\boldsymbol{Q}$. We have the following interesting consequence of *(2)* and *(18.10)*.

(4)[1] *Assume that the elasticity tensor is invertible. Then a necessary and sufficient condition that the mean strain corresponding to an elastic state vanish is that the external force system be in astatic equilibrium.*

It follows as a corollary of *(4)* that *the volume change vanishes when the external force system is in astatic equilibrium.*

30. The reciprocal theorem. Betti's reciprocal theorem is one of the major results of linear elastostatics. In essence, it expresses the fact that the underlying system of field equations is self-adjoint. Further, it is essential in establishing integral representation theorems for elastostatics. In this section we will establish Betti's theorem and discuss some of its consequences. We assume throughout that C is *smooth* on $\overline{B}$.

(1) Betti's reciprocal theorem.[2] *Suppose that the elasticity field* C *is symmetric. Let* $[\boldsymbol{u}, \boldsymbol{E}, \boldsymbol{S}]$ and $[\tilde{\boldsymbol{u}}, \tilde{\boldsymbol{E}}, \tilde{\boldsymbol{S}}]$ *be elastic states corresponding to external force systems* $[\boldsymbol{b}, \boldsymbol{s}]$ *and* $[\tilde{\boldsymbol{b}}, \tilde{\boldsymbol{s}}]$, *respectively. Then*

$$\int_{\partial B} \boldsymbol{s} \cdot \tilde{\boldsymbol{u}}\, da + \int_{B} \boldsymbol{b} \cdot \tilde{\boldsymbol{u}}\, dv = \int_{\partial B} \tilde{\boldsymbol{s}} \cdot \boldsymbol{u}\, da + \int_{B} \tilde{\boldsymbol{b}} \cdot \boldsymbol{u}\, dv = \int_{B} \boldsymbol{S} \cdot \tilde{\boldsymbol{E}}\, dv = \int_{B} \tilde{\boldsymbol{S}} \cdot \boldsymbol{E}\, dv.$$

[1] BERG [1969, *1*].

[2] BETTI [1872, *1*], § 6, [1874, *1*]. See also LÉVY [1888, *3*]. An extension of Betti's theorem to include the possibility of dislocations was established by INDENBOM [1960, *8*]. For homogeneous and isotropic bodies, an interesting generalization of Betti's theorem was given by KUPRADZE [1963, *17*, *18*], § I.1. See also BEATTY [1967, *2*].

Proof. Since C is symmetric, we conclude from the stress-strain relation that

$$\boldsymbol{S}\cdot\tilde{\boldsymbol{E}}=\mathsf{C}[\boldsymbol{E}]\cdot\tilde{\boldsymbol{E}}=\mathsf{C}[\tilde{\boldsymbol{E}}]\cdot\boldsymbol{E}=\tilde{\boldsymbol{S}}\cdot\boldsymbol{E}.$$

Further, it follows from the theorem of work expended *(18.2)* that

$$\int_{\partial B}\mathbf{s}\cdot\tilde{\boldsymbol{u}}\,da+\int_{B}\boldsymbol{b}\cdot\tilde{\boldsymbol{u}}\,dv=\int_{B}\boldsymbol{S}\cdot\tilde{\boldsymbol{E}}\,dv,$$

$$\int_{\partial B}\tilde{\mathbf{s}}\cdot\boldsymbol{u}\,da+\int_{B}\tilde{\boldsymbol{b}}\cdot\boldsymbol{u}\,dv=\int_{B}\tilde{\boldsymbol{S}}\cdot\boldsymbol{E}\,dv,$$

and the proof is complete. □

Betti's theorem asserts that given two elastic states, the work done by the external forces of the first over the displacements of the second equals the work done by the external forces of the second over the displacements of the first. The next theorem shows that the assumption that C be symmetric is necessary for the validity of Betti's theorem.

(2)[1] *Betti's reciprocal theorem is false if C is not symmetric.*

Proof. If C is not symmetric, then there exist symmetric tensors $\boldsymbol{E}$, $\tilde{\boldsymbol{E}}$ such that

$$\boldsymbol{E}\cdot\mathsf{C}[\tilde{\boldsymbol{E}}]\neq\tilde{\boldsymbol{E}}\cdot\mathsf{C}[\boldsymbol{E}].$$

Let

$$\boldsymbol{u}=\boldsymbol{E}\boldsymbol{p},\qquad \boldsymbol{S}=\mathsf{C}[\boldsymbol{E}],\qquad \boldsymbol{b}=-\operatorname{div}\boldsymbol{S},$$

$$\tilde{\boldsymbol{u}}=\tilde{\boldsymbol{E}}\boldsymbol{p},\qquad \tilde{\boldsymbol{S}}=\mathsf{C}[\tilde{\boldsymbol{E}}],\qquad \tilde{\boldsymbol{b}}=-\operatorname{div}\tilde{\boldsymbol{S}}.$$

Then $[\boldsymbol{u},\boldsymbol{E},\boldsymbol{S}]$ and $[\tilde{\boldsymbol{u}},\tilde{\boldsymbol{E}},\tilde{\boldsymbol{S}}]$ are elastic states corresponding to the body force fields $\boldsymbol{b}$ and $\tilde{\boldsymbol{b}}$ and

$$\int_{B}\boldsymbol{S}\cdot\tilde{\boldsymbol{E}}\,dv\neq\int_{B}\tilde{\boldsymbol{S}}\cdot\boldsymbol{E}\,dv.\quad □$$

The following elegant corollary of Betti's theorem is due to SHIELD and ANDERSON.[2]

(3) Suppose that C is symmetric and positive semi-definite. Let $[\boldsymbol{u},\boldsymbol{E},\boldsymbol{S}]$ and $[\tilde{\boldsymbol{u}},\tilde{\boldsymbol{E}},\tilde{\boldsymbol{S}}]$ *be elastic states corresponding to external force systems* $[\boldsymbol{b},\mathbf{s}]$ and $[\tilde{\boldsymbol{b}},\tilde{\mathbf{s}}]$, *respectively. Then*

$$U\{\boldsymbol{E}\}\leqq U\{\tilde{\boldsymbol{E}}\}$$

provided

$$\int_{\partial B}\mathbf{s}\cdot(\boldsymbol{u}-\tilde{\boldsymbol{u}})\,da+\int_{B}\boldsymbol{b}\cdot(\boldsymbol{u}-\tilde{\boldsymbol{u}})\,dv\leqq 0$$

or

$$\int_{\partial B}(\mathbf{s}-\tilde{\mathbf{s}})\cdot\boldsymbol{u}\,da+\int_{B}(\boldsymbol{b}-\tilde{\boldsymbol{b}})\cdot\boldsymbol{u}\,dv\leqq 0.$$

Thus, if $\mathscr{S}_1$ and $\mathscr{S}_2$ are complementary subsets of ∂B,

$$\left.\begin{array}{lll}\boldsymbol{u}=\tilde{\boldsymbol{u}} & \text{on} & \mathscr{S}_1\\ \mathbf{s}=\mathbf{0} & \text{on} & \mathscr{S}_2\\ \boldsymbol{b}=\mathbf{0} & \text{on} & B\end{array}\right\}\Rightarrow U\{\boldsymbol{E}\}\leqq U\{\boldsymbol{E}\},$$

$$\left.\begin{array}{lll}\boldsymbol{u}=\mathbf{0} & \text{on} & \mathscr{S}_1\\ \mathbf{s}=\tilde{\mathbf{s}} & \text{on} & \mathscr{S}_2\\ \boldsymbol{b}=\tilde{\boldsymbol{b}} & \text{on} & B\end{array}\right\}\Rightarrow U\{\boldsymbol{E}\}\leqq U\{\tilde{\boldsymbol{E}}\}.$$

---

[1] TRUESDELL [1963, *24*]. See also BURGATTI [1931, *1*], p. 152.
[2] [1966, *23*].

Proof. By the principle of superposition *(28.2)*, $[\tilde{\boldsymbol{u}}-\boldsymbol{u}, \tilde{\boldsymbol{E}}-\boldsymbol{E}, \tilde{\boldsymbol{S}}-\boldsymbol{S}]$ is an elastic state corresponding to the external forces $[\tilde{\boldsymbol{b}}-\boldsymbol{b}, \tilde{\mathbf{s}}-\mathbf{s}]$. By *(28.1)* with $\tilde{\boldsymbol{E}}$ replaced by $\tilde{\boldsymbol{E}}-\boldsymbol{E}$,

$$U\{\tilde{\boldsymbol{E}}\}=U\{\boldsymbol{E}\}+U\{\tilde{\boldsymbol{E}}-\boldsymbol{E}\}+\int_B \boldsymbol{E}\cdot(\tilde{\boldsymbol{S}}-\boldsymbol{S})\,dv,$$

and by Betti's theorem,

$$\begin{aligned}\int_B \boldsymbol{E}\cdot(\tilde{\boldsymbol{S}}-\boldsymbol{S})\,dv &= \int_B (\tilde{\boldsymbol{b}}-\boldsymbol{b})\cdot\boldsymbol{u}\,dv+\int_{\partial B}(\tilde{\mathbf{s}}-\mathbf{s})\cdot\boldsymbol{u}\,da\\ &=\int_B \boldsymbol{b}\cdot(\tilde{\boldsymbol{u}}-\boldsymbol{u})\,dv+\int_{\partial B}\mathbf{s}\cdot(\tilde{\boldsymbol{u}}-\boldsymbol{u})\,da.\end{aligned}$$

Thus, since $U\{\tilde{\boldsymbol{E}}-\boldsymbol{E}\}\geqq 0$, we have the inequalities

$$U\{\boldsymbol{E}\}\leqq U\{\tilde{\boldsymbol{E}}\}+\int_B(\boldsymbol{b}-\tilde{\boldsymbol{b}})\cdot\boldsymbol{u}\,dv+\int_{\partial B}(\mathbf{s}-\tilde{\mathbf{s}})\cdot\boldsymbol{u}\,da,$$

$$U\{\boldsymbol{E}\}\leqq U\{\tilde{\boldsymbol{E}}\}+\int_B\boldsymbol{b}\cdot(\boldsymbol{u}-\tilde{\boldsymbol{u}})\,dv+\int_{\partial B}\mathbf{s}\cdot(\boldsymbol{u}-\tilde{\boldsymbol{u}})\,da.$$

The remainder of the proof follows from these inequalities. □

Theorem *(3)* may be called a least work principle, since by *(28.3)* $U\{\boldsymbol{E}\}\leqq U\{\tilde{\boldsymbol{E}}\}$ if and only if the work done by the external forces corresponding to $[\boldsymbol{u}, \boldsymbol{E}, \boldsymbol{S}]$ is less than or equal to the work done by the external forces of $[\tilde{\boldsymbol{u}}, \tilde{\boldsymbol{E}}, \tilde{\boldsymbol{S}}]$.

(4)[1] *Suppose that* C *is symmetric and positive semi-definite. Let* $[\boldsymbol{u}, \boldsymbol{E}, \boldsymbol{S}]$ *and* $[\tilde{\boldsymbol{u}}, \tilde{\boldsymbol{E}}, \tilde{\boldsymbol{S}}]$ *be elastic states corresponding to external force systems* $[\boldsymbol{b}, \mathbf{s}]$ *and* $[\tilde{\boldsymbol{b}}, \tilde{\mathbf{s}}]$, *respectively. Let* $\mathscr{S}_0, \mathscr{S}_1, \ldots, \mathscr{S}_N$ *be complementary regular subsurfaces of* ∂B, *and suppose that for each* $n=1, 2, \ldots, N$, *the restriction of* $\boldsymbol{u}$ *to* $\mathscr{S}_n$ *is a rigid displacement of* $\mathscr{S}_n$. *Then*

$$\left.\begin{array}{c}\mathbf{s}=\tilde{\mathbf{s}}\quad on\ \mathscr{S}_0\\ \int_{\mathscr{S}_n}(\mathbf{s}-\tilde{\mathbf{s}})\,da=\int_{\mathscr{S}_n}\mathbf{p}\times(\mathbf{s}-\tilde{\mathbf{s}})\,da=0\\ n=1, 2, \ldots, N\\ \boldsymbol{b}=\tilde{\boldsymbol{b}}\quad on\ B\end{array}\right\}\Rightarrow U\{\boldsymbol{E}\}\leqq U\{\tilde{\boldsymbol{E}}\}.$$

Proof. By *(3)* it is sufficient to show that

$$I=\int_{\partial B}(\mathbf{s}-\tilde{\mathbf{s}})\cdot\boldsymbol{u}\,da+\int_B(\boldsymbol{b}-\tilde{\boldsymbol{b}})\cdot\boldsymbol{u}\,dv=0.$$

Since $\mathbf{s}=\tilde{\mathbf{s}}$ on $\mathscr{S}_0$ and $\boldsymbol{b}=\tilde{\boldsymbol{b}}$ on B,

$$I=\sum_{n=1}^{N} I_n,\qquad I_n=\int_{\mathscr{S}_n}(\mathbf{s}-\tilde{\mathbf{s}})\cdot\boldsymbol{u}\,da.$$

Further, as the restriction of $\boldsymbol{u}$ to each $\mathscr{S}_n$, $n=1, 2, \ldots, N$, is rigid, there exist vectors $\boldsymbol{v}_1, \boldsymbol{v}_2, \ldots, \boldsymbol{v}_n$ and skew tensors $\boldsymbol{W}_1, \boldsymbol{W}_2, \ldots, \boldsymbol{W}_N$ such that

$$\boldsymbol{u}=\boldsymbol{W}_n\mathbf{p}+\boldsymbol{v}_n\qquad\text{on } \mathscr{S}_n.$$

Thus

$$I_n=\int_{\mathscr{S}_n}(\mathbf{s}-\tilde{\mathbf{s}})\cdot\boldsymbol{v}_n\,da+\int_{\mathscr{S}_n}(\mathbf{s}-\tilde{\mathbf{s}})\cdot\boldsymbol{W}_n\mathbf{p}\,da,$$

[1] SHIELD and ANDERSON [1966, 23].

and since $\boldsymbol{v}_n$ and $\boldsymbol{W}_n$ are constant and $(\mathbf{s}-\tilde{\mathbf{s}})\cdot\boldsymbol{W}_n\boldsymbol{p}=\boldsymbol{W}_n\cdot[(\mathbf{s}-\tilde{\mathbf{s}})\otimes\boldsymbol{p}]$, this expression reduces to

$$I_n=\boldsymbol{v}_n\cdot\int_{\mathscr{S}_n}(\mathbf{s}-\tilde{\mathbf{s}})\,da+\boldsymbol{W}_n\cdot\int_{\mathscr{S}_n}(\mathbf{s}-\tilde{\mathbf{s}})\otimes\boldsymbol{p}\,da. \tag{a}$$

But

$$\int_{\mathscr{S}_n}(\mathbf{s}-\tilde{\mathbf{s}})\,da=\int_{\mathscr{S}_n}\boldsymbol{p}\times(\mathbf{s}-\tilde{\mathbf{s}})\,da=\mathbf{0}. \tag{b}$$

The second relation in (b) implies that the tensor

$$\boldsymbol{A}=\int_{\mathscr{S}_n}(\mathbf{s}-\tilde{\mathbf{s}})\otimes\boldsymbol{p}\,da$$

is symmetric. Thus, since $\boldsymbol{W}_n$ is skew, $\boldsymbol{W}_n\cdot\boldsymbol{A}=0$, and it follows from (a) and the first of (b) that $I_n=0$ for each n. Therefore $I=0$. □

Theorem *(29.3)* can also be established using Betti's theorem. Indeed, the volume change $\delta v(B)$ is given by

$$\delta v(B)=\int_B \operatorname{tr}\boldsymbol{E}\,dv.$$

If we assume that B is homogeneous and define

$$\tilde{\boldsymbol{S}}=\mathbf{1},$$
$$\tilde{\boldsymbol{E}}=\mathsf{K}[\mathbf{1}],$$
$$\tilde{\boldsymbol{u}}=\tilde{\boldsymbol{E}}\boldsymbol{p},$$

then $[\tilde{\boldsymbol{u}},\tilde{\boldsymbol{E}},\tilde{\boldsymbol{S}}]$ is an elastic state corresponding to vanishing body forces, and we conclude from Betti's theorem *(1)* that

$$\int_B\tilde{\boldsymbol{S}}\cdot\boldsymbol{E}\,dv=\int_{\partial B}\mathbf{s}\cdot\tilde{\boldsymbol{u}}\,da+\int_B\boldsymbol{b}\cdot\tilde{\boldsymbol{u}}\,dv,$$
$$=\int_{\partial B}\mathbf{s}\cdot\tilde{\boldsymbol{E}}\boldsymbol{p}\,da+\int_B\boldsymbol{b}\cdot\tilde{\boldsymbol{E}}\boldsymbol{p}\,dv.$$

Since

$$\tilde{\boldsymbol{S}}\cdot\boldsymbol{E}=\mathbf{1}\cdot\boldsymbol{E}=\operatorname{tr}\boldsymbol{E},$$
$$\boldsymbol{a}\cdot\tilde{\boldsymbol{E}}\boldsymbol{p}=\tilde{\boldsymbol{E}}\cdot(\boldsymbol{a}\otimes\boldsymbol{p})\quad\text{for any vector }\boldsymbol{a},$$

and $\tilde{\boldsymbol{E}}=\mathsf{K}[\mathbf{1}]$ is constant and symmetric, it follows that

$$\delta v(B)=\mathsf{K}[\mathbf{1}]\cdot\Big\{\int_{\partial B}\boldsymbol{p}\otimes\mathbf{s}\,da+\int_B\boldsymbol{p}\otimes\boldsymbol{b}\,dv\Big\}.$$

The following generalization of Betti's theorem holds even when C is *not* symmetric.

(5) Reciprocal theorem. *Let* $[\boldsymbol{u},\boldsymbol{E},\boldsymbol{S}]$ *and* $[\tilde{\boldsymbol{u}},\tilde{\boldsymbol{E}},\tilde{\boldsymbol{S}}]$ *be elastic states corresponding to external force systems* $[\boldsymbol{b},\mathbf{s}]$ *and* $[\tilde{\boldsymbol{b}},\tilde{\mathbf{s}}]$, *respectively, and let* $[\boldsymbol{u},\boldsymbol{E},\boldsymbol{S}]$ *correspond to the elasticity field* C, $[\tilde{\boldsymbol{u}},\tilde{\boldsymbol{E}},\tilde{\boldsymbol{S}}]$ *to the elasticity field* $\tilde{\mathsf{C}}=\mathsf{C}^T$. *Then*

$$\int_{\partial B}\mathbf{s}\cdot\tilde{\boldsymbol{u}}\,da+\int_B\boldsymbol{b}\cdot\tilde{\boldsymbol{u}}\,dv=\int_{\partial B}\tilde{\mathbf{s}}\cdot\boldsymbol{u}\,da+\int_B\tilde{\boldsymbol{b}}\cdot\boldsymbol{u}\,dv=\int_B\boldsymbol{S}\cdot\tilde{\boldsymbol{E}}\,dv=\int_B\tilde{\boldsymbol{S}}\cdot\boldsymbol{E}\,dv.$$

Proof. We will prove that $\boldsymbol{S}\cdot\tilde{\boldsymbol{E}}=\tilde{\boldsymbol{S}}\cdot\boldsymbol{E}$; the remainder of the proof is identical to the proof of Betti's theorem *(1)*. Since $\tilde{\mathsf{C}}=\mathsf{C}^T$,

$$\boldsymbol{S}\cdot\tilde{\boldsymbol{E}}=\mathsf{C}[\boldsymbol{E}]\cdot\tilde{\boldsymbol{E}}=\boldsymbol{E}\cdot\mathsf{C}^T[\tilde{\boldsymbol{E}}]=\boldsymbol{E}\cdot\tilde{\mathsf{C}}[\tilde{\boldsymbol{E}}]=\boldsymbol{E}\cdot\tilde{\boldsymbol{S}}.\quad\square$$

III. Boundary-value problems. Uniqueness.

31. The boundary-value problems of elastostatics. Throughout the following six sections we assume given an elasticity field $\mathbf{C}$ on B, body forces $\boldsymbol{b}$ on B, surface displacements $\hat{\boldsymbol{u}}$ on $\mathscr{S}_1$, and surface forces $\hat{\mathbf{s}}$ on $\mathscr{S}_2$, where $\mathscr{S}_1$ and $\mathscr{S}_2$ are complementary regular subsurfaces of ∂B. Given the above data, the *mixed problem of elastostatics*[1] is to find an elastic state $[\boldsymbol{u}, \boldsymbol{E}, \boldsymbol{S}]$ that corresponds to $\boldsymbol{b}$ and satisfies the ***displacement condition***

$$\boldsymbol{u} = \hat{\boldsymbol{u}} \quad \text{on } \mathscr{S}_1$$

and the ***traction condition***

$$\mathbf{s} = \boldsymbol{S}\boldsymbol{n} = \hat{\mathbf{s}} \quad \text{on } \mathscr{S}_2.$$

We will call such an elastic state a ***solution of the mixed problem***.[2] When $\mathscr{S}_2$ is empty, so that $\mathscr{S}_1 = \partial B$, the above boundary conditions reduce to

$$\boldsymbol{u} = \hat{\boldsymbol{u}} \quad \text{on } \partial B,$$

and the associated problem is called the ***displacement problem***.[3] If $\mathscr{S}_2 = \partial B$, the boundary conditions become

$$\mathbf{s} = \hat{\mathbf{s}} \quad \text{on } \partial B,$$

and we refer to the resulting problem as the ***traction problem***.[4]

To avoid repeated regularity assumptions we assume that:

(i) $\mathbf{C}$ is smooth on $\bar{B}$;

(ii) $\boldsymbol{b}$ is continuous on $\bar{B}$;

(iii) $\hat{\boldsymbol{u}}$ is continuous on $\mathscr{S}_1$;

(iv) $\hat{\mathbf{s}}$ is piecewise regular on $\mathscr{S}_2$.

---

[1] There are problems of importance in elastostatics not included in this formulation; e.g., the contact problem, studied in Sect. 40, in which the normal component of the displacement and the tangential component of the traction are prescribed over a portion of the boundary. Another example is the SIGNORINI problem [1959, *13*] in which a portion $\mathscr{S}_3$ of the boundary rests on a rigid, frictionless surface, but is allowed to separate from this surface. Thus $\boldsymbol{s}$ is perpendicular to $\mathscr{S}_3$ and at each point of $\mathscr{S}_3$ *either*

$$\boldsymbol{u}\cdot\boldsymbol{n} = 0, \qquad \boldsymbol{s}\cdot\boldsymbol{n} \leqq 0,$$

or

$$\boldsymbol{u}\cdot\boldsymbol{n} < 0, \qquad \boldsymbol{s}\cdot\boldsymbol{n} = 0.$$

A detailed treatment of this difficult problem is given by FICHERA [1963, *6*] and in an article following in this volume.

[2] Our definition of a solution, since it is based on the notion of an elastic state (§ 28), does not cover certain important problems. Indeed, as is well known, there exist genuine mixed problems ($\mathscr{S}_1 \cap \mathscr{S}_2 \neq \emptyset$) and problems involving bodies with reentrant corners for which the corresponding stress field is unbounded. For such problems the classical theorems of elastostatics, as presented here, are vacuous. FICHERA [1971, *1*], Theorem 12.IV has shown that in the case of the genuine mixed problem with C^∞ boundary, C^∞ body force field, C^∞ positive definite and symmetric elasticity field, and null boundary data the displacement field $\boldsymbol{u}$ belongs to $H^1(B)$ and is C^∞ on $B - \mathscr{S}_1 \cap \mathscr{S}_2$. [$H^1(B)$ is the completion of $C^1(\bar{B})$ with respect to the norm defined by $\|\boldsymbol{u}\|^2 = \|\boldsymbol{u}\|_2^2 + \|\nabla\boldsymbol{u}\|_2^2$, where $\|\cdot\|_2$ is the usual L^2 norm.] It is possible to establish minor variants of most of our results assuming only that $\boldsymbol{u} \in H^1(B)$, but this would require analytical machinery much less elementary than that used here. (In this connection see the article by FICHERA [1971, *1*].) Finally, FICHERA [1951, *6*], [1953, *11*] has shown that for the genuine mixed problem $\boldsymbol{u}$ will not be smooth on $\bar{B}$ unless the boundary data obey certain hypotheses.

[3] Sometimes called the first boundary-value problem.

[4] Sometimes called the second boundary-value problem.

Assumptions (ii)–(iv) are necessary for the existence of a solution to the mixed problem. In the definition of a solution $[\boldsymbol{u}, \boldsymbol{E}, \boldsymbol{S}]$, the requirement that $\boldsymbol{S}$ be admissible[1] is redundant; indeed, the required properties of $\boldsymbol{S}$ follow from (i), (ii), the admissibility of $\boldsymbol{u}$, and the field equations.

By a ***displacement field corresponding to a solution of the mixed problem*** we mean a vector field $\boldsymbol{u}$ with the property that there exist fields $\boldsymbol{E}$, $\boldsymbol{S}$ such that $[\boldsymbol{u}, \boldsymbol{E}, \boldsymbol{S}]$ is a solution of the mixed problem. We define a ***stress field corresponding to a solution of the mixed problem*** analogously.

(1) Characterization of the mixed problem in terms of displacements. *Let $\boldsymbol{u}$ be an admissible displacement field. Then $\boldsymbol{u}$ corresponds to a solution of the mixed problem if and only if*

$$\begin{aligned} \operatorname{div} \mathsf{C}[\nabla \boldsymbol{u}] + \boldsymbol{b} &= \boldsymbol{0} && \text{on } B, \\ \boldsymbol{u} &= \hat{\boldsymbol{u}} && \text{on } \mathscr{S}_1, \\ \mathsf{C}[\nabla \boldsymbol{u}]\, \boldsymbol{n} &= \hat{\mathbf{s}} && \text{on } \mathscr{S}_2. \end{aligned}$$

Further, when B is homogeneous and isotropic, these relations take the forms[2]

$$\begin{aligned} \mu \Delta \boldsymbol{u} + (\lambda + \mu)\, \nabla \operatorname{div} \boldsymbol{u} + \boldsymbol{b} &= \boldsymbol{0} && \text{on } B, \\ \boldsymbol{u} &= \hat{\boldsymbol{u}} && \text{on } \mathscr{S}_1, \\ \mu (\nabla \boldsymbol{u} + \nabla \boldsymbol{u}^T)\, \boldsymbol{n} + \lambda (\operatorname{div} \boldsymbol{u})\, \boldsymbol{n} &= \check{\mathbf{s}} && \text{on } \mathscr{S}_2. \end{aligned}$$

Proof. That the above relations are necessary follows from the discussion given at the beginning of Sect. 27. To establish sufficiency assume that $\boldsymbol{u}$ satisfies the above relations. Define $\boldsymbol{E}$ through the strain-displacement relation and $\boldsymbol{S}$ through the stress-strain relation. Then, since $\boldsymbol{u}$ is admissible, $\boldsymbol{E}$ is continuous on $\bar{B}$, and, by assumption (i), $\boldsymbol{S}$ is continuous on $\bar{B}$ and smooth on B. In addition,

$$\begin{aligned} \operatorname{div} \boldsymbol{S} + \boldsymbol{b} &= \boldsymbol{0} && \text{on } B, \\ \boldsymbol{S}\boldsymbol{n} &= \hat{\mathbf{s}} && \text{on } \mathscr{S}_2. \end{aligned} \tag{a}$$

Further, (a) and assumption (ii) imply that $\operatorname{div} \boldsymbol{S}$ is continuous on $\bar{B}$; thus $\boldsymbol{S}$ is an admissible stress field. Therefore $[\boldsymbol{u}, \boldsymbol{E}, \boldsymbol{S}]$ meets all the requirements of a solution to the mixed problem.

When B is homogeneous and isotropic, the last three relations are equivalent to the first three, as is clear from Sect. 27. □

In view of the discussion given at the end of Sect. 27, the boundary condition in *(1)* on $\mathscr{S}_2$ for isotropic bodies may be replaced by

$$2\mu \frac{\partial \boldsymbol{u}}{\partial \boldsymbol{n}} + \mu \boldsymbol{n} \times \operatorname{curl} \boldsymbol{u} + \lambda (\operatorname{div} \boldsymbol{u})\, \boldsymbol{n} = \hat{\mathbf{s}} \qquad \text{on } \mathscr{S}_2.$$

(2) Characterization of the traction problem in terms of stresses. *Suppose that the elasticity field is invertible with class C^2 inverse K on $\bar{B}$. Let B be simply-connected, and let $\boldsymbol{S}$ be an admissible stress field of class C^2 on $\bar{B}$. Then $\boldsymbol{S}$ corresponds to a solution of the traction problem if and only if*

$$\left.\begin{aligned} \operatorname{div} \boldsymbol{S} + \boldsymbol{b} &= \boldsymbol{0} \\ \operatorname{curl}\operatorname{curl} \mathsf{K}[\boldsymbol{S}] &= \boldsymbol{0} \end{aligned}\right\} \quad \text{on } B,$$

$$\boldsymbol{S}\boldsymbol{n} = \hat{\mathbf{s}} \qquad \text{on } \partial B.$$

[1] See p. 59.

[2] LODGE [1955, *9*] has given conditions under which the basic equations of elastostatics for an anisotropic material can be transformed into equations formally identical with those for an isotropic material. See also KACZKOWSKI [1955, *6*].

When B is homogeneous and isotropic, these relations are equivalent to

$$\left.\begin{aligned} \operatorname{div} \boldsymbol{S} + \boldsymbol{b} &= \boldsymbol{0} \\ \Delta \boldsymbol{S} + \frac{1}{1+\nu} \nabla\nabla(\operatorname{tr} \boldsymbol{S}) + 2\hat{\nabla}\boldsymbol{b} + \frac{\nu}{1-\nu} (\operatorname{div} \boldsymbol{b})\, \boldsymbol{1} &= \boldsymbol{0} \end{aligned}\right\} \quad \text{on } B,$$

$$\boldsymbol{S}\boldsymbol{n} = \hat{\mathbf{s}} \qquad \text{on } \partial B.$$

Proof. Let

$$\boldsymbol{E} = \mathbf{K}[\boldsymbol{S}].$$

Then $\boldsymbol{E}$ is of class C^2 on $\bar{B}$ and

$$\operatorname{curl} \operatorname{curl} \boldsymbol{E} = \boldsymbol{0}.$$

Thus, by the compatibility theorem ***(14.2)***, there exists a class C^3 displacement field $\boldsymbol{u}$ on B that satisfies the strain-displacement relation. Moreover, we conclude from ***(14.4)*** that $\boldsymbol{u}$ is of class C^3 on $\bar{B}$.

When B is homogeneous and isotropic, the last three relations are equivalent to the first three, as is clear from the remarks given at the end of Sect. 27. □

32. Uniqueness.[1] In this section we discuss the uniqueness question appropriate to the fundamental boundary-value problems of elastostatics.

We say that two solutions $[\boldsymbol{u}, \boldsymbol{E}, \boldsymbol{S}]$ and $[\tilde{\boldsymbol{u}}, \tilde{\boldsymbol{E}}, \tilde{\boldsymbol{S}}]$ of the mixed problem are ***equal modulo a rigid displacement*** provided

$$[\boldsymbol{u}, \boldsymbol{E}, \boldsymbol{S}] = [\tilde{\boldsymbol{u}} + \boldsymbol{w}, \tilde{\boldsymbol{E}}, \tilde{\boldsymbol{S}}],$$

where $\boldsymbol{w}$ is a rigid displacement field.

(1) Uniqueness theorem for the mixed problem.[2] *Suppose that the elasticity field is positive definite. Then any two solutions of the mixed problem are equal modulo a rigid displacement. Moreover, if $\mathscr{S}_1$ is non-empty, then the mixed problem has at most one solution.*

Proof. Let $\mathfrak{s} = [\boldsymbol{u}, \boldsymbol{E}, \boldsymbol{S}]$ and $\tilde{\mathfrak{s}} = [\tilde{\boldsymbol{u}}, \tilde{\boldsymbol{E}}, \tilde{\boldsymbol{S}}]$ be solutions of the mixed problem, and let

$$[\bar{\boldsymbol{u}}, \bar{\boldsymbol{E}}, \bar{\boldsymbol{S}}] = [, \boldsymbol{E}, \boldsymbol{S}] - [\tilde{\boldsymbol{u}}, \tilde{\boldsymbol{E}}, \tilde{\boldsymbol{S}}].$$

Then, by the principle of superposition ***(28.2)***, $[\bar{\boldsymbol{u}}, \bar{\boldsymbol{E}}, \bar{\boldsymbol{S}}]$ is an elastic state corresponding to zero body forces. Moreover,

$$\bar{\boldsymbol{u}} = \boldsymbol{0} \quad \text{on } \mathscr{S}_1,$$

$$\bar{\mathbf{s}} = \boldsymbol{0} \quad \text{on } \mathscr{S}_2;$$

thus, since $\mathscr{S}_1$ and $\mathscr{S}_2$ are complementary subsets of ∂B,

$$\bar{\mathbf{s}} \cdot \bar{\boldsymbol{u}} = 0 \quad \text{on } \partial B,$$

and we conclude from ***(28.5)*** that $\bar{\boldsymbol{u}}$ is rigid. Thus $\bar{\boldsymbol{E}} = \bar{\boldsymbol{S}} = \boldsymbol{0}$, and the two solutions $\mathfrak{s}$ and $\tilde{\mathfrak{s}}$ are equal modulo a rigid displacement. $\mathscr{S}_1$, since it is a regular surface, if non-empty must contain at least three non-collinear points. Thus, in this instance, since $\bar{\boldsymbol{u}} = \boldsymbol{0}$ on $\mathscr{S}_1$, $\bar{\boldsymbol{u}}$ must vanish identically. □

It is clear from the proof of ***(1)*** that the smoothness assumptions tacit in the definition of a solution are more stringent than required. For the purpose of this theorem it is sufficient to assume that a solution $[\boldsymbol{u}, \boldsymbol{E}, \boldsymbol{S}]$ is a triplet of fields

[1] This subject is discussed at great length in the tract by KNOPS and PAYNE [1971, *2*].

[2] KIRCHHOFF [1859, *1*], [1876, *1*]. See also CLEBSCH [1862, *1*].

on $\bar{B}$ with the following properties: (i) $\boldsymbol{u}$, $\boldsymbol{E}$, and $\boldsymbol{S}$ are continuous on $\bar{B}$; (ii) $\boldsymbol{u}$ and $\boldsymbol{S}$ are differentiable on B; (iii) $\boldsymbol{u}$, $\boldsymbol{E}$, and $\boldsymbol{S}$ satisfy the field equations and boundary conditions.

The last theorem implies that *uniqueness holds for the displacement and traction problems provided the elasticity field is positive definite.* When B is homogeneous, we can prove a much stronger uniqueness theorem for the displacement problem.

(2) Uniqueness theorem for the displacement problem.[1] *Let B be homogeneous, let $\mathbf{C}$ be symmetric, and assume that either $\mathbf{C}$ or $-\mathbf{C}$ is strongly elliptic. Then the displacement problem has at most one solution with smooth displacements on $\bar{B}$.*

Proof. By the principle of superposition, it suffices to show that $\boldsymbol{b} \equiv \mathbf{0}$ on B and

$$\boldsymbol{u} = \mathbf{0} \quad \text{on } \partial B \tag{a}$$

imply $\boldsymbol{u} \equiv \mathbf{0}$ on B. Thus assume that the body forces and surface displacements are zero. Then by ***(28.3)***

$$2U\{\boldsymbol{E}\} = \int_B \boldsymbol{E} \cdot \mathbf{C}[\boldsymbol{E}]\, dv = 0, \tag{b}$$

and, since $\boldsymbol{E} = \hat{\nabla}\boldsymbol{u}$, (ii) of ***(20.1)*** implies

$$2U\{\boldsymbol{E}\} = \int_B \nabla\boldsymbol{u} \cdot \mathbf{C}[\nabla\boldsymbol{u}]\, dv. \tag{c}$$

Let us extend the definition of $\boldsymbol{u}$ from $\bar{B}$ to all of $\mathscr{E}$ by defining

$$\boldsymbol{u} = \mathbf{0} \quad \text{on } \mathscr{E} - \bar{B}. \tag{d}$$

Then by (a) $\boldsymbol{u}$ is continuous and piecewise smooth on $\mathscr{E}$, and $\nabla\boldsymbol{u}$ has discontinuities only on ∂B. Thus $\boldsymbol{u}$ and $\nabla\boldsymbol{u}$ are both absolutely and square integrable over $\mathscr{E}$ and possess the three-dimensional Fourier transforms $\boldsymbol{a}$ and $\boldsymbol{A}$ defined on $\mathscr{E}$ by

$$\boldsymbol{a}(\boldsymbol{x}) = \left(\frac{1}{2\pi}\right)^{\frac{3}{2}} \int_{\mathscr{E}} e^{i\boldsymbol{p}(\boldsymbol{x})\cdot\boldsymbol{p}(\boldsymbol{y})}\, \boldsymbol{u}(\boldsymbol{y})\, dv_{\boldsymbol{y}}, \tag{e}$$

$$\boldsymbol{A}(\boldsymbol{x}) = \left(\frac{1}{2\pi}\right)^{\frac{3}{2}} \int_{\mathscr{E}} e^{i\boldsymbol{p}(\boldsymbol{x})\cdot\boldsymbol{p}(\boldsymbol{y})}\, \nabla\boldsymbol{u}(\boldsymbol{y})\, dv_{\boldsymbol{y}}, \tag{f}$$

where $\boldsymbol{p}(\boldsymbol{x}) = \boldsymbol{x} - \mathbf{0}$. Using (d)–(f) and the divergence theorem, we obtain

$$\boldsymbol{A} = -i\boldsymbol{a} \otimes \boldsymbol{p}. \tag{g}$$

---

[1] E. and F. Cosserat [1898, *1, 2, 3*] proved that strong ellipticity implies uniqueness for the displacement problem in a homogeneous and *isotropic* body. (See also Fredholm [1906, *3*], Boggio [1907, *1*], and Sherman [1938, *2*].) This result can also be inferred from the earlier work of Kelvin [1888, *4*]. For a homogeneous but possibly anisotropic body, theorem ***(2)*** follows from the work of Browder [1954, *4*] and Morrey [1954, *14*]. The proof given here, which is based on the work of Van Hove [1947, *9*], was furnished by R. A. Toupin (private communication in 1961). A slightly more general theorem was established by Hill [1957, *8*], [1961, *12, 13*], who noted that uniqueness holds provided $U\{\boldsymbol{u}\} > 0$ for all class C^2 fields $\boldsymbol{u}$ that vanish on ∂B but are not identically zero. Hill's result is not restricted to homogeneous bodies. Hayes [1966, *9*] established a uniqueness theorem for the displacement problem under a slightly weaker hypothesis on $\mathbf{C}$, which he calls moderate strong ellipticity. It is not difficult to prove that moderate strong ellipticity and strong ellipticity are equivalent notions when the body is isotropic. Additional papers concerned with uniqueness for the displacement problem are Muskhelishvili [1933, *2*], Mişicu [1953, *16*], Ericksen and Toupin [1956, *3*], Ericksen [1957, *5*], Duffin and Noll [1958, *8*], Gurtin and Sternberg [1960, *7*], Gurtin [1963, *10*], Hayes [1963, *12*], Knops [1964, *13*], [1965, *12, 13*], Zorski [1964, *24*], Edelstein and Fosdick [1968, *4*], Mikhlin [1966. *16*].

Since B is homogeneous, a fundamental theorem of Fourier analysis[1] yields

$$\int_{\mathscr{E}} \nabla \boldsymbol{u} \cdot \mathbf{C}[\nabla \boldsymbol{u}]\, dv = C_{ijkl} \int_{\mathscr{E}} u_{i,j}\, u_{k,l}\, dv = C_{ijkl} \int_{\mathscr{E}} A_{ij} \bar{A}_{kl}\, dv = \int_{\mathscr{E}} \boldsymbol{A} \cdot \mathbf{C}[\bar{\boldsymbol{A}}]\, dv, \tag{h}$$

where $\bar{\boldsymbol{A}}$ is the complex conjugate of $\boldsymbol{A}$.

If we write

$$\boldsymbol{A} = \boldsymbol{A}_R + i\boldsymbol{A}_I,$$

$$\boldsymbol{a} = \boldsymbol{a}_R + i\boldsymbol{a}_I,$$

then (g) implies that

$$\boldsymbol{A}_R = \boldsymbol{a}_I \otimes \boldsymbol{p}, \qquad \boldsymbol{A}_I = -\boldsymbol{a}_R \otimes \boldsymbol{p}. \tag{i}$$

On the other hand, we conclude from the symmetry of $\mathbf{C}$ that

$$\boldsymbol{A} \cdot \mathbf{C}[\bar{\boldsymbol{A}}] = \boldsymbol{A}_R \cdot \mathbf{C}[\boldsymbol{A}_R] + \boldsymbol{A}_I \cdot \mathbf{C}[\boldsymbol{A}_I]. \tag{j}$$

By (c), (d), and (h)–(j),

$$2U\{\boldsymbol{E}\} = \int_{\mathscr{E}} (\boldsymbol{a}_R \otimes \boldsymbol{p}) \cdot \mathbf{C}[\boldsymbol{a}_R \otimes \boldsymbol{p}]\, dv + \int_{\mathscr{E}} (\boldsymbol{a}_I \otimes \boldsymbol{p}) \cdot \mathbf{C}[\boldsymbol{a}_I \otimes \boldsymbol{p}]\, dv. \tag{k}$$

Assume that $\boldsymbol{u}$ does not vanish identically on B. Then by the uniqueness of the Fourier transform, $\boldsymbol{a}$ cannot vanish identically on $\mathscr{E}$, and we conclude from (k) that if $\mathbf{C}$ is strongly elliptic,

$$U\{\boldsymbol{E}\} > 0,$$

which contradicts (b). We also arrive at a contradiction if $-\mathbf{C}$ is strongly elliptic. Thus $\boldsymbol{u} \equiv \mathbf{0}$ on B. □

For homogeneous and isotropic bodies, a converse to the uniqueness theorem ***(2)*** is furnished by the

(3) Cosserat-Ericksen-Toupin theorem.[2] *Suppose that the material is homogeneous and isotropic, and that neither the elasticity tensor nor its negative is strongly elliptic. Then there exists a body B and an elastic state $\mathfrak{s} = [\boldsymbol{u}, \boldsymbol{E}, \boldsymbol{S}]$ on B with the following properties: $\mathfrak{s}$ corresponds to zero body forces and satisfies the null boundary condition*

$$\boldsymbol{u} = \mathbf{0} \quad \text{on } \partial B,$$

and $\boldsymbol{u}$ is not identically zero.

Proof. By ***(25.1)*** we have only three cases to consider:

$$\mu = 0, \qquad \lambda + 2\mu = 0, \qquad (\lambda + 2\mu)\,\mu < 0.$$

In view of ***(31.1)***, it is sufficient to exhibit a body B and a non-trivial class C^2 displacement field $\boldsymbol{u}$ on $\bar{B}$ that satisfies

$$\mu \Delta \boldsymbol{u} + (\lambda + \mu)\, \nabla \operatorname{div} \boldsymbol{u} = \mathbf{0}, \tag{a}$$

$$\boldsymbol{u} = \mathbf{0} \quad \text{on } \partial B. \tag{b}$$

(i)[3] Assume $\lambda + 2\mu = 0$. Then (a) is satisfied by any displacement field of the form $\boldsymbol{u} = \nabla \varphi$. Thus we have only to exhibit a body B and a scalar field φ of class C^3 on $\bar{B}$ such that

$$\nabla \varphi = \mathbf{0} \quad \text{on } \partial B.$$

---

[1] See, e.g., GOLDBERG [1961, *9*], Theorem 13E.

[2] COSSERAT [1898, *2*, *3*], ERICKSEN and TOUPIN [1956, *3*], ERICKSEN [1957, *5*]. See also MIKHLIN [1966, *16*].

[3] ERICKSEN and TOUPIN [1956, *3*]. See also HILL [1961, *13*].

We take B equal to the open ball of unit radius centered at the origin $\mathbf{0}$, and

$$\varphi(\boldsymbol{x}) = r^3 (r-1)^3, \qquad r = |\boldsymbol{x} - \mathbf{0}|.$$

(ii)[1] Assume $\mu = 0$. Then it suffices to find a displacement field $\boldsymbol{u}$ that satisfies

$$\operatorname{div} \boldsymbol{u} = 0 \tag{c}$$

and (b). The field $\boldsymbol{u} = \operatorname{curl} \boldsymbol{w}$ obeys (c). Thus it suffices to exhibit a vector field $\boldsymbol{w}$ satisfying

$$\operatorname{curl} \boldsymbol{w} = \mathbf{0} \quad \text{on } \partial B. \tag{d}$$

Let B be as in (i) and take

$$\boldsymbol{w}(\boldsymbol{x}) = r^3 (r-1)^3 \boldsymbol{e},$$

where $\boldsymbol{e}$ is a unit vector. Then it is easily verified that $\boldsymbol{w}$ satisfies (d).

(iii)[2] Assume $(\lambda + 2\mu)\,\mu < 0$. Let

$$u_1(\boldsymbol{x}) = 2\mu\, x_1^2 - (\lambda + 2\mu)(x_2^2 + x_3^2) - \mu,$$
$$u_2(\boldsymbol{x}) = u_3(\boldsymbol{x}) = 0.$$

Then $\boldsymbol{u}$ satisfies (a) everywhere and vanishes on an ellipsoid. □

That *the assumption of homogeneity is necessary for the validity of theorem (2)* is clear from the following counterexample of EDELSTEIN and FOSDICK.[3] They take B to be the spherical shell

$$B = \{\boldsymbol{x} \colon \pi < |\boldsymbol{x} - \mathbf{0}| < 2\pi\}$$

composed of the *inhomogeneous* but isotropic material defined by

$$\mu(r) = \frac{1}{4}\left\{\frac{3}{r^2} + \ln \frac{3\pi}{r}\right\}, \qquad \lambda(r) = \frac{1}{r^2} - 2\mu(r).$$

Since $\mu > 0$ and $\lambda + 2\mu > 0$, the elasticity field is strongly elliptic on $\bar{B}$. They show that the displacement field $\boldsymbol{u}$ defined in spherical coordinates by

$$u_r(r) = \sin r, \qquad u_\theta = u_\gamma = 0,$$

satisfies the displacement equation of equilibrium with zero body forces and vanishes on the boundary of B (i.e. when $r = \pi$ or $r = 2\pi$).

(4) Uniqueness theorem for the traction problem for star-shaped bodies.[4] *Suppose that the body is homogeneous, isotropic, and star-shaped, and let the Lamé moduli μ and λ satisfy*

$$(\lambda + 2\mu)\,\mu < 0.$$

---

[1] ERICKSEN and TOUPIN [1956, *3*]. See also MILLS [1963, *20*].

[2] COSSERAT [1898, *2*]. This solution was discovered independently by ERICKSEN [1957, *5*].

[3] [1968, *4*].

[4] BRAMBLE and PAYNE [1962, *4*]. In terms of Poisson's ratio ν ($\nu \neq \frac{1}{2}$), theorems *(1)* and *(4)* imply that for a star-shaped body uniqueness holds for the surface force problem provided $\mu \neq 0$ and $-1 < \nu < 1$. The question of whether or not one can drop the hypothesis that B be star-shaped is open. A partial answer was supplied by BRAMBLE and PAYNE [1962, *3*], who have shown that uniqueness holds if $-1 < \nu < 1 - \frac{1}{2}K(1+K)^{-1}$, where K is a non-negative constant that depends on the shape of the body. (See also HILL [1961, *13*].) A counterexample by E. and F. COSSERAT [1901, *3*] (see also BRAMBLE and PAYNE [1961, *3*]) establishes lack of uniqueness for $-\infty < \nu < -1$, while counterexamples by E. and F. COSSERAT [1901, *2, 4*] and BRAMBLE and PAYNE [1961, *3*] indicate non-uniqueness for a countable number of discrete values of Poisson's ratio in the interval $[1, \infty)$. See also MIKHLIN [1966, *16*]. For the *plane* traction problem SHERMAN [1938, *2*] has shown that uniqueness holds if $\nu \neq 1$, $\mu \neq 0$. That these conditions are also necessary follows from counterexamples of HILL [1961, *13*] and KNOPS [1965, *12*].

Then any two solutions of the traction problem with class C^2 displacements on $\bar{B}$ are equal modulo a rigid displacement.

Proof. It is sufficient to show that

$$\boldsymbol{b}=\mathbf{0} \quad \text{on } B, \qquad \boldsymbol{s}=\mathbf{0} \quad \text{on } \partial B \tag{a}$$

imply $\boldsymbol{E}\equiv\mathbf{0}$ on B. Thus assume (a) holds. Then ***(24.5)*** and the theorem of work and energy ***(28.3)*** yield

$$U\{\boldsymbol{E}\}=\int_B \left[\mu|\boldsymbol{E}|^2+\frac{\lambda}{2}(\operatorname{tr}\boldsymbol{E})^2\right]dv=0. \tag{b}$$

Recall Navier's equation

$$\begin{gathered}\Delta\boldsymbol{u}+\gamma\nabla\operatorname{div}\boldsymbol{u}=\mathbf{0},\\ \gamma=\frac{\lambda+\mu}{\mu}.\end{gathered} \tag{c}$$

Since B is star-shaped, it follows from ***(5.1)*** that there exists a point in B, which we take to be the origin $\mathbf{0}$, such that

$$\boldsymbol{p}\cdot\boldsymbol{n}>0 \tag{d}$$

at every regular point of ∂B. Consider the identity

$$\begin{aligned}\int_B [\Delta\boldsymbol{u}+\gamma\nabla\operatorname{div}\boldsymbol{u}]\cdot(\nabla\boldsymbol{u})\,\boldsymbol{p}\,dv=\frac{1}{\mu}\int_{\partial B}\boldsymbol{s}\cdot(\nabla\boldsymbol{u})\,\boldsymbol{p}\,da+\frac{1}{\mu}U\{\boldsymbol{E}\}\\ -\int_{\partial B}(\boldsymbol{p}\cdot\boldsymbol{n})\left[|\boldsymbol{E}|^2+\frac{\gamma-1}{2}(\operatorname{div}\boldsymbol{u})^2\right]da,\end{aligned} \tag{e}$$

which can be established with the aid of the divergence theorem, the stress equation of equilibrium $\operatorname{div}\boldsymbol{S}=\mathbf{0}$, and the stress-strain relation in the form

$$\boldsymbol{S}=2\mu\left[\boldsymbol{E}+\frac{\gamma-1}{2}(\operatorname{div}\boldsymbol{u})\,\mathbf{1}\right]. \tag{f}$$

By (a), (b), and (c), (e) reduces to

$$\int_{\partial B}(\boldsymbol{p}\cdot\boldsymbol{n})\left[|\boldsymbol{E}|^2+\frac{\gamma-1}{2}(\operatorname{div}\boldsymbol{u})^2\right]da=0. \tag{g}$$

Since $\boldsymbol{n}\cdot\boldsymbol{S}\boldsymbol{n}=0$ on ∂B, (f) implies

$$(\gamma-1)\operatorname{div}\boldsymbol{u}=-2\boldsymbol{n}\cdot\boldsymbol{E}\boldsymbol{n}.$$

Thus, for $\gamma\neq 1$, (g) becomes

$$\int_{\partial B}(\boldsymbol{p}\cdot\boldsymbol{n})\left[|\boldsymbol{E}|^2+\frac{2}{\gamma-1}(\boldsymbol{n}\cdot\boldsymbol{E}\boldsymbol{n})^2\right]da=0. \tag{h}$$

The inequality $(\lambda+2\mu)\,\mu<0$ implies $\gamma<-1$; hence

$$\frac{\gamma+1}{\gamma-1}>0. \tag{i}$$

Therefore (h) may be rewritten as

$$\int_{\partial B}(\boldsymbol{p}\cdot\boldsymbol{n})\left|\boldsymbol{E}-\left(1+\sqrt{\frac{\gamma+1}{\gamma-1}}\right)(\boldsymbol{n}\cdot\boldsymbol{E}\boldsymbol{n})\,\boldsymbol{n}\otimes\boldsymbol{n}\right|^2da=0,$$

and by (d) this implies

$$\boldsymbol{E} = \left(1 + \sqrt{\frac{\gamma+1}{\gamma-1}}\right) (\boldsymbol{n} \cdot \boldsymbol{E}\boldsymbol{n})\, \boldsymbol{n} \otimes \boldsymbol{n} \quad \text{on } \partial B. \tag{j}$$

In view of (i) and (j),

$$\boldsymbol{n} \cdot \boldsymbol{E}\boldsymbol{n} = 0 \quad \text{on } \partial B,$$

and hence by (j)

$$\boldsymbol{E} = \boldsymbol{0} \quad \text{on } \partial B.$$

This implies that $\operatorname{div} \boldsymbol{u} = \operatorname{tr} \boldsymbol{E} = 0$ on ∂B. But $\operatorname{div} \boldsymbol{u}$ is harmonic;[1] hence $\operatorname{div} \boldsymbol{u} = 0$ on B, and we conclude from (b) that $\boldsymbol{E} \equiv \boldsymbol{0}$. □

33. Nonexistence. For the traction problem a *necessary* condition for the existence of a solution is that the external forces be in equilibrium, i.e. that

$$\int_{\partial B} \hat{\boldsymbol{s}}\, da + \int_B \boldsymbol{b}\, dv = \boldsymbol{0},$$

$$\int_{\partial B} \boldsymbol{p} \times \hat{\boldsymbol{s}}\, da + \int_B \boldsymbol{p} \times \boldsymbol{b}\, dv = \boldsymbol{0}.$$

A deeper result was established by Ericksen[2], who proved that, in general, *lack of uniqueness implies lack of existence*, or equivalently, that *existence implies uniqueness*.[3]

Suppose there were two solutions to a given mixed problem, and that these solutions were not equal modulo a rigid displacement. Then their difference $[\boldsymbol{u}, \boldsymbol{E}, \boldsymbol{S}]$ would have $\boldsymbol{E} \not\equiv \boldsymbol{0}$, would satisfy

$$\boldsymbol{u} = \boldsymbol{0} \quad \text{on } \mathscr{S}_1, \quad \boldsymbol{s} = \boldsymbol{S}\boldsymbol{n} = \boldsymbol{0} \quad \text{on } \mathscr{S}_2, \tag{N}$$

and would correspond to vanishing body forces. We call an elastic state $[\boldsymbol{u}, \boldsymbol{E}, \boldsymbol{S}]$ with the above properties a ***non-trivial solution of the mixed problem with null data.***

(1) Nonexistence theorem.[4] *Let the elasticity field be symmetric. Assume that there exists a non-trivial solution of the mixed problem with null data. Then there exists a continuous body force field* $\boldsymbol{b}$ *on* $\bar{B}$ *of class* C^2 *on* B *with the following property: the mixed problem corresponding to this body force field and to the null boundary condition* (N) *has no solution. Further, if* $\mathscr{S}_1$ *is empty, then* $\boldsymbol{b}$ *can be chosen so as to satisfy*

$$\int_B \boldsymbol{b}\, dv = \boldsymbol{0}, \qquad \int_B \boldsymbol{p} \times \boldsymbol{b}\, dv = \boldsymbol{0}.$$

Proof. Let $\tilde{\mathfrak{s}} = [\tilde{\boldsymbol{u}}, \tilde{\boldsymbol{E}}, \tilde{\boldsymbol{S}}]$ be a non-trivial solution of the mixed problem with null data. Then if $\mathfrak{s} = [\boldsymbol{u}, \boldsymbol{E}, \boldsymbol{S}]$ is any other elastic state with $\boldsymbol{b}$ the corresponding body force field, Betti's theorem ***(30.1)*** implies

$$\int_{\mathscr{S}_1} \tilde{\boldsymbol{s}} \cdot \boldsymbol{u}\, da = \int_{\mathscr{S}_2} \boldsymbol{s} \cdot \tilde{\boldsymbol{u}}\, da + \int_B \boldsymbol{b} \cdot \tilde{\boldsymbol{u}}\, dv,$$

where we have used the fact that $\tilde{\mathfrak{s}}$ corresponds to null data. Now choose $\boldsymbol{b} = \tilde{\boldsymbol{u}}$, and suppose that $\mathfrak{s}$ satisfies the null boundary conditions (N). Then the above relation implies

$$\int_B |\tilde{\boldsymbol{u}}|^2\, dv = 0,$$

[1] This will be proved in Sect. 42.

[2] Ericksen [1963, *4*], [1965, *7*].

[3] This result is a direct analog of the Fredholm alternative for symmetric linear operators on an inner product space.

[4] The basic idea behind this theorem is due to Ericksen [1963, *4*], who established a theorem similar to ***(1)*** for the traction problem and asserted an analogous result for the displacement problem. Cf. Fichera [1971, *1*], Theorem 6.IV (with $\lambda = 0$).

which yields $\tilde{\boldsymbol{u}} \equiv \boldsymbol{0}$, and we have a contradiction. Thus there cannot exist an elastic state $\mathfrak{s}$ with the above properties.

Now assume that $\mathscr{S}_1 = \emptyset$. Clearly, if $[\tilde{\boldsymbol{u}}, \tilde{\boldsymbol{E}}, \tilde{\boldsymbol{S}}]$ is a non-trivial solution of the surface force problem with null data, then $[\tilde{\boldsymbol{u}} + \boldsymbol{w}, \tilde{\boldsymbol{E}}, \tilde{\boldsymbol{S}}]$ also has this property for every rigid displacement field $\boldsymbol{w}$. Thus to complete the proof it suffices to find a rigid displacement $\boldsymbol{w}$ such that

$$\int_B (\tilde{\boldsymbol{u}} + \boldsymbol{w})\, dv = \boldsymbol{0}, \qquad \int_B \boldsymbol{p} \times (\tilde{\boldsymbol{u}} + \boldsymbol{w})\, dv = \boldsymbol{0}. \tag{a}$$

Let $\boldsymbol{c}$ denote the centroid of B and $\boldsymbol{I}$ its centroidal inertia tensor; i.e. if

$$\boldsymbol{p_c}(\boldsymbol{x}) = \boldsymbol{x} - \boldsymbol{c},$$

then

$$\int_B \boldsymbol{p_c}\, dv = \boldsymbol{0},$$

$$\boldsymbol{I} = \int_B [|\boldsymbol{p_c}|^2 \boldsymbol{1} - \boldsymbol{p_c} \otimes \boldsymbol{p_c}]\, dv.$$

Let $\boldsymbol{w}$ denote the rigid displacement field defined by

$$\boldsymbol{w} = \boldsymbol{a} + \boldsymbol{\omega} \times \boldsymbol{p_c},$$

$$\boldsymbol{a} = -\frac{1}{v(B)} \int_B \tilde{\boldsymbol{u}}\, dv, \qquad \boldsymbol{\omega} = -\boldsymbol{I}^{-1} \int_B \boldsymbol{p_c} \times \tilde{\boldsymbol{u}}\, dv.$$

Then a simple computation based on the identity

$$\int_B \boldsymbol{p_c} \times (\boldsymbol{\omega} \times \boldsymbol{p_c})\, dv = \boldsymbol{I}\boldsymbol{\omega}$$

yields

$$\int_B (\tilde{\boldsymbol{u}} + \boldsymbol{w})\, dv = \boldsymbol{0}, \qquad \int_B \boldsymbol{p_c} \times (\tilde{\boldsymbol{u}} + \boldsymbol{w})\, dv = \boldsymbol{0},$$

which is equivalent to (a). □

The non-existence theorem ***(1)*** has the following immediate corollary: *If the mixed problem with null boundary data has a solution whenever the body force field is sufficiently smooth, then there is at most one solution to the mixed problem.*

IV. The variational principles of elastostatics.

34. Minimum principles. In this section we will establish the two classical minimum principles of elastostatics: the principle of minimum potential energy and the principle of minimum complementary energy.[1] These principles completely characterize the solution of the mixed problem discussed previously.

We assume throughout that the data has properties (i)–(iv) of Sect. 31, and, in addition, that

(v) the elasticity field C is *symmetric* and *positive definite* on B.

Let $\boldsymbol{E}$ and $\boldsymbol{S}$ be continuous symmetric tensor fields on $\bar{B}$. For convenience, we now write $U_{\mathsf{C}}\{\boldsymbol{E}\}$, rather than $U\{\boldsymbol{E}\}$, for the ***strain energy***:

$$U_{\mathsf{C}}\{\boldsymbol{E}\} = \tfrac{1}{2} \int_B \boldsymbol{E} \cdot \mathsf{C}[\boldsymbol{E}]\, dv.$$

[1] The treatment of the minimum principles ***(1)*** and ***(3)***, which is based on the notion of an elastic state, was furnished by E. STERNBERG (private communication) in 1959.

Further, we define the ***stress energy*** $U_{\mathbf{K}}\{S\}$ by

$$U_{\mathbf{K}}\{S\} = \tfrac{1}{2}\int_B S \cdot \mathbf{K}[S]\, dv,$$

where $\mathbf{K} = \mathbf{C}^{-1}$ is the compliance tensor. In view of the remarks made on p. 84,

$$U_{\mathbf{K}}\{S\} = U_{\mathbf{C}}\{E\}$$

provided $S = \mathbf{C}[E]$.

By a ***kinematically admissible state*** we mean an admissible state that satisfies the strain-displacement relation, the stress-strain relation, and the displacement boundary condition.

(1) Principle of minimum potential energy.[1] *Let $\mathscr{A}$ denote the set of all kinematically admissible states, and let Φ be the functional on $\mathscr{A}$ defined by*

$$\Phi\{\mathfrak{s}\} = U_{\mathbf{C}}\{E\} - \int_B \mathbf{b} \cdot \mathbf{u}\, dv - \int_{\mathscr{S}_2} \hat{\mathbf{s}} \cdot \mathbf{u}\, da$$

for every $\mathfrak{s} = [\mathbf{u}, E, S] \in \mathscr{A}$. Further, let $\mathfrak{s}$ be a solution of the mixed problem. Then

$$\Phi\{\mathfrak{s}\} \leqq \Phi\{\tilde{\mathfrak{s}}\}$$

for every $\tilde{\mathfrak{s}} \in \mathscr{A}$, and equality holds only if $\tilde{\mathfrak{s}} = \mathfrak{s}$ modulo a rigid displacement.

Proof. Let $\mathfrak{s}, \tilde{\mathfrak{s}} \in \mathscr{A}$ and define

$$\mathfrak{s}' = \tilde{\mathfrak{s}} - \mathfrak{s}. \tag{a}$$

Then $\mathfrak{s}'$ is an admissible state and

$$E' = \tfrac{1}{2}(\nabla \mathbf{u}' + \nabla \mathbf{u}'^T), \tag{b}$$

$$S' = \mathbf{C}[E'], \tag{c}$$

$$\mathbf{u}' = \mathbf{0} \quad \text{on } \mathscr{S}_1. \tag{d}$$

Moreover, $S = \mathbf{C}[E]$, since $\mathfrak{s} \in \mathscr{A}$; thus ***(28.1)*** and (c) imply

$$U_{\mathbf{C}}\{\tilde{E}\} - U_{\mathbf{C}}\{E\} = U_{\mathbf{C}}\{E'\} + \int_B S \cdot E'\, dv. \tag{e}$$

Next, if we apply ***(18.1)*** to S and $\mathbf{u}'$, we conclude, with the aid of (b) and (d), that

$$\int_B S \cdot E'\, dv = \int_{\mathscr{S}_2} \mathbf{s} \cdot \mathbf{u}'\, da - \int_B \mathbf{u}' \cdot \operatorname{div} S\, dv.$$

[1] It is somewhat difficult to trace the history of this and other variational and minimum principles, since in the older work the nature of the allowed variations was often left to be inferred from the result. The basic ideas appear in the work of GREEN [1839, *1*], pp. 253–256, HAUGHTON [1849, *1*], p. 152, KIRCHHOFF [1850, *1*], § 1, KELVIN [1863, *3*], §§ 61–62, DONATI [1894, *2*]. As a proved theorem it appears to have first been given by LOVE [1906, *5*], § 119 for the displacement problem, by HADAMARD [1903, *3*], § 264, COLONNETTI [1912, *1*], and PRANGE [1916, *1*], pp. 51–54 for the traction problem, and by TREFFTZ [1928, *3*], § 18 for the mixed problem. An extension to exterior domains was given by GURTIN and STERNBERG [1961, *11*], Theorem 6.5. Variants of this theorem, which are useful in the determination of bounds on effective elastic moduli for heterogeneous materials, are contained in the work of HASHIN and SHTRIKMAN [1962, *9*], HILL [1963, *13*], HASHIN [1967, *8*], BERAN and MOLYNEUX [1966, *3*], and RUBENFELD and KELLER [1969, *6*].

For the displacement problem, the hypothesis that $\mathbf{C}$ be positive definite is not necessary for the validity of the principle of minimum potential energy. This fact is clear from the work of HILL [1957, *8*], GURTIN and STERNBERG [1960, *7*], and GURTIN [1963, *10*].

For various related references, see PRAGER and SYNGE [1947, *6*], SYNGE [1957, *15*], MIKHLIN [1957, *10*], and ORAVAS and MCLEAN [1966, *19*].

Thus

$$\Phi\{\tilde{\mathfrak{s}}\} - \Phi\{\mathfrak{s}\} = U_{\mathbf{C}}\{\boldsymbol{E}'\} - \int_B (\operatorname{div} \boldsymbol{S} + \boldsymbol{b}) \cdot \boldsymbol{u}' \, dv + \int_{\mathscr{S}_2} (\mathbf{s} - \hat{\mathbf{s}}) \cdot \boldsymbol{u}' \, da, \tag{f}$$

and since $\mathfrak{s}$ is a solution of the mixed problem, (f) implies

$$\Phi\{\tilde{\mathfrak{s}}\} - \Phi\{\mathfrak{s}\} = U_{\mathbf{C}}\{\boldsymbol{E}'\}.$$

Thus, since $\mathbf{C}$ is positive definite,

$$\Phi\{\mathfrak{s}\} \leqq \Phi\{\tilde{\mathfrak{s}}\},$$

$$\Phi\{\mathfrak{s}\} = \Phi\{\tilde{\mathfrak{s}}\} \Leftrightarrow \boldsymbol{E}' = \tilde{\boldsymbol{E}} - \boldsymbol{E} = \boldsymbol{0}.$$

Moreover, since $\mathfrak{s}$ and $\tilde{\mathfrak{s}}$ are kinematically admissible, $\boldsymbol{E} = \tilde{\boldsymbol{E}}$ only when $\mathfrak{s} = \tilde{\mathfrak{s}}$ modulo a rigid displacement. □

In words, the principle of minimum potential energy asserts that the difference between the strain energy and the work done by the body forces and prescribed surface forces assumes a smaller value for the solution of the mixed problem than for any other kinematically admissible state.

The uniqueness theorem ***(32.1)*** for the mixed problem follows as a corollary of the principle of minimum potential energy. Indeed, let $\mathfrak{s}$ and $\tilde{\mathfrak{s}}$ be two solutions. Then this principle tells us that

$$\Phi\{\mathfrak{s}\} \leqq \Phi\{\tilde{\mathfrak{s}}\}, \quad \Phi\{\tilde{\mathfrak{s}}\} \leqq \Phi\{\mathfrak{s}\}.$$

Thus

$$\Phi\{\mathfrak{s}\} = \Phi\{\tilde{\mathfrak{s}}\},$$

and we conclude from ***(1)*** that $\mathfrak{s}$ and $\tilde{\mathfrak{s}}$ must be equal modulo a rigid displacement.

By a ***kinematically admissible displacement field*** we mean an admissible displacement field that satisfies the displacement boundary condition, and for which $\operatorname{div} \mathbf{C}[\nabla \boldsymbol{u}]$ is continuous on $\bar{B}$. The corresponding strain field is then called a ***kinematically admissible strain field***. If $[\boldsymbol{u}, \boldsymbol{E}, \boldsymbol{S}]$ is a kinematically admissible state, then $\boldsymbol{u}$ is a kinematically admissible displacement field. Conversely, the latter assertion implies the former provided $\boldsymbol{E}$ and $\boldsymbol{S}$ are defined through the strain-displacement and stress-strain relations. In view of these equations, $\Phi\{\mathfrak{s}\}$ can be written as a functional $\Phi\{\boldsymbol{u}\}$ of $\boldsymbol{u}$:

$$\Phi\{\boldsymbol{u}\} = \tfrac{1}{2} \int_B \nabla \boldsymbol{u} \cdot \mathbf{C}[\nabla \boldsymbol{u}] \, dv - \int_B \boldsymbol{b} \cdot \boldsymbol{u} \, dv - \int_{\mathscr{S}_2} \hat{\mathbf{s}} \cdot \boldsymbol{u} \, da;$$

therefore we have the following partial restatement of ***(1)***

(2) *Let* $\boldsymbol{u}$ *correspond to a solution of the mixed problem. Then*

$$\Phi\{\boldsymbol{u}\} \leqq \Phi\{\tilde{\boldsymbol{u}}\}$$

for every kinematically admissible displacement field $\tilde{\boldsymbol{u}}$.

By a ***statically admissible stress field*** we mean an admissible stress field that satisfies the equation of equilibrium and the traction boundary condition.

(3) Principle of minimum complementary energy.[1] *Let* $\mathscr{A}$ *denote the set of all statically admissible stress fields, and let* Ψ *be the functional on* $\mathscr{A}$ *defined by*

$$\Psi\{\boldsymbol{S}\} = U_{\mathbf{K}}\{\boldsymbol{S}\} - \int_{\mathscr{S}_1} \mathbf{s} \cdot \hat{\boldsymbol{u}} \, da$$

---

[1] The basic ideas appear in the work of COTTERILL [1865, *1*] and DONATI [1890, *1*], [1894, *2*]. As a proved theorem it seems to have first been given by COLONNETTI [1912, *1*] and PRANGE [1916, *1*] for the surface force problem and by TREFFTZ [1928, *3*], § 19 for the mixed problem. See also DOMKE [1915, *1*].

for every $\boldsymbol{S} \in \mathscr{A}$. *Let* $\boldsymbol{S}$ *be a stress field corresponding to a solution of the mixed problem. Then*

$$\Psi\{\boldsymbol{S}\} \leqq \Psi\{\tilde{\boldsymbol{S}}\}$$

for every $\tilde{\boldsymbol{S}} \in \mathscr{A}$, *and equality holds only if* $\boldsymbol{S} = \tilde{\boldsymbol{S}}$.

Proof. Let $[\boldsymbol{u}, \boldsymbol{E}, \boldsymbol{S}]$ be a solution of the mixed problem, let $\tilde{\boldsymbol{S}} \in \mathscr{A}$, and define

$$\boldsymbol{S}' = \tilde{\boldsymbol{S}} - \boldsymbol{S}. \qquad \text{(a)}$$

Then $\boldsymbol{S}'$ satisfies

$$\begin{aligned} \operatorname{div} \boldsymbol{S}' &= \boldsymbol{0} \quad \text{on } B, \\ \boldsymbol{s}' = \boldsymbol{S}'\boldsymbol{n} &= \boldsymbol{0} \quad \text{on } \mathscr{S}_2. \end{aligned} \qquad \text{(b)}$$

Since $\boldsymbol{E} = \mathsf{K}[\boldsymbol{S}]$, (a) and an obvious analog of *(28.1)* imply

$$U_{\mathsf{K}}\{\tilde{\boldsymbol{S}}\} - U_{\mathsf{K}}\{\boldsymbol{S}\} = U_{\mathsf{K}}\{\boldsymbol{S}'\} + \int_B \boldsymbol{S}' \cdot \boldsymbol{E}\, dv. \qquad \text{(c)}$$

Further, in view of (b) and *(18.1)*,

$$\int_B \boldsymbol{S}' \cdot \boldsymbol{E}\, dv = \int_{\mathscr{S}_1} \boldsymbol{s}' \cdot \boldsymbol{u}\, da.$$

Thus, since $\boldsymbol{u} = \hat{\boldsymbol{u}}$ on $\mathscr{S}_1$,

$$\Psi\{\tilde{\boldsymbol{S}}\} - \Psi\{\boldsymbol{S}\} = U_{\mathsf{K}}\{\boldsymbol{S}'\}.$$

Therefore, since K is positive definite, $\Psi\{\boldsymbol{S}\} \leqq \Psi\{\tilde{\boldsymbol{S}}\}$, and $\Psi\{\boldsymbol{S}\} = \Psi\{\tilde{\boldsymbol{S}}\}$ only if $\boldsymbol{S}' = \tilde{\boldsymbol{S}} - \boldsymbol{S} = \boldsymbol{0}$. □

The principle of minimum complementary energy asserts that of all statically admissible stress fields, the one belonging to a solution of the mixed problem renders a minimum the difference between the stress energy and the work done over the prescribed displacements.

We shall now use the foregoing minimum principles to establish

(4) Upper and lower bounds for the strain energy.[1] *Let* U *be the strain energy associated with a solution of the displacement problem. Assume that the body forces vanish. Then*

$$\int_{\partial B} \bar{\boldsymbol{s}} \cdot \hat{\boldsymbol{u}}\, da - U_{\mathsf{K}}\{\bar{\boldsymbol{S}}\} \leqq U \leqq U_{\mathsf{C}}\{\tilde{\boldsymbol{E}}\},$$

where $\tilde{\boldsymbol{E}}$ *is a kinematically admissible strain field,* $\bar{\boldsymbol{S}}$ *a statically admissible stress field, and* $\bar{\boldsymbol{s}}$ *the corresponding surface traction.*

On the other hand, let U *be the strain energy corresponding to a solution of the traction problem. Then*

$$\int_B \boldsymbol{b} \cdot \tilde{\boldsymbol{u}}\, dv + \int_{\partial B} \hat{\boldsymbol{s}} \cdot \tilde{\boldsymbol{u}}\, da - U_{\mathsf{C}}\{\tilde{\boldsymbol{E}}\} \leqq U \leqq U_{\mathsf{K}}\{\bar{\boldsymbol{S}}\},$$

where $\tilde{\boldsymbol{u}}$ *is a kinematically admissible displacement field,* $\tilde{\boldsymbol{E}}$ *the corresponding strain field, and* $\bar{\boldsymbol{S}}$ *a statically admissible stress field.*

Proof, Let $\mathfrak{s} = [\boldsymbol{u}, \boldsymbol{E}, \boldsymbol{S}]$ be a solution of the displacement problem with $\boldsymbol{b} = \boldsymbol{0}$, and let U be the associated strain energy:

$$U = U_{\mathsf{C}}\{\boldsymbol{E}\} = U_{\mathsf{K}}\{\boldsymbol{S}\}.$$

[1] Cf. Aymerich [1955, *1*] who obtains upper and lower bounds on the strain energy by embedding the body in a larger body.

Then, since $\mathscr{S}_2=\emptyset$, we conclude from the principle of minimum potential energy that

$$U=U_{\mathsf{C}}\{\boldsymbol{E}\}\leqq U_{\mathsf{C}}\{\tilde{\boldsymbol{E}}\}.$$

On the other hand, since $\mathscr{S}_1=\partial B$, it follows from the principle of work and energy *(28.3)* that the last term in the expression for $\Psi\{\boldsymbol{S}\}$ in *(3)* is equal to $-2U$. Thus

$$\Psi\{\boldsymbol{S}\}=-U,$$

and *(3)* implies

$$U\geqq\int_{\partial B}\bar{\mathbf{s}}\cdot\hat{\boldsymbol{u}}\,da-U_{\mathsf{K}}\{\bar{\boldsymbol{S}}\}.$$

Next, let $\mathfrak{s}$ be a solution of the surface force problem. Then $\mathscr{S}_1=\emptyset$, and the principle of minimum complementary energy yields

$$U\leqq U_{\mathsf{K}}\{\bar{\boldsymbol{S}}\}.$$

Moreover, since $\mathscr{S}_2=\partial B$, we conclude from *(28.3)* that the last two terms in the expression for $\Phi\{\mathfrak{s}\}$ in *(1)* are equal to $-2U$. Thus

$$\Phi\{\mathfrak{s}\}=-U,$$

and *(1)* implies

$$U\geqq\int_B\boldsymbol{b}\cdot\tilde{\boldsymbol{u}}\,dv+\int_{\partial B}\hat{\mathbf{s}}\cdot\tilde{\boldsymbol{u}}\,da-U_{\mathsf{C}}\{\tilde{\boldsymbol{E}}\}.\quad\square$$

Using *(18.2)* we can write the inequalities in *(4)* in the following forms:

$$\int_B\bar{\boldsymbol{S}}\cdot\{\tilde{\boldsymbol{E}}-\tfrac{1}{2}\mathsf{K}[\bar{\boldsymbol{S}}]\}\,dv\leqq U\leqq U_{\mathsf{C}}\{\tilde{\boldsymbol{E}}\},$$

$$\int_B\tilde{\boldsymbol{E}}\cdot\{\bar{\boldsymbol{S}}-\tfrac{1}{2}\mathsf{C}[\tilde{\boldsymbol{E}}]\}\,dv\leqq U\leqq U_{\mathsf{K}}\{\bar{\boldsymbol{S}}\}.$$

By a ***statically admissible state*** we mean an admissible state $[\boldsymbol{u},\boldsymbol{E},\boldsymbol{S}]$ with $\boldsymbol{S}$ a statically admissible stress field. For convenience, we now define Ψ on the set of all statically admissible states $\mathfrak{s}=[\boldsymbol{u},\boldsymbol{E},\boldsymbol{S}]$ by writing $\Psi\{\mathfrak{s}\}=\Psi\{\boldsymbol{S}\}$. Clearly, an admissible state $\mathfrak{s}$ is a solution of the mixed problem if and only if $\mathfrak{s}$ is both kinematically and statically admissible. We shall tacitly use this fact in establishing the next theorem, which furnishes an interesting relation between the functionals Φ and Ψ.

(5) *Let $\mathfrak{s}$ be a solution of the mixed problem. Then*

$$\Phi\{\mathfrak{s}\}+\Psi\{\mathfrak{s}\}=0.$$

Proof. In view of the definition of Φ and Ψ,

$$\begin{aligned}\Phi\{\mathfrak{s}\}+\Psi\{\mathfrak{s}\}&=U_{\mathsf{C}}\{\boldsymbol{E}\}+U_{\mathsf{K}}\{\boldsymbol{S}\}-\int_B\boldsymbol{b}\cdot\boldsymbol{u}\,dv-\int_{\mathscr{S}_1}\mathbf{s}\cdot\hat{\boldsymbol{u}}\,da-\int_{\mathscr{S}_2}\hat{\mathbf{s}}\cdot\boldsymbol{u}\,da\\&=2U_{\mathsf{C}}\{\boldsymbol{E}\}-\int_B\boldsymbol{b}\cdot\boldsymbol{u}\,dv-\int_{\partial B}\mathbf{s}\cdot\boldsymbol{u}\,da,\end{aligned}$$

and the desired result follows from the principle of work and energy *(28.3)*. $\square$

If we assume that the mixed problem has a solution, then the preceding theorem has the following corollary.

(6) *Let $\tilde{\mathfrak{s}}$ and $\bar{\mathfrak{s}}$ be admissible states with $\tilde{\mathfrak{s}}$ kinematically admissible and $\bar{\mathfrak{s}}$ statically admissible. Then*

$$\Phi\{\tilde{\mathfrak{s}}\}+\Psi\{\bar{\mathfrak{s}}\}\geqq0.$$

Proof. Let $\mathfrak{s}$ be a solution of the mixed problem. Then the principles of minimum potential and complementary energy imply that

$$\Phi\{\tilde{\mathfrak{s}}\} \geqq \Phi\{\mathfrak{s}\}, \quad \Psi\{\tilde{\mathfrak{s}}\} \geqq \Psi\{\mathfrak{s}\}.$$

Thus

$$\Phi\{\tilde{\mathfrak{s}}\} + \Psi\{\tilde{\mathfrak{s}}\} \geqq \Phi\{\mathfrak{s}\} + \Psi\{\mathfrak{s}\} = 0. \quad \square$$

35. Some extensions of the fundamental lemma. In this section we shall establish three lemmas whose role in elastostatics is analogous to the role of the fundamental lemma in the calculus of variations. Let $\partial B(\boldsymbol{x}_0, h)$ denote the intersection of ∂B with the open ball $\Sigma_h(\boldsymbol{x}_0)$ of radius h and center at $\boldsymbol{x}_0$, and let $\Phi_h(\boldsymbol{x}_0)$ denote the set of functions defined in Sect. 7.

(1) *Let $\mathscr{W}$ be a finite-dimensional inner product space. Let $\boldsymbol{w}: \mathscr{S}_2 \to \mathscr{W}$ be piecewise regular and satisfy*

$$\int_{\mathscr{S}_2} \boldsymbol{w} \cdot \boldsymbol{v} \, da = 0$$

for every class C^∞ function $\boldsymbol{v}: \bar{B} \to \mathscr{W}$ that vanishes near $\mathscr{S}_1$. Then

$$\boldsymbol{w} = \boldsymbol{0} \quad \text{on } \mathscr{S}_2.$$

Proof. Let $\mathbf{e}_1, \mathbf{e}_2, \ldots, \mathbf{e}_n$ be an orthonormal basis for $\mathscr{W}$, and let

$$\boldsymbol{w} = \sum_{i=1}^{n} w_i \boldsymbol{e}_i.$$

Assume that for some regular interior point $\boldsymbol{x}_0$ of $\mathscr{S}_2$ and some k, $w_k(\boldsymbol{x}_0) > 0$ (say). Then there exists an $h > 0$ such that $\partial B(\boldsymbol{x}_0, h)$ is contained in $\mathscr{S}_2$ and $w_k > 0$ on $\partial B(\boldsymbol{x}_0, h)$. Let $\boldsymbol{v} = \varphi \, \boldsymbol{e}_k$ with $\varphi \in \Phi_h(\boldsymbol{x}_0)$. Then $\boldsymbol{v}$ is of class C^∞ and vanishes near $\mathscr{S}_1$. In addition,

$$\int_{\mathscr{S}_2} \boldsymbol{w} \cdot \boldsymbol{v} \, da > 0,$$

which is a contradiction. Thus $\boldsymbol{w}(\boldsymbol{x}_0) = \boldsymbol{0}$ at every regular interior point $\boldsymbol{x}_0$ of $\mathscr{S}_2$. But since $\boldsymbol{w}$ is piecewise regular, this implies $\boldsymbol{w} = \boldsymbol{0}$ on $\mathscr{S}_2$. $\square$

(2) *Let $\boldsymbol{u}$ be a piecewise regular vector field on $\mathscr{S}_1$, and suppose that*

$$\int_{\mathscr{S}_1} (\boldsymbol{S}\boldsymbol{n}) \cdot \boldsymbol{u} \, da = 0$$

for every class C^∞ symmetric tensor field $\boldsymbol{S}$ on $\bar{B}$ that vanishes near $\mathscr{S}_2$. Then

$$\boldsymbol{u} = \boldsymbol{0} \quad \text{on } \mathscr{S}_1.$$

Proof. Since $\boldsymbol{S}$ is symmetric,

$$(\boldsymbol{S}\boldsymbol{n}) \cdot \boldsymbol{u} = \boldsymbol{S} \cdot (\boldsymbol{u} \otimes \boldsymbol{n}) = \boldsymbol{S} \cdot \boldsymbol{T},$$

where $\boldsymbol{T}$ is the symmetric part of $\boldsymbol{u} \otimes \boldsymbol{n}$:

$$\boldsymbol{T} = \tfrac{1}{2}(\boldsymbol{u} \otimes \boldsymbol{n} + \boldsymbol{n} \otimes \boldsymbol{u}).$$

Therefore

$$\int_{\mathscr{S}_1} \boldsymbol{S} \cdot \boldsymbol{T} \, da = 0$$

whenever $\boldsymbol{S}$ satisfies the above hypotheses. Consequently, letting $\mathscr{W}$ in ***(1)*** be the space of all symmetric tensors, we conclude from ***(1)*** that

$$\boldsymbol{T} = \boldsymbol{0} \quad \text{on } \mathscr{S}_1,$$

and hence that

$$2\boldsymbol{T}\boldsymbol{n} = \boldsymbol{u} + \boldsymbol{n}(\boldsymbol{u}\cdot\boldsymbol{n}) = \boldsymbol{0} \quad \text{on } \mathscr{S}_1.$$

Taking the inner product of this relation with $\boldsymbol{n}$, we find that $\boldsymbol{u}\cdot\boldsymbol{n}=0$ on $\mathscr{S}_1$, and this fact and the above relation imply $\boldsymbol{u}=\boldsymbol{0}$ on $\mathscr{S}_1$. □

(3) *Let $\boldsymbol{u}$ be a piecewise regular vector field on $\mathscr{S}_2$, and suppose that*

$$\int_{\mathscr{S}_2} \boldsymbol{u}\cdot \operatorname{div}\boldsymbol{S}\, da = 0$$

for every class C^∞ symmetric tensor field $\boldsymbol{S}$ on $\bar{B}$. Then

$$\boldsymbol{u} = \boldsymbol{0} \quad \textit{on } \mathscr{S}_2.$$

Proof. Let $\boldsymbol{v}$ be an arbitrary class C^∞ vector field on $\bar{B}$ which vanishes near $\mathscr{S}_1$. Then by ***(6.5)*** there exists a class C^∞ vector field $\boldsymbol{g}$ on $\bar{B}$ with the property that

$$\Delta\boldsymbol{g} = \boldsymbol{v}.$$

Let[1]

$$\boldsymbol{S} = \nabla\boldsymbol{g} + \nabla\boldsymbol{g}^T - \boldsymbol{1}\operatorname{div}\boldsymbol{g}.$$

Then $\boldsymbol{S}$ is a class C^∞ symmetric tensor field on $\bar{B}$, and hence

$$\int_{\mathscr{S}_2} \boldsymbol{u}\cdot \operatorname{div}\boldsymbol{S}\, da = 0.$$

But a simple calculation yields

$$\operatorname{div}\boldsymbol{S} = \Delta\boldsymbol{g} = \boldsymbol{v}.$$

Thus

$$\int_{\mathscr{S}_2} \boldsymbol{u}\cdot\boldsymbol{v}\, da = 0$$

for every class C^∞ vector field $\boldsymbol{v}$ on $\bar{B}$ which vanishes near $\mathscr{S}_1$, and the desired result follows from ***(1)***. □

36. Converses to the minimum principles. In this section we shall use the results just established to prove converses to the principles of minimum potential and complementary energy. We continue to assume that hypotheses (i)–(v) of Sects. 31 and 34 hold. Our first theorem shows that if a kinematically admissible state minimizes the functional Φ, then that state is a solution of the mixed problem.

(1) Converse of the principle of minimum potential energy. *Let $\mathfrak{s}$ be a kinematically admissible state, and suppose that*

$$\Phi\{\mathfrak{s}\} \leqq \Phi\{\tilde{\mathfrak{s}}\}$$

for every kinematically admissible state $\tilde{\mathfrak{s}}$. Then $\mathfrak{s}$ is a solution of the mixed problem.

Proof. Let $\boldsymbol{u}'$ be an arbitrary vector field of class C^∞ on $\bar{B}$, and suppose that $\boldsymbol{u}'$ vanishes near $\mathscr{S}_1$. Further, let $\mathfrak{s}' = [\boldsymbol{u}', \boldsymbol{E}', \boldsymbol{S}']$, where

$$\boldsymbol{E}' = \tfrac{1}{2}(\nabla\boldsymbol{u}' + \nabla\boldsymbol{u}'^T),$$
$$\boldsymbol{S}' = \mathsf{C}[\boldsymbol{E}'].$$

Then $\tilde{\mathfrak{s}} = \mathfrak{s} + \mathfrak{s}'$ is kinematically admissible, and it is not difficult to verify that (f) in the proof of ***(34.1)*** also holds in the present circumstances. Thus, since $\Phi\{\mathfrak{s}\} \leqq \Phi\{\tilde{\mathfrak{s}}\}$,

$$0 \leqq U_{\mathsf{C}}\{\boldsymbol{E}'\} - \int_B (\operatorname{div}\boldsymbol{S} + \boldsymbol{b})\cdot\boldsymbol{u}'\, dv + \int_{\mathscr{S}_2} (\mathbf{s} - \hat{\mathbf{s}})\cdot\boldsymbol{u}'\, da.$$

[1] Cf. the remarks following ***(17.8)***.

Clearly, this relation must hold with $\boldsymbol{u}'$ replaced by $\alpha\boldsymbol{u}'$ and $\boldsymbol{E}'$ by $\alpha\boldsymbol{E}'$; hence

$$0 \leqq \alpha^2 U_{\mathbf{C}}\{\boldsymbol{E}'\} - \alpha \int_B (\operatorname{div} \boldsymbol{S} + \boldsymbol{b}) \cdot \boldsymbol{u}'\, dv + \alpha \int_{\mathscr{S}_2} (\mathbf{s} - \hat{\mathbf{s}}) \cdot \boldsymbol{u}'\, da$$

for every scalar α, which implies

$$-\int_B (\operatorname{div} \boldsymbol{S} + \boldsymbol{b}) \cdot \boldsymbol{u}'\, dv + \int_{\mathscr{S}_2} (\mathbf{s} - \hat{\mathbf{s}}) \cdot \boldsymbol{u}'\, da = 0 \tag{a}$$

for every C^∞ vector field $\boldsymbol{u}'$ that vanishes near $\mathscr{S}_1$. If, in addition, $\boldsymbol{u}'$ vanishes near ∂B, then

$$\int_B (\operatorname{div} \boldsymbol{S} + \boldsymbol{b}) \cdot \boldsymbol{u}'\, dv = 0,$$

and we conclude from ***(7.1)*** with $\mathscr{W} = \mathscr{V}$ that

$$\operatorname{div} \boldsymbol{S} + \boldsymbol{b} = \mathbf{0}. \tag{b}$$

By (a) and (b),

$$\int_{\mathscr{S}_2} (\mathbf{s} - \hat{\mathbf{s}}) \cdot \boldsymbol{u}'\, da = 0$$

for every C^∞ field $\boldsymbol{u}'$ that vanishes near $\mathscr{S}_1$, and ***(35.1)*** with $\mathscr{W} = \mathscr{V}$ yields

$$\mathbf{s} = \hat{\mathbf{s}} \quad \text{on } \mathscr{S}_2. \tag{c}$$

Thus $\mathfrak{s}$ is a kinematically admissible state that satisfies (b) and (c); hence $\mathfrak{s}$ is a solution of the mixed problem. □

Note that ***(1)*** does not presuppose the existence of a solution to the mixed problem. If one knows a priori that a solution $\bar{\mathfrak{s}}$ exists,[1] then it follows from ***(34.1)*** that any kinematically admissible state $\mathfrak{s}$ that minimizes Φ must be equal to $\bar{\mathfrak{s}}$, and hence must be a solution. Indeed, if $\bar{\mathfrak{s}}$ minimizes Φ, then

$$\Phi\{\bar{\mathfrak{s}}\} \leqq \Phi\{\mathfrak{s}\};$$

but ***(34.1)*** implies

$$\Phi\{\mathfrak{s}\} \leqq \Phi\{\bar{\mathfrak{s}}\},$$

so that

$$\Phi\{\mathfrak{s}\} = \Phi\{\bar{\mathfrak{s}}\},$$

and we conclude from ***(34.1)*** that $\mathfrak{s} = \bar{\mathfrak{s}}$ (modulo a rigid displacement).

The next theorem yields a converse to the principle of minimum complementary energy. To prove this theorem we need to assume that B is simply-connected and convex with respect to $\mathscr{S}_1$ (when $\mathscr{S}_1 \neq \emptyset$). In view of the discussion given in the preceding paragraph, if B and the boundary data are such that existence holds for the corresponding mixed problem,[1] then this converse follows trivially without the above assumptions concerning B.

(2) Converse of the principle of minimum complementary energy.[2] *Assume that B is simply-connected and convex with respect to $\mathscr{S}_1$ and that $\mathbf{K}$ is of class C^2 on $\bar{B}$. Let $\boldsymbol{S}$ be a statically admissible stress field of class C^2 on B, and suppose that*

$$\Psi\{\boldsymbol{S}\} \leqq \Psi\{\tilde{\boldsymbol{S}}\}$$

for every statically admissible stress field $\tilde{\boldsymbol{S}}$. Then $\boldsymbol{S}$ is a stress field corresponding to a solution of the mixed problem.

Before proving this theorem we shall establish two subsidiary results; these results are not only basic to the proof of ***(2)***, but are also of interest in themselves.

---

[1] Existence theorems for the mixed problem are given, e.g., by FICHERA [1971, *1*], § 12.

[2] The basic ideas behind this theorem are due to COTTERILL [1865, *1*], DONATI [1890, *1*], [1894, *2*], DOMKE [1915, *1*], SOUTHWELL [1936, *5*], [1938, *5*], LOCATELLI [1940, *3, 4*], DORN and SCHILD [1956, *1*]. As a proved theorem it seems to have first been given by SOKOLNIKOFF [1956, *12*] for the traction problem and by GURTIN [1963, *11*] for the mixed problem.

(3)*[1] *Let $\boldsymbol{w}$ be a continuous vector field on a regular subsurface $\mathscr{S}$ of ∂B, and let B be convex with respect to $\mathscr{S}$. Further, suppose that

$$\int_{\mathscr{S}} (\boldsymbol{S}\boldsymbol{n}) \cdot \boldsymbol{w}\, da = 0$$

for every class C^∞ symmetric tensor field $\boldsymbol{S}$ on $\bar{B}$ that vanishes near $\partial B - \mathscr{S}$ and satisfies

$$\operatorname{div} \boldsymbol{S} = \boldsymbol{0}.$$

Then $\boldsymbol{w}$ is a rigid displacement field.

Proof. Let $\hat{\boldsymbol{x}}$ and $\tilde{\boldsymbol{x}}$ be regular points of $\mathring{\mathscr{S}}$, and choose a cartesian coordinate system such that the coordinates of $\hat{\boldsymbol{x}}$ are $(0, 0, 0)$ and those of $\tilde{\boldsymbol{x}}$ are $(0, 0, x_3)$. Let D_ε be the closed disc in the x_1, x_2-plane with radius $\varepsilon > 0$ and center at $(0, 0)$, and let f_ε be a class C^∞ scalar field on the entire x_1, x_2-plane with the following properties:

(α) $f_\varepsilon \geqq 0$;

(β) $f_\varepsilon = 0$ outside D_ε;

(γ) $\int_{D_\varepsilon} f_\varepsilon\, da = 1$.

Such a function is easily constructed using the procedure given in the first paragraph of Sect. 7. Clearly, the symmetric tensor field $\boldsymbol{S}$ on $\bar{B}$ defined by

$$[\boldsymbol{S}(\boldsymbol{x})] = \begin{bmatrix} 0 & 0 & 0 \\ 0 & 0 & 0 \\ 0 & 0 & f_\varepsilon(x_1, x_2) \end{bmatrix}$$

has zero divergence. Let C_ε be the infinite solid circular cylinder whose axis coincides with the x_3-axis and whose cross-section is D_ε. Then the assumed con-

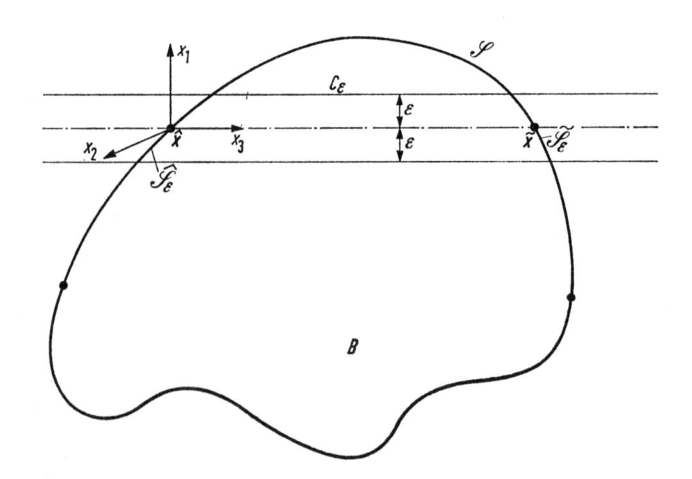

Fig. 10.

[1] GURTIN [1963, *11*].

vexity of B with respect to $\mathscr{S}$ and the regularity of $\mathscr{S}$ imply that for all sufficiently small ε there exist disjoint subregions $\hat{\mathscr{S}}_\varepsilon$ and $\tilde{\mathscr{S}}_\varepsilon$ of $\mathscr{S}$ such that (see Fig. 10)

$$\hat{\boldsymbol{x}} \in \hat{\mathscr{S}}_\varepsilon, \qquad \tilde{\boldsymbol{x}} \in \tilde{\mathscr{S}}_\varepsilon, \qquad C_\varepsilon \cap \partial B = \hat{\mathscr{S}}_\varepsilon \cup \tilde{\mathscr{S}}_\varepsilon.$$

Thus $\boldsymbol{S}$ vanishes near $\partial B - \mathscr{S}$ and

$$0 = \int_{\mathscr{S}} (\boldsymbol{S}\boldsymbol{n}) \cdot \boldsymbol{w}\, da = \int_{\hat{\mathscr{S}}_\varepsilon} f_\varepsilon w_3 n_3\, da + \int_{\tilde{\mathscr{S}}_\varepsilon} f_\varepsilon w_3 n_3\, da. \tag{a}$$

Next, by property (γ) of f_ε,

$$\int_{\hat{\mathscr{S}}_\varepsilon} f_\varepsilon n_3\, da = \int_{D_\varepsilon} f_\varepsilon\, da = 1,$$

$$\int_{\tilde{\mathscr{S}}_\varepsilon} f_\varepsilon n_3\, da = -\int_{D_\varepsilon} f_\varepsilon\, da = -1.$$

Thus if we let $\varepsilon \to 0$ in (a) and use property (α) of f_ε in conjunction with the mean-value theorem of integral calculus, we conclude that

$$w_3(\hat{\boldsymbol{x}}) - w_3(\tilde{\boldsymbol{x}}) = 0,$$

or equivalently that

$$[\boldsymbol{w}(\hat{\boldsymbol{x}}) - \boldsymbol{w}(\tilde{\boldsymbol{x}})] \cdot [\hat{\boldsymbol{x}} - \tilde{\boldsymbol{x}}] = 0.$$

Since $\hat{\boldsymbol{x}}$ and $\tilde{\boldsymbol{x}}$ are arbitrary regular points of $\mathring{\mathscr{S}}$, and since $\boldsymbol{w}$ is continuous on $\mathscr{S}$, $\boldsymbol{w}$ has the projection property on $\mathscr{S}$. Moreover, from our hypotheses it follows that $\mathscr{S}$ is a non-coplanar point set; hence we conclude from ***(13.3)*** that $\boldsymbol{w}$ is rigid. □

(4)[1] *Let $\hat{\boldsymbol{u}}$ be a continuous vector field on a regular subsurface $\mathscr{S}$ of ∂B, and let B be simply-connected and convex with respect to $\mathscr{S}$. Let $\boldsymbol{E}$ be a symmetric tensor field that is continuous on $\bar{B}$ and of class C^2 on B. Further, suppose that*

$$\int_{\mathscr{S}} (\boldsymbol{S}\boldsymbol{n}) \cdot \hat{\boldsymbol{u}}\, da = \int_B \boldsymbol{S} \cdot \boldsymbol{E}\, dv$$

for every class C^∞ symmetric tensor field $\boldsymbol{S}$ on $\bar{B}$ that vanishes near $\partial B - \mathscr{S}$ and satisfies

$$\operatorname{div} \boldsymbol{S} = \boldsymbol{0}.$$

Then there exists an admissible displacement field $\boldsymbol{u}$ such that

$$\boldsymbol{E} = \tfrac{1}{2}(\nabla \boldsymbol{u} + \nabla \boldsymbol{u}^T),$$

$$\boldsymbol{u} = \hat{\boldsymbol{u}} \quad \text{on } \mathscr{S}.$$

Proof. Let $\mathscr{D}$ denote the set of all class C^∞ symmetric tensor fields on $\bar{B}$ that satisfy $\operatorname{div} \boldsymbol{S} = \boldsymbol{0}$. By hypothesis

$$\int_B \boldsymbol{S} \cdot \boldsymbol{E}\, dv = 0$$

for every $\boldsymbol{S} \in \mathscr{D}$ that vanishes near ∂B. Thus we conclude from Donati's theorem ***(18.8)*** that $\boldsymbol{E}$ satisfies the equation of compatibility. In view of the compatibility theorem ***(14.2)***, this result and the fact that B is simply-connected imply the existence of an admissible displacement field $\boldsymbol{u}'$ such that $\boldsymbol{E}$ and $\boldsymbol{u}'$ satisfy the

[1] For the case in which $\mathscr{S} = \partial B$ this theorem is due to Dorn and Schild [1956, *1*]. The present more general case is due to Gurtin [1963, *11*].

strain-displacement relation. Thus we conclude from the present hypotheses and ***(18.1)*** that for every $\boldsymbol{S}\in\mathscr{D}$ which vanishes near $\partial B-\mathscr{S}$,

$$\int_{\mathscr{S}}(\boldsymbol{S}\boldsymbol{n})\cdot\boldsymbol{u}'\,da=\int_{\partial B}(\boldsymbol{S}\boldsymbol{n})\cdot\boldsymbol{u}'\,da=\int_B \boldsymbol{S}\cdot\boldsymbol{E}\,dv=\int_{\mathscr{S}}(\boldsymbol{S}\boldsymbol{n})\cdot\hat{\boldsymbol{u}}\,da.$$

Therefore if we let

$$\boldsymbol{w}=\hat{\boldsymbol{u}}-\boldsymbol{u}',$$

then

$$\int_{\mathscr{S}}(\boldsymbol{S}\boldsymbol{n})\cdot\boldsymbol{w}\,da=0$$

for every $\boldsymbol{S}\in\mathscr{D}$ that vanishes near $\partial B-\mathscr{S}$. Thus we may conclude from ***(3)*** that $\boldsymbol{w}$ is the restriction to $\mathscr{S}$ of a rigid displacement field $\tilde{\boldsymbol{w}}$, and the displacement field $\boldsymbol{u}$ defined on $\bar{B}$ through

$$\boldsymbol{u}=\boldsymbol{u}'+\tilde{\boldsymbol{w}}$$

has all of the desired properties. □

We are now in a position to give the

Proof of ***(2)***. Let $\boldsymbol{S}'\in\mathscr{D}$ vanish near $\mathscr{S}_2$. Then $\tilde{\boldsymbol{S}}=\boldsymbol{S}+\boldsymbol{S}'$ is a statically admissible stress field, and

$$\Psi\{\boldsymbol{S}\}\leqq\Psi\{\tilde{\boldsymbol{S}}\}. \tag{a}$$

Let

$$\boldsymbol{E}=\mathsf{K}[\boldsymbol{S}].$$

Then (c) in the proof of ***(34.3)*** holds in the present circumstances, and we conclude from this result, (a), and the definition of Ψ given in ***(34.3)*** that

$$0\leqq\Psi\{\tilde{\boldsymbol{S}}\}-\Psi\{\boldsymbol{S}\}=U_{\mathsf{K}}\{\boldsymbol{S}'\}+\int_B \boldsymbol{S}'\cdot\boldsymbol{E}\,dv-\int_{\mathscr{S}_1}\boldsymbol{s}'\cdot\hat{\boldsymbol{u}}\,da. \tag{b}$$

The inequality (b) must hold for every $\boldsymbol{S}'\in\mathscr{D}$ that vanishes near $\mathscr{S}_2$. If we replace $\boldsymbol{S}'$ in (b) by $\alpha\boldsymbol{S}'$, where α is a scalar, we find that

$$\int_B \boldsymbol{S}'\cdot\boldsymbol{E}\,dv=\int_{\mathscr{S}_1}\boldsymbol{s}'\cdot\hat{\boldsymbol{u}}\,da$$

for every such field $\boldsymbol{S}'$. Thus we conclude from ***(4)*** that there exists a vector field $\boldsymbol{u}$ such that $[\boldsymbol{u},\boldsymbol{E},\boldsymbol{S}]$ is a solution of the mixed problem. □

37. Maximum principles.[1] The principles of minimum potential and complementary energy can be used to compute *upper* bounds for the "energies" $\Phi\{\mathfrak{s}\}$ and $\Psi\{\mathfrak{s}\}$ corresponding to a solution $\mathfrak{s}$. In this section we shall establish two maximum principles which allow one to compute *lower* bounds for $\Phi\{\mathfrak{s}\}$ and $\Psi\{\mathfrak{s}\}$.

We continue to assume that the hypotheses (i)–(v) of Sects. 31 and 34 hold.

As our first step, we extend the domain of the functionals Φ and Ψ to the set of all admissible states in the obvious manner:

$$\Phi\{\mathfrak{s}\}=U_{\mathsf{C}}\{\boldsymbol{E}\}-\int_B \boldsymbol{b}\cdot\boldsymbol{u}\,dv-\int_{\mathscr{S}_2}\hat{\boldsymbol{s}}\cdot\boldsymbol{u}\,da,$$
$$\Psi\{\mathfrak{s}\}=U_{\mathsf{K}}\{\boldsymbol{S}\}-\int_{\mathscr{S}_1}\boldsymbol{s}\cdot\hat{\boldsymbol{u}}\,da$$

for every admissible state $\mathfrak{s}=[\boldsymbol{u},\boldsymbol{E},\boldsymbol{S}]$.

[1] The basic ideas underlying these principles are due to TREFFTZ [1928, *4*], who established analogous results for boundary-value problems associated with the equations $\Delta u=0$ and $\Delta\Delta u=0$. See also SOKOLNIKOFF [1956, *12*], § 118 and MIKHLIN [1957, *10*], §§ 55, 57, 59.

(1) Principle of maximum potential energy.[1] *Let $\mathfrak{s}$ be a solution of the mixed problem, and let $\tilde{\mathfrak{s}}$ be an elastic state that satisfies*

$$\int_{\mathcal{S}_1} \tilde{\mathbf{s}} \cdot (\hat{\boldsymbol{u}} - \tilde{\boldsymbol{u}})\, da \geqq 0,$$

$$\tilde{\mathbf{s}} = \hat{\mathbf{s}} \qquad on\ \mathcal{S}_2.$$

Then

$$\Phi\{\mathfrak{s}\} \geqq \Phi\{\tilde{\mathfrak{s}}\},$$

and equality holds if and only if $\mathfrak{s} = \tilde{\mathfrak{s}}$ modulo a rigid displacement.

Proof. Let

$$\mathfrak{s}' = \mathfrak{s} - \tilde{\mathfrak{s}}.$$

Then ***(28.1)*** and the fact that $\tilde{\boldsymbol{S}} = \mathbf{C}[\tilde{\boldsymbol{E}}]$ imply

$$U_{\mathbf{C}}\{\boldsymbol{E}\} - U_{\mathbf{C}}\{\tilde{\boldsymbol{E}}\} = U_{\mathbf{C}}\{\boldsymbol{E}'\} + \int_B \tilde{\boldsymbol{S}} \cdot \boldsymbol{E}'\, dv.$$

Next, if we apply the theorem of work expended ***(18.2)*** to $\tilde{\boldsymbol{S}}$ and $\boldsymbol{u}'$, we find that

$$\int_B \tilde{\boldsymbol{S}} \cdot \boldsymbol{E}'\, dv = \int_B \boldsymbol{b} \cdot \boldsymbol{u}'\, dv + \int_{\partial B} \tilde{\mathbf{s}} \cdot \boldsymbol{u}'\, da.$$

Thus, since $\boldsymbol{u} = \hat{\boldsymbol{u}}$ on $\mathcal{S}_1$ and $\tilde{\mathbf{s}} = \hat{\mathbf{s}}$ on $\mathcal{S}_2$,

$$\Phi\{\mathfrak{s}\} - \Phi\{\tilde{\mathfrak{s}}\} = U_{\mathbf{C}}\{\boldsymbol{E}'\} + \int_{\mathcal{S}_1} \tilde{\mathbf{s}} \cdot (\hat{\boldsymbol{u}} - \tilde{\boldsymbol{u}})\, da.$$

This completes the proof, since $\mathbf{C}$ is positive definite and the last term non-negative. □

It is clear from this proof that ***(1)*** continues to hold even when $\mathfrak{s}$ is required only to be kinematically admissible.

Note that in the principle of *minimum* potential energy the admissible states were required to satisfy, in essence, only the displacement boundary condition. On the other hand, the "admissible state" $\tilde{\mathfrak{s}}$ in the principle of *maximum* potential energy is required to satisfy all of the field equations, the traction boundary condition, and a weak form of the displacement boundary condition.

(2) Principle of maximum complementary energy. *Let $\mathfrak{s}$ be a solution of the mixed problem, and let $\tilde{\mathfrak{s}}$ be an elastic state that satisfies*

$$\tilde{\boldsymbol{u}} = \hat{\boldsymbol{u}} \qquad on\ \mathcal{S}_1,$$

$$\int_{\mathcal{S}_2} (\hat{\mathbf{s}} - \tilde{\mathbf{s}}) \cdot \tilde{\boldsymbol{u}}\, da \geqq 0.$$

Then

$$\Psi\{\mathfrak{s}\} \geqq \Psi\{\tilde{\mathfrak{s}}\},$$

and equality holds if and only if $\mathfrak{s} = \tilde{\mathfrak{s}}$ modulo a rigid displacement.

Proof. Let

$$\mathfrak{s}' = \mathfrak{s} - \tilde{\mathfrak{s}};$$

then $\mathfrak{s}'$ is an elastic state corresponding to zero body forces. As before,

$$U_{\mathbf{K}}\{\boldsymbol{S}\} - U_{\mathbf{K}}\{\tilde{\boldsymbol{S}}\} = U_{\mathbf{K}}\{\boldsymbol{S}'\} + \int_B \boldsymbol{S}' \cdot \tilde{\boldsymbol{E}}\, dv,$$

[1] Cooperman [1952, *1*].

and by *(18.1)* and the fact that div $\boldsymbol{S}' = \mathbf{0}$,

$$\int_B \boldsymbol{S}' \cdot \tilde{\boldsymbol{E}}\, dv = \int_{\partial B} \mathbf{s}' \cdot \tilde{\boldsymbol{u}}\, da.$$

Thus, since $\tilde{\boldsymbol{u}} = \hat{\boldsymbol{u}}$ on $\mathscr{S}_1$ and $\mathbf{s} = \hat{\mathbf{s}}$ on $\mathscr{S}_2$,

$$\Psi\{\mathfrak{s}\} - \Psi\{\tilde{\mathfrak{s}}\} = U_{\mathbf{K}}\{\boldsymbol{S}'\} + \int_{\mathscr{S}_2} (\hat{\mathbf{s}} - \tilde{\mathbf{s}}) \cdot \tilde{\boldsymbol{u}}\, da,$$

which, in view of our hypotheses, implies the desired result. □

Note that *(2)* remains valid under the weaker hypothesis that $\mathfrak{s}$ be statically admissible.

38. Variational principles. The admissible states appropriate to the minimum principles of elastostatics are required to meet certain of the field equations and boundary conditions. In some applications it is advantageous to use variational principles in which the admissible states satisfy as few constraints as possible. In this section we establish two such principles.[1]

We continue to assume that the data has properties (i)–(iv) of Sect. 31, but in place of (v) of Sect. 34 we assume that

(v′) the elasticity field $\mathbf{C}$ is *symmetric.*

The two variational principles will be concerned with scalar-valued functionals whose domain of definition is a subset $\mathscr{A}$ of the set of all admissible states. Let Λ be such a functional, let

$$\begin{aligned} &\mathfrak{s} \text{ and } \tilde{\mathfrak{s}} \text{ be admissible states,} \\ &\mathfrak{s} + \lambda\tilde{\mathfrak{s}} \in \mathscr{A} \text{ for every scalar } \lambda, \end{aligned} \tag{a}$$

and formally define the notation

$$\delta_{\tilde{\mathfrak{s}}} \Lambda\{\mathfrak{s}\} = \frac{d}{d\lambda} \Lambda\{\mathfrak{s} + \lambda\tilde{\mathfrak{s}}\}\big|_{\lambda=0}.$$

Then we write

$$\delta\Lambda\{\mathfrak{s}\} = 0$$

if $\delta_{\tilde{\mathfrak{s}}}\Lambda\{\mathfrak{s}\}$ exists and equals zero for every choice of $\tilde{\mathfrak{s}}$ consistent with (a).

We begin with a variational principle in which the admissible states are not required to meet any of the field equations, initial conditions, or boundary conditions.

(1) Hu-Washizu principle.[2] *Let $\mathscr{A}$ denote the set of all admissible states, and let Λ be the functional on $\mathscr{A}$ defined by*

$$\begin{aligned} \Lambda\{\mathfrak{s}\} = U_{\mathbf{C}}\{\boldsymbol{E}\} - \int_B \boldsymbol{S} \cdot \boldsymbol{E}\, dv - \int_B (\operatorname{div} \boldsymbol{S} + \boldsymbol{b}) \cdot \boldsymbol{u}\, dv \\ + \int_{\mathscr{S}_1} \mathbf{s} \cdot \hat{\boldsymbol{u}}\, da + \int_{\mathscr{S}_2} (\mathbf{s} - \hat{\mathbf{s}}) \cdot \boldsymbol{u}\, da \end{aligned}$$

---

[1] The basic ideas behind these principles are contained in the work of BORN [1906, *1*], pp. 91–97, who established analogous results for the plane elastica. BORN was cognizant of the fact that his results applied to elasticity theory, as is clear from his statement (p. 96): „Ich will hier bemerken, daß sich der Vorteil dieser Darstellung eigentlich erst zeigt, wenn man sie auf die allgemeine Elastizitätstheorie anwendet.“ See also ORAVAS and MCLEAN [1966, *19*], pp. 927–929 for a detailed study of the early historical development of these principles.

[2] HU [1955, *5*], WASHIZU [1955, *14*], [1968, *15*], § 2.3. See also DE VEUBEKE [1965, *6*], TONTI [1967, *16*, *17*], and HLÁVAČEK [1967, *6*, *7*]. An extension valid for discontinuous displacement and stress fields was given by PRAGER [1967, *11*], and one in which the stress fields of the admissible states are not required to be symmetric was given by REISSNER [1965, *16*].

for every $\mathfrak{s} = [\boldsymbol{u}, \boldsymbol{E}, \boldsymbol{S}] \in \mathscr{A}$. *Then*

$$\delta \Lambda\{\mathfrak{s}\} = 0$$

at an admissible state $\mathfrak{s}$ *if and only if* $\mathfrak{s}$ *is a solution of the mixed problem.*

Proof. Let $\mathfrak{s} = [\boldsymbol{u}, \boldsymbol{E}, \boldsymbol{S}]$ and $\tilde{\mathfrak{s}} = [\tilde{\boldsymbol{u}}, \tilde{\boldsymbol{E}}, \tilde{\boldsymbol{S}}]$ be admissible states. Then $\mathfrak{s} + \lambda \tilde{\mathfrak{s}} \in \mathscr{A}$ for every scalar λ, and in view of ***(28.1)*** and the symmetry of $\mathbf{C}$,

$$U_{\mathbf{C}}\{\boldsymbol{E} + \lambda \tilde{\boldsymbol{E}}\} = U_{\mathbf{C}}\{\boldsymbol{E}\} + \lambda^2 U_{\mathbf{C}}\{\tilde{\boldsymbol{E}}\} + \lambda \int_B \tilde{\boldsymbol{E}} \cdot \mathbf{C}[\boldsymbol{E}]\, dv.$$

Thus, since

$$\delta_{\tilde{\mathfrak{s}}} \Lambda\{\mathfrak{s}\} = \frac{d}{d\lambda} \Lambda\{\mathfrak{s} + \lambda \tilde{\mathfrak{s}}\}\big|_{\lambda=0},$$

it follows that

$$\delta_{\tilde{\mathfrak{s}}} \Lambda\{\mathfrak{s}\} = \int_B \{(\mathbf{C}[\boldsymbol{E}] - \boldsymbol{S}) \cdot \tilde{\boldsymbol{E}} - (\operatorname{div} \boldsymbol{S} + \boldsymbol{b}) \cdot \tilde{\boldsymbol{u}} - \tilde{\boldsymbol{S}} \cdot \boldsymbol{E} - \boldsymbol{u} \cdot \operatorname{div} \tilde{\boldsymbol{S}}\}\, dv$$
$$+ \int_{\mathscr{S}_1} \tilde{\mathbf{s}} \cdot \hat{\boldsymbol{u}}\, da + \int_{\mathscr{S}_2} \{\tilde{\mathbf{s}} \cdot \boldsymbol{u} + (\mathbf{s} - \hat{\mathbf{s}}) \cdot \tilde{\boldsymbol{u}}\}\, da.$$

If we apply ***(18.1)*** to $\tilde{\boldsymbol{S}}$ and $\boldsymbol{u}$, we find that

$$\int_B \boldsymbol{u} \cdot \operatorname{div} \tilde{\boldsymbol{S}}\, dv = \int_{\partial B} \boldsymbol{u} \cdot \tilde{\mathbf{s}}\, da - \int_B \tfrac{1}{2} (\nabla \boldsymbol{u} + \nabla \boldsymbol{u}^T) \cdot \tilde{\boldsymbol{S}}\, dv;$$

thus

$$\begin{aligned} \delta_{\tilde{\mathfrak{s}}} \Lambda\{\mathfrak{s}\} = & \int_B (\mathbf{C}[\boldsymbol{E}] - \boldsymbol{S}) \cdot \tilde{\boldsymbol{E}}\, dv - \int_B (\operatorname{div} \boldsymbol{S} + \boldsymbol{b}) \cdot \tilde{\boldsymbol{u}}\, dv \\ & + \int_B \{\tfrac{1}{2} (\nabla \boldsymbol{u} + \nabla \boldsymbol{u}^T) - \boldsymbol{E}\} \cdot \tilde{\boldsymbol{S}}\, dv + \int_{\mathscr{S}_1} (\hat{\boldsymbol{u}} - \boldsymbol{u}) \cdot \tilde{\mathbf{s}}\, da \\ & + \int_{\mathscr{S}_2} (\mathbf{s} - \hat{\mathbf{s}}) \cdot \tilde{\boldsymbol{u}}\, da. \end{aligned} \tag{a}$$

If $\mathfrak{s}$ is a solution to the mixed problem, then (a) yields

$$\delta_{\tilde{\mathfrak{s}}} \Lambda\{\mathfrak{s}\} = 0 \qquad \text{for every } \tilde{\mathfrak{s}} \in \mathscr{A}, \tag{b}$$

which implies

$$\delta \Lambda\{\mathfrak{s}\} = 0. \tag{c}$$

To prove the converse assertion assume that (c), and hence (b), holds. If we choose $\tilde{\mathfrak{s}} = [\tilde{\boldsymbol{u}}, \boldsymbol{0}, \boldsymbol{0}]$ and let $\tilde{\boldsymbol{u}}$ vanish near ∂B, then it follows from (a) and (b) that

$$\int_B (\operatorname{div} \boldsymbol{S} + \boldsymbol{b}) \cdot \tilde{\boldsymbol{u}}\, dv = 0.$$

Since this relation must hold for every such $\tilde{\boldsymbol{u}}$ of class C^1 on $\bar{B}$, we conclude from the fundamental lemma ***(7.1)*** that $\operatorname{div} \boldsymbol{S} + \boldsymbol{b} = \boldsymbol{0}$. Next, let $\tilde{\mathfrak{s}} = [\tilde{\boldsymbol{u}}, \boldsymbol{0}, \boldsymbol{0}]$, but this time require only that $\tilde{\boldsymbol{u}}$ vanish near $\mathscr{S}_1$. Then (a) and (b) imply that

$$\int_{\mathscr{S}_2} (\mathbf{s} - \hat{\mathbf{s}}) \cdot \tilde{\boldsymbol{u}}\, da = 0,$$

and we conclude from ***(35.1)*** that $\mathbf{s} = \hat{\mathbf{s}}$ on $\mathscr{S}_1$. Now, let $\tilde{\mathfrak{s}} = [\boldsymbol{0}, \tilde{\boldsymbol{E}}, \boldsymbol{0}]$ and suppose $\tilde{\boldsymbol{E}}$ vanishes near ∂B. Then by (a) and (b),

$$\int_B (\mathbf{C}[\boldsymbol{E}] - \boldsymbol{S}) \cdot \tilde{\boldsymbol{E}}\, dv = 0.$$

Thus, since $\mathbf{C}[\boldsymbol{E}] - \boldsymbol{S}$ and $\tilde{\boldsymbol{E}}$ are symmetric tensor fields, it follows from ***(7.1)*** with $\mathscr{W}$ equal to the set of all symmetric tensors that $\boldsymbol{S} = \mathbf{C}[\boldsymbol{E}]$. In the same manner, choosing $\tilde{\mathfrak{s}}=[0, 0, \tilde{\boldsymbol{S}}]$, where $\tilde{\boldsymbol{S}}$ vanishes near ∂B, we conclude that $\boldsymbol{E}=\frac{1}{2}(\nabla \boldsymbol{u}+\nabla \boldsymbol{u}^T)$. Finally, if we drop the requirement that $\tilde{\boldsymbol{S}}$ vanish near ∂B, we conclude from (a) and (b) that

$$\int_{\mathscr{S}_1} (\hat{\boldsymbol{u}}-\boldsymbol{u}) \cdot (\tilde{\boldsymbol{S}}\boldsymbol{n})\, da=0$$

for every class C^1 symmetric tensor field $\tilde{\boldsymbol{S}}$ on $\bar{B}$, and it follows from ***(35.2)*** that $\boldsymbol{u}=\hat{\boldsymbol{u}}$ on $\mathscr{S}_1$. Thus $\mathfrak{s}=[\boldsymbol{u}, \boldsymbol{E}, \boldsymbol{S}]$ is a solution of the mixed problem. □

(2) Hellinger-Prange-Reissner principle.[1] *Assume that the elasticity field is invertible and that its inverse* K *is smooth on* B. *Let* $\mathscr{A}$ *denote the set of all admissible states that satisfy the strain-displacement relation, and let* Θ *be the functional on* $\mathscr{A}$ *defined by*

$$\Theta\{\mathfrak{s}\}=U_{\mathsf{K}}\{\boldsymbol{S}\}-\int_B \boldsymbol{S}\cdot\boldsymbol{E}\, dv+\int_B \boldsymbol{b}\cdot\boldsymbol{u}\, dv+\int_{\mathscr{S}_1} \mathbf{s}\cdot(\boldsymbol{u}-\hat{\boldsymbol{u}})\, da+\int_{\mathscr{S}_2} \hat{\mathbf{s}}\cdot\boldsymbol{u}\, da$$

for every $\mathfrak{s}=[\boldsymbol{u}, \boldsymbol{E}, \boldsymbol{S}]\in\mathscr{A}$. *Then*

$$\delta\Theta\{\mathfrak{s}\}=0$$

at $\mathfrak{s}\in\mathscr{A}$ *if and only if* $\mathfrak{s}$ *is a solution of the mixed problem.*

Proof. Let $\mathfrak{s}=[\boldsymbol{u}, \boldsymbol{E}, \boldsymbol{S}]$ and $\tilde{\mathfrak{s}}=[\tilde{\boldsymbol{u}}, \tilde{\boldsymbol{E}}, \tilde{\boldsymbol{S}}]$ be admissible states, and suppose that $\mathfrak{s}+\lambda\tilde{\mathfrak{s}}\in\mathscr{A}$ for every scalar λ, or equivalently that $\mathfrak{s}, \tilde{\mathfrak{s}}\in\mathscr{A}$. Then, in view of the symmetry of C, K is symmetric and

$$\delta_{\tilde{\mathfrak{s}}}\, \Theta\{\mathfrak{s}\}=\int_B \{(\mathsf{K}[\boldsymbol{S}]-\boldsymbol{E})\cdot\tilde{\boldsymbol{S}}-\boldsymbol{S}\cdot\tilde{\boldsymbol{E}}+\boldsymbol{b}\cdot\tilde{\boldsymbol{u}}\}\, dv$$
$$+\int_{\mathscr{S}_1} \{\tilde{\mathbf{s}}\cdot(\boldsymbol{u}-\hat{\boldsymbol{u}})+\mathbf{s}\cdot\tilde{\boldsymbol{u}}\}\, da+\int_{\mathscr{S}_2} \hat{\mathbf{s}}\cdot\tilde{\boldsymbol{u}}\, da.$$

If we apply ***(18.1)*** to $\boldsymbol{S}$ and $\tilde{\boldsymbol{u}}$ and use the fact that $\tilde{\boldsymbol{E}}$ and $\tilde{\boldsymbol{u}}$ satisfy the strain-displacement relation, we find that

$$\int_B \boldsymbol{S}\cdot\tilde{\boldsymbol{E}}\, dv=\int_{\partial B} \mathbf{s}\cdot\tilde{\boldsymbol{u}}\, da-\int_B \tilde{\boldsymbol{u}}\cdot\operatorname{div}\boldsymbol{S}\, dv;$$

thus

$$\delta_{\tilde{\mathfrak{s}}}\Theta\{\mathfrak{s}\}=\int_B (\mathsf{K}[\boldsymbol{S}]-\boldsymbol{E})\cdot\tilde{\boldsymbol{S}}\, dv+\int_B (\operatorname{div}\boldsymbol{S}+\boldsymbol{b})\cdot\tilde{\boldsymbol{u}}\, dv$$
$$+\int_{\mathscr{S}_1} (\boldsymbol{u}-\hat{\boldsymbol{u}})\cdot\tilde{\mathbf{s}}\, da+\int_{\mathscr{S}_2} (\hat{\mathbf{s}}-\mathbf{s})\cdot\tilde{\boldsymbol{u}}\, da. \tag{a}$$

If $\mathfrak{s}$ is a solution of the mixed problem, then, clearly,

$$\delta_{\tilde{\mathfrak{s}}}\Theta\{\mathfrak{s}\}=0 \quad \text{for every } \tilde{\mathfrak{s}}\in\mathscr{A}, \tag{b}$$

which implies $\delta\Theta(\mathfrak{s})=0$. On the other hand, (a), (b), ***(7.1)***, ***(35.1)***, ***(35.2)***, and the fact that $\mathfrak{s}$ satisfies the strain-displacement relation imply that $\mathfrak{s}$ is a solution of the mixed problem. □

Let $\mathfrak{s}=[\boldsymbol{u}, \boldsymbol{E}, \boldsymbol{S}]$ be kinematically admissible. Then we conclude from ***(18.1)*** that

$$\int_B \boldsymbol{u}\cdot\operatorname{div}\boldsymbol{S}\, dv=\int_{\mathscr{S}_1} \mathbf{s}\cdot\hat{\boldsymbol{u}}\, da+\int_{\mathscr{S}_2} \mathbf{s}\cdot\boldsymbol{u}\, da-\int_B \boldsymbol{S}\cdot\boldsymbol{E}\, dv.$$

[1] The basic idea is contained in the work of HELLINGER [1914, *1*]. As a proved theorem it was first given by PRANGE [1916, *1*], pp. 54–57 for the traction problem and REISSNER [1950, *10*], [1958, *14*], [1961, *18*] for the mixed problem. See also RÜDIGER [1960, *10*], [1961, *19*], HLÁVAČEK [1967, *6*, *7*], TONTI [1967, *16*, *17*], HORÁK [1968, *8*], SOLOMON [1968, *12*].

Thus, in this instance, $\Lambda\{s\}$, given by *(1)*, reduces to $\Phi\{s\}$, where Φ is the functional of the principle of minimum potential energy *(34.1)*. Also, when s is kinematically admissible,

$$\frac{1}{2}\int_{\mathscr{B}} \boldsymbol{S}\cdot\boldsymbol{E}\,dv = U_{\mathbf{C}}\{\boldsymbol{E}\} = U_{\mathbf{K}}\{\boldsymbol{S}\},$$

and $\Theta\{s\}$ given by *(2)* reduces to $-\Phi\{s\}$. On the other hand, if s is a statically admissible state that satisfies the stress-strain relation, then $\Lambda\{s\}$ reduces to $-\Psi\{s\}$, where Ψ is the functional of the principle of minimum complementary energy *(34.3)*; if s is a statically admissible state that obeys the strain-displacement relation, then $\Theta\{s\}$ reduces to $\Psi\{s\}$.

Table 5 below compares the two variational principles established here with the minimum principles discussed previously.

Table 5.

Principle	Field equations satisfied by the admissible states	Boundary conditions satisfied by the admissible states
Minimum potential energy	Strain-displacement stress-strain	Displacement
Minimum complementary energy	Stress equation of equilibrium	Traction
Maximum potential energy	All	Traction, weak form of displacement
Maximum complementary energy	All	Displacement, weak form of traction
Hu-Washizu	None	None
Hellinger-Prange-Reissner	Strain-displacement	None

39. Convergence of approximate solutions. Recall that the functional of the principle of minimum potential energy *(34.2)* has the form

$$\Phi\{\boldsymbol{u}\} = U\{\boldsymbol{u}\} - \int_B \boldsymbol{b}\cdot\boldsymbol{u}\,dv - \int_{\mathscr{S}_2} \hat{\mathbf{s}}\cdot\boldsymbol{u}\,da, \tag{a}$$

where

$$U\{\boldsymbol{u}\} = \frac{1}{2}\int_B \nabla\boldsymbol{u}\cdot\mathbf{C}[\nabla\boldsymbol{u}]\,dv$$

is the strain energy written, for convenience, as a functional of the displacement field.

The standard method of obtaining an approximate solution to the mixed problem is to minimize the functional Φ over a restricted class of functions.[1] That is, one assumes an approximate solution $\boldsymbol{u}_N$ in the form

$$\boldsymbol{u}_N = \hat{\boldsymbol{u}}_N + \sum_{n=1}^{N} \alpha_n \boldsymbol{f}_n, \tag{b}$$

where $\boldsymbol{f}_1, \boldsymbol{f}_2, \ldots, \boldsymbol{f}_N$ are *given* functions that vanish on $\mathscr{S}_1$, and $\hat{\boldsymbol{u}}_N$ is a function that approximates the boundary data $\hat{\boldsymbol{u}}$ on $\mathscr{S}_1$. Of course, the term $\hat{\boldsymbol{u}}_N$ is omitted when $\mathscr{S}_1$ is empty. The constants $\alpha_1, \alpha_2, \ldots, \alpha_N$ are then chosen so as to render $\Phi\{\boldsymbol{u}_N\}$ a minimum. Indeed, if we write

$$\Phi(\alpha_1, \alpha_2, \ldots, \alpha_N) = \Phi\{\boldsymbol{u}_N\},$$

[1] This idea appears first in the work of Rayleigh (1877) [1945, *6*] and Ritz [1908, *2*]; the method is usually referred to as the Rayleigh-Ritz method.

where $\boldsymbol{u}_N$ is given by (b), then (assuming that $\mathbf{C}$ is symmetric)

$$\Phi(\alpha_1, \alpha_2, \ldots, \alpha_N) = a + \tfrac{1}{2} \sum_{m,n=1}^{N} D_{mn}\, \alpha_m\, \alpha_n + \sum_{n=1}^{N} d_n\, \alpha_n,$$

where

$$D_{mn} = \int_B \nabla \boldsymbol{f}_m \cdot \mathbf{C}[\nabla \boldsymbol{f}_n]\, dv, \tag{c}$$

$$a = U\{\hat{\boldsymbol{u}}_N\} - \int_B \boldsymbol{b} \cdot \hat{\boldsymbol{u}}_N\, dv - \int_{\mathscr{S}_2} \hat{\mathbf{s}} \cdot \hat{\boldsymbol{u}}_N,$$

$$d_n = \int_B \nabla \hat{\boldsymbol{u}}_N \cdot \mathbf{C}[\nabla \boldsymbol{f}_n]\, dv - \int_B \boldsymbol{b} \cdot \boldsymbol{f}_n\, dv - \int_{\mathscr{S}_2} \hat{\mathbf{s}} \cdot \boldsymbol{f}_n\, da. \tag{d}$$

If $\mathscr{S}_1$ is *empty*, the above relations still hold, but with $\hat{\boldsymbol{u}}_N = \mathbf{0}$, so that

$$a = 0,$$

$$d_n = -\int_B \boldsymbol{b} \cdot \boldsymbol{f}_n\, dv - \int_{\partial B} \hat{\mathbf{s}} \cdot \boldsymbol{f}_n\, da. \tag{e}$$

If $\mathbf{C}$ is positive semi-definite, the matrix $[D_{mn}]$ will be positive semi-definite, and $\Phi(\alpha_1, \alpha_2, \ldots, \alpha_N)$ will be a minimum at $\alpha_1 = \hat{\alpha}_1$, $\alpha_2 = \hat{\alpha}_2, \ldots, \alpha_N = \hat{\alpha}_N$ if and only if $\hat{\alpha}_1, \hat{\alpha}_2, \ldots, \hat{\alpha}_N$ is a solution of the following system of equations:

$$\sum_{n=1} D_{mn}\, \hat{\alpha}_n = -d_m \qquad (m = 1, 2, \ldots, N). \tag{f}$$

We will now establish conditions under which solutions of (f) exist, and under which the resulting approximate solutions $\boldsymbol{u}_N$ converge in energy to the actual solution as $N \to \infty$.

We assume for the remainder of this section that $\mathbf{C}$ is *symmetric* and *positive definite*, and that *hypotheses* (i)–(iv) *of* Sect. 31 *hold.*

We write $\mathscr{C}_0$ for the set of all continuous and piecewise smooth vector fields on $\bar{B}$.

(1) Existence of approximate solutions. *Let $\mathscr{F}_N$ be an N-dimensional subspace of $\mathscr{C}_0$ with the property that each $\boldsymbol{f} \in \mathscr{F}_N$ vanishes* on $\mathscr{S}_1$.

(i) *If $\mathscr{S}_1 \neq \emptyset$, let $\hat{\boldsymbol{u}}_N$ be a given function in $\mathscr{C}_0$, and let $\hat{\mathscr{F}}_N$ be the set of all functions $\boldsymbol{f}$ of the form*

$$\boldsymbol{f} = \boldsymbol{g} + \hat{\boldsymbol{u}}_N, \qquad \boldsymbol{g} \in \mathscr{F}_N.$$

(ii) *If $\mathscr{S}_1 = \emptyset$, let $\hat{\mathscr{F}}_N = \mathscr{F}_N$.*

Then there exists a function $\boldsymbol{u}_N \in \hat{\mathscr{F}}_N$ such that

$$\Phi\{\boldsymbol{u}_N\} \leq \Phi\{\boldsymbol{v}\} \qquad \textit{for every } \boldsymbol{v} \in \hat{\mathscr{F}}_N. \tag{g}$$

If $\mathscr{S}_1 \neq \emptyset$, then $\boldsymbol{u}_N$ is unique; if $\mathscr{S}_1 = \emptyset$, then any two solutions of (g) *differ by a rigid displacement. Finally, $\boldsymbol{u}_N$ is optimal in the following sense: If $\boldsymbol{u}$ is the displacement field corresponding to a solution of the mixed problem, then*[1]

$$U\{\boldsymbol{u} - \boldsymbol{u}_N\} = \inf_{\boldsymbol{v} \in \hat{\mathscr{F}}_N} U\{\boldsymbol{u} - \boldsymbol{v}\}. \tag{h}$$

Proof. Let $\boldsymbol{f}_1, \boldsymbol{f}_2, \ldots, \boldsymbol{f}_N$ be a basis for $\mathscr{F}_N$, and let D_{mn} and d_n be defined by (c), (d), and (e). Then, clearly, to establish the existence of a solution $\boldsymbol{u}_N$ of (g) it suffices to establish the existence of a solution $\hat{\alpha}_1, \hat{\alpha}_2, \ldots, \hat{\alpha}_N$ of (f).

---

[1] Cf. SCHULTZ [1969, 7], Theorem 2.4.

Since $\mathbf{C}$ is positive definite, we conclude from (c) that

$$\sum_{m,n=1}^{N} D_{mn}\,\alpha_m\,\alpha_n \geqq 0 \tag{i}$$

for any N-tuple $(\alpha_1, \alpha_2, \ldots, \alpha_N)$. Assume that

$$\sum_{m,n=1}^{N} D_{mn}\,\alpha_m\,\alpha_n = 0, \tag{j}$$

and let

$$\boldsymbol{f} = \sum_{n=1}^{N} \alpha_n \boldsymbol{f}_n. \tag{k}$$

Then it follows from (c) and (j) that

$$\int_B \nabla \boldsymbol{f} \cdot \mathbf{C}[\nabla \boldsymbol{f}]\, dv = 0,$$

and since $\mathbf{C}$ is positive definite, this, in turn, implies that $\boldsymbol{f}$ is rigid. Assume first that $\mathscr{S}_1 \neq \emptyset$. Then $\boldsymbol{f} \equiv \mathbf{0}$, since it is rigid and vanishes on $\mathscr{S}_1$. Consequently, as $\boldsymbol{f}_1, \boldsymbol{f}_2, \ldots, \boldsymbol{f}_N$ is a basis for $\mathscr{F}_N$, this implies that

$$\alpha_1 = \alpha_2 = \cdots = \alpha_N = 0. \tag{l}$$

Therefore, if $\mathscr{S}_1 \neq \emptyset$, then (j) implies (l), and we conclude from (i) that the matrix $[D_{mn}]$ is positive definite; hence in this instance (f) has a unique solution.

Assume next that $\mathscr{S}_1 = \emptyset$. Suppose that

$$\sum_{n=1}^{N} D_{mn}\,\alpha_n = 0; \tag{m}$$

then (j) holds, and hence $\boldsymbol{f}$ defined by (k) is rigid. Thus any two solutions of (g) (if they exist) differ by a rigid displacement. Next, by (e) and (k),

$$\sum_{n=1}^{N} \alpha_n\, d_n = -\int_B \boldsymbol{b} \cdot \boldsymbol{f}\, dv - \int_{\partial B} \hat{\mathbf{s}} \cdot \boldsymbol{f}\, da,$$

and it follows[1] from ***(18.3)*** that

$$\sum_{n=1}^{N} \alpha_n\, d_n = 0.$$

Thus $(d_1, d_2, \ldots, d_N)$ is orthogonal to every solution $(\alpha_1, \alpha_2, \ldots, \alpha_N)$ of the homogeneous equation (m), and we conclude from the Fredholm alternative that the inhomogeneous equation (f) has a solution.

We have only to show that $\boldsymbol{u}_N$ is optimal. To facilitate the proof of this assertion, we now establish the following

(2) Lemma.[2] *Let $[\boldsymbol{u}, \boldsymbol{E}, \boldsymbol{S}]$ be a solution of the mixed problem. Then*

$$\Phi\{\boldsymbol{v}\} - \Phi\{\boldsymbol{u}\} = U\{\boldsymbol{v} - \boldsymbol{u}\} + \int_{\mathscr{S}_1} \mathbf{s} \cdot (\boldsymbol{v} - \hat{\boldsymbol{u}})\, da$$

for every $\boldsymbol{v} \in \mathscr{C}_0$.

Proof. Let

$$\boldsymbol{u}' = \boldsymbol{v} - \boldsymbol{u}, \qquad \boldsymbol{E}' = \hat{\nabla} \boldsymbol{u}'.$$

---

[1] Here we tacitly assume that (when $\mathscr{S}_1 = \emptyset$) the external force system $[\boldsymbol{b}, \hat{\mathbf{s}}]$ is in equilibrium.

[2] Cf. Tong and Pian [1967, *15*], Eq. (2.16).

Then by *(28.1)*,

$$U\{\boldsymbol{v}\} - U\{\boldsymbol{u}\} = U\{\boldsymbol{u}'\} + \int_B \boldsymbol{S}\cdot\boldsymbol{E}'\,dv.$$

Further, since $\operatorname{div}\boldsymbol{S}+\boldsymbol{b}=\boldsymbol{0}$, we conclude from *(18.2)* that

$$\int_B \boldsymbol{S}\cdot\boldsymbol{E}'\,dv = \int_{\partial B} \mathbf{s}\cdot\boldsymbol{u}'\,da + \int_B \boldsymbol{b}\cdot\boldsymbol{u}'\,dv.$$

The last three equations and (a) imply the desired result, since $\boldsymbol{u}=\hat{\boldsymbol{u}}$ on $\mathscr{S}_1$ and $\mathbf{s}=\hat{\mathbf{s}}$ on $\mathscr{S}_2$. □

It is of interest to note that the principle of minimum potential energy in the form *(34.2)* follows as a direct corollary of this lemma. Indeed, if $\boldsymbol{v}$ is kinematically admissible, then $\boldsymbol{v}=\hat{\boldsymbol{u}}$ on $\mathscr{S}_1$, and *(2)* implies that $\Phi\{\boldsymbol{v}\}\geqq\Phi\{\boldsymbol{u}\}$.

We are now in a position to complete the proof of *(1)*. Assume that $\mathscr{S}_1\neq\emptyset$. By hypothesis, every $\boldsymbol{v}\in\hat{\mathscr{F}}_N$ satisfies

$$\boldsymbol{v}=\hat{\boldsymbol{u}}_N \quad \text{on } \mathscr{S}_1.$$

Thus, since $\boldsymbol{u}_N\in\hat{\mathscr{F}}_N$, it follows from *(2)* and (g) that

$$\begin{aligned}U\{\boldsymbol{u}_N-\boldsymbol{u}\} &= \Phi\{\boldsymbol{u}_N\}-\Phi\{\boldsymbol{u}\}-\int_{\mathscr{S}_1}\mathbf{s}\cdot(\hat{\boldsymbol{u}}_N-\hat{\boldsymbol{u}})\,da\\ &\leqq \Phi\{\boldsymbol{v}\}-\Phi\{\boldsymbol{u}\}-\int_{\mathscr{S}_1}\mathbf{s}\cdot(\hat{\boldsymbol{u}}_N-\hat{\boldsymbol{u}})\,da = U\{\boldsymbol{v}-\boldsymbol{u}\}\end{aligned}$$

for every $\boldsymbol{v}\in\hat{\mathscr{F}}_N$. Thus

$$U\{\boldsymbol{u}_N-\boldsymbol{u}\}\leqq U\{\boldsymbol{v}-\boldsymbol{u}\},$$

which implies (h). The proof when $\mathscr{S}_1=\emptyset$ is strictly analogous. □

We now establish the convergence of the approximate solutions established in *(1)*. Thus for each $N=1, 2, \ldots$ let $\hat{\boldsymbol{u}}_N$ and $\hat{\mathscr{F}}_N$ satisfy the hypotheses of *(1)*. We assume that every sufficiently smooth function that satisfies the displacement boundary condition can be approximated arbitrarily closely in *energy* by a sequence of elements of $\hat{\mathscr{F}}_1, \hat{\mathscr{F}}_2, \ldots$. More precisely, we assume that

(A) given any kinematically admissible displacement field $\boldsymbol{u}$, there exists a sequence $\{\boldsymbol{v}_N\}$ with $\boldsymbol{v}_N\in\hat{\mathscr{F}}_N$ such that

$$U\{\boldsymbol{u}-\boldsymbol{v}_N\}\to 0 \quad \text{as} \quad N\to\infty. \tag{n}$$

(3) Approximation theorem.[1] *For each $N=1, 2, \ldots$ let $\hat{\boldsymbol{u}}_N$ and $\hat{\mathscr{F}}_N$ satisfy the hypotheses of **(1)**, and assume that (A) holds. Let $\{\boldsymbol{u}_N\}$ be a sequence of approximate solutions,* i.e. *solutions of* (g), *and let $\boldsymbol{u}$ be the displacement field corresponding to a solution of the mixed problem. Then*

$$U\{\boldsymbol{u}-\boldsymbol{u}_N\}\to 0 \quad \textit{as} \quad N\to\infty. \tag{o}$$

[1] Most of the ideas underlying this theorem are contained in the following works: COURANT and HILBERT [1953, *8*], pp. 175–176; MIKHLIN [1957, *10*], pp. 88–95; FRIEDRICHS and KELLER [1965, *8*]; KEY [1966, *11*]; TONG and PIAN [1967, *15*]. These studies contain general results on the convergence of the Rayleigh-Ritz and finite element methods.

In the usual applications of the Rayleigh-Ritz procedure

$$\mathscr{F}_N\subset\mathscr{F}_{N+1}. \tag{*}$$

The abstract formulation given here also includes the finite element method (see, e.g., ZIENKIEWICZ and CHEUNG [1967, *19*], TONG and PIAN [1967, *15*], ZIENKIEWICZ [1970, *4*]) for which (*) is not necessarily satisfied.

Proof. By (A) there exists a sequence $\{\boldsymbol{v}_N\}$ with $\boldsymbol{v}_N \in \hat{\mathscr{F}}_N$ such that (n) holds. On the other hand, (h) and the fact that C is positive definite imply

$$0 \leqq U\{\boldsymbol{u}-\boldsymbol{u}_N\} \leqq U\{\boldsymbol{u}-\boldsymbol{v}_N\},$$

and the desired result is an immediate consequence of (n). □

V. The general boundary-value problem. The contact problem.

40. Statement of the problem. Uniqueness. In the contact problem of elastostatics[1] the normal component of the displacement and the tangential component of the surface traction are prescribed over a portion $\mathscr{S}_3$ of the boundary:

$$\boldsymbol{u} \cdot \boldsymbol{n} = \hat{u} \quad \text{and} \quad \boldsymbol{s} - (\boldsymbol{s} \cdot \boldsymbol{n})\, \boldsymbol{n} = \hat{\boldsymbol{s}} \quad \text{on } \mathscr{S}_3,$$

where $\hat{u}$ and $\hat{\boldsymbol{s}}$ are prescribed with $\hat{\boldsymbol{s}}$ tangent to $\mathscr{S}_3$. These relations can also be written in the alternative forms:

$$(\boldsymbol{n} \otimes \boldsymbol{n})\, \boldsymbol{u} = \hat{u}\boldsymbol{n} \quad \text{and} \quad (\mathbf{1} - \boldsymbol{n} \otimes \boldsymbol{n})\, \boldsymbol{s} = \hat{\boldsymbol{s}} \quad \text{on } \mathscr{S}_3.$$

We now generalize this boundary condition as follows. We assume given a tensor field $\boldsymbol{P}$ on ∂B whose value $\boldsymbol{P}(\boldsymbol{x})$ at any $\boldsymbol{x} \in \partial B$ is a *perpendicular projection.* The ***generalized boundary condition*** then takes the form:

$$\boldsymbol{P}\boldsymbol{u} = \hat{\boldsymbol{u}} \quad \text{and} \quad (\mathbf{1} - \boldsymbol{P})\, \boldsymbol{s} = \hat{\boldsymbol{s}} \quad \text{on } \partial B,$$

where $\hat{\boldsymbol{u}}$ and $\hat{\boldsymbol{s}}$ are prescribed vector fields. Since[2]

$$(\mathbf{1} - \boldsymbol{P})\boldsymbol{P} = \boldsymbol{P}(\mathbf{1} - \boldsymbol{P}) = \mathbf{0},$$

$\hat{\boldsymbol{u}}$ and $\hat{\boldsymbol{s}}$ must satisfy the ***consistency condition***

$$(\mathbf{1} - \boldsymbol{P})\, \hat{\boldsymbol{u}} = \mathbf{0} \quad \text{and} \quad \boldsymbol{P}\hat{\boldsymbol{s}} = \mathbf{0} \quad \text{on } \partial B.$$

At a point $\boldsymbol{x}$ for which $\boldsymbol{P}(\boldsymbol{x}) = \mathbf{1}$ we have $\boldsymbol{u}(\boldsymbol{x}) = \hat{\boldsymbol{u}}(\boldsymbol{x})$, and no restriction is placed on $\boldsymbol{s}(\boldsymbol{x})$. Thus displacements are prescribed over the subset of ∂B on which $\boldsymbol{P} = \mathbf{1}$; similarly, surface tractions are prescribed over the subset for which $\boldsymbol{P} = \mathbf{0}$. The mixed problem therefore corresponds to

$$\boldsymbol{P} = \mathbf{1} \quad \text{on } \mathscr{S}_1, \qquad \boldsymbol{P} = \mathbf{0} \quad \text{on } \mathscr{S}_2,$$

where $\mathscr{S}_1$ and $\mathscr{S}_2$ are complementary subsets of ∂B.

If $\boldsymbol{P} = \boldsymbol{n} \otimes \boldsymbol{n}$ at a point, then the normal displacement and the tangential traction are prescribed. Thus the contact problem corresponds to situations in which

$$\boldsymbol{P} = \mathbf{1} \quad \text{on } \mathscr{S}_1, \qquad \boldsymbol{P} = \mathbf{0} \quad \text{on } \mathscr{S}_2, \qquad \boldsymbol{P} = \boldsymbol{n} \otimes \boldsymbol{n} \quad \text{on } \mathscr{S}_3,$$

where $\mathscr{S}_1$, $\mathscr{S}_2$, and $\mathscr{S}_3$ are complementary subsets of ∂B.

In view of the preceding discussion, the general problem may be stated as follows: given an elasticity field C and a body force field $\boldsymbol{b}$ on $\bar{B}$ together with $\boldsymbol{P}$, $\hat{\boldsymbol{u}}$, and $\hat{\boldsymbol{s}}$ on ∂B; find an elastic state $[\boldsymbol{u}, \boldsymbol{E}, \boldsymbol{S}]$ that corresponds to $\boldsymbol{b}$ and satisfies the generalized boundary condition. We call such a state a ***solution of the general problem.***

[1] Sometimes called the mixed-mixed problem.
[2] See p. 9.

(1) Uniqueness theorem for the general problem.[1] *Assume that the elasticity field is positive definite. Then any two solutions of the general problem are equal modulo a rigid displacement.*

Proof. Let $\tilde{\mathfrak{s}}$ and $\bar{\mathfrak{s}}$ be solutions of the general problem, and let

$$\mathfrak{s} = [\boldsymbol{u}, \boldsymbol{E}, \boldsymbol{S}] = \bar{\mathfrak{s}} - \tilde{\mathfrak{s}}.$$

Then $\mathfrak{s}$ is an elastic state corresponding to zero body forces. Moreover,

$$\boldsymbol{P}\boldsymbol{u} = (\boldsymbol{1} - \boldsymbol{P})\,\mathbf{s} = \boldsymbol{0} \quad \text{on } \partial B;$$

thus on ∂B

$$\mathbf{s} \cdot \boldsymbol{u} = [(\boldsymbol{1} - \boldsymbol{P} + \boldsymbol{P})\,\boldsymbol{u}] \cdot [(\boldsymbol{1} - \boldsymbol{P} + \boldsymbol{P})\,\mathbf{s}] = [(\boldsymbol{1} - \boldsymbol{P})\,\boldsymbol{u}] \cdot \boldsymbol{P}\mathbf{s} = [\boldsymbol{P}(\boldsymbol{1} - \boldsymbol{P})\,\boldsymbol{u}] \cdot \mathbf{s} = 0.$$

Therefore we conclude from ***(28.5)*** that $\boldsymbol{u}$ is rigid and $\boldsymbol{E} = \boldsymbol{S} = \boldsymbol{0}$. □

Of course, ***(1)*** yields, as a corollary, the uniqueness theorem ***(32.1)*** for the mixed problem.

41. Extension of the minimum principles. The minimum principles established in Sect. 34 are easily extended to the general problem stated in the previous section.[2] For example, if we define a kinematically admissible state to be an admissible state $[\boldsymbol{u}, \boldsymbol{E}, \boldsymbol{S}]$ that satisfies the strain-displacement relation, the stress-strain relation, and the boundary condition

$$\boldsymbol{P}\boldsymbol{u} = \hat{\boldsymbol{u}} \quad \text{on } \partial B, \tag{a}$$

then the principle of minimum potential energy ***(34.1)*** remains valid for the general problem provided we let

$$\Phi\{\mathfrak{s}\} = U_{\boldsymbol{C}}\{\boldsymbol{E}\} - \int_B \boldsymbol{b} \cdot \boldsymbol{u}\, dv - \int_{\partial B} \hat{\mathbf{s}} \cdot \boldsymbol{u}\, da.$$

In view of the consistency condition and the remarks on p. 9,

$$\hat{\mathbf{s}} \cdot \boldsymbol{u} = (\boldsymbol{1} - \boldsymbol{P})\,\hat{\mathbf{s}} \cdot \boldsymbol{u} = (\boldsymbol{1} - \boldsymbol{P})\,\hat{\mathbf{s}} \cdot (\boldsymbol{1} - \boldsymbol{P})\,\boldsymbol{u},$$

and the integral over ∂B can also be written in the form:

$$\int_{\partial B} (\boldsymbol{1} - \boldsymbol{P})\,\hat{\mathbf{s}} \cdot (\boldsymbol{1} - \boldsymbol{P})\,\boldsymbol{u}\, da. \tag{b}$$

For the mixed problem ($\boldsymbol{P} = \boldsymbol{1}$ on $\mathscr{S}_1$, $\boldsymbol{P} = \boldsymbol{0}$ on $\mathscr{S}_2$) the boundary condition (a) reduces to $\boldsymbol{u} = \hat{\boldsymbol{u}}$ on $\mathscr{S}_1$, and the integral (b) reduces to an integral over $\mathscr{S}_2$. Thus in this instance the extended minimum principle is nothing more than the traditional theorem ***(34.1)***.

On the other hand, if we define a statically admissible stress field to be an admissible stress field $\boldsymbol{S}$ that satisfies the equation of equilibrium and the boundary condition

$$(\boldsymbol{1} - \boldsymbol{P})\,\boldsymbol{S}\boldsymbol{n} = \hat{\mathbf{s}} \quad \text{on } \partial B,$$

then the principle of minimum complementary energy ***(34.3)*** remains valid for the general problem[3] provided we let

$$\Psi\{\boldsymbol{S}\} = U_{\boldsymbol{K}}\{\boldsymbol{S}\} - \int_{\partial B} \mathbf{s} \cdot \hat{\boldsymbol{u}}\, da,$$

---

[1] Cf. STERNBERG and KNOWLES [1966, *24*]. Other uniqueness theorems which are not special cases of ***(1)*** are given by BRAMBLE and PAYNE [1961, *5*] and KNOPS and PAYNE [1971, *2*].

[2] Cf. RÜDIGER [1960, *10*], PRAGER [1967, *11*], HLÁVAČEK [1967, *7*].

[3] An extension of the principle of minimum complementary energy to the mixed-mixed problem was given by STERNBERG and KNOWLES [1966, *24*]. See also HLÁVAČEK [1967, *7*]. The result stated here includes, as a special case, the theorem of STERNBERG and KNOWLES.

or equivalently,

$$\Psi\{S\} = U_K\{S\} - \int_{\partial B} \boldsymbol{P s} \cdot \boldsymbol{P} \hat{\boldsymbol{u}}\, da.$$

In a similar manner, the maximum principles discussed in Sect. 37 and the variational principles given in Sect. 38 can be extended to the general problem.

VI. Homogeneous and isotropic bodies.

42. Properties of elastic displacement fields. As we saw in Sect. 27, when the divergence and the curl of the body force field vanish, the displacement field $\boldsymbol{u}$ is biharmonic, while div $\boldsymbol{u}$ and curl $\boldsymbol{u}$ are harmonic. To prove this we tacitly assumed that $\boldsymbol{u}$ is of class C^4. We now show that this somewhat stringent assumption is unnecessary; we will establish the above assertions under the assumption that $\boldsymbol{u}$ is of class C^2.

By an ***elastic displacement field*** corresponding to the body force field $\boldsymbol{b}$ we mean a class C^2 vector field $\boldsymbol{u}$ on B that satisfies the displacement equation of equilibrium

$$\Delta \boldsymbol{u} + \frac{1}{1-2\nu} \nabla \operatorname{div} \boldsymbol{u} + \frac{1}{\mu} \boldsymbol{b} = \boldsymbol{0}.$$

We assume that Poisson's ratio ν is not equal to $\frac{1}{2}$ or 1, and that $\mu \neq 0$.

(1) Analyticity of elastic displacement fields.[1] *Let* $\boldsymbol{u}$ *be an elastic displacement field corresponding to a smooth body force field* $\boldsymbol{b}$ *on* B *that satisfies*

$$\operatorname{div} \boldsymbol{b} = 0, \qquad \operatorname{curl} \boldsymbol{b} = \boldsymbol{0}.$$

Then $\boldsymbol{u}$ *is analytic.*

Proof. The proof is based on the following notion, due to Sobolev[2] and Friedrichs.[3] A *system of mollifiers* is a one-parameter family of functions ϱ^δ, $\delta > 0$, such that for each fixed δ:

(i) ϱ^δ is a C^∞ scalar field on $\mathscr{V}$;[4]

(ii) $\varrho^\delta \geqq 0$ on $\mathscr{V}$;

(iii) $\varrho^\delta(\boldsymbol{v}) = 0$ whenever $|\boldsymbol{v}| \geqq \delta$;

(iv) $\int_{\mathscr{V}} \varrho^\delta\, dv = 1$.

An example is furnished by

$$\varrho^\delta(\boldsymbol{v}) = \begin{cases} A_\delta\, e^{-\frac{1}{\delta^2 - \boldsymbol{v}^2}}, & |\boldsymbol{v}| < \delta \\ 0, & |\boldsymbol{v}| \geqq \delta, \end{cases}$$

with A_δ chosen such that (iv) is satisfied. Let D be a regular region, and let f be a *continuous* scalar, vector, or tensor field on $\bar{D}$. Then the *system of mollified functions* f^δ, $\delta > 0$, is defined on $\mathscr{E}$ by the transformation

$$f^\delta(\boldsymbol{x}) = \int_D \varrho^\delta(\boldsymbol{x} - \boldsymbol{y})\, f(\boldsymbol{y})\, dv_{\boldsymbol{y}}.$$

Clearly,

(v) *each* f^δ *is a class* C^∞ *function on* $\mathscr{E}$.

---

[1] Friedrichs [1947, *1*]. See also Duffin [1956, *2*], whose proof we give here.

[2] [1935, *6*], [1950, *11*].

[3] [1939, *2*], [1944, *2*], [1947, *1*].

[4] Recall that $\mathscr{V}$ is the vector space associated with $\mathscr{E}$.

By (ii), (iii), and (iv),

$$|f^\delta(\boldsymbol{x}) - f(\boldsymbol{x})| = \Big|\int\limits_{\Sigma_\delta(\boldsymbol{x})} \varrho^\delta(\boldsymbol{x}-\boldsymbol{y})\,[f(\boldsymbol{y}) - f(\boldsymbol{x})]\,dv_{\boldsymbol{y}}\Big|$$
$$\leq \sup\{|f(\boldsymbol{y}) - f(\boldsymbol{x})| : \boldsymbol{y}\in\Sigma_\delta(\boldsymbol{x})\},$$

for sufficiently small δ, where $\Sigma_\delta(\boldsymbol{x})$ is the open ball with radius δ and center at $\boldsymbol{x}$. Hence

(vi) *$f^\delta \to f$ as $\delta \to 0$ uniformly on every closed subregion of D.*

Assume now that f is of class C^1 on $\bar{D}$ and, for convenience, that f is scalar-valued. Let

$$D^\delta = \{\boldsymbol{x}\in D : \Sigma_\delta(\boldsymbol{x}) \subset D\}.$$

In view of the divergence theorem,

$$\nabla_{\boldsymbol{x}} f^\delta(\boldsymbol{x}) = \int\limits_D \nabla_{\boldsymbol{x}}\,\varrho^\delta(\boldsymbol{x}-\boldsymbol{y})\,f(\boldsymbol{y})\,dv_{\boldsymbol{y}} = -\int\limits_D [\nabla_{\boldsymbol{y}}\,\varrho^\delta(\boldsymbol{x}-\boldsymbol{y})]\,f(\boldsymbol{y})\,dv_{\boldsymbol{y}}$$
$$= -\int\limits_{\partial D} \varrho^\delta(\boldsymbol{x}-\boldsymbol{y})\,f(\boldsymbol{y})\,\boldsymbol{n}(\boldsymbol{y})\,da_{\boldsymbol{y}} + \int\limits_D \varrho^\delta(\boldsymbol{x}-\boldsymbol{y})\,\nabla_{\boldsymbol{y}} f(\boldsymbol{y})\,dv_{\boldsymbol{y}}.$$

But (iii) implies that

$$\varrho^\delta(\boldsymbol{x}-\boldsymbol{y}) = 0 \qquad \text{for } \boldsymbol{x}\in D^\delta \text{ and } \boldsymbol{y}\in\partial D.$$

Thus we have the following result:

(vii) *$\nabla(f^\delta) = (\nabla f)^\delta$ on D^δ.*

Of course, (vii) also holds when f is a vector or tensor field.

We are now in a position to complete the proof of the theorem. Let $\boldsymbol{u}$ be an elastic displacement field; let D be a regular region with $\bar{D}\subset B$; let ϱ^δ ($\delta>0$) be a system of mollifiers; let $\boldsymbol{u}^\delta$ and $\boldsymbol{b}^\delta$ ($\delta>0$) be the systems of mollified functions corresponding to $\boldsymbol{u}$ and $\boldsymbol{b}$:

$$\boldsymbol{u}^\delta(\boldsymbol{x}) = \int\limits_D \varrho^\delta(\boldsymbol{x}-\boldsymbol{y})\,\boldsymbol{u}(\boldsymbol{y})\,dv_{\boldsymbol{y}},$$
$$\boldsymbol{b}^\delta(\boldsymbol{x}) = \int\limits_D \varrho^\delta(\boldsymbol{x}-\boldsymbol{y})\,\boldsymbol{b}(\boldsymbol{y})\,dv_{\boldsymbol{y}}.$$

Then, by (v), $\boldsymbol{u}^\delta$ and $\boldsymbol{b}^\delta$ are of class C^∞, and, by (vii), $\boldsymbol{u}^\delta$ is an elastic displacement field on D^δ corresponding to $\boldsymbol{b}^\delta$. Further, $\boldsymbol{b}^\delta$ is divergence-free and curl-free on D^δ. Hence we conclude from the results of Sect. 27 (which hold in the present circumstances) that $\boldsymbol{u}^\delta$ is biharmonic. Next, let

$$\boldsymbol{h} = \Delta\boldsymbol{u}.$$

Then (vii) implies that

$$\boldsymbol{h}^\delta = (\Delta\boldsymbol{u})^\delta = \Delta(\boldsymbol{u}^\delta)$$

on D^δ. Therefore, since $\boldsymbol{u}^\delta$ is biharmonic,

$$\Delta\boldsymbol{h}^\delta = 0$$

on D^δ, and we conclude from (vi) that

$$\boldsymbol{h}^\delta \to \boldsymbol{h}$$

uniformly on every closed subregion of D. The last two results and Harnack's convergence theorem *(8.2)* imply that $\boldsymbol{h}$ is harmonic. But

$$\Delta\boldsymbol{h} = \Delta\Delta\boldsymbol{u};$$

thus to prove that $\boldsymbol{u}$ is biharmonic we have only to show that $\boldsymbol{u}$ is of class C^4. But since $\Delta\boldsymbol{u}$ is equal to a harmonic function, *(6.5)* and *(8.1)* imply that $\boldsymbol{u}$ is of class C^∞ on D. Therefore $\boldsymbol{u}$ is biharmonic and, by *(8.1)*, analytic on D. But D was chosen arbitrarily. Thus $\boldsymbol{u}$ is analytic on B. □

In view of the results given in Sect. 27, we have the following immediate corollary of *(1)*.

(2) Properties of elastic states. *Let $[\boldsymbol{u}, \boldsymbol{E}, \boldsymbol{S}]$ be an elastic state corresponding to a smooth body force field $\boldsymbol{b}$ on B that satisfies*

$$\operatorname{div}\boldsymbol{b}=0, \qquad \operatorname{curl}\boldsymbol{b}=\boldsymbol{0}.$$

Then:

(i) *$\boldsymbol{u}$, $\boldsymbol{E}$, and $\boldsymbol{S}$ are biharmonic;*

(ii) $\operatorname{div}\boldsymbol{u}$, $\operatorname{curl}\boldsymbol{u}$, $\operatorname{tr}\boldsymbol{E}$, *and* $\operatorname{tr}\boldsymbol{S}$ *are harmonic.*

The next proposition will be of future use.

(3) *If $\boldsymbol{u}$ is an elastic displacement field corresponding to null body forces, then so also is $\boldsymbol{u}_{,k}=\partial\boldsymbol{u}/\partial x_k$.*

Proof. Since $\boldsymbol{u}$ is an elastic displacement field, $\boldsymbol{u}$ is analytic; therefore

$$\boldsymbol{0}=\left(\Delta\boldsymbol{u}+\frac{1}{1-2\nu}\nabla\operatorname{div}\boldsymbol{u}\right)_{,k}=\Delta(\boldsymbol{u}_{,k})+\frac{1}{1-2\nu}\nabla\operatorname{div}(\boldsymbol{u}_{,k}).$$

Thus $\boldsymbol{u}_{,k}$ is an elastic displacement field. □

Under certain restrictions on the boundary of B, FICHERA[1] has established the following important result: If $\boldsymbol{u}$ is an elastic displacement field that is continuous on $\bar{B}$ and corresponds to zero body forces, then

$$\sup_{B}|\boldsymbol{u}|\leqq H\sup_{\partial B}|\boldsymbol{u}|,$$

where the constant H depends only upon ν and the properties of ∂B. We omit the proof of this result, which is quite difficult. FICHERA[1] has also established the inequality

$$\int_{\partial B}|\boldsymbol{s}|\,da\leqq H\int_{B}|\boldsymbol{b}|\,dv,$$

where $\boldsymbol{s}$ is the surface traction and $\boldsymbol{b}$ the body force field of a smooth elastic displacement field on $\bar{B}$ that vanishes on ∂B.

43. The mean value theorem.[2] The mean value theorem for harmonic functions[3] asserts that the value of a harmonic function at the center of a sphere is equal to the arithmetic mean of its values on the surface of the sphere. In this section we derive similar results for the displacement, strain, and stress fields belonging to an elastic state. In particular, we show that the values of these fields at the center of a sphere are equal to certain weighted averages of the displacement on the surface of the sphere. We also establish a similar result for the stress field in terms of its surface tractions.

---

[1] [1961, *8*]. See also ADLER [1963, *1*], [1964, *1*].
[2] Our presentation follows that of DIAZ and PAYNE [1958, *6*].
[3] See (1) of *(8.3)*.

(1) Mean value theorem.[1] *Assume that the body is homogeneous and isotropic with shear modulus μ and Poisson's ratio ν, and assume that*

$$\mu > 0, \qquad -1 < \nu < \tfrac{1}{2}.$$

Let $[\boldsymbol{u}, \boldsymbol{E}, \boldsymbol{S}]$ be an elastic state corresponding to zero body forces. Then given any ball $\Sigma_\varrho = \Sigma_\varrho(\boldsymbol{y})$ in B,

$$\boldsymbol{u}(\boldsymbol{y}) = \frac{3}{16\pi\varrho^2(2-3\nu)} \int\limits_{\partial\Sigma_\rho} \{5(\boldsymbol{u}\cdot\boldsymbol{n})\,\boldsymbol{n} + (1-4\nu)\,\boldsymbol{u}\}\,da, \tag{1}$$

$$\begin{aligned} \boldsymbol{E}(\boldsymbol{y}) = \frac{3}{8\pi\varrho^3(7-10\nu)} \Bigg[-10\nu \int\limits_{\partial\Sigma_\rho} (\boldsymbol{u}\otimes\boldsymbol{n} + \boldsymbol{n}\otimes\boldsymbol{u})\,da \\ + 7 \int\limits_{\partial\Sigma_\rho} (\boldsymbol{u}\cdot\boldsymbol{n})\{5\boldsymbol{n}\otimes\boldsymbol{n} - \mathbf{1}\}\,da \Bigg], \end{aligned} \tag{2}$$

$$\begin{aligned} \boldsymbol{S}(\boldsymbol{y}) = \frac{3\mu}{4\pi\varrho^3(7-10\nu)} \Bigg[-10\nu \int\limits_{\partial\Sigma_\rho} (\boldsymbol{u}\otimes\boldsymbol{n} + \boldsymbol{n}\otimes\boldsymbol{u})\,da \\ + \int\limits_{\partial\Sigma_\rho} (\boldsymbol{u}\cdot\boldsymbol{n})\{35\boldsymbol{n}\otimes\boldsymbol{n} - \alpha\mathbf{1}\}\,da \Bigg], \end{aligned} \tag{3}$$

$$\alpha = \frac{20\nu^2 - 28\nu + 7}{1-2\nu},$$

$$\boldsymbol{S}(\boldsymbol{y}) = \frac{3}{8\pi\varrho^2(7+5\nu)} \int\limits_{\partial\Sigma_\rho} \{10\nu(\boldsymbol{s}\otimes\boldsymbol{n}) + 7(\boldsymbol{s}\cdot\boldsymbol{n})[5\boldsymbol{n}\otimes\boldsymbol{n} - \mathbf{1}]\}\,da, \tag{4}$$

$$\boldsymbol{s} = \boldsymbol{S}\boldsymbol{n}.$$

Proof.[2] Since the body forces vanish, the displacement equation of equilibrium has the form

$$\Delta\boldsymbol{u} + \gamma\,\nabla\operatorname{div}\boldsymbol{u} = \boldsymbol{0}, \tag{a}$$

where

$$\gamma = \frac{1}{1-2\nu}. \tag{b}$$

Let

$$\boldsymbol{r}(\boldsymbol{x}) = \boldsymbol{x} - \boldsymbol{y}, \qquad r = |\boldsymbol{r}|,$$

and notice that

$$\boldsymbol{n} = \frac{\boldsymbol{r}}{\varrho} \quad \text{on } \partial\Sigma_\varrho. \tag{c}$$

Then (a), (b), (c), and the divergence theorem imply that

$$\begin{aligned} \boldsymbol{0} &= \int\limits_{\Sigma_\rho} r^2[\Delta\boldsymbol{u} + \gamma\nabla\operatorname{div}\boldsymbol{u}]\,dv \\ &= \int\limits_{\partial\Sigma_\rho} \varrho^2[(\nabla\boldsymbol{u})\,\boldsymbol{n} + \gamma(\operatorname{div}\boldsymbol{u})\,\boldsymbol{n}]\,da - 2\int\limits_{\Sigma_\rho} [(\nabla\boldsymbol{u})\,\boldsymbol{r} + \gamma(\operatorname{div}\boldsymbol{u})\,\boldsymbol{r}]\,dv. \end{aligned} \tag{d}$$

[1] Relation (1) is due to AQUARO [1950, *1*], Eq. (13) and SYNGE [1950, *12*], Eq. (2.36); see also SBRANA [1952, *4*], MIKHLIN [1952, *2*], pp. 206–207, GARIBALDI [1957, *7*], DIAZ and PAYNE [1958, *6*, *7*], BRAMBLE and PAYNE [1964, *4*]. Relations (2) and (3) are due to DIAZ and PAYNE [1958, *6*], Eqs. (4.11), (4.16). Relation (4) is due to SYNGE [1950, *12*]; see also DIAZ and PAYNE [1958, *6*], Eq. (5.21). DIAZ and PAYNE [1958, *6*], [1963, *3*] and BRAMBLE and PAYNE [1964, *4*], [1965, *2*] have obtained a number of mean value theorems giving the displacement and stress at the center of a sphere in terms of integrals over the surface of various combinations of normal and tangential displacements and tractions. A mean-value theorem for plane stress was given by FÖPPL [1953, *12*].

[2] The proof given here is that of DIAZ and PAYNE [1958, *6*].

The surface integral in (d) is zero, since

$$\varrho^2 \int_{\partial\Sigma_\varrho} [(\nabla \boldsymbol{u})\, \boldsymbol{n} + \gamma (\operatorname{div} \boldsymbol{u})\, \boldsymbol{n}]\, da = \varrho^2 \int_{\Sigma_\varrho} [\Delta \boldsymbol{u} + \gamma \nabla \operatorname{div} \boldsymbol{u}]\, dv = \boldsymbol{0}.$$

If we use the divergence theorem on the remaining integral in (d), we conclude, with the aid of (c), that

$$\varrho \int_{\partial\Sigma_\varrho} [\boldsymbol{u} + \gamma (\boldsymbol{u} \cdot \boldsymbol{n})\, \boldsymbol{n}]\, da - \int_{\Sigma_\varrho} (3+\gamma)\, \boldsymbol{u}\, dv = \boldsymbol{0}. \tag{e}$$

On the other hand, by (4) of the mean value theorem *(8.3)*,

$$\boldsymbol{u}(\boldsymbol{y}) = \frac{3}{8\pi} \left[\frac{5}{\varrho^3} \int_{\Sigma_\varrho} \boldsymbol{u}\, dv - \frac{1}{\varrho^2} \int_{\partial\Sigma_\varrho} \boldsymbol{u}\, da \right]. \tag{f}$$

If we eliminate the integral of $\boldsymbol{u}$ over Σ_ϱ between (e) and (f), and use (b), we arrive at the first relation in *(1)*.

Our next step is to derive (2). Since $\boldsymbol{u}$ is biharmonic, $\nabla \boldsymbol{u}$ is biharmonic, and we conclude from (4) of *(8.3)* that

$$\nabla \boldsymbol{u}(\boldsymbol{y}) = \frac{3}{8\pi} \left[\frac{5}{\lambda^3} \int_{\Sigma_\lambda} \nabla \boldsymbol{u}\, dv - \frac{1}{\lambda^2} \int_{\partial\Sigma_\lambda} \nabla \boldsymbol{u}\, da \right], \tag{g}$$

where $0 < \lambda \leqq \varrho$. Using the divergence theorem, we can rewrite (g) as follows:

$$\nabla \boldsymbol{u}(\boldsymbol{y}) = \frac{3}{8\pi} \left[\frac{5}{\lambda^3} \int_{\partial\Sigma_\lambda} \boldsymbol{u} \otimes \boldsymbol{n}\, da - \frac{1}{\lambda^2} \int_{\partial\Sigma_\lambda} \nabla \boldsymbol{u}\, da \right].$$

If we multiply this relation by λ^4 and integrate from $\lambda = 0$ to $\lambda = \varrho$, we find that

$$\frac{\varrho^5}{5} \nabla \boldsymbol{u}(\boldsymbol{y}) = \frac{3}{8\pi} \left[5 \int_{\Sigma_\varrho} \boldsymbol{u} \otimes \boldsymbol{r}\, dv - \int_{\Sigma_\varrho} r^2\, \nabla \boldsymbol{u}\, dv \right]. \tag{h}$$

By the divergence theorem,

$$\int_{\Sigma_\varrho} r^2\, \nabla \boldsymbol{u}\, dv = \varrho^2 \int_{\partial\Sigma_\varrho} \boldsymbol{u} \otimes \boldsymbol{n}\, da - 2 \int_{\Sigma_\varrho} \boldsymbol{u} \otimes \boldsymbol{r}\, dv, \tag{i}$$

and (g), (h), and the strain-displacement relation imply that

$$\frac{\varrho^5}{5} \boldsymbol{E}(\boldsymbol{y}) = \frac{3}{16\pi} \left[7 \int_{\Sigma_\varrho} (\boldsymbol{u} \otimes \boldsymbol{r} + \boldsymbol{r} \otimes \boldsymbol{u})\, dv - \varrho^2 \int_{\partial\Sigma_\varrho} (\boldsymbol{u} \otimes \boldsymbol{n} + \boldsymbol{n} \otimes \boldsymbol{u})\, da \right]. \tag{j}$$

Next, we will eliminate the integral in (j) over Σ_ϱ. By *(42.3)*, $\boldsymbol{u}_{,k}$, for each fixed k, satisfies the displacement equation of equilibrium. Thus (e) holds with $\boldsymbol{u}$ replaced by $\boldsymbol{u}_{,k}$, and we are led to the relation

$$\int_{\Sigma_\lambda} \nabla \boldsymbol{u}\, dv = \frac{\lambda}{3+\gamma} \int_{\partial\Sigma_\lambda} [\mathbf{1} + \gamma \boldsymbol{n} \otimes \boldsymbol{n}]\, \nabla \boldsymbol{u}\, da,$$

which can be written, with the aid of the divergence theorem, in the form

$$\int_{\partial\Sigma_\lambda} \boldsymbol{u} \otimes \boldsymbol{n}\, da = \frac{\lambda}{3+\gamma} \int_{\partial\Sigma_\lambda} [\mathbf{1} + \gamma \boldsymbol{n} \otimes \boldsymbol{n}]\, \nabla \boldsymbol{u}\, da.$$

If we multiply this relation by λ, use the fact that $\boldsymbol{n}=\boldsymbol{r}/\lambda$ on $\partial\Sigma_\lambda$, and integrate from $\lambda=0$ to $\lambda=\varrho$, we arrive at

$$\int_{\Sigma_\rho} \boldsymbol{u}\otimes\boldsymbol{r}\,dv=\frac{1}{3+\gamma}\int_{\Sigma_\rho}[r^2\,\mathbf{1}+\gamma\,\boldsymbol{r}\otimes\boldsymbol{r}]\,\nabla\boldsymbol{u}\,dv. \tag{k}$$

Applying the divergence theorem to the right side of (k), we are led to

$$\int_{\Sigma_\rho} \boldsymbol{u}\otimes\boldsymbol{r}\,dv=\frac{1}{3+\gamma}\Big(\varrho^2\int_{\partial\Sigma_\rho}[\boldsymbol{u}\otimes\boldsymbol{n}+\gamma(\boldsymbol{u}\cdot\boldsymbol{n})\,\boldsymbol{n}\otimes\boldsymbol{n}]\,da$$
$$-\int_{\Sigma_\rho}[2\boldsymbol{u}\otimes\boldsymbol{r}+\gamma\boldsymbol{r}\otimes\boldsymbol{u}+\gamma(\boldsymbol{r}\cdot\boldsymbol{u})\,\mathbf{1}]\,dv\Big).$$

If we add this equation to its transpose and rearrange terms, the result is

$$\int_{\Sigma_\rho}(\boldsymbol{u}\otimes\boldsymbol{r}+\boldsymbol{r}\otimes\boldsymbol{u})\,dv$$
$$=\frac{1}{5+2\gamma}\Big(\varrho^2\int_{\partial\Sigma_\rho}[\boldsymbol{u}\otimes\boldsymbol{n}+\boldsymbol{n}\otimes\boldsymbol{u}+2\gamma(\boldsymbol{u}\cdot\boldsymbol{n})\,\boldsymbol{n}\otimes\boldsymbol{n}]\,da-2\gamma\,\mathbf{1}\int_{\Sigma_\rho}(\boldsymbol{r}\cdot\boldsymbol{u})\,dv\Big).$$

Taking the trace of this equation, we obtain

$$\int_{\Sigma_\rho}\boldsymbol{r}\cdot\boldsymbol{u}\,dv=\frac{\varrho^2}{5}\int_{\partial\Sigma_\rho}\boldsymbol{u}\cdot\boldsymbol{n}\,da,$$

and the last two equations yield

$$\int_{\Sigma_\rho}(\boldsymbol{u}\otimes\boldsymbol{r}+\boldsymbol{r}\otimes\boldsymbol{u})\,dv$$
$$=\frac{\varrho^2}{5+2\gamma}\int_{\partial\Sigma_\rho}\Big[\boldsymbol{u}\otimes\boldsymbol{n}+\boldsymbol{n}\otimes\boldsymbol{u}+2\gamma(\boldsymbol{u}\cdot\boldsymbol{n})\Big(\boldsymbol{n}\otimes\boldsymbol{n}-\frac{1}{5}\,\mathbf{1}\Big)\Big]\,da. \tag{l}$$

If we combine (j) and (l) and use (b), the result is (2).

Since div $\boldsymbol{u}$ is harmonic, (2) of *(8.3)* implies that

$$\operatorname{div}\boldsymbol{u}(\boldsymbol{y})=\frac{3}{4\pi\,\varrho^3}\int_{\Sigma_\rho}\operatorname{div}\boldsymbol{u}\,dv=\frac{3}{4\pi\,\varrho^3}\int_{\partial\Sigma_\rho}\boldsymbol{u}\cdot\boldsymbol{n}\,da. \tag{m}$$

The relation (3) follows from (2), (m), and the stress-strain relation in the form

$$\boldsymbol{S}=2\mu\Big[\boldsymbol{E}+\frac{\nu}{1-2\nu}(\operatorname{tr}\boldsymbol{E})\,\mathbf{1}\Big]=2\mu\Big[\boldsymbol{E}+\frac{\nu}{1-2\nu}(\operatorname{div}\boldsymbol{u})\,\mathbf{1}\Big].$$

Since $[\boldsymbol{u},\boldsymbol{E},\boldsymbol{S}]$ is an elastic state corresponding to zero body forces, the stress field $\boldsymbol{S}$ satisfies the equation of equilibrium

$$\operatorname{div}\boldsymbol{S}=\mathbf{0} \tag{n}$$

and the compatibility equation[1]

$$\Delta \boldsymbol{S} + \eta \nabla \nabla (\operatorname{tr} \boldsymbol{S}) = \boldsymbol{0}, \tag{o}$$

where

$$\eta = \frac{1}{1+\nu}. \tag{p}$$

Further, by *(42.2)*, $\boldsymbol{S}$ is biharmonic and $\operatorname{tr} \boldsymbol{S}$ harmonic. Thus we may conclude from the mean value theorem (4) of *(8.3)* that

$$\boldsymbol{S}(\boldsymbol{y}) = \frac{3}{8\pi}\left[\frac{5}{\lambda^3}\int_{\Sigma_\lambda} \boldsymbol{S}\, dv - \frac{1}{\lambda^2}\int_{\partial\Sigma_\lambda} \boldsymbol{S}\, da\right]. \tag{q}$$

By the divergence theorem and (n),

$$\int_{\partial\Sigma_\lambda} \boldsymbol{S}(\boldsymbol{n}\otimes\boldsymbol{n})\, da = \frac{1}{\lambda}\int_{\partial\Sigma_\lambda} \boldsymbol{S}(\boldsymbol{n}\otimes\boldsymbol{r})\, da = \frac{1}{\lambda}\int_{\Sigma_\lambda} \boldsymbol{S}\, dv, \tag{r}$$

and (q), (r) imply that

$$\boldsymbol{S}(\boldsymbol{y}) = \frac{3}{8\pi}\left[\frac{5}{\lambda^2}\int_{\partial\Sigma_\lambda} \boldsymbol{S}(\boldsymbol{n}\otimes\boldsymbol{n})\, da - \frac{1}{\lambda^2}\int_{\partial\Sigma_\lambda} \boldsymbol{S}\, da\right].$$

If we multiply this equation by λ^4, integrate from $\lambda=0$ to $\lambda=\varrho$, and solve for $\boldsymbol{S}(\boldsymbol{y})$, the result is

$$\boldsymbol{S}(\boldsymbol{y}) = \frac{15}{8\pi\varrho^5}\left[5\int_{\Sigma_\varrho} \boldsymbol{S}(\boldsymbol{r}\otimes\boldsymbol{r})\, dv - \int_{\Sigma_\varrho} r^2 \boldsymbol{S}\, dv\right].$$

On the other hand, the divergence theorem and (n) imply

$$\int_{\partial\Sigma_\varrho} \varrho^3 \boldsymbol{S}(\boldsymbol{n}\otimes\boldsymbol{n})\, da = \int_{\partial\Sigma_\varrho} \varrho^2 \boldsymbol{S}(\boldsymbol{n}\otimes\boldsymbol{r})\, da = \int_{\Sigma_\varrho} r^2 \boldsymbol{S}\, dv + 2\int_{\Sigma_\varrho} \boldsymbol{S}(\boldsymbol{r}\otimes\boldsymbol{r})\, dv; \tag{s}$$

thus

$$\boldsymbol{S}(\boldsymbol{y}) = \frac{15}{8\pi\varrho^5}\left[7\int_{\Sigma_\varrho} \boldsymbol{S}(\boldsymbol{r}\otimes\boldsymbol{r})\, dv - \varrho^3\int_{\partial\Sigma_\varrho} \boldsymbol{S}(\boldsymbol{n}\otimes\boldsymbol{n})\, da\right].$$

Further, since

$$\boldsymbol{S}(\boldsymbol{n}\otimes\boldsymbol{n}) = (\boldsymbol{S}\boldsymbol{n})\otimes\boldsymbol{n} = \boldsymbol{s}\otimes\boldsymbol{n}, \tag{t}$$

where

$$\boldsymbol{s} = \boldsymbol{S}\boldsymbol{n},$$

it follows that

$$\boldsymbol{S}(\boldsymbol{y}) = \frac{15}{8\pi\varrho^5}\left[7\int_{\Sigma_\varrho} \boldsymbol{S}(\boldsymbol{r}\otimes\boldsymbol{r})\, dv - \varrho^3\int_{\partial\Sigma_\varrho} \boldsymbol{s}\otimes\boldsymbol{n}\, da\right]. \tag{u}$$

To complete the proof we have only to transform the first integral in (u) into terms involving surface tractions. By the divergence theorem and (o),

$$\int_{\partial\Sigma_\lambda} [(\nabla\boldsymbol{S})\,\boldsymbol{n} + \eta(\nabla \operatorname{tr}\boldsymbol{S})\otimes\boldsymbol{n}]\, da = \boldsymbol{0},$$

and using this indentity, (o), and two applications of the divergence theorem, we see that

$$\begin{aligned} \boldsymbol{0} &= \int_{\Sigma_\lambda} r^2[\Delta\boldsymbol{S} + \eta\nabla\nabla(\operatorname{tr}\boldsymbol{S})]\, dv = -2\int_{\Sigma_\lambda} [(\nabla\boldsymbol{S})\,\boldsymbol{r} + \eta(\nabla\operatorname{tr}\boldsymbol{S})\otimes\boldsymbol{r}]\, dv \\ &= -2\left[\lambda\int_{\partial\Sigma_\lambda} [\boldsymbol{S} + \eta(\operatorname{tr}\boldsymbol{S})\,\boldsymbol{n}\otimes\boldsymbol{n}]\, da - \int_{\Sigma_\lambda} [3\boldsymbol{S} + \eta(\operatorname{tr}\boldsymbol{S})\,\boldsymbol{1}]\, dv\right]. \end{aligned}$$

[1] See p. 92.

Taking the trace of (r), we obtain

$$\int_{\Sigma_\lambda} \operatorname{tr} \boldsymbol{S}\, dv = \lambda \int_{\partial\Sigma_\lambda} \boldsymbol{S} \cdot (\boldsymbol{n} \otimes \boldsymbol{n})\, da,$$

and the last two equations with (r) yield

$$\int_{\partial\Sigma_\lambda} [r^2 \boldsymbol{S} + \eta (\operatorname{tr} \boldsymbol{S})\, \boldsymbol{r} \otimes \boldsymbol{r}]\, da = \int_{\partial\Sigma_\lambda} [3\boldsymbol{S}(\boldsymbol{r} \otimes \boldsymbol{r}) + \eta (\boldsymbol{r} \cdot \boldsymbol{S}\boldsymbol{r})\, \mathbf{1}]\, da.$$

If we integrate this relation from $\lambda = 0$ to $\lambda = \varrho$, we arrive at

$$\int_{\Sigma_\rho} [r^2 \boldsymbol{S} + \eta (\operatorname{tr} \boldsymbol{S})\, \boldsymbol{r} \otimes \boldsymbol{r}]\, dv = \int_{\Sigma_\rho} [3\boldsymbol{S}(\boldsymbol{r} \otimes \boldsymbol{r}) + \eta (\boldsymbol{r} \cdot \boldsymbol{S}\boldsymbol{r})\, \mathbf{1}]\, dv.$$

A simple computation based on (n) and (o) shows that $\boldsymbol{r} \cdot \boldsymbol{S}\boldsymbol{r}$ is biharmonic; thus (4) of *(8.3)* implies

$$\frac{5}{\varrho^3} \int_{\Sigma_\rho} \boldsymbol{r} \cdot \boldsymbol{S}\boldsymbol{r}\, dv = \int_{\partial\Sigma_\rho} \boldsymbol{s} \cdot \boldsymbol{n}\, da.$$

By the last two relations, (s), and (t),

$$5 \int_{\Sigma_\rho} \boldsymbol{S}(\boldsymbol{r} \otimes \boldsymbol{r})\, dv = \varrho^3 \int_{\partial\Sigma_\rho} \left[\boldsymbol{s} \otimes \boldsymbol{n} - \frac{\eta}{5} (\boldsymbol{s} \cdot \boldsymbol{n})\, \mathbf{1}\right] da + \eta \int_{\Sigma_\rho} (\operatorname{tr} \boldsymbol{S})\, \boldsymbol{r} \otimes \boldsymbol{r}\, dv. \tag{v}$$

In view of the identity

$$(S_{kl} r_k r_i r_j)_{,l} = S_{kk} r_i r_j + S_{ki} r_k r_j + S_{kj} r_k r_i$$

and the symmetry of $\boldsymbol{S}$, the volume integral in the right side of (v) takes the form

$$\int_{\Sigma_\rho} (\operatorname{tr} \boldsymbol{S})\, \boldsymbol{r} \otimes \boldsymbol{r}\, dv = \varrho^3 \int_{\partial\Sigma_\rho} (\boldsymbol{s} \cdot \boldsymbol{n})\, \boldsymbol{n} \otimes \boldsymbol{n}\, da - \int_{\Sigma_\rho} [\boldsymbol{S}(\boldsymbol{r} \otimes \boldsymbol{r}) + (\boldsymbol{r} \otimes \boldsymbol{r})\, \boldsymbol{S}]\, dv. \tag{w}$$

On the other hand, if we multiply (r) by λ^2 and integrate with respect to λ from 0 to ϱ, we arrive at

$$\int_{\Sigma_\rho} \boldsymbol{S}(\boldsymbol{r} \otimes \boldsymbol{r})\, dv = \int_0^\varrho \lambda \int_{\Sigma_\lambda} \boldsymbol{S}\, dv\, d\lambda. \tag{x}$$

Since $\boldsymbol{S}$ is symmetric, the left-hand side of (x) is symmetric. Thus the volume integral in the right side of (w) is equal to twice the volume integral in the left side of (v); hence (v) and (w) imply that

$$\int_{\Sigma_\rho} \boldsymbol{S}(\boldsymbol{r} \otimes \boldsymbol{r})\, dv = \frac{\varrho^3}{5 + 2\eta} \int_{\partial\Sigma_\rho} \left[\boldsymbol{s} \otimes \boldsymbol{n} + \eta \left(\boldsymbol{n} \otimes \boldsymbol{n} - \frac{1}{5}\mathbf{1}\right) (\boldsymbol{s} \cdot \boldsymbol{n})\right] da. \tag{y}$$

Eqs. (y), (u), and (p) yield the desired result (4). □

44. Complete solutions of the displacement equation of equilibrium. In this section we study certain general solutions of the displacement equation of equilibrium

$$\Delta \boldsymbol{u} + \frac{1}{1 - 2\nu} \nabla \operatorname{div} \boldsymbol{u} + \frac{1}{\mu} \boldsymbol{b} = \mathbf{0}$$

assuming throughout that the body force $\boldsymbol{b}$, the shear modulus μ, and Poisson's ratio ν are prescribed with $\mu \neq 0$ and $\nu \neq \frac{1}{2}, 1$. Recall that a solution $\boldsymbol{u}$ of class C^2 is referred to as an *elastic displacement field corresponding to* $\boldsymbol{b}$.

(1) Boussinesq-Papkovitch-Neuber solution.[1] *Let*

$$\boldsymbol{u} = \boldsymbol{\psi} - \frac{1}{4(1-\nu)} \nabla(\boldsymbol{p} \cdot \boldsymbol{\psi} + \varphi), \qquad (\mathrm{P}_1)^2$$

where φ *and* $\boldsymbol{\psi}$ *are class* C^3 *fields on* B *that satisfy*

$$\Delta \boldsymbol{\psi} = -\frac{1}{\mu} \boldsymbol{b}, \quad \Delta \varphi = \frac{1}{\mu} \boldsymbol{p} \cdot \boldsymbol{b}. \qquad (\mathrm{P}_2)$$

Then $\boldsymbol{u}$ *is an elastic displacement field corresponding to* $\boldsymbol{b}$.

Proof. Clearly,

$$\Delta \boldsymbol{u} + \frac{1}{1-2\nu} \nabla \operatorname{div} \boldsymbol{u}$$

$$= \Delta \boldsymbol{\psi} + \frac{1}{1-2\nu} \nabla \operatorname{div} \boldsymbol{\psi} - \frac{1}{4(1-\nu)} \left[\Delta \nabla(\boldsymbol{p} \cdot \boldsymbol{\psi} + \varphi) + \frac{1}{1-2\nu} \nabla \Delta (\boldsymbol{p} \cdot \boldsymbol{\psi} + \varphi) \right].$$

Since $\Delta \nabla = \nabla \Delta$ and

$$\Delta(\boldsymbol{p} \cdot \boldsymbol{\psi}) = \boldsymbol{p} \cdot \Delta \boldsymbol{\psi} + 2 \operatorname{div} \boldsymbol{\psi},$$

it follows that

$$\Delta \boldsymbol{u} + \frac{1}{1-2\nu} \nabla \operatorname{div} \boldsymbol{u} = \Delta \boldsymbol{\psi} + \frac{1}{1-2\nu} \nabla \operatorname{div} \boldsymbol{\psi} - \frac{1}{2(1-2\nu)} \nabla(\boldsymbol{p} \cdot \Delta \boldsymbol{\psi} + \Delta \varphi + 2 \operatorname{div} \boldsymbol{\psi}).$$

But

$$\boldsymbol{p} \cdot \Delta \boldsymbol{\psi} = -\Delta \varphi \quad \text{and} \quad \Delta \boldsymbol{\psi} = -\frac{1}{\mu} \boldsymbol{b};$$

thus the above expression reduces to

$$-\frac{1}{\mu} \boldsymbol{b}. \quad \square$$

---

[1] Elements of this solution appear in the work of BOUSSINESQ [1878, *1*], [1885, *1*]. In fact, the difference between the two solutions given by BOUSSINESQ on p. 62 and p. 72 of [1885, *1*] yields the special form to which (P_1) reduces in the case of rotational symmetry. The solution (P_1), (P_2) with $\varphi = 0$ is due to FONTANEAU [1890, *3*], [1892, *3*]. The complete solution (P_1), (P_2) first appears in the work of PAPKOVITCH [1932, *3, 4*]; it was later rediscovered by NEUBER [1934, *4*]. According to PAPKOVITCH [1939, *4*] (footnote on p. 130), (P_1), (P_2) should be attributed to GRODSKI, who arrived at the solution in 1928 (unpublished research). However, it is not clear from the statement of PAPKOVITCH whether or not this work of GRODSKI includes the scalar field φ; in fact, PAPKOVITCH [1937, *5*] claimed that it does not. The complete solution (P_1), (P_2) does appear in GRODSKI'S [1935, *1*] first published research on the subject. Other references concerning this and related solutions and not mentioned elsewhere in this section are: BOUSSINESQ [1888, *1, 2*]. TEDONE [1904, *4*], SERINI [1919, *2*], BURGATTI [1923, *1*], [1926, *1*], TIMPE [1924, *2*], [1948, *6*], [1950, *13*], [1951, *11*], WEBER [1925, *1*], TREFFTZ [1928, *3*], MARGUERRE [1933, *1*], [1935, *3*], [1955, *11*], BIEZENO [1934, *1*], SOBRERO [1934, *6*], [1935, 8], NEUBER [1935, *4*], [1937, *3*], WESTERGAARD [1935, *9*], ZANABONI [1936, *6*], SLOBODYANSKY [1938, *3, 4*], [1959, *15*], PLATRIER [1941, *1*], GUTMAN [1947, *3*], SHAPIRO [1947, *7*], TER-MKRTYCHAN [1947, *8*], KRUTKOV [1949, *5*], BLOKH [1950, *3*], [1958, *2*], [1961, *2*], STERNBERG, EUBANKS, and SADOWSKI [1951, *9*], BRDIČKA [1953, *6*], [1957, *3*], CHURIKOV [1953, *7*], DZHANELIDZE [1953, *9*], HU [1953, *13*], [1954, *8*], SCHAEFER [1953, *18*], TRENIN [1953, *19*], EUBANKS and STERNBERG [1954, *5*], KRÖNER [1954, *11*], LING and YANG [1954, *13*], LUR'E [1955, *10*], DUFFIN [1956, *2*], SOLYANIK-KRASSA [1957, *13*], SOLOMON [1957, *12*], DEEV [1958, *5*], [1959, *3*], HIJAB [1959, *7*], TRUESDELL [1959, *16*], BRAMBLE and PAYNE [1961, *4*], NAGHDI and HSU [1961, *16*], ALEXSANDROV and SOLOVEV [1962, *1*], [1964, *2*], GURTIN [1962, *7*], TANG and SUN [1963, *22*], VOLKOV and KOMISSAROVA [1963, *25*], HERRMANN [1964, *9*], ALEXSANDROV [1965, *1*], STIPPES [1966, *25*], [1967, *14*], SOLOMON [1968, *12*], DE LA PENHA and CHILDS [1969, *3*], YOUNGDAHL [1969, *9*].

[2] Recall that $\boldsymbol{p}(\boldsymbol{x}) = \boldsymbol{x} - \boldsymbol{0}$.

As a corollary of this theorem we have the following result.

(2) *Assume that the body force* $\boldsymbol{b}$ *is continuous on* $\bar{B}$ *and of class* C^N $(N \geqq 2)$ *on* B. *Then*:

(i) *there exists an elastic displacement field* $\boldsymbol{u}$ *of class* C^N *on* B *that corresponds to* $\boldsymbol{b}$;

(ii) *every elastic displacement field that corresponds to* $\boldsymbol{b}$ *is of class* C^N *on* B.

Proof. Let

$$\varphi(\boldsymbol{x}) = -\frac{1}{4\pi\mu}\int_B \frac{\boldsymbol{p}(\boldsymbol{y})\cdot\boldsymbol{b}(\boldsymbol{y})}{|\boldsymbol{x}-\boldsymbol{y}|}\,dv_{\boldsymbol{y}},$$

$$\boldsymbol{\psi}(\boldsymbol{x}) = \frac{1}{4\pi\mu}\int_B \frac{\boldsymbol{b}(\boldsymbol{y})}{|\boldsymbol{x}-\boldsymbol{y}|}\,dv_{\boldsymbol{y}}.$$

Then it is clear from the properties of the Newtonian potential ***(6.5)*** that φ and $\boldsymbol{\psi}$ are of class C^{N+1} on B and satisfy (P_2); thus the field $\boldsymbol{u}$ defined by (P_1) is a class C^N elastic displacement field corresponding to $\boldsymbol{b}$.

To prove (ii) let $\boldsymbol{u}$ be an arbitrary elastic displacement field corresponding to $\boldsymbol{b}$. By (i) there exists a class C^N elastic displacement field $\hat{\boldsymbol{u}}$ that also corresponds to $\boldsymbol{b}$. Let

$$\tilde{\boldsymbol{u}} = \boldsymbol{u} - \hat{\boldsymbol{u}}.$$

It suffices to prove that $\tilde{\boldsymbol{u}}$ is of class C^N. Clearly, $\tilde{\boldsymbol{u}}$ is an elastic displacement field corresponding to *zero* body forces; thus, by ***(42.1)***, $\tilde{\boldsymbol{u}}$ is of class C^∞. □

For the case in which the body force is the gradient of a potential, i.e.

$$\boldsymbol{b} = \nabla\alpha,$$

it is possible to construct a much simpler particular solution. Indeed, the field

$$\boldsymbol{u} = \nabla\varphi$$

satisfies the displacement equation of equilibrium provided

$$\Delta\varphi = -\frac{1}{\lambda+2\mu}\alpha.$$

In the same manner, if

$$\boldsymbol{b} = \operatorname{curl}\boldsymbol{\tau},$$

then

$$\boldsymbol{u} = \operatorname{curl}\boldsymbol{\omega}$$

is a solution provided

$$\Delta\boldsymbol{\omega} = -\frac{1}{\mu}\boldsymbol{\tau}.$$

By Helmholtz's theorem ***(6.7)*** we can always decompose the body force as follows:

$$\boldsymbol{b} = \nabla\alpha + \operatorname{curl}\boldsymbol{\tau}.$$

Thus, combining the previous two displacement fields, we arrive at the solution:[1]

$$\boldsymbol{u} = \nabla\varphi + \operatorname{curl}\boldsymbol{\omega},$$

$$\Delta\varphi = -\frac{1}{\lambda+2\mu}\alpha, \qquad \Delta\boldsymbol{\omega} = -\frac{1}{\mu}\boldsymbol{\tau}.$$

[1] Cf. LAMÉ [1852, *2*], p. 149; BETTI [1872, *1*], § 5; KELVIN and TAIT [1883, *1*], p. 283; VOIGT [1900, *5*], pp. 417, 432, 439; TEDONE and TIMPE [1907, *3*].

This solution, in general, is not complete. Indeed, consider the case in which $\alpha=0$. Then

$$\operatorname{div}\boldsymbol{u}=\Delta\varphi=0;$$

thus, in this instance, the solution can represent at most divergence-free displacement fields.

(3) Boussinesq-Somigliana-Galerkin solution.[1] *Let*

$$\boldsymbol{u}=\Delta\boldsymbol{g}-\frac{1}{2(1-\nu)}\nabla\operatorname{div}\boldsymbol{g}, \tag{G$_1$}$$

where $\boldsymbol{g}$ *is a class* C^4 *vector field on* B *that satisfies*

$$\Delta\Delta\boldsymbol{g}=-\frac{1}{\mu}\boldsymbol{b}. \tag{G$_2$}$$

Then $\boldsymbol{u}$ *is an elastic displacement field corresponding to* $\boldsymbol{b}$.

Proof. Clearly,

$$\begin{aligned}\Delta\boldsymbol{u}&+\frac{1}{1-2\nu}\nabla\operatorname{div}\boldsymbol{u}\\&=\Delta\Delta\boldsymbol{g}-\frac{1}{2(1-\nu)}\Delta\nabla\operatorname{div}\boldsymbol{g}+\frac{1}{1-2\nu}\left[\nabla\operatorname{div}\Delta\boldsymbol{g}-\frac{1}{2(1-\nu)}\nabla\Delta\operatorname{div}\boldsymbol{g}\right].\end{aligned}$$

Since $\Delta\nabla=\nabla\Delta$ and $\operatorname{div}\Delta=\Delta\operatorname{div}$, the right-hand side of the above expression reduces to $\Delta\Delta\boldsymbol{g}$, which by hypothesis is equal to

$$-\frac{1}{\mu}\boldsymbol{b}.\quad\square$$

Our next step is to establish the completeness of the Boussinesq-Papkovitch-Neuber and Boussinesq-Somigliana-Galerkin representations. The notion of completeness is of utmost importance to applications. Indeed, it is concerned with the question: can any (solvable) problem be solved using a given representation; or, in other words, does every sufficiently smooth solution admit such a representation?

(4) Completeness of the Boussinesq-Papkovitch-Neuber and Boussinesq-Somigliana-Galerkin solutions.[2] *Let* $\boldsymbol{u}$ *be an elastic displacement field corresponding to* $\boldsymbol{b}$, *and assume that* $\boldsymbol{u}$ *is continuous on* $\bar{B}$ *and of class* C^4 *on* B. *Then there exists a field* $\boldsymbol{g}$ *of class* C^4 *on* B *that satisfies* (G$_1$), (G$_2$); *and fields* φ, $\boldsymbol{\psi}$ *of class* C^2 *on* B *that satisfy* (P$_1$), (P$_2$).

Proof. We begin by establishing the completeness of the Boussinesq-Somigliana-Galerkin solution. Clearly, (G$_1$) is equivalent to

$$\Delta\boldsymbol{g}+\frac{1}{1-2\hat{\nu}}\nabla\operatorname{div}\boldsymbol{g}+\frac{1}{\mu}\hat{\boldsymbol{b}}=\boldsymbol{0},$$

[1] This solution is due to BOUSSINESQ [1885, *1*], p. 281. It was later rediscovered by SOMIGLIANA [1889, *3*], [1894, *3*] and GALERKIN [1930, *2*, *3*], [1931, *3*] and is usually referred to as GALERKIN's solution, although TEDONE and TIMPE in their classic article [1907, *3*], p. 155 attributed it to BOUSSINESQ. See also IACOVACHE [1949, *4*], IONESCU [1954, *9*], MOISIL [1949, *6*], [1950, *8*], IONESCU-CAZIMIR [1950, *7*], WESTERGAARD [1952, *7*], BRDIČKA [1954, *3*], BERNIKOV [1966, *4*].

[2] PAPKOVITCH [1932, *4*] noticed that if a particular solution to the equation of equilibrium exists for arbitrary values of Poisson's ratio and arbitrary body forces, then the Boussinesq-Somigliana-Galerkin solution is complete. An alternative completeness proof for (G$_1$), (G$_2$) was furnished by NOLL [1957, *11*]. The proof given here, which is based on PAPKOVITCH's argument, is due to STERNBERG and GURTIN [1962, *13*].

where

$$\hat{\boldsymbol{b}} = -\mu \boldsymbol{u}, \quad \hat{\nu} = \tfrac{3}{2} - \nu.$$

Thus the task of finding a class C^4 field $\boldsymbol{g}$ that satisfies (G_1) is reducible to finding a class C^4 particular solution to the displacement equation of equilibrium corresponding to given body forces that are continuous on $\bar{B}$ and of class C^4 on B. This latter objective is met by (i) of *(2)*. Further, using steps identical to those used in the proof of *(3)*, it is easily verified that (G_1) implies

$$\Delta \boldsymbol{u} + \frac{1}{1-2\nu} \nabla \operatorname{div} \boldsymbol{u} = \Delta\Delta \boldsymbol{g}.$$

Thus (G_1) and the displacement equation of equilibrium imply (G_2), and we are entitled to conclude the completeness of the Boussinesq-Somigliana-Galerkin solution (G_1), (G_2). Next, following MINDLIN,[1] we define

$$\varphi = 2 \operatorname{div} \boldsymbol{g} - \boldsymbol{p} \cdot \Delta \boldsymbol{g}, \quad \boldsymbol{\psi} = \Delta \boldsymbol{g}.$$

Then φ and $\boldsymbol{\psi}$ are of class C^2 on B, and (G_2) implies (P_2), (G_1) implies (P_1). Thus the Boussinesq-Papkovitch-Neuber solution is also complete. □

In cartesian coordinates the solutions (P_1), (P_2), and (G_1), (G_2), as well as the formulae for the corresponding stresses, have the following form:

Boussinesq-Papkovitch-Neuber solution

$$\psi_{i,jj} = -\frac{b_i}{\mu}, \quad \varphi_{,jj} = \frac{x_j b_j}{\mu}$$

$$u_i = \psi_i - \frac{1}{4(1-\nu)} (\varphi + x_j \psi_j)_{,i}$$

$$S_{ij} = \frac{2\mu}{4(1-\nu)} \{(1-2\nu)(\psi_{i,j} + \psi_{j,i}) + 2\nu\, \delta_{ij} \psi_{k,k} - x_k \psi_{k,ij} - \varphi_{,ij}\}$$

Boussinesq-Somigliana-Galerkin solution

$$g_{i,jjkk} = -\frac{1}{\mu} b_i$$

$$u_i = g_{i,jj} - \frac{1}{2(1-\nu)} g_{j,ji}$$

$$S_{ij} = \mu \left\{ g_{i,jkk} + g_{j,ikk} + \frac{1}{1-\nu} (\nu \delta_{ij} g_{k,kll} - g_{k,kij}) \right\}.$$

In most applications of elastostatics the Boussinesq-Papkovitch-Neuber solution deserves preference over the Boussinesq-Somigliana-Galerkin solution.[2] Indeed, the transformation (P_1) underlying the former is one degree lower than that of (G_1); secondly, (in the absence of body forces) the stress functions involved in the former are harmonic, while those of the latter are biharmonic; and, finally, (P_1), (P_2) are conveniently transformed into general orthogonal curvilinear coordinates, whereas (G_1), (G_2) give rise to exceedingly cumbersome forms when referred to curvilinear coordinates, with the exception of cylindrical coordinates.

---

[1] [1936, *2*].

[2] Cf. STERNBERG and GURTIN [1962, *13*].

Two other *complete* solutions of the displacement equation of equilibrium (with $\boldsymbol{b}=\mathbf{0}$) are:[1]

$$\boldsymbol{u}=\boldsymbol{\tau}-\frac{1}{2(1-\nu)}\nabla\alpha,$$
$$\Delta\alpha=\operatorname{div}\boldsymbol{\tau},\quad \Delta\boldsymbol{\tau}=\mathbf{0};$$
$$\boldsymbol{u}=\boldsymbol{l}-\frac{1}{1-2\nu}\operatorname{curl}\boldsymbol{\omega},$$
$$\Delta\boldsymbol{\omega}=\operatorname{curl}\boldsymbol{l},\quad \Delta\boldsymbol{l}=\mathbf{0}.$$

Returning to the Boussinesq-Papkovitch-Neuber solution, we note that a given equilibrium displacement field does not uniquely determine its generating stress functions φ and $\boldsymbol{\psi}$. In fact, it is just this lack of uniqueness which, in certain circumstances, allows the elimination of φ or one of the components of $\boldsymbol{\psi}$. The next theorem, which is due to Eubanks and Sternberg,[2] deals with the possibility of setting $\varphi=0$.[3]

(5) Elimination of the scalar potential.[4] *Let B be star-shaped with respect to the origin* $\mathbf{0}$. *Let* $\boldsymbol{u}$ *satisfy the hypotheses of* ***(4)*** *with* $\boldsymbol{b}=\mathbf{0}$. *Further, suppose that* $-1<\nu<\frac{1}{2}$ *with* 4ν *not an integer. Then there exists a harmonic vector field* $\boldsymbol{\psi}$ *on B such that*

$$\boldsymbol{u}=\boldsymbol{\psi}-\frac{1}{4(1-\nu)}\nabla(\boldsymbol{p}\cdot\boldsymbol{\psi}).$$

Proof. By ***(4)*** and since $\boldsymbol{b}=\mathbf{0}$, there exist harmonic fields φ and $\hat{\boldsymbol{\psi}}$ such that

$$\boldsymbol{u}=\hat{\boldsymbol{\psi}}-\frac{1}{4(1-\nu)}\nabla(\boldsymbol{p}\cdot\hat{\boldsymbol{\psi}}+\varphi),$$
$$\Delta\hat{\boldsymbol{\psi}}=\mathbf{0},\quad \Delta\varphi=0.$$

Thus we must show that there exists a harmonic field $\boldsymbol{\psi}$ that satisfies

$$\alpha\boldsymbol{\psi}-\nabla(\boldsymbol{p}\cdot\boldsymbol{\psi})=\alpha\hat{\boldsymbol{\psi}}-\nabla(\boldsymbol{p}\cdot\hat{\boldsymbol{\psi}}+\varphi),$$

where

$$\alpha=4(1-\nu).$$

By hypothesis, we may confine α to the range

$$2<\alpha<8. \tag{a}$$

---

[1] The first solution, which is apparently due to Freiberger [1949, *2*, *3*], is simply a variant of the Boussinesq-Papkovitch-Neuber solution; indeed, take $\boldsymbol{\tau}=\boldsymbol{\psi}$, $2\alpha=\varphi+\boldsymbol{p}\cdot\boldsymbol{\psi}$ in (P_1), (P_2). The second solution is due to Korn [1915, *2*], who used it to prove existence for the displacement problem.

[2] [1956, *4*]. Papkovitch [1932, *3*, *4*], through an erroneous argument, was led to claim that φ may be taken equal to zero without loss in completeness, a claim which is reflected in the title of [1932, *3*]. In a subsequent note [1932, *2*], Papkovitch refers to this argument as "artificial". Neuber [1937, *3*] asserted that φ may be set equal to zero without impairing the generality of the solution. In his proof of this statement he takes for granted the existence of a function satisfying a pair of simultaneous partial differential equations. Therefore, as was pointed out by Sokolnikoff [1956, *12*], p. 331, Neuber's argument is inconclusive. Slobodyansky [1954, *19*] proved that the stress function φ may be eliminated, but as was observed by Sokolnikoff [1956, *12*], p. 331, Slobodyansky's proof is limited to a region interior or exterior to a sphere.

[3] The solution (P_1), (P_2) with $\varphi=0$ appears in the work of Fontaneau [1890, *3*], [1892, *3*], and, according to Papkovitch [1937, *5*], [1939, *4*], p. 130, also in unpublished research of Grodski.

[4] This result has been extended by Stippes [1969, *8*] to an exterior domain whose complement is star-shaped, and to the region between two star-shaped surfaces.

It is sufficient to establish the existence of a harmonic field Φ such that[1]

$$\boldsymbol{p} \cdot \nabla \Phi - \alpha \Phi = \varphi. \tag{b}$$

Indeed, given such a field Φ, the function

$$\boldsymbol{\psi} = \hat{\boldsymbol{\psi}} + \nabla \Phi \tag{c}$$

has all of the desired properties.

Let (r, θ, γ) be spherical coordinates. The function φ, by hypothesis, is harmonic. Hence $\varphi(r, \theta, \gamma)$ admits the following expansion in terms of solid spherical harmonics

$$\varphi(r, \theta, \gamma) = \sum_{n=0}^{\infty} S_n(\theta, \gamma)\, r^n, \tag{d}$$

which is uniformly convergent in any spherical neighborhood of the origin. The functions $S_n(\theta, \gamma)$ are surface harmonics of degree n.

Next, define a function F through

$$F(r, \theta, \gamma) = \varphi(r, \theta, \gamma) - \sum_{n=0}^{[\alpha+1]} S_n(\theta, \gamma)\, r^n, \tag{e}$$

where $[\alpha + 1]$ denotes the largest integer not exceeding $\alpha + 1$. Clearly, F is defined and harmonic on B. Further, by (d) and (e),

$$r^{-\alpha-1} F(r, \theta, \gamma) \to 0 \quad \text{as} \quad r \to 0. \tag{f}$$

In spherical coordinates (b) has the form

$$r \frac{\partial \Phi}{\partial r} - \alpha \Phi = \varphi,$$

or by (e)

$$r^{\alpha+1} \frac{\partial}{\partial r} (r^{-\alpha} \Phi) = F + \sum_{n=0}^{[\alpha+1]} S_n r^n. \tag{g}$$

Consider the function Φ defined by

$$\begin{aligned} \Phi(r, \theta, \gamma) &= G(r, \theta, \gamma) + \sum_{n=0}^{[\alpha+1]} \frac{S_n(\theta, \gamma)\, r^n}{n - \alpha}, \\ G(r, \theta, \gamma) &= r^{\alpha} \int_0^r F(\xi, \theta, \gamma)\, \xi^{-\alpha-1}\, d\xi. \end{aligned} \tag{h}$$

Since B is star-shaped with respect to the origin $\mathbf{0}$, and in view of (a) and (f), the relations (h) define Φ on B. Moreover, Φ is of class C^2 on B and satisfies (g), and hence (b).

It remains to be shown that Φ is harmonic on B. To this end, it suffices to show that G is harmonic. In spherical coordinates the Laplacian operator appears as

$$\begin{aligned} \Delta &= \frac{1}{r^2} \left[\frac{\partial}{\partial r} \left(r^2 \frac{\partial}{\partial r} \right) + \Delta_* \right], \\ \Delta_* &= \operatorname{cosec} \theta \frac{\partial}{\partial \theta} \left(\sin \theta \frac{\partial}{\partial \theta} \right) + \operatorname{cosec}^2 \theta \frac{\partial^2}{\partial \gamma^2}. \end{aligned} \tag{i}$$

[1] NEUBER [1937, *3*] assumed, without proof, the existence of such a Φ, in the absence of any restrictions on B and $\boldsymbol{v}$.

Clearly,

$$\frac{\partial}{\partial r}\left(r^2 \frac{\partial G}{\partial r}\right) = r\frac{\partial F}{\partial r} + (\alpha+1)\,F + \alpha(\alpha+1)\,G,$$
$$\Delta_* G = r^\alpha \int_0^r \Delta_* F(\xi, \theta, \gamma)\, \xi^{-\alpha-1}\, d\xi, \tag{j}$$

and, with the aid of two successive integrations by parts applied to the second of (h), we reach the identity

$$\alpha(\alpha+1)\,G = -(\alpha+1)\,F - r\frac{\partial F}{\partial r} + r^\alpha \int_0^r \frac{\partial}{\partial \xi}\left[\xi^2 \frac{\partial F(\xi,\theta,\gamma)}{\partial \xi}\right] \xi^{-\alpha-1}\, d\xi. \tag{k}$$

Substitution of (k) in (j), and use of (i), now yield

$$r^2 \Delta G = r^\alpha \int_0^r \Delta F(\xi, \theta, \gamma)\, \xi^{-\alpha-1}\, d\xi.$$

But F is harmonic. Hence G and Φ are harmonic on B. □

(6)*[1] *Theorem (5) is false if the hypothesis that 4ν not be an integer is omitted

Proof. Suppose that 4ν is an integer. Then the modulus $\alpha = 4(1-\nu)$ is also an integer with $2<\alpha<8$. Let B be a closed ball centered at $\mathbf{0}$, and let $\boldsymbol{u}$ be the elastic displacement field (corresponding to $\boldsymbol{b}=\mathbf{0}$) defined by

$$\boldsymbol{u} = \frac{1}{\alpha} \nabla \varphi, \tag{a}$$

where φ is the solid spherical harmonic

$$\varphi(r, \theta, \gamma) = r^\alpha S_\alpha(\theta, \gamma). \tag{b}$$

Assume that ***(5)*** remains valid. Then there exists a field $\boldsymbol{\psi}$ that is harmonic on B and satisfies

$$\boldsymbol{u} = \boldsymbol{\psi} - \frac{1}{\alpha} \nabla(\boldsymbol{p}\cdot\boldsymbol{\psi}). \tag{c}$$

By (a) and (c),

$$\nabla(\boldsymbol{p}\cdot\boldsymbol{\psi}) - \alpha\boldsymbol{\psi} = \nabla\varphi, \tag{d}$$

and upon operating on (d) with the curl, we conclude that $\boldsymbol{\psi}$ is irrotational. Since B is star-shaped and hence simply-connected, there exists an analytic field Φ on B such that $\boldsymbol{\psi} = \nabla\Phi$; thus we conclude from (d) that

$$\boldsymbol{p}\cdot\nabla\Phi - \alpha\Phi = \varphi + c_1, \tag{e}$$

where c_1 is a constant. By (e) and (b),

$$\frac{\partial}{\partial r}(r^{-\alpha}\Phi) = r^{-1} S_\alpha(\theta, \gamma) + c_1. \tag{f}$$

The general solution of (f) is

$$\Phi(r, \theta, \gamma) = r^\alpha f(\theta, \gamma) + r^\alpha (\log r)\, S_\alpha(\theta, \gamma) + c_1 r^{\alpha+1}. \tag{g}$$

There is no choice of $f(\theta, \gamma)$ that leaves Φ without a singularity at the origin. But Φ was required to be analytic on B. Hence ***(5)*** cannot be valid in the present circumstances. □

[1] Eubanks and Sternberg [1956, *4*]. See also Slobodyansky [1954, *19*], Solomon [1961, *20*]. Even though the potential φ cannot be eliminated when 4ν is an integer, Stippes [1969, *8*] has shown that φ can be set equal to a solid spherical harmonic of degree $4(1-\nu)$.

The next theorem, which is also due to EUBANKS and STERNBERG,[1] deals with the possibility of eliminating one of the components of $\boldsymbol{\psi}$.

(7) Elimination of a component of the vector potential. *Assume that B satisfies hypotheses* (i) *and* (iv) *of* ***(5.2)*** *with $M=2$ and $N=3$. Let $\boldsymbol{u}$ be an elastic displacement field corresponding to $\boldsymbol{b}=\boldsymbol{0}$, and assume that $\boldsymbol{u}$ is of class C^4 on $\bar{B}$. Then $\boldsymbol{u}$ admits the representation* (P_1) *with φ and $\boldsymbol{\psi}$ harmonic on B and $\psi_3=0$.*

Proof. By ***(4)*** there exist harmonic fields $\hat{\varphi}$ and $\hat{\boldsymbol{\psi}}$ such that

$$\boldsymbol{u}=\hat{\boldsymbol{\psi}}-\frac{1}{\alpha}(\boldsymbol{p}\cdot\hat{\boldsymbol{\psi}}+\hat{\varphi}),$$

where

$$\alpha=4(1-\nu).$$

Thus we must establish the existence of harmonic fields φ and $\boldsymbol{\psi}$ such that

$$\alpha\boldsymbol{\psi}-\nabla(\boldsymbol{p}\cdot\boldsymbol{\psi}+\varphi)=\alpha\hat{\boldsymbol{\psi}}-\nabla(\boldsymbol{p}\cdot\hat{\boldsymbol{\psi}}+\hat{\varphi}), \tag{a}$$

$$\psi_3=0. \tag{b}$$

It suffices to find a harmonic field H that satisfies[2]

$$H_{,3}=\hat{\psi}_3, \tag{c}$$

for then the functions

$$\boldsymbol{\psi}=\hat{\boldsymbol{\psi}}-\nabla H,$$
$$\varphi=\hat{\varphi}+\boldsymbol{p}\cdot\nabla H-\alpha H$$

will be harmonic and will satisfy (a) and (b).

By tracing the steps in the proof of ***(4)***, it is not difficult to verify that $\hat{\boldsymbol{\psi}}$ is of class C^2 on $\bar{B}$, since $\boldsymbol{u}$ is of class C^4 on $\bar{B}$. By (a) of ***(5.2)*** with $M=2$ and $N=3$, there exists a scalar field J of class C^2 on $\bar{B}$ and C^3 on B such that

$$J_{,3}=\hat{\psi}_3.$$

Then

$$(\Delta J)_{,3}=\Delta\hat{\psi}_3=0,$$

and there exists a continuous function h on $\bar{B}_{12}$ that is class C^1 on B_{12} and satisfies

$$\Delta J(x_1,x_2,x_3)=h(x_1,x_2).$$

(Recall that $B_{12}=\{(x_1,x_2):\boldsymbol{x}\in B\}$.) By ***(6.6)*** there exists a class C^2 scalar field K on B_{12} such that

$$\Delta K=-h.$$

Thus if we define H on B through

$$H(x_1,x_2,x_3)=J(x_1,x_2,x_3)+K(x_1,x_2),$$

then H is harmonic and satisfies (c). □

Assume now that the body force $\boldsymbol{b}=\boldsymbol{0}$, and let $\boldsymbol{k}$ be a given unit vector. If in the Boussinesq-Somigliana-Galerkin solution we take

$$\boldsymbol{g}=\boldsymbol{k}\mathscr{X},$$

[1] [1956, *4*].

[2] NEUBER [1934, *4*], without restricting B, assumes the existence of such a function.

then (G_1), (G_2) reduces to ***Love's solution***[1]

$$u = k \Delta \mathscr{X} - \frac{1}{2(1-\nu)} \nabla (k \cdot \nabla \mathscr{X}), \tag{L$_1$}$$

$$\Delta \Delta \mathscr{X} = 0. \tag{L$_2$}$$

On the other hand, if we take

$$\boldsymbol{\psi} = k \psi$$

in the Boussinesq-Papkovitch-Neuber solution (P_1), (P_2), we arrive at ***Boussinesq's solution***:[2]

$$u = k \psi + \frac{1}{4(1-\nu)} \nabla (\varphi + z\psi), \tag{B$_1$}$$

$$\Delta \varphi = 0, \qquad \Delta \psi = 0, \tag{B$_2$}$$

where $z = k \cdot (x - 0)$ is the coordinate in the direction of k. Both of these solutions are extremely useful in applications involving torsionless rotational symmetry.

It is clear from (L_1) and (B_1) that these solutions can at most represent displacement fields that satisfy[3]

$$k \cdot \operatorname{curl} u = 0. \tag{C}$$

The next theorem shows that under certain hypotheses on B, condition (C) is also sufficient for the completeness of the above solutions.

(8) Completeness of Love's and Boussinesq's solutions.[4] *Assume that B satisfies hypotheses* (i)–(iv) *of* ***(5.2)*** *with $M=2$ and $N=5$. Let u be an elastic displacement field corresponding to $b=0$ that satisfies* (C) *with $k=e_3$, and suppose that u is of class C^2 on $\bar{B}$. Then there exist scalar fields $\mathscr{X}$, φ, and ψ on B that satisfy* (L_1), (L_2) *and* (B_1), (B_2).

Proof. By hypothesis and ***(42.1)***, u is of class C^∞ on B. Since $k=e_3$, (C) yields

$$u_{1,2} - u_{2,1} = 0,$$

and we conclude from ***(5.2)*** with $M=2$ and $N=5$ that there exists a scalar field τ of class C^2 on $\bar{B}$ and C^5 on B such that

$$u_\alpha = \tau_{,\alpha}. \tag{a}$$

The displacement equation of equilibrium can be written in the form

$$\Delta u_\alpha + \gamma \theta_{,\alpha} = 0, \tag{b}$$

$$\Delta u_3 + \gamma \theta_{,3} = 0, \tag{c}$$

where

$$\theta = \operatorname{div} u, \tag{d}$$

$$\gamma = \frac{1}{1-2\nu}. \tag{e}$$

---

[1] [1927, *3*], p. 276. Cf. Michell [1900, *4*].

[2] [1885, *1*], p. 62 and p. 72.

[3] Noll [1957, *11*].

[4] The portion of this theorem concerning Love's solution is due to Noll [1957, *11*], Theorem 2. Under the assumption of torsionless rotational symmetry, Love [1927, *3*], pp. 274–276 established the completeness of (L_1), (L_2); under the same assumption Eubanks and Sternberg [1956, *4*] gave a completeness proof for (B_1), (B_2) (see ***(9)***).

Before proceeding any further, recall that

$$B_{12}=\{(x_1, x_2): \boldsymbol{x}\in B\}, \quad B_3=\{x_3: \boldsymbol{x}\in B\}.$$

By (a) and (b),

$$[\Delta \tau+\gamma\theta]_{,\alpha}=0;$$

thus

$$\Delta \tau+\gamma\theta=\lambda, \tag{f}$$

where $\lambda=\lambda(x_3)$ is a class C^3 function on B_3 that is continuous on $\bar{B}_3$. Now let ϱ be a solution of

$$\frac{d^2\varrho}{dx_3^2}=-\lambda, \tag{g}$$

and let

$$\varkappa(x_1, x_2, x_3)=\tau(x_1, x_2, x_3)+\varrho(x_3). \tag{h}$$

Then $\varkappa$ is class C^5 on B and class C^2 on $\bar{B}$. By (h) and (a),

$$u_\alpha=\varkappa_{,\alpha}, \tag{i}$$

and since

$$\Delta\varrho=\varrho_{,33},$$

it follows from (f), (g), and (h) that

$$\Delta\varkappa+\gamma\theta=0. \tag{j}$$

If we differentiate (j) with respect to x_3 and subtract the result from (c), we arrive at

$$\Delta(u_3-\varkappa_{,3})=0. \tag{k}$$

Next, by (d) and (i),

$$\theta=u_{\alpha,\alpha}+u_{3,3}=\varkappa_{,\alpha\alpha}+u_{3,3}=\Delta\varkappa+(u_3-\varkappa_{,3})_{,3}, \tag{l}$$

and (j) and (l) imply

$$(1+\gamma)\Delta\varkappa+\gamma(u_3-\varkappa_{,3})_{,3}=0. \tag{m}$$

In view of ***(5.2)***, there exists a solution Φ^* of

$$\Phi^*_{,3}=\varkappa \tag{n}$$

that is class C^2 on $\bar{B}$ and C^5 on B; thus (m) takes the form

$$[(1+\gamma)\Delta\Phi^*+\gamma(u_3-\varkappa_{,3})]_{,3}=0. \tag{o}$$

Since B is x_3-convex, (o) implies that

$$g^*=(1+\gamma)\Delta\Phi^*+\gamma(u_3-\varkappa_{,3}) \tag{p}$$

is independent of x_3. Moreover, g^* is continuous on $\bar{B}_{12}$ and class C^3 on B_{12}. Thus, in view of the properties of the logarithmic potential ***(6.6)***, there exists a class C^4 function g on B_{12} such that

$$g_{,\alpha\alpha}=\frac{1}{1+\gamma}g^*. \tag{q}$$

If we let

$$\Phi(x_1, x_2, x_3)=\Phi^*(x_1, x_2, x_3)-g(x_1, x_2), \tag{r}$$

then Φ is of class C^4 on B and (p), (q) imply

$$(1+\gamma)\Delta\Phi+\gamma(u_3-\varkappa_{,3})=0. \tag{s}$$

Moreover, (n) and (r) yield

$$\Phi_{,3} = \varkappa. \tag{t}$$

Finally, using (i), (s), and (t)

$$\begin{aligned} \boldsymbol{u} &= u_\alpha \boldsymbol{e}_\alpha + u_3 \boldsymbol{e}_3 = \varkappa_{,\alpha} \boldsymbol{e}_\alpha + u_3 \boldsymbol{e}_3 = \nabla\varkappa + (u_3 - \varkappa_{,3}) \boldsymbol{e}_3 \\ &= \nabla\Phi_{,3} - \frac{1+\gamma}{\gamma} \Delta\Phi \boldsymbol{e}_3, \end{aligned}$$

and letting

$$\mathscr{X} = -\frac{1+\gamma}{\gamma} \Phi,$$

we arrive, with the aid of (e), at (L_1). Since (L_1) and the displacement equation of equilibrium imply (L_2), the completeness of Love's solution is established.[1]

If we now define[2]

$$\begin{aligned} \varphi &= 2\mathscr{X}_{,3} - x_3 \mathscr{X}_{,33}, \\ \psi &= \mathscr{X}_{,33}, \end{aligned}$$

then (B_1), (B_2) follow at once from (L_1), (L_2). □

Let (r, γ, z) be cylindrical coordinates with the z-axis parallel to $\boldsymbol{k}$. We say that a displacement field $\boldsymbol{u}$ is ***torsionless and rotationally symmetric with respect to the z-axis*** if

$$u_r = u_r(r, z), \quad u_\gamma = 0, \quad u_z = u_z(r, z).$$

It is not difficult to show that such displacement fields satisfy (C), and hence, by *(8)*, admit the representation (B_1), (B_2). An alternative proof of this assertion is furnished by the following theorem due to EUBANKS and STERNBERG.[3]

(9) *Suppose that ∂B is a surface of revolution with the z-axis the axis of revolution. Let $\boldsymbol{u}$ be torsionless and rotationally symmetric with respect to the z-axis and satisfy the hypotheses of* ***(4)*** *with $\boldsymbol{b} = \boldsymbol{0}$. Then there exist scalar fields $\varphi = \varphi(r, z)$ and $\psi = \psi(r, z)$ on B such that* (B_1), (B_2) *hold. Moreover, the functions φ and ψ are single-valued and analytic on any simply-connected plane cross-section of B parallel to the z-axis.*

Proof. By *(4)* there exist harmonic fields φ' and $\boldsymbol{\psi}'$ such that (P_1) holds. Let

$$\begin{aligned} f_r &= \frac{\partial \psi_z'}{\partial r} - \frac{\partial \psi_r'}{\partial z}, \\ f_z &= \frac{1}{r} \frac{\partial}{\partial r} (r \psi_r') + \frac{1}{r} \frac{\partial \psi_\gamma'}{\partial \gamma} + \frac{\partial \psi_z'}{\partial z}. \end{aligned}$$

Then an elementary computation based on (P_1) and the fact that $\boldsymbol{\psi}'$ is harmonic and $\boldsymbol{u}$ torsionless and rotationally symmetric confirms that

$$\frac{\partial f_r}{\partial z} - \frac{\partial f_z}{\partial r} = 0.$$

Thus there exists a field ψ such that

$$f_r = \frac{\partial \psi}{\partial r}, \quad f_z = \frac{\partial \psi}{\partial z}.$$

---

[1] The proof up to this point is due to NOLL [1957, *11*].
[2] Cf. the last equation in the proof of *(4)*.
[3] [1956, *4*].

In the same manner, let

$$g_r = \frac{\partial \varphi'}{\partial r} + \frac{\partial}{\partial r}(r\psi_r') + z\frac{\partial \psi_z'}{\partial r} - \alpha\psi_r',$$

$$g_z = \alpha\psi + \frac{\partial \varphi'}{\partial z} + r\frac{\partial \psi_r'}{\partial z} + \frac{\partial}{\partial z}(z\psi_z') - \alpha\psi_z'.$$

Then

$$\frac{\partial g_r}{\partial z} - \frac{\partial g_z}{\partial r} = 0,$$

and there exists a field ω such that

$$g_r = \frac{\partial \omega}{\partial r}, \qquad g_z = \frac{\partial \omega}{\partial z}.$$

Now let

$$\varphi = \omega - z\psi;$$

then a trivial computation reveals that

$$\boldsymbol{u} = \boldsymbol{k}\psi + \frac{1}{\alpha}\nabla(\varphi + z\psi). \quad \square$$

VII. The plane problem.[1]

45. The associated plane strain and generalized plane stress solutions. Throughout this section B is *homogeneous* and *isotropic*[2] with Lamé moduli μ and λ.

Let (x_1, x_2, x_3) be a rectangular coordinate system with origin at **0**, and let B be a (not necessarily circular) cylinder centered at **0** with generators parallel to

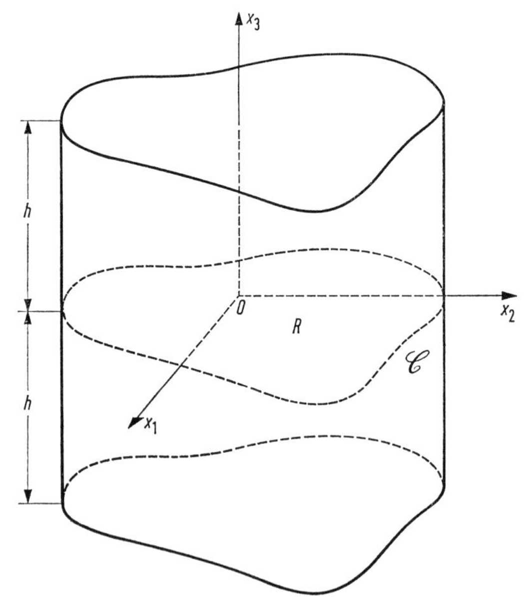

Fig. 11.

[1] Most of this subchapter was taken from the unpublished lecture notes of E. STERNBERG (1959). Extensive reviews of the subject are given by TEODORESCU [1964, *22*], and MUSKHELISHVILI [1965, *15*]. See also the eighth chapter (by BARENBLATT, KALANDYA, and MANDZHAVIDZE) of MUSKHELISHVILI's book [1966, *17*].

[2] The plane problem for anisotropic media is discussed by FRIDMAN [1950, *5*] and KUPRADZE [1963, *17*, *18*], Chap. VIII.

the x_3-axis and with end faces at $x_3 = \pm h$ (Fig. 11). Let $\mathscr{S}_1$ and $\mathscr{S}_2$ be complementary subsets of the set

$$\{\boldsymbol{x} \in \partial B\colon\ x_3 \neq \pm h\}.$$

We assume that $\mathscr{S}_1$ and $\mathscr{S}_2$ are independent of x_3 in the sense that $\overline{\mathscr{S}}_1 \cap \overline{\mathscr{S}}_2$ consists at most of line segments parallel to the x_3-axis and of length $2h$.

Given elastic constants μ and λ, body forces $\boldsymbol{b}$ on B, surface displacements $\hat{\boldsymbol{u}}$ on $\mathscr{S}_1$, and surface tractions $\hat{\boldsymbol{s}}$ on $\mathscr{S}_2$, with $\boldsymbol{b}$, $\hat{\boldsymbol{u}}$, and $\hat{\boldsymbol{s}}$ independent of x_3 and parallel to the x_1, x_2-plane, the ***plane problem of elastostatics*** consists in finding an elastic state $[\boldsymbol{u}, \boldsymbol{E}, \boldsymbol{S}]$ on B that corresponds to the body force field $\boldsymbol{b}$ and satisfies the boundary conditions:

$$\boldsymbol{u} = \hat{\boldsymbol{u}} \quad \text{on } \mathscr{S}_1, \qquad \boldsymbol{s} = \hat{\boldsymbol{s}} \quad \text{on } \mathscr{S}_2,$$
$$\boldsymbol{s} = \boldsymbol{0} \quad \text{when} \quad x_3 = \pm h.$$

Since the surface displacements, surface tractions, and body forces are all independent of x_3 and parallel to the x_1, x_2-plane, one might expect that the displacement field $\boldsymbol{u}$ will also have this property. With this in mind, we introduce the following definition.

Let $[\boldsymbol{u}, \boldsymbol{E}, \boldsymbol{S}]$ be an elastic state. Then $[\boldsymbol{u}, \boldsymbol{E}, \boldsymbol{S}]$ is a ***state of plane strain*** provided

$$u_\alpha = u_\alpha(x_1, x_2), \qquad u_3 = 0.$$

Here and in what follows Greek subscripts range over the integers (1, 2). The above restriction, in conjunction with the strain-displacement and stress-strain relations, implies that $\boldsymbol{E} = \boldsymbol{E}(x_1, x_2)$ and $\boldsymbol{S} = \boldsymbol{S}(x_1, x_2)$. Further,

$$E_{\alpha\beta} = \tfrac{1}{2}(u_{\alpha,\beta} + u_{\beta,\alpha}),$$
$$E_{13} = E_{23} = E_{33} = 0,$$
$$S_{\alpha\beta} = 2\mu\, E_{\alpha\beta} + \lambda\,\delta_{\alpha\beta}\, E_{\gamma\gamma},$$
$$S_{13} = S_{23} = 0, \qquad S_{33} = \lambda E_{\alpha\alpha},$$

and the equation of equilibrium takes the form

$$S_{\alpha\beta,\beta} + b_\alpha = 0.$$

A simple calculation based on the above relations yields the important relation

$$S_{33} = \nu S_{\alpha\alpha},$$

where ν is Poisson's ratio:

$$\nu = \frac{\lambda}{2(\lambda + \mu)}.$$

Thus the complete system of field equations consists of

$$\left.\begin{aligned} S_{\alpha\beta,\beta} + b_\alpha &= 0,\\ S_{\alpha\beta} &= \mu(u_{\alpha,\beta} + u_{\beta,\alpha}) + \lambda\,\delta_{\alpha\beta}\, u_{\gamma,\gamma}, \end{aligned}\right\} \tag{P$_1$}$$

supplemented by

$$S_{13} = S_{23} = u_3 = 0, \qquad S_{33} = \nu S_{\alpha\alpha}; \tag{P$_1^*$}$$

while the boundary conditions are

$$u_\alpha = \hat{u}_\alpha \quad \text{on } \mathscr{C}_1, \qquad S_{\alpha\beta}\, n_\beta = \hat{s}_\alpha \quad \text{on } \mathscr{C}_2, \tag{P$_2$}$$

supplemented by

$$S_{13}=S_{23}=S_{33}=0 \quad \text{when} \quad x_3=\pm h. \tag{P$_2^*$}$$

Here $\mathscr{C}_1$ and $\mathscr{C}_2$ are the intersections, respectively, of $\mathscr{S}_1$ and $\mathscr{S}_2$ with the x_1, x_2-plane.

The system (P_1), (P_2) constitutes a two-dimensional boundary-value problem. This problem has a unique solution which we call the *plane strain solution* associated with the plane problem. As is clear from (P_1^*) and (P_2^*), *this solution actually solves the plane problem only when* $\nu=0$ *or* $S_{11}+S_{22}\equiv 0$. However, the plane strain solution is an actual solution of the modified problem that results if we replace (P_1^*) by the mixed-mixed boundary condition:

$$S_{13}=S_{23}=u_3=0 \quad \text{when} \quad x_3=\pm h.$$

For future use, we note here that in terms of μ and ν the two-dimensional stress-strain relation

$$S_{\alpha\beta}=2\mu\, E_{\alpha\beta}+\lambda\,\delta_{\alpha\beta}\, E_{\gamma\gamma}$$

can be inverted to give

$$2\mu\, E_{\alpha\beta}=S_{\alpha\beta}-\nu\,\delta_{\alpha\beta}\, S_{\gamma\gamma}.$$

We turn next to the associated *generalized plane stress solution*,[1] which characterizes the thickness averages of $\boldsymbol{u}$, $\boldsymbol{E}$, and $\boldsymbol{S}$ under the *approximative assumption*:

$$S_{33}\equiv 0. \tag{A}$$

Thus let $[\boldsymbol{u}, \boldsymbol{E}, \boldsymbol{S}]$ (no longer assumed to be a state of plane strain) be a solution of the plane problem and assume that (A) holds. It follows from the symmetry inherent in the plane problem that

$$u_\alpha(x_1, x_2, x_3)=u_\alpha(x_1, x_2, -x_3), \qquad u_3(x_1, x_2, x_3)=-u_3(x_1, x_2, -x_3),$$

for otherwise the displacement field $\boldsymbol{u}^*$ defined by

$$u_\alpha^*(x_1, x_2, x_3)=u_\alpha(x_1, x_2, -x_3), \qquad u_3^*(x_1, x_2, x_3)=-u_3(x_1, x_2, -x_3)$$

would generate a second solution of the plane problem, a situation ruled out by the uniqueness theorem of elastostatics (at least when the elasticity tensor is positive definite).

Given a function f on B, let $\bar{f}$ denote its *thickness average*:

$$\bar{f}(x_1, x_2)=\frac{1}{2h}\int_{-h}^{+h} f(x_1, x_2, x_3)\, dx_3.$$

Since the thickness average of an x_3-odd function is zero, and since the x_3-derivative of an x_3-even function is x_3-odd, it follows that

$$\bar{u}_3=\bar{E}_{13}=\bar{E}_{23}=\bar{S}_{13}=\bar{S}_{23}=0.$$

Further,

$$\begin{aligned}
\bar{E}_{\alpha\beta}&=\tfrac{1}{2}(\bar{u}_{\alpha,\beta}+\bar{u}_{\beta,\alpha}),\\
\bar{S}_{\alpha\beta}&=2\mu\,\bar{E}_{\alpha\beta}+\lambda\,\delta_{\alpha\beta}(\bar{E}_{\gamma\gamma}+\bar{E}_{33}),\\
\bar{S}_{33}&=2\mu\,\bar{E}_{33}+\lambda(\bar{E}_{\gamma\gamma}+\bar{E}_{33}).
\end{aligned}$$

[1] Thickness averages of solutions to the plane problem under the assumption that S_{33} vanish were first considered by FILON [1903, *2*], who derived the results (a)–(c). See also FILON [1930, *1*] and COKER and FILON [1931, *2*]. The term "generalized plane stress" is due to LOVE [1927, *3*], § 94. The assumption $S_{33}=0$ was studied by MICHELL [1900, *3*].

In view of (A), $\bar{S}_{33}=0$; thus the above relations imply that

$$\bar{S}_{\alpha\beta}=\mu(\bar{u}_{\alpha,\beta}+\bar{u}_{\beta,\alpha})+\bar{\lambda}\,\delta_{\alpha\beta}\,\bar{u}_{\gamma,\gamma}, \tag{a}$$

where

$$\bar{\lambda}=\frac{2\mu\,\lambda}{\lambda+2\mu}.$$

Further, the averaged equation of equilibrium takes the form

$$\bar{S}_{\alpha\beta,\beta}+b_\alpha=0, \tag{b}$$

while the boundary conditions become

$$\bar{u}_\alpha=\hat{u}_\alpha \quad \text{on } \mathscr{C}_1, \qquad \bar{S}_{\alpha\beta}\,n_\beta=\hat{s}_\alpha \quad \text{on } \mathscr{C}_2. \tag{c}$$

The relations (a), (b), and (c) constitute a two dimensional boundary-value problem. This problem has a unique solution called the *generalized plane stress solution* associated with the plane problem. Note that the relations (a), (b), and (c), which constitute the generalized plane stress solution, are *identical* to the relations (P_1) and (P_2) of the plane strain solution provided we replace the Lamé modulus λ by

$$\bar{\lambda}=\frac{2\mu\,\lambda}{\lambda+2\mu}.$$

It must be emphasized that *basic to the derivation of* (a) *is the approximative assumption that S_{33} vanish.*

The stress-strain relation for generalized plane stress is easily inverted; indeed,

$$2\mu\,\bar{E}_{\alpha\beta}=\bar{S}_{\alpha\beta}-\bar{\nu}\,\delta_{\alpha\beta}\,\bar{S}_{\gamma\gamma},$$

where

$$\bar{\nu}=\frac{\nu}{1+\nu}$$

is the appropriate "Poisson's ratio" for generalized plane stress. Fig. 12 shows the graph of $\bar{\nu}/\nu$ as a function of ν.

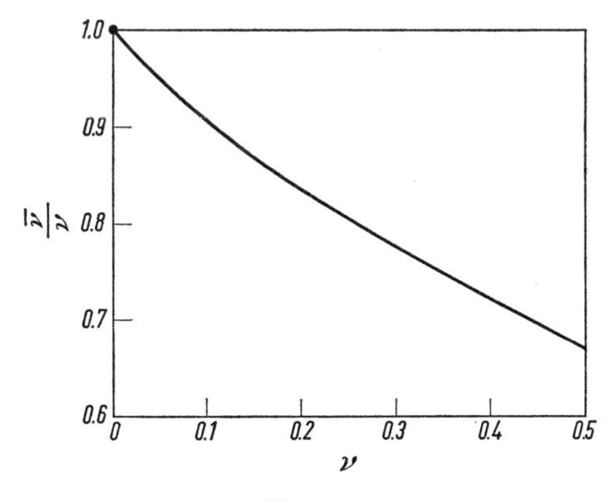

Fig. 12.

Let us agree to call the elastic constants μ and ν *positive definite* if

$$\mu>0, \qquad -1<\nu<\tfrac{1}{2}.$$

This definition is motivated by (iv) of ***(24.5)***. Note that

$$-1<\nu<\tfrac{1}{2} \Leftrightarrow -\infty<\bar{\nu}<\tfrac{1}{3}, \tag{d}$$

so that μ, ν positive definite does *not* imply μ, $\bar{\nu}$ positive definite. Note, however, that

$$0 \leqq \nu < \tfrac{1}{2} \Leftrightarrow 0 \leqq \bar{\nu} < \tfrac{1}{3}.$$

The result (d) will be of importance when we discuss the uniqueness issue appropriate to the generalized plane stress solution.

46. Plane elastic states. In this section we study properties of the plane strain solution associated with the plane problem under the assumption that B is *homogeneous* and *isotropic* with $\mu \neq 0$, $\nu \neq \frac{1}{2}, 1$. In addition, *we assume throughout this section and the next that the body forces are zero.*

We write R for the intersection of B with the x_1, x_2-plane. Further, we let $\mathscr{C}_1$ and $\mathscr{C}_2$ denote complementary regular subcurves[1] of the boundary curve ∂R and write n_α for the components of the outward unit normal to ∂R.

In view of the discussion given in the previous section, to specify a state of plane strain or generalized plane stress it suffices to specify the plane components u_α and $S_{\alpha\beta}$ of the displacement and stress fields. This remark should motivate the following definition. We say that $[u_\alpha, S_{\alpha\beta}]$ is a ***plane elastic state*** corresponding to the elastic constants μ and ν if:

(i) u_α is continuous on $\bar{R}$ and smooth on R;

(ii) $S_{\alpha\beta}$ is continuous on $\bar{R}$ and smooth on R;

(iii) u_α and $S_{\alpha\beta}$ satisfy the field equations

$$S_{\alpha\beta,\beta}=0,$$
$$\mu(u_{\alpha,\beta}+u_{\beta,\alpha})=S_{\alpha\beta}-\nu\,\delta_{\alpha\beta}\,S_{\gamma\gamma}.$$

Since every plane elastic state generates, with the aid of (P_1^*) of Sect. 45, a (three-dimensional) state of plane strain, we have the following direct corollary of ***(42.1)***:

(1) *Let $[u_\alpha, S_{\alpha\beta}]$ be a plane elastic state corresponding to the elastic constants μ and ν. Then u_α and $S_{\alpha\beta}$ are analytic on R.*

Each state of plane strain is an actual solution of the modified plane problem for which the boundary conditions at $x_3=\pm h$ are replaced by the mixed-mixed conditions

$$S_{13}=S_{23}=u_3=0 \quad \text{when} \quad x_3=\pm h.$$

Therefore, as a consequence of the uniqueness theorem ***(40.1)***, we have the

(2) Uniqueness theorem for the plane problem.[2] *Let $[u_\alpha, S_{\alpha\beta}]$ and $[\tilde{u}_\alpha, \tilde{S}_{\alpha\beta}]$ be plane elastic states corresponding to the same elastic constants μ and ν, with μ and ν positive definite, and let both states correspond to the same boundary data, i.e.,*

$$u_\alpha=\tilde{u}_\alpha \ \text{ on } \mathscr{C}_1 \quad \text{and} \quad S_{\alpha\beta}\,n_\beta=\tilde{S}_{\alpha\beta}\,n_\beta \ \text{ on } \mathscr{C}_2.$$

[1] $\mathscr{C}_1$ and $\mathscr{C}_2$ are assumed to have properties strictly analogous to those of $\mathscr{S}_1$ and $\mathscr{S}_2$ given on p. 14.

[2] Trivially, this theorem is valid without change when the body forces do not vanish provided, of course, that both states correspond to the same body forces. A computation of the strain energy of a *plane* elastic state leads, at once, to the conclusion that uniqueness holds under the less restrictive hypotheses: $\mu \neq 0$, $\nu<\frac{1}{2}$. In view of (d), this extension is important as it assures uniqueness when the plane states correspond to generalized plane stress solutions and when the elastic constants μ and $\bar{\nu}=\nu/(1+\nu)$ are such that μ and ν are positive definite.

Then

$$S_{\alpha\beta}=\tilde{S}_{\alpha\beta} \quad \text{on } \overline{R},$$

and hence the two states are equal modulo a plane rigid displacement.

The next theorem shows that for plane states the stress equations of compatibility take a particularly simple form.

(3) Characterization of the stress field. *Let $[u_\alpha, S_{\alpha\beta}]$ be a plane elastic state corresponding to the elastic constants μ and ν. Then $S_{\alpha\beta}$ satisfies the* ***compatibility relation***[1]

$$\Delta S_{\alpha\alpha}=0. \tag{a}$$[2]

Conversely, let $S_{\alpha\beta}$ $(=S_{\beta\alpha})$ be of class C^2 on R and smooth on $\overline{R}$, and suppose that $S_{\alpha\beta}$ satisfies (a) *and the equation of equilibrium*

$$S_{\alpha\beta,\beta}=0. \tag{b}$$

Assume, in addition, that R is simply-connected. Then there exists a field u_α such that $[u_\alpha, S_{\alpha\beta}]$ is a plane elastic state corresponding to μ and ν.

Proof. Let $[u_\alpha, S_{\alpha\beta}]$ be a plane elastic state. By ***(1)*** the strains $E_{\alpha\beta}$ corresponding to u_α are analytic; thus we conclude from ***(14.5)*** that $E_{\alpha\beta}$ satisfies the compatibility relation

$$2E_{12,12}=E_{11,22}+E_{22,11}. \tag{c}$$

The stress-strain relation, in conjunction with (c), yields

$$2S_{12,12}=S_{11,22}+S_{22,11}-\nu\Delta S_{\alpha\alpha}, \tag{d}$$

while the equation of equilibrium implies

$$S_{12,12}=-S_{11,11},$$
$$S_{12,12}=-S_{22,22};$$

hence

$$2S_{12,12}=-S_{11,11}-S_{22,22}. \tag{e}$$

Eqs. (d) and (e) imply (a).

On the other hand, let $S_{\alpha\beta}$ satisfy (a) and (b) and define $E_{\alpha\beta}$ through the stress-strain relation. Then by reversing the above steps, it is a simple matter to verify that (c) holds. Thus ***(14.5)*** implies the existence of a class C^2 displacement field u_α that satisfies the strain-displacement relation, and we conclude from ***(14.4)*** that u_α is continuous on $\overline{R}$. Thus $[u_\alpha, S_{\alpha\beta}]$ has all of the desired properties. □

Theorem ***(3)*** remains valid in the presence of body forces provided (a) and (b) are replaced by

$$\Delta S_{\alpha\alpha}=-\frac{1}{1-\nu}\,b_{\alpha,\alpha}, \qquad S_{\alpha\beta,\beta}+b_\alpha=0.$$

The next theorem asserts that the *stresses corresponding to a solution of the traction problem are independent of the elastic constants*[3] (provided the body is

[1] LÉVY [1898, *7*].

[2] Here Δ is the two-dimensional Laplacian:

$$\Delta=\frac{\partial^2}{\partial x_1^2}+\frac{\partial^2}{\partial x_2^2}.$$

[3] This assertion was first made by LÉVY [1898, *7*]. However, LÉVY did not require that R be simply-connected, and theorem ***(4)*** is false if this restriction is omitted (cf. ***(47.5)*** and ***(47.6)***). The first correct statement is due to MICHELL [1900, *3*], p. 109. Reference should also be made to a remark by MAXWELL [1870, *1*], p. 161, who appears to anticipate LÉVY's claim, but I am unable to follow MAXWELL's argument.

simply-connected). Much of the experimental work in photoelasticity theory is based on this result.

(4) Lévy's theorem. *Assume that R is simply-connected. Let $[u_\alpha, S_{\alpha\beta}]$ and $[\tilde{u}_\alpha, \tilde{S}_{\alpha\beta}]$ be plane elastic states corresponding to positive definite elastic constants μ, ν and $\tilde{\mu}, \tilde{\nu}$, respectively. Suppose, further, that $S_{\alpha\beta}$ and $\tilde{S}_{\alpha\beta}$ are smooth on $\bar{R}$ and*

$$S_{\alpha\beta}\, n_\beta = \tilde{S}_{\alpha\beta}\, n_\beta \quad \text{on } \partial R.$$

Then

$$S_{\alpha\beta} = \tilde{S}_{\alpha\beta} \quad \text{on } \bar{R}.$$

Proof. Clearly, $\tilde{S}_{\alpha\beta}$ satisfies (a) and (b) of ***(3)***. Thus there exists a field $\bar{u}_\alpha$ such that $[\bar{u}_\alpha, \tilde{S}_{\alpha\beta}]$ is a plane elastic state corresponding to μ and ν. Therefore $[u_\alpha, S_{\alpha\beta}]$ and $[\bar{u}_\alpha, \tilde{S}_{\alpha\beta}]$ correspond to the same elastic constants and the same surface forces. Thus we conclude from the uniqueness theorem ***(2)*** with $\mathscr{C}_1 = \emptyset$ and $\mathscr{C}_2 = \partial R$ that $S_{\alpha\beta} = \tilde{S}_{\alpha\beta}$ on $\bar{R}$. □

47. Airy's solution. In this section some of the functions we study will be multi-valued. Therefore, *for this section only, if a function is single-valued we will say so; if nothing is said the function is allowed to be multi-valued.* Thus the assertion "φ is a function of class C^N" means that "φ is a possibly multi-valued function of class C^N". However, we will always assume that the displacement and stress fields corresponding to a plane elastic state are single-valued.

We continue to assume that *the body forces vanish.* Then, as we saw in the previous section, a plane stress field is completely characterized by the following pair of field equations:

$$S_{\alpha\beta,\beta} = 0, \qquad \Delta S_{\alpha\alpha} = 0.$$

The next theorem shows that these equations can be solved with the aid of a single biharmonic function.

(1) Airy's solution. *Let φ be a scalar field of class C^3 on R, and let*

$$S_{11} = \varphi_{,22}, \qquad S_{22} = \varphi_{,11}, \qquad S_{12} = -\varphi_{,12}. \tag{a}$$ [1]

Then

$$S_{\alpha\beta,\beta} = 0. \tag{b}$$

Further,

$$\Delta S_{\alpha\alpha} = 0 \tag{c}$$

if and only if φ is biharmonic.[2]

Conversely, let $S_{\alpha\beta}$ $(= S_{\beta\alpha})$ be single-valued and of class C^N $(N \geqq 1)$ on R, and suppose that (b) *holds. Then there exists a class C^{N+2} function φ on R that satisfies* (a). *Moreover, φ is single-valued if R is simply-connected.*

Proof. Clearly, (a) implies (b), while φ biharmonic is equivalent to (c). To prove the converse assertion assume that (b) holds. Then

$$(S_{11})_{,1} = (-S_{12})_{,2},$$
$$(-S_{12})_{,1} = (S_{22})_{,2},$$

[1] This solution is due to AIRY [1863, *1*]. See also FOX and SOUTHWELL [1945, *1*], SCHAEFER [1956, *11*], TEODORESCU [1958, *19*], [1963, *23*], STRUBECKER [1962, *15*], SCHULER and FOSDICK [1967, *13*]. A generalization for the dynamic plane problem is given by RADOK [1956, *10*]. See also SNEDDON [1952, *5*], [1958, *17*], SVEKLO [1961, *21*].

[2] MAXWELL [1870, *1*] was the first to notice that φ is biharmonic. See also VOIGT [1882, *1*], IBBETSON [1886, *3*], [1887, *2*], pp. 358–359, LÉVY [1898, *7*].

and there exists class C^{N+1} functions g and h on R such that

$$S_{11}=g_{,2}, \quad S_{12}=-g_{,1}, \quad S_{22}=h_{,1}, \quad S_{12}=-h_{,2}. \tag{d}$$

Hence

$$g_{,1}=h_{,2},$$

and there exists a class C^{N+2} function φ such that

$$g=\varphi_{,2}, \quad h=\varphi_{,1}. \tag{e}$$

Eqs. (d) and (e) imply (a). □

Henceforth, by an ***Airy function*** we mean a (possibly multi-valued) *biharmonic* function φ on R with the property that the ***associated stresses***

$$S_{11}=\varphi_{,22}, \quad S_{22}=\varphi_{,11}, \quad S_{12}=-\varphi_{,12}$$

are single-valued. For future use we note that these relations can be written in the concise form:[1]

$$S_{\alpha\beta}=\varepsilon_{\alpha\lambda}\,\varepsilon_{\beta\tau}\,\varphi_{,\lambda\tau},$$

where $\varepsilon_{\alpha\beta}$ is the two-dimensional alternator.

By (a) of ***(1)***, any two Airy functions φ and $\tilde{\varphi}$ generating the same stresses differ by a linear function of position, i.e.

$$\varphi(x)-\tilde{\varphi}(x)=k_\alpha\, x_\alpha+c,$$

where k_α and c are constants and $x=(x_1, x_2)$. Conversely, we can always add a linear function of position to an Airy function without changing the associated stresses. Further, given a point $y=(y_1, y_2)$ in R we can always choose the linear function so that the resulting function φ satisfies

$$\varphi(y)=\varphi_{,\alpha}(y)=0.$$

The (two-dimensional) traction problem for a simply-connected region consists in finding stresses $S_{\alpha\beta}$ that satisfy the field equations

$$S_{\alpha\beta,\beta}=0, \quad \Delta S_{\alpha\alpha}=0,$$

and the boundary condition

$$S_{\alpha\beta}\, n_\beta=\hat{s}_\alpha \quad \text{on } \partial R,$$

where the $\hat{s}_\alpha$ are the prescribed surface tractions. In view of the last theorem, this is equivalent to finding a scalar field φ such that

$$\Delta\Delta\varphi=0,$$
$$\varepsilon_{\alpha\lambda}\,\varepsilon_{\beta\tau}\,\varphi_{,\lambda\tau}\, n_\beta=\hat{s}_\alpha \quad \text{on } \partial R.$$

Thus the traction problem reduces to finding a biharmonic function φ that satisfies the above boundary conditions. Our next theorem, which gives a physical interpretation of the Airy function, allows us to write the boundary condition for φ in a more familiar form. With a view toward a concise statement of this theorem, let

$$\Gamma=\{x(\sigma): 0\leqq\sigma\leqq\sigma_1\}$$

[1] Finzi [1956, 5].

be a piecewise smooth curve in R parametrized by its arc length σ, and let n_α and t_α designate the normal and tangent unit vectors along Γ, i.e.,

$$n_\alpha = \varepsilon_{\alpha\beta}\dot{x}_\beta, \quad t_\alpha = \dot{x}_\alpha,$$

where $\dot{x}_\alpha = dx_\alpha/d\sigma$.

(2) Physical interpretation of the Airy function.[1] *Let φ be an Airy function and write*

$$\varphi(\sigma) = \varphi(x(\sigma)), \quad \frac{\partial\varphi}{\partial n}(\sigma) = \varphi_{,\alpha}(x(\sigma))\, n_\alpha(\sigma)$$

for the values if φ and its normal derivative along the curve Γ. Assume that

$$\varphi(0) = \varphi_{,\alpha}(0) = 0.$$

Then

$$\varphi(\sigma) = m(\sigma),$$

$$\frac{\partial\varphi}{\partial n}(\sigma) = -l(\sigma),$$

where

$$m(\sigma) = \int_0^\sigma \varepsilon_{\alpha\beta}[x_\alpha(\sigma') - x_\alpha(\sigma)]\, s_\beta(\sigma')\, d\sigma',$$

$$l(\sigma) = t_\alpha(\sigma) \int_0^\sigma s_\alpha(\sigma')\, d\sigma',$$

$$s_\alpha(\sigma) = S_{\alpha\beta}(x(\sigma))\, n_\beta(\sigma).$$

Proof. By hypothesis

$$S_{\alpha\beta} = \varepsilon_{\alpha\lambda}\,\varepsilon_{\beta\tau}\,\varphi_{,\lambda\tau}. \tag{a}$$

Thus, in view of the identity

$$\varepsilon_{\alpha\lambda}\,\varepsilon_{\alpha\tau} = \delta_{\lambda\tau}, \tag{b}$$

the surface force field $s_\alpha = S_{\alpha\beta}\, n_\beta$ can be written in the form

$$s_\alpha = \varepsilon_{\alpha\lambda}\,\varepsilon_{\beta\tau}\,\varphi_{,\lambda\tau}\, n_\beta = \varepsilon_{\alpha\lambda}\,\varphi_{,\lambda\tau}\,\dot{x}_\tau.$$

In view of (b),

$$\varphi_{,\lambda\tau}\,\dot{x}_\tau = -\varepsilon_{\lambda\tau}\, s_\tau,$$

and thus

$$\frac{d}{d\sigma}(\varphi_{,\lambda}) = -\varepsilon_{\lambda\tau}\, s_\tau. \tag{c}$$

Therefore, since $\varphi_{,\lambda}(0) = 0$,

$$\varphi_{,\lambda}(\sigma) = f_\lambda(\sigma), \tag{d}$$

where

$$f_\lambda(\sigma) = -\int_0^\sigma \varepsilon_{\lambda\tau}\, s_\tau(\sigma')\, d\sigma'. \tag{e}$$

On the other hand, since $\varphi(0) = 0$,

$$\varphi(\sigma) = \int_0^\sigma \dot{\varphi}(\sigma')\, d\sigma' = \int_0^\sigma \varphi_{,\lambda}(\sigma')\,\dot{x}_\lambda(\sigma')\, d\sigma'. \tag{f}$$

If we integrate (f) by parts and use (d), we arrive at

$$\varphi(\sigma) = f_\lambda\, x_\lambda\big|_0^\sigma - \int_0^\sigma \dot{f}_\lambda(\sigma')\, x_\lambda(\sigma')\, d\sigma'.$$

[1] MICHELL [1900, *3*], p. 107 was the first to notice that φ and $\partial\varphi/\partial n$ are essentially determined by the surface forces along Γ. The relations given here, which follow from a minor modification of MICHELL's results, are due to TREFFTZ [1928, *3*]. See also PHILLIPS [1934, *5*], SOBRERO [1935, *8*], and FINZI [1956, *5*].

By (e), $f_\lambda(0)=0$ and $\dot{f}_\lambda = -\varepsilon_{\lambda\tau} s_\tau$; therefore

$$\varphi(\sigma) = -\int_0^\sigma x_\lambda(\sigma)\,\varepsilon_{\lambda\tau}\, s_\tau(\sigma')\,d\sigma' + \int_0^\sigma x_\lambda(\sigma')\,\varepsilon_{\lambda\tau}\, s_\tau(\sigma')\,d\sigma',$$

or equivalently

$$\varphi(\sigma) = m(\sigma).$$

Next, by definition,

$$\frac{\partial\varphi}{\partial n} = \varphi_{,\alpha} n_\alpha = \varphi_{,\alpha}\,\varepsilon_{\alpha\beta}\,\dot{x}_\beta,$$

and (d), (e), and (b) imply

$$\frac{\partial\varphi}{\partial n}(\sigma) = -l(\sigma). \quad \square$$

If φ is of class C^2 on $\bar{R}$, then the results of *(2)* hold even when Γ is contained in the boundary of R. If R is simply-connected, its boundary consists of a single closed curve; thus taking $\Gamma = \partial R$ in *(2)*, we see that the surface force problem reduces to finding a biharmonic function φ that satisfies the "standard" boundary conditions:[1]

$$\varphi = m \quad \text{and} \quad \frac{\partial\varphi}{\partial n} = -l \quad \text{on } \partial R,$$

where m and l are completely determined by the prescribed surface tractions on ∂R.

If φ and $\varphi_{,\alpha}$ do not vanish at $x(0)$, we can still apply *(2)* to the Airy function

$$\tilde{\varphi}(x) = \varphi(x) - k_\alpha x_\alpha - c,$$

with the constants k_α and c chosen so that

$$\tilde{\varphi}(x(0)) = \tilde{\varphi}_{,\alpha}(x(0)) = 0.$$

Thus, in this instance, the relations for φ and $\partial\varphi/\partial n$ should be replaced by

$$\begin{aligned} \varphi &= m + k_\alpha x_\alpha + c, \\ \frac{\partial\varphi}{\partial n} &= -l + k_\alpha n_\alpha. \end{aligned} \tag{*}$$

If R is multiply-connected, its boundary ∂R is the union of a finite number of closed curves $\mathscr{C}_1, \mathscr{C}_2, \ldots, \mathscr{C}_M$. In this case it is not possible to adjust the Airy function (without changing the stresses) to make φ and $\varphi_{,\alpha}$ vanish at one point on *each* $\mathscr{C}_m$. Thus relations of the form (*) will hold on each $\mathscr{C}_m$; of course, the constants will, in general, be different on each $\mathscr{C}_m$:

$$\varphi = m + k_\alpha^m x_\alpha + c^m \quad \text{and} \quad \frac{\partial\varphi}{\partial n} = -l + k_\alpha^m n_\alpha \quad \text{on } \mathscr{C}_m \quad (m = 1, 2, \ldots, M).$$

In the above relations k_α^m and c^m can be set equal to zero on one of the curves $\mathscr{C}_m$; as we shall see, the remaining constants can be evaluated using the requirement that the displacements be single-valued.

With a view toward deriving a relation for the displacement field in terms of the Airy function, we now establish an important representation theorem for the Airy function in terms of a pair of analytic functions of the complex variable $z = x_1 + i x_2$.

[1] Michell [1900, *3*], p. 108.

(3) Goursat's theorem.[1] *Let φ be biharmonic on R. Then there exist analytic complex functions ψ and χ on R such that*[2]

$$\varphi(x_1, x_2) = \mathrm{Re}\{\bar{z}\,\psi(z) + \chi(z)\},$$

or equivalently,

$$2\varphi(x_1, x_2) = \bar{z}\,\psi(z) + z\,\overline{\psi(z)} + \chi(z) + \overline{\chi(z)}.$$

Conversely, if ψ and χ are analytic functions and φ is defined on R by the above relation, then φ is biharmonic on R.

Proof.[3] Assume that φ is biharmonic on R. Since $\Delta\varphi$ is harmonic, there exists an analytic function f such that $\Delta\varphi$ is the real part of f. Let ψ be a solution of the equation

$$\frac{d\psi}{dz} = \frac{1}{4} f, \tag{a}$$

and let

$$\psi = \tfrac{1}{4}(g_1 + i\, g_2), \tag{b}$$

where g_1 and g_2 are real. Then, since g_α is harmonic,

$$\Delta(\varphi - \tfrac{1}{4} g_\alpha\, x_\alpha) = \Delta\varphi - \tfrac{1}{2} g_{\alpha,\alpha}. \tag{c}$$

By the Cauchy-Riemann equations,

$$g_{1,1} = g_{2,2}, \tag{d}$$

and by (a) and (b),

$$g_{1,1} = \Delta\varphi; \tag{e}$$

thus (c) implies that

$$\Delta(\varphi - \tfrac{1}{4} g_\alpha\, x_\alpha) = 0. \tag{f}$$

Let

$$q_1 = \varphi - \tfrac{1}{4} g_\alpha\, x_\alpha. \tag{g}$$

In view of (f), q_1 is harmonic and hence the real part of an analytic function χ. This fact, (g), and (b) imply that

$$\varphi(x_1, x_2) = \mathrm{Re}\{\bar{z}\,\psi(z) + \chi(z)\}. \tag{h}$$

Conversely, let φ be given by (h) with ψ and χ analytic. Then, by reversing the above steps, it is a simple matter to verify that φ is biharmonic. □

(4) Representation theorem for the displacement field.[4] *Let $[u_\alpha, S_{\alpha\beta}]$ be a plane elastic state corresponding to the elastic constants μ and ν (with $\mu \neq 0$). Further, let φ be an Airy function corresponding to $S_{\alpha\beta}$, and let*

$$4\psi(z) = g_1(x_1, x_2) + i\, g_2(x_1, x_2),$$

*where ψ is the analytic complex function established in **(3)** and g_α is real. Then*

$$u_\alpha = \frac{1}{2\mu}\left[-\varphi_{,\alpha} + (1-\nu)\, g_\alpha\right] + w_\alpha,$$

where w_α is a plane rigid displacement.

Conversely, let φ be an Airy function of class C^2 on $\bar{R}$, let $S_{\alpha\beta}$ be the associated stresses, and let u_α, defined as above, be single-valued. Then $[u_\alpha, S_{\alpha\beta}]$ is a plane elastic state corresponding to μ and ν.

---

[1] GOURSAT [1898, *6*].

[2] Here Re{ } denotes the real part of the quantity { }, and a bar over a letter designates the complex conjugate.

[3] This proof is due to MUSKHELISHVILI [1919, *1*], [1954, *16*], § 31.

[4] LOVE [1927, *3*], § 145. See also CLEBSCH [1862, *1*], § 39.

Proof. Using (a) of ***(1)*** we can write the displacement-stress relations in the form

$$\begin{aligned} 2\mu\, u_{1,1} &= -\varphi_{,11} + (1-\nu)\Delta\varphi, \\ 2\mu\, u_{2,2} &= -\varphi_{,22} + (1-\nu)\Delta\varphi, \\ \mu(u_{1,2}+u_{2,1}) &= -\varphi_{,12}. \end{aligned} \tag{a}$$

Thus we conclude from (d) and (e) in the proof of ***(3)*** that

$$\begin{aligned} 2\mu\, u_{1,1} &= -\varphi_{,11} + (1-\nu)\, g_{1,1}, \\ 2\mu\, u_{2,2} &= -\varphi_{,22} + (1-\nu)\, g_{2,2}; \end{aligned}$$

hence

$$2\mu\, u_\alpha = -\varphi_{,\alpha} + (1-\nu)\, g_\alpha + 2\mu\, w_\alpha, \tag{b}$$

where

$$w_{1,1} = w_{2,2} = 0. \tag{c}$$

Substituting (b) into the third of (a), we find, with the aid of the Cauchy-Riemann equation

$$g_{1,2} = -g_{2,1},$$

that

$$w_{1,2} + w_{2,1} = 0. \tag{d}$$

By (c) and (d), w_α is a plane rigid displacement; thus (b) is the desired representation for u_α.

To prove the converse assertion we simply reverse the steps (a)–(d). □

The next theorem yields conditions that must be satisfied by the Airy function in order to insure that the corresponding displacements be single-valued.

(5) Theorem on single-valued displacements. *Let R be multiply-connected, and suppose that*

$$\partial R = \bigcup_{m=1}^{M} \mathscr{C}_m,$$

where each $\mathscr{C}_m$ is a closed curve. Let φ be an Airy function of class C^3 on $\bar{R}$, and let

$$u_\alpha = \frac{1}{2\mu}\left[-\varphi_{,\alpha} + (1-\nu)\, g_\alpha\right] + w_\alpha,$$

where w_α is rigid and g_α is the function defined in ***(4)***. *Then a necessary and sufficient condition that the displacement field u_α be single-valued on R is that φ satisfy the $3M-3$ conditions:*[1]

$$\int_{\mathscr{C}_m} \frac{\partial}{\partial n}(\Delta\varphi)\, d\sigma = 0,$$

$$\int_{\mathscr{C}_m} \left[x_1 \frac{\partial}{\partial n}(\Delta\varphi) + x_2 \frac{d}{d\sigma}(\Delta\varphi)\right] d\sigma = -\frac{1}{1-\nu} f_1^{(m)},$$

$$\int_{\mathscr{C}_m} \left[x_1 \frac{d}{d\sigma}(\Delta\varphi) - x_2 \frac{\partial}{\partial n}(\Delta\varphi)\right] d\sigma = \frac{1}{1-\nu} f_2^{(m)},$$

[1] These relations are due to Michell [1900, *3*], p. 108. Prager [1946, *4*] proved that for a doubly-connected region these relations are the natural boundary conditions of a variational principle for the Airy function.

for $m=2, 3, \ldots, M$, *where*[1]

$$f_\alpha^{(m)} = \int_{\mathscr{C}_m} S_{\alpha\beta}\, n_\beta\, d\sigma,$$

and $S_{\alpha\beta}$ *is the stress field generated by* φ.

Proof. The field u_α will be single-valued if and only if

$$\int_{\mathscr{C}_m} \frac{d u_\alpha}{d\sigma}\, d\sigma = 0 \qquad (m=2, 3, \ldots, M),$$

or, in view of the definition of u_α, if and only if

$$\int_{\mathscr{C}_m} \frac{d g_\alpha}{d\sigma}\, d\sigma = \frac{1}{1-\nu} \int_{\mathscr{C}_m} \frac{d}{d\sigma} (\varphi_{,\alpha})\, d\sigma \qquad (m=2, 3, \ldots, M). \tag{a}$$

It is a simple matter to verify that (c) in the proof of ***(2)*** is valid in the present circumstances in which φ and $\varphi_{,\alpha}$ do not necessarily vanish at some point on $\mathscr{C}_m$. Thus (a) can be written in the form

$$\int_{\mathscr{C}_m} \frac{d g_\alpha}{d\sigma}\, d\sigma = -\frac{\varepsilon_{\alpha\beta}}{1-\nu} f_\beta^{(m)}, \tag{b}$$

where here and in what follows m is assumed to have the range $(2, 3, \ldots, M)$.

Clearly,

$$\frac{d g_\alpha}{d\sigma} = g_{\alpha,\beta}\, \dot{x}_\beta \tag{c}$$

on $\mathscr{C}_m$, where

$$x(\sigma) = (x_1(\sigma), x_2(\sigma)), \qquad 0 \leqq \sigma \leqq \sigma_1,$$

is the parametrization for $\mathscr{C}_m$. Now let P_1 and P_2 denote the real and imaginary parts of the complex function f established in the proof of ***(3)***:

$$f(z) = P_1(x_1, x_2) + i P_2(x_1, x_2).$$

By (a) and (b) in the proof of ***(3)***,

$$\frac{d}{dz} (g_1 + i g_2) = P_1 + i P_2;$$

thus

$$\begin{aligned} g_{1,1} &= P_1, & g_{2,1} &= P_2, \\ g_{1,2} &= -P_2, & g_{2,2} &= P_1, \end{aligned} \tag{d}$$

and by (c) and (d) the relation (b) is equivalent to

$$\begin{aligned} \int_{\mathscr{C}_m} [P_2 \dot{x}_1 + P_1 \dot{x}_2]\, d\sigma &= \frac{1}{1-\nu} f_1^{(m)}, \\ \int_{\mathscr{C}_m} [P_1 \dot{x}_1 - P_2 \dot{x}_2]\, d\sigma &= -\frac{1}{1-\nu} f_2^{(m)}. \end{aligned} \tag{e}$$

Integrating the left-hand side of (e) by parts, we arrive at

$$\begin{aligned} [P_2 x_1 + P_1 x_2]_{\sigma=0}^{\sigma=\sigma_1} - \int_{\mathscr{C}_m} \left[x_1 \frac{dP_2}{d\sigma} + x_2 \frac{dP_1}{d\sigma} \right] d\sigma &= \frac{1}{1-\nu} f_1^{(m)}, \\ [P_1 x_1 - P_2 x_2]_{\sigma=0}^{\sigma=\sigma_1} - \int_{\mathscr{C}_m} \left[x_1 \frac{dP_1}{d\sigma} - x_2 \frac{dP_2}{d\sigma} \right] d\sigma &= -\frac{1}{1-\nu} f_2^{(m)}. \end{aligned} \tag{f}$$

---

[1] We assume that the parametrization of the $\mathscr{C}_m$ are such that $n_\alpha = \varepsilon_{\alpha\beta}\, \dot{x}_\beta$ coincides with the outward unit normal to ∂R.

It follows from the definition of the Airy function that $P_1 = \Delta\varphi$ is single-valued. Thus for (f) to hold it is *sufficient* that

$$\int_{\mathscr{C}_m} \frac{dP_2}{d\sigma}\, d\sigma = 0,$$

$$\int_{\mathscr{C}_m} \left[x_1 \frac{dP_2}{d\sigma} + x_2 \frac{dP_1}{d\sigma}\right] d\sigma = -\frac{1}{1-\nu} f_1^{(m)}, \tag{g}$$

$$\int_{\mathscr{C}_m} \left[x_1 \frac{dP_1}{d\sigma} - x_2 \frac{dP_2}{d\sigma}\right] d\sigma = \frac{1}{1-\nu} f_2^{(m)}.$$

Indeed, the first of (g) asserts that P_2 is single-valued, and the second and third of (g) are equivalent to (f) when both P_1 and P_2 are single-valued. Next, by (d) and the definition of u_α,

$$P_2 = -\frac{\mu}{1-\nu}\left[u_{1,2} - u_{2,1} - \omega\right], \tag{h}$$

where $\omega = w_{1,2} - w_{2,1}$ is constant (since w_α is rigid). To prove that (g) is also *necessary* for (f) to hold it suffices to show that P_2 is single-valued whenever u_α is single-valued. But this fact follows from (h).[1] Finally, the Cauchy-Riemann equations and the relation $n_\alpha = \varepsilon_{\alpha\beta}\dot{x}_\beta$ imply that

$$\begin{aligned} \frac{dP_2}{d\sigma} &= P_{2,1}\dot{x}_1 + P_{2,2}\dot{x}_2 = -P_{1,2}\dot{x}_1 + P_{1,1}\dot{x}_2 \\ &= P_{1,2} n_2 + P_{1,1} n_1 = \frac{\partial P_1}{\partial n} = \frac{\partial}{\partial n}(\Delta\varphi). \end{aligned} \tag{i}$$

In view of (i) and since $P_1 = \Delta\varphi$, the relations (g) reduce to the desired $3M-3$ conditions. □

The next result extends ***(46.4)*** to multiply-connected regions. It asserts that *the stresses corresponding to a solution of the traction problem with null body forces are independent of the elastic constants if the resultant loading on each boundary curve vanishes.*

(6) Michell's theorem.[2] *Let R be multiply-connected, and suppose that*

$$\partial R = \bigcup_{m=1}^{M} \mathscr{C}_m,$$

where each $\mathscr{C}_m$ is a closed curve. Let $[u_\alpha, S_{\alpha\beta}]$ and $[\tilde{u}_\alpha, \tilde{S}_{\alpha\beta}]$ be plane elastic states corresponding to positive definite elastic constants μ, ν and $\tilde{\mu}, \tilde{\nu}$, respectively, and assume that $S_{\alpha\beta}$ and $\tilde{S}_{\alpha\beta}$ are smooth on $\overline{R}$. Then

$$S_{\alpha\beta} = \tilde{S}_{\alpha\beta} \quad \text{on } \overline{R}$$

provided both states correspond to the same surface tractions and the resultant force on each $\mathscr{C}_m$ vanishes, i.e.,

$$S_{\alpha\beta} n_\beta = \tilde{S}_{\alpha\beta} n_\beta = \hat{s}_\alpha \quad \text{on } \partial R$$

and

$$\int_{\mathscr{C}_m} \hat{s}_\alpha\, d\sigma = 0 \qquad (m = 1, 2, \ldots, M).$$

[1] Since u_α is analytic, $u_{\alpha,\beta}$ is single-valued whenever u_α has this property.
[2] [1900, 3], p. 109.

Proof. By **(1)** there exists an Airy function φ generating $\tilde{S}_{\alpha\beta}$. Moreover, φ is of class C^3 on $\overline{R}$, since $\tilde{S}_{\alpha\beta}$ is smooth there. Since $[\tilde{u}_\alpha, \tilde{S}_{\alpha\beta}]$ is a plane elastic state, we conclude from **(4)** that

$$\tilde{u}_\alpha = \frac{1}{2\tilde{\mu}}\left[-\varphi_{,\alpha} + (1-\tilde{\nu})\, g_\alpha\right] + w_\alpha$$

with w_α rigid. Moreover, since $\tilde{u}_\alpha$ is single-valued, and since the resultant force on each $\mathscr{C}_m$ vanishes, it follows from **(5)** with $f_\alpha^{(m)} = 0$ that

$$\begin{gathered}\int_{\mathscr{C}_m} \frac{\partial}{\partial n}(\Delta\varphi)\, d\sigma = 0,\\ \int_{\mathscr{C}_m}\left[x_1 \frac{\partial}{\partial n}(\Delta\varphi) + x_2 \frac{d}{d\sigma}(\Delta\varphi)\right] d\sigma = 0,\\ \int_{\mathscr{C}_m}\left[x_1 \frac{d}{d\sigma}(\Delta\varphi) - x_2 \frac{\partial}{\partial n}(\Delta\varphi)\right] d\sigma = 0,\end{gathered} \tag{a}$$

for $m = 2, 3, \ldots, M$. Now let

$$\bar{u}_\alpha = \frac{1}{2\mu}\left[-\varphi_{,\alpha} + (1-\nu)\, g_\alpha\right].$$

By (a) and **(5)**, $\bar{u}_\alpha$ is single-valued, and hence the converse assertion in **(4)** implies that $[\bar{u}_\alpha, \tilde{S}_{\alpha\beta}]$ is a plane elastic state corresponding to μ and ν. Therefore $[u_\alpha, S_{\alpha\beta}]$ and $[\bar{u}_\alpha, \tilde{S}_{\alpha\beta}]$ correspond to the same elastic constants and the same surface forces. Thus we conclude from the uniqueness theorem **(46.2)** that $S_{\alpha\beta} = \tilde{S}_{\alpha\beta}$ on $\overline{R}$. □

(7) *Kolosov's theorem.*[1] *Let $[u_\alpha, S_{\alpha\beta}]$ be a plane elastic state corresponding to the elastic constants μ and ν. Then there exist complex analytic functions ψ and χ on R such that*[2]

$$\begin{gathered}u_1(x_1, x_2) + i\, u_2(x_1, x_2) = \frac{1}{2\mu}\left[(3-4\nu)\,\psi(z) - z\,\overline{\psi'(z)} - \overline{\chi'(z)}\right] + w(z),\\ S_{11}(x_1, x_2) + S_{22}(x_1, x_2) = 4\,\mathrm{Re}\{\psi'(z)\},\\ S_{11}(x_1, x_2) - S_{22}(x_1, x_2) + 2i\, S_{12}(x_1, x_2) = 2\left[\bar{z}\,\psi''(z) + \chi'(z)\right],\end{gathered}$$

where w is a complex rigid displacement.

Conversely, let ψ and χ be complex analytic functions on R, and let u_α and $S_{\alpha\beta}$ be defined by the above relations. In addition, assume that u_α is single-valued and u_α and $S_{\alpha\beta}$ continuous on $\overline{R}$. Then $[u_\alpha, S_{\alpha\beta}]$ is an elastic state corresponding to μ and ν.

Proof. Let φ be an Airy function generated by $S_{\alpha\beta}$. By Goursat's theorem **(3)**, there exist analytic complex functions ψ and χ such that

$$\varphi(x_1, x_2) = \mathrm{Re}\{\bar{z}\,\psi(z) + \chi(z)\}. \tag{a}$$

A simple calculation shows that

$$\varphi_{,1}(x_1, x_2) + i\,\varphi_{,2}(x_1, x_2) = \psi(z) + z\,\overline{\psi'(z)} + \overline{\chi'(z)}. \tag{b}$$

[1] KOLOSOV [1909, *2*], [1914, *2*], [1935, *2*]. A related complex representation theorem was established earlier by FILON [1903, *2*]. See also MUSKHELISHVILI [1932, *1*], [1934, *3*], [1954, *16*], § 32; STEVENSON [1943, *2*], [1945, *7*]; PORITSKY [1946, *3*]; SOKOLNIKOFF [1956, *12*], Chap. 5; SOLOMON [1968, *12*], Chap. 6, § 7.

[2] Here $\psi' = d\psi/dz$.

Thus we conclude from ***(4)*** that the desired relation for u_α holds. Further, by (a) of ***(1)***,

$$\begin{aligned} S_{11} + i S_{12} &= -i(\varphi_{,1} + i\varphi_{,2})_{,2}, \\ S_{22} - i S_{12} &= (\varphi_{,1} + i\varphi_{,2})_{,1}, \end{aligned} \tag{c}$$

and (b), (c) imply the desired results concerning $S_{\alpha\beta}$. Conversely, let ψ and χ be complex analytic functions on R, let u_α and $S_{\alpha\beta}$ be defined by the relations in ***(7)***, and let φ be defined by (a) so that (b) holds. Then it is not difficult to verify that $\boldsymbol{u}$ satisfies the relation in ***(4)***, and that $S_{\alpha\beta}$ satisfies (c). Thus $S_{\alpha\beta}$ are the stresses associated with φ, and we conclude from the converse assertion in ***(4)*** that $[u_\alpha, S_{\alpha\beta}]$ is an elastic state correspondign to μ and ν. □

VIII. Exterior domains.

48. Representation of elastic displacement fields in a neighborhood of infinity. The conventional proofs of the fundamental theorems of elastostatics are confined in their validity to bounded domains. Of course, they can be trivially generalized to exterior domains provided one is willing to lay down sufficiently stringent restrictions on the behavior at infinity of the relevant fields, e.g., that the displacement be $O(r^{-1})$ and the stress $O(r^{-2})$ as $r\to\infty$. However, such *a priori* assumptions are quite artificial: the *rate* at which these fields decay is an item of information that one would expect to infer from the solution, rather than a condition to be imposed on a solution in advance.

In this section we shall establish representation theorems for elastostatic fields in the following neighborhood of infinity:

$$\Sigma(\infty) = \{\boldsymbol{x}: r_0 < |\boldsymbol{x} - \boldsymbol{0}| < \infty\},$$

where $r_0 > 0$ is fixed. These representation theorems will be used in subsequent sections to infer the rate at which elastic states decay at infinity, and this, in turn, will be utilized to extend the fundamental theorems of elastostatics to exterior domains.

We assume that the body $B = \Sigma(\infty)$ is *homogeneous* and *isotropic*, and that the elasticity tensor is *positive definite*; thus the Lamé moduli satisfy

$$\mu > 0, \quad 2\mu + 3\lambda > 0.$$

(1) Representation theorem.[1] *Let* $\boldsymbol{u}$ *be an elastic displacement field on* $\Sigma(\infty)$ *corresponding to zero body forces. Then* $\boldsymbol{u}$ *admits the representation*

$$\boldsymbol{u} = \sum_{k=-\infty}^{\infty} \boldsymbol{g}^{(k)} + r^2 \sum_{k=-\infty}^{\infty} \boldsymbol{h}^{(k)}, \tag{a}$$

where $\boldsymbol{g}^{(k)}$ *and* $\boldsymbol{h}^{(k)}$ *are vector solid spherical harmonics of degree* k, *and*

$$\boldsymbol{h}^{(k)} = \frac{-(\lambda+\mu)}{2[(k+1)(\lambda+3\mu)+\mu]} \nabla \operatorname{div} \boldsymbol{g}^{(k+2)}. \tag{b}$$

Both of the series in (a), *as well as the series resulting from any finite number of termwise differentiations, are uniformly convergent on every closed subregion of* $\Sigma(\infty)$.

Proof. First of all, it is not difficult to verify that the denominator in (b) cannot vanish, since the elasticity tensor is positive definite.

[1] Kelvin [1863, *3*]. See also Gurtin and Sternberg [1961, *11*], Theorem 1.2. A generalization of this result to anisotropic bodies is given by Bézier [1967, *3*].

In view of *(42.2)*, $\boldsymbol{u}$ is biharmonic on $\Sigma(\infty)$. Theorem *(8.5)* therefore assures the existence of an expansion of the form (a) which has the asserted convergence properties.

This leaves only (b) to be confirmed. A simple computation shows that

$$\Delta(r^2 \boldsymbol{h}^{(k)}) = 4(\nabla \boldsymbol{h}^{(k)})\,\boldsymbol{p} + 6\boldsymbol{h}^{(k)},$$

and thus the identity

$$(\nabla \boldsymbol{h}^{(k)})\,\boldsymbol{p} = k\,\boldsymbol{h}^{(k)}, \tag{c}$$

which is valid for any vector solid harmonic $\boldsymbol{h}^{(k)}$, implies

$$\Delta(r^2 \boldsymbol{h}^{(k)}) = 2(2k+3)\,\boldsymbol{h}^{(k)}.$$

Hence

$$\Delta \boldsymbol{u} = 2 \sum_{k=-\infty}^{\infty} (2k+3)\,\boldsymbol{h}^{(k)}. \tag{d}$$

Next, by hypothesis and by *(42.2)*,

$$\operatorname{div} \Delta \boldsymbol{u} = 0, \qquad \operatorname{curl} \Delta \boldsymbol{u} = \boldsymbol{0};$$

thus (d) implies

$$\sum_{k=-\infty}^{\infty} (2k+3) \operatorname{div} \boldsymbol{h}^{(k)} = 0, \qquad \sum_{k=-\infty}^{\infty} (2k+3) \operatorname{curl} \boldsymbol{h}^{(k)} = \boldsymbol{0}. \tag{e}$$

But $\operatorname{div} \boldsymbol{h}^{(k)}$ is a solid harmonic of degree $k-1$. Hence the first of (e), by virtue of the uniqueness of the expansion in *(8.5)*, implies

$$\operatorname{div} \boldsymbol{h}^{(k)} = 0. \tag{f}$$

Similarly,

$$\operatorname{curl} \boldsymbol{h}^{(k)} = \boldsymbol{0},$$

or equivalently,

$$\nabla \boldsymbol{h}^{(k)} = [\nabla \boldsymbol{h}^{(k)}]^T. \tag{g}$$

Consequently, (a), (c), (f), and (g) imply

$$\nabla \operatorname{div} \boldsymbol{u} = \sum_{k=-\infty}^{\infty} [\nabla \operatorname{div} \boldsymbol{g}^{(k)} + 2(k+1)\,\boldsymbol{h}^{(k)}]. \tag{h}$$

By hypothesis,

$$\mu \Delta \boldsymbol{u} + (\lambda + \mu) \nabla \operatorname{div} \boldsymbol{u} = \boldsymbol{0}, \tag{i}$$

and substitution of (d) and (h) in (i) yields

$$\sum_{k=-\infty}^{\infty} \{2[(k+1)(\lambda+3\mu) + \mu]\,\boldsymbol{h}^{(k)} + (\lambda+\mu) \nabla \operatorname{div} \boldsymbol{g}^{(k)}\} = 0,$$

which implies the desired result (b). □

The next theorem gives a particular solution of the equations of elastostatics in a neighborhood of infinity.

(2)[1] *Let* $\boldsymbol{b}$ *be a smooth vector field on* $\Sigma(\infty)$, and assume that

$$\operatorname{div} \boldsymbol{b} = 0, \qquad \operatorname{curl} \boldsymbol{b} = \boldsymbol{0}.$$

---

[1] GURTIN and STERNBERG [1961, *11*], Theorem 4.3.

Then:

(i) $\boldsymbol{b}$ *admits the representation*

$$\boldsymbol{b} = \sum_{k=-\infty}^{\infty} \boldsymbol{b}^{(k)},$$

where the $\boldsymbol{b}^{(k)}$ *are divergence-free and curl-free vector solid spherical harmonics of degree* k, *and this infinite series, as well as the series resulting from any finite number of termwise differentiations, is uniformly convergent in every closed subregion of* $\Sigma(\infty)$;

(ii) *the same convergence properties apply to the series*

$$\boldsymbol{u} = -\frac{r^2}{2} \sum_{k=-\infty}^{\infty} \frac{1}{(k+1)(\lambda+3\mu)+\mu} \boldsymbol{b}^{(k)},$$

and the field $\boldsymbol{u}$ *so defined is an elastic displacement field on* $\Sigma(\infty)$ *corresponding to the body force field* $\boldsymbol{b}$;

(iii) *for any integer* n,

$$\boldsymbol{b}(\boldsymbol{x}) = o(r^n) \qquad \text{as } r \to \infty,$$

implies

$$\boldsymbol{u}(\boldsymbol{x}) = O(r^{n+1}) \qquad \text{as } r \to \infty.$$

Proof. By *(8.6)* $\boldsymbol{b}$ is harmonic. Thus *(8.4)* implies part (i) of this theorem. The uniform convergence of the series of (ii), and of the corresponding series resulting from successive differentiations, is apparent from the convergence properties of the series in (i). Thus $\boldsymbol{u}$ is of class C^∞ on $\Sigma(\infty)$. That $\boldsymbol{u}$ satisfies the displacement equation of equilibrium may be verified by direct substitution. This establishes (ii).

To justify (iii) we note from (i) and *(8.4)* that $\boldsymbol{b}(\boldsymbol{x}) = o(r^n)$ implies $\boldsymbol{b}^{(k)}(\boldsymbol{x}) = \boldsymbol{0}$ for $k \geqq n$. Bearing in mind (ii) and appealing once more to *(8.4)*, we see that (iii) holds true. □

49. Behavior of elastic states at infinity. We now use the representation theorems developed in the preceding section to determine the behavior of elastic fields at infinity.

We continue to assume that $B = \Sigma(\infty)$ is *homogeneous* and *isotropic* and that the Lamé moduli obey the inequalities

$$\mu > 0, \qquad 2\mu + 3\lambda > 0.$$

In addition, we assume that the body force field $\boldsymbol{b}$ is *smooth* on $\Sigma(\infty)$, and that

$$\operatorname{div} \boldsymbol{b} = 0, \qquad \operatorname{curl} \boldsymbol{b} = \boldsymbol{0}.$$

(1) Behavior of elastic states with vanishing displacements at infinity.[1] *Let* $[\boldsymbol{u}, \boldsymbol{E}, \boldsymbol{S}]$ *be an elastic state on* $\Sigma(\infty)$ *corresponding to* $\boldsymbol{b}$, *and assume that*

$$\boldsymbol{u}(\boldsymbol{x}) = o(1)$$

as $r \to \infty$. *Then*

$$\boldsymbol{u}(\boldsymbol{x}) = O(r^{-1}), \qquad \boldsymbol{S}(\boldsymbol{x}) = O(r^{-2}),$$
$$\boldsymbol{E}(\boldsymbol{x}) = O(r^{-2}), \qquad \boldsymbol{b}(\boldsymbol{x}) = O(r^{-3})$$

as $r \to \infty$.

[1] Fichera [1950, *4*]. See also Gurtin and Sternberg [1961, *11*], Theorem 5.1. Kupradze [1963, *17*, *18*], § III.3 established the estimate $\boldsymbol{u} = O(r^{-1})$, but only with the additional assumption that $\partial \boldsymbol{u}/\partial r = o(r^{-1})$. Note that if $\boldsymbol{u} = o(1)$, then div $\boldsymbol{u}$ and curl $\boldsymbol{u}$ are both $O(r^{-2})$, a conclusion which Duffin and Noll [1958, *8*] reached by entirely different means.

Proof. By ***(42.2)*** $\boldsymbol{u}$ is biharmonic on $\Sigma(\infty)$. If we apply ***(8.5)*** to $\boldsymbol{u}$, we conclude that $\boldsymbol{u}=O(r^{-1})$. The remaining assertions then follow immediately from (iii) of ***(8.5)*** in conjunction with the strain-displacement relation, the stress-strain relation, and the equation of equilibrium. □

(2) Behavior of elastic states with vanishing stresses at infinity.[1] *Let* $[\boldsymbol{u}, \boldsymbol{E}, \boldsymbol{S}]$ *be an elastic state on* $\Sigma(\infty)$ *corresponding to* $\boldsymbol{b}$, *and assume that*

$$\boldsymbol{S}(\boldsymbol{x}) = o(1), \qquad \boldsymbol{b}(\boldsymbol{x}) = o(r^{-2})$$

as $r\to\infty$. *Then*

$$\boldsymbol{u}(\boldsymbol{x}) = \boldsymbol{w}(\boldsymbol{x}) + O(r^{-1}), \qquad \boldsymbol{S}(\boldsymbol{x}) = O(r^{-2}),$$
$$\boldsymbol{E}(\boldsymbol{x}) = O(r^{-2}), \qquad \boldsymbol{b}(\boldsymbol{x}) = O(r^{-3})$$

as $r\to\infty$, *where* $\boldsymbol{w}$ *is a rigid displacement field.*

Proof. In view of the assumed properties of $\boldsymbol{b}$, we infer from ***(48.2)*** the existence of a particular solution $\boldsymbol{u}$ of the displacement equation of equilibrium that is class C^2 on $\Sigma(\infty)$ and satisfies $\boldsymbol{u}=O(r^{-1})$. Accordingly, by ***(1)***, the elastic state associated with $\boldsymbol{u}$ (i.e., generated by the strain-displacement and stress-strain relations) satisfies $\boldsymbol{E}=O(r^{-2})$ and $\boldsymbol{S}=O(r^{-2})$, while $\boldsymbol{b}$ must conform to $\boldsymbol{b}=O(r^{-3})$. We are thus able to exhibit a *particular* elastic state corresponding to $\boldsymbol{b}$ which possesses the requisite orders of magnitude at infinity. Therefore, since the fundamental system of field equations satisfied by elastic states is linear, it remains to be demonstrated that ***(2)*** is true when $\boldsymbol{b}=\boldsymbol{0}$.

Suppose now that $\boldsymbol{b}=\boldsymbol{0}$. Then, by hypothesis and by ***(48.1)***, $\boldsymbol{u}$ admits the representation

$$\boldsymbol{u} = \mathring{\boldsymbol{u}} + \tilde{\boldsymbol{u}},$$

where

$$\begin{aligned} \mathring{\boldsymbol{u}} &= \sum_{k=0}^{\infty} \boldsymbol{g}^{(k)} + r^2 \sum_{k=-2}^{\infty} \boldsymbol{h}^{(k)}, \\ \tilde{\boldsymbol{u}} &= \sum_{k=-\infty}^{-1} \boldsymbol{g}^{(k)} + r^2 \sum_{k=-\infty}^{-3} \boldsymbol{h}^{(k)}, \end{aligned} \tag{a}$$

and where the solid harmonics $\boldsymbol{g}^{(k)}$, $\boldsymbol{h}^{(k)}$ obey (b) of ***(48.1)***.

Next, define fields $\mathring{\boldsymbol{E}}$ and $\tilde{\boldsymbol{E}}$ through

$$\begin{aligned} \mathring{\boldsymbol{E}} &= \tfrac{1}{2}(\nabla\mathring{\boldsymbol{u}} + \nabla\mathring{\boldsymbol{u}}^T), \\ \tilde{\boldsymbol{E}} &= \tfrac{1}{2}(\nabla\tilde{\boldsymbol{u}} + \nabla\tilde{\boldsymbol{u}}^T), \end{aligned} \tag{b}$$

so that

$$\boldsymbol{E} = \mathring{\boldsymbol{E}} + \tilde{\boldsymbol{E}}. \tag{c}$$

The two infinite series entering the second of (a) each represent a harmonic field on $\Sigma(\infty)$. The second of (b) and part (iii) of ***(8.5)*** therefore imply

$$\tilde{\boldsymbol{u}} = O(r^{-1}), \qquad \tilde{\boldsymbol{E}} = O(r^{-2}), \tag{d}$$

while we conclude from the stress-strain relation that

$$\tilde{\boldsymbol{S}} = O(r^{-2}).$$

Consequently, all we have left to show is that $\mathring{\boldsymbol{u}}$ is rigid; or equivalently that

$$\mathring{\boldsymbol{E}} = \boldsymbol{0}. \tag{e}$$

---

[1] GURTIN and STERNBERG [1961, *11*], Theorem 5.2.

To this end we conclude from (b) in *(48.1)* that

$$\operatorname{div} \boldsymbol{h}^{(k)} = \mathbf{0},$$

and hence reach, with the aid of (a) and (b) above,

$$\operatorname{tr} \mathring{\boldsymbol{E}} = \sum_{k=0}^{\infty} \operatorname{div} \boldsymbol{g}^{(k)} + 2 \sum_{k=-2}^{\infty} \boldsymbol{p} \cdot \boldsymbol{h}^{(k)}, \tag{f}$$

where $\boldsymbol{p}(\boldsymbol{x}) = \boldsymbol{x} - \mathbf{0}$. Next, since $\operatorname{div} \boldsymbol{g}^{(k+2)}$ is a solid harmonic of degree $k+1$,

$$\boldsymbol{p} \cdot \nabla \operatorname{div} \boldsymbol{g}^{(k+2)} = (k+1) \operatorname{div} \boldsymbol{g}^{(k+2)}, \tag{g}$$

and since $\boldsymbol{g}^{(0)}$ is a constant,

$$\nabla \boldsymbol{g}^{(0)} = \mathbf{0}. \tag{h}$$

Thus if we eliminate $\boldsymbol{h}^{(k)}$ from (f), by recourse to (b) of *(48.1)*, and use (g) and (h), we arrive at

$$\operatorname{tr} \mathring{\boldsymbol{E}} = \sum_{k=1}^{\infty} \frac{\mu(2k-1)}{(k-1)(\lambda+3\mu)+\mu} \operatorname{div} \boldsymbol{g}^{(k)}. \tag{i}$$

By hypothesis, $\boldsymbol{S} = o(1)$; thus the stress-strain relation implies

$$\boldsymbol{E} = o(1),$$

and we conclude from (c) and the second of (d) that

$$\mathring{\boldsymbol{E}} = o(1). \tag{j}$$

Thus if we apply *(8.4)* to the harmonic function $\operatorname{tr} \mathring{\boldsymbol{E}}$ given by (i), bearing in mind that $\operatorname{div} \boldsymbol{g}^{(k)}$ is a solid harmonic of degree $k-1$, we find that

$$\operatorname{div} \boldsymbol{g}^{(k)} = 0 \quad \text{for } k \geqq 1,$$

which, together with (b) of *(48.1)* and (h), yields

$$\boldsymbol{h}^{(k)} = \mathbf{0} \quad \text{for } k \geqq -2. \tag{k}$$

Inserting (k) in the first of (a), and using (h) and the first of (b), we obtain

$$\mathring{\boldsymbol{E}} = \sum_{k=1}^{\infty} \widehat{\nabla} \boldsymbol{g}^{(k)}. \tag{l}$$

Since $\nabla \boldsymbol{g}^{(k)}$ is a solid harmonic of degree $k-1$, we conclude from part (iii) of *(8.4)* that (l) is consistent with (j) only if (e) holds. □

50. Extension of the basic theorems in elastostatics to exterior domains. We now use the results established in the last section to extend the basic theorems of elastostatics to exterior domains. Thus for this section we drop the requirement that B be bounded and assume instead that B is an *exterior domain.* We shall continue to assume that B is *homogeneous* and *isotropic* with Lamé moduli μ and λ.

Let $\mathfrak{s} = [\boldsymbol{u}, \boldsymbol{E}, \boldsymbol{S}]$ be an elastic state on B corresponding to the body force field $\boldsymbol{b}$, and suppose that the limits

$$\boldsymbol{f} = \lim_{\varrho \to \infty} \int_{\partial \Sigma_\varrho} \mathbf{s}\, da,$$

$$\boldsymbol{m} = \lim_{\varrho \to \infty} \int_{\partial \Sigma_\varrho} \boldsymbol{p} \times \mathbf{s}\, da$$

exist, where $\mathbf{s}=\boldsymbol{S}\boldsymbol{n}$, $\boldsymbol{n}$ is the outward unit normal to $\partial\Sigma_\varrho$, and Σ_ϱ is the ball of radius ϱ and center at $\mathbf{0}$. Then we call $\boldsymbol{f}$ and $\boldsymbol{m}$ the associated ***force*** and ***moment at infinity***. Since $\boldsymbol{S}$ and $\boldsymbol{b}$ satisfy balance of forces and moments in every part of B,

$$\int_{\partial B}\mathbf{s}\,da+\int_{B_\rho}\boldsymbol{b}\,dv=-\int_{\partial\Sigma_\rho}\mathbf{s}\,da,$$

$$\int_{\partial B}\boldsymbol{p}\times\mathbf{s}\,da+\int_{B_\rho}\boldsymbol{p}\times\boldsymbol{b}\,dv=-\int_{\partial\Sigma_\rho}\boldsymbol{p}\times\mathbf{s}\,da,$$

for sufficiently large ϱ, where $B_\varrho=B\cap\Sigma_\varrho$.[1] Thus, letting $\varrho\to\infty$, we see that $\boldsymbol{f}$ and $\boldsymbol{m}$ exist if and only if the integrals

$$\int_B\boldsymbol{b}\,dv,\quad\int_B\boldsymbol{p}\times\boldsymbol{b}\,dv$$

exist, in which case

$$\boldsymbol{f}=-\int_{\partial B}\mathbf{s}\,da-\int_B\boldsymbol{b}\,dv,$$

$$\boldsymbol{m}=-\int_{\partial B}\boldsymbol{p}\times\mathbf{s}\,da-\int_B\boldsymbol{p}\times\boldsymbol{b}\,dv.$$

(1) Extension of Betti's reciprocal theorem.[2] *Assume that the Lamé moduli obey $\mu>0$, $2\mu+3\lambda>0$. Let $[\boldsymbol{u},\boldsymbol{E},\boldsymbol{S}]$ and $[\tilde{\boldsymbol{u}},\tilde{\boldsymbol{E}},\tilde{\boldsymbol{S}}]$ be elastic states corresponding to the external force systems $[\boldsymbol{b},\mathbf{s}]$ and $[\tilde{\boldsymbol{b}},\tilde{\mathbf{s}}]$, respectively, with $\boldsymbol{b}$ and $\tilde{\boldsymbol{b}}$ divergence-free and curl-free and of class C^2 on B. Further, assume that the associated forces $\boldsymbol{f},\tilde{\boldsymbol{f}}$ and the associated moments $\boldsymbol{m},\tilde{\boldsymbol{m}}$ at infinity exist. Finally, assume that*

$$\boldsymbol{u}(\boldsymbol{x})=\boldsymbol{w}(\boldsymbol{x})+o(1),$$

$$\tilde{\boldsymbol{u}}(\boldsymbol{x})=\tilde{\boldsymbol{w}}(\boldsymbol{x})+o(1),$$

as $r\to\infty$, where $\boldsymbol{w}$ and $\tilde{\boldsymbol{w}}$ are the rigid displacements:

$$\boldsymbol{w}=\boldsymbol{u}_\infty+\boldsymbol{\omega}_\infty\times\boldsymbol{p},$$

$$\tilde{\boldsymbol{w}}=\tilde{\boldsymbol{u}}_\infty+\tilde{\boldsymbol{\omega}}_\infty\times\boldsymbol{p}.$$

Then

$$\int_{\partial B}\mathbf{s}\cdot\tilde{\boldsymbol{u}}\,da+\int_B\boldsymbol{b}\cdot\tilde{\boldsymbol{u}}\,dv+\boldsymbol{f}\cdot\tilde{\boldsymbol{u}}_\infty+\boldsymbol{m}\cdot\tilde{\boldsymbol{\omega}}_\infty$$
$$=\int_{\partial B}\tilde{\mathbf{s}}\cdot\boldsymbol{u}\,da+\int_B\tilde{\boldsymbol{b}}\cdot\boldsymbol{u}\,dv+\tilde{\boldsymbol{f}}\cdot\boldsymbol{u}_\infty+\tilde{\boldsymbol{m}}\cdot\boldsymbol{\omega}_\infty=\int_B\tilde{\boldsymbol{S}}\cdot\boldsymbol{E}\,dv=\int_B\boldsymbol{S}\cdot\tilde{\boldsymbol{E}}\,dv.$$

Proof. Applying Betti's theorem ***(30.1)*** to the region $B_\varrho=B\cap\Sigma_\varrho$, with ϱ large enough so that $\partial B\subset\Sigma_\varrho$, we find that

$$\begin{aligned}&\int_{\partial B}\mathbf{s}\cdot\tilde{\boldsymbol{u}}\,da+\int_{B_\rho}\boldsymbol{b}\cdot\tilde{\boldsymbol{u}}\,dv+\int_{\partial\Sigma_\rho}\mathbf{s}\cdot\tilde{\boldsymbol{u}}\,da\\&\quad=\int_{\partial B}\tilde{\mathbf{s}}\cdot\boldsymbol{u}\,da+\int_{B_\rho}\tilde{\boldsymbol{b}}\cdot\boldsymbol{u}\,dv+\int_{\partial\Sigma_\rho}\tilde{\mathbf{s}}\cdot\boldsymbol{u}\,da=\int_{B_\rho}\tilde{\boldsymbol{S}}\cdot\boldsymbol{E}\,dv=\int_{B_\rho}\boldsymbol{S}\cdot\tilde{\boldsymbol{E}}\,dv.\end{aligned}\tag{a}$$

Clearly,

$$\int_{\partial\Sigma_\rho}\mathbf{s}\cdot\tilde{\boldsymbol{u}}\,da=\int_{\partial\Sigma_\rho}\mathbf{s}\cdot(\tilde{\boldsymbol{u}}-\tilde{\boldsymbol{w}})\,da+\int_{\partial\Sigma_\rho}\mathbf{s}\cdot\tilde{\boldsymbol{w}}\,da\tag{b}$$

---

[1] Note that $\partial B_\varrho=\partial B\cup\partial\Sigma_\varrho$ for ϱ sufficiently large, since B is an exterior domain.

[2] This result for the case in which $\boldsymbol{w}=\tilde{\boldsymbol{w}}=\mathbf{0}$ is due to GURTIN and STERNBERG [1961, *11*], Theorem 6.1.

and

$$\int_{\partial\Sigma_\varrho} \mathbf{s}\cdot\tilde{\boldsymbol{w}}\,da = \int_{\partial\Sigma_\varrho} \mathbf{s}\cdot(\tilde{\boldsymbol{u}}_\infty + \tilde{\boldsymbol{\omega}}_\infty \times \boldsymbol{p})\,da = \boldsymbol{u}_\infty \cdot \int_{\partial\Sigma_\varrho} \mathbf{s}\,da + \boldsymbol{\omega}_\infty \cdot \int_{\partial\Sigma_\varrho} \boldsymbol{p}\times\mathbf{s}\,da. \quad \text{(c)}$$

By hypothesis and by ***(49.1)***,

$$\begin{aligned} &\boldsymbol{u} = \boldsymbol{w} + O(r^{-1}), \quad &&\tilde{\boldsymbol{u}} = \tilde{\boldsymbol{w}} + O(r^{-1}), \\ &\boldsymbol{E} = O(r^{-2}), \quad &&\boldsymbol{S} = O(r^{-2}), \quad &&\boldsymbol{b} = O(r^{-3}), \\ &\tilde{\boldsymbol{E}} = O(r^{-2}), \quad &&\tilde{\boldsymbol{S}} = O(r^{-2}), \quad &&\tilde{\boldsymbol{b}} = O(r^{-3}). \end{aligned} \quad \text{(d)}$$

Thus (b) and (c) imply

$$\int_{\Sigma_\varrho} \mathbf{s}\cdot\tilde{\boldsymbol{u}}\,da \to \boldsymbol{f}\cdot\tilde{\boldsymbol{u}}_\infty + \boldsymbol{m}\cdot\tilde{\boldsymbol{\omega}}_\infty \quad \text{(e)}$$

as $\varrho\to\infty$. Similarly,

$$\int_{\Sigma_\varrho} \tilde{\mathbf{s}}\cdot\boldsymbol{u}\,da \to \tilde{\boldsymbol{f}}\cdot\boldsymbol{u}_\infty + \tilde{\boldsymbol{m}}\cdot\boldsymbol{\omega}_\infty \quad \text{(f)}$$

as $\varrho\to\infty$. The estimates (d) also imply that the volume integrals over B_ϱ in (a) are convergent as $\varrho\to\infty$ and tend to the corresponding integrals over B. Thus if we let $\varrho\to\infty$ in (a) and use (e) and (f), we arrive at the desired conclusion. □

As a direct corollary of this result we have the

(2) Theorem of work and energy for exterior domains.[1] *Let* $[\boldsymbol{u}, \boldsymbol{E}, \boldsymbol{S}]$, $[\boldsymbol{b}, \mathbf{s}]$, $\boldsymbol{f}$, $\boldsymbol{m}$, *as well as* μ *and* λ *obey the same hypotheses as in* ***(1)***. *Then*

$$\int_{\partial B} \mathbf{s}\cdot\boldsymbol{u}\,da + \int_B \boldsymbol{b}\cdot\boldsymbol{u}\,dv + \boldsymbol{f}\cdot\boldsymbol{u}_\infty + \boldsymbol{m}\cdot\boldsymbol{\omega}_\infty = 2U_{\mathsf{C}}\{\boldsymbol{E}\}.$$

This relation differs from the corresponding result for finite regions by the addition of the term $\boldsymbol{f}\cdot\boldsymbol{u}_\infty + \boldsymbol{m}\cdot\boldsymbol{\omega}_\infty$, which represents the work done by the force and moment at infinity.

(3) Uniqueness theorem for exterior domains.[2] *Assume that*

$$\mu > 0, \quad 2\mu + 3\lambda > 0.$$

Let $\mathfrak{s} = [\boldsymbol{u}, \boldsymbol{E}, \boldsymbol{S}]$ *and* $\tilde{\mathfrak{s}} = [\tilde{\boldsymbol{u}}, \tilde{\boldsymbol{E}}, \tilde{\boldsymbol{S}}]$ *be elastic states corresponding to the same body force field and the same boundary data in the sense that*

$$\boldsymbol{u} = \tilde{\boldsymbol{u}} \quad on\ \mathscr{S}_1, \qquad \mathbf{s} = \tilde{\mathbf{s}} \quad on\ \mathscr{S}_2. \quad \text{(a)}$$

Further, assume that either

$$\boldsymbol{u}(\boldsymbol{x}) = \tilde{\boldsymbol{u}}(\boldsymbol{x}) + o(1) \quad as \quad r\to\infty, \quad \text{(b)}$$

or both

$$\boldsymbol{S}(\boldsymbol{x}) = \tilde{\boldsymbol{S}}(\boldsymbol{x}) + o(1) \quad as \quad r\to\infty \quad \text{(c)}$$

and the corresponding traction fields are statically equivalent on ∂B, i.e.

$$\int_{\partial B} \mathbf{s}\,da = \int_{\partial B} \tilde{\mathbf{s}}\,da, \quad \int_{\partial B} \boldsymbol{p}\times\mathbf{s}\,da = \int_{\partial B} \boldsymbol{p}\times\tilde{\mathbf{s}}\,da. \quad \text{(d)}^3$$

Then $\mathfrak{s}$ *and* $\tilde{\mathfrak{s}}$ *are equal modulo a rigid displacement.*

[1] Gurtin and Sternberg [1961, *11*], Theorem 6.2, for the case in which $\boldsymbol{w} = \boldsymbol{0}$.

[2] Fichera [1950, *4*], Theorems II, V, and VI for case (b); Gurtin and Sternberg [1961, *11*] for case (c). See also Duffin and Noll [1958, *8*] and Bézier [1967, *3*]. Analogous uniqueness theorems for two-dimensional elastostatics are given by Tiffen [1952, *6*] and Muskhelishvili [1954, *16*], § 40.

[3] Note that this condition is satisfied automatically when B is finite.

Proof. Let $\mathfrak{s}' = \mathfrak{s} - \tilde{\mathfrak{s}}$. It suffices to prove that

$$\boldsymbol{E}' = \boldsymbol{0}. \tag{e}$$

Clearly, $\mathfrak{s}'$ is an elastic state corresponding to *zero body forces* and *vanishing boundary data*. Assume (b) holds. Then

$$\boldsymbol{u}'(\boldsymbol{x}) = o(1) \quad \text{as} \quad r \to \infty,$$

and *(2)* (with $\boldsymbol{u}'_\infty = \boldsymbol{\omega}'_\infty = \boldsymbol{0}$) yields

$$U_{\mathsf{C}}\{\boldsymbol{E}'\} = 0, \tag{f}$$

which implies (e). Next assume (c) and (d) hold. Then

$$\boldsymbol{S}'(\boldsymbol{x}) = o(1) \quad \text{as} \quad r \to \infty,$$

and we conclude from *(49.2)* that

$$\boldsymbol{u}'(\boldsymbol{x}) = \boldsymbol{w}(\boldsymbol{x}) + O(r^{-1}) \quad \text{as} \quad r \to \infty$$

with $\boldsymbol{w}$ rigid. Further, by (d),

$$\boldsymbol{f}' = -\int_{\partial B} \boldsymbol{s}'\, da = \boldsymbol{0}, \qquad \boldsymbol{m}' = -\int_{\partial B} \boldsymbol{p} \times \boldsymbol{s}'\, da = \boldsymbol{0},$$

and *(2)* again implies (f) and hence (e). □

It should be noted that *(3)* is valid only for unbounded regions with finite boundaries. The uniqueness question associated with problems for general domains whose boundaries extend to infinity is yet to be disposed of satisfactorily.[1]

We emphasize that theorem *(3)*, in contrast to *(1)* and *(2)*, involves no explicit restrictions concerning the body force field. If this field is integrable over B, then (d) can be replaced by the requirement that the forces and moments at infinity coincide:

$$\boldsymbol{f} = \tilde{\boldsymbol{f}}, \qquad \boldsymbol{m} = \tilde{\boldsymbol{m}}.$$

For the surface force problem, (d) *may be omitted because it is implied by* (a). On the other hand, if $\mathscr{S}_1$ is not empty, (d) is an independent hypothesis. This hypothesis, which requires the prescription of the resultant force and moment on ∂B (or equivalently, the force and moment at infinity), appears at first sight artificial, since the surface forces over at least a portion of the boundary are not known beforehand. One is therefore led to ask whether (d) is necessary for the truth of *(3)* when the regularity conditions are taken in the form (c). That this is indeed the case is clear from the next theorem.

(4) *Theorem (3) is false if either of the two hypotheses in* (d) *is omitted.*

Proof. For the purpose at hand, let

$$B = \{\boldsymbol{x} : |\boldsymbol{x} - 0| > 1\},$$

and assume that $\partial B = \mathscr{S}_1$ ($\mathscr{S}_2 = \emptyset$) in *(3)*. It clearly suffices to exhibit two elastic states, both corresponding to zero body forces, both having strains not identically zero, both satisfying

$$\boldsymbol{u} = \boldsymbol{0} \quad \text{on } \partial B, \tag{a}$$

$$\boldsymbol{S} = o(1) \quad \text{as} \quad r \to \infty, \tag{b}$$

[1] For the special case of the displacement and surface force problems appropriate to the half-space, this question was settled by TURTELTAUB and STERNBERG [1967, *18*]. See also KNOPS [1965, *13*].

with one state obeying

$$\int_{\partial B} \mathbf{s}\, da = \mathbf{0}, \tag{c}$$

the other

$$\int_{\partial B} \mathbf{p} \times \mathbf{s}\, da = \mathbf{0}. \tag{d}$$

To this end, let $\boldsymbol{\lambda} \neq \mathbf{0}$ and $\boldsymbol{\omega} \neq \mathbf{0}$ be given vectors and consider the following two displacement fields:[1]

$$\mathbf{u}(\mathbf{x}) = \frac{\boldsymbol{\lambda}}{r} + \frac{r^2 - 1}{2(5-6\nu)} \nabla \operatorname{div}\left(\frac{\boldsymbol{\lambda}}{r}\right) - \boldsymbol{\lambda}, \tag{e}$$

$$\mathbf{u}(\mathbf{x}) = \frac{\boldsymbol{\omega} \times \mathbf{p}}{r^3} + \frac{r^2 - 1}{4(4-5\nu)} \nabla \operatorname{div}\left(\frac{\boldsymbol{\omega} \times \mathbf{p}}{r^3}\right) - \boldsymbol{\omega} \times \mathbf{p}. \tag{f}$$

Both of these displacement fields generate elastic states that correspond to zero body forces and obey (a), (b). Moreover, for (e):

$$\mathbf{f} = \int_{\partial B} \mathbf{s}\, da = 24\pi\mu \frac{1-\nu}{5-6\nu} \boldsymbol{\lambda} \neq \mathbf{0}, \qquad \mathbf{m} = \int_{\partial B} \mathbf{p} \times \mathbf{s}\, da = \mathbf{0};$$

while for (f):

$$\mathbf{f} = \mathbf{0}, \qquad \mathbf{m} = 8\pi\mu\boldsymbol{\omega}. \quad \square$$

The leading two terms in (e) represent the displacement induced in the medium by a rigid translation $\boldsymbol{\lambda}$ applied to the spherical boundary ∂B, the body being constrained against displacements at infinity; the last term in (e) corresponds to a rigid translation $-\boldsymbol{\lambda}$ of the entire medium. On the other hand, the leading two terms in (f) represent the displacement generated by a rotation $\boldsymbol{\omega}$ of ∂B, while the last term corresponds to a rotation of the entire medium.

IX. Basic singular solutions. Concentrated loads. Green's functions.

51. Basic singular solutions. In this section we study the basic singular solutions of elastostatics, assuming throughout that the body is *homogeneous* and *isotropic*, and that $\mu \neq 0, \nu \neq \frac{1}{2}, 1$. We begin with the Kelvin problem,[2] which is concerned with a concentrated load applied at a point of a body occupying the entire space $\mathscr{E}$. With a view toward giving a solution of this problem, we introduce the following notation.

Throughout this section $\mathbf{y}$ is a fixed point of $\mathscr{E}$,

$$\mathbf{r} = \mathbf{x} - \mathbf{y}$$

is the position vector from $\mathbf{y}$, and

$$r = |\mathbf{r}|.$$

Further, we write Σ_η for the open ball $\Sigma_\eta(\mathbf{y})$ with radius η centered at $\mathbf{y}$.

[1] GURTIN and STERNBERG [1961, *11*] attribute these examples to R. T. SHIELD.

[2] The solution to this problem was first given by KELVIN [1848, *3*] (cf. SOMIGLIANA [1885, *4*], BOUSSINESQ [1885, *1*]). It was derived by KELVIN and TAIT [1883, *1*], §§ 730, 731 through a limit process that was made explicit by STERNBERG and EUBANKS [1955, *13*] (cf. LOVE [1927, *3*], § 130; MINDLIN [1936, *1*], [1953, *15*]). The limit formulation presented here follows closely that adopted by STERNBERG and AL-KHOZAI [1964, *20*] and TURTELTAUB and STERNBERG [1968, *14*]. The solution for an anisotropic but homogeneous body was given by FREDHOLM [1900, *2*]. A precise statement of FREDHOLM's result is contained in the work of SÁENZ [1953, *17*]. See also ZEILON [1911, *1*], CARRIER [1944, *1*], LIFSHIC and ROZENCVEIG [1947, *4*], ELLIOT [1948, *2*], SYNGE [1957, *15*], KRÖNER [1953, *14*], BASHELEISHVILI [1957, *1*], MURTAZAEV [1962, *12*], BÉZIER [1967, *3*], BROSS [1968, *1*], CĄKALA, DOMAŃSKI, and MILICER-GRUŻEWSKA [1968, *2*].

By a ***sequence of body force fields tending to a concentrated load $\boldsymbol{l}$ at $\boldsymbol{y}$*** we mean a sequence $\{\boldsymbol{b}_m\}$ of class C^2 vector fields on $\mathscr{E}$ with the following properties:

(i) $\boldsymbol{b}_m = \boldsymbol{0}$ on $\mathscr{E} - \Sigma_{1/m}$;

(ii) $\int_{\mathscr{E}} \boldsymbol{b}_m \, dv \to \boldsymbol{l}$ as $m \to \infty$;

(iii) the sequence $\left\{\int_{\mathscr{E}} |\boldsymbol{b}_m| \, dv\right\}$ is bounded.

(1) Limit definition of the solution to Kelvin's problem.[1] *Let $\{\boldsymbol{b}_m\}$ be a sequence of body force fields tending to a concentrated load $\boldsymbol{l}$ at $\boldsymbol{y}$. Then:*

(i) *For each m there exists a unique elastic state s_m on $\mathscr{E}$ that corresponds to the body force $\boldsymbol{b}_m$ and has uniformly vanishing displacements at infinity.*

(ii) *The sequence $\{s_m\}$ converges (uniformly on closed subsets of $\mathscr{E} - \{\boldsymbol{y}\}$) to an elastic state $s_{\boldsymbol{y}}[\boldsymbol{l}]$ on $\mathscr{E} - \{\boldsymbol{y}\}$.*

(iii) *The limit state $s_{\boldsymbol{y}}[\boldsymbol{l}]$ is independent of the sequence $\{\boldsymbol{b}_m\}$ tending to $\boldsymbol{l}$ and is generated by the displacement field $\boldsymbol{u}_{\boldsymbol{y}}[\boldsymbol{l}]$ with values*

$$\boldsymbol{u}_{\boldsymbol{y}}[\boldsymbol{l}](\boldsymbol{x}) = \frac{1}{c r}\left[\frac{\boldsymbol{r} \otimes \boldsymbol{r}}{r^2} + (3 - 4\nu)\,\mathbf{1}\right]\boldsymbol{l},$$

$$c = 16\pi\mu(1-\nu).$$

We call $s_{\boldsymbol{y}}[\boldsymbol{l}]$ the ***Kelvin state*** *corresponding to a concentrated load $\boldsymbol{l}$ at $\boldsymbol{y}$.*

Proof. Let $\boldsymbol{u}_m$ be defined by

$$\boldsymbol{u}_m = \boldsymbol{\psi}_m - \frac{1}{4(1-\nu)} \nabla(\varphi_m + \boldsymbol{r} \cdot \boldsymbol{\psi}_m),$$

where

$$\boldsymbol{\psi}_m(\boldsymbol{x}) = \alpha \int_{\Sigma_{1/m}} \frac{\boldsymbol{b}_m(\xi)}{|\boldsymbol{x} - \xi|}\, dv_\xi, \qquad \varphi_m(\boldsymbol{x}) = -\alpha \int_{\Sigma_{1/m}} \frac{\boldsymbol{r}(\xi) \cdot \boldsymbol{b}_m(\xi)}{|\boldsymbol{x} - \xi|}\, dv_\xi, \qquad \alpha = \frac{1}{4\pi\mu}. \tag{a}$$

Then, by ***(6.5)*** and ***(44.1)***, $\boldsymbol{u}_m$ is an equilibrium displacement field corresponding to $\boldsymbol{b}_m$. Moreover, it is easy to verify that

$$\boldsymbol{u}_m(\boldsymbol{x}) = o(1) \quad \text{as} \quad |\boldsymbol{x} - \boldsymbol{0}| \to \infty,$$

and, by the uniqueness theorem ***(50.3)*** for exterior domains, there is no other equilibrium displacement field corresponding to $\boldsymbol{b}_m$ with this property. This establishes (i).

Next, a simple calculation shows that the functions defined on $\mathscr{E} - \{\boldsymbol{y}\}$ by

$$\varphi(\boldsymbol{x}) = 0, \qquad \boldsymbol{\psi}(\boldsymbol{x}) = \frac{\alpha \boldsymbol{l}}{|\boldsymbol{x} - \boldsymbol{y}|}, \tag{b}$$

generate, in the sense of the Boussinesq-Papkovitch-Neuber solution ***(44.1)***, the displacement field $\boldsymbol{u}_{\boldsymbol{y}}[\boldsymbol{l}]$ given in (iii). Thus to complete the proof we have only to show that given any closed region $\overline{R}$ in $\mathscr{E} - \{\boldsymbol{y}\}$, $s_m \to s_{\boldsymbol{y}}[\boldsymbol{l}]$ uniformly on $\overline{R}$; or equivalently, that $\varphi_m \to \varphi$ and $\boldsymbol{\psi}_m \to \boldsymbol{\psi}$ uniformly on $\overline{R}$, and that the first and second gradients of φ_m and $\boldsymbol{\psi}_m$ tend to the corresponding gradients of φ and $\boldsymbol{\psi}$ uniformly on $\overline{R}$. Since the argument in each instance is strictly analogous, we shall merely prove that $\boldsymbol{\psi}_m \to \boldsymbol{\psi}$.

---

[1] STERNBERG and EUBANKS [1955, *13*], Theorem 4.2. An analogous theorem for concentrated surface loads was given by TURTELTAUB and STERNBERG [1968, *14*].

Let $\bar{R}$ be a closed subset of $\mathscr{E}-\{\boldsymbol{y}\}$. Then there exists an $m_0>0$ such that

$$\bar{\Sigma}_{1/m_0} \cap \bar{R} = \emptyset;$$

hence

$$\frac{1}{|\boldsymbol{x}-\boldsymbol{y}|} \leqq m_0 \tag{c}$$

for $\boldsymbol{x}\in\bar{R}$ and

$$K = \sup\left\{\frac{1}{|\boldsymbol{x}-\boldsymbol{\xi}|} : \boldsymbol{x}\in\bar{R},\ \boldsymbol{\xi}\in\Sigma_{1/m_0}\right\}$$

is finite. Thus

$$\left|\frac{1}{|\boldsymbol{x}-\boldsymbol{y}|} - \frac{1}{|\boldsymbol{x}-\boldsymbol{\xi}|}\right| = \left|\frac{|\boldsymbol{x}-\boldsymbol{\xi}|-|\boldsymbol{x}-\boldsymbol{y}|}{|\boldsymbol{x}-\boldsymbol{y}|\,|\boldsymbol{x}-\boldsymbol{\xi}|}\right| \leqq \frac{K m_0}{m} \tag{d}$$

for $\boldsymbol{x}\in\bar{R}$, $\boldsymbol{\xi}\in\Sigma_{1/m}$, and $m>m_0$. Next, by (a) and (b),

$$|\boldsymbol{\psi}_m(\boldsymbol{x}) - \boldsymbol{\psi}(\boldsymbol{x})| = |\alpha|\left|\int\limits_{\Sigma_{1/m}} \boldsymbol{b}_m(\boldsymbol{\xi})\left[\frac{1}{|\boldsymbol{x}-\boldsymbol{\xi}|} - \frac{1}{|\boldsymbol{x}-\boldsymbol{y}|}\right] dv_{\boldsymbol{\xi}} + \frac{1}{|\boldsymbol{x}-\boldsymbol{y}|}\left[\int\limits_{\Sigma_{1/m}} \boldsymbol{b}_m\, dv - \boldsymbol{l}\right]\right|;$$

consequently, (c) and (d) imply

$$|\boldsymbol{\psi}_m(\boldsymbol{x}) - \boldsymbol{\psi}(\boldsymbol{x})| \leqq \frac{|\alpha|\, K m_0}{m}\int\limits_{\Sigma_{1/m}} |\boldsymbol{b}_m|\, dv + m_0|\alpha|\left|\int\limits_{\Sigma_{1/m}} \boldsymbol{b}_m\, dv - \boldsymbol{l}\right|$$

for $\boldsymbol{x}\in\bar{R}$ and $m>m_0$. This inequality, when combined with (ii) and (iii) in the definition of a sequence of body force fields tending to a concentrated load, implies $\boldsymbol{\psi}_m\to\boldsymbol{\psi}$ uniformly on $\bar{R}$. □

The need for condition (iii) in the definition of a sequence of body force fields tending to a concentrated load was established by STERNBERG and EUBANKS,[1] who showed by means of a counterexample that conclusions (ii) and (iii) in *(1)* become invalid if this hypothesis is omitted.

We will consistently write

$$\mathfrak{s}_{\boldsymbol{y}}[\boldsymbol{l}] = [\boldsymbol{u}_{\boldsymbol{y}}[\boldsymbol{l}], \boldsymbol{E}_{\boldsymbol{y}}[\boldsymbol{l}], \boldsymbol{S}_{\boldsymbol{y}}[\boldsymbol{l}]]$$

for the Kelvin state corresponding to a concentrated load $\boldsymbol{l}$ at $\boldsymbol{y}$. Clearly, the mapping $\boldsymbol{l}\mapsto\mathfrak{s}_{\boldsymbol{y}}[\boldsymbol{l}]$ is linear on $\mathscr{V}$.

In view of the stress-strain and strain-displacement relations, the stress field $\boldsymbol{S}_{\boldsymbol{y}}[\boldsymbol{l}]$ has the form

$$\boldsymbol{S}_{\boldsymbol{y}}[\boldsymbol{l}](\boldsymbol{x}) = -\frac{2\mu}{c r^3}\left\{\frac{3(\boldsymbol{r}\cdot\boldsymbol{l})\,\boldsymbol{r}\otimes\boldsymbol{r}}{r^2} + (1-2\nu)\,[\boldsymbol{r}\otimes\boldsymbol{l} + \boldsymbol{l}\otimes\boldsymbol{r} - (\boldsymbol{r}\cdot\boldsymbol{l})\,\boldsymbol{1}]\right\}.$$

Let $\boldsymbol{S}_{\boldsymbol{y}}^*[\boldsymbol{l}]$ be the unique tensor field with the property that

$$\boldsymbol{S}_{\boldsymbol{y}}[\boldsymbol{v}]\,\boldsymbol{l} = \boldsymbol{S}_{\boldsymbol{y}}^*[\boldsymbol{l}]^T\boldsymbol{v}$$

for every vector $\boldsymbol{v}$, and let

$$\boldsymbol{s}_{\boldsymbol{y}}^*[\boldsymbol{l}] = \boldsymbol{S}_{\boldsymbol{y}}^*[\boldsymbol{l}]\,\boldsymbol{n}$$

on ∂B. We call $\boldsymbol{S}_{\boldsymbol{y}}^*[\boldsymbol{l}]$ the *adjoint stress field* and $\boldsymbol{s}_{\boldsymbol{y}}^*[\boldsymbol{l}]$ the ***adjoint traction field*** corresponding to $\mathfrak{s}_{\boldsymbol{y}}[\boldsymbol{l}]$. A simple calculation shows that

$$\boldsymbol{S}_{\boldsymbol{y}}^*[\boldsymbol{l}](\boldsymbol{x}) = -\frac{2\mu}{c r^3}\left\{\frac{3(\boldsymbol{r}\cdot\boldsymbol{l})\,\boldsymbol{r}\otimes\boldsymbol{r}}{r^2} + (1-2\nu)\,[\boldsymbol{l}\otimes\boldsymbol{r} - \boldsymbol{r}\otimes\boldsymbol{l} + (\boldsymbol{r}\cdot\boldsymbol{l})\,\boldsymbol{1}]\right\}.$$

Note that $\boldsymbol{S}_{\boldsymbol{y}}^*[\boldsymbol{l}]$ is not symmetric.

[1] [1955, *13*], Theorem 4.3.

(2) Properties of the Kelvin state.[1] *The Kelvin state* $\mathfrak{s}_{\boldsymbol{y}}[\boldsymbol{l}]$ *has the following properties:*

(i) $\mathfrak{s}_{\boldsymbol{y}}[\boldsymbol{l}]$ *is an elastic state on* $\mathscr{E}-\{\boldsymbol{y}\}$ *corresponding to zero body forces.*

(ii) $\boldsymbol{u}_{\boldsymbol{y}}[\boldsymbol{l}](\boldsymbol{x})=O(r^{-1})$ *and* $\boldsymbol{S}_{\boldsymbol{y}}[\boldsymbol{l}](\boldsymbol{x})=O(r^{-2})$ *as* $r\to 0$ *and also as* $r\to\infty$.

(iii) *For all* $\eta>0$,

$$\int_{\partial\Sigma_\eta} \mathbf{s}_{\boldsymbol{y}}[\boldsymbol{l}]\,da=\boldsymbol{l}, \qquad \int_{\partial\Sigma_\eta} \boldsymbol{r}\times\mathbf{s}_{\boldsymbol{y}}[\boldsymbol{l}]\,da=\boldsymbol{0},$$

where

$$\mathbf{s}_{\boldsymbol{y}}[\boldsymbol{l}]=\boldsymbol{S}_{\boldsymbol{y}}[\boldsymbol{l}]\,\boldsymbol{n}$$

on $\partial\Sigma_\eta$ *with* $\boldsymbol{n}$ *the inward unit normal.*

(iv) *For every vector* $\boldsymbol{v}$,

$$\boldsymbol{l}\cdot\boldsymbol{u}_{\boldsymbol{y}}[\boldsymbol{v}]=\boldsymbol{v}\cdot\boldsymbol{u}_{\boldsymbol{y}}[\boldsymbol{l}], \qquad \boldsymbol{l}\cdot\mathbf{s}_{\boldsymbol{y}}[\boldsymbol{v}]=\boldsymbol{v}\cdot\mathbf{s}^*_{\boldsymbol{y}}[\boldsymbol{l}],$$

where $\mathbf{s}_{\boldsymbol{y}}[\boldsymbol{v}]$ *is the traction field on* ∂B *corresponding* to $\mathfrak{s}_{\boldsymbol{y}}[\boldsymbol{v}]$ *and* $\mathbf{s}^*_{\boldsymbol{y}}[\boldsymbol{l}]$ *is the adjoint traction field on* ∂B *corresponding to* $\mathfrak{s}_{\boldsymbol{y}}[\boldsymbol{l}]$.

Proof. Property (i) follows from ***(44.1)*** and the fact that the stress functions φ and $\boldsymbol{\psi}$ defined in (b) of the proof of ***(1)*** are harmonic on $\mathscr{E}-\{\boldsymbol{y}\}$. Properties (ii) and the first of (iv) are established by inspection of the fields $\boldsymbol{u}_{\boldsymbol{y}}[\boldsymbol{l}]$ and $\boldsymbol{S}_{\boldsymbol{y}}[\boldsymbol{l}]$.

Property (iii) can be established by a direct computation based on the form of $\boldsymbol{S}_{\boldsymbol{y}}[\boldsymbol{l}]$. An alternative proof proceeds as follows. Let $\{\mathfrak{s}_m\}$ and $\{\boldsymbol{b}_m\}$ be as in ***(1)***. Then balance of forces and moments imply that

$$-\int_{\partial\Sigma_\eta}\mathbf{s}_m\,da+\int_{\Sigma_\eta}\boldsymbol{b}_m\,dv=\boldsymbol{0},$$

$$-\int_{\partial\Sigma_\eta}\boldsymbol{r}\times\mathbf{s}_m\,da+\int_{\Sigma_\eta}\boldsymbol{r}\times\boldsymbol{b}_m\,dv=\boldsymbol{0},$$

where $\mathbf{s}_m$ is the traction field on that side of $\partial\Sigma_\eta$ facing the point $\boldsymbol{y}$. If we let $m\to\infty$ in the above relations and use (ii) of ***(1)*** and properties (i)–(iii) of $\{\boldsymbol{b}_m\}$, we arrive at the desired relations in (iii).

Finally, to establish the second of (iv) note that, by the symmetry of $\boldsymbol{S}_{\boldsymbol{y}}[\boldsymbol{l}]$ together with the definitions of $\boldsymbol{S}^*_{\boldsymbol{y}}[\boldsymbol{l}]$ and $\mathbf{s}^*_{\boldsymbol{y}}[\boldsymbol{l}]$,

$$\begin{aligned}\boldsymbol{l}\cdot\mathbf{s}_{\boldsymbol{y}}[\boldsymbol{v}]&=\boldsymbol{l}\cdot(\boldsymbol{S}_{\boldsymbol{y}}[\boldsymbol{v}]\,\boldsymbol{n})=(\boldsymbol{S}_{\boldsymbol{y}}[\boldsymbol{v}]\,\boldsymbol{l})\cdot\boldsymbol{n}=(\boldsymbol{S}^*_{\boldsymbol{y}}[\boldsymbol{l}]^T\,\boldsymbol{v})\cdot\boldsymbol{n}\\&=\boldsymbol{v}\cdot(\boldsymbol{S}^*_{\boldsymbol{y}}[\boldsymbol{l}]\,\boldsymbol{n})=\boldsymbol{v}\cdot\mathbf{s}^*_{\boldsymbol{y}}[\boldsymbol{l}].\quad\square\end{aligned}$$

STERNBERG and EUBANKS[2] have shown that the formulation of Kelvin's problem in terms of (i), (iii), and the portion of (ii) concerning the limit as $r\to\infty$, which appears in the literature,[3] is incomplete in view of the existence of elastic states on $\mathscr{E}-\{\boldsymbol{y}\}$ that possess self-equilibrated singularities at $\boldsymbol{y}$.[4] In contrast, properties (i), (ii), and the first of (iii) suffice to characterize the Kelvin state uniquely.[5]

---

[1] Properties (i)–(iii), in the precise form stated here, are taken from STERNBERG and EUBANKS [1955, *13*], Theorem 4.4. Assertion (iii) can be found in LOVE [1927, *3*], § 131; LUR'E [1955, *10*], § 2.1.

[2] [1955, *13*].

[3] See, e.g., TIMOSHENKO and GOODIER [1951, *10*], § 120.

[4] E.g. the center of compression defined on p. 179.

[5] As STERNBERG and EUBANKS [1955, *13*] have remarked, TREFFTZ [1928, *3*], § 32 and BUTTY [1946, *1*], § 350 approached the KELVIN problem on the basis of properties (i), (ii), and the first of (iii) but made the erroneous assertion that (ii) is a consequence of the first of (iii). That this is not true is apparent from the state $\mathfrak{s}_{\boldsymbol{y}}[\boldsymbol{l}]+\alpha\,\mathring{\mathfrak{s}}_{\boldsymbol{y}}$, where α is a scalar and $\mathring{\mathfrak{s}}_{\boldsymbol{y}}$ is the center of compression defined on p. 179. Indeed, this state obeys the first of (iii) without conforming to (ii).

Let $\bar{\boldsymbol{u}}_{\boldsymbol{y}}^i[\boldsymbol{l}]$ be defined on $\mathscr{E}-\{\boldsymbol{y}\}$ by

$$\bar{\boldsymbol{u}}_{\boldsymbol{y}}^i[\boldsymbol{l}](\boldsymbol{x}) = \frac{\partial}{\partial x_i} \boldsymbol{u}_{\boldsymbol{y}}[\boldsymbol{l}](\boldsymbol{x}),$$

so that

$$\bar{\boldsymbol{u}}_{\boldsymbol{y}}^i[\boldsymbol{l}](\boldsymbol{x}) = \frac{1}{c r^3}\left[\boldsymbol{r}\otimes\boldsymbol{e}_i + \boldsymbol{e}_i\otimes\boldsymbol{r} - \frac{3 r_i \boldsymbol{r}\otimes\boldsymbol{r}}{r^2} - (3-4\nu)\, r_i\, \mathbf{1}\right]\boldsymbol{l}.$$

By *(42.3)*, $\bar{\boldsymbol{u}}_{\boldsymbol{y}}^i[\boldsymbol{l}]$ is an elastic displacement field on $\mathscr{E}-\{\boldsymbol{y}\}$ corresponding to null body forces. Let $\bar{\mathfrak{s}}_{\boldsymbol{y}}^i[\boldsymbol{l}]$ be the elastic state on $\mathscr{E}-\{\boldsymbol{y}\}$ generated by $\bar{\boldsymbol{u}}_{\boldsymbol{y}}^i[\boldsymbol{l}]$. We call $\bar{\mathfrak{s}}_{\boldsymbol{y}}^i[\boldsymbol{l}]$ the ***doublet state*** (corresponding to $\boldsymbol{y}$, $\boldsymbol{l}$ and $\boldsymbol{e}_i$).[1]

Note that

$$\bar{\boldsymbol{u}}_{\boldsymbol{y}}^i[\boldsymbol{l}](\boldsymbol{x}) = \lim_{h\to 0}\frac{1}{h}\{\boldsymbol{u}_{\boldsymbol{y}}[\boldsymbol{l}](\boldsymbol{x}) - \boldsymbol{u}_{\boldsymbol{y}}[\boldsymbol{l}](\boldsymbol{x}-h\boldsymbol{e}_i)\}.$$

Thus, since

$$\boldsymbol{u}_{\boldsymbol{y}}[\boldsymbol{l}](\boldsymbol{x}-\boldsymbol{a}) = \boldsymbol{u}_{\boldsymbol{y}+\boldsymbol{a}}[\boldsymbol{l}](\boldsymbol{x}),$$

it follows that

$$\bar{\boldsymbol{u}}_{\boldsymbol{y}}^i[\boldsymbol{l}](\boldsymbol{x}) = \lim_{h\to 0}\left\{\boldsymbol{u}_{\boldsymbol{y}}\left[\frac{1}{h}\boldsymbol{l}\right](\boldsymbol{x}) + \boldsymbol{u}_{\boldsymbol{y}+h\boldsymbol{e}_i}\left[-\frac{1}{h}\boldsymbol{l}\right](\boldsymbol{x})\right\},$$

or equivalently that

$$\bar{\boldsymbol{u}}_{\boldsymbol{y}}^i[\boldsymbol{l}] = \lim_{h\to 0}\left\{\boldsymbol{u}_{\boldsymbol{y}}\left[\frac{1}{h}\boldsymbol{l}\right] + \boldsymbol{u}_{\boldsymbol{y}+h\boldsymbol{e}_i}\left[-\frac{1}{h}\boldsymbol{l}\right]\right\}.$$

Thus $\bar{\boldsymbol{u}}_{\boldsymbol{y}}^i[\boldsymbol{l}]$ is the limit as $h\to 0$ of the sum of two displacement fields. These fields are associated with two Kelvin states: the first corresponds to a concentrated load

$$\frac{1}{h}\boldsymbol{l} \text{ applied at } \boldsymbol{y};$$

the second corresponds to a concentrated load

$$-\frac{1}{h}\boldsymbol{l} \text{ applied at } \boldsymbol{y}+h\boldsymbol{e}_i.$$

The two load systems are shown in Fig. 13. The above relation also yields the important result

$$\bar{\boldsymbol{u}}_{\boldsymbol{y}}^i[\boldsymbol{l}] = -\frac{\partial}{\partial y_i}\boldsymbol{u}_{\boldsymbol{y}}[\boldsymbol{l}].$$

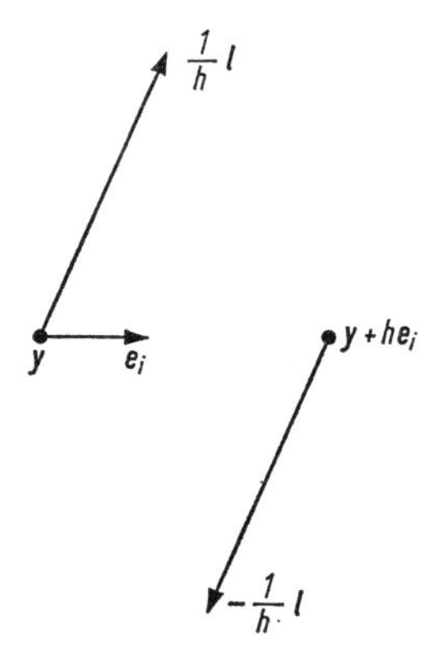

Fig. 13.

[1] These states were introduced by LOVE [1927, *3*], § 132. See also STERNBERG and EUBANKS [1955, *13*].

The next theorem is the analog of *(2)* for doublet states; its proof is strictly analogous and can safely be omitted.

(3) Properties of the doublet states.[1] *The doublet state* $\bar{\mathfrak{s}}^i_{\boldsymbol{y}}[\boldsymbol{l}]$ *has the following properties*:

(i) $\bar{\mathfrak{s}}^i_{\boldsymbol{y}}[\boldsymbol{l}]$ *is an elastic state on* $\mathscr{E}-\{y\}$ *corresponding to zero body forces.*

(ii) $\bar{\boldsymbol{u}}^i_{\boldsymbol{y}}[\boldsymbol{l}](\boldsymbol{x})=O(r^{-2})$ *and* $\bar{\boldsymbol{S}}^i_{\boldsymbol{y}}[\boldsymbol{l}](\boldsymbol{x})=O(r^{-3})$ *as* $r\to 0$ *and also as* $r\to\infty$.

(iii) *For all* $\eta>0$,

$$\int_{\partial\Sigma_\eta}\bar{\boldsymbol{s}}^i_{\boldsymbol{y}}[\boldsymbol{l}]\,da=\boldsymbol{0},\qquad \int_{\partial\Sigma_\eta}\boldsymbol{r}\times\bar{\boldsymbol{s}}^i_{\boldsymbol{y}}[\boldsymbol{l}]\,da=-\boldsymbol{e}_i\times\boldsymbol{l},$$

where

$$\bar{\boldsymbol{s}}^i_{\boldsymbol{y}}[\boldsymbol{l}]=\bar{\boldsymbol{S}}^i_{\boldsymbol{y}}[\boldsymbol{l}]\,\boldsymbol{n}$$

on $\partial\Sigma_\eta$ *with* $\boldsymbol{n}$ *the inward unit normal.*

(iv) *For every vector* $\boldsymbol{v}$,

$$\boldsymbol{l}\cdot\bar{\boldsymbol{u}}^i_{\boldsymbol{y}}[\boldsymbol{v}]=\boldsymbol{v}\cdot\bar{\boldsymbol{u}}^i_{\boldsymbol{y}}[\boldsymbol{l}].$$

(v) *If* $\mathfrak{s}_{\boldsymbol{y}}[\boldsymbol{l}]$ *is the Kelvin state corresponding to the concentrated load* $\boldsymbol{l}$ *at* $\boldsymbol{y}$, *then*

$$\frac{\partial}{\partial y_i}\mathfrak{s}_{\boldsymbol{y}}[\boldsymbol{l}]=-\bar{\mathfrak{s}}^i_{\boldsymbol{y}}[\boldsymbol{l}],$$

i.e.

$$\frac{\partial}{\partial y_i}\boldsymbol{u}_{\boldsymbol{y}}[\boldsymbol{l}]=-\bar{\boldsymbol{u}}^i_{\boldsymbol{y}}[\boldsymbol{l}],\qquad \frac{\partial}{\partial y_i}\boldsymbol{S}_{\boldsymbol{y}}[\boldsymbol{l}]=-\bar{\boldsymbol{S}}^i_{\boldsymbol{y}}[\boldsymbol{l}].$$

Property (iii) could be anticipated intuitively because of the physical meaning attached to the doublet states (cf. Fig. 13). As is clear from (iii), the singularity of $\bar{\mathfrak{s}}^i_{\boldsymbol{y}}[\boldsymbol{l}]$ is statically equivalent to a couple or to null according as

$$\boldsymbol{e}_i\times\boldsymbol{l}\neq\boldsymbol{0}\quad\text{or}\quad \boldsymbol{e}_i\times\boldsymbol{l}=\boldsymbol{0}.$$

In contrast to properties (i)–(iii) of *(2)*, which characterize the Kelvin state uniquely, (i)–(iii) of *(3)*, clearly, do not supply a unique characterization.[2] Indeed, $\bar{\mathfrak{s}}^i_{\boldsymbol{y}}[\boldsymbol{l}+\alpha\,\boldsymbol{e}_i]$ (no sum) has the same properties as $\bar{\mathfrak{s}}^i_{\boldsymbol{y}}[\boldsymbol{l}]$.

We call

$$\mathfrak{s}^i_{\boldsymbol{y}}=\mathfrak{s}_{\boldsymbol{y}}[\boldsymbol{e}_i]$$

the ***unit Kelvin state*** corresponding to a unit load at $\boldsymbol{y}$ in the direction of $\boldsymbol{e}_i$. On the other hand, the ***unit doublet states*** at $\boldsymbol{y}$ are defined by

$$\bar{\mathfrak{s}}^{ij}_{\boldsymbol{y}}=\bar{\mathfrak{s}}^j_{\boldsymbol{y}}[\boldsymbol{e}_i].$$

By *(3)* the singularity of $\bar{\mathfrak{s}}^{ij}_{\boldsymbol{y}}$ at $\boldsymbol{y}$ is statically equivalent to a couple or to null depending on whether $i\neq j$ or $i=j$. We refer to $\bar{\mathfrak{s}}^{ij}_{\boldsymbol{y}}$, $i\neq j$, as the *unit doublet state with moment* $\boldsymbol{e}_i\times\boldsymbol{e}_j$ *at* $\boldsymbol{y}$, and to $\bar{\mathfrak{s}}^{ii}_{\boldsymbol{y}}$ (no sum) as the *unit doublet state corresponding to a force doublet at* $\boldsymbol{y}$ *parallel to the* x_i*-axis*. Clearly,

$$\mathfrak{s}_{\boldsymbol{y}}[\boldsymbol{l}]=l_i\,\mathfrak{s}^i_{\boldsymbol{y}},$$
$$\bar{\mathfrak{s}}^i_{\boldsymbol{y}}[\boldsymbol{l}]=l_j\,\bar{\mathfrak{s}}^{ji}_{\boldsymbol{y}},$$

and the displacement fields belonging to $\mathfrak{s}^i_{\boldsymbol{y}}$ and $\bar{\mathfrak{s}}^{ij}_{\boldsymbol{y}}$ are given by

$$\boldsymbol{u}^i_{\boldsymbol{y}}(\boldsymbol{x})=\frac{1}{c\,r}\left[\frac{r_i\,\boldsymbol{r}}{r^2}+(3-4\nu)\,\boldsymbol{e}_i\right],$$

$$\bar{\boldsymbol{u}}^{ij}_{\boldsymbol{y}}(\boldsymbol{x})=\frac{1}{c\,r^3}\left[\delta_{ij}\,\boldsymbol{r}+r_i\,\boldsymbol{e}_j-\frac{3r_ir_j\,\boldsymbol{r}}{r^2}-(3-4\nu)\,r_j\,\boldsymbol{e}_i\right].$$

[1] STERNBERG and EUBANKS [1955, *13*], Theorem 5.2.
[2] STERNBERG and EUBANKS [1955, *13*], p. 152.

The state

$$\mathring{\mathfrak{s}}_{\boldsymbol{y}} = \bar{\mathfrak{s}}_{\boldsymbol{y}}^{11} + \bar{\mathfrak{s}}_{\boldsymbol{y}}^{22} + \bar{\mathfrak{s}}_{\boldsymbol{y}}^{33}$$

is called the ***center of compression*** at $\boldsymbol{y}$, while

$$\mathring{\mathfrak{s}}_{\boldsymbol{y}}^{i} = \tfrac{1}{2}\varepsilon_{ijk}\,\bar{\mathfrak{s}}_{\boldsymbol{y}}^{jk}$$

is the ***center of rotation*** at $\boldsymbol{y}$ parallel to the x_i-axis.[1] Using the formula for $\bar{\boldsymbol{u}}_{\boldsymbol{y}}^{ij}$, it is a simple matter to verify that the displacement fields $\mathring{\boldsymbol{u}}_{\boldsymbol{y}}$ and $\mathring{\boldsymbol{u}}_{\boldsymbol{y}}^{i}$ associated with $\mathring{\mathfrak{s}}_{\boldsymbol{y}}$ and $\mathring{\mathfrak{s}}_{\boldsymbol{y}}^{i}$ are given by

$$\mathring{\boldsymbol{u}}_{\boldsymbol{y}}(\boldsymbol{x}) = -\frac{2(1-2\nu)\,\boldsymbol{r}}{c\,r^3},$$

$$\mathring{\boldsymbol{u}}_{\boldsymbol{y}}^{i}(\boldsymbol{x}) = -\frac{\boldsymbol{r}\times\boldsymbol{e}_i}{8\pi\mu r^3}.$$

Moreover, it follows from (iii) of *(3)* that

$$\int_{\partial\Sigma_\eta} \mathring{\boldsymbol{s}}_{\boldsymbol{y}}\,da = \int_{\partial\Sigma_\eta} \boldsymbol{r}\times\mathring{\boldsymbol{s}}_{\boldsymbol{y}}\,da = \boldsymbol{0},$$

$$\int_{\partial\Sigma_\eta} \mathring{\boldsymbol{s}}_{\boldsymbol{y}}^{i}\,da = \boldsymbol{0}, \qquad \int_{\partial\Sigma_\eta} \boldsymbol{r}\times\mathring{\boldsymbol{s}}_{\boldsymbol{y}}^{i}\,da = \boldsymbol{e}_i,$$

where

$$\mathring{\boldsymbol{s}}_{\boldsymbol{y}} = \mathring{\boldsymbol{S}}_{\boldsymbol{y}}\boldsymbol{n}, \qquad \mathring{\boldsymbol{s}}_{\boldsymbol{y}}^{i} = \mathring{\boldsymbol{S}}_{\boldsymbol{y}}^{i}\boldsymbol{n},$$

and $\boldsymbol{n}$ is the inward unit normal to $\partial\Sigma_\eta$.

52. Concentrated loads. The reciprocal theorem. By a ***system of concentrated loads*** we mean a vector-valued function $\boldsymbol{l}$ whose domain $D_{\boldsymbol{l}}$ is a finite set of (boundary or interior) points of $\bar{B}$; the vector $\boldsymbol{l}(\boldsymbol{\xi})$ is to be interpreted as a *concentrated load at* $\boldsymbol{\xi}$. If $D_{\boldsymbol{l}}\cap\partial B=\emptyset$, we say that the concentrated loads are ***internal***. Finally, we write $\boldsymbol{l}=\boldsymbol{l}(\boldsymbol{\xi})$ when $D_{\boldsymbol{l}}$ is a singleton $\{\boldsymbol{\xi}\}$.

Our next definition is motivated by *(51.2)* and the ensuing discussion. Let $\boldsymbol{l}$ be a system of concentrated loads. We say that $\mathfrak{s}=[\boldsymbol{u},\boldsymbol{E},\boldsymbol{S}]$ is a ***singular elastic state*** corresponding to the external force system $[\boldsymbol{b},\boldsymbol{s},\boldsymbol{l}]$ if

(i) $\mathfrak{s}$ is a (regular) elastic state on $\bar{B}-D_{\boldsymbol{l}}$ corresponding to the external force system $[\boldsymbol{b},\boldsymbol{s}]$;

(ii) for each $\boldsymbol{\xi}\in D_{\boldsymbol{l}}$, $\boldsymbol{u}(\boldsymbol{x})=O(r^{-1})$ and $\boldsymbol{S}(\boldsymbol{x})=O(r^{-2})$ as $r=|\boldsymbol{x}-\boldsymbol{\xi}|\to 0$;

(iii) for each $\boldsymbol{\xi}\in D_{\boldsymbol{l}}$,

$$\lim_{\varrho\to 0}\int_{\mathscr{S}_\varrho(\boldsymbol{\xi})} \boldsymbol{S}\boldsymbol{n}\,da = \boldsymbol{l}(\boldsymbol{\xi}),$$

where $\mathscr{S}_\varrho(\boldsymbol{\xi})=B\cap\partial\Sigma_\varrho(\boldsymbol{\xi})$ and $\boldsymbol{n}$ is the *inward* unit normal to $\partial\Sigma_\varrho(\boldsymbol{\xi})$.

It follows from *(51.2)* that for $\boldsymbol{\xi}\in\bar{B}$ the Kelvin state corresponding to a concentrated load $\boldsymbol{l}$ at $\boldsymbol{\xi}$ is a singular elastic state on $\bar{B}$ corresponding to $\boldsymbol{l}$ and to zero body forces.

By (ii) and (iii) we have:

(iv) for each $\boldsymbol{\xi}\in D_{\boldsymbol{l}}$,

$$\lim_{\varrho\to 0}\int_{\mathscr{S}_\varrho(\boldsymbol{\xi})} \boldsymbol{p}\times\boldsymbol{S}\boldsymbol{n}\,da = (\boldsymbol{\xi}-\boldsymbol{0})\times\boldsymbol{l}(\boldsymbol{\xi}).$$

[1] These solutions were first introduced by DOUGALL [1898, *4*]. Cf. LOVE [1927, *3*], § 132; LUR'E [1955, *10*], § 2.3; STERNBERG and EUBANKS [1955, *13*].

Indeed, by (iii),

$$\int_{\mathscr{S}_\varrho(\xi)} \boldsymbol{p}\times\boldsymbol{S}\,\boldsymbol{n}\,da - (\xi - 0)\times\boldsymbol{l}(\xi) = \int_{\mathscr{S}_\varrho(\xi)} (\boldsymbol{x}-\xi)\times\boldsymbol{S}(\boldsymbol{x})\,\boldsymbol{n}(\boldsymbol{x})\,da_{\boldsymbol{x}} + o(1)$$

as $\varrho\to 0$, and by (ii) the integral on the right tends to zero in this limit.

The next theorem shows that for internal concentrated loads in a homogeneous and isotropic body, every singular elastic state admits a representation as a regular elastic state plus a sum of Kelvin states.

(1) Decomposition theorem. *Assume that the body is homogeneous and isotropic. Let $\boldsymbol{l}$ be a system of internal concentrated loads, and let $\mathfrak{s}$ be a singular elastic state corresponding to $\boldsymbol{l}$ and to zero body forces. Then $\mathfrak{s}$ admits the decomposition*

$$\mathfrak{s} = \mathfrak{s}_R + \sum_{\xi\in D_{\boldsymbol{l}}} \mathfrak{s}_\xi,$$

where $\mathfrak{s}_R$ is a (regular) elastic state corresponding to vanishing body forces, and for each $\xi\in D_{\boldsymbol{l}}$, $\mathfrak{s}_\xi(=\mathfrak{s}_\xi[\boldsymbol{l}(\xi)])$ is the Kelvin state corresponding to the concentrated load $\boldsymbol{l}(\xi)$ at ξ.

Proof. For each $\xi\in D_{\boldsymbol{l}}$ let

$$\mathring{\Sigma}(\xi) = \Sigma_\varrho(\xi) - \{\xi\},$$

with ϱ sufficiently small that

$$\mathring{\Sigma}(\xi)\subset B, \quad \mathring{\Sigma}(\xi)\cap D_{\boldsymbol{l}} = \emptyset$$

for all $\xi\in D_{\boldsymbol{l}}$. Then, on $\mathring{\Sigma}(\xi)$, $\boldsymbol{u}$ admits the representation[1]

$$\boldsymbol{u} = \sum_{k=-\infty}^{\infty} \boldsymbol{g}_\xi^{(k)} + r_\xi^2 \sum_{k=-\infty}^{\infty} \boldsymbol{h}_\xi^{(k)}, \qquad r_\xi = |\boldsymbol{x}-\xi|, \tag{a}$$

where $\boldsymbol{g}_\xi^{(k)}$ and $\boldsymbol{h}_\xi^{(k)}$ are vector solid spherical harmonics of degree k (ξ being the origin of the spherical coordinate system) and

$$\boldsymbol{h}_\xi^{(k)} = \frac{-(\lambda+\mu)}{2[(k+1)(\lambda+3\mu)+\mu]}\,\nabla\,\mathrm{div}\,\boldsymbol{g}_\xi^{(k)}. \tag{b}$$

Further, by (ii) in the definition of a singular elastic state,[2]

$$\boldsymbol{g}_\xi^{(k)} = \boldsymbol{h}_\xi^{(k-2)} = \boldsymbol{0}, \qquad k \leqq -2, \tag{c}$$

and since $\boldsymbol{g}_\xi^{(k)}$, $k\geqq 0$, is a homogeneous polynomial of degree k, (b) implies

$$\boldsymbol{h}_\xi^{(-2)} = \boldsymbol{h}_\xi^{(-1)} = \boldsymbol{0}. \tag{d}$$

For each $\xi\in D_{\boldsymbol{l}}$ let

$$\boldsymbol{u}_\xi = \boldsymbol{g}_\xi^{(-1)} + r_\xi^2\,\boldsymbol{h}^{(-3)} \tag{e}$$

on $\mathscr{E}-\{\xi\}$; then a simple computation using (b) shows that the state $\mathfrak{s}_\xi$ corresponding to $\boldsymbol{u}_\xi$ (in the sense of the strain-displacement and stress-strain relations) is an elastic state on $\mathscr{E}-\{\xi\}$ corresponding to $\boldsymbol{b}=\boldsymbol{0}$. Thus if we let

$$\boldsymbol{u}_R = \boldsymbol{u} - \sum_{\xi\in D_{\boldsymbol{l}}} \boldsymbol{u}_\xi \tag{f}$$

[1] It is clear that the representation theorem ***(48.1)*** remains valid when the deleted neighborhood of infinity, $\Sigma(\infty)$, is replaced by the deleted neighborhood of ξ, $\mathring{\Sigma}(\xi)$.

[2] This follows from the analog of ***(8.5)*** for $\mathring{\Sigma}(\xi)$.

on $\bar{B} - D_{\boldsymbol{l}}$, then the associated state $\mathfrak{s}_R$ is an elastic state on $\bar{B} - D_{\boldsymbol{l}}$ also corresponding to $\boldsymbol{b} = \mathbf{0}$. It follows from (a) and (c)–(e) that for each $\boldsymbol{\eta} \in D_{\boldsymbol{l}}$,

$$\boldsymbol{u}_R = \sum_{k=0}^{\infty} \boldsymbol{g}_{\boldsymbol{\eta}}^{(k)} + r_{\boldsymbol{\eta}}^2 \sum_{k=0}^{\infty} \boldsymbol{h}_{\boldsymbol{\eta}}^{(k)} - \sum_{\substack{\boldsymbol{\xi} \in D_{\boldsymbol{l}} \\ \boldsymbol{\xi} \neq \boldsymbol{\eta}}} \boldsymbol{u}_{\boldsymbol{\xi}} \tag{g}$$

on $\overset{\circ}{\Sigma}(\boldsymbol{\eta})$; thus $\boldsymbol{u}_R(\boldsymbol{x})$ can be continuously extended to $\boldsymbol{x} = \boldsymbol{\eta}$, and the resulting extension is C^∞ on $\Sigma_\varrho(\boldsymbol{\eta})$. Therefore $\mathfrak{s}_R$ so extended is an elastic state on $\bar{B}$ corresponding to zero body forces.

To complete the proof, it clearly suffices to show that $\mathfrak{s}_{\boldsymbol{\xi}}$ is the Kelvin state corresponding to $\boldsymbol{l}(\boldsymbol{\xi})$. Since

$$\boldsymbol{g}_{\boldsymbol{\xi}}^{(-1)}(\boldsymbol{x}) = \frac{\boldsymbol{\alpha}}{r_{\boldsymbol{\xi}}},$$

where the vector $\boldsymbol{\alpha}$ is a constant, (b) and (e) yield

$$\boldsymbol{u}_{\boldsymbol{\xi}}(\boldsymbol{x}) = \frac{\boldsymbol{\alpha}}{r_{\boldsymbol{\xi}}} + \frac{(\lambda + \mu)}{2(2\lambda + 5\mu)} r_{\boldsymbol{\xi}}^2 \nabla \operatorname{div}\left(\frac{\boldsymbol{\alpha}}{r_{\boldsymbol{\xi}}}\right). \tag{h}$$

By (f) and (g), $\boldsymbol{u} - \boldsymbol{u}_{\boldsymbol{\xi}}$ is C^∞ on $\Sigma_\varrho(\boldsymbol{\xi})$; thus the stress field $\boldsymbol{S}_{\boldsymbol{\xi}}$ corresponding to $\boldsymbol{u}_{\boldsymbol{\xi}}$ satisfies (iii) in the definition of a singular elastic state:

$$\lim_{\delta \to 0} \int_{\partial \Sigma_\delta(\boldsymbol{\xi})} \boldsymbol{S}_{\boldsymbol{\xi}} \boldsymbol{n} \, da = \boldsymbol{l}(\boldsymbol{\xi}). \tag{i}$$

An elementary computation based on (h) and (i) shows that

$$\boldsymbol{\alpha} = \frac{(2\lambda + 5\mu)}{12\pi \mu (\lambda + 2\mu)} \boldsymbol{l}(\boldsymbol{\xi});$$

hence we conclude from (h) and ***(51.1)*** that $\mathfrak{s}_{\boldsymbol{\xi}}$ is the Kelvin state corresponding to $\boldsymbol{l}(\boldsymbol{\xi})$ at $\boldsymbol{\xi}$. □

(2) Balance of forces and moments for singular states. *Let $\boldsymbol{l}$ be a system of concentrated loads, and let $\mathfrak{s}$ be a singular elastic state corresponding to the external force system $[\boldsymbol{b}, \mathbf{s}, \boldsymbol{l}]$ with $\boldsymbol{b}$ continuous on $\bar{B}$. Then*

$$\int_{\partial B} \mathbf{s} \, da + \int_B \boldsymbol{b} \, dv + \sum_{\boldsymbol{\xi} \in D_{\boldsymbol{l}}} \boldsymbol{l}(\boldsymbol{\xi}) = \mathbf{0},$$

$$\int_{\partial B} \boldsymbol{p} \times \mathbf{s} \, da + \int_B \boldsymbol{p} \times \boldsymbol{b} \, dv + \sum_{\boldsymbol{\xi} \in D_{\boldsymbol{l}}} (\boldsymbol{\xi} - 0) \times \boldsymbol{l}(\boldsymbol{\xi}) = \mathbf{0}.$$

Here the surface integrals are to be interpreted as Cauchy principal values; i.e.

$$\int_{\partial B} \mathbf{s} \, da = \lim_{\varrho \to 0} \int_{\mathscr{C}(\varrho)} \mathbf{s} \, da,$$

where

$$\mathscr{C}(\varrho) = \partial B - \bigcup_{n=1}^{N} \Sigma_{\varrho_n}(\boldsymbol{\xi}_n),$$

$$\{\boldsymbol{\xi}_1, \ldots, \boldsymbol{\xi}_N\} = D_{\boldsymbol{l}} \cap \partial B, \qquad \varrho = (\varrho_1, \ldots, \varrho_N).$$

Proof. Let $B_{\boldsymbol{l}} = D_{\boldsymbol{l}} \cap B$,

$$B(\delta, \varrho) = B - \bigcup_{\boldsymbol{\xi} \in B_{\boldsymbol{l}}} \overline{\Sigma_\delta(\boldsymbol{\xi})} - \bigcup_{n=1}^{N} \overline{\Sigma_{\varrho_n}(\boldsymbol{\xi}_n)}, \tag{a}$$

and choose δ and ϱ sufficiently small that the closed balls mentioned in (a) are mutually disjoint. (Note that $\boldsymbol{\xi}_1, \ldots, \boldsymbol{\xi}_N$ are all points of ∂B; while each $\boldsymbol{\xi} \in B_{\boldsymbol{l}}$ is an interior point of B.) Then $\mathfrak{s}$ is a (regular) elastic state on $B(\delta, \varrho)$; hence

$$\int_{\partial B(\delta, \varrho)} \mathbf{s} \, da + \int_{B(\delta, \varrho)} \boldsymbol{b} \, dv = \mathbf{0}. \tag{b}$$

Since $\boldsymbol{b}$ is continuous on $\bar{B}$,

$$\int\limits_{B(\delta,\varrho)} \boldsymbol{b}\,dv \to \int\limits_{B} \boldsymbol{b}\,dv \quad \text{as} \quad \delta\to 0,\ \boldsymbol{\varrho}\to\boldsymbol{0}. \tag{c}$$

Further, by (a), if δ and $\boldsymbol{\varrho}$ are sufficiently small,

$$\int\limits_{\partial B(\delta,\varrho)} = \int\limits_{\mathscr{C}(\varrho)} + \sum_{n=1}^{N} \int\limits_{\mathscr{S}_{\varrho_n}(\xi_n)} + \sum_{\xi\in B_{\boldsymbol{l}}} \int\limits_{\mathscr{S}_\delta(\xi)}, \tag{d}$$

where

$$\mathscr{S}_\eta(\boldsymbol{x}) = B \cap \partial\Sigma_\eta(\boldsymbol{x}). \tag{e}$$

Combining (b) and (d) and letting $\delta\to 0$, $\boldsymbol{\varrho}\to 0$, we conclude, with the aid of (c) and (iii) in the definition of a singular elastic state, that

$$\int\limits_{\partial B} \mathbf{s}\,da$$

exists as a Cauchy principal value, and that the first relation in ***(2)*** holds. The second relation is derived in exactly the same manner using (iv) in place of (iii). □

We now give an extension of Betti's reciprocal theorem ***(30.1)*** which includes concentrated loads. This result, which is due to TURTELTAUB and STERNBERG,[1] will be extremely useful in establishing integral representation theorems for elastic states.

(3) Reciprocal theorem for singular states. *Assume that the elasticity field is symmetric and invertible. Let $\boldsymbol{l}$ and $\tilde{\boldsymbol{l}}$ be systems of concentrated loads with $D_{\boldsymbol{l}}$ and $D_{\tilde{\boldsymbol{l}}}$ disjoint. Further, let $[\boldsymbol{u}, \mathbf{E}, \mathbf{S}]$ and $[\tilde{\boldsymbol{u}}, \tilde{\mathbf{E}}, \tilde{\mathbf{S}}]$ be singular elastic states corresponding to the external force systems $[\boldsymbol{b}, \mathbf{s}, \boldsymbol{l}]$ and $[\tilde{\boldsymbol{b}}, \tilde{\mathbf{s}}, \tilde{\boldsymbol{l}}]$, respectively, with $\boldsymbol{b}$ and $\tilde{\boldsymbol{b}}$ continuous on $\bar{B}$. Then*

$$\begin{aligned}\int\limits_{\partial B} \tilde{\mathbf{s}}\cdot\boldsymbol{u}\,da + \int\limits_{B} \tilde{\boldsymbol{b}}\cdot\boldsymbol{u}\,dv + \sum_{\xi\in D_{\tilde{l}}} \tilde{\boldsymbol{l}}(\xi)\cdot\boldsymbol{u}(\xi)\\ = \int\limits_{\partial B} \mathbf{s}\cdot\tilde{\boldsymbol{u}}\,da + \int\limits_{B} \boldsymbol{b}\cdot\tilde{\boldsymbol{u}}\,dv + \sum_{\xi\in D_{l}} \boldsymbol{l}(\xi)\cdot\tilde{\boldsymbol{u}}(\xi)\\ = \int\limits_{B} \tilde{\mathbf{S}}\cdot\mathbf{E}\,dv = \int\limits_{B} \mathbf{S}\cdot\tilde{\mathbf{E}}\,dv,\end{aligned}$$

where the surface integrals are to be interpreted as Cauchy principal values.

Proof. Let

$$\tilde{\boldsymbol{\delta}}_\xi(\boldsymbol{x}) = \tilde{\boldsymbol{u}}(\boldsymbol{x}) - \tilde{\boldsymbol{u}}(\xi) \tag{a}$$

for each $\boldsymbol{x}\in\bar{B}$ and $\xi\in D_{\boldsymbol{l}}$. Then (iii) in the definition of a singular elastic state implies

$$\int\limits_{\mathscr{S}_\varrho(\xi)} \mathbf{s}\cdot\tilde{\boldsymbol{u}}\,da = \int\limits_{\mathscr{S}_\varrho(\xi)} \mathbf{s}\cdot\tilde{\boldsymbol{\delta}}_\xi\,da + \boldsymbol{l}(\xi)\cdot\tilde{\boldsymbol{u}}(\xi) + o(1) \quad \text{as} \quad \varrho\to 0 \tag{b}$$

for every $\xi\in D_{\boldsymbol{l}}$. Since $\mathfrak{s}$ is an elastic state on $\bar{B} - D_{\boldsymbol{l}}$ and $\tilde{\mathfrak{s}}$ an elastic state on $\bar{B} - D_{\tilde{\boldsymbol{l}}}$, and since $D_{\boldsymbol{l}} \cap D_{\tilde{\boldsymbol{l}}} = \emptyset$, (a), (b), and (ii) in the definition of a singular state yield

$$\int\limits_{\mathscr{S}_\varrho(\xi)} \mathbf{s}\cdot\tilde{\boldsymbol{u}}\,da \to \boldsymbol{l}(\xi)\cdot\tilde{\boldsymbol{u}}(\xi), \qquad \int\limits_{\mathscr{S}_\varrho(\xi)} \tilde{\mathbf{s}}\cdot\boldsymbol{u}\,da \to 0 \tag{c}$$

[1] [1968, *14*].

as $\varrho \to 0$ for each $\xi \in D_{\boldsymbol{l}}$. Similarly,

$$\int\limits_{\mathscr{S}_\varrho(\xi)} \tilde{\mathbf{s}} \cdot \boldsymbol{u}\, da \to \tilde{\boldsymbol{l}}(\xi) \cdot \boldsymbol{u}(\xi), \qquad \int\limits_{\mathscr{S}_\varrho(\xi)} \mathbf{s} \cdot \tilde{\boldsymbol{u}}\, da \to 0 \tag{d}$$

as $\varrho \to 0$ for each $\xi \in D_{\tilde{\boldsymbol{l}}}$.

Now let

$$\begin{aligned} \{\xi_1, \ldots, \xi_N\} &= (D_{\boldsymbol{l}} \cup D_{\tilde{\boldsymbol{l}}}) \cap \partial B, \qquad \varrho = (\varrho_1, \ldots, \varrho_N), \\ \mathscr{C}(\varrho) &= \partial B - \bigcup_{n=1}^{N} \Sigma_{\varrho_n}(\xi_n), \\ B_{\boldsymbol{l}\tilde{\boldsymbol{l}}} &= (D_{\boldsymbol{l}} \cup D_{\tilde{\boldsymbol{l}}}) \cap B, \\ B(\delta, \varrho) &= B - \bigcup_{\xi \in B_{\boldsymbol{l}\tilde{\boldsymbol{l}}}} \overline{\Sigma_\delta(\xi)} - \bigcup_{n=1}^{N} \overline{\Sigma_{\varrho_n}(\xi_n)}. \end{aligned} \tag{e}$$

Then $\mathfrak{s}$ and $\tilde{\mathfrak{s}}$ are (regular) elastic states on $B(\delta, \varrho)$, and Betti's theorem ***(30.1)*** implies

$$\begin{aligned} \int\limits_{\partial B(\delta,\varrho)} \tilde{\mathbf{s}} \cdot \boldsymbol{u}\, da + \int\limits_{B(\delta,\varrho)} \tilde{\boldsymbol{b}} \cdot \boldsymbol{u}\, dv &= \int\limits_{\partial B(\delta,\varrho)} \mathbf{s} \cdot \tilde{\boldsymbol{u}}\, da + \int\limits_{B(\delta,\varrho)} \boldsymbol{b} \cdot \tilde{\boldsymbol{u}}\, dv \\ &= \int\limits_{B(\delta,\varrho)} \tilde{\boldsymbol{S}} \cdot \boldsymbol{E}\, dv = \int\limits_{B(\delta,\varrho)} \boldsymbol{S} \cdot \tilde{\boldsymbol{E}}\, dv. \end{aligned} \tag{f}$$

Further, by (e), if δ and ϱ are sufficiently small,

$$\int\limits_{\partial B(\delta,\varrho)} = \int\limits_{\mathscr{C}(\varrho)} + \sum_{n=1}^{N} \int\limits_{\mathscr{S}_{\varrho_n}(\xi_n)} + \sum_{\xi \in B_{\boldsymbol{l}\tilde{\boldsymbol{l}}}} \int\limits_{\mathscr{S}_\delta(\xi)}, \tag{g}$$

where we have used the notation given in (e) in the proof of ***(2)***.

Since $\mathbf{C}$ is invertible on $\bar{B}$, we conclude from (ii) that $\tilde{\boldsymbol{E}}$ is $O(r^{-2})$ as the distance r from any of its singularities tends to zero. Thus, since $D_{\boldsymbol{l}} \cap D_{\tilde{\boldsymbol{l}}} = \emptyset$, (ii) implies that $\boldsymbol{S} \cdot \tilde{\boldsymbol{E}}$ is also $O(r^{-2})$ near its singularities. Clearly, the same property holds for $\tilde{\boldsymbol{S}} \cdot \boldsymbol{E}$. Further, since $\boldsymbol{b}$ and $\tilde{\boldsymbol{b}}$ are continuous on $\bar{B}$, $\boldsymbol{b} \cdot \tilde{\boldsymbol{u}}$ and $\tilde{\boldsymbol{b}} \cdot \boldsymbol{u}$ are $O(r^{-1})$ near their singularities. Thus as $\delta \to 0$ and $\varrho \to \mathbf{0}$ all of the volume integrals in (f) tend to the corresponding integrals over B. Therefore, if we pass to this limit in (f) and use (g), (c), and (d), we arrive at the desired result. □

(4) Somigliana's theorem.[1] *Assume that the body is homogeneous and isotropic with Lamé moduli that satisfy $\mu \neq 0$, $2\mu + 3\lambda \neq 0$. Let $[\boldsymbol{u}, \boldsymbol{E}, \boldsymbol{S}]$ be a singular elastic state corresponding to the external force system $[\boldsymbol{b}, \mathbf{s}, \boldsymbol{l}]$. Then*[2]

$$\boldsymbol{u}(\boldsymbol{y}) = \int\limits_{\partial B} (\boldsymbol{u}_{\boldsymbol{y}}[\mathbf{s}] - \mathbf{s}^*_{\boldsymbol{y}}[\boldsymbol{u}])\, da + \int\limits_{B} \boldsymbol{u}_{\boldsymbol{y}}[\boldsymbol{b}]\, dv + \sum_{\xi \in D_{\boldsymbol{l}}} \boldsymbol{u}_{\boldsymbol{y}}[\boldsymbol{l}(\xi)](\xi),$$

or equivalently,

$$u_i(\boldsymbol{y}) = \int\limits_{\partial B} (\boldsymbol{u}^i_{\boldsymbol{y}} \cdot \mathbf{s} - \mathbf{s}^i_{\boldsymbol{y}} \cdot \boldsymbol{u})\, da + \int\limits_{B} \boldsymbol{u}^i_{\boldsymbol{y}} \cdot \boldsymbol{b}\, dv + \sum_{\xi \in D_{\boldsymbol{l}}} \boldsymbol{u}^i_{\boldsymbol{y}}(\xi) \cdot \boldsymbol{l}(\xi),$$

---

[1] Somigliana [1885, *4*], [1886, *4*], [1889, *3*] for a regular elastic state. See also Dougall [1898, *4*]. The extension to singular states is due to Turteltaub and Sternberg [1968, *14*], pp. 236–237. Somigliana's theorem was utilized to determine upper and lower bounds on elastic states by Diaz and Greenberg [1948, *1*], Synge [1950, *12*], Washizu [1953, *20*] and Bramble and Payne [1961, *6*]. Kanwal [1969, *4*] used Somigliana's theorem to derive integral equations analogous to those used to study Stokes flow problems in hydrodynamics. See also Maiti and Makan [1971, *3*].

[2] The properties of the Kelvin state in conjunction with this result furnish an alternative proof of the fact that $\boldsymbol{u}$ is of class C^∞ on B.

for $\boldsymbol{y} \in B - D_l$, *where for any vector* $\boldsymbol{v}$, $\boldsymbol{u}_{\boldsymbol{y}}[\boldsymbol{v}]$ *is the displacement field and* $\mathbf{s}^*_{\boldsymbol{y}}[\boldsymbol{v}]$ *the adjoint traction field corresponding to the Kelvin state*[1] $\mathfrak{s}_{\boldsymbol{y}}[\boldsymbol{v}]$, *while* $\boldsymbol{u}^i_{\boldsymbol{y}}$ *and* $\mathbf{s}^i_{\boldsymbol{y}}$ *are the displacement and traction fields for the unit Kelvin state*[2] $\mathfrak{s}^i_{\boldsymbol{y}}$.

Proof. Let $\boldsymbol{v}$ be an arbitrary vector. Then the reciprocal theorem ***(3)*** applied to the states $\mathfrak{s} = [\boldsymbol{u}, \boldsymbol{E}, \boldsymbol{S}]$ and $\mathfrak{s}_{\boldsymbol{y}}[\boldsymbol{v}]$ yields

$$\boldsymbol{u}(\boldsymbol{y}) \cdot \boldsymbol{v} = \int_{\partial B} (\mathbf{s} \cdot \boldsymbol{u}_{\boldsymbol{y}}[\boldsymbol{v}] - \boldsymbol{u} \cdot \mathbf{s}_{\boldsymbol{y}}[\boldsymbol{v}])\, da + \int_B \boldsymbol{b} \cdot \boldsymbol{u}_{\boldsymbol{y}}[\boldsymbol{v}]\, dv + \sum_{\xi \in D_l} \boldsymbol{l}(\xi) \cdot \boldsymbol{u}_{\boldsymbol{y}}[\boldsymbol{v}](\xi).$$

In view of (iv) of ***(51.2)***, this relation implies the first formula in ***(4)***. On the other hand, if we take $\boldsymbol{v} = \boldsymbol{e}_i$ in the above relation, we are led to the second formula. □

As a corollary of Somigliana's theorem we have the

(5) Integral identity for the displacement gradient.[3] *Assume that the hypotheses of **(4)** hold. Then for every* $\boldsymbol{y} \in B - D_l$,

$$u_{i,j}(\boldsymbol{y}) = -\int_{\partial B} (\bar{\boldsymbol{u}}^{ij}_{\boldsymbol{y}} \cdot \mathbf{s} - \bar{\mathbf{s}}^{ij}_{\boldsymbol{y}} \cdot \boldsymbol{u})\, da - \int_B \bar{\boldsymbol{u}}^{ij}_{\boldsymbol{y}} \cdot \boldsymbol{b}\, dv - \sum_{\xi \in D_l} \bar{\boldsymbol{u}}^{ij}_{\boldsymbol{y}}(\xi) \cdot \boldsymbol{l}(\xi),$$

$$\operatorname{div} \boldsymbol{u}(\boldsymbol{y}) = -\int_{\partial B} (\mathring{\boldsymbol{u}}_{\boldsymbol{y}} \cdot \mathbf{s} - \mathring{\mathbf{s}}_{\boldsymbol{y}} \cdot \boldsymbol{u})\, da - \int_B \mathring{\boldsymbol{u}}_{\boldsymbol{y}} \cdot \boldsymbol{b}\, dv - \sum_{\xi \in D_l} \mathring{\boldsymbol{u}}_{\boldsymbol{y}}(\xi) \cdot \boldsymbol{l}(\xi),$$

$$\tfrac{1}{2}[\operatorname{curl} \boldsymbol{u}(\boldsymbol{y})]_i = \int_{\partial B} (\mathring{\boldsymbol{u}}^i_{\boldsymbol{y}} \cdot \mathbf{s} - \mathring{\mathbf{s}}^i_{\boldsymbol{y}} \cdot \boldsymbol{u})\, da + \int_B \mathring{\boldsymbol{u}}^i_{\boldsymbol{y}} \cdot \boldsymbol{b}\, dv + \sum_{\xi \in D_l} \mathring{\boldsymbol{u}}^i_{\boldsymbol{y}}(\xi) \cdot \boldsymbol{l}(\xi),$$

where $\bar{\boldsymbol{u}}^{ij}_{\boldsymbol{y}}$ *and* $\bar{\mathbf{s}}^{ij}_{\boldsymbol{y}}$ *are the displacement and traction fields for the unit doublet state*[4] $\bar{\mathfrak{s}}^{ij}_{\boldsymbol{y}}$, $\mathring{\boldsymbol{u}}_{\boldsymbol{y}}$ *and* $\mathring{\mathbf{s}}_{\boldsymbol{y}}$ *for the center of compression*[5] $\mathring{\mathfrak{s}}_{\boldsymbol{y}}$, *and* $\mathring{\boldsymbol{u}}^i_{\boldsymbol{y}}$ *and* $\mathring{\mathbf{s}}^i_{\boldsymbol{y}}$ *for the center of rotation*[6] $\mathring{\mathfrak{s}}^i_{\boldsymbol{y}}$.

Proof. If we differentiate the second relation in ***(4)*** with respect to y_j, we arrive at[7]

$$u_{i,j}(\boldsymbol{y}) = \int_{\partial B} \left(\frac{\partial}{\partial y_j} \boldsymbol{u}^i_{\boldsymbol{y}} \cdot \mathbf{s} - \frac{\partial}{\partial y_j} \mathbf{s}^i_{\boldsymbol{y}} \cdot \boldsymbol{u}\right) da + \int_B \frac{\partial}{\partial y_j} \boldsymbol{u}^i_{\boldsymbol{y}} \cdot \boldsymbol{b}\, dv + \sum_{\xi \in D_l} \frac{\partial}{\partial y_j} \boldsymbol{u}^i_{\boldsymbol{y}}(\xi) \cdot \boldsymbol{l}(\xi).$$

In view of (v) of ***(51.3)*** and the definition of unit doublet states given on page 178, this relation implies the first formula in ***(5)***. The second and third formulae are immediate consequences of the first. □

It is a simple matter to extend the mean strain and volume change theorems ***(29.2)*** and ***(29.3)*** to include concentrated loads. In particular, the formula for the volume change in a homogeneous and isotropic body becomes

$$\delta v(B) = \frac{1}{3k}\left\{\int_{\partial B} \boldsymbol{p} \cdot \mathbf{s}\, da + \int_B \boldsymbol{p} \cdot \boldsymbol{b}\, dv + \sum_{\xi \in D_l} (\xi - 0) \cdot \boldsymbol{l}(\xi)\right\},$$

where k is the modulus of compression.

---

[1] See p. 174.

[2] See p. 178.

[3] The first identity is apparently due to LOVE [1927, *3*], § 169. The formulae for the divergence and curl of $\boldsymbol{u}$ are due to BETTI [1872, *1*], §§ 8, 9. See also DOUGALL [1898, *4*], LOVE [1927, *3*], §§ 160, 162.

[4] See p. 178.

[5] See p. 179.

[6] See p. 179.

[7] The differentiation under the integral sign is easily justified. Cf. the proof (KELLOGG [1929, *1*], p. 151) of the differentiability under the integral sign of Newtonian potentials of volume distributions.

Suppose now that

$$\mathbf{s} = \mathbf{0} \quad \text{on } \partial B, \qquad \boldsymbol{b} = \mathbf{0} \quad \text{on } B,$$

and that the system $\boldsymbol{l}$ consists of a pair of concentrated loads. If we let $D_{\boldsymbol{l}} = \{\xi_1, \xi_2\}$, then it follows from balance of forces and moments *(2)* that

$$\boldsymbol{l}(\xi_1) = -\boldsymbol{l}(\xi_2) = l\,\boldsymbol{e},$$

$$\boldsymbol{e} = \frac{\xi_1 - \xi_2}{d}, \qquad d = |\xi_1 - \xi_2|;$$

and there results the following elegant formula for the volume change:

$$\delta v(B) = \frac{l d}{3k}.$$

Thus the *volume change due to a pair of concentrated loads is independent of the shape or volume of B; it depends only upon the magnitude of the loads, their distance apart, and the modulus of compression of the material.*

53. Integral representation of solutions to concentrated-load problems. In this section we will use the reciprocal theorem for singular states to establish an integral representation theorem for the mixed problem of elastostatics with concentrated loads included. Thus we assume given a *symmetric* and *invertible* elasticity field C on $\bar{B}$, surface displacements $\hat{\boldsymbol{u}}$ on $\mathscr{S}_1$, surface forces $\hat{\mathbf{s}}$ on $\mathscr{S}_2$, a *continuous* body force field $\boldsymbol{b}$ on $\bar{B}$, and a system of concentrated loads $\boldsymbol{l}$. The *generalized mixed problem* is to find a singular elastic state $[\boldsymbol{u}, \boldsymbol{E}, \boldsymbol{S}]$ that corresponds to the body force field $\boldsymbol{b}$ and to the system of concentrated loads $\boldsymbol{l}$, and that satisfies the boundary conditions

$$\boldsymbol{u} = \hat{\boldsymbol{u}} \quad \text{on } \mathscr{S}_1 - D_{\boldsymbol{l}}, \qquad \mathbf{s} = \hat{\mathbf{s}} \quad \text{on } \mathscr{S}_2 - D_{\boldsymbol{l}}.$$

We will call such a singular elastic state a ***solution of the generalized mixed problem.***

In order to introduce the notion of a Green's state we need the following

(1) Lemma. *Let $\boldsymbol{f}$ and $\boldsymbol{m}$ be given vectors. Then there exists a unique rigid displacement field $\boldsymbol{w}$ such that*

$$\int_{\partial B} \boldsymbol{w}\, da = \boldsymbol{f}, \qquad \int_{\partial B} \mathbf{p_c} \times \boldsymbol{w}\, da = \boldsymbol{m},$$

where $\mathbf{c}$ is the centroid of ∂B and $\mathbf{p_c}(\boldsymbol{x}) = \boldsymbol{x} - \mathbf{c}$. We call $\boldsymbol{w}$ the ***rigid field*** *on ∂B with* ***force*** *$\boldsymbol{f}$ and* ***moment*** *$\boldsymbol{m}$.*

Proof. Suppose such a field $\boldsymbol{w}$ exists and let

$$\boldsymbol{w} = \boldsymbol{a} + \boldsymbol{\omega} \times \mathbf{p_c}. \tag{a}$$

Since $\boldsymbol{c}$ is the centroid,

$$\int_{\partial B} \mathbf{p_c}\, da = \mathbf{0}, \tag{b}$$

and thus

$$\int_{\partial B} \boldsymbol{w}\, da = \alpha\, \boldsymbol{a}, \tag{c}$$

where α is the area of ∂B. Further, (a) and (b) also imply

$$\int_{\partial B} \mathbf{p_c} \times \boldsymbol{w}\, da = \int_{\partial B} \mathbf{p_c} \times (\boldsymbol{\omega} \times \mathbf{p_c})\, da = \boldsymbol{I}\, \boldsymbol{\omega}, \tag{d}$$

where $\boldsymbol{I}$ is the centroidal inertia tensor:

$$\boldsymbol{I} = \int_{\partial B} [|\boldsymbol{p}_c|^2 \mathbf{1} - \boldsymbol{p}_c \otimes \boldsymbol{p}_c] \, da. \tag{e}$$

Since $\boldsymbol{I}$ is invertible, (c) and (d) imply that if such a rigid field $\boldsymbol{w}$ exists, it is unique and

$$\boldsymbol{a} = \frac{\boldsymbol{f}}{\alpha}, \qquad \boldsymbol{\omega} = \boldsymbol{I}^{-1} \boldsymbol{m}. \tag{f}$$

Conversely, it is a simple matter to verify that $\boldsymbol{w}$ defined by (a) and (f) has all of the desired properties. □

Let $\boldsymbol{y} \in B$. We call $\tilde{\mathfrak{s}}_{\boldsymbol{y}}^i = [\tilde{\boldsymbol{u}}_{\boldsymbol{y}}^i, \tilde{\boldsymbol{E}}_{\boldsymbol{y}}^i, \tilde{\boldsymbol{S}}_{\boldsymbol{y}}^i]$ $(i = 1, 2, 3)$ ***Green's states*** at $\boldsymbol{y}$ for the mixed problem provided:

(i) $\tilde{\mathfrak{s}}_{\boldsymbol{y}}^i$ is a singular elastic state corresponding to vanishing body forces and to a concentrated load $\boldsymbol{e}_i$ at $\boldsymbol{y}$;

(ii) if $\mathscr{S}_1$ is not empty, then

$$\tilde{\boldsymbol{u}}_{\boldsymbol{y}}^i = \mathbf{0} \quad \text{on } \mathscr{S}_1, \qquad \tilde{\mathbf{s}}_{\boldsymbol{y}}^i = \mathbf{0} \quad \text{on } \mathscr{S}_2,$$

where $\tilde{\mathbf{s}}_{\boldsymbol{y}}^i$ is the surface traction field of $\tilde{\mathfrak{s}}_{\boldsymbol{y}}^i$;

(iii) if $\mathscr{S}_1$ is empty, then

$$\tilde{\mathbf{s}}_{\boldsymbol{y}}^i = \boldsymbol{w}_{\boldsymbol{y}}^i \qquad \text{on } \partial B,$$

where $\boldsymbol{w}_{\boldsymbol{y}}^i$ is the rigid field with force $-\boldsymbol{e}_i$ and moment $-(\boldsymbol{y} - \boldsymbol{c}) \times \boldsymbol{e}_i$.

The above boundary condition insures that balance of forces and moments are satisfied when $\mathscr{S}_1$ is empty. Indeed, by ***(52.2)*** and (i), the Green's state $\tilde{\mathfrak{s}}_{\boldsymbol{y}}^i$ must satisfy

$$\int_{\partial B} \tilde{\mathbf{s}}_{\boldsymbol{y}}^i \, da + \boldsymbol{e}_i = \mathbf{0},$$

$$\int_{\partial B} \boldsymbol{p}_c \times \tilde{\mathbf{s}}_{\boldsymbol{y}}^i da + (\boldsymbol{y} - \boldsymbol{c}) \times \boldsymbol{e}_i = \mathbf{0};$$

that $\tilde{\mathfrak{s}}_{\boldsymbol{y}}^i$ is consistent with these results follows from ***(1)*** and (iii).

For a homogeneous and isotropic body, the Green's states $\tilde{\mathfrak{s}}_{\boldsymbol{y}}^i$ may be constructed as follows:

$$\tilde{\mathfrak{s}}_{\boldsymbol{y}}^i = \mathfrak{s}_{\boldsymbol{y}}^i + \overset{*}{\mathfrak{s}}{}_{\boldsymbol{y}}^i,$$

where $\mathfrak{s}_{\boldsymbol{y}}^i$ is the unit Kelvin state corresponding to a concentrated load $\boldsymbol{e}_i$ at $\boldsymbol{y}$, and $\overset{*}{\mathfrak{s}}{}_{\boldsymbol{y}}^i$ is a regular elastic state on $\bar{B}$ corresponding to zero body forces and chosen so that the boundary conditions (ii) and (iii) for $\tilde{\mathfrak{s}}_{\boldsymbol{y}}^i$ are satisfied.

We say that an integrable vector field $\boldsymbol{u}$ on ∂B is ***normalized*** if

$$\int_{\partial B} \boldsymbol{u} \, da = \mathbf{0}, \qquad \int_{\partial B} \boldsymbol{p}_c \times \boldsymbol{u} \, da = \mathbf{0}.$$

Given a solution $\boldsymbol{u}$ of the traction problem $(\mathscr{S}_1 = \emptyset)$, the field $\boldsymbol{u} + \boldsymbol{w}$ with $\boldsymbol{w}$ rigid is also a solution. By ***(1)*** there exists a unique rigid $\boldsymbol{w}$ such that $\boldsymbol{u} + \boldsymbol{w}$ is normalized. Thus we may always assume, without loss in generality, that solutions of the traction problem are normalized.

(2) Integral representation theorem.[1] *Let* $[\tilde{\boldsymbol{u}}^i_{\boldsymbol{y}}, \tilde{\boldsymbol{E}}^i_{\boldsymbol{y}}, \tilde{\boldsymbol{S}}^i_{\boldsymbol{y}}]$ $(i=1, 2, 3)$ *be Green's states at* $\boldsymbol{y}\in B$ *for the mixed problem. Further, let* $[\boldsymbol{u}, \boldsymbol{E}, \boldsymbol{S}]$ *be a solution of the generalized mixed problem with* $\boldsymbol{u}$ *normalized if* $\mathscr{S}_1$ *is empty. Then for* $\boldsymbol{y}\notin D_{\boldsymbol{l}}$,

$$u_i(\boldsymbol{y}) = -\int_{\mathscr{S}_1} \tilde{\mathbf{s}}^i_{\boldsymbol{y}} \cdot \hat{\boldsymbol{u}}\, da + \int_{\mathscr{S}_2} \tilde{\boldsymbol{u}}^i_{\boldsymbol{y}} \cdot \hat{\mathbf{s}}\, da + \int_B \tilde{\boldsymbol{u}}^i_{\boldsymbol{y}} \cdot \boldsymbol{b}\, dv + \sum_{\boldsymbol{\xi}\in D_{\boldsymbol{l}}} \tilde{\boldsymbol{u}}^i_{\boldsymbol{y}}(\boldsymbol{\xi}) \cdot \boldsymbol{l}(\boldsymbol{\xi}).$$

Proof. Assume first that $\mathscr{S}_1$ is not empty. If we apply the reciprocal theorem ***(52.3)*** to the states $\mathfrak{s}$ and $\tilde{\mathfrak{s}}^i_{\boldsymbol{y}}$ and use (i) and (ii) in the definition of $\tilde{\mathfrak{s}}^i_{\boldsymbol{y}}$ together with the boundary conditions satisfied by $\mathfrak{s}$, we find that

$$\int_{\mathscr{S}_1} \tilde{\mathbf{s}}^i_{\boldsymbol{y}} \cdot \hat{\boldsymbol{u}}\, da + u_i(\boldsymbol{y}) = \int_{\mathscr{S}_2} \hat{\mathbf{s}} \cdot \tilde{\boldsymbol{u}}^i_{\boldsymbol{y}}\, da + \int_B \boldsymbol{b} \cdot \tilde{\boldsymbol{u}}^i_{\boldsymbol{y}}\, dv + \sum_{\boldsymbol{\xi}\in D_{\boldsymbol{l}}} \boldsymbol{l}(\boldsymbol{\xi}) \cdot \tilde{\boldsymbol{u}}^i_{\boldsymbol{y}}(\boldsymbol{\xi}),$$

which implies the above expression for $u_i(\boldsymbol{y})$.

If $\mathscr{S}_1$ is empty, we conclude from (iii) in the definition of Green's states that there exist vectors $\boldsymbol{a} = \boldsymbol{a}^i_{\boldsymbol{y}}$ and $\boldsymbol{\omega} = \boldsymbol{\omega}^i_{\boldsymbol{y}}$ such that

$$\tilde{\mathbf{s}}^i_{\boldsymbol{y}} = \boldsymbol{a} + \boldsymbol{\omega}\times\boldsymbol{p}_{\boldsymbol{c}} \qquad \text{on } \partial B.$$

Thus, since $\boldsymbol{u}$ is normalized,

$$\int_{\partial B} \tilde{\mathbf{s}}^i_{\boldsymbol{y}} \cdot \boldsymbol{u}\, da = \int_{\partial B} (\boldsymbol{a} + \boldsymbol{\omega}\times\boldsymbol{p}_{\boldsymbol{c}}) \cdot \boldsymbol{u}\, da = \boldsymbol{a} \cdot \int_{\partial B} \boldsymbol{u}\, da + \boldsymbol{\omega} \cdot \int_{\partial B} \boldsymbol{p}_{\boldsymbol{c}}\times\boldsymbol{u}\, da = 0,$$

and the reciprocal theorem again implies the desired result. □

(3) Symmetry of the Green's states. *Let* $\tilde{\boldsymbol{u}}^j_{\boldsymbol{x}}$ *and* $\tilde{\boldsymbol{u}}^i_{\boldsymbol{y}}$ *be displacement fields corresponding to Green's states at* $\boldsymbol{x}, \boldsymbol{y}\in B$ $(\boldsymbol{x}\neq\boldsymbol{y})$ *for the mixed problem. Further, if* $\mathscr{S}_1$ *is empty, assume that* $\tilde{\boldsymbol{u}}^j_{\boldsymbol{x}}$ *and* $\tilde{\boldsymbol{u}}^i_{\boldsymbol{y}}$ *are both normalized. Then*

$$\tilde{\boldsymbol{u}}^j_{\boldsymbol{x}}(\boldsymbol{y}) \cdot \boldsymbol{e}_i = \tilde{\boldsymbol{u}}^i_{\boldsymbol{y}}(\boldsymbol{x}) \cdot \boldsymbol{e}_j.$$

Proof. We simply apply the integral representation theorem ***(2)*** to the state $[\boldsymbol{u}, \boldsymbol{E}, \boldsymbol{S}] = \tilde{\mathfrak{s}}^j_{\boldsymbol{x}}$ and use properties (i)–(iii) of Green's states. □

In the presence of sufficient smoothness one can deduce an integral representation theorem for the displacement *gradient* (and hence also for the strain and stress fields) by differentiating the relations in ***(2)*** under the integral sign. An alternative approach is furnished by the

(4) Representation theorem for the displacement gradient. *Assume that B is homogeneous and isotropic. Choose* $\boldsymbol{y}\in B$ *and let*

$$\tilde{\mathfrak{s}}^{ij}_{\boldsymbol{y}} = \overset{*}{\mathfrak{s}}{}^{ij}_{\boldsymbol{y}} - \bar{\mathfrak{s}}^{ij}_{\boldsymbol{y}},$$

[1] A slightly different version of this theorem for the displacement and surface force problems without concentrated loads was given by LAURICELLA [1895, *1*], who attributed the method to VOLTERRA (cf. the earlier work of SOMIGLIANA discussed in footnote 1 on p. 183). A precise statement of the Lauricella-Volterra theorem is contained in the work of STERNBERG and EUBANKS [1955, *13*]. In [1895, *1*] and [1955, *13*] the equilibration of the concentrated load $\boldsymbol{e}_i$ at $\boldsymbol{y}$ in the Green's state $\tilde{\mathfrak{s}}^i_{\boldsymbol{y}}$ when $\mathscr{S}_1$ is empty is effected through the introduction of a second internal singularity; the normalization of $\boldsymbol{u}$ is achieved by requiring the displacements and rotations to vanish at the location of this added singularity. The method used here to equilibrate the concentrated load and the normalization procedure for $\boldsymbol{u}$ is due to BERGMAN and SCHIFFER [1953, *3*], pp. 223–224 (cf. TURTELTAUB and STERNBERG [1968, *14*]). For the surface force problem with concentrated loads this theorem is due to TURTELTAUB and STERNBERG [1968, *14*]. See also DOUGALL [1898, *4*], [1904, *1*], § 32; LICHTENSTEIN [1924, *1*]; KORN [1927, *2*], § 14; LOVE [1927, *3*], § 169; TREFFTZ [1928, *3*], § 49; ARZHANYH [1950, *2*], [1951, *2*], [1953, *1*], [1954, *2*]; RADZHABOV [1966, *20*]; IDENBOM and ORLOV [1968, *9*].

where:

(i) $\bar{\mathfrak{s}}_{\boldsymbol{y}}^{ij}$ *are the unit doublet states*[1] *at* $\boldsymbol{y}$;

(ii) $\overset{*}{\mathfrak{s}}{}_{\boldsymbol{y}}^{ij}$ *are regular elastic states corresponding to zero body forces*;

(iii) *if* $\mathscr{S}_1$ *is not empty, then*

$$\tilde{\boldsymbol{u}}_{\boldsymbol{y}}^{ij} = \mathbf{0} \quad on\ \mathscr{S}_1, \qquad \tilde{\mathbf{s}}_{\boldsymbol{y}}^{ij} = \mathbf{0} \quad on\ \mathscr{S}_2,$$

where $\tilde{\boldsymbol{u}}_{\boldsymbol{y}}^{ij}$ *is the displacement field and* $\tilde{\mathbf{s}}_{\boldsymbol{y}}^{ij}$ *the traction field of* $\tilde{\mathfrak{s}}_{\boldsymbol{y}}^{ij}$;

(iv) *if* $\mathscr{S}_1$ *is empty, then*

$$\tilde{\mathbf{s}}_{\boldsymbol{y}}^{ij} = \boldsymbol{w}_{\boldsymbol{y}}^{ij} \quad on\ \partial B,$$

where $\boldsymbol{w}_{\boldsymbol{y}}^{ij}$ *is a rigid field.*

Let $[\boldsymbol{u}, \boldsymbol{E}, \boldsymbol{S}]$ *be a solution of the generalized mixed problem with* $\boldsymbol{u}$ *normalized if* $\mathscr{S}_1$ *is empty. Then for* $\boldsymbol{y} \notin D_{\boldsymbol{l}}$,

$$u_{i,j}(\boldsymbol{y}) = -\int_{\mathscr{S}_1} \tilde{\mathbf{s}}_{\boldsymbol{y}}^{ij} \cdot \hat{\boldsymbol{u}}\, da + \int_{\mathscr{S}_2} \tilde{\boldsymbol{u}}_{\boldsymbol{y}}^{ij} \cdot \hat{\mathbf{s}}\, da + \int_B \tilde{\boldsymbol{u}}_{\boldsymbol{y}}^{ij} \cdot \boldsymbol{b}\, dv + \sum_{\boldsymbol{\xi} \in D_{\boldsymbol{l}}} \tilde{\boldsymbol{u}}_{\boldsymbol{y}}^{ij}(\boldsymbol{\xi}) \cdot \boldsymbol{l}(\boldsymbol{\xi}).$$

Proof. By (ii) and the reciprocal theorem ***(52.3)***,

$$\int_{\partial B} (\mathbf{s} \cdot \overset{*}{\boldsymbol{u}}{}_{\boldsymbol{y}}^{ij} - \overset{*}{\mathbf{s}}{}_{\boldsymbol{y}}^{ij} \cdot \boldsymbol{u})\, da + \int_B \boldsymbol{b} \cdot \overset{*}{\boldsymbol{u}}{}_{\boldsymbol{y}}^{ij}\, dv + \sum_{\boldsymbol{\xi} \in D_{\boldsymbol{l}}} \boldsymbol{l}(\boldsymbol{\xi}) \cdot \overset{*}{\boldsymbol{u}}{}_{\boldsymbol{y}}^{ij}(\boldsymbol{\xi}) = 0. \tag{a}$$

Since $\tilde{\mathfrak{s}}_{\boldsymbol{y}}^{ij} = \overset{*}{\mathfrak{s}}{}_{\boldsymbol{y}}^{ij} - \bar{\mathfrak{s}}_{\boldsymbol{y}}^{ij}$, if we add (a) to the right-hand side of the first relation in ***(52.5)***, we find that

$$u_{i,j}(\boldsymbol{y}) = \int_{\partial B} (\mathbf{s} \cdot \tilde{\boldsymbol{u}}_{\boldsymbol{y}}^{ij} - \tilde{\mathbf{s}}_{\boldsymbol{y}}^{ij} \cdot \boldsymbol{u})\, da + \int_B \boldsymbol{b} \cdot \tilde{\boldsymbol{u}}_{\boldsymbol{y}}^{ij}\, dv + \sum_{\boldsymbol{\xi} \in D_{\boldsymbol{l}}} \boldsymbol{l}(\boldsymbol{\xi}) \cdot \tilde{\boldsymbol{u}}_{\boldsymbol{y}}^{ij}(\boldsymbol{\xi}). \tag{b}$$

It follows from (iii) that when $\mathscr{S}_1$ is not empty (b) reduces to the required expression for $u_{i,j}(\boldsymbol{y})$. If $\mathscr{S}_1$ is empty, then an argument identical to that utilized in the proof of ***(2)*** can be used to verify that

$$\int_{\partial B} \tilde{\mathbf{s}}_{\boldsymbol{y}}^{ij} \cdot \boldsymbol{u}\, da = 0; \tag{c}$$

and (b) and (c) yield the desired relation for $u_{i,j}(\boldsymbol{y})$. □

The rigid field $\boldsymbol{w}_{\boldsymbol{y}}^{ij}$ of (iv) is not arbitrary. Indeed, in view of (i) and (ii), $\boldsymbol{w}_{\boldsymbol{y}}^{ij}$ must obey balance of forces and moments on $B - \Sigma_\varrho(\boldsymbol{y})$. If we apply these laws and then let $\varrho \to 0$, we conclude, with the aid of ***(51.3)***, that

$$\int_{\partial B} \tilde{\mathbf{s}}_{\boldsymbol{y}}^{ij}\, da = \mathbf{0},$$

$$\int_{\partial B} \boldsymbol{r} \times \tilde{\mathbf{s}}^{ij}\, da = \boldsymbol{e}_i \times \boldsymbol{e}_j.$$

Thus $\boldsymbol{w}_{\boldsymbol{y}}^{ij}$ *must be the rigid field with force equal to zero and moment equal to* $\boldsymbol{e}_i \times \boldsymbol{e}_j$.

The proof of the next theorem, which we omit, is completely analogous to the proof of ***(4)*** and is based on the relations for $\operatorname{div} \boldsymbol{u}(\boldsymbol{y})$ and $\operatorname{curl} \boldsymbol{u}(\boldsymbol{y})$ given in ***(52.5)***.

(5) Representation theorem for the dilatation and rotation.[2] *Assume that* B *is homogeneous and isotropic. Choose* $\boldsymbol{y} \in B$ *and let*

$$\tilde{\mathfrak{s}}_{\boldsymbol{y}} = \overset{*}{\mathfrak{s}}_{\boldsymbol{y}} - \overset{\circ}{\mathfrak{s}}_{\boldsymbol{y}}, \qquad \tilde{\mathfrak{s}}_{\boldsymbol{y}}^{i} = \overset{*}{\mathfrak{s}}{}_{\boldsymbol{y}}^{i} - \overset{\circ}{\mathfrak{s}}{}_{\boldsymbol{y}}^{i},$$

---

[1] See p. 178.

[2] BETTI [1872, *1*]. See also LOVE [1927, *3*], §§ 161, 163.

where:

(i) $\overset{\circ}{\mathfrak{s}}_{\boldsymbol{y}}$ *is the center of compression at* $\boldsymbol{y}$ *and* $\overset{\circ i}{\mathfrak{s}}_{\boldsymbol{y}}$ *the center of rotation at* $\boldsymbol{y}$ *parallel to the* x_i*-axis;*[1]

(ii) $\overset{*}{\mathfrak{s}}_{\boldsymbol{y}}$ and $\overset{*i}{\mathfrak{s}}_{\boldsymbol{y}}$ *are regular elastic states corresponding to zero body forces;*

(iii) *if* $\mathscr{S}_1$ *is not empty, then*

$$\tilde{\boldsymbol{u}}_{\boldsymbol{y}} = \boldsymbol{0} \quad on\ \mathscr{S}_1, \qquad \tilde{\mathbf{s}}_{\boldsymbol{y}} = \boldsymbol{0} \quad on\ \mathscr{S}_2,$$
$$\tilde{\boldsymbol{u}}^i_{\boldsymbol{y}} = \boldsymbol{0} \quad on\ \mathscr{S}_1, \qquad \tilde{\mathbf{s}}^i_{\boldsymbol{y}} = \boldsymbol{0} \quad on\ \mathscr{S}_2,$$

where $\tilde{\boldsymbol{u}}_{\boldsymbol{y}}$ *and* $\tilde{\boldsymbol{u}}^i_{\boldsymbol{y}}$ *are the displacement fields and* $\tilde{\mathbf{s}}_{\boldsymbol{y}}$ *and* $\tilde{\mathbf{s}}^i_{\boldsymbol{y}}$ *the traction fields of* $\tilde{\mathfrak{s}}_{\boldsymbol{y}}$ *and* $\tilde{\mathfrak{s}}^i_{\boldsymbol{y}}$.

(iv) *if* $\mathscr{S}_1$ *is empty, then*

$$\tilde{\mathbf{s}}_{\boldsymbol{y}} = \boldsymbol{0} \qquad on\ \partial B,$$
$$\tilde{\mathbf{s}}^i_{\boldsymbol{y}} = \boldsymbol{w}^i_{\boldsymbol{y}} \qquad on\ \partial B,$$

where $\boldsymbol{w}^i_{\boldsymbol{y}}$ *is a rigid field.*

Let $[\boldsymbol{u}, \boldsymbol{E}, \boldsymbol{S}]$ *be a solution of the generalized mixed problem with* $\boldsymbol{u}$ *normalized if* $\mathscr{S}_1$ *is empty. Then for* $\boldsymbol{y} \notin D_{\boldsymbol{l}}$,

$$\operatorname{div} \boldsymbol{u}(\boldsymbol{y}) = -\int_{\mathscr{S}_1} \tilde{\mathbf{s}}_{\boldsymbol{y}} \cdot \hat{\boldsymbol{u}}\, da + \int_{\mathscr{S}_2} \tilde{\boldsymbol{u}}_{\boldsymbol{y}} \cdot \hat{\mathbf{s}}\, da + \int_B \tilde{\boldsymbol{u}}_{\boldsymbol{y}} \cdot \boldsymbol{b}\, dv + \sum_{\xi \in D_{\boldsymbol{l}}} \tilde{\boldsymbol{u}}_{\boldsymbol{y}}(\xi) \cdot \boldsymbol{l}(\xi),$$

$$\tfrac{1}{2}[\operatorname{curl} \boldsymbol{u}(\boldsymbol{y})]_i = \int_{\mathscr{S}_1} \tilde{\mathbf{s}}^i_{\boldsymbol{y}} \cdot \hat{\boldsymbol{u}}\, da - \int_{\mathscr{S}_2} \tilde{\boldsymbol{u}}^i_{\boldsymbol{y}} \cdot \hat{\mathbf{s}}\, da - \int_B \tilde{\boldsymbol{u}}^i_{\boldsymbol{y}} \cdot \boldsymbol{b}\, dv - \sum_{\xi \in D_{\boldsymbol{l}}} \tilde{\boldsymbol{u}}^i_{\boldsymbol{y}}(\xi) \cdot \boldsymbol{l}(\xi).$$

As before, the field $\boldsymbol{w}^i_{\boldsymbol{y}}$ of (iv) is not arbitrary; it follows from (i), (ii), and the relations on page 179 that $\boldsymbol{w}^i_{\boldsymbol{y}}$ *must be the rigid field with force equal to zero and moment equal to* $\mathbf{e}_i$.

In view of the last theorem, div $\boldsymbol{u}$ and curl $\boldsymbol{u}$ can be computed, at least in principle, from the data. Further, since the displacement equation of equilibrium is equivalent to the relation[2]

$$\Delta\left[\boldsymbol{u} + \frac{1}{2}\left(1 + \frac{\lambda}{\mu}\right)\mathbf{p} \operatorname{div} \boldsymbol{u}\right] = \boldsymbol{0},$$

the determination of $\boldsymbol{u}$ when div $\boldsymbol{u}$ is known is reduced to a problem in the theory of harmonic functions. Moreover, since the surface force field on ∂B can be written in the form[3]

$$\mathbf{s} = 2\mu \frac{\partial \boldsymbol{u}}{\partial \boldsymbol{n}} + \mu\, \boldsymbol{n} \times \operatorname{curl} \boldsymbol{u} + \lambda (\operatorname{div} \boldsymbol{u})\, \boldsymbol{n},$$

when div $\boldsymbol{u}$ and curl $\boldsymbol{u}$ have been found, the surface value of $\partial \boldsymbol{u}/\partial \boldsymbol{n}$ is known at points at which the surface force field is prescribed. This method of integration is due to BETTI.[4]

ERICKSEN'S theorem ***(33.1)*** extends, at once, to problems involving concentrated loads.

(6) *Assume that there exists a non-trivial solution of the generalized mixed problem with null data* ($\hat{\boldsymbol{u}} = \boldsymbol{0}, \hat{\mathbf{s}} = \boldsymbol{0}, \boldsymbol{b} = \boldsymbol{0}, \boldsymbol{l} = \boldsymbol{0}$). *Then there exists a continuous body force field* $\boldsymbol{b}$ *on* $\bar{B}$ *of class* C^2 *on* B *with the following property: the regular*

[1] See p. 179.
[2] See p. 91.
[3] See p. 93.
[4] [1872, *1*]. See also LOVE [1927, *3*], § 159.

mixed problem corresponding to this body force and to null boundary data ($\hat{\boldsymbol{u}}=\mathbf{0}, \hat{\mathbf{s}}=\mathbf{0}$) *has no solution. If* $\mathscr{S}_1$ *is empty,* $\boldsymbol{b}$ *can be chosen so as to satisfy*

$$\int_B \boldsymbol{b}\, dv = \int_B \boldsymbol{p}\times\boldsymbol{b}\, dv = \mathbf{0}.$$

In addition, there exists a system of concentrated loads $\boldsymbol{l}$ *such that the generalized mixed problem corresponding to* $\boldsymbol{l}$*, vanishing body forces, and null boundary data has no solution. If* $\mathscr{S}_1$ *is empty,* $\boldsymbol{l}$ *can be chosen so as to satisfy*

$$\sum_{\xi\in D_l} \boldsymbol{l}(\xi) = \sum_{\xi\in D_l} (\xi - 0)\times\boldsymbol{l}(\xi) = \mathbf{0}.$$

The proof of this theorem is based on the reciprocal theorem ***(52.3)***; it is strictly analogous to the proof of ***(33.1)*** and can safely be omitted.

As an immediate corollary of ***(6)*** we have the following theorem, due to TURTELTAUB and STERNBERG:[1] *If the regular mixed problem with null boundary data has a solution whenever the body force field is sufficiently smooth, then there is at most one solution to the generalized mixed problem.*

In the case of internal concentrated loads, a uniqueness theorem can be established that does not presuppose existence.

(7) Uniqueness theorem for problems involving internal concentrated loads. *Assume that the body is homogeneous and isotropic with positive definite elasticity tensor. Then any two solutions of the generalized mixed problem with internal concentrated loads are equal modulo a rigid displacement.*

Proof. Let $\mathfrak{s}$ denote the difference between two solutions. Then, by the decomposition theorem ***(52.1)***, $\mathfrak{s}$ is a *regular* elastic state corresponding to null data. Thus the uniqueness theorem ***(32.1)*** implies that $\mathfrak{s}=[\boldsymbol{w}, \mathbf{0}, \mathbf{0}]$ with $\boldsymbol{w}$ rigid. □

X. Saint-Venant's principle.

54. The v. Mises-Sternberg version of Saint-Venant's principle. SAINT-VENANT, in his great memoir[2] on torsion and flexure, studied the deformation of a cylindrical body loaded by surface forces distributed over its plane ends. In order to justify using his results as an approximation in situations involving other end loadings, SAINT-VENANT made a conjecture which may roughly be worded as follows: If two sets of loadings are statically equivalent at each end, then the difference in stress fields is negligible, except possibly near the ends; or equivalently, a system of loads having zero resultant force and moment at each end produces a stress field that is negligible away from the ends. This idea has since been elevated to the status of a general principle bearing the name of SAINT-VENANT. Its first general statement was given by BOUSSINESQ:[3] "An equilibrated system of external forces applied to an elastic body, all of the points of application lying within a given sphere, produces deformations of negligible magnitude at distances from the sphere which are sufficiently large compared to its radius."

As was first pointed out by v. MISES,[4] this formulation of Saint-Venant's principle, which has since become conventional, is in need of clarification, since

---

[1] [1968, *14*], Theorem 5.2. Actually, the TURTELTAUB-STERNBERG theorem is slightly different from the one given above; the idea, however, is the same.

[2] [1855, *1*].

[3] [1885, *1*], p. 298. "Des forces extérieures, qui se font équilibre sur un solide élastique et dont les points d'application se trouvent tous à l'intérieur d'une sphère donnée, ne produisent pas de déformations sensibles à des distances de cette sphère qui sont d'une certaine grandeur par rapport à son rayon."

[4] [1945, *5*].

the system of external forces must necessarily be in equilibrium, at least when the body is finite. Further, as was noted by STERNBERG,[1] in the case of an unbounded body loaded on a finite portion of its surface, the stresses are arbitrarily small at sufficiently large distances from the load region.[2]

The first of these criticisms led v. MISES[3] to suggest the following interpretation of the principle: "If the forces acting upon a body are restricted to several small parts of the surface, each included in a sphere of radius ϱ, then the ... stresses produced in the interior of the body ... are smaller in order of magnitude when the forces for each single part are in equilibrium than when they are not."

It is clear that this is the meaning intended by BOUSSINESQ. Indeed, in order to justify the principle, BOUSSINESQ examined the strains at an interior point of an elastic half-space subjected to *normal* concentrated loads upon its boundary. Assuming the points of application of the loads to lie within a sphere of radius ϱ, he showed that the order of magnitude of the strains is ϱ if the resultant force is zero, and ϱ^2 when the resultant moment also vanishes. Various arguments have since been given in support of the principle.[4]

v. MISES utilized two examples involving *tangential* as well as normal surface loads to demonstrate that the above version of Saint-Venant's principle cannot be valid without qualification. The two examples chosen by v. MISES are the three-dimensional problem of the half-space and the plane problem of the circular disk, each under concentrated surface loads.[5] Guided by these examples, v. MISES[6] conjectured a modified Saint-Venant principle which he stated as follows:

"(a) If a system of loads on an adequately supported body, all applied at surface points within a sphere of diameter ϱ, have the vector sum zero, they produce in an inner point $\boldsymbol{y}$ of the body a strain or stress value s of the order of magnitude ϱ.

(b) If the loads, in addition to having vector sum zero, ... form an equilibrium system within the sphere of diameter ϱ, the s-value produced in the point $\boldsymbol{y}$ will, in general, still be of the order of magnitude ϱ.

(c) If the loads, in addition to being an equilibrium system, ... form a system in astatic equilibrium, then the s-value produced in $\boldsymbol{y}$ will be of the order of magnitude ϱ^2 or smaller. In particular, if loads applied to a small area are parallel to each other and not tangential to the surface and if they form an equilibrium system, they are also in astatic equilibrium and thus lead to an s of the order ϱ^2."

The conjectures of v. MISES concerning concentrated surface loads were analyzed by STERNBERG, whose results are also valid for continuously distributed surface loads. In order to give a concise statement of STERNBERG'S theorem, we introduce the following definitions:

We say that $\mathscr{F}_\varrho$, $0<\varrho\leqq\varrho_0$, is a ***family of load regions on*** ∂B ***contracting to*** $\boldsymbol{z}\in\partial B$ if:[7]

(i) $\mathscr{F}_\varrho=\partial B\cap\Sigma_\varrho(\boldsymbol{z})$ for every $\varrho\in(0,\varrho_0]$;

---

[1] [1954, *20*].

[2] TOUPIN [1965, *20*] gives several counterexamples to the conventional statement of Saint-Venant's principle.

[3] [1945, *5*].

[4] SOUTHWELL [1923, *2*], SUPINO [1931, *5*], GOODIER [1937, *1*, *2*], [1942, *2*], ZANABONI [1937, *6*, *7*, *8*], LOCATELLI [1940, *2*], DOU [1966, *7*].

[5] ERIM [1948, *3*] has applied v. MISES' analysis to the half-plane under concentrated loads.

[6] [1945, *5*].

[7] Cf. STERNBERG and AL-KHOZAI [1964, *20*], Def. 4.1.

(ii) there exists a class C^2 mapping $\hat{\boldsymbol{x}}: \mathscr{P} \to \mathscr{F}_{\varrho_0}$, where $\mathscr{P}$ is an open region in the *plane* containing the origin, such that

$$\boldsymbol{z} = \hat{\boldsymbol{x}}(\boldsymbol{0}), \qquad \frac{\partial \hat{\boldsymbol{x}}(\boldsymbol{\alpha})}{\partial \alpha_1} \times \frac{\partial \hat{\boldsymbol{x}}(\boldsymbol{\alpha})}{\partial \alpha_2}\bigg|_{\boldsymbol{\alpha}=\boldsymbol{0}} \neq \boldsymbol{0}.$$

Note that by (ii) the boundary of B possesses continuous curvatures near $\boldsymbol{z}$; thus this portion of the boundary is required to exhibit a higher degree of smoothness than that automatically assured by the assumption that B be a regular region.

For each $k \in \{1, 2, \ldots, K\}$ let $\mathscr{F}_\varrho(k)$, $0 < \varrho \leq \varrho_0$, be a family of load regions on ∂B contracting to a point $\boldsymbol{z}_k \in \partial B$. We say that $\mathfrak{s}_\varrho = [\boldsymbol{u}_\varrho, \boldsymbol{E}_\varrho, \boldsymbol{S}_\varrho]$, $0 < \varrho < \varrho_0$, is a ***family of singular elastic states corresponding to loads on*** $\mathscr{F}_\varrho(k)$ if given any $\varrho \in (0, \varrho_0)$:[1]

(iii) $\mathfrak{s}_\varrho$ is a singular elastic state corresponding to zero body forces, to surface tractions $\boldsymbol{s}_\varrho$, and to a system $\boldsymbol{l}_\varrho$ of concentrated loads;

(iv) the traction field $\boldsymbol{s}_\varrho$ vanishes on $\partial B - \bigcup_{k=1}^{K} \mathscr{F}_\varrho(k)$;

(v) $D_{\boldsymbol{l}_\varrho} \subset \left(\bigcup_{k=1}^{K} \mathscr{F}_\varrho(k) \right)$;

(vi) $|\boldsymbol{s}_\varrho(\boldsymbol{x})| \leq M_1$ for every $\boldsymbol{x} \in \partial B$ and $|\boldsymbol{l}_\varrho(\boldsymbol{\xi})| \leq M_2$ for every $\boldsymbol{\xi} \in D_{\boldsymbol{l}_\varrho}$, where the constants M_1 and M_2 are independent of ϱ;

(vii) $\boldsymbol{u}_\varrho$ and $\nabla \boldsymbol{u}_\varrho$ admit the representations

$$u_{\varrho i}(\boldsymbol{y}) = \int_{\partial B} \boldsymbol{s}_\varrho \cdot \tilde{\boldsymbol{u}}_{\boldsymbol{y}}^i \, da + \sum_{\boldsymbol{\xi} \in D_{\boldsymbol{l}_\varrho}} \boldsymbol{l}_\varrho(\boldsymbol{\xi}) \cdot \tilde{\boldsymbol{u}}_{\boldsymbol{y}}^i(\boldsymbol{\xi}),$$

$$u_{\varrho i,j}(\boldsymbol{y}) = \int_{\partial B} \boldsymbol{s}_\varrho \cdot \tilde{\boldsymbol{u}}_{\boldsymbol{y}}^{ij} \, da + \sum_{\boldsymbol{\xi} \in D_{\boldsymbol{l}_\varrho}} \boldsymbol{l}_\varrho(\boldsymbol{\xi}) \cdot \tilde{\boldsymbol{u}}_{\boldsymbol{y}}^{ij}(\boldsymbol{\xi}),$$

for every $\boldsymbol{y} \in B$, where for each $\boldsymbol{y} \in B$ the functions $\tilde{\boldsymbol{u}}_{\boldsymbol{y}}^i$ and $\tilde{\boldsymbol{u}}_{\boldsymbol{y}}^{ij}$ are continuous on $\bar{B} - \{\boldsymbol{y}\}$ and of class C^2 on $\mathscr{F}_{\varrho_0}(k)$ for $k = 1, 2, \ldots, K$.

Since $D_{\boldsymbol{l}_\varrho}$ is the set of points at which the concentrated loads of the system $\boldsymbol{l}_\varrho$ are acting, (v) is simply the requirement that there be no concentrated loads applied outside of the K load regions $\mathscr{F}_\varrho(k)$.

In view of (iii), (v), and the integral representation theorem ***(53.2)***, the first representation formula in (vii) is simply the requirement that the Green's states for the surface force problem exist. For a homogeneous and isotropic body, it is clear from the discussion on page 186 that such Green's states will exist if the surface force problem for arbitrary (equilibrated) surface tractions has a solution with sufficiently smooth displacements. In view of the integral representation theorem ***(53.4)***, this also insures that the assertion in (vii) concerning $\nabla \boldsymbol{u}_\varrho$ hold.

For convenience, we adopt the following notation:

$$\begin{aligned}
D_\varrho(k) &= D_{\boldsymbol{l}_\varrho} \cap \mathscr{F}_\varrho(k), \\
\boldsymbol{f}_\varrho(k) &= \int_{\mathscr{F}_\varrho(k)} \boldsymbol{s}_\varrho \, da + \sum_{\boldsymbol{\xi} \in D_\varrho(k)} \boldsymbol{l}_\varrho(\boldsymbol{\xi}), \\
\boldsymbol{m}_\varrho(k) &= \int_{\mathscr{F}_\varrho(k)} \boldsymbol{p} \times \boldsymbol{s}_\varrho \, da + \sum_{\boldsymbol{\xi} \in D_\varrho(k)} (\boldsymbol{\xi} - \boldsymbol{0}) \times \boldsymbol{l}_\varrho(\boldsymbol{\xi}), \\
\boldsymbol{M}_\varrho(k) &= \int_{\mathscr{F}_\varrho(k)} \boldsymbol{p} \otimes \boldsymbol{s}_\varrho \, da + \sum_{\boldsymbol{\xi} \in D_\varrho(k)} (\boldsymbol{\xi} - \boldsymbol{0}) \otimes \boldsymbol{l}_\varrho(\boldsymbol{\xi}),
\end{aligned}$$

[1] Cf. STERNBERG and AL-KHOZAI [1964, *20*], Def. 4.2.

so that $\boldsymbol{f}_\varrho(k)$ and $\boldsymbol{m}_\varrho(k)$ denote the resultant force and resultant moment about the origin of the surface tractions and concentrated loads on $\mathscr{F}_\varrho(k)$. We say that the ***force system is in equilibrium on each load region*** if

$$\boldsymbol{f}_\varrho(k) = \boldsymbol{m}_\varrho(k) = \boldsymbol{0}$$

for $k=1, 2, \ldots, K$ and $0<\varrho<\varrho_0$. Further, using the terminology introduced in Sect. 18, we say that the ***force system is in astatic equilibrium on each load region*** if given any orthogonal tensor $\boldsymbol{Q}$

$$\int_{\mathscr{F}_\varrho(k)} \boldsymbol{Q}\, \boldsymbol{s}_\varrho\, da + \sum_{\xi \in D_\varrho(k)} \boldsymbol{Q}\, \boldsymbol{l}_\varrho(\xi) = \boldsymbol{0},$$

$$\int_{\mathscr{F}_\varrho(k)} \boldsymbol{p} \times \boldsymbol{Q}\, \boldsymbol{s}_\varrho\, da + \sum_{\xi \in D_\varrho(k)} (\xi - 0) \times \boldsymbol{Q}\, \boldsymbol{l}_\varrho(\xi) = \boldsymbol{0}.$$

Taking $\boldsymbol{Q}=\boldsymbol{1}$, we see that astatic equilibrium implies equilibrium. In addition, it follows from theorem ***(18.9)*** that *the force system is in astatic equilibrium on each load region if and only if*

$$\boldsymbol{f}_\varrho(k) = \boldsymbol{0}, \qquad \boldsymbol{M}_\varrho(k) = \boldsymbol{0},$$

for $k=1, 2, \ldots, K$ and $0<\varrho<\varrho_0$. Finally, we say that the ***forces are parallel and non-tangential to the boundary*** if there exist scalar functions φ_ϱ on ∂B and ψ_ϱ on $D_{\boldsymbol{l}_\varrho}$ such that for each k,

$$\boldsymbol{s}_\varrho(\boldsymbol{x}) = \varphi_\varrho(\boldsymbol{x})\, \boldsymbol{a}_k, \qquad \boldsymbol{l}_\varrho(\xi) = \psi_\varrho(\xi)\, \boldsymbol{a}_k$$

for all $\boldsymbol{x} \in \mathscr{F}_\varrho(k)$, $\xi \in D_\varrho(k)$, and $\varrho \in (0, \varrho_0)$, where $\boldsymbol{a}_k$ is a unit vector that is not tangent to ∂B at $\boldsymbol{z}_k$.

(1) Sternberg's theorem.[1] *Let $\boldsymbol{z}_k$, $k=1, 2, \ldots, K$, be K distinct regular points of ∂B, and for each k let $\mathscr{F}_\varrho(k)$, $0<\varrho\leqq\varrho_0$, be a family of load regions on ∂B contracting to $\boldsymbol{z}_k$. Let $\mathfrak{s}_\varrho=[\boldsymbol{u}_\varrho, \boldsymbol{E}_\varrho, \boldsymbol{S}_\varrho]$, $0<\varrho<\varrho_0$, be a family of singular elastic states corresponding to loads on $\mathscr{F}_\varrho(k)$. Then given $\boldsymbol{y} \in B$:*

(a) $\boldsymbol{u}_\varrho(\boldsymbol{y}) = O(\varrho^\delta)$, $\boldsymbol{E}_\varrho(\boldsymbol{y}) = O(\varrho^\delta)$, $\boldsymbol{S}_\varrho(\boldsymbol{y}) = O(\varrho^\delta)$ *as* $\varrho\to 0$, *where* $\delta=0$;

(b) $\delta=1$ *if* $\boldsymbol{f}_\varrho(k)=\boldsymbol{0}$ *for* $k=1, 2, \ldots, K$ *and* $0<\varrho<\varrho_0$;

(c) $\delta=2$ *if on each load region the force system is in equilibrium with forces that are parallel and non-tangential to the boundary;*

(d) $\delta=2$ *if the force system is in astatic equilibrium on each load region.*

Moreover, if the concentrated loads all vanish (i.e. $\boldsymbol{l}_\varrho\equiv\boldsymbol{0}$), *which will be the case if $\mathfrak{s}_\varrho$ is regular for each ϱ, then:*

(a′) $\delta=2$ *(in general);*

(b′) $\delta=3$ *if* $\boldsymbol{f}_\varrho(k)\equiv\boldsymbol{0}$ *for* $k=1, 2, \ldots, K$ *and* $0<\varrho<\varrho_0$;

(c′) $\delta=4$ *if on each load region the force system is in equilibrium with forces that are parallel and non-tangential to the boundary;*

(d′) $\delta=4$ *if the force system is in astatic equilibrium on each load region.*

Proof. Fix $\boldsymbol{y} \in B$. In view of the strain-displacement and stress-strain relations, it clearly suffices to establish the above order of magnitude estimates for $\boldsymbol{u}_\varrho$ and $\nabla\boldsymbol{u}_\varrho$. Let u_ϱ denote either $u_{\varrho i}(\boldsymbol{y})$ or $u_{\varrho i,j}(\boldsymbol{y})$. Then by (vii) u_ϱ admits the representation

$$u_\varrho = \int_{\partial B} \boldsymbol{s}_\varrho \cdot \boldsymbol{g}\, da + \sum_{\xi \in D_{\boldsymbol{l}_\varrho}} \boldsymbol{l}_\varrho(\xi) \cdot \boldsymbol{g}(\xi), \tag{e}$$

[1] [1954, *20*]. See also Schumann [1954, *18*], Boley [1958, *3*], [1960, *2*], Sternberg [1960, *12*], Sternberg and Al-Khozai [1964, *20*], Keller [1965, *11*]. The last author discusses the form which the principle might be expected to take in the case of thin bodies.

where $\boldsymbol{g}$ denotes either $\tilde{\boldsymbol{u}}^i_{\boldsymbol{y}}$ or $\tilde{\boldsymbol{u}}^{ij}_{\boldsymbol{y}}$. Since the $\boldsymbol{z}_k$ are all distinct, there exists a $\varrho_1 \in (0, \varrho_0)$ such that the K sets $\mathscr{F}_\varrho(k)$ are mutually disjoint for each $\varrho \in (0, \varrho_1)$. Thus, in view of (iv) and (v),

$$u_\varrho = \sum_{k=1}^{K} v_\varrho(k), \tag{f}$$

where

$$v_\varrho(k) = \int_{\mathscr{F}_\varrho(k)} \boldsymbol{s}_\varrho \cdot \boldsymbol{g}\, da + \sum_{\boldsymbol{\xi} \in D_\varrho(k)} \boldsymbol{l}_\varrho(\boldsymbol{\xi}) \cdot \boldsymbol{g}(\boldsymbol{\xi}).$$

We now fix the integer k and, for convenience, write

$$v_\varrho = v_\varrho(k), \qquad \mathscr{F}_\varrho = \mathscr{F}_\varrho(k), \qquad D_\varrho = D_\varrho(k), \qquad \boldsymbol{z} = \boldsymbol{z}_k,$$

so that

$$v_\varrho = \int_{\mathscr{F}_\varrho} \boldsymbol{s}_\varrho \cdot \boldsymbol{g}\, da + \sum_{\boldsymbol{\xi} \in D_\varrho} \boldsymbol{l}_\varrho(\boldsymbol{\xi}) \cdot \boldsymbol{g}(\boldsymbol{\xi}). \tag{g}$$

Let $\hat{\boldsymbol{x}}: \mathscr{P} \to \mathscr{F}_{\varrho_0}$ be the function specified in (ii). It follows from the properties of $\hat{\boldsymbol{x}}$ that there exists an open set $\mathring{\mathscr{P}}$ in $\mathscr{P}$ containing the origin, a number $\varrho_2 \in (0, \varrho_1)$, and a mapping $\hat{\boldsymbol{\alpha}} = (\hat{\alpha}_1, \hat{\alpha}_2)$ of $\mathscr{F}_{\varrho_2}$ onto $\mathring{\mathscr{P}}$ such that $\hat{\boldsymbol{\alpha}}$ is the inverse of the restriction of $\boldsymbol{x}$ to $\mathring{\mathscr{P}}$. Further, it follows from the properties of $\hat{\boldsymbol{x}}$ that

$$\hat{\boldsymbol{\alpha}}(\boldsymbol{x}) = O(|\boldsymbol{x} - \boldsymbol{z}|) \quad \text{as} \quad \boldsymbol{x} \to \boldsymbol{z}. \tag{h}$$

Since the composite function $\boldsymbol{g} \circ \hat{\boldsymbol{x}}$ (with values $\boldsymbol{g}(\hat{\boldsymbol{x}}(\boldsymbol{\alpha}))$) is class C^2 on $\mathscr{P}$, if we expand this function in a Taylor series in $\boldsymbol{\alpha}$ about $\boldsymbol{\alpha} = \boldsymbol{0}$ and evaluate the result at $\boldsymbol{\alpha} = \hat{\boldsymbol{\alpha}}(\boldsymbol{x})$, we arrive at

$$\boldsymbol{g}(\boldsymbol{x}) = \boldsymbol{a}_0 + \boldsymbol{a}_1 \hat{\alpha}_1(\boldsymbol{x}) + \boldsymbol{a}_2 \hat{\alpha}_2(\boldsymbol{x}) + \boldsymbol{c}(\boldsymbol{x}) \tag{i}$$

for $\boldsymbol{x} \in \mathscr{F}_{\varrho_2}$, where $\boldsymbol{a}_0$, $\boldsymbol{a}_1$, and $\boldsymbol{a}_2$ are fixed vectors and, in view of (h), $\boldsymbol{c}$ is a class C^2 function on $\mathscr{F}_{\varrho_2}$ that satisfies

$$\boldsymbol{c}(\boldsymbol{x}) = O(|\boldsymbol{x} - \boldsymbol{z}|^2) \quad \text{as} \quad \boldsymbol{x} \to \boldsymbol{z}. \tag{j}$$

By (h)–(j) and (vi),

$$\begin{aligned} \int_{\mathscr{F}_\varrho} \boldsymbol{s}_\varrho \cdot \boldsymbol{g}\, da &= \boldsymbol{a}_0 \cdot \int_{\mathscr{F}_\varrho} \boldsymbol{s}_\varrho\, da + \sum_{\beta=1}^{2} \boldsymbol{a}_\beta \cdot \int_{\mathscr{F}_\varrho} \hat{\alpha}_\beta \boldsymbol{s}_\varrho\, da + O(\varrho^4), \\ &= \boldsymbol{a}_0 \cdot \int_{\mathscr{F}_\varrho} \boldsymbol{s}_\varrho\, da + O(\varrho^3), \\ &= O(\varrho^2), \end{aligned} \tag{k}$$

$$\begin{aligned} \sum_{\boldsymbol{\xi} \in D_\varrho} \boldsymbol{l}_\varrho(\boldsymbol{\xi}) \cdot \boldsymbol{g}(\boldsymbol{\xi}) &= \boldsymbol{a}_0 \cdot \sum_{\boldsymbol{\xi} \in D_\varrho} \boldsymbol{l}_\varrho(\boldsymbol{\xi}) + \sum_{\beta=1}^{2} \boldsymbol{a}_\beta \cdot \sum_{\boldsymbol{\xi} \in D_\varrho} \hat{\alpha}_\beta(\boldsymbol{\xi})\, \boldsymbol{l}_\varrho(\boldsymbol{\xi}) + O(\varrho^2), \\ &= \boldsymbol{a}_0 \cdot \sum_{\boldsymbol{\xi} \in D_\varrho} \boldsymbol{l}_\varrho(\boldsymbol{\xi}) + O(\varrho), \\ &= O(1) \end{aligned} \tag{l}$$

as $\varrho \to 0$. Conclusions (a) and (a') follow from (f), (g), $(\mathrm{k})_3$, and $(\mathrm{l})_3$. Further, it follows from the definition of $\boldsymbol{f}_\varrho = \boldsymbol{f}_\varrho(k)$ in conjunction with (f), (g), $(\mathrm{k})_2$, and $(\mathrm{l})_2$ that conclusions (b) and (b') are also valid.

To establish the remaining results we first expand the function $\hat{\boldsymbol{x}}$ in a Taylor series about $\boldsymbol{\alpha} = \boldsymbol{0}$:

$$\hat{\boldsymbol{x}}(\boldsymbol{\alpha}) = \boldsymbol{z} + \boldsymbol{d}_1 \alpha_1 + \boldsymbol{d}_2 \alpha_2 + \boldsymbol{h}(\boldsymbol{\alpha}), \tag{m}$$

where

$$\boldsymbol{d}_\beta = \left.\frac{\partial \hat{\boldsymbol{x}}(\boldsymbol{\alpha})}{\partial \alpha_\beta}\right|_{\boldsymbol{\alpha}=\boldsymbol{0}} \qquad (\beta = 1, 2), \tag{n}$$

and $\boldsymbol{h}$ is a class C^2 function on $\mathscr{P}$ that obeys

$$\boldsymbol{h}(\boldsymbol{\alpha}) = O(|\boldsymbol{\alpha}|^2) \quad \text{as} \quad \boldsymbol{\alpha} \to \mathbf{0}. \tag{o}$$

Since $\boldsymbol{p}(\boldsymbol{x}) = \boldsymbol{x} - \boldsymbol{0}$, we conclude from (m), (h), and (o) that

$$\begin{aligned}\int_{\mathscr{F}_\varrho} \boldsymbol{p} \otimes \boldsymbol{s}_\varrho \, da &= (\boldsymbol{z} - \boldsymbol{0}) \otimes \int_{\mathscr{F}_\varrho} \boldsymbol{s}_\varrho \, da + \sum_{\beta=1}^{2} \boldsymbol{d}_\beta \otimes \int_{\mathscr{F}_\varrho} \hat{\alpha}_\beta \, \boldsymbol{s}_\varrho \, da + O(\varrho^4), \\ \sum_{\boldsymbol{\xi} \in D_\varrho} (\boldsymbol{\xi} - \boldsymbol{0}) \otimes \boldsymbol{l}_\varrho(\boldsymbol{\xi}) &= (\boldsymbol{z} - \boldsymbol{0}) \otimes \sum_{\boldsymbol{\xi} \in D_\varrho} \boldsymbol{l}_\varrho(\boldsymbol{\xi}) + \sum_{\beta=1}^{2} \boldsymbol{d}_\beta \otimes \sum_{\boldsymbol{\xi} \in D_\varrho} \hat{\alpha}_\beta(\boldsymbol{\xi}) \, \boldsymbol{l}_\varrho(\boldsymbol{\xi}) + O(\varrho^2).\end{aligned} \tag{p}$$

If we assume that the force system is in astatic equilibrium on each load region, then

$$\boldsymbol{f}_\varrho \equiv \mathbf{0}, \qquad \boldsymbol{M}_\varrho \equiv \mathbf{0}, \tag{q}$$

and it follows from (p) and the definition of $\boldsymbol{f}_\varrho$ and $\boldsymbol{M}_\varrho$ that

$$\begin{aligned}\sum_{\beta=1}^{2} \boldsymbol{d}_\beta \otimes \boldsymbol{f}_{\beta\varrho} &= O(\varrho^2), \\ &= O(\varrho^4) \quad \text{if} \quad \boldsymbol{l}_\varrho \equiv \mathbf{0},\end{aligned} \tag{r}$$

where

$$\boldsymbol{f}_{\beta\varrho} = \int_{\mathscr{F}_\varrho} \hat{\alpha}_\beta \, \boldsymbol{s}_\varrho \, da + \sum_{\boldsymbol{\xi} \in D_\varrho} \hat{\alpha}_\beta(\boldsymbol{\xi}) \, \boldsymbol{l}_\varrho(\boldsymbol{\xi}). \tag{s}$$

Each of the relations in (r) may be regarded as a system of inhomogeneous linear equations in the two unknowns $\boldsymbol{f}_{\beta\varrho}$ $(\beta = 1, 2)$. It follows from (n) and the restrictions placed on $\hat{\boldsymbol{x}}$ in (ii) that the coefficient matrix of this system has rank two. Hence (r) implies

$$\begin{aligned}\boldsymbol{f}_{\beta\varrho} &= O(\varrho^2), \\ &= O(\varrho^4) \quad \text{if} \quad \boldsymbol{l}_\varrho \equiv \mathbf{0},\end{aligned} \tag{t}$$

and this result with (f), (g), $(\mathrm{k})_1$, $(\mathrm{l})_1$, and (s) yields conclusions (d) and (d′).

We have only to establish (c) and (c′). Thus suppose, in place of (q), that

$$\boldsymbol{f}_\varrho \equiv \mathbf{0}, \qquad \boldsymbol{m}_\varrho \equiv \mathbf{0}. \tag{u}$$

Clearly, (p) holds in the present circumstances with the operation "$\otimes$" replaced by "$\times$". It follows from this relation in conjunction with (u) and (s) that (r) also holds with "$\otimes$" replaced by "$\times$". Thus if we assume that $\boldsymbol{s}_\varrho(\boldsymbol{x}) = \varphi_\varrho(\boldsymbol{x}) \, \boldsymbol{a}$ and $\boldsymbol{l}_\varrho(\boldsymbol{\xi}) = \psi_\varrho(\boldsymbol{\xi}) \, \boldsymbol{a}$ on $\mathscr{F}_\varrho = \mathscr{F}_\varrho(k)$ and $D_\varrho = D_\varrho(k)$, respectively, then

$$\begin{aligned}\sum_{\beta=1}^{2} \boldsymbol{d}_\beta \times \boldsymbol{a} \left\{ \int_{\mathscr{F}_\varrho} \hat{\alpha}_\beta \, \varphi_\varrho \, da + \sum_{\boldsymbol{\xi} \in D_\varrho} \hat{\alpha}_\beta(\boldsymbol{\xi}) \, \psi_\varrho(\boldsymbol{\xi}) \right\} &= O(\varrho^2), \\ &= O(\varrho^4) \quad \text{if} \quad \boldsymbol{l}_\varrho \equiv \mathbf{0}.\end{aligned}$$

This, in turn, implies that

$$\begin{aligned}[(\boldsymbol{d}_1 \times \boldsymbol{d}_2) \cdot \boldsymbol{a}] \left\{ \int_{\mathscr{F}_\varrho} \hat{\alpha}_\beta \, \varphi_\varrho \, da + \sum_{\boldsymbol{\xi} \in D_\varrho} \hat{\alpha}_\beta(\boldsymbol{\xi}) \, \psi_\varrho(\boldsymbol{\xi}) \right\} &= O(\varrho^2), \\ &= O(\varrho^4) \quad \text{if} \quad \boldsymbol{l}_\varrho \equiv \mathbf{0}.\end{aligned} \tag{v}$$

It follows from (n) and (ii) that $\boldsymbol{d}_1 \times \boldsymbol{d}_2 \neq \mathbf{0}$. Therefore, since $\boldsymbol{d}_1 \times \boldsymbol{d}_2$ is normal to ∂B at $\boldsymbol{z}$, the requirement that $\boldsymbol{a}$ not be tangent to ∂B at $\boldsymbol{z}$ implies $(\boldsymbol{d}_1 \times \boldsymbol{d}_2) \cdot \boldsymbol{a} \neq 0$. Thus, in this instance, (v) implies (t), and, as before, we see that (c) and (c′) are valid. □

On the basis of the traditional statement of Saint-Venant's principle discussed at the beginning of this section, one would expect a reduction from $\delta = 1$

to $\delta = 2$ in the order of magnitude estimates whenever the force system is in equilibrium on each load region, even when the surface tractions are not of the special form specified in (c), i.e. parallel and non-tangential to the boundary. That this is not the case is apparent from counterexamples of v. MISES.[1]

The preceding conclusion has a counterpart in the theory of basic singular states discussed in Sect. 51. Thus consider the unit Kelvin state $\mathfrak{s}^i$ and the unit doublet state $\bar{\mathfrak{s}}^{ij}$, with $\mathbf{0}$ the point of application in both cases. Both of these states are regular on any infinite region B that does not contain $\mathbf{0}$. If the boundary of B is bounded, then

$$\boldsymbol{f}^i \neq \mathbf{0}, \qquad \boldsymbol{m}^i = \mathbf{0},$$
$$\bar{\boldsymbol{f}}^{ij} = \mathbf{0}, \qquad \overline{\boldsymbol{m}}^{ij} \neq \mathbf{0} \quad (i \neq j),$$
$$\bar{\boldsymbol{f}}^{ij} = \mathbf{0}, \qquad \overline{\boldsymbol{m}}^{ij} = \mathbf{0} \quad (i = j),$$

where $\boldsymbol{f}^i$, $\boldsymbol{m}^i$ and $\bar{\boldsymbol{f}}^{ij}$, $\overline{\boldsymbol{m}}^{ij}$ denote the resultant force and moment about the origin of the surface forces on ∂B corresponding to $\mathfrak{s}^i$ and $\bar{\mathfrak{s}}^{ij}$, respectively. Consider the rate of decay of the corresponding stress field at infinity. It is clear from the results on p. 178 that

$$\boldsymbol{S}^i(\boldsymbol{x}) = O(r^{-2}), \qquad \boldsymbol{S}^i(\boldsymbol{x}) \neq O(r^{-3})$$
$$\bar{\boldsymbol{S}}^{ij}(\boldsymbol{x}) = O(r^{-3}), \qquad \bar{\boldsymbol{S}}^{ij}(\boldsymbol{x}) \neq O(r^{-4})$$

as $r = |\boldsymbol{x} - \mathbf{0}| \to \infty$, regardless of whether $i = j$ or $i \neq j$. Consequently, whereas the stress field decays more rapidly when the resultant force on ∂B vanishes than when this is not the case, the additional vanishing of the resultant moment does not produce a further reduction in the order of magnitude of the stresses as $r \to \infty$.[2]

55. Toupin's version of Saint-Venant's principle. Two alternative versions of Saint-Venant's principle were established by TOUPIN[3] and KNOWLES.[4] These authors considered the solution corresponding to a single system of surface loads, and both authors arrived at estimates for the strain energy $U(l)$ of a portion of the body beyond a distance l from the load region.[5] KNOWLES' result, which we will discuss in the next section, applies to the plane problem, while TOUPIN's result is for a semi-infinite cylinder loaded on its end face.[6]

(1) Toupin's theorem.[7] *Let B be a homogeneous semi-infinite cylindrical body with end face $\mathscr{S}_0$, and assume that the elasticity tensor is symmetric and positive definite. Let $[\boldsymbol{u}, \boldsymbol{E}, \boldsymbol{S}]$ be an elastic state corresponding to zero body forces and to a surface force field $\mathbf{s}$ that satisfies:*

(i) $\mathbf{s} = \mathbf{0}$ *on* $\partial B - \mathscr{S}_0$,

(ii) $\int_{\mathscr{S}_0} \mathbf{s}\, da = \int_{\mathscr{S}_0} \mathbf{p} \times \mathbf{s}\, da = \mathbf{0}$,

(iii) $\lim_{l \to \infty} \int_{\mathscr{S}_l} \boldsymbol{u} \cdot \boldsymbol{S}\boldsymbol{n}\, da = 0$,

[1] [1945, *5*].

[2] The conclusions expressed in the last two paragraphs are due to STERNBERG and AL-KHOZAI [1964, *20*], pp. 144–145.

[3] [1965, *21*].

[4] [1966, *12*].

[5] The idea of estimating the energy $U(l)$ of that portion of a circular cylinder a distance l from the loaded end appears in the work of ZANABONI [1937, *8*].

[6] KNOWLES and STERNBERG [1966, *13*] have established a Saint-Venant principle for the torsion of solids of revolution. A result that is valid for more general regions was given by MELNIK [1963, *19*].

[7] [1965, *21*]. See also TOUPIN [1965, *20*].

where $\mathscr{S}_l$ is the intersection with B of a plane perpendicular to the axis $\mathbf{e}$ *of the cylinder and a distance l from $\mathscr{S}_0$. Let $\mathbf{0}\in\mathscr{S}_0$, let*

$$B_l = \{\boldsymbol{x}\in B: (\boldsymbol{x}-\mathbf{0})\cdot\mathbf{e} > l\},$$

and let $U(l)$ denote the strain energy of B_l:

$$U(l) = \tfrac{1}{2}\int_{B_l} \boldsymbol{E}\cdot\mathbf{C}[\boldsymbol{E}]\,dv.$$

Then

$$U(l) \leqq U(0)\, e^{\frac{-(l-t)}{\gamma(t)}} \qquad (l \geqq t)$$

for any $t>0$, where the decay length $\gamma(t)$ is given by

$$\gamma(t) = \sqrt{\frac{\mu_M^2}{\mu_m\,\lambda(t)}}\,.$$

Here μ_M and μ_m are the maximum and minimum elastic moduli, while $\lambda(t)$ is the lowest (non-zero) characteristic value for the free vibrations of a disc with cross section $\mathscr{S}_0$ and thickness t, a disc that is composed of the same material and that has its boundary traction-free.[1]

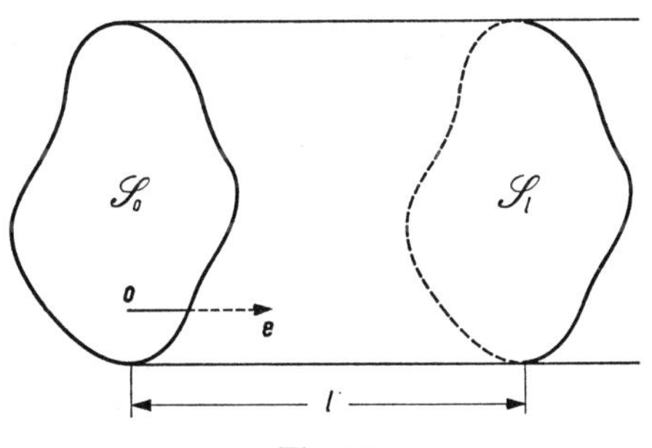

Fig. 14.

Proof. Choose $t, l>0$ and let

$$B(l,t) = B_l - B_{l+t}.$$

Then

$$U(l) = \lim_{t\to\infty} \tfrac{1}{2} \int_{B(l,t)} \boldsymbol{E}\cdot\mathbf{C}[\boldsymbol{E}]\,dv$$

provided the limit exists. By (i) and the theorem of work and energy ***(28.3)***,

$$\int_{\mathscr{S}_l} \boldsymbol{u}\cdot\boldsymbol{S}\boldsymbol{n}\,da + \int_{\mathscr{S}_{l+t}} \boldsymbol{u}\cdot\boldsymbol{S}\boldsymbol{n}\,da = \int_{B(l,t)} \boldsymbol{E}\cdot\mathbf{C}[\boldsymbol{E}]\,dv.$$

Thus if we let $t\to\infty$ and use (iii) we conclude that $U(l)$ exists and obeys the formula

$$U(l) = \tfrac{1}{2}\int_{\mathscr{S}_l} \boldsymbol{u}\cdot\boldsymbol{S}\boldsymbol{n}\,da. \tag{a}$$

Equating to zero the total force and moment on the portion $B(0,l)$ of B between $\mathscr{S}_0$ and $\mathscr{S}_l$, we conclude, with the aid of (i) and (ii), that

$$\int_{\mathscr{S}_l} \boldsymbol{S}\boldsymbol{n}\,da = \int_{\mathscr{S}_l} \mathbf{p}\times(\boldsymbol{S}\boldsymbol{n})\,da = \mathbf{0}. \tag{b}$$

[1] Thus we make the tacit assumption that the characteristic value problem stated in Sect. 76 has a solution, at least for the lowest non-zero characteristic value.

Let $\mathbf{s}'$ be the vector field on $\partial B(0, l)$ defined by

$$\mathbf{s}' = \boldsymbol{S}\boldsymbol{n} \quad \text{on } \mathscr{S}_l, \qquad \mathbf{s}' = \mathbf{0} \quad \text{on } \partial B(0, l) - \mathscr{S}_l.$$

Then (b) implies

$$\int_{\partial B(0,l)} \mathbf{s}'\, da = \int_{\partial B(0,l)} \boldsymbol{p} \times \mathbf{s}'\, da = \mathbf{0},$$

and we conclude from Piola's theorem ***(18.3)*** with $\mathbf{s} = \mathbf{s}'$ and $\boldsymbol{b} = \mathbf{0}$ that

$$\int_{\mathscr{S}_l} \boldsymbol{w} \cdot \boldsymbol{S}\boldsymbol{n}\, da = 0$$

for every rigid displacement $\boldsymbol{w}$. Therefore (a) yields

$$U(l) = \tfrac{1}{2} \int_{\mathscr{S}_l} \bar{\boldsymbol{u}} \cdot \boldsymbol{S}\boldsymbol{n}\, da \tag{c}$$

provided

$$\bar{\boldsymbol{u}} = \boldsymbol{u} + \boldsymbol{w}. \tag{d}$$

By (c) and the Schwarz inequality,

$$U(l) \leqq \tfrac{1}{2} \sqrt{\int_{\mathscr{S}_l} \bar{\boldsymbol{u}}^2\, da \int_{\mathscr{S}_l} |\boldsymbol{S}\boldsymbol{n}|^2\, da}. \tag{e}$$

If we expand $(\alpha\sqrt{a} - \sqrt{b})^2$, where a, b, and α are non-negative scalars with $\alpha > 0$, we arrive at the geometric-arithmetic mean inequality

$$\sqrt{a\,b} \leqq \frac{\alpha a + b/\alpha}{2}.$$

Applying this inequality to (e) yields

$$U(l) \leqq \frac{1}{4}\left\{\alpha \int_{\mathscr{S}_l} |\boldsymbol{S}\boldsymbol{n}|^2\, da + \frac{1}{\alpha} \int_{\mathscr{S}_l} \bar{\boldsymbol{u}}^2\, da\right\}. \tag{f}$$

Since $|\boldsymbol{n}| = 1$ and C is symmetric,

$$|\boldsymbol{S}\boldsymbol{n}|^2 \leqq |\boldsymbol{S}|^2 = \mathsf{C}[\boldsymbol{E}] \cdot \mathsf{C}[\boldsymbol{E}] = \boldsymbol{E} \cdot \mathsf{C}^2[\boldsymbol{E}].$$

Let μ_M and μ_m denote the maximum and minimum elastic moduli for C; then by ***(24.6)***

$$\mu_m |\boldsymbol{E}|^2 \leqq \boldsymbol{E} \cdot \mathsf{C}[\boldsymbol{E}] \leqq \mu_M |\boldsymbol{E}|^2.$$

Further, it follows from the definition of μ_M and μ_m that μ_M^2 and μ_m^2 are the maximum and minimum moduli for C^2; thus

$$\mu_m^2 |\boldsymbol{E}|^2 \leqq \boldsymbol{E} \cdot \mathsf{C}^2[\boldsymbol{E}] \leqq \mu_M^2 |\boldsymbol{E}|^2.$$

The last three inequalities imply that

$$|\boldsymbol{S}\boldsymbol{n}|^2 \leqq \mu_M^2 |\boldsymbol{E}|^2 \leqq \mu^* \boldsymbol{E} \cdot \mathsf{C}[\boldsymbol{E}],$$

where

$$\mu^* = \frac{\mu_M^2}{\mu_m};$$

thus (*f*) implies

$$U(l) \leqq \frac{1}{4}\left\{\alpha \mu^* \int_{\mathscr{S}_l} \boldsymbol{E} \cdot \mathsf{C}[\boldsymbol{E}]\, da + \frac{1}{\alpha} \int_{\mathscr{S}_l} \bar{\boldsymbol{u}}^2\, da\right\}. \tag{g}$$

Next, choose $t > 0$ and let

$$Q(l, t) = \frac{1}{t} \int_{l}^{l+t} U(\lambda)\, d\lambda. \tag{h}$$

If we integrate (g) between l and $l+t$, we arrive at

$$tQ(l,t) \leqq \frac{1}{4} \left\{ \alpha \mu^* \int_{B(l,t)} \boldsymbol{E} \cdot \mathsf{C}[\boldsymbol{E}]\, dv + \frac{1}{\alpha} \int_{B(l,t)} \bar{\boldsymbol{u}}^2\, dv \right\}. \tag{i}$$

Now let $\lambda(t)$ denote the lowest non-zero characteristic value corresponding to the *free* vibrations of the cylindrical disc $B(l, t)$. Then[1]

$$\lambda(t) \leqq \frac{\int_{B(l,t)} \nabla \boldsymbol{v} \cdot \mathsf{C}[\nabla \boldsymbol{v}]\, dv}{\int_{B(l,t)} \boldsymbol{v}^2\, dv}$$

for every admissible displacement field $\boldsymbol{v}$ on $B(l, t)$ that satisfies

$$\int_{B(l,t)} \boldsymbol{v}^2\, dv \neq 0, \qquad \int_{B(l,t)} \boldsymbol{v}\, dv = \int_{B(l,t)} \boldsymbol{p}_c \times \boldsymbol{v}\, dv = \boldsymbol{0}.$$

By use of the same procedure as that used to prove ***(53.1)***, it is not difficult to show that the rigid displacement $\boldsymbol{w}$ in (d) can be chosen so as to satisfy

$$\int_{B(l,t)} \boldsymbol{w}\, dv = -\int_{B(l,t)} \boldsymbol{u}\, dv, \qquad \int_{B(l,t)} \boldsymbol{p}_c \times \boldsymbol{w}\, dv = -\int_{B(l,t)} \boldsymbol{p}_c \times \boldsymbol{u}\, dv,$$

where $\boldsymbol{c}$ is the centroid of $B(l, t)$ and $\boldsymbol{p}_c(\boldsymbol{x}) = \boldsymbol{x} - \boldsymbol{c}$. By (d) when this is the case we have

$$\int_{B(l,t)} \bar{\boldsymbol{u}}\, dv = \int_{B(l,t)} \boldsymbol{p}_c \times \bar{\boldsymbol{u}}\, dv = \boldsymbol{0};$$

thus

$$\int_{B(l,t)} \bar{\boldsymbol{u}}^2\, dv \leqq \frac{1}{\lambda(t)} \int_{B(l,t)} \boldsymbol{E} \cdot \mathsf{C}[\boldsymbol{E}]\, dv, \tag{j}$$

and (i) and (j) yield

$$tQ(l,t) \leqq \frac{1}{4} \left(\frac{1}{\alpha \lambda(t)} + \alpha \mu^* \right) \int_{B(l,t)} \boldsymbol{E} \cdot \mathsf{C}[\boldsymbol{E}]\, dv. \tag{k}$$

If we differentiate (h) with respect to l, we arrive at

$$\frac{dQ(l,t)}{dl} = \frac{1}{t} [U(l+t) - U(l)]. \tag{l}$$

On the other hand,

$$\int_{B(l,t)} \boldsymbol{E} \cdot \mathsf{C}[\boldsymbol{E}]\, dv = \int_{B_l} \boldsymbol{E} \cdot \mathsf{C}[\boldsymbol{E}]\, dv - \int_{B_{l+t}} \boldsymbol{E} \cdot \mathsf{C}[\boldsymbol{E}]\, dv = 2[U(l) - U(l+t)], \tag{m}$$

and (k)–(m) imply

$$\gamma(t) \frac{dQ(l,t)}{dl} + Q(l,t) \leqq 0, \tag{n}$$

where

$$\gamma(t) = \frac{1}{2} \left\{ \frac{1}{\alpha \lambda(t)} + \alpha \mu^* \right\}.$$

[1] See p. 264.

Clearly, $\gamma(t)$ is a minimum when

$$\alpha = \frac{1}{\sqrt{\mu^* \lambda(t)}}.$$

Henceforth we assume that α has this value, so that

$$\gamma(t) = \sqrt{\frac{\mu^*}{\lambda(t)}}.$$

By (n)

$$\frac{d}{dl}\{e^{l/\gamma(t)} Q(l,t)\} = e^{l/\gamma(t)} \frac{dQ(l,t)}{dl} + \frac{1}{\gamma(t)} e^{l/\gamma(t)} Q(l,t) \leqq 0,$$

and thus

$$e^{l_2/\gamma(t)} Q(l_2,t) - e^{l_1/\gamma(t)} Q(l_1,t) \leqq 0 \quad (l_2 \geqq l_1)$$

or

$$\frac{Q(l_2,t)}{Q(l_1,t)} \leqq e^{\frac{-(l_2-l_1)}{\gamma(t)}}. \tag{o}$$

Since $\mathbf{C}$ is positive definite, $U(l)$ is a non-increasing function of l, and by (h) $Q(l,t)$ is the mean value of $U(l)$ in the interval $[l, l+t]$; therefore

$$U(l+t) \leqq Q(l,t) \leqq U(l),$$

and (o) implies

$$\frac{U(l_2+t)}{U(l_1)} \leqq e^{\frac{-(l_2-l_1)}{\gamma(t)}}. \tag{p}$$

Setting $l_1=0$ and $l_2=l-t$ in (p), we obtain the desired energy estimate. □

In view of ***(24.6)***, if the material is homogeneous and isotropic with strictly positive Lamé moduli μ and λ, then the characteristic decay length $\gamma(t)$ has the form

$$\gamma(t) = \frac{2\mu + 3\lambda}{\sqrt{2\mu\, \lambda(t)}}.$$

TOUPIN[1] also derived an estimate for the magnitude of the strain tensor at interior points of the cylinder. A somewhat stronger estimate was established by ROSEMAN[2] for an isotropic cylinder. ROSEMAN proved that

$$|\boldsymbol{S}(\boldsymbol{x})|^2 \leqq K[U(l-a) - U(l+a)]$$

for every $\boldsymbol{x} \in \mathcal{S}_l$, where K is a constant that depends on the Lamé moduli and the cross-section, while a is a constant that depends only on the cross-section. When combined with Toupin's theorem, the above inequality yields pointwise exponential decay for the stress throughout the cylinder.

56. Knowles' version of Saint-Venant's principle. KNOWLES' version[3] of Saint-Venant's principle is concerned with the plane problem of elastostatics for a simply-connected region loaded over a portion of its boundary curve. To state this result concisely we introduce the following notation.

Let R be a *simply-connected* regular region in the x_1, x_2-plane, so that its boundary curve $\mathcal{C}$ is a piecewise smooth simple closed curve. As indicated in Fig. 15,

$$\mathcal{C}_0 = \{x \in \mathcal{C} : x_1 \geqq 0\},$$
$$\mathcal{C}_1 = \mathcal{C} - \mathcal{C}_0,$$

[1] [1965, *21*]. TOUPIN's estimate is based on (h) of Sect. 43, which is due to DIAZ and PAYNE [1958, *6*].

[2] [1966, *22*]. The techniques used by ROSEMAN are based on results of JOHN [1963, *15*], [1965, *10*].

[3] [1966, *12*].

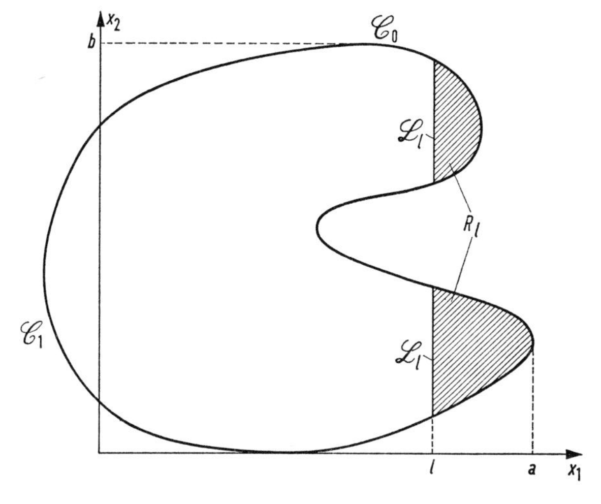

Fig. 15.

and we choose our coordinates so that the minimum value of x_2 on $\mathscr{C}_0$ is zero. $\mathscr{C}_1$ will represent that portion of the boundary curve along which surface tractions are applied. Let

$$a = \sup_{x \in \mathscr{C}_0} x_1, \qquad b = \sup_{x \in \mathscr{C}_0} x_2,$$

$$R_l = \{x \in R \colon x_1 > l\},$$

$$\mathscr{L}_l = \{x \in R \colon x_1 = l\}.$$

We assume that for each $l \in (0, a)$, the set $\mathscr{L}_l$ is the union of a finite number of line segments; we regard $\mathscr{L}_l$ as parametrized by x_2.

Since R is simply-connected, *(46.3)* implies that the stress field $S_{\alpha\beta}$ corresponding to a plane elastic state is completely characterized by the field equations

$$S_{\alpha\beta,\beta} = 0, \qquad \Delta S_{\alpha\alpha} = 0. \tag{a}$$

Knowles' theorem gives a decay estimate for the *norm*

$$\|S\|_l = \sqrt{\int_{R_l} S_{\alpha\beta} S_{\alpha\beta}\, da} \tag{b}$$

of such a stress field over R_l.

(1) Knowles' theorem.[1] *Let $S_{\alpha\beta}$ $(= S_{\beta\alpha})$ be smooth on $\overline{R}_0$ and of class C^2 on R_0. Suppose further that $S_{\alpha\beta}$ satisfies* (a) *on R_0 and*

$$S_{\alpha\beta} n_\beta = 0 \quad \textit{on } \mathscr{C}_0. \tag{c}$$

Then for $0 \leq l \leq a$

$$\|S\|_l \leq \sqrt{2}\, \|S\|_0\, e^{-\frac{kl}{b}},$$

where

$$k = \pi \left(\frac{\sqrt{2} - 1}{2}\right)^{\frac{1}{2}}.$$

[1] [1966, *12*].

With a view toward proving this theorem, we first establish the following

(2) Lemma.[1] *Let f be continuous on $\overline{R}_0$. Then*

$$\frac{d}{dl}\int_{R_l} f\,da = -\int_{\mathscr{L}_l} f\,dx_2.$$

Proof. Consider the difference quotient

$$\frac{F(l+\delta)-F(l)}{\delta}$$

for the function F defined by

$$F(l)=\int_{R_l} f\,da.$$

If $\mathscr{L}_l$ is connected, this difference quotient can be written, for sufficiently small δ, as an iterated integral, and the limit as $\delta\to 0$ can be computed directly to prove ***(2)***. If $\mathscr{L}_l$ is the union of disjoint line segments, the difference quotient is written as a sum of iterated integrals, and ***(2)*** again follows upon letting $\delta\to 0$. □

*Proof of **(1)***. By ***(47.1)*** there exists an Airy function φ for $S_{\alpha\beta}$ on R_0; thus

$$S_{11}=\varphi_{,22},\qquad S_{22}=\varphi_{,11},\qquad S_{12}=-\varphi_{,12},\tag{d}$$

$$\Delta\Delta\varphi=0.\tag{e}$$

Since R_0 is simply-connected, φ is single-valued; further, it follows from (d), (e), and our hypotheses on $S_{\alpha\beta}$ that φ is analytic on R_0 and of class C^3 on $\overline{R}_0$. Moreover, in view of (c), the relation $\varphi(\sigma)=m(\sigma)$ in ***(47.2)***, (d) and (e) in the proof of ***(47.2)***, and the discussion given in the second paragraph following ***(47.1)***, we can assume, without loss in generality, that

$$\varphi=\varphi_{,\alpha}=0\quad\text{ on } \mathscr{C}_0.\tag{f}$$

Let

$$W(l)=\int_{R_l} S_{\alpha\beta}S_{\alpha\beta}\,da,\tag{g}$$

so that

$$\|S\|_l=\sqrt{W(l)}.\tag{h}$$

Then (d) and (g) imply that

$$W(l)=\int_{R_l}\varphi_{,\alpha\beta}\varphi_{,\alpha\beta}\,da.\tag{i}$$

In view of (e), we can write (i) in the form

$$W(l)=\int_{R_l}[(\varphi_{,1}\varphi_{,11}-\varphi\varphi_{,111}+2\varphi_{,2}\varphi_{,12})_{,1} + (\varphi_{,2}\varphi_{,22}-\varphi\varphi_{,222}-2\varphi\varphi_{,112})_{,2}]\,da,$$

where $l\in(0,a)$. Thus (f) and Green's theorem yield the relation

$$W(l)=-\int_{\mathscr{L}_l}(\varphi_{,1}\varphi_{,11}-\varphi\varphi_{,111}+2\varphi_{,2}\varphi_{,12})\,dx_2,$$

and by ***(2)*** this relation, in turn, implies that

$$W(l)=\frac{d}{dl}\int_{R_l}(\varphi_{,1}\varphi_{,11}-\varphi\varphi_{,111}+2\varphi_{,2}\varphi_{,12})\,da.$$

[1] KNOWLES [1966, *12*], Eq. (2.6).

Hence

$$\int_l^a W(\lambda)\,d\lambda = -\int_{R_l} (\varphi_{,1}\varphi_{,11} - \varphi\varphi_{,111} + 2\varphi_{,2}\varphi_{,12})\,da, \tag{j}$$

since the integral on the right vanishes when $l = a$. Finally, since

$$\int_l^a W(\lambda)\,d\lambda = -\int_{R_l} (\varphi_{,1}^2 + \varphi_{,2}^2 - \varphi\varphi_{,11})_{,1}\,da,$$

the integral over R_l in (j) can be transformed as follows by again applying Green's theorem and the boundary condition (f):

$$\int_l^a W(\lambda)\,d\lambda = \int_{\mathcal{L}_l} (\varphi_{,1}^2 + \varphi_{,2}^2 - \varphi\varphi_{,11})\,dx_2. \tag{k}$$

Next, by (i) and *(2)*,

$$\frac{dW(l)}{dl} = -\int_{\mathcal{L}_l} \varphi_{,\alpha\beta}\,\varphi_{,\alpha\beta}\,dx_2. \tag{l}$$

Thus given any scalar $\varkappa$, the formulae (k) and (l) imply the following identity:

$$\begin{aligned}\frac{dW(l)}{dl} &+ 4\varkappa^2\int_l^a W(\lambda)\,d\lambda \\ &= -\int_{\mathcal{L}_l} [\varphi_{,11}^2 + \varphi_{,22}^2 + 2\varphi_{,12}^2 - 4\varkappa^2(\varphi_{,1}^2 + \varphi_{,2}^2 - \varphi\varphi_{,11})]\,dx_2.\end{aligned} \tag{m}$$

Letting

$$F(l) = W(l) + 2\varkappa\int_l^a W(\lambda)\,d\lambda, \tag{n}$$

we can write (m) in the form

$$\begin{aligned}\frac{dF(l)}{dl} + 2\varkappa F(l) = &-\int_{\mathcal{L}_l} ([\varphi_{,11} + 2\varkappa^2\varphi)^2 \\ &+ (2\varphi_{,12}^2 - 4\varkappa^2\varphi_{,1}^2) + (\varphi_{,22}^2 - 4\varkappa^2\varphi_{,2}^2 - 4\varkappa^4\varphi^2)]\,dx_2,\end{aligned} \tag{o}$$

after a rearrangement of terms in the integrand on the right side of (m).

Our next objective is to determine a positive value for $\varkappa$ for which the right side of (o) is non-positive. To do so we make use of the following inequality. If ψ is a smooth function on the closed interval $[t_0, t_1]$, and if $\psi(t_0) = \psi(t_1) = 0$, then

$$\int_{t_0}^{t_1} \dot{\psi}^2\,dt \geq \frac{\pi^2}{(t_1 - t_0)^2}\int_{t_0}^{t_1} \psi^2\,dt, \tag{p}$$ [1]

where $\dot{\psi}$ is the derivative of ψ. This inequality reflects the fact that $\pi^2/(t_1 - t_0)^2$ is the smallest characteristic value associated with the problem of minimizing

$$\int_{t_0}^{t_1} \dot{\psi}^2\,dt,$$

[1] For a discussion of inequalities of this type, reference may be made to § 0.6 of TEMPLE and BICKLEY [1956, *14*]. A direct proof of (p) can be found in HARDY, LITTLEWOOD, and POLYA [1959, *6*].

with the minimum taken over all smooth functions ψ on $[t_0, t_1]$ that vanish at the end points and have

$$\int_{t_0}^{t_1} \psi^2\, dt = 1.$$

We require a slight generalization of (p). If, for fixed $l \in (0, a)$, ψ is a smooth function (of x_2) on $\bar{\mathscr{L}}_l$, and if ψ vanishes at the end points of each line segment that constitutes $\bar{\mathscr{L}}_l$, then

$$\int_{\mathscr{L}_l} \psi_{,2}^2\, dx_2 \geqq \frac{\pi^2}{b^2} \int_{\mathscr{L}_l} \psi^2\, dx_2. \tag{q}$$

This inequality follows upon decomposing the integral on the left in (q) into a sum of integrals over the constituent line segments of $\mathscr{L}_l$, applying (p) to each of these, and observing that the length of each line segment is at most b.

Fix $l \in (0, a)$. Since $\varphi_{,1}$ vanishes on $\mathscr{C}_0$, it vanishes at the end points of the line segments that form $\bar{\mathscr{L}}_l$. Thus, because of the smoothness of φ, we may take ψ in (q) equal to $\varphi_{,1}$ with the result that

$$\int_{\mathscr{L}_l} \varphi_{,12}^2\, dx_2 \geqq \frac{\pi^2}{b^2} \int_{\mathscr{L}_l} \varphi_{,1}^2\, dx_2.$$

In view of this inequality, we can write (o) in the form

$$\frac{dF(l)}{dl} + 2\varkappa F(l) \leqq -\int_{\mathscr{L}_l} \left[\left(\frac{2\pi^2}{b^2} - 4\varkappa^2\right) \varphi_{,1}^2 + \varphi_{,22}^2 - 4\varkappa^2 \varphi_{,2}^2 - 4\varkappa^4 \varphi^2\right] dx_2. \tag{r}$$

Alternatively, we can let $\psi = \varphi_{,2}$ in (q) with the result that

$$\int_{\mathscr{L}_l} \varphi_{,22}^2\, dx_2 \geqq \frac{\pi^2}{b^2} \int_{\mathscr{L}_l} \varphi_{,2}^2\, dx_2,$$

and this result when substituted into (r) yields

$$\frac{dF(l)}{dl} + 2\varkappa F(l) \leqq -\int_{\mathscr{L}_l} \left[\left(\frac{2\pi^2}{b^2} - 4\varkappa^2\right) \varphi_{,1}^2 + \left(\frac{\pi^2}{b^2} - 4\varkappa^2\right) \varphi_{,2}^2 - 4\varkappa^4 \varphi^2\right] dx_2. \tag{s}$$

We now assume that

$$\varkappa \leqq \frac{\pi}{2b}, \tag{t}$$

so that the coefficients of $\varphi_{,1}^2$ and $\varphi_{,2}^2$ in (s) are non-negative. Then we may take $\psi = \varphi$ in (q) and infer from (s) that

$$\frac{dF(l)}{dl} + 2\varkappa F(l) \leqq -\left[\left(\frac{\pi^2}{b^2} - 4\varkappa^2\right) \frac{\pi^2}{b^2} - 4\varkappa^4\right] \int_{\mathscr{L}_l} \varphi^2\, dx_2.$$

The polynomial in $\varkappa$ in the bracket has one positive zero, namely

$$\varkappa = \frac{k}{b},$$

where

$$k = \pi \left(\frac{\sqrt{2} - 1}{2}\right)^{\frac{1}{2}}.$$

This value of $\varkappa$ satisfies (*t*). We now choose this value of $\varkappa$ and conclude that $F(l)$ satisfies the differential inequality

$$\frac{dF(l)}{dl} + 2\varkappa F(l) \leqq 0, \qquad 0 < l < a. \tag{u}$$

The inequality (u) implies that

$$F(l) \leqq F(0)\, e^{-2\varkappa l}, \qquad 0 \leqq l \leqq a. \tag{v}$$

Since $\varkappa \geqq 0$, (i) and (n) imply that

$$F(l) \geqq W(l);$$

thus

$$W(l) \leqq F(0)\, e^{-2\varkappa l}, \qquad 0 \leqq l \leqq a. \tag{w}$$

It remains to compute an upper bound for $F(0)$.

If we insert (n) into the left side of (v), we obtain an inequality which may be written in the form

$$-\frac{d}{dl}\left[e^{-2\varkappa l}\int_l^a W(\lambda)\, d\lambda\right] \leqq F(0)\, e^{-4\varkappa l}.$$

Integrating both sides of this inequality from $l=0$ to $l=a$, we find, with the aid of (n), that

$$\int_0^a W(\lambda)\, d\lambda \leqq \frac{F(0)}{4\varkappa}\,(1-e^{-4\varkappa a}) = \frac{1}{4\varkappa}\left[W(0) + 2\varkappa \int_0^a W(\lambda)\, d\lambda\right](1-e^{-4\varkappa a}).$$

Therefore

$$2\varkappa \int_0^a W(\lambda)\, d\lambda \leqq \left[\frac{1-e^{-4\varkappa a}}{1+e^{-4\varkappa a}}\right] W(0) \leqq W(0),$$

and this result when inserted into (n) with $l=0$ yields the inequality

$$F(0) \leqq 2W(0). \tag{x}$$

The inequalities (w) and (x) imply that

$$W(l) \leqq 2W(0)\, e^{-2\varkappa l}, \qquad 0 \leqq l \leqq a;$$

in view of (i), this is the desired result. □

As was noted by KNOWLES,[1] when R is a rectangular domain subject to self-equilibrated tractions along one edge, the optimum choice of the constant k in the estimate

$$\|S\|_l \leqq \sqrt{2}\,\|S\|_0\, e^{-\frac{kl}{b}}$$

s approximately 4.2.[2] Thus the universal constant k in *(1)* appears, in this particular case, to be conservative by a factor of approximately three.

As is clear from the relation on page 152, the *strain energy* $U(l)$ of R_l is given by

$$U(l) = \frac{1}{2}\int_{R_l} S_{\alpha\beta} E_{\alpha\beta}\, dv$$

$$= \frac{1}{4\mu}\int_R [S_{\alpha\beta} S_{\alpha\beta} - \nu S_{\gamma\gamma}^2]\, dv.$$

---

[1] [1966, *12*], p. 11.

[2] See, e.g., FADLE [1940, *1*], HORVAY [1957, *9*], and BABUŠKA, REKTORYS, and VYČICHLO [1955, *4*]. The discussion by BABUŠKA, REKTORYS, and VYČICHLO of the exponential decay of stresses in the rectangular case appears to be in error in that the roots of smallest modulus of the equation $\sin^2 z - z^2 = 0$ are underestimated by a factor of two (cf. FADLE [1940, *1*]).[3]

[3] KNOWLES [1966, *12*], p. 11, footnote 1.

If we assume that $\mu>0$, $0<\nu<\frac{1}{2}$, then the inequality

$$(1-2\nu)\, S_{\alpha\beta} S_{\alpha\beta} \leqq S_{\alpha\beta} S_{\alpha\beta} - \nu S_{\gamma\gamma}^2 \leqq S_{\alpha\beta} S_{\alpha\beta}$$

is easily derived using the same procedure as was used to establish ***(24.6)***. Thus

$$\frac{1-2\nu}{4\mu}\|S\|_l \leqq U(l) \leqq \frac{1}{4\mu}\|S\|_l,$$

and the inequality in ***(1)*** can be written in the form

$$U(l) \leqq \frac{\sqrt{2}}{1-2\nu}\, U(0)\, e^{-\frac{kl}{b}}.$$

KNOWLES[1] also proved that under the hypotheses of ***(1)*** the stresses satisfy the inequality

$$|S_{\alpha\beta}(x_1, x_2)| \leqq \frac{5}{\delta}\sqrt{\frac{2U(0)}{\pi}}\, e^{-\frac{kx_1}{b}},$$

where δ is the distance from the point (x_1, x_2) to the boundary of R_0 and k is defined in ***(1)***. Note that this inequality is poor at points which are close to the boundary.

56a. The Zanaboni-Robinson version of Saint-Venant's principle. (Added in proof.) This version of the principle is concerned with a one-parameter family $B_l (0<l<\infty)$ of bodies (bounded regular regions) such that

$$B_\tau \subset B_l \qquad (\tau<l)$$

and

$$B(\tau, l) \equiv B_l - B_\tau \qquad (\tau<l)$$

has a regular interior. We assume given a continuous, symmetric, positive definite elasticity field $\mathbf{C}$ on

$$\overline{\bigcup_{l\in[0,\infty)} B_l},$$

a regular surface $\mathscr{S}$ such that

$$\mathscr{S} \subset \partial B_l \qquad (0<l<\infty),$$

and integrable surface tractions $\hat{\mathbf{s}}$ on $\mathscr{S}$ such that

$$\int_{\mathscr{S}} \hat{\mathbf{s}}\, da = \int_{\mathscr{S}} \mathbf{p}\times\hat{\mathbf{s}}\, da = \mathbf{0}.$$

(1) Zanaboni-Robinson theorem.[2] *For each $l\in[0,\infty)$ let $[\boldsymbol{u}_l, \boldsymbol{E}_l, \boldsymbol{S}_l]$ be an elastic state on B_l corresponding to the elasticity field $\mathbf{C}$ (restricted to B_l), to zero body forces, and to the following surface tractions:*

$$\mathbf{s}_l = \hat{\mathbf{s}} \quad \textit{on } \mathscr{S},$$

$$\mathbf{s}_l = \mathbf{0} \quad \textit{on } \partial B_l - \mathscr{S}.$$

Let

$$U(\tau, l) = \tfrac{1}{2}\int_{B(\tau,l)} \boldsymbol{E}_l\cdot\mathbf{C}[\boldsymbol{E}_l]\, dv \qquad (\tau<l).$$

---

[1] [1966, *12*], Theorem 2. See also ROSEMAN [1967, *12*].

[2] The precise version stated here is due to ROBINSON [1966, *21*], § 9.7; the underlying ideas, however, are contained in the work of ZANABONI [1937, *6*, *7*, *8*]. See also SEGENREICH [1971, *4*], who established interesting corollaries of this theorem.

Then

$$\lim_{\substack{\tau, l \to \infty \\ \tau < l}} U(\tau, l) = 0.$$

For the proof of this theorem we refer the reader to the book by ROBINSON.[1]

XI. Miscellaneous results.

57. Some further results for homogeneous and isotropic bodies. In this section we shall establish some results appropriate to *homogeneous* and *isotropic* bodies.

Given a smooth vector field $\boldsymbol{u}$ on $\bar{B}$ and a closed regular surface $\mathscr{S}$ in $\bar{B}$, we define the vector field $\boldsymbol{t}(\boldsymbol{u})$ on $\mathscr{S}$ by[2]

$$\boldsymbol{t}(\boldsymbol{u}) = (\lambda + 2\mu)(\operatorname{div} \boldsymbol{u})\, \boldsymbol{n} + \mu (\operatorname{curl} \boldsymbol{u}) \times \boldsymbol{n},$$

where $\boldsymbol{n}$ is the unit outward normal to $\mathscr{S}$.[3]

The first theorem, which is due to FOSDICK[2], shows that *the vector field* $\boldsymbol{t}(\boldsymbol{u})$ *when integrated over* ∂B *gives the actual load on the boundary.*

(1) *Let* $\mathscr{S}$ *be a closed regular surface in* $\bar{B}$. *Further, let* $\boldsymbol{u}$ *be a smooth vector field on* $\bar{B}$, *and let* $\mathbf{s}$ *be the corresponding surface traction field on* $\mathscr{S}$:

$$\mathbf{s} = \mu (\nabla \boldsymbol{u} + \nabla \boldsymbol{u}^T)\, \boldsymbol{n} + \lambda (\operatorname{div} \boldsymbol{u})\, \boldsymbol{n}.$$

Then

$$\int_{\mathscr{S}} \boldsymbol{t}(\boldsymbol{u})\, da = \int_{\mathscr{S}} \mathbf{s}\, da.$$

Proof. Since

$$(\operatorname{curl} \boldsymbol{u}) \times \boldsymbol{n} = (\nabla \boldsymbol{u} - \nabla \boldsymbol{u}^T)\, \boldsymbol{n},$$

it follows that

$$\mathbf{s} - \boldsymbol{t}(\boldsymbol{u}) = 2\mu [\nabla \boldsymbol{u}^T - (\operatorname{div} \boldsymbol{u})\, \mathbf{1}]\, \boldsymbol{n},$$

and thus that

$$\int_{\mathscr{S}} \mathbf{s}\, da - \int_{\mathscr{S}} \boldsymbol{t}(\boldsymbol{u})\, da = 2\mu \int_{\mathscr{S}} [\nabla \boldsymbol{u}^T - (\operatorname{div} \boldsymbol{u})\, \mathbf{1}]\, \boldsymbol{n}\, da.$$

Letting $\boldsymbol{T}$ denote the skew tensor field whose axial vector is $\boldsymbol{u}$, we conclude from (15) of ***(4.1)*** that

$$\int_{\mathscr{S}} \mathbf{s}\, da - \int_{\mathscr{S}} \boldsymbol{t}(\boldsymbol{u})\, da = -2\mu \int_{\mathscr{S}} (\operatorname{curl} \boldsymbol{T})^T \boldsymbol{n}\, da,$$

and the desired result follows from Stokes' theorem ***(6.2)***. □

Note that since ∂B is the union of *closed* surfaces $\mathscr{S}_1, \ldots, \mathscr{S}_N$, if $\boldsymbol{u}$ is constant on each $\mathscr{S}_n$, then ***(1)*** implies

$$\int_{\partial B} \boldsymbol{t}(\boldsymbol{u}) \cdot \boldsymbol{u}\, da = \int_{\partial B} \mathbf{s} \cdot \boldsymbol{u}\, da;$$

thus, in this instance, "the work done by $\boldsymbol{t}(\boldsymbol{u})$" is equal to the work done by $\mathbf{s}$.

(2) Reciprocal theorem.[4] *Let* $\boldsymbol{u}$ *and* $\tilde{\boldsymbol{u}}$ *be elastic displacement fields of class* C^2 *on* $\bar{B}$ *corresponding to body force fields* $\boldsymbol{b}$ *and* $\tilde{\boldsymbol{b}}$, *respectively. Then*

$$\int_{\partial B} \boldsymbol{t}(\boldsymbol{u}) \cdot \tilde{\boldsymbol{u}}\, da + \int_B \boldsymbol{b} \cdot \tilde{\boldsymbol{u}}\, dv = \int_{\partial B} \boldsymbol{t}(\tilde{\boldsymbol{u}}) \cdot \boldsymbol{u}\, da + \int_B \tilde{\boldsymbol{b}} \cdot \boldsymbol{u}\, dv$$
$$= \int_B [(\lambda + 2\mu)(\operatorname{div} \boldsymbol{u})(\operatorname{div} \tilde{\boldsymbol{u}}) + \mu (\operatorname{curl} \boldsymbol{u}) \cdot (\operatorname{curl} \tilde{\boldsymbol{u}})]\, dv.$$

---

[1] [1966, *21*], § 9.7.
[2] FOSDICK [1968, *5*].
[3] Sometimes we will take $\mathscr{S} = \partial B$. It will be clear from the context when this is the case.
[4] WEYL [1915, *3*], p. 6; FOSDICK [1968, *5*].

Proof Let $I(\boldsymbol{u}, \tilde{\boldsymbol{u}})$ denote the last term in the above relation. It follows from the identity

$$\int_{\partial B} (\tilde{\boldsymbol{u}} \times \operatorname{curl} \boldsymbol{u}) \cdot \boldsymbol{n} \, da = \int_{B} (\operatorname{curl} \boldsymbol{u}) \cdot (\operatorname{curl} \tilde{\boldsymbol{u}}) \, dv - \int_{B} \tilde{\boldsymbol{u}} \cdot \operatorname{curl} \operatorname{curl} \boldsymbol{u} \, dv$$

in conjunction with the definition of $\boldsymbol{t}(\boldsymbol{u})$ that

$$\int_{\partial B} \boldsymbol{t}(\boldsymbol{u}) \cdot \tilde{\boldsymbol{u}} \, da = \int_{\partial B} [(\lambda + 2\mu)(\operatorname{div} \boldsymbol{u}) \, \tilde{\boldsymbol{u}} \cdot \boldsymbol{n} + \mu (\tilde{\boldsymbol{u}} \times \operatorname{curl} \boldsymbol{u}) \cdot \boldsymbol{n}] \, da$$
$$= I(\boldsymbol{u}, \tilde{\boldsymbol{u}}) + \int_{B} [(\lambda + 2\mu) \nabla \operatorname{div} \boldsymbol{u} - \mu \operatorname{curl} \operatorname{curl} \boldsymbol{u}] \cdot \tilde{\boldsymbol{u}} \, dv.$$

But since $\boldsymbol{u}$ is an elastic displacement field,

$$(\lambda + 2\mu) \nabla \operatorname{div} \boldsymbol{u} - \mu \operatorname{curl} \operatorname{curl} \boldsymbol{u} + \boldsymbol{b} = \boldsymbol{0}.$$ [1]

Therefore

$$\int_{\partial B} \boldsymbol{t}(\boldsymbol{u}) \cdot \tilde{\boldsymbol{u}} \, da + \int_{B} \boldsymbol{b} \cdot \tilde{\boldsymbol{u}} \, dv = I(\boldsymbol{u}, \tilde{\boldsymbol{u}}),$$

and, since $I(\boldsymbol{u}, \tilde{\boldsymbol{u}}) = I(\tilde{\boldsymbol{u}}, \boldsymbol{u})$, the proof is complete. □

If we let $\boldsymbol{u} = \tilde{\boldsymbol{u}}$ and $\boldsymbol{b} = \tilde{\boldsymbol{b}}$ in ***(2)***, we arrive at the following result, which is an analog of the theorem of work and energy ***(28.3)***.

(3) Kelvin's theorem.[2] *Let* $\boldsymbol{u}$ *be an elastic displacement field of class* C^2 *on* $\bar{B}$ *corresponding to the body force field* $\boldsymbol{b}$. *Then*

$$\int_{\partial B} \boldsymbol{t}(\boldsymbol{u}) \cdot \boldsymbol{u} \, da + \int_{B} \boldsymbol{b} \cdot \boldsymbol{u} \, dv = 2G\{\boldsymbol{u}\},$$

where

$$G\{\boldsymbol{u}\} = \tfrac{1}{2} \int_{B} [(\lambda + 2\mu)(\operatorname{div} \boldsymbol{u})^2 + \mu |\operatorname{curl} \boldsymbol{u}|^2] \, dv.$$

In view of the remark made following the proof of ***(1)***, when $\boldsymbol{u}$ is constant on each closed surface comprising ∂B, the left-hand sides of the relations in ***(3)*** and ***(28.3)*** coincide; therefore, in this instance, $G\{\boldsymbol{u}\} = U\{\boldsymbol{E}\}$. In particular, it follows that $G\{\boldsymbol{u}\}$ *is equal to the strain energy when the displacement field vanishes on the boundary.*

Theorem ***(3)*** yields a much simpler proof[3] for the uniqueness theorem ***(32.2)*** when B is homogeneous and isotropic. Indeed, if $\boldsymbol{u} = \boldsymbol{0}$ on ∂B and $\boldsymbol{b} = \boldsymbol{0}$ on B, then Kelvin's theorem implies

$$G\{\boldsymbol{u}\} = 0.$$

But if C or $-\mathsf{C}$ is strongly elliptic, we conclude from ***(25.1)*** that $\mu(\lambda + 2\mu) > 0$, and hence that

$$\operatorname{div} \boldsymbol{u} = 0, \qquad \operatorname{curl} \boldsymbol{u} = \boldsymbol{0},$$

which implies $\Delta \boldsymbol{u} = \boldsymbol{0}$. But if $\boldsymbol{u}$ is harmonic and vanishes on ∂B, then $\boldsymbol{u}$ must also vanish on B, and the proof is complete.

Since B is homogeneous and isotropic, the functional Φ of the principle of minimum potential energy ***(34.2)*** has the form

$$\Phi\{\boldsymbol{u}\} = \int_{B} \left[\frac{\mu}{4} |\nabla \boldsymbol{u} + \nabla \boldsymbol{u}^T|^2 + \frac{\lambda}{2} (\operatorname{div} \boldsymbol{u})^2 \right] dv - \int_{B} \boldsymbol{b} \cdot \boldsymbol{u} \, dv - \int_{\mathscr{S}_2} \hat{\boldsymbol{s}} \cdot \boldsymbol{u} \, da,$$

[1] This identity is derived on pp. 90-91.

[2] KELVIN [1888, *4*]. See also WEYL [1915, *3*], p. 6.

[3] This argument is due to ERICKSEN and TOUPIN [1956, 3]. See also DUFFIN and NOLL [1958, *8*], GURTIN and STERNBERG [1960, *7*].

as is clear from **(24.5)**. If we confine our attention to the displacement problem, we can establish a minimum principle much simpler in nature than the principle of minimum potential energy.

(4) Minimum principle for the displacement problem[1]. *Consider the displacement problem for which $\mathscr{S}_1 = \partial B$. Let $\mathscr{A}$ be the set of all kinematically admissible displacement fields, and let $\varkappa$ be the functional on $\mathscr{A}$ defined by*

$$\varkappa\{\boldsymbol{u}\} = G\{\boldsymbol{u}\} - \int_B \boldsymbol{b}\cdot\boldsymbol{u}\,dv$$

for every $\boldsymbol{u}\in\mathscr{A}$. Then given $\boldsymbol{u}\in\mathscr{A}$,

$$\varkappa\{\boldsymbol{u}\} \leqq \varkappa\{\tilde{\boldsymbol{u}}\}$$

for all $\tilde{\boldsymbol{u}}\in\mathscr{A}$ if and only if $\boldsymbol{u}$ is a displacement field corresponding to a solution of the displacement problem.

The proof of **(4)** follows at once from **(34.2)** by virtue of the following theorem, which establishes the connection between the functionals Φ and $\varkappa$.

(5)[2] *Suppose that the hypotheses of* **(4)** *hold. Let $\mathscr{A}$ be the set of all kinematically admissible displacement fields and let*

$$\lambda\{\boldsymbol{u}\} = \varkappa\{\boldsymbol{u}\} - \Phi\{\boldsymbol{u}\}$$

for every $\boldsymbol{u}\in\mathscr{A}$. Then

$$\lambda\{\boldsymbol{u}\} = \lambda\{\boldsymbol{u}'\}$$

for every $\boldsymbol{u}, \boldsymbol{u}'\in\mathscr{A}$; i.e. *$\lambda\{\boldsymbol{u}\}$ is independent of $\boldsymbol{u}$.*

Proof. Let $\boldsymbol{u}\in\mathscr{A}$. Since

$$|\operatorname{curl}\boldsymbol{u}|^2 = \nabla\boldsymbol{u}\cdot(\nabla\boldsymbol{u} - \nabla\boldsymbol{u}^T),$$

$$|\nabla\boldsymbol{u} + \nabla\boldsymbol{u}^T|^2 = 2\nabla\boldsymbol{u}\cdot(\nabla\boldsymbol{u} + \nabla\boldsymbol{u}^T),$$

it follows that

$$\lambda\{\boldsymbol{u}\} = \mu\int_B \{(\operatorname{div}\boldsymbol{u})^2 - \nabla\boldsymbol{u}\cdot\nabla\boldsymbol{u}^T\}\,dv = \mu\int_B \operatorname{div}[\boldsymbol{u}(\operatorname{div}\boldsymbol{u}) - (\nabla\boldsymbol{u})\,\boldsymbol{u}]\,dv.$$

Thus

$$\lambda\{\boldsymbol{u}\} = \mu\int_{\partial B}\eta\,da, \qquad \eta = [\boldsymbol{u}(\operatorname{div}\boldsymbol{u}) - (\nabla\boldsymbol{u})\,\boldsymbol{u}]\cdot\boldsymbol{n}.$$

We now complete the argument by showing that η depends *solely* on the boundary displacement $\hat{\boldsymbol{u}}$, which is common to all members of $\mathscr{A}$. Let the origin $\mathbf{0}$ of the cartesian frame be an arbitrary regular point of ∂B, and choose the frame so that the plane $x_3 = 0$ coincides with the tangent plane of ∂B at $\mathbf{0}$, the x_3 axis pointing in the direction of the inner normal of ∂B at $\mathbf{0}$. For this choice of coordinates,

$$\eta = u_1\frac{\partial u_3}{\partial x_1} + u_2\frac{\partial u_3}{\partial x_2} - u_3\left(\frac{\partial u_1}{\partial x_1} + \frac{\partial u_2}{\partial x_2}\right) \quad \text{at } \mathbf{0}. \tag{a}$$

In a neighborhood of $\mathbf{0}$ the boundary ∂B admits the parametrization

$$x_1 = \xi_1, \quad x_2 = \xi_2, \quad x_3 = h(\boldsymbol{\xi}), \quad \boldsymbol{\xi} = (\xi_1, \xi_2), \tag{b}$$

where

$$\frac{\partial h}{\partial \xi_\alpha}(\mathbf{0}) = 0 \quad (\alpha = 1, 2). \tag{c}$$

Further, if we agree to write $\hat{\boldsymbol{u}}(\boldsymbol{\xi}) = \hat{\boldsymbol{u}}(\xi_1, \xi_2, h(\xi_1, \xi_2))$, then, since $\boldsymbol{u}\in\mathscr{A}$,

$$\boldsymbol{u}(\boldsymbol{x}) = \hat{\boldsymbol{u}}(\boldsymbol{\xi})$$

[1] Love [1906, *5*], § 116. See also Gurtin and Sternberg [1960, *7*], Theorem 2.
[2] Gurtin and Sternberg [1960, *7*], Theorem 3.

in the neighborhood under consideration. Thus by (b)–(c)

$$\frac{\partial \boldsymbol{u}}{\partial x_\alpha}(\mathbf{0}) = \frac{\partial \hat{\boldsymbol{u}}}{\partial \xi_\alpha}(\mathbf{0}) \qquad (\alpha = 1, 2),$$

and we conclude from (a) that

$$\eta = \hat{u}_1 \frac{\partial \hat{u}_3}{\partial \xi_1} + \hat{u}_2 \frac{\partial \hat{u}_3}{\partial \xi_2} - \hat{u}_3 \left(\frac{\partial \hat{u}_1}{\partial \xi_1} + \frac{\partial \hat{u}_2}{\partial \xi_2} \right) \quad \text{at } \mathbf{0}.$$

Therefore η at $\mathbf{0}$ depends exclusively on the boundary data $\hat{\boldsymbol{u}}$. But the point $\mathbf{0}$ was an arbitrarily chosen regular point of ∂B. Thus $\lambda\{\boldsymbol{u}\}$ depends solely on the boundary data $\hat{\boldsymbol{u}}$; hence $\lambda\{\boldsymbol{u}\} = \lambda\{\boldsymbol{u}'\}$ for every $\boldsymbol{u}, \boldsymbol{u}' \in \mathscr{A}$. □

58. Incompressible materials. An ***incompressible body*** is one for which only isochoric motions are possible, or equivalently only motions that satisfy

$$\operatorname{div} \boldsymbol{u} = 0.$$

For a ***linearly elastic incompressible body*** the stress is specified by the strain tensor only up to an arbitrary pressure. Thus

$$\boldsymbol{S} + p\mathbf{1} = \mathsf{C}[\boldsymbol{E}],$$

where

$$p = -\tfrac{1}{3} \operatorname{tr} \boldsymbol{S}$$

and $\mathsf{C} = \mathsf{C}_{\boldsymbol{x}}$ is a linear transformation from the *space of traceless symmetric tensors into itself.*[1] The symmetry group $\mathscr{G}_{\boldsymbol{x}}$ for the material at $\boldsymbol{x}$ is the set of all orthogonal tensors $\boldsymbol{Q}$ such that

$$\boldsymbol{Q}\,\mathsf{C}[\boldsymbol{E}]\,\boldsymbol{Q}^T = \mathsf{C}[\boldsymbol{Q}\boldsymbol{E}\boldsymbol{Q}^T]$$

for every traceless symmetric tensor $\boldsymbol{E}$. This definition is consistent, since

$$\operatorname{tr}(\boldsymbol{Q}\boldsymbol{E}\boldsymbol{Q}^T) = \operatorname{tr}(\boldsymbol{Q}^T\boldsymbol{Q}\boldsymbol{E}) = \operatorname{tr}\boldsymbol{E} = 0$$

when $\boldsymbol{Q}$ is orthogonal.

If the material is isotropic, then (iii) of ***(22.1)*** is still valid; therefore

$$\mathsf{C}[\boldsymbol{E}] = 2\mu\,\boldsymbol{E},$$

and hence

$$\boldsymbol{S} + p\mathbf{1} = 2\mu\,\boldsymbol{E}.$$

Thus *there is but one elastic constant for an isotropic and incompressible material— the shear modulus μ.*

The fundamental field equations for the time-independent behavior of a linearly elastic incompressible body are therefore

$$\begin{gathered} \operatorname{div}\boldsymbol{u} = 0, \\ \boldsymbol{E} = \tfrac{1}{2}(\nabla\boldsymbol{u} + \nabla\boldsymbol{u}^T), \\ \boldsymbol{S} + p\mathbf{1} = \mathsf{C}[\boldsymbol{E}], \quad p = -\tfrac{1}{3}\operatorname{tr}\boldsymbol{S}, \\ \operatorname{div}\boldsymbol{S} + \boldsymbol{b} = \mathbf{0}. \end{gathered} \tag{a}$$

[1] We could also begin by assuming that

$$\boldsymbol{S} + p\mathbf{1} = \mathsf{C}[\nabla\boldsymbol{u}];$$

the steps leading to the reduction $\mathsf{C}[\nabla\boldsymbol{u}] = \mathsf{C}[\boldsymbol{E}]$ would then be exactly the same those given in Sect. 20.

If we combine these relations, we are led to the displacement equation of equilibrium

$$\operatorname{div} \mathsf{C}[\hat{\nabla} \boldsymbol{u}] - \nabla p + \boldsymbol{b} = \mathbf{0},$$

which for a homogeneous and isotropic body reduces to

$$\mu \Delta \boldsymbol{u} - \nabla p + \boldsymbol{b} = \mathbf{0}. \tag{b}$$

Since $\boldsymbol{u}$ is divergence-free,

$$\Delta p = \operatorname{div} \boldsymbol{b}; \tag{c}$$

hence in the absence of body forces the pressure p is *harmonic*. In view of $(a)_1$ and (3) of ***(4.1)***, we can also write (b) in the form

$$\mu \operatorname{curl} \operatorname{curl} \boldsymbol{u} + \nabla p - \boldsymbol{b} = \mathbf{0}.$$

On the other hand, if $\boldsymbol{b} = \mathbf{0}$, then (b) and (c) imply that

$$\Delta \operatorname{curl} \boldsymbol{u} = \mathbf{0},$$
$$\Delta \Delta \boldsymbol{u} = \mathbf{0}.$$

Since $\operatorname{tr} \boldsymbol{E} = 0$, the compatibility equation ***(14.3)*** has the following form for an incompressible body:

$$\Delta \boldsymbol{E} - \nabla \operatorname{div} \boldsymbol{E} - (\nabla \operatorname{div} \boldsymbol{E})^T = \mathbf{0}.$$

If the body is homogeneous and isotropic, then the same procedure used on p. 92 to derive the stress equation of compatibility now yields

$$\Delta \boldsymbol{S} - 2 \nabla \nabla p + 2 \hat{\nabla} \boldsymbol{b} + (\operatorname{div} \boldsymbol{b})\, \mathbf{1} = \mathbf{0},$$

which in the absence of body forces has the simple form

$$\Delta \boldsymbol{S} - 2 \nabla \nabla p = \mathbf{0}.$$

For an isotropic incompressible material the surface traction on the boundary ∂B takes the form

$$\boldsymbol{s} = \boldsymbol{S}\boldsymbol{n} = -p\, \boldsymbol{n} + \mu (\nabla \boldsymbol{u} + \nabla \boldsymbol{u}^T)\, \boldsymbol{n}.$$

Using the steps given on p. 93, we can also write this relation as follows:

$$\boldsymbol{s} = -p\, \boldsymbol{n} + 2\mu \frac{\partial \boldsymbol{u}}{\partial \boldsymbol{n}} + \mu \boldsymbol{n} \times \operatorname{curl} \boldsymbol{u}.$$

We define, for an incompressible elastic body, the notion of an elastic state in the same manner as before,[1] except that the constitutive relation is replaced by $(a)_3$ and the incompressibility condition $(a)_1$ is added. If this is done, the theorem of work and energy ***(28.3)*** remains valid for incompressible bodies, where, as before,

$$U\{\boldsymbol{E}\} = \tfrac{1}{2} \int_B \boldsymbol{E} \cdot \mathsf{C}[\boldsymbol{E}]\, dv,$$

but now, of course, $\boldsymbol{E}$ is required to be traceless. The crucial step in proving this result is to note that if $\boldsymbol{S}$ and $\boldsymbol{E}$ obey $(a)_3$, then

$$\boldsymbol{S} \cdot \boldsymbol{E} = -p \mathbf{1} \cdot \boldsymbol{E} + \boldsymbol{E} \cdot \mathsf{C}[\boldsymbol{E}] = \boldsymbol{E} \cdot \mathsf{C}[\boldsymbol{E}],$$

since $\mathbf{1} \cdot \boldsymbol{E} = \operatorname{tr} \boldsymbol{E} = 0$. Using a similar result, it is a simple matter to verify that Betti's reciprocal theorem ***(30.1)*** also remains valid.

[1] p. 95.

If we define the mixed problem as before,[1] then steps exactly the same as those used to establish *(32.1)* now yield the

(1) Uniqueness theorem for an incompressible body. *Assume that the body is incompressible and that the elasticity field* $\mathbf{C}$ *is positive definite. Let* $\mathfrak{s}=[\boldsymbol{u},\boldsymbol{E},\boldsymbol{S}]$ *and* $\tilde{\mathfrak{s}}=[\tilde{\boldsymbol{u}},\tilde{\boldsymbol{E}},\tilde{\boldsymbol{S}}]$ *be two solutions of the same mixed problem. Then* $\mathfrak{s}$ *and* $\tilde{\mathfrak{s}}$ *are equal modulo a rigid displacement and a uniform pressure,* i.e.

$$[\boldsymbol{u},\boldsymbol{E},\boldsymbol{S}]=[\tilde{\boldsymbol{u}}+\boldsymbol{w},\tilde{\boldsymbol{E}},\tilde{\boldsymbol{S}}+p\,\mathbf{1}],$$

where $\boldsymbol{w}$ *is a rigid displacement field and* p *is a scalar constant. Moreover,* $\boldsymbol{w}=\mathbf{0}$ *if* $\mathscr{S}_1\neq\emptyset$, *while* $p=0$ *if* $\mathscr{S}_2\neq\emptyset$.

The principle of minimum potential energy *(34.1)* remains valid in the present circumstances provided we require that the kinematically admissible states satisfy $\operatorname{div}\boldsymbol{u}=0$ and change the last few words to read: "only if $\tilde{\mathfrak{s}}=\mathfrak{s}$ modulo a rigid displacement and a uniform pressure." On the other hand, the principle of minimum complementary energy *(34.3)* remains valid if we replace $U_{\mathbf{K}}\{\boldsymbol{S}\}$ by $U_{\mathbf{K}}\{\boldsymbol{S}+p\,\mathbf{1}\}$, where $p=-\frac{1}{3}\operatorname{tr}\boldsymbol{S}$ and $\mathbf{K}$ is the inverse of $\mathbf{C}$, and if we change the last few words to read: "only if $\boldsymbol{S}=\tilde{\boldsymbol{S}}$ modulo a uniform pressure."

E. Elastodynamics.

I. The fundamental field equations. Elastic processes. Power and energy. Reciprocity.

59. The fundamental system of field equations. The fundamental system of field equations describing the motion of a linear elastic body consists of the ***strain-displacement relation***

$$\boldsymbol{E}=\tfrac{1}{2}(\nabla\boldsymbol{u}+\nabla\boldsymbol{u}^T),$$

the ***stress-strain relation***

$$\boldsymbol{S}=\mathbf{C}[\boldsymbol{E}],$$

and the ***equation of motion***[2]

$$\operatorname{div}\boldsymbol{S}+\boldsymbol{b}=\varrho\ddot{\boldsymbol{u}}.$$

Here $\boldsymbol{u}$, $\boldsymbol{E}$, $\boldsymbol{S}$, and $\boldsymbol{b}$ are the displacement, strain, stress, and body force fields defined on $B\times(0,t_0)$, where $(0,t_0)$ is a given interval of time, while $\mathbf{C}$ and ϱ are the elasticity and density fields over B.

Since $\mathbf{C}[\boldsymbol{E}]=\mathbf{C}[\nabla\boldsymbol{u}]$, if the displacement field is sufficiently smooth, these equations combine to yield the ***displacement equation of motion***:

$$\operatorname{div}\mathbf{C}[\nabla\boldsymbol{u}]+\boldsymbol{b}=\varrho\ddot{\boldsymbol{u}},$$

which in components has the form

$$(C_{ijkl}u_{k,l})_{,j}+b_i=\varrho\ddot{u}_i.$$

It is a simple matter to verify that, conversely, if $\boldsymbol{u}$ satisfies the displacement equation of motion, and if $\boldsymbol{E}$ and $\boldsymbol{S}$ are *defined* by the strain-displacement and stress-strain relations, then the equation of motion is satisfied.

---

[1] p. 102.

[2] Since the values of $\mathbf{C}_{\boldsymbol{x}}[\cdot]$ are assumed to be symmetric tensors, we need not add the requirement that $\boldsymbol{S}=\boldsymbol{S}^T$.

Assume for the moment that the body is homogeneous and isotropic. Then the procedure used to derive (a) on p. 90 now yields ***Navier's equation of motion***[1]

$$\mu \Delta \boldsymbol{u} + (\lambda + \mu) \nabla \operatorname{div} \boldsymbol{u} + \boldsymbol{b} = \varrho \ddot{\boldsymbol{u}}. \tag{a}$$

By (3) of ***(4.1)***, this equation may be written in the alternative form[2]

$$(\lambda + 2\mu) \nabla \operatorname{div} \boldsymbol{u} - \mu \operatorname{curl} \operatorname{curl} \boldsymbol{u} + \boldsymbol{b} = \varrho \ddot{\boldsymbol{u}}.$$

If we assume that $\lambda + 2\mu$, μ, and ϱ are positive, we can define constants c_1 and c_2 by

$$c_1 = \sqrt{\frac{\lambda + 2\mu}{\varrho}}, \qquad c_2 = \sqrt{\frac{\mu}{\varrho}}; \tag{b}$$

in terms of these constants the above equation becomes

$$c_1^2 \nabla \operatorname{div} \boldsymbol{u} - c_2^2 \operatorname{curl} \operatorname{curl} \boldsymbol{u} + \frac{\boldsymbol{b}}{\varrho} = \ddot{\boldsymbol{u}}. \tag{c}$$

For convenience, we introduce the ***wave operators*** $\square_1$ and $\square_2$ defined by

$$\square_\alpha f = c_\alpha^2 \Delta f - \ddot{f} \qquad (\alpha = 1, 2).$$

Then, since

$$\frac{\lambda + \mu}{\varrho} = c_1^2 - c_2^2,$$

Navier's equation (a) can be written in the form[3]

$$\square_2 \boldsymbol{u} + (c_1^2 - c_2^2) \nabla \operatorname{div} \boldsymbol{u} + \frac{\boldsymbol{b}}{\varrho} = \boldsymbol{0}. \tag{d}$$

Assume now that $\boldsymbol{b} = \boldsymbol{0}$. It is clear from (c) and (3) of ***(4.1)*** that[4]

$$\operatorname{curl} \boldsymbol{u} = \boldsymbol{0} \Rightarrow \square_1 \boldsymbol{u} = \boldsymbol{0},$$
$$\operatorname{div} \boldsymbol{u} = 0 \Rightarrow \square_2 \boldsymbol{u} = \boldsymbol{0}.$$

For this reason, we call c_1 the ***irrotational velocity***, c_2 the ***isochoric velocity***. Operating on Navier's equation, first with the divergence and then with the curl, we find that[5]

$$\begin{aligned} \square_1 \operatorname{div} \boldsymbol{u} &= 0, \\ \square_2 \operatorname{curl} \boldsymbol{u} &= \boldsymbol{0}. \end{aligned} \tag{e}$$

Thus the dilatation $\operatorname{div} \boldsymbol{u}$ satisfies a wave equation with speed c_1, while the rotation $\frac{1}{2} \operatorname{curl} \boldsymbol{u}$ satisfies a wave equation with speed c_2. Clearly,

$$\frac{c_1}{c_2} = \sqrt{\frac{\lambda + 2\mu}{\mu}} = \sqrt{\frac{2(1-\nu)}{1-2\nu}};$$

[1] This relation was first derived by NAVIER [1823, *2*], [1827, *2*] in 1821. NAVIER's theory, which is based on a molecular model, is limited to situations for which $\mu = \lambda$. The general relation involving two elastic constants first appears in the work of CAUCHY [1828, *1*]; cf. POISSON [1829, *2*], LAMÉ and CLAPEYRON [1833, *1*], STOKES [1845, *1*], LAMÉ [1852, *2*], § 26.

[2] LAMÉ and CLAPEYRON [1833, *1*], LAMÉ [1852, *2*], § 26.

[3] CAUCHY [1840, *1*], p. 120.

[4] The first of these results is due to LAMÉ [1852, *2*], § 61–62; the second to CAUCHY [1840, *1*], p. 137.

[5] CAUCHY [1828, *1*], [1840, *1*], pp. 120, 137. See also LAMÉ and CLAPEYRON [1833, *1*], STOKES [1851, *1*], LAMÉ [1852, *2*], § 27.

thus for most materials $c_1 > c_2$, so that dilatational waves travel faster than rotational waves.[1] In fact, $c_1 = 2c_2$ when $\nu = \frac{1}{3}$.

Next, if we apply $\square_1$ to (a) (with $\boldsymbol{b} = \boldsymbol{0}$) and use the first of (e), we arrive at

$$\square_1 (\mu \Delta \boldsymbol{u} - \varrho \ddot{\boldsymbol{u}}) = \boldsymbol{0},$$

or equivalently,[2]

$$\square_1 \square_2 \boldsymbol{u} = \boldsymbol{0}. \tag{f}$$

Eq. (f) and the strain-displacement and stress-strain relations imply that

$$\square_1 \square_2 \boldsymbol{E} = \square_1 \square_2 \boldsymbol{S} = \boldsymbol{0}.$$

Eqs. (e) are counterparts of the harmonicity of $\operatorname{div} \boldsymbol{u}$ and $\operatorname{curl} \boldsymbol{u}$ in the static theory; (f) is the counterpart of the biharmonicity of $\boldsymbol{u}$.

We now drop the assumption that B be homogeneous and isotropic. The quantity

$$e = \tfrac{1}{2} \boldsymbol{E} \cdot \mathsf{C}[\boldsymbol{E}] + \tfrac{1}{2} \varrho \dot{\boldsymbol{u}}^2$$

represents the total energy per unit volume. If C is symmetric, then the stress-strain and strain-displacement relations imply that

$$\dot{e} = \boldsymbol{S} \cdot \nabla \dot{\boldsymbol{u}} + \varrho \ddot{\boldsymbol{u}} \cdot \dot{\boldsymbol{u}}.$$

Thus, if we define

$$\boldsymbol{q} = -\boldsymbol{S} \dot{\boldsymbol{u}},$$

we are led, with the aid of the equation of motion, to the following result:

$$\dot{e} + \operatorname{div} \boldsymbol{q} = \boldsymbol{b} \cdot \dot{\boldsymbol{u}}. \tag{g}$$

The quantity $\boldsymbol{q}$ is called the *energy-flux vector*, and (g) expresses a local form of balance of energy.

Next, operating on the equation of motion with the symmetric gradient $\hat{\nabla}$ and using the strain-displacement relation, we find that

$$\hat{\nabla} \left(\frac{\operatorname{div} \boldsymbol{S} + \boldsymbol{b}}{\varrho} \right) = \ddot{\boldsymbol{E}}.$$

If we assume that C is invertible, so that

$$\boldsymbol{E} = \mathsf{K}[\boldsymbol{S}],$$

where K is the compliance field, we arrive at the ***stress equation of motion***[3]

$$\hat{\nabla} \left(\frac{\operatorname{div} \boldsymbol{S} + \boldsymbol{b}}{\varrho} \right) = \mathsf{K}[\ddot{\boldsymbol{S}}].$$

If ϱ is independent of $\boldsymbol{x}$, this relation takes the form

$$\hat{\nabla} (\operatorname{div} \boldsymbol{S} + \boldsymbol{b}) = \varrho \mathsf{K}[\ddot{\boldsymbol{S}}].$$

If, in addition, the material is isotropic, then the formula on page 78 giving $\boldsymbol{E}$ in terms of $\boldsymbol{S}$ yields

$$\mathsf{K}[\ddot{\boldsymbol{S}}] = \frac{1}{\beta} [(1 + \nu) \ddot{\boldsymbol{S}} - \nu (\operatorname{tr} \ddot{\boldsymbol{S}}) \boldsymbol{1}],$$

---

[1] In seismology waves traveling with speed c_1 are called primary waves (P-waves), those with speed c_2 are called secondary waves (S-waves).

[2] CAUCHY [1840, *1*], p. 137.

[3] IGNACZAK [1963, *14*] generalizing results of VÂLCOVICI [1951, *12*] for a homogeneous, isotropic body and IACOVACHE [1950, *6*] for a homogeneous, isotropic, and incompressible body. See also IGNACZAK [1959, *8*], TEODORESCU [1965, *19*].

where β is Young's modulus and ν is Poisson's ratio; thus

$$\beta(\hat{\nabla} \operatorname{div} \boldsymbol{S} + \hat{\nabla} \boldsymbol{b}) = \varrho[(1+\nu)\ddot{\boldsymbol{S}} - \nu(\operatorname{tr}\ddot{\boldsymbol{S}})\mathbf{1}].$$

Given initial values $\boldsymbol{u}_0$ and $\boldsymbol{v}_0$ for $\boldsymbol{u}$ and $\dot{\boldsymbol{u}}$, the stress equation of motion implies the fundamental system in the following sense: let $\boldsymbol{S}$ satisfy this equation and the initial conditions

$$\boldsymbol{S}(\cdot, 0) = \mathbf{C}[\nabla \boldsymbol{u}_0], \qquad \dot{\boldsymbol{S}}(\cdot, 0) = \mathbf{C}[\nabla \boldsymbol{v}_0];$$

let $\boldsymbol{E}$ be defined by the inverted stress-strain relation; and let $\boldsymbol{u}$ be the unique solution of the equation of motion with the initial values $\boldsymbol{u}_0$ and $\boldsymbol{v}_0$. Then the strain-displacement relation holds, and hence $[\boldsymbol{u}, \boldsymbol{E}, \boldsymbol{S}]$ satisfies the fundamental system of equations. The fact that in elastodynamics the fundamental system of field equations is equivalent to a *single* equation for the stress tensor is most interesting, especially when compared to the equilibrium theory where (for a simply-connected body) the fundamental system is equivalent to a *pair* of equations: the stress equations of compatibility and equilibrium.

60. Elastic processes. Power and energy. By an ***admissible process*** we mean an ordered array $\not{p} = [\boldsymbol{u}, \boldsymbol{E}, \boldsymbol{S}]$ with the following properties:

(i) $\boldsymbol{u}$ is an admissible motion;[1]

(ii) $\boldsymbol{E}$ is a continuous symmetric tensor field on $\bar{B} \times [0, t_0)$;

(iii) $\boldsymbol{S}$ is a time-dependent admissible stress field.[2]

Clearly, the set of all admissible processes is a vector space provided we define addition and scalar multiplication in the natural manner; i.e. as we did for admissible states.[3]

We say that $\not{p} = [\boldsymbol{u}, \boldsymbol{E}, \boldsymbol{S}]$ is an ***elastic process*** (on $\bar{B}$) corresponding to the body force $\boldsymbol{b}$ if $\not{p}$ is an admissible process and

$$\boldsymbol{E} = \tfrac{1}{2}(\nabla \boldsymbol{u} + \nabla \boldsymbol{u}^T),$$
$$\boldsymbol{S} = \mathbf{C}[\boldsymbol{E}],$$
$$\operatorname{div} \boldsymbol{S} + \boldsymbol{b} = \varrho \ddot{\boldsymbol{u}}.$$

As before, the ***surface force field*** $\boldsymbol{s}$ is then defined at every regular point of $\partial B \times [0, t_0)$ by

$$\boldsymbol{s}(\boldsymbol{x}, t) = \boldsymbol{S}(\boldsymbol{x}, t)\, \boldsymbol{n}(\boldsymbol{x}),$$

and we call the pair $[\boldsymbol{b}, \boldsymbol{s}]$ the ***external force system*** for $\not{p}$.

If

$$\boldsymbol{u}(\cdot, 0) = \dot{\boldsymbol{u}}(\cdot, 0) = \mathbf{0} \tag{I}$$

on B, we can extend $\not{p}$ *smoothly* to $\bar{B} \times (-\infty, t_0)$ by defining

$$\boldsymbol{u} = \mathbf{0} \quad \text{and} \quad \boldsymbol{E} = \boldsymbol{S} = \mathbf{0} \quad \text{on } \bar{B} \times (-\infty, 0).$$

When (I) holds we call the resulting *extended* process $\not{p}$ an ***elastic process with a quiescent past.***[4] Further, to be consistent, we also extend the external force system by assuming that $\boldsymbol{b}$ and $\boldsymbol{s}$ vanish for negative time.

---

[1] Cf. p. 43.

[2] Cf. p. 49.

[3] Cf. p. 95.

[4] This terminology is due to WHEELER and STERNBERG [1968, *16*].

When we discuss the basic singular solutions of elastodynamics, we will be forced to deal with processes whose domain is of the form $\mathscr{E}_\xi \times [0, \infty)$, where ξ is a point of $\mathscr{E}$ and

$$\mathscr{E}_\xi = \mathscr{E} - \{\xi\}.$$

We say that $\mathfrak{p} = [\boldsymbol{u}, \boldsymbol{E}, \boldsymbol{S}]$ is an elastic process on $\mathscr{E}_\xi$ corresponding to $\boldsymbol{b}$ if $\boldsymbol{u}, \boldsymbol{E}, \boldsymbol{S}$ and $\boldsymbol{b}$ are functions on $\mathscr{E}_\xi \times [0, \infty)$ and given any closed regular subregion $P \subset \mathscr{E}_\xi$, the restriction of $\mathfrak{p}$ to $P \times [0, \infty)$ is an elastic process on P corresponding to the restriction of $\boldsymbol{b}$ to $P \times [0, \infty)$.

When we omit mention of the domain of definition of an elastic process, it will always be understood to be $\bar{B} \times [0, t_0)$.

The following simple proposition will be extremely useful.

(1) *If* $[\boldsymbol{u}, \boldsymbol{E}, \boldsymbol{S}]$ *is an elastic process corresponding to the body force field* $\boldsymbol{b}$, *then given any* $t \in (0, t_0)$, $[\boldsymbol{u}(\cdot, t), \boldsymbol{E}(\cdot, t), \boldsymbol{S}(\cdot, t)]$ *is an elastic state corresponding to the body force* $\boldsymbol{b} - \varrho \ddot{\boldsymbol{u}}$.

Given an elastic process, we call the function $\mathscr{U}$ on $[0, t_0)$ defined by

$$\mathscr{U} = K + U$$

the ***total energy***. Here K is the ***kinetic energy***:

$$K = \frac{1}{2} \int_B \varrho \dot{\boldsymbol{u}}^2 \, dv;$$

U is the strain energy:

$$U = U_{\mathbf{C}}\{\boldsymbol{E}\} = \frac{1}{2} \int_B \boldsymbol{E} \cdot \mathbf{C}[\boldsymbol{E}] \, dv.$$

If $\mathbf{C}$ is symmetric,

$$\dot{U} = \frac{1}{2} \int_B \{\dot{\boldsymbol{E}} \cdot \mathbf{C}[\boldsymbol{E}] + \boldsymbol{E} \cdot \mathbf{C}[\dot{\boldsymbol{E}}]\} \, dv = \int_B \dot{\boldsymbol{E}} \cdot \mathbf{C}[\boldsymbol{E}] \, dv = \int_B \boldsymbol{S} \cdot \dot{\boldsymbol{E}} \, dv;$$

therefore *the rate of change of total energy equals the rate of change of kinetic energy plus the stress power*.[1] This fact and the theorem of power expended ***(19.1)*** together yield the

(2) Theorem of power and energy. *Assume that* $\mathbf{C}$ *is symmetric. Let* $[\boldsymbol{u}, \boldsymbol{E}, \boldsymbol{S}]$ *be an elastic process corresponding to the external force system* $[\boldsymbol{b}, \boldsymbol{s}]$. *Then*

$$\int_{\partial B} \boldsymbol{s} \cdot \dot{\boldsymbol{u}} \, da + \int_B \boldsymbol{b} \cdot \dot{\boldsymbol{u}} \, dv = \dot{\mathscr{U}},$$

where $\mathscr{U}$ *is the total energy.*

This theorem asserts that the rate at which work is done by the surface and body forces equals the rate of change of total energy. The following proposition, which is an immediate corollary of ***(2)***, will be useful in establishing uniqueness.

(3) *Assume that* $\mathbf{C}$ *is symmetric. Let* $[\boldsymbol{u}, \boldsymbol{E}, \boldsymbol{S}]$ *be an elastic process corresponding to the external force system* $[\boldsymbol{b}, \boldsymbol{s}]$, *and suppose that*

$$\boldsymbol{b} = \boldsymbol{0} \quad \text{on } B, \qquad \boldsymbol{s} \cdot \dot{\boldsymbol{u}} = 0 \quad \text{on } \partial B.$$

Then the total energy is constant, i.e.

$$\mathscr{U}(t) = \mathscr{U}(0), \qquad 0 \leqq t < t_0.$$

[1] Cf. p. 65.

Another interesting corollary of **(2)** may be deduced as follows. By **(2)**

$$\int_0^\tau \int_{\partial B} \mathbf{s} \cdot \dot{\boldsymbol{u}}\, da\, dt + \int_0^\tau \int_B \boldsymbol{b} \cdot \dot{\boldsymbol{u}}\, dv\, dt = \mathscr{U}(\tau) - \mathscr{U}(0).$$

The left-hand side of this expression is the *work done in the time interval* $[0, \tau]$. If the initial state $[\boldsymbol{u}(\cdot, 0), \boldsymbol{E}(\cdot, 0), \boldsymbol{S}(\cdot, 0)]$ is an unstrained state of rest, i.e. if

$$\boldsymbol{E}(\cdot, 0) = \mathbf{0}, \qquad \dot{\boldsymbol{u}}(\cdot, 0) = \mathbf{0},$$

then

$$\mathscr{U}(0) = 0.$$

If, in addition, the elasticity tensor is positive semi-definite and the density positive, then $\mathscr{U}(\tau) \geqq 0$ and

$$\int_0^\tau \int_{\partial B} \mathbf{s} \cdot \dot{\boldsymbol{u}}\, da\, dt + \int_0^\tau \int_B \boldsymbol{b} \cdot \dot{\boldsymbol{u}}\, dv\, dt \geqq 0;$$

in words, *the work done starting from an unstrained rest state is always nonnegative.*[1]

The next theorem, which is of interest in itself, forms the basis of a very strong uniqueness theorem for elastodynamics. For the purpose of this theorem we assume that the underlying time-interval $[0, t_0)$ is equal to $[0, \infty)$.

(4) Brun's theorem.[2] *Assume that* $\mathbf{C}$ *is symmetric. Let* $[\boldsymbol{u}, \boldsymbol{E}, \boldsymbol{S}]$ *be an elastic process corresponding to the external force system* $[\boldsymbol{b}, \mathbf{s}]$, *and suppose that*

$$\dot{\boldsymbol{u}}(\cdot, 0) = \mathbf{0}, \qquad \boldsymbol{E}(\cdot, 0) = \mathbf{0}.$$

Let

$$P(\alpha, \beta) = \int_{\partial B} \mathbf{s}(\boldsymbol{x}, \alpha) \cdot \dot{\boldsymbol{u}}(\boldsymbol{x}, \beta)\, da_{\boldsymbol{x}} + \int_B \boldsymbol{b}(\boldsymbol{x}, \alpha) \cdot \dot{\boldsymbol{u}}(\boldsymbol{x}, \beta)\, dv_{\boldsymbol{x}}$$

for all $\alpha, \beta \in [0, \infty)$. *Then*

$$U(t) - K(t) = \tfrac{1}{2} \int_0^t [P(t+\lambda, t-\lambda) - P(t-\lambda, t+\lambda)]\, d\lambda \qquad (0 \leqq t < \infty),$$

where U *is the strain energy and* K *the kinetic energy.*

Proof. By **(18.2)** with $\boldsymbol{b}$ replaced by $\boldsymbol{b}(\cdot, \alpha) - \varrho \ddot{\boldsymbol{u}}(\cdot, \alpha)$, $\mathbf{s}$ by $\mathbf{s}(\cdot, \alpha)$, $\boldsymbol{u}$ by $\dot{\boldsymbol{u}}(\cdot, \beta)$, and $\boldsymbol{S}$ by $\mathbf{C}[\boldsymbol{E}(\cdot, \alpha)]$,

$$P(\alpha, \beta) = \int_B \{\mathbf{C}[\boldsymbol{E}(\alpha)] \cdot \dot{\boldsymbol{E}}(\beta) + \varrho \ddot{\boldsymbol{u}}(\alpha) \cdot \dot{\boldsymbol{u}}(\beta)\}\, dv, \tag{a}$$

where, for convenience, we have suppressed the argument $\boldsymbol{x}$. Next, in view of the symmetry of $\mathbf{C}$ and the initial vanishing of the strain field,

$$\begin{aligned} &\int_B \int_0^t \{\mathbf{C}[\boldsymbol{E}(t+\lambda)] \cdot \dot{\boldsymbol{E}}(t-\lambda) - \mathbf{C}[\boldsymbol{E}(t-\lambda)] \cdot \dot{\boldsymbol{E}}(t+\lambda)\}\, d\lambda\, dv \\ &\qquad = -\int_B \int_0^t \frac{d}{d\lambda} \{\mathbf{C}[\boldsymbol{E}(t+\lambda)] \cdot \boldsymbol{E}(t-\lambda)\}\, d\lambda\, dv = 2U(t). \end{aligned} \tag{b}$$

Similarly,

$$\int_B \int_0^t \varrho \{\ddot{\boldsymbol{u}}(t+\lambda) \cdot \dot{\boldsymbol{u}}(t-\lambda) - \ddot{\boldsymbol{u}}(t-\lambda) \cdot \dot{\boldsymbol{u}}(t+\lambda)\}\, d\lambda\, dv = -2K(t), \tag{c}$$

and (a), (b), and (c) imply the desired result. □

---

[1] Some related inequalities are given by MARTIN [1964, *14*].

[2] BRUN [1965, *4*], Eq. (7b); [1969, *2*].

61. Graffi's reciprocal theorem. We shall assume throughout this section that the elasticity field is *symmetric*.

Given an elastic process $\mathfrak{p}=[\boldsymbol{u}, \boldsymbol{E}, \boldsymbol{S}]$ corresponding to $\boldsymbol{b}$, we call the pair

$$\boldsymbol{u}_0=\boldsymbol{u}(\cdot, 0), \qquad \boldsymbol{v}_0=\dot{\boldsymbol{u}}(\cdot, 0)$$

the ***initial data*** for $\mathfrak{p}$ and the function $\boldsymbol{f}$ defined by

$$\boldsymbol{f}(\boldsymbol{x}, t)=i*\boldsymbol{b}(\boldsymbol{x}, t)+\varrho(\boldsymbol{x})[\boldsymbol{u}_0(\boldsymbol{x})+t\boldsymbol{v}_0(\boldsymbol{x})]$$

the associated ***pseudo body force field***. Here, as before, i is the function on $[0, t_0)$ defined by

$$i(t)=t.$$

Betti's reciprocal theorem has the following important counterpart in elastodynamics:

(1) Graffi's reciprocal theorem.[1] *Let* $[\boldsymbol{u}, \boldsymbol{E}, \boldsymbol{S}]$ *and* $[\tilde{\boldsymbol{u}}, \tilde{\boldsymbol{E}}, \tilde{\boldsymbol{S}}]$ *be elastic processes corresponding to the external force systems* $[\boldsymbol{b}, \mathbf{s}]$ *and* $[\tilde{\boldsymbol{b}}, \tilde{\mathbf{s}}]$. *Further, let* $[\boldsymbol{u}_0, \boldsymbol{v}_0]$ *and* $[\tilde{\boldsymbol{u}}_0, \tilde{\boldsymbol{v}}_0]$ *be the corresponding initial data,* $\boldsymbol{f}$ *and* $\tilde{\boldsymbol{f}}$ *the corresponding pseudo body force fields. Then*

$$i*\int_{\partial B}\mathbf{s}*\tilde{\boldsymbol{u}}\,da+\int_B\boldsymbol{f}*\tilde{\boldsymbol{u}}\,dv=i*\int_{\partial B}\tilde{\mathbf{s}}*\boldsymbol{u}\,da+\int_B\tilde{\boldsymbol{f}}*\boldsymbol{u}\,dv, \tag{a}$$

$$\int_B\boldsymbol{S}*\tilde{\boldsymbol{E}}\,dv=\int_B\tilde{\boldsymbol{S}}*\boldsymbol{E}\,dv, \tag{b}$$

$$\begin{aligned}&\int_{\partial B}\mathbf{s}*\tilde{\boldsymbol{u}}\,da+\int_B\boldsymbol{b}*\tilde{\boldsymbol{u}}\,dv+\int_B\varrho(\boldsymbol{u}_0\cdot\dot{\tilde{\boldsymbol{u}}}+\boldsymbol{v}_0\cdot\tilde{\boldsymbol{u}})\,dv\\&\quad=\int_{\partial B}\tilde{\mathbf{s}}*\boldsymbol{u}\,da+\int_B\tilde{\boldsymbol{b}}*\boldsymbol{u}\,dv+\int_B\varrho(\tilde{\boldsymbol{u}}_0\cdot\dot{\boldsymbol{u}}+\tilde{\boldsymbol{v}}_0\cdot\boldsymbol{u})\,dv.\end{aligned} \tag{c}$$

Thus if both processes correspond to null initial data, then

$$\int_{\partial B}\mathbf{s}*\tilde{\boldsymbol{u}}\,da+\int_B\boldsymbol{b}*\tilde{\boldsymbol{u}}\,dv=\int_{\partial B}\tilde{\mathbf{s}}*\boldsymbol{u}\,da+\int_B\tilde{\boldsymbol{b}}*\boldsymbol{u}\,dv. \tag{d}$$

Proof. In view of the reciprocal theorem ***(19.5)***, it suffices to show that

$$\boldsymbol{S}*\tilde{\boldsymbol{E}}=\tilde{\boldsymbol{S}}*\boldsymbol{E}.$$

By definition,

$$\boldsymbol{S}*\tilde{\boldsymbol{E}}(t)=\int_0^t\boldsymbol{S}(t-\tau)\cdot\tilde{\boldsymbol{E}}(\tau)\,d\tau,$$

where, for convenience, we have omitted the dependence of these functions on $\boldsymbol{x}$. Since $\mathbf{C}$ is symmetric,

$$\boldsymbol{S}(t-\tau)\cdot\tilde{\boldsymbol{E}}(\tau)=\tilde{\boldsymbol{E}}(\tau)\cdot\mathbf{C}[\boldsymbol{E}(t-\tau)]=\boldsymbol{E}(t-\tau)\cdot\mathbf{C}[\tilde{\boldsymbol{E}}(\tau)]=\boldsymbol{E}(t-\tau)\cdot\tilde{\boldsymbol{S}}(\tau).$$

Thus

$$\boldsymbol{S}*\tilde{\boldsymbol{E}}(t)=\boldsymbol{E}*\tilde{\boldsymbol{S}}(t)=\tilde{\boldsymbol{S}}*\boldsymbol{E}(t).\quad\square$$

---

[1] The result (c) for homogeneous and isotropic bodies is due to GRAFFI [1939, *3*], [1947, *2*], [1954, *6*], whose proof is based on Betti's theorem and systematic use of the Laplace transform. Later, GRAFFI [1963, *8*] gave a direct proof utilizing the properties of the convolution. HU [1958, *10*] noticed that GRAFFI's results extend trivially to inhomogeneous and anisotropic bodies; however, HU gave his results in the Laplace transform domain rather than the time domain. WHEELER and STERNBERG [1968, *16*], assuming homogeneity and isotropy, gave an extension of Graffi's theorem valid for infinite regions; their result was extended to homogeneous but anisotropic bodies by WHEELER [1970, *3*]. AINOLA [1966 *1*], showed that Graffi's theorem can be derived from a variational principle. Some interesting applications of Graffi's theorem are given by DIMAGGIO and BLEICH [1959, *4*] and PAYTON [1964, *16*].

GRAFFI[1] also noticed that if both systems correspond to null external forces, then (c) implies

$$\int_B \varrho(\boldsymbol{u}_0 \cdot \dot{\tilde{\boldsymbol{u}}} + \boldsymbol{v}_0 \cdot \tilde{\boldsymbol{u}})\, dv = \int_B \varrho(\tilde{\boldsymbol{u}}_0 \cdot \dot{\boldsymbol{u}} + \tilde{\boldsymbol{v}}_0 \cdot \boldsymbol{u})\, dv;$$

if, in addition, $\boldsymbol{v}_0 = \tilde{\boldsymbol{v}}_0 = \mathbf{0}$, then

$$\int_B \varrho \boldsymbol{u}_0 \cdot \dot{\tilde{\boldsymbol{u}}}\, dv = \int_B \varrho \tilde{\boldsymbol{u}}_0 \cdot \dot{\boldsymbol{u}}\, dv.$$

This relation can be used to establish the orthogonality of the displacement fields corresponding to distinct modes of free vibration.[2]

For the remainder of this section let $t_0 = \infty$. We say that two external force systems $[\boldsymbol{b}, \mathbf{s}]$ and $[\tilde{\boldsymbol{b}}, \tilde{\mathbf{s}}]$ are ***synchronous*** if there exists a continuous scalar function g on $[0, \infty)$ such that

$$\boldsymbol{b}(\boldsymbol{x}, t) = \boldsymbol{b}_0(\boldsymbol{x})\, g(t), \qquad \tilde{\boldsymbol{b}}(\boldsymbol{x}, t) = \tilde{\boldsymbol{b}}_0(\boldsymbol{x})\, g(t)$$

for all $(\boldsymbol{x}, t) \in B \times [0, \infty)$ and

$$\mathbf{s}(\boldsymbol{x}, t) = \mathbf{s}_0(\boldsymbol{x})\, g(t), \qquad \tilde{\mathbf{s}}(\boldsymbol{x}, t) = \tilde{\mathbf{s}}_0(\boldsymbol{x})\, g(t)$$

for all $(\boldsymbol{x}, t) \in \partial B \times [0, \infty)$.

The next theorem shows that for synchronous external force systems a reciprocal theorem holds having the same form as Betti's theorem in elastostatics.

(2) Reciprocal theorem for synchronous external force systems.[3] *Let* $[\boldsymbol{u}, \boldsymbol{E}, \boldsymbol{S}]$ *and* $[\tilde{\boldsymbol{u}}, \tilde{\boldsymbol{E}}, \tilde{\boldsymbol{S}}]$ *be elastic processes corresponding to null initial data and to synchronous external force systems* $[\boldsymbol{b}, \mathbf{s}]$ *and* $[\tilde{\boldsymbol{b}}, \tilde{\mathbf{s}}]$. *Then*

$$\int_{\partial B} \mathbf{s} \cdot \tilde{\boldsymbol{u}}\, da + \int_B \boldsymbol{b} \cdot \tilde{\boldsymbol{u}}\, dv = \int_{\partial B} \tilde{\mathbf{s}} \cdot \boldsymbol{u}\, da + \int_B \tilde{\boldsymbol{b}} \cdot \boldsymbol{u}\, dv.$$

Proof. By hypothesis and (d) of ***(1)***, $g * \varphi = 0$, where

$$\varphi = \int_{\partial B} \mathbf{s}_0 \cdot \tilde{\boldsymbol{u}}\, da + \int_B \boldsymbol{b}_0 \cdot \tilde{\boldsymbol{u}}\, dv - \int_{\partial B} \tilde{\mathbf{s}}_0 \cdot \boldsymbol{u}\, da - \int_B \tilde{\boldsymbol{b}}_0 \cdot \boldsymbol{u}\, dv,$$

and we conclude from (v) of ***(10.1)*** that either $g = 0$ or $\varphi = 0$. If $g = 0$ the theorem holds trivially. Thus assume $\varphi = 0$. Then $g\varphi = 0$, and the desired result follows. □

II. Boundary-initial-value problems. Uniqueness.

62. The boundary-initial-value problem of elastodynamics. Throughout the following three sections we assume given the following *data*: a time interval $[0, t_0)$, an elasticity field C on B, a density field ϱ on B, body forces $\boldsymbol{b}$ on $B \times (0, t_0)$, initial displacements $\boldsymbol{u}_0$ on B, initial velocities $\boldsymbol{v}_0$ on B, surface displacements $\hat{\boldsymbol{u}}$ on $\mathscr{S}_1 \times (0, t_0)$, and surface tractions $\hat{\mathbf{s}}$ on $\mathscr{S}_2 \times (0, t_0)$. Given the above data, the *mixed problem of elastodynamics* consists in finding an elastic process $[\boldsymbol{u}, \boldsymbol{E}, \boldsymbol{S}]$ that corresponds to $\boldsymbol{b}$ and satisfies the ***initial conditions***

$$\boldsymbol{u}(\cdot, 0) = \boldsymbol{u}_0 \quad \text{and} \quad \dot{\boldsymbol{u}}(\cdot, 0) = \boldsymbol{v}_0 \quad \text{on } B,$$

[1] [1947, *2*].
[2] Cf. (*76. 3*)
[3] GRAFFI [1947, *2*].

the ***displacement condition***

$$\boldsymbol{u} = \hat{\boldsymbol{u}} \quad \text{on } \mathcal{S}_1 \times (0, t_0),$$

and the ***traction condition***

$$\boldsymbol{s} = \hat{\boldsymbol{s}} \quad \text{on } \mathcal{S}_2 \times (0, t_0).$$

We call such an elastic process a ***solution of the mixed problem***.[1] As before, when $\mathcal{S}_1 = \partial B$ ($\mathcal{S}_2 = \emptyset$) we refer to the ***displacement problem***, when $\mathcal{S}_2 = \partial B$ ($\mathcal{S}_1 = \emptyset$) we refer to the ***traction problem***.

To avoid repeated hypotheses we assume once and for all that:

(i) C and ϱ are smooth on $\bar{B}$;

(ii) $\boldsymbol{b}$ is continuous on $\bar{B} \times [0, t_0)$;

(iii) $\boldsymbol{u}_0$ and $\boldsymbol{v}_0$ are continuous on $\bar{B}$;

(iv) $\hat{\boldsymbol{u}}$ is continuous on $\mathcal{S}_1 \times [0, t_0)$;

(v) $\hat{\boldsymbol{s}}$ is continuous in time and piecewise regular on $\mathcal{S}_2 \times [0, t_0)$.

The following proposition is an immediate consequence of the remarks made in Sect. 59; its proof is completely analogous to the proof of ***(31.1)*** and can safely be omitted.

(1) Characterization of the mixed problem in terms of displacements. *Let $\boldsymbol{u}$ be an admissible motion. Then $\boldsymbol{u}$ corresponds to a solution of the mixed problem if and only if*

$$\operatorname{div} \mathsf{C}[\nabla \boldsymbol{u}] + \boldsymbol{b} = \varrho \ddot{\boldsymbol{u}} \quad \text{on } B \times (0, t_0),$$
$$\boldsymbol{u}(\cdot, 0) = \boldsymbol{u}_0 \quad \text{and} \quad \dot{\boldsymbol{u}}(\cdot, 0) = \boldsymbol{v}_0 \quad \text{on } B,$$
$$\boldsymbol{u} = \hat{\boldsymbol{u}} \quad \text{on } \mathcal{S}_1 \times (0, t_0),$$
$$\mathsf{C}[\nabla \boldsymbol{u}]\, \boldsymbol{n} = \hat{\boldsymbol{s}} \quad \text{on } \mathcal{S}_2 \times (0, t_0).$$

Further, when the body is homogeneous and isotropic, the first and last of these conditions may be replaced by

$$\mu \Delta \boldsymbol{u} + (\lambda + \mu) \nabla \operatorname{div} \boldsymbol{u} + \boldsymbol{b} = \varrho \ddot{\boldsymbol{u}} \quad \text{on } B \times (0, t_0),$$
$$\mu (\nabla \boldsymbol{u} + \nabla \boldsymbol{u}^T)\, \boldsymbol{n} + \lambda (\operatorname{div} \boldsymbol{u})\, \boldsymbol{n} = \hat{\boldsymbol{s}} \quad \text{on } \mathcal{S}_2 \times (0, t_0).$$

For the remainder of this section we assume that

$$\varrho > 0.$$

(2) Characterization of the traction problem in terms of stresses.[2] *Assume that:*

(α) *C is invertible on $\bar{B}$;*

(β) *ϱ, $\boldsymbol{u}_0$, and $\boldsymbol{v}_0$ are of class C^2 on B;*

(γ) *$\boldsymbol{b}$ is of class $C^{2,0}$ on $B \times [0, t_0)$;*

(δ) *$\boldsymbol{S}$ is a time-dependent admissible stress field*[3] *of class $C^{3,2}$ on $B \times [0, t_0)$ with $\dot{\boldsymbol{S}}$ continuous on $\bar{B} \times [0, t_0)$.*

[1] See footnote 2 on p. 102.
[2] IGNACZAK [1963, *14*] for the case in which $\boldsymbol{u}_0 = \boldsymbol{v}_0 = \boldsymbol{0}$.
[3] See p. 49.

Then $\boldsymbol{S}$ corresponds to a solution of the traction problem if and only if

$$\hat{\nabla}\left(\frac{\operatorname{div}\boldsymbol{S}+\boldsymbol{b}}{\varrho}\right)=\mathbf{K}[\ddot{\boldsymbol{S}}] \quad \textit{on } B\times(0,t_0),$$

$$\boldsymbol{S}(\cdot,0)=\mathbf{C}[\nabla\boldsymbol{u}_0] \quad \textit{and} \quad \dot{\boldsymbol{S}}(\cdot,0)=\mathbf{C}[\nabla\boldsymbol{v}_0] \quad \textit{on } B,$$

$$\boldsymbol{s}=\hat{\boldsymbol{s}} \quad \textit{on } \partial B\times(0,t_0).$$

Proof. The necessity of the above relations follows from the discussion given at the end of Sect. 59. The sufficiency also follows from this discussion, but we will give the argument in detail. Let $\boldsymbol{S}$ satisfy the above system of equations. Then, since $\boldsymbol{S}$ is a time-dependent admissible stress field,

(1) $\boldsymbol{S}$ and div $\boldsymbol{S}$ are continuous on $\overline{B}\times[0,t_0)$; and, by (i) and ($\alpha$),

(2) $\mathbf{K}=\mathbf{C}^{-1}$ is continuous on $\overline{B}$.

We now define $\boldsymbol{u}$ and $\boldsymbol{E}$ through the relations

$$\boldsymbol{u}=\frac{i*\operatorname{div}\boldsymbol{S}+\boldsymbol{f}}{\varrho}, \qquad \boldsymbol{E}=\mathbf{K}[\boldsymbol{S}], \tag{a}$$

where $\boldsymbol{f}$ is the pseudo body force field

$$\boldsymbol{f}(\boldsymbol{x},t)=i*\boldsymbol{b}(\boldsymbol{x},t)+\varrho(\boldsymbol{x})\,[\boldsymbol{u}_0(\boldsymbol{x})+t\,\boldsymbol{v}_0(\boldsymbol{x})], \tag{b}$$
$$i(t)=t.$$

It then follows that $\boldsymbol{u}$, $\boldsymbol{S}$, and $\boldsymbol{b}$ satisfy the equation of motion,[1] and that $\boldsymbol{u}$ satisfies the initial conditions. Next, by (ii), (iii), (β), (γ), (δ), (1), (2), (a), and (b),

(3) $\boldsymbol{u}$, $\dot{\boldsymbol{u}}$, $\ddot{\boldsymbol{u}}$, $\boldsymbol{E}$, and $\dot{\boldsymbol{E}}$ are continuous on $\overline{B}\times[0,t_0)$, while $\boldsymbol{u}$ is of class C^2 on $B\times[0,t_0)$;

(4) $\boldsymbol{E}$ is of class $C^{0,2}$ on $B\times[0,t_0)$, $\nabla(\ddot{\boldsymbol{u}})$ and $\ddot{\overline{\nabla\boldsymbol{u}}}$ are continuous on $B\times[0,t_0)$, and $\nabla(\ddot{\boldsymbol{u}})=\ddot{\overline{\nabla\boldsymbol{u}}}$.

To complete the proof we have only to show that the strain-displacement relation holds (for then we could conclude from (3) and (4) that $\boldsymbol{u}$ is an admissible motion). It follows from (4), $(\mathrm{a})_2$, the first relation in ***(2)***, and the equation of motion that

$$\ddot{\boldsymbol{E}}=\mathbf{K}[\ddot{\boldsymbol{S}}]=\hat{\nabla}\left(\frac{\operatorname{div}\boldsymbol{S}+\boldsymbol{b}}{\varrho}\right)=\hat{\nabla}(\ddot{\boldsymbol{u}})=\ddot{\overline{\hat{\nabla}\boldsymbol{u}}}. \tag{c}$$

On the other hand, since $\boldsymbol{u}$ satisfies the initial conditions, we conclude from $(\mathrm{a})_2$ and the initial conditions satisfied by $\boldsymbol{S}$ that

$$\boldsymbol{E}(\cdot,0)=\hat{\nabla}\boldsymbol{u}(\cdot,0), \qquad \dot{\boldsymbol{E}}(\cdot,0)=\hat{\nabla}\dot{\boldsymbol{u}}(\cdot,0); \tag{d}$$

thus, by (c) and (d), $\boldsymbol{E}=\hat{\nabla}\boldsymbol{u}$. □

An interesting consequence of theorem ***(62.2)*** and the uniqueness theorem ***(63.1)*** is that *a body force field with $\nabla(\boldsymbol{b}/\varrho)$ skew induces zero stresses in the body*; i.e., if ∂B is traction-free, if $\boldsymbol{u}_0=\boldsymbol{v}_0=\boldsymbol{0}$, and if $\nabla(\boldsymbol{b}/\varrho)$ is skew, then $\boldsymbol{S}\equiv\boldsymbol{0}$ and $\boldsymbol{u}=i*\boldsymbol{b}/\varrho$ is a rigid motion.

The next two theorems give alternative characterizations of the mixed problem in which the initial conditions are incorporated into the field equations. These results will be extremely useful in the derivation of variational principles.

[1] Cf. the proof of ***(19.2)***.

(3)[1] *Let $\mathfrak{p}=[\boldsymbol{u}, \boldsymbol{E}, \boldsymbol{S}]$ be an admissible process. Then $\mathfrak{p}$ is a solution of the mixed problem if and only if*

$$\boldsymbol{E}=\hat{\nabla}\boldsymbol{u},$$
$$\boldsymbol{S}=\mathbf{C}[\boldsymbol{E}],$$
$$i*\operatorname{div}\boldsymbol{S}+\boldsymbol{f}=\varrho\boldsymbol{u}$$

on $B\times[0, t_0)$, and

$$\boldsymbol{u}=\hat{\boldsymbol{u}} \quad on\ \mathscr{S}_1\times(0, t_0),$$
$$\mathbf{s}=\hat{\mathbf{s}} \quad on\ \mathscr{S}_2\times(0, t_0).$$

Proof. This theorem is a direct consequence of ***(19.2)***. □

(4) Characterization of the mixed problem in terms of stresses.[2] *Assume that hypotheses* (α), (β), and (γ) *of **(2)** hold. Let $\boldsymbol{S}$ be a time-dependent admissible stress field of class $C^{3,0}$ on $B\times[0, t_0)$, and suppose that $\dot{\boldsymbol{S}}$ is continuous on $\bar{B}\times[0, t_0)$. Then $\boldsymbol{S}$ corresponds to a solution of the mixed problem if and only if*

$$\hat{\nabla}\left\{\frac{i*\operatorname{div}\boldsymbol{S}+\boldsymbol{f}}{\varrho}\right\}=\mathbf{K}[\boldsymbol{S}] \quad on\ B\times[0, t_0),$$
$$i*\operatorname{div}\boldsymbol{S}=\varrho\hat{\boldsymbol{u}}-\boldsymbol{f} \quad on\ \mathscr{S}_1\times(0, t_0),$$
$$\mathbf{s}=\hat{\mathbf{s}} \quad on\ \mathscr{S}_2\times(0, t_0).$$

Proof. Clearly, (1) and (2) in the proof of ***(2)*** hold. Assume that $\boldsymbol{S}$ satisfies the above relations. If we define

$$\boldsymbol{u}=\frac{i*\operatorname{div}\boldsymbol{S}+\boldsymbol{f}}{\varrho}, \qquad \boldsymbol{E}=\mathbf{K}[\boldsymbol{S}],$$

then (3) of ***(2)*** also holds, and it is not difficult to verify that $\hat{\nabla}\boldsymbol{u}=\boldsymbol{E}$ and $\boldsymbol{u}=\hat{\boldsymbol{u}}$ on $\mathscr{S}_1\times(0, t_0)$. It therefore follows from ***(3)*** that $\mathfrak{p}=[\boldsymbol{u}, \boldsymbol{E}, \boldsymbol{S}]$ is a solution of the mixed problem. Conversely, if $\mathfrak{p}$ is a solution of the mixed problem, then we conclude, with the aid of ***(3)***, that the relations in ***(4)*** hold. □

63. Uniqueness. In this section we discuss the uniqueness question appropriate to the fundamental initial-boundary-value problems discussed in the preceding section. We assume throughout that the density ϱ is *continuous* on $\bar{B}$.

We will first establish the classical uniqueness theorem of NEUMANN. We will then give a more general theorem due to BRUN; we present NEUMANN's theorem separately because its hypotheses are sufficiently general to include most applications, and because its proof is quite simple.

(1) Uniqueness theorem for the mixed problem.[3] *Suppose that the density field is strictly positive and the elasticity field symmetric and positive semi-definite. Then the mixed problem of elastodynamics has at most one solution.*

Proof. Let $\mathfrak{p}=[\boldsymbol{u}, \boldsymbol{E}, \boldsymbol{S}]$ be the difference between two solutions. Then $\mathfrak{p}$ corresponds to vanishing body forces and satisfies

$$\boldsymbol{u}(\cdot, 0)=\dot{\boldsymbol{u}}(\cdot, 0)=\mathbf{0} \quad \text{on } B, \tag{a}$$

$$\boldsymbol{u}=\mathbf{0} \quad \text{on } \mathscr{S}_1\times(0, t_0), \qquad \mathbf{s}=\mathbf{0} \quad \text{on } \mathscr{S}_2\times(0, t_0). \tag{b}$$

By (a)

$$\boldsymbol{E}(\cdot, 0)=\mathbf{0}, \tag{c}$$

[1] GURTIN [1964, *8*], Theorem 3.2.
[2] GURTIN [1964, *8*], Theorem 3.4.
[3] NEUMANN [1885, *3*], § 61. An interesting generalization was given by HILL [1967, *9*].

and (a) and (c) yield

$$\mathscr{U}(0)=0, \tag{d}$$

where $\mathscr{U}$ is the total energy.[1] Moreover, in view of ***(60.3)***, (b) and (d) yield

$$\mathscr{U}(t)=0, \quad 0\leqq t<t_0. \tag{e}$$

By definition $\mathscr{U}$ is the sum of the kinetic and strain energies, and by hypothesis both of these energies are positive; therefore they must both vanish. In particular,

$$\int_B \varrho\, \dot{\boldsymbol{u}}^2\, dv=0, \tag{f}$$

and, since $\varrho>0$,

$$\dot{\boldsymbol{u}}=\mathbf{0} \quad \text{on } B\times(0, t_0).$$

This fact, when combined with the first of (a), yields

$$\boldsymbol{u}=\mathbf{0} \quad \text{on } B\times(0, t_0),$$

and the proof is complete. □

Recently BRUN has established the following major generalization of Neumann's theorem ***(1)***; it is important to note that no assumptions other than symmetry are placed on the elasticity field. We suppose for the purpose of this theorem that $t_0=\infty$, so that the underlying time interval is $[0, \infty)$.

(2) Extended uniqueness theorem.[2] *Suppose that the density field is strictly positive (or negative) and the elasticity field symmetric. Then the mixed problem of elastodynamics has at most one solution.*

Proof. As in the proof of ***(1)***, let $\mathfrak{p}=[\boldsymbol{u}, \boldsymbol{E}, \boldsymbol{S}]$ be the difference between two solutions. Then $\mathfrak{p}$ corresponds to vanishing body forces and satisfies (a), (b), and (c) in the proof of ***(1)***. Thus we conclude from Brun's theorem ***(60.4)*** that

$$U(t)-K(t)=0, \quad 0\leqq t<\infty. \tag{a}$$

On the other hand, by (e) in the proof of ***(1)***, which also holds in the present circumstances,

$$U(t)+K(t)=0, \quad 0\leqq t<\infty. \tag{b}$$

Thus $K(t)\equiv 0$ and the last few steps in ***(1)*** yield the desired conclusion that $\boldsymbol{u}\equiv\mathbf{0}$. □

III. Variational principles.

64. Some further extensions of the fundamental lemma. We now establish, for functions of position and time, counterparts of the lemmas established in Sect. 35. With this in mind, we introduce the following definition. We say that a function f on $B\times[0, t_0)$ ***vanishes near*** $\mathscr{S}\subset\partial B$ if there exists a neighborhood N of $\mathscr{S}$ such that $f=0$ on $(N\cap\bar{B})\times[0, t_0)$.

---

[1] See p. 216.

[2] BRUN [1965, *4*], [1969, *2*]. This theorem was established independently by KNOPS and PAYNE [1968, *11*], using an entirely different method. For the displacement problem appropriate to a homogeneous and isotropic body, GURTIN and STERNBERG [1961, *10*] proved that uniqueness holds provided both wave speeds are positive; i.e. provided ϱ is positive and the elasticity field strongly elliptic. This result was extended to anisotropic, but homogeneous, bodies by GURTIN and TOUPIN [1965, *9*]. HAYES and KNOPS [1968, *7*] showed, without assuming $\mathbf{C}$ symmetric, that uniqueness also holds when $-\mathbf{C}$ is strongly elliptic. Of course, the extended uniqueness theorem ***(2)*** includes as special cases NEUMANN's theorem and all of the above results except that of HAYES and KNOPS. See also KNOPS and PAYNE [1971, *2*], § 8, for a thorough discussion of uniqueness in elastodynamics.

(1)[1] *Let $\mathscr{W}$ be a finite-dimensional inner product space. Let $\boldsymbol{w}: B \times [0, t_0) \to \mathscr{W}$ be continuous and satisfy*

$$\int_B \boldsymbol{w} * \boldsymbol{v}(\boldsymbol{x}, t)\, dv_{\boldsymbol{x}} = 0 \quad (0 \leqq t < t_0)$$

for every class C^∞ function $\boldsymbol{v}: B \times [0, t_0) \to \mathscr{W}$ that vanishes near ∂B. Then

$$\boldsymbol{w} = \boldsymbol{0} \quad \text{on } \bar{B} \times [0, t_0).$$

Proof. Let $I_h(\tau)$ denote the open interval $(\tau - h, \tau + h)$, and for $I_h(\tau) \subset (0, t_0)$ let $\Psi_h(\tau)$ be the set of all C^∞ scalar functions ψ on $[0, t_0)$ with the properties:

(a) $\psi > 0$ on $I_h(\tau)$,

(b) $\psi = 0$ on $[0, t_0) - I_h(\tau)$.

It is clear from the discussion at the beginning of Sect. 7 that $\Psi_h(\tau)$ is not empty.

Let $\boldsymbol{e}_1, \boldsymbol{e}_2, \ldots, \boldsymbol{e}_n$ be an orthonormal basis in $\mathscr{W}$, and let

$$\boldsymbol{w} = \sum_{i=1}^{n} w_i\, \boldsymbol{e}_i.$$

As in the proof of the fundamental lemma ***(7.1)***, we assume that for some $(\boldsymbol{x}_0, \tau) \in B \times (0, t_0)$ and some integer k, $w_k(\boldsymbol{x}_0, \tau) > 0$. Then there exists an $h > 0$ such that $\overline{\Sigma_h(\boldsymbol{x}_0)} \subset B$, $I_h(\tau) \subset (0, t_0)$, and $w_k > 0$ on $\Sigma_h(\boldsymbol{x}_0) \times I_h(\tau)$. Let

$$\boldsymbol{v}(\boldsymbol{x}, s) = \varphi(\boldsymbol{x})\, \psi(s)\, \boldsymbol{e}_k,$$

where[2] $\varphi \in \Phi_h(\boldsymbol{x}_0)$, $\psi \in \Psi_h(t - \tau)$, and $t \in (\tau + h, t_0)$. Then $\boldsymbol{v}$ is of class C^∞ on $\bar{B} \times [0, t_0)$, vanishes near ∂B, and

$$\begin{aligned} \int_B \boldsymbol{w} * \boldsymbol{v}(\boldsymbol{x}, t)\, dv_{\boldsymbol{x}} &= \int_B \int_0^t \boldsymbol{w}(\boldsymbol{x}, t - s) \cdot \boldsymbol{v}(\boldsymbol{x}, s)\, ds\, dv_{\boldsymbol{x}} \\ &= \int_B \int_0^t w_k(\boldsymbol{x}, t - s)\, \varphi(\boldsymbol{x})\, \psi(s)\, ds\, dv_{\boldsymbol{x}} \\ &= \int_{\Sigma_h(x_0)} \int_{t-\tau-h}^{t-\tau+h} w_k(\boldsymbol{x}, t - s)\, \varphi(\boldsymbol{x})\, \psi(s)\, ds\, dv_{\boldsymbol{x}} > 0, \end{aligned}$$

which is a contradiction. Thus $\boldsymbol{w} = \boldsymbol{0}$ on $\bar{B} \times [0, t_0)$. □

The proofs of the next three lemmas can safely be omitted; they are completely analogous, respectively, to the proofs of ***(35.1)***, ***(35.2)***, and ***(35.3)*** and utilize the procedure given in the proof of ***(1)***.

(2)[3] *Let $\mathscr{W}$ be a finite-dimensional inner product space. Let $\boldsymbol{w}: \mathscr{S}_2 \times [0, t_0) \to \mathscr{W}$ be piecewise regular and continuous in time, and assume that*

$$\int_{\mathscr{S}_2} \boldsymbol{w} * \boldsymbol{v}(\boldsymbol{x}, t)\, da_{\boldsymbol{x}} = 0 \quad (0 \leqq t < t_0)$$

for every class C^∞ function $\boldsymbol{v}: \bar{B} \times [0, t_0) \to \mathscr{W}$ that vanishes near $\mathscr{S}_1$. Then

$$\boldsymbol{w} = \boldsymbol{0} \quad \text{on } \mathscr{S}_2 \times [0, t_0).$$

[1] GURTIN [1964, *8*], Lemma 2.1.
[2] $\Phi_h(\boldsymbol{x}_0)$ is defined in Sect. 7.
[3] GURTIN [1964, *8*], Lemma 2.2.

(3)[1] *Let* $\boldsymbol{u}$ *be a vector field on* $\mathscr{S}_1 \times [0, t_0)$ *that is piecewise regular and continuous in time, and suppose that*

$$\int_{\mathscr{S}_1} (\boldsymbol{S}\boldsymbol{n}) * \boldsymbol{u}(\boldsymbol{x}, t)\, d\boldsymbol{v}_{\boldsymbol{x}} = 0 \qquad (0 \leqq t < t_0)$$

for every class C^∞ *symmetric tensor field* $\mathbf{S}$ *on* $\bar{B} \times [0, t_0)$ *which vanishes near* $\mathscr{S}_2$. *Then*

$$\boldsymbol{u} = \boldsymbol{0} \qquad on\ \ \mathscr{S}_1 \times [0, t_0).$$

(4)[2] *Let* $\boldsymbol{u}$ *be a vector field on* $\mathscr{S}_2 \times [0, t_0)$ *that is piecewise regular and continuous in time, and suppose that*

$$\int_{\mathscr{S}_2} \boldsymbol{u} * \operatorname{div} \mathbf{S}(\boldsymbol{x}, t)\, dv_{\boldsymbol{x}} = 0 \qquad (0 \leqq t < t_0)$$

for every class C^∞ *symmetric tensor field* $\mathbf{S}$ *on* $\bar{B} \times [0, t_0)$. *Then*

$$\boldsymbol{u} = \boldsymbol{0} \qquad on\ \ \mathscr{S}_2 \times [0, t_0).$$

65. Variational principles. In this section we will establish several variational principles for elastodynamics. In addition to hypotheses (i)–(v)[3] of Sect. 62 we assume throughout that the elasticity field $\mathbf{C}$ is *symmetric*. We begin by giving KIRCHHOFF's generalization of Hamilton's principle in particle mechanics.

By a ***kinematically admissible process*** we mean an admissible process that satisfies the strain-displacement relation, the stress-strain relation, and the displacement boundary condition.

(1) Hamilton-Kirchhoff principle.[4] *Let* $\mathscr{A}$ *denote the set of all kinematically admissible processes* $\not{p} = [\boldsymbol{u}, \boldsymbol{E}, \boldsymbol{S}]$ *that have* $\ddot{\boldsymbol{u}}$ *and* $\boldsymbol{E}$ *continuous on* $\bar{B} \times [0, t_0]$ *and*

$$\boldsymbol{u}(\cdot, 0) = \boldsymbol{\alpha}, \qquad \boldsymbol{u}(\cdot, t_0) = \boldsymbol{\beta} \qquad on\ \ B,$$

where $\boldsymbol{\alpha}$ *and* $\boldsymbol{\beta}$ *are prescribed vector fields. Let* $\eta\{\cdot\}$ *be the functional on* $\mathscr{A}$ *defined by*

$$\eta\{\not{p}\} = \int_0^{t_0} U\{\boldsymbol{E}\}\, dt - \int_0^{t_0}\int_B \left(\tfrac{1}{2}\varrho \dot{\boldsymbol{u}}^2 + \boldsymbol{b} \cdot \boldsymbol{u}\right) dv\, dt - \int_0^{t_0}\int_{\mathscr{S}_2} \hat{\mathbf{s}} \cdot \boldsymbol{u}\, da\, dt$$

for every $\not{p} = [\boldsymbol{u}, \boldsymbol{E}, \boldsymbol{S}] \in \mathscr{A}$. *Then*

$$\delta\eta\{\not{p}\} = 0$$

at $\not{p} \in \mathscr{A}$ *if and only if* $\not{p}$ *satisfies the equation of motion and the traction boundary condition.*

Proof. Let $\not{p} \in \mathscr{A}$, and let $\tilde{\not{p}} = [\tilde{\boldsymbol{u}}, \widetilde{\boldsymbol{E}}, \widetilde{\boldsymbol{S}}]$ be an admissible process with the following property:

$$\not{p} + \lambda \tilde{\not{p}} \in \mathscr{A} \qquad \text{for every scalar } \lambda. \tag{a}$$

[1] GURTIN [1964, *8*], Lemma 2.3. In 1965 HLAVÁČEK (private communication) pointed out that the proof given by GURTIN [1964, *8*] is incorrect and furnished a valid proof.

[2] GURTIN [1964, *8*], Lemma 2.4.

[3] For the validity of *(1)* we need the stronger assumption that (ii), (iv), and (v) hold with the interval $[0, t_0)$ replaced by $[0, t_0]$.

[4] KIRCHHOFF [1852, *1*], [1859, *1*], [1876, *1*], § 11. See also LOVE [1927, *3*], § 115; LOCATELLI [1945, *3*, *4*]; WASHIZU [1957, *18*]; KNESHKE and RÜDIGER [1962, *10*]; CHEN [1964, *6*]; YU [1964, *23*]; BEN-AMOZ [1966, *2*]; KARNOPP [1966, *10*], [1968, *10*]; KOMKOV [1966, *15*]; TONTI [1966, *26*].

Clearly, (a) holds if and only if $\tilde{p}$ satisfies the strain-displacement relation, the stress-strain relation, the boundary condition

$$\tilde{\boldsymbol{u}}=\mathbf{0} \quad \text{on } \mathscr{S}_1\times[0,t_0], \tag{b}$$

and the end conditions

$$\tilde{\boldsymbol{u}}(\cdot,0)=\tilde{\boldsymbol{u}}(\cdot,t_0)=\mathbf{0} \quad \text{on } B. \tag{c}$$

Next,

$$\begin{aligned}\delta_{\tilde{p}}\,\eta\{p\}&=\frac{d}{d\lambda}\,\eta\{p+\lambda\,\tilde{p}\}|_{\lambda=0}\\ &=\int_0^{t_0}\int_B(\mathbf{S}\cdot\tilde{\boldsymbol{E}}-\varrho\dot{\boldsymbol{u}}\cdot\dot{\tilde{\boldsymbol{u}}}-\boldsymbol{b}\cdot\tilde{\boldsymbol{u}})\,dv\,dt-\int_0^{t_0}\int_{\mathscr{S}_2}\hat{\mathbf{s}}\cdot\tilde{\boldsymbol{u}}\,da\,dt,\end{aligned} \tag{d}$$

where we have used the symmetry of $\mathbf{C}$ and the stress-strain relation. By (c)

$$\int_0^{t_0}\varrho\dot{\boldsymbol{u}}\cdot\dot{\tilde{\boldsymbol{u}}}\,dt=-\int_0^{t_0}\varrho\ddot{\boldsymbol{u}}\cdot\tilde{\boldsymbol{u}}\,dt, \tag{e}$$

and (d), (e), the fact that $\tilde{\boldsymbol{E}}$ and $\tilde{\boldsymbol{u}}$ satisfy the strain-displacement relation, ***(18.1)***, and (b) imply

$$\delta_{\tilde{p}}\,\eta\{p\}=-\int_0^{t_0}\int_B(\operatorname{div}\mathbf{S}+\boldsymbol{b}-\varrho\ddot{\boldsymbol{u}})\cdot\tilde{\boldsymbol{u}}\,dv+\int_0^{t_0}\int_{\mathscr{S}_2}(\mathbf{s}-\hat{\mathbf{s}})\cdot\tilde{\boldsymbol{u}}\,da\,dt. \tag{f}$$

Clearly, $\delta\eta\{p\}=0$ if

$$\operatorname{div}\mathbf{S}+\boldsymbol{b}=\varrho\ddot{\boldsymbol{u}} \quad \text{on } B\times[0,t_0), \qquad \mathbf{s}=\hat{\mathbf{s}} \quad \text{on } \mathscr{S}_2\times[0,t_0). \tag{g}$$

On the other hand, if (f) vanishes for every $\tilde{p}$ consistent with (a), then (f) vanishes for every admissible motion $\tilde{\boldsymbol{u}}$ consistent with (b) and (c), and obvious extensions of ***(7.1)*** and ***(35.1)*** yield (g). □

The Hamilton-Kirchhoff principle is concerned with the variation of a functional over a set of processes that assume a given displacement distribution at an initial as well as at a *later time*. This type of principle clearly does *not* characterize the initial-value problem of elastodynamics, since it fails to take into account the initial velocity distribution and presupposes the knowledge of the displacements at a later time—an item of information not available in advance. We now give several variational principles which do, in fact, characterize this initial-value problem.

We assume for the remainder of this section that the density ϱ is *strictly positive*. Note that, in view of assumptions (i)–(iii) of Sect. 62, the pseudo body force field

$$\boldsymbol{f}(\boldsymbol{x},t)=i*\boldsymbol{b}(\boldsymbol{x},t)+\varrho(\boldsymbol{x})\,[\boldsymbol{u}_0(\boldsymbol{x})+t\boldsymbol{v}_0(\boldsymbol{x})], \quad i(t)=t$$

is continuous on $\bar{B}\times[0,t_0)$.

We begin with a variational principle in which the admissible processes are not required to meet any of the field equations, initial conditions, or boundary conditions; this principle is a counterpart of the Hu-Washizu principle ***(38.1)*** in elastostatics.

(2)[1] *Let $\mathscr{A}$ denote the set of all admissible processes, and for each $t\in[0,t_0)$ define the functional $\Lambda_t\{\cdot\}$ on $\mathscr{A}$ by*

$$\begin{aligned}\Lambda_t\{p\}=&\int_B\{\tfrac{1}{2}i*\boldsymbol{E}*\mathbf{C}[\boldsymbol{E}]+\tfrac{1}{2}\,\varrho\boldsymbol{u}*\boldsymbol{u}-i*\boldsymbol{S}*\boldsymbol{E}-[i*\operatorname{div}\boldsymbol{S}+\boldsymbol{f}]*\boldsymbol{u}\}(\boldsymbol{x},t)\,dv_{\boldsymbol{x}}\\ &+\int_{\mathscr{S}_1}\{i*\mathbf{s}*\hat{\boldsymbol{u}}\}(\boldsymbol{x},t)\,da_{\boldsymbol{x}}+\int_{\mathscr{S}_2}\{i*(\mathbf{s}-\hat{\mathbf{s}})*\boldsymbol{u}\}\,da_{\boldsymbol{x}}\end{aligned}$$

[1] GURTIN [1964, *8*], Theorem 4.1.

for every $\mathfrak{p}=[\boldsymbol{u}, \boldsymbol{E}, \boldsymbol{S}]\in\mathscr{A}$. *Then*

$$\delta\Lambda_t\{\mathfrak{p}\}=0 \qquad (0\leqq t<t_0)$$

at an admissible process $\mathfrak{p}$ *if and only if* $\mathfrak{p}$ *is a solution of the mixed problem.*

Proof. Let $\mathfrak{p}=[\boldsymbol{u}, \boldsymbol{E}, \boldsymbol{S}]$ and $\tilde{\mathfrak{p}}=[\tilde{\boldsymbol{u}}, \tilde{\boldsymbol{E}}, \tilde{\boldsymbol{S}}]$ be admissible processes. Then $\mathfrak{p}+\lambda\tilde{\mathfrak{p}}$ is an admissible process for every scalar λ. Since $\mathbf{C}$ is symmetric,

$$\tilde{\boldsymbol{E}}(t-\tau)\cdot\mathbf{C}[\boldsymbol{E}(\tau)]=\mathbf{C}[\tilde{\boldsymbol{E}}(t-\tau)]\cdot\boldsymbol{E}(\tau),$$

and thus

$$\tilde{\boldsymbol{E}}*\mathbf{C}[\boldsymbol{E}]=\mathbf{C}[\tilde{\boldsymbol{E}}]*\boldsymbol{E}=\boldsymbol{E}*\mathbf{C}[\tilde{\boldsymbol{E}}]. \tag{a}$$

Let

$$U_t\{\boldsymbol{E}\}=\tfrac{1}{2}\int_B i*\boldsymbol{E}*\mathbf{C}[\boldsymbol{E}](\boldsymbol{x}, t)\, dv_{\boldsymbol{x}}.$$

Then (a) implies

$$U_t\{\boldsymbol{E}+\lambda\tilde{\boldsymbol{E}}\}=U_t\{\boldsymbol{E}\}+\lambda^2\, U_t\{\tilde{\boldsymbol{E}}\}+\lambda\int_B i*\tilde{\boldsymbol{E}}*\mathbf{C}[\boldsymbol{E}](\boldsymbol{x}, t)\, dv_{\boldsymbol{x}}. \tag{b}$$

Thus, since

$$\delta_{\tilde{\mathfrak{p}}}\Lambda_t\{\mathfrak{p}\}=\frac{d}{d\lambda}\Lambda_t\{\mathfrak{p}+\lambda\tilde{\mathfrak{p}}\}\big|_{\lambda=0},$$

it follows that

$$\begin{aligned}&\delta_{\tilde{\mathfrak{p}}}\Lambda_t\{\mathfrak{p}\}\\&=\int_B\{i*(\mathbf{C}[\boldsymbol{E}]-\boldsymbol{S})*\tilde{\boldsymbol{E}}-i*\tilde{\boldsymbol{S}}*\boldsymbol{E}-(i*\operatorname{div}\boldsymbol{S}+\boldsymbol{f}-\varrho\,\boldsymbol{u})*\tilde{\boldsymbol{u}}-i*\boldsymbol{u}*\operatorname{div}\tilde{\boldsymbol{S}}\}(\boldsymbol{x}, t)\, dv_{\boldsymbol{x}}\\&\quad+\int_{\mathscr{S}_1}\{i*\tilde{\mathbf{s}}*\hat{\boldsymbol{u}}\}(\boldsymbol{x}, t)\, da_{\boldsymbol{x}}+\int_{\mathscr{S}_2} i*\{\tilde{\mathbf{s}}*\boldsymbol{u}+(\mathbf{s}-\hat{\mathbf{s}})*\tilde{\boldsymbol{u}}\}(\boldsymbol{x}, t)\, da_{\boldsymbol{x}}.\end{aligned}$$

If we apply ***(19.3)*** to $\tilde{\boldsymbol{S}}$ and $\boldsymbol{u}$, we find that

$$i*\int_B \boldsymbol{u}*\operatorname{div}\tilde{\boldsymbol{S}}\, dv=i*\int_{\partial B}\tilde{\mathbf{s}}*\boldsymbol{u}\, da-i*\int_B\tilde{\boldsymbol{S}}*\hat{\nabla}\boldsymbol{u}\, dv;$$

thus

$$\begin{aligned}\delta_{\tilde{\mathfrak{p}}}\Lambda_t\{\mathfrak{p}\}=&\int_B i*(\mathbf{C}[\boldsymbol{E}]-\boldsymbol{S})*\tilde{\boldsymbol{E}}(\boldsymbol{x}, t)\, dv_{\boldsymbol{x}}\\&-\int_B(i*\operatorname{div}\boldsymbol{S}+\boldsymbol{f}-\varrho\boldsymbol{u})*\tilde{\boldsymbol{u}}(\boldsymbol{x}, t)\, dv_{\boldsymbol{x}}\\&+\int_B i*(\hat{\nabla}\boldsymbol{u}-\boldsymbol{E})*\tilde{\boldsymbol{S}}(\boldsymbol{x}, t)\, dv_{\boldsymbol{x}}\\&+\int_{\mathscr{S}_1} i*(\hat{\boldsymbol{u}}-\boldsymbol{u})*\tilde{\mathbf{s}}\,(\boldsymbol{x}, t)\, da_{\boldsymbol{x}}\\&+\int_{\mathscr{S}_2} i*(\mathbf{s}-\hat{\mathbf{s}})*\tilde{\boldsymbol{u}}(\boldsymbol{x}, t)\, da_{\boldsymbol{x}} \qquad (0\leqq t<t_0).\end{aligned} \tag{c}$$

If $\mathfrak{p}$ is a solution of the mixed problem, then we conclude from (c) and ***(62.3)*** that

$$\delta_{\tilde{\mathfrak{p}}}\Lambda_t\{\mathfrak{p}\}=0 \qquad (0\leqq t<t_0) \quad \text{for every} \quad \tilde{\mathfrak{p}}\in\mathscr{A} \tag{d}$$

which implies

$$\delta\Lambda_t\{\mathfrak{p}\}=0 \qquad (0\leqq t<t_0). \tag{e}$$

To prove the converse assertion assume that (e) and hence (d) holds. If we choose $\tilde{\mathfrak{p}}=[\tilde{\boldsymbol{u}}, \boldsymbol{0}, \boldsymbol{0}]$ and let $\tilde{\boldsymbol{u}}$ vanish near ∂B, then it follows from (c) and (d) that

$$\int_B(i*\operatorname{div}\boldsymbol{S}+\boldsymbol{f}-\varrho\boldsymbol{u})*\tilde{\boldsymbol{u}}(\boldsymbol{x}, t)\, dv_{\boldsymbol{x}}=0 \qquad (0\leqq t<t_0).$$

Since this relation must hold for every $\tilde{\boldsymbol{u}}$ of class C^∞ on $\bar{B}\times[0, t_0)$ that vanishes near ∂B, we conclude from ***(64.1)*** that $i*\operatorname{div}\boldsymbol{S}+\boldsymbol{f}=\varrho\boldsymbol{u}$. Next, by choosing $\tilde{p}\in\mathscr{A}$ in the forms $[\mathbf{0}, \tilde{\boldsymbol{E}}, \mathbf{0}]$ and $[\mathbf{0}, \mathbf{0}, \tilde{\boldsymbol{S}}]$, where $\tilde{\boldsymbol{E}}$ and $\tilde{\boldsymbol{S}}$ vanish near ∂B, we conclude from (c), (d), and ***(64.1)*** with $\mathscr{W}$ equal to the set of all symmetric tensors that

$$i*(\mathbf{C}[\boldsymbol{E}]-\boldsymbol{S})=\mathbf{0}, \qquad i*(\hat{\nabla}\boldsymbol{u}-\boldsymbol{E})=\mathbf{0}.$$

Thus, by (iv) of ***(10.1)*** with $\varphi=i$, p satisfies the stress-strain and displacement-strain relations. In view of ***(62.3)***, to complete the proof we have only to show that p satisfies the boundary conditions. By (c), (d), (iv) of ***(10.1)***, and the results established thus far,

$$\int_{\mathscr{S}_1}(\hat{\boldsymbol{u}}-\boldsymbol{u})*(\tilde{\boldsymbol{S}}\boldsymbol{n})\,(\boldsymbol{x},t)\,da_{\boldsymbol{x}}+\int_{\mathscr{S}_2}(\mathbf{s}-\hat{\mathbf{s}})*\tilde{\boldsymbol{u}}(\boldsymbol{x},t)\,da_{\boldsymbol{x}}=0 \qquad (0\leqq t<t_0)$$

for every $\tilde{p}\in\mathscr{A}$. Therefore, if we take $\tilde{p}\in\mathscr{A}$ equal to $[\tilde{\boldsymbol{u}}, \mathbf{0}, \mathbf{0}]$ and $[\mathbf{0}, \mathbf{0}, \tilde{\boldsymbol{S}}]$ and appeal to ***(64.2)*** and ***(64.3)***, we conclude that p satisfies the boundary conditions. □

Our next theorem is an analog of the Hellinger-Prange-Reissner principle ***(38.2)***.

(3)[1] *Assume that the elasticity field is invertible. Let $\mathscr{A}$ denote the set of all admissible processes that satisfy the strain-displacement relation, and for each $t\in(0, t_0)$ define the functional $\Theta_t\{\cdot\}$ on $\mathscr{A}$ by*

$$\Theta_t\{p\}=\int_B\{\tfrac{1}{2}i*\boldsymbol{S}*\mathbf{K}[\boldsymbol{S}]-\tfrac{1}{2}\varrho\boldsymbol{u}*\boldsymbol{u}-i*\boldsymbol{S}*\boldsymbol{E}+\boldsymbol{f}*\boldsymbol{u}\}\,(\boldsymbol{x},t)\,dv_{\boldsymbol{x}}$$
$$+\int_{\mathscr{S}_1}\{i*\mathbf{s}*(\boldsymbol{u}-\hat{\boldsymbol{u}})\}\,(\boldsymbol{x},t)\,da_{\boldsymbol{x}}+\int_{\mathscr{S}_2}\{i*\hat{\mathbf{s}}*\boldsymbol{u}\}\,(\boldsymbol{x},t)\,da_{\boldsymbol{x}}$$

for every $p=[\boldsymbol{u}, \boldsymbol{E}, \boldsymbol{S}]\in\mathscr{A}$. Then

$$\delta\Theta_t\{p\}=0 \qquad (0\leqq t<t_0)$$

at $p\in\mathscr{A}$ if and only if p is a solution of the mixed problem.

Proof. Let p and $\tilde{p}$ be admissible processes, and suppose that $p+\lambda\tilde{p}\in\mathscr{A}$ for every scalar λ, or equivalently, that $p, \tilde{p}\in\mathscr{A}$. In view of the symmetry of $\mathbf{C}$, $\mathbf{K}$ is symmetric. Thus if we proceed as in the proof of ***(2)***, we conclude, with the aid of ***(19.3)***, that[2]

$$\begin{aligned}\delta_{\tilde{p}}\Theta_t\{p\}=&\int_B i*(\mathbf{K}[\boldsymbol{S}]-\boldsymbol{E})*\tilde{\boldsymbol{S}}(\boldsymbol{x},t)\,dv_{\boldsymbol{x}}\\&+\int_B(i*\operatorname{div}\boldsymbol{S}+\boldsymbol{f}-\varrho\boldsymbol{u})*\tilde{\boldsymbol{u}}(\boldsymbol{x},t)\,dv_{\boldsymbol{x}}\\&+\int_{\mathscr{S}_1}i*(\boldsymbol{u}-\hat{\boldsymbol{u}})*\tilde{\mathbf{s}}\,(\boldsymbol{x},t)\,da_{\boldsymbol{x}}\\&+\int_{\mathscr{S}_2}i*(\hat{\mathbf{s}}-\mathbf{s})*\tilde{\boldsymbol{u}}(\boldsymbol{x},t)\,da_{\boldsymbol{x}} \qquad (0\leqq t<t_0).\end{aligned} \tag{a}$$

If p is a solution of the mixed problem, then (a), because of ***(62.3)***, yields

$$\delta\Theta_t\{p\}=0 \qquad (0\leqq t<t_0). \tag{b}$$

On the other hand, (a), (b), ***(64.1)***, ***(64.2)***, ***(64.3)***, (iv) of ***(10.1)***, and ***(62.3)*** imply that p is a solution of the mixed problem. □

[1] GURTIN [1964, *8*], Theorem 4.2.
[2] Cf. the proof of ***(38.2)***.

If $\not p = [\boldsymbol{u}, \boldsymbol{E}, \boldsymbol{S}]$ is a kinematically admissible process, then, by virtue of ***(19.3)***, $\Lambda_t\{\not p\}$ defined in ***(2)*** reduces to $\Phi_t\{\not p\}$, where

$$\Phi_t\{\not p\} = \tfrac{1}{2}\int_B \{i*\boldsymbol{S}*\boldsymbol{E} + \varrho\boldsymbol{u}*\boldsymbol{u} - 2\boldsymbol{f}*\boldsymbol{u}\}(\boldsymbol{x}, t)\, dv_{\boldsymbol{x}} - \int_{\mathscr{S}_2} \{i*\hat{\mathbf{s}}*\boldsymbol{u}\}(\boldsymbol{x}, t)\, da_{\boldsymbol{x}}.$$

Thus we are led to the following:

(4) Analog of the principle of minimum potential energy.[1] *For each $t\in[0, t_0)$ let $\Phi_t\{\cdot\}$ be the functional defined on the set $\mathscr{A}$ of all kinematically admissible processes by the above relation. Then*

$$\delta\Phi_t\{\not p\} = 0 \qquad (0 \leqq t < t_0)$$

at $\not p\in\mathscr{A}$ if and only if $\not p$ is a solution of the mixed problem.

Proof. Let $\tilde{\not p} = [\tilde{\boldsymbol{u}}, \tilde{\boldsymbol{E}}, \tilde{\boldsymbol{S}}]$ be an admissible process, and suppose that

$$\not p + \lambda\tilde{\not p}\in\mathscr{A} \quad \text{for every scalar} \quad \lambda. \tag{a}$$

Condition (a) is equivalent to the requirement that $\tilde{\not p}$ satisfy the strain-displacement relation, the stress-strain relation, and the boundary condition

$$\tilde{\boldsymbol{u}} = \boldsymbol{0} \quad \text{on } \mathscr{S}_1\times[0, t_0). \tag{b}$$

Next, since the restriction of $\Lambda_t\{\cdot\}$ of ***(2)*** to (the present) $\mathscr{A}$ is equal to $\Phi_t\{\cdot\}$, we conclude from (c) in the proof of ***(2)*** and the properties of $\not p$ and $\tilde{\not p}$ that

$$\begin{aligned}\delta_{\tilde{\not p}}\, \Phi_t\{\not p\} = &-\int_B (i*\operatorname{div}\boldsymbol{S} + \boldsymbol{f} - \varrho\boldsymbol{u})*\tilde{\boldsymbol{u}}(\boldsymbol{x}, t)\, dv_{\boldsymbol{x}} \\ &+\int_{\mathscr{S}_2} i*(\mathbf{s} - \hat{\mathbf{s}})*\tilde{\boldsymbol{u}}(\boldsymbol{x}, t)\, da_{\boldsymbol{x}} \qquad (0\leqq t < t_0)\end{aligned} \tag{c}$$

for every admissible motion $\tilde{\boldsymbol{u}}$ that satisfies (b). If $\not p$ is a solution of the mixed problem, then, by virtue of ***(62.3)***, $\delta\Phi_t(\not p) = 0$ $(0\leqq t < t_0)$. Conversely, if the variation of Φ_t vanishes at $\not p$ for all t, then it follows from (c), ***(64.1)***, and ***(64.2)*** that $\not p$ is a solution of the mixed problem. □

It is a simple matter to prove counterparts of theorems ***(34.2)*** and ***(57.4)*** characterizing the displacement field corresponding to a solution of the mixed problem.[2] Rather than do this we shall establish a variational principle which involves only the stress field and which has no counterpart in elastostatics.

In addition to the assumptions made previously, we now assume that:[3]

(A) $\mathbf{C}$ is invertible on $\overline{B}$; ϱ, $\boldsymbol{u}_0$, and $\boldsymbol{v}_0$ are of class C^2 on $\overline{B}$; $\boldsymbol{b}$ is of class $C^{2,0}$ on $\overline{B}\times[0, t_0)$.

(5) First variational principle for the stress field[4]. *Let $\mathscr{A}$ denote the set of all time-dependent admissible stress fields of class $C^{3,0}$ on $B\times(0, t_0)$ with $\dot{\mathbf{S}}$ continuous on $\overline{B}\times[0, t_0)$. For each $t\in[0, t_0)$ let $\Gamma_t\{\cdot\}$ be the functional on $\mathscr{A}$ defined by*

$$\begin{aligned}\Gamma_t\{\boldsymbol{S}\} = \tfrac{1}{2}&\int_B \{i'*\operatorname{div}\boldsymbol{S}*\operatorname{div}\boldsymbol{S} + \boldsymbol{S}*\mathbf{K}[\boldsymbol{S}] - 2\boldsymbol{S}*\nabla\boldsymbol{f}'\}(\boldsymbol{x}, t)\, dv_{\boldsymbol{x}} \\ &+\int_{\mathscr{S}_1} \{(\boldsymbol{f}' - \hat{\boldsymbol{u}})*\mathbf{s}\}(\boldsymbol{x}, t)\, da_{\boldsymbol{x}} + \int_{\mathscr{S}_2} \{i'*(\hat{\mathbf{s}} - \mathbf{s})*\operatorname{div}\boldsymbol{S}\}(\boldsymbol{x}, t)\, da_{\boldsymbol{x}},\end{aligned}$$

where

$$i' = \frac{i}{\varrho}, \qquad \boldsymbol{f}' = \frac{\boldsymbol{f}}{\varrho}.$$

---

[1] GURTIN [1964, *8*], Theorem 4.3.
[2] Cf. GURTIN [1964, *8*], § 5.
[3] These assumptions are slightly more stringent than necessary.
[4] GURTIN [1964, *8*], Theorem 6.1.

Then

$$\delta\Gamma_t\{\mathbf{S}\}=0 \qquad (0\leqq t<t_0)$$

at $\mathbf{S}\in\mathscr{A}$ *if and only if* $\mathbf{S}$ *corresponds to a solution of the mixed problem.*

Proof. Let $\tilde{\mathbf{S}}\in\mathscr{A}$, so that $\mathbf{S}+\lambda\tilde{\mathbf{S}}\in\mathscr{A}$ for every scalar λ. Then, since $\mathbf{K}$ is symmetric, the divergence theorem and the properties of the convolution yield, after some work,

$$\begin{aligned}\delta_{\tilde{\mathbf{S}}}\,\Gamma_t\{\mathbf{S}\} = &-\int_B \{\hat{\nabla}(i'*\operatorname{div}\mathbf{S}+\mathbf{f}')-\mathbf{K}[\mathbf{S}]\}*\tilde{\mathbf{S}}(\mathbf{x},t)\,dv_{\mathbf{x}}\\ &+\int_{\mathscr{S}_1}(i'*\operatorname{div}\mathbf{S}-\hat{\mathbf{u}}+\mathbf{f}')*\tilde{\mathbf{s}}(\mathbf{x},t)\,da_{\mathbf{x}}\\ &+\int_{\mathscr{S}_2} i'*(\hat{\mathbf{s}}-\mathbf{s})*\operatorname{div}\tilde{\mathbf{S}}(\mathbf{x},t)\,da_{\mathbf{x}}\qquad (0\leqq t<t_0)\end{aligned}\tag{a}$$

for every $\tilde{\mathbf{S}}\in\mathscr{A}$. If $\mathbf{S}$ corresponds to a solution of the mixed problem, then ***(62.4)*** implies that

$$\delta\Gamma_t\{\mathbf{S}\}=0 \qquad (0\leqq t<t_0). \tag{b}$$

On the other hand, (a), (b), the symmetry of $\tilde{\mathbf{S}}$, ***(64.1)***, ***(64.3)***, ***(64.4)***, (iv) of ***(10.1)***, and ***(62.4)*** imply that $\mathbf{S}$ corresponds to a solution of the mixed problem. □

Let $\mathscr{A}$ be as defined in ***(5)***. By a ***dynamically admissible stress field*** we mean a tensor field $\mathbf{S}\in\mathscr{A}$ that satisfies the traction boundary condition. For such a stress field, $\Gamma_t\{\mathbf{S}\}$, defined in ***(5)***, reduces to

$$\begin{aligned}\Omega_t\{\mathbf{S}\} = &\tfrac{1}{2}\int_B \{i'*\operatorname{div}\mathbf{S}*\operatorname{div}\mathbf{S}+\mathbf{S}*\mathbf{K}[\mathbf{S}]-2\mathbf{S}*\nabla\mathbf{f}'\}(\mathbf{x},t)\,dv_{\mathbf{x}}\\ &+\int_{\mathscr{S}_1}\{(\mathbf{f}'-\hat{\mathbf{u}})*\mathbf{s}\}(\mathbf{x},t)\,da_{\mathbf{x}}.\end{aligned}$$

Thus we are led to the following theorem, the proof of which is strictly analogous to that given for ***(5)***.

(6) Second variational principle for the stress field.[1] *For each* $t\in[0,t_0)$ *let* $\Omega_t\{\cdot\}$ *be defined on the set of dynamically admissible stress fields by the above relation. Then*

$$\delta\Omega_t\{\mathbf{S}\}=0 \qquad (0\leqq t<t_0)$$

at a dynamically admissible stress field $\mathbf{S}$ *if and only if* $\mathbf{S}$ *corresponds to a solution of the mixed problem.*

If we apply ***(19.3)*** to the term involving $\mathbf{S}*\nabla\mathbf{f}'$, we can rewrite the expression for $\Omega_t\{\mathbf{S}\}$ in the form

$$\begin{aligned}\Omega_t\{\mathbf{S}\} = &\tfrac{1}{2}\int_B \{i'*\operatorname{div}\mathbf{S}*\operatorname{div}\mathbf{S}+\mathbf{S}*\mathbf{K}[\mathbf{S}]+2\operatorname{div}\mathbf{S}*\mathbf{f}'\}(\mathbf{x},t)\,dv_{\mathbf{x}}\\ &-\int_{\mathscr{S}_1}\{\hat{\mathbf{u}}*\mathbf{s}\}(\mathbf{x},t)\,da_{\mathbf{x}}-\int_{\mathscr{S}_2}\{\mathbf{f}'*\hat{\mathbf{s}}\}(\mathbf{x},t)\,da_{\mathbf{x}}.\end{aligned}$$

Of course, the last term may be omitted without destroying the validity of ***(6)***, since it is the same for every dynamically admissible $\mathbf{S}$.

66. Minimum principles. For the variational principles established in the preceding section the relevant functionals were stationary, but not necessarily a minimum, at a solution. On the other hand, the principles of minimum potential

[1] GURTIN [1964, *8*], Theorem 6.2.

energy and minimum complementary energy are true minimum principles. This difference is nontrivial; indeed, the lower bound on the energy provided by a minimum principle is useful in establishing the convergence of numerical solutions. In this section we establish *minimum* principles for the Laplace transforms of the functionals associated with the variational principles **(65.4)** and **(65.6)**.

We let $t_0=\infty$, so that the underlying time interval is $[0, \infty)$. In addition to hypotheses (i)–(v) of Sect. 62 we now assume that:

(vi) C is symmetric and positive definite;

(vii) $\varrho>0$;

(viii) $\boldsymbol{b}$, $\hat{\boldsymbol{u}}$, and $\hat{\boldsymbol{s}}$ possess Laplace transforms.

Let $\mathscr{A}_L$ denote the set of all kinematically admissible processes $\not{p}=[\boldsymbol{u}, \boldsymbol{E}, \boldsymbol{S}]$ such that[1] $\not{p}$ and $\nabla\not{p}$ possess Laplace transforms. Then for $\not{p}\in\mathscr{A}_L$ the function

$$t\mapsto \Phi_t\{\not{p}\}$$

of **(65.4)** has a Laplace transform

$$\bar{\Phi}_\eta\{\not{p}\}=\int_0^\infty e^{-\eta t}\,\Phi_t\{\not{p}\}\,dt \qquad (\eta_0\leqq\eta<\infty)$$

for some $\eta_0\geqq 0$. In fact, since the Laplace transform of $i(t)=t$ is $1/\eta^2$, it follows from (vii) of **(10.1)** and the transformed stress-strain relation $\bar{\boldsymbol{S}}=\mathsf{C}[\bar{\boldsymbol{E}}]$ that

$$\begin{aligned}\bar{\Phi}_\eta\{\not{p}\}=\frac{1}{\eta^2}\,U_{\mathsf{C}}\{\bar{\boldsymbol{E}}(\cdot,\eta)\}+\int_B\left\{\frac{1}{2}\,\varrho(\boldsymbol{x})\,\bar{\boldsymbol{u}}(\boldsymbol{x},\eta)^2-\bar{\boldsymbol{f}}(\boldsymbol{x},\eta)\cdot\bar{\boldsymbol{u}}(\boldsymbol{x},\eta)\right\}dv_{\boldsymbol{x}}\\ -\frac{1}{\eta^2}\int_{\mathscr{S}_2}\bar{\hat{\boldsymbol{s}}}(\boldsymbol{x},\eta)\cdot\bar{\boldsymbol{u}}(\boldsymbol{x}\ \eta)\,da_{\boldsymbol{x}} \qquad (\eta_0\leqq\eta<\infty).\end{aligned} \tag{L}$$

It is interesting to note the similarity between $\bar{\Phi}_\eta\{\cdot\}$ and the functional $\Phi\{\cdot\}$ of the principle of minimum potential energy **(34.1)**. This similarity is emphasized by the

(1) Theorem of minimum transformed energy.[2] *Assume that the elasticity field is positive semi-definite. Let $\not{p}\in\mathscr{A}_L$ be a solution of the mixed problem. Then for each $\tilde{\not{p}}\in\mathscr{A}_L$ there exists an $\eta_0\geqq 0$ such that*

$$\bar{\Phi}_\eta\{\not{p}\}\leqq\bar{\Phi}_\eta\{\tilde{\not{p}}\} \qquad (\eta_0\leqq\eta<\infty).$$

Proof. Since $\not{p}$, $\tilde{\not{p}}\in\mathscr{A}_L$, there exists an $\eta_0\geqq 0$ such that the Laplace transforms of $\not{p}$, $\tilde{\not{p}}$, $\nabla\not{p}$, $\nabla\tilde{\not{p}}$, and $\boldsymbol{b}$ exist on $\bar{B}\times[\eta_0, \infty)$. Let

$$\not{p}'=\tilde{\not{p}}-\not{p}. \tag{a}$$

Then $\not{p}'$ is an admissible process, $\not{p}'$ and $\nabla\not{p}'$ possess Laplace transforms, and

$$\bar{\boldsymbol{E}}'=\tfrac{1}{2}(\nabla\bar{\boldsymbol{u}}'+\nabla\bar{\boldsymbol{u}}'^T), \tag{b}$$

$$\bar{\boldsymbol{S}}'=\mathsf{C}[\bar{\boldsymbol{E}}'], \tag{c}$$

$$\bar{\boldsymbol{u}}'=\boldsymbol{0} \quad \text{on } \mathscr{S}_1. \tag{d}$$

Moreover $\bar{\boldsymbol{S}}=\mathsf{C}[\bar{\boldsymbol{E}}]$, since $\not{p}\in\mathscr{A}_L$; thus **(28.1)** and (a) imply

$$U_{\mathsf{C}}\{\bar{\tilde{\boldsymbol{E}}}\}-U_{\mathsf{C}}\{\bar{\boldsymbol{E}}\}=U_{\mathsf{C}}\{\bar{\boldsymbol{E}}'\}+\int_B\bar{\boldsymbol{S}}\cdot\bar{\boldsymbol{E}}'\,dv. \tag{e}$$

[1] Here $\nabla\not{p}=[\nabla\boldsymbol{u}, \nabla\boldsymbol{E}, \nabla\boldsymbol{S}]$.

[2] Benthien and Gurtin [1970, *1*].

Similarly,

$$\int_B \varrho \bar{\tilde{u}}^2 \, dv - \int_B \varrho \bar{u}^2 \, dv = \int_B \varrho \bar{u}'^2 \, dv + 2 \int_B \varrho \bar{u} \cdot \bar{u}' \, dv. \tag{f}$$

If we apply ***(18.1)*** to $\bar{S}$ and $\bar{u}'$, we conclude, with the aid of (b) and (d), that

$$\int_B \bar{S} \cdot \bar{E}' \, dv = \int_{\mathscr{S}_2} \bar{s} \cdot \bar{u}' \, da - \int_B \bar{u}' \cdot \operatorname{div} \bar{S} \, dv. \tag{g}$$

Further, since $p \in \mathscr{A}_L$ is a solution of the mixed problem, if we take the Laplace transform of the third and fifth relations in ***(62.3)***, we find that

$$\frac{1}{\eta^2} \operatorname{div} \bar{S}(x, \eta) + \bar{f}(x, \eta) = \varrho(x) \, \bar{u}(x, \eta) \tag{h}$$

for every $(x, \eta) \in B \times [\eta_0, \infty)$ and

$$\bar{s} = \bar{\hat{s}} \quad \text{on} \quad \mathscr{S}_2 \times [\eta_0, \infty). \tag{i}$$

By (L) and (e)–(i),

$$\bar{\Phi}_\eta(\tilde{p}) - \bar{\Phi}_\eta(p) = \frac{1}{\eta^2} U_{\mathbf{C}}\{\bar{E}'(\cdot, \eta)\} + \frac{1}{2} \int_B \varrho(x) \, \bar{u}'(x, \eta)^2 \, dv_x \qquad (\eta_0 \leqq \eta < \infty);$$

this implies the desired result, since $\mathbf{C}$ is positive semi-definite and ϱ positive. □

Note that, since $\varrho > 0$, we can strengthen the inequality in ***(1)*** as follows:

$$\bar{\Phi}_\eta\{p\} < \bar{\Phi}_\eta\{\tilde{p}\} \quad \text{whenever} \quad \bar{u}(\cdot, \eta) \neq \bar{\tilde{u}}(\cdot, \eta).$$

We now assume that (A) on p. 229 holds, and, in addition, that ∇b possesses a Laplace transform. Let $\mathscr{D}_L$ denote the set of all dynamically admissible stress fields S such that S and ∇S possess Laplace transforms. Then for $S \in \mathscr{D}_L$ the function $t \mapsto \Omega_t\{S\}$ of ***(65.6)*** has a Laplace transform $\bar{\Omega}_\eta\{S\}$ $(\eta_0 \leqq \eta < \infty)$ for some $\eta_0 > 0$. In fact,

$$\bar{\Omega}_\eta\{S\} = U_{\mathbf{K}}\{\bar{S}(\cdot, \eta)\} + \int_B \left[\frac{[\operatorname{div} \bar{S}(x, \eta)]^2}{2\varrho(x) \, \eta^2} - \bar{S}(x, \eta) \cdot \nabla \bar{f}'(x, \eta) \right] dv_x$$
$$+ \int_{\mathscr{S}_1} [\bar{f}'(x, \eta) - \bar{\hat{u}}(x, \eta)] \cdot \bar{s}(x, \eta) \, da_x \qquad (\eta_0 \leqq \eta < \infty),$$

and we are led to the

(2) Minimum principle for the stress field. *Let $S \in \mathscr{D}_L$ be the stress field corresponding to a solution of the mixed problem. Then for each $\tilde{S} \in \mathscr{D}_L$ there exists an $\eta_0 \geqq 0$ such that*

$$\bar{\Omega}_\eta\{S\} \leqq \bar{\Omega}_\eta\{\tilde{S}\} \qquad (\eta_0 \leqq \eta < \infty).$$

The proof of this theorem is strictly analogous to that of ***(1)*** and can safely be omitted.

IV. Homogeneous and isotropic bodies.

67. Complete solutions of the field equations. In this section we study certain general solutions of the displacement equation of motion[1]

$$c_1^2 \nabla \operatorname{div} u - c_2^2 \operatorname{curl} \operatorname{curl} u + \frac{b}{\varrho} = \ddot{u},$$

[1] Cf. p. 213.

assuming throughout that $c_1^2 > c_2^2 > 0$ and that $\boldsymbol{b}/\varrho$ admits the Helmholtz decomposition[1]

$$\frac{\boldsymbol{b}}{\varrho} = -\nabla \varkappa - \operatorname{curl} \boldsymbol{\gamma}, \qquad \operatorname{div} \boldsymbol{\gamma} = 0,$$

with $\varkappa$ and $\boldsymbol{\gamma}$ of class $C^{2,0}$ on $B \times (0, t_0)$. A solution $\boldsymbol{u}$ of class C^2 on $B \times (0, t_0)$ will be referred to as an ***elastic motion*** corresponding to $\boldsymbol{b}$.

(1) Green-Lamé solution.[2] Let

$$\boldsymbol{u} = \nabla \varphi + \operatorname{curl} \boldsymbol{\psi}, \tag{L$_1$}$$

where φ and $\boldsymbol{\psi}$ are class C^3 fields on $B \times (0, t_0)$ that satisfy[3]

$$\Box_1 \varphi = \varkappa, \qquad \Box_2 \boldsymbol{\psi} = \boldsymbol{\gamma}. \tag{L$_2$}$$

Then $\boldsymbol{u}$ is an elastic motion corresponding to $\boldsymbol{b}$.

Proof. By (L$_1$)

$$c_1^2 \nabla \operatorname{div} \boldsymbol{u} - c_2^2 \operatorname{curl} \operatorname{curl} \boldsymbol{u} - \ddot{\boldsymbol{u}} + \frac{\boldsymbol{b}}{\varrho}$$
$$= \nabla\{c_1^2 \Delta \varphi - \ddot{\varphi} - \varkappa\} + \operatorname{curl}\{- c_2^2 \operatorname{curl} \operatorname{curl} \boldsymbol{\psi} - \ddot{\boldsymbol{\psi}} - \boldsymbol{\gamma}\}.$$

In view of (1) and (3) of ***(4.1)***,

$$\operatorname{curl} \operatorname{curl} \operatorname{curl} \boldsymbol{\psi} = - \operatorname{curl} \Delta \boldsymbol{\psi},$$

and the desired result follows from (L$_2$). □

(2) Completeness of the Green-Lamé solution.[4] *Let $\boldsymbol{u}$ be an elastic motion of class C^3 corresponding to $\boldsymbol{b}$, and assume that $\boldsymbol{u}(\cdot, \tau)$ and $\dot{\boldsymbol{u}}(\cdot, \tau)$ are continuous on $\bar{B}$ at some time $\tau \in (0, t_0)$. Then there exist class C^2 fields φ and $\boldsymbol{\psi}$ on $B \times (0, t_0)$ that satisfy* (L$_1$), (L$_2$). *Moreover,*

$$\operatorname{div} \boldsymbol{\psi} = 0.$$

---

[1] As is clear from the proof of ***(6.7)***, such an expansion is ensured if $\boldsymbol{b}/\varrho$ is continuous on $\bar{B} \times (0, t_0)$ and of class $C^{2,0}$ on $B \times (0, t_0)$.

[2] Lamé [1852, *2*], p. 149. The two-dimensional version was obtained earlier by Green [1839, *1*]. The solution $\boldsymbol{u} = \nabla \varphi$, $\Box_1 \varphi = \varkappa$ appears in the work of Poisson [1829, *2*], p. 404. Related solutions are given by Bondarenko [1960, *4*], Kilchevskii and Levchuk [1967, *10*]. Chadwick and Trowbridge [1967, *5*] show that

$$\boldsymbol{u} = \nabla \varphi - \operatorname{curl} \operatorname{curl} (\boldsymbol{p}\, \psi) - \operatorname{curl} (\boldsymbol{p}\, \omega)$$
$$\Box_1 \varphi = 0, \qquad \Box_2 \psi = 0, \qquad \Box_2 \omega = 0,$$

defines an elastic motion corresponding to $\boldsymbol{b} = \boldsymbol{0}$. Further, they prove that this solution is complete for regions bounded by concentric spheres.

[3] The wave operators $\Box_\alpha$ are defined on p. 213.

[4] Clebsch [1863, *2*] was the first to assert that the Green-Lamé solution is complete. Clebsch also gave a proof which Sternberg [1960, *13*] showed is open to serious objection. (Actually, Clebsch's proof is valid when the boundary of the region consists of a single closed surface.) Another inconclusive proof was given by Kelvin [1904, *5*] in 1884. The first general completeness proof was furnished by Somigliana [1892, *7*]. Completeness proofs were also supplied by Tedone [1897, *1*], Duhem [1898, *5*], and Sternberg and Gurtin [1962, *13*]. An explicit version of Duhem's proof was given by Sternberg [1960, *13*]. The work of Somigliana, Tedone, and Duhem appears to be little known; in fact, Sneddon and Berry [1958, *18*], p. 109 state that the Green-Lamé solution is incomplete if the region occupied by the medium has a boundary. Further, as Sternberg [1960, *13*] has remarked, the completeness proof given by Pearson [1959, *11*] is open to objections.

Proof.[1] If we integrate the equation of motion twice, we arrive at

$$\begin{aligned} \boldsymbol{u}(t) = \boldsymbol{u}(\tau) + \dot{\boldsymbol{u}}(\tau)(t-\tau) + \nabla \int_\tau^t \int_\tau^s [c_1^2 \operatorname{div} \boldsymbol{u}(\lambda) - \varkappa(\lambda)]\, d\lambda\, ds \\ + \operatorname{curl} \int_\tau^t \int_\tau^s [-c_2^2 \operatorname{curl} \boldsymbol{u}(\lambda) - \boldsymbol{\gamma}(\lambda)]\, d\lambda\, ds, \end{aligned} \tag{a}$$

where, for convenience, we have omitted all mention of the variable $\boldsymbol{x}$. If we apply the Helmholtz resolution ***(6.7)*** to $\boldsymbol{u}(\tau)$ and $\dot{\boldsymbol{u}}(\tau)$, we find that

$$\boldsymbol{u}(\tau) + \dot{\boldsymbol{u}}(\tau)(t-\tau) = \nabla\alpha(t) + \operatorname{curl} \boldsymbol{\beta}(t), \tag{b}$$

where α and $\boldsymbol{\beta}$ are class C^2 fields that satisfy

$$\ddot{\alpha} = 0, \qquad \ddot{\boldsymbol{\beta}} = \boldsymbol{0}, \qquad \operatorname{div} \boldsymbol{\beta} = 0. \tag{c}$$

Thus if we define functions φ and $\boldsymbol{\psi}$ on $B \times (0, t_0)$ by

$$\varphi(t) = \alpha(t) + \int_\tau^t \int_\tau^s [c_1^2 \operatorname{div} \boldsymbol{u}(\lambda) - \varkappa(\lambda)]\, d\lambda\, ds, \tag{d}$$

$$\boldsymbol{\psi}(t) = \boldsymbol{\beta}(t) - \int_\tau^t \int_\tau^s [c_2^2 \operatorname{curl} \boldsymbol{u}(\lambda) + \boldsymbol{\gamma}(\lambda)]\, d\lambda\, ds, \tag{e}$$

then

$$\boldsymbol{u} = \nabla\varphi + \operatorname{curl} \boldsymbol{\psi}, \tag{f}$$

and, since $\operatorname{div} \boldsymbol{\gamma} = 0$, (c), (d), and (e) imply

$$\operatorname{div} \boldsymbol{\psi} = 0. \tag{g}$$

Moreover, by (c), (d), and (e),

$$\ddot{\varphi} = c_1^2 \operatorname{div} \boldsymbol{u} - \varkappa, \qquad \ddot{\boldsymbol{\psi}} = -c_2^2 \operatorname{curl} \boldsymbol{u} - \boldsymbol{\gamma}, \tag{h}$$

and (f), (g), and (1)–(3) of ***(4.1)*** yield

$$\operatorname{div} \boldsymbol{u} = \Delta\varphi, \qquad \operatorname{curl} \boldsymbol{u} = -\Delta\boldsymbol{\psi}. \tag{i}$$

Finally, (h) and (i) imply the desired result (L_2). □

It follows from (d), (e), and (i) that φ and $\boldsymbol{\psi}$ are sufficiently smooth that

$$\Box_1 \nabla\varphi = \nabla\Box_1\varphi,$$
$$\Box_2 \operatorname{curl} \boldsymbol{\psi} = \operatorname{curl} \Box_2 \boldsymbol{\psi},$$

and ***(2)*** has the following corollary:

(3) Poisson's decomposition theorem.[2] *Let* $\boldsymbol{u}$ *satisfy the hypotheses of* ***(2)*** *with body force field* $\mathbf{b} = \mathbf{0}$. *Then* $\boldsymbol{u}$ *admits the decomposition*

$$\boldsymbol{u} = \boldsymbol{u}_1 + \boldsymbol{u}_2,$$

where $\boldsymbol{u}_1$ *and* $\boldsymbol{u}_2$ *are class* C^2 *fields on* $B \times (0, t_0)$ *that satisfy*

$$\Box_1 \boldsymbol{u}_1 = \boldsymbol{0}, \qquad \operatorname{curl} \boldsymbol{u}_1 = \boldsymbol{0},$$
$$\Box_2 \boldsymbol{u}_2 = \boldsymbol{0}, \qquad \operatorname{div} \boldsymbol{u}_2 = 0.$$

---

[1] SOMIGLIANA [1892, *7*]. See also BISHOP [1953, *4*].

[2] POISSON [1829, *3*]. Although POISSON's results are based on a molecular model which yields $\mu = \lambda$, his proof in no way depends on this constraint.

(4) Cauchy-Kovalevski-Somigliana solution.[1] *Let*

$$\boldsymbol{u} = \Box_1 \boldsymbol{g} + (c_2^2 - c_1^2)\, \nabla \operatorname{div} \boldsymbol{g}, \tag{S$_1$}$$

where $\boldsymbol{g}$ *is a class* C^4 *vector field on* $B \times (0, t_0)$ *that satisfies*

$$\Box_2 \Box_1 \boldsymbol{g} = -\frac{\boldsymbol{b}}{\varrho}. \tag{S$_2$}$$

Then $\boldsymbol{u}$ *is an elastic motion corresponding to* $\boldsymbol{b}$.

Proof. By (S_1)

$$\Box_2 \boldsymbol{u} + (c_1^2 - c_2^2)\, \nabla \operatorname{div} \boldsymbol{u}$$
$$= \Box_2 \Box_1 \boldsymbol{g} + (c_2^2 - c_1^2)\, \Box_2 \nabla \operatorname{div} \boldsymbol{g} + (c_1^2 - c_2^2)\{\Box_1 \nabla \operatorname{div} \boldsymbol{g} + (c_2^2 - c_1^2)\, \Delta \nabla \operatorname{div} \boldsymbol{g}\}.$$

In view of (S_2) and the identity

$$\Box_2 \boldsymbol{f} - \Box_1 \boldsymbol{f} = (c_2^2 - c_1^2)\, \Delta \boldsymbol{f},$$

the right-hand side of the above relation reduces to $-\boldsymbol{b}/\varrho$; thus we conclude from (d) on p. 213 that $\boldsymbol{u}$ is an elastic motion corresponding to $\boldsymbol{b}$. □

If $\boldsymbol{g}$ is independent of time, then (S_1), (S_2) reduce to

$$\boldsymbol{u} = c_1^2 \Delta \boldsymbol{g} + (c_2^2 - c_1^2)\, \nabla \operatorname{div} \boldsymbol{g},$$
$$\Delta\Delta \boldsymbol{g} = -\frac{\boldsymbol{b}}{\varrho c_1^2 c_2^2},$$

which is, to within a multiplicative constant, the Boussinesq-Somigliana-Galerkin solution ***(44.3)***.

If in the time-dependent Cauchy-Kovalevski-Somigliana solution we define[2]

$$\boldsymbol{\psi} = \Box_1 \boldsymbol{g}, \qquad \varphi = 2 c_1^2 \operatorname{div} \boldsymbol{g} - \boldsymbol{p} \cdot \boldsymbol{\psi},$$

then (S_1), (S_2) reduce to

$$\boldsymbol{u} = \boldsymbol{\psi} + \frac{c_2^2 - c_1^2}{2 c_1^2}\, \nabla(\varphi + \boldsymbol{p} \cdot \boldsymbol{\psi}), \qquad \Box_2 \boldsymbol{\psi} = -\frac{\boldsymbol{b}}{\varrho}, \qquad \Box_1 \varphi = -\boldsymbol{p} \cdot \Box_1 \boldsymbol{\psi},$$

which may be regarded as a dynamic generalization of the Boussinesq-Papkovitch-Neuber solution ***(44.1)***. Since the potentials φ and $\boldsymbol{\psi}$ obey a coupled differential equation, this solution is of no practical interest in elastodynamics.[3]

We now establish the completeness of the Cauchy-Kovalevski-Somigliana solution. For convenience, we shall assume that the body forces are zero.

(5) Completeness of the Cauchy-Kovalevski-Somigliana solution.[4] *Let* $\boldsymbol{u}$ *be a class* C^6 *elastic motion corresponding to zero body forces, and suppose that* $\boldsymbol{u}(\cdot, \tau)$ *and* $\dot{\boldsymbol{u}}(\cdot, \tau)$ *are continuous on* $\bar{B}$ *at some* $\tau \in (0, t_0)$. *Then there exists a class* C^4 *vector field* $\boldsymbol{g}$ *on* $B \times (0, t_0)$ *such that* (S_1), (S_2) *hold.*

---

[1] Cauchy [1840, *1*], pp. 208–209; Kovalevsky [1885, *2*], p. 269; Somigliana [1892, *7*]. This solution was arrived at independently by Iacovache [1949, *4*]; it was extended to include body forces by Sternberg and Eubanks [1957, *14*]. See also Bondarenko [1957, *2*], Predeleanu [1958, *13*], Teodorescu [1960, *14*].

[2] Sternberg and Eubanks [1957, *14*]. Cf. the remarks on p. 142.

[3] Sternberg [1960, *13*].

[4] Somigliana [1892, *7*]. This theorem was arrived at independently by Sternberg and Eubanks [1957, *14*] using an entirely different method of proof and without assuming that $\boldsymbol{b} = \boldsymbol{0}$. A third completeness proof is contained in the work of Sternberg and Gurtin [1962, *13*].

Proof. By hypothesis and **(2)**,

$$c_1^2 \nabla \operatorname{div} \boldsymbol{u} - c_2^2 \operatorname{curl} \operatorname{curl} \boldsymbol{u} = \ddot{\boldsymbol{u}}, \tag{a}$$

$$\boldsymbol{u} = \nabla \varphi + \operatorname{curl} \boldsymbol{\psi}, \tag{b}$$

$$\Box_1 \varphi = 0, \qquad \Box_2 \boldsymbol{\psi} = \boldsymbol{0}. \tag{c}$$

Further, it follows from the proof of **(2)** that φ and $\boldsymbol{\psi}$ are of class C^5. Substituting (b) into (a),

$$\ddot{\boldsymbol{u}} = c_2^2 \Delta \operatorname{curl} \boldsymbol{\psi} - \nabla \operatorname{div} (c_2^2 \operatorname{curl} \boldsymbol{\psi} - c_1^2 \nabla \varphi), \tag{d}$$

where we have used the fact that div curl $=0$. By (c)

$$c_2^2 \Delta \operatorname{curl} \boldsymbol{\psi} = \operatorname{curl} \ddot{\boldsymbol{\psi}} = c_1^2 \Delta \operatorname{curl} \boldsymbol{\psi} - \Box_1 \operatorname{curl} \boldsymbol{\psi},$$

and thus

$$c_2^2 \Delta \operatorname{curl} \boldsymbol{\psi} = \frac{c_2^2}{c_1^2 - c_2^2} \Box_1 \operatorname{curl} \boldsymbol{\psi}. \tag{e}$$

Eqs. (d), (e), and the first of (c) imply that

$$\ddot{\boldsymbol{u}} = \frac{1}{c_1^2 - c_2^2} \Box_1 (c_2^2 \operatorname{curl} \boldsymbol{\psi} - c_1^2 \nabla \varphi) - \nabla \operatorname{div} (c_2^2 \operatorname{curl} \boldsymbol{\psi} - c_1^2 \nabla \varphi),$$

and thus

$$\boldsymbol{u}(t) = \boldsymbol{v}(\tau) + \dot{\boldsymbol{v}}(\tau)(t-\tau) + \Box_1 \boldsymbol{h}(t) - (c_1^2 - c_2^2) \nabla \operatorname{div} \boldsymbol{h}(t),$$

where

$$\boldsymbol{h}(t) = \frac{1}{c_1^2 - c_2^2} \int_\tau^t \int_\tau^s [c_2^2 \operatorname{curl} \boldsymbol{\psi}(\lambda) - c_1^2 \nabla \varphi(\lambda)] \, d\lambda \, ds,$$

$$\boldsymbol{v} = \boldsymbol{u} + \frac{1}{c_1^2 - c_2^2} [c_2^2 \operatorname{curl} \boldsymbol{\psi} - c_1^2 \nabla \varphi].$$

In the above equations we have omitted all mention of the variable $\boldsymbol{x}$.

Next, since the Boussinesq-Somigliana-Galerkin solution **(44.3)** is complete, we can expand $\boldsymbol{v}(\tau) = \boldsymbol{v}(\cdot, \tau)$ and $\dot{\boldsymbol{v}}(\tau) = \dot{\boldsymbol{v}}(\cdot, \tau)$ as follows

$$\boldsymbol{v}(\tau) = c_1^2 \Delta \boldsymbol{g}_1 + (c_2^2 - c_1^2) \nabla \operatorname{div} \boldsymbol{g}_1,$$

$$\dot{\boldsymbol{v}}(\tau) = c_1^2 \Delta \boldsymbol{g}_2 + (c_2^2 - c_1^2) \nabla \operatorname{div} \boldsymbol{g}_2,$$

where $\boldsymbol{g}_1$ and $\boldsymbol{g}_2$ are of class C^4 on B. Thus if we define

$$\boldsymbol{g}(\boldsymbol{x}, t) = \boldsymbol{h}(\boldsymbol{x}, t) + \boldsymbol{g}_1(\boldsymbol{x}) + (t-\tau) \boldsymbol{g}_2(\boldsymbol{x}),$$

then $\boldsymbol{g}$ is of class C^4,

$$\boldsymbol{u} = \Box_1 \boldsymbol{g} + (c_2^2 - c_1^2) \nabla \operatorname{div} \boldsymbol{g},$$

and (S_1) holds. Finally, (S_1) and (a) imply (S_2). $\Box$

In attempting to judge the comparative merits of the Cauchy-Kovalevski-Somigliana solution (S_1), (S_2) and the Green-Lamé solution (L_1), (L_2), the following considerations would appear to be pertinent.[1] The former remains complete in the equilibrium case if the generating potential $\boldsymbol{g}(\boldsymbol{x}, t)$ is taken to be a function of position alone. In contrast, the Green-Lamé solution has no equilibrium counterpart in the foregoing sense. This circumstance reflects the relative economy of the Green-Lamé solution, which is simpler in structure than the available complete solutions to the equations of equilibrium. Further, (S_1) contains space

[1] These observations are due to STERNBERG [1960, *13*].

derivatives of $\boldsymbol{g}$ up to the second order while only first-order derivatives of φ and $\boldsymbol{\psi}$ are seen to enter (L_2). Finally, (L_1), (L_2) are conveniently transformed into general orthogonal curvilinear coordinates, whereas (S_1), (S_2) give rise to exceedingly cumbersome forms when referred to curvilinear coordinates, with the exception of cylindrical coordinates. For all of these reasons, the Green-Lamé solution deserves preference over the Cauchy-Kovalevski-Somigliana solution in applications.

In the absence of body forces the Cauchy-Kovalevski-Somigliana stress function satisfies the repeated wave equation

$$\Box_2 \Box_1 \boldsymbol{g} = \boldsymbol{0}.$$

The next result yields an important decomposition for solutions of this equation.

(6) Boggio's theorem.[1] *Let $\boldsymbol{g}$ be a vector field of class C^4 on $\bar{B}\times(0, t_0)$, and suppose that*

$$\Box_2 \Box_1 \boldsymbol{g} = \boldsymbol{0}.$$

Then

$$\boldsymbol{g} = \boldsymbol{g}_1 + \boldsymbol{g}_2,$$

where $\boldsymbol{g}_1$ and $\boldsymbol{g}_2$ are class C^2 fields on $B\times(0, t_0)$ that satisfy

$$\Box_1 \boldsymbol{g}_1 = \boldsymbol{0}, \qquad \Box_2 \boldsymbol{g}_2 = \boldsymbol{0}.$$

Proof.[2] It suffices to establish the existence of a class C^2 field $\boldsymbol{g}_1$ on $B\times(0, t_0)$ with the following properties:

$$\Box_1 \boldsymbol{g}_1 = \boldsymbol{0}, \qquad \Box_2(\boldsymbol{g} - \boldsymbol{g}_1) = \boldsymbol{0}. \tag{a}$$

Indeed, once such a $\boldsymbol{g}_1$ has been found, we simply define

$$\boldsymbol{g}_2 = \boldsymbol{g} - \boldsymbol{g}_1.$$

Since

$$\Box_\alpha \boldsymbol{g}_1 = c_\alpha^2 \Delta \boldsymbol{g}_1 - \ddot{\boldsymbol{g}}_1 \qquad (\alpha = 1, 2),$$

it follows that if $\Box_1 \boldsymbol{g}_1 = \boldsymbol{0}$, then

$$\Box_2 \boldsymbol{g}_1 = \frac{c_2^2 - c_1^2}{c_1^2} \ddot{\boldsymbol{g}}_1.$$

Thus (a) is satisfied if

$$\Box_1 \boldsymbol{g}_1 = \boldsymbol{0}, \qquad \ddot{\boldsymbol{g}}_1 = \boldsymbol{f}, \tag{b}$$

where

$$\boldsymbol{f} = \frac{c_1^2}{c_2^2 - c_1^2} \Box_2 \boldsymbol{g}.$$

Notice that $\boldsymbol{f}$ is of class C^2 on $\bar{B}\times(0, t_0)$ and obeys

$$\Box_1 \boldsymbol{f} = \boldsymbol{0}. \tag{c}$$

Thus to complete the proof it suffices to exhibit a field $\boldsymbol{g}_1$ satisfying (b), subject to (c).

Let $\tau\in(0, t_0)$ and consider the function $\boldsymbol{g}_1$ defined by

$$\boldsymbol{g}_1(\boldsymbol{x}, t) = \int_\tau^t \int_\tau^s \boldsymbol{f}(\boldsymbol{x}, \lambda)\, d\lambda\, ds + \boldsymbol{h}(\boldsymbol{x}) + t\,\boldsymbol{q}(\boldsymbol{x})$$

---

[1] Boggio [1903, *1*]. This theorem was established independently by Sternberg and Eubanks [1957, *14*].

[2] Sternberg and Eubanks [1957, *14*].

for $(\boldsymbol{x}, t) \in \bar{B} \times (0, t_0)$, where $\boldsymbol{h}$ and $\boldsymbol{q}$ are as of yet unspecified. Then $\boldsymbol{g}_1$ meets the second of (b). We now show that $\boldsymbol{h}$ and $\boldsymbol{q}$ can be chosen such that $\square_1 \boldsymbol{g}_1 = \boldsymbol{0}$. Clearly,

$$\square_1 \boldsymbol{g}_1(\boldsymbol{x}, t) = \int_\tau^t \int_\tau^s c_1^2 \Delta \boldsymbol{f}(\boldsymbol{x}, \lambda)\, d\lambda\, ds - \boldsymbol{f}(\boldsymbol{x}, t) + c_1^2 \Delta \boldsymbol{h}(\boldsymbol{x}) + t\, c_1^2 \Delta \boldsymbol{q}(\boldsymbol{x}),$$

and by (c)

$$c_1^2 \Delta \boldsymbol{f} = \ddot{\boldsymbol{f}}.$$

Thus

$$\square_1 \boldsymbol{g}_1(\boldsymbol{x}, t) = -\boldsymbol{f}(\boldsymbol{x}, \tau) + \tau \dot{\boldsymbol{f}}(\boldsymbol{x}, \tau) - t \dot{\boldsymbol{f}}(\boldsymbol{x}, \tau) + c_1^2 \Delta \boldsymbol{h}(\boldsymbol{x}) + t\, c_1^2 \Delta \boldsymbol{q}(\boldsymbol{x}).$$

Since $\boldsymbol{f}$ is of class C^2 on $\bar{B} \times (0, t_0)$, $\boldsymbol{f}(\cdot, \tau)$ and $\dot{\boldsymbol{f}}(\cdot, \tau)$ are of class C^1 on $\bar{B}$; hence, in view of ***(6.5)***, there exist functions $\boldsymbol{h}$ and $\boldsymbol{q}$ of class C^2 on B such that

$$c_1^2 \Delta \boldsymbol{h}(\boldsymbol{x}) = \boldsymbol{f}(\boldsymbol{x}, \tau) - \tau \dot{\boldsymbol{f}}(\boldsymbol{x}, \tau),$$
$$c_1^2 \Delta \boldsymbol{q}(\boldsymbol{x}) = \dot{\boldsymbol{f}}(\boldsymbol{x}, \tau),$$

which yields $\square_1 \boldsymbol{g}_1 = \boldsymbol{0}$. □

An interesting relation between the Cauchy-Kovalevski-Somigliana solution and the Green-Lamé solution was established by STERNBERG.[1] Assume that the body forces vanish. Let $\boldsymbol{g}$ be the stress function in (S_1), (S_2). Then by (S_2) and Boggio's theorem, there exist functions $\boldsymbol{g}_1$ and $\boldsymbol{g}_2$ such that

$$\boldsymbol{g} = \boldsymbol{g}_1 + \boldsymbol{g}_2, \tag{a}$$

$$\square_1 \boldsymbol{g}_1 = \boldsymbol{0}, \qquad \square_2 \boldsymbol{g}_2 = \boldsymbol{0}. \tag{b}$$

By the second of (b),

$$\square_1 \boldsymbol{g}_2 = (c_1^2 - c_2^2) \Delta \boldsymbol{g}_2; \tag{c}$$

thus, using (a) and (b), we can write (S_1) as

$$\boldsymbol{u} = (c_1^2 - c_2^2)[\Delta \boldsymbol{g}_2 - \nabla \operatorname{div}(\boldsymbol{g}_1 + \boldsymbol{g}_2)], \tag{d}$$

and, since

$$\operatorname{curl}\operatorname{curl} \boldsymbol{g}_2 = \nabla \operatorname{div} \boldsymbol{g}_2 - \Delta \boldsymbol{g}_2,$$

(d) can be written in the form

$$\boldsymbol{u} = (c_2^2 - c_1^2)[\nabla \operatorname{div} \boldsymbol{g}_1 + \operatorname{curl}\operatorname{curl} \boldsymbol{g}_2].$$

If we define functions φ and $\boldsymbol{\psi}$ by

$$\varphi = (c_2^2 - c_1^2) \operatorname{div} \boldsymbol{g}_1, \qquad \boldsymbol{\psi} = (c_2^2 - c_1^2) \operatorname{curl} \boldsymbol{g}_2, \tag{e}$$

then (e) takes the form

$$\boldsymbol{u} = \nabla \varphi + \operatorname{curl} \boldsymbol{\psi}, \qquad \operatorname{div} \boldsymbol{\psi} = 0, \tag{f}$$

while (b) yields

$$\square_1 \varphi = 0, \qquad \square_2 \boldsymbol{\psi} = \boldsymbol{0}. \tag{g}$$

Eqs. (f) and (g) are identical to (L_1) and (L_2). Thus we have STERNBERG's result: The *Cauchy-Kovalevski-Somigliana solution reduces to the Green-Lamé solution provided the field $\boldsymbol{g}$ is subjected to the transformation* (a), (b), *and* (e). In view of the completeness of the Cauchy-Kovalevski-Somigliana solution, this reduction theorem furnishes an alternative proof of the completeness of the Green-Lamé solution.

[1] [1960, *13*].

68. Basic singular solutions. In this section we will discuss the basic singular solutions of elastodynamics, assuming throughout that the body is *homogeneous* and *isotropic* with $c_1>0$, $c_2>0$.

Throughout this section $\boldsymbol{y}$ is a fixed point of $\mathscr{E}$,

$$\boldsymbol{r}=\boldsymbol{x}-\boldsymbol{y}, \qquad r=|\boldsymbol{r}|,$$

and $\Sigma_\eta=\Sigma_\eta(\boldsymbol{y})$ is the open ball with center at $\boldsymbol{y}$ and radius η.

Recall[1] that $\mathscr{L}$ is the set of all *smooth* vector functions on $(-\infty,\infty)$ that vanish on $(-\infty,0)$. For each $\boldsymbol{l}\in\mathscr{L}$ let $\boldsymbol{u}_{\boldsymbol{y}}\{\boldsymbol{l}\}$ be the vector field on $(\mathscr{E}-\{\boldsymbol{y}\})\times(-\infty,\infty)$ defined by

$$\boldsymbol{u}_{\boldsymbol{y}}\{\boldsymbol{l}\}(\boldsymbol{x},t)=\frac{1}{4\pi\varrho r}\Bigg[\left(\frac{3\boldsymbol{r}\otimes\boldsymbol{r}}{r^2}-\mathbf{1}\right)\int\limits_{1/c_1}^{1/c_2}\lambda\boldsymbol{l}(t-\lambda r)\,d\lambda + \frac{\boldsymbol{r}\otimes\boldsymbol{r}}{r^2}\left(\frac{1}{c_1^2}\boldsymbol{l}(t-r/c_1)-\frac{1}{c_2^2}\boldsymbol{l}(t-r/c_2)\right)+\frac{1}{c_2^2}\boldsymbol{l}(t-r/c_2)\Bigg].$$

Further, let $\boldsymbol{E}_{\boldsymbol{y}}\{\boldsymbol{l}\}$ and $\boldsymbol{S}_{\boldsymbol{y}}\{\boldsymbol{l}\}$ be defined through the strain-displacement and stress-strain relations, so that

$$\boldsymbol{S}_{\boldsymbol{y}}\{\boldsymbol{l}\}(\boldsymbol{x},t)=\frac{1}{4\pi r^3}\left[(\boldsymbol{r}\cdot\boldsymbol{f}_1)\frac{\boldsymbol{r}\otimes\boldsymbol{r}}{r^2}-(\boldsymbol{r}\cdot\boldsymbol{f}_2)\,\mathbf{1}-\boldsymbol{r}\otimes\boldsymbol{f}_3-\boldsymbol{f}_3\otimes\boldsymbol{r}\right],$$

where

$$\boldsymbol{f}_i=\boldsymbol{f}_i\{\boldsymbol{l}\}(\boldsymbol{x},t)$$

with

$$\boldsymbol{f}_1\{\boldsymbol{l}\}(\boldsymbol{x},t)=-6c_2^2\int\limits_{1/c_1}^{1/c_2}\lambda\boldsymbol{l}(t-\lambda r)\,d\lambda+12\left[\boldsymbol{l}(t-r/c_2)-\frac{c_2^2}{c_1^2}\boldsymbol{l}(t-r/c_1)\right] + \frac{2r}{c_2}\left[\dot{\boldsymbol{l}}(t-r/c_2)-\frac{c_2^3}{c_1^3}\dot{\boldsymbol{l}}(t-r/c_1)\right],$$

$$\boldsymbol{f}_2\{\boldsymbol{l}\}(\boldsymbol{x},t)=-6c_2^2\int\limits_{1/c_1}^{1/c_2}\lambda\boldsymbol{l}(t-\lambda r)\,d\lambda+2\boldsymbol{l}(t-r/c_2)+\left(1-4\frac{c_2^2}{c_1^2}\right)\boldsymbol{l}(t-r/c_1) + \frac{r}{c_1}\left(1-2\frac{c_2^2}{c_1^2}\right)\dot{\boldsymbol{l}}(t-r/c_1),$$

$$\boldsymbol{f}_3\{\boldsymbol{l}\}(\boldsymbol{x},t)=-6c_2^2\int\limits_{1/c_1}^{1/c_2}\lambda\boldsymbol{l}(t-\lambda r)\,d\lambda+3\boldsymbol{l}(t-r/c_2)-\frac{2c_2^2}{c_1^2}\boldsymbol{l}(t-r/c_1) + \frac{r}{c_2}\dot{\boldsymbol{l}}(t-r/c_1).$$

We write

$$\not{p}_{\boldsymbol{y}}\{\boldsymbol{l}\}=[\boldsymbol{u}_{\boldsymbol{y}}\{\boldsymbol{l}\},\boldsymbol{E}_{\boldsymbol{y}}\{\boldsymbol{l}\},\boldsymbol{S}_{\boldsymbol{y}}\{\boldsymbol{l}\}]$$

and call $\not{p}_{\boldsymbol{y}}\{\boldsymbol{l}\}$ the ***Stokes process***[2] corresponding to a concentrated load $\boldsymbol{l}$ at $\boldsymbol{y}$. Clearly, the mapping $\boldsymbol{l}\mapsto\not{p}_{\boldsymbol{y}}\{\boldsymbol{l}\}$ is linear on $\mathscr{L}$.

---

[1] See p. 26.

[2] This solution is due to STOKES [1851, *1*]. It was deduced by LOVE [1904, *3*], [1927, *3*], § 212 through a limit process based on a sequence of time-dependent body force fields that approaches a concentrated load (cf. ***(51.1)***). The form of STOKES' solution presented here is taken from WHEELER and STERNBERG [1968, *16*]. A generalization of STOKES' solution, valid for inhomogeneous media, is contained in the work of MIKHLIN [1947, *5*]. See also BABICH [1961, *1*], GUTZWILLER [1962, *8*], MURTAZAEV [1962, *12*], NOWAK [1969, *5*].

(1) Let $\boldsymbol{l} \in \mathscr{L}$ *and define* $\mathbf{S}_{\boldsymbol{y}}^{*}\{\boldsymbol{l}\}$ *on* $(\mathscr{E} - \{\boldsymbol{y}\}) \times (-\infty, \infty)$ by

$$\mathbf{S}_{\boldsymbol{y}}^{*}\{\boldsymbol{l}\}(\boldsymbol{x}, t) = \frac{1}{4\pi r^3}\left[\frac{(\boldsymbol{r}\cdot\boldsymbol{f}_1)\,\boldsymbol{r}\otimes\boldsymbol{r}}{r^2} - \boldsymbol{r}\otimes\boldsymbol{f}_2 - \boldsymbol{f}_3\otimes\boldsymbol{r} - (\boldsymbol{r}\cdot\boldsymbol{f}_3)\,\mathbf{1}\right],$$

where $\boldsymbol{f}_i = \boldsymbol{f}_i\{\boldsymbol{l}\}(\boldsymbol{x}, t)$ *is defined above. Then*

$$\mathbf{S}_{\boldsymbol{y}}\{\boldsymbol{k}\} * \boldsymbol{l} = \mathbf{S}_{\boldsymbol{y}}^{*}\{\boldsymbol{l}\}^T * \boldsymbol{k}.$$

Proof. It follows from the definition of $\boldsymbol{f}_i$ and *(10.2)* that

$$\boldsymbol{f}_i\{\boldsymbol{l}\} \# \boldsymbol{k} = \boldsymbol{l} \# \boldsymbol{f}_i\{\boldsymbol{k}\};$$

hence

$$\boldsymbol{f}_i\{\boldsymbol{l}\} * \boldsymbol{k} = \boldsymbol{l} * \boldsymbol{f}_i\{\boldsymbol{k}\}.$$

Thus, in view of the definition of $\mathbf{S}_{\boldsymbol{y}}\{\boldsymbol{k}\}$,

$$\begin{aligned}
4\pi r^3\, \mathbf{S}_{\boldsymbol{y}}\{\boldsymbol{k}\} * \boldsymbol{l} &= (\boldsymbol{r}\cdot\boldsymbol{f}_1\{\boldsymbol{k}\}) * (\boldsymbol{r}\cdot\boldsymbol{l})\,\frac{\boldsymbol{r}}{r^2} - (\boldsymbol{r}\cdot\boldsymbol{f}_2\{\boldsymbol{k}\}) * \boldsymbol{l} - (\boldsymbol{f}_3\{\boldsymbol{k}\} * \boldsymbol{l})\,\boldsymbol{r} - (\boldsymbol{r}\cdot\boldsymbol{l}) * \boldsymbol{f}_3\{\boldsymbol{k}\}\\
&= \left[\left(\frac{\boldsymbol{r}\otimes\boldsymbol{r}}{r^2}\right)\cdot(\boldsymbol{f}_1\{\boldsymbol{k}\} \# \boldsymbol{l})\right]\boldsymbol{r} - (\boldsymbol{l} \# \boldsymbol{f}_2\{\boldsymbol{k}\})\,\boldsymbol{r} - (\boldsymbol{f}_3\{\boldsymbol{k}\} * \boldsymbol{l})\,\boldsymbol{r} - (\boldsymbol{f}_3\{\boldsymbol{k}\} \# \boldsymbol{l})\,\boldsymbol{r}\\
&= \left[\left(\frac{\boldsymbol{r}\otimes\boldsymbol{r}}{r^2}\right)\cdot(\boldsymbol{k} \# \boldsymbol{f}_1\{\boldsymbol{l}\})\right]\boldsymbol{r} - (\boldsymbol{f}_2\{\boldsymbol{l}\} \# \boldsymbol{k})\,\boldsymbol{r} - (\boldsymbol{k} * \boldsymbol{f}_3\{\boldsymbol{l}\})\,\boldsymbol{r} - (\boldsymbol{k} \# \boldsymbol{f}_3\{\boldsymbol{l}\})\,\boldsymbol{r}\\
&= 4\pi r^3\, \mathbf{S}_{\boldsymbol{y}}^{*}\{\boldsymbol{l}\}^T * \boldsymbol{k}. \quad \square
\end{aligned}$$

Let

$$\mathbf{s}_{\boldsymbol{y}}^{*}\{\boldsymbol{l}\} = \mathbf{S}_{\boldsymbol{y}}^{*}\{\boldsymbol{l}\}\,\boldsymbol{n}$$

on $\partial B \times (-\infty, \infty)$, where $\mathbf{S}_{\boldsymbol{y}}^{*}\{\boldsymbol{l}\}$ is the tensor field defined in *(1)*. We call $\mathbf{s}_{\boldsymbol{y}}^{*}\{\boldsymbol{l}\}$ the ***adjoint traction field*** on ∂B corresponding to $\mathscr{p}_{\boldsymbol{y}}\{\boldsymbol{l}\}$.

(2) Properties of the Stokes process.[1] *Let* $\boldsymbol{l} \in \mathscr{L}$. *Then the Stokes process* $\mathscr{p}_{\boldsymbol{y}}\{\boldsymbol{l}\}$ *has the following properties:*

(i) *If* $\boldsymbol{l}$ *is of class* C^2, *then* $\mathscr{p}_{\boldsymbol{y}}\{\boldsymbol{l}\}$ *is an elastic process on* $\mathscr{E} - \{\boldsymbol{y}\}$ *with a quiescent past, and* $\mathscr{p}_{\boldsymbol{y}}\{\boldsymbol{l}\}$ *corresponds to zero body forces.*

(ii) $\boldsymbol{u}_{\boldsymbol{y}}\{\boldsymbol{l}\}(\boldsymbol{x}, \cdot) = O(r^{-1})$ *and* $\mathbf{S}_{\boldsymbol{y}}\{\boldsymbol{l}\}(\boldsymbol{x}, \cdot) = O(r^{-2})$ *as* $r \to 0$ *and also as* $r \to \infty$, *uniformly on every interval of the form* $(-\infty, t]$.

(iii) $\lim\limits_{\eta\to 0} \int\limits_{\partial\Sigma_\eta} \mathbf{s}_{\boldsymbol{y}}\{\boldsymbol{l}\}\, da = \boldsymbol{l}, \quad \lim\limits_{\eta\to 0} \int\limits_{\partial\Sigma_\eta} \boldsymbol{r} \times \mathbf{s}_{\boldsymbol{y}}\{\boldsymbol{l}\}\, da = \mathbf{0},$

where

$$\mathbf{s}_{\boldsymbol{y}}\{\boldsymbol{l}\} = \mathbf{S}_{\boldsymbol{y}}\{\boldsymbol{l}\}\,\boldsymbol{n}$$

on $\partial\Sigma_\eta \times (-\infty, \infty)$ *with* $\boldsymbol{n}$ *the inward unit normal. Moreover, the above limits are attained uniformly on every interval of the form* $(-\infty, t]$.

(iv) *For every* $\boldsymbol{k} \in \mathscr{L}$,

$$\boldsymbol{l} * \boldsymbol{u}_{\boldsymbol{y}}\{\boldsymbol{k}\} = \boldsymbol{k} * \boldsymbol{u}_{\boldsymbol{y}}\{\boldsymbol{l}\}, \quad \boldsymbol{l} * \mathbf{s}_{\boldsymbol{y}}\{\boldsymbol{k}\} = \boldsymbol{k} * \mathbf{s}_{\boldsymbol{y}}^{*}\{\boldsymbol{l}\},$$

where $\mathbf{s}_{\boldsymbol{y}}\{\boldsymbol{k}\}$ *is the traction field on* ∂B *corresponding to* $\mathscr{p}_{\boldsymbol{y}}\{\boldsymbol{k}\}$, *and* $\mathbf{s}_{\boldsymbol{y}}^{*}\{\boldsymbol{l}\}$ *is the adjoint traction field on* ∂B *corresponding to* $\mathscr{p}_{\boldsymbol{y}}\{\boldsymbol{l}\}$.

Proof. Parts (i) and (ii) follow from the assumed properties of $\boldsymbol{l}$ and the formulae for $\boldsymbol{u}_{\boldsymbol{y}}\{\boldsymbol{l}\}$ and $\mathbf{S}_{\boldsymbol{y}}\{\boldsymbol{l}\}$.

---

[1] This theorem, with the exception of the second of (iv), is due to WHEELER and STERNBERG [1968, *16*]. See also LOVE [1927, *3*], who asserts (iii).

Consider now part (iii). A simple computation based on the formula for $\boldsymbol{S_y}\{\boldsymbol{l}\}$ leads to the result

$$\int\limits_{\partial\Sigma_\eta} \boldsymbol{s_y}\{\boldsymbol{l}\}(\boldsymbol{x},\tau)\,da_{\boldsymbol{x}} = \frac{1}{3}\Big[\boldsymbol{l}(\tau-\eta/c_1)+2\boldsymbol{l}(\tau-\eta/c_2) + \frac{\eta}{c_1}\dot{\boldsymbol{l}}(\tau-\eta/c_1)+\frac{2\eta}{c_2}\dot{\boldsymbol{l}}(\tau-\eta/c_2)\Big],$$

which implies the first of (iii), since $\boldsymbol{l}\in\mathscr{L}$. The second of (iii) is an immediate consequence of the order of magnitude estimate for $\boldsymbol{S_y}\{\boldsymbol{l}\}$ in (ii).

The first of property (iv) is easily established with the aid of ***(10.2)***. To prove the second, note that by ***(1)***

$$\begin{aligned}\boldsymbol{l}*\boldsymbol{s_y}\{\boldsymbol{k}\} &= \boldsymbol{l}*(\boldsymbol{S_y}\{\boldsymbol{k}\}\,\boldsymbol{n}) = (\boldsymbol{S_y}\{\boldsymbol{k}\}*\boldsymbol{l})\cdot\boldsymbol{n}\\ &= (\boldsymbol{S}^*_{\boldsymbol{y}}\{\boldsymbol{l}\}^T*\boldsymbol{k})\cdot\boldsymbol{n} = \boldsymbol{k}*(\boldsymbol{S}^*_{\boldsymbol{y}}\{\boldsymbol{l}\}\,\boldsymbol{n}) = \boldsymbol{k}*\boldsymbol{s}^*_{\boldsymbol{y}}\{\boldsymbol{l}\}.\quad\square\end{aligned}$$

Let $\boldsymbol{l}\in\mathscr{L}$ be of class C^2, and let $\bar{\boldsymbol{u}}^i_{\boldsymbol{y}}\{\boldsymbol{l}\}$ be defined on $(\mathscr{E}-\{\boldsymbol{y}\})\times(-\infty,\infty)$ by

$$\bar{\boldsymbol{u}}^i_{\boldsymbol{y}}\{\boldsymbol{l}\}(\boldsymbol{x},t) = \frac{\partial}{\partial x_i}\boldsymbol{u_y}\{\boldsymbol{l}\}(\boldsymbol{x},t).$$

Then

$$\begin{aligned}\bar{\boldsymbol{u}}^i_{\boldsymbol{y}}\{\boldsymbol{l}\}(\boldsymbol{x},t) = -\frac{1}{4\pi\varrho r^3}\Bigg[&3\left(\frac{5r_i\,\boldsymbol{r}\otimes\boldsymbol{r}}{r^2}-r_i\,\mathbf{1}-\boldsymbol{e}_i\otimes\boldsymbol{r}-\boldsymbol{r}\otimes\boldsymbol{e}_i\right)\int\limits_{1/c_1}^{1/c_2}\lambda\boldsymbol{l}(t-\lambda r)\,d\lambda\\ &+\left(\frac{6r_i\,\boldsymbol{r}\otimes\boldsymbol{r}}{r^2}-r_i\,\mathbf{1}-\boldsymbol{e}_i\otimes\boldsymbol{r}-\boldsymbol{r}\otimes\boldsymbol{e}_i\right)\left(\frac{1}{c_1^2}\boldsymbol{l}(t-r/c_1)-\frac{1}{c_2^2}\boldsymbol{l}(t-r/c_2)\right)\\ &+\frac{r_i}{c_2^2}\left(\boldsymbol{l}(t-r/c_2)+\frac{r}{c_2}\dot{\boldsymbol{l}}(t-r/c_2)\right)\\ &+\frac{r_i\,\boldsymbol{r}\otimes\boldsymbol{r}}{r}\left(\frac{1}{c_1^3}\dot{\boldsymbol{l}}(t-r/c_1)-\frac{1}{c_2^3}\dot{\boldsymbol{l}}(t-r/c_2)\right)\Bigg].\end{aligned}$$

We write

$$\bar{\boldsymbol{p}}^i_{\boldsymbol{y}}\{\boldsymbol{l}\} = [\bar{\boldsymbol{u}}^i_{\boldsymbol{y}}\{\boldsymbol{l}\},\bar{\boldsymbol{E}}^i_{\boldsymbol{y}}\{\boldsymbol{l}\},\bar{\boldsymbol{S}}^i_{\boldsymbol{y}}\{\boldsymbol{l}\}],$$

where $\bar{\boldsymbol{E}}^i_{\boldsymbol{y}}\{\boldsymbol{l}\}$ and $\bar{\boldsymbol{S}}^i_{\boldsymbol{y}}\{\boldsymbol{l}\}$ are defined through the strain-displacement and stress-strain relations, and call $\bar{\boldsymbol{p}}^i_{\boldsymbol{y}}\{\boldsymbol{l}\}$ the ***doublet process*** (corresponding to $\boldsymbol{y}$, $\boldsymbol{l}$, and $\boldsymbol{e}_i$).

(3) Properties of the doublet processes.[1] *Let $\boldsymbol{l}\in\mathscr{L}$ be of class C^2. Then the doublet process $\bar{\boldsymbol{p}}^i_{\boldsymbol{y}}\{\boldsymbol{l}\}$ has the following properties:*

(i) *If $\boldsymbol{l}$ is of class C^3, then $\bar{\boldsymbol{p}}^i_{\boldsymbol{y}}\{\boldsymbol{l}\}$ is an elastic process on $\mathscr{E}-\{\boldsymbol{y}\}$ with a quiescent past, and $\bar{\boldsymbol{p}}^i_{\boldsymbol{y}}\{\boldsymbol{l}\}$ corresponds to zero body forces.*

(ii) *$\bar{\boldsymbol{u}}^i_{\boldsymbol{y}}\{\boldsymbol{l}\}(\boldsymbol{x},\cdot)=O(r^{-2})$ and $\bar{\boldsymbol{S}}^i_{\boldsymbol{y}}\{\boldsymbol{l}\}(\boldsymbol{x},\cdot)=O(r^{-3})$ as $r\to 0$ and also as $r\to\infty$, uniformly on every interval of the form $(-\infty,t]$.*

(iii) $\lim\limits_{\eta\to 0}\int\limits_{\partial\Sigma_\eta}\bar{\boldsymbol{s}}^i_{\boldsymbol{y}}\{\boldsymbol{l}\}\,da=\boldsymbol{0},\quad \lim\limits_{\eta\to 0}\int\limits_{\partial\Sigma_\eta}\boldsymbol{r}\times\bar{\boldsymbol{s}}^i_{\boldsymbol{y}}\{\boldsymbol{l}\}\,da=-\boldsymbol{e}_i\times\boldsymbol{l},$

where

$$\bar{\boldsymbol{s}}^i\{\boldsymbol{l}\}=\bar{\boldsymbol{S}}^i\{\boldsymbol{l}\}\,\boldsymbol{n}$$

on $\partial\Sigma_\eta$ with $\boldsymbol{n}$ the inward unit normal.

(iv) *If $\boldsymbol{k}\in\mathscr{L}$ is of class C^2, then*

$$\boldsymbol{k}*\bar{\boldsymbol{u}}_{\boldsymbol{y}}\{\boldsymbol{l}\}=\boldsymbol{l}*\bar{\boldsymbol{u}}_{\boldsymbol{y}}\{\boldsymbol{k}\}.$$

[1] Wheeler and Sternberg [1968, *16*], Theorem 3.2.

(v) *If $\mathfrak{p}_{\boldsymbol{y}}\{\boldsymbol{l}\}$ is the Stokes process corresponding to the concentrated load $\boldsymbol{l}$ at $\boldsymbol{y}$, then*

$$\frac{\partial}{\partial y_i}\,\mathfrak{p}_{\boldsymbol{y}}\{\boldsymbol{l}\} = -\bar{\mathfrak{p}}^i_{\boldsymbol{y}}\{\boldsymbol{l}\}.$$

69. Love's integral identity. In this section we will establish Love's theorem; this theorem is a direct analog of Somigliana's theorem ***(52.4)*** in elastostatics. We continue to assume that B is *homogeneous* and *isotropic* with $c_1>0$, $c_2>0$.

(1) Lemma.[1] *Let $\mathfrak{p}=[\boldsymbol{u}, \boldsymbol{E}, \boldsymbol{S}]$ be an elastic process with a quiescent past, and let $\mathfrak{p}_{\boldsymbol{y}}\{\boldsymbol{l}\}=[\boldsymbol{u}_{\boldsymbol{y}}\{\boldsymbol{l}\}, \boldsymbol{E}_{\boldsymbol{y}}\{\boldsymbol{l}\}, \boldsymbol{S}_{\boldsymbol{y}}\{\boldsymbol{l}\}]$ be the Stokes process corresponding to a concentrated load $\boldsymbol{l}\in\mathscr{L}$ at $\boldsymbol{y}\in B$. Then*

$$\lim_{\eta\to 0}\int_{\partial\Sigma_\eta(\boldsymbol{y})} \boldsymbol{s}*\boldsymbol{u}_{\boldsymbol{y}}\{\boldsymbol{l}\}\,da = 0,$$

$$\lim_{\eta\to 0}\int_{\partial\Sigma_\eta(\boldsymbol{y})} \boldsymbol{u}*\boldsymbol{s}_{\boldsymbol{y}}\{\boldsymbol{l}\}\,da = \boldsymbol{l}*\boldsymbol{u}(\boldsymbol{y},\cdot),$$

where

$$\boldsymbol{s}=\boldsymbol{S}\boldsymbol{n}, \qquad \boldsymbol{s}_{\boldsymbol{y}}\{\boldsymbol{l}\}=\boldsymbol{S}_{\boldsymbol{y}}\{\boldsymbol{l}\}\,\boldsymbol{n}$$

on $\partial\Sigma_\eta(\boldsymbol{y})\times(-\infty,\infty)$ with $\boldsymbol{n}$ the inward unit normal.

Proof. Since both states have quiescent pasts, the above relations hold trivially on $(-\infty, 0]$. Thus hold $t>0$ fixed for the remainder of the argument. Let β be such that $\bar{\Sigma}_\beta(\boldsymbol{y})\subset B$, choose $\eta\in(0,\beta]$, and write $\Sigma_\eta=\Sigma_\eta(\boldsymbol{y})$. Further, let

$$\boldsymbol{v}(\boldsymbol{x},\tau)=\boldsymbol{u}(\boldsymbol{x},\tau)-\boldsymbol{u}(\boldsymbol{y},\tau) \tag{a}$$

for $(\boldsymbol{x},\tau)\in\bar{B}\times[0,\infty)$. Then

$$\begin{aligned}|I(\eta)-\boldsymbol{l}*\boldsymbol{u}(\boldsymbol{y},t)| \leqq &\left|\int_{\partial\Sigma_\eta}\int_0^t \boldsymbol{s}_{\boldsymbol{y}}\{\boldsymbol{l}\}(\boldsymbol{x},t-\tau)\cdot\boldsymbol{v}(\boldsymbol{x},\tau)\,d\tau\,da_{\boldsymbol{x}}\right|\\ &+\left|\int_0^t \boldsymbol{u}(\boldsymbol{y},t-\tau)\cdot\left[\int_{\partial\Sigma_\eta}\boldsymbol{s}_{\boldsymbol{y}}\{\boldsymbol{l}\}(\boldsymbol{x},\tau)\,da_{\boldsymbol{x}}-\boldsymbol{l}(\tau)\right]d\tau\right|,\end{aligned} \tag{b}$$

where

$$I(\eta)=\int_{\partial\Sigma_\eta}\boldsymbol{u}*\boldsymbol{s}_{\boldsymbol{y}}\{\boldsymbol{l}\}(\boldsymbol{x},t)\,da_{\boldsymbol{x}}.$$

The second term in the right-hand member of (b) tends to zero with η, since the limit may be taken under the time-integral[2] and because of (iii) in ***(68.2)***. Consequently,

$$|I(\eta)-\boldsymbol{l}*\boldsymbol{u}(\boldsymbol{y},t)|\leqq 4\pi\eta^2\,t M_3(\eta)\,M_4(\eta)+o(1) \quad\text{as}\quad \eta\to 0, \tag{c}$$

where

$$M_3(\eta)=\sup\{|\boldsymbol{v}(\boldsymbol{x},\tau)|:(\boldsymbol{x},\tau)\in\partial\Sigma_\eta\times[0,t]\},$$
$$M_4(\eta)=\sup\{|\boldsymbol{s}_{\boldsymbol{y}}\{\boldsymbol{l}\}(\boldsymbol{x},\tau)|:(\boldsymbol{x},\tau)\in\partial\Sigma_\eta\times[0,t]\},$$

for $\eta\in(0,\beta]$. From (a) and the continuity of $\boldsymbol{u}$ on $\bar{B}\times[0,\infty)$ follows

$$M_3(\eta)=o(1)\quad\text{as}\quad\eta\to 0.$$

On the other hand, (ii) of ***(68.2)*** yields

$$M_4(\eta)=O(\eta^{-2})\quad\text{as}\quad\eta\to 0.$$

The inequality (c), when combined with the above estimates, implies the second result in ***(1)***. A strictly analogous argument yields the first result in ***(1)***. □

[1] WHEELER and STERNBERG [1968, *16*], Lemma 3.1.
[2] See MIKUSINSKI [1959, *10*], p. 143.

(2) Love's integral identity.[1] *Let $p=[\boldsymbol{u}, \boldsymbol{E}, \boldsymbol{S}]$ be an elastic process on B corresponding to the external force system $[\boldsymbol{b}, \boldsymbol{s}]$, let p have a quiescent past, and suppose that $\boldsymbol{s}(\boldsymbol{x}, \cdot)\in\mathscr{L}$ for every $\boldsymbol{x}\in\partial B$, and that $\boldsymbol{b}(\boldsymbol{x}, \cdot)\in\mathscr{L}$ for every $\boldsymbol{x}\in B$. Then*

$$\boldsymbol{u}(\boldsymbol{y}, \cdot)=\int_{\partial B}[\boldsymbol{u}_{\boldsymbol{y}}\{\boldsymbol{s}\}-\boldsymbol{s}_{\boldsymbol{y}}^{*}\{\boldsymbol{u}\}]\,da+\int_{B}\boldsymbol{u}_{\boldsymbol{y}}\{\boldsymbol{b}\}\,dv$$

for $\boldsymbol{y}\in B$, where, for any $\boldsymbol{l}\in\mathscr{L}$, $\boldsymbol{u}_{\boldsymbol{y}}\{\boldsymbol{l}\}$ is the displacement field and $\boldsymbol{s}_{\boldsymbol{y}}^{*}\{\boldsymbol{l}\}$ the adjoint traction field corresponding to the Stokes process $p_{\boldsymbol{y}}\{\boldsymbol{l}\}$.

Proof.[2] By hypothesis, $\boldsymbol{u}$ is an admissible motion and

$$\boldsymbol{u}(\cdot, 0)=\dot{\boldsymbol{u}}(\cdot, 0)=\boldsymbol{0} \tag{a}$$

on B. Thus $\boldsymbol{u}(\boldsymbol{x}, \cdot)\in\mathscr{L}$ for every $\boldsymbol{x}\in B$. Therefore, since $\boldsymbol{s}(\boldsymbol{x}, \cdot)\in\mathscr{L}$ for every $\boldsymbol{x}\in\partial B$, the above expression holds trivially on $(-\infty, 0]$.

We now establish its validity on $(0, \infty)$. Choose $\alpha>0$ with $\bar{\Sigma}_{\alpha}(\boldsymbol{y})\subset B$, and set

$$B_{\eta}=B-\bar{\Sigma}_{\eta}(\boldsymbol{y}) \tag{b}$$

for $\eta\in(0, \alpha]$. Let $\boldsymbol{l}\in\mathscr{L}$ be of class C^2. By (a) and (i) of ***(68.2)***, p and $p_{\boldsymbol{y}}\{\boldsymbol{l}\}$ both correspond to null initial data, while $p_{\boldsymbol{y}}\{\boldsymbol{l}\}$ corresponds to vanishing body forces. Thus we are entitled to apply (d) of Graffi's reciprocal theorem ***(61.1)*** to p and $p_{\boldsymbol{y}}\{\boldsymbol{l}\}$ on B_{η} with the result that

$$\int_{\partial B_{\eta}}\boldsymbol{s}*\boldsymbol{u}_{\boldsymbol{y}}\{\boldsymbol{l}\}\,da+\int_{B_{\eta}}\boldsymbol{b}*\boldsymbol{u}_{\boldsymbol{y}}\{\boldsymbol{l}\}\,dv=\int_{\partial B_{\eta}}\boldsymbol{s}_{\boldsymbol{y}}\{\boldsymbol{l}\}*\boldsymbol{u}\,da \tag{c}$$

on $(0, \infty)$. If we let $\eta\to 0$ in (c) and use (b) and ***(1)***, we arrive at the result

$$\int_{\partial B}\boldsymbol{s}*\boldsymbol{u}_{\boldsymbol{y}}\{\boldsymbol{l}\}\,da+\int_{B}\boldsymbol{b}*\boldsymbol{u}_{\boldsymbol{y}}\{\boldsymbol{l}\}\,dv=\int_{\partial B}\boldsymbol{s}_{\boldsymbol{y}}\{\boldsymbol{l}\}*\boldsymbol{u}\,da+\boldsymbol{l}*\boldsymbol{u}(\boldsymbol{y}, \cdot). \tag{d}$$

By (iv) of ***(68.2)*** and the properties of the convolution ***(10.1)***, (d) can be written in the form

$$\left[\int_{\partial B}[\boldsymbol{u}_{\boldsymbol{y}}\{\boldsymbol{s}\}-\boldsymbol{s}_{\boldsymbol{y}}^{*}\{\boldsymbol{u}\}]\,da+\int_{B}\boldsymbol{u}_{\boldsymbol{y}}\{\boldsymbol{b}\}\,dv-\boldsymbol{u}(\boldsymbol{y}, \cdot)\right]*\boldsymbol{l}=0 \tag{e}$$

on $[0, \infty)$. Since $\boldsymbol{l}\in\mathscr{L}$ is an arbitrary class C^2 function, (e) and (v) of ***(10.1)*** yield the desired result. □

V. Wave propagation.

70. The acoustic tensor. As we will see in the next few sections, the notion of an acoustic tensor is central to the study of wave propagation. In this section we define this tensor and establish some of its properties.

Fix $\boldsymbol{x}\in B$ and let $\mathbf{C}=\mathbf{C}_{\boldsymbol{x}}$ be the elasticity tensor and $\varrho=\varrho(\boldsymbol{x})>0$ the density at $\boldsymbol{x}$. Let $\boldsymbol{m}$ be a unit vector. Since $\mathbf{C}$ is a linear transformation,

$$\varrho^{-1}\mathbf{C}[\boldsymbol{a}\otimes\boldsymbol{m}]\boldsymbol{m}$$

[1] LOVE [1904, *3*]. A similar result was deduced previously by VOLTERRA [1894, *4*], but VOLTERRA's result is confined to two-dimensional elastostatics. See also SOMIGLIANA [1906, *6*], DE HOOP [1958, *4*], WHEELER and STERNBERG [1968, *16*]. None of the above authors utilizes the adjoint traction field, which considerably simplifies Love's identity. Related integral identities are contained in the work of ARZHANYH [1950, *2*], [1951, *2, 3, 4*], [1954, *1, 2*], [1955, *2, 3*], KNOPOFF [1956, *6*], DOYLE [1966, *8*].

A similar integral identity for the stress field, involving linear combinations of doublet processes, was established by WHEELER and STERNBERG [1968, *16*], Theorem 3.4.

[2] The proof we give here is due to WHEELER and STERNBERG [1968, *16*], pp. 71, 72. LOVE [1904, *3*] based his proof on Betti's reciprocal theorem ***(30.1)***, treating the inertia term as a body force.

is a linear, vector-valued function of the vector $\boldsymbol{a}$ and hence is represented by the operation of a tensor $\boldsymbol{A}(\boldsymbol{m})$ acting on $\boldsymbol{a}$:

$$\boldsymbol{A}(\boldsymbol{m})\,\boldsymbol{a} = \varrho^{-1}\mathsf{C}[\boldsymbol{a}\otimes\boldsymbol{m}]\,\boldsymbol{m} \quad \text{for every vector} \quad \boldsymbol{a}.$$

We call $\boldsymbol{A}(\boldsymbol{m})$ the ***acoustic tensor*** for the direction $\boldsymbol{m}$. In components

$$A_{ik}(\boldsymbol{m}) = \varrho^{-1}\,C_{ijkl}\,m_j\,m_l.$$

(1) Properties of the acoustic tensor.

(i) $\boldsymbol{A}(\boldsymbol{m})$ *is symmetric for every* $\boldsymbol{m}$ *if and only if* C *is symmetric.*

(ii) $\boldsymbol{A}(\boldsymbol{m})$ *is positive definite for every* $\boldsymbol{m}$ *if and only if* C *is strongly elliptic.*

(iii) $\boldsymbol{A}(\boldsymbol{m})$ *is positive definite for every* $\boldsymbol{m}$ *whenever* C *is positive definite.*

(iv) *Given any* $\boldsymbol{Q}$ *in the symmetry group* $\mathcal{G}_{\boldsymbol{x}}$,

$$\boldsymbol{Q}\boldsymbol{A}(\boldsymbol{m})\,\boldsymbol{Q}^T = \boldsymbol{A}(\boldsymbol{Q}\boldsymbol{m})$$

for every $\boldsymbol{m}$.

Proof. Let $\bar{\mathsf{C}} = \varrho^{-1}\mathsf{C}$. Clearly,

$$\boldsymbol{b}\cdot\boldsymbol{A}(\boldsymbol{m})\,\boldsymbol{a} = \boldsymbol{b}\cdot\bar{\mathsf{C}}[\boldsymbol{a}\otimes\boldsymbol{m}]\,\boldsymbol{m} = (\boldsymbol{b}\otimes\boldsymbol{m})\cdot\bar{\mathsf{C}}[\boldsymbol{a}\otimes\boldsymbol{m}]. \tag{a}$$

Properties (i) and (ii) are direct consequences of (a) and (ii) of ***(20.1)***. Further, since C positive definite implies C strongly elliptic, (ii) implies (iii). Finally, since

$$(\boldsymbol{Q}\boldsymbol{a})\otimes(\boldsymbol{Q}\boldsymbol{m}) = \boldsymbol{Q}(\boldsymbol{a}\otimes\boldsymbol{m})\,\boldsymbol{Q}^T,$$

it follows from (f) on p. 71 that

$$\begin{aligned}\boldsymbol{Q}^T\boldsymbol{A}(\boldsymbol{Q}\boldsymbol{m})\,\boldsymbol{Q}\boldsymbol{a} &= \boldsymbol{Q}^T\bar{\mathsf{C}}[\boldsymbol{Q}\boldsymbol{a}\otimes\boldsymbol{Q}\boldsymbol{m}]\,\boldsymbol{Q}\boldsymbol{m} = \boldsymbol{Q}^T\bar{\mathsf{C}}[\boldsymbol{Q}(\boldsymbol{a}\otimes\boldsymbol{m})\,\boldsymbol{Q}^T]\,\boldsymbol{Q}\boldsymbol{m}\\ &= \bar{\mathsf{C}}[\boldsymbol{a}\otimes\boldsymbol{m}]\,\boldsymbol{m} = \boldsymbol{A}(\boldsymbol{m})\,\boldsymbol{a},\end{aligned}$$

which implies (iv). □

(2) Properties of the acoustic tensor when the material is isotropic.[1] *Assume that the material at* $\boldsymbol{x}$ *is isotropic with Lamé moduli* μ *and* λ. *Then*

$$\boldsymbol{A}(\boldsymbol{m}) = c_1^2\,\boldsymbol{m}\otimes\boldsymbol{m} + c_2^2(\mathbf{1}-\boldsymbol{m}\otimes\boldsymbol{m}),$$

where c_1 *and* c_2 *are the wave speeds*:

$$c_1^2 = \frac{2\mu+\lambda}{\varrho}, \qquad c_2^2 = \frac{\mu}{\lambda}.$$

Further, c_1^2 *and* c_2^2 *are the principal values of* $\boldsymbol{A}(\boldsymbol{m})$; *the line spanned by* $\boldsymbol{m}$ *is the characteristic space for* c_1^2, *the plane perpendicular to* $\boldsymbol{m}$ *the characteristic space for* c_2^2.

Proof. By (i) of ***(20.1)*** and ***(22.2)***,

$$\mathsf{C}[\boldsymbol{a}\otimes\boldsymbol{m}] = \mathsf{C}[\operatorname{sym}\boldsymbol{a}\otimes\boldsymbol{m}] = \mu(\boldsymbol{a}\otimes\boldsymbol{m}+\boldsymbol{m}\otimes\boldsymbol{a}) + \lambda(\boldsymbol{a}\cdot\boldsymbol{m})\,\mathbf{1};$$

thus

$$\begin{aligned}\mathsf{C}[\boldsymbol{a}\otimes\boldsymbol{m}]\,\boldsymbol{m} &= \mu\boldsymbol{a} + (\lambda+\mu)\,(\boldsymbol{a}\cdot\boldsymbol{m})\,\boldsymbol{m} = [\mu\,\mathbf{1} + (\lambda+\mu)\,\boldsymbol{m}\otimes\boldsymbol{m}]\,\boldsymbol{a}\\ &= [\mu(\mathbf{1}-\boldsymbol{m}\otimes\boldsymbol{m}) + (\lambda+2\mu)\,\boldsymbol{m}\otimes\boldsymbol{m}]\,\boldsymbol{a},\end{aligned}$$

which yields the above formula for $\boldsymbol{A}(\boldsymbol{m})$. The remainder of the proof follows from the spectral theorem ***(3.2)***. □

[1] The form of the acoustic tensor for various anisotropic materials was given by SAKADI [1941, *2*] and FEDOROV [1965, *5*], § 19. FEDOROV [1963, *5*] has established a method of computing the isotropic acoustic tensor that is closest (in a precise sense) to a given anisotropic acoustic tensor.

When we wish to make explicit the dependence of the acoustic tensor $\boldsymbol{A}(\boldsymbol{m})$ on $\boldsymbol{x} \in \bar{B}$, we write $\boldsymbol{A}(\boldsymbol{x}, \boldsymbol{m})$, i.e.

$$\boldsymbol{A}(\boldsymbol{x}, \boldsymbol{m})\, \boldsymbol{a} = \varrho^{-1}(\boldsymbol{x})\, \mathbf{C}_{\boldsymbol{x}}[\boldsymbol{a} \otimes \boldsymbol{m}]\, \boldsymbol{m}.$$

In the following proposition the region B is completely arbitrary.

(3) Let $\mathbf{C}$ *and* ϱ^{-1} *be bounded on* $\bar{B}$ *with* $\mathbf{C}$ *symmetric and strongly elliptic. For each* $\boldsymbol{x} \in \bar{B}$ *and unit vector* $\boldsymbol{m}$, *let* $\lambda(\boldsymbol{x}, \boldsymbol{m})$ *be the largest characteristic value of the acoustic tensor* $\boldsymbol{A}(\boldsymbol{x}, \boldsymbol{m})$. *Then the number* $c_M > 0$ *defined by*

$$c_M^2 = \sup\{\lambda(\boldsymbol{x}, \boldsymbol{m}) : \boldsymbol{x} \in \bar{B}, |\boldsymbol{m}| = 1\}$$

is finite, and

$$|\boldsymbol{G} \cdot \mathbf{C}_{\boldsymbol{x}}[\boldsymbol{G}]| \leqq \varrho(\boldsymbol{x})\, c_M^2 |\boldsymbol{G}|^2$$

for every $\boldsymbol{x} \in \bar{B}$ *whenever* $\boldsymbol{G}$ *is of the form* $\boldsymbol{G} = \boldsymbol{a} \otimes \boldsymbol{v}$. *We call* c_M *the* ***maximum speed of propagation*** *corresponding to* $\mathbf{C}$ *and* ϱ.

Proof. By hypothesis and (i) and (ii) of *(1)*, $\boldsymbol{A}(\boldsymbol{x}, \boldsymbol{m})$ is always symmetric and positive definite. Thus $\lambda(\boldsymbol{x}, \boldsymbol{m}) > 0$ and

$$\boldsymbol{v} \cdot \boldsymbol{A}(\boldsymbol{x}, \boldsymbol{m})\, \boldsymbol{v} \leqq \lambda(\boldsymbol{x}, \boldsymbol{m}) |\boldsymbol{v}|^2 \tag{a}$$

for every vector $\boldsymbol{v}$. Thus if the supremum of $\lambda(\boldsymbol{x}, \boldsymbol{m})$ is finite, c_M will be well defined. Let $\boldsymbol{e}(\boldsymbol{x}, \boldsymbol{m})$ with $|\boldsymbol{e}(\boldsymbol{x}, \boldsymbol{m})| = 1$ be a characteristic vector corresponding to $\lambda(\boldsymbol{x}, \boldsymbol{m})$. Then

$$\lambda(\boldsymbol{x}, \boldsymbol{m}) = \boldsymbol{e}(\boldsymbol{x}, \boldsymbol{m}) \cdot \boldsymbol{A}(\boldsymbol{x}, \boldsymbol{m})\, \boldsymbol{e}(\boldsymbol{x}, \boldsymbol{m}), \tag{b}$$

and since

$$\begin{aligned} \boldsymbol{v} \cdot \boldsymbol{A}(\boldsymbol{x}, \boldsymbol{m})\, \boldsymbol{v} &= \varrho^{-1}(\boldsymbol{x})\, (\boldsymbol{v} \otimes \boldsymbol{m}) \cdot \mathbf{C}_{\boldsymbol{x}}[\boldsymbol{v} \otimes \boldsymbol{m}] \\ &\leqq \varrho^{-1}(\boldsymbol{x})\, |\mathbf{C}_{\boldsymbol{x}}|\, |\boldsymbol{v} \otimes \boldsymbol{m}|^2 = \varrho^{-1}(\boldsymbol{x})\, |\mathbf{C}_{\boldsymbol{x}}|\, |\boldsymbol{v}|^2 \end{aligned} \tag{c}$$

for every vector $\boldsymbol{v}$, (b) implies that

$$\lambda(\boldsymbol{x}, \boldsymbol{m}) \leqq \varrho^{-1}(\boldsymbol{x})\, |\mathbf{C}_{\boldsymbol{x}}|. \tag{d}$$

Since ϱ^{-1} and $\mathbf{C}$ are bounded on $\bar{B}$, $\varrho^{-1}|\mathbf{C}|$ is bounded, and therefore c_M is finite. The final inequality in *(3)* follows from (a), (c), and the definition of c_M. □

71. Progressive waves. In this section we study plane progressive wave solutions[1] to the displacement equation of motion. These simple solutions are important in that they yield valuable information concerning the propagation characteristics of elastic materials.

We assume that the body is *homogeneous* and the density strictly positive.

By a ***progressive wave*** we mean a function $\boldsymbol{u}$ on $\mathscr{E} \times (-\infty, \infty)$ of the form

$$\boldsymbol{u}(\boldsymbol{x}, t) = \boldsymbol{a}\, \varphi(\boldsymbol{p} \cdot \boldsymbol{m} - c t), \tag{a}$$ [2]

where:

(i) φ is a real-valued function of class C^2 on $(-\infty, \infty)$ with

$$\frac{d^2 \varphi(s)}{d s^2} \not\equiv 0;$$

(ii) $\boldsymbol{a}$ and $\boldsymbol{m}$ are unit vectors called, respectively, the ***direction of motion*** and the ***direction of propagation***;

[1] For studies of steady waves, of which plane progressive waves are a special case, see SÁENZ [1953, *17*] and STROH [1962, *14*].

[2] Recall that $\boldsymbol{p}(\boldsymbol{x}) = \boldsymbol{x} - \boldsymbol{0}$.

(iii) c is a scalar called the ***velocity of propagation.***

We say that the wave is ***longitudinal*** if $\boldsymbol{a}$ and $\boldsymbol{m}$ are linearly dependent; ***transverse*** if $\boldsymbol{a}$ and $\boldsymbol{m}$ are perpendicular; ***elastic*** if $\boldsymbol{u}$ satisfies the displacement equation of motion

$$\operatorname{div} \mathbf{C}[\nabla \boldsymbol{u}] = \varrho \ddot{\boldsymbol{u}} \tag{b}$$

on $\mathscr{E} \times (-\infty, \infty)$. Thus an elastic progressive wave is an elastic motion of the form (a) that corresponds to zero body forces.

Let α denote any given constant, and let π_t denote the plane defined by

$$\pi_t = \{\boldsymbol{x}: (\boldsymbol{x} - \boldsymbol{0}) \cdot \boldsymbol{m} - ct = \alpha\}.$$

Then at any given time t the displacement field $\boldsymbol{u}$ is constant on π_t. This plane is perpendicular to $\boldsymbol{m}$ and, as a function of t, is moving with velocity c in the direction $\boldsymbol{m}$. For this reason $\boldsymbol{u}$ is sometimes referred to as a *plane* progressive wave.

By (a)

$$\nabla \boldsymbol{u} = \varphi' \, \boldsymbol{a} \otimes \boldsymbol{m}, \tag{c}$$

where

$$\varphi'(\boldsymbol{x}, t) = \left. \frac{d\varphi(s)}{ds} \right|_{s = \boldsymbol{p} \cdot \boldsymbol{m} - ct},$$

$$\varphi''(\boldsymbol{x}, t) = \left. \frac{d^2\varphi(s)}{ds^2} \right|_{s = \boldsymbol{p} \cdot \boldsymbol{m} - ct}.$$

Thus

$$\operatorname{div} \boldsymbol{u} = \varphi' \, \boldsymbol{a} \cdot \boldsymbol{m},$$

$$\operatorname{curl} \boldsymbol{u} = \varphi' \, \boldsymbol{m} \times \boldsymbol{a},$$

and, since $\varphi' \not\equiv 0$ (because $\varphi'' \not\equiv 0$), it follows that a progressive wave is:

$$\textit{longitudinal} \Leftrightarrow \operatorname{curl} \boldsymbol{u} = \boldsymbol{0},$$

$$\textit{transverse} \Leftrightarrow \operatorname{div} \boldsymbol{u} = 0.$$

By (c)

$$\mathbf{C}[\nabla \boldsymbol{u}] = \varphi' \, \mathbf{C}[\boldsymbol{a} \otimes \boldsymbol{m}],$$

and, as $\mathbf{C}$ is independent of $\boldsymbol{x}$,

$$\operatorname{div} \mathbf{C}[\nabla \boldsymbol{u}] = \varphi'' \, \mathbf{C}[\boldsymbol{a} \otimes \boldsymbol{m}] \, \boldsymbol{m} = \varrho \, \varphi'' \boldsymbol{A}(\boldsymbol{m}) \boldsymbol{a}, \tag{d}$$

where $\boldsymbol{A}(\boldsymbol{m})$ is the acoustic tensor discussed in the previous section:

$$\boldsymbol{A}(\boldsymbol{m}) \, \boldsymbol{a} = \varrho^{-1} \mathbf{C}[\boldsymbol{a} \otimes \boldsymbol{m}] \, \boldsymbol{m}.$$

Next,

$$\varrho \ddot{\boldsymbol{u}} = \varrho \, c^2 \varphi'' \, \boldsymbol{a}, \tag{e}$$

and (d) and (e) yield the

(1) Propagation condition for progressive waves.[1] *A necessary and sufficient condition that the progressive wave* $\boldsymbol{u}$ *defined in* (a) *be elastic is that the* ***Fresnel-Hadamard propagation condition***

$$\boldsymbol{A}(\boldsymbol{m}) \, \boldsymbol{a} = c^2 \boldsymbol{a}$$

hold.

---

[1] According to TRUESDELL and TOUPIN [1960, *17*], § 300, the ideas behind this theorem can be traced back to the work of FRESNEL. See also CAUCHY [1840, *1*], pp. 42.50; GREEN [1841, *1*]; MUSGRAVE [1954, *15*]; MILLER and MUSGRAVE [1956, *8*]; SYNGE [1956, *13*], [1957, *16*]; FARNELL [1961, *7*]; MUSGRAVE [1961, *15*].

Thus for a progressive wave to propagate in a direction $\boldsymbol{m}$ its amplitude must be a principal vector of the acoustic tensor $\boldsymbol{A}(\boldsymbol{m})$ and the square of the velocity of propagation must be the associated characteristic value. If the elasticity tensor C is symmetric, then, by (i) of ***(70.1)***, $\boldsymbol{A}(\boldsymbol{m})$ is symmetric, and hence has at least three orthogonal principal directions $\boldsymbol{a}_1, \boldsymbol{a}_2, \boldsymbol{a}_3$ and three associated principal values c_1^2, c_2^2, c_3^2 for every $\boldsymbol{m}$. Further, if C is strongly elliptic, then $\boldsymbol{A}(\boldsymbol{m})$ will be positive definite and c_1, c_2, and c_3 will be real. Thus, *if C is symmetric and strongly elliptic, there exist, for every direction $\boldsymbol{m}$, three orthogonal directions of motion and three associated velocities of propagation c_1, c_2, c_3 for progressive waves.*

Theorem ***(1)***, when combined with ***(70.2)***, implies the

(2) Propagation condition for isotropic bodies.[1] *For an isotropic material the progressive wave $\boldsymbol{u}$ defined in* (a) *will be elastic if and only if either*

(i) $c^2 = c_1^2 = \dfrac{2\mu+\lambda}{\varrho}$ *and the wave is longitudinal; or*

(ii) $c^2 = c_2^2 = \mu/\varrho$ *and the wave is transverse.*

Thus for an isotropic material there are but two types of progressive waves: longitudinal and transverse. One can ask whether or not longitudinal and/or transverse waves exist for anisotropic materials. This question is answered by the

(3) Fedorov-Stippes theorem.[2] *There exist longitudinal and transverse elastic progressive waves provided the elasticity tensor is symmetric and strongly elliptic.*

Proof. By ***(1)*** there exists a longitudinal progressive wave in the direction $\boldsymbol{m}$ provided

$$\boldsymbol{A}(\boldsymbol{m})\,\boldsymbol{m} = \lambda\,\boldsymbol{m} \tag{a}$$

for some scalar $\lambda > 0$. Since C is strongly elliptic, $\boldsymbol{A}(\boldsymbol{m})$ is positive definite; thus, given any $\boldsymbol{m}$,

$$\boldsymbol{m}\cdot\boldsymbol{A}(\boldsymbol{m})\,\boldsymbol{m} > 0, \qquad \boldsymbol{A}(\boldsymbol{m})\,\boldsymbol{m} \neq \boldsymbol{0}. \tag{b}$$

Consider the function $\boldsymbol{l}$ defined on the surface of the unit sphere by

$$\boldsymbol{l}(\boldsymbol{m}) = \frac{\boldsymbol{A}(\boldsymbol{m})\,\boldsymbol{m}}{|\boldsymbol{A}(\boldsymbol{m})\,\boldsymbol{m}|}\,. \tag{c}$$

By the first of (b), $\boldsymbol{l}$ is continuous, maps the surface of the unit sphere into itself, and maps no point into its antipode. By a fixed point theorem,[3] any such map has a fixed point. Thus there exists a unit vector $\boldsymbol{m}_1$ such that $\boldsymbol{l}(\boldsymbol{m}_1) = \boldsymbol{m}_1$, or equivalently, by (c),

$$\boldsymbol{A}(\boldsymbol{m}_1)\,\boldsymbol{m}_1 = \lambda\boldsymbol{m}_1, \qquad \lambda = |\boldsymbol{A}(\boldsymbol{m}_1)\,\boldsymbol{m}_1|\,.$$

Thus there exists a longitudinal progressive wave in the direction $\boldsymbol{m}_1$. By the spectral theorem ***(3.2)***, there exist unit vectors $\boldsymbol{m}_2$ and $\boldsymbol{m}_3$ such that $\boldsymbol{m}_1, \boldsymbol{m}_2, \boldsymbol{m}_3$ are orthonormal and $\boldsymbol{m}_2$ and $\boldsymbol{m}_3$ are principal directions for $\boldsymbol{A}(\boldsymbol{m}_1)$. Since $\boldsymbol{A}(\boldsymbol{m}_1)$

[1] Cauchy [1840, *1*], pp. 137–142. See also Lamé [1852, *2*], § 59; Weyrauch [1884, *2*], §§ 98–99; Butty [1946, *1*], §§ 60–62; Pailloux [1956, *9*], p. 35.

[2] This theorem was arrived at independently by Fedorov [1964, *7*] and Stippes [1965, *18*]. See also Truesdell [1966, *27*], [1968, *13*] and Fedorov [1965, *5*], § 18. Kolodner [1966, *14*] has extended the Fedorov-Stippes theorem by showing that there exist three distinct directions along which longitudinal waves propagate. Sadaki [1941, *2*] has calculated the actual directions along which longitudinal and transverse waves propagate for various crystal classes. See also Brugger [1965, *3*]. Waterman [1959, *18*] has established results for waves that are nearly longitudinal or nearly transverse.

[3] See, e.g., Bourgin [1963, *2*], Theorem 8.7 on p. 132.

is positive definite, the corresponding characteristic values c_2^2 and c_3^2 are positive and hence are the squares of positive numbers. Thus there exist progressive waves in the direction $\boldsymbol{m}_1$ whose directions of motion are equal to $\boldsymbol{m}_2$ and $\boldsymbol{m}_3$. But $\boldsymbol{m}_2$ and $\boldsymbol{m}_3$ are orthogonal to $\boldsymbol{m}_1$. Thus these progressive waves are transverse. □

(4)[1] *Assume that the elasticity tensor is symmetric and strongly elliptic. Let* $\boldsymbol{e}$ *be an axis of symmetry for the material.*[2] *Then there exist longitudinal and transverse elastic progressive waves whose direction of propagation is* $\boldsymbol{e}$.

Proof. Clearly, we have only to show that $\boldsymbol{e}$ is a characteristic vector for $\boldsymbol{A}(\boldsymbol{e})$. Since $\boldsymbol{e}$ is an axis of symmetry,

$$\boldsymbol{Q}\boldsymbol{e} = \boldsymbol{e} \tag{a}$$

for some $\boldsymbol{Q} \neq \boldsymbol{1}$ in the symmetry group. Thus, by (iv) of ***(70.1)***,

$$\boldsymbol{Q}\boldsymbol{A}(\boldsymbol{e}) = \boldsymbol{A}(\boldsymbol{e})\,\boldsymbol{Q},$$

and we conclude from the commutation theorem ***(3.3)*** that $\boldsymbol{Q}$ leaves invariant each characteristic space of $\boldsymbol{A}(\boldsymbol{e})$. By (a) $\boldsymbol{Q}$ must be a proper rotation about $\boldsymbol{e}$ through an angle ϑ. If $\vartheta \neq \pi$, then the only spaces left invariant by $\boldsymbol{Q}$ are the one-dimensional space $\mathscr{U}$ spanned by $\boldsymbol{e}$, the plane perpendicular to $\mathscr{U}$, and the entire space $\mathscr{V}$. If $\vartheta = \pi$, then $\boldsymbol{Q}$ leaves invariant, in addition, each one-dimensional space spanned by a vector perpendicular to $\boldsymbol{e}$. In either event, it is clear that at least one characteristic space of $\boldsymbol{A}(\boldsymbol{e})$ must contain $\boldsymbol{e}$. □

It follows from the above proof that if $\vartheta \neq \pi$, elastic transverse waves can propagate in the direction of $\boldsymbol{e}$ with *any* amplitude perpendicular to $\boldsymbol{e}$, and all of these waves propagate with the *same* velocity.

72. Propagating surfaces. Surfaces of discontinuity. In this section we define the notion of a propagating singular surface. Roughly speaking, this is a surface, moving with time, across which some kinematic quantity suffers a jump discontinuity.

Recall that

$$\mathscr{E}^{(4)} = \mathscr{E} \times (-\infty, \infty), \qquad \mathscr{V}^{(4)} = \mathscr{V} \times (-\infty, \infty).$$

By a ***smoothly propagating surface*** we mean a smooth three-dimensional manifold $\mathscr{W}$ in $\mathscr{E}^{(4)}$ with the following property: given any $(\boldsymbol{x}, t) \in \mathscr{W}$, there exists a normal $(\boldsymbol{k}, \varkappa) \in \mathscr{V}^{(4)}$ to $\mathscr{W}$ at $(\boldsymbol{x}, t)$ with $\boldsymbol{k} \neq \boldsymbol{0}$, $\varkappa \neq 0$. This assumption implies the existence of a four-vector field

$$\mathbf{m} = (\boldsymbol{m}, -c)$$

on $\mathscr{W}$ that is normal to $\mathscr{W}$ and has

$$|\boldsymbol{m}| = 1, \qquad c > 0.$$

We call $\boldsymbol{m}$ the ***direction of propagation*** and c the ***speed of propagation.***

For the remainder of this section let $\mathscr{W}$ be a smoothly propagating surface in $B \times (0, t_0)$.

Since $\mathscr{W}$ is a smooth manifold, and since c is strictly positive, $\mathscr{W}$ has the following property: given any point $(\boldsymbol{x}_0, t_0) \in \mathscr{W}$ there exists a neighborhood Σ of $\boldsymbol{x}_0$, a neighborhood T of t_0, and a smooth scalar field ψ on Σ such that

$$t = \psi(\boldsymbol{x}) \quad \text{for all} \quad (\boldsymbol{x}, t) \in N \cap \mathscr{W}, \qquad N = \Sigma \times T.$$

---

[1] FEDOROV [1965, 5], pp. 92–94.

[2] See p. 73.

In terms of ψ:

$$\boldsymbol{m}(\boldsymbol{x}, t) = \frac{\nabla\psi(\boldsymbol{x})}{|\nabla\psi(\boldsymbol{x})|}, \qquad c(\boldsymbol{x}, t) = \frac{1}{|\nabla\psi(\boldsymbol{x})|}$$

for all $(\boldsymbol{x}, t) \in N \cap \mathscr{W}$.

For t fixed,

$$\mathscr{S}_t = \{\boldsymbol{x} \in B : (\boldsymbol{x}, t) \in \mathscr{W}\}$$

is the smooth surface in B occupied by $\mathscr{W}$ at time t;[1] it is not difficult to show that the vector $\boldsymbol{m}(\boldsymbol{x}, t)$ is normal to $\mathscr{S}_t$ at $\boldsymbol{x}$.

Let $\Gamma = \{\mathbf{y}(t) : t \in T\}$ be a smooth curve on $\mathscr{W}$ of the form $\mathbf{y}(t) = (\boldsymbol{y}(t), t)$. We say that Γ is a ***ray*** if $\boldsymbol{y}(\cdot)$ is a solution of the differential equation[2]

$$\dot{\boldsymbol{y}}(t) = c(\boldsymbol{y}(t), t)\, \boldsymbol{m}(\boldsymbol{y}(t), t).$$

Thus $\boldsymbol{y}(\cdot)$ is the path traversed by a particle on $\mathscr{W}$ whose trajectory at any time t is perpendicular to $\mathscr{S}_t$ and whose speed is c. Assume, without loss in generality, that the ray Γ passes through the neighborhood N in which the manifold $\mathscr{W}$ is described by ψ. Let σ denote arc length along Γ:

$$\sigma = \int_0^t |\dot{\boldsymbol{y}}(\tau)|\, d\tau.$$

Since $|\dot{\boldsymbol{y}}|$ never vanishes, σ is an invertible function of t; we write $\bar{t}$ for the inverse function, i.e.

$$t = \bar{t}(\sigma),$$

and we define

$$\begin{aligned}
\bar{\boldsymbol{y}}(\sigma) &= \boldsymbol{y}(\bar{t}(\sigma)),\\
\bar{c}(\sigma) &= c(\bar{\boldsymbol{y}}(\sigma), \bar{t}(\sigma)),\\
\nabla\bar{c}(\sigma) &= \nabla_{\boldsymbol{x}} c(\boldsymbol{x}, \psi(\boldsymbol{x}))\Big|_{\boldsymbol{x}=\bar{\boldsymbol{y}}(\sigma)} = \nabla\left(\frac{1}{|\nabla\psi(\boldsymbol{x})|}\right)\Big|_{\boldsymbol{x}=\bar{\boldsymbol{y}}(\sigma)},\\
\overline{\boldsymbol{m}}(\sigma) &= \overline{\boldsymbol{m}}(\bar{\boldsymbol{y}}(\sigma), \bar{t}(\sigma)).
\end{aligned}$$

Since

$$\frac{d\bar{\boldsymbol{y}}}{d\sigma} = \overline{\boldsymbol{m}}(\sigma) = \bar{c}(\sigma)\, \nabla\psi(\bar{\boldsymbol{y}}(\sigma)), \qquad \left(\frac{1}{\bar{c}(\sigma)}\right)^2 = |\nabla\psi(\bar{\boldsymbol{y}}(\sigma))|^2,$$

it follows that

$$\frac{d}{d\sigma}\left(\frac{1}{\bar{c}}\frac{d\bar{\boldsymbol{y}}}{d\sigma}\right) = [\nabla\nabla\psi(\bar{\boldsymbol{y}}(\sigma))]\frac{d\bar{\boldsymbol{y}}}{d\sigma} = \bar{c}(\sigma)\,[\nabla\nabla\psi(\bar{\boldsymbol{y}}(\sigma))]\,\nabla\psi(\bar{\boldsymbol{y}}(\sigma)) = \frac{\bar{c}}{2}\nabla\left(\frac{1}{\bar{c}^2}\right).$$

Thus we are led to the following *differential equation for the ray*:

$$\frac{1}{\bar{c}}\frac{d^2\bar{\boldsymbol{y}}}{d\sigma^2} - \frac{1}{\bar{c}^2}\left(\frac{d\bar{c}}{d\sigma}\right)\left(\frac{d\bar{\boldsymbol{y}}}{d\sigma}\right) = -\frac{1}{\bar{c}^2}\nabla\bar{c}.$$

It follows from this equation that *if the velocity c is constant on $\mathscr{W}$, then every ray is a straight line.*

Let f be a function on $B \times (0, t_0)$. We call $\mathscr{W}$ a ***singular surface of order zero*** with respect to f if f is continuous on $B \times (0, t_0) - \mathscr{W}$ and f suffers a jump discontinuity across $\mathscr{W}$. The ***jump*** in f across $\mathscr{W}$ is then the function $[f]$ on $\mathscr{W}$

---

[1] In the literature it is customary to define a smoothly propagating surface to be a (smooth) one-parameter family $\mathscr{S}_t$, $0 < t < t_0$, of smooth surfaces $\mathscr{S}_t$ in B. In this instance

$$\mathscr{W} = \{(\boldsymbol{x}, t) : \boldsymbol{x} \in \mathscr{S}_t,\ 0 < t < t_0\}.$$

[2] Since $\dot{\mathbf{y}} \cdot \mathbf{m} = (c\boldsymbol{m}, 1) \cdot (\boldsymbol{m}, -c) = 0$, this definition is consistent in the sense that the tangent to Γ is also tangent to $\mathscr{W}$.

defined by[1]

$$[f]\,(\boldsymbol{x},t)=\lim_{h\to 0+}\{f(\boldsymbol{x},t+h)-f(\boldsymbol{x},t-h)\}.$$

Thus the jump in f is the difference between the values of f just behind and immediately in front of the wave. Since f is continuous on each side of $\mathscr{W}$, we can evaluate $[f]$ by taking limits along the normal $\mathbf{m}$:

$$[f]\,(\boldsymbol{\xi})=\lim_{h\to 0+}\{f(\boldsymbol{\xi}-h\mathbf{m})-f(\boldsymbol{\xi}+h\mathbf{m})\}$$

for $\boldsymbol{\xi}=(\boldsymbol{x},t)\in\mathscr{W}$.

Let $n\geqq 1$ be a fixed integer. We call $\mathscr{W}$ a ***singular surface of order*** n with respect to f if:[2]

(i) f is class C^{n-1} on $B\times(0,t_0)$ and class C^n on $B\times(0,t_0)-\mathscr{W}$;

(ii) the derivatives of f of order n suffer jump discontinuities across $\mathscr{W}$.

The next theorem shows that for a singular surface of order 1, the jumps in the space and time derivatives of the associated function are related.

(1) Maxwell's compatibility theorem.[3] *Let φ be a scalar field, $\boldsymbol{v}$ a vector field, and $\boldsymbol{T}$ a tensor field on $B\times(0,t_0)$, and suppose that $\mathscr{W}$ is a singular surface of order 1 with respect to φ, $\boldsymbol{v}$, and $\boldsymbol{T}$. Then*

$$c\,[\nabla\varphi]=-[\dot{\varphi}]\,\boldsymbol{m},\tag{1}$$

$$c\,[\nabla\boldsymbol{v}]=-[\dot{\boldsymbol{v}}]\otimes\boldsymbol{m},\tag{2}$$

$$c\,[\operatorname{div}\boldsymbol{v}]=-[\dot{\boldsymbol{v}}]\cdot\boldsymbol{m},\tag{3}$$

$$c\,[\operatorname{curl}\boldsymbol{v}]=[\dot{\boldsymbol{v}}]\times\boldsymbol{m},\tag{4}$$

$$c\,[\operatorname{div}\boldsymbol{T}]=-[\dot{\boldsymbol{T}}]\,\boldsymbol{m}.\tag{5}$$

Proof. Choose a point $\boldsymbol{\xi}\in\mathscr{W}$ and a (non-zero) four-vector $\mathbf{a}$ such that

$$\mathbf{a}\cdot\mathbf{m}(\boldsymbol{\xi})=0.$$

Since $\mathscr{W}$ has dimension 3, $\mathbf{a}$ is tangent to $\mathscr{W}$ at $\boldsymbol{\xi}$. Further, since $\mathscr{W}$ is smooth, there exists a smooth curve $\Omega=\{\mathbf{z}(\alpha):\alpha_1\leqq\alpha\leqq\alpha_2\}$ on $\mathscr{W}$ through $\boldsymbol{\xi}$ whose tangent at $\boldsymbol{\xi}$ is $\mathbf{a}$; i.e. for some $\alpha_0\in(\alpha_1,\alpha_2)$

$$\mathbf{z}(\alpha_0)=\boldsymbol{\xi},\qquad \mathbf{z}'(\alpha_0)=\mathbf{a},\tag{a}$$

where

$$\mathbf{z}'=\frac{d\mathbf{z}}{d\alpha}.$$

For $\boldsymbol{\zeta}\in\mathscr{W}$ and any function f on $B\times(0,t_0)$, let

$$f_1(\boldsymbol{\zeta})=\lim_{h\to 0+}f(\boldsymbol{\zeta}-h\mathbf{m}),$$

$$f_2(\boldsymbol{\zeta})=\lim_{h\to 0+}f(\boldsymbol{\zeta}+h\mathbf{m}),$$

so that

$$[f]\,(\boldsymbol{\zeta})=f_1(\boldsymbol{\zeta})-f_2(\boldsymbol{\zeta}).\tag{b}$$

[1] This notation was introduced by CHRISTOFFEL [1877, *1*], § 6.

[2] The notion of a singular surface of order $n\geqq 0$ is due to DUHEM [1900, *1*] and HADAMARD [1901, *5*], § 1, [1903, *3*], § 75. See also TRUESDELL and TOUPIN [1960, *17*], § 187.

[3] MAXWELL [1873, *1*], § 78, [1881, *1*], § 78a. Cf. also WEINGARTEN [1901, *7*], HADAMARD [1903, *3*], § 73, TRUESDELL and TOUPIN [1960, *17*], § 175, HILL [1961, *14*].

Then, by the smoothness properties of φ,

$$\frac{d}{d\alpha}\varphi_1(\mathbf{z}(\alpha)) = \frac{d}{d\alpha}\lim_{h\to 0^+}\varphi(\mathbf{z}(\alpha) - h\boldsymbol{m})$$
$$= \lim_{h\to 0^+}\frac{d}{d\alpha}\varphi(\mathbf{z}(\alpha) - h\boldsymbol{m})$$
$$= \lim_{h\to 0^+}\nabla_{(4)}\varphi(\mathbf{z}(\alpha) - h\boldsymbol{m})\cdot\mathbf{z}'(\alpha).$$

Thus (a) implies[1]

$$\frac{d}{d\alpha}\varphi_1(\mathbf{z}(\alpha))\big|_{\alpha=\alpha_0} = \boldsymbol{a}\cdot(\nabla_{(4)}\varphi)_1(\boldsymbol{\xi}). \tag{c}$$

A similar result holds also for φ_2; if we subtract this result from (c) and use (b), we arrive at the relation

$$\frac{d}{d\alpha}[\varphi](\mathbf{z}(\alpha))\big|_{\alpha=\alpha_0} = \boldsymbol{a}\cdot[\nabla_{(4)}\varphi](\boldsymbol{\xi}). \tag{d}$$

On the other hand, since φ is continuous across $\mathscr{W}$,

$$[\varphi] \equiv 0;$$

hence (d) implies

$$\boldsymbol{a}\cdot[\nabla_{(4)}\varphi](\boldsymbol{\xi}) = 0$$

for every $\boldsymbol{\xi}\in\mathscr{W}$ and every $\boldsymbol{a}$ tangent to $\mathscr{W}$ at $\boldsymbol{\xi}$. Thus $[\nabla_{(4)}\varphi](\boldsymbol{\xi})$ must be parallel to $\boldsymbol{m}(\boldsymbol{\xi})$:

$$[\nabla_{(4)}\varphi](\boldsymbol{\xi}) = \lambda(\boldsymbol{\xi})\,\boldsymbol{m}(\boldsymbol{\xi}); \tag{e}$$

equivalently,

$$[\nabla\varphi] = \lambda\boldsymbol{m}, \qquad [\dot{\varphi}] = -\lambda c,$$

which implies (1).

If we apply (1) to the scalar field $\boldsymbol{v}\cdot\boldsymbol{a}$, where $\boldsymbol{a}$ is an arbitrarily chosen vector, we find that

$$c[\nabla(\boldsymbol{v}\cdot\boldsymbol{a})] = -[\dot{\boldsymbol{v}}\cdot\boldsymbol{a}]\,\boldsymbol{m}.$$

Thus

$$c[\nabla\boldsymbol{v}^T]\,\boldsymbol{a} = -(\boldsymbol{m}\otimes[\dot{\boldsymbol{v}}])\,\boldsymbol{a},$$

which yields (2).

If we take the trace of (2), we arrive at (3); if we take the skew part, we arrive at (4).

Finally, (5) follows from (3) with $\boldsymbol{v} = \boldsymbol{T}^T\boldsymbol{a}$, where $\boldsymbol{a}$ is an arbitrary vector. □

We now sketch an alternative proof of the relation (1) in ***(1)***. By a four-dimensional counterpart of Stokes' theorem,

$$\int_{\mathscr{S}} \operatorname{skw}\{(\nabla_{(4)}\varphi)\otimes\boldsymbol{n}\}\,da = \mathbf{0}$$

for every closed $\mathscr{W}$-regular[2] hypersurface $\mathscr{S}$ in $B\times(0,t_0)$. The same argument used to derive (c) in the proof of ***(73.1)*** here yields

$$\operatorname{skw}\{[\nabla_{(4)}\varphi]\otimes\boldsymbol{m}\} = \mathbf{0}.$$

Operating with this equation on $\boldsymbol{m}$, we are led to the relation (e) in the proof of ***(1)***.

The next theorem shows that for a vector field $\boldsymbol{v}$ with a discontinuity of order 1, the jumps in $\nabla\boldsymbol{v}$ and $\dot{\boldsymbol{v}}$ are completely determined by the jumps in the divergence and curl of $\boldsymbol{v}$.

---

[1] Hadamard [1903, *3*], § 72. Cf. Lichtenstein [1929, *2*], Chap. 1, § 9, and Truesdell and Toupin [1960, *17*], § 174.

[2] This term is defined on p. 252.

(2) Weingarten's theorem.[1] *Let $\mathscr{W}$ be a singular surface of order 1 with respect to a vector field $\boldsymbol{v}$ on $B\times(0, t_0)$, and let*

$$\boldsymbol{a}=[\dot{\boldsymbol{v}}].$$

Then

$$\boldsymbol{a}\cdot\boldsymbol{m}=0 \Leftrightarrow [\operatorname{div}\boldsymbol{v}]=0, \tag{1}$$

$$\boldsymbol{a}\times\boldsymbol{m}=\boldsymbol{0} \Leftrightarrow [\operatorname{curl}\boldsymbol{v}]=\boldsymbol{0}, \tag{2}$$

$$\boldsymbol{a}=-c\boldsymbol{m}[\operatorname{div}\boldsymbol{v}]+c\boldsymbol{m}\times[\operatorname{curl}\boldsymbol{v}]. \tag{3}$$

Proof. The results (1) and (2) follow from (3) and (4) of ***(1)***. To establish (3) we note that the identity

$$\boldsymbol{a}=(\boldsymbol{a}\cdot\boldsymbol{m})\,\boldsymbol{m}+(\boldsymbol{a}\otimes\boldsymbol{m})\,\boldsymbol{m}-(\boldsymbol{m}\otimes\boldsymbol{a})\,\boldsymbol{m}$$

in conjunction with (2) and (3) of ***(1)*** imply

$$\boldsymbol{a}=-c[\operatorname{div}\boldsymbol{v}]\,\boldsymbol{m}-c[\nabla\boldsymbol{v}]\,\boldsymbol{m}+c[\nabla\boldsymbol{v}^T]\,\boldsymbol{m}.$$

This relation yields (3), since

$$[\nabla\boldsymbol{v}-\nabla\boldsymbol{v}^T]\,\boldsymbol{m}=[\operatorname{curl}\boldsymbol{v}]\times\boldsymbol{m}. \quad \square$$

We will occasionally deal with integrals of the form

$$\int_{\mathscr{S}}\mathbf{M}\mathbf{n}\,da,$$

where $\mathscr{S}$ is a closed hypersurface with outward unit normal field $\mathbf{n}$. If $\mathscr{W}$ is a singular surface of order zero with respect to $\mathbf{M}$, then $\mathbf{M}$ will suffer a jump discontinuity across $\mathscr{S}\cap\mathscr{W}$. To insure that the above integral exists and that the value of $\mathbf{M}$ in the integrand at a point of discontinuity equals its limit from the region interior to $\mathscr{S}$, we give the following definitions.

Let $\mathscr{S}$ be a closed regular hypersurface in $B\times(0, t_0)$, i.e. $\mathscr{S}$ is the boundary of a regular region $D_{\mathscr{S}}$ in $B\times(0, t_0)$. A point $\boldsymbol{\xi}\in\mathscr{S}$ is $\mathscr{W}$-regular if either $\boldsymbol{\xi}\notin\mathscr{W}$ or $\boldsymbol{\xi}\in\mathscr{W}$ and there exists an open ball Σ in $\mathscr{E}^{(4)}$ with center at $\boldsymbol{\xi}$ such that $(\mathscr{W}\cap\Sigma)\subset\mathscr{S}$ (see Fig. 16). We say that $\mathscr{S}$ is ***$\mathscr{W}$-regular*** if the set of points of $\mathscr{S}$ that are *not* $\mathscr{W}$-regular has zero area measure.

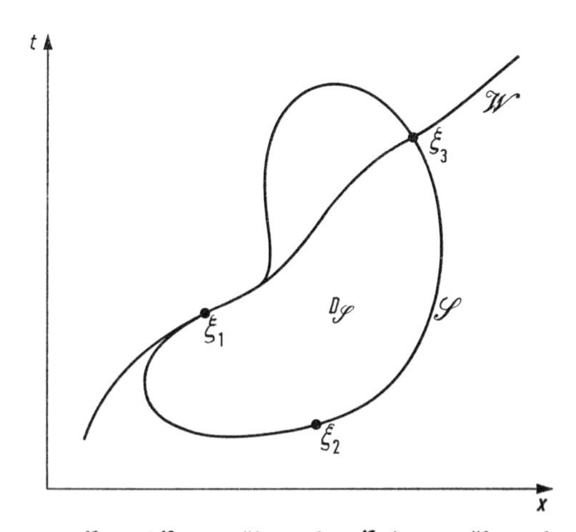

Fig. 16. $\boldsymbol{\xi}_1$ and $\boldsymbol{\xi}_2$ are $\mathscr{W}$-regular; $\boldsymbol{\xi}_3$ is not $\mathscr{W}$-regular.

[1] [1901, *7*]. Cf. also TRUESDELL and TOUPIN [1960, *17*], § 175.

Let $\mathscr{W}$ be a singular surface of order zero with respect to a four-tensor field $\mathbf{M}$ on $B\times(0, t_0)$. Then the limit

$$\hat{\mathbf{M}}(\boldsymbol{\xi}) = \lim_{\substack{\boldsymbol{\zeta}\to\boldsymbol{\xi}\\ \boldsymbol{\zeta}\in D_{\mathscr{S}}}} \mathbf{M}(\boldsymbol{\zeta})$$

exists for every regular $\boldsymbol{\xi}\in\mathscr{S}$. Further, the integral

$$\int_{\mathscr{S}} \hat{\mathbf{M}}\mathbf{n}\, da$$

exists whenever $\mathscr{S}$ is a closed $\mathscr{W}$-regular hypersurface in $B\times(0, t_0)$. When such is the case we define

$$\int_{\mathscr{S}} \mathbf{M}\mathbf{n}\, da = \int_{\mathscr{S}} \hat{\mathbf{M}}\mathbf{n}\, da.$$

73. Shock waves. Acceleration waves. Mild discontinuities. Throughout this section $\mathscr{W}$ is a smoothly propagating surface in $B\times(0, t_0)$ with direction of propagation $\boldsymbol{m}$ and speed of propagation c. Further, we assume that the elasticity field $\mathbf{C}$ and the density field ϱ are *continuous* on B, that ϱ is *strictly positive*, and that the *body force field vanishes*.

Let $n\geqq 1$ be a fixed integer. By a ***wave of order*** n we mean an ordered array $[\boldsymbol{u}, \boldsymbol{S}]$ with properties (i)–(iii) below.

(i) $\boldsymbol{u}$ is a vector field and $\boldsymbol{S}$ a *symmetric* tensor field on $B\times(0, t_0)$.

(ii) $\mathscr{W}$ is a singular surface of order n with respect to $\boldsymbol{u}$ and order $n-1$ with respect to $\boldsymbol{S}$.

(iii) $\boldsymbol{u}$ and $\boldsymbol{S}$ satisfy balance of linear momentum in the following form: given any closed $\mathscr{W}$-regular hypersurface $\mathscr{S}$ in $B\times(0, t_0)$,

$$\int_{\mathscr{S}} \mathbf{M}\mathbf{n}\, da = \mathbf{0},$$

where $\mathbf{M}$ is the stress momentum field[1] corresponding to $\boldsymbol{S}$ and $\boldsymbol{u}$.

Requirement (iii) is motivated by ***(19.6)***. For $n\geqq 2$, (iii) may be replaced by the assumption that

$$\operatorname{div}\boldsymbol{S} = \varrho\ddot{\boldsymbol{u}} \quad \text{on} \quad B\times(0, t_0) - \mathscr{W}. \tag{a}$$

For $n=1$ the discussion given at the end of the last section insures the existence of the integral in (iii); in addition, the value of $\mathbf{M}$ in the integrand is equal to its limit from the region interior to $\mathscr{S}$. When $n=1$, (iii) implies (a), but the converse is not true. However, in this instance (iii) is equivalent to the following requirement:

(iii′) for every one-parameter family P_t $(t_1<t<t_2)$ of parts of B,

$$\int_{\partial P_t} (\boldsymbol{S}\boldsymbol{n})(\boldsymbol{x}, t)\, da_{\boldsymbol{x}} = \frac{d}{dt}\int_{P_t} (\varrho\dot{\boldsymbol{u}})(\boldsymbol{x}, t)\, dv_{\boldsymbol{x}}$$

whenever the integral on the left and the derivative on the right exist. Here the stress $\boldsymbol{S}$ on ∂P_t is to be interpreted as its limit from the interior of P_t.

Given a wave of order n, we call

$$\boldsymbol{a} = [\overset{(n)}{\boldsymbol{u}}]$$

[1] See p. 67.

the ***amplitude***, and we assume, for convenience, that $\boldsymbol{a}$ never vanishes. A wave of order 1 is sometimes referred to as a ***stress wave*** or ***shock wave***; a wave of order 2 is often called an ***acceleration wave***. For a shock wave the amplitude equals the jump in the velocity, while for an acceleration wave $\boldsymbol{a}$ is equal to the jump in the acceleration. We call a wave ***longitudinal*** or ***transverse*** according as $\boldsymbol{a}$ and $\boldsymbol{m}$ are parallel or perpendicular. For a wave of order n, $\mathscr{W}$ is a singular surface of order 1 with respect to $\overset{(n-1)}{\boldsymbol{u}}$; thus Weingarten's theorem ***(72.2)*** implies that $\boldsymbol{a}$ is completely determined by the jumps in the divergence and curl of $\overset{(n-1)}{\boldsymbol{u}}$'s:

$$\boldsymbol{a} = -c\boldsymbol{m}[\operatorname{div}\overset{(n-1)}{\boldsymbol{u}}] + c\boldsymbol{m}\times[\operatorname{curl}\overset{(n-1)}{\boldsymbol{u}}].$$

In addition, we conclude from ***(72.2)*** that the wave is:

$$\textit{longitudinal} \Leftrightarrow [\operatorname{curl}\overset{(n-1)}{\boldsymbol{u}}] = \boldsymbol{0},$$

$$\textit{transverse} \Leftrightarrow [\operatorname{div}\overset{(n-1)}{\boldsymbol{u}}] = 0.$$

For a wave of order n the time-derivative of $\boldsymbol{S}$ of order $n-1$ is allowed to jump across the wave. The next result gives a relation between this jump and the amplitude $\boldsymbol{a}$.

(1) Balance of momentum at the wave.[1] *For a wave of order* $n \geqq 1$,

$$[\overset{(n-1)}{\boldsymbol{S}}]\boldsymbol{m} = -\varrho c\boldsymbol{a}.$$

Proof. Let $n=1$. Let Σ be an open ball in $B\times(0, t_0)$ with center on $\mathscr{W}$. For Σ sufficiently small, $\mathscr{W}$ divides Σ into two regular regions Σ^+ and Σ^- with $\mathscr{W}$-regular boundaries. Let Σ^- be that portion of Σ into which $\boldsymbol{m}$ points (see Fig. 17).

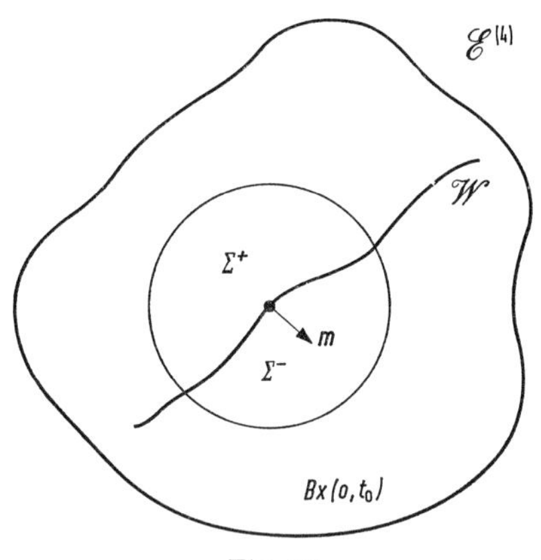

Fig. 17.

[1] Although we have assumed that the body force field $\boldsymbol{b}=\boldsymbol{0}$, the above result remains valid if we assume instead that $\boldsymbol{b}$ is of class C^{n-1} on $B\times(0, t_0)$. For a shock wave ($n=1$) the relation in ***(1)*** has a long history. Hydrodynamical special cases for situations in which $\boldsymbol{S}=-p\boldsymbol{1}$ were obtained by STOKES [1848, *2*], Eq. (3), RIEMANN [1860, *1*], § 5, RANKINE [1870, *2*], §§ 3–5, CHRISTOFFEL [1877, *1*], § 1, HUGONIOT [1887, *1*], § 140, and JOUGUET [1901, *6*]; for general $\boldsymbol{S}$ but linearized acceleration, by CHRISTOFFEL [1877, *2*], § 4; for finite elastic deformation by ZEMPLÉN [1905, *3*], p. 448, [1905, *4*], § 3, whose derivation is based on Hamilton's principle. The general result (for $n=1$) is due to KOTCHINE [1926, *2*], Eq. (18). Cf. also TRUESDELL and TOUPIN [1960, *17*], § 205, JEFFREY [1964, *11*].

By (iii)

$$\int_{\partial\Sigma} \mathbf{M}\mathbf{n}\, da = \int_{\partial\Sigma^+} \mathbf{M}\mathbf{n}\, da = \int_{\partial\Sigma^-} \mathbf{M}\mathbf{n}\, da = \mathbf{0},$$

where $\mathbf{M}$ is the stress momentum field; thus

$$\int_{\partial\Sigma^+} \mathbf{M}\mathbf{n}\, da + \int_{\partial\Sigma^-} \mathbf{M}\mathbf{n}\, da - \int_{\partial\Sigma} \mathbf{M}\mathbf{n}\, da = \mathbf{0}. \tag{a}$$

The boundary of Σ can be decomposed into a portion contained in $\partial\Sigma^+$ and a portion contained in $\partial\Sigma^-$; the remainder of $\partial\Sigma^+$ and the remainder of $\partial\Sigma^-$ are equal, as sets, to $\Sigma \cap \mathscr{W}$; as surfaces, however, they have opposite orientation. Thus (a) reduces to

$$\int_{\Sigma\cap\mathscr{W}} \{\lim_{\boldsymbol{\xi}^+\to\boldsymbol{\xi}} \mathbf{M}(\boldsymbol{\xi}^+)\, \mathbf{m}(\boldsymbol{\xi}) - \lim_{\boldsymbol{\xi}^-\to\boldsymbol{\xi}} \mathbf{M}(\boldsymbol{\xi}^-)\, \mathbf{m}(\boldsymbol{\xi})\}\, da_{\boldsymbol{\xi}} = \mathbf{0}, \tag{b}$$

where $\boldsymbol{\xi}^\pm$ is a point of $\Sigma^\pm$. In deriving (b) we have used the fact that the value of $\mathbf{M}$ in the integral over $\partial\Sigma^\pm$ is equal to its limit from the interior of $\Sigma^\pm$. Since (b) must hold for every sufficiently small open ball Σ with center on $\mathscr{W}$, it follows that

$$[\mathbf{M}]\, \mathbf{m} = \mathbf{0}, \tag{c}$$

or equivalently, using the definitions of $\mathbf{M}$ and $\mathbf{m}$,[1]

$$[\boldsymbol{S}]\, \boldsymbol{m} = -\varrho c\, [\dot{\boldsymbol{u}}] = -\varrho c \boldsymbol{a},$$

which is the desired relation for $n = 1$.[2]

Consider now a wave of order $n \geqq 2$. Choose $(\boldsymbol{x}, t)$ in $B \times (0, t_0) - \mathscr{W}$. Then there exists an open ball Σ in B centered at $\boldsymbol{x}$ and an open time interval $T \subset (t_0, t_1)$ containing t such that $\mathscr{W} \cap (\Sigma \times T) = \emptyset$. Thus $\boldsymbol{S}$ is a time-dependent admissible stress field and $\boldsymbol{u}$ is an admissible motion on $\Sigma \times T$; (iii) and ***(19.6)*** therefore imply

$$\operatorname{div} \boldsymbol{S} = \varrho \ddot{\boldsymbol{u}} \quad \text{on} \quad \Sigma \times T.$$

Since $(\boldsymbol{x}, t)$ in $B \times (0, t_0) - \mathscr{W}$ was arbitrarily chosen, the above relation holds on $B \times (0, t_0) - \mathscr{W}$. If we differentiate $n-2$ times with respect to t and take the jump in the resulting equation, we arrive at

$$[\operatorname{div} \overset{(n-2)}{\boldsymbol{S}}] = \varrho [\overset{(n)}{\boldsymbol{u}}] = \varrho \boldsymbol{a}. \tag{d}$$

Clearly, $\mathscr{W}$ is a singular surface of order 1 with respect to $\overset{(n-2)}{\boldsymbol{S}}$; thus (5) of ***(72.1)*** yields

$$c [\operatorname{div} \overset{(n-2)}{\boldsymbol{S}}] = -[\overset{(n-1)}{\boldsymbol{S}}]\, \boldsymbol{m}. \tag{e}$$

Eqs. (d) and (e) imply the desired result for $n \geqq 2$. □

We say that a wave $[\boldsymbol{u}, \boldsymbol{S}]$ of order $n \geqq 1$ is ***elastic*** if

$$\boldsymbol{S} = \mathbf{C}[\nabla \boldsymbol{u}].$$

We now show that for an elastic wave of any order traveling in the direction $\boldsymbol{m}$, the square of the speed must be a characteristic value of the acoustic tensor $\boldsymbol{A}(\boldsymbol{m})$, and the amplitude must lie in the corresponding characteristic space.

---

[1] See pp. 67, 248.

[2] The proof up to this point is due to Keller [1964, *12*].

(2) Propagation theorem for waves of order n.[1] *Elastic waves of all orders $n \geqq 1$ obey the Fresnel-Hadamard propagation condition*

$$\boldsymbol{A}(\boldsymbol{m})\,\boldsymbol{a} = c^2\,\boldsymbol{a}.$$

Proof. Let $[\boldsymbol{u}, \boldsymbol{S}]$ be an elastic wave of order $n \geqq 1$. Since $\mathscr{W}$ is a singular surface of order 1 with respect to $\boldsymbol{v} = \overset{(n-1)}{\boldsymbol{u}}$, we can apply (2) of ***(72.1)*** with the result:

$$c\,[\nabla \overset{(n-1)}{\boldsymbol{u}}] = -[\overset{(n)}{\boldsymbol{u}}] \otimes \boldsymbol{m} = -\boldsymbol{a} \otimes \boldsymbol{m}. \tag{a}$$

Next, (a) and the stress-strain relation imply

$$c\,[\overset{(n-1)}{\boldsymbol{S}}] = \mathbf{C}[c\,[\nabla \overset{(n-1)}{\boldsymbol{u}}]] = -\mathbf{C}[\boldsymbol{a} \otimes \boldsymbol{m}], \tag{b}$$

and since

$$\boldsymbol{A}(\boldsymbol{m})\,\boldsymbol{a} = \varrho^{-1}\mathbf{C}[\boldsymbol{a} \otimes \boldsymbol{m}]\,\boldsymbol{m},$$

(b) and ***(1)*** imply the desired result. □

More pedantically, the Fresnel-Hadamard condition reads

$$\boldsymbol{A}(\boldsymbol{x}, \boldsymbol{m})\,\boldsymbol{a}(\boldsymbol{x}, t) = c^2(\boldsymbol{x}, t)\,\boldsymbol{a}(\boldsymbol{x}, t)$$

for every $(\boldsymbol{x}, t) \in \mathscr{W}$.

Combining ***(2)*** and ***(70.2)***, we arrive at the

(3) Propagation theorem for isotropic bodies.[2] *If the body is isotropic then a wave of order $n \geqq 1$ is either longitudinal, in which case*

$$c^2 = c_1^2 = \frac{2\mu + \lambda}{\varrho},$$

or transverse, in which case

$$c^2 = c_2^2 = \frac{\mu}{\varrho}.$$

Comparing ***(2)*** and ***(3)*** to ***(71.1)*** and ***(71.2)***, we see that the laws of propagation of waves of discontinuity are the same as those of progressive waves; thus the discussion following ***(71.1)*** and ***(71.2)*** also holds in the present circumstances. In particular, we note that if $\mathbf{C}$ is symmetric and strongly elliptic, there exist, for every direction $\boldsymbol{m}$, three associated velocities of propagation for singular surfaces.

Note that the maximum speed of propagation c_M defined in ***(70.3)*** is the supremum over all possible velocities of propagation of elastic waves of any order.

For a homogeneous and isotropic body the velocity of propagation c is identically constant. Therefore ***(3)*** and the remark made on p. 249 yield the following result.

---

[1] This theorem is essentially due to CHRISTOFFEL [1877, *2*], who gave a proof for $n = 1$. HADAMARD [1903, *3*], §§ 260, 267–268, extended CHRISTOFFEL's result to acceleration waves ($n = 2$). That the Fresnel-Hadamard condition should hold for all n is clear from the results of ERICKSEN [1953, *10*], § 6, and TRUESDELL [1961, *22*], § 3, in finite elasticity theory (cf. TRUESDELL and NOLL [1965, *22*], § 72). See also HUGONIOT [1886, *2*]; DUHEM [1904, *2*], Part 4, Chap. 1, § 5; LOVE [1927, *3*], § 206–209; FINZI [1942, *1*]; PASTORI [1949, *7*]; PETRASHEN [1958, *12*]; BUCHWALD [1959, *1*]; TRUESDELL and TOUPIN [1960, *17*], § 301; NARIBOLI [1966, *18*].

[2] HUGONIOT [1886, *2*]. Elastic waves of order zero (for which $\mathscr{W}$ is a singular surface of order zero with respect to the *displacement* field) in an isotropic but inhomogeneous body are discussed by GVOZDEV [1959, *5*] using the theory of weak solutions. GVOZDEV shows that such waves travel with a velocity equal to c_1 or c_2. Other studies concerned with wave propagation in isotropic elastic media are: LEVIN and RYTOV [1956, *7*], THOMAS [1957, *17*], BABICH and ALEKSEEV [1958, *1*], SKURIDIN and GVOZDEV [1958, *15*], KARAL and KELLER [1959, *9*], SKURIDIN [1959, *14*].

(4) *Suppose that the body is homogeneous and isotropic. Then, for a wave of any order* $n \geqq 1$*, the rays are straight lines.*

74. Domain of influence. Uniqueness for infinite regions. In this section we allow B to be a bounded or *unbounded* regular region. We assume that $\mathbf{C}$ is *positive definite* and *symmetric*, that ϱ is *strictly positive*, and that $\mathbf{C}$ and ϱ^{-1} are *continuous* and *bounded* on $\bar{B}$. It then follows that the *maximum speed of propagation* c_M defined in ***(70.3)*** is finite.

Given an elastic process $[\boldsymbol{u}, \boldsymbol{E}, \boldsymbol{S}]$ and a given time $t \in (0, t_0)$, let $\hat{D}_t$ denote the set of all $\boldsymbol{x} \in \bar{B}$ such that:

(i) if $\boldsymbol{x} \in B$, then

$$\boldsymbol{u}(\boldsymbol{x}, 0) \neq \mathbf{0} \quad \text{or} \quad \dot{\boldsymbol{u}}(\boldsymbol{x}, 0) \neq \mathbf{0} \quad \text{or} \quad \boldsymbol{b}(\boldsymbol{x}, \tau) \neq \mathbf{0} \qquad \text{for some } \tau \in [0, t];$$

(ii) if $\boldsymbol{x} \in \partial B$, then

$$\boldsymbol{s}(\boldsymbol{x}, \tau) \cdot \dot{\boldsymbol{u}}(\boldsymbol{x}, \tau) \neq 0 \qquad \text{for some } \tau \in [0, t].$$

Here $[\boldsymbol{b}, \boldsymbol{s}]$ is the external force system associated with the process. Roughly speaking, $\hat{D}_t$ is the support of the initial and boundary data. Consider next the set D_t of all points of $\bar{B}$ that can be reached by signals propagating from $\hat{D}_t$ with speeds equal to or less than the maximum speed of propagation c_M:

$$D_t = \{\boldsymbol{x} \in \bar{B} : \hat{D}_t \cap \overline{\Sigma_{c_M t}(\boldsymbol{x})} \neq \emptyset\}.$$

We call D_t the ***domain of influence***[1] of the data at time t. The next theorem, which is the main result of this section, shows that on $[0, t]$ the data has no effect on points outside of D_t.

(1) Domain of influence theorem.[2] *Let* $[\boldsymbol{u}, \boldsymbol{E}, \boldsymbol{S}]$ *be an elastic process, and let* D_t *be the domain of influence of its data at time* t*. Then*

$$\boldsymbol{u} = \mathbf{0}, \qquad \boldsymbol{E} = \boldsymbol{S} = \mathbf{0} \quad \text{on } (\bar{B} - D_t) \times [0, t].$$

The proof of this theorem is based on the following lemma. In the statement and proof of this lemma $\varepsilon(\boldsymbol{x}, \boldsymbol{G})$ is the strain energy density at $\boldsymbol{x}$ corresponding to the (not necessarily symmetric) tensor $\boldsymbol{G}$:

$$\varepsilon(\boldsymbol{x}, \boldsymbol{G}) = \tfrac{1}{2} \boldsymbol{G} \cdot \mathbf{C}_{\boldsymbol{x}}[\boldsymbol{G}]. \tag{a}$$

(2) Wheeler-Sternberg lemma.[3] *Let* $[\boldsymbol{u}, \boldsymbol{E}, \boldsymbol{S}]$ *be an elastic state corresponding to the external force system* $[\boldsymbol{b}, \boldsymbol{s}]$*. Further, let* $\tau : \bar{B} \to [0, t_0)$ *be a continuous*

---

[1] Since D_t is based on the maximum speed of propagation, it is, actually, an upper bound on the domain of influence. Cf. DUFF [1960, *5*].

[2] This is a minor modification of a theorem due to WHEELER and STERNBERG [1968, *16*], Lemma 2.2 and WHEELER [1970, *3*], Theorem 1. These authors assume vanishing initial data and do not utilize the notion of a domain of influence. The general idea is due to ZAREMBA [1915, *4*], who established an analogous result for the wave equation. ZAREMBA's scheme was rediscovered independently by RUBINOWICZ [1920, *2*] and FRIEDRICHS and LEWY [1928, *1*]. See also COURANT and HILBERT [1962, *5*], pp. 659–661, BERS, JOHN, and SCHECHTER [1964, *3*], DUFF [1960, *5*].

[3] WHEELER and STERNBERG [1968, *16*], Lemma 2.1 for homogeneous and isotropic media; WHEELER [1970, *3*], Lemma 1, for anisotropic but homogeneous media. Both of the above results assume vanishing initial data.

and piecewise smooth field on $\bar{B}$ with $\{\boldsymbol{x}\in\bar{B}: \tau(\boldsymbol{x})>0\}$ finite. Then

$$\int_{\partial B}\int_0^{\tau(\boldsymbol{x})} \boldsymbol{s}(\boldsymbol{x},t)\cdot\dot{\boldsymbol{u}}(\boldsymbol{x},t)\,dt\,da+\int_B\int_0^{\tau(\boldsymbol{x})} \boldsymbol{b}(\boldsymbol{x},t)\cdot\dot{\boldsymbol{u}}(\boldsymbol{x},t)\,dt\,dv$$

$$=\int_B\Big\{\frac{\varrho(\boldsymbol{x})}{2}\,[\dot{\boldsymbol{u}}^2(\boldsymbol{x},\tau(\boldsymbol{x}))-\dot{\boldsymbol{u}}^2(\boldsymbol{x},0)]-\varepsilon(\boldsymbol{x},\boldsymbol{G}(\boldsymbol{x}))+\varepsilon(\boldsymbol{x},\nabla\boldsymbol{u}(\boldsymbol{x},\tau(\boldsymbol{x}))+\boldsymbol{G}(\boldsymbol{x})) \tag{b}$$

$$-\varepsilon(\boldsymbol{x},\nabla\boldsymbol{u}(\boldsymbol{x},0))\Big\}\,dv$$

where

$$\boldsymbol{G}(\boldsymbol{x})=\dot{\boldsymbol{u}}(\boldsymbol{x},\tau(\boldsymbol{x}))\otimes\nabla\tau(\boldsymbol{x}). \tag{c}$$

Proof of (2). For any function f on $\bar{B}\times[0,t_0)$, let f_τ be the function on $\bar{B}$ defined by

$$f_\tau(\boldsymbol{x})=f(\boldsymbol{x},\tau(\boldsymbol{x})). \tag{d}$$

Further, let

$$\boldsymbol{k}=\boldsymbol{S}\,\dot{\boldsymbol{u}},\qquad \boldsymbol{v}(\boldsymbol{x})=\int_0^{\tau(\boldsymbol{x})}\boldsymbol{k}(\boldsymbol{x},t)\,dt. \tag{e}$$

Then $\boldsymbol{v}$ is continuous on $\bar{B}$ and piecewise smooth on B, and we conclude, with the aid of the equation of motion, that

$$\operatorname{div}\boldsymbol{v}(\boldsymbol{x})=\boldsymbol{k}_\tau(\boldsymbol{x})\cdot\nabla\tau(\boldsymbol{x})+\int_0^{\tau(\boldsymbol{x})}\Big[\boldsymbol{S}\cdot\nabla\dot{\boldsymbol{u}}+\frac{\varrho}{2}\,\overline{(\dot{\boldsymbol{u}}\cdot\dot{\boldsymbol{u}})}^{\boldsymbol{\cdot}}-\boldsymbol{b}\cdot\dot{\boldsymbol{u}}\Big](\boldsymbol{x},t)\,dt. \tag{f}$$

Since

$$\dot{\varepsilon}=\boldsymbol{S}\cdot\nabla\dot{\boldsymbol{u}},$$

(f) takes the form

$$\operatorname{div}\boldsymbol{v}(\boldsymbol{x})=\boldsymbol{k}_\tau(\boldsymbol{x})\cdot\nabla\tau(\boldsymbol{x})+\varepsilon(\boldsymbol{x},\nabla\boldsymbol{u}_\tau(\boldsymbol{x}))-\varepsilon(\boldsymbol{x},\nabla\boldsymbol{u}(\boldsymbol{x},0))$$

$$+\frac{\varrho(\boldsymbol{x})}{2}\,[\dot{\boldsymbol{u}}_\tau^2(\boldsymbol{x})-\dot{\boldsymbol{u}}^2(\boldsymbol{x},0)]-\int_0^{\tau(\boldsymbol{x})}\boldsymbol{b}\cdot\dot{\boldsymbol{u}}(\boldsymbol{x},t)\,dt.$$

A simple calculation using the symmetry of C yields

$$\boldsymbol{k}_\tau(\boldsymbol{x})\cdot\nabla\tau(\boldsymbol{x})=\nabla\tau(\boldsymbol{x})\cdot(\boldsymbol{S}_\tau(\boldsymbol{x})\,\dot{\boldsymbol{u}}_\tau(\boldsymbol{x}))=\boldsymbol{S}_\tau(\boldsymbol{x})\cdot\boldsymbol{G}(\boldsymbol{x})=\boldsymbol{G}(\boldsymbol{x})\cdot\mathsf{C}_{\boldsymbol{x}}[\nabla\boldsymbol{u}_\tau(\boldsymbol{x})]$$

$$=\varepsilon(\boldsymbol{x},\nabla\boldsymbol{u}_\tau(\boldsymbol{x})+\boldsymbol{G}(\boldsymbol{x}))-\varepsilon(\boldsymbol{x},\nabla\boldsymbol{u}_\tau(\boldsymbol{x}))-\varepsilon(\boldsymbol{x},\boldsymbol{G}(\boldsymbol{x}));$$

thus

$$\operatorname{div}\boldsymbol{v}(\boldsymbol{x})=\frac{\varrho(\boldsymbol{x})}{2}\Big[\dot{\boldsymbol{u}}_\tau^2(\boldsymbol{x})-\dot{\boldsymbol{u}}^2(\boldsymbol{x},0)\Big]-\varepsilon(\boldsymbol{x},\boldsymbol{G}(\boldsymbol{x}))+\varepsilon(\boldsymbol{x},\nabla\boldsymbol{u}_\tau(\boldsymbol{x})+\boldsymbol{G}(\boldsymbol{x}))$$

$$-\varepsilon(\boldsymbol{x},\nabla\boldsymbol{u}(\boldsymbol{x},0))-\int_0^{\tau(\boldsymbol{x})}\boldsymbol{b}\cdot\dot{\boldsymbol{u}}(\boldsymbol{x},t)\,dt. \tag{g}$$

Each of the terms on the right hand side of (g) is a piecewise continuous function of $\boldsymbol{x}$ on $\bar{B}$. Moreover, it is clear from (e) and the assumed properties of τ that $\boldsymbol{v}$ has bounded support. Thus div $\boldsymbol{v}$ is integrable on $\bar{B}$, and the divergence theorem together with (g) and the second of (e) imply (b). □

Proof of (1). Let $(\boldsymbol{z},\lambda)\in(B-D_t)\times(0,t)$ be fixed, let

$$\Omega=\bar{B}\cap\overline{\Sigma_{\lambda c_M}(\boldsymbol{z})}, \tag{h}$$

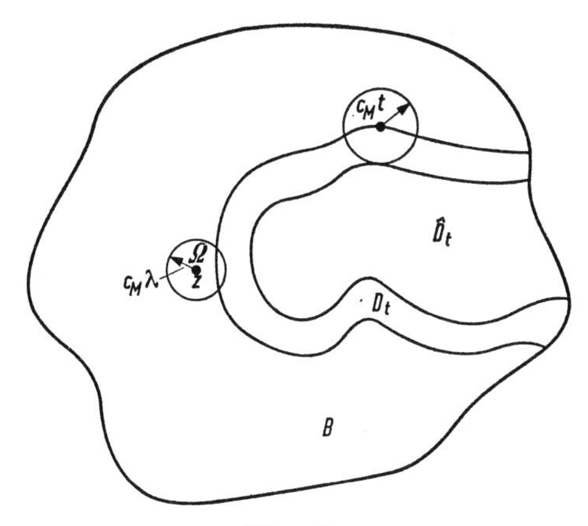

Fig. 18.

and let $\tau: \bar{B} \to [0, t)$ be defined by

$$\tau(\boldsymbol{x}) = \begin{cases} \lambda - \frac{1}{c_M} |\boldsymbol{x} - \boldsymbol{z}|, & \boldsymbol{x} \in \Omega, \\ 0, & \boldsymbol{x} \notin \Omega. \end{cases} \tag{i}$$

Then τ is continuous and piecewise smooth on $\bar{B}$, and $\nabla\tau$ has a weak point discontinuity at $\boldsymbol{x} = \boldsymbol{z}$ and a jump discontinuity across $B \cap \partial\Omega$. Moreover,

$$\begin{aligned} |\nabla\tau| &= \frac{1}{c_M} \quad \text{on } \mathring{\Omega}, \\ \nabla\tau &= \mathbf{0} \quad \text{on } \bar{B} - \Omega. \end{aligned} \tag{j}$$

Since $\lambda < t$, we conclude from (h) and the definition of the domain of influence D_t that

$$\Omega \cap \hat{D}_t = \emptyset;$$

hence

$$\begin{aligned} \boldsymbol{b} &= \mathbf{0} \quad \text{on } \Omega \times [0, t], \\ \boldsymbol{s} \cdot \dot{\boldsymbol{u}} &= 0 \quad \text{on } (\Omega \cap \partial B) \times [0, t], \end{aligned} \tag{k}$$

$$\boldsymbol{u}(\cdot, 0) = \dot{\boldsymbol{u}}(\cdot, 0) = \mathbf{0} \quad \text{on } \Omega. \tag{l}$$

By (i) and (k),

$$\begin{aligned} \int_B \int_0^{\tau(\boldsymbol{x})} \boldsymbol{b}(\boldsymbol{x}, \eta) \cdot \dot{\boldsymbol{u}}(\boldsymbol{x}, \eta)\, d\eta\, dv &= 0, \\ \int_{\partial B} \int_0^{\tau(\boldsymbol{x})} \boldsymbol{s}(\boldsymbol{x}, \eta) \cdot \dot{\boldsymbol{u}}(\boldsymbol{x}, \eta)\, d\eta\, da &= 0. \end{aligned} \tag{m}$$

Next, (i), (j), and (c) imply

$$\dot{\boldsymbol{u}}(\boldsymbol{x}, \tau(\boldsymbol{x})) = \dot{\boldsymbol{u}}(\boldsymbol{x}, 0), \quad \nabla\boldsymbol{u}(\boldsymbol{x}, \tau(\boldsymbol{x})) = \nabla\boldsymbol{u}(\boldsymbol{x}, 0), \quad \boldsymbol{G}(\boldsymbol{x}) = \mathbf{0}, \quad \boldsymbol{x} \notin \Omega. \tag{n}$$

By (a), (n), and (l),

$$\varepsilon(\boldsymbol{x}, \boldsymbol{G}(\boldsymbol{x})) = 0, \quad \boldsymbol{x} \notin \Omega,$$

$$\dot{\boldsymbol{u}}^2(\boldsymbol{x}, \tau(\boldsymbol{x})) - \dot{\boldsymbol{u}}^2(\boldsymbol{x}, 0) = \begin{cases} \dot{\boldsymbol{u}}^2(\boldsymbol{x}, \tau(\boldsymbol{x})), & \boldsymbol{x} \in \Omega, \\ \mathbf{0}, & \boldsymbol{x} \notin \Omega, \end{cases} \tag{o}$$

$$\varepsilon(\boldsymbol{x}, \nabla\boldsymbol{u}(\boldsymbol{x}, \tau(\boldsymbol{x})) + \boldsymbol{G}(\boldsymbol{x})) - \varepsilon(\boldsymbol{x}, \nabla\boldsymbol{u}(\boldsymbol{x}, 0)) = \begin{cases} \varepsilon(\boldsymbol{x}, \nabla\boldsymbol{u}(\boldsymbol{x}, \tau(\boldsymbol{x})) + \boldsymbol{G}(\boldsymbol{x})), & \boldsymbol{x} \in \Omega, \\ 0, & \boldsymbol{x} \notin \Omega. \end{cases}$$

Clearly, τ satisfies the hypotheses of the Wheeler-Sternberg lemma *(2)*. Therefore we may conclude from (b), (m), and (o) that

$$\int_\Omega \left[\frac{\varrho(\boldsymbol{x})}{2}\dot{\boldsymbol{u}}^2(\boldsymbol{x},\tau(\boldsymbol{x})) - \varepsilon(\boldsymbol{x},\boldsymbol{G}(\boldsymbol{x})) + \varepsilon(\boldsymbol{x},\nabla\boldsymbol{u}(\boldsymbol{x},\tau(\boldsymbol{x}))+\boldsymbol{G}(\boldsymbol{x}))\right] dv = 0. \tag{p}$$

Since

$$|\boldsymbol{G}(\boldsymbol{x})|^2 = |\nabla\tau(\boldsymbol{x})|^2\,\dot{\boldsymbol{u}}^2(\boldsymbol{x},\tau(\boldsymbol{x}))$$

for $\boldsymbol{x}\in(B-\{\boldsymbol{z}\})$, it follows from (a), (j) and the inequality in *(70.3)* that

$$\varepsilon(\boldsymbol{x},\boldsymbol{G}(\boldsymbol{x})) \leqq \frac{\varrho(\boldsymbol{x})}{2}\dot{\boldsymbol{u}}^2(\boldsymbol{x},\tau(\boldsymbol{x}));$$

thus

$$\frac{\varrho(\boldsymbol{x})}{2}\dot{\boldsymbol{u}}^2(\boldsymbol{x},\tau(\boldsymbol{x})) - \varepsilon(\boldsymbol{x},\boldsymbol{G}(\boldsymbol{x})) \geqq 0 \tag{q}$$

for $\boldsymbol{x}\in(B-\{\boldsymbol{z}\})$. Since $\mathbf{C}$ is positive definite, the function ε is non-negative, and (q) implies that the integrand in (p) is non-negative on $\Omega-\{\boldsymbol{z}\}$. Thus, since this integrand is also continuous on $\Omega-\{\boldsymbol{z}\}$, we conclude from (p) that it must vanish on this set. This fact, in conjunction with (q), implies

$$\varepsilon(\boldsymbol{x},\nabla\boldsymbol{u}(\boldsymbol{x},\tau(\boldsymbol{x}))+\boldsymbol{G}(\boldsymbol{x})) = 0, \quad \boldsymbol{x}\in(\Omega-\{\boldsymbol{z}\}).$$

Thus, since $\mathbf{C}$ is positive definite, (a) yields

$$\boldsymbol{E}(\boldsymbol{x},\tau(\boldsymbol{x})) + \operatorname{sym}\boldsymbol{G}(\boldsymbol{x}) = \boldsymbol{0}, \quad \boldsymbol{x}\in(\Omega-\{\boldsymbol{z}\}). \tag{r}$$

By (i)

$$\nabla\tau(\boldsymbol{x}) = \frac{\boldsymbol{x}-\boldsymbol{z}}{c_M|\boldsymbol{x}-\boldsymbol{z}|}, \quad \boldsymbol{x}\in(\Omega-\{\boldsymbol{z}\}). \tag{s}$$

In view of (i) and the continuity of $\boldsymbol{E}$ and $\dot{\boldsymbol{u}}$,

$$\boldsymbol{E}(\boldsymbol{x},\tau(\boldsymbol{x})) \to \boldsymbol{E}(\boldsymbol{z},\lambda), \quad \dot{\boldsymbol{u}}(\boldsymbol{x},\tau(\boldsymbol{x})) \to \dot{\boldsymbol{u}}(\boldsymbol{z},\lambda) \quad \text{as} \quad \boldsymbol{x}\to\boldsymbol{z}. \tag{t}$$

If we let $\boldsymbol{x}=\boldsymbol{z}+\delta\boldsymbol{e}$, where $\boldsymbol{e}$ is a fixed unit vector, and take the limit in (r) as $\delta\to 0$, we conclude, with the aid of (c), (s), and (t), that

$$\boldsymbol{E}(\boldsymbol{z},\lambda) + \frac{1}{c_M}\operatorname{sym}\{\dot{\boldsymbol{u}}(\boldsymbol{z},\lambda)\otimes\boldsymbol{e}\} = \boldsymbol{0}. \tag{u}$$

Since (u) must hold for every unit vector $\boldsymbol{e}$, (u) must hold with $\boldsymbol{e}$ replaced by $-\boldsymbol{e}$; thus

$$\boldsymbol{E}(\boldsymbol{z},\lambda) = \boldsymbol{0}.$$

As $(\boldsymbol{z},\lambda)$ was arbitrarily chosen in $(B-D_t)\times(0,t)$, and since $\boldsymbol{E}$ is continuous,

$$\boldsymbol{E} = \boldsymbol{0} \quad \text{on} \quad (\bar{B}-D_t)\times[0,t]. \tag{v}$$

By the definition of D_t,

$$\boldsymbol{b} = \boldsymbol{0} \quad \text{on} \quad (\bar{B}-D_t)\times[0,t], \qquad \boldsymbol{u}(\cdot,0) = \dot{\boldsymbol{u}}(\cdot,0) = \boldsymbol{0} \quad \text{on} \quad \bar{B}-D_t; \tag{w}$$

(v) and the first of (w) together with the stress-strain relation and the equation of motion imply

$$\boldsymbol{S} = \boldsymbol{0}, \quad \ddot{\boldsymbol{u}} = \boldsymbol{0} \quad \text{on} \quad (\bar{B}-D_t)\times[0,t]. \tag{x}$$

The second of (x) and (w) yield

$$\boldsymbol{u} = \boldsymbol{0} \quad \text{on} \quad (\bar{B}-D_t)\times[0,t),$$

and the proof is complete. □

We now use the domain of influence theorem to establish a uniqueness theorem for elastodynamics that is valid for *infinite* domains. In comparing this theorem to the uniqueness theorems of NEUMANN ***(63.1)*** and BRUN ***(63.2)*** it must be remembered that among the assumptions made at the beginning of this section are the assumptions that C be positive definite and symmetric and that ϱ be strictly positive.

(3) Uniqueness theorem.[1] *Under the assumptions made at the beginning of this section the mixed problem of elastodynamics has at most one solution.*

Proof. Let $\mathfrak{p} = [\boldsymbol{u}, \boldsymbol{E}, \boldsymbol{S}]$ be the difference between two solutions. Then $\mathfrak{p}$ corresponds to vanishing body forces and satisfies

$$\boldsymbol{u}(\cdot, 0) = \dot{\boldsymbol{u}}(\cdot, 0) = \mathbf{0} \quad \text{on } B, \qquad \boldsymbol{s} \cdot \dot{\boldsymbol{u}} = 0 \quad \text{on } B \times (0, t_0).$$

Choose t arbitrarily in $(0, t_0)$. Clearly, $\hat{D}_t$ is empty, and thus the domain of influence D_t of the data at time t is empty. Therefore we conclude from ***(1)*** that $\mathfrak{p}$ vanishes on $\bar{B} \times [0, t]$. Since t was arbitrarily chosen, the proof is complete. □

It should be noted that WHEELER and STERNBERG'S uniqueness theorem ***(3)*** is valid for exterior domains and also for regions whose boundaries extend to infinity. An elastodynamic uniqueness theorem valid for infinite regions may alternatively be based on the classical energy identity following NEUMANN'S procedure ***(63.1)***, provided one introduces suitable restrictions on the orders of magnitude of the velocity and stress fields at infinity. The advantage of ***(3)*** is that it does not involve such artificial a priori assumptions. The analogous uniqueness issue in elastostatics, where the equations are elliptic rather than hyperbolic, is considerably more involved. For *exterior* domains elastostatic uniqueness theorems that avoid extraneous order prescriptions at infinity are given in Sect. 50.

VI. The free vibration problem.

75. Basic equations. The free vibration problem[2] is concerned with an elastic body undergoing motions of the form

$$\begin{aligned} \boldsymbol{u}(\boldsymbol{x}, t) &= \boldsymbol{u}(\boldsymbol{x}) \sin(\omega t + \gamma), \\ \boldsymbol{E}(\boldsymbol{x}, t) &= \boldsymbol{E}(\boldsymbol{x}) \sin(\omega t + \gamma), \\ \boldsymbol{S}(\boldsymbol{x}, t) &= \boldsymbol{S}(\boldsymbol{x}) \sin(\omega t + \gamma), \end{aligned} \tag{a}$$

where ω is the *frequency* of vibration and γ is the phase. If we assume null body forces, then these fields will satisfy the fundamental field equations of elastodynamics if and only if the *amplitudes* $\boldsymbol{u}(\boldsymbol{x})$, $\boldsymbol{E}(\boldsymbol{x})$, and $\boldsymbol{S}(\boldsymbol{x})$ obey the relations

$$\begin{aligned} \boldsymbol{E} &= \tfrac{1}{2}(\nabla \boldsymbol{u} + \nabla \boldsymbol{u}^T), \\ \boldsymbol{S} &= \mathsf{C}[\boldsymbol{E}], \\ \operatorname{div} \boldsymbol{S} &+ \varrho \omega^2 \boldsymbol{u} = \mathbf{0}. \end{aligned}$$

Since

$$\mathsf{C}[\boldsymbol{E}] = \mathsf{C}[\nabla \boldsymbol{u}],$$

[1] WHEELER and STERNBERG [1968, *16*] for the case in which B is homogeneous and isotropic; WHEELER [1970, *3*] for the case in which B is homogeneous, but anisotropic.

[2] The vibration problem for infinite domains is treated in detail by KUPRADZE [1963, *17, 18*].

the above relations yield the following equation for the displacement amplitude:

$$\operatorname{div} \mathbf{C}[\nabla \boldsymbol{u}] + \varrho \omega^2 \boldsymbol{u} = \mathbf{0}, \tag{b}$$

which, for a homogeneous and isotropic body, takes the form

$$\mu \Delta \boldsymbol{u} + (\lambda + \mu) \nabla \operatorname{div} \boldsymbol{u} + \varrho \omega^2 \boldsymbol{u} = \mathbf{0}.$$

Also, a simple calculation, similar to that used to deduce (e) of Sect. 59, yields the relations

$$c_1^2 \Delta \operatorname{div} \boldsymbol{u} + \omega^2 \operatorname{div} \boldsymbol{u} = 0,$$
$$c_2^2 \Delta \operatorname{curl} \boldsymbol{u} + \omega^2 \operatorname{curl} \boldsymbol{u} = \mathbf{0},$$

where c_1 and c_2 are the irrotational and isochoric velocities. Of course, these formulae could also have been derived by substituting the relation for $\boldsymbol{u}(\boldsymbol{x}, t)$ in (a) into (e) on p. 213.

Returning to the general inhomogeneous and anisotropic body, we see that the procedure given on p. 214. here yields the following relation for the stress amplitude:

$$\hat{\nabla}\left(\frac{\operatorname{div} \boldsymbol{S}}{\varrho}\right) + \omega^2 \mathbf{K}[\boldsymbol{S}] = \mathbf{0},$$

where $\mathbf{K} = \mathbf{C}^{-1}$ is the compliance tensor.

76. Characteristic solutions. Minimum principles. We assume that the elasticity field has the following properties:

(i) $\mathbf{C}$ is *smooth* on $\bar{B}$;

(ii) $\mathbf{C}$ is *positive definite* and *symmetric.*

Let

$$\mathscr{D} = \text{the set of all admissible displacement fields,}^1$$

and let $\mathscr{L}$ be the operator on $\mathscr{D}$ defined by

$$\mathscr{L}\boldsymbol{u} = \operatorname{div} \mathbf{C}[\nabla \boldsymbol{u}].$$

For convenience, we introduce the following notation:

$$\langle \boldsymbol{u}, \boldsymbol{v} \rangle = \int_B \boldsymbol{u} \cdot \boldsymbol{v} \, dv,$$
$$\|\boldsymbol{u}\| = \sqrt{\langle \boldsymbol{u}, \boldsymbol{u} \rangle},$$
$$\langle \boldsymbol{u}, \boldsymbol{v} \rangle_{\mathbf{C}} = \int_B \nabla \boldsymbol{u} \cdot \mathbf{C}[\nabla \boldsymbol{v}] \, dv.$$

Note that, by (ii) of ***(20.1)***,

$$\langle \boldsymbol{u}, \boldsymbol{v} \rangle_{\mathbf{C}} = \int_B \hat{\nabla} \boldsymbol{u} \cdot \mathbf{C}[\hat{\nabla} \boldsymbol{v}] \, dv.$$

The *free vibration problem* consists in finding combinations of the frequency ω and amplitude $\boldsymbol{u}$ that are possible when a portion $\mathscr{S}_1$ of the boundary is clamped and the remainder $\mathscr{S}_2$ is free. This should motivate the following definition: By a ***characteristic solution*** (for the free vibration problem) we mean an ordered pair $[\lambda, \boldsymbol{u}]$ such that λ is a scalar, $\boldsymbol{u} \in \mathscr{D}$, and

$$\mathscr{L}\boldsymbol{u} + \lambda \boldsymbol{u} = \mathbf{0}, \quad \|\boldsymbol{u}\| = 1, \quad \boldsymbol{u} = \mathbf{0} \text{ on } \mathscr{S}_1, \quad \boldsymbol{s} = \mathbf{0} \text{ on } \mathscr{S}_2;$$

[1] See p. 59.

here $\mathbf{s}$ is the ***traction field*** corresponding to $\boldsymbol{u}$, i.e.

$$\mathbf{s} = \mathbf{C}[\nabla \boldsymbol{u}]\, \boldsymbol{n} \quad \text{on } \partial B.$$

We call λ a ***characteristic value***, $\boldsymbol{u}$ an associated ***characteristic displacement***. Given λ, the corresponding *frequency of vibration* is given by

$$\omega = \sqrt{\frac{\lambda}{\varrho}}\,;$$

as we will soon see, λ is always non-negative. Finally, we say that the ***entire boundary is clamped*** if $\mathscr{S}_1 = \partial B$ ($\mathscr{S}_2 = \emptyset$), ***free*** if $\mathscr{S}_2 = \partial B$ ($\mathscr{S}_1 = \emptyset$).

(1) Lemma. *Assume that:*

(i) $\boldsymbol{u} \in \mathscr{D}$ *with* $\mathscr{L}\boldsymbol{u}$ *continuous on* $\bar{B}$;

(ii) $\boldsymbol{v}$ *is continuous on* $\bar{B}$ *and piecewise smooth on* B, *while* $\hat{\nabla}\boldsymbol{v}$ *is piecewise continuous on* $\bar{B}$.

Then

$$\langle \mathscr{L}\boldsymbol{u}, \boldsymbol{v}\rangle = -\langle \boldsymbol{u}, \boldsymbol{v}\rangle_{\mathbf{C}} + \int_{\partial B} \mathbf{s}\cdot \boldsymbol{v}\, da,$$

where $\mathbf{s}$ *is the traction field corresponding to* $\boldsymbol{u}$.

Proof. Let $\boldsymbol{S} = \mathbf{C}[\nabla \boldsymbol{u}]$, so that $\mathbf{s} = \boldsymbol{S}\boldsymbol{n}$ on ∂B. Then

$$\begin{aligned}\int_B (\mathscr{L}\boldsymbol{u})\cdot \boldsymbol{v}\, dv &= \int_B (\operatorname{div} \boldsymbol{S})\cdot \boldsymbol{v}\, dv = \int_{\partial B} \mathbf{s}\cdot \boldsymbol{v}\, da - \int_B \boldsymbol{S}\cdot \nabla \boldsymbol{v}\, dv\\ &= \int_{\partial B} \mathbf{s}\cdot \boldsymbol{v}\, da - \langle \boldsymbol{u}, \boldsymbol{v}\rangle_{\mathbf{C}}. \quad \square\end{aligned}$$

(2) Properties of characteristic values.[1] *Let* $[\lambda, \boldsymbol{u}]$ *be a characteristic solution. Then*

$$\lambda = \langle \boldsymbol{u}, \boldsymbol{u}\rangle_{\mathbf{C}},$$

so that

$$\lambda \geqq 0.$$

In fact, $\lambda = 0$ *if and only if* $\boldsymbol{u}$ *is rigid. Moreover, a characteristic solution* $[\lambda, \boldsymbol{u}]$ *with* $\lambda = 0$ *exists only if the entire boundary is free, in which case every rigid displacement* $\boldsymbol{u}$ *with* $\|\boldsymbol{u}\| = 1$ *is a characteristic vector corresponding to* $\lambda = 0$.

Proof. Since $[\lambda, \boldsymbol{u}]$ is a characteristic solution,

$$\mathscr{L}\boldsymbol{u} + \lambda \boldsymbol{u} = \boldsymbol{0}, \qquad \|\boldsymbol{u}\| = 1, \tag{a}$$

and

$$\mathbf{s}\cdot \boldsymbol{u} = 0 \quad \text{on } \partial B, \tag{b}$$

where $\mathbf{s}$ is the corresponding traction field. By (a)

$$\langle \mathscr{L}\boldsymbol{u}, \boldsymbol{u}\rangle = -\lambda \langle \boldsymbol{u}, \boldsymbol{u}\rangle = -\lambda;$$

therefore we conclude from (b) and ***(1)*** that

$$\lambda = \langle \boldsymbol{u}, \boldsymbol{u}\rangle_{\mathbf{C}}.$$

Moreover, since $\mathbf{C}$ is positive definite,

$$\langle \boldsymbol{u}, \boldsymbol{u}\rangle_{\mathbf{C}} \geqq 0,$$

[1] Clebsch [1862, *1*], § 20, generalizing a result of Poisson [1829, *2*] for the vibrations of an elastic sphere. See also Ibbetson [1887, *2*], § 256, Cesàro [1894, *1*], p. 58.

Azhmudinov and Mamatordiev [1967, *1*] show that when the boundary is clamped λ is given by a certain integral over ∂B.

with equality holding if and only if the associated strain field vanishes, that is, if and only if $\boldsymbol{u}$ is rigid. Further, since $\boldsymbol{u}$ vanishes on $\mathscr{S}_1$, a rigid $\boldsymbol{u}$ is possible only if $\mathscr{S}_1=\emptyset$. (Since $\mathscr{S}_1$ is a regular surface, if non-empty it must contain at least three non-collinear points.) On the other hand, if $\mathscr{S}_1=\emptyset$, it is a simple matter to verify that $[0, \boldsymbol{u}]$ is a characteristic solution whenever $\boldsymbol{u}$ is rigid and $\|\boldsymbol{u}\|=1$. □

(3) Orthogonality of characteristic displacements.[1] *Let* $[\lambda, \boldsymbol{u}]$ *and* $[\tilde{\lambda}, \tilde{\boldsymbol{u}}]$ *be characteristic solutions with* $\lambda \neq \tilde{\lambda}$. *Then*

$$\langle \boldsymbol{u}, \tilde{\boldsymbol{u}}\rangle = \langle \boldsymbol{u}, \tilde{\boldsymbol{u}}\rangle_{\mathbf{C}} = \langle \mathscr{L}\boldsymbol{u}, \tilde{\boldsymbol{u}}\rangle = 0.$$

Proof. By hypothesis

$$\mathscr{L}\boldsymbol{u}+\lambda\boldsymbol{u}=\mathbf{0}, \tag{a}$$

$$\mathscr{L}\tilde{\boldsymbol{u}}+\tilde{\lambda}\tilde{\boldsymbol{u}}=\mathbf{0}, \tag{b}$$

$$\tilde{\mathbf{s}}\cdot\boldsymbol{u}=\mathbf{s}\cdot\tilde{\boldsymbol{u}}=0 \quad \text{on } \partial B, \tag{c}$$

where $\mathbf{s}$ and $\tilde{\mathbf{s}}$ are the traction fields of $\boldsymbol{u}$ and $\tilde{\boldsymbol{u}}$. By (a) and (b)

$$\left.\begin{aligned} &\langle \mathscr{L}\boldsymbol{u}, \tilde{\boldsymbol{u}}\rangle + \lambda\langle \boldsymbol{u}, \tilde{\boldsymbol{u}}\rangle = 0,\\ &\langle \mathscr{L}\tilde{\boldsymbol{u}}, \boldsymbol{u}\rangle + \tilde{\lambda}\langle \tilde{\boldsymbol{u}}, \boldsymbol{u}\rangle = 0. \end{aligned}\right\} \tag{d}$$

On the other hand, (c), *(1)*, and the symmetry of $\mathbf{C}$ imply that

$$\langle \mathscr{L}\boldsymbol{u}, \tilde{\boldsymbol{u}}\rangle = -\langle \boldsymbol{u}, \tilde{\boldsymbol{u}}\rangle_{\mathbf{C}} = -\langle \tilde{\boldsymbol{u}}, \boldsymbol{u}\rangle_{\mathbf{C}} = \langle \mathscr{L}\tilde{\boldsymbol{u}}, \boldsymbol{u}\rangle. \tag{e}$$

By (d) and (e)

$$(\lambda-\tilde{\lambda})\langle \boldsymbol{u}, \tilde{\boldsymbol{u}}\rangle = 0;$$

hence $\langle \boldsymbol{u}, \tilde{\boldsymbol{u}}\rangle=0$ if $\lambda\neq\tilde{\lambda}$. The remaining results follow from (d) and (e). □

Theorem *(3)* asserts that characteristic displacements corresponding to distinct characteristic values are orthogonal. This is not necessarily true if the characteristic values are equal. Indeed, if $\boldsymbol{u}_1, \boldsymbol{u}_2, \ldots, \boldsymbol{u}_n$ are characteristic displacements corresponding to the *same* characteristic value λ, then it is a simple matter to verify that any field $\boldsymbol{u}$ with $\|\boldsymbol{u}\|=1$ lying in the span of $\{\boldsymbol{u}_1, \boldsymbol{u}_2, \ldots, \boldsymbol{u}_n\}$ is also a characteristic displacement corresponding to λ.

Henceforth $\mathscr{K}$ denotes the set of all vector fields $\boldsymbol{v}$ with the following properties:

(i) $\boldsymbol{v}$ is continuous on $\bar{B}$ and piecewise smooth on B;
(ii) $\hat{\nabla}\boldsymbol{v}$ is piecewise continuous on $\bar{B}$;
(iii) $\|\boldsymbol{v}\|=1$;
(iv) $\boldsymbol{v}=\mathbf{0}$ on $\mathscr{S}_1$.

(4) Minimum principle for the lowest characteristic value.[2] *Let* $\boldsymbol{u}_1\in\mathscr{D}\cap\mathscr{K}$ *and assume that*

$$\langle \boldsymbol{u}_1, \boldsymbol{u}_1\rangle_{\mathbf{C}} \leq \langle \boldsymbol{v}, \boldsymbol{v}\rangle_{\mathbf{C}} \quad \textit{for every } \boldsymbol{v}\in\mathscr{K}.$$

Define

$$\lambda_1 = \langle \boldsymbol{u}_1, \boldsymbol{u}_1\rangle_{\mathbf{C}}.$$

Then $[\lambda_1, \boldsymbol{u}_1]$ *is a characteristic solution, and* λ_1 *is the lowest characteristic value.*

---

[1] See footnote 1 on p. 263. See also POISSON [1826, *1*].

[2] POINCARÉ [1890, *2*], p. 211, [1892, *6*], p. 103. See also TEDONE [1907, *2*], § 15d; COURANT [1920, *1*]; COURANT and HILBERT [1953, *8*], p. 339; HU [1958, *11*].

Proof. Let $\alpha \neq \pm 1$ be an arbitrary real number and choose $\boldsymbol{v} \in \mathscr{K}$. Then $\|\boldsymbol{u}_1 + \alpha \boldsymbol{v}\| \neq 0$, and

$$\boldsymbol{w} = \frac{1}{\|\boldsymbol{u}_1 + \alpha \boldsymbol{v}\|} (\boldsymbol{u}_1 + \alpha \boldsymbol{v})$$

belongs to $\mathscr{K}$; thus, by hypothesis,

$$\lambda_1 = \langle \boldsymbol{u}_1, \boldsymbol{u}_1 \rangle_{\mathbf{C}} \leqq \langle \boldsymbol{w}, \boldsymbol{w} \rangle_{\mathbf{C}} = \frac{1}{\|\boldsymbol{u}_1 + \alpha \boldsymbol{v}\|^2} \langle \boldsymbol{u}_1 + \alpha \boldsymbol{v}, \boldsymbol{u}_1 + \alpha \boldsymbol{v} \rangle_{\mathbf{C}},$$

and hence

$$\lambda_1 \langle \boldsymbol{u}_1 + \alpha \boldsymbol{v}, \boldsymbol{u}_1 + \alpha \boldsymbol{v} \rangle \leqq \langle \boldsymbol{u}_1 + \alpha \boldsymbol{v}, \boldsymbol{u}_1 + \alpha \boldsymbol{v} \rangle_{\mathbf{C}}.$$

If we expand both bilinear forms and use the relations $\|\boldsymbol{v}\| = \|\boldsymbol{u}_1\| = 1$ and $\lambda_1 = \langle \boldsymbol{u}_1, \boldsymbol{u}_1 \rangle_{\mathbf{C}}$, we arrive at

$$2\alpha [\lambda_1 \langle \boldsymbol{u}_1, \boldsymbol{v} \rangle - \langle \boldsymbol{u}_1, \boldsymbol{v} \rangle_{\mathbf{C}}] + \alpha^2 [\lambda_1 - \langle \boldsymbol{v}, \boldsymbol{v} \rangle_{\mathbf{C}}] \leqq 0.$$

Since this inequality must hold for every $\alpha \neq \pm 1$, we must have

$$\lambda_1 \langle \boldsymbol{u}_1, \boldsymbol{v} \rangle - \langle \boldsymbol{u}_1, \boldsymbol{v} \rangle_{\mathbf{C}} = 0. \tag{a}$$

As $\boldsymbol{v} = \mathbf{0}$ on $\mathscr{S}_1$, ***(1)*** implies that

$$\langle \boldsymbol{u}_1, \boldsymbol{v} \rangle_{\mathbf{C}} = -\langle \mathscr{L} \boldsymbol{u}_1, \boldsymbol{v} \rangle + \int_{\mathscr{S}_2} \mathbf{s}_1 \cdot \boldsymbol{v} \, da,$$

where $\mathbf{s}_1$ is the traction field corresponding to $\boldsymbol{u}_1$. In view of the last two relations,

$$\int_B (\mathscr{L} \boldsymbol{u}_1 + \lambda_1 \boldsymbol{u}_1) \cdot \boldsymbol{v} \, dv - \int_{\mathscr{S}_2} \mathbf{s}_1 \cdot \boldsymbol{v} \, da = 0$$

for *every* continuous and piecewise smooth vector field $\boldsymbol{v}$ with $\boldsymbol{v} = \mathbf{0}$ on $\mathscr{S}_1$ and $\|\boldsymbol{v}\| = 1$. Since this expression is linear in $\boldsymbol{v}$, we can drop the requirement that $\|\boldsymbol{v}\| = 1$. Therefore it follows from ***(7.1)*** and ***(35.1)*** that

$$\mathscr{L} \boldsymbol{u}_1 + \lambda_1 \boldsymbol{u}_1 = \mathbf{0} \quad \text{on } B,$$
$$\mathbf{s}_1 = \mathbf{0} \quad \text{on } \mathscr{S}_2.$$

Thus, since $\boldsymbol{u}_1 \in \mathscr{K}$, $[\lambda_1, \boldsymbol{u}_1]$ is a characteristic solution.

If $[\lambda, \boldsymbol{u}]$ is another characteristic solution, then ***(2)*** implies that

$$\lambda = \langle \boldsymbol{u}, \boldsymbol{u} \rangle_{\mathbf{C}}.$$

Thus we conclude from our present hypotheses that

$$\lambda_1 \leqq \lambda,$$

and hence λ_1 is the lowest characteristic value. □

Theorem ***(4)*** asserts that the minimum (if it exists) of the functional

$$\langle \boldsymbol{v}, \boldsymbol{v} \rangle_{\mathbf{C}} \tag{a}$$

under the supplementary condition

$$\|\boldsymbol{v}\| = 1 \tag{b}$$

is the lowest characteristic value. We now show that the determination of the n-th characteristic value can be reduced to the variational problem of finding a minimum of the functional (a) under the supplementary condition (b) and the constraint

$$\langle \boldsymbol{v}, \boldsymbol{u}_r \rangle = 0, \qquad r = 1, 2, \ldots, n-1,$$

where $\boldsymbol{u}_1, \boldsymbol{u}_2, \ldots, \boldsymbol{u}_{n-1}$ are the characteristic vectors for the first $n-1$ characteristic values.

Let $\boldsymbol{u}_1, \boldsymbol{u}_2, \ldots, \boldsymbol{u}_n$ be piecewise continuous vector fields on $\bar{B}$. Then we write

$$\{\boldsymbol{u}_1, \boldsymbol{u}_2, \ldots, \boldsymbol{u}_n\}^{\perp}$$

for the set of all piecewise continuous vector fields $\boldsymbol{v}$ on $\bar{B}$ such that

$$\langle \boldsymbol{v}, \boldsymbol{u}_r \rangle = 0, \qquad r = 1, 2, \ldots, n.$$

(5) Minimum principle.[1] *Let $\boldsymbol{u}_1, \boldsymbol{u}_2, \ldots$ belong to $\mathscr{D} \cap \mathscr{K}$, and let*

$$\mathscr{K}_1 = \mathscr{K},$$
$$\mathscr{K}_r = \mathscr{K} \cap \{\boldsymbol{u}_1, \boldsymbol{u}_2, \ldots, \boldsymbol{u}_{r-1}\}^{\perp}, \qquad r \geqq 2.$$

Assume that, for each $r \geqq 1$, $\boldsymbol{u}_r \in \mathscr{K}_r$ and

$$\langle \boldsymbol{u}_r, \boldsymbol{u}_r \rangle_{\mathsf{C}} \leqq \langle \boldsymbol{v}, \boldsymbol{v} \rangle_{\mathsf{C}} \quad \textit{for every } \boldsymbol{v} \in \mathscr{K}_r.$$

Define λ_r by

$$\lambda_r = \langle \boldsymbol{u}_r, \boldsymbol{u}_r \rangle_{\mathsf{C}}.$$

Then each $[\lambda_r, \boldsymbol{u}_r]$ is a characteristic solution and

$$0 \leqq \lambda_1 \leqq \lambda_2 \leqq \cdots.$$

Proof. By ***(4)***, $[\lambda_1, \boldsymbol{u}_1]$ is a characteristic solution. We now proceed by induction. Thus we assume that $n \geqq 2$ and that $[\lambda_r, \boldsymbol{u}_r]$ is a characteristic solution for $1 \leqq r \leqq n-1$. Since $\boldsymbol{u}_n \in \mathscr{K}_n$,

$$\langle \boldsymbol{u}_n, \boldsymbol{u}_i \rangle = 0, \qquad i = 1, 2, \ldots, n-1. \tag{a}$$

On the other hand, our induction hypothesis and ***(1)*** imply that

$$\begin{aligned} 0 = \lambda_i \langle \boldsymbol{u}_n, \boldsymbol{u}_i \rangle &= -\langle \boldsymbol{u}_n, \mathscr{L} \boldsymbol{u}_i \rangle \\ &= \langle \boldsymbol{u}_n, \boldsymbol{u}_i \rangle_{\mathsf{C}} - \int_{\partial B} \mathbf{s}_i \cdot \boldsymbol{u}_n \, da, \qquad i = 1, 2, \ldots, n-1. \end{aligned}$$

But $\boldsymbol{u}_n = \mathbf{0}$ on $\mathscr{S}_1$ and $\mathbf{s}_i = \mathbf{0}$ on $\mathscr{S}_2$; therefore

$$\langle \boldsymbol{u}_n, \boldsymbol{u}_i \rangle_{\mathsf{C}} = 0, \qquad i = 1, 2, \ldots, n-1. \tag{b}$$

Now choose $\boldsymbol{v} \in \mathscr{K}_n$ and let

$$\boldsymbol{w} = \frac{1}{\|\boldsymbol{u}_n + \alpha \boldsymbol{v}\|} (\boldsymbol{u}_n + \alpha \boldsymbol{v}),$$

where $\alpha \neq \pm 1$. Then $\boldsymbol{w} \in \mathscr{K}_n$, and steps directly analogous to those used to establish (a) in the proof of ***(4)*** now lead to the result that

$$\lambda_n \langle \boldsymbol{u}_n, \boldsymbol{v} \rangle - \langle \boldsymbol{u}_n, \boldsymbol{v} \rangle_{\mathsf{C}} = 0 \tag{c}$$

for every $\boldsymbol{v} \in \mathscr{K}_n$.

Let $\bar{\boldsymbol{v}} \in \mathscr{K}$ be arbitrarily chosen. If $\bar{\boldsymbol{v}}$ lies in the span of $\boldsymbol{u}_1, \boldsymbol{u}_2, \ldots, \boldsymbol{u}_{n-1}$, then (a) and (b) imply that

$$\lambda_n \langle \boldsymbol{u}_n, \bar{\boldsymbol{v}} \rangle - \langle \boldsymbol{u}_n, \bar{\boldsymbol{v}} \rangle_{\mathsf{C}} = 0. \tag{d}$$

Assume that $\boldsymbol{v}$ does not lie in the span of $\boldsymbol{u}_1, \boldsymbol{u}_2, \ldots, \boldsymbol{u}_{n-1}$. Then there exist scalars $\beta, t_1, t_2, \ldots, t_{n-1}$ such that

$$\boldsymbol{v} = \beta \left(\bar{\boldsymbol{v}} + \sum_{i=1}^{n-1} t_i \boldsymbol{u}_i \right) \tag{e}$$

[1] See footnote 2 on p. 264.

belongs to $\mathcal{K}_n$. Indeed, let

$$t_i = -\langle \bar{\boldsymbol{v}}, \boldsymbol{u}_i \rangle,$$

and $\left(\text{since } \bar{\boldsymbol{v}} \neq -\sum_{i=1}^{n-1} t_i \boldsymbol{u}_i\right)$ choose β such that $\|\boldsymbol{v}\| = 1$. By (c) and (e),

$$\lambda_n \langle \boldsymbol{u}_n, \bar{\boldsymbol{v}} \rangle - \langle \boldsymbol{u}_n, \bar{\boldsymbol{v}} \rangle_{\mathbf{C}} + \sum_{i=1}^{n-1} t_i [\lambda_n \langle \boldsymbol{u}_n, \boldsymbol{u}_i \rangle - \langle \boldsymbol{u}_n, \boldsymbol{u}_i \rangle_{\mathbf{C}}] = 0. \tag{f}$$

Thus we conclude from (a) and (b) that (d) remains valid in the present circumstances. Hence (d) holds for every $\bar{\boldsymbol{v}} \in \mathcal{K}$, and, as in the proof of ***(4)***, this implies that $[\lambda_n, \boldsymbol{u}_n]$ is a characteristic solution. □

We now show that the minimum principle ***(5)*** characterizes *all* of the solutions of the free vibration problem provided $\lambda_n \to \infty$.[1]

(6) Completeness of the minimum principle. *Let $\lambda_1, \lambda_2, \ldots$ and $\boldsymbol{u}_1, \boldsymbol{u}_2, \ldots$ obey the hypotheses of the minimum principle* ***(5)*** *and assume, in addition, that $\lambda_n \to \infty$ as $n \to \infty$. Let $[\lambda, \boldsymbol{u}]$ be a characteristic solution. Then $\lambda = \lambda_r$ for some integer r, and if*

$$\lambda_{r-1} < \lambda_r = \lambda_{r+1} = \cdots = \lambda_{r+n} < \lambda_{r+n+1},$$

then $\boldsymbol{u}$ is contained in the span of $\boldsymbol{u}_r, \boldsymbol{u}_{r+1}, \ldots, \boldsymbol{u}_{r+n}$. In particular, if $n = 0$ so that λ_r is distinct, then $\boldsymbol{u} = \pm \boldsymbol{u}_r$.

Proof. Assume that λ is not equal to any of the λ_r's. By ***(4)***, $\lambda > \lambda_1$. Thus, since $\lambda_n \to \infty$, there exists an $m \geqq 2$ such that

$$\lambda_{m-1} < \lambda < \lambda_m. \tag{a}$$

Since $[\lambda_r, \boldsymbol{u}_r]$ is a characteristic value for every r, we conclude from ***(3)*** that

$$\boldsymbol{u} \in \{\boldsymbol{u}_1, \boldsymbol{u}_2, \ldots, \boldsymbol{u}_{m-1}\}^{\perp}.$$

Thus, as the hypotheses of ***(5)*** hold,

$$\langle \boldsymbol{u}_m, \boldsymbol{u}_m \rangle_{\mathbf{C}} \leqq \langle \boldsymbol{u}, \boldsymbol{u} \rangle_{\mathbf{C}},$$

or equivalently, by ***(2)***,

$$\lambda_m \leqq \lambda,$$

which contradicts (a). Thus $\lambda = \lambda_r$ for some r.

Assume now that $\lambda_{r-1} < \lambda_r = \lambda_{r+1} = \cdots = \lambda_{r+n} < \lambda_{r+n+1}$, and let

$$\mathcal{U} = \{\boldsymbol{u}_r, \boldsymbol{u}_{r+1}, \ldots, \boldsymbol{u}_{r+n}\}.$$

Since $\lambda \neq \lambda_j$ for $j < r$, it follows from ***(3)*** that

$$\boldsymbol{u} \in \{\boldsymbol{u}_1, \boldsymbol{u}_2, \ldots, \boldsymbol{u}_{r-1}\}^{\perp}. \tag{b}$$

We decompose $\boldsymbol{u}$ as follows:

$$\boldsymbol{u} = \boldsymbol{u}_0 + \boldsymbol{u}_{\perp}, \tag{c}$$

where $\boldsymbol{u}_0$ belongs to the span of $\mathcal{U}$ and $\boldsymbol{u}_{\perp} \in \mathcal{U}^{\perp}$. Indeed, we simply take

$$\boldsymbol{u}_0 = \sum_{j=r}^{r+n} \langle \boldsymbol{u}, \boldsymbol{u}_j \rangle \boldsymbol{u}_j.$$

To complete the proof it suffices to show that $\boldsymbol{u}_{\perp} = \mathbf{0}$. If $\boldsymbol{u}_0 = \mathbf{0}$, then $\boldsymbol{u} = \boldsymbol{u}_{\perp}$, and it follows from (b) and the fact that $\boldsymbol{u}_{\perp} \in \mathcal{U}^{\perp}$ that

$$\boldsymbol{u} \in \{\boldsymbol{u}_1, \boldsymbol{u}_2, \ldots, \boldsymbol{u}_{r+n}\}^{\perp}. \tag{d}$$

---

[1] See ***(78.1)***.

Thus, as before,

$$\lambda_{r+n+1} \leqq \lambda,$$

and we have a contradiction. Thus $\boldsymbol{u}_0 \neq \boldsymbol{0}$. Assume now that $\boldsymbol{u}_\perp \neq \boldsymbol{0}$ and let

$$\bar{\boldsymbol{u}}_0 = \frac{\boldsymbol{u}_0}{|\boldsymbol{u}_0|}, \qquad \bar{\boldsymbol{u}}_\perp = \frac{\boldsymbol{u}_\perp}{|\boldsymbol{u}_\perp|}.$$

Since $\bar{\boldsymbol{u}}_0$ is a linear combination of characteristic displacements corresponding to the same characteristic value λ, $[\lambda, \bar{\boldsymbol{u}}_0]$ is a characteristic solution; this fact and (c) imply that $[\lambda, \bar{\boldsymbol{u}}_\perp]$ is also a characteristic solution. Thus, by ***(3)***, $\bar{\boldsymbol{u}}_\perp$ is orthogonal to every $\boldsymbol{u}_j \notin \mathscr{U}$, and hence (d) holds with $\boldsymbol{u}$ replaced by $\bar{\boldsymbol{u}}_\perp$. Thus we are again led to a contradiction; consequently

$$\boldsymbol{u}_\perp = \boldsymbol{0}. \quad \square$$

77. The minimax principle and its consequences. We now establish an alternative characterization of the free vibration problem. We continue to assume that $\mathbf{C}$ is *symmetric, positive definite,* and *smooth* on $\bar{B}$. For convenience, we introduce the following notation:

$\mathscr{W}$ = the set of all piecewise continuous vector fields on $\bar{B}$,

$$\mathscr{W}^n = \mathscr{W} \times \mathscr{W} \times \cdots \times \mathscr{W} \quad (n \text{ times}).$$

Recall that $\mathscr{K}$ is the set of all vector fields on $\bar{B}$ with properties (i)–(iv) listed on p. 264.

(1) Minimax principle.[1] *Let $[\lambda_r, \boldsymbol{u}_r]$ satisfy the hypotheses of* ***(76.5)*** *for $1 \leqq r \leqq n+1$. Define φ on $\mathscr{W}^n$ by*

$$\varphi(\boldsymbol{w}_1, \boldsymbol{w}_2, \ldots, \boldsymbol{w}_n) = \inf\{\langle \boldsymbol{v}, \boldsymbol{v}\rangle_{\mathbf{C}} \colon \boldsymbol{v} \in \mathscr{K} \cap \{\boldsymbol{w}_1, \boldsymbol{w}_2, \ldots, \boldsymbol{w}_n\}^\perp\}.$$

Then

$$\lambda_{n+1} = \varphi(\boldsymbol{u}_1, \boldsymbol{u}_2, \ldots, \boldsymbol{u}_n) = \sup\{\varphi(\boldsymbol{w}_1, \boldsymbol{w}_2, \ldots, \boldsymbol{w}_n) \colon (\boldsymbol{w}_1, \boldsymbol{w}_2, \ldots, \boldsymbol{w}_n) \in \mathscr{W}^n\}.$$

Proof. By hypothesis,

$$\lambda_{n+1} = \varphi(\boldsymbol{u}_1, \boldsymbol{u}_2, \ldots, \boldsymbol{u}_n) = \langle \boldsymbol{u}_{n+1}, \boldsymbol{u}_{n+1}\rangle_{\mathbf{C}}. \tag{a}$$

Thus it suffices to prove that

$$\lambda_{n+1} \geqq \varphi(\boldsymbol{w}_1, \boldsymbol{w}_2, \ldots, \boldsymbol{w}_n) \quad \text{for every } (\boldsymbol{w}_1, \boldsymbol{w}_2, \ldots, \boldsymbol{w}_n) \in \mathscr{W}^n. \tag{b}$$

To accomplish this we will show that for every $(\boldsymbol{w}_1, \boldsymbol{w}_2, \ldots, \boldsymbol{w}_n) \in \mathscr{W}^n$ there exists a $\boldsymbol{v} \in \mathscr{K} \cap \{\boldsymbol{w}_1, \boldsymbol{w}_2, \ldots, \boldsymbol{w}_n\}^\perp$ such that

$$\langle \boldsymbol{v}, \boldsymbol{v}\rangle_{\mathbf{C}} \leqq \lambda_{n+1}; \tag{c}$$

for then (b) would follow from the inequality

$$\varphi(\boldsymbol{w}_1, \boldsymbol{w}_2, \ldots, \boldsymbol{w}_n) \leqq \langle \boldsymbol{v}, \boldsymbol{v}\rangle_{\mathbf{C}} \leqq \lambda_{n+1}.$$

Choose $(\boldsymbol{w}_1, \boldsymbol{w}_2, \ldots, \boldsymbol{w}_n) \in \mathscr{W}^n$. Let

$$\boldsymbol{v} = \sum_{i=1}^{n+1} \alpha_i \boldsymbol{u}_i, \tag{d}$$

where the α_i's are as yet undetermined. Clearly, $\boldsymbol{v} \in \mathscr{K} \cap \{\boldsymbol{w}_1, \boldsymbol{w}_2, \ldots, \boldsymbol{w}_n\}^\perp$ if

$$\sum_{i=1}^{n+1} \alpha_i^2 = 1 \tag{e}$$

[1] COURANT [1920, *1*]. See also COURANT and HILBERT [1953, *8*], p. 406.

and

$$\sum_{i=1}^{n+1} \alpha_i \langle \boldsymbol{u}_i, \boldsymbol{w}_r \rangle = 0 \qquad (r = 1, 2, \ldots, n). \tag{f}$$

The system (f) consists of n equations in $n+1$ unknowns and can always be solved for $\alpha_1, \alpha_2, \ldots, \alpha_{n+1}$. Moreover, since the system is homogeneous, the requirement (e) can also be met. We have only to show that the function $\boldsymbol{v}$ so defined satisfies (c). Since[1]

$$\langle \boldsymbol{u}_i, \boldsymbol{u}_j \rangle_{\mathbf{C}} = \lambda_j \, \delta_{ij} \quad \text{(no sum)} \quad i, j = 1, 2, \ldots, n+1,$$

(d) implies

$$\langle \boldsymbol{v}, \boldsymbol{v} \rangle_{\mathbf{C}} = \sum_{i=1}^{n+1} \alpha_i^2 \, \lambda_i. \tag{g}$$

Further, (e) and the fact that

$$\lambda_i \leqq \lambda_{n+1}, \qquad i \leqq n+1$$

imply

$$\sum_{i=1}^{n+1} \alpha_i^2 \, \lambda_i \leqq \lambda_{n+1}, \tag{h}$$

and (g), (h) yield the desired inequality (c). □

Let $\lambda_1, \lambda_2, \ldots$ and $\boldsymbol{u}_1, \boldsymbol{u}_2, \ldots$ obey the hypotheses of the minimum principle ***(76.5)***. We call the ordered array

$$\{\lambda_1, \lambda_2, \ldots\}$$

the ***spectrum*** corresponding to $\mathbf{C}$, B, $\mathscr{S}_1$, and $\mathscr{S}_2$, and we say that $\boldsymbol{u}_1, \boldsymbol{u}_2, \ldots$ is an associated system of ***normal modes.*** By ***(76.6)***, ***(78.1)***, and Footnote 2 on p. 270, the spectrum contains every characteristic value, and hence it is independent of the choice of $\boldsymbol{u}_1, \boldsymbol{u}_2, \ldots$ (as long as this choice is consistent with ***(76.5)***).

We now consider a second body $\tilde{B}$ composed of an elastic material with elasticity field $\tilde{\mathbf{C}}$. We will discuss the vibration problem for such a body taking as boundary conditions

$$\boldsymbol{u} = \mathbf{0} \quad \text{on } \tilde{\mathscr{S}}_1 \quad \text{and} \quad \boldsymbol{s} = \mathbf{0} \quad \text{on } \tilde{\mathscr{S}}_2,$$

where $\tilde{\mathscr{S}}_1$ and $\tilde{\mathscr{S}}_2$ are complementary subsets of $\partial \tilde{B}$. In particular, we compare the spectrum corresponding to this problem with the spectrum for the problem discussed previously (assuming, of course that both spectrums exist).

We assume that $\tilde{\mathbf{C}}$ is *positive definite, symmetric, and smooth on* $\bar{\tilde{B}}$. We write

$$\tilde{\mathbf{C}} \leqq \mathbf{C}$$

if

$$\boldsymbol{E} \cdot \tilde{\mathbf{C}}_{\boldsymbol{x}}[\boldsymbol{E}] \leqq \boldsymbol{E} \cdot \mathbf{C}_{\boldsymbol{x}}[\boldsymbol{E}]$$

for every symmetric tensor $\boldsymbol{E}$ and every $\boldsymbol{x} \in B \cap \tilde{B}$.

(2) Comparison theorem.[2] *Let* $\{\lambda_1, \lambda_2, \ldots\}$ *be the spectrum corresponding to* $\mathbf{C}$, B, $\mathscr{S}_1$, *and* $\mathscr{S}_2$; $\{\tilde{\lambda}_1, \tilde{\lambda}_2, \ldots\}$, *the spectrum corresponding to* $\tilde{\mathbf{C}}$, $\tilde{B}$, $\tilde{\mathscr{S}}_1$, *and* $\tilde{\mathscr{S}}_2$. *Assume that either*

---

[1] This follows from (d) and (e) in the proof of ***(16.3)*** and the fact that

$$\langle \boldsymbol{u}_i, \boldsymbol{u}_j \rangle = \delta_{ij} \qquad (i, j = 1, 2, \ldots, n+1).$$

[2] Cf. Courant and Hilbert [1953, *8*], pp. 407–412.

(i) $B=\tilde{B}$, $\tilde{\mathscr{S}}_1 \subset \mathscr{S}_1$, $\tilde{\mathbf{C}} \leqq \mathbf{C}$; *or*
(ii) *the entire boundary of B is clamped, and*

$$B \subset \tilde{B}, \quad \tilde{\mathbf{C}} \leqq \mathbf{C}.$$

Then

$$\tilde{\lambda}_r \leqq \lambda_r, \quad r=1, 2, \ldots$$

Proof. Let $\mathscr{K}$ be as before, and let $\tilde{\mathscr{K}}$ be the corresponding set for $\tilde{B}$ and $\tilde{\mathscr{S}}_1$. Assume that (i) holds. Then, since $B=\tilde{B}$ and $\tilde{\mathscr{S}}_1 \subset \mathscr{S}_1$,

$$\mathscr{K} \subset \tilde{\mathscr{K}}. \tag{a}$$

Further, since $\tilde{\mathbf{C}} \leqq \mathbf{C}$,

$$\langle \boldsymbol{v}, \boldsymbol{v} \rangle_{\tilde{\mathbf{C}}} \leqq \langle \boldsymbol{v}, \boldsymbol{v} \rangle_{\mathbf{C}}. \tag{b}$$

Let $n \geqq 1$ be a fixed integer, let φ be as defined in ***(1)***, and let $\tilde{\varphi}$ be defined on $\mathscr{W}^n$ by

$$\tilde{\varphi}(\boldsymbol{w}_1, \boldsymbol{w}_2, \ldots, \boldsymbol{w}_n) = \inf\{\langle \boldsymbol{v}, \boldsymbol{v} \rangle_{\tilde{\mathbf{C}}} : \boldsymbol{v} \in \tilde{\mathscr{K}} \cap \{\boldsymbol{w}_1, \boldsymbol{w}_2, \ldots, \boldsymbol{w}_n\}^{\perp}\}.$$

Then by (a) and (b)

$$\tilde{\varphi} \leqq \varphi,$$

and we conclude from the minimax principle ***(1)*** that

$$\tilde{\lambda}_n \leqq \lambda_n.$$

The proof under hypothesis (ii) is strictly analogous. □

78. Completeness of the characteristic solutions. In this section we show that the characteristic values λ_n become infinite as $n \to \infty$. This implies, in particular, that each characteristic value has only a finite multiplicity. More important, however, is that the unboundedness of the characteristic values yields the completeness of the characteristic displacement fields. Thus we assume that the spectrum $\{\lambda_1, \lambda_2, \ldots\}$ exists; in addition, we suppose that $\mathbf{C}$ is *positive definite, symmetric,* and *smooth* on $\bar{B}$.

(1) Infinite growth of the characteristic values.[1] *Assume that B is properly regular and that $\mathscr{S}_1 \neq \emptyset$.*[2] *Then the spectrum $\{\lambda_1, \lambda_2, \ldots\}$ has the following property:*

$$\lim_{n \to \infty} \lambda_n = \infty.$$

Proof. Assume that there exists a real number M such that

$$\lambda_n \leqq M \tag{a}$$

for all n. Then by ***(76.2)***

$$\langle \boldsymbol{u}_n, \boldsymbol{u}_n \rangle_{\mathbf{C}} \leqq M,$$

where $\boldsymbol{u}_1, \boldsymbol{u}_2, \ldots$ is an associated system of normal modes. Thus, since

$$\langle \boldsymbol{u}_n, \boldsymbol{u}_n \rangle_{\mathbf{C}} = 2U\{\boldsymbol{E}_n\},$$

[1] WEYL [1915, *3*]. See also COURANT and HILBERT [1953, *8*], p. 412 and FICHERA [1971, *1*], Theorem 6.V.

[2] Note that we do not include the important case in which the entire boundary is free. That the theorem also holds in that instance is clear from the work of WEYL [1915, *3*] and FICHERA [1971, *1*], Theorem 6.V. In fact, WEYL [1915, *3*] proves that (for a homogeneous and isotropic body) λ_n goes to infinity as $n^{\frac{2}{3}}$. FICHERA (private communication, 1971) has remarked that WEYL's result is easily extended to include inhomogeneous and anisotropic bodies (with symmetric and positive definite $\mathbf{C}$).

where $\boldsymbol{E}_n$ is the strain field corresponding to $\boldsymbol{u}_n$, we conclude from Korn's inequality ***(13.9)*** that there exists a constant K such that

$$\int_B |\nabla \boldsymbol{u}_n|^2 \, dv \leqq K$$

for all n. Consequently, since

$$\|\boldsymbol{u}_n\|^2 = 1, \tag{b}$$

we conclude from Rellich's lemma ***(7.2)*** that there exists a subsequence $\{\boldsymbol{u}_{n_k}\}$ such that

$$\lim_{j,\, k \to \infty} \|\boldsymbol{u}_{n_j} - \boldsymbol{u}_{n_k}\| = 0. \tag{c}$$

But in view of the method of construction of $\boldsymbol{u}_1, \boldsymbol{u}_2, \ldots$ in ***(76.5)***,

$$\langle \boldsymbol{u}_{n_j}, \boldsymbol{u}_{n_k} \rangle = 0, \qquad n_j \neq n_k. \tag{d}$$

By (b) and (d),

$$\|\boldsymbol{u}_{n_j} - \boldsymbol{u}_{n_k}\|^2 = \|\boldsymbol{u}_{n_j}\|^2 + \|\boldsymbol{u}_{n_k}\|^2 = 2,$$

which contradicts (c). Thus there cannot exist a number M that satisfies (a); hence $\lambda_n \to \infty$ as $n \to \infty$. □

The next theorem shows that every vector field $\boldsymbol{f}$ with property (A) below admits an expansion in terms of normal modes.

(A) $\boldsymbol{f}$ is continuous on $\bar{B}$, piecewise smooth on B, and vanishes on $\mathscr{S}_1$; while $\hat{\nabla} \boldsymbol{f}$ is piecewise continuous on $\bar{B}$.

(2) Completeness of the normal modes.[1] *Let $\boldsymbol{u}_1, \boldsymbol{u}_2, \ldots$ be a system of normal modes corresponding to the spectrum $\{\lambda_1, \lambda_2, \ldots\}$, and assume that*

$$\lim_{n \to \infty} \lambda_n = \infty.$$

Let $\boldsymbol{f}$ satisfy property (A). *Then*

$$\lim_{N \to \infty} \Big\| \boldsymbol{f} - \sum_{n=1}^{N} c_n \boldsymbol{u}_n \Big\| = 0,$$

where the c_n are the Fourier coefficients

$$c_n = \langle \boldsymbol{f}, \boldsymbol{u}_n \rangle.$$

Moreover, the c_n satisfy the completeness relation

$$\sum_{n=1}^{\infty} c_n^2 = \|\boldsymbol{f}\|^2.$$

Proof. Let

$$\boldsymbol{g}_N = \boldsymbol{f} - \sum_{n=1}^{N} c_n \boldsymbol{u}_n, \tag{a}$$

with

$$c_n = \langle \boldsymbol{f}, \boldsymbol{u}_n \rangle.$$

Then, since

$$\langle \boldsymbol{u}_m, \boldsymbol{u}_n \rangle = \delta_{mn}, \tag{b}$$

it follows that

$$0 \leqq \|\boldsymbol{g}_N\|^2 = \|\boldsymbol{f}\|^2 - \sum_{n=1}^{N} c_n^2. \tag{c}$$

[1] Cf. COURANT and HILBERT [1953, *8*], p. 424.

Thus the series $\sum_{n=1}^{\infty} c_n^2$ converges and satisfies Bessel's inequality

$$\sum_{n=1}^{\infty} c_n^2 \leqq \|f\|^2. \tag{d}$$

By (a) and (c), to complete the proof it suffices to prove that

$$\|\boldsymbol{g}_N\| \to 0 \quad \text{as} \quad N \to \infty. \tag{e}$$

Suppose that for some M, $\|\boldsymbol{g}_M\|=0$. Then (c) would imply that $c_{M+n}=0$, $n=1, 2, \ldots$, and hence that $\|\boldsymbol{g}_{M+n}\|=0$, $n=1, 2, \ldots$. Thus, in this instance, (e) holds trivially.

Assume now that $\|\boldsymbol{g}_N\|$ never vanishes and let

$$\bar{\boldsymbol{g}}_N = \frac{\boldsymbol{g}_N}{\|\boldsymbol{g}_N\|}, \qquad N=1, 2, \ldots \tag{f}$$

By (a), (b), and (f),

$$\langle \bar{\boldsymbol{g}}_N, \boldsymbol{u}_n \rangle = 0, \qquad n=1, 2, \ldots, N, \tag{g}$$

and we conclude from the properties of $\boldsymbol{u}_n$ and $\boldsymbol{f}$ that

$$\bar{\boldsymbol{g}}_N \in \mathscr{K}_{N+1}, \tag{h}$$

where $\mathscr{K}_{N+1}$ is defined in *(76.5)*. Hence

$$\lambda_{N+1} \leqq \langle \bar{\boldsymbol{g}}_N, \bar{\boldsymbol{g}}_N \rangle_{\mathbf{C}} = \frac{1}{\|\boldsymbol{g}_N\|^2} \langle \boldsymbol{g}_N, \boldsymbol{g}_N \rangle_{\mathbf{C}},$$

or equivalently,

$$\|\boldsymbol{g}_N\|^2 \leqq \frac{\langle \boldsymbol{g}_N, \boldsymbol{g}_N \rangle_{\mathbf{C}}}{\lambda_{N+1}}.$$

Since $\lambda_{N+1} \to \infty$ as $N \to \infty$, to complete the proof it suffices to show that $\langle \boldsymbol{g}_N, \boldsymbol{g}_N \rangle_{\mathbf{C}}$ remains bounded in this limit.

It follows from *(76.1)*, the fact that $[\lambda_n, \boldsymbol{u}_n]$ is a characteristic solution, (g), and (h) that

$$\langle \boldsymbol{g}_N, \boldsymbol{u}_n \rangle_{\mathbf{C}} = 0, \qquad n=1, 2, \ldots, N.$$

Thus, since

$$\langle \boldsymbol{u}_n, \boldsymbol{u}_m \rangle_{\mathbf{C}} = \lambda_n \delta_{mn} \qquad \text{(no sum)},$$

we conclude from (a) that

$$\langle \boldsymbol{f}, \boldsymbol{f} \rangle_{\mathbf{C}} = \langle \boldsymbol{g}_N + \sum_{n=1}^{N} c_n \boldsymbol{u}_n, \boldsymbol{g}_N + \sum_{n=1}^{N} c_n \boldsymbol{u}_n \rangle_{\mathbf{C}}$$

$$= \langle \boldsymbol{g}_N, \boldsymbol{g}_N \rangle_{\mathbf{C}} + \sum_{n=1}^{N} \lambda_n c_n^2.$$

Consequently, as $\lambda_n \geqq 0$,

$$\langle \boldsymbol{g}_N, \boldsymbol{g}_N \rangle_{\mathbf{C}} \leqq \langle \boldsymbol{f}, \boldsymbol{f} \rangle_{\mathbf{C}};$$

hence $\langle \boldsymbol{g}_N, \boldsymbol{g}_N \rangle_{\mathbf{C}}$ remains bounded as $N \to \infty$. □

Under reasonable assumptions on B and on $\mathscr{S}_1$, the set of vector fields with property (A) is dense (with respect to the norm $\|\cdot\|$) in the space of continuous functions on $\bar{B}$. Thus, under such assumptions, *(2)* will hold for functions $\boldsymbol{f}$ that are continuous on $\bar{B}$.

References.

(*Italic* numbers in parentheses following the reference indicate the sections where the work is mentioned.)

1780 *1.* EULER, L.: Determinatio onerum, quae columnae gestare valent. Acta Acad. Sci. Petrop. **2** (1778), 121–145. To be reprinted in L. Euleri Opera Omnia (2) **17**. (*22*)

1822 *1.* FOURIER, J.: Théorie Analytique de la Chaleur. Paris: Didot = Œuvres **1**. (*15*)

1823 *1.* CAUCHY, A.-L.: Recherches sur l'équilibre et le mouvement intérieur des corps solides ou fluides, élastiques ou non élastiques. Bull. Soc. Philomath. 9–13 = Œuvres (2) **2**, 300–304. (*15, 22*)

2. NAVIER, C.-L.-M.-H.: Sur les lois de l'équilibre et du mouvement des corps solides élastiques (1821). Bull. Soc. Philomath. 177–181. (Abstract of [1827, *2*].) (*22, 27, 59*)

1826 *1.* POISSON, S.-D.: Notes sur les racines des équations transcendantes. Bull. Soc. Philomath. 145–148. (*76*)

1827 *1.* CAUCHY, A.-L.: De la pression ou tension dans un corps solide (1822). Ex. de Math. **2**, 42–56 = Œuvres (2) **7**, 60–78. (*15*)

2. NAVIER, C.-L.-M.-H.: Mémoire sur les lois de l'équilibre et du mouvement des corps solides élastiques (1821). Mém. Acad. Sci. Inst. France (2) **7**, 375–393. (*22, 27, 59*)

1828 *1.* CAUCHY, A.-L.: Sur les équations qui expriment les conditions d'équilibre ou les lois du mouvement intérieur d'un corps solide, élastique ou non élastique (1822). Ex. de Math. **3**, 160–187 = Œuvres (2) **8**, 195–226. (*22, 27, 59*)

1829 *1.* CAUCHY, A.-L.: Sur l'équilibre et le mouvement intérieur des corps considérés comme des masses continues. Ex. de Math. **4**, 293–319 = Œuvres (2) **9**, 343–369. (*20, 22*)

2. POISSON, S.-D.: Mémoire sur l'équilibre et le mouvement des corps élastiques (1828). Mém. Acad. Sci. Inst. France (2) **8**, 357–570. (*15, 22, 27, 59, 67, 76*)

3. — Addition au mémoire sur l'équilibre et le mouvement des corps élastiques. Mém. Acad. Sci. Inst. France. (2) **8**, 623–627. (*67*)

1830 *1.* CAUCHY, A.-L.: Sur les diverses méthodes à l'aide desquelles on peut établir les équations qui représentent les lois d'équilibre, ou le mouvement intérieur des corps solides ou fluides. Bull. Sci. Math. Soc. Prop. Conn. **13**, 169–176. (*22*)

1831 *1.* POISSON, S.-D.: Mémoire sur les équations générales de l'équilibre et du mouvement des corps solides élastiques et des fluides (1829). J. École Polytech. **13**, cahier 20, 1–174. (*20, 22*)

1833 *1.* LAMÉ, G., and E. CLAPEYRON: Mémoire sur l'équilibre des corps solides homogènes. Mém. Divers Savants Acad. Sci. Paris (1828) (2) **4**, 465–562. (*27, 59*)

2. PIOLA, G.: La meccanica dei corpi naturalmente estesi trattata col calcolo delle variazioni. Opusc. Mat. Fis. di Diversi Autori. Milano: Giusti **1**, 201–236. (*18*)

1839 *1.* GREEN, G.: On the laws of reflection and refraction of light at the common surface of two non-crystallized media (1837). Trans. Cambridge Phil. Soc. **7** (1838–1842), 1–24 = Papers, 245–269. (*24, 34, 67*)

1840 *1.* CAUCHY, A. L.: Exercises d'Analyse et de Physique Mathématique **1**. Paris: Bachelier = Œuvres (2) **12** (*59, 67, 71*)

1841 *1.* GREEN, G.: On the propagation of light in crystallized media (1839). Trans. Cambridge Phil. Soc. **7** (1838–1842), 121–140 = Papers 293–311. (*24, 71*)

2. — Supplement to a memoir on the reflexion and refraction of light. Trans. Cambridge Phil. Soc. **7**, (1838–1842), 113–120 = Papers 283–290.

1845 *1.* STOKES, G. G.: On the theories of the internal friction of fluids in motion, and of the equilibrium and motion of elastic solids. Trans. Cambridge Phil. Soc. **8** (1844–1849), 287–319 = Papers 1, 75–129. (*22, 27, 59*)

1848 *1.* PIOLA, G.: Intorno alle equazioni fondamentali del movimento di corpi qualsivogliono, considerati secondo la naturale loro forma e costituzione (1845). Mem. Mat. Fis. Soc. Ital. Modena **24**[1], 1–186. (*18*)

2. STOKES, G. G.: On a difficulty in the theory of sound. Phil. Mag. **23**, 349–356. A drastically condensed version, with an added note, appears in Papers **2**, 51–55. (*73*)

3. THOMSON, W. (Lord KELVIN): On the equations of equilibrium of an elastic solid. Cambr. Dubl. Math. J. **3**, 87–89. (*51*)

1849 *1.* HAUGHTON, S.: On the equilibrium and motion of solid and fluid bodies (1846). Trans. Roy. Irish Acad. **21**, 151–198. (*34*)

1850 *1.* KIRCHHOFF, G.: Über das Gleichgewicht und die Bewegung einer elastischen Scheibe. J. Reine Angew. Math. **40**, 51–88 = Ges. Abh. 237–279. (*34*)

1851 *1.* STOKES, G. G.: On the dynamical theory of diffraction (1849). Trans. Cambridge Phil. Soc. **9**, 1–62 = Papers **2**, 243–328. (*59, 68*)

2. — On the effect of the internal friction of fluids on the motion of pendulums (1850). Trans. Cambridge Phil. Soc. **9**^{2}, 8–106 = Papers **3**, 1–141. (*19*)

1852 *1.* KIRCHHOFF, G.: Über die Gleichungen des Gleichgewichts eines elastischen Körpers bei nicht unendlich kleinen Verschiebungen seiner Theile. Sitzgsber. Akad. Wiss. Wien **9**, 762–773. (Not reprinted in Ges. Abh.) (*65*)

2. LAMÉ, G.: Leçons sur la Théorie Mathématique de l'Élasticité. Paris: Bachelier. Also 2nd ed., Paris 1866. (*22, 27, 28, 44, 59, 67, 71*)

1855 *1.* ST. VENANT, A.-J.-C. B., DE: Mémoire sur la torsion des prismes ... (1853). Mém. Divers Savants Acad. Sci. Paris **14**, 233–560. (*54*)

1859 *1.* KIRCHHOFF, G.: Über das Gleichgewicht und die Bewegung eines unendlich dünnen elastischen Stabes. J. Reine Angew. Math. **56**, 285–313 = Ges. Abh., 285–316. (*13, 14, 32, 65*)

1860 *1.* RIEMANN, B.: Über die Fortpflanzung ebener Luftwellen von endlicher Schwingungsweite. Gött. Abh., Math. Cl. **8** (1858–1859), 43–65 = Werke, 145–164. (*73*)

1862 *1.* CLEBSCH, A.: Theorie der Elasticität fester Körper. Leipzig: Teubner. There is an expanded version in French by B. DE SAINT VENANT and A. FLAMANT, Paris 1883. (*32, 47, 76*)

1863 *1.* AIRY, G. B.: On the strains in the interior of beams. Phil. Trans. Roy. Soc. Lond. **153**, 49–80. Abstract in Rep. Brit. Assn. 1862, 82–86 (1863). (*17, 47*)

2. CLEBSCH, A.: Über die Reflexion an einer Kugelfläche. J. Reine Angew. Math. **61**, 195–262. (*67*)

3. THOMSON, W. (Lord KELVIN): Dynamical problems regarding elastic spheroidal shells and spheroids of incompressible liquid. Phil. Trans. Roy. Soc. Lond. **153**, 583–616 = Papers **3**, 351–394. (*34, 48*)

1864 *1.* ST. VENANT, A.-J.-C. B. DE: Établissement élémentaire des formules et équations générales de la théorie de l'élasticité des corps solides. Appendix in: Résumé des Leçons données à l'École des Ponts et Chaussées sur l'Application de la Mécanique, première partie, première section, De la Résistance des Corps Solides, par C.-L.-M.-H. NAVIER, 3rd ed. Paris. (*14*)

1865 *1.* COTTERILL, J. H.: Further application of the principle of least action. Phil. Mag. (4), **29**, 430–436. (*18, 34, 36*)

1866 *1.* VERDET, E.: Introduction, Œuvres de **Fresnel 1**, IX–XCIX. (*22*)

1868 *1.* FRESNEL, A.: Second supplément au mémoire sur la double réfraction (1822). Œuvres **2**, 369–442. (*15*)

2. MAXWELL, J. C.: On reciprocal diagrams in space, and their relation to Airy's function of stress. Proc. Lond. Math. Soc. (1) **2** (1865–1969), 58–60 = Papers **2**, 102–104. (*17*)

1870 *1.* MAXWELL, J. C.: On reciprocal figures, frames, and diagrams of forces. Trans. Roy. Soc. Edinburgh **26** (1869–1872), 1–40 = Papers **2**, 161–207. (*17, 46, 47*)

2. RANKINE, W. J. M.: On the thermodynamic theory of waves of finite longitudinal disturbance. Phil. Trans. Roy. Soc. Lond. **160**, 277–288 = Papers, 530–543. (*73*)

1871 *1.* BOUSSINESQ, J.: Étude nouvelle sur l'équilibre et le mouvement des corps solides élastiques dont certaines dimensions sont très petites par rapport à d'autres. Premier Mémoire. J. Math. Pures Appl. (2) **16**, 125–240. (*14*)

1872 *1.* BETTI, E.: Teoria dell'Elasticità. Nuovo Cim. (2) **7–8**, 5–21, 69–97, 158–180; **9**, 34–43; **10** (1873), 58–84 = Opere **2**, 291–390. (*27, 29, 30, 44, 52, 53*)

1873 *1.* MAXWELL, J. C.: A Treatise on Electricity and Magnetism. 2 vols. Oxford. Cf. [1881, **1**]. (*72*)

1874 *1.* BETTI, E.: Sopra l'equazioni di equilibrio dei corpi solidi elastici. Ann. Mat. (2) **6**, 101–111 = Opere **2**, 379–390. (*30*)

2. UMOV, N.: Ableitung der Bewegungsgleichungen der Energie in continuirlichen Körpern. Z. Math. Phys. **19**, 418–431. (*19*)

1876 *1.* KIRCHHOFF, G.: Vorlesungen über mathematische Physik: Mechanik. Leipzig. 2nd ed., 1877; 3rd ed., 1883. (*14, 26, 32, 65*)

1877 *1.* CHRISTOFFEL, E. B.: Untersuchungen über die mit dem Fortbestehen linearer partieller Differentialgleichungen verträglichen Unstetigkeiten. Ann. Mat. (2) **8**, 81–113 = Abh. **2**, 51–80. (*72, 73*)

2. — Über die Fortpflanzung von Stößen durch elastische feste Körper. Ann. Mat. (2) **8**, 193–244 = Abh. **1**, 81–126. (*73*)

1878 *1.* BOUSSINESQ, J.: Équilibre d'élasticité d'un solide isotrope sans pesanteur, supportant différents poids. C. R. Acad. Sci., Paris **86**, 1260–1263. (*44*)

1881 *1.* MAXWELL, J. C.: 2nd ed. of [1873, *1*]. (*72*)

1882 *1.* Voigt, W.: Allgemeine Formeln für die Bestimmung der Elasticitätsconstanten von Krystallen durch die Beobachtung der Biegung und Drillung von Prismen. Ann. Phys. (2) **16**, 273–321, 398–416. (*26, 47*)

1883 *1.* Thomson, W. (Lord Kelvin), and P. G. Tait: Treatise on Natural Philosophy, Part 2. Cambridge. (*44, 51*)

1884 *1.* Minnigerode, B.: Untersuchungen über die Symmetrieverhältnisse und die Elasticität der Krystalle. Nach Königl. Ges. Wiss. Göttingen, 195–226, 374–384, 488–492. (*26*)

2. Weyrauch, J. J.: Theorie elastischer Körper. Leipzig: Teubner. (*71*)

1885 *1.* Boussinesq, J.: Application des potentiels à l'étude de l'équilibre et des mouvements des solides élastiques. Paris: Gauthier-Villars. (*44, 51, 54*)

2. Kovalevski, S.: Über die Brechung des Lichtes in cristallinischen Mitteln. Acta Math. **6**, 249–304. (*67*)

3. Neumann, F.: Vorlesungen über die Theorie der Elastizität. Leipzig: Teubner. (*26, 63*)

4. Somigliana, C.: Sopra l'equilibrio di un corpo elastico isotropo. Nuovo Cimento (3) **17**, 140–148, 272–276; **18**, 161–166. (*51, 52*)

1886 *1.* Beltrami, E.: Sull'interpretazione meccanica delle formole di Maxwell. Mem. Accad. Sci. Bologna (4) **7**, 1–38 = Opere **4**, 190–223. (*14*)

2. Hugoniot, H.: Sur un théorème général relatif à la propagation du mouvement. C.R. Acad. Sci. Paris **102**, 858–860. (*73*)

3. Ibbetson, W. J.: On the Airy-Maxwell solution of the equations of equilibrium of an isotropic elastic solid, under conservative forces. Proc. Lond. Math. Soc. **17**, 296–309. (*47*)

4. Somigliana, C.: Sopra l'equilibrio di un corpo elastico isotropo. Nuovo Cimento (3) **19**, 84–90, 278–282; **20**, 181–185. (*52*)

1887 *1.* Hugoniot, H.: Mémoire sur la propagation du mouvement dans les corps et spécialement dans les gaz parfaits. J. École Polytech. **57**, 3–97; **58**, 1–125 (1889). The date of the memoir is 26 October 1885. (*73*)

2. Ibbetson, W. J.: An Elementary Treatise on the Mathematical Theory of Perfectly Elastic Solids with a Short Account of Viscous Fluids. London: MacMillan. (*47, 76*)

3. Voigt, W.: Theoretische Studien über die Elasticitätsverhältnisse der Krystalle. Abh. Ges. Wiss. Göttingen **34**, 100 pp. (*26*)

1888 *1.* Boussinesq, J.: Équilibre d'élasticité d'un solide sans pesanteur, homogène et isotrope, dont les parties profondes sont maintenues fixes, pendant que sa surface éprouve des pressions ou des déplacements connus, s'annulant hors d'une région restreinte où ils sont arbitraires. C.R. Acad. Sci., Paris **106**, 1043–1048. (**44**)

2. — Équilibre d'élasticité d'un solide sans pesanteur, homogène et isotrope, dont les parties profondes sont maintenues fixes, pendant que sa surface éprouve des pressions ou des déplacements connus, s'annulant hors d'une région restreinte où ils sont arbitraires – cas où les donées sont mixtes, c'est-à-dire relatives en partie aux pressions et en partie aux déplacements. C.R. Acad. Sci., Paris **106**, 1119–1123. (*44*)

3. Lévy, M.: Sur une propriété générale des corps solides élastiques. C.R. Acad. Sci. Paris **107**, 414–416. (*30*)

4. Thomson, W. (Lord Kelvin): On the reflection and refraction of light. Phil. Mag. (5) **26**, 414–425. Reprinted in part as Sections 107–111 of Baltimore Lectures on Molecular Dynamics and the Wave Theory of Light. London: Clay and Sons 1904. (*32, 57*)

1889 *1.* Beltrami, E.: Sur la théorie de la déformation infiniment petite d'un milieu. C.R. Acad. Sci., Paris **108**, 502–504 = Opere **4**, 344–347. (*14*)

2. Padova, E.: Sulle deformazioni infinitesimi. Atti Accad. Lincei Rend. (4) $\mathbf{5}_2$, 174–178. (*14*)

3. Somigliana, C.: Sulle equazioni della elasticità. Ann. Mat. (2) **17**, 37–64. (*27, 44, 52*)

1890 *1.* Donati, L.: Illustrazione al teorema del Menabrea. Mem. Accad. Sci. Bologna (4) **10**, 267–274. (*18, 34, 36*)

2. Poincaré, H.: Sur les équations aux dérivées partielles de la physique mathématique. Am. J. Math. **12**, 211–294. (*76*)

3. Fontaneau, M. E.: Sur l'équilibre d'élasticité d'une enveloppe sphérique. Nouv. Ann. de Math. (3) **9**, 455–471. (*44*)

1891 *1.* Schoenfliess, A.: Kristallsysteme und Kristallstruktur. Leipzig. (*21*)

1892 *1.* Beltrami, E.: Osservazioni sulla nota precedente. Atti Accad. Lincei Rend. (5) $\mathbf{1}_1$, 141–142 = Opere **4**, 510–512. (*17, 27*)

2. Chree, C.: Changes in the dimensions of elastic solids due to given systems of forces. Cambridge Phil. Soc. Trans. **15**, 313–337. (*18, 29*)

3. FONTANEAU, M. E.: Sur la déformation des corps isotropes en équilibre d'élasticité. Assoc. Fran. Avan. Sci., C. R. 21st Sess. (2), 190–207. (*44*)
4. MORERA, G.: Soluzione generale delle equazioni indefinite dell'equilibrio di un corpo continuo. Atti Accad. Lincei Rend. (5) $\mathbf{1}_1$, 137–141. (*17*)
5. MORERA, G.: Appendice alla Nota: Sulla soluzione più generale delle equazioni indefinite dell'equilibrio di un corpo continuo. Atti Accad. Lincei Rend. (5) $\mathbf{1}_1$, 233–234. (*17*)
6. POINCARÉ, H.: Leçons sur la théorie de l'élasticité. Paris. (*76*)
7. SOMIGLIANA, C.: Sulle espressioni analitiche generali dei movimenti oscillatori. Atti Accad. Lincei Rend. (5) $\mathbf{1}^1$, 111–119. (*67*)

1894 1. CESÀRO, E.: Introduzione alla teoria matematica della elasticità. Torino: Fratelli, Bocca. (*76*)
2. DONATI, L.: Ulteriori osservazioni intorno al teorema del Menabrea. Mem. Accad. Sci. Bologna (5) **4**, 449–474. (*18, 27, 34, 36*)
3. SOMIGLIANA, C.: Sopra gli integrali delle equazioni della isotropia elastica. Nuovo Cimento (3) **36**, 28–39. (*44*)
4. VOLTERRA, V.: Sur les vibrations des corps élastiques isotropes. Acta Math. **18**, 161–232. (*69*)

1895 1. LAURICELLA, G.: Equilibrio dei corpi elastici isotropi. Ann. Scuola Norm. Pisa **7**, 1–120. (*53*)

1896 1. COSSERAT, E., et F.: Sur la théorie de l'élasticité. Ann. Fac. Sci. Toulouse **10**, 1–116. (*14*)

1897 1. TEDONE, O.: Sulle vibrazioni dei corpi solidi, omogenei ed isotropi. Mem. Accad. Sci. Torino (2) **47**, 181–258. (*67*)

1898 1. COSSERAT, E., et F.: Sur la déformation infiniment petite d'une ellipsoïde élastique. C. R. Acad. Sci., Paris **127**, 315–318. (*32*)
2. — Sur les équations de la théorie de l'élasticité. C. R. Acad. Sci., Paris **126**, 1089–1091. (*32*)
3. — Sur les fonctions potentielles de la théorie de l'élasticité. C. R. Acad. Sci., Paris **126**, 1129–1132. (*32*)
4. DOUGALL, J.: A general method of solving the equations of elasticity. Edinburgh Math. Soc. Proc. **16**, 82–98. (*51, 52, 53*)
5. DUHEM, P.: Sur l'intégrale des équations des petits mouvements d'un solide isotrope. Mém. Soc. Sci. Bordeaux (5) **3**, 317–329. (*67*)
6. GOURSAT, É.: Sur l'équation $\Delta\Delta u = 0$. Bull. Soc. Math. France **26**, 236–237. (*47*)
7. LÉVY, M.: Sur la légitimité de la règle dite du trapèze dans l'étude de la résistance des barrages en maçonnerie. C. R. Acad. Sci., Paris **126**, 1235. (*46, 47*)

1899 1. POINCARÉ, H.: Théorie du potentiel Newtonien. Paris: Gauthier-Villars. (*8*)

1900 1. DUHEM, P.: Sur le théorème d'Hugoniot et quelques théorèmes analogues. C. R. Acad. Sci., Paris **131**, 1171–1173. (*72*)
2. FREDHOLM, I.: Sur les équations de l'équilibre d'un corps solide élastique. Acta Math. **23**, 1–42. (*51*)
3. MICHELL, J. H.: On the direct determination of stress in an elastic solid, with applications to the theory of plates. Proc. Lond. Math. Soc. **31** (1899), 100–124. (*27, 45, 46, 47*)
4. — The uniform torsion and flexure of incomplete tores, with application to helical springs. Proc. Lond. Math. Soc. **31**, 130–146. (*44*)
5. VOIGT, W.: Elementare Mechanik. Leipzig: Veit. (*44*)
6. — L'Élasticité des Cristaux. Rapports Congrès Int. Phys. **1**. Paris: Gauthier-Villars, 277–347. (*26*)

1901 1. ABRAHAM, M.: Geometrische Grundbegriffe. In: Vol. 4_3 of the Encyklopädie der Mathematischen Wissenschaften, edited by F. KLEIN and C. MÜLLER. Leipzig: Teubner. (*14*)
2. COSSERAT, E., et F.: Sur la déformation infiniment petite d'un corps élastique soumis à des forces données. C. R. Acad. Sci., Paris **133**, 271–273, 400. (*32*)
3. — Sur la déformation infiniment petite d'un ellipsoïde élastique soumis à des efforts donnés sur la frontière. C. R. Acad. Sci., Paris **133**, 361–364. (*32*)
4. — Sur la déformation infiniment petite d'une enveloppe sphérique élastique. C. R. Acad. Sci., Paris **133**, 326–329. (*32*)
5. HADAMARD, J.: Sur la propagation des ondes. Bull. Soc. Math. France **29**, 50–60. (*72*)
6. JOUGUET, E.: Sur la propagation des discontinuités dans les fluides. C. R. Acad. Sci., Paris **132**, 673–676. (*73*)
7. WEINGARTEN, G.: Sulle superficie di discontinuità nella teoria della elasticità dei corpi solidi. Atti Accad. Lincei Rend. (5) $\mathbf{10}^1$, 57–60. (*14, 72*)

1903 *1.* Boggio, T.: Sull'integrazione di alcune equazioni lineari alle derivate parziali. Ann. Mat. (3) **8**, 181–231. (*67*)
2. Filon, L. N. G.: On an approximative solution for the bending of a beam of rectangular cross-section under any system of load, with special reference to points of concentrated or discontinuous loading. Phil. Trans. Roy Soc. Lond., Ser. A **201**, 63–155. (*45, 47*)
3. Hadamard, J.: Leçons sur la Propagation des Ondes et les Équations de l'Hydrodynamique. Paris: Hermann. (*34, 72, 73*)
4. Tedone, O.: Saggio di una teoria generale delle equazioni dell'equilibrio elastico per un corpo isotropo. Ann. Mat. (3) **8**, 129–180. (*27*)
1904 *1.* Dougall, J.: The equilibrium of an isotropic elastic plate. Trans. Roy. Soc. Edinburgh **41**, 129–228. (*53*)
2. Duhem, P.: Recherches sur l'élasticité. Ann. École Norm. (3) **21**, 99–139, 375–414; **22** (1905), 143–217; *23* (1906), 169–223 = repr. separ., Paris 1906. (*73*)
3. Love, A. E. H.: The propagation of wave-motion in an isotropic elastic solid medium. Proc. Lond. Math. Soc. (2) **1**, 291–344. (*68, 69*)
4. Tedone, O.: Saggio di una teoria generale delle equazioni dell'equilibrio elastico per un corpo isotropo. Ann. Mat. (3) **10**, 13–64. (*27, 44*)
5. Thomson, W. (Lord Kelvin): Baltimore Lectures on Molecular Dynamics and the Wave Theory of Light. London: Clay and Sons. (*67*)
1905 *1.* Klein, F., and K. Wieghardt: Über Spannungsflächen und reziproke Diagramme, mit besonderer Berücksichtigung der Maxwellschen Arbeiten. Arch. Math. Phys. (3) **8**, 1–10, 95–119. (*17*)
2. Timpe, A.: Probleme der Spannungsverteilung in ebenen Systemen einfach gelöst mit Hilfe der Airyschen Funktion. Dissertation. Göttingen. (*14*)
3. Zemplén, G.: Kriterien für die physikalische Bedeutung der unstetigen Lösungen der hydrodynamischen Bewegungsgleichungen. Math. Ann. **61**, 437–449. (*73*)
4. — Besondere Ausführungen über unstetige Bewegungen in Flüssigkeiten. In: Vol. 4^4 of the Encyklopädie der Mathematischen Wissenschaften, edited by F. Klein and C. Müller. Leipzig: Teubner. (*73*)
1906 *1.* Born, M.: Untersuchungen über die Stabilität der elastischen Linie in Ebene und Raum unter verschiedenen Grenzbedingungen. Dissertation. Göttingen. (*38*)
2. Cesàro, E.: Sulle formole del Volterra, fondamentali nella teoria delle distorsioni elastiche. Rend. Napoli (3a) **12**, 311–321. (*14*)
3. Fredholm, I.: Solution d'un problème fondamental de la théorie de l'élasticité. Ark. Mat. Astr. Fys. **2**, 1–8. (*32*)
4. Korn, A.: Die Eigenschwingungen eines elastischen Körpers mit ruhender Oberfläche. Akad. Wiss., München, Math. phys. Kl. Sitz. **36**, 351–402. (*13*)
5. Love, A. E. H.: A Treatise on the Mathematical Theory of Elasticity, 2nd ed. Cambridge. (*17, 34, 57*)
6. Somigliana, C.: Sopra alcune formole fondamentali della dinamica dei mezzi isotropi. Atti Accad. Sci. Torino **41**, 869–885, 1070–1090. (*69*)
1907 *1.* Boggio T.: Nuova risoluzione di un problema fondamentale della teoria dell' elasticità. Atti Accad. Lincei Rend. **16**, 248–255. (*32*)
2. Tedone, O.: Allgemeine Theoreme der mathematischen Elastizitätslehre (Integrationstheorie). In Vol. 4 of the Encyklopädie der Mathematischen Wissenschaften, edited by F. Klein and C. Müller. Leipzig: Teubner. (*76*)
3. —, and A. Timpe: Spezielle Ausführungen zur Statik elastischer Körper. In Vol. 4 of the Encyklopädie der Mathematischen Wissenschaften, edited by F. Klein and C. Müller. Leipzig: Teubner. (*44*)
4. Volterra, V.: Sur l'équilibre des corps élastiques multiplement connexes. Ann. École Norm. (3) **24**, 401–517. (*14*)
1908 *1.* Korn, A.: Solution générale du problème d'équilibre dans la théorie de l'élasticité, dans le cas où les efforts sont donnés à la surface. Ann. Fac. Sci. Toulouse (2) **10**, 165–269. (*13*)
2. Ritz, W.: Über eine neue Methode zur Lösung gewisser Variationsprobleme der mathematischen Physik. J. Reine Angew. Math. **135**, 1–61 = Œuvres 192–264. (*39*)
3. Routh, E. T.: A Treatise on Analytical Statics, 2nd ed. Cambridge. (*18*)
1909 *1.* Mises, R. v.: Theorie der Wasserräder. Z. Math. Phys. **57**, 1–120. (*19*)
2. Kolosov, G. V.: On an application of complex function theory to a plane problem in the mathematical theory of elasticity (in Russian). Dissertation at Dorpat (Yuriev) University. (*47*)
3. Korn, A.: Über einige Ungleichungen, welche in der Theorie der elastischen und elektrischen Schwingungen eine Rolle spielen. Akad. Umiejęt. Krakow Bulletin Int. 705–724. (*13*)

1910 *1.* VOIGT, W.: Lehrbuch der Kristallphysik. Leipzig: Teubner. (*21, 26*)

1911 *1.* ZEILON, N.: Das Fundamentalintegral der allgemeinen partiellen linearen Differentialgleichung mit konstanten Koeffizienten. Ark. Mat. Ast. Fys. **6**, 38, 1–32. (*51*)

1912 *1.* COLONNETTI, G.: L'equilibrio elastico dal punto di vista energetico. Mem. Accad. Sci. Torino (2) **62**, 479–495. (*34*)

2. GWYTHER, R. F.: The formal specification of the elements of stress in cartesian, and in cylindrical and spherical polar coordinates. Mem. Manchester Lit. Phil. Soc. **56**, No. 10 (13 pp.). (*17*)

1913 *1.* GWYTHER, R. F.: The specification of the elements of stress. Part II. A simplification of the specification given in Part I. Mem. Manchester Lit. Phil. Soc. **57**, No. 5 (4 pp.) (*17*)

1914 *1.* HELLINGER, E.: Die allgemeinen Ansätze der Mechanik der Kontinua. In: Vol. 4^4 of the Encyklopädie der Mathematischen Wissenschaften, edited by F. KLEIN and C. MÜLLER. Leipzig: Teubner. (*38*)

2. KOLOSOV, G. V.: Über einige Eigenschaften des ebenen Problems der Elastizitätstheorie. Z. Math. Phys. **62**, 383–409. (*47*)

1915 *1.* DOMKE, O.: Über Variationsprinzipien in der Elastizitätslehre nebst Anwendungen auf die technische Statik. Z. Math. Phys. **63**, 174–192. (*34, 36*)

2. KORN, A.: Über die beiden bisher zur Lösung der ersten Randwertaufgabe der Elastizitätstheorie eingeschlagenen Wege. Annaes. Acad. Poly. Porto **10**, 129–156. (*44*)

3. WEYL, H.: Das asymptotische Verteilungsgesetz der Eigenschwingungen eines beliebig gestalteten elastischen Körpers. Rend. Circ. Mat. Palermo **39**, 1–50. (*57, 78*)

4. ZAREMBA, S.: Sopra un teorema d'unicitá relativo alla equazione delle onde sferiche. Atti Accad. Lincei Rend. (5) **24**, 904–908. (**74**)

1916 *1.* PRANGE, G.: Die Variations- und Minimalprinzipe der Statik der Baukonstruktionen. Habilitationsschrift. Tech. Univ. Hannover. (*34, 38*)

1919 *1.* MUSKHELISHVILI, N. I.: Sur l'intégration de l'équation biharmonique. Izv. Akad. Nauk SSSR. 663–686. (*47*)

2. SERINI, R.: Deformazioni simmetriche dei corpi elastici. Atti Accad. Lincei Rend. (5) **28**, 343–347. (*44*)

1920 *1.* COURANT, R.: Über die Eigenwerte bei den Differentialgleichungen der mathematischen Physik. Math. Z. **7**, 1–57. (*76, 77*)

2. RUBINOWICZ, A.: Herstellung von Lösungen gemischter Randwertprobleme bei hyperbolischen Differentialgleichungen zweiter Ordnung durch Zusammenstückelung aus Lösungen einfacher gemischter Randwertaufgaben. Mon. Math. Phys. **30**, 65–79. (*74*)

1923 *1.* BURGATTI, P.: Sopra una soluzione molto generale dell'equazioni dell'equilibrio elastico. Atti Accad. Bologna Rend. (2) **27**, 66–73. (*44*)

2. SOUTHWELL, R.: On Castigliano's theorem of least work and the principle of Saint-Venant. Phil. Mag. (6) **45**, 193–212. (*54*)

1924 *1.* LICHTENSTEIN, L.: Über die erste Randwertaufgabe der Elastizitätstheorie. Math. Z. **20**, 21–28. (*27, 53*)

2. TIMPE, A.: Achsensymmetrische Deformation von Umdrehungskörpern. Z. Angew. Math. Mech. **4**, 361–376. (*44*)

1925 *1.* WEBER, C.: Achsensymmetrische Deformation von Umdrehungskörpern. Z. Angew. Math. Mech. **5**, 477–468. (*44*)

1926 *1.* BURGATTI, P.: Sopra due utili forme dell'integrale generale dell'equazione per l'equilibrio dei solidi elastici isotropi. Mem. Accad. Sci. Bologna (8) **3**, 63–67. (*44*)

2. KOTCHINE, N. E.: Sur la théorie des ondes de choc dans un fluide. Rend. Circ. Mat. Palermo **50**, 305–344. (*73*)

1927 *1.* AUERBACH, F.: Elastizität der Kristalle. Handbuch der Physikalischen und Technischen Mechanik **3**, 239–282. Leipzig: Barth. (*26*)

2. KORN, A.: Allgemeine Theorie der Elastizität. Handbuch der Physikalischen und Technischen Mechanik **3**, 1–47. Leipzig: Barth. (*27, 53*)

3. LOVE, A. E. H.: A Treatise on the Mathematical Theory of Elasticity, 4th ed. Cambridge. (*18, 26, 27, 29, 44, 45, 47, 51, 52, 53, 54, 65, 68, 73*)

1928 *1.* FRIEDRICHS, K. O., and H. LEWY: Über die Eindeutigkeit und das Abhängigkeitsgebiet der Lösungen beim Anfangswertproblem linearer hyperbolischer Differentialgleichungen. Math. Ann. **98**, 192–204. (*74*)

2. GECKELER, J. W.: Elastizitätstheorie anisotroper Körper (Kristallelastizität). In: Volume VI of the Handbuch der Physik, edited by R. GRAMMEL. Berlin: Springer. (*26*)

3. TREFFTZ, E.: Mathematische Elastizitätstheorie. In: Volume VI of the Handbuch der Physik, edited by R. GRAMMEL. Berlin: Springer. (*34, 44, 47, 51, 53*)

4. — Konvergenz und Fehlerschätzung beim Ritzschen Verfahren. Math. Ann. **100**, 503–521. (*37*)

1929 *1.* Kellogg, O. D.: Foundations of Potential Theory. Berlin: Springer. (*5, 6, 8, 52*)

2. Lichtenstein, L.: Grundlagen der Hydromechanik. Berlin: Springer. (*6, 72*)

1930 *1.* Filon, L. N. G.: On the relation between corresponding problems in plane stress and in generalized plane stress. Q. J. Appl. Math. **1**, 289–299. (*45*)

2. Galerkin, B.: On an investigation of stresses and deformations in elastic isotropic solids [in Russian]. Dokl. Akad. Nauk SSSR, 353–358. (*44*)

3. — Contribution à la solution générale du problème de la théorie de l'élasticité dans le cas de trois dimensions. C. R. Acad. Sci., Paris **190**, 1047–1048. (*44*)

4. Rellich, F.: Ein Satz über mittlere Konvergenz. Nachr. Ges. Wissen. Göttingen. Math.-Phys. Kl., 30–35. (*7*)

1931 *1.* Burgatti, P.: Teoria Matematica della Elasticità. Bologna: Zanichelli. (*30*)

2. Coker, E. G., and L. N. G. Filon: A Treatise on Photoelasticity. Cambridge. (*45*)

3. Galerkin, B.: On the general solution of a problem in the theory of elasticity in three dimensions by means of stress and displacement functions [in Russian]. Dokl. Akad. Nauk SSSR, 281–286. (*44*)

4. Hobson, E. W.: The Theory of Spherical and Ellipsoidal Harmonies. Cambridge. (*8*)

5. Supino, G.: Sopra alcune limitazioni per la sollecitazione elastica e sopra la dimostrazione del principio del De Saint Venant. Ann. Mat. (4) **9**, 91–119. (*54*)

1932 *1.* Muskhelishvili, N. I.: Recherches sur les problèmes aux limites relatifs à l'équation biharmonique et aux équations de l'élasticité à deux dimensions. Math. Ann. **107**, 282–312. (*47*)

2. Papkovitch, P. F.: Expressions générales des composantes des tensions, ne renfermant comme fonctions arbitraires que des fonctions harmoniques. C.R. Acad. Sci., Paris **195**, 754–756. (*44*)

3. — Solution générale des équations différentielles fondamentales d'élasticité, exprimée par trois fonctions harmoniques. C. R. Acad. Sci., Paris **195**, 513–515. (*44*)

4. — The representation of the general integral of the fundamental equations of the theory of elasticity in terms of harmonic functions [in Russian]. Izv. Akad. Nauk SSSR, Fiz.-Mat. Ser. **10**, 1425–1435. (*44*)

5. Signorini, A.: Sollecitazioni iperastatiche. Rend. Ist. Lombardo (2) **65**, 1–7. (*18*)

1933 *1.* Marguerre, K.: Ebenes und achsensymmetrisches Problem der Elastizitätstheorie. Z. Angew. Math. Mech. **13**, 437–438. (*44*)

2. Muskhelishvili, N. I.: Recherches sur les problèmes aux limites relatifs à l'équation biharmonique et aux équations de l'élasticité à deux dimensions. Math. Ann. **107**, 282–312. (*32*)

3. Phillips, H. B.: Vector Analysis. New York: Wiley. (*6*)

4. Signorini, A.: Sopra alcune questioni di statica dei sistemi continui. Ann. Scuola Norm. Pisa (2) **2**, 231–257. (*18*)

1934 *1.* Biezeno, C.: Über die Marguerresche Spannungsfunktion. Ing.-Arch. **5**, 120–124. (*44*)

2. Finzi, B.: Integrazione delle equazioni indefinite della meccanica dei sistemi continui. Atti Accad. Lincei Rend. (6) **19**, 578–584, 620–623. (*17*)

3. Muskhelishvili, N. I.: A new general method of solution of the fundamental boundary problems of the plane theory of elasticity [in Russian]. Dokl. Akad. Nauk SSSR **3**, 7–9. (*47*)

4. Neuber, H.: Ein neuer Ansatz zur Lösung räumlicher Probleme der Elastizitätstheorie. Der Hohlkegel unter Einzellast als Beispiel. Z. Angew. Math. Mech. **14**, 203–212. (*44*)

5. Phillips, H. B.: Stress functions. J. Math. Phys. **13**, 421–425. (*47*)

6. Sobrero, L.: Nuovo metodo per lo studio dei problemi di elasticità con applicazione al problema della piastra forata. Ric. Ingegneria **2**, 255–264. (*44*)

1935 *1.* Grodski, G. D.: The integration of the general equations of equilibrium of an isotropic elastic body by means of the Newtonian potential and harmonic functions [in Russian]. Izv. Akad. Nauk SSSR Otdel Mat. **4**, 587–614. (*44*)

2. Kolosov, G. V.: An Application of the Complex Variable in the Theory of Elasticity [in Russian]. Moscow-Leningrad: Gostehizdat. (*47*)

3. Marguerre, K.: Ebenes und achsensymmetrisches Problem der Elastizitätstheorie. Z. Angew. Math. Mech. **13**, 437–438. (*44*)

4. Neuber, H.: Der räumliche Spannungszustand in Umdrehungskerben. Ing. Arch. **6**, 133–156. (*44*)

5. Nielsen, T.: Elementare Mechanik. Berlin: Springer. (*13*)

6. Sobolev, S. L.: General theory of the diffraction of waves on Riemann surfaces. Trudy Mat. Inst. Steklov **9**, 39–105. (*42*)

7. SOBRERO, L.: Del significato meccanico della funzione di Airy. Ric. Ingegneria **3**, 77–80 = Atti. Accad. Lincei Rend. (6) **21**, 264–269. (*17*)
8. — Delle funzioni analoghe al potentiale intervenienti nella fisica-matematica. Atti. Accad. Lincei. Rend. (6) **21**, 448–454. (*44, 47*)
9. WESTERGAARD, H.: General solution of the problem of elastostatics of an n-dimensional homogeneous isotropic solid in an n-dimensional space. Bull. Amer. Math. Soc. **41**, 695–699. (*44*)

1936 *1.* MINDLIN, R. D.: Force at a point in the interior of a semi-infinite solid. Phys. **7**, 195–202. (*51*)
2. MINDLIN R.: Note on the Galerkin and Papkovitch stress functions. Bull. Amer. Math. Soc. **42**, 373–376. (*44*)
3. NICOLESCO, M.: Les fonctions polyharmoniques. Actualités Sci. Ind. No. 331. (*8*)
4. PICONE, M.: Nuovi indirizzi di ricerca nella teoria e nel calcolo delle soluzioni di talune equazioni lineari alle derivate parziali della Fisica Matematica. Ann. Scuola Norm. Pisa (2) **5**, 213–288 (*8*)
5. SOUTHWELL, R. V.: Castigliano's principle of minimum strain energy. Proc. Roy. Soc. Lond. (A) **154**, 4–21. (*18, 36*)
6. ZANABONI, O.: Il problema della funzione delle tensioni in un sistema spaziale isotropo. Bull. Un. Mat. Ital. **15**, 71–76. (*44*)

1937 *1.* GOODIER, J. N.: A general proof of Saint-Venant's principle. Phil. Mag. (7) **23**, 607–609. (*54*)
2. — Supplementary note on "A general proof of Saint-Venant's principle". Phil. Mag. (7) **24**, 325. (*54*)
3. NEUBER, H.: Kerbspannungslehre. Berlin: Springer. (*44*)
4. ODQVIST, F.: Équations de compatibilité pour un système de coordonnées triples orthogonaux quelconques. C. R. Acad. Sci., Paris **205**, 202–204. (*14*)
5. PAPKOVITCH, P. F.: Survey of some general solutions of the fundamental differential equations of equilibrium of an isotropic elastic body [in Russian]. Prikl. Mat. Meh. **1**, 117–132. (*44*)
6. ZANABONI, O.: Dimostrazione generale del principio del De Saint-Venant. Atti Accad. Lincei Rend. **25**, 117–121. (*54, 56a*)
7. — Valutazione dell'errore massimo cui da luogo l'applicazione del principio del Saint-Venant. Atti Accad. Lincei Rend. **25**, 595–601. (*54, 56a*)
8. — Sull'approssimazione dovuta al principio del De Saint-Venant nei solidi prismatici isotropi. Atti Accad. Lincei Rend. **26**, 340–345. (*54, 55, 56a*)

1938 *1.* BLINCHIKOV, T. N.: The differential equations of equilibrium of the theory of elasticity in curvilinear coordinate systems [in Russian]. Prikl. Mat. Meh. **2**, 407–413. (*14*)
2. SHERMAN, D. J.: On the distribution of eigenvalues of integral equations in the plane theory of elasticity [in Russian]. Trudy Seismol. Inst. Akad. Nauk SSSR, No. 82. (*32*)
3. SLOBODYANSKY, M.: Expression of the solution of the diferential equations of elasticity by means of one, two, and three functions and proof of the generality of these solutions [in Russian]. Uch. Zap. Mosk. Gos. Univ. **24**, 191–202. (*44*)
4. — Stress functions for spatial problems in the theory of elasticity [in Russian]. Uch. Zap. Mosk. Gos. Univ. **24**, 181–190. (*44*)
5. SOUTHWELL, R. V.: Castigliano's principle of minimum strain-energy, and the conditions of compatibility for strain. Stephen Timoshenko 60th Aniv. Vol. New York: MacMillan. (*18, 36*)

1939 *1.* BURGERS, J. M.: Some considerations on the fields of stress connected with dislocations in a regular crystal lattice. Proc. Koninkl. Ned. Akad. Wetenschap. **42**, 293–325, 378–399. (*14*)
2. FRIEDRICHS, K. O.: On differential operators in Hilbert spaces. Am. J. Math. **61**, 523–544. (*42*)
3. GRAFFI, D.: Sui teoremi di reciprocità nei fenomeni dipendenti dal tempo. Ann. Mat. (4) **18**, 173–200. (*61*)
4. PAPKOVITCH, P. F.: Theory of Elasticity [in Russian]. Moscow. (*44*)

1940 *1.* FADLE, J.: Die Selbstspannungs-Eigenwertfunktionen der quadratischen Scheibe. Ing. Arch. **11**, 125–149. (*56*)
2. LOCATELLI, P.: Estensione del principio di St. Venant a corpi non perfettamente elastici. Atti Accad. Sci. Torino **75**, 502–510. (54)
3. — Sul principio di Menabrea. Boll. Un. Nat. Ital. (2) **2**, 342–347. (*18, 36*)
4. — Sulla congruenza delle deformazioni. Rend. Ist. Lombardo **73** = (3) **4** (1939–1940), 457–464. (*18, 36*)

1941 *1.* PLATRIER, C.: Sur l'intégration des équations indéfinies de l'équilibre élastique. C. R. Acad. Sci., Paris **212**, 749–751. (*44*)

2. SAKADI, Z.: Elastic waves in crystals. Proc. Phys. Math. Soc. Japan (3) **23**, 539–547. (*70, 71*)

1942 *1.* FINZI, B.: Propagazione ondosa nei continui anisotropi. Rend. Ist. Lombardo, **75** = (3) **6** (1941/42) 630–640. (**73**)

2. GOODIER, J. N.: An extension of Saint-Venant's principle, with applications. J. Appl. Phys. **13**, 167–171. (*54*)

1943 *1.* FRANK, P., and R. v. MISES: Die Differential- und Integralgleichungen der Mechanik und Physik, Vol. 1. New York: Rosenberg. (*8*)

2. STEVENSON, A. C.: Some boundary problems of two-dimensional elasticity. Phil. Mag. (7) **34**, 766–793. (*47*)

3. UDESCHINI, P.: Sull'energia di deformazione. Rend. Ist. Lombardo **76** = (3) **7**, 25–34. (*24*)

1944 *1.* CARRIER, G. F.: The thermal stress and body-force problems of the infinite orthotropic solid. Q. Appl. Math. **2**, 31–36. (*51*)

2. FRIEDRICHS, K. O.: The identity of weak and strong extensions of differential operators. Trans. Am. Math. Soc. **55**, 132–151. (*42*)

3. VLASOV, V. Z.: Equations of compatibility of deformation in curvilinear co-ordinates [in Russian]. Prikl. Mat. Meh. **8**, 301–306. (*14*)

1945 *1.* FOX, L., and R. V. SOUTHWELL: Relaxation methods applied to engineering problems. VII A. Biharmonic analysis as applied to the flexure and extension of flat elastic plates. Phil. Trans. Roy. Soc. Lond. Ser. A **239**, 419–460. (*47*)

2. KUZMIN, R. O.: On Maxwell's and Morera's formulae in the theory of elasticity. Dokl. Akad. Nauk SSSR **49**, 326–328. (*17*)

3. LOCATELLI, P.: Nuove espressioni variazionali nella dinamica elastica. Rend. Ist. Lombardo **78** = (3) **9**, 247–257. (*65*)

4. — Sull'interpretazione di principi variazionali dinamica nella statica elastica. Rend. Ist. Lombardo **78** = (3) **9**, 301–306. (*65*)

5. MISES, R. v.: On Saint-Venant's principle. Bull. Amer. Math. Soc. **51**, 555–562. (*54*)

6. RAYLEIGH, J. W. S.: The Theory of Sound, Vol. 1. New York: Dover. (*39*)

7. STEVENSON, A. C.: Complex potentials in two-dimensional elasticity. Proc. Roy. Soc. Lond. (A) **184**, 129–179, 218–229. (*47*)

1946 *1.* BUTTY, E.: Tratado de elasticidad teorico-technica. Buenos Aires: Centro Estudiantes de Ingenieria. (*51, 71*)

2. HEARMON, R. F. S.: The elastic constants of anisotropic materials. Rev. Mod. Phys. **18**, 409–440. (*26*)

3. PORITSKY, H.: Application of analytic functions to two-dimensional biharmonic analysis. Trans. Am. Math. Soc. **59**, 248–279. (*47*)

4. PRAGER, W.: On plane elastic strain in doubly-connected domains. Q. Appl. Math. **3**, 377–380. (*47*)

1947 *1.* FRIEDRICHS, K. O.: On the boundary-value problems of the theory of elasticity and Korn's inequality. Ann. Math. **48**, 441–471. (*13, 42*)

2. GRAFFI, D.: Sul teorema di reciprocità nella dinamica dei corpi elastici. Mem. Accad. Sci. Bologna **18**, 103–109. (*19, 61*)

3. GUTMAN, C. G.: General solution of a problem in the theory of elasticity in generalized cylindrical coordinates [in Russian]. Dokl. Akad. Nauk SSSR **58**, 993–996. (*44*)

4. LIFSHIC, I. M., and L. N. ROZENCVEIG: On the construction of the Green's tensor for the fundamental equations of the theory of elasticity in the case of an unrestricted anisotropic elastic medium [in Russian]. Akad. Nauk SSSR. Zhurnal Eksper. Teoret. Fiz. **17**, 783–791. (*51*)

5. MIKHLIN, S. G.: Fundamental solutions of the dynamic equations of the theory of elasticity for inhomogeneous media [in Russian]. Prikl. Mat. Meh. **11**, 423–432. (*68*)

6. PRAGER, W., and J. L. SYNGE: Approximations in elasticity based on the concept of function space. Q. Appl. Math. **5**, 241–269. (*34*)

7. SHAPIRO, G.: Les fonctions des tensions dans un système arbitraire de coordonnées curvilignes. Dokl. Acad. Sci. URSS **55**, 693–695. (*44*)

8. TER-MKRTYCHAN, L. N.: On general solutions of the problems of the theory of elasticity [in Russian]. Trudy Leningr. Politekhn. Akad. in-ta, No. 4. (*44*)

9. VAN HOVE, L.: Sur l'extension de la condition de Legendre du calcul des variations aux intégrales multiples à plusieurs fonctions inconnues. Proc. Koninkl. Ned. Akad. Wetenschap. **50**, 18–23. (*32*)

1948 *1.* DIAZ, J. B., and H. J. GREENBERG: Upper and lower bounds for the solution of the first boundary value problem of elasticity. Q. Appl. Math. **6**, 326–331. (*52*)

2. ELLIOT, H. A.: Three-dimensional stress distributions in hexagonal aeolotropic crystals. Proc. Cambridge Phil. Soc. **44**, 522–533. (*51*)
3. ERIM, K.: Sur le principe de Saint-Venant. Proc. Seventh Int. Cong. Appl. Mech. London. (*54*)
4. FLINT, E. E.: A Practical Handbook on Geometrical Crystallography [in Russian]. Moscow: Gosgeolizdat. (*21*)
5. MORIGUTI, S.: On Castigliano's theorem in three-dimensional elastostatics [in Japanese]. J. Soc. Appl. Mech. Japan **1**, 175–180. (*18*)
6. TIMPE, A.: Torsionsfreie achsensymmetrische Deformation von Umdrehungskörpern und ihre Inversion. Z. Angew. Math. Mech. **28**, 161–166. (*44*)
7. WEBER, C.: Spannungsfunktionen des dreidimensionalen Kontinuums. Z. Angew. Math. Mech. **28**, 193–197. (*17*)

1949 1. FINZI, B., and M. PASTORI: Calcolo Tensoriale e Applicazioni. Bologna: Zanichelli. (*17*)
2. FREIBERGER, W.: On the solution of the equilibrium equations of elasticity in general curvilinear coordinates. Australian J. Sci. A **2**, 483–492. (*44*)
3. — The uniform torsion of an incomplete tore. Australian J. Sci. A **2**, 354–375. (*44*)
4. IACOVACHE, M.: O extindere a metodei lui Galerkin pentru sistemul ecuaţiilor elasticităţii. Bul. St. Acad. R. P. Române **1**, 593–596. (*44, 67*)
5. KRUTKOV, Y. A.: Tensor Stress Functions and General Solutions in the Static Theory of Elasticity [in Russian]. Moscow: Akademkniga. (*17, 44*)
6. MOISIL, G.: Asupra formulelor lui Galerkin in teoria elasticităţii. Bul. St. Acad. R. P. Române **1**, 587–592. (*44*)
7. PASTORI, M.: Propagazione ondosa nei continui anisotropi e corrispondenti direzioni principali. Nuovo Cimento (9) **6**, 187–193. (*73*)
8. PERETTI, G.: Significato del tensore arbitrario che interviene nell'integrale generale delle equazioni della statica dei continui. Atti Sem. Mat. Fis. Univ. Modena **3**, 77–82. (*17*)

1950 1. AQUARO, G.: Un teorema di media per le equazioni dell'elasticità. Riv. Mat. Univ. Parma **1**, 419–424. (*43*)
2. ARZHANYH, I. S.: On the theory of integration of the dynamical equations of an isotropic elastic body [in Russian]. Dokl. Akad. Nauk SSSR (N.S.) **73**, 41–44. (*53, 69*)
3. BLOKH, V.: Stress functions in the theory of elasticity [in Russian]. Prikl. Mat. Meh. **14**, 415–422. (*17, 44*)
4. FICHERA, G.: Sull'esistenza e sul calcolo delle soluzioni dei problemi al contorno, relativi all'equilibrio di un corpo elastico. Ann. Scuola Norm. Pisa (3) **4**, 35–99. (*49, 50*)
5. FRIDMAN, M. M.: The mathematical theory of elasticity of anisotropic media [in Russian]. Prikl. Mat. Meh. **14**, 321–340. (*45*)
6. IACOVACHE, M.: Asupra relatiilor dintre tensiuni intr'un corp elastic in miscare. Bul. St. Acad. R. P. Române **2**, 699–705. (*59*)
7. IONESCU-CAZIMIR, V.: Asupra ecuatiilor echilibrului termoelastic. I. Analogul vectorului lui Galerkin. Bul. St. Acad. R. P. Române **2**, 589–595. (*44*)
8. MOISIL, G.: Asupra unui vector analog vectorului lui Galerkin pentru echilibrul corpurilor elastice cu isotropie transversa. Bul. St. Acad. R. P. Române **2**, 207–210 (*44*)
9. MORINAGA, K., and T. NÔNO: On stress-functions in general coordinates. J. sci. Hiroshima Univ. Ser. A **14**, 181–194. (*17*)
10. REISSNER, E.: On a variational theorem in elasticity. J. Math. Phys. **29**, 90–95. (*38*)
11. SOBOLEV, S. L.: Some Applications of Functional Analysis in Mathematical Physics [in Russian]. Leningrad: University Press. (I have seen this work only in the English translation by F. E. BROWDER, Providence: Am. Math. Soc. 1963.) (*42*)
12. SYNGE, J. L.: Upper and lower bounds for solutions of problems in elasticity. Proc. Roy. Irish Acad. Sect. A **53**, 41–64. (*43, 52*)
13. TIMPE, A.: Spannungsfunktionen für die von Kugel- und Kegelflächen begrenzten Körper- und Kuppelprobleme. Z. Angew. Math. Mech. **30**, 50–61. (*44*)

1951 1. ALBRECHT, F.: L'équilibre élastique des cristaux du système cubique. Com. Acad. R. P. Române **1**, 403–408. (*27*)
2. ARZHANYH, I. S.: The integral equations of the dynamics of an elastic body [in Russian]. Dokl. Akad. Nauk SSSR (N. S.) **76**, 501–503. (*53, 69*)
3. — Integral equations for the representation of the vector of translation, spatial dilatation, and rotation of an elastic body [in Russian]. Prikl. Mat. Meh. **15**, 387–391. (*69*)
4. — The fundamental integral equations of the dynamics of an elastic body [in Russian]. Dokl. Akad. Nauk SSSR (N. S.) **81**, 513–516. (*69*)

5. Eidus, D. M.: On the mixed problem of the theory of elasticity [in Russian]. Dokl. Akad. Nauk SSSR. **76**, 181–184. (*13*)
6. Fichera, G.: Über eine Möglichkeit zur Kontrolle der physikalischen Widerspruchsfreiheit der Gleichungen der mathematischen Elastizitätstheorie. Z. Angew. Math. Mech. **31**, 268–270. (*31*)
7. Filonenko-Borodich, M. M.: The problem of the equilibrium of an elastic parallelepiped subject to assigned loads on its boundaries [in Russian]. Prikl. Mat. Meh. **15**, 137–148. (*17*)
8. — Two problems on the equilibrium of an elastic parallelepiped [in Russian]. Prikl. Mat. Meh. **15**, 563–574. (*17*)
9. Sternberg, E., R. A. Eubanks and M. A. Sadowski: On the stress function approaches of Boussinesq and Timpe to the axisymmetric problem of elasticity theory. J. Appl. Phys. **22**, 1121–1124. (44)
10. Timoshenko, S., and J. N. Goodier: Theory of Elasticity, 2nd ed. New York: McGraw Hill. (*51*)
11. Timpe, A.: Spannungsfunktionen achsensymmetrischer Deformationen in Zylinderkoordinaten. Z. Angew. Math. Mech. **31**, 220–224. (*44*)
12. Vâlcovici, V.: Sur les relations entre les tensions. Com. Acad. R. P. Române **1**, 337–339. (*59*)

1952 *1.* Cooperman, P.: An extension of the method of Trefftz for finding local bounds on the solutions of boundary-value problems, and on their derivatives. Q. Appl. Math. **10**, 359–373. (*37*)
2. Mikhlin, S. G.: The Problem of the Minimum of a Quadratic Functional [in Russian]. Moscow. (I have seen this work only in the English translation by A. Feinstein, San Francisco: Holden Day 1965.) (*13, 43*)
3. Moisil, A.: Les relations entre les tensions pour les corps élastiques à isotropie transverse. Bul. St. Acad. R. P. Române **3**, 473–480. (*27*)
4. Sbrana, F.: Una proprietá caratteristica delle equazioni dell'elasticità. Atti Accad. Ligure Sci. Lett. **9**, 84–88. (*43*)
5. Sneddon, I. N.: The stress produced by a pulse of pressure moving along the surface of a semi-infinite solid. Rend. Circ. Mat. Palermo (2) **1**, 57–62. (*47*)
6. Tiffen, R.: Uniqueness theorems of two-dimensional elasticity theory. Q. J. Mech. Appl. Math. **5**, 237–252. (*50*)
7. Westergaard, H.: Theory of Elasticity and Plasticity. Cambridge. (*44*)

1953 *1.* Arzhanyh, I. S.: Construction of the integral equations of statics in the theory of elasticity by means of Green's functions [in Russian]. Akad. Nauk Uzbek. SSR. Trudy Inst. Mat. Meh. **10**, 5–25. (*53*)
2. — Stress tensor functions for the dynamics of an elastic body [in Russian]. Dokl. Akad. Nauk Uzbek. SSR., No. 7, 3–4. (*17*)
3. Bergman, S., and M. Schiffer: Kernel functions and elliptic differential equations in mathematical physics. New York: Academic. (*53*)
4. Bishop, R. E. D.: On dynamical problems of plane stress and plane strain. Q. J. Mech. Appl. Math. **6**, 250–254. (*67*)
5. Bland, D. R.: Mean displacements on the boundary of an elastic solid. Q. J. Mech. Appl. Math. **6**, 379–384. (*29*)
6. Brdička, M.: The equations of compatibility and stress functions in tensor form [in Russian]. Czech. J. Phys. **3**, 36–52. (*44*)
7. Churikov, F.: On a form of the general solution of the equilibrium equations for the displacements in the theory of elasticity [in Russian]. Prikl. Mat. Meh. **17**, 751–754. (*44*)
8. Courant, R., and D. Hilbert: Methods of Mathematical Physics, Vol. 1. New York: Interscience. (*39, 76, 77, 78*)
9. Dzhanelidze, G. I.: The general solution of the equations of the theory of elasticity in arbitrary linear coordinates [in Russian]. Dokl. Akad. Nauk SSSR **88**, 423–425. (*44*)
10. Ericksen, J. L.: On the propagation of waves in isotropic incompressible perfectly elastic materials. J. Rational Mech. Anal. **2**, 329–337. Reprinted in Problems of Non-linear Elasticity. Intl. Sci. Rev. Ser. New York: Gordon and Breach 1965. (*73*)
11. Fichera, G.: Condizioni perchè sia compatibile il problema principale della statica elastica. Atti Accad. Lincei Rend. (8) **14**, 397–400. (*31*)
12. Föppl, L.: Ein Mittelwertsatz der ebenen Elastizitätstheorie. Sitzber. Math.-Naturw. Kl. Bayer Akad. Wiss. München **1952**, 215–217. (*43*)
13. Hu, H.: On the three-dimensional problems of the theory of elasticity of a transversely isotropic body. Sci. Sinica (Peking) **2**, 145–151. (*44*)
14. Kröner, E.: Das Fundamentalintegral der anisotropen elastischen Differentialgleichungen. Z. Physik **136**, 402–410. (*51*)

15. MINDLIN, R. D.: Force at a point in the interior of a semi-infinite solid. Proc. First Midwestern Conf. Solid Mech. (*51*)
16. MIŞICU, M.: Echilibrul mediilor continue cu deformări mari. Stud. Cercet. Mec. Metal. **4**, 31–53. (*32*)
17. SÁENZ, A. W.: Uniformly moving dislocations in anisotropic media. J. Rational Mech. Anal. **2**, 83–98. (*51, 71*)
18. SCHAEFER, H.: Die Spannungsfunktionen des dreidimensionalen Kontinuums und des elastischen Körpers. Z. Angew. Math. Mech. **33**, 356—362 (*17, 18, 44*).
19. TRENIN, S.: On the solutions of the equilibrium equations of an axisymmetrical problem in the theory of elasticity [in Russian]. Vestn. Mosk. Univ. Ser. Fiz., Mat. **8**, 7–13. (*44*)
20. WASHIZU, K.: Bounds for solutions of boundary-value problems in elasticity. J. Math. Phys. **32**, 117–128. (*52*)

1954 *1.* ARZHANYH, I. S.: Dynamical potentials of the theory of elasticity [in Russian]. Akad. Nauk Uzbek. SSR. Trudy Inst. Mat. Meh. **13**, 3–17. (*69*)
2. — Integral Equations of Basic Problems in the Theory of Vector Fields and in Elasticity [in Russian]. Tashkent: Izdat. Akad. Nauk Uzbek. SSR. (*53, 69*)
3. BRDIČKA, M.: Covariant form of the general solution of the Galerkin equations of elastic equilibrium [in Russian]. Czech. J. Phys. **4**, 246. (*44*)
4. BROWDER, F. E.: Strongly elliptic systems of differential equations. Contrib. Th. Partial Diff. Eqns. Annals. of Math. Studies No. 33, 15–51. (*32*)
5. EUBANKS, R., and E. STERNBERG: On the axisymmetric problem of elasticity theory for a medium with transverse isotropy. J. Rational Mech. Anal. **3**, 89–101. (*44*)
6. GRAFFI, D.: Über den Reziprozitätssatz in der Dynamik der elastischen Körper. Ing. Arch. **22**, 45–46. (*19, 61*)
7. GÜNTHER, W.: Spannungsfunktionen und Verträglichkeitsbedingungen der Kontinuumsmechanik. Abhandl. Braunschweig. Wiss. Ges. **6**, 207–219. (17)
8. HU, H.: On the general theory of elasticity for a spherically isotropic medium. Sci. Sinica (Peking) **3**, 247–260. (*44*)
9. IONESCU, D.: Asupra vectorului lui Galerkin în teoria elasticităţii şi în hidrodinamica fluidelor vîscoase. Bul. Şt. Acad. R. P. Române **6**, 555–571 (*44*)
10. KLYUSHNIKOV, V. D.: Derivation of the Beltrami-Michell equations from a variational principle [in Russian]. Prikl. Mat. Meh. **18**, 250–252. (*18*)
11. KRÖNER, E.: Die Spannungsfunktionen der dreidimensionalen isotropen Elastizitätstheorie. Z. Physik **139**, 175–188. Correction, Z. Phys. **143**, 175 (1955). (*17, 44*)
12. LANGHAAR, H., and M. STIPPES: Three-dimensional stress functions. J. Franklin Inst. **258**, 371–382. (*17*)
13. LING, C. B., and K. L. YANG: On symmetric strains in solids of revolution in curvilinear coordinates. Ann. Acad. Sinica Taipei **1**, 507–516. (*44*)
14. MORREY, C. B., JR.: Second order elliptic systems of differential equations. Contrib. Th. Partial Diff. Eqns. Annals of Math. Studies No. 33, 101–159. (*32*)
15. MUSGRAVE, M. J. P.: On the propagation of elastic waves in aeolotropic media. Proc. Roy. Soc. Lond., Ser. (A) **226**, 339–366. (*71*)
16. MUSKHELISHVILI, N. I.: Some Basic Problems of the Mathematical Theory of Elasticity [in Russian], 4th Ed. Moscow. (I have seen this work only in the English translation by J. R. M. RADOK, Groningen: Noordhoff 1963.) (*47, 50*).
17. ORNSTEIN, W.: Stress functions of Maxwell and Morera. Q. Appl. Math. **12**, 198–201. (*17*)
18. SCHUMANN, W.: Sur différentes formes du principe de B. de Saint-Venant. C.R. Acad. Sci., Paris **238**, 988–990. (*54*)
19. SLOBODYANSKY, M. G.: The general form of solutions of the equations of elasticity for simply connected and multiply connected domains expressed by harmonic functions [in Russian]. Prikl. Mat. Meh. **18**, 55–74. (*44*)
20. STERNBERG, E.: On Saint-Venant's principle. Q. Appl. Math. **11**, 393–402. (*54*)

1955 *1.* AYMERICH, G.: Una proprietà dell'energia elastica. Boll. Un. Mat. Ital. (3) **10**, 332–336. (*34*)
2. ARZHANYH, I. S.: Regular integral equations of the dynamics of an elastic body [in Russian]. Akad. Nauk Uzbek. SSR. Trudy Inst. Mat. Meh. **15**, 79–85. (*69*)
3. — Retarded potentials of the dynamics of an elastic body [in Russian]. Akad. Nauk Uzbek. SSR. Trudy Inst. Mat. Meh. **16**, 5–22. (*69*)
4. BABUŠKA, I., K. REKTORYS, and F. VYČICHLO: Mathematical Theory of Elasticity for the Plane Problem [in Czechoslovakian]. Prague. (I have seen this work only in the German translation by W. HEINRICH, Berlin: Akademie 1960.) (*56*)
5. HU, H.: On some variational principles in the theory of elasticity and the theory of plasticity. Sci. Sinica (Peking) **4**, 33–54. (*38*)

6. KACZKOWSKI, Z.: The conjugate directions in an anisotropic body [in Polish]. Arch. Mech. Stos. **7**, 52–86. (*31*)

7. KRÖNER, E.: Die Spannungsfunktionen der dreidimensionalen anisotropen Elastizitätstheorie. Z. Physik **141**, 386–398. (*17*)

8. — Die inneren Spannungen und der Inkompatibilitätstensor in der Elastizitätstheorie. Z. Angew. Phys. **7**, 249–257. (*17*)

9. LODGE, A. S.: The transformation to isotropic form of the equilibrium equations for a class of anisotropic elastic solids. Q. J. Mech. Appl. Math. **8**, 211–225. (*31*)

10. LUR'E, A. I.: Three-dimensional problems in the theory of elasticity [in Russian]. Moscow: Gostekhizdat. (I have seen this work only in the English translation by D. B. MCVEAN, New York: Interscience 1964.) (*17, 44, 51*)

11. MARGUERRE, K.: Ansätze zur Lösung der Grundgleichungen der Elastizitätstheorie. Z. Angew. Math. Mech. **35**, 241–263. (*17, 44*)

12. SCHAEFER, H.: Die Spannungsfunktion einer Dyname. Abhandl. Braunschweig. Wiss. Ges. **7**, 107–112. (*17*)

13. STERNBERG, E., and R. A. EUBANKS: On the concept of concentrated loads and an extension of the uniqueness theorem in the linear theory of elasticity. J. Rational Mech. Anal. **4**, 135–168. (*28, 51, 53*)

14. WASHIZU, K.: On the variational principles of elasticity and plasticity. Rept. 25–18, Cont. Nsori-07833 Massachusetts Institute of Technology, March. (*38*)

1956 *1.* DORN, W. S., and A. SCHILD: A converse to the virtual work theorem for deformable solids. Q. Appl. Math. **14**, 209–213. (*17, 36*)

2. DUFFIN, R. J.: Analytic continuation in elasticity. J. Rational Mech. Anal. **5**, 939–950. (*42, 44*)

3. ERICKSEN, J. L., and R. A. TOUPIN: Implications of Hadamard's condition for elastic stability with respect to uniqueness theorems. Canad. J. Math. **8**, 432–436. Reprinted in Foundations of Elasticity Theory. Intl. Sci. Rev. Ser. New York: Gordon and Breach 1965. (*32, 57*)

4. EUBANKS, R. A., and E. STERNBERG: On the completeness of the Boussinesq-Papkovitch stress functions. J. Rational Mech. Anal. **5**, 735–746. (*44*)

5. FINZI, L.: Legame fra equilibrio e congruenza e suo significato fisico. Atti Accad. Lincei Rend. (8) **20**, 205–211. (*47*)

6. KNOPOFF, L.: Diffraction of elastic waves. J. Acoust. Soc. Amer. **28**, 217–229. (*69*)

7. LEVIN, M. L., and S. M. RYTOV: Transition to geometrical approximations in elasticity theory [in Russian], Soviet Phys. Acoust. **2**, 179–184. (*73*)

8. MILLER, G. F., and M. J. P. MUSGRAVE: On the propagation of elastic waves in aeolotropic media. III. Media of cubic symmetry. Proc. Roy. Soc. Lond. (A) **236**, 352–383. (*71*)

9. PAILLOUX, M. H.: Élasticité. Mem. Sci. Math. Acad. Sci. Paris **132**. (*71*)

10. RADOK, J. R. M.: On the solutions of problems of dynamic plane elasticity. Q. Appl. Math. **14**, 289–298. (*47*)

11. SCHAEFER, H.: Die drei Spannungsfunktionen des zweidimensionalen ebenen Kontinuums. Österreich. Ing. Arch. **10**, 267–277. (*47*)

12. SOKOLNIKOFF, I. S.: Mathematical Theory of Elasticity. 2nd Ed. New York: McGraw-Hill. (*14, 27, 36, 37, 44, 47*)

13. SYNGE, J. L.: Flux of energy for elastic waves in anisotropic media. Proc. Roy. Irish Acad., Sect. A **58**, 13–21. (*71*)

14. TEMPLE, G., and W. G. BICKLEY: Rayleigh's principle and its applications to engineering. New York: Dover. (*56*)

1957 *1.* BASHELEISHVILI, M. O.: On fundamental solutions of the differential equations of an anisotropic elastic body [in Russian]. Soobšč. Akad. Nauk Gruzin. SSR. **19**, No. 4, 393–400. (*51*)

2. BONDARENKO, B. A.: On a class of solutions of dynamical equations in the theory of elasticity [in Russian]. Akad. Nauk Uzbek. SSR. Trudy Inst. Mat. Meh. **21**, 41–49. (*67*)

3. BRDIČKA, M.: On the general form of the Beltrami equation and Papkovitch's solution of the axially symmetric problem of the classical theory of elasticity. Czech. J. Phys. **7**, 262–274. (*17, 44*)

4. COURANT, R.: Differential and Integral Calculus, Vol. 2. New York: Interscience. (*6*)

5. ERICKSEN, J. L.: On the Dirichlet problem for linear differential equations. Proc. Amer. Math. Soc. **8**, 521–522. (*32*)

6. FILONENKO-BORODICH, M. M.: On the problem of Lamé for the parallelepiped in the general case of surface loads [in Russian]. Prikl. Mat. Meh. **21**, 550–559. (*17*)

7. GARABALDI, A. C.: Su una proprietà di media caratteristica per l'equazione dell'elastostatica. Rend. Sem. Mat. Univ. Padova **27**, 306–318. (*43*)

8. HILL, R.: On uniqueness and stability in the theory of finite elastic strain. J. Mech. Phys. Solids **5**, 229–241. (*32, 34*)
9. HORVAY, G.: Saint-Venant's principle: a biharmonic eigenvalue problem. J. Appl. Mech. **24**, 381–386. (*56*)
10. MIKHLIN, S. G.: Variational Methods in Mathematical Physics [in Russian]. Moscow. (I have seen this work only in the English translation by T. BODDINGTON, New York: Pergamon 1964.) (*34, 37, 39*).
11. NOLL, W.: Verschiebungsfunktionen für elastische Schwingungsprobleme. Z. Angew. Math. Mech. **37**, 81–87. (*44*)
12. SOLOMON, L.: Displacement potentials for the equations of elastostatics [in Roumanian]. Buletinul Stiintific Ac. R.P.R., **9**, 415–431. (*44*)
13. SOLYANIK-KRASSA, K. V.: Stress functions of an axisymmetric problem in the theory of elasticity [in Russian]. Prikl. Mat. Meh. **21**, 285–286. (*44*)
14. STERNBERG, E., and R. A. EUBANKS: On stress functions for elastokinetics and the integrations of the repeated wave equation. Q. Appl. Math. **15**, 149–153. (*67*)
15. SYNGE, J. L.: The Hypercircle in Mathematical Physics; a Method for the Approximate Solution of Boundary-Value Problems. Cambridge. (*34, 51*)
16. — Elastic waves in anisotropic media. J. Math. Phys. **35**, 323–334. (*71*)
17. THOMAS, T. Y.: The decay of waves in elastic solids. J. Math. Mech. **6**, 759–768. (*73*)
18. WASHIZU, K.: On the variational principles applied to dynamic problems of elastic bodies. ASRL TR 25–23. Massachusetts Institute of Technology. (*65*)

1958 *1.* BABICH, V. M., and A. S. ALEKSEEV: A ray method of computing wave front intensities [in Russian], Izv. Akad. Nauk SSSR, Ser. Geofiz. 17–31. (*73*)
2. BLOKH, V. I.: On the representation of the general solution of the basic equations of the static theory of elasticity for an isotropic body with the aid of harmonic functions [in Russian]. Prikl. Mat. Meh. **22**, 473–479. (*44*)
3. BOLEY, B.: Some observations on Saint-Venant's principle. Proc. Third U.S. National Congress Appl. Mech. 259–264. (*54*)
4. DE HOOP, A. T.: Representation theorems for the displacement in an elastic solid and their application to elastodynamic diffraction theory. Doctoral Dissertation, Technische Hogeschool, Delft. (*69*)
5. DEEV, V. M.: The solution of the spatial problem in the theory of elasticity [in Ukrainian]. Dopovidi Acad. Nauk Ukr. RSR 29–32. (*44*)
6. DIAZ, J. B., and L. E. PAYNE: Mean value theorems in the theory of elasticity. Proc. Third U. S. Natl. Congress Appl. Mech. 293–303. (*8, 43, 55*)
7. — — On a mean value theorem and its converse for the displacements in the theory of elasticity. Port. Mat. **17**, 123–126. (*43*)
8. DUFFIN, R. J., and W. NOLL: On exterior boundary value problems in elasticity. Arch. Rational Mech. Anal. **2**, 191–196. (*32, 49, 50, 57*)
9. HALMOS, P.: Finite-Dimensional Vector Spaces. New York: Van Nostrand. (*3, 23*)
10. HU, H.: On reciprocal theorems in the dynamics of elastic bodies and some applications. Sci. Sinica (Peking) **7**, 137–150. (*61*)
11. — On two variational principles about the natural frequencies of elastic bodies. Sci. Sinica (Peking) **7**, 293–312. (*76*)
12. PETRASHEN, G. I.: The investigations of the propagation of elastic waves [in Russian]. Vestn. Leningrad Univ. **13**, No. 22, 119–136. (*73*)
13. PREDELEANU, M.: Über die Verschiebungsfunktionen für das achsensymmetrische Problem der Elastodynamik. Z. Angew. Math. Mech. **38**, 402–405. (*67*)
14. REISSNER, E.: On variational principles in elasticity. Proc. Symp. Appl. Math. Vol. 8, Calculus of Variations and its Applications. New York: McGraw Hill. (*38*)
15. SKURIDIN, G. A., and A. A. GVOZDEV: Boundary conditions for the jumps of discontinuous solutions of the dynamical equations of the theory of elasticity [in Russian]. Izv. Akad. Nauk SSSR. Ser. Geofiz. 145–156. (*73*)
16. SMITH, G. F., and R. S. RIVLIN: The strain-energy function for anisotropic elastic materials. Trans. Amer. Math. Soc. **88**, 175–193. (*21, 26*)
17. SNEDDON, I. N.: Note on a paper by J. R. M. Radok. Q. Appl. Math. **16**, 197. (*47*)
18. — and D. S. BERRY: The classical theory of elasticity. Handbuch der Physik **6**. Berlin-Göttingen-Heidelberg: Springer. (*67*)
19. TEODORESCU, P. P.: A plane problem of the theory of elasticity with arbitrary body forces [in Russian]. Rev. Méc. Appl. **3**, 101–108. (*47*)
20. WASHIZU, K.: A note on the conditions of compatibility. J. Math. Phys. **36**, 306–312. (*18*)

1959 *1.* BUCHWALD, V. T.: Elastic waves in anisotropic media. Proc. Roy. Soc. Lond. (A) **253**, 563–580. (*73*)

2. Dana, J. S.: Dana's manual of mineralogy 17th ed., revised by C. S. Hurlbut, Jr. New York: John Wiley. (*21*)
3. Deev, V. M.: On the form of the general solution of the three-dimensional problem of the theory of elasticity expressed with the aid of harmonic functions [in Russian]. Prikl. Math. Meh. **23**, 1132–1133; translated as J. Appl. Math. Mech. **23**, 1619–1622. (*44*)
4. DiMaggio, F. L., and H. H. Bleich: An application of a dynamic reciprocal theorem. J. Appl. Mech. **26**, 678–679. (*61*)
5. Gvozdev, A. A.: On the conditions at elastic wave fronts propagating in a non-homogeneous medium [in Russian]. Prikl. Math. Meh. **23**, 395–397; translated as J. Appl. Math. Mech. **23**, 556–561. (**73**)
6. Hardy, G. H., J. E. Littlewood, and G. Polya: Inequalities. Cambridge. (*56*)
7. Hijab, W. A.: Application of Papkovitch functions to three-dimensional problems of elasticity. Proc. Math. Phys. Soc. U.A.R. No. 23, 99–115. (*44*)
8. Ignaczak, J.: Direct determination of stresses from the stress equations of motion in elasticity. Arch. Mech. Stos. **11**, 671–678. (*59*)
9. Karal, F. C., and J. B. Keller: Elastic wave propagation in homogeneous and inhomogeneous media. J. Acoust. Soc. Am. **31**, 694–705. (*73*)
10. Mikusinski, J.: Operational Calculus. New York: Pergamon. (*10, 69*)
11. Pearson, C. E.: Theoretical Elasticity. Cambridge: Harvard University Press. (*67*)
12. Schaefer, H.: Die Spannungsfunktionen des dreidimensionalen Kontinuums; statische Deutung und Randwerte. Ing. Arch. **28**, 291–306. (*17*)
13. Signorini, A.: Questioni di elasticità non linearizzata e semilinearizzata. Rend. Mat. e Appl. **18**, 95–139. (*31*)
14. Skuridin, G. A.: Duhamel's principle and asymptotic solutions of dynamical equations in the theory of elasticity. II [in Russian]. Izv. Akad. Nauk SSSR. Ser. Geofiz. 337–343. (*73*)
15. Slobodyansky, M. G.: On the general and complete form of solutions of the equations of elasticity [in Russian]. Prikl. Math. Meh. **23**, 468–482; translated as J. Appl. Math. Mech. **23**, 666–685. (*44*)
16. Truesdell, C.: Invariant and complete stress functions for general continua. Arch. Rational Mech. Anal. **3**, 1–29. (*17, 44*)
17. — The rational mechanics of materials–past, present, future. Applied Mech. Rev. **12**, 75–80. Corrected reprint, Applied Mechanics Surveys. Washington: Spartan Books. 1965. (*22*)
18. Waterman, P. C.: Orientation dependence of elastic waves in single crystals. Phys. Rev. (2) *113*, 1240–1253. (*71*)

1960 *1.* Bernstein, B., and R. Toupin: Korn inequalities for the sphere and for the circle. Arch. Rational Mech. Anal. **6**, 51–64. (*13*)
2. Boley, B. A.: On a dynamical Saint-Venant principle. J. Appl. Mech. *27*, 74–78. (*54*) (*54*)
3. — and J. H. Weiner: Theory of Thermal Stresses. New York: Wiley. (*14*)
4. Bondarenko, B. A.: Gradient and curl solutions of the dynamical equations of elasticity theory [in Russian]. Issled. Mat. Analizu Mehanike Uzbek. Izdat. Akad. Nauk Uzbek. SSR., Tashkent, 17–29. (*67*)
5. Duff, G. F. D.: The Cauchy problem for elastic waves in an anisotropic medium. Trans. Roy. Soc. Lond. (A) **252**, 249–273. (*74*)
6. Ericksen, J. L.: Tensor fields. Appendix to [1960, 17]. (*27*)
7. Gurtin, M. E., and E. Sternberg: On the first boundary-value problem of linear elastostatics. Arch. Rational Mech. Anal. **3**, 177–187. (*32, 34, 57*)
8. Indenbom, V. L.: Reciprocity theorems and influence functions for the dislocation density tensor and the deformation incompatibility tensor [in Russian]. Dokl. Akad. Nauk SSSR **128**, 906–909; translated as Soviet Phys. Doklady **4**, 1125–1128. (*30*)
9. Rieder, G.: Topologische Fragen in der Theorie der Spannungsfunktionen. Abhandl. Braunschweig. Wiss. Ges. **7**, 4–65. (*17*)
10. Rüdiger, D.: Eine Verallgemeinerung des Prinzips vom Minimum der potentiellen Energie elastischer Körper. Ing. Arch. **27**, 421–428. (*38, 41*)
11. Sirotin, Y. I.: Group tensor spaces [in Russian]. Kristallografiya **5**, 171–179; translated as Soviet Phys.-Cryst. **5**, 157–165. (*26*)
12. Sternberg, E.: On some recent developments in the linear theory of elasticity. In Structural Mechanics. New York: Pergamon Press. (*17, 54*)
13. — On the integration of the equations of motion in the classical theory of elasticity. Arch. Rational Mech. Anal. **6**, 34–50. (*67*)

14. TEODORESCU, P. P.: Sur une représentation par potentiels dans le problème tridimensionnel de l'élastodynamique. C. R. Acad. Sci., Paris **250**, 1792–1794. (*67*)

15. TRUESDELL, C.: A program toward rediscovering the rational mechanics of the age of reason. Arch. Hist. Exact Sc. **1**, 3–36. Corrected reprint in Essays in the History of Mechanics. Berlin-Heidelberg-New York: Springer-Verlag 1968. (*22*)

16. — The Rational Mechanics of Flexible or Elastic Bodies, 1638–1788. L. Euleri Opera Omnia (2) **11**^{2}. Zurich: Füssli. (*22*)

17. — and R. TOUPIN: The classical field theories. In Vol. 3^{1} of the Handbuch der Physik, edited by S. FLÜGGE. Berlin-Göttingen-Heidelberg: Springer-Verlag. (*12, 15, 16, 17, 18, 19, 71, 72, 73*)

1961 *1.* BABICH, V. M.: Fundamental solutions of the dynamical equations of elasticity for nonhomogeneous media [in Russian]. Prikl. Mat. Meh. **25**, 38–45; translated as J. Appl. Math. Mech. **25**, 49–60. (*68*)

2. BLOKH, V. I.: Stress functions and displacement functions [in Russian]. Dopovidi Akad. Nauk. Ukr. RSR 455–458. (*17, 44*)

3. BRAMBLE, J. H., and L. E. PAYNE: An analogue of the spherical harmonics for the equation of elasticity. J. Math. Phys. **15**, 163–171. (*32*)

4. — — A new decomposition formula in the theory of elasticity. J. Res. Nat. Bur. Std. **65 B**, 151–156. (*44*)

5. — — On some new continuation formulas and uniqueness theorems in the theory of elasticity. J. Math. Anal. Appl. **3**, 1–17. (*40*)

6. — — A priori bounds in the first boundary-value problem in elasticity. J. Res. Nat. Bur. Std. **65 B**, 269–276. (*52*)

7. FARNELL, G. W.: Elastic waves in trigonal crystals. Can. J. Phys. **39**, 65–80. (*71*)

8. FICHERA, G.: Il teorema del massimo modulo per l'equazione dell'elastostatica. Arch. Rational Mech. Anal. **7**, 373–387. (*42*)

9. GOLDBERG, R. R.: Fourier Transforms. Cambridge. (*32*)

10. GURTIN, M. E., and E. STERNBERG: A note on uniqueness in classical elastodynamics. Q. Appl. Math. **19**, 169–171. (*63*)

11. — — Theorems in linear elastostatics for exterior domains. Arch. Rational Mech. Anal. **8**, 99–119. (*8, 34, 48, 49, 50*)

12. HILL, R.: Bifurcation and uniqueness in non-linear mechanics of continua. Problems Contin. Mech. (Muskhelisvili Anniv. Vol.), 155–164. Phila.: Soc. Indust. Appl. Math. (*32*)

13. — Uniqueness in general boundary-value problems for elastic or inelastic solids. J. Mech. Phys. Solids **9**, 114–130. (*32*)

14. — Discontinuity relations in mechanics of solids. Progress in Solid Mechanics. Vol. II, 245–276. Amsterdam: North-Holland. (*72*)

15. MUSGRAVE, M. J. P.: Elastic waves in anisotropic media. Progress in Solid Mechanics. Vol. II, 61–85. (*71*)

16. NAGHDI, P. M., and C. S. HSU: On a representation of displacements in linear elasticity in terms of three stress functions. J. Math. Mech. **10**, 233–245. (*44*)

17. PAYNE, L. E., and H. F. WEINBERGER: On Korn's inequality. Arch. Rational Mech. Anal. **8**, 89–98. (*13*)

18. REISSNER, E.: On some variational theorems in elasticity. Problems of Continuum Mechanics (Muskhelishvili Anniversary Volume), 370–381. Phila.: Soc. Indust. Appl. Math. (*38*)

19. RÜDIGER, D.: Ein neues Variationsprinzip in der Elastizitätstheorie. Ing. Arch. **30**, 220–224. (*38*)

20. SOLOMON, L.: On the degree of arbitrariness in the determination of the functions of Papkovitch [in Roumanian]. Comun. Acad. RPR., **11**, 1039–1045. (*44*)

21. SVEKLO, V. A.: On the solution of dynamic problems in the plane theory of elasticity for anisotropic media [in Russian]. Prikl. Mat. Meh. **25**, 885–896; translated as J. Appl. Math. Mech. **25**, 1324–1339. (*47*)

22. TRUESDELL, C.: General and exact theory of waves in finite elastic strain. Arch. Rational Mech. Anal. **8**, 263–296. Reprinted in Problems of Non-linear Elasticity. Intl. Sci. Rev. Ser. New York: Gordon and Breach 1965 and also in Wave Propagation in Dissipative Materials. Berlin-Heidelberg-New York: Springer 1965. (*73*)

1962 *1.* ALEXANDROV, A. Y., and I. I. SOLOVEV: One form of solution of three-dimensional axisymmetric problems of elasticity theory by means of functions of a complex variable and the solution of these problems for the sphere [in Russian]. Prikl. Mat. Mech. **26**, 138–145; translated as J. Appl. Math. Mech. **26**, 188–198. (*44*)

2. BLOKH, V. I.: On some consequences of two variational principles in the mechanics of solid deformable bodies. Akad. Nauk. Ukr. RSR Prikl. Meh. **8**, 56–62. (*18*)

3. BRAMBLE, J. H., and L. E. PAYNE: On the uniqueness problem in the second boundary-value problem in elasticity. Proc. Fourth U. S. National Congress Appl. Mech. **1**, 469–473. (*32*)
4. — — Some uniqueness theorems in the theory of elasticity. Arch. Rational Mech. Anal. **9**, 319–328. (*32*)
5. COURANT, R.: Partial Differential Equations. Vol. 2 of Methods of Mathematical Physics by R. COURANT and D. HILBERT. New York: Interscience. (*6, 74*)
6. GOBERT, J.: Une inégalité fondamentale de la théorie de l'élasticité. Bull. Soc. Roy Sci. Liège **31**, 182–191. (*13*)
7. GURTIN, M.: On Helmholtz's theorem and the completeness of the Papkovitch-Neuber stress functions for infinite domains. Arch. Rational Mech. Anal. **9**, 225–233. (*44*)
8. GUTZWILLER, M. C.: Note on Green's function in anisotropic elasticity. Q. Appl. Math. **20**, 249–256. (*68*)
9. HASHIN, Z., and S. SHTRIKMAN: On some variational principles in anisotropic and nonhomogeneous elasticity. J. Mech. Phys. Solids **10**, 335–352. (*34*)
10. KNESHKE, A., u. D. RÜDIGER: Eine Erweiterung des Hamiltonschen Prinzips zur Integration mechanischer Anfangs-Randwertaufgaben. Ing. Arch. **31**, 101–112. (*65*)
11. MINAGAWA, S.: Riemannian three-dimensional stress-function space. RAAG Mem. **3**, 69–81. (*17*)
12. MURTAZAEV, D.: Fundamental solutions of systems of equations in elasticity theory [in Russian]. Izv. Vysš. Učebn. Zaved. Matematika No. 1 (26) 109–117. (*51, 68*)
13. STERNBERG, E., and M. GURTIN: On the completeness of certain stress functions in the linear theory of elasticity. Proc. Fourth U. S. Nat. Cong. Appl. Mech. 793–797. (*44, 67*)
14. STROH, A. N.: Steady state problems in anisotropic elasticity. J. Math. Phys. **41**, 77–103. (*71*)
15. STRUBECKER, K.: Airysche Spannungsfunktion und isotrope Differentialgeometrie. Math. Z. **78**, 189–198. (*47*)

1963 *1.* ADLER, G.: Maggiorazione delle tensioni in un corpo elastico mediante gli spostamenti superficiali. Atti Accad. Lincei Rend. (8) **34**, 369–371. (*42*)
2. BOURGIN, D. G.: Modern Algebraic Topology. New York: Macmillan. (*71*)
3. DIAZ, J. B., and L. E. PAYNE: New mean value theorems in the linear theory of elasticity. Contrib. Diff. Eqts. **1**, 29–38. (*43*)
4. ERICKSEN, J. L.: Non-existence theorems in linear elasticity theory. Arch. Rational Mech. Anal. **14**, 180–183. (*33*)
5. FEDOROV, F. I.: The theory of elastic waves in crystals. Comparison with isotropic medium [in Russian]. Kristallografiya **8**, 213–220; translated as Soviet Phys.-Cryst. **8**, 159–163. (*70*)
6. FICHERA, G.: Problemi elastostatici con vincoli unilaterali: Il problema di Signorini con ambigue condizioni al contorno. Atti Accad. Naz. Lincei Mem. (8) **7**, 91–140. English translation in: Estratto dai Seminari dell'Istituto Nazionale di Alta Matematica 1962–1963. Rome: Edizioni Cremonese 1964. (*31*)
7. GAZIS, D. C., I. TADJBAKHSH, and R. A. TOUPIN: The elasticity tensor of a given symmetry nearest to an anisotropic elastic tensor. Acta Cryst. **16**, 917–922. (*22*)
8. GRAFFI, D.: Sui teoremi di reciprocità nei fenomeni non stazionari. Atti Accad. Sci. Bologna (11) **10**, 33–40. (*19, 61*)
9. GURTIN, M. E.: A generalization of the Beltrami stress functions in continuum mechanics. Arch. Rational Mech. Anal. **13**, 321–329. (*17*)
10. — A note on the principle of minimum potential energy for linear anisotropic elastic solids. Q. Appl. Math. **20**, 379–382. (*32, 34*)
11. — Variational principles in the linear theory of viscoelasticity. Arch. Rational Mech. Anal. **13**, 179–191. (*36*)
12. HAYES, M.: Wave propagation and uniqueness in prestressed elastic solids. Proc. Roy. Soc. Lond. (A) **274**, 500–506. (*32*)
13. HILL, R.: New derivations of some elastic extremum principles. Progress in Applied Mechanics, the Prager Anniversary Volume. New York: Macmillan. 99–106. (*34*)
14. IGNACZAK, J.: A completeness problem for the stress equation of motion in the linear theory of elasticity. Arch. Mech. Stosow. **15**, 225–234. (*59, 62*)
15. JOHN, F.: Estimates for the error in the equations of nonlinear plate theory. New York University, Courant Institute of Math. Sc. Rept. IMM-NYU 308. (*55*)
16. KAWATATE, K.: A note on the stress function. Rep. Res. Inst. Appl. Mech. Kyushi Univ. **11**, 21–27. (*17*)

17. Kupradze, V. D.: Potential Methods in the Theory of Elasticity [in Russian]. Moscow. (I have seen this work only in the English translation by H. Gutfreund, Jerusalem: Monson 1965.) (*27, 30, 45, 49, 75*).

18. — Dynamical problems in elasticity. Vol. III of Progress in Solid Mechanics. New York: Wiley. (*27, 30, 45, 49, 75*)

19. Melnik, S. I.: Estimates for the St. Venant principle [in Russian]. Perm. Gos. Univ. Učen. Zap. Mat. **103**, 178–180. (*55*)

20. Mills, N.: Uniqueness in classical linear isotropic elasticity theory. M. Sci. Thesis, University of Newcastle upon Tyne. (*32*)

21. Pastori, M.: Equilibrio e congruenza nei sistemi continui. Aspetto formale e geometrico. Aspetto energetico ed estensioni. Confer. Sem. Mat. Univ. Bari, 95–96. (*17*)

22. Tang, L.-M., and H.-C. Sun: Three-dimensional elasticity problems solved by complex variable method. Sci. Sinica (Peking) **12**, 1627–1649. (*44*)

23. Teodorescu, P. P.: Sur l'utilisation des potentiels réels dans le problème plan de la théorie de l'élasticité. Bull. Math. Soc. Sci. Math. Phys. R.P. Romaine (N.S.) **7** (55), 77–111. (*47*)

24. Truesdell, C.: The meaning of Betti's reciprocal theorem. J. Res. Nat. Bur. Std. B **67**, 85–86. (*30*)

25. Volkov, S. D., and M. L. Komissarova: Certain representations of the general solutions of the boundary-value problems of elasticity theory [in Russian]. Inžh. Žh. **3**, 86–92. (*44*)

1964 *1.* Adler, G.: Majoration des tensions dans un corps élastique à l'aide des déplacements superficiels. Arch. Rational Mech. Anal. **16**, 345–372. (*42*)

2. Aleksandrov, A. Y., and I. I. Solovev: On a generalization of a method of solution of axially symmetric problems in the theory of elasticity by means of analytic functions to spatial problems without axial symmetry [in Russian]. Dokl. Akad. Nauk SSSR **154**, 294–297; translated as Soviet Phys. Dokl. **9**, 99–102. (*44*)

3. Bers, L., F. John, and M. Schechter: Partial Differential Equations. New York: Interscience. (*74*)

4. Bramble, J. H., and L. E. Payne: Some mean value theorems in elastostatics. SIAM J. **12**, 105–114. (*43*)

5. Coleman, B. D., and W. Noll: Material symmetry and thermostatic inequalities in finite elastic deformations. Arch. Rational Mech. Anal. **15**, 87–111. (*21*)

6. Chen, Y.: Remarks on variational principles in elastodynamics. J. Franklin Inst. **278**, 1–7. (*65*)

7. Fedorov, F. I.: On the theory of waves in crystals. [in Russian]. Mosk. Univ. Vestn. Fiz. Ast. (3) **6**, 36–40. (*71*)

8. Gurtin, M. E.: Variational principles in linear elastodynamics. Arch. Rational Mech. Anal. **16**, 34–50. (*19, 62, 64, 65*)

9. Herrmann, L. R.: Stress functions for the asymmetric, orthotropic elasticity equations. AIAA J. **2**, 1822–1824. (*44*)

10. Ionov, V. N., and G. A. Vvodenskiĭ: On the possible forms of the general solution of the equilibrium equations in curvilinear coordinates [in Russian]. Izv. Vysš. Učebn. Zaved. Mat. (43) **6**, 59–66. (*17*)

11. Jeffrey, A.: A note on the derivation of the discontinuity conditions across contact discontinuities, shocks, and phase fronts. Z. Angew. Math. Phys. **15**, 68–71. (*73*)

12. Keller, H. B.: Propagation of stress discontinuities in inhomogeneous elastic media. SIAM Rev. **6**, 356–382. (*73*)

13. Knops, R.: Uniqueness for the whole space in classical elasticity. J. Lond. Math. Soc. **39**, 708–712. (*32*)

14. Martin, J.: A displacement bound technique for elastic continua subjected to a certain class of dynamic loading. J. Mech. Phys Solids **12**, 165–175. (*60*)

15. Nôno, T.: Sur les champs de tensions. J. Sci. Hiroshima Univ. Ser. A–I. **28**, 209–221. (*17*)

16. Payton, R. G.: An application of the dynamic Betti-Rayleigh reciprocal theorem to moving point loads in elastic media. Q. Appl. Math. **21**, 299–313. (*61*)

17. Rieder, G.: Die Berechnung des Spannungsfeldes von Einzelkräften mit Hilfe räumlicher Spannungsfunktionen und ihre Anwendung zur quellenmäßigen Darstellung der Verschiebung bei Eigenspannungszuständen. Österr. Ing. Arch. **18**, 173–201. (*17*)

18. — Die Randbedingungen für den Spannungsfunktionentensor an ebenen und gekrümmten belasteten Oberflächen. Österr. Ing. Arch. **18**, 208–243. (*17*)

19. — Über eine Spezialisierung des Schaeferschen Spannungsfunktionenansatzes in der räumlichen Elastizitätstheorie. Z. Angew. Math. Mech. **44**, 329–330. (*17*)

20. STERNBERG, E., and S. AL-KHOZAI: On Green's functions and Saint Venant's principle in the linear theory of viscoelasticity. Arch. Rational Mech. Anal. **15**, 112–146. (*51, 54*)

21. STICKFORTH, J.: Zur Anwendung des Castiglianoschen Prinzips und der Beltramischen Spannungsfunktionen bei mehrfach zusammenhängenden elastischen Körpern unter Berücksichtigung von Eigenspannungen. Tech. Mitt. Krupp Forsch.-Ber. **22**, 83–92. (*18*)

22. TEODORESCU, P. P.: A hundred years of research on the plane problem of the theory of elasticity [in Romanian]. Stud. Cerc. Mat. **15**, 355–367. (*45*)

23. YU, Y.-Y.: Generalized Hamilton's principle and variational equation of motion in nonlinear elasticity theory, with application to plate theory. J. Acoust. Soc. Am. *36*, 111–120. (*65*)

24. ZORSKI, H.: On the equations describing small deformations superposed on finite deformations. Proc. Intl. Sympos. Second-order Effects, Haifa 1962, 109–128. (*32*)

1965 *1.* ALEXSANDROV, I. Y.: Solution of axisymmetric and other three-dimensional problems in the theory of elasticity by means of analytical functions. Appl. Theory of Functions in Continuum Mechanics (Proc. Int. Sympos. Tbilisi. 1963) Vol. I, 97–118. Moscow: Izdat. Nauka. (*44*)

2. BRAMBLE, J. H., and L. E. PAYNE: Some converses of mean value theorems in the theory of elasticity. J. Math. Anal. Appl. **10**, 553–567. (*43*)

3. BRUGGER, K.: Pure modes for elastic waves in crystals. J. Appl. Phys. **36**, 759–768. (*71*)

4. BRUN, L.: Sur l'unicité en thermoélasticité dynamique et diverses expressions analogues à la formule de Clapeyron. C. R. Acad. Sci., Paris **261**, 2584–2587. (*60, 63*)

5. FEDOROV, F. I.: Theory of Elastic Waves in Crystals [in Russian]. Moscow: Nauka Press (I have seen this work only in the English translation by J. E. S. BRADLEY, New York: Plenum 1968.) (*70, 71*)

6. DE VEUBEKE, B. F.: Displacement and equilibrium models in the finite element method. Stress Analysis. New York: Wiley. (*38*)

7. ERICKSEN, J. L.: Nonexistence theorems in linearized elastostatics. J. Diff. Eqts. **1**, 446–451. (*33*)

8. FRIEDRICHS, K. O., and H. B. KELLER: A finite difference scheme for generalized Neumann problems. Numerical Solution of Partial Differential Equations. New York: Academic. (*39*)

9. GURTIN, M. E., and R. A. TOUPIN: A uniqueness theorem for the displacement boundary-value problem of linear elastodynamics. Q. Appl. Math. **23**, 79–81. (*63*)

10. JOHN, F.: Estimates for the derivatives of the stresses in a thin shell and interior shell equations. Comm. Pure Appl. Math. **18**, 235–267. (*55*)

11. KELLER, H. B.: Saint-Venant's procedure and Saint Venant's principle. Q. Appl. Math. **22**, 293–304. (*54*)

12. KNOPS, R. J.: Uniqueness of axisymmetric elastostatic problems for finite regions. Arch. Rational Mech. Anal. **18**, 107–116. (*32*)

13. — Uniqueness of the displacement boundary-value problem for the classical elastic half-space. Arch. Rational Mech. Anal. **20**, 373–377. (*32, 50*)

14. MINAGAWA, S.: On Riemannian and non-Riemannian stress function spaces with reference to two-dimensional stress functions. Reports of the Research Institute for Strength and Fracture of Materials. Tohuku Univ. **1**, 15–31. (*17*)

15. MUSKHELISHVILI, N. I.: Applications of the theory of functions of a complex variable to the theory of elasticity. Appl. Theory of Functions in Continuum Mechanics (Proc. Int. Sympos. Tbilisi, 1963) Vol. I, 32–55 (Russian); 56–75 (English) Moscow: Izdat Nauka. (*45*)

16. REISSNER, E.: A note on variational principles in elasticity. Int. J. Solids Structures **1**, 93–95. (*38*)

17. STICKFORTH, J.: On the derivation of the conditions of compatibility from Castigliano's principle by means of three-dimensional stress functions. J. Math. Phys. **44**, 214–226. (*18*)

18. STIPPES, M.: Steady state waves in anisotropic media. Ann. Meeting Soc. Eng. Sci., Nov., 1965 (unpublished). (*71*)

19. TEODORESCU, P. P.: Über das dreidimensionale Problem der Elastokinetik. Z. Angew. Math. Mech. **45**, 513–523. (*59*)

20. TOUPIN, R. A.: Saint-Venant and a matter of principle. Trans. N.Y. Acad. Sci. **28**, 221–232. (*54, 55*)

21. — Saint-Venant's principle. Arch. Rational Mech. Anal. **18**, 83–96. (*24, 55*)

22. TRUESDELL, C., and W. NOLL: The non-linear field theories of mechanics. In Vol. 3^3 of the Handbuch der Physik, edited by S. FLÜGGE. Berlin-Heidelberg-New York: Springer Verlag. (*12, 16, 19, 20, 21, 23, 73*)

1966 *1.* AINOLA, L. I.: A reciprocal theorem for dynamic problems of the theory of elasticity [in Russian]. Prikl. Math. Meh. **31**, 176–177; translated as J. Appl. Math. Mech. **31**, 190–192. (*61*)

2. BEN-AMOZ, M.: Variational principles in anisotropic and non-homogeneous elastokinetics. Q. Appl. Math. **24**, 82–86. (*65*)

3. BERAN, M., and J. MOLYNEUX: Use of classical variational principles to determine bounds for the effective bulk modulus in heterogeneous media. Q. Appl. Math. **24**, 107–118. (*34*)

4. BERNIKOV, H. B.: Generalization of the Boussinesq-Galerkin solution to anisotropic bodies [in Russian]. Dokl. Akad. Nauk. Tadžh. SSR **9**, No. 6, 3–6. (*44*)

5. BURCKHARDT, J. J.: Die Bewegungsgruppen der Kristallographie. Basel und Stuttgart: Birkhäuser Verlag. (*21*)

6. CARLSON, D.: Completeness of the Beltrami stress functions in continuum mechanics. J. Math. Anal. Appl. **15**, 311–315. (*17*)

7. DOU, A.: Upper estimate of the potential elastic energy of a cylinder. Comm. Pure Appl. Math. **19**, 83–93. (*54*)

8. DOYLE, J. M.: Integration of the Laplace transformed equations of classical elastokinetics. J. Math. Anal. Appl. **13**, 118–131. (*69*)

9. HAYES, M.: On the displacement boundary-value problem in linear elastostatics. Q. J. Mech. Appl. Math. **19**, 151–155. (*32*)

10. KARNOPP, B. H.: Complementary variational principles in linear elastodynamics. Tensor (N.S.) **17**, 300–307. (*65*)

11. KEY, S.: A convergence investigation of the direct stiffness method. Ph. D. Thesis. Dept. Aero. Eng. University of Washington. (*39*)

12. KNOWLES, J. K.: On Saint Venant's principle in the two-dimensional linear theory of elasticity. Arch. Rational Mech. Anal. **21**, 1–22. (*55, 56*)

13. — and E. STERNBERG: On Saint Venant's principle and the torsion of solids of revolution. Arch. Rational Mech. Anal. **22**, 100–120. (*55*)

14. KOLODNER, I.: Existence of longitudinal waves in anisotropic media. J. Acoust. Soc. Am. **40**, 730–731. (*71*)

15. KOMKOV, V.: Application of Rall's theorem to classical elastodynamics. J. Math. Anal. Appl. **14**, 511–521. (*65*)

16. MIKHLIN, S. G.: On Cosserat functions. Problems in Mathematical Analysis. Vol. 1. Boundary Value Problems and Integral Equations [in Russian]. Leningrad University Press. (I have seen this work only in the English translation. New York: Consultant's Bureau 1968, 55–64.) (*32*)

17. MUSKHELISHVILI: Some Basic Problems of the Mathematical Theory of Elasticity [in Russian] 5th Ed. Moscow. (I have not seen this work.) (*45*)

18. NARIBOLI, G. A.: Wave propagation in anisotropic elasticity. J. Math. Anal. Appl. **16**, 108–122. (*73*)

19. ORAVAS, G., and L. MCLEAN: Historical developments of energetical principles in elastomechanics. Appl. Mech. Rev. **19**, 647–656, 919–933. (*34, 38*)

20. RADZHABOV, R. R.: The construction of the Green's tensor for the first boundary-value problem of the theory of elasticity [in Russian]. Voprosy Kibernet. Vychisl. Mat. Vyp. **6**, 96–111. (*53*)

21. ROBINSON, A.: Non-Standard Analysis. Amsterdam: North-Holland. (*56a*)

22. ROSEMAN, J. J.: A pointwise estimate for the stress in a cylinder and its application to Saint-Venant's principle. Arch. Rational Mech. Anal. **21**, 23–48. (*55*)

23. SHIELD, R. T., and C. A. ANDERSON: Some least work principles for elastic bodies. Z. Angew. Math. Phys. **17**, 663–676. (*30*)

24. STERNBERG, E., and J. K. KNOWLES: Minimum energy characterizations of Saint Venant's solution to the relaxed Saint-Venant problem. Arch. Rational Mech. Anal. **21**, 89–107. (*40, 41*)

25. STIPPES, M.: On stress functions in classical elasticity. Q. Appl. Math. **24**, 119–125. (*17, 44*)

26. TONTI, E.: Condizioni iniziali nei principi variazionali. Rend. Ist. Lombardo (A) **100**, 982–988. (*65*)

27. TRUESDELL, C.: Existence of longitudinal waves. J. Acoust. Soc. Am. **40**, 729–730. (*71*)

28. — The Elements of Continuum Mechanics. Berlin-Heidelberg-New York: Springer. (*27*)

1967 *1.* AZHYMUDINOV, T., and G. MAMATURDIEV: Representation of eigenvalues of the first boundary-value problem of elasticity theory by means of a contour-integral. Izv. Akad. Nauk UzSSR Ser. Fiz.-Mat. Nauk **11**, No. 3, 3–9. (*76*)

2. Beatty, M. F.: A reciprocal theorem in the linearized theory of couple-stresses. Acta Mech. **3**, 154–166. (*30*)
3. Bézier, P.: Sur quelques propriétés des solutions de problèmes de l'élastostatique linéaire. C. R. Acad. Sci., Paris (A) **265**, 365–367. (*48, 50, 51*)
4. Carlson, D.: A note on the Beltrami stress functions. Z. Angew. Math. Mech. **47**, 206–207. (*17*)
5. Chadwick, P., and E. A. Trowbridge: Elastic wave fields generated by scalar wave functions. Proc. Cambridge Phil. Soc. **63**, 1177–1187. (*67*)
6. Hlaváček, I.: Derivation of non-classical variational principles in the theory of elasticity. Apl. Mat. **12**, 15–29. (*38*)
7. — Variational principles in the linear theory of elasticity for general boundary conditions. Apl. Mat. **12**, 425–448. (*38, 41*)
8. Hashin, Z.: Variational principles of elasticity in terms of the polarization tensor. Int. J. Eng. Sci. **5**, 213–223. (*34*)
9. Hill, R.: Eigenmodal deformations in elastic/plastic continua. J. Mech. Phys. Solids **15**, 371–386. (*63*)
10. Kilchevskii, M. O., and O. F. Levchuk: One of the forms of the general solution of the elastodynamic equations [in Ukrainian]. Dopovidi Akad. Nauk. Ukr. RSR (A) 801–804. (*67*)
11. Prager, W.: Variational principles of linear elastostatics for discontinuous displacements, strains and stresses. Recent Progress in Applied Mechanics. The Folke Odqvist Volume. New York: Wiley. 463–474. (*38, 41*)
12. Roseman, J. J.: The principle of Saint Venant in linear and non-linear plane elasticity. Arch. Rational Mech. Anal. **26**, 142–162. (*56*)
13. Schuler, K. W., and R. L. Fosdick: Generalized Beltrami stress functions. Dept. Mech. Rept., Illinois Inst. Tech. (*17, 47*)
14. Stippes, M.: A note on stress functions. Int. J. Solids Structures **3**, 705–711. (*17, 44*)
15. Tong, P., and T. H. H. Pian: The convergence of finite element method in solving linear elastic problems. Int. J. Solids Structures **3**, 865–879. (*39*)
16. Tonti, E.: Principii variazionali nell'elastostatica. Atti Accad. Lincei Rend. (8) **42**, 390–394. (*18, 38*)
17. — Variational principles in elastostatics. Meccanica **2**, 201–208. (*18, 38*)
18. Turteltaub, M. J., and E. Sternberg: Elastostatic uniqueness in the half-space. Arch. Rational Mech. Anal. **24**, 233–242. (*50*)
19. Zienkiewicz, O. C., and Y. K. Cheung: The Finite Element Method in Structural and Continuum Mechanics. New York: McGraw-Hill. (*39*)

1968 *1.* Bross, H.: Das Fundamentalintegral der Elastizitätstheorie für kubische Medien. Z. Angew. Math. Phys. **19**, 434–446. (*51*)
2. Cąkala, S., Z. Domański, and H. Milicer-Grużewska: Construction of the fundamental solution of the system of equations of the static theory of elasticity and the solution of the first boundary-value problem for a half space [in Polish]. Zeszyty Nauk. Politech. Warszaw. Mat. No. 14, 151–164. (*51*)
3. Dafermos, C. M.: Some remarks on Korn's inequality. Z. Angew. Math. Phys. **19**, 913–920. (*13*)
4. Edelstein, W. S., and R. L. Fosdick: A note on non-uniqueness in linear elasticity theory. Z. Angew. Math. Phys. **19**, 906–912. (*32*)
5. Fosdick, R. L.: On the displacement boundary-value problem of static linear elasticity theory. Z. Angew. Math. Phys. **19**, 219–233. (*57*)
6. Gurtin, M. E., V. J. Mizel, and W. O. Williams: A note on Cauchy's stress theorem. J. Math. Anal. Appl. **22**, 398–401. (*15*)
7. Hayes, M., and R. J. Knops: On the displacement boundary-value problem of linear elastodynamics. Q. Appl. Math. **26**, 291–293. (*63*)
8. Horák, V.: Inverse Variationsprinzipien der Mechanik fester Körper. Z. Angew. Math. Mech. **48**, 143–146. (*38*)
9. Indenbom, V. L., and S. S. Orlov: Construction of Green's function in terms of Green's function of lower dimensions [in Russian]. Prikl. Math. Meh. **32**, 414–420; translated as J. Appl. Math. Mech. 414–420. (*53*)
10. Karnopp, B. H.: Reissner's variational principle and elastodynamics. Acta Mech. **6**, 158–164. (*65*)
11. Knops, R. J., and L. E. Payne: Uniqueness in classical elastodynamics. Arch. Rational Mech. Anal. **27**, 349–355. (*63*)
12. Solomon, L.: Élasticité Linéaire. Paris: Masson. (*27, 38, 44, 47*)
13. Truesdell, C.: Comment on longitudinal waves. J. Acoust. Soc. Am. **43**, 170. (*71*)

14. TURTELTAUB, M. J., and E. STERNBERG: On concentrated loads and Green's functions in elastostatics. Arch. Rational Mech. Anal. **29**, 193–240. (*51, 52, 53*)
15. WASHIZU, K.: Variational methods in elasticity and plasticity. New York: Pergamon. (*38*)
16. WHEELER, L. T., and E. STERNBERG: Some theorems in classical elastodynamics. Arch. Rational Mech. Anal. **31**, 51–90. (*60, 61, 68, 69, 74*)

1969 *1.* BERG, C. A.: The physical meaning of astatic equilibrium in Saint Venant's principle for linear elasticity. J. Appl. Mech. **36**, 392–396. (*18, 29*)
2. BRUN, L.: Méthodes énergétiques dans les systèmes évolutifs linéaires. J. Mécanique **8**, 125–191. (*60, 63*)
3. DE LA PENHA, G., and S. B. CHILDS: A note on Love's stress function. AIAA J. **7**, 540. (*44*)
4. KANWAL, R. P.: Integral equations formulation of classical elasticity. Q. Appl. Math. **27**, 57–65. (*52*)
5. NOWAK, M.: Equations of motion of an elastic free-free body. Bull. Acad. Polon. Sci., Ser. Sci. Tech. **17**, 31–36. (*68*)
6. RUBENFELD, L. A., and J. B. KELLER: Bounds on elastic moduli of composite media. SIAM J. Appl. Math. **17**, 495–510. (*34*)
7. SCHULTZ, M. H.: Error bounds on the Rayleigh-Ritz-Galerkin method. J. Math. Anal. Appl. **27**, 524–533. (*39*)
8. STIPPES, M.: Completeness of the Papkovitch potentials. Q. Appl. Math. **26**, 477–483. (*44*)
9. YOUNGDAHL, C. K.: On the completeness of a set of stress functions appropriate to the solution of elasticity problems in general cylindrical coordinates. Int. J. Eng. Sci. **7**, 61–79. (*44*)

1970 *1.* BENTHIEN, G., and M. E. GURTIN: A principle of minimum transformed energy in elastodynamics. J. Appl. Mech. **31**, 1147–1149. (*66*)
2. HLAVÁČEK, I., and J. NEČAS: On inequalities of Korn's type. I. Boundary-value problems for elliptic systems of partial differential equations. II. Applications to linear elasticity. Arch. Rational Mech. Anal. **36**, 305–334. (*28*)
3. WHEELER, L. T.: Some results in the linear dynamical theory of anisotropic elastic solids. Q. Appl. Math. **28**, 91–101. (*61, 74*)
4. ZIENKIEWICZ, O. C.: The finite element method; from intuition to generality. Appl. Mech. Rev. **23**, 249–256. (*39*)

1971 *1.* FICHERA, G.: Existence theorems in elasticity. Below in Volume VIa/2 of the Handbuch der Physik, edited by C. TRUESDELL. Berlin-Heidelberg-New York: Springer. (*5, 7, 13, 31, 33, 36, 78*)
2. KNOPS, R. J., and L. E. PAYNE: Uniqueness theorems in linear elasticity. In: Springer Tracts in Natural Philosophy (ed. B. D. COLEMAN), Vol. 19. Berlin-Heidelberg-New York: Springer. (*32, 40, 63*)
3. MAITI, M., and G. R. MAKAN: On an integral equation approach to displacement problems of classical elasticity. Q. Appl. Math. To appear. (*52*)
4. SEGENREICH, S. A.: On the strain energy distribution in self equilibrated cylinders.
5. Thesis. COPPE, Federal Univ. Rio de Janeiro.

Addendum.

The following additional papers would have been referred to, had I seen them in time.

1956 TELEMAN, S.: The method of orthogonal projection in the theory of elasticity. Rev. Math. Pures Appl. **1**, 49–66.

1957 ROSEAU, M.: Sur un théorème d'unicité applicable à certains problèmes de diffraction d'ondes élastiques. C.R. Acad. Sci., Paris **245**, 1780–1782.

1959 CAMPANATO, S.: Sui problemi al contorno per sistemi de equazioni differenziali lineari del tipo dell'elasticità. Ann. Scuola Norm. Pisa (3) **13**, 223–258, 275–302.

1961 BAĬKUZIEV, K.: Some methods of solving the Cauchy problem of the mathematical theory of elasticity [in Russian]. Izv. Akad. Nauk USSR Ser. Fiz.-Mat. 3–12.

1962 GEGELIA, T. G.: Some fundamental boundary-value problems in elasticity theory in space [in Russian]. Akad. Nauk Gruzin. SSSR Trudy Tbiliss. Mat. Inst. Razmadze **28**, 53–72.
KUPRADZE, V. D.: Singular integral equations and boundary-value problems of elasticity theory [in Russian]. Tbiliss. Gos. Univ. Trudy Ser. Meh.-Mat. Nauk. **84**, 63–75.
RIEDER, G.: Iterationsverfahren und Operatorgleichungen in der Elastizitätstheorie. Abhandl. Braunschweig. Wiss. Ges. **14**, 109–343.

1963 DEEV, V. M.: Representation of the general solution of a three-dimensional problem of elasticity theory by particular solutions of Lamé's equations [in Ukrainian]. Dopovidi Akad. Nauk. Ukr. RSR 1464–1467.

VOLKOV, S. D., and M. L. KOMISSAROVA: Certain representations of the general solutions of the boundary-value problems of elasticity theory [in Russian]. Inzh. Zh. **3**, 86–92.

1964 KURLANDZKI, J.: Mathematical formulation of a certain method for solving boundary problems of mechanics. Proc. Vibration Problems **5**, 117–124.

1965 NIEMEYER, H.: Über die elastischen Eigenschwingungen endlicher Körper. Arch. Rational Mech. Anal. **19**, 24–61.

1966 BARR, A. D. S.: An extension of the Hu-Washizu variational principle in linear elasticity for dynamic problems. J. Appl. Mech. **88**, 465.

MIKHLIN, S. G.: Numerical Realization of Variational Methods [in Russian]. Moscow. (I have not seen this work.)

1967 COLAUTTI, M. P.: Sui problemi variazionali di un corpo elastico incompressibile. Mem. Accad. Lincei (8) **8**, 291–343.

LAI, P. T.: Potentiels élastiques: Tenseurs de Green et de Neumann. J. Mécan. **6**, 211–242.

1968 CAKALA, S., Z. DOMAŃSKI, and H. MILICER-GRUŻEWSKA: Construction of the fundamental solution of the system of equations of the static theory of elasticity and the solution of the first boundary value problem for a half space [in Polish]. Zeszyty Nauk. Politech. Warszaw. Mat. No. **14**, 151–164.

KNOPS, R. J., and L. E. PAYNE: Stability in linear elasticity. Int. J. Solids Structures **4**, 1233–1242.

LAI, P. T.: Elastic potentials and Green's tensor. Part I. Mémorial de l'Artillerie Française **42**, 23–96.

1970 BUFLER, H.: Erweiterung des Prinzips der virtuellen Verschiebungen und des Prinzips der virtuellen Kräfte. Z. Angew. Math. Mech. **50**, 104–108.

1971 HLAVÁČEK, I.: On Reissner's variational theorem for boundary values in linear elasticity. Apl. Mat. **16**, 109–124.

HORGAN, C. O., and J. K. KNOWLES: Eigenvalue problems associated with Korn's inequalities. Arch. Rational Mech. Anal. **40**, 384–402.

MAISONNEUVE, O.: Sur le principe de Saint-Venant. Thèse, Université de Poitiers.

1972 DUVAUT, G., and J. L. LIONS: Sur les Inéquations en Mécanique et en Physique. Paris: Dunod.

Linear Thermoelasticity.

By

DONALD E. CARLSON.[1]

A. Introduction.

1. The nature of this article. It is the purpose of this article to present a short account of the linear theory of thermoelasticity using the style and notation of GURTIN'S treatise, *The Linear Theory of Elasticity*[2], preceding in this volume. While it would probably be possible to put together a thermoelastic counterpart to each of the results given there, considerations of space and time have caused extensive selection to be made.

In Chap. B, the linear theory of thermoelasticity is derived by starting with the nonlinear theory. The exposition of nonlinear thermoelasticity given here is in no sense complete; only enough is presented to provide a base for the linear theory.

Chaps. C and D contain the equilibrium and dynamic theories, respectively. They may be read independently of each other and, for those already somewhat familiar with the subject, of Chap. B. The basic equations and smoothness hypotheses are given in the first section of each chapter. The treatment is fully three-dimensional. No special results associated with geometrical symmetries are considered.

2. Notation. Following LTE, *direct notation* is employed. Thus, *scalars* appear as italic light face letters, *vectors* and *points* as italic bold face minuscules, *second-order tensors* as italic bold face majuscules, and *fourth-order tensors* as sans-serif bold face majuscules. Of course, the same scheme is used for functions whose values are such quantities. Reference may be made to Chap. B of LTE for more details on this notational scheme and also for standard mathematical results utilized throughout.

---

[1] *Acknowledgment*. This article was planned jointly with M. E. GURTIN as an extension of his treatise, *The Linear Theory of Elasticity*, preceding in this volume. Unfortunately, the completion of that work kept him from participating as fully as he had intended. Nonetheless, the text of Chap. B was essentially written by GURTIN, and I am deeply grateful to him for his detailed criticism of the manuscript as well. I would also like to thank M. C. STIPPES for numerous discussions throughout the course of the writing. Finally, I wish to express my gratitude to the U.S. National Science Foundation for their support through a research grant to the University of Illinois.

[2] Referenced hereafter as "LTE".

Index of frequently used symbols. Only symbols used in more than one section are listed.

Symbol	Name	Section of first occurrence
$\boldsymbol{A}$	Thermal expansion tensor	7
B	Body	3
∂B	Boundary of B	3
$\mathbf{\mathsf{C}}$	Elasticity tensor	7
$\mathscr{C}$	Reference configuration	3
$\boldsymbol{D}$	Finite strain tensor	5
$\boldsymbol{E}$	Infinitesimal strain tensor	7
$\boldsymbol{E}_0$	Initial strain	21
$\boldsymbol{F}$	Deformation gradient	3
$\boldsymbol{K}$	Conductivity tensor	6
$\mathbf{\mathsf{K}}$	Compliance tensor	7
$\boldsymbol{M}$	Stress-temperature tensor	7
$\mathbf{0}$	Origin, zero vector, zero tensor	3
P	Part of B	3
$\mathbf{S}$	Stress tensor	3
$\mathbf{S}'$	Reduced stress	11
$\mathscr{S}_1, \mathscr{S}_2; \mathscr{S}_3, \mathscr{S}_4$	Complementary Subsets of ∂B	14, 22
$\mathscr{U}$	Total energy	21
$\boldsymbol{b}$	Non-inertial body force	7
$\mathscr{b}$	Pseudo body force	21
$\boldsymbol{b}'$	Reduced body force	11
c	Specific heat	4
e	Internal energy	3
$\boldsymbol{f}$	Inertial body force	3
$\boldsymbol{g}$	Temperature gradient	3
i	Function with values $i(t)=t$	21
	$\sqrt{-1}$	25
k	Conductivity	8
m	Stress-temperature modulus	8
$\boldsymbol{n}$	Unit outward normal to ∂B	3
$\boldsymbol{p}$	Position vector	3
$\boldsymbol{q}$	Heat flux vector	3
q	Heat flux	18
$\hat{q}$	Prescribed heat flux on boundary	21
r	Heat supply	3
$\mathscr{r}$	Pseudo heat supply	21
$\mathbf{s}$	Surface traction	9
$\mathbf{s}'$	Reduced traction	11
$\hat{\mathbf{s}}$	Prescribed traction on boundary	14
t	Time	3
$\boldsymbol{u}$	Displacement vector	3
$\hat{\boldsymbol{u}}$	Prescribed displacement on boundary	14
$\boldsymbol{u}_0$	Initial displacement	21
$\boldsymbol{v}_0$	Initial velocity	21
$\boldsymbol{x}$	Point in B	3
α	Coefficient of thermal expansion	8

Symbol	Name	Section of first occurrence
η	Entropy	3
θ	Absolute temperature	3
θ_0	Reference temperature	7
ϑ	Temperature difference	9
ϑ_0	Initial temperature	21
$\hat{\vartheta}$	Prescribed temperature on boundary	22
λ	Lamé modulus	8
μ	Shear modulus	8
ψ	Free energy	3
ϱ	Density	3
$\mathbf{1}$	Unit tensor	3
∇	Gradient	3
Δ	Laplacian	12
$\otimes$	Tensor product	10
$*$	Convolution	21
$\dot{(\ )}$	Time derivative	3
$(\ )^T$	Transpose of a tensor	3
curl	Curl	12
div	Divergence	3
sym	Symmetric part of a tensor	9
tr	Trace of a tensor	8

B. The foundations of the linear theory of thermoelasticity.

3. The basic laws of mechanics and thermodynamics. We identify the ***body***[3] B with the bounded *regular region*[4] of space it occupies in a fixed ***reference configuration*** $\mathscr{C}$. Points $\boldsymbol{x} \in B$ will be referred to as ***material points***.[5]

A ***motion*** of the body B is a class C^2 vector field $\boldsymbol{u}$ on $B \times (0, t_0)$, where $(0, t_0)$ is a fixed open interval of time. Given $\boldsymbol{x} \in B$ and $t \in (0, t_0)$, the vector $\boldsymbol{u}(\boldsymbol{x}, t)$ represents the ***displacement*** of the material point $\boldsymbol{x}$ at the time t; thus, the point $\boldsymbol{x} + \boldsymbol{u}(\boldsymbol{x}, t)$ is the position in space of $\boldsymbol{x}$ at time t. The spatial gradient $\boldsymbol{F}$ of the function $\boldsymbol{x} \rightarrow \boldsymbol{x} + \boldsymbol{u}(\boldsymbol{x}, t)$ is called the ***deformation gradient***; clearly,

$$\boldsymbol{F} = \mathbf{1} + \nabla \boldsymbol{u}. \tag{3.1}$$

As is usual, *we assume that for each t the mapping* $\boldsymbol{x} \rightarrow \boldsymbol{x} + \boldsymbol{u}(\boldsymbol{x}, t)$ *is one-to-one on* B *and that its inverse is smooth*; thus, $\det \boldsymbol{F} \neq 0$.

If $\boldsymbol{S}(\boldsymbol{x}, \boldsymbol{t})$ denotes the ***first Piola-Kirchhoff stress tensor***[6] measured per unit surface area in the reference configuration $\mathscr{C}$, and if $\boldsymbol{f}(\boldsymbol{x}, t)$ is the ***body force***[7]

[3] A precise definition of a body has been given by NOLL [1958, *3*]. See also TRUESDELL and NOLL [1965, *5*, Sect. 15].

[4] See LTE, Sect. 5.

[5] More commonly in the treatise literature of continuum mechanics, the letter X is used to designate an arbitrary material point; while $\boldsymbol{x}$ is the position in space it occupies at some time in a motion.

[6] See TRUESDELL and NOLL [1965, *5*, Sect. 43A].

[7] For convenience the body force $\boldsymbol{f}$ *includes the inertial body force* $-\varrho \ddot{\boldsymbol{u}}$, where ϱ is the density in the reference configuration.

per unit volume in $\mathscr{C}$, then the ***laws of balance of forces and moments*** take the forms

$$\int_{\partial P} \boldsymbol{S}\boldsymbol{n}\, da + \int_P \boldsymbol{f}\, dv = \boldsymbol{0},$$

$$\int_{\partial P} (\boldsymbol{p}+\boldsymbol{u}) \times (\boldsymbol{S}\boldsymbol{n})\, da + \int_P (\boldsymbol{p}+\boldsymbol{u}) \times \boldsymbol{f}\, dv = \boldsymbol{0}$$

for every part[8] P of B and every time t, where

$$\boldsymbol{p} = \boldsymbol{x} - \boldsymbol{0}$$

is the *position vector* of $\boldsymbol{x}$ relative to the origin $\boldsymbol{0}$ and $\boldsymbol{n}$ is the unit outward normal to ∂P. If $\boldsymbol{S}$ is of class $C^{1,0}$ and $\boldsymbol{f}$ continuous on $B \times (0, t_0)$, then these laws together are equivalent to

$$\operatorname{div} \boldsymbol{S} + \boldsymbol{f} = \boldsymbol{0}, \tag{3.2}$$ [9]

$$\boldsymbol{S}\boldsymbol{F}^T = \boldsymbol{F}\boldsymbol{S}^T. \tag{3.3}$$

The ***rate of working*** of a part P at time t is defined by

$$W(P) = \int_{\partial P} (\boldsymbol{S}\boldsymbol{n}) \cdot \dot{\boldsymbol{u}}\, da + \int_P \boldsymbol{f} \cdot \dot{\boldsymbol{u}}\, dv. \tag{3.4}$$ [10]

Since

$$\dot{\boldsymbol{F}} = \nabla \dot{\boldsymbol{u}},$$

it follows from the divergence theorem that

$$\int_{\partial P} (\boldsymbol{S}\boldsymbol{n}) \cdot \dot{\boldsymbol{u}}\, da = \int_P \boldsymbol{S} \cdot \dot{\boldsymbol{F}}\, dv + \int_P \dot{\boldsymbol{u}} \cdot \operatorname{div} \boldsymbol{S}\, dv;$$

and hence (3.2) and (3.4) imply

$$W(P) = \int_P \boldsymbol{S} \cdot \dot{\boldsymbol{F}}\, dv. \tag{3.5}$$

The quantity $\boldsymbol{S} \cdot \dot{\boldsymbol{F}}$ is called the ***stress power***.

The ***first law of thermodynamics*** is the assertion that

$$\frac{d}{dt} \int_P e\, dv = W(P) - \int_{\partial P} \boldsymbol{q} \cdot \boldsymbol{n}\, da + \int_P r\, dv$$

for every part P and every time t. Here $e(\boldsymbol{x}, t)$ is the ***internal energy*** per unit volume in $\mathscr{C}$; $\boldsymbol{q}(\boldsymbol{x}, t)$ is the ***heat flux vector*** measured per unit surface area in $\mathscr{C}$; and $r(\boldsymbol{x}, t)$ is the ***heat supply*** per unit volume in $\mathscr{C}$ from the external world. The quantity

$$\int_P e\, dv$$

is the internal energy of P;

$$-\int_{\partial P} \boldsymbol{q} \cdot \boldsymbol{n}\, da$$

is the total heat flux into P by conduction across ∂P; and

$$\int_P r\, dv$$

---

[8] A regular subregion of B.

[9] The differential operator div, like ∇, is with respect to the *material point* $\boldsymbol{x} \in B$.

[10] Note that since $\boldsymbol{f}$ includes the inertial body force, $W(P)$ includes the term $-\int_P \varrho\, \ddot{\boldsymbol{u}} \cdot \dot{\boldsymbol{u}}\, dv$, which is the negative of the rate of change of the kinetic energy.

represents the total heat supplied into the interior of P from the external world. If e is of class $C^{0,1}$, $\boldsymbol{q}$ of class $C^{1,0}$, and r continuous on $B\times(0, t_0)$, then, in view of (3.5), the first law has the local equivalent

$$\dot{e} = \boldsymbol{S}\cdot\dot{\boldsymbol{F}} - \operatorname{div}\boldsymbol{q} + r. \tag{3.6}$$

The ***second law of thermodynamics*** is the assertion that

$$\frac{d}{dt}\int_P \eta\, dv \geqq -\int_{\partial P}\frac{\boldsymbol{q}\cdot\boldsymbol{n}}{\theta}\, da + \int_P \frac{r}{\theta}\, dv$$

for every part P and every time t. Here $\eta(\boldsymbol{x}, t)$ is the ***entropy*** per unit volume in $\mathscr{C}$, and $\theta(\boldsymbol{x}, t) > 0$ is the ***absolute temperature***. Hence,

$$\int_P \eta\, dv$$

is the entropy of P;

$$\int_{\partial P}\frac{\boldsymbol{q}\cdot\boldsymbol{n}}{\theta}\, da$$

is the total entropy flux across ∂P due to conduction; and

$$\int_P \frac{r}{\theta}\, dv$$

represents the entropy supplied into the interior of P from the external world. If, in addition to the previous assumptions, η is of class $C^{0,1}$ and θ of class $C^{1,0}$ on $B\times(0, t_0)$, the second law is equivalent to

$$\dot{\eta} \geqq -\operatorname{div}\left(\frac{\boldsymbol{q}}{\theta}\right) + \frac{r}{\theta}. \tag{3.7}$$

Introducing the ***free energy***

$$\psi = e - \eta\theta \tag{3.8}$$

and combining (3.6) and (3.7), we arrive at the ***local dissipation inequality***

$$\dot{\psi} + \eta\dot{\theta} - \boldsymbol{S}\cdot\dot{\boldsymbol{F}} + \frac{1}{\theta}\boldsymbol{q}\cdot\boldsymbol{g} \leqq 0. \tag{3.9}$$

Here $\boldsymbol{g}$ is the ***temperature gradient***

$$\boldsymbol{g} = \nabla\theta. \tag{3.10}$$

Granted (3.6) and (3.8), then (3.7) and (3.9) are equivalent.

4. Elastic materials. Consequences of the second law. An ***elastic material*** is defined by *constitutive equations* giving the free energy ψ, the stress $\boldsymbol{S}$, the entropy η, and the heat flux $\boldsymbol{q}$ at each material point $\boldsymbol{x}$ whenever the deformation gradient $\boldsymbol{F}$, the temperature θ, and the temperature gradient $\boldsymbol{g}$ are known at $\boldsymbol{x}$:

$$\begin{aligned} \psi &= \hat{\psi}(\boldsymbol{F}, \theta, \boldsymbol{g}, \boldsymbol{x}), \\ \boldsymbol{S} &= \hat{\boldsymbol{S}}(\boldsymbol{F}, \theta, \boldsymbol{g}, \boldsymbol{x}), \\ \eta &= \hat{\eta}(\boldsymbol{F}, \theta, \boldsymbol{g}, \boldsymbol{x}), \\ \boldsymbol{q} &= \hat{\boldsymbol{q}}(\boldsymbol{F}, \theta, \boldsymbol{g}, \boldsymbol{x}). \end{aligned} \tag{4.1}$$

We assume, once and for all, that the ***response functions*** *$\hat{\psi}$, $\hat{\boldsymbol{S}}$, $\hat{\eta}$, and $\hat{\mathrm{g}}$ are smooth on their domain $\mathscr{D}$,* which is the set of all $(\boldsymbol{F}, \theta, \boldsymbol{g}, \boldsymbol{x})$ where $\boldsymbol{F}$ is a tensor with

$\det \boldsymbol{F} \neq 0$, θ is a strictly positive scalar, $\boldsymbol{g}$ is a vector, and $\boldsymbol{x} \in B$. Further, *we assume that $\hat{\boldsymbol{S}}$ is compatible with balance of moments in the sense of* (3.3):

$$\hat{\boldsymbol{S}}(\boldsymbol{F}, \theta, \boldsymbol{g}, \boldsymbol{x})\, \boldsymbol{F}^T = \boldsymbol{F} \hat{\boldsymbol{S}}(\boldsymbol{F}, \theta, \boldsymbol{g}, \boldsymbol{x})^T. \tag{4.2}$$

For convenience we will generally suppress the argument $\boldsymbol{x}$ in (4.1).

By an ***admissible thermodynamic process*** we mean an ordered array

$$[\boldsymbol{u}, \theta, \psi, \boldsymbol{S}, \eta, \boldsymbol{q}]$$

with the following properties:

(i) All of the functions have a common domain of the form $P \times T$, where P is a part of B and T is an open (time) interval of $(0, t_0)$;

(ii) $\boldsymbol{u}$ is a motion on $P \times T$, and θ is a strictly positive class $C^{2,1}$ scalar field on $P \times T$;

(iii) ψ, $\boldsymbol{S}$, η, and $\boldsymbol{q}$ are defined on $P \times T$ through the constitutive relations (4.1) with

$$\boldsymbol{F} = \boldsymbol{1} + \nabla \boldsymbol{u}, \qquad \boldsymbol{g} = \nabla \theta.$$

We call $P \times T$ the *domain of the process*; this domain may vary from process to process.

Given an admissible thermodynamic process, balance of forces (3.2) and balance of energy (3.6) yield the body force $\boldsymbol{f}$ and heat supply r that must be applied to support the process:

$$\boldsymbol{f} = -\operatorname{div} \boldsymbol{S}, \qquad r = \dot{e} - \boldsymbol{S} \cdot \dot{\boldsymbol{F}} + \operatorname{div} \boldsymbol{q}.$$

However, the local dissipation inequality will imply certain restrictions on the response functions.

(1) Restrictions placed on elastic materials by the second law.[11] *A necessary and sufficient condition that every admissible thermodynamic process obey the local dissipation inequality* (3.9) *is that the following three statements hold*:

(i) *The response functions $\hat{\psi}$, $\hat{\boldsymbol{S}}$, and $\hat{\eta}$ are independent of the temperature gradient $\boldsymbol{g}$*:

$$\psi = \hat{\psi}(\boldsymbol{F}, \theta), \qquad \boldsymbol{S} = \hat{\boldsymbol{S}}(\boldsymbol{F}, \theta), \qquad \eta = \hat{\eta}(\boldsymbol{F}, \theta);$$

(ii) *$\hat{\psi}$ determines $\hat{\boldsymbol{S}}$ through the* **stress relation**

$$\hat{\boldsymbol{S}}(\boldsymbol{F}, \theta) = \partial_{\boldsymbol{F}} \hat{\psi}(\boldsymbol{F}, \theta)$$

---

[11] COLEMAN and NOLL [1963, *1*] established this theorem for the case in which the temperature gradient is omitted in the constitutive relations for the stress, free energy, and entropy. COLEMAN and MIZEL [1964, *1*] proved that the second law rules out such a dependence, even if it is assumed from the outset. Actually, some of the underlying ideas are contained in the treatise of GREEN and ADKINS [1960, *3*, Sect. 8.3], who established the stress relation and the heat conduction inequality as consequences of the local dissipation inequality under the following assumptions: the entropy relation holds (an assumption not crucial to their argument); the free energy depends only on the strain and temperature; the heat flux and stress do not depend on the time derivative of the strain and temperature. However, GREEN and ADKINS do not include the heat supply term in their statement of the first law; the inclusion of this term allows one to specify arbitrarily the mechanical and thermodynamic variables without violating balance of energy. Evidently, COLEMAN and NOLL [1963, *1*] were the first to notice this fact. TRUESDELL and TOUPIN [1960, *6*, Sects. 240, 241, 256–268] appear to be the first to write the first two laws with the heat supply terms indicated.

Theorem ***(1)*** was generalized by GURTIN [1965, *3*, *4*] [1967, *2*], who showed that if, in place of (4.1), ψ, $\boldsymbol{S}$, η, and $\boldsymbol{q}$ are allowed to be *functionals* of the displacement and temperature *fields* over the entire body, then the local dissipation inequality implies that such a constitutive assumption must necessarily reduce to the forms given in (i) of ***(1)***.

and $\hat{\eta}$ *through the* ***entropy relation***

$$\hat{\eta}(\boldsymbol{F}, \theta) = -\partial_\theta \hat{\psi}(\boldsymbol{F}, \theta);$$

(iii) $\hat{\boldsymbol{q}}$ *obeys the* ***heat conduction inequality***

$$\hat{\boldsymbol{q}}(\boldsymbol{F}, \theta, \boldsymbol{g}) \cdot \boldsymbol{g} \leqq 0.$$

Proof. In view of (4.1), an admissible thermodynamic process is compatible with the local dissipation inequality if and only if

$$\begin{aligned}[\partial_{\boldsymbol{F}} \hat{\psi}(\boldsymbol{F}, \theta, \boldsymbol{g}) - \hat{\boldsymbol{S}}(\boldsymbol{F}, \theta, \boldsymbol{g})] \cdot \dot{\boldsymbol{F}} + [\partial_\theta \hat{\psi}(\boldsymbol{F}, \theta, \boldsymbol{g}) + \hat{\eta}(\boldsymbol{F}, \theta, \boldsymbol{g})]\, \dot{\theta} \\ + \partial_{\boldsymbol{g}} \hat{\psi}(\boldsymbol{F}, \theta, \boldsymbol{g}) \cdot \dot{\boldsymbol{g}} + \frac{1}{\theta} \hat{\boldsymbol{q}}(\boldsymbol{F}, \theta, \boldsymbol{g}) \cdot \boldsymbol{g} \leqq 0\end{aligned} \tag{4.3}$$

on its domain. The *sufficiency* of (i)–(iii) follows from (4.3). With a view toward demonstrating the *necessity* of these conditions, we first establish the following

(2) Lemma. *Given* $(\boldsymbol{F}^*, \theta^*, \boldsymbol{g}^*, \boldsymbol{x}^*) \in \mathscr{D}$, $t^* \in (0, t_0)$, *a tensor* $\boldsymbol{A}$, *a scalar* α, *and a vector* $\boldsymbol{a}$, *there exists an admissible thermodynamic process with*

$$\begin{aligned} \boldsymbol{F}(\boldsymbol{x}^*, t^*) &= \boldsymbol{F}^*, & \dot{\boldsymbol{F}}(\boldsymbol{x}^*, t^*) &= \boldsymbol{A}, \\ \theta(\boldsymbol{x}^*, t^*) &= \theta^*, & \dot{\theta}(\boldsymbol{x}^*, t^*) &= \alpha, \\ \boldsymbol{g}(\boldsymbol{x}^*, t^*) &= \boldsymbol{g}^*, & \dot{\boldsymbol{g}}(\boldsymbol{x}^*, t^*) &= \boldsymbol{a}. \end{aligned}$$

Proof of ***(2)***. Clearly, there exists an open ball $P \subset B$ about $\boldsymbol{x}^*$ and an open interval $T \subset (0, t_0)$ containing t^* such that

$$\begin{gathered} \det[\boldsymbol{F}^* + (t - t^*)\, \boldsymbol{A}] \neq 0, \\ \theta^* + \alpha(t - t^*) + [\boldsymbol{g}^* + (t - t^*)\, \boldsymbol{a}] \cdot (\boldsymbol{x} - \boldsymbol{x}^*) > 0 \end{gathered}$$

for every $(\boldsymbol{x}, t) \in P \times T$. Let $\boldsymbol{u}$ and θ be defined on $P \times T$ by

$$\begin{aligned} \boldsymbol{u}(\boldsymbol{x}, t) &= [-\boldsymbol{1} + \boldsymbol{F}^* + (t - t^*)\, \boldsymbol{A}]\, (\boldsymbol{x} - \boldsymbol{x}^*), \\ \theta(\boldsymbol{x}, t) &= \theta^* + \alpha(t - t^*) + [\boldsymbol{g}^* + (t - t^*)\, \boldsymbol{a}] \cdot (\boldsymbol{x} - \boldsymbol{x}^*). \end{aligned}$$

The corresponding admissible thermodynamic process has all of the desired properties. □

We now continue with the proof of ***(1)***. Applying the inequality (4.3) at $\boldsymbol{x} = \boldsymbol{x}^*$ and $t = t^*$ to the process constructed in ***(2)*** yields

$$\begin{aligned}[\partial_{\boldsymbol{F}} \hat{\psi}(\boldsymbol{F}^*, \theta^*, \boldsymbol{g}^*) - \hat{\boldsymbol{S}}(\boldsymbol{F}^*, \theta^*, \boldsymbol{g}^*)] \cdot \boldsymbol{A} + [\partial_\theta \hat{\psi}(\boldsymbol{F}^*, \theta^*, \boldsymbol{g}^*) + \hat{\eta}(\boldsymbol{F}^*, \theta^*, \boldsymbol{g}^*)]\, \alpha \\ + \partial_{\boldsymbol{g}} \hat{\psi}(\boldsymbol{F}^*, \theta^*, \boldsymbol{g}^*) \cdot \boldsymbol{a} + \frac{1}{\theta^*} \hat{\boldsymbol{q}}(\boldsymbol{F}^*, \theta^*, \boldsymbol{g}^*) \cdot \boldsymbol{g}^* \leqq 0,\end{aligned}$$

which implies the necessity of (i)–(iii) since $\boldsymbol{F}^*$, θ^*, $\boldsymbol{g}^*$, $\boldsymbol{x}^*$, t^*, $\boldsymbol{A}$, α, and $\boldsymbol{a}$ are arbitrary. □

Henceforth, we assume that the conditions (i)–(iii) *of* ***(1)*** *hold.*

By (3.8),

$$\dot{e} = \dot{\psi} + \theta \dot{\eta} + \eta \dot{\theta},$$

while (i) and (ii) of ***(1)*** imply that

$$\dot{\psi} = \boldsymbol{S} \cdot \dot{\boldsymbol{F}} - \eta \dot{\theta};$$

thus,

$$\dot{e} = \boldsymbol{S} \cdot \dot{\boldsymbol{F}} + \theta \dot{\eta},$$

and we have the following result.

(3) *In an admissible thermodynamic process, the energy equation* (3.6) *takes the form*

$$\theta \dot{\eta} = -\operatorname{div} \boldsymbol{q} + r. \tag{4.4}$$

Thus, an admissible thermodynamic process is ***adiabatic*** ($-\operatorname{div} \boldsymbol{q} + r \equiv 0$) *if and only if it is* ***isentropic*** ($\dot{\eta} \equiv 0$).

As another important consequence of ***(1)***, we have

(4) *In an admissible thermodynamic process, the response functions for stress and entropy satisfy the* ***Maxwell relation***

$$\partial_\theta \hat{\boldsymbol{S}}(\boldsymbol{F}, \theta) = -\partial_{\boldsymbol{F}} \hat{\eta}(\boldsymbol{F}, \theta). \tag{4.5}$$

Proof. Since $\hat{\boldsymbol{S}}$ and $\hat{\eta}$ are smooth by hypothesis, it follows from (ii) of ***(1)*** that $\hat{\psi}$ is of class C^2. Therefore,

$$\partial_\theta \, \partial_{\boldsymbol{F}} \, \hat{\psi}(\boldsymbol{F}, \theta) = \partial_{\boldsymbol{F}} \, \partial_\theta \, \hat{\psi}(\boldsymbol{F}, \theta),$$

and (4.5) is implied by (ii) of ***(1)***. □

In view of (3.8), (i) of ***(1)***, and (4.1), the internal energy obeys a constitutive relation of the form

$$e = \hat{e}(\boldsymbol{F}, \theta), \tag{4.6}$$

where

$$\hat{e}(\boldsymbol{F}, \theta) = \hat{\psi}(\boldsymbol{F}, \theta) + \theta \hat{\eta}(\boldsymbol{F}, \theta). \tag{4.7}$$

The number

$$c(\boldsymbol{F}, \theta) = \partial_\theta \, \hat{e}(\boldsymbol{F}, \theta) \tag{4.8}$$

is called the ***specific heat***. By (4.7) and (4.8),

$$c(\boldsymbol{F}, \theta) = \partial_\theta \, \hat{\psi}(\boldsymbol{F}, \theta) + \hat{\eta}(\boldsymbol{F}, \theta) + \theta \, \partial_\theta \hat{\eta}\,(\boldsymbol{F}, \theta),$$

and thus (ii) of ***(1)*** implies

$$c(\boldsymbol{F}, \theta) = \theta \, \partial_\theta \, \hat{\eta}(\boldsymbol{F}, \theta). \tag{4.9}$$

For the remainder of this section, we assume that

$$c(\boldsymbol{F}, \theta) > 0 \tag{4.10}$$

for all $(\boldsymbol{F}, \theta)$. Since $\theta > 0$, (4.9) and (4.10) imply that $\hat{\eta}(\boldsymbol{F}, \theta)$ is smoothly invertible in θ for each choice of $\boldsymbol{F}$; therefore, we may rewrite our basic constitutive assumption (4.1) in the form

$$\begin{aligned} e &= \bar{e}(\boldsymbol{F}, \eta), & \boldsymbol{S} &= \bar{\boldsymbol{S}}(\boldsymbol{F}, \eta), \\ \theta &= \bar{\theta}(\boldsymbol{F}, \eta), & \boldsymbol{q} &= \bar{\boldsymbol{q}}(\boldsymbol{F}, \eta, \boldsymbol{g}). \end{aligned} \tag{4.11}$$

For given $\boldsymbol{F}$, the function $\eta \to \bar{\theta}(\boldsymbol{F}, \eta)$ in (4.11) is just the inverse of the function $\theta \to \hat{\eta}(\boldsymbol{F}, \theta)$ in ***(1)***; while the function $\bar{e}$, e.g., is given by

$$\bar{e}(\boldsymbol{F}, \eta) = \hat{e}\big(\boldsymbol{F}, \bar{\theta}(\boldsymbol{F}, \eta)\big). \tag{4.12}$$

It follows from (4.7) and (4.12) that

$$\bar{e}(\boldsymbol{F}, \eta) = \hat{\psi}\big(\boldsymbol{F}, \bar{\theta}(\boldsymbol{F}, \eta)\big) + \eta \bar{\theta}(\boldsymbol{F}, \eta);$$

therefore,

$$\partial_{\boldsymbol{F}}\, \bar{e} = \partial_{\boldsymbol{F}}\, \hat{\psi} + (\partial_\theta\, \hat{\psi})\, \partial_{\boldsymbol{F}}\, \bar{\theta} + \eta\, \partial_{\boldsymbol{F}}\, \bar{\theta},$$
$$\partial_\eta\, \bar{e} = (\partial_\theta\, \hat{\psi})\, \partial_\eta\, \bar{\theta} + \bar{\theta} + \eta\, \partial_\eta\, \bar{\theta}$$

and we conclude from (ii) of *(1)* that

$$\begin{aligned} \bar{\boldsymbol{S}}(\boldsymbol{F}, \eta) &= \partial_{\boldsymbol{F}}\, \bar{e}(\boldsymbol{F}, \eta), \\ \bar{\theta}(\boldsymbol{F}, \eta) &= \partial_\eta\, \bar{e}(\boldsymbol{F}, \eta). \end{aligned} \tag{4.13}$$

Eqs. (4.13) are the ***stress*** and ***temperature relations*** when entropy is taken as the independent thermodynamic variable.

5. The principle of material frame-indifference. The ***principle of material frame-indifference*** asserts that the constitutive assumption (4.1) be independent of the observer.[12] Under a change of observer the relevant mechanical and thermodynamic quantities transform as follows:[13]

$$\begin{aligned} &\boldsymbol{F} \to \boldsymbol{Q}\boldsymbol{F}, \\ &\boldsymbol{S} \to \boldsymbol{Q}\boldsymbol{S}, \\ &\theta \to \theta, \\ &\boldsymbol{g} \to \boldsymbol{g}, \\ &\psi \to \psi, \\ &\eta \to \eta, \\ &\boldsymbol{q} \to \boldsymbol{q}, \end{aligned}$$

where $\boldsymbol{Q}$ is the orthogonal tensor corresponding to this change. Since (4.1) is to be invariant under all such changes, it must satisfy

$$\begin{aligned} \hat{\psi}(\boldsymbol{F}, \theta) &= \hat{\psi}(\boldsymbol{Q}\boldsymbol{F}, \theta), \\ \hat{\boldsymbol{S}}(\boldsymbol{F}, \theta) &= \boldsymbol{Q}^T \hat{\boldsymbol{S}}(\boldsymbol{Q}\boldsymbol{F}, \theta), \\ \hat{\eta}(\boldsymbol{F}, \theta) &= \hat{\eta}(\boldsymbol{Q}\boldsymbol{F}, \theta), \\ \hat{\boldsymbol{q}}(\boldsymbol{F}, \theta, \boldsymbol{g}) &= \hat{\boldsymbol{q}}(\boldsymbol{Q}\boldsymbol{F}, \theta, \boldsymbol{g}) \end{aligned} \tag{5.1}$$

for every orthogonal tensor $\boldsymbol{Q}$ and for all $(\boldsymbol{F}, \theta, \boldsymbol{g}) \in \mathscr{D}$.

By the polar decomposition theorem,

$$\boldsymbol{F} = \boldsymbol{R}\boldsymbol{U} \tag{5.2}$$

where $\boldsymbol{R}$ is orthogonal and $\boldsymbol{U}$ is the positive definite square root of $\boldsymbol{F}^T\boldsymbol{F}$. Choosing $\boldsymbol{Q} = \boldsymbol{R}^T$ in (5.1) and using (5.2), we arrive at the relations

$$\begin{aligned} \psi &= \hat{\psi}(\boldsymbol{U}, \theta), \\ \boldsymbol{S} &= \boldsymbol{F}\boldsymbol{U}^{-1}\hat{\boldsymbol{S}}(\boldsymbol{U}, \theta), \\ \eta &= \hat{\eta}(\boldsymbol{U}, \theta), \\ \boldsymbol{q} &= \hat{\ }(\boldsymbol{U}, \theta, \boldsymbol{g}). \end{aligned} \tag{5.3}$$

[12] See, e.g., TRUESDELL and NOLL [1965, *5*, Sects. 17–19A].

[13] The transformation rule for $\boldsymbol{F}$ is a consequence of Eq. $(19.2)_1$ of TRUESDELL and NOLL [1965, *5*]. The rule for the first Piola-Kirchhoff stress $\boldsymbol{S}$ (their $\boldsymbol{T}_R$) follows from the transformation law for the Cauchy stress $\boldsymbol{T}$ (their Eq. $(19.2)_2$) and their Eq. (43A.2) relating $\boldsymbol{S}$ to $\boldsymbol{T}$ and $\boldsymbol{F}$. The transformation rules for θ, ψ, and η are immediate. The temperature gradient $\boldsymbol{g}$ is invariant since it is the gradient relative to the reference configuration $\mathscr{C}$. The heat flux $\boldsymbol{q}$ is invariant since it is measured per unit area in $\mathscr{C}$; the details may be worked out as for $\boldsymbol{S}$.

Defining the ***finite strain tensor D*** through

$$\boldsymbol{D} = \tfrac{1}{2}(\boldsymbol{U}^2 - \boldsymbol{1}) = \tfrac{1}{2}(\boldsymbol{F}^T\boldsymbol{F} - \boldsymbol{1}), \tag{5.4}$$

we are led to

(1) Consequences of material frame-indifference.[14] *The constitutive equations* (4.1) *satisfy the principle of material frame-indifference if and only if they can be written in the* ***reduced forms***

$$\begin{aligned} \psi &= \tilde{\psi}(\boldsymbol{D}, \theta) \\ \boldsymbol{S} &= \boldsymbol{F}\tilde{\boldsymbol{S}}(\boldsymbol{D}, \theta), \\ \eta &= \tilde{\eta}(\boldsymbol{D}, \theta), \\ \boldsymbol{q} &= \tilde{\boldsymbol{q}}(\boldsymbol{D}, \theta, \boldsymbol{g}). \end{aligned} \tag{5.5}$$

Moreover, the stress and entropy relations of ***(4.1)*** *reduce to*

$$\begin{aligned} \tilde{\boldsymbol{S}}(\boldsymbol{D}, \theta) &= \partial_{\boldsymbol{D}}\tilde{\psi}(\boldsymbol{D}, \theta), \\ \tilde{\eta}(\boldsymbol{D}, \theta) &= -\partial_\theta \tilde{\psi}(\boldsymbol{D}, \theta). \end{aligned} \tag{5.6}$$

Proof. The necessity of (5.5) follows from (5.3) and (5.4) with the function $\tilde{\psi}$, e.g., given by

$$\tilde{\psi}(\boldsymbol{D}, \theta) = \hat{\psi}\left((2\boldsymbol{D} + \boldsymbol{1})^{\frac{1}{2}}, \theta\right).$$

That the reduced relations (5.5) satisfy (5.1) is a simple calculation.

The derivation of $(5.6)_1$ is carried out most readily in terms of components. From the chain rule,

$$\frac{\partial\hat{\psi}}{\partial F_{kl}} = \frac{\partial\tilde{\psi}}{\partial D_{ij}} \frac{\partial D_{ij}}{\partial F_{kl}}.$$

By (5.4),

$$D_{ij} = \tfrac{1}{2}(F_{mi}F_{mj} - \delta_{ij}),$$

and thus

$$\frac{\partial D_{ij}}{\partial F_{kl}} = \frac{1}{2}(F_{ki}\delta_{lj} + F_{kj}\delta_{lj}).$$

Therefore, since

$$\frac{\partial\tilde{\psi}}{\partial D_{ij}} = \frac{\partial\tilde{\psi}}{\partial D_{ji}},$$

we conclude that

$$\frac{\partial\hat{\psi}}{\partial F_{kl}} = F_{ki}\frac{\partial\tilde{\psi}}{\partial D_{il}};$$

or, equivalently,

$$\partial_{\boldsymbol{F}}\hat{\psi} = \boldsymbol{F}\,\partial_{\boldsymbol{D}}\tilde{\psi}.$$

This result, the stress relation of ***(4.1)***, and $(5.5)_2$ imply $(5.6)_1$.

The second of (5.6) follows immediately from the entropy relation of ***(4.1)***. □

It is interesting to note that

$$\boldsymbol{S} = \boldsymbol{F}\,\partial_{\boldsymbol{D}}\tilde{\psi}(\boldsymbol{D}, \theta)$$

automatically satisfies

$$\boldsymbol{S}\boldsymbol{F}^T = \boldsymbol{F}\boldsymbol{S}^T.$$

Thus, *in thermoelasticity, balance of moments is a consequence of balance of forces, the two laws of thermodynamics, and material frame-indifference.*

[14] RICHTER [1952, *2*] was apparently the first to employ the polar decomposition theorem to find the implications of material frame-indifference.

6. Consequences of the heat conduction inequality. In view of (iii) of *(4.1)*, the reduced response function $\tilde{\boldsymbol{q}}$ must obey the inequality

$$\tilde{\boldsymbol{q}}(\boldsymbol{D}, \theta, \boldsymbol{g}) \cdot \boldsymbol{g} \leqq 0. \tag{6.1}$$

In this section, we obtain some important implications of this inequality.

We call

$$\boldsymbol{K}(\boldsymbol{D}, \theta) = -\partial_{\boldsymbol{g}} \boldsymbol{q}(\boldsymbol{D}, \theta, \boldsymbol{g})|_{\boldsymbol{g}=\boldsymbol{0}} \tag{6.2}$$

the ***conductivity tensor.*** Fixing $\boldsymbol{D}$ and θ and expanding

$$\tilde{\boldsymbol{q}}(\boldsymbol{g}) \equiv \tilde{\boldsymbol{q}}(\boldsymbol{D}, \theta, \boldsymbol{g})$$

in a Taylor series about $\boldsymbol{g} = \boldsymbol{0}$ yields

$$\tilde{\boldsymbol{q}}(\boldsymbol{g}) = \tilde{\boldsymbol{q}}(\boldsymbol{0}) - \boldsymbol{K}\boldsymbol{g} + o(|\boldsymbol{g}|) \quad \text{as} \quad \boldsymbol{g} \to \boldsymbol{0}. \tag{6.3}$$

By (6.1) and (6.3),

$$\tilde{\boldsymbol{q}}(\boldsymbol{0}) \cdot \boldsymbol{g} - \boldsymbol{g} \cdot \boldsymbol{K}\boldsymbol{g} + o(|\boldsymbol{g}|^2) \leqq 0 \quad \text{as} \quad \boldsymbol{g} \to \boldsymbol{0}.$$

This inequality holds for all $\boldsymbol{g}$ only if $\tilde{\boldsymbol{q}}(0) = \boldsymbol{0}$ and $\boldsymbol{g} \cdot \boldsymbol{K}\boldsymbol{g} \geqq 0$ for every $\boldsymbol{g}$. Thus we have the following result.

(1) Consequences of the heat conduction inequality.[15] *The heat flux vanishes whenever the temperature gradient vanishes,* i.e.,

$$\tilde{\boldsymbol{q}}(\boldsymbol{D}, \theta, \boldsymbol{0}) = \boldsymbol{0}; \tag{6.4}$$

and the conductivity tensor $\boldsymbol{K}(\boldsymbol{D}, \theta)$ *is positive semi-definite.*

From (6.4),

$$\partial_{\boldsymbol{D}} \tilde{\boldsymbol{q}}(\boldsymbol{D}, \theta, \boldsymbol{0}) = \boldsymbol{0}, \qquad \partial_{\theta} \tilde{\boldsymbol{q}}(\boldsymbol{D}, \theta, \boldsymbol{0}) = \boldsymbol{0}; \tag{6.5}$$

these results will prove useful in the linearization of the next section.

7. Derivation of the linear theory. The complete system of field equations for nonlinear thermoelasticity theory was derived in the preceding sections; it consists in the balance laws

$$\operatorname{div} \boldsymbol{S} + \boldsymbol{f} = \boldsymbol{0}, \qquad \theta \dot{\eta} = -\operatorname{div} \boldsymbol{q} + r \tag{7.1}$$

together with the constitutive equations

$$\begin{aligned} \psi &= \tilde{\psi}(\boldsymbol{D}, \theta), \\ \boldsymbol{S} &= \boldsymbol{F}\tilde{\boldsymbol{S}}(\boldsymbol{D}, \theta), \\ \eta &= \tilde{\eta}(\boldsymbol{D}, \theta), \\ \boldsymbol{q} &= \tilde{\boldsymbol{q}}(\boldsymbol{D}, \theta, \boldsymbol{g}), \end{aligned} \tag{7.2}$$

where

$$\begin{aligned} \boldsymbol{F} &= \boldsymbol{1} + \nabla \boldsymbol{u}, \\ \boldsymbol{D} &= \tfrac{1}{2}(\boldsymbol{F}^T\boldsymbol{F} - \boldsymbol{1}), \\ \boldsymbol{g} &= \nabla \theta. \end{aligned} \tag{7.3}$$

Of course, these constitutive relations are subject to the thermodynamic restrictions

$$\begin{aligned} \tilde{\boldsymbol{S}}(\boldsymbol{D}, \theta) &= \partial_{\boldsymbol{D}} \tilde{\psi}(\boldsymbol{D}, \theta), \\ \tilde{\eta}(\boldsymbol{D}, \theta) &= -\partial_{\theta} \tilde{\psi}(\boldsymbol{D}, \theta), \\ \tilde{\boldsymbol{q}}(\boldsymbol{D}, \theta, \boldsymbol{g}) \cdot \boldsymbol{g} &\leqq 0. \end{aligned} \tag{7.4}$$

---

[15] Pipkin and Rivlin [1958, *5*]. See also Coleman and Noll [1963, *1*], Coleman and Mizel [1963, *2*], and Gurtin [1965, *3*].

In this section, we determine the linear approximation to this system under the following assumptions: the displacement gradient and its rate of change are small; the temperature field is nearly equal to a given *uniform* field θ_0 called the ***reference temperature***; the temperature rate and the temperature gradient are small. Thus, *we assume that*

$$|\nabla \boldsymbol{u}|, \quad |\nabla \dot{\boldsymbol{u}}|, \quad |\theta - \theta_0|, \quad |\dot{\theta}|, \quad |\boldsymbol{g}| \leqq \delta, \tag{7.5}$$

where δ is small.

By $(7.3)_1$ and (7.5),

$$\boldsymbol{F} = \mathbf{1} + O(\delta), \tag{7.6}$$

where here and in what follows all order of magnitude statements are as $\delta \to 0$. The tensor

$$\boldsymbol{E} = \tfrac{1}{2}(\nabla \boldsymbol{u} + \nabla \boldsymbol{u}^T) \tag{7.7}$$

is called the ***infinitesimal strain tensor***; by (7.5),

$$\left.\begin{aligned} \boldsymbol{E} &= O(\delta), \\ \dot{\boldsymbol{E}} &= O(\delta). \end{aligned}\right\} \tag{7.8}$$

From $(7.3)_1$, $(7.3)_2$, (7.5), (7.7), and (7.8),

$$\left.\begin{aligned} \boldsymbol{D} &= \boldsymbol{E} + \tfrac{1}{2}\nabla \boldsymbol{u}^T \nabla \boldsymbol{u}, \\ &= \boldsymbol{E} + O(\delta^2), \\ &= O(\delta); \end{aligned}\right\} \tag{7.9}$$

while $(7.9)_1$ and (7.5) imply that

$$\left.\begin{aligned} \dot{\boldsymbol{D}} &= \dot{\boldsymbol{E}} + \tfrac{1}{2}(\nabla \boldsymbol{u}^T \nabla \dot{\boldsymbol{u}} + \nabla \dot{\boldsymbol{u}}^T \nabla \boldsymbol{u}), \\ &= \dot{\boldsymbol{E}} + O(\delta^2). \end{aligned}\right\} \tag{7.10}$$

When $\boldsymbol{F} = \mathbf{1}$ and $\theta = \theta_0$, $\boldsymbol{D} = \mathbf{0}$ and

$$\boldsymbol{S} = \tilde{\boldsymbol{S}}(\mathbf{0}, \theta_0).$$

Thus $\tilde{\boldsymbol{S}}(\mathbf{0}, \theta_0)$ represents the *residual stress* at the reference temperature, i.e., the stress that the body would experience if it were held in the reference configuration at the uniform temperature θ_0. *We assume that*

$$\tilde{\boldsymbol{S}}(\mathbf{0}, \theta_0) = \mathbf{0}. \tag{7.11}$$

The assumptions of zero residual stress and uniform reference temperature are crucial to the development of the *classical* theory.

By the assumed differentiability of the function $\tilde{\boldsymbol{S}}$ in conjunction with (7.5), $(7.9)_2$, and (7.11),

$$\tilde{\boldsymbol{S}}(\boldsymbol{D}, \theta) = \mathsf{C}[\boldsymbol{E}] + (\theta - \theta_0)\,\boldsymbol{M} + o(\delta), \tag{7.12}$$

where

$$\left.\begin{aligned} \mathsf{C} &= \partial_{\boldsymbol{D}} \boldsymbol{S}(\boldsymbol{D}, \theta)\big|_{\substack{\boldsymbol{D}=\mathbf{0},\\ \theta=\theta_0}} \\ \boldsymbol{M} &= \partial_\theta \tilde{\boldsymbol{S}}(\boldsymbol{D}, \theta)\big|_{\substack{\boldsymbol{D}=\mathbf{0}.\\ \theta=\theta_0}} \end{aligned}\right\} \tag{7.13}$$

Substituting (7.12) into $(7.2)_2$, we find, with the aid of (7.5), (7.6), and (7.8), that

$$\boldsymbol{S} = \mathsf{C}[\boldsymbol{E}] + (\theta - \theta_0)\,\boldsymbol{M} + o(\delta). \tag{7.14}$$

The fourth-order tensor C is called the ***elasticity tensor***; it is a linear transformation of the space of all symmetric (second-order) tensors into itself.[16] By $(7.4)_1$ and $(7.13)_1$,

$$\mathsf{C} = \partial_{\boldsymbol{D}}^2 \tilde{\psi}(\boldsymbol{D}, \theta)\big|_{\substack{\boldsymbol{D}=\boldsymbol{0}\\ \theta=\theta_0}},$$

and we conclude[17] that for any pair of symmetric tensors $\boldsymbol{G}$ and $\boldsymbol{H}$

$$\boldsymbol{G} \cdot \mathsf{C}[\boldsymbol{H}] = \frac{\partial^2}{\partial\alpha\,\partial\beta}\tilde{\psi}(\alpha\boldsymbol{G}+\beta\boldsymbol{H}, \theta_0)\big|_{\alpha=\beta=0} = \boldsymbol{H} \cdot \mathsf{C}[\boldsymbol{G}];$$

thus, *the elasticity tensor is symmetric*. The (second-order) tensor $\boldsymbol{M}$ is called the ***stress-temperature tensor***; again since $\tilde{\boldsymbol{S}} = \tilde{\boldsymbol{S}}^T$, *the stress-temperature tensor is symmetric*.

Expansion of $\boldsymbol{q} = \tilde{\boldsymbol{q}}(\boldsymbol{D}, \theta, \boldsymbol{g})$ in a Taylor series about $(\boldsymbol{0}, \theta_0, \boldsymbol{0})$ yields, with the aid of (6.4), (6.5), (7.5) and $(7.9)_3$,

$$\boldsymbol{q} = -\boldsymbol{K}\boldsymbol{g} + o(\delta), \tag{7.15}$$

where $\boldsymbol{K}$ is the *conductivity tensor* [cf. (6.2)] corresponding to $\boldsymbol{D} = \boldsymbol{0}$ and $\theta = \theta_0$, i.e.,

$$\boldsymbol{K} = -\partial_{\boldsymbol{g}}\tilde{\boldsymbol{q}}(\boldsymbol{0}, \theta_0, \boldsymbol{g})\big|_{\boldsymbol{g}=\boldsymbol{0}}. \tag{7.16}$$

Eq. (7.15) asserts that *to within an error of $o(\delta)$ the heat flux depends linearly on the temperature gradient and is independent of the strain and the temperature.* It is important to note that this result followed from the heat conduction inequality; no arguments concerning symmetry were involved. Furthermore, we do *not* make the common *assumption* that $\boldsymbol{K}$ is symmetric.[18] From *(6.1)*, $\boldsymbol{K}$ is positive semi-definite.

The asymptotic form for the law of balance of forces is simply $(7.1)_1$. If we introduce the ***non-inertial body force*** $\boldsymbol{b}$, then

$$\boldsymbol{f} = \boldsymbol{b} - \varrho\ddot{\boldsymbol{u}}, \tag{7.17}$$

where $\varrho(\boldsymbol{x})$ is the ***density*** in the reference configuration, and $(7.1)_1$ reduces to

$$\operatorname{div}\boldsymbol{S} + \boldsymbol{b} = \varrho\ddot{\boldsymbol{u}}. \tag{7.18}$$

To complete our development of the classical theory, we have only to derive the linearized form of the energy equation $(7.1)_2$. By $(7.2)_3$,

$$\dot{\eta} = \partial_{\boldsymbol{D}}\eta(\boldsymbol{D}, \theta) \cdot \dot{\boldsymbol{D}} + \partial_\theta\tilde{\eta}(\boldsymbol{D}, \theta)\,\dot{\theta}. \tag{7.19}$$

Since $\tilde{\eta}$ is smooth,

$$\begin{aligned}\partial_{\boldsymbol{D}}\tilde{\eta}(\boldsymbol{D}, \theta) &= \partial_{\boldsymbol{D}}\tilde{\eta}(\boldsymbol{D}, \theta)\big|_{\substack{\boldsymbol{D}=\boldsymbol{0}\\ \theta=\theta_0}} + o(1),\\ \partial_\theta\tilde{\eta}(\boldsymbol{D}, \theta) &= \partial_\theta\tilde{\eta}(\boldsymbol{D}, \theta)\big|_{\substack{\boldsymbol{D}=\boldsymbol{0}\\ \theta=\theta_0}} + o(1),\end{aligned} \tag{7.20}$$

where we have made use of (7.5) and $(7.9)_3$. From $(7.4)_1$ and $(7.4)_2$, we see that

$$\partial_{\boldsymbol{D}}\tilde{\eta} = -\partial_\theta\tilde{\boldsymbol{S}}.$$

---

[16] The values of C are symmetric since $\tilde{\boldsymbol{S}} = \tilde{\boldsymbol{S}}^T$ by $(7.4)_1$.

[17] Cf. LTE, Sect. 4.

[18] The history of this assumption goes back to STOKES [1851, *1*]. An extensive discussion has been provided recently by TRUESDELL [1969, *3*, Sect. 7]; to his list of references we add the note by LESSEN [1956, *2*]. See also DAY and GURTIN [1969, *2*]. I. MÜLLER, in results presented to the Society for Natural Philosophy in November, 1970, has constructed a theory with a symmetric conductivity tensor by inserting $\dot{\theta}$ as a variable in the constitutive equations (4.1).

This reduced form of the Maxwell relation (4.5) and $(7.13)_2$ imply that

$$\partial_{\boldsymbol{D}}\tilde{\eta}(\boldsymbol{D},\theta)\big|_{\substack{\boldsymbol{D}=\boldsymbol{0}\\ \theta=\theta_0}} = -\boldsymbol{M}. \tag{7.21}$$

By (4.9) and (5.5), the number

$$c = \theta_0\, \partial_\theta \tilde{\eta}(\boldsymbol{D},\theta)\big|_{\substack{\boldsymbol{D}=\boldsymbol{0}\\ \theta=\theta_0}} \tag{7.22}$$

is the *specific heat* corresponding to $\boldsymbol{D}=\boldsymbol{0}$ and $\theta=\theta_0$. Combining (7.19)–(7.22) and using (7.5), $(7.10)_2$, and $(7.8)_2$ yields

$$\theta\dot{\eta} = -\theta_0\,\boldsymbol{M}\cdot\dot{\boldsymbol{E}} + c\,\dot{\theta} + o(\delta).$$

Substituting this result into $(7.1)_2$, we are led to the following asymptotic form for the energy equation:

$$-\operatorname{div}\boldsymbol{q} + \theta_0\,\boldsymbol{M}\cdot\dot{\boldsymbol{E}} + r + o(\delta) = c\,\dot{\theta}. \tag{7.23}$$

Eqs. $(7.3)_3$, (7.7), (7.14), (7.15), (7.18), and (7.23) with the terms of $o(\delta)$ neglected are the ***basic equations of linearized thermoelasticity theory***:[19]

$$\begin{gathered}
\boldsymbol{E} = \tfrac{1}{2}(\nabla\boldsymbol{u} + \nabla\boldsymbol{u}^T),\\
\operatorname{div}\boldsymbol{S} + \boldsymbol{b} = \varrho\ddot{\boldsymbol{u}},\\
-\operatorname{div}\boldsymbol{q} + \theta_0\,\boldsymbol{M}\cdot\dot{\boldsymbol{E}} + r = c\,\dot{\theta},\\
\boldsymbol{S} = \mathsf{C}[\boldsymbol{E}] + (\theta-\theta_0)\,\boldsymbol{M},\\
\boldsymbol{q} = -\boldsymbol{K}\,\nabla\theta.
\end{gathered} \tag{7.24}$$

For the remainder of this article, we limit our discussion to the linearized system (7.24).

Since $\boldsymbol{M}$ and the values of C are symmetric, we see from $(7.24)_4$ that $\boldsymbol{S}=\boldsymbol{S}^T$; i.e., *the stress tensor is symmetric in the linear theory.*

According to $(7.24)_4$, the stress-temperature tensor $\boldsymbol{M}$ gives the stress resulting from a given temperature distribution when the strain vanishes:

$$\boldsymbol{S} = (\theta-\theta_0)\,\boldsymbol{M} \quad \text{when} \quad \boldsymbol{E}=\boldsymbol{0}.$$

When the elasticity tensor C is invertible,[20] $(7.24)_4$ can be solved for $\boldsymbol{E}$:

$$\boldsymbol{E} = \mathsf{K}[\boldsymbol{S}] + (\theta-\theta_0)\boldsymbol{A}, \tag{7.25}$$

where

$$\begin{aligned}
\mathsf{K} &= \mathsf{C}^{-1},\\
\boldsymbol{A} &= -\mathsf{K}[\boldsymbol{M}]
\end{aligned} \tag{7.26}$$

are called the ***compliance tensor*** and the ***thermal expansion tensor***, respectively. According to (7.25), the thermal expansion tensor gives the strain resulting from a given temperature distribution when the stress vanishes:

$$\boldsymbol{E} = (\theta-\theta_0)\,\boldsymbol{A} \quad \text{when} \quad \boldsymbol{S}=\boldsymbol{0}.$$

---

[19] In the isotropic case, these equations were postulated directly by DUHAMEL [1837, *1*] and NEUMANN [1885, *1*]. The earlier work of DUHAMEL [1838, *1*] and NEUMANN [1841, *1*] did not include the strain rate term in the energy equation. Attempts to provide a thermodynamic basis for them were later made by VOIGT [1895, *1*, Pt. 3, Chap. 1], JEFFREYS [1930, *1*], and LESSEN and DUKE [1953, *1*]. They were produced on the basis of "linear irreversible thermodynamics" by BIOT [1956, *1*]. A derivation founded on modern continuum thermomechanics similar to the one presented here has been given by ERINGEN [1967, *1*, Chap. 8].

Note added in proof. C. TRUESDELL has pointed out to me the surprising fact that DUHAMEL's equations even preceded the enunciation of the general principle of equivalence of heat and work. DUHAMEL's success is evidently attributable to the principle of superposition of effects inherent in a linear theory and to good fortune. These matters will be fully discussed in a treatise on the history of thermodynamics under preparation by TRUESDELL.

[20] Recall that the domain of C has been restricted to the space of all symmetric tensors.

It is important to realize that while the *material functions* $\mathbf{C}$, $\boldsymbol{M}$, $\boldsymbol{K}$, c, and ϱ will generally depend on the reference temperature θ_0, they do not depend on the temperature θ. To allow them to be θ-dependent would be inconsistent with the assumptions leading to the linearized energy equation $(7.24)_3$. Of course, unless the body is ***homogeneous***, the material functions will depend on the position $\boldsymbol{x}$ in B.

Two simplifying approximations are often made to facilitate the solution of actual problems. The first is to neglect the term $\theta_0 \boldsymbol{M} \cdot \dot{\boldsymbol{E}}$ in the energy equation $(7.24)_3$; the resulting theory is referred to as being ***uncoupled***. The second, resulting in the ***quasi-static theory***, is to neglect the term $\varrho \ddot{\boldsymbol{u}}$ in the momentum equation $(7.24)_2$. In practice, these two approximations are usually made together. They are thoroughly discussed in the treatise of BOLEY and WEINER.[21]

8. Isotropy. The effect of material symmetry on the form of the elasticity tensor $\mathbf{C}$ has been discussed in detail by GURTIN.[22] His formulation of the notion of symmetry may be carried over without difficulty to the stress-temperature tensor $\boldsymbol{M}$ and the conductivity tensor $\boldsymbol{K}$.[23]

When the material is *isotropic*.

$$\begin{aligned} \mathbf{C}[\boldsymbol{E}] &= 2\mu\, \boldsymbol{E} + \lambda (\operatorname{tr} \boldsymbol{E})\, \mathbf{1}, \\ \boldsymbol{M} &= m\, \mathbf{1}, \\ \boldsymbol{K} &= k\, \mathbf{1}. \end{aligned} \tag{8.1}$$

The scalars μ and λ are called the ***Lamé moduli***; μ is the ***shear modulus***; m is the ***stress-temperature modulus***; and k is the ***conductivity***. Combining the relations (8.1) with (7.24) leads to the ***basic equations of linear thermoelasticity for an isotropic body***:

$$\begin{aligned} \boldsymbol{E} &= \tfrac{1}{2}(\nabla \boldsymbol{u} + \nabla \boldsymbol{u}^T), \\ \operatorname{div} \boldsymbol{S} + \boldsymbol{b} &= \varrho \ddot{\boldsymbol{u}}, \\ -\operatorname{div} \boldsymbol{q} + m\, \theta_0 \operatorname{tr} \dot{\boldsymbol{E}} + r &= c\, \dot{\theta}, \\ \boldsymbol{S} &= 2\mu\, \boldsymbol{E} + \lambda (\operatorname{tr} \boldsymbol{E})\, \mathbf{1} + m(\theta - \theta_0)\, \mathbf{1}, \\ \boldsymbol{q} &= -k\, \nabla \theta. \end{aligned} \tag{8.2}$$

It follows from $(8.2)_5$ that in an isotropic body, the stress is equal to a *pressure* when the strain vanishes:

$$\boldsymbol{S} = m(\theta - \theta_0)\, \mathbf{1} \quad \text{when} \quad \boldsymbol{E} = \mathbf{0}.$$

If $\mu \neq 0$ and $3\lambda + 2\mu \neq 0$, $(8.2)_5$ can be inverted to yield

$$\boldsymbol{E} = \frac{1}{2\mu} \boldsymbol{S} - \frac{\lambda}{2\mu(3\lambda + 2\mu)} (\operatorname{tr} \boldsymbol{S})\, \mathbf{1} + \alpha(\theta - \theta_0)\, \mathbf{1}, \tag{8.3}$$

where

$$\alpha = -\frac{m}{3\lambda + 2\mu} \tag{8.4}$$

is called the ***coefficient of thermal expansion***. On comparing (8.3) with (7.25) we see that, in the present circumstances,

$$\boldsymbol{A} = \alpha \mathbf{1}. \tag{8.5}$$

[21] [1960, *1*, Chap. 2].
[22] LTE, Sects. 21, 22, 26.
[23] See also COLEMAN and MIZEL [1963, *2*], and WANG's appendix to TRUESDELL [1969, *3*].

According to (8.3), the strain is equal to a *dilatation* when the stress vanishes:

$$\boldsymbol{E} = \alpha(\theta - \theta_0)\,\mathbf{1} \quad \text{when} \quad \boldsymbol{S} = \mathbf{0}.$$

C. Equilibrium theory.

9. Basic equations. Thermoelastic states. In this chapter, we consider the ***equilibrium theory*** of linear elastic heat conductors. In the absence of time dependence, the fundamental system of field equations (7.24) reduces to the ***strain-displacement relation***

$$\boldsymbol{E} = \tfrac{1}{2}(\nabla \boldsymbol{u} + \nabla \boldsymbol{u}^T),$$

the ***equation of equilibrium***

$$\operatorname{div} \boldsymbol{S} + \boldsymbol{b} = \mathbf{0},$$

the ***equilibrium energy equation***

$$-\operatorname{div} \boldsymbol{q} + r = 0,$$

the ***stress-strain-temperature relation***

$$\boldsymbol{S} = \mathsf{C}[\boldsymbol{E}] + (\theta - \theta_0)\,\boldsymbol{M},$$

and the ***heat conduction equation***

$$\boldsymbol{q} = -\boldsymbol{K}\,\nabla\,\theta,$$

here $\boldsymbol{u}$, $\boldsymbol{E}$, $\boldsymbol{S}$, θ, θ_0, $\boldsymbol{q}$, $\boldsymbol{b}$, and r are the displacement, strain, stress, temperature reference temperature, heat flux, body force, and heat supply fields, respectively; while C, $\boldsymbol{M}$, and $\boldsymbol{K}$ are the elasticity, stress-temperature, and conductivity fields, respectively.

We note that the above system is *uncoupled* in the sense that the temperature field can be found by solving the heat flow problem associated with the heat conduction and energy equations. Generally, in this chapter, we shall treat the temperature field as having already been so determined.

In what follows, we shall find it convenient to work with the ***temperature difference field***

$$\vartheta = \theta - \theta_0.$$

We presume that the uniform reference temperature θ_0 has been fixed once and for all. It is important to observe that ϑ can have negative values. In terms of ϑ, the stress-strain-temperature relation and the heat conduction equation become

$$\boldsymbol{S} = \mathsf{C}[\boldsymbol{E}] + \vartheta\,\boldsymbol{M}$$

$$\boldsymbol{q} = -\boldsymbol{K}\,\nabla\vartheta,$$

respectively.

Following LTE,[24] we say that a vector field $\boldsymbol{u}$ is an ***admissible displacement field*** provided

(i) $\boldsymbol{u}$ is of class C^2 on B;

(ii) $\boldsymbol{u}$ and sym $\nabla\boldsymbol{u}$ are continuous on $\bar{B}$.

---

[24] Sect. 18.

An ***admissible stress field*** $\boldsymbol{S}$ is a symmetric tensor field with the properties:

(i) $\boldsymbol{S}$ is smooth on B;

(ii) $\boldsymbol{S}$ and div $\boldsymbol{S}$ are continuous on $\bar{B}$.

The associated ***surface traction*** $\mathbf{s}$ is then defined at every regular point $\boldsymbol{x} \in \partial B$ by

$$\mathbf{s}(\boldsymbol{x}) = \boldsymbol{S}(\boldsymbol{x})\,\boldsymbol{n}(\boldsymbol{x}),$$

where $\boldsymbol{n}(\boldsymbol{x})$ is the unit outward normal to ∂B at $\boldsymbol{x}$.

By an ***admissible state*** we mean an ordered array $[\boldsymbol{u}, \boldsymbol{E}, \boldsymbol{S}]$ with the properties:

(i) $\boldsymbol{u}$ is an admissible displacement field;

(ii) $\boldsymbol{E}$ is a continuous symmetric tensor field on $\bar{B}$;

(iii) $\boldsymbol{S}$ is an admissible stress field.

The fields $\boldsymbol{u}$, $\boldsymbol{E}$, and $\boldsymbol{S}$ need not be related. The set of all admissible states is a vector space provided addition and scalar multiplication are defined in the natural manner:

$$[\boldsymbol{u}, \boldsymbol{E}, \boldsymbol{S}] + [\tilde{\boldsymbol{u}}, \tilde{\boldsymbol{E}}, \tilde{\boldsymbol{S}}] = [\boldsymbol{u} + \tilde{\boldsymbol{u}}, \boldsymbol{E} + \tilde{\boldsymbol{E}}, \boldsymbol{S} + \tilde{\boldsymbol{S}}],$$

$$a[\boldsymbol{u}, \boldsymbol{E}, \boldsymbol{S}] = [a\boldsymbol{u}, a\boldsymbol{E}, a\boldsymbol{S}].$$

We assume for the rest of Chap. C that the elasticity tensor $\mathbf{C}$ *and the stress-temperature tensor* $\boldsymbol{M}$ *are given and that each is smooth on* $\bar{B}$.[25] *Furthermore,* $\mathbf{C}$ *and* $\boldsymbol{M}$ *are symmetric, and the values of* $\mathbf{C}[\cdot]$ *are symmetric.*[26]

By a ***thermoelastic state corresponding to the body force field b and the temperature difference field*** ϑ we mean an admissible state $[\boldsymbol{u}, \boldsymbol{E}, \boldsymbol{S}]$ that meets the field equations

$$\boldsymbol{E} = \tfrac{1}{2}(\nabla\boldsymbol{u} + \nabla\boldsymbol{u}^T),$$

$$\boldsymbol{S} = \mathbf{C}[\boldsymbol{E}] + \vartheta\,\boldsymbol{M},$$

$$\operatorname{div}\boldsymbol{S} + \boldsymbol{b} = \boldsymbol{0}.$$

Of course, this implies that $\boldsymbol{b}$ is continuous and ϑ is smooth on $\bar{B}$. Note that ϑ is not required to satisfy the heat conduction and energy equations, even though this will normally be the case. The ordered array $[\boldsymbol{b}, \mathbf{s}, \vartheta]$ is called the ***external force-temperature system*** for the thermoelastic state $[\boldsymbol{u}, \boldsymbol{E}, \boldsymbol{S}]$.

The following two theorems are immediate consequences of the above definitions and the linearity of the field equations.

(1) Principle of superposition of thermoelastic states. *Let* $[\boldsymbol{u}, \boldsymbol{E}, \boldsymbol{S}]$ *and* $[\tilde{\boldsymbol{u}}, \tilde{\boldsymbol{E}}, \tilde{\boldsymbol{S}}]$ *be thermoelastic states corresponding to the external force-temperature systems* $[\boldsymbol{b}, \mathbf{s}, \vartheta]$ *and* $[\tilde{\boldsymbol{b}}, \tilde{\mathbf{s}}, \tilde{\vartheta}]$, *respectively. Then if* a *and* $\tilde{a}$ *are scalars,* $[a\boldsymbol{u} + \tilde{a}\tilde{\boldsymbol{u}}, a\boldsymbol{E} + \tilde{a}\tilde{\boldsymbol{E}}, a\boldsymbol{S} + \tilde{a}\tilde{\boldsymbol{S}}]$ *is a thermoelastic state corresponding to the external force-temperature system* $[a\boldsymbol{b} + \tilde{a}\tilde{\boldsymbol{b}}, a\mathbf{s} + \tilde{a}\tilde{\mathbf{s}}, a\vartheta + \tilde{a}\tilde{\vartheta}]$.

(2) Principle of decomposition of thermoelastic states. *Let* $[\boldsymbol{u}, \boldsymbol{E}, \boldsymbol{S}]$ *be a thermoelastic state corresponding to the external force-temperature system* $[\boldsymbol{b}, \mathbf{s}, \vartheta]$. *Then* $[\boldsymbol{u}, \boldsymbol{E}, \boldsymbol{S}]$ *admits the decomposition.*

$$[\boldsymbol{u}, \boldsymbol{E}, \boldsymbol{S}] = [\tilde{\boldsymbol{u}}, \tilde{\boldsymbol{E}}, \tilde{\boldsymbol{S}}] + [\hat{\boldsymbol{u}}, \hat{\boldsymbol{E}}, \hat{\boldsymbol{S}}],$$

[25] LTE assumes only that $\mathbf{C}$ is continuous on $\bar{B}$.

[26] See Sect. 7.

where $[\tilde{\boldsymbol{u}}, \tilde{\boldsymbol{E}}, \tilde{\boldsymbol{S}}]$ *corresponds to the purely mechanical external force-temperature system* $[\boldsymbol{b}, \mathbf{s}, 0]$ *and* $[\hat{\boldsymbol{u}}, \hat{\boldsymbol{E}}, \hat{\boldsymbol{S}}]$ *corresponds to the purely thermal system* $[\mathbf{0}, \mathbf{0}, \vartheta]$.

10. Mean strain and mean stress. Volume change. Throughout this section, the ***mean value*** of a continuous function f on $\overline{B}$ is denoted by

$$\bar{f}(B) = \frac{1}{v(B)} \int_B f \, dv,$$

where $v(B)$ is the volume of B.

Our first theorem is especially useful in the displacement problem.[27]

(1) Mean stress and strain in terms of surface displacement and mean temperature. *Let the vector field* $\boldsymbol{u}$ *and the corresponding strain field* $\boldsymbol{E} = \operatorname{sym} \nabla \boldsymbol{u}$ *be continuous on* $\overline{B}$. *Then the mean strain* $\overline{\boldsymbol{E}}(B)$ *and the associated volume change*[28] $\delta v(B)$ *depend only on the boundary values of* $\boldsymbol{u}$ *and are given by*

$$\overline{\boldsymbol{E}}(B) = \frac{1}{v(B)} \int_{\partial B} \operatorname{sym} \boldsymbol{u} \otimes \boldsymbol{n} \, da$$

and

$$\delta v(B) = \int_{\partial B} \boldsymbol{u} \cdot \boldsymbol{n} \, da.$$

Suppose, in addition, that ϑ *is a continuous scalar field on* $\overline{B}$ *and that* C *and* $\boldsymbol{M}$ *are independent of* $\boldsymbol{x}$. *Then the corresponding mean stress*[29] $\overline{\boldsymbol{S}}(B)$ *depends only on the boundary values of* $\boldsymbol{u}$ *and the mean temperature difference* $\bar{\vartheta}(B)$ *and is given by*

$$\overline{\boldsymbol{S}}(B) = \frac{1}{v(B)} \mathsf{C}\left[\int_{\partial B} \boldsymbol{u} \otimes \boldsymbol{n} \, da\right] + \bar{\vartheta}(B) \, \boldsymbol{M}.$$

Therefore, if the boundary is ***clamped*** *in the sense that*

$$\boldsymbol{u} = \mathbf{0} \quad \text{on } \partial B,$$

then,

$$\overline{\boldsymbol{E}}(B) = \mathbf{0},$$
$$\delta v(B) = 0,$$
$$\overline{\boldsymbol{S}}(B) = \bar{\vartheta}(B) \, \boldsymbol{M}.$$

Proof. The first two assertions are the mean strain theorem[30] and the definition of infinitesimal volume change.[31] Since B is homogeneous,

$$\overline{\boldsymbol{S}}(B) = \mathsf{C}[\overline{\boldsymbol{E}}(B)] + \bar{\vartheta}(B) \, \boldsymbol{M},$$

and by the symmetry properties of C,

$$\mathsf{C}\left[\int_{\partial B} \operatorname{sym} \boldsymbol{u} \otimes \boldsymbol{n} \, da\right] = \mathsf{C}\left[\int_{\partial B} \boldsymbol{u} \otimes \boldsymbol{n} \, da\right]. \quad \square$$

Note that for an isotropic and homogeneous body,

$$\overline{\boldsymbol{S}}(B) = \frac{1}{v(B)} \int_{\partial B} [\mu(\boldsymbol{u} \otimes \boldsymbol{n} + \boldsymbol{n} \otimes \boldsymbol{u}) + \lambda(\boldsymbol{u} \cdot \boldsymbol{n}) \, \mathbf{1}] \, da + m \, \bar{\vartheta}(B) \, \mathbf{1}.$$

---

[27] See Sect. 14.
[28] LTE, Sect. 12.
[29] We define $\boldsymbol{S}$ through the stress-strain-temperature relation $\boldsymbol{S} = \mathsf{C}[\boldsymbol{E}] + \vartheta \boldsymbol{M}$.
[30] LTE, *(13.8)*.
[31] LTE, Sect. 12.

The next result is of interest when the surface traction is prescribed on ∂B.[32]

(2) Mean stress and strain in terms of the external force-temperature system. *Let* $[\boldsymbol{u}, \boldsymbol{E}, \boldsymbol{S}]$ *be a thermoelastic state corresponding to the external force-temperature system* $[\boldsymbol{b}, \boldsymbol{s}, \vartheta]$. *Then the mean stress* $\bar{\boldsymbol{S}}(B)$ *depends only on the surface traction and body force and is given by*

$$\bar{\boldsymbol{S}}(B) = \frac{1}{v(B)}\left(\int_{\partial B} \boldsymbol{p}\otimes\boldsymbol{s}\, da + \int_B \boldsymbol{p}\otimes\boldsymbol{b}\, dv\right),$$

where $\boldsymbol{p} = \boldsymbol{x} - \boldsymbol{x}_0$ *with* $\boldsymbol{x}_0$ *an arbitrarily chosen point. Suppose, in addition, that* C *is invertible so that* $\boldsymbol{E} = \mathsf{K}[\boldsymbol{S}] + \vartheta \boldsymbol{A}$ *and that* K *and* $\boldsymbol{A}$ *are independent of* $\boldsymbol{x}$. *Then the mean strain* $\bar{\boldsymbol{E}}(B)$ *and the volume change* $\delta v(B)$ *depend only on the external force system and the mean temperature difference* $\bar{\vartheta}(B)$ *and are given by*

$$\bar{\boldsymbol{E}}(B) = \frac{1}{v(B)}\mathsf{K}\left[\int_{\partial B} \boldsymbol{p}\otimes\boldsymbol{s}\, da + \int_B \boldsymbol{p}\otimes\boldsymbol{b}\, dv\right] + \bar{\vartheta}(B)\,\boldsymbol{A}$$

and

$$\delta v(B) = \mathsf{K}[\boldsymbol{1}]\cdot\left(\int_{\partial B} \boldsymbol{p}\otimes\boldsymbol{s}\, da + \int_B \boldsymbol{p}\otimes\boldsymbol{b}\, dv\right) + v(B)\,\bar{\vartheta}(B)\operatorname{tr}\boldsymbol{A}.$$ [33]

Therefore, if the body is ***free*** *in the sense that*

$$\boldsymbol{s} = \boldsymbol{0} \quad \text{on } \partial B, \qquad \boldsymbol{b} = \boldsymbol{0} \quad \text{on } B,$$

then

$$\bar{\boldsymbol{S}}(B) = \boldsymbol{0},$$ [34]

$$\bar{\boldsymbol{E}}(B) = \bar{\vartheta}(B)\,\boldsymbol{A},$$ [35]

$$\delta v(B) = v(B)\,\bar{\vartheta}(B)\operatorname{tr}\boldsymbol{A}.$$ [36]

Proof. The first assertion is the mean stress theorem.[37] Since B is homogeneous,

$$\bar{\boldsymbol{E}}(B) = \mathsf{K}[\bar{\boldsymbol{S}}(B)] + \bar{\vartheta}(B)\,\boldsymbol{A}.$$

Finally, it follows from ***(1)*** that

$$\delta v(B) = \int_{\partial B} \boldsymbol{u}\cdot\boldsymbol{n}\, da = v(B)\operatorname{tr}\bar{\boldsymbol{E}}(B). \quad \square$$

On the assumption that B is isotropic as well as homogeneous, the above results reduce to

$$\bar{\boldsymbol{E}}(B) = \frac{1}{v(B)}\left[\frac{1}{2\mu}\left(\int_{\partial B} \boldsymbol{p}\otimes\boldsymbol{s}\, da + \int_B \boldsymbol{p}\otimes\boldsymbol{b}\, dv\right) - \frac{2}{2\mu(3\lambda+2\mu)}\left(\int_{\partial B} \boldsymbol{p}\cdot\boldsymbol{s}\, da + \int_B \boldsymbol{p}\cdot\boldsymbol{b}\, dv\right)\right] + \alpha\bar{\vartheta}(B)\,\boldsymbol{1}$$

and

$$\delta v(B) = \frac{1}{3\lambda+2\mu}\left(\int_{\partial B} \boldsymbol{p}\cdot\boldsymbol{s}\, da + \int_B \boldsymbol{p}\cdot\boldsymbol{b}\, dv\right) + 3\alpha\, v(B)\,\bar{\vartheta}(B).$$

[32] See Sect. 14.
[33] GURTIN, private communication, 1969.
[34] GURTIN, Lecture Notes, Brown University, 1963.
[35] NOWACKI [1962, *1*].
[36] NOWACKI [1954, *1*], HIEKE [1955, *1*].
[37] LTE, ***(18.5)***.

Therefore, if B is free,

$$\delta v(B) = 3\alpha\, v(B)\, \bar{\vartheta}(B).$$

This last formula provides a method for the measurement of the coefficient of thermal expansion α.

11. The body force analogy. The following theorem allows us to bring all of the results of isothermal elastostatics to bear on the equilibrium theory of thermoelasticity.

(1) Body force analogy.[38] *Let $[\boldsymbol{u}, \boldsymbol{E}, \boldsymbol{S}]$ be a thermoelastic state corresponding to the body force $\boldsymbol{b}$ and the temperature difference ϑ. Define the **reduced stress** $\boldsymbol{S}'$ by*

$$\boldsymbol{S}' = \boldsymbol{S} - \vartheta \boldsymbol{M} \tag{11.1}$$

*and the **reduced body force** $\boldsymbol{b}'$ by*

$$\boldsymbol{b}' - \boldsymbol{b} + \operatorname{div}(\vartheta \boldsymbol{M}). \tag{11.2}$$

Then $\boldsymbol{u}$, $\boldsymbol{E}$, and $\boldsymbol{S}'$ satisfy the elasticity equations

$$\begin{aligned} \boldsymbol{E} &= \tfrac{1}{2}(\nabla \boldsymbol{u} + \nabla \boldsymbol{u}^T), \\ \boldsymbol{S}' &= \mathbf{C}[\boldsymbol{E}], \\ \operatorname{div} \boldsymbol{S}' &+ \boldsymbol{b}' = \boldsymbol{0}. \end{aligned} \tag{11.3}$$

Conversely, let $[\boldsymbol{u}, \boldsymbol{E}, \boldsymbol{S}']$ be an admissible state that satisfies the elasticity equations (11.3). *Define $\boldsymbol{S}$ and $\boldsymbol{b}$ through* (11.1) *and* (11.2), *respectively. Then $[\boldsymbol{u}, \boldsymbol{E}, \boldsymbol{S}]$ is a thermoelastic state corresponding to the body force $\boldsymbol{b}$ and the temperature difference ϑ.*

Proof. Suppose that $[\boldsymbol{u}, \boldsymbol{E}, \boldsymbol{S}]$ is a thermoelastic state corresponding to $\boldsymbol{b}$ and ϑ. Then,

$$\begin{gathered} \boldsymbol{E} = \tfrac{1}{2}(\nabla \boldsymbol{u} + \nabla \boldsymbol{u}^T), \\ \boldsymbol{S} - \vartheta \boldsymbol{M} = \mathbf{C}[\boldsymbol{E}], \\ \operatorname{div}(\boldsymbol{S} - \vartheta \boldsymbol{M}) + \boldsymbol{b} + \operatorname{div}(\vartheta \boldsymbol{M}) = \boldsymbol{0}. \end{gathered}$$

Setting $\boldsymbol{S}' = \boldsymbol{S} - \vartheta \boldsymbol{M}$ and $\boldsymbol{b}' = \boldsymbol{b} + \operatorname{div}(\vartheta \boldsymbol{M})$ yields the elasticity equations (11.3).

The converse assertion follows by reversing the above steps. □

The body force analogy asserts that $[\boldsymbol{u}, \boldsymbol{E}, \boldsymbol{S}]$ is a thermoelastic state corresponding to the body force $\boldsymbol{b}$ and the temperature difference ϑ if and only if $[\boldsymbol{u}, \boldsymbol{E}, \boldsymbol{S}' = \boldsymbol{S} - \vartheta \boldsymbol{M}]$ is an *elastic state*[39] corresponding to the reduced body force $\boldsymbol{b}' = \boldsymbol{b} + \operatorname{div}(\vartheta \boldsymbol{M})$. Thus, for every theorem in linear elastostatics, there is an associated result in the equilibrium theory of linear thermoelasticity. Since LTE is necessarily at our fingertips, we shall make full use of the body force analogy in subsequent sections of this part, where we present some of its more interesting and less obvious consequences.

Note that the reduced surface traction associated with the reduced stress $\boldsymbol{S}' = \boldsymbol{S} - \vartheta \boldsymbol{M}$ is given by

$$\boldsymbol{s}' = \boldsymbol{S}' \boldsymbol{n} = \boldsymbol{s} - \vartheta \boldsymbol{M} \boldsymbol{n}, \tag{11.4}$$

where $\boldsymbol{s}$ is the traction field corresponding to $\boldsymbol{S}$.

---

[38] DUHAMEL [1838, *1*].

[39] LTE, Sect. 28.

12. Special results for homogeneous and isotropic bodies. In this section, we assume that the body B is *homogeneous* and *isotropic*[40] so that

$$\mathsf{C}[\boldsymbol{E}] = 2\mu\,\boldsymbol{E} + \lambda(\operatorname{tr}\boldsymbol{E})\,\mathbf{1},$$
$$\boldsymbol{M} = m\mathbf{1},$$
$$\boldsymbol{K} = k\mathbf{1},$$

where the Lamé moduli λ and μ, the stress-temperature modulus m, and the conductivity k are constants. Then the basic field equations of the equilibrium theory reduce to

$$\begin{aligned} \boldsymbol{E} &= \tfrac{1}{2}(\nabla\boldsymbol{u} + \nabla\boldsymbol{u}^T),\\ \operatorname{div}\boldsymbol{S} + \boldsymbol{b} &= \mathbf{0},\\ -\operatorname{div}\boldsymbol{q} + r &= 0,\\ \boldsymbol{S} &= 2\mu\,\boldsymbol{E} + \lambda(\operatorname{tr}\boldsymbol{E})\,\mathbf{1} + m\vartheta\mathbf{1},\\ \boldsymbol{q} &= -k\nabla\vartheta. \end{aligned} \tag{12.1}$$

Substitution of $(12.1)_1$, $(12.1)_4$ into $(12.1)_2$ and $(12.1)_5$ into $(12.1)_3$ yields the ***displacement-temperature equation of equilibrium*** and the ***equilibrium heat equation***:

$$\mu\,\Delta\boldsymbol{u} + (\lambda+\mu)\,\nabla\operatorname{div}\boldsymbol{u} + m\,\nabla\vartheta + \boldsymbol{b} = \mathbf{0} \tag{12.2}$$

and

$$k\,\Delta\vartheta + r = 0. \tag{12.3}$$

In reaching (12.2) and (12.3), we have assumed that $\boldsymbol{u}$ and ϑ are of class C^2. If, in addition, we assume that $k \neq 0$,[41] $\boldsymbol{b}$ is smooth, and $\boldsymbol{u}$ is of class C^3, we can operate on (12.2) with the divergence and curl and use (12.3) to obtain,

$$\begin{aligned} (\lambda+2\mu)\,\Delta\operatorname{div}\boldsymbol{u} &= -\operatorname{div}\boldsymbol{b} + \frac{m}{k}\,r,\\ \mu\,\Delta\operatorname{curl}\boldsymbol{u} &= -\operatorname{curl}\boldsymbol{b}. \end{aligned} \tag{12.4}$$

If $(\lambda+2\mu) \neq 0$, $\mu \neq 0$, $\operatorname{div}\boldsymbol{b} = \frac{m}{k}\,r$, and $\operatorname{curl}\boldsymbol{b} = \mathbf{0}$, (12.4) becomes

$$\operatorname{div}\Delta\boldsymbol{u} = 0, \qquad \operatorname{curl}\Delta\boldsymbol{u} = \mathbf{0}.$$

Thus, if $\boldsymbol{u}$ is of class C^4, we can conclude[42] that $\boldsymbol{u}$ is biharmonic and therefore analytic on B.

As in the isothermal case, the hypothesis that $\boldsymbol{u}$ is of class C^4 can be weakened to class C^2. To obtain this result, we recall the body force analogy ***(11.1)*** which asserts, in the present context, that the equations

$$\begin{aligned} \boldsymbol{E} &= \tfrac{1}{2}(\nabla\boldsymbol{u} + \nabla\boldsymbol{u}^T),\\ \operatorname{div}\boldsymbol{S}' + \boldsymbol{b}' &= \mathbf{0},\\ \boldsymbol{S}' &= 2\mu\,\boldsymbol{E} + \lambda(\operatorname{tr}\boldsymbol{E})\,\mathbf{1} \end{aligned} \tag{12.5}$$

are equivalent to $(12.1)_1$, $(12.1)_2$, $(12.1)_4$ provided

$$\begin{aligned} \boldsymbol{S}' &= \boldsymbol{S} - m\,\vartheta\mathbf{1},\\ \boldsymbol{b}' &= \boldsymbol{b} + m\,\nabla\vartheta. \end{aligned} \tag{12.6}$$

---

[40] See Sect. 8.

[41] The second law of thermodynamics [see ***(6.1)*** and (7.16)] yields only that the conductivity tensor $\boldsymbol{K} = k\,\mathbf{1}$ must be positive semi-definite, i.e., $k \geqq 0$.

[42] LTE, Sect. 27.

By ***(42.1)*** and ***(42.2)*** of LTE, if $\boldsymbol{u}$ is of class C^2 and if $\operatorname{div} \boldsymbol{b}' = 0$ and $\operatorname{curl} \boldsymbol{b}' = \boldsymbol{0}$, then $\boldsymbol{u}$ is analytic and $\Delta \operatorname{div} \boldsymbol{u} = 0$, $\Delta \operatorname{curl} \boldsymbol{u} = \boldsymbol{0}$, $\Delta\Delta \boldsymbol{u} = \boldsymbol{0}$. From (12.6) and (12.3),

$$\operatorname{div} \boldsymbol{b}' = \operatorname{div} \boldsymbol{b} + m\,\Delta\vartheta = \operatorname{div} \boldsymbol{b} - \frac{m}{k} r,$$
$$\operatorname{curl} \boldsymbol{b}' = \operatorname{curl} \boldsymbol{b}.$$

It is unlikely that the condition $\operatorname{div} \boldsymbol{b}' = \operatorname{div} \boldsymbol{b} - \frac{m}{k} r = 0$ would be met except in the important special case $\boldsymbol{b} = \boldsymbol{0}$ and $r = 0$, and we state our main result accordingly.

(1) Analyticity of thermoelastic displacement and temperature fields. *Let $\boldsymbol{u}$ and ϑ be of class C^2 on B and satisfy the displacement-temperature equation of equilibrium*

$$\mu\,\Delta\boldsymbol{u} + (\lambda + \mu)\,\nabla \operatorname{div} \boldsymbol{u} + m\,\nabla\vartheta = \boldsymbol{0}$$

and the equilibrium heat equation

$$\Delta\vartheta = 0.$$

Then $\boldsymbol{u}$ and ϑ are analytic on B and

$$\Delta \operatorname{div} \boldsymbol{u} = 0, \qquad \Delta \operatorname{curl} \boldsymbol{u} = \boldsymbol{0}, \qquad \Delta\Delta\boldsymbol{u} = \boldsymbol{0}.$$

In the present case ($\boldsymbol{b} = \boldsymbol{0}$ and $r = 0$), (1), $(12.1)_1$, and $(12.1)_4$ imply that

$$\Delta \operatorname{tr} \boldsymbol{E} = \Delta \operatorname{tr} \boldsymbol{S} = 0,$$
$$\Delta\Delta\boldsymbol{E} = \Delta\Delta\boldsymbol{S} = \boldsymbol{0}.$$

The body force analogy can also be used to write the *compatibility equation*[43]

$$\operatorname{curl}\operatorname{curl} \boldsymbol{E} = \boldsymbol{0}$$

in terms of the stress and temperature. The *stress equation of compatibility*[44] for the isothermal system (12.5) is

$$\Delta\boldsymbol{S}' + \frac{1}{1+\nu}\,\nabla\nabla \operatorname{tr} \boldsymbol{S}' + \nabla\boldsymbol{b}' + \nabla\boldsymbol{b}'^T + \frac{\nu}{1-\nu}\,(\operatorname{div} \boldsymbol{b}')\,\boldsymbol{1} = \boldsymbol{0} \tag{12.7}$$

where

$$\nu = \frac{\lambda}{2(\lambda+\mu)}$$

is ***Poisson's ratio***.[45] In deriving (12.7) it is assumed that $\mu \neq 0$ and $\nu \neq \pm 1$. Using (12.6) and (12.3), (12.7) becomes

$$\Delta\boldsymbol{S} + \frac{1}{1+\nu}\,\nabla\nabla \operatorname{tr} \boldsymbol{S} - m\,\frac{1-2\nu}{1+\nu}\,\nabla\nabla\vartheta$$
$$+ \nabla\boldsymbol{b} + \nabla\boldsymbol{b}^T + \frac{\nu}{1-\nu}\,(\operatorname{div} \boldsymbol{b})\,\boldsymbol{1} + \left(\frac{1-2\nu}{1-\nu}\right)\left(\frac{m}{k}\right) r\,\boldsymbol{1} = \boldsymbol{0};$$

thus, we have the

(2) Stress-temperature equation of compatibility.[46] *Let the symmetric tensor $\boldsymbol{E}$ be of class C^2 on B and define the stress through the inverted stress-strain-temperature relation ($\mu \neq 0$, $\nu \neq -1$)*

$$\boldsymbol{E} = \frac{1}{2\mu}\left[\boldsymbol{S} - \frac{\nu}{1+\nu}\,(\operatorname{tr} \boldsymbol{S})\,\boldsymbol{1} - m\,\frac{1-2\nu}{1+\nu}\,\vartheta\boldsymbol{1}\right],$$

[43] LTE, Sect. 14.
[44] LTE, Sect. 27.
[45] LTE, Sect. 22.
[46] See I. S. and E. S. SOKOLNIKOFF [1939, *1*] for a similar result. See ***(14.2)*** for the nonhomogeneous and anisotropic case.

where the temperature difference ϑ is of class C^2 and meets the equilibrium heat equation

$$k\,\Delta\vartheta + r = 0.$$

Assume, in addition, that $k \neq 0$ and $\nu \neq 1$. Then the equilibrium equation

$$\operatorname{div} \mathbf{S} + \boldsymbol{b} = \mathbf{0}$$

and the compatibility equation

$$\operatorname{curl}\operatorname{curl} \boldsymbol{E} = \mathbf{0}$$

imply the ***stress-temperature equation of compatibility***

$$\Delta\mathbf{S} + \frac{1}{1+\nu}\nabla\nabla\operatorname{tr}\mathbf{S} - m\frac{1-2\nu}{1+\nu}\nabla\nabla\vartheta + \nabla\boldsymbol{b} + \nabla\boldsymbol{b}^T$$
$$+ \frac{1+\nu}{\nu}(\operatorname{div}\boldsymbol{b})\,\mathbf{1} + \left(\frac{1-2\nu}{1-\nu}\right)\left(\frac{m}{k}\right) r\,\mathbf{1} = \mathbf{0}.$$

Conversely, if $\mathbf{S}$ is a symmetric tensor field of class C^2 on B that satisfies the equation of equilibrium and the stress-temperature equation of compatibility, then the corresponding strain $\mathbf{E}$ defined by the inverted stress-strain-temperature relation satisfies the equation of compatibility. Accordingly, if B is simply connected, there exists a displacement field $\boldsymbol{u}$ that meets the strain-displacement relation

$$\boldsymbol{E} = \tfrac{1}{2}(\nabla\boldsymbol{u} + \nabla\boldsymbol{u}^T).$$

From the body force analogy and the *Boussinesq-Papkovitch-Neuber solution*[47] of the isothermal theory, we see that a complete solution of the displacement-temperature equation of equilibrium is provided by

$$\boldsymbol{u} = \boldsymbol{\psi} - \frac{1}{4(1-\nu)}\nabla(\boldsymbol{p}\cdot\boldsymbol{\Psi} + \varphi)$$

where

$$\Delta\boldsymbol{\psi} = -\frac{1}{\mu}(\boldsymbol{b} + m\nabla\vartheta), \qquad \Delta\varphi = \frac{1}{\mu}\boldsymbol{p}\cdot(\boldsymbol{b} + m\nabla\vartheta)$$

and $\boldsymbol{p} = \boldsymbol{x} - \mathbf{0}$.

Finally, we recall that, in the isothermal case, the surface traction can be expressed in terms of the displacement by

$$\mathbf{s}' = 2\mu\frac{\partial\boldsymbol{u}}{\partial\boldsymbol{n}} + \mu\,\boldsymbol{n}\times\operatorname{curl}\boldsymbol{u} + \lambda(\operatorname{div}\boldsymbol{u})\,\boldsymbol{n},$$ [48]

where

$$\frac{\partial\boldsymbol{u}}{\partial\boldsymbol{n}} = (\nabla\boldsymbol{u})\,\boldsymbol{n}.$$

Therefore, by (11.4),

$$\mathbf{s} = 2\mu\frac{\partial\boldsymbol{u}}{\partial\boldsymbol{n}} + \mu\,\boldsymbol{n}\times\operatorname{curl}\boldsymbol{u} + \lambda(\operatorname{div}\boldsymbol{u})\,\boldsymbol{n} + m\,\vartheta\boldsymbol{n}.$$

13. The theorem of work and energy. The reciprocal theorem. While the two results presented in this section can be obtained by applying the body force analogy to their counterparts in the isothermal theory, it is actually quicker to proceed directly.

---

[47] LTE, Sect. 44.
[48] LTE, Sect. 27.

The theorem of work expended[49] states that

$$\int_{\partial B} \mathbf{s} \cdot \boldsymbol{u}\, da + \int_B \boldsymbol{b} \cdot \boldsymbol{u}\, dv = \int_B \boldsymbol{S} \cdot \boldsymbol{E}\, dv$$

provided $\boldsymbol{S}$ is an admissible stress field that satisfies the equation of equilibrium and $\boldsymbol{u}$ is an admissible displacement field with strain field $\boldsymbol{E}$. Hence, we have the

(1) Theorem of work and energy. *Let* $[\boldsymbol{u}, \boldsymbol{E}, \boldsymbol{S}]$ *be a thermoelastic state corresponding to the external force-temperature system* $[\boldsymbol{b}, \mathbf{s}, \vartheta]$. *Then,*

$$\int_{\partial B} \mathbf{s} \cdot \boldsymbol{u}\, da + \int_B \boldsymbol{b} \cdot \boldsymbol{u}\, dv - \int_B \vartheta \boldsymbol{M} \cdot \boldsymbol{E}\, dv = \int_B \boldsymbol{E} \cdot \mathsf{C}[\boldsymbol{E}]\, dv.$$

The left-hand side of the above equation may be interpreted as the work done by the external force-temperature system, while the right-hand side is twice the strain energy.

Now suppose that $[\boldsymbol{u}, \boldsymbol{E}, \boldsymbol{S}]$ and $[\tilde{\boldsymbol{u}}, \tilde{\boldsymbol{E}}, \tilde{\boldsymbol{S}}]$ are thermoelastic states corresponding to the external force-temperature systems $[\boldsymbol{b}, \mathbf{s}, \vartheta]$ and $[\tilde{\boldsymbol{b}}, \tilde{\mathbf{s}}, \tilde{\vartheta}]$, respectively. Then we can use the theorem of work expended to write

$$\int_{\partial B} \mathbf{s} \cdot \tilde{\boldsymbol{u}}\, da + \int_B \boldsymbol{b} \cdot \tilde{\boldsymbol{u}}\, dv = \int_B \boldsymbol{S} \cdot \tilde{\boldsymbol{E}}\, dv = \int_B \vartheta \boldsymbol{M} \cdot \tilde{\boldsymbol{E}}\, dv + \int_B \tilde{\boldsymbol{E}} \cdot \mathsf{C}[E]\, dv$$

and

$$\int_{\partial B} \tilde{\mathbf{s}} \cdot \boldsymbol{u}\, da + \int_B \tilde{\boldsymbol{b}} \cdot \boldsymbol{u}\, dv = \int_B \boldsymbol{S} \cdot \boldsymbol{E}\, dv = \int_B \tilde{\vartheta} \boldsymbol{M} \cdot \boldsymbol{E}\, dv + \int_B \boldsymbol{E} \cdot \mathsf{C}[\tilde{\boldsymbol{E}}]\, dv.$$

Since C is symmetric,

$$\tilde{\boldsymbol{E}} \cdot \mathsf{C}[\boldsymbol{E}] = \boldsymbol{E} \cdot \mathsf{C}[\tilde{\boldsymbol{E}}],$$

and we have the

(2) Reciprocal theorem.[50] *Let* $[\boldsymbol{u}, \boldsymbol{E}, \boldsymbol{S}]$ *and* $[\tilde{\boldsymbol{u}}, \tilde{\boldsymbol{E}}, \tilde{\boldsymbol{S}}]$ *be thermoelastic states corresponding to the external force-temperature systems* $[\boldsymbol{b}, \mathbf{s}, \vartheta]$ *and* $[\tilde{\boldsymbol{b}}, \tilde{\mathbf{s}}, \tilde{\vartheta}]$, *respectively. Then*

$$\int_{\partial B} \mathbf{s} \cdot \tilde{\boldsymbol{u}}\, da + \int_B \boldsymbol{b} \cdot \tilde{\boldsymbol{u}}\, dv - \int_B \vartheta \boldsymbol{M} \cdot \tilde{\boldsymbol{E}}\, dv = \int_B \tilde{\boldsymbol{E}} \cdot \mathsf{C}[\boldsymbol{E}]\, dv$$
$$= \int_{\partial B} \tilde{\mathbf{s}} \cdot \boldsymbol{u}\, da + \int_B \tilde{\boldsymbol{b}} \cdot \boldsymbol{u}\, dv - \int_B \tilde{\vartheta} \boldsymbol{M} \cdot \boldsymbol{E}\, dv = \int_B \boldsymbol{E} \cdot \mathsf{C}[\tilde{\boldsymbol{E}}]\, dv.$$

The reciprocal theorem asserts that given two thermoelastic states, the work done by the external force-temperature system of the first over the displacement field of the second equals the work done by the external force-temperature system of the second over the displacement field of the first.

When the elasticity tensor C is invertible, we have $\boldsymbol{E} = \mathsf{K}[\boldsymbol{S}] + \vartheta \boldsymbol{A}$, where $\mathsf{K} = \mathsf{C}^{-1}$ is the compliance tensor and $\boldsymbol{A} = -\mathsf{K}[\boldsymbol{M}]$ is the thermal expansion tensor. In this case, the reciprocal theorem can be written in the form

$$\int_{\partial B} \mathbf{s} \cdot \tilde{\boldsymbol{u}}\, da + \int_B \boldsymbol{b} \cdot \tilde{\boldsymbol{u}}\, dv - \int_B \boldsymbol{S} \cdot \tilde{\vartheta} \boldsymbol{A}\, dv = \int_B \boldsymbol{S} \cdot \mathsf{K}[\tilde{\boldsymbol{S}}]\, dv$$
$$= \int_{\partial B} \tilde{\mathbf{s}} \cdot \boldsymbol{u}\, da + \int_B \tilde{\boldsymbol{b}} \cdot \boldsymbol{u}\, dv - \int_B \tilde{\boldsymbol{S}} \cdot \vartheta \boldsymbol{A}\, dv = \int_B \tilde{\boldsymbol{S}} \cdot \mathsf{K}[\boldsymbol{S}]\, dv.$$

14. The boundary-value problems of the equilibrium theory. Uniqueness. *Throughout this section, we assume that, in addition to* C *and* $\boldsymbol{M}$,[51] *a smooth temperature difference field* ϑ *on* $\bar{B}$, *a continuous body force field* $\boldsymbol{b}$ *on* $\bar{B}$, *a continuous*

[49] LTE, *(18.2)*.
[50] MAYSEL [1941, *1*].
[51] See Sect. 9.

surface displacement field $\hat{\boldsymbol{u}}$ *on* $\mathscr{S}_1$, *and a piecewise regular*[52] *surface traction field* $\hat{\mathbf{s}}$ *on* $\mathscr{S}_2$ *are prescribed.*[53] Then the ***mixed problem of the equilibrium theory of thermoelasticity*** is to find a thermoelastic state $[\boldsymbol{u}, \boldsymbol{E}, \boldsymbol{S}]$ corresponding to $\boldsymbol{b}$ and ϑ that satisfies the ***displacement condition***

$$\boldsymbol{u} = \hat{\boldsymbol{u}} \quad \text{on } \mathscr{S}_1$$

and the ***traction condition***

$$\mathbf{s} = \boldsymbol{S}\boldsymbol{n} = \hat{\mathbf{s}} \quad \text{on } \mathscr{S}_2.$$

If such a state exists, it is called a ***solution of the mixed problem.***

When $\mathscr{S}_1 = \partial B$ (i.e., $\mathscr{S}_2$ is empty), the above boundary conditions reduce to

$$\boldsymbol{u} = \hat{\boldsymbol{u}} \quad \text{on } \partial B,$$

and the resulting problem is called the ***displacement problem.*** If $\mathscr{S}_2 = \partial B$, the boundary conditions become

$$\mathbf{s} = \hat{\mathbf{s}} \quad \text{on } \partial B,$$

and we have the ***traction problem.*** It is, of course, possible to set up more complicated (but yet physically meaningful) boundary-value problems than those considered here.[54]

The body force analogy allows us to read off the following two theorems from Sect. 31 of LTE.

(1) Characterization of the mixed problem in terms of the displacement. *An admissible displacement field* $\boldsymbol{u}$ *corresponds to a solution of the mixed problem if and only if*

$$\begin{aligned} \operatorname{div} \mathbf{C}[\nabla \boldsymbol{u}] + \operatorname{div}(\vartheta \boldsymbol{M}) + \boldsymbol{b} &= \mathbf{0} \quad \textit{on } B,^{55} \\ \boldsymbol{u} &= \hat{\boldsymbol{u}} \quad \textit{on } \mathscr{S}_1 \\ \mathbf{C}[\nabla \boldsymbol{u}]\,\boldsymbol{n} &= \hat{\mathbf{s}} - \vartheta \boldsymbol{M}\boldsymbol{n} \quad \textit{on } \mathscr{S}_2; \end{aligned}$$

or equivalently, if B *is homogeneous and isotropic,*

$$\begin{aligned} \mu\,\Delta \boldsymbol{u} + (\lambda + \mu)\,\nabla \operatorname{div} \boldsymbol{u} + m\,\nabla \vartheta + \boldsymbol{b} &= \mathbf{0} \quad \textit{on } B, \\ \boldsymbol{u} &= \hat{\boldsymbol{u}} \quad \textit{on } \mathscr{S}_1 \\ \mu(\nabla \boldsymbol{u} + \nabla \boldsymbol{u}^T) + \lambda(\operatorname{div} \boldsymbol{u})\,\boldsymbol{n} &= \hat{\mathbf{s}} - m\,\vartheta\,\boldsymbol{n} \quad \textit{on } \mathscr{S}_2.^{56} \end{aligned}$$

(2) Characterization of the traction problem in terms of stress. *Suppose that the elasticity tensor* $\mathbf{C}$ *is invertible and that the compliance tensor* $\mathbf{K} = \mathbf{C}^{-1}$, *the stress-temperature tensor* $\boldsymbol{M}$, *and the temperature difference* ϑ *are each of class* C^2 *on* $\bar{B}$. *Let* B *be simply-connected. Then a symmetric tensor field* $\boldsymbol{S}$ *that is of class* C^2 *on* $\bar{B}$ *is a stress field corresponding to a solution of the traction problem if and only if*

$$\begin{aligned} \operatorname{div} \boldsymbol{S} + \boldsymbol{b} &= \mathbf{0} \quad \textit{on } B, \\ \operatorname{curl}\operatorname{curl} \mathbf{K}[\boldsymbol{S} - \vartheta \boldsymbol{M}] &= \mathbf{0} \quad \textit{on } B,^{57} \\ \boldsymbol{S}\boldsymbol{n} &= \hat{\mathbf{s}} \quad \textit{on } \partial B; \end{aligned}$$

---

[52] LTE, Sect. 5.

[53] $\mathscr{S}_1$ and $\mathscr{S}_2$ are complementary regular subsurfaces of ∂B, see LTE, Sect. 5.

[54] Cf. LTE, Sect. 40.

[55] We refer to this equation as the ***displacement-temperature equation of equilibrium.***

[56] This boundary condition can be written as

$$2\mu \frac{\partial \boldsymbol{u}}{\partial \boldsymbol{n}} + \mu\,\boldsymbol{n} \times \operatorname{curl} \boldsymbol{u} + \lambda(\operatorname{div} \boldsymbol{u})\,\boldsymbol{n} = \hat{\mathbf{s}} - m\,\vartheta\boldsymbol{n} \quad \text{on } \mathscr{S}_2,$$

see the end of Sect. 12.

[57] We refer to this equation as the ***stress-temperature equation of compatibility.***

or equivalently, if B is homogeneous and isotropic,

$$\operatorname{div} \boldsymbol{S} + \boldsymbol{b} = \boldsymbol{0} \quad \text{on } B,$$

$$\Delta \boldsymbol{S} + \frac{1}{1+\nu} \nabla\nabla \operatorname{tr} \boldsymbol{S} - m \frac{1-2\nu}{1+\nu} \nabla\nabla\vartheta + \nabla \boldsymbol{b} + \nabla \boldsymbol{b}^T + \frac{\nu}{1+\nu} (\operatorname{div} \boldsymbol{b}) \boldsymbol{1}$$
$$+ \left(\frac{1-2\nu}{1-\nu}\right)\left(\frac{m}{k}\right) r \boldsymbol{1} = \boldsymbol{0} \quad \text{on } B,$$ [58]

$$\boldsymbol{S}\boldsymbol{n} = \hat{\boldsymbol{s}} \quad \text{on } \partial B.$$

The body force analogy implies uniqueness theorems for the boundary-value problems discussed above. We give only one such theorem here, but more detailed results may be read off from Sect. 32 of LTE.

(3) Uniqueness theorem for the mixed problem. *Let the elasticity tensor C be positive definite. Then any two solutions of the mixed problem are equal modulo a rigid displacement. Moreover, if $\mathscr{S}_1$ is non-empty, the mixed problem has at most one solution.*

15. Temperature fields that induce displacement free and stress free states. Throughout this section, we assume that the body B is *homogeneous*, the elasticity tensor C is *positive definite*, and the stress-temperature tensor $\boldsymbol{M}$ is *invertible.*

We first suppose that the boundary of B is clamped ($\boldsymbol{u} = \boldsymbol{0}$ on ∂B) and seek a temperature difference field ϑ that induces zero displacement in B. Accordingly, we assume that $\boldsymbol{u} = \boldsymbol{0}$ on $\bar{B}$. Then the stress-strain-temperature relation reduces to $\boldsymbol{S} = \vartheta \boldsymbol{M}$, and the equilibrium equation requires that

$$\operatorname{div}(\vartheta \boldsymbol{M}) + \boldsymbol{b} = \boldsymbol{0}. \tag{15.1}$$

Since $\boldsymbol{M}$ is constant and invertible, (15.1) implies

$$\nabla\vartheta = -\boldsymbol{M}^{-1} \boldsymbol{b};$$

and assuming that the body force $\boldsymbol{b}$ is constant we may integrate this equation so as to obtain

$$\vartheta = -(\boldsymbol{M}^{-1} \boldsymbol{b}) \cdot \boldsymbol{p} + \vartheta_0, \tag{15.2}$$

where ϑ_0 is a constant and $\boldsymbol{p} = \boldsymbol{x} - \boldsymbol{x}_0$ with $\boldsymbol{x}_0$ a fixed point.

Conversely, if ϑ is given by (15.2), then the state

$$[\boldsymbol{u}, \boldsymbol{E}, \boldsymbol{S}] = [\boldsymbol{0}, \boldsymbol{0}, \vartheta \boldsymbol{M}]$$

is a solution of the displacement problem defined by $\boldsymbol{u} = \boldsymbol{0}$ on ∂B. By the uniqueness theorem ***(14.3)***, it is the only solution. Thus we have the following theorem:

(1) Temperature fields that induce displacement free states.[59]

Let $[\boldsymbol{u}, \boldsymbol{E}, \boldsymbol{S}]$ be the thermoelastic state corresponding to the uniform body force $\boldsymbol{b}$, the temperature difference ϑ, and the clamped condition $\boldsymbol{u} = \boldsymbol{0}$ on ∂B. Then $\boldsymbol{u} = \boldsymbol{0}$ on B if and only if

$$\vartheta = -(\boldsymbol{M}^{-1} \boldsymbol{b}) \cdot \boldsymbol{p} + \vartheta_0$$

where $\boldsymbol{p} = \boldsymbol{x} - \boldsymbol{x}_0$ and $\boldsymbol{x}_0$ and ϑ_0 are arbitrary constants; in this case,

$$\boldsymbol{S} = \vartheta \boldsymbol{M}.$$

In particular, if $\boldsymbol{b} = \boldsymbol{0}$, then $\vartheta = \vartheta_0$.

---

[58] Here we are assuming, in addition, that $\nu \neq 1$, $k \neq 0$, and that ϑ satisfies the equilibrium heat equation $k \Delta\vartheta + r = 0$, see ***(12.2)***.

[59] VOIGT [1895, *1*, Pt. 3, Sect. 7] for the case $\boldsymbol{b} = \boldsymbol{0}$. See also VOIGT [1910, *2*, Sect. 385].

For the remainder of this section, we suppose that B is free ($\mathbf{s}=\mathbf{0}$ on ∂B and $\boldsymbol{b}=\mathbf{0}$ on $\bar{B}$) and seek a temperature difference field ϑ that induces a stress free state in B. We assume now that B is *isotropic* as well as homogeneous. Accordingly, we set $\boldsymbol{S}=\mathbf{0}$ on $\bar{B}$. Then the stress -temperature equation of compatibility ***(12.2)*** reduces to

$$\nabla\nabla\vartheta=\mathbf{0};^{60}$$

consequently,

$$\vartheta=\boldsymbol{a}\cdot\boldsymbol{p}+\vartheta_0, \tag{15.3}$$

where $\boldsymbol{a}$ and ϑ_0 are constants. Then, by the inverted stress-strain-temperature relation (8.3),

$$\boldsymbol{E}=\alpha(\boldsymbol{a}\cdot\boldsymbol{p}+\vartheta_0)\,\mathbf{1}. \tag{15.4}$$

The corresponding displacement field is

$$\boldsymbol{u}=\alpha[(\boldsymbol{a}\cdot\boldsymbol{p}+\vartheta_0)\,\boldsymbol{p}-\tfrac{1}{2}\,\boldsymbol{a}(\boldsymbol{p}\cdot\boldsymbol{p})]+\mathring{\boldsymbol{u}}, \tag{15.5}$$

where $\mathring{\boldsymbol{u}}$ is an arbitrary rigid displacement field.[61]

Conversely, if ϑ is given by (15.3), then the state $[\boldsymbol{u},\boldsymbol{E},\boldsymbol{S}]$ with $\boldsymbol{u}$ and $\boldsymbol{E}$ given by (15.5) and (15.4), respectively, and $\boldsymbol{S}=\mathbf{0}$ is a solution of the traction problem defined by $\mathbf{s}=\mathbf{0}$ on ∂B. By the uniqueness theorem ***(14.3)***, it is the only solution. Thus, we have established the following result:

(2) Temperature fields that induce stress free states.[62]

Let $[\boldsymbol{u},\boldsymbol{E},\boldsymbol{S}]$ be a thermoelastic state corresponding to zero body force, a temperature difference ϑ that is of class C^2 on B and satisfies the equilibrium heat equation $\Delta\vartheta=0$, and the traction free condition $\mathbf{s}=\mathbf{0}$ on ∂B. Assume that B is isotropic and that the conductivity $k\neq 0$. Then $\boldsymbol{S}=\mathbf{0}$ on $\bar{B}$ if and only if

$$\vartheta=\boldsymbol{a}\cdot\boldsymbol{p}+\vartheta_0,$$

where $\mathbf{p}=\boldsymbol{x}-\boldsymbol{x}_0$ and $\boldsymbol{a}$, $\boldsymbol{x}_0$, and ϑ_0 are arbitrary constants; in this case

$$\boldsymbol{u}=\alpha[(\boldsymbol{a}\cdot\boldsymbol{p}+\vartheta_0)\,\boldsymbol{p}-\tfrac{1}{2}\,\boldsymbol{a}(\boldsymbol{p}\cdot\boldsymbol{p})]+\mathring{\boldsymbol{u}},$$

where α is the coefficient of thermal expansion and $\mathring{\boldsymbol{u}}$ is an arbitrary rigid displacement field.

16. Minimum principles. In this section, we use the body force analogy to write the thermoelastic versions of the theorems of minimum potential energy and minimum complementary energy from linear elastostatics.[63] These theorems completely characterize the solution of the mixed problem. We recall that the boundary conditions for the mixed problem are the displacement condition

$$\boldsymbol{u}=\hat{\boldsymbol{u}} \quad \text{on } \mathscr{S}_1$$

and the traction condition

$$\mathbf{s}=\hat{\mathbf{s}} \quad \text{on } \mathscr{S}_2.$$

[60] We assume that ϑ is of class C^2 and satisfies the equilibrium heat equation $k\,\Delta\vartheta+r=0$ for $r=0$ and that $k\neq 0$. The conditions that $\mu\neq 0$, $\nu\neq 1,\tfrac{1}{2}$ are met by the positive definiteness of $\mathbf{C}$ (LTE, ***(24.5)***), and $m\neq 0$ since $\boldsymbol{M}$ is invertible. *Note added in proof.* Eq. (8.3) with $\boldsymbol{S}=\mathbf{0}$ and curl curl $\boldsymbol{E}=\mathbf{0}$ imply $\nabla\nabla\vartheta=\mathbf{0}$ (and hence $\Delta\vartheta=0$) without the assumptions that $k\neq 0$ and $\Delta\vartheta=0$.

[61] LTE, ***(13.2)***.

[62] VOIGT [1910, *2*, Sect. 384]. Partial results are contained in VOIGT [1895, *1*, Pt. 3, Sect. 7].

[63] LTE, Sects. 34, 36.

Throughout this section, we assume that the data have the smoothness properties demanded in Sect. 14 and that, in addition, the elasticity tensor C is *positive definite.*

If $\boldsymbol{E}$ and $\boldsymbol{S}$ are continuous symmetric tensor fields on $\bar{B}$, we define the ***strain energy***

$$U_{\mathsf{C}}\{\boldsymbol{E}\} = \tfrac{1}{2}\int_B \boldsymbol{E}\cdot\mathsf{C}[\boldsymbol{E}]\,dv$$

and the ***stress energy***

$$U_{\mathsf{K}}\{\boldsymbol{S}\} = \tfrac{1}{2}\int_B \boldsymbol{S}\cdot\mathsf{K}[\boldsymbol{S}]\,dv,$$

where $\mathsf{K} = \mathsf{C}^{-1}$ is the compliance tensor. Generally, in thermoelasticity,

$$U_{\mathsf{K}}\{\boldsymbol{S}\} \neq U_{\mathsf{C}}\{\boldsymbol{E}\}$$

because we have $\boldsymbol{S} = \mathsf{C}[\boldsymbol{E}]$ if and only if $\vartheta = 0$. However, since $\boldsymbol{M} = -\mathsf{C}[\boldsymbol{A}]$, where $\boldsymbol{A}$ is the thermal expansion tensor and $\boldsymbol{M}$ is stress-temperature tensor,

$$U_{\mathsf{K}}\{\boldsymbol{M}\} = U_{\mathsf{C}}\{\boldsymbol{A}\}.$$

By a ***kinematically admissible state*** we mean an admissible state that satisfies the strain-displacement relation, the stress-strain-temperature relation, and the displacement boundary condition.

Using $\mathfrak{s} = [\boldsymbol{u}, \boldsymbol{E}, \boldsymbol{S}]$ to denote a kinematically admissible state, we see from the body force analogy ***(11.1)*** that the *isothermal principle of minimum potential energy*[64] and its converse[65] assert that the functional

$$\Phi\{\mathfrak{s}\} = U_{\mathsf{C}}\{\boldsymbol{E}\} - \int_B \boldsymbol{b}'\cdot\boldsymbol{u}\,dv - \int_{\mathscr{S}_2} \hat{\boldsymbol{s}}'\cdot\boldsymbol{u}\,da$$

assumes a minimum at and only at the solution of the mixed problem. Here we are using the previously adopted notation[66]

$$\boldsymbol{b}' = \boldsymbol{b} + \operatorname{div}(\vartheta\boldsymbol{M}),$$
$$\boldsymbol{S}' = \boldsymbol{S} - \vartheta\boldsymbol{M}, \qquad \boldsymbol{s}' = \boldsymbol{s} - \vartheta\boldsymbol{M}\boldsymbol{n}.$$

With these last relations, the expression for Φ can be written as

$$\begin{aligned}\Phi\{\mathfrak{s}\} = U_{\mathsf{C}}\{\boldsymbol{E}\} &- \int_B \boldsymbol{b}\cdot\boldsymbol{u}\,dv + \int_B \vartheta\boldsymbol{M}\cdot\boldsymbol{E}\,dv \\ &- \int_{\mathscr{S}_1} (\vartheta\boldsymbol{M}\boldsymbol{n})\cdot\hat{\boldsymbol{u}}\,da - \int_{\mathscr{S}_2} \hat{\boldsymbol{s}}\cdot\boldsymbol{u}\,da.\end{aligned} \tag{16.1}$$

Thus, we have the

(1) Theorem of minimum potential energy. *Let $\mathscr{A}$ be the set of all kinematically admissible states, and let Φ be the functional defined on $\mathscr{A}$ by* (16.1). *Then $\mathfrak{s}$ is a solution of the mixed problem if and only if*

$$\Phi\{\mathfrak{s}\} \leqq \Phi\{\tilde{\mathfrak{s}}\}$$

for every $\tilde{\mathfrak{s}} \in \mathscr{A}$, and equality holds only if $\tilde{\mathfrak{s}} = \mathfrak{s}$ modulo a rigid displacement.

By a ***statically admissible stress field*** we mean an admissible stress field that satisfies the equation of equilibrium and the traction boundary condition.

---

[64] LTE, ***(34.1)***.
[65] LTE, ***(36.1)***.
[66] Sect. 11.

Again employing the body force analogy, the isothermal *principle of minimum complementary energy*[67] and its converse[68] assert that the functional

$$\Psi\{S\} = U_{\mathsf{K}}\{S'\} - \int_{\mathscr{S}_2} s' \cdot \hat{u}\, da$$

defined on the set of all statically admissible stress fields assumes a minimum at and (with certain additional restrictions) only at the solution of the mixed problem. Using

$$S' = S - \vartheta M, \qquad s' = s - \vartheta M n$$

and the symmetry of K, we obtain

$$\Psi\{S\} = U_{\mathsf{K}}\{S\} + U_{\mathsf{K}}\{\vartheta M\} + \int_B S \cdot \vartheta A\, dv - \int_{\mathscr{S}_1} (s - \vartheta M n) \cdot \hat{u}\, da. \tag{16.2}$$

Consequently, we have the

(2) Theorem of minimum complementary energy. *Let $\mathscr{A}$ be the set of all statically admissible stress fields, and let Ψ be the fucntional defined on $\mathscr{A}$ by* (16.2). *Let S be the stress field corresponding to the solution of the mixed problem. Then*

$$\Psi\{S\} \leqq \Psi\{\tilde{S}\}$$

for every $\tilde{S} \in \mathscr{A}$, and equality holds only if $\tilde{S} = S$.

Conversely, assume that B is simply-connected and convex with respect to $\mathscr{S}_1$ and that K is of class C^2 on B. Let $S \in \mathscr{A}$ be of class C^2 on B, and suppose that

$$\Psi\{S\} \leqq \Psi\{\tilde{S}\}$$

for every $\tilde{S} \in \mathscr{A}$. Then S is the stress field corresponding to the solution of the mixed problem.

We close this section with the observation that since the temperature difference field ϑ is viewed as having been previously determined, the terms

$$\int_{\mathscr{S}_1} (\vartheta M n) \cdot \hat{u}\, da$$

and

$$U_{\mathsf{K}}\{\vartheta M\} + \int_{\mathscr{S}_1} (\vartheta M n) \cdot \hat{u}\, da$$

may be dropped from (16.1) and (16.2), respectively.

17. The uncoupled-quasi-static theory. The basic equations of the approximate ***uncoupled-quasi-static theory***[69] are

$$\begin{gathered} E = \tfrac{1}{2}(\nabla u + \nabla u^T), \\ \operatorname{div} S + b = 0, \\ -\operatorname{div} q + r = c\dot{\theta}, \\ S = \mathsf{C}[E] + (\theta - \theta_0)\, M, \\ q = -K \nabla \theta. \end{gathered}$$

Except for the energy equation, these equations are identical to those of the exact (within the context of the linear theory) equilibrium theory. As before, the

[67] LTE, *(34.3)*.
[68] LTE, *(36.2)*.
[69] See the end of Sect. 7.

theory is uncoupled, but now the temperature field will be time dependent. Treating time as a parameter, we can carry over nearly all of the results of the equilibrium theory. The few exceptions are those places where we assumed that the temperature satisfied the equilibrium heat equation.

D. Dynamic theory.

18. Basic equations. Thermoelastic processes. In this chapter, we treat the coupled dynamic theory of linear elastic heat conductors. The fundamental system of field equations consists of the ***strain-displacement relation***

$$\boldsymbol{E} = \tfrac{1}{2}\left(\nabla \boldsymbol{u} + \nabla \boldsymbol{u}^T\right),$$

the ***thermal gradient-temperature relation***

$$\boldsymbol{g} = \nabla \vartheta,$$

the ***equation of motion***

$$\operatorname{div} \boldsymbol{S} + \boldsymbol{b} = \varrho \ddot{\boldsymbol{u}},$$

the ***energy equation***

$$-\operatorname{div} \boldsymbol{q} + \theta_0 \boldsymbol{M} \cdot \dot{\boldsymbol{E}} + r = c \dot{\vartheta},$$

the ***stress-strain-temperature relation***

$$\boldsymbol{S} = \mathsf{C}[\boldsymbol{E}] + \vartheta \boldsymbol{M},$$

and the ***heat conduction equation***

$$\boldsymbol{q} = -\boldsymbol{K} \boldsymbol{g}.$$

Here $\boldsymbol{u}$, $\boldsymbol{E}$, $\boldsymbol{S}$, ϑ, $\boldsymbol{g}$, $\boldsymbol{q}$, $\boldsymbol{b}$, and r are the displacement, strain, stress, temperature difference,[70] thermal gradient, heat flux, body force, and heat supply fields, respectively; while ϱ, C, $\boldsymbol{M}$, $\boldsymbol{K}$, and c are the density, elasticity, stress-temperature, conductivity, and specific heat fields, respectively; finally θ_0 is the fixed uniform reference temperature. Of course, a body B and a fixed time interval $(0, t_0)$ are presumed given.

Again following LTE,[71] we say that $\boldsymbol{u}$ is an ***admissible motion*** provided

(i) $\boldsymbol{u}$ is of class C^2 on $B \times (0, t_0)$;

(ii) $\boldsymbol{u}$, $\dot{\boldsymbol{u}}$, $\ddot{\boldsymbol{u}}$, sym $\nabla \boldsymbol{u}$, and sym $\nabla \dot{\boldsymbol{u}}$ are continuous on $\bar{B} \times [0, t_0)$.

A ***time-dependent admissible stress field*** $\boldsymbol{S}$ is a symmetric tensor field with the properties:

(i) $\boldsymbol{S}$ is of class $C^{1,0}$ on $B \times (0, t_0)$;

(ii) $\boldsymbol{S}$ and div $\boldsymbol{S}$ are continuous on $\bar{B} \times [0, t_0)$.

A ***time-dependent admissible temperature difference field*** ϑ is defined by the following conditions:

(i) ϑ is of class $C^{2,1}$ on $B \times (0, t_0)$;

(ii) ϑ, $\nabla \vartheta$, and $\dot{\vartheta}$ are continuous on $\bar{B} \times [0, t_0)$.

We say that $\boldsymbol{q}$ is a ***time-dependent admissible heat flux field*** if

(i) $\boldsymbol{q}$ is of class $C^{1,0}$ on $B \times (0, t_0)$;

(ii) $\boldsymbol{q}$ and div $\boldsymbol{q}$ are continuous on $\bar{B} \times [0, t_0)$.

---

[70] See Sect. 9.

[71] Sects. 15, 60.

By an ***admissible process*** we mean an ordered array $[\boldsymbol{u}, \boldsymbol{E}, \boldsymbol{S}, \vartheta, \boldsymbol{g}, \boldsymbol{q}]$ with the following properties:

(i) $\boldsymbol{u}$ is an admissible motion;
(ii) $\boldsymbol{E}$ is a continuous symmetric tensor field on $\bar{B}\times[0, t_0)$;
(iii) $\boldsymbol{S}$ is a time-dependent admissible stress field;
(iv) ϑ is a time-dependent admissible temperature difference field;
(v) $\boldsymbol{g}$ is a continuous vector field on $\bar{B}\times[0, t_0)$;
(vi) $\boldsymbol{q}$ is a time-dependent admissible heat flux field.

The fields $\boldsymbol{u}, \boldsymbol{E}, \boldsymbol{S}, \vartheta, \boldsymbol{g}$, and $\boldsymbol{q}$ need not be related, and the set of all admissible processes is a vector space provided addition and scalar multiplication are defined in the natural manner.[72]

We assume for the rest of Chap. D that the elasticity tensor C, *the stress-temperature tensor* $\boldsymbol{M}$, *the conductivity tensor* $\boldsymbol{K}$, *the specific heat* c, *and the density* ϱ *are prescribed and that* C, $\boldsymbol{M}$, *and* $\boldsymbol{K}$ *are smooth on* $\bar{B}$ *while* c *and* ϱ *are continuous on* $\bar{B}$. *Furthermore,* C *and its values are symmetric,* $\boldsymbol{M}$ *is symmetric,* $\boldsymbol{K}$ *is positive semi-definite,*[73] *and* ϱ *is strictly positive.*

By a ***thermoelastic process corresponding to the body force b and the heat supply*** r we mean an admissible process $[\boldsymbol{u}, \boldsymbol{E}, \boldsymbol{S}, \vartheta, \boldsymbol{g}, \boldsymbol{q}]$ that meets the fundamental system of field equations listed at the beginning of this section. Of course, this implies that $\boldsymbol{b}$ and r are continuous on $\bar{B}\times[0, t_0)$. The corresponding ***surface traction s*** and ***heat flux*** q are defined at every regular point of $\partial B\times[0, t_0)$ by

$$\boldsymbol{s}(\boldsymbol{x}, t) = \boldsymbol{S}(\boldsymbol{x}, t)\, \boldsymbol{n}(\boldsymbol{x})$$

and

$$q(\boldsymbol{x}, t) = \boldsymbol{q}(\boldsymbol{x}, t)\cdot \boldsymbol{n}(\boldsymbol{x}),$$

where $\boldsymbol{n}(\boldsymbol{x})$ is the unit outward normal to ∂B at $\boldsymbol{x}$. We call $[\boldsymbol{b}, \boldsymbol{s}]$ the ***external force system*** and $[r, q]$ the ***external thermal system*** for the thermoelastic process $[\boldsymbol{u}, \boldsymbol{E}, \boldsymbol{S}, \vartheta, \boldsymbol{g}, \boldsymbol{q}]$.

As in the equilibrium theory, superposition holds due to the linearity of the field equations.[74]

19. Special results for homogeneous and isotropic bodies. In this section, we assume that the body is *homogeneous* and *isotropic*[75] so that the basic field equations reduce to

$$\begin{aligned}
\boldsymbol{E} &= \tfrac{1}{2}(\nabla\boldsymbol{u} + \nabla\boldsymbol{u}^T),\\
\boldsymbol{g} &= \nabla\vartheta,\\
\operatorname{div}\boldsymbol{S} + \boldsymbol{b} &= \varrho\ddot{\boldsymbol{u}},\\
-\operatorname{div}\boldsymbol{q} + m\,\theta_0 \operatorname{tr}\dot{\boldsymbol{E}} + r &= c\dot{\vartheta},\\
\boldsymbol{S} &= 2\mu\,\boldsymbol{E} + \lambda(\operatorname{tr}\boldsymbol{E})\,\mathbf{1} + m\,\vartheta\mathbf{1},\\
\boldsymbol{q} &= -k\boldsymbol{g},
\end{aligned} \tag{19.1}$$

where the density ϱ, the stress-temperature modulus m, the specific heat c, the Lamé moduli λ and μ, and the conductivity k are constants.

---

[72] See Sect. 9.
[73] See Sect. 7.
[74] Cf. ***(9.1)***.
[75] See Sect. 8.

Suppose that $[\boldsymbol{u}, \boldsymbol{E}, \boldsymbol{S}, \vartheta, \boldsymbol{g}, \boldsymbol{q}]$ is a thermoelastic process corresponding to $\boldsymbol{b}$ and r; then substitution of $(19.1)_1$, $(19.1)_5$ into $(19.1)_3$ and $(19.1)_1$, $(19.1)_2$, $(19.1)_6$ into $(19.1)_4$ yields the ***displacement-temperature equation of motion***[76]

$$\mu\,\Delta\boldsymbol{u} + (\lambda+\mu)\,\nabla\operatorname{div}\boldsymbol{u} + m\,\nabla\vartheta + \boldsymbol{b} = \varrho\ddot{\boldsymbol{u}}$$

and the ***coupled heat equation***[76]

$$k\,\Delta\vartheta + m\,\theta_0\operatorname{div}\dot{\boldsymbol{u}} + r = c\,\dot{\vartheta}.$$

The statement of the following theorem is facilitated by the introduction of the ***wave operators***[77] $\Box_1$ and $\Box_2$ defined by

$$\Box_\alpha f = c_\alpha^2\,\Delta f - \ddot{f} \qquad (\alpha = 1, 2),$$

where

$$c_1 = \sqrt{\frac{\lambda+2\mu}{\varrho}}, \qquad c_2 = \sqrt{\frac{\mu}{\varrho}}.$$ [78]

(1) Dilatational and rotational wave equations. *Let* $\boldsymbol{u}$ *of class* $C^{3,2}$, ϑ *of class* $C^{2,1}$, *and* $\boldsymbol{b}$ *of class* $C^{1,0}$ *on* $B\times(0, t_0)$ *satisfy the displacement-temperature equation of motion. Then* $\boldsymbol{u}$, ϑ, *and* $\boldsymbol{b}$ *obey the* **dilatational wave equation**[79]

$$\Box_1\operatorname{div}\boldsymbol{u} = -\frac{1}{\varrho}\,(m\,\Delta\vartheta + \operatorname{div}\boldsymbol{b})$$

and the **rotational wave equation**[80]

$$\Box_2\operatorname{curl}\boldsymbol{u} = -\frac{1}{\varrho}\operatorname{curl}\boldsymbol{b}.$$

Proof. Operate on the displacement-temperature equation of motion with the divergence and the curl. □

Note that the rotational wave equation is independent of the temperature. The next theorem provides an equation for the temperature that is independent of the displacement.

(2) Temperature equation.[81] *Let* $\boldsymbol{u}$ *of class* C^3, ϑ *of class* $C^{4,3}$, $\boldsymbol{b}$ *of class* C^1, *and* r *of class* C^2 *on* $B\times(0, t_0)$ *satisfy the displacement-temperature equation of motion and the coupled heat equation. Then* ϑ, $\boldsymbol{b}$, *and* r *obey the* **temperature equation**

$$k\,c_1^2\,\Delta\,\Delta\vartheta - \left(\frac{m^2\,\theta_0}{\varrho} + c\,c_1^2\right)\Delta\dot{\vartheta} - k\,\Delta\ddot{\vartheta} + c\,\dddot{\vartheta} - \frac{m\,\theta_0}{\varrho}\operatorname{div}\dot{\boldsymbol{b}} + \Box_1 r = 0.$$

Proof. Operate on the coupled heat equation with $\Box_1$ and use ***(1)***. □

On eliminating the temperature between the coupled heat equation and the dilatational wave equation, VOIGT[82] found that *the dilatation* $\operatorname{div}\boldsymbol{u}$ *also satisfies the temperature equation.*

[76] See ***(22.1)*** for the nonhomogeneous and anisotropic case.
[77] Cf. LTE, Sect. 59.
[78] Note that the wave operators are well defined even if $\lambda + 2\mu$ and μ are negative.
[79] VOIGT [1895, *1*, p. 554].
[80] CRISTEA [1952, *1*].
[81] The temperature equation was obtained by CRISTEA [1952, *1*] through the formal technique of associated matrices. Rigorous deductions were given by LESSEN and DUKE [1953, *1*] in the one-dimensional case, by CHADWICK and SNEDDON [1958, *1*] in the three-dimensional case, and by PARIA [1958, *4*] (for the entropy) in the axisymmetric case.
[82] [1895, *1*, p. 554].

Next let

$$\sigma = \frac{m^2\,\theta_0}{\varrho} + c\,c_1^2,$$

$$\varkappa_1 = \frac{k\,c_1^2}{\sigma},$$

$$\varkappa_2 = \frac{k}{c},$$

and define the ***heat operators*** $\bigcirc_1$ and $\bigcirc_2$ by

$$\bigcirc_\alpha f = \varkappa_\alpha\,\Delta f - \dot f \qquad (\alpha = 1, 2).$$

Then the temperature equation in ***(2)*** can be written in the following forms (assuming $\boldsymbol{b} = \mathbf{0}$, $r = 0$):

$$\left(\sigma\,\Delta\,\bigcirc_1 - c\,\frac{\partial^2}{\partial t^2}\,\bigcirc_2\right)\vartheta = 0,$$

$$\left(c\,\bigcirc_2\,\Box_1 - \frac{m^2\,\theta_0}{\varrho}\,\frac{\partial}{\partial t}\,\Delta\right)\vartheta = 0.$$

Since the dilatation satisfies the temperature equation, we can operate on the displacement-temperature equation of motion with the above operator to obtain

$$\Box_2\left(\sigma\,\Delta\,\bigcirc_1 - c\,\frac{\partial^2}{\partial t^2}\,\bigcirc_2\right)\boldsymbol{u} = \mathbf{0}\ \text{[83]}$$

(assuming $\boldsymbol{b} = \mathbf{0}$, $r = 0$ and sufficient smoothness). This is the counterpart of the equation $\Box_1\,\Box_2\,\boldsymbol{u} = \mathbf{0}$[84] in the isothermal theory.

In the next section, we will uncouple the field equations in a different manner.

20. Complete solutions of the field equations. Here we study a general solution of the displacement-temperature equation of motion

$$\mu\,\Delta\boldsymbol{u} + (\lambda+\mu)\,\nabla\,\mathrm{div}\,\boldsymbol{u} + m\,\nabla\vartheta + \boldsymbol{b} = \varrho\,\ddot{\boldsymbol{u}}$$

and the coupled heat equation

$$k\,\Delta\vartheta + m\,\theta_0\,\mathrm{div}\,\dot{\boldsymbol{u}} + r = c\,\dot\vartheta$$

that is analogous to the *Green-Lamé solution*[85] in the isothermal case.

We assume initially that the body force $\boldsymbol{b}$ is continuous on $\overline{B}\times(0, t_0)$ and of class $C^{2,1}$ on $B\times(0, t_0)$. Then we have the *Helmholtz resolution*[86]

$$\frac{\boldsymbol{b}}{\varrho} = -\nabla\varkappa - \mathrm{curl}\,\boldsymbol{\gamma},$$

where $\varkappa$ and $\boldsymbol{\gamma}$ are of class $C^{2,1}$ on $B\times(0, t_0)$ and $\mathrm{div}\,\boldsymbol{\gamma} = 0$. We also assume that the stress-temperature modulus $m \neq 0$; otherwise the mechanical and thermal aspects of the theory are completely uncoupled.

Proceeding formally, we substitute the Helmholtz decomposition

$$\boldsymbol{u} = \nabla\varphi + \mathrm{curl}\,\boldsymbol{\psi}$$

into the displacement-temperature equation of motion to get[87]

$$\nabla\left(\Box_1\,\varphi + \frac{m}{\varrho}\,\vartheta - \varkappa\right) + \mathrm{curl}\,(\Box_2\,\boldsymbol{\psi} - \boldsymbol{\gamma}) = \mathbf{0}.$$

---

[83] CRISTEA [1952, *1*].

[84] LTE, (59.6).

[85] LTE, ***(67.1)***.

[86] LTE, ***(6.7)***.

[87] The wave operators $\Box_1$ and $\Box_2$ are defined in Sect. 19.

Thus, we can satisfy the equation of motion by taking

$$\Box_1 \varphi = \varkappa - \frac{m}{\varrho} \vartheta,$$
$$\Box_2 \psi = \gamma.$$

From the first of these,

$$\vartheta = \frac{\varrho}{m} (\varkappa - \Box_1 \varphi),$$

and substitution of this and $\boldsymbol{u} = \nabla \varphi + \operatorname{curl} \psi$ into the coupled heat equation yields [88]

$$c \bigcirc_2 \Box_1 \varphi - \frac{m^2 \theta_0}{\varrho} \Delta \dot{\varphi} = c \bigcirc_2 \varkappa + \frac{m}{\varrho} r.$$

Hence, we have the

(1) Deresiewicz-Zorski solution.[89] *Let*

$$\boldsymbol{u} = \nabla \varphi + \operatorname{curl} \psi$$

and

$$\vartheta = \frac{\varrho}{m} (\varkappa - \Box_1 \varphi),$$

where the fields φ of class $C^{4,3}$ and ψ of class $C^{3,2}$ on $B \times (0, t_0)$ satisfy

$$\left(c \bigcirc_2 \Box_1 - \frac{m^2 \theta_0}{\varrho} \frac{\partial}{\partial t} \Delta\right) \varphi = c \bigcirc_2 \varkappa + \frac{m}{\varrho} r$$

and

$$\Box_2 \psi = \gamma.$$

Then $\boldsymbol{u}$ and ϑ meet the displacement-temperature equation of motion and the coupled heat equation.

It is interesting to note that the scalar potential φ satisfies the temperature equation if $\boldsymbol{b} = \boldsymbol{0}$, $r = 0$.[90]

The next theorem establishes that the Deresiewicz-Zorski solution provides a general solution of the equations of motion and energy.

(2) Completeness of the Deresiewicz-Zorski solution.[91] *Let $\boldsymbol{u}$ of class $C^{M,N}$ on $B \times (0, t_0)$ and continuous on $\bar{B} \times (0, t_0)$, where $M \geqq 3$, $N \geqq 2$, and ϑ of class $C^{P,Q}$ on $B \times (0, t_0)$, where $P \geqq 2$, $Q \geqq 0$, meet the displacement-temperature equation of motion for $\boldsymbol{b}/\varrho = -\nabla \varkappa - \operatorname{curl} \gamma$ with $\varkappa$ and γ of class $C^{P,Q}$ on $B \times (0, t_0)$ and* $\operatorname{div} \gamma = 0$. *Then there exist fields φ and ψ of class $C^{M,R}$ on $B \times (0, t_0)$, where*

$$R = \min \{N + 2, Q + 2\},$$

such that

$$\boldsymbol{u} = \nabla \varphi + \operatorname{curl} \psi,$$

$$\vartheta = \frac{m}{\varrho} (\varkappa - \Box_1 \varphi),$$

$$\Box_2 \psi = \gamma;$$

moreover,

$$\operatorname{div} \psi = \boldsymbol{0}.$$

---

[88] The heat operator $\bigcirc_2$ is defined in Sect. 19.

[89] DERESIEWICZ [1958, *2*], ZORSKI [1958, *6*].

[90] CHADWICK [1960, *2*, p. 278].

[91] STERNBERG [1960, *5*] indicated that DUHEM's completeness proof for the Green-Lamé solution of the isothermal theory could be easily extended to the present case; actually SOMIGLIANA's proof is more readily extended. See also NOWACKI [1967, *3*].

If, in addition, $M \geqq 4$, $P \geqq 2$, and $Q \geqq 1$, and if $\boldsymbol{u}$ and ϑ meet the coupled heat equation, then φ satisfies

$$\left(c \bigtriangleup_2 \Box_1 - \frac{m^2 \theta_0}{\varrho} \frac{\partial}{\partial t} \Delta\right) \varphi = c \bigtriangleup_2 \varkappa + \frac{m}{\varrho} r.$$

Proof. The first assertion follows from the completeness proof given in LTE[92] for the Green-Lamé solution of elastodynamics provided we replace the body force there by $\boldsymbol{b} + m \nabla \vartheta$. The equation for φ is obtained by substituting the representations of $\boldsymbol{u}$ and ϑ into the coupled heat equation. □

It is interesting to observe that on eliminating φ between the representation $\vartheta = \varrho/m(\varkappa - \Box_1 \varphi)$ and the coupled heat equation with div $\dot{\boldsymbol{u}}$ replaced by $\Delta \dot{\varphi}$, we reach the temperature equation.[93] However, the direct derivation of the temperature equation given in ***(19.2)*** is preferable in that it requires less severe smoothness assumptions.

A counterpart of the *Cauchy-Kovalevski-Somigliana solution*[94] of the isothermal theory has been given by PODSTRIGACH,[95] NOWACKI,[96] and RÜDIGER.[97] NOWACKI[98] has established its uniqueness and its relation to the Deresiewicz-Zorski solution. Since it does not appear to be a particularly useful solution,[99] we do not elaborate on it here.

21. The theorem of power and energy. The reciprocal theorem. The *theorem of power expended*[100] asserts that

$$\int_{\partial B} \boldsymbol{s} \cdot \dot{\boldsymbol{u}}\, da + \int_B \boldsymbol{b} \cdot \dot{\boldsymbol{u}}\, dv = \int_B \boldsymbol{S} \cdot \dot{\boldsymbol{E}}\, dv + \frac{1}{2} \frac{d}{dt} \int_B \dot{\boldsymbol{u}} \cdot \dot{\boldsymbol{u}}\, dv$$

provided $\boldsymbol{S}$ is a time-dependent admissible stress field,[101] $\boldsymbol{u}$ is an admissible motion[101] with strain field $\boldsymbol{E}$, and the equation of motion is satisfied.

By the stress-strain-temperature relation,

$$\int_B \boldsymbol{S} \cdot \dot{\boldsymbol{E}}\, dv = \int_B \mathsf{C}[\boldsymbol{E}] \cdot \dot{\boldsymbol{E}}\, dv + \int_B \vartheta \boldsymbol{M} \cdot \dot{\boldsymbol{E}}\, dv.$$

Since C is symmetric,

$$\mathsf{C}[\boldsymbol{E}] \cdot \dot{\boldsymbol{E}} = \frac{1}{2} \frac{d}{dt} (\boldsymbol{E} \cdot \mathsf{C}[\boldsymbol{E}]).$$

From the energy, thermal gradient-temperature, and heat conduction equations

$$\vartheta \boldsymbol{M} \cdot \dot{\boldsymbol{E}} = \frac{c}{2\theta_0} \frac{\partial}{\partial t} (\vartheta^2) + \frac{1}{\theta_0} [\operatorname{div}(\vartheta \boldsymbol{q}) - \vartheta r + \boldsymbol{g} \cdot \boldsymbol{K} \boldsymbol{g}].$$

Therefore, the theorem of power expended implies the

(1) Theorem of power and energy.[102] *Let $[\boldsymbol{u}, \boldsymbol{E}, \boldsymbol{S}, \vartheta, \boldsymbol{g}, \boldsymbol{q}]$ be a thermoelastic process corresponding to the external force and thermal systems $[\boldsymbol{b}, \boldsymbol{s}]$ and*

[92] ***(67.2)***.
[93] CHADWICK [1960, *2*, p. 278].
[94] LTE, ***(67.4)***.
[95] [1960, *4*].
[96] [1964, *5*].
[97] [1964, *6*].
[98] [1967, *3*].
[99] Cf. GURTIN's remarks on the relative merits of Green-Lame and the Cauchy-Kovalevski-Somigliana solutions in elastodynamics, LTE, Sect. 67.
[100] LTE, ***(60.2)***.
[101] See Sect. 18.
[102] In the case of homogeneous data, WEINER [1957, *3*] for isotropic bodies and IONESCU-CAZIMIR [1964, *2*] for anisotropic bodies. See also BRUN [1965, *1*].

$[r, q]$. *Then*

$$\int_{\partial B} \mathbf{s} \cdot \dot{\boldsymbol{u}}\, da + \int_B \boldsymbol{b} \cdot \dot{\boldsymbol{u}}\, dv - \frac{1}{\theta_0} \int_{\partial B} q\vartheta\, da + \frac{1}{\theta_0} \int_B r\vartheta\, dv - \frac{1}{\theta_0} \int_B \boldsymbol{g} \cdot \boldsymbol{K}\boldsymbol{g}\, dv = \dot{\mathscr{U}}$$

where $\mathscr{U}$ *is the* ***total energy***:

$$\mathscr{U} = \frac{1}{2} \int_B \left(\varrho \dot{\boldsymbol{u}} \cdot \dot{\boldsymbol{u}} + \boldsymbol{E} \cdot \mathbf{C}[\boldsymbol{E}] + \frac{c}{\theta_0} \vartheta^2\right) dv.$$

Since the conductivity tensor $\boldsymbol{K}$ is positive semi-definite, an immediate consequence of ***(1)*** is the inequality

$$\int_{\partial B} \mathbf{s} \cdot \dot{\boldsymbol{u}}\, da + \int_B \boldsymbol{b} \cdot \dot{\boldsymbol{u}}\, dv - \frac{1}{\theta_0} \int_{\partial B} q\vartheta\, da + \frac{1}{\theta_0} \int_B r\vartheta\, dv - \dot{\mathscr{U}} \geqq 0.$$

Thus, we can state the following corollary of ***(1)***, which will be the key to uniqueness.

(2) *Let* $[\boldsymbol{u}, \boldsymbol{E}, \boldsymbol{S}, \vartheta, \boldsymbol{g}, \boldsymbol{q}]$ *be a thermoelastic process corresponding to the external force and thermal systems* $[\boldsymbol{b}, \mathbf{s}]$ *and* $[r, q]$, *and suppose that* $\mathbf{s} \cdot \dot{\boldsymbol{u}} = q\vartheta = 0$ *on* $\partial B \times [0, t_0)$ *and* $\boldsymbol{b} = \boldsymbol{0}$, $r = 0$, *on* $\bar{B} \times [0, t_0)$; *then*

$$\mathscr{U}(t) \leqq \mathscr{U}(0), \quad 0 \leqq t < t_0.$$

Roughly speaking, the above inequality asserts that the total energy of an isolated body decreases in time.

With a view toward establishing a counterpart of *Graffi's reciprocal theorem*[103] in the isothermal theory, we introduce the following notation. Let $[\boldsymbol{u}, \boldsymbol{E}, \mathbf{s}, \vartheta, \boldsymbol{g}, \boldsymbol{q}]$ be a thermoelastic process corresponding to the external force and thermal systems $[\boldsymbol{b}, \mathbf{s}]$ and $[r, q]$. Then the associated ***pseudo body force field*** $\mathscr{b}$[104] and the ***pseudo heat supply field*** $\mathscr{r}$ are defined on $\bar{B} \times [0, t_0)$ by

$$\mathscr{b}(\boldsymbol{x}, t) = i * \boldsymbol{b}(\boldsymbol{x}, t) + \varrho(x)[\boldsymbol{u}_0(\boldsymbol{x}) + t\, \boldsymbol{v}_0(\boldsymbol{x})],$$
$$\mathscr{r}(\boldsymbol{x}, t) = 1 * r(\boldsymbol{x}, t) + c(x)\, \vartheta_0(\boldsymbol{x}) - \theta_0 \boldsymbol{M}(\boldsymbol{x}) \cdot \boldsymbol{E}_0(\boldsymbol{x}).$$

Here

$$\boldsymbol{u}_0 = \boldsymbol{u}(\cdot, 0), \quad \boldsymbol{v}_0 = \dot{\boldsymbol{u}}(\cdot, 0), \quad \boldsymbol{E}_0 = \boldsymbol{E}(\cdot, 0), \quad \text{and} \quad \vartheta_0 = \vartheta(\cdot, 0)$$

are the initial values of the displacement, velocity, strain, and temperature difference fields, $*$ denotes the operation of *convolution*,[105] and i and 1 are the functions on $[0, t_0)$ defined by

$$i(t) \equiv t, \quad 1(t) \equiv 1.$$

Thus, for any function f,

$$1 * f(x, t) = \int_0^t f(x, t)\, dt.$$

Further, we write j for the function

$$j(t) \equiv \frac{t^2}{2}.$$

(3) Reciprocal theorem.[106] *Assume that the conductivity tensor* $\boldsymbol{K}$ *is symmetric. Let* $[\boldsymbol{u}, \boldsymbol{E}, \boldsymbol{S}, \vartheta, \boldsymbol{g}, \boldsymbol{q}]$ *and* $[\tilde{\boldsymbol{u}}, \tilde{\boldsymbol{E}}, \tilde{\boldsymbol{S}}, \tilde{\vartheta}, \tilde{\boldsymbol{g}}, \tilde{\boldsymbol{q}}]$ *be thermoelastic processes corre-*

[103] LTE, ***(61.1)***.

[104] Cf. LTE, Sect. 61, where the notation $\boldsymbol{f}$ is used.

[105] See LTE, Sect. 10.

[106] IONESCU-CAZMIR [1964, *2*] for null initial data. The formulation given here was communicated to be my M. E. GURTIN in 1970. A different type of reciprocal theorem has been given by BRUN [1965, *1*, Eq. (6)].

sponding to the external force and thermal systems $[\mathbf{b}, \mathbf{s}]$, $[r, q]$ *and* $[\tilde{\mathbf{b}}, \tilde{\mathbf{s}}]$, $[\tilde{r}, \tilde{q}]$, *respectively. Further, let* $\boldsymbol{u}_0$, $\boldsymbol{v}_0$, $\boldsymbol{E}_0$, ϑ_0 *and* $\tilde{\boldsymbol{u}}_0$, $\tilde{\boldsymbol{v}}_0$, $\tilde{\boldsymbol{E}}_0$, $\tilde{\vartheta}_0$ *be the corresponding initial displacement, velocity, strain, and temperature difference fields; and let* $\mathfrak{b}$, $\mathfrak{r}$ *and* $\tilde{\mathfrak{b}}$, $\tilde{\mathfrak{r}}$ *be the respective pseudo body force and pseudo heat supply fields. Then*

$$i*\int_{\partial B} \mathbf{s}*\tilde{\boldsymbol{u}}\,da+\int_B \mathfrak{b}*\tilde{\boldsymbol{u}}\,dv+\frac{j}{\theta_0}*\int_{\partial B} q*\tilde{\vartheta}\,da-\frac{i}{\theta_0}*\int_B \mathfrak{r}*\vartheta\,dv$$
$$=i*\int_{\partial B} \tilde{\mathbf{s}}*\boldsymbol{u}\,da+\int_B \tilde{\mathfrak{b}}*\boldsymbol{u}\,dv+\frac{j}{\theta_0}*\int_{\partial B} \tilde{q}*\vartheta\,da-\frac{i}{\theta_0}\int_B \tilde{\mathfrak{r}}*\vartheta\,dv$$

and

$$\int_{\partial B} \mathbf{s}*\tilde{\boldsymbol{u}}\,da+\int_B \boldsymbol{b}*\tilde{\boldsymbol{u}}\,dv+\int_B \varrho(\boldsymbol{u}_0\cdot\dot{\tilde{\boldsymbol{u}}}+\boldsymbol{v}_0\cdot\tilde{\boldsymbol{u}})\,dv$$
$$-\frac{1}{\theta_0}*\left[-\int_{\partial B} q*\tilde{\vartheta}\,da+\int_B r*\tilde{\vartheta}\,dv+\int_B (c\vartheta_0-\theta_0\boldsymbol{M}\cdot\boldsymbol{E}_0)\,\tilde{\vartheta}\,dv\right]$$
$$=\int_{\partial B} \tilde{\mathbf{s}}*\boldsymbol{u}\,da+\int_B \tilde{\boldsymbol{b}}*\boldsymbol{u}\,dv+\int_B \varrho(\boldsymbol{u}_0\cdot\dot{\boldsymbol{u}}+\boldsymbol{v}_0\cdot\boldsymbol{u})\,dv$$
$$-\frac{1}{\theta_0}*\left[-\int_{\partial B} \tilde{q}*\vartheta\,da+\int_B \tilde{r}*\vartheta\,dv+\int_B (c\tilde{\vartheta}_0-\theta_0\boldsymbol{M}\cdot\tilde{\boldsymbol{E}}_0)\,\vartheta\,dv\right].$$

In particular, if both processes correspond to null initial displacement, velocity, and temperature difference fields, then

$$\int_{\partial B} \mathbf{s}*\tilde{\boldsymbol{u}}\,da+\int_B \boldsymbol{b}*\tilde{\boldsymbol{u}}\,dv-\frac{1}{\theta_0}*\left(-\int_{\partial B} q*\tilde{\vartheta}\,da+\int_B r*\tilde{\vartheta}\,dv\right)$$
$$=\int_{\partial B} \tilde{\mathbf{s}}*\boldsymbol{u}\,da+\int_B \tilde{\boldsymbol{b}}*\boldsymbol{u}\,dv-\frac{1}{\theta_0}*\left(-\int_{\partial B} \tilde{q}*\vartheta\,da+\int_B \tilde{r}*\vartheta\,dv\right).$$

Proof. It follows from a general dynamical reciprocal theorem[107] that

$$i*\int_{\partial B} \mathbf{s}*\tilde{\boldsymbol{u}}\,da+\int_B \mathfrak{b}*\tilde{\boldsymbol{u}}\,dv-i*\int_B \boldsymbol{S}*\tilde{\boldsymbol{E}}\,dv$$
$$=i*\int_{\partial B} \tilde{\mathbf{s}}*\boldsymbol{u}\,da+\int_B \tilde{\mathfrak{b}}*\boldsymbol{u}\,dv-i*\int_B \tilde{\boldsymbol{S}}*\boldsymbol{E}\,dv,$$
$$\int_{\partial B} \mathbf{s}*\tilde{\boldsymbol{u}}\,da+\int_B \boldsymbol{b}*\tilde{\boldsymbol{u}}\,dv-\int_B \boldsymbol{S}*\tilde{\boldsymbol{E}}\,dv+\int_B \varrho(\boldsymbol{u}_0\cdot\dot{\tilde{\boldsymbol{u}}}+\boldsymbol{v}_0\cdot\tilde{\boldsymbol{u}})\,dv$$
$$=\int_{\partial B} \tilde{\mathbf{s}}*\boldsymbol{u}\,da+\int_B \tilde{\boldsymbol{b}}*\boldsymbol{u}\,dv-\int_B \tilde{\boldsymbol{S}}*\boldsymbol{E}\,dv+\int_B \varrho(\tilde{\boldsymbol{u}}_0\cdot\dot{\boldsymbol{u}}+\tilde{\boldsymbol{v}}_0\cdot\boldsymbol{u})\,dv.$$

By the stress-strain-temperature relation,

$$\boldsymbol{S}*\tilde{\boldsymbol{E}}=\mathsf{C}[\boldsymbol{E}]*\tilde{\boldsymbol{E}}+(\vartheta\boldsymbol{M})*\tilde{\boldsymbol{E}},$$
$$\tilde{\boldsymbol{S}}*\boldsymbol{E}=\mathsf{C}[\tilde{\boldsymbol{E}}]*\boldsymbol{E}+(\tilde{\vartheta}\boldsymbol{M})*\boldsymbol{E}.$$

From the symmetry of C and the definition of convolution, it follows[108] that

$$\mathsf{C}[\boldsymbol{E}]*\tilde{\boldsymbol{E}}=\mathsf{C}[\tilde{\boldsymbol{E}}]*\boldsymbol{E}.$$

---

[107] LTE, ***(19.5)***.

[108] See LTE, Sect. 61.

Consequently,

$$
\begin{aligned}
i * \int_{\partial B} \mathbf{s} * \tilde{\boldsymbol{u}}\, da + \int_B \boldsymbol{b} * \tilde{\boldsymbol{u}}\, dv - i * \int_B (\vartheta \boldsymbol{M}) * \tilde{\boldsymbol{E}}\, dv \\
= i * \int_{\partial B} \tilde{\mathbf{s}} * \boldsymbol{u}\, da + \int_B \tilde{\boldsymbol{b}} * \boldsymbol{u}\, dv - i * \int_B (\tilde{\vartheta} \boldsymbol{M}) * \boldsymbol{E}\, dv,
\end{aligned} \tag{21.1}
$$

$$
\begin{aligned}
\int_{\partial B} \mathbf{s} * \tilde{\boldsymbol{u}}\, da + \int_B \boldsymbol{b} * \tilde{\boldsymbol{u}}\, dv - \int_B (\vartheta \boldsymbol{M}) * \tilde{\boldsymbol{E}}\, dv + \int_B \varrho (\boldsymbol{u}_0 \cdot \dot{\tilde{\boldsymbol{u}}} + \boldsymbol{v}_0 \cdot \tilde{\boldsymbol{u}})\, dv \\
= \int_{\partial B} \tilde{\mathbf{s}} * \boldsymbol{u}\, da + \int_B \tilde{\boldsymbol{b}} * \boldsymbol{u}\, dv - \int_B (\tilde{\vartheta} \boldsymbol{M}) * \boldsymbol{E}\, dv + \int_B \varrho (\tilde{\boldsymbol{u}}_0 \cdot \dot{\boldsymbol{u}} + \tilde{\boldsymbol{v}}_0 \cdot \boldsymbol{u})\, dv.
\end{aligned} \tag{21.2}
$$

Next, if we integrate the energy equation

$$
-\operatorname{div} \boldsymbol{q} + \theta_0 \boldsymbol{M} \cdot \dot{\boldsymbol{E}} + r = c \dot{\vartheta}
$$

from 0 to t, we arrive at

$$
-1 * \operatorname{div} \boldsymbol{q} + \theta_0 \boldsymbol{M} \cdot \boldsymbol{E} + \mathscr{r} = c \vartheta .
$$

Taking the convolution of this equation with $\tilde{\vartheta}$ and integrating over B, we conclude, with the aid of the divergence theorem, that

$$
\begin{aligned}
-1 * \int_{\partial B} q * \tilde{\vartheta}\, da + 1 * \int_B \boldsymbol{q} * \nabla \tilde{\vartheta}\, dv + \theta_0 \int_B \boldsymbol{E} * (\tilde{\vartheta} \boldsymbol{M})\, dv + \int_B \mathscr{r} * \tilde{\vartheta}\, dv \\
= \int_B c \vartheta * \tilde{\vartheta}\, dv .
\end{aligned} \tag{21.3}
$$

Similarly,

$$
\begin{aligned}
-1 * \int_{\partial B} \tilde{q} * \vartheta\, da + 1 * \int_B \tilde{\boldsymbol{q}} * \nabla \vartheta\, dv + \theta_0 \int_B \tilde{\boldsymbol{E}} * (\vartheta \boldsymbol{M})\, dv + \int_B \tilde{\mathscr{r}} * \vartheta\, dv \\
= \int_B c \tilde{\vartheta} * \vartheta\, dv .
\end{aligned} \tag{21.4}
$$

By the commutativity of the convolution,

$$
\vartheta * \tilde{\vartheta} = \tilde{\vartheta} * \vartheta . \tag{21.5}
$$

Further, in view of the heat conduction equation and the symmetry of $\boldsymbol{K}$,

$$
\boldsymbol{q} * \nabla \tilde{\vartheta} = \tilde{\boldsymbol{q}} * \nabla \vartheta . \tag{21.6}
$$

Indeed, suppressing the argument $\boldsymbol{x}$ and writing $\boldsymbol{g} = \nabla \vartheta$, $\tilde{\boldsymbol{g}} = \nabla \tilde{\vartheta}$,

$$
\begin{aligned}
\boldsymbol{q} * \nabla \tilde{\vartheta}(t) &= \int_0^t [\boldsymbol{K} \boldsymbol{g}(t-\tau)] \cdot \tilde{\boldsymbol{g}}(\tau)\, d\tau = \int_0^t \boldsymbol{g}(t-\tau) \cdot \boldsymbol{K} \tilde{\boldsymbol{g}}(\tau)\, dt \\
&= \int_0^t [\boldsymbol{K} \tilde{\boldsymbol{g}}(t-\tau)] \cdot \boldsymbol{g}(\tau)\, dt = \tilde{\boldsymbol{q}} * \nabla \vartheta(t).
\end{aligned}
$$

Finally, if we solve (21.3) and (21.4) for

$$
\int_B \boldsymbol{E} * (\tilde{\vartheta} \boldsymbol{M})\, dv \quad \text{and} \quad \int_B \tilde{\boldsymbol{E}} * (\vartheta \boldsymbol{M})\, dv,
$$

substitute the results into (21.1) and (21.2), and use (21.5) and (21.6), we are led to the first two relations in *(3)*. The third result in *(3)* is an obvious consequence of the second. □

22. The boundary-initial-value problems of the dynamic theory. In this section, in addition to the specification of the body B and its material properties,[109] we assume that the following ***data*** are given:

(i) the body force $\boldsymbol{b}$ and heat supply r both continuous on $\bar{B}\times[0, t_0)$;

(ii) the initial displacement, velocity, and temperature $\boldsymbol{u}_0$, $\boldsymbol{v}_0$, and ϑ_0 all continuous on $\bar{B}$;

(iii) the surface displacement $\hat{\boldsymbol{u}}$ continuous on $\mathscr{S}_1\times[0, t_0)$;

(iv) the surface traction $\hat{\mathbf{s}}$ piecewise regular on $\mathscr{S}_2\times[0, t_0)$ and continuous in time;

(v) the surface temperature $\hat{\vartheta}$ continuous on $\mathscr{S}_3\times[0, t_0)$;

(vi) the surface heat flux $\hat{q}$ piecewise regular on $\mathscr{S}_4\times[0, t_0)$ and continuous in time.[110]

Then the ***mixed problem of the dynamic theory of thermoelasticity*** is to find a thermoelastic process $[\boldsymbol{u}, \boldsymbol{E}, \boldsymbol{S}, \vartheta, \boldsymbol{g}, \boldsymbol{q}]$ corresponding to $\boldsymbol{b}$ and r that satisfies the ***initial conditions***

$$\boldsymbol{u}(\cdot, 0)=\boldsymbol{u}_0, \quad \dot{\boldsymbol{u}}(\cdot, 0)=\boldsymbol{v}_0, \quad \vartheta(\cdot, 0)=\vartheta_0 \quad \text{on } \bar{B},$$

the ***displacement condition***

$$\boldsymbol{u}=\hat{\boldsymbol{u}} \quad \text{on} \quad \mathscr{S}_1\times[0, t_0),$$

the ***traction condition***

$$\mathbf{s}=\boldsymbol{S}\boldsymbol{n}=\hat{\mathbf{s}} \quad \text{on} \quad \mathscr{S}_2\times[0, t_0),$$

the ***temperature condition***

$$\vartheta=\hat{\vartheta} \quad \text{on} \quad \mathscr{S}_3\times[0, t_0),$$

and the ***heat flux condition***

$$q=\boldsymbol{q}\cdot\boldsymbol{n}=\hat{q} \quad \text{on} \quad \mathscr{S}_4\times[0, t_0).$$

If such a thermoelastic process exists, it is called a ***solution of the mixed problem.***

We refrain from naming the numerous special cases of the mixed problem that arise when one or more of the subsurfaces $\mathscr{S}_1-\mathscr{S}_4$ is empty. As in the equilibrium theory, more complicated boundary-initial-value problems can be considered.

(1) Characterization of the mixed problem in terms of the displacement and the temperature. *An admissible motion $\boldsymbol{u}$ and a time-dependent admissible temperature field ϑ correspond to a solution of the mixed problem if and only if*

$$\operatorname{div}\mathbf{C}[\nabla\boldsymbol{u}]+\operatorname{div}(\vartheta\boldsymbol{M})+\boldsymbol{b}=\varrho\ddot{\boldsymbol{u}} \quad \textit{on} \quad B\times(0, t_0),$$ [111]

$$\operatorname{div}(\boldsymbol{K}\nabla\vartheta)+\theta_0\boldsymbol{M}\cdot\nabla\dot{\boldsymbol{u}}+r=c\dot{\vartheta} \quad \textit{on} \quad B\times(0, t_0),$$ [112]

$$\boldsymbol{u}(\cdot, 0)=\boldsymbol{u}_0, \quad \dot{\boldsymbol{u}}(\cdot, 0)=\boldsymbol{v}_0, \quad \vartheta(\cdot, 0)=\vartheta_0 \quad \textit{on } \bar{B},$$

$$\boldsymbol{u}=\hat{\boldsymbol{u}} \quad \textit{on} \quad \mathscr{S}_1\times[0, t_0),$$

$$(\mathbf{C}[\nabla\boldsymbol{u}]+\vartheta\boldsymbol{M})\boldsymbol{n}=\hat{\mathbf{s}} \quad \textit{on} \quad \mathscr{S}_2\times[0, t_0),$$

$$\vartheta=\hat{\vartheta} \quad \textit{on} \quad \mathscr{S}_3\times[0, t_0),$$

$$-(\boldsymbol{K}\nabla\vartheta)\cdot\boldsymbol{n}=\hat{q} \quad \textit{on} \quad \mathscr{S}_4\times[0, t_0).$$

[109] See Sect. 18.
[110] $\mathscr{S}_1$ and $\mathscr{S}_2$ are complementary subsurfaces of ∂B as are $\mathscr{S}_3$ and $\mathscr{S}_4$, see LTE, Sect. 5.
[111] We refer to this equation as the ***displacment-temperature equation of motion.***
[112] This equation is known as the ***coupled heat equation.***

When B is homogeneous and isotropic, the first, second, fifth, and seventh of these conditions reduce to

$$\begin{aligned} \mu\,\Delta\boldsymbol{u}+(\lambda+\mu)\,\nabla\operatorname{div}\boldsymbol{u}+m\,\nabla\vartheta+\boldsymbol{b}&=\varrho\ddot{\boldsymbol{u}} && \text{on} \quad B\times(0,t_0),\\ k\,\Delta\vartheta+m\theta_0\operatorname{div}\dot{\boldsymbol{u}}+r&=c\dot{\vartheta} && \text{on} \quad B\times(0,t_0),\\ \mu(\nabla\boldsymbol{u}+\nabla\boldsymbol{u}^T)\,\boldsymbol{n}+\lambda(\operatorname{div}\boldsymbol{u})\,\boldsymbol{n}+m\,\vartheta\boldsymbol{n}&=\hat{\mathbf{s}} && \text{on} \quad \mathscr{S}_2\times[0,t_0),\text{[113]}\\ -k\,\nabla\vartheta\cdot\boldsymbol{n}&=\hat{q} && \text{on} \quad \mathscr{S}_4\times[0,t_0). \end{aligned}$$

Proof. Suppose that $[\boldsymbol{u},\boldsymbol{E},\boldsymbol{S},\vartheta,\boldsymbol{g},\boldsymbol{q}]$ is a solution of the mixed problem. Then substitution of the stress-strain-temperature relation and the strain-displacement relation into the equation of motion yields the displacement-temperature equation of motion. Similarly, the heat conduction equation, the thermal gradient-temperature relation, the strain-displacement relation, and the energy equation imply the coupled heat equation. With the stress-strain-temperature relation and the strain-displacement relation the traction condition becomes

$$\mathbf{C}[\nabla\boldsymbol{u}]\,\boldsymbol{n}+\vartheta\boldsymbol{M}\boldsymbol{n}=\hat{\mathbf{s}} \quad \text{on} \quad \mathscr{S}_2\times[0,t_0),$$

while the heat conduction equation, the thermal gradient-temperature relation, and the heat flux condition give

$$-(\boldsymbol{K}\,\nabla\vartheta)\cdot\boldsymbol{n}=\hat{q} \quad \text{on} \quad \mathscr{S}_4\times[0,t_0).$$

The remaining conditions in ***(1)*** are identical to ones appearing in the statement of the mixed problem.

Conversely, suppose that an admissible motion $\boldsymbol{u}$ and a time-dependent admissible temperature field ϑ satisfy the above conditions. Define $\boldsymbol{E}$ through the strain-displacement relation, $\boldsymbol{g}$ through the thermal gradient-temperature relation, $\boldsymbol{S}$ through the stress-strain temperature relation, and $\boldsymbol{q}$ through the heat conduction equation. Then

$$\begin{aligned} \operatorname{div}\boldsymbol{S}+\boldsymbol{b}&=\varrho\ddot{\boldsymbol{u}} && \text{on} \quad B\times(0,t_0),\\ -\operatorname{div}\boldsymbol{q}+\theta_0\boldsymbol{M}\cdot\dot{\boldsymbol{E}}+r&=c\dot{\vartheta} && \text{on} \quad B\times(0,t_0),\\ \boldsymbol{S}\boldsymbol{n}&=\hat{\mathbf{s}} && \text{on} \quad \mathscr{S}_2[0,t_0),\\ \boldsymbol{q}\cdot\boldsymbol{n}&=\hat{q} && \text{on} \quad \mathscr{S}_4[0,t_0). \end{aligned}$$

Thus, all that remains to be shown is that $[\boldsymbol{u},\boldsymbol{E},\boldsymbol{S},\vartheta,\boldsymbol{g},\boldsymbol{q}]$ is an admissible process. Since $\boldsymbol{u}$ is an admissible motion and ϑ is a time-dependent admissible temperature difference field, $\boldsymbol{E}$ and $\boldsymbol{g}$ are continuous on $\bar{B}\times[0,t_0)$. From the smoothness of $\mathbf{C}$, $\boldsymbol{M}$, and $\boldsymbol{K}$ and the admissibility of $\boldsymbol{u}$ and ϑ, $\boldsymbol{S}$ and $\boldsymbol{q}$ are continuous on $\bar{B}\times[0,t_0)$ and of class $C^{1,0}$ on $B\times(0,t_0)$. The continuity of $\operatorname{div}\boldsymbol{S}$ and $\operatorname{div}\boldsymbol{q}$ on $\bar{B}\times[0,t_0)$ follows from the continuity of $\boldsymbol{b}$, r, ϱ, and c and the equations of motion and energy. □

When some of the subsurfaces $\mathscr{S}_1$–$\mathscr{S}_4$ are empty, it is easy to construct alternative formulations analogous to IGNACZAK'S[114] characterization of the traction

[113] This boundary condition can be written as

$$2\mu\frac{\partial\boldsymbol{u}}{\partial\boldsymbol{n}}+\mu\boldsymbol{n}\times\operatorname{curl}\boldsymbol{u}+\lambda(\operatorname{div}\boldsymbol{u})\,\boldsymbol{n}+m\,\vartheta\boldsymbol{n}=\hat{\mathbf{s}} \quad \text{on} \quad \mathscr{S}_2\times[0,t_0),$$

see the end of Sect. 12.

[114] [1963, *3*]. See also LTE, ***(62.2)***.

problem of elastodynamics in terms of the stress. An incomplete formulation in terms of $\boldsymbol{S}$ and ϑ has been given by IGNACZAK[115] for the case where $\mathscr{S}_1 = \mathscr{S}_4 = \emptyset$.

With an eye toward the derivation of variational principles in Sect. 24, we next give a characterization of the mixed problem in which the initial conditions are incorporated into the field equations. First, we recall[116] that the pseudo body field $\mathscr{b}$ and the pseudo heat supply $\mathscr{r}$ are defined on $\bar{B} \times [0, t_0)$ by

$$\mathscr{b}(\boldsymbol{x}, t) = i * \boldsymbol{b}(\boldsymbol{x}, t) + \varrho(\boldsymbol{x})[\boldsymbol{u}_0(\boldsymbol{x}) + t\,\boldsymbol{v}_0(\boldsymbol{x})],$$

$$\mathscr{r}(\boldsymbol{x}, t) = 1 * r(\boldsymbol{x}, t) + c(\boldsymbol{x})\,\vartheta_0(\boldsymbol{x}) - \theta_0 \boldsymbol{M}(\boldsymbol{x}) \cdot \boldsymbol{E}_0(\boldsymbol{x}),$$

where

$$i(t) \equiv t, \qquad 1(t) \equiv 1,$$

and $*$ denotes the operation of convolution.[117]

(2)[118] *An admissible process* $[\boldsymbol{u}, \boldsymbol{E}, \boldsymbol{S}, \vartheta, \boldsymbol{g}, \boldsymbol{q}]$ *is a solution of the mixed problem if and only if*

$$\boldsymbol{E} = \operatorname{sym} \nabla \boldsymbol{u}, \tag{22.1}$$

$$\boldsymbol{g} = \nabla \vartheta, \tag{22.2}$$

$$\boldsymbol{S} = \mathbf{C}[\boldsymbol{E}] + \vartheta \boldsymbol{M}, \tag{22.3}$$

$$\boldsymbol{q} = -\boldsymbol{K}\boldsymbol{g}, \tag{22.4}$$

$$i * \operatorname{div} \boldsymbol{S} + \mathscr{b} = \varrho \boldsymbol{u}, \tag{22.5}$$

$$-1 * \operatorname{div} \boldsymbol{q} + \theta_0 \boldsymbol{M} \cdot \boldsymbol{E} + \mathscr{r} = c\vartheta, \tag{22.6}$$

on $B \times [0, t_0)$, *and*

$$\boldsymbol{u} = \hat{\boldsymbol{u}} \quad on \quad \mathscr{S}_1 \times [0, t_0),$$

$$\mathbf{s} = \hat{\mathbf{s}} \quad \text{on} \quad \mathscr{S}_2 \times [0, t_0),$$

$$\vartheta = \hat{\vartheta} \quad on \quad \mathscr{S}_3 \times [0, t_0),$$

$$q = \hat{q} \quad on \quad \mathscr{S}_4 \times [0, t_0).$$

Proof. That

$$i * \operatorname{div} \boldsymbol{S} + \mathscr{b} = \varrho \boldsymbol{u}$$

is equivalent to the equation of motion and the mechanical initial conditions is ***(19.2)*** of LTE. We saw in the proof of ***(21.3)*** that the energy equation and $\vartheta(\cdot, 0) = \vartheta_0$ imply

$$-1 * \operatorname{div} \boldsymbol{q} + \theta_0 \boldsymbol{M} \cdot \boldsymbol{E} + \mathscr{r} = c\vartheta.$$

Conversely, differentiation of this equation with respect to time yields the energy equation, and $\vartheta(\cdot, 0) = \vartheta_0$ is recovered by setting $t = 0$. □

Clearly ***(2)*** can be written to give a characterization of the mixed problem in terms of $\boldsymbol{u}$ and ϑ which incorporates the initial conditions into the field equations. Actually such a characterization in terms of any pair of mechanical and thermal variables is possible.[119]

23. Uniqueness.

Next, we establish a uniqueness theorem for the mixed problem that is a direct counterpart of *Neumann's theorem*[120] in the isothermal case. Reference to more general results is made at the end of the section.

---

[115] [1963, *4*].

[116] See Sect. 21.

[117] See LTE, Sect. 10.

[118] IESAN [1966, *1*], NICKELL and SACKMAN [1968, *2*].

[119] See IESAN [1966, *1*] for $\boldsymbol{S}$, ϑ and $\boldsymbol{u}$, $\boldsymbol{q}$ formulations, and NICKELL and SACKMAN [1968, *2*] for $\boldsymbol{u}$, ϑ and $\boldsymbol{S}$, $\boldsymbol{q}$ formulations.

[120] See LTE, ***(63.1)***.

(1) Uniqueness theorem for the mixed problem.[121] *Let the elasticity tensor* $\mathbf{C}$ *be positive semi-definite and the specific heat c be strictly positive.*[122] *Then the mixed problem has at most one solution.*

Proof. Suppose that there are two solutions. Then their difference $[\boldsymbol{u}, \boldsymbol{E}, \boldsymbol{S}, \vartheta, \boldsymbol{g}, \boldsymbol{q}]$ corresponds to null data, i.e.,

$$\boldsymbol{u}(\cdot, 0) = \dot{\boldsymbol{u}}(\cdot, 0) = \mathbf{0}, \quad \vartheta(\cdot, 0) = 0, \tag{23.1}$$

and

$$\boldsymbol{s} \cdot \dot{\boldsymbol{u}} = q\vartheta = 0 \quad \text{on} \quad \partial B \times [0, t_0), \quad \boldsymbol{b} = \mathbf{0}, \quad r = 0. \tag{23.2}$$

By (23.2) and *(21.2)*,

$$\mathscr{U}(t) = \frac{1}{2} \int_B \left(\varrho \dot{\boldsymbol{u}} \cdot \dot{\boldsymbol{u}} + \boldsymbol{E} \cdot \mathbf{C}[\boldsymbol{E}] + \frac{c}{\theta_0} \vartheta^2\right) dv \leqq \mathscr{U}(0), \quad 0 \leqq t < t_0.$$

By (23.1),

$$\mathscr{U}(0) = 0;$$

and, therefore, since $\varrho > 0$, $\boldsymbol{E} \cdot \mathbf{C}[\boldsymbol{E}] \geqq 0$, and $\frac{c}{\theta_0} > 0$,

$$\mathscr{U}(t) = 0, \quad 0 \leqq t < t_0.$$

Hence,

$$\dot{\boldsymbol{u}} = \mathbf{0} \quad \text{and} \quad \vartheta = 0 \quad \text{on} \quad \bar{B} \times [0, t_0).$$

But $\boldsymbol{u}$ vanishes initially; thus

$$\boldsymbol{u} = \mathbf{0} \quad \textit{on} \quad \bar{B} \times [0, t_0).$$

Consequently,

$$[\boldsymbol{u}, \boldsymbol{E}, \boldsymbol{S}, \vartheta, \boldsymbol{g}, \boldsymbol{q}] = [\mathbf{0}, \mathbf{0}, \mathbf{0}, 0, \mathbf{0}, \mathbf{0}]. \quad \square$$

BRUN[123] has been able to prove uniqueness without the hypothesis that $\mathbf{C}$ be positive semi-definite. KNOPS and PAYNE[124] have reached a similar conclusion and established continuous dependence of the solution on the initial data. Their results have been generalized by LEVINE,[125] but the generalization does not appear to be relevant to the theory presented here. DAFERMOS[126] has provided existence, uniqueness, and asymptotic stability theorems for the mixed problem.

24. Variational principles. Here we derive variational principles which are direct counterparts of those given by GURTIN[127] for elastodynamics. In addition to the hypotheses (i)–(vi) of Sect. 22 on the data and the conditions of Sect. 18 on the material properties, *we assume that the conductivity* $\boldsymbol{K}$ *is symmetric.*

We will be concerned with scalar-valued functionals whose domain is a subset $\mathscr{A}$ of the space of all admissible processes. Letting Λ be such a functional, we introduce the formal notation

$$\delta_{\tilde{p}} \Lambda\{p\} = \frac{d}{d\lambda} \Lambda\{p + \lambda \tilde{p}\}\big|_{\lambda=0},$$

---

121 WEINER [1957, *3*] for homogeneous and isotropic bodies. IONESCU-CAZIMIR [1964, *3*] for the general case. See also IONESCU-CAZIMIR [1964, *4*] and BRUN [1965, *1*]. ROSENBLATT [1910, *1*] attributes uniqueness to VOIGT [1895, *1*], but I have been unable to locate such a theorem in VOIGT's treatise.

122 We have already assumed that the conductivity tensor $\boldsymbol{K}$ is positive semi-definite and that the density ϱ is strictly positive; see Sect. 18.

123 [1969, *1*].

124 [1970, *1*].

125 [1970, *2*].

126 [1968, *1*].

127 LTE, Sect. 65.

where $\not p + \lambda \tilde{\not p} \in \mathscr{A}$ for every scalar λ. If $\delta_{\tilde{\not p}} \Lambda\{\not p\}$ exists and equals zero for every $\tilde{\not p}$ meeting the above requirement, then we write

$$\delta \Lambda\{\not p\} = 0.$$

We recall[128] that the pseudo body force field $\not b$ and the pseudo heat supply field $\not r$ are defined on $\bar{B} \times [0, t_0)$ by

$$\not b(\boldsymbol{x}, t) = i * \mathbf{b}(\boldsymbol{x}, t) + \varrho(\boldsymbol{x})[\boldsymbol{u}_0(\boldsymbol{x}) + t\,\boldsymbol{v}_0(\boldsymbol{x})],$$
$$\not r(\boldsymbol{x}, t) = \varrho * r(\boldsymbol{x}, t) + c(\boldsymbol{x})\,\vartheta_0(\boldsymbol{x}) - \theta_0 \boldsymbol{M}(\boldsymbol{x}) \cdot \boldsymbol{E}_0(\boldsymbol{x}),$$

where

$$i(t) \equiv t, \qquad 1(t) \equiv 1,$$

and $*$ denotes the operation of convolution.[129]

In the first variational principle, the admissible processes are not required to meet any of the field equations, initial conditions, or boundary conditions.

(1)[130] *Let $\mathscr{A}$ denote the set of all admissible processes, and for each $t \in [0, t_0)$, define the functional $\Lambda_t\{\cdot\}$ on $\mathscr{A}$ by*

$$\begin{aligned}
\Lambda_t\{\not p\} = \int_B \Big[\tfrac{1}{2}\, i * \boldsymbol{E} * \mathbf{C}[\boldsymbol{E}] + \tfrac{1}{2}\,\varrho \boldsymbol{u} * \boldsymbol{u} - i * \boldsymbol{S} * \boldsymbol{E} - (i * \operatorname{div} \boldsymbol{S} + \not b) * \boldsymbol{u}\Big]\, dv \\
- \frac{1}{\theta_0} \int_B i * \Big[\tfrac{1}{2} * \boldsymbol{g} * \boldsymbol{K} \boldsymbol{g} + \tfrac{1}{2}\, c\vartheta * \vartheta + 1 * \boldsymbol{q} * \boldsymbol{g} \\
- (-1 * \operatorname{div} \boldsymbol{q} + \theta_0 \boldsymbol{M} \cdot \boldsymbol{E} + \not r) * \vartheta\Big]\, dv + \int_{\mathscr{S}_1} i * \mathbf{s} * \hat{\boldsymbol{u}}\, da + \int_{\mathscr{S}_2} i * (\mathbf{s} - \hat{\mathbf{s}}) * \boldsymbol{u}\, da \\
+ \frac{1}{\theta_0} \int_{\mathscr{S}_3} i * 1 * q * \hat{\vartheta}\, da + \frac{1}{\theta_0} \int_{\mathscr{S}_4} i * 1 * (q - \hat{q}) * \vartheta\, da
\end{aligned}$$

for every $\not p = [\boldsymbol{u}, \boldsymbol{E}, \boldsymbol{S}, \vartheta, \boldsymbol{g}, \boldsymbol{q}] \in \mathscr{A}$. Then

$$\delta \Lambda_t\{\not p\} = 0 \qquad (0 \le t < t_0)$$

at an admissible process $\not p$ if and only if $\not p$ is a solution of the mixed problem.

Proof. Let $\not p, \tilde{\not p} \in \mathscr{A}$; then $\not p + \lambda \tilde{\not p} \in \mathscr{A}$. A straightforward calculation[131] yields

$$\begin{aligned}
\delta_{\tilde{\not p}} \Lambda_t\{\not p\} = \int_B i * (\mathbf{C}[\boldsymbol{E}] + \vartheta \boldsymbol{M} - \boldsymbol{S}) * \tilde{\boldsymbol{E}}\, dv - \int_B (i * \operatorname{div} \boldsymbol{S} + \not b - \varrho \boldsymbol{u}) * \tilde{\boldsymbol{u}}\, dv \\
+ \int_B i * (\operatorname{sym} \nabla \boldsymbol{u} - \boldsymbol{E}) * \tilde{\boldsymbol{S}}\, dv - \frac{1}{\theta_0} \int_B i * 1 * (\boldsymbol{K} \boldsymbol{g} + \boldsymbol{q}) * \tilde{\boldsymbol{g}}\, dv \\
+ \frac{1}{\theta_0} \int_B i * (-1 * \operatorname{div} \boldsymbol{q} + \theta_0 \boldsymbol{M} \cdot \boldsymbol{E} + \not r - c\vartheta) * \tilde{\vartheta}\, dv \\
+ \frac{1}{\theta_0} \int_B i * 1 * (\nabla \vartheta - \boldsymbol{g}) * \tilde{\boldsymbol{q}}\, dv + \int_{\mathscr{S}_1} i * (\hat{\boldsymbol{u}} - \boldsymbol{u}) * \tilde{\mathbf{s}}\, da + \int_{\mathscr{S}_2} i * (\mathbf{s} - \hat{\mathbf{s}}) * \tilde{\boldsymbol{u}}\, da \\
+ \frac{1}{\theta_0} \int_{\mathscr{S}_3} i * 1 * (\hat{\vartheta} - \vartheta) * \tilde{q}\, da + \frac{1}{\theta_0} \int_{\mathscr{S}_4} i * 1 * (q - \hat{q}) * \tilde{\vartheta}\, da
\end{aligned} \tag{24.1}$$

for $0 \le t < t_0$.

[128] See Sect. 21.

[129] See LTE, Sect. 10.

[130] IESAN [1966, *1*], NICKELL and SACKMAN [1968, *2*]. Their results appear somewhat different because they do not eliminate the entropy from the basic equations. See also RAFALSKI [1968, *3*]. These papers reference several other variational principles for thermoelasticity.

[131] Cf. the proof of ***(65.1)*** of LTE.

If $\mathscr{p}$ is a solution of the mixed problem, it follows from (24.1) and ***(22.2)*** that

$$\delta_{\tilde{\mathscr{p}}} \Lambda_t\{\mathscr{p}\} = 0 \quad (0 \leqq t < t_0) \quad \text{for every } \tilde{\mathscr{p}} \in \mathscr{A}, \tag{24.2}$$

and therefore,

$$\delta \Lambda_t\{\mathscr{p}\} = 0 \quad (0 \leqq t < t_0). \tag{24.3}$$

Conversely, suppose that (24.3) and hence (24.2) holds. Choose

$$\tilde{\mathscr{p}} = [\tilde{\boldsymbol{u}}, \mathbf{0}, \mathbf{0}, 0, \mathbf{0}, \mathbf{0}]$$

and let $\tilde{\boldsymbol{u}}$ *vanish near*[132] ∂B to get from (24.1) and (24.2) that

$$\int_B (i * \operatorname{div} \boldsymbol{S} + \boldsymbol{b} - \varrho \boldsymbol{u}) * \tilde{\boldsymbol{u}} \, dv = 0 \quad (0 \leqq t < t_0).$$

Since this must hold for every $\boldsymbol{u}$ of class C^∞ on $\bar{B} \times [0, t_0)$ that vanishes near ∂B, we conclude from ***(64.1)*** of LTE that

$$i * \operatorname{div} \boldsymbol{S} + \boldsymbol{b} = \varrho \boldsymbol{u}.$$

Similarly, by choosing $\tilde{\mathscr{p}}$ in the forms $[\mathbf{0}, \tilde{\boldsymbol{E}}, \mathbf{0}, 0, \mathbf{0}, \mathbf{0}]$, $[\mathbf{0}, \mathbf{0}, \tilde{\boldsymbol{S}}, 0, \mathbf{0}, \mathbf{0}]$, $[\mathbf{0}, \mathbf{0}, \mathbf{0}, \tilde{\vartheta}, \mathbf{0}, \mathbf{0}]$, $[\mathbf{0}, \mathbf{0}, \mathbf{0}, 0, \tilde{\boldsymbol{g}}, \mathbf{0}]$, and $[\mathbf{0}, \mathbf{0}, \mathbf{0}, 0, \mathbf{0}, \tilde{\boldsymbol{q}}]$, where $\tilde{\boldsymbol{E}}$, $\tilde{\boldsymbol{S}}$, $\tilde{\vartheta}$, $\tilde{\boldsymbol{g}}$, and $\tilde{\boldsymbol{q}}$ vanish near ∂B, we conclude from (24.1), (24.2) and ***(64.1)*** of LTE that

$$\begin{aligned} i * (\mathsf{C}[\boldsymbol{E}] + \vartheta \boldsymbol{M} - \boldsymbol{S}) &= \mathbf{0}, \\ i * (\operatorname{sym} \nabla \boldsymbol{u} - \boldsymbol{E}) &= \mathbf{0}, \\ i * 1 * (\boldsymbol{K} \boldsymbol{g} + \boldsymbol{q}) &= \mathbf{0}, \\ i * (-1 * \operatorname{div} \boldsymbol{q} + \theta_0 \boldsymbol{M} \cdot \boldsymbol{E} + r - c \vartheta) &= 0, \\ i * 1 * (\nabla \vartheta - \boldsymbol{g}) &= \mathbf{0}. \end{aligned}$$

Hence, by the cancellation property[133] of convolutions and ***(22.2)***, $\mathscr{p}$ satisfies ail the field equations and initial conditions of the mixed problem. Thus (24.1) and (24.2) imply

$$\int_{\mathscr{S}_1} (\hat{\boldsymbol{u}} - \boldsymbol{u}) * \tilde{\boldsymbol{S}} \boldsymbol{n} \, da + \int_{\mathscr{S}_2} (\boldsymbol{s} - \hat{\boldsymbol{s}}) * \tilde{\boldsymbol{u}} \, da$$
$$+ \frac{1}{\theta_0} \int_{\mathscr{S}_3} 1 * (\hat{\vartheta} - \vartheta) * \tilde{\boldsymbol{q}} \cdot \boldsymbol{n} \, da + \frac{1}{\theta_0} \int_{\mathscr{S}_4} 1 * (q - \hat{q}) * \tilde{\vartheta} \, da = 0$$

for every $0 \leqq t < t_0$ and every $\tilde{\mathscr{p}} \in \mathscr{A}$. Therefore, on making obvious choices of $\tilde{\mathscr{p}}$ and appealing to Lemmas ***(64.2)*** and ***(64.3)*** of LTE, we see that $\mathscr{p}$ meets the boundary conditions.[134] Thus $\mathscr{p}$ is a solution of the mixed problem. □

The functional whose variation vanishes at a solution can be simplified by putting restrictions on its domain.

(2)[135] *Assume that the elasticity tensor* C *and the conductivity tensor* $\boldsymbol{K}$ *are invertible. Let* $\mathscr{A}$ *denote the set of all admissible processes that satisfy the strain-*

[132] I.e., there exists a neighborhood N of ∂B such that $\tilde{\boldsymbol{u}} = \mathbf{0}$ on $(N \cap \bar{B}) \times [0, t_0)$. See LTE, Sect. 64.
[133] See (iv) of ***(10.1)*** of LTE.
[134] Actually to obtain the temperature condition, we need a scalar version of ***(64.3)*** of LTE.
[135] IESAN [1966, *1*], NICKELL and SACKMAN [1968, *2*].

displacement and thermal gradient-temperature relations, and for each $t \in [0, t_0)$ *define the functional* $\Theta_t\{\cdot\}$ *on* $\mathscr{A}$ *by*

$$\Theta_t\{p\} = \int_B \left(\frac{1}{2} i * \boldsymbol{S} * \mathbf{K}[\boldsymbol{S}] - \frac{1}{2} \varrho \boldsymbol{u} * \boldsymbol{u} - i * \boldsymbol{S} * \boldsymbol{E} + \boldsymbol{b} * \boldsymbol{u}\right) dv$$
$$- \frac{1}{\theta_0} \int_B i * \left(\frac{1}{2} 1 * \boldsymbol{q} * \boldsymbol{K}^{-1} \boldsymbol{q} - \frac{1}{2} c' \vartheta * \vartheta\right.$$
$$\left. - \theta_0 \boldsymbol{A} \cdot \boldsymbol{S} * \vartheta + 1 * \boldsymbol{q} * \boldsymbol{g} + r * \vartheta\right) dv$$
$$+ \int_{\mathscr{S}_1} i * \mathbf{s} * (\boldsymbol{u} - \hat{\boldsymbol{u}})\, da + \int_{\mathscr{S}_2} i * \hat{\mathbf{s}} * \boldsymbol{u}\, da$$
$$+ \frac{1}{\theta_0} \int_{\mathscr{S}_3} i * 1 * q * (\vartheta - \hat{\vartheta})\, da + \frac{1}{\theta_0} \int_{\mathscr{S}_4} i * 1 * \hat{q} * \vartheta\, da$$

for every $p = [\boldsymbol{u}, \boldsymbol{E}, \boldsymbol{S}, \vartheta, \boldsymbol{g}, \boldsymbol{q}] \in \mathscr{A}$, *where*

$$\mathbf{K} = \mathbf{C}^{-1}, \quad \boldsymbol{A} = -\mathbf{K}[\boldsymbol{M}], \quad c' = c - \theta_0 \boldsymbol{M} \cdot \boldsymbol{A}.$$

Then

$$\delta \Theta_t\{p\} = 0 \qquad (0 \leqq t < t_0)$$

at $p \in \mathscr{A}$ *if and only if* p *is a solution of the mixed problem.*

Proof. Let $p, \tilde{p} \in \mathscr{A}$ so that $p + \lambda \tilde{p} \in \mathscr{A}$. Again an easy calculation[136] leads to

$$\delta_{\tilde{p}} \Theta_t\{p\} = \int_B i * (\mathbf{K}[\boldsymbol{S}] + \vartheta \boldsymbol{A} - \boldsymbol{E}) * \tilde{\boldsymbol{S}}\, dv + \int_B (i * \operatorname{div} \boldsymbol{S} + \boldsymbol{b} - \varrho \boldsymbol{u}) * \tilde{\boldsymbol{u}}\, dv$$
$$- \frac{1}{\theta_0} \int_B i * 1 * (\boldsymbol{K}^{-1} \boldsymbol{q} + \boldsymbol{g}) * \tilde{\boldsymbol{q}}\, dv$$
$$- \frac{1}{\theta_0} \int_B i * (-1 * \operatorname{div} \boldsymbol{q} + \theta_0 \boldsymbol{M} \cdot \mathbf{K}[\boldsymbol{S}] + r - c' \vartheta) * \tilde{\vartheta}\, dv$$
$$+ \int_{\mathscr{S}_1} i * (\boldsymbol{u} - \hat{\boldsymbol{u}}) * \tilde{\mathbf{s}}\, da + \int_{\mathscr{S}_2} i * (\hat{\mathbf{s}} - \mathbf{s}) * \tilde{\boldsymbol{u}}\, da$$
$$+ \frac{1}{\theta_0} \int_{\mathscr{S}_3} i * 1 * (\vartheta - \hat{\vartheta}) * \tilde{q}\, da + \frac{1}{\theta_0} \int_{\mathscr{S}_4} i * 1 * (\hat{q} - q) * \tilde{\vartheta}\, da.$$

On recognizing that the pair of equations

$$\boldsymbol{E} = \mathbf{K}[\boldsymbol{S}] + \vartheta A,$$
$$-1 * \operatorname{div} \boldsymbol{q} + \theta_0 \boldsymbol{M} \cdot \mathbf{K}[\boldsymbol{S}] + r = c' \vartheta$$

is equivalent to the pair (24.3), (24.6) of ***(22.2)***, the proof is completed as in ***(1)***. □

By a ***kinematically and thermally admissible process*** we mean an admissible process that satisfies the strain-displacement and thermal gradient-temperature relations, the stress-strain-temperature and heat conduction equations, and the displacement and temperature boundary conditions. If we restrict $\mathscr{A}$ in ***(1)*** to the set of kinematically and thermally admissible processes, we are led to the following analog of the *principle of minimum potential energy*:[137]

---

[136] Cf. the proof of ***(65.2)*** of LTE.
[137] ***(16.1)***.

(3)[138] *Let $\mathscr{A}$ denote the set of all kinematically and thermally admissible processes, and for each $t \in [0, t_0)$, define the functional $\Phi_t\{\cdot\}$ on $\mathscr{A}$ by*

$$\Phi_t\{p\} = \int_B \left(\frac{1}{2} i*\boldsymbol{S}*\boldsymbol{E} + \frac{1}{2} \varrho \boldsymbol{u}*\boldsymbol{u} - \boldsymbol{b}*\boldsymbol{u}\right) dv$$
$$- \frac{1}{\theta_0} \int_B i*\left(-\frac{1}{2} *\boldsymbol{q}*\boldsymbol{g} + \frac{1}{2} c\vartheta*\vartheta - \frac{1}{2} \theta_0 \boldsymbol{M}\cdot\boldsymbol{E}*\vartheta - r*\vartheta\right) dv$$
$$- \int_{\mathscr{S}_2} i*\hat{\mathbf{s}}*\boldsymbol{u}\, da - \frac{1}{\theta_0} \int_{\mathscr{S}_4} i*1*\hat{q}*\vartheta\, da$$

for every $p = [\boldsymbol{u}, \boldsymbol{E}, \boldsymbol{S}, \vartheta, \boldsymbol{g}, \boldsymbol{q}] \in \mathscr{A}$. Then,

$$\delta\Phi_t\{p\} = 0 \qquad (0 \leqq t < t_0)$$

at $p \in \mathscr{A}$ if and only if p is a solution of the mixed problem.

Proof. Let $\tilde{p}$ be an admissible process, and suppose that $p + \lambda\tilde{p} \in \mathscr{A}$ for every scalar λ. Then $\tilde{p}$ must satisfy the strain-displacement and thermal gradient-temperature relations, the stress-strain-temperature and heat conduction equations, and the boundary conditions

$$\tilde{\boldsymbol{u}} = \boldsymbol{0} \quad \text{on} \quad \mathscr{S}_1 \times [0, t_0), \qquad \tilde{\vartheta} = 0 \quad \text{on} \quad \mathscr{S}_3 \times [0, t_0).$$

Since $\Phi_t\{\cdot\}$ is the restriction of $\Lambda_t\{\cdot\}$ of ***(1)*** to the set of all kinematically and thermally admissible processes, we conclude from (24.1) of the proof of ***(1)*** that

$$\delta_{\tilde{p}} \Phi_t\{p\} = -\int_B (i*\operatorname{div} \boldsymbol{S} + \boldsymbol{b} - \varrho \boldsymbol{u}) * \tilde{\boldsymbol{u}}\, dv$$
$$+ \frac{1}{\theta_0} \int_B i*(-1*\operatorname{div} \boldsymbol{q} + \theta_0 \boldsymbol{M}\cdot\boldsymbol{E} + r - c\vartheta) * \tilde{\vartheta}\, dv$$
$$+ \int_{\mathscr{S}_2} i*(\mathbf{s} - \mathbf{s}) * \tilde{\boldsymbol{u}}\, da + \frac{1}{\theta_0} \int_{\mathscr{S}_4} i*1*(q - \hat{q}) * \tilde{\vartheta}\, da.$$

The desired conclusions follow as in the proof of ***(1)***. □

The papers of IESAN[139] and NICKELL and SACKMAN[140] may be consulted for variational principles formulated in terms of various pairs of mechanical and thermal variables.

25. Progressive waves. In this section, we briefly consider ***harmonic plane progressive wave*** solutions[141] for *homogeneous* and *isotropic* bodies with zero body force and heat supply. Thus, we seek solutions of the displacement-temperature equation of motion[142]

$$\mu\, \Delta \boldsymbol{u} + (\lambda + \mu)\, \nabla \operatorname{div} \boldsymbol{u} + m\, \nabla\vartheta = \varrho \ddot{\boldsymbol{u}} \tag{25.1}$$

and the coupled heat equation[142]

$$k\, \Delta\vartheta + m\, \theta_0 \operatorname{div} \dot{\boldsymbol{u}} = c\dot{\vartheta} \tag{25.2}$$

[138] NICKELL and SACKMAN [1968, *2*]. IESAN [1966, *1*] obtains a similar result by requiring that the heat flux boundary condition be satisfied instead of the temperature boundary condition.

[139] [1966, *1*].

[140] [1968, *2*].

[141] CHADWICK and POWDRILL [1965, *2*] have treated the problem of waves as propagating surfaces of discontinuity in some detail.

[142] See Sect. 19.

in the form

$$\begin{aligned} \boldsymbol{u} &= \boldsymbol{a}\, A \exp[i(\gamma \boldsymbol{m}\cdot\boldsymbol{p} - \xi t)], \\ \vartheta &= B \exp[i(\gamma \boldsymbol{m}\cdot\boldsymbol{p} - \xi t)]. \end{aligned} \tag{25.3}$$

Here $\boldsymbol{a}$ and $\boldsymbol{m}$ are unit vectors called the ***direction of displacement*** and the ***direction of propagation***, respectively; $i=\sqrt{-1}$; $\boldsymbol{p}=\boldsymbol{x}-\boldsymbol{0}$ is the position vector; A and B are real constants; the constants γ and ξ are allowed to be complex; and $\mathrm{Re}(\xi/\gamma)$ is the ***speed of propagation***[143].

Substitution of (25.3) into (25.1) and (25.2) yields

$$\begin{aligned} (\varrho\xi^2 - \mu\gamma^2) A\, \boldsymbol{a} - [(\lambda+\mu)(\boldsymbol{a}\cdot\boldsymbol{m})\gamma^2 A - i m \gamma B]\, \boldsymbol{m} &= 0, \\ m\theta_0 \gamma\xi(\boldsymbol{a}\cdot\boldsymbol{m}) A + (ic\xi - k\gamma^2) B &= 0. \end{aligned} \tag{25.4}$$

In the case of a ***transverse wave*** (i.e., $\boldsymbol{a}\cdot\boldsymbol{m}=0$), (25.4) is equivalent to

$$\begin{aligned} (\varrho\xi^2 - \mu\gamma^2) A &= 0, \\ i m \gamma B &= 0, \\ (ic\xi - k\gamma^2) B &= 0. \end{aligned}$$

Hence,

$$B = 0,\text{[144]}$$

$$\varrho\xi^2 - \mu\gamma^2 = 0,\text{[145]}$$

and we conclude that *transverse waves are independent of thermal effects, and they propagate with speed* $\sqrt{\mu/\varrho}$.[146]

When the wave is ***longitudinal*** (i.e., $\boldsymbol{a}\cdot\boldsymbol{m}=1$), (25.4) is satisfied if and only if

$$\begin{aligned} [\varrho\xi^2 - (\lambda+2\mu)\gamma^2] A + i m \gamma B &= 0, \\ m\theta_0 \gamma \xi A + (ic\xi - k\gamma^2) B &= 0. \end{aligned}$$

A necessary and sufficient condition for this system to have a nontrivial solution for A and B is

$$[\varrho\xi^2 - (\lambda+2\mu)\gamma^2](ic\xi - k\gamma^2) - i m^2 \theta_0 \gamma^2 \xi = 0.\text{[147]}$$

This equation has been solved for γ (given real ξ) and for ξ (given real γ) by CHADWICK.[148] *For given ξ, the waves are* **attenuated** *in that γ is complex and* **dispersed** *since the speed of propagation is a function of ξ. For given γ, the waves are* **damped** *in that ξ is complex and* **dispersed** *because the speed of propagation depends on γ.*

List of works cited.

This list is not a complete bibliography of linear thermoelasticity. I have been interested primarily in establishing priorities. Italic numbers in parentheses following the reference indicate the sections in which it is cited.

1837 *1.* DUHAMEL, J. M. C.: Second mémoire sur les phénomènes thermomécaniques. J. École Polytechn. **15**, 1–57. (*7*)

[143] See LTE, Sect. 11 for motivation.

[144] If $m\gamma=0$, either the theory is uncoupled or we do not have a progressive wave.

[145] If $A=0$, we are left with the trivial solution.

[146] LESSEN and DUKE [1953, *1*], LESSEN [1957, *2*], DERESIEWICZ [1957, *1*], CHADWICK and SNEDDON [1958, *1*].

[147] LESSEN and DUKE [1953, *1*], LESSEN [1957, *2*], DERESIEWICZ [1957, *1*], CHADWICK and SNEDDON [1958, *1*].

[148] [1960, *2*].

1838 *1.* — Mémoire sur le calcul des actions moléculaires développées par les changements de température dans les corps solides. Mémoires par Divers Savants **5**. 440–498. (*7, 11*)

1841 *1.* NEUMANN, F. E.: Die Gesetze der Doppelbrechung des Lichts in comprimirten oder ungleichförmig erwärmten unkrystallinischen Körpern. Abhandl. k. Akad. Wiss. zu Berlin, 2. Theil, 1–254. (*7*)

1851 *1.* STOKES, G. G.: On the conduction of heat in crystals. Cambr. Dubl. Math. J. **6**, 215–238 = Papers 67–101. (*7*)

1885 *1.* NEUMANN, F.: Vorlesungen über die Theorie der Elasticität. Leipzig: Teubner. (*7*)

1895 *1.* VOIGT, W.: Kompendium der theoretischen Physik, vol. 1. Leipzig: Von Veit & Comp. (*7, 15, 19, 23*)

1910 *1.* ROSENBLATT, A.: Über das allgemeine thermoelastische Problem. Rend. Circolo Mat. Palermo **29**, 324–328. (*23*)

2. VOIGT, W.: Lehrbuch der Kristallphysik. Leipzig: Teubner. (*15*)

1930 *1.* JEFFREYS, H.: The thermodynamics of an elastic solid. Proc. Cambridge Phil. Soc. **26**, 101–106. (*7*)

1939 *1.* SOKOLNIKOFF, I. S., and E. S. SOKOLNIKOFF: Thermal stress in elastic plates. Trans. Am. Math. Soc. **45**, 235–255. (*12*)

1941 *1.* MAYSEL, V. M.: A generalization of the Betti-Maxwell theorem to the case of thermal stresses and some of its application [in Russian]. Dokl. Acad. Sci. USSR **30**, 115–118. (*13*)

1952 *1.* CRISTEA, M.: Asupra ecuatiilor micilor miscari termoelastice. Rev. Univ. "C. I. Parhon" Ploiteh. Bucuresti **1**, 72–76. (*19*)

2. RICHTER, H.: Zur Elastizitätstheorie endlicher Verformungen. Math. Nachr. **8**, 65–73. English transl. in Foundations of Elasticity Theory. Intl. Sci. Rev. Ser. New York: Gordon & Breach 1965. (*5*)

1953 *1.* LESSEN, M., and C. E. DUKE: On the motion of an elastic thermally conducting solid. Proc. 1st Midwest Conf. on Solid Mech., Univ. of Illinois, 14–18. (*7, 19, 25*)

1954 *1.* NOWACKI, W.: Thermal stress in anisotropic bodies (I) [in Polish]. Arch. Mech. Stos. **6**, 481–492. (*10*)

1955 *1.* HIEKE, M.: Eine indirekte Bestimmung der Airyschen Fläche bei unstetigen Wärmespannungen. Z. Angew. Math. Mech. **35**, 285–294. (*10*)

1956 *1.* BIOT, M. A.: Thermoelasticity and irreversible thermodynamics. J. Appl. Phys. **27**, 240–253. (*7*)

2. LESSEN, M.: Note on the symmetrical property of the thermal conductivity tensor. Quart. Appl. Math. **14**, 208–209. (*7*)

1957 *1.* DERESIEWICZ, H.: Plane waves in a thermoelastic solid. J. Acoust. Soc. Am. **29**, 204–209. (*25*)

2. LESSEN, M.: The motion of a thermoelastic solid. Quart. Appl. Math. **15**, 105–108. (*25*)

3. WEINER, J. H.: A uniqueness theorem for the coupled thermoelastic problem. Quart. Appl. Math. **15**, 102–105. (*21, 23*)

1958 *1.* CHADWICK, P., and I. N. SNEDDON: Plane waves in an elastic solid conducting heat. J. Mech. Phys. Solids **6**, 223–230. (*19, 25*)

2. DERESIEWICZ, H.: Solution of the equations of thermoelasticity. Proc. 3rd U.S. National Congr. Appl. Mech., Brown University, 287–291. (*20*)

3. NOLL, W.: A mathematical theory of the mechanical behavior of continuous media. Arch. Rational Mech. Anal. **2**, 197–226 (1958/59). Reprinted in Rational mechanics of materials, edit. by C. TRUESDELL, Intl. Sci. Rev. Ser. New York: Gordon and Breach, 1965, and in Continuum theory of inhomogeneities in simple bodies. Berlin-Heidelberg-New York: Springer 1968. (*3*)

4. PARIA, G.: Coupling of elastic and thermal deformations I. Appl. Sci. Res., Sect. A **7**, 463–475. (*19*)

5. PIPKIN, A. C., and R. S. RIVLIN: The formulation of constitutive equations in continuum physics. Div. Appl. Math., Brown University Report. September. (*6*)

6. ZORSKI, H.: Singular solutions for thermoelastic media. Bull. Acad. Polon. Sci., Sér. Sci. Tech. **6**, 331–339. (*20*)

1960 *1.* BOLEY, B. A., and J. H. WEINER: Theory of Thermal Stresses. New York: John Wiley & Sons. (*7*)

2. CHADWICK, P.: Thermoelasticity. The dynamical theory. In: Vol. I of Progress in solid mechanics, edit. by I. N. SNEDDON and R. HILL. Amsterdam: North-Holland. (*20, 25*)

3. GREEN, A. E., and J. E. ADKINS: Large Elastic Deformations and Non-linear Continuum Mechanics. Oxford: Clarendon Press. (*4*)

4. Podstrigach, Y. S.: General solution of the non-steady thermoelastic problem [in Ukrainian]. Prykladnaya Mat. Mekh. **6**, 215–219. (*20*)
5. Sternberg, E.: On the integration of the equations of motion in the classical theory of elasticity. Arch. Rational Mech. Anal. **6**, 34–50. (*20*)
6. Truesdell, C., and R. Toupin: The classical field theories. In: Vol. III/1 of the Handbuch der Physik, edit. by S. Flügge. Berlin-Göttingen-Heidelberg: Springer. (*4*)

1962 *1.* Nowacki, W.: Thermoelasticity. Reading, Mass.: Addison-Wesley. (*10*)

1963 *1.* Coleman, B. D., and W. Noll: The thermodynamics of elastic materials with heat conduction and viscosity. Arch. Rational Mech. Anal. **13**, 167–178. (*4*)
2. —, and V. Mizel: Thermodynamics and departures from Fourier's law of heat conduction. Arch. Rational Mech. Anal. **13**, 245–261. (*6, 8*)
3. Ignaczak, J.: A completeness problem for the stress equations of motion in the linear theory of elasticity. Arch. Mech. Stosow. **15**, 225–234. (*22*)
4. — On the stress equations of motion in the linear thermoelasticity. Arch. Mech. Stosow. **15**, 691–695. (*22*)

1964 *1.* Coleman, B. D., and V. J. Mizel: Existence of caloric equations of state in thermodynamics. J. Chem. Phys. **40**, 1116–1125. (*4*)
2. Ionescu-Cazimir, V.: Problem of linear coupled thermoelasticity. Theorems on reciprocity for the dynamic problem of coupled thermoelasticity. I. Bull. Acad. Polon. Sci., Sér. Sci. Tech. **12**, 473–480. (*21*)
3. — Problem of linear coupled thermoelasticity. III. Uniqueness theorem. Bull. Acad. Polon. Sci., Sér. Sci. Tech. **12**, 565–573. (*21, 23*)
4. — Problem of linear coupled thermoelasticity. IV. Uniqueness theorem. Bull. Acad. Polon. Sci. Sér. Sci. Tech. **12**, 575–579. (*23*)
5. Nowacki, W.: Green functions for the thermoelastic medium. II. Bull. Acad. Polon. Sci. Sér. Sci. Tech. **12**, 465–472. (*20*)
6. Rüdiger, D.: Bemerkung zur Integration der thermoelastischen Grundgleichungen. Österr. Ing.-Arch. **18**, 121–122. (*20*)

1965 *1.* Brun, L.: Sur l'unicité en thermoélasticité dynamique et diverses expressions analogues à la formule de Clapeyron. Compt. Rend. **261**, 2584–2587. (*21, 23*)
2. Chadwick, P., and B. Powdrill: Singular surfaces in linear thermoelasticity. Int. J. Engng. Sci. **3**, 561–595. (*25*)
3. Gurtin, M. E.: Thermodynamics and the possibility of spatial interaction in rigid heat conductors. Arch. Rational Mech. Anal. **18**, 335–342. (*4*)
4. — Thermodynamics and the possibility of spatial interaction in elastic materials. Arch. Rational Mech. Anal. **19**, 339–352. (*4*)
5. Truesdell, C., and W. Noll: The non-linear field theories of mechanics. In: Vol. III/3 of the Handbuch der Physik, edit. by S. Flügge. Berlin-Heidelberg-New York: Springer. (*3, 23*)

1966 *1.* Iesan, D.: Principes variationnels dans la théorie de la thermoélasticité couplée. Analele Stiint. Univ. "A. I. Cuza" Iasi, Sect. I, Matematică **12**, 439–456. (*22, 24*)

1967 *1.* Eringen, A. C.: Mechanics of Continua. New York: John Wiley & Sons. (*7*)
2. Gurtin, M. E.: On the thermodynamics of elastic materials. J. Math. Anal. Appl. **18**, 39–44. (*4*)
3. Nowacki, W.: On the completeness of stress functions in thermoelasticity. Bull. Acad. Polon. Sci., Sér. Sci. Tech. **15**, 583–591 (*20*)

1968 *1.* Dafermos, C. M.: On the existence and asymptotic stability of solutions to the equations of linear thermoelasticity. Arch. Rational Mech. Anal. **29**, 241–271. (*23*)
2. Nickell, R. E., and J. L. Sackman: Variational principles for linear coupled thermoelasticity. Quart. Appl. Math. **26**, 11–26. (*22, 24*)
3. Rafalski, P.: A variational principle for the coupled thermoelastic problem. Int. J. Engng. Sci. **6**, 465–471. (*24*)

1969 *1.* Brun, L.: Méthodes énergétiques dans les systèmes evolutifs linéaires. Deuxième partie: Théorèmes d'unicité. J. Mécanique **8**, 167–192. (*23*)
2. Day, W. A., and M. E. Gurtin: On the symmetries of the conductivity tensor and other restrictions in the nonlinear theory of heat conduction. Arch. Rational Mech. Anal. **33**, 26–32. (*1*)
3. Truesdell, C.: Rational Thermodynamics. New York: McGraw-Hill. (*7, 8*).

1970 *1.* Knops, R. J., and L. E. Payne: On uniqueness and continuous dependence in dynamical problems of linear thermoelasticity. Int. J. Solids Structures **6**, 1173–1184. (*23*)
2. Levine, H. A.: On a theorem of Knops and Payne in dynamical linear thermoelasticity. Arch. Rational Mech. Anal. **38**, 290–307. (*23*)

Existence Theorems in Elasticity.

By

GAETANO FICHERA.

The subject to be developed in this article covers a very large field of existence theory for linear and nonlinear partial differential equations. Indeed, problems of static elasticity, of the propagation of waves in elastic media, and of the thermodynamics of continua require existence theorems for elliptic, hyperbolic and parabolic equations both linear and nonlinear. Even if one restricts oneself to linear elasticity, there are several kinds of partial differential equations to be considered. In static problems we encounter second order systems, either with constant or with variable coefficients (homogeneous and non-homogeneous bodies), scalar second order equations (for instance either in the St. Venant torsion problems or in the membrane theory), fourth order equations (equilibrium of thin plates), eighth order equations (equilibrium of shells). Each case must be considered with several kinds of boundary conditions, corresponding to different physical situations. On the other hand, to every problem of static elasticity corresponds a dynamical one, connected with the study of vibrations in the elastic system under consideration. Moreover, problems of thermodynamics require the study of certain diffusion problems of parabolic type. In addition to that, the study of materials with memory requires existence theorems for certain integro-differential equations, first considered by VOLTERRA.

I confess having had great difficulty in attempting to fit all these subjects into the space allotted. Thus I have not tried to include nonlinear problems, which, on the other hand, as far as existence theory is concerned, remain far from a definitive settlement. The existence theory for problems of elasticity defined by unilateral constraints, although started recently (see [*5*, *6*]) is nowadays so important that I have devoted to it a separate article, which follows this one.

Concerning the linear theory, I have found it more convenient, instead of considering separately different specific cases concerning elasticity, to incorporate all of them within the theory of strongly elliptic linear systems. Of course it would have been preferable to develop the more general theory of elliptic systems, considering strong ellipticity as a particular case of ellipticity, but obviously such a program could not be carried through in a relatively short article. On the other hand, strongly elliptic systems provide enough generality to cover the most significant applications. In connection with these systems, propagation and diffusion problems as well as integro-differential equations have been considered. For all of them the existence theorems have been given for the main cases of interest. Among the many applications of the general theory, let us mention here the existence theorem for the nonstandard boundary value problem connected with the equilibrium of heterogeneous elastic media. Of course some of the problems which arise in the applications are not included here as, for instance, exterior boundary value problems. However I believe that the reader should obtain from the general theories developed in this paper enough information to be able to handle by himself some specific cases which are not considered here.

1. Prerequisites and notations. Throughout this article we shall use basic concepts of LEBESGUE integration theory and of functional analysis. In order to provide the reader with a concrete basis, it is worthwhile to mention that the concepts and the theories which we shall use in this paper can be found (i) in the first six chapters of the book [*31*], (ii) in the first two chapters of [*16*], (iii) in the first chapter of [*14*]. For some additional items which we shall use in the course of the paper, we shall give specific references.

The real r-dimensional cartesian space will be denoted by X^r, i.e. X^r is the real vector space of all the ordered r-tuples of real numbers $x \equiv (x_1, \ldots, x_r)$, where the distance between two points $x \equiv (x_1, \ldots, x_r)$ and $y \equiv (y_1, \ldots, y_r)$ is defined as:

$$|x-y| = \left[\sum_{i=1}^{r} (x_i - y_i)^2\right]^{\frac{1}{2}}.$$

A *domain* of X^r is a connected open set of X^r. A *multiindex* p is an r-vector with integer components $p \equiv (p_1, \ldots, p_r)$. The norm of a multi-index p will be denoted by $|p|$, where either

$$|p| = \left(\sum_{i=1}^{r} p_i^2\right)^{\frac{1}{2}} \text{ or } |p| = \sum_{i=1}^{r} |p|,$$

the choice being clear from the context. In general for an n-vector $u \equiv (u_1, \ldots, u_n)$ with n complex components, we shall define its norm in the standard way

$$|u| = \left(\sum_{i=1}^{n} |u_i|^2\right)^{\frac{1}{2}}.$$

The scalar product of two n-vectors $u \equiv (u_1, \ldots, u_n)$ and $v \equiv (v_1, \ldots, v_n)$ (both with complex components) will be indicated by uv, which must be understood as

$$uv = \sum_{i=1}^{n} u_i \bar{v}_i$$

or, in terms of the extended summation convention $uv = u_i \bar{v}_i$. If ξ is a real r-vector and p a multi-index with non-negative components, we shall use the following abridged notations:

$$\xi^p = \xi_1^{p_1} \ldots \xi_r^{p_r} \quad (\xi_i^{p_i} = 1 \text{ if } \xi_i = p_i = 0) \quad \text{and} \quad D^p = \frac{\partial^{|p|}}{\partial x_1^{p_1} \ldots \partial x_r^{p_r}}.$$

If $z = x + iy$ is a complex number, $\bar{z}$ denotes its conjugate, i.e. $\bar{z} = x - iy$. Let $a = \{a_{ij}\}$ $(i = 1, \ldots, m;\ j = 1, \ldots, n)$ be an $m \times n$-matrix with complex entries and u a complex n-vector; by au we mean the m-vector with components $(a_{ij} u_j, \ldots, a_{mj} u_j)$. We denote by $\bar{a}$ the adjoint matrix of a, i.e. the $n \times m$-matrix $\{\alpha_{ji}\}$ $(j = 1, \ldots, n;\ i = 1, \ldots, m)$ with $\alpha_{ji} = \bar{a}_{ij}$. If v is an m-vector, by auv and vau we mean

$$auv = (au)\, v = a_{ij} u_j \bar{v}_i; \qquad vau = v\,(au) = v_i \bar{a}_{ij} \bar{u}_j.$$

It is easy to see that $auv = u\bar{a}v$.

If B is any point-set of X^r with interior points, we shall denote by $C^k(B)$ the class of all the complex n-vector valued functions $u(x)$ possessing continuous derivatives up to the order k $(k \geq 0)$ in B. This means that every derivative of u

of order $\leqq k$ exists at all interior points of B and coincides with a function which is continuous in the whole set B. Let us remark that it would be more proper to use the symbol $C_n^k(B)$, instead of $C^k(B)$, showing explicitly the number of components of the vector $u(x)$. However, since we shall consider the integer n as fixed, no confusion should arise. The reader will understand from the context what class $C^k(B)$ we are dealing with. If u is a function defined in the set B of X^r, we denote by *spt* u (support of u) the closure of the set where $|u(x)|>0$. Let A be a domain. By $\overset{\circ}{C}{}^k(A)$ we shall denote the subclass of $C^k(A)$ formed by all the functions u with a bounded support and such that *spt* $u \subset A$. We denote by C^k the class $C^k(X^r)$. By $\overset{\circ}{C}{}^k$ we denote the subclass of C^k formed by all the functions with bounded support. The intersection of all the classes $C^k(A)$ will be denoted by $C^\infty(A)$. By $\overset{\circ}{C}{}^\infty(A)$ we denote the intersection of all the classes $\overset{\circ}{C}{}^k(A)$. In the case when $A=X^r$, we simply write C^∞ and $\overset{\circ}{C}{}^\infty$. Let B and E be two point-sets of X^r and let B be a proper subset of E. Suppose that $u(x)$ is a function defined in B. Throughout this paper we shall use the convention that, whenever we consider $u(x)$ as a function defined in E, we suppose that $u(x)$ is continued in the whole set E by assuming $u(x)=0$ for $x \in E-B$. A *regular domain* is any bounded domain whose boundary ∂A is piecewise smooth [i.e. decomposable into a finite number of non-overlapping differentiable $(r-1)$-cells] such that for any $u \in C^1(\bar{A})$ the Gauss-Green formula holds:

$$\int_A \frac{\partial u}{\partial x_i}\, dx - \int_{\partial A} u \nu_i\, d\sigma = 0$$

[$dx = dx_1 \ldots dx_r$, $d\sigma$ = measure of the hypersurface element of ∂A, $\nu \equiv (\nu_1, \ldots, \nu_r)$ inward normal at any regular point of ∂A].

Additional hypotheses concerning the smoothness of the boundary of a regular domain will be considered in the course of this article.

2. The function spaces $\overset{\circ}{\boldsymbol{H}}_{\boldsymbol{m}}$ and $\boldsymbol{H}_{\boldsymbol{m}}$. Let A be a bounded domain of the real cartesian space X^r. Without any loss in generality we may assume from now on that $\bar{A}$ be contained in the open square Q: $|x_k|<\pi$ $(k=1, \ldots, r)$, since in general a suitable change of coordinates will bring this about. Let us consider the vector space $\overset{\circ}{C}{}^m(A)$ [the vector space $C^m(\bar{A})$] of complex n-vector valued functions. For any pair of vectors of this space let us define the following scalar product:

$$(u, v)_m = \int_A D^s u\, D^s v\, dx \qquad (0 \leqq |s| \leqq m). \tag{2.1}$$

We define as the space $\overset{\circ}{H}_m(A)$ [as the space $H_m(A)$] (or, simply, $\overset{\circ}{H}_m$ [H_m] when no confusion need arise) the Hilbert space obtained by functional completion of $\overset{\circ}{C}{}^m(A)$ [of $C^m(\bar{A})$] with respect to the scalar product (2.1). It is obvious that for $m=0$ we have $\overset{\circ}{H}_0(A)=H_0(A)=\mathscr{L}^2(A)$. A function u will belong to $\overset{\circ}{H}_m(A)$ [to $H_m(A)$] if and only if there is a sequence of functions of $\overset{\circ}{C}{}^m(A)$ [of $C^m(\bar{A})$]—say $\{v_k(x)\}$—such that $\lim\limits_{k\to\infty} \int_A |v_k(x)-u(x)|^2\, dx = 0$ and, moreover, there are functions $\varphi^s(x)$ $(0 \leqq |s| \leqq m)$ of $\mathscr{L}^2(A)$ such that $\lim\limits_{k\to\infty} \int_A |D^s v_k(x) - \varphi^s(x)|^2\, dx = 0$. The function $\varphi^s(x)$—which obviously does not depend on the sequence $\{v_k\}$—is called the *strong derivative* $D^s u$ of the function u of $\overset{\circ}{H}_m(A)$ [of $H_m(A)$], or, simply, the *s-derivative* of u. For any $u \in \mathring{H}_m$ and any $w \in H_m$ the integration by parts formula holds:

$$\int_A u D^s w \, dx = (-1)^{|s|} \int_A (D^s u) \, w \, dx, \qquad (0 \leqq |s| \leqq m). \tag{2.2}$$

In fact (2.2) holds if we replace u by v_k and w by z_k, where $v_k \in \mathring{C}^m(A)$ and $z_k \in C^m(\bar{A})$. Since we may assume that v_k converges to u in the metrics of $\mathring{H}_m$ and z_k converges to w in the metrics of H_m, by making k tend to infinity we get (2.2).

Let us remark that if $w \in \mathring{H}_m$ and φ^s is the s-derivative of w as a function of $\mathring{H}_m$, then φ^s is also the s-derivative of w as a function of H_m. That follows obviously from (2.2).

Set $c_k = (2\pi)^{-r/2} \int_A u e^{-ikx} \, dx$. Any function u of H_m can be developed in a Fourier trigonometrical series

$$u(x) = (2\pi)^{-r/2} \sum_{-\infty}^{+\infty}{}_k \, c_k \, e^{ikx}, \tag{2.3}$$

the development being convergent in $\mathscr{L}^2(A)$. For $|s| \leqq m$ we have also

$$D^s u(x) = (2\pi)^{-r} \sum_{-\infty}^{+\infty}{}_k \, e^{ikx} \int_A (D^s u) \, e^{-ikx} \, dx$$

with convergence in $\mathscr{L}^2(A)$. Supposing $u \in \mathring{H}_m$ and using (2.2), we get

$$D^s u(x) = (2\pi)^{-r/2} \sum_{-\infty}^{+\infty} (i)^{|s|} k^s \, c_k \, e^{ikx}. \tag{2.4}$$

This means that for any $u \in \mathring{H}_m$, the Fourier series (2.3) can be differentiated term by term, provided the order of the differentiation does not exceed m. For $u \in \mathring{H}_m$, we have

$$\int_A |D^s u|^2 \, dx = \sum_{-\infty}^{+\infty}{}_k (k^s)^2 |c_k|^2.$$

For any fixed ν, there exist two positive numbers $p_0^{(\nu)}$ and $p_1^{(\nu)}$ such that

$$p_0^{(\nu)} |k|^{2\nu} \leqq \sum_{|s|=\nu} |k^s|^2 \leqq p_1^{(\nu)} |k|^{2\nu} \quad [|k| = (k_1^2 + \cdots + k_r^2)^{\frac{1}{2}}].$$

Hence

$$p_0^{(\nu)} \sum_{-\infty}^{+\infty} |k|^{2\nu} \, |c_k|^2 \leqq \sum_{|s|=\nu} \int_A |D^s u|^2 \, dx \leqq p_1^{(\nu)} \sum_{-\infty}^{+\infty}{}_k |k|^{2\nu} \, |c_k|^2. \tag{2.5}$$

We shall denote by $\|u\|_{m,A}$ the norm of an element of $H_m(A)$ or, in particular, of $\mathring{H}_m(A)$. When there is no ambiguity about the domain, we simply write $\|u\|_m$. The following lemma is called the *Poincaré inequality*:

2.I *For any* $u \in \mathring{H}_m(A)$

$$\|u\|_m^2 \leqq c \sum_{|s|=m} \int_A |D^s u|^2 \, dx. \tag{2.6}$$

The constant c depends only on A (and on m).

It is evident that it suffices to prove (2.6) for $m=1$; then, by induction, the general case follows. Let us denote by Du the $r\times n$-matrix $\{\partial u_k/\partial x_h\}$ $(h=1,\ldots,r;$ $k=1,\ldots,n)$. We have

$$|c_0|^2=(2\pi)^{-r}\left|\int_A u\,dx\right|^2=(2\pi)^{-r}r^{-2}\left|\int_A (Du)\,x\,dx\right|^2$$
$$\leqq(2\pi)^{-r}r^{-2}\int_A |Du|^2\,dx\int_A |x|^2\,dx.$$

Hence

$$\|u\|_1^2=\int_A |u|^2\,dx+\int_A |Du|^2\,dx\leqq\sum_{-\infty}^{+\infty}{}_k|c_k|^2+p_1^{(1)}\sum_{-\infty}^{+\infty}{}_k|k|^2\,|c_k|^2$$
$$\leqq\left[(2\pi)^{-r}r^{-2}\int_A |x|^2\,dx+\frac{1+p^{(1)}}{p_0^{(1)}}\right]\int_A |Du|^2\,dx.$$

From (2.6) and (2.5) it follows that

2.II *The norms $\|u\|_m^2$ and $\sum_{-\infty}^{+\infty}{}_k|k|^{2m}|c_k|^2$ are isomorphic in the space $\mathring{H}_m$.*

Let us now consider, for any $u\in\mathring{H}_m$ $(m>0)$ the operator $\mathscr{I}_{m,l}$ with domain $\mathring{H}_m$ and range $\mathring{H}_l$ (with $l<m$) which associates with u the same u, but considered as a vector in $\mathring{H}_l$. It is obvious that $\mathscr{I}_{m,l}$ is a bounded linear transformation of $\mathring{H}_m$ into $\mathring{H}_l$. This transformation will be called the *embedding of $\mathring{H}_m$ into $\mathring{H}_l$*.

2.III *The embedding $\mathscr{I}_{m,l}$ of $\mathring{H}_m$ into $\mathring{H}_l$ is compact.*

Let U be a bounded set of $\mathring{H}_m$. There is a positive constant L such that

$$\sum_{-\infty}^{+\infty}{}_k|k|^{2m}|c_k|^2<L$$

for any $u\in U$. It follows that

$$\sum_{-\infty}^{+\infty}{}_k|k|^{2l}|c_k|^2<L$$

and that the latter series is uniformly convergent when u varies in U. Since $\mathscr{I}_{m,l}$ takes a bounded set into a set with a compact closure, compactness of $\mathscr{I}_{m,l}$ follows[1].

Let A be a regular domain. The function u is said to belong to $D^m(A)$ $(m>0)$ whenever it belongs to $C^{m-1}(\bar{A})$ and, corresponding to u, there exists a decomposition of A into a finite set of non-overlapping regular domains $A_1,\ldots,A_s$ [2] such that $u\in C^m(\bar{A}_i)$ $(i=1,\ldots,s)$. When $u\in D^m(A)$, we say that u has piecewise continuous m-th derivatives. It is possible to prove that $D^m(A)\subset H_m(A)$. If we denote by $\mathring{D}^m(A)$ the subset of $D^m(A)$ consisting of those functions such that *spt* $u\subset A$, then $\mathring{D}^m(A)\subset\mathring{H}_m(A)$.

A domain A is said to be a *properly regular domain* if the following conditions are satisfied: α) A is regular; β) for any $x^0\in\partial A$ there exists a neighborhood I of x^0 (open set containing x^0) such that the set $J=\bar{I}\cap\bar{A}$ is homeomorphic to the closed

[1] We have been using the following lemma: *If S is a separable Hilbert space and $\{v_k\}$ a complete orthonormal system in it, the subset U of S has a compact closure if and only if the series $\sum_k|(u,v_k)|^2$ is uniformly bounded and uniformly convergent in U.* (See [4], pp. 231–232]).

[2] By "decomposition of A into a finite set of non-overlapping domains" we mean that $\bar{A}=\bar{A}_1\cup\cdots\cup\bar{A}_s$ and $A_i\cap A_j=\emptyset$ for $i\neq j$.

semiball B^+: $y_r \geqq 0$, $|y| \leqq 1$ of the cartesian space Y^r. In this homeomorphism the set $\partial A \cap \tilde{I}$ is mapped onto the set $y_r = 0$, $y_1^2 + \cdots + y_{r-1}^2 \leqq 1$; γ) the vector-valued function $y = y(x)$ which maps homeomorphically J onto B^+ has piecewise continuous first derivatives [i.e. $y(x) \in D^1(J - \partial J)$] and the jacobian matrix $\partial y / \partial x$ has a positive determinant which is bounded away from zero in the whole J. It follows that the inverse function $x = x(y)$ which maps B^+ onto J has properties analogous to those of $y = y(x)$.

2.IV *(Rellich selection principle). If A is properly regular, the embedding $\mathscr{I}_{m,l}$ of H_m into H_l $(m > l)$ is compact.*

It is sufficient to prove the theorem in the case $m = 1$, $l = 0$. Let $I_1, \ldots, I_q$ be a finite set of neighborhoods, like those mentioned in the condition β), which cover the boundary ∂A of A. Set $\tilde{I} = I_1 \cup \cdots \cup I_q$ and suppose that the set $A - \tilde{I}$ be non-empty (otherwise the proof would be simpler). The closed set $A - \tilde{I}$ is contained in A. Let I_0 be any open subset of A containing $A - \tilde{I}$. The sets I_0, $I_1, \ldots, I_q$ form an *open covering of* $\bar{A}$.

Let $\sum_{h=0}^{q} \varphi_h(x) = 1$ be a partition of unity with non-negative C^∞ functions such that $spt\ \varphi_h(x) \subset I_h$ $(h = 0, \ldots, q)$.

Let $u \in H_1(A)$; we have $u = \sum_{h=0}^{q} \varphi_h u$ and

$$\|\varphi_h u\|_1^2 \leqq \int_A |\varphi_h u|^2 dx + 2 \int_A |u|^2 \sum_{k=1}^{r} \left|\frac{\partial \varphi_h}{\partial x_k}\right|^2 dx + 2 \int_A |\varphi_h|^2 \sum_{k=1}^{r} \left|\frac{\partial u}{\partial x_k}\right|^2 dx \leqq c \|u\|_1^2.$$

If U is a bounded set of $H_1(A)$, then the set U_h of functions $\varphi_h u$ $(u \in U)$ is a bounded set of $H_1(I_h \cap A)$. For $h = 0$ we have $U_0 \subset \mathring{H}_1(I_0)$. Hence boundedness of U implies compactness of $\overline{U}_0$ in $H_0(I_0)$ (see Theorem 2.III). Suppose $h > 0$. For simplicity set $I_h = I$. For any $u \in C^1(\overline{I \cap A})$ set $\tilde{u}(y) = u[x(y)]$. Almost everywhere on B^+ we have

$$\frac{\partial \tilde{u}}{\partial y_j} = \frac{\partial u}{\partial x_h} \frac{\partial x_h}{\partial y_j}.$$

It follows that

$$\int_{B^+} |\tilde{u}|^2 dy + \sum_{j=1}^{r} \int_{B^+} \left|\frac{\partial \tilde{u}}{\partial y_j}\right|^2 dy = \int_J |u|^2 \left|\frac{\partial y}{\partial x}\right| dx$$
$$+ \sum_{j=1}^{n} \int_J \frac{\partial x_h}{\partial y_j} \frac{\partial x_k}{\partial y_j} \frac{\partial u}{\partial x_h} \frac{\partial u}{\partial x_k} \left|\frac{\partial y}{\partial x}\right| dx \leqq c_1 \|u\|_{1,J}^2.$$

Let us now assume $\tilde{u}(y) = \varphi_h[x(y)]\, u[x(y)]$, where u is any function of the subset U which, with no loss in generality, we suppose contained in $C^1(\bar{A})$, bounded in $H_1(A)$. Let us define in the closed ball B: $|y| \leqq 1$ the following function:

$$\tilde{u}^*(y) = \begin{cases} \tilde{u}(y_1, \ldots, y_{r-1}, y_r) & \text{for } y_r \geqq 0, \\ \tilde{u}(y_1, \ldots, y_{r-1}, -y_r) & \text{for } y_r < 0. \end{cases}$$

The function $\tilde{u}^*(y)$ belongs to $\mathring{H}_1(B - \partial B)$. We have $\|\tilde{u}^*\|_{1,B}^2 = 2\|\tilde{u}\|_{1,B^+}^2 \leqq 2c_1 c \|u\|_1^2$. Then we can extract from any sequence $\{u_n\} \in U$ a subsequence such that $\{\tilde{u}_{nk}^*\}$ is convergent in $H_0(B)$. This means that $\{\tilde{u}_{n_k}\}$ is convergent in $H_0(B^+)$ and therefore $\{\varphi_h u_{n_k}\}$ is convergent in $H_0(I \cap A) = H_0(I_h \cap A)$. Of course we may suppose $\{n_k\}$

independent of h. Thus we deduce that

$$u_{n_k} = \sum_{h=0}^{q} \varphi_h \, u_{n_k}$$

is convergent in $H_0(A)$. We have proven that the operator $\mathscr{I}_{1,0}$, when restricted to $C^1(\bar{A})$, is compact. Since $C^1(\bar{A})$ is dense in $H_1(A)$, the proof of the theorem follows.

Let A be a properly regular domain of X^r. Let τ denote the linear transformation which to any $u \in C^m(\bar{A})$ associates its boundary values τu on ∂A. Set

$$\|\tau u\|^2_{m-1,\partial A} = \sum_{|k|=0}^{m-1} \int_{\partial A} |D^k u|^2 \, d\sigma, \qquad m \geqq 1.$$

We wish to prove that

$$\|\tau u\|^2_{m-1,\partial A} \leqq c \, \|u\|^2_m \tag{2.7}$$

where c is a constant depending only on A (and on m). It is sufficient to prove (2.7) in the case $m=1$. To this end, let us consider the open covering $I_0, I_1, \ldots, I_q$ introduced in the proof of Theorem 2.IV and the corresponding partition of unity $\sum_{u=0}^{q} \varphi_h(x) = 1$. Inequality (2.7) (for $m=1$) will follow from the inequalities

$$\int_{\partial A} |\varphi_h u|^2 \, d\sigma \leqq c \left\{ \int_A |\varphi_h u|^2 \, dx + \sum_{k=0}^{r} \int_A \left| \frac{\partial \varphi_h u}{\partial x_k} \right|^2 dx \right\} \qquad (h=1, \ldots, q). \tag{2.8}$$

As before, we set $I_h = I$ $(h>0)$ and consider the homeomorphism which maps $J = \bar{I} \cap \bar{A}$ onto the semiball B^+ of the cartesian (y, t)-space defined by $t \geqq 0$, $y_1^2 + \cdots + y_{r-1}^2 + t^2 \leqq 1$. Let D be the $(r-1)$-ball: $t=0$, $|y|^2 = \sum_{k=1}^{r-1} y_k^2 \leqq 1$ i.e. the image of $\bar{I} \cap \partial A$ under the homeomorphism of J onto B^+. Let $\tilde{u}(y, t)$ be the function obtained from $\varphi_h(x) u(x)$ by the above-mentioned homeomorphism. We continue $\tilde{u}(y, t)$ outside of B^+ in the half-space $t \geqq 0$ by assuming $\tilde{u}(y, t) \equiv 0$ for $t \geqq 0$, $|y|^2 + t^2 > 1$. We have for every $t > 0$

$$\tilde{u}(y, 0) = \tilde{u}(y, t) + \int_t^0 \tilde{u}_s(y, s) \, ds.$$

Hence

$$\int_D |\tilde{u}(y, 0)|^2 \, dy \leqq 2 \int_D |\tilde{u}(y, t)|^2 \, dy + 2 \int_0^1 \int_D |\tilde{u}_t(y, t)|^2 \, dy \, dt.$$

It follows that

$$\int_D |\tilde{u}(y, 0)|^2 \, dy \leqq 2 \iint_{B^+} |\tilde{u}(y, t)|^2 \, dy \, dt + 2 \iint_{B^+} |D\tilde{u}|^2 \, dy \, dt.$$

From this inequality, by transformation back to the $x_1, \ldots, x_r$ coordinates, we deduce inequality (2.8).

From (2.7) it follows that the operator τ can be continuously extended to the whole space $H_m(A)$. Then for any u a set of vectors $\tau u \equiv \{\varphi^p\}$ $(0 \leqq |p| \leqq m-1)$ is determined such that $\varphi^p \in \mathscr{L}^2(\partial A)$ and $\varphi^p \equiv D^p u$ if $u \in C^m(\bar{A})$. The operator τ will be called the *trace operator*, and φ^p will be considered as the "boundary value" of $D^p u$ in a generalized sense. It is obvious that $u \in \overset{\circ}{H}_m(A)$ implies $\tau u = 0$, i.e. any function of $\overset{\circ}{H}_m(A)$ vanishes (in a generalized sense) on ∂A together with any derivative of order $\leqq m-1$.

2.V *If* $u \in C^{m-1}(\bar{A}) \cap \overset{\circ}{H}_m(A)$, *then* $D^p u = 0$ *on* ∂A $(0 \leqq |p| \leqq m-1)$ *in the classical sense.*

A proof is easy to construct.

If u and v are two functions of $C^m(\bar{A})$ and $0 < |p| \leqq m$, the Gauss-Green integral formula holds:

$$\int_A u\, D^p v\, dx + (-1)^{|p|} \int_A (D^p u)\, v\, dx = \int_{\partial A} M(u, v)\, d\sigma, \tag{2.9}$$

where $M(u, v)$ is a bilinear differential operator of order $|p|-1$ both in u and in v. From (2.7) it follows that (2.9) still holds if u and v are functions of $H_m(A)$, provided the boundary values of $D^q u$ and $D^q v$ $(0 \leqq |q| < |p|)$ are understood in the generalized sense introduced above.

Let x^0 be any point of X^r and Σ the unit sphere $|x| = 1$. Let Γ be a set of positive measure on the unit sphere Σ and let R be a positive number. The set $C_{x^0}(\Gamma, R)$ of all x such that $(x - x^0)|x - x^0|^{-1} \in \Gamma$, $0 < |x - x^0| \leqq R$ will be called a *cone*; x^0 is the *vertex* of the cone. The domain A of X^r is said to satisfy the "*cone hypothesis*" if for every x^0 of $\bar{A}$ there exists a cone $C_{x^0}(\Gamma_{x^0}, R)$ of vertex x^0 which is contained in A and is such that R is independent of x^0 and Γ_{x^0} is congruent to a fixed set Γ on Σ.

It is not very difficult to prove that a properly regular domain satisfies the cone hypothesis.

2.VI *(Sobolev lemma.) If A is a bounded domain of X^r satisfying the cone hypothesis, the functions u of any space $H_m(A)$ with $m > r/2$ are continuous in $\bar{A}$, and*

$$\max_{\bar{A}} |u| \leqq c \|u\|_m, \tag{2.10}$$

where c depends only on A (and on m).

It is enough to prove inequality (2.10) for any $u \in C^m(\bar{A})$ in order to get the proof of the lemma. Let x^0 be any point of $\bar{A}$. We denote by $\varphi(x)$ a C^∞ real-valued function which satisfies the following condition: $\varphi(x) = 1$ for $|x - x^0| < R/2$, $\varphi(x) = 0$ for $|x - x^0| > R$. We have

$$u(x^0) = -\int_0^R \frac{\partial \varphi u}{\partial \varrho}\, d\varrho = \frac{(-1)^m}{(m-1)!} \int_0^R \varrho^{m-1} \frac{\partial^m \varphi u}{\partial \varrho^m}\, d\varrho.$$

After integration over Γ_{x^0}, on the unit sphere, we get

$$\begin{aligned} |u(x^0) \operatorname{meas} \Gamma| &= \frac{1}{(m-1)!} \left| \int_{C_{x^0}(\Gamma_{x^0}, R)} \varrho^{m-r} \frac{\partial^m \varphi u}{\partial \varrho^m}\, dx \right| \\ &\leqq \frac{1}{(m-1)!} \left(\int_{C_{x^0}(\Gamma_{x^0}, R)} \left| \frac{\partial^m \varphi u}{\partial \varrho^m} \right|^2 dx \right)^{\frac{1}{2}} \left(\int_{C_{x^0}(\Gamma, R)} \varrho^{2(m-r)}\, dx \right)^{\frac{1}{2}} \leqq c^1 \|u\|_m, \end{aligned}$$

which implies (2.10).

2.VII *(First Ehrling lemma.) Let A be any bounded domain of X^r. For any $\varepsilon > 0$ there exists a positive constant $c(\varepsilon)$ (depending only on ε, A, m) such that for any $u \in \overset{\circ}{H}_m(A)$ the following inequality holds*

$$\|u\|_{m-1} \leqq \varepsilon \|u\|_m + c(\varepsilon) \|u\|_0. \tag{2.11}$$

2.VIII *(Second Ehrling lemma.) Let A be any properly regular domain of X^r. For any $\varepsilon > 0$ there exists a positive constant $c(\varepsilon)$ (depending only on ε, A, m) such that for any $u \in H_m(A)$, inequality (2.11) holds.*

Let B_1, B_2, B_3 be three complex Banach spaces, and assume that, as vector spaces, they satisfy the following inclusion conditions: $B_1 \subset B_2 \subset B_3$. Let us assume that the embedding $\mathscr{I}_{12}$ of B_1 into B_2 be compact and the embedding $\mathscr{I}_{23}$ of B_2 into B_3 be continuous. For any given $\varepsilon > 0$ there exists a positive constant $c(\varepsilon)$ such that

$$\|u\|_{B_2} \leqq \varepsilon \|u\|_{B_1} + c(\varepsilon)\|u\|_{B_3}. \tag{2.12}$$

Suppose (2.12) false for some $\varepsilon > 0$. Then for any positive integer n, there must exist some u_n such that $\|u_n\|_{B_2} > \varepsilon \|u_n\|_{B_1} + n\|u_n\|_{B_3}$. Set $v_n = \|u_n\|_{B_1}^{-1} u_n$. We have $\|v_n\|_{B_2} > \varepsilon + n\|v_n\|_{B_3}$, $\|v_n\|_{B_1} = 1$. The latter implies $\|v_n\|_{B_2} < L$ with L independent of n; the former $\lim_{n\to\infty} \|v_n\|_{B_3} = 0$. Since the sequence $\{v_n\}$ is compact in B_2, we can extract a subsequence $\{v_{n_k}\}$ converging in B_2 to some v. Because of the continuity of $\mathscr{I}_{23}$, $v_{n_k} \to v$ also in B_3. Hence $v = 0$, which makes impossible the inequality $\|v_{n_k}\|_{B_2} > \varepsilon$.

Taking $B_1 = \mathring{H}_m$, $B_2 = \mathring{H}_{m-1}$, $B_3 = H_0$, we have the proof of the first Ehrling lemma; taking $B_1 = H_m$, $B_2 = H_{m-1}$, $B_3 = H_0$, the proof of the second one.

3. Elliptic linear systems. Interior regularity. Let A be a domain of X^r. Suppose that the $n \times n$ complex matrices $a_s(x)$ $(0 \leqq |s| \leqq \nu)$ be defined in A. Consider the linear matrix differential operator $Lu \equiv a_s(x) D^s u$. This operator is said to be an *elliptic operator* in A if, for any real non-zero r-vector ξ the following condition is satisfied:

$$\det \sum_{|s|=\nu} a_s(x) \xi^s \neq 0$$

at every point $x \in A$.

From now on we shall suppose, for simplicity, that the matrices $a_s(x)$ are defined in the whole X^r and $a_s(x) \in C^\infty(X^r)$. We define as *adjoint operator* of the operator L the operator

$$L^* u \equiv \sum_{|s|=0}^{\nu} (-1)^{|s|} D^s(\bar{a}_s(x)\, u).$$

The function $u \in \mathscr{L}^2(A)$ is said to be a *weak solution* in A of the differential system $Lu = f$, where $f \in \mathscr{L}^2(A)$, if for any function $v \in \mathring{C}^\infty(A)$ we have

$$\int_A u\, L^* v\, dx = \int_A f v\, dx. \tag{3.1}$$

From now on we shall suppose that A is bounded [and $\bar{A}$ contained in the square Q: $|x_k| < \pi$ $(k = 1, \ldots, r)$] and L elliptic in A.

3.I *Let u be a weak solution in A of the elliptic differential system $Lu = f$. Let f be a function belonging to $H_m(A)$. If x^0 is any point of A, there exists a $\delta > 0$ such that $u \in H_{m+\nu}(\Gamma_\delta)$ and*

$$\|u\|^2_{m+\nu, \Gamma_\delta} \leqq c\left(\|f\|^2_{m, \Gamma_{2\delta}} + \|u\|^2_{m+\nu-1, \Gamma_{2\delta}}\right),$$

where Γ_ϱ is the ball of center x^0 and radius ϱ and c is a positive constant depending only on L and x^0.

Set

$$L_0(x, D) = \sum_{|s|=\nu} a_s(x)\, D^s,$$

$$L^*(x, D) = \sum_{|s|=0}^{\nu} (-1)^{|s|} D^s\, \overline{a_s(x)},$$

$$L_0^*(x, D) = \sum_{|s|=\nu} \bar{a}_s(x)\, D^s.$$

Let x^0 be any fixed point of A. Let $\Gamma_{2\delta}$ be such that $\overline{\Gamma}_{2\delta} \subset A$. We denote by $\varphi(x)$ a real valued function which is equal to 1 at every point of Γ_δ and vanishes identically outside of $\Gamma_{2\delta}$. Let γ be an arbitrary complex n-vector. Assume $v = \varphi(x)\, e^{ikx}\gamma$, where k is a non-zero vector with integer components. Then

$$L^*(x, D)\, v = \varphi(x)\, L_0^*(x, ik)\, e^{ikx}\gamma + \sum_{|s|=0}^{\nu-1} \Phi_s(x)\, k^s\, e^{ikx}\gamma.$$

The $n\times n$ matrices $\Phi_s(x)$ have their supports contained in $\Gamma_{2\delta}$. From (3.1), because of the arbitrariness of γ, we obtain

$$L_0(x^0, ik)\int_A \varphi u e^{-ikx}\, dx = -\sum_{|s|=0}^{\nu-1} k^s \int_A \overline{\Phi}_s\, u e^{-ikx}\, dx + \int_A \varphi f e^{-ikx}\, dx + \int_A [L_0(x^0, ik) - L_0(x, ik)]\, \varphi u e^{-ikx}\, dx.$$

Since $\det L_0(x^0, ik) \neq 0$ (L is elliptic!),

$$\int_A \varphi u e^{-ikx}\, dx = -\sum_{|s|=0}^{\nu-1} k^s [L_0(x^0, ik)]^{-1} \int_A \overline{\Phi}_s(x)\, u e^{-ikx}\, dx + [L(x^0, ik)]^{-1}\int_A \varphi f e^{-ikx}\, dx + [L_0(x^0, ik)]^{-1}\int_A [L_0(x^0, ik) - L_0(x, ik)]\, \varphi u e^{-ikx}\, dx.$$

Set $[L_0(x^0, ik) - L_0(x, ik)] = \sum_{|s|=\nu} \alpha_s(x)\, k^s$, $U = \varphi u$ in $\Gamma_{2\delta}$, $U = 0$ outside of $\Gamma_{2\delta}$. Let h be any non-vanishing constant real r-vector. Then

$$\begin{aligned}\frac{e^{ikh}-1}{|h|}\int_A U e^{-ikx}\, x = &-\sum_{|s|=0}^{\nu-1} k^s\, [L_0(x^0, ik)]^{-1}\frac{e^{ikh}-1}{|h|}\int_A \overline{\Phi}_s\, u e^{-ikx}\, dx \\ &+ [L_0(x^0, ik)]^{-1}\frac{e^{ikh}-1}{|h|}\int_A \varphi f e^{-ikx}\, dx \\ &+ \sum_{|s|=\nu} k^s [L_0(x^0, ik)]^{-1}\frac{e^{ikh}-1}{|h|}\int_A \alpha_s(x)\, U(x)\, e^{-ikx}\, dx.\end{aligned} \tag{3.2}$$

Let $|h|$ be less than the distance between $\Gamma_{2\delta}$ and ∂A. Then

$$\begin{aligned}\frac{e^{ikh}-1}{|h|}\int_A \alpha_s(x)\, U(x)\, e^{-ikx}\, dx &= \int_Q \frac{\alpha_s(x+h)\, U(x+h) - \alpha_s(x)\, U(x)}{|h|}\, e^{-ikx}\, dx \\ &= \int_Q \alpha_s(x+h)\,\frac{U(x+h) - U(x)}{|h|}\, e^{-ikx}\, dx + \int_Q \frac{\alpha_s(x+h) - \alpha_s(x)}{|h|}\, U(x)\, e^{-ikx}\, dx.\end{aligned}$$

Let us denote by c_k the Fourier coefficient of U and suppose that $u \in H_l(\Gamma_{2\delta})$. From (3.2) we get (using the Landau symbol $\mathcal{O}$)

$$\begin{aligned}\sum_{|k|<m} |k|^{2l}\left|\frac{e^{ikh}-1}{|h|}\right|^2 |c_k|^2 = &\,\mathcal{O}(\|u\|^2_{l,\Gamma_{2\delta}}) + \mathcal{O}\left(\sum_{k\neq 0}^{-\infty,+\infty} |k|^{2(1-\nu+l)}\left|\int_Q \varphi f e^{-ikx}\, dx\right|^2\right) \\ &+ \sum_{|s|=\nu}\left\{\mathcal{O}\left(\left\|\alpha_s(x+h)\,\frac{U(x+h)-U(x)}{|h|}\right\|_l^2\right) + \mathcal{O}\left(\left\|\frac{\alpha_s(x+h)-\alpha_s(x)}{|h|}\, U(x)\right\|_l^2\right)\right\}.\end{aligned} \tag{3.3}$$

Since the $(l+1)$-th derivatives of $\alpha_s(x)$ are bounded ($\alpha_s \in C^\infty$!),

$$\left\|\frac{\alpha_s(x+h) - \alpha_s(x)}{|h|}\, U(x)\right\|_l^2 = \mathcal{O}(\|u(x)\|^2_{l,\Gamma_{2\delta}}).$$

Denoting by c' a suitable positive constant, we have

$$\left\|\alpha_s(x+h)\frac{U(x+h)-U(x)}{|h|}\right\|_l^2 \leq c'\left\{\sum_{|k|=l}\int_Q\left|\alpha_s(x+h)D^k\frac{U(x+h)-U(x)}{|h|}\right|^2 dx + \sum_{|k|<l}\int_Q\left|\beta_{sk}(x+h)D^k\frac{U(x+h)-U(x)}{|h|}\right|^2 dx\right\}. \tag{3.4}$$

Given $\varepsilon>0$, we take σ_ε such that for any s $(|s|=\nu)$: $\max_{|x+h-x^0|\leq 2\sigma_\varepsilon}|\alpha_s(x+h)|<\varepsilon$. Assume $2\delta<\sigma_\varepsilon$ and $|h|<\sigma_\varepsilon$. Because of (3.4) we may take ε such that

$$\left\|\alpha_s(x+h)\frac{U(x+h)-U(x)}{|h|}\right\|_l^2 \leq \frac{\eta}{(2\pi)^r}\sum_{-\infty}^{+\infty}{}_k|k|^{2l}\left|\int_Q\frac{U(x+h)-U(x)}{|h|}e^{-ikx}dx\right|^2 + c''\sum_{-\infty}^{+\infty}{}_k|k|^{2l-2}\left|\int_Q\frac{U(x+h)-U(x)}{|h|}e^{-ikx}dx\right|^2,$$

where η is an arbitrarily given positive number and c'' a positive constant. Then, from (3.2) and (3.3) we deduce that, for δ and $|h|$ sufficiently small,

$$\sum_{-\infty}^{+\infty}{}_h|k|^{2l}\left|\frac{e^{ikh}-1}{|h|}\right|^2|c_k|^2 = \mathcal{O}(\|u(x)\|_{l,\Gamma_{2\delta}}^2) + \mathcal{O}\left(\sum_{k\neq 0}^{-\infty,+\infty}|k|^{2(1-\nu+l)}\left|\int_Q\varphi f e^{-ikx}dx\right|^2\right).$$

By considering the following successive choices of h: $h\equiv(t,0,\ldots,0)$, $h\equiv(0,t,\ldots,0),\ldots,h\equiv(0,0,\ldots,t)$ and making $t\to 0$, we get

$$\sum_{-\infty}^{+\infty}{}_h|k|^{2(l+1)}|c_k|^2 = \mathcal{O}(\|u(x)\|_{l,\Gamma_{2\delta}}^2) + \mathcal{O}\left(\sum_{k\neq 0}^{-\infty,+\infty}|k|^{2(1-\nu+l)}\left|\int_Q\varphi f e^{-ikx}dx\right|^2\right).$$

It follows that $u\in H_{l+1}(\Gamma_\delta)$, and the theorem is proved.

3.II *If u is a weak solution in A of the elliptic system $Lu=f$ and if $f\in H_m(A)$ with $m>r/2$, then u is a solution in the classical sense*[3]. *In particular, if $f\in C^\infty(A)$, then $u\in C^\infty(A)$.*

In fact, by the Sobolev lemma 2.VI and by the above theorem u belongs to $C^\nu(A)$. Then, by the Gauss-Green identity,

$$\int_A uL^*v\,dx = \int_A Lu\,v\,dx$$

for any $v\in\overset{\circ}{C}{}^\infty(A)$. By subtracting from (3.1) and using the arbitrariness of v, one deduces $Lu=f$. If $f\in C^\infty(A)$, then for every $x^0\in A$, u belongs to $H_l(\Gamma_\delta)$ for any l; that implies $u\in C^\infty(A)$.

4. Results preparatory to the regularization at the boundary. Let u be a function of $\mathcal{L}^2(A)$ (A bounded domain of X^r). We say that u has the *$\mathcal{L}^2$-strong derivative* $D^p u$ whenever a sequence of functions $\{v_k\}$ of $C^m(\bar{A})$ $(m=|p|)$ exists such that $\{v_k\}$ converges in $\mathcal{L}^2(A)$ towards u and $D^p v_k$ converges in $\mathcal{L}^2(A)$ to some function ψ^p, which we define to be the $\mathcal{L}^2$-strong derivative $D^p u$ of u. Since

$$\int_A u\,D^p v\,dx = (-1)^{|p|}\int_A \psi^p v\,dx, \tag{4.1}$$

[3] The condition $f\in H_m(A)$ with $m>r/2$ is too strong for insuring that $u\in C^\nu(A)$. By using methods of potential theory one may show that Hölder continuity of f is sufficient for having $u\in C^\nu(A)$.

for any $v \in \overset{\circ}{C}^m(A)$, it follows that ψ^p does not depend on the particular sequence $\{v_k\}$. If $u \in H_m(A)$ $(m>0)$, then u has all $\mathcal{L}^2$-strong derivative $D^p u$ with $0<|p|\leqq m$. If K is a closed subset of A and u vanishes in $A-K$, then, if u has the $\mathcal{L}^2$-strong derivative $D^p u$, this derivative vanishes in $A-K$, as readily follows from (4.1). In this case (4.1) holds for any $v \in C^m(A)$. Then, provided $\bar{A} \subset Q$, the Fourier development of $D^p u$ is obtained from the Fourier development of u by formally differentiating it term-wise by means of D^p.

In the X^r space for any point $x \equiv (x_1, \ldots, x_r)$ we shall denote by y_k the coordinate x_k $(k=1, \ldots, r-1)$ and by t the coordinate x_r. The point $(y_1, \ldots, y_{r-1})$ of the $(r-1)$-dimensional cartesian space will be denoted by y and the point x of X^r, when convenient, by (y, t). Let R be the open interval of X^r defined by: $-\pi<y_k<\pi$ $(k=1, \ldots, r-1)$, $0<t<\pi$.

We say that the function v belongs to $C_\sigma^m(R)$ (σ is any fixed positive number such that $0<\sigma<\pi$) if: (i) $v \in C^m$; (ii) $v \equiv 0$, when at least one of the following conditions is satisfied: $|y|>\sigma$, $t>\sigma$. The closure of $C_\sigma^m(R)$ in the space $H_m(R)$ will be denoted by $H_m^\sigma(R)$. Let $v(x)$ be a function of $C_\sigma^m(R)$. In X^r we define the following function:

$$v^*(y,t)=\begin{cases} v(y, t) & \text{for } t \geqq 0, \\ \sum_{j=1}^{m+1} \lambda_j v(y, -jt) & \text{for } t<0, \end{cases} \tag{4.2}$$

where the scalars λ_j are the solutions of the following algebraic system

$$\sum_{j=1}^{m+1} (-j)^k \lambda_j = 1 \qquad (k=0, \ldots, m). \tag{4.3}$$

It is easily seen that the function $v^*(y, t)$ belongs to C^m and that its support is contained in the square Q: $|x_k|<\pi$ $(k=1, \ldots, r)$. It is also very easy to verify that

$$\int_Q |D^p v^*|^2 dx \leqq c_p \int_R |D^p v|^2 dx \qquad 0 \leqq |p| \leqq m, \tag{4.4}$$

where c_p is a constant which does not depend on v. If $u \in H_m^\sigma(R)$, the function u^*, defined by means of (4.2), (4.3), belongs to $\overset{\circ}{H}_m(Q)$. In fact, if $\{v_k\} \in C_\sigma^m(R)$ and v_k converges in $H_m(R)$ to u, then $\{v_k^*\}$ converges, because of (4.4), in the space $\overset{\circ}{H}_m(Q)$, and its limit is u^*. Hence, from (4.4)

$$\|u^*\|_{m,Q}^2 \leqq c \|u\|_{m,R}^2, \tag{4.5}$$

with c independent on u.

4.I *If $u \in H_0^\sigma(R)$ and u has all $\mathcal{L}^2$-strong derivatives of order m, then $u \in H_m^{\sigma'}(R)$ for any σ' such that $\sigma<\sigma'<\pi$.*

From (4.4) it follows that u^* has the $\mathcal{L}^2$-strong derivative $D^p u^*$ $(|p|=m)$, Since u^* vanishes outside of a closed subset K of Q, if u^* has the following Fourier development:

$$u^* = (2\pi)^{-r/2} \sum_{-\infty}^{+\infty}{}_k\, c_k e^{ikx}; \tag{4.6}$$

$D^p u^*$ has the following one:

$$D^p u^* = (2\pi)^{-r/2} \sum_{-\infty}^{+\infty}{}_k\, i^{|p|} k^p c_k e^{ikx}. \tag{4.7}$$

It follows that $u^* \in H_m(Q)$. Since u^* vanishes outside of the cylinder $|y|<\sigma$, $|t|<\sigma$, it is easy to construct a sequence of C^m functions $\{v_k\}$ converging towards

u^* in $H_m(Q)$ and such that $spt\, v_k$ is contained in the cylinder $|y|<\sigma'$, $|t|<\sigma'$. Then $u\in H_m^{\sigma'}(R)$.

4.II *If $u\in H_0^\sigma(R)$ has the $\mathscr{L}^2$-strong derivative $D^p u=\varphi$ and φ has the $\mathscr{L}^2$-strong derivative $D^q\varphi=\psi$, then u has the $\mathscr{L}^2$-strong derivative $D^q D^p u=D^p D^q u=\psi$.*

The proof is a trivial consequence of (4.6) and (4.7).

Whenever a function $u\in\mathscr{L}^2(A)$ has in A some $\mathscr{L}^2$-strong derivative, say $D^p u$, we simply write $D^p u\in\mathscr{L}^2(A)$. If $u_1,\ldots,u_n$ are the components of the n-vector valued function u, then by $D_y u$ we mean the $n(r-1)$-vector valued function whose components are $\partial u_h/\partial y_k$ $(h=1,\ldots,n;\ k=1,\ldots,r-1)$. It is evident that $D_y u^*=(D_y u)^*$.

4.III *Let $u\in H_m^\sigma(R)$. A necessary and sufficient condition for $D_y u\in H_m(R)$ is that for any non-zero real r-vector $h\equiv(h_1,\ldots,h_{r-1},0)$ such that $|h|<\pi-\sigma$, the following inequality be satisfied:*

$$\left\|\frac{u(x+h)-u(x)}{|h|}\right\|_{m,R}^2\leqq c_1, \tag{4.8}$$

where c_1 is a constant independent of h. If (4.8) is satisfied, then

$$\left\|\frac{u(x+h)-u(x)}{|h|}\right\|_{m,R}^2=\mathcal{O}(\|D_y u\|_{m,R}^2). \tag{4.9}$$

Let us first prove sufficiency. From (4.5), (4.8) it follows that

$$\left\|\frac{u^*(x+h)-u^*(x)}{|h|}\right\|_{m,Q}^2\leqq c\,c_1. \tag{4.10}$$

Let us consider the Fourier development (4.6) of u^* in Q. From (4.10) we deduce

$$\sum_{|p|=0}^{m}\sum_{-\infty}^{+\infty}{}_k|k|^{2|p|}\,|c_k|^2\left|\frac{e^{ikh}-1}{|h|}\right|^2<c_2 \tag{4.11}$$

where c_2 is independent of h.

Arguing as in the proof of Theorem 3.I, we deduce $\|D_y u\|_{m,R}^2<+\infty$. Proof of (4.9) follows from the inequality $|h|^{-2}|e^{ikh}-1|^2\leqq k_1^2+\cdots+k_{r-1}^2$. Let us now assume $D_y u\in H_m(R)$. Then $D_y u^*\subset H_m(Q)$. Since u^* vanishes in a strip near ∂Q, any derivative $D^p D_{y_i} u^*$ of u^* $(|p|\leqq m)$ is given by the development obtained by merely differentiating termwise (4.6). Hence the series $\sum_{|p|=0}^{m}\sum_{-\infty}^{+\infty}{}_k|k|^{2|p|}|c_k|^2$ $(k_1^2+\cdots+k_{r-1}^2)$ is convergent. That implies (4.11). From (4.11), inequality (4.8) follows.

4.IV *Let $u\in H_m^\sigma(R)$ and $D_y u\in H_m(R)$. Set $h=(0,\ldots,h_i,\ldots,0)$, $0<|h_i|<\sigma$, $1\leqq i<r$. Then*

$$\lim_{h\to 0}\left\|\frac{u(x+h)-u(x)}{h_i}-\frac{\partial u}{\partial y_i}\right\|_{m,R}=0.$$

The proof is easily obtained by using the u^* continuation of u and the Fourier development (4.6).

4.V *Let $\psi\in H_0^\sigma(R)$ and $w_{kp}\in H_0^\sigma(R)$ $(k+|p|\leqq\nu,\ \nu\geqq 0)$; let w_{kp} be an $n(r-1)$-vector valued function. Assume that $D_y\psi\in\mathscr{L}^2(R)$ and that for every $v\in\mathring{C}^{\nu+1}(R)$:*

$$E v\equiv\int_R\psi\frac{\partial^{\nu+1}v}{\partial t^{\nu+1}}\,dx+\int_R\Big(\sum_{k+|p|\leqq\nu}w_{kp}\frac{\partial^k}{\partial t^k}D_y^p v\Big)dx=0.$$

Then $\frac{\partial\psi}{\partial t}\in\mathcal{L}^2(R)$ *and* $\left\|\frac{\partial\psi}{\partial t}\right\|^2_{0,R}=O\left(\sum_{k,p}\|w_{k,p}\|^2_{0,R}+\|D_y\psi\|^2_{0,R}\right)$.

Let m be any integer not less than $\nu+1$. Set

$$\psi^*(y,t)=\begin{cases}\psi(y,t) & \text{for } t>0,\\ \sum_{j=1}^{m+1}\lambda_j\psi(y,-jt) & \text{for } t<0,\end{cases}$$

$$w^*_{kp}(y,t)=\begin{cases}w_{kp}(y,t) & \text{for } t>0,\\ \sum_{j=1}^{m+1}\lambda_j(-j)^{\nu+1-k}\,w_{kp}(y,-jt) & \text{for } t<0,\end{cases}$$

where the λ_j's are such that $\sum_{j=1}^{m+1}(-j)^{\nu-k}\lambda_j=1$ $(k=0,1,\dots,\nu+1)$.

Let $f\in\mathring{C}^\infty(Q)$. We claim that

$$\tilde{E}f\equiv\int_Q\psi^*\frac{\partial^{\nu+1}f}{\partial t^{\nu+1}}\,dx+\int_Q\left(\sum_{k+|p|\le\nu}w^*_{kp}\frac{\partial^k}{\partial t^k}D^p_y f\right)dx=0. \tag{4.12}$$

In fact we have:

$$\begin{aligned}\tilde{E}f&=Ef+\sum_{j=1}^{m+1}\lambda_j\int_R\psi(y,jt)\,f_{t^{\nu+1}}(y,-t)\,dx\\&\quad+\sum_{k+|p|\le\nu}\sum_{j=1}^{m+1}\lambda_j(-j)^{\nu+1-k}\int_R w_{kp}(y,jt)\,D^p_y f_{t^k}(y,-t)\,dx\\&=Ef+\sum_{j=1}^{m+1}\lambda_j j^{-1}\int_R\psi(y,t)\,f_{t^{\nu+1}}(y,-tj^{-1})\,dx\\&\quad+\sum_{k+|p|\le\nu}\sum_{j=1}^{m+1}\lambda_j j^{-1}(-j)^{\nu+1-k}\int_R w_{kp}(y,t)\,D^p_y f_{t^k}(y,-tj^{-1})\,dx.\end{aligned}$$

Set $f_0(y,t)=-\sum_{j=1}^{m+1}\lambda_j(-j)^\nu f(y,-tj^{-1})$, $v=f+f_0$. Then $\tilde{E}f=E(f+f_0)=Ev$, and $[v_{t^k}(y,t)]_{t=0}=0$ $(k=0,1,\dots,\nu+1)$. For every positive integer s, set $v_s(y,t)=\eta(t)\,v(y,t-s^{-1})$ for $t\ge s^{-1}$, $v_s(y,t)=0$ for $t<s^{-1}$, where $\eta(t)$ is a C^∞ scalar function which equals 1 in $(0,\sigma)$ and vanishes identically in a left neighborhood of π. Since $v_s\in\mathring{C}^{\nu+1}(R)$, we have $Ev_s=0$. On the other hand $\lim_{s\to\infty}Ev_s=Ev$. Hence (4.12) is proven. Let us now assume $f=i^{\nu+1}\varphi(x)\,e^{ikx}\gamma$, where φ is a real-valued function belonging to $\mathring{C}^\infty(Q)$ and identically equal to 1 in the cylinder $|y|<\sigma$, $|t|<\sigma$, γ is an arbitrary constant n-vector and $k\equiv(l_1,\dots,l_{r-1},\tau)$ is different from zero. From Eq. (4.12) we get

$$\tau^{\nu+1}\int_Q\psi^*e^{-ikx}\,dx=\sum_{0\le|\alpha|\le\nu}\int_Q W_\alpha(x)\,k^\alpha e^{-ikx}\,dx,$$

where $W_\alpha(x)$ is a function belonging to $\mathcal{L}^2(Q)$. Let c_k and $\gamma_{\alpha k}$ be the Fourier coefficients of ψ^* and W_α, respectively. We have

$$\begin{aligned}|\tau c_k|&\le 2^{\nu+1}\frac{|\tau|^{2\nu+3}}{|k|^{2\nu+2}}|c_k|+2^{\nu+1}\frac{|\tau|\,|l|^{2\nu+2}}{|k|^{2\nu+2}}|c_k|\\&\le\frac{2^{\nu+1}}{|k|^\nu}\Sigma_\alpha|k|^\nu|\gamma_{\alpha k}|+2^{\nu+1}|l|\,|c_k|.\end{aligned} \tag{4.13}$$

Since $D_y \psi^* \in \mathcal{L}^2(Q)$, we deduce from (4.13) that $\sum_k^{+\infty}{}_{-\infty} \tau^2 |c_k|^2 < +\infty$, i.e. $\psi_t^* \in \mathcal{L}^2(Q)$. From (4.13) we also deduce $\left\| \frac{\partial \psi^*}{\partial t} \right\|_{0,Q}^2 = \mathcal{O}(\|W_\alpha\|_{0,Q}^2 + \|D_y \psi^*\|_{0,Q}^2)$ which enables us to complete the proof of the theorem.

For any pair of r-multi-indices p and q with $0 \leqq |p| \leqq m$, $0 \leqq |q| \leqq m$ $(m \geqq 1)$ let us suppose that the $n \times n$ complex matrix $\alpha_{pq}(x)$ be defined in X^r and that $\alpha_{pq}(x) \in C^\infty$. If A is a bounded domain of X^r for $u \in H_m(A)$ and $v \in H_m(A)$ we shall consider the following bilinear integro-differential form:

$$B(u, v) = \int_A \alpha_{pq}(x) D^q u \, D^p v \, dx = \int_A D^q u \, \bar{\alpha}_{pq}(x) \, D^p v \, dx.$$

In this section we take as domain A the semiball Σ^+: $|y|^2 + t^2 < 1$, $t > 0$. Let us now introduce a function class W enjoying the following properties:

(1) W is the closure in the space $H_m(\Sigma^+)$ of a set constituted by functions v_0 belonging to $C^\infty(\bar{R})$ and such that for each of them a ϱ exists such that $0 < \varrho < 1$ and $v_0 \equiv 0$ for $|x| > \varrho$.

(2) If $v \in W$ there exists a positive ε such that, if the vector $h \equiv (h_1, \ldots, h_{r-1}, 0)$ satisfies the condition $0 < |h| < \varepsilon$, then $|h|^{-1}[v(x+h) - v(x)] \in W$.

(3) Given arbitrary δ and σ such that $0 < \delta < \sigma < 1$, for any real-valued C^∞ function $\varphi(x)$ such that $\varphi(x) \equiv 1$ for $|x| < \delta$ and $\varphi(x) \equiv 0$ for $|x| > \sigma$, we have $\varphi v \in W$ if $v \in W$.

(4) There exists a positive constant c_0 such that for any $v \in W$ the following inequality holds:

$$\mathcal{R}e\, B(v, v) = \mathcal{R}e \int_{\Sigma^+} \sum_{|p|,|q|}^{0,m} \alpha_{pq}(x) \, D^q v \, D^p v \, dx \geqq c_0 \|v\|_{m,\Sigma^+}^2. \tag{4.14}$$

Let u and f satisfy the following hypotheses.

(i) $u \in H_m(\Sigma^+)$, $f \in H_\nu(\Sigma^+)$, $(\nu \geqq 0)$.

(ii) If φ is the function introduced above $\varphi u \in W$;

(iii) For any $v \in W$ we have

$$B(u, v) = (f, v)_0.$$

4.VI *Let W satisfy conditions* (1), (2), (3), (4) *and let u and f satisfy hypotheses* (i), (ii), (iii). *Denote by δ an arbitrary positive number less than* 1 *and by Γ_δ^+ the semi-ball $|y|^2 + t^2 < \delta^2$, $t > 0$. Let $s \equiv (s_1, \ldots, s_{r-1})$ be an arbitrary $(r-1)$-multi-index such that $|s| \leqq \nu + m$. The function u is such that $D_y^s u \in H_m(\Gamma_\delta^+)$.*[4] *Moreover*

$$\|D_y^s u\|_{m,\Gamma_\delta^+}^2 \leqq c_1 (\|f\|_{|s|-m,\Sigma^+}^2 + \|u\|_{m,\Sigma^+}^2),^5 \tag{4.15}$$

where c_1 for any given s depends only on the α_{pq}'s and on δ.

The theorem is true for $|s| = 0$. We shall suppose it to be true for $|s| \leqq k < \nu + m$ and shall prove it for $|s| = k+1$. *Let D_y^k denote any y-partial derivative of order k.* Since φu belongs to W [hypothesis (ii)], by using repeatedly condition (2) on W, Lemma 4.IV and the fact that W is a closed subset of $H_m(\Sigma^+)$, we see that $D_y^k \varphi u \in W$. Set $U(x) = D_y^k \varphi u$. We have for $v \in W \cap C^\infty(\bar{R})$ and any real $h \equiv (h_1, \ldots, h_{r-1}, 0)$

[4] By the symbol D_y^s we mean $\dfrac{\partial^{|s|}}{\partial y_1^{s_1} \ldots \partial y_{r-1}^{s_{r-1}}}$.

[5] From now on, when in the norm $\|f\|_{h,A}$ we have a negative h, then the norm must be understood as $\|f\|_{0,A}$, since we do not need to introduce here negative norms.

such that $0<|h|<1-\sigma$

$$B\left(\frac{U(x+h)-U(x)}{|h|}, v\right)=(-1)^k\int_{\Sigma^+} D_q \frac{\varphi(x+h)u(x+h)-\varphi(x)u(x)}{|h|} D_y^k(\bar{\alpha}_{pq} D^p v)\,dx$$

$$=(-1)^k\int_{\Sigma^+}\alpha_{pq} D^q \frac{\varphi(x+h)u(x+h)-\varphi(x)u(x)}{|h|} D_y^k D^p v\,dx$$

$$+\sum_{0\leqq|j|\leqq k-1}\int_{\Sigma^+} D_y^j\left[\beta_{pq}^j D^q \frac{\varphi(x+h)u(x+h)-\varphi(x)u(x)}{|h|}\right] D^q v\,dx.$$

The meaning of the symbol β_{pq}^j is obvious. Due to the induction hypothesis and to lemma 4.III the last integral is of order

$$\mathcal{O}\left(\sum_{|s|\leqq k}\|D_y^s u\|_{m,\Gamma_\sigma^+}\|v\|_m\right).[6]$$

Assuming $0<|h|<\frac{1}{2}(1-\sigma)$ and $\sigma'=\frac{1}{2}(1+\sigma)$, we have:

$$(-1)^k\int_{\Sigma^+}\alpha_{pq} D^q \frac{\varphi(x+h)u(x+h)-\varphi(x)u(x)}{|h|} D_y^k D^p v\,dx$$

$$=(-1)^k\int_{\Sigma^+}\alpha_{pq}\frac{\varphi(x+h)D^q u(x+h)-\varphi(x)D^q u(x)}{|h|} D^p D_y^k v\,dx$$

$$+\int_{\Sigma^+} D_y^k\left[\sum_{\substack{|q'|\leqq m\\|q''|<m}}\tilde{\alpha}_{p,q',q''}\frac{D^{q'}\varphi(x+h)D^{q''}u(x+h)-D^{q'}\varphi(x)D^{q''}u(x)}{|h|}\right] D^p v\,dx$$

$$=(-1)^k\int_{\Sigma^+}\frac{\alpha_{pq}(x+h)\varphi(x+h)D^q u(x+h)-\alpha_{pq}(x)\varphi(x)D^q u(x)}{|h|} D^p D_y^k v\,dx$$

$$-\int_{\Sigma^+} D_y^k\left[\frac{\alpha_{pq}(x+h)-\alpha_{pq}(x)}{|h|}\varphi(x+h)D^q u(x+h)\right] D^p v\,dx$$

$$+\mathcal{O}\left(\sum_{|s|\leqq k}\|D_y^s u\|_{m-1,\Gamma_\sigma^+}\|v\|_m\right)$$

$$=(-1)^k\int_{\Sigma^+}\alpha_{pq} D^q u(x)\varphi(x) D^p D_y^k\frac{v(x-h)-v(x)}{|h|}\,dx$$

$$+\mathcal{O}\left(\sum_{|s|\leqq k}\|D_k^s u\|_{m,\Gamma_{\sigma'}^+}\|v\|_m\right)$$

$$=(-1)^k\int_{\Sigma^+}\alpha_{pq}(x) D^q u(x) D^p\left[\varphi(x) D_y^k\frac{v(x-h)-v(x)}{|h|}\right]dx$$

$$+\mathcal{O}\left(\sum_{|s|\leqq k}\|D_y^s u\|_{m,\Gamma_{\sigma'}^+}\|v\|_m\right).$$

We have proved thus that

$$B\left(\frac{U(x+h)-U(x)}{|h|}, v\right)=(-1)^k B\left(u, \varphi(x) D_y^k\frac{v(x-h)-v(x)}{|h|}\right)$$

$$+\mathcal{O}\left(\sum_{|s|\leqq k}\|D_y^s u\|_{m,\Gamma_{\sigma'}^+}\|v\|_m\right).$$

[6] By simply writing $\|\ \|_m$ we mean the norm over Σ^+.

Since $D_y^k \dfrac{v(x-h)-v(x)}{|h|} \in W$, from hypothesis (iii) we get

$$B\left(u, \varphi(x) D_y^k \frac{v(x-h)-v(x)}{|h|}\right) = \int_{\Sigma^+} \varphi f D_y^k \frac{v(x-h)-v(x)}{|h|}\, dx$$

$$= \mathcal{O}\left(\|f\|_{k-m+1} \left\| \frac{v(x-h)-v(x)}{|h|} \right\|_{m-1}\right) = \mathcal{O}(\|f\|_{k-m-1} \|v\|_m).$$

Then, for any $v \in W$, we have

$$B\left(\frac{U(x+h)-U(x)}{|h|}, v\right) = \mathcal{O}\left[\left(\sum_{|s|\leq k} \|D_y^s u\|_{m, \Gamma_\sigma^+} + \|f\|_{k-m+1}\right) \|v\|_m\right].$$

By assuming $v = \dfrac{U(x+h)-U(x)}{|h|}$ we get from (4.14)

$$\left\| \frac{U(x+h)-U(x)}{|h|} \right\|_m = \mathcal{O}\left(\sum_{|s|\leq k} \|D_y^s u\|_{m, \Gamma_\sigma^+} + \|f\|_{k-m+1}\right).$$

From this estimate, by a usual argument, the proof follows.

Let us now consider for the class W the further condition:

(5) Every function v of $\mathring{C}^\infty(\Sigma^+)$ belongs to W. Set $\tilde{p} = (0, \ldots, 0, m)$.

4.VII *If inequality* (4.14) *holds, for any* $v \in \mathring{C}^\infty(\Sigma^+)$, *then* $\det \alpha_{\tilde{p}\tilde{p}}(x) \neq 0$ *for any* $x \in \bar{\Sigma}^+$.

We shall give later the proof of this lemma as a consequence of a stronger result.

4.VIII *Let W satisfy conditions* (1), (2), (3), (4), (5) *and u and f hypotheses* (i), (ii), (iii). *For every s and every i such that $|s| + i \leq \nu + m$, the function u has the $\mathscr{L}^2$-strong derivative $\dfrac{\partial^{m+i}}{\partial t^{m+i}} D_y^s u$ in any Γ_δ^+ $(0 < \delta < 1)$. There is a constant c_2, depending only on the α_{pq}'s and on δ, such that*

$$\left\| \frac{\partial^{m+i}}{\partial t^{m+i}} D_y^s u \right\|_{0, \Gamma_\delta^+}^2 \leq c_2 (\|f\|_{|s|+i-m, \Sigma^+}^2 + \|u\|_{m, \Sigma^+}^2).$$

The plan of the proof is the following. Take s and i such that $|s| + i \leq \nu + m$. The statement of the theorem is true if $i = 0$, $|s| \leq \nu + m$. Then we take $i > 0$ and assume the following induction hypothesis:

(h$_1$) *The statement is true for $|s| + j \leq \nu + m$ and $j < i$*, and we prove that the theorem is true for $|s| = 0$, $j = i$.

Then we assume, together with (h$_1$), this new induction hypothesis:

(h$_2$) *The statement is true for $|s| + i \leq \nu + m$ and $|s| \leq k$*, and we prove that the theorem is true for $|s| + i \leq \nu + m$ and $|s| = k + 1$.

Assume that (h$_1$) holds. With the usual meaning for $\varphi(x)$, set $U(x) = \varphi(x) u(x)$. For any $v \in \mathring{C}^\infty(\Sigma^+)$ we have, with an obvious meaning for β_{pq}^h,

$$B\left(U, \frac{\partial^{i-1}}{\partial t^{i-1}} v\right) = (-1)^{i-1} \int_{\Sigma^+} \alpha_{pq} \frac{\partial^{i-1}}{\partial t^{i-1}} D^q U D^p v\, dx + \sum_{h=0}^{i-2} \int_{\Sigma^+} \beta_{pq}^h \frac{\partial^h}{\partial t^h} D^q U D^p v\, dx$$

$$= (-1)^{i-1} \int_{\Sigma^+} \alpha_{pq} \frac{\partial^{i-1}}{\partial t^{i-1}} D^q U D^p v\, dx + \mathcal{O}\left(\sum_{j+|l|\leq m+i-1} \left\| \frac{\partial^j}{\partial t^j} D_y^l u \right\|_{0, \Gamma_\sigma^+} \|v\|_{m-1}\right).$$

We have also, using symbols whose meaning is self-explanatory,

$$B\left(U, \frac{\partial^{i-1}}{\partial t^{i-1}} v\right) = \int_{\Sigma^+} \alpha_{pq} D^q u\, D^p \left(\varphi \frac{\partial^{i-1}}{\partial t^{i-1}} v\right) dx + \sum_{\substack{|p| \leq m \\ |k|<m}} \int_{\Sigma^+} \beta_{kq} D^k u\, D^p \frac{\partial^{i-1}}{\partial t^{i-1}} v\, dx$$

$$+ \sum_{\substack{|k|<m \\ |q| \leq m}} \int_{\Sigma^+} \gamma_{kq} D^q u\, D^k \frac{\partial^{i-1}}{\partial t^{i-1}} v\, dx$$

$$= \left(f, \varphi \frac{\partial^{i-1}}{\partial t^{i-1}} v\right)_0 + \mathcal{O}\left(\sum_{j+|l| \leq m+i-1} \left\| \frac{\partial^j}{\partial t^j} D_y^l u \right\|_{0, \Gamma_\sigma^+} \|v\|_{m-1}\right).$$

With the meaning introduced above for $\tilde{p}$, we denote by $\sum_{p,q}^{(\tilde{p})}$ a summation over every pair of multi-indices p, q $(0 \leq |p| \leq m,\ 0 \leq |q| \leq m)$ except the pair $\tilde{p}, \tilde{p}$.

$$\int_{\Sigma^+} \alpha_{pq} \frac{\partial^{i-1}}{\partial t^{i-1}} D^q U\, D^p v\, dx = \int_{\Sigma^+} \alpha_{\tilde{p}\tilde{p}} \frac{\partial^{m+i-1}}{\partial t^{m+i-1}} U \frac{\partial^m v}{\partial t^m} dx$$

$$+ \sum_{p,q}^{(\tilde{p})} \int_{\Sigma^+} \alpha_{pq} \frac{\partial^{i-1}}{\partial t^{i-1}} D^q U\, D^p v\, dx = \int_{\Sigma^+} \alpha_{\tilde{p}\tilde{p}} \frac{\partial^{m+i-1}}{\partial t^{m+i-1}} U \frac{\partial^m v}{\partial t^m} dx$$

$$+ \mathcal{O}\left(\sum_{\substack{j+|l| \leq m+i \\ j \leq m+i-1}} \left\| \frac{\partial^j}{\partial t^j} D_y^l u \right\|_{0, \Gamma_\sigma^+} \|v\|_{m-1}\right).$$

From the above estimates we deduce that

$$\int_{\Sigma^+} \alpha_{\tilde{p}\tilde{p}} \frac{\partial^{m+i-1}}{\partial t^{m+i-1}} U \frac{\partial^m v}{\partial t^m} dx = \mathcal{O}\left(\left\{\sum_{\substack{j+|l| \leq m+i \\ j \leq m+i-1}} \left\| \frac{\partial^j}{\partial t^j} D_y^l u \right\|_{0, \Gamma_\sigma^+} + \|f\|_{i-m}\right\} \|v\|_{m-1}\right).$$

Using Lemma 4.VII and very simple arguments, we see that we are in position to use Theorem 4.V. Then $U_{t^{m+i}} \in \mathcal{L}^2(\Sigma^+)$ and

$$\left\| \frac{\partial^{m+i} v}{\partial t^{m+i}} \right\|_{0, \Gamma_\delta^+} = \mathcal{O}\left(\sum_{\substack{j+|l| \leq m+i \\ j \leq m+i-1}} \left\| \frac{\partial^j}{\partial t^j} D_y^l u \right\|_{0, \Gamma_\sigma^+} + \|f\|_{i-m}^2\right).$$

Now let us suppose that we have, together with (h_1), the induction hypothesis (h_2). We have just proven that (h_2) is true for $k=0$. Let us prove the statement for $|s| = k+1$. We have for any D_y^s such that $|s| = k+1$

$$B\left(U, \frac{\partial^{i-1}}{\partial t^{i-1}} D_y^s v\right) = (-1)^{k+i} \int_{\Sigma^+} \alpha_{pq} \frac{\partial^{i-1}}{\partial t^{i-1}} D_y^s D^q U\, D^p v\, dx$$

$$+ \sum_{\substack{j+|h|<k+i \\ j \leq i-1}} \int_{\Sigma^+} \beta_{pq}^{jh} \frac{\partial^j}{\partial t^j} D_y^h D^q U\, D^p v\, dx$$

$$= (-1)^{k+i} \int_{\Sigma^+} \alpha_{pq} \frac{\partial^{i-1}}{\partial t^{i-1}} D_y^s D^q u\, D^p v\, dx$$

$$+ \mathcal{O}\left(\sum_{\substack{j+|l| \leq m+k+i \\ j \leq m+i}} \left\| \frac{\partial^j}{\partial t^j} D_y^l u \right\|_{0, \Gamma_\sigma^+} \|v\|_{m-1}\right)$$

and also, using the same arguments as above,

$$B\left(U, \frac{\partial^{i-1}}{\partial t^{i-1}} D_y^s v\right) = \left(f, \varphi \frac{\partial^{i-1}}{\partial t^{i-1}} D_y^s v\right)_0 + \mathcal{O}\left(\sum_{\substack{j+|l| \leq m+k+i \\ j \leq m+i-1}} \left\| \frac{\partial^j}{\partial t^j} D_y^l u \right\|_{0, \Gamma_\sigma^+} \|v\|_{m-1}\right);$$

$$\int_{\Sigma^+} \alpha_{pq} \frac{\partial^{i-1}}{\partial t^{i-1}} D_y^s D^q U D^p v \, dx = \int_{\Sigma^+} \alpha_{\tilde{p}\tilde{p}} \frac{\partial^{m+i-1}}{\partial t^{m+i-1}} D_y^s U \frac{\partial^m v}{\partial t^m} dx$$
$$+ \mathcal{O}\left(\sum_{\substack{j+|l| \leq m+k+1+i \\ j \leq m+i-1}} \left\| \frac{\partial^j}{\partial t^j} D_y^l u \right\|_{0, \Gamma_\sigma^+} \|v\|_{m-1}\right).$$

Arguing as before, we deduce that $\frac{\partial^{m+i}}{\partial t^{m+i}} D_y^s U \in \mathcal{L}^2(\Sigma^+)$ and

$$\left\| \frac{\partial^{m+i}}{\partial t^{m+i}} D_y^s u \right\|_{0, \Gamma_\sigma^+}^2 = \mathcal{O}\left(\sum_{\substack{j+|l| \leq m+k+i \\ j \leq m+i}} \left\| \frac{\partial^j}{\partial t^j} D_y^l u \right\|_{0, \Gamma_\sigma^+} \right.$$
$$\left. + \sum_{\substack{j+|l| \leq m+k+1+i \\ j \leq m+i-1}} \left\| \frac{\partial^j}{\partial t^j} D_y^l u \right\|_{0, \Gamma_\sigma^+}^2 + \|f\|_{|s|+i-m}^2\right).$$

That proves the theorem.

As corollary of Theorems 4.VI, 4.VIII we have

4.IX *Under the hypotheses of Theorem* 4.VIII,

$$\|u\|_{2m+\nu, \Gamma_\delta^+}^2 \leq c\left(\|f\|_{\nu, \Sigma^+}^2 + \|u\|_{m, \Sigma^+}^2\right)$$

where, for a given ν, c depends only on the α_{pq}'s and on δ.

5. Strongly elliptic systems. The matrix differential operator $L(x, D) \equiv a_s(x) D^s$ $(0 \leq |s| \leq \nu)$ is said to be a *strongly elliptic operator* in the domain A if, for every real non-zero r-vector ξ and for every non-zero complex n-vector η, we have

$$\mathcal{R}e\left(\sum_{s=|\nu|} a_s(x) \xi^s \eta \eta\right) \neq 0.$$

It is evident that strong ellipticity implies ellipticity. The converse is not true, as the example of the operator $\frac{\partial}{\partial x_1} + i \frac{\partial}{\partial x_2}$ in X^2 proves.

Let us consider in the bounded domain A the $n \times n$ matrices $a_{pq}(x)$ $(0 \leq |p| \leq m,$ $0 \leq |q| \leq m)$ and set

$$Q(x, \xi, \eta) = \mathcal{R}e\left(\sum_{|p|, |q|=m} a_{pq}(x) \xi^p \xi^q \eta \eta\right)$$

where ξ is a non-zero real r-vector and η a non-zero complex n-vector. Suppose $a_{pq}(x)$ $(|p|=m, |q|=m)$ continuous in $\bar{A}$. Then the condition $Q(x, \xi, \eta) > 0$ for any $x \in \bar{A}$ is equivalent to the following one:

$$\mathcal{R}e\left(\sum_{|p|, |q|=m} a_{pq}(x) \xi^p \xi^q \eta \eta\right) \geq p_0 |\xi|^{2m} |\eta|^2 \qquad (x \in \bar{A}) \tag{5.1}$$

where p_0 is a positive constant (i.e. the minimum of $Q(x, \xi, \eta)$ in the compact set: $x \in \bar{A}$, $|\xi| = 1$, $|\eta| = 1$).

Suppose that $a_{pq}(x)$ (for $|p| + |q| < 2m$) be bounded and measurable in A. For any $u, v \in H_m(A)$, set

$$B(u, v) = \sum_{|p|, |q|}^{0, m} (-1)^{|p|} \int_A a_{pq} D^q u D^p v \, dx.$$

Under the above assumptions on the a_{pq}'s, if (5.1) is satisfied, the following theorem holds. It is called the theorem of the *Gärding inequality.*

5.I *There exist two constants γ_0 and λ_0 ($\gamma_0>0$, $\lambda_0\geqq 0$) such that for any $v\in \mathring{H}_m(A)$*

$$(-1)^m \mathscr{R}e\, B(v,v) \geqq \gamma_0 \|v\|_m^2 - \lambda_0 \|v\|_0^2. \tag{5.2}$$

Set $\beta_{ls}=(-1)^{|l|+m}\,\alpha_{ls}$ for $|l|+|s|<2m$. We may write

$$(-1)^m B(u,v) = \int_A a_{pq} D^q u\, D^p v\, dx + \int_A \beta_{ls} D^s u\, D^l v\, dx.$$

In the first integral the summation must be understood to be restricted to $|p|=|q|=m$ in the second integral to $|l|+|s|<2m$, $0\leqq|l|\leqq m$, $0\leqq|s|\leqq m$. From now on, in the course of this proof, we write a_{pq} only in the case when $|p|=|q|=m$. The proof is obtained by considering three cases:

1st case: a_{pq}=constant, $\beta_{ls}=0$. We suppose, as usual, $\bar{A}\subset Q$. By using (2.3) and (2.4) we have, for $v\in\mathring{H}_m(A)$,

$$D^p v = (2\pi)^{-r/2} \sum_{k=-\infty}^{+\infty} i^{|p|}\, k^p\, c_k\, e^{ikx}, \qquad a_{pq} D^q v = (2\pi)^{-r/2} \sum_{k=-\infty}^{+\infty} i^{|q|}\, k^q\, a_{pq}\, c_k\, e^{ikx}.$$

Then by the Parseval theorem, inequality (5.1), Theorem 2.II, we have

$$(-1)^m \mathscr{R}e\, B(v,v) = \sum_{k=-\infty}^{+\infty} \mathscr{R}e (a_{pq} k^p k^q c_k \bar{c}_k) \geqq p_0 \sum_{k=-\infty}^{+\infty} |k|^{2m} |c_k|^2 \geqq \gamma_0 \|v\|_m^2.$$

In this particular case (5.2) holds with $\lambda_0=0$ and $\gamma_0=p_0\, c$, where c is a constant depending only on A.

2nd case: diameter $spt\, v<\delta_0$ (where δ_0 is a positive number). Let $x^0\in spt\, v$. We have (with $\tilde{\gamma}_0$ independent of x^0)

$$\begin{aligned}
\tilde{\gamma}_0 \|v\|_m^2 &\leqq \mathscr{R}e \int_A a_{pq}(x^0)\, D^q v\, \overline{D^p v} = \mathscr{R}e \int_A a_{pq}(x)\, D^q v\, \overline{D^p v}\, dx \\
&+ \mathscr{R}e \int_A \beta_{ls}(x)\, D^s v\, \overline{D^l v}\, dx + \mathscr{R}e \int_A [a_{pq}(x^0) - a_{pq}(x)]\, D^q v\, \overline{D^p v}\, dx \\
&- \mathscr{R}e \int_A \beta_{ls}(x)\, D^s v\, \overline{D^l v}\, dx \leqq (-1)^m \mathscr{R}e\, B(v,v) \\
&+ \max_{x\in spt\, v} \sum_{p,q} |a_{pq}(x^0) - a_{pq}(x)|\, \|v\|_m^2 + \gamma_1 \|v\|_m \|v\|_{m-1}.
\end{aligned}$$[7]

Assume δ_0 so small that

$$\max_{x\in spt\, v} \sum_{p,q} |a_{pq}(x^0) - a_{pq}(x)| < \tfrac{1}{2}\tilde{\gamma}_0.$$

Then we have $2^{-1}\tilde{\gamma}_0 \|v\|_m^2 \leqq (-1)^m \mathscr{R}e\, B(v,v) + \gamma_1 \|v\|_m \|v\|_{m-1}$. From this inequality, by use of Lemma 2.VII, inequality (5.2) follows.

3rd case: general case. Let us consider the following partition of unity in

$$\bar{A}: \sum_{h=1}^{\nu} \varphi_h^2(x) = 1$$

[7] In the proofs of this theorem and the next, $\gamma_1, \gamma_2, \gamma_3, \ldots$ denote positive constants.

with $\varphi_h(x) \in C^\infty$ and diameter $spt\ \varphi_h(x) < \delta_0$; δ_0 is such that for $|x^1 - x^2| < \delta_0$ one has: $\sum_{p,q} |a_{pq}(x^1) - a_{pq}(x^2)| < \frac{1}{2}\tilde{\gamma}_0$. We have

$$(-1)^m B(v, v) = \sum_{h=1}^{\nu} \int_A \varphi_h^2 a_{pq} D^q v\, D^p v\, dx + \int_A \beta_{ls} D^s v\, D^l v\, dx$$
$$= \sum_{h=1}^{\nu} \int_A a_{pq} D^q \varphi_h v\, D^p \varphi_h v\, dx + \mathcal{O}(\|v\|_m \|v\|_{m-1}).$$

From the 2nd case we have

$$\|\varphi_h v\|_m^2 \leqq \gamma_2 \mathcal{Re} \int_A a_{pq} D^q \varphi_h v\, D^p \varphi_h v\, dx + \gamma_3 \|\varphi_h v\|_0^2.$$

Then

$$(-1)^m \mathcal{Re}\, B(v, v) \geqq \gamma_4 \|v\|_m^2 - \gamma_5 \|v\|_0^2 - \gamma_6 \|v\|_m \|v\|_{m-1}.$$

Arguing as in the 2nd case, we get (5.2).

The theorem can be inverted, and we have

5.II *If* (5.2) *holds for any* $v \in \mathring{H}_m(A)$, *then for any* $x \in \bar{A}$, (5.1) *holds.*

Suppose (5.1) false. Then there exist $x^0 \in \bar{A}$, a real $\xi \neq 0$ and $\eta \neq 0$ such that

$$\mathcal{Re}\Big(\sum_{|p|,|q|=m} a_{pq}(x^0)\, \xi^p \xi^q \eta \bar{\eta}\Big) \leqq 0. \tag{5.3}$$

For $v \in \mathring{H}_m(A)$ we have

$$\begin{aligned} \gamma_0 \|v\|_m^2 &\leqq (-1)^m \mathcal{Re}\, B(v, v) + \lambda_0 \|v\|_0^2 \leqq \mathcal{Re} \int_A a_{pq}(x^0) D^q v\, D^p v\, dx \\ &+ \mathcal{Re} \int_A [a_{pq}(x) - a_{pq}(x^0)] D^q v\, D^p v\, dx + \gamma_1 \|v\|_m \|v\|_{m-1} + \lambda_0 \|v\|_0^2. \end{aligned} \tag{5.4}$$

Let δ_0 be such that for $|x - x^0| < \delta_0$: $\sum_{pq} |a_{pq}(x) - a_{pq}(x^0)| < \frac{1}{2}\gamma_0$. Let I_0 be the ball $|x - x^0| < \delta_0$. If $spt\ v \in A \cap I_0$, from (5.4) and Lemma 2.VII we deduce

$$\|v\|_m^2 \leqq \gamma_7 \mathcal{Re} \int_A a_{pq}(x^0) D^q v\, D^p v\, dx + \gamma_8 \|v\|_0^2. \tag{5.5}$$

Let $\varphi(x)$ be a C^∞ real function which is not identically zero and such that $spt\ \varphi \in A \cap I_0$. Set $v = \varphi(x)\, e^{i\lambda\xi x} \eta$ with λ real. We have

$$\|v\|_m^2 = \lambda^{2m} |\eta|^2 \sum_{|p|=m} (\xi^p)^2 \int_A |\varphi(x)|^2 dx + \mathcal{O}(\lambda^{2m-1})$$
$$\mathcal{Re} \int_A a_{pq}(x^0) D^q v\, D^p v\, dx = \lambda^{2m} \left(\mathcal{Re}\, a_{pq}(x^0)\, \xi^p \xi^q \eta \bar{\eta}\right) \int_A |\varphi|^2 dx + \mathcal{O}(\lambda^{2m-1}).$$

Then (5.5) is in contradiction with (5.3)

Remark. From this theorem it follows that if (5.2) holds, then, in particular, one must have

$$\det \sum_{|p|,|q|=m} a_{pq}(x)\, \xi^p \xi^q \neq 0$$

for any $x \in \bar{A}$ and any non-zero ξ. Then, from (4.14), it follows

$$\det \sum_{|p|,|q|=m} \alpha_{pq}(x)\, \xi^p \xi^q \neq 0$$

for $x \in \Sigma^+$, $\xi \neq 0$. Assuming $\xi_1 = \cdots = \xi_{r-1} = 0$, $\xi_r = 1$, we have the proof of Lemma 4.VII.

6. General existence theorems. Let us now consider in X^r the differential operator $L(x, D) = D^p a_{pq}(x) D^q$ $(0 \leqq |p| \leqq m,\ 0 \leqq |q| \leqq m)$ and suppose, for simplicity, that the $n \times n$ complex matrices $a_{pq}(x)$ belong to C^∞. Let A be a bounded domain of X^r. Let us consider a function space V and assume the following hypotheses:

(I) V is a closed linear subspace of $H_m(A)$;

(II) $V \supset \mathring{H}_m(A)$;

(III) for every $v \in V$, the following inequality holds:

$$(-1)^m \mathcal{R}e\, B(v, v) \geqq \gamma_0 \|v\|_m^2 \qquad (\gamma_0 > 0). \tag{6.1}$$

If L satisfies (5.1) and we assume $V = \mathring{H}_m(A)$, hypotheses (I), (II), (III) are satisfied, provided we replace the operator L by the operator $L_0 = L + (-1)^m \lambda_0 I$, where λ_0 is the constant which enters in (5.2) and I the identity operator.

6.I *Hypotheses* (I), (II), (III) *be satisfied; given* $f \in H_0(A)$, *there exists one and only one function* $u \in V$ *such that*

$$B(u, v) = \int_A f v\, dx,$$

for every $v \in V$. *This function satisfies the inequality* $\|u\|_m \leqq \gamma_0^{-1} \|f\|_0$.

Since the bilinear form $B(w, v)$ is continuous in $V \times V$, there exists a linear bounded transformation T from V into itself such that $B(w, v) = (w, Tv)_m$ for any $w, v \in V$. Since $|(v, Tv)_m| = |B(v, v)| \geqq \gamma_0 \|v\|_m^2$, we have: $\|Tv\|_m \geqq \gamma_0 \|v\|_m$. Let $T(V)$ be the range of T. For any $w \in T(V)$, set $\psi(w) = (f, v)_0$, where v is the only function of V such that $w = Tv$. We have $|\psi(w)| \leqq \|f\|_0 \|v\|_m \leqq \gamma_0^{-1} \|f\|_0 \|w\|_m$. Then $\psi(w)$ is an anti-linear bounded functional defined over $T(V)$, such that $\|\psi\| \leqq \gamma_0^{-1} \|f\|_0$. We can continue ψ in the whole space V, in such a way that the continued functional Ψ shall have the same norm. On the other hand, since V is a Hilbert space, there exists $u \in V$ such that $\Psi(v) = (u, v)_m$, for any $v \in V$, and $\|u\|_m = \|\Psi\| \leqq \gamma_0^{-1} \|f\|_0$. The uniqueness of u is obvious since $B(u, v) = 0$, for any $v \in V$, implies $0 = |B(u, u)| \geqq \gamma_0 \|u\|_m^2$.[8]

Let us now define the linear space $\mathcal{U}_V$ which enjoys the following properties:

(1) $\mathcal{U}_V$ is a linear subspace of V;

(2) if $u \in \mathcal{U}_V$, then $u \in H_{2m}(B)$ for any domain B such that $\bar{B} \subset A$ and $Lu \in H_0(A)$;

(3) for any $v \in V$ and any $u \in \mathcal{U}_V$, one has $B(u, v) = (Lu, v)_0$.

Let us consider the following problem:

$$\text{(P)} \qquad Lu = f \quad (f \in H_0(A)), \quad u \in \mathcal{U}_V.$$

6.II *If the hypotheses of Theorem* 6.I *are satisfied, there exists one and only one solution of problem* (P).

It is obvious that if u is a solution of problem (P), then u is a solution of the problem considered in Theorem 6.I.

Conversely, if u is the solution of such a problem, since u is a weak solution of $Lu = f$ (as it is easily seen through integration by parts) then u is a solution of problem (P) (Theorem 3.I).

[8] It must be remarked that in proving the theorem only hypotheses (I) and (III) were used. Hypothesis (II) will be used later.

6.III *Let $B(u, v)$ be symmetric on V (i.e. $B(u, v) = \overline{B(v, u)}$ for $u, v \in V$). Under the hypotheses of Theorem* 6.I, *the functional $I(v) = \frac{1}{2} B(v, v) - \mathcal{R}e(f, v)_0$ has an absolute minimum in V, and the minimizing function is the solution of problem* (P).

If u is the solution of problem (P) and v any function of V, we have $I(v) - I(u) = \frac{1}{2} B(v - u, v - u)$, which gives the proof.

A sufficient condition in order $B(u, v)$ be symmetric in V is that L be *formally self-adjoint*. By saying that, we mean $a_{pq}(x) \equiv (-1)^{|p|+|q|} \bar{a}_{qp}(x)$ for any pair p, q of multi-indices. In this case we have $L^*(x, D) = L(x, D)$.

Let us now consider the problem:

$$(\mathrm{P}_\lambda) \qquad Lu - (-1)^m \lambda u = f \quad (f \in H_0(A)), \quad u \in \mathcal{U}_V$$

where λ is any given complex number. *Suppose hypotheses* (I), (II), (III) *satisfied.* For every $\varphi \in H_0(A)$ denote by $G\varphi$ the solution of problem (P) with $f = (-1)^m \varphi$; G is a linear transformation with domain $H_0(A)$ and range in V. Suppose φ is a solution of the following problem:

$$\varphi - \lambda G \varphi = (-1)^m f, \quad \varphi \in H_0(A). \tag{6.2}$$

Then $u = G\varphi$ is a solution of (P_λ). Conversely, if u is a solution of (P_λ), then $\varphi = (-1)^m Lu$ is a solution of (6.2). The operator G considered as an operator from $H_0(A)$ into $H_0(A)$ is compact, *if we suppose that A is properly regular*[9] (see Theorem 2.IV). Then we can apply to problem (6.2) the Riesz-Fredholm theory of compact operators. Thus

6.IV *Problem (P_λ) has a discrete (i.e. empty, finite or countable) set of eigenvalues for the parameter λ, each with a finite geometric multiplicity and with no finite limit-points. If λ is not an eigenvalue, problem (P_λ) has a unique solution for any given $f \in H_0(A)$. If λ is an eigenvalue, then a solution exists when and only when $(f, \varphi_k)_0 = 0$ $(k = 1, \ldots, l)$, where $\varphi_1, \ldots, \varphi_l$ are a complete set of eigenfunctions of the equation $\varphi - \bar{\lambda} G^* \varphi = 0$; G^* is the adjoint transformation of G considered as a mapping from $H_0(A)$ into $H_0(A)$.*

It must be remarked that if λ is an eigenvalue, the eigenfunctions of (P_λ) coincide with the eigenfunctions of (6.2). If $B(u, v)$ is symmetric on V, for φ and ψ in $H_0(A)$, we have $(\varphi, G\psi)_0 = (-1)^m B(G\varphi, G\psi) = (-1)^m \overline{B(G\psi, G\varphi)} = (G\varphi, \psi)_0$, i.e. G is a symmetric operator in $H_0(A)$. Moreover from hypothesis (III) we get $(\varphi, G\varphi)_0 = (-1)^m B(G\varphi, G\varphi) \geqq \gamma_0 \|G\varphi\|_m^2$; thus G is a positive operator[10]; and G is strictly positive since $G\varphi = 0$ implies $LG\varphi = (-1)^m \varphi = 0$.

6.V *If $B(u, v)$ is symmetric on V, problem (P_λ) has an increasing sequence of positive eigenvalues converging to $+\infty$. A complete set of eigensolutions of problem (P_λ) constitutes a complete set in the space $H_0(A)$.*

We say that the domain A is *C^ν-smooth at the point x^0 of ∂A* if a neighborhood I of x^0 exists with the following properties:

(i) There exists a C^ν homeomorphism which maps the set $J = \bar{I} \cap \bar{A}$ onto the closed semiball $\bar{\Sigma}^+$: $|y|^2 + t^2 \leqq 1$, $t \geqq 0$ of the r-dimensional (y, t) space;

(ii) the set $\bar{I} \cap \partial A$ is mapped onto the $(r-1)$-dimensional ball $t = 0$, $|y| \leqq 1$. A is called *C^ν-smooth* if it is C^ν-smooth at every point of its boundary.

---

[9] This hypothesis on A is not needed if $V = \mathring{H}_m(A)$.

[10] Since we consider $H_0(A)$ as a complex Hilbert space, we could derive the symmetry of G from its positiveness.

Suppose that A is C^∞-smooth at the point x^0 of ∂A and suppose that, in addition to hypotheses (I), (II), (III), the space V satisfies the following further condition:

$(\mathrm{IV})_{x^0}$ Given any C^∞ real function $\psi(x)$, with support in I and any $v \in V$, the function $\psi(x)\,v(x)$ belongs to V, and the class W of all the functions $\psi[x(\xi)]\,v[x(\xi)]$ $\{\psi \in \overset{\circ}{C}{}^\infty(I), v \in V, x = x(\xi)$ C^∞-homeomorphism from $\bar{\Sigma}^+$ to $J\}$ enjoys the properties (1), (2), considered in Sect. 4.

The class W which we have now defined enjoys property (3) of Sect. 4 as one can easily prove. Moreover if we take as the functional $B(v, v)$ which appears in (4.14) the one which is obtained from the $B(v, v)$ of (6.1) by transforming this last functional, for any v of V with support in I, by the C^∞-homeomorphism $x = x(\xi)$, then we see that W satisfies also condition (4). Condition (5) is satisfied since $V \supset \overset{\circ}{H}_m(A)$.

It is not difficult to prove that, assuming either $V = \overset{\circ}{H}_m(A)$ or $V = H_m(A)$, condition $(\mathrm{IV})_{x^0}$ is satisfied.

From the regularization theory developed in Sects. 3 and 4, we deduce the following regularization theorems:

6.VI *Let A be C^∞-smooth at $x^0 \in \partial A$. If V satisfies hypotheses* (I), (II), (III), $(\mathrm{IV})_{x^0}$ *and $f \in H_{\nu-2m}(A)$,*[11] *then the solution of problem* (P) *belongs to $H_\nu(B)$ for any domain B such that $\bar{B} \subset \bar{A} \cap (A \cup I)$; a positive c exists such that:*

$$\|u\|^2_{\nu,B} \leq c\,(\|f\|^2_{\nu-2m,A} + \|u\|^2_{m,A}).$$

If A is C^∞-smooth and $(\mathrm{IV})_{x_0}$ is satisfied for every $x^0 \in \partial A$, then $u \in H_\nu(A)$ and a positive c exists such that: $\|u\|^2_{\nu,A} \leq c\,\|f\|^2_{\nu-2m,A}$. If A is C^∞-smooth, $(\mathrm{IV})_{x^0}$ is satisfied for every $x^0 \in \partial A$ and $f \in C^\infty(\bar{A})$, then $u \in C^\infty(\bar{A})$.

6.VII *If A is C^∞-smooth $(\mathrm{IV})_{x_0}$ is satisfied for every $x^0 \in \partial A$ and u is an eigenfunction of problem (P_λ) (i.e. $Lu - (-1)^m \lambda u = 0$, $u \in \mathrm{U}_V$), then $u \in C^\infty(\bar{A})$.*

In fact, since $Lu \in H_m(A)$, $u \in H_{3m}(A)$, then $Lu \in H_{3m}(A)$ and $u \in H_{5m}(A)$, etc.

If $V = \overset{\circ}{H}_m(A)$, A is C^∞-smooth, $f \in H_{\nu-2m}(A)$, with $\nu > \frac{r}{2} + m - 1$, then the functions u considered in Theorems 6.VI and 6.VII belong to $C^{m-1}(\bar{A})$ (Theorem 2.VI) and satisfy in the classical sense the *Dirichlet boundary conditions* $D^p u = 0$ $(0 \leq |p| \leq m-1)$.

If $f \in H_{\nu-2m}(A)$ with $\nu > \frac{r}{2} + 2m - 1$, A is C^∞-smooth and $(\mathrm{IV})_{x^0}$ is satisfied for every $x^0 \in \partial A$, then $u \in C^{2m-1}(\bar{A})$. In this case the function u satisfies the Green identity for every $v \in H_m(A)$:

$$B(u, v) = \int_A (Lu)\,v\,dx + \int_{\partial A} M(u, v)\,d\sigma$$

where $M(u, v)$ is a certain bilinear differential operator of order $2m-1$ in u and of order $m-1$ in v. The fact that $u \in \mathscr{U}_V$ implies

$$\int_{\partial A} M(u, v)\,d\sigma = 0 \tag{6.3}$$

for every $v \in V$. In many applications Eqs. (6.3) enable us to characterize the boundary conditions satisfied by u.

Let us now suppose A properly regular and consider the space $\mathscr{T}_{m-1}(\partial A)$ spanned by the vector τv, when v spans $H_m(A)$. The norm in $\mathscr{T}_{m-1}(\partial A)$ is the

[11] $H_{\nu-2m}(A)$ stands for $H_0(A)$ if $\nu - 2m < 0$.

one introduced in Sect. 2. Let $D(w, z)$ be a bilinear continuous functional in $\mathcal{T}_{m-1}(\partial A) \times \mathcal{T}_{m-1}(\partial A)$. Let γ be a linear bounded operator with domain $H_m(A)$ and range in $\mathcal{T}_{m-1}(\partial A)$. Set $\Phi(u, v) = D(\gamma u, \tau v) + B(u, v)$. Let V be the subspace of $H_m(A)$ satisfying the hypotheses (I), (II). Let us suppose that D and B are such that

(III′) $$(-1)^m \mathcal{R}e\,[D(\gamma v, \tau v) + B(v, v)] \geqq \gamma_0' \|v\|_m^2 \qquad (\gamma_0' > 0).$$

It is easy to check that the theory developed in Theorems 6.I, 6.II, 6.III, 6.IV, 6.V holds unchanged if we replace, in the statements and in the definitions, the bilinear form $B(u, v)$ by $\Phi(u, v)$.

As far as the boundary regularization theory is concerned, some additional hypotheses on D are required in order to carry out, in this more general connection, the theory developed in Sect. 4. Looking back at the proofs of Theorems 4.VI and 4.VIII, the reader may formulate hypotheses to be made on D in order to get the same results in this more general case.

7. Propagation problems. Let us now consider, in the $(r+1)$-dimensional cartesian space of the variables $x_1, \ldots, x_r, t$, the cylinder $E = A \times I$ where A is a domain of X^r and I the open interval $(0, T)$. We shall consider initial-boundary value problems for the differential equation $(-1)^{m-1} Lu - u_{tt} = f$ in E, where L is the operator considered in the preceding section, with coefficients $a_{pq}(x)$ which depend on x. We shall consider only a "C^∞-theory"; yielding the maximum regularization on the unknown function u, since this theory seems to be, in general, the most useful in applications. However, by using the theory of strongly elliptic operators developed in the preceding sections, it should be possible to get a theory for propagation problems with more general domains and data.

Let A be a bounded C^∞-smooth domain of X^r and L be the matrix differential operator with C^∞ coefficients, introduced in Sect. 6. Let V be the subspace considered in that section satisfying hypotheses (I), (II), (III), $(\text{IV})_{x^0}$ (the last one with respect to every $x^0 \in \partial A$)[12]. Let us in addition suppose that

(V) The bilinear form $B(u, v)$ is symmetric on V.

We shall consider the following problem:

(P_{tt}) *Given* $f(x, t) \in C^\infty(\bar{E})$, $g_0(x) \in C^\infty(\bar{A})$, $g_1(x) \in C^\infty(\bar{A})$, *find a function* $u(x, t) \in C^\infty(\bar{E})$ *satisfying the differential equation* $(-1)^{m-1} Lu - u_{tt} = f(x, t)$ *in* E, *belonging to* $\mathcal{U}_V$, *for every* $t \in \bar{I}$, *and such that* $u(x, 0) = g_0(x)$, $u_t(x, 0) = g_1(x)$.

The fact that $u(x, t) \in \mathcal{U}_V$ for every $t \in \bar{I}$, is equivalent to imposing condition (6.3) for every $t \in \bar{I}$ and $v \in V$.

Let us denote by $\lambda_1^2 \leqq \lambda_2^2 \leqq \cdots \leqq \lambda_k^2 \leqq \cdots$ the sequence of the eigenvalues of problem (P_λ), each eigenvalue being repeated according to its multiplicity. Let $\{v_k(x)\}$ be a corresponding complete and orthonormal set of eigenfunctions.

7.I *The solution of problem* (P_{tt}) *is unique. Necessary and sufficient conditions for the existence of the solution are the following*:

$$\begin{aligned} &g_0(x) \in \mathcal{U}_V, \quad g_1(x) \in \mathcal{U}_V, \\ &L^j g_0(x) - \sum_{k=1}^{j} (-1)^{k(m-1)} L^{j-k} f_{t^{2k-2}}(x, 0) \in \mathcal{U}_V, \\ &L^j g_1(x) - \sum_{k=1}^{j} (-1)^{k(m-1)} L^{j-k} f_{t^{2k-1}}(x, 0) \in \mathcal{U}_V \qquad (j = 1, 2, \ldots). \end{aligned} \tag{7.1}$$

[12] We must remark that this is a severely restrictive condition on V; for instance, it excludes "mixed boundary conditions". However, these more general problems could be handled in a "non-C^∞-theory".

If (7.1) *are satisfied, the solution is given by the following development:*

$$u(x,t)=\sum_{k=1}^{\infty}\Big\{(g_0,v_k)_0\cos\lambda_k t+\lambda_k^{-1}(g_1,v_k)\sin\lambda_k t - \int_0^t\frac{\sin\lambda_k(t-\tau)}{\lambda_k}\,d\tau\int_A f(\xi,\tau)\,v_k(\xi)\,d\xi\Big\}\,v_k(x). \tag{7.2}$$

Set $U_k(t)=\big(u(x,t),v_k(x)\big)_0$, $F_k(t)=\big(f(x,t),v_k(x)\big)_0$. If u is a solution of problem (P_{tt}), we have: $(Lu,v_k)_0=B(u,v_k)=\overline{B(v_k,u)}=(u,Lv_k)_0=(-1)^m\lambda_k^2 U_k(t)$. Then $U_k''(t)+\lambda_k^2U_k(t)=-F_k(t)$, $U_k(0)=(g_0,v_k)_0$, $U_k'(0)=(g_1,v_k)_0$. That proves uniqueness of the solution and moreover, if the solution exists, is given by (7.2). From (6.3) we deduce that, if $u(x,t)$ is a solution of problem (P_{tt}), then $u_{t^s}(x,t)$, for any s and any $t\in\bar{I}$, belongs to $\mathscr{U}_V$. Then $g_0\in\mathscr{U}_V$, $g_1\in\mathscr{U}_V$. We have

$$u_{tt}=(-1)^{m-1}Lu-f(x,t).$$

By successive differentiations we get

$$u_{t^{2j}}=(-1)^{j(m-1)}L^ju-\sum_{k=1}^{j}(-1)^{(j-k)(m-1)}L^{j-k}f_{t^{2k-2}}.$$

Hence

$$L^ju-\sum_{k=1}^{j}(-1)^{k(m-1)}L^{j-k}f_{t^{2k-2}}\quad\text{and}\quad L^ju_t-\sum_{k=1}^{t}(-1)^{k(m-1)}L^{j-k}f_{t^{2k-1}}$$

belong to $\mathscr{U}_V$ for every $t\in I$. Assuming $t=0$ we get conditions (7.1). Suppose that (7.1) are satisfied. We claim that (7.2) gives the solution of problem (P_{tt}). Let G be the operator introduced in Sect. 6 and ϱ any, arbitrarily fixed, positive integer. By elementary computations and using an induction argument, we recognize that the development (7.2) can be written as follows:

$$u(x,t)=\sum_{k=1}^{\infty}W_{k\varrho}(t)\,v_k(x)+\sum_{h=1}^{\varrho}(-1)^hG^h\,[f_{t^{2h-2}}(\xi,t)],$$

where

$$\begin{aligned}W_{k\varrho}(t)&=(-1)^{m(\varrho+1)}\Big(L^{\varrho+1}g_0-\sum_{h=1}^{\varrho}(-1)^{h(m-1)}L^{\varrho+1-h}f_{t^{2h-2}}(\xi,0),\,v_k(\xi)\Big)_0\frac{\cos\lambda_k t}{\lambda_k^{2\varrho+2}}\\&+(-1)^{m\varrho}\Big(L^\varrho g_1-\sum_{h=1}^{\varrho-1}(-1)^{h(m-1)}L^{\varrho-h}f_{t^{2h-1}}(\xi,0),v_k(\xi)\Big)_0\frac{\sin\lambda_k t}{\lambda_k^{2\varrho+1}}\\&+(-1)^{\varrho+1}\int_0^t\frac{\cos\lambda_k(t-\tau)}{\lambda_k^{2\varrho}}F_k^{(2\varrho-1)}(\tau)\,d\tau.\end{aligned}$$

Let s and q, respectively, be an index and a multi-index such that $s+|q|\leqq\varrho$. For any l such that $\lambda_l^{-2}<1$ (denoting by c and c_1 positive constants),

$$\begin{aligned}\int_0^T\Big\|\sum_{k=l+1}^{l+p}D_t^sW_{k\varrho}(t)\,D_x^qv_k(x)\Big\|_{0,A}^2dt&\leqq c\int_0^T\Big\|\sum_{k=l+1}^{l+p}D_t^sW_{k\varrho}(t)\,L^{\varrho-s}v_k(x)\Big\|_{0,A}^2dt\\&=c\int_0^T\sum_{k=l+1}^{l+p}\lambda_k^{4\varrho-4s}|D_t^sW_{k\varrho}(t)|^2dt\\&\leqq c_1\sum_{k=l+1}^{l+p}\lambda_k^{-2s}\Big[\Big|\Big(L^{\varrho+1}g_0-\sum_{h=1}^{\varrho}(-1)^{h(m-1)}L^{\varrho+1-h}f_{t^{2h-2}}(\xi,0),v_k(\xi)\Big)_0\Big|^2\\&+\Big|\Big(L^\varrho g_1-\sum_{h=1}^{\varrho-1}(-1)^{h(m-1)}L^{\varrho-h}f_{t^{2h-1}}(\xi,0),v_k(\xi)\Big)_0\Big|^2\\&+\sum_{h=0}^{s-1}\int_0^T|F_k^{(2\varrho-1+h)}(t)|^2dt\Big].\end{aligned}$$

Thus u belongs to $H_\varrho(E)$ for any ϱ and, in consequence, $u \in C^\infty(\bar{E})$. We are now in position to recognize, by inspection, that $u(x,t)$ as given by (7.2) is the solution of problem (P_{tt}).

By the methods we have used, it is possible to derive estimates which prove the continuous dependence of the solution of problem (P_{tt}), on the data. We write down explicitly only the simplest one:

$$\frac{1}{2}\int_A |u(x,t)|^2\,dx \leqq \int_A |g_0(x)|^2\,dx + \lambda_1^{-2}\int_A |g_1(x)|^2\,dx + \lambda_1^{-2}\,t\int_0^t dt\int_A |f(x,t)|^2\,dx,$$

which readily follows from (7.2).

8. Diffusion problems. We assume for A, L, B and V the same hypotheses as in the preceding section. Now we consider the following problem:

(P_t) *Given* $f(x,t) \in C^\infty(\bar{E})$, $g_0(x) \in C^\infty(\bar{A})$, *find a function* $u(x,t) \in C^\infty(\bar{E})$ *satisfying the differential equation* $(-1)^{m-1}Lu - u_t = f(x,t)$ *in* E, *belonging to* $\mathcal{U}_V$, *for every* $t \in \bar{I}$, *and such that* $u(x,0) = g_0(x)$.

8.I *The solution of problem* (P_t) *is unique. Necessary and sufficient conditions for the existence of the solution are the following*:

$$\begin{gathered} g_0(x) \in \mathcal{U}_V, \\ L^j g_0(x) - \sum_{k=1}^{j} (-1)^{k(m-1)} L^{j-k} f_{t^{k-1}}(x,0) \in \mathcal{U}_V \qquad (j = 1, 2, \ldots). \end{gathered} \tag{8.1}$$

If (8.1) *are satisfied, the solution is given by the following development*:

$$u(x,t) = \sum_{k=1}^{\infty}\left\{(g_0, v_k)_0\, e^{-\lambda_k^2 t} - \int_0^t e^{-\lambda_k^2(t-\tau)} F_k(\tau)\,d\tau\right\} v_k(x). \tag{8.2}$$

Since the proof exactly parallels the one of Theorem 7.I, we need not repeat it here. We only note the following estimate which readily follows from (8.2):

$$\frac{1}{2}\int_A |u(x,t)|^2\,dx \leqq \int_A |g_0(x)|^2\,dx + t\int_0^t dt\int_A |f(x,t)|^2\,dx.$$

9. Integro-differential equations. In the context of general strongly elliptic operators, we shall now consider integro-differential equations, which arise, for instance, in theories of materials with "memory". VOLTERRA, about sixty years ago, was the first to consider problems of this kind, especially in relation with classical elasticity.

We assume on A, L, B, and V the same hypotheses as in Sects. 7 and 8, but we drop hypothesis (V), since what we are going to say holds also when $B(v,v)$ is not symmetric on V.

We shall denote by $W(E)$ the class constituted by the (complex n-vector valued) functions $w(x,t)$ defined in $\bar{E}$, such that:

(1) for every $t \in I$, $w(x,t)$ belongs to $C^\infty(\bar{A})$;

(2) $D_x^p w(x,t)$ $(|p| \geqq 0)$ belongs to $C^0(\bar{E})$.

Let us observe that w belongs to $W(E)$ when and only when, for any given ν, we have $w(x,t) \in H_\nu(A)$, for every $t \in \bar{I}$, and $\lim_{t \to t_0} \|w(x,t) - w(x,t_0)\|_{\nu,A} = 0$. In fact, it is obvious that, if $w \in W(E)$, then the above conditions are satisfied. Conversely, if $w(x,t) \in H_\nu(A)$, for any ν, then $w(x,t) \in C^\infty(\bar{A})$. Let l be an integer such that $l > r/2$. We have (Theorem 2.VI):

$$|(D^p w)_{x,t} - (D^p w)_{x^0,t_0}| \leqq c\,\|w(x,t) - w(x,t_0)\|_{p+l,A} + |(D^p w)_{x,t_0} - (D^p w)_{x^0,t_0}|,$$

which proves the continuity of $D^p w$ in $\bar{E}$.

Let $K_\alpha(t, \tau)$ $(0 \leqq |\alpha| \leqq 2m)$ be a complex $n \times n$ matrix, continuous in $\bar{I} \times \bar{I}$. We shall consider the following problem:

(P_K) *Given $f(x, t) \in W(E)$, find a function $u(x, t) \in W(E)$ satisfying the integro-differential equation*

$$L u(x, t) + \sum_{|\alpha|}^{0,2m} \int_0^t K_\alpha(t, \tau)\, D^\alpha u(x, \tau)\, d\tau = f(x, t)$$

in E and such that, for every $t \in \bar{I}$, $u(x, t) \in \mathscr{U}_V$.

The proof of existence and uniqueness of a solution of problem (P_K) is a consequence of the theory of Volterra integral equations in Banach spaces. Let S be a complex Banach space and $K(t, \tau)$ a function whose values are linear bounded operators from S into S. Let $K(t, \tau)$ be continuous in $\bar{I} \times \bar{I}$. Let $v(t)$ and $h(t)$ be functions with values in S and continuous in $\bar{I}$. The Volterra integral equation

$$v(t) = \int_0^t K(t, \tau)\, v(\tau)\, d\tau + h(t) \text{ [13]} \tag{9.1}$$

has one and only one solution u continuous in $\bar{I}$; v is the limit of the sequence $\{v_k(t)\}$, uniformly convergent in $\bar{I}$, obtained by the method of successive approximation:

$$v_0(t) = h(t), \qquad v_k(t) = \int_0^t K(t, \tau)\, v_{k-1}(\tau)\, d\tau + h(t).$$

The proof of the convergence of $\{v_k(t)\}$ and of the uniqueness of the solution proceeds exactly the same as in the classical scalar case.

We can apply the general results just stated to the proof of the following theorem:

9.I *Problem* (P_K) *has one and only one solution.*

Let G be the operator already considered in the previous sections. We claim that problem (P_K) has a solution when and only when there exists $\varphi(x, t) \in W(E)$ satisfying the equation

$$(-1)^m \varphi(x, t) + \sum_{|\alpha|}^{0,2m} \int_0^t K_\alpha(t, \tau)\, D^\alpha G[\varphi(\xi, \tau)]\, d\tau = f(x, t). \tag{9.2}$$

In fact, if u is a solution of problem (P_K), then $\varphi(x, t) = (-1)^m L u$ satisfies (9.2). Conversely, if $\varphi(x, t)$ is a solution of (9.2), then $u(x, t) = G[\varphi(\xi, t)]$ is a solution of problem (P_K). Consider any space $H_\nu(A)$ with $\nu \geqq 2m$ and set $S = H_\nu(A)$, $v(t) = \varphi(x, t)$, $h(t) = (-1)^m f(x, t)$,

$$K(t, \tau)\, v(\tau) = (-1)^m \sum_{|\alpha|}^{0,2m} K_\alpha(t, \tau)\, D^\alpha G[\varphi(\xi, \tau)].$$

Thus (9.2) reduces to (9.1). Hence (9.2) has one and only one solution belonging to $H_\nu(A)$, for every $t \in \bar{I}$, and such that $\lim_{t \to t_0} \|\varphi(x, t) - \varphi(x, t_0)\|_{\nu, A} = 0$. Since ν is arbitrary, $\varphi \in W(E)$.

10. Classical boundary value problems for a scalar 2nd order elliptic operator. In this section we shall consider only real-valued functions. Let L be a 2nd order linear elliptic operator (with real coefficients)

$$L u \equiv \frac{\partial}{\partial x_i}\left[a_{ij}(x) \frac{\partial u}{\partial x_j}\right] + b_i(x) \frac{\partial u}{\partial x_i} + c u \qquad [a_{ij}(x) \equiv a_{ji}(x);\ a_{ij}(x)\, \xi_i \xi_j > 0].$$

[13] For the definition and the properties of the integral of a continuous function with values in a Banach space, see [*16*, Chap. III, Sect. 1].

We assume that the coefficients a_{ij}, b_i, c belong to C^∞. Let A be a bounded domain, which we suppose C^∞-smooth. We shall consider for L some classical boundary value problems. We restrict ourselves to list only the results of the "C^∞-theory" leaving to the reader to derive, from the general statements of Sect. 6, other particular cases, concerning this particular operator L.

(I) *Dirichlet problem: Given $f(x)\in C^\infty(\bar{A})$, find $u\in C^\infty(\bar{A})$ such that $Lu=f$ in A, $u=0$ on ∂A.*

Let us consider problem (P) of Sect. 6, assuming as V the space $\overset{\circ}{H}_1(A)$. In order to apply the results of Theorems 6.II and 6.VI we need only to prove inequality (6.1), i.e.

$$-B(v,v)=\int_A\left[a_{ij}\frac{\partial v}{\partial x_i}\frac{\partial v}{\partial x_j}+\left(\frac{1}{2}\frac{\partial b_i}{\partial x_i}-c\right)v^2\right]dx\geqq\gamma_0\|v\|_1^2 \tag{10.1}$$

for every $v\in\overset{\circ}{H}_1(A)$. Inequality (10.1) holds if

$$\frac{1}{2}\frac{\partial b_i}{\partial x_i}-c\geqq 0. \tag{10.2}$$

Then, because of Theorem 6.IV, if there is a uniqueness theorem for the Dirichlet problem, there is an existence theorem. We have uniqueness not only when (10.2) is satisfied but also when $c(x)\leqq 0$ in A. In fact, in that case we have, for any solution of the homogeneous equation $Lu=0$: $|u(x)|\leqq\max_{x\in\partial A}|u(x)|$ *(maximum principle)*, which implies the uniqueness of the solution of the Dirichlet problem[14].

(II) *Neumann problem: Given $f(x)\in C^\infty(\bar{A})$, find $u\in C^\infty(\bar{A})$ such that $Lu=f$ in A, $a_{ij}\nu_i u_{x_j}=0$ on ∂A [$\nu\equiv(\nu_1,\ldots,\nu_r)$ being the inward unit normal to ∂A].*

Now we consider problem (P), assuming as V the space $H_1(A)$. Assume

$$B(u,v)=\int_A\left[-a_{ij}\frac{\partial v}{\partial x_i}\frac{\partial v}{\partial x_j}+v b_i\frac{\partial u}{\partial x_i}+cuv\right]dx.$$

Suppose that for $x\in\bar{A}$: $a_{ij}\xi_i\xi_j>p_0|\xi|^2$, $|b_i|\leqq p_1$ $(i=1,\ldots,r)$, $-c\geqq p_2$. Then for any $v\in H_1(A)$ and p_2 large enough,

$$-B(v,v)\geqq\int_A\left(p_0\sum_{i=1}^r\left(\frac{\partial v}{\partial x_i}\right)^2-p_1|v|\sum_{i=1}^r\left|\frac{\partial v}{\partial x_i}\right|+p_2|v|^2\right)dx\geqq c_0\|v\|_1^2.$$

As in the preceding case, we may conclude existence is implied by uniqueness, Eq. (6.3) in this particular case becomes

$$\int_{\partial A} v\, a_{ij}\nu_i u_{x_j}\,d\sigma=0$$

which must be satisfied for every $v\in H_1(A)$. Then $a_{ij}\nu_i u_{x_j}=0$ on ∂A.

Let us consider the particular case $L\equiv\Delta_2\equiv\sum_{i=1}^r\frac{\partial^2}{\partial x_i^2}$. In this case $B(v,v)=-\int_A v_{x_i}v_{x_i}\,dx$ is symmetric. Hence the solution exists when and only when $(f,\varphi)_0=0$, where φ is a solution of the homogeneous problem $\Delta_2\varphi=0$ in A. $\varphi_{x_i}\nu_i=0$ on ∂A. Hence, in the case of the Laplace operator Δ_2, the Neumann problem has a solution if and only if $(f,1)_0=0$.

[14] See [*26*, p. 5].

Let $\lambda \equiv (\lambda_1, \ldots, \lambda_r)$ be a real unit vector defined on ∂A and which is C^∞ when considered as a function of the point x varying on ∂A. The following boundary value problem is known as the *oblique derivative problem* for the operator L:

$$Lu = f \quad \text{in} \quad A, \tag{10.3}$$

$$\frac{\partial u}{\partial \lambda} = 0 \quad \text{on} \quad \partial A. \tag{10.4}$$

Under the further assumption $\lambda \nu > 0$ the problem is known as the

(III) *Regular oblique derivative problem*: Let ϱ_{ij} be arbitrary functions, belonging to $C^\infty(\bar{A})$, such that $\varrho_{ij} = -\varrho_{ji}$. The operator L can be written as follows:

$$Lu = \frac{\partial}{\partial x_i}\left[\alpha_{ij}\frac{\partial u}{\partial x_j}\right] + \beta_i \frac{\partial u}{\partial x_i} + cu,$$

where $\alpha_{ij} = a_{ij} + \varrho_{ij}$, $\beta_i = b_i + \dfrac{\partial \varrho_{ij}}{\partial x_j}$. It is possible to choose the functions ϱ_{ij} in such a way that

$$\alpha_{ij}\nu_i = \varrho \lambda_j \qquad (j = 1, \ldots, r), \tag{10.5}$$

ϱ being a positive C^∞ function defined on ∂A.

Let us assume on ∂A

$$\varrho_{ij} = \frac{1}{\lambda \nu} a_{hk}(\nu_i \lambda_j - \nu_j \lambda_i)\nu_h \nu_k - a_{hj}\nu_h \nu_i + a_{hi}\nu_h \nu_j \text{ [15]}$$

and continue this function throughout $\bar{A}$ in such a way that the continued function belongs to $C^\infty(\bar{A})$ and $\varrho_{ij} = -\varrho_{ji}$. Condition (10.5) is satisfied with $\varrho = (\lambda\nu)^{-1} a_{ij}\nu_i\nu_j$

Let us now consider for $u, v \in H_1(A)$ the bilinear form

$$B(u, v) = \int_A \left\{-\alpha_{ij}\frac{\partial v}{\partial x_i}\frac{\partial u}{\partial x_j} + \beta_i v \frac{\partial u}{\partial x_i} + cvu\right\} dx.$$

Assume that $c < 0$ in $\bar{A}$. Let p_0, p_1, p_2 be positive numbers such that for every $x \in \bar{A}$: $\alpha_{ij}(x)\xi_i\xi_j \geqq p_0|\xi|^2$, $|\beta_i| \leqq p_1$ $(i = 1, \ldots, r)$, $-c \geqq p_2$. Then:

$$-B(v, v) \geqq \int_A \left(p_0 \sum_{i=1}^r \left|\frac{\partial v}{\partial x_i}\right|^2 - p_1|v| \sum_{i=1}^r \left|\frac{\partial v}{\partial x_i}\right| + p_2|v|^2\right) dx.$$

Hence, *if p_2 is large enough*, $-B(v, v) \geqq c_0\|v\|_1^2$, for every $v \in H_1(A)$. It follows that a function $u \in H_1(A)$ exists which satisfies the equation $B(u, v) = (f, v)_0$ for any $v \in V \equiv H_1(A)$. Since we can apply the regularization theory, we deduce that $u \in C^\infty(\bar{A})$. Because of (10.5) we have

$$B(u, v) = \int_A (Lu)\, v\, dx + \int_{\partial A} v\varrho \frac{\partial u}{\partial \lambda} d\sigma.$$

Hence, due to the arbitrariness of v, the function u is a solution of (10.3), (10.4). On the other hand, a solution of $u \in C^\infty(\bar{A})$ of the regular oblique derivative problem exists if uniqueness holds. We have such uniqueness if $c(x) < 0$ in $\bar{A}$. In fact, if $Lu = 0$ in A and $\partial u/\partial \lambda = 0$ on ∂A, the function u, because of a theorem of GIRAUD and HOPF[16], cannot have any maximum or minimum on ∂A unless

[15] This definition of ϱ_{ij} which simplifies the one considered in [7], was suggested to me by H. WEINBERGER.

[16] See [26, p. 5].

it is a constant. Since

$$|u(x)| \leq \max_{x \in \partial A} |u|, \quad c < 0,$$

it must be $u \equiv 0$ in A.

Let us now suppose that $\partial_1 A$ and $\partial_2 A$ are two disjoint subsets of ∂A such that if x^0 is any point of $\partial_k A$ $(k = 1, 2)$ the neighborhood I of x^0 considered for insuring the C^∞-smoothness of ∂A at x^0 can be choosen in such a way that $I \cap \partial A \subset \partial_k A$. Moreover we assume that $\overline{\partial_1 A} = \partial A - \partial_2 A$. Let λ be the C^∞ unit vector defined on ∂A (already introduced in the oblique derivative problem) which satisfies the regularity condition $\lambda \nu > 0$ on ∂A. The following problem is known as the

(IV) *Mixed boundary value problem*:

$$Lu = f \quad in\ A, \qquad u = 0 \quad on\ \partial_1 A, \qquad \frac{\partial u}{\partial \lambda} = 0 \quad on\ \partial_2 A.$$

Now, as the space V we take the closure in the space $H_1(A)$ of the subclass of $C^1(\bar{A})$ constituted by the functions which vanish on $\overline{\partial_1 A}$. For any x^0 lying either in $\partial_1 A$ or in $\partial_2 A$, the condition $(\mathrm{IV})_{x^0}$ of Sect. 6 is satisfied with $\nu = \infty$.

We use for Lu the same representation as in the oblique derivative problem with the α_{ij}'s satisfying (10.5). By the argument used in the oblique derivative problem, we see that $-B(v, v) \geqq c_0 \|v\|_1^2$ for any $v \in V$, when $-c \geqq p_2$ with p_2 large enough. Then we have a unique solution of the boundary value problem corresponding to the present choice of V. The solution of the problem can be regularized in the neighborhood of any point either of $\partial_1 A$ or of $\partial_2 A$. Hence the solution u has the following regularity properties:

(i) u belongs to $C^\infty(A) \cap H_1(A)$;

(ii) u belongs to $C^\infty(A \cup \partial_1 A \cup \partial_2 A)$.

If $\partial A = \partial_1 A \cup \partial_2 A$, then u is C^∞ in $\bar{A}$; otherwise the only points where u could fail to be C^∞ are in the set $\partial A - (\partial_1 A \cup \partial_2 A)$. By standard arguments we see that the boundary conditions are satisfied in $\partial_1 A$ and in $\partial_2 A$ respectively.

We leave it to the reader to state results concerning propagation problems, diffusion problems and hereditary problems connected with a scalar second order elliptic operator, by specialization from the general theory developed in Sects. 7, 8, 9.

11. Equilibrium of a thin plate. The classical theory of the equilibrium of a thin plate requires the solution of certain boundary value problems for the iterated Laplace operator in two real variables x, y:

$$\Delta_4 u \equiv \Delta_2 \Delta_2 u = \frac{\partial^4 u}{\partial x^4} + 2 \frac{\partial^4 u}{\partial x^2 \partial y^2} + \frac{\partial^4 u}{\partial y^4}$$

with several kinds of boundary conditions. Let us suppose that A be a C^∞-smooth bounded plane domain. The theory of thin plates considers the following boundary conditions on ∂A:

$$u = 0, \tag{11.1}$$

$$\frac{\partial u}{\partial \nu} = 0, \tag{11.2}$$

$$\frac{\partial^2 u}{\partial \nu^2} + \sigma \left(\frac{\partial^2 u}{\partial s^2} - \frac{1}{\varrho} \frac{\partial u}{\partial \nu} \right) = 0, \tag{11.3}$$

$$\frac{\partial}{\partial \nu} \Delta_2 u + (1 - \sigma) \frac{\partial}{\partial s} \frac{\partial^2 u}{\partial \nu \partial s} = 0. \tag{11.4}$$

Here ν is the unit inward normal to ∂A; $\partial/\partial s$ denotes differentiation with respect to the arc (increasing counter clock-wise); $1/\varrho$ is the curvature of ∂A; σ is a constant such that $-1<\sigma<1$. The differential equation to be considered is the following:

$$\Delta_4 u = f \tag{11.5}$$

where u and f are real-valued functions. The boundary value problem (11.5), (11.1), (11.2) corresponds to the equilibrium problem for a plate clamped along its boundary. The boundary conditions (11.1) and (11.3) express the fact that the plate is supported along its edge. The boundary conditions (11.3) and (11.4) mean that the part of the boundary where these conditions are satisfied is free. We restrict ourselves to the consideration of the boundary value problem (11.5), (11.1), (11.2) and the mixed boundary value problem for a partially clamped plate, i.e. when (11.1) and (11.2) are satisfied on a part $\partial_1 A$ of ∂A and (11.3), (11.4) on a part $\partial_2 A$, where $\partial_1 A$ and $\partial_2 A$ are the subsets of ∂A considered in Sect. 10. We leave it to the reader to apply the general theory so as to derive results concerning further boundary conditions, for instance the ones corresponding to a plate partially clamped on ∂A, partially supported and partially free. As a bilinear form for the boundary value problem (11.5), (11.1), (11.2), we take

$$B(u,v) = \int_A \left(\frac{\partial^2 u}{\partial x^2}\frac{\partial^2 v}{\partial x^2} + 2\frac{\partial^2 u}{\partial x\,\partial y}\frac{\partial^2 v}{\partial x\,\partial y} + \frac{\partial^2 u}{\partial y^2}\frac{\partial^2 v}{\partial y^2}\right) dx\,dy.$$

The subspace V of $H_2(A)$ to be considered in $\mathring{H}_2(A)$. In this case inequality (6.1) reduces to inequality (2.6) for $m=2$. Thus $B(v,v) \geqq c_0 \|v\|_2^2$ $(v \in \mathring{H}_2(A))$. All the hypotheses (I), (II), (III), $(\mathrm{IV})_{x^0}$ (for any $x^0 \in \partial A$) of Sect. 6 are satisfied, so

11.I *If* $f \in C^\infty(\bar{A})$, *the boundary value problem* (11.5), (11.1), (11.2) *has one and only one solution belonging to* $C^\infty(\bar{A})$.

In order to consider the above-mentioned mixed boundary value problem [i.e. conditions (11.1), (11.2) on $\partial_1 A$ and conditions (11.3), (11.4) on $\partial_2 A$] it is helpful to remark that, for u and v belonging to $C^\infty(\bar{A})$, we have

$$\begin{aligned}
&\int_{\partial A} v\left[\frac{\partial}{\partial \nu}\Delta_2 u + (1-\sigma)\frac{\partial}{\partial s}\frac{\partial^2 u}{\partial \nu\,\partial s}\right] ds - \int_{\partial A} \frac{\partial v}{\partial \nu}\left[\frac{\partial^2 u}{\partial \nu^2} + \sigma\left(\frac{\partial^2 u}{\partial s^2} - \frac{1}{\varrho}\frac{\partial u}{\partial \nu}\right)\right] ds \\
&= \int_A \left\{\frac{\partial^2 u}{\partial x^2}\frac{\partial^2 v}{\partial x^2} + (2-2\sigma)\frac{\partial^2 u}{\partial x\,\partial y}\frac{\partial^2 v}{\partial x\,\partial y} + \frac{\partial^2 u}{\partial y^2}\frac{\partial^2 v}{\partial y^2}\right. \\
&\quad \left. + \sigma\left[\frac{\partial^2 u}{\partial x^2}\frac{\partial^2 v}{\partial y^2} + \frac{\partial^2 u}{\partial y^2}\frac{\partial^2 v}{\partial x^2}\right]\right\} dx\,dy - \int_A v\,\Delta_4 u\,dx\,dy.
\end{aligned} \tag{11.6}$$

Let us now take as the space V the subspace of $H_2(A)$ composed by the functions which satisfy conditions (11.1), (11.2) on $\partial_1 A$. Set

$$\begin{aligned}
B(u,v) = \int_A &\left\{\frac{\partial^2 u}{\partial x^2}\frac{\partial^2 v}{\partial x^2} + (2-2\sigma)\frac{\partial^2 u}{\partial x\,\partial y}\frac{\partial^2 v}{\partial x\,\partial y} + \frac{\partial^2 u}{\partial y^2}\frac{\partial^2 v}{\partial y^2}\right. \\
&\left. + \sigma\left[\frac{\partial^2 u}{\partial x^2}\frac{\partial^2 v}{\partial y^2} + \frac{\partial^2 u}{\partial y^2}\frac{\partial^2 v}{\partial x^2}\right]\right\} dx\,dy.
\end{aligned}$$

Because of the assumption $-1<\sigma<1$, we have

$$B(v,v) \geqq c(\sigma) \sum_{|p|=2} \int_A |D^p v|^2\,dx\,dy.$$

The constant $c(\sigma)$ depends only on σ. In order to prove (6.1) we need only to show that there exists $c_1 > 0$ such that, for any $v \in V$

$$\sum_{|p|=2} \int_A |D^p v|^2 \, dx \, dy \geqq c_1 \|v\|_2^2. \tag{11.7}$$

Suppose (11.7) to be false. Then there exists $\{v_n\} \in V$ such that

$$\|v_n\|_2 = 1, \tag{11.8}$$

$$\sum_{|p|=2} \int_A |D^p v_n|^2 \, dx \, dy < \frac{1}{n}. \tag{11.9}$$

We can suppose that $\{v_n\}$ converges in $H_1(A)$ (Theorem 2.IV). Then, because of (11.9), $\{v_n\}$ converges in $H_2(A)$, and the limit function has strong second derivatives vanishing on A. It follows readily that v is a polynomial of degree 1.[17]

Since v belongs to V, $v \equiv 0$ in A. That contradicts (11.8). Since (6.1) has been proved, and hypotheses (I), (II), (III) of Sect. 6 are satisfied, there exists one and only one solution u of the equations $B(u, v) = (f, v)_0$, for $v \in V$, belonging to V. Since hypothesis $(\text{IV})_{x^0}$ is satisfied for $x^0 \in \partial_1 A$ and $x^0 \in \partial_2 A$, u belongs to $C^\infty(A \cup \partial_1 A \cup \partial_2 A)$. By using (11.6), we deduce that u is the solution of our mixed boundary value problem.

11.II *The mixed boundary value problem* (11.5), (11.1), (11.2) *on* $\partial_1 A$; (11.3), (11.4) *on* $\partial_2 A$, *with* $f \in C^\infty(\bar{A})$, *has one and only one solution belonging to*

$$C^\infty(A \cup \partial_1 A \cup \partial_2 A) \cap H_2(A) \cap C^0(\bar{A}).$$

The fact that u belongs to $C^0(\bar{A})$ is a consequence of Lemma 2.VI.

We leave it to the reader to state the results concerning the particular cases of problems (P_λ), (P_{tt}), (P_t) and (P_K), connected, for instance, with a clamped plate.

---

[17] Let be $u \in H_m(A)$ and $D^s u = 0$ for $|s| = m$. For any $w \in \overset{\circ}{C}{}^\infty(A)$ we have

$$\int_A D^s u \, D^s w \, dx = (-1)^m \int_A u \, D^s D^s w \, dx = 0.$$

Since the operator $L(w) = D^s D^s w$ is elliptic, from Theorem 3.II we deduce that $u \in C^\infty(A)$ and, in consequence, u must be a polynomial of degree $m - 1$.

Let us note the following lemma, which will be used in the sequel:

Let A be a properly regular domain. For any $u \in H_m(A)$ $(m \geqq 1)$ such that:

$$\int D^s u \, dx = 0, \qquad 0 \leqq |s| \leqq m-1,$$

the following inequality holds:

$$\|u\|_{m,A}^2 \leqq c \sum_{|s|=m} \int_A |D^s u|^2 \, dx \qquad (c > 0). \tag{*}$$

Suppose the inequality not true. Then there exists a sequence $\{u^{(n)}\}$ of functions of $H_m(A)$ such that

$$\|u^{(n)}\|_{m,A} = 1, \quad \lim_{n \to \infty} \sum_{|s|=m} \int_A |D^s u^{(n)}|^2 \, dx = 0.$$

We may assume that $u^{(n)}$ converges in $H_{m-1}(A)$ (Theorem 2.IV). Let u be the limit of $\{u^{(n)}\}$ in $H_{m-1}(A)$. We have

$$\lim_{n \to \infty} \left(\|u^{(n)} - u\|_{m-1,A}^2 + \sum_{|s|=m} \int_A |D^s u^{(n)}|^2 \, dx \right) = 0.$$

Then $u^{(n)}$ converges to u in $H_m(A)$, and $D^s u = 0$ for $|s| = m$. It follows that u is a polynomial of degree $m - 1$ which satisfies the conditions

$$\|u\|_{m-1,A}^2 = 1, \qquad \int_A D^s u \, dx = 0, \qquad |s| < m.$$

That is impossible.

12. Boundary value problems of equilibrium in linear elasticity. We shall now consider the classical boundary value problems of linear elasticity in the case of an inhomogeneous anisotropic body. In order to include both the cases of physical interest, of plane and 3-dimensional elasticity, it is convenient to study these problems in the space X^r. In this section we shall consider r-vector valued functions $u, v, f, \ldots$ with real components. Set

$$\varepsilon_{ih} = 2^{-1}\left(\frac{\partial u_i}{\partial x_h} + \frac{\partial u_h}{\partial x_i}\right) \; (i, h = 1, \ldots, r).$$

Let us consider the *elastic potential*

$$W(x, \varepsilon) = \sum_{i \leqq h}^{1,r} \sum_{j \leqq k}^{1,r} \alpha_{ih,jk}(x)\, \varepsilon_{ih}\, \varepsilon_{jk}.$$

The (real valued) functions $\alpha_{ih,jk}$ are supposed to belong to C^∞, and the quadratic form $W(x, \varepsilon)$ is supposed to be positive definite in the $\frac{r(r+1)}{2}$ variables ε_{ih} $(1 \leqq i \leqq h \leqq r)$, for any $x \in X^r$. We can assume that $\alpha_{ih,jk}(x) \equiv \alpha_{jk,ih}(x)$. Let us now define for i, h, j, k arbitrary values of the indeces $1, \ldots, r$

$$a_{ihk}(x)_{,j} = \begin{cases} \alpha_{ih,jk}(x) & \text{for } i \leqq h,\ j < k;\ i < h,\ j \leqq k; \\ \alpha_{hi,jk}(x) & \text{for } i > h,\ j \leqq k; \\ \alpha_{ih,kj}(x) & \text{for } i \leqq h,\ j > k; \\ \alpha_{hi,kj}(x) & \text{for } i > h,\ j > k; \\ 2\alpha_{ih,jk}(x) & \text{for } i = h,\ j = k. \end{cases}$$

We have $a_{ih,jk}(x) \equiv a_{jk,ih}(x) \equiv a_{hi,jk}(x) \equiv a_{ih,kj}(x)$. Moreover

$$W(x, \varepsilon) = \frac{1}{2}\, a_{ih,jk}\, \varepsilon_{ih}\, \varepsilon_{jk} = \frac{1}{2}\, a_{ih,jk} \frac{\partial u_i}{\partial x_h} \frac{\partial u_j}{\partial x_k}. \tag{12.1}$$

It must be pointed out that the quadratic form $a_{ih,jk}\,\eta_{ih}\,\eta_{jk}$ is *not* positive; rather, as a function of the r^2 real variables η_{ih} $(i, h = 1, \ldots, r)$ it is only non-negative. It is positive, in general, only on the subspace of the r^2-dimensional space of the η_{ih}'s defined by the conditions $\eta_{ih} = \eta_{hi}$.

Of particular interest is the case when

$$a_{ih,jk}(x) \equiv \delta_{ij}\, \delta_{hk} + \delta_{ik}\, \delta_{hj} + (\nu - 1)\, \delta_{ih}\, \delta_{jk}, \tag{12.2}$$

where ν is a constant. This is the case of classical linear elasticity for a homogeneous isotropic body. Since here we have $W(\varepsilon) = \varepsilon_{ih}\, \varepsilon_{ih} + 2^{-1}(\nu - 1)\, \varepsilon_{ii}\, \varepsilon_{hh}$, W is positive if and only if $\nu > r^{-1}(r-2)$.

Let A be a bounded domain of X^r, which we suppose C^∞-smooth. The equations of equilibrium for an elastic body whose *natural configuration* coincides with A are the following

$$\frac{\partial}{\partial x_h} \frac{\partial}{\partial \varepsilon_{ih}} W(x, \varepsilon) + f_i = 0 \quad \text{in } A. \tag{12.3}$$

We have three kinds of boundary conditions corresponding to the three main problems of elasticity. We consider here only homogeneous boundary conditions.

1st Boundary Value Problem (body fixed along its boundary)

$$u = 0 \quad on\ \partial A. \tag{12.4}$$

2nd Boundary Value Problem (body free along its boundary)

$$t_i(u) \equiv \nu_h \frac{\partial}{\partial \varepsilon_{ih}} W(x, \varepsilon) = 0 \quad on\ \partial A \tag{12.5}$$

(ν is the unit inward normal to ∂A).

3rd Boundary Value Problem (mixed boundary value problem)

$$u = 0 \quad on\ \partial_1 A, \tag{12.6}$$

$$t(u) = 0 \quad on\ \partial_2 A, \tag{12.7}$$

where $\partial_1 A$ and $\partial_2 A$ are the subsets of ∂A already introduced in Sect. 10.

Other boundary value problems could be considered. For instance, the ones assigning p components of u and $r-p$ components of $t(u)$ on ∂A. However we shall restrict ourselves to the three above-considered cases just stated and leave it as an exercise for the reader to study other boundary value problems for (12.3). Eq. (12.3) can be written in the form

$$\frac{\partial}{\partial x_h}\left[a_{ih,jk}(x)\frac{\partial u_j}{\partial x_k}\right] + f_i = 0. \tag{12.3'}$$

In order to prove the existence and the uniqueness of the solution of the boundary value problem (12.3), (12.4), we need to prove inequality (6.1), for any $v \in \mathring{H}_1(A)$. Inequality (6.1) in this particular case becomes

$$\int_A a_{ih,jk}(x)\frac{\partial v_i}{\partial x_h}\frac{\partial x_h}{\partial v_i}\,dx \geqq c_0 \|v\|_1^2 \quad (v \in \mathring{H}_1(A)). \tag{12.8}$$

Inequality (12.8), because of (12.1), is equivalent to the following one:

$$\int_A \sum_{i,h}^{1,r}\left(\frac{\partial v_i}{\partial x_h} + \frac{\partial v_h}{\partial x_i}\right)^2 dx \geqq c_1 \|v\|_1^2 \quad (v \in \mathring{H}_1(A)), \tag{12.9}$$

which is known as *Korn's first inequality*. Inequality (12.9) is immediately obtained by using the Fourier developments of the functions v_i and Parseval's theorem. Thus we get the following theorem.

12.I *Given $f \in C^\infty(\bar{A})$, there exists one and only one solution of the boundary value problem* (12.3), (12.4), *which belongs to $C^\infty(\bar{A})$.*

Let us remark that as consequence of (12.8), we have for any $x \in \bar{A}$, any non-zero real ξ and any non-zero real η: $a_{ih,jk}(x)\,\xi_h\,\xi_k\,\eta_i\,\eta_j > 0$ (see Theorem 5.II). Thus, as a consequence of the positiveness of the elastic potential $W(x, \varepsilon)$, the operator of elasticity is strongly elliptic.

In order to prove the existence theorem for the boundary value problem (12.3), (12.5), let us consider the system

$$\frac{\partial}{\partial x_h}\left[a_{ih,jk}(x)\frac{\partial u_j}{\partial x_k}\right] - p_0\,u_i + f_i = 0, \tag{12.10}$$

where p_0 is an arbitrarily fixed positive constant. We wish first to give an existence theorem for the problem (12.10), (12.5). It is easily seen that the inequality to be proven in this case is the following *(Korn's second inequality)*:

$$\int_A \sum_{i,h}^{1,r}\left(\frac{\partial v_i}{\partial x_h} + \frac{\partial v_h}{\partial x_i}\right)^2 dx + \int_A |v|^2\,dx \geqq c_2 \|v\|_1^2, \tag{12.11}$$

for any $v \in H_1(A)$. The proof of this inequality is anything but trivial. We shall present here a reasonably simple proof. To this end, let us suppose that the bounded domain A satisfies the following "restricted cone-hypothesis": *let $C_0(\Gamma_s, R)$ $(s=1, \ldots, q)$ be cones whose vertex is the origin 0 and let Γ_s be open sets relative to the sphere $|x|=1$. There exists an open covering of $\bar{A}$ (by open sets $I_1, \ldots, I_q$) such that, for any $x \in \bar{A} \cap I_s$, the cone $C_x(\Gamma_s, R)$ is contained in A.* We have the following theorem regarding Korn's second inequality.

12.II *If the bounded domain A satisfies the restricted cone-hypothesis, inequality* (12.11) *holds for any $v \in H_1(A)$.*

Let $\{J\}$ be the set of all the open balls such that:

(i) the center x of J is in $\bar{A}$;

(ii) the radius of J is less than $\frac{1}{2}R$;

(iii) J is contained in some open set I_s.

Let $J_1, J_2, \ldots, J_m$ be a finite covering of $\bar{A}$ extracted from $\{J\}$. Let $\sum_{h=1}^{m} \varphi_h^2(x) = 1$ be a partition of unity with $\varphi_h(x) \in C^\infty$ and $spt\ \varphi_h \subset J_h$ $(h=1, \ldots, m)$. Denoting by $/k$ differentiation with respect to x_k, we have

$$\begin{aligned}\int_A v_{i/k}\, v_{i/k}\, dx &= \sum_{h=0} \int_A \varphi_h^2\, v_{i/k}\, v_{i/k}\, dx \\ &= \int_A (\varphi_h v_i)_{/k} (\varphi_h v_i)_{/k}\, dx = \int_A \varphi_{h/k} v_i\, \varphi_{h/k} v_i\, dx - 2\int_A \varphi_h\, \varphi_{h/k}\, v_i\, v_{i/k}\, dx.\end{aligned} \tag{12.12}$$

Suppose that for any $v \in H_1(A)$ and any h we have proven that a constant $c_4 > 0$ exists such that

$$\int_A (\varphi_h v_i)_{/k} (\varphi_h v_i)_{/k}\, dx \leqq c_4 \sum_{i,k}^{1,r} \int_A [(\varphi_h v_i)_{/k} + (\varphi_h v_k)_{/i}]^2\, dx. \tag{12.13}$$

If we set:

$$|||v^2||| = \sum_{i,h}^{1,r} \int_A (v_{i/h} + v_{h/i})^2\, dx + \int_A |v|^2\, dx,$$

we deduce from (12.12) and (12.13) that $\|v\|_1^2 \leqq c_5\, |||v^2|||$, which is the inequality (12.11) to be proved. Set $\varphi_h v_i = u_i$. Since, under the assumed hypotheses on A, we have $H_1(A) = \overline{C^\infty(\bar{A})}$ [the closure of $C^\infty(\bar{A})$ being understood in the topology of $H_1(A)$][18], we may assume that $u_i \in C^\infty(\bar{A})$. Set $J_h = J$, $\Gamma_s = \Gamma$. For any $x \in J \cap \bar{A}$,

$$u(x) = \int_0^R u_{\varrho\varrho}(\varrho, \omega)\, \varrho\, d\varrho = \int_0^R u_{y_h y_k}(y)\, \frac{(y_h - x_h)(y_k - x_k)}{|y-x|}\, d\varrho, \tag{12.14}$$

where ω is any fixed point on the set Γ of the unit sphere and $\varrho = |y-x|$. Let $\psi(\omega)$ be a real function, which, as a function of the point ω of the unit sphere Ω, be of class C^∞ on Ω and be such that

$$spt\, \psi \in \Gamma, \qquad \int_\Omega \psi(\omega)\, d\omega = 1.$$

From (12.14), by multiplying for $\psi(\omega)$ and integrating over Γ, we deduce

$$u(x) = \int_{C_x(\Gamma, R)} u_{y_h y_k}(y)\, \frac{(y_h - x_h)(y_k - x_k)}{|y-x|^r}\, \psi\left(\frac{y-x}{|y-x|}\right) dy.$$

[18] See [*1*, p. 11].

Since

$$2u_{l/hk} = (u_{l/k} + u_{k/l})_{/h} + (u_{l/h} + u_{h/l})_{k} - (u_{h/k} + u_{k/h})_{/l},$$

we get

$$u_l(x) = \int_{C_x(\Gamma, R)} [\varepsilon_{lk/h}(y) + \varepsilon_{lh/k}(y) - \varepsilon_{hk/l}(y)] \frac{(y_h - x_h)(y_k - x_k)}{|y-x|^r} \psi\left(\frac{y-x}{|y-x|}\right) dy.$$

Set

$$M_{hk}(y-x) = \frac{(y_h - x_h)(y_k - x_k)}{|y-x|^r} \psi\left(\frac{y-x}{|y-x|}\right); \qquad \alpha_l(x) = \int_{C_x(\Gamma, R)} \varepsilon_{lk/h}(y)\, M_{hk}(y-x)\, dy$$

$$\beta_l(x) = \int_{C_x(\Gamma, R)} \varepsilon_{lh/k}(y)\, M_{hk}(y-x)\, dy; \qquad \gamma_l(x) = \int_{C_x(\Gamma, R)} \varepsilon_{hk/l}(y)\, M_{hk}(y-x)\, dy;$$

$$H_{jhk}(y-x) = \frac{\partial}{\partial x_j} M_{hk}(y-x); \qquad \frac{\partial}{\partial y_h} H_{jhk}(y-x) = S_{jk}(y-x).$$

Then

$$\frac{\partial}{\partial x_j} \alpha_l(x) = \int_{C_x(\Gamma, R)} \varepsilon_{lk/h}(y)\, H_{jhk}(y-x)\, dy = \lim_{\varepsilon \to 0} \int_{C_x(\Gamma, R) - C_x(\Gamma, \varepsilon)} \varepsilon_{lk/h}(y)\, H_{jhk}(y-x)\, dy$$

$$= -\lim_{\varepsilon \to 0} \int_{C_x(\Gamma, R) - C_x(\Gamma, \varepsilon)} \varepsilon_{lk}(y) \frac{\partial}{\partial y_h} H_{jhk}(y-x)\, dy$$

$$-\lim_{\varepsilon \to 0} \int_{\Gamma} \varepsilon_{lk}(x + \varepsilon\omega)\, H_{jhk}(\omega)\, \omega_h\, d\omega$$

$$= -\lim_{\varepsilon \to 0} \int_{C_x(\Gamma, R) - C_x(\Gamma, \varepsilon)} \varepsilon_{lk}(y)\, S_{jk}(y-x)\, dy - \varepsilon_{lk}(x) \int_{\Gamma} H_{jhk}(\omega)\, \omega_h\, d\omega.$$

Set $S_{jk}^{(\varepsilon)}(t) = S_{jk}(t)$ for $\varepsilon \leqq |t| \leqq R$; set $S_{jk}^{(\varepsilon)} = 0$ either for $0 \leqq |t| < \varepsilon$ or for $|t| > R$. We suppose $\bar{A}$ contained in the square $|x_k| < \frac{\pi}{2}$ $(k = 1, \ldots, r)$. We have, assuming $\varepsilon_{ik}(y) = 0$ for y outside of $\bar{A}$,

$$F_{lj}^{(\varepsilon)}(x) = \int_{C_x(\Gamma, R) - C_x(\Gamma, \varepsilon)} \varepsilon_{lk}(y)\, S_{jk}(y-x)\, dy = \int_{Q} \varepsilon_{lk}(y)\, S_{jk}^{(\varepsilon)}(y-x)\, dy.$$

Then

$$\int_Q F_{lj}^{(\varepsilon)}(x)\, e^{isx}\, dx = \int_Q \varepsilon_{lk}(y)\, e^{isy}\, dy \int_Q S_{jk}^{(\varepsilon)}(t)\, e^{-ist}\, dt.$$

On the other hand,

$$\int_Q S_{jk}^{(\varepsilon)}(t)\, e^{-ist}\, dt = \int_\Gamma \tau_{jk}(\omega)\, d\omega \int_\varepsilon^R \frac{e^{-i\varrho s\omega}}{\varrho}\, d\varrho$$

where $\tau_{jk}(\omega)$ is a C^∞ function on Ω such that

$$spt\, \tau_{jk} \in \Gamma, \qquad \int_\Gamma \tau_{jk}(\omega)\, d\omega = 0.$$

So, we are permitted to write, for $s \neq 0$,

$$\int_Q S_{jk}^{(\varepsilon)}(t)\, e^{-ist}\, dt = \int_\Gamma \tau_{jk}(\omega)\, d\omega \int_\varepsilon^R \frac{e^{-i\varrho|s|\alpha} - e^{-\varrho|s|}}{\varrho}\, d\varrho,$$

where $\alpha = |s|^{-1} s\omega$. By elementary arguments we prove that

$$\left| \int_\varepsilon^R \frac{e^{-i\varrho|s|\alpha} - e^{-\varrho|s|}}{\varrho}\, d\varrho \right| \leqq \log \frac{c_6}{|\alpha|} \text{ [19]}$$

[19] See [25, p. 95].

with c_6 independent of ε and of s. It follows that

$$\left|\int_Q S_{jk}^{(\varepsilon)}(t)\, e^{-ist}\, dt \leqq c\right| \leqq c_7$$

with c_7 independent of s. Then

$$\int_J |F_{lj}^{(\varepsilon)}(x)|^2\, dx \leqq \frac{c_8}{(2\pi)^r} \sum_{k=1}^{r} \sum_{-\infty}^{+\infty}{}_s \left|\int_Q \varepsilon_{lk}(y)\, e^{-isy}\, dy\right|^2. \tag{12.15}$$

Since, for any $x \in J \cap \bar{A}$,

$$\lim_{\varepsilon\to\infty} F_{lj}^{(\varepsilon)}(x) = -\alpha_{l/j}(x) - \varepsilon_{lk}(x) \int_\Gamma H_{jhk}(\omega)\, \omega_h\, d\omega,$$

from (12.15), by a well known theorem of integration theory (Fatou's lemma) [20] we deduce

$$\|\alpha_{l/j}\|_{0,J\cap\bar{A}}^2 \leqq c_9 \sum_{h,k} \|\varepsilon_{hk}\|_{0,A}^2.$$

Arguing in the same manner for β_l and γ_l, we get the proof of (12.11).

Remark. The original second Korn inequality is

$$\int_A v_{i/k}\, v_{i/k}\, dx \leqq c_0 \int_A (v_{i/k} + v_{k/i})(v_{i/k} + v_{k/i})\, dx \tag{12.16}$$

for any $v \in H_1(A)$ such that

$$\int_A (v_{i/k} - v_{k/i})\, dx = 0. \tag{12.17}$$

Let us briefly prove that (12.16), with the side condition (12.17), is equivalent to (12.11). For the sake of simplicity we assume, as further hypothesis, that A is properly regular. Let us suppose (12.11) to hold and assume that there exists a sequence $v^{(n)} \in H_1(A)$ such that (12.17) is satisfied, but such that

$$\int_A v_{i/k}^{(n)} v_{i/k}^{(n)}\, dx = 1, \qquad \lim_{n\to\infty} \int_A (v_{i/k}^{(n)} + v_{i/k}^{(n)})(v_{k/i}^{(n)} + v_{k/i}^{(n)})\, dx = 0.$$

Since we may assume $\int_A v^{(n)}\, dx = 0$, we are permitted to suppose that $v^{(n)}$ converges in the space $H_0(A)$ to a vector v (see footnote [17] and Theorem 2.IV). Set $\varepsilon_{ik}^{(n)} = 2^{-1}(v_{i/k}^{(n)} + v_{k/i}^{(n)})$. We have

$$\|v^{(n+p)} - v^{(n)}\|_1^2 \leqq c_2^{-1} \left(4 \sum_{i,k}^{1,r} \|\varepsilon_{ik}^{(n+p)} - \varepsilon_{ik}^{(n)}\|_0^2 + \|v^{(n+p)} - v^{(n)}\|_0^2\right).$$

Thus $v^{(n)}$ converges to v in $H_1(A)$. Since $v_{i/k} + v_{k/i} = 0$, we must have $v = a + Bx$ where a is a constant vector and B a skew-symmetric $r \times r$ matrix.[21] Conditions (12.17) imply $B \equiv 0$ and, in consequence, $v = 0$, which contradicts $\sum_{i,k}^{1,r} \|v_{i/k}\|_0^2 = 1$. Let us now suppose that (12.16) holds when (12.17) is satisfied.

---

[20] See [*15*, p. 113], [*4*, p. 362].

[21] Let w be a vector of $\overset{\circ}{C}{}^\infty(A)$. Since $v_{i/k} + v_{k/i} = 0$,

$$\int_A v_i (w_{i/kk} + w_{k/ii})\, dx = 0.$$

Since the operator $L_i(w) = w_{i/kk} + w_{k/ii}$ $(i = 1, \ldots, r)$ is elliptic, from Theorem 3.II it follows that $v_i \in {}^\infty C(A)$. Then $v_{i/hk} + v_{k/hi} = 0$, which implies $v_{i/hk} = 0$. Then $v_i = a_i + b_{ih} x_h$. The conditions $v_{i/k} + v_{k/i} = 0$ imply $b_{ih} = -b_{hi}$.

Denote by $\tilde{H}_1(A)$ the subspace of $H_1(A)$ definedby the condition $\int_A v\,dx=0$. Assume as norm in $\tilde{H}_1(A)$: $|v|_1^2=(v_{i/k}, v_{i/k})_0$. Let R be the orthogonal complement of the subspace V_0 of $\tilde{H}_1(A)$ defined by (12.17). For any vector $v\in\tilde{H}_1(A)$ we have $v=v_0+\varrho$ with $v_0\in V_0$ and $\varrho\in R$. It is easily seen that $\varrho=a+Bx$, where B is a skew-symmetric $r\times r$ matrix. Suppose that there exists a sequence in $\tilde{H}_1(A)$ such that

$$|v^{(n)}|_1=1, \quad \lim_{n\to\infty}\left(\sum_{i,k}^{1,r}\|\varepsilon_{ik}^{(n)}\|_0^2+\|v^{(n)}\|_0^2\right)=0.$$

We have $|v_0^{(n)}|_1^2+|\varrho^{(n)}|_1^2=1$, $\varrho^{(n)}=v^{(n)}-v_0^{(n)}$, then $\|\varrho^{(n)}\|_0^2\leq 2\,\|v^{(n)}\|_0^2+2\,\|v_0^{(n)}\|_0^2$, which implies $\|\varrho^{(n)}\|_0<c_{10}$, with c_{10} independent of n. On the other hand we have $\|\varrho^{(n)}\|_1\leq c_{11}\|\varrho^{(n)}\|_0$, with c_{11} independent of n. So we may assume that $\{\varrho^{(n)}\}$ converges in $H_1(A)$. By using (12.16) we see that $\{v_0^n\}$ converges in $H_1(A)$. Hence $\{v^n\}$ converges in $H_1(A)$. The limit function v must satisfy the conditions

$$|v|_1=1, \sum_{i,k}^{1,r}\|\varepsilon_{ik}\|_0^2+\|v\|_0^2=0.$$

That is impossible.

Let us now turn back to the boundary value problem (12.10), (12.5). Since, as is easily seen, a C^1-smooth domain satisfies the restricted cone-hypothesis, our boundary value problem has only one solution which is C^∞ in $\bar{A}$. The differential operator is formally self-adjoint. Then a C^∞ solution in $\bar{A}$ of the following differential system:

$$\frac{\partial}{\partial x_h}\left[a_{ih,jk}(x)\,\frac{\partial u_j}{\partial x_k}\right]-p_0\,u_i+\lambda u_i+f_i=0, \tag{12.18}$$

with the boundary conditions (12.5), exists when and only when

$$\int_A f_i\,\varrho_i\,dx=0, \tag{12.19}$$

where ϱ is any solution belonging to $C^\infty(\bar{A})$ of (12.18) with $f_i\equiv 0$. In the case when $\lambda=p_0$ the only $C^\infty(\bar{A})$ solutions of the homogeneous system are

$$\varrho_i=a_i+b_{ih}\,x_h, \tag{12.20}$$

where a_i and b_{ih} are arbitrary constants such that $b_{ih}=-b_{hi}$.

12.III *Let f be a function of $C^\infty(\bar{A})$. The boundary value problem* (12.3), (12.5) *has solutions belonging to $C^\infty(\bar{A})$ if and only if f satisfies the compatibility conditions* (12.19) *with ϱ given by* (12.20).

So as to prove the existence and uniqueness of the solution of the mixed boundary value problem (12.3), (12.6), (12.7), we take as V the subspace of $H_1(A)$ constituted by the functions of $H_1(A)$ vanishing on $\partial_1 A$. We can consider Korn's second inequality for any $v\in V$ and argue as in the case of the previous boundary value problem. In the present case we obtain the following theorem:

12.IV *For $f\in C^\infty(\bar{A})$ there is one and only one solution of the mixed boundary value problem* (12.3), (12.6), (12.7) *which belongs to $C^\infty(A\cup\partial_1 A\cup\partial_2 A)\cap H_1(A)$.*

It is worthwile to remark that the solution of the first, second, and thus boundary value problems which we have obtained are the ones required by the mathematical theory of elasticity, since they minimize the *energy functional*

$$\int_A [W(x,\varepsilon)-uf]\,dx$$

in the classes $\mathring{H}_1(A)$, $H_1(A)$ and V, respectively (see Theorem 6.III). Let us also observe that existence theorems relative to problems (P_λ), (P_{tt}), (P_t), (P_K) provide existence theorems connected with the eigenvalue problems, propagation of elastic waves, diffusion problems and hereditary problems (visco-elasticity) connected with linear elasticity.

13. Equilibrium problems for heterogeneous media. Let A be a C^1-smooth domain and let Σ_1 and Σ_2 be two disjoint subsets of ∂A such that $\partial A = \bar{\Sigma}_1 \cup \bar{\Sigma}_2$. Let Σ be a set contained in A satisfying the following condition: $\bar{\Sigma}_k \cup \bar{\Sigma}$ $(k=1, 2)$ is the boundary of a regular domain A_k contained in A such that $A_1 \cap A_2 = \emptyset$. Suppose that A be C^∞-smooth for any $x^0 \in \Sigma_1 \cup \Sigma_2$ and, moreover, the neighborhood I of $x^0 \in \Sigma_1$ (of $x^0 \in \Sigma_2$) considered for insuring the C^∞-smoothness of A at x^0 be such that $\bar{I} \cap \bar{\Sigma}_2 = \emptyset$ $(\bar{I} \cap \bar{\Sigma}_1 = \emptyset)$ and $\bar{I} \cap \bar{\Sigma} = \emptyset$. Let us also suppose that for any $x^0 \in \Sigma$ there exists a neighborhood J of x^0 such that:

(1) $\bar{J}$ is disjoint from $\bar{\Sigma}_1 \cup \bar{\Sigma}_2$;

(2) the set $\bar{J} \cap \bar{A}_k$ $(k=1, 2)$ is C^∞-homeomorphic to the closed semiball $\bar{B}^+ : |y|^2 + t^2 \leqq 1$ of the r-dimensional (y, t) space;

(3) the set $\bar{J} \cap \partial A_k$ $(k=1, 2)$ is mapped onto the $(r-1)$-dimensional bal $t=0$, $|y| \leqq 1$;

(4) if $\xi = \omega_k(x)$ $[\xi \equiv (y, t)]$ is the homeomorphism which maps $\bar{J} \cap \bar{A}_k$ onto $\bar{B}^+$, we have $\bar{J} \cap \partial A_1 \equiv \bar{J} \cap \partial A_2$ and the restriction of $\omega_1(x)$ on $\bar{J} \cap \partial A_1$ must coincide with the restriction of $\omega_2(x)$ to $\bar{J} \cap \partial A_2$.

Suppose that A_1 and A_2 are occupied by two elastic bodies in their natural configurations. Let $W^{(k)}(x, \varepsilon)$ $(k=1, 2)$ be the elastic potential corresponding to the body which occupies A_k. *We assume that $W^{(k)}(x, \varepsilon)$ satisfies the hypotheses laid down for $W(x, \varepsilon)$ in the previous section.* Let $a_{ih,js}^{(k)}(x)$ be C^∞-functions representing the coefficients relative to A_k. We can imagine A as being a hetereogeneous elastic medium in its natural configuration. The elastic coefficients of A are, in general, discontinuous along Σ. The equilibrium problems for the hetereogeneous medium are the following.

(HP) *Given $f^{(k)} \in C^\infty(\bar{A}_k)$ $(k=1, 2)$, find a function u satisfying the following conditions:*

(i) $u \in H_1(A) \cap C^\infty(A_1 \cup \Sigma_1 \cup \Sigma) \cap C^\infty(A_2 \cup \Sigma_2 \cup \Sigma) \cap C^0(A)$;

(ii) *u satisfies the differential equations*

$$\frac{\partial}{\partial x_h} \frac{\partial}{\partial \varepsilon_{ih}} W^{(k)}(x, \varepsilon) + f_i^{(k)} = 0 \quad in\ A_k \qquad (k=1, 2);$$

(iii) *u satisfies on Σ_1 and on Σ_2 boundary conditions defining the first, second, or third boundary value problem considered in Sect. 12;*

(iv) *if $\nu^{(k)}$ is the inward unit normal to A_k at a point of Σ, u satisfies, at each point of Σ the boundary conditions*

$$\frac{\partial W^{(1)}}{\partial \varepsilon_{ih}} \nu_h^{(1)} + \frac{\partial W^{(2)}}{\partial \varepsilon_{ih}} \nu_h^{(2)} = 0 \qquad (i=1, \dots, r). \tag{13.1}$$

For the sake of concreteness we assume that the boundary conditions to be satisfied by u on Σ_1 and Σ_2 are

$$\nu_h \frac{\partial}{\partial \varepsilon_{ih}} W^{(1)}(x, \varepsilon) = 0 \quad \text{on } \Sigma_1, \qquad \nu_h \frac{\partial}{\partial \varepsilon_{ih}} W^{(2)}(x, \varepsilon) = 0 \quad \text{on } \Sigma_2. \tag{13.2}$$

The reader will notice by himself how the theory must be modified for other boundary conditions on $\Sigma_1 \cup \Sigma_2$.

Set $\varepsilon_{ik}(u) = 2^{-1}(u_{i/k} + u_{k/i})$ and for $u, v \in H_1(A)$ set

$$B(u, v) = \int_{A_1} a^{(1)}_{ih,jk}(x)\, \varepsilon_{ih}(u)\, \varepsilon_{jk}(v)\, dx + \int_{A_2} a^{(2)}_{ih,jk}(x)\, \varepsilon_{ih}(u)\, \varepsilon_{jk}(v)\, dx.$$

Let p_0 be a positive constant. From Korn's second inequality it follows that for any $v \in H_1(A)$

$$B(v, v) + p_0 \|v\|_0^2 \geqq \gamma_0 \|v\|_1^2 \quad (\gamma_0 > 0). \tag{13.3}$$

Then there is one and only one $u \in H_1(A)$ such that

$$B(u, v) + p_0(u, v)_0 = (f^{(1)}, v)_{0, A_1} + (f^{(2)}, v)_{0, A_2} \tag{13.4}$$

(see the proof of Theorem 6.I). From the general theory developed in Sects. 3 and 6, it follows that u belongs to $C^\infty(A_1 \cup \Sigma_1) \cap C^\infty(A_2 \cup \Sigma_2)$, satisfies the differential equations

$$\frac{\partial}{\partial x_h} \frac{\partial}{\partial \varepsilon_{ih}} W^{(k)}(x, \varepsilon) - p_0 u_i + f_i^{(k)} = 0 \quad \text{in } A_k \qquad (k = 1, 2)$$

and, on Σ_1 and on Σ_2, the boundary conditions (13.2).

Let x^0 be a point of Σ and v any r-vector belonging to $\mathring{C}^\infty(J)$. Denote by $w(\xi)$ the $2r$-vector valued function defined in B^+ whose components are $v_1[\omega_1^{-1}(\xi)], \ldots, v_r[\omega_1^{-1}(\xi)], v_1[\omega_2^{-1}(\xi)], \ldots, v_r[\omega_2^{-1}(\xi)]$. We have for $v \in \mathring{C}^\infty(J)$

$$B(v, v) + p_0 \|v\|_0^2 = \int_{B^+} \sum_{i,j}^{1,2r} \sum_{h,k}^{1,r} \alpha_{ih,jk}(\xi)\, w_{i/h}\, w_{j/k}\, d\xi + \int_{B^+} \sum_{i,j}^{1,2r} \beta_{ij}\, w_i\, w_j\, d\xi \equiv \tilde{B}(w, w),$$

with $\alpha_{ih,jk}, \beta_{ij}$ being C^∞-functions. From (13.3), (13.4) we deduce

$$\tilde{B}(w, w) \geqq \tilde{\gamma}_0 \|w\|_{1, B^+}^2$$
$$\tilde{B}(\tilde{u}, w) = (\tilde{f}, w)_{0, B^+} \qquad (w \in W),$$

where $\tilde{u}$ and $\tilde{f}$ have an obvious definition and W is the closure in $H_1(B^+)$ of the manifold spanned by w when v spans $\mathring{C}^\infty(J)$. It is readily seen that W satisfies hypotheses (1), (2), (3), (4), (5) of Sect. 4. From the theory developed in that section, we deduce that $\tilde{u}$ is C^∞ in every semiball $t \geqq 0$, $|y|^2 + t^2 \leqq \delta$ $(0 < \delta < 1)$. Coming back to u, we see that u belongs to $C^\infty(A_1 \cup \Sigma)$, to $C^\infty(A_2 \cup \Sigma)$ and to $C^0(A)$. Using Green's identity and the arbitrariness of v, from (13.4) we deduce that u satisfies the boundary condition (13.1) on Σ. Arguing as in the proof of the existence theorem for the second boundary value problem of the previous section, we get the following result.

13.I *Problem* (HP) *has a solution when and only when*

$$\int_{A_1} f_i^{(1)} \varrho_i\, dx + \int_{A_2} f_i^{(2)} \varrho_i\, dx = 0,$$

where the ϱ_i*'s are given by* (12.20); *the* ϱ_i*'s are the only solutions of the homogeneous problem* (*i.e.* $f^{(1)} \equiv 0$, $f^{(2)} \equiv 0$).

Theorem 13.I is easily extended to the case when A is composed of n elastic bodies with different elasticities. It is also trivial to find the variational principle corresponding to problem (HP).

Bibliography

[1] AGMON, S.: Lectures on elliptic boundary value problems. Princeton, N. J.-Toronto-New York-London: D. V. Nostrand Co. Inc. 1965.
[2] EHRLING, G.: On a type of eigenvalue problem for certain elliptic differential operators. Math. Scand. 2 (1954).
[3] FICHERA, G.: Sull'esistenza e sul calcolo delle soluzioni dei problemi al contorno, relativi all'equilibrio di un corpo elastico. Ann. Scuola Norm. Sup. Pisa, s. III 4 (1950).
[4] — Lezioni sulle trasformazioni lineari. Ist. Mat. Univ. Trieste (Edit. VESCHI) (1954).
[5] — The Signorini elastostatics problem with ambiguous boundary conditions. Proc. of the Int. Symp. "Applications of the theory of functions in continuum mechanics", vol. I, Tbilisi, Sept. 1963.
[6] — Problemi elastostatici con vincoli unilaterali: il problema di Signorini con ambigue condizioni al contorno. Mem. Acc. Naz. Lincei, s. VIII 7, fasc. 5 (1964).
[7] — Linear elliptic differential systems and eigenvalue problems. Lecture notes in mathematics, vol. 8. Berlin-Heidelberg-New York: Springer 1965.
[8] FREDHOLM, J.: Solution d'un problème fondamental de la théorie de l'élasticité. Ark. Mat. Astron. Fys. 2, 28 (1906).
[9] FRIEDRICHS, K. O.: Die Randwert- und Eigenwert-Probleme aus der Theorie der elastischen Platten. Math. Annalen, Bd. 98 (1928).
[10] — On the boundary value problems of the theory of elasticity and Korn's inequality. Ann. of Math. 48 (1947).
[11] FUBINI, G.: Il principio di minimo e i teoremi di esistenza per i problemi al contorno relativi alle equazioni alle derivate parziali di ordine pari. Rend. Circ. Mat. Palermo (1907).
[12] GÅRDING, L.: Dirichlet's problem for linear elliptic partial differential equations. Math. Scand. 1 (1953).
[13] GOBERT, J.: Une inégalité fondamentale de la théorie de l'élasticité .Bull. Soc. Roy. Sci. Liège 3-4 (1962).
[14] HALMOS, P. R.: Introduction to Hilbert space and the theory of spectral multiplicity. New York, N. Y.: Chelsea Publ. Comp. 1951.
[15] — Measure theory. Princeton, N. J.-Toronto-New York-London: D. V. Nostrand Co. Inc. 1950.
[16] HILLE, E., and R. PHILLIPS: Functional analysis and semi-groups. Amer. Math. Soc. Coll. Publ. 31 (1957).
[17] JOHN, F.: Plane waves and spherical means applied to partial differential equations. Intersc. Tracts in Pure and Appl. Math. No. 2 (1955).
[18] KORN, A.: Solution générale du problème d'équilibre dans la théorie de l'élasticité dans le cas où les efforts sont donnés à la surface. Ann. Université Toulouse (1908).
[19] — Über einige Ungleichungen, welche in der Theorie der elastischen und elektrischen Schwingungen eine Rolle spielen. Bull. Intern., Cracov. Akad. umiejet (Classe Sci. Math. nat.) (1909).
[20] KUPRADZE, V. D.: Potential methods in the theory of elasticity. Israel Program for Scientific Translations, Jerusalem (1965).
[21] LAURICELLA, G.: Alcune applicazioni della teoria delle equazioni funzionali alla Fisica-Matematica. Nuovo Cimento, s. V 13 (1907).
[22] LAX, P.: On Cauchy's problem for hyperbolic equations and the differentiability of solutions of elliptic equations. Comm. Pure Appl. Math. 8 (1955).
[23] LICHTENSTEIN, L.: Über die erste Randwertaufgabe der Elastizitätstheorie. Math. Z. 20 (1924).
[24] MARCOLONGO, R.: La teoria delle equazioni integrali e le sue applicazioni alla Fisica-Matematica. Rend. Accad. Naz. Lincei, s. V 16, 1 (1907).
[25] MICHLIN, S. G.: Multidimensional singular integrals and integral equations. Oxford: Pergamon Press 1965.
[26] MIRANDA, C.: Equazioni alle derivate parziali di tipo ellittico. Ergeb. der Mathem., H. 2. Berlin-Göttingen-Heidelberg: Springer 1955.
[27] NIRENBERG, L.: Remarks on strongly elliptic partial differential equations. Comm. Pure Appl. Math. 8 (1955).
[28] — On elliptic partial differential equations. Ann. Scuola Norm. Sup. Pisa, s. III 13 (1959).
[29] PAYNE, L. E., and H. F. WEINBERGER: On Korn's inequality. Arch. Rational Mech. Anal. 8 (1961).
[30] PICONE, M.: Sur un problème nouveau pour l'équation linéaire aux dérivées partielles de la théorie mathématique classique de l'élasticité. Colloque sur les équations aux dérivées partielles, CBRM, Bruxelles, May 1954.

[31] RIESZ, F., et B. SZ. NAGY: Leçons d'Analyse Fonctionelle, 3me éd. Acad. des Sciences de Hongrie, 1955.
[32] SCHECHTER, M.: A generalization of the problem of transmission. Ann. Scuola Norm. Sup. Pisa, s. III **15** (1960).
[33] SOBOLEV, S. L.: Applications of functional analysis in mathematical physics. Trans. Math. Monogr. **7**, Amer. Math. Soc. (1963).
[34] VOLTERRA, V.: Sulle equazioni integro-differenziali della teoria dell'elasticità. Rend. Accad. Naz. Lincei, s. V **18** (1909).
[35] WEYL, H.: Das asymptotische Verteilungsgesetz der Eigenschwingungen eines beliebig gestalteten elastischen Körpers. Rend. Circ. Mat. Palermo **39** (1915).

Boundary Value Problems of Elasticity with Unilateral Constraints.

By

GAETANO FICHERA.

In the preceding article "Existence Theorems in Elasticity", which henceforth will be cited as E.T.E.[1], I have treated boundary value problems of Elasticity in the case when the side conditions to be associated with the differential equations of equilibrium correspond to bilateral constraints imposed upon the elastic body. In this article I will treat the analytical problems which arise when unilateral constraints are imposed. In Sect. 6 of E.T.E. it is shown that the "bilateral problems", as far as the existence theory is concerned, are founded on the solution of a system of equations of the following type

$$B(u, v) = F(v), \qquad u \in V, \; \forall v \in V, \tag{1}$$

where $B(u, v)$ is a bounded bilinear form defined in $H \times H$ ($H =$ Hilbert space), F is a linear functional and V is a closed linear subspace of H. These equations, in the case when B is symmetric and the space H real, are easily obtained by imposing upon the *energy functional*

$$\mathscr{I}(v) = \tfrac{1}{2} B(v, v) - F(v)$$

the condition of attaining a minimum on V.

In the case of unilateral constraints, the manifold V is *not* a closed linear subspace of H but a closed convex set of H. The condition for $\mathscr{I}(v)$ to have a minimum in the point u of V leads, in this case, to the inequalities

$$B(u, v-u) \geqq F(v-u), \qquad u \in V, \; \forall v \in V \tag{2}$$

which easily reduce to (1) if V is a linear subspace of H.

In order to provide, as in the case of bilateral constraints, satisfactory foundations for unilateral problems of elasticity, we shall develop in the first two sections of this article an abstract theory of functional inequalities (2), considering not only the case of a symmetric bilinear form $B(u, v)$ but also the non symmetric case. We shall denote these problems as *abstract unilateral problems*.

1. Abstract unilateral problems: the symmetric case. Let H be a real Hilbert space. We denote by (u, v) the scalar product in H and by $\|u\|$ the norm of the vector u of H. Let $B(u, v)$ be a symmetric bounded bilinear form defined on $H \times H$. From the elementary theory of Hilbert spaces it is well known that there exists one and only one symmetric, bounded linear operator T from H into H such that $B(u, v) = (Tu, v)$ for any u and v of H. Let $N(T)$ be the kernel of the operator T, i.e.

$$N(T) \equiv \{v;\, T(v) = 0\}.$$

[1] See this volume, pp. 347–389.

The linear subspace $N(T)$ is also the kernel of the nonnegative quadratic form $B(v, v)$, i.e.

$$N(T) \equiv \{v;\ B(v, v) = 0\}.$$

Let Q be the orthogonal projector of H into the kernel of $B(v, v)$ and set $P = I - Q$ (I = identity operator). We shall assume the following hypotheses.

(I) *Semi-coerciveness hypothesis:*

$$B(v, v) \geqq c\|Pv\|^2 \qquad \forall v \in H, \tag{1.1}$$

where c is a positive constant independent on v.

(II) *The kernel of the quadratic form $B(v, v)$ is finite dimensional.*

Let $F(v)$ be a bounded linear functional defined on H and V a closed convex set of H.[2]

1.I *Let us consider the functional*

$$\mathscr{I}(v) = \tfrac{1}{2} B(v, v) - F(v).$$

There exists in V a vector u which minimizes $\mathscr{I}(v)$ in V, if and only if there exists a solution of the unilateral problem.

$$B(u, v-u) \geqq F(v-u), \qquad u \in V,\ \forall v \in V. \tag{1.2}$$

If $\mathscr{I}(v)$ has a minimum for $v = u$, given any $v \in V$, the function of t $(0 \leqq t \leqq 1)$ $g(t) = \mathscr{I}[u + t(v-u)]$ has a minimum for $t = 0$. The condition $g'(0) \geqq 0$ exactly coincides with (1.2). Conversely if (1.2) is satisfied for any $v \in V$, we have

$$\mathscr{I}(v) - \mathscr{I}(u) = \mathscr{I}[u + (v-u)] - \mathscr{I}(u) = B(u, v-u) - F(v-u) + \tfrac{1}{2} B(v-u, v-u) \geqq 0.$$

Remark that in the proof of this lemma only the condition $B(v, v) \geqq 0$ has been used.

Let U be a point set of H containing some $u \neq 0$. Let us consider for any $u \in U$, $u \neq 0$, the set of nonnegative numbers t such that $t\|u\|^{-1}u$ is contained in U. We shall denote by $p(u, U)$ the supremum of this set, i.e.

$$p(u, U) = \sup\{t;\ u \in U,\ u \neq 0,\ t\|u\|^{-1} u \in U\}.$$

If T is a mapping from H into H and U a set of H, by $T[U]$ we shall indicate the image of U under the mapping T.

Let $N(F)$ be the kernel of the functional F, i.e.

$$N(F) = \{v;\ F(v) = 0\}.$$

Set

$$L = N(F) \cap N(T),$$

$$N(T) = L \oplus L_1.$$

Let $\tilde{Q}$ be the orthogonal projector of H onto L. Set $\tilde{P} = I - \tilde{Q}$. Let Q_1, be the orthogonal projector of H onto L_1.

1.II *The functional $\mathscr{I}(v)$ has an absolute minimum in V if there exists $u_0 \in V$ such that the following conditions are satisfied*

(i) *$F(\varrho) < 0$ for $\varrho \in N(T) \cap V(u_0)$, $p\{Q_1 \varrho, Q_1[N(T) \cap V(u_0)]\} = +\infty$.*

(ii) *The set $\tilde{P}[V(u_0)]$ is closed.*

---

[2] When we say that V is a convex set, we mean that $v_1 \in V$, $v_2 \in V$ imply $t_1 v_1 + t_2 v_2 \in V$ for any t_1, t_2 such that $0 \leqq t_i \leqq 1$ $(i = 1, 2)$, $t_1 + t_2 = 1$. By $V(u_0)$ we denote the set of all v such that $v + u_0 \in V$.

Set $Q=I-P$. Since L is a subspace of $N(T)$, we have $\widetilde{Q}<Q$ and therefore $\widetilde{P}>P$. Let i be the infimum (finite or not) of $\mathscr{I}(v)$ in V.

Let $\{u_n\}$ be a *minimizing sequence* for $\mathscr{I}(v)$ in V, i.e.

$$\lim_{n\to\infty}\mathscr{I}(u_n)=i, \qquad u_n\in V.$$

Set $z_n=u_n-u_0$. The sequence $\{z_n\}$ is minimizing for the functional $\mathscr{I}(w+u_0)$ for $w\in V(u_0)$. Suppose we have proved that $\{\widetilde{P}z_n\}$ contains a bounded subsequence. Then we may extract a subsequence, which we still denote by $\{z_n\}$, such that $\{\widetilde{P}z_n\}$ converges weakly to some limit. This limit must belong to $\widetilde{P}[V(u_0)]$, since this set is closed and convex and, in consequence, weakly closed. Let $\widetilde{P}z$ be the weak limit of $\{\widetilde{P}z_n\}$. Since $F(\widetilde{Q}\,v)=0$, we have

$$\mathscr{I}(z_n+u_0)=\mathscr{I}(\widetilde{P}z_n+u_0).$$

On the other hand, since $B(v,v)$ is lower semicontinuous with respect to weak convergence[3] and we may suppose $z\in V(u_0)$, we have

$$i=\lim_{n\to\infty}\mathscr{I}(\widetilde{P}z_n+u_0)\geqq\mathscr{I}(\widetilde{P}z+u_0)=\mathscr{I}(z+u_0)\geqq i.$$

Assuming $u=z+u_0$, we have $u\in V$ and $\mathscr{I}(u)=i$. Thus we have to prove that, from the sequence $\{\widetilde{P}z_n\}$, we may extract a bounded subsequence. Set $\sigma_n=\|\widetilde{P}z_n\|$ and suppose $\lim\limits_{n\to\infty}\sigma_n=+\infty$. We have, using (1.1)

$$c\|Pz_n\|^2\leqq B(z_n,z_n)=2\mathscr{I}(u_n)-2\mathscr{I}(u_0)+2F(z_n)-2B(u_0,z_n). \tag{1.3}$$

Set $w_n=\sigma_n^{-1}z_n$. We have, denoting by c_1 a suitable constant,

$$c\,\sigma_n^2\|Pw_n\|^2\leqq c_1+2(\|F\|+\|T\|\,\|u_0\|)\,\sigma_n\|\widetilde{P}w_n\|$$

which implies $\lim\limits_{n\to\infty}\|Pw_n\|=0$. We have $Q_1=\widetilde{P}-P$. Q_1 is a projector which is orthogonal to P. Hence

$$\|\widetilde{P}w_n\|^2=\|Pw_n\|^2+\|Q_1w_n\|^2=1.$$

Since $Q_1w_n\in N(T)$, we may extract (because of Hypothesis (II)) from $\{w_n\}$ a subsequence, which we still denote by $\{w_n\}$, such that $\{Q_1w_n\}$ converges (strongly) to some vector ϱ belonging to $Q_1[N(T)]$ and such that $\|\varrho\|=1$. We have $z_n\in V(u_0)$. Moreover $\{0\}\in V(u_0)$. Hence, for any $t>0$ and n large enough, $t\,w_n\in V(u_0)$. Then $t\widetilde{P}w_n\in\widetilde{P}[V(u_0)]$ and, for condition ii),

$$t\,\varrho=\lim_{n\to\infty}t\widetilde{P}w_n\in\widetilde{P}[V(u_0)].$$

Let ϱ' be a vector of $V(u_0)$ such that $\widetilde{P}\varrho'=\varrho$. We have $\varrho'=\varrho+(\varrho'-\varrho)$, $\varrho\in Q_1[N(T)]$, $\varrho'-\varrho\in\widetilde{Q}[N(T)]$; thus $\varrho'\in N(T)\cap V(u_0)$ and $F(\varrho)=F(\varrho)+F(\varrho'-\varrho)=F(\varrho')<0$.

[3] Suppose that $B(u,v)$ is a symmetric bounded bilinear form and $B(v,v)\geqq 0$. Suppose that $\{w_n\}$ converges weakly to the limit w. We have $B(w_n,w_n)=B(w,w)+2B(w,w_n-w)+B(w_n-w,w_n-w)$. Since $B(w_n-w,w_n-w)\geqq 0$, $\lim\limits_{n\to\infty}B(w,w_n-w)=0$, then

$$\operatorname*{minlim}_{n\to\infty}B(w_n,w_n)\geqq B(w,w).$$

Of course the hypothesis of the symmetry of $B(u,v)$ is not restrictive. In fact, if $B(u,v)$ is not symmetric, we can apply the argument to $\frac{1}{2}\{B(u,v)+B(v,u)\}$.

From (1.3) we deduce the inequality

$$c\,\sigma_n \|P w_n\|^2 \leqq \frac{2}{\sigma_n}\,[\mathcal{I}(u_n) - \mathcal{I}(u_0)] + 2F(P w_n) + 2F(Q_1 w_n) - 2B(u_0, P w_n),$$

which is absurd since the *minlim* of the left hand side is nonnegative while the *max lim* of the right hand side is negative.

Remark. If the set $N(T) \cap V(u_0)$ is bounded [and ii) holds], the existence of the minimum holds for any given F. In particular this condition is satisfied if V is bounded. In this case the condition (1.1) is not needed, but only $B(v, v) \geqq 0$. In fact, if V is bounded, any minimizing sequence is bounded and, in consequence, weakly compact.

1.III *If $\mathcal{I}(v)$ has a minimum in V, for any $u_0 \in V$ and any ϱ such that*

$$\varrho \in N(T) \cap V(u_0), \qquad p\{Q_1 \varrho, Q_1[N(T) \cap V(u_0)]\} = +\infty$$

one must have $F(\varrho) < 0$.

Let $\varrho(t)$ be such that $Q_1 \varrho(t) = t\, Q_1 \varrho$, $\varrho(t) \in N(T) \cap V(u_0)$ $t > 0$. From (1.2) we have: $B(u, \varrho(t) + u_0 - u) \geqq F[\varrho(t) + u_0 - u]$. Since $F[\varrho(t)] = t\,F(\varrho) + 0$, we have $t\,F(\varrho) \leqq B(u, u_0) - B(u, u) + F(u - u_0)$, which implies $F(\varrho) < 0$.

This theorem proves that condition i) of Theorem 1.III is necessary for the existence of the minimum. Let us prove, by an example, that, if condition ii) does not hold, the theorem could fail to be true.

Let H be a two dimensional space, and let v_1 and v_2 be two orthogonal unit vectors of H. Assume $B = (u, v_1)(v, v_1)$. Let V be the convex set defined by the conditions

$$0 \leqq (v, v_1) < 1 \qquad (v, v_2)[1 - (v, v_1)] \geqq (v, v_1).$$

Assume that

$$F(v) = a(v, v_1) + b(v, v_2)$$

with $a \geqq 1$. The kernel $N(T)$ is defined by the condition $(v, v_1) = 0$. For any choice of u_0 the set $N(T) \cap V(u_0)$ is formed by the vectors ϱ such that $(\varrho, v_1) = 0$, $(\varrho, v_2) \geqq \sigma$, where σ is a constant depending on u_0. If $b \neq 0$ the condition $F(\varrho) < 0$ can be satisfied only if we assume

$$(u_0, v_2)[1 - (u_0, v_1)] = (u_0, v_1).$$

In this case $F(\varrho) < 0$ is equivalent to $b < 0$. In the case when $b < 0$ the space L coincides with $\{0\}$, and the functional $\mathcal{I}(v)$ has a minimum in V, as is easily checked by elementary arguments. If $b = 0$, the space L_M is the space $(v, v_1) = 0$. Then the set $\tilde{P}[V(u_0)]$ is defined by the conditions

$$(v, v_2) = 0, \qquad -\alpha^2 \leqq (v, v_1) < 1 - \alpha^2$$

and it is not closed, i.e. condition ii) of Theorem 1.III is violated. It is easily seen that in this case $\mathcal{I}(v)$ has no minimum in V.

1.IV *If for any $w \in P[H]$ there exists a $\varrho \in N(T)$ such that $w + \varrho \in V$, and if $F(\varrho) = 0$ for every $\varrho \in N(T)$, then $\mathcal{I}(v)$ has a minimum in V and the vector u which minimizes $\mathcal{I}(v)$ in V minimizes $\mathcal{I}(v)$ in the whole H.*

The existence of the vectors u and u_0 minimizing $\mathcal{I}(v)$ respectively in V and in H, follows from Theorem 1.II.

Let $\varrho \in N(T)$ and $P u_0 + \varrho \in V$. We have

$$\mathcal{I}(u_0) = \mathcal{I}(P u_0) = \mathcal{I}(P u_0 + \varrho) \geqq \mathcal{I}(u);$$

hence $\mathcal{I}(u) = \mathcal{I}(u_0)$.

1.V *If u minimizes $\mathscr{I}(v)$ in V, the vector Pu is uniquely determined by F. Any other vector u' minimizing $\mathscr{I}(v)$ in V is given by $u'=u+\varrho$, where ϱ is a vector of $N(T)$ such that $F(\varrho)=0$, $u+\varrho\in V$.*

Let u be a solution of the unilateral problem (1.2), i.e. a vector minimizing $\mathscr{I}(v)$ in V. Let u' be a solution of the unilateral problem

$$B(u', v-u')\geqq F'(v-u'), \qquad u'\in V,\ \forall v\in V. \tag{1.4}$$

F' is a linear bounded functional. From (1.2), (1.4) we deduce

$$B(u, u'-u)\geqq F(u'-u)$$
$$B(u', u-u')\geqq F'(u-u');$$

hence

$$B(u-u', u-u')\leqq F(u-u')-F'(u-u')$$

and, by (1.1),

$$c\,\|Pu-Pu'\|^2\leqq \|F-F'\|\,\|u-u'\|, \tag{1.5}$$

which, for $F=F'$, implies $Pu=Pu'$. If u' minimizes $\mathscr{I}(v)$ in V, one has $Pu=Pu'$ and $u'=u+\varrho$. Since $\mathscr{I}(u')=\mathscr{I}(u)-F(\varrho)$, it follows that $F(\varrho)=0$. Conversely, if one has $u'=u+\varrho$ and ϱ satisfies the conditions stated in the theorem, u' minimizes $\mathscr{I}(v)$ in V.

2. Abstract unilateral problems: the nonsymmetric case. We maintain in this section all the hypotheses stated for $B(u,v)$, for V and for F, in Sect. 1, but we no longer assume $B(u,v)$ to be symmetric. As in Sect. 1 we denote by Q the orthogonal projector of H onto the kernel of the quadratic form $B(v,v)$ and set $P=I-Q$. Thus $B(u,v)$ is a bounded bilinear form defined on $H\times H$ and satisfying Hypotheses I and II of Sect. 1. V is a closed convex set of H. F is a bounded linear functional defined in H.

We have $B(u,v)=(Tu,v)$ where T is a bounded linear operator from H into H. We shall denote by T^* the adjoint operator of T, i.e. the bounded linear operator defined by the condition

$$(Tu, v)=(u, T^*v), \qquad \forall u, v\in H.$$

We have

$$B(v,v)=(Tv,v)=(T^*v,v)=\tfrac{1}{2}\left((T+T^*)\,v, v\right).$$

Since $B(v,v)$ is nonnegative, its kernel coincides with the kernel $N(T+T^*)$ of the symmetric operator $T+T^*$, and moreover

$$N(T)\equiv N(T^*)\subset N(T+T^*).$$

We shall consider the unilateral problem (1.1) under these more general hypotheses on B.

Let us denote by $K(T)$ the orthogonal complement of $N(T)$ with respect to $N(T+T^*)$, i.e.

$$N(T+T^*)=N(T)\oplus K(T).$$

The following theorem generalizes Theorem 1.III to a nonsymmetric form $B(u,v)$. As in Sect. 1 we set

$$L=N(F)\cap N(T), \qquad N(T)=L\oplus L_1,$$

$\tilde{Q}$, $\tilde{P}$, Q_1 have the same meaning as in Sect. 1.

2.I *The unilateral problem*

$$B(u, v-u) \geqq F(v-u), \quad u \in V, \ \forall v \in V \tag{2.1}$$

for the (not necessarily symmetric) bilinear form $B(u, v)$ has a solution u if there exists a $u_0 \in V$ such that the following conditions are satisfied:

(i) $F(\varrho) < 0$ *for* $\varrho \in N(T) \cap V(u_0)$, $p\{Q_1 \varrho, Q_1[N(T) \cap V(u_0)]\} = +\infty$.

(ii) *The set* $\tilde{P}[V(u_0)]$ *is closed.*

(iii) *Let Q_0 be the orthogonal projector of H onto $K(T)$. For every ϱ satisfying the conditions* $Q_0 \varrho \neq 0$, $\varrho \in N(T+T^*) \cap V(u_0)$, $p\{(Q_0+Q_1)\varrho, (Q_0+Q_1)[N(T+T^*) \cap V(u_0)]\} = +\infty$ *there exists a vector* $v_\varrho \in V$ *such that*

$$F(\varrho) + B(\varrho, v_\varrho) < 0.$$

Let f be a vector of H such that

$$F(v) = (f, v), \quad \forall v \in H.$$

Let us consider the unilateral problem

$$(u, w-u) \geqq (f, w-u), \quad u \in \tilde{P}[V(u_0)], \ \forall w \in \tilde{P}[V(u_0)]. \tag{2.2}$$

This is a particular case of the problem solved by Theorem 1.II, when we assume $B(u, v) = (u, v)$, as Hilbert space the space $\tilde{P}[H]$ and as convex set $\tilde{P}[V(u_0)]$. Since in this case the kernel of the quadratic form is $\{0\}$, we have a solution for any given f. This solution is unique and, in this particular case, (1.5) gives

$$\|u-x\| \leqq \|f-g\|; \tag{2.3}$$

x is the solution of the unilateral problem

$$(x, w-x) \geqq (g, w-x), \quad x \in \tilde{P}[V(u_0)], \ \forall w \in \tilde{P}[V(u_0)].$$

Let us denote by Rf the solution u of the unilateral problem (2.2).

For every positive integer n set $T_n = T + n^{-1} I$. Consider the unilateral problem

$$(T_n \zeta, w-\zeta) \geqq (f - T u_0, w-\zeta), \quad \zeta \in \tilde{P}[V(u_0)], \ \forall w \in \tilde{P}[V(u_0)]. \tag{2.4}$$

Let λ be a positive constant. We can write (2.4) as follows

$$(\zeta, w-\zeta) \geqq (\zeta - \lambda T_n \zeta + \lambda f - \lambda T u_0, w-\zeta), \quad \zeta \in \tilde{P}[V(u_0)], \ \forall w \in \tilde{P}[V(u_0)].$$

Then a solution of (2.4) exists if and only if there exists a solution ζ of the equation

$$\zeta = R(\zeta - \lambda T_n \zeta + \lambda f - \lambda T u_0), \quad \zeta \in \tilde{P}[V(u_0)]. \tag{2.5}$$

Set $S_n \zeta = R(\zeta - \lambda T_n \zeta + \lambda f - \lambda T u_0)$ and consider S_n as a mapping from $\tilde{P}[V(u_0)]$ into $\tilde{P}[V(u_0)]$. We have, because of (2.3)

$$\|S_n \zeta - S_n \varkappa\| \leqq \|(I - \lambda T_n)(\zeta - \varkappa)\| \leqq \|I - \lambda T_n\| \, \|\zeta - \varkappa\|.$$

It is easily seen that, by assuming $\lambda < 2n^{-1} \|T_n\|^{-2}$, we have $\|I - \lambda T_n\| < 1$. Thus S_n is a contraction, and (2.5) has one and only one solution. Let us denote by z_n the unique solution of (2.4). Suppose we have proved that $\{z_n\}$ contains a bounded subsequence. We may extract from $\{z_n\}$ a weakly convergent subsequence, which we still denote by $\{z_n\}$. Let $\tilde{P} z$ be the weak limit of $\{z_n\}$. We may suppose $z \in V(u_0)$. Set $u = z + u_0$. Keeping in mind that (Tv, v) is lower semicontinuous with respect to weak convergence and setting $\widetilde{P}(v-u_0)=w$ $(v\in V)$, we have

$$\begin{aligned}(Tu, v-u) &= \big(Tz, (v-u_0)-(u-u_0)\big)+(Tu_0, v-u)\\ &= (T\widetilde{P}z, w-\widetilde{P}z)+(Tu_0, v-u) \geqq -\operatorname*{minlim}_{n\to\infty}(Tz_n, z_n-w)+(Tu_0, v-u)\\ &\geqq -\lim_{n\to\infty}\left\{(f-Tu_0, z_n-w)-\frac{1}{n}(z_n, z_n-w)\right\}+(Tu_0, v-u)\\ &= (f-Tu_0, w-\widetilde{P}z)+(Tu_0, v-u)\\ &= (f-Tu_0, v-u_0-z)+(Tu_0, v-u)=(f, v-u).\end{aligned}$$

Thus we have to prove that $\{z_n\}$ contains a bounded subsequence. Set $\sigma_n=\|z_n\|$ and suppose that $\lim\limits_{n\to\infty}\sigma_n=+\infty$.

From (2.4), assuming $w=\widetilde{P}(v-u_0)$ $(v\in V)$, we deduce that

$$c\,\|Pz_n\|^2 \leqq (z_n, T^*v-T^*u_0)+\frac{1}{n}(z_n, v-u_0)+(f-Tu_0, z_n)-(f-Tu_0, v-u_0).$$

Set $x_n=\sigma_n^{-1}z_n$. Then

$$\begin{aligned}c\,\sigma_n^2\,\|Px_n\|^2 \leqq \sigma_n(x_n, T^*v-T^*u_0)+\frac{\sigma_n}{n}(x_n, v-u_0)\\ +\sigma_n(f-Tu_0, x_n)-(f-Tu_0, v-u_0).\end{aligned} \tag{2.6}$$

Hence $\lim\limits_{n\to\infty}\|Px_n\|=0$.

We have $\widetilde{P}=P+Q_0+Q_1$ and P, Q_0, Q_1 are mutually orthogonal projectors. Hence

$$\|Px_n\|^2+\|Q_0x_n\|^2+\|Q_1x_n\|^2=1.$$

Thus we may extract from $\{x_n\}$ a subsequence, which we still denote by $\{x_n\}$, such that

$$\begin{gathered}\lim_{n\to\infty}Q_0x_n=\varrho_0\in K(T), \qquad \lim_{n\to\infty}Q_1x_n=\varrho_1\in L_1,\\ \|\varrho_0\|^2+\|\varrho_1\|^2=1.\end{gathered} \tag{2.7}$$

Set $\varrho=\varrho_0+\varrho_1$. Since

$$t\,\varrho=\lim_{n\to\infty}t\,x_n \qquad (t>0)$$

and, for n large enough, $t\,x_n\in\widetilde{P}[V(u_0)]$, then $t\,\varrho\in\widetilde{P}[V(u_0)]$. Let ϱ' be a vector of $V(u_0)$ such that $\widetilde{P}\varrho'=\varrho$. We have $\varrho'=\varrho+(\varrho'-\varrho)$. Since $\varrho\in L_1\oplus K(T)$ and $\varrho'-\varrho\in L$, we have $\varrho'\in N(T+T^*)\cap V(u_0)$. If $\varrho_0=0$, then $\varrho'\in N(T)\cap V(u_0)$. Eq. (2.7) implies $\varrho_1\neq 0$ and, in consequence, $F(\varrho_1)=F(\varrho_1)+F(\varrho'-\varrho_1)=F(\varrho')<0$. Then

$$\lim_{n\to\infty}(f+T^*v, Q_0x_n+Q_1x_n)<0. \tag{2.8}$$

If $\varrho_0\neq 0$, then $Q_0\varrho'\neq 0$ and, assuming $v=v_{\varrho'}$, we have $(f, \varrho')+(T^*v_{\varrho'}, \varrho')<0$. Hence

$$\begin{aligned}(f, \varrho)+(T^*v_{\varrho'}, \varrho) &= (f, \varrho)+(T^*v_{\varrho'}, \varrho)+(f, \varrho'-\varrho)\\ &\quad+\big(v_{\varrho'}, T(\varrho'-\varrho)\big)=(f, \varrho')+(T^*v_{\varrho'}, \varrho')<0,\end{aligned}$$

and (2.8) holds also in this case, provided we assume $v=v_{\varrho'}$.

From (2.6), where v is any fixed vector of V if $\varrho_0=0$ and $v=v_{\varrho'}$ if $\varrho_0\neq 0$, we deduce that

$$\begin{aligned}c\,\sigma_n\,\|Px_n\|^2 &\leqq (Px_n, T^*v-T^*u_0)+\frac{1}{n}(x_n, v-u_0)\\ &\quad+(Px_n, f-Tu_0)+(Q_0x_n+Q_1x_n, f+T^*v)-\frac{1}{\sigma_n}(f-Tu_0, v-u_0),\end{aligned}$$

which is absurd since the *minlim* of the left hand side is nonnegative while the limit of the right hand side is negative.

Remark. If the set $N(T+T^*)\cap V(u_0)$ is bounded [and (ii) holds], the unilateral problem (2.1) has a solution for any given F. In particular this condition is satisfied if V is bounded. In this case the condition (1.1) is not needed but only $B(v, v)\geqq 0$. In fact, if V is bounded, the sequence $\{z_n\}$ is weakly compact.

The following theorem proves the necessity of condition i) of Theorem 2.I.

2.II *If the unilateral problem* (2.1) *has a solution u, for any $u_0\in V$ and any ϱ such that*

$$\varrho\in N(T)\cap V(u_0), \quad p\{Q_1\varrho, Q_1[N(T)\cap V(u_0)]\}=+\infty,$$

one must have $F(\varrho)<0$.

The proof is exactly the same as for Theorem 1.III.

Concerning condition ii), the same example as given in the symmetric case proves that if condition ii) is dropped the theorem may fail to be true.

With respect to condition iii) we can only prove the following

2.III *If for every $u_0\in V$ and for some ϱ satisfying the conditions*

$$\begin{gathered} Q_0\varrho\neq 0, \quad \varrho\in N(T+T^*)\cap V(u_0), \\ p\{(Q_0+Q_1)\varrho, (Q_0+Q_1)[N(T+T^*)\cap V(u_0)]\}=+\infty, \end{gathered} \tag{2.9}$$

we have

$$F(\varrho)+B(\varrho, v)>0 \tag{2.10}$$

for every $v\in V$, then the unilateral problem (2.1) *has no solution.*

Suppose, contrarywise, that problem (2.1) has a solution u. Let us have, for every $t>0$, $(Q_0+Q_1)\varrho(t)=t(Q_0+Q_1)\varrho$, $\varrho(t)\in N(T+T^*)\cap V(u_0)$. Then

$$(Tu+T^*u, \varrho(t)+u_0-u)\geqq t(f+T^*u, \varrho(t))+(f+T^*u, u_0-u).$$

Since $T\varrho(t)+T^*\varrho(t)=0$, $(f+T^*u, \varrho(t))=t(f+T^*u, \varrho)$, we conclude that

$$(Tu, u_0-u)-(f, u_0-u)\geqq t(f+T^*u, \varrho)$$

which implies

$$(f+T^*u, \varrho)=F(\varrho)+B(\varrho, u)\leqq 0$$

in contradiction with (2.10).

2.IV *If u is a solution of the unilateral problem* (2.1) *the vector Pu is uniquely determined by F.*

Arguing as in the proof of Theorem 1.V, we get (1.5) and the proof of Theorem 2.IV.

2.V *If $B(v, v)$ is coercive on H, i.e.*

$$c\|v\|^2\leqq B(v, v) \qquad \forall v\in H,$$

the unilateral problem (2.1) *has only one solution for any given F. If we denote by $G(F)$ the solution of the unilateral problem* (2.1), *G is a Lipschitz-continuous mapping from H^* (dual space of H) into V.*

Under the assumed hypotheses, problem (2.1) has a solution for every given $F\in H^*$. In this case, since $P\equiv I$, the inequality (1.4) gives

$$c\|u-u'\|\leqq\|F-F'\|,$$

which proves uniqueness and, moreover,

$$\|G(F)-G(F')\| \leqq c^{-1}\|F-F'\|.$$

Remark. For the uniqueness of the solution of the unilateral problem (2.1), it is sufficient the quadratic form $B(v, v)$ be strictly positive on H. In fact, if u and u' are two solutions of (2.1), we have $B(u, u'-u) \geqq F(u'-u)$, $B(u', u-u') \geqq F(u-u')$. Hence $B(u-u', u-u') \leqq 0$, which implies $u=u'$.

3. Unilateral problems for elliptic operators. Let A be a bounded domain of the cartesian space X^r, and let $B(u, v)$ be the bilinear form considered in Sect. 5 of E.T.E.

$$B(u, v)=(-1)^m \sum_{|p|,|q|}^{c,m} (-1)^{|p|} \int_A a_{pq}(x)\, D^q u\, D^p v\, dx.$$

We now assume that the $n\times n$ matrices a_{pq} have real entries which are bounded and measurable functions in A; u and v are real n-vector valued functions of $H_m(A)$ which we now consider as a real Hilbert space.[4]

Let H be a closed, linear subspace of $H_m(A)$, and suppose that the bilinear form, when restricted to $H\times H$, satisfies Hypotheses I and II considered in Sects. 1 and 2.

(I) *Denote by P the projector of H onto the orthogonal complement of the kernel of $B(v, v)$ (restricted to H). A positive constant c exists such that*

$$B(v, v) \geqq c\,\|P v\|_m, \qquad \forall v\in H.$$

(II) *The kernel of $B(v, v)$ is finite dimensional.*

Let V be a closed convex set of H. We can apply the theory of Sects. 1 and 2 to this particular case and give existence theorems for the unilateral problem

$$(-1)^m \sum_{|p|,|q|}^{0,m} (-1)^{|p|} \int_A a_{pq}(x)\, D^q u\, D^p (v-u)\, dx \geqq F(v-u), \qquad u\in V,\ \forall v\in V, \tag{3.1}$$

where F is a given bounded linear functional defined on H.

Suppose that the following further hypotheses are satisfied.

(III) *The convex set V contains $\mathring{H}_m(A)$.*

(IV) *The differential operator*

$$L(x, D)=(-1)^m D^p a_{pq} D^q, \qquad (0\leqq|p|\leqq m),\ (0\leqq|q|\leqq m)$$

is elliptic for every $x\in A$.

In addition assume that $a_{pq}(x)\in C^\infty(A)$ and $F(v)=(f, v)_0$[5] for any $v\in\mathring{H}_m(A)$, where f is a function of $H_\nu(A)$. We have the following theorems:

3.I *Under the assumed hypotheses, if u is a solution of the unilateral problem* (3.1), *then u belongs to $H_{2m+\nu}(B)$, where B is any domain such that $\overline{B}\subset A$, and u is a solution in A of the differential system $Lu=f$.*

If $v\in\mathring{C}^\infty(A)$ and t is an arbitrary real constant, from (3.1) we deduce $tB(u, v)-B(u, u)\geqq t(f, v)_0-(f, u)_0$. Hence, since t is arbitrary, $B(u, v)=(f, v)_0$, and after integrations by parts we obtain

$$\int_A u\, L^* v\, dx=\int_A f\, v\, dx.$$

[4] We use in this paper concepts and notations already introduced in E.T.E.

[5] $(\cdot,\cdot)_0$ denotes scalar product in $H_0(A)\equiv\mathscr{L}^2(A)$.

From Theorem 3.I of E.T.E. we get the proof.

3.II *The space $N(T)$ consists in all the functions ϱ such that*

$$L\varrho=0, \qquad \varrho\in\mathcal{U}_H.^{6} \tag{3.2}$$

We obtain the proof of this theorem by observing that $\varrho\in N(T)$ if and only if

$$B(\varrho, v)=0, \qquad \varrho\in H, \ \forall v\in H.$$

From the results of Sect. 6 of E.T.E. it follows that $\varrho\in N(T)$ when and only when ϱ satisfies (3.2).

We have a similar result for characterizing the functions ϱ which belong to $N(T+T^*)$. In fact we have to use the same argument but referred to the symmetric bilinear form $B(u, v)+B(v, u)$. As a consequence we deduce that $N(T+T^*)$ is formed by all the solutions of the equation $Lu+L^*u=0$ that belong to a certain subspace $\mathcal{U}'_H$ of H; L^* is the operator $L^*=(-1)^m D^p \alpha_{pq} D^q$ where $\alpha_{pq}=(-1)^{|p|+|q|}\bar{a}_{qp}$.

Let A' be an open subset of A and consider the function space $\mathring{H}_m(A')$. We may consider $\mathring{H}_m(A')$ as a subspace of $\mathring{H}_m(A)$ if we suppose that each function u of $\mathring{H}_m(A')$ is continued into the whole set A by assuming $u=0$ in $A-A'$.

Suppose we have, instead of (III) the weaker hypothesis

(III′) *The convex set V contains $\mathring{H}_m(A')$.*

If we retain unchanged the other hypotheses on the ellipticity of L and on the smoothness of the coefficients a_{pq} and of f (or, at most, we restrict them to A'), Theorem 3.I still holds, provided we read A' instead of A in its statement.

Let us now suppose that the domain A is C^∞-smooth at the point x^0 of ∂A (see E.T.E., Sect. 6). Let I_0 be a neighborhood of x^0 containing the neighborhood I of x^0 which is involved in the definition of C^∞-smoothness. Let V_0 be a class of functions v which are defined on the whole space X^r and which have their supports in I_0. If B is any set of X^r, we shall denote by $V_0(B)$ the class of functions on B obtained by taking the restriction to B of every $v\in V_0$. Let us suppose that: (i) $V_0(A)$ is a closed linear subspace of $H_m(A)$; (ii) $V_0(A\cap I_0)\supset\mathring{H}_m(A\cap I_0)$; (iii) for every $v\in V_0(A\cap I_0)$, the following inequality holds

$$B(v, v)\geqq\gamma_0\|v\|^2_{m,A} \qquad (\gamma_0>0); \tag{3.3}$$

(iv) Hypothesis $(\mathrm{IV})_{x_0}$ of Sect. 6 of E.T.E. is satisfied for the class $V_0(A)$.

Let us now assume that

(III″) *The convex set V contains the subspace $V_0(A)$.*

Let us remark that Hypothesis (III″) implies Hypothesis (III′) (with $A'=A\cap I_0$), because of ii), and implies Hypothesis (IV) (restricted to $A\cap I_0$), because of inequality (3.3) (see E.T.E. Theorem 5.II).

Let us assume that $a_{pq}\in C^\infty(X^r)$ and that for every $v\in V_0(A)$ we have $F(v)=(f, v)_0$ where f is a function of $H_\nu(A)$. The following theorem holds.

3.III *Under the assumed hypotheses, every solution u of the unilateral problem* (3.1) *belongs to $H_{2m+\nu}(B)$ for any domain B such that $\bar{B}\subset\bar{A}\cap(A\cup I)$.*

Arguing as in the proof of Theorem 3.I we see that $B(u, v)=(f, v)_0$ for every $v\in V_0(A)$. Then the proof of our theorem follows from Theorem 6.VI of E.T.E.

[6] For the definition of $\mathcal{U}_H$ see E.T.E., Sect. 6 (definition of problem (P)).

Theorem 3.I and the analogous theorem relative to Hypothesis (III'), give results concerning the *interior regularity* of u, while Theorem 3.III concerns the *boundary regularization* of u. Of course these theorems are obtained under the strong assumption that V contains linear subspaces such as to permit us to carry over the results of the regularization theory developed for linear problems. When this assumption is not satisfied, the regularization theory becomes much more difficult, as we shall see later, and the solution has, in general, only a mild degree of regularity even if the data are very smooth.

Remark. Suppose that a solution u of (3.1) exists such that $B(u, u)=F(u)$. Then Theorem 3.I [3.III] still holds for such a solution if we substitute for Hypothesis (III) [III'] the weaker one: *the convex set V contains the closed ball $\|w\| \leqq \varepsilon$ $(\varepsilon > 0)$ of $\mathring{H}_m(A)$ $[V_0(A)]$.*

The proof is readily obtained by an easy modification of the proof of Theorem 3.I.

4. General definition for the convex set V. Let us now consider three definitions for the convex set V that include the unilateral problems which generally arise in the theory of elasticity.

Let $\Phi_h(x, z_0, z_1, \ldots, z_s)$ $(h=1, \ldots, l)$ be a real valued function defined for $x \in X^r$ and for every choice of the vectors $z_0, \ldots, z_s$. We suppose that $z_0, \ldots, z_s$ are such that for every n-vector valued function $v \in C^\infty(X^r)$ we may consider the functions

$$\Phi_h[x, v(x), \ldots, D^p v(x), \ldots]$$

for $0 \leqq |p| \leqq m$.

We suppose that every Φ_h depends continuously on the variables $x, z_0, \ldots, z_s$. Moreover we suppose that, for $t_1 \geqq 0$, $t_2 \geqq 0$ and $t_1+t_2=1$ and for any choice of $x, z_0^{(1)}, \ldots, z_s^{(1)}, z_0^{(2)}, \ldots, z_s^{(2)}$,

$$\begin{aligned} \Phi_h(x, t_1 z_0^{(1)}+t_2 z_0^{(2)}, \ldots, t_1 z_s^{(1)}+t_2 z_s^{(2)}) \leqq t_1 \Phi_h(x, z_0^{(1)}, \ldots, z_s^{(1)}) \\ +t_2 \Phi_h(x, z_0^{(2)}, \ldots, z_s^{(2)}) \qquad \Phi_h(x, 0, \ldots, 0) \leqq 0. \end{aligned} \tag{4.1}$$

Let H be a closed linear subspace of $H_m(A)$, and let A_h be a measurable subset of A $(h=1, \ldots, l)$.

Let $\varphi_h(x)$ $(h=1, \ldots, l)$ be a real valued function, nonnegative and measurable in A_h.

We define V as follows.

α) *V is the set of all the functions v of H such that almost everywhere on A_h*

$$\Phi_h[x, v(x), \ldots, D^p v(x), \ldots] \leqq \varphi_h(x) \qquad (h=1, \ldots, l). \tag{4.2}$$

From (4.1) we deduce that V is not empty and is convex. Let $\{v_n\}$ be a converging sequence (in the topology of $H_m(A)$) of functions belonging to V. Let v be the limit of this sequence. Since from $\{v_n\}$ we can extract a subsequence, which we still denote by $\{v_n\}$, such that $\{D^p v_n\}$ converges to $D^p v$ almost everywhere on A, we see that the function v satisfies condition (4.2), i.e. belongs to V. Thus we have proved that V is closed.

Thus we may apply to the bilinear form $B(u, v)$ and to the convex set V the theory developed in the preceding sections. In particular if we choose $l=1$,

$$\Phi_1(x, v, \ldots, D^p v, \ldots) = \sum_{|p|}^{0, m} |D^p v|^2,$$

take $A_1 \equiv A$ and suppose that $\varphi_1(x)$ is bounded in A, we have a bounded closed convex set V. In this case the unilateral problem (3.1) has a solution for any given F.

Let us now consider a second general definition for the convex set V. Let us suppose that A is properly regular (see E.T.E. Sect. 2) and let $\Psi_h(x, z)$ $(h=1, \ldots, l')$ be a real valued function defined for $x \in \partial A$ and for any choice of the vector z. We suppose that z has as many components as needed for considering the function

$$\Psi_h[x, \tau v(x)] \qquad (h=1, \ldots, l')$$

where $v(x)$ is an n-vector valued function belonging to $C^m(\bar{A})$ and τ is the "boundary value operator" defined in Sect. 2 of E.T.E. We suppose that the boundary ∂A is decomposable into a finite number of non-overlapping differentiable $(r-1)$-cells[7] and that $\Psi_h(x, z)$ is continuous when x varies in any of these cells and z varies arbitrarily. Moreover for $t_1 \geqq 0$, $t_2 \geqq 0$, $t_1+t_2=1$ and any x, $z^{(1)}$, $z^{(2)}$, we have

$$\Psi_h(x, t_1 z^{(1)} + t_2 z^{(2)}) \leq t_1 \Psi_h(x, z^{(1)}) + t_2 \Psi_h(x, z^{(2)}) \qquad \Psi_h(x, 0) \leq 0.$$

Let H be a closed linear subspace of $H_m(A)$ and Σ_h $(h=1, \ldots, l')$ a subset of ∂A composed by non-overlapping differentiable $(r-1)$-cells. Let $\psi_h(x)$ be a real valued function nonnegative and measurable on Σ_h.

We now define V as follows.

β) *V is the set of all the functions v of H such that almost everywhere on Σ_h we have*

$$\Psi[x, \tau v(x)] \leqq \psi_h(x) \qquad (h=1, \ldots, l').^8 \tag{4.3}$$

As for the definition α) it is easy to prove that V is non-empty, closed and convex.

As third definition we take the one corresponding to the set which is the intersection of the two convex sets defined, respectively, by α) and by β).

γ) *V is the set of all the functions of H satisfying conditions* (4.2) and (4.3).

Since the intersection of two closed convex sets is closed and convex, so is the set V defined by γ).

5. Unilateral problems for an elastic body. Let A be a bounded domain of X^r which we suppose properly regular and with a boundary ∂A that can be decomposed into a finite number of non-overlapping $(r-1)$-cells of class C^n $(n \geqq 1)$. Let A represent the natural configuration of an r-dimensional elastic body, which we denote by the same letter A. Using the notations of Sect. 12 of E.T.E., for every pair of r-vectors u, v of $H_1(A)$ we consider the bilinear form

$$B(u, v) = \int_A a_{ih,jk}(x)\, u_{i/h}\, u_{j/k}\, dx,$$

[7] It is convenient to recall here the definition of a differentiable $(r-1)$-cell. Let $x = x(t) \equiv [x_1(t), \ldots, x_r(t)]$ be a r-vector valued function defined in the closed domain T^{r-1}, $0 \leqq t_i \leqq 1$ $(i=1, \ldots, r-1)$ $0 \leqq t_1 + \cdots + t_{r-1} \leqq 1$ of the $(r-1)$-dimensional space, satisfying the following conditions: i) $x = x(t)$ belongs to the class $C^n(T^{r-1})$ $(n \geqq 1)$; ii) the jacobian matrix $\partial x/\partial t$ has rank $r-1$ at every point of T^{r-1}; iii) the function $x = x(t)$ is univalent in T^{r-1}. The range Γ of the function $x = x(t)$ is, by definition, a *differentiable $(r-1)$-cell of class C^n*. The set of all the points of Γ, which correspond to the boundary ∂T^{r-1} of T^{r-1}, is the *border* $\partial\Gamma$ of Γ. When we say that two $(r-1)$-cells are *non-overlapping*, we mean that they have in common, at most, points of their borders.

[8] It is evident that now the concepts "measurable function" and "almost everywhere" must be understood with respect to the hypersurface measure on ∂A.

where the functions $a_{ih,jk}(x)$ are the ones introduced in Sect. 12 of E.T.E. and satisfy the hypotheses stated in that section. Let f be a given r-vector belonging to $H_0(A)$ and g a given r-vector belonging to $\mathscr{L}^2(\partial A)$. Let the *energy functional* to be associated with the body A be the following

$$\mathscr{I}(v)=\tfrac{1}{2}B(v,v)-F(v),$$

where

$$F(v)=\int_A f_i v_i\,dx+\int_{\partial A} g_i v_i\,d\sigma.$$

Let $\Phi_h(x;z_0;z_1)$ $(h=1,\ldots,l)$ be a real valued function depending on the r-vector x, the r-vector z_0, and the r^2-vector z_1, defined in the whole $(2r+r^2)$-dimensional cartesian space and continuous at every point $(x;z_0;z_1)$.

For the functions Φ_h we assume that condition (4.1) is satisfied, i.e. for $t_1\geqq 0$, $t_2\geqq 0$, $t_1+t_2=1$ and any choice of x, $z_0^{(1)}$, $z_1^{(1)}$, $z_0^{(2)}$, $z_1^{(2)}$, we have

$$\Phi_h(x;t_1 z_0^{(1)}+t_2 z_0^{(2)};t_1 z_1^{(1)}+t_2 z_1^{(2)})\leqq t_1\,\Phi_h(x;z_0^{(1)};z_1^{(1)})+t_2\,\Phi_h(x;z_0^{(2)};z_1^{(2)})$$
$$\Phi_h(x;0;0)\leqq 0.$$

Let $\varphi_h(x)$ be a nonnegative, bounded measurable function in the subdomain A_h of A.

We suppose that the elastic body is subjected to the following *internal constraints*

$$\Phi_h(x;v;v_{/1},\ldots,v_{/r})\leqq\varphi_h(x) \tag{5.1}$$

almost everywhere in A_h $(h=1,\ldots,l)$.

Let $\Psi_h(x,z)$ $(h=1,\ldots,l')$ be a real valued function defined for $x\in\partial A$ and for every r-vector z. Let the function $\Psi_h(x,z)$ be continuous when x varies in each $(r-1)$-cell of ∂A and z in the r-dimensional cartesian space. Moreover for $t_1\geqq 0$, $t_2\geqq 0$, $t_1+t_2=1$, for each $x\in\partial A$ and every choice of $z^{(1)}$ and $z^{(2)}$

$$\Psi_h(x,t_1 z^{(1)}+t_2 z^{(2)})\leqq t_1\,\Psi_h(x,z^{(1)})+t_2\,\Psi_h(x,z^{(2)})\qquad \Psi_h(x,0)\leqq 0.$$

Let Σ_h $(h=1,\ldots,l')$ be a subset of ∂A formed by non-overlapping differentiable $(r-1)$-cells of ∂A. Let $\psi_h(x)$ be a nonnegative bounded measurable function in Σ_h.

The elastic body be subjected to the following *boundary constraints*

$$\Psi_h(x,\tau v)\leqq\psi_h(x) \tag{5.2}$$

almost everywhere on Σ_h $(h=1,\ldots,l')$.

Let H be a closed linear subspace of $H_1(A)$.

The following one is a very general problem in the classical theory of elasticity. *To minimize the functional $\mathscr{I}(v)$ in the subset V of H formed by all the functions v that satisfy the conditions* (5.1) *and* (5.2).

The problem consists in finding the equilibrium configuration of an elastic body subjected to the body forces determined by f, to the surface forces determined by g, and subjected to the *bilateral constraints*, imposed upon the admissible displacements by requiring that they belong to H, and to the *unilateral constraints* represented by (5.1) and (5.2).

It must be remarked that either (5.1) or (5.2), which are stated here as unilateral constraints, may turn out, in some particular case, to be bilateral, as we shall see by an interesting illustration.

The problem stated, although very general in elasticity, is a particular case of the theory developed in the previous sections of this article. The convex set V is the one determined by condition γ) of the preceding section. Of course, since we do not exclude either the case when $l'=0$ or the case when $l=0$, we may have convex sets either of the kind α) or of the kind β). If we take $l=l'=0$, we have the problems discussed in Sect. 12 of E.T.E.

Of particular relevance is the case when $l=0$, $l'=1$; $\Psi_1(x,z)=-\nu(x)\,z$ [$\nu(x)$ is the unit inward normal to ∂A]; $\Sigma_1\equiv\Sigma$ [subset of ∂A composed by non-overlapping $(r-1)$-cells]; $\psi_1(x)\equiv 0$; $g\equiv 0$ on Σ; $H=H_1(A)$. The unilateral constraint is now the following

$$v_i\,\nu_i\geqq 0 \qquad \text{a.e. on } \Sigma.$$

This particular case defines what is nowadays known in the literature as the *Signorini problem*. It corresponds to the equilibrium problem of an elastic body, which in its natural configuration is supported by a rigid frictionless surface Σ. We shall study this problem in full detail in the sequel.

Let us now assume $l=2$, $l'=0$. Let Σ be a subset of ∂A formed by non-overlapping $(r-1)$-cells. Set $g\equiv 0$ on Σ. Let H be the subspace of $H_1(A)$ formed by the functions of $H_1(A)$ vanishing on ∂A. Assume

$$\Phi_1(x;v;v_{/1},\ldots,v_{/r})\equiv-\Phi_2(x;v;v_{/1},\ldots,v_{/r})\equiv v_{i/i},$$

$A_1=A_2=A$ and $\varphi_1(x)\equiv\varphi_2(x)\equiv 0$. In this case the two unilateral constraints given by (5.1) are equivalent to the unique bilateral constraint $v_{i/i}=0$ in A. This is the *incompressibility condition* for the elastic body A. The equilibrium problem concerns now an incompressible elastic body, fixed along the part Σ of its boundary, subjected to given surface forces φ on the remaining part of the boundary and to given body forces f on A. Theorem 1.III gives readily an existence theorem for the minimum of $\mathscr{I}(v)$ in this particular case.

6. Other examples of unilateral problems.

(I) *The Signorini problem for a scalar 2nd order elliptic operator.* Let A be the domain considered in Sect. 10 of E.T.E. and let $B(u,v)$ be the bilinear form defined in the example III of that section. Under the hypotheses there assumed we have $-B(v,v)\geqq c_0\,\|v\|_1^2$. Let $H=H_1(A)$ and V the convex set defined by the condition $v\geqq 0$ on ∂A. For any given $f\in H_0(A)$ there exists one and only one solution of the unilateral problem

$$\begin{gathered}\int_A\left\{\alpha_{ij}\frac{\partial(v-u)}{\partial x_i}\frac{\partial u}{\partial x_j}-\beta_i(v-u)\frac{\partial u}{\partial x_i}-c\,(v-u)\,u\right\}dx\\ \geqq\int_A f(v-u)\,dx\qquad u\in V,\ \forall v\in V.\end{gathered}\tag{6.1}$$

In this case the convex set V satisfies the Hypothesis (III) of Sect. 3. Then $u\in H_2(B)$ for any domain B such that $\overline{B}\subset A$ and satisfies in A the differential equation

$$\frac{\partial}{\partial x_i}\left[\alpha_{ij}\frac{\partial u}{\partial x_j}\right]+\beta_i\frac{\partial u}{\partial x_i}+c\,u+f=0.$$

Let us now suppose that $u\in H_2(A)$.[9] From (6.1) we easily deduce $-B(u,u)=(f,u)_0$, $-B(u,v)\geqq(f,v)_0$. Moreover we have (see E.T.E., Sect. 10, III)

$$B(u,v)=-\int_A f\,v\,dx+\int_{\partial A}v\,\varrho\,\frac{\partial u}{\partial\lambda}\,d\sigma.$$

[9] This will turn out to be true from the analysis we shall develop in Sects. 8 and 9.

Then

$$\int_{\partial A} v\,\varrho \frac{\partial u}{\partial \lambda}\, d\sigma \leqq 0, \qquad \int_{\partial A} u\,\varrho \frac{\partial u}{\partial \lambda}\, d\sigma = 0.$$

By the arbitrariness of v we deduce that u satisfies almost everywhere on ∂A the conditions:

$$u \geqq 0, \qquad \frac{\partial u}{\partial \lambda} \leqq 0, \qquad u \frac{\partial u}{\partial \lambda} = 0,$$

which—provided they hold in the whole of ∂A—lead to the *ambiguous boundary conditions*

$$\text{either} \begin{cases} u > 0 \\ \dfrac{\partial u}{\partial \lambda} = 0 \end{cases} \qquad \text{or} \begin{cases} u = 0 \\ \dfrac{\partial u}{\partial \lambda} \leqq 0. \end{cases}$$ [10]

(II) *Membrane fixed along its boundary and stretched over an obstacle.* Let A be a bounded domain of X^r. Assume

$$B(u, v) = \int_A u_{/i} v_{/i}\, dx \qquad (u, v \text{ real valued functions}). \tag{6.2}$$

Let $\varphi(x)$ be a continuous function defined in $\bar{A}$ which takes nonpositive values on ∂A. Assume $H = \mathring{H}_1(A)$, and let the closed convex set V be defined by the condition $u \geqq \varphi$ in A.[11]

There exists one and only one solution u of the unilateral problem

$$\int_A u_{/i}(v_{/i} - u_{/i})\, dx \geqq 0.$$

The function u minimizes in V the Dirichlet integral

$$\int_A v_{/i} v_{/i}\, dx$$

in the set V. In the case when $r=2$, u gives the equilibrium configuration of a membrane (coinciding with the domain A in its natural configuration) which is stretched over an obstacle represented by the function $\varphi(x)$.

(III) *Elastic plastic torsion problem.* This problem leads to a unilateral problem relative to a bounded domain A of X^r. Also in this case, the bilinear form $B(u,v)$ is given by (6.2). H is again $\mathring{H}_1(A)$. The closed convex set V is defined by the condition

$$|u_{/i} u_{/i}| \leqq a$$

where a is a positive constant. From the general theory we know that there exists one and only one solution of the unilateral problem.

$$\int_A u_{/i}(v_{/i} - u_{/i})\, dx \geqq b \int_A (v - u)\, dx$$

[10] Boundary conditions of this kind were indicated for the first time by SIGNORINI in his problem in elasticity. He proposed the term "ambiguous boundary conditions" because it is not known a priori what set of conditions at a given point of the boundary is satisfied by the solution u of the problem.

[11] For a more general formulation of this problem see [*26*].

(b given constant). The function u minimizes in V the functional

$$\frac{1}{2}\int_A v_{/i}\, v_{/i}\, dx - b\int_A v\, dx.$$

(IV) *Clamped plate, partially supported on a subdomain.* Let A be a bounded domain of the plane x, y, and let $B(u, v)$ be the bilinear form defined in Sect. 11 of E.T.E. which is used for proving Theorem 11.I. Let $H \equiv \mathring{H}_2(A)$. Let A_0 be a subdomain of A, and let V be the closed convex set defined by the condition $v \geqq 0$ in A_0. There exists one and only one solution u of the unilateral problem

$$\int_A \left(\frac{\partial^2 u}{\partial x^2}\,\frac{\partial^2 (v-u)}{\partial x^2} + 2\,\frac{\partial^2 u}{\partial x\,\partial y}\,\frac{\partial^2 (v-u)}{\partial x\,\partial y} + \frac{\partial^2 u}{\partial y^2}\,\frac{\partial^2 (v-u)}{\partial y^2}\right) dx\, dy \geqq \int_A f(v-u)\, dx\, dy$$

(f is a function of $H_0(A)$).

The function u minimizes in V the functional

$$\frac{1}{2}\int_A \left\{\left(\frac{\partial^2 v}{\partial x^2}\right)^2 + 2\left(\frac{\partial^2 v}{\partial x\,\partial y}\right)^2 + \left(\frac{\partial^2 v}{\partial y^2}\right)^2\right\} dx\, dy - \int_A f\, v\, dx\, dy.$$

Let us observe that in this example, if A_0 is a proper subdomain of A, the convex set V satisfies the condition (III′) of Sect. 3 for every domain $A' \subset A - \bar{A}_0$ (if there is any). Then $u \in H_{4+\nu}(A')$ if $f \in H_\nu(A')$ and $\Delta_4 v = f$ in A'. Moreover if ∂A is C^∞-smooth in x^0, Hypothesis (III″) of Sect. 3 is satisfied by assuming as V_0 the space $\mathring{H}_2(A \cap I_0)$[12] provided we can take I_0 at a positive distance from A_0. It follows that u is smooth in the neighborhood of x^0 according to the smoothness of f.

(V) *Elastic, perfectly locking body.* A is the domain considered in Sect. 5 and $B(u, v)$ the bilinear form defined in that section. H is one of the following three spaces: $\mathring{H}_1(A)$, $H_1(A)$, $H_\Sigma \equiv \{v;\, v \in H_1(A),\, v=0 \text{ on } \Sigma\}$ (Σ is a subset of ∂A formed by $(r-1)$-cells of ∂A).

Let $\varphi(\varepsilon)$ be a function depending on the symmetric tensor $\varepsilon \equiv \{\varepsilon_{ij}\}$ which is continuous for every ε, is a convex function of ε and is such that $\varphi(0) < 0$.

Let V be the closed convex subset of H formed by all the functions v such that

$$\varphi\left[\ldots, 2^{-1}(v_{j/i} + v_{i/j}), \ldots\right] \leqq 0 \qquad \text{a.e. in } A.$$

The kernel $N(T)$ of the quadratic form $B(v, v)$ is constituted by all the rigid displacements $\varrho = a + Cx$ (a constant r-vector, C skew-symmetric constant $r \times r$-matrix) belonging to H. From the Korn 2nd inequality, which we suppose to hold in A, it follows that Hypothesis (I of Sect. 1 is satisfied.[13]

If R is the space of all the rigid displacements we have $N(T) = R \cap H$. Define $F(v)$ as in Sect. 5, taking $g \equiv 0$ on ∂A if $H = \mathring{H}_1(A)$, and $g \equiv 0$ on $\partial A - \Sigma$ if $H = H_\Sigma$. A necessary condition for the existence of the minimum of $\mathscr{I}(v)$ in V is $F(\varrho) = 0$ for every $\varrho \in R \cap H$.

If L, $\tilde{Q}$ and $\tilde{P}$ are defined as in Sect. 1, and if we assume $u_0 = 0$, we see that $L = R \cap H$ and $\tilde{P}[V] \subset V$. Hence, from Theorem 1.II, it follows that $\mathscr{I}(v)$ has a minimum in V. The minimizing function is determined up to a vector of $R \cap H$.

It must be remarked that $R \cap H \neq \{0\}$ only in the case when $H = H_1(A)$.

---

[12] Of course we suppose, as usual, that each function of $\mathring{H}_2(A \cap I_0)$ is continued in the whole space and coincides with 0 outside of $A \cap I_0$.

[13] We shall develop that in more detail in the next section.

7. Existence theorem for the generalized Signorini problem. Let us now consider the problem stated in Sect. 5 under more specific assumptions on the functions Φ_h and Ψ_h, φ_h, ψ_h. We shall suppose that

(i) $\Phi_h(x; z_0; z_1)$ is a continuous function of $(x; z_0; z_1)$ for every choice of these variables;

(ii) $\Phi_h(x; z_0; z_1)$ for any $x \in X^r$ is a convex function of $(z_0; z_1)$;

(iii) $\Phi_h(x; z_0; z_1)$ for any $x \in X^r$ is a homogeneous function of degree 1 of $(z_0; z_1)$.

It follows that for $t_1 \geqq 0$, $t_2 \geqq 0$ we have

$$\Phi_h(x; t_1 z_0^{(1)} + t_2 z_0^{(2)}; t_1 z_1^{(1)} + t_2 z_1^{(2)}) \leqq t_1 \Phi_h(x; z_0^{(1)}; z_1^{(1)}) + t_2 \Phi_h(x; z_0^{(2)}; z_1^{(2)}) \tag{7.1}$$

and moreover

$$\Phi_h(x; 0; 0) = 0.$$

We assume similar hypotheses on the $\Psi_h(x; z)$:

(i) $\Psi_h(x; z)$ is a continuous function of $(x; z)$ when x varies in each $(r-1)$-cell of ∂A and z is an arbitrary r-vector;

(ii) $\Psi_h(x; z)$ for any $x \in \partial A$ is a convex function of z;

(iii) $\Psi_h(x; z)$ for any $x \in \partial A$ is a homogeneous function of degree 1 of z.

We have for $t_1 \geqq 0$, $t_2 \geqq 0$

$$\Psi_h(x, t_1 z^{(1)} + t_2 z^{(2)}) \leqq t_1 \Psi(x, z^{(1)}) + t_2 \Psi(x, z^{(2)}) \qquad \Psi(x, 0) = 0. \tag{7.2}$$

Let H be the space $H_1(A)$, and let Σ be the subset of ∂A considered in stating the Signorini problem in Sect. 5.

Let A_h $(h=1, \ldots, l)$ be a subdomain of A and Σ_h a subset of Σ composed by $(r-1)$-cells of ∂A.

We shall define the closed convex set V as the set of all the functions of $H_1(A)$ such that

$$\Phi_h(x, v; v_{/1}, \ldots, v_{/r}) \leqq 0 \qquad \text{a.e. in } A_h \ (h=1, \ldots, l), \tag{7.3}$$

$$\Psi_h(x, \tau v) \leqq 0 \qquad \text{a.e. on } \Sigma_h \ (h=1, \ldots, l'). \tag{7.4}$$

Let $B(u, v)$ be the bilinear form of elasticity already considered in Sect. 5. We have

$$B(v, v) \geqq c_1 \int_A (v_{i/j} + v_{j/i})(v_{i/j} + v_{j/i})\, dx \qquad (c_1 > 0).$$

The kernel of $B(v, v)$ is the space R of the rigid displacements. Let Q be the projector of $H_1(A)$ onto R. For any vector $v \in H_1(A)$ such that $Q v = 0$, we have, as it is easily seen,

$$\int_A v\, dx = 0, \qquad \int_A (v_{i/j} - v_{j/i})\, dx = 0.$$

Then, if we assume that Korn's second inequality holds in A,[14] from the remark contained in Sect. 12 of E.T.E. and from the footnote[17] of that paper it follows that $B(v, v)$ satisfies Hypothesis (I) of Sect. 1. Since R is finite dimensional, Hypothesis (II) is satisfied too.

Let us now consider the subset $R' = R \cap V$. Let R^* be the subset of R' formed by all the vectors of R' which are bilateral, i.e. R^* is defined as follows

$$R^* \equiv \{\varrho; \varrho \in R', \varrho \in R \Rightarrow -\varrho \in R\}.$$

---

[14] Actually from the hypotheses assumed on A, it is possible to prove that A satisfies a restricted cone hypothesis.

It is easy to prove that R^* is a linear space. Let

$$f \in H_0(A), \quad g \in \mathscr{L}^2(\Sigma^*) \quad [\Sigma^* = \partial A - \Sigma].$$

By the *generalized Signorini problem* we shall mean the following unilateral problem

$$B(u, v-u) \geqq \int_A f(v-u)\,dx + \int_{\Sigma^*} g(v-u)\,d\sigma, \qquad u \in V,\ \forall v \in V. \tag{7.5}$$

It is equivalent to minimize

$$\mathscr{I}(v) = \tfrac{1}{2} B(v, v) - \int_A f v\,dx - \int_{\Sigma^*} g v\,d\sigma$$

in V. The Signorini problem is a particular case of this more general problem, as is easily seen.

7.I *If for every* $\varrho \in R'$

$$F(\varrho) \equiv \int_A f \varrho\,dx + \int_{\Sigma^*} g \varrho\,d\sigma \leqq 0 \tag{7.6}$$

and if the sign = holds when and only when $\varrho \in R^*$, *the generalized Signorini problem has a solution. If u is a solution of* (7.5), *any other solution* u' *is given by* $u' = u + \varrho$, *where* ϱ *is a rigid displacement such that* $F(\varrho) = 0$, $u + \varrho \in V$.

We may apply Theorem 1.II, assuming $u_0 = 0$. We need only to show that, if $\tilde{Q}$ is the projector of $H_1(A)$ onto R^* and $\tilde{P} = I - \tilde{Q}$, then $\tilde{P}[V]$ is closed. If $v \in V$, from (7.1) we have

$$\begin{aligned}\Phi_h(x; \tilde{P}v; (\tilde{P}v)_{/1}, \ldots, (\tilde{P}v)_{/r}) &\leqq \Phi_h(x; v; v_{/1}, \ldots, v_{/r}) \\ &+ \Phi_h(x; -\tilde{Q}v; (-\tilde{Q}v)_{/1}, \ldots, (-\tilde{Q}v)_{/r}) \leqq 0 \quad \text{a.e. in } A_h \ (h=1, \ldots, l)\end{aligned}$$

and, analogously, from (7.2) we deduce that

$$\Psi_h(x; \tau \tilde{P} v) \leqq 0 \quad \text{a.e. on } \Sigma_h \ (h=1, \ldots, l').$$

This means $\tilde{P}[V] \subset V$. Hence $\tilde{P}[V]$ is closed. The statement about uniqueness follows from 1.VII.

Let us explicitly note the following particular case of Theorem 1.IV. Let $P = I - Q$.

7.II *If for every* $w \in P[H_1(A)]$ *there exists a* $\varrho \in R$ *such that* $w + \varrho \in V$, *and if* $F(\varrho) = 0$ *for every* $\varrho \in R$, *there exists a solution of the unilateral problem* (7.5).

8. Regularization theorem: interior regularity. We consider now a regularity theorem which, under suitable hypotheses, is able to guarantee a certain regularity to a solution of a unilateral problem in the neighborhood of an interior point x^0 of A. We consider a bilinear form of the following kind

$$B(u, v) = \int_A \{\alpha_{hk}(x)\, u_{/h} v_{/k} + \beta_h(x)\, u_{/h} v + \beta'_h(x)\, u\, v_{/h} + \gamma(x)\, u\, v\}\,dx, \tag{8.1}$$

where A is a bounded domain of X^r, and assume that the $n \times n$ matrices α_{hk} belong to $C^2(\bar{A})$, the matrices β'_h belong to $C^1(\bar{A})$[15] and the matrices β_h, γ are bounded and measurable in A. Moreover

$$\alpha_{hk}(x) \equiv \bar{\alpha}_{kh}(x).$$

[15] Actually we shall see that the theorem, we are going to prove, still holds, with the same proof, under less restrictive assumptions on these matrices.

Let us suppose that there exists a solution u of the unilateral problem

$$B(u, v-u) \geqq F(v-u), \quad u \in V, \ \forall v \in V, \tag{8.2}$$

that V is a closed convex set of the space $H_1(A)$ (of the n-vector valued functions) and that $F(v)$ is a bounded linear functional in $H_1(A)$ such that, for every $v \in \mathring{H}_1(A)$,

$$F(v) = \int_A f\, v\, dx$$

with $f \in H_0(A)$.

Let us assume the following hypotheses:

1) If v is any function of V and φ a nonnegative scalar function of C^∞, the function $\varphi\, v$ belongs to V;

2) There exists a subdomain E of A such that

$$c\,\|v\|^2 \leqq B(v, v), \quad \forall v \in \mathring{H}_1(E), \quad (c > 0).$$

3) Let $v \in V$ and $spt\, v \in E$. Let y be such that $spt\, v(x+y) \subset E$. Then $v(x+y) \in V$.

Observe that these conditions are satisfied in the generalized Signorini problem if we assume the further hypothesis that the functions Φ_h do not depend on x and on z_1 but are functions of the variable z_0 alone.

8.I *The solution u of the unilateral problem* (8.2) *belongs to $H_2(I)$ for any I such that $\bar{I} \subset E$. Given $x_0 \in E$, there exists a $\delta > 0$ such that $\bar{\Gamma}_\delta \in E$ and*

$$\|u\|^2_{2, \Gamma_\delta} \leqq c_0 \left(\|f\|^2_{0, \Gamma_{2\delta}} + \|u\|^2_{1, \Gamma_{2\delta}}\right)$$

where Γ_ϱ is the ball of center $x^0 \in E$ and radius ϱ and where c_0 is a positive constant depending only on the coefficients of the bilinear form (8.1) *and on x^0.*

Let us first remark that, because of the hypotheses we have assumed on V, we have $B(u, u) = F(u)$, $B(u, v) \geqq F(v)$ $(\forall v \in V)$. If ψ is a scalar function of C^∞ such that $0 \leqq \psi \leqq 1$, we have $B(u, \psi\, u) \geqq F(\psi\, u)$, $B(u, (1-\psi)\, u) \geqq F(u) - F(\psi\, u)$. Hence $B(u, \psi\, u) = F(\psi\, u)$. Fixed $x^0 \in E$, let $\delta > 0$ be such that $\bar{\Gamma}_{2\delta} \in E$. Let $\varphi(x)$ be a scalar function of C^∞ such that $\varphi(x) \geqq 0$, $\varphi(x) \equiv 1$ for $|x - x_0| \leqq \delta$, $\varphi(x) \equiv 0$ for $|x - x_0| \geqq (3/2)\, \delta$. Set $U(x) = \varphi(x)\, u(x)$. Using arguments employed in E.T.E., it is easy to see that for $0 < |y| \leqq \delta/2$

$$\left\| \frac{2U(x) - U(x-y) - U(x+y)}{|y|^2} \right\|_{0, A} \leqq c_1 \left\| \frac{U(x+y) - U(x)}{|y|} \right\|_{1, A} \quad (c_1 > 0).$$

Moreover, if $g \in C^2(\bar{A})$

$$\left\| \frac{2g(x)\, U(x) - g(x-y)\, U(x-y) - g(x+y)\, U(x+y)}{|y|^2} \right\|_{0, A} \leqq c_2 \left\| \frac{2U(x) - U(x-y) - U(x+y)}{|y|^2} \right\|_{0, A}$$

where c_2 depends on the C^2 norm of $g(x)$.

Let us recall that, if the function of y

$$\left\| \frac{U(x+y) - U(x)}{|y|} \right\|_{1, A}$$

is bounded by c_3, then $\|U\|_{2,A}$ is finite and $\|U\|_{2,A} \leqq c_4 c_3$. We have for $0 \leqq |y| \leqq \delta/2$

$$
\begin{aligned}
B[U(x+y)-U(x),\, U(x+y)-U(x)] = {} & 2B[u(x),\, \varphi(x)\, U(x)] \\
& -2B[u(x),\, \varphi(x)\, U(x-y)] \\
& +\int_A \{\alpha_{hk}(x)\,[U_{/h}(x+y)-U_{/h}(x)]\,[U_{/k}(x+y)-U_{/k}(x)] \\
& -2\alpha_{hk}(x)\, u_{/h}(x)\,[\varphi(x)\, U(x)]_{/k} + 2\alpha_{hk}(x)\, u_{/h}(x)\,[\varphi(x)\, U(x-y)]_{/k} \\
& +\beta_h(x)\,[U_{/h}(x+y)-U_{/h}(x)]\,[U(x+y)-U(x)] \\
& -2\beta_h(x)\, u_{/h}(x)\, \varphi(x)\, U(x) + 2\beta_h(x)\, u_{/h}(x)\, \varphi(x)\, U(x-y) \\
& +\beta'_h(x)\,[U(x+y)-U(x)]\,[U_{/h}(x+y)\, U_{/h}(x)] \\
& -2\beta'_h(x)\, u(x)\,[\varphi(x)\, U(x)]_{/h} + 2\beta'_h(x)\, u(x)\,[\varphi(x)\, U(x-y)]_{/h} \\
& +\gamma(x)\,[U(x+y)-U(x)]\,[U(x+y)-U(x)] \\
& -2\gamma(x)\, U(x)\, U(x) + 2\gamma(x)\, U(x)\, U(x-y)\}\, dx.
\end{aligned} \tag{8.3}
$$

Set

$$\lambda(y) = B[U(x+y)-U(x),\, U(x+y)-U(x)].$$

From (8.3), keeping in mind that

$$
\begin{aligned}
B[u(x),\, \varphi^2(x)\, u(x)] = {} & (\varphi(x)\, f(x),\, U(x))_0 \\
& - B[u(x), \varphi(x)\, \varphi(x-y)\, u(x-y)] \leqq -(\varphi(x)\, f(x), U(x-y))_0,
\end{aligned}
$$

we deduce an inequality which we write briefly as $\lambda(y) \leqq \mu(y)$. On the other hand we have $\lambda(y) = \lambda(-y) + \sigma(y)$, where

$$
\begin{aligned}
\sigma(y) = \int_A \{ & [\alpha_{hk}(x-y)-\alpha_{hk}(x)]\,[U(x-y)-U(x)]_{/h}\,[U(x-y)-U(x)]_{/k} \\
& +[\beta_h(x-y)-\beta_h(x)]\,[U(x-y)-U(x)]_{/h}\,[U(x-y)-U(x)] \\
& +[\beta'_h(x-y)-\beta'_h(x)]\,[U(x-y)-U(x)]\,[U(x-y)-U(x)]_{/h} \\
& +[\gamma(x-y)-\gamma(x)]\,[U(x-y)-U(x)]\,[U(x-y)-U(x)]\}\, dx.
\end{aligned}
$$

Hence $\lambda(y) \leqq 2^{-1}[\mu(y)+\mu(-y)+\sigma(y)]$. If we set $\Phi(x) = \varphi(x)\, f(x)$, this last inequality, after simple transformations, gives

$$
\begin{aligned}
B[U(x+y)-U(x),\, U(x+y)-U(x)] \leqq {} & (\Phi(x),\, 2U(x)-U(x-y)-U(x+y))_0 \\
& +\int_A \{[\varphi_{/h}(x+y)\, u(x+y)-\varphi_{/h}(x)\, u(x)]\,[\bar\alpha_{hk}(x+y)\, U_{/k}(x+y)-\bar\alpha_{hk}(x)\, U_{/k}(x)] \\
& +\varphi_{/k}(x)\, u_{/h}(x)\,[\bar\alpha_{hk}(x-y)\, U(x-y)+\bar\alpha_{hk}(x+y)\, U(x+y)-2\bar\alpha_{hk}(x)\, U(x)] \\
& -[\alpha_{hk}(x-y)-\alpha_{hk}(x)]\, u_{/h}(x)\,([\varphi(x)\, U(x-y)]_{/k}-[\varphi(x)\, U(x)]_{/k}) \\
& -[\alpha_{hk}(x+y)-\alpha_{hk}(x)]\, u_{/h}(x)\,([\varphi(x)\, U(x+y)]_{/k}-[\varphi(x)\, U(x)]_{/k}) \\
& -[\alpha_{hk}(x-y)+\alpha_{hk}(x+y)-2\alpha_{hk}(x)]\, u_{/h}(x)\,[\varphi(x)\, U(x)]_{/k} \\
& +2^{-1}[\alpha_{hk}(x-y)+\alpha_{hk}(x+y)-2\alpha_{hk}(x)]\, U_{/h}(x)\, U_{/k}(x) \\
& +2^{-1}\beta_h(x)\,[U_{/h}(x+y)-U_{/h}(x)]\,[U(x+y)-U(x)] \\
& +2^{-1}\beta_h(x)\,[U_{/h}(x-y)-U_{/h}(x)]\,[U(x-y)-U(x)] \\
& +\beta_h(x)\, \varphi(x)\, u_{/h}(x)\,[U(x-y)+U(x+y)-2U(x)] \\
& +2^{-1}\beta'_h(x)\,[U(x+y)-U(x)]\,[U_{/h}(x+y)-U_{/h}(x)] \\
& +2^{-1}\beta'_h(x)\,[U(x-y)-U(x)]\,[U_{/h}(x-y)-U_{/h}(x)] \\
& -[\beta'_h(x+y)\, U(x+y)-\beta'_h(x)\, U(x)]\,[U_{/h}(x+y)-U_{/h}(x)] \\
& +\beta'_h(x)\, \varphi_{/h}(x)\, u(x)\,[U(x-y)+U(x+y)-2U(x)] \\
& +2^{-1}\gamma(x)\,[U(x+y)-U(x)]\,[U(x+y)-U(x)] \\
& +2^{-1}\gamma(x)\,[U(x-y)-U(x)]\,[U(x-y)-U(x)] \\
& +\gamma(x)\, U(x)\,[U(x-y)+U(x+y)-2U(x)]\}\, dx + 2^{-1}\sigma(y).
\end{aligned}
$$

Denoting by $p_1, p_2, \ldots$ positive constants and by $\| \|$ norms over A, we have for $0<|y|\leqq\delta/2$

$$c\left\|\frac{U(x+y)-U(x)}{|y|}\right\|_1^2\leqq p_1\|\Phi\|\left\|\frac{U(x+y)-U(x)}{|y|}\right\|_1$$
$$+p_2\|u\|_{1,\Gamma_{2\delta}}\left\|\frac{U(x+y)-U(x)}{|y|}\right\|_1+p_3\|u\|^2_{1,\Gamma_{2\delta}} \tag{8.4}$$
$$+\sum_{h,k}^{1,r}\sup_{x,\xi\in\Gamma_{2\delta}}|a_{hk}(x)-a_{hk}(\xi)|\left\|\frac{U(x+y)-U(x)}{|y|}\right\|_1^2.$$

We can choose δ such that

$$\sum_{h,k}^{1,r}\sup_{x,\xi\in\Gamma_{2\delta}}|a_{hk}(x)-a_{hk}(\xi)|<c.$$

Then the proof of the theorem is an obvious consequence of (8.4).

9. Regularization theorem: regularity near the boundary. The technique used in the preceding section can be extended so as to get analogous results concerning regularization near the regular points of the boundary of A.

We consider the bilinear form $B(u,v)$ given by (8.1) and impose on the matrices $\alpha_{hk}, \beta_h, \beta'_h, \gamma_h$ the hypotheses of Sect. 8.

Let x^0 be a point of ∂A, and suppose that A is C^3-smooth in x^0. This means that there exists a neighborhood I of x^0 such that the set $J=\bar{I}\cap\bar{A}$ can be mapped C^3-homeomorphically onto the closed semiball $\bar{\Sigma}^+$: $|y|^2+t^2\leqq 1$, $t\geqq 0$ of the r-dimensional (y,t)-space in such a way that the set $\bar{I}\cap\partial A$ is mapped onto the $(r-1)$-dimensional ball $t=0$, $|y|\leqq 1$.

As in Sect. 8, we assume that there exists a solution u of the unilateral problem (8.2). We suppose that the functional $F(v)$ for every $v\in H_1(A)$ such that *spt* $v\in J$, admits the following representation

$$F(v)=\int_A f\,v\,dx$$

with $f\in H_0(A)$.

We now assume the following hypotheses:

1) If v is any function of V and φ a nonnegative scalar function of C^∞, the function φv belongs to V.

2) We have

$$c\|v\|^2\leqq B(v,v), \quad \forall v\in H_1(A),\ spt\, v\in J \quad (c>0).$$

3) Let $x=x(\xi)\equiv x(y,t)$ the C^3-homeomorphism from $\bar{\Sigma}^+$ to J. Let E be the image of the semiball Σ^+: $|y|^2+t^2<1$, $t\geqq 0$ under this homeomorphism. Let $v\in V$ and *spt* $v\in E$. Let y be such that *spt* $v[x(\xi+y)]\subset\Sigma^+$. Then $v[x(\xi+y)]\in V$.

4) The solution u of the unilateral problem (8.2) satisfies in $A\cap I$ the differential equation

$$(\alpha_{hk}u_{/h})_{/k}+\beta_h u_{/h}-(\beta'_h u)_{/h}+\gamma u+f=0. \tag{9.1}$$

If U is a subset of $\bar{\Sigma}^+$ we shall denote by $x(U)$ its image under the homeomorphism $x=x(\xi)$.

9.I *There exists* $\delta>0$ such that $\Gamma_\delta^+\subset\Sigma^+$, $u\in H_2[x(\Gamma_\delta^+)]$ *and*

$$\|u\|^2_{2,x(\Gamma_\delta^+)}\leqq c_0(\|f\|^2_{0,x(\Gamma_{2\delta}^+)}+\|u\|^2_{1,x(\Gamma_{2\delta}^+)});$$

Γ_δ^+ is the semiball $|y|^2+t^2 \leq \delta^2$, $t>0$ and c_0 a positive constant only depending on the coefficients of the bilinear form (8.1) *and on x^0.*

The proof parallels the one of Theorem 8.I. We have first to transform the bilinear form $B(u,v)$, for any v such that $spt\, v \in E$, by using the homeomorphism $x=x(\xi)$. We get a bilinear form which we indicate as follows

$$\tilde{B}(u,v)=\int_{\Sigma^+} (\tilde{\alpha}_{hk}\, u_{/h}\, v_{/k}+\tilde{\beta}_h\, u_{/h}\, v+\tilde{\beta}'_h\, u\, v_{/h}+\tilde{\gamma}\, u\, v)\, d\xi .$$

From now on we exactly follow the proof of Theorem 8.I with the only difference that now y must be parallel to the hyperplane $t=0$. Thus we arrive at inequality (8.4). However the boundedness of

$$\left\| \frac{U(x+y)-U(x)}{|y|} \right\|_1^2$$

is only able, in this case, to insure that all second derivatives of U except U_{tt} belong to $\mathscr{L}^2(\Sigma^+)$. Since u satisfies the differential equation (9.1), after writing this equation in the variables y,t, we deduce, since, for Hypothesis 2), the equation is elliptic, that $U_{tt} \in \mathscr{L}^2(\Sigma^+)$ and, in consequence, the proof of the theorem.

Remark. In the proof of Theorem 8.I we considered in the integrand of the bilinear form (8.1) the presence of the term $\beta'_h\, u\, v_{/h}$. This could have been avoided by an integration by parts. However since this term cannot be eliminated in the proof of Theorem 9.I, we have retained it also in the proof of the earlier theorem.

In the case of the generalized Signorini problem, assuming that Φ_h does not depend on z_i, the hypotheses assumed in this section are satisfied if we suppose that: i) the domains A_h ($h=1, \ldots, l$) (see Sect. 7) are all disjoint from J; ii) for each h ($h=1, \ldots, l'$) either the subset Σ_h is disjoint from $\bar{I} \cap \partial A$ or the function $\Psi_h(x, z)$ does not depend on x (for $x \in \bar{I} \cap \partial A$).

In the original Signorini problem there is no question about the Φ_h's since $l=0$; the condition ii) concerning the Ψ_h's is not satisfied if $x^0 \in \Sigma$ since $l'=1$, $\Psi_1 = -\nu(x)\, z$, $\Sigma_1 \equiv \Sigma$. However, we may overcome this difficulty by introducing in the set E an orthogonal system of unit-vectors $\nu_1(x), \ldots, \nu_r(x)$ such that: 1) $\nu_i(x) \in C^2(\bar{E})$ ($i=1, \ldots, r$); 2) $\nu_r(x) \equiv \nu(x)$ for $x \in \bar{I} \cap \partial A$. If ∂A is C^3-smooth in x^0 the construction of this set of vectors is trivial. For each $v \in H_1(A)$ and for $x \in E$, set $v(x) = \tilde{v}_i(x)\, \nu_i(x)$. Let

$$a_{hk} \equiv \{a_{hk}^{ij}\} \equiv \{a_{jh,ik}\} \qquad (i,j=1, \ldots, r)$$

be the matrices considered in Sect. 12 of E.T.E. in connection with problems of elasticity and employed in Sect. 5.[16] If $spt\, v \in E$ we have

$$a_{hk}\, u_{/h}\, v_{/k} = a_{hk}\, \nu_i\, \tilde{u}_{i/h}\, \nu_j\, \tilde{v}_{j/k} + a_{hk}\, \nu_i\, \tilde{u}_{i/h}\, \nu_{j/k}\, \tilde{v}_j + a_{hk}\, \nu_{i/h} \tilde{u}_i\, \nu_j\, \tilde{v}_{j/k} + a_{hk}\, \nu_{i/h} \tilde{u}_i\, \nu_{j/k} \tilde{u}_j .$$

Let us define the matrices α_{hk}, β_h, β'_h, γ as follows:

$$\alpha_{hk} \equiv \{a_{hk}\, \nu_i\, \nu_j\}, \qquad \beta_h \equiv \{a_{hk}\, \nu_i\, \nu_{j/k}\}, \qquad \beta'_h \equiv \{a_{kh}\, \nu_{i/k}\, \nu_j\}, \qquad \gamma \equiv \{a_{hk}\, \nu_{i/h}\, \nu_{j/k}\}.$$

[16] By writing $\{a_{hk}^{ij}\}$ we mean the matrix a_{hk} which has a_{hk}^{ij} as entry in the i-th row and in the j-th column. Thus we assume $a_{hk}^{ij} = a_{jh,\, ik}$.

If we denote by $\tilde{u}$ and $\tilde{v}$ the r-vectors having as components $\tilde{u}_1, \ldots, \tilde{u}_r$ and $\tilde{v}_1, \ldots, \tilde{v}_r$, we have, if $spt\; v \in E$,

$$B(u, v) = \int_A (\alpha_{hk}\, u_{/h}\, v_{/k} + \beta_h\, u_{/h}\, v + \beta'_h\, u\, v_{/h} + \gamma\, u\, v)\, dx$$

and the constraint of the Signorini problem is expressed by $-\tilde{v}_r(x) \leqq 0$ $(x \in \bar{I} \cap \partial A)$. Then the above condition ii) is satisfied and we can apply the regularization theory developed in this section.

10. Analysis of the Signorini problem. On the regular bounded domain A we shall assume the following more specific hypotheses:

1) ∂A is decomposable into a finite set of non-overlapping differentiable $(r-1)$-cells of class $C^\infty: \Gamma_1, \ldots, \Gamma_q$.

2) It is possible to define a unit vector $\mu(x)$ which is a function of the point x variable on ∂A, which is a continuous function of x on ∂A and which always points inside A, while $-\mu(x)$ points into the complement of $\bar{A}$. Denoting by ω the angle (between 0 and π) which $\mu(x)$ forms with the inner unit normal vector $\nu(x)$, we always have $0 \leqq \omega \leqq \omega_0 < \pi/2$ in every point x of $\Gamma_k - \partial\Gamma_k$ $(k = 1, \ldots, q;$ $\partial\Gamma_k \equiv$ border of $\Gamma_k)$.

3) Denoting by $x = x^k(t)$ the parametric representation of Γ_k on T^{r-1} (see Footnote [7], p. 402) we suppose that $\mu[x^k(t)] \in C^\infty(T^{r-1})$.

4) There exists a positive number λ_0 such that for every λ such that $0 < \lambda \leqq \lambda_0$ the range described by $y = x + \lambda\, \mu(x)$ as x varies on ∂A is entirely contained in A and is in one-to-one correspondence with ∂A.

It is easy to prove that conditions 1), 2), 3), 4), imply that A is properly regular.

We shall consider in the domain A the Signorini problem. Assume

$$\Sigma = \Gamma_1 \cup \Gamma_2 \cup \ldots \cup \Gamma_{q'} \qquad (q' \leqq q).$$

Let

$$f \in H_0(A), \quad g \in \mathscr{L}^2(\Sigma^*) \quad (\Sigma^* = \partial A - \Sigma).$$

From the theory we have developed in the preceding sections we derive this theorem

10.I *Set*

$$B(u, v) = \int_A a_{hk}(x)\, u_{/h}\, v_{/k}\, dx, \qquad F(v) = \int_A f\, v\, dx + \int_{\Sigma^*} g\, v\, d\sigma,$$

where $\{a^{ij}_{hk}\} \equiv \{a_{jh,ik}\}$ *are the matrices of elasticity* (see E.T.E., Sect. 12). *Let V be the convex set formed by all the functions of $H_1(A)$ such that $v(x)\, \nu(x) \geqq 0$ on ∂A.*[17] *Let $R' = R \cap V$ and R^* be the subset of R' formed by all the bilateral displacements. A necessary condition for the existence of a minimum of the functional $\mathscr{I}(v) = 2^{-1} B(v, v) - F(v)$ in V is*

$$F(\varrho) \leqq 0 \quad \forall\, \varrho \in R'. \tag{10.1}$$

If this condition is satisfied in the strong sense, i.e. if the sign $=$ holds when and only when $\varrho \in R^*$, then *$\mathscr{I}(v)$ has an absolute minimum in V. If u is a minimizing function, u is a solution in A of the differential system*

$$(a_{hk}\, u_{/h})_{/k} + f = 0. \tag{10.2}$$

[17] Of course the inequality $v(x)\, \nu(x) \geqq 0$ must be understood in the sense of the functions of $H_1(A)$, i.e. almost everywhere.

If $f \in C^\infty(A)$, *then* $u \in C^\infty(A)$. *If* $q' < q$, $x^0 \in \Gamma_k - \partial\Gamma_k$ $(k = q'+1, \ldots, q)$ *and* $g[x^k(x)] \in C^\infty(T^{r-1})$. *Then* $u \in C^\infty(\bar{B})$, *for any domain B such that* $\bar{B} \subset \bar{A} \cap (A \cup I)$, *where I is a suitable neighborhood of* x^0,[18] *and u satisfies in* $\partial A \cap I$ *the boundary condition*

$$\sigma_{ih}(u)\,\nu_h = g_i \qquad [\sigma_{ih}(u) = -a_{ih,jk}\,u_{j/k}].$$

If $x^0 \in \Gamma_k - \partial\Gamma_k$ $(k = 1, \ldots, q')$ *then* $u \in H_2(A \cap I)$,[19] *where I is a suitable neighborhood of* x^0, *and u satisfies in* $\partial A \cap I$ *the "ambiguous boundary conditions"*

$$\text{either}\begin{cases} u_i\nu_i = 0 \\ \sigma_{ih}(u)\,\nu_i\nu_h \geqq 0 \\ \sigma_{ih}(u)\,\nu_i\,\tau_h = 0 \end{cases} \quad (10.3) \qquad \text{or}\begin{cases} u_i\nu_i > 0 \\ \sigma_{ih}(u)\,\nu_i\nu_h = 0 \\ \sigma_{ih}(u)\,\nu_i\,\tau_h = 0 \end{cases} \quad (10.4)$$

where τ *is any vector tangent to* Σ *in the points of* $\partial A \cap I$.

The conditions (10.3) express the fact that at the point of $\partial A \cap I$ under consideration the elastic body in its equilibrium configuration rests on Σ, and therefore, that the reaction of the constraints has a nonnegative component along the inward normal. Any tangential component of such a reaction is null since the surface Σ is supposed frictionless. On the other hand, if the conditions (10.4) are satisfied, this means that in coming to equilibrium the body has left the supporting surface, which therefore no longer reacts on the body.

Evidently it is not known a priori which of the two sets of conditions (10.3) and (10.4) is to be satisfied at a given point of Σ. Hence the use of the name proposed by Signorini of "ambiguous boundary conditions".

All the facts considered in the statement of the theorem are evident consequences of the theory developed in the preceding sections. For proving that the ambiguous conditions (10.3) (10.4) are satisfied in $\partial A \cap I$ (I being a suitable neighborhood of a *regular* point x^0 of Σ), one only needs to repeat, with easy generalizations, the arguments used in the example I of Sect. 6. Thus one shows that, almost everywhere on $\partial A \cap I$, the following conditions are satisfied:

$$u_i\nu_i \geqq 0, \qquad \sigma_{ik}\nu_i\nu_k \geqq 0, \qquad \sigma_{ik}\nu_i\nu_k u_j\nu_j = 0, \qquad \sigma_{ik}\nu_i\tau_k = 0. \tag{10.5}$$

Since the functions u_i, σ_{ik} are defined almost everywhere, we may assume, by suitably defining these functions on sets of measure zero,[20] that (10.5) hold at every point of $\partial A \cap I$. Thus we deduce that in $\partial A \cap I$ the ambiguous conditions (10.3) (10.4) hold.

The following theorem gives information about the global nature of the reaction that the supporting surface Σ exerts on the elastic body.

10.II *A real-valued nonnegative measure* $\gamma(B)$ *is defined in the* σ*-ring* $\{B\}_\Sigma$ *of the Borel sets B contained in* Σ *such that, if u minimizes* $\mathscr{I}(v)$ *in V, we have for every* $v \in H_1(A) \cap C^0(A \cup \Sigma)$

$$\int_\Sigma v_i\nu_i\,d\gamma = B(u, v) - F(v). \tag{10.6}$$

[18] Actually, in the case when $g \equiv 0$, this is consequence of Theorem 3.III. For the general case we refer the reader to the paper [6] (see Bibliography of E.T.E.).

[19] In the case $r \leqq 3$ this implies that u is continuous in $\overline{A \cup I}$ (see E.T.E., 2.VI).

[20] Actually, in the case $r \leqq 3$, we have to take care only of the functions σ_{ik}.

The singular set of the measure γ is contained in

$$\bigcup_{k=1}^{q'} \partial \Gamma_k$$

and the Lebesgue derivative of γ is the function $\sigma_{ih}(u)\,\nu_i\,\nu_h$.[21]

We denote by W the linear manifold that consists of the real valued functions w which are defined almost everywhere on Σ and are such that, for every w, and almost everywhere on Σ, we have $w = v_i\,\nu_i$ with $v \in H_1(A) \cap C^0(A \cup \Sigma)$. We put

$$\Phi(w) = B(u, v) - F(v). \tag{10.7}$$

The linear functional Φ is defined without ambiguity by (10.7) for every $w \in W$, because, if $\tilde{v} \in H_1(A) \cap C^0(A \cup \Sigma)$ and $w = \tilde{v}_i\,\nu_i$ a.e. on Σ, then, since $v - \tilde{v}$ is a bilateral displacement of V (i.e. both $v - \tilde{v}$ and $\tilde{v} - v$ belong to V), we have $B(u, v) - F(v) = B(u, \tilde{v}) - F(\tilde{v})$.

Let A_λ $(0 < \lambda \leqq \lambda_0)$ be the domain bounded by the range of $x = \xi + \lambda\,\mu(\xi)$ as ξ describes ∂A. Let $\psi(x)$ be a real valued function of C^∞ which is null in A_{λ_0} and is equal to 1 in the exterior of $A_{\lambda_0/2}$. Set

$$v^0(x) = \begin{cases} \mu(\xi)\,\psi(x) & \text{for} \quad x = \xi + \lambda\,\mu(\xi), \quad \xi \in \partial A, \quad 0 \leqq \lambda \leqq \lambda_0, \\ 0 & \text{for} \quad x \in A_{\lambda_0}. \end{cases}$$

If λ_0 is sufficiently small the function $v^0(x)$ is defined everywhere in $\bar{A}$ and belongs to $H_1(A) \cap C^0(A \cup \Sigma)$. Moreover we have on ∂A: $v_i^0\,\nu_i \geqq \cos\omega_0 > 0$.

We can assume that every $w \in W$ coincides in Γ_k $(k = 1, \dots, q')$ with a continuous function w_k. Let us use the symbol $\max|w_k|$ for the maximum of $|w_k|$ in Γ_k and put $\mathcal{M}(w) = \max(\max|w_1|, \dots, \max|w_{q'}|)$. Consider the Banach space $\overline{W}$ obtained by functional completion from W by means of the norm $\mathcal{M}(w)$. On every $\Gamma_k (k = 1, \dots, q')$ we have for any $w \in W$

$$-\frac{\mathcal{M}(w)}{\cos w_0}\,v_i^0(x)\,\nu_i(x) \leqq w(x) \leqq \frac{\mathcal{M}(w)}{\cos w_0}\,v_i^0(x)\,\nu_i(x)$$

and therefore

$$|\Phi(w)| \leqq \frac{1}{\cos w_0}\,[B(u, v^0) - F(v^0)]\,\mathcal{M}(w).$$

Thus Φ is a continuous functional in $\overline{W}$. On the other hand, when $w \in W$ and $w \geqq 0$ on Σ, we have, because of (10.7), $\Phi(w) \geqq 0$. From classical theorems of Functional Analysis we deduce the existence of the non-negative measure γ such that (10.6) holds.

---

[21] A σ-ring of sets is a family of sets which is closed with respect to union and intersection either of a finite or of a countable collection of sets of the family. The σ-ring $\{B\}_\Sigma$ is the intersection of all the σ-rings formed by subsets of Σ. A *measure* defined on $\{B\}_\Sigma$ is any real valued function of B such that $\gamma(B) = \sum_k \gamma(B_k)$ if $B = B_1 \cup B_2 \cup \cdots \cup B_k \dots$ and $B_1, \dots, B_k, \dots$ is either a finite or a countable collection of mutually disjoint sets of $\{B\}_\Sigma$. A measure γ^* is said to be *singular* if, for any $B \in \{B\}_\Sigma$, $\gamma^*(B) = \gamma^*(B \cap N)$, where N is a set of zero Lebesgue measure on Σ. Hence $\gamma^*(B) = 0$ if $B \cap N = \phi$. For every measure $\gamma(B)$ defined on $\{B\}_\Sigma$ there exist, and are uniquely determined, a singular measure $\gamma^*(B)$ and a function $\varphi(x) \in \mathcal{L}^1(\Sigma)$ such that the *Lebesgue decomposition* holds

$$\gamma(B) = \gamma^*(B) + \int_B \varphi(x)\,d\sigma.$$

The above mentioned set N is called the *singular set* of the measure γ and the function $\varphi(x)$ is called the *Lebesgue derivative* of the measure γ.

Let $x^0 \in \Gamma_k - \partial\Gamma_k$ and let I be a suitable small neighborhood of x^0 such that u belongs to $H_2(A \cap I)$. Let v be a function belonging to $H_1(A) \cap \mathring{C}(I)$. For the w which corresponds to this particular v we have by (10.2)

$$\Phi(w) = B(u, v) - F(v) = -\int_{\partial A} (a_{hk}\, u_{/h}\, \nu_k)\, v\, d\sigma = \int_{\partial A} (v_j \nu_j)\, [\sigma_{ih}(u)\, \nu_i \nu_h]\, d\sigma.$$

That implies that the singular set of γ must be contained in $\bigcup_{k=1}^{q'} \partial\Gamma_k$ and that the Lebesgue derivative of γ is $\sigma_{ih}(u)\, \nu_i \nu_h$.

The mechanical meaning of the measure γ is evident: $\gamma(B)$ *represents the intensity of the global reaction exerted by the constraint of support on Σ over the whole set B. This reaction may have concentrated stresses only on point-sets formed by singular points of Σ; no concentration can occur in the neighborhood of any regular point of Σ.*[22]

We wish now to discuss another delicate question concerning the Signorini problem. We saw that the condition (10.1) is sufficient for the existence of the solution u of the Signorini problem *provided it is satisfied in the strong sense,* i.e. the sign $=$ can occur in (10.1) when and only when $\varrho \in R^*$. We shall prove, considering a particular case of paramount mechanical interest, that the *strong condition is necessary* for the existence of the solution. That will lead us to a remarkable mechanical interpretation of this condition.

We take $r=3$ and we suppose that the supporting surface Σ is planar and connected. We are permitted to suppose that Σ is a bounded closed region of the plane $x_3=0$. Moreover we suppose that $\bar{A} - \Sigma$ is contained in the half-space $x_3 > 0$. The linear space of the rigid displacements is formed by the vectors ϱ such that

$$\varrho_1 = a_1 + b_2 x_3 - b_3 x_2, \qquad \varrho_2 = a_2 + b_3 x_1 - b_1 x_3, \qquad \varrho_3 = a_3 + b_1 x_2 - b_2 x_1,$$

where $a_1, a_2, a_3, b_1, b_2, b_3$ are constants. The vector ϱ belongs to R' if and only if

$$a_3 + b_1 x_2 - b_2 x_1 \geqq 0 \qquad \text{for } (x_1, x_2) \in \Sigma. \tag{10.8}$$

We are permitted to suppose that the x_3-axis intersects $\Sigma - \partial\Sigma$. That implies $a_3 \geqq 0$. For any integrable real valued function w, set

$$F_h(w) = \int_A f_h\, w\, dx + \int_{\Sigma^*} g_h\, w\, d\sigma \qquad (h = 1, 2, 3).$$

Condition (10.1) is equivalent to the following conditions

$$\begin{gathered} F_1(1) = F_2(1) = 0, \qquad F_2(x_1) - F_1(x_2) = 0, \\ a_3 F_3(1) - b_2 [F_3(x_1) - F_1(x_3)] + b_1 [F_3(x_2) - F_2(x_3)] \leqq 0. \end{gathered} \tag{10.9}$$

The last inequality is to be taken for a_3, b_2 and b_1 satisfying (10.7). It follows that

$$F_3(1) \leqq 0.$$

If the sign $=$ held, then from (10.8) (10.9) we could deduce that $F_3(x_1) - F_1(x_3) = 0$, $F_3(x_2) - F_2(x_3) = 0$, i.e. the system of the applied forces would be equilibrated. Then the problem has a solution, since it reduces to the classical one which consists in assigning the body forces and the surface forces everywhere on ∂A, on the assumption that the given surface forces vanish on Σ (see Theorems 1.VI and 7.II).

[22] This circumstance was conjectured by the soviet mathematician G. I. BARENBLATT, during a seminar that the author held at Moscow University in 1969.

From now on we shall exclude equilibrated systems of applied forces.

Then we must have

$$F_3(1) < 0.$$

From (10.9) we deduce the relation

$$F_1(1)\,[F_3(x_2) - F_2(x_3)] + F_2(1)\,[F_1(x_3) - F_3(x_1)] + F_3(1)\,[F_2(x_1) - F_1(x_2)] = 0.$$

Since the vector $\{F_1(1), F_2(1), F_3(1)\}$ does not vanish, we know from elementary mechanics[23] that the system of applied forces is equivalent to a single force orthogonal to the plane $x_3=0$, directed downwards and applied in any point of the *central axis* of the system which is the straight line $x_1 = x_1^0$, $x_2 = x_2^0$ with

$$x_1^0 = \frac{F_3(x_1) - F_1(x_3)}{F_3(1)}, \qquad x_2^0 = \frac{F_3(x_2) - F_2(x_3)}{F_3(1)}.$$

We shall call the set obtained as the intersection of all the closed half-planes which contain Σ, the *convex hull* of Σ and shall denote it by $K(\Sigma)$. $K(\Sigma)$ is obviously a closed convex set. A half-plane having as its origin the straight line $a_3 + b_1 x_2 - b_2 x_1 = 0$, with $a_3 > 0$, contains Σ if and only if a_3, b_1, b_2 satisfy condition (10.8). Thus the last of (10.9) expresses the fact that $K(\Sigma)$ contains (x_1^0, x_2^0). Vice-versa, if $K(\Sigma)$ contains (x_1^0, x_2^0) and if the system of applied forces is equivalent to a single force, orthogonal to the plane $x_3=0$ and directed downwards, conditions (10.9) are satisfied and thus also (10.1).

We shall now prove that

10.III *A necessary and sufficient condition in order that, given* (10.1), *it hold in the strong sense, is that the central axis of the system of applied forces meet* $K(\Sigma)$ *at an internal point.*

If (x_1^0, x_2^0) is internal to $K(\Sigma)$, it will be internal to every half-plane which contains Σ. This implies that the last of (10.9) is satisfied in the strong sense for a_3, b_1, b_2 satisfying (10.8), and $a_3 > 0$. Thus (10.1) is satisfied in the strong sense. Vice-versa, if (10.1) is satisfied and if we had $(x_1^0, x_2^0) \in \partial K(\Sigma)$, we could consider a straight line $a_3 + b_1 x_2 - b_2 x_1 = 0$ $(a_3 > 0)$ passing through (x_1^0, x_2^0) and such that one of the half-planes which admit this straight line origin (indeed the half-plane $a_3 + b_1 x_2 - b_2 x_1 \geqq 0$) contains $K(\Sigma)$.[24] The rigid displacement $\varrho_1 = b_2 x_3$, $\varrho_2 = -b_1 x_3$, $\varrho_3 = a_3 + b_1 x_2 - b_2 x_1$ belongs to R' and is unilateral. However, the sign $=$ holds in (10.1) for this displacement.

From this lemma it follows that the case in which (10.1) is satisfied, but not in the strong sense, presents itself as a limiting case in which the central axis meets $K(\Sigma)$ on the boundary.

We shall prove that in this limiting case *a solution of the Signorini problem does not exist.* This proves the *necessity of the strong condition for the existence of a solution of the Signorini problem.*

---

[23] See [*25*, pp. 36–37].

[24] *If K is a planar convex set and x a point of ∂K, there exists a straight line λ passing through x such that K lies in one of the two closed half-planes of origin λ.*

Let us exclude the trivial case $K = \{x\}$. Consider a system of polar coordinates with pole x. If there existed two numbers α, β such that $\beta - \alpha \geqq \pi$ and such that for $\alpha < \theta < \beta$, the half axis originating in x of argument θ did not contain any point of K different from x, the assertion would be proved. Assume these numbers do not exist. Then there will be three values θ_1, θ_2, θ_3 for which $0 < \theta_2 - \theta_1 < \pi$, $0 < \theta_3 - \theta_2 < \pi$, $0 < \theta_1 + 2\pi - \theta_3 < \pi$ and such that the three half axes, determined by them, each contains a point of K different from x. Let x^1, x^2, x^3 be these three points. The triangle $x^1 x^2 x^3$ is contained in K and thus x cannot be a point of ∂K.

10.IV *If* (10.1) *is satisfied, but if the central axis of the system of applied forces meets* $K(\Sigma)$ *at a point of its boundary, then* $\mathscr{I}(v)$ *has no minimum in* V.

Let $\varrho_1^0 = a_1^0 + b_2^0 x_3 - b_3 x_2$, $\varrho_2^0 = a_2^0 + b_3^0 x_1 - b_1^0 x_3$, $\varrho_3^0 = a_3^0 + b_1^0 x_2 - b_2^0 x_1$ be a displacement of R' which is unilateral and such that

$$\int_A f_i \varrho_i^0 \, dx + \int_{\Sigma^*} g_i \varrho_i^0 \, d\sigma = 0.$$

Let λ be the straight line of the plane $x_3 = 0$ given by the equation $\varrho_3^0 = 0$. Then, recalling (10.9), $(x_1^0, x_2^0) \in \lambda$. Note that the intersection of λ with $K(\Sigma)$ is not empty.

Suppose there exists a solution u of the Signorini problem. Then also $u^0 = u + \varrho^0$ will be a solution of the problem (see Theorem 7.I). Since $\varrho_3^0 > 0$ in $\Sigma - (\lambda \cap \partial \Sigma)$, we have $u_3^0 > 0$ a.e. in $\Sigma - (\lambda \cap \partial \Sigma)$. Let z be a point of $\Sigma - (\lambda \cap \partial \Sigma)$ and let $D_\delta(z)$ be a 3-dimensional ball of center z and radius δ at positive distance from $\lambda \cap \partial \Sigma$. Let $w \in \mathring{C}^1[D_\delta(z)]$. We can determine a positive number $t(\delta, z)$ such that for $|t| < t(\delta, z)$ the function $u^0 + t w$ belongs to V. It follows that $I(u^0 + t w)$ has a minimum for $t = 0$. Then $B(u, w) - F(w) = 0$. This implies that the measure γ defined by Theorem 10.II is such that $\gamma(B) = 0$ for each Borel set B of Σ contained in $D_\delta(z)$. It follows that the measure γ is singular and the singular set of γ must be contained in $\lambda \cap \partial \Sigma$.

Observe that γ cannot be identically null, since, from (10.6), assuming $v_1 = 0$, $v_2 = 0$, $v_3 = 1$, we deduce

$$\gamma(\lambda \cap \partial \Sigma) = - F_3(1) > 0.$$

Thus: *if there exists a solution of the Signorini problem, the reaction of the constraint is concentrated on the linear set* $\lambda \cap \partial \Sigma$.

For every $v \in H_1(A) \cap C^0(A \cup \Sigma)$ we put

$$\mathscr{I}_0(v) = \tfrac{1}{2} B(v, v) - F(v) - \int_{\lambda \cap \partial \Sigma} v_3 \, d\gamma.$$

Since we have $B(u, u) = F(u)$ and, because of (10.6),

$$\int_{\lambda \cap \delta \Gamma} v_3 \, d\gamma = B(u, v) - F(v),$$

we get

$$\mathscr{I}_0(v) - \mathscr{I}(u) \geqq B(u - v, u - v) \geqq 0.$$

Let p be a positive constant greater than the diameter of A. We put

$$v_1^\delta = 0, \qquad v_2^\delta = 0, \qquad v_3^\delta = \log \log \frac{(|a^0| + |b^0| p)^2 + p^2 + 1}{|\varrho_3^0(x)|^2 + x_3^2 + \delta}.$$

For $0 < \delta < 1$ we have $v^\delta \in C^\infty(\bar{A})$. On the other hand we have

$$\lim_{\delta \to 0} \mathscr{I}_0[v^\delta] = -\infty, \qquad \mathscr{I}_0[v^\delta] \geqq \mathscr{I}(u).$$

This proves that no function u such as to minimize $\mathscr{I}(v)$ in V exists.

11. Historical and bibliographical remarks concerning Existence Theorems in Elasticity. Existence theorems are of prominent interest in problems of mechanics and physics, since they provide a rational tool for proving, independently of any physical plausibility and experimental evidence, the consistency of a theory which brings into a mathematical scheme facts and phenomena of the physical world. Unfortunately, they very often constitute the most difficult part of the theory.

Concerning the classical linear elasticity of homogeneous isotropic bodies, the first existence theorem was given for the 1st Boundary Value Problem by FREDHOLM **[8]**[25] as an application of the discovery of his fundamental theorems concerning integral equations. The same problem was also considered, always using the Fredholm integral equations, by LAURICELLA **[21]**, MARCOLONGO **[24]** and, later, by LICHTENSTEIN **[23]**.

The use of integral equations in the 1st Boundary Value Problem (Dirichlet's Problem) is possible by considering the classical differential system of elasticity as homogeneous, with non-homogeneous the boundary values and by representing the solution as a "potential of a double layer" relative to the *fundamental solution matrix*, which was given for the differential system in question by Lord KELVIN [14] and by SOMIGLIANA [40].

The approach exactly parallels the one which is used in potential theory for getting the existence theorem for the Dirichlet problem relative to harmonic functions [13], [12]. Real difficulties arise when one tries to use the same method for the 2nd Boundary Value Problem of classical elasticity. Following the analogy with harmonic functions one would represent the solution as a "potential of a simple layer", just as is done in the 2nd Boundary Value Problem for harmonic functions (Neumann's Problem) [13], [12].

Unfortunately the system of integral equations which one gets is not a Fredholm system. In fact the "kernels" of these integral equations are *not* absolutely integrable, and the corresponding integrals have a meaning only if they are interpreted as "Cauchy singular integrals".

This has misled some authors, who have taken them to be Fredholm integral equations. Let us spend a few words on this phenomenon.

Let us consider the differential operator of classical elasticity in the case $r=3$, which we write as follows (see E.T.E., Sect. 12).

$$L_i u = u_{i/kk} + \sigma u_{k/ki}$$

where σ is a constant $>\frac{1}{3}$. To this operator we associate the boundary operator tu which represents the forces on the boundary, corresponding to the displacement u:

$$t_i u = u_{i/k} \nu_k + \sigma u_{k/k} \nu_i + (u_{k/i} \nu_k - u_{k/k} \nu_i).$$

In the bounded regular domain A the Betti reciprocity theorem holds:

$$\int_{\partial A} (u_i t_i v - v_i t_i u)\, d\sigma + \int_A (u_i L_i v - v_i L_i u)\, dx = 0, \tag{11.1}$$

which is analogous to the Green formula for the Laplace operator

$$\int_{\partial A} \left(u \frac{\partial v}{\partial \nu} - v \frac{\partial u}{\partial \nu}\right) d\sigma + \int_A (u \Delta_2 v - v \Delta_2 u)\, dx = 0.$$

This has induced several researchers to believe that the operator tu plays in elasticity the same role as the operator $\partial u/\partial \nu$ plays in potential theory. But it must be remarked that if we consider instead of tu, the more general operator $t(u;\lambda)$

$$t_i(u;\lambda) = u_{i/k} \nu_k + \sigma u_{k/k} \nu_i + \lambda (u_{k/i} \nu_k - u_{k/k} \nu_i),$$

where λ is an arbitrary constant, the reciprocity relation (11.1) still holds

$$\int_{\partial A} [u_i t_i(v;\lambda) - v_i t_i(u;\lambda)]\, d\sigma + \int_A (u_i L_i v - v_i L_i u)\, dx = 0.$$

Then $t(u;\lambda)$ has the same right as tu to play in elasticity the role of the normal derivative in potential theory. With respect to the theory of Fredholm integral equations, the operator $t(u;\lambda)$ which behaves like the normal derivative in potential theory, is one that, ∂A being supposed a Liapounov boundary (i.e. with a Hölder continuous normal field), when it operates on the Somigliana fundamental matrix

$$\left\{\frac{1}{x-y} - \delta_{ij} \frac{\sigma}{2(1+\sigma)} \frac{\partial^2 |x-y|}{\partial x_i \partial x_j}\right\} \quad (i,j=1,2,3),$$

produces kernels $k_{ij}(x,y)$ which have a Fredholm singularity, i.e. are such that $k_{ij}(x,y) = \mathcal{O}(|x-y|^{\alpha-2})$ $0<\alpha\leqq 1$. It is not difficult to see that this happens if and only if $\lambda = \sigma(2+\sigma)^{-1}$.

[25] Numbers in boldface square brackets refer to the Bibliography at the end of E.T.E. Numbers in lightface square brackets refer to the Bibliography at the end of the present paper.

Unfortunately the case concerning elasticity is $\lambda = 1$. It follows that the Fredholm method can be applied to the 2nd Boundary Value Problem provided one considers as boundary operator $t[u; \sigma(2+\sigma)^{-1}]$ which has no physical meaning. Actually this was done by LAURICELLA [21], who called $t[u; \sigma(2+\sigma)^{-1}]$ the *pseudo-tensions operator*.

The real 2nd Boundary Value Problem of elasticity was studied by KORN [18] in a very long paper by a very complicated method which uses integral equations and for the first time (see also [19]) introduces the inequalities, nowadays known as Korn's inequalities (see E.T.E., Sect. 12). FRIEDRICHS [10], citing the work of KORN writes: "The author of the present paper has been unable to verify KORN's proof for the second case". BERNSTEIN and TOUPIN in their paper on KORN's inequalities, after quoting FRIEDRICHS' statement, write: "With him we confess unability to follow KORN's original treatment".

A few years later H. WEYL [35] tried to study the 2nd Boundary Value Problem of elasticity by using Fredholm integral equations obtained by means of the so-called antenna potential. However a certain hypothesis, which he assumes for carrying out his approach, has not been proved to hold in general.

It must be remarked that, in any case, H. WEYL's paper is of fundamental interest in elasticity for the analysis of the asymptotic distribution of eigenvalues in problems of elasticity.

For long time the theory of Boundary Value Problems in elasticity made no substantial progress in the case when $r > 2$. On the other hand, in the case when $r = 2$, many important achievements were obtained, mainly by MUSKHELISHVILI, I. N. VEKUA and the Georgian school, using complex methods and the theory of singular integral equations on a curve (for extensive bibliography see [34]).

In 1947 FRIEDRICHS published an important paper [10] on r-dimensional problems of elasticity. He gives the first acceptable proof of KORN's second inequality (the proof of the first one is almost trivial) and new proofs of the existence theorems for the 1st and the 2nd Boundary Value Problems of classical elasticity and for the related eigenvalue problems. His method is founded on the variational approach (by the same method FRIEDRICHS had given in 1928 the existence theorem for a clamped plate [9]) and he succeeds in proving, by employing his technique of "mollifiers", the interior regularity of the solutions.

In 1950 appeared paper [3] in which, by use of methods of functional analysis, new proofs of the existence theorems for the 1st and the 2nd Boundary Value Problems of elasticity were given and, for the first time, the existence theorem for the 3rd B.V.P. (mixed B.V.P.) was obtained. It is worthwhile to recall here briefly the method used in [3] for the proof of this theorem. We shall consider, for simplicity, the case of the Laplace operator. We shall also introduce some simplifications and shall use language more modern than that of [3].

Let A be a bounded domain of X^3 with a Liapounov boundary ∂A, which is decomposed into two open hypersurfaces Σ_1 and Σ_2 which have a common border $\partial\Sigma_1 = \partial\Sigma_2$ and no other point in common. We shall consider Σ_i ($i = 1, 2$) as an open set respect to ∂A. Let us suppose that there exists a domain A' with a Liapounov boundary $\partial A'$ and such that $A' \supset A$, $\partial A' \cap \partial A = \bar{\Sigma}_1$. Let H_{Σ_1} be the subspace of $H_1(A)$ obtained as closure of the linear manifold of all the real valued functions v such that $v \in C^1(\bar{A})$, $spt\; v \cap \bar{\Sigma}_1 = \phi$. Let δ be a function Hölder continuous on $\bar{\Sigma}_2$. We wish to prove that there exists one and only one function u such that

$$u \in H_{\Sigma_1} \cap C^2(A) \cap C^0(\bar{A} - \partial\Sigma_1) \cap C^1(A \cup \Sigma_2); \quad \Delta_2 u = 0 \quad \text{in } A, \quad \frac{\partial u}{\partial \nu} = \delta \quad \text{on } \Sigma_2.$$

Let $\tilde{u}$ be a function such that $\tilde{u} \in C^2(A) \cap C^1(\bar{A})$, $\Delta_2 \tilde{u} = 0$ in A, $\tilde{u} = \delta$ on Σ_2. Such a function is easily obtained by suitably continuing δ on ∂A and solving the corresponding Neumann problem in A. Let us introduce in $H_1(A)$ the new scalar product

$$((u, v)) = \int_A u_{/i}\, v_{/i}\, dx$$

and, identifying two functions of $H_1(A)$ which differ by a constant, let us denote by $\mathscr{H}$ the corresponding Hilbert space. Considering H_{Σ_1} as a subspace of $\mathscr{H}$, let u be the orthogonal projection of $\tilde{u}$ on H_{Σ_1}. We have for any $v \in H_{\Sigma_1}$

$$\int_A u_{/i}\, v_{/i}\, dx = \int_A \tilde{u}_{/i}\, v_{/i}\, dx = -\int_{\Sigma_2} v\, \delta\, d\sigma. \tag{11.2}$$

If we take any $\varphi \in \overset{\circ}{C}{}^\infty(A')$ and put

$$v(x) = \int_{A'} G(x, y)\, \varphi(y)\, dy, \tag{11.3}$$

where $G(x, y)$ is the Green function of the Dirichlet problem for Δ_2, relative to A', we have

$$-\int_{\Sigma_2} v\,\delta\,d\sigma = \int_A u_{/i}\,v_{/i}\,dx = -\int_{\Sigma_2} u\,\frac{\partial v}{\partial \nu}\,d\sigma - \int_A u\,\varphi\,dx.$$

By the arbitrariness of φ we easily deduce that

$$\theta(y)\,u(y) = \int_{\Sigma_2} \delta(x)\,G(x, y)\,d\sigma_x - \int_{\Sigma_2} u(x)\,\frac{\partial}{\partial \nu_x} G(x, y)\,d\sigma_x \tag{11.4}$$

with $\theta(y) = 1$ if $y \in A$, $\theta(y) = 0$ if $y \in A' - \bar{A}$. From (11.4) it is easy to deduce, using standard arguments of potential theory (jump relations) that u is a solution of the problem. If u^0 is another solution of the problem, since, for any $v \in H_{\Sigma_1}$, we have

$$\int_A u^0_{/i}\,v_{/i}\,dx = -\int_{\Sigma_2} v\,\delta\,d\sigma,$$

from (11.2) we see that $u^0 \equiv u$.[26]

This approach, since only with difficulty could it be extended to higher order elliptic systems with variable coefficients, was not followed in E.T.E. Nevertheless, over the approach used there the earlier one has the advantage of not requiring such severe restrictions upon ∂A.

One year after the publication of **[3]** a note of Eidus [*7*] appeared on the mixed problem of elasticity. It must be remarked that Soviet mathematicians have been very active in the field of the existence theory for classical elasticity. Besides the above cited contributions of the Georgian school, let us quote here the papers of S. L. Sobolev [*39*], Michlin [*31*], S. Y. Kogan [*15*], E. N. Nikolsky [*35*] on the extension to elasticity of the Schwarz alternating method and, in particular, the work of Michlin and Kupradze. The former has considered elasticity problems from several points of view. In his monograph [*32*] he considers, besides the above quoted alternating method, the variational approach and he reviews the results obtained in this field by Soviet mathematicians. However he does not seem to be aware of some of the work done in the western world. In the monograph **[*25*]** Michlin applies his theory of multidimensional singular integral equations to problems of elasticity. This theory, initiated by Tricomi [*43*] and Giraud [*10*], has been concluded by Michlin and is the starting point of the modern theory of pseudo-differential operators. Michlin in **[*25*]** is able to solve the system of singular integral equations, to which the 2nd B.V.P. of classical elasticity gives rise when the solution is represented by simple layer potentials. Similar results were obtained by Kupradze **[*20*]**, almost at the same time. He also uses Michlin's theory to solve the same system of singular integral equations.

Let us mention here, besides other relevant contributions of Kupradze, his work on dynamic problems and on problems for heterogeneous media **[*20*]**, [*15–22*].

The equilibrium problem for a heterogeneous elastic medium (an elastic body composed by two homogeneous isotropic bodies with different Lamè constants) was first posed by

---

[26] Let us remark that the use of the domain A' and of the corresponding Green's function $G(x, y)$, according to the procedure used in **[3]** (which was followed by Kupradze **[*20*]** for the analysis of mixed problems) can be avoided and a further simplification in the proof of the existence theorem introduced. To this end, instead of using $v(x)$ as given by (11.3), let us assume

$$v(x) = \psi(x) \int_A s(x, y)\,\varphi(y)\,dy$$

where

$$s(x, y) = \begin{cases} (2\pi)^{-1} \log|x - y| & \text{for } r = 2, \\ [(2 - r)\,\omega_r]^{-1}\,|x - y|^{2-r} & \text{for } r > 2 \end{cases}$$

(ω_r is the hypersurface measure of the unit sphere of X^r), $\varphi(x) \in \mathring{C}^\infty$ and $\psi(x)$ is a C^∞ function such that, given arbitrarily x^0 disjoint from $\bar{\Sigma}_1$, $\psi(x) \equiv 1$ for $|x - x^0| \leqq \varepsilon$, $\psi(x) \equiv 0$ for $|x - x^0| \geqq 2\varepsilon$; ε is a positive number such that the ball $\bar{\Gamma}_{2\varepsilon}$: $|x - x^0| \leqq 2\varepsilon$ is disjoint from $\bar{\Sigma}_1$. It is not difficult to see that for $y \in \Gamma_\varepsilon$ one gets the representation

$$\theta(y)\,u(y) = \int_{\Sigma_2 \cap \Gamma_\varepsilon} \delta(x)\,s(x, y)\,d\sigma_x - \int_{\Sigma_2 \cap \Gamma_\varepsilon} u(x)\,\frac{\partial}{\partial \nu_x} s(x, y)\,d\sigma_x + h(y), \tag{11.5}$$

where $h(y)$ is a harmonic function in Γ_ε. By using (11.5) in place of (11.4) one gets the same conclusions.

Picone [30], who proposed a method for numerical solution. Papers by Lions [28] and by Campanato [2] are concerned with this problem. The case of anisotropic bodies, with general elasticities, is for the first time considered in E.T.E. In that article the modern approach to boundary value problems for strongly elliptic operators is followed, which has made it possible, for the first time, to treat with great generality dynamic problems as well as diffusion problems and integro-differential problems. There can be no doubt that the analytical investigations of problems of elasticity have greatly contributed to the modern development of the theory of partial differential equations. For instance researches concerning the Korn inequalities are among the first examples of investigations connected with the concept of "coerciveness" now very important. In this respect, besides the papers of Friedrichs and of Bernstein and Toupin, already cited, let us mention here the work of Campanato [3], [4], which is mainly interesting for the examples in which Korn's inequalities fail, and a remarkable paper by Payne and Weinberger [29], where the best estimate for the 2nd Korn inequality, in the case of a sphere, is obtained and a new proof of this inequality provided for a class of domains. Unfortunately the results of the paper do not have so large a range of validity as the authors claim. The paper [13] of Gobert also concerns Korn's inequality.

The first author to consider a unilateral problem for elasticity was Signorini [37] early in 1933. He presented again his theory, in more complete fashion, in 1959 [38]. In this paper the problem nowadays known as the Signorini problem is proposed. This problem was investigated and solved in the paper [6], which was submitted for publication in September, 1963, and which appeared in 1964. The results of paper [6] had been announced in [8] (February 1963), in [5] and in [9]. The results of [6], although relative to a specific problem of elasticity, are immediately extensible—as far as the abstract theory is concerned—to an abstract unilateral problem relative to a symmetric semi-coercive bilinear form considered in a cone.

The existence and uniqueness theorem given in [6] is the first example of an existence and uniqueness theorem for a unilateral problem connected with a differential operator.

More than one year later Stampacchia [41] considered unilateral problems relative to nonsymmetric coercive bilinear forms. The case considered by Stampacchia because of the coerciveness hypothesis he assumes cannot cover that considered in [6]. In fact—as has been shown in the present paper—it is just the absence of the coerciveness condition which makes the problem a complicated one, since, in this case, one has to face the delicate question of the compatibility conditions. When the bilinear form is coercive, the problem is almost trivial in the symmetric case. The non symmetric case is easily reduced to the symmetric one by a simple argument, shown in [30], which makes use of a suitable contraction mapping. We have used this argument at the beginning of the proof of Theorem 2.I of the present paper. The authors of [30] consider also the case of non-coercive non-symmetric bilinear forms. In the abstract scheme which they assume, they exactly reproduce the situation which arises in [6] in connection with elasticity and give a theorem which imitates the results of [6], transferring them into their abstract setting. The proof of the boundedness of the sequence which furnishes the solution of their unilateral problem is strongly inspired by the proof of the boundedness of the analogous sequence, presented in [6]. Unfortunately, imitation of the results of [6] (which relate to a particular convex set) in the case of the general convex set which they introduce, without considering the extremely more general geometric nature of an arbitrary convex set, lead them to state unacceptable results. As a matter of fact, if one uses results of [30] for solving unilateral problems for non-coercive forms, one has to impose compatibility conditions on the data, even when that is not necessary.

Nowadays the literature on unilateral problems grows more and more.[27] The abstract point of view has especially considered extensions to non linear operators. These researches mainly interest pure mathematics. However, a number of concrete specific problems have also been investigated and the associated questions of regularization studied. While I believe that the abstract theory, developed in this paper, is able to cover with sufficient generality the unilateral problems connected with linear operators of applied mathematics, I realize that the regularization theory is still very far from the generality that it has in the case of bilateral problems. The regularization results stated in the present paper, as far as the relevant differential operator is concerned, are the most general known up to date for linear operators.

With specific reference to elasticity, let us mention that one of the first unilateral problems in elasticity after Signorini's was formulated by Prager [36]. It has been considered in Sect. 6 (example V) of the present paper. A unilateral problem for a membrane was investigated by H. Lewy [26] and later by Lewy and Stampacchia [27]. Elastic plastic torsion problems have been studied by T. W. Ting [42], Duvaut and Lanchon [24] and Lanchon [23]. The theory of elastic-plastic torsion problem is explained in detail in the

[27] For a bibliography on this subject we refer to the book [29].

article by TING in volume VI a/3 of this Encyclopedia. DUVAUT has considered the Signorini problem in visco-elasticity [5] (i.e. hereditary elasticity); GRIOLI [11] and DUVAUT [6] elastic problems when the bounding surface has friction. The paper [33] of MOREAU concerns unilateral problems of elastostatics; he has considered also a number of abstract and concrete topics connected with unilateral problems.

So far we have considered problems connected with linear elasticity. Although the unilateral problems do not belong to the domain of linear analysis and many questions are still not settled, it has been possible to put them in a general scheme. That is not possible for nonlinear elasticity. As far as existence theory is concerned, there are particular results, for special problems but nothing sufficient to give concrete foundations to a general theory. That probably will be one of the tasks of applied analysts in the next years.

Bibliography.

[1] BERNSTEIN, B., and R. A. TOUPIN: Korn inequalities for the sphere and circle. Arch. Rational Mech. Anal. **6** (1960).

[2] CAMPANATO, S.: Sul problema di M. Picone relative all'equilibrio di un corpo elastico incastrato. Ricerche di Matem. **VI** (1957).

[3] — Sui problemi al contorno per sistemi di equazioni differenziali lineari del tipo dell'elasticità. Parte I and Parte II. Ann. Scuola Norm. Sup. Pisa, s. III, **13** (1959).

[4] — Proprietà di taluni spazi di Banach connessi con la teoria dell'elasticità. Ann. Scuola Norm. Sup. Pisa, s. III, **16** (1962).

[5] DUVAUT, G.: Problème de Signorini en viscoélasticité linéaire. Compt. Rend. **268** (1969).

[6] — Problèmes unilatéraux en mécanique des milieux continus. Proceedings of the International Congress of Mathematicians, Nice (1970).

[7] EIDUS, D. M.: On a mixed problem of elasticity theory [in Russian]. Dokl. Akad. Nauk USSR **76** (1971).

[8] FICHERA, G.: Sul problema elastostatico di Signorini con ambigue condizioni al contorno. Rend. Accad. Naz. Lincei, s. VIII, **34** (1963).

[9] — Un teorema generale di semicontinuità per gli integrali multipli e sue applicazioni alla fisica-matematica. Atti del Convegno Lagrangiano, Acc. Sci. Torino (1963).

[10] GIRAUD, G.: Equations à intégrales principales. Ann. Sci. École Norm. Supér. **51** (1934).

[11] GRIOLI, G.: Problemi d'integrazione e formulazione integrale del problema fondamentale dell'elastostatica. Simposio intern. sulle Applicazioni della Analisi alla Fisica Matematica, Cagliari-Sassari (1964). Roma: Edit. Cremonese 1965.

[12] GÜNTER, N. M.: Die Potentialtheorie und ihre Anwendung auf Grundaufgaben der Mathematischen Physik. Leipzig: B. G. Teubner 1957.

[13] KELLOGG, O. D.: Foundations of potential theory. Berlin: Springer 1929; New York: F. Ungar 1946.

[14] KELVIN, Lord (W. Thomson): Note on the integration of the equations of equilibrium of an elastic solid. Cambridge & Dublin Math. J. (1848).

[15] KOGAN, S. Y.: On the resolution of the threedimensional problem of elasticity by means of the Schwarz alternating method [in Russian]. Izv. Akad. Nauk (Geophys. Ser.) **3** (1956).

[16] KUPRADZE, V. D.: Boundary value problems of the theory of forced elastic oscillations [in Russian]. Usp. Matem. Nauk **8** (1953).

[17] — Some new theorems on oscillation equations and their application to boundary value problems [in Russian]. Tr. Tbilisk. Universiteta **25**a (1944).

[18] — Boundary value problems of the theory of elasticity for piecewise-homogeneous elastic bodies [in Russian]. Soobshch. Akad. Nauk. Gruz. **22** (1959).

[19] — On the theory of boundary value problems for inhomogeneous elastic bodies [in Russian]. Soobshch. Akad. Nauk Gruz. **22** (1959).

[20] — On boundary value problems of the theory of elasticity for piecewise-homogeneous bodies [in Russian]. Soobshch. Akad. Nauk Gruz. **22** (1959).

[21] — The boundary value problems of the oscillation theory and their integral equations [in Russian]. Moscow: Gostekhizdat (1950).

[22] — Dynamical problems in elasticity (from Progress in solid mechanics), vol. 3. Amsterdam: North Holland (1963).

[23] LANCHON, H.: Solution du problème de torsion élastoplastique d'une barre cylindrique de section quelconque. Compt. Rend. **269** (1969).

[24] —, et G. DUVAUT: Sur la solution du problème de torsion élastoplastique d'une barre cylindrique de section quelconque. Compt. Rend. **264** (1967).

[25] LEVI CIVITA, T., e U. AMALDI: Compendio di Meccanica razionale. II Ediz. N. Zanichelli (1938).

[26] Lewy, H.: On a variational problem with inequalities on the boundary. J. Math. & Mech. **17** (1968).
[27] —, and G. Stampacchia: On the regularity of the solution of a variational inequality. Comm. Pure Appl. Math. **22** (1969).
[28] Lions, J. L.: Contribution à un problème de M. Picone. Ann. Matem. pura e appl., s. IV, **16** (1955).
[29] — Quelques méthodes de résolution des problèmes aux limites non linéaires. Paris: Dunod Gauthier-Villars (1969).
[30] —, and G. Stampacchia: Variational inequalities. Comm. Pure Appl. Math. **20** (1967).
[31] Michlin, S. G.: On the Schwarz algorithm [in Russian]. Dokl. Akad. Nauk USSR **77** (1951).
[32] — The problem of the minimum of a quadratic functional. S. Francisco-London-Amsterdam: Holden-Day 1965.
[33] Moreau, J. J.: La notion de sur-potential et les liasons unilatérales en élastostatique. Proc. of the 12th Inter. Congress of Appl. Mech. Stanford (1968).
[34] Muskhelishvili, N. I.: Singular integral equation, 2nd ed. Groningen: Nordhoff 1959, and 3rd ed. [in Russian] published by the State Publ. House for Physics-Mathematics Literature, Moscow (1968).
[35] Nikolski, E. N.: The Schwarz algorithm in the problem of tensions in the theory of elasticity [in Russian]. Dokl. Akad. Nauk USSR **135** (1960).
[36] Prager, W.: Unilateral constraints in mechanics of continua. Atti del Convegno Lagrangiano, Acc. Sci. Torino (1963).
[37] Signorini, A.: Sopra alcune questioni di elastostatica. Atti Soc. Ital. per il Progesso delle Scienze (1933).
[38] — Questioni di elasticità non linearizzata o semi-linearizzata. Rend. di Matem. e delle sue appl. **18** (1959).
[39] Sobolev, S. L.: The Schwarz algorithm in elasticity theory. Dokl. Akad. Nauk USSR **4** (1936).
[40] Somigliana, C.: Sopra l'equilibrio di un corpo elastico isotropo. Nuovo Cimento **3** (1885).
[41] Stampacchia, G.: Formes bilinéaires coercitives sur les ensembles convexes. Compt. Rend. **258** (1964).
[42] Ting, T. W.: Elastic-plastic torsion problem. II Arch. Rat. Mech. Anal. **25** (1967); III Arch. Rat. Mech. Anal. **34** (1969).
[43] Tricomi, F.: Equazioni integrali contenenti il valor principale di un integrale doppio. Math. Z. **27** (1928).

The Theory of Shells and Plates.

By

P. M. Naghdi.

With 2 Figures.

A. Introduction.[1]

1. Preliminary remarks. A plate and more generally a shell is a special three-dimensional body whose boundary surface has special features. Although we defer defining a shell-like body in precise terms until Sect. 4, for the purpose of these preliminary remarks consider a surface—called a reference surface—and imagine material filaments from above and below surrounding the surface along the normal at each point of the reference surface. Suppose further that the bounding surfaces formed by the end points of the material filaments are equidistant from the reference surface. Such a three-dimensional body is called a shell if the dimension of the body along the normals, called the thickness, is *small*. A shell is said to be *thin* if its thickness is much *smaller* than a certain characteristic length of the reference surface, e.g., the minimum radius of the curvature of the reference surface for initially curved shells.[2]

Interest in the construction of a linear theory for the extensional and flexural deformation of plates (from the three-dimensional equations of linear elasticity) dates back to the early part of the nineteenth century. Following a short period of controversy (especially with regard to the nature of boundary conditions), the complete theory for bending of thin elastic plates, under certain special assumptions, was finally derived in 1850 by Kirchhoff.[3] This theory, now classical (and occasionally referred to as Poisson-Kirchhoff theory for bending of plates), remains virtually unchanged even in its details. The corresponding development for shells from the three-dimensional equations of linear elasticity was given some thirty-eight years later in a pioneering paper by Love (cited below). This paper, containing the first complete linear bending theory for *thin* shells, employs certain special assumptions analogous to Kirchhoff's; in the current literature on

---

[1] *Acknowledgement.* The support of the U. S. Office of Naval Research, during preparation of this article, under Contract N 00014-69-A-0200-1008 with the University of California, Berkeley is gratefully acknowledged.

Special thanks are due to A. E. Green with whom I have had a number of discussions on the subject of shell theory since 1964 and with whom I have collaborated on a number of papers pertaining to the subject of this article. I am deeply indepted to Drs. J. A. Trapp and M. L. Wenner who read the manuscript, independently checked all formulae and offered helpful criticisms and suggestions. The responsibility for any remaining errors is, of course, mine; although I hope that these have been reduced to a minimum or to a superficial level.

I should like to express my appreciation also to M. M. Carroll, J. L. Ericksen, R. P. Nordgren and C. Truesdell, who offered helpful comments on the final form of the manuscript.

[2] In the case of initially flat plates, the characteristic length is taken to be the smallest dimension of the reference plane.

[3] Kirchhoff [1850, *1*]. See also the *Historical Introduction* in Love's Treatise [1892, *1*] and [1893, *2*] or its subsequent editions, e.g., [1927, *1*] or [1944, *4*].

shell theory, these assumptions are sometimes referred to as Kirchhoff-Love assumptions. This theory, which has come to be known as Love's first approximation, despite shortcomings, has since occupied a position of prominence.[4] The rather large number of papers in recent years devoted to re-examinations, re-derivations, extensions and generalizations of the equations of the linear theory of thin shells, in itself may be a sufficient indication of not only the shortcomings of LOVE's theory, but also that the foundations of the theory (even linear theory) as derived from the three-dimensional equations is not as yet firmly established. Indeed, there have been continual efforts by a number of investigators (especially during the last two decades) to rigorize and systematize the derivation of the (two-dimensional) equations of the classical linear theory of thin shells; and these efforts have resulted in considerable improvements toward a more satisfactory derivation of an approximate linear theory of shells, particularly the constitutive equations, from the three-dimensional equations. Although these improvements, achieved by a variety of means (e.g., use of variational theorems of classical elasticity, asymptotic expansion technique or consideration of error estimates, etc.), have contributed to our understanding of the subject, *alas* too slowly, a fully satisfactory derivation of an approximate system of constitutive equations along with an appraisal of their accuracy as compared with the three-dimensional equations still remains an open issue. Further remarks on this point are made in Sect. 4 and again in Sect. 20.

Historically, interest in the construction of theories of shells and plates grew from the desire to treat vibrations of plates and shells, aimed at deducing the tones of vibrating bells.[5] However, the motivating factors for the development of a general bending theory of sufficiently thin shells stem mainly from the following considerations: (i) The reduction of an otherwise intractable (at least analytically) three-dimensional problem to one characterized by two-dimensional equations; and (ii) the need to treat shell-like bodies when bending effects are prominent. Such bending effects arise in the open shell which has an edge but they may also be significant in both open and closed shells due to purely geometrical or load discontinuities. Thus, the purpose of a theory of shells (and plates) is to provide appropriate two-dimensional equations applicable to shell-like bodies. Examples of bodies to which shell theory is applicable are numerous and range from machine parts, electronic devices, domes, variety of ship and aerospace structures to components of physiological systems such as arteries, the cornea of the eye and the periodontal membrane, to name only a few.

Because of the considerable difficulties associated with the derivation of shell theory from the three-dimensional equations mentioned above, the possibility of employing a (two-dimensional) *model* for a thin shell presents itself in a natural way. Indeed, such an approach for thin shells was conceived of and dealt with (although at the time with some limitations) by the brothers E. and F. COSSERAT but remained largely unknown or unnoticed until fairly recently.[6] The idea of representing a (three-dimensional) thin shell by a two-dimensional continuum

---

[4] LOVE [1888, *1*]. The reference to LOVE's first approximation in contemporary literature is often confusing, since different versions bearing his name can be found. The difference between these and their origin is briefly discussed in Sect. 21 A.

[5] Interestingly enough in the Summary of his [1888, *1*] paper, LOVE wrote: "This paper is really an attempt to construct a theory of the vibrations of bells." For historical background on the subject, see LOVE's [1944, *4*] *Historical Introduction* and TODHUNTER and PEARSON's history of elasticity [1886, *1*], [1893, *3*].

[6] [1908, *1*] and [1909, *1*]. The monograph by E. and F. COSSERAT [1909, *1*] and a number of recent papers bearing on the matter are cited and elaborated upon at some length in Sect. 4.

which would then permit the development of shell theory by direct approach (rather than from the three-dimensional equations) is not as strange as may at first appear to some. The notion of a model for an idealized body, a system or even a universe permeates the structure of classical physics; and is, in fact, the cornerstone of all field theories. Before continuing our discussion of the derivation of shell theory by direct approach, it is desirable to enumerate briefly some examples pertaining to different ways in which the idea of a model is used in the classical (non-polar) continuum mechanics.

To begin with the *continuum* itself is a model representing an idealized body in some sense. Roughly speaking, we may recall that the continuum model (in classical mechanics) is intended to represent phenomena in nature which appear at a scale larger than the interatomic distances. From such intuitive notions and by assigning a (continuous) mass density to the continuum model (corresponding to that of the macroscopic behavior of the general medium in question) we put forth a well-defined model for which the classical field theories of mechanics can be constructed. The plausibility of such an idealized model depends, of course, on its relevance; but this hardly needs emphasis here, in view of the success of the continuum theories and wide and extensive demonstrations of their usefulness and relevance to phenomena in the physical world for nearly two centuries. To continue, the idea of the use of a model in the classical three-dimensional (non-polar) theory is not limited to the concept of the continuum itself and is, in fact, adopted also for a different purpose. To elaborate, we recall that since the field equations in the classical continuum mechanics hold for every medium, it is only the constitutive equations which differ from one medium to another. Now constitutive equations (even those which embrace considerable generality) are always developed with a view toward a particular *model.*[7] For example, constitutive equations which define Newtonian viscous fluids describe the behavior of a class of fluids in all motions. Moreover, the success and the prominence of the theory of Newtonian viscous fluids is simply due to its usefulness for certain purposes and not because it is a valid theory for all fluid media; indeed, for another fluid medium (having a different behavior from Newtonian viscous fluid) we require a different set of constitutive equations. We have come to accept the point of view of describing the behavior of different materials through different sets of constitutive equations which, in turn, represent characterization of a particular model we have in mind.

In the classical theories of the type just referred to, it is the equations representing the material behavior which distinguish one theory from another; i.e., it is the characterization of constitutive equations which is intended to represent the particular behavior of a phenomenon in nature, whenever the classical theory can be assumed to hold. But, as in the case of the continuum itself, we need not confine the idea of a model to the sole purpose of characterizing the behavior of materials by constitutive equations; and, in fact, we may appeal to a suitable model for different purposes whenever such notions are conceptually helpful.[8]

Returning to our earlier discussion regarding the foundations and formulation of the general theory of shells (and plates), we observe here that the two-dimen-

[7] Our use of the term model here is intended to reflect the nature of the behavior of materials and should not be regarded as synonymous with other usages of the term, e.g., with reference to combinations of springs and dashpots.

[8] Indeed it was a point of view of this kind that prompted DUHEM [1893, *1*] to propose the concept of *oriented* media, which was subsequently adopted by the COSSERATS. Additional remarks, in this connection, are made in Sect. 4.

sional field equations for shells often have been derived by direct procedures (rather than from the three-dimensional equations). For example as noted in Sect. 12A, almost from the very early developments in shell theory, the derivation of the (two-dimensional) equations of equilibrium by LOVE and others was accomplished by considering a portion of a reference surface (embedded in a Euclidean 3-space) subjected to load resultants acting on the reference surface and various stress-resultants and stress-couples on the edge curves of the reference surface (corresponding to the middle surface of the shell in a reference configuration).[9] Similarly, in some of the literature on the linear shell theory devoted to derivations from the three-dimensional equations, a (two-dimensional) virtual work principle in terms of two-dimensional variables is stated *ab initio* and is assumed to be valid without any previous appeal to its derivation from the corresponding virtual work principle in the three-dimensional theory. The justification for such an approach (which is not uncommon even in some of the recent or current literature) is of course based on the fact that the two-dimensional principle is *postulated* to be valid on the middle surface of the shell. As another example, consider the membrane theory of elastic shells. Whenever this special theory is regarded to be applicable, it is tacitly assumed that the behavior of the (three-dimensional) shell can be represented by a membrane with only the gradient of the position vector of a material surface (corresponding to, say, the initial middle surface of the shell) in the deformed configuration as its kinematic ingredient. In this connection, it may be recalled that the idea of a *membrane* is not limited to a special theory of elastic shells and in a wider sense pertains to a model which reflects only extensional properties of a thin shell-like body.

The foregoing remarks are intended not only to illustrate the usefulness of the direct approach, but also to remind the reader of the fact that the direct approach has been known and utilized all along in shell theory. Thus, in addition to the definition of the (three-dimensional) shell-like body, we also motivate and introduce in Sect. 4 a (two-dimensional) model which portrays a thin shell. This model [described in detail in Sect. 4, following (4.34)], called a *Cosserat surface*, consists of a surface with a single director (i.e., a deformable vector) assigned to every point of the surface. As will become evident in subsequent chapters, such a *directed* two-dimensional continuum is conceptually simple and provides a fruitful means for characterization and direct development of a general theory for shells. In Chaps. B to D we pursue the construction of a general theory of shells and plates, both by direct approach (based mainly on the concept of a Cosserat surface) and from the three-dimensional equations of the classical (non-polar) continuum mechanics. Such parallel developments are illuminating and provide at the same time a basis of comparison between the various results (especially the field equations) from the direct approach and those emerging from a derivation via the three-dimensional theory. Inasmuch as considerable difficulties remain in the derivation of an *approximate* system of (two-dimensional) equations for thin shells (especially with regard to an approximate set of constitutive equations) from the three-dimensional theory, the alternative development by the direct approach offers a great deal of appeal and is relatively simple. We emphasize the latter approach in Chap. E, which is confined to the linear theory, with the aim of demonstrating the relevance and applicability of the theory of Cosserat surface to shells and plates. The Cosserat surface, as emphasized in Sect. 4, is not a two-dimensional surface alone and can be regarded as represent-

[9] The derivation of the equations of equilibrium for shells from the three-dimensional equations is of a more recent origin. An account of the history of the derivations of the equations of equilibrium for shells is given in Sect. 12A.

ing a model for a thin shell. It will become evident in Chaps. D and E that the director is an effective part of the model reflecting the three-dimensional effects in thin shells and plates.

2. Scope and contents. This monograph is concerned mainly with the foundations of the general theory of shells and plates. Our point of view and approach to the subject are motivated and spelled out in some detail at the beginning of Chap. B (Sect. 4). The preliminary remarks and various definitions in Sect. 4 should be kept in mind in connection with the remaining developments. An effort is made to provide a systematic treatment of the subject in the context of the nonlinear theory, both by direct approach and also from the three-dimensional equations of the classical continuum mechanics. While the equations of the linear theories of shells and plates are included and fully discussed, generally these are obtained by a systematic linearization of the results from the nonlinear theory. On the whole the subject matter is directed toward recent developments, and all aspects of the theory pertaining to the mechanical and thermodynamical foundations of the subject are treated. The problem of stability, however, is not considered. A number of results included here have not appeared or been discussed previously in the literature.

No attempt is made to provide a complete list of works on shells and plates. Our guideline in compiling the bibliography has been to select those which directly bear on our treatment, those which are pertinent to various discussions and provide a source for further related references and a few of the historical papers on the subject.[10] Neither the contents, nor the list of works cited are exhaustive. Nevertheless, it is hoped that the developments presented reflect accurately the state of knowledge in the foundations of the general theory. Some familiarity with the classical three-dimensional (non-polar) continuum mechanics is assumed. In this connection and for background and additional information, we frequently refer to TRUESDELL and TOUPIN and to TRUESDELL and NOLL.[11]

The kinematics of the subject and primitive concepts associated with the basic principles are developed and emphasized only to the extent that they are needed in our treatment of the subject. Moreover, we have made an effort not to take the reader through an extended excursion of the most primitive notions and ideas which are familiar from their counterparts in the three-dimensional theory. For example, the idea of force and couple vectors, each per unit length of a curve on a surface, and associated results are introduced as rapidly as possible (without sacrifice to clarity) but not such details as the notion of the stress vector, the stress tensor, etc., which are discussed in a treatise on continuum mechanics. Similarly, some of the formulae such as those for linearized kinematic measures can be put in a variety of forms by straightforward (although on occasion lengthy) manipulations. Whenever possible such lengthy formulae are conveniently catalogued in separate tables. This should enable a reader not interested in detailed linear kinematic formulae to pass over such special topics easily. A short appendix is included at the end (Chap. F), where for convenience selected formulae from the differential geometry of a surface and related results are collected.

The kinematics of shells and plates, both by direct approach and from the three-dimensional theory, are discussed in Chap. B (Sects. 4–7) at some length. In Sect. 4, we first define a shell-like body and then elaborate in some detail

[10] Our bibliography includes several papers, cited in the text, which became available when this work was almost completed. Consequently, we have not had an opportunity to examine fully these papers which either appeared in recen: periodicals or were kindly sent by their authors in manuscript form.

[11] [1960, *14*] and [1965, *9*].

about the nature of shell theory and motivate the introduction of a Cosserat surface as a continuum portraying a *thin* shell. This background information should be kept in mind in connection with all subsequent developments. Sects. 5 and 6 are concerned with kinematical results by direct approach while their counterparts from the three-dimensional theory are developed in Sect. 7.

Chap. C (Sects. 8–12A) deals with basic principles for shells and plates and shell-like bodies and derivations of the field equations both by direct approach (Sects. 8–10) and from the three-dimensional equations (Sects. 11–12). The basic principles and conservation laws for a Cosserat surface are discussed in detail in Sect. 8 and the local field equations are deduced in Sect. 9, where the field equations appropriate to the linearized theory are also obtained by specialization. Sect. 10 contains a derivation of the field equations for a restricted theory by direct approach, i.e., a theory in which the director is not admitted as an independent primitive kinematic ingredient. The contents of Sects. 11–12 are concerned with the construction of shell theory from the three-dimensional equations of the classical nonlinear continuum mechanics and deal chiefly with the derivation of the local field equations for shell-like bodies. In addition, the nature of the results which may be deduced for an approximate system of field equations for *thin* shells and their relationship with corresponding known results in the classical linear theory of shells and plates are elaborated upon in some detail. Also included in this chapter is a sketch of the history of the derivations of equations of equilibrium for shells (Sect. 12A).

Chap. D (Sects. 13–22) is concerned mainly with constitutive equations for elastic shells and plates, both by direct approach (Sects. 13–16) and from the three-dimensional theory (Sects. 17–21). This chapter also includes a number of related results pertaining to the initial boundary-value problem of the dynamical theory (or the boundary-value problem of the equilibrium theory), special cases of the general theory such as the membrane theory and the inextensional theory (Sect. 14), as well as the complete restricted theory by direct method (Sect. 15) which bears on the classical bending theory of shells. A fairly detailed development of the constitutive equations for an elastic Cosserat surface is given in Sect. 13 and their subsequent linearization is considered in Sect. 16. In the derivation of these results, particular attention is paid to the role played by the director and the manner in which the director (and its gradient) affect the structure of the constitutive equations. The corresponding developments from the three-dimensional theory, including certain approximation schemes for the purpose of obtaining an approximate system of (two-dimensional) equations for thin shells and plates, are discussed in Sects. 17–20. The contents of Sect. 20 (together with part of the remarks made in Sect. 21) provide an account of the classical results and their generalizations, as well as the nature of recent efforts, in the approximate linear theories of shells and plates obtained from the three-dimensional equations. In addition, a sketch of the history of the derivation of linear constitutive equations for elastic shells is provided in Sect. 21A. Also included in Chap. D is a general discussion concerning the relationship and correspondence between the results derived from the three-dimensional theory and those in the theory of Cosserat surface (Sect. 22).

Chap. E (Sects. 23–26) deals exclusively with the linear isothermal theory of elastic plates and shells. It represents a culmination of the point of view expressed in Sect. 4 and begins with a system of equations for the complete linear theory derived by direct method in the previous chapters. Since the constitutive coefficients are arbitrary (in the developments by direct approach) and are not predetermined from an approximate expression for the three-dimensional strain energy density function, a major portion of this chapter (Sects. 24–25) is devoted to the determination of the constitutive coefficients for an isotropic Cosserat surface and the corresponding results in the restricted theory. Chap. E also includes a uniqueness theorem, as well as some remarks on the general theorems in the linear theory of elastic shells.

Apart from the present introductory chapter, the contents of the remaining four Chaps. B to E are so arranged that to a large extent the various sections can be read independently of each other. As should be clear from the table of contents, the developments in each of the three Chaps. B to D are carried out both by direct approach and from the three-dimensional theory. By the time a reader has reached the end of Chap. D, he should be convinced of the following: (1) The development of kinematical results and derivation of field equations can be pursued in a systematic manner from both approaches; (2) the various results for thin shells, in particular the field equations and general aspects of the constitutive equations (in terms of a thermodynamic potential or a strain energy density function), are formally equivalent; and (3) while explicit forms of constitutive equations by direct approach can be dealt with systematically (and free from *ad hoc* assumptions), in general a great deal of difficulty is encountered when explicit constitutive equations are sought from the three-dimensional equations. It is partly here that the theory of a Cosserat surface can be put into fruitful and effective use, as brought out in Chap. E for the linear theory.

3. Notation and a list of symbols used. General convected curvilinear coordinates θ^i $(i=1, 2, 3)$ are used to identify a particle of a body. Similarly, a particle on a surface is identified by convected curvilinear coordinates θ^α $(\alpha=1, 2)$. Throughout this work Latin indices (subscripts or superscripts) have the range 1, 2, 3, Greek indices have the range 1, 2 and the usual summation convention is employed. We use a vertical bar $(|)$ for covariant differentiation with respect to the first fundamental form of a surface, a comma for partial differentiation with respect to surface coordinates θ^α and a superposed dot for material time derivative, i.e., differentiation with respect to time, holding the material coordinates (either θ^α or θ^i) fixed.

Fields and functions which are defined throughout a three-dimensional body are clearly distinguished from the corresponding fields and functions defined over a two-dimensional manifold. Whenever the same symbol is used in both cases, an asterisk is added to the symbol representing a field or a function in a three-dimensional body. Symbols standing for quantities associated with superposed rigid body motions are distinguished by placing a plus sign $(+)$ on the upper right-hand side of the symbol.

While the basic developments are carried out (to a large extent) in an invariant vector notation, the various vector quantities which occur in the field equations and constitutive equations are also expressed in terms of their tensor components. Boldface symbols are employed to designate vector fields or functions defined either throughout a region of a Euclidean 3-space or on a two-dimensional manifold. In general, the boldface symbols represent three-dimensional vector fields or functions; but occasionally they are also used for tangential vector fields defined over a surface. Greek lower case letters are used (although not exclusively) for local thermodynamic scalar functions or variables.

Since the various developments in Chaps. B to D are pursued both by direct approach and also from the three-dimensional theory, often by choice we employ the same symbols in the two developments. This choice of notations for similar quantities is suggestive and will not be confusing, since the two developments are carried out in parallel and entirely separately from one another. Moreover, in order to emphasize the separate nature of these two parallel developments, sometimes formulae of the same type and forms are repeated in a section pertaining to a derivation from the three-dimensional theory even though similar formulae have been already recorded in an earlier section concerned with direct approach.

The notations and formulae given in the Appendix (Chap. F) concerning the geometry of a surface and related results are used throughout and are particularly helpful in early sections. Although all symbols are defined when first introduced, a list of frequently used symbols is provided in a table below. It has not been possible to maintain a complete uniformity in notations and on occasions we have found it necessary to deviate from the scheme of the table and use the same symbols in different senses and in entirely different contexts.

Table of frequently used symbols

Symbol	Name or description	Place of definition or first occurrence
$\boldsymbol{a}_\alpha, \boldsymbol{A}_\alpha$	Base vectors of a surface in the present and reference configurations	(4.10)
$\boldsymbol{a}_3, \boldsymbol{A}_3$	Unit normal vector to a surface in the present and reference configurations	(4.11)
$a_{\alpha\beta}, A_{\alpha\beta}$	First fundamental form of a surface in the present and reference configurations	(4.12)
B	Flexural rigidity	(20.13)
$b_{\alpha\beta}, B_{\alpha\beta}$	Second fundamental form of a surface in the present and reference configurations	(4.13)
$\mathscr{B}$	Body	Sect. 4
c	A curve on the surface $\mathfrak{s}$	Sect. 8
$\mathbf{c}, c_i$	Acceleration vector for a Cosserat surface	(9.46)
C	Extensional rigidity	(20.13)
$\mathscr{C}$	Cosserat surface	Sect. 4
$\mathbf{d}, \boldsymbol{D}$	Director of the continuum $\mathscr{C}$ in the present and reference configurations	Sect. 4; $(5.1)_2$ and $(5.2)_2$
D	Component D_3 of $\boldsymbol{D}$	(4.35)
d_i, D_i	Components of director $\mathbf{d}, \boldsymbol{D}$	(5.25), $(5.34)_2$
$\boldsymbol{d}_N, \boldsymbol{D}_N$	Vector fields occuring in a representation of $\boldsymbol{p}, \boldsymbol{P}$	(7.1)–(7.2)
d_{Ni}, D_{Ni}	Components of $\boldsymbol{d}_N, \boldsymbol{D}_N$	(7.13)
$e_{\alpha\beta}$	A surface kinematic measure	(5.31)
E	Young's modulus of elasticity	(19.3)
$\mathscr{E}$	Internal energy of a Cosserat surface	(8.9)
$\boldsymbol{f}$	Assigned force per unit mass of a surface	(8.6)

Symbol	Name or description	Place of definition or first occurrence
$\bar{\boldsymbol{f}}$	Difference of $\boldsymbol{f}$ and acceleration vector	$(8.19)_1$
$\boldsymbol{f}^*$	External body force density	(8.26)
f^i, F^i	Components of $\boldsymbol{f}$ in the present and reference configurations	(9.38)
$\boldsymbol{g}_i, \boldsymbol{g}^i, g_{ij}, g^{ij}$	Base vectors, metric tensor and conjugate tensor referred to coordinates θ^i in the present configuration	(4.7)
$\boldsymbol{G}_i, \boldsymbol{G}^i, G_{ij}, G^{ij}$	Base vectors, metric tensor and conjugate tensor referred to coordinates θ^i in the reference configuration	(4.23)
$\boldsymbol{G}'_i, G'_{ij}$	Base vectors and metric tensor referred to a normal coordinate system in the reference configuration	(4.29)
h	Initial thickness of (three-dimensional) shell or plate	(4.30)
k, k^N	Scalar functions of position defined by mass density and determinant of the metric tensor	(4.21), (11.28)
$\mathcal{K}$	Kinetic energy of a Cosserat surface	(8.11)
$\boldsymbol{l}$	Assigned director couple per unit mass of a surface	(8.6)
$\grave{\boldsymbol{l}}$	Assigned couple per unit mass of a surface (a tangential vector field)	(10.1)
$\bar{\boldsymbol{l}}$	Difference of $\boldsymbol{l}$ and the inertia term due to director acceleration	$(8.19)_2$
$\boldsymbol{l}^N$ $(N=0, 1, 2, \ldots)$	Body force resultants	(11.29)–(11.30)
$l^i, \bar{l}^i, L^i, \bar{L}^i$	Components of $\boldsymbol{l}$ and $\bar{\boldsymbol{l}}$ in the present and reference configurations	(9.38), (9.57)
$\grave{l}^\alpha, \grave{L}^\alpha$	Components of $\grave{\boldsymbol{l}}$ in the present and reference configurations	(10.20), $(10.27)_2$
$\boldsymbol{m}, {}_R\boldsymbol{m}$	Surface director couple measured per unit area in the present and reference configurations	(8.16), $(8.58)_3$
m^i	Components of the surface director couple $\boldsymbol{m}$	(9.39)
$\boldsymbol{m}^N, m^{Ni}$ $(N=0, 1, 2, \ldots)$	Shear stress- and normal stress-resultants	(11.55), (12.11)
$\boldsymbol{M}, {}_R\boldsymbol{M}$	Contact director couple measured per unit length of a curve on a surface in the present and reference configurations	(8.4), (8.56)
$\grave{\boldsymbol{M}}$	Contact couple per unit length of a curve on a surface (a tangential vector field)	(10.1)

Symbol	Name or description	Place of definition of first occurrence
$\boldsymbol{M}^N$ $(N=0, 1, 2, \ldots)$ $\boldsymbol{M}^{N\alpha}$, $M^{N\alpha i}$ $(N=0, 1, 2, \ldots)$	Stress-resultants and stress-couples	(11.34), (11.36), (12.10)
$M^{\alpha i}$	Components of the contact director couple $\boldsymbol{M}$	(9.42)–(9.43)
$\grave{M}^{\alpha\gamma}$	Components of the contact couple $\grave{\boldsymbol{M}}$	(10.20)
$\boldsymbol{n}$	Outward unit normal vector to a surface in a body	Sect. 8
$\boldsymbol{N}$, ${}_R\boldsymbol{N}$	Contact force measured per unit length of a curve on a surface in the present and reference configurations	(8.4), (8.56)
$N^{\alpha i}$	Components of the contact force $\boldsymbol{N}$	(9.40)–(9.41)
$N'^{\alpha\beta}$	Certain combination of the components $N^{\alpha\beta}$, m^α and $M^{\alpha\beta}$	(9.31), (9.53)
$\grave{N}^{\alpha\beta}$	Certain combination of the components $N^{\alpha\beta}$ and $\grave{M}^{\alpha\beta}$	(10.26)
$\boldsymbol{p}$, $\boldsymbol{P}$	Position vector of the place in a body occupied by the material point in the present and reference configurations	(4.5)–(4.6)
P, ${}_RP$	Scalar quantities representing mechanical power	(9.28), (9.83)
$\mathscr{P}$	Part of a surface	Sect. 4; (4.43)
$\mathscr{P}^*$	Part of a body containing the corresponding part of the surface $\xi=0$	Sect. 11
$\overline{\mathscr{P}}$	Part of a body (not necessarily the same as $\mathscr{P}^*$)	Sect. 11
$\boldsymbol{q}$, ${}_R\boldsymbol{q}$	Heat flux vector per unit time for a Cosserat surface measured per unit length in the present and reference configurations	(9.27), (9.76)
$\boldsymbol{q}^*$, q^{*k}	Heat flux vector per unit area per unit time	(11.11), (11.15)
q_α	Components of $\boldsymbol{q}$ per unit length in the present configuration	(9.26)–(9.27)
${}_Rq_\alpha$	Components of ${}_R\boldsymbol{q}$ per unit length in the reference configuration	$(9.76)_3$
Q_α	Components of $\boldsymbol{q}$ per unit length in the reference configuration	(9.59)
Q^i_j, Q	Proper orthogonal tensor	(4.33), (5.37)
r	Heat supply function per unit mass per unit time for a Cosserat surface	(8.10)
r^*	Heat supply function per unit mass per unit time	$(11.11)_2$

Symbol	Name or description	Place of definition or first occurrence
R	Rate of work by contact and assigned forces and couples for a Cosserat surface	(8.8)
$\boldsymbol{r}$, $\boldsymbol{R}$	Position vector of the place on a surface occupied by the material point in the present and reference configurations	Sect. 4; $(5.1)_1$ and $(5.2)_1$
s, s_α	Relative kinematic measures for a Cosserat surface arising from normal components of director and its gradient	(13.58)
$\mathscr{s}$, $\mathscr{S}$	Surface of the continuum $\mathscr{C}$ in the present and reference configurations	Sect. 4
$\mathfrak{s}$, $\mathfrak{S}$	Surface $\xi=0$ in a body $\mathscr{B}$	Sect. 4
t	Time	(4.6)
$\boldsymbol{t}$	Stress vector	(8.26)
$\boldsymbol{u}$, u^i	Infinitesimal displacement vector of a surface	(6.1)
$\boldsymbol{u}^*$, u^{*i}	Infinitesimal displacement vector of a shell-like body	(7.50)–(7.51)
$\boldsymbol{v}$, v^i	Velocity vector of a surface at time t	$(5.3)_1$, (5.7)
$\boldsymbol{v}^*$	Velocity vector of a shell-like body at time t	(7.4)–(7.5)
V^α, V^3	Combination of certain components of m^i and $M^{\alpha i}$ for a Cosserat surface	(9.62), (9.67) and (9.72)
$V_{i\alpha}$, $V^i_{.\alpha}$	Components of the gradient of a vector field $\boldsymbol{V}$	(5.5)
$\boldsymbol{w}$, w_i	Director velocity vector of a Cosserat surface at time t	$(5.3)_2$, (5.26)
$\grave{\boldsymbol{w}}$	Angular velocity of the unit normal to a surface at time t	$(5.61)_1$
W_{ki}	Spin tensor (a space tensor)	(5.13)–(5.14)
x_i, x^i	Rectangular Cartesian coordinates	(4.1)
α	Coefficient of director inertia	(8.11)
$\alpha_1, \alpha_2, \alpha_3, \ldots, \alpha_{13}$	Constitutive coefficients in the linear theory (direct approach)	(16.21), (16.33)
$\beta_1, \beta_2, \beta_5, \beta_6$	Constitutive coefficients in the restricted linear theory (direct approach)	(16.40)
$\boldsymbol{\beta}$, β_α	Infinitesimal kinematic measure arising from rotation of the unit normal to a surface	(6.6)
γ	Specific entropy production for a Cosserat surface	(8.22)
γ_i	A relative surface kinematic measure due to components of director	(5.33)

Symbol	Name or description	Place of definition or first occurrence
γ_{Ni}	A kinematic measure in the three-dimensional theory	(7.19)
γ^*_{ij}	Strain tensor	(7.27)
$\boldsymbol{\Gamma}, \boldsymbol{\Gamma}_{:\alpha}$	Measures of rate of deformation arising from director velocity and director velocity gradient	(5.26), (5.29)
$\boldsymbol{\delta}, \delta_i$	Infinitesimal director displacement vector	(6.2)
$\delta^i_j, \delta^\alpha_\beta$	Components of a unit tensor	(4.7), (4.12)
ε	Specific internal energy of a Cosserat surface	(8.9)
ε^*	Specific internal energy	$(11.11)_1$
ε^n $(n=0, 1, 2, \ldots)$	Specific internal energy resultants	(11.43)
$\varepsilon_{kim}, \varepsilon^{kim}$; $\varepsilon_{\alpha\beta}, \varepsilon^{\alpha\beta}$; $\bar{\varepsilon}_{\alpha\beta}, \bar{\varepsilon}^{\alpha\beta}$	Absolute permutation symbols	(5.43), (5.63), (6.30)
ζ	Third coordinate in a normal coordinate system $[\theta^\alpha, \zeta]$	(4.25)
η	Specific entropy for a Cosserat surface	(8.21)
η^*	Specific entropy	Sect. 11; (11.12)
η^n $(n=0, 1, 2, \ldots)$	Specific entropy resultants	(11.44)
$\eta_{\alpha\beta}, \boldsymbol{\eta}_\alpha$	Surface rate of deformation	(5.15), (5.19)
θ	Temperature field defined on a surface	(8.20)
θ^*	Temperature field in a body	(11.12)
θ^i	Material coordinates in a body	(4.1) and (4.4)
θ^α	Material coordinates on a surface	Sect. 4
$\varkappa_{i\alpha}$	A relative surface kinematic measure due to components of the director gradients	(5.32)
$\bar{\varkappa}_{\beta\alpha}$	A surface kinematic measure (linear theory)	Sect. 6
$\varkappa_{Ni\alpha}$	A kinematic measure in the three-dimensional theory	(7.19)
$\boldsymbol{\lambda}, \lambda^\alpha$	Unit tangent vector to a curve on a surface in the present configuration	(8.1)
$\lambda_{i\alpha}, \Lambda_{i\alpha}$	Components of the director gradient $\boldsymbol{d}_{,\alpha}$ and $\boldsymbol{D}_{,\alpha}$	(5.28) and (5.34)
$\lambda_{Ni\alpha}, \Lambda_{Ni\alpha}$	Kinematic measures in the three-dimensional theory	$(7.13)_4$, (7.14), $(7.20)_3$
μ	Shear modulus of elasticity	(20.13)
ν	Poisson's ratio	(19.3)

Symbol	Name or description	Place of definition or first occurrence
$\boldsymbol{\nu}$, ν^α	Unit normal to a curve on a surface in the present configuration	(8.2)
${}_0\boldsymbol{\nu}$, ${}_0\nu^\alpha$	Unit normal to a curve on a surface in the reference configuration	(8.55)
ξ	Third coordinate in a general convected coordinate system θ^i	(4.4)
ϱ, ϱ_0	Mass densities (per unit area) of a surface	Sect. 4; (4.17), (4.21), (4.38)
ϱ^*, ϱ_0^*	Mass densities of a body	Sect. 4; (4.16), (4.36)
$\varrho_{i\alpha}$, $\bar{\varrho}_{\alpha\beta}$	Surface kinematic measures (linear theory)	(6.24), (20.37), (25.16)
σ, σ_α	Certain kinematic variables for a Cosserat surface arising from normal components of director and its gradient	$(13.33)_{1,2}$
σ_{ij}	Cartesian components of the stress tensor	(24.1)
Σ	Strain energy density for a Cosserat surface	(14.3)–(14.4)
τ_{ij}	Symmetric stress tensor of Cauchy in coordinates θ^i	(11.7)
ψ	Specific Helmholtz free energy for a Cosserat surface	(9.34)
ψ^*	Specific Helmholtz free energy	(11.18)
ψ^n $(n=0, 1, 2, \ldots)$	Specific Helmholtz free energy resultants	(11.44)
φ	Specific Gibbs free energy for a Cosserat surface (linear theory)	(16.12)
φ^*	Specific Gibbs free energy	(19.1)
φ_N $(N=0, 1, \ldots)$	Scalar functions in a representation for temperature θ^*	(11.42)
$\boldsymbol{\omega}$, ω^i Ω, Ω_{ki}	Vector and skew-symmetric space tensor representing rigid body angular velocity	(5.39)–(5.42)
${}_0\bar{\boldsymbol{\omega}}$	Infinitesimal rigid body rotation	(6.42)
dv	Element of volume	$(4.19)_1$
$d\sigma$, $d\Sigma$	Element of area	$(4.19)_2$, (4.39)
ds, dS	Line element	(8.4), (8.56)
$(\ )_{,\alpha}$	Partial differentiation with respect to surface coordinates	(4.13)
$(\ )_{\mid\alpha}$	Covariant differentiation with respect to first fundamental form of a surface	(4.13)

Symbol	Name or description	Place of definition or first occurrence
$(\ )_{\Vert i}$	Covariant differentiation with respect to the metric tensor g_{ij}	(11.15), (A.1.27)
$\dot{(\ )}$	Material time derivative	(5.3)
$[L]$	Physical dimension of length	(4.30)
$[M]$	Physical dimension of mass	(4.37)
$[T]$	Physical dimension of time	(8.5)

B. Kinematics of shells and plates.

This chapter is concerned with the kinematics of shells and initially flat plates both by direct approach and from the three-dimensional theory of classical continuum mechanics. Definition of a shell-like body and motivation for introducing another continuum, namely a Cosserat surface, as a *model* reflecting the main features of a *thin* shell are included in Sect. 4. This preliminary and background material should be kept in mind in relation to the remaining sections of this chapter, as well as most of the subsequent developments in Chaps. C and D.

4. Coordinate systems. Definitions. Preliminary remarks. A shell or a plate is a three-dimensional body whose boundary surface enjoys special features. Before describing such a body in precise terms, it is convenient to introduce suitable coordinate systems. Let the points of a region $\mathscr{R}$ in a Euclidean 3-space be referred to a fixed right-handed rectangular Cartesian coordinate system x_i $(i=1, 2, 3)$ and let $\{\theta^1, \theta^2, \theta^3\}$ be a general *convected* curvilinear system defined by the transformation relations[1]

$$x_i = x_i(\theta^1, \theta^2, \theta^3). \tag{4.1}$$

We assume

$$\det(\partial x_i/\partial\theta^j) \neq 0, \tag{4.2}$$

so that (4.1) is nonsingular in $\mathscr{R}$ and has a unique inverse

$$\theta^i = \theta^i(x_1, x_2, x_3). \tag{4.3}$$

In what follows, all Latin indices (subscripts or superscripts) take the values 1, 2, 3 and Greek indices (subscripts or superscripts) take the values 1, 2. Also, for later convenience, we set $\theta^3 = \xi$ and adopt the notation

$$\theta^i = \{\theta^\alpha, \xi\}. \tag{4.4}$$

Consider now a three-dimensional body $\mathscr{B}$, embedded in the region $\mathscr{R}$ of the Euclidean 3-space, and let the particles of $\mathscr{B}$ be identified by a general convected coordinate system (4.3). Let $\boldsymbol{P}$ denote the position vector, relative to a fixed

[1] Often our notation and particularly the choice of convected (or moving) curvilinear coordinates is patterned after GREEN and ZERNA [1954, *1*] or [1968, *9*]. Although the use of a convected coordinate system is by no means essential, it is particularly suited in studies of special bodies (such as shells, plates and rods) and often results in simplification of intermediate steps in the development of the subject.

origin, of a typical particle of $\mathscr{B}$ in a reference configuration. Then,[2]

$$\boldsymbol{P} = \boldsymbol{P}(\theta^\alpha, \xi) \tag{4.5}$$

which can also be expressed as a function of x_i, in view of (4.2). We denote the position vector, relative to the same fixed origin, of a typical particle of $\mathscr{B}$ in the deformed configuration at time t by

$$\boldsymbol{p} = \boldsymbol{p}(\theta^\alpha, \xi, t). \tag{4.6}$$

Thus, (4.5) specifies the place occupied by the material point θ^i in a reference configuration while the place occupied by the material point θ^i in the deformed configuration is given by (4.6). The region of space at time t into which the body is mapped by the vector function $\boldsymbol{p}$ in (4.6) is the region occupied by the body in a given configuration. We assume that the vector function $\boldsymbol{p}$—a 1-parameter family of configurations with t as the real parameter—which describes the motion of the body $\mathscr{B}$, is sufficiently smooth in the sense that it is differentiable with respect to θ^α, ξ and t as many times as may be needed. We recall the formulae[3]

$$\begin{aligned} \boldsymbol{g}_i &= \frac{\partial \boldsymbol{p}}{\partial \theta^i}, \quad g_{ij} = \boldsymbol{g}_i \cdot \boldsymbol{g}_j, \quad g = \det(g_{ij}), \\ \boldsymbol{g}^i &= g^{ij}\, \boldsymbol{g}_j, \quad \boldsymbol{g}^i \cdot \boldsymbol{g}^j = g^{ij}, \quad \boldsymbol{g}^i \cdot \boldsymbol{g}_j = \delta^i_j, \end{aligned} \tag{4.7}$$

and further assume that

$$g^{\frac{1}{2}} = [\boldsymbol{g}_1\, \boldsymbol{g}_2\, \boldsymbol{g}_3] > 0 \tag{4.8}$$

for physically possible motions.[4] In (4.7), $\boldsymbol{g}_i$ and $\boldsymbol{g}^i$ are the covariant and the contravariant base vectors at time t, respectively, g_{ij} is the metric tensor, g^{ij} is its conjugate and δ^i_j is the Kronecker symbol in 3-space. Formulae analogous to those in (4.7), valid in a reference configuration, can be deduced from (4.5) but we postpone recording such results.

A material surface in $\mathscr{B}$ can be defined by the equation $\xi = \xi(\theta^\alpha)$; the equations resulting from (4.5) and (4.6) with $\xi = \xi(\theta^\alpha)$ represent the parametric forms of this surface in the reference and deformed configurations. In particular, with reference to (4.6), $\xi = 0$ defines a 1-parameter family of surfaces in space each of which we assume to be smooth and non-intersecting. We refer to the surface $\xi = 0$ at time t (i.e., in the deformed configuration) by $\mathfrak{s}$. Any point of this surface is specified by the position vector $\boldsymbol{r}$, relative to the same fixed origin to which $\boldsymbol{p}$ is referred, where

$$\boldsymbol{r} = \boldsymbol{r}(\theta^\alpha, t) = \boldsymbol{p}(\theta^\alpha, 0, t). \tag{4.9}$$

Let $\boldsymbol{a}_\alpha$ denote the base vectors along the θ^α-curves on the surface $\mathfrak{s}$. By (4.9) and $(4.7)_1$,

$$\boldsymbol{a}_\alpha = \frac{\partial \boldsymbol{r}}{\partial \theta^\alpha} = \boldsymbol{g}_\alpha(\theta^\gamma, 0, t), \tag{4.10}$$

---

[2] We may recall that when the particles of the body are referred to a convected (or moving) coordinate system, the numerical values of the coordinates associated with each material point remain the same for all time.

In (4.5) and in most of the developments that follow, the same symbol can be used for a function and its value without confusion. Only on occasions and wherever it serves clarity, the symbol for a function will be distinguished from that of its value. When a function is first introduced we either state or exhibit its arguments; but, subsequently we often employ the symbol designating the function without an explicit indication of the arguments as this will be clear from the context.

[3] For details see Sect. A.1 of the Appendix (Chap. F).

[4] Strictly speaking, for physically possible motions we only need to assume that $g^{\frac{1}{2}} \neq 0$ with the understanding that in any given motion $[\boldsymbol{g}_1\, \boldsymbol{g}_2\, \boldsymbol{g}_3]$ is either > 0 or < 0. The condition (4.8) also requires that θ^i be a right-handed coordinate system.

and the unit normal $\boldsymbol{a}_3$ to $\mathfrak{s}$ may be defined by

$$\boldsymbol{a}_\alpha \cdot \boldsymbol{a}_3 = 0, \quad \boldsymbol{a}_3 \cdot \boldsymbol{a}_3 = 1, \quad \boldsymbol{a}_3 = \boldsymbol{a}^3, \quad [\boldsymbol{a}_1\, \boldsymbol{a}_2\, \boldsymbol{a}_3] > 0. \tag{4.11}$$

We also recall the formulae[5]

$$\begin{gathered} a_{\alpha\beta} = \boldsymbol{a}_\alpha \cdot \boldsymbol{a}_\beta, \qquad a = \det(a_{\alpha\beta}), \\ \boldsymbol{a}^\alpha = a^{\alpha\beta}\, \boldsymbol{a}_\beta, \qquad \boldsymbol{a}^\alpha \cdot \boldsymbol{a}^\beta = a^{\alpha\beta}, \qquad a^{\alpha\gamma}\, a_{\gamma\beta} = \delta^\alpha_\beta, \qquad a_{\alpha\gamma}\, a^{\gamma\beta} = \delta^\beta_\alpha, \end{gathered} \tag{4.12}$$

$$\begin{gathered} b_{\alpha\beta} = b_{\beta\alpha} = -\boldsymbol{a}_\alpha \cdot \boldsymbol{a}_{3,\beta} = \boldsymbol{a}_3 \cdot \boldsymbol{a}_{\alpha,\beta}, \\ \boldsymbol{a}_{\alpha|\beta} = b_{\alpha\beta}\, \boldsymbol{a}_3, \qquad \boldsymbol{a}_{3,\alpha} = -b^\gamma_\alpha\, \boldsymbol{a}_\gamma, \qquad b_{\alpha\beta|\gamma} = b_{\alpha\gamma|\beta}, \end{gathered} \tag{4.13}$$

where $\boldsymbol{a}^\alpha$ denote the reciprocal base vectors of the surface $\mathfrak{s}$, $a_{\alpha\beta}$ and $b_{\alpha\beta}$ are its first and its second fundamental forms, a comma denotes partial differentiation with respect to the surface coordinates θ^α, a vertical bar stands for covariant differentiation with respect to $a_{\alpha\beta}$ and δ^α_β is the Kronecker symbol in 2-space.

Let $\partial\mathscr{B}$, the boundary of the body $\mathscr{B}$, be specified by (i) the material surfaces

$$\xi = \alpha(\theta^\alpha), \qquad \xi = \beta(\theta^\alpha), \qquad \alpha < 0 < \beta, \tag{4.14}$$

with the surface $\xi = 0$ lying entirely between them; and (ii) a material surface

$$f(\theta^1, \theta^2) = 0, \tag{4.15}$$

which is such that $\xi =$ const. are closed smooth curves on the surface (4.15).[6] Since the convected coordinates (4.4) are identified as material coordinates, the material surfaces $(4.14)_{1,2}$ have the same parametric representations in all configurations; and, in general α and β are functions of the surface coordinates θ^α but in special cases they may be constants. We observe that by virtue of (4.8) and $(4.14)_3$, the material surfaces $(4.14)_{1,2}$ do not intersect themselves, each other, or the surface $\xi = 0$; and in keeping with the designation of the surface $\xi = 0$ at time t by $\mathfrak{s}$, we refer to the surfaces $(4.14)_{1,2}$ in the deformed configuration as $\mathfrak{s}^-$ and $\mathfrak{s}^+$, respectively. The surface $\mathfrak{s}$ is not necessarily midway between the bounding surfaces $\mathfrak{s}^-$ and $\mathfrak{s}^+$ and the middle surface is arbitrarily situated with respect to the boundary surfaces $(4.14)_{1,2}$; however, a reference configuration may be chosen in which the initial middle surface is midway between the surfaces defined by $(4.14)_{1,2}$.

Let $\varrho^*(\theta^\alpha, \xi, t)$ and $\varrho_0^*(\theta^\alpha, \xi)$ be the mass densities of $\mathscr{B}$ in the deformed and reference configurations, respectively. Then, the (local) equation of conservation of mass is

$$\varrho^*\, g^{\frac{1}{2}} = \varrho_0^*\, G^{\frac{1}{2}}, \tag{4.16}$$

where G is the dual of g in a reference configuration. We define a mass per unit area of $\mathfrak{s}$ at time t, namely $\varrho(\theta^\alpha, t)$, by the formula

$$\varrho\, a^{\frac{1}{2}} = \int_\alpha^\beta \varrho^*\, g^{\frac{1}{2}}\, d\xi, \tag{4.17}$$

where a and g are given by $(4.12)_2$ and $(4.7)_3$. Since θ^i are convected coordinates and since $\varrho^* g^{\frac{1}{2}}$ is independent of time, it follows from (4.17) that $\varrho a^{\frac{1}{2}}$ is also independent of time, although both ϱ and a may depend on t. The mass of an

[5] See Sects. A.2 and A.3 of the Appendix (Chap. F).

[6] In place of (4.15) we can specify a more general boundary surface of the form $\bar{f}(\theta^\alpha, \xi) = 0$ such that $\xi =$ const are closed smooth curves on this surface. However, (4.15) will suffice for our present purpose.

arbitrary portion of $\mathscr{B}$ bounded by the surfaces (4.14) and a surface of the form (4.15) may be expressed as

$$\iiint \varrho^* g^{\frac{1}{2}}\, d\theta^1\, d\theta^2\, d\theta^3 = \int \varrho^*\, dv = \iint \varrho\, a^{\frac{1}{2}}\, d\theta^1\, d\theta^2 = \int \varrho\, d\sigma, \tag{4.18}$$

where the ranges of integration in (4.18) are clear and dv and $d\sigma$ given by

$$\begin{aligned} dv &= (\boldsymbol{g}_1 \times \boldsymbol{g}_2 \cdot \boldsymbol{g}_3)\, d\theta^1\, d\theta^2\, d\theta^3 = g^{\frac{1}{2}}\, d\theta^1\, d\theta^2\, d\theta^3, \\ d\sigma &= (\boldsymbol{a}_1 \times \boldsymbol{a}_2 \cdot \boldsymbol{a}_3)\, d\theta^1\, d\theta^2 = a^{\frac{1}{2}}\, d\theta^1\, d\theta^2 \end{aligned} \tag{4.19}$$

are an element of volume of the body $\mathscr{B}$ in the deformed configuration and an element of area of the surface $\mathfrak{s}$, respectively.

The relation of the surface $\xi = 0$ at time t, i.e., the surface $\mathfrak{s}$, to the bounding surfaces $\mathfrak{s}^-$ and $\mathfrak{s}^+$ can be fixed by imposing the condition[7]

$$\int_\alpha^\beta \varrho^* g^{\frac{1}{2}}\, \xi\, d\xi = \int_\alpha^\beta k\, \xi\, d\xi = 0, \tag{4.20}$$

where we have put for later convenience

$$k = k(\theta^\alpha, \xi) = \varrho^* g^{\frac{1}{2}} = \varrho_0^* G^{\frac{1}{2}}. \tag{4.21}$$

The condition (4.20) is independent of time, so that once the relative position of $\mathfrak{s}$ (i.e., relative to $\mathfrak{s}^-$ and $\mathfrak{s}^+$) is determined by such an equation (in, say, a reference configuration) it remains so determined. This completes the description of a shell-like body, i.e., a three-dimensional continuum bounded by the surfaces (4.14) and (4.15).

In order to continue our preliminary remarks, it is desirable to dispose of some further notation. Henceforth, let the initial configuration of $\mathscr{B}$ be taken as the reference configuration. Then, by (4.5), the initial position vector of the body $\mathscr{B}$, at time $t = 0$, is specified by

$$\boldsymbol{P} = \boldsymbol{P}(\theta^\alpha, \xi) = \boldsymbol{p}(\theta^\alpha, \xi, 0). \tag{4.22}$$

The initial values of $\boldsymbol{g}_i$, $\boldsymbol{g}^i$, g_{ij} will be designated by $\boldsymbol{G}_i$, $\boldsymbol{G}^i$, G_{ij} and we have the formulae

$$\begin{aligned} &\boldsymbol{G}_i = \frac{\partial \boldsymbol{P}}{\partial \theta^i}, \qquad G_{ij} = \boldsymbol{G}_i \cdot \boldsymbol{G}_j, \qquad G = \det(G_{ij}), \\ &\boldsymbol{G}^i = G^{ij}\, \boldsymbol{G}_j, \qquad \boldsymbol{G}^i \cdot \boldsymbol{G}^j = G^{ij}, \qquad \boldsymbol{G}^i \cdot \boldsymbol{G}_j = \delta^i_j, \end{aligned} \tag{4.23}$$

as the duals of those in (4.7). The initial surfaces in the initial configuration of $\mathscr{B}$, which become the surfaces $\mathfrak{s}$, $\mathfrak{s}^-$, $\mathfrak{s}^+$ at time t, will be referred to as $\mathfrak{S}$, $\mathfrak{S}^-$, $\mathfrak{S}^+$, respectively. Similarly, we write the initial position vector of $\mathfrak{S}$ as

$$\boldsymbol{R} = \boldsymbol{R}(\theta^\alpha) = \boldsymbol{P}(\theta^\alpha, 0) = \boldsymbol{r}(\theta^\alpha, 0) \tag{4.24}$$

and designate the initial values of $\boldsymbol{a}_i$, $a_{\alpha\beta}$, $b_{\alpha\beta}$, a by $\boldsymbol{A}_i$, $A_{\alpha\beta}$, $B_{\alpha\beta}$, A, respectively, and we note that formulae of the type (4.10) to (4.13) hold also for the surface $\mathfrak{S}$.

We have already defined a general shell-like body. But, in order to fix ideas, it is desirable to describe a shell in somewhat less general and more familiar terms. For this purpose, consider a surface (embedded in a Euclidean 3-space) which

[7] This restriction is not severe and imposes only a minor loss of generality. The formula (4.17) defines the mass, per unit area of $\mathfrak{s}$, for an arbitrary portion of $\mathscr{B}$ (above and below the surface $\xi = 0$). The relation (4.20), involving essentially the "moment" of $\varrho^* g^{\frac{1}{2}}$, is in effect a condition on the distribution of mass above and below the surface $\xi = 0$. Condition (4.20) was introduced by GREEN, LAWS and NAGHDI [1968, *4*] in their derivation of two-dimensional shell equations from the three-dimensional equations of classical continuum mechanics.

we temporarily designate by $\mathfrak{S}'$. Let the position vector of any point on this surface be denoted by $\boldsymbol{R}'(y^1, y^2)$ with $\{y^1, y^2\}$ being simply parameters as yet unrelated to the coordinates θ^α. With formulae of the type (4.10) and (4.11) in mind, let $\boldsymbol{A}'_\alpha(y^1, y^2)$ denote the base vectors of $\mathfrak{S}'$ and $\boldsymbol{A}'_3(y^1, y^2)$ its unit normal such that $[\boldsymbol{A}'_1\, \boldsymbol{A}'_2\, \boldsymbol{A}'_3] > 0$. Assume the existence of a neighborhood $\mathfrak{N}(\mathfrak{S}')$ in which points in space lie along one and only one normal to $\mathfrak{S}'$. Let y^3, a parameter measured to the scale of the (rectangular Cartesian coordinates) x_i along the positive direction of the uniquely defined normal $\boldsymbol{A}'_3(y^1, y^2)$ from $\mathfrak{S}'$, denote the distance to any point in $\mathfrak{N}(\mathfrak{S}')$. Then, any point in $\mathfrak{N}(\mathfrak{S}')$ can be located by $\boldsymbol{R}'(y^1, y^2) + y^3 \boldsymbol{A}'_3(y^1, y^2)$. If we now identify $\mathfrak{S}'$ with $\mathfrak{S}$, the parameters $\{y^1, y^2\}$ with θ^α and put for later convenience $y^3 = \zeta$, then

$$y^i = \{\theta^\alpha, \zeta\} \tag{4.25}$$

may be regarded as a *normal* coordinate system in which y^α coincides with the convected θ^α on $\mathfrak{S}$ and y^3 is normal to $\mathfrak{S}$.

Now in a reference configuration of $\mathscr{B}$, which again we take to be the initial configuration, the convected general coordinates θ^i can always be related to y^i with ζ a specified function of θ^α and ξ. For simplicity and to avoid undue complications, we specify ζ in the form[8]

$$\zeta = \zeta'(\theta^\alpha)\, \xi, \tag{4.26}$$

where ζ' is a function of θ^α only. In the special case that $\zeta'(\theta^\alpha) = 1, \zeta = \xi$, the coordinates (4.4) become coincident with (4.25) in the reference configuration. Thus, with ζ specified by (4.26), the position vector of the body $\mathscr{B}$ in the initial configuration referred to the normal coordinate system y^i is

$$\boldsymbol{P} = \boldsymbol{P}(\theta^\alpha, \xi) = \boldsymbol{P}'(\theta^\alpha, \zeta) = \boldsymbol{R}(\theta^\alpha) + \zeta \boldsymbol{A}_3(\theta^\alpha) \tag{4.27}$$

and (4.27) represents the transformation relations between the normal coordinates (4.25) and x_i, the rectangular Cartesian components of $\boldsymbol{P}$. The existence of a neighborhood $\mathfrak{N}(\mathfrak{S})$ in which every point in the initial configuration of $\mathscr{B}$ is uniquely located by (4.27) may be verified in the manner discussed in Sect. A.3 of the Appendix [between (A.3.14)–(A.3.17) in Chap. F]. Indeed, if R_1 and R_2 are the principal radii of curvature of the surface $\mathfrak{S}$ and since $A = \det(A_{\alpha\beta}) \neq 0$, we need only choose

$$\mathfrak{N}(\mathfrak{S}) = \{(\theta^\alpha, \zeta) : |\zeta| < R\}, \qquad R = \min(|R_1|, |R_2|) \neq 0, \tag{4.28}$$

to ensure that a condition of the form (4.3) between the coordinates y^i and x_i is always satisfied in the (initial) reference configuration. With this choice of $\mathfrak{N}(\mathfrak{S})$, (4.27) is nonsingular and hence the duals of the formulae of the type (4.10)–(4.13) remain valid. Let the base vectors and the metric tensor in the reference configuration referred to the coordinates y^i, with ζ given by (4.26), be denoted by $\boldsymbol{G}'_i(\theta^\alpha, \zeta)$ and $G'_{ij}(\theta^\alpha, \zeta)$. These and other results of the type (4.23) can be calculated using (4.27). In particular, on the surface $\zeta = 0$, the metric tensor G'_{ij}, its conjugate G'^{ij} and the determinant G' reduce to

$$\begin{gathered} G'_{\alpha\beta}(\theta^\gamma, 0) = A_{\alpha\beta}, \qquad G'_{\alpha 3}(\theta^\gamma, 0) = 0, \qquad G'_{33}(\theta^\gamma, 0) = 1, \\ G'^{\alpha\beta}(\theta^\gamma, 0) = A^{\alpha\beta}, \qquad G'^{\alpha 3}(\theta^\gamma, 0) = 0, \qquad G'^{33}(\theta^\gamma, 0) = 1, \\ G'(\theta^\alpha, 0) = A, \end{gathered} \tag{4.29}$$

respectively.

[8] In place of (4.26) we can write $\zeta = \bar{\zeta}(\theta^\alpha, \xi)$, $\bar{\zeta}$ being a function of θ^α and ξ; but this generality is not needed in our subsequent developments.

Consider again the part of $\partial\mathscr{B}$ referred to previously by $\mathfrak{S}^-$ and $\mathfrak{S}^+$ in the reference configuration. Recalling (4.27), let these bounding surfaces be specified by[9]

$$\begin{aligned} &\zeta = h_1(\theta^\alpha), \quad \zeta = h_2(\theta^\alpha), \qquad h_1 < 0 < h_2, \\ &h = h(\theta^\alpha) = h_2(\theta^\alpha) - h_1(\theta^\alpha), \quad \text{phys. dim. } h = [L], \end{aligned} \tag{4.30}$$

where the functions h_1 and h_2, as well as h, have the physical dimension of length designated by $[L]$ and $\zeta = 0$ lies entirely between the surfaces $(4.30)_{1,2}$. Then, with reference to the initial configuration of the body and (4.27)–(4.30), a shell may be defined as a region of space $h_1 < \zeta < h_2$, $\max(|h_1|, |h_2|) < R$ bounded by the two surfaces $(4.30)_{2,1}$, i.e., $\mathfrak{S}^+$ and $\mathfrak{S}^-$ (called the upper and lower surfaces or faces) which are situated above and below a surface $\mathfrak{S}$ (specified by $\zeta = 0$) and a lateral surface [or an edge boundary, i.e., the surface corresponding to (4.15) in the reference configuration], the intersection of which with surfaces $\zeta =$ constant are closed smooth curves. By (4.27) and (4.30) the distance between $\mathfrak{S}^-$ and $\mathfrak{S}^+$, measured along $\boldsymbol{A}_3$ is h and is called the (initial) thickness of the shell. If h is constant, the shell is said to be of uniform thickness, otherwise of variable thickness.

It is clear from either of the above descriptions that a shell (or a plate) is a three-dimensional continuum whose boundary surface has special features, as remarked at the beginning of this section. If *full* information is desired regarding the motion and deformation of such bodies in the context of the classical three-dimensional theory of continuum mechanics, then there would be no point to the present article or the extensive and continuing efforts on the foundations of the theories of shells and plates which have spanned the literature during the twentieth century. In this connection, it is worth recalling the well-known fact that as a problem in the three-dimensional theory, the closed shell is often amenable to analytical treatment at least within the scope of linear elastostatics, whereas an exact analysis of the open shell (i.e., one with an edge boundary) would lead to a formidable task.

Suppose, instead, we are content with only partial information (in some sense) for a sufficiently *thin* shell, i.e., when the thickness h is much smaller than the minimum radius of the curvature R defined in (4.28) or equivalently when[10]

$$\frac{h}{R} \ll 1. \tag{4.31}$$

---

[9] Instead of $(4.30)_{1,2}$, it is tempting to specify $\mathfrak{S}^-$ and $\mathfrak{S}^+$ by the symmetric conditions

$$\zeta = \mp \frac{h}{2}$$

in which case $\zeta = 0$ will be the middle surface. Because of (4.20), our specification of $\mathfrak{S}^-$ and $\mathfrak{S}^+$ in the forms $(4.30)_{1,2}$ is simply motivated by the fact that we do not wish to exclude an important class of shell-like bodies whose initial mass density ϱ_0^* is independent of ζ. In general, the symmetric specification

$$\zeta = \mp \frac{h}{2}$$

with $\zeta = 0$ as the middle surface is possible only for initially flat plates; but it will not be consistent with the condition corresponding to (4.20) in the (initial) reference configuration of a curved shell, unless the mass density ϱ_0^* is dependent on ζ. Further remarks on this point, including the consideration of (4.20) in the context of an approximate theory for a shell with its initial mass density independent of ζ, is made in Sect. 7 [Subsect. β) between (7.42)–(7.48)].

[10] The criterion (4.31) is generally used to define a *thin* shell and is also invoked in the development of approximate theories of thin shells from the three-dimensional equation. In the case of an initially flat plate (for which R is infinite), in place of (4.31), it is assumed that h is much smaller than the smallest dimension of its middle plane.

By partial information we mean, for example, information concerning quantities which can be regarded as representing the response of the surface $\zeta=0$ (or its neighborhood) as a consequence of the (three-dimensional) motion of the body $\mathscr{B}$ or the determination of certain averages of quantities resulting from the (three-dimensional) motion of $\mathscr{B}$. Indeed, the desire for such partial or limited information is the basic motivation for the construction of a two-dimensional theory for a thin shell as defined above, with the aim of providing a simpler theory for the partial or limited information sought. A useful two-dimensional theory of this type is necessarily approximate and is referred to as shell theory, in order to distinguish it from the three-dimensional theory. In fact, the main problem of the general theory of thin shells (and plates) may be stated as follows:

(a) The development of a two-dimensional theory—an approximate theory relative to the three-dimensional theory—so constructed as to be capable of supplying the partial information mentioned above.
(b) The development of a scheme or a systematic procedure for estimating the "error" involved in the use of the (approximate) two-dimensional theory in comparison with that of the full three-dimensional theory. Alternatively, we may ask under what circumstances do the equations of shell theory supply an approximate solution to the three-dimensional equations and how "close" is this approximate solution to the exact solution? (4.32)

With reference to (b) in (4.32), of course, an idea of the range of applicability of shell theory often can be had *a priori* through intuitive reasoning and for certain specific problems (especially in the linear theory) it is possible to support such intuitive reasoning by analysis or similar considerations.[11] In fact, historically speaking, such *a priori* considerations have been partially relied upon by various investigators in developing specialized or more general theories of shells and plates. However, an explicit answer to (b) is not available at present and is not attempted here.[12] Rather, the main developments of this article are concerned with (a) and associated topics. Even here there are considerable difficulties when the complete theory (i.e., all field equations and constitutive equations) is deduced from the full three-dimensional equations. The main difficulties stem chiefly from the fact that at some stage in the development of shell theory, from the three-dimensional equations, approximations must be introduced and that the nature of validity of such approximations probably cannot be entirely divorced from the question posed under (b) in (4.32). Again, with reference to (a) in (4.32) and a general development of the linear theory of thin elastic shells, we may note that there remained unsettled questions for almost three-quarters of a century be-

[11] In addition to intuitive notions, such a priori knowledge may be based on available exact or asymptotic solutions of specific problems obtained from special linear theories of shells such as plate theory, membrane theory or other special cases of the general theory. Also, for certain simple problems (such as pure bending of a rectangular plate, torsion of a rectangular plate and torsion of a circular cylindrical shell) a solution via a general linear theory of shells may be expected to agree (exactly or very nearly) with the corresponding known exact solutions from the three-dimensional theory.

[12] The problem posed under (b) in (4.32) is a formidable one and has not been solved with finality even in the case of the linear theory of bending of elastic plates. Recently, however, some effort in this direction has been made by JOHN [1965, *5*] and [1969, *4*] in connection with the v. Kármán theory of plates.

ginning with the pioneering work of LOVE.[13] From a practical point of view, there seems to be a reasonable agreement among a number of recent and different derivations of an approximate linear bending theory of shells starting from the three-dimensional equations; but still these derivations employ a number of approximations or special assumptions (sometimes in an *ad hoc* manner) which, when taken collectively, are not particularly appealing and often leave something to be desired. Again, the nature of the difficulties here is not unrelated to (b) in (4.32).

With reference to (a) in (4.32), two aspects of the above remarks or observations are worth recapitulating: Firstly, shell theory (being an approximate two-dimensional theory) can neither be expected to, nor can it be capable of, predicting full and exact information (in the sense of three-dimensional theory), except possibly in very special circumstances; and secondly the derivation of a theory for thin shells from the three-dimensional equations involve considerable difficulties which are largely mathematical and have to do with the nature of approximations and special assumptions introduced.[14] It then seems natural to ask if it is possible to replace the continuum characterizing the body $\mathscr{B}$ with another continuum, a *model* which would reflect the main features of a thin shell and which would then permit the development of an exact theory without recourse to the approximations or special assumptions mentioned above? To this end and preliminary to the description of the alternative model, we recall that a motion of a body in classical (three-dimensional) continuum mechanics is said to be *rigid* if and only if the rectangular Cartesian components of the position of every material point θ^i at time t are related to the rectangular Cartesian components $x^i(\theta^k)$ in a reference position by a relation of the form[15]

$$x^{+i}(\theta^k, t) = C^i(t) + Q^i_j(t)\, x^j(\theta^k). \tag{4.33}$$

In (4.33), C^i is some vector-valued function of time and Q^i_j are the components of a proper orthogonal tensor (or matrix) function of time which may be interpreted as the rotation tensor. Consider now a body regarded as consisting not only of material particles identified with material points θ^i, but also of directions associated with the material points. For such a model of *oriented media*, the directions are characterized by deformable vector fields—called the *directors*—which are capable of rotation and stretches independently of the deformation of materials points.[16] In a three-dimensional theory of an oriented medium with a

---

[13] [1888, *1*].

[14] One of the main obstacles in the development of a general theory of thin shells (from the three-dimensional equations), even in the case of linear theory, lies in the difficulty of rendering the notion of "thinness" precise. What is generally invoked is the criterion (4.31), often supplemented by other special assumptions.

[15] Although subscripts and superscripts are employed in writing (4.33), we recall that no distinction between the character of these indices is necessary in a rectangular Cartesian coordinate system.

[16] Historically the concept of "directed" or "oriented" media was originated by DUHEM [1893, *1*] and a first systematic development of theories of oriented media in one, two and three dimensions (the first two being motivated by rods and shells) was carried out by E. and F. COSSERAT [1909, *1*]. In their work (see also [1968, *3*]—translation of the original [1909, *1*]), the COSSERATS represented the orientation of each point of their continuum by a set of mutually perpendicular rigid vectors. A general development of the kinematics of oriented media, in the presence of n stretchable directors (in n-dimensional space), has been given more recently by ERICKSEN and TRUESDELL [1958, *1*] who also introduced the terminology of *directors*. An exposition of the kinematics of the theory of oriented bodies, together with references to other contributions on the subject prior to 1960, may be found in TRUESDELL and TOUPIN [1960, *14*].

single deformable director[17] $\boldsymbol{d}$ at every material point of the body, the above definition of a rigid motion displayed through (4.33), is supplemented by

$$\bar{d}^{+i}(\theta^k, t) = Q^i_j(t)\,\bar{d}^j(\theta^k), \tag{4.34}$$

where $\bar{d}^{+i}$ and $\bar{d}^i$ are the rectangular Cartesian components of the director field $\boldsymbol{d}$ at x^{+i} and x^i, respectively. The condition (4.34) relates the components $\bar{d}^{+i}$ at every material point θ^i at time t to the reference values $\bar{d}^i$ under rigid motions. It can be readily verified from (4.34) that the director has the property that it remains unaltered in magnitude under rigid motions of the continuum.

Returning to our objective of an alternative model for a thin shell, consider a body $\mathscr{C}$ consisting of a surface embedded in a Euclidean 3-space together with a single deformable director assigned to every point of the surface. Let the particles on the surface of $\mathscr{C}$ be identified by the convected coordinates θ^α ($\alpha = 1, 2$) and let $\mathscr{S}$ and $\mathfrak{s}$ refer to the surface of $\mathscr{C}$ in the reference and deformed configurations, respectively. Let[18] $\boldsymbol{r} = \boldsymbol{r}(\theta^\alpha, t)$ denote the position vector of $\mathfrak{s}$, relative to a fixed origin (say relative to the same fixed origin used previously), which specifies the place occupied by the material point θ^α in the deformed configuration at time t. Let $\boldsymbol{a}_3 = \boldsymbol{a}_3(\theta^\alpha, t)$ be the unit normal to $\mathfrak{s}$ and let $\boldsymbol{d} = \boldsymbol{d}(\theta^\alpha, t)$ stand for the deformable director assigned to every point of $\mathfrak{s}$. The three-dimensional vector field $\boldsymbol{d}$ at $\boldsymbol{r}$ which we specify to be dimensionless[19] is not necessarily along the normal to $\mathfrak{s}$ and, as already noted, has the property that it remains invariant in magnitude under rigid motions of the continuum. Henceforth (except when noted otherwise) we identify the reference surface $\mathscr{S}$ (i.e., the surface which becomes the surface $\mathfrak{s}$ at time t) with the initial surface and denote the initial values, at time $t = 0$, of the position vector, the unit normal to $\mathscr{S}$ and the initial director as $\boldsymbol{R}$, $\boldsymbol{A}_3$ and $\boldsymbol{D}$, respectively. Also, we denote the base vectors of $\mathfrak{s}$ and $\mathscr{S}$ by $\boldsymbol{a}_\alpha$ and $\boldsymbol{A}_\alpha$, respectively, and note that formulae of the type $(4.10)_1$ and (4.11)–(4.13) and their duals hold also for the surface of $\mathscr{C}$ in the deformed and reference configurations.

The continuum just described is called a *Cosserat surface*.[20] It may be emphasized that this continuum is not just a two-dimensional surface alone; it consists of a surface with a director assigned to every point of the surface. The

[17] Fairly general nonlinear theories of this type, developed in different contexts, have been given by ERICKSEN [1961, *1*] and GREEN, NAGHDI and RIVLIN [1965, *3*]. Also, TOUPIN [1964, *8*] has discussed, among other developments, a mechanical theory of couple-stress with directors for elastic materials. References to related works on the subject may be found in the above papers and in TRUESDELL and NOLL [1965, *9*].

[18] Although the symbols for the position vector and the unit normal of the surface $\mathfrak{s}$ are the same as those used for the surface $\bar{\mathfrak{s}}$, this need not give rise to confusion. If desired, the two surfaces may be identified; but we postpone such identifications until later.

[19] Our specification of $\boldsymbol{d}$ as a dimensionless vector field is for later convenience and in anticipation of later interpretations. For other directed media it may be more convenient to regard $\boldsymbol{d}$ as having the dimension of length.

[20] Such a surface is also referred to as a directed or an oriented surface in the recent literature. The idea of representing a (three-dimensional) shell by such a model was initially conceived by E. and F. COSSERAT [1909, *1*]; see also [1968, *3*]. In their work, however, the COSSERATS had considered a surface with a triad of rigid directors assigned to every point of the surface. A development of the kinematics of such directed or oriented surfaces with n deformable directors is contained in the paper of ERICKSEN and TRUESDELL [1958, *1*]. Here, as in the paper of GREEN, NAGHDI and WAINWRIGHT [1965, *4*], we use a single deformable director. A triad of deformable directors assigned to every point of the surface, based on the kinematics of ERICKSEN and TRUESDELL, has been employed by COHEN and DESILVA [1966, *2*] in their study of directed surfaces. With reference to shells and plates, however, at present it appears that the use of a single director should be sufficient to model a shell, as has been remarked also by TOUPIN [1964, *8*].

assigned director is intended to portray the "thickening" about the surface $\xi=0$ (or $\zeta=0$) of the three-dimensional shell and its component along the unit normal to the surface can be regarded as representing the thickness of the three-dimensional shell. More specifically, in the initial configuration of a Cosserat surface, we may specify the initial director $\boldsymbol{D}$ (which is dimensionless) to be directed along $\boldsymbol{A}_3$, i.e., $\boldsymbol{D}=D\boldsymbol{A}_3$; and we may then regard the magnitude of $\boldsymbol{D}$, namely D, as representing the initial thickness of the (three-dimensional) shell:

$$D \propto h \quad \text{or} \quad D=\frac{h}{h_R}, \tag{4.35}$$

where h_R designates a reference value of the thickness h and may be specified by the maximum value of h. For shells and plates of constant thickness, since $h_R=h$, $D=1$.

To complete the specification of the continuum $\mathscr{C}$, we need to specify its mass. Let $\mathscr{P}_{\mathscr{C}}$ refer to an arbitrary region of the material surface of $\mathscr{C}$ which is mapped into a part $\mathscr{P}_0$ of $\mathscr{S}$ in the reference configuration and into a corresponding part $\mathscr{P}$ of $\mathscr{s}$ in the deformed configuration at time t. Then, the mass $m(\mathscr{P}_{\mathscr{C}})$ for each part $\mathscr{P}_{\mathscr{C}}$ of the Cosserat surface can be defined by a non-negative scalar measure[21]

$$m(\mathscr{P}_{\mathscr{C}})=\int_{\mathscr{P}} \varrho\, d\sigma=\int_{\mathscr{P}_0} \varrho_0\, d\Sigma. \tag{4.36}$$

In (4.36), $\varrho=\varrho(\theta^\alpha, t)$ is the mass density at time t having the physical dimension

$$\text{phys. dim. } \varrho=[M\, L^{-2}], \tag{4.37}$$

with the symbol $[M]$ standing for the physical dimension of mass, and

$$\varrho_0=\varrho_0(\theta^\alpha)=\varrho(\theta^\alpha, 0) \tag{4.38}$$

is the initial mass density. Also, the element of area $d\sigma$ of $\mathscr{s}$ is given by a formula in the form $(4.19)_2$ and

$$d\Sigma=(\boldsymbol{A}_1\times\boldsymbol{A}_2\cdot\boldsymbol{A}_3)\, d\theta^1\, d\theta^2=A^{\frac{1}{2}}\, d\theta^1\, d\theta^2, \tag{4.39}$$

the dual of $(4.19)_2$ is an element of area of $\mathscr{S}$. It is easily verified from $(4.19)_2$ and (4.39) that the area elements $d\sigma$ and $d\Sigma$ are related by the formula

$$d\sigma=J\, d\Sigma, \tag{4.40}$$

where for later reference we have introduced the notation

$$J=\left(\frac{a}{A}\right)^{\frac{1}{2}}. \tag{4.41}$$

The formula (4.40) relates the elements of area in the deformed and undeformed configurations of the surface. By (4.40), an immediate consequence of (4.36) is the relation

$$\varrho=J^{-1}\varrho_0 \quad \text{or} \quad \varrho_0=J\varrho, \tag{4.42}$$

---

[21] For background information regarding the concept of mass in continuum mechanics, see TRUESDELL and TOUPIN [1960, *14*]. Note also that the physical dimension of ϱ given by (4.37) differs from the physical dimension of the mass density ϱ^* (of the three-dimensional body), but agrees with that defined by (4.17).

as one form of the equation of continuity.[22] Other forms of the continuity equation for the Cosserat surface will be given in Chap. C.

While the mass $m(\mathscr{P}_{\mathscr{C}})$ is a part of the specification of the continuum—here the Cosserat surface $\mathscr{C}$—the mass density depends on the particular configuration which $\mathscr{C}$ will occupy and its value in different configurations is determined by the motion of the continuum. In subsequent developments we need to introduce certain physical entities for a part $\mathscr{P}_{\mathscr{C}}$ of the Cosserat surface which are defined by means of integrals whose range of integration is over $\mathscr{P}$ in the present configuration. To be specific, let $f(\theta^\alpha, t)$ be any scalar-valued or vector-valued function denoting a physical quantity per unit mass ϱ and consider an integral of the type

$$\int_{\mathscr{P}} \varrho f \, d\sigma. \tag{4.43}$$

The above integral is associated with the motion of the continuum and represents a physical entity (say F) defined for a part of the Cosserat surface which occupies the region $\mathscr{P}$ in the present configuration. Strictly speaking an integral of the type (4.43) should be designated as $F(\mathscr{P}_{\mathscr{C}})$; however, in order to avoid cumbersome notations, we shall write $F(\mathscr{P})$ in place of $F(\mathscr{P}_{\mathscr{C}})$ when defining expressions of the type

$$F(\mathscr{P}) = \int_{\mathscr{P}} \varrho f \, d\sigma \tag{4.44}$$

for each part $\mathscr{P}$ of the Cosserat surface in the present configuration.[23] No confusion should arise from the above abbreviated notation, since it will be clear from the particular context [e.g., the presence of $d\sigma$ on the right-hand side of (4.44)] which region of Euclidean space $\mathscr{C}$ occupies. Whenever we need to emphasize in the same equation the distinction between the region of integration over a part of $\mathscr{C}$ in the present configuration and the corresponding part in the reference configuration, we employ the designation $\mathscr{P}_0$ for the latter.

As will become apparent later, a general and exact theory of a Cosserat surface can be constructed systematically and in the same spirit as currently enjoyed by the classical theory of continuum mechanics. This approach, which we call a *direct* approach, will not have the difficulties involving the approximations when shell theory is developed from the three-dimensional equations of the classical theory.[24] On the other hand, the direct approach (via a Cosserat surface) requires additional and sometimes difficult considerations, namely the interpretations of the results of the complete theory (at least in special cases), the identification of the constitutive coefficients and the demonstration of the relevance and the applicability of the theory of a Cosserat surface to shells and plates (regarded as three-dimensional bodies). Either of the two approaches, when carried to completion, has difficulties of its own. Nevertheless, much can be gained by studying both, with the ultimate goal of showing the relationship between them. With this point of view, in this and the next two chapters the subject is discussed both from a direct approach and also from the three-dimensional theory. In each

[22] This is a two-dimensional analogue of (4.16) which expresses conservation of mass in the three-dimensional theory.

[23] Our abbreviated notation here is similar (but not identical) to that in Sect. 15 of TRUESDELL and NOLL [1965, *9*].

[24] It will become evident by the end of Chap. D that considerable difficulties are associated with the derivation of the constitutive equations from the three-dimensional theory, even in the case of the linear theory of thin elastic shells; and it is partly for this reason that the direct approach is emphasized in Chap. E. However, these remarks and those made earlier at the end of Sect. 2 should not be construed as minimizing or discouraging the efforts directed toward the solution of the problem posed under (b) in (4.32).

chapter the two approaches are considered separately so that no confusion should arise when sometimes the same symbols are used for a Cosserat surface and in the theory developed from the three-dimensional equations.

5. Kinematics of shells: I. Direct approach. We develop in this section the basic kinematical results for a Cosserat surface already defined in Sect. 4. For ease of reference, however, we repeat here once more that a Cosserat surface is a body $\mathscr{C}$ comprising a surface (embedded in a Euclidean 3-space) and a single deformable director attached to every point of the surface. Moreover, the directors which are not necessarily along the unit normals to the surface have, in particular, the property that they remain unaltered in magnitude under superposed rigid body motions.

α) *General kinematical results.* Let the particles (or the material points) of the material surface of $\mathscr{C}$ be identified with the convected coordinates θ^α. In a reference configuration of the Cosserat surface $\mathscr{C}$ which we take to be the initial configuration, let the reference surface be referred to by $\mathscr{S}$, let $\boldsymbol{R}$ be the position vector of $\mathscr{S}$ and $\boldsymbol{D}$ the reference value of the director. Further, let the surface occupied by the material surface of $\mathscr{C}$ in the present configuration at time t be referred to by $\mathfrak{s}$, let $\boldsymbol{r}$ be the position vector of $\mathfrak{s}$ and $\boldsymbol{d}$ the director at $\boldsymbol{r}$. Then, the motion of a Cosserat surface is defined by

$$\boldsymbol{r}=\boldsymbol{r}(\theta^\alpha, t), \qquad \boldsymbol{d}=\boldsymbol{d}(\theta^\alpha, t), \qquad [\boldsymbol{a}_1\, \boldsymbol{a}_2\, \boldsymbol{d}]>0, \tag{5.1}$$

where the condition $(5.1)_3$ ensures that $\boldsymbol{d}$ is not tangent to the surface $\mathfrak{s}$. Alternatively, instead of $(5.1)_3$, it will suffice to assume that $[\boldsymbol{a}_1\, \boldsymbol{a}_2\, \boldsymbol{d}]\neq 0$ with the understanding that in any given motion $[\boldsymbol{a}_1\, \boldsymbol{a}_2\, \boldsymbol{d}]$ is either >0 or <0. In the reference configuration, the initial position vector $\boldsymbol{R}$ of the surface $\mathscr{S}$ and the initial director $\boldsymbol{D}$ at $\boldsymbol{R}$ are:

$$\boldsymbol{R}=\boldsymbol{R}(\theta^\alpha)=\boldsymbol{r}(\theta^\alpha, 0), \qquad \boldsymbol{D}=\boldsymbol{D}(\theta^\alpha)=\boldsymbol{d}(\theta^\alpha, 0). \tag{5.2}$$

At the risk of being repetitious, we observe that $(5.2)_1$ specifies the place occupied by the material point θ^α of the surface of $\mathscr{C}$ in the reference configuration while the place occupied by the material point of the surface of $\mathscr{C}$ in the deformed configuration at time t is given by $(5.1)_1$. We also note that the base vectors $\boldsymbol{a}_\alpha$, the unit normal $\boldsymbol{a}_3$ and the first and the second fundamental forms of $\mathfrak{s}$, as well as their duals associated with the reference surface $\mathscr{S}$, satisfy the relations of the forms (4.10)–(4.13).

The vector functions $\boldsymbol{r}$ and $\boldsymbol{d}$, jointly represent a 1-parameter family of configurations which describe the motion of the Cosserat surface. We assume that these vector functions are sufficiently smooth in the sense that they are differentiable with respect to θ^α and t as many times as required. Let $\boldsymbol{v}$ and $\boldsymbol{w}$, each a three-dimensional vector field, denote the velocity of a point of $\mathfrak{s}$ and the director velocity at time t. Then

$$\boldsymbol{v}=\frac{d}{dt}\boldsymbol{r}(\theta^\alpha, t)=\dot{\boldsymbol{r}}(\theta^\alpha, t), \qquad w=\frac{d}{dt}\boldsymbol{d}(\theta^\alpha, t)=\dot{\boldsymbol{d}}(\theta^\alpha, t), \tag{5.3}$$

where a superposed dot stands for the material time derivative with respect to t, holding θ^α fixed.

In what follows, we frequently encounter the partial derivatives of three-dimensional vector fields or functions with respect to the surface coordinates. In this connection, let $\boldsymbol{V}$ be a three-dimensional vector field defined on $\mathfrak{s}$ and let V^i be the components of $\boldsymbol{V}$ referred to the base vectors $\boldsymbol{a}_i=\{\boldsymbol{a}_\alpha, \boldsymbol{a}_3\}$. Then, $\boldsymbol{V}$ can be expressed as

$$\boldsymbol{V}=V^i\,\boldsymbol{a}_i=V^\alpha\,\boldsymbol{a}_\alpha+V^3\,\boldsymbol{a}_3=V_i\,\boldsymbol{a}^i=V_\alpha\,\boldsymbol{a}^\alpha+V_3\,\boldsymbol{a}^3, \tag{5.4}$$

and we recall that the gradient of $\boldsymbol{V}$ and its components are[25]

$$
\begin{aligned}
&\boldsymbol{V}_{,\alpha}=\boldsymbol{V}_{|\alpha}=V_{i\alpha}\,\boldsymbol{a}^i=V^i_{.\alpha}\,\boldsymbol{a}_i,\\
&V_{i\alpha}=\boldsymbol{a}_i\cdot\boldsymbol{V}_{,\alpha}, \qquad V^i_{.\alpha}=\boldsymbol{a}^i\cdot\boldsymbol{V}_{,\alpha},\\
&V_{\lambda\alpha}=V_{\lambda|\alpha}-b_{\alpha\lambda}V_3, \qquad V_{3\alpha}=V_{3,\alpha}+b^\lambda_\alpha V_\lambda,\\
&V^\lambda_{.\alpha}=V^\lambda{}_{|\alpha}-b^\lambda_\alpha V_3, \qquad V^3_{.\alpha}=V^3{}_{,\alpha}+b_{\lambda\alpha}V^\lambda,
\end{aligned}
\tag{5.5}
$$

where a vertical bar stands for covariant differentiation with respect to $a_{\alpha\beta}$ and $V^\alpha\,\boldsymbol{a}_\alpha$ and $V_\alpha\,\boldsymbol{a}^\alpha$ are surface vectors with contravariant and covariant components V^α and V_α, respectively. We also note here that since $\boldsymbol{a}_3$ is a unit normal to $\mathscr{s}$ and satisfies $(4.11)_1$, the lowering and raising of superscripts and subscripts of space tensor functions such as V^i in (5.4) and $V_{i\alpha}$ in (5.5) can be accomplished by using a space metric tensor g_{ij} defined by

$$g_{\alpha\beta}=a_{\alpha\beta}, \qquad g_{\alpha 3}=0, \qquad g_{33}=1. \tag{5.6}$$

Consider now the velocity vector $\boldsymbol{v}$ which can be written in the form

$$\boldsymbol{v}=v^i\,\boldsymbol{a}_i=v^\alpha\,\boldsymbol{a}_\alpha+v^3\,\boldsymbol{a}_3=v_i\,\boldsymbol{a}^i. \tag{5.7}$$

Since the coordinate curves on $\mathscr{s}$ are convected, it follows that

$$\dot{\boldsymbol{a}}_\alpha=\boldsymbol{v}_{,\alpha}=\boldsymbol{v}_{|\alpha}, \tag{5.8}$$

and by (5.5) we have

$$
\begin{aligned}
&\boldsymbol{v}_{|\alpha}=v_{i\alpha}\,\boldsymbol{a}^i, \qquad v_{i\alpha}=\boldsymbol{a}_i\cdot\boldsymbol{v}_{,\alpha},\\
&v_{\lambda\alpha}=\boldsymbol{a}_\lambda\cdot\boldsymbol{v}_{,\alpha}=v_{\lambda|\alpha}-b_{\alpha\lambda}v_3,\\
&v_{3\alpha}=\boldsymbol{a}_3\cdot\boldsymbol{v}_{,\alpha}=v_{3,\alpha}+b^\lambda_\alpha v_\lambda.
\end{aligned}
\tag{5.9}
$$

From $(4.11)_{1,2}$ and (5.9), it can be easily shown that

$$\dot{\boldsymbol{a}}_3=-(\boldsymbol{a}_3\cdot\boldsymbol{v}_{,\alpha})\,\boldsymbol{a}^\alpha=-v_{3\alpha}\,\boldsymbol{a}^\alpha=-(v_{3,\alpha}+b^\lambda_\alpha v_\lambda)\,\boldsymbol{a}^\alpha. \tag{5.10}$$

Since each of the vectors $\dot{\boldsymbol{a}}_i$ $(i=1,2,3)$, may be expressed as a linear combination of $\boldsymbol{a}^i$, we may write

$$\dot{\boldsymbol{a}}_i=c_{ki}\,\boldsymbol{a}^k, \qquad c_{ki}=\boldsymbol{a}_k\cdot\dot{\boldsymbol{a}}_i. \tag{5.11}$$

Let $T_{(ik)}$ and $T_{[ik]}$ stand, respectively, for the symmetric and the skew-symmetric parts of a second order tensor T_{ik}, i.e.,

$$T_{ik}=T_{(ik)}+T_{[ik]}, \qquad T_{(ik)}=\tfrac{1}{2}(T_{ik}+T_{ki}), \qquad T_{[ik]}=\tfrac{1}{2}(T_{ik}-T_{ki}). \tag{5.12}$$

Then, with the notation

$$\eta_{ki}=c_{(ki)}, \qquad W_{ki}=c_{[ki]}, \tag{5.13}$$

it follows from (5.11) that

$$
\begin{aligned}
&c_{ki}=\eta_{ki}+W_{ki},\\
&2\eta_{ki}=\boldsymbol{a}_k\cdot\dot{\boldsymbol{a}}_i+\boldsymbol{a}_i\cdot\dot{\boldsymbol{a}}_k=\overline{\boldsymbol{a}_k\cdot\boldsymbol{a}_i}^{\,\boldsymbol{\cdot}}=\dot{a}_{ki}=2\eta_{ik},\\
&2W_{ki}=\boldsymbol{a}_k\cdot\dot{\boldsymbol{a}}_i-\boldsymbol{a}_i\cdot\dot{\boldsymbol{a}}_k=-2W_{ik}.
\end{aligned}
\tag{5.14}
$$

[25] See Eqs. (A.2.54) in Chap. F.

From $(5.14)_{2,3}$, together with (5.5) and $(4.11)_{1,2}$, we obtain

$$\begin{aligned} 2\eta_{\alpha\beta} &= v_{\alpha|\beta} + v_{\beta|\alpha} - 2b_{\alpha\beta}\, v_3, \\ \eta_{3\alpha} &= \eta_{\alpha 3} = 0, \qquad \eta_{33} = 0, \end{aligned} \tag{5.15}$$

$$\begin{aligned} 2W_{\alpha\beta} &= -2W_{\beta\alpha} = v_{\alpha|\beta} - v_{\beta|\alpha}, \\ W_{\alpha 3} &= -W_{3\alpha} = -\boldsymbol{a}_3 \cdot \boldsymbol{v}_{,\alpha} = -(v_{3,\alpha} + b_\alpha^\beta\, v_\beta), \\ W_{33} &= 0. \end{aligned} \tag{5.16}$$

In view of (5.15) and (5.16), the components of the velocity gradient $\boldsymbol{v}_{,\alpha}$ are

$$v_{\lambda\alpha} = \eta_{\lambda\alpha} + W_{\lambda\alpha}, \qquad v_{3\alpha} = W_{3\alpha} \tag{5.17}$$

and we may express $\dot{\boldsymbol{a}}_i$ in the form

$$\dot{\boldsymbol{a}}_i = (\eta_{ki} + W_{ki})\, \boldsymbol{a}^k. \tag{5.18}$$

It is clear from (5.15)–(5.18) that $\eta_{\alpha\beta}$ and $W_{\alpha\beta}$ (a subtensor of W_{ki}) are *surface tensors* whereas W_{ki} is a *space tensor*. The functions $\eta_{\alpha\beta}$ and $W_{\alpha\beta}$ may be called the *surface rate of deformation* tensor and the *surface spin* tensor, respectively. For later reference, we introduce the notation [26]

$$\boldsymbol{\eta}_\alpha = \eta_{k\alpha}\, \boldsymbol{a}^k \tag{5.19}$$

and also record here the time rate of change of the determinant of $a_{\alpha\beta}$. Thus, from (4.12) and $(5.14)_2$,

$$\begin{aligned} \dot{a} &= \overline{\det(a_{\alpha\beta})}^{\,\boldsymbol{\cdot}} = \frac{\partial}{\partial a_{\lambda\nu}} [\det(a_{\alpha\beta})]\, \dot{a}_{\lambda\nu} \\ &= a\, a^{\alpha\beta}\, \dot{a}_{\alpha\beta} = 2a\, \eta_\alpha^\alpha \\ &= 2a\, (v^\alpha{}_{|\alpha} - b_\alpha^\alpha\, v_3). \end{aligned} \tag{5.20}$$

By (5.20), the time rate of change of J defined by (4.41) is

$$\dot{J} = \tfrac{1}{2} J a^{-1} \dot{a} = J \eta_\alpha^\alpha. \tag{5.21}$$

Before proceeding further, we need suitable expressions for the time rate of change of the reciprocal base vectors. From differentiation of $(4.12)_5$ follows

$$\dot{a}^{\alpha\beta} = -a^{\alpha\lambda}\, a^{\beta\nu}\, \dot{a}_{\lambda\nu}. \tag{5.22}$$

Recalling $(4.12)_3$ and using (5.14), (5.18) and (5.11), we obtain

$$\begin{aligned} \dot{\boldsymbol{a}}^\alpha &= \overline{a^{\alpha\beta}\, \boldsymbol{a}_\beta}^{\,\boldsymbol{\cdot}} \\ &= a^{\alpha\beta} (\eta_{k\beta} + W_{k\beta})\, \boldsymbol{a}^k - 2a^{\alpha\beta}\, a^{\lambda\nu}\, \eta_{\beta\nu}\, \boldsymbol{a}_\lambda \\ &= a^{\alpha\beta} (W_{k\beta} - \eta_{k\beta})\, \boldsymbol{a}^k \end{aligned} \tag{5.23}$$

and from $(4.11)_3$, (5.10) and (5.16), we have

$$\dot{\boldsymbol{a}}^3 = \dot{\boldsymbol{a}}_3 = W_{\alpha 3}\, \boldsymbol{a}^\alpha = -W_{3\alpha}\, \boldsymbol{a}^\alpha = -v_{3\alpha}\, \boldsymbol{a}^\alpha. \tag{5.24}$$

We now introduce additional kinematical results in terms of the director and its derivatives. Let $\boldsymbol{d}$ be referred to the reciprocal base vectors $\boldsymbol{a}^i$. Then,

$$\boldsymbol{d} = d_i\, \boldsymbol{a}^i = d_\alpha\, \boldsymbol{a}^\alpha + d_3\, \boldsymbol{a}^3, \qquad d_i = \boldsymbol{a}_i \cdot \boldsymbol{d} \tag{5.25}$$

---

[26] The vector function $\boldsymbol{\eta}_\alpha$ as defined in (5.19) is the negative of the corresponding quantity in the paper of Green, Naghdi and Wainwright [1965, *4*]. We note that $\boldsymbol{\eta}_\alpha$ is a vector tangent to the surface (as $\eta_{3\alpha} = 0$).

and by $(5.3)_2$, (5.23) and (5.24), the director velocity $\boldsymbol{w}$ can be put in the forms

$$\begin{aligned}\boldsymbol{w} &= w_k\, \boldsymbol{a}^k \\ &= \boldsymbol{\Gamma} + d_i\, \dot{\boldsymbol{a}}^i = \boldsymbol{\Gamma} + d^i\, W_{ki}\, \boldsymbol{a}^k - d^\alpha\, \boldsymbol{\eta}_\alpha \\ &= \boldsymbol{\Gamma} + d^\alpha (\boldsymbol{v}_{,\alpha} - 2\boldsymbol{\eta}_\alpha) + d^3\, W_{k3}\, \boldsymbol{a}^k \\ &= [\dot{d}_k + d^i (W_{ki} - \eta_{ki})]\, \boldsymbol{a}^k,\end{aligned} \tag{5.26}$$

where

$$\boldsymbol{\Gamma} = \dot{d}_i\, \boldsymbol{a}^i \tag{5.27}$$

and $\boldsymbol{\eta}_\alpha$ is defined by (5.19). The gradient of the director $\boldsymbol{d}$, with the help of (5.5), can be written as

$$\begin{aligned} &\boldsymbol{d}_{,\alpha} = \lambda_{i\alpha}\, \boldsymbol{a}^i = \lambda^i_{.\alpha}\, \boldsymbol{a}_i, && \lambda_{i\alpha} = \boldsymbol{a}_i \cdot \boldsymbol{d}_{,\alpha}, \\ &\lambda_{\beta\alpha} = d_{\beta|\alpha} - b_{\alpha\beta}\, d_3, && \lambda_{3\alpha} = d_{3,\alpha} + b^\beta_\alpha\, d_\beta, \\ &\lambda^\beta_{.\alpha} = a^{\beta\gamma}\, \lambda_{\gamma\alpha}, && \lambda^3_{.\alpha} = \lambda_{3\alpha}. \end{aligned} \tag{5.28}$$

Also, from (5.26), the gradient of the director velocity is

$$\begin{aligned}\boldsymbol{w}_{,\alpha} &= \boldsymbol{\Gamma}_{:\alpha} + \lambda^i_{.\alpha}\, W_{ki}\, \boldsymbol{a}^k - \lambda^\beta_{.\alpha}\, \boldsymbol{\eta}_\beta \\ &= [\dot{\lambda}_{k\alpha} + \lambda^\beta_{.\alpha} (W_{k\beta} - \eta_{k\beta}) + \lambda^3_{.\alpha}\, W_{k3}]\, \boldsymbol{a}^k,\end{aligned} \tag{5.29}$$

where

$$\dot{\lambda}_{i\alpha} = \boldsymbol{a}_i \cdot \boldsymbol{\Gamma}_{:\alpha}. \tag{5.30}$$

The kinematic quantities introduced above involve mainly $a_{\alpha\beta}$, $\lambda_{i\alpha}$, d_i and their rates.[27] Often, it is convenient to employ the alternative kinematic measures

$$e_{\alpha\beta} = \tfrac{1}{2} (a_{\alpha\beta} - A_{\alpha\beta}), \tag{5.31}$$

$$\varkappa_{\gamma\alpha} = \lambda_{\gamma\alpha} - \Lambda_{\gamma\alpha}, \qquad \varkappa_{3\alpha} = \lambda_{3\alpha} - \Lambda_{3\alpha}, \tag{5.32}$$

$$\gamma_\alpha = d_\alpha - D_\alpha, \qquad \gamma_3 = d_3 - D_3, \tag{5.33}$$

where

$$\Lambda_{i\alpha} = \boldsymbol{A}_i \cdot \boldsymbol{D}_{,\alpha}, \qquad D_i = \boldsymbol{A}_i \cdot \boldsymbol{D} \tag{5.34}$$

are the initial values of $\lambda_{i\alpha}$ and d_i. We note that

$$\dot{e}_{\alpha\beta} = \tfrac{1}{2}\, \dot{a}_{\alpha\beta} = \eta_{\alpha\beta}, \qquad \dot{\varkappa}_{i\alpha} = \dot{\lambda}_{i\alpha}, \qquad \dot{\gamma}_i = \dot{d}_i. \tag{5.35}$$

β) *Superposed rigid body motions.* For later considerations, we need to determine whether or not the above kinematical quantities remain invariant under superposed rigid body motions. For this purpose, we consider a motion of the Cosserat surface which differs from the previous motion, defined by (5.1), only by superposed rigid body motions of the whole continuum at different times. Suppose that under such superposed rigid body motions (since the surface $\mathfrak{s}$ now assumes a new orientation in space) the position $\boldsymbol{r}$ and the director $\boldsymbol{d}$ at $\boldsymbol{r}$ are displaced to the position $\boldsymbol{r}^+$ and the director $\boldsymbol{d}^+$ at $\boldsymbol{r}^+$. Then

$$\begin{aligned}\boldsymbol{r}^+ &= \boldsymbol{r}^+(\theta^\alpha, t') = \boldsymbol{r}_0^+(t') + Q(t)\, [\boldsymbol{r}(\theta^\alpha, t) - \boldsymbol{r}_0(t)], \\ \boldsymbol{d}^+ &= \boldsymbol{d}^+(\theta^\alpha, t') = Q(t)\, \boldsymbol{d}(\theta^\alpha, t),\end{aligned} \tag{5.36}$$

[27] These kinematical results were given by GREEN, NAGHDI and WAINWRIGHT [1965, *4*]. Apart from differences in notation, these results can be brought into correspondence with the kinematical results of COHEN and DESILVA [1966, *2*] when their theory is properly specialized to a single director. As already mentioned in Sect. 4, COHEN and DESILVA employ a triad of deformable directors and their analysis is based on ERICKSEN and TRUESDELL's general kinematics of oriented media in n-dimensional space [1958, *1*].

where $\boldsymbol{r}_0^+$ and $\boldsymbol{r}_0$ are vector-valued functions of t' and t, respectively, $t'=t+a'$, a' being an arbitrary constant, and Q is a proper orthogonal tensor-valued function of t. The tensor Q, a second order space tensor, satisfies the conditions

$$Q Q^T = Q^T Q = I, \quad \det(Q) = 1, \tag{5.37}$$

where I stands for the unit tensor and Q^T denotes the transpose of Q. In what follows, we designate the quantities associated with the motion (5.36) by the same symbols to which we also attach a plus sign (+). Thus, let the base vectors of the surface associated with $(5.36)_1$ be denoted by $\boldsymbol{a}_i^+$. Then, from (4.10), (4.11) and (5.36), we have

$$\boldsymbol{a}_\alpha^+ = Q\,\boldsymbol{a}_\alpha, \quad \boldsymbol{a}_3^+ = Q\,\boldsymbol{a}_3, \quad \boldsymbol{d}_{,\alpha}^+ = Q\,\boldsymbol{d}_{,\alpha}. \tag{5.38}$$

Let Ω be a second order space tensor-valued function of time defined by

$$\Omega = \Omega(t) = \dot{Q}(t)\,Q(t)^T. \tag{5.39}$$

Then, by $(5.37)_1$,

$$\dot{Q} = \Omega Q, \quad \Omega = -\Omega^T, \tag{5.40}$$

so that Ω is a skew-symmetric tensor. Hence, there exists a vector-valued function $\boldsymbol{\omega}$ such that for any vector $\boldsymbol{V}$

$$\Omega \boldsymbol{V} = \boldsymbol{\omega} \times \boldsymbol{V}. \tag{5.41}$$

In particular,

$$\begin{gathered} \boldsymbol{\omega} \times \boldsymbol{a}_\alpha = -\varepsilon_{k\alpha m}\,\omega^m\,\boldsymbol{a}^k = -\Omega_{k\alpha}\,\boldsymbol{a}^k, \\ \omega^m = \boldsymbol{\omega} \cdot \boldsymbol{a}^m, \quad \Omega_{ki} = \varepsilon_{kim}\,\omega^m = a^{\frac{1}{2}}\,e_{kim}\,\omega^m = -\Omega_{ik}, \end{gathered} \tag{5.42}$$

where the ε-system is related to the permutation symbols e_{kim}, e^{kim} through

$$\varepsilon_{kim} = a^{\frac{1}{2}}\,e_{kim}, \quad \varepsilon^{kim} = a^{-\frac{1}{2}}\,e^{kim}. \tag{5.43}$$

Using (5.40), it follows from $(5.36)_1$ that

$$\Omega(\boldsymbol{r}^+ - \boldsymbol{r}_0^+) = \dot{Q}(\boldsymbol{r} - \boldsymbol{r}_0). \tag{5.44}$$

The velocity vector $\boldsymbol{v}^+$, obtained from $(5.36)_1$, can then be written in the forms

$$\begin{aligned} \boldsymbol{v}^+ = \dot{\boldsymbol{r}}^+ &= \dot{\boldsymbol{r}}_0^+ + Q(\boldsymbol{v} - \dot{\boldsymbol{r}}_0) + \Omega(\boldsymbol{r}^+ - \boldsymbol{r}_0^+) \\ &= [\dot{\boldsymbol{r}}_0^+ - Q\,\dot{\boldsymbol{r}}_0 - \dot{Q}\boldsymbol{r}_0] + Q\,\boldsymbol{v} + \dot{Q}\boldsymbol{r}, \end{aligned} \tag{5.45}$$

where the quantity in the square bracket on the right-hand side of $(5.45)_2$ is a function of time only and we note that the material time derivative operator $\dot{\overline{(\ )}}$ is unaltered under superposed rigid body motions $(5.36)_1$.

Before proceeding further, it is instructive to consider a special case of (5.36) for which the function Ω (and therefore $\boldsymbol{\omega}$) is constant for all time. To this end we consider a motion of the type $(5.36)_1$ such that for a given time the function Q is specified by a special value. For later convenience we take t to be the given time and specify Q by

$$Q(\tau) = \exp[\Omega_0(\tau - t)], \quad \Omega_0 = -\Omega_0^T = \text{const}, \tag{5.46}$$

τ being real. Then,

$$\dot{Q}(\tau) = \Omega_0\,Q(\tau) \tag{5.47}$$

and

$$Q(t) = I, \quad \dot{Q}(t) = \Omega_0. \tag{5.48}$$

But, since (5.40) holds for all time, from comparison of $(5.40)_1$ and (5.47) we have $\Omega(\tau)=\Omega_0$. Hence, for the special motion with Q specified by (5.46), Ω is constant for all time and (5.45) can be reduced to

$$\begin{aligned} \boldsymbol{v}^+(\tau) &= \boldsymbol{b}(\tau) + Q(\tau)\,\boldsymbol{v}(\tau) + \Omega_0\, Q(\tau)\,\boldsymbol{r}(\tau), \\ \boldsymbol{b}(\tau) &= \dot{\boldsymbol{r}}_0^+(\tau) - Q(\tau)\,\dot{\boldsymbol{r}}_0(\tau) - \Omega_0\, Q(\tau)\,\boldsymbol{r}_0(\tau), \end{aligned} \tag{5.49}$$

or for time t to

$$\boldsymbol{v}^+(t) = \boldsymbol{v}(t) + [\boldsymbol{b}(t) + \boldsymbol{\omega}_0 \times \boldsymbol{r}(t)]. \tag{5.50}$$

The square bracket on the right-hand side of (5.50) is due to superposed rigid body motion, $\boldsymbol{\omega}_0$ is a uniform rigid body angular velocity and $b(t)$ may be interpreted as a uniform rigid body translatory velocity at time t.

Returning to (5.38), we have

$$\dot{\boldsymbol{a}}_i^+ = Q\,\dot{\boldsymbol{a}}_i + \dot{Q}\,\boldsymbol{a}_i = Q\,[\dot{\boldsymbol{a}}_i + Q^T \Omega\, Q\,\boldsymbol{a}_i] \tag{5.51}$$

which also provides the expression for the velocity gradient $\boldsymbol{v}^+_{,\alpha} = \dot{\boldsymbol{a}}^+_\alpha$. In addition, the director velocity and its gradient associated with the motion (5.36) are

$$\begin{aligned} \boldsymbol{w}^+ &= Q\,\boldsymbol{w} + \dot{Q}\,\boldsymbol{d} = Q[\boldsymbol{w} + Q^T \Omega\, Q\,\boldsymbol{d}], \\ \boldsymbol{w}^+_{,\alpha} &= Q\,\boldsymbol{w}_{,\alpha} + \dot{Q}\,\boldsymbol{d}_{,\alpha} = Q[\boldsymbol{w}_{,\alpha} + Q^T \Omega\, Q\,\boldsymbol{d}_{,\alpha}]. \end{aligned} \tag{5.52}$$

Recalling the expressions for the first and the second fundamental forms of the surface, as well as $(5.28)_2$, from (5.38) and (5.36) and using the relation

$$\boldsymbol{U} \cdot Q\boldsymbol{V} = Q^T \boldsymbol{U} \cdot \boldsymbol{V} \tag{5.53}$$

with $\boldsymbol{U}$ and $\boldsymbol{V}$ being any two vectors, we have

$$\begin{aligned} a^+_{\alpha\beta} &= a_{\alpha\beta}, \qquad & b^+_{\alpha\beta} &= b_{\alpha\beta}, \\ d^+_i &= d_i, & \lambda^+_{i\alpha} &= \lambda_{i\alpha}, \end{aligned} \tag{5.54}$$

for all proper orthogonal Q. Similarly, with the use of (5.51), it can be readily verified that

$$\begin{aligned} &\eta^+_{ki} = \eta_{ki}, \qquad \dot{a}^+ = \dot{a}, \qquad \boldsymbol{\eta}^+_\alpha = Q\,\boldsymbol{\eta}_\alpha, \\ &W^+_{ki} = W_{ki} + \tfrac{1}{2}[Q^T \dot{Q}\,\boldsymbol{a}_i \cdot \boldsymbol{a}_k - Q^T \dot{Q}\,\boldsymbol{a}_k \cdot \boldsymbol{a}_i] \end{aligned} \tag{5.55}$$

and

$$\begin{aligned} \dot{d}^+_i &= \dot{d}_i, \qquad & \dot{\lambda}^+_{i\alpha} &= \dot{\lambda}_{i\alpha}, \\ \boldsymbol{\Gamma}^+ &= Q\boldsymbol{\Gamma}, & \boldsymbol{\Gamma}^+_{:\alpha} &= Q\boldsymbol{\Gamma}_{:\alpha}. \end{aligned} \tag{5.56}$$

The foregoing results, except for (5.46)–(5.50), are valid for every proper orthogonal Q and for all t. In the special case of (5.36) in which Q has the value specified by (5.46), some of the formulae simplify and assume a more revealing form. In particular, with $b(\tau)=0$ in $(5.49)_2$ and with Q and $\dot{Q}$ given by (5.48) at time t, the superposed velocity and the superposed velocity gradient at time t become

$$\begin{aligned} \boldsymbol{v}^+ &= \boldsymbol{v} + \boldsymbol{\omega}_0 \times \boldsymbol{r}, \\ \boldsymbol{v}^+_{,\alpha} &= \dot{\boldsymbol{a}}^+_\alpha = \boldsymbol{v}_{,\alpha} + \boldsymbol{\omega}_0 \times \boldsymbol{a}_\alpha = \boldsymbol{v}_{,\alpha} - \Omega^0_{k\alpha}\,\boldsymbol{a}_k \end{aligned} \tag{5.57}$$

and (5.51)–(5.52) reduce to

$$\dot{\boldsymbol{a}}^+_i = \dot{\boldsymbol{a}}_i + \boldsymbol{\omega}_0 \times \boldsymbol{a}_i \tag{5.58}$$

and

$$\begin{aligned} \boldsymbol{w}^+ &= \boldsymbol{w} + \boldsymbol{\omega}_0 \times \boldsymbol{d} = \boldsymbol{w} + d^i\, \Omega^0_{ik}\,\boldsymbol{a}^k, \\ \boldsymbol{w}^+_{,\alpha} &= \boldsymbol{w}_{,\alpha} + \boldsymbol{\omega}_0 \times \boldsymbol{d}_{,\alpha} = \boldsymbol{w}_{,\alpha} + \lambda^i_{.\alpha}\, \Omega^0_{ik}\,\boldsymbol{a}^k, \end{aligned} \tag{5.59}$$

where Ω^0_{ki} are related to the components of $\boldsymbol{\omega}_0$ by (5.42) and we have omitted the argument t [corresponding to the given time in (5.46)] from the various functions in (5.57)–(5.59). Similarly, in view of $(5.46)_1$, most of the remaining expressions have an obvious simplification in this case; in particular $(5.55)_4$ becomes

$$W^+_{ki} = W_{ki} - \Omega^0_{ki}. \tag{5.60}$$

We emphasize that the special results (5.57)–(5.60) are obtained corresponding to (5.46) for a given time t and with Q and $\dot{Q}$ specified by (5.48).

γ) *Additional kinematics.* In the remainder of this section we consider some additional (but unrelated) kinematics which will be used subsequently.

The preceding developments in this section represent kinematic results by direct approach appropriate to the theory of a Cosserat surface. However, other developments by direct approach, in the absence of the director field and less general than the earlier results [in Subsect. α)], are possible; and we discuss now one such possibility suitable for a theory which we call a *restricted theory*. Briefly, consider a material surface and identify the material points of the surface with convected coordinates θ^α. Adopting the previous notation and terminology, we continue to refer to the (initial) reference surface (with position vector $\boldsymbol{R}$) by $\mathscr{S}$ and to the surface in the present configuration at time t (with position vector $\boldsymbol{r}$) by $\mathfrak{s}$. Since we do not admit a director, the motion of the surface is simply characterized by $(5.1)_1$ and instead of $(5.1)_3$ we have $(4.11)_4$. The velocity vector $\boldsymbol{v}$ of $\mathfrak{s}$ at time t is defined by $(5.3)_1$. In anticipation of results to be derived for the restricted theory (in Sect. 10) and in order to easily contrast these with those of the more general theory of the Cosserat surface (in Sects. 8–9), we introduce the notations

$$\dot{\boldsymbol{w}} = \dot{\boldsymbol{a}}_3, \qquad \dot{\boldsymbol{w}}_{,\alpha} = \dot{\boldsymbol{a}}_{3,\alpha}, \qquad \lambda_{\alpha\beta} = \lambda_{\beta\alpha} = -b_{\alpha\beta}, \qquad \dot{\lambda}_{\alpha\beta} = -\dot{b}_{\alpha\beta}, \tag{5.61}$$

where $\dot{\boldsymbol{w}}$ is the angular velocity of the surface $\mathfrak{s}$. Then, by $(4.13)_3$, (5.18) and (5.23), we may write

$$\begin{aligned} \dot{\boldsymbol{w}}_{,\alpha} &= -\overline{(b^\gamma_\alpha \boldsymbol{a}_\gamma)}^{\boldsymbol{\cdot}} = \dot{\lambda}^\gamma_\alpha \boldsymbol{a}_\gamma + \lambda^\gamma_\alpha (\eta_{k\gamma} + W_{k\gamma})\, \boldsymbol{a}^k \\ &= \dot{\lambda}_{\gamma\alpha} \boldsymbol{a}^\gamma + \lambda^\gamma_\alpha (W_{k\gamma} - \eta_{k\gamma})\, \boldsymbol{a}^k. \end{aligned} \tag{5.62}$$

This completes our brief discussion of the kinematics of the restricted theory. We note, however, that earlier formulae of this section which do not involve the director or its gradient (including those under superposed rigid body motions) remain valid in the restricted theory. For later reference, we also recall here the formulae which relate the two-dimensional ε-system $\varepsilon_{\alpha\beta}$, $\varepsilon^{\alpha\beta}$ to the two-dimensional permutation symbols:

$$\begin{aligned} &\varepsilon_{\alpha\beta} = \varepsilon_{\alpha\beta3} = a^{\frac{1}{2}}\, e_{\alpha\beta}, &&\varepsilon^{\alpha\beta} = \varepsilon^{\alpha\beta3} = a^{-\frac{1}{2}}\, e^{\alpha\beta}, \\ &e_{11} = e_{22} = e^{11} = e^{22} = 0, &&e_{12} = -e_{21} = e^{12} = -e^{21} = 1. \end{aligned} \tag{5.63}$$

Next, we consider the kinematics of a surface integral and deduce an integral formula which will be utilized in the next chapter. Let $\varphi(\theta^\alpha, t)$ stand for a (sufficiently smooth) scalar-valued or vector-valued function of position and time and define the integral

$$\int_{\mathscr{P}} \varphi \, d\sigma \tag{5.64}$$

over $\mathscr{P}$ in the present configuration. Since the above integral is a function of time, its derivative with respect to t can be calculated as follows:

$$\frac{d}{dt}\int_{\mathscr{P}} \varphi\, d\sigma = \frac{d}{dt}\int_{\mathscr{P}_0} J\varphi\, d\Sigma = \int_{\mathscr{P}_0} \overline{J\varphi}^{\boldsymbol{\cdot}}\, d\Sigma = \int_{\mathscr{P}} (\dot{\varphi} + J^{-1}\dot{J}\varphi)\, d\sigma, \tag{5.65}$$

where J is defined by (4.41) and the region of integration of the last integral is again over $\mathscr{P}$. But $J^{-1}\dot{J}=\eta^\alpha_\alpha$ by (5.21). Hence, from $(5.65)_3$ we have

$$\frac{d}{dt}\int\limits_{\mathscr{P}}\varphi\,d\sigma=\int\limits_{\mathscr{P}}(\dot{\varphi}+\eta^\alpha_\alpha\,\varphi)\,d\sigma \tag{5.66}$$

which is the desired result. The last formula is essentially the two-dimensional analogue of the transport theorem in the three-dimensional theory.

6. Kinematics of shells continued (linear theory): I. Direct approach. This section is devoted to linearized kinematics for shells and plates by direct approach. In particular, we deduce the linearized kinematic measures for a Cosserat surface with infinitesimal displacements and infinitesimal director displacements as a special case of the general results in Sect. 5.

δ) *Linearized kinematics.* Let

$$\boldsymbol{r}=\boldsymbol{R}+\varepsilon\,\boldsymbol{u},\qquad \boldsymbol{u}=u^i\boldsymbol{A}_i,\qquad \boldsymbol{v}=\varepsilon\,\dot{\boldsymbol{u}}, \tag{6.1}$$

$$\boldsymbol{d}=\boldsymbol{D}+\varepsilon\,\boldsymbol{\delta},\qquad \boldsymbol{\delta}=\delta_i\boldsymbol{A}^i,\qquad \boldsymbol{w}=\varepsilon\,\dot{\boldsymbol{\delta}}, \tag{6.2}$$

where ε is a non-dimensional parameter. We say the motion of a Cosserat surface characterized by $(6.1)_1$ and $(6.2)_1$ describes infinitesimal deformation if the magnitudes of $\boldsymbol{u}$, $\boldsymbol{\delta}$ and all their derivatives are bounded by 1 and if

$$\varepsilon\ll 1. \tag{6.3}$$

We shall be concerned in the following developments with (scalar, vector or tensor) functions of position and time, determined by $\varepsilon\,\boldsymbol{u}$, $\varepsilon\,\boldsymbol{\delta}$ and their surface and time derivatives. We denote these functions by the customary order symbol $O(\varepsilon^n)$ if there exists a real number C, independent of ε, $\boldsymbol{u}$, $\boldsymbol{\delta}$ and their derivatives, such that

$$|O(\varepsilon^n)|<C\,\varepsilon^n, \tag{6.4}$$

as $\varepsilon\to 0$.

We emphasize that the infinitesimal theory which we wish to obtain as a special case of the results in Sect. 5 and in the sense of (6.3) is such that all kinematical quantities (including the displacement $\boldsymbol{u}$, the director displacement $\boldsymbol{\delta}$ and such measures as $e_{\alpha\beta}$, $\varkappa_{i\alpha}$ and γ_i, as well as their derivatives with respect to the surface coordinates and t) are of $O(\varepsilon)$. Moreover, throughout this section, we again use a vertical bar to denote covariant differentiation; this, however, should not be confusing. The designation of a vertical bar in the present section (or whenever we are concerned with linearized measures) is for covariant differentiation with respect to $A_{\alpha\beta}$ of the undeformed surface, in contrast to the meaning of a vertical bar in Sect. 5 and also in parts of later sections.[28]

From $(6.1)_1$ and $(6.2)_1$, we have

$$\boldsymbol{a}_\alpha=\boldsymbol{A}_\alpha+\varepsilon\,\boldsymbol{u}_{,\alpha},\qquad \boldsymbol{d}_{,\alpha}=\boldsymbol{D}_{,\alpha}+\varepsilon\,\boldsymbol{\delta}_{,\alpha} \tag{6.5}$$

and by $(6.5)_1$ and $(4.11)_{1,2}$ we can show that

$$\boldsymbol{a}_3=\boldsymbol{A}_3+\varepsilon\,\boldsymbol{\beta}+O(\varepsilon^2),\qquad \boldsymbol{\beta}=\beta^\alpha\boldsymbol{A}_\alpha=\beta_\alpha\boldsymbol{A}^\alpha,\qquad \beta_\alpha=\boldsymbol{\beta}\cdot\boldsymbol{A}_\alpha=-\boldsymbol{u}_{,\alpha}\cdot\boldsymbol{A}_3. \tag{6.6}$$

[28] It will be clear from the particular context, in later sections, whenever a vertical bar denotes covariant differentiation with respect to $a_{\alpha\beta}$ or $A_{\alpha\beta}$.

With the use of (6.5) and $(6.6)_1$ and recalling $(4.12)_1$, (4.13), $(5.25)_2$ and $(5.28)_2$, the expressions for $a_{\alpha\beta}$, $b_{\alpha\beta}$, d_i and $\lambda_{i\alpha}$ can be written as

$$\begin{aligned} a_{\alpha\beta} &= \boldsymbol{a}_\alpha \cdot \boldsymbol{a}_\beta = A_{\alpha\beta} + \varepsilon(\boldsymbol{u}_{,\alpha} \cdot \boldsymbol{A}_\beta + \boldsymbol{u}_{,\beta} \cdot \boldsymbol{A}_\alpha) + O(\varepsilon^2),\\ b_{\alpha\beta} &= -\boldsymbol{a}_\alpha \cdot \boldsymbol{a}_{3,\beta} = B_{\alpha\beta} - \varepsilon(\boldsymbol{u}_{,\alpha} \cdot \boldsymbol{A}_{3,\beta} + \boldsymbol{A}_\alpha \cdot \boldsymbol{\beta}_{,\beta}) + O(\varepsilon^2), \end{aligned} \tag{6.7}$$

$$\begin{aligned} d_\alpha &= \boldsymbol{a}_\alpha \cdot \boldsymbol{d} = D_\alpha + \varepsilon(\boldsymbol{A}_\alpha \cdot \boldsymbol{\delta} + \boldsymbol{u}_{,\alpha} \cdot \boldsymbol{D}) + O(\varepsilon^2),\\ d_3 &= \boldsymbol{a}_3 \cdot \boldsymbol{d} = D_3 + \varepsilon(\boldsymbol{A}_3 \cdot \boldsymbol{\delta} + \boldsymbol{\beta} \cdot \boldsymbol{D}) + O(\varepsilon^2), \end{aligned} \tag{6.8}$$

$$\begin{aligned} \lambda_{\beta\alpha} &= \boldsymbol{a}_\beta \cdot \boldsymbol{d}_{,\alpha} = \Lambda_{\beta\alpha} + \varepsilon(\boldsymbol{A}_\beta \cdot \boldsymbol{\delta}_{,\alpha} + \boldsymbol{u}_{,\beta} \cdot \boldsymbol{D}_{,\alpha}) + O(\varepsilon^2),\\ \lambda_{3\alpha} &= \boldsymbol{a}_3 \cdot \boldsymbol{d}_{,\alpha} = \Lambda_{3\alpha} + \varepsilon(\boldsymbol{A}_3 \cdot \boldsymbol{\delta}_{,\alpha} + \boldsymbol{\beta} \cdot \boldsymbol{D}_{,\alpha}) + O(\varepsilon^2), \end{aligned} \tag{6.9}$$

where $A_{\alpha\beta}$, $B_{\alpha\beta}$, D_i and $\Lambda_{i\alpha}$ are the initial reference values of the functions given by (6.7)–(6.9). These results can now be used to obtain the appropriate expressions for $e_{\alpha\beta}$, $\varkappa_{i\alpha}$ and γ_i in (5.29) to (5.31) in terms of $\varepsilon\,\boldsymbol{u}$, $\varepsilon\,\boldsymbol{\delta}$ and their derivatives.

It is desirable at this stage to elaborate on the manner in which the process of linearization may be accomplished. For this purpose, let $\boldsymbol{u}'$ and $\boldsymbol{\delta}'$ be vector functions defined by

$$\begin{aligned} \boldsymbol{u}' &= \varepsilon\,\boldsymbol{u} = O(\varepsilon), & u'^i &= \boldsymbol{A}^i \cdot \boldsymbol{u}' = O(\varepsilon),\\ \boldsymbol{\delta}' &= \varepsilon\,\boldsymbol{\delta} = O(\varepsilon), & \delta'_i &= \boldsymbol{A}_i \cdot \boldsymbol{\delta}' = O(\varepsilon), \end{aligned} \tag{6.10}$$

which can be used to express all kinematical quantities in terms of $\boldsymbol{u}'$, $\boldsymbol{\delta}'$ and their derivatives. For example, if we define $e'_{\alpha\beta}$ and $\varkappa'_{\beta\alpha}$ by

$$e'_{\alpha\beta} = \tfrac{1}{2}(\boldsymbol{u}'_{,\alpha} \cdot \boldsymbol{A}_\beta + \boldsymbol{u}'_{,\beta} \cdot \boldsymbol{A}_\alpha), \qquad \varkappa'_{\beta\alpha} = \boldsymbol{A}_\beta \cdot \boldsymbol{\delta}'_{,\alpha} + \boldsymbol{u}'_{,\beta} \cdot \boldsymbol{D}_{,\alpha}, \tag{6.11}$$

each of which is of $O(\varepsilon)$, we can then write (5.31)–(5.32) and the ratio a/A in (4.41) as

$$\begin{aligned} e_{\alpha\beta} &= e'_{\alpha\beta} + O(\varepsilon^2) = O(\varepsilon), & \varkappa_{\beta\alpha} &= \varkappa'_{\beta\alpha} + O(\varepsilon^2) = O(\varepsilon),\\ \frac{a}{A} &= 1 + 2A^{\alpha\beta}\,e'_{\alpha\beta} + O(\varepsilon^2), & \left(\frac{a}{A}\right)^{\frac{1}{2}} &= 1 + e'^{\alpha}_{\alpha} + O(\varepsilon^2) \end{aligned} \tag{6.12}$$

and other kinematical quantities can be expressed similarly. A straightforward procedure is now to retain only terms of $O(\varepsilon)$ in such expressions as (6.12), hence approximate $e_{\alpha\beta}$ and $\varkappa_{\beta\alpha}$ by $e'_{\alpha\beta}$ and $\varkappa'_{\beta\alpha}$, etc., and complete the linearization in this manner. However, in order to avoid the introduction of unnecessary additional notations, we may proceed with the linearization from (6.7)–(6.9) by retaining only terms of $O(\varepsilon)$ and after the approximations, without loss in generality, we set $\varepsilon = 1$. In what follows, we adopt this latter procedure. Thus, the kinematic measures (5.31)–(5.33), after linearization, reduce to

$$\begin{aligned} e_{\alpha\gamma} &= \tfrac{1}{2}(\boldsymbol{u}_{,\alpha} \cdot \boldsymbol{A}_\gamma + \boldsymbol{u}_{,\gamma} \cdot \boldsymbol{A}_\alpha),\\ \varkappa_{\gamma\alpha} &= \boldsymbol{A}_\gamma \cdot \boldsymbol{\delta}_{,\alpha} + \boldsymbol{u}_{,\gamma} \cdot \boldsymbol{D}_{,\alpha}, & \varkappa_{3\alpha} &= \boldsymbol{A}_3 \cdot \boldsymbol{\delta}_{,\alpha} + \boldsymbol{\beta} \cdot \boldsymbol{D}_{,\alpha},\\ \gamma_\alpha &= \boldsymbol{A}_\alpha \cdot \boldsymbol{\delta} + \boldsymbol{u}_{,\alpha} \cdot \boldsymbol{D}, & \gamma_3 &= \boldsymbol{A}_3 \cdot \boldsymbol{\delta} + \boldsymbol{\beta} \cdot \boldsymbol{D} \end{aligned} \tag{6.13}$$

and (4.42) becomes

$$\varrho = \varrho_0(1 - e^\alpha_\alpha), \tag{6.14}$$

in view of (6.7)–(6.9) and $(6.12)_4$. In (6.13), $\boldsymbol{\beta}$ is defined by $(6.6)_3$ and the partial derivatives of various vector functions can be calculated by using a formula of the type (5.5). In particular, we record here the following formulae

$$\begin{aligned} &\boldsymbol{u}_{,\alpha}=u_{i\alpha}\boldsymbol{A}^i=u^i_{.\alpha}\boldsymbol{A}_i, \quad u^\lambda_{.\alpha}=A^{\lambda\nu}u_{\nu\alpha}, \quad u^3_{.\alpha}=u_{3\alpha}, \\ &u_{\gamma\alpha}=\boldsymbol{A}_\gamma\cdot\boldsymbol{u}_{,\alpha}=u_{\gamma|\alpha}-B_{\alpha\gamma}u_3, \quad u_{3\alpha}=\boldsymbol{A}_3\cdot\boldsymbol{u}_{,\alpha}=u_{3,\alpha}+B^\gamma_\alpha u_\gamma \end{aligned} \tag{6.15}$$

and

$$\begin{aligned} &\boldsymbol{A}_\gamma\cdot\boldsymbol{\delta}_{,\alpha}=\delta_{\gamma|\alpha}-B_{\alpha\gamma}\delta_3, \quad \boldsymbol{A}_3\cdot\boldsymbol{\delta}_{,\alpha}=\delta_{3,\alpha}+B^\gamma_\alpha\delta_\gamma, \\ &\boldsymbol{u}_{,\gamma}\cdot\boldsymbol{D}_{,\alpha}=\Lambda_{\nu\alpha}(u^\nu{}_{|\gamma}-B^\nu_\gamma u_3)+\Lambda_{3\alpha}(u_{3,\gamma}+B^\nu_\gamma u_\nu), \\ &\boldsymbol{\beta}\cdot\boldsymbol{D}_{,\alpha}=\Lambda_{\nu\alpha}\beta^\nu, \quad \boldsymbol{u}_{,\alpha}\cdot\boldsymbol{D}=D^i u_{i\alpha}=D_i u^i_{.\alpha}, \quad \boldsymbol{\beta}\cdot\boldsymbol{D}=\beta^\alpha D_\alpha, \end{aligned} \tag{6.16}$$

where

$$\Lambda_{\nu\alpha}=\boldsymbol{A}_\nu\cdot\boldsymbol{D}_{,\alpha}=D_{\nu|\alpha}-B_{\alpha\nu}D_3, \quad \Lambda_{3\alpha}=\boldsymbol{A}_3\cdot\boldsymbol{D}_{,\alpha}=D_{3,\alpha}+B^\nu_\alpha D_\nu, \tag{6.17}$$

by $(5.34)_1$. Also, in (6.15)–(6.17) and throughout the present section, a vertical bar denotes covariant differentiation with respect to $A_{\alpha\beta}$ of the initial undeformed surface. Introducing the above results in (6.13), we finally obtain the following expressions for the kinematic measures and their time rates: [29]

$$\begin{aligned} &e_{\alpha\gamma}=\tfrac{1}{2}(u_{\alpha|\gamma}+u_{\gamma|\alpha})-B_{\alpha\gamma}u_3, \\ &\varkappa_{\gamma\alpha}=\delta_{\gamma|\alpha}-B_{\alpha\gamma}\delta_3+\Lambda_{\nu\alpha}u^\nu_{.\gamma}+\Lambda_{3\alpha}u^3_{.\gamma}, \\ &\varkappa_{3\alpha}=\delta_{3,\alpha}+B^\gamma_\alpha\delta_\gamma+\Lambda_{\nu\alpha}\beta^\nu, \quad \beta_\alpha=-u^3_{.\alpha}=-(u_{3,\alpha}+B^\gamma_\alpha u_\gamma), \\ &\gamma_\alpha=\delta_\alpha+D_\lambda u^\lambda_{.\alpha}-D_3\beta_\alpha=\delta_\alpha+D_i u^i_{.\alpha}, \quad \gamma_3=\delta_3+D_\alpha\beta^\alpha \end{aligned} \tag{6.18}$$

and

$$\begin{aligned} &\dot{e}_{\alpha\beta}=\tfrac{1}{2}(v_{\alpha|\gamma}+v_{\gamma|\alpha})-B_{\alpha\gamma}v_3, \\ &\dot{\varkappa}_{\gamma\alpha}=w_{\gamma|\alpha}-B_{\alpha\gamma}w_3+\Lambda_{\nu\alpha}v^\nu_{.\gamma}+\Lambda_{3\alpha}v^3_{.\gamma}, \\ &\dot{\varkappa}_{3\alpha}=w_{3,\alpha}+B^\gamma_\alpha w_\gamma+\Lambda_{\nu\alpha}\dot{\beta}^\nu, \\ &\dot{\gamma}_\alpha=w_\alpha+D_\lambda v^\lambda_{.\alpha}-D_3\dot{\beta}_\alpha, \quad \dot{\gamma}_3=w_3+D_\alpha\dot{\beta}^\alpha, \end{aligned} \tag{6.19}$$

where u_i and δ_i are components of the infinitesimal displacement and the infinitesimal director displacement and where

$$v_i=\dot{u}_i, \quad w_i=\dot{\delta}_i, \quad v^\lambda_{.\alpha}=v^\lambda{}_{|\alpha}-B^\lambda_\alpha v_3, \quad v^3_{.\alpha}=-\dot{\beta}_\alpha=v_{3,\alpha}+B^\lambda_\alpha v_\lambda. \tag{6.20}$$

ε) *A catalogue of linear kinematic measures.* The linearized kinematic measures (6.18)–(6.19) are valid for a variable initial director. Apart from (6.14) and $(6.18)_1$ which do not depend on $\boldsymbol{D}$, some of the kinematical results simplify if the initial director $\boldsymbol{D}$ is along the initial unit normal $\boldsymbol{A}_3$, i.e., if

$$\boldsymbol{D}=D\boldsymbol{A}_3, \quad D_\alpha=0, \quad D_3=D, \tag{6.21}$$

where the notation $D_3=D$ is introduced for convenience. Below we collect a catalogue of formulae for linearized kinematic measures when $\boldsymbol{D}$ is of the form (6.21) or a more specialized case in which $\boldsymbol{D}=\boldsymbol{A}_3$. The resulting expressions, which can be expressed in a variety of forms, are of particular interest in connection with kinematic measures in the classical linear theories of shells and plates.

Formulae A

$$\begin{aligned} &\boldsymbol{D}=D\boldsymbol{A}_3, \quad \Lambda_{\nu\alpha}=-B_{\nu\alpha}D, \quad \Lambda_{3\alpha}=D_{,\alpha}, \\ &e_{\alpha\gamma}=\tfrac{1}{2}(u_{\alpha|\gamma}+u_{\gamma|\alpha})-B_{\alpha\gamma}u_3, \\ &\delta_\alpha=\gamma_\alpha+D\beta_\alpha, \quad \delta_3=\gamma_3, \quad \beta_\alpha=-(u_{3,\alpha}+B^\nu_\alpha u_\nu), \\ &\varkappa_{\gamma\alpha}=\gamma_{\gamma|\alpha}-B_{\alpha\gamma}\gamma_3-D(B_{\nu\alpha}u^\nu_{.\gamma}-\beta_{\gamma|\alpha})=\varrho_{\gamma\alpha}-B_{\alpha\gamma}\gamma_3, \\ &\varrho_{\gamma\alpha}=\gamma_{\gamma|\alpha}-D(B_{\nu\alpha}u^\nu_{.\gamma}-\beta_{\gamma|\alpha}), \\ &\varkappa_{3\alpha}=\gamma_{3,\alpha}+B^\nu_\alpha\gamma_\nu=\varrho_{3\alpha}+B^\nu_\alpha\gamma_\nu, \quad \varrho_{3\alpha}=\gamma_{3,\alpha}. \end{aligned} \tag{6.22}$$

[29] GREEN, NAGHDI and WAINWRIGHT [1965, 4].

Formulae B

$$
\begin{gathered}
\boldsymbol{D}=D\boldsymbol{A}_3, \quad \Lambda_{\nu\alpha}=-B_{\nu\alpha}D, \quad \Lambda_{3\alpha}=D_{,\alpha}, \\
\hat{\boldsymbol{\delta}}=\frac{1}{D}\boldsymbol{\delta}, \quad \hat{\delta}_i=\frac{1}{D}\delta_i, \quad \hat{\gamma}_i=\frac{1}{D}\gamma_i, \quad \hat{\varkappa}_{i\alpha}=\frac{1}{D}\varkappa_{i\alpha}, \quad \hat{\varrho}_{i\alpha}=\frac{1}{D}\varrho_{i\alpha}, \\
e_{\alpha\gamma}=\tfrac{1}{2}(u_{\alpha|\gamma}+u_{\gamma|\alpha})-B_{\alpha\gamma}u_3, \\
\hat{\delta}_\alpha=\hat{\gamma}_\alpha+\beta_\alpha, \quad \hat{\delta}_3=\hat{\gamma}_3, \quad \beta_\alpha=-(u_{3,\alpha}+B^\nu_\alpha u_\nu), \\
\hat{\varkappa}_{\gamma\alpha}=\hat{\varrho}_{\gamma\alpha}-B_{\alpha\gamma}\hat{\gamma}_3, \quad \hat{\varrho}_{\gamma\alpha}=\hat{\gamma}_{\gamma|\alpha}+\frac{D_{,\alpha}}{D}\hat{\gamma}_\gamma-(B_{\nu\alpha}u^\nu_{.\gamma}-\beta_{\gamma|\alpha}), \\
\hat{\varkappa}_{3\alpha}=\hat{\varrho}_{3\alpha}+B^\nu_\alpha\hat{\gamma}_\nu, \quad \hat{\varrho}_{3\alpha}=\hat{\gamma}_{3,\alpha}+\frac{D_{,\alpha}}{D}\hat{\gamma}_3.
\end{gathered} \tag{6.23}
$$

Formulae C

$$
\begin{gathered}
\boldsymbol{D}=\boldsymbol{A}_3, \quad \Lambda_{\nu\alpha}=-B_{\nu\alpha}, \quad \Lambda_{3\alpha}=0, \\
e_{\alpha\gamma}=\tfrac{1}{2}(u_{\alpha|\gamma}+u_{\gamma|\alpha})-B_{\alpha\gamma}u_3, \\
\delta_\alpha=\gamma_\alpha+\beta_\alpha, \quad \delta_3=\gamma_3, \quad \beta_\alpha=-(u_{3,\alpha}+B^\nu_\alpha u_\nu), \\
\varkappa_{\gamma\alpha}=\varkappa_{(\gamma\alpha)}+\varkappa_{[\gamma\alpha]}=\varrho_{\gamma\alpha}-B_{\alpha\gamma}\gamma_3, \\
\varkappa_{3\alpha}=\varrho_{3\alpha}+B^\gamma_\alpha\gamma_\gamma, \\
\varrho_{\gamma\alpha}=\gamma_{\gamma|\alpha}-\bar{\varkappa}_{\gamma\alpha}, \quad \varrho_{3\alpha}=\gamma_{3,\alpha}, \\
\bar{\varkappa}_{\gamma\alpha}=\bar{\varkappa}_{\alpha\gamma}=u_{3|\gamma\alpha}+B^\nu_{\gamma|\alpha}u_\nu+B^\nu_\alpha u_{\nu|\gamma}+B^\nu_\gamma u_{\nu|\alpha}-B^\nu_\alpha B_{\nu\gamma}u_3=(B_{\nu\alpha}u^\nu_{.\gamma}-\beta_{\gamma|\alpha}), \\
\varkappa_{(\gamma\alpha)}=\varrho_{(\gamma\alpha)}-B_{\alpha\gamma}\gamma_3=\tfrac{1}{2}(\gamma_{\gamma|\alpha}+\gamma_{\alpha|\gamma})-\bar{\varkappa}_{\alpha\gamma}-B_{\alpha\gamma}\gamma_3, \\
\varkappa_{[\gamma\alpha]}=\varrho_{[\gamma\alpha]}=\tfrac{1}{2}(\gamma_{\gamma|\alpha}-\gamma_{\alpha|\gamma}).
\end{gathered} \tag{6.24}
$$

Formulae D

$$
\begin{gathered}
\boldsymbol{D}=\boldsymbol{A}_3, \quad B_{\alpha\beta}=0, \\
e_{\alpha\gamma}=\tfrac{1}{2}(u_{\alpha|\gamma}+u_{\gamma|\alpha}), \quad \varkappa_{3\alpha}=\varrho_{3\alpha}=\gamma_{3,\alpha}, \quad \gamma_3=\delta_3, \\
\varkappa_{\gamma\alpha}=\varrho_{\gamma\alpha}=\varkappa_{(\gamma\alpha)}+\varkappa_{[\gamma\alpha]}, \quad \gamma_\alpha=\delta_\alpha-\beta_\alpha=\delta_\alpha+u_{3,\alpha}, \\
\varkappa_{(\gamma\alpha)}=\tfrac{1}{2}(\gamma_{\gamma|\alpha}+\gamma_{\alpha|\gamma})-u_{3|\alpha\gamma}, \quad \varkappa_{[\gamma\alpha]}=\tfrac{1}{2}(\gamma_{\gamma|\alpha}-\gamma_{\alpha|\gamma}).
\end{gathered} \tag{6.25}
$$

We briefly elaborate on the nature of the above catalogue of formulae for linearized kinematic measures.[30] The relative simplicity of Formulae A, in

---

[30] The linearized kinematic measures (6.18)–(6.19) were obtained by GREEN, NAGHDI and WAINWRIGHT [1965, *4*]. The linearized measures in Formulae D and C were employed by GREEN and NAGHDI in a number of studies concerned with linear constitutive equations for a Cosserat surface and their applicability to thin elastic shells and plates: [1967, *4*], [1968, *6*], [1969, *3*]. Formulae B involving the variables $\hat{\gamma}_i$ and $\hat{\varkappa}_{i\alpha}$ have been used by GREEN, NAGHDI and WENNER [1971, *6*] in connection with application of the linear theory of a Cosserat surface (with variable initial director) to plates of variable thickness.

Linearized kinematics for a surface with a single director are given also by GÜNTHER [1961, *4*] who, however, allows the director to undergo infinitesimal rotation without stretch. Thus GÜNTHER's kinematical results are more restrictive than those given by Formulae C. GÜNTHER's paper [1961, *4*] seems to be the first attempt in recent years to construct a complete linear mechanical theory for shells based on the concept of oriented media. However, apart from his somewhat restrictive kinematics, his paper has another undesirable feature: After the development of his kinematics and the equations of motion for the linear theory by direct approach, the rest of his developments are obtained from the three-dimensional equations and he considers the question of constitutive relations on the basis of the generalized Hooke's law (in the three-dimensional theory of non-polar linear elasticity). It is difficult to assess the nature of the final results in his paper which, in addition to an awkward notation for kinematic quantities, involves approximations of the type often used in the development of shell theory from three-dimensional equations.

comparison with (6.18), is mainly due to the simpler expressions for $\Lambda_{\nu\alpha}$ and $\Lambda_{3\alpha}$ in $(6.22)_{2,3}$ which follow from (6.17) when $\boldsymbol{D}$ is specified by (6.21). In Formulae B the infinitesimal director displacement is put in non-dimensional form and an alternative set of kinematic variables are introduced which consist of $e_{\alpha\beta}$ and the new variables $\hat{\gamma}_i$ and $\hat{\varkappa}_{i\alpha}$ in place of γ_i and $\varkappa_{i\alpha}$. The expressions for $\varrho_{\gamma\alpha}$ in (6.22) and $\hat{\varrho}_{\gamma\alpha}$ in (6.23) may also be expressed in terms of $\bar{\varkappa}_{\gamma\alpha}$ in (6.24). Formulae C are special cases of (6.18) in which the initial director is of constant length and coincident with the unit normal to the initial surface. The results in (6.24) can also be obtained as special cases of those in Formulae B, since with $D=1$ the distinctions between $\hat{\delta}_i$ and δ_i, $\hat{\gamma}_i$ and γ_i, $\hat{\varkappa}_{i\alpha}$ and $\varkappa_{i\alpha}$ disappear. Most of the above results simplify in the case of an initially flat Cosserat surface for which

$$B_{\alpha\beta}=0. \tag{6.26}$$

In particular, Formulae D represent the linearized kinematic measures for an initially flat Cosserat surface as special cases of the results in (6.24). Evidently, when (6.26) holds, the kinematic measures separate into two sets given by $(6.25)_{3,4,5}$ and $(6.25)_{6,7}$, respectively. The kinematic variables in the former set, namely $e_{\alpha\gamma}$, $\varkappa_{3\alpha}$ and γ_3 arise from u_γ and δ_3 and characterize the *extensional motion* (or the stretching) of the Cosserat surface. The kinematic variables $\varkappa_{\gamma\alpha}$ and γ_α in the latter set, on the other hand, are specified in terms of u_3 and δ_α which represent the *flexural motion* (or the bending) of the Cosserat surface.

The foregoing kinematic measures in Formulae A to D are appropriate to a linear *direct* theory for the infinitesimal deformation of a Cosserat surface. The primitive relative kinematic measures of this direct theory are the displacement $\boldsymbol{u}$ and the director displacement $\boldsymbol{\delta}$ defined by (6.1)–(6.2)[31] and the resulting kinematic measures (in Formulae A to D) include several features which should be noted. Among these,[32] with reference to Formulae C, we mention the presence of: (i) The component $\delta_3=\gamma_3$ of the director representing extensibility in the direction of the unit normal to $\mathscr{S}$; (ii) the components $\gamma_\alpha=\delta_\alpha-\beta_\alpha$, i.e., the difference in components of rotation of the director relative to the unit normal and the components of the angular displacement β_α, which may be regarded as representing the effect of "transverse shear deformation"; and (iii) the anti-symmetric $\varkappa_{[\gamma\alpha]}$, as well as the components $\varkappa_{3\alpha}$. Parallel observations can be made in the case of other kinematic measures and, in particular, for those in Formulae D.

It is possible to construct by direct approach a *restricted* theory in which the director is not admitted. In a restricted theory of this type (appropriate for a deformable surface embedded in a Euclidean 3-space), the primitive kinematic quantities may be specified by the displacement $\boldsymbol{u}$ and the angular displacement $\boldsymbol{\beta}$ in (6.1) and (6.6). A special set of kinematic measures which emerges for such a restricted theory is summarized below:[33]

---

[31] Inasmuch as the unit normal to the surface is determined by the surface base vectors, the angular displacement $\boldsymbol{\beta}$ in (6.6) is not a primitive kinematic measure.

[32] These features are ordinarily absent or are accounted for only approximately in the existing derivations of the linear theory of shells from the three-dimensional equations.

[33] Although the special set of kinematic measures given by Formulae E resembles that sometimes used in the literature on shell theory, it should not be confused with the set of kinematic measures obtained from the three-dimensional equations under special assumptions such as Kirchhoff-Love hypothesis. In this connection, see the remarks in Sect. 7 preceding (7.76).

Formulae E

$$\begin{aligned}
&\boldsymbol{u}=u^i\boldsymbol{A}_i=u_i\boldsymbol{A}^i, && \boldsymbol{\beta}=\beta^\gamma\boldsymbol{A}_\gamma=\beta_\gamma\boldsymbol{A}^\gamma,\\
&\boldsymbol{u}_{,\alpha}=u^i_{.\alpha}\boldsymbol{A}_i=u_{\gamma\alpha}\boldsymbol{A}^\gamma-\beta_\alpha\boldsymbol{A}^3, && \boldsymbol{\beta}_{,\alpha}=\beta_{\gamma|\alpha}\boldsymbol{A}^\gamma+B^\gamma_\alpha\beta_\gamma\boldsymbol{A}^3,\\
&e_{\gamma\alpha}=\tfrac{1}{2}(u_{\gamma|\alpha}+u_{\alpha|\gamma})-B_{\alpha\gamma}u_3, && \beta_\alpha=-u^3_{.\alpha}=-(u_{3,\alpha}+B^\nu_\alpha u_\nu),\\
&\varkappa_{\gamma\alpha}=\varrho_{\gamma\alpha}=B_{\nu\alpha}u^\nu_{.\gamma}-\beta_{\gamma|\alpha}=-\bar{\varkappa}_{\gamma\alpha}=-\bar{\varkappa}_{\alpha\gamma}, && \bar{\varkappa}_{\alpha\gamma}\text{ defined in (6.24)}.
\end{aligned}\tag{6.27}$$

The above special kinematic measures can also be obtained as a special case of Formulae C if we put $\gamma_i=0$. In contrast to certain features of Formulae C noted above, Formulae E contain only a symmetric $\varkappa_{\gamma\alpha}=\varrho_{\gamma\alpha}=-\bar{\varkappa}_{\gamma\alpha}$ and do not contain (i) a measure for the extensibility along the normal to the surface or (ii) a measure representing the "transverse shear deformation."

ζ) *Additional linear kinematic formulae.* Some of the expressions in Formulae C can be expressed in slightly different forms but the interrelations are not always immediately apparent. To facilitate such comparisons and in order to record some additional formulae for later use, we introduce the notations

$$\begin{aligned}
&u_{\lambda|\alpha}=\gamma_{\lambda\alpha}=\gamma_{(\lambda\alpha)}+\gamma_{[\lambda\alpha]},\\
&\gamma_{(\lambda\alpha)}=\tfrac{1}{2}(u_{\lambda|\alpha}+u_{\alpha|\lambda}),\quad \gamma_{[\lambda\alpha]}=\tfrac{1}{2}(u_{\lambda|\alpha}-u_{\alpha|\lambda})=-\gamma_{[\alpha\lambda]},
\end{aligned}\tag{6.28}$$

where $\gamma_{(\lambda\alpha)}$ is the part of the strain measure $e_{\lambda\alpha}$ resulting from the displacement gradient $(6.28)_1$ and $\gamma_{[\lambda\alpha]}$ can be interpreted as the infinitesimal rotation at a point about the unit normal to the surface $\mathscr{S}$. Keeping this interpretation in mind and remembering that $\boldsymbol{\beta}$ (with components $\beta_\alpha=\boldsymbol{\beta}\cdot\boldsymbol{A}_\alpha$) in (6.6) is a measure of the infinitesimal rotation of the unit normal to the surface, we introduce a three-dimensional vector field $\bar{\boldsymbol{\omega}}$ defined by

$$\begin{aligned}
&\bar{\boldsymbol{\omega}}=\bar{\omega}^i\boldsymbol{A}_i=\bar{\omega}_i\boldsymbol{A}^i,\quad \gamma_{[\lambda\alpha]}=-\bar{\varepsilon}_{\lambda\alpha}\bar{\omega}^3,\\
&u_{3\alpha}=-\beta_\alpha=\bar{\varepsilon}_{\lambda\alpha}\bar{\omega}^\lambda,\quad \bar{\omega}^3=-\tfrac{1}{2}\bar{\varepsilon}^{\lambda\alpha}\gamma_{[\lambda\alpha]},\quad \bar{\omega}^\gamma=-\bar{\varepsilon}^{\gamma\alpha}\beta_\alpha,
\end{aligned}\tag{6.29}$$

where $\bar{\varepsilon}_{\alpha\beta}$, $\bar{\varepsilon}^{\alpha\beta}$ are the two-dimensional ε-system for the surface $\mathscr{S}$ defined similarly to those in (5.63) but with $a^{\frac{1}{2}}$ replaced by $A^{\frac{1}{2}}$:

$$\bar{\varepsilon}_{\alpha\beta}=A^{\frac{1}{2}}e_{\alpha\beta},\quad \bar{\varepsilon}^{\alpha\beta}=A^{-\frac{1}{2}}e^{\alpha\beta}.\tag{6.30}$$

We note here the identity

$$\bar{\boldsymbol{\omega}}\times\boldsymbol{A}_\alpha=\gamma_{[\lambda\alpha]}\boldsymbol{A}^\lambda+u_{3\alpha}\boldsymbol{A}^3,\tag{6.31}$$

which may be used to express the displacement gradient $\boldsymbol{u}_{,\alpha}$ in terms of $e_{\lambda\alpha}\boldsymbol{A}^\lambda$ and the rotation $\bar{\boldsymbol{\omega}}\times\boldsymbol{A}_\alpha$.

For later reference, we record below the expression resulting from covariant derivative of $\gamma_{[\lambda\nu]}$ given by $(6.28)_3$. Thus

$$\begin{aligned}
\gamma_{[\lambda\nu]|\alpha}&=\tfrac{1}{2}(u_{\lambda|\alpha\nu}-u_{\nu|\alpha\lambda})+\tfrac{1}{2}(R^\gamma_{.\lambda\nu\alpha}-R^\gamma_{.\nu\lambda\alpha})u_\gamma\\
&=\tfrac{1}{2}(u_{\lambda|\alpha\nu}-u_{\nu|\alpha\lambda})+\tfrac{1}{2}(u_{\alpha|\lambda\nu}-u_{\alpha|\lambda\nu})+\tfrac{1}{2}(R^\gamma_{.\lambda\nu\alpha}-R^\gamma_{.\nu\lambda\alpha})u_\gamma\\
&=\tfrac{1}{2}(u_{\lambda|\alpha\nu}+u_{\alpha|\lambda\nu})-\tfrac{1}{2}(u_{\nu|\alpha\lambda}+u_{\alpha|\nu\lambda})+\tfrac{1}{2}[R^\gamma_{.\lambda\nu\alpha}-R^\gamma_{.\nu\lambda\alpha}-R^\gamma_{.\alpha\lambda\nu}]u_\gamma\\
&=(e_{\lambda\alpha}+B_{\lambda\alpha}u_3)_{|\nu}-(e_{\nu\alpha}+B_{\nu\alpha}u_3)_{|\lambda}-R^\gamma_{.\alpha\lambda\nu}u_\gamma\\
&=(e_{\lambda\alpha|\nu}-e_{\nu\alpha|\lambda})-(B_{\lambda\alpha}\beta_\nu-B_{\nu\alpha}\beta_\lambda),
\end{aligned}\tag{6.32}$$

where

$$R^\gamma_{.\lambda\nu\alpha}=-R^\gamma_{.\lambda\alpha\nu}=B_{\lambda\alpha}B^\gamma_\nu-B_{\lambda\nu}B^\gamma_\alpha\tag{6.33}$$

is the Riemann-Christoffel surface tensor. We also collect below a list of formulae which hold when $\boldsymbol{D}=\boldsymbol{A}_3$ (as in Formulae C) and which will be useful in our subsequent derivation of the compatibility equations later in this section:

Formulae F

$$
\begin{aligned}
&u_{\lambda|\alpha}-B_{\alpha\lambda}\,u_3=e_{\lambda\alpha}+\gamma_{[\lambda\alpha]}, \qquad \beta_{\lambda|\alpha}=-\bar{\varkappa}_{\alpha\lambda}+B_\alpha^\nu(e_{\nu\lambda}+\gamma_{[\nu\lambda]}),\\
&B_\beta^\lambda\,\beta_{\lambda|\alpha}=-B_\beta^\lambda\,\bar{\varkappa}_{\lambda\alpha}+B_\beta^\lambda\,B_\alpha^\nu(e_{\nu\lambda}+\gamma_{[\nu\lambda]}),\\
&\beta_{\lambda|\alpha\beta}=-\bar{\varkappa}_{\alpha\lambda|\beta}+B^\nu_{\alpha|\beta}(e_{\nu\lambda}+\gamma_{[\nu\lambda]})+B_\alpha^\nu(e_{\nu\lambda|\beta}+\gamma_{[\nu\lambda]|\beta}),\\
&\boldsymbol{u}_{,\alpha}=(e_{\lambda\alpha}+\gamma_{[\lambda\alpha]})\,\boldsymbol{A}^\lambda-\beta_\alpha\,\boldsymbol{A}^3=e_{\lambda\alpha}\,\boldsymbol{A}^\lambda+\overline{\boldsymbol{\omega}}\times\boldsymbol{A}_\alpha,\\
&\boldsymbol{\delta}_{,\alpha}=[\varkappa_{\lambda\alpha}+B_\alpha^\nu(e_{\nu\lambda}+\gamma_{[\nu\lambda]})]\,\boldsymbol{A}^\lambda+[\varkappa_{3\alpha}+B_\alpha^\nu\,\beta_\nu]\,\boldsymbol{A}^3=\boldsymbol{J}_\alpha-B_\alpha^\nu\,\overline{\boldsymbol{\omega}}\times\boldsymbol{A}_\nu,\\
&\boldsymbol{J}_\alpha=(\varkappa_{\lambda\alpha}+B_\alpha^\nu\,e_{\nu\lambda})\,\boldsymbol{A}^\lambda+\varkappa_{3\alpha}\,\boldsymbol{A}^3,\\
&\boldsymbol{J}_{\alpha|\beta}=[\varkappa_{\lambda\alpha|\beta}+B^\nu_{\alpha|\beta}\,e_{\nu\lambda}+B_\alpha^\nu\,e_{\nu\lambda|\beta}-B_{\lambda\beta}\,\varkappa_{3\alpha}]\,\boldsymbol{A}^\lambda\\
&\qquad\qquad+[\varkappa_{3\alpha|\beta}+B_\beta^\lambda\,\varkappa_{\lambda\alpha}+B_\beta^\lambda\,B_\alpha^\nu\,e_{\nu\lambda}]\,\boldsymbol{A}^3,\\
&\bar{\varepsilon}^{\alpha\beta}\{\boldsymbol{J}_{\alpha|\beta}-B_\beta^\nu\,\boldsymbol{A}_\nu\times\overline{\boldsymbol{\omega}}_{,\alpha}\}=\bar{\varepsilon}^{\alpha\beta}\{[\varkappa_{\lambda\alpha|\beta}+B_\alpha^\nu\,e_{\nu\lambda|\beta}-B_{\lambda\beta}\,\varkappa_{3\alpha}+B_\beta^\nu\,e_{\lambda\alpha|\nu}-B_\beta^\nu\,e_{\nu\alpha|\lambda}]\,\boldsymbol{A}^\lambda\\
&\qquad\qquad+[\varkappa_{3\alpha|\beta}-B_\alpha^\nu\,\gamma_{\nu|\beta}]\,\boldsymbol{A}^3\}.
\end{aligned}
\tag{6.34}
$$

Expressions for $\overline{\boldsymbol{\omega}}$ and $\overline{\boldsymbol{\omega}}_{,\alpha}$ are listed in (6.35).

Formulae G

$$
\begin{aligned}
&\overline{\boldsymbol{\omega}}=-\tfrac{1}{2}\,\bar{\varepsilon}^{\lambda\nu}\,\gamma_{[\lambda\nu]}\,\boldsymbol{A}_3-\bar{\varepsilon}^{\lambda\nu}\,\beta_\nu\,\boldsymbol{A}_\lambda,\\
&\overline{\boldsymbol{\omega}}_{,\alpha}=[-\tfrac{1}{2}\,\bar{\varepsilon}^{\lambda\nu}\,\gamma_{[\lambda\nu]|\alpha}-\bar{\varepsilon}^{\lambda\nu}\,\beta_\nu\,B_{\lambda\alpha}]\,\boldsymbol{A}_3+[+\tfrac{1}{2}\,\bar{\varepsilon}^{\lambda\nu}\,\gamma_{[\lambda\nu]}\,B_\alpha^\gamma-\bar{\varepsilon}^{\gamma\nu}\,\beta_{\nu|\alpha}]\,\boldsymbol{A}_\gamma\\
&\qquad=\bar{\varepsilon}^{\nu\lambda}\,e_{\lambda\alpha|\nu}\,\boldsymbol{A}_3+\bar{\varepsilon}^{\gamma\nu}(\gamma_{\nu|\alpha}-\varrho_{\nu\alpha}-B_\alpha^\sigma\,e_{\sigma\nu})\,\boldsymbol{A}_\gamma\\
&\qquad=\bar{\varepsilon}^{\nu\lambda}\,e_{\lambda\alpha|\nu}\,\boldsymbol{A}_3+\bar{\varepsilon}^{\gamma\nu}(\gamma_{\nu|\alpha}-\varkappa_{\nu\alpha}-B_{\nu\alpha}\,\gamma_3-B_\alpha^\sigma\,e_{\sigma\nu})\,\boldsymbol{A}_\gamma,\\
&\bar{\varepsilon}^{\alpha\beta}\,\overline{\boldsymbol{\omega}}_{|\alpha\beta}=\bar{\varepsilon}^{\alpha\beta}\{\bar{\varepsilon}^{\nu\lambda}\,e_{\lambda\alpha|\nu\beta}+B_{\tau\beta}\,\bar{\varepsilon}^{\tau\nu}(\gamma_{\nu|\alpha}-\varkappa_{\nu\alpha}-B_{\nu\alpha}\,\gamma_3-B_\alpha^\sigma\,e_{\sigma\nu})\}\,\boldsymbol{A}_3\\
&\qquad+\bar{\varepsilon}^{\alpha\beta}\{\bar{\varepsilon}^{\tau\nu}[\gamma_{\nu|\alpha\beta}-\varkappa_{\nu\alpha|\beta}-B_{\nu\alpha}\,\gamma_{3|\beta}-B_\alpha^\sigma\,e_{\sigma\nu|\beta}]-\bar{\varepsilon}^{\nu\lambda}\,B_\beta^\tau\,e_{\lambda\alpha|\nu}\}\,\boldsymbol{A}_\tau.
\end{aligned}
\tag{6.35}
$$

Before turning our attention to a derivation of compatibility equations, we make one further observation regarding the kinematic measures obtained in this section. The strain measures (6.18) or an equivalent set given by (6.24) when $\boldsymbol{D}=\boldsymbol{A}_3$, should be unaffected by infinitesimal rigid body displacement of the Cosserat surface. To show this, we first introduce the notations

$$
{}_0\boldsymbol{R}=\boldsymbol{R}({}_0\theta^\alpha), \qquad {}_0\boldsymbol{D}=\boldsymbol{D}({}_0\theta^\alpha), \qquad {}_0\boldsymbol{u}=\boldsymbol{u}({}_0\boldsymbol{R}), \qquad {}_0\boldsymbol{\delta}=\boldsymbol{\delta}({}_0\boldsymbol{R}),
\tag{6.36}
$$

where ${}_0\boldsymbol{R}$ and ${}_0\boldsymbol{D}$ stand for the position vector and the director of an arbitrary reference point of $\mathscr{S}$ while $(6.36)_{3,4}$ are abbreviations for the infinitesimal displacement and the infinitesimal director displacement at ${}_0\boldsymbol{R}$. Now, in order to obtain the appropriate expressions for $\boldsymbol{u}=\boldsymbol{u}(\boldsymbol{R})$ and $\boldsymbol{\delta}=\boldsymbol{\delta}(\boldsymbol{R})$ due to purely *infinitesimal rigid body displacements* of the Cosserat surface, we only need to consider a special case of (5.36) in which the reference values of position and director are specified to be $\boldsymbol{R}$ and $\boldsymbol{D}$. With this proviso, it follows from (5.36) that under rigid body displacement alone the position $\boldsymbol{r}$ and the director $\boldsymbol{d}$ can be written as

$$
\boldsymbol{r}=\boldsymbol{C}+Q\boldsymbol{R}, \qquad \boldsymbol{d}=Q\boldsymbol{D},
\tag{6.37}
$$

where the contributions corresponding to $\boldsymbol{r}_0^+$ and $\boldsymbol{r}_0$ in (5.36) have been absorbed into $\boldsymbol{C}$. Recalling $(6.1)_1$–$(6.2)_1$ and the earlier linearization procedure adopted in this section, we have to $O(\varepsilon)$:

$$\boldsymbol{u} = \boldsymbol{r} - \boldsymbol{R} = \boldsymbol{C} + {}_0Q\boldsymbol{R}, \qquad \boldsymbol{\delta} = {}_0Q\boldsymbol{D}, \tag{6.38}$$

$$\boldsymbol{C} = O(\varepsilon), \qquad {}_0Q = Q - I = O(\varepsilon). \tag{6.39}$$

Using the notations of $(6.36)_{3,4}$, the expressions (6.38) when evaluated at the reference point ${}_0\boldsymbol{R}$ yield

$${}_0\boldsymbol{u} = \boldsymbol{C} + {}_0Q\,{}_0\boldsymbol{R}, \qquad {}_0\boldsymbol{\delta} = {}_0Q\,{}_0\boldsymbol{D}. \tag{6.40}$$

By subtraction, from (6.38) and (6.40) follow

$$\boldsymbol{u} = {}_0\boldsymbol{u} + {}_0Q(\boldsymbol{R} - {}_0\boldsymbol{R}), \qquad \boldsymbol{\delta} = {}_0\boldsymbol{\delta} + {}_0Q(\boldsymbol{D} - {}_0\boldsymbol{D}), \tag{6.41}$$

as the displacement and the director displacement relative to those at ${}_0\boldsymbol{R}$. In (6.38)–(6.41), the vector $\boldsymbol{C}$ is a uniform infinitesimal rigid body translatory displacement while the second order tensor ${}_0Q$ represents a uniform infinitesimal rotation. In order to write (6.38) and (6.41) in alternative forms, we express the orthogonality condition $(5.37)_1$ in terms of ${}_0Q$ defined by $(6.39)_2$. Since $(I + {}_0Q)^T = I + {}_0Q^T$,

$$QQ^T = (I + {}_0Q)(I + {}_0Q^T) = I + {}_0Q + {}_0Q^T + O(\varepsilon^2) = I.$$

It is easily seen from the last result that $(5.37)_1$ is satisfied to $O(\varepsilon)$ if ${}_0Q$ is a skew-symmetric tensor; hence, there exists an infinitesimal vector-valued function ${}_0\overline{\boldsymbol{\omega}}$ such that for any vector $\boldsymbol{V}$

$${}_0Q\boldsymbol{V} = {}_0\overline{\boldsymbol{\omega}} \times \boldsymbol{V}, \qquad {}_0Q = -{}_0Q^T \tag{6.42}$$

and ${}_0\overline{\boldsymbol{\omega}}$ can be interpreted as a uniform infinitesimal rigid body angular displacement vector. Using $(6.42)_1$, (6.38) and (6.41) can be written in more familiar forms

$$\boldsymbol{u} = \boldsymbol{C} + {}_0\overline{\boldsymbol{\omega}} \times \boldsymbol{R}, \qquad \boldsymbol{\delta} = {}_0\overline{\boldsymbol{\omega}} \times \boldsymbol{D} \tag{6.43}$$

and

$$\boldsymbol{u} = {}_0\boldsymbol{u} + {}_0\overline{\boldsymbol{\omega}} \times (\boldsymbol{R} - {}_0\boldsymbol{R}), \qquad \boldsymbol{\delta} = {}_0\boldsymbol{\delta} + {}_0\overline{\boldsymbol{\omega}} \times (\boldsymbol{D} - {}_0\boldsymbol{D}), \tag{6.44}$$

respectively. Moreover, since ${}_0\boldsymbol{u}$, ${}_0\boldsymbol{\delta}$ and ${}_0\overline{\boldsymbol{\omega}}$ are independent of surface coordinates (but may be functions of time), from (6.43) or (6.44) and (6.6) we have

$$\boldsymbol{u}_{,\alpha} = {}_0\overline{\boldsymbol{\omega}} \times \boldsymbol{A}_\alpha, \qquad \boldsymbol{\delta}_{,\alpha} = {}_0\overline{\boldsymbol{\omega}} \times \boldsymbol{D}_{,\alpha}, \qquad \boldsymbol{\beta} = {}_0\overline{\boldsymbol{\omega}} \times \boldsymbol{A}_3, \tag{6.45}$$

where in obtaining $(6.45)_3$ we have used the identity

$$[(\boldsymbol{V} \times \boldsymbol{A}_\lambda) \cdot \boldsymbol{A}_3]\,\boldsymbol{A}^\lambda = -[(\boldsymbol{V} \times \boldsymbol{A}_3) \cdot \boldsymbol{A}_\lambda]\,\boldsymbol{A}^\lambda = -\boldsymbol{V} \times \boldsymbol{A}_3 \tag{6.46}$$

which holds for any vector $\boldsymbol{V}$. It can be easily verified, with the help of (6.13) and (6.45), that the rigid body displacements (6.44) have no effect on the kinematic measures $e_{\alpha\beta}$, γ_i, $\varkappa_{i\alpha}$ in (6.24).[34]

η) *Compatibility equations.* We include here a relatively simple derivation of compatibility equations which provides both necessary and sufficient conditions for the existence of single-valued displacements $\boldsymbol{u}$ and $\boldsymbol{\delta}$. For the purpose of the

[34] It may be of interest to note here that in the older literature of linear shell theory (developed from the three-dimensional equations) some of the kinematic measures are not invariant under infinitesimal rigid body displacements and this, of course, affects the constitutive relations. Further remark on this is made in Sect. 21 A.

derivation at hand, we assume $\mathscr{S}$ to be a simply connected surface and suppose that the displacement ${}_0\boldsymbol{u}=\boldsymbol{u}({}_0\boldsymbol{R})$, ${}_0\boldsymbol{\delta}=\boldsymbol{\delta}({}_0\boldsymbol{R})$ and the rotation ${}_0\overline{\boldsymbol{\omega}}=\overline{\boldsymbol{\omega}}({}_0\boldsymbol{R})$ are known at some point ${}_0\boldsymbol{R}$ of this surface. We wish to determine the displacement $\boldsymbol{u}$ and the director displacement $\boldsymbol{\delta}$ at any other point $R'=R(\theta'^{\alpha})$ of $\mathscr{S}$ in terms of the known kinematic measures $e_{\alpha\beta}$, γ_i, $\varkappa_{i\alpha}$ (assumed to be at least twice continuously differentiable) by means of the line integrals

$$\boldsymbol{u}(\boldsymbol{R}')={}_0\boldsymbol{u}+\int_{{}_0\boldsymbol{R}}^{\boldsymbol{R}'}\frac{\partial\boldsymbol{u}}{\partial\theta^{\alpha}}\,d\theta^{\alpha},\qquad \boldsymbol{\delta}(\boldsymbol{R}')={}_0\boldsymbol{\delta}+\int_{{}_0\boldsymbol{R}}^{\boldsymbol{R}'}\frac{\partial\boldsymbol{\delta}}{\partial\theta^{\alpha}}\,d\theta^{\alpha} \tag{6.47}$$

over a continuous curve joining the points ${}_0\boldsymbol{R}$ and $\boldsymbol{R}'$. For simplicity, we carry out the derivation with reference to the kinematic measures in Formulae C which hold when the initial director[35] $\boldsymbol{D}=\boldsymbol{A}_3$.

Consider first $(6.47)_1$ which, with the use of $(6.34)_5$, can be written as

$$\boldsymbol{u}(\boldsymbol{R}')={}_0\boldsymbol{u}+\int_{{}_0\boldsymbol{R}}^{\boldsymbol{R}'}e_{\lambda\alpha}\boldsymbol{A}^{\lambda}\,d\theta^{\alpha}+\int_{{}_0\boldsymbol{R}}^{\boldsymbol{R}'}\overline{\boldsymbol{\omega}}\times\boldsymbol{A}_{\alpha}\,d\theta^{\alpha}. \tag{6.48}$$

The last integral in (6.48) after an integration by parts gives

$$\begin{aligned}\int_{{}_0\boldsymbol{R}}^{\boldsymbol{R}'}\overline{\boldsymbol{\omega}}\times\boldsymbol{A}_{\alpha}\,d\theta^{\alpha}&=\int_{{}_0\boldsymbol{R}}^{\boldsymbol{R}'}\overline{\boldsymbol{\omega}}\times d\boldsymbol{R}=\int_{{}_0\boldsymbol{R}}^{\boldsymbol{R}'}\overline{\boldsymbol{\omega}}\times d(\boldsymbol{R}-\boldsymbol{R}')\\&={}_0\overline{\boldsymbol{\omega}}\times(\boldsymbol{R}'-{}_0\boldsymbol{R})+\int_{{}_0\boldsymbol{R}}^{\boldsymbol{R}'}(\boldsymbol{R}-\boldsymbol{R}')\times\overline{\boldsymbol{\omega}}_{,\alpha}\,d\theta^{\alpha}.\end{aligned} \tag{6.49}$$

Hence, from (6.48)–(6.49),

$$\boldsymbol{u}(\boldsymbol{R}')={}_0\boldsymbol{u}+{}_0\overline{\boldsymbol{\omega}}\times(\boldsymbol{R}'-{}_0\boldsymbol{R})+\int_{{}_0\boldsymbol{R}}^{\boldsymbol{R}'}\boldsymbol{U}_{\alpha}\,d\theta^{\alpha} \tag{6.50}$$

and we have put

$$\boldsymbol{U}_{\alpha}=e_{\lambda\alpha}\boldsymbol{A}^{\lambda}+(\boldsymbol{R}-\boldsymbol{R}')\times\overline{\boldsymbol{\omega}}_{,\alpha}, \tag{6.51}$$

which is a known function of the kinematic measures [see $(6.35)_2$]. Since the displacement $\boldsymbol{u}$ must be independent of the path of integration for simply connected surfaces, the integrand $\boldsymbol{U}_{\alpha}\,d\theta^{\alpha}$ must be an exact differential. A necessary and sufficient condition that the integrand in (6.50) be an exact differential is $\boldsymbol{U}_{1,2}=\boldsymbol{U}_{2,1}$. The latter condition can equivalently be stated as

$$\bar{\varepsilon}^{\alpha\beta}\,\boldsymbol{U}_{\alpha|\beta}=0, \tag{6.52}$$

since in the expression for the covariant derivative of $\boldsymbol{U}_{\alpha|\beta}$ the Christoffel symbol is symmetric in α, β. Further, since

$$\begin{aligned}\boldsymbol{A}_{\beta}\times\overline{\boldsymbol{\omega}}_{,\alpha}&=(\boldsymbol{A}_{\beta}\times\overline{\boldsymbol{\omega}})_{|\alpha}-B_{\beta\alpha}\boldsymbol{A}_3\times\overline{\boldsymbol{\omega}}\\&=-[\boldsymbol{u}_{,\beta}-e_{\lambda\beta}\boldsymbol{A}^{\lambda}]_{|\alpha}-B_{\beta\alpha}\boldsymbol{A}_3\times\overline{\boldsymbol{\omega}},\end{aligned}$$

[35] Our derivation of the compatibility equations differs from similar previous developments in the literature. A discussion of compatibility equations by direct approach is contained in GÜNTHER's paper [1961, *4*] which, as noted earlier, employs a restrictive kinematic measure. Compatibility equations (in terms of the kinematic measures in Formulae C), as necessary conditions for the existence of single-valued displacements $\boldsymbol{u}$ and $\boldsymbol{\delta}$, have been considered also by CROCHET [1967, *2*] but his development appears to be incomplete.

Compatibility equations are sometimes useful in the formulation of a class of boundary-value problems for shells with infinitesimal deformation. Their utility in the linear theory of elastic shells is somewhat similar to the use of compatibility equations in the formulation of two-dimensional problems (in terms of a stress function) in the classical theory of linear elasticity.

from (6.51) we have

$$\begin{aligned}\boldsymbol{U}_{\alpha|\beta} &= (e_{\lambda\alpha}\boldsymbol{A}^{\lambda})_{|\beta} + \boldsymbol{A}_{\beta}\times\overline{\boldsymbol{\omega}}_{|\alpha} + (\boldsymbol{R}-\boldsymbol{R}')\times\overline{\boldsymbol{\omega}}_{|\alpha\beta}\\ &= (e_{\lambda\alpha}\boldsymbol{A}^{\lambda})_{|\beta} + (e_{\lambda\beta}\boldsymbol{A}^{\lambda})_{|\alpha} - \boldsymbol{u}_{|\beta\alpha} - B_{\beta\alpha}\boldsymbol{A}_3\times\overline{\boldsymbol{\omega}} + (\boldsymbol{R}-\boldsymbol{R}')\times\overline{\boldsymbol{\omega}}_{|\alpha\beta}.\end{aligned}$$

Each of the third and fourth terms in the last equation, as well as the combination of the first and second term, is symmetric in α, β and vanishes identically when multiplied by $\bar{\varepsilon}^{\alpha\beta}$. Hence, application of (6.52) to the above expression for $\boldsymbol{U}_{\alpha|\beta}$ yields

$$(\boldsymbol{R}-\boldsymbol{R}')\times\bar{\varepsilon}^{\alpha\beta}\,\overline{\boldsymbol{\omega}}_{|\alpha\beta}=0.$$

But, since the last equation must hold for an arbitrary choice of $\boldsymbol{R}'$ and therefore $(\boldsymbol{R}-\boldsymbol{R}')$, it follows that

$$\bar{\varepsilon}^{\alpha\beta}\,\overline{\boldsymbol{\omega}}_{|\alpha\beta}=0. \tag{6.53}$$

We postpone examining the further implication of (6.53) and now turn our attention to $(6.47)_2$ which, with the use of $(6.34)_4$, can be written in the form

$$\boldsymbol{\delta}(\boldsymbol{R}') = {}_0\boldsymbol{\delta} + \int_{{}_0\boldsymbol{R}}^{\boldsymbol{R}'} \boldsymbol{J}_{\alpha}\,d\theta^{\alpha} + \int_{{}_0\boldsymbol{R}}^{\boldsymbol{R}'} \overline{\boldsymbol{\omega}}\times\boldsymbol{D}_{,\alpha}\,d\theta^{\alpha}, \tag{6.54}$$

where we have written $\boldsymbol{D}_{,\alpha}$ $(=\boldsymbol{A}_{3,\alpha})$ in place of $-B_{\alpha}^{\nu}\boldsymbol{A}_{\nu}$ for clarity. Again, by an integration by parts of the last integral in (6.54) and using the notation $\boldsymbol{D}'=\boldsymbol{D}(\boldsymbol{R}')$,

$$\begin{aligned}\int_{{}_0\boldsymbol{R}}^{\boldsymbol{R}'} \overline{\boldsymbol{\omega}}\times\boldsymbol{D}_{,\alpha}\,d\theta^{\alpha} &= \int_{{}_0\boldsymbol{R}}^{\boldsymbol{R}'} \overline{\boldsymbol{\omega}}\times d\boldsymbol{D} = \int_{{}_0\boldsymbol{R}}^{\boldsymbol{R}'} \overline{\boldsymbol{\omega}}\times d(\boldsymbol{D}-\boldsymbol{D}')\\ &= {}_0\overline{\boldsymbol{\omega}}\times(\boldsymbol{D}'-{}_0\boldsymbol{D}) + \int_{{}_0\boldsymbol{R}}^{\boldsymbol{R}'} (\boldsymbol{D}-\boldsymbol{D}')\times\overline{\boldsymbol{\omega}}_{,\alpha}\,d\theta^{\alpha}\end{aligned} \tag{6.55}$$

and hence (6.54) becomes

$$\boldsymbol{\delta}(\boldsymbol{R}') = {}_0\boldsymbol{\delta} + {}_0\overline{\boldsymbol{\omega}}\times(\boldsymbol{D}'-{}_0\boldsymbol{D}) + \int_{{}_0\boldsymbol{R}}^{\boldsymbol{R}'} \boldsymbol{V}_{\alpha}\,d\theta^{\alpha}, \tag{6.56}$$

where we have put

$$\boldsymbol{V}_{\alpha} = \boldsymbol{J}_{\alpha} + (\boldsymbol{D}-\boldsymbol{D}')\times\overline{\boldsymbol{\omega}}_{,\alpha}. \tag{6.57}$$

Again, since the director displacement must be independent of the path of integration for a simply connected surface, the integrand $\boldsymbol{V}_{\alpha}\,d\theta^{\alpha}$ must be an exact differential. Hence, $\boldsymbol{V}_{1,2}=\boldsymbol{V}_{2,1}$ which is a necessary and sufficient condition for the integrand in (6.56) to be an exact differential. But the latter condition, similar to (6.52), can also be written as

$$\bar{\varepsilon}^{\alpha\beta}\,\boldsymbol{V}_{\alpha|\beta}=0. \tag{6.58}$$

From (6.57),

$$\boldsymbol{V}_{\alpha|\beta} = \boldsymbol{J}_{\alpha|\beta} - B_{\beta}^{\nu}\boldsymbol{A}_{\nu}\times\overline{\boldsymbol{\omega}}_{,\alpha} + (\boldsymbol{D}-\boldsymbol{D}')\times\overline{\boldsymbol{\omega}}_{|\alpha\beta}.$$

But, in view of (6.53), application of (6.58) to the last result yields

$$\bar{\varepsilon}^{\alpha\beta}\{\boldsymbol{J}_{\alpha|\beta} - B_{\beta}^{\nu}\boldsymbol{A}_{\nu}\times\boldsymbol{\omega}_{,\alpha}\}=0. \tag{6.59}$$

The conditions (6.53) and (6.59) are primitive forms of equations of compatibility. By taking the scalar products of these two vector equations with $\boldsymbol{A}^{\gamma}$ and $\boldsymbol{A}_3$ and using the appropriate expressions in (6.34)–(6.35), we readily deduce the following equations:

$$\begin{aligned}&\bar{\varepsilon}^{\alpha\beta}[\bar{\varepsilon}^{\gamma\nu}(\varkappa_{\nu\alpha|\beta} + B_{\alpha}^{\sigma}e_{\sigma\nu|\beta} - \gamma_{\nu|\alpha\beta} + B_{\nu\alpha}\gamma_{3|\beta}) + \bar{\varepsilon}^{\nu\lambda}B_{\beta}^{\gamma}e_{\lambda\alpha|\nu}]=0,\\ &\bar{\varepsilon}^{\alpha\beta}\bar{\varepsilon}^{\lambda\nu}[e_{\alpha\lambda|\nu\beta} - B_{\nu\beta}(\varkappa_{\lambda\alpha} + B_{\alpha}^{\sigma}e_{\sigma\lambda} - \gamma_{\lambda|\alpha} + B_{\lambda\alpha}\gamma_3=0])\end{aligned} \tag{6.60}$$

and

$$\bar{\varepsilon}^{\alpha\beta}[\gamma_{\nu|\alpha\beta} + B_{\nu\alpha} B^{\sigma}_{\beta} \gamma_{\sigma}] = 0, \qquad \bar{\varepsilon}^{\alpha\beta}(\varkappa_{3\alpha|\beta} - B^{\nu}_{\alpha} \gamma_{\nu|\beta}) = 0, \tag{6.61}$$

where $(6.61)_1$ is obtained with the help of $(6.60)_1$ and is equivalent to that resulting from (6.59) and (6.34). The two sets of compatibility equations $(6.60)_{1,2}$ and $(6.61)_{1,2}$ consist of six equations expressed in terms of the kinematic measures $e_{\lambda\alpha}$, $\varkappa_{i\lambda}$, γ_i. Alternatively, they can also be expressed in terms of the measures $e_{\lambda\alpha}$, $\varrho_{i\lambda}$ and γ_i [see (6.24)] but we do not record these.

We examine now the reduction of (6.60)–(6.61) for the restricted theory whose kinematic measures are summarized in (6.27). Thus, if we put $\gamma_i = 0$, the two equations in (6.61) vanish identically and $(6.60)_{1,2}$ reduce to

$$\begin{aligned} \bar{\varepsilon}^{\alpha\beta}[\bar{\varepsilon}^{\gamma\nu}(\varrho_{(\nu\alpha)|\beta} + B^{\sigma}_{\alpha} e_{\sigma\nu|\beta}) + \bar{\varepsilon}^{\nu\lambda} B^{\gamma}_{\nu} e_{\lambda\alpha|\beta}] = 0, \\ \bar{\varepsilon}^{\alpha\beta} \bar{\varepsilon}^{\lambda\nu}[e_{\alpha\lambda|\nu\beta} - B_{\nu\beta}(\varrho_{(\lambda\alpha)} + B^{\sigma}_{\alpha} e_{\sigma\lambda})] = 0. \end{aligned} \tag{6.62}$$

The fact that $\varrho_{\nu\alpha}$ is symmetric in the restricted theory has been explicitly indicated in (6.62).[36] In $(6.62)_1$, we have also used the identity $\bar{\varepsilon}^{\alpha\beta} \bar{\varepsilon}^{\nu\lambda} B^{\gamma}_{\nu} e_{\lambda\alpha|\beta} = \bar{\varepsilon}^{\alpha\beta} \bar{\varepsilon}^{\nu\lambda} B^{\gamma}_{\beta} e_{\lambda\alpha|\nu}$.

The foregoing derivation of the compatibility equations provides both necessary and sufficient conditions for the existence of single-valued displacements $\boldsymbol{u}$ and $\boldsymbol{\delta}$: Given the strain measures $e_{\alpha\gamma}$, γ_i, $\varkappa_{i\alpha}$ satisfying the conditions (6.52) and (6.58) or equivalently (6.60) and (6.61), then (6.50) and (6.56) determine the displacements $\boldsymbol{u}$ and $\boldsymbol{\delta}$ (corresponding to the given strain measures) at any point $\boldsymbol{R}'$ of the surface $\mathscr{S}$ uniquely to within rigid displacements of the forms (6.44). On the other hand, if the functions $e_{\alpha\gamma}$, γ_i, $\varkappa_{i\alpha}$ satisfy the differential equations in u_i and δ_i given by (6.24), then (6.52) and (6.58) or (6.60) and (6.61) are necessary conditions for the existence of single-valued displacements $\boldsymbol{u}$ and $\boldsymbol{\delta}$.

7. Kinematics of shells: II. Developments from the three-dimensional theory. We have already defined a shell-like body in Sect. 4. Here we derive the kinematics of such three-dimensional continua from the three-dimensional theory. Often, we employ (by choice) the same symbols which have been used previously in Sects. 5 and 6; but this need not be confusing, since the contents of this and the two previous sections are developed independently of each other.

α) *General kinematical results.* We begin our development of the kinematical results from the three-dimensional equations by assuming that the position vector $\boldsymbol{p}(\theta^{\alpha}, \xi, t)$ of a material point in the deformed shell is an analytic function of ξ in the region $\alpha < \xi < \beta$. Thus, recalling (4.6) and (4.9), $\boldsymbol{p}$ can be represented as[37]

$$\boldsymbol{p} = \boldsymbol{r}(\theta^{\alpha}, t) + \sum_{N=1}^{\infty} \xi^N \boldsymbol{d}_N(\theta^{\alpha}, t) \tag{7.1}$$

and its dual in a reference configuration is

$$\boldsymbol{P} = \boldsymbol{R}(\theta^{\alpha}) + \sum_{N=1}^{\infty} \xi^N \boldsymbol{D}_N(\theta^{\alpha}), \tag{7.2}$$

where $\boldsymbol{r}$ and $\boldsymbol{R}$, defined by (4.9) and (4.24), are the position vectors of the surface $\xi = 0$ in the deformed and the reference configurations, respectively, $\boldsymbol{d}_N$ are

[36] We may observe that these equations are of the same form as the corresponding compatibility equations in the classical theory of shells derived from the three-dimensional equations. Compare, for example, with Eq. (3.4) in [1963, 7].

[37] The representation (7.1), along with the interpretation for $\boldsymbol{d}_N$ stated after (7.8), was introduced by GREEN, LAWS and NAGHDI [1968, 4].

vector functions of θ^α, t and their reference values are denoted by $\boldsymbol{D}_N$, i.e.,

$$\boldsymbol{D}_N(\theta^\alpha) = \boldsymbol{d}_N(\theta^\alpha, 0). \tag{7.3}$$

We assume that the two series (7.1)–(7.2) may be differentiated as many times as required with respect to any of their variables, at least in the open region $\alpha < \xi < \beta$.

The velocity vector $\boldsymbol{v}^*$, of the three-dimensional continuum, at time t is given by

$$\boldsymbol{v}^* = \frac{d\boldsymbol{p}(\theta^\alpha, \xi, t)}{dt} = \dot{\boldsymbol{p}}(\theta^\alpha, \xi, t), \tag{7.4}$$

where a superposed dot denotes the material time derivative, holding $\theta^i = \{\theta^\alpha, \xi\}$ fixed. From (7.4) and (7.1), we have

$$\boldsymbol{v}^* = \boldsymbol{v} + \sum_{N=1}^{\infty} \xi^N \boldsymbol{w}_N, \tag{7.5}$$

where

$$\boldsymbol{v} = \dot{\boldsymbol{r}}, \qquad \boldsymbol{w}_N = \dot{\boldsymbol{d}}_N. \tag{7.6}$$

We recall that when the motion of $\mathscr{B}$ differs from (4.6) only by superposed rigid body motions, the position $\boldsymbol{p}^+$ (using a by now familiar notation) has the form

$$\overset{+}{\boldsymbol{p}} = \boldsymbol{p}^+(\theta^i, t') = \overset{+}{\boldsymbol{p}}_0(t') + Q(t)\,[\boldsymbol{p}(\theta^i, t) - \boldsymbol{p}_0(t)], \tag{7.7}$$

where Q is a proper orthogonal tensor function of time which satisfies the conditions (5.37). Since under superposed rigid body motions the position vector $\boldsymbol{r}^+$ of the surface $\mathfrak{s}$ transforms by a formula of the form $(5.36)_1$ and since

$$\boldsymbol{p}^+ - \boldsymbol{r}^+ = \sum_{N=1}^{\infty} \xi^N \overset{+}{\boldsymbol{d}}_N$$

by (7.1), it follows that the vector functions $\overset{+}{\boldsymbol{d}}_N$ must transform according to

$$\overset{+}{\boldsymbol{d}}_N(\theta^\alpha, t) = Q(t)\, \boldsymbol{d}_N(\theta^\alpha, t) \tag{7.8}$$

and hence remain unchanged in magnitude. We may, therefore, call the vectors $\boldsymbol{d}_N$ *directors* and $\boldsymbol{w}_N$ *director velocities*. Also, for reasons that will become apparent later, we introduce the notations

$$\boldsymbol{d} = \boldsymbol{d}_1, \qquad \boldsymbol{D} = \boldsymbol{D}_1, \qquad \boldsymbol{w} = \boldsymbol{w}_1. \tag{7.9}$$

By $(4.7)_1$ and (7.1), the base vectors $\boldsymbol{g}_i$ can be written as

$$\boldsymbol{g}_\alpha = \boldsymbol{a}_\alpha + \sum_{N=1}^{\infty} \xi^N \frac{\partial \boldsymbol{d}_N}{\partial \theta^\alpha}, \qquad \boldsymbol{g}_3 = \sum_{N=1}^{\infty} N\, \xi^{N-1} \boldsymbol{d}_N, \tag{7.10}$$

where $\boldsymbol{a}_\alpha$ are the base vectors of the surface $\mathfrak{s}$ (i.e., $\xi = 0$ in the deformed configuration) defined by (4.10). The unit normal $\boldsymbol{a}_3$ to $\mathfrak{s}$ and the first and the second fundamental forms of $\mathfrak{s}$ are given by (4.11)–(4.13). The base vectors $\boldsymbol{g}_i(\theta^\alpha, \xi, t)$ when evaluated on the surface $\mathfrak{s}$ reduce to

$$\boldsymbol{g}_\alpha(\theta^\gamma, 0, t) = \boldsymbol{a}_\alpha(\theta^\gamma, t), \qquad \boldsymbol{g}_3(\theta^\gamma, 0, t) = \boldsymbol{d}(\theta^\gamma, t), \tag{7.11}$$

where $\boldsymbol{d}$ is defined by $(7.9)_1$. The restriction (4.8) holds for all time and all values of θ^i. In particular, it is valid for $\xi = 0$ so that by (7.11) we also have

$$[\boldsymbol{a}_1\, \boldsymbol{a}_2\, \boldsymbol{d}] > 0. \tag{7.12}$$

We now introduce some additional kinematical quantities. Let the three-dimensional vector fields $\boldsymbol{d}_N$ and their partial derivatives with respect to θ^α be referred to the base vectors $\boldsymbol{a}_i$. Then,[38]

$$\begin{aligned} &\boldsymbol{d}_N = d_{Ni}\,\boldsymbol{a}^i = d_{N.}^{\ i}\,\boldsymbol{a}_i, \quad d_{N.}^{\ \gamma} = a^{\gamma\beta}\,d_{N\beta},\\ &d_{N.}^{\ 3} = d_{N3}, \qquad \frac{\partial \boldsymbol{d}_N}{\partial\theta^\alpha} = \lambda_{Ni\alpha}\,\boldsymbol{a}^i \qquad (N \geqq 2),\\ &\boldsymbol{d}_1 = \boldsymbol{d} = d_i\,\boldsymbol{a}^i = d^i\,\boldsymbol{a}_i, \quad d^\gamma = a^{\gamma\beta}\,d_\beta,\\ &d^3 = d_3, \qquad \frac{\partial \boldsymbol{d}}{\partial\theta^\alpha} = \lambda_{1i\alpha}\,\boldsymbol{a}^i = \lambda_{i\alpha}\,\boldsymbol{a}^i. \end{aligned} \tag{7.13}$$

By application of the general formula (5.5), the components $\lambda_{Ni\alpha}$ are

$$\begin{aligned} &\lambda_{N\gamma\alpha} = \boldsymbol{a}_\gamma\cdot\boldsymbol{d}_{N,\alpha} = d_{N\gamma|\alpha} - b_{\gamma\alpha}\,d_{N3}, \qquad \lambda_{N.\alpha}^{\ \gamma} = a^{\gamma\beta}\,\lambda_{N\beta\alpha},\\ &\lambda_{N3\alpha} = \boldsymbol{a}_3\cdot\boldsymbol{d}_{N,\alpha} = d_{N3,\alpha} + b_\alpha^\gamma\,d_{N\gamma}, \qquad \lambda_{N.\alpha}^{\ 3} = \lambda_{N3\alpha} \end{aligned} \tag{7.14}$$

and

$$\begin{aligned} &\lambda_{\gamma\alpha} = \lambda_{1\gamma\alpha} = d_{\gamma|\alpha} - b_{\gamma\alpha}\,d_3, \qquad \lambda_{.\alpha}^{\gamma} = a^{\gamma\beta}\,\lambda_{\beta\alpha},\\ &\lambda_{3\alpha} = \lambda_{13\alpha} = d_{3,\alpha} + b_\alpha^\gamma\,d_\gamma, \qquad \lambda_{.\alpha}^{3} = \lambda_{3\alpha} \end{aligned} \tag{7.15}$$

for $N=1$, where a vertical bar denotes covariant differentiation with respect to $a_{\alpha\beta}$. Also, in anticipation of certain results to be obtained presently, we record the expressions

$$\begin{aligned} &\boldsymbol{d}_N\cdot\boldsymbol{d}_M = d_{N.}^{\ \gamma}\,d_{M\gamma} + \sigma_{NM}, \qquad \sigma_{NM} = d_{N.}^{\ 3}\,d_{M3},\\ &\boldsymbol{d}_N\cdot\boldsymbol{d}_{M,\alpha} = d_{N.}^{\ \gamma}\,\lambda_{M\gamma\alpha} + \sigma_{NM\alpha}, \qquad \sigma_{NM\alpha} = d_{N.}^{\ 3}\,\lambda_{M3\alpha},\\ &\boldsymbol{d}_{N,\alpha}\cdot\boldsymbol{d}_{M,\beta} = \lambda_{N.\alpha}^{\ \gamma}\,\lambda_{M\gamma\beta} + \sigma_{NM\alpha\beta}, \qquad \sigma_{NM\alpha\beta} = \lambda_{N.\alpha}^{\ 3}\,\lambda_{M3\beta} \end{aligned} \tag{7.16}$$

and

$$\begin{aligned} &\boldsymbol{d}\cdot\boldsymbol{d} = \boldsymbol{d}_1\cdot\boldsymbol{d}_1 = d^\gamma\,d_\gamma + \sigma, \qquad \sigma = (d_3)^2,\\ &\boldsymbol{d}\cdot\boldsymbol{d}_{,\alpha} = \boldsymbol{d}_1\cdot\boldsymbol{d}_{1,\alpha} = d^\gamma\,\lambda_{\gamma\alpha} + \sigma_\alpha, \qquad \sigma_\alpha = d^3\,\lambda_{3\alpha},\\ &\boldsymbol{d}_{,\alpha}\cdot\boldsymbol{d}_{,\beta} = \boldsymbol{d}_{1,\alpha}\cdot\boldsymbol{d}_{1,\beta} = \lambda_{.\alpha}^{\gamma}\,\lambda_{\gamma\beta} + \sigma_{\alpha\beta}, \qquad \sigma_{\alpha\beta} = \lambda_{.\alpha}^{3}\,\lambda_{3\beta} \end{aligned} \tag{7.17}$$

for $N, M=1$.

We further introduce the kinematic variables

$$2e_{\alpha\beta} = a_{\alpha\beta} - A_{\alpha\beta}, \tag{7.18}$$

$$\begin{aligned} &\varkappa_{Ni\alpha} = \lambda_{Ni\alpha} - \Lambda_{Ni\alpha}, \qquad \gamma_{Ni} = d_{Ni} - D_{Ni},\\ &\varkappa_{i\alpha} = \varkappa_{1i\alpha} = \lambda_{i\alpha} - \Lambda_{i\alpha}, \qquad \gamma_i = \gamma_{1i} = d_i - D_i, \end{aligned} \tag{7.19}$$

where D_{Ni} and $\Lambda_{Ni\alpha}$, the reference values of d_{Ni} and $\lambda_{Ni\alpha}$, are given by

$$\begin{aligned} &D_{Ni} = \boldsymbol{A}_i\cdot\boldsymbol{D}_N, \qquad D_i = D_{1i} = \boldsymbol{A}_i\cdot\boldsymbol{D},\\ &\Lambda_{Ni\alpha} = \boldsymbol{A}_i\cdot\frac{\partial\boldsymbol{D}_N}{\partial\theta^\alpha}, \qquad \Lambda_{i\alpha} = \Lambda_{1i\alpha} = \boldsymbol{A}_i\cdot\frac{\partial\boldsymbol{D}}{\partial\theta^\alpha}, \end{aligned} \tag{7.20}$$

$\boldsymbol{D}$ is defined by $(7.9)_2$ and $\boldsymbol{A}_i = \{\boldsymbol{A}_\gamma, \boldsymbol{A}_3\}$ are the base vectors and the unit normal of the surface $\mathfrak{S}$ (i.e., $\xi=0$ in the reference configuration). We observe that the kinematic quantities d_{Ni} and $\lambda_{Ni\alpha}$, as well as those in (7.18)–(7.19), remain unaltered under superposed rigid body motions. This can be easily verified in a

[38] The kinematic variables (7.13)–(7.15) were employed by GREEN, LAWS and NAGHDI [1968, 4] and GREEN and NAGHDI [1970, 2] but the variables (7.16)–(7.17) were not explicitly introduced in these papers. The latter variables, which appear in (7.25)–(7.26), will subsequently bear on the structure of the constitutive equations derived in Sect. 17 from the three-dimensional theory.

manner similar to the invariance conditions expressed in (5.54). We record here the relative measures

$$\begin{aligned} \boldsymbol{d}_N \cdot \boldsymbol{d}_M - \boldsymbol{D}_N \cdot \boldsymbol{D}_M &= d_{N.}^{\gamma}\, d_{M\gamma} - D_{N.}^{\gamma}\, D_{M\gamma} + s_{NM}, \\ \boldsymbol{d}_N \cdot \boldsymbol{d}_{M,\alpha} - \boldsymbol{D}_N \cdot \boldsymbol{D}_{M,\alpha} &= d_{N.}^{\gamma}\, \lambda_{M\gamma\alpha} - D_{N.}^{\gamma}\, \Lambda_{M\gamma\alpha} + s_{NM\alpha}, \\ \boldsymbol{d}_{N,\alpha} \cdot \boldsymbol{d}_{M,\beta} - \boldsymbol{D}_{N,\alpha} \cdot \boldsymbol{D}_{M,\beta} &= \lambda_{N.\alpha}^{\gamma}\, \lambda_{M\gamma\beta} - \Lambda_{N.\alpha}^{\gamma}\, \Lambda_{M\gamma\beta} + s_{NM\alpha\beta}, \end{aligned} \tag{7.21}$$

where

$$\begin{gathered} s_{NM} = d_{N.}^{3}\, d_{M3} - D_{N.}^{3}\, D_{M3}, \quad s_{NM\alpha} = d_{N.}^{3}\, \lambda_{M3\alpha} - D_{N.}^{3}\, \Lambda_{M3\alpha}, \\ s_{NM\alpha\beta} = \lambda_{N.\alpha}^{3}\, \lambda_{M3\beta} - \Lambda_{N.\alpha}^{3}\, \Lambda_{M3\beta} \end{gathered} \tag{7.22}$$

and note that the expressions corresponding to (7.21)–(7.22) for $N, M = 1$, in the notations of $(7.9)_1$, are:

$$\begin{aligned} \boldsymbol{d} \cdot \boldsymbol{d} - \boldsymbol{D} \cdot \boldsymbol{D} &= (d^{\gamma}\, d_{\gamma} - D^{\gamma}\, D_{\gamma}) + s, \\ \boldsymbol{d} \cdot \boldsymbol{d}_{,\alpha} - \boldsymbol{D} \cdot \boldsymbol{D}_{,\alpha} &= (d^{\gamma}\, \lambda_{\gamma\alpha} - D^{\gamma}\, \Lambda_{\gamma\alpha}) + s_{\alpha}, \\ \boldsymbol{d}_{,\alpha} \cdot \boldsymbol{d}_{,\beta} - \boldsymbol{D}_{,\alpha} \cdot \boldsymbol{D}_{,\beta} &= (\lambda^{\gamma}_{.\alpha}\, \lambda_{\gamma\beta} - \Lambda^{\gamma}_{.\alpha}\, \Lambda_{\gamma\beta}) + s_{\alpha\beta} \end{aligned} \tag{7.23}$$

and

$$s = (d_3)^2 - (D_3)^2, \quad s_{\alpha} = d^3\, \lambda_{3\alpha} - D^3\, \Lambda_{3\alpha}, \quad s_{\alpha\beta} = \lambda^{3}_{.\alpha}\, \lambda_{3\beta} - \Lambda^{3}_{.\alpha}\, \Lambda_{3\beta}. \tag{7.24}$$

For later reference, it is desirable to calculate the components of a three-dimensional strain measure in terms of the variables (7.13) referred to the base vectors $\boldsymbol{a}^i$. For this purpose, using (7.10), we first record the expressions for the components g_{ij} as follows:

$$\begin{aligned} g_{\alpha\beta} &= \left(\boldsymbol{a}_{\alpha} + \sum_{N=1}^{\infty} \xi^N \frac{\partial \boldsymbol{d}_N}{\partial \theta^{\alpha}}\right) \cdot \left(\boldsymbol{a}_{\beta} + \sum_{M=1}^{\infty} \xi^M \frac{\partial \boldsymbol{d}_M}{\partial \theta^{\beta}}\right) \\ &= a_{\alpha\beta} + \sum_{N=1}^{\infty} \xi^N \left(\boldsymbol{a}_{\beta} \cdot \frac{\partial \boldsymbol{d}_N}{\partial \theta^{\alpha}} + \boldsymbol{a}_{\alpha} \cdot \frac{\partial \boldsymbol{d}_N}{\partial \theta^{\beta}}\right) + \sum_{P=2}^{\infty} \xi^P \sum_{M=1}^{P-1} \frac{\partial \boldsymbol{d}_{P-M}}{\partial \theta^{\alpha}} \cdot \frac{\partial \boldsymbol{d}_M}{\partial \theta^{\beta}} \\ &= a_{\alpha\beta} + \xi \left(\boldsymbol{a}_{\beta} \cdot \frac{\partial \boldsymbol{d}_1}{\partial \theta^{\alpha}} + \boldsymbol{a}_{\alpha} \cdot \frac{\partial \boldsymbol{d}_1}{\partial \theta^{\beta}}\right) \\ &\quad + \sum_{P=2}^{\infty} \xi^P \left[\boldsymbol{a}_{\beta} \cdot \frac{\partial \boldsymbol{d}_P}{\partial \theta^{\alpha}} + \boldsymbol{a}_{\alpha} \cdot \frac{\partial \boldsymbol{d}_P}{\partial \theta^{\beta}} + \sum_{M=1}^{P-1} \left(\frac{\partial \boldsymbol{d}_{P-M}}{\partial \theta^{\alpha}} \cdot \frac{\partial \boldsymbol{d}_M}{\partial \theta^{\beta}}\right)\right], \\ g_{\alpha 3} &= \left(\boldsymbol{a}_{\alpha} + \sum_{N=1}^{\infty} \xi^N \frac{\partial \boldsymbol{d}_N}{\partial \theta^{\alpha}}\right) \cdot \left(\sum_{M=1}^{\infty} M\, \xi^{M-1}\, \boldsymbol{d}_M\right) \\ &= \boldsymbol{a}_{\alpha} \cdot \boldsymbol{d}_1 + \sum_{P=2}^{\infty} \xi^{P-1} \left[P\, \boldsymbol{a}_{\alpha} \cdot \boldsymbol{d}_P + \sum_{M=1}^{P-1} M\, \frac{\partial \boldsymbol{d}_{P-M}}{\partial \theta^{\alpha}} \cdot \boldsymbol{d}_M\right], \\ g_{33} &= \left(\sum_{N=1}^{\infty} N\, \xi^{N-1}\, \boldsymbol{d}_N\right) \cdot \left(\sum_{M=1}^{\infty} M\, \xi^{M-1}\, \boldsymbol{d}_M\right) \\ &= \sum_{P=2}^{\infty} \xi^{P-2} \left(\sum_{M=1}^{P-1} (P-M)\, M\, \boldsymbol{d}_{P-M} \cdot \boldsymbol{d}_M\right). \end{aligned} \tag{7.25}$$

A close examination of the series in each of the second expressions of $(7.25)_{1,2}$ reflects a particular form for the dependence of $g_{\alpha\beta}$ and $g_{\alpha 3}$ on $\boldsymbol{d}_N$ and $\partial \boldsymbol{d}_N / \partial \theta^{\alpha}$. To see this more easily, we write the right-hand side of each of $(7.25)_{1,2,3}$ in expanded form indicating explicitly all terms involving only $\boldsymbol{d}_1$ and $\partial \boldsymbol{d}_1 / \partial \theta^{\alpha}$.

Thus, with the notations of $(7.9)_1$, (7.15) and (7.17):

$$\begin{aligned} g_{\alpha\beta} &= a_{\alpha\beta} + \xi\left(\boldsymbol{a}_\beta \cdot \frac{\partial \boldsymbol{d}}{\partial \theta^\alpha} + \boldsymbol{a}_\alpha \cdot \frac{\partial \boldsymbol{d}}{\partial \theta^\beta}\right) + \xi^2 \frac{\partial \boldsymbol{d}}{\partial \theta^\alpha} \cdot \frac{\partial \boldsymbol{d}}{\partial \theta^\beta} + \cdots \\ &= a_{\alpha\beta} + \xi(\lambda_{\beta\alpha} + \lambda_{\alpha\beta}) + \xi^2(\lambda^{\nu}_{.\alpha}\,\lambda_{\nu\beta} + \sigma_{\alpha\beta}) + \cdots, \\ g_{\alpha 3} &= \boldsymbol{a}_\alpha \cdot \boldsymbol{d} + \xi\, \boldsymbol{d} \cdot \frac{\partial \boldsymbol{d}}{\partial \theta^\alpha} + \cdots \\ &= d_\alpha + \xi(d^\nu \lambda_{\nu\alpha} + \sigma_\alpha) + \cdots, \\ g_{33} &= \boldsymbol{d} \cdot \boldsymbol{d} + \cdots = d^\gamma d_\gamma + \sigma + \cdots, \end{aligned} \tag{7.26}$$

which readily reveal the manner that the components of $\boldsymbol{d} = \boldsymbol{d}_1$ and $\boldsymbol{d}_{,\alpha}$ contribute to the components of g_{ij}. Results similar to (7.25) for the components G_{ij} can be calculated with the help of (7.2). We now recall the formula[39]

$$\gamma^*_{ij} = \tfrac{1}{2}(\boldsymbol{g}_i \cdot \boldsymbol{g}_j - \boldsymbol{G}_i \cdot \boldsymbol{G}_j) = \tfrac{1}{2}(g_{ij} - G_{ij}), \tag{7.27}$$

where γ^*_{ij} are the covariant components of a strain measure in the three-dimensional (nonlinear) theory. Using (7.25) and the corresponding expressions for G_{ij}, the components of γ^*_{ij} may be expressed in the following forms:

$$\begin{aligned} 2\gamma^*_{\alpha\beta} &= 2e_{\alpha\beta} + \xi(\varkappa_{\beta\alpha} + \varkappa_{\alpha\beta}) \\ &\quad + \sum_{P=2}^{\infty} \xi^P \left[\varkappa_{P\beta\alpha} + \varkappa_{P\alpha\beta} + \sum_{M=1}^{P-1} (\boldsymbol{d}_{(P-M),\alpha} \cdot \boldsymbol{d}_{M,\beta} - \boldsymbol{D}_{(P-M),\alpha} \cdot \boldsymbol{D}_{M,\beta})\right], \\ 2\gamma^*_{\alpha 3} &= \gamma_\alpha + \sum_{P=2}^{\infty} \xi^{P-1}\left[P\gamma_{P\alpha} + \sum_{M=1}^{P-1} M(\boldsymbol{d}_{(P-M),\alpha} \cdot \boldsymbol{d}_M - \boldsymbol{D}_{(P-M),\alpha} \cdot \boldsymbol{D}_M)\right], \\ 2\gamma^*_{33} &= \sum_{P=2}^{\infty} \xi^{P-2}\left[\sum_{M=1}^{P-1} (P-M)\, M(\boldsymbol{d}_{(P-M)} \cdot \boldsymbol{d}_M - \boldsymbol{D}_{(P-M)} \cdot \boldsymbol{D}_M)\right]. \end{aligned} \tag{7.28}$$

Again, in order to give an idea regarding the manner of dependence of γ^*_{ij} on the components of $\boldsymbol{d}_N$ and $\partial \boldsymbol{d}_N/\partial\theta^\alpha$, we rewrite (7.28) in expanded form similar to (7.26) indicating explicitly only terms due to $\boldsymbol{d} = \boldsymbol{d}_1$, $\boldsymbol{d}_{1,\alpha} = \boldsymbol{d}_{,\alpha}$ and their reference values. Thus

$$\begin{aligned} 2\gamma^*_{\alpha\beta} &= 2e_{\alpha\beta} + \xi(\varkappa_{\beta\alpha} + \varkappa_{\alpha\beta}) + \xi^2[(\lambda^{\nu}_{.\alpha}\,\lambda_{\nu\beta} - \Lambda^{\nu}_{.\alpha}\,\Lambda_{\nu\beta}) + s_{\alpha\beta}] + \cdots, \\ 2\gamma^*_{\alpha 3} &= \gamma_\alpha + \xi[(d^\nu \lambda_{\nu\alpha} - D^\nu \Lambda_{\nu\alpha}) + s_\alpha] + \cdots, \\ 2\gamma^*_{33} &= (d^\gamma d_\gamma - D^\gamma D_\gamma) + s + \cdots, \end{aligned} \tag{7.29}$$

where all terms in (7.29) not explicitly recorded arise from $\boldsymbol{d}_N$ and $\boldsymbol{D}_N$ for $N \geqq 2$.

The kinematic variables (7.18)–(7.19) and (7.21)–(7.24) which occur in (7.28) are independent of ξ and thus can be referred to as two-dimensional variables. Given (7.1), the two-dimensional kinematic quantities may be regarded as an exact characterization of the kinematics of shells; however, (7.18)–(7.19) and (7.21)–(7.22) form an infinite set of variables and this is an undesirable feature of such characterizations. It is therefore clear that the introduction of suitable approximations is necessary in order to obtain useful measures of deformation for shells in line with the objective stated under (a) in (4.32). This requires additional elaboration but we postpone further comments and take up the question of suitable approximation later in this section and again in Chap. D.

[39] See, e.g., GREEN and ZERNA [1968, 9].

β) *Some results valid in a reference configuration.* Before proceeding further, we dispose of some additional results which are independent of linearization, although their utility in a linear theory is particularly significant. First, we observe that the convected coordinates θ^i can always be so chosen in the reference configuration that $\boldsymbol{D}_N = 0$ for $N \geqq 2$. Hence, instead of (7.2), without loss in generality we may write the position vector in the reference configuration of $\mathscr{B}$ as

$$\boldsymbol{P} = \boldsymbol{R}(\theta^\alpha) + \xi\, \boldsymbol{D}(\theta^\alpha), \tag{7.30}$$

with $\boldsymbol{D}$ specified by

$$D_\alpha = 0, \quad D = D_3, \quad \boldsymbol{D} = D\boldsymbol{A}_3. \tag{7.31}$$

From (4.23) and (7.30)–(7.31)$_3$, it follows that the base vectors and the metric tensor in the (initial) reference configuration are

$$\begin{gathered} \boldsymbol{G}_\alpha = \boldsymbol{A}_\alpha + \xi\, \boldsymbol{D}_{,\alpha} = \nu_\alpha^\gamma \boldsymbol{A}_\gamma + \xi\, D_{,\alpha} \boldsymbol{A}_3, \quad \boldsymbol{G}_3 = D\boldsymbol{A}_3, \\ G_{\alpha\beta} = \nu_\alpha^\gamma\, \nu_\beta^\delta\, A_{\gamma\delta} + \xi^2\, D_{,\alpha}\, D_{,\beta}, \quad G_{\alpha 3} = \xi D D_{,\alpha} = \tfrac{1}{2}\, \xi\, (D^2)_{,\alpha}, \quad G_{33} = D^2, \end{gathered} \tag{7.32}$$

where

$$\nu_\alpha^\gamma = \delta_\alpha^\gamma - \xi\, D\, B_\alpha^\gamma. \tag{7.33}$$

We note for later reference that

$$\nu = D \det(\nu_\alpha^\gamma) = \left(\frac{G}{A}\right)^{\frac{1}{2}} = D[1 - 2\xi\, DH + \xi^2\, D^2\, K], \tag{7.34}$$

with H and K being the mean curvature and the Gaussian curvature of the surface $\xi = 0$ in the reference configuration defined by

$$\begin{aligned} H &= \tfrac{1}{2} B_\alpha^\alpha, \\ K &= \det(B_\beta^\alpha) = A^{-1} \det(B_{\alpha\beta}) = B_1^1\, B_2^2 - B_2^1\, B_1^2. \end{aligned} \tag{7.35}$$

In view of (7.31) and our choice (7.30), the functions $\Lambda_{Ni\alpha}$ in (7.20) reduce to

$$\Lambda_{\beta\alpha} = \Lambda_{1\beta\alpha} = -D\, B_{\alpha\beta}, \quad \Lambda_{3\alpha} = \Lambda_{13\alpha} = D_{,\alpha}, \quad \Lambda_{Ni\alpha} = 0 \quad (N \geqq 2). \tag{7.36}$$

Since the convected coordinates θ^i may be identified with the normal coordinates (4.25) in the reference configuration of $\mathscr{B}$, the initial position vector can also be taken in the form (4.27) instead of (7.30). Recalling our notations for the base vectors and the metric tensor associated with the coordinates (4.25), we have

$$\begin{aligned} \boldsymbol{G}'_\alpha &= \mu_\alpha^\gamma\, \boldsymbol{A}_\gamma, & \boldsymbol{G}'_3 &= \boldsymbol{A}_3, \\ G'_{\alpha\beta} &= \mu_\alpha^\gamma\, \mu_\beta^\delta\, A_{\gamma\delta}, & G'_{\alpha 3} &= 0, \quad G'_{33} = 1, \end{aligned} \tag{7.37}$$

where (4.27) has been used and where

$$\mu_\alpha^\gamma = \delta_\alpha^\gamma - \zeta\, B_\alpha^\gamma \tag{7.38}$$

and

$$\mu = \det(\mu_\alpha^\gamma) = \left(\frac{G'}{A}\right)^{\frac{1}{2}} = 1 - 2\zeta H + \zeta^2 K. \tag{7.39}$$

From comparison of (4.27) and (7.30) with $\boldsymbol{D}$ specified by (7.31), we have

$$y^\alpha = \theta^\alpha, \quad \zeta = D\, \xi \tag{7.40}$$

as the transformation relations between the coordinate system y^i or (4.25) and θ^i in the reference configuration. Moreover, under the transformations (7.40), from (7.37)–(7.39) and (7.32)–(7.34) follow the relations

$$\mu^{\alpha}_{\beta} = \nu^{\alpha}_{\beta}, \quad \mu = \frac{\nu}{D}. \tag{7.41}$$

It may be noted here that the metric tensors G'_{ij} and G_{ij} become identical when evaluated on the surface $\zeta = 0$ (or $\xi = 0$) in the reference configuration and both reduce to (4.29).

With reference to the condition (4.20), recall that k is independent of time and suppose further that k is also independent of ξ. Then, for $k = k(\theta^{\alpha})$, (4.20) gives

$$\int_{\alpha}^{\beta} k\,\xi\,d\xi = \frac{k}{2}(\beta^2 - \alpha^2) = 0 \Rightarrow \alpha = -\beta, \tag{7.42}$$

since $\alpha < 0$. By (4.21) and (7.34), the mass density in the reference configuration can be written in the form

$$\varrho_0^* = k\,G^{-\frac{1}{2}} = \frac{k}{D A^{\frac{1}{2}}[1 - 2\xi\, D H + \xi^2 D^2 K]}, \qquad k = k(\theta^{\alpha}). \tag{7.43}$$

In the case of an initially flat plate, for which both H and K vanish, the expression $(7.43)_1$ is independent of ξ and satisfies (4.20) and therefore (7.42) exactly. For initially curved shells, on the other hand, the mass density ϱ_0^* depends on ξ or ζ in view of $(7.40)_2$. In terms of the principal radii of curvature of the surface $\xi = 0$ in the (initial) reference configuration,

$$H = -\frac{1}{2}\left(\frac{1}{R_1} + \frac{1}{R_2}\right), \qquad K = \frac{1}{R_1 R_2} \tag{7.44}$$

and $(7.43)_1$ can be expressed as

$$\varrho_0^* = \frac{k}{D A^{\frac{1}{2}}\left[1 + \left(\frac{\zeta}{R_1} + \frac{\zeta}{R_2}\right) + \left(\frac{\zeta}{R_1}\right)\left(\frac{\zeta}{R_2}\right)\right]}, \tag{7.45}$$

where we have also used $(7.40)_2$. Recalling the notation of $(4.28)_2$, we may invoke (4.31) for sufficiently *thin* shells and neglect ζ/R in comparison with unity in the denominator in (7.45). Thus, if we approximate (7.45) by

$$\varrho_0^* \cong \frac{k}{D A^{\frac{1}{2}}}, \tag{7.46}$$

ϱ_0^* is independent of ζ (or ξ) and the condition (7.42) is also satisfied to the order of approximation used in the sense of (4.31).

With the above approximation for the initial mass density and with $\boldsymbol{P}$ specified by (7.30)–(7.31), it follows from $(7.42)_2$ that the boundary surfaces (4.14) can be regarded as symmetrically situated [approximately in the sense of (4.31)] about the surface $\xi = 0$ in the reference configuration of the initially curved shells. Hence, the limits of integration in such integrals as (4.17) and (4.20) can be replaced by $-\beta, \beta$. Also, in view of (4.29), from $(7.40)_2$ and $(7.42)_2$ follows

$$D\beta = \frac{h}{2}, \tag{7.47}$$

which relates β to D and h. In particular if $D = 1$ so that $(7.31)_3$ becomes $\boldsymbol{D} = \boldsymbol{A}_3$, the distinction between ξ and ζ in (7.30) and (4.27) disappears and we have

$$\begin{gathered} D = 1, \quad \xi = \zeta, \quad \beta = -\alpha = \frac{h}{2}, \\ \nu^{\gamma}_{\alpha} = \mu^{\gamma}_{\alpha}, \quad \mu = \nu. \end{gathered} \tag{7.48}$$

γ) *Linearized kinematics.* We now proceed to obtain the linearized version of the foregoing kinematical results. Since the linearization procedure is analogous to that employed in Sect. 6, details will not be given; but we note that a vertical bar, in the linearized expressions, will denote covariant differentiation with respect to $A_{\alpha\beta}$. Also, in the remainder of this section, instead of (7.2) we specify the position vector in the reference configuration of $\mathscr{B}$ by (7.30)–(7.31) so that the reference values $(7.20)_{1,3}$ for $N \geqq 2$ become[40]

$$D_{Ni} = 0, \qquad A_{Ni\alpha} = 0 \qquad (N \geqq 2). \tag{7.49}$$

Let

$$\boldsymbol{p} = \boldsymbol{P} + \varepsilon\, \boldsymbol{u}^*, \qquad \boldsymbol{u}^* = u^{*i} \boldsymbol{A}_i, \qquad \boldsymbol{v}^* = \varepsilon\, \dot{\boldsymbol{u}}^* \tag{7.50}$$

and put

$$\boldsymbol{u}^* = \boldsymbol{u}^*(\theta^\alpha, \xi, t) = \boldsymbol{u}(\theta^\alpha, t) + \sum_{N=1}^{\infty} \xi^N \boldsymbol{\delta}_N(\theta^\alpha, t). \tag{7.51}$$

Introduction of (7.51) into (7.50), together with (7.1) and (7.30), results in

$$\begin{aligned} &\boldsymbol{r} = \boldsymbol{R} + \varepsilon\, \boldsymbol{u}, && \boldsymbol{u} = u_i \boldsymbol{A}^i = u^i \boldsymbol{A}_i, && \boldsymbol{v} = \varepsilon\, \dot{\boldsymbol{u}}, \\ &\boldsymbol{d} = \boldsymbol{d}_1 = \boldsymbol{D} + \varepsilon\, \boldsymbol{\delta}, && \boldsymbol{\delta} = \boldsymbol{\delta}_1 = \delta_i \boldsymbol{A}^i, && \boldsymbol{w} = \boldsymbol{w}_1 = \varepsilon\, \dot{\boldsymbol{\delta}}, \\ &\boldsymbol{d}_N = \varepsilon\, \boldsymbol{\delta}_N, && \boldsymbol{\delta}_N = \delta_{Ni} \boldsymbol{A}^i, && \boldsymbol{w}_N = \varepsilon\, \dot{\boldsymbol{\delta}}_N \quad (N \geqq 2). \end{aligned} \tag{7.52}$$

The base vectors $\boldsymbol{G}_i$ are given by $(7.32)_{1,2}$ and the base vectors $\boldsymbol{g}_i$, by $(4.7)_1$, $(4.23)_1$ and (7.52), can be expressed as

$$\begin{aligned} \boldsymbol{g}_i &= \boldsymbol{G}_i + \varepsilon\, \boldsymbol{u}^*_{,i}, \\ \boldsymbol{g}_\alpha &= (\boldsymbol{A}_\alpha + \varepsilon\, \boldsymbol{u}_{,\alpha}) + \xi \frac{\partial}{\partial \theta^\alpha} (\boldsymbol{D} + \varepsilon\, \boldsymbol{\delta}) + \sum_{N=2}^{\infty} \xi^N \frac{\partial}{\partial \theta^\alpha} (\varepsilon\, \boldsymbol{\delta}_N), \\ \boldsymbol{g}_3 &= (\boldsymbol{D} + \varepsilon\, \boldsymbol{\delta}) + \sum_{N=2}^{\infty} N \xi^{N-1} (\varepsilon\, \boldsymbol{\delta}_N). \end{aligned} \tag{7.53}$$

The above formulae on the surface $\xi = 0$ become

$$\begin{aligned} \boldsymbol{a}_\alpha = \boldsymbol{g}_\alpha(\theta^\gamma, 0, t) &= \boldsymbol{G}_\alpha(\theta^\gamma, 0) + \varepsilon\, \boldsymbol{u}^*_{,\alpha}(\theta^\gamma, 0, t) \\ &= \boldsymbol{A}_\alpha(\theta^\gamma) + \varepsilon\, \boldsymbol{u}_{,\alpha}(\theta^\gamma, t), \\ \boldsymbol{d} = \boldsymbol{g}_3(\theta^\gamma, 0, t) &= \boldsymbol{G}_3(\theta^\gamma, 0) + \varepsilon\, \boldsymbol{u}^*_{,3}(\theta^\gamma, 0, t) \\ &= \boldsymbol{D}(\theta^\gamma) + \varepsilon\, \boldsymbol{\delta}(\theta^\gamma, t), \end{aligned} \tag{7.54}$$

in view of (4.9) and (7.11).

We say $(7.50)_1$–(7.51) characterize the infinitesimal motion of the three-dimensional continuum $\mathscr{B}$ if the magnitude of $\boldsymbol{u}^*$ and all its derivatives are bounded by 1 and if (6.3) holds. Also, we use the order symbol $O(\varepsilon^n)$ in the sense of (6.4). The components of the vector functions $\boldsymbol{u}$ and $\boldsymbol{\delta}_N$ in (7.52) are referred to the base vectors $\boldsymbol{A}_i$ or $\boldsymbol{A}^i$ and subsequently the components of $\boldsymbol{u}_{,\alpha}$ and $\boldsymbol{\delta}_{N,\alpha}$ will be defined with reference to $\boldsymbol{A}_i$. On the other hand, the components of the vector functions $\boldsymbol{d}_N$ and $\boldsymbol{d}_{N,\alpha}$ are defined with respect to the base vectors $\boldsymbol{a}_i$ or $\boldsymbol{a}^i$ [see (7.13)]. For this reason, in the sequel, we need to have the approximate expressions for $\boldsymbol{a}_i$ to $O(\varepsilon)$. We have already obtained the desired approximation for $\boldsymbol{a}_\alpha$ to $O(\varepsilon)$ in (7.54) and the corresponding expression for the unit normal

[40] As noted already, there is no loss of generality in writing the reference position vector in the form (7.30). Later, in order to simplify some of the calculations, we also take $D = 1$ but the various formulae between (7.50)–(7.57) hold with $\boldsymbol{D}$ specified by (7.31).

$\boldsymbol{a}_3$ to $\mathfrak{s}$ can be deduced from $(7.54)_1$ and $(4.11)_{1,2}$. Thus,

$$\begin{gathered}\boldsymbol{a}_3=\boldsymbol{A}_3+\varepsilon\boldsymbol{\beta}+O(\varepsilon^2),\\ \boldsymbol{\beta}=\beta_\alpha\boldsymbol{A}^\alpha=\beta^\alpha\boldsymbol{A}_\alpha,\quad \beta_\alpha=-(u_{3,\alpha}+B_\alpha^\lambda u_\lambda),\end{gathered}\tag{7.55}$$

which together with $(7.54)_1$ can be used to rewrite (7.18)–(7.19) in terms of $\varepsilon\boldsymbol{u}$, $\varepsilon\boldsymbol{\beta}$, $\varepsilon\boldsymbol{\delta}_N$ and their derivatives. For example, the functions γ_{Ni} in $(7.19)_{2,4}$ can be written as

$$\begin{gathered}\gamma_\alpha=\varepsilon(\boldsymbol{A}_\alpha\cdot\boldsymbol{\delta}+\boldsymbol{u}_{,\alpha}\cdot\boldsymbol{D})+O(\varepsilon^2)=O(\varepsilon),\\ \gamma_3=\varepsilon(\boldsymbol{A}_3\cdot\boldsymbol{\delta}+\boldsymbol{\beta}\cdot\boldsymbol{D})+O(\varepsilon^2)=O(\varepsilon),\\ \gamma_{Ni}=\varepsilon(\boldsymbol{A}_i\cdot\boldsymbol{\delta}_N)+O(\varepsilon^2)=O(\varepsilon)\quad(N\geqq 2).\end{gathered}\tag{7.56}$$

Also, the variables $\varkappa_{i\alpha}$ and those in (7.24) upon linearization become

$$\begin{aligned}\varkappa_{\beta\alpha}&=\varepsilon(\boldsymbol{u}_{,\beta}\cdot\boldsymbol{D}_{,\alpha}+\boldsymbol{A}_\beta\cdot\boldsymbol{\delta}_{,\alpha})+O(\varepsilon^2)=O(\varepsilon),\\ \varkappa_{3\alpha}&=\varepsilon(\boldsymbol{A}_3\cdot\boldsymbol{\delta}_{,\alpha}+\boldsymbol{\beta}\cdot\boldsymbol{D}_{,\alpha})+O(\varepsilon^2)=O(\varepsilon),\\ s&=2\varepsilon[D(\boldsymbol{A}_3\cdot\boldsymbol{\delta}+\boldsymbol{\beta}\cdot\boldsymbol{D})]+O(\varepsilon^2)=O(\varepsilon),\\ s_\alpha&=\varepsilon[D(\boldsymbol{A}_3\cdot\boldsymbol{\delta}_{,\alpha}+\boldsymbol{\beta}\cdot\boldsymbol{D}_{,\alpha})+\Lambda_{3\alpha}(\boldsymbol{A}_3\cdot\boldsymbol{\delta}+\boldsymbol{\beta}\cdot\boldsymbol{D})+O(\varepsilon^2)=O(\varepsilon),\\ s_{\alpha\beta}&=\varepsilon[\Lambda_{3\alpha}(\boldsymbol{A}_3\cdot\boldsymbol{\delta}_{,\beta}+\boldsymbol{\beta}\cdot\boldsymbol{D}_{,\beta})+\Lambda_{3\beta}(\boldsymbol{A}_3\cdot\boldsymbol{\delta}_{,\alpha}+\boldsymbol{\beta}\cdot\boldsymbol{D}_{,\alpha})]+O(\varepsilon^2)=O(\varepsilon)\end{aligned}\tag{7.57}$$

and the expressions for other quantities in (7.21)–(7.23) may be put in similar forms.

In what follows, in order to achieve some simplification in otherwise lengthy and elaborate calculations, we confine attention to the case corresponding to $D=1$ as in (7.48). Thus, with $\boldsymbol{D}=\boldsymbol{A}_3$, in addition to the reference values (7.49) we now also have

$$D_\alpha=0,\quad D=1,\quad \Lambda_{\beta\alpha}=-B_{\beta\alpha},\quad \Lambda_{3\alpha}=0.\tag{7.58}$$

The simplification resulting from (7.58) is rather substantial. For example, the first term in $(7.56)_2$ reduces to $\varepsilon(\boldsymbol{A}_3\cdot\boldsymbol{\delta})$. Similarly, in view of $(7.58)_{3,4}$, the last three of (7.57) now become $s=2\gamma_3$, $s_\alpha=\varkappa_{3\alpha}$ and $s_{\alpha\beta}=0$, respectively. Now remembering the remarks made in Sect. 6 [between (6.10) and (6.13)] concerning the linearization process, again we retain only terms of $O(\varepsilon)$; and after the approximations, without loss of generality, we set $\varepsilon=1$ in order to avoid the introduction of additional notations. In this way, the linearized kinematic measures resulting from (7.18)–(7.19) are

$$\begin{gathered}e_{\alpha\beta}=\tfrac{1}{2}(u_{\alpha|\beta}+u_{\beta|\alpha})-B_{\alpha\beta}u_3,\\ \gamma_\alpha=\delta_\alpha-\beta_\alpha,\quad \gamma_3=\delta_3,\\ \varkappa_{\beta\alpha}=\delta_{\beta|\alpha}-B_{\alpha\beta}\delta_3-B_\alpha^\nu(u_{\nu|\beta}-B_{\nu\beta}u_3),\\ \varkappa_{3\alpha}=\delta_{3,\alpha}+B_\alpha^\nu(\delta_\nu-\beta_\nu),\\ \gamma_{N\alpha}=\delta_{N\alpha},\quad \gamma_{N3}=\delta_{N3}\quad(N\geqq 2),\\ \varkappa_{N\beta\alpha}=\delta_{N\beta|\alpha}-B_{\beta\alpha}\delta_{N3},\quad \varkappa_{N3\alpha}=\delta_{N3,\alpha}+B_\alpha^\beta\delta_{N\beta}\quad(N\geqq 2),\end{gathered}\tag{7.59}$$

where the vertical bar now stands for covariant differentiation with respect to $A_{\alpha\beta}$.

The infinitesimal (three-dimensional) strain tensor, resulting from linearization of (7.27), is given by

$$\gamma_{ij}^*=\tfrac{1}{2}(\boldsymbol{G}_i\cdot\boldsymbol{u}_{,j}^*+\boldsymbol{G}_j\cdot\boldsymbol{u}_{,i}^*).\tag{7.60}$$

The relationship between the two-dimensional kinematic variables (7.59) and the infinitesimal (three-dimensional) strain tensor can be found either by linearization of (7.28), remembering also (7.49) and (7.58), or directly from (7.60) and (7.51). Here, we record the final results:

$$
\begin{aligned}
2\gamma^*_{\alpha\beta} &= 2e_{\alpha\beta} + \xi(\varkappa_{\alpha\beta} + \varkappa_{\beta\alpha}) - 2\xi^2 B^\nu_\beta B^\mu_\alpha e_{\mu\nu} + \sum_{N=2}^{\infty} \xi^N (\varkappa_{N\alpha\beta} + \varkappa_{N\beta\alpha}) \\
&\quad - \sum_{N=1}^{\infty} \xi^{N+1} [B^\gamma_\beta \varkappa_{N\gamma\alpha} + B^\gamma_\alpha \varkappa_{N\gamma\beta}], \\
2\gamma^*_{\alpha 3} &= 2\gamma^*_{3\alpha} = \gamma_\alpha + \xi(\varkappa_{3\alpha} - B^\nu_\alpha \gamma_\nu) \\
&\quad + \sum_{N=2}^{\infty} \xi^{N-1} [N \gamma_{N\alpha} + \xi \varkappa_{N3\alpha} - N\xi B^\nu_\alpha \gamma_{N\nu}], \\
\gamma^*_{33} &= \gamma_3 + \sum_{N=2}^{\infty} N \xi^{N-1} \gamma_{N3}.
\end{aligned}
\tag{7.61}
$$

The complexity of the expressions (7.61) for γ^*_{ij} is perhaps suggestive of the degree of the difficulties that can be encountered even in the development of the linear theory. However, prior to consideration of an approximation scheme, it is instructive to specialize (7.59) and (7.61) to the case of an initially flat plate for which (6.26) holds. Thus, for an initially flat plate, the kinematic measures (7.59) reduce to

$$
\begin{gathered}
e_{\alpha\beta} = \tfrac{1}{2}(u_{\alpha|\beta} + u_{\beta|\alpha}), \\
\gamma_\alpha = \delta_\alpha - \beta_\alpha, \quad \gamma_3 = \delta_3, \quad \beta_\alpha = -u_{3,\alpha}, \\
\varkappa_{\beta\alpha} = \gamma_{\beta|\alpha} - u_{3|\alpha\beta}, \quad \varkappa_{3\alpha} = \gamma_{3,\alpha}, \\
\gamma_{N\alpha} = \delta_{N\alpha}, \quad \gamma_{N3} = \delta_{N3} \quad (N \geqq 2), \\
\varkappa_{N\beta\alpha} = \gamma_{N\beta|\alpha}, \quad \varkappa_{N3\alpha} = \gamma_{N3,\alpha} \quad (N \geqq 2)
\end{gathered}
\tag{7.62}
$$

and the expressions in (7.61) assume the forms

$$
\begin{aligned}
\gamma^*_{\alpha\beta} &= e_{\alpha\beta} + \xi \varkappa_{(\alpha\beta)} + \tfrac{1}{2} \sum_{N=2}^{\infty} \xi^N (\varkappa_{N\alpha\beta} + \varkappa_{N\beta\alpha}), \\
\gamma^*_{\alpha 3} &= \gamma^*_{3\alpha} = \tfrac{1}{2}\left[\gamma_\alpha + \sum_{N=2}^{\infty} N \xi^{N-1} \gamma_{N\alpha} + \sum_{N=1}^{\infty} \xi^N \varkappa_{N3\alpha}\right] \\
&= \tfrac{1}{2}\left[\sum_{N=1}^{\infty} \xi^{N-1} (N \gamma_{N\alpha} + \xi \varkappa_{N3\alpha})\right], \\
\gamma^*_{33} &= \gamma_3 + \sum_{N=2}^{\infty} N \xi^{N-1} \gamma_{N3} = \sum_{N=1}^{\infty} N \xi^{N-1} \gamma_{N3},
\end{aligned}
\tag{7.63}
$$

where the notation $\varkappa_{(\alpha\beta)}$ stands for

$$
\varkappa_{(\alpha\beta)} = \tfrac{1}{2}(\varkappa_{\alpha\beta} + \varkappa_{\beta\alpha}) = \tfrac{1}{2}(\gamma_{\alpha|\beta} + \gamma_{\beta|\alpha}) - u_{3|\alpha\beta}
\tag{7.64}
$$

and conforms to that introduced in (5.12).

The above kinematical results for initially flat plates, although simpler than (7.61), still involve two infinite sets of variables δ_{Ni} and $\varkappa_{Ni\alpha}$. Nevertheless, some observations may be made regarding the structure of γ^*_{ij} in (7.63). The components $\gamma^*_{\alpha\beta}$ are independent of δ_{Ni} and $\varkappa_{N3\alpha}$ while the components γ^*_{i3} are independent of $\varkappa_{N\beta\alpha}$. This uncoupling of the effects of δ_{Ni}, $\varkappa_{N3\alpha}$ and $\varkappa_{N\beta\alpha}$ immediately suggests consideration of an approximate theory for thin plates in which an approximate expression for $\gamma^*_{\alpha\beta}$, obtained from $(7.63)_1$ by ignoring $\varkappa_{N\beta\alpha}(N \geqq 2)$, plays a dominant role.[41] We return to this problem of approximations later.

δ) *Approximate linearized kinematic measures.* The complexity of the foregoing linearized kinematic measures, especially for shells, clearly indicates the need for a *suitable* approximative scheme which would make possible the development of a complete theory which is both useful and manageable. However, in general, the introduction of an approximation for the kinematic quantities alone (and separate from the rest of the theory) is difficult to justify, since any approximation introduced here must be compatible with the entire theory including all field equations and the constitutive relations.[42] Thus, strictly speaking, the introduction of any approximative scheme should be postponed until a complete theory (which includes all field equations and constitutive relations) from the three-dimensional equations is developed. On the other hand, this may require too much patience from the reader. For this reason and in anticipation of certain results, whose range of validity and limitations will be spelled out eventually, we discuss here a set of kinematic measures which are obtained by an approximation from (7.61).

With reference to (7.59) and (7.61), suppose now that γ_{Ni} and $\varkappa_{Ni\alpha}$ vanish for $N \geqq 2$, i.e.,

$$\gamma_{Ni} = 0, \quad \varkappa_{Ni\alpha} = 0 \quad (N \geqq 2). \tag{7.65}$$

Then, in view of (7.65), by $(7.59)_{6,7}$ $\delta_{Ni} = 0$ for $N \geqq 2$. Hence

$$\boldsymbol{\delta}_N = 0 \quad (N \geqq 2), \tag{7.66}$$

and (7.51) assumes the simple form

$$\boldsymbol{u}^* = \boldsymbol{u} + \xi\, \boldsymbol{\delta}. \tag{7.67}$$

We expect (7.67) to be a valid approximation for *thin* shells in the region $\alpha < \xi < \beta$ We postpone the justification of (7.65) or the limitations under which it may be assumed; but, given (7.65), the simple expression (7.67) follows without further assumption.[43]

---

[41] The simple form of such an approximate expression for $\gamma^*_{\alpha\beta}$, corresponding to the first two terms on the right-hand side of $(7.63)_1$, is partially the reason for the success in the construction of an approximate classical theory of plates as compared to that for shells. We postpone further remarks on this until later in this section [Subsect. ε)] and again in Sect. 20. Here, however, we note that the development of the approximate theory just referred to usually begins with a set of displacements (or special assumptions for the three-dimensional strain tensor) which amounts to ignoring γ_{Ni} $(N \geqq 1)$, $\varkappa_{N3\alpha}$ $(N \geqq 1)$ and $\varkappa_{N\beta\alpha}$ $(N \geqq 2)$ or equivalently $\gamma^*_{\alpha 3}$ and $\varkappa_{N\beta\alpha}$ $(N \geqq 2)$ in the kinematical results, but later include (through an additional assumption) the effect of γ^*_{33} in the constitutive relations.

[42] Some of the difficulties and inconsistencies in the past developments of shell theory may be traced directly to the fact that approximations for kinematics were introduced at the outset and without due consideration for their effects in the rest of the theory. To compensate for these difficulties in the derivations of the linear theory of elastic shells, sometimes in recent years use has been made of variational theorems such as those of the Hellinger-Reissner, Hu-Washizu or variants thereof: HELLINGER [1914, *1*], REISSNER [1950, *5*] and [1953, *4*], HU [1955, *4*], WASHIZU [1955, *7*], REISSNER [1964, *7*], NAGHDI [1964, *6*] and REISSNER [1965, *7*]. What these (three-dimensional) variational theorems have in common, apart from the boundary conditions, is one form of the constitutive relations for linear elasticity as part of their Euler equations; they differ from one another mainly in the degree of generality provided by their additional Euler equations for the variational problem. For derivations of an approximate shell theory by means of variational theorems (in the three-dimensional theory) of the type mentioned above, see, e.g., [1963, *6*] and [1964, *6*] which contain additional related references.

[43] Alternatively, we may assume (7.67) together with (7.66) and conclude (7.65) but this appears to be more difficult to justify. An assumption for the displacement vector in the form (7.67) or a more specialized version of it is made in nearly all developments of the linear theory of thin shells at the outset. In this connection, see also Subsect. ε) of this section.

With the approximation (7.65), the kinematic measures (7.59) can be written as

$$\begin{aligned}
e_{\alpha\beta} &= \tfrac{1}{2}(u_{\alpha|\beta} + u_{\beta|\alpha}) - B_{\alpha\beta}\, u_3, \\
\gamma_\alpha &= \delta_\alpha - \beta_\alpha, \quad \gamma_3 = \delta_3, \quad \beta_\alpha = -(u_{3,\alpha} + B^\lambda_\alpha\, u_\lambda), \\
\varkappa_{\beta\alpha} &= \varrho_{\beta\alpha} - B_{\alpha\beta}\,\gamma_3, \quad \varkappa_{3\alpha} = \varrho_{3\alpha} + B^\lambda_\alpha\,\gamma_\lambda, \\
\varrho_{\beta\alpha} &= \gamma_{\beta|\alpha} - \bar{\varkappa}_{\beta\alpha}, \quad \varrho_{3\alpha} = \gamma_{3,\alpha}, \\
\bar{\varkappa}_{\beta\alpha} = \bar{\varkappa}_{\alpha\beta} &= [u_{3|\beta\alpha} + B^\nu_{\beta|\alpha}\, u_\nu + B^\nu_\alpha\, u_{\nu|\beta} + B^\nu_\beta\, u_{\nu|\alpha} - B^\nu_\alpha\, B_{\nu\beta}\, u_3] \\
&= -\tfrac{1}{2}(\beta_{\alpha|\beta} + \beta_{\beta|\alpha}) + \tfrac{1}{2}[B^\nu_\alpha(e_{\nu\beta} + \gamma_{[\nu\beta]}) + B^\nu_\beta(e_{\nu\alpha} + \gamma_{[\nu\alpha]})], \\
\gamma_{[\alpha\beta]} &= \tfrac{1}{2}(u_{\alpha|\beta} - u_{\beta|\alpha}),
\end{aligned} \tag{7.68}$$

where $\varrho_{i\alpha}$ defined by the last three of (7.68) are introduced for later reference. The components of γ^*_{ij} in (7.61) now simplify considerably and reduce to

$$\begin{aligned}
\gamma^*_{\alpha\beta} &= e_{\alpha\beta} + \xi\,\varkappa_{(\alpha\beta)} - \xi^2\,\chi_{\alpha\beta}, \\
\gamma^*_{\alpha 3} &= \gamma^*_{3\alpha} = \gamma_\alpha + \xi\,\varrho_{3\alpha}, \quad \gamma^*_{33} = \gamma_3,
\end{aligned} \tag{7.69}$$

where $\varkappa_{(\alpha\beta)}$ is the symmetric part of $\varkappa_{\alpha\beta}$ and where we have put

$$\begin{aligned}
\chi_{\alpha\beta} &= \bar{\chi}_{\alpha\beta} + B^\lambda_\alpha\, B^\nu_\beta\, e_{\lambda\nu} = \chi_{\beta\alpha}, \\
\bar{\chi}_{\alpha\beta} &= \tfrac{1}{2}(B^\gamma_\alpha\,\varkappa_{\gamma\beta} + B^\gamma_\beta\,\varkappa_{\gamma\alpha}).
\end{aligned} \tag{7.70}$$

It is not difficult to see that the kinematic measures (7.68) are formally equivalent to those given in Sect. 6 for a Cosserat surface [compare with (6.24)], if the two surfaces $\mathfrak{S}$ and $\mathscr{S}$ are identified. However, we postpone such identifications until later. Also, as might be expected, only the symmetric part of $\varkappa_{\beta\alpha}$ occurs in $(7.69)_1$.

ε) *Other kinematic approximations in the linear theory.* The literature on the linear theory, especially for shells, abounds with a variety of kinematic approximations. Most of these are variants, or a special case, of (7.67) and often constitute the starting point in the development of complete approximate theories.[44] In order to indicate here the nature of these approximations and at the same time call attention to a certain kinematic approximation which is generally adopted for classical theories of plates and shells, it is expedient to consider the case of a flat plate.

Referred to the base vectors $\boldsymbol{A}_i$, the approximate expression (7.67) in component form reads

$$u^*_\alpha = u_\alpha + \xi\,\delta_\alpha, \quad u^*_3 = u_3 + \xi\,\delta_3. \tag{7.71}$$

For initially flat plates, in view of (6.26) and the assumptions (7.65), the approximate kinematic measures (7.68) simplify and reduce to the first six relations in (7.62). It is then easily seen that the linearized kinematic variables for initially flat plates separate into two parts, namely

$$\begin{aligned}
E &= \{u_\alpha, \delta_3, e_{\alpha\gamma}, \gamma_3, \varkappa_{3\alpha}\}, \\
F &= \{u_3, \delta_\alpha, \varkappa_{\gamma\alpha}, \gamma_\alpha\}.
\end{aligned} \tag{7.72}$$

The displacements u_α, δ_3 in the former set E characterize the extensional motion while u_3, δ_α in the latter set F represent the displacements associated with the flexural motion of a plate.[45] Moreover, with reference to (7.71), it is clear that the tangential and normal components of $\boldsymbol{u}^*$ are, respectively, even and odd in ξ for the extensional motion while they are odd and even in ξ for the flexural motion.

[44] An account of such kinematic approximations can be found in [1963, *6*].

[45] This observation parallels that made in Sect. 6 [following (6.26)] for a Cosserat surface.

Special cases of the kinematic variables listed in (7.72) correspond to those employed in the classical theory of plates.[46] The classical extensional case does not include δ_3 (and hence the components $\gamma_3, \varkappa_{3\alpha}$) and only admits the tangential displacements u_α. Such a kinematic assumption implies, in turn, inextensibility along the normal to the plate. The kinematic variables for the classical flexural (or bending) theory is obtained from the set F in (7.72), if we put $\gamma_\alpha=0$. Then, the relevant kinematic variables reduce to

$$\delta_\alpha=\beta_\alpha=-u_{3,\alpha}, \qquad \varkappa_{\beta\alpha}=-u_{3|\alpha\beta}. \tag{7.73}$$

The above results for the classical theory of plates is summarized in the displacement assumption

$$u_\alpha^*=u_\alpha+\xi\,\beta_\alpha, \qquad u_3^*=u_3, \tag{7.74}$$

which is known as the Kirchhoff hypothesis.[47] It is clear that the approximate displacements (7.74) imply inextensibility (in the extensional case) along the normal to the plate[48] and also include only the angular displacements β_α, so that the effect of transverse shear deformation is ignored (in the case of bending theory).

The kinematic assumptions in the derivations of the classical theory of shells, beginning with LOVE's paper[49]—known as Love's first approximation—are similar to those in the classical theory of plates. In particular, it is assumed that (i) normals to the undeformed middle surface remain normals and (ii) suffer no extension. From these assumptions, sometimes referred to as Kirchhoff-Love hypothesis (especially in the Russian literature), it follows that the components of displacements must have the form (7.74) or equivalently can be obtained from[50]

$$\boldsymbol{u}^*=\boldsymbol{u}+\xi\,\boldsymbol{\beta}, \tag{7.75}$$

where $\boldsymbol{\beta}$ is defined by $(7.55)_{2,3}$. With the displacement vector given by (7.75), the components $\gamma_{\alpha3}^*$ and γ_{33}^* (for transverse shear and transverse normal strain) vanish[51] and since now $\gamma_i=0$ the variables (7.68) reduce to the set

$$e_{\alpha\beta}, \quad \varkappa_{\beta\alpha}=\varrho_{\alpha\beta}=-\bar{\varkappa}_{\alpha\beta}, \quad \delta_\alpha=\beta_\alpha, \tag{7.76}$$

where the expressions for $e_{\alpha\beta}, \bar{\varkappa}_{\alpha\beta}, \beta_\alpha$ in terms of u_α, u_3 and their derivatives are those recorded in (7.68).[52]

The above remarks pertain to kinematic approximations in the classical theories of plates and shells. Other displacement assumptions or numerous other

---

[46] KIRCHHOFF [1850, *1*].

[47] Despite the implication of (7.74) regarding inextensibility along the normal, the effect of the component of the strain tensor γ_{33}^* is accounted for in the constitutive equations. This seemingly inconsistent set of assumptions, however, leads to the correct linear constitutive relations.

[48] See also Sect. 20.

[49] LOVE [1888, *1*].

[50] Given the kinematic assumptions (i) and (ii) above, the form (7.75) follows even without the limitation to smallness of $\boldsymbol{u}^*$ or its components. See Sect. 4 of [1963, *6*].

[51] Even though (7.75) imply the inextensibility along the normal to the middle surface, the effect of γ_{33}^* is nevertheless included in the constitutive relations. This is similar to the case of plate noted above.

[52] The expressions for these kinematic variables have the same forms as those given by (6.27) for a restricted theory by direct approach.

ad hoc schemes have continuously appeared in the literature but we do not elaborate on these here.[53]

A derivation of compatibility equations, in terms of the kinematic measures resulting from an approximate expression for the displacement $\boldsymbol{u}^*$ in the form (7.67) or the more restrictive assumption (7.75), may be accomplished in a manner similar to the derivation by direct method given in Sect. 6 [Subsect. η)]. We do not include here such a derivation but note that the resulting compatibility equations will be of the same forms as those given by (6.60)–(6.61) and (6.62). In the literature on the classical linear shell theory, compatibility equations [corresponding to those in (6.62)] are generally obtained using the equations of Gauss and Mainardi-Codazzi as the starting point. A derivation of this kind is readily available elsewhere and was first given by GOL'DENVEIZER.[54]

C. Basic principles for shells and plates.

This chapter is devoted to the basic principles, derivations of the appropriate field equations and related results for shells and plates. Again various developments are pursued both by direct approach (Sects. 8–10) and from the three-dimensional theory of (non-polar) classical continuum mechanics (Sects. 11–12).

8. Basic principles for shells: I. Direct approach. This section is concerned with basic principles for a Cosserat surface, including conservation laws and invariance requirements under superposed rigid body motions. Our developments are in the context of a general thermodynamical theory. But a reader who prefers to confine himself to the purely mechanical theory should be able to do so without difficulty, although some adjustment and omission [especially that of Subsect. β)] will then be necessary.

The ideas of force, couple, stress vector, etc., are familiar from the three-dimensional theory. Here, we need to define the corresponding quantities for a Cosserat surface; and, it is not difficult to see that they may be introduced in a manner which parallels the concepts of the external body force and the contact force in the three-dimensional theory. However, prior to a statement of conservation laws, it is more enlightening (and perhaps even economical) in the present development to introduce the notion of force and related quantities for a Cosserat surface through their rate of work expressions.

α) *Conservation laws*. Let $\mathscr{P}$ be a part of $\mathscr{s}$ occupied by the arbitrary part $\mathscr{P}_{\mathscr{C}}$ of the material region of the Cosserat surface $\mathscr{C}$ (defined in Sect. 4) in the deformed configuration at time t. Let c be any curve on $\mathscr{s}$ which may also be taken as the

[53] The nature of LOVE's second approximation in which instead of (7.75) terms involving ξ^2 are also retained in the expansion of $\boldsymbol{u}^*$ was examined and explored by HILDEBRAND, REISSNER and THOMAS [1949, *4*]. A further account of such assumptions for displacements in the linear theory of shells may be found in [1963, *6*]. Similar and other types of kinematic approximations are used in the papers of PARKUS [1950, *4*], RÜDIGER [1959, *5*], DUDDECK [1962, *1*], ZERNA [1962, *8*] and ZERNA [1968, *14*].

[54] GOL'DENVEIZER [1940, *1*]. See also his book [1961, *3*]. The earliest attempt to derive the compatibility equations in shell theory appears to be by ODQVIST [1937, *1*]. A short account of compatibility equations in the context of the classical theory may be found also in the paper of GOL'DENVEIZER and LUR'E [1947, *3*]. A derivation in lines of curvature coordinates is given in NOVOZHILOV's book [1959, *3*]. Another derivation in general coordinates, involving kinematic variables based on (7.75), can be found in [1963, *6*] and its generalization is considered by KOLLMAN [1966, *5*]. Compatibility equations in lines of curvature coordinates and in terms of physical components of the variables (7.76) are given by SANDERS [1959, *6*]. A shorter version in general coordinates is included in [1963, *7*] and another (in coordinate free notation) is contained in a recent paper by STEELE [1971, *9*].

boundary $\partial\mathscr{P}$ of $\mathscr{P}$. As the boundary of $\mathscr{P}$, c will be a closed curve defined for the points $\boldsymbol{r}$ in $\partial\mathscr{P}$ of the deformed configuration at time t. Let $\theta^\alpha = \theta^\alpha(s)$ be the parametric equations of the curve c, with s as the arc parameter; and let $\boldsymbol{\lambda}$ denote the unit tangent vector to the curve c defined for the points $\boldsymbol{r}(\theta^\alpha(s), t)$ on c. Then,

$$\boldsymbol{\lambda} = \frac{\partial \boldsymbol{r}}{\partial s} = \lambda^\alpha \boldsymbol{a}_\alpha, \qquad \lambda^\alpha = \frac{d\theta^\alpha(s)}{ds} \tag{8.1}$$

and the outward unit normal $\boldsymbol{\nu}$ to c lying in the surface is given by

$$\boldsymbol{\nu} = \boldsymbol{\lambda} \times \boldsymbol{a}_3 = \nu^\alpha \boldsymbol{a}_\alpha = \nu_\alpha \boldsymbol{a}^\alpha = \varepsilon_{\alpha\beta} \lambda^\beta \boldsymbol{a}^\alpha, \tag{8.2}$$

where $\varepsilon_{\alpha\beta}$ is defined in (5.63). For later reference, we also note that $\boldsymbol{\lambda}$ can be expressed as

$$\boldsymbol{\lambda} = \boldsymbol{a}_3 \times \boldsymbol{\nu} = \boldsymbol{a}_3 \times \nu_\alpha \boldsymbol{a}^\alpha = \varepsilon^{\alpha\beta} \nu_\alpha \boldsymbol{a}_\beta. \tag{8.3}$$

Let[1] $\boldsymbol{N} = \boldsymbol{N}(\theta^\alpha, t; \boldsymbol{\nu})$ and $\boldsymbol{M} = \boldsymbol{M}(\theta^\alpha, t; \boldsymbol{\nu})$, each a three-dimensional vector field, be defined for points $\boldsymbol{r}$ on the boundary curve c of $\mathscr{P}$ as follows: If for all arbitrary velocity fields $\boldsymbol{v}$, the scalar $\boldsymbol{N} \cdot \boldsymbol{v}$ is a rate of work per unit length of c, then $\boldsymbol{N}$ is called a *contact force* (or a curve force) vector per unit length of c. Similarly, if the scalar $\boldsymbol{M} \cdot \boldsymbol{w}$ is a rate of work per unit length of c for all arbitrary director velocities $\boldsymbol{w}$, then $\boldsymbol{M}$ is called a *contact director couple* (or a curve director couple) per unit length of c. It is clear that the definition for $\boldsymbol{M}$ parallels that for $\boldsymbol{N}$ and can be obtained from the sentence preceding the last if we replace the symbols $\boldsymbol{N}$ and $\boldsymbol{v}$ by $\boldsymbol{M}$ and $\boldsymbol{w}$ and the words velocity and force by director velocity and director couple, respectively. The resultant contact force $\boldsymbol{F}_c(\mathscr{P})$ and the resultant contact director couple $\boldsymbol{G}_c(\mathscr{P})$ exerted on the part $\mathscr{P}$ of the Cosserat surface $\mathscr{C}$ at time t are defined by the line integrals

$$\boldsymbol{F}_c(\mathscr{P}) = \int_{\partial\mathscr{P}} \boldsymbol{N}\, ds, \qquad \boldsymbol{G}_c(\mathscr{P}) = \int_{\partial\mathscr{P}} \boldsymbol{M}\, ds \tag{8.4}$$

over the boundary $\partial\mathscr{P}$ of $\mathscr{P}$ in the present configuration.[2] Moreover, it is clear from the above definitions and the physical dimensions of $\boldsymbol{v}$ and $\boldsymbol{w}$ that $\boldsymbol{N}$ and $\boldsymbol{M}$ have, respectively, the physical dimensions of force and couple per unit length, namely

$$\begin{aligned} \text{phys. dim. } \boldsymbol{N} &= \left[\frac{MLT^{-2}}{L}\right] = [MT^{-2}], \\ \text{phys. dim. } \boldsymbol{M} &= \left[\frac{ML^2T^{-2}}{L}\right] = [MLT^{-2}], \end{aligned} \tag{8.5}$$

where the symbol $[T]$ designates the dimension of time and the symbols $[M]$ and $[L]$ for the dimensions of mass and length were introduced previously.

The vector fields $\boldsymbol{N}$ and $\boldsymbol{M}$ act across any curve on $\mathfrak{s}$. In an analogous fashion we may define $\boldsymbol{f} = \boldsymbol{f}(\theta^\alpha, t)$ and $\boldsymbol{l} = \boldsymbol{l}(\theta^\alpha, t)$, each a three-dimensional vector field per unit mass, for points $\boldsymbol{r}$ on the part $\mathscr{P}$ of $\mathfrak{s}$: If the scalar $\boldsymbol{f} \cdot \boldsymbol{v}$ is a rate of work per unit mass for all arbitrary velocities $\boldsymbol{v}$, then $\boldsymbol{f}$ is called an *assigned force* vector per unit mass of $\mathfrak{s}$. The definition for the *assigned director couple* $\boldsymbol{l}$ parallels that for $\boldsymbol{f}$ and can be obtained from the preceding sentence, if we replace the symbols $\boldsymbol{f}$ and $\boldsymbol{v}$ by $\boldsymbol{l}$ and $\boldsymbol{w}$ and the words velocities and assigned force by director veloc-

[1] To emphasize the dependence of such vector fields as $\boldsymbol{N}$ and $\boldsymbol{M}$ on the unit normal $\boldsymbol{\nu}$, it is customary to write them as $\boldsymbol{N}_{(\boldsymbol{\nu})}$ and $\boldsymbol{M}_{(\boldsymbol{\nu})}$. Here we omit the subscript $(\boldsymbol{\nu})$ from $\boldsymbol{N}_{(\boldsymbol{\nu})}$ and $\boldsymbol{M}_{(\boldsymbol{\nu})}$, except on one or two occasions which may serve clarity.

[2] In (8.4), (8.6) and other definitions in this chapter, the abbreviated notation introduced in Sect. 4 [following (4.43)] is used. The designation of the left-hand sides of $(8.4)_{1,2}$ by $\boldsymbol{F}_c(\mathscr{P})$ and $\boldsymbol{G}_c(\mathscr{P})$, instead of $\boldsymbol{F}_c(\mathscr{P}_\mathscr{C})$ and $\boldsymbol{G}_c(\mathscr{P}_\mathscr{C})$, is in accord with our previous notational agreement.

ities and assigned director couple, respectively. The *resultant assigned force* $\boldsymbol{F}_b(\mathscr{P})$ and the *resultant assigned director couple* $\boldsymbol{G}_b(\mathscr{P})$ acting on the part $\mathscr{P}$ of the Cosserat surface $\mathscr{C}$ at time t are defined by the surface integrals

$$\boldsymbol{F}_b(\mathscr{P}) = \int_{\mathscr{P}} \varrho \boldsymbol{f}\, d\sigma, \qquad \boldsymbol{G}_b(\mathscr{P}) = \int_{\mathscr{P}} \varrho\, \boldsymbol{l}\, d\sigma \tag{8.6}$$

over $\mathscr{P}$ in the present configuration, where $\varrho = \varrho(\theta^\alpha, t)$ is the mass density of $\mathfrak{s}$ defined by (4.36) and the area element $d\sigma$ is given by $(4.19)_2$. In view of (4.37) it is evident that the vector fields $\boldsymbol{f}$ and $\boldsymbol{l}$ have, respectively, the physical dimensions of force and couple per unit mass, namely

$$\text{phys. dim.}\, \boldsymbol{f} = \left[\frac{MLT^{-2}}{M}\right] = [LT^{-2}], \qquad \text{phys. dim.}\, \boldsymbol{l} = \left[\frac{ML^2T^{-2}}{M}\right] = [L^2T^{-2}]. \tag{8.7}$$

We assume that all forces and couples are continuously distributed.[3] The assigned force $\boldsymbol{f}$ and the assigned director couple $\boldsymbol{l}$, each per unit mass, act throughout an arbitrary part $\mathscr{P}$ of $\mathscr{C}$ in the present configuration; and the contact force $\boldsymbol{N}$ and the contact director couple $\boldsymbol{M}$, each per unit length, act across the boundary $\partial\mathscr{P}$ of $\mathscr{P}$ in the present configuration at time t. It is convenient to record here the rate of work by these contact and assigned forces and couples in the form

$$\begin{gathered} R(\mathscr{P}) = R_c(\mathscr{P}) + R_b(\mathscr{P}), \\ R_c(\mathscr{P}) = \int_{\partial\mathscr{P}} (\boldsymbol{N}\cdot\boldsymbol{v} + \boldsymbol{M}\cdot\boldsymbol{w})\, ds, \qquad R_b(\mathscr{P}) = \int_{\mathscr{P}} \varrho(\boldsymbol{f}\cdot\boldsymbol{v} + \boldsymbol{l}\cdot\boldsymbol{w})\, d\sigma. \end{gathered} \tag{8.8}$$

We now introduce some additional quantities which we associate with the motion of the Cosserat surface $\mathscr{C}$. We assume the existence of a scalar potential function per unit mass $\varepsilon = \varepsilon(\theta^\alpha, t)$, called the specific internal energy. The surface integral

$$\mathscr{E}(\mathscr{P}) = \int_{\mathscr{P}} \varrho\, \varepsilon\, d\sigma \tag{8.9}$$

defines the *internal energy* for each part $\mathscr{P}$ in the present configuration. We also introduce a scalar field $r = r(\theta^\alpha, t)$ per unit mass per unit time, called the specific heat supply (or heat absorption); and the heat flux, across a curve with the unit normal $\boldsymbol{\nu}$, by the scalar[4] $h = h(\theta^\alpha, t; \boldsymbol{\nu})$ per unit length per unit time. The integral[5]

$$H(\mathscr{P}) = \int_{\mathscr{P}} \varrho\, r\, d\sigma - \int_{\partial\mathscr{P}} h\, ds, \tag{8.10}$$

where $\partial\mathscr{P}$ is the boundary of $\mathscr{P}$, defines the heat per unit time entering the part $\mathscr{P}$ of $\mathscr{C}$ in the present configuration. The first term on the right-hand side of (8.10) represents the heat transmitted into the surface by radiation and the second term the heat entering the surface by conduction.

The kinetic energy $\mathscr{K}(\mathscr{P})$ for each part $\mathscr{P}$ of $\mathscr{C}$ in the present configuration is defined by

$$\mathscr{K}(\mathscr{P}) = \int_{\mathscr{P}} \tfrac{1}{2}\varrho(\boldsymbol{v}\cdot\boldsymbol{v} + \alpha\, \boldsymbol{w}\cdot\boldsymbol{w})\, d\sigma \tag{8.11}$$

which includes a contribution $\frac{1}{2}\alpha\, \boldsymbol{w}\cdot\boldsymbol{w}$ per unit mass due to director velocity $\boldsymbol{w}$, the coefficient $\alpha = \alpha(\theta^\gamma)$ being independent of time but possibly a function of surface coordinates. We also define, for each part $\mathscr{P}$ of $\mathscr{C}$ in the present con-

[3] The terminology of *director force* and *assigned director force* was used in [1965, *4*] for $\boldsymbol{M}$ and $\boldsymbol{l}$, respectively, in place of *director couple* and *assigned director couple*. The latter terminology, adopted here, seems to be more suggestive.

[4] The use of the symbol h here is temporary and need not be confused with that in (4.30).

[5] Alternatively we can introduce the heat flux by a vector field $\boldsymbol{q} = \boldsymbol{q}(\theta^\alpha, t)$ such that $\boldsymbol{q}\cdot\boldsymbol{\nu} = h$. The negative sign in (8.10) is in accord with the usual convention, since $\boldsymbol{q}\cdot\boldsymbol{\nu} < 0$ at points where heat is entering the surface through the boundary curve. Here, our r and $\boldsymbol{q}$ are the two-dimensional counterparts of q and $-\boldsymbol{h}$ (in a three-dimensional theory) employed by some writers; see, e.g., Sect. 79 of TRUESDELL and NOLL [1965, *9*].

figuration, the integrals[6]

$$\begin{aligned}
&\text{the linear momentum:} && \mathcal{L}(\mathcal{P}) = \int_{\mathcal{P}} \varrho\, \boldsymbol{v}\, d\sigma,\\
&\text{the director momentum:} && \mathcal{D}(\mathcal{P}) = \int_{\mathcal{P}} \varrho\, \alpha\, \boldsymbol{w}\, d\sigma,\\
&\text{the moment of momentum:} && \mathcal{M}(\mathcal{P}) = \int_{\mathcal{P}} \varrho\, \mathcal{A}\, d\sigma,
\end{aligned} \tag{8.12}$$

where

$$\mathcal{A} = \boldsymbol{r} \times \boldsymbol{v} + \boldsymbol{d} \times \alpha\, \boldsymbol{w}. \tag{8.13}$$

Further, let $\boldsymbol{A}(\mathcal{P})$ represent the sum of the resultants of the supply of moment of momentum $\boldsymbol{A}_b(\mathcal{P})$ due to $\boldsymbol{f}$, $\boldsymbol{l}$ and the flux of moment of momentum $\boldsymbol{A}_c(\mathcal{P})$ due to $\boldsymbol{N}$, $\boldsymbol{M}$. Then,

$$\begin{gathered}
\boldsymbol{A}(\mathcal{P}) = \boldsymbol{A}_b(\mathcal{P}) + \boldsymbol{A}_c(\mathcal{P}),\\
\boldsymbol{A}_b(\mathcal{P}) = \int_{\mathcal{P}} [\boldsymbol{r} \times \varrho \boldsymbol{f} + \boldsymbol{d} \times \varrho\, \boldsymbol{l}]\, d\sigma, \quad \boldsymbol{A}_c(\mathcal{P}) = \int_{\partial\mathcal{P}} (\boldsymbol{r} \times \boldsymbol{N} + \boldsymbol{d} \times \boldsymbol{M})\, ds.
\end{gathered} \tag{8.14}$$

We also admit the existence of a vector field[7] $\boldsymbol{m} = \boldsymbol{m}(\theta^\alpha, t)$, an *intrinsic* (or *surface*) *director couple* per unit area of $\mathfrak{s}$ which makes no contribution to the supply of moment of momentum; the physical dimension of $\boldsymbol{m}$ is

$$\text{phys. dim. } \boldsymbol{m} = \left[\frac{ML^2T^{-2}}{L^2}\right] = [MT^{-2}], \tag{8.15}$$

i.e., a physical dimension of couple per unit area. Recalling $(8.6)_2$, we define the combined resultant director couple due to $\varrho\, \boldsymbol{l}$ and $\boldsymbol{m}$ for each part $\mathcal{P}$ of $\mathcal{C}$ in the present configuration as

$$\boldsymbol{G}_b'(\mathcal{P}) = \int_{\mathcal{P}} (\varrho\, \boldsymbol{l} - \boldsymbol{m})\, d\sigma = \boldsymbol{G}_b(\mathcal{P}) - \int_{\mathcal{P}} \boldsymbol{m}\, d\sigma. \tag{8.16}$$

Having disposed of the foregoing preliminaries, we are now in a position to state the conservation laws (or principles) for a Cosserat surface. With reference to the present configuration, these conservation laws may be stated in the forms[8]

$$\begin{aligned}
&\frac{d}{dt}\int_{\mathcal{P}} \varrho\, d\sigma = 0,\\
&\frac{d}{dt}\int_{\mathcal{P}} \varrho\, \boldsymbol{v}\, d\sigma = \boldsymbol{F}_b(\mathcal{P}) + \boldsymbol{F}_c(\mathcal{P}),\\
&\frac{d}{dt}\int_{\mathcal{P}} \varrho\, \alpha\, \boldsymbol{w}\, d\sigma = \boldsymbol{G}_b'(\mathcal{P}) + \boldsymbol{G}_c(\mathcal{P}),\\
&\frac{d}{dt}\int_{\mathcal{P}} \varrho\, \mathcal{A}\, d\sigma = \boldsymbol{A}_b(\mathcal{P}) + \boldsymbol{A}_c(\mathcal{P}),\\
&\frac{d}{dt}\int_{\mathcal{P}} \varrho\left[\varepsilon + \frac{1}{2}(\boldsymbol{v}\cdot\boldsymbol{v} + \alpha\, \boldsymbol{w}\cdot\boldsymbol{w})\right] d\sigma = R(\mathcal{P}) + H(\mathcal{P}).
\end{aligned} \tag{8.17}$$

---

[6] The director momentum $\mathcal{D}(\mathcal{P})$ in $(8.12)_2$ has the dimension of moment of momentum, since $\boldsymbol{d}$ was specified to be dimensionless. In other theories of directed media, where $\boldsymbol{d}$ more conveniently may be defined to have the dimension of length, the expression corresponding to $\mathcal{D}(\mathcal{P})$ would have the dimension of linear momentum.

[7] In contrast to the contact force $\boldsymbol{N}$ and the contact couple $\boldsymbol{M}$, $\boldsymbol{m}$ does not depend on the unit normal $\boldsymbol{\nu}$.

[8] Apart from minor variations in form, these conservation laws may be regarded as the two-dimensional counterparts of those in the three-dimensional theory of directed media with a single deformable director: ERICKSEN [1961, *1*], GREEN, NAGHDI and RIVLIN [1965, *3*]. See also, in this connection, Sect. 127 of TRUESDELL and NOLL [1965, *9*] which contains an interesting approach to the conservation laws for linear momentum and linear director momentum in the context of ERICKSEN's theory of liquid crystals [1961, *1*].

The first of (8.17) is a mathematical statement of conservation of mass, the second that of the linear momentum principle, the third that of the director momentum, the fourth is the moment of momentum and the fifth represents the balance of energy for a Cosserat surface. The left-hand sides of the last four in (8.17) represent, respectively, the rate of increase of the linear momentum, the director momentum, the moment of momentum (including ordinary momentum and director momentum) and the total energy, i.e., the sum of internal and kinetic energies.

Assuming that ϱ is continuously differentiable and recalling (5.66), from $(8.17)_1$ we obtain

$$\int_{\mathscr{P}} (\dot{\varrho} + \varrho\, \eta^{\alpha}_{\alpha})\, d\sigma = 0,$$

which holds for each part $\mathscr{P}$ in the present configuration. Hence follows

$$\dot{\varrho} + \varrho\, \eta^{\alpha}_{\alpha} = \dot{\varrho} + \varrho\, (v^{\alpha}{}_{|\alpha} - b^{\alpha}_{\alpha}\, v_3) = 0 \tag{8.18}$$

as the spatial form of the continuity equation for the Cosserat surface in contrast to the material form of the continuity equation expressed by (4.42). We postpone the derivations of other local field equations as consequences of the remaining conservation laws in (8.17) and consider first some additional thermodynamical preliminaries. However, in anticipation of certain future results, we introduce here vector fields $\bar{\boldsymbol{f}}$ and $\bar{\boldsymbol{l}}$ defined by

$$\bar{\boldsymbol{f}} = \boldsymbol{f} - \dot{\boldsymbol{v}}, \qquad \bar{\boldsymbol{l}} = \boldsymbol{l} - \alpha\, \dot{\boldsymbol{w}}. \tag{8.19}$$

In (8.19), $\bar{\boldsymbol{f}}$ is the difference of the assigned force $\boldsymbol{f}$ and the acceleration vector $\dot{\boldsymbol{v}}$ of the surface $\mathscr{s}$ and $\bar{\boldsymbol{l}}$ is the difference of the assigned director couple $\boldsymbol{l}$ and the inertia term due to the director velocity.

β) *Entropy production.* We introduce now further thermodynamic preliminaries. Let the temperature field be denoted by $\theta = \theta(\theta^{\alpha}, t)$ which we assume to have positive values, i.e.,

$$\theta > 0 \tag{8.20}$$

and define a scalar field per unit mass $\eta = \eta(\theta^{\alpha}, t)$, called the *specific entropy.* The surface integral

$$\mathscr{H}(\mathscr{P}) = \int_{\mathscr{P}} \varrho\, \eta\, d\sigma \tag{8.21}$$

defines the entropy of a part $\mathscr{P}$ in the present configuration and we define the *production of entropy* per unit time in a part $\mathscr{P}$ (in the present configuration) at time t by

$$\begin{aligned} \Gamma(\mathscr{P}) &= \frac{d}{dt} \int_{\mathscr{P}} \varrho\, \eta\, d\sigma - \left[\int_{\mathscr{P}} \varrho\, \frac{r}{\theta}\, d\sigma - \int_{\partial\mathscr{P}} \frac{h}{\theta}\, ds \right], \\ \Gamma(\mathscr{P}) &= \int_{\mathscr{P}} \varrho\, \gamma\, d\sigma, \end{aligned} \tag{8.22}$$

where the scalar γ per unit mass is the *specific production of entropy.* The quantity r/θ in $(8.22)_1$ is the entropy due to radiation entering $\mathscr{P}$ and

$$-\frac{h}{\theta}$$

is the flux of entropy due to conduction entering $\mathscr{P}$ through the boundary $\partial\mathscr{P}$.

For a given Cosserat surface $\mathscr{C}$, it is convenient to speak of a *thermodynamic process* or simply a *process* if the set of twelve functions consisting of the vector functions $\boldsymbol{r}$ and $\boldsymbol{d}$ in (5.1), the scalar function η and the mechanical and thermal fields $\mathfrak{M}$, $\mathfrak{T}$, $\mathfrak{B}$, namely

$$\mathfrak{M}=\{\boldsymbol{N}, \boldsymbol{M}, \boldsymbol{m}\}, \quad \mathfrak{T}=\{\varepsilon, \theta, h\}, \quad \mathfrak{B}=\{\boldsymbol{f}, \boldsymbol{l}, r\} \tag{8.23}$$

as functions of θ^α and t and with $\{\boldsymbol{N}, \boldsymbol{M}, h\}$ being also dependent on $\boldsymbol{v}$, satisfy all conservation laws in (8.17). We speak of an *admissible* process if the set of functions $\mathfrak{M}$ and $\mathfrak{T}$ in $(8.23)_{1,2}$ are specified by constitutive equations (characterizing the thermo-mechanical behavior of a given material) which hold at each material point θ^α and at all times. We return to the above terminology in Sect. 9 and amplify the remarks concerning a process and an admissible process.

Recalling $(8.22)_1$, we now postulate the inequality

$$\Gamma \geqq 0 \tag{8.24}$$

which we require to be valid for every admissible process describing the motion and the thermo-mechanical behavior of the Cosserat surface.[9] The entropy production inequality (8.24) is the two-dimensional counterpart of the Clausius-Duhem inequality (in the three-dimensional theory) and is a statement of the second law of thermodynamics.[10] The inequality (8.24) will be viewed as a condition to be satisfied identically by every admissible process which describes the motion and the material behavior of the continuum. Thus, it will narrow the class of all admissible processes and will place restrictions on the functions in $(8.23)_{1,2}$ to be specified by constitutive equations.

γ) *Invariance conditions.* Previously, in Sect. 5, we have considered motions of a Cosserat surface which differ from (5.1) only by superposed rigid body motions and have also obtained the transformation relations for various kinematical quantities under such superposed rigid body motions. We consider now important invariance conditions to be satisfied by various dynamic and thermodynamic quantities when the position vector $\boldsymbol{r}^+$ and the director $\boldsymbol{d}^+$ are specified by (5.36).

Preliminary to our consideration of invariance conditions, we make certain observations regarding $(5.36)_1$ and its characteristic features. It is clear from $(5.36)_1$ that $\{\boldsymbol{r}^+, t^+\}$ and $\{\boldsymbol{r}, t\}$ are related by rigid transformations combined with a time shift. An immediate consequence of $(5.36)_1$ is that for each time t and corresponding to any two material points (say θ^α and ${}_0\theta^\alpha$) of the Cosserat surface $\mathscr{C}$, the magnitude of the relative displacement $|\boldsymbol{r}(\theta^\alpha) - \boldsymbol{r}({}_0\theta^\alpha)|$ remains unaltered:

$$|\boldsymbol{r}^+(\theta^\alpha) - \boldsymbol{r}^+({}_0\theta^\alpha)| = |\boldsymbol{r}(\theta^\alpha) - \boldsymbol{r}({}_0\theta^\alpha)|.$$

Hence, $(5.36)_1$ is distance (or length) preserving and it follows that the element of area of the surface $\mathscr{s}$ and therefore its mass density ϱ remains unaltered under the transformation. We have already seen in Sect. 5 that the relationship $(5.36)_1$

[9] The point of view adopted here is that currently used in the three-dimensional theory of continuum mechanics. For a more detailed treatment of the basic concepts in thermodynamics of deformable media, we refer the reader to Chap. E of TRUESDELL and TOUPIN [1960, *14*] and to Sects. 79–81 of TRUESDELL and NOLL [1965, *9*].

[10] There is some discussion in the current literature regarding the limitation of (8.24) or a more general form for the entropy inequality, but these need not affect our limited use of the inequality (8.24) here. In later sections, we shall appeal to (8.24) only in the context of elastic materials. Moreover, although our subsequent developments are carried out within the framework of a thermodynamic theory, most of the results without much effort can be specialized to the *isothermal* or the purely mechanical theory. Thus a reader who prefers to avoid the use of the inequality (8.24) should be able to do so easily.

induces certain transformations on various kinematic scalar or vector fields and that some of these transform according to formulae of the type

$$V^+ = V, \qquad \boldsymbol{V}^+ = Q(t)\,\boldsymbol{V}, \tag{8.25}$$

where V and $\boldsymbol{V}$ stand for a scalar and a vector field and $Q(t)$ is the proper orthogonal tensor appearing in (5.36).[11] However, not all kinematic quantities transform according to (8.25) and this is plainly evident from (5.45), (5.51) and $(5.55)_4$. The foregoing remarks parallel corresponding observations that can be made regarding the characteristic features of superposed rigid body motions in the (non-polar) three-dimensional theory. In particular, since the relationship (7.7) between the position $\boldsymbol{p}^+$ and $\boldsymbol{p}$ is distance preserving, volume elements and therefore the mass density ϱ^* remain unaltered under the transformation (7.7).

In order to motivate the invariance conditions sought, we deviate briefly from our main task and consider the nature of invariance conditions under superposed rigid body motions in the context of the classical three-dimensional continuum mechanics. Confining attention for simplicity to the purely mechanical theory, we recall that the mechanical fields which enter the linear momentum and the moment of momentum principles in the three-dimensional theory are the stress vector $\boldsymbol{t} = \boldsymbol{t}(\theta^i, t; \boldsymbol{n})$ and the body force $\boldsymbol{f}^* = \boldsymbol{f}^*(\theta^i, t)$ per unit mass; the latter is defined throughout a region of space occupied at time t by a corresponding material region of the body $\mathscr{B}$ while the former is defined over the boundary surface of the region in question at time t, with $\boldsymbol{n}$ as its outward unit normal vector. Consider now a second motion of $\mathscr{B}$ which differs from the given one at time t only by superposed rigid body motions. The second motion imparts a change in the orientation of the body, so that the outward unit normal vector to the same material surface (in the present configuration) becomes $\boldsymbol{n}^+$ under superposed rigid body motions; and, by virtue of (7.7), $\boldsymbol{n}^+$ is related to $\boldsymbol{n}$ through $\boldsymbol{n}^+ = Q\,\boldsymbol{n}$. Further, let $\boldsymbol{t}^+ = \boldsymbol{t}^+(\theta^i, t; \boldsymbol{n}^+)$ denote the stress vector in the second motion acting on the surface (in the present configuration) whose unit normal is $\boldsymbol{n}^+$. Because the transformation (7.7) is distance preserving, we expect the stress vector $\boldsymbol{t}^+$ (i) to have the same magnitude as $\boldsymbol{t}$ (which acts on the surface whose unit normal is $\boldsymbol{n}$) and (ii) to have the same orientation relative to $\boldsymbol{n}^+$ as $\boldsymbol{t}$ has relative to $\boldsymbol{n}$. These remarks, in mathematical terms, can be stated as $\boldsymbol{t}^+(\theta^i, t; \boldsymbol{n}^+) = Q(t)\,\boldsymbol{t}(\theta^i, t; \boldsymbol{n})$; and hence, under superposed rigid body motions, the stress vector transforms according to $(8.25)_2$.

To examine the manner in which $\boldsymbol{f}^*$ transforms under superposed rigid body motions, we make use of the linear momentum principle (from the three-dimensional theory) which holds for every motion and for each part $\mathscr{P}_{\mathscr{B}}$ of the body $\mathscr{B}$; and, in particular, compare a statement of this principle for two motions, one specified by $\boldsymbol{p}$ in (4.6) and another by $\boldsymbol{p}^+$ in (7.7) at time t. We observe that these two motions differ by only superposed rigid body motions and that the comparison in question is sought with reference to the same region of space (in the present configuration) at time t. For this purpose let $\mathscr{P}^1_{\mathscr{B}}$ and $\mathscr{P}^2_{\mathscr{B}}$ refer to two different materials rgions of the body $\mathscr{B}$ with at least one material point in common. Further let $\mathscr{P}^1_{\mathscr{B}}$ and $\mathscr{P}^2_{\mathscr{B}}$ occupy the same region of space at time t designated by[12] $\bar{\mathscr{P}}$ with the common material point coincident in the present configuration at

[11] In addition to the mass density noted above, examples of kinematic quantities which transform according to (8.25) under superposed rigid body motions are a, $\dot{a}$, $\boldsymbol{a}_\alpha$ and η_α. See $(5.54)_1$, $(5.55)_{2,3}$ and $(5.38)_1$.

[12] Our notation here for $\mathscr{P}^1_{\mathscr{B}}$ and $\mathscr{P}^2_{\mathscr{B}}$ is temporary and in line with that adopted in Sect. 4 [following (4.42)].

time t. The linear momentum principle corresponding to the motion (4.6) and for the part $\mathscr{P}^1_{\mathscr{B}}$ in the present configuration may be stated in the form[13]

$$\int_{\bar{\mathscr{P}}} \varrho^* \dot{\boldsymbol{v}}^* \, dv = \int_{\bar{\mathscr{P}}} \varrho^* \boldsymbol{f}^* \, dv + \int_{\partial\bar{\mathscr{P}}} \boldsymbol{t} \, d\sigma. \tag{8.26}$$

Similarly, we may write

$$\int_{\bar{\mathscr{P}}} \varrho^{*+} \dot{\boldsymbol{v}}^{*+} \, dv = \int_{\bar{\mathscr{P}}} \varrho^{*+} \boldsymbol{f}^{*+} \, dv + \int_{\partial\bar{\mathscr{P}}} \boldsymbol{t}^+ \, d\sigma, \tag{8.27}$$

corresponding to the motion (7.7) and for the part $\mathscr{P}^2_{\mathscr{B}}$ in the present configuration. But, since $\boldsymbol{t}^+$ transforms by $(8.25)_2$, the last integral in (8.27) can be written as

$$\int_{\partial\bar{\mathscr{P}}} \boldsymbol{t}^+ \, d\sigma = Q \int_{\partial\bar{\mathscr{P}}} \boldsymbol{t} \, d\sigma = Q \int_{\bar{\mathscr{P}}} \varrho^* (\dot{\boldsymbol{v}}^* - \boldsymbol{f}^*) \, dv, \tag{8.28}$$

where (8.26) has been used. Substitution of (8.28) into (8.27) yields

$$Q \int_{\bar{\mathscr{P}}} \varrho^* (\dot{\boldsymbol{v}}^* - \boldsymbol{f}^*) \, dv = \int_{\bar{\mathscr{P}}} \varrho^{*+} (\dot{\boldsymbol{v}}^{*+} - \boldsymbol{f}^{*+}) \, dv \Rightarrow Q(\dot{\boldsymbol{v}}^* - \boldsymbol{f}^*) = \dot{\boldsymbol{v}}^{*+} - \boldsymbol{f}^{*+}, \tag{8.29}$$

since $(8.29)_1$ holds for each part $\bar{\mathscr{P}}$ and since ϱ^* transforms according to $(8.25)_1$. Thus, $(8.29)_2$ provides the relation between $\boldsymbol{f}^{*+}$ and $\boldsymbol{f}^*$ under the distance preserving transformation (7.7).

Returning to our main objective, we first observe that the mechanical fields (corresponding to the stress vector and the body force per unit mass of the three-dimensional theory) in the theory of a Cosserat surface under consideration are specified by $\{\boldsymbol{N}, \boldsymbol{M}\}$ and $\{\boldsymbol{f}, \boldsymbol{l}\}$. Each of the vector fields $\boldsymbol{N}$ and $\boldsymbol{M}$ depends on the unit normal vector $\boldsymbol{\nu}$ [defined by (8.2)] which transforms by $(8.25)_2$ under superposed rigid body motions. By an argument which entirely parallels that for the stress vector in the paragraph preceding the last, we can motivate the manner in which the curve force vector $\boldsymbol{N}$ and the curve director couple $\boldsymbol{M}$ are affected when a second motion of $\mathscr{C}$ differs from (5.1) only by superposed rigid body motions. In this way, we arrive at the conclusion that both $\boldsymbol{N}$ and $\boldsymbol{M}$ must transform according to $(8.25)_2$ under the transformation $(5.36)_1$. Consider next the conservation law $(8.17)_2$. With the help of (5.66) and after invoking (8.18), $(8.17)_2$ can be written in the form

$$\int_{\mathscr{P}} \varrho \, \dot{\boldsymbol{v}} \, d\sigma = \int_{\mathscr{P}} \varrho \, \boldsymbol{f} \, d\sigma + \int_{\partial\mathscr{P}} \boldsymbol{N} \, ds,$$

which is the analogue of (8.26) for the Cosserat surface. By repeating the steps between (8.26)–(8.29) but now applied to the last equation, we arrive at the conclusion that $\bar{\boldsymbol{f}}$ defined by $(8.19)_1$ must transform according to $(8.25)_2$. In a similar manner, by considering the conservation laws $(8.17)_{3,4}$ and remembering also that the director transforms by $(5.36)_2$, we can motivate the manner in which the vector fields $\boldsymbol{m}$ and $\bar{\boldsymbol{l}}$ defined by $(8.19)_2$ must transform under superposed rigid body motions (5.36). The above background consideration for the invariance requirements can be extended to thermal fields. For example (as in the three-dimensional theory), we expect such scalar fields as the specific internal energy, entropy, temperature and the heat flux h to remain unaffected by superposed rigid body motions.

[13] In this form the conservation of mass is already invoked.

The foregoing discussion was intended to serve as background and motivation for invariance conditions under superposed rigid body motions which we now postulate:

$$\varrho^+ = \varrho, \quad h^+ = h, \tag{8.30}$$

$$\varepsilon^+ = \varepsilon, \quad \eta^+ = \eta, \quad \theta^+ = \theta, \tag{8.31}$$

$$\boldsymbol{N}^+ = Q(t)\,\boldsymbol{N}, \quad \boldsymbol{M}^+ = Q(t)\,\boldsymbol{M}, \quad \boldsymbol{m}^+ = Q(t)\,\boldsymbol{m} \tag{8.32}$$

and

$$r^+ = r, \quad \bar{\boldsymbol{f}}^+ = Q(t)\,\bar{\boldsymbol{f}}, \quad \bar{\boldsymbol{l}}^+ = Q(t)\,\bar{\boldsymbol{l}}. \tag{8.33}$$

These conditions may be viewed as physical requirements—consistent with the conservation laws—imposed on certain physical quantities when the motion of the Cosserat surface differs from (5.1) only by superposed rigid body motions. In subsequent developments, we shall frequently encounter functions and fields [such as those in (8.30)–(8.33)] whose values are scalars and vectors and which obey transformation laws of the type (8.25). For brevity we may refer to such functions and fields as *objective*.[14]

It is clear that the requirements (8.30)–(8.33) impose restrictions on the class of admissible functions characterizing the thermo-mechanical response of the material; however, we have imposed no restrictions relating to any symmetry that the material may possess. Finally, we may adopt the terminology of *equivalent processes* often employed in the three-dimensional theory of continuum mechanics. Thus, recalling that a process is specified here by the motion of the Cosserat surface, the entropy η and the functions $\mathfrak{M}$, $\mathfrak{T}$ and $\mathfrak{B}$ in (8.23), we say two processes are *equivalent* if the motions are related by (5.36) and if the temperature and the various functions in (8.23) are related by the transformations (8.30)–(8.33).

δ) *An alternative statement of the conservation laws.* Starting with the balance of energy $(8.17)_5$ which holds for every motion of the Cosserat surface, we include here a derivation which shows that three of the conservation laws, namely $(8.17)_{1,2,4}$, can be deduced from the energy balance together with the postulated invariance requirements (8.30)–(8.33) under superposed rigid body motions.[15] Such a derivation, apart from any intrinsic value that it may have, is often useful in the construction of a dynamical theory of the type under consideration (or for other generalized continua) and provides some insight into the nature of the conservation laws.

Consider first a special motion of the type (5.36), namely one for which

$$Q(t) = I, \quad \dot{Q}(t) = 0. \tag{8.34}$$

Recalling (5.45) and $(5.52)_1$, we see that the superposed velocity $\boldsymbol{v}^+$ and the superposed director velocity $\boldsymbol{w}^+$ in the above special motion include those which may be obtained from the linear transformations of the forms

$$\begin{aligned} &\boldsymbol{v} \to \boldsymbol{v} + \boldsymbol{b}, \quad \boldsymbol{b} = \text{const}, \\ &\boldsymbol{w} \to \boldsymbol{w}, \end{aligned} \tag{8.35}$$

[14] The use of the term *objective* here is different from the corresponding usage by many who appeal to the *principle of material frame-indifference* and thus allow Q to be an orthogonal tensor and regard $(5.36)_1$ as a change of frame. For a discussion of differences between invariance requirements under superposed rigid body motions and the requirements demanded by the principle of material frame-indifference, see TRUESDELL and NOLL [1965, *9*].

[15] The invariance requirements (8.30)–(8.32) must be utilized in the development of constitutive equations for all materials and are not additional assumptions in any complete theory.

the first of which corresponds to a superposed *uniform* rigid body translational velocity of the continuum while the director velocity $\boldsymbol{w}$ remains unaltered. Moreover, the fields

$$\varrho, r, h, \varepsilon, \boldsymbol{N}, \boldsymbol{M}, \bar{\boldsymbol{f}}, \bar{\boldsymbol{l}} \tag{8.36}$$

all remain unchanged under the above transformation, in view of (8.30)–(8.33). On the other hand, the kinetic energy (8.11) and the rate of work expressions $(8.8)_{2,3}$ become[16]

$$\begin{aligned} &\mathcal{K} \to \mathcal{K} + \boldsymbol{b} \cdot \int_{\mathcal{P}} \varrho \boldsymbol{v}\, d\sigma + \tfrac{1}{2}(\boldsymbol{b} \cdot \boldsymbol{b}) \int_{\mathcal{P}} \varrho\, d\sigma, \\ &R_b \to R_b + \boldsymbol{b} \cdot \int_{\mathcal{P}} \varrho \boldsymbol{f}\, d\sigma, \\ &R_c \to R_c + \boldsymbol{b} \cdot \int_{\partial\mathcal{P}} \boldsymbol{N}\, ds, \end{aligned} \tag{8.37}$$

since $\boldsymbol{b}$ is a constant vector. Also,

$$\begin{aligned} &\frac{d\mathcal{E}}{dt} \to \frac{d\mathcal{E}}{dt}, \\ &\left(\frac{d\mathcal{K}}{dt} - R_b\right) \to \left(\frac{d\mathcal{K}}{dt} - R_b\right) + \boldsymbol{b} \cdot \left[\frac{d}{dt} \int_{\mathcal{P}} \varrho \boldsymbol{v}\, d\sigma - \int_{\mathcal{P}} \varrho \boldsymbol{f}\, d\sigma\right] \\ &\qquad + \frac{1}{2} (\boldsymbol{b} \cdot \boldsymbol{b}) \frac{d}{dt} \int_{\mathcal{P}} \varrho\, d\sigma. \end{aligned} \tag{8.38}$$

Thus, if in the balance of energy $(8.17)_5$ we replace $\boldsymbol{v}$ and $\boldsymbol{w}$ by the transformations (8.35) and keep in mind (8.37)–(8.38), then after subtraction we deduce

$$\boldsymbol{b} \cdot \left\{\frac{d}{dt} \int_{\mathcal{P}} \varrho \boldsymbol{v}\, d\sigma - \int_{\mathcal{P}} \varrho \boldsymbol{f}\, d\sigma - \int_{\partial\mathcal{P}} \boldsymbol{N}\, ds\right\} + \frac{1}{2} (\boldsymbol{b} \cdot \boldsymbol{b}) \frac{d}{dt} \int_{\mathcal{P}} \varrho\, d\sigma = 0, \tag{8.39}$$

which holds for all arbitrary $\boldsymbol{b}$. By replacing $\boldsymbol{b}$ by $\beta\, \boldsymbol{b}$, β being an arbitrary scalar, from (8.39) we obtain the Eq. $(8.17)_1$ for conservation of mass and

$$\frac{d}{dt} \int_{\mathcal{P}} \varrho \boldsymbol{v}\, d\sigma - \int_{\mathcal{P}} \varrho \boldsymbol{f}\, d\sigma - \int_{\partial\mathcal{P}} \boldsymbol{N}\, ds = 0, \tag{8.40}$$

which is the conservation law $(8.17)_2$.

Next, consider a motion of the type (5.36) in which the position $\boldsymbol{r}^+$ is related to $\boldsymbol{r}$ by rotation alone while Ω in (5.39) is a constant tensor, i.e.,

$$\boldsymbol{r}^+ = Q(t)\, \boldsymbol{r}, \quad \boldsymbol{d}^+ = Q(t)\, \boldsymbol{d}, \quad \Omega = \Omega_0 = \text{const}. \tag{8.41}$$

In this case, corresponding to (5.47),

$$\dot{Q}(t) = \Omega_0\, Q(t), \quad \dot{Q}(t)^T = Q(t)^T\, \Omega_0^T = -Q(t)^T\, \Omega_0, \quad \ddot{Q}(t) = \Omega_0^2\, Q(t), \tag{8.42}$$

$$\begin{aligned} \boldsymbol{v}^+ &= Q\, \boldsymbol{v} + \dot{Q}\, \boldsymbol{r} = Q\, \boldsymbol{v} + \Omega_0\, Q\, \boldsymbol{r}, \\ \dot{\boldsymbol{v}}^+ &= Q\, \dot{\boldsymbol{v}} + 2\Omega_0\, Q\, \boldsymbol{v} + \Omega_0^2\, Q\, \boldsymbol{r}, \end{aligned} \tag{8.43}$$

and the kinetic energy per unit mass due to velocity $\boldsymbol{v}^+$ is

$$\begin{aligned} \tfrac{1}{2} \boldsymbol{v}^+ \cdot \boldsymbol{v}^+ &= \tfrac{1}{2}(Q\, \boldsymbol{v} + \Omega_0\, Q\, \boldsymbol{r}) \cdot (Q\, \boldsymbol{v} + \Omega_0\, Q\, \boldsymbol{r}) \\ &= \tfrac{1}{2} \boldsymbol{v} \cdot \boldsymbol{v} + \boldsymbol{v} \cdot Q^T\, \Omega_0\, Q\, \boldsymbol{r} + \tfrac{1}{2}(\Omega_0\, Q\, \boldsymbol{r}) \cdot (\Omega_0\, Q\, \boldsymbol{r}) \end{aligned} \tag{8.44}$$

[16] In (8.37)–(8.38) and again in (8.52)–(8.53), for brevity we have used the symbols $\mathcal{E}$ and $\mathcal{K}$ in place of the integrals in (8.9) and (8.11).

with similar expressions for $\boldsymbol{w}^+$, $\dot{\boldsymbol{w}}^+$ and $\frac{1}{2}\alpha\,\boldsymbol{w}^+\cdot\boldsymbol{w}^+$ which we do not record here. We now observe that

$$\frac{d}{dt}Q^T\Omega_0 Q=\dot{Q}^T\Omega_0 Q+Q^T\Omega_0\dot{Q}=-Q^T\Omega_0^2 Q+Q^T\Omega_0^2 Q=0. \tag{8.45}$$

Hence the quantity $Q^T\Omega_0 Q$ which occurs in the second term on the right-hand side of (8.44) is independent of time and is equal to its initial value Ω_0 for all time. For later reference, we record here the identities

$$\Omega_0 Q\,\boldsymbol{v}\cdot Q\,\boldsymbol{v}=Q^T\Omega_0 Q\,\boldsymbol{v}\cdot\boldsymbol{v}=\Omega_0\boldsymbol{v}\cdot\boldsymbol{v}=0 \tag{8.46}$$

and

$$\begin{aligned}2\Omega_0 Q\,\boldsymbol{v}\cdot\Omega_0 Q\,\boldsymbol{r}+\Omega_0^2 Q\,\boldsymbol{r}\cdot Q\,\boldsymbol{v}&=2\Omega_0 Q\,\boldsymbol{v}\cdot\Omega_0 Q\,\boldsymbol{r}+\Omega_0 Q\,\boldsymbol{r}\cdot\Omega_0^T Q\,\boldsymbol{v}\\&=\Omega_0 Q\,\boldsymbol{v}\cdot\Omega_0 Q\,\boldsymbol{r},\end{aligned} \tag{8.47}$$

where in obtaining (8.46) we have made use of (8.45). By $(8.33)_2$, (8.43) and (8.46)–(8.47),

$$\begin{aligned}\boldsymbol{f}^+&=Q(\boldsymbol{f}-\dot{\boldsymbol{v}})+\dot{\boldsymbol{v}}^+=Q\,\boldsymbol{f}+2\Omega_0 Q\,\boldsymbol{v}+\Omega_0^2 Q\,\boldsymbol{r},\\ \boldsymbol{f}^+\cdot\boldsymbol{v}^+&=\boldsymbol{f}\cdot\boldsymbol{v}+\boldsymbol{f}\cdot Q^T\Omega_0 Q\,\boldsymbol{r}+[(\Omega_0 Q\,\boldsymbol{v})\cdot(\Omega_0 Q\,\boldsymbol{r})+(\Omega_0^2\,Q\,\boldsymbol{r})\cdot(\Omega_0 Q\,\boldsymbol{r})]\end{aligned} \tag{8.48}$$

with similar expressions for $\boldsymbol{l}^+$ and $\boldsymbol{l}^+\cdot\boldsymbol{w}^+$. Also, in view of (8.18) and recalling (5.66), we have

$$\begin{aligned}&\frac{d}{dt}\int_{\mathscr{P}}\varrho\left[\frac{1}{2}(\Omega_0 Q\,\boldsymbol{r})\cdot(\Omega_0 Q\,\boldsymbol{r})\right]d\sigma\\&\qquad=\int_{\mathscr{P}}\varrho[(\Omega_0 Q\,\boldsymbol{v})\cdot(\Omega_0 Q\,\boldsymbol{r})+(\Omega_0^2 Q\,\boldsymbol{r})\cdot(\Omega_0 Q\,\boldsymbol{r})]\,d\sigma,\end{aligned} \tag{8.49}$$

where the quantity in the square bracket on the right-hand side of (8.49) is the material time derivative of the third term on the right-hand side of (8.44).

Consider now a special case of the motion (8.41) with Q corresponding to (5.46) so that (5.48) holds. Then, the superposed velocity $\boldsymbol{v}^+$ and the superposed director velocity $\boldsymbol{w}^+$ of the special motion under consideration are given by the linear transformations $(5.57)_1$ and $(5.59)_1$ which correspond to a superposed *uniform* rigid body angular velocity, the continuum occupying the same position at time t. Under these transformations, the fields (8.36) all remain unchanged, in view of (8.30)–(8.33). Keeping the results (8.44)–(8.49) in mind, with the use of identities of the type

$$(\boldsymbol{\omega}_0\times\boldsymbol{r})\cdot\boldsymbol{v}=\boldsymbol{\omega}_0\cdot(\boldsymbol{r}\times\boldsymbol{v}),\qquad \boldsymbol{v}\times\boldsymbol{v}\equiv 0 \tag{8.50}$$

and the temporary notation

$$\begin{aligned}\mathscr{W}=&(\boldsymbol{\omega}_0\times\boldsymbol{v})\cdot(\boldsymbol{\omega}_0\times\boldsymbol{r})+(\boldsymbol{\omega}_0\times\boldsymbol{\omega}_0\times\boldsymbol{r})\cdot(\boldsymbol{\omega}_0\times\boldsymbol{r})\\&+\alpha[(\boldsymbol{\omega}_0\times\boldsymbol{w})\cdot(\boldsymbol{\omega}_0\times\boldsymbol{d})+(\boldsymbol{\omega}_0\times\boldsymbol{\omega}_0\times\boldsymbol{d})\cdot(\boldsymbol{\omega}_0\times\boldsymbol{d}),\end{aligned} \tag{8.51}$$

we see that the rate of the kinetic energy and the expressions for R_b and R_c transform according to

$$\begin{aligned}\frac{d\mathscr{K}}{dt}&\rightarrow\frac{d\mathscr{K}}{dt}+\boldsymbol{\omega}_0\cdot\frac{d}{dt}\int_{\mathscr{P}}\varrho\,\mathscr{A}\,d\sigma+\int_{\mathscr{P}}\varrho\,\mathscr{W}\,d\sigma,\\ R_b&\rightarrow R_b+\boldsymbol{\omega}_0\cdot\int_{\mathscr{P}}\varrho[\boldsymbol{r}\times\boldsymbol{f}+\boldsymbol{d}\times\boldsymbol{l}]\,d\sigma+\int_{\mathscr{P}}\varrho\,\mathscr{W}\,d\sigma,\\ R_c&\rightarrow R_c+\boldsymbol{\omega}_0\cdot\int_{\partial\mathscr{P}}(\boldsymbol{r}\times\boldsymbol{N}+\boldsymbol{d}\times\boldsymbol{M})\,ds,\end{aligned} \tag{8.52}$$

since $\boldsymbol{\omega}_0$ is a constant vector. Also,

$$\frac{d\mathscr{E}}{dt} \to \frac{d\mathscr{E}}{dt},$$
$$\left(\frac{d\mathscr{K}}{dt} - R_b\right) \to \left(\frac{d\mathscr{K}}{dt} - R_b\right) + \boldsymbol{\omega}_0 \cdot \left\{\frac{d}{dt}\int_{\mathscr{P}} \varrho\, \mathscr{A}\, d\sigma - \boldsymbol{A}_b \cdot\right\}. \tag{8.53}$$

Thus, if in the balance of energy $(8.17)_5$, we replace $\boldsymbol{v}$ and $\boldsymbol{w}$ by the transformations $(5.57)_1$ and $(5.59)_1$ and use (8.52)–(8.53), then after subtraction and since $\boldsymbol{\omega}_0$ is a constant vector we arrive at the conservation law $(8.17)_4$ or equivalently

$$\frac{d}{dt}\int_{\mathscr{P}} \varrho\, \mathscr{A}\, d\sigma - \int_{\mathscr{P}} \varrho\,[\boldsymbol{r}\times\boldsymbol{f} + \boldsymbol{d}\times\boldsymbol{l}]\, d\sigma - \int_{\partial\mathscr{P}} (\boldsymbol{r}\times\boldsymbol{N} + \boldsymbol{d}\times\boldsymbol{M})\, ds = 0, \tag{8.54}$$

where $\mathscr{A}$ is defined by (8.13) and we have used (8.14) in writing the right-hand side of (8.54).

In the foregoing derivation, the conservation of mass in the form $(8.17)_1$, (8.40) and (8.54) have been deduced with the use of the invariance requirements (8.30)–(8.33) and the balance of energy $(8.17)_5$. This clearly shows that the four conservation laws $(8.17)_{1,2,4,5}$ are *equivalent* to the balance of energy and the invariance requirements under superposed rigid body motions.[17]

ε) *Conservation laws in terms of field quantities in a reference state.* The conservation laws (8.17) have been stated with reference to the present configuration and in terms of field quantities which are measured per unit length, per unit area or per unit mass of $\mathscr{s}$ in the present configuration. For certain purposes, it is more convenient to have available a statement of these conservation laws in terms of field quantities which are measured per unit length, per unit area or per unit mass of the material surface of $\mathscr{C}$ in a reference configuration.

Although the basic structure of the conservation laws (8.17) remains unaltered, certain modifications are necessary in the definitions of some of the field quantities. We assume, in what follows, that a reference configuration of the Cosserat surface is specified by (5.2) and do not insist that it be necessarily an initial configuration. However, in order to simplify the notation as much as possible, we continue (i) to refer to each part $\mathscr{P}_{\mathscr{C}}$ of the Cosserat surface by $\mathscr{P}_0$ as in (4.36), (ii) to identify the material surface of $\mathscr{C}$ in the reference configuration by $\mathscr{S}$ and (iii) to designate the mass density in the reference configuration by ϱ_0. Now, let C be a closed curve on the reference surface $\mathscr{S}$ which becomes a curve c on $\mathscr{s}$ in the present configuration, let $\theta^\alpha = \theta^\alpha(S)$ be the parametric equations of

---

[17] The idea of this manner of obtaining the conservation laws from the balance of energy and the invariance requirements under superposed rigid body motions was evidently known to ERICKSEN (see the final remark in his [1961, *1*] paper on liquid crystals). An explicit derivation of this kind in the context of the three-dimensional theory of continuum mechanics was first given by GREEN and RIVLIN [1964, *3*], where Cauchy's equations of motion and the local equation for conservation of mass are derived from the balance of energy and the invariance requirements under superposed rigid body motions. Although the spirit of our development in Subsect. δ) is patterned after that given by GREEN, NAGHDI and WAINWRIGHT [1965, *4*] for a Cosserat surface, the method of derivation differs somewhat from that in [1964, *3*] and [1965, *4*].

The conservation law for the director momentum in the form $(8.17)_3$ cannot be deduced from the balance of energy and the invariance requirements under superposed rigid body motions. In their derivation of the field equations via the energy balance and the invariance requirements, GREEN, NAGHDI and WAINWRIGHT [1965, *4*] did not explicitly state a separate postulate for the integral form of the director momentum principle but assumed a local form of the equations of motion for the director couple which can be deduced from $(8.17)_3$.

the curve C with S as the arc parameter and let

$$ {}_0\boldsymbol{\nu} = {}_0\nu_\alpha \boldsymbol{A}^\alpha = {}_0\nu^\alpha \boldsymbol{A}_\alpha \tag{8.55} $$

be the outward unit normal to C lying in the surface $\mathscr{S}$. Formulae analogous to (8.1)–(8.3) hold also for ${}_0\boldsymbol{\nu}$ and the unit tangent vector ${}_0\boldsymbol{\lambda}$ to C on $\mathscr{S}$.

We introduce here a contact force and a contact director couple, each of which acts across c on $\mathfrak{s}$ but is measured per unit length of C on $\mathscr{S}$. Thus, the three-dimensional vector field ${}_R\boldsymbol{N} = {}_R\boldsymbol{N}(\theta^\alpha, t; {}_0\boldsymbol{\nu})$ defined for points $\boldsymbol{r}$ on c and measured per unit length of C will be called a contact force if the scalar ${}_R\boldsymbol{N}\cdot\boldsymbol{v}$ is a rate of work per unit length of C for all arbitrary velocity fields $\boldsymbol{v}$. Similarly, the three-dimensional vector field ${}_R\boldsymbol{M} = {}_R\boldsymbol{M}(\theta^\alpha, t; {}_0\boldsymbol{\nu})$ defined for points $\boldsymbol{r}$ on c and measured per unit length of C will be called a contact director couple if the scalar ${}_R\boldsymbol{M}\cdot\boldsymbol{w}$ is a rate of work per unit length of C for all arbitrary director velocity $\boldsymbol{w}$. It is clear that the above definitions for ${}_R\boldsymbol{N}$ and ${}_R\boldsymbol{M}$ are analogous to that of the Piola-Kirchhoff stress vector in the classical three-dimensional theory. In terms of ${}_R\boldsymbol{N}$ and ${}_R\boldsymbol{M}$ and the notation of (4.36), the resultant contact force and the resultant contact director couple exerted on the part $\mathscr{P}_{\mathscr{C}}$ of the Cosserat surface at time t are defined by the line integrals

$$ \int_{\partial\mathscr{P}_0} {}_R\boldsymbol{N}\, dS, \quad \int_{\partial\mathscr{P}_0} {}_R\boldsymbol{M}\, dS, \tag{8.56} $$

the integration being over the boundary $\partial\mathscr{P}_0$ of $\mathscr{P}_0$ in the reference configuration. Also, let ${}_Rh = {}_Rh(\theta^\alpha, t; {}_0\boldsymbol{\nu})$ be the heat flux per unit time, acting across a curve c on $\mathfrak{s}$ but measured per unit length of C in the reference configuration. The line integral

$$ \int_{\partial\mathscr{P}_0} {}_Rh\, dS, \tag{8.57} $$

where $\partial\mathscr{P}_0$ is the boundary of $\mathscr{P}_0$, defines the heat per unit time entering the surface by conduction.

The expressions (8.56) and (8.57) in terms of ${}_R\boldsymbol{N}$, ${}_R\boldsymbol{M}$, ${}_Rh$ define resultants which parallel the line integrals in (8.4) and (8.10). In a like manner, we introduce now a vector field ${}_R\boldsymbol{m} = {}_R\boldsymbol{m}(\theta^\alpha, t)$ as an intrinsic (surface) director couple per unit area of $\mathscr{S}$. The remaining field quantities were defined previously [Sect. 9, Subsect. α)] per unit mass and hence require no new definition. The only modification occurs in the surface integrals such as those in (8.6) and $(8.8)_3$, where now $\varrho\, d\sigma$ is replaced by $\varrho_0\, d\Sigma$ and the integration is over $\mathscr{P}_0$ in the reference configuration. With the above background, the conservation laws (8.17) can be easily rewritten and expressed in terms of field quantities in the reference state as follows:

$$ \begin{gathered} \frac{d}{dt}\int_{\mathscr{P}_0} \varrho_0\, d\Sigma = 0, \quad \frac{d}{dt}\int_{\mathscr{P}_0} \varrho_0 \boldsymbol{v}\, d\Sigma = \int_{\mathscr{P}_0} \varrho_0 \boldsymbol{f}\, d\Sigma + \int_{\partial\mathscr{P}_0} {}_R\boldsymbol{N}\, dS, \\ \frac{d}{dt}\int_{\mathscr{P}_0} \varrho_0\, \alpha\, \boldsymbol{w}\, d\Sigma = \int_{\mathscr{P}_0} (\varrho_0 \boldsymbol{l} - {}_R\boldsymbol{m})\, d\Sigma + \int_{\partial\mathscr{P}_0} {}_R\boldsymbol{M}\, dS, \\ \frac{d}{dt}\int_{\mathscr{P}_0} \varrho_0\, \mathscr{A}\, d\Sigma = \int_{\mathscr{P}_0} (\boldsymbol{r}\times\varrho_0\boldsymbol{f} + \boldsymbol{d}\times\varrho_0\boldsymbol{l})\, d\Sigma + \int_{\partial\mathscr{P}_0} (\boldsymbol{r}\times{}_R\boldsymbol{N} + \boldsymbol{d}\times{}_R\boldsymbol{M})\, dS, \\ \frac{d}{dt}\int_{\mathscr{P}_0} \varrho_0 \left[\varepsilon + \frac{1}{2}(\boldsymbol{v}\cdot\boldsymbol{v} + \alpha\boldsymbol{w}\cdot\boldsymbol{w})\right] d\Sigma = \int_{\mathscr{P}_0} \varrho_0 (\boldsymbol{f}\cdot\boldsymbol{v} + \boldsymbol{l}\cdot\boldsymbol{w} + r)\, d\Sigma \\ + \int_{\partial\mathscr{P}_0} ({}_R\boldsymbol{N}\cdot\boldsymbol{v} + {}_R\boldsymbol{M}\cdot\boldsymbol{w} - {}_Rh)\, dS. \end{gathered} \tag{8.58} $$

The conservation of mass expressed by $(8.58)_1$ is equivalent to (4.36). The remaining four conservation laws in (8.58), in the order listed, correspond to the last four of (8.17).

For completeness, we also record here the entropy production corresponding to (8.22) in terms of the field quantities in the reference state. Thus, the production of entropy per unit time in a part of $\mathscr{C}$ at time t and measured relative to the reference configuration is defined by

$$\Gamma = \int_{\mathscr{P}_0} \varrho_0 \gamma \, d\Sigma = \frac{d}{dt} \int_{\mathscr{P}_0} \varrho_0 \eta \, d\Sigma - \left[\int_{\mathscr{P}_0} \varrho_0 \frac{r}{\theta} \, d\Sigma - \int_{\partial \mathscr{P}_0} \frac{R^h}{\theta} \, dS \right] \tag{8.59}$$

and satisfies the inequality (8.24).

9. Derivation of the basic field equations for shells: I. Direct approach. We derive in this section the basic field equations for a Cosserat surface from the conservation laws of the previous section. As the local form of the conservation of mass has been already given by (8.18), we shall be concerned with the remaining four conservation laws, namely (8.40), (8.54), $(8.17)_3$ and $(8.17)_5$. First we deduce the basic field equations in vector form, using an invariant vector notation, but subsequently we record these equations in terms of tensor components. In these general results, a vertical bar stands for covariant differentiation with respect to the surface metric tensor $a_{\alpha\beta}$; however, in the latter part of this section dealing with linearized field equations and the basic equations in a reference state, a vertical bar denotes covariant differentiation with respect to $A_{\alpha\beta}$ of the reference surface.

α) *General field equations in vector forms.* Consider an arbitrary part of the material region of the surface of $\mathscr{C}$ which is mapped into a part $\mathscr{P}$ of the surface $\mathscr{s}$ in the present configuration at time t. Let $\mathscr{P}$ be divided into two regions $\mathscr{P}_1$, $\mathscr{P}_2$ separated by a curve $\mathscr{p}$ on $\mathscr{s}$ in the present configuration (see Fig. 1). Further let $\partial\mathscr{P}_1$, $\partial\mathscr{P}_2$ refer to the boundaries of $\mathscr{P}_1$, $\mathscr{P}_2$, respectively; and let $\partial\mathscr{P}'$, $\partial\mathscr{P}''$ be the portions of the boundaries of $\mathscr{P}_1$, $\mathscr{P}_2$ such that $\partial\mathscr{P}' = \partial\mathscr{P}_1 \cap \partial\mathscr{P}$, $\partial\mathscr{P}'' = \partial\mathscr{P}_2 \cap \partial\mathscr{P}$. The above description can be summarized as follows:

$$\mathscr{P} = \mathscr{P}_1 \cup \mathscr{P}_2, \qquad \partial\mathscr{P} = \partial\mathscr{P}' \cup \partial\mathscr{P}'', \qquad \partial\mathscr{P}_1 = \partial\mathscr{P}' \cup \mathscr{p}, \qquad \partial\mathscr{P}_2 = \partial\mathscr{P}'' \cup \mathscr{p}. \tag{9.1}$$

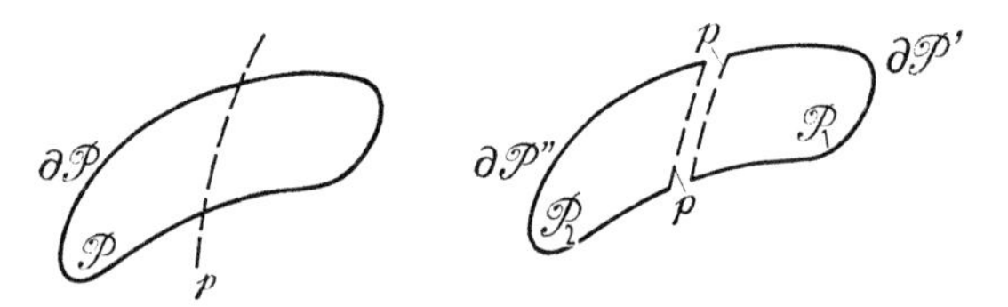

Fig. 1. A part of the surface $\mathscr{s}$ divided into two regions separated by a curve $\mathscr{p}$

Application of the principle of linear momentum (8.40) separately to the parts $\mathscr{P}_1$, $\mathscr{P}_2$ and again to $\mathscr{P}_1 \cup \mathscr{P}_2$ in the present configuration yields

$$\begin{aligned} \frac{d}{dt} \int_{\mathscr{P}_1} \varrho \boldsymbol{v} \, d\sigma - \int_{\mathscr{P}_1} \varrho \boldsymbol{f} \, d\sigma - \int_{\partial\mathscr{P}_1} \boldsymbol{N} \, ds = 0, \\ \frac{d}{dt} \int_{\mathscr{P}_2} \varrho \boldsymbol{v} \, d\sigma - \int_{\mathscr{P}_2} \varrho \boldsymbol{f} \, d\sigma - \int_{\partial\mathscr{P}_2} \boldsymbol{N} \, ds = 0 \end{aligned} \tag{9.2}$$

and

$$\frac{d}{dt}\int\limits_{\mathscr{P}_1\cup\mathscr{P}_2} \varrho\, \boldsymbol{v}\, d\sigma - \int\limits_{\mathscr{P}_1\cup\mathscr{P}_2} \varrho\, \boldsymbol{f}\, d\sigma - \int\limits_{\partial\mathscr{P}'\cup\partial\mathscr{P}''} \boldsymbol{N}\, ds = 0. \tag{9.3}$$

The curve force vector $\boldsymbol{N}$ in $(9.2)_1$ acting over the boundary $\partial\mathscr{P}_1$ (in the present configuration) is due to forces exerted *by* the material on one side of the boundary (exterior to $\mathscr{P}_1$) *on* the material of the other side ($\mathscr{P}_1$). Parallel remarks can be made for the curve force vector in $(9.2)_2$ and (9.3). Let $\boldsymbol{\nu}$ denote the outward unit normal at a point on $\mathscr{p}$ when $\mathscr{p}$ is a portion of $\partial\mathscr{P}_1$. Then, the outward unit normal at the same point on $\mathscr{p}$ when $\mathscr{p}$ is a portion of $\partial\mathscr{P}_2$ is $-\boldsymbol{\nu}$. Keeping this in mind and recalling $(9.1)_{3,4}$, from combination of $(9.2)_{1,2}$ we obtain an equation which must hold also for $\mathscr{P}_1\cup\mathscr{P}_2$. Comparison of the latter result with (9.3) gives[18]

$$\int\limits_{\mathscr{p}} [\boldsymbol{N}_{(\boldsymbol{\nu})} + \boldsymbol{N}_{(-\boldsymbol{\nu})}]\, ds = 0 \tag{9.4}$$

over the arbitrary curve $\mathscr{p}$ of $\mathscr{P}$ in the present configuration. Assuming that $\boldsymbol{N}_{(\boldsymbol{\nu})}$ is a continuous function of position and $\boldsymbol{\nu}$, from (9.4) we conclude that

$$\boldsymbol{N}_{(\boldsymbol{\nu})} = -\boldsymbol{N}_{(-\boldsymbol{\nu})}, \tag{9.5}$$

an analogue of a familiar result in classical continuum mechanics. According to (9.5), the curve force vectors acting on opposite sides of the same curve at a given point are equal in magnitude and opposite in direction.

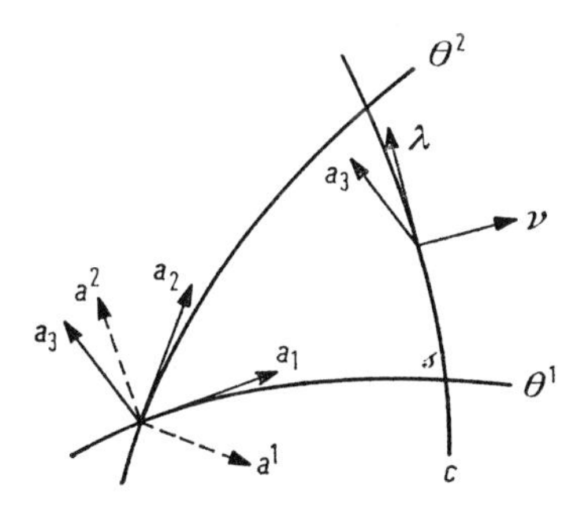

Fig. 2. An elementary curvilinear triangle on $\mathscr{s}$ bounded by coordinate curves and a curve c

Next, consider an elementary curvilinear triangle on $\mathscr{s}$ (see Fig. 2) bounded by the coordinate curves θ^1, θ^2 and a curve c with a unit tangent $\boldsymbol{\lambda}$ and an outward unit normal $\boldsymbol{\nu}$ defined by (8.1)–(8.2). Let ds and ds_α denote, respectively, the element of arc length of c and each of the coordinate curves θ^α. Then,

$$ds_1 = (a_{11})^{\frac{1}{2}}\, d\theta^1, \qquad ds_2 = (a_{22})^{\frac{1}{2}}\, d\theta^2 \tag{9.6}$$

on θ^1 and θ^2 curves, respectively. The unit tangent vectors to θ^1 and θ^2 coordinate curves are[19] $\boldsymbol{\lambda}^{(1)} = \boldsymbol{a}_1 (a_{11})^{-\frac{1}{2}}$ and $\boldsymbol{\lambda}^{(2)} = -\boldsymbol{a}_2 (a_{22})^{-\frac{1}{2}}$. Using (8.2), the outward unit normal vectors to θ^1 and θ^2 curves are given by

$$\boldsymbol{\nu}^{(1)} = -\frac{a^{\frac{1}{2}}\,\boldsymbol{a}^2}{(a_{11})^{\frac{1}{2}}} = -\frac{\boldsymbol{a}^2}{(a^{22})^{\frac{1}{2}}}, \qquad \boldsymbol{\nu}^{(2)} = -\frac{a^{\frac{1}{2}}\,\boldsymbol{a}^1}{(a_{22})^{\frac{1}{2}}} = -\frac{\boldsymbol{a}^1}{(a^{11})^{\frac{1}{2}}}, \tag{9.7}$$

---

[18] Temporarily we use $\boldsymbol{N}_{(\boldsymbol{\nu})}$ in place of $\boldsymbol{N}$, in order to emphasize its dependence on $\boldsymbol{\nu}$.

[19] Here the sign of the unit tangent vector is chosen in accord with the convention that if one proceeds along a curve in a closed circuit, the area is always kept to the left. This is consistent with $\{\boldsymbol{\nu}, \boldsymbol{\lambda}, \boldsymbol{a}_3\}$ being a right-handed triad.

respectively (see Fig. 2). Also,

$$\begin{aligned}\boldsymbol{\lambda}\, ds &= -\boldsymbol{a}_1 (a_{11})^{-\frac{1}{2}}\, ds_1 + \boldsymbol{a}_2 (a_{22})^{-\frac{1}{2}}\, ds_2 \\ &= \varepsilon^{\alpha\beta}\, \nu_\alpha\, \boldsymbol{a}_\beta\, ds = \left[\sum_{\alpha,\beta} e^{\alpha\beta} \frac{\boldsymbol{a}_\beta}{(a_{\beta\beta})^{\frac{1}{2}}}\, \nu_\alpha (a^{\alpha\alpha})^{\frac{1}{2}}\right] ds,\end{aligned} \tag{9.8}$$

where the expression in the square bracket on the right-hand side of $(9.8)_2$ is written with the help of (8.3) and $(5.63)_2$. From $(9.8)_1$ and $(9.8)_2$, we obtain the formulae

$$ds_1 = (a^{22})^{\frac{1}{2}}\, \nu_2\, ds, \qquad ds_2 = (a^{11})^{\frac{1}{2}}\, \nu_1\, ds \tag{9.9}$$

relating ds_α to ds.

We apply (8.40) to the elementary curvilinear triangle described above. Over the curve c with the outward unit normal $\boldsymbol{\nu}$ the physical curve force vector is $\boldsymbol{N}_{(\boldsymbol{\nu})}$. Let $\boldsymbol{n}^{(1)}$ and $\boldsymbol{n}^{(2)}$ denote the physical force vectors acting on the sides of the coordinate curves with the outward unit normals $\boldsymbol{a}^1/(a^{11})^{\frac{1}{2}}$ and $\boldsymbol{a}^2/(a^{22})^{\frac{1}{2}}$, respectively, so that by (9.5) the physical force vectors on the opposite sides of the coordinate curves [corresponding to the outward unit normals (9.7)] are $-\boldsymbol{n}^{(1)}$ and $-\boldsymbol{n}^{(2)}$. Assume that the fields $\varrho\, \dot{\boldsymbol{v}}$ and $\varrho\, \boldsymbol{f}$ are bounded and that $\boldsymbol{N}$ is a continuous function of position and $\boldsymbol{\nu}$. Next, estimate the surface integrals in (8.40) and apply the mean-value theorem to the line integral. Then, in the limit, as the point under consideration (the vertex of the curvilinear triangle in Fig. 2) approaches the boundary curve[20] c:

$$\boldsymbol{N}_{(\boldsymbol{\nu})}\, ds - \boldsymbol{n}^{(2)}\, ds_1 - \boldsymbol{n}^{(1)}\, ds_2 = 0. \tag{9.10}$$

From (9.10), after use of (9.9), follows the relation

$$\begin{aligned}\boldsymbol{N} = \boldsymbol{N}_{(\boldsymbol{\nu})} &= \sum_\alpha \nu_\alpha\, \boldsymbol{n}^{(\alpha)} (a^{\alpha\alpha})^{\frac{1}{2}} = \boldsymbol{N}^\alpha\, \nu_\alpha, \\ \boldsymbol{N}^\alpha &= \boldsymbol{n}^{(\alpha)} (a^{\alpha\alpha})^{\frac{1}{2}} \quad \text{(no sum on } \alpha\text{)},\end{aligned} \tag{9.11}$$

which also indicates that $\boldsymbol{N}^\alpha$ transforms as a contravariant surface vector. In obtaining the above results, the point under consideration was taken to be an interior point on $\mathscr{s}$. But the argument can easily be extended to hold if the point is on the bounding curve with a continuous tangent.

Assume now that $\boldsymbol{N}$ is continuously differentiable, $\varrho\, \dot{\boldsymbol{v}}$ and $\varrho\, \boldsymbol{f}$ are continuous and substitute $(9.11)_1$ into (8.40). Then, recalling (5.66), using (8.18) and applying Stokes' theorem to the line integral, we obtain

$$\int_{\mathscr{P}} [\varrho (\dot{\boldsymbol{v}} - \boldsymbol{f}) - \boldsymbol{N}^\alpha{}_{|\alpha}]\, d\sigma = 0, \tag{9.12}$$

which holds for each part $\mathscr{P}$ of the Cosserat surface. Hence, from the vanishing of the integrand follows the local equations of motion

$$\boldsymbol{N}^\alpha{}_{|\alpha} + \varrho\, \boldsymbol{f} = \varrho\, \dot{\boldsymbol{v}}, \tag{9.13}$$

where a vertical bar stands for covariant differentiation with respect to $a_{\alpha\beta}$.

Next, turning to (8.54), with the help of $(9.11)_1$ we write the line integral containing the term $\boldsymbol{r} \times \boldsymbol{N}$ as

$$\begin{aligned}\int_{\partial\mathscr{P}} (\boldsymbol{r} \times \boldsymbol{N})\, ds &= \int_{\partial\mathscr{P}} (\boldsymbol{r} \times \boldsymbol{N}^\alpha\, \nu_\alpha)\, ds = \int_{\mathscr{P}} (\boldsymbol{r} \times \boldsymbol{N}^\alpha)_{|\alpha}\, d\sigma \\ &= \int_{\mathscr{P}} (\boldsymbol{r} \times \boldsymbol{N}^\alpha{}_{|\alpha} + \boldsymbol{a}_\alpha \times \boldsymbol{N}^\alpha)\, d\sigma.\end{aligned} \tag{9.14}$$

---

[20] The argument in obtaining (9.10) parallels that in three-dimensional continuum mechanics. See, for example, Sect. 203 of TRUESDELL and TOUPIN [1960, *14*].

By use of (9.14) and (9.13), as well as (8.18), (8.54) can be reduced to

$$\int_{\mathscr{P}} (\boldsymbol{a}_\alpha \times \boldsymbol{N}^\alpha + \varrho\, \boldsymbol{d} \times \bar{\boldsymbol{l}})\, d\sigma + \int_{\partial\mathscr{P}} \overline{\boldsymbol{M}}\, ds = 0, \tag{9.15}$$

where we have used $(8.19)_2$ and where we have introduced the field $\overline{\boldsymbol{M}}$, which has the physical dimension of couple per unit length:

$$\overline{\boldsymbol{M}} = \boldsymbol{d} \times \boldsymbol{M}, \qquad \text{phys. dim. } \overline{\boldsymbol{M}} = \left[\frac{M L^2 T^{-2}}{L}\right] = [M L T^{-2}]. \tag{9.16}$$

Recalling (9.1), $(8.4)_2$ and (8.16), from the application of the director momentum principle $(8.17)_3$ to the parts $\mathscr{P}_1$, $\mathscr{P}_2$ and again to $\mathscr{P}_1 \cup \mathscr{P}_2$ we can deduce the result[21]

$$\boldsymbol{M}_{(\boldsymbol{\nu})} = -\boldsymbol{M}_{(-\boldsymbol{\nu})}. \tag{9.17}$$

Since the contact director couple $\boldsymbol{M}_{(\boldsymbol{\nu})}$ depends on the unit normal $\boldsymbol{\nu}$, the vector $\overline{\boldsymbol{M}}$ in $(9.16)_1$ also depends on $\boldsymbol{\nu}$. Moreover, as the director field $\boldsymbol{d}$ does not depend on $\boldsymbol{\nu}$, it follows that

$$\overline{\boldsymbol{M}}_{(\boldsymbol{\nu})} = -\overline{\boldsymbol{M}}_{(-\boldsymbol{\nu})}. \tag{9.18}$$

Let $\boldsymbol{m}^{(1)}$, $\boldsymbol{m}^{(2)}$ denote the physical director couple vectors (associated with $\boldsymbol{M} = \boldsymbol{M}_{(\boldsymbol{\nu})}$) acting on the sides of the coordinate curves whose outward unit normal vectors are $\boldsymbol{a}^1/(a^{11})^{\frac{1}{2}}$, $\boldsymbol{a}^2/(a^{22})^{\frac{1}{2}}$; and let $\overline{\boldsymbol{m}}^{(1)} = \boldsymbol{d} \times \boldsymbol{m}^{(1)}$, $\overline{\boldsymbol{m}}^{(2)} = \boldsymbol{d} \times \boldsymbol{m}^{(2)}$ be the corresponding physical quantities associated with $\overline{\boldsymbol{M}} = \overline{\boldsymbol{M}}_{(\boldsymbol{\nu})}$. Assume $\boldsymbol{d} \times \varrho\, \bar{\boldsymbol{l}}$ is bounded and that $\boldsymbol{d} \times \boldsymbol{M}$ (and therefore $\overline{\boldsymbol{M}}$) is a continuous function of position and $\boldsymbol{\nu}$. Then, with the help of (9.17) and by an argument similar to that which led to (9.11), the application of $(8.17)_3$ to an elementary curvilinear triangle yields

$$\boldsymbol{M} = \boldsymbol{M}_{(\boldsymbol{\nu})} = \boldsymbol{M}^\alpha \nu_\alpha, \qquad \boldsymbol{M}^\alpha = \boldsymbol{m}^{(\alpha)} (a^{\alpha\alpha})^{\frac{1}{2}} \quad (\text{no sum on } \alpha), \tag{9.19}$$

and we also conclude that

$$\begin{aligned} \overline{\boldsymbol{M}} &= \overline{\boldsymbol{M}}_{(\boldsymbol{\nu})} = \overline{\boldsymbol{M}}^\alpha \nu_\alpha = (\boldsymbol{d} \times \boldsymbol{M}^\alpha)\, \nu_\alpha, \\ \overline{\boldsymbol{M}}^\alpha &= \overline{\boldsymbol{m}}^{(\alpha)} (a^{\alpha\alpha})^{\frac{1}{2}} = \boldsymbol{d} \times \boldsymbol{M}^\alpha = \boldsymbol{d} \times \boldsymbol{m}^{(\alpha)} (a^{\alpha\alpha})^{\frac{1}{2}} \quad (\text{no sum on } \alpha), \end{aligned} \tag{9.20}$$

in view of $(9.16)_1$. Assume now that $\boldsymbol{M}$ is continuously differentiable and that $\boldsymbol{d} \times \varrho\, \bar{\boldsymbol{l}}$ is continuous. After introducing (9.20) into (9.15) and transforming the line integral into a surface integral by Stokes' theorem, there follows the equation

$$\boldsymbol{a}_\alpha \times \boldsymbol{N}^\alpha + \boldsymbol{d} \times \varrho\, \bar{\boldsymbol{l}} + (\boldsymbol{d} \times \boldsymbol{M}^\alpha)_{|\alpha} = 0, \tag{9.21}$$

as a consequence of the moment of momentum principle. Similarly, after introducing $(9.19)_1$ in $(8.17)_3$, recalling (5.66) and making suitable continuity assumptions, transforming the line integral into a surface integral and using also (8.18), we deduce in the usual manner the second set of equations of motion:

$$\boldsymbol{M}^\alpha{}_{|\alpha} + \varrho\, \bar{\boldsymbol{l}} = \boldsymbol{m}, \tag{9.22}$$

as a consequence of the director momentum principle. In view of (9.22), (9.21) can also be written as

$$\boldsymbol{a}_\alpha \times \boldsymbol{N}^\alpha + \boldsymbol{d} \times \boldsymbol{m} + \boldsymbol{d}_{,\alpha} \times \boldsymbol{M}^\alpha = 0. \tag{9.23}$$

[21] The argument leading to (9.17) is entirely similar to that used in obtaining (9.5). It should be evident that a statement paralleling that made following (9.5) is applicable also to (9.17) if we replace the word "force" by "director couple".

We consider next the energy balance $(8.17)_5$. Recalling (5.66), using $(9.11)_1$ and $(9.19)_1$, and after invoking (8.18), (9.13), (9.22) and rearranging the result, we reduce the energy balance to

$$\int_{\mathscr{P}} [\varrho r - \varrho \dot{\varepsilon} + \boldsymbol{N}^\alpha \cdot \boldsymbol{v}_{,\alpha} + \boldsymbol{m} \cdot \boldsymbol{w} + \boldsymbol{M}^\alpha \cdot \boldsymbol{w}_{,\alpha}]\, d\sigma - \int_{\partial\mathscr{P}} h\, ds = 0. \tag{9.24}$$

Assuming that the integrand of the surface integral is bounded and that h is a continuous function of position and $\boldsymbol{\nu}$, by repeated application of (9.24), first to $\mathscr{P}_1$, $\mathscr{P}_2$ and $\mathscr{P}_1 \cup \mathscr{P}_2$ defined in (9.1) and then to an elementary curvilinear triangle on $\mathfrak{s}$, we get[22]

$$h = h_{(\boldsymbol{\nu})} = -h_{(-\boldsymbol{\nu})}, \tag{9.25}$$

together with

$$h = q^\alpha \nu_\alpha = \boldsymbol{q} \cdot \boldsymbol{\nu}, \qquad q^\alpha = h^{(\alpha)} (a^{\alpha\alpha})^{\frac{1}{2}}, \tag{9.26}$$

where $h^{(1)}$, $h^{(2)}$ are the values of the flux of heat across the coordinate curves whose outward unit normals are $\boldsymbol{a}^1/(a^{11})^{\frac{1}{2}}$, $\boldsymbol{a}^2/(a^{22})^{\frac{1}{2}}$ and $q^\alpha = q^\alpha(\theta^\gamma, t)$ are the contravariant components of the heat flux vector

$$\boldsymbol{q} = q^\alpha \boldsymbol{a}_\alpha. \tag{9.27}$$

Substituting $(9.26)_1$ into (9.24), making the appropriate continuity assumptions and transforming the line integral into a surface integral, we finally obtain

$$\varrho r - q^\alpha{}_{|\alpha} - \varrho \dot{\varepsilon} + P = 0, \qquad P = \boldsymbol{N}^\alpha \cdot \boldsymbol{v}_{,\alpha} + \boldsymbol{m} \cdot \boldsymbol{w} + \boldsymbol{M}^\alpha \cdot \boldsymbol{w}_{,\alpha}. \tag{9.28}$$

The scalar P in (9.28) may be called the *mechanical power* and corresponds to the *stress power* in the three-dimensional theory of continuum mechanics.

As the above energy equation is not in invariant form, we consider now the previous argument about *uniform* superposed rigid body angular velocity, the continuum occupying the same position at time t. From $(9.26)_1$–(9.27) and $(8.30)_2$, it follows that under superposed rigid body motions $\boldsymbol{q}$ and q^α transform according to $(8.25)_2$ and $(8.25)_1$, respectively. Keeping this in mind and recalling (8.30)–(8.32), then under superposed rigid body motions at time t in which the values of Q and $\dot{Q}$ are specified by (5.48), we again obtain (9.23) and the local energy equation is st ll given by $(9.28)_1$ but with P in the invariant form

$$\begin{aligned} P &= \boldsymbol{N}^\alpha \cdot \boldsymbol{\eta}_\alpha + \boldsymbol{m} \cdot (\boldsymbol{\Gamma} - d^\alpha \boldsymbol{\eta}_\alpha) + \boldsymbol{M}^\alpha \cdot (\boldsymbol{\Gamma}_{:\alpha} - \lambda^\beta_{\cdot\alpha} \boldsymbol{\eta}_\beta) \\ &= \boldsymbol{N}'^\alpha \cdot \boldsymbol{\eta}_\alpha + \boldsymbol{m} \cdot \boldsymbol{\Gamma} + \boldsymbol{M}^\alpha \cdot \boldsymbol{\Gamma}_{:\alpha}, \end{aligned} \tag{9.29}$$

where various kinematic results of Sect. 5 [including $(5.57)_2$ and (5.59)] have been used and where we have introduced the notation

$$\boldsymbol{N}'^\alpha = \boldsymbol{N}^\alpha - d^\alpha \boldsymbol{m} - \lambda^\alpha_{\cdot\gamma} \boldsymbol{M}^\gamma. \tag{9.30}$$

While the above expression for $\boldsymbol{N}'^\alpha$ is a convenient notation, it should be noted that the only values of (9.30) which occur in $(9.29)_2$, namely in the term

$$\boldsymbol{N}'^\alpha \cdot \boldsymbol{\eta}_\alpha = \boldsymbol{N}'^\alpha \cdot \eta_{\beta\alpha} \boldsymbol{a}^\beta,$$

are the symmetric components

$$N'^{\alpha\beta} = \tfrac{1}{2} (\boldsymbol{N}'^\alpha \cdot \boldsymbol{a}^\beta + \boldsymbol{N}'^\beta \cdot \boldsymbol{a}^\alpha). \tag{9.31}$$

The conservation of mass (8.18), the equations of motion (9.13) and (9.22), the Eq. (9.23) [or equivalently (9.21)] and the energy equation $(9.28)_1$ are the

[22] The arguments here again parallel those used previously in obtaining (9.5) and (9.11).

basic field equations for a Cosserat surface. Each of these five sets of equations is a necessary and sufficient condition for the respective conservation laws stated in (8.17). For completeness, we also obtain here a local entropy production inequality. Recalling (8.24), we introduce $(9.26)_1$ into (8.24) and after transforming the line integral into a surface integral, we obtain the local inequality

$$\varrho\,\theta\,\gamma = \varrho\,\theta\,\dot{\eta} - \varrho\,r + q^{\alpha}{}_{|\alpha} - q^{\alpha}\frac{\theta_{,\alpha}}{\theta} \geqq 0, \tag{9.32}$$

which is deduced with the help of (8.18). From combination of (9.32) and the energy equation $(9.28)_1$ follows the inequality

$$-\varrho\,\dot{\varepsilon} + \varrho\,\theta\,\dot{\eta} + P - q^{\alpha}\frac{\theta_{,\alpha}}{\theta} \geqq 0. \tag{9.33}$$

With reference to the terminology of a process (or a thermodynamic process) introduced in Sect. 8, we now observe that to define a process it is sufficient to prescribe the nine functions $\boldsymbol{r}$, $\boldsymbol{d}$, η and those in $(8.23)_{1,2}$ such that (9.23) is satisfied. The remaining three functions $(8.23)_3$ can then be determined from (9.13), (9.22) and $(9.28)_1$. Also, in order to specify an admissible process it is sufficient to prescribe $\boldsymbol{r}$, $\boldsymbol{d}$ and η (as functions of position and time) and to specify the fields $(8.23)_{1,2}$ by constitutive equations consistent with (9.23). Then, to each such choice of $\boldsymbol{r}$, $\boldsymbol{d}$ and η, there corresponds a unique admissible process since the functions $(8.23)_3$ can be so chosen that the field equations (9.13), (9.22) and $(9.28)_1$ are satisfied. Moreover, since (8.24) was postulated to hold for every admissible process, it follows that the inequality (9.33) will serve to restrict the class of admissible processes or equivalently the constitutive equations which describe the thermo-mechanical behavior of the material. In this connection, we note that (9.23) is regarded also as a restriction on the constitutive equations.

The energy equation $(9.28)_1$, which involves the specific internal energy ε, can be written in terms of an alternative thermodynamic potential, namely the specific *Helmholtz free energy*. For this purpose, we introduce a (two-dimensional) Helmholtz function per unit mass $\psi = \psi(\theta^{\alpha}, t)$ by

$$\psi = \varepsilon - \eta\,\theta. \tag{9.34}$$

By use of (9.34), the energy equation $(9.28)_1$ can be expressed as

$$\varrho\,r - q^{\alpha}{}_{|\alpha} - \varrho(\theta\,\dot{\eta} + \dot{\theta}\,\eta) - \varrho\,\dot{\psi} + P = 0 \tag{9.35}$$

and when this is combined with (9.32), we obtain the inequality

$$-\varrho\,\dot{\psi} - \varrho\,\eta\,\dot{\theta} + P - q^{\alpha}\frac{\theta_{,\alpha}}{\theta} \geqq 0, \tag{9.36}$$

where P is given by $(9.28)_2$ or the invariant form $(9.29)_2$. The remarks made following (9.33) refer to the energy equation $(9.28)_1$ and the entropy inequality in the form (9.33) and when the entropy η is regarded as an independent thermodynamic variable. If, on the other hand, the temperature θ (as a function of position and time) is regarded as the independent thermodynamic variable (in place of η) and the energy equation (9.35) and the inequality (9.36) are employed instead of $(9.28)_1$ and (9.33), the previous remarks [following (9.33)] require only a minor modification. In particular, in order to specify an admissible process, it will be sufficient to prescribe $\boldsymbol{r}$, $\boldsymbol{d}$ and θ (as functions of position and time) and to specify the mechanical fields $(8.23)_1$, as well as ψ, η and the components of heat flux q^α, by constitutive equations consistent with (9.23).[23]

β) *Alternative forms of the field equations.* The elegance and the simplicity of the foregoing derivation does not display the relative complexity of the results. For this reason and for future use, we now deduce the basic field equations in tensor components. The contact force $\boldsymbol{N}$ and the contact couple $\boldsymbol{M}$, when referred to the base vectors $\boldsymbol{a}_i$, can be written as

$$\boldsymbol{N} = N^i\,\boldsymbol{a}_i = N^\alpha\,\boldsymbol{a}_\alpha + N^3\,\boldsymbol{a}_3, \qquad \boldsymbol{M} = M^i\,\boldsymbol{a}_i = M^\alpha\,\boldsymbol{a}_\alpha + M^3\,\boldsymbol{a}_3. \tag{9.37}$$

Similarly, the assigned force $\boldsymbol{f}$, the assigned couple $\boldsymbol{l}$ and the surface couple $\boldsymbol{m}$ can be expressed as

$$\boldsymbol{f} = f^i\,\boldsymbol{a}_i, \qquad \boldsymbol{l} = l^i\,\boldsymbol{a}_i \tag{9.38}$$

and

$$\boldsymbol{m} = m^i\,\boldsymbol{a}_i. \tag{9.39}$$

Since $\boldsymbol{N}^\alpha$ in $(9.11)_1$ transforms as a contravariant surface vector, we can put[24]

$$\boldsymbol{N}^\alpha = N^{\alpha i}\,\boldsymbol{a}_i = N^{\alpha\gamma}\,\boldsymbol{a}_\gamma + N^{\alpha 3}\,\boldsymbol{a}_3. \tag{9.40}$$

Also, by $(9.37)_1$,

$$N^i = N^{\alpha\gamma}\,\nu_\alpha, \qquad N^\gamma = N^{\alpha\gamma}\,\nu_\alpha, \qquad N^3 = N^{\alpha 3}\,\nu_\alpha, \tag{9.41}$$

where $N^{\alpha\gamma}$ and $N^{\alpha 3}$ are surface tensors under the transformation of surface coordinates. In a similar manner, since $\boldsymbol{M}^\alpha$ in $(9.19)_1$ transforms as a contravariant surface vector, we can set

$$\boldsymbol{M}^\alpha = M^{\alpha i}\,\boldsymbol{a}_i = M^{\alpha\gamma}\,\boldsymbol{a}_\gamma + M^{\alpha 3}\,\boldsymbol{a}_3 \tag{9.42}$$

and by $(9.37)_2$

$$M^i = M^{\alpha i}\,\nu_\alpha, \qquad M^\gamma = M^{\alpha\gamma}\,\nu_\alpha, \qquad M^3 = M^{\alpha 3}\,\nu_\alpha, \tag{9.43}$$

where $M^{\alpha\gamma}$ and $M^{\alpha 3}$ are surface tensors under the transformation of surface coordinates. It is also clear that $\overline{\boldsymbol{M}}$ defined by (9.16) and $\overline{\boldsymbol{M}}^\alpha$ in $(9.20)_1$ may be expressed as

$$\overline{\boldsymbol{M}} = \overline{M}^i\,\boldsymbol{a}_i, \qquad \overline{\boldsymbol{M}}^\alpha = \overline{M}^{\alpha i}\,\boldsymbol{a}_i, \qquad \overline{M}^i = \overline{M}^{\alpha i}\,\nu_\alpha, \tag{9.44}$$

the components $\overline{M}^{\alpha\gamma}$ and $\overline{M}^{\alpha 3}$ being surface tensors under the transformation of surface coordinates.

To obtain the equations of motion (9.13) in component form, consider the scalar product of (9.13) with $\boldsymbol{a}^\beta$ and again with $\boldsymbol{a}^3$ and deduce

$$\begin{aligned} (\boldsymbol{a}^\beta \cdot \boldsymbol{N}^\alpha)_{|\alpha} - \boldsymbol{N}^\alpha \cdot \boldsymbol{a}^\beta{}_{|\alpha} + \varrho\,(\boldsymbol{f} - \dot{\boldsymbol{v}}) \cdot \boldsymbol{a}^\beta = 0, \\ (\boldsymbol{a}_3 \cdot \boldsymbol{N}^\alpha)_{|\alpha} - \boldsymbol{N}^\alpha \cdot \boldsymbol{a}_{3|\alpha} + \varrho\,(\boldsymbol{f} - \dot{\boldsymbol{v}}) \cdot \boldsymbol{a}_3 = 0. \end{aligned} \tag{9.45}$$

Introducing c^i for the components of the acceleration vector by

$$\boldsymbol{c} = \dot{\boldsymbol{v}} = c^i\,\boldsymbol{a}_i \tag{9.46}$$

and using (9.40) and $(4.13)_{2,3}$, from (9.45) we deduce

$$N^{\alpha\beta}{}_{|\alpha} - b^\beta_\alpha\,N^{\alpha 3} + \varrho\,f^\beta = \varrho\,c^\beta, \qquad N^{\alpha 3}{}_{|\alpha} + b_{\alpha\beta}\,N^{\alpha\beta} + \varrho\,f^3 = \varrho\,c^3. \tag{9.47}$$

[23] The results in Subsect. α) can be easily specialized to the isothermal or the purely mechanical theory. For the isothermal theory with the heat flux $q^\alpha = 0$, r in $(9.28)_1$ or (9.35) is a specified function in order to balance the energy equation.

[24] Our notation for the order of superscripts αi in $N^{\alpha i}$ (and also in $M^{\alpha i}$ to be introduced presently) differs from that used by GREEN, NAGHDI and WAINWRIGHT [1965, *4*]. The above notation also differs from that used by GREEN and NAGHDI [1967, *4*] but is in agreement with a number of subsequent papers, e.g., [1968, *7*] and [1971, *6*].

In an entirely similar manner, using (9.42) and (9.39), we obtain from (9.22) the equations

$$M^{\alpha\beta}{}_{|\alpha} - b^{\beta}_{\alpha} M^{\alpha 3} + \varrho \bar{l}^{\beta} = m^{\beta}, \qquad M^{\alpha 3}{}_{|\alpha} + b_{\alpha\beta} M^{\alpha\beta} + \varrho \bar{l}^{3} = m^{3}, \tag{9.48}$$

where $\bar{l}^i$ are the components of $\bar{\boldsymbol{l}}$ in $(8.19)_2$ referred to the base vectors $\boldsymbol{a}_i$, i.e.,

$$\bar{\boldsymbol{l}} = \bar{l}^i \boldsymbol{a}_i. \tag{9.49}$$

In an analogous manner, from (9.23) we have

$$\varepsilon_{jim}[\delta^j_\alpha N^{\alpha i} + d^j m^i + \lambda^j_{.\alpha} M^{\alpha i}] = 0, \tag{9.50}$$

or equivalently

$$\begin{gathered} \varepsilon_{\beta\alpha}[N^{\alpha\beta} + m^\beta d^\alpha + M^{\gamma\beta} \lambda^{\alpha}_{.\gamma}] = 0, \\ N^{\alpha 3} + (m^3 d^\alpha - m^\alpha d^3) + M^{\gamma 3} \lambda^{\alpha}_{.\gamma} - M^{\gamma\alpha} \lambda^{3}_{.\gamma} = 0, \end{gathered} \tag{9.51}$$

where ε_{jim} and $\varepsilon_{\beta\alpha}$ are defined by (5.43) and (5.63) and the components of $\lambda^i_{.\alpha}$ are given by (5.28). In order to obtain the energy equation in terms of the tensor components of the various quantities in $(9.28)_1$ or (9.35), it will suffice to record here the expression for P in the form

$$P = N'^{\alpha\beta} \eta_{\alpha\beta} + m^i \dot{d}_i + M^{\alpha i} \dot{\lambda}_{i\alpha}, \tag{9.52}$$

where $N'^{\alpha\beta}$ defined by (9.31) can be written as

$$N'^{\alpha\beta} = N'^{\beta\alpha} = N^{\alpha\beta} - m^\alpha d^\beta - M^{\gamma\alpha} \lambda^{\beta}_{.\gamma}. \tag{9.53}$$

The above expression is equivalent to $(9.51)_1$.

The various field equations derived in this section, including the component forms (9.47)–(9.48) and (9.51) of the equations of motion, were obtained previously.[25] Although GREEN, NAGHDI and WAINWRIGHT did not postulate a director momentum principle corresponding to $(8.17)_3$, in the course of derivation they assumed the local equation (9.22) from which $(9.48)_{1,2}$ result. Without such an assumption, the right-hand sides of $(9.48)_{1,2}$ would be zero and the resulting theory would be too restrictive for a Cosserat surface. Earlier in Sect. 4, we called attention to a paper of ERICKSEN and TRUESDELL which contains a general development of the kinematics of oriented media; this paper also includes a derivation of equilibrium equations for shells by direct approach.[26] In the context of an oriented surface, the equilibrium equations given by ERICKSEN and TRUESDELL are incomplete or restrictive;[27] and this is mainly because their basic principles do not include a director momentum principle corresponding to $(8.17)_3$. However, if their basic principles are supplemented by $(8.17)_3$ [or equivalently (9.22)], then their equilibrium equations can be shown to be of the same form as (9.47)–(9.48) and (9.51).[28]

[25] GREEN, NAGHDI and WAINWRIGHT [1965, *4*]. The equations of equilibrium contained in the paper of COHEN and DE SILVA [1966, *2*], apart from a generality resulting from the fact that they admit a triad of deformable directors at each material point of their directed surface, are similar to (9.47)–(9.48) and (9.51).

[26] Sects. 24–27 of ERICKSEN and TRUESDELL [1958, *1*]. This derivation of equilibrium equations for shells may also be found in Sect. 212 of TRUESDELL and TOUPIN [1960, *14*].

[27] The equilibrium equations of ERICKSEN and TRUESDELL are of the type obtained in Sect. 10 for the restricted theory.

[28] In this connection and with reference to Sect. 26 of the paper by ERICKSEN and TRUESDELL [1958, *1*], we may note that their notations $\boldsymbol{S}, \boldsymbol{F}, \boldsymbol{M}, \boldsymbol{L}$ correspond, respectively, to our $\boldsymbol{N}, \varrho \boldsymbol{f}, \boldsymbol{d} \times \boldsymbol{M}, \boldsymbol{d} \times \varrho \boldsymbol{l}$.

γ) *Linearized field equations.* Previously, with reference to the linearization of the kinematical results for a Cosserat surface (Sect. 6), it was assumed that all kinematic measures such as $e_{\alpha\beta}$, $\varkappa_{i\alpha}$ and γ_i, as well as their derivatives with respect to the surface coordinates and time, are of $O(\varepsilon)$. These must now be supplemented by additional assumptions in a complete infinitesimal theory. Thus, let θ_0 and η_0 refer to a standard temperature and entropy in the (initial) reference configuration and also assume that the vector fields $\boldsymbol{N}^\alpha$, $\boldsymbol{M}^\alpha$, $\boldsymbol{m}$ are all zero in the reference configuration. We further assume that $\boldsymbol{N}^\alpha$, $\boldsymbol{M}^\alpha$, $\boldsymbol{m}$ (or their components) when expressed in suitable non-dimensional forms, as well as their derivatives with respect to the surface coordinates, are of $O(\varepsilon)$; and that $(\theta-\theta_0)/\theta_0$ and $(\eta-\eta_0)/\eta_0$ and their derivatives are $O(\varepsilon)$.

Recalling the linearization procedures of Sect. 6, we again avoid introducing additional notations but now regard $N^{\alpha i}$, $M^{\alpha i}$, m^i, etc., as well as θ and η (measured from their reference values), as infinitesimal quantities of $O(\varepsilon)$. As a result of linearization, all tensors are now referred to the initial undeformed surface and covariant differentiation is with respect to $A_{\alpha\beta}$. It follows that in the Eqs. (9.47)–(9.51) each term is $O(\varepsilon)$ and that $b_{\alpha\beta}$, d_i, $\lambda_{i\alpha}$ must be replaced to the order ε by $B_{\alpha\beta}$, D_i, $\Lambda_{i\alpha}$, respectively. We omit the details which involve a straightforward calculation and merely record the linearized versions of (9.47)–(9.51) as follows:

$$N^{\alpha\beta}{}_{|\alpha}-B^{\beta}_{\alpha}\,N^{\alpha 3}+\varrho_0\,\bar{F}^{\beta}=0,\qquad N^{\alpha 3}{}_{|\alpha}+B_{\alpha\beta}\,N^{\alpha\beta}+\varrho_0\,\bar{F}^{3}=0, \tag{9.54}$$

$$M^{\alpha\beta}{}_{|\alpha}-B^{\beta}_{\alpha}\,M^{\alpha 3}+\varrho_0\,\bar{L}^{\beta}=m^{\beta},\qquad M^{\alpha 3}{}_{|\alpha}+B_{\alpha\beta}\,M^{\alpha\beta}+\varrho_0\,\bar{L}^{3}=m^{3} \tag{9.55}$$

and

$$\begin{gathered}\bar{\varepsilon}_{\beta\alpha}\left[N^{\alpha\beta}+m^{\beta}\,D^{\alpha}+M^{\gamma\beta}\,\Lambda^{\alpha}_{\cdot\gamma}\right]=0,\\ N^{\alpha 3}+(m^{3}\,D^{\alpha}-m^{\alpha}\,D^{3})+M^{\gamma 3}\,\Lambda^{\alpha}_{\cdot\gamma}-M^{\gamma\alpha}\,\Lambda^{3}_{\cdot\gamma}=0,\end{gathered} \tag{9.56}$$

where $\bar{\varepsilon}_{\beta\alpha}$ was introduced previously in (6.30), the vertical bar in (9.54)–(9.55) and in the rest of this subsection stands for covariant differentiation with respect to $A_{\alpha\beta}$, all quantities are now referred to the base vectors $\boldsymbol{A}_i$ of the initial Cosserat surface and the notations $\bar{F}^i$ and $\bar{L}^i$ stand for the components of the vector fields $\bar{\boldsymbol{f}}$ and $\bar{\boldsymbol{l}}$ in (8.19) referred to the base vectors $\boldsymbol{A}_i$, i.e.,

$$\bar{F}^i=\bar{\boldsymbol{f}}\cdot\boldsymbol{A}^i,\qquad \bar{L}^i=\bar{\boldsymbol{l}}\cdot\boldsymbol{A}^i. \tag{9.57}$$

Also, upon linearization, (9.53) becomes

$$N'^{\alpha\beta}=N'^{\beta\alpha}=N^{\alpha\beta}-m^{\alpha}\,D^{\beta}-M^{\gamma\alpha}\,\Lambda^{\beta}_{\cdot\gamma}. \tag{9.58}$$

Before recording the linearized energy equation, we note that the heat flux h in the infinitesimal theory assumes the form

$$h={}_0\nu^{\alpha}\,Q_{\alpha}={}_0\nu_{\alpha}\,Q^{\alpha},\qquad Q^{\alpha}=\boldsymbol{q}\cdot\boldsymbol{A}^{\alpha}, \tag{9.59}$$

where Q_α defined by $(9.59)_2$ are the components of the heat flux vector per unit length (in the reference configuration) per unit time and ${}_0\nu^\alpha$ are the components of the outward unit normal ${}_0\boldsymbol{\nu}={}_0\nu^\alpha\boldsymbol{A}_\alpha$ to the θ^α-curves on the reference surface $\mathscr{S}$. By virtue of the assumptions stated above and in view of (9.52), the energy equation (9.35) becomes

$$\begin{gathered}\varrho_0\,r-Q^{\alpha}{}_{|\alpha}-\varrho_0\,(\theta\,\dot{\eta}+\eta\,\dot{\theta})-\varrho_0\,\dot{\psi}+P=0,\\ P=N'^{\alpha\beta}\,\dot{e}_{\alpha\beta}+m^{i}\,\dot{\gamma}_{i}+M^{\alpha i}\,\dot{\varkappa}_{i\alpha},\end{gathered} \tag{9.60}$$

where $N'^{\alpha\beta}$ is given by (9.58) and the infinitesimal kinematical quantities in $(9.60)_2$ are defined by (6.18). At this point the energy equation $(9.60)_1$ still contains terms of up to $O(\varepsilon^2)$ such as $\eta\dot{\theta}$ and those in P. However, as will become apparent later (in Chap. D), these terms will cancel others resulting from $\dot{\psi}$; and the final linearized (residual) energy equation, after linearization of $\theta\dot{\eta}$ term, will only contain terms of $O(\varepsilon)$ in line with the linearized versions of the other field equations.

It is instructive to discuss briefly special cases of Eqs. (9.54)–(9.58). Consider first the special case in which the initial director is along the unit normal $\boldsymbol{A}_3$ specified by (6.21). In this case, $\Lambda_{i\alpha}$ are given by $(6.22)_{2,3}$ and corresponding to the alternative set of the kinematic measures (6.22)–(6.23), we introduce the variables[29]

$$\hat{N}^{\alpha\beta} = N^{\alpha\beta}, \quad \hat{M}^{\alpha i} = D M^{\alpha i}, \tag{9.61}$$

$$\begin{aligned} V^\alpha &= N^{\alpha 3} = D m^\alpha + D B^\alpha_\gamma M^{\gamma 3} + M^{\beta\alpha} D_{,\beta}, \\ V^3 &= D m^3 - D B_{\alpha\beta} M^{\alpha\beta} + D_{,\alpha} M^{\alpha 3}. \end{aligned} \tag{9.62}$$

The first of (9.62) satisfies $(9.56)_2$ while in terms of the above variables (9.58), which satisfies $(9.56)_1$, becomes

$$N'^{\alpha\beta} = N'^{\beta\alpha} = \hat{N}^{\alpha\beta} + \hat{M}^{\gamma\alpha} B^\beta_\gamma. \tag{9.63}$$

The equations of motion (9.54)–(9.55) can then be written in the more compact forms

$$\hat{N}^{\alpha\beta}{}_{|\alpha} - B^\beta_\alpha V^\alpha + \varrho_0 \bar{F}^\beta = 0, \qquad V^\alpha{}_{|\alpha} + B_{\alpha\beta} \hat{N}^{\alpha\beta} + \varrho_0 \bar{F}^3 = 0, \tag{9.64}$$

$$\hat{M}^{\alpha\beta}{}_{|\alpha} + \varrho_0 \hat{L}^\beta = V^\beta, \qquad \hat{M}^{\alpha 3}{}_{|\alpha} + \varrho_0 \hat{L}^3 = V^3, \tag{9.65}$$

where we have put

$$\hat{L}^i = D \bar{L}^i. \tag{9.66}$$

We now further specialize the above alternative forms of the equations of motion to the case in which the initial director $\boldsymbol{D}$ is of constant magnitude and coincident with the unit normal $\boldsymbol{A}_3$. With $D=1$, the distinction between $\hat{M}^{\alpha i}$ and $M^{\alpha i}$, $\hat{L}^i$ and $\bar{L}^i$ disappears while (9.62)–(9.63) and $(9.56)_1$ reduce to

$$\begin{aligned} V^\alpha = N^{\alpha 3} = m^\alpha + B^\alpha_\gamma M^{\gamma 3}, \quad & V^3 = m^3 - B_{\alpha\beta} M^{\alpha\beta}, \\ N'^{\alpha\beta} = N'^{\beta\alpha} &= N^{\alpha\beta} + M^{\gamma\alpha} B^\beta_\gamma \end{aligned} \tag{9.67}$$

and

$$\varepsilon_{\beta\alpha} [N^{\alpha\beta} - B^\alpha_\gamma M^{\gamma\beta}] = 0. \tag{9.68}$$

Then, the equations of motion (9.64)–(9.65) can be expressed as[30]

$$N^{\alpha\beta}{}_{|\alpha} - B^\beta_\alpha V^\alpha + \varrho_0 \bar{F}^\beta = 0, \qquad V^\alpha{}_{|\alpha} + B_{\alpha\beta} N^{\alpha\beta} + \varrho_0 \bar{F}^3 = 0, \tag{9.69}$$

$$M^{\alpha\beta}{}_{|\alpha} + \varrho_0 \bar{L}^\beta = V^\beta, \qquad M^{\alpha 3}{}_{|\alpha} + \varrho_0 \bar{L}^3 = V^3. \tag{9.70}$$

Also, with reference to the energy equation (9.60), we observe that the expression for P can be expressed in an alternative form when $\boldsymbol{D} = \boldsymbol{A}_3$. Thus, in terms of the kinematic variables $e_{\alpha\beta}$, $\varrho_{i\alpha}$, γ_i in (6.24), P becomes

$$P = N'^{\alpha\beta} \dot{e}_{\alpha\beta} + M^{\alpha i} \dot{\varrho}_{i\alpha} + V^i \dot{\gamma}_i, \tag{9.71}$$

where we have used the definitions (9.67).

---

[29] [1968, *6*] and [1970, *7*]. The variables (9.61)–(9.62) are of special significance in regard to the classical linear theories of shells and plates. At this stage the forms of these new variables can be easily motivated from the energy equation (9.60) with the help of the alternative kinematic measures (6.22)–(6.23).

[30] The equations of motion (or equilibrium) in the linear theory of a Cosserat surface, either in the general forms (9.69)–(9.70) or their special cases, have been used in a number of recent papers, e.g., [1967, *4*], [1967, *6*] and [1968, *6*]. A special case of the variables (9.61)–(9.62) and the Eqs. (9.64)–(9.65) with $B_{\alpha\beta} = 0$ have been employed in the application of the theory with initial variable directors to plates of variable thickness [1971, *6*].

We close this subsection by making an observation regarding a special case of (9.69)–(9.70) in the linearized theory of an initially flat Cosserat plate for which (6.26) holds. In this case (9.67)–(9.68) reduce to

$$V^\alpha = N^{\alpha 3} = m^\alpha, \qquad V^3 = m^3, \qquad N'^{\alpha\beta} = N^{\alpha\beta} = N^{\beta\alpha}. \tag{9.72}$$

Hence, for a flat Cosserat surface, $N^{\alpha 3}$ and m^α are identical, $N^{\alpha\beta}$ is symmetric and the equations of motion (9.69)–(9.70) become

$$N^{\alpha\beta}{}_{|\alpha} + \varrho_0 \bar{F}^\beta = 0, \qquad M^{\alpha 3}{}_{|\alpha} + \varrho_0 \bar{L}^3 = V^3, \tag{9.73}$$

$$M^{\alpha\beta}{}_{|\alpha} + \varrho_0 \bar{L}^\beta = V^\beta, \qquad V^\alpha{}_{|\alpha} + \varrho_0 \bar{F}^3 = 0. \tag{9.74}$$

Evidently the variables occurring in the above equations separate into two sets, namely $\{N^{\alpha\beta}, V^3, M^{\alpha 3}; \bar{F}^\beta, \bar{L}^3\}$ and $\{M^{\alpha\beta}, V^\beta; \bar{F}^3, \bar{L}^\beta\}$, and this is analogous to the separation of the kinematic variables in (6.25), as discussed following (6.26). In fact, as will become apparent later, the former set governed by the equations of motion (9.73) is associated with the extensional motion while the latter set of variables governed by the equations of motion (9.74) is associated with the flexural motion of the Cosserat plate.

δ) *The basic field equations in terms of a reference state.* The various forms of the field equations in this section deduced from the conservation laws (8.17) involve field quantities which are measured in the present configuration. We now proceed to obtain the counterparts of these results from the conservation laws (8.58) which involve field quantities measured in the reference configuration.

First, we observe that by arguments similar to those used earlier in this section [Subsect. α)], we can readily deduce the results

$$_R\boldsymbol{N} = {}_R\boldsymbol{N}^\alpha\, {}_0\nu_\alpha, \qquad {}_R\boldsymbol{M} = {}_R\boldsymbol{M}^\alpha\, {}_0\nu_\alpha, \tag{9.75}$$

$$_Rh = {}_Rq^\alpha\, {}_0\nu_\alpha = {}_R\boldsymbol{q} \cdot {}_0\boldsymbol{\nu}, \qquad {}_R\boldsymbol{q} = {}_Rq^\alpha \boldsymbol{A}_\alpha, \tag{9.76}$$

where ${}_Rq^\alpha = {}_Rq^\alpha(\theta^\gamma; t)$ are the contravariant components of the heat flux vector ${}_R\boldsymbol{q}$ defined by $(9.76)_3$. The relation $(9.75)_1$ is the counterpart of (9.11), while $(9.75)_2$ and (9.76) follow from the conservation laws (8.58) at a stage comparable to that where (9.19) and (9.26) were obtained from those in (8.17). Next, making smoothness assumptions similar to those made in obtaining (9.13) and using $(9.75)_1$, from $(8.58)_2$ we obtain the local equations of motion in terms of ${}_R\boldsymbol{N}^\alpha$:

$$_R\boldsymbol{N}^\alpha{}_{|\alpha} + \varrho_0 \boldsymbol{f} = \varrho_0 \dot{\boldsymbol{v}}, \tag{9.77}$$

where now the vertical bar stands for covariant differentiation with respect to $A_{\alpha\beta}$ of the reference surface. The remaining field equations are obtained in a similar manner from the last three of (8.58). Thus, in place of (9.22), (9.23) and (9.28) we now have

$$_R\boldsymbol{M}^\alpha{}_{|\alpha} + \varrho_0 \boldsymbol{l} = {}_R\boldsymbol{m}, \tag{9.78}$$

$$\boldsymbol{a}_\alpha \times {}_R\boldsymbol{N}^\alpha + \boldsymbol{d} \times {}_R\boldsymbol{m} + \boldsymbol{d}_{,\alpha} \times {}_R\boldsymbol{M}^\alpha = 0, \tag{9.79}$$

$$\varrho_0\, r - {}_Rq^\alpha{}_{|\alpha} - \varrho_0\, \dot{\varepsilon} + {}_R\boldsymbol{N}^\alpha \cdot \boldsymbol{v}_{,\alpha} + {}_R\boldsymbol{m} \cdot \boldsymbol{w} + {}_R\boldsymbol{M}^\alpha \cdot \boldsymbol{w}_{,\alpha} = 0 \tag{9.80}$$

and we emphasize that the vertical bar throughout the present subsection stands for covariant differentiation with respect to $A_{\alpha\beta}$. Also, from (8.59) and (8.24), the entropy production inequality in terms of the field quantities in the reference state is given by

$$\varrho_0\, \theta\gamma = \varrho_0\, \theta\dot{\eta} - \varrho_0\, r + {}_Rq^\alpha{}_{|\alpha} - {}_Rq^\alpha \frac{\theta_{,\alpha}}{\theta} \geq 0. \tag{9.81}$$

By use of (9.34), from combination of (9.80) and (9.81) follows the inequality

$$-\varrho_0\,\dot{\psi}-\varrho_0\,\eta\,\dot{\theta}+{}_R P-{}_R q^\alpha\,\frac{\theta_{,\alpha}}{\theta}\geqq 0, \tag{9.82}$$

where

$${}_R P={}_R\boldsymbol{N}^\alpha\cdot\boldsymbol{v}_{,\alpha}+{}_R\boldsymbol{m}\cdot\boldsymbol{w}+{}_R\boldsymbol{M}^\alpha\cdot\boldsymbol{w}_{,\alpha}. \tag{9.83}$$

The inequality (9.82) is the counterpart of (9.36) and the mechanical power ${}_R P$ is the counterpart of $(9.28)_2$ in terms of the reference state.

The structure of (9.77)–(9.80) is entirely analogous to that of the field equations (9.13), (9.22), (9.23) and (9.28). The field equations in terms of a reference state and in tensor components can be obtained from (9.77)–(9.80) but we do not consider these in detail here. We note, however, that if we put

$${}_R\boldsymbol{N}^\alpha={}_R N^{\alpha i}\,\boldsymbol{A}_i,\qquad {}_R\boldsymbol{m}={}_R m^i\,\boldsymbol{A}_i,\qquad {}_R\boldsymbol{M}^\alpha={}_R M^{\alpha i}\,\boldsymbol{A}_i \tag{9.84}$$

and also refer $\boldsymbol{f}$, $\dot{\boldsymbol{v}}$, $\boldsymbol{l}$ to the reference base vectors $\boldsymbol{A}_i$, then the equations of motion in tensor components resulting from (9.77)–(9.78) will be similar in forms to the equations of motion (9.47)–(9.48).

It is of some interest to obtain the relations between the field variables ${}_R\boldsymbol{N}^\alpha$, ${}_R\boldsymbol{M}^\alpha$, ${}_R h$ and $\boldsymbol{N}^\alpha$, $\boldsymbol{M}^\alpha$, h which are measured in the present configuration. For this purpose, recall the expression for the resultant contact force as defined by the line integrals $(8.4)_1$ and $(8.56)_1$. Their integrands, by use of (9.11), $(9.75)_1$ and formulae of the type (8.1)–(8.2), can be expressed as

$$\begin{aligned}\boldsymbol{N}\,ds&=\boldsymbol{N}^\alpha\,\nu_\alpha\,ds=a^{\frac{1}{2}}\,e_{\alpha\beta}\,\boldsymbol{N}^\alpha\,\lambda^\beta\,ds=a^{\frac{1}{2}}\,e_{\alpha\beta}\,\boldsymbol{N}^\alpha\,d\theta^\beta,\\ {}_R\boldsymbol{N}\,dS&={}_R\boldsymbol{N}^\alpha\,{}_0\nu_\alpha\,dS=A^{\frac{1}{2}}\,e_{\alpha\beta}\,{}_R\boldsymbol{N}^\alpha\,{}_0\lambda^\beta\,dS=A^{\frac{1}{2}}\,e_{\alpha\beta}\,{}_R\boldsymbol{N}^\alpha\,d\theta^\beta,\end{aligned} \tag{9.85}$$

where $e_{\alpha\beta}$ is the permutation symbol defined in (5.63) and where ${}_0\lambda^\beta$ are the components of the unit tangent vector to the curve C lying in the reference surface. Since for a given Cosserat surface in the present configuration, the resultants $(8.4)_1$ and $(8.56)_1$ must be the same, it follows that $\boldsymbol{N}\,ds={}_R\boldsymbol{N}\,dS$ and hence by (9.85) and (4.41) we have

$$a^{\frac{1}{2}}\,\boldsymbol{N}^\alpha=A^{\frac{1}{2}}\,{}_R\boldsymbol{N}^\alpha\quad\text{or}\quad {}_R\boldsymbol{N}^\alpha=J\,\boldsymbol{N}^\alpha. \tag{9.86}$$

Similar results hold between $\boldsymbol{M}^\alpha$ and ${}_R\boldsymbol{M}^\alpha$, between q^α and ${}_R q^\alpha$ and between $\boldsymbol{m}$ and ${}_R\boldsymbol{m}$. The latter follows from the fact that $\boldsymbol{m}\,d\sigma={}_R\boldsymbol{m}\,d\Sigma$. For later reference, we summarize the last results as

$$\{{}_R\boldsymbol{N}^\alpha,\,{}_R\boldsymbol{m},\,{}_R\boldsymbol{M}^\alpha,\,{}_R q^\alpha\}=\frac{\varrho_0}{\varrho}\,\{\boldsymbol{N}^\alpha,\,\boldsymbol{m},\,\boldsymbol{M}^\alpha,\,q^\alpha\}, \tag{9.87}$$

where we have also used (4.42).

10. Derivation of the basic field equations of a restricted theory: I. Direct approach. Our derivation of the field equations by a direct approach in Sects. 8–9 is founded on the concept of a Cosserat surface whose basic kinematic variables are the position vector $\boldsymbol{r}$ and the director $\boldsymbol{d}$. As already mentioned in Sect. 5 [Subsect. γ)], other developments by direct approach in which a director is not admitted are possible. For example, we may consider a material surface embedded in a Euclidean 3-space and construct a theory in which the basic kinematic ingredients are the position vector of the surface and suitable first and higher order gradients (with respect to the surface coordinates) of the position vector.[31] Although we do not undertake a general development of this type in this section or even a special case in which the basic kinematic variables are the position vector $\boldsymbol{r}$ and the gradient[32] $\boldsymbol{r}_{,\alpha}$ (giving rise to velocity and velocity gradient $\boldsymbol{v}_{,\alpha}=v_{i\alpha}\boldsymbol{a}^i$), we consider a restricted theory which bears on the classical theory of shells.[33]

In the developments of the restricted theory we retain, of course, the principles $(8.17)_{1,2}$ but replace the remaining conservation laws with a different set. For this purpose, we need to introduce certain quantities not defined previously. Let a (tangential) vector field $\grave{\boldsymbol{M}}=\grave{\boldsymbol{M}}(\theta^\alpha, t; \boldsymbol{\nu})$ with components $\grave{M}^\gamma=\grave{\boldsymbol{M}}\cdot\boldsymbol{a}^\gamma$ be defined for points $\boldsymbol{r}$ on the boundary curve c of the part $\mathscr{P}$ of $\mathfrak{s}$. If for all arbitrary angular velocity fields $\grave{\boldsymbol{w}}$ [defined in (5.61)], the scalar $\grave{\boldsymbol{M}}\cdot\grave{\boldsymbol{w}}$ is a rate of work per unit length, then $\grave{\boldsymbol{M}}$ is called a *contact couple* (or a *curve couple*) vector per unit length of[34] c. Further, let a (tangential) vector field $\grave{\boldsymbol{l}}=\grave{\boldsymbol{l}}(\theta^\alpha, t)$ per unit mass with components $\grave{l}^\gamma=\grave{\boldsymbol{l}}\cdot\boldsymbol{a}^\gamma$ be defined for points $\boldsymbol{r}$ on the part $\mathscr{P}$ of $\mathfrak{s}$. Then, if the scalar $\grave{\boldsymbol{l}}\cdot\grave{\boldsymbol{w}}$ is a rate of work per unit mass for all arbitrary angular velocities $\grave{\boldsymbol{w}}$, $\grave{\boldsymbol{l}}$ is called an *assigned couple* per unit mass of $\mathfrak{s}$. The *resultant contact couple* $\grave{\boldsymbol{G}}_c(\mathscr{P})$ and the *resultant assigned* couple $\grave{\boldsymbol{G}}_b$ over a part $\mathscr{P}$ in the present configuration are defined by

$$\grave{\boldsymbol{G}}_c=\int_{\partial\mathscr{P}}\grave{\boldsymbol{M}}\,ds, \quad \grave{\boldsymbol{G}}_b=\int_{\mathscr{P}}\varrho\,\grave{\boldsymbol{l}}\,d\sigma. \tag{10.1}$$

Similarly, corresponding to (8.14), the sum $\grave{\boldsymbol{A}}(\mathscr{P})$ of the supply of moment of momentum $\grave{\boldsymbol{A}}_b(\mathscr{P})$ due to the assigned force and the assigned couple, each per unit mass, and the flux of moment of momentum $\grave{\boldsymbol{A}}_c(\mathscr{P})$ due to the contact force and the contact couple, each per unit length, is defined by

$$\begin{gathered}\grave{\boldsymbol{A}}(\mathscr{P})=\grave{\boldsymbol{A}}_b(\mathscr{P})+\grave{\boldsymbol{A}}_c(\mathscr{P}),\\ \grave{\boldsymbol{A}}_b(\mathscr{P})=\int_{\mathscr{P}}[\boldsymbol{r}\times\varrho\boldsymbol{f}+\boldsymbol{a}_3\times\varrho\,\grave{\boldsymbol{l}}]\,d\sigma, \quad \grave{\boldsymbol{A}}_c(\mathscr{P})=\int_{\partial\mathscr{P}}(\boldsymbol{r}\times\boldsymbol{N}+\boldsymbol{a}_3\times\grave{\boldsymbol{M}})\,ds.\end{gathered} \tag{10.2}$$

We also record below the expression for the total rate of work by the contact force and the contact couple and by the assigned force and the assigned couple in the form

$$\begin{gathered}\grave{R}(\mathscr{P})=\grave{R}_c(\mathscr{P})+\grave{R}_b(\mathscr{P}),\\ \grave{R}_c(\mathscr{P})=\int_{\partial\mathscr{P}}(\boldsymbol{N}\cdot\boldsymbol{v}+\grave{\boldsymbol{M}}\cdot\grave{\boldsymbol{w}})\,ds, \quad \grave{R}_b(\mathscr{P})=\int_{\mathscr{P}}\varrho(\boldsymbol{f}\cdot\boldsymbol{v}+\grave{\boldsymbol{l}}\cdot\grave{\boldsymbol{w}})\,d\sigma.\end{gathered} \tag{10.3}$$

---

[31] A theory of this type, concerned with a deformable surface with simple force multipoles in which $\boldsymbol{r}$ and its first and second gradients ($\boldsymbol{r}_{,\alpha}, \boldsymbol{r}_{,\alpha\beta}$) are taken as the basic kinematic variables, has been developed by BALABAN, GREEN and NAGHDI [1967, *1*]. A similar theory, but less general than that in [1967, *1*], is given by COHEN and DE SILVA [1966, *1*] who have subsequently modified their analysis [1968, *2*]. The work of COHEN and DE SILVA in [1968, 2] may be compared with a special case of the results in [1967, *1*], called the restricted theory of simple force dipoles. Also, SERBIN [1963, *13*] has considered an exact linear (isothermal) theory of an elastic surface by direct approach; but, in early stages of his analysis, he assumes a strain energy function for the linear theory which is too restrictive.

[32] A direct theory in which $\boldsymbol{r}$ and $\boldsymbol{r}_{,\alpha}$ are regarded as the basic kinematic variables will obviously have some overlapping features with that of a Cosserat surface; but, in general the two theories are different in character.

[33] In the existing literature, however, the classical theory is often derived from the linearized three-dimensional equations.

[34] The use of the term contact couple is justified in view of the physical dimension of $\grave{\boldsymbol{w}}$.

It is clear from the above definitions and the physical dimension of $\grave{\boldsymbol{w}}$ that $\grave{\boldsymbol{M}}$ and $\grave{\boldsymbol{l}}$ have, respectively, the physical dimensions of couple per unit length and couple per unit mass, namely

$$\begin{aligned} \text{phys. dim. } \grave{\boldsymbol{M}} &= \left[\frac{ML^2\,T^{-2}}{L}\right] = [MLT^{-2}], \\ \text{phys. dim. } \grave{\boldsymbol{l}} &= \left[\frac{ML^2\,T^{-2}}{M}\right] = [L^2\,T^{-2}]. \end{aligned} \tag{10.4}$$

Having disposed of the above preliminaries, we begin the development of the restricted theory under consideration by adopting $(8.17)_{1,2}$ supplemented by

$$\begin{aligned} \frac{d}{dt}\int_{\mathscr{P}} \varrho(\boldsymbol{r}\times\boldsymbol{v})\,d\sigma &= \grave{\boldsymbol{A}}_b(\mathscr{P}) + \grave{\boldsymbol{A}}_c(\mathscr{P}), \\ \frac{d}{dt}\int_{\mathscr{P}} \left(\varepsilon + \frac{1}{2}\boldsymbol{v}\cdot\boldsymbol{v}\right) d\sigma &= \grave{R}(\mathscr{P}) + H(\mathscr{P}), \end{aligned} \tag{10.5}$$

which are stated with reference to the present configuration of the surface. The conservation laws $(10.5)_{1,2}$ represent, respectively, the balance of moment of momentum and the balance of energy in the restricted theory.[35] They should be contrasted with $(8.17)_{4,5}$; and we note that $(8.17)_3$ has no counterpart here, since the director is not admitted in the restricted theory.

In place of the invariance condition $(8.32)_2$ we now require that $\grave{\boldsymbol{M}}$ be objective. Similarly, since moment of momentum due to angular velocity is excluded, instead of $(8.33)_3$ we require that $\grave{\boldsymbol{l}}$ be objective. To summarize, it follows that the invariance conditions under superposed rigid body motions in the restricted theory consist of (8.30)–(8.31), together with

$$\boldsymbol{N}^+ = Q(t)\,\boldsymbol{N}, \qquad \grave{\boldsymbol{M}}^+ = Q(t)\,\grave{\boldsymbol{M}} \tag{10.6}$$

and

$$r^+ = r, \qquad \bar{\boldsymbol{f}}^+ = Q(t)\,\bar{\boldsymbol{f}}, \qquad \grave{\boldsymbol{l}}^+ = Q(t)\,\grave{\boldsymbol{l}}. \tag{10.7}$$

By (5.66), (8.18) and (9.13)–(9.14), $(10.5)_1$ can be reduced to

$$\int_{\mathscr{P}} (\boldsymbol{a}_\alpha \times \boldsymbol{N}^\alpha + \varrho\,\boldsymbol{a}_3 \times \grave{\boldsymbol{l}})\,d\sigma + \int_{\partial\mathscr{P}} \boldsymbol{a}_3 \times \grave{\boldsymbol{M}}\,ds = 0. \tag{10.8}$$

Using the temporary notation $\grave{\boldsymbol{M}}_{(\boldsymbol{\nu})}$ in place of $\grave{\boldsymbol{M}}$ (in order to emphasize its dependence on $\boldsymbol{\nu}$), in the same manner that (9.5) and (9.17) were obtained, we can deduce the result $\boldsymbol{a}_3 \times [\grave{\boldsymbol{M}}_{(\boldsymbol{\nu})} + \grave{\boldsymbol{M}}_{(-\boldsymbol{\nu})}] = 0$. But since $\grave{\boldsymbol{M}}_{(\boldsymbol{\nu})}$ is a tangential vector field, it follows that

$$\grave{\boldsymbol{M}}_{(\boldsymbol{\nu})} = -\grave{\boldsymbol{M}}_{(-\boldsymbol{\nu})}. \tag{10.9}$$

Let $\grave{\boldsymbol{m}}^{(1)}$, $\grave{\boldsymbol{m}}^{(2)}$ denote the physical couple vectors acting on the sides of the coordinate curves whose unit normal vectors are $\boldsymbol{a}^1/(a^{11})^{\frac{1}{2}}$, $\boldsymbol{a}^2/(a^{22})^{\frac{1}{2}}$, respectively. Then, with the help of (10.9), application of (10.8) to an elementary curvilinear triangle on $\mathscr{s}$ (Fig. 2) yields

$$\grave{\boldsymbol{M}} = \grave{\boldsymbol{M}}_{(\boldsymbol{\nu})} = \grave{\boldsymbol{M}}^\alpha\,\nu_\alpha, \qquad \grave{\boldsymbol{M}}^\alpha = \grave{\boldsymbol{m}}^{(\alpha)}\,(a^{\alpha\alpha})^{\frac{1}{2}} \quad (\text{no sum on } \alpha). \tag{10.10}$$

[35] In writing (10.5), the contributions of the moment of momentum and of the kinetic energy due to the angular velocity have been excluded. These contributions are usually absent in the classical (approximate) theories derived from the three-dimensional equations. Their omission here does not lead to an essential limitation; they can be easily included if desired.

With the use of (10.10) and under suitable continuity assumptions, from (10.8) we obtain the equation

$$\boldsymbol{a}_\alpha \times \boldsymbol{N}^\alpha + \boldsymbol{a}_3 \times \varrho\, \dot{\boldsymbol{l}} + (\boldsymbol{a}_3 \times \dot{\boldsymbol{M}}^\alpha)_{|\alpha} = 0 \tag{10.11}$$

or equivalently

$$\boldsymbol{a}_\alpha \times (\boldsymbol{N}^\alpha - b^\alpha_\gamma\, \dot{\boldsymbol{M}}^\gamma) + \boldsymbol{a}_3 \times (\dot{\boldsymbol{M}}^\alpha{}_{|\alpha} + \varrho\, \dot{\boldsymbol{l}}) = 0, \tag{10.12}$$

as a consequence of the moment of momentum principle $(10.5)_1$.

We now turn to the energy balance $(10.5)_2$ which, with the help of (8.8) and (9.13), can be reduced to

$$\begin{aligned} \int_{\mathscr{P}} [\varrho\, r - \varrho\, \dot{\varepsilon} + \boldsymbol{N}^\alpha \cdot \boldsymbol{v}_{,\alpha} + \dot{\boldsymbol{M}}^\alpha \cdot \dot{\boldsymbol{w}}_{,\alpha}]\, d\sigma \\ + \int_{\mathscr{P}} (\dot{\boldsymbol{M}}^\alpha{}_{|\alpha} + \varrho\, \dot{\boldsymbol{l}}) \cdot \dot{\boldsymbol{w}}\, d\sigma - \int_{\partial\mathscr{P}} h\, ds = 0. \end{aligned} \tag{10.13}$$

Before proceeding further, we indicate a simplification which can be effected in (10.13). To this end, consider the scalar product of (10.11) with $\boldsymbol{a}_3 \times \dot{\boldsymbol{w}}$ and obtain

$$[(\boldsymbol{a}_3 \times \dot{\boldsymbol{M}}^\alpha)_{|\alpha} + (\boldsymbol{a}_3 \times \varrho\, \dot{\boldsymbol{l}})] \cdot (\boldsymbol{a}_3 \times \dot{\boldsymbol{w}}) = -(\boldsymbol{a}_\alpha \times \boldsymbol{N}^\alpha) \cdot (\boldsymbol{a}_3 \times \dot{\boldsymbol{w}}). \tag{10.14}$$

By use of the formulae for scalar triple product and vector triple product of vectors, namely

$$\boldsymbol{U} \cdot (\boldsymbol{V} \times \boldsymbol{W}) = (\boldsymbol{U} \times \boldsymbol{V}) \cdot \boldsymbol{W}, \qquad (\boldsymbol{U} \times \boldsymbol{V}) \times \boldsymbol{W} = \boldsymbol{V}(\boldsymbol{U} \cdot \boldsymbol{W}) - \boldsymbol{U}(\boldsymbol{V} \cdot \boldsymbol{W}), \tag{10.15}$$

the various terms in (10.14) can be reduced as follows:

$$\begin{aligned} (\boldsymbol{a}_3 \times \dot{\boldsymbol{M}}^\alpha)_{|\alpha} \cdot (\boldsymbol{a}_3 \times \dot{\boldsymbol{w}}) &= [(\boldsymbol{a}_{3,\alpha} \times \dot{\boldsymbol{M}}^\alpha + \boldsymbol{a}_3 \times \dot{\boldsymbol{M}}^\alpha{}_{|\alpha}) \times \boldsymbol{a}_3] \cdot \dot{\boldsymbol{a}}_3 \\ &= [\dot{\boldsymbol{M}}^\alpha (\boldsymbol{a}_{3,\alpha} \cdot \boldsymbol{a}_3) - \boldsymbol{a}_{3,\alpha} (\dot{\boldsymbol{M}}^\alpha \cdot \boldsymbol{a}_3) \\ &\quad + \dot{\boldsymbol{M}}^\alpha{}_{|\alpha} (\boldsymbol{a}_3 \cdot \boldsymbol{a}_3) - \boldsymbol{a}_3 (\dot{\boldsymbol{M}}^\alpha{}_{|\alpha} \cdot \boldsymbol{a}_3] \cdot \dot{\boldsymbol{a}}_3 \\ &= \dot{\boldsymbol{M}}^\alpha{}_{|\alpha} \cdot \dot{\boldsymbol{w}}, \end{aligned} \tag{10.16}$$

$$\begin{aligned} (\boldsymbol{a}_3 \times \varrho\, \dot{\boldsymbol{l}}) \cdot (\boldsymbol{a}_3 \times \dot{\boldsymbol{w}}) &= \varrho [\dot{\boldsymbol{l}} (\boldsymbol{a}_3 \cdot \boldsymbol{a}_3) - \boldsymbol{a}_3 (\dot{\boldsymbol{l}} \cdot \boldsymbol{a}_3)] \cdot \dot{\boldsymbol{a}}_3 = \varrho\, \dot{\boldsymbol{l}} \cdot \dot{\boldsymbol{w}}, \\ (\boldsymbol{a}_\alpha \times \boldsymbol{N}^\alpha) \cdot (\boldsymbol{a}_3 \times \dot{\boldsymbol{w}}) &= [\boldsymbol{N}^\alpha (\boldsymbol{a}_\alpha \cdot \boldsymbol{a}_3) - \boldsymbol{a}_\alpha (\boldsymbol{N}^\alpha \cdot \boldsymbol{a}_3)] \cdot \dot{\boldsymbol{a}}_3 \\ &= -(\boldsymbol{N}^\alpha \cdot \boldsymbol{a}_3)(\boldsymbol{a}_\alpha \cdot \dot{\boldsymbol{a}}_3) = \boldsymbol{N}^\alpha \cdot (v_{3\alpha}\, \boldsymbol{a}_3), \end{aligned}$$

where temporarily we have recalled (5.61) and we have used $(4.11)_{1,2}$ and (5.10). Substitution of (10.16) into (10.14) results in

$$(\dot{\boldsymbol{M}}^\alpha{}_{|\alpha} + \varrho\, \dot{\boldsymbol{l}}) \cdot \dot{\boldsymbol{w}} = -\boldsymbol{N}^\alpha \cdot (v_{3\alpha}\, \boldsymbol{a}_3). \tag{10.17}$$

The combination of the term $\boldsymbol{N}^\alpha \cdot \boldsymbol{v}_{,\alpha}$ in the first integral of (10.13) and the terms in the second integral, in view of (10.17), yield $\boldsymbol{N}^\alpha \cdot v_{\gamma\alpha}\, \boldsymbol{a}^\gamma$ and the energy balance becomes

$$\int_{\mathscr{P}} [\varrho r - \varrho\, \dot{\varepsilon} + \boldsymbol{N}^\alpha \cdot v_{\gamma\alpha}\, \boldsymbol{a}^\gamma + \dot{\boldsymbol{M}}^\alpha \cdot \dot{\boldsymbol{w}}_{,\alpha}]\, d\sigma - \int_{\partial\mathscr{P}} h\, ds = 0, \tag{10.18}$$

which is the counterpart of (9.24) in the present development. The results (9.25)–(9.27) still hold and, after making the appropriate continuity assumptions, from (10.18) we obtain

$$\varrho r - q^\alpha{}_{|\alpha} - \varrho\, \dot{\varepsilon} + \boldsymbol{N}^\alpha \cdot v_{\gamma\alpha}\, \boldsymbol{a}^\gamma + \dot{\boldsymbol{M}}^\alpha \cdot \dot{\boldsymbol{w}}_{,\alpha} = 0. \tag{10.19}$$

It is convenient at this point to write (10.11) in component form. Since $\grave{\boldsymbol{M}}$ and $\grave{\boldsymbol{l}}$ are tangential vector fields and since $\grave{\boldsymbol{M}}^\alpha$ by (10.10) transforms as a contravariant surface vector, we have

$$\grave{\boldsymbol{M}} = \grave{M}^\gamma \boldsymbol{a}_\gamma, \qquad \grave{\boldsymbol{l}} = \grave{l}^\gamma \boldsymbol{a}_\gamma, \qquad \grave{\boldsymbol{M}}^\alpha = \grave{M}^{\alpha\gamma} \boldsymbol{a}_\gamma, \qquad \grave{M}^\gamma = \grave{M}^{\alpha\gamma} \nu_\alpha. \tag{10.20}$$

Then, recalling (9.40), from (10.11) we obtain the equations of motion

$$\grave{M}^{\gamma\alpha}{}_{|\gamma} - N^{\alpha 3} + \varrho\, \grave{l}^\alpha = 0, \qquad \varepsilon_{\beta\alpha} [N^{\alpha\beta} - \grave{M}^{\gamma\beta} b_\gamma^\alpha] = 0. \tag{10.21}$$

The set of Eqs. (10.21) are the counterparts of $(9.48)_1$ and $(9.51)_1$ in the restricted theory.[36] We also observe here that the differential equations of motion (9.47) and $(10.21)_1$ can be easily written in alternative forms by elimination of $N^{\alpha 3}$. Thus, substituting for $N^{\alpha 3}$ from $(10.21)_1$ into $(9.47)_{1,2}$, we obtain

$$N^{\alpha\beta}{}_{|\alpha} - b_\alpha^\beta \grave{M}^{\gamma\alpha}{}_{|\gamma} + \varrho\, \bar{f}^\beta = 0, \qquad \grave{M}^{\gamma\alpha}{}_{|\gamma\alpha} + b_{\alpha\beta} N^{\alpha\beta} + \varrho\, \bar{f}^3 = 0, \tag{10.22}$$

where we have set

$$\varrho\, \bar{f}^\beta = \varrho\, (f^\beta - c^\beta + b_\alpha^\beta\, \grave{l}^\alpha), \qquad \varrho\, \bar{f}^3 = \varrho\, (f^3 - c^3) + (\varrho\, \grave{l}^\alpha)_{|\alpha}. \tag{10.23}$$

Returning (10.19), as in Sect. 9, we now use the argument about uniform superposed rigid body angular velocity, the continuum occupying the same position at time t. Recalling that the fields which occur in (8.30)–(8.31) and (10.6)–(10.7) [as well as q^α] are unaltered under such superposed rigid body motions, with the use of $(5.17)_1$ and (5.62), we again obtain $(10.21)_2$ and the local energy equation in the invariant form

$$\begin{aligned} &\varrho\, r - q^\alpha{}_{|\alpha} - \varrho\, \dot{\varepsilon} + \grave{P} = 0, \\ &\grave{P} = \grave{\boldsymbol{N}}^\alpha \cdot \boldsymbol{\eta}_\alpha + \grave{\boldsymbol{M}}^\alpha \cdot \grave{\lambda}_{\gamma\alpha}\, \boldsymbol{a}^\lambda \\ &\quad = \grave{N}^{\alpha\beta} \eta_{\alpha\beta} - \grave{M}^{\alpha\gamma} \dot{b}_{\gamma\alpha}, \end{aligned} \tag{10.24}$$

where we have introduced the notation

$$\grave{\boldsymbol{N}}^\alpha = \boldsymbol{N}^\alpha + b_\gamma^\alpha \grave{\boldsymbol{M}}^\gamma \tag{10.25}$$

and where $\grave{N}^{\alpha\beta}$ which satisfies $(10.21)_2$ is given by

$$\begin{aligned} \grave{N}^{\alpha\beta} &= \grave{N}^{\beta\alpha} = \tfrac{1}{2} (\grave{\boldsymbol{N}}^\alpha \cdot \boldsymbol{a}^\beta + \grave{\boldsymbol{N}}^\beta \cdot \boldsymbol{a}^\alpha) \\ &= N^{\alpha\beta} + \grave{M}^{\gamma\alpha} b_\gamma^\beta = N^{\beta\alpha} + \grave{M}^{\gamma\beta} b_\gamma^\alpha. \end{aligned} \tag{10.26}$$

It is clear that the only values of $N^{\alpha\beta} = \boldsymbol{N}^\alpha \cdot \boldsymbol{a}^\beta$ which are present in $(10.24)_2$ are the symmetric components $N^{(\alpha\beta)}$. Also, the components $N^{\alpha 3} = \boldsymbol{N}^\alpha \cdot \boldsymbol{a}_3$ do not occur in the energy equation and will be absent from an inequality resulting from combination of $(10.24)_1$ and (9.32). Moreover, since $b_{\gamma\alpha}$ is symmetric, it follows that only the symmetric part of $\grave{M}^{\alpha\gamma}$ occurs in the energy equation $(10.24)_1$. We may therefore replace $\grave{M}^{\alpha\gamma}$ in (10.24) by $\grave{M}^{(\alpha\gamma)}$. This completes our derivation of the basic field equations of the restricted theory. They consist of the continuity equation (8.18), the equations of motion (9.47) and (10.21) [or equivalently (10.22)

[36] Eqs. $(9.48)_2$ and $(9.51)_2$ have no counterparts in the restricted theory. Had we included the moment of momentum due to the angular velocity in (10.5), instead of $\varrho\, \grave{\boldsymbol{l}}$ in (10.11) we should have $\varrho\, (\grave{\boldsymbol{l}} - \dot{\boldsymbol{w}})$ representing the difference of the assigned couple vector per unit mass and the inertia term due to $\dot{\boldsymbol{w}}$.

and (10.21)] and the energy equation (10.24) which can also be expressed in terms of the specific Helmholtz free energy.[37]

Before closing this section we observe that the linearized version of the above field equations can be obtained in the same manner as the linearized field equations of a Cosserat surface in Sect. 9 [Subsect. γ)]. In particular, all quantities are referred to the base vectors $\boldsymbol{A}_i$, covariant differentiation is with respect to $A_{\alpha\beta}$ and the components of $\grave{\boldsymbol{M}}^\alpha$ and $\grave{\boldsymbol{l}}$ are now defined through

$$M^\gamma = \grave{\boldsymbol{M}} \cdot \boldsymbol{A}_\gamma, \qquad \grave{L}^\gamma = \grave{\boldsymbol{l}} \cdot \boldsymbol{A}_\gamma, \qquad \grave{\boldsymbol{M}}^\alpha = \grave{M}^{\alpha\gamma} \boldsymbol{A}_\gamma. \tag{10.27}$$

The linearized versions of (10.21) then become

$$\grave{M}^{\alpha\gamma}{}_{|\gamma} - N^{\alpha 3} + \varrho_0 \grave{L}^\alpha = 0, \tag{10.28}$$

$$\bar{\varepsilon}_{\beta\alpha}[N^{\alpha\beta} - \grave{M}^{\gamma\beta} B^\alpha_\gamma] = 0, \tag{10.29}$$

where the notation $\bar{\varepsilon}_{\alpha\beta} = A^{\frac{1}{2}} e_{\alpha\beta}$ was introduced previously in (6.30). Similarly, the linearized equations of motion corresponding to (10.22) can be obtained from (10.28) and (9.54). Also, the energy equation in the restricted linear theory can be written as

$$\varrho_0 r - Q^\alpha{}_{|\alpha} - \varrho_0 \dot{\varepsilon} + \tilde{N}^{\alpha\beta} \dot{e}_{\alpha\beta} + \grave{M}^{\alpha\gamma} \dot{\bar{\varkappa}}_{\gamma\alpha} = 0. \tag{10.30}$$

In (10.30), Q^α is defined in (9.59), $\tilde{N}^{\alpha\beta}$ is of the form (10.26) with b^β_γ replaced by B^β_γ and $\bar{\varkappa}_{\gamma\alpha}$ [which can be obtained from the linearization of $(b_{\alpha\beta} - B_{\alpha\beta})$] is given by the symmetric expression in (6.24). Again, since $\bar{\varkappa}_{\gamma\alpha} = \bar{\varkappa}_{\alpha\gamma}$, we may replace $\grave{M}^{\alpha\gamma}$ in (10.30) by $\grave{M}^{(\alpha\gamma)}$.

11. Basic field equations for shells: II. Derivation from the three-dimensional theory. Prior to our derivation of the basic field equations for shell-like bodies from the three-dimensional equations of the classical continuum mechanics, we need to recall some preliminary results from the three-dimensional theory for non-polar media and also define certain resultants, including stress-resultants and stress-couples. Since our developments are carried out in the context of a thermodynamic theory, we also recall the Clausius-Duhem inequality and obtain its analogue for shells.[38]

α) *Some preliminary results.* A (three-dimensional) shell-like body $\mathscr{B}$ with its boundary surface $\partial\mathscr{B}$ [specified by (4.14)–(4.15)] is defined in Sect. 4, where the relation of the material surface $\xi = 0$ to the bounding surfaces $(4.14)_{1,2}$ is fixed by the condition (4.20). Let[39] $\mathscr{P}_{\mathscr{B}}$ refer to an arbitrary region of the material

[37] These field equations can be brought into correspondence with those of the restricted theory of simple force dipoles; see Sect. 8 of the paper by BALABAN, GREEN and NAGHDI [1967, *1*].

[38] Our developments in this and the next section follow largely GREEN, LAWS and NAGHDI [1968, *4*] and GREEN and NAGHDI [1970, *2*] and to some extent NAGHDI [1964, *4*] and GREEN, NAGHDI and WENNER [1971, *6*]. The thermodynamic results are mainly those obtained in [1964, *4*] and [1970, *2*].

We again remark that there is some discussion in the current literature regarding the limitations of the Clausius-Duhem inequality or a more general form for the entropy inequality, but these need not affect our limited use of the inequality (11.20) here. Later, we shall appeal to (11.20) only in the context of elastic materials. We further remark here that while our subsequent developments are carried out within the framework of a thermodynamic theory, most of the results can be easily specialized to the isothermal or the purely mechanical theory.

[39] Although the symbol $\mathscr{P}$ designating a part of the surface $\xi = 0$ in the present configuration at time t is the same as that used previously [in Sects. 4, 8–9] to designate a part which the Cosserat surface $\mathscr{C}$ occupies at time t, this need not give rise to confusion. If desired, the two surfaces may be identified; but we postpone such identifications until later.

surface $\xi = 0$ which is mapped into a part $\mathscr{P}$ in the present configuration and let $\partial\mathscr{P}$ denote the boundary of $\mathscr{P}$. Further let $\mathscr{P}^*_{\mathscr{B}}$ refer to an arbitrary part of $\mathscr{B}$ which is mapped into a part $\mathscr{P}^*$ in the present configuration such that (i) $\mathscr{P}^*$ contains $\mathscr{P}$; (ii) the boundary $\partial\mathscr{P}^*$ of $\mathscr{P}^*$ consists of portions of the surfaces $(4.14)_{1,2}$ and a surface of the form (4.15) at time t; and (iii) the boundary $\partial\mathscr{P}^*$ coincides with $\partial\mathscr{P}$ on $\xi = 0$. Also, for later reference, let $\partial\mathscr{P}^*_n$ refer to the part of $\partial\mathscr{P}^*$ specified by a normal surface of the form (4.15) so that $\partial\mathscr{P}^*_n = \partial\mathscr{P}^* = \partial\mathscr{P}$ on $\xi = 0$, and let $\partial\mathscr{P}^{*c}_n = \partial\mathscr{P}^* - \partial\mathscr{P}^*_n$ refer to the complement of $\partial\mathscr{P}^*_n$ in $\partial\mathscr{P}^*$.

Recalling that ϱ is a mass per unit area of $\mathfrak{s}$ (the surface $\xi = 0$ at time t) given by (4.17), we define the mass of any part $\mathscr{P}_{\mathscr{B}}$ by

$$m(\mathscr{P}_{\mathscr{B}}) = \int_{\mathscr{P}} \varrho \, d\sigma, \tag{11.1}$$

where the surface integral is over the part $\mathscr{P}$ of the surface $\xi = 0$ at time t. Let $m^*(\mathscr{P}^*_{\mathscr{B}})$ denote the mass of a part $\mathscr{P}^*_{\mathscr{B}}$ of the shell-like body $\mathscr{B}$. The mass density of $\mathscr{B}$ in different configurations is determined by the motion of $\mathscr{B}$; it is a relation between the physical mass $m^*(\mathscr{P}^*_{\mathscr{B}})$ and the volume of the region of the Euclidean space occupied by $\mathscr{P}^*_{\mathscr{B}}$ in a given configuration as defined, for example, by the volume integral in (4.18). From (4.18) and (11.1) follows the relation

$$m^*(\mathscr{P}^*_{\mathscr{B}}) = m(\mathscr{P}_{\mathscr{B}}). \tag{11.2}$$

A local form of the (three-dimensional) continuity equation has been stated previously by (4.16). Integration of this equation with respect to ξ, between the limits $\xi = \alpha, \beta$, results in

$$\varrho \, a^{\frac{1}{2}} = \varrho_0 \, A^{\frac{1}{2}}. \tag{11.3}$$

The right-hand side of (11.3) is the reference value of (4.17) and

$$\varrho_0 = \varrho_0(\theta^\alpha) = \varrho(\theta^\alpha, 0)$$

is a mass per unit area of $\mathfrak{S}$ (the surface $\xi = 0$ in the initial reference configuration).[40]

Assume that a shell-like body $\mathscr{B}$ and a motion of $\mathscr{B}$ are given. Let $\overline{\mathscr{P}}$, not necessarily the same as $\mathscr{P}^*$, refer to an arbitrary part of $\mathscr{B}$ in the present configuration and let $\partial\overline{\mathscr{P}}$ denote the boundary of $\overline{\mathscr{P}}$. Then, within the scope of non-polar continuum mechanics, the system of forces acting over any part $\overline{\mathscr{P}}_{\mathscr{B}}$ of the body $\mathscr{B}$ in motion consist of the sum of two types of forces $\boldsymbol{F}^*_b$ and $\boldsymbol{F}^*_c$ defined as follows: Let $\boldsymbol{f}^* = \boldsymbol{f}^*(\theta^i, t)$ be a vector field, per unit mass, defined for material points in the region occupied by $\mathscr{B}$ at time t; it is called the *external body force* or simply the *body force*. The *resultant body force* exerted on the part $\overline{\mathscr{P}}$ at time t is defined by the volume integral[41]

$$\boldsymbol{F}^*_b(\overline{\mathscr{P}}) = \int_{\overline{\mathscr{P}}} \varrho^* \boldsymbol{f}^* \, dv \tag{11.4}$$

[40] Despite similarity and our use of the same symbols in (11.3) and the continuity equation (4.42) for the Cosserat surface, we emphasize that these two equations have been deduced on the basis of different concepts. Clearly, they may be brought into one-to-one correspondence, but we postpone such identifications until later.

[41] Strictly speaking, the left-hand sides of (11.4) and (11.6) should be denoted as $\boldsymbol{F}^*_b(\overline{\mathscr{P}}_{\mathscr{B}})$ and $\boldsymbol{F}^*_c(\overline{\mathscr{P}}_{\mathscr{B}})$, respectively; however, in order to avoid cumbersome notation, we adopt the simpler designations on the left-hand sides of (11.4), (11.6) and elsewhere in this section. Our abbreviated notation which is similar to that in (4.44) should not cause confusion, since it will be clear from the particular context [e.g., the presence of dv on the right-hand side of (11.4)] which region of Euclidean space $\mathscr{B}$ occupies.

over $\overline{\mathscr{P}}$ in the present configuration and the element of volume dv is given by $(4.19)_1$. Further, let

$$\boldsymbol{n} = n_i \boldsymbol{g}^i = n^i \boldsymbol{g}_i \tag{11.5}$$

be the outward unit normal vector at the material point on the boundary $\partial\overline{\mathscr{P}}$ at time t and let $\boldsymbol{t} = \boldsymbol{t}(\theta^i, t; \boldsymbol{n})$ be defined for the material points on the boundary $\partial\overline{\mathscr{P}}$ at time t. The vector field[42] $\boldsymbol{t}$ is called the *contact force* or the *stress vector* acting on the part $\overline{\mathscr{P}}$ of $\mathscr{B}$. The *resultant contact force* exerted on the part $\overline{\mathscr{P}}$ at time t is defined by the surface integral

$$\boldsymbol{F}_c^*(\overline{\mathscr{P}}) = \int_{\partial\overline{\mathscr{P}}} \boldsymbol{t}(\theta^i, t; \boldsymbol{n})\, da \tag{11.6}$$

over the boundary $\partial\overline{\mathscr{P}}$ of $\overline{\mathscr{P}}$ in the present configuration, where da is an element of area whose outward unit normal is $\boldsymbol{n}$.

Under suitable continuity assumptions, the principles of linear momentum and moment of momentum for non-polar media imply[43] (i) the existence of a tensor field $\tau^{ij} = \tau^{ij}(\theta^k, t)$ such that

$$\boldsymbol{t} = \frac{\boldsymbol{T}^i n_i}{g^{\frac{1}{2}}} = \tau^{ij} n_i \boldsymbol{g}_j, \qquad \boldsymbol{T}^i = g^{\frac{1}{2}} \tau^{ij} \boldsymbol{g}_j = g^{\frac{1}{2}} \tau^i_j \boldsymbol{g}^j, \tag{11.7}$$

and (ii) the Cauchy equations of motion, which may be expressed as

$$\boldsymbol{T}^i{}_{,i} + \varrho^* \boldsymbol{f}^* g^{\frac{1}{2}} = \varrho^* \dot{\boldsymbol{v}}^* g^{\frac{1}{2}}, \qquad \boldsymbol{g}_i \times \boldsymbol{T}^i = 0. \tag{11.8}$$

In (11.7), τ^{ij} and τ^i_j are the contravariant and the mixed components of the stress tensor and it follows from $(11.7)_2$ and $(11.8)_2$ that τ^{ij} is symmetric, i.e.,

$$\tau^{ij} = \tau^{ji}. \tag{11.9}$$

We observe that if $\boldsymbol{n}$ is the outward unit normal to $\partial\mathscr{B}$, then $\boldsymbol{t}(\theta^i, t; \boldsymbol{n})$ reduces to a function of position for points on $\partial\mathscr{B}$, i.e.,

$$\boldsymbol{t} = \bar{\boldsymbol{t}}(\theta^i, t), \tag{11.10}$$

where $\bar{\boldsymbol{t}}$ is the surface traction prescribed on the boundary $\partial\mathscr{B}$.

We introduce the following additional five quantities which we associate with a motion of the body:

The specific *internal energy* $\varepsilon^* = \varepsilon^*(\theta^i, t)$.

The *heat flux* vector $\boldsymbol{q}^* = \boldsymbol{q}^*(\theta^i, t)$.

The *heat supply* or *heat absorption* $r^* = r^*(\theta^i, t)$.

The specific *entropy* $\eta^* = \eta^*(\theta^i, t)$.

The local temperature $\theta^* = \theta^*(\theta^i, t)$ which is assumed to be always positive, i.e., $\theta^* > 0$.

Consider the eight functions consisting of the vector function $\boldsymbol{p}$ in (4.6) which describes the motion of the body, the body force $\boldsymbol{f}^*$, the surface force $\boldsymbol{T}^i$ (or equivalently the stress tensor τ^{ij}) and the five functions defined above. It is convenient to speak of a *thermodynamic process* or simply a *process* if the eight functions are so prescribed that the conservation laws, namely the principles of

[42] Often $\boldsymbol{t}$ is denoted by $\boldsymbol{t}_{(\boldsymbol{n})}$ in order to emphasize its dependence on the unit normal $\boldsymbol{n}$, but we omit the subscript $\boldsymbol{n}$ here.

[43] GREEN and ZERNA [1968, 9, Chap. 2]. We may observe here that the *equivalence* of the conservation laws for mass, linear momentum, moment of momentum and energy *and* the balance of energy together with invariance requirements under superposed rigid body motions can be demonstrated in a manner similar to that in Sect. 8.

linear and moment of momentum and balance of energy, are satisfied for every part $\overline{\mathscr{P}}$ of the body $\mathscr{B}$. The internal energy $\mathscr{E}^*(\overline{\mathscr{P}})$ of a part $\overline{\mathscr{P}}$ and the heat $H^*(\overline{\mathscr{P}})$ entering the part $\overline{\mathscr{P}}$ per unit time are defined by[44]

$$\mathscr{E}^*(\overline{\mathscr{P}}) = \int_{\overline{\mathscr{P}}} \varrho^* \, \varepsilon^* \, dv, \qquad H^*(\overline{\mathscr{P}}) = \int_{\overline{\mathscr{P}}} \varrho^* \, r^* \, dv - \int_{\partial\overline{\mathscr{P}}} \boldsymbol{q}^* \cdot \boldsymbol{n} \, da. \tag{11.11}$$

Also, the *entropy* $\mathscr{H}^*(\overline{\mathscr{P}})$ of a part $\overline{\mathscr{P}}$ of $\mathscr{B}$ and the *production of entropy* Γ^* per unit time in a part $\overline{\mathscr{P}}$ are defined by

$$\begin{aligned} \mathscr{H}^*(\overline{\mathscr{P}}) &= \int_{\overline{\mathscr{P}}} \varrho^* \, \eta^* \, dv, \\ \Gamma^*(\overline{\mathscr{P}}) = \int_{\overline{\mathscr{P}}} \varrho^* \, \gamma^* \, dv &= \frac{d}{dt} \int_{\overline{\mathscr{P}}} \varrho^* \, \eta^* \, dv - \left[\int_{\overline{\mathscr{P}}} \varrho^* \frac{r^*}{\theta^*} \, dv - \int_{\partial\overline{\mathscr{P}}} \frac{\boldsymbol{q}^* \cdot \boldsymbol{n}}{\theta^*} \, da \right], \end{aligned} \tag{11.12}$$

where γ^* is the *specific production of entropy*, r^*/θ^* in (11.12) is the entropy due to radiation entering $\overline{\mathscr{P}}$ and $-(\boldsymbol{q}^* \cdot \boldsymbol{n})/\theta^*$ is the flux of entropy due to conduction entering $\overline{\mathscr{P}}$ through the boundary $\partial\overline{\mathscr{P}}$.

For later convenience, we recall the principle of balance of energy (for non-polar media) over any part $\overline{\mathscr{P}}$ in the form

$$\frac{d}{dt} \int_{\overline{\mathscr{P}}} \left(\varepsilon^* + \frac{1}{2} \boldsymbol{v}^* \cdot \boldsymbol{v}^* \right) \varrho^* \, dv = \int_{\overline{\mathscr{P}}} (r^* + \boldsymbol{f}^* \cdot \boldsymbol{v}^*) \, \varrho^* \, dv + \int_{\partial\overline{\mathscr{P}}} (\boldsymbol{t} \cdot \boldsymbol{v}^* - h^*) \, da, \tag{11.13}$$

the left-hand side of which represents the rate of increase of the sum of internal and kinetic energies, $\boldsymbol{t} \cdot \boldsymbol{v}^*$ and $\boldsymbol{f}^* \cdot \boldsymbol{v}^*$ are rate of work by surface and body forces,

$$h^* = \boldsymbol{q}^* \cdot \boldsymbol{n} \tag{11.14}$$

is the flux of heat entering $\overline{\mathscr{P}}$ through the boundary $\partial\overline{\mathscr{P}}$ and all other quantities have been defined previously. Making the appropriate continuity assumptions, recalling (7.27) and using (11.8), we can write the local balance of energy as

$$\varrho^* \, r^* - \varrho \, \dot{\varepsilon}^* + \tau^{ij} \, \dot{\gamma}^*_{ij} - q^{*k}{}_{\|k} = 0, \tag{11.15}$$

where q^{*k} and γ_{ij} are defined by

$$\boldsymbol{q}^* = q^{*k} \, \boldsymbol{g}_k, \qquad \dot{\gamma}^*_{ij} = \tfrac{1}{2} \, \dot{g}_{ij}, \tag{11.16}$$

and the double vertical line in (11.15) denotes covariant differentiation with respect to the metric tensor g_{ij}. For later use, we record the local energy equation in the alternative form

$$\varrho^* \, r^* - \varrho^* \, \dot{\varepsilon}^* + g^{-\frac{1}{2}} [\boldsymbol{T}^i \cdot \boldsymbol{v}^*_{,i} - (q^{*k} \, g^{\frac{1}{2}})_{,k}] = 0, \tag{11.17}$$

and note that this equation and (11.15) can also be expressed in terms of the (three-dimensional) Helmholtz free energy function defined by

$$\psi^* = \varepsilon^* - \theta^* \, \eta^*. \tag{11.18}$$

---

[44] The negative sign in $(11.11)_2$ is in accord with the usual convention, since $\boldsymbol{q}^* \cdot \boldsymbol{n} < 0$ at points where heat is entering the surface. The sign convention in (11.11) agrees with that used by some writers (see, e.g., Sects. 2.7–2.8 of GREEN and ZERNA [1968, *9*]) but differs from that employed by others (see, e.g., Sect. 79 of TRUESDELL and NOLL [1965, *9*]). Our r^* and $\boldsymbol{q}^*$ correspond to q and $-\boldsymbol{h}$ in [1965, *9*].

Before stating the Clausius-Duhem entropy production inequality, we briefly recall certain additional results from the (three-dimensional) theory of the classical continuum mechanics. The invariance conditions under superposed rigid body motions are assumed to be the requirements that physical quantities

$$\varrho^*, r^*, h^* \text{ (or } \boldsymbol{q}^*), \quad \varepsilon^*, \eta^*, \theta^*, \psi^*, \quad \boldsymbol{T}_i, \tau_{ij}, (\boldsymbol{f}^* - \dot{\boldsymbol{v}}^*) \tag{11.19}$$

transform according to (8.25) and are therefore *objective.*[45] With reference to the terminology introduced preceding (11.11), a process is said to be *admissible* if the set of five functions τ^{ij} (or $\boldsymbol{T}^i$,) ε^*, θ^*, η^* and h^* (or $\boldsymbol{q}^*$) are specified by constitutive equations which hold at each material point and for all time. If the vector function $\boldsymbol{p}$ in (4.6) and either θ^* and η^* are specified, then the body force $\boldsymbol{f}^*$ and the heat supply r^* are uniquely determined from $(11.8)_1$ and (11.15), respectively, while $(11.8)_2$ [or the symmetry condition (11.9)] is assumed to be satisfied identically.

The Clausius-Duhem inequality (in the three-dimensional theory) is specified by

$$\Gamma^*(\overline{\mathscr{P}}) \geqq 0, \tag{11.20}$$

which we require to be valid for every admissible process describing the motion and the thermo-mechanical behavior of the body. From (11.20) and $(11.12)_2$, after introducing the appropriate continuity assumptions, follows the local inequality

$$\varrho^* \theta^* \dot{\eta}^* - \varrho^* r^* + \theta^* g^{-\frac{1}{2}} \left(\frac{q^{*k} g^{\frac{1}{2}}}{\theta^*}\right)_{,k} \geqq 0. \tag{11.21}$$

Elimination of r^* and $(q^{*k} g^{\frac{1}{2}})_{,k}$ between (11.15) and (11.21) yields the inequality

$$\varrho (\theta^* \dot{\eta}^* - \dot{\varepsilon}^*) + \tau^{ij} \dot{\gamma}_{ij} - \frac{q^{*k}}{\theta^*} \theta^*_{,k} \geqq 0, \tag{11.22}$$

which in terms of the Helmholtz function ψ^* becomes

$$-\varrho (\eta^* \dot{\theta}^* + \dot{\psi}^*) + \tau^{ij} \dot{\gamma}_{ij} - \frac{q^{*k}}{\theta^*} \theta_{,k} \geqq 0. \tag{11.23}$$

We observe that the left-hand sides of the energy equation (11.15) and the inequality (11.22), as well as the left-hand side of (11.23), remain unaltered under superposed rigid body motions in view of the invariance requirements noted above. Following a current procedure in continuum mechanics, we regard (11.21) as a condition to be satisfied identically by every admissible process describing the motion and the thermo-mechanical behavior of the medium. Thus, (11.22) or (11.23) will narrow the class of all admissible processes and will place restrictions on the constitutive equations to be specified for the five functions τ^{ij}, ε^* (or ψ^*), η^*, θ^* and q^{*k}.

β) *Stress-resultants, stress-couples and other resultants for shells.* We introduce here the definitions of various resultants of the field quantities which occur in the three-dimensional equations by suitable integration with respect to ξ between the limits $\xi = \alpha(\theta^\alpha)$ and $\xi = \beta(\theta^\alpha)$. First, however, we need the expressions for the

[45] See in this connection the earlier remarks made following (8.25). Our use of the term *objective* is different from the corresponding usage by many who appeal to the *principle of material frame-indifference* which is concerned with invariance under a change of frame employing orthogonal transformations. For a discussion of differences between invariance requirements demanded by the principle of material frame-indifference and those under superposed rigid body motions, see TRUESDELL and NOLL [1965, 9].

element of area of the surfaces $(4.14)_{1,2}$. The outward unit normal vector to the surfaces $(4.14)_{1,2}$ are:

$$\begin{aligned} \boldsymbol{n} &= \mathfrak{f}_{(\alpha)}^{-1}[\alpha_{,1}\,\boldsymbol{g}^1+\alpha_{,2}\,\boldsymbol{g}^2-\boldsymbol{g}^3] \quad \text{on} \quad \xi=\alpha(\theta^\alpha), \\ \boldsymbol{n} &= \mathfrak{f}_{(\beta)}^{-1}[-\beta_{,1}\,\boldsymbol{g}^1-\beta_{,2}\,\boldsymbol{g}^2+\boldsymbol{g}^3] \quad \text{on} \quad \xi=\beta(\theta^\alpha), \end{aligned} \tag{11.24}$$

where a comma denotes partial differentiation with respect to surface coordinates and where

$$\mathfrak{f}_{(\alpha)} = [(\alpha_{,1})^2 g^{11} + (\alpha_{,2})^2 g^{22} + g^{33} + 2(\alpha_{,1}\,\alpha_{,2}\,g^{12} - \alpha_{,1}\,g^{13} - \alpha_{,2}\,g^{23})]^{\frac{1}{2}} \tag{11.25}$$

with an analogous expression for $\mathfrak{f}_{(\beta)}$. The elements of area on the surfaces $(4.14)_{1,2}$ are then given by

$$\begin{aligned} da &= \mathfrak{f}_{(\alpha)}\, g^{\frac{1}{2}}\, d\theta^1\, d\theta^2 \quad \text{on} \quad \xi=\alpha(\theta^\alpha), \\ da &= \mathfrak{f}_{(\beta)}\, g^{\frac{1}{2}}\, d\theta^1\, d\theta^2 \quad \text{on} \quad \xi=\beta(\theta^\alpha). \end{aligned} \tag{11.26}$$

In each of the above expressions, $\boldsymbol{g}^i$, g^{ij}, g are evaluated either at $\xi=\alpha(\theta^\alpha)$ or at $\xi=\beta(\theta^\alpha)$ as indicated. If α and β are constants, then both expressions in (11.26) reduce to

$$da = (g\,g^{33})^{\frac{1}{2}}\, d\theta^1\, d\theta^2 \quad (\text{for } \alpha,\, \beta \text{ constants}). \tag{11.27}$$

Recalling (4.17) and (4.21), we put

$$\varrho\, a^{\frac{1}{2}} = \int_\alpha^\beta k\, d\xi, \qquad \varrho\, k^N a^{\frac{1}{2}} = \int_\alpha^\beta \xi^N k\, d\xi \quad (N \geqq 2) \tag{11.28}$$

and note that α, β are not necessarily constants and an integral corresponding to the right-hand side of $(11.28)_2$ with $N=1$ has been introduced previously in (4.20). We define two-dimensional body forces and a two-dimensional heat supply by

$$\begin{aligned} \varrho\, \boldsymbol{f}\, a^{\frac{1}{2}} = \varrho\, \boldsymbol{l}^0\, a^{\frac{1}{2}} &= \int_\alpha^\beta \varrho^* \boldsymbol{f}^* g^{\frac{1}{2}}\, d\xi + [\bar{\boldsymbol{t}}\, g^{\frac{1}{2}}\, \mathfrak{f}_{(\beta)}]_{\xi=\beta} + [\bar{\boldsymbol{t}}\, g^{\frac{1}{2}}\, \mathfrak{f}_{(\alpha)}]_{\xi=\alpha} \\ &= \int_\alpha^\beta \varrho^* \boldsymbol{f}^* g^{\frac{1}{2}}\, d\xi + [-\beta_{,1}\,\boldsymbol{T}^1 - \beta_{,2}\,\boldsymbol{T}^2 + \boldsymbol{T}^3]_{\xi=\beta} \\ &\quad - [-\alpha_{,1}\,\boldsymbol{T}^1 - \alpha_{,2}\,\boldsymbol{T}^2 + \boldsymbol{T}^3]_{\xi=\alpha}, \end{aligned} \tag{11.29}$$

$$\begin{aligned} \varrho\, \boldsymbol{l}^N a^{\frac{1}{2}} &= \int_\alpha^\beta \varrho^* \boldsymbol{f}^* g^{\frac{1}{2}}\, \xi^N\, d\xi + [\bar{\boldsymbol{t}}\, \xi^N g^{\frac{1}{2}}\, \mathfrak{f}_{(\beta)}]_{\xi=\beta} + [\bar{\boldsymbol{t}}\, \xi^N g^{\frac{1}{2}}\, \mathfrak{f}_{(\alpha)}]_{\xi=\alpha} \\ &= \int_\alpha^\beta \varrho^* \boldsymbol{f}^* g^{\frac{1}{2}}\, \xi^N\, d\xi + [\xi^N(-\beta_{,1}\,\boldsymbol{T}^1 - \beta_{,2}\,\boldsymbol{T}^2 + \boldsymbol{T}^3)]_{\xi=\beta} \\ &\quad - [\xi^N(-\alpha_{,1}\,\boldsymbol{T}^1 - \alpha_{,2}\,\boldsymbol{T}^2 + \boldsymbol{T}^3)]_{\xi=\alpha} \end{aligned} \tag{11.30}$$

and

$$\varrho\, r\, a^{\frac{1}{2}} = \int_\alpha^\beta \varrho^* r^* g^{\frac{1}{2}}\, d\xi - [h^* g^{\frac{1}{2}}\, \mathfrak{f}_{(\beta)}]_{\xi=\beta} - [h^* g^{\frac{1}{2}}\, \mathfrak{f}_{(\alpha)}]_{\xi=\alpha}, \tag{11.31}$$

where $\bar{\boldsymbol{t}}$ in (11.29)–(11.30) is the prescribed surface load on the surfaces $(4.14)_{1,2}$.

Expressions for the stress-resultant and the stress-couples, in terms of the stress tensor τ^{ij} or alternatively the stress vector $\boldsymbol{t}$, can be defined directly using (11.7). First, however, we need certain preliminaries. Remembering the boundary surface (4.15), let $\boldsymbol{\nu}$ be the outward unit normal in the surface $\mathfrak{s}$ (i.e., the surface $\xi=0$ at time t) to a curve of the form $f(\theta^1, \theta^2)=0$, $\xi=0$ and let ds denote the element of arc length of this curve. Then,

$$\boldsymbol{\nu} = \nu_\alpha\, \boldsymbol{a}^\alpha, \qquad \nu_1\, ds = a^{\frac{1}{2}}\, d\theta^2, \qquad \nu_2\, ds = -a^{\frac{1}{2}}\, d\theta^1. \tag{11.32}$$

Also, an element of surface area in a surface of the form (4.15) with outward unit normal $\boldsymbol{n} = n^i \boldsymbol{g}_i = n_i \boldsymbol{g}^i$ is given by

$$\begin{gathered} n_1\, da = g^{\frac{1}{2}}\, d\theta^2\, d\theta^3, \qquad n_2\, da = -g^{\frac{1}{2}}\, d\theta^1\, d\theta^3, \\ da = (n^1\, d\theta^2 - n^2\, d\theta^1)\, g^{\frac{1}{2}}\, d\theta^3. \end{gathered} \tag{11.33}$$

Let $\boldsymbol{N} = \boldsymbol{N}(\theta^\alpha, t; \boldsymbol{\nu})$ and $\boldsymbol{M}^N = \boldsymbol{M}^N(\theta^\alpha, t; \boldsymbol{\nu})$ denote, respectively, the resultant force and the resultant couple vectors of order N ($N = 1, 2, \ldots$), each per unit length of a curve c on $\partial\mathscr{P}$ in the deformed configuration. These resultants are defined by the conditions

$$\int_{\partial\mathscr{P}} \boldsymbol{N}\, ds = \int_{\partial\mathscr{P}_n^*} \boldsymbol{t}\, da, \qquad \int_{\partial\mathscr{P}} \boldsymbol{M}^N\, ds = \int_{\partial\mathscr{P}_n^*} \boldsymbol{t}\, \xi^N\, da, \tag{11.34}$$

the integration on the right-hand sides of $(11.34)_{1,2}$ being over a surface of the form (4.15) between $\xi = \alpha$ and $\xi = \beta$. The above conditions stipulate that the action of $\boldsymbol{N}$ and $\boldsymbol{M}^N$ on a portion of the curve c is equipollent to the action of the stress vector $\boldsymbol{t}$ upon a corresponding portion of the normal surface $\partial\mathscr{P}_n^*$ (which as defined at the beginning of this section coincides with $\partial\mathscr{P}$ on $\xi = 0$). Similarly, the resultant heat flux can be introduced through the surface integral involving h^* in (11.13). Thus, the resultant heat flux $h = h(\theta^\alpha, t; \boldsymbol{\nu})$ per unit time entering $\mathscr{P}$ through the boundary $\partial\mathscr{P}$ is defined by the condition

$$\int_{\partial\mathscr{P}} h\, ds = \int_{\partial\mathscr{P}_n^*} h^*\, da, \tag{11.35}$$

which stipulates that the effect of the resultant heat flux h entering $\mathscr{P}$ is equivalent to the effect of the flux of heat h^* entering through a corresponding portion of the normal surface $\partial\mathscr{P}_n^*$. We also define the resultants $\boldsymbol{N}^\alpha$ and $\boldsymbol{M}^{N\alpha}$ ($N = 1, 2, \ldots$), as well as q^α, by

$$\boldsymbol{N}^\alpha a^{\frac{1}{2}} = \int_\alpha^\beta \boldsymbol{T}^\alpha\, d\xi, \qquad \boldsymbol{M}^{N\alpha} a^{\frac{1}{2}} = \int_\alpha^\beta \xi^N\, \boldsymbol{T}^\alpha\, d\xi, \qquad \boldsymbol{M}^{0\alpha} = \boldsymbol{N}^\alpha, \tag{11.36}$$

$$q^\alpha a^{\frac{1}{2}} = \int_\alpha^\beta q^{*\alpha} g^{\frac{1}{2}}\, d\xi. \tag{11.37}$$

Consider now $(11.34)_1$ and substitute for $\boldsymbol{t}$ from $(11.7)_1$ to obtain

$$\begin{aligned} \int_{\partial\mathscr{P}} \boldsymbol{N}\, ds &= \int_{\partial\mathscr{P}_n^*} g^{-\frac{1}{2}}\, \boldsymbol{T}^i\, n_i\, da = \int_{\partial\mathscr{P}} \int_\alpha^\beta (\boldsymbol{T}^1\, d\theta^2 - \boldsymbol{T}^2\, d\theta^1)\, d\xi \\ &= \int_{\partial\mathscr{P}} a^{\frac{1}{2}} (\boldsymbol{N}^1\, d\theta^2 - \boldsymbol{N}^2\, d\theta^1) = \int_{\partial\mathscr{P}} \boldsymbol{N}^\alpha\, \nu_\alpha\, ds, \end{aligned} \tag{11.38}$$

where (11.33), $(11.36)_1$ and (11.32) have been used. From (11.38) follows the relation[46]

$$\boldsymbol{N} = \boldsymbol{N}^\alpha\, \nu_\alpha. \tag{11.39}$$

In a similar manner, we can deduce the results

$$\boldsymbol{M}^N = \boldsymbol{M}^{N\alpha}\, \nu_\alpha, \qquad h = q^\alpha\, \nu_\alpha. \tag{11.40}$$

---

[46] This corresponds to $(9.11)_1$, but here (11.39) is obtained from the three-dimensional theory through the condition $(11.34)_1$ and the definition of the resultant $\boldsymbol{N}^\alpha$ in $(11.36)_1$.

As in (7.1), we assume that the temperature function $\theta^*(\theta^\alpha, \xi, t)$ at a material point in the deformed shell is an analytic function of ξ in the region $\alpha < \xi < \beta$. Thus, with the notation

$$\varphi_0 = \theta = \theta(\theta^\alpha, t) = \theta^*(\theta^\alpha, 0, t) \tag{11.41}$$

we write[47]

$$\theta^* = \varphi_0 + \sum_{N=1}^{\infty} \xi^N \varphi_N, \tag{11.42}$$

where φ_N are scalar functions of the surface coordinates θ^α and t. We also assume that the series (11.42) may be differentiated as many times as required with respect to any of its variables, at least in the open region $\alpha < \xi < \beta$. We complete our definitions of resultants by introducing the functions $\varepsilon^n = \varepsilon^n(\theta^\alpha, t)$, $(n = 0, 1, 2, \ldots)$ as

$$\begin{aligned} \varrho\, \varepsilon\, a^{\frac{1}{2}} &= \int_\alpha^\beta \varepsilon^*\, k\, d\xi, \\ \varrho\, \varepsilon^n\, a^{\frac{1}{2}} &= \int_\alpha^\beta \varepsilon^*\, k\, \xi^n\, d\xi \quad (n = 0, 1, 2, \ldots), \qquad \varepsilon^0 = \varepsilon, \end{aligned} \tag{11.43}$$

together with the functions $\eta^n = \eta^n(\theta^\alpha, t)$ and $\psi^n = \psi^n(\theta^\alpha, t)$, $(n = 0, 1, 2, \ldots)$, through

$$\begin{aligned} \varrho\, \eta^n\, a^{\frac{1}{2}} &= \int_\alpha^\beta \eta^*\, k\, \xi^n\, d\xi, \qquad \eta^0 = \eta, \\ \varrho\, \psi^n\, a^{\frac{1}{2}} &= \int_\alpha^\beta \psi^*\, k\, \xi^n\, d\xi, \qquad \psi^0 = \psi. \end{aligned} \tag{11.44}$$

The relation

$$\psi^n = \varepsilon^n - \sum_{N=0}^{\infty} \eta^{N+n} \varphi_N \tag{11.45}$$

follows from (11.18) and (11.41)–(11.43). Our motivation for definitions (11.43)–(11.45) will become apparent presently.

γ) *Developments from the energy equation. Entropy inequalities.* First, with the help of the resultants (11.29)–(11.31), (11.34)–(11.35) and $(11.43)_1$, we deduce a two-dimensional energy balance from (11.13). The expression for the kinetic energy of the shell over $\mathscr{P}^*$, using (7.5), (11.28) and (11.2) becomes

$$\tfrac{1}{2} \int_{\mathscr{P}^*} \varrho^*\, \boldsymbol{v}^* \cdot \boldsymbol{v}^*\, dv = \tfrac{1}{2} \int_{\mathscr{P}} \varrho \left[\boldsymbol{v} \cdot \boldsymbol{v} + 2 \sum_{N=2}^{\infty} k^N\, \boldsymbol{v} \cdot \boldsymbol{w}_N + \sum_{M,N=1}^{\infty} k^{M+N}\, \boldsymbol{w}_M \cdot \boldsymbol{w}_N \right] d\sigma, \tag{11.46}$$

where k^N is given by $(11.28)_2$ and

$$\varrho\, k^{M+N}\, a^{\frac{1}{2}} = \int_\alpha^\beta \xi^{M+N}\, k\, d\xi \qquad (M \geqq 1,\ N \geqq 1). \tag{11.47}$$

We note that k^N, k^{M+N} and k defined by (4.21) are functions of θ^1, θ^2 but are independent of t. The surface integral in (11.13) with $\partial\overline{\mathscr{P}}$ identified with $\partial\mathscr{P}^*_n$,

[47] This representation and associated definition for η^n in (11.44) were introduced in [1964, 4].

using (7.5), (11.36)–(11.37) and (11.39)–(11.40), can be reduced as follows:

$$
\begin{aligned}
\int_{\partial\mathscr{P}_n^*} (\boldsymbol{t}\cdot\boldsymbol{v}^* - h^*)\, da &= \int_{\partial\mathscr{P}_n^*} \boldsymbol{t}\cdot\left(\boldsymbol{v} + \sum_{N=1}^{\infty} \xi^N \boldsymbol{w}_N\right) da - \int_{\partial\mathscr{P}_n^*} h^*\, da,\\
&= \iint_{\partial\mathscr{P}_n^*} (\boldsymbol{T}^1\, d\theta^2 - \boldsymbol{T}^2\, d\theta^1)\cdot\left(\boldsymbol{v} + \sum_{N=1}^{\infty} \xi^N \boldsymbol{w}_N\right) d\xi\\
&\quad - \iint_{\partial\mathscr{P}_n^*} h^*(n^1\, d\theta^2 - n^2\, d\theta^1)\, g^{\frac{1}{2}}\, d\xi\\
&= \int_{\partial\mathscr{P}} \left(\boldsymbol{N}^\alpha\cdot\boldsymbol{v} + \sum_{N=1}^{\infty} \boldsymbol{M}^{N\alpha}\cdot\boldsymbol{w}_N\right) \nu_\alpha\, ds - \int_{\partial\mathscr{P}} a^{\frac{1}{2}}(q^1\, d\theta^2 - q^2\, d\theta^1)\\
&= \int_{\partial\mathscr{P}} \left(\boldsymbol{N}\cdot\boldsymbol{v} + \sum_{N=1}^{\infty} \boldsymbol{M}^N\cdot\boldsymbol{w}_N\right) ds - \int_{\partial\mathscr{P}} q^\alpha \nu_\alpha\, ds\\
&= \int_{\partial\mathscr{P}} \left(\boldsymbol{N}\cdot\boldsymbol{v} + \sum_{N=1}^{\infty} \boldsymbol{M}^N\cdot\boldsymbol{w}_N - h\right) ds.
\end{aligned} \tag{11.48}
$$

Also, with the help of (7.5) and (11.29)–(11.31), it can be readily shown that the volume integral on the right-hand side of (11.13) with $\overline{\mathscr{P}}$ identified with $\mathscr{P}^*$ and contributions due to the surface integral in (11.13) evaluated on the surfaces $(4.14)_{1,2}$ may be written as

$$
\int_{\mathscr{P}^*} \varrho^*(r^* + \boldsymbol{f}^*\cdot\boldsymbol{v}^*)\, dv + \int_{\partial\mathscr{P}_n^{*c}} (\boldsymbol{t}\cdot\boldsymbol{v}^* - h^*)\, da = \int_{\mathscr{P}} \varrho\left[r + \boldsymbol{f}\cdot\boldsymbol{v} + \sum_{N=1}^{\infty} \boldsymbol{l}^N\cdot\boldsymbol{w}_N\right] d\sigma. \tag{11.49}
$$

By use of the results (11.46)–(11.49), as well as $(11.43)_1$, the (three-dimensional) balance of energy (11.13) reduces to

$$
\begin{aligned}
&\frac{d}{dt}\int_{\mathscr{P}} \varrho\left(\varepsilon + \frac{1}{2}\boldsymbol{v}\cdot\boldsymbol{v} + \sum_{N=2}^{\infty} k^N \boldsymbol{w}_N\cdot\boldsymbol{v} + \frac{1}{2}\sum_{M,N=1}^{\infty} k^{M+N}\boldsymbol{w}_M\cdot\boldsymbol{w}_N\right) d\sigma\\
&\quad = \int_{\mathscr{P}} \varrho\left(r + \boldsymbol{f}\cdot\boldsymbol{v} + \sum_{N=1}^{\infty} \boldsymbol{l}^N\cdot\boldsymbol{w}_N\right) d\sigma + \int_{\partial\mathscr{P}} \left(\boldsymbol{N}\cdot\boldsymbol{v} + \sum_{N=1}^{\infty} \boldsymbol{M}^N\cdot\boldsymbol{w}_N - h\right) ds
\end{aligned} \tag{11.50}
$$

over $\mathscr{P}$ in the present configuration.[48]

In what follows, we require generalized forms of the energy equation and the entropy production inequality and these can be obtained from (11.17) and (11.21). Let $\Phi = \Phi(\theta^i)$ be a scalar function of the coordinates θ^i. For example, a useful form of Φ used below in the development of shell theory is $\Phi = \xi^n$ $(n = 0, 1, 2, \ldots)$. Now multiply (11.17) by Φ and integrate over an arbitrary part $\mathscr{P}^*$ (not $\overline{\mathscr{P}}$) in the present configuration. After some straightforward manipulation and using (11.8), we obtain

$$
\begin{aligned}
\frac{d}{dt}\int_{\mathscr{P}^*} \left(\varepsilon^* + \frac{1}{2}\boldsymbol{v}^*\cdot\boldsymbol{v}^*\right) \varrho^*\, \Phi\, dv &= \int_{\mathscr{P}^*} (r^* + \boldsymbol{f}^*\cdot\boldsymbol{v}^*)\, \varrho^*\, \Phi\, dv\\
&\quad + \int_{\partial\mathscr{P}^*} (\boldsymbol{t}\cdot\boldsymbol{v}^* - \boldsymbol{q}^*\cdot\boldsymbol{n})\, \Phi\, da\\
&\quad - \int_{\mathscr{P}^*} [\boldsymbol{T}^i\cdot\boldsymbol{v}^* - q^{*i} g^{\frac{1}{2}}]\, g^{-\frac{1}{2}} \frac{\partial\Phi}{\partial\theta^i}\, dv.
\end{aligned} \tag{11.51}
$$

[48] The two-dimensional energy balance (11.50) was obtained in [1968, *4*]. A less general result was given previously in [1964, *4*]. The derivation in the rest of this section closely follows that of GREEN and NAGHDI [1970, *2*].

Again let $\Psi = \Psi(\theta^i)$, a non-negative scalar function of coordinates, be defined over a part $\mathscr{P}^*$ of $\mathscr{B}$:

$$\Psi \geqq 0. \tag{11.52}$$

Multiply (11.21) by Ψ/θ^* and integrate over a part $\mathscr{P}^*$ to obtain

$$\begin{aligned}\frac{d}{dt}\int\limits_{\mathscr{P}^*} \varrho^* \eta^* \Psi\, dv - \int\limits_{\mathscr{P}^*} \varrho^* \frac{r^*}{\theta^*} \Psi\, dv + \int\limits_{\partial\mathscr{P}^*} \frac{\boldsymbol{q}^* \cdot \boldsymbol{n}}{\theta^*} \Psi\, da \\ - \int\limits_{\mathscr{P}^*} \frac{q^{*i}}{\theta^*} \frac{\partial \Psi}{\partial \theta^i}\, dv \geqq 0.\end{aligned} \tag{11.53}$$

It is clear that when Φ and Ψ are constants, (11.51) and (11.53) reduce to the integral forms of balance of energy and the Clausius-Duhem inequality over a part $\mathscr{P}^*$ of $\mathscr{B}$ in the present configuration.

We now substitute $\Phi = \xi^n$ $(n = 1, 2, \ldots)$ in (11.51) and by a procedure similar to that used in obtaining (11.50) we deduce

$$\begin{aligned}\frac{d}{dt}\int\limits_{\mathscr{P}} \varrho \left[\varepsilon^n + \frac{1}{2} k^n \boldsymbol{v} \cdot \boldsymbol{v} + \sum_{N=1}^{\infty} k^{N+n} \boldsymbol{w}_N \cdot \boldsymbol{v} + \frac{1}{2} \sum_{M,N=1}^{\infty} k^{M+N+n} \boldsymbol{w}_M \cdot \boldsymbol{w}_N\right] d\sigma \\ = \int\limits_{\mathscr{P}} \cdot \varrho \left[\left(r^n + R^n + \boldsymbol{l}^n \cdot \boldsymbol{v} + \sum_{N=1}^{\infty} \boldsymbol{l}^{N+n} \cdot \boldsymbol{w}_N\right) - \boldsymbol{m}^n \cdot \boldsymbol{v} \right. \\ \left. - \sum_{N=1}^{\infty} \frac{n}{N+n} \boldsymbol{m}^{N+n} \cdot \boldsymbol{w}_N\right] d\sigma \\ + \int\limits_{\partial\mathscr{P}} \left[\boldsymbol{M}^n \cdot \boldsymbol{v} + \sum_{N=1}^{\infty} \boldsymbol{M}^{N+n} \cdot \boldsymbol{w}_N - h^n\right] ds,\end{aligned} \tag{11.54}$$

where we have introduced

$$\boldsymbol{m}^N a^{\frac{1}{2}} = N \int\limits_{\alpha}^{\beta} \xi^{N-1} \boldsymbol{T}^3\, d\xi \qquad (N \geqq 0) \tag{11.55}$$

and

$$\varrho\, R^n a^{\frac{1}{2}} = n \int\limits_{\alpha}^{\beta} q^{*3} \xi^{n-1} g^{\frac{1}{2}}\, d\xi, \tag{11.56}$$

$$q^{n\alpha} a^{\frac{1}{2}} = \int\limits_{\alpha}^{\beta} q^{*\alpha} \xi^n g^{\frac{1}{2}}\, d\xi, \qquad h^n = q^{n\alpha} \nu_\alpha, \tag{11.57}$$

$$\varrho\, r^n a^{\frac{1}{2}} = \int\limits_{\alpha}^{\beta} r^* k\, \xi^n\, d\xi - [\xi^n h^* g^{\frac{1}{2}} \mathfrak{f}_{(\beta)}]_{\xi=\beta} - [\xi^n h^* g^{\frac{1}{2}} \mathfrak{f}_{(\alpha)}]_{\xi=\alpha}, r^0 = r, \tag{11.58}$$

in addition to resultants defined previously. Under suitable continuity assumptions (11.54) can be reduced to

$$\begin{aligned}\varrho\, r^n + \varrho\, R^n - \varrho\, \dot{\varepsilon}^n + (\boldsymbol{M}^{n\alpha}{}_{|\alpha} + \varrho\, \bar{\boldsymbol{l}}^n - \boldsymbol{m}^n) \cdot \boldsymbol{v} + \boldsymbol{M}^{n\alpha} \cdot \boldsymbol{v}_{,\alpha} \\ + \sum_{N=1}^{\infty} \left[\boldsymbol{M}^{(N+n)\alpha}{}_{|\alpha} + \varrho\, \bar{\boldsymbol{l}}^{N+n} - \frac{n}{N+n} \boldsymbol{m}^{N+n}\right] \cdot \boldsymbol{w}_N \\ + \sum_{N=2}^{\infty} \boldsymbol{M}^{(N+n)\alpha} \cdot \boldsymbol{w}_{N,\alpha} - q^{n\alpha}{}_{|\alpha} = 0 \qquad (n = 0, 1, 2, \ldots),\end{aligned} \tag{11.59}$$

where we have put

$$\left.\begin{aligned}&\bar{\boldsymbol{f}}=\bar{\boldsymbol{l}}^0=\boldsymbol{f}-\dot{\boldsymbol{v}}-\sum_{N=1}^{\infty}k^N\dot{\boldsymbol{w}}_N, \qquad \bar{\boldsymbol{l}}=\bar{\boldsymbol{l}}^1=\boldsymbol{l}-\sum_{M=1}^{\infty}k^{M+1}\dot{\boldsymbol{w}}_M,\\&\bar{\boldsymbol{l}}^N=\boldsymbol{l}^N-k^N\dot{\boldsymbol{v}}-\sum_{M=1}^{\infty}k^{M+N}\dot{\boldsymbol{w}}_M \qquad (N\geqq 2).\end{aligned}\right\}\tag{11.60}$$

In (11.59), for clarity, parentheses have been placed around the number $N+n$ which occurs as a superscript in $\boldsymbol{M}^{(N+n)\alpha}$. Henceforth, however, we shall omit the parentheses and will write $\boldsymbol{M}^{(N+n)\alpha}$ as $\boldsymbol{M}^{N+n\alpha}$.

We postpone further consideration of (11.59) and turn our attention to the inequality (11.53). Recalling (11.42) and the fact that $\theta^*>0$, we introduce

$$\Phi^*=\frac{1}{\theta^*}=\sum_{N=0}^{\infty}\xi^N\Phi_N, \qquad \Phi_N=\Phi_N(\theta^1,\theta^2,t), \tag{11.61}$$

so that

$$\varphi_0\,\Phi_0=1, \qquad \sum_{N=0}^{r}\varphi_N\,\Phi_{r-N}=0 \qquad (r-1,2,\ldots), \tag{11.62}$$

in view of (11.42). Also, remembering (11.52), we set

$$\Psi=(-\alpha+\xi)^n \qquad (\alpha\leqq\xi\leqq\beta). \tag{11.63}$$

Substituting (11.61) and (11.63) into (11.53) and making the appropriate continuity assumptions, we obtain the inequalities

$$\left.\begin{aligned}&\varrho\sum_{r=0}^{n}\binom{n}{r}(-\alpha)^{n-r}\left[\dot{\eta}^r-\sum_{N=0}^{\infty}\gamma^{r+N}\Phi_N\right]\\&\quad+\sum_{r=0}^{n}\binom{n}{r}(-\alpha)^{n-r}\sum_{N=0}^{\infty}(q^{r+N\alpha}|_\alpha\,\Phi_N+q^{r+N\alpha}\,\Phi_{N,\alpha})\\&\quad-\varrho\sum_{r=0}^{n-1}\binom{n-1}{r}(-\alpha)^{n-r-1}\sum_{N=0}^{\infty}\frac{n}{r+N+1}R^{r+N+1}\Phi_N\geqq 0\\&\quad(n=0,1,2,\ldots).\end{aligned}\right\}\tag{11.64}$$

Since φ_N and Φ_N are not defined for negative values of N, we may write $(11.62)_2$ as

$$\sum_{N=0}^{\infty}\varphi_N\,\Phi_{r-N}=0 \quad (r=1,2,\ldots), \qquad \varphi_M=\Phi_M=0 \quad \text{for} \quad M<0. \tag{11.65}$$

Also, for later reference, we record here the identity

$$\sum_{N=0}^{\infty}\sum_{M=0}^{\infty}\Phi_N\,\dot{\eta}^{M+N+r}\varphi_M=\dot{\eta}^r. \tag{11.66}$$

To verify the last result, let the left-hand side of (11.66) be denoted by $\mathfrak{g}$ and write

$$\begin{aligned}\mathfrak{g}&=\sum_{N=0}^{\infty}\sum_{M=0}^{\infty}\Phi_N\,\dot{\eta}^{M+N+r}\,\varphi_M\\&=\sum_{n'=N'}^{\infty}\sum_{N'=0}^{\infty}\Phi_{n'-N'}\,\varphi_{N'}\,\dot{\eta}^{n'+r}\\&=\sum_{n=0}^{\infty}\sum_{N=0}^{\infty}\Phi_{n-N}\,\varphi_N\,\dot{\eta}^{n+r},\end{aligned}$$

where in the last of the above the first summation sign $\sum_{n=N}^{\infty}$ has been replaced with $\sum_{n=0}^{\infty}$ since $\Phi_{n-N}=0$ for $N>n$. Now the last expression can be written as

$$\mathfrak{g}=\sum_{N=0}^{\infty}\Phi_{-N}\,\varphi_N\,\dot{\eta}^r+\sum_{n=1}^{\infty}\left[\sum_{N=0}^{\infty}\Phi_{n-N}\,\varphi_N\right]\dot{\eta}^{n+r}. \tag{11.67}$$

But the term in the square bracket vanishes by virtue of (11.65) and the first term on the right-hand side of (11.67) is $\Phi_0\,\varphi_0\,\dot{\eta}^r=\dot{\eta}^r$ by $(11.62)_1$ and $(11.65)_2$. Hence, (11.67) reduces to (11.66).

12. Basic field equations for shells continued: II. Derivation from the three-dimensional theory. We continue here the development of basic field equations for shells from the three-dimensional theory. After the derivation of the local field equations, several related aspects of the subject are discussed including the linearization of the field equations and their relations to the corresponding results in the classical linear theories of shells and plates.

δ) *General field equations.* The derivation of the equations of motion for shells can be accomplished by direct integration of $(11.8)_{1,2}$ after their multiplication by $\xi^N\,(N\geqq 0)$ and with the use of the resultants $(11.36)_{1,2}$, (11.55) and (11.29)–(11.30). Here, however, we deduce the equations of motion for shells from the energy equations (11.59) together with the invariance requirements under superposed rigid body motions. From the fact that the quantities listed in (11.19) are objective in the sense of (8.25) and the definitions of the various resultants, it follows that

$$\begin{aligned}&\boldsymbol{M}^{n\alpha},\ \boldsymbol{m}^n,\ \boldsymbol{l}^n\quad(n\geqq 0),\\&\varepsilon^n,\ \eta^n,\ \varphi^n,\ \psi^n\quad(n\geqq 0),\\&r^n,\ R^n,\ q^{n\alpha}\quad(n\geqq 0)\end{aligned} \tag{12.1}$$

are also objective and hence transform according to (8.25) under superposed rigid body motions.

Consider now two special motions corresponding to those under superposed rigid body motions when the body occupies the same position at time t. Then, with Q and $\dot{Q}$ given by (5.48) at time t, the superposed velocities are given, respectively, by the linear transformations

$$\boldsymbol{v}^*\to\boldsymbol{v}^*+\boldsymbol{b},\qquad \boldsymbol{b}=\text{const.}, \tag{12.2}$$

$$\boldsymbol{v}^*\to\boldsymbol{v}^*+\boldsymbol{\omega}_0\times\boldsymbol{p},\qquad \boldsymbol{\omega}_0=\text{const.}, \tag{12.3}$$

and these, in turn, imply the transformations

$$\boldsymbol{v}\to\boldsymbol{v}+\boldsymbol{b},\qquad \boldsymbol{w}_N\to\boldsymbol{w}_N, \tag{12.4}$$

$$\boldsymbol{v}\to\boldsymbol{v}+\boldsymbol{\omega}_0\times\boldsymbol{r},\qquad \boldsymbol{w}_N\to\boldsymbol{w}_N+\boldsymbol{\omega}_0\times\boldsymbol{d}_N, \tag{12.5}$$

in view of (7.1) and (7.5). The linear transformations (12.2) or (12.4) correspond to superposed uniform rigid body translational velocity while (12.3) or (12.5) represent uniform rigid body angular velocity, the (three-dimensional) continuum occupying the same position at time t.

Since the energy equations (11.59) hold for every motion of the shell, by a familiar argument and using (12.4)–(12.5), the equations of motion can quickly be deduced from (11.59). Thus, if we replace $\boldsymbol{v}$ and $\boldsymbol{w}_N$ in (11.59) by the transformations (12.4) and keep in mind that the functions listed in (12.1) remain unaltered under superposed rigid body motions, then after subtraction we obtain

$$\boldsymbol{N}^{\alpha}{}_{|\alpha} + \varrho \bar{\boldsymbol{f}} = 0, \qquad \boldsymbol{M}^{n\alpha}{}_{|\alpha} + \varrho \bar{\boldsymbol{l}}^{n} - \boldsymbol{m}^{n} = 0 \quad (n = 1, 2, \ldots). \tag{12.6}$$

With the use of (12.6) and recalling the remark (concerning notation) following (11.60), the energy equations (11.59) can be reduced to

$$\begin{aligned} &\varrho r^{n} + \varrho R^{n} - \varrho \dot{\varepsilon}^{n} + \boldsymbol{M}^{n\alpha} \cdot \boldsymbol{v}_{,\alpha} + \sum_{N=1}^{\infty} \frac{N}{N+n} \boldsymbol{m}^{N+n} \cdot \boldsymbol{w}_{N} \\ &+ \sum_{N=1}^{\infty} \boldsymbol{M}^{N+n\alpha} \cdot \boldsymbol{w}_{N,\alpha} - q^{n\alpha}{}_{|\alpha} = 0 \quad (n = 0, 1, 2, \ldots). \end{aligned} \tag{12.7}$$

Next, considering a motion of the shell corresponding to (12.5) and again keeping in mind that the functions listed in (12.1) remain unaltered under such superposed rigid body motions, from (12.7) we deduce the equations

$$\boldsymbol{M}^{n\alpha} \times \boldsymbol{a}_{\alpha} + \sum_{N=1}^{\infty} \left(\frac{N}{N+n} \boldsymbol{m}^{N+n} \times \boldsymbol{d}_{N} + \boldsymbol{M}^{N+n\alpha} \times \boldsymbol{d}_{N,\alpha} \right) = 0 \quad (n = 0, 1, 2, \ldots). \tag{12.8}$$

For $n = 0$, (12.8) can be written in a more transparent form, namely

$$\boldsymbol{N}^{\alpha} \times \boldsymbol{a}_{\alpha} + \sum_{N=1}^{\infty} (\boldsymbol{m}^{N} \times \boldsymbol{d}_{N} + \boldsymbol{M}^{N\alpha} \times \boldsymbol{d}_{N,\alpha}) = 0. \tag{12.9}$$

As noted earlier, the equations of motion (12.6) can also be obtained directly from $(11.8)_1$ while (12.8) can be deduced from $(11.8)_2$ and is therefore a consequence of the symmetry of the stress tensor.[49]

The Eqs. (12.6) and (12.8) are the equations of motion for shells in vector form. For later reference, it is desirable to record these equations also in tensor components. By (11.39)–$(11.40)_1$, the resultants $\boldsymbol{N}^{\alpha}$ and $\boldsymbol{M}^{N\alpha}$ transform as contravariant surface vectors. When referred to the base vectors $\boldsymbol{a}_i$, these vector fields and the resultants $\boldsymbol{m}^{N}$ can be written as[50]

$$\boldsymbol{N}^{\alpha} = N^{\alpha i} \boldsymbol{a}_i, \qquad \boldsymbol{M}^{N\alpha} = M^{N\alpha i} \boldsymbol{a}_i \tag{12.10}$$

and

$$\boldsymbol{m}^{N} = m^{N i} \boldsymbol{a}_i. \tag{12.11}$$

Similarly, the resultants in (11.60) can be expressed as

$$\bar{\boldsymbol{f}} = \bar{\boldsymbol{l}}^{0} = \bar{f}^{i} \boldsymbol{a}_i, \qquad \bar{\boldsymbol{l}} = \bar{\boldsymbol{l}}^{1} = \bar{l}^{i} \boldsymbol{a}_i, \qquad \bar{\boldsymbol{l}}^{N} = \bar{l}^{N i} \boldsymbol{a}_i \quad (N \geqq 2). \tag{12.12}$$

---

[49] The equations of motion (12.6) and (12.8) in vector forms were derived by GREEN and NAGHDI [1970, *2*] and the component forms (12.13)–(12.15) were obtained earlier by GREEN, LAWS and NAGHDI [1968, *4*]. Two-dimensional equations of motion (or equilibrium) of this type, limited to the linear theory of flat plates, can be found in the papers of TIFFEN and LOWE [1963, *15*] and [1965, *8*]. We also mention here an extensive work of CHIEN [1944, *2*], where his equations of equilibrium (corresponding to the equations of the restricted theory in Sect. 10) are those which were derived in [1941, *2*] by direct approach; however, in the remainder of the series of papers [1944, *2*], the stress tensor and the body force are expanded in powers of the thickness coordinate. This procedure is not the same as that in which resultants of the type (11.29)–(11.30) and (11.36) are employed.

[50] The order of indices αi in $N^{\alpha i}$ and $M^{N\alpha i}$ differs from those in [1968, *4*] and [1970, *2*] but agrees with that in [1971, *6*] and corresponds to the usual notation in shell theory.

With the help of (12.10)–(12.12), the equations of motion (12.6) in component form are[51]

$$N^{\alpha\beta}{}_{|\alpha} - b^{\alpha}_{\beta} N^{\alpha 3} + \varrho \bar{f}^{\beta} = 0, \qquad N^{\alpha 3}{}_{|\alpha} + b_{\alpha\beta} N^{\alpha\beta} + \varrho \bar{f}^{3} = 0, \tag{12.13}$$

$$\begin{aligned} &M^{N\alpha\beta}{}_{|\alpha} - b^{\beta}_{\alpha} M^{N\alpha 3} + \varrho \bar{l}^{N\beta} = m^{N\beta}, \\ &M^{N\alpha 3}{}_{|\alpha} + b_{\alpha\beta} M^{N\alpha\beta} + \varrho \bar{l}^{N3} = m^{N3} \qquad (N = 1, 2, \ldots). \end{aligned} \tag{12.14}$$

In a similar manner and by recalling the kinematical results (7.13)–(7.15), from (12.8) we deduce

$$\begin{aligned} &N'^{\beta\alpha} = N'^{\alpha\beta} = N^{\alpha\beta} - \sum_{N=1}^{\infty} (m^{N\alpha} d_{N\cdot}^{\ \beta} + M^{N\gamma\alpha} \lambda_{N\cdot\gamma}^{\ \beta}), \\ &N^{\alpha 3} + \sum_{N=1}^{\infty} (m^{N3} d_{N\cdot}^{\ \alpha} - m^{N\alpha} d_{N\cdot}^{\ 3}) + \sum_{N=1}^{\infty} (M^{N\gamma 3} \lambda_{N\cdot\gamma}^{\ \alpha} - M^{N\gamma\alpha} \lambda_{N\cdot\gamma}^{\ 3}) = 0 \end{aligned} \tag{12.15}$$

and

$$\begin{aligned} &M'^{n\alpha\beta} = M'^{n\beta\alpha} = M^{n\alpha\beta} - \sum_{N=1}^{\infty} \left(\frac{N}{N+n} m^{N+n\alpha} d_{N\cdot}^{\ \beta} + M^{N+n\gamma\alpha} \lambda_{N\cdot\gamma}^{\ \beta} \right) \qquad (n = 1, 2, \ldots), \\ &M^{n\alpha 3} + \sum_{N=1}^{\infty} \frac{N}{N+n} (m^{N+n3} d_{N\cdot}^{\ \alpha} - m^{N+n\alpha} d_{N\cdot}^{\ 3}) \\ &\quad + \sum_{N=1}^{\infty} (M^{N+n\gamma 3} \lambda_{N\cdot\gamma}^{\ \alpha} - M^{N+n\gamma\alpha} \lambda_{N\cdot\gamma}^{\ 3}) = 0 \qquad (n = 1, 2, \ldots). \end{aligned} \tag{12.16}$$

Eqs. (12.15) are the component form of (12.9) or (12.8) for $n = 0$. It is clear that the conditions (12.15)–(12.16), which result from the symmetry of the stress tensor, are identities in the exact theory.

Recalling the kinematic variables (7.13)–(7.16) and using (12.8), the energy equations (12.7) reduce to

$$\varrho r^n + \varrho R^n - \varrho \dot{\varepsilon}^n + P^n - q^{n\alpha}{}_{|\alpha} = 0, \tag{12.17}$$

where

$$P^n = M'^{n\alpha\beta} \eta_{\alpha\beta} + \sum_{N=1}^{\infty} \left(\frac{N}{N+n} m^{N+ni} \dot{d}_{Ni} + M^{N+n\alpha i} \dot{\lambda}_{Ni\alpha} \right), \tag{12.18}$$

$M'^{n\alpha\beta}$ is given by $(12.16)_1$ and we have introduced the notation

$$\eta_{\alpha\beta} = \dot{e}_{\alpha\beta} = \tfrac{1}{2} \dot{a}_{\alpha\beta} \tag{12.19}$$

for the time rate of the kinematic measure (7.18). In terms of the resultants ψ^n defined by $(11.44)_3$, the energy equations (12.17) can be expressed as

$$\varrho r^n + \varrho R^n - \varrho \left[\dot{\psi}^n + \sum_{N=0}^{\infty} (\dot{\eta}^{N+n} \varphi_N + \eta^{N+n} \dot{\varphi}_N) \right] + P^n - q^{n\alpha}{}_{|\alpha} = 0. \tag{12.20}$$

For completeness, we also record here the inequalities which can be deduced from the combination of (11.64) and (12.20). In view of (11.66), we have

$$\begin{aligned} &\dot{\eta}^r - \sum_{N=0}^{\infty} \Phi_N \left[\dot{\psi}^{r+N} + \sum_{M=0}^{\infty} \overline{\eta^{M+N+r} \varphi_M}^{\,\cdot} \right] \\ &\quad = \dot{\eta}^r - \sum_{N=0}^{\infty} \Phi_N \left[\dot{\psi}^{r+N} + \sum_{M=0}^{\infty} \eta^{M+N+r} \dot{\varphi}_M \right] - \sum_{n=0}^{\infty} \sum_{N=0}^{\infty} \dot{\eta}^{n+r} \Phi_{n-N} \varphi_N \\ &\quad = - \sum_{N=0}^{\infty} \Phi_N \left[\dot{\psi}^{r+N} + \sum_{M=0}^{\infty} \eta^{M+N+r} \dot{\varphi}_M \right]. \end{aligned} \tag{12.21}$$

[51] These are deduced in a manner similar to (9.42) by considering the scalar product of each of (12.6) with $\boldsymbol{a}^{\beta}$ and again with $\boldsymbol{a}^{3}$.

Substituting for r^n from (12.20) into (11.64) and using (12.21), we finally obtain the inequalities[52]

$$\sum_{r=0}^{n}\binom{n}{r}(-\alpha)^{n-r}\Bigg\{\sum_{N=0}^{\infty}\Phi_N\Bigg[-\varrho\Big(\dot\psi^{r+N}+\sum_{M=0}^{\infty}\eta^{M+N+r}\,\dot\varphi_M\Big)+M'^{r+N\alpha\beta}\,\eta_{\alpha\beta}$$
$$+\sum_{M=1}^{\infty}\frac{M}{M+N+r}\,m^{M+N+ri}\,\dot d_{Mi}+\sum_{M=1}^{\infty}M^{M+N+r\alpha i}\,\dot\lambda_{Mi\alpha}+\frac{N}{N+r}\,\varrho\,R^{N+r}\Bigg] \tag{12.22}$$
$$+\sum_{N=0}^{\infty}q^{N+r\alpha}\,\Phi_{N,\alpha}\Bigg\}\geqq 0\qquad (n=0,1,2,\ldots).$$

ε) *An approximate system of equations of motion.* Although the equations of motion (12.6) and (12.8) or (12.13)–(12.16) are exact, they consist in an infinite system of equations in an infinite number of unknowns. The complexity of this system and a need for a *suitable* approximation are evident. As remarked in Sect. 7, the introduction of an approximative scheme is premature at this stage and strictly speaking should be postponed until the complete theory (including the constitutive equations) has been developed. However, in order to give an indication of the nature of the field equations of an approximate theory (derived from the three-dimensional equations), we consider briefly a system of approximate equations of motion but postpone its justification.[53]

Suppose that[54]

$$M^{N\alpha i}=0,\qquad m^{Ni}=0\qquad (N\geqq 2). \tag{12.23}$$

Then, the equations of motion (12.14) for $N\geqq 2$ are satisfied if we specify

$$\bar l^{Ni}=0\qquad (N\geqq 2). \tag{12.24}$$

Eqs. (12.24) are usually satisfied approximately, since the quantities $\bar l^{Ni}$ $(N\geqq 2)$ involve moments (of order greater than one across the thickness of the shell) of body forces, applied surface loads and inertia terms. Recalling (11.60) and the notation in (7.9), consistent with the above approximation, we also take

$$\bar{\boldsymbol f}=\boldsymbol f-\dot{\boldsymbol v},\qquad \bar{\boldsymbol l}=\bar{\boldsymbol l}^1=\boldsymbol l-k^{11}\,\dot{\boldsymbol w}, \tag{12.25}$$

where

$$\varrho\,k^{11}\,a^{\frac12}=\int_\alpha^\beta \varrho^*\,g^{\frac12}\,\xi^2\,d\xi. \tag{12.26}$$

With the notations

$$M^{1\alpha i}=M^{\alpha i},\qquad m^{1i}=m^i, \tag{12.27}$$

as a consequence of the above approximation specified by (12.23)–(12.25), the equations of motion (12.13)–(12.14) reduce to

$$N^{\alpha\beta}{}_{|\alpha}-b^\beta_\alpha\,N^{\alpha 3}+\varrho\,\bar f^\beta=0,\qquad N^{\alpha 3}{}_{|\alpha}+b_{\alpha\beta}\,N^{\alpha\beta}+\varrho\,\bar f^3=0, \tag{12.28}$$

$$M^{\alpha\beta}{}_{|\alpha}-b^\beta_\alpha\,M^{\alpha 3}+\varrho\,\bar l^\beta=m^\beta,\qquad M^{\alpha 3}{}_{|\alpha}+b_{\alpha\beta}\,M^{\alpha\beta}+\varrho\,\bar l^3=m^3, \tag{12.29}$$

---

[52] GREEN and NAGHDI [1970, *2*].

[53] The approximations under which the system of approximate equations of motion are obtained were employed by GREEN, LAWS and NAGHDI [1968, *4*] and GREEN and NAGHDI [1970, *2*].

[54] In view of the generality of our development leading to (12.13)–(12.16), conditions (12.23) represent an approximation. On the other hand, in most of the existing literature on shell theory confined to a less general development, the introduction of the resultants $M^{N\alpha i}$ and m^{Ni} for $N\geqq 2$ is avoided and thereby the question of approximation (12.23) does not arise.

while Eqs. $(12.15)_{1,2}$ become

$$\begin{aligned} N'^{\alpha\beta} = N'^{\beta\alpha} = N^{\alpha\beta} - m^\alpha d^\beta - M^{\gamma\alpha} \lambda^\beta_{.\gamma}, \\ N^{\alpha 3} + m^3 d^\alpha - m^\alpha d^3 + M^{\gamma 3} \lambda^\alpha_{.\gamma} - M^{\gamma\alpha} \lambda^3_{.\gamma} = 0. \end{aligned} \tag{12.30}$$

Also, the Eqs. $(12.16)_{1,2}$ reduce to

$$\begin{aligned} M'^{1\alpha\beta} = M^{\alpha\beta} = M^{\beta\alpha}, \\ M^{1\alpha 3} = M^{\alpha 3} = 0. \end{aligned} \tag{12.31}$$

Although $(12.16)_{1,2}$ are identities in an exact theory, in general we cannot expect that they be satisfied in an approximate theory. Indeed, from the restrictive nature of (12.31), it can be inferred that they could be satisfied only in special cases of an approximate theory. The identities $(12.16)_{1,2}$ and the question of their subsequent approximations do not arise in most of the existing literature on the classical theory of shells developed from the three-dimensional equations. This is simply due to the fact that the introduction of certain resultants such as $M^{N\alpha i}$ and m^{Ni} for $N \geqq 2$ is avoided *ab initio.*

As our purpose of describing the above approximation at this stage is merely to give an indication of the nature of a system of approximate equations of motion, we do not dwell on the corresponding approximation for the temperature functions in (11.61) and their subsequent effect, as well as those in (12.23), on the energy equations (12.20) and the inequalities (12.22). However, it is worth observing that if the surfaces $\mathfrak{s}$ and $\mathfrak{z}$ are identified, the set of Eqs. (12.28)–(12.30) are formally the same as the equations of motion for a Cosserat surface derived in Sect. 9.

ζ) *Linearized field equations.* Previously in Sect. 7, where the linearization of the kinematic quantities was considered, the (three-dimensional) infinitesimal strain tensor was expressed in terms of linearized two-dimensional kinematic measures. In carrying out the linearization procedure, it was assumed that all (two-dimensional) kinematic measures such as those in (7.59), as well as their derivatives with respect to the surface coordinates and time, are of $O(\varepsilon)$ in the sense of (6.4). We now suppose that the continuum is initially stress-free and let θ_0^* and η_0^* refer to a standard temperature and entropy in the (initial) reference configuration. We further recall that in a complete linearized (three-dimensional) theory, $\boldsymbol{T}^i$ or the stress tensor τ^{ij} when expressed in suitable non-dimensional form, as well as its derivative, is of $O(\varepsilon)$; $(\theta^* - \theta_0^*)/\theta_0^*$ and $(\eta^* - \eta_0^*)/\eta_0^*$ and their derivatives are of $O(\varepsilon)$; and that the internal energy ε^* includes terms of $O(\varepsilon)$ and $O(\varepsilon^2)$ while the free energy ψ^* is of $O(\varepsilon^2)$. Keeping this in mind, the order of magnitude of the various resultants defined in Sect. 11 is clear in a linearized theory. In particular, the vector fields $\boldsymbol{N}^\alpha$, $\boldsymbol{M}^{N\alpha}$, $\boldsymbol{m}^N$ $(N \geqq 1)$ or their tensor components when expressed in suitable non-dimensional forms, as well as their derivatives with respect to the surface coordinates, are of $O(\varepsilon)$.

Thus, in obtaining the linearized version of the basic field equations, all tensors in these equations are referred to the initial undeformed surface $\mathfrak{S}$, covariant differentiation is with respect to $A_{\alpha\beta}$ and in Eqs. (12.13)–(12.20), as well as the inequality (12.22), $b_{\alpha\beta}$, d_{Ni}, $\lambda_{Ni\alpha}$, ϱ must be replaced to order ε by their initial values $B_{\alpha\beta}$, D_{Ni}, $\Lambda_{Ni\alpha}$, ϱ_0, respectively. Similarly, in (12.10)–(12.11) and in the definitions of the various resultants [e.g., (11.36)–(11.37), (11.55), (11.43)], $\boldsymbol{g}_i$, $\boldsymbol{a}_i$, g and a may now be replaced with their initial values. We omit the details of the linearization of all field equations but in the rest of this subsection confine our attention to the linearized versions of the approximate equations of motion (12.28)–(12.30). Let $\bar{F}^i$ and $\bar{L}^i$ denote the components of $\bar{\boldsymbol{f}}$ and $\bar{\boldsymbol{l}}$ in (12.25) referred to the base vectors $\boldsymbol{A}_i$, i.e.,

$$\bar{F}^i = \bar{\boldsymbol{f}} \cdot \boldsymbol{A}^i, \quad \bar{L}^i = \bar{\boldsymbol{l}} \cdot \boldsymbol{A}^i. \tag{12.32}$$

Then, upon linearization, (12.28)–(12.30) become

$$N^{\alpha\beta}{}_{|\alpha} - B^\beta_\alpha N^{\alpha 3} + \varrho_0 \bar{F}^\beta = 0, \qquad N^{\alpha 3}{}_{|\alpha} + B_{\alpha\beta} N^{\alpha\beta} + \varrho_0 \bar{F}^3 = 0, \tag{12.33}$$

$$M^{\alpha\beta}{}_{|\alpha} - B^\beta_\alpha M^{\alpha 3} + \varrho_0 \bar{L}^\beta = m^\beta, \qquad M^{\alpha 3}{}_{|\alpha} + B_{\alpha\beta} M^{\alpha\beta} + \varrho_0 \bar{L}^3 = m^3, \tag{12.34}$$

and

$$\begin{gathered} N'^{\alpha\beta} = N'^{\beta\alpha} = N^{\alpha\beta} - m^\alpha D^\beta - M^{\gamma\alpha} \Lambda^\beta_{\cdot\gamma}, \\ N^{\alpha 3} + m^3 D^\alpha - m^\alpha D^3 + M^{\gamma 3} \Lambda^\alpha_{\cdot\gamma} - M^{\gamma\alpha} \Lambda^3_{\cdot\gamma} = 0. \end{gathered} \tag{12.35}$$

The above equations are the linearized equations of motion, obtained under the linearized version of the approximations (12.23)–(12.25).[55] The Eqs. (12.33)–(12.35) are formally equivalent to the linearized equations of motion for a Cosserat surface given by (9.47)–(9.51).

η) *Relationship with results in the classical linear theory of thin shells and plates.* In our description of a shell-like body in Sect. 4 it was emphasized that the surface $\xi = 0$ is not necessarily midway between the surfaces $(4.14)_{1,2}$, although a reference configuration could be chosen in which $\xi = 0$ is the initial middle surface. The various resultants which occur in the field equations of this section, e.g., (12.13)–(12.16) or (12.28)–(12.30), are defined between the limits α, β and in terms of $\boldsymbol{T}^i$ (or equivalently the symmetric stress tensor τ^{ij}) in the present configuration.

In what follows, we confine our attention to the field equations of the linearized theory and assume that the position vector in the initial reference configuration is given by (7.30) with $\boldsymbol{D}$ specified by $(7.31)_3$. For sufficiently *thin* shells, we may also adopt the criterion (4.31) which is generally assumed in the classical linear theories of shells developed from the three-dimensional equations. Then, the mass density is given by the approximate expression (7.46) and the limits of integration in the various resultants may be replaced by[56] $-\beta, \beta$. Moreover, in the special case in which $\boldsymbol{D} = \boldsymbol{A}_3$, by (7.48) the limits of integration may be replaced by

$$-\frac{h}{2}, \frac{h}{2}.$$

As noted above, within the scope of the linear theory, $\boldsymbol{g}_i$, $\boldsymbol{a}_i$, g and a may be replaced by their initial values in the definitions of the various resultants. For example, the resultants which occur in the linearized equations (12.33)–(12.35) can be written in the forms

$$\begin{aligned} N^{\alpha\beta} &= \int_{-\beta}^{\beta} \nu\, \tau^{\alpha\gamma} \nu^\beta_\gamma \, d\xi, & N^{\alpha 3} &= \int_{-\beta}^{\beta} \nu (\xi D_{,\beta} \tau^{\alpha\beta} + D \tau^{\alpha 3}) \, d\xi, \\ M^{\alpha\beta} &= \int_{-\beta}^{\beta} \nu\, \tau^{\alpha\gamma} \nu^\beta_\gamma \, \xi \, d\xi, & M^{\alpha 3} &= \int_{-\beta}^{\beta} \nu (\xi D_{,\beta} \tau^{\alpha\beta} + D \tau^{\alpha 3}) \, \xi \, d\xi, \\ m^\alpha &= \int_{-\beta}^{\beta} \nu\, \tau^{3\gamma} \nu^\alpha_\gamma \, d\xi, & m^3 &= \int_{-\beta}^{\beta} \nu (\xi D_{,\alpha} \tau^{3\alpha} + D \tau^{33}) \, d\xi. \end{aligned} \tag{12.36}$$

[55] The equations of motion (12.33)–(12.35) are obtained by ignoring $(12.31)_{1,2}$ which result from $(12.16)_{1,2}$ by approximation. As was noted earlier, the latter conditions do not arise if (for thin shells) resultants of the type $M^{N\alpha i}$ and m^{Ni} for $N \geqq 2$ are not admitted. A further remark on this point will be made later in this section.

[56] In this connection, recall the remarks made following (7.46).

The above expressions for thin shells have been obtained from the definitions (11.36) and (11.55) with $N=1$ referred to the base vectors $\boldsymbol{G}_i$ given by $(7.32)_{1,2}$.

The resultants (12.36) are defined in terms of the stress tensor τ^{ij} referred to the convected coordinates θ^i. Let $\hat{\tau}^{ij}$ denote the contravariant stress tensor referred to the normal coordinates y^i defined by the transformation relations (7.40). The relationships between the components of τ^{ij} and $\hat{\tau}^{ij}$ can be readily found from (7.40) and the transformation law between two second order tensors. Thus

$$\begin{gathered}\hat{\tau}^{\alpha\beta}=\tau^{\alpha\beta}, \qquad \hat{\tau}^{\alpha 3}=\xi D_{,\beta}\,\tau^{\alpha\beta}+D\,\tau^{\alpha 3},\\ \hat{\tau}^{33}=\xi^2 D_{,\alpha} D_{,\beta}\,\tau^{\alpha\beta}+2\xi D D_{,\alpha}\,\tau^{\alpha 3}+D^2\,\tau^{33}.\end{gathered} \tag{12.37}$$

We define a new set of stress-resultants and stress-couples in terms of $\hat{\tau}^{ij}$ by

$$\begin{gathered}\hat{N}^{\alpha\beta}=\int_{-h/2}^{h/2}\mu\,\hat{\tau}^{\alpha\gamma}\mu_\gamma^\beta\,d\zeta, \qquad V^\alpha=\int_{-h/2}^{h/2}\mu\,\hat{\tau}^{\alpha 3}\,d\zeta,\\ \hat{M}^{\alpha\beta}=\int_{-h/2}^{h/2}\mu\,\hat{\tau}^{\alpha\gamma}\mu_\gamma^\beta\,\zeta\,d\zeta, \qquad \hat{M}^{\alpha 3}=\int_{-h/2}^{h/2}\mu\,\hat{\tau}^{\alpha 3}\,\zeta\,d\zeta,\\ V^3=\int_{-h/2}^{h/2}\mu\,(\hat{\tau}^{33}-B_{\alpha\beta}\,\hat{\tau}^{\alpha\gamma}\mu_\gamma^\beta\,\zeta)\,d\zeta,\end{gathered} \tag{12.38}$$

where μ_γ^β and μ are given by (7.38)–(7.39). Using (7.41), (7.47) and (12.37) in (12.38) and recalling (12.36), we obtain

$$\begin{aligned}&\hat{N}^{\alpha\beta}=N^{\alpha\beta}, \qquad V^\alpha=N^{\alpha 3}=D\,m^\alpha+D\,B_\gamma^\alpha\,M^{\gamma 3}+M^{\beta\alpha}\,D_{,\beta},\\ &\hat{M}^{\alpha i}=D\,M^{\alpha i}, \qquad V^3=D\,m^3-D\,B_{\alpha\beta}\,M^{\alpha\beta}+D_{,\alpha}\,M^{\alpha 3},\end{aligned} \tag{12.39}$$

which relate the definitions (12.36) and (12.38). The definitions (12.38) are those usually employed in the classical theories of shells.[57] The corresponding expressions for initially flat plates are obtained by putting $B_{\alpha\beta}=0$, $\mu=1$ and $\mu_\beta^\alpha=\delta_\beta^\alpha$. It is clear from (7.48) that with $D=1$ (corresponding to $\boldsymbol{D}=\boldsymbol{A}_3$), the distinction between $M^{\alpha i}$ and $\hat{M}^{\alpha i}$ disappears and that the two sets of resultants are equivalent, apart from the expressions which involve τ^{33} or $\hat{\tau}^{33}$.

The linearized field equations and the linearized version of (12.22) can be expressed in different forms depending on the choice of the definitions of the type (12.36) or (12.38). We illustrate this with reference to the approximate equations of motion (12.33)–(12.35) in which, along with the definitions (12.36) and (12.38), corresponding expressions for the resultants F^i, L^i (and $\bar{F}^i$, $\bar{L}^i$) with the limits of integration

$$-\beta,\ \beta \text{ and } \frac{h}{2},\ \frac{h}{2}$$

are used.[58] Below a catalogue of formulae is provided which, in particular, are of interest in connection with the definitions of resultants and the equations of motion in the classical linear theories of shells and plates. With reference to the equations of motion (12.28)–(12.31), it was remarked earlier in this section that the question of approximation does not arise and that these equations are exact

[57] The expressions corresponding to $(12.38)_{4,5}$ are not defined in most of the literature on the linear theory of shells. The results (12.39) relating the two sets of definitions (12.36) and (12.38) were noted by Green, Naghdi and Wenner [1971, *6*].

[58] These expressions are not listed here since they can be easily obtained from (11.29), (11.30) with $N=1$ after linearization and use of (12.32).

if the resultants corresponding to $M^{N\alpha i}$ and m^{Ni} for $N \geqq 2$ are not admitted in the development of the theory. We now observe specifically that if (as in the classical shell theory) only the resultants of the type (12.38) are admitted and $\boldsymbol{D}$ is identified with $\boldsymbol{A}_3$, then the resulting equations of motion recorded in Formulae C below are exact.[59]

Formulae A

$$\boldsymbol{D} = D\boldsymbol{A}_3, \quad \Lambda_{\nu\alpha} = -B_{\nu\alpha} D, \quad \Lambda_{3\alpha} = D_{,\alpha}.$$

Stress-resultants and stress-couples defined by (12.36),

$$\begin{gathered} N^{\alpha\beta}{}_{|\alpha} - B^{\beta}_{\alpha} N^{\alpha 3} + \varrho_0 \bar{F}^{\beta} = 0, \qquad N^{\alpha 3}{}_{|\alpha} + B_{\alpha\beta} N^{\alpha\beta} + \varrho_0 \bar{F}^3 = 0, \\ M^{\alpha\beta}{}_{|\alpha} - B^{\beta}_{\alpha} M^{\alpha 3} + \varrho_0 \bar{L}^{\beta} = m^{\beta}, \qquad M^{\alpha 3}{}_{|\alpha} + B_{\alpha\beta} M^{\alpha\beta} + \varrho_0 \bar{L}^3 = m^3, \\ N'^{\alpha\beta} = N'^{\beta\alpha} = N^{\alpha\beta} + D\, M^{\gamma\alpha} B^{\beta}_{\gamma}, \\ N^{\alpha 3} - D\, m^{\alpha} - D\, M^{\gamma 3} B^{\alpha}_{\gamma} - M^{\gamma\alpha} D_{,\gamma} = 0. \end{gathered} \tag{12.40}$$

Formulae B

$$\boldsymbol{D} = D\boldsymbol{A}_3, \quad \Lambda_{\nu\alpha} = -B_{\nu\alpha} D, \quad \Lambda_{3\alpha} = D_{,\alpha}.$$

Stress-resultants and stress-couples defined by (12.38),

$$\begin{gathered} \hat{N}^{\alpha\beta}{}_{|\alpha} - B^{\beta}_{\alpha} V^{\alpha} + \varrho_0 \bar{F}^{\beta} = 0, \qquad V^{\alpha}{}_{|\alpha} + B_{\alpha\beta} \hat{N}^{\alpha\beta} + \varrho_0 \bar{F}^3 = 0, \\ \hat{M}^{\alpha\beta}{}_{|\alpha} + \varrho_0 \hat{\bar{L}}^{\beta} = V^{\beta}, \qquad \hat{M}^{\alpha 3}{}_{|\alpha} + \varrho_0 \bar{L}^3 = V^3, \\ N'^{\alpha\beta} = N'^{\beta\alpha} = \hat{N}^{\alpha\beta} + \hat{M}^{\gamma\alpha} B^{\beta}_{\gamma}, \\ V^{\alpha} = D\, m^{\alpha} + B^{\alpha}_{\gamma} \hat{M}^{\gamma 3} + \hat{M}^{\beta\alpha} \frac{D_{,\beta}}{D}, \qquad V^3 = D\, m^3 - B_{\alpha\beta} \hat{M}^{\alpha\beta} + \hat{M}^{\alpha 3} \frac{D_{,\alpha}}{D}, \\ \hat{\bar{L}}^{\beta} = D\, \bar{L}^{\beta}, \quad \hat{\bar{L}}^3 = D\, \bar{L}^3. \end{gathered} \tag{12.41}$$

Formulae C

$$\begin{gathered} \boldsymbol{D} = \boldsymbol{A}_3, \quad \Lambda_{\beta\alpha} = -B_{\alpha\beta}, \quad \Lambda_{3\alpha} = 0, \\ N^{\alpha\beta} = \int_{-h/2}^{h/2} \mu\, \tau^{\alpha\gamma} \mu^{\beta}_{\gamma}\, d\zeta, \qquad V^{\alpha} = \int_{-h/2}^{h/2} \mu\, \tau^{\alpha 3}\, d\zeta, \\ M^{\alpha\beta} = \int_{-h/2}^{h/2} \mu\, \tau^{\alpha\gamma} \mu^{\beta}_{\gamma} \zeta\, d\zeta, \qquad M^{\alpha 3} = \int_{-h/2}^{h/2} \mu\, \tau^{\alpha 3} \zeta\, d\zeta, \\ V^3 = \int_{-h/2}^{h/2} \mu\, (\tau^{33} - B_{\alpha\beta}\, \tau^{\alpha\gamma} \mu^{\beta}_{\gamma} \zeta)\, d\zeta, \\ N^{\alpha\beta}{}_{|\alpha} - B^{\beta}_{\alpha} V^{\alpha} + \varrho_0 \bar{F}^{\beta} = 0, \qquad V^{\alpha}{}_{|\alpha} + B_{\alpha\beta} N^{\alpha\beta} + \varrho_0 \bar{F}^3 = 0, \\ M^{\alpha\beta}{}_{|\alpha} + \varrho_0 \bar{L}^{\beta} = V^{\beta}, \qquad M^{\alpha 3}{}_{|\alpha} + \varrho_0 \bar{L}^3 = V^3, \\ N'^{\alpha\beta} = N'^{\beta\alpha} = N^{\alpha\beta} + M^{\gamma\alpha} B^{\beta}_{\gamma}, \\ V^{\alpha} = N^{\alpha 3} = m^{\alpha} + B^{\alpha}_{\gamma} M^{\gamma 3}, \qquad V^3 = m^3 - B_{\alpha\beta} M^{\alpha\beta}. \end{gathered} \tag{12.42}$$

[59] A derivation leading to the equations of motion in (12.42) from the three-dimensional stress equations of motion and with the use of the resultants $(12.38)_{1,2,3}$ is given in Sect. 5 of [1963, *6*].

Formulae D

$$\boldsymbol{D}=\boldsymbol{A}_3, \qquad B_{\alpha\beta}=0,$$

$$N^{\alpha\beta}=\int_{-h/2}^{h/2}\tau^{\alpha\beta}\,d\zeta, \qquad V^{\alpha}=\int_{-h/2}^{h/2}\tau^{\alpha 3}\,d\zeta,$$

$$M^{\alpha\beta}=\int_{-h/2}^{h/2}\tau^{\alpha\beta}\,\zeta\,d\zeta, \qquad M^{\alpha 3}=\int_{-h/2}^{h/2}\tau^{\alpha 3}\,\zeta\,d\zeta,$$

$$V^{3}=\int_{-h/2}^{h/2}\tau^{33}\,d\zeta, \tag{12.43}$$

$$N^{\alpha\beta}{}_{|\alpha}+\varrho_0\bar{F}^{\beta}=0, \qquad M^{\alpha 3}{}_{|\alpha}+\varrho_0\bar{L}^{3}=V^{3},$$

$$M^{\alpha\beta}{}_{|\alpha}+\varrho_0\bar{L}^{\beta}=V^{\beta}, \qquad V^{\alpha}{}_{|\alpha}+\varrho_0\bar{F}^{3}=0,$$

$$N'^{\alpha\beta}=N^{\alpha\beta}, \qquad V^{\alpha}=N^{\alpha 3}=m^{\alpha}, \qquad V^{3}=m^{3}.$$

12A. Appendix on the history of derivations of the equations of equilibrium for shells. A method of derivation of equations of equilibrium for shells which is only partly direct was originated by LOVE; it has long received acceptance and has been widely practiced and reproduced in books on the subject. Briefly, this method consists in two parts: (i) First the stress-resultants and the stress-couples are defined by integrals of the type in (12.42), together with similar definitions for load resultants [see (12.32) and (11.29)–(11.30)]; and then (ii) the equilibrium equations for shells are derived not from the three-dimensional equations but by consideration of the equilibrium of an element of the curved shell (or effectively its middle surface) under the action of the stress-resultants and the stress-couples (each per unit length of curves on the middle surface), as well as load resultant (per unit area of the middle surface). As a whole, this method is neither direct nor one in which the equations of equilibrium are derived fully from the three-dimensional equations. It is the second part of the procedure which is direct; and, hence this manner of obtaining the equilibrium equations for shells may properly be regarded as a derivation by direct approach.

The derivation of the equilibrium equations for shells via the direct method (and in lines of curvature coordinates) was given by LOVE.[60] In effect, this derivation is equivalent to that of our restricted theory when the equilibrium equations corresponding to those in Sect. 10 are specialized to lines of curvature coordinates and are also expressed in terms of physical components. A neat vectorial treatment of LOVE's derivation (again in lines of curvature coordinates) was given by REISSNER.[61] The corresponding derivation in general coordinates was supplied by SYNGE and CHIEN.[62] A derivation of equilibrium equations by direct method was also considered in 1958 by ERICKSEN and TRUESDELL to which reference was made in Sect. 9 [Subsect. β)]. This derivation of ERICKSEN and TRUESDELL is shorter and more appealing than that of SYNGE and CHIEN or other earlier

[60] Sect. 340 of LOVE [1893, *2*] and subsequent editions of his treatise, e.g., Sect. 331 of [1944, *4*]. Although LOVE defines the stress-resultants and the stress-couples, these are not used *per se* in his subsequent derivation of equilibrium equations according to the procedure mentioned above. With the help of formulae of the type (A.4.11)–(A.4.12) in Chap. F, the equilibrium equations in lines of curvature coordinates can easily be obtained from (12.28)–(12.30) or from (9.47)–(9.48) and (9.51). The corresponding results for the restricted theory can be obtained from (9.47) and (10.21). See also Eqs. (A.4.13)–(A.4.14) in Chap. F.

[61] REISSNER [1941, *1*].

[62] SYNGE and CHIEN [1941, *2*]. Derivations of this type by direct approach and in terms of tensor components are contained also in a paper by ZERNA [1949, *6*] and Chap. 10 of GREEN and ZERNA [1954, *1*].

derivations by direct approach. References to other recent direct derivations of equations of motion or equilibrium (after 1958) have already been cited in Sects. 9–10 and need not be repeated here.

The definitions of stress-resultants and stress-couples in terms of the three-dimensional stress tensor, corresponding to the resultants in (12.42) but in lines of curvature coordinates and for physical components of $N^{\alpha\beta}$, $M^{\alpha\beta}$, V^{α}, are due to LOVE.[63] The definitions of the resultants in general coordinates corresponding to $N^{\alpha\beta}$, $M^{\alpha\beta}$, V^{α} in (12.42) were given by ZERNA.[64] The earliest derivation which can be regarded as fully derived from the three-dimensional equations appears to be due to NOVOZHILOV and to NOVOZHILOV and FINKELSTEIN.[65] This derivation, carried out in lines of curvature coordinates and in terms of physical components of the resultants, is accomplished by integration of the (three-dimensional) differential equations of equilibrium across the thickness of the shell; the derivation is independent of any kinematic assumption and no approximation is involved. An exposition of the derivation of NOVOZHILOV and of NOVOZHILOV and FINKELSTEIN (again in lines of curvature coordinates) is given by TRUESDELL and TOUPIN.[66] A derivation in general coordinates (from the three-dimensional theory), resulting in the equilibrium equations in (12.42) which involve $N^{\alpha\beta}$, $M^{\alpha\beta}$, V^{α}, was given by NAGHDI.[67] References to more recent derivations of equations of motion for shells (from the three-dimensional equations) in terms of more general definitions for resultants [such as those in (11.36)] are cited in Sects. 11–12.

D. Elastic shells.

This chapter is concerned mainly with the development of constitutive equations for elastic shells, both by direct approach and from the three-dimensional equations.[1] Nonlinear and linearized constitutive relations are discussed and the complete theory is recapitulated. While we confine our attention here to elastic shells, it may be noted that the previous developments (in Chaps. B and C) are not limited to elastic materials.[2]

13. Constitutive equations for elastic shells (nonlinear theory): I. Direct approach. In this section, we introduce nonlinear constitutive equations for thermo-

---

[63] Sect. 339 of LOVE [1893, *2*] and subsequent editions of his treatise, e.g., Sect. 328 of [1944, *4*]. The main ingredient for the definitions of the resultants of the type in (12.42), namely the presence of the curvature factor, was noted by LAMB [1890, *2*] and BASSET [1890, *1*]. In his paper of [1888, *1*], LOVE defines the resultants approximately (without the curvature factor) and obtains his equations of motion by integration of the (three-dimensional) virtual work principle and in terms of displacements of the middle surface.

[64] ZERNA [1949, *6*]. See also GREEN and ZERNA [1954, *1*] and [1968, *9*].

[65] NOVOZHILOV [1943, *1*] and NOVOZHILOV and FINKELSTEIN [1943, *2*]. A less general derivation of this kind from the three-dimensional equations was given independently by TRUESDELL [1945, *3*] in a paper which deals chiefly with the membrane theory of shells of revolution.

[66] Sect. 213 of TRUESDELL and TOUPIN [1960, *14*] which also contains some historical remarks concerning the definitions of resultants and derivation of equilibrium equations for shells from the three-dimensional equations.

[67] NAGHDI [1963, *6*]. See also Chap. 10 (in 2nd edition) of GREEN and ZERNA [1968, *9*], where the equilibrium equations in general coordinates are derived from the three-dimensional equations.

[1] What we regard here as *elastic materials* are often called *hyperelastic materials* in the recent literature on continuum mechanics. For an extensive account of the three-dimensional (non-polar) theory of hyperelastic materials, together with thermodynamic aspects of the subject, see TRUESDELL and NOLL [1965, *9*].

[2] With reference to elastic-plastic shells and plates, mention may be made of a paper by GREEN, NAGHDI and OSBORN [1968, *8*] which deals with an elastic-plastic Cosserat surface.

mechanical behavior of an elastic Cosserat surface and then deduce the restrictions placed on them by the invariance requirements under superposed rigid body motions, by the entropy inequality and by material symmetries.

α) *General considerations. Thermodynamical results.* We recall that a material is defined by a *constitutive assumption* which characterizes the thermo-mechanical behavior of the medium; the constitutive assumption places a restriction on the processes which are admissible in a body—here the Cosserat surface $\mathscr{C}$. Preliminary to the introduction of the constitutive assumption for an elastic Cosserat surface, we recall the invariance requirements under superposed rigid body motions and examine their implications regarding the tensor components of the various vector fields in (8.30)–(8.32). Consider $(8.32)_1$ according to which $\boldsymbol{N}$ transforms as

$$\boldsymbol{N} \to Q\boldsymbol{N},$$

where Q is a proper orthogonal tensor defined by (5.37). By (9.11) and $(5.38)_1$ and since $\boldsymbol{v}$ transforms as $\boldsymbol{v} \to Q\,\boldsymbol{v}$, $v_\alpha \to v_\alpha$, we readily deduce the transformations

$$\boldsymbol{N}^\alpha \to Q\boldsymbol{N}^\alpha, \qquad N^{\alpha i} \to N^{\alpha i}, \tag{13.1}$$

for $\boldsymbol{N}^\alpha$ and $N^{\alpha i}$ under superposed rigid body motions.[3] Similar objective transformations can be obtained for $\boldsymbol{m}, \boldsymbol{M}^\alpha, \boldsymbol{q}$ or their components $m^i, M^{\alpha i}, q^\alpha$, as well as the temperature gradient $\theta_{,\alpha}$, all of which follow from the invariance conditions (8.30)–(8.32). Collecting these results, in addition to (13.1), we have[4]

$$\boldsymbol{m} \to Q\,\boldsymbol{m}, \qquad \boldsymbol{M}^\alpha \to Q\boldsymbol{M}^\alpha, \qquad \boldsymbol{N}'^\alpha \to Q\boldsymbol{N}'^\alpha, \qquad \boldsymbol{q} \to Q\boldsymbol{q}, \tag{13.2}$$

or

$$m^i \to m^i, \qquad M^{\alpha i} \to M^{\alpha i}, \qquad N'^{\alpha\beta} \to N'^{\alpha\beta}, \qquad q^\alpha \to q^\alpha, \tag{13.3}$$

as well as

$$\theta_{,\alpha} \to \theta_{,\alpha}. \tag{13.4}$$

An elastic Cosserat surface is defined by a set of response functions which depend on appropriate kinematic and thermal variables. For example, if in addition to the kinematic variables, the temperature and the temperature gradient are also taken as the independent variables, then the response functions consist of the six functions[5]

$$\overline{\boldsymbol{N}}^\alpha \text{ (or } \overline{\boldsymbol{N}}'^\alpha), \ \overline{\boldsymbol{m}}, \ \overline{\boldsymbol{M}}^\alpha, \ \bar{\psi}, \ \bar{\eta}, \ \bar{\boldsymbol{q}}, \tag{13.5}$$

or an equivalent set

$$\bar{N}^{\alpha i}, \ \bar{m}^i, \ \bar{M}^{\alpha i}, \ \bar{\psi}, \ \bar{\eta}, \ \bar{q}^\alpha. \tag{13.6}$$

We introduce constitutive equations which must hold at each material point and for all t in terms of the response functions (13.5).[6] In this connection, we

[3] While $\boldsymbol{N}^\alpha$ transforms as a vector under superposed rigid body motions, $N^{\alpha i}$ (with six components) transform as six scalars; this is because of our use of convected coordinates.

[4] Recalling the definitions (9.30)–(9.31), the transformations for $\boldsymbol{N}'^\alpha$ and $N'^{\alpha\beta}$ follow from (13.1), $(13.2)_{1,2}$, $(13.3)_{1,2}$ and (5.54).

[5] The overbar in (13.5) is introduced temporarily for added clarity and in order to distinguish a function from its value. The notation $\overline{\boldsymbol{M}}^\alpha$ in (13.5) should not be confused with the previous use of the same symbol in (9.20) and (9.44).

[6] Often, in the three-dimensional theory of (non-polar) continuum mechanics, the material point in a body is identified with its position in a reference configuration and the constitutive equations are then introduced relative to the reference configuration. Although such a procedure is also possible here, it is simpler to introduce the constitutive equations at each material point of $\mathscr{C}$ relative to a material coordinate system, rather than in terms of the response functions relative to a reference configuration. Thus, our developments in the first part of this section [between (13.8)–(13.57)] are independent of any reference configuration; and, until further notice in this section, we disregard the fact that the coordinates representing the particles (or the material points) of $\mathscr{C}$ were identified previously (Sect. 4) with the convected coordinates on the surface in the (initial) reference configuration.

recall that the displacement function $\boldsymbol{r}$ in $(5.1)_1$ is the place occupied by the material point θ^μ (representing a typical surface particle of $\mathscr{C}$) in the present configuration; and, similarly, the function $\boldsymbol{d}$ in $(5.1)_2$ is the director at the material point in the present configuration. Thus, apart from temperature and its gradient, the local state of an elastic Cosserat surface $\mathscr{C}$ can be defined by the functions in $(5.1)_{1,2}$ and their gradients at each material point in the present configuration, namely

$$\boldsymbol{r},\ \boldsymbol{r}_{,\alpha},\ \boldsymbol{d},\ \boldsymbol{d}_{,\gamma}.$$

But, since the response functions must remain unaltered under superposed rigid body translational displacement, dependence of the functions (13.5) on $\boldsymbol{r}$ must be excluded. Hence, the last three in the above set, or equivalently

$$\boldsymbol{a}_\alpha,\ \boldsymbol{d},\ \boldsymbol{d}_{,\gamma}, \tag{13.7}$$

can be regarded as the primitive kinematic ingredients which define the local state and which occur in the constitutive equations. Keeping this in mind, we assume that the constitutive equations for an elastic Cosserat surface depend on the kinematic variables (13.7), as well as θ and $\theta_{,\alpha}$. We therefore write[7]

$$\begin{aligned}
\psi &= \overline{\psi}(\boldsymbol{a}_\alpha, \boldsymbol{d}, \boldsymbol{d}_{,\gamma}, \theta, \theta_{,\alpha};\ \theta^\mu),\\
\eta &= \overline{\eta}(\boldsymbol{a}_\alpha, \boldsymbol{d}, \boldsymbol{d}_{,\gamma}, \theta, \theta_{,\alpha};\ \theta^\mu),\\
\boldsymbol{N}^\alpha &= \overline{\boldsymbol{N}}^\alpha(\boldsymbol{a}_\alpha, \boldsymbol{d}, \boldsymbol{d}_{,\gamma}, \theta, \theta_{,\alpha};\ \theta^\mu),\\
\boldsymbol{m} &= \overline{\boldsymbol{m}}(\boldsymbol{a}_\alpha, \boldsymbol{d}, \boldsymbol{d}_{,\gamma}, \theta, \theta_{,\alpha};\ \theta^\mu),\\
\boldsymbol{M}^\alpha &= \overline{\boldsymbol{M}}^\alpha(\boldsymbol{a}_\alpha, \boldsymbol{d}, \boldsymbol{d}_{,\gamma}, \theta, \theta_{,\alpha};\ \theta^\mu),
\end{aligned} \tag{13.8}$$

and

$$\boldsymbol{q} = \overline{\boldsymbol{q}}(\boldsymbol{a}_\alpha, \boldsymbol{d}, \boldsymbol{d}_{,\gamma}, \theta, \theta_{,\alpha};\ \theta^\mu). \tag{13.9}$$

The above constitutive equations, which characterize the thermo-mechanical response of the medium, are assumed to hold at each particle θ^μ of the Cosserat surface and for all times t. Any dependence of the response functions on inhomogeneity, anisotropy or a (physically) preferred reference state is indicated through the argument θ^μ. Moreover, the constitutive equations (13.8)–(13.9) represent the response of the medium relative to a material coordinate system and would be different relative to another material coordinate system. This latter has not been explicitly exhibited in (13.8)–(13.9); and, strictly speaking, a constitutive equation such as $(13.8)_1$ should be recorded as

$$\psi = \overline{\psi}_{(\theta^\mu)}(\boldsymbol{a}_\alpha, \boldsymbol{d}, \boldsymbol{d}_{,\gamma}, \theta, \theta_{,\alpha};\ \theta^\mu),$$

where the subscript (θ^μ) attached to $\overline{\psi}$ indicates the choice of coordinates in contrast to the argument θ^μ which signifies the choice of particle. To elaborate, let θ'^μ be any other material coordinate system related to θ^μ by

$$\theta'^\mu = \bar{\theta}'^\mu(\theta^\nu) \quad \text{and} \quad \theta^\mu = \bar{\theta}^\mu(\theta'^\nu),$$

where the function $\bar{\theta}^\mu$ is the inverse of $\bar{\theta}'^\mu$. Further, temporarily let the surface base vectors and the gradients of the director and the temperature relative to the

[7] These constitutive equations satisfy *equipresence,* which appears to be viewed by some writers as a physical principle. Here, we regard the notion of equipresence as a convenient mathematical procedure. A statement of equipresence, as currently understood, is as follows: An independent variable present in one constitutive equation should be so present in all, unless its presence is contradicted by the conservation laws, the entropy inequality or a rule of invariance. For further background information on equipresence, see Sect. 96 of TRUESDELL and NOLL [1965, *9*].

primed coordinate system be written as

$$\frac{\partial \boldsymbol{r}}{\partial \theta'^{\alpha}}, \quad \frac{\partial \boldsymbol{d}}{\partial \theta'^{\gamma}}, \quad \frac{\partial \theta}{\partial \theta'^{\alpha}},$$

respectively. Then, under the coordinate transformation, the response function for the free energy relative to the primed coordinate system is related to $\overline{\psi}_{(\theta^\mu)}$ by

$$\begin{aligned}\overline{\psi}_{(\theta'^\mu)}&\left(\frac{\partial \boldsymbol{r}}{\partial \theta'^{\alpha}}, \boldsymbol{d}, \frac{\partial \boldsymbol{d}}{\partial \theta'^{\gamma}}, \theta, \frac{\partial \theta}{\partial \theta'^{\alpha}}; \theta'^{\mu}\right)\\ &=\overline{\psi}_{(\theta^\mu)}\left(\frac{\partial \boldsymbol{r}}{\partial \theta'^{\mu}}\frac{\partial \theta'^{\mu}}{\partial \theta^{\alpha}}, \boldsymbol{d}, \frac{\partial \boldsymbol{d}}{\partial \theta'^{\mu}}\frac{\partial \theta'^{\mu}}{\partial \theta^{\gamma}}, \theta, \frac{\partial \theta}{\partial \theta'^{\mu}}\frac{\partial \theta'^{\mu}}{\partial \theta^{\alpha}}; \theta^{\mu}\right),\end{aligned}$$

where the argument θ^μ refers to the particle identified by $\theta^\mu=\bar{\theta}^\mu(\theta'^\nu)$. According to the last relation, from the knowledge of the function $\psi_{(\theta^\mu)}$ and the coordinate transformation we can readily calculate the response function in any other coordinate system. Having discussed the nature of the dependence of the response functions on the material coordinates, in order to avoid cumbersome notation in what follows we do not explicitly display this dependence, and we continue to write the response functions, as in (13.8)–(13.9), without the subscript (θ^μ) attached to them.

Before proceeding further, we first effect a certain simplification in our constitutive assumptions (13.8). We had previously postulated that the inequality (8.24) must hold for every admissible process. We now introduce the constitutive equations (13.8)–(13.9) into the inequality (9.36) which [since it is obtained from (8.24)] must hold for every admissible process. From $(13.8)_1$, we have

$$\begin{aligned}\dot{\overline{\psi}}&=\frac{\partial \overline{\psi}}{\partial \boldsymbol{a}_\alpha}\cdot\dot{\boldsymbol{a}}_\alpha+\frac{\partial \overline{\psi}}{\partial \boldsymbol{d}}\cdot\dot{\boldsymbol{d}}+\frac{\partial \overline{\psi}}{\partial \boldsymbol{d}_{,\gamma}}\cdot\dot{\boldsymbol{d}}_{,\gamma}+\frac{\partial \overline{\psi}}{\partial \theta}\dot{\theta}+\frac{\partial \overline{\psi}}{\partial \theta_{,\alpha}}\dot{\theta}_{,\alpha}\\ &=\left(\frac{\partial \overline{\psi}}{\partial \boldsymbol{a}_\alpha}-d^\alpha\frac{\partial \overline{\psi}}{\partial \boldsymbol{d}}-\lambda^{\alpha}_{\cdot\gamma}\frac{\partial \overline{\psi}}{\partial \boldsymbol{d}_{,\gamma}}\right)\cdot\boldsymbol{\eta}_\alpha\\ &\quad+\frac{\partial \overline{\psi}}{\partial \boldsymbol{d}}\cdot\boldsymbol{\Gamma}+\frac{\partial \overline{\psi}}{\partial \boldsymbol{d}_{,\gamma}}\cdot\boldsymbol{\Gamma}_{:\gamma}+\frac{\partial \overline{\psi}}{\partial \theta}\dot{\theta}+\frac{\partial \overline{\psi}}{\partial \theta_{,\alpha}}\dot{\theta}_{,\alpha}\\ &\quad+W_{ki}\,\boldsymbol{a}^k\cdot\left[\delta^i_\alpha\frac{\partial \overline{\psi}}{\partial \boldsymbol{a}_\alpha}+d^i\frac{\partial \overline{\psi}}{\partial \boldsymbol{d}}+\lambda^{i}_{\cdot\gamma}\frac{\partial \overline{\psi}}{\partial \boldsymbol{d}_{,\gamma}}\right],\end{aligned}\tag{13.10}$$

where (5.19), (5.26) and (5.29) have been used. After introducing the constitutive assumption into (9.36) with P given by $(9.29)_2$, we obtain an inequality which must hold for all arbitrary values of[8]

$$\dot{\boldsymbol{a}}_\alpha,\ \dot{\boldsymbol{d}},\ \dot{\boldsymbol{d}}_{,\alpha},\ \dot{\theta},\ \dot{\theta}_{,\alpha}.\tag{13.11}$$

The coefficients of the above rate quantities in the inequality are independent of the variables (13.11) and, as functions of time, can be chosen arbitrarily. It follows that for a given time (say at the present time t), there exist admissible

[8] For convenience we postpone the use of invariance requirements under superposed rigid body motions which places a restriction on the function $\overline{\psi}$ in (13.8) and therefore $\dot{\overline{\psi}}$ which appears in the inequality. Operators of the form $\partial f/\partial \boldsymbol{x}$, where f is a scalar function whose arguments include the vector $\boldsymbol{x}$, occur in (13.10) and elsewhere in this chapter. By an operator of this type, whenever it exists, we mean the partial derivative with respect to $\boldsymbol{x}$, which satisfies

$$\lim_{\beta\to 0}\frac{f(\boldsymbol{x}+\beta\boldsymbol{V})-f(\boldsymbol{x})}{\beta}=\frac{\partial f}{\partial \boldsymbol{x}}\cdot\boldsymbol{V}$$

for all values of the arbitrary vector $\boldsymbol{V}$.

processes such that all coefficient functions in the inequality and all rate quantities in (13.11) can be assigned arbitrarily. Consider now a process in which all kinematic and thermal variables in the argument of the response functions in (13.8)–(13.9) at time t, as well as all rate quantities in (13.11) at time t, except $\dot{\theta}_{,\alpha}$ are prescribed. Then, all terms in the resulting inequality are fixed except $\dot{\theta}_{,\alpha}$. In order that the inequality holds for all $\dot{\theta}_{,\alpha}$, the coefficient of the term before last in $(13.10)_2$ must vanish:

$$\frac{\partial \bar{\psi}}{\partial \theta_{,\alpha}} = 0. \tag{13.12}$$

Hence $\bar{\psi}$ must be independent of $\theta_{,\alpha}$ and $(13.8)_1$ reduces to[9]

$$\psi = \bar{\psi}(\boldsymbol{a}_\alpha, \boldsymbol{d}, \boldsymbol{d}_{,\gamma}, \theta; \theta^\mu), \tag{13.13}$$

where $\bar{\psi}$ is now a different function from that in $(13.8)_1$.

Having shown that the constitutive equation $(13.8)_1$ can be reduced to (13.13), with the use of the kinematic results in Sect. 5 and the remaining constitutive equations in (13.8)–(13.9), we write the inequality (9.36) in the form

$$\begin{aligned} &-\varrho\left(\bar{\eta} + \frac{\partial \bar{\psi}}{\partial \theta}\right)\dot{\theta} + \left[\boldsymbol{N}'^\alpha - \varrho\left(\frac{\partial \bar{\psi}}{\partial \boldsymbol{a}_\alpha} - d^\alpha \frac{\partial \bar{\psi}}{\partial \boldsymbol{d}} - \lambda^\alpha_{\cdot\gamma} \frac{\partial \bar{\psi}}{\partial \boldsymbol{d}_{,\gamma}}\right)\right] \cdot \boldsymbol{\eta}_\alpha \\ &\quad + \left[\boldsymbol{m} - \varrho \frac{\partial \bar{\psi}}{\partial \boldsymbol{d}}\right] \cdot \boldsymbol{\Gamma} + \left[\boldsymbol{M}^\gamma - \varrho \frac{\partial \bar{\psi}}{\partial \boldsymbol{d}_{,\gamma}}\right] \cdot \boldsymbol{\Gamma}_{:\gamma} - (\bar{\boldsymbol{q}} \cdot \boldsymbol{a}^\alpha) \frac{\theta_{,\alpha}}{\theta} \geqq 0, \end{aligned} \tag{13.14}$$

where the function $\bar{\psi}$ defined in (13.13) must satisfy

$$\varepsilon_{jkm}\, \boldsymbol{a}^k \cdot \left[\delta^j_\alpha \frac{\partial \bar{\psi}}{\partial \boldsymbol{a}_\alpha} + d^j \frac{\partial \bar{\psi}}{\partial \boldsymbol{d}} + \lambda^j_{\cdot\gamma} \frac{\partial \bar{\psi}}{\partial \boldsymbol{d}_{,\gamma}}\right] = 0. \tag{13.15}$$

The condition (13.15) arises from the fact that the inequality (9.36) must remain unaffected by superposed rigid body motions. It is not difficult to see that it is deduced from the vanishing of the square bracket in $(13.10)_2$ if, in the notation of (5.42), the skew-symmetric W_{ki} is expressed as $\varepsilon_{kim}\,\omega^m$. As will become apparent later in Subsect. β), (13.15) will be satisfied identically after restriction arising from invariance conditions under superposed rigid body motions is placed on the free energy ψ.

The inequality (13.14) holds for all admissible processes and, in particular, for all arbitrary values of $\dot{\theta}$, $\boldsymbol{\eta}_\alpha$, $\boldsymbol{\Gamma}$, $\boldsymbol{\Gamma}_{:\alpha}$. Consider now an admissible process such that all kinematic and thermal variables which occur in the response functions (including $\theta_{,\alpha}$ at time t), as well as $\boldsymbol{\eta}_\alpha$, $\boldsymbol{\Gamma}$, $\boldsymbol{\Gamma}_{:\alpha}$ at time t, are prescribed. Then, all quantities in (13.14) are fixed except $\dot{\theta}$, which may assume arbitrary values. Hence, in order that (13.14) hold for all $\dot{\theta}$, its coefficient must vanish and we must have

$$\bar{\eta} = -\frac{\partial \bar{\psi}}{\partial \theta}. \tag{13.16}$$

Next, consider an admissible process at time t corresponding to which (13.7), the thermal variables θ and $\theta_{,\alpha}$ and all rate quantities, except $\boldsymbol{\eta}_\alpha$, are prescribed; but, since $\boldsymbol{a}_\beta$ (and therefore $\boldsymbol{a}^\beta$) are tangent vectors at each point of $\mathscr{s}$, $\boldsymbol{\eta}_\alpha = \eta_{\beta\alpha}\, \boldsymbol{a}^\beta$

[9] The argument here and those leading to (13.16)–(13.19) parallels similar arguments by COLEMAN and NOLL [1963, 2] in the (three-dimensional) theory of nonlinear elastic materials with heat conduction and viscosity. We recall that the justification for such arguments necessarily rests on the notion that the functions $(8.23)_3$ may be arbitrarily chosen and are not assigned *a priori*.

can only be varied arbitrarily in the tangent plane and only the components $\eta_{\beta\alpha}$ may assume arbitrary values. Keeping this in mind and by repeating similar arguments, from (13.14) we obtain the further relations

$$\begin{aligned}\bar{N}'^{\alpha\beta} = \bar{N}'^{\beta\alpha} &= \frac{1}{2}\left(\bar{\boldsymbol{N}}'^{\alpha}\cdot\boldsymbol{a}^{\beta} + \bar{\boldsymbol{N}}'^{\beta}\cdot\boldsymbol{a}^{\alpha}\right)\\ &= \frac{1}{2}\varrho\left\{\left(\frac{\partial\bar{\psi}}{\partial\boldsymbol{a}_{\alpha}} - d^{\alpha}\frac{\partial\bar{\psi}}{\partial\boldsymbol{d}} - \lambda^{\alpha}_{\cdot\gamma}\frac{\partial\bar{\psi}}{\partial\boldsymbol{d}_{,\gamma}}\right)\cdot\boldsymbol{a}^{\beta}\right.\\ &\quad\left.+\left(\frac{\partial\bar{\psi}}{\partial\boldsymbol{a}_{\beta}} - d^{\beta}\frac{\partial\bar{\psi}}{\partial\boldsymbol{d}} - \lambda^{\beta}_{\cdot\gamma}\frac{\partial\bar{\psi}}{\partial\boldsymbol{d}_{,\gamma}}\right)\cdot\boldsymbol{a}^{\alpha}\right\},\end{aligned} \tag{13.17}$$

$$\bar{\boldsymbol{m}} = \varrho\frac{\partial\bar{\psi}}{\partial\boldsymbol{d}}, \qquad \bar{\boldsymbol{M}}^{\gamma} = \varrho\frac{\partial\bar{\psi}}{\partial\boldsymbol{d}_{,\gamma}} \tag{13.18}$$

and the inequality

$$-\bar{q}^{\alpha}\,\theta_{,\alpha} \geqq 0, \tag{13.19}$$

since $\theta > 0$. Also, in view of (9.30), from (13.17)–(13.18) follows the expression for the symmetric part of $\bar{\boldsymbol{N}}^{\alpha}\cdot\boldsymbol{a}^{\beta}$:

$$\bar{N}^{(\alpha\beta)} = \frac{1}{2}\left(\bar{N}^{\alpha\beta} + \bar{N}^{\beta\alpha}\right) = \frac{1}{2}\varrho\left(\frac{\partial\bar{\psi}}{\partial\boldsymbol{a}_{\alpha}}\cdot\boldsymbol{a}^{\beta} + \frac{\partial\bar{\psi}}{\partial\boldsymbol{a}_{\beta}}\cdot\boldsymbol{a}^{\alpha}\right). \tag{13.20}$$

It is clear that the relations (13.20) and (13.18) are equivalent to the set (13.17)–(13.18).

Previously, we had shown that $\bar{\psi}$ must be independent of the temperature gradient $\theta_{,\alpha}$, but the remaining response functions in (13.8) are still dependent on $\theta_{,\alpha}$. The relations (13.16)–(13.18) now show that the response functions $\bar{\eta}$, $\bar{N}'^{\alpha\beta}$, $\bar{\boldsymbol{m}}$, $\bar{\boldsymbol{M}}^{\gamma}$, as well as $\bar{N}^{(\alpha\beta)}$, which are determined by the partial derivatives of $\bar{\psi}$ are also independent of $\theta_{,\alpha}$. Hence, $\bar{\boldsymbol{q}}$ (or $\bar{q}^{\alpha}$) is the only response function in (13.9) which remains dependent on the temperature gradient. It follows from these results and (13.12) that the five constitutive equations in (13.8) have been reduced to (13.13) and

$$\eta = -\frac{\partial\bar{\psi}}{\partial\theta}, \tag{13.21}$$

$$\begin{aligned}N'^{\alpha\beta} = N'^{\beta\alpha} &= \frac{1}{2}\varrho\left\{\left(\frac{\partial\bar{\psi}}{\partial\boldsymbol{a}_{\alpha}} - d^{\alpha}\frac{\partial\bar{\psi}}{\partial\boldsymbol{d}} - \lambda^{\alpha}_{\cdot\gamma}\frac{\partial\bar{\psi}}{\partial\boldsymbol{d}_{,\gamma}}\right)\cdot\boldsymbol{a}^{\beta}\right.\\ &\quad\left.+\left(\frac{\partial\bar{\psi}}{\partial\boldsymbol{a}_{\beta}} - d^{\beta}\frac{\partial\bar{\psi}}{\partial\boldsymbol{d}} - \lambda^{\beta}_{\cdot\gamma}\frac{\partial\bar{\psi}}{\partial\boldsymbol{d}_{,\gamma}}\right)\cdot\boldsymbol{a}^{\alpha}\right\},\end{aligned} \tag{13.22}$$

$$\boldsymbol{m} = \varrho\frac{\partial\bar{\psi}}{\partial\boldsymbol{d}}, \qquad \boldsymbol{M}^{\gamma} = \varrho\frac{\partial\bar{\psi}}{\partial\boldsymbol{d}_{,\gamma}}. \tag{13.23}$$

Also, the constitutive relation for $N^{(\alpha\beta)}$ is now given by the right-hand side of (13.20):

$$N^{(\alpha\beta)} = \frac{1}{2}\varrho\left(\frac{\partial\bar{\psi}}{\partial\boldsymbol{a}_{\alpha}}\cdot\boldsymbol{a}^{\beta} + \frac{\partial\bar{\psi}}{\partial\boldsymbol{a}_{\beta}}\cdot\boldsymbol{a}^{\alpha}\right).$$

The foregoing results have been deduced from (13.14) or equivalently from the entropy inequality (9.36) with the mechanical power P given by $(9.29)_2$. Alternatively, we could have used the inequality (9.36) with P defined by $(9.28)_2$. If the latter inequality is employed, then with the help of $\dot{\bar{\psi}}$ given by $(13.10)_1$ we should obtain slightly different results at this stage of our derivation. In particular, with the use of (9.36) with P given by $(9.28)_2$, an expression corresponding to (13.15) does not arise and while the results (13.21), (13.23) and (13.19) are again recovered, instead of (13.22) we have

$$\boldsymbol{N}^{\alpha} = \varrho \frac{\partial \overline{\psi}}{\partial \boldsymbol{a}_{\alpha}}. \tag{13.24}$$

Recalling (9.40), from (13.24) we can readily calculate the expressions for $N^{\alpha 3} = \boldsymbol{N}^{\alpha} \cdot \boldsymbol{a}^{3}$ and $N^{\alpha\beta} = \boldsymbol{N}^{\alpha} \cdot \boldsymbol{a}^{\beta}$. The symmetric part of the latter is easily seen to be the expression for $N^{(\alpha\beta)}$ noted above and at this point it appears that we also have constitutive equations for $N^{\alpha 3}$ and the skew-symmetric part of $N^{\alpha\beta}$. However, after allowance is made for the restriction placed on the free energy ψ by the invariance conditions under superposed rigid body motions, it will become evident in Subsect. β) [see (13.38)–(13.39)] that the expressions for $N^{\alpha 3}$ and $N^{[\alpha\beta]}$ calculated from (13.24) are identical with those resulting from (9.51). In this connection, it is worth observing that substitution of (13.23) and (13.24) into (13.15) at once yields (9.50). We further observe that in obtaining the relations (13.16)–(13.18) or (13.21)–(13.24), as well as the inequality (13.19), we have employed the procedure that the entropy inequality be identically satisfied for every admissible process defined by the constitutive equations. Thus (13.16)–(13.19) are necessary conditions for the validity of the inequality (9.36) and it is easily seen that they are also sufficient.

A different set of constitutive equations can also be obtained in terms of the specific internal energy ε. Through (13.21), θ can be expressed as a function of η and the set of the kinematic variables (13.7). Recalling (9.34) and (13.13), it follows that ε has the form

$$\varepsilon = \overline{\varepsilon}(\boldsymbol{a}_{\alpha}, \boldsymbol{d}, \boldsymbol{d}_{,\gamma}, \eta;\ \theta^{\mu}) \tag{13.25}$$

and instead of (13.21)–(13.23) we obtain

$$\theta = \frac{\partial \overline{\varepsilon}}{\partial \eta}, \tag{13.26}$$

$$\begin{gathered} N'^{\alpha\beta} = N'^{\beta\alpha} = \frac{1}{2} \varrho \left\{ \left(\frac{\partial \overline{\varepsilon}}{\partial \boldsymbol{a}_{\alpha}} - d^{\alpha} \frac{\partial \overline{\varepsilon}}{\partial \boldsymbol{d}} - \lambda^{\alpha}_{\cdot\gamma} \frac{\partial \overline{\varepsilon}}{\partial \boldsymbol{d}_{,\gamma}} \right) \cdot \boldsymbol{a}^{\beta} \right. \\ \left. + \left(\frac{\partial \overline{\varepsilon}}{\partial \boldsymbol{a}_{\beta}} - d^{\beta} \frac{\partial \overline{\varepsilon}}{\partial \boldsymbol{d}} - \lambda^{\beta}_{\cdot\gamma} \frac{\partial \overline{\varepsilon}}{\partial \boldsymbol{d}_{,\gamma}} \right) \cdot \boldsymbol{a}^{\alpha} \right\}, \\ \boldsymbol{m} = \varrho \frac{\partial \overline{\varepsilon}}{\partial \boldsymbol{d}}, \qquad \boldsymbol{M}^{\gamma} = \varrho \frac{\partial \overline{\varepsilon}}{\partial \boldsymbol{d}_{,\gamma}}. \end{gathered} \tag{13.27}$$

The results (13.26)–(13.27) with $\overline{\varepsilon}$ given by the response function in (13.25), together with the inequality (13.19), may be obtained directly from the inequality (9.33) by a procedure similar to that used earlier in this section and after introducing a set of constitutive assumptions similar to those in (13.8)–(13.9).

β) *Reduction of the constitutive equations under superposed rigid body motions.* As remarked previously, the local state of the Cosserat surface is defined by the kinematic variables (13.7), together with θ and $\theta_{,\alpha}$. These kinematic and thermal variables which occur in the function $\overline{\psi}$ and other response functions are objective, in view of $(5.36)_2$, $(5.38)_{1,3}$ and (13.4). Moreover, by virtue of (8.30)–(8.32) and (13.1)–(13.3), η, θ, $\theta_{,\alpha}$, ε (and therefore ψ), $\boldsymbol{N}^{\alpha}$, $\boldsymbol{m}$, $\boldsymbol{M}^{\alpha}$, $\boldsymbol{N}'^{\alpha}$, $\boldsymbol{q}$ or their components, all are objective under superposed rigid body motions.

In what follows, let a typical constitutive equation such as (13.13), which is an objective scalar, be written in the form

$$f = \overline{f}(\boldsymbol{a}_{\alpha}, \boldsymbol{d}, \boldsymbol{d}_{,\gamma}), \tag{13.28}$$

where for brevity dependence on θ and the material point θ^μ is temporarily suppressed. Now a constitutive equation in the form (13.28) which holds for an admissible process must also hold for a motion differing from the given one only by superposed rigid body motions. This requirement is fulfilled if and only if the response function $\bar{f}$ satisfies the identity

$$\bar{f}(\boldsymbol{a}_\alpha, \boldsymbol{d}, \boldsymbol{d}_{,\gamma}) = \bar{f}(Q\boldsymbol{a}_\alpha, Q\boldsymbol{d}, Q\boldsymbol{d}_{,\gamma}) \tag{13.29}$$

for all values of the arguments in the domain of $\bar{f}$ and for all proper orthogonal Q. From Cauchy's representation theorem on isotropic functions, $\bar{f}$ may be expressed as a (different) function of the inner products and the scalar triple products of its arguments.[10] The inner products of the arguments of $\bar{f}$ in (13.28) are the set

$$\boldsymbol{a}_\alpha \cdot \boldsymbol{a}_\beta = a_{\alpha\beta}, \quad \boldsymbol{a}_\alpha \cdot \boldsymbol{d} = d_\alpha, \quad \boldsymbol{a}_\gamma \cdot \boldsymbol{d}_{,\alpha} = \lambda_{\gamma\alpha}, \tag{13.30}$$

and

$$\boldsymbol{d} \cdot \boldsymbol{d}, \quad \boldsymbol{d} \cdot \boldsymbol{d}_{,\gamma}, \quad \boldsymbol{d}_{,\alpha} \cdot \boldsymbol{d}_{,\gamma}, \tag{13.31}$$

while the independent scalar triple products in question are formed from the vectors $\boldsymbol{a}_\alpha$, $\boldsymbol{d}$ and $\boldsymbol{d}_{,\gamma}$. However, before considering these scalar triple products, it is expedient to examine further the inner products (13.30)–(13.31).

Let each of the two vectors $\boldsymbol{d}$ and $\boldsymbol{d}_{,\alpha}$ be resolved into their respective tangential and normal components. Thus, we write[11]

$$\begin{aligned} &\boldsymbol{d} = \boldsymbol{d}^T + \boldsymbol{d}^N, && \boldsymbol{d}^T = d^\alpha \boldsymbol{a}_\alpha, && \boldsymbol{d}^N = d^3 \boldsymbol{a}_3, \\ &\boldsymbol{d}_{,\alpha} = (\boldsymbol{d}_{,\alpha})^T + (\boldsymbol{d}_{,\alpha})^N, && (\boldsymbol{d}_{,\alpha})^T = \lambda^\gamma_{.\alpha} \boldsymbol{a}_\gamma, && (\boldsymbol{d}_{,\alpha})^N = \lambda^3_{.\alpha} \boldsymbol{a}_3, \end{aligned} \tag{13.32}$$

where $\boldsymbol{d}^T$ and $\boldsymbol{d}^N$ designate the tangential and normal components of $\boldsymbol{d}$ while $(\boldsymbol{d}_{,\alpha})^T$ and $(\boldsymbol{d}_{,\alpha})^N$ stand for the tangential and normal components of $\boldsymbol{d}_{,\alpha}$. According to (13.28), f is determined by a response function which depends on the variables characterizing the local state of the Cosserat surface; and we have noted that $\bar{f}$ is expressible as a different function of (13.30)–(13.31) and the scalar triple products. But the latter function already depends on the tangential components $\boldsymbol{d}^T$ and $(\boldsymbol{d}_{,\alpha})^T$, as is evident from $(13.30)_{2,3}$. Hence, without loss of generality, the set (13.31) may be replaced by the set of variables $\sigma, \sigma_\alpha, \sigma_{\alpha\gamma}$ defined by

$$\begin{aligned} \sigma = \boldsymbol{d}^N \cdot \boldsymbol{d}^N = (d^3)^2, \quad \sigma_\alpha = \boldsymbol{d}^N \cdot (\boldsymbol{d}_{,\alpha})^N = d^3 \lambda_{3\alpha}, \\ \sigma_{\alpha\gamma} = (\boldsymbol{d}_{,\alpha})^N \cdot (\boldsymbol{d}_{,\gamma})^N = \lambda_{3\alpha} \lambda_{3\gamma}. \end{aligned} \tag{13.33}$$

Before proceeding further, we note that $(5.25)_1$ can be solved for

$$\boldsymbol{a}_3 = \frac{\boldsymbol{d} - d^\alpha \boldsymbol{a}_\alpha}{d^3}$$

and then $(5.28)_1$ can be expressed in the form

$$\boldsymbol{d}_{,\gamma} = \left(\lambda^\nu_{.\gamma} - \frac{d^\nu \sigma_\gamma}{\sigma}\right) \boldsymbol{a}_\nu + \frac{\sigma_\gamma}{\sigma} \boldsymbol{d}. \tag{13.34}$$

As noted above [following (13.31)], the independent scalar triple products are formed from the appropriate combination of the vectors $\boldsymbol{a}_1$, $\boldsymbol{a}_2$, $\boldsymbol{d}$, $\boldsymbol{d}_{,1}$ and $\boldsymbol{d}_{,2}$.

---

[10] For a proof of CAUCHY's representation theorem, see TRUESDELL and NOLL [1965, 9]. The scalar triple products must be included since the identity (13.29) is required to hold only for proper orthogonal Q.

[11] The temporary notations for tangential and normal components in (13.32)–(13.33) should not be confused, respectively, with the use of superscript T for transpose of a second order tensor in Sect. 5 and the use of the index N in formulae of Sect. 7.

One of these, namely $[\boldsymbol{a}_1\, \boldsymbol{a}_2\, \boldsymbol{d}]$ can be expressed in terms of $(13.30)_{1,2}$, $(13.33)_1$ and τ defined by

$$\tau = \operatorname{sgn}\,[\boldsymbol{a}_1\, \boldsymbol{a}_2\, \boldsymbol{d}]. \tag{13.35}$$

Clearly the magnitude of $[\boldsymbol{a}_1\, \boldsymbol{a}_2\, \boldsymbol{d}]$, but not its sign, is determined by $(13.30)_{1,2}$ and $(13.33)_1$. With the help of (13.34), it can be easily verified that the remaining scalar triple products can be expressed in terms of (13.30), $(13.33)_{1,2}$ and (13.35). In addition, it is easily seen from (13.34) that the scalar $\boldsymbol{d}_{,\alpha} \cdot \boldsymbol{d}_{,\gamma}$ can be expressed in terms of the arguments[12] (13.30) and $(13.33)_{1,2}$; hence $\sigma_{\alpha\gamma}$ may be suppressed from the arguments of the response function $\bar{f}$. Thus, from the results between (13.29)–(13.35), it follows that $\bar{f}$ may be expressed as a function of (13.35) and the set of variables

$$\begin{aligned} \mathscr{V}:\ & a_{\alpha\beta},\ d_\alpha,\ \lambda_{\gamma\alpha}, \\ & \sigma,\quad \sigma_\alpha. \end{aligned} \tag{13.36}$$

However, since we have considered only motions of a Cosserat surface which are consistent with $(5.1)_3$, τ will always have the positive value 1. With this restriction, (13.35) may be eliminated from the domain of the function $\bar{f}$ which can be replaced by a (different) function of (13.36):[13]

$$\bar{f}\,(\boldsymbol{a}_\alpha,\boldsymbol{d},\boldsymbol{d}_{,\gamma}) = \tilde{f}(\mathscr{V}). \tag{13.37}$$

By application of the result (13.37) to the function $\bar{\psi}$, the constitutive equation (13.13), apart from its dependence on the material point θ^μ and temperature θ, can alternatively be expressed in terms of a different function of (13.36):

$$\begin{aligned} \psi &= \bar{\psi}\,(\boldsymbol{a}_\alpha, \boldsymbol{d}, \boldsymbol{d}_{,\gamma}, \theta;\ \theta^\mu) \\ &= \tilde{\psi}\,(a_{\alpha\beta}, d_\alpha, \lambda_{\beta\alpha}, \sigma, \sigma_\alpha, \theta;\ \theta^\mu). \end{aligned} \tag{13.38}$$

In order to obtain an alternative form for the constitutive equations (13.21)–(13.23) in tensor components and in terms of the function $\tilde{\psi}$, we first calculate the partial derivatives which occur in (13.22)–(13.23) in terms of the partial derivatives of $\tilde{\psi}$. Thus, by chain rule differentiation,

$$\begin{aligned} \frac{\partial\bar{\psi}}{\partial \boldsymbol{a}_\beta} &= \left(\frac{\partial\tilde{\psi}}{\partial a_{\beta\gamma}}\boldsymbol{a}_\gamma + \frac{\partial\tilde{\psi}}{\partial a_{\gamma\beta}}\boldsymbol{a}_\gamma\right) + \frac{\partial\tilde{\psi}}{\partial d_\beta}\boldsymbol{d} + \frac{\partial\tilde{\psi}}{\partial\lambda_{\beta\gamma}}\boldsymbol{d}_{,\gamma} \\ &\quad - \left[2 d^3 d^\beta \frac{\partial\tilde{\psi}}{\partial\sigma} + (d^\beta \lambda^3_{\cdot\alpha} + d^3 \lambda^\beta_{\cdot\alpha})\frac{\partial\tilde{\psi}}{\partial\sigma_\alpha}\right]\boldsymbol{a}_3, \\ \frac{\partial\bar{\psi}}{\partial\boldsymbol{d}} &= \frac{\partial\tilde{\psi}}{\partial d_\gamma}\boldsymbol{a}_\gamma + 2\frac{\partial\tilde{\psi}}{\partial\sigma} d^3 \boldsymbol{a}_3 + \frac{\partial\tilde{\psi}}{\partial\sigma_\alpha}\lambda^3_{\cdot\alpha}\boldsymbol{a}_3, \\ \frac{\partial\bar{\psi}}{\partial\boldsymbol{d}_{,\gamma}} &= \frac{\partial\tilde{\psi}}{\partial\lambda_{\alpha\gamma}}\boldsymbol{a}_\alpha + \frac{\partial\tilde{\psi}}{\partial\sigma_\gamma} d^3\boldsymbol{a}_3. \end{aligned} \tag{13.39}$$

Then, with the help of (13.39) and recalling $(5.25)_2$, $(5.28)_2$, (9.39), (9.40) and (9.42), we can express the constitutive relations (13.21)–(13.23) in tensor compo-

[12] This further reduction resulting in the suppression of $\boldsymbol{d}_{,\alpha} \cdot \boldsymbol{d}_{,\gamma}$ was noted by Dr. S. L. PASSMAN. I express here my appreciation to him for bringing it to my attention.

[13] Such a reduction would result eventually for any material symmetry restriction characterized by an improper symmetry transformation, i.e., one in which (13.36) remain invariant and $\tau \to -\tau$, even if we had assumed $[\boldsymbol{a}_1\, \boldsymbol{a}_2\, \boldsymbol{d}] \neq 0$ instead of $(5.1)_3$.

nents as follows:[14]

$$\eta = -\frac{\partial \tilde{\psi}}{\partial \theta} \tag{13.40}$$

and

$$N'^{\alpha\beta} = N'^{\beta\alpha} = 2\varrho \frac{\partial \tilde{\psi}}{\partial a_{\alpha\beta}}, \tag{13.41}$$

$$\begin{aligned} m^\alpha &= \varrho \frac{\partial \tilde{\psi}}{\partial d_\alpha}, & m^3 &= \varrho \left(2 d^3 \frac{\partial \tilde{\psi}}{\partial \sigma} + \lambda^3_{\cdot\,\alpha} \frac{\partial \tilde{\psi}}{\partial \sigma_\alpha}\right), \\ M^{\alpha\gamma} &= \varrho \frac{\partial \tilde{\psi}}{\partial \lambda_{\gamma\alpha}}, & M^{\alpha 3} &= \varrho d^3 \frac{\partial \tilde{\psi}}{\partial \sigma_\alpha}. \end{aligned} \tag{13.42}$$

It is clear from the above that only the symmetric part of $N^{\alpha\beta}$ or equivalently $N'^{\alpha\beta}$ is determined by a constitutive equation in terms of $\tilde{\psi}$. Also, using (13.39), (13.24) and (13.41)–(13.42), the expressions for $N^{\alpha 3}$ and the skew-symmetric part of $N^{\alpha\beta}$ are found from the restriction (13.15). These expressions are the same as those provided by (9.51); and this is in line with earlier observations in this section and the remarks in Sect. 9 [Subsect. α)] that (9.23) and therefore (9.51) is regarded as a restriction on the constitutive equations.

For completeness, we also record here a reduced form of the constitutive relation for the components q^α of the heat flux vector. Recalling the transformations $(13.3)_4$–(13.4) under superposed rigid body motions, we see easily that the conclusion (13.37) is also applicable to the response functions $\bar{q}^\alpha = \bar{\boldsymbol{q}} \cdot \boldsymbol{a}^\alpha$ in (13.9). Hence, the constitutive relation (13.9) reduces to

$$q^\alpha = \tilde{q}^\alpha (a_{\gamma\delta}, d_\gamma, \lambda_{\delta\gamma}, \sigma, \sigma_\gamma, \theta, \theta_{,\gamma};\ \theta^\mu), \tag{13.43}$$

and is subject to the restriction (13.19) with $\bar{q}^\alpha$ replaced by $\tilde{q}^\alpha$.

The foregoing nonlinear constitutive equations, namely (13.38) and (13.40)–(13.43), are valid for an elastic Cosserat surface which may be anisotropic with reference to preferred directions associated with the material points of $\mathscr{C}$. Since the response functions for η, $N'^{\alpha\beta}$, m^i and $M^{\alpha i}$ are fully determined from the knowledge of the response functions for the specific Helmholtz free energy, in any discussion of the effect of material symmetry it will suffice to consider only the response functions for the specific free energy and the heat flux vector. In subsequent developments, however, we shall be largely concerned with the isothermal case; and, in the discussion of material symmetries which serve to restrict the form of the constitutive equations, we shall mainly consider the symmetries associated with the function $\bar{\psi}$ or $\tilde{\psi}$.

γ) *Material symmetry restrictions.* According to a primitive notion of *material symmetry*, the isotropy group (also called the symmetry group) $\mathscr{G}$ of a material is the group of density-preserving transformations of the material coordinates which leave the response of the material unaltered.[15] Before describing the

[14] Although we often write the partial derivative of a function with respect to a symmetric tensor such as $a_{\alpha\beta}$ in the form indicated in (13.41), the partial derivative $\partial\tilde{\psi}/\partial a_{\alpha\beta}$ is understood to have the symmetric form

$$\frac{1}{2}\left(\frac{\partial \tilde{\psi}}{\partial a_{\alpha\beta}} + \frac{\partial \tilde{\psi}}{\partial a_{\beta\alpha}}\right).$$

A parallel remark applies to all similar partial derivatives with respect to symmetric tensors elsewhere in this chapter.

[15] In general, the isotropy group consists of density-preserving transformations; but, since we are concerned with elastic solids, we only need to consider length-preserving transformations. A clear development of the basic concepts of material symmetry in the three-dimensional theory can be found in a paper by Noll [1958, *4*]. See also Sect. 293 of Truesdell and Toupin [1960, *14*].

consequences of such restrictions in mathematical terms, for background information, it is desirable to elaborate briefly (in descriptive terms) on the primitive notion of material symmetry with reference to the Cosserat surface $\mathscr{C}$. Consider the tangent plane and the normal at each material point of $\mathscr{C}$, which point may be regarded as the origin of a Cartesian coordinate system x^m. We associate with x^m a set of orthonormal basis vectors $\boldsymbol{E}_m$ such that $\boldsymbol{E}_\alpha$ are in the tangent plane and $\boldsymbol{E}_3$ is directed along the normal. If a typical vector $\boldsymbol{V}$ is decomposed in such a coordinate system along the unit vectors $\boldsymbol{E}_m$ into tangential components and a normal component, then under a change from one coordinate system in the tangent plane to another the tangential components transform like the components of a 2-vector under the orthogonal group and the third component remains invariant. The symmetry of the material, defined by preferred directions in the body manifold (i.e., the Cosserat surface) is then characterized by $\boldsymbol{E}_\alpha$ and the appropriate group of (two-dimensional) transformations which specify the equivalent positions of the vectors $\boldsymbol{E}_\alpha$ from one system to another and the constitutive relations must then be form-invariant under each transformation of this group. We consider here symmetries for which the associated group of transformations is a subgroup of the full orthogonal group.

Since the body manifold—here the Cosserat surface $\mathscr{C}$—has no metric property, we begin our discussion of material symmetry by assigning a metric tensor to the material surface of $\mathscr{C}$; this is conveniently realized by assigning the (reference) metric tensor $A_{\alpha\beta}$ at each material point[16] θ^μ. Next, we introduce another coordinate system $\bar{\theta}^\mu = \{\bar{\theta}^1, \bar{\theta}^2\}$ on the material surface of $\mathscr{C}$ by the length-preserving transformation

$$\theta^\alpha = \theta^\alpha(\bar{\theta}^1, \bar{\theta}^2), \tag{13.44}$$

such that

$$A_{\gamma\delta} = H^\alpha_\gamma H^\beta_\delta A_{\alpha\beta}, \qquad H^\alpha_\gamma = \frac{\partial\theta^\alpha}{\partial\bar{\theta}^\gamma}. \tag{13.45}$$

In other words, (13.44) is so restricted that at a given material point the components H^α_γ satisfy $(13.45)_1$.

The length-preserving coordinate transformation (13.44) relates two curvilinear coordinate systems on the material surface of $\mathscr{C}$; and it is not necessary, for our present purpose, to be more specific than this. However, often it is simpler in applications to introduce the length-preserving transformation by means of two coordinate systems which are locally Cartesian at the material point in question of the body manifold. Thus, let x^α and $\bar{x}^\gamma$ refer to two locally rectangular Cartesian coordinate systems and let the origin of the former be identified with the material point in question. Then, in line with earlier remarks (in this subsection) and instead of (13.44)–(13.45), we may introduce the length-preserving transformation by

$$x^\alpha = \bar{H}^\alpha_\gamma \bar{x}^\gamma + \bar{H}^\alpha, \qquad \bar{H}\,\bar{H}^T = \bar{H}^T\,\bar{H} = I, \tag{13.46}$$

such that $\boldsymbol{E}_\alpha$ are the basis in the x^α coordinates and $\bar{H}$ (with components $\bar{H}^\alpha_\gamma$) is an orthogonal matrix which satisfies $(13.46)_2$.

---

[16] Since the assigned metric tensor is that of the reference configuration, in reality we are considering symmetries relative to the reference configuration of $\mathscr{C}$. As we are concerned with material symmetries possessed by the body manifold relative to a *local* reference configuration, we have assigned the values of $A_{\alpha\beta}$ to each material point θ^μ in order to render the notion of distance on the material surface of $\mathscr{C}$ meaningful.

For convenience, in what follows, we again employ (13.28) as a typical constitutive equation and study first the effect of material symmetry on the response functions $\bar{f}$ and $\tilde{f}$ in (13.37). If a response function for an elastic Cosserat surface [such as $\bar{f}$ in (13.28)] is form-invariant under a subgroup of the distance-preserving transformations (13.44), then the material is said to have the symmetry represented by that subgroup. Hence, for each element in the symmetry group of the material, we must have[17]

$$\bar{f}\left(\frac{\partial \boldsymbol{r}}{\partial \theta^\alpha}, \boldsymbol{d}, \frac{\partial \boldsymbol{d}}{\partial \theta^\gamma}; \theta^\mu\right) = \bar{f}\left(\frac{\partial \boldsymbol{r}}{\partial \bar{\theta}^\alpha}, \boldsymbol{d}, \frac{\partial \boldsymbol{d}}{\partial \bar{\theta}^\gamma}; \theta^\mu\right), \tag{13.47}$$

where $\partial \boldsymbol{r}/\partial \bar{\theta}^\alpha$ and $\partial \boldsymbol{d}/\partial \bar{\theta}^\alpha$ by (13.44)–(13.45) are:

$$\frac{\partial \boldsymbol{r}}{\partial \bar{\theta}^\alpha} = H_\alpha^\gamma \boldsymbol{a}_\gamma, \qquad \frac{\partial \boldsymbol{d}}{\partial \bar{\theta}^\alpha} = H_\alpha^\gamma \boldsymbol{d}_{,\gamma}. \tag{13.48}$$

Evidently, under the transformation (13.44), a scalar V or a vector field $\boldsymbol{V}$ remains unaffected but tangential derivatives (such as $\partial V/\partial \bar{\theta}^\alpha$, $\partial \boldsymbol{V}/\partial \bar{\theta}^\alpha$) transform according to formulae of the type (13.48).

Under the symmetry transformation (13.48), (13.47) becomes

$$\bar{f}(\boldsymbol{a}_\alpha, \boldsymbol{d}, \boldsymbol{d}_{,\beta}; \theta^\mu) = \bar{f}(H_\alpha^\gamma \boldsymbol{a}_\gamma, \boldsymbol{d}, H_\beta^\delta \boldsymbol{d}_{,\delta}; \theta^\mu). \tag{13.49}$$

Similarly, recalling (13.37) and the variables $\mathscr{V}$ in (13.36), we see that the material symmetry group of the response function $\tilde{f}$ is the set of all length-preserving transformations such that

$$\tilde{f}(\mathscr{V}; \theta^\mu) = \tilde{f}(\bar{\mathscr{V}}; \theta^\mu), \tag{13.50}$$

where $\bar{\mathscr{V}}$ stands for the set of variables

$$\begin{gathered} \bar{\mathscr{V}}\colon H_\alpha^\tau H_\gamma^\nu a_{\tau\nu}, \; H_\alpha^\tau d_\tau, \; H_\gamma^\nu H_\alpha^\tau \lambda_{\nu\tau}, \\ \sigma, \; H_\alpha^\tau \sigma_\tau \end{gathered} \tag{13.51}$$

resulting from the set $\mathscr{V}$ under the symmetry transformations (13.48).

We can now apply the result (13.49) to $\bar{\psi}$ or (13.50) to $\tilde{\psi}$ in (13.38). Considering only the latter, we have

$$\tilde{\psi}(\mathscr{V}, \theta; \theta^\mu) = \tilde{\psi}(\bar{\mathscr{V}}, \theta; \theta^\mu), \tag{13.52}$$

which must be satisfied by the response function $\tilde{\psi}$ for all H_γ^α in $\mathscr{G}$. It should be apparent that a result similar to (13.52) can also be recorded for the response function $\tilde{q}^\alpha$ in (13.43), except that $H_\alpha^\tau \theta_{,\tau}$ should be added to the variables $\bar{\mathscr{V}}$. Given a group of symmetry transformations (13.44) for an elastic material, the identity (13.52) serves to restrict the response function $\tilde{\psi}$ and hence the constitutive relations $(13.38)_2$ and (13.40)–(13.42). In particular, if the symmetry group $\mathscr{G}$ consists of the set of all length-preserving transformations (13.44), the elastic material is said to be isotropic with a center of symmetry. This completes our discussion of material symmetry restrictions. However, independently of material symmetries, a further restriction may be imposed and this is discussed next.

The material symmetries discussed above pertain to a given Cosserat surface $\mathscr{C}$ and are valid irrespective of the choice of direction associated with the director

[17] Previously in (13.28) the dependence of $\bar{f}$ on the material point θ^μ was suppressed; it is included in (13.47) for clarity.

at each material point of $\mathscr{C}$. Suppose we require that the response functions be also independent of the particular orientation of the director at each material point of the surface which, in turn, implies that the constitutive equations remain unaltered also under the reflection[18]

$$\boldsymbol{d} \to -\boldsymbol{d} \tag{13.53}$$

and hence

$$\boldsymbol{d}_{,\alpha} \to -\boldsymbol{d}_{,\alpha}. \tag{13.54}$$

Then, for a given Cosserat surface, $\boldsymbol{a}_\alpha$ and therefore $\boldsymbol{a}_3$ remain unaltered but the kinematic variables d_i and $\lambda_{i\alpha}$ transform according to

$$\begin{aligned} d_\alpha &\to -d_\alpha, & d_3 &\to -d_3, \\ \lambda_{\beta\alpha} &\to -\lambda_{\beta\alpha}, & \lambda_{3\alpha} &\to -\lambda_{3\alpha}. \end{aligned} \tag{13.55}$$

It is noteworthy that the variables σ and σ_α defined by $(13.33)_{1,2}$ remain unaffected by the results (13.55). To summarize, under the added restriction (13.53), the set of variables $\mathscr{V}$ in (13.36) becomes

$$\begin{gathered} \mathscr{V}': \; a_{\alpha\gamma}, \; -d_\alpha, \; -\lambda_{\gamma\alpha}, \\ \sigma, \; \sigma_\alpha \end{gathered} \tag{13.56}$$

and hence

$$\tilde{\psi}(\mathscr{V}, \theta; \theta^\mu) = \tilde{\psi}(\mathscr{V}', \theta; \theta^\mu). \tag{13.57}$$

The restriction imposed by (13.57) requires that $\tilde{\psi}$ be an even function of the ordered pair $(\lambda_{\beta\alpha}, d_\gamma)$. This must be kept in mind, if the restriction (13.53) is introduced prior to considerations of material symmetry.

δ) *Alternative forms of the constitutive equations.* It is sometimes desirable to express the response functions in terms of *relative* kinematic measures such as those in (5.31)–(5.33), which are obtained relative to a fixed reference configuration. To this end and in view of the dependence of $\tilde{\psi}$ and q^α in (13.38) and (13.43) on the last two of the set (13.36), we introduce the relative measures

$$\begin{aligned} s &= \boldsymbol{d}^N \cdot \boldsymbol{d}^N - \boldsymbol{D}^N \cdot \boldsymbol{D}^N = (d^3)^2 - (D^3)^2, \\ s_\alpha &= \boldsymbol{d}^N \cdot (\boldsymbol{d}_{,\alpha})^N - \boldsymbol{D}^N \cdot (\boldsymbol{D}_{,\alpha})^N = d^3 \lambda_{3\alpha} - D^3 \Lambda_{3\alpha}, \end{aligned} \tag{13.58}$$

where in line with (13.32)–(13.33) we have used the notations

$$\boldsymbol{D}^N = D^3 \boldsymbol{A}_3, \qquad (\boldsymbol{D}_{,\alpha})^N = \Lambda^3_{.\alpha} \boldsymbol{A}_3. \tag{13.59}$$

The variable $e_{\alpha\beta}$ defined by (5.31) is a kinematic measure in which $a_{\alpha\beta}$ is computed relative to its reference value $A_{\alpha\beta}$. Hence, a function of $a_{\alpha\beta}$ can be expressed as a different function of $e_{\alpha\beta}$ and $A_{\alpha\beta}$; but it should be kept in mind that a constitutive equation in terms of the latter function is now one that holds relative to a reference configuration, with the material point identified with its position in the reference configuration. Parallel remarks apply to more general response functions.

Henceforth, we identify the material point with its position in the reference configuration and also restrict attention to materials which are homogeneous in the reference configuration. It follows from the above remarks that a response

[18] To give an interpretation of (13.53), we first recall the definition of the three-dimensional shell and the observation in Sect. 4 that the magnitude of the director (say directed along the normal to the material surface prior to any deformation) can be regarded as representing the thickness of the (three-dimensional) shell. Then, the condition (13.53) can be interpreted as reflecting the fact that a material filament above and below the surface $\xi = 0$ of the (three-dimensional) shell possesses no intrinsic positive or negative direction.

function such as $\tilde{\psi}$ in (13.38) may be replaced by a (different) function of temperature, the variables

$$\mathscr{U}: \; e_{\alpha\beta}, \; \gamma_\alpha, \; \varkappa_{\beta\alpha}, \\ s, \; s_\alpha \tag{13.60}$$

and the reference values

$$\mathscr{U}_R: \; A_{\alpha\beta}, \; D_\alpha, \; \Lambda_{\beta\alpha}, \\ (D^3)^2, \; D^3 \Lambda_{3\alpha}. \tag{13.61}$$

In particular, the constitutive equation $(13.38)_2$ for the specific free energy may be replaced by

$$\psi = \hat{\psi}(e_{\alpha\beta}, \gamma_\alpha, \varkappa_{\beta\alpha}, s, s_\alpha, \theta; \mathscr{U}_R). \tag{13.62}$$

The above constitutive equation holds relative to a fixed homogeneous reference configuration which may be taken to be the initial configuration. It is now a straightforward matter to obtain the constitutive equations in tensor components and in terms of the function $\hat{\psi}$ from (13.40)–(13.42). These are given by[19]

$$\eta = -\frac{\partial \hat{\psi}}{\partial \theta} \tag{13.63}$$

and

$$N'^{\alpha\beta} = N'^{\beta\alpha} = \varrho \frac{\partial \hat{\psi}}{\partial e_{\alpha\beta}}, \tag{13.64}$$

$$m^\alpha = \varrho \frac{\partial \hat{\psi}}{\partial \gamma_\alpha}, \qquad m^3 = \varrho \left(2 d^3 \frac{\partial \hat{\psi}}{\partial s} + \lambda^3_{\cdot\alpha} \frac{\partial \hat{\psi}}{\partial s_\alpha}\right), \\ M^{\alpha\gamma} = \varrho \frac{\partial \hat{\psi}}{\partial \varkappa_{\gamma\alpha}}, \qquad M^{\alpha 3} = \varrho\, d^3 \frac{\partial \hat{\psi}}{\partial s_\alpha}. \tag{13.65}$$

Also, corresponding to (13.43), we have

$$q^\alpha = \hat{q}^\alpha(e_{\alpha\beta}, \gamma_\alpha, \varkappa_{\beta\alpha}, s, s_\alpha, \theta, \theta_{,\alpha}; \mathscr{U}_R). \tag{13.66}$$

Both sets of constitutive equations $(13.38)_2$ and (13.40)–(13.43) and (13.62)–(13.66) hold at each material point of the Cosserat surface $\mathscr{C}$; however, in contrast to the former set, the constitutive equations (13.62)–(13.66) involve kinematic variables which are measured relative to a fixed reference configuration. Moreover, the response functions in (13.62)–(13.66) depend on the choice of the reference configuration as indicated explicitly by the presence of $\mathscr{U}_R$ in the arguments of $\hat{\psi}$ and $\hat{q}^\alpha$. We also observe that the response function $\hat{q}^\alpha$ must satisfy an inequality of the form (13.19) but with $\bar{q}^\alpha$ replaced by $\hat{q}^\alpha$. Also, with the help of (13.62)–(13.65), the residual energy equation can be obtained from (9.35) and (9.52).

The material symmetry restrictions discussed above [Subsect. γ)] may be readily extended to the response functions $\hat{\psi}$ and $\hat{q}^\alpha$. Recalling that the material point is now identified with its position in the reference configuration, it can be shown that $\partial \boldsymbol{R}/\partial \bar{\theta}^\gamma$ and $\partial \boldsymbol{D}/\partial \bar{\theta}^\gamma$ also transform as in (13.48) under the length-preserving transformation (13.44). Moreover, the material isotropy group $\mathscr{G}$ now becomes identical to the isotropy group relative to the reference configuration.[20]

[19] The partial derivative $\partial \hat{\psi}/\partial e_{\alpha\beta}$ in (13.64) is understood to have the symmetric form

$$\frac{1}{2}\left(\frac{\partial \hat{\psi}}{\partial e_{\alpha\beta}} + \frac{\partial \hat{\psi}}{\partial e_{\beta\alpha}}\right).$$

[20] We may recall here that according to Noll [1958, 4], the isotropy groups corresponding to two different reference configurations of the same material can be brought into one-to-one correspondence preserving the group structure or that, in more precise terms, they are conjugate and hence isomorphic.

It then follows that the response function $\hat{\psi}$ must satisfy the identity

$$\hat{\psi}(\mathcal{U}, \theta; \mathcal{U}_R) = \hat{\psi}(\overline{\mathcal{U}}, \theta; \overline{\mathcal{U}}_R), \tag{13.67}$$

where $\overline{\mathcal{U}}$ and $\overline{\mathcal{U}}_R$ stand for

$$\begin{gathered} \overline{\mathcal{U}}: \; H^{\tau}_{\alpha} H^{\nu}_{\gamma} e_{\tau\nu}, \; H^{\tau}_{\alpha} \gamma_{\tau}, \; H^{\nu}_{\gamma} H^{\tau}_{\alpha} \varkappa_{\nu\tau}, \\ s, \; H^{\tau}_{\alpha} s_{\tau} \end{gathered} \tag{13.68}$$

and

$$\begin{gathered} \overline{\mathcal{U}}_R: \; H^{\tau}_{\alpha} H^{\nu}_{\gamma} A_{\tau\nu}, \; H^{\tau}_{\alpha} D_{\tau}, \; H^{\nu}_{\gamma} H^{\tau}_{\alpha} \Lambda_{\nu\tau}, \\ (D^3)^2, \; H^{\tau}_{\alpha}(D^3 \Lambda_{3\tau}). \end{gathered} \tag{13.69}$$

An identity similar to (13.67) holds also for $\hat{q}^{\alpha}$. The symmetry restriction (13.67) serves to restrict the response function $\hat{\psi}$ and hence the constitutive equation (13.62). The previous remarks concerning the symmetry group $\mathcal{G}$ made with reference to (13.52) apply also to (13.67).

Now let the director in the (initial) reference configuration be specified along the unit normal to $\mathcal{S}$ so that

$$\begin{gathered} \boldsymbol{D} = D\,\boldsymbol{A}_3, \quad D_{\alpha} = 0, \quad D_3 = D, \\ \Lambda_{\beta\alpha} = -D\,B_{\alpha\beta}, \quad \Lambda_{3\alpha} = D_{,\alpha}, \end{gathered} \tag{13.70}$$

where the notation $(13.70)_3$ is introduced for convenience. Further, if $\boldsymbol{D}$ is of constant magnitude and coincident with the unit normal $\boldsymbol{A}_3$, we have

$$\begin{gathered} \boldsymbol{D} = \boldsymbol{A}_3, \quad D_{\alpha} = 0, \quad D_3 = D = 1, \\ \Lambda_{\beta\alpha} = -B_{\alpha\beta}, \quad \Lambda_{3\alpha} = 0. \end{gathered} \tag{13.71}$$

Then, corresponding to (13.71), the reference values (13.61) become simply

$$\mathcal{U}'_R: \; A_{\alpha\beta}, \; -B_{\alpha\beta} \tag{13.72}$$

and the constitutive equation (13.62) reduces to

$$\psi = \hat{\psi}(e_{\alpha\beta}, \gamma_{\alpha}, \varkappa_{\beta\alpha}, s, s_{\alpha}, \theta; \mathcal{U}'_R), \tag{13.73}$$

where $\hat{\psi}$ is a different function from that in (13.62). In our further developments of the constitutive equations (both nonlinear and linear) by direct approach, we shall restrict attention mainly to a Cosserat surface whose (initial) reference director is specified by (13.71).

Previously [in Subsect. γ)], with reference to $(13.38)_2$ and (13.40)–(13.43), we also examined the consequences of an added restriction (13.53). This restriction, which is separate from material symmetry restrictions, requires that the response of the Cosserat surface be independent of the particular orientation of the director at each point of the surface. Since the constitutive equations (13.62)–(13.66) hold relative to the (initial) reference configuration, the added restriction requires (13.53)–(13.54) together with

$$\boldsymbol{D} \to -\boldsymbol{D} \tag{13.74}$$

and hence

$$\boldsymbol{D}_{,\alpha} \to -\boldsymbol{D}_{,\alpha}. \tag{13.75}$$

The conditions (13.74)–(13.75), in turn, imply

$$\begin{gathered} D_{\alpha} \to -D_{\alpha}, \quad D_3 \to -D_3, \\ \Lambda_{\beta\alpha} \to -\Lambda_{\beta\alpha}, \quad \Lambda_{3\alpha} \to -\Lambda_{3\alpha}. \end{gathered} \tag{13.76}$$

Under the additional restriction (13.53) and (13.74), the set of variables $\mathscr{U}$ and $\mathscr{U}_R$ in (13.60)–(13.61) become

$$\begin{gathered}\mathscr{U}'' \colon e_{\alpha\gamma}, \ -\gamma_\alpha, \ -\varkappa_{\gamma\alpha}, \\ s, \ s_\alpha\end{gathered} \tag{13.77}$$

and

$$\begin{gathered}\mathscr{U}_R'' \colon A_{\alpha\beta}, \ -D_\alpha, \ -\Lambda_{\gamma\alpha}, \\ (D^3)^2, \ D^3 \Lambda_{3\alpha}\end{gathered} \tag{13.78}$$

and hence the response function $\hat{\psi}$ must also satisfy

$$\hat{\psi}(\mathscr{U}, \theta; \mathscr{U}_R) = \hat{\psi}(\mathscr{U}'', \theta; \mathscr{U}_R''), \tag{13.79}$$

in addition to (13.67) resulting from material symmetry restrictions. It should be clear that the restriction specified by (13.53) and (13.74) could have been introduced prior to any consideration of material symmetries.

The constitutive relations for $N'^{\alpha\beta}$, m^i and $M^{\alpha i}$ can be expressed in somewhat simpler forms than those in (13.64)–(13.65); but, this can be accomplished at the expense of a loss in the representation for the function $\tilde{\psi}$ or $\hat{\psi}$. To elaborate, consider the constitutive relations given by (13.62)–(13.65) which reflect, in particular, the manner in which $\hat{\psi}$ depends on the last two variables in each of (13.60)–(13.61). Suppose, instead of (13.62), the constitutive equation for ψ is assumed to depend on the temperature, the kinematic variables (5.31)–(5.33) and the reference values $A_{\alpha\beta}$, D_i, $\Lambda_{i\alpha}$:

$$\psi = \psi'(e_{\alpha\beta}, \gamma_i, \varkappa_{i\alpha}, \theta; A_{\alpha\beta}, D_i, \Lambda_{i\alpha}). \tag{13.80}$$

Then, from (9.36) with P given by (9.52) and by a procedure similar to that used earlier, we can deduce the expressions[21]

$$\eta = -\frac{\partial \psi'}{\partial \theta}, \tag{13.81}$$

$$N'^{\alpha\beta} = N'^{\beta\alpha} = \varrho \frac{\partial \psi'}{\partial e_{\alpha\beta}}, \qquad m^i = \varrho \frac{\partial \psi'}{\partial \gamma_i}, \qquad M^{\alpha i} = \varrho \frac{\partial \psi'}{\partial \varkappa_{i\alpha}}. \tag{13.82}$$

Mathematically, the constitutive equations (13.80)–(13.82) and a similar constitutive equation for q^α are as general as those in (13.62)–(13.66). But the simpler forms (13.82) are gained at a loss in the representation for the specific free energy, since the function ψ' is not endowed with the property that its arguments γ_3, $\varkappa_{3\alpha}$ and their reference values appear according to the forms of the last two variables in each of (13.60)–(13.61). Of course, the constitutive equations (13.80)–(13.82) can be supplemented by separate conditions reflecting the effects of (13.58) and their initial values in the response functions.[22]

We close this section with a remark concerning the nature of constitutive equations in an alternative formulation of the theory of a Cosserat surface in terms of a reference state [Sect. 8, Subsect. ε) and Sect. 9, Subsect. δ)]. We recall that the basic field equations in terms of a reference state involve, in particular, the

[21] The constitutive equations in the form (13.80)–(13.82) were derived by GREEN, NAGHDI and WAINWRIGHT [1965, *4*]. For a purely mechanical theory, results similar in forms to (13.82) and in terms of a strain energy function were also obtained by COHEN and DE SILVA [1966, *2*] who admit three independent directors at each point of their directed surfaces.

[22] In the context of the linear theory and on the basis of a different argument, a restriction corresponding to (13.79) was introduced by GREEN and NAGHDI [1967, *4*], [1968, *6*] and imposed by them on the linearized versions of the constitutive equations (13.80)–(13.82). In this connection, see Sect. 16 [Subsect. γ)].

variables

$$\{{}_R\boldsymbol{N}^\alpha,\ {}_R\boldsymbol{m},\ {}_R\boldsymbol{M}^\alpha,\ {}_R q^\alpha\}, \tag{13.83}$$

which are measured relative to a reference configuration. The constitutive relations and related thermodynamical results for an elastic Cosserat surface in terms of the variables (13.83) can be obtained from (9.82) in the same manner that various results in this section [Subsect. α)] were deduced in terms of the variables $\{\boldsymbol{N}^\alpha, \boldsymbol{m}, \boldsymbol{M}^\alpha, q^\alpha\}$ from the inequality (9.36). However, any additional discussion of this kind is unnecessary, in view of the relations (9.87) between the two sets of variables. In fact, apart from a factor of ϱ_0/ϱ, it is clear that the constitutive equations for the variables (13.83) are also given by (13.24), (13.23) and (13.9).

14. The complete theory. Special results: I. Direct approach. We recapitulate here the complete thermodynamical theory of an elastic Cosserat surface and then separately discuss the constitutive equations of the purely mechanical theory of an elastic Cosserat surface. Also, we briefly indicate the nature of some special theories and some available special results and solutions within the scope of the general isothermal (or mechanical) theory.

α) *The boundary-value problem in the general theory.* The basic field equations of the nonlinear theory consist in the equations of motion (9.47)–(9.48), (9.51) and the energy equation $(9.28)_1$ with P given by $(9.28)_2$ or (9.52). The constitutive equations for an elastic Cosserat surface are specified by $(13.38)_2$, (13.40)–(13.43) or any of the alternative forms discussed in the previous section. Also the heat flux q^α, whose constitutive equation has the form (13.43), must satisfy an inequality of the form (13.19).

The above field equations and constitutive relations characterize the initial boundary-value problem in the nonlinear theory of an elastic Cosserat surface. The nature of the boundary conditions in this theory is clear from the rate of work expression $R_c(\mathscr{P})$ in (8.8) and the line integral in (8.10). In particular, the force and the director couple boundary conditions are given by (9.41) and (9.43) and these hold pointwise on the boundary c of the Cosserat surface with outward unit normal ν_α.

The rather difficult question of existence and nonuniqueness for elastostatic boundary-value problems has been considered very recently by ANTMAN, whose analysis is confined to axisymmetric deformations in the isothermal theory of a Cosserat surface.[23] Employing the direct methods of the calculus of variations and imposing fairly mild restrictions on the material response, ANTMAN's paper deals with existence, regularity and nonuniqueness of solutions for equilibrium problems of shells of revolution (under a wide class of boundary conditions) subjected to hydrostatic pressure and distributed axial load.

β) *Constitutive equations in a mechanical theory.* It is desirable to indicate briefly one way in which the constitutive relations of the purely mechanical theory can be expressed in terms of an *elastic potential*. For this purpose, we first recall the expression for the rate of work by contact force and contact director couple for each part $\mathscr{P}$ of $\mathscr{C}$ in the present configuration. Thus, after introducing (9.11) and (9.20) into $(8.8)_2$ and transforming the line integral into a surface integral, the expression for $R_c(\mathscr{P})$ becomes

$$\begin{aligned} R_c(\mathscr{P}) &= \int_{\mathscr{P}} [(\boldsymbol{N}^\alpha\cdot\boldsymbol{v})_{|\alpha} + (\boldsymbol{M}^\alpha\cdot\boldsymbol{w})_{|\alpha}]\, d\sigma \\ &= -\int_{\mathscr{P}} \varrho[\bar{\boldsymbol{f}}\cdot\boldsymbol{v} + \bar{\boldsymbol{l}}\cdot\boldsymbol{w}]\, d\sigma + \int_{\mathscr{P}} [\boldsymbol{N}^\alpha\cdot\boldsymbol{v}_{,\alpha} + \boldsymbol{m}\cdot\boldsymbol{w} + \boldsymbol{M}^\alpha\cdot\boldsymbol{w}_{,\alpha}]\, d\sigma, \end{aligned} \tag{14.1}$$

[23] ANTMAN [1971, *1*].

where in obtaining $(14.1)_2$ use has been made of (9.13) and (9.22). Consider next the reduction of the rate of work of the contact and the assigned forces and couples in $(8.8)_1$. Combining (14.1) and $(8.8)_3$, recalling the kinematical results (5.18)–(5.19), (5.26), (5.29) and using (9.23), we finally obtain

$$\begin{aligned} R(\mathscr{P}) &= \frac{d}{dt}\mathscr{K}(\mathscr{P}) + \int_{\mathscr{P}} [\boldsymbol{N}'^{\alpha}\cdot\boldsymbol{\eta}_{\alpha} + \boldsymbol{m}\cdot\boldsymbol{\Gamma} + \boldsymbol{M}^{\alpha}\cdot\boldsymbol{\Gamma}_{:\alpha}]\, d\sigma \\ &= \frac{d}{dt}\mathscr{K}(\mathscr{P}) + \int_{\mathscr{P}} [N'^{\alpha\beta}\eta_{\alpha\beta} + m^i\dot{d}_i + M^{\alpha i}\dot{\lambda}_{i\alpha}]\, d\sigma, \end{aligned} \tag{14.2}$$

where the kinetic energy $\mathscr{K}(\mathscr{P})$ is defined by (8.11) and the integrand in the second surface integral is the component form of the integrand in the first integral.

We assume the existence of a *strain energy* or a *stored energy* per unit mass $\Sigma(\theta^{\alpha}, t)$ such that

$$\boldsymbol{N}'^{\alpha}\cdot\boldsymbol{\eta}_{\alpha} + \boldsymbol{m}\cdot\boldsymbol{\Gamma} + \boldsymbol{M}^{\alpha}\cdot\boldsymbol{\Gamma}_{:\alpha} = N'^{\alpha\beta}\eta_{\alpha\beta} + m^i\dot{d}_i + M^{\alpha i}\dot{\lambda}_{i\alpha} = \varrho\dot{\Sigma} \tag{14.3}$$

and we define the strain energy for each part $\mathscr{P}$ in the present configuration at time t by the surface integral

$$U(\mathscr{P}) = \int_{\mathscr{P}} \varrho\,\Sigma\, d\sigma. \tag{14.4}$$

From (14.2)–(14.4), (5.66) and (8.18) follows

$$R(\mathscr{P}) = \frac{d}{dt}[\mathscr{K}(\mathscr{P}) + U(\mathscr{P})], \tag{14.5}$$

which is the analogue of a familiar result in the three-dimensional theory. According to (14.5) the rate of work by the contact and the assigned forces and couples $R(\mathscr{P})$ is equal to the sum of the rate of the kinetic energy $\mathscr{K}(\mathscr{P})$ and the rate of the strain energy $U(\mathscr{P})$.

Returning to (14.3), in parallel with the developments of Sect. 13 [Subsect. α)], we assume that for an elastic Cosserat surface the strain energy density Σ at each material point of $\mathscr{C}$ and for all t is specified by a response function which depends on the kinematic variables (13.7):

$$\Sigma = \bar{\Sigma}(\boldsymbol{a}_{\alpha}, \boldsymbol{d}, \boldsymbol{d}_{,\gamma}; \theta^{\mu}). \tag{14.6}$$

Since

$$\dot{\Sigma} = \frac{\partial\bar{\Sigma}}{\partial\boldsymbol{a}_{\alpha}}\cdot\dot{\boldsymbol{a}}_{\alpha} + \frac{\partial\bar{\Sigma}}{\partial\boldsymbol{d}}\cdot\dot{\boldsymbol{d}} + \frac{\partial\bar{\Sigma}}{\partial\boldsymbol{d}_{,\gamma}}\cdot\dot{\boldsymbol{d}}_{,\gamma}$$

and since (14.3) is defined to hold for all arbitrary values of $(13.11)_{1,2,3}$ or equivalently $\eta_{\alpha\beta}$, $\boldsymbol{\Gamma}$, $\boldsymbol{\Gamma}_{:\alpha}$, from (14.3) we can deduce a condition of the form (13.15) together with constitutive equations of the forms (13.22)–(13.23) but with $\bar{\psi}$ replaced by $\bar{\Sigma}$. Further, by an argument entirely similar to that which led to (13.38) and (13.41)–(13.42), the constitutive equations can be reduced to

$$\Sigma = \tilde{\Sigma}(a_{\alpha\beta}, d_{\alpha}, \lambda_{\beta\alpha}, \sigma, \sigma_{\alpha}; \theta^{\mu}) \tag{14.7}$$

and

$$N'^{\alpha\beta} = N'^{\beta\alpha} = 2\varrho\frac{\partial\tilde{\Sigma}}{\partial a_{\alpha\beta}}, \tag{14.8}$$

$$\begin{aligned} m^{\alpha} &= \varrho\frac{\partial\tilde{\Sigma}}{\partial d_{\alpha}}, \qquad & m^3 &= \varrho\left(2d^3\frac{\partial\tilde{\Sigma}}{\partial\sigma} + \lambda^3_{\cdot\alpha}\frac{\partial\tilde{\Sigma}}{\partial\sigma_{\alpha}}\right), \\ M^{\alpha\gamma} &= \varrho\frac{\partial\tilde{\Sigma}}{\partial\lambda_{\gamma\alpha}}, \qquad & M^{\alpha 3} &= \varrho\, d^3\frac{\partial\tilde{\Sigma}}{\partial\sigma_{\alpha}}. \end{aligned} \tag{14.9}$$

Alternative forms of the constitutive equations in which the strain energy density depends on the relative kinematic measures (13.60) and their reference values (13.61), the material symmetry, as well as the restriction imposed by the condition (13.53), can be discussed as in Sect. 13.

The above results, in terms of the response function $\tilde{\Sigma}$, represent the constitutive equations of the purely mechanical theory of an elastic Cosserat surface. They can be regarded as a special case of the formulae (13.38) and (13.41)–(13.42) when the temperature is constant. Alternatively, they can be regarded as a special case of those resulting from (13.25)–(13.27) when the entropy is constant. We note that the constitutive relations (14.8)–(14.9) may also be deduced from a virtual work principle.

γ) *Some special results.* We include here some remarks concerning certain special results, including a class of large deformation solutions. Based on the above theory and when the constitutive equations are restricted to correspond to an elastic Cosserat surface which is initially isotropic with a center of symmetry, a class of static and dynamic solutions for finite isothermal deformation has been given by CROCHET and NAGHDI.[24] The problems considered by CROCHET and NAGHDI are concerned mainly with a sector of a circular cylindrical surface, problems of a closed circular cylindrical surface and spherically symmetric deformation of a spherical surface. The solutions are obtained without specializing the form of the specific free energy (or the strain energy density) function and are such that in all cases (i) the initial director is taken to be coincident with the unit normal to the initial undeformed surface and (ii) the displacement gradients and the director displacements are independent of surface coordinates, but are either constants or functions of time. These solutions may be regarded as the surface counterpart of a number of existing exact solutions in the three-dimensional (non-polar) finite elasticity, often called controllable solutions in the recent literature.[25] We also note here that the theory of infinitesimal deformation superimposed upon a given deformation of an elastic Cosserat surface can be constructed by employing concepts similar to the corresponding developments in the three-dimensional theory.[26]

Within the scope of the purely mechanical theory of an elastic Cosserat surface and with the limitation to elastostatic problems, recently ERICKSEN has posed the question of what is meant by a uniform state for shells and has provided conditions for the existence of uniform states.[27] In essence, ERICKSEN defines a local state by the kinematic variables (13.7) and proceeds to obtain an equivalence relation between local states at different points of the same surface at time t. His motivation is based on the idea that easily analyzed problems in shell theory will involve such uniform states.

δ) *Special theories.* The theory of a Cosserat surface summarized above [Sect. 14, Subsect. α)] includes, or is more general than, a number of special or restrictive theories of shells. Notable among these is the restricted theory whose developments began in Sect. 10 and will be completed in the next section. Here, we discuss two well-known special theories or special cases of the general theory, namely the *membrane theory* and the *inextensional theory*.

---

[24] CROCHET and NAGHDI [1969, *1*].

[25] A detailed account of such exact solutions in the three-dimensional theory can be found in GREEN and ADKINS [1960, *5*] and in TRUESDELL and NOLL [1965, *9*].

[26] General results concerning superposed small deformations on a large deformation of an elastic Cosserat surface are contained in a paper by GREEN and NAGHDI [1971, *5*].

[27] ERICKSEN [1970, *1*].

The membrane theory of shells can be obtained by returning to the general principles of Sect. 8 and omitting the vector $\boldsymbol{d}$. Alternatively, the membrane theory follows from the general theory of an elastic Cosserat surface if we set $\bar{\boldsymbol{l}}=0$ and assume that the free energy function in (13.38) is independent of $\boldsymbol{d}$ and $\boldsymbol{d}_{,\gamma}$. Then, $\boldsymbol{M}^\alpha$ and $\boldsymbol{m}$ vanish by (13.23) and from (9.51) and (9.53) we obtain

$$N^{\alpha\beta}=N^{\beta\alpha}=N'^{\alpha\beta}, \qquad N^{\alpha 3}=0. \tag{14.10}$$

Hence, in the membrane theory, $N^{\alpha\beta}$ is a symmetric tensor and the equations of motion (9.47) reduce to

$$N^{\alpha\beta}{}_{|\alpha}+\varrho f^\beta=\varrho c^\beta, \qquad b_{\alpha\beta} N^{\alpha\beta}+\varrho f^3=\varrho c^3, \qquad \varepsilon_{\beta\alpha} N^{\alpha\beta}=0. \tag{14.11}$$

The energy equation and the entropy inequality are still of the forms (9.35)–(9.36) but with $P=N^{\alpha\beta}\eta_{\alpha\beta}$. For completeness we record below the constitutive equations for an elastic membrane which follow from (13.62)–(13.66):

$$\psi=\hat{\psi}(e_{\alpha\beta}, \theta; A_{\alpha\beta}), \qquad \eta=-\frac{\partial\hat{\psi}}{\partial\theta}, \qquad N^{\alpha\beta}=N^{\beta\alpha}=\varrho\frac{\partial\hat{\psi}}{\partial e_{\alpha\beta}} \tag{14.12}$$

and

$$q^\alpha=\hat{q}^\alpha(e_{\gamma\delta}, \theta, \theta_{,\gamma}; A_{\gamma\delta}), \tag{14.13}$$

where the response functions $\hat{\psi}$, $\hat{q}^\alpha$ are now different functions from those in (13.62) and (13.66) and $\hat{q}^\alpha$ must satisfy an inequality of the form (13.19). Alternatively, the constitutive equations for an elastic membrane may be obtained from specialization of (13.38) and (13.40)–(13.43), in terms of the kinematic measure $a_{\alpha\beta}$ rather than $e_{\alpha\beta}$, but we do not record these here. The material symmetry restrictions discussed in Sect. 13 [Subsects. γ) and δ)] with an obvious modification (in the absence of $\boldsymbol{d}$, $\boldsymbol{d}_{,\gamma}$) apply also to the response functions in (14.12)–(14.13). In this connection, we observe that the appropriate material symmetries could be used in application to thin shells reinforced with cords.[28]

It is perhaps worth observing here that such phenomenon as *surface tension* can be described as a special case of the foregoing membrane theory. The classical theory of surface tension may be summarized by the statement that the total free energy is proportional to the surface area. In the context of the three-dimensional theory, for a *thin* film, the reference (or middle) surface has about the same area as the lower and upper surfaces so that adding their contributions results in an energy proportional to the area of the reference surface. Within the framework of the above membrane theory, the free energy can contain such a contribution. Indeed, in view of (4.40), if the response function $\tilde{\psi}$ is assumed to depend on $a^{\frac{1}{2}}$ only or equivalently if $\psi=\tilde{\psi}(a^{\frac{1}{2}})$, then the expressions for $N^{\alpha\beta}$ appropriate to surface tension can be calculated from $(14.10)_1$ and (13.41). In the special case that $\tilde{\psi}$ is quadratic in $a^{\frac{1}{2}}$; i.e., $\tilde{\psi}=\beta a$, where β is a constant, the results in lines of curvature coordinates will correspond to those usually given in the context of the classical theory of surface tension.

We consider now a special theory, called the inextensional theory, wherein the length of each element of the surface $\mathscr{C}$ is assumed to remain constant throughout all motions. This requirement can be expressed by the three equations

$$a_{\alpha\beta}=A_{\alpha\beta}=\text{const.}, \tag{14.14}$$

[28] In the context of the purely mechanical theory, RIVLIN [1959, *4*] has discussed the deformation of a net of inextensible cords regarded as a membrane of fabric with two directions of inextensibility at each point of the surface. Related problems of sheets reinforced with perfectly flexible but inextensible cords can be found in Chap. 7 of GREEN and ADKINS [1960, *5*].

which by (5.31) is equivalent to $e_{\alpha\beta}=0$. The condition (14.14) implies that the surface area is "preserved" for all times and it follows that $\varrho=\varrho_0$, in view of the continuity equation (4.42).[29] It is clear that (14.14) is a restriction on the class of possible motions and hence represents an internal constraint. We can therefore deduce the inextensional theory from the general theory of a Cosserat surface only by the introduction of additional assumptions appropriate to a theory with internal constraints. Although we do not consider here a general development of internal constraints for shells, the development of the theory in the presence of mechanical constraints of the type (14.14) is fairly straightforward and may be patterned after analogous treatments of internal constraints in the three-dimensional (non-polar) continuum mechanics.[30]

From differentiation of (14.14) with respect to t follow the three constraint equations $\dot{a}_{\alpha\beta}=0$. These constraint equations remain unaltered under superposed rigid body motions, and by (5.14) may be written as

$$\eta_{\alpha\beta}=0. \tag{14.15}$$

From the energy equation (9.29) and the conditions (14.15), it is at once apparent that $\boldsymbol{N}'^{\alpha}$ can no longer be determined entirely by constitutive equations in the sense that a part of $\boldsymbol{N}'^{\alpha}$ which occurs in

$$\boldsymbol{N}'^{\alpha}\cdot\boldsymbol{\eta}_{\alpha}=(\boldsymbol{N}'^{\alpha}\cdot\boldsymbol{a}^{\beta})\,\eta_{\alpha\beta}$$

is *workless*. It is evident that the indeterminate part of $\boldsymbol{N}'^{\alpha}$, i.e., the part not specified by constitutive equations, pertains to the tangential components $N'^{\alpha\beta}$ defined by (9.31) and not to the normal component $\boldsymbol{N}'^{\alpha}\cdot\boldsymbol{a}^3$. The indeterminate part of $\boldsymbol{N}'^{\alpha}$ may be thought of as the "forces required to maintain the constraint" and for brevity may be referred to as the *constraint response*.[31]

Keeping the above in mind, in the development of the inextensional theory under consideration, we make the following assumptions: (i) The vector functions $\boldsymbol{N}'^{\alpha}$ are determined to within an additive constraint response in the form

$$\boldsymbol{N}'^{\alpha}=\mathcal{N}^{\alpha}+{}_E\boldsymbol{N}'^{\alpha}, \tag{14.16}$$

where ${}_E\boldsymbol{N}'^{\alpha}$ are specified by constitutive equations, while the constraint response $\mathcal{N}^{\alpha}$ are functions of position and time but are independent of the rate quantities (13.11); (ii) the indeterminate forces $\mathcal{N}^{\alpha}$ are workless in all motions consistent with (14.15), i.e.,[32]

$$\mathcal{N}^{\alpha}\cdot\boldsymbol{\eta}_{\alpha}=0, \tag{14.17}$$

where $\boldsymbol{\eta}_{\alpha}$ is defined by (5.19) and (5.15). Now by any of a number of standard procedures, from (14.15) and (14.17) we may conclude that $\mathcal{N}^{\alpha}\cdot\boldsymbol{a}^{\beta}=\lambda^{\alpha\beta}$ or equivalently

$$\mathcal{N}^{\alpha}=\lambda^{\alpha\beta}\,\boldsymbol{a}_{\beta},\qquad \lambda^{\alpha\beta}=\lambda^{\beta\alpha}, \tag{14.18}$$

---

[29] The inextensional deformation specified by (14.14) implies $\varrho=$ const (and hence "surface incompressibility"), but the converse is not true.

[30] For an account of mechanical constraints in the three-dimensional theory, see Sect. 30 of TRUESDELL and NOLL [1965, *9*]. A thermodynamical theory of a continuum in the presence of thermo-mechanical constraints has been given recently by GREEN, NAGHDI and TRAPP [1970, *3*]. A development of the inextensional theory by direct approach, based on the theory of a Cosserat surface subject to the conditions (14.14), is contained in a paper by CROCHET [1971, *3*].

[31] This terminology is used in [1970, *3*].

[32] In the context of the thermodynamical treatment of constraints in [1970, *3*], the requirement (14.17) is equivalent to the assumption that the local production of entropy due to $\mathcal{N}^{\alpha}$ is zero.

where $\lambda^{\alpha\beta}$ is an arbitrary symmetric tensor function of position and time.[33] It may be noted that since only the tangential components of $\boldsymbol{N}'^{\alpha}$ are affected by the constraint conditions (14.15), in the case of the inextensional theory we could absorb the tangential components of ${}_E\boldsymbol{N}'^{\alpha}$ into $\mathcal{N}^{\alpha}$ and replace (14.16) by $N'^{\alpha\beta} = N'^{\beta\alpha} = \mathcal{N}^{\alpha\beta}$. In this way we would again arrive at the conclusion $\mathcal{N}^{\alpha\beta} = \lambda^{\alpha\beta}$.

We summarize now the field equations and the constitutive equations for an inextensible Cosserat surface. The symmetric tensor $\lambda^{\alpha\beta}$ in (14.18) is governed by the equations of motion (9.13) or equivalently[34] $(9.47)_{1,2}$. The equations of motion of the inextensional theory are given by (9.48) and the components $N^{[\alpha\beta]}$ and $N^{\alpha 3}$ are still determined from (9.51). The energy equation and the entropy inequality are still of the forms (9.35)–(9.36) but instead of P in $(9.29)_2$ or (9.52) we now have

$$\begin{aligned} P &= \boldsymbol{m}\cdot\boldsymbol{\Gamma} + \boldsymbol{M}^{\alpha}\cdot\boldsymbol{\Gamma}_{:\alpha} \\ &= m^i\,\dot{d}_i + M^{\alpha i}\,\dot{\lambda}_{i\alpha}. \end{aligned} \tag{14.19}$$

The developments of the constitutive equations for an elastic inextensional Cosserat surface now parallels that given in Sect. 13. Thus, except for $\boldsymbol{N}^{\alpha}$, we may begin by introducing constitutive assumptions similar to those in (13.8) in terms of response functions which, aside from their dependence on the material point θ^{μ}, are functions of

$$\boldsymbol{d},\ \boldsymbol{d}_{,\gamma},\ \theta \tag{14.20}$$

and the temperature gradient $\theta_{,\alpha}$. Moreover, because of the constraint conditions (14.15), instead of the inequality (13.14) we now have

$$-\varrho\left(\bar{\eta} + \frac{\partial\bar{\psi}}{\partial\theta}\right)\dot{\theta} + \left(\overline{\boldsymbol{m}} - \varrho\frac{\partial\bar{\psi}}{\partial\boldsymbol{d}}\right)\cdot\boldsymbol{\Gamma} + \left(\overline{\boldsymbol{M}}^{\gamma} - \varrho\frac{\partial\bar{\psi}}{\partial\boldsymbol{d}_{,\gamma}}\right)\cdot\boldsymbol{\Gamma}_{:\gamma} - (\bar{\boldsymbol{q}}\cdot\boldsymbol{a}^{\alpha})\frac{\theta_{,\alpha}}{\theta} \geqq 0, \tag{14.21}$$

where $\bar{\psi}$ is a different function from that in (13.14) and depends on the variables (14.20). The rest of the developments, with obvious modifications, are entirely similar to those in Sect. 13.[35]

15. The complete restricted theory: I. Direct approach. Previously, the field equations of the restricted theory in which the motion is characterized by $(5.1)_1$ alone were derived in Sect. 10. Here we complete this theory for an elastic surface in a manner similar to the more general developments of Sect. 13. Preliminary to the introduction of the constitutive assumption appropriate to the restricted theory, we observe that under superposed rigid body motions the results (13.1) still hold but instead of (13.2) we now have the transformations

$$\check{\boldsymbol{M}}^{\alpha} \to Q\,\check{\boldsymbol{M}}^{\alpha}, \qquad \check{\boldsymbol{N}}^{\alpha} \to Q\,\check{\boldsymbol{N}}^{\alpha}, \qquad \boldsymbol{q} \to Q\,\boldsymbol{q} \tag{15.1}$$

under superposed rigid body motions. It follows that the transformation (13.4) still holds but those in (13.3) are now replaced by

$$\check{M}^{\alpha\beta} \to \check{M}^{\alpha\beta}, \qquad \check{N}^{\alpha\beta} \to \check{N}^{\alpha\beta}, \qquad q^{\alpha} \to q^{\alpha}, \tag{15.2}$$

where $\check{N}^{\alpha\beta}$ and $\check{M}^{\alpha\beta}$ are defined by (10.26) and $(10.20)_3$.

---

[33] It is clear that $\lambda^{\alpha\beta}$ play the role of Lagrange multipliers.

[34] These equations may not always uniquely determine $\lambda^{\alpha\beta}$ throughout the surface of $\mathscr{C}$.

[35] Inasmuch as the variables $\boldsymbol{a}_{\alpha}$ are absent from the argument of the response functions, the inequality (14.21) is unaffected by the constraint equations (14.15). In the presence of more general constraints of the type $\boldsymbol{\gamma}^{\alpha}\cdot\dot{\boldsymbol{a}}_{\beta} = 0$, where $\boldsymbol{\gamma}^{\alpha}$ are vector functions of $\boldsymbol{a}_{\nu}$ and the variables (14.20), the inequality corresponding to (14.21) will be subject to the constraint equations and this must also be taken into account when obtaining results similar to (13.18). See, in this connection, Sect. 4 of [1970, *3*] for a discussion of constitutive equations for an elastic continuum subject to a general thermo-mechanical constraint.

Recalling the general remarks concerning constitutive equations at the beginning of Sect. 13 [between (13.4) and (13.8)], within the scope of the restricted theory, we define an elastic surface by the following five response functions:

$$\overset{\backprime}{\boldsymbol{N}}^{\alpha}\,(\text{or } \bar{\boldsymbol{N}}^{\alpha}),\ \overset{\backprime}{\boldsymbol{M}}^{\alpha},\ \bar{\psi},\ \bar{\eta},\ \bar{q}^{\alpha}. \tag{15.3}$$

We introduce constitutive equations, in terms of the response functions (15.3), at each material point θ^{μ} of the surface in the present configuration; and we assume that the response functions depend on the kinematic variables

$$\boldsymbol{a}_{\alpha},\ \boldsymbol{a}_{3,\gamma}, \tag{15.4}$$

as well as the temperature and the temperature gradient.[36] For example, the constitutive equation for the specific Helmholtz free energy will have the form

$$\psi = \bar{\psi}(\boldsymbol{a}_{\alpha}, \boldsymbol{a}_{3,\gamma}, \theta, \theta_{,\alpha}; \theta^{\mu}) \tag{15.5}$$

with similar constitutive equations for $\overset{\backprime}{\boldsymbol{N}}^{\alpha}$, $\overset{\backprime}{\boldsymbol{M}}^{\alpha}$, η and q^{α} in terms of the response functions in (15.3). These constitutive equations are assumed to hold at each material point θ^{μ} and for all times t.

From combination of the entropy inequality (9.32) and the energy equation $(10.24)_1$ in terms of ψ in (9.34), follows the inequality

$$-\varrho\,\dot{\psi} - \varrho\,\eta\,\dot{\theta} + \overset{\backprime}{P} - q^{\alpha}\frac{\theta_{,\alpha}}{\theta} \geqq 0, \tag{15.6}$$

which must hold for every admissible process and where $\overset{\backprime}{P}$ is given by $(10.24)_2$. After introducing the above constitutive assumptions for the restricted theory in (15.6) and following the procedure of Sect. 13, we can show that $\bar{\psi}$ in (15.5) must be independent of $\theta_{,\alpha}$. Hence, ψ can be expressed in terms of a (different) response function of θ and the variables (15.4):

$$\psi = \bar{\psi}(\boldsymbol{a}_{\alpha}, \boldsymbol{a}_{3,\gamma}, \theta; \theta^{\mu}). \tag{15.7}$$

Moreover, the inequality (15.6) can then be written in the form

$$\begin{aligned} -\varrho\left(\bar{\eta} + \frac{\partial\bar{\psi}}{\partial\theta}\right)\dot{\theta} + \left[\overset{\backprime}{\boldsymbol{N}}^{\alpha} - \varrho\left(\frac{\partial\bar{\psi}}{\partial\boldsymbol{a}_{\alpha}} + b_{\gamma}^{\alpha}\frac{\partial\bar{\psi}}{\partial\boldsymbol{a}_{3,\gamma}}\right)\right]\cdot\boldsymbol{\eta}_{\alpha} \\ + \left[\overset{\backprime}{\boldsymbol{M}}^{\gamma} - \varrho\frac{\partial\bar{\psi}}{\partial\boldsymbol{a}_{3,\gamma}}\right]\cdot\dot{\lambda}_{\alpha\gamma}\,\boldsymbol{a}^{\alpha} - \bar{q}^{\alpha}\frac{\theta_{,\alpha}}{\theta} \geqq 0, \end{aligned} \tag{15.8}$$

where the function $\bar{\psi}$ in (15.7) must satisfy

$$\varepsilon_{\gamma\alpha}\,\boldsymbol{a}^{\gamma}\cdot\left[\frac{\partial\bar{\psi}}{\partial\boldsymbol{a}_{\alpha}} - b_{\lambda}^{\alpha}\frac{\partial\bar{\psi}}{\partial\boldsymbol{a}_{3,\lambda}}\right] = 0 \tag{15.9}$$

and $\lambda_{\alpha\gamma}$ is defined by $(5.61)_3$. The condition (15.9) arises from the fact that the inequality (15.6) must remain unaffected by superposed rigid body motions.[37] Using arguments similar to those in Sect. 13 [Subsect. α)] and recalling our

[36] Apart from the temperature and its gradient, the variables (15.4) define the local state in the restricted theory. They may be contrasted with the kinematic variables (13.7) which define the local state in the theory of an elastic Cosserat surface. In fact, since the unit normal to a surface is determined by the surface base vectors, the variables (15.4) representing the local state in the restricted theory follow from (13.7) with $\boldsymbol{d} = \boldsymbol{a}_3$.

[37] The inequality (15.8) and the condition (15.9) are, respectively, the counterparts of (13.14) and (13.15) in the restricted theory.

constitutive assumptions, from (15.8) we deduce the relations

$$\eta = -\frac{\partial \overline{\psi}}{\partial \theta}, \tag{15.10}$$

$$\begin{aligned} \overset{\backprime}{N}^{\alpha\beta} = \overset{\backprime}{N}^{\beta\alpha} &= \frac{1}{2}\varrho\left\{\left(\frac{\partial\overline{\psi}}{\partial \boldsymbol{a}_\alpha} + b^\alpha_\gamma \frac{\partial\overline{\psi}}{\partial \boldsymbol{a}_{3,\gamma}}\right)\cdot \boldsymbol{a}^\beta + \left(\frac{\partial\overline{\psi}}{\partial \boldsymbol{a}_\beta} + b^\beta_\gamma \frac{\partial\overline{\psi}}{\partial \boldsymbol{a}_{3,\gamma}}\right)\cdot \boldsymbol{a}^\alpha\right\}, \\ \overset{\backprime}{M}^{(\gamma\alpha)} &= \frac{1}{2}\varrho\left\{\frac{\partial\overline{\psi}}{\partial \boldsymbol{a}_{3,\gamma}}\cdot \boldsymbol{a}^\alpha + \frac{\partial\overline{\psi}}{\partial \boldsymbol{a}_{3,\alpha}}\cdot \boldsymbol{a}^\gamma\right\} \end{aligned} \tag{15.11}$$

and an inequality in the form (13.19) but with $\overline{q}^\alpha$ now a function of θ, $\theta_{,\alpha}$ and the variables (15.4).

Next, we consider the reduction of (15.7) under superposed rigid body motions. As in Sect. 13 [Subsect. β)], we observe that a constitutive equation in the form (15.7) which holds for an admissible process must also hold for a motion which differs from $(5.1)_1$ only by superposed rigid body motions. This requirement is fulfilled if and only if a response function such as $\overline{\psi}$ satisfies the identity

$$\overline{\psi}(\boldsymbol{a}_\alpha, \boldsymbol{a}_{3,\gamma}, \theta; \theta^\mu) = \overline{\psi}(Q\,\boldsymbol{a}_\alpha, Q\,\boldsymbol{a}_{3,\gamma}, \theta; \theta^\mu), \tag{15.12}$$

for all proper orthogonal Q. An identity similar to (15.12) holds for $\overline{q}^\alpha$. We now use Cauchy's representation theorem for isotropic functions and express $\overline{\psi}$ as a (different) function in the form

$$\psi = \tilde{\psi}(a_{\alpha\beta}, \overset{\backprime}{\lambda}_{\alpha\beta}, \theta; \theta^\mu), \tag{15.13}$$

where $(4.12)_1$, $(4.13)_1$ and the definition $(5.61)_3$ for $\overset{\backprime}{\lambda}_{\alpha\beta}$ have been used. With the help of (15.7) and (15.13) and using the chain rule for differentiation, the constitutive equations (15.10)–(15.11) can be reduced to[38]

$$\eta = -\frac{\partial\tilde{\psi}}{\partial\theta}, \tag{15.14}$$

$$\overset{\backprime}{N}^{\alpha\beta} = \overset{\backprime}{N}^{\beta\alpha} = 2\varrho\frac{\partial\tilde{\psi}}{\partial a_{\alpha\beta}}, \qquad \overset{\backprime}{M}^{(\alpha\gamma)} = \varrho\frac{\partial\tilde{\psi}}{\partial\overset{\backprime}{\lambda}_{\alpha\beta}}. \tag{15.15}$$

Also, the constitutive equation for the heat flux can be expressed as

$$q^\alpha = \tilde{q}^\alpha(a_{\gamma\delta}, \overset{\backprime}{\lambda}_{\gamma\delta}, \theta, \theta_{,\gamma}; \theta^\mu). \tag{15.16}$$

The discussion of material symmetry restrictions and other considerations for constitutive equations can be pursued in a manner similar to that in Sect. 13. We can also write the constitutive equations in terms of relative kinematic measures and the reference values (13.72) but we do not record these.

The components $N^{\alpha 3}$ are absent from the energy equation and also from the inequality (15.6), as was noted in Sect. 10. Hence, the components $N^{\alpha 3}$ cannot be specified by constitutive equations and may be determined from the equations of motion $(10.21)_1$. Moreover, as is evident from (15.15), the restricted theory provides constitutive equations for only the symmetric $\overset{\backprime}{N}^{\alpha\beta}$ and the symmetric part of $\overset{\backprime}{M}^{\alpha\beta}$ or equivalently for $N^{(\alpha\beta)}$ and $\overset{\backprime}{M}^{(\alpha\beta)}$. The skew-symmetric part of $N^{\alpha\beta}$ is determined from $(10.21)_2$ but the latter also involves $\overset{\backprime}{M}^{\alpha\beta}$. Thus, in order to obtain a determinate theory, we now assume that

$$\overset{\backprime}{M}^{[\alpha\beta]} = 0. \tag{15.17}$$

[38] Once more we recall that in evaluating partial derivatives with respect to symmetric tensors such as $\partial\tilde{\psi}/\partial\overset{\backprime}{\lambda}_{\alpha\beta}$, the tensor $\overset{\backprime}{\lambda}_{\alpha\beta}$ is understood to stand for $\frac{1}{2}(\overset{\backprime}{\lambda}_{\alpha\beta} + \overset{\backprime}{\lambda}_{\beta\alpha})$.

Then, from $(10.21)_2$, the skew-symmetric part of $N^{\alpha\beta}$ is given by

$$N^{[\alpha\beta]} = \tfrac{1}{2}\{b^\alpha_\gamma \grave{M}^{(\gamma\beta)} - b^\beta_\gamma \grave{M}^{(\gamma\alpha)}\}. \tag{15.18}$$

Further, with the help of (15.18), the differential equations of motion (10.22) can be written in the alternative form

$$\begin{aligned} N^{(\alpha\beta)}{}_{|\alpha} - \tfrac{1}{2}[b^\beta_\gamma \grave{M}^{(\gamma\alpha)}]_{|\alpha} + \tfrac{1}{2}[b^\alpha_\gamma \grave{M}^{(\gamma\beta)}]_{|\alpha} - b^\beta_\alpha \grave{M}^{(\gamma\alpha)}{}_{|\gamma} + \varrho\, \bar{f}^\beta &= 0, \\ \grave{M}^{(\alpha\beta)}{}_{|\alpha\beta} + b_{\alpha\beta}\, N^{(\alpha\beta)} + \varrho\, \bar{f}^3 &= 0. \end{aligned} \tag{15.19}$$

The above differential equations involve only the symmetric $N^{(\alpha\beta)}$ and $\grave{M}^{(\alpha\beta)}$. They can also be expressed in terms of $\tilde{N}^{\alpha\beta}$ and $\grave{M}^{(\alpha\beta)}$ as follows:

$$\begin{aligned} \tilde{N}^{\alpha\beta}{}_{|\alpha} - b^\beta_{\gamma|\alpha}\, \grave{M}^{(\gamma\alpha)} - 2 b^\beta_\alpha \grave{M}^{(\gamma\alpha)}{}_{|\gamma} + \varrho\, \bar{f}^\beta &= 0, \\ \grave{M}^{(\alpha\beta)}{}_{|\alpha\beta} + b_{\alpha\beta}\, \tilde{N}^{\alpha\beta} - b_{\alpha\beta}\, b^\beta_\gamma \grave{M}^{(\gamma\alpha)} + \varrho\, \bar{f}^3 &= 0. \end{aligned} \tag{15.20}$$

To summarize, the basic field equations and the constitutive relations for an elastic surface in the restricted theory consist of the equations of motion (9.47) and (10.21) together with (15.17) [or any of the alternative sets recorded above], the energy equation (10.24) and the constitutive equations (15.13)–(15.16). Also the response function $\tilde{q}^\alpha$ must satisfy an inequality of the form (13.19).[39] It remains to consider the nature of the boundary conditions in the restricted theory. Recalling the expression for the rate of work of contact force and couple in (10.3) and remembering $(10.20)_3$, (15.17), $(5.61)_1$ and (5.24), we have

$$\begin{aligned} \grave{R}_c(\mathscr{P}) &= \int_{\partial\mathscr{P}} (\boldsymbol{N}\cdot\boldsymbol{v} + \grave{\boldsymbol{M}}\cdot\grave{\boldsymbol{w}})\, ds \\ &= \int_{\partial\mathscr{P}} \nu_\alpha \{N^{\alpha i}\, v_i - \grave{M}^{(\alpha\gamma)}\, b^\beta_\gamma\, v_\beta - \grave{M}^{(\alpha\gamma)}\, v_{3,\gamma}\}\, ds. \end{aligned} \tag{15.21}$$

Let $\partial/\partial\nu$ stand for the directional derivative along the unit normal $\boldsymbol{\nu}$ to the boundary curve c of the surface $\mathscr{s}$ in the present configuration. Then

$$v_{3,\gamma} = \nu_\gamma \frac{\partial v_3}{\partial \nu} - \varepsilon_{\gamma\alpha}\, \nu^\alpha \frac{\partial v_3}{\partial s}, \tag{15.22}$$

where $\partial/\partial s$ is the directional derivative along the tangent to c introduced in (8.1). Provided the quantities in (15.21) are single-valued on a (sufficiently smooth) closed curve c, with the use of (15.22) and an integration by parts, (15.21) can be reduced to

$$R_c(\mathscr{P}) = \int_{\partial\mathscr{P}} \left\{P^\beta_{\scriptscriptstyle\blacksquare}\, v_\beta + P^3\, v_3 - G \frac{\partial v_3}{\partial \nu}\right\} ds, \tag{15.23}$$

---

[39] These results, including the constitutive relations of the restricted theory for an elastic surface, can be brought into correspondence with those of the restricted theory of simple force dipole contained in the paper of BALABAN, GREEN and NAGHDI [1967, *1*] and referred to earlier in Sect. 10.

In the context of the purely mechanical theory, there exist in the literature derivations (with various degrees of approximation) from the three-dimensional equations which are aimed at the types of results supplied by the restricted theory. Such developments from the three-dimensional theory which employ special or restrictive kinematic assumptions (e.g., NAGHDI and NORDGREN [1963, *8*]) are necessarily approximate in character; and although they often contain formulae similar in form to those of the restricted theory [including those corresponding to (15.15) with ψ regarded as the strain energy density], they should not be confused with the latter. The essential point here to be born in mind is that although the two sets of formulae are similar, the constitutive coefficients in (15.15) are arbitrary and are not predetermined from an (approximate) expression of the three-dimensional specific free energy or strain energy density.

where we have put

$$P^{\beta} = \nu_{\alpha}[N^{\alpha\beta} - \tilde{M}^{(\alpha\gamma)} b^{\beta}_{\gamma}], \qquad G = \tilde{M}^{(\alpha\gamma)} \nu_{\alpha} \nu_{\gamma}, \tag{15.24}$$

$$P^3 = \nu_{\alpha}[\tilde{M}^{(\beta\alpha)}{}_{|\beta} + \varrho\, \tilde{l}^{\alpha}] - \frac{\partial}{\partial s}[\varepsilon_{\beta\gamma} \tilde{M}^{(\alpha\beta)} \nu_{\alpha} \nu^{\gamma}], \tag{15.25}$$

and in obtaining (15.25) use has been made of $(10.21)_1$. The nature of the boundary conditions in the restricted theory is now clear from (15.23). In particular, the modified force and couple boundary conditions which hold pointwise on c are given by (15.24)–(15.25).[40]

As remarked earlier, the above restricted theory bears on the classical theory of shells.[41] This will become apparent in Sect. 25, where the complete restricted linear theory is discussed. We note here that the membrane theory follows also from the above restricted theory by setting $\tilde{l}^{\alpha} = 0$ and assuming that $\tilde{\psi}$ in (5.13) is independent of $\grave{\lambda}_{\alpha\beta}$. Similarly, by introducing additional assumptions, an inextensional theory—less general than that given in Sect. 14 [Subsect. δ)]—can be derived from the restricted theory subject to the constraint equations (14.15).

16. Linear constitutive equations: I. Direct approach. Previously in Sects. 6 and 9, we considered linearization of the kinematic variables and the field equations in a systematic manner. Here, we complete the linearization procedure in a complete theory of an elastic Cosserat surface and obtain explicit expressions for the constitutive equations.

α) *General considerations.* We begin our general considerations of constitutive equations of the linear theory with the thermodynamical results (13.62)–(13.66). However, as in $(13.70)_1$, we take the initial director $\boldsymbol{D}$ to be along the normal to the reference surface $\mathscr{S}$ so that the set of variables $\mathscr{U}_R$ are still given by (13.61) but with values D_i and $\Lambda_{i\alpha}$ in (13.70).

Since the kinematic variables γ_3, $\varkappa_{3\alpha}$ and their reference values occur in the arguments of the response functions according to the last two variables in each of (13.60)–(13.61), in addition to the linear kinematic variables of Sect. 6, we need also the expressions resulting from linearization of (13.58). Thus, by the procedure of Sect. 6 and recalling $(6.8)_2$, $(6.9)_2$ and (6.13), we readily find

$$\begin{aligned} s &= 2\varepsilon(D^3\gamma_3) + O(\varepsilon^2) = O(\varepsilon), \\ s_{\alpha} &= \varepsilon(D^3\varkappa_{3\alpha} + \Lambda^3_{.\alpha}\gamma_3) + O(\varepsilon^2) = O(\varepsilon). \end{aligned} \tag{16.1}$$

Recalling the earlier remarks concerning the linear theory [Sect. 9, Subsect. γ)], the response function $\hat{\psi}$ in (13.62) now becomes a quadratic function of the appropriate kinematic variables either in the forms (6.22) or (6.23) together with

$$s = 2D^3\gamma_3, \qquad s_{\alpha} = D^3\varkappa_{3\alpha} + \Lambda^3_{.\alpha}\gamma_3, \tag{16.2}$$

which are obtained from (16.1) in accordance with the linearization procedure of Sect. 6. Moreover, since the left-hand side of each constitutive equation in (13.63)–(13.66) is now of $O(\varepsilon)$, it follows that on the right-hand sides ϱ, d^3, $\lambda^3_{.\alpha}$ must be replaced to the order ε by ϱ_0, D^3, $\Lambda^3_{.\alpha}$, respectively. It is easily seen that the linearized constitutive equations for η, $N'^{\alpha\beta}$, m^{α}, $M^{\alpha\gamma}$ will have the same

[40] These results were given in [1968, *7*]. They correspond to similar formulae obtained in the literature on the classical linear theory of shells; see, e.g., [1963, *7*].

[41] A special case of the theory of Cosserat surface resulting in a system of equations analogous to that of the restricted theory has been discussed by Green and Naghdi [1968, *5*]. The constitutive equations used in [1968, *5*] are, however, those in the form (13.80)–(13.82).

forms as those in (13.63)–(13.65) but the linear constitutive equations for m^3 and $M^{\alpha 3}$ are

$$m^3 = \varrho_0 \left(2D^3 \frac{\partial \hat{\psi}}{\partial s} + \Lambda^3_{\cdot\alpha} \frac{\partial \hat{\psi}}{\partial s_\alpha}\right), \qquad M^{\alpha 3} = \varrho_0 D^3 \frac{\partial \hat{\psi}}{\partial s_\alpha}, \tag{16.3}$$

where $\hat{\psi}$ is a different function from that in (13.62) and is quadratic in the arguments $e_{\alpha\gamma}$, γ_α, $\varkappa_{\gamma\alpha}$, s, s_α and θ.

The foregoing results are obtained from (13.62)–(13.65) for a Cosserat surface whose (initial) reference director is specified by $(13.70)_1$. Consider now a Cosserat surface with its (initial) reference director coincident with the unit normal to $\mathscr{S}$ given by $(13.71)_1$. For a Cosserat surface $\mathscr{C}$ so specified in its reference configuration, in view of $(13.71)_{3,5}$, the linearized expressions (16.2) to $O(\varepsilon)$ become

$$s = 2\gamma_3, \qquad s_\alpha = \varkappa_{3\alpha}. \tag{16.4}$$

Then, the right-hand sides of $(16.3)_{1,2}$ reduce to

$$\varrho_0 \frac{\partial \hat{\psi}}{\partial \gamma_3} \quad \text{and} \quad \varrho_0 \frac{\partial \hat{\psi}}{\partial \varkappa_{3\alpha}},$$

respectively; and the constitutive equations of the linear theory may be expressed as[42]

$$\psi = \psi(e_{\alpha\beta}, \gamma_i, \varkappa_{i\alpha}, \theta; \mathscr{U}'_R) \tag{16.5}$$

and

$$\eta = -\frac{\partial \psi}{\partial \theta}, \tag{16.6}$$

$$N'^{\alpha\beta} = N'^{\beta\alpha} = \varrho_0 \frac{\partial \psi}{\partial e_{\alpha\beta}}, \qquad m^i = \varrho_0 \frac{\partial \psi}{\partial \gamma_i}, \qquad M^{\alpha i} = \varrho_0 \frac{\partial \psi}{\partial \varkappa_{i\alpha}}. \tag{16.7}$$

In (16.5)–(16.7) θ and η are now the temperature difference and the entropy difference, each of $O(\varepsilon)$ measured from a standard temperature θ_0 and entropy η_0 in the initial undeformed surface; and the response function ψ is quadratic in the temperature θ and the kinematic variables $e_{\alpha\beta}$, γ_i, $\varkappa_{i\alpha}$. Also, instead of (13.66), we now have a constitutive equation for the linearized heat flux Q^α defined by $(9.59)_2$. The response function for the heat flux, apart from its dependence on $\mathscr{U}'_R$, is now a linear function of the remaining arguments in (16.5) in addition to $\theta_{,\alpha}$. Moreover, instead of (13.19), in the linear theory we have the inequality

$$-Q^\alpha \theta_{,\alpha} \geqq 0, \tag{16.8}$$

which places restriction on the response function Q^α. It then follows that Q^α must be linear in the temperature gradient and can depend in addition only on the reference values $\mathscr{U}'_R$. We defer recording an explicit constitutive equation for Q^α but note here that the residual energy equation in the linear theory becomes

$$\varrho_0 r - \varrho_0 \theta_0 \dot{\eta} - Q^\alpha{}_{|\alpha} = 0, \tag{16.9}$$

where the vertical bar denotes covariant differentiation with respect to $A_{\alpha\beta}$. The above residual energy equation is obtained from $(9.60)_1$ with the help of (16.5)–(16.7). Each term in (16.9) is of $O(\varepsilon)$ consistent with our linearization procedure and in accord with the remarks made in Sect. 9 [Subsect. γ)].

In the remainder of this section and in our further considerations of the linear theory by direct approach, we confine attention to an elastic Cosserat surface

[42] Here and in most of the developments that follow we return to our earlier notation and often employ the same symbol for a function and its value without confusion.

with its (initial) reference director specified by $(13.71)_1$. The constitutive relations are then given by (16.5)–(16.7), apart from an equation for heat flux.[43] Moreover, we observe here that the results (16.5)–(16.7) can also be deduced by linearization of (13.80)–(13.82) with the reference values (13.71). It is rather striking that the distinction between the two sets of constitutive equations (13.62)–(13.65) and (13.80)–(13.82) disappears for the linear theory and with $\boldsymbol{D}$ specified by $(13.71)_1$.

If we recall the energy equation $(9.60)_1$ with P in the form (9.71), it becomes evident that thermodynamical results corresponding to (16.7) can be deduced also for $N'^{\alpha\beta}$, $M^{\alpha i}$ and V^i. Thus, in terms of the kinematic variables $e_{\alpha\beta}, \varrho_{i\alpha}$ and γ_i, the specific free energy in the linear theory can alternatively be expressed as[44]

$$\psi = \overline{\psi}(e_{\alpha\beta}, \gamma_i, \varrho_{i\alpha}, \theta; \mathscr{U}'_R) \tag{16.10}$$

and we can deduce the results

$$N'^{\alpha\beta} = N'^{\beta\alpha} = \varrho_0 \frac{\partial \overline{\psi}}{\partial e_{\alpha\beta}}, \qquad V^i = \varrho_0 \frac{\partial \overline{\psi}}{\partial \gamma_i}, \qquad M^{\alpha i} = \varrho_0 \frac{\partial \overline{\psi}}{\partial \varrho_{i\alpha}} \tag{16.11}$$

in place of (16.7). The relation for entropy is still of the form (16.6) but with the function ψ replaced by $\overline{\psi}$.

The constitutive equations of the linear theory may also be expressed in terms of the specific internal energy. Such results can be deduced in a manner similar to those in Sect. 13 [between (13.25)–(13.27)] but will not be recorded here. Instead we consider next a different set of constitutive equations for the linear theory in terms of another thermodynamic potential, namely the specific *Gibbs free energy function* φ. Occasionally, it is useful to be able to express the kinematic variables such as $e_{\alpha\beta}, \gamma_i, \varkappa_{i\alpha}$ in terms of θ and $N'^{\alpha\beta}$, m^i, $M^{\alpha i}$, regarded as the independent variables. This can be achieved by introduction of the specific Gibbs function φ defined by

$$\varphi = \varphi(N'^{\alpha\beta}, m^i, M^{\alpha i}, \theta; \mathscr{U}'_R) = \psi - \frac{1}{\varrho_0}\left[N'^{\alpha\beta} e_{\alpha\beta} + m^i \gamma_i + M^{\alpha i} \varkappa_{i\alpha}\right], \tag{16.12}$$

where the specific Helmholtz free energy function ψ is of the form (16.5). Then, we have

$$\eta = -\frac{\partial \varphi}{\partial \theta}, \tag{16.13}$$

$$e_{\alpha\beta} = -\varrho_0 \frac{\partial \varphi}{\partial N'^{\alpha\beta}}, \qquad \gamma_i = -\varrho_0 \frac{\partial \varphi}{\partial m^i}, \qquad \varkappa_{i\alpha} = -\varrho_0 \frac{\partial \varphi}{\partial M^{\alpha i}}, \tag{16.14}$$

instead of the relations (16.6)–(16.7).

β) *Explicit results for linear constitutive equations.* Let the surface $\mathscr{S}$ in the initial configuration of the Cosserat surface $\mathscr{C}$ be homogeneous, free from curve force and director couple and in the state of rest at a constant temperature θ_0 and entropy η_0. Then, to the order of approximation considered, it is sufficient to express ψ in (16.5) as a quadratic function of $e_{\alpha\beta}, \gamma_i, \varkappa_{i\alpha}$ and θ. Thus, if we put

---

[43] The constitutive relations (16.5)–(16.7), being valid for a Cosserat surface whose initial director $\boldsymbol{D}$ is of constant magnitude, are intended for shells and plates of uniform thickness.

[44] The use of the symbol $\overline{\psi}$ here should not be confused with a similar notation in Sect. 13. In (16.10), $\overline{\psi}$ is a quadratic function of temperature and the kinematic variables $e_{\alpha\beta}, \gamma_i, \varrho_{i\alpha}$.

$\eta_0 = 0$, we have[45]

$$
\begin{aligned}
\varrho_0 \psi = {}& {}_1C^{\alpha\beta\gamma\delta} e_{\alpha\beta} e_{\gamma\delta} + {}_2C^{\alpha\beta\gamma\delta} \varkappa_{\alpha\beta} \varkappa_{\gamma\delta} + {}_3C^{\alpha\beta\gamma\delta} e_{\alpha\beta} \varkappa_{\gamma\delta} \\
& + {}_1C^{\alpha\beta\gamma} \varkappa_{3\alpha} \varkappa_{\beta\gamma} + {}_2C^{\alpha\beta\gamma} e_{\alpha\beta} \gamma_\gamma \\
& + {}_3C^{\alpha\beta\gamma} e_{\alpha\beta} \varkappa_{3\gamma} + {}_4C^{\alpha\beta\gamma} \gamma_\alpha \varkappa_{\beta\gamma} \\
& + {}_1C^{\alpha\beta} \gamma_\alpha \gamma_\beta + {}_2C^{\alpha\beta} \varkappa_{3\alpha} \varkappa_{3\beta} + {}_3C^{\alpha\beta} \gamma_\alpha \varkappa_{3\beta} \\
& + {}_4C^{\alpha\beta} e_{\alpha\beta} \gamma_3 + {}_5C^{\alpha\beta} \varkappa_{\alpha\beta} \gamma_3 + {}_4C'^{\alpha\beta} e_{\alpha\beta} \theta + {}_5C'^{\alpha\beta} \varkappa_{\alpha\beta} \theta \\
& + {}_1C^{\alpha} \gamma_\alpha \gamma_3 + {}_2C^{\alpha} \varkappa_{3\alpha} \gamma_3 + {}_1C'^{\alpha} \gamma_\alpha \theta + {}_2C'^{\alpha} \varkappa_{3\alpha} \theta \\
& + C (\gamma_3)^2 + C' \theta^2 + C'' \gamma_3 \theta ,
\end{aligned}
\tag{16.15}
$$

where some of the coefficients satisfy certain symmetry conditions, e.g.,

$$
\begin{gathered}
{}_1C^{\alpha\beta\gamma\delta} = {}_1C^{\beta\alpha\gamma\delta} = {}_1C^{\alpha\beta\delta\gamma} = {}_1C^{\gamma\delta\alpha\beta}, \\
{}_2C^{\alpha\beta\gamma\delta} = {}_2C^{\gamma\delta\alpha\beta}, \quad {}_3C^{\alpha\beta\gamma\delta} = {}_3C^{\beta\alpha\gamma\delta}, \quad {}_2C^{\alpha\beta\gamma} = {}_2C^{\beta\alpha\gamma}.
\end{gathered}
\tag{16.16}
$$

From (16.15), we can readily calculate explicit relations for η, $N'^{\alpha\beta}$, m^i, $M^{\alpha i}$ which would be valid for an anisotropic Cosserat surface, but we do not record these here. For completeness we also record an explicit constitutive relation for Q^α. Recalling the remark made following (16.8), we write[46]

$$
Q^\alpha = -\varkappa^{\alpha\beta} \theta_{,\beta}, \qquad \varkappa^{\alpha\beta} = \varkappa^{\beta\alpha}, \tag{16.17}
$$

where the conductivity coefficients $\varkappa^{\alpha\beta}$ (assumed to be symmetric) depend on the reference values $\mathscr{U}_R'$ and satisfy the restrictions

$$
\varkappa^{11} \geqq 0, \qquad \varkappa^{22} \geqq 0, \qquad \varkappa^{11} \varkappa^{22} - \varkappa^{12} \varkappa^{21} \geqq 0 \tag{16.18}
$$

imposed by the inequality (16.8). We emphasize here that the free energy function (16.15) [and also the response function $(16.17)_1$] is recorded for a Cosserat surface with its (initial) reference director specified by $(13.71)_1$; and hence, in line with the remark made in Sect. 4 [following (4.35)], is appropriate for a shell which is of uniform thickness in the initial reference configuration. For a shell which is of variable thickness in the reference configuration, the free energy is still of the form (16.15) except that the various coefficients are now functions of the reference values (13.61) [instead of (13.72)] and also the kinematic variables γ_3 and $\varkappa_{3\alpha}$ in (16.15) must be replaced with s and s_α, respectively. Moreover, in this case, the constitutive relations for m^3 and $M^{\alpha 3}$ are found from (16.3) instead of the corresponding expressions in (16.7).

Henceforth, with reference to (16.15) and (16.17), we restrict our attention to an elastic Cosserat surface possessing holohedral isotropy (i.e., isotropy with a center of symmetry). In this case, a tensor basis is given by $A^{\alpha\beta}$ and since there are no holohedral isotropic tensors of odd order, it follows that all odd order coefficients in (16.15) must vanish:

$$
{}_1C^{\alpha\beta\gamma} = {}_2C^{\alpha\beta\gamma} = {}_3C^{\alpha\beta\gamma} = {}_4C^{\alpha\beta\gamma} = 0, \qquad {}_1C^{\alpha} = {}_2C^{\alpha} = {}_1C'^{\alpha} = {}_2C'^{\alpha} = 0. \tag{16.19}
$$

---

[45] Our results between (16.15)–(16.25) follow GREEN, NAGHDI and WAINWRIGHT [1965, *4*]. The various coefficients in (16.15) are functions of the reference values $\mathscr{U}_R'$ in (13.72).

Since for simplicity we have specified the initial director to be $\boldsymbol{D} = \boldsymbol{A}_3$ (with $D^3 = 1$), the dependence of the various coefficients in (16.15) on $(D^3)^2$ and therefore the magnitude of $\boldsymbol{D}$ is not explicit. [Recall the reference values (13.61).] Indeed, instead of specifying $D^3 = 1$, we could leave $(D^3)^2$ as an arbitrary parameter in the various coefficients of (16.15).

[46] This is simply the two-dimensional analogue of the Fourier law in the linear three-dimensional theory.

Moreover, the remaining coefficients in (16.15) must be homogeneous linear functions of products of $A^{\alpha\beta}$ so that, for example, ${}_1C^{\alpha\beta\gamma\delta}$ may be written as

$$2\,{}_1C^{\alpha\beta\gamma\delta} = \alpha_1 A^{\alpha\beta} A^{\gamma\delta} + \alpha_2 A^{\alpha\gamma} A^{\beta\delta} + \alpha_3 A^{\alpha\delta} A^{\beta\gamma}. \tag{16.20}$$

But by $(16.16)_1$, $\alpha_2 = \alpha_3$ and the same conclusions can be reached by $(16.16)_2$. Similar arguments can be applied to other coefficients in (16.15). In particular, it can be shown that ${}_3C^{\alpha\beta\gamma\delta}$, as ${}_1C^{\alpha\beta\gamma\delta}$, has only two independent scalar coefficients but ${}_2C^{\alpha\beta\gamma\delta}$ involves three independent scalar coefficients.

Keeping the above in mind, we can finally write the specific free energy function ψ in the form

$$\begin{aligned}
\varrho_0 \psi = {} & \tfrac{1}{2}[\alpha_1 A^{\alpha\beta} A^{\gamma\delta} + \alpha_2 (A^{\alpha\gamma} A^{\beta\delta} + A^{\alpha\delta} A^{\beta\gamma})]\, e_{\alpha\beta}\, e_{\gamma\delta} \\
& + \tfrac{1}{2}\alpha_3 A^{\alpha\beta} \gamma_\alpha \gamma_\beta + \tfrac{1}{2}\alpha_4 (\gamma_3)^2 + \alpha_4' \gamma_3\, \theta + \tfrac{1}{2}\alpha_4'' \theta^2 \\
& + \tfrac{1}{2}[\alpha_5 A^{\alpha\beta} A^{\gamma\delta} + \alpha_6 A^{\alpha\gamma} A^{\beta\delta} + \alpha_7 A^{\alpha\delta} A^{\beta\gamma}]\, \varkappa_{\alpha\beta}\, \varkappa_{\gamma\delta} \\
& + \tfrac{1}{2}\alpha_8 A^{\alpha\beta} \varkappa_{3\alpha} \varkappa_{3\beta} + \alpha_9 A^{\alpha\beta} e_{\alpha\beta} \gamma_3 + \alpha_9' A^{\alpha\beta} e_{\alpha\beta}\, \theta \\
& + [\alpha_{10} A^{\alpha\beta} A^{\gamma\delta} + \alpha_{11} (A^{\alpha\gamma} A^{\beta\delta} + A^{\alpha\delta} A^{\beta\gamma})]\, e_{\alpha\beta}\, \varkappa_{\gamma\delta} \\
& + \alpha_{12} A^{\alpha\beta} \varkappa_{\alpha\beta} \gamma_3 + \alpha_{12}' A^{\alpha\beta} \varkappa_{\alpha\beta}\, \theta \\
& + \alpha_{13} A^{\alpha\beta} \gamma_\alpha \varkappa_{3\beta},
\end{aligned} \tag{16.21}$$

and by (16.7) we also have

$$\begin{aligned}
N'^{\alpha\beta} = {} & [\alpha_1 A^{\alpha\beta} A^{\gamma\delta} + \alpha_2 (A^{\alpha\gamma} A^{\beta\delta} + A^{\alpha\delta} A^{\beta\gamma})]\, e_{\gamma\delta} \\
& + [\alpha_{10} A^{\alpha\beta} A^{\gamma\delta} + \alpha_{11} (A^{\alpha\gamma} A^{\beta\delta} + A^{\alpha\delta} A^{\beta\gamma})]\, \varkappa_{\gamma\delta} \\
& + \alpha_9 A^{\alpha\beta} \gamma_3 + \alpha_9' A^{\alpha\beta}\, \theta,
\end{aligned} \tag{16.22}$$

$$\begin{aligned}
m^\alpha &= \alpha_3 A^{\alpha\gamma} \gamma_\gamma + \alpha_{13} A^{\alpha\gamma} \varkappa_{3\gamma}, \\
m^3 &= \alpha_4 \gamma_3 + \alpha_4' \theta + \alpha_9 A^{\alpha\beta} e_{\alpha\beta} + \alpha_{12} A^{\alpha\beta} \varkappa_{\alpha\beta},
\end{aligned} \tag{16.23}$$

and

$$\begin{aligned}
M^{\beta\alpha} = {} & [\alpha_5 A^{\alpha\beta} A^{\gamma\delta} + \alpha_6 A^{\alpha\gamma} A^{\beta\delta} + \alpha_7 A^{\alpha\delta} A^{\beta\gamma}]\, \varkappa_{\gamma\delta} \\
& + [\alpha_{10} A^{\alpha\beta} A^{\gamma\delta} + \alpha_{11} (A^{\alpha\gamma} A^{\beta\delta} + A^{\alpha\delta} A^{\beta\gamma})]\, e_{\gamma\delta} \\
& + \alpha_{12} A^{\alpha\beta} \gamma_3 + \alpha_{12}' A^{\alpha\beta}\, \theta, \\
M^{\alpha 3} = {} & \alpha_8 A^{\alpha\gamma} \varkappa_{3\gamma} + \alpha_{13} A^{\alpha\gamma} \gamma_\gamma,
\end{aligned} \tag{16.24}$$

where the coefficients $\alpha_1, \ldots, \alpha_{13}, \alpha_4', \alpha_4'', \alpha_9'$ and α_{12}' depend on the initial values $A_{\alpha\beta}$, $B_{\alpha\beta}$. We also note that for an isotropic material with a center of symmetry the constitutive relation for Q^α in (16.17) reduces to

$$Q^\alpha = -\varkappa A^{\alpha\beta} \theta_{,\beta}, \qquad \varkappa \geqq 0. \tag{16.25}$$

Apart from dependence of the coefficients $\alpha_1, \ldots, \alpha_{13}, \alpha_4', \ldots, \alpha_{12}', \alpha_4''$ on the reference values $\mathscr{U}_R'$ in (13.72), the foregoing results for an elastic Cosserat surface which is initially homogeneous and possesses holohedral isotropy are not particularly simple even in the case of an initially flat Cosserat surface for which (6.26) holds. For this reason and for later reference we consider next an additional restriction corresponding to (13.53) and (13.74) and examine its consequences.

γ) *A restricted form of the constitutive equations for an isotropic material.* We consider now a further restriction on the constitutive equations introduced previously in Sect. 13. According to this restriction, which is separate from material symmetries, for a given Cosserat surface the constitutive equations are required to remain unaltered under the reflection (13.53) and hence (13.54) or equivalently

$$\boldsymbol{\delta} \to -\boldsymbol{\delta}, \qquad \boldsymbol{\delta}_{,\alpha} \to -\boldsymbol{\delta}_{,\alpha}, \tag{16.26}$$

together with (13.74)–(13.75).[47] In order to obtain the appropriate transformation relations for the kinematic variables (which occur in the constitutive equations) under the restriction $(16.26)_1$, we should return to (16.2) and examine the effect of $(16.26)_1$ prior to the specification of $\boldsymbol{D}=\boldsymbol{A}_3$. For this purpose and with reference to the constitutive equations resulting from linearization of (13.63)–(13.65) with $\boldsymbol{D}$ along the normal to $\mathscr{S}$ [such as the constitutive equation (16.3)], suppose that the initial director along the normal is of constant magnitude specified by $\boldsymbol{D}=\bar{D}\boldsymbol{A}_3$ with $\bar{D}$ being constant. Then, $D^3=\bar{D}=$ const, $\Lambda_{3\alpha}=0$ and the linearized variables (16.2) reduce to

$$s=2\bar{D}\,\gamma_3, \qquad s_\alpha=\bar{D}\,\varkappa_{3\alpha}. \tag{16.27}$$

Keeping the above in mind, we see that the kinematic variables $\gamma_\alpha, \varkappa_{\gamma\alpha}, s, s_\alpha$ which occur in the linear constitutive equations (with $\boldsymbol{D}=\bar{D}\boldsymbol{A}_3$) transform according to

$$\begin{aligned} &\gamma_\alpha \to -\gamma_\alpha, \qquad \varkappa_{\gamma\alpha} \to -\varkappa_{\gamma\alpha},\\ &s \to s, \qquad s_\alpha \to s_\alpha, \end{aligned} \tag{16.28}$$

under the conditions (16.26). Now, without loss in generality, we put $\bar{D}=1$ (or $\boldsymbol{D}=\boldsymbol{A}_3$) and conclude that under the restriction $(16.26)_1$ the kinematic variables $\gamma_i, \varkappa_{i\alpha}$ [which occur in the constitutive equations (16.5)–(16.7)] must transform according to

$$\begin{aligned} &\gamma_\alpha \to -\gamma_\alpha, \qquad \varkappa_{\gamma\alpha} \to -\varkappa_{\gamma\alpha},\\ &\gamma_3 \to \gamma_3, \qquad \varkappa_{3\alpha} \to \varkappa_{3\alpha}, \end{aligned} \tag{16.29}$$

in view of the relations (16.4) between[48] s, γ_3 and $s_\alpha, \varkappa_{3\alpha}$.

We restrict attention in the remainder of this section to the isothermal theory, although this is by no means essential, and we return to our objective of obtaining a restricted form of the constitutive equations (16.21)–(16.24). It is instructive to consider first the case of an initially flat Cosserat plate for which the specific free energy (or the strain energy density, since $\theta=$ const.) takes the form[49]

$$\psi=\psi(e_{\alpha\beta}, \varkappa_{i\alpha}, \gamma_i, A_{\alpha\beta}), \tag{16.30}$$

where the kinematic variables in the argument of ψ are given by (6.25). It is clear that the constitutive equations obtained from (16.30) and (16.7), for an isotropic material with a center of symmetry, will have the same form as (16.21)–(16.24) except that the coefficients $\alpha_1, \dots, \alpha_{13}$ are now constants. We now introduce the further restriction that ψ be invariant under the reflection $(16.26)_1$ which, in

[47] See also the remarks made preceding both (13.53) and (13.74). The interpretation which may be associated with $(16.26)_1$ parallels that given in Sect. 13 with reference to (13.53).

[48] Note that with $\bar{D}=1$, the expressions (16.27) reduce to (16.4) and the previous conclusion regarding the reduction of the constitutive equations (16.3) to corresponding expressions in (16.7) holds.

[49] This follows from (16.5) for an isothermal theory of an initially flat Cosserat plate for which (6.26) holds.

view of (16.29), can be stated as[50]

$$\psi(e_{\alpha\beta}, \varkappa_{\gamma\alpha}, \varkappa_{3\alpha}, \gamma_\alpha, \gamma_3, A_{\alpha\beta}) = \psi(e_{\alpha\beta}, -\varkappa_{\gamma\alpha}, \varkappa_{3\alpha}, -\gamma_\alpha, \gamma_3, A_{\alpha\beta}). \quad (16.31)$$

The condition (16.31) is a further restriction on the free energy function ψ, separate from that arising from holohedral isotropy (or isotropy with a center of symmetry), and results in simpler expressions for ψ, $N'^{\alpha\beta}$, m^i, $M^{\alpha i}$. These expressions can be obtained as special cases of (16.21)–(16.24), apart from the terms involving θ, if we put

$$\alpha_{10} = \alpha_{11} = \alpha_{12} = \alpha_{13} = 0. \quad (16.32)$$

We postpone recording the explicit form of these constitutive relations, but note that the simplification leading to (16.32) can also be achieved in the case of (16.10)–(16.11) by specializing (16.10) for the isothermal theory of an initially flat Cosserat plate. For later convenience, we record the resulting expression for $\bar{\psi}$ in the form

$$\begin{aligned}
\bar{\psi} &= \bar{\psi}_e + \bar{\psi}_b, \\
2\varrho_0 \bar{\psi}_e &= [\alpha_1 A^{\alpha\beta} A^{\gamma\delta} + \alpha_2 (A^{\alpha\gamma} A^{\beta\delta} + A^{\alpha\delta} A^{\beta\gamma})]\, e_{\alpha\beta}\, e_{\gamma\delta} \\
&\quad + \alpha_4 (\gamma_3)^2 + \alpha_8 A^{\alpha\beta} \varrho_{3\alpha} \varrho_{3\beta} + 2\alpha_9 A^{\alpha\beta} e_{\alpha\beta} \gamma_3, \\
2\varrho_0 \bar{\psi}_b &= [\alpha_5 A^{\alpha\beta} A^{\gamma\delta} + \alpha_6 A^{\alpha\gamma} A^{\beta\delta} + \alpha_7 A^{\alpha\delta} A^{\beta\gamma}]\, \varrho_{\alpha\beta}\, \varrho_{\gamma\delta} \\
&\quad + \alpha_3 A^{\alpha\beta} \gamma_\alpha \gamma_\beta.
\end{aligned} \quad (16.33)$$

We may note that the variable $\varrho_{\gamma\alpha}$ in (16.33) may be replaced with $\varkappa_{\gamma\alpha}$, in view of $(6.25)_6$.

We now turn to the more general case (in which $B_{\alpha\beta} \neq 0$) but discuss the additional restriction with reference to the constitutive relations (16.10)–(16.11). Again, limiting the discussion to the isothermal theory, we see that the free energy as a function of the kinematic variables (6.24) takes the form

$$\psi = \bar{\psi}(e_{\alpha\beta}, \varrho_{i\alpha}, \gamma_i; \mathscr{U}'_R), \quad (16.34)$$

where $\bar{\psi}$ is now a function different from that in (16.10). The additional restriction that $\bar{\psi}$ remain unaltered under the transformations $(16.26)_1$ and (13.74), in view of (16.29) and (13.78) with $\boldsymbol{D} = \boldsymbol{A}_3$, implies that[51]

$$\begin{aligned}
&\bar{\psi}(e_{\alpha\beta}, \varrho_{\alpha\beta}, \varrho_{3\alpha}, \gamma_\alpha, \gamma_3; A_{\alpha\beta}, -B_{\alpha\beta}) \\
&\quad = \bar{\psi}(e_{\alpha\beta}, -\varrho_{\alpha\beta}, \varrho_{3\alpha}, -\gamma_\alpha, \gamma_3; A_{\alpha\beta}, B_{\alpha\beta}).
\end{aligned} \quad (16.35)$$

The restriction (16.35), together with that due to isotropy with a center of symmetry, enables us to write down the complete form for $\bar{\psi}$ which is quadratic in $e_{\alpha\beta}$, $\varrho_{i\alpha}$, γ_i. The resulting expression will contain terms of the type in (16.21) with $\varrho_{i\alpha}$ defined in (6.24), in addition to terms involving $B_{\alpha\beta}$. As the latter terms (which vanish in the case of an initially flat surface) are rather unwieldy, we may regard the free energy to have the form (16.33) but with the kinematic variables defined by (6.24). This is equivalent to the specification that the response function

[50] The restriction (16.31) was introduced by GREEN and NAGHDI [1967, *4*], [1968, *6*] on the basis of a different argument than that given here, but the results and the conclusions remain the same.

[51] The restriction (16.35) was introduced by GREEN and NAGHDI [1968, *6*], [1969, *3*] on the basis of an argument different from that given here, but the results and conclusions remain the same.

$\bar{\psi}$ be independent of $B_{\alpha\beta}$. It should be emphasized that this is *not* an approximation but merely a particular choice for $\bar{\psi}$ as a special case of the general theory.[52]

δ) *Constitutive equations of the restricted linear theory.* The results in Subsect. γ) for a restricted form of constitutive equations of an elastic Cosserat surface should not be confused with the corresponding set of linear constitutive equations appropriate to the restricted theory (Sects. 10 and 15). It is not difficult to see that the latter set (with different constitutive coefficients) may be obtained in a similar manner by linearization of the nonlinear constitutive equations of the restricted theory in Sect. 15. In particular, with the limitation to isothermal deformations, the specific free energy in the restricted linear theory is a quadratic function of the kinematic variables $e_{\alpha\beta}$, $\varrho_{(\beta\alpha)}$ defined in (6.27). Hence, in parallel with (16.34), we may write

$$\psi = \bar{\psi}(e_{\alpha\beta}, \varrho_{(\beta\alpha)}; \mathscr{U}'_R), \tag{16.36}$$

where $\bar{\psi}$ is a different function from that in (16.34).

We may also impose a condition similar to (16.35) on the response function $\bar{\psi}$ in (16.36) by requiring that the function $\bar{\psi}$ remain unaltered under the reflection of the unit normal and hence its gradient:[53]

$$\boldsymbol{A}_3 \to -\boldsymbol{A}_3, \quad \boldsymbol{A}_{3,\alpha} \to -\boldsymbol{A}_{3,\alpha}. \tag{16.37}$$

Under the transformations (16.37), the linearized kinematic formulae of the restricted theory [collected in (6.27)] transform according to

$$\begin{gathered} u_\alpha \to u_\alpha, \quad u_3 \to -u_3, \quad B_{\alpha\beta} \to -B_{\alpha\beta}, \\ e_{\alpha\beta} \to e_{\alpha\beta}, \quad \varrho_{(\beta\alpha)} \to -\varrho_{(\beta\alpha)}. \end{gathered} \tag{16.38}$$

Hence, the added requirement that $\bar{\psi}$ in (16.36) remain unaltered under the reflection (16.37) implies that

$$\bar{\psi}(e_{\alpha\beta}, \varrho_{(\beta\alpha)}; A_{\alpha\beta}, -B_{\alpha\beta}) = \bar{\psi}(e_{\alpha\beta}, -\varrho_{(\beta\alpha)}; A_{\alpha\beta}, B_{\alpha\beta}). \tag{16.39}$$

In line with the remarks made following (16.35), we now regard the response function $\bar{\psi}$ in (16.39) to be independent of $B_{\alpha\beta}$ but the kinematic variables in its argument being those in (6.27). It then follows that for an isotropic material with a center of symmetry, the constitutive equation for ψ [in parallel to (16.33)] can be written in the form

$$\begin{aligned} \psi &= \bar{\psi}_e + \bar{\psi}_b, \\ 2\varrho_0 \bar{\psi}_e &= [\beta_1 A^{\alpha\beta} A^{\gamma\delta} + \beta_2 (A^{\alpha\gamma} A^{\beta\delta} + A^{\alpha\delta} A^{\beta\gamma})]\, e_{\alpha\beta}\, e_{\gamma\delta}, \\ 2\varrho_0 \bar{\psi}_b &= [\beta_5 A^{\alpha\beta} A^{\gamma\delta} + \beta_6 (A^{\alpha\gamma} A^{\beta\delta} + A^{\alpha\delta} A^{\beta\gamma})]\, \varrho_{(\alpha\beta)}\, \varrho_{(\gamma\delta)}, \end{aligned} \tag{16.40}$$

where $\beta_1, \beta_2, \beta_5$ and β_6 are arbitrary constants and do not necessarily have the same values as the corresponding coefficients in (16.33). We postpone recording

[52] The ultimate plausibility for such a special choice of the specific free energy (or the strain energy) function depends, of course, on its usefulness and a demonstration of its relevance to shell theory. Indeed, as remarked by GREEN and NAGHDI [1968, *6*], the problem of choice of the strain energy function here parallels the corresponding problem for a suitable choice of the strain energy function in the three-dimensional theory of nonlinear or linear elasticity and is a common feature of all general theories. Moreover, it is worth noting that in all existing developments for a linear shell theory from the three-dimensional equations (beginning with the paper of LOVE [1888, *1*]), an assumption that the strain energy density (after an integration with respect to the thickness coordinate) has the same form as the (integrated) strain energy density for a flat plate is implicit or is implied by other assumptions. This corresponds to the above special choice for $\bar{\psi}$.

[53] The conditions (16.37) are the counterparts of (13.74)–(13.75) in the restricted theory. We emphasize that (16.37) are conditions which should be imposed only on the arguments of the response functions.

the constitutive equations for $\grave{N}^{\alpha\beta}$ and $\grave{M}^{(\alpha\beta)}$ which can be obtained from (16.40) and the linearized version of (15.15).

17. The complete theory for thermoelastic shells: II. Derivation from the three-dimensional theory. On the basis of the three-dimensional equations of the classical continuum mechanics, two-dimensional kinematical results, field equations, entropy inequalities and related aspects of the subject were developed above for shell-like bodies in Sects. 7 and 11–12. Here we first obtain the corresponding two-dimensional constitutive relations and thermodynamical results for elastic materials and then briefly summarize the basic equations of the complete theory for thermo-elastic shells.

α) *Constitutive equations in terms of two-dimensional variables. Thermodynamical results.* We recall that in the three-dimensional theory of (non-polar) elastic materials the constitutive relations for the specific free energy, the entropy and the stress tensor can be expressed in the forms[54]

$$\psi^* = \hat{\psi}^*(\gamma^*_{ij}, \theta^*), \tag{17.1}$$

$$\eta^* = -\frac{\partial \hat{\psi}^*}{\partial \theta^*}, \tag{17.2}$$

$$\tau^{ij} = \varrho^* \frac{\partial \hat{\psi}^*}{\partial \gamma^*_{ij}}. \tag{17.3}$$

In addition, the constitutive equation for the heat flux vector has the form

$$q^{*k} = \hat{q}^{*k}(\gamma^*_{ij}, \theta^*, \theta^*_{,m}) \tag{17.4}$$

and the response function $\hat{q}^{*k}$ is restricted by the inequality

$$-\hat{q}^{*k}\,\theta^*_{,k} \geqq 0. \tag{17.5}$$

The results (17.2)–(17.3) and (17.5) are deduced from (11.23) and hold for every admissible process.[55] Also, with the help of (17.2)–(17.3), the residual energy equation is obtained from (11.15) and (11.18):

$$\varrho^* r^* - q^{*k}{}_{\|k} - \varrho^* \theta^* \dot{\eta}^* = 0. \tag{17.6}$$

We now proceed to deduce the two-dimensional counterparts of the above results in terms of variables defined in Sects. 11–12. To indicate the nature of the reduction, using the chain rule we first observe the relations:

$$\frac{\partial \psi^*}{\partial \boldsymbol{g}_k} = \frac{\partial \hat{\psi}^*}{\partial \gamma^*_{ij}} \frac{\partial \gamma^*_{ij}}{\partial \boldsymbol{g}_k} = \frac{\partial \hat{\psi}^*}{\partial \gamma^*_{ij}} \left[\frac{1}{2}(\delta^k_i \boldsymbol{g}_j + \delta^k_j \boldsymbol{g}_i)\right] = \frac{\partial \hat{\psi}^*}{\partial \gamma^*_{kj}} \boldsymbol{g}_j \tag{17.7}$$

and

$$\begin{gathered}\frac{\partial \psi^*}{\partial \boldsymbol{a}_\alpha} = \frac{\partial \hat{\psi}^*}{\partial \gamma^*_{ij}} \frac{\partial \gamma^*_{ij}}{\partial \boldsymbol{g}_k} \frac{\partial \boldsymbol{g}_k}{\partial \boldsymbol{a}_\alpha} = \frac{\partial \hat{\psi}^*}{\partial \gamma^*_{kj}} \boldsymbol{g}_j\,\delta^k_\alpha = \frac{\partial \hat{\psi}^*}{\partial \gamma^*_{\alpha j}} \boldsymbol{g}_j,\\ \frac{\partial \psi^*}{\partial \boldsymbol{d}_{N,\alpha}} = \frac{\partial \hat{\psi}^*}{\partial \gamma^*_{\alpha j}} \boldsymbol{g}_j\,\xi^N, \quad \frac{\partial \psi^*}{\partial \boldsymbol{d}_N} = \frac{\partial \hat{\psi}^*}{\partial \gamma^*_{3j}} \boldsymbol{g}_j (N\,\xi^{N-1}),\\ \frac{\partial \psi^*}{\partial \varphi_M} = \frac{\partial \hat{\psi}^*}{\partial \theta^*} \xi^M.\end{gathered} \tag{17.8}$$

[54] The functions $\hat{\psi}^*$ in (17.1) and $\hat{q}^{*k}$ in (17.4) depend also on the reference values G_{ij}, although this is not exhibited in (17.1), (17.4) and elsewhere in this section. The partial derivative of a function with respect to a symmetric tensor such as that in (17.3) is understood to have the symmetric form

$$\frac{1}{2}\left(\frac{\partial \hat{\psi}^*}{\partial \gamma^*_{ij}} + \frac{\partial \hat{\psi}^*}{\partial \gamma^*_{ji}}\right).$$

[55] See Sects. 79–82 of TRUESDELL and NOLL [1965, 9].

The relation (17.7) follows directly from (17.1) when the argument γ^*_{ij} is expressed in terms of the base vectors $\boldsymbol{g}_i$ and $\boldsymbol{G}_i$ as in $(7.27)_1$ while those in (17.8) are obtained with the help of (7.10) and (11.42). Consider now the constitutive equations for the components $\tau^{\alpha j}$ in (17.3). Multiply both sides by $g^{\frac{1}{2}}\,\boldsymbol{g}_j$ and integrate the resulting expression with respect to ξ between the limits α, β to obtain

$$\int\limits_{\alpha}^{\beta} \boldsymbol{T}^{\alpha}\, d\xi = \int\limits_{\alpha}^{\beta} \varrho^*\, g^{\frac{1}{2}} \frac{\partial \hat{\psi}^*}{\partial \gamma^*_{\alpha j}}\, \boldsymbol{g}_j\, d\xi, \tag{17.9}$$

where by $(11.7)_2$ we have used $\boldsymbol{T}^{\alpha} = g^{\frac{1}{2}}\, \tau^{\alpha j}\, \boldsymbol{g}_j$. By $(17.8)_1$, (4.21) and $(11.44)_{3,4}$, the right-hand side of (17.9) can be reduced as follows:

$$\int\limits_{\alpha}^{\beta} \varrho^*\, g^{\frac{1}{2}} \frac{\partial \hat{\psi}^*}{\partial \gamma^*_{\alpha j}}\, \boldsymbol{g}_j\, d\xi = \int\limits_{\alpha}^{\beta} k\, \frac{\partial \hat{\psi}^*}{\partial \boldsymbol{a}_{\alpha}}\, d\xi = \frac{\partial}{\partial \boldsymbol{a}_{\alpha}} \int\limits_{\alpha}^{\beta} k\, \hat{\psi}^*\, d\xi = \varrho\, a^{\frac{1}{2}} \frac{\partial \overline{\psi}}{\partial \boldsymbol{a}_{\alpha}}. \tag{17.10}$$

From $(11.36)_1$ and (17.9)–(17.10), it is easily seen that the stress-resultants $\boldsymbol{N}^{\alpha}$ are related to the partial derivatives $\partial \hat{\psi}^* / \partial \boldsymbol{a}_{\alpha}$ and similar relations can be deduced for other resultants. The relation

$$\overline{\psi} = \frac{1}{\varrho(\theta^{\alpha})\, a^{\frac{1}{2}}(\theta^{\alpha})} \int\limits_{\alpha}^{\beta} k(\theta^{\alpha}, \xi)\, \hat{\psi}^*(\gamma^*_{\alpha\beta}, \gamma^*_{\alpha 3}, \gamma^*_{33}, \theta^*)\, d\xi, \tag{17.11}$$

where the arguments of $\hat{\psi}^*$ have the structure indicated by (7.28) and (11.42), defines the function $\overline{\psi}$ which occurs in (17.10).[56] Clearly, with the help of $(11.44)_{3,4}$, (17.11) may also be obtained from (17.1) and $\overline{\psi}$ can be regarded as a function of the variables

$$\boldsymbol{a}_{\alpha},\ \boldsymbol{d}_N,\ \boldsymbol{d}_{N,\alpha},\ \varphi_M \qquad (N = 1, 2, \ldots;\ M = 0, 1, 2, \ldots).$$

In a manner similar to that indicated above, from (7.28), (11.42), (11.28), (11.36), (11.44), (11.55), (17.7)–(17.8) and (17.1)–(17.3), by direct calculations we deduce the following two-dimensional constitutive equations:[57]

$$\psi = \overline{\psi}(\boldsymbol{a}_{\alpha}, \boldsymbol{d}_N, \boldsymbol{d}_{N,\gamma}, \varphi_M) \qquad (N = 1, 2, \ldots;\ M = 0, 1, 2, \ldots), \tag{17.12}$$

$$\eta^M = -\frac{\partial \overline{\psi}}{\partial \varphi_M} \qquad (M = 0, 1, 2, \ldots), \tag{17.13}$$

$$\left.\begin{aligned} \boldsymbol{N}^{\alpha} &= \varrho \frac{\partial \overline{\psi}}{\partial \boldsymbol{a}_{\alpha}}, \\ \boldsymbol{m}^N = \varrho \frac{\partial \overline{\psi}}{\partial \boldsymbol{d}_N}, \quad \boldsymbol{M}^{N\alpha} &= \varrho \frac{\partial \overline{\psi}}{\partial \boldsymbol{d}_{N,\alpha}} \quad (N = 1, 2, \ldots). \end{aligned}\right\} \tag{17.14}$$

The above constitutive equations involve only the specific free energy ψ corresponding to $n = 0$ in the definition $(11.44)_3$ and an explanation is necessary as to why we have not used the expressions involving ψ^n $(n = 1, 2, \ldots)$. To elaborate, consider for example $(17.14)_3$ which is obtained from the constitutive relation for $\tau^{\alpha j}$ in (17.3) after multiplication of both sides of the equation by $\xi^N g^{\frac{1}{2}}\, \boldsymbol{g}_j$. However, if instead of $\xi^N g^{\frac{1}{2}}\, \boldsymbol{g}_j$ both sides of the equation are multiplied by $\xi^{N+n} g^{\frac{1}{2}}\, \boldsymbol{g}_j$ prior to integration, then

$$\boldsymbol{M}^{(N+n)\alpha} = \varrho \frac{\partial \overline{\psi}^n}{\partial \boldsymbol{d}_{N,\alpha}} \qquad (N = 1, 2, \ldots;\ n = 0, 1, 2, \ldots)$$

---

[56] The function $\hat{\psi}^*$ in (17.11) depends also on the reference values $G_{\alpha\beta}$, $G_{\alpha 3}$, G_{33}.

[57] The function $\overline{\psi}$ in (17.12) depends also on the reference values $\boldsymbol{A}_{\alpha}$, $\boldsymbol{D}_N$, $\boldsymbol{D}_{N,\gamma}$. These reference values, although not exhibited in (17.12), arise from the duals of (7.25) for the reference values $G_{\alpha\beta}$, $G_{\alpha 3}$, G_{33} in the arguments of $\hat{\psi}^*$ in (17.11).

will result instead of $(17.14)_3$, where $\overline{\psi}^n$ depends on the variables which occur on the right-hand side of (17.12) and where for clarity we have again used the notation $M^{(N+n)\alpha}$ [as in (11.59)] instead of $M^{N+n\alpha}$ in (12.7). Similar results can be obtained for

$$\eta^{(n+M)}, \quad \frac{N}{n+N}\,\boldsymbol{m}^{(N+n)\alpha}, \quad \boldsymbol{N}^{n\alpha} \quad \begin{pmatrix} n=0,1,2,\ldots \\ N=1,2,\ldots;\ M=0,1,2,\ldots \end{pmatrix}$$

in terms of $\overline{\psi}^n$. It is clear that these constitutive relations for $n \geqq 1$ are redundant since those in (17.13)–(17.14) already hold for all integer values of N and M. Hence, it will suffice to consider only the set (17.12)–(17.14) as the appropriate two-dimensional constitutive equations.

The constitutive equations (17.1)–(17.4) are expressed in terms of response functions which depend, in particular, on the relative kinematic measures γ_{ij}^* and are specially useful when we obtain their linearized counterparts later in this chapter. Alternatively, the constitutive equations in nonlinear thermo-elasticity may be expressed as functions of g_{ij} (rather than γ_{ij}^*), apart from their dependence on thermal variables and on the reference values G_{ij}. Using a response function for the specific free energy which depends on g_{ij} (instead of γ_{ij}^*), we can readily record a relation similar to (17.11). In this latter relation, the arguments of $\hat{\psi}^*$ (now a different function) have the structure indicated by (7.25) and (11.42); and the response function for the two-dimensional specific free energy can be regarded as a different function of φ_M $(M=0, 1, 2, \ldots)$ and the variables

$$a_{\alpha\beta},\ d_{N\alpha},\ \lambda_{N\beta\alpha},\ \sigma_{NK},\ \sigma_{NK\alpha},\ \sigma_{NK\alpha\beta}, \qquad (N, K=1, 2, \ldots), \tag{17.15}$$

where the last five of the above are defined in (7.13)–(7.17). Introducing the notations

$$\sigma_N = \sigma_{1N} = d^3\, d_{N3}, \qquad \sigma_{N\alpha} = \sigma_{1N\alpha} = d^3\, \lambda_{N3\alpha} \qquad (N=1, 2, \ldots) \tag{17.16}$$

and using the expression for $\boldsymbol{a}_3$ in terms of $\boldsymbol{d}(=\boldsymbol{d}_1)$ and its components [see the equation preceding (13.34)], we can write $\boldsymbol{d}_N$ and its partial derivatives as

$$\boldsymbol{d}_N = d_{N\cdot}^{\ \alpha}\, \boldsymbol{a}_\alpha + d_{N\cdot}^{\ 3}\, \boldsymbol{a}_3 = \left(d_{N\cdot}^{\ \nu} - \frac{d^\nu \sigma_N}{\sigma}\right)\boldsymbol{a}_\nu + \frac{\sigma_N}{\sigma}\,\boldsymbol{d},$$

$$\frac{\partial \boldsymbol{d}_N}{\partial \theta^\alpha} = \lambda_{N\cdot\alpha}^{\ \nu}\, \boldsymbol{a}_\nu + \lambda_{N\cdot\alpha}^{\ 3}\, \boldsymbol{a}_3 = \left(\lambda_{N\cdot\alpha}^{\ \nu} - \frac{d^\nu \sigma_{N\alpha}}{\sigma}\right)\boldsymbol{a}_\nu + \frac{\sigma_{N\alpha}}{\sigma}\,\boldsymbol{d},$$

where $\sigma(=\sigma_1)$ is defined by $(7.17)_2$. We now make an observation parallel to that in Sect. 13 [between (13.32)–(13.33)] regarding the tangential components of $\boldsymbol{d}_N$ and $\boldsymbol{d}_{N,\alpha}$; and also, from the last two expressions, we note that the last three sets of variables in (17.15) can be expressed in terms of (17.16) and the first three of (17.15). It follows that σ_{NK} (for $N \geqq 2$), $\sigma_{NK\alpha}$ (for $N \geqq 2$) and $\sigma_{KN\alpha\beta}$ (for $N, K \geqq 1$) may be suppressed from the arguments of the (two-dimensional free energy) response function and that, instead of (17.15), the relevant variables are

$$\mathscr{V}_N\colon\ a_{\alpha\beta},\ d_{N\alpha},\ \lambda_{N\beta\alpha},\ \sigma_N,\ \sigma_{N\alpha} \qquad (N=1, 2, \ldots). \tag{17.17}$$

Hence, we write[58]

$$\begin{aligned} \psi &= \overline{\psi}(\boldsymbol{a}_\alpha, \boldsymbol{d}_N, \boldsymbol{d}_{N,\gamma}, \varphi_M) \\ &= \tilde{\psi}(\mathscr{V}_N, \varphi_M) \qquad (N=1, 2, \ldots;\ M=0, 1, 2, \ldots), \end{aligned} \tag{17.18}$$

where $\mathscr{V}_N$ stand for the set of kinematic variables (17.17). As in (13.40)–(13.42), using the chain rule for differentiation, the constitutive equations (17.13)–

[58] As already noted with reference to (17.12), the function $\overline{\psi}$ depends also on the reference values $\boldsymbol{A}_\alpha$, $\boldsymbol{D}_N$, $\boldsymbol{D}_{N,\gamma}$. Similarly, the function $\tilde{\psi}$ in $(17.18)_2$ depends also on the reference values which are the duals of those in (17.17).

(17.14) can be reduced to a set in terms of $\tilde{\psi}$ and the tensor components of the resultants defined by (12.10)–(12.11) and (12.15).

Before proceeding further, we make an observation regarding the representation of the two-dimensional specific free energy function. As long as no kinematic approximation is introduced in the argument of $\hat{\psi}^*$ in (17.11), the two-dimensional function $\bar{\psi}$ (and therefore $\tilde{\psi}$) is exact. However, if instead of the exact expressions (7.28) the kinematic variables in the argument of the response function $\hat{\psi}^*$ are replaced by a set of approximate expressions, then neither $\bar{\psi}$ nor $\tilde{\psi}$ will necessarily reflect the structure of the original three-dimensional response function.[59] Because of this, if an approximation is contemplated, it is equally acceptable to obtain an approximate two-dimensional specific free energy from a function of the kinematic variables $a_{\alpha\beta}$, d_{Ni}, $\lambda_{Ni\alpha}$ in the form[60]

$$\psi = \psi'(a_{\alpha\beta}, d_{Ni}, \lambda_{Ni\alpha}, \varphi_M) \qquad (N=1, 2, \ldots; M=0, 1, 2, \ldots). \tag{17.19}$$

However, even though an approximation procedure will be considered in the next section, in what follows we retain the constitutive equation for ψ in the form (17.18) since the variables $\mathscr{V}_N$ in the argument of $\tilde{\psi}$ do reflect the structure of the leading terms in the expansion of g_{ij} and hence γ^*_{ij} [see (7.25)–(7.26) and (7.28)–(7.29)].

Returning to our main task in this section, instead of recording the two-dimensional constitutive equations in terms of $\tilde{\psi}$, for later convenience we express the two-dimensional specific free energy as a function of the relative kinematic measures

$$\mathscr{U}_N:\ e_{\alpha\beta},\ \gamma_{N\alpha},\ \varkappa_{N\beta\alpha},\ s_N,\ s_{N\alpha} \qquad (N=1, 2, \ldots) \tag{17.20}$$

and the reference values

$$\begin{aligned} {}_R\mathscr{U}_N:\ & A_{\alpha\beta},\ D_{N\alpha},\ \Lambda_{N\beta\alpha}, \\ & D^3\, D_{N3},\ D^3\, \Lambda_{N3\alpha} \qquad (N=1, 2, \ldots). \end{aligned} \tag{17.21}$$

The last two relative measures in (17.20) correspond to those in (17.16) and are given by

$$s_N = s_{1\,N} = \sigma_N - D^3\, D_{N3}, \qquad s_{N\alpha} = s_{1\,N\alpha} = \sigma_{N\alpha} - D^3\, \Lambda_{N3\alpha},$$

where $s_{1\,N}$, $s_{1\,N\alpha}$ and $s_1 (=s)$, $s_{1\alpha} (=s_\alpha)$ are defined by $(7.22)_{1,2}$ and $(7.24)_{1,2}$. Thus, with

$$\psi = \hat{\psi}(\mathscr{U}_N, \varphi_M; {}_R\mathscr{U}_N) \qquad (N=1, 2, \ldots; M=0, 1, 2, \ldots), \tag{17.22}$$

the constitutive equations corresponding to (17.13)–(17.14) are:

$$\eta^M = -\frac{\partial\psi}{\partial\varphi_M} \qquad (M=0, 1, 2, \ldots), \tag{17.23}$$

$$N'^{\alpha\beta} = N'^{\beta\alpha} = \varrho\,\frac{\partial\hat{\psi}}{\partial e_{\alpha\beta}}, \tag{17.24}$$

$$\begin{aligned} m^{N\alpha} &= \varrho\,\frac{\partial\hat{\psi}}{\partial\gamma_{N\alpha}}, \qquad m^{N3} = \varrho\left(d^3\,\frac{\partial\hat{\psi}}{\partial s_N} + \delta_1^N \sum_{M=1}^{\infty}\left\{\frac{\partial\hat{\psi}}{\partial s_M}\, d_{M.}^{\ 3} + \frac{\partial\hat{\psi}}{\partial s_{M\alpha}}\,\lambda_{M.\alpha}^{\ 3}\right\}\right), \\ M^{N\alpha\beta} &= \varrho\,\frac{\partial\hat{\psi}}{\partial\varkappa_{N\beta\alpha}}, \qquad M^{N\alpha 3} = \varrho\, d^3\,\frac{\partial\hat{\psi}}{\partial s_{N\alpha}} \qquad (N=1, 2, \ldots). \end{aligned} \tag{17.25}$$

---

[59] In effect, the approximation will partially mask the manner of dependence of the two-dimensional free energy function on the detailed structure of γ^*_{ij} in (7.28).

[60] The form (17.19) is used in [1970, 2]. Two-dimensional constitutive equations of the type (17.13)–(17.14), either in terms of $\tilde{\psi}$ or ψ', may also be obtained from combination of (11.59) and (11.64) or from (12.22). This manner of obtaining the two-dimensional constitutive equations will also include those in terms of $\tilde{\psi}^n$ or ψ'^n $(n \geqq 1)$; but, as noted above, the expressions involving $\tilde{\psi}^n$ or ψ'^n $(n \geqq 1)$ are redundant.

The results (17.23)–(17.25) are obtained from (17.22) and (17.18) by using the chain rule and the tensor components of the resultants defined by (12.10)–(12.11) and (12.15).

To complete our development, we now proceed to obtain the two-dimensional counterparts of the inequality (17.5) and the residual energy equation (17.6). Use of (11.42) and (11.56)–(11.57) followed by integration of (17.5) yields

$$-\sum_{r=0}^{n}\binom{n}{r}(-\alpha)^{n-r}\sum_{N=0}^{\infty}\left[\frac{N}{r+N}\,\varrho\,R^{r+N}\,\varphi_N+q^{r+N\alpha}\,\varphi_{N,\alpha}\right]\geqq 0 \qquad (n=0,1,2,\ldots), \tag{17.26}$$

where R^n and $q^{n\alpha}$, which require constitutive equations, depend on the variables in the argument of the function $\hat{\psi}$ in (17.22) or $\tilde{\psi}$ in (17.18) and $\varphi_{N,\alpha}$. Alternatively, observing that the inequality (17.5) is equivalent to

$$q^{*k}\,\Phi^{*}{}_{,k}\geqq 0,$$

instead of (17.26), we can deduce the equivalent inequalities

$$\sum_{r=0}^{n}\binom{n}{r}(-\alpha)^{n-r}\sum_{N=0}^{\infty}\left[\frac{N}{r+N}\,\varrho\,R^{r+N}\,\Phi_N+q^{r+N\alpha}\,\Phi_{N,\alpha}\right]\geqq 0 \qquad (n=0,1,2,\ldots). \tag{17.27}$$

In order to obtain the two-dimensional residual energy equations, we first multiply (17.6) by ξ^n and then integrate over a part $\mathscr{P}^*$. The resulting equation, after use of (11.42), $(11.44)_1$ and (11.56)–(11.58), becomes

$$\int_{\mathscr{P}}\varrho\,(r^n+R^n)\,d\sigma-\int_{\mathscr{P}}\varrho\sum_{M=0}^{\infty}\dot{\eta}^{n+M}\,\varphi_M\,d\sigma-\int_{\partial\mathscr{P}}h^n\,ds=0, \tag{17.28}$$

from which follow the residual energy equations

$$\varrho\,(r^n+R^n)-\varrho\sum_{M=0}^{\infty}\varphi_M\,\dot{\eta}^{n+M}-q^{n\alpha}{}_{|\alpha}=0 \qquad (n=0,1,2,\ldots). \tag{17.29}$$

The foregoing two-dimensional results are obtained by direct integration of (17.1)–(17.6) and with the use of various resultants defined in Sects. 11–12. Alternatively, by introducing suitable constitutive assumptions, the results (17.22)–(17.25), (17.27) and (17.29) can be deduced from the inequalities (12.22) and the energy equations (12.20).

β) *Summary of the basic equations in a complete theory.* In addition to the constitutive relations and thermodynamical results (17.22)–(17.29) or an equivalent set of results, the complete theory includes the equations of motion (12.13)–(12.14), Eqs. $(12.15)_{1,2}$ which give values for $N^{\beta\alpha}$ and $N^{\alpha 3}$ and the set of Eqs.[61] $(12.16)_{1,2}$. Earlier we observed that constitutive equations can also be deduced in terms of the functions $\bar{\psi}^n$ or $\tilde{\psi}^n$ $(n\geqq 1)$ [and therefore also in terms of $\hat{\psi}^n$ $(n\geqq 1)$], obtained from $(11.44)_3$; however, such results were not recorded since they are redundant. Keeping this in mind, we see that the set of Eqs. (12.16) is satisfied identically. As a result, there is some redundancy in the system of two-dimensional equations obtained for shell-like bodies and we summarize below the essential results.

The equations of motion are given by (12.13)–(12.15). The constitutive equations in terms of the function $\hat{\psi}$ have the forms (17.22)–(17.25). The constitutive

[61] Eqs. (12.15)–(12.16) arise from the symmetry of the stress tensor.

equations for $q^{0\alpha}$ and $q^{n\alpha}$ and R^n, $(n=1, 2, \ldots)$, depend on $\varphi_{M,\alpha}$ and the variables in the argument of the function $\hat{\psi}$ in (17.22). The functions R^n and $q^{n\alpha}$ satisfy the inequalities (17.26) and the residual energy equations are given by (17.29).

Special cases of the above development can be discussed in a manner somewhat analogous to the special theory considered in Sect. 15 [Subsect. α)]. In particular, the membrane theory can be obtained by assuming that the function $\hat{\psi}$ in (17.22) depends only on $e_{\alpha\beta}$ and φ_0 [or θ in the notation of (11.41)] and by introducing other appropriate specializations. The resulting theory, under isothermal conditions, will be of the same form as the nonlinear theory of elastic membranes discussed by GREEN and ADKINS.[62]

18. Approximation for thin shells: II. Developments from the three-dimensional theory. While the two-dimensional equations summarized in Sect. 17 [Subsect. β)] have been obtained systematically from the corresponding three-dimensional equations of thermoelasticity, they consist in an infinite set of equations for an infinite number of unknowns. The desirability of an approximation procedure for *thin* shells is, therefore, self evident. Indeed, the need for a suitable approximative scheme in conjunction with the exact two-dimensional results deduced from the three-dimensional equations is already indicated in Sect. 7 [Subsects. α), δ)] and in Sect. 12 [Subsect. ε)]. Here, we first outline an approximation procedure suggested by GREEN and NAGHDI[63] and then make some remarks pertaining to the resulting approximate theory and other approximations.

α) *An approximation procedure.* Before introducing an approximation procedure, we recall that the initial position vector (without loss in generality) can always be taken in the form (7.30)–(7.31) so that, as indicated in (7.49), D_{Ni} and $\Lambda_{Ni\alpha}$ vanish for $N \geqq 2$. In addition, henceforth we specify $D_3 = D = 1$. Then, (7.48) and (7.58) hold and the reference values (17.21) reduce to

$$A_{\alpha\beta}, \quad -B_{\alpha\beta}. \tag{18.1}$$

We now assume that the free energy function $\hat{\psi}$ in (17.22) for sufficiently *thin* shells, can be represented by an approximate expression in terms of the kinematic variables

$$\begin{gathered} e_{\alpha\beta}, \ \gamma_\alpha, \ \varkappa_{(\beta\alpha)}, \\ s, \ s_\alpha, \end{gathered} \tag{18.2}$$

the reference values (18.1) and φ_N $(N=0, 1, 2, \ldots)$ only.[64] Thus, we set[65]

$$\psi = \psi(e_{\alpha\beta}, \gamma_\alpha, \varkappa_{(\beta\alpha)}, s, s_\alpha, \varphi_N) \qquad (N=0, 1, 2, \ldots) \tag{18.3}$$

approximately. The problem of how to determine the approximate form of ψ in (18.3) from $\hat{\psi}$ in (17.22) and therefore from (17.11) is not considered here.[66]

---

[62] GREEN and ADKINS [1960, *5*].

[63] [1970, *2*].

[64] At this stage it is not necessary to introduce an approximation for the temperature, which can be considered separately. The kinematic variables (18.2) correspond to those in (17.20) for $N=1$ only; however, in introducing the approximation, we also write $\varkappa_{(\beta\alpha)}$ in place of $\varkappa_{\beta\alpha} = \varkappa_{1\beta\alpha}$. The latter is motivated by the fact that the leading terms of $\gamma^*_{\alpha\beta}$ in the argument of $\hat{\psi}^*$ in (17.11) contain only the symmetric $\varkappa_{(\beta\alpha)}$.

[65] In (18.3) and throughout this section, we again use the same symbol for a function and its value. Although not explicitly exhibited, the dependence of the function ψ in (18.3) on the (initial) reference values (18.1) is understood. In view of (7.58), the variables γ_α and s_α in the argument of ψ can be replaced, respectively, by d_α and σ_α defined in (7.17). But we retain the set of variables (18.2).

[66] Except possibly in very special cases, it appears to be exceedingly difficult to calculate ψ in (18.3) from the free energy function $\hat{\psi}^*$ or $\hat{\psi}$ in (17.22).

It follows from (18.3) and (17.25) that

$$M^{N\alpha i}=0, \quad m^{Ni}=0 \quad (N \geqq 2),$$

approximately.[67] The equations of motion (12.14) for $N \geqq 2$ are then satisfied if we also specify (12.24). The remaining equations of motion are given by (12.28)–(12.30). The constitutive relation for η^N is still of the form (17.23) but with the function $\hat{\psi}$ replaced by that in (18.3). However, in place of (17.24)–(17.25) we now have

$$N'^{\alpha\beta}=N'^{\beta\alpha}=\varrho\frac{\partial\psi}{\partial e_{\alpha\beta}}, \quad m^{\alpha}=\varrho\frac{\partial\psi}{\partial\gamma_{\alpha}}, \quad M^{\alpha\beta}=M^{(\alpha\beta)}=\varrho\frac{\partial\psi}{\partial\varkappa_{(\beta\alpha)}}, \tag{18.4}$$

$$m^3=\varrho\left(2d^3\frac{\partial\psi}{\partial s}+\lambda^3_{.\alpha}\frac{\partial\psi}{\partial s_{\alpha}}\right), \quad M^{\alpha 3}=\varrho d^3\frac{\partial\psi}{\partial s_{\alpha}}, \tag{18.5}$$

where the notations of (12.27) have been used. As we have made no approximation so far about the temperature, the residual energy equations are still given by (17.29) and constitutive equations are required for $q^{n\alpha}$ and R^n. These constitutive equations must satisfy the inequalities (17.26).

Consider now the question of approximation for the temperature and the remaining thermal variables. In view of (11.41)–(11.42), one possibility is to make the approximation

$$\theta^*(\theta^{\alpha}, \xi, t)=\varphi_0(\theta^{\alpha}, t)+\xi\,\varphi_1(\theta^{\alpha}, t), \tag{18.6}$$

or adopt the more specialized approximation[68]

$$\theta^*(\theta^{\alpha}, \xi, t)=\theta(\theta^{\alpha}, t)=\varphi_0(\theta^{\alpha}, t). \tag{18.7}$$

In the former case we also assume that $q^{n\alpha}=0$, $R^n=0$ for $n \geqq 2$. Then, the approximate expression for the free energy ψ in (18.3) will depend only on φ_0 and φ_1 and the complete theory will involve only two energy equations and two entropy inequalities corresponding to $n=0, 1$ in (17.29) and (17.26). On the other hand, if the approximation (18.7) is adopted, we also assume that $q^{n\alpha}=0$, $R^n=0$ for $n \geqq 1$. The approximate expression for the free energy will now depend only on θ (in place of φ_N) and it follows from a relation of the form (17.23) that $\eta^N=0$ for $N \geqq 1$ and

$$\eta=-\frac{\partial\psi}{\partial\theta}, \tag{18.8}$$

where we use the notation of $(11.44)_2$. Moreover, in this case $r^n=0$ for $n \geqq 1$ and we have the single residual energy equation

$$\varrho r-\varrho\,\theta\,\dot{\eta}-q^{\alpha}{}_{|\alpha}=0 \tag{18.9}$$

for the determination of temperature and the single inequality

$$-q^{\alpha}\,\theta_{,\alpha} \geqq 0, \tag{18.10}$$

in place of (17.29) and (17.26). It is worth observing that if the approximation (18.7) was adopted at an earlier stage in place of (11.42), then there would have been no need to introduce the thermal resultants $\varepsilon^n, \eta^n, \psi^n, R^n$ for $n>0$; and, as a result, we would obtain a single energy equation and a single entropy in-

[67] These results, which were stated in (12.23), hold approximately, since they are obtained with the use of (18.3).

[68] The approximation (18.6) allows for temperature variation through the thickness of the shell while the approximation (18.7) accounts only for temperature variation on the surface $\xi=0$.

equality corresponding to (11.59) and (11.64) for $n=0$ only.[69] A similar remark can be made if (18.6) was adopted earlier in place of (11.42).

Apart from the approximation for the temperature, the constitutive equations of the above approximate theory are functions of the kinematic variables (18.2) and the reference values (18.1). We now adopt the approximation (18.7) for the temperature; and, for later reference and subsequent linearization, record the two-dimensional free energy as

$$\psi=\psi(e_{\alpha\beta}, \gamma_\alpha, \varkappa_{(\beta\alpha)}, s, s_\alpha, \theta), \tag{18.11}$$

where ψ in (18.11) is a different function from that in (18.3) and depends also on the reference values (18.1). The constitutive equations for η, $N'^{\alpha\beta}$, m^α, $M^{(\alpha\beta)}$, m^3, $M^{\alpha 3}$ are still of the forms in (18.4)–(18.5) and (18.8) but with ψ now given by that in (18.11). Also the constitutive equation for q^α depends on $\theta, \theta_{,\alpha}$, the variables (18.2) and the reference values (18.1).[70] One further remark should be made regarding the nature of the equation of motion in the above approximation procedure, either with ψ given by the approximate expression (18.3) or (18.11). In considering the approximate equations of motion [Sect. 12, Subsect. ε)], it was observed that while $(12.16)_{1,2}$ are identities in an exact theory, they cannot in general be satisfied in an approximate theory. In addition to the remarks made following (12.31), we note that since the expression (18.3) [or (18.11)] for the free energy is no longer exact, we expect that some of the equations in $(12.16)_{1,2}$ may not be satisfied by the approximation procedure used in obtaining (12.28)–(12.30), (18.4)–(18.5) and related results. In fact, in view of (12.23) which follow from the constitutive relations and the approximate expression for the free energy, only equations corresponding to $n=1$ in $(12.16)_2$ are violated and these reduce to $(12.31)_2$.

β) *Approximation in the linear theory.* We summarize here the main results of the linear theory for thermoelastic shells based on the above approximation procedure but confine our attention to the case in which the free energy is specified by (18.11). Some aspects of linearized kinematics and the forms of the linearized field equations were discussed previously in Sect. 7 [Subsect. γ)] and Sect. 12 [Subsect. ζ)]. In particular, we recall that when the reference values D_i, $\Lambda_{i\alpha}$ are specified by (7.58), the variables s and s_α to $O(\varepsilon)$ become $2\gamma_3$ and $\varkappa_{3\alpha}$, respectively, as was noted in Sect. 7 [see the remarks following (7.58)]. Thus, when the components of $\boldsymbol{D}$ are specified by $(7.58)_{1,2}$, the distinction between the linearized kinematic variables corresponding to those in the argument of ψ in (18.11) and the set defined by (7.68), namely

$$e_{\alpha\beta}, \gamma_i, \varkappa_{(\beta\alpha)}, \varkappa_{3\alpha} \tag{18.12}$$

disappears. Also, the temperature θ is now measured from the reference temperature θ_0^* and is of $O(\varepsilon)$ [see Sect. 12, Subsect. ζ)].

The field equations and constitutive relations of the approximate theory can be linearized in the same way as their counterparts in the exact theory. In particular, the specific free energy of the approximate linear theory has the form[71]

$$\psi=\psi(e_{\alpha\beta}, \gamma_i, \varkappa_{i\alpha}, \theta), \tag{18.13}$$

[69] The approximation (18.7) for the temperature is employed by GREEN, LAWS and NAGHDI [1968, *4*] prior to consideration of constitutive equations.

[70] These constitutive equations of the approximate theory with the free energy specified by (18.11) are of the same form as those in (13.62)–(13.66), apart from an extra generality for the skew-symmetric part of $M^{\alpha\gamma}$ in $(13.65)_3$.

[71] In (18.13) and subsequent results for the approximate linear theory, we no longer emphasize the symmetry of $\varkappa_{\beta\alpha}$ and, for convenience, write $\varkappa_{i\alpha}$ for $\{\varkappa_{\beta\alpha}, \varkappa_{3\alpha}\}$. However, it is understood that $\varkappa_{\beta\alpha}$, as well as $M^{\alpha\beta}$ in the constitutive relation $(18.4)_3$ are symmetric.

where ψ is quadratic in the variables (18.12) and the temperature θ (measured from the reference temperature) and depends also on the reference values (18.1). It can be readily seen that the remaining constitutive equations upon linearization of (18.8), (18.4)–(18.5) and by virtue of the reduction of s, s_α mentioned above reduce to

$$\eta = -\frac{\partial \psi}{\partial \theta}, \tag{18.14}$$

$$N'^{\alpha\beta} = N'^{\beta\alpha} = \varrho_0 \frac{\partial \psi}{\partial e_{\alpha\beta}}, \qquad m^i = \varrho_0 \frac{\partial \psi}{\partial \gamma_i}, \qquad M^{\alpha i} = \varrho_0 \frac{\partial \psi}{\partial \varkappa_{i\alpha}}, \tag{18.15}$$

where η in (18.14) is measured from the reference entropy η_0^* and is of $O(\varepsilon)$ [see Sect. 12, Subsect. ζ)]. The approximate linear constitutive equations (18.13)–(18.15) have the same forms as (16.5)–(16.7), apart from an extra generality for the skew-symmetric part of $M^{\alpha\beta}$ in $(16.7)_3$.

Recalling (7.37)–(7.39) and the linearized version of (11.37), the definition of the heat flux resultant is now given by

$$Q^\alpha = \boldsymbol{q} \cdot \boldsymbol{A}^\alpha, \qquad Q^\alpha = \int_{-h/2}^{h/2} \mu\, q^{*\alpha}\, d\zeta, \tag{18.16}$$

where Q^α are the contravariant components of the heat flux resultant $\boldsymbol{q}$ and are measured per unit length (in the reference configuration) per unit time. The stress-resultants, the stress couples and the equations of motion of the approximate linear theory are given by (12.42). The residual energy equation (18.9) now reduces to

$$\varrho_0\, r - \varrho_0\, \theta_0^*\, \dot{\eta} - Q^\alpha{}_{|\alpha} = 0, \tag{18.17}$$

where the vertical bar denotes covariant differentiation with respect to $A_{\alpha\beta}$. The constitutive relation for Q^α will be of a form similar to that stated in Sect. 16 and its coefficients will be restricted by an inequality in the form (18.10) but with q^α replaced by Q^α. Our above discussion of the approximate linear theory utilizes the approximation (18.7) for the temperature. A more general development can be pursued in the presence of the approximation (18.6) and has been given in some detail by GREEN and NAGHDI.[72]

19. An alternative approximation procedure in the linear theory: II. Developments from the three-dimensional theory. The central point in the approximation procedure outlined in Sect. 18 is the assumption that the specific free energy can be approximated by an expression of the form (18.11) or in the case of the linear theory by a function which is quadratic in the infinitesimal temperature θ and the variables (18.12). However, even in the linear theory an explicit form of the function ψ in (18.13) was not obtained from the full three-dimensional expression for the free energy so that the constitutive coefficients (of the linear theory) still remain arbitrary and unrelated to the elastic constants in the three-dimensional

[72] GREEN and NAGHDI [1970, *2*]. In the older literature on the linear theories of elastic shells and plates, thermal effects were generally confined to thermal stresses without full thermodynamical considerations. Within the scope of the classical plate theory and in the presence of steady state temperature distribution of the form (18.6), thermal stresses in plates have been discussed by MARGUERRE [1935, *1*] and by MELAN and PARKUS [1953, *1*]. A more general derivation of heat conduction equations in the linear theory of shells is given by BOLOTIN [1960, *1*] and is used in [1962, *6*] in connection with a formulation of non-isothermal elastokinetic problems of shallow shells; BOLOTIN's approximate equations do not account for thermo-mechanical coupling effects. The results given in [1970, *2*] with the use of approximation (18.6) for the temperature and upon the neglect of thermo-mechanical coupling effects reduce to those obtained by BOLOTIN [1960, *1*].

theory. In order to provide explicit constitutive relations (for the approximate linear theory) in which the coefficients are related to the elastic constants in the three-dimensional theory, we outline in this section an approximation procedure in terms of the specific *Gibbs free energy function*. This approximation procedure will be helpful in effecting an explicit derivation of the constitutive relations for thin plates and shells from the three-dimensional equations.[73]

Our developments in this and the next section are carried out in the context of elastostatic theories of shells and plates and are also confined to isothermal deformation and to isotropic materials. However, the latter limitations are not essential and are adopted in order to focus attention on the main features of the approximation procedure.

We recall that the constitutive equations of the isothermal linear theory of elasticity may be expressed in terms of the (three-dimensional) specific Gibbs free energy function φ^* in the form[74]

$$\gamma_{ij}^* = -\varrho_0^* \frac{\partial \varphi^*}{\partial \tau^{ij}}, \tag{19.1}$$

where the infinitesimal strain γ_{ij}^* is given by (7.60),

$$\varphi^* = \varphi^*(\tau^{ij}) = \psi^*(\gamma_{ij}^*) - \frac{1}{\varrho_0^*} \tau^{ij} \gamma_{ij}^* \tag{19.2}$$

and φ^* and ψ^* are quadratic functions of their arguments and both also depend upon the reference values G_{ij}. The Gibbs function φ^* (or the complementary energy function in the isothermal theory) for an initially homogeneous and isotropic material can be expressed as

$$\varrho_0^* \varphi^* = \left[-\frac{1+\nu}{2E} G_{im} G_{jn} + \frac{\nu}{2E} G_{ij} G_{mn} \right] \tau^{ij} \tau^{mn}, \tag{19.3}$$

where G_{ij} is the initial metric tensor defined in (4.23), E is Young's modulus of elasticity and ν is Poisson's ratio.[75]

Within the scope of the linear theory and remembering the remarks made following (7.46), corresponding to the resultant (11.43) and $(11.44)_3$ with $n=1$, we define

$$\varrho_0 \varphi A^{\frac{1}{2}} = \int_{-\frac{h}{2}}^{\frac{h}{2}} \varrho_0^* \varphi^* G^{\frac{1}{2}} d\xi, \tag{19.4}$$

where φ is a two-dimensional Gibbs free energy (or a complementary energy in the isothermal theory). With the use of the linearized versions of $(11.7)_2$, (11.36), (11.55), (12.10)–(12.11), as well as (19.1)–(19.2) and (19.4), in a manner similar to the development in Sect. 17 we can show that the constitutive equations of

[73] The approximation procedure in terms of the Gibbs function is proposed by GREEN, NAGHDI and WENNER [1971, *6*] and is used by them to obtain an approximate theory for plates of variable thickness from the three-dimensional equations. This procedure has some features in common with an approximation procedure employing a variational theorem in which assumptions are admitted simultaneously for both stresses and displacements.

[74] The partial derivative $\partial\varphi^*/\partial\tau^{ij}$ is understood to have the symmetric form

$$\frac{1}{2}\left(\frac{\partial \varphi^*}{\partial \tau^{ij}} + \frac{\partial \varphi^*}{\partial \tau^{ji}}\right).$$

[75] The use of the letter ν in (19.3) and other constitutive relations should not be confused with that in (7.34).

the linear isothermal theory may also be expressed in the forms

$$\begin{aligned}\varrho_0\,\varphi &= \varrho_0\,\overline{\varphi}\,(N'^{\alpha\beta}, M^{N\alpha i}, m^{Ni})\\ &= \varrho_0\,\overline{\psi} - \left\{N'^{\alpha\beta}\,e_{\alpha\beta} + \sum_{N=1}^{\infty}(M^{N\alpha i}\,\varkappa_{Ni\alpha} + m^{Ni}\,\gamma_{Ni})\right\},\end{aligned} \tag{19.5}$$

$$e_{\alpha\beta} = -\varrho_0\,\frac{\partial\overline{\varphi}}{\partial N'^{\alpha\beta}}, \qquad \varkappa_{Ni\alpha} = -\varrho_0\,\frac{\partial\overline{\varphi}}{\partial M^{N\alpha i}}, \qquad \gamma_{Ni} = -\varrho_0\,\frac{\partial\overline{\varphi}}{\partial m^{Ni}}. \tag{19.6}$$

In (19.5), $\overline{\varphi}$ is a quadratic function of its arguments, $\overline{\psi}$ is a quadratic function of $e_{\alpha\beta}, \varkappa_{Ni\alpha}, \gamma_{Ni}$ and both $\overline{\psi}$ and $\overline{\varphi}$ depend also on the reference values (18.1).

The constitutive equations (19.5)–(19.6), together with the linearized version of the equations of motion (12.13)–(12.16), form a system of infinite equations in an infinite number of unknowns. In parallel to the approximation procedure of Sect. 18, we now assume that the Gibbs free energy function $\overline{\varphi}$ can be represented by an approximate expression which is independent of $M^{N\alpha i}$, m^{Ni} for $N \geqq 2$. Thus, using the notations of (12.27), we set

$$\varphi = \tilde{\varphi}(N'^{\alpha\beta}, M^{\alpha i}, m^i) \tag{19.7}$$

approximately, where $\tilde{\varphi}$ is a different function from that in (19.5). Using (19.7) it follows from $(19.6)_{2,3}$ that

$$\gamma_{Ni} = 0, \qquad \varkappa_{Ni\alpha} = 0 \qquad (N \geqq 2),$$

approximately.[76] We also assume that in the linearized equations of equilibrium, $M^{N\alpha i}$, m^{Ni} and L^{Ni} (for $N \geqq 2$) can be neglected so that these equations reduce to (12.33)–(12.35) with D_i, $\Lambda_{i\alpha}$ given by (7.58) and with F^i, L^i in place of $\overline{F}^i$, $\overline{L}^i$, respectively. Moreover, as in the approximation procedure of Sect. 18, the linearized versions of the conditions (12.16) are satisfied approximately, except those for $n=1$. These reduce to the forms (12.31), which may be violated in an approximate theory and about which we have already remarked in Sect. 18 and in Sect. 12 following (12.31). By means of the above procedure, the constitutive relations of the approximate theory are easily seen to be[77]

$$e_{\alpha\beta} = -\varrho_0\,\frac{\partial\tilde{\varphi}}{\partial N'^{\alpha\beta}}, \qquad \varkappa_{i\alpha} = -\varrho_0\,\frac{\partial\tilde{\varphi}}{\partial M^{\alpha i}}, \qquad \gamma_i = -\varrho_0\,\frac{\partial\tilde{\varphi}}{\partial m^i}, \tag{19.8}$$

where $\tilde{\varphi}$ is the function in (19.7). Alternatively, the constitutive equations (19.8) can be expressed in terms of the variables $e_{\alpha\beta}, \varrho_{(\beta\alpha)}, \varrho_{3\alpha}, \gamma_i$ [see (7.68)] and those in (12.42). Thus, writing $\varrho_{i\alpha}$ for $\{\varrho_{(\beta\alpha)}, \varrho_{3\alpha}\}$, we have

$$e_{\alpha\beta} = -\varrho_0\,\frac{\partial\hat{\varphi}}{\partial N'^{\alpha\beta}}, \qquad \varrho_{i\alpha} = -\varrho_0\,\frac{\partial\hat{\varphi}}{\partial M^{\alpha i}}, \qquad \gamma_i = -\varrho_0\,\frac{\partial\hat{\varphi}}{\partial V^i}, \tag{19.9}$$

where $\hat{\varphi}$ is a function of $N'^{\alpha\beta}$, $M^{\alpha i}$, V^i [and also depends on the reference values (18.1)]:

$$\varphi = \hat{\varphi}(N'^{\alpha\beta}, M^{\alpha i}, V^i). \tag{19.10}$$

Further conclusions may be obtained from the above procedure without additional assumptions. We recall again that the initial position vector can always be taken

---

[76] These results which were stated in (7.65) hold approximately, since they are obtained with the use of (19.7).

[77] Although in (19.7)–(19.8) and elsewhere in this section we have written $\varkappa_{i\alpha}$ for $\{\varkappa_{\beta\alpha}, \varkappa_{3\alpha}\}$ and $M^{\alpha i}$ for $\{M^{\alpha\beta}, M^{\alpha 3}\}$, it is understood that $\varkappa_{\beta\alpha}$ and $M^{\alpha\beta}$ in the constitutive relations $(19.8)_2$ and (19.7) are symmetric tensors. This is analogous to the notation adopted in (18.13) and (18.15).

in the form (7.30), with the initial values of D_i and $\Lambda_{i\alpha}$ specified by (7.58). Then, as discussed in Sect. 12 [Subsect. δ)], it follows from (7.65) and (7.59) that consistently with the above approximation procedure the displacement $\boldsymbol{u}^*$ has the form (7.67) and by (7.68) the relevant (two-dimensional) kinematic measures are either the set $e_{\alpha\beta}, \varkappa_{i\alpha}, \gamma_i$ or $e_{\alpha\beta}, \varrho_{i\alpha}, \gamma_i$.

20. Explicit constitutive equations for approximate linear theories of plates and shells: II. Developments from the three-dimensional theory. Using the approximation procedure of Sect. 19, in this section we obtain explicit forms for constitutive equations in approximate linear theories of thin plates and shells of uniform thickness h. Although our derivations are carried out in the context of elastostatic theories, some remarks regarding dynamical problems are also included. Moreover, we confine attention to isotropic materials but the extension to anisotropic materials will be essentially similar.

α) *Approximate constitutive equations for plates.* Prior to the calculation of an explicit form for the approximate expression (19.10), we need to dispose of certain preliminaries. For this purpose, recall (7.48) and let the surface loads on the bounding surfaces $\zeta = \pm h/2$ of the initially flat plate be specified by

$$\begin{aligned} \boldsymbol{P}'' &= p''^{\alpha} \boldsymbol{A}_\alpha - p'' \boldsymbol{A}_3 \quad \text{on} \quad \zeta = \frac{h}{2}, \\ \boldsymbol{P}' &= p'^{\alpha} \boldsymbol{A}_\alpha + p' \boldsymbol{A}_3 \quad \text{on} \quad \zeta = -\frac{h}{2}. \end{aligned} \tag{20.1}$$

Then, in the absence of three-dimensional body force $\boldsymbol{f}^*$, from resultants of the type (11.29)–(11.30) we find

$$\begin{aligned} \varrho_0 F^\alpha &= p''^{\alpha} + p'^{\alpha}, \qquad & \varrho_0 F^3 &= -p'' + p', \\ \varrho_0 L^\alpha &= \frac{h}{2}(p''^{\alpha} - p'^{\alpha}), \qquad & \varrho_0 L^3 &= -\frac{h}{2}(p'' + p'). \end{aligned} \tag{20.2}$$

The outward unit normals to the surfaces $\zeta = \pm h/2$ have the components $\{n_\alpha = 0, n_3 = \pm 1\}$. Hence, by (11.10) and $(11.7)_1$, the surface tractions on $\zeta = \pm h/2$ are

$$\begin{aligned} \tau^{3\alpha} &= \begin{Bmatrix} p''^{\alpha} \\ -p'^{\alpha} \end{Bmatrix} \quad \text{on} \quad \zeta = \pm\frac{h}{2}, \\ \tau^{33} &= \begin{Bmatrix} -p'' \\ -p' \end{Bmatrix} \quad \text{on} \quad \zeta = \pm\frac{h}{2}. \end{aligned} \tag{20.3}$$

In order to calculate an approximate expression for $\hat{\varphi}$ in the form (19.10), we need to introduce *suitable* assumptions for the stresses τ^{ij} in terms of the resultants $N^{\alpha\beta}$, $M^{\alpha i}$, V^i defined in (12.43). Also, in the calculation of $\hat{\varphi}$, we shall assume that the effect of surface loads is negligible and attempt to satisfy the following conditions as closely as possible: The stresses must satisfy (i) the definitions of the resultants in (12.43) and (ii) the boundary conditions (20.3) with $p', p'', p'^{\alpha}, p''^{\alpha}$ all zero.[78]

Since the two-dimensional equations governing the behavior of isotropic plates separate into those for bending and extensional theories, it is instructive

[78] We could delete (ii) and satisfy the boundary conditions in the presence of $p', p'', p'_\alpha, p''_\alpha$; but then the effect of the surface loads will appear in the constitutive relations. It may be noted that the effect of the surface loads is sometimes retained in the constitutive equations of the approximate linear theories of shells and plates. See, e.g., Sect. 7.7 of GREEN and ZERNA [1968, 9].

to carry out the calculation for $\hat{\varphi}$ in (19.10) in two parts. Thus, employing again the same symbol for a function and its value (in line with our earlier notation), we write

$$\varphi = \varphi_e + \varphi_b, \tag{20.4}$$

where φ_e is associated with the extensional theory and φ_b with the bending theory, and calculate separately the approximate expressions for $\varrho_0 \varphi_e$ and $\varrho_0 \varphi_b$.

Consider first the case of bending and introduce the following approximate expressions for stresses:

$$\tau^{\alpha\beta} = \frac{6 M^{\alpha\beta}}{h^2} \frac{\zeta}{h/2}, \qquad \tau^{\alpha 3} = \tau^{3\alpha} = \frac{3}{2h} V^\alpha \left[1 - \left(\frac{\zeta}{h/2}\right)^2\right], \qquad \tau^{33} = 0. \tag{20.5}$$

The expressions (20.5) meet the conditions (i) and (ii) stated above.[79] Introducing (20.5) into (19.3) and using the resulting expression in (19.4), we obtain

$$\begin{aligned} \varrho_0 \varphi_b = & \frac{6}{E h^3} \left[\nu A_{\alpha\beta} A_{\gamma\delta} - \frac{1+\nu}{2} (A_{\alpha\gamma} A_{\beta\delta} + A_{\alpha\delta} A_{\beta\gamma})\right] M^{(\alpha\beta)} M^{(\gamma\delta)} \\ & - \frac{6(1+\nu)}{5 E h} A_{\alpha\beta} V^\alpha V^\beta, \end{aligned} \tag{20.6}$$

where we have written $M^{(\alpha\beta)}$ for $M^{\alpha\beta}$ in order to emphasize that $M^{\alpha\beta}$ in (20.6) is symmetric. The constitutive equations for $\varrho_{(\alpha\beta)}$ and γ_α can be obtained from (19.9) and (20.6). Thus,

$$\begin{aligned} \varrho_{(\alpha\beta)} &= \frac{12}{E h^3} \left[-\nu A_{\alpha\beta} A_{\gamma\delta} + (1+\nu) A_{\alpha\gamma} A_{\beta\delta}\right] M^{(\gamma\delta)}, \\ \gamma_\alpha &= \frac{12(1+\nu)}{5 E h} A_{\alpha\beta} V^\beta \end{aligned} \tag{20.7}$$

and there is no constitutive relation for $\varrho_{[\alpha\beta]}$ since the symmetric $M^{(\alpha\beta)}$ occurs in (20.6).

For the extensional theory of plates, we again assume that the effect of surface loads (in the constitutive relations) are negligible and write the following approximate expressions for stresses:

$$\tau^{\alpha\beta} = \frac{N^{\alpha\beta}}{h}, \qquad \tau^{\alpha 3} = \tau^{3\alpha} = \frac{15}{h^2} M^{\alpha 3} \left[\frac{\zeta}{h/2} - \left(\frac{\zeta}{h/2}\right)^3\right], \qquad \tau^{33} = \frac{V^3}{h}. \tag{20.8}$$

The expressions (20.8) meet the condition (i) above and also satisfy $(20.3)_1$ with $p'^\alpha = p''^\alpha = 0$. The approximate expression for φ_e can now be obtained from (19.3)–(19.4) and (20.8). Thus,

$$\begin{aligned} \varrho_0 \varphi_e = & \frac{1}{2 E h} \left[\nu A_{\alpha\beta} A_{\gamma\delta} - \frac{1+\nu}{2} (A_{\alpha\gamma} A_{\beta\delta} + A_{\alpha\delta} A_{\beta\gamma})\right] N^{\alpha\beta} N^{\gamma\delta} \\ & + \frac{\nu}{E h} A_{\alpha\beta} N^{\alpha\beta} V^3 - \frac{1}{2 E h} (V^3)^2 \\ & - \frac{120}{7} \frac{1+\nu}{E h^3} A_{\alpha\beta} M^{\alpha 3} M^{\beta 3} \end{aligned} \tag{20.9}$$

---

[79] Our assumption for τ^{33} in (20.5) could be easily modified so as to satisfy $(20.3)_2$. But, as already mentioned, this would result in the effect of the surface loads appearing in the constitutive relations. We also note here that $(20.5)_{1,2}$ satisfy the local forms of the first two of the three-dimensional equations of equilibrium $\boldsymbol{T}^i{}_{,i} = 0$, rather than merely the integrated forms of equilibrium equations.

and the constitutive relations for $e_{\alpha\beta}$, $\varrho_{3\alpha}$, γ_3 are found from (19.9) and (20.9):

$$
\begin{aligned}
e_{\alpha\beta} &= \frac{1}{Eh}\left[-\nu A_{\alpha\beta} A_{\gamma\delta} + (1+\nu) A_{\alpha\gamma} A_{\beta\delta}\right] N^{\gamma\delta} - \frac{\nu}{Eh} A_{\alpha\beta} V^3, \\
\gamma_3 &= \frac{1}{Eh} V^3 - \frac{\nu}{Eh} A_{\alpha\beta} N^{\alpha\beta}, \\
\varrho_{3\alpha} &= \frac{240}{7} \frac{1+\nu}{Eh^3} A_{\alpha\beta} M^{\beta 3}.
\end{aligned}
\tag{20.10}
$$

The constitutive equations (20.10) and (20.7) with the help of the dual of $(4.12)_6$, i.e., $A_{\alpha\gamma} A^{\gamma\beta} = \delta^\beta_\alpha$, can be easily inverted to give the relations

$$
\begin{aligned}
N^{\alpha\beta} &= (1-\nu)\, C \left\{\left[\frac{\nu}{1-2\nu} A^{\alpha\beta} A^{\gamma\delta} + A^{\alpha\gamma} A^{\beta\delta}\right] e_{\gamma\delta} + \frac{\nu}{1-2\nu} A^{\alpha\beta} \gamma_3\right\}, \\
V^3 &= \frac{1-\nu}{1-2\nu} C\left[(1-\nu)\gamma_3 + \nu A^{\alpha\beta} e_{\alpha\beta}\right], \\
M^{\alpha 3} &= \frac{7}{20} (1-\nu)\, B A^{\alpha\beta} \varrho_{3\beta}
\end{aligned}
\tag{20.11}
$$

for the extensional theory and the relations

$$
\begin{aligned}
M^{(\alpha\beta)} &= B\left[\nu A^{\alpha\beta} A^{\gamma\delta} + (1-\nu) A^{\alpha\gamma} A^{\beta\delta}\right] \varrho_{(\gamma\delta)}, \\
V^\alpha &= \frac{5}{6} \mu h A^{\alpha\beta} \gamma_\beta
\end{aligned}
\tag{20.12}
$$

for the bending theory. In (20.11)–(20.12), the shear modulus of elasticity μ and the coefficients C and B are given by

$$
\mu = \frac{E}{2(1+\nu)}, \qquad C = \frac{Eh}{1-\nu^2}, \qquad B = \frac{Eh^3}{12(1-\nu^2)}.
\tag{20.13}
$$

Also, the approximate Helmholtz free energy function corresponding to $\hat{\varphi}$ in (19.10) can be calculated with the help of (19.5), (7.65), (20.9), (20.6) and (20.11)–(20.12). Thus

$$
\begin{aligned}
\psi &= \psi_e + \psi_b, \\
\varrho_0 \psi_e &= \frac{1-\nu}{2} C \left\{\left[\frac{\nu}{1-2\nu} A^{\alpha\beta} A^{\gamma\delta} + \frac{1}{2}\left(A^{\alpha\gamma} A^{\beta\delta} + A^{\alpha\delta} A^{\beta\gamma}\right)\right] e_{\alpha\beta} e_{\gamma\delta}\right. \\
&\quad \left. + \frac{1-\nu}{1-2\nu} (\gamma_3)^2 + \frac{2\nu}{1-2\nu} A^{\alpha\beta} e_{\alpha\beta} \gamma_3\right\} + \frac{7}{40}(1-\nu)\, B A^{\alpha\beta} \varrho_{3\alpha} \varrho_{3\beta}, \\
\varrho_0 \psi_b &= \frac{1}{2} B \left[\nu A^{\alpha\beta} A^{\gamma\delta} + \frac{1-\nu}{2}\left(A^{\alpha\gamma} A^{\beta\delta} + A^{\alpha\delta} A^{\beta\gamma}\right)\right] \varrho_{(\alpha\beta)} \varrho_{(\gamma\delta)} \\
&\quad + \frac{5}{12} \mu h A^{\alpha\beta} \gamma_\alpha \gamma_\beta.
\end{aligned}
\tag{20.14}
$$

The constitutive relations (20.11), apart from the presence of V^3 and $M^{\alpha 3}$, correspond to the classical result for the extensional theory of isotropic plates.[80] The constitutive Eqs. (20.12) for equilibrium problems in the bending theory of

[80] The relation between $(20.11)_1$ and the constitutive equation for $N^{\alpha\beta}$ in the classical extensional theory (i.e., for generalized plane stress) will become apparent below in Subsect. β). The value of $\frac{7}{20}(1-\nu) B$ for the coefficient of $(20.11)_3$ is a consequence of our approximation (20.8) and was also obtained in [1971, *6*]. An explicit relation of the form $(20.11)_3$ has not been discussed previously in the literature and there is no evidence available at present regarding the reasonableness of $(20.11)_3$ as an approximate relation for $M^{\alpha 3}$.

plates were first derived by REISSNER, using a variational theorem.[81] This theory includes the effect of transverse shear deformation γ_α in the constitutive equations and requires the specification of three boundary conditions at each edge of the plate. These are either the displacement conditions[82]

$$\gamma_\alpha, \; u_3 \text{ on } \partial\mathfrak{S} \tag{20.15}$$

or the boundary conditions for the resultants (in the bending theory)

$${}_0\nu_\alpha M^{\alpha\beta}, \; {}_0\nu_\alpha V^\alpha \quad \text{on } \partial\mathfrak{S}, \tag{20.16}$$

where $\partial\mathfrak{S}$ refers to the boundary $\mathfrak{S}$ of the middle surface $\xi=0$ of the plate and ${}_0\nu_\alpha$ are the components of the outward unit normal ${}_0\boldsymbol{\nu}$ to $\partial\mathfrak{S}$. This is in contrast to the classical bending theory of plates which, as will be seen below, requires only two boundary conditions at each edge of the plate.

A theory similar to REISSNER's but for flexural vibrations of plates is developed by MINDLIN.[83] In MINDLIN's development, the constitutive equation for the shear-stress resultant corresponding to $(20.12)_2$ is obtained as follows: In the early stage of his derivation he assumes a constitutive relation in the form

$$V^\alpha = \mu' \, h A^{\alpha\beta} \gamma_\beta, \qquad \mu' = \varkappa^2 \mu, \tag{20.17}$$

where $\varkappa^2$ (and therefore μ') is a constant but unspecified. Then, at the completion of his derivation by comparing the solution for the circular frequency of the thickness-shear vibration with the corresponding exact three-dimensional solution due to LAMB,[84] MINDLIN makes the identification $\varkappa^2 = \pi^2/12$ [a value which is very close to REISSNER's $5/6$ in $(20.12)_2$].

β) *The classical plate theory. Additional remarks.* We recall that the constitutive relations in the three-dimensional theory of elasticity, referred to normal coordinates (4.25) with metric tensor components of the type in (7.37), may be expressed in the form

$$\begin{aligned}
\tau_{\alpha\beta} &= 2\mu \left[\frac{\nu}{1-\nu} \gamma^{*\lambda}_{\;\lambda} G_{\alpha\beta} + \gamma^*_{\alpha\beta}\right] + \frac{\nu}{1-\nu} \tau_{33} G_{\alpha\beta}, \\
\tau_{\alpha 3} &= 2\mu \, \gamma^*_{\alpha 3}, \\
\gamma^*_{33} &= -\frac{\nu}{1-\nu} \gamma^{*\lambda}_{\;\lambda} + \frac{1-2\nu}{1-\nu} \frac{1}{2\mu} \tau_{33}.
\end{aligned} \tag{20.18}$$

The constitutive equations of the classical theory of plates is generally obtained from integration of (20.18) or from the counterparts of (17.1) and (17.3) in the linear isothermal theory with $\hat{\psi}^*$ being a quadratic function of the linearized γ^*_{ij},

---

[81] REISSNER [1945, *2*]; see also [1944, *5*], [1947, *5*] and [1950, *5*]. A derivation of REISSNER's theory of bending of plates from the three-dimensional equations of classical elasticity is given by GREEN [1949, *2*] and may also be found in GREEN and ZERNA [1968, *9*]. Some related additional references on the subject include the papers by BOLLE [1947, *1*], SCHÄFER [1952, *3*], NAGHDI and ROWLEY [1953, *2*] and ESSENBURG and NAGHDI [1958, *2*]. The last contains a derivation of constitutive equations [corresponding to (20.12)] for plates of variable thickness.

[82] The nature of the boundary conditions for the bending of plates in REISSNER's theory, as well as those for the extensional theory, is clear from an examination of the linearized balance of energy or a corresponding virtual work principle.

[83] MINDLIN [1951, *1*]. A previous development for flexural vibrations of plates was given by UFLYAND [1948, *1*]. Related aspects of the subject are discussed in several papers by MINDLIN: E.g., [1951, *2*], [1955, *6*] and [1960, *8*]. For extensional vibrations of plates at moderate or high frequencies, reference may be made to the papers by MINDLIN [1960, *8*], [1961, *8*], [1963, *5*] where additional references are cited.

[84] LAMB [1917, *1*].

after introduction of the kinematic approximation (7.74) and certain additional assumptions. These assumptions pertain to (a) the neglect of the effect of "transverse shear deformation" due to $\gamma^*_{\alpha 3}$ and (b) the neglect of transverse normal stress τ_{33} in the constitutive equations. From the displacement approximation (7.74) and the assumption (a) follows the expression $(7.73)_1$ for β_α. In view of assumption (a) and $(20.18)_2$, it is clear that there will be no constitutive equation for the shear stress-resultant V^α defined in (12.43). Further, by assumption (b) and the definitions for $N^{\alpha\beta}$ and $M^{\alpha\beta}$ in (12.43), constitutive equations of the classical theory of plates can easily be obtained from integration of $(20.18)_1$.

Alternatively, the constitutive equations of the classical theory for $N^{\alpha\beta}$ and $M^{\alpha\beta}$ in (12.43) may be deduced in the following manner: The expression for the three-dimensional response function appropriate to the linear theory (mentioned above) can be calculated from $\varrho_0^* \hat{\psi}^*(\gamma^*_{ij}) = \frac{1}{2} \tau^{ij} \gamma^*_{ij}$, where $\hat{\psi}^*$ is a quadratic function of the infinitesimal strain tensor. Substitute for τ_{33} from $(20.18)_3$ into $(20.18)_1$ and introduce the resulting expression along with $(20.18)_2$ on the right-hand side of $\hat{\psi}^*$. Then, after invoking the assumptions (a) and (b) and using the displacement approximation (7.74), integration of $\hat{\psi}^*$ by means of the linearized version of a formula of the type (17.11) yields a two-dimensional function $\overline{\psi}$ which is quadratic in $e_{\alpha\beta}$ and $\varrho_{(\alpha\beta)}$ calculated from (7.74). Remembering the procedure by which the approximate constitutive equations of the linear theory were obtained in Sect. 18 and observing that the resultant V^α will not occur in the energy equation [as a consequence of assumption (a)], we are led to constitutive equations of the type (18.15) with ψ replaced by $\overline{\psi}$ for $N'^{\alpha\beta} = N^{\alpha\beta}$ and $M^{(\alpha\beta)}$.

The discussion in the preceding two paragraphs should indicate the manner in which the constitutive equations of the classical theory of plates are usually derived. Here, in keeping with our development in Subsect. α), we obtain the constitutive equations of the classical plate theory by the approximation method of Sect. 19. For this purpose, however, we must return to (20.5), (20.8) and modify the assumptions for stresses. With reference to constitutive equations of the type (19.9) in terms of a two-dimensional Gibbs free energy function (or a complementary energy function in the isothermal theory) and in view of the remarks made following (7.74), since $\gamma_i = 0$ in the classical theory, it follows from $(19.9)_{2,3}$ that $\hat{\varphi}$ in (19.10) cannot depend on V^α, $M^{\alpha 3}$ and V^3. Hence, consistently with this observation, we introduce the following assumptions for stresses:[85]

$$\tau^{\alpha\beta} = \frac{N^{\alpha\beta}}{h} + \frac{6 M^{\alpha\beta}}{h^2} \frac{\zeta}{h/2}, \qquad \tau^{\alpha 3} = 0, \quad \tau^{33} = 0. \tag{20.19}$$

Using the assumptions (20.19), from (19.4) we calculate the approximate expressions

$$\varphi = \varphi_e + \varphi_b,$$

$$\varrho_0 \varphi_e = \frac{1}{2Eh} \left[\nu A_{\alpha\beta} A_{\gamma\delta} - \frac{1+\nu}{2} (A_{\alpha\gamma} A_{\beta\delta} + A_{\alpha\delta} A_{\beta\gamma})\right] N^{\alpha\beta} N^{\gamma\delta}, \tag{20.20}$$

$$\varrho_0 \varphi_b = \frac{6}{E h^3} \left[\nu A_{\alpha\beta} A_{\gamma\delta} - \frac{1+\nu}{2} (A_{\alpha\gamma} A_{\beta\delta} + A_{\alpha\delta} A_{\beta\gamma})\right] M^{(\alpha\beta)} M^{(\gamma\delta)}$$

[85] The approximate expression $(20.19)_1$ represents combined contribution to both stretching and bending of a plate. The first and the second term on the right-hand side of $(20.19)_1$ are the same as $(20.8)_1$ and $(20.5)_1$, respectively.

and also obtain

$$e_{\alpha\beta} = \frac{1}{E h} \left[-\nu A_{\alpha\beta} A_{\gamma\delta} + (1+\nu) A_{\alpha\gamma} A_{\beta\delta} \right] N^{\gamma\delta},$$
$$\varrho_{(\alpha\beta)} = \frac{12}{E h^3} \left[-\nu A_{\alpha\beta} A_{\gamma\delta} + (1+\nu) A_{\alpha\gamma} A_{\beta\delta} \right] M^{(\gamma\delta)}. \tag{20.21}$$

The above results can be easily inverted to yield

$$N^{\alpha\beta} = C \left[\nu A^{\alpha\beta} A^{\gamma\delta} + (1-\nu) A^{\alpha\gamma} A^{\beta\delta} \right] e_{\gamma\delta},$$
$$M^{(\alpha\beta)} = B \left[\nu A^{\alpha\beta} A^{\gamma\delta} + (1-\nu) A^{\alpha\gamma} A^{\beta\delta} \right] \varrho_{(\gamma\delta)}. \tag{20.22}$$

Similarly, the approximate expression for the Helmholtz free energy function in the classical theory can be calculated from (19.5) and in a manner similar to that in (20.14). Thus, for the classical theory,

$$\psi = \psi_e + \psi_b,$$
$$\varrho_0 \psi_e = \frac{1}{2} C \left[\nu A^{\alpha\beta} A^{\gamma\delta} + \frac{1-\nu}{2} (A^{\alpha\gamma} A^{\beta\delta} + A^{\alpha\delta} A^{\beta\gamma}) \right] e_{\alpha\beta} e_{\gamma\delta},$$
$$\varrho_0 \psi_b = \frac{1}{2} B \left[\nu A^{\alpha\beta} A^{\gamma\delta} + \frac{1-\nu}{2} (A^{\alpha\gamma} A^{\beta\delta} + A^{\alpha\delta} A^{\beta\gamma}) \right] \varrho_{(\alpha\beta)} \varrho_{(\gamma\delta)}. \tag{20.23}$$

The constitutive equations $(20.21)_{1,2}$ or equivalently $(20.22)_{1,2}$ are those for the classical extensional theory (corresponding to generalized plane stress) and the classical bending theory of plates. The boundary conditions in the classical theory are four at each edge of the plate, two for the extensional theory and two for the bending theory. For the extensional theory they are specified by[86]

$$u_\alpha \quad \text{or} \quad {}_0\nu_\alpha N^{\alpha\beta} \quad \text{on } \partial\mathfrak{S} \tag{20.24}$$

and for the bending theory by

$$u_3, \quad \frac{\partial u_3}{\partial \nu_0}$$

or

$$M^{(\alpha\gamma)} {}_0\nu_\alpha {}_0\nu_\gamma, \quad {}_0\nu_\alpha M^{(\beta\alpha)}{}_{|\beta} - \frac{\partial}{\partial s_0} \left[\bar{\varepsilon}_{\beta\gamma} M^{(\alpha\beta)} {}_0\nu_\alpha {}_0\nu^\gamma \right] \quad \text{on } \partial\mathfrak{S}, \tag{20.25}$$

where $\partial\mathfrak{S}$ refers to the boundary of the middle surface $\xi = 0$ of the plate and $\partial/\partial\nu_0$, $\partial/\partial s_0$ denote the directional derivatives along the normal and the tangent to the boundary curve $\partial\mathfrak{S}$, respectively. The Kirchhoff boundary conditions $(20.25)_{3,4}$ should be compared with the boundary conditions (20.16) in REISSNER'S theory.[87] In this connection, we may recall that the differential equations of the classical bending theory (which requires the specification of two boundary conditions at each edge of the plate) can be reduced to the fourth-order partial differential equation

$$\nabla^4 u_3 = \frac{p}{B}, \tag{20.26}$$

---

[86] These boundary conditions may be derived in a manner similar to those for the restricted theory in Sect. 15.

[87] The classical theory of plates, prior to KIRCHHOFF'S paper [1850, *1*], was considered by POISSON (1828) and by CAUCHY (1828). References to these as well as some remarks on the history of the subject may be found in LOVE'S [1944, *4*] *Historical Introduction*. An account of the controversy over POISSON'S boundary conditions, KIRCHHOFF'S derivation of plate theory with correct boundary conditions and KELVIN and TAIT'S interpretation (1867) of KIRCHHOFF'S boundary conditions can be found in TODHUNTER and PEARSON [1886, *1*], Sects. 488–489 and [1893, *3*], Sects. 1238–1239.

where ∇^2 is the two-dimensional Laplacian and where we have used the notation $p=\varrho_0 F^3$ in place of that in $(20.2)_2$.[88] In contrast to (20.26), the differential equations of REISSNER's theory (which requires the specification of three boundary conditions at each edge of the plate) are equivalent to a system of sixth-order partial differential equations.

With a view toward an assessment of the effect of transverse shear deformation, we note that the system of differential equations in REISSNER's theory of bending of plates is characterized by the appropriate equilibrium equations in (12.43) and the constitutive equations (20.12). While this system of equations after elimination of V^α includes as a special case (with $\gamma_\alpha=0$) the corresponding system of differential equations in KIRCHHOFF's classical theory or equivalently the fourth-order differential equation (20.26), the practical significance of the former should not be exaggerated. Indeed, it is well known that for most purposes the solution of the simpler Eq. (20.26) compares favorably with the solution for the displacement u_3 of the corresponding equilibrium boundary-value problem in REISSNER's theory. On the other hand, the additional ingredients in REISSNER's theory (including the presence of three boundary conditions instead of two) may be significant in certain circumstances. Roughly speaking, the effect of transverse shear deformation may give rise to a noticeable contribution, if the equilibrium boundary-value problem in question involves a geometrical parameter which has the dimension of length and which is smaller than any other characteristic length on the middle plane of the plate but not so small as to be of the same order of magnitude as the plate thickness. Such a parameter may arise due to the presence of a hole in an infinite plate, from a distributed load acting over a small area on the middle plane of the plate or when a small portion of the surface is constrained against normal displacement. Some idea along these lines can be gained from a number of existing solutions in the literature;[89] and, in particular, from REISSNER's solution of the "stress concentration" (or rather stress-couple concentration) problem of an infinite plate with a circular hole subjected to a uniform state of plain bending (and twist) at infinity.[90]

The above discussion regarding the effect of transverse shear deformation is limited to elastostatic flexural problems. For dynamical problems of plates, where a parameter corresponding to wavelength is always present, the significance of the effect of transverse shear deformation (and rotatory inertia) for moderate frequencies has been brought out in extensive studies by MINDLIN and others.[91]

γ) *Approximate constitutive relations for thin shells.* We obtain here a system of constitutive relations for thin elastic shells using the approximation procedure

[88] An account of the classical bending theory of plates can be found in LOVE's treatise [1944, *4*] or in the books by NADAI [1925, *1*] and TIMOSHENKO and WOINOWSKY-KRIEGER [1959, *7*].

[89] Among such solutions we mention those in [1945, *2*], [1953, *2*], [1956, *1*] and [1962, *2*].

[90] REISSNER [1945, *2*]. For a ratio of hole radius to plate thickness $a/h<5$, the results (especially the stress-couple concentration factor) differ substantially from the solution of the corresponding problem according to the classical theory by GOODIER [1936, *1*]. In this connection, the discussions of the solution in [1945, *2*] by DRUCKER [1946, *1*] and by GOODIER [1946, *2*] are particularly interesting. The solution for the *stress-couple concentration* problem in [1945, *2*] should not be confused with the solution of the corresponding *stress concentration* problem within the scope of the three-dimensional theory. The latter three-dimensional problem has been considered by ALBLAS [1957, *1*] on the basis of an exact formulation of plate theory given by GREEN [1949, *3*]. The work of ALBLAS also includes a comparison with REISSNER's solution.

[91] For example, MINDLIN [1951, *1*], [1960, *8*], MINDLIN and ONOE [1957, *3*] and MIKLOWITZ [1960, *7*]. Additional references may be found in [1960, *8*] and in TIERSTEN's monograph [1969, *5*] on vibrations of piezoelectric plates.

of Sect. 19. Again our derivation is limited to elastostatic problems and, for simplicity, attention is confined to isotropic materials. As in Subsect. α), let the surface loads on

$$\zeta = \pm \frac{h}{2}$$

be specified by (20.1). Then, F^i, L^i are given by (20.2) and the boundary conditions on

$$\zeta = \pm \frac{h}{2}$$

are specified by (20.3).

Again, we calculate an approximate expression for φ in the form (19.10) and this requires the introduction of *suitable* assumptions for the stresses τ^{ij} in terms of the resultants $N^{\alpha\beta}$, $M^{\alpha i}$, V^i in (12.42). In this connection, we assume that the effect of surface load is negligible in the evaluation of $\hat{\varphi}$ and attempt to satisfy the conditions (i) and (ii) stated above [in Subsect. α)] as closely as possible. With this background and remembering the definitions of the stress-resultants and the stress-couples in (12.42), we write the following approximate expressions for the stresses:

$$\begin{aligned} \mu\, \tau^{\alpha\gamma}\, \mu_\gamma^\beta &= \frac{N^{\alpha\beta}}{h} + \frac{6M^{\alpha\beta}}{h^2}\,\frac{\zeta}{h/2}, \\ \mu\, \tau^{\alpha 3} = \mu\, \tau^{3\alpha} &= \frac{3\,V^\alpha}{2h}\left[1 - \left(\frac{\zeta}{h/2}\right)^2\right] + \frac{15}{h^2}\, M^{\alpha 3}\left[\frac{\zeta}{h/2} - \left(\frac{\zeta}{h/2}\right)^3\right], \\ \mu\, \tau^{33} &= \frac{V^3}{h} + \zeta\, B_{\alpha\beta}\left(\frac{N^{\alpha\beta}}{h} + \frac{6M^{\alpha\beta}}{h^2}\,\frac{\zeta}{h/2}\right). \end{aligned} \tag{20.27}$$

The expressions (20.27) meet the condition (i) above, and they satisfy the boundary conditions $(20.3)_1$ with $p'^\alpha = p''^\alpha = 0$. However, as in the case of flat plates, the boundary conditions $(20.3)_2$ with $p' = p'' = 0$ are not satisfied.[92] In view of the expression for $N'^{\alpha\beta}$ in (12.42), $(20.27)_1$ becomes

$$\mu\, \tau^{\alpha\gamma}\, \mu_\gamma^\beta = \frac{N'^{\alpha\beta}}{h} - \frac{M^{\gamma\alpha}}{h}\, B_\gamma^\beta + \frac{6M^{\alpha\beta}}{h^2}\,\frac{\zeta}{h/2}. \tag{20.28}$$

Since $D = 1$ in the present development, $\zeta = \xi$ by (7.40). Hence, use of (7.37)–(7.39) in (19.3) yields

$$\begin{aligned} \mu\, \varrho_0^*\, \varphi^* = &-\frac{1+\nu}{2E}\,\mu^{-1}\big[A_{\delta\nu}\, A_{\theta\varrho}\, \mu_\alpha^\delta\, \mu_\beta^\nu (\mu\, \tau^{\beta\gamma}\, \mu_\gamma^\varrho)\, (\mu\, \tau^{\alpha\lambda}\, \mu_\lambda^\theta) \\ &+ 2A_{\delta\nu}\, \mu_\alpha^\delta\, \mu_\beta^\nu (\mu\, \tau^{\alpha 3})\, (\mu\, \tau^{\beta 3}) + (\mu\, \tau^{33})^2\big] \\ &+ \frac{\nu}{2E}\,\mu^{-1}\big[A_{\delta\nu}\, \mu_\alpha^\delta (\mu\, \tau^{\alpha\beta}\, \mu_\beta^\nu) + \mu\, \tau^{33}\big]^2. \end{aligned} \tag{20.29}$$

In writing (20.29) no assumption is introduced beyond those already made in (20.27). However, before calculating φ from (19.4), we invoke the condition (4.31) and also make the following approximations in (20.29):

$$\begin{aligned} &\frac{1}{\mu} \cong 1, \quad \mu_\beta^\alpha \cong \delta_\beta^\alpha, \\ &\mu\, \tau^{\alpha\gamma}\, \mu_\gamma^\beta \cong \frac{N'^{\alpha\beta}}{h} + \frac{6M^{(\alpha\beta)}}{h^2}\,\frac{\zeta}{h/2}, \\ &\mu\, \tau^{33} \cong \frac{V^3}{h}, \end{aligned} \tag{20.30}$$

[92] We can modify $(20.27)_3$ so as to satisfy the boundary conditions $(20.3)_2$ but then the effect of surface loads will appear in the constitutive relations. See [1957, 4].

where in obtaining $(20.30)_3$ the term

$$\frac{M^{\gamma\alpha}}{h} B^{\beta}_{\gamma}, \text{ which is } O[(h\, B^{\gamma}_{\beta})] \text{ times } \frac{6\, M^{\alpha\beta}}{h^2}\, \frac{\zeta}{h/2},$$

has been neglected in view of (4.31). Also, since by $(20.30)_{1,2}$ the left-hand side of $(20.30)_3$ is now $\mu\, \tau^{\alpha\gamma}\, \mu^{\beta}_{\gamma} \cong \tau^{\alpha\beta} = \tau^{\beta\alpha}$, we have replaced $M^{\alpha\beta}$ in (20.30) by its symmetric part so that

$$M^{[\alpha\beta]} = 0. \tag{20.31}$$

The approximation (20.30) need not be made at this stage and could be postponed: Since μ^{-1} can be expressed as a convergent series in[93] ζ, (20.29) can be integrated without the introduction of the approximation (20.30). This is however a lengthy procedure and since approximations are eventually introduced after the integration, the development is tantamount to adopting (20.30).[94]

Using (20.29)–(20.30) and $(20.27)_2$ in (19.4) yields the approximate expression

$$\begin{aligned} 2\varrho_0\, \varphi = {} & \frac{1}{E h} [\nu A_{\alpha\beta} A_{\gamma\delta} - (1+\nu) A_{\alpha\delta} A_{\beta\gamma}]\, N'^{\alpha\beta} N'^{\gamma\delta} \\ & + \frac{2\nu}{E h} A_{\alpha\beta} N'^{\alpha\beta} V^3 - \frac{1}{E h} (V^3)^2 - \frac{240(1+\nu)}{7 E h^3} A_{\alpha\beta} M^{\alpha 3} M^{\beta 3} \\ & + \frac{12}{E h^3} [\nu A_{\alpha\beta} A_{\gamma\delta} - (1+\nu) A_{\alpha\delta} A_{\beta\gamma}]\, M^{(\alpha\beta)} M^{(\gamma\delta)} - \frac{6}{5\mu h} A_{\alpha\beta} V^{\alpha} V^{\beta}. \end{aligned} \tag{20.32}$$

The remaining constitutive relations, obtained by direct calculation from (19.9) and (20.32), are:

$$\begin{aligned} e_{\alpha\beta} &= \frac{1}{E h} \{[-\nu A_{\alpha\beta} A_{\gamma\delta} + (1+\nu) A_{\alpha\gamma} A_{\beta\delta}]\, N'^{\gamma\delta} - \nu A_{\alpha\beta} V^3\}, \\ \gamma_{\alpha} &= \frac{6}{5\mu h} A_{\alpha\beta} V^{\beta}, \qquad \gamma_3 = \frac{1}{E h} [V^3 - \nu A_{\alpha\beta} N'^{\alpha\beta}], \\ \varrho_{(\alpha\beta)} &= \frac{12}{E h^3} [-\nu A_{\alpha\beta} A_{\gamma\delta} + (1+\nu) A_{\alpha\gamma} A_{\beta\delta}]\, M^{(\gamma\delta)}, \\ \varrho_{3\alpha} &= \frac{240(1+\nu)}{7 E h^3} A_{\alpha\beta} M^{\beta 3}. \end{aligned} \tag{20.33}$$

From comparison of (20.32)–(20.33) with corresponding results in the case of flat plates, it is easily seen that the approximate expression for φ in (20.32) is of the same form as the combination of φ_e and φ_b in (20.9) and (20.6) and that the constitutive equations in (20.33) are of the same forms as those in (20.10) and (20.7). It follows that the inverted constitutive relations (20.33) will be of the same form as those in (20.11)–(20.12) but with $N^{\alpha\beta}$ replaced by $N'^{\alpha\beta}$. Also, the approximate expression for the Helmholtz free energy ψ obtained from (19.5) and (20.32) will be of the same form as (20.14). We postpone further remarks on the constitutive relations (20.33) or their inverted forms and consider next a derivation of the constitutive equations of the classical theory.

δ) *Classical shell theory. Additional remarks.* It was indicated in Sect. 7 [Subsect. ε)] that the displacement vector usually employed in the derivations of the

[93] Such convergent series representation for μ^{-1} and $\mu^{-1}{}^{\alpha}_{\beta}$ are given by NAGHDI [1963, *6*].

[94] In essence the approximation (20.30) will yield an energy function without coupling terms, i.e., one which has the same form as that for a flat plate. If the lengthier procedure is adopted and if the approximation of the type (20.30) is adopted after the integration, then the consequences of the two derivations will be the same.

classical theory (under Kirchhoff-Love assumptions) has the form (7.75). In order to obtain the constitutive relations of the classical theory by the procedure of this section, we must return to (19.4) and introduce a different set of approximate expressions for stresses. Since the classical theory ignores the effect of transverse shear deformation and the transverse normal stress in the constitutive equations, instead of (20.27) we now write

$$\mu\,\tau^{\alpha\gamma}\,\mu_\gamma^\beta = \frac{N^{\alpha\beta}}{h} + \frac{6M^{\alpha\beta}}{h^2}\,\frac{\zeta}{h/2}, \qquad \tau^{\alpha 3} = \tau^{33} = 0. \tag{20.34}$$

We can then calculate the expression for $\hat{\varphi}$ in a manner analogous to that in Subsect. β): First $\mu\,\varrho_0\,\varphi^*$ can be calculated using (19.3), (7.37)–(7.39) and (7.48). Next we introduce the assumptions (20.34) and use the approximation $(20.30)_{1,2}$ to obtain the function φ for the classical theory. This function will have the same form as that in (20.20) but with $N^{\alpha\beta}$ replaced by $N'^{\alpha\beta}$ and $\varrho_{(\alpha\beta)}$ by that for shells in (7.68). For ease of reference, we record below these constitutive relations in the form

$$\begin{aligned} N'^{\alpha\beta} &= C\left[\nu A^{\alpha\beta} A^{\gamma\delta} + (1-\nu)\, A^{\alpha\gamma} A^{\beta\delta}\right] e_{\gamma\delta}, \\ M^{(\alpha\beta)} &= B\left[\nu A^{\alpha\beta} A^{\gamma\delta} + (1-\nu)\, A^{\alpha\gamma} A^{\beta\delta}\right] \varrho_{(\gamma\delta)}, \\ e_{\gamma\delta} &= \tfrac{1}{2}(u_{\gamma|\delta} + u_{\delta|\gamma}) - B_{\gamma\delta}\, u_3, \qquad \varrho_{(\gamma\delta)} = -\bar{\varkappa}_{\gamma\delta}, \end{aligned} \tag{20.35}$$

where $\bar{\varkappa}_{\gamma\delta}$ is given by $(7.68)_9$.

The above linear constitutive equations of the classical theory involve the kinematic variables $e_{\alpha\beta}$ and $\varrho_{(\beta\alpha)}$. They can also be deduced from expressions of the type (18.15), namely

$$N'^{\alpha\beta} = N'^{\beta\alpha} = \varrho_0 \frac{\partial \psi}{\partial e_{\alpha\beta}}, \qquad M^{(\alpha\beta)} = \varrho_0 \frac{\partial \psi}{\partial \varrho_{(\beta\alpha)}}, \tag{20.36}$$

where ψ is a function different from that in (18.15). In fact, as in (20.23), ψ in (20.36) is a quadratic function of $e_{\alpha\beta}$ and $\varrho_{(\beta\alpha)}$. We consider now an alternative form of the constitutive equations of the classical theory employing a slightly different set of kinematic variables. For this purpose, put

$$\bar{\varrho}_{(\alpha\beta)} = \varrho_{(\alpha\beta)} + \tfrac{1}{2}(B_\alpha^\nu\, e_{\nu\beta} + B_\beta^\nu\, e_{\nu\alpha}). \tag{20.37}$$

Assuming now for ψ the form

$$\psi = \bar{\psi}(e_{\alpha\beta}, \bar{\varrho}_{(\beta\alpha)}), \tag{20.38}$$

we can then deduce the results

$$N^{(\alpha\beta)} = \varrho_0 \frac{\partial \bar{\psi}}{\partial e_{\alpha\beta}}, \qquad M^{(\alpha\beta)} = \varrho_0 \frac{\partial \bar{\psi}}{\partial \bar{\varrho}_{(\beta\alpha)}}, \tag{20.39}$$

where

$$\begin{aligned} N^{\alpha\beta} &= N^{(\alpha\beta)} + N^{[\alpha\beta]}, \\ N^{(\alpha\beta)} &= N'^{\alpha\beta} - \tfrac{1}{2}(M^{\lambda\alpha}\, B_\lambda^\beta + M^{\lambda\beta}\, B_\lambda^\alpha), \\ N^{[\alpha\beta]} &= \tfrac{1}{2}\left[B_\lambda^\alpha\, M^{\lambda\beta} - B_\lambda^\beta\, M^{\lambda\alpha}\right]. \end{aligned} \tag{20.40}$$

Because of the approximations of the type (20.30) already used in the calculation from the three-dimensional expressions for φ^*, i.e., the neglect of all terms of $O(h/R)$ or smaller compared with those in $\bar{\psi}$ [or a function φ in the form (20.20)], to the same order of approximation we can neglect the second and the third terms on the right-hand side of $(20.40)_2$ and write

$$N^{(\alpha\beta)} = N'^{\alpha\beta}. \tag{20.41}$$

It should be evident from the above development [between (20.34)–(20.41)] that the two sets of constitutive relations (20.36) and (20.39) are equivalent. These equations are effectively of the type originally given by LOVE (in lines of curvature coordinates) but with added recent improvements.[95]

To summarize the system of differential equations which characterize the (classical) bending theory of shells with the approximation (7.75) for the displacement, we first recall that resultants corresponding to $(12.38)_{4,5}$ are not admitted in the classical theory. Moreover, in line with the approximations used in the derivation of the constitutive equations and in order to render the theory determinate, the skew-symmetric part of $M^{\alpha\beta}$ is specified by (20.31). Then, $M^{\alpha\beta}$ can be replaced by $M^{(\alpha\beta)}$ also in the equations of equilibrium (or motion) while the skew-symmetric part of $N^{\alpha\beta}$, calculated from $(12.35)_1$ or the corresponding expression in (12.42), is given by

$$N^{[\alpha\beta]} = \tfrac{1}{2}\{B^{\alpha}_{\gamma} M^{(\gamma\beta)} - B^{\beta}_{\gamma} M^{(\gamma\alpha)}\}. \tag{20.42}$$

It follows that the equilibrium equations in the classical bending theory are given by the first three differential equations in (12.42) in the absence of the inertia terms and with $L^i = 0$. These equations involve the resultants V^α, $N^{\alpha\beta}$ and the symmetric $M^{(\alpha\beta)}$ by virtue of (20.31). The shear stress-resultant V^α may be eliminated from the equilibrium equations in (12.42) in a manner similar to that in Sect. 15 [see (15.19)–(15.20)]. Thus, from the three equilibrium equations in (12.42), we obtain either the set[96]

$$\begin{aligned} N'^{\alpha\beta}{}_{|\alpha} - B^{\beta}_{\gamma|\alpha} M^{(\gamma\alpha)} - 2B^{\beta}_{\gamma} M^{(\gamma\alpha)}{}_{|\alpha} + \varrho_0 F^\beta = 0, \\ M^{(\alpha\beta)}{}_{|\alpha\beta} + B_{\alpha\beta} N'^{\alpha\beta} - B_{\alpha\beta} B^{\beta}_{\gamma} M^{(\gamma\alpha)} + \varrho_0 F^3 = 0 \end{aligned} \tag{20.43}$$

or the set

$$\begin{aligned} N^{(\alpha\beta)}{}_{|\alpha} - \tfrac{1}{2}[B^{\beta}_{\gamma} M^{(\gamma\alpha)}]_{|\alpha} + \tfrac{1}{2}[B^{\alpha}_{\gamma} M^{(\gamma\beta)}]_{|\alpha} - B^{\beta}_{\gamma} M^{(\gamma\alpha)}{}_{|\alpha} + \varrho_0 F^\beta = 0, \\ M^{(\alpha\beta)}{}_{|\alpha\beta} + B_{\alpha\beta} N^{(\alpha\beta)} + \varrho_0 F^3 = 0. \end{aligned} \tag{20.44}$$

---

[95] The constitutive equations (20.35) or (20.36), as well as (20.39), possess a certain symmetry in structure and the kinematic variables $\varrho_{(\alpha\beta)}$ and $\bar{\varrho}_{(\alpha\beta)}$ are unaffected by infinitesimal rigid body displacements. This is in contrast to corresponding kinematic quantities in various versions of LOVE's first approximation (see Sect. 21 A). The set of constitutive equations (20.39) when expressed in a form similar to (20.35) and specialized to lines of curvature coordinates have features which are essentially those contained in a derivation first given by NOVOZHILOV [1946, *3*]. (For further details, see Sect. 21 A.) The kinematic measure $\bar{\varrho}_{\alpha\beta}$ was introduced by SANDERS [1959, *6*] in lines of curvature coordinates and by KOITER [1960, *6*], [1961, *5*] in a form which can be expressed as in (20.37). The choice of $\bar{\varrho}_{(\alpha\beta)}$ by SANDERS is motivated mainly from his consideration of a (two-dimensional) virtual work principle for shells. KOITER [1960, *6*], who does not introduce a displacement assumption such as (7.75), obtains his expression for $\bar{\varrho}_{\alpha\beta}$ by approximation and begins his calculation of a strain energy function by assuming that the state of stress as in (20.34) is approximately plane so that transverse shear stress and transverse normal stress may be neglected in the (three-dimensional) strain energy density; and he subsequently defines a modified stress-resultant and obtains his equilibrium equations by a variational method. An account of SANDER's and KOITER's contributions may be found in [1963, *6*]. Our derivation of (20.35) or the relations that follow from (20.39) is patterned after that given by NAGHDI [1963, *7*] and [1964, *5*], who used a virtual work principle for shells derived from the corresponding principle in the three-dimensional theory. See also NAGHDI [1966, *7*].

[96] The Eqs. (20.43) can be deduced from the original differential equations in (12.42) even without the specification of (20.31). See Eqs. (3.8) in [1963, *7*].

Both sets of equations are remarkably free from the skew-symmetric $N^{[\alpha\beta]}$. The former involves only the symmetric $N'^{\alpha\beta}$ and $M^{(\alpha\beta)}$ while the latter is expressed in terms of the symmetric part of $N^{\alpha\beta}$ and $M^{(\alpha\beta)}$.

It is now clear that the classical theory may be characterized by two entirely equivalent systems of differential equations: Either by the system of Eqs. (20.43) and (20.36) or by (20.44) and (20.39).[97] The boundary conditions appropriate to the classical theory may be obtained by considering the linearized (three-dimensional) rate of work expression. After integrating with respect to ζ between the limits

$$-\frac{h}{2}, \frac{h}{2}$$

and using the definitions of the resultants in (12.42), the procedure to be followed is similar to that in Sect. 15 [between (15.21)–(15.25)]. For equilibrium problems, these boundary conditions are either[98]

$$u_\beta, u_3, \frac{\partial u_3}{\partial \nu_0} \quad \text{on } \partial\mathfrak{S} \tag{20.45}$$

or

$$\begin{gathered} {}_0\nu_\alpha [N'^{\alpha\beta} - 2 B^\beta_\gamma M^{(\alpha\gamma)}], \; M^{(\alpha\gamma)} {}_0\nu_\alpha {}_0\nu_\gamma, \\ {}_0\nu_\alpha M^{(\beta\alpha)}{}_{|\beta} - \frac{\partial}{\partial s_0} (\bar{\varepsilon}_{\beta\gamma} M^{(\alpha\beta)} {}_0\nu_\alpha {}_0\nu^\gamma) \quad \text{on } \partial\mathfrak{S} \end{gathered} \tag{20.46}$$

and hold pointwise on the boundary $\partial\mathfrak{S}$. The boundary condition $(20.46)_1$ can also be expressed in terms of $N^{(\alpha\beta)}$.

The derivation of the approximate constitutive equations in Subsect. γ) is based on the assumption (20.27) and leads to (20.33) or their inverted forms, including a constitutive relation for $M^{\alpha 3}$. These constitutive equations are appropriate for a bending theory whose equilibrium equations are given by those which can be obtained from (12.42) in the absence of inertia terms. Also, it is clear that the number of the independent boundary conditions in this theory are six at each edge of the shell. This is in contrast to the four boundary conditions in the classical theory given by (20.45)–(20.46).

We consider now briefly a (slightly less general) system of equations of the linear theory in which the resultant $M^{\alpha 3}$ is not admitted. The equations of equilibrium are then given by those associated with the first three differential equations in (12.42) and $V^3 = \varrho_0 L^3$. The derivation of constitutive equations in this theory can be accomplished in just the way that led to (20.33) but with a modification in the initial assumption for stresses. In the present development, we retain $(20.27)_{1,3}$ but (since $M^{\alpha 3}$ is not admitted) replace $(20.27)_2$ by

$$\mu\,\tau^{\alpha 3} = \mu\,\tau^{3\alpha} = \frac{3\,V^\alpha}{2h}\left[1 - \left(\frac{\zeta}{h/2}\right)^2\right].$$

The above expression will still meet the boundary conditions $(20.3)_1$ with $p'^\alpha = p''^\alpha = 0$. The rest of the development parallels that in Subsect. γ) with obvious modifications, in view of the absence of $M^{\alpha 3}$. In this manner we finally obtain a system of constitutive equations which are the same as the first four of

[97] In the recent literature, sometimes preference is indicated in favor of the constitutive equations (20.36) or (20.39) together with (20.41) [or a variant of (20.40)] and with $\bar{\varrho}_{(\alpha\beta)}$ expressed in terms of the components of rotation about the normal to the middle surface. In view of the equivalence of the two systems of equations indicated above, such a preference can only be argued on the basis of convenience or a possible simplification resulting with reference to a specific problem. Budiansky and Sanders [1963, *1*] have argued in favor of a set of constitutive equations corresponding to (20.39); the label of "the 'best' first order linear shell theory" adopted by them and repeated by others is misleading.

[98] The boundary conditions (20.46) can also be derived (from the three-dimensional theory) even without the specification of (20.31). See Eqs. (4.20) in [1963, *7*].

(20.33). For later reference, we record below the inverted forms of these constitutive equations:

$$\begin{aligned} N'^{\alpha\beta} &= (1-\nu)\, C\left\{\left[\frac{\nu}{1-2\nu} A^{\alpha\beta} A^{\gamma\delta} + A^{\alpha\gamma} A^{\beta\delta}\right] e_{\gamma\delta} + \frac{\nu}{1-2\nu} A^{\alpha\beta} \gamma_3\right\}, \\ M^{(\alpha\beta)} &= B\left[\nu A^{\alpha\beta} A^{\gamma\delta} + (1-\nu) A^{\alpha\gamma} A^{\beta\delta}\right] \varrho_{(\gamma\delta)}, \\ V^{\alpha} &= \frac{5}{6}\mu h A^{\alpha\beta} \gamma_{\beta}, \\ V^{3} &= \frac{1-\nu}{1-2\nu} C\left[(1-\nu)\gamma_3 + \nu A^{\alpha\beta} e_{\alpha\beta}\right]. \end{aligned} \tag{20.47}$$

The above constitutive equations include the effect of transverse shear deformation and transverse normal stress.[99] The boundary conditions in the above linear theory with constitutive equations (20.47) are either the displacement boundary conditions

$$u_\alpha,\ u_3,\ \gamma_\alpha \quad \text{on } \partial\mathfrak{S} \tag{20.48}$$

or the resultant boundary conditions

$${}_0\nu_\alpha N^{\alpha\beta},\ {}_0\nu_\alpha M^{\alpha\beta},\ {}_0\nu_\alpha V^{\alpha} \quad \text{on } \partial\mathfrak{S}. \tag{20.49}$$

The boundary-value problem of the linear bending theory characterized by the appropriate equations of equilibrium (noted earlier in this paragraph), as well as (20.47)–(20.49), include certain features in common with REISSNER'S plate theory. Remarks similar to those made at the end of Subsect. β) can also be made here.

Before closing this section some additional remarks should be made regarding the nature of the foregoing derivations of constitutive equations for shells and plates from the three-dimensional theory. Although an effort was made to be as systematic as possible, the shortcomings of the derivations associated with the approximations are self-evident. Moreover, no attempt has been made here to provide clear-cut justifications (which can be supported by analysis) for the approximations or to obtain an estimate of the "error" involved in the use of the approximate constitutive equations. As remarked previously (Sect. 4), the problem posed under (b) in (4.32) regarding a scheme for estimating the "error" involved in the use of the (approximate) two-dimensional equations of the classical theory of shells or more general linear theories, as compared to the three-dimensional equations of linear elasticity, has not been solved with finality even in the case of the linear bending theory of plates.[100]

[99] Constitutive equations of this type with various degrees of generality and from different points of view have been given previously: HILDEBRAND, REISSNER and THOMAS [1949, *4*], GREEN and ZERNA [1950, *2*], REISSNER [1952, *2*], [1964, *7*] and NAGHDI [1957, *4*], [1963, *6*], [1964, *6*]. These papers (except for [1964, *6*]), however, employ a kinematic measure different from $\varrho_{(\gamma\delta)}$ in $(20.47)_2$.

[100] With reference to the classical bending theory of plates, however, MORGENSTERN [1959, *2*] has shown that the stresses and strains obtained from a solution in plate theory converge in a mean square sense to a solution in elasticity theory as the plate thickness approaches zero. A recent related paper by NORDGREN [1971, *7*] contains an explicit estimate of the mean square error for the stresses obtained from a solution in plate theory with respect to the exact solution of the corresponding problem in the three-dimensional theory. By constructing a somewhat more elaborate three-dimensional displacement field than that in [1971, *7*], an improved estimate for the error in the classical bending theory of plates has been obtained by SIMMONDS [1971, *8*]. A more ambitious effort toward the problem posed under (b) in (4.32) was undertaken by JOHN ([1965, *5*] and [1969, *4*]) in the context of a nonlinear theory, as was noted earlier. A comparison of a linear shell theory with the exact three-dimensional linear theory, based to some extent on the work of JOHN [1965, *5*], is given by SENSENIG [1968, *11*]. The question of "error" estimate in the linear theory of shells has been considered recently by KOITER [1970, *4*]. KOITER'S analysis is carried out in the spirit of earlier works of MORGENSTERN [1959, *2*] for the bending of flat plates and MORGENSTERN and SZABO [1961, *9*] for generalized plane stress.

21. Further remarks on the approximate linear and nonlinear theories developed from the three-dimensional equations. In this section, we first supplement the previous remarks on the classical linear theories of shells and plates and then briefly discuss the nature of some of the approximate nonlinear theories obtained from the three-dimensional equations. Preliminary to the discussion which follows, we recall that certain mathematical rules or procedures (such as those pertaining to dimensional invariance, coordinate invariance, consistency, equipresence, etc.) are observed or employed directly in the construction of the constitutive equations.[101] In particular, we recall here that the constitutive equations must fulfill the following requirements: (a) They must be consistent with all conservation laws and equations resulting therefrom—hence they must be consistent with the equations of motion, the energy equation and all energetic theorems and related results; (b) they must remain unaffected by rigid body motions;[102] and (c) they must remain invariant under the transformation of coordinate systems or, in the present context, under the transformation of the middle surface coordinates.[103]

Ordinarily, most of the rules of invariance or procedures referred to above are not explicitly stated; and often, especially in the case of linear theories of continuum mechanics, it is taken for granted that the derived constitutive equations meet all of the above requirements. Yet, as the history of the subject shows,[104] the above requirements have not always been satisfied in the case of the constitutive equations of the linear theory of elastic shells. These shortcomings stem largely from the trend in which the bending theory of elastic shells has been historically developed—often piecemeal and in an *ad hoc* fashion, in contrast to the developments of both the three-dimensional and two-dimensional (plane) linear elasticity theories—but also from the difficult nature of the topic, as well as the use of special coordinates (e.g., lines of curvature coordinates) or special shapes (e.g., cylindrical and spherical shells).[105]

The constitutive equations derived in Sect. 20 [Subsects. γ) and δ)] meet all invariance requirements mentioned above. However, some of the linear constitutive equations employed in the current literature on shell theory and even recent books on the subject violate one or more of the invariance requirements (see Sect. 21 A). Although the extent to which the latter constitutive equations violate the requirements of the type (a) to (c) may be insignificant in most practical applications, such shortcomings are nevertheless undesirable from a theoretical standpoint.[106]

---

[101] For an account of these rules, see Sect. 293 of TRUESDELL and TOUPIN [1960, *14*].

[102] In the case of linear theory, this requirement implies that the constitutive equations must remain unaffected by infinitesimal rigid body motions.

[103] This requirement guarantees that the response of the material is unaltered by a different coordinate description; in the case of shell theory, it is easily fulfilled by deriving the equations in tensorial or similar general forms.

[104] A brief history of the derivation of the constitutive equations of the linear theory of elastic shells, beginning with the work of LOVE [1888, *1*], is given in Sect. 21 A.

[105] Some of the trends have surprisingly persisted throughout the years; even in very recent and current literature, it is not uncommon to find the use of lines of curvature coordinates in discussions intended to reveal features of a general theory.

[106] The foregoing remarks are not intended to detract from most of the specific results and solutions in the literature which have been obtained in the past with the use of somewhat simpler sets of constitutive equations [such as the set which includes (21 A.7) and described in Sect. 21 A], despite their shortcomings. As noted by KOITER [1960, *6*], if the nonvanishing physical component of the rotation [corresponding to that in (21 A.8)] is *small* or of the same order of magnitude when compared with the physical components of $e_{\alpha\beta}$, then from a practical point of view the use of the simpler set of constitutive equations may be justified. On the other hand, as brought out by KNOWLES and REISSNER [1958, *3*] and by COHEN's [1960, *2*] analysis of a helicoidal shell, if the physical component of rotation is *large* in comparison with components of $e_{\alpha\beta}$, then the use of the simpler set of constitutive equations [e.g., the set which includes (21 A.7)] can lead to serious errors.

Having disposed of the foregoing preliminaries, we return once more to the discussion of an approximate linear theory of elastic shells obtained from the three-dimensional equations. Our derivation of linear constitutive equations in Sect. 20 is carried out in the spirit of earlier developments (from the three-dimensional equations) in Chap. D. Alternatively (and with some adjustments in the initial assumptions), the same results can be deduced from the generalized Hooke's law or with the use of the (three-dimensional) virtual work principle or any of the variational theorems of three-dimensional linear elasticity. The latter approach has been popular over the years beginning with a paper of TREFFTZ.[107]

An entirely different approach to the linear theories of plates and shells, based on the use of asymptotic expansion techniques, has attracted considerable attention in recent years. This approach which begins with the three-dimensional equations of linear elasticity, roughly speaking, may be described as follows: With the initial position vector of a material point specified by (4.27), first *suitable* scaling of the coordinates and, in turn of the displacements u_i^* and stresses τ^{ij} are introduced. This scaling is such that it gives distinction between the tangential components u_α^*, $\tau^{\alpha\gamma}$ and the components u_3^*, τ^{i3} along the normal to the middle surface of the shell. Next, with a view toward separate consideration of the "interior" problem (i.e., for the region away from the edge of the shell or plate) and that of the boundary-layer region, all equations are expanded in powers of a *small* parameter (say $\varepsilon = h/R$)[108] and successive approximations to the three-dimensional equations are deduced for the interior and the boundary-layer regions. Inasmuch as a fairly detailed and recent account of the asymptotic expansion procedure for shells is available elsewhere (Chap. 16 of GREEN and ZERNA),[109] we confine ourselves here to only a few remarks in order to give an indication of the nature of such developments.

In the case of flat plates, the earliest derivation by an asymptotic procedure is included in a paper by GOODIER.[110] The subject was considered anew by FRIEDRICHS and DRESSLER with particular reference to the boundary-layer equations for elastic plates.[111] In essence, the lowest order approximation to the "interior" problem leads to the classical Kirchhoff theories for stretching and bending of plates. However, the interior equations are not uniformly valid in the immediate region of the edge surface of the plate in the sense that it is not possible to satisfy arbitrary specified boundary conditions on the edge surface with the interior equations and the boundary-layer problem must be considered. From the latter and by a method first used by GREEN,[112] the usual Kirchhoff boundary conditions

---

[107] TREFFTZ [1935, *2*]. The use of variational theorems of a virtual work principle are prominent in a number of papers cited in Sect. 20 and are also employed in most books on the subject: E.g., NOVOZHILOV [1959, *3*], GOL'DENVEIZER [1961, *3*] and AMBARTSUMIAN [1964, *1*].

[108] It is usual in the case of shells to introduce two parameters, one given by ε and another associated with the deformation pattern on the middle surface which implies dependence on the nature of the solutions of the differential equations sought. However, this is not essential and can be avoided as indicated by GREEN and NAGHDI [1965, *2*].

[109] GREEN and ZERNA [1968, *9*].

[110] GOODIER [1938, *1*].

[111] FRIEDRICHS and DRESSLER [1961, *2*]. For related papers on asymptotic expansions and boundary-layer equations for elastic plates, see REISS and LOCKE [1961, *11*], GOL'DENVEIZER [1962, *3*], FOX [1964, *2*], GOL'DENVEIZER and KOLOS [1965, *1*] and LAWS [1966, *6*].

The above papers are all concerned with derivations of classical plate theory by asymptotic expansions of the equations of (non-polar) linear elasticity. GREEN and NAGHDI [1967, *5*] have shown that the equations of the linear theory of an elastic Cosserat plate can be obtained as a first approximation to an asymptotic expansion of an exact linearized three-dimensional theory of a generalized continuum which admits a director. See also a related paper by ERINGEN [1967, *3*] concerning the derivation of plate equations from a theory of generalized continua.

[112] GREEN [1962, *5*]; see also GREEN and LAWS [1966, *3*].

can then be deduced and these must be applied to the major terms of the interior stresses.[113]

As might be expected, the derivations by asymptotic methods for elastic shells are both more complex and laborious than that for flat plates, but the essence of the ideas (though not the details of the method and results) are similar. Asymptotic derivations of this type have been discussed by JOHNSON and REISSNER, REISS, GREEN and others.[114] Roughly speaking, four sets of approximate system of equations for shells are obtained by the asymptotic procedure and these correspond to (1) membrane approximation, (2) inextensional approximation, (3) a simplified system of equations for bending and (4) boundary-layer equations.[115] The four sets of Eqs. (1)–(4) do not constitute a complete system of shell equations in the sense of the classical shell theory obtained in Sect. 20 [Subsect. δ)], but they may be regarded as a basis for obtaining approximate solutions to shell problems. In fact, in the context of the asymptotic procedure used, the Eqs. (20.39) of the classical shell theory contain a significant contribution of an order higher than the first as was noted by GREEN and NAGHDI.[116] In any case, the derivations by asymptotic expansion techniques for shells (and plates) do not appear to be conclusive, especially with regard to the problem posed under (b) in Sect. 4.[117] The derivations by asymptotic expansion techniques, at first sight, may appear to be free from *ad hoc* assumptions; but, in fact, this is not the case. The scaling of stresses and displacement is tantamount to *a priori* special assumptions regarding the transverse components u_3^* and τ^{i3}, although subsequent developments are carried out systematically. Moreover, no proof is available that the expansions obtained are asymptotic and of uniform validity over the entire region of the shell or plate.

We close this section with some remarks concerning the nature of the existing approximate nonlinear theories of elastic plates and shells derived from the three-dimensional equations. In nearly all of these approximate theories, the underlying kinematic assumptions are such that the strains are small while the rotation may be *large* or moderately large[118] and linear constitutive equations are assumed to be

[113] Boundary conditions for plates have been discussed by FRIEDRICHS [1949, *1*], [1950, *1*], FRIEDRICHS and DRESSLER [1961, *2*], REISSNER [1963, *9*] and others. Some of these papers use a "matching" procedure between the interior and the boundary-layer stresses, but the simpler method of GREEN [1962, *5*] is used by LAWS [1966, *6*] to obtain the classical Kirchhoff boundary conditions.

[114] The earliest papers are by JOHNSON and REISSNER [1959, *1*] and REISS [1960, *12*] who confined themselves to symmetric deformations of cylindrical shells. Subsequently, the case of axisymmetric deformation of shells of revolution was considered by REISSNER [1960, *13*] and the unsymmetric deformation of cylindrical shells was dealt with by REISS [1962, *7*]. General derivations of the interior and the boundary-layer equations for shells were given by GREEN [1962, *4*], [1962, *5*]; see also GREEN and LAWS [1966, *3*]. Some simplifications in the procedure, as well as extension of the results to dynamical problems, were noted by GREEN and NAGHDI [1965, *2*]. For other related references, see the papers by GOL'DENVEIZER [1963, *3*], JOHNSON [1963, *4*] and REISSNER [1963, *10*].

[115] The system of equations in each of the categories (1)–(4) represents an approximation to the three-dimensional equations. Sometimes the approximations (1) and (2) are referred to as the membrane theory and the inextensional theory, respectively. But the latter terminologies may possibly detract from clarity here, since what is sought in an asymptotic development are approximations to the three-dimensional equations. While the approximate system of equations in each of the categories (1)–(4) may violate one or more of the requirements stated under (a) to (c) above, it should be noted that this is simply a consequence of the particular approximation used.

[116] [1965, *2*].

[117] This is perhaps partly evident from GOL'DENVEIZER's recent effort [1969, *2*] with regard to the interaction of boundary layer and interior stresses for thin shells.

[118] A discussion of such kinematic assumptions is given in Sects. 48–49 of NOVOZHILOV [1953, *3*]. See also the review of [1953, *3*] by TRUESDELL [1953, *5*].

valid. In general, however, a systematic development of such approximate theories is not available; and even those few contributions which have striven toward a more satisfactory derivation either employ assumptions which are too special or else still contain a number of *ad hoc* approximations. Nevertheless, it should be noted that these approximate developments have been largely motivated by applications and have served a useful purpose over the years. The origin of such approximate theories for initially flat plates goes back to papers of FÖPPL (in which bending effects are ignored) and VON KÁRMÁN.[119] The latter contains the well-known von Kármán equations for nonlinear bending of thin plates characterized by a system of coupled nonlinear partial differential equations in terms of a stress function and the normal displacement.[120]

With reference to initially curved shells, a general derivation of the nonlinear membrane theory (in the context of nonlinear elasticity) is given in GREEN and ADKINS.[121] This derivation does not contain approximations for the tangential components of strain measures and employs nonlinear constitutive equations. A more special nonlinear membrane theory in which the strains (but not the displacements) are assumed to be small was given earlier by BROMBERG and STOKER.[122] A number of approximate nonlinear theories have been developed with reference to special geometries. Notable among these are the papers by DONNELL for cylindrical shells,[123] MARGUERRE for shallow shells[124] and by REISSNER for axisymmetric deformation of shells of revolution.[125] Finally, we note that more general approximate nonlinear theories have been obtained in recent years with different aims and varying degrees of generality.[126]

---

[119] FÖPPL [1907, *1*], VON KÁRMÁN [1910, *1*].

[120] An account of VON KÁRMÁN's equations can be found in STOKER's [1968, *12*] monograph and in TIMOSHENKO and WOINOWSKY-KREIGER [1959, *7*]. Related and more general derivations are contained in the papers by REISSNER [1953, *4*] and ERINGEN [1955, *1*]. Large inextensional deformations of plates are discussed by FUNG and WITRICK [1955, *2*], MANSFIELD [1955, *5*] and ASHWELL [1957, *2*].

[121] GREEN and ADKINS [1960, *5*].

[122] BROMBERG and STOKER [1945, *1*]. Related results pertaining to nonlinear membrane theory are contained in the papers of STOKER [1963, *14*], ERINGEN [1952, *1*] and BUDIANSKY [1968, *1*].

[123] DONNELL [1933, *1*]. A linearized version of DONNELL's equations, because of their relative simplicity, has been extensively employed in applications but not without concern regarding their accuracy. In this connection, see a paper by HOFF [1955, *3*] which contains a discussion of the accuracy of DONNELL's equations as compared to those of FLÜGGE [1932, *1*].

[124] MARGUERRE [1938, *2*]. The papers of MARGUERRE and DONNELL are also significant because of the linearized versions of their developments in [1938, *2*] and [1933, *1*]. In fact, MARGUERRE's paper contains, as a special case, the first satisfactory derivation of the linear theory of shallow shells.

Equations equivalent to those given by DONNELL [1933, *1*] and MARGUERRE [1938, *2*] or variants thereof were independently discovered and utilized by others, e.g., VLASOV and MUSHTARI. For references, see VLASOV [1958, *5*], NOVOZHILOV [1959, *3*], ONIASHVILI [1960, *11*] and MUSHTARI and GALIMOV [1961, *10*]. The effect of nonlinearity in [1933, *1*] and [1938, *2*] is accounted for in a manner similar to that in VON KÁRMÁN's equations. A linearized version of DONNELL's equations may also be deduced as a special case of the differential equations for bending of circular cylindrical shells in GREEN and ZERNA [1968, *9*]. Similarly, a linearized version of MARGUERRE's theory of shallow shells may be found in GREEN and ZERNA [1968, *9*].

[125] REISSNER [1950, *6*], [1960, *13*] and [1963, *11*].

[126] We mention, in particular, the papers by LEONARD [1961, *6*], RÜDIGER [1961, *12*], NAGHDI and NORDGREN [1963, *8*], SANDERS [1963, *12*], KOITER [1966, *4*] and BIRICIKOGLU and KALNINS [1971, *2*]. All of these papers invoke the Kirchhoff-Love hypothesis or a similar equivalent set of assumptions and most (but not all) of them employ linear constitutive equations. However, a nonlinear theory in which the strains are small but the constitutive equations (through their coefficients) contain nonlinear effects has been also discussed by ZERNA [1960, *15*] and WAINWRIGHT [1963, *16*].

21 A. Appendix on the history of the derivation of linear constitutive equations for thin elastic shells. We begin our account of the history of the derivation of linear constitutive equations for elastic shells with that of LOVE's first approximation (LOVE's theory of 1888). However, as remarked in Sect. 1, the reference to LOVE's first approximation in the contemporary literature is often confusing, as there are at least three different versions which bear his name.[127] In order to distinguish between these and also provide a basis for their comparison with our results of the classical shell theory (Sect. 20), it is expedient to record the constitutive equations of the classical theory corresponding to one of the alternative forms (20.36) or (20.39) in lines of curvature coordinates and in terms of physical components. Here we choose the latter, since the constitutive equations of LOVE's first approximation involve resultants which correspond to $N^{(\alpha\beta)}$ and $M^{(\alpha\beta)}$ instead of $N'^{\alpha\beta}$ and $M^{(\alpha\beta)}$.

Thus, we refer the surface coordinates in the reference configuration to lines of curvature coordinates so that $A_{12}=B_{12}=0$ and also introduce the notations

$$\begin{gathered}(A_1)^2=A_{11}=|\boldsymbol{A}_1|^2, \quad (A_2)^2=A_{22}=|\boldsymbol{A}_2|^2,\\ B_1^1=-\frac{1}{R_1}, \quad B_2^2=-\frac{1}{R_2}, \quad B_{11}=-\frac{(A_1)^2}{R_1}, \quad B_{22}=-\frac{(A_2)^2}{R_2},\end{gathered} \tag{21 A.1}$$

where R_1 and R_2 are the principal radii of curvature of the initial reference surface and the notations A_1, A_2 are introduced for convenience. Further, let the physical components of tensors of the type $e_{\alpha\beta}$, $N^{\alpha\beta}$ be designated as $e_{\langle\alpha\beta\rangle}$, $N_{\langle\alpha\beta\rangle}$. Then[128]

$$\begin{aligned} e_{\langle\alpha\beta\rangle}&=\frac{e_{\alpha\beta}}{A_\alpha A_\beta} && \text{(no summation over } \alpha,\beta),\\ e_{\alpha\beta}&=A_\alpha A_\beta\, e_{\langle\alpha\beta\rangle} && \text{(no summation over } \alpha,\beta),\\ N_{\langle\alpha\beta\rangle}&=A_\alpha A_\beta N^{\alpha\beta} && \text{(no summation over } \alpha,\beta),\\ N^{\alpha\beta}&=\frac{N_{\langle\alpha\beta\rangle}}{A_\alpha A_\beta} && \text{(no summation over } \alpha,\beta). \end{aligned} \tag{21 A.2}$$

We record below the physical components of the symmetric strain measures $e_{\alpha\beta}$ and $\bar{\varrho}_{\alpha\beta}$, namely

$$\begin{aligned} e_{\langle 11\rangle}&=\frac{1}{A_1}\left[u_{\langle 1\rangle,1}+\frac{A_{1,2}}{A_2}u_{\langle 2\rangle}\right]+\frac{u_3}{R_1},\\ e_{\langle 22\rangle}&=\frac{1}{A_2}\left[u_{\langle 2\rangle,2}+\frac{A_{2,1}}{A_1}u_{\langle 1\rangle}\right]+\frac{u_3}{R_2},\\ e_{\langle 12\rangle}=e_{\langle 21\rangle}&=\frac{1}{2}\left\{\frac{A_1}{A_2}\left[\frac{u_{\langle 1\rangle}}{A_1}\right]_{,2}+\frac{A_2}{A_1}\left[\frac{u_{\langle 2\rangle}}{A_2}\right]_{,1}\right\} \end{aligned} \tag{21 A.3}$$

and

$$\begin{aligned} \bar{\varrho}_{\langle 11\rangle}&=-\frac{1}{A_1}\left[\frac{u_{3,1}}{A_1}-\frac{u_{\langle 1\rangle}}{R_1}\right]_{,1}-\frac{A_{1,2}}{A_1A_2}\left[\frac{u_{3,2}}{A_2}-\frac{u_{\langle 2\rangle}}{R_1}\right],\\ \bar{\varrho}_{\langle 22\rangle}&=-\frac{1}{A_2}\left[\frac{u_{3,2}}{A_2}-\frac{u_{\langle 2\rangle}}{R_2}\right]_{,2}-\frac{A_{2,1}}{A_1A_2}\left[\frac{u_{3,1}}{A_1}-\frac{u_{\langle 1\rangle}}{R_2}\right],\\ 2\bar{\varrho}_{\langle 12\rangle}=2\bar{\varrho}_{\langle 21\rangle}&=\left\{\frac{1}{A_1A_2}\left[(A_1)^2\left(\frac{u_{\langle 1\rangle}}{A_1R_1}\right)_{,2}+(A_2)^2\left(\frac{u_{\langle 2\rangle}}{A_2R_2}\right)_{,1}\right]\right.\\ &\quad\left.+\frac{2}{A_1A_2}\left[-u_{3,12}+\frac{A_{1,2}}{A_1}u_{3,1}+\frac{A_{2,1}}{A_2}u_{3,2}\right]\right\}\\ &\quad+\left(\frac{1}{R_1}-\frac{1}{R_2}\right)\left[\frac{1}{2A_1A_2}\left(\frac{1}{(A_1)^2}\{(A_1)^3u_{\langle 1\rangle}\}_{,2}-\frac{1}{(A_2)^2}\{(A_2)^3u_{\langle 2\rangle}\}_{,1}\right)\right]. \end{aligned} \tag{21 A.4}$$

---

[127] LOVE's original derivations [1888, *1*] and [1893, *2*], as well as related subsequent developments, were carried out in lines of curvature coordinates and in terms of physical components of the stress-resultants and the stress-couples.

[128] See Sect. A.4. of the Appendix (Chap. F).

The physical constitutive equations for $N^{(\alpha\beta)}$ and $M^{(\alpha\beta)}$ which follow from (20.39) in the case of initially homogeneous and isotropic elastic materials are then given by relations of the type

$$\begin{aligned} N_{\langle 11\rangle} &= C\,[e_{\langle 11\rangle} + \nu\, e_{\langle 22\rangle}], \\ N_{\langle(12)\rangle} &= N_{\langle(21)\rangle} = (1-\nu)\, C\, e_{\langle 12\rangle}, \\ M_{\langle 11\rangle} &= B\,[\bar{\varrho}_{\langle 11\rangle} + \nu\, \bar{\varrho}_{\langle 22\rangle}], \\ M_{\langle 12\rangle} &= M_{\langle 21\rangle} = (1-\nu)\, B\, \bar{\varrho}_{\langle 12\rangle}, \end{aligned} \tag{21 A.5}$$

where the kinematic measures are those in (21 A.3)–(21 A.4). Also, the physical equation for the skew-symmetric part of $N^{\alpha\beta}$ becomes

$$N_{\langle 12\rangle} - N_{\langle 21\rangle} = \left(\frac{1}{R_2} - \frac{1}{R_1}\right) M_{\langle 12\rangle}. \tag{21 A.6}$$

Returning to LOVE's first approximation, we first observe that a derivation of LOVE's constitutive equations in lines of curvature coordinates is contained in a paper by REISSNER.[129] This formulation of LOVE's first approximation, as will become apparent presently, is different from that in LOVE's treatise or the version originally given by LOVE in 1888. The essential difference between the physical constitutive equations (21 A.5) and the version of LOVE's equations as derived by REISSNER is in the expression for $M_{\langle 12\rangle}$, namely[130]

$$\begin{aligned} M_{\langle 12\rangle} &= M_{\langle 21\rangle} = \frac{1-\nu}{2}\, B\, \tau, \\ \tau &= \left(\frac{A_2}{A_1}\right)\left[\frac{\beta_{\langle 2\rangle}}{A_2}\right]_{,1} + \left(\frac{A_1}{A_2}\right)\left[\frac{\beta_{\langle 1\rangle}}{A_1}\right]_{,2}, \end{aligned} \tag{21 A.7}$$

where $\beta_{\langle 1\rangle}$, $\beta_{\langle 2\rangle}$ are the physical components of β_α given by $(7.68)_4$. The remaining constitutive equations of LOVE's theory (as derived by REISSNER) are the same as the corresponding equations in (21 A.5). Thus, the difference between the two sets of constitutive equations lies in the kinematic measures which occur in (21 A.7) and $(21\,\text{A}.5)_4$, respectively. An easy comparison between τ and $\bar{\varrho}_{\langle 12\rangle}$ can be made if, with the use of $(20.35)_4$ and the second expression for $\bar{\varkappa}_{\alpha\beta}$ in (7.68), we write (20.37) in the form

$$\bar{\varrho}_{\alpha\beta} = \tfrac{1}{2}(\beta_{\alpha|\beta} + \beta_{\beta|\alpha}) - \tfrac{1}{2}(B^{\nu}_{\alpha}\, \gamma_{[\nu\beta]} + B^{\nu}_{\beta}\, \gamma_{[\nu\alpha]}). \tag{21 A.8}$$

It is then easily seen that the expression for τ in $(21\,\text{A}.7)_2$ arises only from the first parenthesis on the right-hand side of (21 A.8). As noted above, the version of LOVE's theory as given in his treatise differs from that which includes (21 A.7) and has also been used extensively in the literature (at least until the early 1950's). The version of the constitutive equations in LOVE's treatise differs from the set derived by REISSNER only in the relation for $M_{\langle 12\rangle}$, where the expression corresponding to the kinematic measure τ is not symmetric in $u_{\langle 1\rangle}$ and $u_{\langle 2\rangle}$; and hence, in this sense, the constitutive equations in LOVE's treatise are considered unsatisfactory.[131]

---

[129] REISSNER [1941, *1*]. A similar derivation is included in [1949, *4*].

[130] The difference of the factor of $\frac{1}{2}$ in the coefficients of $(21\,\text{A}.5)_4$ and $(21\,\text{A}.7)_1$ is due to the definition of τ in $(21\,\text{A}.7)_2$.

[131] See Sect. 329 of LOVE [1944, *4*] or the corresponding sections of earlier editions of his treatise. Still, shortly after his [1941, *1*] paper, another version was proposed by REISSNER [1942, *1*]. The constitutive equations in [1942, *1*] contain higher-order terms in h/R in comparison with the set of equations which include (21 A.7); and, in particular, those corresponding to $M_{\langle 12\rangle}$ and $M_{\langle 21\rangle}$ are not the same.

Almost immediately after its appearance, LOVE'S work of 1888 drew criticism from Lord RAYLEIGH[132] who, favoring his own inextensional theory, objected to the application of LOVE'S equations to extensional vibrations of shells.[133] Shortly thereafter, LAMB[134] published a paper in which he derived by another approach equations essentially equivalent to those of LOVE. But despite LAMB'S praise of LOVE'S work, at the time LOVE appears to have been pessimistic about his own 1888 derivation of the general bending theory of thin shells.[135] Responding to RAYLEIGH'S criticism,[136] LOVE after again pointing out that the inextensional theory violates the boundary conditions and agreeing that the extensional theory will not predict the lowest frequency of free vibrations,[137] also notes that it may be necessary to retain higher-order terms in h/R as was done by BASSETT in special cases.[138] In this connection, it may be noted that LOVE'S second approximation[139] does contain some higher order terms in h/R.

During the past two or three decades, when speaking of LOVE'S first approximation, some authors have in mind the version of LOVE'S theory with $M_{\langle 12\rangle} = M_{\langle 21\rangle}$ given by (21 A.7) while others (especially the Russian investigators) have reference to the equations originally supplied by LOVE or those in his treatise mentioned above. Neither of the two sets of constitutive equations of LOVE'S first approximation (i) satisfies the equilibrium equation arising from the symmetry of the stress tensor or equivalently (21 A.6). Nor are they (ii) invariant under infinitesimal rigid body displacement.[140] Objections to LOVE'S first approximation (in the form given originally by LOVE) were first raised by VLASOV[141] who, in addition to the above shortcomings (i) and (ii), also pointed out that LOVE'S equations (iii) violate a reciprocity theorem, i.e., a two-dimensional analogue of the (BETTI-RAYLEIGH) reciprocity theorem in three-dimensional linear elasticity. The version of LOVE'S first approximation as derived by REISSNER also has the shortcomings (i) and (ii) but not (iii); and, moreover, REISSNER'S version can be put in a tensorially invariant form, i.e., in a form which remains unaltered under the transformation of the middle surface coordinates.[142]

Subsequent attempts to remedy the unsatisfactory state of the subject that existed around 1940 were largely confined to derivations within the framework

---

[132] RAYLEIGH [1888, *2*].

[133] In his paper of [1888, *1*], after deriving the equations of the bending theory, LOVE considers the special cases of the extensional vibrations of cylindrical and spherical shells.

[134] LAMB [1890, 2].

[135] LOVE [1891, *1*].

[136] In his response to RAYLEIGH'S criticism and with reference to his own theory of [1888, *1*], LOVE writes ([1891, *1*]) "it would most probably be sufficiently exact for the application of a method of approximation."

[137] However, the argument given by Lord RAYLEIGH [1888, *2*] is faulty since the inextensional and the extensional displacements both violate the boundary conditions; and, therefore, RAYLEIGH'S principle as used in the argument does not apply.

[138] BASSETT [1890, *1*].

[139] See Sect. 330 of [1944, *4*]. A discussion of LOVE'S second approximation is included also in [1949, *4*].

[140] In the case of (21 A.7), this can be easily verified with the use of expressions of the type $(6.44)_1$ and $(6.45)_3$ for infinitesimal rigid body displacements of the shell. See also Sect. 6 of [1963, *6*].

[141] VLASOV [1944, *6*]. See also his book [1958, *5*].

[142] With reference to REISSNER'S version of LOVE'S first approximation [1941, *1*] the shortcomings (i) and (ii) above were also mentioned by KNOWLES and REISSNER [1958, *3*]. The fact that this version admits a reciprocity theorem was noted in [1960, *10*] and the observation that it can be put in tensorially invariant form was made in [1963, *6*].

of the Kirchhoff-Love assumption.[143] In this connection, we now indicate the nature of a system of constitutive equations, derived under the Kirchhoff-Love assumption [see Sect. 7, Subsect. ε)], where higher-order terms in the thickness coordinate (i.e., in h/R) are retained. For this purpose, we introduce the notations

$$\begin{aligned} \gamma_1^0 &= \frac{1}{A_1}\left[u_{\langle 2\rangle,1} - \frac{u_{\langle 1\rangle}}{A_2} A_{1,2}\right], & \gamma_2^0 &= \frac{1}{A_2}\left[u_{\langle 1\rangle,2} - \frac{u_{\langle 2\rangle}}{A_1} A_{2,1}\right], \\ \tau_1 &= \frac{1}{A_1}\left[\beta_{\langle 2\rangle,1} - \frac{\beta_{\langle 1\rangle}}{A_2} A_{1,2}\right], & \tau_2 &= \frac{1}{A_2}\left[\beta_{\langle 1\rangle,2} - \frac{\beta_{\langle 2\rangle}}{A_1} A_{2,1}\right] \end{aligned} \tag{21 A.9}$$

and also observe that the above quantities are such that $\gamma_1^0 + \gamma_2^0 = 2e_{\langle 12\rangle}$ and $\tau_1 + \tau_2 = \tau$. Then, referred to lines of curvature coordinates and in terms of physical components, the constitutive equations under consideration can be written in the form[144]

$$\begin{aligned} N_{\langle 11\rangle} &= C\left\{[e_{\langle 11\rangle} + \nu\, e_{\langle 22\rangle}] + \frac{h^2}{12}\left(\frac{1}{R_1} - \frac{1}{R_2}\right)\left(\frac{e_{\langle 11\rangle}}{R_1} - \bar{\varrho}_{\langle 11\rangle}\right)\right\}, \\ N_{\langle 22\rangle} &= C\left\{[e_{\langle 22\rangle} + \nu\, e_{\langle 11\rangle}] + \frac{h^2}{12}\left(\frac{1}{R_2} - \frac{1}{R_1}\right)\left(\frac{e_{\langle 22\rangle}}{R_2} - \bar{\varrho}_{\langle 22\rangle}\right)\right\}, \\ N_{\langle 12\rangle} &= \frac{1-\nu}{2}\, C\left\{2\, e_{\langle 12\rangle} + \frac{h^2}{12}\left(\frac{1}{R_1} - \frac{1}{R_2}\right)\left(\frac{\gamma_1^0}{R_1} - \tau_1\right)\right\}, \\ N_{\langle 21\rangle} &= \frac{1-\nu}{2}\, C\left\{2\, e_{\langle 12\rangle} + \frac{h^2}{12}\left(\frac{1}{R_2} - \frac{1}{R_1}\right)\left(\frac{\gamma_2^0}{R_2} - \tau_2\right)\right\} \end{aligned} \tag{21 A.10}$$

and

$$\begin{aligned} M_{\langle 11\rangle} &= B\left\{[\bar{\varrho}_{\langle 11\rangle} + \nu\, \bar{\varrho}_{\langle 22\rangle}] - \left(\frac{1}{R_1} - \frac{1}{R_2}\right) e_{\langle 11\rangle}\right\}, \\ M_{\langle 22\rangle} &= B\left\{[\bar{\varrho}_{\langle 22\rangle} + \nu\, \bar{\varrho}_{\langle 11\rangle}] - \left(\frac{1}{R_2} - \frac{1}{R_1}\right) e_{\langle 22\rangle}\right\}, \\ M_{\langle 12\rangle} &= \frac{1-\nu}{2}\, B\left\{\tau - \left(\frac{1}{R_1} - \frac{1}{R_2}\right)\gamma_1^0\right\}, \\ M_{\langle 21\rangle} &= \frac{1-\nu}{2}\, B\left\{\tau - \left(\frac{1}{R_2} - \frac{1}{R_1}\right)\gamma_2^0\right\}. \end{aligned} \tag{21 A.11}$$

---

[143] An exception is the work of CHIEN [1944, *2*] which, although quite general in scope, does not appear to have directed itself toward the resolution of the issues that existed at the time. Criticisms of CHIEN's work can be found in the papers of GOL'DENVEIZER and LUR'E [1947, *3*] and GREEN and ZERNA [1950, *2*].

[144] In the literature on shell theory, (21 A.10)–(21 A.11) are often referred to as the Flügge-Lur'e-Byrne equations. These constitutive relations were originally derived in the case of cylindrical shells by FLÜGGE [1932, *1*] and in general form by LUR'E [1940, *2*]. Similar developments were constructed independently in general form by BYRNE (1941) and for shells of revolution by TRUESDELL (1943). The paper by LUR'E [1940, *2*] did not become known until after World War II and the works of BYRNE and TRUESDELL were evidently delayed in publication (BYRNE [1944, 1] and TRUESDELL [1945, *3*]); see also, in this regard, the remarks by TRUESDELL in the last paragraph of [1953, *5*], where references to the original works of BYRNE (1941) and TRUESDELL (1943) are cited. In view of this background, the relations (21 A. 10)–(21 A. 11) may be referred to as the FLÜGGE-LUR'E-BYRNE-TRUESDELL equations. For accounts of constitutive equations (21 A.10)–(21 A.11) see also FLÜGGE's books [1934, *1*], [1960, *4*] and "Notes to Chapter I" at the end of the book by LUR'E [1947, *4*]. A general derivation of (21 A.10)–(21 A.11) can be found in [1963, *6*].

In recording the constitutive equations (21 A.10)–(21 A.11), so far as possible, we have employed the same notations as in (21 A.5). As should be apparent from (21 A.8) when specialized to lines of curvature coordinates, the measures $\bar{\varrho}_{\langle 11\rangle}$ and $\bar{\varrho}_{\langle 22\rangle}$ in (21 A.10)–(21 A.11) are equivalent to the physical components of $\beta_{\alpha|\beta}$ for values of $\alpha = \beta$. It is the latter variables that occur in the constitutive equations (21 A.10)–(21 A.11) as used in the literature on shell theory.

If the second set of terms on the right-hand sides of the above equations, i.e., those involving

$$\pm\left(\frac{1}{R_1}-\frac{1}{R_2}\right)$$

are neglected, then we recover the version of LOVE's first approximation which includes (21 A.7). The constitutive equations (21 A.10)–(21 A.11) do not have any of the shortcomings associated with different versions of LOVE's first approximation and, in particular, are invariant under infinitesimal rigid body displacement and under transformation of the middle surface coordinates.[145] However, it is quite clear that the additional terms in (21 A.10)–(21 A.11) which involve

$$\pm\left(\frac{1}{R_1}-\frac{1}{R_2}\right)$$

are of $O(h/R)$ and thus these equations represent more than a first approximation in the linear theory. Mention should be made here of several related derivations which include the effect of transverse shear deformation and sometimes also the effect of transverse normal stress.[146] Most of these results are such that upon the neglect of the latter effects, they reduce to (21 A.10)–(21 A.11) or variants thereof.

A noteworthy contribution toward a satisfactory derivation of the constitutive equations of a first approximation theory was published in 1946 by NOVOZHILOV.[147] This derivation, which is carried out in lines of curvature coordinates, remained virtually unknown in the Western hemisphere until the early 1960's. NOVOZHILOV's constitutive equations have certain features which (in lines of curvature coordinates) appear to be close to those in (21 A.5).[148] Whereas NOVOZHILOV's equations remain invariant under infinitesimal rigid body displacement, satisfy the equilibrium equation resulting from the symmetry of the stress tensor and also admit two-dimensional energetic theorems (including a reciprocity theorem), nevertheless they are not entirely satisfactory. For example, these equations do not attain a simple form for spherical shells and NOVOZHILOV himself recommends the use of the constitutive equations (21 A.10)–(21 A.11) in this case.[149] Moreover, NOVOZHILOV's equations are not invariant under the transformation of the middle surface coordinates. This shortcoming no doubt is because approximations were made in lines of curvature coordinates rather than in a derivation carried out in general coordinates; however, if this deficiency is remedied and the set of equations given by NOVOZHILOV is put in an invariant form, then part of the simplicity is lost and some of the equations will contain additional terms of $O(h/R)$ similar to those in (21 A.10)–(21 A.11).[150] References to other and more recent contributions concerning the derivation of constitutive equations of a first approximation theory are cited in Sects. 20–21 and need not be repeated here.

---

[145] For details on these and related discussions concerning the constitutive equations (21 A.10)–(21 A.11), see [1963, *6*] and [1966, *8*].

[146] Among these we mention the derivations by HILDEBRAND, REISSNER and THOMAS [1949, *4*], GREEN and ZERNA [1950, *2*], REISSNER [1952, *2*] and NAGHDI [1957, *4*], [1963, *6*]. A criticism by GOL'DENVEIZER (cited in [1960, *9*]) to the effect that the constitutive equations in [1957, *4*] are not invariant under infinitesimal rigid body displacements has no basis. This was acknowledged by him later, as was noted in [1960, *9*].

[147] NOVOZHILOV [1946, *3*]. See also the English translation of his book [1959, *3*]. According to NOVOZHILOV [1959, *3*, p. 54] similar approximate constitutive equations were obtained independently by L. I. BALABUKH. The reference to the work of BALABUKH (1946) is cited on p. 83 of [1961, *3*].

[148] See NOVOZHILOV's equations (10.16) in [1959, *3*].

[149] See the remark on p. 54 of [1959, *3*].

[150] A more detailed discussion of NOVOZHILOV's equations can be found in Sects. 6.3 and 6.4 of [1963, *6*].

22. Relationship of results from the three-dimensional theory and the theory of Cosserat surface. We return to the field equations and the constitutive relations of the approximate nonlinear theory (derived from the three-dimensional equations) and consider their comparison with the corresponding results derived by direct approach (and summarized in Sect. 14). A close examination of the field equations (9.47)–(9.48) and (9.51) readily reveals that these equations are of the same form as (12.28)–(12.30). Moreover, the set of constitutive equations (13.64)–(13.65) for the Cosserat surface has the same form as (18.4)–(18.5), apart from an extra generality in Sect. 18 for the temperature. If we also adopt the approximation (18.7) for the temperature, identify the surface $\mathfrak{s}$ of the (three-dimensional) shell-like body with $\mathscr{s}$ and (11.1) with (4.36), then the field equations and the constitutive equations for elastic shells in the two developments are formally equivalent.

In the theory of a Cosserat surface, the skew-symmetric part of $M^{\alpha\beta}$ is specified by a constitutive equation which has no counterpart in the approximate development from the three-dimensional equations. The latter is due to the nature of the approximation adopted for the (two-dimensional) specific free energy ψ in the derivation from the three-dimensional theory. Our reasons for this choice of an approximate expression for ψ were discussed in Sects. 17–18. Alternatively, a specification of a different (and equally acceptable) approximate expression for ψ could result in a constitutive equation for $M^{\alpha\beta}$ rather than $M^{(\alpha\beta)}$.

Although the constitutive equations in Sect. 18 are obtained from the three-dimensional equations, they involve an approximate function for the free energy and, as already remarked, in general it is difficult to evaluate this approximate expression from the full three-dimensional expression for the free energy. On the other hand, given the Cosserat surface $\mathscr{C}$ as a model for a thin shell and the balance principles stated in Sect. 8, the resulting theory is exact; but, it requires the additional (and sometimes difficult) considerations regarding the interpretations of the results and identification of the constitutive coefficients.

The contact force $\boldsymbol{N}$ and the contact director couple $\boldsymbol{M}$ in the theory of a Cosserat surface, as well as $\boldsymbol{N}^\alpha$ in (9.11) and $\boldsymbol{M}^\alpha$ in (9.19), have respectively the physical dimensions of force per unit length and couple per unit length, as indicated in (8.5). Moreover, these vector fields have the same physical dimensions as the stress-resultants and the stress-couples defined by (11.36). Hence, in view of the formal equivalence of the two developments noted above, we may identify $\boldsymbol{N}^\alpha$ and $\boldsymbol{M}^\alpha$ (in the theory of a Cosserat surface) with the corresponding stress-resultants and stress-couples. Similar identifications can be made for other quantities, including the identification of the assigned force $\boldsymbol{f}$ and the assigned director couple $\boldsymbol{l}$ with the corresponding load resultants in (11.30) with $N=0, 1$. Further, recalling (8.19), we may identify $f^i - c^i = \bar{f}^i$ and $\bar{l}^i$ in (9.47)–(9.48) with the corresponding quantities in (12.28)–(12.29). The identification of $\bar{l}^i$ in the two sets of equations entails also that we put the director inertia coefficient α in $(8.19)_2$ equal to k^{11} in $(12.25)_2$, i.e.,

$$\alpha = k^{11}. \tag{22.1}$$

With (22.1) the identification is complete and the equations of motion (9.47)–(9.48) correspond exactly to (12.28)–(12.29).

While a one-to-one correspondence can be established between the various quantities in our two developments, namely that by direct approach and from the approximate three-dimensional theory, the relationship of the latter to the results in the classical theory of shells needs further elaboration. In the classical (approximate) linear theory of shells developed from the three-dimensional equations, where the initial position vector is of the form (4.27), the stress-couple resultant is defined by a tangential vector field[151]

$$\begin{aligned}\mathscr{M}^\alpha &= \int_{-h/2}^{h/2} (\boldsymbol{A}_3 \times \boldsymbol{T}^\alpha)\, \zeta\, d\zeta, \\ &= M^{\alpha\beta}\, \boldsymbol{A}_3 \times \boldsymbol{A}_\beta.\end{aligned} \tag{22.2}$$

On the other hand, in the linearized version of the field equations as derived (from the three-dimensional equations) in Sects. 11–12 and with the initial position vector specified by (7.30)–(7.31), it is the quantity $\boldsymbol{D} \times \boldsymbol{M}^\alpha$ or

$$\begin{aligned}\boldsymbol{D} \times \boldsymbol{M}^\alpha &= D\, \boldsymbol{A}_3 \times M^{\alpha i} \boldsymbol{A}_i, \\ &= D\, M^{\alpha\beta} \boldsymbol{A}_3 \times \boldsymbol{A}_\beta,\end{aligned} \tag{22.3}$$

which corresponds to (22.2). In view of earlier remarks, a similar comparison also holds between $\boldsymbol{D} \times \boldsymbol{M}^\alpha$ (in the theory of a Cosserat surface) and $\mathscr{M}^\alpha$ as defined by (22.2). It is therefore clear that in the comparison with the classical theory (where $M^{\alpha 3}$ is not defined) it is the quantity $D M^{\alpha\beta}$ [or $\boldsymbol{D} \times \boldsymbol{M}^\alpha$] which must be identified with $M^{\alpha\beta}$ in (22.2). This is simply due to the fact that in the classical theory of shells $M^{\alpha\beta}$ is defined through (22.2), or through the stress-couples in $(12.42)_6$, and *not* by $\mathscr{M}^\alpha = M^{\alpha i} \boldsymbol{A}_i$. In this connection, see also the remarks made in Sect. 12 [Subsect. η)].

Before closing this section, we calculate explicitly the value of the inertia coefficient k^{11} in (22.1) appropriate to the linear theory and when the position vector of the shell-like body in the initial reference configuration is given by (7.30) with $\boldsymbol{D} = \boldsymbol{A}_3$. Recalling (4.21), (7.46) and $(7.48)_{1,3}$, from (12.26) we readily obtain

$$\varrho\, k^{11}\, a^{\frac{1}{2}} = \varrho_0^*\, A^{\frac{1}{2}} \frac{h^3}{12}. \tag{22.4}$$

But $\varrho_0 = \varrho_0^*\, h$ by (4.17) and (7.46) and $\varrho\, a^{\frac{1}{2}} = \varrho_0\, A^{\frac{1}{2}}$ by (4.41)–(4.42). Using these results, in (22.4) yields $k^{11} = h^2/12$. Hence, the director inertia coefficient α which occurs in $\bar{L}^i$ of the linearized equations of motion (9.55) is given by

$$\alpha = \frac{h^2}{12}, \tag{22.5}$$

in view of the identification (22.1).

E. Linear theory of elastic plates and shells.

This concluding chapter, though limited only to the linear theory, is a culmination of the point of view adopted earlier in Sect. 4 regarding shells and plates. Our starting point here is a system of equations for the linear theory derived in the previous chapters by direct approach. After providing for ease of reference a brief summary of the field equations and the constitutive relations, the remainder of the chapter is devoted to the detailed considerations of the linear theory. Special attention is given to the determination of the constitutive coefficients, proof of a uniqueness theorem and certain results pertaining to the classical linear theory of shells and plates.

In order to concentrate attention on the main aspects of the subject, especially in regard to the identification of the constitutive coefficients, we limit ourselves to the isothermal theory. We note, however, that this limitation is not essential. Moreover, it should be clear that the developments of this chapter can be regarded

[151] See for example Chap. 10 of GREEN and ZERNA [1968, 9].

as appropriate for the purely mechanical theory in line with the remarks made in Sect. 14 [Subsect. β)].

23. The boundary-value problem in the linear theory. In this section, we summarize the field equations and a system of constitutive relations in the linear isothermal theory of a Cosserat surface which characterize the initial boundary-value problem of elastic shells and plates. However, the case of a flat plate is considered separately from that of the initially curved shell, as this will be more illuminating.

α) *Elastic plates.* It is convenient to recall here the basic equations for elastic plates with reference to rectangular Cartesian coordinates. Thus, let the surface coordinates on the initial surface $\mathscr{S}$ be identified with the rectangular Cartesian coordinates x_α. Then, referred to rectangular Cartesian coordinates, the equations of motion (9.73)–(9.74) can be written as

$$N_{\alpha\beta,\alpha}+\varrho_0\bar{F}_\beta=0, \qquad M_{\alpha 3,\alpha}+\varrho_0\bar{L}_3=V_3, \tag{23.1}$$

$$M_{\alpha\beta,\alpha}+\varrho_0\bar{L}_\beta=V_\beta, \qquad V_{\alpha,\alpha}+\varrho_0\bar{F}_3=0, \tag{23.2}$$

where a comma denotes partial differentiation with respect to x_α and all quantities in (23.1)–(23.2) are now referred to rectangular Cartesian coordinates so that no distinction between the position of the indices (subscripts and superscripts) in such quantities as $N^{\alpha\beta}$, $M^{\alpha i}$, $\bar{L}^i$ is necessary. The above equations of motion include the effect of inertia due to both displacement $\boldsymbol{u}$ and director displacement $\boldsymbol{\delta}$. The latter, which corresponds to "rotatory inertia", occurs through $\bar{L}_i$ defined by $(9.57)_2$ and $(8.19)_2$ with the director inertia coefficient α given by (22.5).

The constitutive relations of the linear theory considered here are those for an isotropic Cosserat plate obtained under the restriction (16.31). Recalling (16.33) and (9.72), from (16.11) referred to rectangular Cartesian coordinates, we have the following constitutive relations for initially flat elastic Cosserat plates:

$$\begin{gathered} N_{\alpha\beta}=N_{\beta\alpha}=\alpha_1\,\delta_{\alpha\beta}\,e_{\gamma\gamma}+2\alpha_2\,e_{\alpha\beta}+\alpha_9\,\delta_{\alpha\beta}\,\gamma_3, \\ V_3=\alpha_4\,\gamma_3+\alpha_9\,e_{\gamma\gamma}, \qquad M_{\alpha 3}=\alpha_8\,\varkappa_{3\alpha} \end{gathered} \tag{23.3}$$

and

$$M_{\alpha\beta}=\alpha_5\,\delta_{\alpha\beta}\,\varkappa_{\gamma\gamma}+\alpha_6\,\varkappa_{\beta\alpha}+\alpha_7\,\varkappa_{\alpha\beta}, \qquad V_\alpha=\alpha_3\,\gamma_\alpha. \tag{23.4}$$

The kinematic measures in (23.3)–(23.4) are those in (6.25) referred to rectangular Cartesian coordinates so that now covariant differentiation in (6.25) is replaced with partial differentiation.

An examination of the equations of motion (23.1)–(23.2) and the constitutive equations (23.3)–(23.4) readily reveals that the differential equations of the linear theory of an isotropic Cosserat plate separate into two systems of uncoupled equations: One system, consisting of (23.1), (23.3) and $(6.25)_{3,4,5}$, represents the stretching (or the extensional motion) of the plate while the other, given by (23.2), (23.4) and $(6.25)_{6,7}$, characterizes the bending (or the flexural motion) of the plate. Among the features of the extensional theory, we note the presence of five constitutive coefficients in (23.3), a constitutive equation for the normal force V_3 (corresponding to that for the normal stress-resultant) and a constitutive equation for $M_{\alpha 3}$ (corresponding to that for the shear stress-couple). Similarly, the bending theory[1] contains four constitutive coefficients in (23.4) and includes a constitutive

[1] Apart from the presence of a constitutive equation for $M_{[\alpha\beta]}$, the equations of the bending theory are of the same form as those in REISSNER's plate theory discussed in Sect. 20.

equation for V^α (corresponding to that for the shear stress-resultant) in terms of the kinematic variable γ_α; the latter can be regarded as representing the effect of "transverse shear deformation".

The nature of the boundary conditions in the above linear theory is clear from the linearized version of the rate of work expression (8.8). In particular, the force and the director couple boundary conditions are given by[2]

$$N_\gamma = N_{\alpha\gamma}\,{}_0\nu_\alpha, \quad M_3 = M_{\alpha 3}\,{}_0\nu_\alpha, \quad N_3 = V_\alpha\,{}_0\nu_\alpha, \quad M_\gamma = M_{\alpha\gamma}\,{}_0\nu_\alpha, \tag{23.5}$$

where ${}_0\nu_\alpha$ in (23.5) are the components of the outward unit normal to the boundary curve of the initial surface $\mathscr{S}$ referred to rectangular Cartesian coordinates. The boundary conditions (23.5) hold pointwise on the boundary $\partial\mathscr{S}$.

β) *Elastic shells.* We recall the equations of motion for shells in terms of the variables $N^{\alpha\beta}$, $M^{\alpha i}$, V^i and in the forms (9.68)–(9.70) which correspond to those in (12.42). The equations of motion (9.70) include the inertia terms in $\bar{L}^i$ arising from the director displacement. These terms, with director inertia coefficient α given by (22.5), represent the effect of "rotatory inertia".

The constitutive equations considered here are those for an isotropic material which are obtained under the restriction (16.35) with the further stipulation that $\bar{\psi}$ be independent of $B_{\alpha\beta}$. Thus, in line with the discussion following (16.35), we assume that $\bar{\psi}$ is specified by (16.33) and from (16.11) obtain the desired constitutive relations for an elastic Cosserat surface. These constitutive relations, which are similar to those in (16.22)–(16.24), can be expressed in the form

$$\begin{aligned} N'^{\alpha\beta} &= [\alpha_1 A^{\alpha\beta} A^{\gamma\delta} + \alpha_2 (A^{\alpha\gamma} A^{\beta\delta} + A^{\alpha\delta} A^{\beta\gamma})]\, e_{\gamma\delta} + \alpha_9 A^{\alpha\beta} \gamma_3, \\ M^{\beta\alpha} &= [\alpha_5 A^{\alpha\beta} A^{\gamma\delta} + \alpha_6 A^{\alpha\gamma} A^{\beta\delta} + \alpha_7 A^{\alpha\delta} A^{\beta\gamma}]\, \varrho_{\gamma\delta} \end{aligned} \tag{23.6}$$

$$V^\alpha = \alpha_3 A^{\alpha\gamma} \gamma_\gamma, \qquad V^3 = \alpha_4 \gamma_3 + \alpha_9 A^{\alpha\beta} e_{\alpha\beta}, \tag{23.7}$$

$$M^{\alpha 3} = \alpha_8 A^{\alpha\gamma} \varrho_{3\gamma}, \tag{23.8}$$

where the kinematic measures are defined in (6.24). The above constitutive relations have been expressed in terms of the variables $N'^{\alpha\beta}$, $M^{\alpha i}$, V^i and the kinematic measures $e_{\alpha\beta}$, $\varrho_{i\alpha}$, γ_i. Alternatively, we could adopt the set $e_{\alpha\beta}$, $\varkappa_{i\alpha}$, γ_i as the kinematic variables for the linear theory and record the constitutive relations corresponding to (23.6) in terms of $N'^{\alpha\beta}$, $M^{\alpha i}$, m^i. We note that upon specialization and with $B_{\alpha\beta}=0$, (23.6)–(23.8) reduce to (23.3)–(23.4), since for flat plates $\varrho_{i\alpha} = \varkappa_{i\alpha}$.

We again observe that the nature of the boundary conditions is clear from the linearized version of the rate of work expression (8.8). In particular, with reference to the above forms of the equations of motion and constitutive equations, the force and the director couple boundary conditions for shells are given by

$$N^\gamma = N^{\alpha\gamma}\,{}_0\nu_\alpha, \qquad N^3 = V^\alpha\,{}_0\nu_\alpha, \qquad M^i = M^{\alpha i}\,{}_0\nu_\alpha, \tag{23.9}$$

where ${}_0\nu_\alpha$ are the components of the outward unit normal ${}_0\boldsymbol{\nu}$ to the boundary curve $\partial\mathscr{S}$ of the initial surface $\mathscr{S}$ defined previously in Sect. 9 [Subsect. γ)]. It should be noted that the components $N^{\alpha\gamma}$ in $(23.9)_1$ involve both the symmetric components $N^{(\alpha\gamma)}$ and the skew-symmetric part $N^{[\alpha\gamma]}$. The former is calculated from $(23.6)_1$ and $(9.67)_3$ while the latter is given by[3]

$$N^{[\alpha\gamma]} = \tfrac{1}{2} (M^{\beta\gamma} B^\alpha_\beta - M^{\beta\alpha} B^\gamma_\beta). \tag{23.10}$$

[2] These follow from linearized versions of (9.41) and (9.43), as well as (9.72).

[3] Recall that only the symmetric part of $N^{\alpha\beta}$ has a constitutive equation. The expression (23.10) follows from (9.68).

Before closing this section, we briefly indicate the manner in which a (slightly less general) system of equations for the linear theory can be obtained as a special case of those stated above.[4] Let

$$\alpha_8 = 0 \tag{23.11}$$

and hence[5]

$$M^{\alpha 3} = 0. \tag{23.12}$$

Then, the differential equations of motion are given by (9.69)–$(9.70)_1$ and in place of $(9.70)_2$ we have

$$V^3 = \varrho_0 L^3. \tag{23.13}$$

The constitutive equations are given by (23.6)–(23.7) and the specification of five (instead of six) boundary conditions are required at each edge of the shell. These boundary conditions are either the displacement boundary conditions

$$u_\alpha,\ u_3,\ \delta_\alpha \text{ (or } \gamma_\alpha) \tag{23.14}$$

or the force boundary conditions $(23.9)_{1,2}$ and

$$M^\gamma = M^{\alpha\gamma}\ {}_0\nu_\alpha. \tag{23.15}$$

Among other features of this theory we note the presence of: (i) a constitutive equation for V^α representing the effect of "transverse shear deformation"; (ii) a constitutive equation for the normal force V^3 per unit length (corresponding to a normal stress-resultant); (iii) a constitutive equation for the skew-symmetric director couple $M^{[\beta\alpha]}$; and (iv) eight constitutive coefficients in (23.6)–(23.7).

24. Determination of the constitutive coefficients. The relationship and the correspondence between the theory of a Cosserat surface and the theory of shells and plates obtained from the general three-dimensional equations have been already brought out. Moreover, in view of the observations made in Sect. 22, we may identify $N^{\alpha\beta}$ and $N^{\alpha 3}$ (or V^α) as stress-resultants and shear stress-resultants, $M^{\alpha\beta}$ and $M^{\alpha 3}$ as stress-couples and shear stress-couples and V^3 may be called a normal stress-resultant. In the constitutive relations (23.6)–(23.8), as well as in (23.3)–(23.4), the coefficients $\alpha_1, \ldots, \alpha_9$ are so far arbitrary; however, as will become evident presently, most of these coefficients can be identified by comparison of certain simple solutions with corresponding exact solutions in the three-dimensional theory of linear elasticity. In what follows, we first consider the determination of the constitutive coefficients for initially flat plates and then discuss the constitutive coefficients for shells separately.

α) *The constitutive coefficients for plates.* Before proceeding with the identification of the constitutive coefficients, we elaborate once more on the relationship between the variables $N^{\alpha\beta}$, $M^{\alpha i}$, V^i and the stress-resultants and the stress-couples defined in the course of our derivation from the three-dimensional equations. For this purpose, consider a (three-dimensional) plate of uniform thickness h. Let the plate, referred to a system of rectangular Cartesian coordinates x_i, be defined by its middle plane $x_3 = 0$ and the region

$$-\frac{h}{2} \leqq x_3 \leqq \frac{h}{2}.$$

---

[4] The system of equations discussed below [between (23.11)–(23.15)] correspond to those in Sect. 20 with constitutive equations (20.47).

[5] It should be noted in obtaining (23.12) no restriction is placed on δ_3 or $\varrho_{3\gamma}$.

Further, let σ_{ij} denote the Cartesian components of the symmetric stress tensor and consider the resultants

$$\int_{-h/2}^{h/2} \sigma_{\alpha\beta}\, dx_3, \quad \int_{-h/2}^{h/2} \sigma_{\alpha 3}\, dx_3, \quad \int_{-h/2}^{h/2} \sigma_{33}\, dx_3, \quad \int_{-h/2}^{h/2} \sigma_{\alpha\beta}\, x_3\, dx_3, \quad \int_{-h/2}^{h/2} \sigma_{\alpha 3}\, x_3\, dx_3. \tag{24.1}$$

It follows from (12.43) and the remarks made in Sect. 22 that the resultants (24.1), in the order listed, may be identified with $N_{\alpha\beta}, V_\alpha, V_3, M_{(\alpha\beta)}, M_{\alpha 3}$ in the linear theory of a Cosserat plate.[6]

We now proceed to identify the constitutive coefficients in (23.3)–(23.4). We consider, in particular, three simple elastostatic problems, namely pure bending of a plate, extensional deformation in the plane of the plate and a plate under a uniform hydrostatic pressure. From a comparison of solutions of these simple examples with the corresponding exact solutions in linear three-dimensional elasticity, we determine seven of the nine coefficients in (23.3)–(23.4). In addition, we also make some remarks pertaining to the specification of the remaining coefficients. In this connection and for later reference, we recall here the three-dimensional linear constitutive relations for transversely isotropic elastic materials. These constitutive relations which involve five independent coefficients are given by[7]

$$\begin{aligned} \sigma_{11} &= c_{11}\, e^*_{11} + c_{12}\, e^*_{22} + c_{13}\, e^*_{33}, & \sigma_{23} &= 2c_{44}\, e^*_{23}, \\ \sigma_{22} &= c_{12}\, e^*_{11} + c_{11}\, e^*_{22} + c_{13}\, e^*_{33}, & \sigma_{31} &= 2c_{44}\, e^*_{31}, \\ \sigma_{33} &= c_{13}(e^*_{11} + e^*_{22}) + c_{33}\, e^*_{33}, & \sigma_{12} &= (c_{11} - c_{12})\, e^*_{12}, \end{aligned} \tag{24.2}$$

where e^*_{ij} are the Cartesian components of the three-dimensional strain tensor defined by (7.60) which is now referred to rectangular Cartesian coordinates and the coefficients $c_{11}, \ldots, c_{44}$ in (24.2) are the elasticity constants. In the case of an isotropic material, the coefficients $c_{11}, \ldots, c_{44}$ assume the values

$$c_{11} = c_{33} = \lambda + 2\mu, \qquad c_{12} = c_{13} = \lambda, \qquad c_{44} = \mu, \tag{24.3}$$

where λ and μ are the Lamé constants.

(i) *Pure bending of a plate.* Consider a rectangular Cosserat plate in equilibrium, bounded by the lines (or edges) $x_1 = a_1, a_2$, $x_2 = b_1, b_2$, and subjected to uniform couples of constant magnitude M_1 and M_2 along the edges $x_1 = \text{const.}$ and $x_2 = \text{const.}$, respectively. The appropriate boundary conditions in this case are[8]

$$\begin{aligned} M_{11} &= M_1, & M_{12} &= V_1 = 0 \quad \text{on} \quad x_1 = a_1, a_2, \\ M_{22} &= M_2, & M_{21} &= V_2 = 0 \quad \text{on} \quad x_2 = b_1, b_2. \end{aligned} \tag{24.4}$$

The differential equations for elastostatic bending problems of a Cosserat plate subjected to edge tractions alone are given by (23.2) and (23.4) in the absence of $\bar{L}_\beta$ and $\bar{F}_3$. We seek a solution of these equations for which

$$\begin{gathered} \gamma_\alpha = 0, \\ \varkappa_{11} = -u_{3,11} = \text{const.}, \qquad \varkappa_{22} = -u_{3,22} = \text{const.}, \qquad \varkappa_{12} = -u_{3,12} = 0. \end{gathered} \tag{24.5}$$

[6] It should be recalled that in the linear theory of a Cosserat plate characterized by (23.1)–(23.4), the initial director is of constant length and coincident with the unit normal to the initial surface (i.e., $D_\alpha = 0$, $D_3 = D = 1$). Also, since $\sigma_{\alpha\beta}$ is symmetric, only the symmetric part of $M_{\alpha\beta}$, namely $M_{(\alpha\beta)}$, can be identified with $(24.1)_4$.

[7] See Green and Zerna [1968, *9*, p. 178].

[8] Since we are concerned with bending of an isotropic plate, it is not necessary to consider the equations of the extensional theory. However, for the example under consideration, we note that a zero solution for $\{N_{\alpha\beta}, V_3, M_{\alpha 3}\}$ identically satisfies (23.1) and (23.3) in the absence of $\bar{F}_\beta$ and $\bar{L}_3$.

It then follows that all equilibrium equations in (23.2) with $\bar{L}_\beta = \bar{F}_3 = 0$ are identically satisfied and the only non-identically vanishing constitutive relations are those for M_{11} and M_{22} from which we obtain

$$-u_{3,11} = \varkappa_{11} = \frac{M_1 - \beta M_2}{\alpha(1-\beta^2)}, \qquad -u_{3,22} = \varkappa_{22} = \frac{M_2 - \beta M_1}{\alpha(1-\beta^2)}, \tag{24.6}$$

where we have put

$$\alpha = \alpha_5 + \alpha_6 + \alpha_7, \qquad \beta = \frac{\alpha_5}{\alpha}. \tag{24.7}$$

The solution of the system of differential Eqs. $(24.5)_4$ and (24.6) is

$$u_3 = -\tfrac{1}{2}(\varkappa_{11}\, x_1^2 + \varkappa_{22}\, x_2^2), \tag{24.8}$$

where the constants $\varkappa_{11}$ and $\varkappa_{22}$ are given by the right-hand sides of each of the expressions in $(24.6)_{1,2}$ and where the arbitrary constants of integration have been set equal to zero without loss in generality. The solution (24.8) includes those for bending of a Cosserat plate into a spherical surface ($\varkappa_{11} = \varkappa_{22}$, $M_{11} = M_{22}$) or into an anticlastic surface ($\varkappa_{11} = -\varkappa_{22}$, $M_{11} = -M_{22}$).

Consider now a three-dimensional rectangular plate of uniform thickness h which initially is transversely isotropic with respect to the normals of the plate and let the plate be subjected to uniform bending couples of magnitude M_1 and M_2, each per unit length, along the edges parallel to x_2 and x_1 directions, respectively. From an elementary result in three-dimensional elasticity, the solution of this example for $\varkappa_{11}$ and $\varkappa_{22}$ has the same form as those given by the right-hand members in each of the expressions (24.6) and the solution for u_3 (on the middle plane only) is of the same form as (24.8).[9] Hence, by comparison, we set[10]

$$\begin{aligned} \alpha_6 + \alpha_7 &= \frac{h^3}{12}(c_{11} - c_{12}), \qquad \alpha_6 - \alpha_7 = 0, \\ \alpha_5 &= \frac{h^3}{12}\,\frac{c_{12}\,c_{33} - c_{13}^2}{c_{33}}, \end{aligned} \tag{24.9}$$

where $c_{11}, c_{12}, c_{13}, c_{33}$ are the elastic coefficients for a plate which is transversely isotropic with respect to its normals. In the case of an isotropic plate, these coefficients are given by (24.3) and the coefficients (24.9) reduce to

$$\alpha_5 = \nu B, \qquad \alpha_6 = \alpha_7 = \tfrac{1}{2}(1-\nu)\, B, \tag{24.10}$$

where the flexural rigidity B is defined by $(20.13)_3$ and ν is Poisson's ratio.

(ii) *Extensional deformation of a plate.* Consider a rectangular Cosserat plate in equilibrium, bounded by the lines $x_1 = a_1, a_2$, $x_2 = b_1, b_2$ and subjected to uniform forces per unit length (in the plane of the plate) of magnitude N_1 and N_2 along the edges $x_1 =$ const. and $x_2 =$ const., respectively. The appropriate boundary conditions in this case are[11]

$$\begin{aligned} N_{11} &= N_1, \quad N_{12} = M_{13} = 0 \quad \text{on} \quad x_1 = a_1, a_2, \\ N_{22} &= N_2, \quad N_{21} = M_{23} = 0 \quad \text{on} \quad x_2 = b_1, b_2. \end{aligned} \tag{24.11}$$

[9] In the case of transversely isotropic materials for which the components σ_{i3} of the stress tensor vanish, the results can be easily deduced using the equations given in GREEN and ZERNA [1968, *9*, p. 178]. The corresponding solution for an isotropic plate may be found in LOVE [1944, *4*, p. 132], or in TIMOSHENKO and GOODIER [1951, *3*, p. 255].

[10] In making such comparisons we put $\alpha_6 - \alpha_7 = 0$, since the coefficient $(\alpha_6 - \alpha_7)$ has no counterpart in (non-polar) three-dimensional linear elasticity. The results (24.9) and (24.10) were given previously by GREEN and NAGHDI [1967, *4*] through consideration of pure bending of a plate as discussed here.

[11] Since we are concerned with extensional deformation of an isotropic plate, it is not necessary to consider the equations of the bending theory. However, for the example under consideration, we note that a zero solution for $\{M_{\alpha\beta}, V_\alpha\}$ identically satisfies (23.2) and (23.4) in the absence of $\bar{L}_\beta$ and $\bar{F}_3$.

The differential equations for elastostatic extensional problems of a Cosserat plate subjected to edge tractions alone are given by (23.1) and (23.3) in the absence of $\overline{F}_\beta$ and $\overline{L}_3$. Here we seek a solution of these equations in the form

$$\gamma_3 = \text{const.}, \qquad e_{11} = u_{1,1} = \text{const.}, \quad e_{22} = u_{2,2} = \text{const.}, \quad e_{12} = u_{1,2} = 0. \tag{24.12}$$

From $(23.3)_3$ we have $M_{\alpha 3} = 0$ and, in the absence of $\overline{L}_3$, $(23.1)_2$ yields $V_3 = 0$ which along with $(23.3)_2$ gives

$$\gamma_3 = -\frac{\alpha_9}{\alpha_4} e_{\gamma\gamma}. \tag{24.13}$$

The equilibrium equation $(23.1)_1$ with $\overline{F}_\beta = 0$ is identically satisfied and the only remaining non-identically vanishing constitutive relations are those for N_{11} and N_{22} from which we obtain

$$\begin{aligned} u_{1,1} &= \frac{\bar{\beta} N_2 - \bar{\alpha} N_1}{\bar{\beta}^2 - \bar{\alpha}^2} = \text{const.} = \lambda_1 \text{ (say)}, \\ u_{2,2} &= \frac{\bar{\beta} N_1 - \bar{\alpha} N_2}{\bar{\beta}^2 - \bar{\alpha}^2} = \text{const.} = \lambda_2 \text{ (say)}, \end{aligned} \tag{24.14}$$

where

$$\bar{\alpha} = 2\alpha_2 + \bar{\beta}, \quad \bar{\beta} = \alpha_1 - \frac{\alpha_9^2}{\alpha_4} = \alpha_1 - \alpha_9 \bar{\gamma}, \quad \bar{\gamma} = \frac{\alpha_9}{\alpha_4} \tag{24.15}$$

and we have introduced $(24.15)_3$ for later convenience. The solution of the system of differential equations (24.14) and $(24.12)_4$ is

$$u_1 = \lambda_1 x_1, \qquad u_2 = \lambda_2 x_2, \tag{24.16}$$

where the constants λ_1 and λ_2 are those in $(24.14)_{1,2}$. Also, by (24.13) and (24.16), γ_3 can be written as

$$\gamma_3 = -\bar{\gamma}(\lambda_1 + \lambda_2). \tag{24.17}$$

Consider now a three-dimensional rectangular plate of uniform thickness h, as defined earlier in this section. The plate is initially transversely isotropic with respect to its normals and is subjected to uniform tractions N_1 and N_2, each per unit length, along the edges parallel to the x_1 and x_2 directions. The boundary conditions in this case can be written as

$$\begin{aligned} &\sigma_{11} = \frac{N_1}{h}, \quad \sigma_{12} = \sigma_{13} = 0 \quad \text{on} \quad x_1 = a_1, a_2, \\ &\sigma_{22} = \frac{N_2}{h}, \quad \sigma_{21} = \sigma_{23} = 0 \quad \text{on} \quad x_2 = b_1, b_2, \\ &\qquad\qquad \sigma_{3i} = 0 \quad \text{on} \quad x_3 = \pm\frac{h}{2}. \end{aligned} \tag{24.18}$$

From an elementary result in three-dimensional elasticity for extensional deformation of a plate, the (three-dimensional) displacements are

$$u_1^* = \lambda_1 x_1, \qquad u_2^* = \lambda_2 x_2, \qquad u_3^* = \lambda_3 x_3. \tag{24.19}$$

With the use of (24.19), (24.2) and the boundary conditions (24.18), the solutions for e_{11}^*, e_{22}^* and e_{33}^* are easily found to have the same form as those given by the right-hand members in each of (24.14) and (24.17). Hence, by comparison, we may set

$$\bar{\alpha}=\alpha_1+2\alpha_2-\frac{\alpha_9^2}{\alpha_4}=h\left(c_{11}-\frac{c_{13}^2}{c_{33}}\right),\quad \bar{\beta}=\alpha_1-\frac{\alpha_9^2}{\alpha_4}=h\left(c_{12}-\frac{c_{13}^2}{c_{33}}\right),$$
$$\bar{\gamma}=\frac{\alpha_9}{\alpha_4}=\frac{c_{13}}{c_{33}}. \tag{24.20}$$

The results in (24.20) provide only three relations involving the four coefficients $\alpha_1, \alpha_2, \alpha_4, \alpha_9$; and, before the coefficients $\alpha_1, \ldots, \alpha_9$ can be solved in terms of the elastic constants $c_{11}, c_{12}, c_{13}, c_{33}$, we need a fourth relation which is obtained next.

(iii) *A plate under a uniform hydrostatic pressure.* Consider a plate of uniform thickness h subjected to a uniform hydrostatic pressure and recall first its solution as an elastostatic problem in the three-dimensional theory. The solution for the stresses is simply

$$\sigma_{ij}=-p\,\delta_{ij}, \tag{24.21}$$

where p is a positive constant. The above solution satisfies all stress boundary conditions and the (three-dimensional) equilibrium equations in the absence of body forces. Using (24.21), the constitutive equations (24.2) yield

$$e_{11}^*=e_{22}^*=-\left[\frac{c_{13}-c_{33}}{2c_{13}^2-(c_{11}+c_{12})\,c_{33}}\right]p=\lambda \quad \text{(say)},$$
$$e_{33}^*=-\left[\frac{2c_{13}-(c_{11}+c_{12})}{2c_{13}^2-(c_{11}+c_{12})\,c_{33}}\right]p=\bar{\lambda} \quad \text{(say)}.$$
$$\text{All other } e_{ij}^*=0. \tag{24.22}$$

From (24.22) and the strain-displacements relations, we readily obtain the displacements

$$u_1^*=\lambda\,x_1,\quad u_2^*=\lambda\,x_2,\quad u_3^*=\bar{\lambda}\,x_3, \tag{24.23}$$

for a body under a uniform hydrostatic pressure.

Consider now the same example within the scope of the theory of Cosserat surface. Thus, for a Cosserat plate in equilibrium bounded by the lines (or edges) $x_1=a_1, a_2$, $x_2=b_1, b_2$, we specify the boundary conditions by

$$N_{11}=-h\,p,\quad N_{12}=M_{13}=0 \quad \text{on} \quad x_1=a_1, a_2,$$
$$N_{22}=-h\,p,\quad N_{21}=M_{23}=0 \quad \text{on} \quad x_2=b_1, b_2, \tag{24.24}$$

and put[12]

$$\varrho_0\,L_3=-h\,p. \tag{24.25}$$

The relevant differential equations in this case are given by (23.3) and (23.1) in the absence of $\bar{F}_\beta$ and we seek a solution of these equations[13] in the form

$$\gamma_3=\text{const.},\quad e_{11}=e_{22}=\text{const.},\quad e_{12}=0. \tag{24.26}$$

Then, from (23.3), we have $M_{\alpha 3}=0$, $N_{12}=0$ and we also conclude that the components N_{11}, N_{22}, V_3 given by

$$N_{11}=N_{22}=2(\alpha_1+\alpha_2)\,e_{11}+\alpha_9\,\gamma_3,\quad V_3=\alpha_4\,\gamma_3+2\alpha_9\,e_{11} \tag{24.27}$$

[12] The specification of (24.25) is suggested by the definitions $(12.32)_2$ and (11.30) for $N=1$.

[13] If one considers the remaining Eqs. (23.4) and (23.2), it is easily seen that these are identically satisfied with $\bar{L}_\beta=0$, $\bar{F}_3=0$.

are constants. It follows that $(23.1)_1$ is identically satisfied and $(23.1)_2$ yields

$$V_3 = \varrho_0 L_3. \tag{24.28}$$

From (24.27)–(24.28) and (24.24)–(24.25) we obtain

$$2(\alpha_1 + \alpha_2) e_{11} + \alpha_9 \gamma_3 = -h p, \quad \alpha_4 \gamma_3 + 2\alpha_9 e_{11} = -h p. \tag{24.29}$$

Introducing the notation

$$\bar{\delta} = \frac{2(\alpha_1 + \alpha_2)}{\alpha_4} \tag{24.30}$$

and solving for e_{11} and γ_3 from (24.29) we finally obtain the expressions

$$e_{11} = e_{22} = -\frac{1-\bar{\gamma}}{\bar{\alpha}+\bar{\beta}} h p, \quad \gamma_3 = -\frac{\bar{\delta} - 2\bar{\gamma}}{\bar{\alpha}+\bar{\beta}} h p, \tag{24.31}$$

where $\bar{\alpha}$, $\bar{\beta}$, $\bar{\gamma}$ are defined by (24.15). By comparison of (24.31) and (24.22), we may set

$$\frac{h(1-\bar{\gamma})}{\bar{\alpha}+\bar{\beta}} = \frac{c_{13} - c_{33}}{2c_{13}^2 - (c_{11} + c_{12}) c_{33}}, \quad \frac{h(\bar{\delta} - 2\bar{\gamma})}{\bar{\alpha}+\bar{\beta}} = \frac{2c_{13} - (c_{11} + c_{12})}{2c_{13}^2 - (c_{11} + c_{12}) c_{33}}. \tag{24.32}$$

The previous expressions for $\bar{\alpha}$, $\bar{\beta}$, $\bar{\gamma}$ given by (24.20) identically satisfy $(24.32)_1$ and when used in $(24.32)_2$ result in

$$\bar{\delta} = \frac{2(\alpha_1 + \alpha_2)}{\alpha_4} = \frac{c_{11} + c_{12}}{c_{33}}. \tag{24.33}$$

The four relations in (24.20) and (24.33) can be solved for α_1, α_2, α_4, α_9 as follows:

$$\alpha_1 = h c_{12}, \quad 2\alpha_2 = h(c_{11} - c_{12}), \quad \alpha_4 = h c_{33}, \quad \alpha_9 = h c_{13}. \tag{24.34}$$

For an isotropic plate, using (24.3), the above coefficients become

$$\alpha_1 = \frac{\nu(1-\nu)}{1-2\nu} C, \quad \alpha_2 = \frac{1-\nu}{2} C, \quad \alpha_4 = \frac{(1-\nu)^2}{1-2\nu} C, \quad \alpha_9 = \frac{\nu(1-\nu)}{1-2\nu} C, \tag{24.35}$$

where C is defined by $(20.13)_2$ and ν is Poisson's ratio.

This completes the determination of seven of the constitutive coefficients in (23.3)–(23.4) for initially flat Cosserat plates.[14] In the theory of a Cosserat plate characterized by (23.1)–(23.4), the coefficients α_6 and α_7 or equivalently $(\alpha_6 + \alpha_7)$ and $(\alpha_6 - \alpha_7)$ are arbitrary and there is a constitutive relation for $M_{[\alpha\beta]}$ which involves the coefficient $(\alpha_6 - \alpha_7)$. However, since we have undertaken to determine the coefficients by direct comparison with the exact three-dimensional solutions, then we should put $\alpha_6 = \alpha_7$ in view of the symmetry of the stress-tensor in the three-dimensional theory.

Two other coefficients, namely α_3 and α_8 which occur in $(23.4)_2$ and $(23.3)_3$, remain arbitrary and we note that these coefficients have the orders of magnitude

$$\alpha_3 = O(C), \quad \alpha_8 = O(B). \tag{24.36}$$

The coefficients α_1, α_2, α_4, α_5, α_6, α_7 and α_9 in (24.10) and (24.35), apart from their dependence on the thickness h, are expressed in terms of material properties which are constants in the three-dimensional theory. By contrast, a little reflection will reveal that the coefficients α_3 and α_8 in (24.36) cannot be determined as

[14] The determination of the constitutive coefficients (24.10) and (24.35) has been discussed in a number of papers: [1967, *4*], [1967, *6*], [1968, *6*], [1969, *3*] and [1970, *2*]. However, the identification (24.35) was achieved by different procedures in these papers.

constants from comparison with the three-dimensional solutions.[15] Even for a slightly less general system of equations [discussed following (23.10)], where the coefficient $\alpha_8=0$ and $M_{\alpha 3}=0$, we still have α_3 unspecified. Instead of assigning a definite approximate value to α_3, it seems preferable to allow α_3 (and also α_8 when a constitutive equation for $M_{\alpha 3}$ is present) to have different possible values depending on the particular context in which the theory of Cosserat surface is used. This requires elaboration but first we consider below a further example which will be helpful in our subsequent discussion.

(iv) *Torsion of a rectangular plate.* Consider a rectangular Cosserat plate in equilibrium and let the boundary lines (or edges)

$$x_1=\pm\frac{a}{2}$$

be traction free while the boundaries $x_2=\pm l$ rotate and are free from normal tractions. These boundary conditions can be stated as

$$M_{11}=M_{12}=V_1=0 \quad \text{on} \quad x_1=\pm\frac{a}{2}, \tag{24.37}$$

$$M_{22}=0, \quad \delta_1=\mp\bar{\theta}\, l, \quad u_3=\pm\bar{\theta}\, x_1\, l \quad \text{on} \quad x_2=\pm l, \tag{24.38}$$

where $\bar{\theta}$ is the angle of twist per unit length. We assume $M_{\alpha\beta}$, V_α to be independent of x_2 so that the kinematic variables $\varrho_{\alpha\beta}$ and γ_α are also functions of x_1 only. In the absence of $\bar{L}_\beta$ and $\bar{F}_3$, Eqs. (23.2) reduce to

$$M_{11,1}=V_1, \quad M_{12,1}=V_2, \quad V_{1,1}=0. \tag{24.39}$$

Keeping the boundary conditions $(24.38)_{2,3}$ in mind, we assume the displacement solutions in the form

$$u_3=\bar{\theta}\, x_1\, x_2, \quad \delta_1=-\bar{\theta}\, x_2, \quad \delta_2=\delta_2(x_1). \tag{24.40}$$

From $(24.39)_{1,3}$, (23.4) and the kinematic results (6.25), as well as the boundary conditions $(24.37)_{1,3}$ and $(24.38)_1$, it follows that

$$\begin{gathered} \gamma_1=0, \quad \gamma_2=\gamma_2(x_1)=\delta_2+\bar{\theta}\, x_1, \\ \varrho_{11}=\varrho_{22}=0, \quad \varrho_{12}=-\bar{\theta}, \quad \varrho_{21}=\gamma_{2,1}-\bar{\theta}, \\ M_{11}=0, \quad M_{22}=0, \quad V_1=0. \end{gathered} \tag{24.41}$$

By use of the expressions for γ_2 and ϱ_{12} in (24.41), the remaining constitutive equations in (23.4) become[16]

$$M_{(12)}=\alpha_6(\gamma_{2,1}-2\,\bar{\theta}), \quad V_2=\alpha_3\,\gamma_2. \tag{24.42}$$

A differential equation for γ_2 can be obtained by substituting (24.42) into $(24.39)_2$. From the solution of this equation for γ_2 (which we expect to be an odd function of x_1) and the use of the boundary condition $(24.37)_2$, we readily deduce

$$\gamma_2=A\,\mathrm{Sinh}\frac{x_1}{\lambda}, \quad \lambda^2=\frac{\alpha_6}{\alpha_3}, \quad A=\frac{2\lambda\,\bar{\theta}}{\mathrm{Cosh}\dfrac{a}{2\lambda}}. \tag{24.43}$$

[15] This will become evident also in the example (iv) considered below.

[16] Since we have already determined the coefficients α_6 and α_7 according to (24.10), only the symmetric part of $M_{\alpha\beta}$ is given by a constitutive equation and $M_{[\alpha\beta]}=0$.

We do not record here the final expressions for $M_{(12)}$ and V_2 but note that the resultant torque is given by

$$T = \int_{-a/2}^{a/2} [x_1 V_2 - M_{(12)}]\, dx_1 = -2 \int_{-a/2}^{a/2} M_{(12)}\, dx_1, \tag{24.44}$$

in view of $(24.39)_2$ and the edge conditions (24.37). For a rectangular strip of breadth a and thickness h, the expression for the torsional rigidity resulting from (24.44) is easily found to be

$$\mu \frac{h^3 a}{3} \left[1 - \frac{2\lambda}{a} \tanh \frac{a}{2\lambda}\right], \tag{24.45}$$

where μ is defined by $(20.13)_1$ and we have also used the value of α_6 given by $(24.10)_2$. The above expression, for a wide range of a/h, will be in remarkably close agreement with the prediction of the corresponding result in the Saint-Venant theory of torsion if we choose $\lambda^2 = h^2/10$ or equivalently

$$\alpha_3 = \tfrac{5}{6} \mu h. \tag{24.46}$$

This completes our consideration of torsion of a rectangular plate.[17]

We return now to our previous discussion concerning the coefficients α_3, α_8 and also recall the remarks made following (24.36). As far as $(24.36)_2$ is concerned, at present we have no definite evidence regarding a suitable approximate value for [18] α_8. However, in a theory which is only slightly less general than that characterized by (23.1)–(23.4), $M_{\alpha 3}$ is absent and the coefficient $\alpha_8 = 0$ or does not arise.[19]

With reference to $(24.36)_1$, in view of the conclusion reached in example (iv) above, we may specify α_3 by the approximate value (24.46) for elastostatic problems. On the other hand, for dynamical problems, we can determine α_3 by comparison of an appropriate elastodynamic solution of the system of equations of the bending theory [i.e., the system of Eqs. (23.2) and (23.4) with values (24.10)] with the prediction of the corresponding exact (three-dimensional) solution due to LAMB[20] for the circular frequency of the first anti-symmetric mode of thickness-shear

---

[17] The approximate value for α_3 is the same as that in REISSNER's plate theory discussed in Sect. 20. Our above solution for torsion of a rectangular plate [between (24.37)–(24.44)] parallels that contained in REISSNER's [1945, *2*] paper. It is perhaps of interest to indicate here the nature of a corresponding result for torsion of a rectangular plate of variable thickness, with the thickness being dependent on one coordinate (say x_1), as given by ESSENBURG and NAGHDI [1958, *2*]. Using constitutive equations appropriate to plates of variable thickness (derived from the three-dimensional equations), they have shown that when the cross-section is an ellipse or an equilateral triangle the results agree exactly with the corresponding solutions in the Saint-Venant theory of torsion. A solution parallel to that in [1958, *2*] is included in the paper of GREEN, NAGHDI and WENNER [1971, *6*] who employ the theory of a Cosserat surface with the initial director in the form (6.21) and with D as a function of x_1.

[18] A value for α_8 is suggested by the coefficient in $(20.11)_3$. This value is a consequence of the approximation $(20.8)_2$ and there is no evidence in the existing literature as to its being a reasonable approximate value, as remarked also following (20.14). However, it is possible to specify an approximate value for α_8 by a suitable comparison of an extensional solution of (23.1) and (23.3) with the corresponding solution of a simple two-dimensional problem in elasticity theory.

[19] The nature of this linear theory, in the absence of $M_{\alpha 3}$, was discussed following (23.10). This type of linear theory already includes all existing linear theories of shells and plates currently employed in the literature.

[20] LAMB [1917, *1*].

vibration in a plate. Such a comparison leads to the value[21]

$$\alpha_3 = \frac{\pi^2}{12} \mu h,$$

which is very close to (24.46). In subsequent developments in this chapter, we shall not record (or utilize) the constitutive equations for V_α and $M_{\alpha 3}$ with specific values assigned to the coefficients (24.36); but the foregoing discussion regarding the specification of α_8 and α_3 will be understood.

β) The constitutive coefficients for shells. It is clear from the developments of Sect. 16 that the constitutive coefficients in the linear theory of shells depend, in general, on $B_{\alpha\beta}$. However, the constitutive relations (23.6)–(23.8) are deduced by assuming a special form for the Helmholtz free energy function which does not depend explicitly on[22] $B_{\alpha\beta}$. Moreover, since these constitutive relations must reduce to those appropriate for flat plates, we adopt the values of the coefficients (24.10) and (24.35). Hence, the constitutive relations (23.6)–(23.8) can be written as

$$\begin{aligned} N'^{\alpha\beta} &= N'^{\beta\alpha} = C H^{\alpha\beta\gamma\delta} e_{\gamma\delta} + \frac{\nu}{1-\nu} A^{\alpha\beta} V^3, \\ M^{\alpha\beta} &= M^{\beta\alpha} = B H^{\alpha\beta\gamma\delta} \varrho_{\gamma\delta}, \\ V^\alpha &= N^{\alpha 3} = \alpha_3 A^{\alpha\beta} \gamma_\beta, \\ V^3 &= C \frac{(1-\nu)^2}{1-2\nu} \gamma_3 + C \frac{\nu(1-\nu)}{1-2\nu} A^{\alpha\beta} e_{\alpha\beta}, \\ M^{\alpha 3} &= \alpha_8 A^{\alpha\gamma} \varrho_{3\gamma}, \end{aligned} \tag{24.47}$$

where

$$\begin{aligned} H^{\alpha\beta\gamma\delta} &= \tfrac{1}{2}\{A^{\alpha\gamma} A^{\beta\delta} + A^{\alpha\delta} A^{\beta\gamma} + \nu[2A^{\alpha\beta} A^{\gamma\delta} - A^{\alpha\gamma} A^{\beta\delta} - A^{\alpha\delta} A^{\beta\gamma}]\} \\ &= \tfrac{1}{2}\{A^{\alpha\gamma} A^{\beta\delta} + A^{\alpha\delta} A^{\beta\gamma} + \nu(\bar{\varepsilon}^{\alpha\gamma} \bar{\varepsilon}^{\beta\delta} + \bar{\varepsilon}^{\alpha\delta} \bar{\varepsilon}^{\beta\gamma})\} \end{aligned} \tag{24.48}$$

and $\bar{\varepsilon}^{\alpha\beta}$ is an ε-symbol defined previously in (6.30).

For reasons stated earlier in this section, we have left the coefficients α_3 and α_8 unspecified in (24.47). However, we emphasize that the discussion at the end of Subsect. α) is also pertinent to the constitutive coefficients in $(24.47)_{3,5}$. In particular, the values suggested for α_3 can also be used for shells. We further observe that some of the remaining coefficients in (24.47) can be determined by means of solutions of special examples in the theory of Cosserat surface with $B_{\alpha\beta} \neq 0$ (rather than those for initially flat plates). For example, the coefficients α_2, α_6 and α_7 with values the same as those in (24.35) and (24.10) can be deter-

[21] This manner of determining the value of α_3 for dynamical problems is similar to that used by MINDLIN [1951, *1*] and mentioned previously in Sect. 20 [Subsect. α)]. As noted with reference to (20.17), in the early stage of his derivation of plate equations for dynamical problems (from the three-dimensional equations), MINDLIN assumes the coefficient corresponding to α_3 to be unspecified; but subsequently he determines this coefficient by a comparison with LAMB's solution [1917, *1*]. This manner of determining the constitutive coefficients is in accord with our point of view in this chapter; in fact, the use of the equations of the Cosserat surface (whose constitutive coefficients are not predetermined) supplies a justification for MINDLIN's procedure.

[22] As already remarked in Sect. 16, the ultimate plausibility for such a special choice of the free energy (or the strain energy) function depends, of course, on its usefulness. In the context of the three-dimensional theory, it is clear from comparison of (16.33) and (20.14) that the relations (23.6)–(23.8) correspond to the neglect of terms of $O(h/R)$ or smaller compared with those in the special free energy function (16.33).

mined from a comparison of the solution for torsion of a cylindrical Cosserat surface with the corresponding results in the Saint-Venant theory of torsion.[23]

25. The boundary-value problem of the restricted linear theory. For convenience, we first summarize below the linearized field equations of the restricted isothermal theory. Thus, it follows from the linearized versions of the results in Sects. 10 and 15 that the equations of motion are given by (9.69) and

$$\grave{M}^{(\gamma\alpha)}{}_{|\gamma} = V^\alpha, \qquad \bar{\varepsilon}_{\alpha\beta}[N^{\alpha\beta} - \grave{M}^{(\gamma\beta)} B^\alpha_\gamma] = 0. \tag{25.1}$$

The above equations are the same as those in (10.28)–(10.29), together with (15.17), except that in (25.1) we have used V^α in place of $N^{\alpha 3}$ for ready comparison with $(9.70)_1$ and we have also put[24] $\grave{L}^\alpha = 0$. We recall here that in the restricted theory only $\grave{M}^{(\gamma\alpha)}$ and the linearized expression corresponding to (10.26), namely the symmetric

$$\grave{N}^{\alpha\beta} = \grave{N}^{\beta\alpha} = N^{\alpha\beta} + \grave{M}^{(\gamma\alpha)} B^\beta_\gamma, \tag{25.2}$$

are specified by constitutive equations; and that V^α, which is not specified by a constitutive equation, is determined from $(25.1)_1$. Remembering the brief remarks included in Sect. 16 [Subsect. δ)] concerning the linear constitutive equations of the restricted isothermal theory, from (16.40) we obtain the following constitutive equation for an isotropic elastic material:

$$\begin{aligned} \grave{N}^{\alpha\beta} &= [\beta_1 A^{\alpha\beta} A^{\gamma\delta} + \beta_2 (A^{\alpha\gamma} A^{\beta\delta} + A^{\alpha\delta} A^{\beta\gamma})]\, e_{\gamma\delta}, \\ \grave{M}^{(\alpha\beta)} &= [\beta_5 A^{\alpha\beta} A^{\gamma\delta} + \beta_6 (A^{\alpha\gamma} A^{\beta\delta} + A^{\alpha\delta} A^{\beta\gamma})]\, \varrho_{\gamma\delta}. \end{aligned} \tag{25.3}$$

The kinematic measures in (25.3) are defined in (6.27) and the coefficients β_1, β_2, β_5 and β_6 are arbitrary constants and do not necessarily have the values obtained previously in Sect. 24 for the corresponding coefficients in (23.6).

We proceed to identify the constitutive coefficients in (25.3) in a manner entirely similar to that employed in Sect. 24. On this occasion we only need to consider two simple elastostatic problems, namely those for flexural and extensional deformations of a plate, discussed in examples (i) and (ii) in Sect. 24. Thus, by comparisons of these solutions with corresponding exact solutions in the three-dimensional theory, we determine all four coefficients in (25.3) as follows:

$$\beta_1 = \nu\, C, \quad \beta_2 = \frac{1-\nu}{2} C, \quad \beta_5 = \nu\, B, \quad \beta_6 = \frac{1-\nu}{2} B. \tag{25.4}$$

With the above coefficients, the constitutive equations (25.3) can be expressed as

$$\grave{N}^{\alpha\beta} = C\, H^{\alpha\beta\gamma\delta} e_{\gamma\delta}, \qquad \grave{M}^{(\alpha\beta)} = B\, H^{\alpha\beta\gamma\delta} \varrho_{\gamma\delta}, \tag{25.5}$$

where $H^{\alpha\beta\gamma\delta}$ is given by (24.48).

The boundary conditions appropriate to the above restricted linear theory can be obtained from the linearized version of the corresponding results in Sect. 15.

[23] The determination of α_2, α_6 and α_7 in this manner can be found in [1967, *6*]. The value of α_1 appearing in Eq. (3.20) of this paper contains an error and should conform to that in (24.35); however, this does not affect the remaining discussion in [1967, *6*] concerning torsion of a circular cylindrical Cosserat surface. Related results for torsion of a cylindrical Cosserat surface, in which the edge curve perpendicular to the generator is not necessarily circular, are given by Wenner [1968, *13*].

[24] This restriction is not essential. It is introduced here simply because a quantity corresponding to $\grave{L}^\alpha$ is usually absent or is neglected in the derivations of the classical theory of shells and plates from the three-dimensional equations.

These conditions require the specification of either the displacement boundary conditions

$$u_\gamma, \quad u_3, \quad \frac{\partial u_3}{\partial \nu_0} \tag{25.6}$$

or the force and couple boundary conditions

$$\begin{aligned} {}_0P^\beta &= {}_0\nu_\alpha [N^{\alpha\beta} - \grave{M}^{(\alpha\gamma)} B^\beta_\gamma], \qquad {}_0G = \grave{M}^{(\alpha\gamma)}\, {}_0\nu_\alpha\, {}_0\nu_\gamma, \\ {}_0P^3 &= {}_0\nu_\alpha (\grave{M}^{(\beta\alpha)}{}_{|\beta}) - \frac{\partial}{\partial s_0} [\bar{\varepsilon}_{\beta\gamma}\, \grave{M}^{(\alpha\beta)}\, {}_0\nu_\alpha\, {}_0\nu^\gamma], \end{aligned} \tag{25.7}$$

where ${}_0\nu_\alpha$ are the components of the outward unit normal ${}_0\boldsymbol{\nu}$ to the boundary curve on the reference surface $\mathscr{S}$ and $\partial/\partial\nu_0$, $\partial/\partial s_0$ denote the directional derivatives along the normal and the tangent to the boundary curve on $\mathscr{S}$.

The system of equations and boundary conditions in the above restricted theory correspond to those of the classical theory of shells and plates obtained (from the three-dimensional equations) under the Kirchhoff (or the Kirchhoff-Love) hypothesis or any other equivalent set of assumptions. The derivation of the restricted theory (by direct approach) is, however, free of the inconsistencies present in the usual derivations of the classical theory.[25] In this connection, we emphasize that the constitutive coefficients in (25.3) are arbitrary and can be assigned the specified values (25.4) without violating any of the kinematic assumptions appropriate to the restricted theory. This is in contrast to the usual derivations of the classical theory in which an additional assumption is introduced (beyond that of the kinematic assumptions), in order to compensate for the fact that the constitutive coefficients in an approximate expression for the three-dimensional strain energy density are predetermined.

The system of equations of the restricted theory can also be obtained as a special case of those in the linear theory of a Cosserat surface [Sect. 23, Subsect. β)] under suitable constraints. To see this, with reference to (23.6)–(23.8), let

$$\alpha_8 = 0, \qquad \frac{\alpha_3}{\alpha_5} \to \infty, \qquad \gamma_\alpha \to 0, \tag{25.8}$$

in such a manner that

$$V^\alpha \to \text{finite limit}. \tag{25.9}$$

Then, $M^{\alpha 3} = 0$ and V^α is not determined by a constitutive equation. In addition, under the condition $(25.8)_3$, the kinematic measure $\varrho_{\gamma\alpha}$ in (6.24) becomes symmetric in the indices γ, α [see (6.27)] and hence the skew-symmetric part of $M^{\beta\alpha}$ in $(23.6)_2$ vanishes, i.e., $M^{[\beta\alpha]} = 0$. We also assume that the assigned director couple and the inertia terms due to director velocity are either zero or negligible so that $\bar{L}^\beta = 0$, $\bar{L}^3 = 0$. Then, from (9.70) and (25.9), we have

$$M^{(\gamma\alpha)}{}_{|\gamma} = V^\alpha, \qquad V^3 = 0, \tag{25.10}$$

while the equations of motion (9.69) remain unchanged. In view of $(25.10)_2$, the constitutive relation $(24.47)_1$ reduces to

$$N'^{\alpha\beta} = N'^{\beta\alpha} = C H^{\alpha\beta\gamma\delta} e_{\gamma\delta} \tag{25.11}$$

and $(24.47)_4$ yields

$$\gamma_3 = -\frac{\nu}{1-\nu} A^{\alpha\beta} e_{\alpha\beta}. \tag{25.12}$$

[25] These inconsistencies refer, on the one hand, to the kinematic assumptions in the classical theory of shells and plates [discussed in Sect. 7, Subsect. ε)]; and, on the other hand, to the manner in which the constitutive equations with appropriate constitutive coefficients are derived.

Further, by (25.8) and in view of the expressions for the symmetric $\bar{\varkappa}_{\gamma\delta}$ in (6.24), $\varrho_{\gamma\delta}$ is now given by the last expression in (6.27) and $(24.47)_2$ becomes

$$M^{(\alpha\beta)} = B\, H^{\alpha\beta\gamma\delta}\, \varrho_{\gamma\delta}, \qquad \varrho_{\gamma\delta} = -\bar{\varkappa}_{\gamma\delta}. \tag{25.13}$$

Also, the remaining Eq. (9.68) takes the form

$$N'^{\alpha\beta} = N'^{\beta\alpha} = N^{\alpha\beta} + M^{(\gamma\alpha)}\, B^{\beta}_{\gamma}. \tag{25.14}$$

The boundary conditions for the above special theory can be obtained from the more general boundary conditions using (25.8)–(25.9) and other results between (25.10)–(25.14). These reduced boundary conditions will be similar in form to those in (25.6)–(25.7) and hence will not be recorded. It should be clear that the system of equations of the restricted theory characterized by (9.69), (25.1)–(25.2) and (25.5) are formally equivalent to those of the special theory [discussed between (25.8)–(25.14)] if we identify $\grave{M}^{(\alpha\beta)}$ with $M^{(\alpha\beta)}$ and $\grave{N}^{\alpha\beta}$ with $N'^{\alpha\beta}$, apart from (25.12) in the special theory.[26]

Returning to the restricted theory, we recall from Sect. 15 that the equations of motion (after elimination of V^{α}) can be expressed in either one of the alternative forms resulting from linearization of (15.19) and (15.20). The linearized equations of motion in terms of $\grave{N}^{\alpha\beta}$ and $\grave{M}^{(\alpha\beta)}$ obtained by linearization of $(15.20)_{1,2}$, together with the constitutive equations (25.3) in terms of the kinematic variables $e_{\alpha\beta}$ and $\varrho_{\alpha\beta}$, constitute a determinate system. However, in obtaining solutions to boundary-value problems (or initial boundary-value problems), it is also possible to employ the equations of motion which result from the linearization of $(15.19)_{1,2}$, namely

$$\begin{aligned} N^{(\alpha\beta)}{}_{|\alpha} - \tfrac{1}{2}[B^{\beta}_{\gamma}\, \grave{M}^{(\gamma\alpha)}]_{|\alpha} + \tfrac{1}{2}[B^{\alpha}_{\gamma}\, \grave{M}^{(\gamma\beta)}]_{|\alpha} - B^{\beta}_{\alpha}\, \grave{M}^{(\gamma\alpha)}{}_{|\gamma} + \varrho_0 \bar{F}^{\beta} = 0, \\ \grave{M}^{(\alpha\beta)}{}_{|\alpha\beta} + B_{\alpha\beta}\, N^{(\alpha\beta)} + \varrho_0 \bar{F}^{3} = 0, \end{aligned} \tag{25.15}$$

where the components $\bar{F}^{i}$ are defined by $(9.57)_1$ instead of the linearized version of (10.23) since we have put $\grave{L}^{\alpha} = 0$. Although $\grave{M}^{(\alpha\beta)}$ in (25.15) can be identified with $M^{(\alpha\beta)}$ in view of the equivalence of the restricted theory and a special case of the general theory noted above, we continue our use of the notations $\grave{N}^{\alpha\beta}$ and $\grave{M}^{(\alpha\beta)}$ in order to distinguish between the restricted theory and the general theory.

The equations of motion (25.15) involve $N^{(\alpha\beta)}$, $\grave{M}^{(\alpha\beta)}$ rather than $\grave{N}^{\alpha\beta}$, $\grave{M}^{(\alpha\beta)}$. Hence, it is more convenient to express the constitutive equations also in terms of the former set whenever (25.15) is employed. To this end, we introduce the symmetric kinematic variable

$$\bar{\varrho}_{\alpha\beta} = \varrho_{\alpha\beta} + \tfrac{1}{2}(B^{\nu}_{\alpha}\, e_{\nu\beta} + B^{\nu}_{\beta}\, e_{\nu\alpha}), \tag{25.16}$$

where $\varrho_{\alpha\beta}$, $e_{\alpha\beta}$ are defined in (6.27).[27] Assuming now for the specific free energy (or the strain energy density) the form

$$\psi = \grave{\psi}\,(e_{\alpha\beta}, \bar{\varrho}_{\alpha\beta};\, A_{\alpha\beta}, -B_{\alpha\beta}), \tag{25.17}$$

we can then deduce the results[28]

$$N^{(\alpha\beta)} = \varrho_0 \frac{\partial \grave{\psi}}{\partial e_{\alpha\beta}}, \qquad \grave{M}^{(\alpha\beta)} = \varrho_0 \frac{\partial \grave{\psi}}{\partial \bar{\varrho}_{\alpha\beta}}, \tag{25.18}$$

[26] The reduction of the general linear theory to the special theory as discussed above is contained in a paper by Green and Naghdi [1968, 6] and is patterned after that discussed by them [1968, 5] in a more general context.

[27] The kinematic variable in (25.16) corresponds to that defined by (20.37) in an alternative formulation of the classical shell theory [Sect. 20, Subsect. δ)].

[28] The expressions (25.18) may be compared with (20.39) in the classical linear theory of shells, obtained from the three-dimensional equations by approximation.

where $\grave{\psi}$ is a quadratic function of $e_{\alpha\beta}$ and $\bar{\varrho}_{\alpha\beta}$. Further, by imposing a restriction on the function $\grave{\psi}$ similar to (16.39) and for an isotropic material with a center of symmetry, we can write the constitutive equation for ψ in a form similar to (16.40) but in terms of $e_{\alpha\beta}$ and $\bar{\varrho}_{\alpha\beta}$. The linear constitutive equations for $N^{(\alpha\beta)}$ and $\grave{M}^{(\alpha\beta)}$ in terms of $e_{\alpha\beta}$ and $\bar{\varrho}_{\alpha\beta}$ then follow but we do not pursue the matter further. We note, however, the relationship between $N^{(\alpha\beta)}$ and $\grave{N}^{\alpha\beta}$, namely

$$N^{(\alpha\beta)} = \grave{N}^{\alpha\beta} - \tfrac{1}{2}(B^{\alpha}_{\lambda}\,\grave{M}^{(\lambda\beta)} + B^{\beta}_{\lambda}\,\grave{M}^{(\lambda\alpha)}), \tag{25.19}$$

which is obtained from (25.2). Also, from (10.29), the skew-symmetric part of $N^{\alpha\beta}$ is given by

$$N^{[\alpha\beta]} = \tfrac{1}{2}[B^{\alpha}_{\lambda}\,\grave{M}^{(\lambda\beta)} - B^{\beta}_{\lambda}\,\grave{M}^{(\lambda\alpha)}]. \tag{25.20}$$

26. A uniqueness theorem. Remarks on the general theorems. This concluding section is concerned mainly with a uniqueness theorem for solutions of the initial (isothermal) boundary-value problems of elastic shells and plates. In addition, some remarks are included here concerning the nature of the general theorems which can be obtained in the linear theory. Although we establish a uniqueness theorem below for the initial mixed boundary-value problems of the dynamical theory, a parallel development can also be given for equilibrium mixed boundary-value problems. We first obtain the conditions *sufficient* for uniqueness, using the constitutive equations (16.11) with a quadratic specific free energy (or the strain energy density) function $\bar{\psi}$ in the form (16.10);[29] and then we further examine the restrictions on the constitutive coefficients required by uniqueness when $\bar{\psi}$ has the special form (16.33).

Let $\mathscr{S}$ be a bounded regular region of two-dimensional space occupied by the initial elastic Cosserat surface. (By a regular region we mean one to which the divergence theorem is applicable.) Let $\partial\mathscr{S}$ be the boundary and let $\mathscr{S}^0$ denote the interior of $\mathscr{S}$ and introduce the regions of space-time by

$$\begin{aligned} \mathscr{R} &= \{(\theta^{\alpha}, t) : \theta^{\alpha} \in \mathscr{S}, t \geqq 0\}, \\ \mathscr{R}^0 &= \{(\theta^{\alpha}, t) : \theta^{\alpha} \in \mathscr{S}^0, t \geqq 0\}. \end{aligned} \tag{26.1}$$

Let ${}_0\boldsymbol{v} = {}_0\nu_{\alpha}\,\boldsymbol{A}^{\alpha}$ be the outward unit normal to $\partial\mathscr{S}$ and $\partial\mathscr{S}_1$, $\partial\mathscr{S}_2$ be arbitrary disjoint subsets of $\partial\mathscr{S}$ such that $\partial\mathscr{S}_1 \cup \partial\mathscr{S}_2 = \partial\mathscr{S}$. Then, the initial boundary-value problem of the isothermal linear theory of elastic shells is characterized by the equations of motion (9.68)–(9.70) and the constitutive relations (16.11) with the function $\bar{\psi}$ given by that in (16.10). We require that the free energy (or the strain energy density) function $\bar{\psi}$ be nonnegative, i.e.,

$$\bar{\psi} \geqq 0 \quad \text{for} \quad t \geqq 0. \tag{26.2}$$

To the above field equations and constitutive relations, we supplement the boundary conditions

$$\begin{aligned} \boldsymbol{v} &= \boldsymbol{v}', & \boldsymbol{w} &= \boldsymbol{w}' & &\text{on } \partial\mathscr{S}_1 \quad (\text{for } t \geqq 0), \\ \boldsymbol{N} &= \boldsymbol{N}', & \boldsymbol{M} &= \boldsymbol{M}' & &\text{on } \partial\mathscr{S}_2 \quad (\text{for } t \geqq 0) \end{aligned} \tag{26.3}$$

and the initial conditions

$$\begin{aligned} \boldsymbol{u} &= \boldsymbol{u}^0, & \boldsymbol{\delta} &= \boldsymbol{\delta}^0 & &\text{on } \mathscr{S} \quad (\text{for } t = 0), \\ \boldsymbol{v} &= \boldsymbol{v}^0, & \boldsymbol{w} &= \boldsymbol{w}^0 & &\text{on } \mathscr{S} \quad (\text{for } t = 0). \end{aligned} \tag{26.4}$$

[29] Since we are concerned with the isothermal theory, θ in the argument of $\bar{\psi}$ in (16.10) is now a constant.

Here $\boldsymbol{u}^0$, $\boldsymbol{\delta}^0$, $\boldsymbol{v}^0$, $\boldsymbol{w}^0$, $\boldsymbol{v}'$, $\boldsymbol{w}'$, $\boldsymbol{N}'$, $\boldsymbol{M}'$ are prescribed functions on the appropriate domains and the mass density ϱ_0 is a strictly positive constant.

We now state the following uniqueness theorem: Let $\boldsymbol{u}$, $\boldsymbol{\delta}$ be the displacement vector and the director displacement vector which satisfy the above mentioned field equations on $\mathscr{R}$; and let the assigned surface force $\boldsymbol{f}$ and the assigned director couple $\boldsymbol{l}$ be prescribed. Then, provided (26.2) holds and that the director inertia coefficient $\alpha>0$, there exists at most one set of functions $\boldsymbol{u}$, $\boldsymbol{\delta}$ satisfying (26.3)–(26.4) such that u_α, δ_i are of class C^1 and u_3 is of class C^2 on $\mathscr{R}$ while u_α, δ_i are of class C^2 and u_3 is of class C^3 on $\mathscr{R}^0$.

Our method of proof is similar to the classical uniqueness proofs in three-dimensional linear elasticity.[30] In essence, we consider the difference of two possible solutions and make use of an equation for the mechanical balance of energy, i.e., an equation in which the rate of increase of the sum of kinetic energy and internal energy is equal to the rate of work by the contact and the assigned forces and couples. Thus, assume that there are two sets of solutions to the initial boundary-value problem under consideration, namely

$$\begin{gathered} {}^{(1)}u_i,\ {}^{(1)}\delta_i,\ {}^{(1)}N'^{\alpha\beta},\ {}^{(1)}M^{\alpha i},\ {}^{(1)}V^{\alpha}, \\ {}^{(2)}u_i,\ {}^{(2)}\delta_i,\ {}^{(2)}N'^{\alpha\beta},\ {}^{(2)}M^{\alpha i},\ {}^{(2)}V^{\alpha}, \end{gathered} \tag{26.5}$$

and denote the difference of the two sets of solutions by

$$\begin{gathered} \bar{u}_i = {}^{(1)}u_i - {}^{(2)}u_i, \qquad \bar{\delta}_i = {}^{(1)}\delta_i - {}^{(2)}\delta_i, \\ \bar{N}'^{\alpha\beta} = {}^{(1)}N'^{\alpha\beta} - {}^{(2)}N'^{\alpha\beta}, \qquad \bar{M}^{\alpha i} = {}^{(1)}M^{\alpha i} - {}^{(2)}M^{\alpha i}, \qquad \bar{V}^{\alpha} = {}^{(1)}V^{\alpha} - {}^{(2)}V^{\alpha}. \end{gathered} \tag{26.6}$$

Because of the linear character of all field equations and constitutive relations, it is clear that the set of functions $\bar{u}_i$, $\bar{\delta}_i$, etc., defined by (26.6) satisfy equations of the forms (9.68)–(9.70) in the absence of the assigned force and the assigned director couple F^i, L^i [defined as in (9.57)] and also equations of the forms (16.11) with $\bar{\psi}$ given by that in (16.10) as a quadratic function of its arguments. Moreover, since each of the two sets of solutions in (26.5) satisfies the boundary conditions, we have

$$\begin{aligned} \bar{\boldsymbol{v}} &= 0, \quad \bar{\boldsymbol{w}} = 0 \quad \text{on } \partial\mathscr{S}_1 \quad (\text{for } t \geqq 0), \\ \bar{\boldsymbol{N}} &= 0, \quad \bar{\boldsymbol{M}} = 0 \quad \text{on } \partial\mathscr{S}_2 \quad (\text{for } t \geqq 0). \end{aligned} \tag{26.7}$$

We keep these in mind in what follows and for simplicity delete the overbar from the difference functions in (26.6).

Since F_i and L_i are prescribed, from the linearized versions of (8.8) and (14.2) applied to the difference of the two solutions, we obtain

$$\frac{d}{dt}\int_{\mathscr{S}} \frac{1}{2}\varrho_0(\boldsymbol{v}\cdot\boldsymbol{v} + \alpha\,\boldsymbol{w}\cdot\boldsymbol{w})\,d\Sigma + \int_{\mathscr{S}} [N'^{\alpha\beta}\,\dot{e}_{\alpha\beta} + V^i\dot{\gamma}_i + M^{\alpha i}\dot{\varrho}_{i\alpha}]\,d\Sigma = \int_{\partial\mathscr{S}} (\boldsymbol{N}\cdot\boldsymbol{v} + \boldsymbol{M}\cdot\boldsymbol{w})\,dS, \tag{26.8}$$

where the element of area $d\Sigma$ is defined by (4.39), dS is an element of length of the boundary curve with outward unit normal ${}_0\boldsymbol{\nu}$ and we have omitted the overbar from the difference of the two sets of functions. Using (16.11) and the bound-

[30] For the classical uniqueness theorems in three-dimensional elasticity, see Love [1944, *4*, p. 170 and p. 176] or Sokolnikoff [1956, *2*, p. 86]. The requirement that the free energy function $\bar{\psi}$ be nonnegative can probably be relaxed as in the paper of Knops and Payne [1968, *10*], which is concerned with uniqueness in three-dimensional elastodynamics.

ary conditions (26.7), (26.8) leads to

$$\frac{d}{dt}\int_{\mathscr{S}}\left\{\frac{1}{2}\varrho_0(\boldsymbol{v}\cdot\boldsymbol{v}+\alpha\boldsymbol{w}\cdot\boldsymbol{w})+\varrho_0\bar{\psi}\right\}d\Sigma=0. \tag{26.9}$$

Since $\bar{\psi}$ is nonnegative and $\alpha>0$, it follows from (26.9) that the difference of the functions for velocity and director velocity must vanish:

$$\boldsymbol{v}=0, \quad \boldsymbol{w}=0. \tag{26.10}$$

Moreover, in view of the initial conditions (26.4), we also have

$$\boldsymbol{u}=0, \quad \boldsymbol{\delta}=0, \tag{26.11}$$

which proves uniqueness.

The above uniqueness proof is valid for a general linear theory of shells and is not limited to isotropic materials. Obviously, it is valid also for the initial mixed boundary-value problem characterized by the equations of the restricted linear theory [Sect. 16, Subsect. δ) and Sect. 25].[31] It is straightforward to obtain corresponding sufficient conditions for uniqueness of the equilibrium mixed boundary-value problem in which case instead of (26.2) we require that $\bar{\psi}$ be positive definite, i.e.,

$$\bar{\psi}>0 \tag{26.12}$$

for all non-zero values of the kinematic variables; however, in view of the close similarity of the proof with that given above, it will not be included here.

The above theorem provides sufficient conditions for uniqueness without detailed specification of the quadratic function $\bar{\psi}$ in (16.10). We now briefly examine the restrictions on the constitutive coefficients (required by our uniqueness proof) when $\bar{\psi}$ is specified by (16.33)[32] for an isotropic material and further limit ourselves to the case of flat plates. Since the differential equations characterizing the streching of the plate and those for the bending of the plate separate into two distinct sets (Sect. 23), the nature of the restrictions on the constitutive coefficients can be examined separately. Consider first the extensional case and introduce $(16.33)_2$ in the condition (26.2). It then follows that

$$\alpha_1+\alpha_2\geqq 0, \quad \alpha_2\geqq 0, \quad \alpha_4\geqq 0, \quad (\alpha_1+\alpha_2)\alpha_4\geqq\alpha_9^2 \tag{26.13}$$

and

$$\alpha_8\geqq 0. \tag{26.14}$$

Similarly, for the flexure of the plate, the condition (26.2) with $\bar{\psi}$ given by $(16.33)_3$ yields

$$\alpha_6\geqq 0, \quad \alpha_6\pm\alpha_7\geqq 0, \quad 2\alpha_5+\alpha_6+\alpha_7\geqq 0 \tag{26.15}$$

and

$$\alpha_3\geqq 0. \tag{26.16}$$

[31] In the case of the classical theory of plates (corresponding to equations for flat plates resulting from the restricted theory), the uniqueness proof was given by KIRCHHOFF [1850, *1*]. The earliest proof of uniqueness for shells, based on the constitutive relations (21 A.10)–(21 A.11), is contained in a paper by BYRNE [1944, *1*]. A uniqueness theorem is also given by GOL'DENVEIZER [1944, *3*]; see also, GOL'DENVEIZER and LUR'E [1947, *3*]. Uniqueness of solution in the context of thermoelastic shells (using a system of differential equations which include those recorded in Sect. 23), along with some related results pertaining to the nature of the restrictions placed on the constitutive coefficients, is discussed in [1971, *4*].

[32] Recall that the constitutive equations for shells and plates in Sect. 23 are those which are obtained from $\bar{\psi}$ in (16.33).

If we now make use of the values (24.10) and (24.35), the inequalities (26.13) imply

$$\mu \geqq 0, \qquad -1 \leqq \nu < \tfrac{1}{2} \tag{26.17}$$

while those in (26.15) give

$$\mu \geqq 0, \qquad -1 \leqq \nu < 1. \tag{26.18}$$

The conditions (26.17) are the same as those found (by energy arguments) for the initial boundary-value problem in the three-dimensional theory of elasticity for isotropic materials. This is in contrast to the results of the classical theory of generalized plane stress for the stretching of the plate, where the corresponding conditions deduced by energy arguments are $\mu \geqq 0$, $-1 \leqq \nu < 1$. These latter conditions can be obtained by putting

$$\alpha_4 \gamma_3 + 2\alpha_9 A^{\alpha\beta} e_{\alpha\beta} = 0$$

in $(16.33)_2$ before using $(26.2)_1$. Returning to (26.18), we note that these conditions are less restrictive than the corresponding conditions (26.17) for the stretching of the plate. The conditions (26.18) are the same as those found for the bending of the plate according to the classical thin plate theory [Sect. 20, Subsect. β)] using energy arguments. This completes our discussion of the restrictions on the constitutive coefficients required by the above uniqueness proof.

Before closing this section, we make a few remarks concerning the nature of the general theorems for shells. It is not difficult to see that most of the general theorems in the three-dimensional linear theory of elasticity have analogues in shell theory. We have already used an analogue of a three-dimensional result (in the above uniqueness proof), namely the linearized versions of (8.8) and (14.2) according to which the rate of kinetic energy and internal energy is equal to the rate of work by $\boldsymbol{N}$, $\boldsymbol{M}$ and $\boldsymbol{f}$, $\boldsymbol{l}$. From this result and the structure of the linear constitutive equations for shells (Sect. 23), it should be apparent that one can easily obtain analogues of all variational theorems of classical elasticity (including the minimum potential energy and complementary energy theorems), the analogue of BETTI's (also known as the BETTI-RAYLEIGH) reciprocity theorem, as well as a number of related representation theorems.[33]

All of the general theorems mentioned above may be established using the constitutive equations (16.11), with a quadratic function $\overline{\psi}$ of the form (16.10) specialized for isothermal theory, and thus their validity is not limited to isotropic materials. Some of these theorems are analogues of those in the three-dimensional dynamical theory and others represent the counterparts of the corresponding theorems in the three-dimensional elastostatic theory. There exists, however, a correspondence theorem for elastostatic problems of shells which has no direct counterpart in the classical three-dimensional theory. This correspondence theorem, known as *static-geometric analogy*, was introduced simultaneously and independently by GOL'DENVEIZER and LUR'E in the context of the classical theory of shells.[34] According to this analogy, in the absence of the assigned surface

[33] Theorems of this kind, over the years, have been obtained for shells with the use of a variety of constitutive equations which are less general than those in Sect. 23. In particular, an analogue of Betti's reciprocity theorem was first derived by GOL'DENVEIZER [1944, *3*]. For this and analogues of related results, see also [1947, *3*], [1960, *10*] and [1963, *6*].

[34] The static-geometric analogy was introduced by GOL'DENVEIZER [1940, *1*] and LUR'E [1940, *2*], within the scope of the classical shell theory under the Kirchhoff-Love hypothesis. Since the classical theory does not provide a constitutive equation for the shear-stress resultants V^{α}, the analogy in its most convenient form makes use of the equations of equilibrium corresponding to (20.43) or (20.44). Related to this is the idea of reducing the number of the stress-resultants and the stress-couples on the one hand, and the number of the independent strain measures on the other hand, which was apparently first suggested by LUR'E [1950,*3*].

forces $\{F^\beta, F^3\}$, a one-to-one correspondence exists between the equations of equilibrium and the compatibility equations.[35] Although the static-geometric analogy can be discussed with reference to the linear theory summarized in Sect. 23, we confine attention here to the restricted theory since the analogy is particularly useful in the latter framework.

In order to show (within the scope of the restricted theory) the truth of the assertion concerning the static-geometric analogy stated above, it will suffice to consider one of the alternative forms of the equilibrium equations resulting from the linearization of (15.19) or (15.20). We consider here the former or equivalently the equilibrium equations associated with (25.15). For this purpose, recall the compatibility equations $(6.62)_{1,2}$ for the restricted theory and write these in terms of the kinematic variable $\bar{\varrho}_{\alpha\beta}$ defined by (25.16). Thus

$$\begin{aligned} \bar{\varepsilon}^{\alpha\beta}\{\bar{\varepsilon}^{\gamma\nu}[\bar{\varrho}_{\nu\alpha|\beta} + \tfrac{1}{2}(B^\lambda_\alpha e_{\lambda\nu})_{|\beta} - \tfrac{1}{2}(B^\lambda_\nu e_{\lambda\alpha})_{|\beta}] + \bar{\varepsilon}^{\nu\lambda} B^\gamma_\nu e_{\lambda\alpha|\beta}\} = 0, \\ \bar{\varepsilon}^{\alpha\beta}\bar{\varepsilon}^{\lambda\nu}[e_{\alpha\lambda|\nu\beta} - B_{\nu\beta}\bar{\varrho}_{\lambda\alpha}] = 0, \end{aligned} \tag{26.19}$$

where $(4.13)_4$ has been used. Introducing now the correspondence

$$N^{(\alpha\beta)} \Leftrightarrow -\bar{\varepsilon}^{\alpha\sigma}\bar{\varepsilon}^{\beta\nu}\bar{\varrho}_{\sigma\nu}, \qquad \overset{*}{M}{}^{(\alpha\beta)} \Leftrightarrow \bar{\varepsilon}^{\alpha\sigma}\bar{\varepsilon}^{\beta\nu} e_{\sigma\nu}, \tag{26.20}$$

we can verify easily that substitution of (26.20) into the equilibrium equations (25.15) with $F^\beta = F^3 = 0$ yields the compatibility equations (26.19) and this establishes the correspondence sought.[36] Moreover, by virtue of the correspondence (26.20), the equilibrium and the compatibility equations can be combined into a single system of complex differential equations. For this purpose, put

$$P^{\alpha\beta} = N^{(\alpha\beta)} - i K\bar{\varepsilon}^{\alpha\sigma}\bar{\varepsilon}^{\beta\nu}\bar{\varrho}_{\sigma\nu}, \qquad Q^{\alpha\beta} = \overset{*}{M}{}^{(\alpha\beta)} + i K\bar{\varepsilon}^{\alpha\sigma}\bar{\varepsilon}^{\beta\nu} e_{\sigma\nu}, \tag{26.21}$$

where $i = (-1)^{\frac{1}{2}}$ and K is an arbitrary (real) constant. Using (26.21), by a suitable combination of (26.19) and (25.15) with $\bar{F}^i$ replaced with F^i we obtain

$$\begin{aligned} P^{\alpha\beta}{}_{|\beta} - \tfrac{1}{2}[B^\alpha_\lambda Q^{\beta\lambda}]_{|\beta} + \tfrac{1}{2}[B^\beta_\lambda Q^{\lambda\alpha}]_{|\beta} - B^\alpha_\lambda Q^{\beta\lambda}{}_{|\beta} + \varrho_0 F^\alpha = 0, \\ Q^{\alpha\beta}{}_{|\alpha\beta} + B_{\alpha\beta} P^{\alpha\beta} + \varrho_0 F^3 = 0, \end{aligned} \tag{26.22}$$

the real and imaginary parts of which yield the equations of equilibrium and compatibility, respectively. The complex differential equations (26.22) can be further reduced in the case of isotropic shells but we do not pursue the matter here.[37]

---

[35] The static-geometric analogy, in lines of curvature coordinates, is discussed in the books by NOVOZHILOV [1959, *3*] and GOL'DENVEIZER [1961, *3*] and was also noted independently by SANDERS [1959, *6*]. A related analogy has been discussed more recently by LUR'E [1961, *7*]. An account of these results can be found in [1963, *6*]. Within the scope of the classical shell theory and in terms of the variables employed in Sect. 20 [Subsect. δ)], different versions of the analogy can be deduced. See, in this connection, [1963, *7*], [1964, *5*] and [1966, *7*].

[36] As noted earlier, the correspondence theorem has no direct counterpart in the classical (non-polar) three-dimensional theory. However, a *static-geometric analogue* can be shown to hold in the three-dimensional linear theory with couple stresses. An observation of this kind is made in [1965, *6*], where it is also shown that the analogy in the non-polar case merely degenerates to the representation of the stress tensor in terms of a Beltrami-Gwyther-Finzi stress function; for the latter, see TRUESDELL and TOUPIN [1960, *14*].

[37] The complex differential equations (26.22) and their further reduction in the case of isotropic shells were obtained by NAGHDI [1966, *7*]. Similar complex differential equations, together with their reduction for isotropic shells, were previously discussed by NOVOZHILOV [1959, *3*] in lines of curvature coordinates. However, in effecting the reduction, NOVOZHILOV makes use of a different set of constitutive equations along with a further approximation which renders his derivation rigorous only if Poisson's ratio is zero. The complex differential equations (26.22) and the results in [1966, *7*] have been utilized in a recent study by STEELE [1971, *9*].

F. Appendix: Geometry of a surface and related results.

This Appendix contains various formulae from tensor calculus and selected results from the differential geometry of a surface which are essential in our development of the theory of shells and plates (Chaps. B to E). Some familiarity with tensor calculus and elementary differential geometry is assumed. Our short exposition is intended to serve mainly as background and to facilitate and shorten some of the developments in Chaps. B to E. More elaborate accounts and detailed proofs may be found in standard books on tensor calculus and differential geometry.[1]

Our notations in this Appendix generally correspond to that of the main text; in particular, we use the same symbol for a function and its value without confusion. Throughout the Appendix, all Latin indices (subscripts or superscripts) take the values 1, 2, 3, Greek indices (subscripts or superscripts) have the range 1, 2, and the usual summation convention over a repeated index (one subscript and one superscript) is employed. We use a comma for partial differentiation with respect to coordinates, a single vertical line ($|$) for covariant differentiation with respect to the metric tensor of the (two-dimensional) surface and double vertical lines ($\|$) to designate covariant differentiation with respect to the metric tensor of a Euclidean 3-space. Also, in general, whenever the same letter is used for a quantity defined on a (two-dimensional) surface and a corresponding quantity in a Euclidean 3-space, for clarity the latter is distinguished by an added asterisk.

The contents of the Appendix are arranged in four sections as follows: In Sect. A.1, we have summarized the essential results and formulae from tensor calculus for a Euclidean 3-space with real coordinates. The restriction to three-dimensional space is not essential and can be easily dropped. Although the results in Sect. A.1 are collected with reference to Euclidean space, many of the formulae hold in fact in Riemannian space. Sect. A.2 contains selected results from differential geometry pertaining to *intrinsic* and *non-intrinsic* properties of a (two-dimensional) surface embedded in a Euclidean 3-space. In contrast to the developments in Sect. A.2, we consider in Sect. A.3 the geometrical properties of a surface embedded in a Euclidean 3-space covered by *normal coordinates*. We also indicate in Sect. A.3 the relationships between the components of space tensors and their surface counterparts. Finally, in Sect. A.4 we briefly discuss the nature of physical components of surface tensors and their tensor derivatives referred to lines of curvature coordinates on a surface.

A.1. Geometry of Euclidean space. Let x^i $(i=1, 2, 3)$ refer to a fixed right-handed orthogonal Cartesian coordinate system in a Euclidean 3-space and let θ^i denote an arbitrary (real) curvilinear coordinate system defined by the transformation

$$x^i = x^i(\theta^1, \theta^2, \theta^3), \qquad \det\left(\frac{\partial x^i}{\partial \theta^j}\right) \neq 0. \tag{A.1.1}$$

The condition $(A.1.1)_2$ ensures the existence of a unique inverse of $(A.1.1)_1$ so that

$$\theta^i = \theta^i(x^1, x^2, x^3). \tag{A.1.2}$$

Let $\bar{\theta}^i$ be a new set of curvilinear coordinates and consider the invertible transformation

$$\bar{\theta}^i = \bar{\theta}^i(\theta^1, \theta^2, \theta^3) \tag{A.1.3}$$

[1] A large number of books and monographs pertaining to the subject are available: For example, McConnell [1931, *1*], Eisenhart [1947, *2*], Synge and Schild [1949, *5*], Ch. 1 of Green and Zerna [1954, *1*] (see also [1968, *9*]), Willmore [1959, *8*] and Ericksen [1960, *3*].

and its inverse defined by

$$\theta^i = \theta^i(\bar\theta^1, \bar\theta^2, \bar\theta^3). \tag{A.1.4}$$

We assume that the functions describing the coordinate transformations (A.1.3)–(A.1.4) are single-valued and possess as many partial derivatives as required. From (A.1.3)–(A.1.4), the transformation of differentials $d\theta^i$, $d\bar\theta^i$ are:

$$d\theta^i = \frac{\partial\theta^i}{\partial\bar\theta^j}\, d\bar\theta^j, \qquad d\bar\theta^i = \frac{\partial\bar\theta^i}{\partial\theta^j}\, d\theta^j. \tag{A.1.5}$$

It is clear from (A.1.5)$_1$ that (A.1.4) induces a homogeneous linear transformation on the differentials $d\theta^i$ and $d\bar\theta^i$, the coefficients of transformation being functions of the coordinates.

Let $T^i = \{T^1, T^2, T^3\}$ and $T_i = \{T_1, T_2, T_3\}$ be two distinct sets of 3 real numbers associated with a point P of the Euclidean 3-space with coordinates θ^i. Similarly, let there be associated with the same point of space two sets of real numbers $\bar T^i = \{\bar T^1, \bar T^2, \bar T^3\}$ and $\bar T_i = \{\bar T_1, \bar T_2, \bar T_3\}$ defined with respect to a coordinate system $\bar\theta^i$ obtained from (A.1.3). Then, $T^i (T_i)$ are said to be components of a contravariant (covariant) tensor of order 1 (in the coordinates θ^i) or simply contravariant (covariant) components of a vector at P if the numbers T^i and $\bar T^i$ (T_i and $\bar T_i$) satisfy the relations

$$\bar T^i = \frac{\partial\bar\theta^i}{\partial\theta^j}\, T^j, \qquad T^i = \frac{\partial\theta^i}{\partial\bar\theta^j}\, \bar T^j, \tag{A.1.6}$$

$$\bar T_i = \frac{\partial\theta^j}{\partial\bar\theta^i}\, T_j, \qquad T_i = \frac{\partial\bar\theta^j}{\partial\theta^i}\, \bar T_j. \tag{A.1.7}$$

Consider next a system of 3^2 real numbers, associated with the point P of the Euclidean 3-space, which we designate by T^{ij} and $\bar T^{ij}$ $(i, j = 1, 2, 3)$ in the coordinates θ^i and $\bar\theta^i$, respectively. These numbers are called the components of a contravariant tensor of order 2 (in their respective coordinates) if they are related by the transformation laws

$$\begin{aligned} T^{ij} &= \frac{\partial\theta^i}{\partial\bar\theta^k}\,\frac{\partial\theta^j}{\partial\bar\theta^l}\, \bar T^{kl}, \\ \bar T^{ij} &= \frac{\partial\bar\theta^i}{\partial\theta^k}\,\frac{\partial\bar\theta^j}{\partial\theta^l}\, T^{kl}. \end{aligned} \tag{A.1.8}$$

The set of numbers T^{ij} are the components of the contravariant tensor at P in the coordinates θ^i and similarly $\bar T^{ij}$ are the components of the same tensor at P in the coordinates $\bar\theta^i$. The transformation laws for the components of a covariant tensor of order 2 can be defined analogously. Let T_{ij} and $\bar T_{ij}$ denote a system of 3^2 real numbers (associated with the point P) in the coordinates θ^i and $\bar\theta^i$, respectively. These numbers are called the components of a covariant tensor of order 2 if, under transformation of coordinates, T_{ij} and $\bar T_{ij}$ are related by

$$\begin{aligned} T_{ij} &= \frac{\partial\bar\theta^k}{\partial\theta^i}\,\frac{\partial\bar\theta^l}{\partial\theta^j}\, \bar T_{kl}, \\ \bar T_{ij} &= \frac{\partial\theta^k}{\partial\bar\theta^i}\,\frac{\partial\theta^l}{\partial\bar\theta^j}\, T_{kl}. \end{aligned} \tag{A.1.9}$$

More generally, let $T_{i_1\ldots i_r\cdot}{}^{j_1\ldots j_s}$ and $\bar T_{i_1\ldots i_r\cdot}{}^{j_1\ldots j_s}$ $(i_1, \ldots, i_r, j_1, \ldots, j_s = 1, 2, 3)$ be two systems of 3^{r+s} real numbers associated with a point P in the coordinates

θ^i and $\bar{\theta}^i$, respectively. We say these numbers (in their respective coordinates) represent the components of a mixed tensor of order $r+s$ at P, covariant of order r and contravariant of order s, if they transform according to the laws

$$
\begin{aligned}
T_{i_1 \ldots i_r}{}^{j_1 \ldots j_s}_{\cdot \quad \cdot} &= \frac{\partial \bar{\theta}^{k_1}}{\partial \theta^{i_1}} \cdots \frac{\partial \bar{\theta}^{k_r}}{\partial \theta^{i_r}} \frac{\partial \theta^{j_1}}{\partial \bar{\theta}^{l_1}} \cdots \frac{\partial \theta^{j_s}}{\partial \bar{\theta}^{l_s}} \bar{T}_{k_1 \ldots k_r}{}^{l_1 \ldots l_s}_{\cdot \quad \cdot}, \\
\bar{T}_{i_1 \ldots i_r}{}^{j_1 \ldots j_s}_{\cdot \quad \cdot} &= \frac{\partial \theta^{k_1}}{\partial \bar{\theta}^{i_1}} \cdots \frac{\partial \theta^{k_r}}{\partial \bar{\theta}^{i_r}} \frac{\partial \bar{\theta}^{j_1}}{\partial \theta^{l_1}} \cdots \frac{\partial \bar{\theta}^{j_s}}{\partial \theta^{l_s}} T_{k_1 \ldots k_r}{}^{l_1 \ldots l_s}_{\cdot \quad \cdot}.
\end{aligned}
\tag{A.1.10}
$$

For reasons that will become apparent later, in writing a mixed tensor such as $T_{i\cdot k}^{\ j}$ or those in (A.1.10), we refrain from placing two indices above each other (along the same vertical line). This is easily effected by reserving a vacant space indicated by a dot.

The above definitions and transformation laws are those appropriate for absolute tensors and require modifications in the case of the so-called relative tensors. Definitions (A.1.6)–(A.1.9) may be regarded as special cases of (A.1.10). A scalar, i.e., a quantity which remains invariant with respect to any coordinate transformation, is called a tensor of order zero while a tensor of order 1 is also known as a vector. A tensor is called *covariant* or *contravariant* if it has only covariant indices (subscripts) or contravariant indices (superscripts), respectively; otherwise the tensor is called *mixed.* If the relative order of two indices—either both subscripts or both superscripts—of a tensor is immaterial, the tensor is called *symmetric with respect to these indices.* A tensor, either covariant or contravariant, is said to be *symmetric* if it is symmetric with respect to all pairs of indices. If two components of a given tensor (either covariant or contravariant) can be obtained from one another by the interchange of two particular indices (either both subscripts or both superscripts) and a change in sign, the tensor is said to be *skew-symmetric (or alternating) with respect to these indices.* A tensor, either covariant or contravariant, is said to be *skew-symmetric* if it is skew-symmetric with respect to all pairs of indices. Any covariant or contravariant tensor of order 2 can be represented uniquely as a sum of symmetric and a skew-symmetric tensor. Thus

$$
\begin{gathered}
T_{ij} = T_{(ij)} + T_{[ij]}, \\
T_{(ij)} = \tfrac{1}{2}(T_{ij} + T_{ji}), \qquad T_{[ij]} = \tfrac{1}{2}(T_{ij} - T_{ji}) = -T_{[ji]},
\end{gathered}
\tag{A.1.11}
$$

where the notations $T_{(ij)}$, $T_{[ij]}$ stand, respectively, for the symmetric and the skew-symmetric part of T_{ij}. Results analogous to (A.1.11) can be recorded also for T^{ij}.

Two tensors are said to be *of the same type* if they are covariant of the same order and contravariant of the same order. By adding (at the same point in space) the corresponding components of two tensors of the same type (say in the coordinates θ^i), we obtain a third tensor of the same type (in the coordinates θ^i). If we multiply every component of a tensor by a scalar, a tensor of the same type will result. From this multiplication and the addition of two tensors of the same type, it follows that the totality of tensors of the same type form a vector space. If every component of a tensor such as $A_{ij}{}^{kl}_{\cdot\cdot}$ is multiplied by every component of another arbitrary tensor such as $B_{mn}{}^{p}_{\cdot}$, there results

$$
T_{ijmn}{}^{klp}_{\cdot\cdot\cdot} = A_{ij}{}^{kl}_{\cdot\cdot} B_{mn}{}^{p}_{\cdot}, \tag{A.1.12}
$$

called the *outer product* of the tensor A and B. By (A.1.10), the outer product of any two tensors is a tensor whose character is indicated by the position of its indices; the number of covariant and contravariant indices of the resulting tensor T is equal to the sum of the numbers of covariant or contravariant indices of the tensors A and B. Thus, the left-hand side of (A.1.12) are components of a mixed tensor, covariant of order 4 and contravariant of order 3. If a subscript and a superscript of a mixed tensor are identified, so that the same index is repeated with the implied summation, the process is called *contraction*. For example, contraction of the indices j and k of the tensor $T_{ijmn\dots}^{klp}$ in (A.1.12) results in the mixed tensor $T_{ijmn\dots}^{jlp}$ which is of order 1 less than the original tensor in both covariant and contravariant indices. In particular, for a mixed tensor of order 2 such as $T^i_{.j}$, contraction results in a scalar $T^i_{.i}$. Multiplication of two tensors accompanied by a contraction results in a tensor called an *inner product*. Contraction may be applied more than once and the number of inner products formed from any two tensors depends on the number of indices involved. For example, from the tensors $A_{i.k}^{\,j}$ and $B^m_{.n}$ different inner products $A_{i.k}^{\,j} B^k_{.n}$, $A_{i.k}^{\,j} B^m_{.j}$, $A_{i.k}^{\,j} B^k_{.j}$, etc., can be obtained.

Let

$$\boldsymbol{p} - \boldsymbol{p}(\theta^i) \tag{A.1.13}$$

denote the position vector of a typical point (with coordinates θ^i) in a region $\mathscr{R}$ of the Euclidean space. Then, the square of a line element is given by

$$ds^{*2} = d\boldsymbol{p} \cdot d\boldsymbol{p} = g_{ij}\, d\theta^i\, d\theta^j, \tag{A.1.14}$$

where

$$\boldsymbol{g}_i = \frac{\partial \boldsymbol{p}}{\partial \theta^i}, \qquad g_{ij} = \boldsymbol{g}_i \cdot \boldsymbol{g}_j \tag{A.1.15}$$

are, respectively, the (covariant) base vectors and the metric tensor while a comma denotes partial differentiation with respect to θ^i. The reciprocal (contravariant) base vectors $\boldsymbol{g}^i$ and the conjugate tensor g^{ij} are defined by

$$\begin{gathered} \boldsymbol{g}^i = g^{ij}\, \boldsymbol{g}_j, \qquad g^{ij} = g^{ji} = \boldsymbol{g}^i \cdot \boldsymbol{g}^j = \frac{\mathscr{G}^{ij}}{g}, \\ g = \det(g_{ij}), \end{gathered} \tag{A.1.16}$$

where $\mathscr{G}^{ij}$ are the cofactors of g_{ij} in the expansion of the determinant g. The base vectors as well as the metric and the conjugate tensors, in addition to those indicated by $(A.1.15)_2$ and $(A.1.16)_2$, also satisfy the relations

$$\boldsymbol{g}^i \cdot \boldsymbol{g}_j = \delta^i_j, \qquad g^{ik}\, g_{kj} = g_{jk}\, g^{ki} = \delta^i_j. \tag{A.1.17}$$

The vector products of the base vectors yield

$$\begin{aligned} \boldsymbol{g}_i \times \boldsymbol{g}_j &= \varepsilon_{ijk}\, \boldsymbol{g}^k, \\ \boldsymbol{g}^i \times \boldsymbol{g}^j &= \varepsilon^{ijk}\, \boldsymbol{g}_k, \end{aligned} \tag{A.1.18}$$

where δ^i_j stands for Kronecker delta and the ε-system is related to the permutation symbols e_{ijk}, e^{ijk}, through

$$\varepsilon_{ijk} = g^{\frac{1}{2}}\, e_{ijk}, \qquad \varepsilon^{ijk} = g^{-\frac{1}{2}}\, e^{ijk}. \tag{A.1.19}$$

In contrast to ε_{ijk}, ε^{ijk} which are absolute tensors and transform according to (A.1.10), the permutation symbols are relative tensors and have different transformation laws. We do not elaborate here in detail on relative tensors but note that the permutation symbols e_{ijk} and e^{ijk} are relative tensors of weights -1 and

$+1$, respectively, which transform according to

$$\begin{aligned} e_{ijk} &= J \frac{\partial \bar{\theta}^m}{\partial \theta^i} \frac{\partial \bar{\theta}^n}{\partial \theta^j} \frac{\partial \bar{\theta}^p}{\partial \theta^k} \bar{e}_{mnp}, \\ e^{ijk} &= J^{-1} \frac{\partial \theta^i}{\partial \bar{\theta}^m} \frac{\partial \theta^j}{\partial \bar{\theta}^n} \frac{\partial \theta^k}{\partial \bar{\theta}^p} \bar{e}^{mnp}, \end{aligned} \tag{A.1.20}$$

where $J = \det\left(\frac{\partial \theta^i}{\partial \bar{\theta}^j}\right)$ is the Jacobian of the transformation.

An index of a tensor can be raised or lowered through inner multiplication by g^{ij} or g_{ij}. To illustrate this process, consider a tensor T_{hjk} which may be generated from $T^{i}_{.jk}$ through inner multiplication by g_{hi}:

$$T_{hjk} = g_{hi}\, T^{i}_{.jk}. \tag{A.1.21}$$

It is clear from the right-hand side of (A.1.21) that through inner multiplication by g_{hi}, the superscript i is lowered into the vacant space indicated by a dot in $T^{i}_{.jk}$. Consider next the raising of an index by inner multiplication of g^{ih} and T_{hjk}. Thus, using (A.1.21) and $(\text{A.1.17})_2$, we have

$$\begin{aligned} g^{ih}\, T_{hjk} &= g^{ih} g_{hm}\, T^{m}_{.jk} \\ &= \delta^i_m\, T^{m}_{.jk} = T^{i}_{.jk}, \end{aligned} \tag{A.1.22}$$

which also shows that the process of raising and lowering indices is reversible. However, in general, the expressions

$$g^{ki}\, T_{ij} = T^{k}_{.j}, \qquad g^{ki}\, T_{ji} = T_{j.}^{k} \tag{A1.23}$$

are different. They become identical if $T_{ij} = T_{ji}$.

The Christoffel symbols of the first and second kinds are defined by

$$\overset{*}{\Gamma}_{ijk} = g_{mk}\, \overset{*}{\Gamma}{}^m_{ij} = \tfrac{1}{2}[g_{ik,j} + g_{jk,i} - g_{ij,k}], \qquad \overset{*}{\Gamma}_{ijk} = \overset{*}{\Gamma}_{jik}, \qquad \overset{*}{\Gamma}{}^k_{ij} = \overset{*}{\Gamma}{}^k_{ji}. \tag{A.1.24}$$

In view of (A.1.15) and since $\boldsymbol{g}_{i,j} = \boldsymbol{g}_{j,i}$, they can also be expressed as

$$\overset{*}{\Gamma}_{ijk} = \overset{*}{\Gamma}_{jik} = \boldsymbol{g}_k \cdot \boldsymbol{g}_{i,j}, \qquad \overset{*}{\Gamma}{}^k_{ij} = \overset{*}{\Gamma}{}^k_{ji} = \boldsymbol{g}^k \cdot \boldsymbol{g}_{i,j} = -\boldsymbol{g}_i \cdot \boldsymbol{g}^k{}_{,j} \tag{A.1.25}$$

and hence

$$\boldsymbol{g}_{i,j} = \overset{*}{\Gamma}{}^k_{ij}\, \boldsymbol{g}_k, \qquad \boldsymbol{g}^i{}_{,j} = -\overset{*}{\Gamma}{}^i_{jk}\, \boldsymbol{g}^k. \tag{A.1.26}$$

To indicate the nature of covariant differentiation of a tensor function, we record below the expressions for covariant derivatives of a covariant tensor T_i, a contravariant tensor T^i, a mixed second order tensor $T^{i}_{.j}$ and a contravariant second order tensor T^{ij}. Thus, designating covariant differentiation with respect to g_{ij} by double vertical lines ($\|$), we have:

$$\begin{aligned} T_{i\|k} &= T_{i,k} - \overset{*}{\Gamma}{}^l_{ik}\, T_l, \\ T^i{}_{\|k} &= T^i{}_{,k} + \overset{*}{\Gamma}{}^i_{lk}\, T^l, \end{aligned} \tag{A.1.27}$$

$$\begin{aligned} T^{i}_{.j\|k} &= T^{i}_{.j,k} + \overset{*}{\Gamma}{}^i_{lk}\, T^{l}_{.j} - \overset{*}{\Gamma}{}^l_{jk}\, T^{i}_{.l}, \\ T^{ij}{}_{\|k} &= T^{ij}{}_{,k} + \overset{*}{\Gamma}{}^i_{lk}\, T^{lj} + \overset{*}{\Gamma}{}^j_{lk}\, T^{il}. \end{aligned} \tag{A.1.28}$$

The covariant derivative $T^i{}_{\|k}$ (and also $T_{i\|k}$) has a simple geometrical meaning. To see this, let T^i and T_i be the contravariant and the covariant components of a vector $\boldsymbol{T}$ so that

$$\boldsymbol{T} = T^i \boldsymbol{g}_i = T_i \boldsymbol{g}^i. \tag{A.1.29}$$

Then, with the help of (A.1.25)–(A.1.26), from $(A.1.29)_1$ we can readily obtain the expression

$$\frac{\partial \boldsymbol{T}}{\partial \theta^k} = \frac{\partial T^i}{\partial \theta^k} \boldsymbol{g}_i + \frac{\partial \boldsymbol{g}_i}{\partial \theta^k} T^i = \{T^i{}_{,k} + \overset{*}{\Gamma}{}^i_{lk} T^l\} \boldsymbol{g}_i = T^i{}_{\|k} \boldsymbol{g}_i. \tag{A.1.30}$$

Similarly, from $(A.1.29)_2$, we obtain

$$\frac{\partial \boldsymbol{T}}{\partial \theta^k} = T_{i\|k} \boldsymbol{g}^i. \tag{A.1.31}$$

The expressions (A.1.30)–(A.1.31) show that the covariant derivatives $T^i{}_{\|k}$ $(T_{i\|k})$ are components of the partial derivative $\partial \boldsymbol{T}/\partial \theta^k$ referred to the base vectors $\boldsymbol{g}_i (\boldsymbol{g}^i)$. We note here that the base vectors $\boldsymbol{g}_i$, $\boldsymbol{g}^i$, the metric tensor g_{ij} and its conjugate g^{ij}, as well as the tensor derivatives defined by (A.1.27)–(A.1.28), are tensors whose covariant and contravariant characters are indicated by the position of their indices. The Christoffel symbols do not transform according to (A.1.10) and hence are not tensors.

We examine now the nature of two successive covariant derivatives of tensor functions. For this purpose, it will suffice to consider the tensors T_i and T_{ij} whose characters are indicated by the position of their respective indices. Then

$$T_{i\|jk} - T_{i\|kj} = \overset{*}{R}{}^m{}_{.ijk} T_m, \qquad T_{ij\|kl} - T_{ij\|lk} = \overset{*}{R}{}^m{}_{.ikl} T_{mj} + \overset{*}{R}{}^m{}_{.jkl} T_{im}, \tag{A.1.32}$$

where the mixed curvature tensor (also known as the mixed Riemann-Christoffel tensor) is given by

$$\overset{*}{R}{}^m{}_{.ijk} = \overset{*}{\Gamma}{}^m_{ik,j} - \overset{*}{\Gamma}{}^m_{ij,k} + \overset{*}{\Gamma}{}^p_{ik} \overset{*}{\Gamma}{}^m_{pj} - \overset{*}{\Gamma}{}^p_{ij} \overset{*}{\Gamma}{}^m_{pk}. \tag{A.1.33}$$

According to the relations (A.1.32), the order of covariant differentiation does not commute unless $\overset{*}{R}{}^m{}_{.ijk} = 0$ which is both a necessary and sufficient condition for the space to be Euclidean. Hence, in a Euclidean space, the order of covariant differentiation is immaterial.

The covariant curvature tensor, namely

$$\overset{*}{R}_{mijk} = g_{mp} \overset{*}{R}{}^p{}_{.ijk}$$

satisfies the relations

$$\overset{*}{R}_{pijk} = -\overset{*}{R}_{ipjk} = -\overset{*}{R}_{pikj} = \overset{*}{R}_{jkpi}, \tag{A.1.34}$$

$$\overset{*}{R}_{mijk} + \overset{*}{R}_{mjki} + \overset{*}{R}_{mkij} = 0, \tag{A.1.35}$$

and the number of independent components of $\overset{*}{R}_{mijk}$ which depends on the dimension of the space, with the aid of (A.1.34)–(A.1.35), may be shown to be[2]

$$\tfrac{1}{12} N^2 (N^2 - 1), \tag{A.1.36}$$

N being the dimension of the space. For a 2-space, the covariant curvature tensor has only one independent component as all of its components either vanish or are

[2] See, e.g., p. 87 of SYNGE and SCHILD [1949, 5].

expressible in terms of $\overset{*}{R}_{1212}$. For $N=3$, it follows from (A.1.36) that the number of independent components of the covariant curvature tensor is six, which can be shown to be

$$\overset{*}{R}_{1212}, \overset{*}{R}_{3112}, \overset{*}{R}_{3221}, \overset{*}{R}_{1313}, \overset{*}{R}_{2323}, \overset{*}{R}_{1323}. \tag{A.1.37}$$

A.2. Some results from the differential geometry of a surface. This section contains a heterogeneous collection of certain results and formulae pertaining to (local) intrinsic and non-intrinsic properties of a surface $\mathscr{s}$ embedded in a Euclidean 3-space.

α) *Definition of a surface. Preliminaries.* A surface $\mathscr{s}$ may be specified in terms of parametric equations of the form

$$x_i = f_i(\theta^1, \theta^2), \tag{A.2.1}$$

where $x_i (i=1, 2, 3)$ are rectangular Cartesian coordinates of a point P on the surface $\mathscr{s}$ and $\theta^\alpha (\alpha=1, 2)$ are parameters, as yet unrelated to the curvilinear coordinates θ^i utilized in Sect. A.1. We assume that the functions f_i are single-valued and continuous and also require that they possess as many derivatives as may be required. The position vector $\boldsymbol{r}$ (with rectangular Cartesian components x_i) at P, in view of (A.2.1), may be expressed as a function of the parameters[3] θ^α. Thus, we write

$$\boldsymbol{r} = \boldsymbol{r}(\theta^1, \theta^2), \tag{A.2.2}$$

where the function $\boldsymbol{r}$ in (A.2.2) has the same continuity and differentiability properties as f_i. With any point P of the surface, we can associate a set of base vectors $\boldsymbol{a}_1$, $\boldsymbol{a}_2$ defined by

$$\boldsymbol{a}_\alpha = \boldsymbol{r}_{,\alpha}, \qquad \boldsymbol{a}_1 \times \boldsymbol{a}_2 \neq 0, \tag{A.2.3}$$

where a comma denotes partial differentiation with respect to θ^α-coordinates. As will become evident presently, the condition $(A.2.3)_2$ on the vector product of the base vectors ensures the existence of a unique normal at each *ordinary point*[4] on $\mathscr{s}$.

We introduce new coordinates $\bar{\theta}^1$, $\bar{\theta}^2$ on the surface $\mathscr{s}$ by means of the coordinate transformation

$$\theta^\alpha = \theta^\alpha(\bar{\theta}^1, \bar{\theta}^2), \qquad \det\left(\frac{\partial \theta^\alpha}{\partial \bar{\theta}^\gamma}\right) \neq 0, \tag{A.2.4}$$

so that the inverse transformation

$$\bar{\theta}^\alpha = \bar{\theta}^\alpha(\theta^1, \theta^2) \tag{A.2.5}$$

exists. From (A.2.4)–(A.2.5), the transformation of the differentials $d\theta^\alpha$ and $d\bar{\theta}^\alpha$ are:

$$d\theta^\alpha = \frac{\partial \theta^\alpha}{\partial \bar{\theta}^\gamma} d\bar{\theta}^\gamma, \qquad d\bar{\theta}^\alpha = \frac{\partial \bar{\theta}^\alpha}{\partial \theta^\gamma} d\theta^\gamma. \tag{A.2.6}$$

In what follows an overbar will be placed on quantities at P associated with coordinates $\bar{\theta}^\alpha$ in order to distinguish these from the same quantities at P associated with coordinates θ^α. Thus, for example the base vectors at P associated

[3] These parameters are usually called curvilinear coordinates on the surface.

[4] An ordinary point is defined by the condition $(A.2.3)_2$. A point which is not ordinary, such as the vertex of a cone, is called a *singularity*. Here we restrict the domain of θ^α such that every point of $\mathscr{s}$ is ordinary.

with $\bar{\theta}^\alpha$ will be designated as $\bar{\boldsymbol{a}}_\alpha$. It can be readily verified that the property which defines an ordinary point is unaltered under the coordinate transformation (A.2.4), so that

$$\bar{\boldsymbol{a}}_1 \times \bar{\boldsymbol{a}}_2 = \frac{\partial \boldsymbol{r}}{\partial \bar{\theta}^1} \times \frac{\partial \boldsymbol{r}}{\partial \bar{\theta}^2} \neq 0$$

[$\boldsymbol{r}$ being now a different function from that in (A.2.2)].

Definitions of covariant, contravariant and mixed tensors at a point P on s are similar to those given in Sect. A.1. In fact, with an obvious modification, various tensor transformation laws between (A.1.6)–(A.1.10) and the subsequent discussion until the end of the paragraph containing (A.1.12) hold for the Riemannian space under consideration. For example, let a system of 2^{r+s} real numbers $T_{\alpha_1 \ldots \alpha_r .}{}^{\gamma_1 \ldots \gamma_s}$ $(\alpha_1, \ldots, \alpha_r, \gamma_1, \ldots, \gamma_s = 1, 2)$ be associated with a point P on s (a Riemannian 2-space) with coordinates θ^α and let $\overline{T}_{\alpha_1 \ldots \alpha_r .}{}^{\gamma_1 \ldots \gamma_s}$ be the corresponding 2^{r+s} real numbers at P in the coordinates $\bar{\theta}^\alpha$ defined by (A.2.5). We say that these numbers represent a mixed tensor of order $r+s$ at P, covariant of order r and contravariant of order s, if they are related by the transformation laws

$$\begin{aligned} T_{\alpha_1 \ldots \alpha_r .}{}^{\gamma_1 \ldots \gamma_s} &= \frac{\partial \bar{\theta}^{\beta_1}}{\partial \theta^{\alpha_1}} \cdots \frac{\partial \bar{\theta}^{\beta_r}}{\partial \theta^{\alpha_r}} \frac{\partial \theta^{\gamma_1}}{\partial \bar{\theta}^{\delta_1}} \cdots \frac{\partial \theta^{\gamma_s}}{\partial \bar{\theta}^{\delta_s}} \overline{T}_{\beta_1 \ldots \beta_r .}{}^{\delta_1 \ldots \delta_s}, \\ \overline{T}_{\alpha_1 \ldots \alpha_r .}{}^{\gamma_1 \ldots \gamma_s} &= \frac{\partial \theta^{\beta_1}}{\partial \bar{\theta}^{\alpha_1}} \cdots \frac{\partial \theta^{\beta_r}}{\partial \bar{\theta}^{\alpha_r}} \frac{\partial \bar{\theta}^{\gamma_1}}{\partial \theta^{\delta_1}} \cdots \frac{d \bar{\theta}^{\gamma_s}}{d \theta^{\delta_s}} T_{\beta_1 \ldots \beta_r .}{}^{\delta_1 \ldots \delta_s}. \end{aligned} \tag{A.2.7}$$

A curve on a surface s can be determined by a sufficiently smooth parametric representation $\theta^\alpha = \theta^\alpha(u)$, where u is a real variable. The direction of the tangent to a curve on s is determined by the vector

$$\frac{d\boldsymbol{r}}{du} = \frac{\partial \boldsymbol{r}}{\partial \theta^\alpha} \frac{d\theta^\alpha}{du} = \boldsymbol{a}_\alpha \frac{d\theta^\alpha}{du}, \tag{A.2.8}$$

which, in general, depends on u. Since $\boldsymbol{a}_\alpha$ (which are tangent to the coordinate curves) are non-zero and independent, the tangent to any curve on s through a point P lies in a plane which contains the two vectors $\boldsymbol{a}_1$, $\boldsymbol{a}_2$ at P; this plane is the *tangent plane* at P. A vector $\boldsymbol{U}$ (in the Euclidean 3-space at P) which lies in the tangent plane to s at P is called a *tangent vector* or a *vector in the surface*. Any such vector (since it lies in the tangent plane at P) can be expressed as a linear combination of $\boldsymbol{a}_1$, $\boldsymbol{a}_2$:

$$\boldsymbol{U} = U^\alpha \boldsymbol{a}_\alpha . \tag{A.2.9}$$

The base vectors $\boldsymbol{a}_1$, $\boldsymbol{a}_2$ defined by (A.2.3) are linearly independent, are tangent to the θ^α-curves through a point P on s and span the tangent plane to s at P. The normal to s at P is the normal to the tangent plane at P and is therefore perpendicular to $\boldsymbol{a}_1$, $\boldsymbol{a}_2$. Let $\boldsymbol{a}_3$ denote the unit normal to s at P, the sense of the unit normal being fixed by the convention that $\boldsymbol{a}_1$, $\boldsymbol{a}_2$, $\boldsymbol{a}_3$ form a right-handed triad. It then follows that

$$\boldsymbol{a}_3 = \frac{\boldsymbol{a}_1 \times \boldsymbol{a}_2}{|\boldsymbol{a}_1 \times \boldsymbol{a}_2|}, \qquad |\boldsymbol{a}_1 \times \boldsymbol{a}_2| \neq 0 \tag{A.2.10}$$

and

$$\boldsymbol{a}_3 \cdot \boldsymbol{a}_3 = 1, \qquad \boldsymbol{a}_3 \cdot \boldsymbol{a}_\alpha = 0, \qquad \boldsymbol{a}_3 \cdot \boldsymbol{a}_{3,\alpha} = 0. \tag{A.2.11}$$

The notation $|\boldsymbol{V}|$ in (A.2.10) stands for the magnitude of $\boldsymbol{V}$ and the last of (A.2.11) follows from differentiation of $(\text{A.2.11})_1$.

β) *First and second fundamental forms.* The square of a line element of the surface is given by

$$ds^2 = d\mathbf{r}\cdot d\mathbf{r} = a_{\alpha\beta}\, d\theta^\alpha\, d\theta^\beta, \tag{A.2.12}$$

where

$$a_{\alpha\beta} = \mathbf{a}_\alpha \cdot \mathbf{a}_\beta \tag{A.2.13}$$

and the base vectors $\mathbf{a}_\alpha$ are defined by (A.2.3)$_1$. The quadratic form (A.2.12) is called the *first fundamental form* of the surface. The reciprocals of (A.2.3)$_1$ and (A.2.13) are denoted by $\mathbf{a}^\alpha$ and $a^{\alpha\beta}$, respectively. They are defined for all ordinary points of the surface, i.e., points for which[5]

$$a = \det(a_{\alpha\beta}) \neq 0, \tag{A.2.14}$$

and are given by

$$\mathbf{a}^\alpha = a^{\alpha\beta}\, \mathbf{a}_\beta, \qquad a^{\alpha\beta} = a^{\beta\alpha} = \mathbf{a}^\alpha \cdot \mathbf{a}^\beta = \frac{\mathscr{A}^{\alpha\beta}}{a}, \tag{A.2.15}$$

where $\mathscr{A}^{\alpha\beta}$ are the cofactors of $a_{\alpha\beta}$ in the expansion of the determinant a. The (covariant) base vectors $\mathbf{a}_\alpha$, the (contravariant) reciprocal base vectors $\mathbf{a}^\alpha$, the symmetric tensor $a_{\alpha\beta}$ and its conjugate $a^{\alpha\beta}$ all satisfy the appropriate laws for tensor transformations of the type (A.2.7). These vectors and tensors also satisfy the relations

$$\mathbf{a}^\alpha \cdot \mathbf{a}_\beta = \delta^\alpha_\beta, \qquad a^{\alpha\lambda}\, a_{\lambda\beta} = a_{\beta\lambda}\, a^{\lambda\alpha} = \delta^\alpha_\beta. \tag{A.2.16}$$

The raising and the lowering of indices is accomplished here with the use of $a^{\alpha\beta}$ and $a_{\alpha\beta}$; the process is similar to that discussed in Sect. A.1 [between (A.1.21)–(A.1.22)].

The vector products involving $\mathbf{a}_1, \mathbf{a}_2, \mathbf{a}_3$ are

$$\begin{aligned} \mathbf{a}_\alpha \times \mathbf{a}_\beta &= \varepsilon_{\alpha\beta}\, \mathbf{a}_3, & \mathbf{a}^\alpha \times \mathbf{a}^\beta &= \varepsilon^{\alpha\beta}\, \mathbf{a}_3, \\ \mathbf{a}_3 \times \mathbf{a}_\beta &= \varepsilon_{\beta\lambda}\, \mathbf{a}^\lambda, & \mathbf{a}_3 \times \mathbf{a}^\beta &= \varepsilon^{\beta\lambda}\, \mathbf{a}_\lambda, \end{aligned} \tag{A.2.17}$$

where $\varepsilon_{\alpha\beta}$, $\varepsilon^{\alpha\beta}$ are the ε-system for the surface defined by

$$\varepsilon_{\alpha\beta} = a^{\frac{1}{2}}\, e_{\alpha\beta}, \qquad \varepsilon^{\alpha\beta} = a^{-\frac{1}{2}}\, e^{\alpha\beta} \tag{A.2.18}$$

and

$$e_{11} = e_{22} = e^{11} = e^{22} = 0, \qquad e_{12} = e^{12} = 1, \qquad e_{21} = e^{21} = -1. \tag{A.2.19}$$

Let $\theta^\alpha = \theta^\alpha(s)$ be the parametric equations of a curve c, where the parameter s is the arc length; and let $\boldsymbol{\lambda}$ denote the unit tangent vector to c defined for points $\mathbf{r}(\theta^\alpha(s))$ on c. Then,

$$\boldsymbol{\lambda} = \frac{d\mathbf{r}}{ds} = \lambda^\alpha\, \mathbf{a}_\alpha, \qquad \lambda^\alpha = \frac{d\theta^\alpha(s)}{ds}, \tag{A.2.20}$$

which follows also from (A.2.8) if the parameter u is identified with the arc length s. The outward unit normal $\boldsymbol{\nu}$ to a curve c on $\mathscr{s}$, through a point P, is a tangent vector whose sense is fixed by the convention that $\boldsymbol{\nu}$, $\boldsymbol{\lambda}$, $\mathbf{a}_3$ form a right-handed triad. Thus, using (A.2.17),

$$\begin{aligned} \boldsymbol{\nu} &= \boldsymbol{\lambda} \times \mathbf{a}_3 = \nu^\alpha\, \mathbf{a}_\alpha = \nu_\alpha\, \mathbf{a}^\alpha = \varepsilon_{\alpha\beta}\, \lambda^\beta\, \mathbf{a}^\alpha, \\ \boldsymbol{\lambda} &= \mathbf{a}_3 \times \boldsymbol{\nu} = \varepsilon^{\alpha\beta}\, \nu_\alpha\, \mathbf{a}_\beta. \end{aligned} \tag{A.2.21}$$

We define the generalized Kronecker delta for the surface by

$$\varepsilon^{\alpha\beta}\, \varepsilon_{\lambda\mu} = \delta^{\alpha\beta}_{\lambda\mu} \tag{A.2.22}$$

[5] The condition (A.2.14) is equivalent to (A.2.3)$_2$.

and note that

$$\delta^{\alpha\beta}_{\lambda\beta} = \varepsilon^{\alpha\beta}\,\varepsilon_{\lambda\beta} = a_{\lambda\mu}\,a_{\beta\nu}\,\varepsilon^{\alpha\beta}\,\varepsilon^{\mu\nu} = a^{\alpha\mu}\,a_{\lambda\mu} = \delta^{\alpha}_{\lambda}, \qquad \delta^{\alpha\beta}_{\alpha\beta} = \delta^{\alpha}_{\alpha} = 2. \tag{A.2.23}$$

Let $T^{\alpha\beta}$ (not necessarily in the surface) denote the components of a second-order tensor defined at a point on the surface $\mathscr{s}$. Then,

$$T^{\alpha\beta} - T^{\beta\alpha} = \delta^{\alpha\beta}_{\lambda\mu}\,T^{\lambda\mu}. \tag{A.2.24}$$

Similarly, with the help of (A.2.22) and (A.2.19), it can be easily verified that

$$\delta^{\alpha\eta}_{\lambda\nu}\,T^{\lambda}_{\alpha} = \delta^{\eta}_{\nu}\,T^{\alpha}_{\alpha} - T^{\eta}_{\nu}. \tag{A.2.25}$$

The magnitude or the length of a vector in the surface such as $\boldsymbol{U}$ in (A.2.9) is easily seen to be $|\boldsymbol{U}| = [a_{\alpha\beta}\,U^{\alpha}\,U^{\beta}]^{\frac{1}{2}} = [a^{\alpha\beta}\,U_{\alpha}\,U_{\beta}]^{\frac{1}{2}}$. We recall that the element of area of the surface (by definition) is $d\sigma = a^{\frac{1}{2}}\,d\theta^1\,d\theta^2$ and that the expression for the angle between two vectors in the surface involves $a_{\alpha\beta}$. It is, therefore, clear that the coefficients of the first fundamental form determine lengths, angles and areas on a surface. This justifies the use of the term *metric tensor of the surface* for $a_{\alpha\beta}$.

We now turn to the *second fundamental form* of the surface, defined by the scalar product

$$-d\boldsymbol{r}\cdot d\boldsymbol{a}_3 = b_{\alpha\beta}\,d\theta^{\alpha}\,d\theta^{\beta}, \tag{A.2.26}$$

where by $(\text{A.2.3})_1$ and $(\text{A.2.11})_2$,

$$b_{\alpha\beta} = b_{\beta\alpha} = -\boldsymbol{a}_{\alpha}\cdot\boldsymbol{a}_{3,\beta} = \boldsymbol{a}_3\cdot\boldsymbol{a}_{\alpha,\beta}. \tag{A.2.27}$$

The *mean curvature* of the surface is defined by

$$H = \tfrac{1}{2}\,b^{\alpha}_{\alpha} \tag{A.2.28}$$

and the *Gaussian curvature* of the surface by

$$K = \det(b^{\alpha}_{\beta}) = a^{-1}\det(b_{\alpha\beta}) = b^1_1\,b^2_2 - b^1_2\,b^2_1. \tag{A.2.29}$$

Recalling the well-known formula for the expansion of a determinant, namely

$$\varepsilon^{\lambda\mu}\det(b^{\alpha}_{\beta}) = \varepsilon^{\alpha\beta}\,b^{\lambda}_{\alpha}\,b^{\mu}_{\beta}, \tag{A.2.30}$$

then after multiplication by $\varepsilon_{\lambda\mu}$ and using (A.2.22) and $(\text{A.2.23})_5$, we see that (A.2.29) may alternatively be written in the form

$$K = \tfrac{1}{2}\,\delta^{\alpha\beta}_{\lambda\mu}\,b^{\lambda}_{\alpha}\,b^{\mu}_{\beta}. \tag{A.2.31}$$

It is clear from (A.2.28) and (A.2.31) that both H and K are surface invariants which depend only on the coefficients of the second fundamental form $b_{\alpha\beta}$ of the surface. However, as will be seen presently, the Gaussian curvature can also be expressed entirely in terms of the coefficients of the first fundamental form $a_{\alpha\beta}$ and its derivatives[6].

γ) *Covariant derivatives. The curvature tensor.* The operator representing the covariant derivative with respect to the metric tensor of the surface will be designated by a vertical line (|). In order to indicate the nature of covariant differentiation of a tensor function defined on the surface, we record below the

[6] See Eq. (A.2.52).

expressions for covariant derivatives, with respect to $a_{\alpha\beta}$, of a covariant tensor T_α, a contravariant tensor T^α, a mixed second order tensor $T^\alpha_{.\beta}$ and a contravariant second order tensor $T^{\alpha\beta}$:

$$\begin{aligned} &T_{\alpha|\gamma} = T_{\alpha,\gamma} - \Gamma^\lambda_{\alpha\gamma} T_\lambda, \qquad T^\alpha{}_{|\gamma} = T^\alpha{}_{,\gamma} + \Gamma^\alpha_{\lambda\gamma} T^\lambda, \\ &T^\alpha_{.\beta|\gamma} = T^\alpha_{.\beta,\gamma} + \Gamma^\alpha_{\lambda\gamma} T^\lambda_{.\beta} - \Gamma^\lambda_{\beta\gamma} T^\alpha_{.\lambda}, \\ &T^{\alpha\beta}{}_{|\gamma} = T^{\alpha\beta}{}_{,\gamma} + \Gamma^\alpha_{\lambda\gamma} T^{\lambda\beta} + \Gamma^\beta_{\lambda\gamma} T^{\alpha\lambda}, \end{aligned} \tag{A.2.32}$$

where the surface Christoffel symbols are given by

$$\begin{aligned} &\Gamma_{\alpha\beta\gamma} = a_{\mu\gamma} \Gamma^\mu_{\alpha\beta} = \tfrac{1}{2}(a_{\alpha\gamma,\beta} + a_{\beta\gamma,\alpha} - a_{\alpha\beta,\gamma}), \\ &\Gamma_{\alpha\beta\gamma} = \Gamma_{\beta\alpha\gamma} = \boldsymbol{a}_\gamma \cdot \boldsymbol{a}_{\alpha,\beta}, \\ &\Gamma^\gamma_{\alpha\beta} = \Gamma^\gamma_{\beta\alpha} = \boldsymbol{a}^\gamma \cdot \boldsymbol{a}_{\alpha,\beta} = -\boldsymbol{a}_\alpha \cdot \boldsymbol{a}^\gamma{}_{,\beta}. \end{aligned} \tag{A.2.33}$$

We note that application of (A.2.32) to (A.2.13), (A.2.15) and (A.2.18) results in

$$a_{\alpha\beta|\gamma} = a^{\alpha\beta}{}_{|\gamma} = 0, \qquad \varepsilon_{\alpha\beta|\gamma} = \varepsilon^{\alpha\beta}{}_{|\gamma} = 0. \tag{A.2.34}$$

By considering the second covariant derivatives of surface vectors and tensors, we find the surface analogues of the expressions (A.1.32). For example

$$\begin{aligned} &T_{\alpha|\beta\gamma} - T_{\alpha|\gamma\beta} = R^\lambda_{.\alpha\beta\gamma} T_\lambda, \\ &T_{\alpha\beta|\gamma\delta} - T_{\alpha\beta|\delta\gamma} = R^\lambda_{.\alpha\gamma\delta} T_{\lambda\beta} + R^\lambda_{.\beta\gamma\delta} T_{\alpha\lambda}, \\ &T^{\alpha\beta}{}_{|\gamma\delta} - T^{\alpha\beta}{}_{|\delta\gamma} = -R^\alpha_{.\lambda\gamma\delta} T^{\lambda\beta} - R^\beta_{.\lambda\gamma\delta} T^{\alpha\lambda}, \end{aligned} \tag{A.2.35}$$

where the curvature tensor (also called the Riemann-Christoffel tensor) for the surface is defined analogously to (A.1.33):

$$R^\lambda_{.\alpha\beta\gamma} = \Gamma^\lambda_{\alpha\gamma,\beta} - \Gamma^\lambda_{\alpha\beta,\gamma} + \Gamma^\mu_{\alpha\gamma} \Gamma^\lambda_{\mu\beta} - \Gamma^\mu_{\alpha\beta} \Gamma^\lambda_{\mu\gamma}. \tag{A.2.36}$$

The covariant curvature tensor for the surface is given by

$$R_{\lambda\alpha\beta\gamma} = a_{\lambda\mu} R^\mu_{.\alpha\beta\gamma} \tag{A.2.37}$$

and satisfies the identities

$$\begin{aligned} &R_{\alpha\beta\gamma\delta} = -R_{\beta\alpha\gamma\delta}, \qquad R_{\alpha\beta\gamma\delta} = -R_{\alpha\beta\delta\gamma}, \qquad R_{\alpha\beta\gamma\delta} = R_{\gamma\delta\alpha\beta}, \\ &R_{\alpha\beta\gamma\delta} + R_{\alpha\gamma\delta\beta} + R_{\alpha\delta\beta\gamma} = 0. \end{aligned} \tag{A.2.38}$$

It follows from (A.2.38) that in a Riemannian 2-space all components of $R_{\lambda\alpha\beta\gamma}$ either vanish or are expressible in terms of R_{1212}. We return to this later in this section. It is clear from (A.2.35) that the successive covariant differentiations do not commute in a Riemannian space. However, we note here the useful identity

$$T^{\alpha\beta}{}_{|\alpha\beta} = T^{\alpha\beta}{}_{|\beta\alpha}, \tag{A.2.39}$$

where $T^{\alpha\beta}$ is not necessarily symmetric.

δ) *Formulae of Weingarten and Gauss. Integrability conditions.* At every point P of the surface $\mathfrak{s}$, we have a set of base vectors $\boldsymbol{a}_i = \{\boldsymbol{a}_1, \boldsymbol{a}_2, \boldsymbol{a}_3\}$ defined by $(A.2.3)_1$ and (A.2.10). We examine now the partial derivatives, with respect to θ^α, of these vectors which can be represented as a linear combination of $\boldsymbol{a}_i$.

Consider first the partial derivatives $\boldsymbol{a}_{3,1}$ and $\boldsymbol{a}_{3,2}$ which by (A.2.11) lie in the tangent plane to the surface spanned by $\boldsymbol{a}_1, \boldsymbol{a}_2$. Hence, we put

$$\boldsymbol{a}_{3,\alpha} = k^\lambda_\alpha \boldsymbol{a}_\lambda. \tag{A.2.40}$$

The coefficient k_α^λ can be determined by taking the scalar product of both sides of (A.2.40) with $\boldsymbol{a}_\beta$:

$$\boldsymbol{a}_{3,\alpha}\cdot\boldsymbol{a}_\beta = k_\alpha^\lambda\,\boldsymbol{a}_\lambda\cdot\boldsymbol{a}_\beta = a_{\lambda\beta}\,k_\alpha^\lambda.$$

It follows from the last expression and (A.2.27) that $k_\alpha^\lambda = -b_\alpha^\lambda$ and we have

$$\boldsymbol{a}_{3,\alpha} = \boldsymbol{a}_{3|\alpha} = -b_\alpha^\gamma\,\boldsymbol{a}_\gamma. \tag{A.2.41}$$

The relations (A.2.41) are known as the *formulae of Weingarten.*

Next, we write the partial derivatives of $\boldsymbol{a}_1$, $\boldsymbol{a}_2$ as linear combinations of $\boldsymbol{a}_i$ in the form

$$\boldsymbol{a}_{\alpha,\beta} = c_{\alpha\beta}^\gamma\,\boldsymbol{a}_\gamma + c_{\alpha\beta}\,\boldsymbol{a}_3 \tag{A.2.42}$$

and determine the coefficients $c_{\alpha\beta}^\gamma$, $c_{\alpha\beta}$ by taking the scalar product of both sides of (A.2.42) first with $\boldsymbol{a}_\lambda$ and then with $\boldsymbol{a}_3$. In this way, using (A.2.27) and (A.2.33), we obtain

$$c_{\alpha\beta}^\gamma\,a_{\gamma\lambda} = c_{\alpha\beta\lambda} = \boldsymbol{a}_\lambda\cdot\boldsymbol{a}_{\alpha,\beta} = \Gamma_{\alpha\beta\lambda},$$
$$c_{\alpha\beta}\,\boldsymbol{a}_3\cdot\boldsymbol{a}_3 = c_{\alpha\beta} = \boldsymbol{a}_{\alpha,\beta}\cdot\boldsymbol{a}_3 = b_{\alpha\beta}.$$

Hence, the relations (A.2.42) become

$$\boldsymbol{a}_{\alpha,\beta} = \Gamma_{\alpha\beta}^\gamma\,\boldsymbol{a}_\gamma + b_{\alpha\beta}\,\boldsymbol{a}_3 \tag{A.2.43}$$

and are known as the *formulae of Gauss*. In view of $(A.2.32)_1$, the last result may also be written in the useful form

$$\boldsymbol{a}_{\alpha|\beta} = b_{\alpha\beta}\,\boldsymbol{a}_3. \tag{A.2.44}$$

Weingarten's formula may be used to define the *third fundamental form* of the surface, namely

$$d\boldsymbol{a}_3\cdot d\boldsymbol{a}_3 = b_\alpha^\lambda\,b_{\lambda\beta}\,d\theta^\alpha\,d\theta^\beta. \tag{A.2.45}$$

We now proceed to obtain certain integrability conditions which ensure that the partial differential equations (A.2.41) and (A.2.43) [or equivalently (A.2.44)] always have solutions. The conditions we derive are necessary but they can also be shown to be sufficient for the existence of solutions of (A.2.41) and (A.2.43).[7] Assuming that (A.2.2) is at least of class C^3, we must have

$$\frac{\partial\boldsymbol{a}_{\alpha,\beta}}{\partial\theta^\lambda} = \frac{\partial\boldsymbol{a}_{\alpha,\lambda}}{\partial\theta^\beta}. \tag{A.2.46}$$

After calculating the partial derivatives of (A.2.43) and substituting the results in (A.2.46), we may rewrite the latter in the form

$$[R_{\delta\alpha\beta\gamma} - (b_{\alpha\gamma}\,b_{\beta\delta} - b_{\alpha\beta}\,b_{\gamma\delta})]\,\boldsymbol{a}^\delta - (b_{\alpha\beta|\gamma} - b_{\alpha\gamma|\beta})\,\boldsymbol{a}^3 = 0, \tag{A.2.47}$$

where $R_{\delta\alpha\beta\gamma}$ is defined by (A.2.37).[8] The scalar product of (A.2.47) with $\boldsymbol{a}_3$ yields

$$b_{\alpha\beta|\gamma} = b_{\alpha\gamma|\beta}, \tag{A.2.48}$$

which are known as the *Mainardi-Codazzi equations*. For $\beta=\gamma$ these equations are identically satisfied, while for $\beta\neq\gamma$ they give the relations

$$b_{\alpha1|2} = b_{\alpha2|1}. \tag{A.2.49}$$

---

[7] See, in this connection, p. 119 of WILLMORE [1959, *8*].

[8] The result (A.2.47) can also be obtained from application of $(A.2.35)_1$ to $\boldsymbol{a}_\alpha$ and the use of (A.2.44).

Also, from the scalar product of (A.2.47) with $\boldsymbol{a}_\mu$, we obtain *the equations of Gauss*:

$$R_{\delta\alpha\beta\gamma} = b_{\alpha\gamma} b_{\beta\delta} - b_{\alpha\beta} b_{\gamma\delta}. \tag{A.2.50}$$

It follows from the symmetry conditions (A.2.38) that the non-zero components of $R_{\delta\alpha\beta\gamma}$ are those for which $\delta \neq \alpha$ and $\beta \neq \gamma$. Hence, there are only four non-zero components of the curvature tensor for the surface and these are given by

$$\begin{aligned} R_{2112} &= R_{1221} = -R_{1212} = -R_{2121}, \\ R_{1212} &= b_{22} b_{11} - (b_{12})^2 = \det(b_{\alpha\beta}). \end{aligned} \tag{A.2.51}$$

The Mainardi-Codazzi equations (A.2.48) and the expression (A.2.50) constitute the integrability conditions for the formulae of Weingarten and Gauss. From $(\text{A.2.51})_5$ and the definition (A.2.29) follows the well-known theorem of Gauss, namely

$$K = \frac{R_{1212}}{a}, \tag{A.2.52}$$

according to which the Gaussian curvature of a surface is independent of the coefficients of the second fundamental form but depends only on the coefficients $a_{\alpha\beta}$ of the first fundamental form and their first and second partial derivatives. It is clear from (A.2.35) and (A.2.52) that, in general, the interchange of the order of covariant differentiation is not permissible unless the Gaussian curvature vanishes. In the special case when the surface (A.2.2) is a *plane*, the unit vector $\boldsymbol{a}_3$ is a constant and $\boldsymbol{a}_{3,\alpha} = 0$. Hence, $b_{\alpha\beta} = 0$, $R_{1212} = 0$ and the order of covariant differentiation is immaterial in the case of a plane.

The formulae of Weingarten and Gauss can be used to calculate the partial derivatives of a vector which is not necessarily in the surface. To show this, consider a three-dimensional vector field $\boldsymbol{V}$ defined on $\mathcal{s}$. Referred to the base vectors $\boldsymbol{a}_i$, $\boldsymbol{V}$ may be expressed in the form

$$\boldsymbol{V} = V^\alpha \boldsymbol{a}_\alpha + V^3 \boldsymbol{a}_3 = V_\alpha \boldsymbol{a}^\alpha + V_3 \boldsymbol{a}^3. \tag{A.2.53}$$

Using (A.2.41) and (A.2.44), the partial derivatives of $\boldsymbol{V}$ are found to be

$$\begin{aligned} \boldsymbol{V}_{,\alpha} \equiv \boldsymbol{V}_{|\alpha} &= (V^\lambda{}_{|\alpha} - b^\lambda_\alpha V^3)\, \boldsymbol{a}_\lambda + (V^3{}_{,\alpha} + b_{\lambda\alpha} V^\lambda)\, \boldsymbol{a}_3 \\ &= (V_{\lambda|\alpha} - b_{\lambda\alpha} V_3)\, \boldsymbol{a}^\lambda + (V_{3,\alpha} + b^\lambda_\alpha V_\lambda)\, \boldsymbol{a}^3. \end{aligned} \tag{A.2.54}$$

ε) *Principal curvatures. Lines of curvature.* The unit tangent vector $\boldsymbol{\lambda}$ to a curve on the surface at the point $r = r(\theta^\alpha(s))$ is given by $(\text{A.2.21})_2$. Let $\varkappa(s)$, where s is the arc parameter, be the *curvature* of the curve and let $\boldsymbol{\mu}$ denote the (space) unit *principal normal* to the curve at the point $\boldsymbol{r} = \boldsymbol{r}(\theta^\alpha(s))$. Then, the *curvature vector* $\boldsymbol{\kappa}$ defined by

$$\boldsymbol{\kappa} = \varkappa \boldsymbol{\mu} = \frac{d\boldsymbol{\lambda}}{ds} \tag{A.2.55}$$

can, with the help of (A.2.44), be written as

$$\boldsymbol{\kappa} = \left(\frac{d\lambda^\mu}{ds} + \Gamma^\mu_{\alpha\beta} \lambda^\alpha \lambda^\beta\right) \boldsymbol{a}_\mu + b_{\alpha\beta} \lambda^\alpha \lambda^\beta \boldsymbol{a}_3. \tag{A.2.56}$$

It should be recalled that the unit principal normal vector $\boldsymbol{\mu}$ is independent of the orientation of a curve while the sense of the unit tangent vector $\boldsymbol{\lambda}$ depends on the orientation of the curve.

The *normal curvature* of a curve at the point $\boldsymbol{r}=\boldsymbol{r}(\theta^\alpha(s))$ is obtained from (A.2.56) through

$$\varkappa_{(n)}=\boldsymbol{\kappa}\cdot\boldsymbol{a}_3=b_{\alpha\beta}\,\lambda^\alpha\,\lambda^\beta, \tag{A.2.57}$$

which, in general, assumes an infinity of values corresponding to an infinity of directions at each point of the surface. Except at points where $\varkappa_{(n)}$ is constant, i.e., at *umbilics*, the directions for which the normal curvature $\varkappa_{(n)}$ assumes an extremal value may be determined by the method of Lagrangian multiplier. This leads to the following set of homogeneous equations:[9]

$$(b_{\alpha\beta}-\varkappa_{(n)}\,a_{\alpha\beta})\,\lambda^\beta=0, \tag{A.2.58}$$

which has a nontrivial solution if and only if the values of $\varkappa_{(n)}$ are the roots of

$$\det(b_{\alpha\beta}-\varphi\,a_{\alpha\beta})=0, \tag{A.2.59}$$

or equivalently

$$\varphi^2-2H\,\varphi+K=0, \tag{A.2.60}$$

where H and K are given by (A.2.28) and (A.2.29). The roots $\varkappa_{(1)}=\varkappa_1$, $\varkappa_{(2)}=\varkappa_2$ are the *principal curvatures* of the surface at the point in question and the corresponding directions are called the *principal directions*. It can be shown that the roots of (A.2.60) are always real and that at any point which is not umbilic the principal directions $\lambda^\alpha_{(1)}$, $\lambda^\alpha_{(2)}$ are orthogonal. The principal directions $\lambda^\alpha_{(\nu)}$ can be determined in the usual manner from (A.2.58). At an umbilic, since $\varkappa_{(n)}=\text{const}$, $\varkappa_1=\varkappa_2$ and every direction is a principal direction.

We define the principal radii of curvature r_1 and r_2 by

$$r_1=-\frac{1}{\varkappa_1},\qquad r_2=-\frac{1}{\varkappa_2}, \tag{A.2.61}$$

where

$$\varkappa_1+\varkappa_2=2H,\qquad \varkappa_1\,\varkappa_2=K. \tag{A.2.62}$$

For later reference, we note that (A.2.61) are the solutions of

$$\det(\delta^\alpha_\beta-\varphi^{-1}\,b^\alpha_\beta)=0. \tag{A.2.63}$$

The sign convention in (A.2.61) is in accord with the rule that r_1 and r_2 are positive (negative) if the unit normal vector to the surface is directed away from (toward) the center of curvature.[10]

A curve on a surface whose tangent at each point is along a principal direction is called a *line of curvature*. If $\varkappa_1=\varkappa_2$ at every point, then all coordinate lines are lines of curvature (the only cases being the spherical surface and the plane). For $\varkappa_1\neq\varkappa_2$ it can be shown that the conditions necessary and sufficient for the coordinate curves on $\mathscr{s}$ to be lines of curvature are

$$a_{12}=b_{12}=0. \tag{A.2.64}$$

A.3. Geometry of a surface in a Euclidean space covered by normal coordinates. We begin the development of this section by identifying the curvilinear coordinates θ^i of Sect. A.1 with a system of normal coordinates as follows:[11] Consider a

---

[9] See, e.g., p. 210 of MCCONNELL [1931, *1*] or p. 224 of EISENHART [1947, *2*].

[10] It should be noted that the sign convention in (A.2.61) is opposite to that usually employed in differential geometry, but is in accord with that adopted in most investigations dealing with shell theory. Also some authors prefer to define the second fundamental form as negative of that in (A.2.26) and this in turn influences the signs of $\varkappa_1$ and $\varkappa_2$.

[11] For a general discussion of normal coordinate system, see p. 62 of SYNGE and SCHILD [1949, *5*].

singly infinite family of (two-dimensional) surfaces in the Euclidean 3-space such that θ^3, regarded as a parameter, is a constant over each surface. Let $\theta^\alpha\,(\alpha=1,2)$ be the coordinate system on one of the surfaces, say the surface $\theta^3=0$ which we can call the *reference surface*. Further, let the position vector and the unit normal to the reference surface ($\theta^3=0$) be denoted by $\boldsymbol{r}=\boldsymbol{r}(\theta^\alpha)$ and $\boldsymbol{a}_3=\boldsymbol{a}_3(\theta^\alpha)$, respectively.[12] Then, adopting the notation $\theta^3=\zeta$, we may represent the position vector of any point of the space as

$$\boldsymbol{p}=\boldsymbol{r}+\zeta\,\boldsymbol{a}_3, \tag{A.3.1}$$

where $\boldsymbol{r}$ and $\boldsymbol{a}_3$ are functions of θ^1, θ^2 only and satisfy the restriction

$$\boldsymbol{a}_\alpha\cdot\boldsymbol{a}_3=0, \qquad \boldsymbol{a}_\alpha=\boldsymbol{r}_{,\alpha}. \tag{A.3.2}$$

It should be clear that various formulae of the previous section such as (A.2.3), (A.2.11), (A.2.27), (A.2.41), and (A.2.43) which involve $\boldsymbol{r}$, $\boldsymbol{a}_3$ and their partial derivatives hold also here for the reference surface $\boldsymbol{r}=\boldsymbol{p}(\theta^\alpha, 0)$.

In the remainder of this section we consider the geometrical properties of a surface, i.e., the reference surface ($\zeta=0$) defined above, embedded in the space (A.3.1) and also briefly discuss the relationships between the components of the space tensors and their surface counterparts defined on $\zeta=0$. Thus, in view of (A.1.15), the covariant base vectors and the metric tensor for the space (A.3.1) are

$$\boldsymbol{g}_\alpha=\boldsymbol{a}_\alpha+\zeta\,\boldsymbol{a}_{3,\alpha}=\mu_\alpha^\gamma\,\boldsymbol{a}_\gamma, \qquad \boldsymbol{g}_3=\boldsymbol{a}_3 \tag{A.3.3}$$

and

$$g_{\alpha\beta}=\mu_\alpha^\nu\,\mu_\beta^\lambda\,a_{\nu\lambda}, \qquad g_{\alpha 3}=0, \qquad g_{33}=1, \tag{A.3.4}$$

where $a_{\nu\lambda}=\boldsymbol{a}_\nu\cdot\boldsymbol{a}_\lambda$ is the metric tensor of the reference surface and where

$$\mu_\alpha^\gamma=\delta_\alpha^\gamma-\zeta\,b_\alpha^\gamma, \tag{A.3.5}$$

b_α^γ being the coefficients of the second fundamental form. The value of $g_{\alpha 3}$ given by $(\text{A.3.4})_2$ is the characteristic property of normal coordinates and is a consequence of $(\text{A.3.2})_1$. The nature of the reduction of the above formulae when evaluated on the reference surface ($\zeta=0$) is evident. In particular, we note that

$$\boldsymbol{g}_\alpha(\theta^\gamma, 0)=\boldsymbol{a}_\alpha, \qquad g_{\alpha\beta}(\theta^\gamma, 0)=a_{\alpha\beta} \tag{A.3.6}$$

and

$$g(\theta^\alpha, 0)=\det(g_{\alpha\beta})\big|_{\zeta=0}=\det(a_{\alpha\beta})=a. \tag{A.3.7}$$

Let $\varepsilon_{\alpha\beta}$, $\varepsilon^{\alpha\beta}$ denote the ε-system for the reference surface. Then, the components $\varepsilon_{\alpha\beta 3}$, $\varepsilon^{\alpha\beta 3}$ of the space ε-system in (A.1.19) when evaluated on $\zeta=0$ become

$$\begin{aligned}\varepsilon_{\alpha\beta}&=\varepsilon_{\alpha\beta 3}\big|_{\zeta=0}=a^{\frac{1}{2}}\,e_{\alpha\beta},\\ \varepsilon^{\alpha\beta}&=\varepsilon^{\alpha\beta 3}\big|_{\zeta=0}=a^{-\frac{1}{2}}\,e^{\alpha\beta},\end{aligned} \tag{A.3.8}$$

where we have set $e_{\alpha\beta 3}=e_{\alpha\beta}$, $e^{\alpha\beta 3}=e^{\alpha\beta}$ and the components of the permutation symbols $e_{\alpha\beta}$, $e^{\alpha\beta}$ are defined in (A.2.19). We observe that

$$\varepsilon_{\alpha\beta 3}=\left(\frac{g}{a}\right)^{\frac{1}{2}}\varepsilon_{\alpha\beta}, \qquad \varepsilon^{\alpha\beta 3}=\left(\frac{a}{g}\right)^{\frac{1}{2}}\varepsilon^{\alpha\beta}, \tag{A.3.9}$$

which follow from the comparison of (A.1.19), (A.3.8) and (A.2.19). Also, from (A.3.5),

$$\mu=\det(\mu_\beta^\alpha)=\left(\frac{g}{a}\right)^{\frac{1}{2}}=1-2\zeta H+\zeta^2 K, \tag{A.3.10}$$

---

[12] The designation of the position vector and the unit normal of the reference surface by the same symbols as those used for a surface in Sect. A.2 is made in anticipation of later identification of the two surfaces.

where H and K are defined by (A.2.28)–(A.2.29). Using (A.3.10), we may rewrite (A.3.9) in the form

$$\varepsilon_{\alpha\beta 3} = \mu^{\nu}_{\alpha}\,\mu^{\lambda}_{\beta}\,\varepsilon_{\nu\lambda} = \mu\,\varepsilon_{\alpha\beta} \tag{A.3.11}$$

and a similar expression holds for $\varepsilon^{\alpha\beta 3}$.

The Christoffel symbols with respect to the reference surface are found by evaluating (A.1.24) on $\zeta = 0$ and by using (A.3.3) and (A.1.25):

$$\begin{aligned}
&\overset{*}{\Gamma}_{\alpha\beta\gamma}\big|_{\zeta=0} = \Gamma_{\alpha\beta\gamma}, \qquad \overset{*}{\Gamma}{}^{\alpha}_{\beta\gamma}\big|_{\zeta=0} = \Gamma^{\alpha}_{\beta\gamma} = a^{\alpha\lambda}\,\Gamma_{\beta\gamma\lambda},\\
&\overset{*}{\Gamma}{}^{\alpha}_{\beta 3}\big|_{\zeta=0} = \boldsymbol{a}^{\alpha}\cdot\boldsymbol{a}_{3,\beta} = -\boldsymbol{a}_3\cdot\boldsymbol{a}^{\alpha}{}_{,\beta} = -b^{\alpha}_{\beta},\\
&\overset{*}{\Gamma}{}^{3}_{\alpha\beta}\big|_{\zeta=0} = \boldsymbol{a}^{3}\cdot\boldsymbol{a}_{\alpha,\beta} = -\boldsymbol{a}_{\beta}\cdot\boldsymbol{a}^{3}{}_{,\alpha} = b_{\alpha\beta},\\
&\overset{*}{\Gamma}{}^{\alpha}_{33}\big|_{\zeta=0} = 0, \qquad \overset{*}{\Gamma}{}^{3}_{33}\big|_{\zeta=0} = 0, \qquad \overset{*}{\Gamma}{}^{3}_{3\alpha}\big|_{\zeta=0}
\end{aligned} \tag{A.3.12}$$

where $\Gamma_{\alpha\beta\gamma}$ is defined by (A.2.33). Using the values (A.3.12), the components of the Riemann-Christoffel tensor (A.1.33) when evaluated on the surface $\zeta = 0$ can be put in the forms

$$\overset{*}{R}{}^{\lambda}_{.\alpha\beta\gamma}\big|_{\zeta=0} = R^{\lambda}_{.\alpha\beta\gamma} - b_{\alpha\gamma}\,b^{\lambda}_{\beta} + b_{\alpha\beta}\,b^{\lambda}_{\gamma}, \tag{A.3.13}$$

$$\overset{*}{R}{}^{3}_{.\alpha\beta\gamma}\big|_{\zeta=0} = b_{\alpha\gamma,\beta} - b_{\alpha\beta,\gamma} + \Gamma^{\lambda}_{\alpha\gamma}\,b_{\lambda\beta} - \Gamma^{\lambda}_{\alpha\beta}\,b_{\lambda\gamma}, \tag{A.3.14}$$

where $R^{\lambda}_{.\alpha\beta\gamma}$ is defined by (A.2.36). Since $\overset{*}{R}{}^{m}_{.ijk}$ vanishes in a Euclidean space, the left-hand sides of each of the last two equations are zero and we again obtain formulae of the forms (A.2.50) and (A.2.48) for the reference surface $\zeta = 0$.

In terms of general coordinates θ^i of Sect. A.1, the volume element (by definition) is $dv = g^{\frac{1}{2}}\,d\theta^1\,d\theta^2\,d\theta^3$. In a region of space covered by normal coordinates, the volume element can be expressed in terms of the element of area $d\sigma = a^{\frac{1}{2}}\,d\theta^1\,d\theta^2$ of the reference surface ($\zeta = 0$) in the form $dv = \mu\,d\sigma\,d\zeta$ which also involves the determinant μ defined by (A.3.10). In this connection, we ask what restriction must be placed on the space (A.3.1) in order to ensure the existence of nonvanishing μ for all ζ. For this puspose, we set

$$\mu = \det(\delta^{\alpha}_{\beta} - \zeta\,b^{\alpha}_{\beta}) = 0 \tag{A.3.15}$$

and observe that this equation has the same structure as (A.2.63). It follows from (A.2.59)–(A.2.62) and (A.3.10) that the solutions of (A.3.15) in terms of the principal radii of curvature of the reference surface are

$$\zeta = \{r_1, r_2\}. \tag{A.3.16}$$

Evidently, in view of (A.3.16), for $\mu \neq 0$ it is sufficient to require that[13]

$$|\zeta| < r, \qquad r = \min(|r_1|, |r_2|) \neq 0. \tag{A.3.17}$$

The restriction $(\text{A.3.17})_1$ also ensures that the tensor μ^{γ}_{α} is nonsingular and, therefore, possesses a unique inverse $(\mu^{-1})^{\gamma}_{\alpha}$ such that

$$\mu^{\nu}_{\alpha}\,(\mu^{-1})^{\gamma}_{\nu} = \delta^{\gamma}_{\alpha}. \tag{A.3.18}$$

[13] Of course the restriction $(\text{A.3.17})_1$ is not a necessary condition for the existence of nonvanishing μ. We remark here that (A.3.17) is indeed a weak restriction from the point of view of shell theory.

To obtain an expression for the inverse $(\mu^{-1})^{\gamma}_{\alpha}$, multiply (A.3.11)$_2$ by $\varepsilon^{\gamma\beta}$, use (A.2.22) and put the resulting equation in the form

$$\mu^{\nu}_{\alpha}\left[\frac{1}{\mu}\,\delta^{\gamma\beta}_{\nu\lambda}\,\mu^{\lambda}_{\beta}\right]=\delta^{\gamma}_{\alpha}.$$

Then, comparison of this last result and (A.3.18) yields the expression

$$(\mu^{-1})^{\gamma}_{\alpha}=\frac{1}{\mu}\,\delta^{\gamma\beta}_{\alpha\lambda}\,\mu^{\lambda}_{\beta}. \tag{A.3.19}$$

In a region of space covered by normal coordinates, the relationships between the components of space tensors and their surface counterparts involve the determinant μ, as well as μ^{γ}_{α} and its inverse. This is partly evident from (A.3.3)–(A.3.4) and (A.3.11) which involve μ^{γ}_{α} and μ. Other transformation relations between space tensors and their surface counterparts, such as those between $\boldsymbol{g}^{\alpha}$ and $\boldsymbol{a}^{\gamma}$ and between $g^{\alpha\beta}$ and $a^{\gamma\lambda}$, can be shown to involve the inverse $(\mu^{-1})^{\gamma}_{\alpha}$ but we do not elaborate on these here.[14]

A.4. Physical components of surface tensors in lines of curvature coordinates. We consider here briefly the physical components of tensors defined over a surface and referred to lines of curvature coordinates, i.e., when the conditions (A.2.64) hold. We also obtain explicit forms of the physical components of certain tensor derivatives which will be useful in the discussion of the equations of shell theory in lines of curvature coordinates.

From (A.2.58), the nonvanishing components of b^{α}_{β} in lines of curvature coordinates are

$$\varkappa_1=b^1_1=\frac{b_{11}}{(a_1)^2}=-\frac{1}{r_1},\qquad \varkappa_2=b^2_2=\frac{b_{22}}{(a_2)^2}=-\frac{1}{r_2}, \tag{A.4.1}$$

where r_1 and r_2 are the principal radii of curvature defined by (A.2.61) and where for convenience we have introduced the notations

$$(a_1)^2=a_{11}=|\boldsymbol{a}_1|^2,\qquad (a_2)^2=a_{22}=|\boldsymbol{a}_2|^2. \tag{A.4.2}$$

When the coordinate curves on $\mathscr{s}$ are orthogonal ($a_{12}=0$), the components of the Christoffel symbols in (A.2.33) are given by

$$\begin{aligned}\Gamma^{\alpha}_{\beta\beta}&=-\frac{1}{2a_{\alpha\alpha}}\frac{\partial a_{\beta\beta}}{\partial\theta^{\alpha}}=-\frac{a_{\beta\beta}}{a_{\alpha\alpha}}\frac{\partial}{\partial\theta^{\alpha}}\log\sqrt{a_{\beta\beta}}=-\frac{a_{\beta}}{(a_{\alpha})^2}\,a_{\beta,\alpha}\\&\qquad(\alpha\neq\beta,\ \text{no sum over }\alpha,\beta),\\ \Gamma^{\alpha}_{\alpha\beta}&=\Gamma^{\alpha}_{\beta\alpha}=\frac{1}{2a_{\alpha\alpha}}\frac{\partial a_{\alpha\alpha}}{\partial\theta^{\beta}}=\frac{\partial}{\partial\theta^{\beta}}\log\sqrt{a_{\alpha\alpha}}=\frac{1}{a_{\alpha}}\,a_{\alpha,\beta}\\&\qquad(\text{no sum over }\alpha),\end{aligned} \tag{A.4.3}$$

where we have recorded the formulae both in terms of the metric tensor $a_{\alpha\beta}$ and also in terms of a_{α} defined by (A.4.2).

The physical components of tensors in an orthogonal basis are defined by[15]

$$\begin{aligned}T_{\langle\alpha\ldots\gamma\beta\ldots\delta\rangle}&=\left[\frac{a_{\alpha\alpha}\cdots a_{\gamma\gamma}}{a_{\beta\beta}\cdots a_{\delta\delta}}\right]^{\frac{1}{2}}T^{\alpha\ldots\gamma}{}_{\beta\ldots\delta}\\&=(a_{\alpha\alpha}\cdots a_{\delta\delta})^{\frac{1}{2}}\,T^{\alpha\ldots\delta}=(a_{\alpha\alpha}\cdots a_{\delta\delta})^{-\frac{1}{2}}\,T_{\alpha\ldots\delta},\end{aligned} \tag{A.4.4}$$

---

[14] A general discussion of transformation relations between the components of space tensors and their surface counterparts is given in Sect. 3 of [1963, *6*].

[15] A discussion of the physical components of tensors in orthogonal curvilinear coordinates can be found in McConnell [1931, *1*]. For a more general account of physical components of tensors, see the papers by Truesdell [1953, *6*], [1954, *2*] and Sects. 11–14 of Ericksen [1960, *3*].

where $T_{\langle\alpha\ldots\gamma\beta\ldots\delta\rangle}$ is the symbol designating the physical components. In particular, using the notation of (A.4.2), the physical components of the first and second-order tensors are

$$\begin{aligned} T_{\langle\alpha\rangle} &= \frac{T_\alpha}{a_\alpha} = a_\alpha T^\alpha && \text{(no summation over } \alpha), \\ T_{\langle\alpha\beta\rangle} &= \frac{T_{\alpha\beta}}{a_\alpha a_\beta} = a_\alpha a_\beta T^{\alpha\beta} && \text{(no summation over } \alpha, \beta), \end{aligned} \tag{A.4.5}$$

and

$$\begin{aligned} T_\alpha &= a_\alpha T_{\langle\alpha\rangle}, & T^\alpha &= \frac{T_{\langle\alpha\rangle}}{a_\alpha} && \text{(no summation over } \alpha), \\ T_{\alpha\beta} &= a_\alpha a_\beta T_{\langle\alpha\beta\rangle}, & T^{\alpha\beta} &= \frac{T_{\langle\alpha\beta\rangle}}{a_\alpha a_\beta} && \text{(no summation over } \alpha, \beta). \end{aligned} \tag{A.4.6}$$

Next, introducing the notations

$$\Gamma^{\langle\alpha\beta\gamma\rangle} = \frac{\sqrt{a_{\gamma\gamma}}}{\sqrt{a_{\alpha\alpha}}\sqrt{a_{\beta\beta}}}\,\Gamma^\gamma_{\alpha\beta} = \frac{a_\gamma}{a_\alpha a_\beta}\,\Gamma^\gamma_{\alpha\beta} \tag{A.4.7}$$

and

$$\frac{\partial}{\partial s_\beta} = \frac{1}{\sqrt{a_{\beta\beta}}}\frac{\partial}{\partial\theta^\beta} = \frac{1}{a_\beta}\frac{\partial}{\partial\theta^\beta} \qquad \text{(no sum over } \beta), \tag{A.4.8}$$

from (A.4.3) we obtain

$$\begin{aligned} \Gamma^{\langle\beta\beta\alpha\rangle} &= -\frac{\partial}{\partial s_\alpha}\log\sqrt{a_{\beta\beta}} = -\frac{\partial}{\partial s_\alpha}\log a_\beta && (\alpha \neq \beta, \text{ no sum over } \beta), \\ \Gamma^{\langle\alpha\beta\alpha\rangle} &= \frac{\partial}{\partial s_\beta}\log\sqrt{a_{\alpha\alpha}} = \frac{\partial}{\partial s_\beta}\log a_\alpha && \text{(no sum over } \alpha). \end{aligned} \tag{A.4.9}$$

The physical components of tensor derivatives can be readily obtained by direct substitution from the definitions (A.4.4) and (A.4.7)–(A.4.9) into expressions of the type (A.2.32). Thus, the physical components of $T^{\alpha\ldots\gamma}{}_{.\beta\ldots\delta|\sigma}$ can be calculated from[16]

$$\begin{aligned} T_{\langle\alpha\ldots\gamma\beta\ldots\delta|\sigma\rangle} = {} & \frac{\partial}{\partial s_\sigma} T_{\langle\alpha\ldots\gamma\beta\ldots\delta\rangle} - T_{\langle\alpha\ldots\gamma\beta\ldots\delta\rangle}\frac{\partial}{\partial s_\sigma}\log\frac{a_\alpha\ldots a_\gamma}{a_\beta\ldots a_\delta} \\ & + \Gamma^{\langle\lambda\sigma\alpha\rangle} T_{\langle\lambda\ldots\gamma\beta\ldots\delta\rangle} + \cdots + \Gamma^{\langle\lambda\sigma\gamma\rangle} T_{\langle\alpha\ldots\lambda\beta\ldots\delta\rangle} \\ & - \Gamma^{\langle\beta\sigma\lambda\rangle} T_{\langle\alpha\ldots\gamma\lambda\ldots\delta\rangle} - \cdots - \Gamma^{\langle\delta\sigma\lambda\rangle} T_{\langle\alpha\ldots\gamma\beta\ldots\lambda\rangle}, \end{aligned} \tag{A.4.10}$$

where use has been made of the fact that

$$\frac{a_\beta\ldots a_\delta}{a_\alpha\ldots a_\gamma}\frac{\partial}{\partial s_\sigma}\left[\frac{a_\alpha\ldots a_\gamma}{a_\beta\ldots a_\delta}\right] = \frac{\partial}{\partial s_\sigma}\log\frac{a_\alpha\ldots a_\gamma}{a_\beta\ldots a_\delta}.$$

We record below explicit expressions for the physical components of two tensor derivatives, namely $T^\alpha{}_{|\alpha}$ and $T^{\alpha\beta}{}_{|\alpha}$, which frequently occur in the field equations of shell theory. Thus, for the physical components of $T^\alpha{}_{|\alpha}$ we have

$$\begin{aligned} T_{\langle 1|1\rangle} &= \frac{1}{a_1}\left[T_{\langle 1\rangle,1} + \frac{a_{1,2}}{a_2} T_{\langle 2\rangle}\right], \\ T_{\langle 2|2\rangle} &= \frac{1}{a_2}\left[T_{\langle 2\rangle,2} + \frac{a_{2,1}}{a_1} T_{\langle 1\rangle}\right]. \end{aligned} \tag{A.4.11}$$

[16] The rule (A.4.10) is a special case of a more general formula given by TRUESDELL [1953, *6*]; see Sect. 12 of his paper.

Similarly, the physical components of $T^{\alpha\beta}{}_{|\alpha}$ are given by

$$
\begin{aligned}
T_{\langle 11|1\rangle} &= \frac{1}{a_1}\left[T_{\langle 11\rangle,1} + \frac{a_{1,2}}{a_2}\left(T_{\langle 12\rangle} + T_{\langle 21\rangle}\right)\right], \\
T_{\langle 22|2\rangle} &= \frac{1}{a_2}\left[T_{\langle 22\rangle,2} + \frac{a_{2,1}}{a_1}\left(T_{\langle 12\rangle} + T_{\langle 21\rangle}\right)\right], \\
T_{\langle 21|2\rangle} &= \frac{1}{a_2}\left[T_{\langle 21\rangle,2} + \frac{a_{2,1}}{a_1}\left(T_{\langle 11\rangle} - T_{\langle 22\rangle}\right)\right], \\
T_{\langle 12|1\rangle} &= \frac{1}{a_1}\left[T_{\langle 12\rangle,1} + \frac{a_{1,2}}{a_2}\left(T_{\langle 22\rangle} - T_{\langle 11\rangle}\right)\right].
\end{aligned}
\tag{A.4.12}
$$

In obtaining (A.4.4) to (A.4.12), only the first of the two conditions in (A.2.64), namely $a_{12}=0$, has been invoked. With the use of (A.4.11)–(A.4.12) and (A.4.6), as well as (A.4.1), it is now a simple matter to write the various field equations, kinematical results and the constitutive relations of shell theory in lines of curvature coordinates and in terms of physical components. Thus, for example, the equations of equilibrium obtained from $(9.47)_{1,2}$ in lines of curvature coordinates and in terms of physical components can be written as

$$
\begin{aligned}
N_{\langle 11|1\rangle} + N_{\langle 21|2\rangle} + \frac{N_{\langle 13\rangle}}{r_1} + \varrho f_{\langle 1\rangle} &= 0, \\
N_{\langle 12|1\rangle} + N_{\langle 22|2\rangle} + \frac{N_{\langle 23\rangle}}{r_2} + \varrho f_{\langle 2\rangle} &= 0, \\
N_{\langle 13|1\rangle} + N_{\langle 23|2\rangle} - \left(\frac{N_{\langle 11\rangle}}{r_1} + \frac{N_{\langle 22\rangle}}{r_2}\right) + \varrho f_3 &= 0.
\end{aligned}
\tag{A.4.13}
$$

Here, r_1 and r_2 are the principal radii of curvature of the surface $(5.1)_1$ and the components of $N_{\langle\alpha 3|\alpha\rangle}$ are given by expressions of the type (A.4.11). If we substitute formulae of the forms (A.4.11)–(A.4.12) into (A.4.13), then after a slight rearrangement more familiar forms of the equations of equilibrium in lines of curvature coordinates follow:

$$
\begin{aligned}
&\frac{1}{a_1 a_2}\{(a_2 N_{\langle 11\rangle})_{,1} + (a_1 N_{\langle 21\rangle})_{,2} + a_{1,2} N_{\langle 12\rangle} - a_{2,1} N_{\langle 22\rangle}\} + \frac{N_{\langle 13\rangle}}{r_1} + \varrho f_{\langle 1\rangle} = 0, \\
&\frac{1}{a_1 a_2}\{(a_2 N_{\langle 12\rangle})_{,1} + (a_1 N_{\langle 22\rangle})_{,2} - a_{1,2} N_{\langle 11\rangle} + a_{2,1} N_{\langle 21\rangle}\} \\
&\qquad + \frac{N_{\langle 23\rangle}}{r_2} + \varrho f_{\langle 2\rangle} = 0, \\
&\frac{1}{a_1 a_2}\{(a_2 N_{\langle 13\rangle})_{,1} + (a_1 N_{\langle 23\rangle})_{,2}\} - \left(\frac{N_{\langle 11\rangle}}{r_1} + \frac{N_{\langle 22\rangle}}{r_2}\right) + \varrho f_3 = 0.
\end{aligned}
\tag{A.4.14}
$$

Results similar to (A.4.13) and (A.4.14) hold also for the equations of equilibrium obtained from $(9.48)_{1,2}$ which have essentially the same structure as those in $(9.47)_{1,2}$.

References.

The list below is not a complete bibliography on the subject and is limited to works referred to in the text. Entries in parentheses following each reference indicate the sections or places in which it has been cited.[1]

1850 *1.* Kirchhoff, G.: Über das Gleichgewicht und die Bewegung einer elastischen Scheibe. Crelles J. **40**, 51–88 = Ges. Abh. 237–279. (*1, 7, 20, 26*)

[1] The titles of several papers (in Russian) published in Prikl. Mat. Mekh. are listed in English. The English versions of these titles are taken from a Complete Bibliography of Vols. 1–21 in J. Appl. Math. Mech. (Transl. of PMM) **22** (No. 6), pp. i–xxxix.

1886 *1.* TODHUNTER, I., and K. PEARSON: A history of the theory of elasticity, Vol. I. Cambridge: University Press = reprint, Dover Publications. (*1, 20*)

1888 *1.* LOVE, A. E. H.: The small free vibrations and deformation of a thin elastic shell. Phil. Trans. Roy. Soc. London, Ser. A **179**, 491–546. (*1, 4, 7, 12A, 16, 21, 21A*)

2. RAYLEIGH, (Lord): On the bending and vibration of thin elastic shells, especially of cylindrical form. Proc. Roy. Soc. London, Ser. A **45**, 105–123. (*21A*)

1890 *1.* BASSET, A. B.: On the extension and flexure of cylindrical and spherical thin elastic shells. Phil. Trans. Roy. Soc. London, Ser. A **181**, 433–480. (*12A, 21A*)

2. LAMB, H.: On the deformation of an elastic shell. Proc. London Math. Soc. **21**, 119–146. (*12A, 21A*)

1891 *1.* LOVE, A. E. H.: Note on the present state of the theory of thin elastic shells. Proc. Roy. Soc. London, Ser. A **49**, 100–102. (*21A*)

1892 *1.* LOVE, A. E. H.: A treatise on the mathematical theory of elasticity **1**. Cambridge. ix + 354 pp. (*1*)

1893 *1.* DUHEM, P.: Le potentiel thermodynamique et la pression hydrostatique. Ann. Ecole Norm. (3) **10**, 187–230. (*1, 4*)

2. LOVE, A. E. H.: A treatise on the mathematical theory of elasticity **2**. Cambridge. xi + 327 pp. (*1, 12A, 21A*)

3. TODHUNTER, I., and K. PEARSON: A history of the theory of elasticity, Vol. II (Parts I and II). Cambridge: University Press = reprint, Dover Publications. (*1, 20*)

1907 *1.* FÖPPL, A.: Vorlesungen über Technische Mechanik, Vol. 5. München: R. Oldenbourg. (*21*)

1908 *1.* COSSERAT, E., et F.: Sur la théorie des corps minces. Compt. Rend. **146**, 169–172. (*1*)

1909 *1.* COSSERAT, E., et F.: Théorie des Corps Déformables. Paris. vi + 226 pp. = Appendix, pp. 953–1173, of Chwolson's Traité de Physique, 2nd ed. Paris. (*1, 4*)

1910 *1.* KÁRMÁN, TH. v.: Festigkeitsprobleme im Maschinenbau. Encyklopädie der Mathematischen Wissenschaften, Vol. 4/4, pp. 311–385 = Collected Works, **1**, 141–207. (*21*)

1914 *1.* HELLINGER, E.: Die allgemeinen Ansätze der Mechanik der Kontinua. Encyklopädie der Mathematischen Wissenschaften, Vol. 4/4, pp. 601–694. (*7*)

1917 *1.* LAMB, H.: On waves in an elastic plate. Proc. Roy. Soc. London, Ser. A **93**, 114–128. (*20, 24*)

1925 *1.* NADAI, A.: Die elastischen Platten. Berlin: Springer. viii + 326 pp. (*20*)

1927 *1.* LOVE, A. E. H.: 4th ed. of [1892, *1*] and [1893, *2*]. (*1*)

1931 *1.* MCCONNELL, A. J.: Applications of the absolute differential calculus. London and Glasgow: Blackie & Son. xii + 318 pp. (Appendix, *A.2, A.4*)

1932 *1.* FLÜGGE, W.: Die Stabilität der Kreiszylinderschale. Ing.-Arch. **3**, 463–506. (*21, 21A*)

1933 *1.* DONNELL, L. H.: Stability of thin-walled tubes under torsion. NACA Tech. Rept. No. 479. (*21*)

1934 *1.* FLÜGGE, W.: Statik und Dynamik der Schalen. Berlin: Springer. vii + 240 pp. (*21A*)

1935 *1.* MARGUERRE, K.: Thermo-elastische Plattengleichungen. Z. Angew. Math. Mech. **15**, 369–372. (*18*)

2. TREFFTZ, E.: Ableitung der Schalenbiegungsgleichungen mit dem Castiglianoschen Prinzip. Z. Angew. Math. Mech. **15**, 101–108. (*21*)

1936 *1.* GOODIER, J. N.: The influence of circular and elliptical holes on the transverse flexure of elastic plates. Phil. Mag., Ser. VII **22**, 69–80. (*20*)

1937 *1.* ODQVIST, F.: Equations complètes de l'équilibre des couches minces élastiques gauches. Compt. Rend. **205**, 271–273. (*7*)

1938 *1.* GOODIER, J. N.: On the problems of the beam and the plate in the theory of elasticity. Trans. Roy. Soc. Can., Ser. III **32**, 65–88. (*21*)

2. MARGUERRE, K.: Zur Theorie der gekrümmten Platte großer Formänderung. Proc. Fifth Intern. Congr. Appl. Mech. (Cambridge, Mass.), pp. 93–101. (*21*)

1940 *1.* GOL'DENVEIZER, A. L.: Equations of the theory of thin shells [in Russian]. Prikl. Mat. Mekh. **4**, 35–42. (*7, 26*)

2. LUR'E, A. I.: General theory of thin elastic shells [in Russian]. Prikl. Mat. Mekh. **4**, 7–34. (*21A, 26*)

1941 *1.* REISSNER, E.: A new derivation of the equations for the deformation of elastic shells. Amer. J. Math. **63**, 177–184. (*12A, 21A*)

2. SYNGE, J. L., and W. Z. CHIEN: The intrinsic theory of elastic shells and plates. Th. von Kármán Anniv. Vol., pp. 103–120. (*12, 12A*)

1942 *1.* REISSNER, E.: Note on the expressions for the strains in a bent thin shell. Amer. J. Math. **64**, 768–772. (*21A*)

1943 *1.* NOVOZHILOV, V.: On an error in a hypothesis of the theory of shells. Compt. Rend. Acad. Sci. SSSR (Doklady) (N.S.) **38**, 160–164. (*12A*)

2. NOVOZHILOV V., and R. FINKEL'SHTEIN: On the incorrectness of Kirchhoff's hypotheses in the theory of shells [in Russian]. Prikl. Mat. Mekh. **7**, 331–340. (*12A*)

1944 *1.* BYRNE, R.: Theory of small deformations of a thin elastic shell. Seminar Reports in Mathematics. Univ. of Calif. (Los Angeles) Publ. in Math. (N.S.) **2**, 103–152. (*21A, 26*)

2. CHIEN, W. Z.: The intrinsic theory of thin shells and plates (Parts I, II and III). Quart. Appl. Math. **1**, 297–327; **2**, 43–59 and 120–135. (*12, 21A*)

3. GOL'DENVEIZER, A. L.: On the applicability of general theorems of the theory of elasticity to thin shells [in Russian]. Prikl. Mat. Mekh. **8**, 3–14. (*26*)

4. LOVE, A. E. H.: A treatise on the mathematical theory of elasticity (reprint of [1927, *1*]). New York: Dover Publications. xviii + 643 pp. (*1, 12A, 20, 21A, 26*)

5. REISSNER, E.: On the theory of bending of elastic plates. J. Math. Phys. **23**, 184–191. (*20*)

6. VLASOV, V. Z.: Basic differential equations in general theory of elastic shells [in Russian]. Prikl. Mat. Mekh. **8**, 109–140. [English Transl.: NACA TM 1241 (1951) 58 pp.] (*21A*)

1945 *1.* BROMBERG, E., and J. J. STOKER: Non-linear theory of curved elastic sheets. Quart. Appl. Math. **3**, 246–265. (*21*)

2. REISSNER, E.: The effect of transverse shear deformation on the bending of elastic plates. J. Appl. Mech. **12**, A-69–77. (*20, 24*)

3. TRUESDELL, C.: The membrane theory of shells of revolution. Trans. Am. Math. Soc. **58**, 96–166. (*12A, 21A*)

1946 *1.* DRUCKER, D. C.: Discussion of [1945, *2*]. J. Appl. Mech. **13**, A-250. (*20*)

2. GOODIER, J. N.: Discussion of [1945, *2*]. J. Appl. Mech. **13**, A-251. (*20*)

3. NOVOZHILOV, V. V.: A new method for the analysis of thin shells [in Russian]. Izv. Akad. Nauk SSR Otd. Tekhn. Nauk, No. 1. (*20, 21A*)

1947 *1.* BOLLE, L.: Contribution au problème linéaire de flexion d'une plaque élastique (Parts 1 and 2). Bull. Tech. Suisse Romande **73**, 281–285 and 293–298. (*20*)

2. EISENHART, L. P.: An introduction to differential geometry. Princeton Univ. Press. x + 304 pp. (Appendix, *A.2*)

3. GOL'DENVEIZER, A. L., and A. I. LUR'E: On the mathematical theory of the equilibrium of elastic shells [in Russian]. (Review of works published in the U.S.S.R.). Prikl. Mat. Mekh. **11**, 565–592. (*7, 21A, 26*)

4. LUR'E, A. I.: Statics of thin-walled elastic shells [transl. from 1947 Russian ed.]. AEC-tr-3798 (U.S. Atomic Energy Commission, Tech. Info. Service). iv + 210 pp. (*21A*)

5. REISSNER, E.: On bending of elastic plates. Quart. Appl. Math. **5**, 55–68. (*20*)

1948 *1.* UFLYAND, IA. S.: Wave propagation with transverse vibrations of bars and plates [in Russian]. Prikl. Mat. Mekh. **12**, 287–300. (*20*)

1949 *1.* FRIEDRICHS, K. O.: The edge effect in the bending of plates. H. Reissner Anniversary Volume, pp. 197–210. Ann Arbor, Mich.: J. W. Edwards. (*21*)

2. GREEN, A. E.: On Reissner's theory of bending of elastic plates. Quart. Appl. Math. **7**, 223–228. (*20*)

3. — The elastic equilibrium of isotropic plates and cylinders. Proc. Roy. Soc. London, Ser. A **195**, 533–552. (*20*)

4. HILDEBRAND, F. B., E. REISSNER, and G. B. THOMAS: Notes on the foundations of the theory of small displacements of orthotropic shells. NACA TN 1833. (*7, 20, 21A*)

5. SYNGE, J. L., and A. SCHILD: Tensor calculus. Toronto: Univ. of Toronto Press. xi + 324 pp. (Appendix, *A.1, A.3*)

6. ZERNA, W.: Beitrag zur allgemeinen Schalenbiegetheorie. Ing.-Arch. **17**, 149–164. (*12A*)

1950 *1.* FRIEDRICHS, K. O.: Kirchhoff's boundary conditions and the edge effect for elastic plates. Proc. Symp. Appl. Math. **3**, 117–124. (*21*)

2. GREEN, A. E., and W. ZERNA: The equilibrium of thin elastic shells. Quart. J. Mech. Appl. Math. **3**, 9–22. (*20, 21A*)

3. LUR'E, A. I.: On the equations of the general theory of elastic shells [in Russian]. Prikl. Mat. Mekh. **14**, 558–560. (*26*)

4. PARKUS, H.: Die Grundgleichungen der Schalentheorie in allgemeinen Koordinaten. Oesterr. Ing.-Arch. **4**, 160–174 and **6** (1952), 72. (*7*)

5. REISSNER, E.: On a variational theorem in elasticity. J. Math. Phys. **29**, 90–95. (*7, 20*)

6. — On axisymmetrical deformations of thin shells of revolution. Proc. Symp. Appl. Math. **3**, 27–52. (*21*)

1951 *1.* MINDLIN, R. D.: Influence of rotatory inertia and shear on flexural vibrations of isotropic, elastic plates. J. Appl. Mech. **18**, 31–38. (*20, 24*)

2. — Thickness-shear and flexural vibrations of crystal plates. J. Appl. Phys. **22**, 316–323. (*20*)

3. TIMOSHENKO, S., and J. N. GOODIER: Theory of elasticity, 2nd ed. McGraw-Hill. xviii + 506 pp. (*24*)

1952 *1.* ERINGEN, A. C.: On the non-linear vibration of circular membrane. Proc. First U.S. Nat'l. Congr. Appl. Mech. (Chicago, 1951), pp. 139–145. (*21*)

2. REISSNER, E.: Stress-strain relations in the theory of thin elastic shells. J. Math. Phys. **31**, 109–119. (*20, 21A*)

3. SCHÄFER, M.: Über eine Verfeinerung der klassischen Theorie dünner schwach gebogener Platten. Z. Angew. Math. Mech. **32**, 161–171. (*20*)

1953 *1.* MELAN, E., and H. PARKUS: Wärmespannungen infolge stationärer Temperaturfelder. Wien: Springer. v + 114 pp. (*18*)

2. NAGHDI, P. M., and J. C. ROWLEY: On the bending of axially symmetric plates on elastic foundations. Proc. First Midwestern Conf. on Solid Mechanics (Univ. of Illinois), pp. 119–123. (*20*)

3. NOVOZHILOV, V. V.: Foundations of the nonlinear theory of elasticity [transl. from the 1947 Russian ed.]. Rochester, N.Y.: Graylock Press. vi + 233 pp. (*21*)

4. REISSNER, E.: On a variational theorem for finite elastic deformations. J. Math. Phys. **32**, 129–135. (*7, 21*)

5. TRUESDELL, C.: Review of [1953, *3*]. Bull. Am. Math. Soc. **59**, 467–473. (*21, 21A*)

6. — The physical components of vectors and tensors. Z. Angew. Math. Mech. **33**, 345–356. (*A.4*)

1954 *1.* GREEN, A. E., and W. ZERNA: Theoretical elasticity. Oxford: Clarendon Press. xiii + 442 pp. (*4, 12A*)

2. TRUESDELL, C.: Remarks on the paper "The physical components of vectors and tensors." Z. Angew. Math. Mech. **34**, 69–70. (*A.4*)

1955 *1.* ERINGEN, A. C.: On the nonlinear oscillations of viscoelastic plates. J. Appl. Mech. **22**, 563–567. (*21*)

2. FUNG, Y. C., and W. H. WITRICK: A boundary layer phenomenon in the large deflexion of thin plates. Quart. J. Mech. Appl. Math. **8**, 191–210. (*21*)

3. HOFF, N. J.: The accuracy of Donnell's equations. J. Appl. Mech. **22**, 329–334. (*21*)

4. HU, H.-C.: On some variational principles in the theory of elasticity and plasticity. Scientia Sinica **4**, 33–54. (*7*)

5. MANSFIELD, E. H.: The inextensional theory for thin flat plates. Quart. J. Mech. Appl. Math. **8**, 338–352. (*21*)

6. MINDLIN, R. D.: An introduction to the mathematical theory of vibration of elastic plates. U.S. Army Signal Corps Engineering Laboratories Monograph, Ft. Monmouth, N.J. (*20*)

7. WASHIZU, K.: On the variational principles of elasticity and plasticity. Tech. Rept. 25-18, Contract N 50ri-07833, Mass. Inst. of Tech. (*7*)

1956 *1.* FREDERICK, D.: On some problems in bending of thick circular plates on an elastic foundation. J. Appl. Mech. **23**, 195–200. (*20*)

2. SOKOLNIKOFF, I. S.: Mathematical theory of elasticity, 2nd ed. McGraw-Hill. xi + 476 pp. (*26*)

1957 *1.* ALBLAS, J. B.: Theorie van De Driedimensionale Spanningstoestand in een Doorboorde Plaat. Amsterdam: H. J. Paris. viii + 127 pp. (*20*)

2. ASHWELL, D. G.: The equilibrium equations of the inextensional theory for thin flat plates. Quart. J. Mech. Appl. Math. **10**, 169–182. (*21*)

3. MINDLIN, R. D., and M. ONOE: Mathematical theory of vibrations of elastic plates. Proc. 11th Ann. Symp. on Frequency Control (U.S. Army Signal Engineering Laboratories, Fort Monmouth, N.J.), pp. 17–40. (*20*)

4. NAGHDI, P. M.: On the theory of thin elastic shells. Quart. Appl. Math. **14**, 369–380. (*20, 21A*)

1958 *1.* ERICKSEN, J. L., and C. TRUESDELL: Exact theory of stress and strain in rods and shells. Arch. Rational Mech. Anal. **1**, 295–323. (*4, 5, 9, 12A*)

2. ESSENBURG, F., and P. M. NAGHDI: On elastic plates of variable thickness. Proc. 3rd U.S. Nat'l. Congr. Appl. Mech. (Providence, R.I.), pp. 313–319. (*20, 24*)

3. KNOWLES, J. K., and E. REISSNER: Notes on stress-strain relations for thin elastic shells. J. Math. Phys. **37**, 269–282. (*21, 21A*)

4. NOLL, W.: A mathematical theory of the mechanical behavior of continuous media. Arch. Rational Mech. Anal. **2**, 197–226. (*13*)

5. VLASOV, V. Z.: Allgemeine Schalentheorie und ihre Anwendung in der Technik [transl. from the 1949 Russian ed.]. Berlin: Akademie-Verlag. xi + 661 pp. (*21, 21A*)

1959 *1.* Johnson, M. W., and E. Reissner: On the foundations of the theory of thin elastic shells. J. Math. Phys. **37**, 371–392. (*21*)

2. Morgenstern, D.: Herleitung der Plattentheorie aus der dreidimensionalen Elastizitätstheorie. Arch. Rational Mech. Anal. **4**, 411–417. (*20*)

3. Novozhilov, V. V.: The theory of thin shells [transl. from the 1951 Russian ed.]. Groningen: P. Noordhoof Ltd. vi + 376 pp. (*7, 21, 21A, 26*)

4. Rivlin, R. S.: The deformation of a membrane formed by inextensible cords. Arch. Rational Mech. Anal. **2**, 447–476. (*14*)

5. Rüdiger, D.: Zur Theorie elastischer Schalen. Ing.-Arch. **28**, 281–288. (*7*)

6. Sanders, J. L. Jr.: An improved first-approximation theory for thin shells. NASA Tech. Rept. R-24. 11 pp. (*7, 20, 26*)

7. Timoshenko, S., and W. Woinowsky-Kreiger: Theory of plates and shells, 2nd ed. New York: McGraw-Hill. xiv + 580 pp. (*20, 21*)

8. Willmore, T. J.: An introduction to differential geometry. Oxford: Clarendon Press. x + 317 pp. (Appendix, *A.2*)

1960 *1.* Bolotin, V. V.: Equation for the non-stationary temperature fields in thin shells in the presence of sources of heat. J. Appl. Math. Mech. (Transl. of PMM) **24**, 515–519. (*18*)

2. Cohen, J. W.: The inadequacy of the classical stress-strain relations for the right helicoidal shell. Proc. IUTAM Symposium on the theory of thin elastic shells (Delft, 1959), pp. 415–433. (21)

3. Ericksen, J. L.: Appendix. Tensor fields. Handbuch der Physik, Vol. III/1 (edit. by S. Flügge), pp. 794–858. Berlin-Göttingen-Heidelberg: Springer. (Appendix, *A.4*)

4. Flügge, W.: Stresses in shells. Berlin-Göttingen-Heidelberg: Springer. xi + 499 pp. (*21A*)

5. Green, A. E., and J. E. Adkins: Large elastic deformations. Oxford: Clarendon Press. vii + 348 pp. (*14, 17, 21*)

6. Koiter, W. T.: A consistent first approximation in the general theory of thin elastic shells. Proc. IUTAM Symposium on the theory of thin elastic shells (Delft, 1959), pp. 12–33. (*20, 21*)

7. Miklowitz, J.: Flexural stress waves in an infinite elastic plate due to suddenly applied concentrated transverse load. J. Appl. Mech. **27**, 681–689. (*20*)

8. Mindlin, R. D.: Waves and vibrations in isotropic, elastic plates. Proc. First Symp. on Naval Structural Mechanics (Stanford, Calif., 1958), p. 199–232. (*20*)

9. Naghdi, P. M.: A note on rigid body displacements in the theory of thin elastic shells. Quart. Appl. Math. **18**, 296–298. (*21A*)

10. — On Saint Venant's principle: elastic shells and plates. J. Appl. Mech. **27**, 417–422. (*21A, 26*)

11. Oniashvili, O. D.: Analysis of shells and other thin-walled spatial structures. Chap. 5 of Structural mechanics in the U.S.S.R., 1917–1957, pp. 223–275. Oxford: Pergamon Press. (*21*)

12. Reiss, E. L.: A theory for the small rotationally symmetric deformations of cylindrical shells. Comm. Pure Appl. Math. **13**, 531–550. (*21*)

13. Reissner, E.: On some problems in shell theory. Proc. First Symp. on Naval Structural Mechanics (Stanford, Calif., 1958), pp. 74–114. (*21*)

14. Truesdell, C., and R. Toupin: The classical field theories. Handbuch der Physik, Vol. III/1 (edit. by S. Flügge), pp. 226–793. Berlin-Göttingen-Heidelberg: Springer. (*2, 4, 8, 9, 12A, 13, 21, 26*)

15. Zerna, W.: Über eine nichtlineare allgemeine Theorie der Schalen. Proc. IUTAM symposium on the theory of thin elastic shells (Delft, 1959), pp. 34–42. (*21*)

1961 *1.* Ericksen, J. L.: Conservation laws for liquid crystals. Trans. Soc. Rheol. **5**, 23–34. (*4, 8*)

2. Friedrichs, K. O., and R. F. Dressler: A boundary-layer theory for elastic plates. Comm. Pure Appl. Math. **14**, 1–33. (*21*)

3. Gol'denveizer, A. L.: Theory of elastic thin shells [transl. from the 1953 Russian ed.]. Oxford: Pergamon Press. xxi + 658 pp. (*7, 21, 21A, 26*)

4. Günther, W.: Analoge Systeme von Schalengleichungen. Ing.-Arch. **30**, 160–186. (*6*)

5. Koiter, W. T.: A systematic simplification of the general equations in the linear theory of thin shells. Proc. Koninkl. Ned. Akad. Wetenschap., Ser. B **64**, 612–619. (*20*)

6. Leonard, R. W.: Nonlinear first approximation thin shell and membrane theory. (National Aeronautics and Space Administration, Langley Field, Va.) (*21*)

7. Lur'e, A. I.: On the static geometric analogue of shell theory. Problems of continuum mechanics (N. I. Muskhelishvili Anniv. Vol.), pp. 267–274. Phila.: S.I.A.M. (*26*)

8. MINDLIN, R. D.: High frequency vibrations of crystal plates. Quart. Appl. Math. **19**, 51–61. (*20*)
9. MORGENSTERN, D., and I. SZABÒ: Vorlesungen über theoretische Mechanik. Berlin-Göttingen-Heidelberg: Springer. (*20*)
10. MUSHTARI, KH. M., and K. Z. GALIMOV: Non-linear theory of thin elastic shells [transl. from the 1957 Russian ed.]. Israel Program for Scientific Translations. 374 pp. (*21*)
11. REISS, E. L., and S. LOCKE: On the theory of plane stress. Quart. Appl. Math. **19**, 195–203. (*21*)
12. RÜDIGER, D.: Eine geometrisch nichtlineare Schalentheorie. Z. Angew. Math. Mech. **41**, 198–207. (*21*)

1962 *1.* DUDDECK, H.: Das Randstörungsproblem der technischen Biegetheorie dünner Schalen in drei korrespondierenden Darstellungen. Oester. Ing.-Arch. **17**, 32–57. (*7*)
2. ESSENBURG, F.: On surface constraints in plate problems. J. Appl. Mech. **29**, 340–344. (*20*)
3. GOL'DENVEIZER, A. L.: Derivation of an approximate theory of bending of a plate by the method of asymptotic integration of the equations of the theory of elasticity. J. Appl. Math. Mech. (Transl. of PMM) **26**, 1000–1025. (*21*)
4. GREEN, A. E.: On the linear theory of thin elastic shells. Proc. Roy. Soc. London, Ser. A **266**, 143–160. (*21*)
5. — Boundary-layer equations in the theory of thin elastic shells. Proc. Roy. Soc. London, Ser. A **269**, 481–491. (*21*)
6. NORDGREN, R. P., and P. M. NAGHDI: Propagation of thermoelastic waves in an unlimited shallow spherical shell under heating. Proc. 4th U.S. Nat'l. Congr. Appl. Mech. (Berkeley, Calif.), pp. 311–324. (*18*)
7. REISS, E. L.: On the theory of cylindrical shells. Quart. J. Mech. Appl. Math. **15**, 324–338. (*21*)
8. ZERNA, W.: Mathematisch strenge Theorie elastischer Schalen. Z. Angew. Math. Mech. **42**, 333–341. (*7*)

1963 *1.* BUDIANSKY, B., and J. L. SANDERS, JR.: On the "best" first-order linear shell theory. Progress in applied mechanics (The Prager Anniv. Vol.), pp. 129–140. (*20*)
2. COLEMAN, B. D., and W. NOLL: The thermodynamics of elastic materials with heat conduction and viscosity. Arch. Rational Mech. Anal. **13**, 167–178. (*13*)
3. GOL'DENVEIZER, A. L.: Derivation of an approximate theory of shells by means of asymptotic integration of the equations of the theory of elasticity. J. Appl. Math. Mech. (Transl. of PMM) **27**, 903–924. (*21*)
4. JOHNSON, M. W.: A boundary layer theory of unsymmetric deformations of circular cylindrical elastic shells. J. Math. Phys. **42**, 167–188. (*21*)
5. MINDLIN, R. D.: High frequency vibrations of plated, crystal plates. Progress in applied mechanics (The Prager Anniv. Vol.), pp. 73–84. (*20*)
6. NAGHDI, P. M.: Foundations of elastic shell theory. Progress in solid mech., Vol. 4 (edit. by I. N. SNEDDON and R. HILL), pp. 1–90. Amsterdam: North-Holland Publ. Co. = Tech. Rept. No. 15, Contract Nonr-222 (69), Univ. of Calif., Berkeley (Jan. 1962). (*7, 12, 12A, 20, 21A, 26, A.3*)
7. — A new derivation of the general equations of elastic shells. Int. J. Engng. Sci. **1**, 509–522. (*6, 7, 15, 20, 26*)
8. —, and R. P. NORDGREN: On the nonlinear theory of elastic shells under the Kirchhoff hypothesis. Quart. Appl. Math. **21**, 49–59. (*15, 21*)
9. REISSNER, E.: On the derivation of boundary conditions for plate theory. Proc. Roy. Soc. London, Ser. A **276**, 178–186. (*21*)
10. — On the derivation of the theory of thin elastic shells. J. Math. Phys. **42**, 263–277. (*21*)
11. — On the equations for finite symmetrical deflections of thin shells of revolution. Progress in applied mechanics (The Prager Anniv. Vol.), pp. 171–178. (*21*)
12. SANDERS, J. L., JR.: Nonlinear theories for thin shells. Quart. Appl. Math. **21**, 21–36. (*21*)
13. SERBIN, H.: Quadratic invariants of surface deformations and the strain energy of thin elastic shells. J. Math. Phys. **4**, 838–851. (*10*)
14. STOKER, J. J.: Elastic deformations of thin cylindrical sheets. Progress in applied mechanics (The Prager Anniv. Vol.), pp. 179–188. (*21*)
15. TIFFEN, R., and P. G. LOWE: An exact theory of generally loaded elastic plates in terms of moments of the fundamental equations. Proc. London Math. Soc. **13**, 653–671. (*12*)
16. WAINWRIGHT, W. L.: On a nonlinear theory of elastic shells. Int. J. Engng. Sci. **1**, 339–358. (*21*)

1964 *1.* AMBARTSUMIAN, S. A.: Theory of anisotropic shells [transl. from a 1961 Russian ed.]. NASA Technical Translation TTF-118. vi + 395 pp. (*21*)
2. FOX, N.: On asymptotic expansions in plate theory. Proc. Roy. Soc. London, Ser. A **278**, 228–233. (*21*)
3. GREEN, A. E., and R. S. RIVLIN: On Cauchy's equations of motion. Z. Angew. Math. Phys. **15**, 290–292. (*8*)
4. NAGHDI, P. M.: On the nonlinear thermoelastic theory of shells. Proc. IASS Symposium on non-classical shell problems (Warsaw, 1963), pp. 5–26. (*11*)
5. — Further results in the derivation of the general equations of elastic shells. Int. J. Engng. Sci. **2**, 269–273. (*20, 26*)
6. — On a variational theorem in elasticity and its application to shell theory. J. Appl. Mech. **31**, 647–653. (*7, 20*)
7. REISSNER, E.: On the form of variationally derived shell equations. J. Appl. Mech. *31*, 233–238. (*7, 20*)
8. TOUPIN, R. A.: Theories of elasticity with couple-stress. Arch. Rational Mech. Anal. **17**, 85–112. (*4*)
1965 *1.* GOL'DENVEIZER, A., and A. V. KOLOS: On the derivation of two-dimensional equations in the theory of thin elastic plates. J. Appl. Math. Mech. (Transl. of PMM) **29**, 151–166. (*21*)
2. GREEN, A. E., and P. M. NAGHDI: Some remarks on the linear theory of shells. Quart. J. Mech. Appl. Math. **18**, 257–276 and **20**, 527 (1967). (*21*)
3. — —, and R. S. RIVLIN: Directors and multipolar displacements in continuum mechanics. Int. J. Engng. Sci. **2**, 611–620. (*4, 8*)
4. — —, and W. L. WAINWRIGHT: A general theory of a Cosserat surface. Arch. Rational Mech. Anal. **20**, 287–308. (*4, 5, 6, 8, 9, 13, 16*)
5. JOHN, F.: Estimates for the derivatives of the stresses in a thin shell and interior shell equations. Comm. Pure Appl. Math. **18**, 235–267. (*4, 20*)
6. NAGHDI, P. M.: A static-geometric analogue in the theory of couple-stress. Proc. Koninkl. Ned. Akad. Wetenschap., Ser. B **68**, 29–32. (26)
7. REISSNER, E.: A note on variational principles in elasticity. Int. J. Solids Structures **1**, 93–95 and 351. (*7*)
8. TIFFEN, R., and P. G. LOWE: An exact theory of plane stress. J. London Math. Soc. **40**, 72–86. (*12*)
9. TRUESDELL, C., and W. NOLL: The non-linear field theories of mechanics. In Handbuch der Physik, Vol. III/3 (edit. by S. FLÜGGE). Berlin-Heidelberg-New York: Springer viii + 602 pp. (*2, 4, 8, 11,* Chap. D—preliminary remarks, *13, 14, 17*)
1966 *1.* COHEN, H., and C. N. DE SILVA: Nonlinear theory of elastic surfaces. J. Math. Phys. **7**, 246–253. (*10*)
2. — — Theory of directed surfaces. J. Math. Phys. **7**, 960–966. (*4, 5, 9, 13*)
3. GREEN, A. E., and N. LAWS: Further remarks on the boundary conditions for thin elastic shells. Proc. Roy. Soc. London, Ser. A **289**, 171–176. (*21*)
4. KOITER, W. T.: On the nonlinear theory of thin elastic shells. Proc. Koninkl. Ned. Akad. Wetenschap., Ser. B **69**, 1–54. (*21*)
5. KOLLMAN, F. G.: Eine Ableitung der Kompatibilitätsbedingungen in der linearen Schalentheorie mit Hilfe des Riemannschen Krümmungstensors. Ing.-Arch. **35**, 10–16. (*7*)
6. LAWS, N.: A boundary-layer theory for plates with initial stress. Proc. Cambridge Phil. Soc. **62**, 313–327. (*21*)
7. NAGHDI, P. M.: On the differential equations of the linear theory of elastic shells. Proc. Eleventh Int. Congr. Appl. Mech. (Munich, 1964), pp. 262–269. (*20, 26*)
8. REISSNER, E.: On the foundations of the linear theory of elastic shells. Proc. Eleventh Int. Congr. Appl. Mech. (Munich, 1964), pp. 20–30. (*21A*)
1967 *1.* BALABAN, M. M., A. E. GREEN, and P. M. NAGHDI: Simple force multipoles in the theory of deformable surfaces. J. Math. Phys. **8**, 1026–1036. (*10, 15*)
2. CROCHET, M. J.: Compatibility equations for a Cosserat surface. J. Mécanique **6**, 593–600 and **9**, 600 (1970). (*6*)
3. ERINGEN, A. C.: Theory of micropolar plates. Z. Angew. Math. Phys. **18**, 12–30. (*21*)
4. GREEN, A. E., and P. M. NAGHDI: The linear theory of an elastic Cosserat plate. Proc. Cambridge Phil. Soc. **63**, 537–550 and 922. (*6, 7, 9, 13, 16, 24*)
5. — — Micropolar and director theories of plates. Quart. J. Mech. Appl. Math. **20**, 183–199. (*21*)
6. NAGHDI, P. M.: A theory of deformable surface and elastic shell theory. Proc. Symposium on the theory of shells to honor L. H. DONNELL (Houston, 1966), pp. 25–43. (*9, 24*)
1968 *1.* BUDIANSKY, B.: Notes on nonlinear shell theory. J. Appl. Mech. **35**, 393–401. (*21*)

2. COHEN, H., and C. N. DE SILVA: On a nonlinear theory of elastic shells. J. Mécanique **7**, 459–464. (*10*)
3. COSSERAT, E., and F.: Theory of deformable bodies. Translation of [1909, *1*]. NASA TT F-11, 561 (Washington, D.C.). Clearinghouse for U.S. Federal Scientific and Technical Information, Springfield, Virginia. (*4*)
4. GREEN, A. E., N. LAWS, and P. M. NAGHDI: Rods, plates and shells. Proc. Cambridge Phil. Soc. **64**, 895–913. (*4, 7, 11, 12, 18*)
5. —, and P. M. NAGHDI: A note on the Cosserat surface. Quart. J. Mech. Appl. Math. **21**, 135–139. (*15, 25*)
6. — — The linear elastic Cosserat surface and shell theory. Int. J. Solids and Structures **4**, 585–592. (*6, 9, 13, 16, 24, 25*)
7. — — The Cosserat surface. Proc. IUTAM Symposium on mechanics of generalized continua (Freudenstadt and Stuttgart, 1967), pp. 36–48. (*9, 15*)
8. — —, and R. B. OSBORN: Theory of an elastic-plastic Cosserat surface. Int. J. Solids and Structures **4**, 907–927. (Chap. D—preliminary remarks)
9. —, and W. ZERNA: 2nd ed. of [1954, *1*]. ix + 457 pp. (*4, 7, 11, 12A, 20, 21, 22, 24,* Appendix)
10. KNOPS, R. J., and L. E. PAYNE: Uniqueness in classical elasto-dynamics. Arch. Rational Mech. Anal. **27**, 349–355. (*26*)
11. SENSENIG, C. B.: A shell theory compared with the exact three dimensional theory of elasticity. Int. J. Engng. Sci. **6**, 435–464. (*20*)
12. STOKER, J. J.: Nonlinear elasticity. New York: Gordon and Breach. xi + 130 pp. (*21*)
13. WENNER, M. L.: On torsion of an elastic cylindrical Cosserat surface. Int. J. Solids and Structures **4**, 769–776. (*24*)
14. ZERNA, W.: A new formulation of the theory of elastic shells. IASS Bull. **37**, 61–76. (*7*)

1969 *1.* CROCHET, M. J., and P. M. NAGHDI: Large deformation solutions for an elastic Cosserat surface. Int. J. Engng. Sci. **7**, 309–335. (*14*)
2. GOL'DENVEIZER, A.: Boundary layer and its interaction with the interior state of stress of an elastic thin shell. J. Appl. Math. Mech. (Transl. of PMM) **33**, 971–1001 and **34**, 548, (1970). (*21*)
3. GREEN, A. E., and P. M. NAGHDI: Shells in the light of generalized continua. Proc. Second IUTAM Symposium on the theory of thin shells (Copenhagen, 1967), pp. 39–58. (*6, 16, 24*)
4. JOHN, F.: Refined interior shell equations. Proc. Second IUTAM Symposium on the theory of thin shells (Copenhagen, 1967), pp. 1–14. (*4, 20*)
5. TIERSTEN, H. F.: Linear piezoelectric plate vibrations. New York: Plenum Press. xv + 212 pp. (*20*)

1970 *1.* ERICKSEN, J. L.: Uniformity in shells. Arch. Rational Mech. Anal. **37**, 73–84. (*14*)
2. GREEN, A. E., and P. M. NAGHDI: Non-isothermal theory of rods, plates and shells. Int. J. Solids and Structures **6**, 209–244 and **7**, 127 (1971). (*7, 11, 12, 17, 18, 24*)
3. — —, and J. A. TRAPP: Thermodynamics of a continuum with internal constraints. Int. J. Engng. Sci. **8**, 891–908. (*14*)
4. KOITER, W. T.: On the foundations of the linear theory of thin elastic shells, I and II. Proc. Koninkl. Ned. Akad. Wetenschap., Ser. B **73**, 169–195. (*20*)

1971 *1.* ANTMAN, S. S.: Existence and nonuniqueness of axisymmetric equilibrium states of nonlinearly elastic shells. Arch. Rational. Mech. Anal. **40**, 329–372. (*14*)
2. BIRICIKOGLU, V., and A. KALNINS: Large elastic deformations of shells with the inclusion of transverse normalstra in. Int. J. Solids and Structures **7**, 431–444. (*21*)
3. CROCHET, M. J.: Finite deformations of inextensible Cosserat surface. Int. J. Solids and Structures **7**, 383–397. (*14*)
4. GREEN, A. E., and P. M. NAGHDI: On uniqueness in the linear theory of elastic shells and plates. J. Mécanique **10**, 251–261. (*26*)
5. — — On superposed small deformations on a large deformation of an elastic Cosserat surface. J. Elasticity **1**, 1–17. (*14*)
6. — —, and M. L. WENNER: Linear theory of Cosserat surface and elastic plates of variable thickness. Proc. Cambridge Phil. Soc. **69**, 227–254. (*6, 9, 11, 12, 19, 20, 24*)
7. NORDGREN, R. P.: A bound on the error in plate theory. Quart. Appl Math. **28**, 587–595. (*20*)
8. SIMMONDS, J. G.: An improved estimate for the error in the classical, linear theory of plate bending. Quart. Appl. Math. **29**, 439–447. (*20*)
9. STEELE, C. R.: A geometric optics solution for the thin shell equation. Int. J. Engng. Sci. **9**, 681–704. (*7, 26*)

The Theory of Rods.

By

STUART S. ANTMAN.

With 5 Figures.

A. Introduction.[1]

1. Definition and purpose of rod theories. Nature of this article. A *theory of rods*[2] or, equivalently, a *one-dimensional theory of solids* is a characterization of the behavior of slender three-dimensional solid bodies by a set of equations having the parameter of a certain curve and the time as the only independent variables. The formidable mathematical obstacles presented by three-dimensional continuum theories dictate the need for tractable, accurate, and illuminating one-dimensional models.

We regard the study of rods as a mathematical science that must meet high standards of rigor, logic, and precision. We concentrate on problems that help clarify the structure of nonlinear continuum mechanics, employing linear theories only for illustration. We do not treat technical theories of rods, which are developed in a voluminous literature. Although many technical theories have achieved marked success in applications to structural mechanics, the special nature of the hypotheses underlying these theories tends to obscure the features common to all one-dimensional continua. Our work however will indicate how linear technical theories can be developed without the inconsistencies that arise from time to time in elementary "derivations" and how such theories can be assigned a position within a hierarchy of rationally constructed rod theories. We cannot do the same for a great number of nonlinear technical theories that are based upon ad hoc assumptions leading to the neglect of certain kinematic terms regarded as small. These theories enjoy neither the accuracy and the generality of the complete nonlinear theory nor the analytic simplicity of the linear theory. As often as not, the equations of such approximate nonlinear theories are no easier to study than the equations of the complete nonlinear theory itself.[3]

[1] *Acknowledgment.* This article was begun while the author, on leave of absence from New York University, was a Senior Visiting Fellow of the Science Research Council of Great Britain at the University of Oxford. It was completed at New York University under grant GP-27209 from the National Science Foundation. The author is grateful to these organizations for their support.

[2] We use "rod" as a generic name for "arch", "bar", "beam", "column", "ring", "shaft", etc. We employ "rod" in both the intuitive sense of a slender body and in several precise mathematical senses. The meaning will be clear from the context.

[3] Such approximate nonlinear theories are developed by a procedure analogous to replacing the nonlinear eigenvalue problem $y'' + \lambda \sin y = 0$, $y(0) = y(1) = 0$ by the approximate problem $y'' + \lambda(y - y^3/6) = 0$, $y(0) = y(1) = 0$. The study of the latter is no easier than that of the former, but the behavior of solutions for large y is markedly different in each case. The behavior near the trivial solution $y \equiv 0$ is nearly the same, so no advantage is gained here. Indeed, a perturbation analysis of the first problem (which can be justified mathematically in this case) leads to a sequence of linear problems whose solutions provide a more accurate picture of the local behavior of the first problem than does an exact solution of the second problem.

This article consists of two main parts. In Chap. B, we examine several rational developments of rod theories and the corresponding diverse interpretations of the role and scope of such theories. We are able to bring a certain unity to these various approaches by showing that *each of several distinct constructions yields governing equations of exactly the same form, the variables entering these equations enjoying the greatest latitude of physical interpretation.* We can thereby resolve some of the controversies that have beset our subject. In Chap. C we treat a variety of problems in the theory of nonlinearly elastic rods. The different topics in this chapter are substantially independent. A detailed study of Chap. B is unnecessary for the understanding of Chap. C.

2. Notation. Our notational scheme is in the spirit of that described by TRUESDELL and NOLL in Sect. 6 of *The Non-linear Field Theories of Mechanics* in vol. III/3 of this Encyclopedia. Lower case Latin indices have the range 1, 2, 3 and lower case Greek indices have the range 1, 2. These indices obey the summation convention. German indices are enumerative. Let $\boldsymbol{x}=(x^1, x^2, x^3)$ be a system of curvilinear coordinates for (a region of) the Euclidean 3-space $\mathscr{E}^3$ and let $\boldsymbol{r}(\boldsymbol{x})$ denote the position vector to the point with coordinates $\boldsymbol{x}$. We define the base vectors $\boldsymbol{g}_p$, their duals $\boldsymbol{g}^p$, and the components g_{pq} and g^{pq} of the metric tensor $\mathbf{1}$ by

$$\boldsymbol{g}_p=\frac{\partial \boldsymbol{r}}{\partial x^p}, \quad \boldsymbol{g}^p\cdot\boldsymbol{g}_q=\delta^p_q, \quad g_{pq}=\boldsymbol{g}_p\cdot\boldsymbol{g}_q, \quad g^{pq}=\boldsymbol{g}^p\cdot\boldsymbol{g}^q, \tag{2.1}$$

where the dot denotes the inner product and δ^p_q is the Kronecker delta. The cross product of two base vectors is given by

$$\boldsymbol{g}^p\times\boldsymbol{g}^p=\sqrt{g}\, e_{pqs}\,\boldsymbol{g}^s, \tag{2.2}$$

where e_{pqs} is the permutation symbol and

$$g=\det(g_{pq}), \quad \sqrt{g}=(\boldsymbol{g}_1\times\boldsymbol{g}_2)\cdot\boldsymbol{g}_3. \tag{2.3}$$

We denote vectors in $\mathscr{E}^3$ by lower case bold-face symbols. Any vector $\boldsymbol{v}$ can be written as a linear combination of the base vectors:

$$\boldsymbol{v}=v^p\,\boldsymbol{g}_p=v_p\,\boldsymbol{g}^p, \tag{2.4}$$

with $v^p=\boldsymbol{v}\cdot\boldsymbol{g}^p$ and $v_p=\boldsymbol{v}\cdot\boldsymbol{g}_p$ being the contravariant and covariant components of $\boldsymbol{v}$, respectively. Second order tensors in $\mathscr{E}^3$ are denoted by upper case bold-face symbols. A second order tensor $\boldsymbol{A}$ with contravariant components A^{pq} is a linear transformation that assigns to each vector $\boldsymbol{u}$ the vector

$$\boldsymbol{v}=\boldsymbol{A}\boldsymbol{u} \tag{2.5}$$

with contravariant components

$$v^p=A^{pq}\,u_q. \tag{2.6}$$

The transpose of $\boldsymbol{A}$ is denoted $\boldsymbol{A}^T$.

The product $\boldsymbol{A}\boldsymbol{B}$ of the second order tensors $\boldsymbol{A}$ and $\boldsymbol{B}$ is the second order tensor with contravariant components $A^p{}_s\,B^{sq}$. The dyadic product $\boldsymbol{u}\otimes\boldsymbol{v}$ of two vectors $\boldsymbol{u}$ and $\boldsymbol{v}$ is the second order tensor with contravariant components $u^p\,v^q$. If ϕ is a differentiable scalar function of the vector $\boldsymbol{v}$, then $\partial\phi/\partial\boldsymbol{v}$ denotes the vector with covariant components $\partial\phi/\partial v^p$ and if ψ is a differentiable scalar function of the tensor $\boldsymbol{A}$, then $\partial\psi/\partial\boldsymbol{A}$ is the tensor with covariant components $\partial\psi/\partial A^{pq}$. Generalizations are immediate.

We denote sets by upper case script symbols. We employ the standard notation of set theory. In particular, the boundary of a set $\mathscr{A}$ is denoted $\partial\mathscr{A}$. If f maps a set $\mathscr{A}$ into a set $\mathscr{B}$, we define the range of $\mathscr{A}$ under f to be

$$f(\mathscr{A})=\{b\in\mathscr{B}: b=f(a),\ a\in\mathscr{A}\}. \tag{2.7}$$

When necessary we carefully distinguish between a function and its values, occasionally denoting the function with values $f(a)$ by $f(\cdot)$.

In Chap. B, where we formulate the equations of rod theories, we make the usual assumptions that variables whose derivatives are exhibited actually possess such derivatives and that the boundaries of regions are regular enough for the integral theorems that are used to be valid. In Chap. C, where we discuss solutions of the equations, we precisely delimit the notions of smoothness used.

3. Background. The reader of this article is presumed familiar with the fundamental concepts of three-dimensional continuum physics, detailed expositions of which may be found in the treatises *The Classical Field Theories* (hereafter abbreviated "CFT") by TRUESDELL and TOUPIN (1960) in vol. III/1 of this encyclopedia and *The Non-linear Field Theories of Mechanics* (hereafter abbreviated "NFTM") by TRUESDELL and NOLL (1965) in vol. III/3 of this encyclopedia. In this section we summarize the principal relations to be used in Chap. B.

Let a body $\mathscr{B}$ (NFTM, Sect. 15) consist of particles X. The smooth invertible mapping $\hat{\boldsymbol{r}}$ associates with each particle X its position $\boldsymbol{r}=\hat{\boldsymbol{r}}(X)$ in the body's reference configuration in $\mathscr{E}^3$. $\hat{\boldsymbol{r}}(\mathscr{B})$ accordingly denotes the region of $\mathscr{E}^3$ occupied by $\mathscr{B}$ in its reference configuration. The position of X in the configuration occupied by $\mathscr{B}$ at time t is denoted $\boldsymbol{r}=\hat{\boldsymbol{r}}(X, t)$. Hence the region of $\mathscr{E}^3$ occupied by $\mathscr{B}$ at time t is $\hat{\boldsymbol{r}}(\mathscr{B}, t)$. $\hat{\boldsymbol{r}}(\mathscr{B})$ is assigned a system of curvilinear material coordinates $\boldsymbol{X}=(X^1, X^2, X^3)$ by the smooth invertible mapping $\hat{\boldsymbol{X}}$ that associates with each particle $X\in\mathscr{B}$ the triad $\boldsymbol{X}=\hat{\boldsymbol{X}}(X)$ in the region $\hat{\boldsymbol{X}}(\mathscr{B})$ of the real number space $\mathscr{R}^3$. The smooth invertible mapping of $\hat{\boldsymbol{X}}(\mathscr{B})$ onto $\hat{\boldsymbol{r}}(\mathscr{B})$ given by $\boldsymbol{r}=\hat{\boldsymbol{r}}(\hat{\boldsymbol{X}}^{-1}(\boldsymbol{X}))$ is denoted $\boldsymbol{r}=\boldsymbol{r}(\boldsymbol{X})$ and the smooth invertible mapping of $\hat{\boldsymbol{X}}(\mathscr{B})$ onto $\hat{\boldsymbol{r}}(\mathscr{B}, t)$ given by $\boldsymbol{r}=\hat{\boldsymbol{r}}(\hat{\boldsymbol{X}}^{-1}(\boldsymbol{X}), t)$ is denoted $\boldsymbol{r}=\boldsymbol{r}(\boldsymbol{X}, t)$.

If f is a function of $\mathbf{A}, \boldsymbol{y}, s$ with values $f(\mathbf{A}, \boldsymbol{y}, s)$, then we define

$$\begin{gathered}
f_{,p}\big(\mathbf{B}(\boldsymbol{X},t),\boldsymbol{X},t\big)=\frac{\partial f}{\partial \mathbf{A}}\big(\mathbf{B}(\boldsymbol{X},t),\boldsymbol{X},t\big)\frac{\partial \mathbf{B}}{\partial X^p}(\boldsymbol{X},t)+\frac{\partial f}{\partial y^p}\big(\mathbf{B}(\boldsymbol{X},t),\boldsymbol{X},t\big),\\
\dot{f}\big(\mathbf{B}(\boldsymbol{X},t),\boldsymbol{X},t\big)=\frac{\partial f}{\partial \mathbf{A}}\big(\mathbf{B}(\boldsymbol{X},t)\boldsymbol{X},t\big)\frac{\partial \mathbf{B}}{\partial t}(\boldsymbol{X},t)+\frac{\partial f}{\partial s}\big(\mathbf{B}(\boldsymbol{X},t)\boldsymbol{X},t\big),\\
\frac{\partial f}{\partial X^p}\big(\mathbf{B}(\boldsymbol{X},t),\boldsymbol{X},t\big)=\frac{\partial f}{\partial y^p}\big(\mathbf{B}(\boldsymbol{X},t),\boldsymbol{X},t\big),\quad \frac{\partial f}{\partial t}\big(\mathbf{B}(\boldsymbol{X},t)\,\boldsymbol{X},t\big)=\frac{\partial f}{\partial s}\big(\mathbf{B}(\boldsymbol{X},t),\boldsymbol{X},t\big)\cdot
\end{gathered}\tag{3.1}$$

Thus $f_{,p}$ and $\dot{f}$ represent "total partial derivatives". ($\dot{f}$ is the material time derivative of f.) If f depends only upon $\boldsymbol{X}, t$, *the history of* f *up to time* t is the function $f^{(t)}$ defined by

$$f^{(t)}(\boldsymbol{X}, s)=f(\boldsymbol{X}, t-s).\tag{3.2}$$

The position, velocity, and acceleration of the particle with material coordinates $\boldsymbol{X}$ at time t are respectively denoted

$$\boldsymbol{r}=\boldsymbol{r}(\boldsymbol{X},t),\qquad \dot{\boldsymbol{r}}=\frac{\partial \boldsymbol{r}}{\partial t}(\boldsymbol{X},t),\qquad \ddot{\boldsymbol{r}}=\frac{\partial^2 \boldsymbol{r}}{\partial t^2}(\boldsymbol{X},t).\tag{3.3}$$

The right Cauchy-Green deformation tensor $\boldsymbol{C}$ is defined by

$$C_{pq}=\boldsymbol{r}_{,p}\cdot\boldsymbol{r}_{,q}.\tag{3.4}$$

Since the governing field equations for solids are naturally formulated in material coordinates, there is no restriction and there is considerable convenience in choosing the set of spatial coordinate surfaces $x^p=$ const to be the material surfaces whose particles form the surfaces $\hat{X}^p(X)=$ const in the reference configuration. The coordinates X^p are termed *convected* material coordinates.[4] These coordinates are used throughout this article. We can specialize expressions involving an arbitrary system of spatial coordinates $\boldsymbol{x}=(x^1, x^2, x^3)$ to convected material coordinates by formally identifying $x^p=X^p$. Thus the base vectors and the covariant components of the metric tensor, defined in (2.1), are given for each configuration of the body by

$$\boldsymbol{g}_p=\boldsymbol{r}_{,p},\qquad g_{pq}=\boldsymbol{r}_{,p}\cdot\boldsymbol{r}_{,q}.\tag{3.5}$$

Let G denote the value of g in the reference configuration. If $\varrho_0(\boldsymbol{X})>0$ is the given density at the particle $X=\hat{\boldsymbol{X}}^{-1}(\boldsymbol{X})$ in the reference configuration, then the density $\varrho(\boldsymbol{X}, t)$ at the same particle in the configuration occupied by the body at time t is *defined* by the continuity equation

$$\varrho\sqrt{g}=\varrho_0\sqrt{G}.\tag{3.6}$$

The *continuity condition*, expressing the impossibility of compressing a positive volume of material into a zero volume, can be written as

$$g/G>0.\tag{3.7}$$

[4] These coordinates are discussed in CFT, Sect. 66b; NFTM, Sect. 15; and in the references cited therein. They are particularly suitable for rods, being inherent in the concepts of reference curve and cross-section.

The somewhat stronger requirement that no fiber of positive length can be compressed to zero length is equivalent to the positive-definiteness of $\boldsymbol{C}$:

$$C_{pq}\, v^p\, v^q > 0 \tag{3.8}$$

for all $\boldsymbol{v} \neq \boldsymbol{0}$. We term this the *strong continuity condition.*

We denote the stress vector acting across the material surface $X^p = \text{const}$ by

$$\boldsymbol{t}^p = t^{pq}\, \boldsymbol{g}_q, \tag{3.9}$$

where t^{pq} are the components of the (Cauchy) stress tensor. We introduce convected representations of the Piola-Kirchhoff stress vector $G^{-\frac{1}{2}}\boldsymbol{\tau}^p$ and stress tensor $G^{-\frac{1}{2}}\boldsymbol{T}$ (CFT, Sect. 210; NFTM, Sect. 43A) by

$$\boldsymbol{\tau}^p = \sqrt{g}\,\boldsymbol{t}^p, \qquad T^{pq} = \sqrt{g}\, t^{pq}, \qquad \boldsymbol{\tau}^p = T^{pq}\,\boldsymbol{g}_p, \qquad \boldsymbol{\tau}^p \otimes \boldsymbol{g}_p = \boldsymbol{T}. \tag{3.10}$$

Let $\boldsymbol{\nu}$ be the unit normal to a material surface in its reference configuration. We set

$$\boldsymbol{\tau}_{(\boldsymbol{\nu})} = \boldsymbol{\tau}^p\, \nu_p = T^{pq}\, \nu_p\, \boldsymbol{g}_q. \tag{3.11}$$

The equations of motion of an arbitrary nonpolar medium may be written as

$$\boldsymbol{\tau}^p_{,p} + \varrho_0 \sqrt{G}\boldsymbol{f} = \varrho_0\sqrt{G}\,\ddot{\boldsymbol{r}}, \tag{3.12}$$

$$\boldsymbol{g}_p \times \boldsymbol{\tau}^p = \boldsymbol{0}, \quad \text{or equivalently,} \quad \boldsymbol{T} = \boldsymbol{T}^T, \tag{3.13}$$

where $\boldsymbol{f}$ is the body force per unit mass.

Let e^λ denote the temperature and let ε, $e^{-\lambda}\zeta$, ψ respectively denote the densities per unit mass of the internal energy, entropy, and free energy. We call λ the *logarithmic temperature* and ζ the *temperature entropy function*. These variables are related by

$$\varepsilon = \psi + \zeta. \tag{3.14}$$

Our representation for the temperature ensures its positivity whenever $\lambda > -\infty$. The use of the variables λ and ζ simplifies a number of formulae that follow.

Let $G^{-\frac{1}{2}}\boldsymbol{q}$ denote the heat flux vector measured per unit *reference* area of a material surface through which the heat passes. (In convected coordinates, the usual heat flux vector per unit actual surface area is given by $\boldsymbol{q}/\sqrt{g}$.) $\boldsymbol{q}\cdot\boldsymbol{\nu}$ is positive when heat flows in the $\boldsymbol{\nu}$-direction. Let h denote the rate of heat supply per unit mass. The integral equation of energy balance is

$$\begin{aligned}\frac{d}{dt}\int\limits_{\hat{\boldsymbol{X}}(\mathscr{B})}\left(\varepsilon + \frac{1}{2}\dot{\boldsymbol{r}}\cdot\dot{\boldsymbol{r}}\right)\varrho_0\sqrt{G}\,dX^1\,dX^2\,dX^3 &= \int\limits_{\hat{\boldsymbol{X}}(\mathscr{B})}(\boldsymbol{f}\cdot\dot{\boldsymbol{r}} + h)\,\varrho_0\sqrt{G}\,dX^1\,dX^2\,dX^3 \\ &\quad + \int\limits_{\partial\hat{\boldsymbol{r}}(\mathscr{B})} G^{-\frac{1}{2}}(\boldsymbol{\tau}_{(\boldsymbol{\nu})}\cdot\dot{\boldsymbol{r}} - \boldsymbol{q}\cdot\boldsymbol{\nu})\,dA,\end{aligned} \tag{3.15}$$

where dA denotes the differential surface area of $\partial\hat{\boldsymbol{r}}(\mathscr{B})$. The use of (3.12) reduces (3.15) to the local form[5]

$$\varrho_0\sqrt{G}\,\dot{\varepsilon} \equiv \varrho_0\sqrt{G}\,(\dot{\psi} + \dot{\zeta}) = \boldsymbol{\tau}^p\cdot\dot{\boldsymbol{r}}_{,p} - q^p_{,p} + \varrho_0\sqrt{G}\,h. \tag{3.16}$$

With

$$\boldsymbol{\gamma} = \operatorname{grad}\lambda, \qquad (\gamma_p = \lambda_{,p}) \tag{3.17}$$

the Clausius-Duhem entropy inequality[6] can be written as

$$\varrho_0\sqrt{G}\,(\dot{\zeta} - \zeta\dot{\lambda}) + q^p_{,p} - \boldsymbol{q}\cdot\boldsymbol{\gamma} - \varrho_0\sqrt{G}\,h \geqq 0. \tag{3.18}$$

The substitution of (3.16) into (3.18) reduces the latter to

$$-\varrho_0\sqrt{G}\,(\dot{\psi} + \zeta\dot{\lambda}) + \boldsymbol{\tau}^p\cdot\dot{\boldsymbol{r}}_{,p} - \boldsymbol{q}\cdot\boldsymbol{\gamma} \geqq 0. \tag{3.19}$$

[5] In our convected coordinates, $2\boldsymbol{\tau}^p\cdot\dot{\boldsymbol{r}}_{,p} = T^{pq}\,\dot{C}_{pq}$.

[6] There remains some controversy over the range of validity of this relation. Its replacement by some other thermodynamic inequality would not affect our general methods for constructing rod theories, since the role of such an inequality as a restriction on the constitutive relations would be unchanged.

We restrict our considerations to simple solids (NFTM, Sects. 28, 29, 33) which have constitutive relations of the form

$$\begin{aligned}
\boldsymbol{\tau}^p(\boldsymbol{X}, t) &= \check{\boldsymbol{\tau}}^p\big(\boldsymbol{r}_{,q}^{(t)}(\boldsymbol{X}, \cdot), \lambda^{(t)}(\boldsymbol{X}, \cdot), \gamma^{(t)}(\boldsymbol{X}, \cdot), \boldsymbol{X}\big),\\
\boldsymbol{q}(\boldsymbol{X}, t) &= \check{\boldsymbol{q}}\big(\boldsymbol{r}_{,q}^{(t)}(\boldsymbol{X}, \cdot), \lambda^{(t)}(\boldsymbol{X}, \cdot), \gamma^{(t)}(\boldsymbol{X}, \cdot), \boldsymbol{X}\big),\\
\zeta(\boldsymbol{X}, t) &= \check{\zeta}\big(\boldsymbol{r}_{,q}^{(t)}(\boldsymbol{X}, \cdot), \lambda^{(t)}(\boldsymbol{X}, \cdot), \gamma^{(t)}(\boldsymbol{X}, \cdot), \boldsymbol{X}\big),\\
\psi(\boldsymbol{X}, t) &= \check{\psi}\big(\boldsymbol{r}_{,q}^{(t)}(\boldsymbol{X}, \cdot), \lambda^{(t)}(\boldsymbol{X}, \cdot), \gamma^{(t)}(\boldsymbol{X}, \cdot), \boldsymbol{X}\big).
\end{aligned} \tag{3.20}$$

Here $\check{\boldsymbol{\tau}}^p, \check{\boldsymbol{q}}, \check{\zeta}, \check{\psi}$ are suitably valued functionals of the indicated histories[7] and of the real variables $\boldsymbol{X}$. The application of the principle of material frame indifference to (3.20) yields the reduced constitutive relations (NFTM, Sects. 29, 96 bis) of the form

$$\begin{aligned}
\boldsymbol{T}(\boldsymbol{X}, t) &= \hat{\boldsymbol{T}}\big(\boldsymbol{C}^{(t)}(\boldsymbol{X}, \cdot), \lambda^{(t)}(\boldsymbol{X}, \cdot), \gamma^{(t)}(\boldsymbol{X}, \cdot), \boldsymbol{X}\big),\\
\boldsymbol{q}(\boldsymbol{X}, t) &= \hat{\boldsymbol{q}}\big(\boldsymbol{C}^{(t)}(\boldsymbol{X}, \cdot), \lambda^{(t)}(\boldsymbol{X}, \cdot), \gamma^{(t)}(\boldsymbol{X}, \cdot), \boldsymbol{X}\big),\\
\zeta(\boldsymbol{X}, t) &= \hat{\zeta}\big(\boldsymbol{C}^{(t)}(\boldsymbol{X}, \cdot), \lambda^{(t)}(\boldsymbol{X}, \cdot), \gamma^{(t)}(\boldsymbol{X}, \cdot), \boldsymbol{X}\big),\\
\psi(\boldsymbol{X}, t) &= \hat{\psi}\big(\boldsymbol{C}^{(t)}(\boldsymbol{X}, \cdot), \lambda^{(t)}(\boldsymbol{X}, \cdot), \gamma^{(t)}(\boldsymbol{X}, \cdot), \boldsymbol{X}\big).
\end{aligned} \tag{3.21}$$

Whenever we employ the less specific form (3.20) for its formal simplicity, we assume that it is nevertheless frame-indifferent. These constitutive relations must also be consistent with the symmetry condition (3.13), the entropy inequality (3.18) or (3.19), and the strengthened continuity condition (3.8). These relations discharge their responsibilities to continuum mechanics by being incorporated into the constitutive equations[8]. They therefore play no active role in the formulation of boundary value problems. Indeed, we can replace $\boldsymbol{T}, \boldsymbol{q}, \zeta, \psi$ in (3.12) and (3.16) by their representations (3.20) or (3.21) in terms of $\boldsymbol{r}$ and λ. Eqs. (3.12) and (3.16) thereby become a set of four equations for the unknown functions $\boldsymbol{r} = \boldsymbol{r}(\boldsymbol{X}, t)$ and $\lambda = \lambda(\boldsymbol{X}, t)$. We regard $\boldsymbol{r}$ and λ as the fundamental dependent variables and the Eqs. (3.12) and (3.16) as the fundamental equations of continuum physics.

In general, we prescribe

$$\boldsymbol{f} = \hat{\boldsymbol{f}}(\boldsymbol{r}^{(t)}, \lambda^{(t)}, \boldsymbol{X}, t), \qquad h = \hat{h}(\boldsymbol{r}^{(t)}, \lambda^{(t)}, \boldsymbol{X}, t). \tag{3.22}$$

Let $\{\boldsymbol{e}_p(\boldsymbol{X}, t)\}$ for $\boldsymbol{X} \in \partial\hat{\boldsymbol{X}}(\mathscr{B})$ be a given set of three independent vectors and let $\{\boldsymbol{e}^p(\boldsymbol{X}, t)\}$ be their duals: $\boldsymbol{e}_p \cdot \boldsymbol{e}^q = \delta_p^q$. Let $\boldsymbol{v}$ denote the unit outer normal to $\partial\hat{\boldsymbol{r}}(\mathscr{B})$. For each $\boldsymbol{X} \in \partial\hat{\boldsymbol{X}}(\mathscr{B})$ and for each p, we specify either the functional $\hat{\tau}^p_{(\boldsymbol{v})}$

$$\boldsymbol{\tau}_{(\boldsymbol{v})} \cdot \boldsymbol{e}^p = \hat{\tau}^p_{(\boldsymbol{v})}(\boldsymbol{r}^{(t)}, \lambda^{(t)}, \boldsymbol{X}, t) \tag{3.23}$$

or else the components of $\boldsymbol{r} = \boldsymbol{r}(\boldsymbol{X}, t)$ such that

$$\dot{\boldsymbol{r}} \cdot \boldsymbol{e}_p = \frac{\partial \boldsymbol{r}}{\partial t}(\boldsymbol{X}, t) \cdot \boldsymbol{e}_p(\boldsymbol{X}, t) \tag{3.24}$$

[7] The replacement of $\lambda^{(t)}$ by the history of some other thermodynamic variable as an independent variable in (3.20) and the consequent change in dependent variables causes no difficulty. The form (3.20) is most convenient for our purposes.

[8] The treatment of the strict inequality (3.8), (or of (3.7)) necessarily differs from that for the entropy inequality. We adopt the view that real materials cannot violate (3.8) for finite values of $\boldsymbol{T}, \boldsymbol{q}, \zeta, \psi$ (cf. ANTMAN, 1970b, 1971). We can ensure this by requiring that at least one of the dependent constitutive variables $\boldsymbol{T}, \boldsymbol{q}, \zeta, \psi$ becomes unbounded as any eigenvalue of $\boldsymbol{C}$ approaches zero.

Had we employed the temperature rather than the logarithmic temperature as an independent constitutive variable, we would have had to postulate that the temperature be positive. The condition that at least one of the dependent constitutive variables becomes unbounded as the temperature approaches zero is equivalent to the condition that one of these dependent variables becomes unbounded as the logarithmic temperature approaches $-\infty$. It is clear that these conditions are closely connected to questions of regularity: If the constitutive relations are such that reasonable problems have bounded solutions λ, then the temperature must be positive. The analogous treatment of (3.8) by the introduction of a new tensorial measure of strain whose components take values on $(-\infty, \infty)$ awkwardly complicates the kinematic relations. As we shall see from the treatment of the corresponding one-dimensional problem in Sects. 19–22, the choice of an appropriate new strain measure is strongly influenced by the requirements of a well-set boundary value problem.

is determined. Conditions (3.23) and (3.24) ensure that

$$\oint_{\partial \hat{\boldsymbol{r}}(\mathscr{B})} G^{-\frac{1}{2}} (\boldsymbol{\tau}_{(\boldsymbol{v})} - \overline{\boldsymbol{\tau}}_{(\boldsymbol{v})}) \cdot (\dot{\boldsymbol{r}} - \dot{\overline{\boldsymbol{r}}})\, dA = 0, \tag{3.25}$$

where $\overline{\boldsymbol{\tau}}_{(\boldsymbol{v})}$ is any vector satisfying whichever of the conditions (3.23) are prescribed and where $\dot{\overline{\boldsymbol{r}}}$ is any vector satisfying whichever of the conditions (3.24) are prescribed. We also specify either

$$\boldsymbol{q} \cdot \boldsymbol{v} = \hat{q}_{(\boldsymbol{v})} (\boldsymbol{r}^{(t)}, \lambda^{(t)}, \boldsymbol{X}, t) \tag{3.26}$$

or

$$\lambda = \lambda(\boldsymbol{X}, t) \tag{3.27}$$

for each $\boldsymbol{X} \in \partial \hat{\boldsymbol{X}}(\mathscr{B})$. These boundary conditions generalize accepted conditions for elastic bodies.[9]

B. Formation of rod theories.

I. Approximation of three-dimensional equations.

4. Nature of the approximation process. We are confronted with a straightforward mathematical problem: To construct a rational scheme for approximating the system of field equations of Sect. 3, having the four independent variables $\boldsymbol{X}, t$, by a system in just the two independent variables of a space coordinate and t. Such a scheme would yield a sequence of one-dimensional problems of increasing complexity, the solutions of these problems converging in some appropriate sense to a solution of the three-dimensional problem (provided the latter exists). Our development must also be consistent with the physical objectives of rod theories in that the resulting approximating equations be useful and accurate and that the variables entering these equations have relatively simple physical interpretations. We meet these requirements by using a generalized form of projection methods,[10] which is explained in the rest of this subchapter. In Sect. 13, we discuss asymptotic methods.

Our development is far more general than necessary for any practical purpose. We employ it because it explains the form invariance enjoyed by rod theories constructed by apparently dissimilar processes. Moreover, it contains as special cases the most general constructions heretofore proposed.

For simplicity, we restrict our attention to simple thermo-mechanical materials. It is clear that our projection methods readily handle materials with additional kinematic and physical structure.

Those bodies to which we can apply our approximation scheme are termed *rods* in this subchapter. Precisely, a *rod* is a pair $(\mathscr{B}, \hat{\boldsymbol{X}})$ consisting of a connected (but not necessarily simply-connected) solid body $\mathscr{B}$ and a smooth invertible mapping $\hat{\boldsymbol{X}}$ that assigns material coordinates to $\mathscr{B}$ such that (i) $\hat{X}^1(\mathscr{B})$ and $\hat{X}^2(\mathscr{B})$ are bounded sets of real numbers, and (ii) there is a real number m such that $G(\boldsymbol{X}) < m$ for each $\boldsymbol{X} \in \hat{\boldsymbol{X}}(\mathscr{B})$. These requirements ensure that rods have finite thickness.

This concept of rods clearly includes bodies that are not slender. For such bodies, the utility of the resulting one-dimensional equations may be seriously impaired, although the approximation procedure may still converge. We comment further on this question in Sect. 11.

---

[9] A definitive listing of appropriate boundary conditions for a given body must await the verdict of existence theorems. The conditions given here are based on plausible physical assumptions and are consistent with those for variational problems.

[10] Special cases of such processes are usually associated with the names of RITZ, GALERKIN, and KANTOROVICH. See Sect. 11 for references.

Note that we do not require the X^3-axis to be a curve of centroids for the reference configuration of the body. For segments of a cylinder and for other bodies of simple reference geometry, it is clear that there exists a curve of centroids. But for arbitrary bodies, it is not evident how to define this curve conveniently and unambiguously. Were such a definition possible and were a given body to possess a curve of centroids, the actual construction of the curve from geometrical data would remain a formidable practical task. Moreover, for a given problem, the curve need not be the most natural choice for the X^3-axis.

We frequently denote the distinguished coordinate X^3 by S and use the subscript ",S" in place of ",3". We set

$$\inf \hat{X}^3(\mathscr{B}) = S_1, \qquad \sup \hat{X}^3(\mathscr{B}) = S_2. \tag{4.1}$$

We permit S_1 and S_2 to be infinite. $S=S_1$ and $S=S_2$ are called the *ends* of the rod. We set

$$\begin{aligned} \mathscr{S} &= \partial\hat{\boldsymbol{X}}(\mathscr{B}) \cap \{\boldsymbol{X}: S_1 < S < S_2\}, \\ \mathscr{A}(T) &= \hat{\boldsymbol{X}}(\mathscr{B}) \cap \{\boldsymbol{X}: S = T\}. \end{aligned} \tag{4.2}$$

The *lateral surface* of the rod is $\boldsymbol{r}(\mathscr{S}, t)$ and the *cross-sectional surface* at $S=T$ is $\boldsymbol{r}\left(\mathscr{A}(T), t\right)$.

We introduce surface parameters U, V for $\boldsymbol{r}(\mathscr{S})$ by

$$X^\alpha = X^\alpha(U, V), \qquad S = S(V). \tag{4.3}$$

If $\boldsymbol{\nu}$ denotes the unit outer normal to $\boldsymbol{r}(\mathscr{S})$, then

$$\nu_\alpha \frac{\partial X^\alpha}{\partial U} = 0, \qquad \nu_p \frac{\partial X^p}{\partial V} = 0. \tag{4.4}$$

The differential surface area on $\boldsymbol{r}(\mathscr{S})$ is given by

$$dA = \frac{\sqrt{G}}{1 - \nu^3 \nu_3} \left(\frac{\partial X^2}{\partial U} \nu^1 - \frac{\partial X^1}{\partial U} \nu^2 \right) dU\, dS. \tag{4.5}$$

5. Representation of position and logarithmic temperature. Let $\mathbf{r}_\mathfrak{K}$ denote an ordered set of $\mathfrak{K}+1$ vectors and $\lambda_\mathfrak{L}$, an ordered set of $\mathfrak{L}+1$ scalars:

$$\mathbf{r}_\mathfrak{K} = \{\boldsymbol{r}_\mathfrak{k}, \mathfrak{k} = 0, \ldots, \mathfrak{K}\}, \qquad \lambda_\mathfrak{L} = \{\lambda_\mathfrak{l}, \mathfrak{l} = 0, \ldots, \mathfrak{L}\}. \tag{5.1}$$

We seek to approximate $\boldsymbol{r}(\boldsymbol{X}, t)$ and $\lambda(\boldsymbol{X}, t)$, the fundamental dependent variables of the three-dimensional theory, by representations of the form

$$\boldsymbol{r}(\boldsymbol{X}, t) \sim \boldsymbol{b}\left(\mathbf{r}_\mathfrak{K}(S, t), \boldsymbol{X}, t\right), \tag{5.2}$$

$$\lambda(\boldsymbol{X}, t) \sim \beta\left(\lambda_\mathfrak{L}(S, t), \boldsymbol{X}, t\right), \tag{5.3}$$

where the sets of functions $\mathbf{r}_\mathfrak{K}(S, t)$ and $\lambda_\mathfrak{L}(S, t)$ are to be determined. Here $\boldsymbol{b}(\mathbf{r}_\mathfrak{K}, \boldsymbol{X}, t)$ is a given vector function on $\mathscr{R}^{3(\mathfrak{K}+1)} \times \hat{\boldsymbol{X}}(\mathscr{B}) \times \mathscr{R}$ and $\beta(\lambda_\mathfrak{L}, X, t)$ is a given scalar function on $\mathscr{R}^{\mathfrak{L}+1} \times \hat{\boldsymbol{X}}(\mathscr{B}) \times \mathscr{R}$. It is often useful and illuminating to regard the representations (5.2) and (5.3), not merely as approximations for $\boldsymbol{r}$ and λ, but as exact expressions of permissible forms of $\boldsymbol{r}$ and λ for a class of constrained materials.[11]

Let $\{\boldsymbol{e}_p(\boldsymbol{X}, t)\}$ for $\boldsymbol{X} \in \mathscr{S}$ be a prescribed set of three independent vectors (cf. Sect. 3). If any of the four quantities $\{\dot{\boldsymbol{r}} \cdot \boldsymbol{e}_p\}$, λ is prescribed on parts of $\mathscr{S}$ for some intervals of time,

[11] This interpretation was used by Volterra (1955, 1956, 1961). It was developed further by Antman and Warner (1966). We can interpret the work of Green, Laws, and Naghdi (1968) and of Green and Naghdi (1970) in this context. See the discussion in Sect. 11.

we may employ the following process to construct representations so that $\{\dot{\boldsymbol{b}}\cdot\boldsymbol{e}_p\}$, β have the same specified values on the same parts of $\mathscr{S}$ for the same intervals of time. Suppose that

$$\begin{aligned}\boldsymbol{r}\cdot\boldsymbol{e}_p &= u_p(\boldsymbol{X},t) \quad\text{for}\quad \boldsymbol{X}\in\mathscr{S}_p(t)\subset\mathscr{S},\\ \lambda &= u_0(\boldsymbol{X},t) \quad\text{for}\quad \boldsymbol{X}\in\mathscr{S}_0(t)\subset\mathscr{S},\end{aligned}\tag{5.4}$$

where $u_{\mathfrak{q}}$, $\mathfrak{q}=0, 1, 2, 3$, are prescribed functions of $\boldsymbol{X}, t$ on the time-dependent subsets $\mathscr{S}_{\mathfrak{q}}(t)$ of $\mathscr{S}$. Let $\{\phi^{\mathfrak{a}}_{\mathfrak{q}}(\boldsymbol{X},t)\}$ be a family of sufficiently smooth functions for which $\phi^{\mathfrak{a}}_{\mathfrak{q}}(\boldsymbol{X},t)=0$ is the equation of any surface containing $\mathscr{S}_{\mathfrak{q}}(t)$. (Note that $\dot{\phi}^{\mathfrak{a}}_{\mathfrak{q}}(\boldsymbol{X},t)=0$ for $\boldsymbol{X}\in\mathscr{S}_{\mathfrak{q}}(t)$. See Fig. 1.) By choosing a vector function $\boldsymbol{b}^*(\boldsymbol{X},t)$ and a scalar function $\beta^*(\boldsymbol{X},t)$ such that

$$\dot{\boldsymbol{b}}^*\cdot\boldsymbol{e}_p = u_p \quad\text{for}\quad \boldsymbol{X}\in\mathscr{S}_p(t), \qquad \beta^* = u_0 \quad\text{for}\quad \boldsymbol{X}\in\mathscr{S}_0,\tag{5.5}$$

and by choosing $\boldsymbol{b}-\boldsymbol{b}^*$ and $\beta-\beta^*$ such that

$$\begin{aligned}(\boldsymbol{b}-\boldsymbol{b}^*)\cdot\boldsymbol{e}_p &= \sum_{\mathfrak{a}}\phi^{\mathfrak{a}}_p(\boldsymbol{X},t)\,k^{\mathfrak{a}}(\boldsymbol{r}_{\mathfrak{K}},\boldsymbol{X},t),\\ \beta-\beta^* &= \sum_{\mathfrak{a}}\phi^{\mathfrak{a}}_0(\boldsymbol{X},t)\,\varkappa^{\mathfrak{a}}(\lambda_{\mathfrak{L}},\boldsymbol{X},t),\end{aligned}\tag{5.6}$$

where $k^{\mathfrak{a}}$, $\varkappa^{\mathfrak{a}}$ are sufficiently smooth functions of their indicated arguments, we ensure that

$$\dot{\boldsymbol{b}}\cdot\boldsymbol{e}_p = u_p \quad\text{for}\quad \boldsymbol{X}\in\mathscr{S}_p(t), \qquad \beta = u_0 \quad\text{for}\quad \boldsymbol{X}\in\mathscr{S}_0(t).\tag{5.7}$$

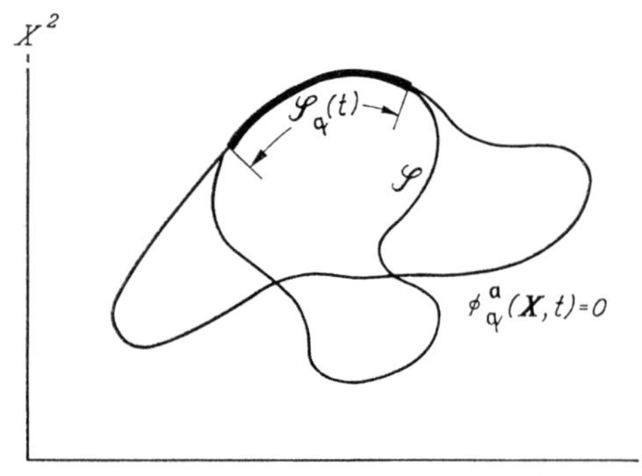

Fig. 1. The intersection (in $\boldsymbol{X}$-space) of the surfaces $\mathscr{S}$, $\mathscr{S}_{\mathfrak{q}}(t)$, and $\phi^{\mathfrak{a}}_{\mathfrak{q}}(\boldsymbol{X},t)=0$ with the plane $X^3=\text{const.}$

If (5.4) is time-independent, there is no need for $\boldsymbol{b}$ and β to depend explicitly on t.[12]

As special choices for $\boldsymbol{b}$ and β we may take the linear relations

$$\boldsymbol{b}=\boldsymbol{b}^*(\boldsymbol{X},t)+\sum_{\mathfrak{k}=0}^{\mathfrak{K}}\boldsymbol{B}^{\mathfrak{k}}(\boldsymbol{X},t)\,\boldsymbol{r}_{\mathfrak{k}}(S,t), \qquad \beta=\beta^*(\boldsymbol{X},t)+\sum_{\mathfrak{l}=0}^{\mathfrak{L}}\beta^{\mathfrak{l}}(\boldsymbol{X},t)\,\lambda_{\mathfrak{l}}(S,t),\tag{5.8}$$

where $\boldsymbol{b}^*$, $\{\boldsymbol{B}^{\mathfrak{k}}\}$, β^*, $\{\beta^{\mathfrak{l}}\}$ are a fixed set of differentiable functions of $\boldsymbol{X}, t$. The generality afforded by the use of tensors $\{\boldsymbol{B}^{\mathfrak{k}}\}$ instead of just scalars proves convenient in the treatment of some important problems (see Sect. 11). It allows each component of $\boldsymbol{r}$ to have an individual representation like that for λ in $(5.8)_2$. If position boundary conditions are not prescribed on $\mathscr{S}$, we may choose $\boldsymbol{b}^*$ to be the position vector to the particles in the reference configuration so that $\boldsymbol{b}-\boldsymbol{b}^*$ represents the displacement from the reference configuration. Moreover, we may assume that $\boldsymbol{b}^*$ itself is approximated by an expansion of the form $\sum_0^{\mathfrak{K}}\boldsymbol{B}^{\mathfrak{k}}\boldsymbol{r}^*_{\mathfrak{k}}$. Then the representation for $\boldsymbol{b}$ would have the same form as if $\boldsymbol{b}^*=\boldsymbol{0}$.

If (5.4) holds and if (5.8) is used, we can ensure (5.7) by choosing $\boldsymbol{b}^*$ and β^* as before and by taking

$$\boldsymbol{B}^{\mathfrak{k}T}\boldsymbol{e}_p=\phi^{\mathfrak{k}}_p(\boldsymbol{X},t)\,\boldsymbol{w}^{\mathfrak{k}}(\boldsymbol{X},t), \qquad \beta^{\mathfrak{l}}=\phi^{\mathfrak{l}}_0(\boldsymbol{X},t)\,\omega^{\mathfrak{l}}(\boldsymbol{X},t),\tag{5.9}$$

where $\boldsymbol{w}^{\mathfrak{k}}$ and $\omega^{\mathfrak{l}}$ are sufficiently smooth functions of $\boldsymbol{X}, t$. Note that eigenfunction expansions in the transverse coordinates have this form.

[12] The explicit presence of t in (5.4) produces difficulties like those caused by rheonomic constraints in rigid body mechanics. In the case of elastic rods, this analogy is most apt.

For given infinite sequences $\{\boldsymbol{B}^{\mathfrak{k}}\}$, $\{\beta^{\mathrm{I}}\}$, the assumption that $\boldsymbol{r}$ and λ equal the infinite series representations obtained formally by setting $\mathfrak{K}=\infty$, $\mathfrak{L}=\infty$ in (5.8) is of course a much sharper restriction on the class of positions $\boldsymbol{r}(\boldsymbol{X},t)$ and logarithmic temperatures $\lambda(\boldsymbol{X},t)$ than the assumption that the representations (5.8) converge to $\boldsymbol{r}$ and λ as $\mathfrak{K}\to\infty$, $\mathfrak{L}\to\infty$. In the absence of regularity theorems from the three-dimensional theory, caution must be exercized in postulating infinite series representations for these variables.[13]

The imposition of a set of constraints on a theory characterized by (5.8) may destroy the linearity of the dependence of $\boldsymbol{b}$ and β upon the reduced set of kinematic and thermal unknowns. This possibility is accounted for in (5.2), (5.3).

In the sequel we shall employ the notation

$$\boldsymbol{B}^{\mathfrak{k}}=\frac{\partial\boldsymbol{b}}{\partial\boldsymbol{r}_{\mathfrak{k}}},\quad \beta^{\mathrm{I}}=\frac{\partial\beta}{\partial\lambda_{\mathrm{I}}}. \tag{5.10}$$

This is consistent with (5.8).

In accord with (5.2) and (5.10), the base vectors $\boldsymbol{g}_p=\boldsymbol{r}_{,p}$ are represented by

$$\boldsymbol{g}_p\sim\boldsymbol{b}_{,p}=\delta_p^3\sum_{\mathfrak{k}=0}^{\mathfrak{K}}\boldsymbol{B}^{\mathfrak{k}}\,\boldsymbol{r}_{\mathfrak{k},S}+\frac{\partial\boldsymbol{b}}{\partial X^p} \tag{5.11}$$

and the Cauchy-Green tensor $\boldsymbol{C}$ is represented by

$$C_{pq}\sim\boldsymbol{b}_{,p}\cdot\boldsymbol{b}_{,q}. \tag{5.12}$$

Similarly,

$$\gamma_p=\lambda_{,p}\sim\beta_{,p}+\delta_p^3\sum_{\mathrm{I}=0}^{\mathfrak{L}}\beta^{\mathrm{I}}\lambda_{\mathrm{I},S}+\frac{\partial\beta}{\partial X^p}. \tag{5.13}$$

We approximate (3.7) by

$$(\boldsymbol{b}_{,1}\times\boldsymbol{b}_{,2})\cdot\boldsymbol{b}_{,3}>0 \tag{5.14}$$

wherever the reference value of the determinant G is positive and we approximate (3.8) by

$$\boldsymbol{b}_{,p}\cdot\boldsymbol{b}_{,q}\,v^p\,v^q>0 \tag{5.15}$$

for all $\boldsymbol{v}\neq\boldsymbol{0}$[14]. The acceleration $\ddot{\boldsymbol{r}}$ is represented by $\ddot{\boldsymbol{b}}$, which may be expanded in the manner of (5.11) by the chain rule.

6. Moments of the fundamental equations. Let $\{\boldsymbol{A}^{\mathfrak{k}}(\boldsymbol{r}_{\mathfrak{K}},\boldsymbol{X},t)\}$ be a fixed sequence of independent differentiable second order tensors defined on $\mathscr{R}^{3(\mathfrak{K}+1)}\times\hat{\boldsymbol{X}}(\mathscr{B})\times\mathscr{R}$ and let $\{\alpha^{\mathrm{I}}(\lambda_{\mathfrak{L}},\boldsymbol{X},t)\}$ be a fixed sequence of independent differentiable scalars defined on $\mathscr{R}^{\mathfrak{L}+1}\times\hat{\boldsymbol{X}}(\mathscr{B})\times\mathscr{R}$. We define the following moments of the dependent constitutive variables:

$$\begin{aligned}
\boldsymbol{\sigma}^{\mathfrak{k}}(S,t)&=\int_{\mathscr{A}(S)}\boldsymbol{A}^{\mathfrak{k}}(\boldsymbol{r}_{\mathfrak{K}}(S,t),\boldsymbol{X},t)\,\boldsymbol{\tau}^3(\boldsymbol{X},t)\,dX^1\,dX^2,\\
\bar{\boldsymbol{\sigma}}^{\mathfrak{k}}(S,t)&=\int_{\mathscr{A}(S)}\boldsymbol{A}^{\mathfrak{k}}_{,p}(\boldsymbol{r}_{\mathfrak{K}}(S,t),\boldsymbol{X},t)\,\boldsymbol{\tau}^p(\boldsymbol{X},t)\,dX^1\,dX^2,\\
q^{\mathrm{I}}(S,t)&=\int_{\mathscr{A}(S)}\alpha^{\mathrm{I}}(\lambda_{\mathfrak{L}}(S,t),\boldsymbol{X},t)\,q^3(\boldsymbol{X},t)\,dX^1\,dX^2,\\
\bar{q}^{\mathrm{I}}(S,t)&=\int_{\mathscr{A}(S)}\alpha^{\mathrm{I}}_{,p}(\lambda_{\mathfrak{L}}(S,t),\boldsymbol{X},t)\,q^p(\boldsymbol{X},t)\,dX^1\,dX^2,\\
\zeta^{\mathrm{I}}(S,t)&=\int_{\mathscr{A}(S)}\alpha^{\mathrm{I}}(\lambda_{\mathfrak{L}}(S,t),\boldsymbol{X},t)\,\zeta(\boldsymbol{X},t)\,\varrho_0(\boldsymbol{X})\sqrt{G(\boldsymbol{X})}\,dX^1\,dX^2,\\
\mathring{\zeta}^{\mathrm{I}}(S,t)&=\int_{\mathscr{A}(S)}\dot{\alpha}^{\mathrm{I}}(\lambda_{\mathfrak{L}}(S,t),\boldsymbol{X},t)\,\zeta(\boldsymbol{X},t)\,\varrho_0(\boldsymbol{X})\sqrt{G(\boldsymbol{X})}\,dX^1\,dX^2,\\
\psi^{\mathrm{I}}(S,t)&=\int_{\mathscr{A}(S)}\alpha^{\mathrm{I}}(\lambda_{\mathfrak{L}}(S,t),\boldsymbol{X},t)\,\psi(\boldsymbol{X},t)\,\varrho_0(\boldsymbol{X})\sqrt{G(\boldsymbol{X})}\,dX^1\,dX^2,\\
\mathring{\psi}^{\mathrm{I}}(S,t)&=\int_{\mathscr{A}(S)}\dot{\alpha}^{\mathrm{I}}(\lambda_{\mathfrak{L}}(S,t),\boldsymbol{X},t)\,\psi(\boldsymbol{X},t)\,\varrho_0(\boldsymbol{X})\sqrt{G(\boldsymbol{X})}\,dX^1\,dX^2.
\end{aligned} \tag{6.1}$$

[13] Formal power series in X^1, X^2 appear in the work of GREEN (1959), ANTMAN and WARNER (1966), GREEN, LAWS, and NAGHDI (1968), and GREEN and NAGHDI (1970), but their use is not essential to the developments in these papers as we shall show.

[14] There is no ambiguity in retaining the strict inequalities in (5.14), (5.15), provided the position field is regarded as constrained in the sense described above.

The special cases

$$\boldsymbol{A}^{\mathfrak{k}} = \boldsymbol{B}^{\mathfrak{k}T}, \qquad \alpha^{\mathfrak{l}} = \beta^{\mathfrak{l}} \tag{6.2}$$

lead to the most natural formulation of boundary value problems. In this case we denote the stress moments $\boldsymbol{\sigma}^{\mathfrak{k}}$ and $\overline{\boldsymbol{\sigma}}^{\mathfrak{k}}$ by $\boldsymbol{\mu}^{\mathfrak{k}}$ and $\overline{\boldsymbol{\mu}}^{\mathfrak{k}}$ and the temperature-entropy moments $\zeta^{\mathfrak{l}}$ and $\mathring{\zeta}^{\mathfrak{l}}$ by $\eta^{\mathfrak{l}}$ and $\mathring{\eta}^{\mathfrak{l}}$:

$$\begin{aligned}
\boldsymbol{\mu}^{\mathfrak{k}} &= \int_{\mathscr{A}} \boldsymbol{B}^{\mathfrak{k}T} \boldsymbol{\tau}^3 \, dX^1 \, dX^2, \\
\overline{\boldsymbol{\mu}}^{\mathfrak{k}} &= \int_{\mathscr{A}} \boldsymbol{B}^{\mathfrak{k}T}_{,p} \boldsymbol{\tau}^p \, dX^1 \, dX^2, \\
\eta^{\mathfrak{l}} &= \int_{\mathscr{A}} \beta^{\mathfrak{l}} \zeta \, \varrho_0 \sqrt{G} \, dX^1 \, dX^2, \\
\mathring{\eta}^{\mathfrak{l}} &= \int_{\mathscr{A}} \dot{\beta}^{\mathfrak{l}} \zeta \, \varrho_0 \sqrt{G} \, dX^1 \, dX^2.
\end{aligned} \tag{6.3}$$

It may still be convenient to retain the original level of generality so that $\boldsymbol{\sigma}^{\mathfrak{k}}, \overline{\boldsymbol{\sigma}}^{\mathfrak{k}}, \zeta^{\mathfrak{l}}, \mathring{\zeta}^{\mathfrak{l}}$ need not equal $\boldsymbol{\mu}^{\mathfrak{k}}, \overline{\boldsymbol{\mu}}^{\mathfrak{k}}, \eta^{\mathfrak{l}}, \mathring{\eta}^{\mathfrak{l}}$, respectively. In particular, we may wish to identify the resultant force and moment at a section as members of $\{\boldsymbol{\sigma}^{\mathfrak{k}}\}$ while retaining the kinemantic versatility inherent in (6.1). We can express $\boldsymbol{\sigma}^{\mathfrak{k}}, \overline{\boldsymbol{\sigma}}^{\mathfrak{k}}, \zeta^{\mathfrak{l}}, \mathring{\zeta}^{\mathfrak{l}}$ exactly in terms of $\{\boldsymbol{\mu}^{\mathfrak{k}}, \overline{\boldsymbol{\mu}}^{\mathfrak{k}}, \eta^{\mathfrak{l}}, \mathring{\eta}^{\mathfrak{l}}\}$ if there are tensors $\boldsymbol{K}^{\mathfrak{k}}_{\mathfrak{m}}(\boldsymbol{r}_{\mathfrak{K}}, S, t)$ and scalars $k^{\mathfrak{l}}_{\mathfrak{n}}(\lambda_{\mathfrak{L}}, S, t)$ such that

$$\boldsymbol{A}^{\mathfrak{k}} = \sum_{\mathfrak{m}=0}^{\mathfrak{K}} \boldsymbol{K}^{\mathfrak{k}}_{\mathfrak{m}} \boldsymbol{B}^{\mathfrak{m}T}, \qquad \alpha^{\mathfrak{l}} = \sum_{\mathfrak{n}=0}^{\mathfrak{L}} k^{\mathfrak{l}}_{\mathfrak{n}} \beta^{\mathfrak{n}}, \tag{6.4}$$

for then

$$\begin{aligned}
\boldsymbol{\sigma}^{\mathfrak{k}} &= \sum_{\mathfrak{m}=0}^{\mathfrak{K}} K^{\mathfrak{l}}_{\mathfrak{m}} \boldsymbol{\mu}^{\mathfrak{m}}, \\
\overline{\boldsymbol{\sigma}}^{\mathfrak{k}} &= \sum_{\mathfrak{m}=0}^{\mathfrak{K}} (\boldsymbol{K}^{\mathfrak{k}}_{\mathfrak{m},S} \boldsymbol{\mu}^{\mathfrak{m}} + \boldsymbol{K}^{\mathfrak{k}}_{\mathfrak{m}} \overline{\boldsymbol{\mu}}^{\mathfrak{m}}), \\
\zeta^{\mathfrak{l}} &= \sum_{\mathfrak{n}=0}^{\mathfrak{L}} k^{\mathfrak{l}}_{\mathfrak{n}} \eta^{\mathfrak{n}}, \\
\mathring{\zeta}^{\mathfrak{l}} &= \sum_{\mathfrak{n}=0}^{\mathfrak{L}} (\dot{k}^{\mathfrak{l}}_{\mathfrak{n}} \eta^{\mathfrak{n}} + k^{\mathfrak{l}}_{\mathfrak{n}} \mathring{\eta}^{\mathfrak{n}}).
\end{aligned} \tag{6.5}$$

[If (6.4) is just approximate, then (6.5) is also.] We observe that

$$\sum_{\mathfrak{m}=0}^{\mathfrak{K}} \boldsymbol{K}^{\mathfrak{k}}_{\mathfrak{m}} (\boldsymbol{\mu}^{\mathfrak{m}}_{,S} - \overline{\boldsymbol{\mu}}^{\mathfrak{m}}) = \boldsymbol{\sigma}^{\mathfrak{k}}_{,S} - \overline{\boldsymbol{\sigma}}^{\mathfrak{k}}. \tag{6.6}$$

In Sect. 11, we shall illustrate the construction of $\{\boldsymbol{K}^{\mathfrak{k}}_{\mathfrak{m}}\}$ and $\{k^{\mathfrak{l}}_{\mathfrak{n}}\}$ for problems in linear elasticity.

The differential force and differential heat flux across $\boldsymbol{r}(\mathscr{S})$ are

$$G^{-\frac{1}{2}} \boldsymbol{\tau}_{(\boldsymbol{v})} \, dA = \left[\boldsymbol{\tau}^1 \frac{\partial X^2}{\partial U} - \boldsymbol{\tau}^2 \frac{\partial X^1}{\partial U} + \frac{\boldsymbol{\tau}^3 \nu_3}{1 - \nu^3 \nu_3} \left(\frac{\partial X^2}{\partial U} \nu^1 - \frac{\partial X^1}{\partial U} \nu^2\right)\right] dU \, dS, \tag{6.7}$$

$$G^{-\frac{1}{2}} \boldsymbol{q} \cdot \boldsymbol{v} \, dA = \left[q^1 \frac{\partial X^2}{\partial U} - q^2 \frac{\partial X^1}{\partial U} + \frac{q^3 \nu_3}{1 - \nu^3 \nu_3} \left(\frac{\partial X^2}{\partial U} \nu^1 - \frac{\partial X^1}{\partial U} \nu^2\right)\right] dU \, dS. \tag{6.8}$$

We define the combined moment contributions per unit of S of the body force and the tractions on the lateral surface

$$\begin{aligned}
\boldsymbol{f}^{\mathfrak{k}} &= \int_{\mathscr{A}} \boldsymbol{A}^{\mathfrak{k}} \boldsymbol{f} \varrho_0 \sqrt{G} \, dX^1 \, dX^2 + \oint_{\partial\mathscr{A}} \boldsymbol{A}^{\mathfrak{k}} [\boldsymbol{\tau}^1 \, dX^2 - \boldsymbol{\tau}^2 \, dX^1 \\
&\quad + \boldsymbol{\tau}^3 \nu_3 (1 - \nu^3 \nu_3)^{-1} (\nu^1 \, dX^2 - \nu^2 \, dX^1)] \\
&= \int_{\mathscr{A}} \boldsymbol{A}^{\mathfrak{k}} \boldsymbol{f} \varrho_0 \sqrt{G} \, dX^1 \, dX^2 + \oint_{\partial\mathscr{A}} \frac{\boldsymbol{A}^{\mathfrak{k}} \boldsymbol{\tau}_{(\boldsymbol{v})} (\nu^1 \, \partial X^2 - \nu^2 \, dX^1)}{1 - \nu^3 \nu_3}
\end{aligned} \tag{6.9}$$

and the combined moment contributions per unit of S of the heat source and the heat flux through the lateral surface

$$h^{\mathrm{I}} = \int_{\mathscr{A}} \alpha^{\mathrm{I}} h \varrho_0 \sqrt{G}\, dX^1 dX^2 - \oint_{\partial\mathscr{A}} \alpha^{\mathrm{I}} [q^1\, dX^2 - q^2\, dX^1 + q^3 \nu_3 (1 - \nu^3 \nu_3)^{-1} (\nu^1\, dX^2 - \nu^2\, dX^1)] = \int_{\mathscr{A}} \alpha^{\mathrm{I}} h \varrho_0 \sqrt{G}\, dX^1 dX^2 - \oint_{\partial\mathscr{A}} \frac{\alpha^{\mathrm{I}} \boldsymbol{q}\cdot\boldsymbol{\nu} (\nu^1\, dX^2 - \nu^2\, dX^1)}{1 - \nu^3 \nu_3}. \tag{6.10}$$

If the traction boundary conditions are not prescribed over the entire lateral surface, then, in general, $\boldsymbol{f}^{\mathfrak{k}}$ is not a well-defined functional of $\boldsymbol{r}^{(t)}$, $\lambda^{(t)}$, $\boldsymbol{X}$, t [see (3.23), (3.24)] and if $\boldsymbol{q}\cdot\boldsymbol{\nu}$ is not prescribed everywhere on the lateral surface, h^{I} is likewise not well-defined [see (3.26), (3.27)]. We can remove this difficulty by judicious choices of $\{\boldsymbol{A}^{\mathfrak{k}}\}$ and $\{\alpha^{\mathrm{I}}\}$. If (5.4) holds, then $\boldsymbol{\tau}_{(\boldsymbol{\nu})}\cdot\boldsymbol{e}^p$ is not known on $\mathscr{S}_p(t)$ and $\boldsymbol{q}\cdot\boldsymbol{\nu}$ is not known on $\mathscr{S}_0(t)$. By choosing $\{\boldsymbol{A}^{\mathfrak{k}}\}$, $\{\alpha^{\mathrm{I}}\}$ such that

$$\boldsymbol{A}^{\mathfrak{k}} \boldsymbol{e}_p = \phi^{\mathfrak{k}}_p(\boldsymbol{X}, t)\, w^{\mathfrak{k}}(\boldsymbol{r}_{\mathfrak{R}}, \boldsymbol{X}, t), \qquad \alpha^{\mathrm{I}} = \phi^{\mathrm{I}}_0(\boldsymbol{X}, t)\, \omega^{\mathrm{I}}(\lambda_{\mathfrak{L}}, \boldsymbol{X}, t), \tag{6.11}$$

where $\boldsymbol{w}^{\mathfrak{k}}$ and ω^{I} are smooth enough functions of their arguments, we ensure that there are zero contributions to the line integrals of (6.9) and (6.10) from wherever $\boldsymbol{\tau}_{(\boldsymbol{\nu})}\cdot\boldsymbol{e}^p$, $\boldsymbol{q}\cdot\boldsymbol{\nu}$ are not prescribed. If (6.2) holds, the treatment of position and boundary conditions of Sect. 5 automatically includes this treatment since (5.6) and (6.2) imply (6.11). [Also cf. (5.9).]

To obtain useful consequence of the equation of motion (3.12), we multiply it by $\boldsymbol{A}^{\mathfrak{k}}$ and integrate the resulting expression over $\mathscr{A}$. Since

$$\boldsymbol{\sigma}^{\mathfrak{k}}_{,3} = \int_{\mathscr{A}} (\boldsymbol{A}^{\mathfrak{k}} \boldsymbol{\tau}^3)_{,3}\, dX^1 dX^2 - \oint_{\partial\mathscr{A}} \boldsymbol{A}^{\mathfrak{k}} \boldsymbol{\tau}^3 \nu_3 (1 - \nu^3\nu_3)^{-1} (\nu^1\, dX^2 - \nu^2\, dX^1), \tag{6.12}$$

we then obtain from KELVIN's transformation (STOKES' theorem) for the plane that these weighted integrals of (3.12) can be written in the form[15]

$$\boldsymbol{\sigma}^{\mathfrak{k}}_{,S} - \bar{\boldsymbol{\sigma}}^{\mathfrak{k}} + \boldsymbol{f}^{\mathfrak{k}} = \int_{\mathscr{A}} \boldsymbol{A}^{\mathfrak{k}} \ddot{\boldsymbol{r}} \varrho_0 \sqrt{G}\, dX^1 dX^2. \tag{6.13}$$

By multiplying the energy equation (3.16) by α^{I} and integrating the resulting expression over $\mathscr{A}$, we likewise obtain

$$q^{\mathrm{I}}_{,S} - \bar{q}^{\mathrm{I}} + \dot{\psi}^{\mathrm{I}} + \dot{\zeta}^{\mathrm{I}} - \mathring{\psi}^{\mathrm{I}} - \mathring{\zeta}^{\mathrm{I}} - h^{\mathrm{I}} - \int_{\mathscr{A}} \alpha^{\mathrm{I}} \boldsymbol{\tau}^p \cdot \dot{\boldsymbol{r}}_{,p}\, dX^1 dX^2 = 0. \tag{6.14}$$

These equations are exact.[16] We shall use them to generate explicit one-dimensional approximations.

For the purposes of comparison, it is useful to have the equations of motion for the stress resultant vector

$$\boldsymbol{n} = \int_{\mathscr{A}} \boldsymbol{\tau}^3\, dX^1 dX^2 \tag{6.15}$$

and the couple resultant vector

$$\boldsymbol{m} = \int_{\mathscr{A}} (\boldsymbol{r} - \boldsymbol{p}) \times \boldsymbol{\tau}^3\, dX^1 dX^2, \tag{6.16}$$

where $\boldsymbol{p}(S, t) = \boldsymbol{r}(0, 0, S, t)$. By setting $\boldsymbol{A}^k = \boldsymbol{1}$ in (6.12), we get

$$\boldsymbol{n}_{,S} + \boldsymbol{f}_0 = \int_{\mathscr{A}} \ddot{\boldsymbol{r}} \varrho_0 \sqrt{G}\, dX^1 dX^2, \tag{6.17}$$

---

[15] If (6.2) holds then $\boldsymbol{\mu}^{\mathfrak{k}}$ and $\bar{\boldsymbol{\mu}}^{\mathfrak{k}}$ satisfy (6.13). If (6.4) holds we can get Eq. (6.13) for the $\boldsymbol{\sigma}$'s from the equation for the $\boldsymbol{\mu}$'s by multiplying the latter by $\boldsymbol{K}^{\mathfrak{k}}_{\mathfrak{m}}$, summing over $\mathfrak{m}$, and using (6.6). Eq. (6.6) is thus seen to be an explicit manifestation of the form invariance of (6.13) for any choice of $\{\boldsymbol{A}^{\mathfrak{k}}\}$, $\{\alpha^{\mathrm{I}}\}$.

[16] Eqs. (6.13), (6.14) generalize results of GREEN (1959), ANTMAN and WARNER (1966), GREEN, LAWS, and NAGHDI (1968), and GREEN and NAGHDI (1970).

where $\boldsymbol{f}_0$ is the value of $\boldsymbol{f}^{\mathfrak{k}}$ when $\boldsymbol{A}^{\mathfrak{k}} = \boldsymbol{1}$. The integration of the angular momentum balance

$$(\boldsymbol{r} \times \boldsymbol{\tau}^p)_{,p} + \varrho_0 \sqrt{G}\, \boldsymbol{r} \times \boldsymbol{f} = \varrho_0 \sqrt{G}\, \boldsymbol{r} \times \ddot{\boldsymbol{r}} \tag{6.18}$$

over $\mathscr{A}$, the use of KELVIN's transformation, and the use of (6.17) yields[17]

$$\boldsymbol{m}_{,S} + \boldsymbol{p}_{,S} \times \boldsymbol{n} + \boldsymbol{l} = \int_{\mathscr{A}} (\boldsymbol{r} - \boldsymbol{p}) \times \ddot{\boldsymbol{r}}\, \varrho_0 \sqrt{G}\, dX^1\, dX^2, \tag{6.19}$$

where

$$\begin{aligned} \boldsymbol{l} = & \int_{\mathscr{A}} (\boldsymbol{r} - \boldsymbol{p}) \times \boldsymbol{f}\, \varrho_0 \sqrt{G}\, dX^1\, dX^2 \\ & + \oint_{\partial \mathscr{A}} (\boldsymbol{r} - \boldsymbol{p}) \times [\boldsymbol{\tau}^1\, dX^2 - \boldsymbol{\tau}^2\, dX^1 + \boldsymbol{\tau}^3\, \nu_3 (1 - \nu^3 \nu_3)^{-1} (\nu^1\, dX^2 - \nu^2\, dX^1)]. \end{aligned} \tag{6.20}$$

Note that (6.19) may be obtained by choosing $\boldsymbol{A}^{\mathfrak{k}} = (\boldsymbol{r} - \boldsymbol{p}) \times$ and using (3.13).

The symmetry condition (3.13) and the entropy inequality (3.18) or (3.19), which are accounted for in the constitutive relations (3.20) or (3.21) and therefore play no essential role in boundary value problems, are nevertheless fundamental in describing the structure of the equations for continua. The one-dimensional analogs of these relations, which have the same purpose, will be obtained from the exact relations

$$\int_{\mathscr{A}} \boldsymbol{A} (\boldsymbol{g}_p \times \boldsymbol{\tau}^p)\, dX^1\, dX^2 = \boldsymbol{0}, \tag{6.21}$$

$$\int_{\mathscr{A}} \alpha [\varrho_0 \sqrt{G} (\dot{\zeta} - \zeta \dot{\lambda} - h) + q^p_{,p} - \boldsymbol{q} \cdot \boldsymbol{\gamma}]\, dX^1\, dX^2 \geqq 0, \tag{6.22}$$

or

$$\int_{\mathscr{A}} \alpha [-\varrho_0 \sqrt{G} (\dot{\psi} + \zeta \dot{\lambda}) + \boldsymbol{\tau}^p \cdot \dot{\boldsymbol{r}}_{,p} - \boldsymbol{q} \cdot \boldsymbol{\gamma}]\, dX^1\, dX^2 \geqq 0, \tag{6.23}$$

where $\boldsymbol{A} = \boldsymbol{A}(\boldsymbol{X}, t)$ is any second order tensor and $\alpha = \alpha(\boldsymbol{X}, t)$ is any non-negative scalar.

7. Approximation of the fundamental equations. To obtain approximations for the governing equations, we replace $\boldsymbol{r}$ and λ by $\boldsymbol{b}$ and β. Thus we approximate the acceleration terms on the right side of (6.13):

$$\int_{\mathscr{A}} \boldsymbol{A}^{\mathfrak{k}} \ddot{\boldsymbol{r}}\, \varrho_0 \sqrt{G}\, dX^1\, dX^2 \sim \boldsymbol{a}^{\mathfrak{k}} (\boldsymbol{r}_{\mathfrak{K}}, \dot{\boldsymbol{r}}_{\mathfrak{K}}, \ddot{\boldsymbol{r}}_{\mathfrak{K}}, S, t) \equiv \int_{\mathscr{A}} \boldsymbol{A}^{\mathfrak{k}} \ddot{\boldsymbol{b}}\, \varrho_0 \sqrt{G}\, dX^1\, dX^2. \tag{7.1}$$

If (5.8) is used, then

$$\boldsymbol{a}^{\mathfrak{k}} = \boldsymbol{j}^{\mathfrak{k}} + \sum_{\mathfrak{m}=0}^{\mathfrak{K}} [\boldsymbol{J}^{\mathfrak{k}\mathfrak{m}}\, \ddot{\boldsymbol{r}}^{\mathfrak{m}} + 2 \boldsymbol{J}^{\mathfrak{k}\mathfrak{m}}_{(1)}\, \dot{\boldsymbol{r}}^{\mathfrak{m}} + \boldsymbol{J}^{\mathfrak{k}\mathfrak{m}}_{(2)}\, \boldsymbol{r}^{\mathfrak{m}}], \tag{7.2}$$

where the vectors $\boldsymbol{j}^{\mathfrak{k}}$ and the inertia tensors $\boldsymbol{J}^{\mathfrak{k}\mathfrak{m}}$, $\boldsymbol{J}^{\mathfrak{k}\mathfrak{m}}_{(1)}$, $\boldsymbol{J}^{\mathfrak{k}\mathfrak{m}}_{(2)}$ are computed from the reference data by

$$\begin{aligned} \boldsymbol{j}^{\mathfrak{k}} (S, t) &= \int_{\mathscr{A}} \boldsymbol{A}^{\mathfrak{k}}\, \ddot{\boldsymbol{r}}^*\, \varrho_0 \sqrt{G}\, dX^1\, dX^2, \\ \boldsymbol{J}^{\mathfrak{k}\mathfrak{m}} (S, t) &= \int_{\mathscr{A}} \boldsymbol{A}^{\mathfrak{k}}\, \boldsymbol{B}^{\mathfrak{m}}\, \varrho_0 \sqrt{G}\, dX^1\, dX^2, \\ \boldsymbol{J}^{\mathfrak{k}\mathfrak{m}}_{(1)} (S, t) &= \int_{\mathscr{A}} \boldsymbol{A}^{\mathfrak{k}}\, \dot{\boldsymbol{B}}^{\mathfrak{m}}\, \varrho_0 \sqrt{G}\, dX^1\, dX^2, \\ \boldsymbol{J}^{\mathfrak{k}\mathfrak{m}}_{(2)} (S, t) &= \int_{\mathscr{A}} \boldsymbol{A}^{\mathfrak{k}}\, \ddot{\boldsymbol{B}}^{\mathfrak{m}}\, \varrho_0 \sqrt{G}\, dX^1\, dX^2. \end{aligned} \tag{7.3}$$

---

[17] Cf. GREEN (1959).

We replace $\boldsymbol{r}^{(t)}$, $\lambda^{(t)}$ by $\boldsymbol{b}^{(t)}$, $\beta^{(t)}$ in (3.22), (3.23), (3.26), then substitute the resulting expression into (6.9), (6.10) to obtain the approximate relations

$$\begin{aligned} \boldsymbol{f}^{\mathfrak{k}} \sim \hat{\boldsymbol{f}}^{\mathfrak{k}}(\mathbf{r}_{\mathfrak{K}}^{(t)}, \lambda_{\mathfrak{L}}^{(t)} S, t) \equiv & \int_{\mathscr{A}} \boldsymbol{A}^{\mathfrak{k}} \hat{\boldsymbol{f}}(\boldsymbol{b}^{(t)}, \beta^{(t)}, X, t)\, \varrho_0 \sqrt{G}\, dX^1\, dX^2 \\ & + \oint_{\partial\mathscr{A}} \frac{\boldsymbol{A}^{\mathfrak{k}} \boldsymbol{e}_p\, \hat{\tau}^p_{(\boldsymbol{\nu})}(\boldsymbol{b}^{(t)}, \beta^{(t)}, \boldsymbol{X}, t)\,(\nu^1\, dX^2 - \nu^2\, dX^1)}{1 - \nu^3\, \nu_3}, \end{aligned} \tag{7.4}$$

$$\begin{aligned} h^{\mathrm{I}} \sim \hat{h}^{\mathrm{I}}(\mathbf{r}_{\mathfrak{K}}^{(t)}, \lambda_{\mathfrak{L}}^{(t)}, S, t) \equiv & \int_{\mathscr{A}} \alpha^{\mathrm{I}}\, \hat{h}(\boldsymbol{b}^{(t)}, \beta^{(t)}, \boldsymbol{X}, t)\, \varrho_0 \sqrt{G}\, dX^1\, dX^2 \\ & + \int_{\partial\mathscr{A}} \frac{\alpha^{\mathrm{I}}\, \hat{q}_{(\boldsymbol{\nu})}(\boldsymbol{b}^{(t)}, \beta^{(t)}, \boldsymbol{X}, t)\,(\nu^1\, dX^2 - \nu^2\, dX^1)}{1 - \nu^3\, \nu_3}. \end{aligned} \tag{7.5}$$

The development given in Sect. 6 shows that the functionals $\hat{\boldsymbol{f}}^{\mathfrak{k}}$ and $\hat{h}^{\mathrm{I}}$ are well-defined if $\boldsymbol{A}^{\mathfrak{k}}$ and α^{I} are chosen appropriately. Using (7.1)–(7.4), we obtain the approximate equations of motion

$$\boldsymbol{\sigma}^{\mathfrak{k}}_{,S} - \bar{\boldsymbol{\sigma}}^{\mathfrak{k}} + \hat{\boldsymbol{f}}^{\mathfrak{k}}(\mathbf{r}_{\mathfrak{K}}^{(t)}, \lambda_{\mathfrak{L}}^{(t)}, S, t) \sim \boldsymbol{a}^{\mathfrak{k}}(\mathbf{r}_{\mathfrak{K}}, \dot{\boldsymbol{r}}_{\mathfrak{K}}, \ddot{\boldsymbol{r}}_{\mathfrak{K}}, S, t). \tag{7.6}$$

To get analogous results for the energy Eq. (6.14), we first define the additional set of moments

$$\begin{aligned} \sigma^{\mathrm{I}} &= \int_{\mathscr{A}} \alpha^{\mathrm{I}} \left(\frac{\partial \boldsymbol{b}}{\partial t}\right)_{,p} \cdot \boldsymbol{\tau}^p\, dX^1\, dX^2, \\ \boldsymbol{\sigma}^{\mathrm{I}\mathfrak{k}} &= \int_{\mathscr{A}} \alpha^{\mathrm{I}}\, \boldsymbol{B}^{\mathfrak{k}\mathsf{T}}\, \boldsymbol{\tau}^3\, dX^1\, dX^2, \\ \bar{\boldsymbol{\sigma}}^{\mathrm{I}\mathfrak{k}} &= \int_{\mathscr{A}} \alpha^{\mathrm{I}}\, \boldsymbol{B}^{\mathfrak{k}\mathsf{T}}_{,p}\, \boldsymbol{\tau}^p\, dX^1\, dX^2. \end{aligned} \tag{7.7}$$

Then

$$\int_{\mathscr{A}} \alpha^{\mathrm{I}}\, \boldsymbol{\tau}^p \cdot \dot{\boldsymbol{r}}_{,p}\, dX^1\, dX^2 \sim \sigma^{\mathrm{I}} + \sum_{\mathfrak{k}=0}^{\mathfrak{K}} (\boldsymbol{\sigma}^{\mathrm{I}\mathfrak{k}} \cdot \dot{\boldsymbol{r}}_{\mathfrak{k},S} + \bar{\boldsymbol{\sigma}}^{\mathrm{I}\mathfrak{k}} \cdot \dot{\boldsymbol{r}}_{\mathfrak{k}}). \tag{7.8}$$

The substitution of (7.5), (7.8) into the energy Eq. (6.14) yields the approximations

$$\begin{aligned} & q^{\mathrm{I}}_{,S} - \bar{q}^{\mathrm{I}} + \dot{\psi}^{\mathrm{I}} + \dot{\zeta}^{\mathrm{I}} - \mathring{\psi}^{\mathrm{I}} - \mathring{\zeta}^{\mathrm{I}} - \hat{h}^{\mathrm{I}}(\mathbf{r}_{\mathfrak{K}}^{(t)}, \lambda_{\mathfrak{L}}^{(t)}, S, t) \\ & \quad \sim \sigma^{\mathrm{I}} + \sum_{\mathfrak{k}=0}^{\mathfrak{K}} (\boldsymbol{\sigma}^{\mathrm{I}\mathfrak{k}} \cdot \dot{\boldsymbol{r}}_{\mathfrak{k},S} + \bar{\boldsymbol{\sigma}}^{\mathrm{I}\mathfrak{k}} \cdot \dot{\boldsymbol{r}}_{\mathfrak{k}}). \end{aligned} \tag{7.9}$$

We supplement (6.3) with

$$\begin{aligned} \mu &= \int_{\mathscr{A}} \left(\frac{\partial \boldsymbol{b}}{\partial t}\right)_{,p} \cdot \boldsymbol{\tau}^p\, dX^1\, dX^2, \\ \eta &= \int_{\mathscr{A}} \frac{\partial \beta}{\partial t}\, \zeta\, \varrho_0 \sqrt{G}\, dX^1\, dX^2. \end{aligned} \tag{7.10}$$

By $(7.10)_1$ we can approximate the total stress power over a section by

$$\int_{\mathscr{A}} \boldsymbol{\tau}^p \cdot \dot{\boldsymbol{r}}_{,p}\, dX^1\, dX^2 \sim \mu + \sum_{\mathfrak{k}=0}^{\mathfrak{K}} (\boldsymbol{\mu}^{\mathfrak{k}} \cdot \dot{\boldsymbol{r}}_{\mathfrak{k},S} + \bar{\boldsymbol{\mu}}^{\mathfrak{k}} \cdot \dot{\boldsymbol{r}}_{\mathfrak{k}}). \tag{7.11}$$

We may similarly generate approximations to the symmetry condition (6.21) and the entropy inequalities (6.22), (6.23), but these may contain integrals not reducible to combinations of the moments (6.1), (7.7) or their specializations (6.3), (7.10) already defined. Because of the subsidiary role played by these conditions, we do not exhibit general forms for them. We merely observe that for special choices of $\boldsymbol{b}$, β, $\boldsymbol{A}^{\mathfrak{k}}$, $\alpha^{\mathfrak{l}}$ these relations can be expressed in relatively simple form. Thus if $\boldsymbol{A}^{\mathfrak{k}}=\boldsymbol{B}^{\mathfrak{k}}=B^{\mathfrak{k}}\,\mathbf{1}$, where $\{B^{\mathfrak{k}}\}$ are scalars, and if $\boldsymbol{A}=\mathbf{1}$, then the symmetry condition (6.21) is approximated by

$$\int\limits_{\mathscr{A}} \frac{\partial \boldsymbol{b}}{\partial X^p} \times \boldsymbol{\tau}^3\, dX^1\, dX^2 + \sum_{\mathfrak{k}=0}^{\mathfrak{K}} \boldsymbol{r}_{\mathfrak{k},S} \times \boldsymbol{\mu}^{\mathfrak{k}} \sim \mathbf{0}. \tag{7.12}$$

If (5.8) is used, then this reduces to the form

$$\int\limits_{\mathscr{A}} \boldsymbol{b}^*_{,p} \times \boldsymbol{\tau}^p\, dX^1\, dX^2 + \sum_{\mathfrak{k}=0}^{\mathfrak{K}} (\boldsymbol{r}_{\mathfrak{k},S} \times \boldsymbol{\mu}^{\mathfrak{k}} + \boldsymbol{r}_{\mathfrak{k}} \times \bar{\boldsymbol{\mu}}^{\mathfrak{k}}) \sim \mathbf{0}. \tag{7.13}$$

If $\alpha=\alpha^0=1$, then the entropy inequalities (6.22), (6.23) for $\mathfrak{l}=0$ are approximated[18] by

$$\dot{\zeta}^0 - \sum_{\mathfrak{l}=0}^{\mathfrak{L}} \eta^{\mathfrak{l}} \dot{\lambda}_{\mathfrak{l}} - \eta - \hat{h}^0(\mathbf{r}^{(t)}_{\mathfrak{K}}, \boldsymbol{\lambda}^{(t)}_{\mathfrak{L}}, S, t) + q^0_{,S} - \bar{q}^0 - \sum_{\mathfrak{l}=0}^{\mathfrak{L}} \pi^{\mathfrak{l}} \lambda_{\mathfrak{l},S} - \pi \geqq 0, \tag{7.14}$$

$$-\dot{\psi}_0 - \sum_{\mathfrak{l}=0}^{\mathfrak{L}} \eta^{\mathfrak{l}} \dot{\lambda}_{\mathfrak{l}} - \eta + \mu + \sum_{\mathfrak{k}=0}^{\mathfrak{K}} (\boldsymbol{\mu}^{\mathfrak{k}} \cdot \dot{\boldsymbol{r}}_{\mathfrak{k},S} + \bar{\boldsymbol{\mu}}^{\mathfrak{k}} \cdot \dot{\boldsymbol{r}}_{\mathfrak{k}}) - \sum_{\mathfrak{l}=0}^{\mathfrak{L}} \pi^{\mathfrak{l}} \lambda_{\mathfrak{l},S} - \pi \geqq 0, \tag{7.15}$$

where

$$\pi^{\mathfrak{l}} = \int\limits_{\mathscr{A}} \beta^{\mathfrak{l}} q^3\, dX^1\, dX^2, \qquad \pi = \int\limits_{\mathscr{A}} \frac{\partial \beta}{\partial X^p} q^p\, dX^1\, dX^2. \tag{7.16}$$

Further simplications attend the use of (5.8).

If $\mathfrak{K}=\mathfrak{L}=(\mathfrak{M}+1)(\mathfrak{M}+2)/2$, where $\mathfrak{M}$ is a non-negative integer, and if

$$\{E^{\mathfrak{k}}(\mathbf{X}),\ \mathfrak{k}=1,\ldots,\mathfrak{K}\} = \{(X^1)^{\mathfrak{m}}(X^2)^{\mathfrak{n}},\ 0<\mathfrak{m},\ \mathfrak{n}<\mathfrak{M},\ 0<\mathfrak{m}+\mathfrak{n}<\mathfrak{M}\}, \tag{7.17}$$

then the assumption that $\boldsymbol{A}^{\mathfrak{k}}=\boldsymbol{B}^{\mathfrak{k}}=E^{\mathfrak{k}}\,\mathbf{1}$, $\alpha^{\mathfrak{l}}=\beta^{\mathfrak{l}}=E^{\mathfrak{l}}$, yields relations of especially simple form.[19] In particular, all the sets of stress moments can be written in terms of a single set. Similar simplifications arise if $\{E^{\mathfrak{k}}\}$ is taken to be a suitable set of trigonometric polynomials, or, in fact, any set of functions closed under multiplication.

8. Constitutive relations. Let $\mathbf{M}^{\mathfrak{K}\mathfrak{L}}$ denote the collection of moments (6.1), (7.7) for $\mathfrak{k}=0,\ldots,\mathfrak{K}$, $\mathfrak{l}=0,\ldots,\mathfrak{L}$. To obtain approximate representations for the constitutive relations for $\mathbf{M}^{\mathfrak{K}\mathfrak{L}}$, we first substitute the unreduced constitutive equations (3.20) into (6.1), (7.7) and then replace $\boldsymbol{r}_{,p}$, λ, γ by their approximants (5.2), (5.3), (5.13). We get

$$\mathbf{M}^{\mathfrak{K}\mathfrak{L}} \sim \check{\mathbf{M}}^{\mathfrak{K}\mathfrak{L}}(\mathbf{r}^{(t)}_{\mathfrak{K}}, \mathbf{r}^{(t)}_{\mathfrak{K},S}, \boldsymbol{\lambda}^{(t)}_{\mathfrak{L}}, \boldsymbol{\lambda}^{(t)}_{\mathfrak{L},S}, S, t), \tag{8.1}$$

which is an abbreviation for

$$\begin{aligned} \boldsymbol{\sigma}^{\mathfrak{k}} &\sim \check{\boldsymbol{\sigma}}^{\mathfrak{k}}(\mathbf{r}^{(t)}_{\mathfrak{K}}, \mathbf{r}^{(t)}_{\mathfrak{K},S}, \boldsymbol{\lambda}^{(t)}_{\mathfrak{L}}, \boldsymbol{\lambda}^{(t)}_{\mathfrak{L},S}, S, t) \\ &\equiv \int\limits_{\mathscr{A}} \boldsymbol{A}^{\mathfrak{k}}\, \boldsymbol{\tau}^3(\boldsymbol{b}^{(t)}_{,p}, \beta^{(t)}, \beta^{(t)}_{,p}, \mathbf{X})\, dX^1\, dX^2, \qquad \mathfrak{k}=0,\ldots,\mathfrak{K}, \quad \text{etc.} \end{aligned} \tag{8.2}$$

[18] Again, the interpretation of the fields (5.2) (5.3) as constrained in the sense described above removes the ambiguity in the inequalities (7.14), (7.15).

[19] Cf. ANTMAN and WARNER (1966), GREEN, LAWS, and NAGHDI (1968), and GREEN and NAGHDI (1970).

Similarly, we may substitute the reduced constitutive Eqs. (3.21) into (6.1), (7.7), and replace $\boldsymbol{C}$, λ, γ by their approximants (5.12), (5.3), (5.13) to get

$$\mathbf{M}^{\mathfrak{K}\mathfrak{L}} \sim \hat{\mathbf{M}}^{\mathfrak{K}\mathfrak{L}}(\mathbf{C}_{\mathfrak{K}}^{(t)}, \lambda_{\mathfrak{L}}^{(t)}, \lambda_{\mathfrak{L},S}^{(t)}, S, t), \tag{8.3}$$

an abbreviation for

$$\begin{aligned}\boldsymbol{\sigma}^{\mathfrak{k}} \cdot \tilde{\boldsymbol{b}}_{,p} &\sim \hat{\boldsymbol{\sigma}}_p^{\mathfrak{k}}(\mathbf{C}_{\mathfrak{K}}^{(t)}, \lambda_{\mathfrak{L}}^{(t)}, \lambda_{\mathfrak{L},S}^{(t)}, S, t)\\ &\equiv \tilde{\boldsymbol{b}}_{,p} \cdot \int_{\mathscr{A}} A^{\mathfrak{k}} \boldsymbol{b}_{,q}\, \hat{T}^{3q}((\boldsymbol{b}_{,i} \cdot \boldsymbol{b}_{,j})^{(t)}, \beta^{(t)}, \beta_{,i}^{(t)}, \boldsymbol{X})\, dX^1\, dX^2, \quad \mathfrak{k}=0, \ldots, \mathfrak{K}, \text{ etc.,}\end{aligned} \tag{8.4}$$

where $\tilde{\boldsymbol{b}}_{,p}(S, t)$ is the value of $\boldsymbol{b}_{,p}$ at any fixed X^1, X^2, say $X^1 = X^2 = 0$, and where $\mathbf{C}_{\mathfrak{K}}$ denotes the set of algebraically independent combinations of $\mathbf{r}_{\mathfrak{K}}$ and $\mathbf{r}_{\mathfrak{K},S}$ whose components are found in $\boldsymbol{b}_{,i} \cdot \boldsymbol{b}_{,j}$. E.g., if (5.8) is used then

$$\mathbf{C}_{\mathfrak{K}} = \{\boldsymbol{r}_{\mathfrak{k}}, \boldsymbol{r}_{\mathfrak{k},S}, \boldsymbol{r}_{\mathfrak{k}} \otimes \boldsymbol{r}_{\mathfrak{m}}, \boldsymbol{r}_{\mathfrak{k}} \otimes \boldsymbol{r}_{\mathfrak{m},S}, \boldsymbol{r}_{\mathfrak{k},S} \otimes \boldsymbol{r}_{\mathfrak{m},S} \colon \mathfrak{k}, \mathfrak{m} = 0, \ldots, \mathfrak{K}\}. \tag{8.5}$$

The constitutive functionals (3.20) are assumed to be actually in frame-indifferent form and the constitutive functionals (3.21) are in frame-indifferent form. Hence the constitutive functionals $\check{\mathbf{M}}^{\mathfrak{K}\mathfrak{L}}$ and $\hat{\mathbf{M}}^{\mathfrak{K}\mathfrak{L}}$ of (8.1)-(8.4) are approximately frame-indifferent if (5.2) is regarded as an approximation for $\boldsymbol{r}$ and are (exactly) frame-indifferent if (5.2) is regarded as an exact expression for $\boldsymbol{r}$ for a class of constrained materials. Since it is $\boldsymbol{b}$, rather that the vectors $\mathbf{r}_{\mathfrak{K}}$, $\mathbf{r}_{\mathfrak{K},S}$ themselves, that has a well-defined behavior under change of frame, we cannot in general expect these constitutive functionals to have a simple reduced dependence upon these vectors.

This question is clarified by noting that it is possible to replace the variables $\mathbf{r}_{\mathfrak{K}}$ with another set $\mathbf{p}_{\mathfrak{K}} = \{\boldsymbol{p}_{\mathfrak{k}}, \mathfrak{k} = 0, \ldots, \mathfrak{K}\}$ that has simple transformation properties under change of frame. We have much latitude in choosing the members of $\mathbf{p}_{\mathfrak{K}}$. We take

$$\boldsymbol{p}_0(S, t) = \boldsymbol{b}(\mathbf{r}_{\mathfrak{K}}(S, t), 0, 0, S, t) \tag{8.6}$$

and we may take

$$\boldsymbol{p}_{\mathfrak{k}}(S, t) = \boldsymbol{b}(\mathbf{r}_{\mathfrak{K}}(S, t), X_{\mathfrak{k}}^1, X_{\mathfrak{k}}^2, S, t) - \boldsymbol{p}_0(S, t), \mathfrak{k} = 1, \ldots, \mathfrak{K}, \tag{8.7}$$

or

$$\boldsymbol{p}_{\mathfrak{k}_\alpha}(S, t) = \boldsymbol{b}_{,\alpha}(\mathbf{r}_{\mathfrak{K}}(S, t), X_{\mathfrak{k}_\alpha}^1, X_{\mathfrak{k}_\alpha}^2, S, t), \quad \mathfrak{k}_\alpha = 1, \ldots, \mathfrak{K}. \tag{8.8}$$

In these cases

$$\boldsymbol{p}_0 \to \boldsymbol{c}(t) + \boldsymbol{Q}(t)\, \boldsymbol{p}_0, \quad \boldsymbol{p}_{\mathfrak{k}} \to \boldsymbol{Q}(t)\, \boldsymbol{p}_{\mathfrak{k}}, \quad \mathfrak{k} = 1, \ldots, \mathfrak{K} \tag{8.9}$$

under the change of frame

$$\boldsymbol{r} \to \boldsymbol{c}(t) + \boldsymbol{Q}(t)\, \boldsymbol{r}, \tag{8.10}$$

where $\boldsymbol{Q}(t)$ is orthogonal or proper-orthogonal depending on the version of frame-indifference used. From relations such as (8.6)–(8.8), we obtain a system of the form

$$\boldsymbol{p}_{\mathfrak{k}} = \hat{\boldsymbol{p}}_{\mathfrak{k}}(\mathbf{r}_{\mathfrak{K}}, S, t), \quad \mathfrak{k} = 0, \ldots, \mathfrak{K}. \tag{8.11}$$

Provided that

$$\det\left(\frac{\partial \hat{\boldsymbol{p}}_{\mathfrak{k}}}{\partial \boldsymbol{r}_{\mathfrak{m}}}\right) \neq 0, \tag{8.12}$$

we can solve (8.11) for $\mathbf{r}_{\mathfrak{K}}$ as functions of $\mathbf{p}_{\mathfrak{K}}$ and substitute these representations into (5.2) to obtain an expression of the form

$$\boldsymbol{b} = \tilde{\boldsymbol{b}}(\mathbf{p}_{\mathfrak{K}}, \boldsymbol{X}, t). \tag{8.13}$$

If the linear representation (5.8) is used, then in place of (8.13) we obtain an expression of the form

$$\boldsymbol{b} = \boldsymbol{d}^*(\boldsymbol{X}, t) + \sum_{\mathfrak{k}=0}^{\mathfrak{K}} \boldsymbol{D}^{\mathfrak{k}}(X, t)\, \boldsymbol{p}_{\mathfrak{k}}(S, t). \tag{8.14}$$

These representations (8.13), (8.14) meet all the position boundary conditions satisfied by (5.2), (5.8) respectively.[20] Since $\boldsymbol{b}$ itself transforms like (8.10), it is easy to show that the members of (8.14) have the transformations

$$\boldsymbol{D}^{\mathfrak{k}} \to \boldsymbol{Q}\,\boldsymbol{D}^{\mathfrak{k}}\,\boldsymbol{Q}^T, \quad \mathfrak{k}=0,\ldots,\mathfrak{K}, \quad \boldsymbol{d}^* \to (\boldsymbol{1}-\boldsymbol{Q}\,\boldsymbol{D}^0\,\boldsymbol{Q}^T)\,\boldsymbol{c}+\boldsymbol{Q}\,\boldsymbol{d}^* \tag{8.15}$$

under the change of frame (8.10).

The advantage of the variables $\mathbf{p}_{\mathfrak{K}}$ is slight: In general, we can no more infer the desirable conclusion that suitable components of $\mathbf{M}^{\mathfrak{K}\mathfrak{L}}$ depend only upon the inner products of

$$\boldsymbol{p}_{0,S}^{(t)},\, \boldsymbol{p}_{\mathfrak{k}}^{(t)},\, \boldsymbol{p}_{\mathfrak{k},S}^{(t)}, \quad \mathfrak{k}=1,\ldots,\mathfrak{K}, \tag{8.16}$$

(and upon $\lambda_{\mathfrak{L}}^{(t)}, \lambda_{\mathfrak{L},S}^{(t)}$) than we could when the representation (5.2) is used, for $\mathbf{M}^{\mathfrak{K}\mathfrak{L}}$ must be regarded not only as a functional of $\mathbf{p}_{\mathfrak{K}}^{(t)}, \mathbf{p}_{\mathfrak{K},S}^{(t)}, \lambda_{\mathfrak{L}}^{(t)}, \lambda_{\mathfrak{L},S}^{(t)}$ as in (8.1), (8.3), but also as a functional of the suppressed arguments which are the functions $\boldsymbol{b}$ and β. The possible explicit dependence of $\mathbf{M}^{\mathfrak{K}\mathfrak{L}}$ on t is through the histories of the suppressed arguments. In particular, if (8.14) and $(5.8)_2$ is used, the suppressed arguments are the histories of

$$\boldsymbol{d}^*,\, \boldsymbol{d}^*_{,q},\, \boldsymbol{D}^{\mathfrak{k}},\, \boldsymbol{D}^{\mathfrak{k}}_{,q},\, \beta^*,\, \beta^*_{,q},\, \beta^{\mathfrak{l}},\, \beta^{\mathfrak{l}}_{,q}.$$

In this case, we conclude that these suitable components of $\mathbf{M}^{\mathfrak{K}\mathfrak{L}}$ can be reduced to a functional of the inner products of (8.16) and of $\lambda_{\mathfrak{L}}^{(t)}, \lambda_{\mathfrak{L},S}^{(t)}$ and a function of S, t only when

$$\boldsymbol{d}^*=\boldsymbol{0}, \quad \boldsymbol{D}^{\mathfrak{k}}=D^{\mathfrak{k}}\,\boldsymbol{1}, \quad \mathfrak{k}=0,\ldots,\mathfrak{K}, \tag{8.17}$$

where $\{D^{\mathfrak{k}}\}$ are scalars. If in addition, $D^0=1$, then the substitution of (8.14), (8.17) into (8.3), (8.4) gives $\mathbf{M}^{\mathfrak{K}\mathfrak{L}}$ directly as a functional of the inner products of (8.16). Purely algebraic considerations may then be invoked to permit the elimination of $\{\boldsymbol{p}_{\mathfrak{k},S}\cdot\boldsymbol{p}_{\mathfrak{m},S},\, \mathfrak{k},\mathfrak{m}=1,\ldots,\mathfrak{K}\}$ as arguments of $\hat{\mathbf{M}}^{\mathfrak{K}\mathfrak{L}}$.

For any particular class of materials, the one-dimensional constitutive equations reflect the structure of their three-dimensional counterparts. We illustrate this in the next section with the constitutive equations for nonlinear thermo-elasticity.

9. Thermo-elastic rods. The constitutive relations for three-dimensional thermo-elastic materials in unreduced form[21] are given in terms of two functions $\check{\psi}(\boldsymbol{r}_{,p},\lambda,\boldsymbol{X})$ and $\check{\boldsymbol{q}}(\boldsymbol{r}_{,p},\lambda,\gamma,\boldsymbol{X})$ by

$$\begin{aligned}
\boldsymbol{\tau}^p(\boldsymbol{X},t) &= \varrho_0(\boldsymbol{X})\sqrt{G(\boldsymbol{X})}\,\frac{\partial\check{\psi}}{\partial\boldsymbol{r}_{,p}}\left(\boldsymbol{r}_{,q}(\boldsymbol{X},t),\lambda(\boldsymbol{X},t),\boldsymbol{X}\right),\\
\boldsymbol{q}(\boldsymbol{X},t) &= \check{\boldsymbol{q}}\left(\boldsymbol{r}_{,q}(\boldsymbol{X},t),\lambda(\boldsymbol{X},t),\gamma(\boldsymbol{X},t),\boldsymbol{X}\right),\\
\zeta(\boldsymbol{X},t) &= -\frac{\partial\check{\psi}}{\partial\lambda}\left(\boldsymbol{r}_{,q}(\boldsymbol{X},t),\lambda(\boldsymbol{X},t),\boldsymbol{X}\right),\\
\psi(\boldsymbol{X},t) &= \check{\psi}\left(\boldsymbol{r}_{,q}(\boldsymbol{X},t)\,\lambda(\boldsymbol{X},t),\boldsymbol{X}\right),
\end{aligned} \tag{9.1}$$

with $\check{\psi}$ satisfying the symmetry condition

$$\boldsymbol{r}_{,p}\times\frac{\partial\check{\psi}}{\partial\boldsymbol{r}_{,p}}(\boldsymbol{r}_{,q},\lambda,\boldsymbol{X})=\boldsymbol{0} \tag{9.2}$$

and $\check{\boldsymbol{q}}$ satisfying the heat conduction inequality

$$\check{\boldsymbol{q}}(\boldsymbol{r}_{,q},\lambda,\gamma,\boldsymbol{X})\cdot\gamma\leqq 0. \tag{9.3}$$

We define the functions $\tilde{\psi}$ and $\tilde{\boldsymbol{q}}$ by

$$\begin{aligned}
\tilde{\psi}(\mathbf{r}_{\mathfrak{K}},\mathbf{r}_{\mathfrak{K},S},\lambda_{\mathfrak{L}},\boldsymbol{X},t) &= \check{\psi}(\boldsymbol{b}_{,p},\beta,\boldsymbol{X}),\\
\tilde{\boldsymbol{q}}(\mathbf{r}_{\mathfrak{K}},\mathbf{r}_{\mathfrak{K},S},\lambda_{\mathfrak{L}},\lambda_{\mathfrak{L},S},\boldsymbol{X},t) &= \check{\boldsymbol{q}}(\boldsymbol{b}_{,p},\beta,\beta_{,p},\boldsymbol{X}).
\end{aligned} \tag{9.4}$$

---

[20] Note that $\boldsymbol{d}^*(0,0,S,t)=\boldsymbol{0}$, $\boldsymbol{D}^0(0,0,S,t)=\boldsymbol{1}$, $\boldsymbol{D}^{\mathfrak{k}}(0,0,S,t)=\boldsymbol{0}$, $\mathfrak{k}=1,\ldots,\mathfrak{K}$.

[21] Cf. NFTM, Sect. 96, and GREEN and ZERNA (1968).

From (5.11) we then obtain

$$\frac{\partial \tilde{\psi}}{\partial \mathbf{r}_{\mathfrak{k},S}} = \mathbf{B}^{\mathfrak{k}T} \frac{\partial \check{\psi}}{\partial \mathbf{r}_{,S}}, \qquad \frac{\partial \check{\psi}}{\partial \mathbf{r}_{\mathfrak{k}}} = \mathbf{B}^{\mathfrak{k}T}_{,p} \frac{\partial \check{\psi}}{\partial \mathbf{r}_{,p}}, \qquad \frac{\partial \psi}{\partial \lambda_{\mathrm{I}}} = \beta^{\mathrm{I}} \frac{\partial \check{\psi}}{\partial \lambda}. \tag{9.5}$$

We also define the following approximate representation for the total free energy:

$$\Psi(\mathbf{r}_{\mathfrak{K}}, \mathbf{r}_{\mathfrak{K},S}, \lambda_{\mathfrak{L}}, S, t) = \int_{\mathscr{A}} \tilde{\psi}(\mathbf{r}_{\mathfrak{K}}, \mathbf{r}_{\mathfrak{K},S}, \lambda_{\mathfrak{L}}, \mathbf{X}, t)\, \varrho_0 \sqrt{G}\, dX^1\, dX^2. \tag{9.6}$$

We can now obtain simple constitutive relations for some stress and temperature-entropy moments as derivatives of Ψ. We substitute $(9.1)_1$ into $(6.3)_1$ and use $(9.5)_1$, (9.6) to get[22]

$$\begin{aligned} \boldsymbol{\mu}^{\mathfrak{k}} \sim \check{\boldsymbol{\mu}}^{\mathfrak{k}}(\mathbf{r}_{\mathfrak{K}}, \mathbf{r}_{\mathfrak{K},S}, \lambda_{\mathfrak{L}}, S, t) &\equiv \int_{\mathscr{A}} \mathbf{B}^{\mathfrak{k}T} \frac{\partial \check{\psi}}{\partial \mathbf{r}_{,S}}\, \varrho_0 \sqrt{G}\, dX^1\, dX^2 \\ &= \int_{\mathscr{A}} \frac{\partial \tilde{\psi}}{\partial \mathbf{r}_{\mathfrak{k},S}}\, \varrho_0 \sqrt{G}\, dX^1\, dX^2 = \frac{\partial \Psi}{\partial \mathbf{r}_{\mathfrak{k},S}}. \end{aligned} \tag{9.7}$$

We similarly find

$$\bar{\boldsymbol{\mu}}^{\mathfrak{k}} \sim \check{\bar{\boldsymbol{\mu}}}^{\mathfrak{k}}(\mathbf{r}_{\mathfrak{K}}, \mathbf{r}_{\mathfrak{K},S}, \lambda_{\mathfrak{L}}, S, t) = \frac{\partial \Psi}{\partial \mathbf{r}_{\mathfrak{k}}}, \tag{9.8}$$

$$\eta^{\mathrm{I}} \sim \check{\eta}^{\mathrm{I}}(\mathbf{r}_{\mathfrak{K}}, \mathbf{r}_{\mathfrak{K},S}, \lambda_{\mathfrak{L}}, S, t) = -\frac{\partial \Psi}{\partial \lambda_{\mathrm{I}}}, \tag{9.9}$$

$$\mu - \eta \sim \check{\mu}(\mathbf{r}_{\mathfrak{K}}, \mathbf{r}_{\mathfrak{K},S}, \lambda_{\mathfrak{L}}, S, t) - \check{\eta}(\mathbf{r}_{\mathfrak{K}}, \mathbf{r}_{\mathfrak{K},S}, \lambda_{\mathfrak{L}}, S, t) = \frac{\partial \Psi}{\partial t}. \tag{9.10}$$

By introducing various moments of the free energy function, we can represent the remaining moments of (6.1), (6.3), (7.7), other than the moments of $\boldsymbol{q}$, as derivatives of these new free energy moments. E.g., from the definition

$$\Psi^{\mathrm{I}}(\mathbf{r}_{\mathfrak{K}}, \mathbf{r}_{\mathfrak{K},S}, \lambda_{\mathfrak{L}}, S, t) = \int_{\mathscr{A}} \alpha^{\mathrm{I}} \tilde{\psi}(\mathbf{r}_{\mathfrak{K}}, \mathbf{r}_{\mathfrak{K},S}, \lambda_{\mathfrak{L}}, X, t)\, \varrho_0 \sqrt{G}\, dX^1\, dX^2, \tag{9.11}$$

we obtain

$$\psi^{\mathrm{I}} \sim \Psi^{\mathrm{I}}(\mathbf{r}_{\mathfrak{K}}, \mathbf{r}_{\mathfrak{K},S}, \lambda_{\mathfrak{L}}, S, t), \tag{9.12}$$

$$\boldsymbol{\sigma}^{\mathrm{I}\mathfrak{k}} \sim \check{\boldsymbol{\sigma}}^{\mathrm{I}\mathfrak{k}}(\mathbf{r}_{\mathfrak{K}}, \mathbf{r}_{\mathfrak{K},S}, \lambda_{\mathfrak{L}}, S, t) = \frac{\partial \Psi^{\mathrm{I}}}{\partial \mathbf{r}_{\mathfrak{k},S}}, \tag{9.13}$$

$$\bar{\boldsymbol{\sigma}}^{\mathrm{I}\mathfrak{k}} \sim \check{\bar{\boldsymbol{\sigma}}}^{\mathrm{I}\mathfrak{k}}(\mathbf{r}_{\mathfrak{K}}, \mathbf{r}_{\mathfrak{K},S}, \lambda_{\mathfrak{L}}, S, t) = \frac{\partial \Psi^{\mathrm{I}}}{\partial \mathbf{r}_{\mathfrak{k}}}. \tag{9.14}$$

The approximate constitutive relations for q^{I} and $\bar{q}^{\mathrm{I}}$ are found directly from $(6.1)_{3,4}$ and $(9.1)_2$:

$$q^{\mathrm{I}} \sim \check{q}^{\mathrm{I}}(\mathbf{r}_{\mathfrak{K}}, \mathbf{r}_{\mathfrak{K},S}, \lambda_{\mathfrak{L}}, \lambda_{\mathfrak{L},S}, S, t) \equiv \int_{\mathscr{A}} \alpha^{\mathrm{I}} \check{q}^3(\mathbf{b}_{,p}, \beta, \beta_{,p}, \mathbf{X})\, dX^1\, dX^2, \tag{9.15}$$

$$\bar{q}^{\mathrm{I}} \sim \check{\bar{q}}^{\mathrm{I}}(\mathbf{r}_{\mathfrak{K}}, \mathbf{r}_{\mathfrak{K},S}, \lambda_{\mathfrak{L}}, \lambda_{\mathfrak{L},S}, S, t) \equiv \int_{\mathscr{A}} \alpha^{\mathrm{I}}_{,i} \check{q}^i(\mathbf{b}_{,p}, \beta, \beta_{,p}, \mathbf{X})\, dX^1\, dX^2. \tag{9.16}$$

Special choices for (5.8) reduce many of these relations to especially simple form.

We note that the constitutive relations (9.1) reduce the energy Eq. (3.16) to the thermal energy equation

$$q^p_{,p} - \varrho_0 \sqrt{G}\,(\dot{\lambda}\zeta - \dot{\xi} + h) = 0. \tag{9.17}$$

We could use moments of this equation in place of (7.9).

[22] A result of this form was first obtained by Naghdi and Nordgren (1963).

The one-dimensional constitutive relations obtained directly from the reduced three-dimensional constitutive equations[23]

$$\boldsymbol{T}(\boldsymbol{X}, t) = 2\varrho_0(\boldsymbol{X})\sqrt{G(\boldsymbol{X})}\,\frac{\partial\hat{\psi}}{\partial \boldsymbol{C}}\,(\boldsymbol{C}(\boldsymbol{X}, t), \lambda(\boldsymbol{X}, t), \boldsymbol{X}), \quad \text{etc.}, \tag{9.18}$$

are formally more complicated. We restrict our consideration to the linear representations of (5.8). Following the same procedure as before, we define the new functions ω and Ω by

$$\omega(\mathbf{C}_{\mathfrak{K}}, \lambda_{\mathfrak{L}}, \boldsymbol{X}, t) = \hat{\psi}(\boldsymbol{b}_{,p}\cdot\boldsymbol{b}_{,q}, \beta, \boldsymbol{X}), \tag{9.19}$$

$$\Omega(\mathbf{C}_{\mathfrak{K}}, \lambda_{\mathfrak{L}}, S, t) = \int_{\mathscr{A}} \omega(\mathbf{C}_{\mathfrak{K}}, \lambda_{\mathfrak{L}}, \boldsymbol{X}, t)\,\varrho_0\sqrt{G}\,dX^1\,dX^2. \tag{9.20}$$

Since the elements of $\mathbf{C}_{\mathfrak{K}}$ are regarded as independent, we readily find

$$\begin{aligned}
\boldsymbol{\mu}^{\mathfrak{k}} &\sim \hat{\boldsymbol{\mu}}^{\mathfrak{k}}(\mathbf{C}_{\mathfrak{K}}, \lambda_{\mathfrak{L}}, S, t) = \frac{\partial\Omega}{\partial \boldsymbol{r}_{\mathfrak{k},S}} + \sum_{\mathfrak{m}=0}^{\mathfrak{K}}\left[\frac{\partial\Omega}{\partial(\boldsymbol{r}_{\mathfrak{k}}\otimes\boldsymbol{r}_{\mathfrak{m},S})}\,\boldsymbol{r}_{\mathfrak{m}} + 2\,\frac{\partial\Omega}{\partial(\boldsymbol{r}_{\mathfrak{k},S}\otimes\boldsymbol{r}_{\mathfrak{m},S})}\,\boldsymbol{r}_{\mathfrak{m},S}\right],\\
\bar{\boldsymbol{\mu}}^{\mathfrak{k}} &\sim \hat{\bar{\boldsymbol{\mu}}}^{\mathfrak{k}}(\mathbf{C}_{\mathfrak{K}}, \lambda_{\mathfrak{L}}, S, t) = \frac{\partial\Omega}{\partial \boldsymbol{r}_{\mathfrak{k}}} + \sum_{\mathfrak{m}=0}^{\mathfrak{K}}\left[2\,\frac{\partial\Omega}{\partial(\boldsymbol{r}_{\mathfrak{k}}\otimes\boldsymbol{r}_{\mathfrak{m}})}\,\boldsymbol{r}_{\mathfrak{m}} + \frac{\partial\Omega}{\partial(\boldsymbol{r}_{\mathfrak{k}}\otimes\boldsymbol{r}_{\mathfrak{m},S})}\,\boldsymbol{r}_{\mathfrak{m},S}\right], \quad \text{etc.}
\end{aligned} \tag{9.21}$$

If we use the representation (8.14) with $\boldsymbol{d}^* = \boldsymbol{0}$, $\boldsymbol{D}^{\mathfrak{k}} = D^{\mathfrak{k}}\,\boldsymbol{1}$, $D^0 = 1$, then $\mathbf{C}_{\mathfrak{K}}$ reduces to the set of inner products

$$\boldsymbol{p}_{0,S}\cdot\boldsymbol{p}_{0,S},\quad \boldsymbol{p}_{0,S}\cdot\boldsymbol{p}_{\mathfrak{k}},\quad \boldsymbol{p}_{0,S}\cdot\boldsymbol{p}_{\mathfrak{k},S},\quad \boldsymbol{p}_{\mathfrak{k}}\cdot\boldsymbol{p}_{\mathfrak{m}},\quad \boldsymbol{p}_{\mathfrak{k}}\cdot\boldsymbol{p}_{\mathfrak{m},S},\quad \boldsymbol{p}_{\mathfrak{k},S}\cdot\boldsymbol{p}_{\mathfrak{m},S},\quad \mathfrak{k}, \mathfrak{m} = 1, \ldots, \mathfrak{K}, \tag{9.22}$$

and we can replace (9.21) by the simpler and obviously frame-indifferent relations[24]

$$\begin{aligned}
\boldsymbol{\mu}^{\mathfrak{k}} &\sim \hat{\boldsymbol{\mu}}(\mathbf{C}_{\mathfrak{K}}, S, t) = \sum_{\mathfrak{m}=0}^{\mathfrak{K}}\left[\frac{\partial\Omega}{\partial(\boldsymbol{p}_{\mathfrak{k}}\cdot\boldsymbol{p}_{\mathfrak{m},S})}\,\boldsymbol{p}_{\mathfrak{m}} + 2\,\frac{\partial\Omega}{\partial(\boldsymbol{p}_{\mathfrak{k},S}\cdot\boldsymbol{p}_{\mathfrak{m},S})}\,\boldsymbol{p}_{\mathfrak{m},S}\right],\\
\bar{\boldsymbol{\mu}}^{\mathfrak{k}} &\sim \hat{\bar{\boldsymbol{\mu}}}^{\mathfrak{k}}(\mathbf{C}_{\mathfrak{K}}, S, t) = \sum_{\mathfrak{m}=0}^{\mathfrak{K}}\left[2\,\frac{\partial\Omega}{\partial(\boldsymbol{p}_{\mathfrak{k}}\cdot\boldsymbol{p}_{\mathfrak{m}})}\,\boldsymbol{p}_{\mathfrak{m}} + \frac{\partial\Omega}{\partial(\boldsymbol{p}_{\mathfrak{k}}\cdot\boldsymbol{p}_{\mathfrak{m},S})}\,\boldsymbol{p}_{\mathfrak{m},S}\right],
\end{aligned}$$

in which the summation does not extend to terms formally including $\boldsymbol{p}_0$. (Thus $\bar{\boldsymbol{\mu}}^0 = \boldsymbol{0}$.)

The one-dimensional constitutive equations for COLEMAN's materials with fading memory (NFTM, Sect. 96bis) are found in a similar fashion. Roughly speaking, we merely replace the partial derivatives of Ψ that occur in these one-dimensional constitutive equations of elasticity by Fréchet derivatives with respect to the corresponding histories.

10. Statement of the boundary value problems. The governing equations of the approximate or constrained theory of order $(\mathfrak{K}, \mathfrak{L})$ are obtained by replacing the approximation relation "$\sim$" wherever it appears in Sects. 5–9 by equality "$=$". From (7.6) we thus obtain

$$\boldsymbol{\sigma}^{\mathfrak{k}}_{,S} - \bar{\boldsymbol{\sigma}}^{\mathfrak{k}} + \hat{\boldsymbol{f}}^{\mathfrak{k}}(\boldsymbol{r}_{\mathfrak{K}}^{(t)}, \lambda_{\mathfrak{L}}^{(t)}, S, t) = \boldsymbol{a}^{\mathfrak{k}}(\boldsymbol{r}_{\mathfrak{K}}, \dot{\boldsymbol{r}}_{\mathfrak{K}}, \ddot{\boldsymbol{r}}_{\mathfrak{K}}, S, t), \quad \mathfrak{k} = 0, \ldots, \mathfrak{K}, \tag{10.1}$$

where the functionals $\hat{\boldsymbol{f}}^{\mathfrak{k}}$ and the functions $\boldsymbol{a}^{\mathfrak{k}}$ are prescribed. From (7.9) we get

$$\begin{aligned}
q^{\mathfrak{l}}_{,S} - \bar{q}^{\mathfrak{l}} + \dot{\psi}^{\mathfrak{l}} + \dot{\zeta}^{\mathfrak{l}} - \mathring{\psi}^{\mathfrak{l}} - \mathring{\zeta}^{\mathfrak{l}} - \hat{h}^{\mathfrak{l}}(\boldsymbol{r}_{\mathfrak{K}}^{(t)}, \lambda_{\mathfrak{L}}^{(t)}, S, t)\\
= \sigma^{\mathfrak{l}} + \sum_{\mathfrak{k}=0}^{\mathfrak{K}}(\boldsymbol{\sigma}^{\mathfrak{l}\mathfrak{k}}\cdot\dot{\boldsymbol{r}}_{\mathfrak{k},S} + \bar{\boldsymbol{\sigma}}^{\mathfrak{l}\mathfrak{k}}\cdot\dot{\boldsymbol{r}}_{\mathfrak{k}}), \quad \mathfrak{l} = 0, \ldots, \mathfrak{L},
\end{aligned} \tag{10.2}$$

---

[23] Since $\boldsymbol{C}$ is symmetric, the tensor $\dfrac{\partial\hat{\psi}}{\partial\boldsymbol{C}}$ has contravariant components $\dfrac{1}{2}\left(\dfrac{\partial\hat{\psi}}{\partial C_{pq}} + \dfrac{\partial\hat{\psi}}{\partial C_{pq}}\right)$ where $\hat{\psi}$ is treated as a function of the nine components of $\boldsymbol{C}$ regarded as independent (cf. NFTM, Sect. 9).

[24] Explicit forms of the equations when $\{D^{\mathfrak{k}}\}$ are homogeneous polynomials in X^1 and X^2 are given by ANTMAN and WARNER (1966), GREEN, LAWS, and NAGHDI (1968), and GREEN and NAGHDI (1970).

where the functionals $\hat{h}^{\mathrm{I}}$ are prescribed. The constitutive relations are

$$\mathbf{M}^{\mathfrak{K}\mathfrak{L}} = \check{\mathbf{M}}^{\mathfrak{K}\mathfrak{L}}(\mathbf{r}_{\mathfrak{K}}^{(t)}, \mathbf{r}_{\mathfrak{K},S}^{(t)}, \lambda_{\mathfrak{L}}^{(t)}, \lambda_{\mathfrak{L},S}^{(t)}, S, t)$$

or (10.3)

$$\mathbf{M}^{\mathfrak{K}\mathfrak{L}} = \hat{\mathbf{M}}^{\mathfrak{K}\mathfrak{L}}(\mathbf{C}_{\mathfrak{K}}^{(t)}, \lambda_{\mathfrak{L}}^{(t)}, \lambda_{\mathfrak{L},S}^{(t)}, S, t).$$

The elements of $\mathbf{M}^{\mathfrak{K}\mathfrak{L}}$ are listed in (6.1), (7.7) and the forms of $\check{\mathbf{M}}^{\mathfrak{K}\mathfrak{L}}$ and $\hat{\mathbf{M}}^{\mathfrak{K}\mathfrak{L}}$ are described in Sect. 8.

Our boundary value problem is to solve (10.1)–(10.3)[25] for $\mathbf{r}_{\mathfrak{K}}$ and $\lambda_{\mathfrak{L}}$ subject to boundary conditions at $S=S_1, S_2$ and to an appropriate specification of temporal behavior necessary to ensure a well-set problem. We do not discuss the nature of suitable temporal conditions,[26] since they depend strongly on the particular form of (10.3). We limit our consideration to boundary conditions at the ends $S=S_1, S_2$, the boundary conditions on the lateral surface having been incorporated into the formulation of (10.1)–(10.3).

We wish to establish the one-dimensional analog of the rule embodied in (3.25). By a judicious choice of the variables entering (5.5), (5.6), we can ensure that $\frac{\partial \boldsymbol{b}}{\partial t} = \dot{\boldsymbol{b}}^*$ at the ends of the rod. The power contribution of the tractions acting at the ends

$$\int_{\mathscr{A}(S_\alpha)} \boldsymbol{\tau}^3 \cdot \dot{\boldsymbol{r}}\, dX^1\, dX^2 \tag{10.4}$$

can then be approximated by

$$\int_{\mathscr{A}(S_\alpha)} \boldsymbol{\tau}^3 \cdot \dot{\boldsymbol{b}}^*\, dX^1\, dX^2 + \sum_{\mathfrak{k}=0}^{\mathfrak{K}} \boldsymbol{\mu}^{\mathfrak{k}} \cdot \dot{\boldsymbol{r}}_{\mathfrak{k}} \big|_{S=S_\alpha}. \tag{10.5}$$

We need not worry about the integral of (10.5) because $\boldsymbol{b}^*$ is a prescribed function of $\boldsymbol{X}$, t.

If the rod is a ring, i.e., if the ends consist of the same particles, we assume that all the given functions of S (namely, reference geometry, loads, and constitutive functionals) can be smoothly extended to the entire S-axis with period $S_2 - S_1$ in S. Then we require the unknowns of the problem to satisfy the periodicity conditions

$$\boldsymbol{r}_{\mathfrak{k}}(S_1, t) = \boldsymbol{r}_{\mathfrak{k}}(S_2, t), \quad \boldsymbol{\mu}^{\mathfrak{k}}(S_1, t) = \boldsymbol{\mu}^{\mathfrak{k}}(S_2, t), \quad \text{etc.} \tag{10.6}$$

Otherwise we proceed as follows. For each $\mathfrak{k}$ let $\{\boldsymbol{e}_p^{\mathfrak{k}}(S_\alpha, t)\}$ be a given set of three independent vectors and let $\{\boldsymbol{e}_{\mathfrak{k}}^p(S_\alpha, t)\}$ be their duals: $\boldsymbol{e}_p^{\mathfrak{k}} \cdot \boldsymbol{e}_{\mathfrak{k}}^q = \delta_p^q$. Then for each $\mathfrak{k}$, each p, and each α, we specify either

$$\boldsymbol{\mu}^{\mathfrak{k}}(S_\alpha, t) \cdot \boldsymbol{e}_{\mathfrak{k}}^p(S_\alpha, t) = \mu^{\mathfrak{k}p}(\boldsymbol{r}_{\mathfrak{K}}^{(t)}(S_\alpha, \cdot), \lambda_{\mathfrak{L}}^{(t)}(S_\alpha, \cdot), S_\alpha, t), \tag{10.7}$$

where $\mu^{\mathfrak{k}p}$ is a given functional of its arguments, or else we specify

$$\dot{\boldsymbol{r}}_{\mathfrak{k}}(S_\alpha, t) \cdot \boldsymbol{e}_p^{\mathfrak{k}}(S_\alpha, t). \tag{10.8}$$

[25] Component forms of (10.1)–(10.3) are readily obtained by taking the inner products of the equations with a suitable set of independent vectors. The base vectors $\boldsymbol{g}_p$ evaluated at $X^1 = X^2 = 0$ naturally suggest themselves for this purpose, but in fact they are not nearly as useful for the analysis and physics as an appropriate system of orthogonal unit vectors, which generate physical components (cf. Sect. 17). Tensorial forms of these equations may be found in the work of ANTMAN and WARNER (1966), GREEN, LAWS, and NAGHDI (1968), and GREEN and NAGHDI (1970).

[26] In elasticity such conditions would include initial and periodicity conditions.

The expressions (10.4), (10.5) show that in each of the cases (10.6)–(10.8) the ends make (approximately) zero contribution to the integral in (3.25).

We refrain from developing formal strategies for dealing with the most general cases of mixed boundary conditions on the ends [see (3.23)-(3.25)]. If position conditions are prescribed on the ends, we simply choose $\boldsymbol{r}_{\mathfrak{R}}$ so that $\boldsymbol{b}(\boldsymbol{r}_{\mathfrak{R}}(S_\alpha, t), X^1, X^2, S_\alpha, t)$ gives the best approximation to $\boldsymbol{r}(S_\alpha, t)$. If traction conditions are prescribed at the ends, we prescribe $\{\boldsymbol{\mu}^{\mathfrak{k}}(S_\alpha, t)\}$. We also leave open the question whether we can replace boundary conditions (10.7) on $\boldsymbol{\mu}^{\mathfrak{k}}$ by analogous conditions on $\boldsymbol{\sigma}^{\mathfrak{k}}$, when these moments are not equivalent by (6.4), (6.5). This can only be settled by existence theorems. We note however that (10.7) are well-defined conditions even when the stress moments employed are $\{\boldsymbol{\sigma}^{\mathfrak{k}}\}$.

The thermal boundary conditions are analogous. In the ring case we specify periodicity conditions of the form (10.6). Otherwise for each $\mathfrak{l}$ and each α we specify either

$$q^{\mathfrak{l}} = \hat{q}^{\mathfrak{l}}(\boldsymbol{r}_{\mathfrak{R}}^{(t)}(S_\alpha, \cdot), \lambda_{\mathfrak{L}}^{(t)}(S_\alpha, \cdot), S_\alpha, t) \tag{10.9}$$

where $\hat{q}^{\mathfrak{l}}$ is a given functional of its arguments, or else we specify

$$\lambda_{\mathfrak{l}} = \lambda_{\mathfrak{l}}(S_\alpha, t). \tag{10.10}$$

11. Validity of the projection methods. The projection methods described above produce well-defined one-dimensional boundary value problems for the fundamental dependent variables $\boldsymbol{r}_{\mathfrak{R}}, \lambda_{\mathfrak{L}}$. The substitution of a solution $\boldsymbol{r}_{\mathfrak{R}}(S, t)$, $\lambda_{\mathfrak{L}}(S, t)$ into $\boldsymbol{b}(\boldsymbol{r}_{\mathfrak{R}}, S, t), \beta(\lambda_{\mathfrak{L}}, S, t)$ yields approximate representations for $\boldsymbol{r}$ and λ. These may be substituted into the three-dimensional constitutive Eqs. (3.20) or (3.21) to generate approximations for the dependent constitutive variables. (Alternatively, we may construct approximations for these variables directly from the moments $\boldsymbol{M}^{\mathfrak{R}\mathfrak{L}}$.) The resulting collection of approximations for $\boldsymbol{r}$, λ, $\boldsymbol{T}$, $\boldsymbol{q}$, ζ, ψ is called a $(\mathfrak{R}, \mathfrak{L})$-approximation. For each $(\mathfrak{R}, \mathfrak{L})$ we let $\boldsymbol{U}_{\mathfrak{R}\mathfrak{L}}$ denote the set of all $(\mathfrak{R}, \mathfrak{L})$-approximations. The number of elements of $\boldsymbol{U}_{\mathfrak{R}\mathfrak{L}}$ may be zero, one, finite, or infinite. We let $\boldsymbol{U}$ denote the set of all solutions to the exact three-dimensional problem. It too may have any number of elements.

The fundamental problem underlying our projection methods is to determine how accurately $\boldsymbol{U}$ is described by $\boldsymbol{U}_{\mathfrak{R}\mathfrak{L}}$ for a given class of problems and for a given set of functions $\boldsymbol{b}$, β, $\{\boldsymbol{A}^{\mathfrak{k}}, \mathfrak{k}=0, \ldots, \mathfrak{R}\}$, $\{\alpha^{\mathfrak{l}}, \mathfrak{l}=0, \ldots, \mathfrak{L}\}$. (The class of problems is defined by restrictions on material response, reference geometry, loads, and boundary conditions. The efficiency of the projection methods may be considerably enhanced by a choice of functions $\boldsymbol{b}$, β, $\{\boldsymbol{A}^{\mathfrak{k}}\}$, $\{\alpha^{\mathfrak{l}}\}$ reflecting the structure of the class of problems under study.) To pose this problem precisely, we must first define the concept of solution by interpreting the governing equations as certain operator equations on a function space containing $\boldsymbol{U}$ and $\boldsymbol{U}_{\mathfrak{R}\mathfrak{L}}$. The size of the members of a class of functionals depending upon $\boldsymbol{U}$ and $\boldsymbol{U}_{\mathfrak{R}\mathfrak{L}}$ serves as a criterion for accuracy. The function space and the functionals defined thereon should be suited for describing the physical and mathematical structure of the three-dimensional equations. Two less useful but more accessible problems are to show that the elements of $\boldsymbol{U}_{\mathfrak{R}\mathfrak{L}}$ converge to those of $\boldsymbol{U}$ and to determine the rate of covergence.

Projection methods have been justified for many types of equations, the body of results for linear equations being far more extensive than that for nonlinear equations.[27] Unfortunately the results have not yet been extended to systems

[27] See KANTOROVICH and KRYLOV (1952), KANTOROVICH and AKILOV (1959), KRASNOSEL'SKII, VAINIKKO *et al.* (1969), LIONS (1969).

with the complexity of the three-dimensional equations of continuum mechanics.[28] Nevertheless, the available proofs for simpler problems and the body of comparisons of projection approximations with exact solutions suggest that we may expect some sort of weak convergence. Less frequently, stronger modes of convergence may obtain. Weak convergence is appropriate because rod theories are developed to deliver the position of a reference curve and the values of the stress resultants, rather than the pointwise behavior of the fields entering the three-dimensional problem. The lack of uniformity in weak solutions is typical also of the boundary layer behavior to be expected at the ends where the solution form might have to change rapidly to accomodate the specified boundary conditions. This question of uniformity is closely related to the validity of ST. VENANT's principle, which is more naturally treated in a three-dimensional context by means of appropriate estimates or by asymptotic methods.[29]

Rod problems from linear elasticity. An examination of classical problem of linear elasticity indicates the manner in which our projection methods work. We first study the St. Venant problem. Let $\boldsymbol{X}=(X, Y, S)$ be a set of material Cartesian coordinates with corresponding unit base vectors $\boldsymbol{I}, \boldsymbol{J}, \boldsymbol{K}$. The reference configuration of a homogeneous, isotropic, linearly elastic rod occupies the region bounded by the cylinder $\phi(X, Y)=0$ and by the two planes $S=0$, $S=L$. We assume that the S-axis coincides with the line of centroids of this region and that the X and Y axes are along the principal axes of inertia of the cross sections. The resultant force and moment over an end are

$$\begin{aligned} \boldsymbol{n} &= E\,I_1\,Q_1\,\boldsymbol{I}+E\,I_2\,Q_2\,\boldsymbol{J}+E\,A\,P\,\boldsymbol{K}, \\ \boldsymbol{m} &= E\,I_1\,M_1\,\boldsymbol{I}+E\,I_2\,M_2\,\boldsymbol{J}+D\,T\,\boldsymbol{K}, \end{aligned} \tag{11.1}$$

where EI_1, EI_2, EA, D are given constants depending upon the material and the cross-sectional geometry. For zero traction on the lateral surface and for certain specified tractions on the ends having $\boldsymbol{n}$ and $\boldsymbol{m}$ as resultant force and moment, this body undergoes the displacement given (to within a rigid body motion) by

$$\sum_{\mathfrak{k}=0}^{3} \boldsymbol{B}^{\mathfrak{k}}(X, Y)\,\boldsymbol{r}_{\mathfrak{k}}(S), \tag{11.2}$$

where

$$\begin{aligned} &\boldsymbol{B}^0=1, \quad \boldsymbol{B}^1=\begin{pmatrix} X & Y & 0 \\ Y & -X & 0 \\ 0 & 0 & X \end{pmatrix}, \quad \boldsymbol{B}^2=\begin{pmatrix} \frac{1}{2}(X^2-Y^2) & XY & 0 \\ XY & \frac{1}{2}(Y^2-X^2) & 0 \\ 0 & 0 & Y \end{pmatrix}, \\ &\boldsymbol{B}^3=\begin{pmatrix} 0 & 0 & 0 \\ 0 & 0 & 0 \\ \Phi_1(X, Y) & \Phi_2(X, Y) & \Phi(X, Y) \end{pmatrix}, \end{aligned} \tag{11.3}$$

$$\begin{aligned} &\boldsymbol{r}_0=\begin{pmatrix} \frac{1}{2}\left[Q_1\left(L-\frac{S}{3}\right)+M_2\right]S^2 \\ \frac{1}{2}\left[Q_2\left(L-\frac{S}{3}\right)-M_1\right]S^2 \\ PS \end{pmatrix}, \quad \boldsymbol{r}_1=-\begin{pmatrix} \nu P \\ TS \\ \left[Q_1\left(L-\frac{S}{2}\right)+M_2\right]S \end{pmatrix}, \\ &\boldsymbol{r}_2=\begin{pmatrix} \nu\,[Q_1(L-S)+M_2] \\ \nu\,[Q_2(L-S)-M_1] \\ -\left[Q_2\left(L-\frac{S}{2}\right)-M_1\right]S \end{pmatrix}, \quad \boldsymbol{r}_3=\begin{pmatrix} Q_1 \\ Q_2 \\ T \end{pmatrix}. \end{aligned} \tag{11.4}$$

---

[28] One of the difficulties is that physically reasonable conditions that ensure the existence of solutions to the three-dimensional equations are generally not known. These conditions may be quite delicate (cf. Sects. 19–22). Once existence has been established, it may then be possible to employ these conditions to establish constructive procedures (in particular projection methods) to obtain it. E.g., cf. BROWDER (1967), PETRYSHYN (1967).

[29] See Sect. 13.

Here ν is POISSON's ratio and Φ_1, Φ_2, Φ are functions determined from the cross-sectional geometry with Φ_1, Φ_2 depending on ν.

We observe that the displacement has at least a cubic dependence on X and Y. This is not surprising: A quadratic dependence would produce a stress field linear in X and Y and such a stress field could not generally meet the boundary conditions of zero traction on the lateral surface without causing a degeneracy such as the vanishing of ν. If we seek to approximate solutions to a class of problems by polynomials in X and Y (i.e., by a representation in the form (5.8) with $\boldsymbol{B}^{\mathfrak{k}} = E^{\mathfrak{k}}\mathbf{1}$ where $E^{\mathfrak{k}}$ is defined in (7.17)), and if we demand as a criterion of accuracy that these approximate representations reduce exactly to the St. Venant solution (11.2) when the data is that of the St. Venant problem, then (11.2)–(11.4) require that there be at least nine members in the collection $\{\boldsymbol{r}_{\mathfrak{k}}(S, t)\}$ of vectors to be determined. Consequently, the resulting theory is quite complicated. On the other hand such polynomial approximations uniformly approximate the solutions of arbitrary St. Venant problems and yield exact solutions when Φ_1, Φ_2, Φ are polynomials.

We may seek to represent the displacements of a class of problems by (11.2), (11.3). In this case, the class of approximations reduces exactly to the St. Venant problem and there are but four vectors, $\boldsymbol{r}_0$, $\boldsymbol{r}_1$, $\boldsymbol{r}_2$, $\boldsymbol{r}_3$, to be determined. It is easy to see that it is impossible to reduce the number of unknown vectors to fewer than four if the approximation must contain the St. Venant solutions, if $\boldsymbol{B}^0 = \mathbf{1}$, and if $\boldsymbol{B}^1, \boldsymbol{B}^2, \boldsymbol{B}^2$ depend only on X, Y and not on P, Q_1, Q_2, T, M_1, M_2. We also note that with (11.2) we have $\boldsymbol{\mu}^0 = \boldsymbol{n}$. Since

$$(X\boldsymbol{I} + Y\boldsymbol{J}) \times = \begin{pmatrix} 0 & 0 & 0 \\ 0 & 0 & -1 \\ 0 & -1 & 0 \end{pmatrix} \boldsymbol{B}^1 + \begin{pmatrix} 0 & 0 & 1 \\ 0 & 0 & 0 \\ 0 & 0 & 0 \end{pmatrix} \boldsymbol{B}^2, \tag{11.5}$$

we find for linear problems that

$$\boldsymbol{m} = \begin{pmatrix} 0 & 0 & 0 \\ 0 & 0 & -1 \\ 0 & -1 & 0 \end{pmatrix} \boldsymbol{\mu}^1 + \begin{pmatrix} 0 & 0 & 1 \\ 0 & 0 & 0 \\ 0 & 0 & 0 \end{pmatrix} \boldsymbol{\mu}^2. \tag{11.6}$$

If we relax these restrictions, then there are an infinite number of ways of embedding ST. VENANT's problem in a larger class of approximations. The simplest such approach is to represent the displacement by

$$\boldsymbol{u}^*(\boldsymbol{X}) + \boldsymbol{c}(\mathbf{r}_{\mathfrak{K}}, \boldsymbol{X}, t)$$

where $\boldsymbol{u}^*(\boldsymbol{X})$ is the St. Venant displacement field given by (11.2)–(11.4), $\boldsymbol{c}$ is a given function of its arguments with $\boldsymbol{c}(\mathbf{0}, \boldsymbol{X}, t) = \mathbf{0}$, and $\mathbf{r}_{\mathfrak{K}}$ is to be determined.

The comparison of approximate solutions with the St. Venant solutions as a test of accuracy was promoted by NOVOZHILOV (1948). We regard this criterion merely as a reasonable expedient in the absence of sharp estimates. Such estimates would be free of dependence upon closed form solutions.[30] If we lift this comparison criterion, we may still construct approximations that meet the traction boundary conditions on the lateral surface by using a device analogous to that of Sect. 6. In particular, either we may exploit the fact that if $\phi(\boldsymbol{X}) = 0$ is the equation of the lateral surface, then $[\phi^2(\boldsymbol{X})]_{,p} = 0$, or else we may expand $\boldsymbol{r}$ in suitable eigenfunctions in X and Y.

Technical theories of rods may be characterized as approximations for linear theories in which the displacement varies linearly in X and Y. (Generally the displacement is subject to further constraints, such as the Kirchhoff hypotheses.) In cases of technical interest, these theories yield good results for displacements and stress resultants, but, as noted above, furnish an inadequate description of shear [cf. LOVE (1927) for further discussion].

All possible rod theories for linear elasticity based upon the linear representation (5.8) have the same form for fixed $\mathfrak{K}$. (This is a consequence of the general result for any nonlinear material.) Appearing in such theories are a number of functions of S that represent geometric and material properties. The values of the functions depend upon the choice of $\{\boldsymbol{B}^{\mathfrak{k}}\}$. We may regard the one-dimensional equations as typifying all linear elastic rod theories. We are then free to choose these functions to fit the data of a class of problems. This interpretation

[30] This question is clarified by a consideration of the analogous problem of approximating a given function $f(x)$ on a finite interval by a finite linear combination $\sum_{0}^{\mathfrak{K}} r_{\mathfrak{k}} B_{\mathfrak{k}}(x)$ with the coefficients $\{r_{\mathfrak{k}}\}$ to be determined. In general, the requirement that the curve pass through $\mathfrak{K}+1$ prescribed points does not produce the best solution. ("Best" is defined by the choice of metric.) These prescribed points correspond to the St. Venant solutions.

underlies the director theories discussed in Subchapter III. [Also of cf. GREEN, LAWS, and NAGHDI (1967).]

The treatment of dynamical problems is similar to that of static problems. The canonical problem replacing the St. Venant problem is the Pochhammer-Chree problem of wave propagation along an infinite rod with circular cross-section. While classical technical theories give good results for low frequency waves, they prove inadequate for high frequency waves. MINDLIN and HERRMANN (1952), MINDLIN and MCNIVEN (1960), MEDICK (1966), and others used special projection methods to obtain improved models for wave propagation in straight rods. To have their results agree with the Pochhammer-Chree solution, these authors found it expedient to alter some of the constants determined by the projection method. Such alterations, justified by pragmatism, can be traced back to TIMOSHENKO'S (1921) treatment of shear deformation in rods. We prefer to justify these alterations by invoking the form invariance of theories for linearly elastic rods obtained by projection methods. The projection methods developed by VOLTERRA (1955, 1956, 1961) have been applied by him and his collaborators to treat a variety of linear dynamical problems. NOVOZHILOV and SLEPIAN (1965) examined the validity of a dynamic St. Venant principle for theories obtained by projection methods.

12. History of the use of projection methods for the construction of rod theories. The idea of constructing one-dimensional rod theories from three-dimensional models by averaging the stress over a cross-section was introduced by LEIBNIZ in 1684 and was further developed in the next two centuries by JAS. BERNOULLI, EULER, CAUCHY, POISSON, KIRCHHOFF, CLEBSCH, LOVE, and others. A comprehensive historical analysis of the work on rods up to 1788 is given by TRUESDELL (1960). No comparable study of subsequent work is available. Guidance to the research of the nineteenth century may be obtained from CFT, Sects. 63A, 214, TODHUNTER and PEARSON (1886, 1893), LOVE (1927), TIMOSHENKO (1953), TRUESDELL (1959).

Modern work has been devoted to the systemization of the derivation of rod theories from the equations of three-dimensional continuum mechanics. HAY (1942) used expansions in a thickness parameter to obtain a precise description of strain in a rod. His methods are discussed in the following section. NOVOZHILOV (1948) developed nonlinear models for small strain. MINDLIN and HERRMANN (1952), VOLTERRA (1955, 1956), MINDLIN and MCNIVEN (1960), VOLTERRA (1961), MEDICK (1966), and others represented the displacement by a polynomial in the transverse coordinates to generate linearly elastic rod theories. These theories were subsequently used to treat problems of vibrations and wave propagation (see Sect. 11).

GREEN (1959) obtained the exact equilibrium equations for resultant force and moment by integrating the three-dimensional equilibrium equations over a cross-section. ANTMAN and WARNER (1966) represented the position by polynomials in transverse coordinates and took moments of the equations of motion with powers of these coordinates to obtain a hierarchy of theories for hyperelastic rods. They examined several possible approximation techniques. GREEN, LAWS, and NAGHDI (1968) and GREEN and NAGHDI (1970), while retaining some of the apparatus of ANTMAN and WARNER, attacked the problem from a different approach. They sought to construct thermodynamic rod theories that are independent of the choice of representation for position and temperature. This aim was partly motivated by the inadequacy, described in Sect. 11, of theories based upon a representation for position that is linear in the transverse coordinates.[31] These authors represented position and temperature by formal power series in the transverse coordinates and took moments of the equations of motion, energy, and the entropy inequality to generate a theory with infinitely many fundamental dependent variables: $\boldsymbol{r}_{\infty}, \lambda_{\infty}$. They constructed theories in just a finite number of variables by replacing the constitutive functionals depending on $\boldsymbol{r}_{\infty}, \lambda_{\infty}$ by new constitutive functionals depending only upon $\boldsymbol{r}_{\mathfrak{K}}, \lambda_{\mathfrak{L}}$. The replacement process was left unspecified. They retained unchanged an appropriate finite number of field equations. We interpret this procedure as the imposition of an infinite number of constraints on $\boldsymbol{r}_{\infty}$ and λ_{∞}. This view is incorporated in our representations (5.2), (5.3) for $\boldsymbol{r}$ and λ. Thus our projection methods generalize these diverse approaches.

[31] GREEN, LAWS, and NAGHDI (1967, 1968) noted that the theory of VOLTERRA (1955, 1956) and the first order theory of ANTMAN and WARNER (1966), which are of this form, fail to provide results in agreement with those for the St. Venant flexure problem. On the other hand, the theory of VOLTERRA (1961) and the higher order theories of ANTMAN and WARNER, which are correspondingly more complicated, do not suffer from this defect (cf. Sect. 11). The chief disadvantage of these refined theories is their complexity, not their lack of accuracy. Tacit in the work of GREEN, LAWS, and NAGHDI is the concept that a simple theory should suffice to describe the typical problems of rod theory.

Work on the theory of rods is closely related to that for the theory of shells. Indeed, there is a formal duality between these theories: If Y^1 and Y^2 represent convected material coordinates for the reference surface of a shell and if Y^3 is a convected material coordinate in a transverse direction, then the interchange

$$X^\alpha \leftrightarrow Y^3, \quad X^3 \leftrightarrow Y^\alpha$$

effectively converts the equations for rod theories into those for shell theories and vice-versa. The reader is referred to NAGHDI's article on shells in this volume.

13. Asymptotic methods. The construction of rod theories by the asymptotic expansion of the variables of the three-dimensional theory in a small thickness parameter has advantages not enjoyed by other methods: The role of thinness is explicit, boundary layer effects are readily accounted for, and the procedure for correcting a given order of approximation requires the solution of just linear equations. On the other hand, the elegant structure of the nonlinear theory obtained by projection methods [as exemplified by (9.7)-(9.10)] is suppressed and the corrections may require the solution of equations in more than one variable. There are numerous ways of representing the solution by a formal[32] asymptotic expansion. Their variety reflects the range of possible interrelations between the thickness parameter and the other parameters supplied in the data.[33]

HAY (1942) treated the equilibrium of curved nonlinearly elastic rods of uniform cross-section subject only to tractions at the ends.[34] He employed a special nonlinear stress-strain law. He scaled the transverse normal coordinates X^α by

$$X^\alpha = \varepsilon \xi^\alpha, \tag{13.1}$$

where ε is a ratio of thickness to length. He assumed that the variables entering the equations have power series expansions in X^1, X^2. The substitution of (13.1) into these power series generates expansions in powers of ε and the substitution of these ε-series into the governing equations yields equations for the unknown coefficients of powers of ε. Using these in conjunction with equations for the resultants, HAY found the orders in ε that the external load must have in order to produce small displacements, small strains but large displacements, and large strains. These results depend strongly on the stress-strain law used. HAY did not treat boundary layer effects.

NARIBOLI (1969) treated axisymmetric longitudinal waves in a circular, linearly elastic rod. Using (13.1), he represented a typical variable u, regarded as a function of ξ^α, S, t by the interior expansion

$$u(\xi^\alpha, S, t, \varepsilon) \sim \sum_{k=0}^{\infty} u^{(k)}(\xi^\alpha, S, t)\, \varepsilon^k. \tag{13.2}$$

Substituting such expansions into the governing equations, he obtained a sequence of equations for the coefficients of the powers of ε. He showed that the leading term in the expansion for the longitudinal displacement satisfies the classical wave equation and he examined the question of recovering the correction of RAYLEIGH [cf. LOVE (1927, Sect. 278)]. He did not study the boundary layer.

[32] There are but few proofs that a solution of a partial differential equation has a given asymptotic character.

[33] If the thickness parameter is chosen to be the ratio of thickness to length, other dimensionless parameters can be constructed from radii of curvature and torsion, wavelengths, size of solutions, amplitude of applied forces, frequency of applied forces, etc. An important goal of asymptotic methods is to determine, if possible, the asymptotic representations that lead naturally to various classical and engineering theories and thereby assign a precise asymptotic meaning to these theories.

[34] His work generalizes that of GOODIER (1938) for straight, linearly elastic rods.

REISS (1972) used the method of matched asymptotic expansions[35] to obtain both the interior and boundary layer behavior for the planar flexural motion of a linearly elastic beam with square cross-section. Using the scaling (13.1) and setting $\tau=\varepsilon t$, he sought a (formally) uniform asymptotic expansion of a typical variable u, regarded as a function of ξ^α, S, τ, in the form

$$u(\xi^\alpha, S, \tau, \varepsilon) \sim \sum_{k=0}^{\infty} u^{(k)}(\xi^\alpha, S, \tau)\, \varepsilon^k + \sum_{k=0}^{\infty}\left[U_1^{(k)}\left(\xi^\alpha, \frac{S-S_1}{\varepsilon}, \tau\right) + U_2^{(k)}\left(\xi^\alpha, \frac{S_2-S}{\varepsilon}, \tau\right)\right]\varepsilon^k. \tag{13.3}$$

The second sum accounts for boundary layer behavior. Substituting (13.3) into the governing equations, he obtained equations for the coefficients of powers of ε in (13.3). He employed matching conditions to relate the interior and boundary layer coefficients. He found that the leading term in the interior expansion for the transverse displacement satisfies the classical equation for flexural motion. The usual boundary conditions were found directly from the matching conditions. The leading term in the boundary layer expansion is the solution of the three-dimensional equilibrium problem (depending parametrically upon τ) for a semi-infinite rod with a stress-free lateral surface. A similar analysis for longitudinal motion shows that the leading term of the interior expansion for the longitudinal displacement satisfies the classical wave equation.

II. Director theories of rods.

14. Definition of a Cosserat rod. We now adopt an alternative concept of a rod as a material curve c in $\mathscr{E}^3$ together with a collection of vectors assigned to each particle of c that deform independently of c. Precisely, a *Cosserat rod*[36] is a set of vector fields

$$\mathbf{p}_{\mathfrak{K}}(S,t) = \{\boldsymbol{p}_{\mathfrak{k}}(S,t),\ S_1 \leqq S \leqq S_2,\ \mathfrak{k}=0,\ldots,\mathfrak{K}\} \tag{14.1}$$

that transform according to the rules

$$\boldsymbol{p}_0(S,t) \to \boldsymbol{c}(t) + \boldsymbol{Q}(t)\,\boldsymbol{p}_0(S,t), \tag{14.2}$$

$$\boldsymbol{p}_{\mathfrak{k}}(S,t) \to \boldsymbol{Q}(t)\,\boldsymbol{p}_{\mathfrak{k}}(S,t), \qquad \mathfrak{k}=1,\ldots,\mathfrak{K} \tag{14.3}$$

under the change of frame

$$\boldsymbol{r} \to \boldsymbol{c}(t) + \boldsymbol{Q}(t)\,\boldsymbol{r}, \tag{14.4}$$

where $\boldsymbol{Q}$ is orthogonal or proper orthogonal depending on the version of frame-indifference used.[37] The relevant kinematic significance of $\mathbf{p}_{\mathfrak{K}}$ is embodied in (14.2), (14.3): Relation (14.2) indicates that $\boldsymbol{r}=\boldsymbol{p}_0(S,t)$ defines the position of a material curve in $\mathscr{E}^3$, which we denote c, and (14.3) indicates that $\{\boldsymbol{p}_{\mathfrak{k}},\ \mathfrak{k}=1,\ldots,\mathfrak{K}\}$

[35] This method is essentially due to FRIEDRICHS (1949, 1950).

[36] Such models of continua were introduced by DUHEM (1893) and by E. and F. COSSERAT (1907, 1909). A resurgence of interest followed in the wake of ERICKSEN and TRUESDELL'S (1958) paper, an account of which appears in CFT, Sects. 63, 214.

[37] Our development in this subchapter is logically independent of the preceding work. For heuristic purposes we nevertheless retain some notations identical to those used above. The discussion in Sect. 8 indicates that (14.2)–(14.4) represent a rather sharp restriction on the kinematic versatility of the theory.

behave like differences between position vectors. The members of $\{\boldsymbol{p}_{\mathfrak{k}}, \mathfrak{k}=1, \ldots, \mathfrak{K}\}$ are called *directors*. The thermal state of a Cosserat rod is characterized by a set of scalar fields

$$\lambda_{\mathfrak{L}}(S,t) = \{\lambda_{\mathfrak{l}}(S,t), \mathfrak{l}=0, \ldots, \mathfrak{L}\}. \tag{14.5}$$

For simplicity, we restrict our attention to the important special case[38] $\mathfrak{K}=2$, $\mathfrak{L}=0$. We require that the vectors $\boldsymbol{p}_{0,S}, \boldsymbol{p}_1, \boldsymbol{p}_2$ be independent:

$$\boldsymbol{p}_{0,S} \cdot (\boldsymbol{p}_1 \times \boldsymbol{p}_2) > 0. \tag{14.6}$$

We term (14.6) the continuity condition.[39]

15. Field equations. Having defined the kinematic and temperature variables for a one-dimensional theory, we must now supply the remaining ingredients of a general continuum theory: equations of motion and energy and an entropy inequality. Since Cosserat rods are not three-dimensional bodies, the appropriate forms for these relations are not altogether obvious if no appeal is made to a three-dimensional model. Below we examine several intrinsically one-dimensional formulations of these relations. Rather than attempting to maintain the level of generality of Subchapter I, we pursue the more modest course of studying the restricted models that have been developed in the literature. By comparing the forms of the resulting equations with those of Subchapter I, we are able to classify precisely the nature of the special assumptions underlying these Cosserat theories. Here and in the rest of this article we frequently use a prime in place of "$,S$".

α) *Direct postulation of the relations in integral forms.* The most straightforward way of obtaining the one-dimensional balance laws and entropy inequality is simply to postulate them: There are fields $\boldsymbol{\mu}^0(S,t) \equiv \boldsymbol{n}(S,t)$, $\boldsymbol{\mu}^\alpha(S,t)$, $\bar{\boldsymbol{\mu}}^\alpha(S,t)$, $q^0(S,t)$, $\eta^0(S,t)$, $\psi^0(S,t)$ satisfying

the balance of linear momentum

$$[\boldsymbol{n}]_U^V + \int_U^V \boldsymbol{f}^0 \, dS = \frac{d}{dt} \int_U^V \sum_{\mathfrak{m}=0}^{2} J^{0\mathfrak{m}} \dot{\boldsymbol{p}}_{\mathfrak{m}} \, dS, \tag{15.1}$$

the balance of director momentum

$$[\boldsymbol{\mu}^\alpha]_U^V + \int_U^V (\boldsymbol{f}^\alpha - \bar{\boldsymbol{\mu}}^\alpha) \, dS = \frac{d}{dt} \int_U^V \sum_{\mathfrak{m}=0}^{2} J^{\alpha\mathfrak{m}} \dot{\boldsymbol{p}}_{\mathfrak{m}} \, dS, \tag{15.2}$$

the balance of angular momentum

$$[\boldsymbol{p}_0 \times \boldsymbol{n} + \boldsymbol{p}_\alpha \times \boldsymbol{\mu}^\alpha]_U^V + \int_U^V (\boldsymbol{p}_0 \times \boldsymbol{f}^0 + \boldsymbol{p}_\alpha \times \boldsymbol{f}^\alpha) \, dS = \frac{d}{dt} \int_U^V \sum_{\mathfrak{k},\mathfrak{m}=0}^{2} J^{\mathfrak{k}\mathfrak{m}} \boldsymbol{p}_{\mathfrak{k}} \times \dot{\boldsymbol{p}}_{\mathfrak{m}} \, dS, \tag{15.3}$$

[38] This model has sufficient kinematic structure not only to account for flexure, torsion, and axial extension, which characterize the simplest rod theories, but also to account for the shear of three pairs of perpendicular directions and for transverse extensions. In short, this model encompasses all the kinematic phenomena that have names. The case $\mathfrak{K}=3$ has also been developed extensively in the literature, although the only particular virtue that we can attribute to it is that it permits the representation for $\boldsymbol{r}$ based upon (11.3). The case $\mathfrak{K}=2$ is obtained naturally from $\mathfrak{K}=3$ by means of the constraint $\boldsymbol{p}_3 = \boldsymbol{p}_{0,3}$. By taking $\mathfrak{L}=0$, we allow just the usual one-dimensional heat flow.

[39] The nature of the restrictions that (14.6) imposes upon $\boldsymbol{p}_2$ may be ascertained by comparing (5.14), (8.6)–(8.17). We shall discuss the analog of the strong continuity condition (5.15) in Sect. 17.

the balance of energy

$$[\boldsymbol{n}\cdot\dot{\boldsymbol{p}}_0+\boldsymbol{\mu}^\alpha\cdot\dot{\boldsymbol{p}}_\alpha-q^0]_U^V+\int_U^V(h^0+\boldsymbol{f}^0\cdot\dot{\boldsymbol{p}}_0+\boldsymbol{f}^\alpha\cdot\dot{\boldsymbol{p}}_\alpha)\,dS$$
$$=\frac{d}{dt}\int_U^V\left(\psi^0+\eta^0+\frac{1}{2}\sum_{\mathfrak{k},\mathfrak{m}=0}^{2}J^{\mathfrak{k}\mathfrak{m}}\dot{\boldsymbol{p}}_{\mathfrak{k}}\cdot\dot{\boldsymbol{p}}_{\mathfrak{m}}\right)dS\,, \tag{15.4}$$

and *the entropy inequality*

$$\frac{d}{dt}\int_U^V e^{-\lambda_0}\eta^0\,dS-\int_U^V e^{-\lambda_0}h^0\,dS+[e^{-\lambda_0}q^0]_U^V\geqq 0\,, \tag{15.5}$$

for all intervals $[U, V]$ in $[S_1, S_2]$. Here $\{J^{\mathfrak{k}\mathfrak{m}}=J^{\mathfrak{m}\mathfrak{k}}\}$ are given functions of S, and $\{\boldsymbol{f}^{\mathfrak{k}}, h^0\}$ are given functionals of $\mathbf{p}_2^{(t)}$, $\lambda_0^{(t)}$, S, t. We need not postulate conservation of mass since we employ a strictly material description.

The reduction of (15.1), (15.2), (15.4), (15.5) to local form[40] yields special cases of (10.1) with $\mathfrak{k}=0$, (10.1) with $\mathfrak{k}=\alpha$, (7.9) with $\mathfrak{l}=0$, and (7.14), respectively. To obtain the local consequences of (15.3), we set

$$\boldsymbol{m}=\boldsymbol{p}_\alpha\times\boldsymbol{\mu}^\alpha,\qquad \boldsymbol{l}=\boldsymbol{p}_\alpha\times\boldsymbol{f}^\alpha. \tag{15.6}$$

Then from the local forms of (15.1) and (15.3) we readily obtain

$$\boldsymbol{m}'+\boldsymbol{p}_0'\times\boldsymbol{n}+\boldsymbol{l}=\sum_{\mathfrak{m}=0}^{2}J^{\alpha\mathfrak{m}}\boldsymbol{p}_\alpha\times\ddot{\boldsymbol{p}}_{\mathfrak{m}}\,, \tag{15.7}$$

which corresponds to the $\mathfrak{K}=2$ approximation of (6.19). From (15.7) and the local form of (15.2) we get

$$\boldsymbol{p}_0'\times\boldsymbol{n}+\boldsymbol{p}_\alpha'\times\boldsymbol{\mu}^\alpha+\boldsymbol{p}_\alpha\times\bar{\boldsymbol{\mu}}^\alpha=\boldsymbol{0}\,, \tag{15.8}$$

which is a special case of (7.13). We regard both (15.5) and (15.8) as constitutive restrictions.

We can expose the tacit assumptions underlying the postulates (15.1)-(15.6) by examining the special choices of $\boldsymbol{b}$, β, $\{\boldsymbol{A}^{\mathfrak{k}}\}$, $\{\alpha^{\mathfrak{l}}\}$ used to generate the corresponding equations in Subchapter I. We find that

$$\boldsymbol{b}=\boldsymbol{p}_0+D^\alpha(\boldsymbol{X})\,\boldsymbol{p}_\alpha,\qquad \beta=\lambda_0,\qquad \boldsymbol{A}^0=\mathbf{1},\qquad \boldsymbol{A}^\alpha=D^\alpha\,\mathbf{1},\qquad \alpha^0=1. \tag{15.9}$$

It is clear that the greatest restrictions on the generality of Subchapter I occur in the acceleration terms. (In the literature, $J^{0\alpha}$ is usually taken to be zero. This can be effected by choosing D^α appropriately, which corresponds to locating the curve $\boldsymbol{c}$ appropriately in a cross-section.)

From (15.4) we can read off suitable boundary conditions at the ends $S=S_1, S_2$. These are identical with those given in Sect. 10. In particular, if the rod is a ring, we have periodicity conditions. Otherwise, for $\mathfrak{k}=0, 1, 2$, let $\{\boldsymbol{e}_p^{\mathfrak{k}}(S_\alpha, t)\}$ be a given set of three independent vectors and let $\{\boldsymbol{e}_{\mathfrak{k}}^p(S_\alpha, t)\}$ be their duals. For each $\mathfrak{k}$, p, α we specify either

$$\boldsymbol{\mu}^{\mathfrak{k}}(S_\alpha, t)\cdot\boldsymbol{e}_{\mathfrak{k}}^p(S_\alpha, t)=\mu^{\mathfrak{k}p}\left(\mathbf{p}_2^{(t)}(S_\alpha,\cdot),\,\lambda_0^{(t)}(S_\alpha,\cdot)\,S_\alpha,t\right), \tag{15.10}$$

where $\mu^{\mathfrak{k}p}$ is a given functional of its arguments, or else we specify

$$\dot{\boldsymbol{p}}_{\mathfrak{k}}(S_\alpha, t)\cdot\boldsymbol{e}_p^{\mathfrak{k}}(S_\alpha, t). \tag{15.11}$$

---

[40] The integral forms (15.1)–(15.5) not only imply the corresponding differential equations, but also jump conditions at discontinuities. We shall not use these.

The differential equations corresponding to (15.1) and (15.3) for the equilibrium of rods in a plane were laid down by EULER. A penetrating analysis of the precedents of this work is given in TRUESDELL'S (1960) historical treatise. CLEBSCH, following KIRCHHOFF (cf. LOVE, 1927), provided the differential equilibrium equations for (15.1) and (15.3) for rods in space. A careful modern postulation of these equations was given by ERICKSEN and TRUESDELL (1958) and reproduced in CFT, Sect. 214, where further historical details are supplied.

To render determinate the rod theories based upon these six scalar equations, it was found necessary to restrict the number of geometric variables to six or fewer. This reduction was usually effected by the use of "KIRCHHOFF'S hypotheses" (cf. LOVE, 1927), which require plane cross-sections of the rod normal to a given material curve in the reference configuration to remain plane, undeformed, and normal to the same material curve in an arbitrary configuration. (Our previous work shows that this same reduction can be achieved by introducing any of a large class of constraints in place of KIRCHHOFF'S hypotheses. In the nonlinear theory, the mathematical structure of the resulting problems is invariant under the choice of constraints from this class.) To retain the full kinematic generality of the case $\mathfrak{K}=2$, it is necessary to supplement the six scalar equations of motion (15.1), (15.3) by three others, or equivalently to replace (15.3) by six scalar equations and regard (15.3) essentially as a constitutive restriction. In confronting this problem, GREEN and LAWS (1966) first obtained the differential equations (15.1), (15.3) by the method discussed below. They then obtained the differential equations corresponding to (15.2) by effectively postulating them, although their treatment obscures this. Rather than stating (15.2) directly, in their equation (3.17) they defined (in our notation)

$$\bar{\boldsymbol{\mu}}^{\alpha} = \boldsymbol{\mu}^{\alpha}_{,S} + \boldsymbol{f}^{\alpha} - \frac{d}{dt} \sum_{\mathfrak{m}=0}^{2} J^{\alpha\mathfrak{m}}\, \dot{\boldsymbol{p}}_{\mathfrak{m}},$$

the variables on the right side of this equation having been introduced previously. They then prescribed constitutive relations for $\boldsymbol{\mu}^{\alpha}$ and $\bar{\boldsymbol{\mu}}^{\alpha}$, thereby elevating this definition of $\bar{\boldsymbol{\mu}}^{\alpha}$ to a set of fundamental differential equations for the problem. In addition, GREEN and LAWS laid down integral laws equivalent to (15.4), (15.5). DESILVA and WHITMAN (1971) elaborated these results, giving a complete postulational formulation equivalent to (15.1)–(15.5).

β) *Invariance of energy balance under superposed rigid motions.* From the postulate that the energy balance (15.4) be invariant under the change of frame (14.2)–(14.4) with $\boldsymbol{c}=\boldsymbol{0}$, $\ddot{\boldsymbol{c}}=\boldsymbol{0}$, $\boldsymbol{Q}=\boldsymbol{1}$, $\ddot{\boldsymbol{Q}}=\boldsymbol{0}$, GREEN and LAWS (1966) obtained (15.1), (15.3). DESILVA and WHITMAN (1971) refined their analysis by showing that such invariance forces $J^{\mathfrak{k}\mathfrak{m}}$ to be time independent. LAWS (1967a) used this approach in his treatment of the relationship of dipolar rod theories to planar problems.

The advantage of this method is that the single scalar law (15.4) yields consistent forms of (15.1), (15.3). The energy balance is easy to construct because the physical concepts entering it have standard mathematical forms: Kinetic energy is a quadratic form in the generalized velocities, mechanical power terms are bilinear forms in the generalized forces and velocities, etc.

The disadvantage of this method is that it does not yield (15.2). The remedy for this difficulty is implicit in the three-dimensional treatment of GREEN and NAGHDI (1970): Two additional laws of energy balance must be provided. The requirement that these be invariant under change of frame yields (15.2). From the viewpoint of Subchapter I, it appears however that this approach works only when products of $\{D^{\mathfrak{k}}\}$ can be expressed as linear combinations of $\{D^{\mathfrak{k}}\}$ as would be the case if $\{D^{\mathfrak{k}}\}$ were taken to be homogeneous polynomials in X^1, X^2.

γ) *Variational methods.*[41] That variational principles could yield the equations for elastic rods was recognized by D. BERNOULLI, proved by EULER, and further developed by LAGRANGE (cf. TRUESDELL, 1960). KIRCHHOFF initiated the use of variational formulations as a device to ensure the consistency of boundary value

[41] The role of variational principles in continuum mechanics in general, and in elasticity in particular, is analyzed in CFT, Sects. 231–238; NFTM, Sects. 88, 89. These discussions do not treat the virtues of direct variational methods in analysis.

problems for rods. This program was continued by E. and F. COSSERAT (1909). The development by variational principles of theories for nonlinearly elastic Cosserat rods was carried out by MEISSONNIER (1965), TADJBAKHSH (1966), COHEN (1966, 1967), WHITMAN and DE SILVA (1969). The exploitation of the analytic advantages of variational methods has begun in the work of TADJBAKHSH and ODEH (1967), ANTMAN (1968c, 1970b) (cf. Sects. 19–22).

16. Constitutive equations. We employ the entropy principle as interpreted by COLEMAN and NOLL (cf. NFTM, Sect. 96) to obtain the constitutive relations for thermo-elastic Cosserat rods.[41a] We define a thermo-elastic rod by its constitutive equations

$$\begin{aligned}
\boldsymbol{n}(S,t) &= \check{\boldsymbol{n}}\left(\boldsymbol{p}_\alpha(S,t), \boldsymbol{p}_0'(S,t), \boldsymbol{p}_\alpha'(S,t), \lambda_0(S,t), \lambda_0'(S,t), S\right),\\
\boldsymbol{\mu}^\alpha(S,t) &= \check{\boldsymbol{\mu}}^\alpha\left(\boldsymbol{p}_\alpha(S,t), \boldsymbol{p}_0'(S,t), \boldsymbol{p}_\alpha'(S,t), \lambda_0(S,t), \lambda_0'(S,t), S\right),\\
\bar{\boldsymbol{\mu}}^\alpha(S,t) &= \check{\bar{\boldsymbol{\mu}}}^\alpha\left(\boldsymbol{p}_\alpha(S,t), \boldsymbol{p}_0'(S,t), \boldsymbol{p}_\alpha'(S,t), \lambda_0(S,t), \lambda_0'(S,t), S\right),\\
q^0(S,t) &= \check{q}^0\left(\boldsymbol{p}_\alpha(S,t), \boldsymbol{p}_0'(S,t), \boldsymbol{p}_\alpha'(S,t), \lambda_0(S,t), \lambda_0'(S,t), S\right),\\
\eta^0(S,t) &= \check{\eta}^0\left(\boldsymbol{p}_\alpha(S,t), \boldsymbol{p}_0'(S,t), \boldsymbol{p}_\alpha'(S,t), \lambda_0(S,t), \lambda_0'(S,t), S\right),\\
\psi^0(S,t) &= \Psi\left(\boldsymbol{p}_\alpha(S,t), \boldsymbol{p}_0'(S,t), \boldsymbol{p}_\alpha'(S,t), \lambda_0(S,t), \lambda_0'(S,t), S\right).
\end{aligned} \tag{16.1}$$

We simplify the local form of (15.4) by using (15.1), (15.2), and then substitute the resulting equation and (16.1) into the local form of (15.5). We get

$$\begin{aligned}
&\left(\check{\boldsymbol{n}} - \frac{\partial \Psi}{\partial \boldsymbol{p}_0'}\right)\cdot \dot{\boldsymbol{p}}_0' + \left(\check{\boldsymbol{\mu}}^\alpha - \frac{\partial \Psi}{\partial \boldsymbol{p}_\alpha'}\right)\cdot \dot{\boldsymbol{p}}_\alpha' + \left(\check{\bar{\boldsymbol{\mu}}}^\alpha - \frac{\partial \Psi}{\partial \boldsymbol{p}_\alpha}\right)\cdot \dot{\boldsymbol{p}}_\alpha\\
&\qquad - \left(\check{\eta}^0 + \frac{\partial \Psi}{\partial \lambda_0}\right)\dot{\lambda}_0 - \frac{\partial \Psi}{\partial \lambda_0'}\dot{\lambda}_0' - \check{q}^0 \lambda_0' \geqq 0.
\end{aligned} \tag{16.2}$$

Since this inequality must hold for arbitrary thermodynamic processes, we have

$$\frac{\partial \Psi}{\partial \lambda_0'} = 0, \qquad \check{q}^0 \lambda_0' \leqq 0 \tag{16.3}$$

and

$$\check{\boldsymbol{n}} = \frac{\partial \Psi}{\partial \boldsymbol{p}_0'}, \qquad \check{\boldsymbol{\mu}}^\alpha = \frac{\partial \Psi}{\partial \boldsymbol{p}_\alpha'}, \qquad \check{\bar{\boldsymbol{\mu}}}^\alpha = \frac{\partial \Psi}{\partial \boldsymbol{p}_\alpha}, \qquad \check{\eta}^0 = -\frac{\partial \Psi}{\partial \lambda_0}. \tag{16.4}$$

Frame-indifference requires that Ψ and $\check{q}^0$ have the forms

$$\Psi(\boldsymbol{p}_\alpha, \boldsymbol{p}_0', \boldsymbol{p}_\alpha', \lambda_0, S) = \Omega(\boldsymbol{p}_\alpha\cdot\boldsymbol{p}_\beta, \boldsymbol{p}_0'\cdot\boldsymbol{p}_\beta, \boldsymbol{p}_0'\cdot\boldsymbol{p}_0', \boldsymbol{p}_\alpha'\cdot\boldsymbol{p}_0', \lambda_0, S), \tag{16.5}$$

$$\check{q}^0(\boldsymbol{p}_\alpha, \boldsymbol{p}_0', \boldsymbol{p}_\alpha', \lambda_0, \lambda_0', S) = \hat{q}^0(\boldsymbol{p}_\alpha\cdot\boldsymbol{p}_\beta, \boldsymbol{p}_0'\cdot\boldsymbol{p}_\beta, \boldsymbol{p}_0'\cdot\boldsymbol{p}_0', \boldsymbol{p}_\alpha'\cdot\boldsymbol{p}_0', \lambda_0, \lambda_0', S). \tag{16.6}$$

It is easy to show that (15.8) is automatically satisfied by (16.1), (16.5).

The constitutive relations for Cosserat rods of other materials have been obtained. GREEN, LAWS, and NAGHDI (1967) have treated linearly elastic rods; GREEN, KNOPS, and LAWS (1968), elastic rods linearized about a large deformation; LAWS (1967b), elasto-plastic rods; SHACK (1970), viscoelastic rods; and DESILVA and WHITMAN (1971), rods of simple material with no memory. GREEN, LAWS, and NAGHDI (1967) and GREEN, KNOPS, and LAWS (1968) studied the question of material symmetries.

[41a] Our analysis is equivalent to that of GREEN and LAWS (1966).

III. Planar problems.

17. The governing equations. We formulate the equations of planar rod problems in terms of physical components that seem most suitable for applications.[42] We restrict our considerations to purely mechanical theories of the Cosserat rods defined in Subchapter II by assuming that the constitutive relations for $\boldsymbol{n}$, $\boldsymbol{\mu}^{\alpha}$, $\bar{\boldsymbol{\mu}}^{\alpha}$ are independent of λ_0 and λ_0'.

Let x, y, z denote Cartesian coordinates for $\mathscr{E}^3$ and let $\boldsymbol{i}, \boldsymbol{j}, \boldsymbol{k}$ denote the corresponding unit base vectors. The position vector to any point in space is

$$\boldsymbol{r} = x\boldsymbol{i} + y\boldsymbol{j} + z\boldsymbol{k}. \tag{17.1}$$

A problem is *planar* if the equations are invariant under reflection through a plane. Taking the x, y-plane as the locus of $\mathscr{c}$, we can ensure this result by restricting the forms of the kinematic variables, the mechanical variables, the external forces, the boundary conditions, and the material response.

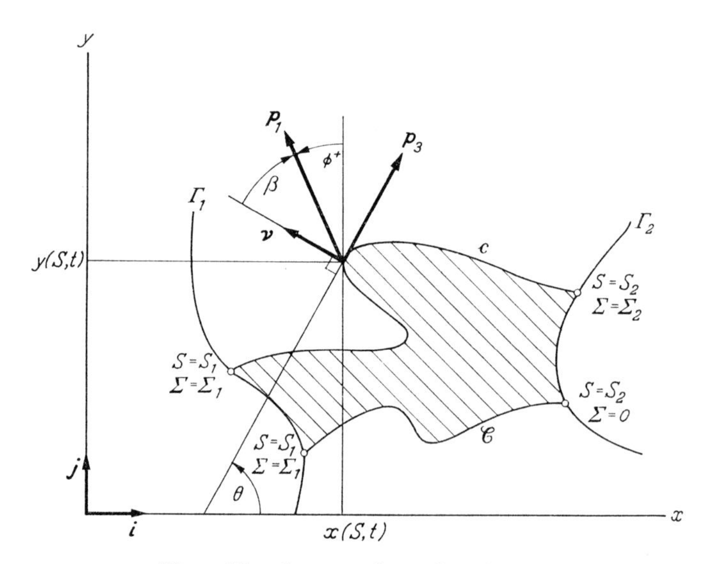

Fig. 2. The planar configuration of a rod.

We begin by requiring the vectors $\boldsymbol{p}_0, \boldsymbol{p}_1, \boldsymbol{p}_2$ to have the form

$$\boldsymbol{p}_0(S,t) = x(S,t)\,\boldsymbol{i} + y(S,t)\,\boldsymbol{j}, \tag{17.2}$$

$$\boldsymbol{p}_1(S,t) = \varrho_1^+(S,t)\,[-\sin\phi^+(S,t)\,\boldsymbol{i} + \cos\phi^+(S,t)\,\boldsymbol{j}] \tag{17.3}$$

$$\boldsymbol{p}_2(S,t) = \varrho_2^+(S,t)\,\boldsymbol{k}, \tag{17.4}$$

with

$$S_1 \leqq S \leqq S_2. \tag{17.5}$$

Let $\theta(S,t)$ denote the tangent angle to the curve (17.2):

$$\theta = \arctan(y'/x'). \tag{17.6}$$

[42] Not enough analytic results are available for motion in space of Cosserat rods to determine canonical forms for the governing equations.

With

$$\varrho_3^+ = [(x')^2 + (y')^2]^{\frac{1}{2}}, \tag{17.7}$$

we may set

$$\begin{aligned} \boldsymbol{p}_3(S, t) &\equiv \boldsymbol{p}_0'(S, t) = x'(S, t)\,\boldsymbol{i} + y'(S, t)\,\boldsymbol{j} \\ &= \varrho_3^+(S, t)\,[\cos\theta(S, t)\,\boldsymbol{i} + \sin\theta(S, t)\,\boldsymbol{j}]. \end{aligned} \tag{17.8}$$

Let $\boldsymbol{\nu}$ denote the unit normal to (17.2)

$$\boldsymbol{\nu} = \boldsymbol{k} \times \frac{\boldsymbol{p}_3}{\varrho_3^+} = -\sin\theta\,\boldsymbol{i} + \cos\theta\,\boldsymbol{j} \tag{17.9}$$

and let β be the angle between $\boldsymbol{\nu}$ and $\boldsymbol{p}_1$ with the orientation shown in Fig. 2. Then

$$\theta = \beta + \phi^+. \tag{17.10}$$

Since the reference configuration need not be stress-free, there is no loss of generality in taking

$$\varrho_k^+ = 1, \quad \beta = 0 \tag{17.11}$$

in this configuration. The reference configuration is thus defined by

$$\boldsymbol{p}_0 = X(S)\,\boldsymbol{i} + Y(S)\,\boldsymbol{j}, \tag{17.12}$$

$$\boldsymbol{p}_1 = -\sin\Phi(S)\,\boldsymbol{i} + \cos\Phi(S)\,\boldsymbol{j}, \tag{17.13}$$

$$\boldsymbol{p}_2 = \boldsymbol{k}, \tag{17.14}$$

with

$$\Phi(S) = \arctan[Y'(S)/X'(S)] \tag{17.15}$$

for S in (17.5). Here $X(S)$, $Y(S)$, and thus $\Phi(S)$ are prescribed. Thus S is the arc length parameter for the curve $\mathscr{C}$ given by (17.12) and the vector $\boldsymbol{p}_1$ of (17.13) is the unit normal to $\mathscr{C}$. Relation (17.11) implies that ϱ_k^+ are stretches. We set

$$\varrho_k^+ = 1 + \varrho_k, \quad \phi^+ = \phi + \Phi, \tag{17.16}$$

so that the reference configuration is characterized by the vanishing of ϱ_k, ϕ. We require

$$\varrho_k > -1, \quad -\pi/2 < \beta < \pi/2. \tag{17.17}$$

This ensures the continuity condition (14.6) and in fact is a strong continuity condition in the sense of the definition of Sect. 3.

The assumption of planarity reduces the strains

$$\boldsymbol{p}_k \cdot \boldsymbol{p}_m, \quad \boldsymbol{p}_\alpha' \cdot \boldsymbol{p}_m, \quad k, m = 1, 2, 3 \tag{17.18}$$

to functions of the septuple

$$\boldsymbol{w} = (\varrho_1, \varrho_2, \varrho_3, \beta, \varrho_1', \varrho_2', \phi'). \tag{17.19}$$

As a second condition of planarity we require the stresses and loads to have the following forms:

$$\boldsymbol{n} = N\left(\frac{\boldsymbol{p}_3}{\varrho_3^+}\right) + Q\boldsymbol{v}, \tag{17.20}$$

$$\boldsymbol{\mu}^1 = G\left(\frac{\boldsymbol{p}_1}{\varrho_1^+}\right) + H\left(\frac{\boldsymbol{k}\times\boldsymbol{p}_1}{\varrho_1^+}\right), \tag{17.21}$$

$$\overline{\boldsymbol{\mu}}^1 = \overline{G}\left(\frac{\boldsymbol{p}_1}{\varrho_1^+}\right) + \overline{H}\left(\frac{\boldsymbol{k}\times\boldsymbol{p}_1}{\varrho_1^+}\right), \tag{17.22}$$

$$\boldsymbol{\mu}^2 = K\,\boldsymbol{k}, \tag{17.23}$$

$$\overline{\boldsymbol{\mu}}^2 = \overline{K}\,\boldsymbol{k}, \tag{17.24}$$

$$\boldsymbol{m} \equiv \boldsymbol{p}_\alpha \times \boldsymbol{\mu}^\alpha = M\,\boldsymbol{k} = \varrho_1^+\,H\,\boldsymbol{k}, \tag{17.25}$$

$$\boldsymbol{f}^0 = f_N\left(\frac{\boldsymbol{p}_3}{\varrho_3^+}\right) + f_Q\,\boldsymbol{v}, \tag{17.26}$$

$$\boldsymbol{f}^1 = f_G\left(\frac{\boldsymbol{p}_1}{\varrho_1^+}\right) + f_H\left(\frac{\boldsymbol{k}\times\boldsymbol{p}_1}{\varrho_1^+}\right), \tag{17.27}$$

$$\boldsymbol{f}^2 = f_K\,\boldsymbol{k}, \tag{17.28}$$

$$\boldsymbol{l} \equiv \boldsymbol{p}_\alpha \times \boldsymbol{f}^\alpha = l\,\boldsymbol{k} = \varrho_1^+\,f_H\,\boldsymbol{k}. \tag{17.29}$$

We also restrict the material properties[43] of the rod by requiring

$$J^{02} = J^{12} = 0. \tag{17.30}$$

For simplicity, we follow the customary practice of choosing $\mathscr{C}$ such that

$$J^{01} = 0. \tag{17.31}$$

We set

$$J^{00} = A, \quad J^{11} = I, \quad J^{22} = J. \tag{17.32}$$

The substitution of (17.20)–(17.32) into the local forms of (15.1), (15.2) and into (15.7), (15.8) yields the nontrivial equations of motion for the planar problem:

$$\begin{aligned} N' - \theta' Q + f_N = {} & \ddot{x}(S_0, t)\cos\theta + \ddot{y}(S_0, t)\sin\theta \\ & + A\int_{S_0}^{S} \{[\ddot{\varrho}_3(\overline{S}, t) - \varrho_3^+(\overline{S}, t)\,\dot{\theta}^2(\overline{S}, t)]\cos[\theta(S, t) - \theta(\overline{S}, t)] \\ & + [2\dot{\varrho}_3(\overline{S}, t)\,\dot{\theta} + \varrho_3^+(\overline{S}, t)\,\ddot{\theta}(\overline{S}, t)]\sin[\theta(S, t) - \theta(\overline{S}, t)]\}\,d\overline{S} \\ \equiv {} & A(\ddot{x}\cos\theta + \ddot{y}\sin\theta), \end{aligned} \tag{17.33}$$

$$\begin{aligned} Q' + \theta' N + f_Q = {} & -\ddot{x}(S_0, t)\sin\theta + \ddot{y}(S_0, t)\cos\theta \\ & + A\int_{S_0}^{S} \{-[\ddot{\varrho}_3(\overline{S}, t) - \varrho_3^+(\overline{S}, \dot{t})\,\theta^2(\overline{S}, t)]\sin[\theta(S, t) - \theta(\overline{S}, t)] \\ & + [2\dot{\varrho}_3(\overline{S}, t)\,\dot{\theta} + \varrho_3^+(\overline{S}, t)\,\ddot{\theta}(\overline{S}, t)]\cos[\theta(S, t) - \theta(\overline{S}, t)]\}\,d\overline{S} \\ \equiv {} & A(-\ddot{x}\sin\theta + \ddot{y}\cos\theta), \end{aligned} \tag{17.34}$$

$$G' - (\phi^+)' H - \overline{G} + f_G = I(\ddot{\varrho}_1 - \varrho_1^+\,\dot{\phi}^2), \tag{17.35}$$

$$H' + (\phi^+)' G - \overline{H} + f_H = I(2\dot{\varrho}_1\,\dot{\phi} + \varrho_1^+\,\ddot{\phi}), \tag{17.36}$$

$$K' - \overline{K} + f_K = J\,\ddot{\varrho}_2, \tag{17.37}$$

$$M' + \varrho_3^+\,Q + l = I\,\varrho_1^+\,(2\dot{\varrho}_1\,\dot{\phi} + \varrho_1^+\,\ddot{\phi}), \tag{17.38}$$

$$\varrho_3^+\,Q + \varrho_1'\,H - \varrho_1^+\,(\phi^+)' G + \varrho_1^+\,\overline{H} = 0. \tag{17.39}$$

---

[43] Note that when such material properties are interpreted in a three-dimensional setting, they include not only the material properties but also the geometric properties of the three-dimensional body. This requirement implies a symmetry in the mass distribution about the plane of motion.

Note the redundancy in (17.36), (17.38), (17.39) (cf. Sect. 15). We regard (17.39) [corresponding to (15.8)] as a constitutive restriction. We are then free to choose whichever of the other two equations is more convenient.

Our final requirement for planarity is that the constitutive equations ensure the consistency of (17.2)–(17.4) with (17.20)–(17.25). The general constitutive equations for purely mechanical theories of rods in space are of the form

$$\begin{aligned}\boldsymbol{n}=\hat{\boldsymbol{n}}\big(&(\boldsymbol{p}_1\cdot\boldsymbol{p}_1)^{(t)},(\boldsymbol{p}_1\cdot\boldsymbol{p}_2)^{(t)},(\boldsymbol{p}_1\cdot\boldsymbol{p}_3)^{(t)},(\boldsymbol{p}_2\cdot\boldsymbol{p}_2)^{(t)},(\boldsymbol{p}_2\cdot\boldsymbol{p}_3)^{(t)},(\boldsymbol{p}_3\cdot\boldsymbol{p}_3)^{(t)},\\&(\boldsymbol{p}_1'\cdot\boldsymbol{p}_1)^{(t)},(\boldsymbol{p}_1'\cdot\boldsymbol{p}_2)^{(t)},(\boldsymbol{p}_1'\cdot\boldsymbol{p}_3)^{(t)},(\boldsymbol{p}_2'\cdot\boldsymbol{p}_1)^{(t)},(\boldsymbol{p}_2'\cdot\boldsymbol{p}_2)^{(t)},(\boldsymbol{p}_2'\cdot\boldsymbol{p}_3)^{(t)},S\big),\quad\text{etc.}\end{aligned}\tag{17.40}$$

For (17.20) to be consistent with (17.2)–(17.4) we require $\hat{\boldsymbol{n}}$ to satisfy

$$0=\boldsymbol{k}\cdot\hat{\boldsymbol{n}}(\alpha_{11}^{(t)},0,\alpha_{13}^{(t)},\alpha_{22}^{(t)},0,\alpha_{33}^{(t)},\beta_{11}^{(t)},0,\beta_{13}^{(t)},0,\beta_{22}^{(t)},0,S)\tag{17.41}$$

for arbitrary $\alpha_{11}^{(t)},\alpha_{13}^{(t)},\alpha_{22}^{(t)},\alpha_{33}^{(t)},\beta_{11}^{(t)},\beta_{13}^{(t)},\beta_{22}^{(t)}$ in the domain of $\hat{\boldsymbol{n}}$. We treat the other constitutive functionals analogously.

Temperature independence reduces the constitutive equations for thermoelastic rods of Sect. 16 to those of hyperelastic rods. In this case the material response is characterized by the strain-energy (free-energy) functions

$$\Psi(\boldsymbol{p}_1,\boldsymbol{p}_2,\boldsymbol{p}_3,\boldsymbol{p}_1',\boldsymbol{p}_2',S)=\Omega(\boldsymbol{p}_k\cdot\boldsymbol{p}_m,\boldsymbol{p}_\alpha'\cdot\boldsymbol{p}_m,S).\tag{17.42}$$

When (17.2)–(17.4) hold, Ω reduces to

$$\begin{aligned}&\Omega\big((\varrho_1^+)^2,0,\varrho_1^+\varrho_3^+\sin\beta,(\varrho_2^+)^2,0,(\varrho_3^+)^2,\\&\quad\varrho_1'\varrho_1^+,0,\varrho_3^+(\varrho_1'\sin\beta-\varrho_1^+\phi'\cos\beta),0,\varrho_2'\varrho_2^+,0,S\big).\end{aligned}\tag{17.43}$$

We guarantee that (17.2)–(17.4) is consistent with (17.20)–(17.25) by requiring

$$\begin{aligned}&\frac{\partial\Omega}{\partial(\boldsymbol{p}_1\cdot\boldsymbol{p}_2)}=0,\quad\frac{\partial\Omega}{\partial(\boldsymbol{p}_2\cdot\boldsymbol{p}_3)}=0,\quad\frac{\partial\Omega}{\partial(\boldsymbol{p}_1'\cdot\boldsymbol{p}_2)}=0,\\&\frac{\partial\Omega}{\partial(\boldsymbol{p}_2'\cdot\boldsymbol{p}_1)}=0,\quad\frac{\partial\Omega}{\partial(\boldsymbol{p}_2'\cdot\boldsymbol{p}_3)}=0,\end{aligned}\tag{17.44}$$

whenever

$$\boldsymbol{p}_1\cdot\boldsymbol{p}_2=\boldsymbol{p}_2\cdot\boldsymbol{p}_3=\boldsymbol{p}_1'\cdot\boldsymbol{p}_2=\boldsymbol{p}_2'\cdot\boldsymbol{p}_1=\boldsymbol{p}_2'\cdot\boldsymbol{p}_3=0.\tag{17.45}$$

This assumption on Ω means that any aeolotropy is consistent with planarity. Since we seek only planar solutions, we introduce a new strain energy function with values

$$W(\varrho_1,\varrho_2,\varrho_3,\beta,\varrho_1',\varrho_2',\phi',S).\tag{17.46}$$

We define (17.46) to equal (17.42).

Using (16.4), (17.2)–(17.4), (17.8), (17.42), we find

$$\frac{\partial W}{\partial\varrho_3}=\frac{\partial\Psi}{\partial\boldsymbol{p}_3}\cdot\frac{\partial\boldsymbol{p}_3}{\partial\varrho_3}=\boldsymbol{n}\cdot\frac{\boldsymbol{p}_3}{\varrho_3^+}=N.\tag{17.47}$$

In this way we obtain the constitutive relations[44]

$$N=\frac{\partial W}{\partial\varrho_3},\quad\varrho_3^+Q=\frac{\partial W}{\partial\beta},\quad M=\frac{\partial W}{\partial\phi'},\tag{17.48}$$

$$G=\frac{\partial W}{\partial\varrho_1'},\quad\overline{G}+(\phi^+)'H=\frac{\partial W}{\partial\varrho_1},\quad K=\frac{\partial W}{\partial\varrho_2'},\quad\overline{K}=\frac{\partial W}{\partial\varrho_2}.\tag{17.49}$$

[44] Relations equivalent to $(17.48)_{1,3}$ were found by TADJBAKHSH (1966). The simple forms shown here are due to ANTMAN (1968a).

A constitutive equation for $\bar{H}$ may be determined from (17.39). On the other hand, we may simply ignore (17.36) and (17.39) because of their equivalence to (17.38).

Special planar theories were developed by TADJBAKHSH (1966), ANTMAN (1968a). LAWS (1967a) obtained a theory with the generality of this section by showing that a special class of Cosserat theories was physically meaningful in the planar case. Also cf. GREEN and LAWS (1966), ANTMAN and WARNER (1966).

18. Boundary conditions. Because this Cosserat theory has more kinematic structure than classical rod theories, we cannot expect to describe boundary conditions unambiguously by such terms as "hinged", "cantilevered", etc. Here we investigate the nature of the simpler boundary conditions (15.10), (15.11) for planar problems.

If the rod is a planar ring so that the ends consist of the same particles, we smoothly extend the functions defining the data of the problem to the entire S-axis with period S_2-S_1 in S (cf. Sect. 10). We then require the variables

$$x,\ y,\ \varrho_1,\ \varrho_2,\ \varrho_3,\ \beta,\ \phi,\ \varrho_1',\ \varrho_2',\ \phi',\ N,\ Q,\ G,\ H,\ K,\ M,\ \bar{G},\ \bar{H},\ \bar{K} \tag{18.1}$$

to satisfy the periodicity conditions

$$x(S_1)=x(S_2), \quad \text{etc.} \tag{18.2}$$

For the conditions on x, y, ϕ to hold in the reference configuration, (17.11), (17.12), (17.16) require

$$\int_{S_1}^{S_2} \exp i\,\Phi(S)\,dS=0. \tag{18.3}$$

If the rod is not a ring, we treat the boundary conditions in three groups corresponding to the separate terms in the power of the end tractions in (15.4). One set of boundary conditions from each of the groups I, II, III must hold.

I. (i) The position of the end is prescribed:

$$x(S_\alpha,t)=\bar{x}_\alpha(t), \qquad y(S_\alpha,t)=\bar{y}_\alpha(t). \tag{18.4}$$

(ii) The end is constrained to move on a smooth time dependent curve

$$\Gamma_\alpha(x,y,t)=0, \tag{18.5}$$

given parametrically by

$$x=x_\alpha(\Sigma,t), \qquad y=y_\alpha(\Sigma,t) \tag{18.6}$$

(cf. Fig. 2), and the component of $\boldsymbol{n}$ tangent to the direction of motion is prescribed. Thus the boundary conditions are

$$\begin{gathered} x(S_\alpha,t)=x_\alpha(\Sigma(t),t), \qquad y(S_\alpha,t)=y_\alpha(\Sigma(t),t), \\ [N(S_\alpha,t)\cos\theta(S_\alpha,t)-Q(S_\alpha,t)\sin\theta(S_\alpha,t)]\,\dot{x}_\alpha(\Sigma(t),t) \\ +[N(S_\alpha,t)\sin\theta(S_\alpha,t)+Q(S_\alpha,t)\cos\theta(S_\alpha,t)]\,\dot{y}_\alpha(\Sigma(t),t) \\ =B_\alpha(\Sigma(t),\dot{\Sigma}(t),t), \end{gathered} \tag{18.7}$$

where B_α is a given function. If Γ_α, x_α, y_α do not depend explicitly upon t, we fix Σ by setting

$$x_\alpha(0)=X(S_\alpha), \qquad y_\alpha(0)=Y(S_\alpha). \tag{18.8}$$

[Thus we are requiring the curve (18.5) to pass through the reference position of the end.] In the static case we replace $\dot{x}_\alpha$ and $\dot{y}_\alpha$ by $\frac{\partial x_\alpha}{\partial \Sigma}$ and $\frac{\partial y_\alpha}{\partial \Sigma}$.

(iii) The end is free of geometric restraint. The stress resultants at the end are prescribed:

$$N(S_\alpha, t) = N_\alpha(t), \quad Q(S_\alpha, t) = Q_\alpha(t). \tag{18.9}$$

II. (i) The director $\boldsymbol{p}_1$ is prescribed:

$$\varrho_1(S_\alpha, t) = \bar{\varrho}_{1\alpha}(t), \quad \phi(S_\alpha, t) = \bar{\phi}_\alpha(t). \tag{18.10}$$

(ii) The director $\boldsymbol{p}_1$ at S_α is required to describe a smooth time dependent curve

$$\varrho_1 = \varrho_{1\alpha}(\Sigma, t), \quad \phi = \phi_\alpha(\Sigma, t) \tag{18.11}$$

and the component of $\boldsymbol{\mu}^1$ tangent to this motion is prescribed. Thus the boundary conditions are

$$\begin{gathered}\varrho_1(S_\alpha, t) = \varrho_{1\alpha}(\Sigma(t), t), \quad \phi(S_\alpha, t) = \phi_\alpha(\Sigma(t), t), \\ G(S_\alpha, t)\, \dot{\varrho}_{1\alpha}(\Sigma(t), t) + M(S_\alpha, t)\, \dot{\phi}_\alpha(\Sigma(t), t) = C_\alpha(\Sigma(t), \dot{\Sigma}(t), t),\end{gathered} \tag{18.12}$$

where C_α is a given function. In the static case we replace $\dot{\varrho}_{1\alpha}$ and $\dot{\phi}_\alpha$ by $\frac{\partial \varrho_{1\alpha}}{\partial \Sigma}(\Sigma)$ and $\frac{\partial \phi_\alpha}{\partial \Sigma}(\Sigma)$.

(iii) The director $\boldsymbol{p}_1$ is free of geometric restraint. We prescribe

$$G(S_\alpha, t) = G_\alpha(t), \quad M(S_\alpha, t) = M_\alpha(t). \tag{18.13}$$

III. (i) The director $\boldsymbol{p}_2$ is prescribed:

$$\varrho_2(S_\alpha, t) = \bar{\varrho}_{2\alpha}(t). \tag{18.14}$$

(iii) The director $\boldsymbol{p}_2$ is free of geometric restraint. We prescribe

$$K(S_\alpha, t) = K_\alpha(t). \tag{18.15}$$

It is important to note that nowhere do we prescribe $\beta(S_\alpha)$. (This is consistent with the absence of the history of β' from the arguments of the constitutive functionals.) Consequently we cannot prescribe $\theta(S_\alpha)$, as we could for classical rod theories. Rather than being a defect of the theory, this fact indicates how the displacement boundary conditions for a three-dimensional body can be more closely approximated by those for our Cosserat theory than by those for a classical rod theory. This is illustrated in Fig. 3. (We use the geometric interpretation of directors suggested by our work in Subchapter I.) In Fig. 3a, we show a neighborhood of the end $S = 0$ of a rod in its reference configuration with $X(0) = Y(0) = \Phi(0) = 0$. In Fig. 3b, we show an arbitrary configuration of the neighborhood of this end satisfying

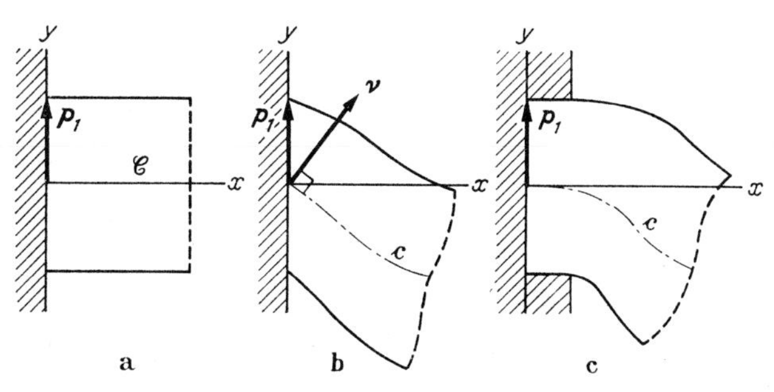

Fig. 3a–c. Comparison of welded and clamped end conditions.

the boundary conditions $x(0,t)=y(0,t)=\varrho_1(0,t)=\phi(0,t)=0$. In Fig. 3c, we show a configuration with $x(0,t)=y(0,t)=\varrho_1(0,t)=\theta(0,t)=0$. The boundary conditions for Fig. 3b, which are special cases of I (i), II (i), correspond to the welding of the end to a rigid wall, a condition typical of the three-dimensional theory. On the other hand, to cause $\theta(0,t)$ to vanish as in Fig. 3c, we must not only weld the end to a rigid wall, but must also clamp the lateral surface of the rod near this end. To attain this situation in both our Cosserat theory and in the three-dimensional theory, we must impose an additional geometric constraint corresponding to this clamping and leave free some combination of surface tractions and body forces to accomodate this constraint. In classical theories, there is insufficient geometric structure to distinguish between Figs. 3b and 3c. The introduction of the material constraint $\beta=0$ into our theory would cause $(18.10)_2$ to degenerate into $\theta(S_\alpha,t)=\bar{\phi}_\alpha(t)$.

Simple geometric models involving rollers and hinges can be made for each of possible combinations of boundary conditions described above. We could introduce various couplings in the specification of boundary conditions I, II, III by requiring the functions $\bar{x}_\alpha, \bar{y}_\alpha, \ldots$ appearing on the right sides to be functionals of the past history of other geometric variables. This would correspond to the introduction of nonlinear viscoelastic springs, the use of follower forces, etc.

C. Problems for nonlinearly elastic rods.

19. Existence. Existence theorems for the equations of linear elasticity are of a simple form: For reasonable data, there exists a unique classical solution. The inherent complexity of three-dimensional nonlinear elasticity has so far prevented the proof of global existence theorems. Still open is the problem of determining physically consistent restrictions on the strain energy function that ensure the existence of solutions in appropriate circumstances.[45] By the direct methods of the calculus of variations, we can however prove a number of global existence theorems for planar equilibrium problems for elastic Cosserat rods, thereby solving this open problem in this one-dimensional case. The delicate nature of parts of our work emphasizes the difficulties posed by the three-dimensional problems.

By limiting our treatment to planar problems, we naturally simplify some details of the work. But this restriction also enables us to treat the physically interesting problem of hydrostatic loads. The resulting existence theorems serve to illuminate the physics.

20. Variational formulation of the equilibrium problems. We introduce appropriate variables that enable us to express clearly the hypotheses that both describe the physics and ensure the success of the mathematics. We define the complex strain variable

$$\omega^+ = 1+\omega = \omega_3^+ + i\omega_4 = \varrho_3^+ \exp i\beta, \qquad \omega_3^+ = 1+\omega_3. \tag{20.1}$$

ω_3^+ is a weighted measure of the stretch ϱ_3^+; ω_4 is a weighted measure of director rotation. From (17.17) we obtain

$$\omega_3 > -1. \tag{20.2}$$

The strict inequalities (17.17), (20.2) cause difficulties for the analysis. (A convergent sequence of approximating solutions satisfying these strict unilateral constraints need not have a limit satisfying them.) To remedy this, we introduce in place of these constrained variables, a new set of strains that can assume values on $(-\infty,\infty)$. We define a class of one-to-one mappings $h_1, h_2, h_3 \equiv h$ of

[45] To appreciate the difficulty of this "main open problem of nonlinear elasticity" (NFTM, Sect. 20), one need only compare the account of physically motivated restrictions on the strain-energy function in NFTM, Sect. 87, with the account of mathematically motivated assumptions for variational problems given by MORREY (1966). It is apparent that freedom in selecting appropriate hypotheses is sharply curtailed.

$(-\infty, \infty)$ onto $(-1, \infty)$ by

$$\varrho = h_k(\gamma), \qquad -\infty < \gamma < \infty, \tag{20.3}$$

where h_k is twice continuously differentiable on $(-\infty, \infty)$,

$$\begin{gathered} h_k'(\gamma) > 0, \quad h_k(0) = 0, \quad h_k'(0) = 1, \\ h_k(\gamma) \to -1 \text{ as } \gamma \to -\infty, \quad h_k(\gamma) \to \infty \text{ as } \gamma \to \infty, \end{gathered} \tag{20.4}$$

$$\gamma h''(\gamma) \leqq 0. \tag{20.5}$$

If $\varrho_1(S)$, $\varrho_2(S)$, $\omega_3(S)$ are measurable, then the composite functions

$$\gamma_1(S) = h_1^{-1}(\varrho_1(S)), \quad \gamma_2(S) = h_2^{-1}(\varrho_2(S)), \quad \gamma_3 = h^{-1}(\omega_3(S)) \tag{20.6}$$

are also measurable [cf. VAINBERG (1956, Chap. 6) or KRASNOSEL'SKII (1956, Chap. 1)]. The functions γ_k are just new strain variables with the desirable property that if they are bounded functions of S, then ϱ_1, ϱ_2, ω_3 are not only bounded, but also satisfy the inequalities (17.17), (20.2).

We let $\boldsymbol{u}$ and $\boldsymbol{v}$ denote the septuples of real variables

$$\boldsymbol{u} = (\gamma_1, \gamma_2, \omega_3, \omega_4, \gamma_1', \gamma_2', \phi'), \tag{20.7}$$

$$\boldsymbol{v} = (\gamma_1, \gamma_2, \gamma_3, \omega_4, \gamma_1', \gamma_2', \phi'). \tag{20.8}$$

These differ only in the third argument. We denote by

$$\boldsymbol{u} = \boldsymbol{h}(\boldsymbol{v}) \tag{20.9}$$

the mapping of seven-dimensional real space $\mathscr{R}^7$ onto $\mathscr{R}^2 \times (-1, \infty) \times \mathscr{R}^4$ induced by h. We introduce new strain-energy functions depending on $\boldsymbol{u}$ and $\boldsymbol{v}$ in terms of the strain-energy function W of (17.46) depending on the septuple $\boldsymbol{w}$ of (17.19) by

$$W(\boldsymbol{w}, S) = U(\boldsymbol{u}, S) = V(\boldsymbol{v}, S), \tag{20.10}$$

with

$$U(\boldsymbol{h}(\boldsymbol{v}), S) = V(\boldsymbol{v}, S), \quad \text{etc.} \tag{20.11}$$

We assume that U, V, W are twice continuously differentiable functions of their eight arguments on the respective domains of these arguments.

Replacing the coordinate functions $x(S)$, $y(S)$ by their indefinite integrals obtained from (17.8), (17.10), we can reduce the set of unknowns to any of the three sets

$$\begin{aligned} \mathfrak{w} &= (\varrho_1, \varrho_2, \varrho_3, \beta, \phi, e_6[x(S_1) - X(S_1)], e_6[y(S_1) - Y(S_1)]), \\ \mathfrak{u} &= (\gamma_1, \gamma_2, \omega_3, \omega_4, \phi, e_6[x(S_1) - X(S_1)], e_6[y(S_1) - Y(S_1)]), \\ \mathfrak{v} &= (\gamma_1, \gamma_2, \gamma_3, \omega_4, \phi, e_6[x(S_1) - X(S_1)], e_6[y(S_1) - Y(S_1)]), \end{aligned} \tag{20.12}$$

where $e_6 = 0$ if the end $S = S_1$ is fixed and $e_6 = 1$ otherwise. We denote the mapping of measurable $\mathfrak{v}$ into measurable $\mathfrak{u}$ induced by (20.9) by

$$\mathfrak{u} = \mathfrak{h}(\mathfrak{v}) \tag{20.13}$$

and its inverse by $\mathfrak{h}^{-1}$.

We explicitly represent the coordinate functions $x(S)$, $y(S)$ by

$$\begin{aligned} x(S) + i y(S) &= \mathfrak{x}[\mathfrak{u}, S] + i \mathfrak{y}[\mathfrak{u}, S] \\ &\equiv x(S_1) + i y(S_1) + \int_{S_1}^{S} \omega^+(T) \exp i\phi^+(T)\, dT. \end{aligned} \tag{20.14}$$

The external distributed loads (body forces) for our problem are taken to have the form

$$f^0 = -\lambda_2 \varrho_3^+ \boldsymbol{v} + \lambda_3 F_x(S)\,\boldsymbol{i} + \lambda_4 F_y(S)\,\boldsymbol{j}, \quad \boldsymbol{f}^1 = \boldsymbol{f}^2 = \boldsymbol{l} = \boldsymbol{0}. \tag{20.15}$$

[cf. (17.26)–(17.29)]. The term $-\lambda_2 \varrho_3^+ \boldsymbol{v}$ represents a hydrostatic pressure. λ_2 measures normal force per unit length of c. The term $\lambda_3 F_x \boldsymbol{i} + \lambda_4 F_y \boldsymbol{j}$ represents a dead loading.[46] λ_3 and λ_4 are just numerical parameters.

We treat several cases of the boundary conditions of Sect. 18. We first consider the conditions of class I. If the position of the end $S = S_1$ is fixed, then

$$x(S_1) = X(S_1), \qquad y(S_1) = Y(S_1). \tag{20.16}$$

If the position of the end $S = S_2$ is also fixed then

$$\mathfrak{x}[\mathfrak{u}, S_2] = X(S_2), \qquad \mathfrak{y}[\mathfrak{u}, S_2] = Y(S_2). \tag{20.17}$$

As a special case of (20.17) we have the ring boundary conditions for which $X(S_1) = X(S_2)$, $Y(S_1) = Y(S_2)$. When the ring conditions hold, we require the body forces to be self-equilibrated:

$$\lambda_3 \int_{S_1}^{S_2} F_x(S)\, dS = 0, \qquad \lambda_4 \int_{S_1}^{S_2} F_y(S)\, dS = 0. \tag{20.18}$$

(The hydrostatic pressure is always self-equilibrated on a ring.)

If the position of only one end is fixed, we take it to be the end $S = S_1$. If the end $S = S_2$ moves on the curve $\Gamma_2(x, y) = 0$ of (18.5) then

$$\Gamma_2(\mathfrak{x}[\mathfrak{u}, S_2], \mathfrak{y}[\mathfrak{u}, S_2]) = 0 \tag{20.19}$$

and we may prescribe an external concentrated dead load at the end $S = S_2$:

$$\lambda_7 \boldsymbol{i} + \lambda_8 \boldsymbol{j}. \tag{20.20}$$

This load plus the normal reaction of the curve $\Gamma_2(x, y) = 0$ gives the total concentrated force $\boldsymbol{n}(S_2)$ at this end. If the end $S = S_1$ also moves on a curve $\Gamma_1(x, y) = 0$, then

$$\Gamma_1(x(S_1), y(S_1)) = 0 \tag{20.21}$$

and we may also prescribe an external concentrated dead load at $S = S_1$:

$$\lambda_5 \boldsymbol{i} + \lambda_6 \boldsymbol{j}. \tag{20.22}$$

If both (20.19) and (20.21) hold, then $x(S_1)$ and $y(S_1)$ are unknowns: $e_6 = 1$. In this case we must restrict the form of the curves $\Gamma_1(x, y) = 0$, $\Gamma_2(x, y) = 0$ to exclude problems in which equilibrium can be lost by a degenerate process as would happen if the curves $\Gamma_1(x_1, y) = 0$, $\Gamma_2(x, y) = 0$ were parallel lines: the application of a hydrostatic pressure to the rod would cause it to accelerate along the lines. To ensure that the curves $\Gamma_\alpha(x, y) = 0$ exert sufficient restraint on possible deformations, we require

Hypothesis G. *If both ends move on curves and if $x_1(\Sigma) + i y_1(\Sigma)$ is unbounded, then*

$$|x_2(\Sigma_2) + i y_2(\Sigma_2) - x_1(\Sigma) - i y_1(\Sigma)| \to \infty \quad \textit{as} \quad |x_1(\Sigma) + i y_1(\Sigma)| \to \infty$$

for arbitrary fixed $x(S_2) + i y(S_2) = x_2(\Sigma_2) + i y_2(\Sigma_2)$.

[46] A dead loading is a system of force functions depending solely upon the material coordinates. It is independent of time and of the configuration occupied by the body.

If the end $S=S_1$ is fixed and the end $S=S_2$ is free, then we require $\lambda_2=0$. (If not, the problem would fail to be variational.) In this case we prescribe the dead load

$$\boldsymbol{n}(S_1)=\lambda_7\boldsymbol{i}+\lambda_8\boldsymbol{j}. \tag{20.23}$$

We assume that the boundary conditions of classes II, III are all homogeneous. This causes no loss of generality since a problem with nonhomogeneous boundary conditions is readily transformed into one with homogeneous boundary conditions and nonhomogeneous equations. There is no inconsistency since we do not require the reference configuration to be stress-free. The homogeneous geometric boundary conditions for $\mathfrak{w}$ are members of the set

$$\begin{aligned}&\varrho_1(S_1)=\varrho_1(S_2), &&\varrho_2(S_1)=\varrho_2(S_2), &&\phi(S_1)=\phi(S_2),\\ &\varrho_1(S_\alpha)=0, &&\varrho_2(S_\alpha)=0, &&\phi(S_\alpha)=0,\end{aligned} \tag{20.24}$$

and correspondingly, the homogeneous boundary conditions for $\mathfrak{u}$ and $\mathfrak{v}$ are members of the set

$$\begin{aligned}&\gamma_1(S_1)=\gamma_1(S_2), &&\gamma_2(S_1)=\gamma_2(S_2), &&\phi(S_1)=\phi(S_2),\\ &\gamma_1(S_\alpha)=0, &&\gamma_2(S_\alpha)=0, &&\phi(S_\alpha)=0.\end{aligned} \tag{20.25}$$

We now define the functionals that arise in our variational problem. The tota strain energy is

$$\mathfrak{U}_1[\mathfrak{u}]=\int_{S_1}^{S_2} U(\boldsymbol{u}(S), S)\,dS. \tag{20.26}$$

If each end is either fixed or constrained to move along a curve, then the area swept out by $\boldsymbol{c}$ in a deformation from the reference configuration to an arbitrary configuration is well-defined. It is the area of the shaded region of Fig. 2 and is given by

$$\begin{aligned}\mathfrak{U}_2[\mathfrak{u}]&=\mathfrak{U}_{2_1}[\mathfrak{u}]+\mathfrak{U}_{2_2}[\mathfrak{u}]-C,\\ \mathfrak{U}_{2_1}[\mathfrak{u}]&=\int_{S_1}^{S_2}\mathfrak{y}[\mathfrak{u},S][\omega_3^+(S)\cos\phi^+(S)-\omega_4(S)\sin\phi^+(S)]\,dS,\\ \mathfrak{U}_{2_2}[\mathfrak{u}]&=\int_0^{\Sigma_1}y_1(\Sigma)\,x_1'(\Sigma)\,d\Sigma-\int_0^{\Sigma_2}y_2(\Sigma)\,x_2'(\Sigma)\,d\Sigma,\\ C&=\int_{S_1}^{S_2}Y(S)\,X'(S)\,dS,\end{aligned} \tag{20.27}$$

with Σ_1, Σ_2 determined by

$$x(S_\alpha)=x_\alpha(\Sigma_\alpha),\qquad y(S_\alpha)=y_\alpha(\Sigma_\alpha). \tag{20.28}$$

Note the C is a given number. We also define

$$\mathfrak{U}_3[\mathfrak{u}]=\int_{S_1}^{S_2}\mathfrak{x}[\mathfrak{u},S]\,F_x(S)\,dS,\qquad \mathfrak{U}_4[\mathfrak{u}]=\int_{S_1}^{S_2}\mathfrak{y}[\mathfrak{u},S]\,F_y(S)\,dS, \tag{20.29}$$

$$\mathfrak{U}_5[\mathfrak{u}]=x(S_1)-X(S_1),\qquad \mathfrak{U}_6[\mathfrak{u}]=y(S_1)-Y(S_1), \tag{20.30}$$

$$\mathfrak{U}_7[\mathfrak{u}]=\mathfrak{x}[\mathfrak{u},S_2]-X(S_2),\qquad \mathfrak{U}_8[\mathfrak{u}]=\mathfrak{y}[\mathfrak{u},S_2]-Y(S_2), \tag{20.31}$$

$$\mathfrak{U}_9[\mathfrak{u}]=\Gamma_1(x(S_1),y(S_1)),\qquad \mathfrak{U}_{10}[\mathfrak{u}]=\Gamma_2(\mathfrak{x}[\mathfrak{u},S_2],\mathfrak{y}[\mathfrak{u},S_2]). \tag{20.32}$$

The potential energy function is given by

$$\mathfrak{U}[\mathfrak{u}] = \sum_{j=1}^{8} \lambda_j \mathfrak{U}_j[\mathfrak{u}] \tag{20.33}$$

where $\lambda_1 \equiv 1$.

We shall consider the problem of extremizing $\mathfrak{U}[\mathfrak{u}]$ subject to any of the following sets of constraints

$$\mathfrak{U}_j[\mathfrak{u}] = 0, \quad j \in m \equiv \{5, 6, 7, 8\} \quad (e_6 = 0), \tag{20.34}$$

$$\mathfrak{U}_j[\mathfrak{u}] = 0, \quad j \in m \equiv \{5, 6, 10\} \quad (e_6 = 0), \tag{20.35}$$

$$\mathfrak{U}_j[\mathfrak{u}] = 0, \quad j \in m \equiv \{9, 10\} \quad (e_6 = 1), \tag{20.36}$$

$$\mathfrak{U}_j[\mathfrak{u}] = 0, \quad j \in m \equiv \{5, 6\}, \quad \lambda_2 = 0, \quad (e_6 = 0) \tag{20.37}$$

and subject to whichever of the homogeneous boundary conditions (20.25) are prescribed. We also treat the reciprocal variational problems that have Euler equations of the same form. We shall show that the Euler equations for the variational problems are equivalent to the complete boundary value problems for the equilibrium of planar elastic rods given in Sects. 18, 19.

We define the functionals

$$\mathfrak{V}[\mathfrak{v}] = \mathfrak{U}[\mathfrak{h}(\mathfrak{v})], \quad \mathfrak{V}_j[\mathfrak{v}] = \mathfrak{U}_j[\mathfrak{h}(\mathfrak{v})], \quad j = 1, \ldots, 10, \tag{20.38}$$

which will play a fundamental role in the subsequent analysis.

21. Statement of theorems. We introduce a set of physically reasonable restrictions on the strain energy function V.

Hypothesis H_1. *There are constants $\varkappa > 0$, $\alpha_j > 1$, $j = 1, 2, 3, 4, 5$, with $\alpha_3 = \alpha_4$ and there is a function $V_0(S)$ integrable on $[S_1, S_2]$ such that*

$$V(\gamma_1, \gamma_2, \gamma_3, \omega_4, \gamma_1', \gamma_2', \phi', S) \geqq \varkappa (e_1|\gamma_1|^{\alpha_1} + e_2|\gamma_2|^{\alpha_2} + |\gamma_3|^{\alpha_3} + |\omega_4|^{\alpha_4} + |\gamma_1'|^{\alpha_1} + |\gamma_2'|^{\alpha_2} + |\phi'|^{\alpha_5}) + V_0(S),$$

where $e_\alpha = 0$ if $\gamma_\alpha(S_1) = 0$ or $\gamma_\alpha(S_2) = 0$ and $e_\alpha = 1$ otherwise.

This condition implies that the rate of growth of V as a function of $\boldsymbol{v}$ is greater than linear for large values of the arguments. This suffices to ensure that $\mathfrak{V}_1$ must get large as the strains get large in a sense to be made precise.

Hypothesis H_2. *The strain energy function $V(\boldsymbol{v}, S)$ is strictly convex in the variables $\gamma_3, \omega_4, \gamma_1', \gamma_2', \phi'$ for fixed values of the remaining variables.* (Since V is assumed to be twice continuously differentiable in its arguments, $\boldsymbol{v}$, S, the Hessian matrix of second partial derivatives of V with respect to $\gamma_3, \omega_4, \gamma_1', \gamma_2', \phi'$, is positive definite.)

This hypothesis is analogous to those proposed by COLEMAN and NOLL (see NFTM, Sect. 87). It ensures that the stress resultants corresponding to the strains $\gamma_3, \omega_4, \gamma_1', \gamma_2', \phi'$ are monotonically increasing functions of these strains for fixed values of the remaining arguments and that the stress-strain relations for these variables can be inverted.

The properties of the mapping h listed in (20.4), (20.5) enable us to show immediately that H_2 implies the strict convexity of U with respect to its arguments $\omega_3, \omega_4, \gamma_1', \gamma_2', \phi'$. Moreover, convexity of U with respect to ω_3, ω_4 implies the convexity of W with respect to ϱ_3 [see (20.10)].

Hypothesis H₃. *The strain energy function $V(\boldsymbol{v}, S)$ satisfies*

$$\sum_{j=1}^{7}\left|\frac{\partial V}{\partial v_j}(\boldsymbol{v}, S)\right| \leq k(1+|\gamma_1|^{\alpha_1}+|\gamma_2|^{\alpha_2}+|\gamma_3|^{\alpha_3}+|\omega_4|^{\alpha_4}+|\gamma_1'|^{\alpha_1}+|\gamma_2'|^{\alpha_2}+|\phi'|^{\alpha_5})$$

where k is a positive constant and v_j is the j-th component of $\boldsymbol{v}$ given in (20.8).

We now define the admissible class of functions for our variational problems. We denote by $\|w\|_p$ the (Lebesgue) $\mathscr{L}_p$ norm of the function w on (S_1, S_2):

$$\|w\|_p = \left[\int_{S_1}^{S_2} |w(S)|^p dS\right]^{1/p} \tag{21.1}$$

and by $\|w\|_{1,p}$ the (Sobolev) $\mathscr{W}_p^1$ norm of $w(S)$ on (S_1, S_2):

$$\|w\|_{1,p} = \|w\|_p + \|w'\|_p. \tag{21.2}$$

Here w' denotes the distributional derivative of w. We assume that

$$\gamma_1 \in \mathscr{W}_{\alpha_1}^1, \quad \gamma_2 \in \mathscr{W}_{\alpha_2}^1, \quad \gamma_3 \in \mathscr{L}_{\alpha_3}, \quad \omega_4 \in \mathscr{L}_{\alpha_4}, \quad \phi \in \mathscr{W}_{\alpha_5}^1. \tag{21.3}$$

It is understood that the functions in these spaces are defined on the interval (S_1, S_2). We define the norm of $\mathfrak{v}$ [see $(20.12)_3$] to be

$$\begin{aligned}\|\mathfrak{v}\| = \|\gamma_1\|_{1,\alpha_1} + \|\gamma_2\|_{1,\alpha_2} + \|\gamma_3\|_{\alpha_3} + \|\omega_4\|_{\alpha_4} + \|\phi\|_{1,\alpha_5} \\ + e_6(|x(S_1) - X(S_1)| + |y(S_1) - Y(S_1)|).\end{aligned} \tag{21.4}$$

We denote by $\mathscr{B}$ the real Banach space of elements $\mathfrak{v}$ defined by (21.3), (21.4) that satisfy whichever of the homogeneous linear boundary conditions (20.25) are prescribed for the given problem. This space is well-defined, i.e. these boundary conditions are meaningful, because a special case of the Sobolev embedding theorem ensures that $\mathscr{W}_p^1$ can be identified with a subspace of the continuous functions provided the elements of $\mathscr{W}_p^1$ are defined only on a finite interval of the real line.

Let

$$\mathscr{E}_k(c_k) = \{\mathfrak{v} \in \mathscr{B} : \mathfrak{B}_k[\mathfrak{v}] = c_k\}. \tag{21.5}$$

The domain of the functionals $\mathfrak{B}_k$, $k = 1, \ldots, 8$ is $\bigcap_{k \in m} \mathscr{E}_k(0)$ where m is defined for each type of boundary condition in (20.34)–(20.37). The boundary conditions not included in this specification of domain are produced as natural boundary conditions by the variational process.

Let s denote the set of those integers $1, \ldots, 8$ not belonging to m. Let e denote any subset of s. e may be the empty set. The complement of e in s is denoted e^c. Let

$$\mathscr{E}_{em} = \left[\bigcap_{k \in e} \mathscr{E}_k(c_k)\right] \cap \left[\bigcap_{k \in m} \mathscr{E}_k(0)\right]. \tag{21.6}$$

We can now state our results.[47]

Theorem 1. *Let hypotheses G, H_1, H_2 hold. Let $F_x, F_y \in \mathscr{L}_1$.*

(i) *Let $1 \in e^c$. If $\alpha_3 = \alpha_4 > 2$ or if $2 \in e$, then for arbitrary fixed values of $\{\lambda_k, k \in e^c, k \neq 1\}$ and for arbitrary fixed values of $\{c_k, k \in e\}$, there exists an element $\mathfrak{v}$*

[47] These theorems generalize those of ANTMAN (1970b) and are closest in spirit to the results for shells given by ANTMAN (1971). Earlier investigations along these lines were made by TADJBAKHSH and ODEH (1967), ANTMAN (1968c). TADJBAKHSH and ODEH showed the advantages of replacing the coordinate functions $x(S)$, $y(S)$ by indefinite integrals.

that minimizes $\sum_{k\in e^c} \lambda_k \mathfrak{V}_k$ *on* $\mathscr{E}_{em}$ *and there is a corresponding element* $\mathfrak{u}=\mathfrak{h}(\mathfrak{v})$ *that minimizes* $\sum_{k\in e^c} \lambda_k \mathfrak{U}_k$ *on* $\mathfrak{h}(\mathscr{E}_{em})$.

(ii) *Let* $1\in e$. *For arbitrary fixed values of* $\{\lambda_k, k\in e^c\}$ *and for arbitrary fixed values of* $\{c_k, k\in e\}$ *there exist elemenst that* $\mathfrak{v}$, $\tilde{\mathfrak{v}}$ *respectively minimize and maximize* $\sum_{k\in e^c} \lambda_k \mathfrak{V}_k$ *on* $\mathscr{E}_{em}$ *and there are corresponding elements* $\mathfrak{u}=\mathfrak{h}(\mathfrak{v})$, $\tilde{\mathfrak{u}}=\mathfrak{h}(\tilde{\mathfrak{v}})$ *that respectively maximize and minimize* $\sum_{k\in e^c} \lambda_k \mathfrak{U}_k$ *on* $\mathfrak{h}(\mathscr{E}_{em})$.

Theorem 2. *Let hypotheses G,* H_1, H_2, H_3 *hold. Then the extremizers of Theorem 1 are solutions of the corresponding boundary value problem and have as much smoothness as the smoothness of* F_x *and* F_y *permits. In particular, if* F_x *and* F_y *are continuous, then the solutions are classical solutions. Moreover,* $\varrho_1>-1$, $\varrho_2>-1$, $\omega_3>-1$ *everywhere.*

Remarks. These theorems show that there are solutions of all sizes. But the existence of solutions for all values of the pressure λ_2 is assured only when $\alpha_3=\alpha_4$ exceeds the critical value 2. In Sect. 23 we show why this is essentially a best possible condition. We note that the analysis leading to the determination of this critical value is quite elementary (see Lemmas 1 and 2 in Sect. 22).

This result holds important lessons for the construction of special nonlinear theories of elastic rods. It has been a common practice to retain true nonlinear strain-displacement relations, but to employ a linear relation between stress resultants and the particular strain measure chosen. Thus if the measure of extensional strain is chosen to be the extension ϱ_3 the corresponding constitutive relation is usually taken to be

$$N=EA\,\varrho_3, \tag{21.7}$$

whereas if the strain is taken to be the material strain $\varrho_3+\frac{1}{2}(\varrho_3)^2$ the corresponding constitutive relation is usually taken to be

$$N=EA\left[\varrho_3+\tfrac{1}{2}(\varrho_3)^2\right]. \tag{21.8}$$

Here EA is a material constant. These two laws clearly agree for ϱ_3 small. But for ϱ_3 large and positive, relation (21.7) corresponds to $\alpha_3=2$ and (21.8) to $\alpha_3=3$. Thus these two commonly used measures of extensional strain produce strikingly different qualitative behavior in the large. These relations are also unacceptable for large deformations because only a finite force N is needed to compress a positive length of rod to zero length. Needless to say, these same comments apply to three-dimensional and shell theories.

Our analysis in the next section shows the essential role played by the complex strain ω: the replacement of the convexity of U with respect to $\omega_3, \omega_4, \gamma_1', \gamma_2', \phi'$, which is a consequence of H_2, by the convexity of another strain energy function (say W) with respect to a different set of arguments (say $\varrho_3, \beta, \varrho_1', \varrho_2', \phi'$) would destroy the proof that the solutions of the boundary value problems are regular.

22. Proofs of the theorems. The principal tool used in the proof of Theorem 1 is the ***Minimization Theorem:*** *A (sequentially) weakly lower semicontinuous functional on a bounded (sequentially) weakly closed nonempty subset of a reflexive Banach space attains its minimum there.* [Proof of this result and relevant definitions are given by VAINBERG (1956), ROTHE (1968), and others.] For the spaces we use, sequential weakness is equivalent to weakness. We prove Theorem 1 by proving a sequence of lemmas that ensure that the hypotheses of this Minimization Theorem are met. We adopt the convention that C represents a positive constant, specific estimates for which are available but unnecessary for our purposes. If C appears more than once in any expression, it need not have the same meaning in each place.

Lemma 1. *If hypotheses G and H_1 hold, then $\mathfrak{B}_1[\mathfrak{v}]\to\infty$ as $\|\mathfrak{v}\|\to\infty$.*

Proof. $\|\mathfrak{v}\|\to\infty$ if and only if at least one of the terms on the right side of (21.4) approaches infinity. Hypothesis H_1 immediately implies that $\mathfrak{B}_1[\mathfrak{v}]\to\infty$ as any one of the norms

$$e_1\|\gamma_1\|_{\alpha_1},\quad e_2\|\gamma_2\|_{\alpha_2},\quad \|\gamma_3\|_{\alpha_3},\quad \|\omega_4\|_{\alpha_4},\quad \|\gamma_1'\|_{\alpha_1},\quad \|\gamma_2'\|_{\alpha_2},\quad \|\phi'\|_{\alpha_5}$$

approaches infinity. That $\mathfrak{B}_1\to\infty$ as $\|\gamma_1\|_{\alpha_1}$ (when $e_1=0$) or $\|\gamma_2\|_{\alpha_2}$ (when $e_2=0$) approaches infinity, is an immediate consequence of the ***Poincaré inequality:*** *If $w(S)$ is a function in $\mathscr{W}_p^1$ with $w(S_0)=0$ for some $S_0\in[S_1, S_2]$, then there is a positive number C such that $\|w\|_p< C\|w'\|_p$.* Similarly, if $\phi(S_1)=0$ or $\phi(S_2)=0$, then $\mathfrak{B}_1\to\infty$ as $\|\phi\|_{\alpha_5}\to\infty$. If ϕ is not required to satisfy one of these boundary conditions, then we observe that the problem is invariant under the addition of multiples of 2π to ϕ. We then define the branch of ϕ for any given configuration to have mean value satisfying

$$0\leqq(S_2-S_1)^{-1}\int_{S_1}^{S_2}\phi(S)\,dS<2\pi.$$

Then $\phi-(S_2-S_2)^{-1}\int_{S_1}^{S_2}\phi(S)\,dS$ satisfies the Poincaré inequality. Thus $\|\phi'\|_{\alpha_5}$ and therefore $\mathfrak{B}_1$ approach infinity as $\|\phi\|_{\alpha_5}\to\infty$. Finally, if $|x(S_1)-X(S_1)|+|y(S_1)-Y(S_1)|$ can approach infinity, then hypothesis G and an elementary estimate ensure that $\|\gamma_3\|_{\alpha_3}+\|\omega_4\|_{\alpha_4}\to\infty$ so that $\mathfrak{B}_1\to\infty$. □[48]

Lemma 2. *Let $F_x, F_y\in\mathscr{L}_1$ and let hypotheses G and H_1 hold. Let $1\in e^c$. If $\alpha_3=\alpha_4>2$ or if $2\in e$, then for arbitrary fixed values of $\{\lambda_k, k\in e^c, k\neq 1\}$, $\sum_{k\in e^c}\lambda_k\mathfrak{B}_k\to\infty$ as $\|\mathfrak{v}\|\to\infty$.*

Proof. The inequalities

$$\begin{gathered}|\mathfrak{B}_2|\leqq C[1+\|\gamma_3\|^2_{\alpha_3}+\|\omega_4\|^2_{\alpha_4}],\\ \sum_{k=2}^{8}|\mathfrak{B}_k|\leqq C[1+\|\gamma_3\|_{\alpha_3}+\|\omega_4\|_{\alpha_4}],\end{gathered}\tag{22.1}$$

obtained by the Hölder inequality and the hypothesis H_1, ensure the result. □

Lemma 3. *Let $F_x, F_y\in\mathscr{L}_1$. The functionals $\mathfrak{U}_k$, $k=2,\dots,10$, are (sequentially) weakly continuous on $\mathfrak{h}(\mathscr{B})\subset\mathscr{B}$.*

Proof. The (sequential) weak continuity of $\mathfrak{U}_k$ means that

$$\mathfrak{U}_k[\mathfrak{u}_j]-\mathfrak{U}_k[\mathfrak{u}]\to 0\quad\text{as}\quad \mathfrak{u}_j\to\mathfrak{u}\quad\text{weakly.}\tag{22.2}$$

The Riesz Representation Theorem implies that a sequence $\{w_j\}$ in $\mathscr{L}_p$ $(p>1)$ converges weakly to w if and only if

$$\int_{S_1}^{S_2}[w_j(S)-w(S)]\,g(S)\,dS\to 0\tag{22.3}$$

for arbitrary $g\in\mathscr{L}_{p*}$ where $(1/p)+(1/p^*)=1$ and that a sequence $\{w_j\}$ in $\mathscr{W}_p^1$ $(p>1)$ converges weakly to w if and only if

$$\int_{S_1}^{S_2}\{[w_j(S)-w(S)]\,g(S)+[w_j'(S)-w'(S)]\,g'(S)\}\,dS\to 0\tag{22.4}$$

for arbitrary $g\in\mathscr{W}^1_{p*}$, where again $(1/p)+(1/p^*)=1$. The weak convergence of $x_j(S_1)$ to $x(S_1)$ and of $y_j(S_1)$ to $y(S_1)$ is the usual numerical convergence. Thus $\mathfrak{u}_j\to\mathfrak{u}$ weakly when the components of $\mathfrak{u}_j$ converge weakly to the components of $\mathfrak{u}$ in their respective spaces.

Now let $\mathfrak{u}_j\to\mathfrak{u}$ weakly as $j\to\infty$. From (20.14) and (20.31) we have

$$\begin{aligned}\mathfrak{U}_7[\mathfrak{u}_j]-\mathfrak{U}_7[\mathfrak{u}]+i(\mathfrak{U}_8[\mathfrak{u}_j]-\mathfrak{U}_8[\mathfrak{u}])&=x_j(S_1)-x(S_1)+i[y_j(S_1)-y(S_1)]\\ &\quad+\int_{S_1}^{S_2}(\omega_j-\omega)\exp i\phi^+\,dS+\int_{S_1}^{S_2}\omega_j^+(\exp\phi_j^+-\exp i\phi^+)\,dS.\end{aligned}\tag{22.5}$$

The first integral on the right side of (22.5) approaches zero by (22.3). Since $\mathscr{W}_p^1$ can be compactly embedded in the space $\mathscr{C}[S_1, S_2]$ of continuous functions with the maximum norm,

[48] The symbol □ indicates the termination of a proof.

the elements of the sequence $\{\phi_j\}$ are (equivalence classes) containing continuous functions converging strongly (i.e. uniformly) to ϕ in $\mathscr{C}[S_1, S_2]$. Since the sequence $\{\omega_j^+\}$ is weakly convergent in $\mathscr{L}_{\alpha_3}$, it is bounded in norm. Thus the second integral of (22.5) is less than

$$\left(\int_{S_1}^{S_2} |\omega_j^+|\, dS\right) \max_{[S_1, S_2]} (\exp i\,\phi_j^+ - \exp i\,\phi^+) \to 0.$$

By the hypothesis of weak convergence, $x_j(S_1) \to x(S_1)$, $y_j(S_1) \to y(S_1)$. Thus $\mathfrak{U}_7$ and $\mathfrak{U}_8$ are weakly continuous. The same reasoning yields the weak continuity of the rest of the functionals $\mathfrak{U}_k$, $k = 3, \ldots, 10$, the proofs for $\mathfrak{U}_3$ and $\mathfrak{U}_4$ being expedited by a change in the order of integration.

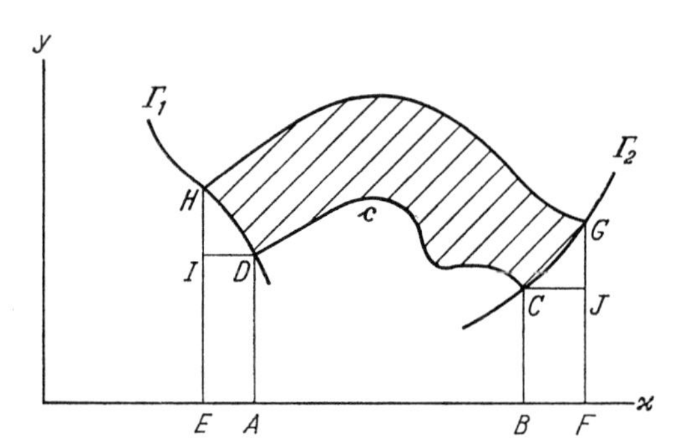

Fig. 4. The area $\Delta_{DCGH} = \mathfrak{U}_2[\mathfrak{u}_j] - \mathfrak{U}_2[\mathfrak{u}]$.

The proof of the weak continuity of $\mathfrak{U}_2$ is more delicate. We first examine Fig. 4. Here the curve DC represents $\mathscr{C}$ and the curve HG represents the curve determined by $\mathfrak{u}_j$. The area of the quadrilateral $DCGH$ is denoted Δ_{DCGH}. We use an analogous notation for other areas. Thus

$$\mathfrak{U}_2[\mathfrak{u}_j] - \mathfrak{U}_2[\mathfrak{u}] = \Delta_{DCGH} = \Delta_{EFGH} - \Delta_{ABCD} - \Delta_{EADI} - \Delta_{BFJC} - \Delta_{IDH} - \Delta_{CJG}, \tag{22.6}$$

where

$$\begin{aligned}
\Delta_{EFGH} &= \mathfrak{U}_{2_1}[\mathfrak{u}_j] + \tfrac{1}{2}\{\mathfrak{x}[\mathfrak{u}_j, S_2]\, \mathfrak{y}[\mathfrak{u}_j, S_2] - x_j(S_1)\, y_j(S_1)\},\\
\Delta_{ABCD} &= \mathfrak{U}_{2_1}[\mathfrak{u}] + \tfrac{1}{2}\{\mathfrak{x}[\mathfrak{u}, S_2]\, \mathfrak{y}[\mathfrak{u}, S_2] - x(S_1)\, y(S_1)\},\\
\Delta_{EADI} &= -y(S_1)\,[x_j(S_1) - x(S_1)],\\
\Delta_{BFJC} &= \mathfrak{y}[\mathfrak{u}, S_2]\, \{\mathfrak{x}[\mathfrak{u}_j, S_2] - \mathfrak{x}[\mathfrak{u}, S_2]\},\\
\Delta_{IDH} &= O\big(|x_j(S_1) - x(S_1)| \cdot |y_j(S_1) - y(S_1)|\big),\\
\Delta_{CJG} &= O\big(|\mathfrak{x}[\mathfrak{u}_j, S_2] - \mathfrak{x}[\mathfrak{u}, S_2]| \cdot |\mathfrak{y}[\mathfrak{u}_j, S_2] - \mathfrak{y}[\mathfrak{u}, S_2]|\big).
\end{aligned} \tag{22.7}$$

From (22.6), (22.7), and from the weak continuity of $\mathfrak{U}_7$, $\mathfrak{U}_8$, we find that $\mathfrak{U}_{2_1}[\mathfrak{u}_j] - \mathfrak{U}_{2_1}[\mathfrak{u}] \to 0$ as $j \to \infty$, so that $\mathfrak{U}_{2_2}$ is weakly continuous.

Now $(20.27)_2$ implies that

$$\mathfrak{U}_{2_1}[\mathfrak{u}_j] - \mathfrak{U}_{2_1}[\mathfrak{u}] = -\operatorname{Im} \int_{S_1}^{S_2} (\omega_j^+ \exp i\,\phi_j^+ - \omega^+ \exp i\,\phi^+)\, \xi\, dS + \int_{S_1}^{S_2} (\xi_j - \xi)\, \omega_j^+ \exp i\,\phi_j^+\, dS, \tag{22.8}$$

where

$$\xi(S) = \mathfrak{x}[\mathfrak{u}, S] - i\mathfrak{y}[\mathfrak{u}, S], \qquad \xi_j(S) = \mathfrak{x}[\mathfrak{u}_j, S] - i\mathfrak{y}[\mathfrak{u}_j, S]. \tag{22.9}$$

Our previous arguments show that the first integral on the right of (22.8) approaches zero and that $\xi_j(S) - \xi(S)$ approaches zero pointwise as $j \to \infty$. The uniform boundedness of $\|\omega_j^+\|_{\alpha_3}$ and the Hölder inequality enable us to show that the sequence of continuous functions $\{\xi_j\}$ is uniformly bounded and equicontinuous and therefore has a uniformly convergent subsequence by the Arzelà-Ascoli Theorem. The entire sequence $\{\xi_j\}$ itself must converge uniformly to zero (or else we could extract a subsequence violating the pointwise convergence to zero). Thus the second integral of (22.8) is bounded above by

$$C \max_{[S_1, S_2]} |\xi_j - \xi| \to 0.$$

Hence $\mathfrak{U}_{2_1}$ is weakly continuous. □

Lemma 4. *Let $\mathcal{K}$ denote the closed ball $\{\mathfrak{v}\in\mathcal{B}: \|\mathfrak{v}\|\leq a\}$ in $\mathcal{B}$. The closure of $\mathfrak{h}(\mathcal{K})$, denoted* $\mathrm{Cl}\,\mathfrak{h}(\mathcal{K})$, *is weakly closed.*

Proof. The convexity of the function $|h^{-1}|$ defined in (20.4), (20.5) enables us to show that $\mathfrak{h}(\mathcal{K})$ is midpoint convex. The lemma follows from the results that the closure of a midpoint convex set is convex and that a strongly closed convex subset of a Banach space is weakly closed. □

We require the lower semicontinuity theorem of FICHERA (1967): *Let $F(\boldsymbol{a}, \boldsymbol{b}, S)$ be a continuous real valued function bounded from below for $\boldsymbol{a}\in\mathcal{R}^p$, $\boldsymbol{b}\in\mathcal{R}^q$, $S\in[S_1, S_2]$ and let F be convex in the components of $\boldsymbol{a}$ for fixed $\boldsymbol{b}$, S. Let $\mathfrak{l}$ be a bounded linear map of a Banach space $\mathcal{B}$ into $[\mathcal{L}_1]^p$ and let $\mathfrak{m}$ be a compact linear map of $\mathcal{B}$ into $[\mathcal{L}_1]^q$. Then the functional*

$$\mathfrak{F}[\mathfrak{f}] = \int_{S_1}^{S_2} F\big((\mathfrak{l}\mathfrak{f})(S), (\mathfrak{m}\mathfrak{f})(S), S\big)\, dS$$

is sequentially weakly lower semicontinuous on the set of all $\mathfrak{f}\in\mathcal{B}$ for which $\mathfrak{F}[\mathfrak{f}]$ is defined. We can now prove

Lemma 5. *Let hypotheses H_1 and H_2 hold. Then $\mathfrak{U}_1$ is (sequentially) weakly lower semicontinuous wherever it is defined on $\mathcal{B}$.*

Proof. We merely apply FICHERA's theorem to $\mathfrak{U}_1$, identifying

$$\mathfrak{l}\mathfrak{u} = (\omega_3, \omega_4, \gamma_1', \gamma_2', \phi'), \qquad \mathfrak{m}\mathfrak{u} = (\gamma_1, \gamma_2),$$

and noting that the requisite convexity of U is an immediate consequence of H_2. The Sobolev embedding theorem ensures the compactness of $\mathfrak{m}$. □

Proof of Part (i) *of Theorem 1.* Lemmas 1 and 2 imply that $\sum_{k\in e^c}\lambda_k\mathfrak{V}_k\to\infty$ as $\|\mathfrak{v}\|\to\infty$. Therefore we need only seek minima for $\sum_{k\in e^c}\lambda_k\mathfrak{V}_k$ on $\mathcal{K}\cap\mathcal{E}_{em}$ where $\mathcal{K}$ is some closed ball of radius a centered at the origin of $\mathcal{B}$. An elementary geometric construction ensures that $\mathcal{K}\cap\mathcal{E}_{em}$ is not empty. Since the functionals $\mathfrak{V}_k$ have the same values as the functionals $\mathfrak{U}_k$ for corresponding arguments, we study the minimization problem for $\sum_{k\in e^c}\lambda_k\mathfrak{U}_k$ on $\mathfrak{h}(\mathcal{K}\cap\mathcal{E}_{em})$. By Lemmas 3 and 5, $\sum_{k\in e^c}\lambda_k\mathfrak{U}_k$ is weakly lower semicontinuous on $\mathrm{Cl}\,\mathfrak{h}(\mathcal{K})$. By Lemmas 3 and 4, $\mathrm{Cl}\,\mathfrak{h}(\mathcal{K}\cap\mathcal{E}_{em})$ is weakly closed. The Minimization Theorem implies that $\sum_{k\in e^c}\lambda_k\mathfrak{U}_k$ is minimized on $\mathrm{Cl}\,\mathfrak{h}(\mathcal{K}\cap\mathcal{E}_{em})$. Thus $\sum_{k\in e^c}\lambda_k\mathfrak{V}_k$ is minimized on the pre-image of $\mathrm{Cl}\,\mathfrak{h}(\mathcal{K}\cap\mathcal{E}_{em})$ (which consists of all $\mathfrak{v}\in\mathcal{B}$ that are taken into $\mathrm{Cl}\,\mathfrak{h}(\mathcal{K}\cap\mathcal{E}_{em})$ by $\mathfrak{h}$ and which may be an unbounded set). But we have shown that if $\sum_{k\in e^c}\lambda_k\mathfrak{V}_k$ is minimized on $\mathcal{E}_{em}$ it must in fact be minimized on $\mathcal{K}\cap\mathcal{E}_{em}$ itself. □

We need some additional results for the proof of Part (ii) of Theorem 1.

Lemma 6. *Let $\delta\mathfrak{v}\in\mathcal{B}$. The functionals $\mathfrak{V}_k$, $k=2, \ldots, 10$, have Gâteaux differentials (first variations)*

$$\mathfrak{V}_k'[\mathfrak{v}]\,\delta\mathfrak{v} = \frac{d}{d\varepsilon}\mathfrak{V}_k[\mathfrak{v}+\varepsilon\delta\mathfrak{v}]\big|_{\varepsilon=0} \tag{22.10}$$

at $\mathfrak{v}$ in the direction $\delta\mathfrak{v}$ and the functionals $\mathfrak{U}_k$, $k=2, \ldots, 10$ have Gâteaux differentials $\mathfrak{U}_k'[\mathfrak{u}]\,\delta\mathfrak{u}$ at $\mathfrak{u}=\mathfrak{h}(\mathfrak{v})$ in the direction $\delta\mathfrak{u}=\mathfrak{h}'(\mathfrak{v})\,\delta\mathfrak{v}$, where $\mathfrak{h}'(\mathfrak{v})$ is the Gâteaux derivative of $\mathfrak{h}$.

Proof. We use the mean value theorem to represent the integrands of

$$\varepsilon^{-1}\{\mathfrak{V}_k[\mathfrak{v}+\varepsilon\delta\mathfrak{v}]-\mathfrak{V}_k[\mathfrak{v}]\}, \qquad k=2_1, 3, 4, 7, 8$$

in the form $\boldsymbol{b}(\varepsilon)\cdot\delta\boldsymbol{v}$, where $\boldsymbol{b}(\varepsilon)$ depends on septuples $\bar{\boldsymbol{v}}$ intermediate to $\boldsymbol{v}$ and $\boldsymbol{v}\varepsilon+\delta\boldsymbol{v}$. These integrands are uniformly bounded by integrable functions for $|\varepsilon|\leq 1$. This follows for $\mathfrak{V}_{2_1}, \mathfrak{V}_3, \mathfrak{V}_4$ by the presence of indefinite integrals in their integrands and for $\mathfrak{V}_7, \mathfrak{V}_8$ by standard arguments. Since the integrands converge pointwise to their obvious limits as $\varepsilon\to 0$, the Lebesgue Dominated Convergence Theorem implies that the limits of these integrals, namely $\mathfrak{V}_k'[\mathfrak{v}]\,\delta\mathfrak{v}$, $k=2_1, 3, 4, 7, 8$, exist. The differentiability of Γ_2 as a function of its arguments x, y then guarantees the existence of $\mathfrak{V}_{10}'[\mathfrak{v}]\,\delta\mathfrak{v}$. $\mathfrak{V}_{2_2}$ is treated as in (22.6)–(22.8). The Gâteaux differentials of $\mathfrak{V}_5, \mathfrak{V}_6, \mathfrak{V}_9$ are just ordinary differentials. The existence of $\mathfrak{U}_k'[\mathfrak{u}]\,\delta\mathfrak{u}$ follows from the chain rule for Gâteaux derivatives. □

These Gâteaux derivatives are readily calculated by the usual procedures of variational calculus.

Lemma 7. *If* $\sum_{k=2}^{12} \bar{\lambda}_k \mathfrak{V}_k'[\mathfrak{v}]\,\delta\mathfrak{v} = 0$ *for arbitrary* $\delta\mathfrak{v}$ *in some dense subset of* $\mathcal{B}$, *then* $\bar{\lambda}_3 F_x = 0$, $\bar{\lambda}_4 F_y = 0$ *and* $\bar{\lambda}_k = 0$, $k = 2, 5, 6, \ldots, 10$.

Proof. If $\bar{\lambda}_2 = 0$, the use of the explicit representations for the Gâteaux differentials and of the arbitrariness of $\delta\mathfrak{v}$ leads to the conclusion of the lemma. If $\bar{\lambda}_2 \neq 0$, a calculation shows that $1 + h(\gamma_3) = 0$ everywhere, implying that $\gamma_3 = -\infty$ everywhere, which is absurd. $\square$

Proof of Part (ii) *of Theorem 1.* Let

$$\mathcal{G} = \{\mathfrak{v} \in \mathcal{B} : \mathfrak{V}_1[\mathfrak{v}] \leqq c_1\} \tag{22.11}$$

so that $\partial\mathcal{G} = \mathcal{E}_1(c_1)$. $\mathcal{G}$ is bounded, for if not, there would be a sequence $\{\mathfrak{v}_k\} \in \mathcal{G}$ with $\|\mathfrak{v}_k\| \to \infty$. But Lemma 1 would then imply that $\mathfrak{V}_1[\mathfrak{v}_k] \to \infty$, violating the definition of $\mathcal{G}$. The properties of $\mathfrak{h}$ ensure that $\mathfrak{h}(\mathcal{G})$ is likewise bounded. Since $\mathfrak{U}_1$ is weakly lower semicontinuous wherever it is defined on $\mathcal{B}$, it follows that $\mathfrak{h}(\mathcal{G})$ is weakly closed and thus $\mathcal{H} = \mathfrak{h}(\mathcal{G} \cap \mathcal{E}_{em})$ is weakly closed. It is readily shown that this set is nonempty. Thus by Lemma 3, $\sum_{k \in e^c} \lambda_k \mathfrak{U}_k$ is weakly continuous on the bounded weakly closed nonempty subset $\mathcal{H}$ of $\mathcal{B}$. The Minimization Theorem then implies that $\sum_{k \in e^c} \lambda_k \mathfrak{U}_k$ attains both its maximum and minimum on $\mathcal{H}$.

We wish to show that these extrema are actually attained on $\partial\mathcal{H}$. If such an extremum were attained at an interior point of $\mathcal{H}$, then $\sum_{k \in e^c} \lambda_k \mathfrak{U}_k'[\mathfrak{u}]\,\delta\mathfrak{u} = 0$ at that point. By Lemma 7 and the chain rule for Gâteaux differentials, $\lambda_k = 0$, $k \in e^c$ so that $\sum_{k \in e^c} \lambda_k \mathfrak{U}_k'[\mathfrak{u}]\,\delta\mathfrak{u}$ would be identically zero on $\mathcal{H}$ and trivially maximized on the boundary. Since $\mathfrak{h}$ is continuous, the pre-image of $\partial\mathcal{H}$ is the closed set $\mathcal{E}_{em}$, so these results immediately carry over to $\sum_{k \in e^c} \lambda_k \mathfrak{V}_k$ which has the same values on $\mathcal{G}$ as $\sum_{k \in e^c} \lambda_k \mathfrak{u}_k$ has on $\mathfrak{h}(\mathcal{G})$. $\square$

For the proof of Theorem 2, we require

Lemma 8. *Let* $e_6\,\delta x(S_1)$, $e_6\,\delta y(S_1)$ *be arbitrary fixed real numbers, let* $\delta\gamma_3(S)$, $\delta\omega_4(S)$ *be arbitrary fixed piecewise continuous functions, and let* $\delta\gamma_1(S)$, $\delta\gamma_2(S)$, $\delta\phi(S)$ *be arbitrary fixed piecewise smooth functions in* $\mathcal{B}$. *Let* $\delta\mathfrak{v}$ *denote the element of* $\mathcal{B}$ *consisting of these components. If hypothesis* H_3 *holds, then* $\mathfrak{V}_1$ *has a Gâteaux differential* $\mathfrak{V}_1'[\mathfrak{v}]\,\delta\mathfrak{v}$ *at* $\mathfrak{v}$ *in the direction of this* $\delta\mathfrak{v}$, *and* $\mathfrak{U}_1$ *has a Gâteaux differential* $\mathfrak{U}_1'[\mathfrak{u}]\,\delta\mathfrak{u}$ *at* $\mathfrak{u} = \mathfrak{h}(\mathfrak{v})$ *in the direction* $\delta\mathfrak{u} = \mathfrak{h}'(\mathfrak{v})\,\delta\mathfrak{v}$.

Proof. The proof imitates that of Lemma 6, using H_3 to obtain a uniform bound on the integrand of $\varepsilon^{-1}[\mathfrak{V}_1[\mathfrak{v} + \varepsilon\,\delta\mathfrak{v}] - \mathfrak{V}_1[\mathfrak{v}]]$ by an integrable function for $|\varepsilon| \leqq 1$. $\square$

Proof of Theorem 2.[49] We just prove the result for the case in which $e^c = \{1\}$. The proofs for the other extremizers are identical. The Multiplier Rule implies there are real constants $\bar{\lambda}_k$, $k = 1, \ldots, 10$, not all zero, such that

$$\sum_{k=1}^{10} \bar{\lambda}_k \mathfrak{V}_k'[\mathfrak{v}]\,\delta\mathfrak{v} = \sum_{k=1}^{10} \bar{\lambda}_k \mathfrak{U}_k'[\mathfrak{u}]\,\delta\mathfrak{u} = 0 \tag{22.12}$$

for arbitrary $\delta\mathfrak{v}$ of the type described in Lemma 8 and for $\delta\mathfrak{u} = \mathfrak{h}'(\mathfrak{v})\,\delta\mathfrak{v}$, where the $\bar{\lambda}$'s are determined by a normalizing condition and the side conditions

$$\mathfrak{V}_k[\mathfrak{v}] = \mathfrak{U}_k[\mathfrak{u}] = 0 \quad \text{for } k \in m; \qquad \mathfrak{V}_k[\mathfrak{v}] = \mathfrak{U}_k[\mathfrak{u}] = c_k \quad \text{for } k \in e. \tag{22.13}$$

Now $\bar{\lambda}_1$ does not vanish, for if it did, Lemma 7 would imply that the remaining $\bar{\lambda}$'s vanish, in violation of the Multiplier Rule. We therefore normalize the $\bar{\lambda}$'s by taking $\bar{\lambda}_1 = 1$.

We now invoke the Fundamental Lemma of the Calculus of Variations in the following two forms: *Let* $f(S) \in \mathcal{L}_1$. *If* $\int_{S_1}^{S_2} f(S)\,g'(S)\,dS = 0$, *for all piecewise smooth* $g(S)$ *with compact support on* $[S_1, S_2]$, *then* $f(S)$ = constant function a.e.; *if* $\int_{S_1}^{S_2} f(S)\,g(S) = 0$ *for all piecewise*

[49] Our proof modifies the standard regularity theory of TONELLI for single integral variational problems in order to account for the analytical difficulty of our problem. Cf. AKHIEZER (1955), FUNK (1962), MORREY (1966, Theorem 1.10).

continuous $g(S)$ *with compact support on* $[S_1, S_2]$, *then* $f(S)=0$, a.e. We change the order of integrations in some terms of $(22.12)_2$ and integrate some terms by parts in such a way that the terms remaining in the integrands have one of the functions $\delta\omega_3(S)$, $\delta\omega_4(S)$, $\delta\gamma_1'(S)$, $\delta\gamma_2'(S)$, $\delta\phi'(S)$ as factors. We then apply the Fundamental Lemma and exploit the possible arbitrariness of $\delta\gamma_1(S_1)$, $\delta\gamma_2(S_1)$, $\delta\phi(S_1)$, $\delta x(S_1)$, $\delta y(S_1)$ to obtain

$$\frac{\partial U}{\partial\gamma_\alpha'}(\boldsymbol{u}(S), S) = \int_{S_1}^{S} \frac{\partial U}{\partial\gamma_\alpha}(\boldsymbol{u}(T), T)\, dT + \text{const}, \tag{22.14}$$

$$\frac{\partial U}{\partial\omega_3}(\boldsymbol{u}(S), S) = -A(S)\cos\phi^+(S) - B(S)\sin\phi^+(S), \tag{22.15}$$

$$\frac{\partial U}{\partial\omega_4}(\boldsymbol{u}(S), S) = -B(S)\cos\phi^+(S) + A(S)\sin\phi^+(S), \tag{22.16}$$

$$\begin{aligned}\frac{\partial U}{\partial\phi'}(\boldsymbol{u}(S), S) = &\int_{S_1}^{S} [\omega_3^+(T)\cos\phi^+(T) - \omega_4(T)\sin\phi^+(T)]\,B(T)\, dT \\ &- \int_{S_1}^{S} [\omega_3^+(T)\sin\phi^+(T) + \omega_4(T)\cos\phi^+(T)]\,A(T)\, dT + \text{const},\end{aligned} \tag{22.17}$$

where

$$\begin{aligned}A(S) \equiv &-\bar\lambda_2 \int_S^{S_2} [\omega_3^+(T)\sin\phi^+(T) + \omega_4(T)\cos\phi^+(T)]\, dT \\ &+ \bar\lambda_3 \int_S^{S_2} F_x(T)\, dT + \bar\lambda_7 + \bar\lambda_{10}\frac{\partial\Gamma_2}{\partial x}(\mathfrak{x}[\mathfrak{u}, S_2], \mathfrak{y}[\mathfrak{u}, S_2]), \\ B(S) \equiv &\,\bar\lambda_2 \int_S^{S_2} [\omega_3^+(T)\cos\phi^+(T) - \omega_4(T)\sin\phi^+(T)]\, dT \\ &+ \bar\lambda_4 \int_S^{S_2} F_y(T)\, dT + \bar\lambda_8 + \bar\lambda_{10}\frac{\partial\Gamma_2}{\partial y}(\mathfrak{x}[\mathfrak{u}, S_2], \mathfrak{y}[\mathfrak{u}, S_2]).\end{aligned} \tag{22.18}$$

Eqs. (22.14)–(22.17) hold a.e. We also obtain the following alternatives for boundary conditions:

Either

$$\gamma_\alpha(S_1) = 0 \quad \text{or} \quad \int_{S_1}^{S_2} \frac{\partial U}{\partial\gamma_\alpha}(\boldsymbol{u}(S), S)\, dS = 0. \tag{22.19}$$

Either

$$\begin{aligned}\phi(S_1) = 0 \quad \text{or} \quad &\int_{S_1}^{S} [\omega_3^+(T)\cos\phi^+(T) - \omega_4(T)\sin\phi^+(T)]\, B(T)\, dT \\ &- \int_{S_1}^{S_2} [\omega_3^+(T)\sin\phi^+(T) + \omega_4(T)\cos\phi^+(T)]\,A(T)\, dT = 0.\end{aligned} \tag{22.20}$$

Either

$$x(S_1) = X(S_1) \quad \text{and} \quad y(S_1) = Y(S_1),$$

or

$$\begin{aligned}A(S_1) + \bar\lambda_5 + \bar\lambda_9 \frac{\partial\Gamma_1}{\partial x}(x(S_1), y(S_1)) &= 0, \\ B(S_1) + \bar\lambda_6 + \bar\lambda_9 \frac{\partial\Gamma_1}{\partial y}(x(S_1), y(S_1)) &= 0.\end{aligned} \tag{22.21}$$

Since $\phi \in \mathscr{W}_{\alpha_8}^1$, it is continuous. Thus the right sides of (22.14)–(22.17) represent continuous functions of S. The strict convexity of U with respect to ω_3, ω_4, γ_1', λ_2', ϕ', which is a consequence of H_2, allows us to apply the Implicit Function Theorem to (22.14)–(22.17) and conclude that we can solve these equations for the $\omega_3, \omega_4, \gamma_1', \gamma_2', \phi'$ that appear as arguments of the derivatives of U in terms of the other members of the equations:

$$(\omega_3(S), \omega_4(S), \gamma_1'(S), \gamma_2'(S), \phi'(S)) = \boldsymbol{G}(S, \gamma_1(S), \gamma_2(S), \boldsymbol{P}(S)), \tag{22.22}$$

where $\boldsymbol{G}$ is a differentiable function of its arguments and $\boldsymbol{P}(S)$ represents the right sides of (22.14)–(22.17). Since the arguments of $\boldsymbol{G}$ are continuous functions of S, it follows that $\omega_3, \omega_4, \gamma_1', \gamma_2', \phi'$ are continuous functions of S. (The uniqueness of the distributional derivative implies that γ_α', ϕ' are classical derivatives of γ_α, ϕ.) If F_x, F_y are continuous, this last result implies that the arguments of $\boldsymbol{G}$ are then continuously differentiable functions of S so that ω_3, ω_4 are continuously differentiable and γ_α, ϕ are twice continuously differentiable. This smoothness of $\mathfrak{u}$ enables us to differentiate (22.14)–(22.17) with respect to S. We obtain

$$\begin{gathered}
\frac{d}{dS}\left(\frac{\partial U}{\partial \gamma_\alpha'}\right) = \frac{\partial U}{\partial \gamma_\alpha}, \\
\frac{d}{dS}\left(\frac{\partial U}{\partial \omega_3}\cos\phi^+ - \frac{\partial U}{\partial \omega_4}\sin\phi^+\right) = -\bar{\lambda}_2(\omega_3^+\sin\phi^+ + \omega_4\cos\phi^+) + \bar{\lambda}_3 F_x, \\
\frac{d}{dS}\left(\frac{\partial U}{\partial \omega_3}\sin\phi^+ + \frac{\partial U}{\partial \omega_4}\cos\phi^+\right) = \bar{\lambda}_2(\omega_3\cos\phi^+ - \omega_4\sin\phi^+) + \bar{\lambda}_4 F_y, \\
\frac{d}{dS}\left(\frac{\partial U}{d\phi'}\right) = \omega_4\frac{\partial U}{\partial \omega_3} - \omega_3^+\frac{\partial U}{\partial \omega_4}.
\end{gathered} \tag{22.23}$$

These are readily shown to be equivalent to the equations of Sect. 18.

By (20.6), the variables ϱ_1, ϱ_2 are twice differentiable images under h_1, h_2 of the twice differentiable functions γ_1, γ_2. The continuity of γ_1, γ_2 ensures that $\varrho_1 > -1$, $\varrho_2 > -1$ everywhere. Since ω_3 is a continuous function and the image of the $\mathscr{L}_{\alpha_3}$ function γ_3 under h, it follows that $\omega_3 > -1$, a.e. To show that $\omega_3 > -1$ everywhere, we proceed thus: We use $(22.12)_1$, thereby obtaining in place of (22.14)–(22.17) a similar system with derivatives of V on the left and with $h'(\gamma_3(S))$ among the expressions on the right. Hypothesis H_2 enables us to use the Implicit Function Theorem to express $\gamma_2(S)$ as a continuous function of S, $\gamma_\alpha(S)$, and $\tilde{\boldsymbol{P}}(S)$, the right side of this system. Since $h'(\gamma_3(S))$ appears in $\tilde{\boldsymbol{P}}(S)$, $\tilde{\boldsymbol{P}}(S)$ need not be continuous. But the inequality $|h'| < 1$ implies that $\tilde{\boldsymbol{P}}(S)$ is bounded, whence it follows that $\gamma_3(S)$ is also bounded. The properties of h then imply that $\omega_3(S) = h(\gamma_3(S))$ is bounded below by a number greater than -1.

We now turn to the study of natural boundary conditions. Following the standard procedure, we integrate by parts the expression

$$\int_{S_1}^{S_2} \frac{\partial U}{\partial \gamma_\alpha'}\,\delta\gamma_\alpha'\,dS = \left[\frac{\partial U}{\partial \gamma_\alpha'}\,\delta\gamma_\alpha\right]_{S_1}^{S_2} - \int_{S_1}^{S_2} \frac{d}{dS}\left(\frac{\partial U}{\partial \gamma_\alpha'}\right)\delta\gamma_\alpha\,dS \tag{22.24}$$

appearing in $(22.12)_2$ and use the arbitrariness of $\delta\gamma_\alpha$ to conclude that if $\gamma_\alpha(S_\nu) \neq 0$, then

$$\frac{\partial U}{\partial \gamma_\alpha'}(\boldsymbol{u}(S_\nu), S_\nu) = 0, \tag{22.25}$$

and if $\gamma_\alpha(S_1) = \gamma_\alpha(S_2)$, then

$$\frac{\partial U}{\partial \gamma_\alpha'}(u(S_1), S_1) = \frac{\partial U}{\partial \gamma_\alpha'}(u(S_2), S_2). \tag{22.26}$$

Similarly, we find that if $\phi(S_\nu) \neq 0$, then

$$\frac{\partial U}{\partial \phi'}(u(S_\nu), S_\nu) = 0, \tag{22.27}$$

and if $\phi(S_1) = \phi(S_2)$, then

$$\frac{\partial U}{\partial \phi'}(u(S_1), S_1) = \frac{\partial U}{\partial \phi'}(u(S_2), S_2). \tag{22.28}$$

Setting $S=S_2$ in (22.15), (22.16), we find

$$\begin{aligned}&\frac{\partial U}{\partial\omega_3}\big(\boldsymbol{u}(S_2),S_2\big)\cos\phi^+(S_2)-\frac{\partial U}{\partial\omega_4}\big(\boldsymbol{u}(S_2),S_2\big)\sin\phi^+(S_2)\\&\qquad=-\bar\lambda_7-\bar\lambda_{10}\frac{\partial\Gamma_2}{\partial x}\big(\mathfrak{x}[\mathfrak{u},S_2],\mathfrak{y}[\mathfrak{u},S_2]\big),\end{aligned}\tag{22.29}$$

$$\begin{aligned}&\frac{\partial U}{\partial\omega_3}\big(\boldsymbol{u}(S_2),S_2\big)\sin\phi^+(S_2)+\frac{\partial U}{\partial\omega_4}\big(\boldsymbol{u}(S_2),S_2\big)\cos\phi^+(S_2)\\&\qquad=-\bar\lambda_8-\bar\lambda_{10}\frac{\partial\Gamma_2}{\partial y}\big(\mathfrak{x}[\mathfrak{u},S_2],\mathfrak{y}[\mathfrak{u},S_2]\big).\end{aligned}\tag{22.30}$$

Now we remove the restriction that $e^c=\{1\}$ and consider the whole range of possibilities for boundary conditions at the end $S=S_2$: (i) If the position of this end is free, then we set $\bar\lambda_{10}=0$ because the boundary condition (20.19) is not operative. If the horizontal component λ_7 of end load is prescribed, we replace $\bar\lambda_7$ by λ_7. Eq. (22.29) then becomes a boundary condition. On the other hand, if we prescribe the horizontal component of displacement for this end, then $\mathfrak{u}_7=c_7$ and (22.29) is an equation giving the force $\bar\lambda_7$ necessary to maintain such a displacement in terms of $\mathfrak{u}$ and c_7. Analogous results hold for (22.30) with respect to vertical components of load and displacement. (ii) If the position of this end is fixed: $\mathfrak{u}_7=c_7$, $\mathfrak{u}_8=c_8$, then we again set $\bar\lambda_{10}=0$. Eqs. (22.29), (22.30) determine the reactions $\bar\lambda_7$, $\bar\lambda_8$ necessary to maintain this state. (iii) If this end moves on the curve $\Gamma_2(x,y)=0$ then (20.19) holds. If we prescribe the horizontal component of displacement at this end, but leave the vertical component free, then the boundary conditions are (20.19) and $\mathfrak{u}_7=c_7$, with (22.29), (22.30) used to determine the reactions $\bar\lambda_7+\bar\lambda_{10}\frac{\partial\Gamma_2}{\partial x}$, $\bar\lambda_8+\bar\lambda_{10}\frac{\partial\Gamma_2}{\partial y}$ needed to maintain this state. If we further prescribe the vertical component λ_8 of the external end load, we merely replace $\bar\lambda_8$ by λ_8. Anaolguos results hold when the vertical component of displacement is prescribed and the horizontal component is left free. If λ_7 and λ_8 are prescribed, then we replace $\bar\lambda_7$, $\bar\lambda_8$ by λ_7, λ_8 and eliminate $\bar\lambda_{10}$ from (22.29), (22.30) to obtain

$$\begin{aligned}&\left(\frac{\partial\Gamma_2}{\partial y}\cos\phi^+-\frac{\partial\Gamma_2}{\partial x}\sin\phi^+\right)\frac{\partial U}{\partial\omega_3}-\left(\frac{\partial\Gamma_2}{\partial y}\sin\phi^++\frac{\partial\Gamma_2}{\partial x}\cos\phi^+\right)\frac{\partial U}{\partial\omega_4}\\&\qquad=-\lambda_7\frac{\partial\Gamma_2}{\partial y}+\lambda_8\frac{\partial\Gamma_2}{\partial x}\quad\text{at}\quad S=S_2.\end{aligned}\tag{22.31}$$

The boundary conditions are then (20.19) and (22.31), with (22.29) or (22.30) used to determine the reaction $\bar\lambda_{10}$.

The boundary conditions at $S=S_1$ are treated analogously. Eqs. (22.21) correspond to (22.29), (22.30). If λ_5 and λ_6 are prescribed, we can eliminate $\bar\lambda_9$ from these equations to obtain a boundary condition corresponding to (22.31).

It is readily shown that all these conditions reduce to those of Sect. 18 when the variables of $\boldsymbol{w}$, given in (17.19), are used.

We finally turn to the ring which has the periodicity conditions $(20.25)_{1,2,3}$, (22.26), (22.28) and

$$\mathfrak{x}[\mathfrak{u},S_2]=x(S_1),\qquad\mathfrak{y}[\mathfrak{u},S_2]=y(S_1).\tag{22.32}$$

[See remarks following (20.17).] Substituting (22.32) into (22.15), (22.16), (22.18), we get

$$\frac{\partial U}{\partial\omega_3}\big(\boldsymbol{u}(S_2),S_2\big)=\frac{\partial U}{\partial\omega_3}\big(\boldsymbol{u}(S_1),S_1\big),\qquad\frac{\partial U}{\partial\omega_4}\big(\boldsymbol{u}(S_2),S_2\big)=\frac{\partial U}{\partial\omega_4}\big(\boldsymbol{u}(S_1),S_1\big).\tag{22.33}$$

Since U has period S_2-S_1 in S, we can write the five Eqs. (22.26), (22.28), (22.33) in the form

$$\frac{\partial U}{\partial\gamma'_\alpha}\big(\boldsymbol{u}(S_2),S_1\big)=\frac{\partial U}{\partial\gamma'_\alpha}\big(\boldsymbol{u}(S_1),S_1\big),\quad\text{etc.},\tag{22.34}$$

which we regard as a system for determining $\gamma_1'(S_2)$, $\gamma_2'(S_2)$, $\omega_3(S_2)$, $\omega_4(S_2)$, $\phi(S_2)$ in terms of $\gamma_1'(S_1)$, $\gamma_2'(S_1)$, $\omega_3(S_1)$, $\omega_4(S_1)$, $\phi(S_1)$. Since the Jacobian of this system is just the nonzero

Hessian determinant of U with respect to $\gamma_1', \gamma_2', \omega_3, \omega_4, \phi$, this system has a unique solution which must be

$$\gamma_\alpha'(S_2) = \gamma_\alpha'(S_1), \quad \text{etc.} \tag{22.35}$$

Thus we can recover all the periodicity conditions (18.2).

This completes the proof of Theorem 2. □

23. Straight and circular rods. The planar equilibrium problems of a straight rod under axially directed end tractions and a circular ring or arch under hydrostatic pressure throw light on the theorems of Sect. 21. The reference configuration for these rods is defined by

$$\Phi'(S) = \Omega\,(\text{const}). \tag{23.1}$$

For a straight rod we take

$$\Omega = 0, \quad X(S) = S, \quad Y(S) = 0, \quad \lambda_k = 0 \quad \text{for} \quad k \neq 1, 7, \tag{23.2}$$

and we prescribe either λ_7 or c_7. For a circular rod we take

$$\Omega = 1, \quad X(S) + i\,Y(S) = \exp iS, \quad \lambda_k = 0 \quad \text{for} \quad k \neq 1, 2, \tag{23.3}$$

and we prescribe either λ_2 or c_2. We assume that the rod is homogeneous, i.e., W does not depend explicitly on S. We also require

$$\frac{\partial W}{\partial \beta} = 0, \quad \frac{\partial W}{\partial \varrho_\alpha'} = 0, \quad \frac{\partial W}{\partial \phi'} = 0 \quad \text{if} \quad \boldsymbol{w} = (\boldsymbol{\varrho}, \boldsymbol{0}) \equiv (\varrho_1, \varrho_2, \varrho_3, 0, 0, 0, 0). \tag{23.4}$$

We seek solutions of the equilibrium equations of the form

$$\boldsymbol{w} = (\mathring{\boldsymbol{\varrho}}, \boldsymbol{0}) \equiv (\mathring{\varrho}_1, \mathring{\varrho}_2, \mathring{\varrho}_3, 0, 0, 0, 0), \tag{23.5}$$

where $\{\mathring{\varrho}_k\}$ are constants. Under these assumptions the nontrivial equilibrium equations from (17.33)–(17.35), (17.37), (17.38) reduce to

$$N' = 0, \quad \Omega N = \lambda_2\, \varrho_3^+, \quad \bar{G} = 0, \quad \bar{K} = 0. \tag{23.6}$$

If the rod is straight $(23.6)_2$ is an identity. We append a set of boundary conditions consistent with (23.5).[50]

There need not be solutions of the algebraic equations (23.6) even though the hypotheses of the existence theory are met because (23.5) may exclude the only solutions that exist. We must therefore impose an additional restriction on the material response to ensure the existence of "trivial" solutions of the form (23.5). We require the pair of algebraic equations

$$\frac{\partial W}{\partial \varrho_\alpha}(\boldsymbol{\varrho}, \boldsymbol{0}) = 0 \tag{23.7}$$

to have solutions

$$\varrho_\alpha = \hat{\varrho}_\alpha(\varrho_3) \tag{23.8}$$

for each $\varrho_3 > -1$.

Using this property and the constitutive equations (17.48), (17.49), we can reduce (23.6) for straight rods to

$$\frac{\partial W}{\partial \varrho_3}\left(\hat{\varrho}_1(\mathring{\varrho}_3), \hat{\varrho}_2(\mathring{\varrho}_3), \mathring{\varrho}_3, \boldsymbol{0}\right) = \mathring{N}\,(\text{const}). \tag{23.9}$$

[50] Note that if $G = 0$, then $\varrho_1' = 0$, if $K = 0$ then $\varrho_2' = 0$, if $M = 0$, then $\phi' = 0$. This follows from (23.4), hypothesis H_2, and (17.48), (17.49).

If c_7 is prescribed, then $\mathring{\varrho}_3 = c_7 (S_2 - S_1)^{-1}$ (provided $x(S_1) = S_1$). The axial stress resultant $\mathring{N} = \bar{\lambda}_7$ is then determined from (23.9). On the other hand if we prescribe

$$N(S_2) \cos \theta(S_2) - Q(S_2) \sin \theta(S_2) = \lambda_7, \tag{23.10}$$

which (23.5) reduces to

$$N(S_2) = \lambda_7, \tag{23.11}$$

then $\mathring{N} = \lambda_7$. Now hypotheses H_1 and H_2 imply that $\frac{\partial W}{\partial \varrho_3} \to -\infty$ as $\varrho_3 \to -1$, $\frac{\partial W}{\partial \varrho_3} \to \infty$ as $\varrho_3 \to \infty$. Since $\frac{\partial W}{\partial \varrho_3}$ is continuous, (23.9) has a solution for each λ_7.

The situation for circular rods subject to hydrostatic pressure is more delicate. The condition specified by (23.7), (23.8) reduces (23.6) to

$$\frac{\partial W}{\partial \varrho_3} \left(\hat{\varrho}_1(\mathring{\varrho}_3), \hat{\varrho}_2(\mathring{\varrho}_3), \mathring{\varrho}_3, \mathbf{0}\right) = \lambda_2 (1 + \mathring{\varrho}_3). \tag{23.12}$$

If c_2 is prescribed, $\mathring{\varrho}_3$ is determined and the pressure λ_2 necessary to effect the deformation is given by (23.12). Thus there are solutions of all sizes. If λ_2 is prescribed then the situation is more complicated. In Fig. 5 we plot a typical $\frac{\partial W}{\partial \varrho_3}$ and the family of rays $\lambda_2(1 + \mathring{\varrho}_3)$ as functions of $1 + \mathring{\varrho}_3$. Since $\frac{\partial W}{\partial \varrho_3} \to -\infty$ as $\varrho_3 \to -\infty$ and since $\frac{\partial^2 W}{\partial (\varrho_3)^2} > 0$, we find

(i) For all $\lambda_2 \leqq 0$, there is at least one solution $\mathring{\varrho}_3$ of (23.12).

(ii) If there are constants $\varkappa > 0$, $\alpha > 2$ such that $\frac{\partial W}{\partial \varrho} > \varkappa [(\varrho_3)^{\alpha-1} - 1]$ for $\varrho_3 > 0$, then there is at least one circular solution for each λ_2.

(iii) If there are constants $\varkappa > 0$, $\alpha \leqq 2$ such that $\frac{\partial W}{\partial \varrho_3} \leqq \varkappa [(\varrho_3)^{\alpha-1} + 1]$ for $\varrho_3 > 0$, then there is a positive value of λ_2 above which there are no circular solutions.

These results illustrate the significance of the hypotheses $H_1 - H_3$ underlying our existence theorems.

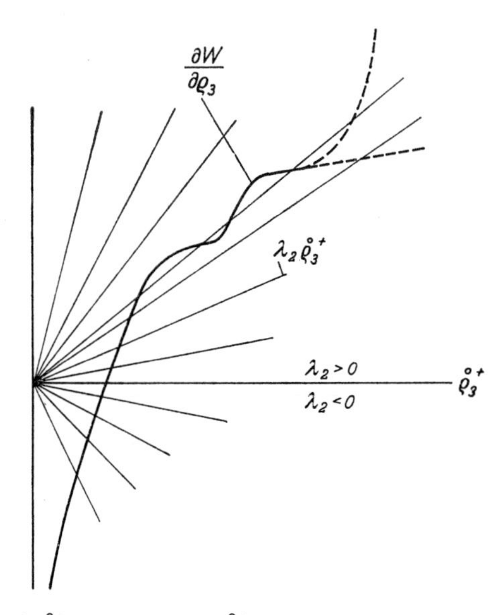

Fig. 5. $\partial W / \partial \varrho_3$ and $\lambda_2 \mathring{\varrho}_3^+$ as functions of $\mathring{\varrho}_3^+$ for circular rods under hydrostatic pressure.

24. Uniqueness theorems. We present a sampling of uniqueness theorems. The hypotheses for these theorems, which are far more stringent than those necessary to ensure existence, reflect that unqualified uniqueness in nonlinear elastostatics is exceptional. We first consider the planar equilibrium problem for the straight rod defined by (23.2) and the boundary conditions

$$\begin{gathered} x(0)=0, \quad y(0)=y(S_2)=0, \quad M(0)=M(S_2)=0, \\ N(S_2)\cos\theta(S_2)-Q(S_2)\sin\theta(S_2)=\lambda_7. \end{gathered} \tag{24.1}$$

We allow any conditions on the director amplitudes. We assume that

$$\frac{\partial W}{\partial\beta}=0 \quad \text{if} \quad \beta=0, \qquad \frac{\partial W}{\partial\phi'}=0 \quad \text{if} \quad \phi'=0 \tag{24.2}$$

and that

$$\frac{\partial^2 W}{\partial\beta^2}>0. \tag{24.3}$$

Theorem.[51] *If hypothesis H_2 and relations* (23.2), (24.1)–(24.3) *hold, if $\lambda_7 \geqq 0$, and if $-\pi<\theta<\pi$, then the classical solutions of the planar equilibrium problem satisfy*

$$y(S)\equiv 0, \quad \beta(S)\equiv 0, \quad \phi(S)\equiv 0 \quad \textit{for} \quad 0\leqq S\leqq S_2. \tag{24.4}$$

Proof. Since the only external force is the tensile force λ_7, we can replace the equilibrium equations for (17.33), (17.34), (17.38) and the constitutive relations (17.48) by

$$\frac{\partial W}{\partial\varrho_3}=\lambda_7\cos\theta, \qquad \frac{\partial W}{\partial\beta}=-\lambda_7\,\varrho_3^+\sin\theta, \qquad \frac{\partial W}{\partial\phi'}=y\,\lambda_7, \tag{24.5}$$

with y satisfying

$$y'=\varrho_3^+\sin\theta. \tag{24.6}$$

If $\lambda_7=0$, then H_2, (24.2), (24.3), $(24.5)_{2,3}$ imply that $\beta\equiv 0$, $\phi'\equiv 0$. Thus $\theta'\equiv 0$. Relations $(24.1)_{2,3}$, (24.6) then imply that $\theta\equiv 0$, $y\equiv 0$. If $\lambda_7>0$, we assume for contradiction that $y\not\equiv 0$. Conditions $(24.1)_{2,3}$ then require y or $-y$ to have a positive interior maximum. We need only treat the case in which y is maximized at $S_0\in(0, S_2)$:

$$y(S_0)>0, \quad y'(S_0)=0, \quad y''(S_0)\leqq 0. \tag{24.7}$$

Eqs. (24.6), (24.7) and the positivity of ϱ_3^+ imply that $\sin\theta(S_0)=0$. The restriction of θ to the range $(-\pi, \pi)$ then requires $\theta(S_0)=0$. Differentiating (24.6), we obtain

$$y''(S_0)=\varrho_3^+(S_0)\,\theta'(S_0). \tag{24.8}$$

The inequality $(24.7)_3$ then implies

$$\theta(S_0)\leqq 0. \tag{24.9}$$

Evaluating the derivative of $(24.5)_2$ at S_0, we obtain

$$W_{\beta\beta}\,\beta'(S_0)=-\lambda_7\,\varrho_3^+(S_0)\,\theta'(S_0)=-\lambda_7\,\varrho_3^+(S_0)\,[\beta'(S_0)+\phi'(S_0)], \tag{24.10}$$

where $W_{\beta\beta}$ denotes the value of $\partial^2 W/\partial\beta^2$ at S_0. We solve this for $\beta'(S_0)$ in terms of $\phi'(S_0)$. We then get

$$\theta'(S_0)=\beta'(S_0)+\phi'(S_0)=\frac{W_{\beta\beta}}{W_{\beta\beta}+\varrho_3^+(S_0)\,\lambda_7}\,\phi'(S_0). \tag{24.11}$$

[51] The theorem and its method of proof generalize results of GREENBERG (1967).

Now H_2, $(24.2)_2$, $(24.5)_3$, $(24.7)_1$ imply that $\phi'(S_0)>0$, so that $\theta'(S_0)>0$ in contradiction to (24.9). □

Some other boundary conditions can be treated similarly. The restriction that $\theta\in(-\pi,\pi)$ prevents solutions with loops that are known to exist for the elastica (cf. LOVE, 1927). There is no reason to expect uniqueness to hold for ϱ_α. Indeed, the absence of such uniqueness may correspond to modes of extensional failure, such as "necking down".

Sharper theorems for more special classes of materials are available. These ensure uniqueness for a range of negative values of λ_7 corresponding to compressive thrusts. REISS (1969) gave a simple proof of uniqueness for the classical elastica for $-\Lambda<\lambda_7<0$, where Λ is the smallest eigenvalue of the problem linearized about the trivial solution, by exploiting the minimizing properties of Λ. OLMSTEAD (1971) obtained uniqueness for a range of compressive thrusts for a material with constitutive relations

$$\beta=0, \qquad M=\hat{M}(\phi'), \qquad N=\hat{N}(\varrho_3).$$

His techniques may be applicable to more general classes of materials. MEISSONNIER (1965) has treated uniqueness for some very special forms of W.

The follower load problem. We now consider the planar equilibrium problem for the straight rod with reference configuration defined by

$$\Phi(S)\equiv 0, \qquad 0\leqq X(S)\equiv S\leqq S_2, \qquad Y(S)\equiv 0, \tag{24.12}$$

subject to boundary conditions

$$\begin{aligned} x(0)&=0, & y(0)&=0, & \phi(0)&=0, \\ N(S_2)&=P, & Q(S_2)&=0, & M(S_2)&=0. \end{aligned} \tag{24.13}$$

Thus the load applied at S_2 remains tangent to c at S_2 and is accordingly termed a *follower* load. This load and the reactions at $S=0$ are the only external forces applied to the rod. We allow any conditions on the director amplitudes. We assume that

$$\frac{\partial W}{\partial\beta}=0 \text{ if and only if } \beta=0, \qquad \frac{\partial W}{\partial\phi'}=0 \text{ if } \phi'=0. \tag{24.14}$$

Theorem. *If* (24.14) *and H_2 hold, then classical solutions to this follower lower load problem satisfy*

$$y(S)\equiv 0, \qquad \beta(S)\equiv 0, \qquad \phi(S)\equiv 0 \quad \textit{for} \quad 0\leqq S\leqq S_2.$$

Proof. From (17.34), (17.35), (24.13), $(17.48)_2$ we get

$$\frac{\partial W}{\partial\beta}=P\varrho_3^+\sin[\theta-\theta(S_2)]. \tag{24.15}$$

Relations (24.14) and (24.15) imply that $\beta(S_2)=0$. Now we substitute (17.48) into the equilibrium equations for (17.34) and (17.38) to get

$$\begin{aligned} &\left(\varrho_3^+\frac{\partial W}{\partial\varrho_3}+\frac{\partial^2 W}{\partial\beta^2}\right)\beta'+\frac{\partial^2 W}{\partial\beta\,\partial\phi'}\phi'' \\ &\qquad=\frac{\partial W}{\partial\beta}\frac{\varrho_3'}{\varrho_3^+}-\frac{\partial^2 W}{\partial\beta\,\partial\varrho_k}\varrho_k'-\frac{\partial^2 W}{\partial\beta\,\partial\varrho_\alpha'}\varrho_\alpha''-\frac{\partial W}{\partial\varrho_3}\phi' \\ &\frac{\partial^2 W}{\partial\phi'\,\partial\beta}\beta'+\frac{\partial^2 W}{\partial(\phi')^2}\phi''=-\frac{\partial^2 W}{\partial\phi'\,\partial\varrho_k}\varrho_k'-\frac{\partial^2 W}{\partial\phi'\,\partial\varrho_\alpha'}\varrho_\alpha''-\frac{\partial^2 W}{\partial\phi'\,\partial S}-\frac{\partial W}{\partial\beta}. \end{aligned} \tag{24.16}$$

By letting ϱ_3^+ be any positive continuously differentiable function and letting ϱ_1^+, ϱ_2^+ be positive twice continuously differentiable functions, we may regard (24.16) as a third order system for β and ϕ. We examine this system subject to initial conditions

$$\beta(S_2)=0, \quad \phi(S_2)=\theta(S_2), \quad \phi'(S_2)=0. \tag{24.17}$$

It is easy to show that H_2 implies that the determinant of coefficients of β' and ϕ'' is positive. The smoothness of W then ensures that the solutions of (24.16), (24.17) are unique and (24.14) requires that these solutions be given by $\beta(S)\equiv 0$, $\phi(S)\equiv\theta(S_2)$. Condition $(24.13)_3$ then implies that $\theta(S_2)=0$, $\phi(S)\equiv 0$ and (24.6) implies that $y(S)\equiv 0$. □

The follower load problem is a celebrated example of a nonconservative problem in elasticity. Since the equilibrium equations for this problem in the classical linear beam theory also have just the unique trivial solution, it was erroneously concluded that the straight state is "stable". BECK (1952) showed however that solutions to the linear dynamic problem become unbounded when P assumes certain critical values. (This still does not ensure non-uniqueness of the straight state for nonlinear equations.) Further examples and historical commentary on related problems are given by BOLOTIN (1961), PANOVKO and GUBANOVA (1967), HERRMANN (1967), VOL'MIR (1967), LEIPHOLZ (1968), ZIEGLER (1968), NEMAT-NASSER (1970).

25. Buckled states. If a boundary value problem depending on several parameters has a simple, explicit solution, called a *trivial* solution, for a range of the parameters, then the study of multiplicity of solutions for this parameter range is greatly simplified. In Sect. 23, we studied straight solutions of initially straight rods under end thrust and circular solutions of initially circular rods under hydrostatic pressure. These solutions may be regarded as trivial; the parameters involved are the end thrust $-\lambda_7$ and the hydrostatic pressure λ_2. In this section, we survey methods by which the existence and properties of nontrivial equilibrium states, called *buckled* states, can be determined for rods.

If a problem known to have a solution that minimizes a variational functional admits a trivial solution,[52] then the simple device of showing that the trivial solution does not furnish the variational functional with even a local minimum ensures that the minimizing solution and the trivial solution are distinct. This is accomplished by exhibiting an admissible variation for which the second variation of the variational functional is negative at the trivial solution.[53] By this process we can determine a range of load parameters for which there is a nontrivial solution. Results of this sort have physical content because local minimizers can be characterized as stable and other extremizers as unstable according to a number of well-defined criteria of stability. (See NFTM, Sect. 89, and the article by KNOPS and WILKES in this volume.)

TADJBAKHSH and ODEH (1967) gave an elegant application of this approach to the problem of the inextensible ring elastica under hydrostatic pressure. They showed that the minimizing solution is nontrivial for all external pressures exceeding the lowest eigenvalue of the problem linearized about the trivial solution. ANTMAN (1970b) used this method to study the buckling of an extensible, nonlinearly elastic circular ring under hydrostatic pressure. The ring's material response is characterized by the strain energy function

$$W=W(\varrho_3, \phi') \quad \text{with} \quad \beta\equiv 0, \quad \varrho_3>-1. \tag{25.1}$$

ANTMAN gave sufficient conditions on W to ensure the existence of nontrivial solutions for certain ranges of pressure. For this problem it may happen that

[52] Cf. Sects. 19–23.

[53] See FUNK's (1962) treatment of the elastica.

there is a maximum external pressure above which the only solution is the trivial solution. This possibility arises because the uniqueness of the trivial solutions for large deformations depends upon the relative strengths of the ring in resisting compression and flexure.

The existence and behavior of other branches of nontrivial solutions may be studied locally or globally. Such branches may represent stable configurations. The goals of a local study are to show that bifurcation actually occurs at the eigenvalues of the problem linearized about the trivial solution[54] and to determine the relationship between the solution and the parameter. If, as in Sect. 23, the trivial solution is not an identically zero solution, then the characteristic equation for the eigenvalues of the linearized problem may have a very complicated structure. For the trivial solutions of Sect. 23, the characteristic equations reduce to a set of nonlinear algebraic equations involving various derivatives of W at $\mathring{\boldsymbol{\varrho}}, \mathbf{0}$. The solutions of these equations given critical values of $\mathring{\boldsymbol{\varrho}}$ from which the buckling loads (eigenvalues) can be determined. No general conclusion about the number or ordering of these eigenvalues can be inferred. In fact this algebraic system need not have any solution. This variety of behavior was found by GREENBERG (1967) for straight rods with constitutive relations[55]

$$M=\hat{M}\left(\frac{\phi'}{\varrho_3^+}\right), \quad N=\hat{N}(\varrho_3), \quad \beta\equiv 0, \quad \varrho_3>-1 \tag{25.2}$$

and by ANTMAN (1970a, b) for rings with constitutive relations (25.1). The constructive local study can be effected by the methods of POINCARÉ (cf. KELLER, 1969) or of LYAPUNOV and SCHMIDT (cf. VAINBERG and TRENOGIN 1962, 1969). POINCARÉ's method has been used by ODEH and TADJBAKHSH (1965) to study the steady states of a rotating straight elastica, by TADJBAKHSH and ODEH (1967) to study the buckling of a ring elastica under hydrostatic pressure, by KELLER (1969) to study the bucking of a nonhomogeneous straight elastica under end thrust, and by ANTMAN (1970a) to study the buckling of a circular ring of material (25.1) under hydrostatic pressure. The method of LYAPUNOV and SCHMIDT was used by ODEH and TADJBAKHSH (1965) in the work just described and by BAZLEY and ZWAHLEN (1967). STAKGOLD (1971) illustrated both these methods for the straight elastica under end thrust. In cases when the computational effort for these constructive techniques becomes excessive, either because the system is of high order or because the eigenvalue has high multiplicity, the existence and certain local behavior of bifurcating branches may be determined by topological methods (cf. KRASNOSEL'SKIĬ, 1956). The most useful result for conservative problems in elasticity is the Theorem of KRASNOSEL'SKIĬ (1956, Chap. 2, Theorem 2.2) that for sufficiently well-behaved variational problems, bifurcation always occurs at the eigenvalues of the problem linearized about a trivial solution. The Leray-Schauder degree theory (cf. KRASNOSEL'SKIĬ, 1956), suitable for non-variational problems, has been applied by ODEH and TADJBAKHSH (1965) and by GREENBERG (1967) in the work described above.

The problem of describing the global behavior of branches is much deeper. The few results for rod problems have been obtained by techniques relying on

[54] There are nonlinear problems for which bifurcation does not occur at the eigenvalues of the problem linearized about the trivial solution. See the references cited at the end of this section.

[55] Note that $\phi'/\varrho_3^+ = \theta'/\varrho_3^+$ is the curvature. Because of the special uncoupling employed, these equations are not of the variational form (25.1). The methods used by GREENBERG can be extended to special cases of the general constitutive relations: $M=\hat{M}(\varrho_3, \phi')$, $N=\hat{N}(\varrho_3, \phi')$, $\beta\equiv 0$.

the availability of first integrals to the governing equations. These are discussed in Sect. 26. There are however a number of techniques including shooting[56] and topological methods that have proved effective for other kinds of equations and that should be useful for the more complicated equations of rod theories. For further details on this subject, consult PRODI (1967), KELLER and ANTMAN (1969), PIMBLEY (1969), VAINBERG and TRENOGIN (1969), STAKGOLD (1971), and the references cited therein.

26. Integrals of the equilibrium equations. Qualitative behavior of solutions. We obtain certain integrals of the equilibrium equations for circular and straight rods and indicate how these may be used to determine the qualitative behavior of the collection of solutions.

α) *Rods in space.* We consider the equilibrium equations corresponding to (15.1)–(15.3) and (15.7):

$$\boldsymbol{n}'+\boldsymbol{f}^0=\boldsymbol{0}, \qquad \boldsymbol{m}'+\boldsymbol{p}_0'\times\boldsymbol{n}+\boldsymbol{l}=\boldsymbol{0}, \qquad (\boldsymbol{\mu}^\alpha)'-\bar{\boldsymbol{\mu}}^\alpha+\boldsymbol{f}^\alpha=\boldsymbol{0}. \tag{26.1}$$

If $\boldsymbol{f}^0=\boldsymbol{0}$, $\boldsymbol{l}=\boldsymbol{0}$ then from $(26.1)_{1,2}$, (15.6), (16.4), we get

$$\frac{\partial\Psi}{\partial\boldsymbol{p}_0'}=\boldsymbol{a}\ (\text{const}), \qquad \boldsymbol{p}_\alpha\times\frac{\partial\Psi}{\partial\boldsymbol{p}_\alpha'}+\boldsymbol{p}_0\times\boldsymbol{a}=\boldsymbol{b}\ (\text{const}). \tag{26.2}$$

More generally, if Ψ does not depend explicitly on S and if there is a function Υ depending only on $\boldsymbol{p}_0$, $\boldsymbol{p}_\alpha$ such that

$$\boldsymbol{f}^0=\frac{\partial\Upsilon}{\partial\boldsymbol{p}_0}, \qquad \boldsymbol{f}^\alpha=\frac{\partial\Upsilon}{\partial\boldsymbol{p}_\alpha}, \tag{26.3}$$

then from (16.4), $(26.1)_{1,3}$, (26.3) we get

$$\boldsymbol{p}_0'\cdot\left(\frac{\partial\Psi}{\partial\boldsymbol{p}_0'}\right)'+\boldsymbol{p}_\alpha'\cdot\left(\frac{\partial\Psi}{\partial\boldsymbol{p}_\alpha}\right)'-\boldsymbol{p}_\alpha'\cdot\frac{\partial\Psi}{\partial\boldsymbol{p}_\alpha}+\boldsymbol{p}_0'\cdot\frac{\partial\Upsilon}{\partial\boldsymbol{p}_0}+\boldsymbol{p}_\alpha'\cdot\frac{\partial\Upsilon}{\partial\boldsymbol{p}_\alpha}=0, \tag{26.4}$$

which can be integrated to yield[57]

$$\boldsymbol{p}_0'\cdot\frac{\partial\Psi}{\partial\boldsymbol{p}_0'}+\boldsymbol{p}_\alpha'\cdot\frac{\partial\Psi}{\partial\boldsymbol{p}_\alpha'}-\Psi+\Upsilon=\text{const}. \tag{26.5}$$

β) *Rods in the plane.* We study the equilibrium problem for (17.33)–(17.38) with constitutive equations (17.48), (17.49). We assume that the only distributed loading is a hydrostatic pressure. Thus $f_Q=-\lambda_2\,\varrho_3^+$ and all other load components vanish. We eliminate Q from (17.33), (17.34), (17.38) to get

$$\left(\frac{M'}{\varrho_3^+}\right)'-\theta'\,N+\lambda_2\,\varrho_3^+=0, \qquad \theta'\,M'+\varrho_3^+\,N'=0. \tag{26.6}$$

Multiplying $(26.6)_1$ by M'/ϱ_3^+ and using $(26.6)_2$, we get

$$\frac{1}{2}\left(\frac{M'}{\varrho_3^+}\right)^2+\frac{1}{2}\,N^2+\lambda_2\,M=a\ (\text{const}). \tag{26.7}$$

From (17.38) we then obtain

$$\tfrac{1}{2}(Q^2+N^2)+\lambda_2\,M=a. \tag{26.8}$$

[56] A beautiful application of the shooting method was given by KOLODNER (1955).
[57] A result like (26.5) was found by ERICKSEN (1970).

If the rod is homogeneous so that W does not depend explicitly upon S and if the reference state is straight or circular so that $\Phi' = \text{const}$, then from $(26.6)_2$, (17.35), (17.37), (17.38), (17.48), (17.49), we get

$$(\phi' + \Phi')\left(\frac{\partial W}{\partial \phi'}\right)' - \beta' \frac{\partial W}{\partial \beta} + \varrho_3^+ \left(\frac{\partial W}{\partial \varrho_3}\right)' + \varrho_\alpha' \left(\frac{\partial W}{\partial \varrho_\alpha'}\right)' - \varrho_\alpha' \frac{\partial W}{\partial \varrho_\alpha} = 0, \tag{26.9}$$

which can be integrated to yield[58]

$$(\phi' + \Phi') \frac{\partial W}{\partial \phi'} + \varrho_3' \frac{\partial W}{\partial \varrho_3} - W = \text{const}. \tag{26.10}$$

γ) *Qualitative behavior.* If the strain energy function W is independent of $\{\varrho_\alpha, \varrho_\alpha'\}$, then we can reduce the entire problem to quadratures, and more importantly, determine some of the qualitative behavior of the solutions. If $\lambda_2 = 0$, then the planar form of (26.2) gives

$$\begin{aligned} &\frac{\partial W}{\partial \varrho_3} = \lambda_7 \cos\theta + \lambda_8 \sin\theta, & &\varrho_3^+ \frac{\partial W}{\partial \beta} = -\lambda_7 \sin\theta + \lambda_8 \cos\theta, \\ &\frac{\partial W}{\partial \phi'} = y\,\lambda_7 - x\,\lambda_8 + \text{const}, \quad \text{or} & &\left(\frac{\partial W}{\partial \phi'}\right)' = \varrho_3' [\lambda_7 \sin\theta - \lambda_8 \cos\theta]. \end{aligned} \tag{26.11}$$

Eqs. $(26.11)_{1,2}$ are just algebraic relations between ϱ_3, β, ϕ', S. If these equations can be solved uniquely for ϱ_3 and β in terms of ϕ' and S (for some range of λ_7, λ_8) and if these results are substituted into $(26.11)_4$, then $(26.11)_4$ becomes a nonlinear second order ordinary differential equation for ϕ. This represents a considerable simplification of the original system. In particular, if W does not depend explicitly on S and if the reference configuration is straight or circular, then the equation for ϕ is autonomous and can be studied by phase plane methods. STAKGOLD (1971) has carried out this program in detail for the compression of a straight hinged rod with invertible constitutive equations of the form

$$\varrho_3 = \hat{\varrho}_3(N), \qquad \phi' = \hat{\mu}(M), \qquad \beta = 0. \tag{26.12}$$

[These equations are a special case of (25.1).] STAKGOLD showed the solutions have substantially the same qualitative behavior as those of the elastica. If $(26.11)_{1,2}$ has a multiple-valued solution for ϱ_3 and β in terms of ϕ', S, then the resulting equation for ϕ' is also multiple-valued but still susceptible of analysis.

If $\lambda_2 \neq 0$, the approach is similar. If $\frac{\partial W}{\partial S} = 0$ and if the reference state is straight or circular, then (26.8), (26.10) are a pair of algebraic equations for ϱ_3, β, ϕ' yielding single- or multiple-valued representations for ϱ_3 and β as functions of ϕ'. The substitution of such representations into (26.7) yields a (single- or multiple-valued) first order autonomous equation for ϕ' that is in a form immediately available for a phase plane analysis. ANTMAN (1970a) investigated the qualitative behavior of a ring of material (25.1). He showed that the algebraic equation (26.10) has unique solutions of ϱ_3 in terms of ϕ' for sufficiently small ϕ' and obtained estimates for the range of ϕ' in terms of W for which the solution is unique. [If the material (26.12) is used, then this solution is unique for all ϕ'.] Using this result he showed that the entire qualitative structure of solutions inherited from the linearized problem is preserved as long as (26.10) has unique solutions. In particular, for this range each solution has at least two axes of symmetry. Earlier results of this form for the ring elastica were obtained by TADJBAKHSH and ODEH (1967) and by ANTMAN (1968b).

[58] Eqs. (26.7), (26.10) were derived by ANTMAN (1968a).

δ) *The elastica.* The planar inextensible elastica theory, characterized by the constitutive equations

$$M = B(S)\,\phi', \qquad \varrho_3 = 0, \qquad \beta = 0, \tag{26.13}$$

was formulated by JAS. BERNOULLI in 1691. EULER (1744) determined the qualitative behavior of the solutions. The extension of this theory to inextensible rods in space was achieved by KIRCHHOFF and CLEBSCH (cf. LOVE, 1927). Compendia of solutions for elastica problems are given by POPOV (1948) and by FRISCH-FAY (1962). These works and the survey of EISLEY (1963) have extensive bibliographies. The variational theory of the elastica in plane and space is given by FUNK (1962). KOVÁRI (1969) has treated some buckling problems for the elastica in space.

27. Problems of design. In design problems for rods, certain properties of the solution are prescribed, and certain properties of the reference configuration and the constitutive relations are left to be determined. Simple examples of such problems were treated in Sects. 17 and 23, where we obtained conditions on the strain energy function to ensure the possibility of planar solutions for initially planar rods and of circular solutions for initially circular rods.

EULER (1744) determined the curved reference configuration that an elastica must have in order to be straightened by an assigned terminal load. TRUESDELL (1954) treated an analogous problem for the elastica under hydrostatic pressure. These problems have practical significance for the design of load-bearing members. TRUESDELL observed that such inverse problems are simpler than the corresponding boundary value problem.

Classes of problems that admit trivial solutions are suitable for bifurcation analysis. The determination of such classes is a design problem. CARRIER (1947) showed that a homogeneous $(B(S) = \text{const})$, inextensible, elastica ring of arbitrary reference shape has a trivial solution provided

$$\frac{f_Q(S)}{\Phi'(S)} = -f_N(S). \tag{27.1}$$

For the hydrostatic loading of planar rods with a strain energy function $W(\varrho_3, \phi', S)$ with $\beta = 0$, ANTMAN (1970b) showed that if $\phi(S) = 0$ is to be a solution of the equilibrium problem, then

$$\Phi'(S) > 0, \qquad W(\varrho_3, \phi', S) = \Phi'(S)\, H\left(\frac{\varrho_3^+}{\phi'(S)}, a(S)\,\phi'\right), \tag{27.2}$$

where the functions H and a are arbitrary.

For a material characterized by a strain energy function

$$\Psi = \Psi(\mathbf{p}_0', \mathbf{p}_0'', \mathbf{p}_\alpha, \mathbf{p}_\alpha', S), \tag{27.3}$$

which is a generalization of our two-director theory, ERICKSEN (1970) has examined the circumstances for which there are *uniform* states. These are solutions $\mathbf{p}_0, \mathbf{p}_\alpha$ of the equilibrium equations that satisfy

$$\begin{aligned} \mathbf{p}_0'(S) &= \mathbf{Q}(S)\,\mathbf{p}_0'(S_1), & \mathbf{p}_0''(S) &= \mathbf{Q}(S)\,\mathbf{p}_0''(S_1), \\ \mathbf{p}_\alpha(S) &= \mathbf{Q}(S)\,\mathbf{p}_\alpha(S_1), & \mathbf{p}_\alpha'(S) &= \mathbf{Q}(S)\,\mathbf{p}_\alpha'(S_1), \end{aligned} \tag{27.4}$$

where $\mathbf{Q}$ is an orthogonal tensor, $\mathbf{Q}(S_1) = \mathbf{1}$.[59] By employing the integrals of Sect. 26α, ERICKSEN was able to describe the nature of solutions, showing that $\mathbf{p}_0(S)$ must

[59] The set of uniform states is not enlarged by dropping the requirement $(27.4)_2$.

represent a helix, possibly degenerate. When $\frac{\partial \Psi}{\partial S} = 0$, he reduced the problem to an algebraic one, obtaining restrictions on the reference state.

The problem of optimal design of a rod has long received much attention. It yields a variational problem with isoperimetric and unilateral constraints. TADJBAKHSH (1968) has treated the problem of determining $B(S)$ [cf. 26.13)] that minimizes the maximum deflection of a cantilevered elastica subject to a transverse end load. Here $\int_{S_1}^{S_2} B(S)\, dS$, which represents the volume, is fixed. This is apparently the only optimal design problem in the nonlinear theory of rods to have been treated. There is a large body of results for linear theories. We just mention the work of KELLER (1960), TADJBAKHSH and KELLER (1962), NIORDSEN (1965), KELLER and NIORDSEN (1966) on maximizing the minimum eigenvalue for linear rod problems. We refer to the comprehensive surveys of WASIUTYŃSKI and BRANDT (1963) and of SHEU and PRAGER (1968) for further details.

28. Dynamical problems. The governing equations for the dynamics of rods form a system of nonlinear equations with just two independent variables S, t. For nonlinear elastic rods these are partial differential equations. The advantage of having just one space variable is partly offset by the high order of the system. The problem of developing methods for treating the difficulties presented by such a system lies on the frontier of present day research. Rigorous results have been obtained chiefly for second order equations. Although such one-dimensional models serve as useful constitutive prototypes, they are far too special for inclusion here.[60] We refer to the review literature for further details.

There are but few results ensuring periodic solutions for nonlinear second order wave equations, scarcely any for nonlinear hyperbolic systems. Even in the degenerate case of planar "breathing" vibrations of a homogeneous circular ring under hydrostatic pressure, in which

$$\varrho_k = \varrho_k(t), \qquad \beta \equiv 0, \qquad \phi \equiv 0, \tag{28.1}$$

and for which (23.3), (23.4) hold, we are confronted with the still formidable system of ordinary differential equations

$$I\ddot{\varrho}_1 + \frac{\partial W}{\partial \varrho_1} = 0, \quad J\ddot{\varrho}_2 + \frac{\partial W}{\partial \varrho_2} = 0, \quad A\ddot{\varrho}_3 + \frac{\partial W}{\partial \varrho_3} + \lambda_2(1+\varrho_3) = 0, \tag{28.2}$$

obtained from (17.33)–(17.35), (17.37), (17.48), (17.49). Only when (25.1), $(24.2)_2$ are used does the problem reduce to a tractable second order ordinary differential equation for ϱ_3. The periodicity of solutions is then assured by the convexity requirement $\frac{\partial^2 W}{\partial \varrho_3^2} > 0$ (cf. TADJBAKHSH, 1966).

Formal perturbation methods, which generalize rigorous procedures for ordinary differential equations, can be used to determine the local behavior of oscillatory solutions near an equilibrium solution. The mathematical apparatus of these methods has been carefully described by KELLER and TING (1966). Perturbation solutions for the vibrating elastica have been found by ERINGEN (1952), KELLER and TING (1966), and WOODALL (1966). Others have similarly treated nonlinear technical theories. These methods are also effective for forced vibrations.

---

[60] We merely note that the solutions of the second order hyperbolic systems that can describe the longitudinal motion of a nonlinear elastic rod manifest a typical lack of regularity. E.g., see MACCAMY and MIZEL (1967).

Both local and global behavior can be studied by the projection method of KANTOROVICH in which the spatial dependence is averaged out and the partial differential equations are approximated by a finite system of ordinary differential equations. For further details consult BOLOTIN (1956, 1961), EVAN-IWANOWSKI (1965), HERRMANN (1967), LEIPHOLZ (1968), METTLER (1968), NEMAT-NASSER (1970) and the accompanying bibliographies.

DICKEY (1970) and REISS and MATKOWSKY (1971) examined the initial value problem for special nonlinear models of a compressed beam. They sought a dynamic description of the buckling process. DICKEY proved the existence and uniqueness of solutions to the coupled infinite system of Duffing equations obtained by separating the original partial differential equation. For sufficiently large compressive loads, he showed that the solutions would move about either one or both the stable buckled states depending upon the initial conditions.[61] REISS and MATKOWSKY employed a two-time asymptotic expansion to examine the behavior of a beam subject to a load slightly larger than critical. Assuming small damping, they determined the manner in which the solutions spiral down on the buckled equilibrium states.

Stability of motion for elastic Cosserat rods has been treated by GREEN, KNOPS, and LAWS (1968) and by WHITMAN and DESILVA (1970). They employed LYAPUNOV'S second method adapted for partial differential equations. (Cf. the article by KNOPS and WILKES in this volume.) DUPUIS (1969) has also examined this question.

References.

This list of works cited reflects neither the historical development of the theory of rods nor contemporary research motivated by engineering applications. Italic numbers in parentheses following the reference indicate the sections in which it has been cited.

AKHIEZER, N. I.: Lectures on the calculus of variations [in Russian]. Moscow: Gostekhteorizdat. (English translation by A. H. FRINK, 1962. New York: Blaisdell.) 1955. (*22*).

ANTMAN, S. S.: General solutions for plane extensible elasticae having nonlinear stress-strain laws. Quart. Appl. Math. **26**, 35–47 (1968a). (*17, 26*).

— A note on a paper of TADJBAKHSH and ODEH. J. Math. Anal. Appl. **21**, 132–135 (1968b). (*26*).

— Equilibrium states of nonlinearly elastic rods. J. Math. Anal. Appl. **23**, 459–470 (1968c). (*15, 21*).

— The shape of buckled nonlinearly elastic rings. Z. Angew. Math. Phys. **21**, 422–438 (1970a). (*25, 26*).

— Existence of solutions of the equilibrium equations for nonlinearly elastic rings and arches. Indiana Univ. Math. J. (J. Math. Mech.) **20**, 281–302 (1970b). (*3, 15, 21, 25, 27*).

— Existence and nonuniqueness of axisymmetric equilibrium states of nonlinearly elastic shells. Arch. Rational Mech. Anal. **40**, 329–372 (1971). (*3, 21*).

—, and W. H. WARNER: Dynamical theory of hyperelastic rods. Arch. Rational Mech. Anal. **23**, 135–162 (1966). (*5, 6, 7, 9, 10, 12, 17*).

BALL, J. M.: Initial-boundary value problems for an extensible beam. J. Math. Anal. Appl. (1972) (to appear). (*28*)

BAZLEY, N., and B. ZWAHLEN: Remarks on the bifurcation of solutions of a nonlinear eigenvalue problem. Arch. Rational Mech. Anal. **28**, 51–58 (1967). (*25*).

BECK, M.: Die Knicklast des einseitig eingespannten, tangential gedrückten Stabes. Z. Angew. Math. Phys. **3**, 225–228, 476–477 (1952). (*24*).

BOLOTIN, V. V.: Dynamic stability of elastic systems [in Russian]. Moscow: Gostekhteorizdat. [English translation (1964). San Francisco: Holden-Day.] 1956. (*28*).

— Nonconservative problems of the theory of elastic stability. Moscow: Fizmatgiz. (English translation by T. K. LUSHER, 1963. Oxford: Pergamon.) 1961. (*24, 28*).

BROWDER, F. E.: Approximation-solvability of nonlinear functional equations in normed linear spaces. Arch. Rational Mech. Anal. **26**, 33–42 (1967). (*11*).

CARRIER, G. F.: On the buckling of elastic rings. J. Math. Phys. **26**, 94–103 (1947). (*27*).

[61] BALL (1972) has obtained existence and regularity theorems for the actual partial differential equation for this problem.

COHEN, H.: A non-linear theory of elastic directed curves. Int. J. Engng. Sci. 4, 511–524 (1966). (*15*).
— The elastic fiber strengthened string—first order theory. Rév. Roum. Sci. Techn.—Mec. Appl. 12, 439–449 (1967). (*15*).
COSSERAT, E. et F.: Sur la statique de la ligne déformable. Compt. Rend. 145, 1409–1412 (1907). (*14*).
— Théorie des Corps Déformables. Paris: Hermann 1909. (*14, 15*).
DE SILVA, C. N., and A. B. WHITMAN: A thermodynamical theory of directed curves. (To appear 1971). (*15, 16*).
DICKEY, R. W.: Free vibrations and dynamic buckling of the extensible beam. J. Math. Anal. Appl. 29, 443–454 (1970). (*28*).
DUHEM, P.: Le potentiel thermodynamique et la pression hydrostatique. Ann. École Norm. (3) 10, 187–230 (1893). (*14*).
DUPUIS, G.: Stabilité élastique des structures unidimensionelles. Z. Angew. Math. Phys. 20, 94–106 (1969). (*28*).
EISLEY, J. G.: Nonlinear deformation of elastic beams, rings, and strings. Appl. Mech. Rev. 16, 677–680 (1963). (*26*).
ERICKSEN, J. L.: Simpler static problems in nonlinear theories of rods. Int. J. Solids Structures 6, 371–377 (1970). (*26, 27*).
—, and C. TRUESDELL: Exact theory of stress and strain in rods and shells. Arch. Rational Mech. Anal. 1, 295–323 (1958). (*14, 15*).
ERINGEN, A. C.: On the nonlinear vibrations of elastic bars. Quart. Appl. Math. 9, 361–369 (1952). (*28*).
EULER, L.: Additamentum I de curvis elasticis, Methodus inveniendi lineas curvas maximi minimive proprietate gaudentes. Lausanne = Opera Omnia I 24, 231–297 (1744). (*26, 27*).
EVAN-IWANOWSKI, R. M.: On the parametric response of structures. Appl. Mech. Rev. 18, 699–702 (1965). (*28*).
FICHERA, G.: Semicontinuità ed esistenza del minimo per una classe di integrali multipli. Rév. Roum. Mat. Pures et Appl. 12, 1217–1220 (1967). (*22*).
FRIEDRICHS, K. O.: The edge effect of transverse shear deformation on the boundary of plates. Reissner Anniversary Volume, p. 197–210. Ann Arbor: J. W. Edwards 1949. (*13*).
— Kirchhoff's boundary condition and the edge effect for elastic plates. Proc. Symp. Appl. Math. 3, p. 117–124. New York: McGraw-Hill 1950. (*13*).
FRISCH-FAY, R.: Flexible bars. London: Butterworths 1962. (*26*).
FUNK, P.: Variationsrechnung und ihre Anwendung in Physik und Technik. Berlin-Göttingen-Heidelberg: Springer 1962. (*22, 25, 26*).
GOODIER, J. N.: On the problem of the beam and the plate in the theory of elasticity. Trans. Roy. Soc. Can., Sect. III, 32, 65–88 (1938). (*13*).
GREEN, A. E.: The equilibrium of rods. Arch. Rational Mech. Anal. 3, 417–421 (1959). (*5, 6, 12*).
—, R. J. KNOPS, and N. LAWS: Large deformations, superposed small deformations and stability of elastic rods. Int. J. Solids Structures 4, 555–577 (1968). (*16, 28*).
—, and N. LAWS: A general theory of rods. Proc. Roy. Soc. (London) 293, 145–155 (1966). (*15, 16, 17*).
—, —, and P. M. NAGHDI: A linear theory of straight elastic rods. Arch. Rational Mech. Anal. 25, 285–298 (1967). (*11, 12, 16*).
— Rods, plates and shells. Proc. Cambridge Phil. Soc. 64, 895–913 (1968). (*5, 6, 7, 9, 10, 12*).
—, and P. M. NAGHDI: Non-isothermal theory of rods, plates, and shells. Int. J. Solids Structures 6, 209–244 (1970). (*5, 6, 7, 9, 10, 12, 15*).
—, and W. ZERNA: Theoretical elasticity, eds. 1, 2. Oxford: Clarendon Press 1954, 1968. (*9*).
GREENBERG, J. M.: On the equilibrium configurations of compressible slender bars. Arch. Rational Mech. Anal. 27, 181–194 (1967). (*24, 25*).
HAY, G. E.: The finite displacement of thin rods. Trans. Am. Math. Soc. 51, 65—102 (1942). (*12, 13*).
HERRMANN, G.: Stability of equilibrium of elastic systems subject to nonconservative forces. Appl. Mech. Rev. 20, 103–108 (1967). (*24, 28*).
KANTOROVICH, L. V., and G. P. AKILOV: Functional analysis in normed spaces [in Russian]. Moscow: Fizmatgiz. (English translation by D. E. BROWN, 1964. Oxford: Pergamon.) 1959. (*11*).
—, and KRYLOV: Approximate methods of higher analysis [in Russian], eds. 1–5. Moscow: Fizmatgiz. (English translation of 3rd ed. by C. D. BENSTER, 1958. New York: Interscience.) 1952–1962. (*11*).
KELLER, J. B.: The shape of the strongest column. Arch. Rational Mech. Anal. 5, 275–285 (1960). (*27*).

KELLER, J. B.: Bifurcation theory for ordinary differential equations. In: Bifurcation theory and nonlinear eigenvalue problems, ed. by J. B. KELLER and S. ANTMAN, p. 17–48. New York: Benjamin 1969. (*25*).

—, and S. ANTMAN (editors): Bifurcation theory and nonlinear eigenvalue problems. New York: Benjamin 1969. (*25*).

—, and F. I. NIORDSEN: The tallest column. J. Math. Mech. **16**, 433–446 (1966). (*27*).

—, and L. TING: Periodic vibrations of systems governed by nonlinear partial differential equations. Comm. Pure Appl. Math. **19**, 371–420 (1966). (*28*).

KOLODNER, I. I.: Heavy rotating string—a nonlinear eigenvalue problem. Comm. Pure Appl. Math. **8**, 395–408 (1955). (*25*).

KOVÁRI, K.: Räumliche Verzweigungsprobleme des dünnen elastischen Stabes mit endlichen Verformungen. Ing.-Arch. **37**, 393–416 (1969). (*26*).

KRASNOSEL'SKIĬ, M. A.: Topological methods in the theory of nonlinear integral equations [in Russian]. Moscow: Gostekhteorizdat. (English transl. by A. H. ARMSTRONG, 1964. Oxford: Pergamon.) 1956. (*20, 25*).

—, G. M. VAĬNIKKO, P. P. ZABREĬKO, YA. B. RUTITSKIĬ, and V. YA. STETSENKO: Approximate solutions of operator equations [in Russian]. Moscow: Izdat. Nauka 1969. (*11*).

LAWS, N.: A simple dipolar curve. Int. J. Engng. Sci. **5**, 653–661 (1967a). (*17*).

— A theory of elastic-plastic rods. Quart. J. Mech. Appl. Math. **20**, 167–181 (1967b). (*16*).

LEIPHOLZ, H.: Stabilitätstheorie. Stuttgart: Teubner. (English translation by Scientific Translation Service, 1970: Stability theory. New York: Academic Press.) 1968. (*24, 28*).

LIONS, J. L.: Quelques méthodes de résolution des problèmes aux limites non linéaires. Paris: Dunod, Gauthier-Villars 1969. (*11*).

LOVE, A. E. H.: A treatise on the mathematical theory of elasticity, eds. 1–4. Cambridge: University Press. [Reprinted (1944). New York: Dover Publications.] 1892–1927. (*11, 12, 13, 15, 24, 26*).

MACCAMY, R. C., and V. J. MIZEL: Existence and nonexistence in the large of solutions of quasilinear wave equations. Arch. Rational Mech. Anal. **25**, 299–320 (1967). (*28*).

MEDICK, M. A.: One dimensional theories of wave propagation and vibrations in elastic bars of rectangular cross-section. J. Appl. Mech. **33**, 489–495 (1966). (*11, 12*).

MEISSONNIER, B.: Étude de quelques figures d'équilibre d'une ligne hyperélastique et considérée comme un milieu orienté. J. Mécanique **4**, 423–437 (1965). (*15, 24*).

METTLER, E.: Combination resonances in mechanical systems under harmonic excitation. Proc. Fourth Conf. on Nonlinear Oscillations, Prague, 1967, p. 51–70. Prague: Academia 1968. (*28*).

MINDLIN, R. D., and G. HERRMANN: A one-dimensional theory of compressional waves in an elastic rod. Proc. First U.S. Natl. Cong. Appl. Mech. Chicago, 1951, p. 187–191. New York: A.S.M.E. 1952. (*11, 12*).

—, and H. D. MCNIVEN: Axially symmetric waves in elastic rods. J. Appl. Mech. **27**, 145–151 (1960). (*11, 12*).

MORREY, C. B.: Multiple integrals in the calculus of variations. Berlin-Heidelberg-New York: Springer 1966. (*19, 22*).

NAGHDI, P. M., and R. P. NORDGREN: On the nonlinear theory of elastic shells under the Kirchhoff hypothesis. Quart. Appl. Math. **21**, 49–59 (1963). (*9*).

NARIBOLI, G. A.: Asymptotic theory of wave motion in rods (longitudinal wave-motion). Z. Angew. Math. Mech. **49**, 525–531 (1969). (*13*).

NEMAT-NASSER, S.: Thermoelastic stability under general loads. Appl. Mech. Revs. **23**, 615–624 (1970). (*24, 28*).

NIORDSEN, F. I.: On the optimal design of a vibrating beam. Quart. Appl. Math. **23**, 47–53 (1965). (*27*).

NOVOZHILOV, V. V.: Foundations of the nonlinear theory of elasticity [in Russian]. Moscow: Gostekhizdat. (English translation by F. BAGEMIHL, H. KOMM, and W. SEIDEL, 1953. Rochester: Graylock.) 1948. (*11, 12*).

—, and L. I. SLEPIAN: On St. Venant's principle in the dynamics of beams. Prikl. Mat. Mekh. **29**, 261—281. English translation in J. Appl. Math. Mech. **29**, 293–315 (1965). (*11*).

ODEH, F., and I. TADJBAKHSH: A nonlinear eigenvalue problem for rotating rods. Arch. Rational Mech. Anal. **20**, 81–94 (1965). (*25*).

OLMSTEAD, W. E.: On the preclusion of buckling of a nonlinearly elastic rod. To appear (1971). (*24*).

PANOVKO, YA. G., and I. I. GUBANOVA: Stability and vibrations of elastic systems, eds. 1, 2 [in Russian]. Moscow: Izdat. Nauka. (English translation by C. V. LARRICK, 1965. New York: Consultants Bureau.) 1964–1967. (*24*).

PETRYSHYN, W. V.: Remarks on approximation-solvability of nonlinear functional equations. Arch. Rational Mech. Anal. **26**, 43–49 (1967). (*11*).

PIMBLEY, G. H.: Eigenfunction branches of nonlinear operators and their bifurcations. Berlin-Heidelberg-New York: Springer 1969. (*25*).

POPOV, E. P.: Nonlinear problems of the statics of thin rods [in Russian]. Leningrad: Gostekhteorizdat 1948. (*26*).

PRODI, G.: Problemi di diramazione per equazioni funzionali. Boll. Un. Mat. Ital. (3) **22**, 413–433 (1967). (*25*).

REISS, E. L.: Column buckling—an elementary example of bifurcation. In: Bifurcation theory and nonlinear eigenvalue problems, ed. by J. B. KELLER and S. ANTMAN, p. 1–16. New York: Benjamin 1969. (*24*).

— To appear (1972). (*13*).

—, and B. J. MATKOWSKY: Nonlinear dynamic buckling of a compressed elastic column. Quart. Appl. Math. (1971) (to appear). (*28*).

ROTHE, E.: Weak topology and the calculus of variations. In: C.I.M.E. course, Calculus of variations, classical and modern, 1966. Rome: Cremonese 1968. (*22*).

SHACK, W. J.: On linear viscoelastic rods. Int. J. Solids Structures **6**, 1–20 (1970). (*16*).

SHEU, C. Y., and W. PRAGER: Recent developments in optimal structural design. Appl. Mech. Rev. **21**, 985–992 (1968). (*27*).

STAKGOLD, I.: Branching of solutions of nonlinear equations. S.I.A.M. Rev. (1971) (to appear). (*25, 26*).

TADJBAKHSH, I.: The variational theory of the plane motion of the extensible elastica. Int. J. Engng. Sci. **4**, 433–450 (1966). (*15, 17, 28*).

— An optimal design problem for the nonlinear elastica. S.I.A.M. J. Appl. Math. **16**, 964–972 (1968). (*27*).

—, and J. B. KELLER: Strongest columns and isoperimetric inequalities for eigenvalues. J. Appl. Mech. **29**, 157–164 (1962). (*27*).

—, and F. ODEH: Equilibrium states of elastic rings. J. Math. Anal. Appl. **18**, 59–74 (1967). (*15, 21, 25, 26*).

TIMOSHENKO, S. P.: On the correction for shear of the differential equations for transverse vibrations of prismatic bars. Phil. Mag. (6) **41**, 744–746 (1921). (*11*).

— History of the strength of materials with a brief account of the history of the theory of elasticity and the theory of structures. New York: McGraw-Hill 1953. (*12*).

TODHUNTER, I., and K. PEARSON: A history of the theory of elasticity and of the strength of materials. Cambridge: University Press. 1886, 1893. [Reprinted (1960). New York: Dover.] (*12*).

TRUESDELL, C.: A new chapter in the theory of the elastica. Proc. First Midwestern Conf. on Solid Mech. 1953, 52–55 (1954). (*24*).

— The rational mechanics of materials—past, present, future. Appl. Mech. Rev. **12**, 75–80 (1959). Corrected reprint in: Applied Mechanics Surveys. Washington: Spartan Books 1966. (*12*).

— The rational mechanics of flexible or elastic bodies, 1638–1788. L. Euleri Opera Omnia (2) $\mathbf{11}_2$. Zürich: Füssli 1960. (*12, 15*).

VAĬNBERG, M. M.: Variational methods for the study of nonlinear operators [in Russian]. Moscow: Gostekhteorizdat. (English translation by A. FEINSTEIN, 1964. San Francisco: Holden-Day.) 1956. (*20, 22*).

—, and V. A. TRENOGIN: The methods of LYAPUNOV and SCHMIDT in the theory of nonlinear equations and their further development. Usp. Mat. Nauk **17** (2), 13–75. [English transl. (1962) in Russian Math. Surveys. **17**, 1–60.] (1962). (*25*).

— The theory of branching of solutions of nonlinear equations [in Russian]. Moscow: Izdat. Nauka 1969. (*25*).

VOL'MIR, A. S.: Stability of deformable systems, eds. 1, 2 [in Russian]. Moscow: Izdat. Nauka 1963–1967. (*24, 28*).

VOLTERRA, E.: Equations of motion for curved elastic bars by the use of the "method of internal constraints". Ing.-Arch. **23**, 402–409 (1955). (*5, 11, 12*).

— Equations of motion for curved and twisted elastic bars deduced by the use of the "method of internal constraints". Ing.-Arch. **24**, 392–400 (1956). (*5, 11, 12*).

— Second approximation of the method of internal constraints and its applications. Int. J. Mech. Sci. **3**, 47–67 (1961). (*5, 11, 12*).

WASIUTYŃSKI, Z., and A. BRANDT: The present state of knowledge in the field of optimal design of structures. Appl. Mech. Rev. **16**, 341–350 (1963). (*27*).

WOODALL, S. R.: On the large amplitude oscillations of a thin elastic beam. Int. J. Nonlin. Mech. **1**, 217–238 (1966). (*28*).

WHITMAN, A. B., and C. N. DE SILVA: A dynamical theory of elastic directed curves. Z. Angew. Math. Phys. **20**, 200–212 (1969). (*15*).

— Dynamics and stability of elastic Cosserat curves. Int. J. Solids Structures **6**, 411–422 (1970). (*28*).

ZIEGLER, H.: Principles of structural stability. Waltham: Ginn/Blaisdell 1968. (*24*).

Namenverzeichnis. — Author Index.

Sachverzeichnis.

(Deutsch-Englisch.)

Bei gleicher Schreibweise in beiden Sprachen sind die Stichwörter nur einmal aufgeführt.

abstrakte einseitige Probleme, symmetrischer Fall, *abstract unilateral problems, symmetric* 391.
— — —, unsymmetrischer Fall, *nonsymmetric* 395.
adiabatisch, *adiabatic* 304.
adjungierte, dem Kelvin-Zustand entsprechende Felder, *adjoint fields corresponding to Kelvin state* 175.
— Transformation 369.
adjungierter Operator, *adjoint operator* 355.
adjungiertes, dem Stokesschen Prozeß entsprechendes Zugkraftfeld, *adjoint traction field corresponding to Stokes process* 240.
Airy-Funktion, *Airy function* 157.
Airysche Lösung, *Airy's solution* 54, 156.
akustischer Tensor, *acoustic tensor* 243–248.
Amplitude, Welle der Ordnung n, *amplitude, wave of order* n 254.
Analogon zum Prinzip der minimalen potentiellen Energie, *analog of principle of minimum potential energy* 229.
analytisches Feld, *analytic field* 11.
Analytizität, elastische Verschiebungsfelder, *analyticity, elastic displacement fields* 131.
—, thermoelastische Verschiebungs- und Temperaturfelder, *thermoelastic displacement and temperature fields* 318.
anfänglich flache Cosserat-Fläche, *initially flat Cosserat surface* 460.
Anfangsbedingungen, *initial conditions* 219, 335.
Anfangswerte, *initial data* 218.
Anfangswertproblem für Stäbe, *initial-value problem for rods* 700.
anisotropes Material, *anisotropic material* 7, 87–89, 530.
Antennenpotential, *antenna potential* 420.
Antwortfunktion in der Elastizitätstheorie, *response function in elasticity* 80.
— — — Schalentheorie, *shell theory* 529, 531.
— — — Thermoelastizität, *thermoelasticity* 301, 302, 304.
Äquipräsenz, *equipresence* 530.
äquivalente Prozesse, *equivalent processes* 487.
— Systeme von Gleichungen, klassische lineare Schalentheorie, *systems of equations, classical linear theory of shells* 583.
Arbeit, Definition, *work, definition* 60.
—, geleistete, beginnend im Ruhezustand, *done, starting from rest* 217.
—, —, für einen Dehnungsverlauf, *along a strain path* 82.
—, —, von einem isotropen Körper, *by an isotropic body* 207.
Arbeitsleistung durch Kontaktkraft und Direktorkräftepaar, *rate of work by contact force and director couple* 60, 65, 544.
— — Kontaktkräfte und Kräftepaare sowie Volumkräfte und Kräftepaare, *contact and assigned forces and couples* 481.
Arbeitssätze in der Elastizitätstheorie, *work theorems in elasticity* 95, 208, 211.
— — — Thermoelastizität, *thermoelasticity* 320.
astatisches Gleichgewicht, *astatic equilibrium* 63, 64, 98, 193.
asymptotische Entwicklungsmethoden für Schalen, *asymptotic expansion techniques for shells* 586, 587.
— Methoden für Stäbe, *methods for rods* 664.
— Verteilung von Eigenwerten, *distribution of eigenvalues* 420.
aufgebrachte Leistung, Satz, *theorem of power expended* 65.
Aufteilung, *partition* 27.
Ausbreitung elastischer Wellen, *propagation of elastic waves* 386.
Ausbreitungsbedingung für fortschreitende Wellen, *propagation condition for progressive waves* 246.
— — isotrope Körper, *isotropic bodies* 247.
Ausbreitungsgeschwindigkeit, *speed of propagation* 246, 248, 343.
Ausbreitungsprobleme, *propagation problems* 371, 377.
Ausbreitungssätze, *propagation theorems* 256.
Ausdehnbarkeit in Richtung der Einheitsnormalen zur Bezugsfläche, *extensibility in the direction of the unit normal to a reference surface* 460.
Ausdehnungsbewegung, *extensional motion* 460.
—, Deformation einer Platte, *deformation of a plate* 600.
—, Schwingungen von Platten, *vibration of plates* 575.

Ausdehnungsbewegung, Theorie von Platten, klassisch, *extensional motion, theory of plates, classical* 577.
Ausgangswertprobleme, *initial-boundary value problems* 371.
Auslegungsprobleme für Stäbe, *design problems for rods* 698.
äußere Körperkraft, *external body force* 509.
äußeres Gebiet, *exterior domain* 14, 165, 261.
— Kraftsystem, *external force system* 95, 215, 327.
— Kraft-Temperatur-System, *force-temperature system* 313.
— Produkt von Tensoren, *outer product of tensors* 617.
— thermisches System, *thermal system* 327.
axialer Vektor, *axial vector* 6.

Basisvektoren, allgemein, *base vectors in general* 439, 618.
— auf einer Oberfläche, *on a surface* 621.
Beltrami-Lösung, *Beltrami's solution* 54.
Beltrami-Schaefer-Lösung, *Beltrami-Schaefer solution* 58.
Beschleunigung, Definition, *acceleration, definition* 43.
— in Schalentheorien, *in shell theories* 494.
— in Stabtheorien, *in rod theories* 649, 652.
Beschleunigungswelle, *acceleration wave* 253, 254.
beschränktes reguläres Gebiet, *bounded regular region* 13.
Betrag eines Tensors, *magnitude of a tensor* 7.
Bettischer Reziprozitätssatz, *Betti's reciprocal theorem* 98, 99, 170, 211, 419.
Beugebewegung, *flexural motion* 460.
Beugeschwingungen von Platten, *flexural vibrations of plates* 575.
bevorzugter Bezugszustand, *preferred reference state* 530.
Bewegung einer Cosserat-Fläche, *motion of Cosserat surface* 449.
— eines Körpers, *of a body* 43, 299.
Bewegungsgleichungen in Elastizitätstheorie, *equations of motion, elasticity* 45, 64, 212.
— — Schalen, abgeleitet aus der 3-dimensionalen Theorie, *shells, derived from 3-dimensional theory* 520.
— — —, direkter Zugang, *direct approach* 494.
— — —, Katalog, *catalogue* 526.
— — —, eingeschränkte Theorie, *restricted theory* 552.
— — Stäbe, *rods* 653, 658, 672.
— — Thermoelastizität, *thermoelasticity* 326.
Bezeichnungen, Liste, *notation, tables* 2–5, 298, 299, 432–438.
Bezugsdirektor, *reference director* 447.
Bezugsfläche, *reference surface* 425.
Bezugskonfiguration, *reference configuration* 299.
Bezugstemperatur, *reference temperature* 308, 312.
Biegungstheorie, dünne Schalen, *bending theory, shells, thin* 425.
—, Platten, klassische Theorie, *plates, classical theory* 577.
—, —, v. Kármáns nichtlineare Theorie, *v. Karman's nonlinear theory* 588.
—, Schalen, klassische Theorie, *of shells, classical* 582.
biharmonisches Feld, *biharmonic field* 20.
Bilanzprinzipien (s. auch Erhaltungssätze) für Schalen, *balance principles (see also conservation laws) for shells* 483.
— — Stäbe, Direktorimpuls, *rods, director momentum* 666.
— — —, Drehimpuls, *angular momentum* 666.
— — —, Energie, *energy* 667.
— — —, Impuls, *linear momentum* 666.
— — 3-dimensionale Theorien, Energie, *3-dimensional theories, energy* 511.
— — — —, Impuls, *momentum* 42.
— — — —, Impuls einer Welle, *momentum of a wave* 254.
— — — —, Impuls in Raum-Zeit, *momentum in space-time* 67.
— — — —, Impuls und Drehimpuls, *linear and angular momentum* 44, 52.
— — — —, Kräfte, für singuläre Zustände, *forces, for singular states* 181.
— — — —, Momente, *moments* 306.
— — — —, Kräfte und Momente, *forces and moments* 300.
bilineare Integrodifferentialform, *bilinear integro-differential form* 361.
Boggioscher Satz, *Boggio's theorem* 237.
Boussinesq-Papkovitch-Neuber-Lösung in der Elastizitätstheorie, allgemeine Form, *Boussinesq-Papkovitch-Neuber solution in elasticity, general form* 139–142, 235.
— — — —, Elimination des skalaren Potentials, *elimination of scalar potential* 143.
— — — —, Elimination einer Komponente des Vektorpotentials, *elimination of component of vector potential* 146.
Boussinesq-Somigliana-Galerkin-Lösung in der Elastizitätstheorie, *Boussinesq-Somigliana-Galerkin solution in elasticity* 141, 142, 235, 236.
Boussinesqs Lösung in der Elastizitätstheorie, *Boussinesq's solution in elasticity* 147.
Brunscher Satz, *Brun's theorem* 217, 223.

Cartesisches Koordinatensystem, *Cartesian coordinate system* 438.
Cauchy-Green-Tensor, Darstellung in der Stabtheorie, *Cauchy-Green tensor, representation in rod theories* 649.
Cauchy-Kovalevski-Somigliana-Lösung der Elastizitätstheorie, *Cauchy-Kovalevski-Somigliana solution in elasticity* 235–237.

Cauchy-Poissonscher Satz, *Cauchy-Poisson theorem* 44.
CAUCHYs Bewegungsgleichungen, *Cauchy equations of motion* 510.
— Darstellungstheorem für isotrope Funktionen, *representation theorem for isotropic functions* 535, 551.
— Hauptwert eines Oberflächenintegrals, *principal value of a surface integral* 181.
— Reziprozitätssatz, *reciprocal theorem* 45.
— singuläre Integrale, *singular integrals* 419.
— Spannungstensor, *stress tensor* 51, 53, 305.
C^∞-glatt, C^∞-*smooth* 370.
C^ν-glatt, C^ν-*smooth* 369.
charakteristische Lösung für das freie Schwingungsproblem, *characteristic solution for free vibration problem* 262.
— Verschiebung für das Schwingungsproblem, *displacement for vibration problem* 262.
— Werte des Elastizitätstensors, *values, elasticity tensor* 76.
— — — Schwingungsproblems, *vibration problem* 263.
charakteristischer Raum, *characteristic space* 8.
Christoffel-Symbole erster und zweiter Art, *Christoffel symbols of the first and second kinds* 619.
Clausius-Duhem-Ungleichung, *Clausius-Duhem inequality* 301, 484, 512, 532.
Cosserat-Fläche, allgemein, *Cosserat surface, in general* 430, 446, 449.
—, Beziehung zur 3-dimensionalen Theorie, *relation to 3-dimensional theory* 594.
Cosserat-Stab, *Cosserat rod* 665.
curl 11.

Darstellung des elastischen Verschiebungsfeldes im Unendlichen, *representation of elastic displacement field at infinity* 165.
— — Verschiebungsfeldes, *displacement fields* 160.
— — Gradienten, *gradient* 187.
— harmonischer Felder im Unendlichen, *harmonic fields at infinity* 22.
— von biharmonischen Feldern im Unendlichen, *biharmonic fields at infinity* 22.
— — Dilatation und Rotation, *dilatation and rotation* 188.
Deformationsgradient, *deformation gradient* 29, 299.
Deformation, Definition 29.
—, infinitesimal 28, 456.
Dehnungsenergie (gesamte gespeicherte Energie) in der Elastizitätstheorie, *strain energy (total stored energy), elasticity* 94, 110, 208, 216.
— — — —, Schranken, *bounds* 113.
— in Schalen, *in shells* 545.
— in der Thermoelastizität, *in thermoelasticity* 324.
Dehnungsfeld, *strain field* 31, 39, 43, 461.
Dehnungsrate, *strain rate* 43.
Dehnungs-Temperatur-Beziehung, *stress-strain temperature relation* 312, 326.
Dehnungsverlauf, *strain path* 82.
Dehnungs-Verschiebungs-Relation, *strain-displacement relation* 31, 89, 212, 312, 326.
Deresiewicz-Zorski-Lösung, *Deresiewicz-Zorski solution* 330.
Determinante, *determinant* 6.
Dichte, *density* 43, 299, 309.
Dicke einer Schale, *thickness of shell* 443.
Dicke-Scherschwingung, *thickness-shear vibration* 575.
differenzierbare $(r-1)$-Zelle der Klasse C^n, *differentiable* $(r-1)$-*cell of Class* C^n 402.
Diffusionsprobleme, *diffusion problems* 373, 377, 386.
Dilatation, Definition 31.
—, gleichförming, *uniform* 34, 77.
Dilatationswelle, Gleichung, *dilatational wave equation* 213, 328.
direkte Bezeichnung, *direct notation* 297.
direkte Ableitung der Schalentheorie, *direct approach to shell theory* 448.
Direktor, Definition, *director, definition* 446, 467, 666.
Direktorgeschwindigkeit, *director velocity* 449, 467.
Direktorimpuls, *director momentum* 482, 483.
Direktorkräftepaar, vorgeschriebenes, *director couple, assigned* 480.
Direktorträgheitskoeffizient, *director inertia coefficient* 594–596.
DIRICHLETs Problem, *Dirichlet problem* 375, 449.
— Randbedingungen, *boundary conditions* 370.
dispersierte Wellen, *dispersed waves* 343.
Divergenz, Definition, *divergence, definition* 11.
—, Satz, *theorem* 16, 28.
Doublet-Prozeß, *doublet process* 241.
Doublet-Zustand, *doublet state* 177, 178.
Drehimpuls siehe Impuls, *angular momentum, see under "momentum"*.
dreidimensionaler Körper *B*, *three-dimensional body B* 438.
Druck, gleichmäßig, *uniform pressure* 50, 77.
Dualität von Stäben und Schalen, *duality between rods and shells* 664.
dünne Schalen und Platten, *thin shells and plates* 438, 524.
dünner Film, *thin film* 547.
dynamisch zulässiges Spannungsfeld, *dynamically admissible stress field* 230.
dynamischer Prozeß, *dynamical process* 44, 49.
dynamisches Problem für Stäbe, *dynamical problem for rods* 699.

ebene Dehnung, *plane strain* 151.
— Dehnungslösung, *strain solution* 152.

ebene Dehnung, konvexe Menge, *plane strain, convex set* 417.
— Probleme für Stäbe, *problems for rods* 670.
— Spannung, verallgemeinert, *stress, generalized* 152, 577.
— starre Verschiebung, *rigid displacement* 39.
— Verschiebung, *displacement* 39.
ebener elastischer Zustand, *plane elastic state* 154.
ebenes Problem, *plane problem* 150, 151.
Ehrling-Lemma 354.
Eigenfunktionen, vollständiger Satz, *eigenfunctions, complete set* 369.
Eigenschwingungen, *normal modes* 269, 271.
eigentlich regulär, *properly regular* 14, 38, 351.
eigentliche orthogonale Gruppe, *proper orthogonal group* 7.
Eigenwert, Definition, *eigenvalue, definition* 369.
Eigenwertprobleme, *eigenvalue problems* 386.
Einbettung, *embedding* 35, 352.
Eindeutigkeitssatz, allgemeines Problem, *uniqueness theorem, general problem* 130.
—, äußeres Gebiet, *exterior domain* 171.
— der Thermoelastizität, dynamisch, *thermoelasticity, dynamic* 338.
— — —, statisch, *static* 322.
—, drittes oder gemischtes Problem, *mixed ("third") problem* 104, 112.
—, ebenes Problem, *plane problem* 154.
—, Elastodynamik, *elastodynamics* 222, 261.
— für Schalen, *for shells* 610.
— — Stäbe, *rods* 692.
—, inkompressibler Körper, *incompressible body* 212.
—, innere konzentrierte Last, *internal concentrated load* 190.
einfach zusammenhängend, *simply-connected* 13.
einfache Ausdehnung, *simple extension* 34.
— Scherung, *shear* 34, 77.
eingeklemmte Platte, unterstützt in einem Teilgebiet, *clamped plate, partially supported on a subdomain* 406.
eingeklemmter Rand, *clamped boundary* 314
eingeschränkte Form der Materialgleichungen für ein isotropes Material, *restricted form of the constitutive equations for an isotropic material* 557.
— Kegelhypothese, *cone-hypothesis* 382.
— Theorie, *theory* 455.
Einheits-Doublet-Zustand, *unit doublet state* 178.
Einheitshauptnormale einer Kurve, *unit principal normal to the curve* 627.
Einheits-Kelvin-Zustand, *unit Kelvin state* 178.
Einheitsnormale einer Fläche, *unit normal to a surface* 440, 622.
Einheitsnormale einer Kurve, *unit normal to a curve* 623, 627.
Einheitstangente einer Kurve, *unit tangent of a curve* 623.
— und Einheitsnormale einer Kurve in der Fläche, *tangent and unit normal to a curve lying in the surface* 623.
einschränkende Bedingung an Antwortfunktionen bei Spiegelung des Direktor, *restriction of response functions under reflection of direction* 540, 542.
— — für die Spannung bei gleichförmiger Dilatation, *on stress due to uniform dilatation* 73.
— Bedingungen für elastische Materialien aus dem zweiten Hauptsatz, *restrictions placed on elastic materials by the second law* 302.
einseitige Zwangsbedingung, *unilateral constraints* 403.
einseitiges Problem, abstrakt, *unilateral problem, abstract* 39, 395.
— —, symmetrische koerzive bilineare Form, *symmetric coercive bilinear form* 422.
— —, elliptische Operatoren, *elliptic operators* 399.
— —, Membran, *membrane* 422.
— —, elastischer Körper, *elastic body* 402.
— —, Elastostatik, *elastostatics* 423.
— —, lineare Operatoren der angewandten Mathematik, *linear operators of applied mathematics* 422.
— —, Differentialoperator, *differential operator* 422.
Elastika, *elastica* 697, 698.
elastische Bewegung, *elastic motion* 233.
— fortschreitende Welle, *progressive wave* 246.
— Koeffizienten, Elastizitätsfeld, *elasticities, elasticity field* 68, 75.
— Verschiebungsfelder, *displacement fields* 131.
— Welle der Ordnung *n*, *wave of order n* 255.
elastischer Prozeß, *process* 215.
— Zustand, *state* 95, 102.
— — mit im Unendlichen verschwindender Spannung, *with stress vanishing at infinity* 168.
— — — im Unendlichen verschwindender Verschiebung, *displacement vanishing at infinity* 167.
— —, Eigenschaften, *properties* 133.
elastisches Material, *elastic material* 301.
— Potential 380, 386, 544.
elastisch-plastisches Torsionsproblem, *elastic-plastic torsion problem* 405, 422.
Elastizitätsmodul, *Young's modulus* 78, 570.

Elastizitätstensor, allgemein, *elasticity tensor, general* 67–69, 309, 313.
— für isotrope Materialien, *isotropic materials* 75, 317.
elementares krummliniges Dreieck, *elementary curvilinear triangle* 493.
Elemente von Fläche und Volumen, *elements of area and volume* 441.
elliptischer Operator, *elliptic operator* 355, 374.
endliche Deformationen, Definition, *finite deformation, definition* 28.
— — in der Elastizitätstheorie, *in elasticity* 80.
— starre Deformation, *rigid deformation* 30.
endlicher Dehnungstensor, *finite strain tensor* 29, 80, 306.
Energie (s. auch „Leistung“ und „Arbeit“), *energy (see also "power" and "work").*
—, freie, *free* 497, 511, 555, 570.
—, innere, *internal* 481, 510.
Energieflußvektor, *energy-flux vector* 214.
Energiefunktional, *energy functional* 385, 391, 403.
Energiegleichung, allgemein, *energy equation, general* 326.
— im Gleichgewicht, *in equilibrium* 312.
— für Schalen, abgeleitet aus der 3-dimensionalen Theorie, *shells, derived from 3-dimensional theory* 515.
— — —, direkte Ableitung, *direct approach* 481, 482.
— — Stäbe, *rods* 653, 658.
Entropiebeziehungen, *entropy relations* 303, 306.
Entropieprinzip für Materialgleichungen, allgemein, *entropy principle for constitutive relations, general* 302.
— — Schalen, *shells* 512.
— — Stäbe, *rods* 669.
Entropieproduktion, *entropy production* 483, 492, 511.
Entropieungleichung (Clausius-Duhem), allgemein, *entropy inequality (Clausius-Duhem), general* 301, 510, 511.
— für Schalen, abgeleitet aus der 3-dimensionalen Theorie, *for shells, derived from 3-dimensional theory* 616.
— — —, direkte Ableitung, *direct approach* 483.
— — Stäbe, *rods* 667.
ε-System 618.
Erhaltungssätze (s. auch Bilanzgleichungen), *conservation laws (see also balance principles).*
—, ausgedrückt im Bezugszustand, *in terms of a reference state* 490.
— der Masse, *of mass* 440, 483.
— für eine Cosserat-Fläche, *for a Cosserat surface* 479, 482, 487.
erste und zweite Fundamentalform einer Oberfläche, *first and second fundamental forms of a surface* 623.
erster Hauptsatz der Thermodynamik, *first law of thermodynamics* 300.
Erweiterungen des Fundamentallemmas, *extensions of the fundamental lemma* 115, 116, 223.
Existenz- und Eindeutigkeitssatz für das gemischte Problem, *existence and uniqueness theorem for the mixed problem* 223.
Existenzsätze, allgemein, *existence theorems, general* 368–387.
— für das verallgemeinerte Signorini-Problem, *generalized Signorini problem* 407.
— — Stäbe, *rods* 680.
— und Nichteindeutigkeit für elastostatische Randwertprobleme, *and nonuniqueness for elastostatic boundary-value problems* 544.
—, Näherungslösungen, *approximate solutions* 126.

Faltung, *convolution* 25, 332.
Familie singulärer elastischer Zustände, *family of singular elastic states* 192.
Familie von Belastungsgebieten, die sich auf einen Randpunkt zusammenziehen, *family of load regions contracting to a boundary point* 191.
Fatous Lemma, *Fatou's lemma* 384.
Federov-Stippesscher Satz, *Federov-Stippes theorem* 247.
Fehlerabschätzung in den linearen Theorien von Platten und Schalen, *error estimate in the linear theories of plates and shells* 584.
Feldgleichungen für Elastizitätstheorie, *field equations of elasticity* 89–94, 212–215.
— — nichtlineare Thermoelastizität, *nonlinear thermoelasticity* 307.
— — Schalen, abgeleitet aus der 3-dimensionalen Theorie, *shells, derived from 3-dimensional theory* 508, 519.
— — —, ausgedrückt durch den Bezugszustand, *in terms of reference state* 502.
— — —, direkte Ableitung, *direct approach* 492, 498.
— — —, eingeschränkte Theorie, *restricted theory* 503.
— — —, linearisiert, *linearized* 500, 523.
Flächenbasisvektoren und reziproke Basisvektoren, *surface base vectors and their reciprocals* 623.
Flächendeformationsrate, *surface rate of deformation* 451.
Flächengeometrie, *surface geometry* 621.
Flächenkraftfeld, *surface force field* 59, 215.
Flächenspannung, *surface tension* 547.
Flächenspintensor, *surface spin tensor* 451.
Flächenvektoren, *surface vectors* 450.
Flächenwärmeflußmomente für Stäbe, *surface heat flux moments for rods* 651.
Flächenzugkraft, *surface traction* 44, 93, 95, 211, 313, 319, 327.

Flügge-Lur'e-Byrne-Truesdell-Gleichungen in der linearen Schalentheorie, *Flügge-Lur'e-Byrne-Truesdell equations in the linear theory of shells* 592.
Folgelastproblem, *follower load problem* 693.
fortschreitende Welle, *progressive wave* 245.
Fredholmsche Integralgleichungen, *Fredholm integral equations* 419.
freie Energie in Schalen, elastische Cosserat-Fläche, free energy in *shells, elastic Cosserat surface* 557.
— — — Schalen, lineare Theorie, *shells, linear theory* 556.
— — — Stäben, *rods* 649, 657.
— —, dreidimensional, *three-dimensional* 301.
freier Körper, *free body* 315.
Frequenz einer Schwingung, *frequency of vibration* 261.
Fresnel-Hadamard-Ausbreitungsbedingung, *Fresnel-Hadamard propagation condition* 246, 256.
fundamentale Lösungsmatrix, *fundamental solution of matrix* 419.
fundamentales Lemma, *fundamental lemma* 20.
— System von Feldgleichungen, *system of field equations* 89.
Fundamentalform einer Fläche, dritte, *fundamental form of a surface, third* 626.
— — —, erste und zweite, *first and second* 623, 624.
Funktionaldeterminante, *Jacobian of transformation* 619.

Gauss, Gleichungen von, *Gauss, equations of* 46, 627.
Gaußsche Formel, *Gauss' formulae* 625, 626.
— Krümmung einer Oberfläche, *Gaussian curvature of surface* 624.
Gebiet, Definition, *domain, definition* 348.
—, Einfluß-, *of influence* 257.
gedämpft, *damped, attenuated* 343.
geknickte Zustände für Stäbe, *buckled states for rods* 694.
gekoppelte Wärmeleitungsgleichung, *coupled heat equation* 328, 335.
gemischter Tensor, *mixed tensor* 617, 622.
gemischtes Randwertproblem für die dynamische Elastizitätstheorie, *mixed boundary value problem of dynamic elasticity* 219.
— — — — — Thermoelastizität, *thermoelasticity* 335.
— — — — statische Elastizitätstheorie, *static elasticity* 102, 377, 379, 385.
— — — — — Thermoelastizität, *thermoelasticity* 321.
Geometrie einer Fläche im Euklidischen Raum mit Normalkoordinaten, *geometry of a surface in an Euclidean space covered by normal coordinates* 628.
geometrische Multiplizität, *geometric multiplicity* 369.
— Symmetrie, *geometrical symmetries* 297.
gerade Stäbe, *straight rods* 690.
Gesamtenergie, *total energy* 216, 332.
gesamte Randfläche, eingeklemmt, *entire boundary, clamped* 263.
Gesamtimpuls, *total moment* 44.
Gesamtkraft, *total force* 44.
Gesamtmoment, *total moment* 44.
Geschichte der Schalentheorie, *history of shell theory* 527, 589.
geschlossene konvexe Menge, *closed convex set* 391.
— reguläre Oberfläche, *regular surface* 13.
geschlossener Dehnungsverlauf, *closed strain path* 82.
Geschwindigkeit, Definition, *velocity, definition* 43, 467.
gespeicherte Energie (Dehnungsenergie), Funktion in der Elastizitätstheorie, *stored energy (strain energy), function in elasticity* 82, 94.
— —, — — Schalen, *shells* 515.
— —, — — Stäben, *rods* 673, 677, 680.
glatt ausbreitende Welle, *smoothly propagating surface* 248.
— in der Zeit, *smooth in time* 24.
glattes Feld, *smooth field* 11.
gleich bis auf ein starres Verschiebungsfeld, *equal modulo a rigid displacement* 104.
gleichförmige Dilatation, *uniform dilatation* 34, 77.
Gleichgewicht in jedem belasteten Gebiet, *equilibrium in each load region* 193.
Gleichgewichtsgleichungen, allgemein, *equilibrium, equations of, general* 49, 59, 90, 312.
— für Stäbe, Integrale, *for rods, integrals* 696.
— in rechtwinkligen, zylindrischen und Kugelkoordinaten, *in rectangular, cylindrical, and polar coordinates* 93.
—, Lösung, *solution* 53.
— mit Krümmungslinien als Koordinaten, *in lines of curvature coordinates* 633.
Gleichgewichtsproblem in dünnen Platten, *equilibrium problem in thin plates* 377.
— — heterogenen elastischen Medien, *heterogeneous elastic medium* 386, 421.
— — der Elastizitätstheorie, *in elasticity* 380.
— astatisch, *astatic* 63, 64, 98, 193.
— von Kräften, *of forces* 63.
—, Wärmeleitungsgleichung, *heat equation* 317.
Gleichgewichts-Spannungsfeld, *self-equilibrated stress field* 55, 59.
gleichmäßiger Druck, *uniform pressure* 50, 77.
Gleichungen für das Innere, *interior equations* 586.
globale Reaktion auf die Stützungsbedingung, *global reaction exerted by the constraint of support* 416.
Goursatscher Satz, *Goursat's theorem* 160.
Gradient der Deformation, *gradient deformation* 29, 299.

Gradient der Direktorgeschwindigkeit, *gradient director velocity* 452.
— der Verschiebung, *displacement* 29.
— des Direktors, *director* 452.
Gradientenfeld, *gradient field* 10.
Graffischer Reziprozitätssatz, *Graffi's reciprocal theorem* 218, 243.
Green-Lamé-Lösung, *Green-Lamé solution* 233, 236, 237.
GREENs Formel für den Laplace-Operator, *Green's formula for the Laplace operator* 419.
— Funktion des Dirichlet-Problems, *function of the Dirichlet problem* 421.
Greensche Zustände, *Green's states* 186.
Grenze, *border* 402.
Grenzschichtgleichungen, *boundary-layer equations* 586.
Grenzwertdefinition der Lösung von KELVINs Problem, *limit definition of solution to Kelvin's problem* 174.
große Deformation, Lösungen für, *large deformation solutions* 546.
grundlegende Gleichungen, lineare Thermoelastizität, *basic equations, linear thermoelasticity* 310, 311.
— —, Schalentheorie, *shell theory* 565.
— Prinzipien für Schalen, *principles for shells* 479.
— singuläre Lösungen, Elastodynamik, *singular solutions, elastodynamics* 239.
— —, Elastostatik, *elastostatics* 173.

halbkoerzive bilineare Form, *semi-coercive bilinear form* 422.
Halbkoerzivität, Hypothese, *semi-coerciveness hypothesis* 392.
halbstetig bez. schwacher Konvergenz, *lower semi-continuous with respect to weak convergence* 393.
Hamilton-Kirchhoff-Prinzip, *Hamilton-Kirchhoff principle* 225, 226.
harmonische ebene fortschreitende Welle, *harmonic plane progressive wave* 342.
harmonisches Feld, *harmonic field* 20.
Harnackscher Konvergenzsatz, *Harnack's convergence theorem* 21.
Hauptdehnungen, *principal strains* 35.
Hauptkrümmung einer Fläche, *principal curvatures of a surface* 628.
Hauptrichtung, *principal direction* 8, 628.
— der Dehnung, *of strain* 35.
— — Spannung, *stress* 50.
Hauptspannungen, *principal stresses* 50.
Hauptwerte, *principal values* 8.
Hellinger-Prange-Reissner-Prinzip, *Hellinger-Prange-Reissner principle* 124, 125, 228.
Helmholtzscher Satz, *Helmholtz's theorem* 19.
heterogene Medien, *heterogeneous media* 421, 422.
homogener Körper, elastisch, *homogeneous body, elastic* 68.
— —, isotrop-elastisch, *isotropic elastic* 380.
— —, isotrop-thermoelastisch, *isotropic thermoelastic* 317, 327.
— —, thermoelastisch, *thermoelastic* 327.
homogenes Verschiebungsfeld, *homogeneous displacement field* 33.
Hu-Washizu-Prinzip, *Hu-Washizu principle* 122, 125, 226.
hydrostatischer Druck, *hydrostatic pressure* 678, 690.
hyperelastische Stäbe, *hyperelastic rods* 673.

Impuls, *momentum, linear* 43, 44, 51, 254, 482, 483.
—, Direktorimpuls und Drehimpuls in Stäben, *in rods (angular, director, linear momentum)* 666.
—, Drehimpuls, *angular momentum* 44, 52.
— in Raum-Zeit, *momentum, space-time* 67.
— in Schalen, *shells* 483.
infinitesimale Bewegung eines 3-dimensionalen Kontinuums, *infinitesimal motion of 3-dimensional continuum* 473.
— Deformation, *deformation* 28, 456.
— Direktor-Verschiebung, *director displacement* 456.
— Rotation, *rotation* 461.
— starre Verschiebungen, *rigid-body displacements* 30, 462, 463.
— Volumenänderung, *volume change* 30.
infinitesimaler Dehnungstensor, *infinitesimal strain tensor* 461.
Inhomogenität, *inhomogeneity* 530.
Inkompressibilitätsbedingung, *incompressibility condition* 404.
inkompressible Körper, Materialien, *incompressible body, material* 210.
innere Energie, *internal energy* 300, 481.
— konzentrierte Belastungen, *concentrated loads* 179.
— Zwangsbedingungen, *constraints* 403, 548.
inneres Direktorkräftepaar, *intrinsic director couple* 482, 491.
— Produkt aus Tensoren, *inner product of tensors* 618.
— — — Tensoren zweiter Ordnung, *second-order tensors* 7.
— — — Vektoren, *vectors* 5.
— — — Vierer-Vektoren, *four-vectors* 27.
Integral-Darstellungssatz für Lösungen der gemischten Probleme, *integral representation theorem for solutions of mixed problem* 187.
Integrale der Gleichgewichtsgleichungen für Stäbe, *integrals of the equilibrium equations for rods* 696.
Integral-Identität für den Verschiebungsgradienten, *integral identity for the displacement gradient* 184.
Integrodifferentialgleichungen, *integrodifferential equations* 373.

Invarianzbedingungen, *invariance conditions* 484, 485, 487.
isentropisch, *isentropic* 304.
isochore Bewegung, *isochoric motion* 210.
— Geschwindigkeit, *velocity* 213.
isotrope Materialien und Körper, elastisch, *isotropic materials and bodies, elastic* 71, 74, 85, 86.
— — — —, elastische Platten, *elastic plates* 573.
— — — —, thermoelastisch, *thermoelastic* 311, 317.
Isotropie, transversale, *isotropy, transverse* 72.
iterierter Laplace-Operator, *iterated Laplace operator* 377.

Kataloge der Schalentheorie, Bezeichnungen, *catalogues for shell theory, notations* 432–438.
— — —, lineare kinematische Maße für eine Cosserat-Fläche, *linear kinematic measures for Cosserat surface* 458.
— — —, Resultierende und Bewegungsgleichungen, *resultants and equations of motion* 526.
Kegelhypothese, *cone hypothesis* 354.
Kelvin-Problem, *Kelvin problem* 173, 176.
Kelvinscher Satz, *Kelvin theorem* 208.
Kelvin-Zustand, *Kelvin state* 174, 176, 178–180.
Kern der nichtnegativen quadratischen Form, *kernel of the nonnegative quadratic form* 392.
— des Funktionals F, *of the functional* F 392.
— des Operators T, *of the operator* T 391.
Kinematik der Cosserat-Fläche, *kinematics of Cosserat surface* 449.
— — orientierten Medien, *oriented media* 445.
— — Schalen, abgeleitet aus der 3-dimensionalen Theorie, *shells, derived from 3-dimensional theory* 466.
— — —, eingeschränkte Theorie, *restricted theory* 455.
kinematisch zulässiges Dehnungsfeld, *kinematically admissible strain field* 112.
— zulässiger Prozeß, *admissible process* 225.
— — Zustand, *state* 111, 212, 324.
— zulässiges Verschiebungsfeld, *admissible displacement field* 112.
kinematisch und thermisch zulässiger Prozeß, *kinematically and thermally admissible process* 341.
kinematische Maße für eine Cosserat-Fläche, *kinematic measures for Cosserat surface* 452.
— — — Schalen, abgeleitet aus der 3-dimensionalen Theorie, *shells, derived from 3-dimensional theory* 468.
— — — —, eingeschränkte Theorie, *restricted theory* 460.
kinematische Maße für Schalen linearisiert, abgeleitet aus der 3-dimensionalen Theorie, *kinematic measures for shells, linearized, derived from 3-dimensional theory* 474.
— — — —, Katalog, *catalogue* 458.
— Näherungen in Platten und Schalen, *approximations in plates and shells* 478.
— Resultate für Schalen, *results for shells* 471.
kinetische Energie, *kinetic energy* 65, 216, 481.
Kirchhoffsche Randbedingungen, *Kirchhoff boundary conditions* 577, 586.
— Theorien für Dehnung und Biegung von Platten, *theories for stretching and bending of plates* 586.
Kirchhoffscher Satz, *Kirchhoff's theorem* 32.
Klasse $C^{M,N}$, *Class* $C^{M,N}$ 24.
— C^N 11.
klassische Theorien der Platten und Schalen, *classical theories of plates and shells* 575–577, 582, 585.
Knowlesscher Satz zum Saint-Venant-Prinzip, *Knowles' theorem on Saint-Venant's principle* 200–206.
Koerzivität, *coerciveness* 422.
Kofaktoren, *cofactors* 618.
Kolosovscher Satz, *Kolosov's theorem* 164.
Kompatibilität, *compatibility* 39.
Kompatibilitätsgleichungen für ebene elastische Zustände, *compatibility equations for plane elastic states* 155.
— — Schalen, *shells* 463, 479.
— — — in der eingeschränkten Theorie, *restricted theory* 466.
— — 3-dimensionale Körper, *3-dimensional bodies* 39, 40, 41, 62, 63, 211, 318.
Kompatibilitätssatz für ebene Verschiebungen, *compatibility theorem for plane displacements* 41.
— — dreidimensionale Deformationen, *three-dimensional deformations* 40.
komplementäre Energiefunktion, *complementary energy function* 570.
komplexe starre Verschiebung, *complex rigid displacement* 39.
Kompressionsmodul (s. auch Elastizität), *modulus of compression (see also elasticities)* 74, 97.
Kompressionszentrum, Rotationszentrum, *center of compression or rotation* 179.
konjugierter Tensor, *conjugate tensor* 618.
Konsistenzbedingung, *consistency condition* 129.
Kontaktdirektorkräftepaar, *contact director couple* 480, 491.
Kontaktkraft, *contact force* 480, 491, 510.
Kontaktkräftepaar, *contact couple* 504.
Kontaktproblem, *contact problem* 129.
Kontraktionsabbildung, *contraction mapping* 396, 422, 618.

kontravariante Tensoren, *contravariant tensors* 617, 622.
konvektierte Koordinaten, *convected coordinates* 431, 439, 449.
Konvergenz von Näherungslösungen, *convergence of approximate solutions* 125.
konvexe Hülle, *convex hull* 417.
— Menge, *set* 392, 417.
konvexes Gebiet bez. $\mathscr{S}$, *convex region with respect to* $\mathscr{S}$ 16.
konzentrierte Last, *concentrated load* 174, 179.
— Spannungen, *stresses* 416.
konzentriertes Kräftesystem, *system of forces* 43.
— Lastsystem, *concentrated loads* 179.
Koordinatensysteme, *coordinate systems* 438.
Koordinatentransformationen, *coordinate transformation* 531, 615.
Körper, Definitionen, *body, definitions* 12, 14, 299.
—, Kraft, Analogon in der Thermoelastizität, *body force analogy in thermoelasticity* 316.
—, —, Definition 44, 299, 309, 485.
—, —, Momente für Stäbe, *moments for rods* 650.
—, —, zweidimensionale, *two-dimensional* 513.
kovariante Ableitung bez. $a_{\alpha\beta}$, *covariant derivative with respect to* $a_{\alpha\beta}$ 450.
— — — A 456.
— — — erster Fundamentalform (metrischer Tensor), *first fundamental form (metric tensor)* 431, 624.
— — — Euklidscher Metrik, *Euclidean metric* 619.
— Tensoren, *tensors* 617, 622.
Kraft im Unendlichen, *force at infinity* 170.
Kräfte, parallel und nichttangential zur Randfläche, *forces, parallel and nontangential to the boundary* 193.
—, System 43.
—, Direktorkräftepaar, vorgeschriebenes, *director couple, assigned* 480.
Kräftepaar, resultierender Vektor, *couple resultant vector* 651.
—, vorgeschriebenes, *assigned* 504.
Kraftvektor, vorgeschrieben, *force vector, assigned* 480.
kreisförmige Stäbe, *circular rods* 690.
Kristallklasse, *crystal class* 72.
Kronecker-Delta, *Kronecker delta* 618, 623.
Krümmung der Oberfläche, Gaußsche, *curvature of surface, Gaussian* 471.
— — —, mittlere, *mean* 471, 624.
— — Randfläche, *of boundary* 378.
Krümmungstensor, *curvature tensor* 620, 625.
Kugelflächenfunktionen, *surface spherical harmonic* 21.
Kugelfunktion, *spherical harmonic* 22.
Kurve auf der Fläche, *curve on the surface* 622.

Lage, Darstellung in Stabtheorien, *position, representation in rod theories* 647.
Lagevektor, Feld, *position-vector field* 5, 439.
Lamé-Koeffizienten, *Lamé moduli* 76, 311.
Landau-Symbol, *Landau symbol* 356.
längenerhaltende Transformationen, *length-preserving transformations* 538.
Laplace-Operator, *Laplace operator, Laplacian* 11, 375.
Laplace-transformierte, *Laplace transform* 25, 231.
Lebesgue-Ableitung, *Lebesgue derivative* 415.
Lebesgue-Zerlegung, *Lebesgue decomposition* 415.
Leistung, *rate of working* 300.
— und Energie, Satz von, *theorem of power and energy* 216, 331.
Leitfähigkeit in isotropen Materialien, *conductivity, isotropic materials* 311.
Leitfähigkeitstensor, *conductivity tensor* 307, 309.
Lévyscher Satz, *Lévy's theorem* 156.
Liapounov-Randfläche, *Liapounov boundary* 419.
lineare Elastizität, Stabprobleme, *linear elasticity, rod problems* 661.
— Materialgleichungen für Schalen, *constitutive equations for shells* 553, 555, 556, 571.
— Theorie für elastische Platten, *theory for elastic plates* 585, 596.
— — — elastische Schalen, *elastic shells* 595, 597.
— — — — —, abgeleitet aus der 3-dimensionalen Theorie, *derived from 3-dimensional theory* 578.
— — — — —, Bemerkungen, *remarks on* 613.
linear-elastischer inkompressibler Körper, *linearly elastic incompressible body* 210.
linearisierte Feldgleichungen für Schalen, *linearized field equations for shells* 500, 523.
— kinematische Maße für Cosserat-Flächen und -Schalen, *kinematic measures for Cosserat surfaces and shells* 456, 460, 474.
Listen der häufig benutzten Symbole, *indexes of frequently used symbols* 2–5, 298, 299, 432–438.
Lösung der Gleichung der Elastizitätstheorie, *solution of equation of elasticity* 102.
— des allgemeinen Problems der Elastizitätstheorie, *of general problems of elasticity* 129.
— — gemischten Problems, *mixed problems* 102, 220, 321, 335.
Lösung des verallgemeinerten gemischten Problems, *solution of generalized mixed problem* 185.
Loves erste Näherung, *Love's first approximation* 478, 582, 589, 590.

LOVES Integralidentität in der Elastizitätstheorie, *Love's integral identity in elasticity* 243.
— Lösung in der Elastizitätstheorie, *solution in elasticity* 147.

Mainardi-Codazzi-Gleichungen, *Mainardi-Codazzi equations* 626.
Maß, *measure* 414, 415.
Massendichte einer Bezugskonfiguration, *mass density in reference configuration* 440, 472.
— — Cosserat-Fläche, *Cosserat surface* 447.
— — deformierten Konfiguration, *deformed configuration* 440.
Materialgleichungen für elastische Materialien, *constitutive equations, elastic materials* 67.
— — Schalen und Platten, ausgedrückt durch materielle Koordinaten, *shells and plates in terms of material coordinates* 529.
— — — — —, ausgedrückt durch relative Maße, *in terms of relative measures* 540.
— — — — —, ausgedrückt im Bezugszustand, *in terms of references state* 544.
— — — — —, Cosserat-Fläche, elastisch, *Cosserat surfaces, elastic* 528.
— — — — —, Cosserat-Fläche, unausdehnbar, *Cosserat-surface, inextensible* 549.
— — — — —, eingeschränkte lineare Theorie, *restricted linear theory* 560.
— — — — —, elastische Schalen, *elastic shells* 528, 533, 537.
— — — — —, elastische Schalen, aus der 3-dimensionalen Theorie, *elastic shells, from 3-dimensional theory* 561, 564.
— — — — —, linear, ausgedrückt durch die Gibbs-Funktion, *linear, in terms of Gibbs function* 571.
— — — — —, linear, für Cosserat-Fläche, *linear, for Cosserat surface* 553.
— — — — —, linear, für Schalen einheitlicher Dicke, *linear, for shells of uniform thickness* 556.
— — — — —, mechanische Theorie, *mechanical theory* 544.
— — Stäbe, *rods* 654, 657, 659, 669, 673.
Materialien mit „Gedächtnis", *materials with "memory"* 373.
Materialkoeffizienten in der linearen Theorie der Cosserat-Flächen, *constitutive coefficients in linear theory of Cosserat surfaces* 598, 606.
— — — — — — Cosserat-Platten, *Cosserat plates* 600, 603.
— — — — — — Schalen (eingeschränkte Theorie), *shells, restricted* 607.
Materialrelationen s. Materialgleichungen, *constitutive relations, see constitutive equations.*
materielle Fläche, *material surface* 439.
— Symmetrie in der dreidimensionalen Theorie, *symmetry in three-dimensional theory* 69–74, 87–89.
— — in Schalen, *in shells* 537, 539, 541.
materielle Systemunabhängigkeit, *material frame-indifference* 80, 305, 487.
— Zeitableitung, *time-derivative* 431, 449, 453.
materieller Punkt, *material point* 229, 439, 446, 530.
maximale Ausbreitungsgeschwindigkeit, *maximum speed of propagation* 245, 256, 257.
— komplementäre Energie, *complementary energy* 125.
— potentielle Energie, *potential energy* 125.
maximaler elastischer Koeffizient, *maximum elastic modulus* 85.
Maximalprinzipien, *maximum principles* 120, 375.
Maxwell-Beziehung, *Maxwell relation* 304, 310.
Maxwellsche Lösung in der Elastizitätstheorie, *Maxwell's solution in elasticity* 54.
Maxwellscher Kompatibilitätssatz, *Maxwell's compatibility theorem* 250.
mechanische Leistung, *mechanical power* 496.
Membran, am Rand festgehalten und über ein Hindernis gespannt, *membrane fixed along boundary and stretched over obstacle* 405.
—, einseitiges Problem, *unilateral constraint* 422.
Membrantheorie von Schalen, *membrane theory of shells* 547.
metrischer Tensor für den Raum, *metric tensor of space* 618.
— — — die Fläche, *surface* 624.
Michellscher Satz, *Michell's theorem* 163.
minimale komplementäre Energie, *minimum complementary energy* 125, 324, 325.
— potentielle Energie, *potential energy* 111, 125.
— — —, Umkehrung, *converse* 116.
— transformierte Energie, *transformed energy* 231.
minimaler elastischer Koeffizient, *minimum elastic modulus* 85.
Minimalprinzip für das Schwingungsproblem, *minimum principle for the vibration problem* 262, 266, 267.
— — — Spannungsfeld, *stress field* 232.
— — — Verschiebungsproblem, *displacement problem* 209.
— — den kleinsten charakteristischen Wert, *lowest characteristic value* 264.
— — die Elastodynamik, *elastodynamics* 230.
Minimax-Prinzip, *minimax principle* 268.
Mittelwert einer Funktion, *mean value of a function* 314.
Mittelwertsatz für elastische Zustände, *mean value theorem for elastic states* 133–138.
— — harmonische und biharmonische Felder, *harmonic and biharmonic fields* 21.

mittlere Dehnung, *mean strain* 37, 38, 97, 98.
— Krümmung, *curvature* 471, 624.
— Spannung, *stress* 96, 97.
— — und Dehnung, ausgedrückt durch äußere Kraft und Temperatur, *and strain in terms of external force-temperature* 315.
Moment-Drehimpuls, *moment of momentum* 482.
Moment im Unendlichen, *moment at infinity* 170.
Momente der Materialgrößen in Stabtheorien, *moments of constitutive variables in rod theories* 649.
Momentenbilanz, *balance of moments* 306.
MORERAs Lösung in der Elastizitätstheorie, *Morera's solution in elasticity* 55.
Multiplizität, *multiplicity* 369.

Nabelpunkte, *umbilics* 628.
Nachgiebigkeitstensor, *compliance tensor* 69, 71, 77, 310.
Nachwirkungselastizität, *hereditary elasticity* 423.
Nachwirkungsprobleme, *hereditary problems* 377, 386.
Näherungen in der Schalen- und Plattentheorie, *approximations in shell and plate theory* 476, 522, 566–568, 570, 587.
Näherungssatz, *approximation theorem* 128.
natürliche Konfiguration, *natural configuration* 380.
NAVIERs Gleichung der Elastizitätstheorie, dynamisch, *Navier's equation of elasticity, dynamic* 213.
— — — —, statisch, *static* 90.
Neumann-Problem, *Neumann problem* 375, 419.
Newtonsches Potential, *Newtonian potential* 19.
Nichtausdehnbarkeitstheorie von Schalen, *inextensional theory of shells* 546, 548.
nichteindeutige Randbedingungen, *ambiguous boundary conditions* 405, 414.
Nichtexistenzsatz für Elastizitätstheorie, *nonexistence theorem of elasticity* 109.
nichtkoerzive, unsymmetrische bilineare Formen, *non-coercive non-symmetric bilinear forms* 422.
nichtlineare Materialgleichungen für Schalen, *nonlinear constitutive equations for shells* 528.
— — — —, eingeschränkte Theorie, *restricted theory* 549.
— — — —, Membrantheorie, *membrane theory* 566, 588.
— — — Stäbe, *rods* 654.
— — — thermoelastische Materialien, *thermoelastic materials* 297.
nichtsymmetrische Form, *nonsymmetric form* 395.
nichttriviale Lösung des gemischten Problems mit Nullwerten, *non-trivial solution of the mixed problem with null data* 109.

nichtüberlappend, *nonoverlapping* 402.
Norm 348.
Normalkoordinatensystem, *normal coordinate system* 442, 628.
Normalkrümmung einer Kurve, *normal curvature of a curve* 628.
Normalspannung, transversale, *transverse normal stress* 584.
normiertes Vektorfeld, *normalized vector field* 186.
Novozhilovsche Ableitung der klassischen linearen Schalentheorie, *Novozhilov's derivation of classical linear theory of shells* 593.

objektiv (s. auch systemunabhängig), *objective (see also frame-indifference)* 487.
offene Kugel, *open ball* 12.
optimale Auslegung eines Stabes, *optimal design of a rod* 699.
orientierte Medien, *oriented media* 445.
orthogonal 7.
orthogonale Gruppe, *orthogonal group* 7.
— Projektion, *perpendicular projection* 9, 129.
orthogonaler Projektor, *orthogonal projector* 392.
orthogonales Komplement, *orthogonal complement* 385, 395.
Orthogonalität charakteristischer Verschiebungen, *orthogonality of characteristic displacements* 264.
orthotrope Materialien, *orthotropic material* 72.

Parsevalscher Satz, *Parseval theorem* 366, 381.
partielle Ableitung nach Oberflächenkoordinaten θ^α, *partial differentiation with respect to surface coordinates* θ^α 431.
partikuläre Lösung der Gleichgewichtsgleichung, *particular solution of the equation of equilibrium* 59.
Permutationssymbole, *permutation symbols* 453, 455, 618.
Photoelastizität, *photoelasticity* 156.
physikalische Interpretation der Airy-Funktion, *physical interpretation of Airy function* 158.
— Komponenten von Tensoren, *components of tensors* 631.
Piola-Kirchhoff-Spannung, *Piola-Kirchhoff stress* 51, 53, 229, 305, 491.
Piolascher Satz, *Piola's theorem* 60.
Platte unter hydrostatischem Druck, *plate under hydrostatic pressure* 602.
POINCARÉs Ungleichung, *Poincaré inequality* 350.
Poisson-Kirchhoff-Theorie für die Biegung von Platten, *Poisson-Kirchhoff theory for bending of plates* 524.
Poissonscher Quotient, *Poisson's ratio* 78, 318, 570.
— — für verallgemeinerte ebene Spannung, *for generalized plane stress* 153.

Poissonscher Zerlegungssatz, *Poisson's decomposition theorem* 234.
positiv definite elastische Konstanten, *positive definite elastic constants* 153.
positive Arbeit, Satz von der, *theory of positive work* 95.
positiver Elastizitätstensor, *positive elasticity tensor* 69, 85–87.
— halbdefiniter Elastizitätstensor, *semi-definite elasticity tensor* 69.
— — Tensor 84.
— Operator 369.
positives Elastizitätsfeld, *positive elasticity field* 104.
Potential von Einfach- und Doppelschichten, *potential of double and simple layers* 419.
Prinzip der Äquipräsenz, *principle of equipresence* 530.
— — maximalen komplementären Energie, *maximum complementary energy* 121.
— — minimalen komplementären Energie, *minimum complementary energy* 112, 130, 212.
— — — potentiellen Energie, *potential energy* 111, 112, 121, 130, 212, 231.
— — Systemunabhängigkeit, *material frame-indifference* 80, 305, 487.
— — Zerlegung thermoelastischer Zustände, *decomposition of thermoelastic states* 313.
— des Impulses, *linear momentum* 483.
Problem „der schiefen Ableitung", *oblique derivative problem* 376.
— freier Schwingungen, *free vibration problem* 261.
Produkt zweier Tensoren, *product of two tensors* 6.
Produktion von Entropie, *production of entropy* 483, 511.
Projektion, orthogonale, *perpendicular projection* 9, 129.
Projektionseigenschaft, *projection property* 31, 32.
Projektionsmethoden für Stabtheorien, *projection methods for rod theories* 663.
Prozeß, dynamischer, *process, dynamical* 44, 49.
—, thermodynamischer, *thermodynamic* 484, 497, 510.
—, thermoelastischer, *thermoelastic* 327.
Pseudo-Differentialoperatoren, *pseudo-differential operators* 421.
Pseudo-Volumenkraftfeld, *pseudo body force field* 66, 218, 332.
Pseudo-Wärmezufuhrfeld, *pseudo heat supply field* 332.
Pseudo-Zugkraftoperator, *pseudo-tension operator* 420.

qualitatives Verhalten von Stäben, *qualitative behavior for rods* 697.
quasistatische Theorie, *quasi-static theory* 311.

Randbedingungen, eingeschränkte Schalentheorie, *boundary conditions, shell theory, restricted* 553.
—, Elastizitätstheorie, *elasticity* 102, 129, 314, 403, 405, 414, 420.
—, Einklemmung, *clamped* 263, 314.
—, lineare Schalentheorie, *shell theory, linear* 583.
—, Stabtheorien, *rod theories* 648, 659, 667, 674, 678.
—, Thermoelastizität, *thermoelasticity* 314, 321.
—, Zwangsbedingungen, *constraints* 403.
Randfläche eines Körpers, *boundary of body* 102, 440.
Rand- und Anfangswertprobleme in der Elastizitätstheorie, *boundary-initial-value problems of elasticity* 335.
Randwertoperator, *boundary-value operator* 402.
Randwertprobleme für Elastizitätstheorie, allgemein, *boundary value problems of elasticity, general* 129, 185.
— — —, gemischt („drittes"), *mixed ("third")* 102, 381, 420.
— — —, nicht-eindeutig, *ambiguous* 405, 414.
— — —, Verschiebung („erstes"), *displacement ("first")* 103, 220, 380.
— — —, Zugkraft („zweites"), *traction ("second")* 103, 220, 381.
— — harmonische Funktionen, *harmonic functions* 419.
— — Schalen, allgemeine Theorie, *shells, general theory* 550.
— — —, eingeschränkte lineare Theorie, *restricted linear theory* 607, 609.
— — —, lineare Biegung, *linear bending* 584.
— — Stäbe, *rods* 658.
— — Theorie der Thermoelastizität, Verschiebung, *thermoelasticity, displacement* 321.
— — — — —, Verschiebung und Temperatur, *displacement and temperature* 335.
— — — — —, Zugkraft, *traction* 321.
Räume $H_m(A)$ und $H_m^0(A)$, *spaces* $H_m(A)$ *and* $H_m^0(A)$ 349.
räumliche Kugelfunktion, *solid spherical harmonic* 22.
Raum-Zeit, *space-time* 27.
Rayleigh-Ritz-Methode, *Rayleigh-Ritz procedure* 125–129.
Reaktion eines isotropen Materials, *response of isotropic materials* 74.
Reduktion von Materialgleichungen durch Überlagerung starrer Bewegungen, *reduction of constitutive equations under superposed rigid body motions* 534.
reduzierte Formen, *reduced forms* 306.
— Spannung, *stress* 316.
— Volumenkraft, *body force* 316.
reguläre Fläche, *regular surface* 13.

reguläre Hyperfläche, *regular hypersurface* 252.
— Unterfläche, *subsurface* 14.
regulärer Punkt der Randfläche, *regular point of boundary* 14.
reguläres Gebiet, *regular domain* 349.
— Problem der „schiefen Ableitung", *oblique derivative problem* 376.
Regularisierung, Regularität, allgemein, *regularization, regularity in general* 370, 422.
— auf oder in der Umgebung der Randfläche, *at or near the boundary* 357, 401, 411.
— bei Stäben, *of rods* 680.
— im Innern, *interior* 408.
Regularität im Innern, *interior regularity* 355, 401.
reine Biegung einer Platte, *pure bending of a plate* 599.
— Dehnung, *strain* 33.
— Scherung, *shear* 50, 77.
reiner Zug, *pure tension* 50, 78.
REISSNERs Plattentheorie, *Reissner's plate theory* 575, 578, 584.
relative Tensoren, *relative tensors* 618.
Rellich-Lemma, *Rellich's lemma* 20.
Rellichsches Auswahlprinzip, *Rellich selection principle* 352.
Restspannung, *residual stress* 8, 308.
Resultierende für spezifische innere Energie und Entropie, *resultants, specific internal energy and entropy* 515.
—, Katalog der Definitionen, *catalogue of definitions* 526.
resultierende Kontaktkraft, *resultant contact force* 480, 510.
— Kraft, *force* 514.
— Kräftepaarvektoren, *couple vectors* 514.
— Volumenkraft, *body force* 509.
— vorgeschriebene Kraft, *assigned force* 481.
resultierendes Kontakt-Direktorkräftepaar, *resultant contact director couple* 580.
— Kontaktkräftepaar, *contact couple* 504.
— vorgeschriebenes Direktorkräftepaar, *assigned director couple* 481.
— vorgeschriebenes Kräftepaar, *assigned couple* 504.
reziproke Basisvektoren, *reciprocal base vectors* 618.
Reziprozitätssatz für Elastodynamik, *reciprocal theorem for elastodynamics* 66.
— — Elastostatik, *elastostatics* 101, 207.
— — singuläre Zustände, *singular states* 182.
— — synchrone äußere Kraft, *synchronous external force* 219.
— — Thermoelastizität, *thermoelasticity* 320, 332.
Richtung der Ausbreitung, *direction of propagation* 245, 248, 343.
Richtung der Bewegung (oder Verschiebung) einer fortschreitenden Welle, *direction of motion (or displacement) of a progressive wave* 245, 343.
— — Verschiebung, *displacement* 343.
Riemann-Christoffel-Tensor einer Fläche, *Riemann-Christoffel tensor of a surface* 462.
— im Raum, *in space* 620.
Riesz-Fredholm-Theorie kompakter Operatoren, *Riesz-Fredholm theory of compact operators* 369.
Ring (Stabtheorie), *ring (rod)* 659, 697.
Ring, σ 415.
Rotationsfeld, *rotation field* 31.
Rotationsträgheit, *rotary inertia* 578, 596.
Rotationsvektor, *rotation vector* 31, 39.
Rotationswellengleichung, *rotational wave equation* 328.

s-Ableitung, *s-derivative* 350.
SAINT-VENANTs Prinzip, *Saint-Venant's principle* 190–207.
Saint-Venant-Problem, *Saint-Venant problem* 661.
Schalen, verstärkt durch Fasern, *shells reinforced with cords* 547.
schalenartiger Körper, *shell-like body* 411, 425, 438, 440, 442.
Schalendicke, *shell thickness* 443, 447.
Schalentheorie, *shell theory* 444, 445.
Scherdeformation, transversal, *shear deformation, transverse* 460, 575, 578, 584.
Schermodul, *shear modulus* 74, 76, 210, 311, 574.
Scherspannung, *shear stress* 50, 51.
Scherung, einfache, *simple shear* 34, 36, 77.
schief, *skew* 6, 617.
schiefsymmetrische $r \times r$ Matrix, *skew-symmetric $r \times r$ matrix* 384.
Schockwelle, *shock wave* 253, 254.
Schranken für Dehnungsenergie, *bounds for strain energy* 113.
schwach geschlossen, *weakly closed* 393.
— kompakt, *compact* 394, 398.
schwache Lösung, *weak solution* 355.
— Unstetigkeiten, *mild discontinuities* 253.
Schwarzsche alternierende Methode, *Schwarz' alternating method* 421.
Schwingungsamplitude, *amplitude, vibration* 261.
selbstadjungiert, *self-adjoint* 385.
σ-Ring 415.
Signorini-Problem, Definition, *Signorini problem, definition* 404, 413.
— in der Theorie der Viskoelastizität, *in viscoelasticity* 423.
—, verallgemeinert, *generalized* 404, 408.
Signorinischer Satz, *Signorini's theorem* 61.
singuläre Fläche der Ordnung n, *singular surface of order n* 250.
— — — — Null, *zero* 249.
— Integralgleichungen auf einer Kurve, *integral equations on a curve* 420.

singuläre Menge vom Maß γ, *singular set of the measure* γ 415.
singulärer elastischer Zustand, *singular elastic state* 179.
skalare räumliche Kugelfunktion, *scalar solid spherical harmonic* 21.
skalares Feld, *scalar field* 10.
— Produkt, *product* 399.
Sobolev-Lemma, *Sobolev lemma* 354.
Somigliana-Fundamentalmatrix, *Somigliana fundamental matrix* 419.
Somiglianascher Satz, *Somigliana's theorem* 183.
Spannungen, unabhängig von den elastischen Konstanten, *stress, independent of elastic constants* 163.
—, Bewegungsgleichung, *stress equations of motion* 214.
—, Kompatibilitätsbeziehung, *stress equation of compatibility* 92.
Spannungsenergie, *stress energy* 11, 324.
Spannungsfeld, *stress field* 45–64.
— für die Lösung des gemischten Problems, *for solution of mixed problem* 103.
spannungsfreier Zustand, *stress-free state* 323.
Spannungsfunktionen, *stress functions* 53–59.
Spannungsimpulsfeld, *stress-momentum field* 67, 253.
Spannungskräftepaare, *stress-couples* 512, 513, 595.
Spannungsleistung, *stress power* 65, 216, 300, 496.
Spannungsmomente für Stäbe, *stress moments for rods* 649.
Spannungs- oder Spannungs-Dehnungs-Beziehung, *stress or stress-strain relation* 89, 212, 302, 305, 306.
Spannungsresultierende, *stress resultant* 512, 513, 651.
Spannungs-Temperatur-Gleichung (Kompatibilitätsbeziehung), *stress-temperature equation of compatibility* 318, 319, 321.
Spannungs-Temperatur-Modul, *stress-temperature modulus* 311.
Spannungs-Temperatur-Tensor, *stress-temperature tensor* 309, 313.
Spannungstensor, *stress tensor* 45–64, 513.
Spannungsvektor, *stress vector* 43, 485, 510.
Spannungswelle, *stress wave* 254.
spektrale Zerlegung und Theorem, *spectral decomposition and theorem* 8.
Spektrum, *spectrum* 269.
spezielle Schalentheorien, *special theories of shells* 546.
spezifische Entropie, *specific entropy* 483, 510.
— Entropieproduktion, *production of entropy* 483, 511.
— Gibbssche freie Energie, *Gibbs free energy* 555, 570.
— Helmholtzsche freie Energie, *Helmholtz free energy* 497, 511.
spezifische innere Energie, *specific internal energy* 481, 510.
— Wärme, *heat* 304.
— Wärmezufuhr, *heat supply* 481.
Sprung an einer singulären Fläche, *jump across a singular surface* 249, 421.
Spur, *trace* 6, 353.
Stab, Definition, *definition of rod* 646.
Stabtheorie, Definition, *theory of rods, definition* 641.
stark elliptischer Elastizitätstensor, *strongly elliptic elasticity tensor* 86, 87, 247, 256.
— — Operator 365, 422.
— elliptisches elastisches Feld, *elliptic elasticity field* 105.
starke Ableitung, *strong derivative* 349, 357.
— Bedingung für die Existenz der Lösung des Signorini-Problems, *condition for existence of solution of Signorini problem* 417.
starken Sinne, im, *strong sense* 416.
starre Verschiebung, *rigid displacement* 31–33, 61, 68, 104.
starres Feld auf ∂B mit Kraft und Moment, *rigid field on* ∂B *with force and moment* 185.
statisch-geometrische Analogie, *static-geometric analogy* 613.
statisch zulässiger Zustand, *statically admissible state* 114.
Sternbergscher Satz zu St. Venants Prinzip, *Sternberg's theorem on St. Venant's principle* 193.
sternförmiges Gebiet, *star-shaped region* 14.
stetig in der Zeit, *continuous in time* 24.
Stetigkeitsgleichung, *continuity, equation of* 448.
Stokes-Prozeß, *Stokes process* 239, 240.
Stokesscher Satz, *Stokes' theorem* 17.
Strahl, *ray* 249, 257.
stückweise glatt, *piecewise smooth* 14.
— kontinuierliche Funktion, *continuous function* 14.
— regulär, *regular* 14, 25.
Superpositionsprinzip für elastische Zustände, *superposition principle for elastic states* 95.
— — thermoelastische Zustände, *thermoelastic states* 313.
Symmetrie, geometrische, *symmetry, geometrical* 297.
— von Green-Zuständen, *of Green's states* 187.
Symmetrieachse, *axis of symmetry* 73, 248.
Symmetriebeziehung für Stäbe, *symmetry condition for rods* 654.
Symmetriegruppe, *symmetry group* 70, 210.
Symmetrietransformation, *symmetry transformation* 70.

symmetrische, beschränkte bilineare Form, *symmetric-bounded bilinear form* 391.
symmetrischer Elastizitätstensor, *symmetric elasticity tensor* 69, 84, 85, 247, 256.
— Gradient 11.
— Operator 369.
— Tensor 6, 617.
synchrone äußere Kraftsysteme, *synchronous external force systems* 219.
Systemunabhängigkeit, materielle (s. auch „Objektivität"), in der Stabtheorie, *frame-indifference, material (see also "objective"), rods* 655, 669.
—, —, dreidimensional, *three-dimensional* 80, 305, 306.

Tangentialebene, *tangent plane* 622.
Tangential- und Normalkomponenten des Direktors und des Direktorgradienten, *tangential and normal components of director and of director gradient* 535.
Teil, *part* 14, 300.
Teilflächen $\mathcal{S}_1$ und $\mathcal{S}_2$, *subsurfaces* $\mathcal{S}_1$ *and* $\mathcal{S}_2$ 12.
Temperatur, absolute, *temperature, absolute* 301, 483, 510.
—, Bedingung, *condition* 335.
—, Beziehungen, *relations* 305.
—, Darstellung in Stabtheorien, *representation in rod theories* 649.
—, Differenzfeld, *difference field* 312.
—, Gleichung, *equation* 328.
—, Gradient 301.
Temperatur-Entropie-Momente für Stäbe, *temperature-entropy moment for rods* 649.
Temperaturgradient-Temperaturbeziehung, *thermal gradient-temperature relation* 326.
Tensor, Art, *tensor, type* 617.
—, Definition 6.
—, kovariant, kontravariant, gemischt, *covariant, contravariant, mixed* 617, 622.
—, relativer, *relative* 618.
—, Transformationsgesetze, *transformation laws* 616.
— vierter Stufe, *fourth-order tensor* 9.
Tensorfelder, *tensor fields* 10.
Tensorfläche, *tensor surface* 631.
Tensorkomponenten, *tensor components* 498.
Tensorprodukt, *tensor product* 7.
thermischer Ausdehnungstensor, *thermal expansion tensor* 310.
— Ausdehnungskoeffizient, *coefficient of thermal expansion* 311, 316.
thermodynamischer Prozeß, *thermodynamic process* 484, 497, 510.
thermoelastische Schale, *thermoelastic shell* 561.
thermoelastischer Prozeß, *thermoelastic process* 327.
— Stab, *rod* 656, 669.
— Zustand, *state* 313.
thermomechanisches Verhalten von elastischen Cosserat-Flächen, *thermo-mechanical behavior (response of), elastic Cosserat surfaces* 529.
— — — Körpern, *bodies* 512.
— — — Schalen, *shells* 497.
Topologie von $H_m(A)$, *topology of* $H_m(A)$ 401.
Torsion einer rechteckigen Platte, *torsion of rectangular plate* 604.
Toupinscher Satz von St. Venants Prinzip, *Toupin's theorem on St. Venant's principle* 196.
Träger von *u*, *support of u* 349.
Transponierte eines Tensors vierter Ordnung, *transpose of a tensor, fourth-order* 10.
— — — zweiter Ordnung, *second-order* 6.
Transporttheorem, zweidimensional, *transport theorem, two-dimensional* 456.
transversale Isotropie, *transverse isotropy* 72.
— Normalspannung, *normal stress* 584.
— Scherdeformation, *shear deformation* 460, 575, 578, 584.
— Welle, *wave* 246–248, 254, 343.
typisches Oberflächenteilchen, *typical surface particle* 530.

überlagerte starre Bewegungen, *superposed rigid body motions* 452, 532.
Umgebung des Unendlichen, *neighborhood of infinity* 165.
Umkehrungsprinzip der minimalen potentiellen Energie, *converse principle of minimum potential energy* 116.
— — — Komplementarenergie, *complementary energy* 117.
unbeschränktes reguläres Gebiet, *unbounded regular region* 13, 257.
ungekoppelt, *uncoupled* 311, 312.
ungekoppelte quasistatische Theorie, *uncoupled quasi-static theory* 325.
uniforme Zustände, *uniform states* 546.
— — für Stäbe, *for rods* 698.
universelles Verschiebungsfeld, *universal displacement field* 91.
Unstetigkeitsflächen, *surfaces of discontinuity* 248.

Variationsmethoden, elastische Stabtheorie, *variational methods, elastic rod theory* 668.
—, dreidimensionale Elastizitätstheorie, *three-dimensional elasticity* 125–129.
Variationsprinzipien und -sätze, Elastizitätstheorie, *variational principles and theorems, elasticity* 122–125, 223–232.
— — —, hyperelastische Stäbe, *hyperelastic rods* 676.
— — —, Schalentheorie, *shell theory* 476.
Vektor in einer Fläche, *vector in a surface* 622.
Vektorfeld, *vector field* 10.
vektorielle räumliche Kugelfunktion, *vector solid spherical harmonic* 22.

Vektorprodukt, *vector product* 5.
Vergleichssatz der Elastizitätstheorie, *comparison theorem of elasticity* 269.
Verschiebung, Bedingung, *displacement, condition* 102, 220, 321, 335.
—, Definition 29, 43, 299.
—, Feld, *field* 31.
—, Feld für gemischtes Problem, *field solving mixed problem* 103.
—, Gleichung für Bewegung, *equation of motion* 212.
—, — — Gleichgewicht, *equilibrium* 90.
—, Gradient 29.
—, Problem 102, 220, 231.
Verschiebungen, Satz für eindeutige, *theorem on single-valued* 161.
verschiebungsfreie Zustände, *displacement-free states* 322.
Verschiebungs-Temperatur-Gleichung bei Bewegung, *displacement-temperature equation of motion* 328, 335.
— für Gleichgewicht, *of equilibrium* 317, 321.
Verschwinden bei einer Fläche, *vanishing near a surface* 20, 223.
Vertauschungssatz, *commutation theorem* 9.
vieldimensionale singuläre Integralgleichung, *multidimensional singular integral equation* 421.
vielfacher Index, *multiindex* 348.
Vierer-Tensor, *four-tensor* 27.
Vierer-Vektor, *four-vector* 27.
Viskoelastizität, *visco-elasticity* 386.
Vollständigkeit der Eigenschwingungen, *completeness of normal modes* 271.
—, Minimalprinzip für das Schwingungsproblem, *minimum principle for the vibration problem* 267.
—, Lösungen von BELTRAMI, *solutions of Beltrami* 56.
—, — — BELTRAMI-SCHAEFER 58.
—, — — BOUSSINESQ-PAPKOVITCH-NEUBER und BOUSSINESQ-SOMIGLIANA-GALERKIN 141.
—, — — CAUCHY-KOVALEVSKI-SOMIGLIANA 235.
—, — — GREEN-LAMÉ 233.
—, — — LOVE-BOUSSINESQ 147, 149.
—, vollständige Lösungen, Elastizitätstheorie, *complete solutions, elasticity* 138.
—, — —, Elastodynamik, *elastodynamics* 232.
—, — —, Thermoelastizität, *thermoelasticity* 329.
Volterra-Integralgleichung im Banach-Raum, *Volterra integral equations in Banach spaces* 374.
Volumenänderung, *volume change* 30, 31, 37, 97.
v. Kármán-Gleichung für nichtlineare Biegung dünner Platten, *v. Kármán equations for nonlinear bending of thin plates* 588.
v. Mises-Sternberg-Version des Prinzips von SAINT-VENANT, *v. Mises-Sternberg version of Saint-Venant's principle* 190.
Voraussetzungen der Theorie der Thermoelastizität, *assumed data of thermoelasticity* 335.

Wachstum charakteristischer Werte ins Unendliche, *infinite growth of the characteristic values* 270.
Wärme, übertragen durch Strahlung und Leitung, *heat transmitted by radiation and by conduction* 581.
Wärmeabsorption, *heat absorption* 510.
Wärmefluß, Bedingung, *heat flux condition* 335.
—, Definition 327.
Wärmeflußmoment für Stäbe, *heat flux moments for rods* 649.
Wärmeflußvektor, *heat flux vector* 300, 510.
Wärmeleitungsgleichung, *heat conduction equation* 312, 326.
Wärmeleitungsungleichung, *heat conduction inequality* 303, 307.
Wärmeoperatoren, *heat operators* 329.
Wärmequellenmomente für Stäbe, *heat source moments for rods* 651.
Wärmezufuhr, *heat supply* 300, 510, 513.
WEINGARTENs Formel, *Weingarten's formulae* 625, 626.
Weingartenscher Satz über singuläre Flächen, *Weingarten's theorem on singular surfaces* 252.
Welle der Ordnung n, *wave of order n* 253.
—, Impulsbilanz, *balance of momentum* 254.
Wellenausbreitung in geraden Stäben, *wave propagation in straight rods* 663.
— in 3-dimensionalen Körpern, *3-dimensional bodies* 243–261.
Wellengleichungen, *wave equations* 328.
Wellenoperatoren, *wave operators* 213, 328.
Wheeler-Sternberg-Lemma, *Wheeler-Sternberg lemma* 257.
wirbelfreie Felder, *irrotational fields* 17.
— Geschwindigkeit, *velocity* 213.

x_i-konvexes Gebiet, *x_i-convex region* 15.

Zanaboni-Robinsonscher Satz zu SAINT-VENANTs Prinzip, *Zanaboni-Robinson theorem on Saint-Venant's principle* 206.
Zeitintervall, *time-interval* 23.
Zentralchse, *central axis* 417.
Zerlegungssatz für einfache Scherung, *decomposition theorem for simple shears* 36
— — reine Dehnung, *pure strains* 35.
— — — —, isochor, *isochoric* 36.
— — singuläre Zustände, *singular states* 180.
— — Spannungsfelder, *stress fields* 59.
Zufuhr von Drehimpuls, *supply of moment of momentum* 482.
Zugkraftbedingung, *traction condition* 102, 220, 321, 335.

Zugkraftproblem, *traction problem* 102, 220, 321.
zulässige Bewegung, *admissible motion* 43, 326.
zulässiger Prozeß, *admissible process* 215, 327, 484, 512, 531, 532.
— thermodynamischer Prozeß, *process, thermodynamic* 302, 304.
zulässiges Spannungsfeld, *admissible stress field* 59, 313.
— zeitabhängiges Feld der Temperaturdifferenz, *temperature difference field, time-dependent* 326.
— — Spannungsfeld, *stress field, time-dependent* 49, 326.
— — Wärmeflußfeld, *heat flux field, time-dependent* 326.
— Verschiebungsfeld, *displacement field* 59, 312.
Zwangsbedingung, Reaktion, *constraint response* 548.
Zwangsbedingungen, einseitig, *constraints, unilateral* 403.
zweidimensionale Kräfte, *two-dimensional body forces* 513.
— Wärmezufuhr, *heat supply* 513.
zweidimensionales ε-System, *two-dimensional ε-system* 455.
— Transporttheorem, *transport theorem* 456.
— Vertauschungssymbole, *permutation symbols* 455.
zweiseitige Zwangsbedingungen, *bilateral constraints* 403.
zweiter Hauptsatz der Thermodynamik (Clausius-Duhem-Ungleichung), *second law of thermodynamics (Clausius-Duhem inequality)* 301, 484, 512, 532.

Subject Index.

(English-German.)

Where English and German spelling of a word is identical the German version is omitted.

abstract unilateral problems, non-symmetric, *abstrakte einseitige Probleme, unsymmetrischer Fall* 395.
— — —, symmetric, *symmetrischer Fall* 391.
acceleration, definition, *Beschleunigung, Definition* 43.
— in rod theories, *in Stabtheorien* 649, 652.
— — shell theories, *Schalentheorien* 494.
— waves, *Beschleunigungswellen* 253, 254.
acoustic tensor, *akustischer Tensor* 243–248.
adiabatic, *adiabatisch* 304.
adjoint operator, *adjungierter Operator* 355.
— fields corresponding to Kelvin state, *adjungierte, dem Kelvin-Zustand entsprechende Felder* 175.
— traction field corresponding to Stokes process, *adjungiertes, dem Stokesschen Prozeß entsprechendes Zugkraftfeld* 240.
— transformation, *adjungierte Transformation* 369.
admissible displacement field, *zulässiges Verschiebungsfeld* 59, 312.
— heat flux field, time-dependent, *zeitabhängiges Wärmeflußfeld* 326.
— motion, *zulässige Bewegung* 43, 326.
— process, *zulässiger Prozeß* 215, 327, 484, 512, 531, 532.
— —, thermodynamic, *thermodynamisch* 302, 304.
— state, *Zustand* 94, 313.
— stress field, *zulässiges Spannungsfeld* 59, 313.
— — —, time-dependent, *zeitabhängig* 49, 326.
— temperature difference field, time-dependent, *zeitabhängiges Feld der Temperaturdifferenz* 326.
Airy function, *Airy-Funktion* 157.
Airy's solution, *Airysche Lösung* 54, 156.
ambiguous boundary conditions, *nichteindeutige Randbedingungen* 405, 414.
amplitude, vibration, *Schwingungsamplitude* 261.
—, wave of order *n*, *Amplitude, Welle der Ordnung n* 254.
analog of principle of minimum potential energy, *Analogon zum Prinzip der minimalen potentiellen Energie* 229.
analytic field, *analytisches Feld* 11.
analyticity, elastic displacement fields, *Analytizität, elastische Verschiebungsfelder* 131.
analyticity, thermoelastic displacement and temperature fields, *Analytizität, thermoelastische Verschiebungs- und Temperaturfelder* 318.
angular momentum, see under "momentum" *Drehimpuls s. Impuls.*
anistropic material, *anisotropes Material* 7, 87–89, 530.
antenna potential, *Antennenpotential* 420.
approximation theorem, *Näherungssatz* 128.
approximations in shell and plate theory, *Näherungen in der Schalen- und Plattentheorie* 476, 522, 566–568, 570, 587.
astatic equilibrium, *astatisches Gleichgewicht* 63, 64, 98, 193.
asymptotic distribution of eigenvalues, *asymptotische Verteilung von Eigenwerten* 420.
— expansion techniques for shells, *Entwicklungsmethoden für Schalen* 586, 587.
— methods for rods, *Methoden für Stäbe* 664.
attenuated, *gedämpft* 343.
axial vector, *axialer Vektor* 6.
axis of symmetry, *Symmetrieachse* 73, 248.

balance principles (see also conservation laws) for 3-dimensional theories, energy, *Bilanzprinzipien (s. auch Erhaltungssätze) für 3-dimensionale Theorien, Energie* 511.
— — — — —, forces, for singular states, *Kräfte, für singuläre Zustände* 181.
— — — — —, forces and moments, *Kräfte und Momente* 300.
— — — — —, linear and angular momentum, *Impuls und Drehimpuls* 44, 52.
— — — — —, moments, *Momente* 306.
— — — — —, momentum, *Impuls* 42, 51.
— — — — —, momentum in space-time, *Impuls in Raum-Zeit* 67.
— — — — —, momentum of a wave, *Impuls einer Welle* 254.
— — for rods, angular momentum, *für Stäbe, Drehimpuls* 666.
— — — —, director momentum, *Direktorimpuls* 666.
— — — —, energy, *Energie* 667.
— — — —, linear momentum, *Impuls* 666.
— — for shells, *für Schalen* 483.
base vectors, in general, *Basisvektoren, allgemein* 439, 618.

base vectors on a surface, *Basisvektoren auf einer Oberfläche* 621.
basic equations, shell theory, *grundlegende Gleichungen, Schalentheorie* 565.
— —, linear thermoelasticity, *lineare Thermoelastizität* 310, 311.
— principles for shells, *Prinzipien für Schalen* 479.
— singular solutions, elastodynamics, *singuläre Lösungen, Elastodynamik* 239.
— — —, elastostatics, *Elastostatik* 173.
Beltrami-Schaefer solution, *Beltrami-Schaefer-Lösung* 58.
BELTRAMI's solution, *Beltramische Lösung* 54.
bending theory, plates, classical theory, *Biegungstheorie, Platten, klassische Theorie* 577.
— —, plates, v. KÁRMÁN's nonlinear theory, *Platten, v. Kármán's nichtlineare Theorie* 588.
— — of shells, classical, *Schalen, klassische Theorie* 582.
— —, shells, thin, *dünne Schalen* 425.
BETTI's reciprocal theorem, *Bettischer Reziprozitätssatz* 98, 99, 170, 211, 419.
biharmonic field, *biharmonisches Feld* 20.
bilateral constraints, *zweiseitige Zwangsbedingungen* 403.
bilinear integro-differential form, *bilineare Integrodifferentialform* 361.
body, definitions, *Körper, Definitionen* 12, 14, 299.
— force, analogy in thermoelasticity, *Kraft, Analogon in der Thermoelastizität* 316.
— —, definition 44, 299, 309, 485.
— —, moments for rods, *Momente für Stäbe* 650.
— —, two-dimensional, *zweidimensional* 513.
BOGGIO's theorem, *Boggioscher Satz* 237.
border, *Grenze* 402.
boundary of body, *Randfläche eines Körpers* 102, 440.
— conditions, elasticity, *Randbedingungen, Elastizitätstheorie* 102, 129, 314, 403, 405, 414, 420.
— —, rod theories, *Stabtheorien* 648, 659, 667, 674, 678.
— —, shell theory, linear, *lineare Schalentheorie* 583.
— —, shell theory, restricted, *eingeschränkte Schalentheorie* 553.
— —, thermoelasticity, *Thermoelastizität* 314, 321.
— —, clamped, *Einklemmung* 263, 314.
— —, constraints, *Zwangsbedingungen* 403.
— —, regularization, *Regularisierung* 401.
boundary-initial-value problems of elasticity, *Rand- und Anfangswertprobleme in der Elastizitätstheorie* 335.
boundary-layer equations, *Grenzschichtgleichungen* 586.
boundary-value operator, *Randwertoperator* 402.
boundary-value problems of elasticity, ambiguous, *Randwertprobleme für Elastizitätstheorie, nicht eindeutig* 405, 414.
— — — —, displacement ("first"), *Verschiebung („erste")* 103, 220, 380.
— — — —, general, *allgemein* 129, 185.
— — — —, mixed ("third"), *gemischt („dritte")* 102, 381, 420.
— — — —, traction ("second"), *Zugkraft („zweite")* 103, 220, 381.
— —, harmonic functions, *harmonische Funktionen* 419.
— —, rods, *Stäbe* 658.
— —, shells, general theory, *Schalen, allgemeine Theorie* 550.
— —, —, linear bending, *lineare Biegung* 584.
— —, —, restricted linear theory, *eingeschränkte lineare Theorie* 607, 609.
— —, thermoelasticity, displacement, *Theorie der Thermoelastizität, Verschiebung* 321.
— —, displacement and temperature, *Verschiebung und Temperatur* 335.
— —, —, traction, *Zugkraft* 321.
bounded regular region, *beschränktes reguläres Gebiet* 13.
bounds for strain energy, *Schranken für Dehnungsenergie* 113.
Boussinesq-Papkovitch-Neuber solution in elasticity, elimination of component of vector potential, *Boussinesq-Papkovitch-Neuber-Lösung in der Elastizitätstheorie, Elimination einer Komponente des Vektorpotentials* 146.
— —, elimination of scalar potential, *Elimination des skalaren Potentials* 143.
— —, general form, *allgemeine Form* 139–142, 235.
Boussinesq-Somigliana-Galerkin solution in elasticity, *Boussinesq-Somigliana-Galerkin-Lösung in der Elastizitätstheorie* 141, 142, 235, 236.
BOUSSINESQ's solution in elasticity, *Boussinesqs Lösung in der Elastizitätstheorie* 147.
BRUN's theorem, *Brunscher Satz* 217, 223.
buckled states for rods, *geknickte Zustände für Stäbe* 694.

C$^\infty$-smooth, *C*$^\infty$*-glatt* 370.
C^ν-smooth, C^ν*-glatt* 369.
Cartesian coordinate system, *Cartesisches Koordinatensystem* 438.
catalogues for shell theory, linear kinematic measures for Cosserat surface, *Kataloge der Schalentheorie, lineare kinematische Maße für eine Cosserat-Fläche* 458.
— — — —, notations, *Bezeichnungen* 432–438.
— — — —, resultants and equations of motion, *Resultierende und Bewegungsgleichungen* 526.

Cauchy equations of motion, *Cauchys Bewegungsgleichungen* 510.
— principal value of a surface integral, *Hauptwert eines Oberflächenintegrals* 181.
— reciprocal theorem, *Reziprozitätssatz* 45.
— representation theorem for isotropic functions, *Darstellungstheorem für isotrope Funktionen* 535, 551.
— singular integrals, *singuläre Integrale* 419.
— stress tensor, *Spannungstensor* 51, 53, 305.
Cauchy-Green tensor, representation in rod theories, *Cauchy-Green-Tensor, Darstellung in der Stabtheorie* 649.
Cauchy-Kovalevski-Somigliana solution in elasticity, *Cauchy-Kovalevski-Somigliana-Lösung in der Elastizitätstheorie* 235–237.
Cauchy-Poisson theorem, *Cauchy-Poissonscher Satz* 44.
center of compression or rotation, *Kompressionszentrum oder Rotationszentrum* 179.
central axis, *Zentralachse* 417.
characteristic displacement for vibration problem, *charakteristische Verschiebung für das Schwingungsproblem* 262.
— solution for free vibration problem, *Lösung für das freie Schwingungsproblem* 262.
— space, *charakteristischer Raum* 8.
— values, elasticity tensor, *charakteristische Werte des Elastizitätstensors* 76.
— —, vibration problem, *des Schwingungsproblems* 263.
Christoffel symbols of the first and second kinds, *Christoffel-Symbole erster und zweiter Art* 619.
circular rods, *kreisförmige Stäbe* 690.
clamped boundary, *eingeklemmter Rand* 314.
— plate, partially supported on a subdomain *eingeklemmte Platte, unterstützt in einem Teilgebiet* 406.
class $C^{M,N}$, *Klasse* $C^{M,N}$ 24.
— C^N 11.
classical theories of plates and shells, *klassische Theorien der Platten und Schalen* 575–577, 582, 585.
Clausius-Duhem inequality, *Clausius-Duhem-Ungleichung* 301, 484, 512, 532.
closed convex set, *geschlossene konvexe Menge* 391.
— regular surface, *reguläre Oberfläche* 13.
— strain path, *geschlossener Dehnungsverlauf* 82.
coefficient of thermal expansion, *thermischer Ausdehnungskoeffizient* 311, 316.
coerciveness, *Koerzivität* 422.
cofactors, *Kofaktoren* 618.
commutation theorem, *Vertauschungssatz* 9.
comparison theorem of elasticity, *Vergleichssatz der Elastizitätstheorie* 269.
compatibility, *Kompatibilität* 39.
— equations for plane elastic states, *Kompatibilitätsgleichungen für ebene elastische Zustände* 155.
— —, shells, *Schalen* 463, 479.
— —, —, restricted theory, *in der eingeschränkten Theorie* 466.
— —, 3-dimensional bodies, *3-dimensionale Körper* 39–41, 62, 63, 211, 318.
— theorem for plane displacements, *Kompatibilitätssatz für ebene Verschiebungen* 41.
complementary energy function, *komplementäre Energiefunktion* 570.
— regular subsurfaces, *reguläre Teilflächen* 14.
complete solutions, elasticity, *vollständige Lösungen, Elastizitätstheorie* 138.
— —, elastodynamics, *Elastodynamik* 232.
— —, thermoelasticity, *Thermoelastizität* 329.
completeness, minimum principle for the vibration problem, *Vollständigkeit, Minimalprinzip für das Schwingungsproblem* 267.
—, normal modes, *Eigenschwingungen* 271.
—, solutions of BELTRAMI, *Lösungen von Beltrami* 56.
—, — — BELTRAMI-SCHAEFER 58.
—, — — BOUSSINESQ-PAPKOVITCH-NEUBER and BOUSSINESQ-SOMIGLIANA-GALERKIN 141.
—, — — CAUCHY-KOVALEVSKI-SOMIGLIANA 235.
—, — — GREEN-LAMÉ 233.
—, — — LOVE-BOUSSINESQ 147, 149.
complex rigid displacement, *komplexe starre Verschiebung* 39.
compliance tensor, *Nachgiebigkeitstensor* 69, 71, 77, 310.
concentrated load, *konzentrierte Last* 174, 179.
— stresses, *Spannungen* 416.
conductivity, isotropic material, *Leitfähigkeit, isotrope Materialien* 311.
— tensor 307, 309.
cone hypothesis, *Kegelhypothese* 354.
conjugate tensor, *konjugierter Tensor* 618.
conservation laws (see also balance principles) for a Cosserat surface, *Erhaltungssätze (s. auch Bilanzgleichungen) für eine Cosserat-Fläche* 479, 482, 487.
— — in terms of a reference state, *ausgedrückt im Bezugszustand* 490.
— — of mass, *der Masse* 440, 483.
consistency condition, *Konsistenzbedingung* 129.
constitutive coefficients in linear theory of Cosserat plates, *Materialkoeffizienten in der linearen Theorie der Cosserat-Platten* 600, 603.
— — — — — — Cosserat surfaces, *Cosserat-Flächen* 598, 606.
— — — — — — shells, restricted, *Schalen (eingeschränkte Theorie)* 607.

constitutive equations for elastic materials, *Materialgleichungen für elastische Materialien* 67.
— — — rods, *Stäbe* 654, 657, 659, 669, 673.
— — — shells and plates, Cosserat surface, elastic, *Schalen und Platten, Cosserat-Fläche, elastisch* 528.
— — — — — —, Cosserat surface, inextensible, *Cosserat-Fläche, unausdehnbar* 549.
— — — — — —, elastic shells, *elastische Schalen* 528, 533, 537.
— — — — — —, elastic shells, from 3-dimensional theory, *elastische Schalen, aus der 3-dimensionalen Theorie* 561, 564.
— — — — — —, in terms of material coordinates, *ausgedrückt durch materielle Koordinaten* 529.
— — — — — —, in terms of reference state, *ausgedrückt im Bezugszustand* 544.
— — — — — —, in terms of relative measures, *ausgedrückt durch relative Maße* 540.
— — — — — —, linear, for Cosserat surface, *linear, für Cosserat-Fläche* 553.
— — — — — —, linear, for shells of uniform thickness, *linear, für Schalen einheitlicher Dicke* 556.
— — — — — —, linear, in terms of Gibbs function, *linear, ausgedrückt durch die Gibbsfunktion* 571.
— — — — — —, mechanical theory, *mechanische Theorie* 544.
— — — — — —, restricted linear theory, *eingeschränkte lineare Theorie* 560.
— relations, see constitutive equations, *Materialrelationen, s. Materialgleichungen.*
constraint response, *Zwangsbedingung, Reaktion* 548.
constraints, unilateral, *Zwangsbedingungen, einseitig* 403.
contact couple, *Kontaktkräftepaar* 504.
— director couple, *Kontaktdirektorkräftepaar* 480, 491.
— force, *Kontaktkraft* 480, 491, 510.
— problem, *Kontaktproblem* 129.
continuity, equation of, *Stetigkeitsgleichung* 448.
continuous in time, *stetig in der Zeit* 24.
contraction mapping, *Kontraktionsabbildung* 396, 422, 618.
contravariant tensors, *kontravariante Tensoren* 617, 622.
convected coordinates, *konvektive Koordinaten* 431, 439, 449.
convergence of approximate solutions, *Konvergenz von Näherungslösungen* 125.
converse principle of minimum complementary energy, *Umkehrung des Prinzips der minimalen Komplementärenergie* 117.
— — — — potential energy, *potentiellen Energie* 116.
convex hull, *konvexe Hülle* 417.
— region with respect to $\mathscr{S}$, *konvexes Gebiet bezüglich* $\mathscr{S}$ 16.
— set, *konvexe Menge* 392, 417.
convolution, *Faltung* 25, 332.
coordinate systems, *Koordinatensysteme* 438.
— transformation, *Koordinatentransformation* 531, 615.
Cosserat rod, *Cosserat-Stab* 665.
— surface, in general, *Cosserat-Fläche, allgemein* 430, 446, 449.
— —, relation to 3-dimensional theory, *Beziehung zur 3-dimensionalen Theorie* 594.
Cosserat-Ericksen-Toupin theorem, *Cosserat-Ericksen-Toupinscher Satz* 106.
couple assigned, *Kräftepaar, vorgeschriebenes* 504.
— director, assigned, *Direktorkräftepaar, vorgeschriebenes* 480.
— resultant vector, *resultierender Vektor* 651.
coupled heat equation, *gekoppelte Wärmeleitungsgleichung* 328, 335.
covariant derivative with respect to $a_{\alpha\beta}$, *kovariante Ableitung bezüglich* $a_{\alpha\beta}$ 450.
— — — — — A 456.
— — — — — Euclidean metric, *Euklidische Metrik* 619.
— — — — — first fundamental form (metric tensor), *erster Fundamentalform (metrischer Tensor)* 431, 624.
— tensors, *Tensoren* 617, 622.
crystal class, *Kristallklasse* 72.
curl 11.
curvature of boundary, *Krümmung der Randfläche* 378.
— — surface, mean, *Oberfläche, mittlere* 471, 624.
— — —, Gaussian, *Gaußsche* 471.
— tensor, *Krümmungstensor* 620, 625.
curve on the surface, *Kurve auf der Fläche* 622.

damped, *gedämpft* 343.
data of thermoelasticity, *Voraussetzungen der Theorie der Thermoelastizität* 335.
decomposition theorem, pure strains, *Zerlegungssatz für reine Dehnung* 35.
— —, — —, isochoric, *isochor* 36.
— —, simple shears, *einfache Scherung* 36.
— —, singular states, *singuläre Zustände* 180.
— —, stress fields, *Spannungsfelder* 59.
deformation, definition 29.
—, infinitesimal 28, 456.
—, gradient 29, 299.
density, *Dichte* 43, 299, 309.
Deresiewicz-Zorski solution, *Deresiewicz-Zorski-Lösung* 330.
design problems for rods, *Auslegungsproblem für Stäbe* 698.
determinant, *Determinante* 6.

differentiable $(r-1)$-cell of Class C^n, *differenzierbare $(r-1)$-Zelle der Klasse C^n* 402.
diffusion problems, *Diffusionsprobleme* 373, 377, 386.
dilatation, definition 31.
—, uniform, *gleichförmig* 34, 77.
dilatational wave equation, *Dilatationswelle, Gleichung* 213, 328.
direct approach to shell theory, *direkte Ableitung der Schalentheorie* 448.
— notation, *Bezeichnung* 297.
direction of displacement, *Richtung der Verschiebung* 343.
— — motion of a progressive wave, *Bewegung einer fortschreitenden Welle* 245.
— — propagation, *Ausbreitung* 245, 248, 343.
director couple, assigned, *Direktorkräftepaar, vorgeschriebenes* 480.
—, definition, *Direktor, Definition* 446, 467, 666.
— inertia coefficient, *Direktorträgheitskoeffizient* 594–596.
— momentum, *Direktorimpuls* 482, 483.
— velocity, *Direktorgeschwindigkeit* 449, 467.
Dirichlet boundary conditions, *Dirichletsche Randbedingungen* 370.
— problem, *Dirichletsches Problem* 375, 449.
dispersed waves, *dispersierte Wellen* 343.
displacement condition, *Verschiebungsbedingung* 102, 220, 321, 335.
—, definition, *Verschiebung, Definition* 29, 43, 299.
—, equation of equilibrium, *Gleichung für Gleichgewicht* 90.
—, — — motion, *Bewegung* 212.
— field, *Verschiebungsfeld* 31.
— — solving mixed problem, *für gemischtes Problem* 103.
— gradient, *Verschiebungsgradient* 29.
— problem, *Verschiebungsproblem* 102, 220, 321.
—, theorem on single-valued, *Verschiebung, Satz für eindeutige* 161.
displacement-free states, *verschiebungsfreie Zustände* 322.
displacement-temperature equation of equilibrium, *Verschiebungs-Temperatur-Gleichung im Gleichgewicht* 317, 321.
— — — motion, *bei Bewegung* 328, 335.
divergence, definition, *Divergenz, Definition* 11.
— theorem, *Divergenzsatz* 16, 28.
domain, definition, *Gebiet, Definition* 348.
— of influence, *Einflußgebiet* 257.
doublet process, *Doublet-Prozeß* 241.
— state, *Doublet-Zustand* 177, 178.
duality between rods and shells, *Dualität von Stäben und Schalen* 664.
dynamical problem for rods, *dynamisches Problem für Stäbe* 699.
— process, *dynamischer Prozeß* 44, 49.
dynamically admissible stress field, *dynamisch zulässiges Spannungsfeld* 230.

Ehrling lemma 354.
eigenfunctions, complete set of, *Eigenfunktionen, vollständiger Satz von* 369.
eigenvalue, definition, *Eigenwert, Definition* 369.
— problems, *Eigenwertprobleme* 386.
elastic displacement fields, *elastische Verschiebungsfelder* 131.
— material, *elastisches Material* 301.
— motion, *elastische Bewegung* 233.
— potential, *elastisches Potential* 380, 386, 544.
— process, *elastischer Prozeß* 215.
— progressive wave, *elastische fortschreitende Welle* 246.
— state, *elastischer Zustand* 95, 102.
— —, properties, *Eigenschaften* 133.
— — with displacement vanishing at infinity, *mit im Unendlichen verschwindender Verschiebung* 167.
— — — stress vanishing at infinity, *im Unendlichen verschwindender Spannung* 168.
— wave of order n, *elastische Welle der Ordnung n* 255.
elastica, *Elastika* 697, 698.
elastic-plastic torsion problem, *elastisch-plastisches Torsionsproblem* 405, 422.
elasticities, elasticity field, *elastische Koeffizienten, Elastizitätsfeld* 68, 75.
elasticity tensor, general, *Elastizitätstensor, allgemein* 67–69, 309, 313.
— —, isotropic materials, *isotrope Materialien* 75, 317.
elementary curvilinear triangle, *elementares krummliniges Dreieck* 493.
elements of area and volume, *Elemente von Fläche und Volumen* 441.
elliptic operator, *elliptischer Operator* 355, 374.
embedding, *Einbettung* 35, 352.
energy (see also "power" and "work"), *Energie (s. auch „Leistung" und „Arbeit").*
— equation, general, *Energiegleichung, allgemein* 326.
— —, rods, *Stäbe* 653, 658.
— —, shells, derived from 3-dimensional theory, *Schalen, abgeleitet aus der 3-dimensionalen Theorie* 515.
— —, shells, direct approach, *Schalen, direkte Ableitung* 481, 482.
—, free, *Energie, freie* 497, 511, 555, 570.
— functional, *Energiefunktional* 385, 391, 403.
—, internal, *Energie, innere* 481, 510.
energy-flux vector, *Energieflußvektor* 214.
entire boundary, clamped, *gesamte Randfläche, eingeklemmt* 263.

entropy inequality (CLAUSIUS-DUHEM), general, *Entropieungleichung (Clausius-Duhem), allgemein* 301, 510, 511.
— —, rods, *Stäbe* 667.
— —, shells, derived from 3-dimensional theory, *Schalen, abgeleitet aus der 3-dimensionalen Theorie* 515.
— —, shells, direct, *Schalen, direkte Ableitung* 483.
— principle for constitutive relations, general, *Entropieprinzip für Materialgleichungen, allgemein* 302.
— — — — —, rods, *Stäbe* 669.
— — — — —, shells, *Schalen* 512.
— production, *Entropieproduktion* 483, 492, 511.
— relations, *Entropiebeziehungen* 303, 306.
ε-system 618.
equal modulo a rigid displacement, *gleich bis auf ein starkes Verschiebungsfeld* 104.
equations of motion, elasticity, *Bewegungsgleichungen, Elastizitätstheorie* 45, 64, 212.
— — —, rods, *Stäbe* 653, 658, 672.
— — —, shells, catalogue of, *Schalen, Katalog der* 526.
— — —, shells, derived from 3-dimensional theory, *Schalen, abgeleitet aus der 3-dimensionalen Theorie* 520.
— — —, shells, direct approach, *Schalen, direkte Ableitung* 494.
— — —, shells, restricted theory, *Schalen, eingeschränkte Theorie* 552.
— — —, thermoelasticity, *Thermoelastizität* 326.
equilibrium, astatic, *Gleichgewicht, astatisch* 63, 64, 98, 193.
—, energy equation, *Energiegleichung* 312.
— equations, general, *Gleichungen, allgemein* 49, 59, 90, 312.
— —, in lines of curvature coordinates, *mit Krümmungslinien als Koordinaten* 633.
— —, in rectangular, cylindrical, and polar coordinates, *in rechtwinkligen, zylindrischen und Kugelkoordinaten* 93.
— —, rods, integrals, *für Stäbe, Integrale* 696.
— —, solution, *Lösung* 53.
— heat equation, *Wärmeleitungsgleichung* 317.
— in each load region, *in jedem belasteten Gebiet* 193.
— of forces, *von Kräften* 63.
— problem, elasticity, *Problem in der Elastizitätstheorie* 380.
— —, heterogeneous elastic medium, *heterogenen elastischen Medien* 386, 421.
— —, thin plate, *dünnen Platten* 377.
equipresence, *Äquipräsenz* 530.
equivalent processes, *äquivalente Prozesse* 487.
— systems of equations, classical linear theory of shells, *Systeme von Gleichungen, klassische lineare Schalentheorie* 583.
error estimate in the linear theories of plates and shells, *Fehlerabschätzung in den linearen Theorien von Platten und Schalen* 584.
existence and nonuniqueness for elastostatic boundary-value problems, *Existenzsätze und Nichteindeutigkeit für elastostatische Randwertprobleme* 544.
— of approximate solutions, *von Näherungslösungen* 126.
— theorems, general, *Theoreme, allgemein* 368–387.
— theorem, generalized Signorini problem, *Theorem für das verallgemeinerte Signorini-Problem* 407.
— — for rods, *für Stäbe* 680.
extended uniqueness theorem for the mixed problem, *erweiterter Existenzsatz für das gemischte Problem* 223.
extensibility in the direction of the unit normal to a reference surface, *Ausdehnbarkeit in Richtung der Einheitsnormalen zur Bezugsfläche* 460.
extensional deformation of a plate, *Ausdehnungsdeformation einer Platte* 600.
— motion, *Ausdehnungsbewegung* 460.
— theory of plates, classical, *Ausdehnungstheorie von Platten, klassisch* 577.
— vibrations of plates, *Ausdehnungsschwingungen von Platten* 575.
extensions of the fundamental lemma, *Erweiterungen des grundlegenden Lemmas* 115, 116, 223.
exterior domain, *äußeres Gebiet* 14, 165, 261.
external body force, *äußere Körperkraft* 509.
— force system, *äußeres Kraft-System* 95, 215, 327.
— force-temperature system, *äußeres Kraft-Temperatur-System* 313.
— thermal system, *äußeres thermisches System* 327.

family of load regions contracting to a boundary point, *Familie von Belastungsgebieten, die sich auf einen Randpunkt zusammenziehen* 191.
— — singular elastic states, *singulärer elastischer Zustände* 192.
FATOU's lemma 384.
Federov-Stippes theorem, *Federov-Stippesscher Satz* 247.
field equations, elasticity, *Feldgleichungen, Elastizitätstheorie* 89–94, 212–215.
— —, nonlinear thermoelasticity, *nichtlineare Thermoelastizität* 307.
— —, shells, direct approach, *Schalen, direkte Ableitung* 492, 498.
— —, —, derived from 3-dimensional theory, *abgeleitet aus der 3-dimensionalen Theorie* 508, 519.
— —, —, in terms of reference state, *ausgedrückt durch den Bezugszustand* 502.
— —, —, linearized, *linearisiert* 500, 523.
— —, —, restricted theory, *eingeschränkte Theorie* 503.

finite deformations, *endliche Deformationen* 28.
— elasticity, *Theorie endlicher Deformationen* 80.
— rigid deformation, *endliche starre Deformationen* 30.
— strain tensor, *endlicher Dehnungstensor* 29, 80, 306.
first and second fundamental forms of a surface, *erste und zweite Fundamentalform einer Oberfläche* 623.
first law of thermodynamics, *erster Hauptsatz der Thermodynamik* 300.
flexural motion, *Beugebewegung* 460.
— vibrations of plates, *Beugeschwingungen von Platten* 575.
Flügge-Lur'e-Byrne-Truesdell equations in the linear theory of shells, *Flügge-Lur'e-Byrne-Truesdellsche Gleichungen in der linearen Schalentheorie* 592.
follower load problem, *Folgelastproblem* 693.
force at infinity, *Kraft im Unendlichen* 170.
— vector, assigned, *Kraftvektor, zugeordnet* 480.
forces, parallel and non-tangential to the boundary, *Kräfte, parallel und nicht-tangential zur Randfläche* 193.
—, system 43.
four-tensor, *Vierer-Tensor* 27.
four-vector, *Vierer-Vektor* 27.
fourth-order tensor, *Tensor vierter Ordnung* 9.
frame-indifference, material (see also "objective"), 3-dimensional, *Systemunabhängigkeit, materielle (s. auch „Objektivität"), 3-dimensional* 80, 305, 306.
—, —, rods, *Stabtheorie* 655, 669.
Fredholm integral equations, *Fredholmsche Integralgleichungen* 419.
free body, *freier Körper* 315.
— energy, 3-dimensional, *Energie, 3-dimensional* 301.
— —, rods, *in Stäben* 649, 657.
— —, shells, elastic Cosserat surface, *in Schalen, elastische Cosserat-Fläche* 557.
— —, shells, linear theory, *in Schalen, lineare Theorie* 556.
— vibration problem, *Problem freier Schwingungen* 261.
frequency of vibration, *Frequenz einer Schwingung* 261.
Fresnel-Hadamard propagation condition, *Fresnel-Hadamardsche Ausbreitungsbedingung* 246, 256.
fundamental form of a surface, first and second, *Fundamentalform einer Fläche, erste und zweite* 623, 624.
— — — — —, third, *dritte* 626.
— lemma, *fundamentales Lemma* 20.
— solution matrix, *fundamentale Lösungsmatrix* 419.
— system of field equations, *fundamentales System von Feldgleichungen* 89.

GAUSS, equations of, *Gauß, Gleichungen von* 46, 627.
GAUSS' formulae, *Gaußsche Formel* 625, 626.
Gaussian curvature of surface, *Gaußsche Krümmung einer Oberfläche* 624.
geometric multiplicity, *geometrische Multiplizität* 369.
geometrical symmetries, *geometrische Symmetrie* 297.
geometry of a surface in a Euclidean space covered by normal coordinates, *Geometrie einer Fläche im Euklidischen Raum mit Normalkoordinaten* 628.
global reaction exerted by the constraint of support, *globale Reaktion auf die Stützungsbedingung* 416.
GOURSAT's theorem, *Goursatscher Satz* 160.
gradient, deformation 29, 299.
—, director, *Direktor* 452.
—, director velocity, *Direktorgeschwindigkeit* 452.
—, displacement, *Verschiebung* 29.
— field, *Feld* 10.
GRAFFI's reciprocal theorem, *Graffischer Reziprozitätssatz* 218, 243.
Green-Lamé solution, *Green-Lamé-Lösung* 233, 236, 237.
GREEN's formula for the Laplace operator, *Greensche Formel für den Laplace-Operator* 419.
— function of the Dirichlet problem, *Funktion des Dirichlet-Problems* 421.
— states, *Greensche Zustände* 186.

Hamilton-Kirchhoff principle, *Hamilton-Kirchhoffsches Prinzip* 255, 226.
harmonic field, *harmonisches Feld* 20.
— plane progressive wave, *harmonische ebene fortschreitende Welle* 342
HARNACK's convergence theorem, *Harnackscher Konvergenzsatz* 21.
heat absorption, *Wärmeabsorption* 510.
— conduction equation, *Wärmeleitungsgleichung* 312, 326.
— — inequality, *Wärmeleitungsungleichung* 303, 307.
— flux condition, *Wärmeflußbedingung* 335.
— —, definition, *Wärmefluß, Definition* 327.
— — moments for rods, *Wärmeflußmoment für Stäbe* 649.
— — vector, *Wärmeflußvektor* 300, 510.
— operators, *Wärmeoperatoren* 329.
— source moments for rods, *Quellenmomente für Stäbe* 651.
— supply, *Wärmezufuhr* 300, 510, 513.
—, transmitted by radiation and by conduction, *Wärme, übertragen durch Strahlung und Leitung* 581.
Hellinger-Prange-Reissner principle, *Hellinger-Prange-Reissnersches Prinzip* 124, 125, 228.
HELMHOLTZ's theorem, *Helmholtzscher Satz* 19.

hereditary elasticity, *Nachwirkungs-elastizität* 423.
— problems, *Nachwirkungsprobleme* 377, 386.
heterogeneous media, *heterogene Medien* 421, 422.
history of shell theory, *Geschichte der Schalentheorie* 527, 589.
homogeneous body, elastic, *homogener Körper, elastisch* 68.
— —, isotropic elastic, *isotrop elastisch* 380.
— —, isotropic thermoelastic, *isotrop thermoelastisch* 317, 327.
— —, thermoelastic, *thermoelastisch* 327.
— displacement field, *homogenes Verschiebungsfeld* 33.
Hu-Washizu principle, *Hu-Washizusches Prinzip* 122, 125, 226.
hydrostatic pressure, *hydrostatischer Druck* 678, 690.
hyperelastic rods, *hyperelastische Stäbe* 673.

incompressibility condition, *Inkompressibilitätsbedingung* 404.
incompressible body, material, *inkompressible Körper, Materialien* 210.
indexes of frequently used symbols, *Listen der häufig benutzten Symbole* 2–5, 298, 299, 432–438.
inextensional theory of shells, *Nichtausdehnbarkeitstheorie von Schalen* 546, 548.
infinite growth of the characteristic values, *Wachstum charakteristischer Werte ins Unendliche* 270.
infinitesimal deformation, *infinitesimale Deformation* 28, 456.
— director displacement, *Direktorverschiebung* 456.
— motion of 3-dimensional continuum, *Bewegung eines 3-dimensionalen* Kontinuums 473.
— rigid-body displacements, *starre Verschiebungen* 30, 462, 463.
— rotation, *Rotation* 461.
— strain tensor, *infinitesimaler Dehnungstensor* 461.
— volume change, *infinitesimale Volumenänderung* 30.
inhomogeneity, *Inhomogenität* 530.
initial-boundary value problems, *Anfangs- und Randwertprobleme* 371.
initial conditions, *Anfangsbedingungen* 219, 335.
— data, *Anfangswerte* 218.
initial-value problem for rods, *Anfangswertproblem für Stäbe* 700.
initially flat Cosserat surface, *anfänglich flache Cosserat-Fläche* 460.
inner product of four-vectors, *inneres Produkt aus Vierer-Vektoren* 27.
— — — second-order tensors, *Tensoren zweiter Ordnung* 7.
— — — tensors, *Tensoren* 618.
— — — vectors, *Vektoren* 5.
integral identity for the displacement gradient, *Integralidentität für den Verschiebungsgradienten* 184.
— representation theorem for solutions of mixed problem, *Integraldarstellungssatz für Lösungen des gemischten Problems* 187.
integrals of the equilibrium equations for rods, *Integrale der Gleichgewichtsgleichungen für Stäbe* 696.
integro-differential equations, *Integrodifferentialgleichungen* 373.
interior equations, *Gleichungen für das Innere* 586.
— regularity, *Regularität im Innern* 355, 401.
internal concentrated loads, *innere konzentrierte Belastungen* 179.
— constraints, *Zwangsbedingungen* 403, 548.
— energy, *Energie* 300, 481.
intrinsic director couple, *inneres Direktorkräftepaar* 482, 491.
invariance conditions, *Invarianzbedingungen* 484, 485, 487.
irrotational fields, *wirbelfreie Felder* 17.
— velocity, *Geschwindigkeit* 231.
isentropic, *isentropisch* 304.
isochoric motion, *isochore Bewegung* 210.
— velocity, *Geschwindigkeit* 213.
isotropic materials and bodies, elastic, *isotrope Materialien und Körper, elastisch* 71, 74, 85, 86.
— — — —, elastic plates, *elastische Platten* 573.
— — — —, thermoelastic, *thermoelastisch* 311, 317.
isotropy, transverse, *Isotropie, transversale* 72.
iterated Laplace operator, *iterierter Laplace-Operator* 377.

Jacobian of transformation, *Funktionaldeterminante* 619.
jump across a singular surface, *Sprung an einer singulären Fläche* 249, 421.

Kelvin problem, *Kelvin-Problem* 173, 176.
— state, *Kelvin-Zustand* 174, 176, 178–180.
— theorem, *Kelvinscher Satz* 208.
kernel of the functional F, *Kern des Funktionals F* 392.
— — — nonnegative quadratic form, *der nichtnegtiven quadratischen Form* 392.
— — — operator T, *des Operators T* 391.
kinematic approximations in plates and shells, *kinematische Näherungen in Platten und Schalen* 478.
— measures for Cosserat surface, *Maße für eine Cosserat-Fläche* 452.
— — — shells, catalogue, *Schalen, Katalog* 458.
— — — —, derived from 3-dimensional theory, *abgeleitet aus der 3-dimensionalen Theorie* 468.

kinematic measures for shells, linearized, derived from 3-dimensional theory, *kinematische Maße für Schalen, linearisiert, abgeleitet aus der 3-dimensionalen Theorie* 474.
— — — —, restricted theory, *eingeschränkte Theorie* 460.
— results for shells, *Resultate für Schalen* 471.
kinematically admissible displacement field, *kinematisch zulässiges Verschiebungsfeld* 112.
— — process, *zulässiger Prozeß* 225.
— — state, *zulässiger Zustand* 111, 212, 324.
— — strain field, *zulässiges Dehnungsfeld* 112.
— and thermally admissible process, *und thermisch zulässiger Prozeß* 341.
kinematics of Cosserat surface, *Kinematik der Cosserat-Fläche* 449.
— — oriented media, *orientierten Medien* 445.
— — shells, derived from 3-dimensional theory, *Schalen, abgeleitet aus der 3-dimensionalen Theorie* 466.
— — —, restricted theory, *eingeschränkte Theorie* 455.
kinetic energy, *kinetische Energie* 65, 216, 481.
Kirchhoff boundary conditions, *Kirchhoffsche Randbedingungen* 577, 586.
KIRCHHOFF's theories for stretching and bending of plates, *Kirchhoffsche Theorien für Dehnung und Biegung von Platten* 586.
— theorem, *Kirchhoffscher Satz* 32.
KNOWLES' theorem on SAINT-VENANT's principle, *Knowlesscher Satz zum Saint-Venant-Prinzip* 200–206.
KOLOSOV's theorem, *Kolosovscher Satz* 164.
KORN's inequality, alternative, *Kornsche Ungleichung, Alternativform* 96.
— —, *first*, erste 38, 381, 384.
— —, second, *zweite* 381, 384.
Kronecker delta, *Kronecker-Delta* 618, 623.

Lamé moduli, *Lamé-Koeffizienten* 76, 311.
Landau symbol, *Landau-Symbol* 356.
Laplace operator Δ_2, *Laplace-Operator* Δ_2 375.
— transform, *-transformierte* 25, 231.
Laplacian, *Laplace-Operator* 11.
large deformation solutions, *große Deformation, Lösungen für* 546.
Lebesgue decomposition, *Lebesgue-Zerlegung* 415.
— derivative, *Lebesgue-Ableitung* 415.
length-preserving transformations, *längenerhaltende Transformationen* 538.
LÉVY's theorem, *Lévyscher Satz* 156.
Liapounov boundary, *Liapounovsche Randfläche* 419.
limit definition of solution to Kelvin's problem, *Grenzwertdefinition der Lösung von Kelvins Problem* 174.
linear constitutive equations for shells, *lineare Materialgleichungen für Schalen* 553, 555, 556, 571.
— elasticity, rod problems, *Elastizität, Stabprobleme* 661.
— momentum, *linearer Impuls* 43, 51, 482, 483.
— theory, elastic plates, *lineare Theorie für elastische Platten* 585, 596.
— —, — shells, *Schalen* 595, 597.
— —, — —, derived from 3-dimensional theory, *abgeleitet aus der 3-dimensionalen Theorie* 578.
— —, — —, remarks on, *Bemerkungen* 613.
linearized field equations for shells, *linearisierte Feldgleichungen für Schalen* 500, 523.
— kinematic measures for Cosserat surfaces and shells, *kinematische Maße für Cosserat-Flächen und -Schalen* 456, 460, 474.
linearly elastic incompressible body, *linearelastischer inkompressibler Körper* 210.
LOVE's first approximation, *Loves erste Näherung* 478, 582, 589, 590.
— integral identity in elasticity, *Integralidentität in der Elastizitätstheorie* 243.
— solution in elasticity, *Lösung in der Elastizitätstheorie* 147.
lower semi-continuous with respect to weak convergence, *halbstetig bezüglich schwacher Konvergenz* 393.

magnitude of a tensor, *Betrag eines Tensors* 7.
Mainardi-Codazzi equations, *Mainardi-Codazzi-Gleichungen* 626.
mass density, Cosserat surface, *Massendichte einer Cosserat-Fläche* 447.
— —, deformed configuration, *deformierten Konfiguration* 440.
— —, reference configuration, *Bezugskonfiguration* 440, 472.
material frame-indifference, *materielle Systemunabhängigkeit* 80, 305, 487.
— point, *materieller Punkt* 229, 439, 446, 530.
— surface, *materielle Fläche* 439.
— symmetry in shells, *Symmetrie in Schalen* 537, 539, 541.
— — — 3-dimensional theory, *der 3-dimensionalen Theorie* 69–74, 87–89.
— time derivative, *Zeitableitung* 431, 449, 453.
materials with "memory", *Materialien mit „Gedächtnis"* 373.
maximum complementary energy, *maximale komplementäre Energie* 125.
— elastic modulus, *maximaler elastischer Koeffizient* 85.
— potential energy, *maximale potentielle Energie* 125.
— principles, *Maximalprinzipien* 120, 375.

maximum speed of propagation, *maximale Ausbreitungsgeschwindigkeit* 245, 256, 257.
Maxwell relation, *Maxwell-Beziehung* 304, 310.
MAXWELL's compatibility theorem, *Maxwellscher Kompatibilitätssatz* 250.
— solution in elasticity, *Maxwellsche Lösung in der Elastizitätstheorie* 54.
mean curvature, *mittlere Krümmung* 471, 624.
— strain, *Dehnung* 37, 38, 97, 98.
— stress, *Spannung* 96, 97.
— — and strain in terms of external force-temperature, *Spannung und Dehnung, ausgedrückt durch äußere Kraft und Temperatur* 315.
— value of a function, *Mittelwert einer Funktion* 314.
— — theorem for elastic states, *Mittelwertsatz für elastische Zustände* 133–138.
— — — — harmonic and biharmonic fields, *harmonische und biharmonische Felder* 21.
measure, *Maß* 414, 415.
mechanical power, *mechanische Leistung* 496.
membrane, fixed along boundary and stretched over obstacle, *Membran, am Rand festgehalten und über ein Hindernis gespannt* 405.
— theory of shells, *Membrantheorie von Schalen* 547.
—, unilateral constraint, *Membran, einseitiges Problem* 422.
metric tensor of space, *metrischer Tensor für den Raum* 618.
— — — surface, *die Fläche* 624.
MICHELL's theorem, *Michellscher Satz* 163.
mild discontinuities, *schwache Unstetigkeiten* 253.
minimax principle, *Minimax-Prinzip* 268.
minimizing sequence, *Minimalisierungsfolge* 393.
minimum complementary energy, *minimale komplementäre Energie* 125, 324, 325.
— elastic modulus, *minimaler elastischer Koeffizient* 85.
— potential energy, *minimale potentielle Energie* 111, 125.
— — —, converse, *Umkehrung* 116.
— principle, displacement problem, *Minimalprinzip für das Verschiebungsproblem* 209.
— —, elastodynamics, *die Elastodynamik* 230.
— —, lowest characteristic value, *den kleinsten charakteristischen Wert* 264.
— —, stress field, *das Spannungsfeld* 232.
— —, vibration problem, *das Schwingungsproblem* 262, 266, 267.
— transformed energy, *minimale transformierte Energie* 231.
mixed boundary-value problem of elasticity, dynamic, *gemischtes Randwertproblem für die dynamische Elastizitätstheorie* 219.
— — — — —, static, *statische Elastizitätstheorie* 102, 377, 379, 385.
— — — — thermoelasticity, dynamic, *dynamische Thermoelastizität* 335.
— — — — —, static, *statische Thermoelastizität* 321.
— tensor, *gemischter Tensor* 617, 622.
modulus of compression (see also elasticities), *Kompressionsmodul (s. auch elastische Koeffizienten)* 74, 97.
mollifiers 420.
moment at infinity, *Moment im Unendlichen* 170.
— of momentum, *Drehimpuls* 482.
moments, balance of, *Momentenbilanz* 306.
— of constitutive variables in rod theories, *Momente der Materialgrößen in Stabtheorien* 649.
momentum, angular, *Impuls, Drehimpuls* 44, 52.
—, linear, *Impuls* 43, 44, 51, 254, 482, 483.
—, rods (angular, director, linear), *Impuls, Direktorimpuls und Drehimpuls in Stäben* 666.
—, shells, *in Schalen* 483.
—, space-time, *in Raum-Zeit* 67.
MORERA's solution in elasticity, *Moreras Lösung in der Elastizitätstheorie* 55.
motion of body, *Bewegung eines Körpers* 43, 299.
— — Cosserat surface, *einer Cosserat-Fläche* 449.
multidimensional singular integral equation, *vieldimensionale singuläre Integralgleichung* 421.
multiindex, *vielfacher Index* 348.
multiplicity, *Multiplizität* 369.

natural configuration, *natürliche Konfiguration* 380.
NAVIER's equation of elasticity, dynamic, *Navier-Gleichung der Elastizitätstheorie, dynamisch* 213.
— — — —, static, *statisch* 90.
neighborhood of infinity, *Umgebung des Unendlichen* 165.
Neumann problem, *Neumann-Problem* 375, 419.
Newtonian potential, *Newtonsches Potential* 19.
non-coercive non-symmetric bilinear forms, *nichtkoerzive, unsymmetrische bilineare Formen* 422.
nonexistence theorem of elasticity, *Nichtexistenzsatz der Elastizitätstheorie* 109.
nonlinear constitutive equations, rods, *nichtlineare Materialgleichungen für Stäbe* 654.
— — —, shells, Schalen 528.
— — —, —, membrane theory, *Membrantheorie* 566, 588.
— — —, —, restricted theory, *eingeschränkte Theorie* 549.

nonlinear constitutive equations, thermoelastic materials, *thermoelastische Materialien* 297.
nonoverlapping, *nichtüberlappend* 402.
nonsymmetric form, *nichtsymmetrische Form* 395.
non-trivial solution of the mixed problem with null data, *nichttriviale Lösung des gemischten Problems mit Nullwerten* 109.
norm 348.
normal coordinate system, *Normalkoordinatensystem* 442, 628.
— curvature of a curve, *Normalkrümmung einer Kurve* 628.
— modes, *Eigenschwingungen* 269, 271.
— unit, of curve, *Einheitsnormale einer Kurve* 623, 627.
— —, — surface, *Fläche* 440, 622.
normalized vector field, *normiertes Vektorfeld* 186.
notation, tables of, *Bezeichnungen, Liste der* 2–5, 298, 299, 432–438.
NOVOZHILOV's derivation of classical linear theory of shells, *Novozhilovs Ableitung der klassischen linearen Schalentheorie* 593.

objective (see also frame-indifference), *objektiv (s. auch systemunabhängig)* 487.
oblique derivative problem, *Problem „der schiefen Ableitung"* 376.
open ball, *offene Kugel* 12.
optimal design of a rod, *optimale Auslegung eines Stabes* 699.
oriented media, *orientierte Medien* 445.
orthogonal 7.
— complement, *Komplement* 385, 395.
— group, *orthogonale Gruppe* 7.
— projector, *orthogonaler Projektor* 392.
orthogonality of characteristic displacements, *Orthogonalität charakteristischer Verschiebungen* 264.
orthotropic material, *orthotrope Materialien* 72.
outer product of tensors, *äußeres Produkt von Tensoren* 617.

Parseval theorem, *Parsevalscher Satz* 366, 381.
part, Teil 14, 300.
partial differentiation with respect to surface coordinates θ^α, *partielle Ableitung nach Oberflächenkoordinaten* θ^α 431.
particular solution of the equation of equilibrium, *partikuläre Lösung der Gleichgewichtsgleichung* 59.
partition, *Aufteilung* 27.
permutation symbols, *Vertauschungssymbole* 453, 455, 618.
perpendicular projection, *orthogonale Projektion* 9, 129.
photoelasticity, *Photoelastizität* 156.
physical components of tensors, *physikalische Komponenten von Tensoren* 631.
— interpretation of Airy function, *Interpretation der Airy-Funktion* 158.
piecewise continuous function, *stückweise kontinuierliche Funktion* 14.
— regular, *regulär* 14, 25.
— smooth, *glatt* 14.
Piola-Kirchhoff stress, *Piola-Kirchhoffsche Spannung* 51, 53, 229, 305, 491.
PIOLA's theorem, *Piolascher Satz* 60.
plane convex set, *ebene konvexe Menge* 417.
— displacement, *Verschiebung* 39.
— elastic state, *ebener elastischer Zustand* 154.
— problem, *ebenes Problem* 150, 151.
— problems for rods, *ebene Probleme für Stäbe* 670.
— rigid displacement, *starre Verschiebung* 39.
— strain, *Dehnung* 151.
— strain solution, *Dehnungslösung* 152.
— stress, generalized, *Spannung, verallgemeinert* 152, 577.
plate under hydrostatic pressure, *Platte unter hydrostatischem Druck* 602.
Poincaré inequality, *Poincarés Ungleichung* 350.
Poisson-Kirchhoff theory for bending of plates, *Poisson-Kirchhoffsche Theorie für die Biegung von Platten* 524.
POISSON's decomposition theorem, *Poissonscher Zerlegungssatz* 234.
— ratio, *Quotient* 78, 318, 570.
— — for generalized plane stress, *Quotient für verallgemeinerte ebene Spannung* 153.
position, representation in rod theories, *Lage, Darstellung in Stabtheorien* 647.
position-vector field, *Lagevektor, Feld des* 5, 439.
positive definite elastic constants, *positiv definite elastische Konstanten* 153.
— — elasticity field, *definites Elastizitätsfeld* 104.
— — elasticity tensor, *definiter Elastizitätstensor* 69, 85–87.
— operator, *positiver Operator* 369.
— semi-definite elasticity tensor, *positiv halbdefiniter Elastizitätstensor* 69.
— — tensor 84.
— work, theorem of, *positive Arbeit, Satz von der* 95.
potential of double and simple layers, *Potential von Einfachschichten und Doppelschichten* 419.
power and energy, theorem of, *Leistung und Energie, Satz von* 216, 331.
— expended, theorem of, *aufgebrachte Leistung, Satz von* 65.
preferred reference state, *bevorzugter Bezugszustand* 530.
pressure, uniform, *Druck, gleichförmig* 50, 77.
principal curvatures of a surface, *Hauptkrümmung einer Fläche* 628.
— direction, *Hauptrichtung* 8, 628.
— — of strain, *Hauptrichtung der Dehnung* 35.
— — of stress, *Hauptrichtung der Spannung* 50.

principal strains, *Hauptdehnungen* 35.
— stresses, *Hauptspannungen* 50.
— value, *Hauptwert* 8.
principle of decomposition of thermoelastic states, *Prinzip der Zerlegung thermoelastischer Zustände* 313.
— — equipresence, *der Äquipräsenz* 530.
— — linear momentum, *des Impulses* 483.
— — material frame-indifference, *der Systemunabhängigkeit* 80, 305, 487.
— — maximum complementary energy, *der maximalen komplementäre Energie* 121.
— — — potential energy, *potentiellen Energie* 121.
— — minimum complementary energy, *der minimalen komplementären Energie* 112, 130, 212.
— — — potential energy, *potentiellen Energie* 111, 112, 130, 212, 231.
— — superposition for elastic states, *Superpositionsprinzip für elastische Zustände* 95.
— — — — thermoelastic states, *thermoelastische Zustände* 313.
process, dynamical, *Prozeß, dynamischer* 44, 49.
—, thermodynamic, *thermodynamischer* 484, 497, 510.
—, thermoelastic, *thermoelastischer* 327.
product of two tensors, *Produkt zweier Tensoren* 6.
production of entropy, *Produktion von Entropie* 483, 511.
progressive wave, *fortschreitende Welle* 245.
projection methods for rod theories, *Projektionsmethoden für Stabtheorien* 663.
—, perpendicular, *Projektion, orthogonale* 9, 129.
— property, *Projektionseigenschaft* 31, 32.
propagation condition for isotropic bodies, *Ausbreitungsbedingung für isotrope Körper* 247.
— — — progressive waves, *fortschreitende Wellen* 246.
— of elastic waves, *Ausbreitung elastischer Wellen* 386.
— problems, *Ausbreitungsprobleme* 371, 377.
— theorems, *Ausbreitungssätze* 256.
proper orthogonal group, *eigentliche orthogonale Gruppe* 7.
properly regular, *eigentlich regulär* 14, 38, 351.
pseudo body force field, *Pseudo-Volumenkraftfeld* 66, 218, 332.
pseudo-differential operators, *Pseudo-Differentialoperatoren* 421.
pseudo heat supply field, *Pseudo-Wärmezufuhrfeld* 332.
pseudo-tension operator, *Pseudo-Zugkraftoperator* 420.
pure bending of a plate, *reine Biegung einer Platte* 599.
— shear, *Scherung* 50, 77.
— strain, *Dehnung* 33.
— tension, *reiner Zug* 50, 78.
qualitative behavior for rods, *qualitatives Verhalten von Stäben* 697.
quasi-static theory, *quasistatische Theorie* 311.

rate of work by contact and assigned forces and couples, *Arbeitsleistung durch Kontaktkräfte und Kräftepaare sowie Volumkräfte und Kräftepaare* 481.
— — — — contact force and director couple, *Kontaktkraft und Direktorkräftepaar* 544.
rate of working, *Leistung* 300.
ray, *Strahl* 249, 257.
Rayleigh-Ritz procedure, *Rayleigh-Ritz-Methode* 125–129.
reciprocal base vectors, *reziproke Basisvektoren* 618.
— theorem for elastodynamics, *Reziprozitätssatz für Elastodynamik* 66.
— — — elastostatics, *Elastostatik* 101, 207.
— — — singular states, *singuläre Zustände* 182.
— — — synchronous external force, *synchrone äußere Kraft* 219.
— — — thermoelasticity, *Thermoelastizität* 320, 332.
reduced body force, *reduzierte Volumenkraft* 316.
— forms, *Formen* 306.
— stress, *Spannung* 316.
reduction of constitutive equations under superposed rigid body motions, *Reduktion von Materialgleichungen durch Überlagerung starrer Bewegungen* 534.
reference configuration, *Bezugskonfiguration* 299.
— director, *Bezugsdirektor* 447.
— surface, *Bezugsfläche* 425.
— temperature, *Bezugstemperatur* 308, 312.
regular domain, *reguläres Gebiet* 349.
— hypersurface, *reguläre Hyperfläche* 252.
— oblique derivative problem, *reguläres Problem „der schiefen Ableitung"* 376.
— point of boundary, *regulärer Punkt der Randfläche* 14.
— subsurface, *reguläre Unterfläche* 14.
— surface, *reguläre Fläche* 13.
regularization, regularity at or near the boundary, *Regularisierung, Regularität auf oder in der Umgebung der Randfläche* 357, 401, 411.
—, — in general, *allgemein* 370, 422.
—, — interior, *im Innern* 408.
—, — rods, *bei Stäben* 680.
REISSNER's plate theory, *Reissnersche Plattentheorie* 575, 578, 584.
relative tensors, *relative Tensoren* 618.
Rellich selection principle, *Rellichsches Auswahlprinzip* 352.
RELLICH's lemma 20.
representation of biharmonic fields at infinity, *Darstellung von biharmonischen Feldern im Unendlichen* 22.

representation of dilatation and rotation, *Darstellung von Dilatation und Rotation* 188.
— — displacement field, *des Verschiebungsfeldes* 160.
— — — gradient, *Verschiebungsgradienten* 187.
— — elastic displacement fields at infinity, *des elastischen Verschiebungsfeldes im Unendlichen* 165.
— — harmonic fields at infinity, *harmonischer Felder im Unendlichen* 22.
residual stress, *Restspannung* 81, 308.
response function in elasticity, *Antwortfunktion in der Elastizitätstheorie* 80.
— — — shell theory, *Schalentheorie* 529, 531.
— — — thermoelasticity, *Thermoelastizität* 301, 302, 304.
response of isotropic materials, *Reaktion eines isotropen Materials* 74.
restricted cone-hypothesis, *eingeschränkte Kegelhypothese* 382.
— form of the constitutive equations for an isotropic material, *Form der Materialgleichungen für ein isotropes Material* 557.
— theory (kinematic) of shells, *Theorie (kinematisch) der Schalen* 455.
restriction of response functions under reflection of director, *einschränkende Bedingung an Antwortfunktionen bei Spiegelung des Direktors* 540, 542.
— on stress due to uniform dilatation, *für die Spannung bei gleichförmiger Dilatation* 73.
restrictions placed on elastic materials by the second law, *einschränkende Bedingungen für elastische Materialien aus dem zweiten Hauptsatz* 302.
resultant assigned couple, *resultierendes vorgeschriebenes Kräftepaar* 504.
— — director couple, *Direktorkräftepaar* 481.
— — force, *resultierende vorgeschriebene Kraft* 481
— body force, *Volumenkraft* 509.
— contact couple, *resultierendes Kontaktkräftepaar* 504.
— — director couple, *Kontaktdirektorkräftepaar* 480.
— — force, *resultierende Kontaktkraft* 480, 510.
— couple vectors, *Kräftepaarvektoren* 514.
— force, *Kraft* 514.
resultants, catalogue of definitions, *Resultierende, Katalog der Definitionen der* 526.
—, specific internal energy and entropy, *für spezifische innere Energie und Entropie* 515.
Riemann-Christoffel tensor, *Riemann-Christoffel-Tensor* 620.
— — in space, *im Raum* 620.
— — of a surface, *auf einer Fläche* 462.
Riesz-Fredholm theory of compact operators, *Riesz-Fredholm-Theorie kompakter Operatoren* 369.
rigid displacement, *starre Verschiebung* 31–33, 61, 68, 104.
— field on ∂B with force and moment, *starres Feld auf* ∂B *mit Kraft und Moment* 185.
ring (rod), *Ring (Stabtheorie)* 659, 697.
ring, σ 415.
rod, definition, *Stab, Definition* 646.
—, definition of theory, *Stabtheorie, Definition* 641.
rotation field, *Rotationsfeld* 31.
— vector, *Rotationsvektor* 31, 39.
rotational wave equation, *Rotationswellengleichung* 328.
rotatory inertia, *Rotationsträgheit* 578, 596.

Saint-Venant problem, *Saint-Venantsches Problem* 661.
SAINT-VENANT's principle, *Saint-Venants Prinzip* 190–207.
scalar field, *skalares Feld* 10.
— product, *Produkt* 399.
— solid spherical harmonic, *skalare räumliche Kugelfunktion* 21.
Schwarz alternating method, *Schwarzsche alternierende Methode* 421.
s-derivative, *s-Ableitung* 350.
second law of thermodynamics (Clausius-Duhem inequality), *zweiter Hauptsatz der Thermodynamik (Clausius-Duhem-Ungleichung)* 301, 484, 512, 532.
self-adjoint, *selbstadjungiert* 385.
self-equilibrated stress field, *Gleichgewichts-Spannungsfeld* 55, 59.
semi-coercive bilinear form, *halbkoerzive bilineare Form* 422.
semi-coerciveness hypothesis, *Halbkoerzivität, Hypothese* 392.
shear deformation, transverse, *Scherdeformation, transversal* 460, 575, 578, 584.
— modulus, *Schermodul* 74, 76, 210, 311, 574.
—, simple, *Scherung, einfache* 34, 36, 77.
— stress, *Scherspannung* 50, 51.
shell-like body, *schalenartiger Körper* 411, 425, 438, 440, 442.
shell theory, *Schalentheorie* 444, 445.
— thickness, *Schalendicke* 443, 447.
shells reinforced with cords, *Schalen, verstärkt durch Fasern* 547.
shock wave, *Schockwelle* 253, 254.
σ-ring 415.
Signorini problem, definition 404, 413.
—, generalized, *verallgemeinert* 404, 408.
— in viscoelasticity, *in der Theorie der Viskoelastizität* 423.
SIGNORINI's theorem, *Signorinischer Satz* 61.
simple extension, *einfache Ausdehnung* 34.
— shear, *einfache Scherung* 34, 77.
simply-connected, *einfach zusammenhängend* 13.
singular elastic state, *singulärer elastischer Zustand* 179.

singular integral equations on a curve, *singuläre Integralgleichungen auf einer Kurve* 420.
— set of the measure γ, *Menge vom Maß* γ 415.
— surface of order n, *Fläche der Ordnung* n 250.
— — — — zero, *Null* 249.
skew, *schief* 6, 617.
skew-symmetric $r \times r$ matrix, *schiefsymmetrische* $r \times r$ *Matrix* 384.
smooth field, *glattes Feld* 11.
— in time, *glatt in der Zeit* 24.
smoothly propagating surface, *glatt sich ausbreitende Welle* 248.
Sobolev lemma 354.
solenoidal fields *solonoidales Feld* 18.
solid spherical harmonic, *räumliche Kugelfunktion* 22.
solution of equation of elasticity, *Lösung der Gleichung der Elastizitätstheorie* 102.
— — general problem of elasticity, *des allgemeinen Problems der Elastizitätstheorie* 129.
— — generalized mixed problem, *des verallgemeinerten gemischten Problems* 185.
— — mixed problem, *des gemischten Problems* 102, 220, 321, 335.
Somigliana fundamental matrix, *Somiglianasche Fundamentalmatrix* 419.
SOMIGLIANA's theorem, *Somiglianascher Satz* 183.
spaces $H_m(A)$ and $\overset{\circ}{H}_m(A)$, *Räume* $H_m(A)$ *und* $\overset{\circ}{H}_m(A)$ 349.
space-time, *Raum-Zeit* 27.
special theories of shells, *spezielle Schalentheorien* 546.
specific entropy, *spezifische Entropie* 483, 510.
— Gibbs free energy, *Gibbssche freie Energie* 555, 570.
— heat, *Wärme* 304.
— — supply, *Wärmezufuhr* 481.
— Helmholtz free energy, *Helmholtzsche freie Energie* 497, 511.
— internal energy, *innere Energie* 481, 510.
— production of entropy, *Entropieproduktion* 483, 511.
spectral decomposition and theorem, *spektrale Zerlegung und Spektralsatz* 8.
spectrum, *Spektrum* 269.
speed of propagation, *Ausbreitungsgeschwindigkeit* 248, 343.
spherical harmonic, *Kugelfunktion* 22.
star-shaped region, *sternförmiges Gebiet* 14.
static-geometric analogy, *statisch-geometrische Analogie* 613.
statically admissible state, *statisch zulässiger Zustand* 114.
— — stress field, *zulässiges Spannungsfeld* 112, 324.
STERNBERG's theorem on ST. VENANT's principle, *Sternbergscher Satz zu St. Venants Prinzip* 193.
Stokes process, *Stokesscher Prozeß* 239, 240.
STOKES' theorem, *Stokesscher Satz* 17.
stored energy (strain energy), function in elasticity, *gespeicherte Energie (Dehnungsenergie), Funktion in der Elastizitätstheorie* 82, 94.
— —, — — rods, *Stäben* 673, 677, 680.
— —, — — shells, *Schalen* 545.
straight rods, *gerade Stäbe* 690.
strain energy (total stored energy), elasticity, *Dehnungsenergie (gesamte gespeicherte Energie), in der Elastizitätstheorie* 94, 110, 208, 216.
— —, elasticity, bounds for, *Elastizitätstheorie, Schranken für* 113.
— —, shells, *Schalen* 545.
— —, thermoelasticity, *Thermoelastizität* 324.
— field, *Dehnungsfeld* 31, 39, 43, 461.
— path, *Dehnungsverlauf* 82.
— rate, *Dehnungsrate* 43.
strain-displacement relation, *Dehnungs-Verschiebungs-Relation* 31, 89, 212, 312, 326.
stress-couples, *Spannungskräftepaare* 512, 513, 595.
stress energy, *Spannungsenergie* 111, 324.
— equation of compatibility, *Spannung, Kompatibilitätsbeziehung* 92.
— — — motion, *Bewegungsgleichung* 214.
— field, *Spannungsfeld* 45–64.
— — for solution of mixed problem, *für die Lösung des gemischten Problems* 103.
— functions, *Spannungsfunktionen* 53–59.
—, independent of elastic constants, *Spannungen, unabhängig von den elastischen Konstanten* 163.
— moments for rods, *Spannungsmomente für Stäbe* 649.
— power, *Leistung* 65, 216, 300, 496.
— relation, *Spannungs-Dehnungs-Beziehung* 89, 212, 302, 305, 306.
— resultant, *Spannung, Resultierende* 512, 513, 651.
— tensor, *Spannungstensor* 45–64, 513.
—, transverse normal, *transverse Normalspannung* 584.
— vector, *Spannungsvektor* 43, 485, 510.
— wave, *Spannungswelle* 254.
stress-free state, *spannungsfreier Zustand* 323.
stress-momentum field, *Spannungs-Impuls-Feld* 67, 253.
stress-strain relation (see also stress relation), *Spannungs-Dehnungs-Beziehung (s. auch Spannungsbeziehung)* 89, 212.
stress-strain temperature relation, *Dehnungs-Temperatur-Beziehung* 312, 326.
stress-temperature equation of compatibility, *Spannungs-Temperatur-Gleichung (Kompatibilitätsbeziehung)* 318, 219, 321.
— modulus, *Spannungs-Temperatur-Modul* 311.
— tensor, *Spannungs-Temperatur-Tensor* 309, 313.

strong condition for existence of solution of Signorini problem, *starke Bedingung für die Existenz der Lösung des Signorini-Problems* 417.
— derivative, *Ableitung* 349, 357.
— sense, *im starken Sinne* 416.
strongly elliptic elasticity field, *stark elliptisches Elastizitätsfeld* 105.
— — elasticity tensor, *elliptischer Elastizitätstensor* 86, 87, 247, 256.
— — operator 365, 422.
subsurfaces $\mathscr{S}_1$ and $\mathscr{S}_2$, *Teilflächen* $\mathscr{S}_1$ und $\mathscr{S}_2$ 12.
superposed rigid body motions, *überlagerte starre Bewegungen* 452, 532.
supply of moment of momentum, *Zufuhr von Drehimpuls* 482.
support of u, *Träger von* u 349.
surface base vectors and their reciprocals, *Flächenbasisvektoren und reziproke Basisvektoren* 623.
— force field, *Flächenkraftfeld* 59, 215.
— geometry, *Flächengeometrie* 621.
— heat flux moments for rods, *Flächenwärmeflußmomente für Stäbe* 651.
— of discontinuity, *Unstetigkeitsflächen* 248.
— rate of deformation, *Flächendeformationsrate* 451.
— spin tensor, *Flächenspintensor* 451.
— spherical harmonic, *Kugelflächenfunktionen* 21.
— tension, *Flächenspannung* 547.
— traction, *Flächenzugkraft* 44, 93, 95, 211, 313, 319, 327.
— vectors, *Flächenvektoren* 450.
symmetric bounded bilinear form, *symmetrische beschränkte bilineare Form* 391.
— elasticity tensor, *symmetrischer Elastizitätstensor* 69, 84, 85, 247, 256.
— gradient 11.
— operator 369.
— tensor 6, 617.
symmetry condition for rods, *Symmetrie Beziehung für Stäbe* 654.
—, geometrical, *geometrische* 297.
— group, *Symmetriegruppe* 70, 210.
— of Green's states, *Symmetrie von Greens Zuständen* 187.
— transformation, *Transformation* 70.
synchronous external force systems, *synchrones äußeres Kraftsystem* 219.
system of concentrated loads, *konzentriertes Lastsystem* 179.
— — forces, *Kraftsystem* 43.

tangent plane, *Tangentialebene* 622.
—, unit of a curve, *Einheitstangente einer Kurve* 623.
tangential and normal components of director and of director gradient, *Tangential- und Normalkomponenten des Direktors und des Direktorgradienten* 535.
temperature, absolute, *Temperatur, absolute* 301, 483, 510.
temperature condition, *Temperaturbedingung* 335.
— difference field, *Temperaturdifferenzfeld* 312.
— equation, *Temperaturgleichung* 328.
— gradient, *Temperaturgradient* 301.
— relations, *Temperaturbeziehungen* 305.
—, representation in rod theories, *Temperatur, Darstellung in Stabtheorien* 649.
temperature-entropy moment for rods, *Temperatur-Entropie-Moment für Stäbe* 649.
tensor components, *Tensorkomponenten* 498.
—, covariant, contravariant, mixed, *Tensor, kovarianter, kontravarianter, gemischter* 617, 622.
—, definition 6.
— fields, *Tensorfelder* 10.
— product, *Tensorprodukt* 7.
—, relative, *Tensor, relativer* 618.
—, surface, *Tensor in einer Fläche* 631.
— transformation laws, *Tensor-Transformationsgesetze* 616.
— type, *Art* 617.
thermal expansion tensor, *thermische Ausdehnung, Tensor der* 310.
thermal gradient-temperature relation, *Temperaturgradient-Temperatur-Beziehung* 326.
thermodynamic process, *thermodynamischer Prozeß* 484, 497, 510.
thermoelastic process, *thermoelastischer Prozeß* 327.
— rod, *thermoelastischer Stab* 656, 669.
— shell, *thermoelastische Schale* 561.
— state, *thermoelastischer Zustand* 313.
thermo-mechanical behavior (response) of bodies, *thermo-mechanisches Verhalten von Körpern* 512.
— — — elastic Cosserat surface, *elastischen Cosserat-Flächen* 529.
— — — shells, *Schalen* 497.
thickness of shell, *Dicke einer Schale* 443.
thickness-shear vibration, *Dicke-Scherschwingung* 575.
thin film, *dünner Film* 547.
— shells and plates, *dünne Schalen und Platten* 438, 524.
three-dimensional body B, *dreidimensionaler Körper* B 438.
time-interval, *Zeitintervall* 23.
topology of $H_m(A)$, *Topologie von* $H_m(A)$ 401.
torsion of rectangular plate, *Torsion einer rechteckigen Platte* 604.
total energy, *Gesamtenergie* 216, 332.
— force, *Gesamtkraft* 44.
— moment, *Gesamtimpuls* 44.
TOUPIN's theorem on ST. VENANT's principle, *Toupinscher Satz zu St. Venants Prinzip* 196.
trace, *Spur* 6, 353.
traction condition, *Zugkraftbedingung* 102, 220, 321, 335.
— problem, *Zugkraftproblem* 102, 220, 321.

transport theorem, two-dimensional, *Transporttheorem, zweidimensional* 456.
transpose of a tensor, fourth-order, *Transponierte eines Tensors vierter Ordnung* 10.
— — — —, second-order, *zweiter Ordnung* 6.
transverse isotropy, *transversale Isotropie* 72.
— normal stress, *Normalspannung* 584.
— shear deformation, *Scherdeformation* 460, 575, 578, 584.
— wave, *Welle* 246–248, 254, 343.
two-dimensional body forces, *zweidimensionale Kräfte* 513.
— ε-system, *zweidimensionales ε-System* 455.
— heat supply, *zweidimensionale Wärmezufuhr* 513.
— permutation symbols, *zweidimensionale Vertauschungssymbole* 455.
— transport theorem, *Transporttheorem* 456.
typical surface particle, *typisches Oberflächenteilchen* 530.

umbilics, *Nabelpunkte* 628.
unbounded regular region, *unbeschränktes reguläres Gebiet* 13, 257.
uncoupled, *ungekoppelt* 311, 312.
— quasi-static theory, *ungekoppelte quasistatische Theorie* 325.
uniform dilatation, *gleichförmige Dilatation* 34, 77.
— pressure, *gleichmäßiger Druck* 50, 77.
— states, *uniforme Zustände* 546.
— — for rods, *für Stäbe* 698.
unilateral constraints, *einseitige Zwangsbedingungen* 403.
— problem, abstract, *einseitiges Problem, abstrakt* 39, 395.
— —, differential operator, *verknüpft mit Differentialoperator* 422.
— —, elastic body, *im elastischen Körper* 402.
— —, elastostatics, *in der Elastostatik* 423.
— —, elliptic operators, *für elliptische Operatoren* 399.
— —, linear operators of applied mathematics, *verknüpft mit linearen Operatoren der angewandten Mathematik* 422.
— —, membrane, *für eine Membran* 422.
— —, symmetric coercive bilinear form, *bzgl. symmetrischer koerziver bilinearer Form* 422.
uniqueness theorem, elasticity, displacement ("first") problem, *Eindeutigkeitssatz der Elastizitätstheorie, erstes oder Verschiebungs-Problem* 105.
— —, —, elastodynamics, *Elastodynamik* 222, 261.
— —, —, exterior domain, *äußeres Gebiet* 171.
— —, —, general problem, *allgemeines Problem* 130.
uniqueness theorem, elasticity, incompressible body, *Eindeutigkeitssatz der Elastizitätstheorie, inkompressibler Körper* 212.
— —, —, internal concentrated loads, *innere konzentrierte Lasten* 190.
— —, —, mixed ("third") problem, *drittes oder gemischtes Problem* 104, 112.
— —, —, plane problem, *ebenes Problem* 154.
— —, rods, *für Stäbe* 692.
— —, shells, *für Schalen* 610.
— —, thermoelasticity, dynamic, *der Thermoelastizität, dynamisch* 338.
— —, —, static, *statisch* 322.
unit doublet state, *Einheits-Doublet-Zustand* 178.
— Kelvin state, *Einheits-Kelvin-Zustand* 178.
— normal to a surface, *Einheitsnormale einer Fläche* 440, 622.
— principal normal to a curve, *Einheitsnormale einer Kurve* 627.
— tangent and unit normal to a curve lying in the surface, *Einheitstangente und Einheitsnormale einer Kurve in der Fläche* 623.
universal displacement field, *universelles Verschiebungsfeld* 91.
upper and lower bounds for strain energy, *obere und untere Schranke für die Dehnungsenergie* 113.

vanishing near a surface, *Verschwinden bei einer Fläche* 20, 223.
variational methods, elastic rod theory, *Variationsmethoden, elastische Stabtheorie* 668.
— —, 3-dimensional elasticity, *3-dimensionale Elastizitätstheorie* 125–129.
— principles and theorems, elasticity, *Variations-Prinzipien und -Sätze, Elastizitätstheorie* 122–125, 223–232.
— — — —, hyperelastic rods, *hyperelastische Stäbe* 676.
— — — —, shell theory, *Schalentheorie* 476.
vector field, *Vektorfeld* 10.
— in a surface, *Vektor in einer Fläche* 622.
— product, *Vektorprodukt* 5.
— solid spherical harmonic, *vektorielle räumliche Kugelfunktion* 22.
velocity, definition, *Geschwindigkeit, Definition* 43, 467.
—, director, *Direktorgeschwindigkeit* 449.
— of propagation of a wave, *Ausbreitungsgeschwindigkeit einer Welle* 246.
visco-elasticity, *Viskoelastizität* 386.
Volterra dislocation, *Volterra-Versetzung* 42.
— integral equations in Banach spaces, *Volterra-Integralgleichung in Banach-Räumen* 374.
volume change, *Volumenänderung* 30, 31, 37, 97.

v. Kármán equations for nonlinear bending, of thin plates, *v. Kármán-Gleichung für nichtlineare Biegung dünner Platten* 588.
v. Mises-Sternberg version of Saint-Venant's principle, *v. Mises-Sternberg-Version des Prinzips von Saint-Venant* 190.

wave, acceleration, *Beschleunigungswelle* 253, 254.
—, balance of momentum for, *Welle, Impulsbilanz für* 254.
—, dispersed, *Wellen, dispersierte* 343.
— equations, *Wellengleichung* 328.
— of order n, *Welle der Ordnung n* 253.
— operators, *Wellenoperator* 213, 328.
— propagation, straight rods, *Wellenausbreitung in geraden Stäben* 663.
— —, 3-dimensional bodies, *in 3-dimensionalen Körpern* 243–261.
—, transverse, *Welle, transversal* 246–248, 254, 343.
weak solution, *schwache Lösung* 355.
weakly closed, *schwach geschlossen* 393.
— compact, *kompakt* 394, 398.

WEINGARTEN's formulae, *Weingartens Formeln* 625, 626.
— theorem on singular surfaces, *Satz über singuläre Flächen* 252.
Wheeler-Sternberg lemma 257.
work, definition, *Arbeit, Definition* 60.
— done along a strain path, *geleistete, für einen Dehnungsverlauf* 82.
— — by an isotropic body, *von einem isotropen Körper* 207.
— — starting from rest, *beginnend im Ruhezustand* 217.
— expended, *Arbeitsleistung* 60.
— theorem in thermoelasticity, *Arbeitssatz in der Thermoelastizität* 320.
— theorems in elasticity, *Arbeitssätze in der Elastizitätstheorie* 95, 208, 211.

x_i-convex region, *x_i-konvexes Gebiet* 15.

YOUNG's modulus, *Elastizitätsmodul* 78, 570.

Zanaboni-Robinson theorem on SAINT-VENANT's principle, *Zanaboni-Robinsonscher Satz zu Saint-Venants Prinzip* 206.